DISEASES AND THE ORGANISMS THAT CAUSE THEM (*Continued*)

VIRAL DISEASES

Disease	Virus	Reservoir	Page
AIDS, ARC	human immunodeficiency virus (HIV)	humans	9, 270, 404, 426, 431, 432, 464, 530, 535–544
aplastic crisis in sickle cell anemia	parvovirus	humans	657
bronchitis, rhinitis	parainfluenza	humans, some other mammals	581
Burkitt's lymphoma	Epstein-Barr	humans	274, 655–656
chickenpox	varicella-zoster	humans	5, 270, 274, 401, 411, 554–556, 561
cold, common	rhinovirus	humans	579–580, 581
colds	coronavirus	humans	464, 580, 581
cytomegalic inclusion disease	cytomegalovirus	humans	711–713
Dengue fever	Dengue	humans	271, 317, 653–654, 662
encephalitis	bunyavirus	mammals	272, 656
	Colorado tick fever	mammals	317, 657
	Eastern equine encephalitis	birds	413, 673–674, 685
	St. Louis encephalitis	birds	411, 413, 673–674, 685
	Venezuelan equine encephalitis	rodents	673–674, 685
	Western equine encephalitis	birds	413, 673–674, 685
enteritis (acute infantile gastroenteritis)	rotavirus	humans	616–618, 629
hemorrhagic fever, Bolivian	arenavirus	rodents and humans	657, 662
hemorrhagic fever, Korean	bunyavirus (Hantaan)	rodents	272, 656, 662
hemorrhagic fever	Ebola virus (filovirus)	humans	656, 662
hemorrhagic fever	Marburg virus (filovirus)	humans	656, 662
hepatitis, infectious	hepatitis A	humans	5, 618–619, 629, 737
hepatitis, serum	hepatitis B	humans	5, 401, 618–620, 629
hepatitis, non-A, non-B	hepatitis C	humans	619–620
herpes, oral	usually herpes simplex type 1, sometimes type 2	humans	270, 274, 463, 706–707
herpes, genital	usually herpes simplex type 2, sometimes type 1	humans	270, 274, 464, 706–710

Disease	Org[illegible]	[illegible]	Page
infectious mononucleosis	E[illegible]	[illegible]	[illegible]274, 401, [illegible]655, 662
influenza	in[illegible]	[illegible] humans (type B) humans (type C)	[illegible]70, 385, [illegible], 412, 464, 591–593, 597–599
keratoconjunctivitis	adenovirus	humans	562
Lassa fever	arenavirus	rodents	272, 657, 662
measles (rubeola)	measles	humans	270, 401, 411, 422
meningoencephalitis	herpes	humans	274, 495, 553–554
molluscum contagiosum	pox virus group	humans	556–557
mumps	paramyxovirus	humans	5, 270, 401, 411, 495, 607–608
muscle damage	Coxsackie	humans, mice	271
pneumonia	respiratory syncytial virus	humans	593–594
poliomyelitis	poliovirus	humans	5, 270, 411, 494–496, 680–683, 685, 737
rabies	rabies	all warm-blooded animals	5, 270, 285, 401, 495, 671–673, 685
respiratory infections	adenovirus	humans	694
respiratory infections	polyomavirus	none	674–675
Rift Valley fever	bunyavirus (phlebovirus)	humans, sheep, cattle	656
rubella (German or 3-day measles)	rubella	humans	271, 287, 392, 401, 495, 552–553, 561
shingles	varicella-zoster	humans	270, 274, 539, 554–555, 561
smallpox	variola (major and minor)	humans	3, 15, 270, 274, 561, 570–571
warts, common (papillomas)	human papillomavirus	humans	556–557, 561
warts, genital (condylomas)	human papillomavirus	humans	556–557, 711, 713
yellow fever	yellow fever	monkeys	5, 17, 270, 271, 317, 414, 654, 662

FUNGAL DISEASES

Disease	Organism	Page
aspergillosis	*Aspergillus* sp.	305, 560, 561, 622
blastomycosis	*Blastomyces dermatitidis*	558, 559, 561, 597
candidiasis	*Candida albicans*	559, 561, 608, 697
coccidioidomycosis (valley fever)	*Coccidioides immitis*	594–595, 597
cryptococcosis	*Cryptococcus neoformans*	596, 597
ergotism	*Claviceps purpura*	306, 622–623
histoplasmosis	*Histoplasma capsulatum*	595, 597
pneumocystic pneumonia	*Pneumocystis carinii*	539, 596, 597
ringworm (tinea)	various species of *Epidermophyton, Trichophyton, Microsporium*	558
sporotrichosis	*Sporothrix schenckii*	558, 561
zygomycosis	*Rhizopus* sp., *Mucor* sp.	560, 561

The table of parasitic diseases appears on the following page.

DISEASES AND THE ORGANISMS THAT CAUSE THEM (*Concluded*)

PARASITIC DISEASES

Disease	Organism	Type*	Page
Acanthamoeba keratitis	*Acanthamoeba culbertsoni*	protozoan	566
amoebic dysentery	*Entamoeba histolytica*	protozoan	298, 621, 629, 737
ascariasis	*Ascaris lumbricoides*	roundworm	626–627, 629
babesiosis	*Babesia microti*	protozoan	662
balantidiasis	*Balantidium coli*	protozoan	621, 629, 737
body lice (pediculosis)	*Pediculus humanus*	louse	568–569
Chagas' disease	*Trypanosoma cruzi*	protozoan	317, 684, 685
chigger dermatitis	*Trombicula*	mite	568–569
chigger infestation	*Tunga penetrans*	sandflea	568–569
Chinese liver fluke	*Clonorchis sinensis*	flatworm	309, 624, 629
crab lice	*Phthirus pubis*	louse	568–569
cryptosporidiosis	*Cryptosporidium* sp.	protozoan	622, 629
elephantiasis (wuchereriasis)	*Wuchereria bancrofti*	roundworm	314, 640–641
fasciolopsiasis	*Fasciolopsis buski*	flatworm	624
giardiasis	*Giardia intestinalis*	protozoan	398, 620–621, 629, 737
Guinea worm (dracunculiasis)	*Dracunculus medinensis*	roundworm	312–314, 561
heart worm disease	*Dirofilaria immitis*	roundworm	640
hookworm	*Ancylostoma duodenale* (Old World hookworm)	roundworm	311, 626, 629
	Necator americanus (New World hookworm)	roundworm	311, 401, 626, 629
leishmaniasis	*Leishmania braziliensis*	protozoan	652, 658
kala-azar	*L. donovani*		317, 658, 662
oriental sore	*L. tropica*		652, 658
liver/lung fluke (paragonimiasis)	*Paragonimus westermani*	flatworm	308, 309, 596, 597
loaisis	*Loa Loa*	roundworm	317, 566
malaria	*Plasmodium* sp.	protozoan	659, 660, 662
pinworm (enterobiasis)	*Enterobius vermicularis*	roundworm	401, 628, 629
river blindness (onchocerciasis)	*Onchocerca volvulvus*	roundworm	563, 566
scabies	*Sarcoptes scabeii*	mite	568, 569
schistosomiasis	*Schistosoma* sp.	flatworm	637–640, 641
sheep liver fluke (fascioliasis)	*Fasciola hepatica*	flatworm	308, 623, 624
sleeping sickness (trypanosomiasis)	*Trypanosoma gambiense*, *Trypanosoma rhodesiense*	protozoan	683–684, 685
strongyloidiasis	*Strongyloides stercoralis*	roundworm	628, 629
swimmer's itch	*Schistosoma* sp.	flatworm	560, 561
tapeworm infestation (taeniasis)	*Hymenolepsis nana* (dwarf tapeworm)	flatworm	624, 629
	Taenia saginata (beef tapeworm)	flatworm	310, 624, 629
	Taenia solium (pork tapeworm)	flatworm	624, 629
	Diphyllobothrium latum (fish tapeworm)	flatworm	624–625, 629
	Echinococcus granulosis (dog tapeworm)	flatworm	624, 629
toxoplasmosis	*Toxoplasma gondii*	protozoan	660–662
trichinosis	*Trichinella spiralis*	roundworm	5, 311, 625–626, 629, 751
trichomoniasis	*Trichomonas vaginalis*	protozoan	696–697, 713
visceral larva migrans	*Toxocara* sp.	roundworm	627
whipworm (trichuriasis)	*Trichuris trichiura*	roundworm	627–628, 629

PATHOGENS AND THE DISEASES THEY CAUSE (*Continued*)

BACTERIA

Organism	Gram Stain	Basic Morphology	Diseases	Page
Actinomadura sp.	+	rod, some filamentous forms	Madura foot (actinomycetoma)	560
Acintomyces israelii	+	filamentous, diphtheroid, & coccal	Actinomycosis, mouth & other lesions	257, 259
Bacillus anthracis	+	rod, encapsulated	anthrax	256, 395, 642, 649
Bacillus cereus	+	rod, encapsulated	food poisoning	255, 395, 609, 756
Bacteroides sp.	–	small rod	mouth lesions, septicemia, abscesses, Vincent's angina	252, 607
Bartonella bacilliformis	–	curved or coccoid	Oroya fever (systemic form), verruga peruana (cutaneous form)	253, 653, 654
Bordetella pertussis	–	coccobacillus	whooping cough	581–583, 597
Borrelia burgdorferi	–	spiral	Lyme disease	248, 315, 316, 647–649
Borrelia recurrentis	–	large spiral	epidemic relapsing fever	248, 316, 646–647, 649
Brucella sp.	–	coccobacillus	Brucellosis (undulant fever or Malta fever)	644, 649
Calymmatobacterium granulomatis	–	rod, encapsulated	granuloma inguinale (Donovanosis)	706, 713
Campylobacter sp.	–	rod	gastroenteritis	248, 614, 617, 737, 756
Chlamydia trachomatis	NA	coccoid, very tiny	conjunctivitis, trachoma, genital tract infection (nongonococcal urethritis), infant pneumonitis, lymphogranuloma venereum	253, 562, 566, 704–706, 713
Chlamydia psittaci	NA	coccoid, very tiny	ornithosis (parrot fever)	253, 585–590, 597
Clostridium difficile	+	rod	pseudomembraneous colitis	366, 379
Clostridium botulinum	+	rod	food poisoning (botulism), wound infections, infant botulism	256, 395, 396, 609, 679–680, 685, 756
Clostridium perfringens	+	rod	gas gangrene, food poisoning	256, 395, 567, 609
Clostridium tetani	+	rod	tetanus	256, 395, 678, 685
Corynebacterium diphtheriae	+	rod, club-shaped, pleomorphic, forms palisades	diphtheria: pharyngeal, laryngeal & cutaneous	256, 279, 395, 577–578
Coxiella burnetii	NA	coccobacillus	Q fever, pneumonia	253, 317, 590, 597
Escherichia coli	–	rod	urinary tract infections, "traveler's diarrhea", nosocomial infections	395, 613–614 (see index)
Francisella tularensis	–	small rod (coccobacillus)	tularemia	249, 316, 645–646, 649
Gardnerella vaginalis	–	small rod	bacterial vaginitis (nonspecific), urethritis	695–696, 713
Haemophilus aegyptius	–	coccobacillus	conjunctivitis	252, 562
Haemophilus ducreyi	–	slender rod (coccobacillus)	chancroid	252, 704, 713
Haemophilus influenzae	–	coccobacillus, some strains form capsules	meningitis in children under 5, epiglottitis, eye infections, pneumonia in elderly or compromised patients	252, 365, 496, 576–579, 670–671
Helicobacter pylori		curved rod	chronic gastritis	615–616, 617
Klebsiella pneumoniae	–	rod, encapsulated	pneumonia, infant diarrhea, urinary tract infections	250, 417, 583–584, 725, 755

Organism	Gram Stain	Basic Morphology	Disease	Page
Legionella pneumophila	–	coccoid rod	Legionnaire's disease (pneumonia)	249, 539, 585, 597
Leptospira interrogans	–	spiral	leptospirosis	248, 695, 713
Listeria monocytogenes	+	rod	listeriosis, meningitis, abortion	256, 671, 685
Mycoplasma pneumoniae	NA	too small to be visualized by light microscope	primary atypical bacterial pneumonia	255, 577, 584, 705, 713
Mycobacterium avium	A-F	rod	chronic pulmonary disease, opportunistic infections in immunosuppressed patients	587
Mycobacterium leprae	A-F	rod	Hansen's disease (leprosy)	67, 257, 390, 584, 675
Mycobacterium tuberculosis	A-F	rod, branching forms	tuberculosis	257, 387, 539, 585–588, 597
Neisseria gonorrhoeae	–	cocci in pairs	gonorrhea, ophthalmia neonatorum, meningitis, arthritis, keratitis	156, 249, 423, 561–562, 566, 697, 700, 713
Neisseria meningitidis	–	cocci in pairs; capsules formed in young cells	meningitis, Waterhouse-Friderichson syndrome	249, 670–671
Nocardia sp.	+	rod, some filamentous forms	nocardiosis, Madura foot (mycetoma)	67, 209, 257, 560, 590–591, 597
Propionibacterium acnes	+	rod	acne	257, 366, 446, 551, 561
Providencia stuartii	–	rod	urinary tract infections, wound infections	251, 435–436, 552
Pseudomonas aeruginosa	–	rod	urinary tract infections, skin lesions, eye & ear infections, septicemia in immunocompromised patients	249, 395, 417, 434, 552, 565, 579, 693, 713
Rickettsia akari	NA	coccobacillus	rickettsialpox	651–654
Rickettsia prowazekii	NA	coccobacillus	epidemic typhus	316, 650, 654
Rickettsia typhi	NA	coccobacillus	endemic or murine typhus	316, 650–651, 654
Rickettsia tsutsugamushi	NA	coccobacillus	tsutsugamushi fever	253, 316, 651, 654
Rickettsia rickettsii	NA	coccobacillus	Rocky Mountain spotted fever	317, 651, 654
Rochalimaea quintana	NA	coccobacillus	trench fever	253, 317, 653, 654
Salmonella typhimurium, Salmonella enteritidis, Salmonella paratyphi	–	rod	salmonellosis (food poisoning)	610, 617, 752, 753, 756, 758
Salmonella typhi	–	rod	typhoid fever	155, 325, 611, 737
Serratia marcescens	–	rod	urinary tract infections, hospital epidemics, septicemia, peritonitis, arthritis, respiratory	251, 552, 565
Shigella boydii, Shigella dysenteriae, Shigella flexneri, Shigella sonnei	–	rod generally single	shigellosis (bacterial dysentery)	395, 611–612, 617, 737, 756
Spirillum minor	–	spiral	sodoku	568, 569
Staphylococcus aureus	+	cocci in clusters	skin lesions, abscesses, boils, scalded skin syndrome, impetigo, toxic shock syndrome, food poisoning, pericarditis	120, 255, 325, 395, 430, 550, 561, 562, 577, 579, 597, 609, 617, 637, 696, 713, 754, 756

PATHOGENS AND THE DISEASES THEY CAUSE

VIRUSES

Virus	Group Family	Disease	Page
adenovirus	Adenoviridae	acute upper & lower respiratory tract distress, pharyngitis, pneumonia, follicular conjunctivitis, epidemic keratoconjunctivitis	268, 270, 273, 274, 397, 562, 566, 580
arenavirus	Arenaviridae	Lassa fever	270, 272, 657
arenavirus	Arenaviridae	Bolivian hemorrhagic fever	270, 657
bunyavirus	Bunyaviridae	encephalitis	272, 656
Rift Valley fever	Bunyaviridae	fever & hemorrhage	657
hantaan	Bunyaviridae	Korean hemorrhagic fever	657
Colorado tick fever	Reoviridae	encephalitis	317, 657
coronavirus	Coronaviridae	colds, GI disturbances	269, 580
coxsackie	Piconaviridae	common cold syndrome & pharyngitis; severe systemic illness of newborn; muscle pain & damage; diabetes; meningoencephalitis	270, 271, 657
cytomegalovirus	Herpesviridae	mononucleosis, congenital cytomegalic inclusion disease, severe birth defects	268, 269, 287, 711–713
dengue	Flaviviridae	dengue fever (break-bone fever)	271, 316, 317, 653–654, 662
Ebola	Filoviridae	hemorrhagic fever	656
Marburg	Filoviridae	hemorrhagic fever	656
Eastern equine encephalitis	Togaviridae	encephalitis	673–674, 685
Venezuelan equine encephalitis	Togaviridae	encephalitis	674, 685
Western equine encephalitis	Togaviridae	encephalitis	317, 673–674, 685
St. Louis encephalitis	Flaviviridae	encephalitis	317, 411, 674, 685
enterovirus	Picornaviridae	acute hemorrhagic conjunctivitis	271, 398, 563, 566
Epstein-Barr	Herpesviridae	Burkitt's lymphoma, infectious mononucleosis, nasopharyngeal carcinoma	274, 500, 539, 654–656, 662
hepatitis A	Picornaviridae	infectious hepatitis	5, 618, 629, 737
hepatitis B	Picornaviridae	serum hepatitis	5, 618–620, 629
herpes virus	Herpesviridae	meningoencephalitis	5, 268, 270, 273–274, 397, 674, 685
herpes simplex type 1	Herpesviridae	oral herpes, gingivostomatitis, herpes labialis (cold sores), keratoconjunctivitis, herpetic whitlow	270, 274, 287, 705–706, 713
herpes simplex type 2	Herpesviridae	genital herpes, oral & whitlow	270, 274, 287, 707–708
human immunodeficiency (HIV)	Retroviridae	AIDS, ARC	273, 540–544
influenza	Orthomyxoviridae	influenza (flu)	5, 9, 270, 271, 272, 592
measles	Paramyxoviridae	rubeola, sometimes subacute sclerosing panencephalitis (SSPE)	270, 271, 500
human papillomavirus	Papovaviridae	common warts (papillomas), genital warts (condylomas); associated with cervical cancer	274, 556–557, 561, 711, 713
parainfluenza	Paramyxoviridae	rhinitis, pharyngitis, bronchitis, pneumonia, croup	397, 500, 579, 580
paramyxovirus (mumps)	Paramyxoviridae	mumps	268, 270, 271, 397, 607–608
parvovirus	Parvoviridae	aplastic crisis in sickle cell anemia	268, 270, 271, 274, 657
canine parvovirus	Parvoviridae	severe vomiting & diarrhea	657
feline panleukopenia	Parvoviridae	decreased number of WBC with fever	657
poliovirus	Picornoviradae (enterovirus)	poliomyelitis	5, 271, 395, 500, 662, 680–682, 685, 737
polyomavirus: JK	Papoviridae	mild respiratory illness	675
polyomavirus: BK	Papoviridae	associated with renal transplant infection, immunosuppressed patients	675
pox virus group (unclassified)	?	molluscum contagiosum	270, 273, 274
rabies	Rhabdoviridae	rabies	5, 271, 397, 671–673, 685
respiratory syncytial	Paramyxoviridae	pneumonia in children under age 1, upper respiratory infection in older children & adults	593–594, 597
rhinovirus (cold virus)	Picornavirdae	common cold	271, 580
rotavirus	Reoviridae	enteritis	272, 616–618, 629
rubella	Togaviridae	German measles, 3-day measles	271
varicella-zoster	Herpesviridae	chickenpox, shingles	274, 411, 539, 554–555
yellow fever	Flaviviridae	yellow fever	5, 271, 317, 654, 662

FUNGI

Organism	Disease	Page
Aspergillus sp.	aspergillosis, pneumonia in compromised patients, skin infections in burn patients, corneal & external ear infections	305, 560, 561, 622
Blastomyces dermatitidis	blastomycosis	558–559, 561, 597
Candida albicans	candidiasis	305, 539, 559
Claviceps purpura	ergotism	306, 559, 622
Coccidioides immitis	coccidioidomycosis (valley fever)	594–595, 597
Cryptococcus neoformans	cryptococcosis	539, 596, 597
Epidermophyton sp.	ringworm (tinea)	557
Histoplasma capsulatum	histoplasmosis	539, 595, 597
Microsporum sp.	ringworm (tinea)	557
Mucor sp.	zygomycosis	560, 561, 751
Philophora verrucosa	corneal ulcers	564
Pneumocystis carinii	Pneumocystis pneumonia	539, 596, 597
Rhizopus sp.	zygomycosis	560, 561
Sporothrix schenckii	sporotrichosis	546, 548, 558, 561
Trichophyton sp.	ringworm (tinea)	294, 304, 305, 545, 557

Microbiology

Jacquelyn G. Black

Marymount University, Arlington, Virginia

Jacquelyn Black received her B.A., B.S., and M.S. from the University of Chicago and her Ph.D. from Catholic University of America. She has been teaching microbiology to undergraduates since 1970. She is a member of the American Society for Microbiology, and she has received grants for conducting teacher-training programs.

In addition to her extensive teaching experience, Dr. Black has engaged in fieldwork and studies throughout the globe. Her travels have taken her from the interior of Iceland to Belgium and Portugal to the barrier reef of Belize.

Dr. Black, pictured at left with her daughter, Laura, describes herself as an "incorrigible snoop" who is interested in all the various aspects and applications of microbiology. This natural curiosity, coupled with her classroom and laboratory experience, make her uniquely qualified to author an introductory microbiology textbook. This book conveys her sense of excitement for microbiology and offers the most current information on developments and applications within the field.

Dear Reader,

Being an author is a joyous thing. It gives you a sort of "license to snoop." You can call up people and say, "I'm writing a book, and I need to know all about . . . ," or "I'd like to come to visit and see. . . ." Suddenly closed doors swing open and I'm off on a wonderful adventure, talking to fascinating people, learning things that will stretch my mind out to a dimension from which it will never return. That's my definition of education. I wish I could have taken you all with me when I went to the U.S. Army Medical Research Institute of Infectious Diseases (USAMRIID)—our former germ warfare center) to interview Dr. John Huggins for a Microbiologist's Notebook. Once every 2 or 3 years, they "break down" a maximum containment laboratory suite, sterilizing it and getting ready to start a new project in it. It can be visited only on the 2 days after sterilization, before it will be closed off for the next several years again. I had been, with my nose and camera pressed up against the glass portholes to these rooms, trying to see everything inside. Then they invited me to come back on one of the 2 days and go inside. Your heart would have pounded a little, too, I think, if you were beside me walking where only scientists dressed in plastic spacesuits had gone up until the day before. You might not have liked falling into the drainage ditch with me at the mushroom farm, but you would have enjoyed the visit to the Carter Foundation in Atlanta. Indeed, had you been able to come with me and see everything yourself, and talk with all the people I did (Dr. Watterson and I had a wonderful conversation about the 2 years of work that went into producing the photo of bacterioform gold that is the Chapter 27 opener), we might not have needed this book. But since you couldn't, I hope that I have been able to transmit in the pages of this book a sense of the excitement that is microbiology. The longer I am a microbiologist, the more excited I become. It never grows old or boring—it is constantly new and vital. I hope you will love it too.

Sincerely,
Jacquelyn Black

MICROBIOLOGY
Principles and Applications

SECOND EDITION

Jacquelyn G. Black

Marymount University

PRENTICE HALL Englewood Cliffs, New Jersey 07632

Library of Congress Cataloging-in-Publication Data

Black, Jacquelyn G.
Microbiology: principles and applications / Jacquelyn G. Black. –
– 2nd ed.
p. cm.
Includes bibliographical references and index.
ISBN 0-13-582917-8
1. Medical microbiology. I. Title.
QR46.C877 1993
616'.01—dc20 92-35694
CIP

Acquisitions Editor: David Kendric Brake
Development Editor: Robert J. Weiss
Design Director: Florence Dara Silverman
Interior Designer and Page Make-up Artist: Lee Goldstein
Cover Designer: Bruce Kenselaar
Pre-press Buyer: Paula Massenaro
Manufacturing Buyer: Lori Bulwin
Photo Researcher: Yvonne Gerin
Photo Editor: Lori Morris-Nantz
Illustrators: William C. Ober, M.D.; Claire W. Garrison, R.N.; Vantage Art, Inc.
Supplements Editor: Mary Hornby

Photo credits begin on page PC1
and constitute a continuation of this copyright page.

Printed in the United States of America

10 9 8 7 6 5 4 3 2 1

ISBN 0-13-582917-8

Prentice-Hall International (UK) Limited, *London*
Prentice-Hall of Australia Pty. Limited, *Sydney*
Prentice-Hall of Canada Inc., *Toronto*
Prentice-Hall Hispanoamericana, S.A., *Mexico*
Prentice-Hall of India Private Limited, *New Delhi*
Prentice-Hall of Japan, Inc., *Tokyo*
Simon & Schuster Asia Pte. Ltd., *Singapore*
Editora Prentice-Hall do Brasil, Ltda., *Rio de Janeiro*

Contents in Brief

Contents

UNIT II MICROBIAL METABOLISM, GROWTH, AND GENETICS

UNIT III THE ROSTER OF MICROBES AND MULTICELLULAR PARASITES

UNIT IV CONTROL OF MICROORGANISMS

UNIT V HOST-MICROBE INTERACTIONS

List of Boxes

CLOSE-UP BOXES

PUBLIC HEALTH BOXES

APPLICATIONS BOXES

BIOTECHNOLOGY BOXES

TRY IT BOXES

Preface

We are surrounded by a world of microorganisms. They inhabit a range of environments from mountains and volcanoes to deep-sea vents. They can be found in the air we breathe, the food we eat, and even within our own bodies. They play a critical role in various processes that provide energy and make life possible. As long as human beings exist, we will continue to interact with countless numbers of microorganisms on a daily basis.

For these reasons, microbiology is and will remain a critical discipline. The knowledge and ability to recognize, coexist with, and manipulate microorganisms is one of the great challenges currently confronting humanity. Although much microbiological research is conducted in laboratories, the subjects and consequences of this research affect our everyday lives. Indeed, many important news items involve microbiology: major epidemics such as AIDS and drug-resistant tuberculosis; the eradication of major diseases such as smallpox; bacteria as causative agents of ulcers; technologies designed to increase food production; genetic engineering; and the Human Genome Project, which will identify all the genetic material within the human body.

A central theme that permeates this entire book, then, is that microbiology is a current, relevant, exciting central science that affects us all. In countless areas, from agriculture to evolution, from ecology to dentistry, microbiology is both contributing to scientific knowledge and solving human problems. This was true in past centuries, when Louis Pasteur was developing the first modern vaccines and Anton van Leeuwenhoek was making simple microscopes and glimpsing the microbial world for the first time, and it remains true today. As you read this text, you will get a sense of the history of this science, its methodology, its many contributions to humanity, and the many ways in which it continues to be on the cutting edge of scientific advancement.

Style, Organization, and Currency

This book is written chiefly for students of the health sciences, but it contains more than enough information to meet the needs of students majoring in biology. We have designed the book to serve the needs of both audiences by using an abundance of clinically important information to amplify and illustrate a thorough treatment of the general principles of microbiology.

We have made this text truly "user-friendly," in the belief that students who enjoy a course retain far more of its content for a longer period of time than those who take the course like a dose of medicine. We want students to experience microbiology as an exciting, dynamic, rapidly changing field that is important to human welfare. The development of microbiology—from Leeuwenhoek's astonished observations of "animalcules," to Pasteur's first use of rabies vaccine on a human being, to Fleming's discovery of penicillin, to today's race to develop an AIDS vaccine—has been one of the most dramatic stories in the history of science. The growth of our knowledge in recent years has done nothing to diminish the drama; microbiology is as intriguing now as it has ever been. There is no reason for a book to be any less interesting than its subject.

This book has deliberately been written in a simple, straightforward, functional style. Our aim is to make information as accessible as possible to students, not to dazzle or intimidate them. Throughout, we have emphasized the connection between microbiological knowledge and the students' personal experiences and career goals. It is this aim, as much as the demands of the microbiology curriculum, that accounts for the special attention we have given to the *clinical aspects* of microbiology and to public health issues.

In a field that changes so quickly—with new research, new drugs, and (unfortunately) new diseases—it is essential that a text be as up to date as possible. We have tried to incorporate the latest information, not just on clinical practice, but on all aspects of microbiology. Special attention has been paid to such important, rapidly evolving topics as genetic engineering (Chapter 8), drug resistance, especially in tuberculosis (Chapter 14), and nosocomial infections (Chapter 16). Chapter 19 contains a newly expanded Essay on AIDS, and topics related to this disease are treated in several other chapters as well.

The organization of this text is designed to combine logic with flexibility. The chapter sequence will be useful in most microbiology courses as they are usually taught. Nevertheless, it is not essential that the chapters be assigned in their present order; thus, it should be possible to use this book in courses organized along quite different lines. The first part of the book (Chapters 1–4) provides the basic information—on the nature of the microbiology, on chemistry, on microscopy, and on cells—that underlies the rest of the course. The second part (Chapters 5–8) deals with the metabolism, growth, and genetics of microorganisms. Including two chapters on genetics allows for a thorough treatment of such important topics as mutation and genetic engineering, plus new techniques such as polymerase chain reaction (PCR). The third part (Chapters 9–12) comprises a survey of the major types of microorganisms, together with an introductory discussion of how they are classified.

The fourth part of the text (Chapters 13 and 14), devoted to control of microorganisms, includes material on physical control (sterilization and disinfection) and antimicrobial chemotherapy. The fifth part (Chapters 15–19) covers all aspects of the relationship between host and microorganism, including the disease process, epidemiology and nosocomial infections, nonspecific body defense mechanisms, and the immune system. Two chapters are allocated to immunology to permit a comprehensive discussion of immunization, immunologic tests, autoimmune disorders, and immunodeficiency diseases, including AIDS. The sixth part (Chapters 20–25) consists of a survey of infectious diseases, organized by the affected organ or system. Finally, the seventh part (Chapters 26 and 27) deals with environmental microbiology (including water pollution, water purification, and sewage treatment) and applied microbiology (including the microbiology of food).

The Art Program

Clear, attractive drawings and carefully chosen photographs can often contribute as much to the student's understanding of a scientific subject as the written text itself. We have been extremely fortunate once again to have the services of an extraordinary artist, William C. Ober, M.D., who, in collaboration with Claire Garrison, has executed many of the illustrations for this text. Bill and Claire combine technical skill with scientific understanding to a degree rarely found in even the best biological illustrators. In fact, they have been awarded nearly every important prize for biomedical illustration, including the Association of Medical Illustrators Award of Excellence, the American Institute of Graphic Arts Certificate of Excellence, and the Chicago Book Clinic award for art and design. The fact that all the drawings in this book are in full color has enabled us to make the maximum use of Bill and Claire's talents.

The use of color in the first edition was widely acclaimed by instructors and students. We have added many new color photos and line drawings to the second edition, responding particularly to reviewer and user comments requesting additions and changes. Throughout, color has been used, not just decoratively, but for its pedagogical value. For example, as in the first edition, similar molecules or cellular structures are colored the same way each time they appear, making them easier to recognize. The availability of color also makes it possible to present many subjects, such as staining reactions, as they are actually seen in the laboratory.

Special Features

The book includes three special features designed to broaden the scope of its coverage and maintain a high level of student interest and involvement with the material.

Each chapter contains a number of **Boxes** that deal with a wide variety of subjects. In the second edition, all boxes are placed in one of five categories, each of which is distinguished by an icon and a color banner.

Biotechnology Boxes report on recent developments in this exciting field, such as triple-stranded DNA, human hemoglobin produced by transgenic pigs, and the box reproduced on p. xxi on roach traps in which a fungus is the active agent.

Try It Boxes encourage students to make the transition from the textbook to their world through hands-on activities that enable them to experience the excitement of discovering things for themselves. Examples include culturing magnetotactic bacteria from local swamps and investigating which organisms grow in the cosmetic and health products found in students' own bathrooms.

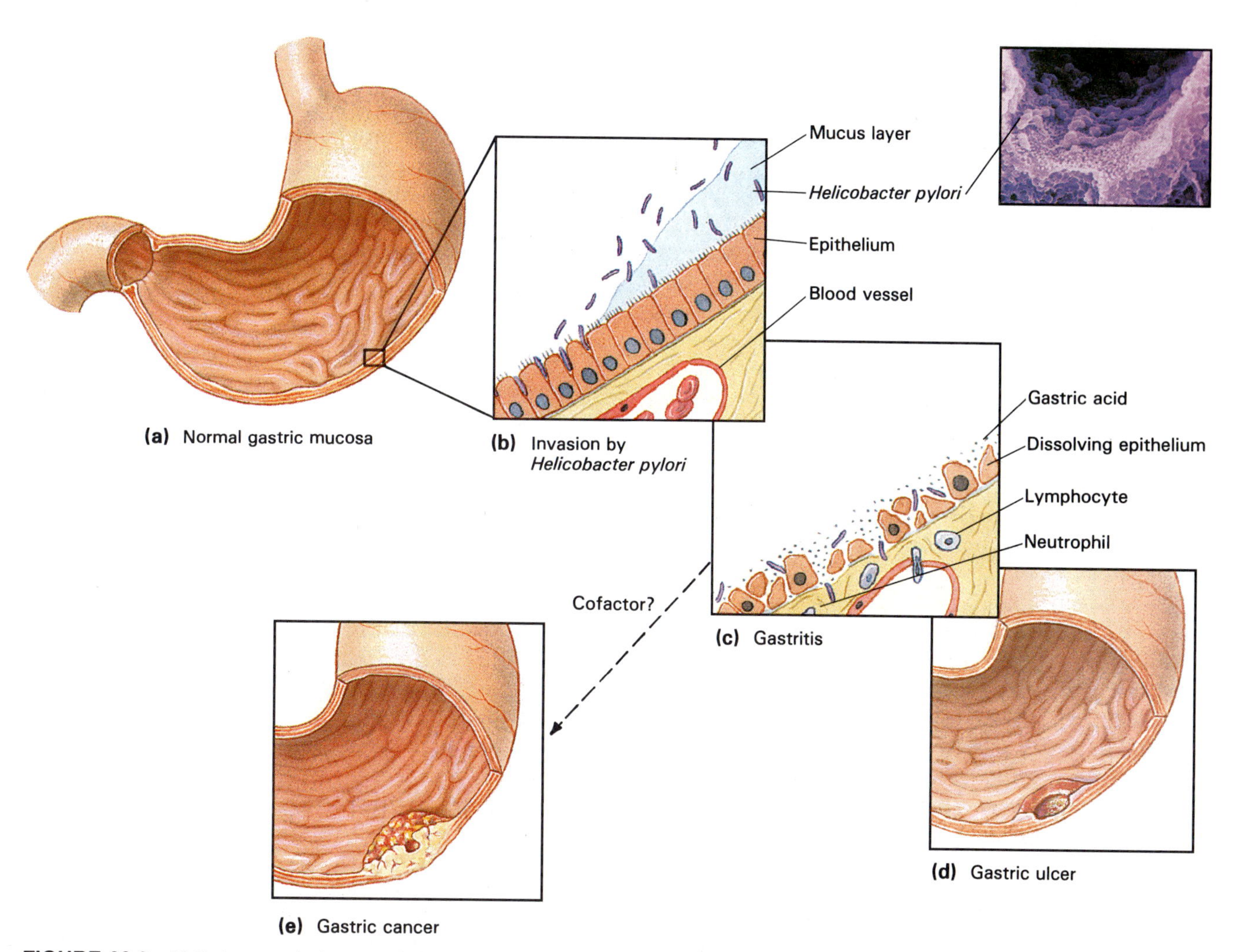

FIGURE 22.9 *Helicobacter pylori*, a spiral-shaped bacterium, has recently been recognized as the cause of peptic ulcers and chronic gastritis, and is involved in stomach cancer. Antibiotic treatment can lead to permanent cure of ulcers, so long as reinfection does not occur.

Public Health Boxes deal with issues such as the polio vaccine controversy, microbial contamination of foods, and HIV testing of patients and health-care providers.

Applications Boxes demonstrate the role of microbiology in the "real world." Many of these applications are clinical; others deal with processes such as fermentation, cleaning up toxic materials, and biodegrading junked autos.

Close-Up Boxes contain interesting anecdotes, historical background, quotations from distinguished writers on microbiology, and information about unusual subjects such as the recently discovered fungus that is one of the world's oldest and largest organisms.

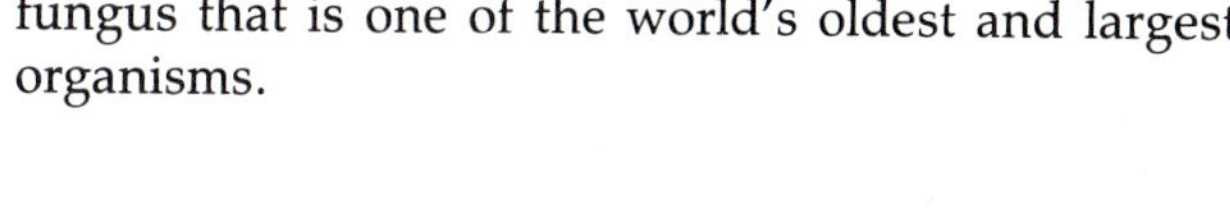

Roach Trap Uses Fungus

A company specializing in pest control using naturally occurring microbes has recently patented a baited roach trap lined with the soil fungus *Metarhizium anisopliae*. As roaches crawl through the trap the fungus sticks to their back and within 12 hours has penetrated the body wall, grown throughout the interior, and is digesting the tissue. A single roach can spread the fungus throughout a colony, as roaches groom each other by licking. The fungus attacks only roaches and is safer than chemical pesticides.

The text of each chapter ends with an **Essay,** a discussion of a supplemental topic that is both interesting in itself and relevant to the content of the chapter. Some instructors may elect to treat these essays as integral parts of the text, whereas others may prefer to assign them as extra reading or make them optional for their students. Many of the essays deal with clinical subjects or public health issues ("Performing Multiple Diagnostic Tests," "Cholera on Our Doorstep"); others develop chapter topics in greater detail ("*Yersinia pestis*"); still others focus on current research or unsolved problems in microbiology ("Mysterious Brain Infections"). Chapter 19 contains a special extended essay on AIDS that examines many aspects of the disease, including its epidemiology, the disease process and its clinical manifestations, HIV-less AIDS, the prospects for treatment and prevention, and the social, ethical, and economic problems associated with this disease.

The **Microbiologist's Notebooks** invite the student into the world of the working microbiologist. They are designed to provide insight into the backgrounds and motivations of typical microbiologists, as well as into the nature of their activities, the varied settings in which they work, the types of problems they investigate, and the experiences they have. Our subjects occupy very diverse careers in a range of microbiological fields and disciplines. They include nurses, zookeepers, mushroom growers, scientific researchers, physicians, and former President Jimmy Carter. These stories have been selected, not only because they are highly interesting in themselves, but also because they are timely and relevant to topics discussed in the text. For example, they examine the control of nosocomial infections and the use of genetically altered microorganisms. The Carter Foundation's Global 2000 Project to eliminate the guinea worm will see the eradication of this scourge during the lifetime of this edition.

The Endpaper Tables

We have made extensive use of summary tables in many of the chapters to convey large quantities of useful information in compact and easily accessible form. In addition, this text has a unique feature: six pages of **endpaper tables** listing virtually all the medically important microorganisms and diseases discussed in the text.

It is not uncommon for someone studying microbiology to remember the name of an infectious organism, but not the diseases associated with it; or to remember the name of a disease, but not that of the causative organism. Moreover, because of the way in which microbiology courses are organized, the same organism or disease may be discussed in several different chapters. Bubonic plague, for example, is mentioned in Chapter 1 (Scope and History of Microbiology), in Chapter 10 (Bacteria), in Chapter 14 (Antimicrobial Therapy), in Chapter 16 (Epidemiology and Nosocomial Infections), in Chapter 23 (Cardiovascular, Lymphatic, and Systemic Diseases), and in several other places as well.

The endpaper tables do two things:

- They reinforce the association between diseases and their causative organisms.
- They make it easy for a student to find the various sections of the text that deal with a particular organism or disease.

The front table lists all the major infectious diseases discussed in the text, their special features, the organisms that cause them, and the text pages on which each is discussed. The back table lists all the major organisms discussed in the text, classified by type (bacteria, viruses, etc.), with their distinguishing characteristics, the diseases that they produce, and the relevant text pages. These tables are a valuable learning tool in their own right, while also enhancing the usefulness of the text for students and instructors.

Pedagogical Apparatus

This book is designed, not simply as a vehicle for transmitting information, but also as a tool for learning. To that end, we have incorporated a set of coordinated pedagogical features.

- Each chapter begins with a list of **Focus Questions** that define the scope of the chapter and offer the student a preview of the major topics that will be covered.
- Within the text of each chapter, **important terms** are highlighted in boldface type.
- The **derivations** of terms are given whenever they are of special interest or will help students to understand and remember the term.
- Unfamiliar terms are accompanied by an easy-to-use **pronunciation guide,** based on *Stedman's Medical Dictionary*. (See "A Note on Pronunciation" at the end of this Preface.)
- At the end of each chapter, there is a concise **Chapter Summary.** Inclusion of the first- and second-level headings in the summary provides students with a quick overview of the chapter's structure, facilitating review.
- The summary also makes it easy to review the essential terminology introduced in the chapter. Terms that are boldfaced in the chapter are boldfaced when they appear in the summary. Those terms that are not included in the summary itself appear next to the relevant summary passage in the column headed **Related Key Terms.** Thus, all of a chapter's core vocab-

ulary is repeated in boldface at the end of the chapter, visible at a single glance for easy review.

- Definitions of boldfaced terms are assembled in the alphabetical **Glossary** at the end of the text.

At the end of each chapter are two sections of exercises and a selection of supplemental readings.

- The **Questions for Review**—more than 700 in all, an average of more than 26 per chapter—are grouped into sections that correspond to the Focus Questions at the start of each chapter. They test the student's mastery of the factual information presented in the chapter.
- The **Problems for Investigation** typically call for more extended answers, often in the form of an oral report or an essay. They demand more thought and generally require the synthesis of information from several parts of the chapter, from several chapters, or sometimes from sources outside the text. Some chapters contain questions that describe a series of symptoms and then ask the students to identify the disease and the causative agent. The answers to these questions are listed in Appendix E.
- The section called **Some Interesting Reading** lists books and articles that provide additional information about topics discussed in the chapter.

Five **Appendices** are provided to assist the student:

- a concise guide to the **metric system** and **scientific notation;**
- a summary of the standard **classification** of bacteria based on *Bergey's Manual*, plus a summary of the classification of viruses that affect humans and other vertebrates;
- a listing of **word roots** important for the student of microbiology, together with their meanings;
- guidelines for the safe collection and handling of **microbiological specimens;**
- answers to case study questions in the chapter-end Problems for Investigation.

Instructor's Edition

A specially augmented **Instructor's Edition** of this text is available to enhance the usefulness of the text as a teaching tool. For each chapter, the Instructor's Edition provides a wide range of supplemental material, designed (1) to assist the instructor in teaching the text, and (2) to allow for deeper and/or more extensive treatment of various topics should time and student interest permit.

Included in the material for each chapter are the following:

1. Chapter Overview The overview explains the organization of the chapter and summarizes its main themes. It is not a recapitulation of facts; rather, it is intended to place the material of the chapter into a broader context. Its purpose is to focus attention on the forest rather than on the trees.

2. Chapter Objectives These objectives are related to the Focus Questions at the start of each chapter, but they are more numerous and detailed. They are couched in behavioral terms, thus directing the instructor's attention to what students ought to be able to *do* after successfully completing each chapter, as well as what they should know. (This feature is also a convenience for instructors who must prepare a course syllabus with specific chapter-by-chapter goals.)

3. Chapter Outline This is a complete outline that includes all levels of headings. It also indicates the location of all Boxes, Figures, and Tables in the text for easy reference.

4. Instructional Suggestions A diversity of material is included in this extensive section:

- *Demonstrations* for class use;
- *Teaching Tips*, including suggestions for interesting examples and analogies, points to emphasize, questions and exercises for class use, effective ways to explain topics that students find especially difficult, and the like;
- *Discussion Topics* to stimulate student interest;
- *Library Assignments* suitable for student papers or oral reports;
- *Laboratory Correlations*, a list of those exercises from the Laboratory Manual that are relevant to the chapter material.

5. Review This section contains complete and detailed answers to all end-of-chapter Questions for Review and Problems for Investigation.

Supplements

To supplement both student and instructor needs, the following materials are available to qualified adopters.

For the Instructor

- **Instructor's Edition** with the Instructor's Manual bound in as an integral part of the book, preceding the main text. Written by William Matthai of Tarrant County Junior College, the Instructor's Manual includes outlines, objectives, suggestions, tips, activities, and answers to all test questions.
- **Prentice Hall Microbiology Laser Disc** contains more than 2000 images, including full-color micrographs, photographs, illustrations, and animations

for use in either a lecture or lab setting. All images are indexed and accessible with or without a bar code scanner.

- **Test Item File,** authored by Denise Friedman of Hudson Valley Community College, contains more than 1000 questions, all referenced to chapter number and section.
- **Prentice Hall Test Manager** allows you both to edit and add test questions to the Test Item File and assemble and save tests both manually and randomly. Additionally, two test-scrambling options combine to give you a virtually unlimited number of versions of your tests. Available in both IBM and Apple versions.
- **Prentice Hall Grade Manager** is an electronic gradebook that merges flexibility, power, and ease of use. With it you can easily maintain and update class records, compute class statistics, print graphs, average grades, and sort by student name or grade. Available in IBM version.
- **Telephone Testing Service** allows you to select questions from the Test Item File, call a toll-free number, and have the test prepared in-house at Prentice Hall with no additional charge. Within 48 hours, a professionally prepared test will arrive at your school, along with an answer key and answer sheets for your students. Two versions of a test can be furnished.
- **Instructor's Edition to the Laboratory Manual** with answers to all the exercises.
- **Transparency Pack** with 150 full-color acetates; or, 150 **full-color slides** from the book.

For the Student

- ***The New York Times* Themes of the Times** consists of selected articles from *The New York Times* dealing with topics related to microbiology. This supplement is updated annually and is available free to adopters, who can order as many copies as the number of new texts that are purchased.
- **Laboratory Manual** to accompany the text contains practically every exercise that might be used in an intro lab (close to 30) and provides such extensive, clearly written background material and helpful pedagogical aids for each chapter that students do not have to consult outside sources during their lab work.
- **Study Guide,** by William Matthai, with chapter overviews, self-tests, key terms, and case studies.
- **Interactive Glossary,** which drills students on vocabulary and word roots covered in the microbiology course. Software is available on 3.5″ and 5.25″ disks for IBM computers.
- **Pronunciation Tapes** feature a review of key terms with their pronunciation. These tapes also drill students on definitions of key terms.

A Note on Pronunciation

The scheme used for pronunciations is simple:

′ is used for the main accent in a word;

″ is used for the secondary accent, if any;

any vowel not followed by a consonant is assumed to be long;

any vowel followed by a consonant is assumed to be short unless it has a macron (bar) over it, in which case it is long;

syllables are separated by either a hyphen or an accent mark.

Acknowledgments

Many generous and talented people helped in the preparation of this text, and I would like to express my appreciation for their efforts. First, I must thank all those persons who helped to make the first edition such a great success. Then to Tim Bozik, Prentice Hall's Editor in Chief for math and science, many thanks for his patient support and encouragement. David Brake, Biology Editor, expertly guided this book through stormy and calm seas, always seeking excellence. I also owe much to Ray Mullaney, Editor in Chief of College Book Editorial Development, for his commitment to helping create the best book possible. I am deeply grateful for the time and effort invested in the refinement of this text by Robert Weiss.

Designer Lee Goldstein worked swiftly and with an enthusiasm that was contagious in designing and dummying the text. Florence Silverman, Design Director, also worked on the design, and I am thankful for the lovely results.

I acknowledge a debt of gratitude to managing editor Jeanne Hoeting for smoothing the inevitable stresses and strains of a long and intricate process, and above all to production editor Barbara DeVries, whose diligence, expertise, keen eye, and prodigious capacity for hard work under sometimes difficult conditions did so much to pilot this book to safe harbor.

Art and photographs are an integral part of this text. We are thus extremely grateful for the efforts of Yvonne Gerin and Lori Morris-Nantz for resourcefulness and perseverance in helping to procure the best photos available. Personnel at the Centers for Disease Control were especially helpful in opening their collection of slides to the author. And once again I must mention Bill Ober; there is no way I can adequately express my admiration for his work or thank him sufficiently for his contribution.

I would also like to express my gratitude to the people who shared their experiences with us in the Essay and Microbiologist's Notebook features, particularly Jim Angelucci of Phillips Mushroom Farms, as well as personnel at the Carter Foundation.

I appreciate also the contributions of Denise Friedman, Bobbie Pettreiss, and Warren Silver in the late stages of manuscript preparation. Their keen eyes caught many things. To my student assistant Liane Benton, who faithfully helped in so many ways, heartfelt thanks. In addition, I take pleasure in thanking Bill Matthai, not only for being a most helpful reviewer of the entire manuscript, but for his invaluable work on the Instructor's Edition of this text.

This project has consumed large amounts of time and energy over the past three years and has required patience and understanding on the part of family members. Special thanks, therefore, to Laura Black for having again shared her mother with "the book" during this further part of her childhood. Finally, the author gratefully acknowledges the support of Marymount University, and especially of Provost Alice S. Mandanis and Dean Robert A. Draghi.

I wish to acknowledge the Prentice Hall reviewers for their thoughtful suggestions.

First edition reviewers:

Kimberley Pearlstein, *Adelphi University*
Donald G. Lehman, *Wright State University*
William C. Matthai, *Tarrant County Junior College*
Dennis J. Russell, *Seattle Pacific University*
Gordon D. Schrank, *St. Cloud State University*
Lawrence W. Hinck, *Arkansas State University*
Raymond B. Otero, *Eastern Kentucky University*
Robert A. Pollack, *Nassau Community College*
Alan J. Sexstone, *West Virginia University*
Deborah Simon-Eaton, *Santa Fe Community College*
Robert E. Sjogren, *University of Vermont*
Keith Bancroft, *Southeastern Louisiana University*
Michael R. Yeaman, *University of New Mexico*
Oswald G. Baca, *University of New Mexico*
David L. Balkwill, *Florida State University*
Wallis L. Jones, *De Kalb College*
Monica A. Devanas, *Rutgers University*
Thomas R. Corner, *Michigan State University*
Larry Stearns, *Central Piedmont Community College*

Second edition reviewers:

D. Andy Anderson, *Utah State University*
Dan C. DeBorde, *University of Montana*
Monica A. Devanas, *Rutgers University*
David L. Filmer, *Purdue University*
Denise Y. Friedman, *Hudson Valley Community College*
Eugene Flaumenhaft, *University of Akron*
William R. Gibbons, *South Dakota State University*
Ronald E. Hurlbert, *Washington State University*
Robert J. Janssen, *University of Arizona*
Thomas R. Jewell, *University of Wisconsin-Eau Claire*
Harvey Liftin, *Broward Community College*
Russell A. Normand, *Northeast Louisiana University*
Joseph M. Sobek, *University of Southwestern Louisiana*
Bernice C. Stewart, *Prince George's Community College*
James Urban, *Kansas State University*
Von Dunn, *Tarrant County Junior College*
John C. Zak, *Texas Tech University*
Thomas E. Zettle, *Illinois Central College*

Comments and suggestions about the book are most welcome.

JACQUELYN BLACK
Arlington, Virginia

Anton van Leeuwenhoek exhibiting his microscopes for Queen Catherine of England. The unseen world of microbes is opened to us all now.

1 Scope and History of Microbiology

This chapter focuses on the following questions:

A. Why is the study of microbiology important?

B. What is the scope of microbiology?

C. What are some major events in the early history of microbiology?

D. What is the germ theory of disease, and what historical developments led to its formulation?

E. What events mark the emergence of immunology, virology, chemotherapy, genetics, and molecular biology as branches of microbiology?

It's just some 'bug' going around." You have heard that from others or said it yourself when you have been ill for a day or two. Indeed, the little unidentified illnesses we all have from time to time and attribute to a "bug" are probably caused by viruses, the tiniest of all microbes. Other groups of **microorganisms**—bacteria, fungi, and protozoa—also have disease-causing members. Before studying microbiology, therefore, we are likely to think of microbes as germs that cause disease. Health scientists are concerned with just such microbes and with treating and preventing the diseases they cause. Yet less than 1 percent of known microorganisms cause disease, so focusing our study of microbes exclusively on disease gives us too narrow a view of microbiology.

BIOTECHNOLOGY

We Are Not Alone

"We are outnumbered. The average human contains about ten trillion cells. On that average human are about 10 times as many microorganisms, or 100 trillion microscopic beings. . . . As long as they stay in balance and where they belong, [they] do us no harm. . . . In fact, many of them provide some important services to us. [But] most are opportunists, who if given the opportunity of increasing growth or invading new territory, will cause infection."

—Robert J. Sullivan, 1989

WHY STUDY MICROBIOLOGY?

If you were to dust your desk and shake your dust cloth over the surface of a medium designed for growing microorganisms, after a day or so you would find a variety of organisms growing on that medium. If you were to cough onto such a medium or make fingerprints on it, you would later find a different assortment of microorganisms growing on the medium. When you have a sore throat and your physician orders a throat culture, a variety of organisms will be present in the culture—perhaps including the one that is causing your sore throat. Thus, microorganisms have a close association with humans. They are in us, on us, and nearly everywhere about us (Figure 1.1). One reason for studying microbiology is that *microorganisms are part of the human environment and are therefore important to human health.*

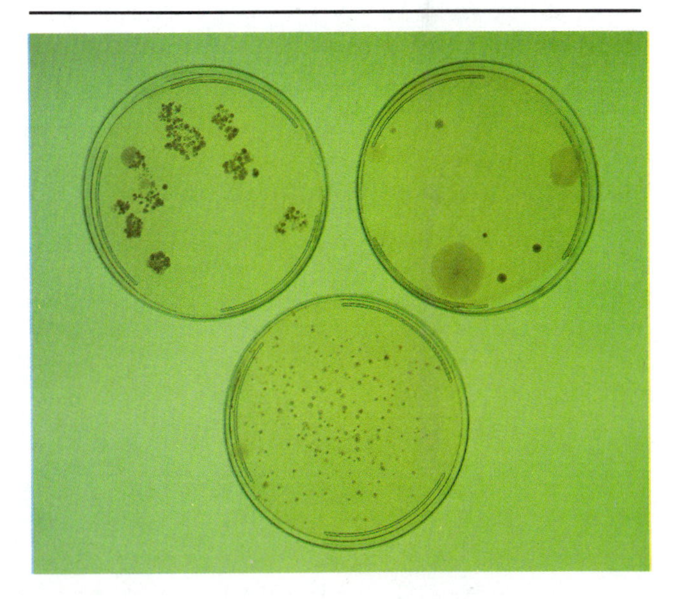

FIGURE 1.1 A simple experiment shows that microorganisms are almost everywhere in our environment. Nutrient agar, a culture medium, was exposed to the air (dish at upper right), touched by fingers (upper left), and coughed on (bottom). After 3 days of incubation under favorable conditions, abundent microbial growth is easily visible in all three dishes.

Microorganisms are essential to the web of life in every environment. Many microorganisms in the ocean and in bodies of fresh water capture energy from light and store it in molecules that other organisms use as food. They decompose dead organisms and waste material from living organisms, and they can decompose some kinds of industrial wastes. They make nitrogen available to plants. These are but a few of many examples of how microorganisms interact with other organisms and help to maintain the balance of nature. The vast majority of microorganisms are directly or indirectly beneficial, not only to other organisms, but also to humans. They form essential links in many food chains that produce plants and animals that humans eat. Aquatic microbes serve as food for small macroscopic animals that, in turn, serve as food for fish and shellfish that humans eat. Certain microorganisms live in the digestive tracts of grazing animals such as cattle and sheep and aid in their digestive processes. Without these microbes, cows could not digest grass, and horses would get no nourishment from hay. Humans occasionally eat microbes, such as some algae, directly. Biochemical reactions carried out by microbes also are used by the food industry to make pickles, sauerkraut, yogurt and other dairy products, fructose used in soft drinks, and the artificial sweetener aspartame. Fermentation reactions in microorganisms are used in the brewing industry to make beer and wine.

One of the most significant benefits of microorganisms for humans is their ability to synthesize *antibiotics,* substances derived from one microorganism that kill or restrict the growth of other microorganisms. In a sense, therefore, microorganisms cure diseases as well as cause them. Finally, microorganisms are the major tools of genetic engineering. Several products important to humans, such as interferon and growth hormones, can now be produced economically by microbes because of genetic engineering.

While only a few microbes cause disease, learning how such diseases are transmitted and how to diagnose, treat, and prevent them is of great importance in a health-science career. Such knowledge will help you care for patients and avoid becoming infected yourself.

Another reason for studying microbiology is that such study *provides insight into life processes in all life forms.* Biologists in many different disciplines use ideas from microbiology and use the organisms themselves. Ecologists draw on principles of microbiology to understand how matter is decomposed and made available for continuous recycling. Biochemists use microbes to study metabolic pathways—sequences of chemical reactions in living organisms. Geneticists use microbes to study how hereditary information is transferred and how such information controls the structure and functions of organisms.

Microorganisms are especially useful in research for at least three reasons:

1. Compared to other organisms, microorganisms have relatively simple structures. It is easier to study most life processes in simple unicellular organisms than in complex multicellular organisms.
2. Large numbers of microorganisms can be used in an experiment to obtain reliable results at reasonable costs. Growing a billion bacteria costs less than maintaining 10 rats. Results from experiments with large numbers of microorganisms are more reliable than are those with small numbers of organisms with individual variations.
3. Because microorganisms reproduce quickly, they are especially suitable for studies involving transmission of genetic information. Some bacteria can undergo three divisions in an hour, so the effects of genetic transmission can be followed through many generations quickly.

Through the study of microbes scientists already have achieved remarkable success in understanding life processes and disease control. For example, within the last quarter century, vaccines have nearly eradicated several dreaded childhood diseases—including measles, German measles, and mumps—in developed countries. Smallpox, which once accounted for 10 percent of all deaths in Europe, has not been reported anywhere on earth since 1978. Much also has been learned about genetic changes that lead to antibiotic resistance and how to manipulate genetic information in bacteria. Much more remains to be learned. For example, how can vaccines be made available on a worldwide basis? How can the development of new antibiotics keep pace with genetic changes in microorganisms? How will increased jet-age world travel affect the spread of infections? Can a vaccine or an effective treatment for AIDS be made available? Therein lie the challenges for the next generation of biologists and health scientists.

SCOPE OF MICROBIOLOGY

Microbiology is the study of **microbes,** very small organisms, so small that a microscope is needed to study them. We consider two dimensions of the scope of microbiology: (1) the variety of kinds of microbes and (2) the kinds of work microbiologists do.

The Microbes

The major groups of organisms studied in microbiology are bacteria, algae, fungi, viruses, and protozoa (Figure 1.2). All are widely distributed in nature. For example, a recent study of bee bread (a pollen-derived nutrient eaten by worker bees) showed it to contain 188 kinds of fungi and 29 kinds of bacteria. Most microbes consist of a single cell. (Cells are the basic units of structure and function in living things; they are discussed in Chapter 4.) Viruses, tiny acellular entities on the borderline between living and nonliving, behave like living organisms when they gain entry to cells. They, too, are studied in microbiology. Microbes range in size from small viruses 15 nm in diameter to large protozoans 5 mm or more in diameter. In other words, the largest microbes are as much as 100,000 times as large as the smallest ones! (Refer to Appendix A for a review of metric units.)

Among the great variety of microorganisms that have been identified, bacteria probably have been the most thoroughly studied. **Bacteria** (singular: *bacterium*) are usually single-celled organisms with spherical, rod, or spiral shapes, but a few form filaments. Most are so small they can be seen with a light microscope only under the highest magnification. Although bacteria are cellular, they do not have a nucleus and they lack the membrane-bound intracellular structures found in most other cells. Many bacteria absorb nutrients from their environment, but some make their own nutrients by photosynthesis or other synthetic processes. Some are stationary, and others move about. Bacteria are widely distributed in nature, for example, in aquatic environments and in decaying matter. They occasionally cause diseases.

In contrast to bacteria, several groups of microorganisms consist of larger, more complex cells that have a nucleus. They include algae, fungi, and protozoa, all of which can be seen easily with a light microscope.

Many **algae** (al'je) are single-celled microscopic organisms, but some marine algae are large, relatively complex, multicellular organisms. Unlike bacteria, al-

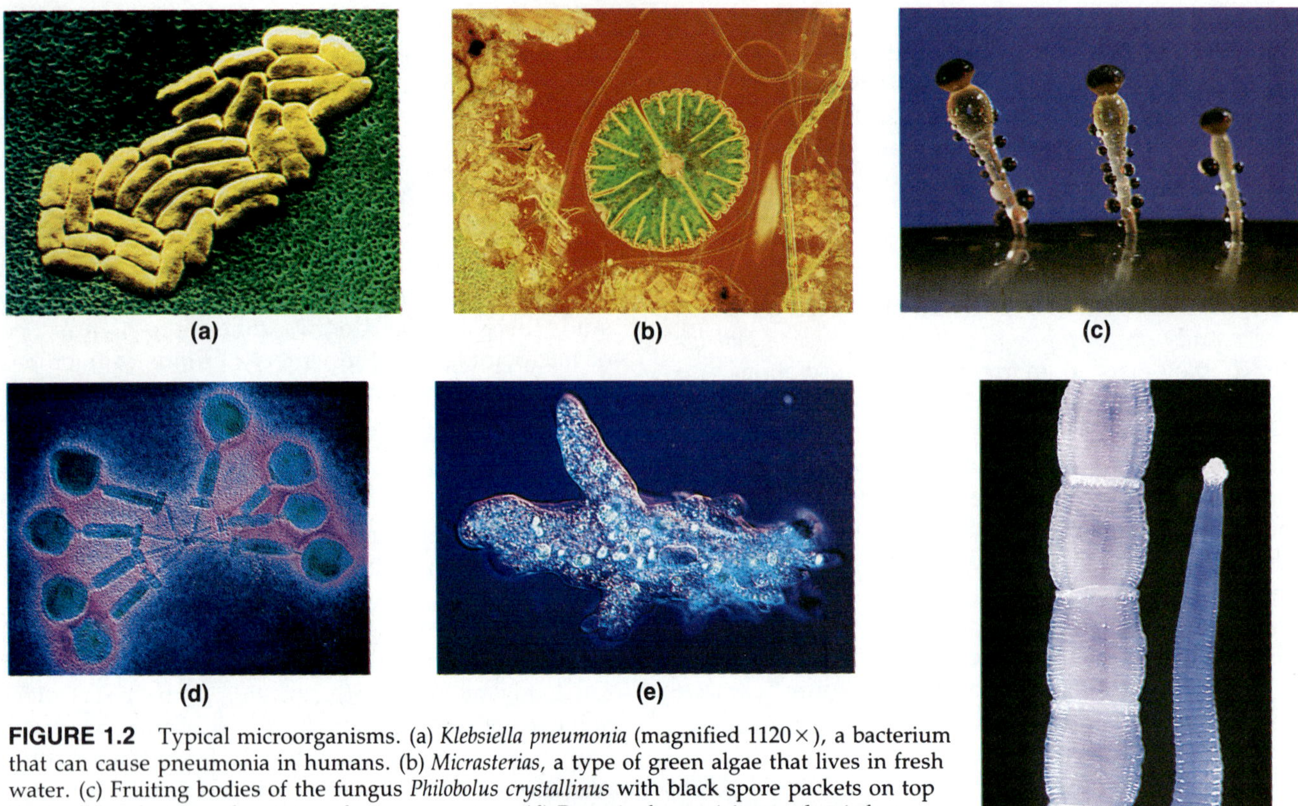

FIGURE 1.2 Typical microorganisms. (a) *Klebsiella pneumonia* (magnified 1120×), a bacterium that can cause pneumonia in humans. (b) *Micrasterias*, a type of green algae that lives in fresh water. (c) Fruiting bodies of the fungus *Philobolus crystallinus* with black spore packets on top that will be shot into the air to colonize new areas. (d) Bacteriophages (viruses that infect bacteria). (e) *Amoeba*, a protozoan. (f) Portions of the tapeworm *Taenia*. Its head and younger segments are at the top of the right portion, while the left portion shows older mature segments.

gae (singular: *alga*) have a clearly defined nucleus and numerous membrane-bound intracellular structures. All algae photosynthesize their own food as plants do, and many can move about. Algae are widely distributed in both fresh water and oceans. Because they are so numerous and because they capture energy from sunlight in the food they make, algae are an important source of food for other organisms. Algae are of little medical importance; only one species has been found to cause disease in humans.

Like algae, many **fungi** (fun'ji), such as yeasts and some molds, are single-celled microscopic organisms. Some, such as mushrooms, are multicellular, macroscopic organisms. Fungi (singular: *fungus*) also have a nucleus and intracellular structures. All fungi absorb ready-made nutrients from their environment. Fungi sometimes form extensive networks of branching filaments, but the organisms themselves generally do not move. Fungi are widely distributed in water and soil as decomposers of dead organisms. Some also are important in medicine either as agents of disease or as sources of antibiotics.

Viruses are acellular entities too small to be seen with a light microscope. They are relatively simple structures composed of specific chemical substances—a nucleic acid and a few proteins (Chapter 2). Indeed, some viruses are so simple that they can be crystallized and stored in a container, but they retain the capacity to invade cells. Viruses replicate themselves and display other properties of living organisms only when they have invaded cells. Many viruses can invade human cells and cause disease.

Protozoa (pro-to-zo'ah) also are single-celled, microscopic organisms with at least one nucleus and numerous intracellular structures. A few species of amoebae are large enough to be seen with the naked eye, but we can study their structure only with a microscope. Many protozoa (singular: *protozoan*) obtain food by engulfing or ingesting smaller microorganisms. Most protozoa can move, but a few, especially those that cause human disease, cannot. Protozoa are found in a variety of water and soil environments.

In addition to organisms properly in the domain of microbiology, in this text we consider some macroscopic *helminths* (worms) and *arthropods* (insects and similar organisms). The helminths have microscopic stages in their life cycles that can cause disease, and the arthropods transmit these microscopic forms.

We will learn more about the classification of microorganisms in Chapter 9. For now it is important to know only that cellular organisms are referred to by

TABLE 1.1 Reportable disease caused by microorganisms and parasites[a]

Bacterial Diseases	Bacterial Diseases (cont.)	Viral Diseases	Algal Diseases
Anthrax	Plague	AIDS (symptomatic cases)	None
Asiatic cholera	Psittacosis	Arbovirus infection	
Bacillary dysentery	Q fever	Aseptic meningitis	**Fungal Diseases**
Bacterial meningitis	Relapsing fever	Chickenpox	None
Botulism	Rheumatic fever	Encephalitis	
Brucellosis	Rocky Mountain spotted fever	Hepatitis A	**Protozoan Diseases**
Chancroid	*Salmonella* infections (exclusive of typhoid fever)	Hepatitis B	Amebiasis
Diphtheria	Scarlet fever	Hepatitis C	Intestinal parasites
Food poisoning	*Shigella* infections	Hepatitis, unspecified	Malaria
Gonorrhea	Syphilis	Influenza	
Granuloma inguinale	Tetanus	Measles (rubeola)	**Helminth Diseases**
Legionnaire's disease	Toxic shock syndrome	Mumps	Intestinal parasites
Leprosy	Trachoma	Poliomyelitis	Trichinosis
Leptospirosis	Tuberculosis	Rabies (in animals)	
Lymphogranuloma venereum	Tularemia	Rubella	
Meningitis	Typyhoid fever	Smallpox	
Paratyphoid fever	Typhus fever	Yellow fever	
Pertussis			

[a] Reportable means the diseases are recommended to be reported to the U.S. Centers for Disease Control.

two names, their genus and species names. For example, a bacterium commonly found in the human gut is called *Escherichia coli,* and a protozoan sometimes responsible for severe diarrhea is called *Giardia intestinalis.* The naming of viruses is less precise. Some viruses, such as herpesviruses, are named for the group to which they belong, and others, such as polioviruses, are named for the disease they cause.

Disease-causing organisms and the human diseases they cause are discussed in detail in Chapters 20–25. Hundreds of infectious diseases are known to medicine. Some of the most important—those diseases that physicians should report to the U.S. Centers for Disease Control (CDC)—are listed in Table 1.1 according to the kind of causative organism. The CDC is a federal agency concerned with collecting information about diseases and developing ways to control them.

The Microbiologists

Microbiologists study many kinds of problems that involve microbes. Some study microbes mainly to find out more about a particular kind of organism—the life stages of a particular fungus, for example. Others are interested in a particular kind of function, such as the metabolism of a certain sugar or the action of a specific gene. Still others focus directly on practical problems, such as how to purify or synthesize a new antibiotic or how to make a vaccine against a particular disease. Quite often the findings from one project are useful in another, as when agricultural scientists use information from microbiologists to control pests and improve crop yields, or when environmentalists attempt to maintain natural food chains and prevent damage to the environment. Some fields of microbiology are described in Table 1.2.

Microbiologists work in a variety of settings (Figure 1.3). Some work in universities, where they are likely to spend some time teaching, some time doing research, and some time teaching students to do research. Microbiologists in both university and commercial laboratories are helping to develop the microorganisms used in genetic engineering. Law firms are beginning to hire microbiologists to help with the complexities of patenting new genetically engineered organisms. Many microbiologists work in health-related positions. Some work in clinical laboratories, performing tests to diagnose diseases or determining which antibiotics will cure a particular disease. A few develop new clinical tests. Others work in industrial laboratories to develop or manufacture antibiotics, vaccines, and similar biological products. Still others, concerned with controlling the spread of infections and related public health matters, work in hospitals and government labs.

From the point of view of health scientists, today's research is the source of tomorrow's new technologies. Research in *immunology* is greatly increasing our knowledge of how microbes elicit host responses and how the microbes escape these responses. It also is contributing to the development of new vaccines and

TABLE 1.2 Fields of microbiology

Field (pronunciation)	What Is Studied
Microbial taxonomy	Classification of microorganisms
Fields according to organisms studied	
Bacteriology (bak"te-re-ol'o-je)	Bacteria
Phycology (fi-kol'o-je)	Algae (phyco, seaweed)
Mycology (mi-kol'o-je)	Fungi (myco, a fungus)
Protozoology (pro"to-zo-ol'o-je)	Protozoa (proto, first; zoo, animal)
Virology (vir-ol'o-je)	Viruses
Parasitology (par"ah-si-tol'o-je)	Parasites
Fields according to processes or functions studied	
Microbial metabolism	Chemical reactions that occur in microbes
Microbial genetics	Transmission and action of genetic information in microorganisms
Microbial ecology	Relationships of microbes with each other and with the environment
Health-related fields	
Immunology (im"mu-nol'o-je)	How host organisms defend themselves against infection by microorganisms
Epidemiology (ep-e-dem-e-ol'o-je)	Frequency and distribution of diseases
Etiology (e-te-ol'o-je)	Causes of disease
Infection control	How to control the spread of nosocomial (no-so-kom'e-al), or hospital-acquired, infections
Chemotherapy	The development and use of chemical substances to treat diseases
Fields according to applications of knowledge	
Food technology	How to protect humans from disease organisms in fresh and preserved foods
Environmental microbiology	How to maintain safe drinking water, dispose of wastes, and control environmental pollution
Industrial microbiology	How to apply knowledge of microorganisms to the manufacture of fermented foods and other products of microorganisms
Pharmaceutical microbiology	How to manufacture antibiotics, vaccines, and other health products
Genetic engineering	How to use microorganisms to synthesize products useful to humans

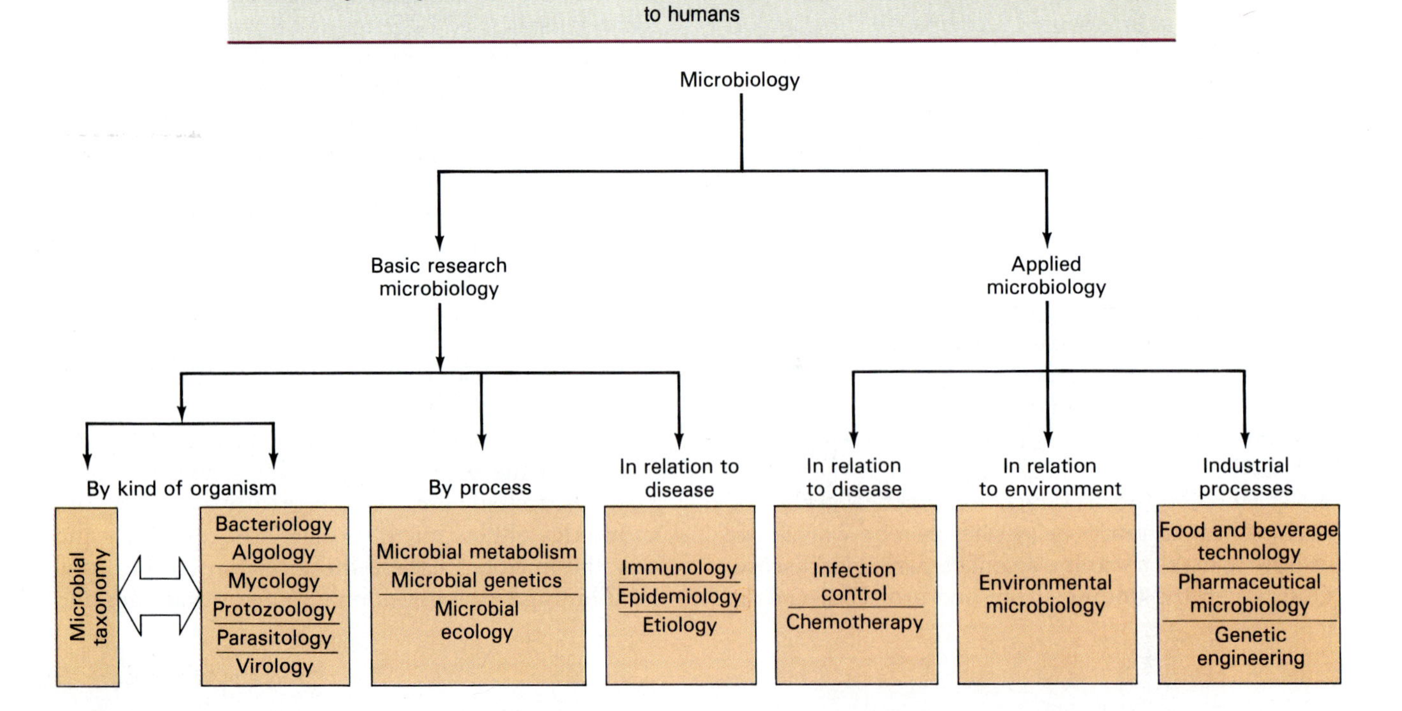

(a) (b) (c)

(d)

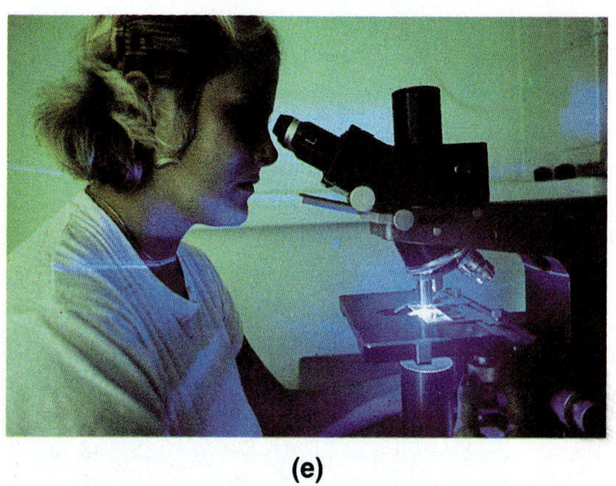

(e)

(f)

FIGURE 1.3 Microbiology is used in careers as diverse as (a) using genetically engineered bacteria to investigate how diet influences risk of developing cancer; (b) inspecting plastics made with as much as 40 percent starch (pieces inside baskets) for signs that aquatic microbes are degrading them; (c) assessing the effectiveness of aerial spraying of a virus deadly to cotton bollworm and the tobacco budworm along field edges, borders, and treelines in Mississippi crop lands; (d) using beating nets to survey for ticks that can spread disease to livestock and humans; (e) examining specimens from a U.S. quarantine station to prevent pathogens from entering the country; (f) keeping our pets and domestic animals healthy, as well as improving their productivity, by means of advances in veterinary science.

to the treatment of immunologic disorders. Research in *virology* is improving our understanding of how viruses cause infections and how they are involved in cancer. Research in *chemotherapy* is increasing the number of drugs available to treat infections and is also improving our knowledge of how these drugs work. Finally, research in genetics is providing new information about the transfer of genetic information and, especially, about how genetic information acts at the molecular level.

HISTORICAL ROOTS

Many of the ancient Mosaic laws found in the Bible about basic sanitation have been used through the centuries and still contribute to our practices of preventive medicine. In Deuteronomy, Chapter 13, Moses instructed the soldiers to carry spades and bury solid waste matter. The Bible also refers to leprosy and to the isolation of lepers. Though in those days the term "leprosy" probably included other infectious and noninfectious diseases, isolation did limit the spread of the infectious diseases.

The Greeks anticipated microbiology as they did so many things. The Greek physician Hippocrates, who lived around 400 B.C., set forth ethical standards for the practice of medicine that are still in use today. Hippocrates was not only wise in human relations, he was also a shrewd observer. He associated particular signs and symptoms with certain illnesses and realized that diseases could be transmitted from one person to another by clothing or other objects. At about the same time the Greek historian Thucydides observed that people who had recovered from the plague could take care of plague victims without danger of getting the disease again.

The Romans also contributed to microbiology as early as the first century B.C. The scholar and writer Varro proposed that tiny invisible animals entered the body through the mouth and nose to cause disease. Lucretius, a philosophical poet, mentioned "seeds" of disease in his major work, *De Rerum Natura* (*On the Nature of Things*).

Bubonic plague, also called the Black Death, appeared in the Mediterranean region around A.D. 542, where it reached epidemic proportions and killed millions. In 1347 the plague invaded Europe by way of the caravan routes and sea lanes from central Asia, affecting Italy first, then France, England, and finally northern Europe. Though no accurate records were kept at that time, it is estimated that tens of millions of people in Europe died during this and successive waves of plague over the next 300 years. The Black Death was a great leveler—it killed rich and poor alike (Figure 1.4). The wealthy fled to isolated summer homes but carried plague-infected fleas with them in unwashed hair and clothing. One group that escaped devastation from the plague was the Jewish population. Jewish laws regarding sanitation offered some protection to those who practiced them. The relatively clean ghettos harbored fewer rats to spread the disease. When Jews did fall ill, they were carefully nursed and treated with herbal remedies rather than by strenuous purging or excessive bleedings with dirty instruments. As a result, a smaller proportion of Jews than gentiles died of the disease. Ironically, some gentiles regarded the Jews' higher survival rates as proof that Jews were the source of the epidemic.

In his *Diary*, the English writer Samuel Pepys gave a vivid, firsthand account of the plague in London in the 1660s.

FIGURE 1.4 A portion of "The Triumph of Death" by Pieter Brueghel the Elder. The picture, painted in the mid-sixteenth century, a time when outbreaks of plague were still common in many parts of Europe, dramatizes the swiftness and inescapability of death for people of all social and economic classes.

PUBLIC HEALTH

AIDS

Suppose several close relatives, two of your best friends, and many of your neighbors are suffering from painful illnesses and will soon die. There are no available beds at the hospital, and most of the doctors and nurses have quit work because they fear becoming infected. The local television station has started broadcasting the latest figures on deaths and new outbreaks on a daily basis, like the stock market prices or the weather report. Nearly every time you go to the shopping center, you meet a funeral procession.

Over the centuries many people have found themselves in analogous situations. Again and again, large proportions of the human population have been devastated by infectious diseases such as typhus, smallpox, and bubonic plague. In the mid-fourteenth century, plague alone wiped out 25 million people—one-fourth of the population of Europe and neighboring regions—in just 5 years. (The opening scenario is an imaginary version of a similarly devastating epidemic in modern times.)

We who live in technologically advanced countries tend to think that outbreaks of this sort are a thing of the past. Yet following World War I, within the memory of many people now living, worldwide outbreaks of influenza claimed 20 million lives. And now, acquired immune deficiency syndrome (AIDS) threatens to kill great numbers of people after they have suffered a long and painful illness. Is AIDS in any way comparable to the great killer diseases of the past? More alarmingly, does its appearance mean that all of our "triumphs" over infectious disease were an illusion? Will the past few decades, during which vaccines and antibiotics have largely kept contagious disease in check and even have wiped out certain ancient curses such as smallpox, prove to have been just a brief and atypical episode in an endless war that can't be won?

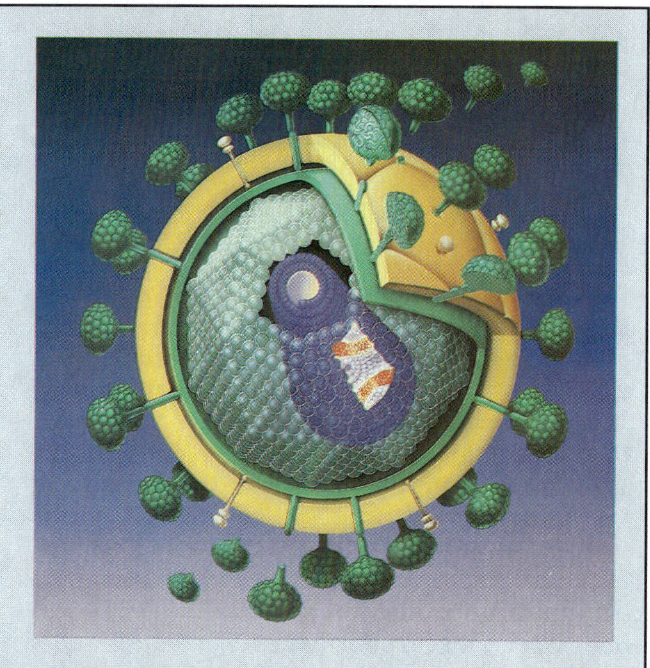

A model of the AIDS virus.

The AIDS quilt, a memorial to victims of this disease, on display in Washington, D.C.

Despite the fears AIDS has aroused, there are many differences between AIDS and the epidemic diseases mentioned above. For one thing, AIDS is not nearly so easily communicable as the epidemic killers of the past. It is not spread by casual contact but largely by certain behaviors, most of which people can learn to avoid (see Chapter 19).

More important, however, is the state of our knowledge. During past epidemics, people were terrified and demoralized because they had no idea of how the disease was caused and spread or what they could do to fight it. Today, we understand far more about the nature of the enemy. The study of AIDS has occupied virologists, chemotherapists, and immunologists from among the world's most talented microbiologists. These scientists have shown that the virus infects cells of the body's immune system, altering genetic information so that instead of fighting the infection, infected cells make viruses and then die. They have determined the precise structure of many of the components of the virus, and they have learned how it attaches itself to its target cells. They have developed some drugs that are being tested to treat AIDS, and they are working on a vaccine. In spite of these efforts, however, AIDS remains one of the most threatening infectious diseases and one of the greatest challenges microbiologists have ever faced.

> The streets mighty empty all the way now even in London, which is a sad sight. . . . Poor Will, that used to sell us ale, . . . his wife and three children died, all I think in a day. . . . home to draw over anew my will, which I had bound myself by oath to dispatch by tomorrow night, the town growing so unhealthy that a man cannot depend upon living two days to an end. In the City died this week 7,496, and of them 6,102 of the plague. But it is feared that the true number of the dead is near 10,000; partly from the poor that cannot be taken notice of through the greatness of the number. . . . I saw a dead corps in a coffin lie in the Close unburied; and a watch is constantly kept there night and day to keep the people in, the plague making us cruel as doggs one to another.

Until the seventeenth century, the advance of microbiology was hampered by the lack of appropriate tools to observe microbes. Around 1665, the English scientist Robert Hooke built a compound microscope (one in which light passes through two lenses) and used it to observe thin slices of cork. He coined the term *cell* to describe the orderly arrangement of small boxes that he saw because they reminded him of the cells (small, bare rooms) of monks. However, it was Anton van Leeuwenhoek (Figure 1.5), a Dutch clothes merchant and amateur lens grinder, who first made and used lenses to observe living microorganisms.

FIGURE 1.5 Anton van Leeuwenhoek (1632–1723), shown holding one of his microscopes.

The lenses Leeuwenhoek made were of excellent quality; some gave magnifications up to 300X and were remarkably free of distortion. Making these lenses and looking through them were the passions of his life. Everywhere he looked he found what he called "animalcules." He found them in stagnant water, in sick people, and even in his own mouth.

Over the years Leeuwenhoek observed all the major kinds of microorganisms—protozoa, algae, yeast, fungi, and bacteria in spherical, rod, and spiral forms. He once wrote, "For my part I judge, from myself (howbeit I clean my mouth like I've already said), that all the people living in our United Netherlands are not as many as the living animals that I carry in my own mouth this very day." Starting in the 1670s he wrote numerous letters to the Royal Society in London and pursued his studies until his death in 1723, at the age of 91. Leeuwenhoek refused to sell his microscopes to others and so failed to foster the development of microbiology as much as he might have.

After Leeuwenhoek's death microbiology failed to advance for more than a century. Eventually microscopes became more widely available, and progress resumed. Several workers discovered ways to stain microorganisms with dyes to make them more visible. Carolus Linnaeus developed a general classification system for all living organisms. Botanist Matthias Schleiden and zoologist Theodor Schwann formulated the **cell theory,** which states that cells are the fundamental units of life and carry out all the basic functions of living things. This theory still applies today to all cellular organisms, but not to viruses.

THE GERM THEORY OF DISEASE

The **germ theory of disease** states that microorganisms (germs) can invade other organisms and cause disease. Although this is a simple idea and is generally accepted today, it was not widely accepted when formulated in the mid-nineteenth century. Many people believed that broth, left standing, turned cloudy because of something about the broth itself. Even after it was shown that microorganisms in the broth caused it to turn cloudy, people believed that the microorganisms, like the "worms" (fly larvae, or maggots) in rotting meat, arose from nonliving things, a concept known as **spontaneous generation.** Widespread belief in spontaneous generation, even among scientists, hampered further development of the science of microbiology and the acceptance of the germ theory of disease. As long as scientists believed that microorganisms could arise from nonliving substances, they saw no purpose in considering how diseases were transmitted or how they could be controlled. Dispelling the belief in spontaneous generation took years of painstaking effort.

Early Studies

For as long as humans have existed, some probably have believed that living things somehow originated spontaneously from nonliving matter. Aristotle's theories about his four "elements"—fire, earth, air, and water—seem to have suggested that nonliving forces somehow contributed to the generation of life. Even some naturalists believed that rodents arose from moist grain, beetles from dust, and worms and frogs from mud. As late as the nineteenth century it seemed obvious to most people that rotting meat gave rise to "worms."

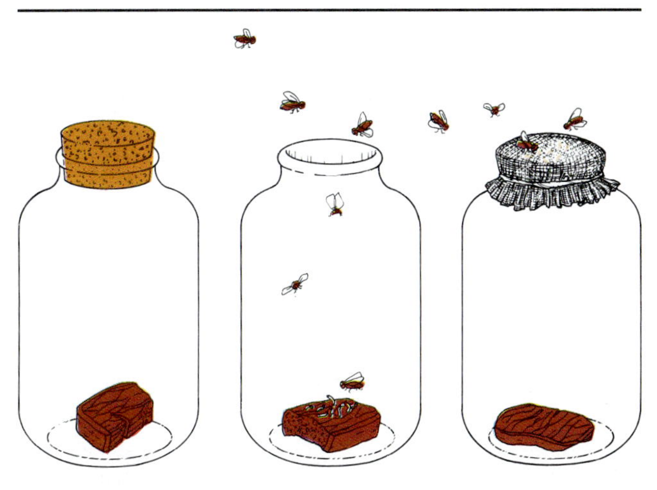

FIGURE 1.6 Redi's experiments refuting the spontaneous generation of maggots in meat. When meat is exposed in an open jar, flies lay their eggs on it, and the eggs hatch into maggots (fly larvae). In a sealed jar, however, no maggots appear. If the jar is covered with gauze, maggots hatch from eggs that the flies lay on top of the gauze, but still no maggots appear in the meat.

In the late seventeenth century the Italian physician Francesco Redi devised a set of experiments to demonstrate that if pieces of meat were covered with gauze so that flies could not reach them, no "worms" appeared in the meat, no matter how rotten it was (Figure 1.6). Maggots did, however, hatch from fly eggs laid on top of the gauze. In spite of proof that maggots did not arise spontaneously, some scientists, such as the British clergyman John Needham, still believed in spontaneous generation—at least of microorganisms. Lazzaro Spallanzani, an Italian cleric and scientist, was more skeptical. He boiled broth infusions containing organic (living or previously living) matter and sealed the flasks to demonstrate that no organisms would develop spontaneously in them. Critics did not accept this as disproof of spontaneous generation; they argued that boiling drove off oxygen, and sealing the flasks prevented its return.

Several scientists tried different ways of introducing air to counter this criticism. Schwann heated air before introducing it into flasks, and others filtered air through chemicals or cotton plugs. All these methods prevented the growth of microorganisms in the flasks. But the critics still argued that altering the air prevented spontaneous generation.

Even nineteenth-century scientists of some stature continued to argue vociferously in favor of spontaneous generation. They believed that an organic compound previously formed by living organisms contained a "vital force" from which life sprang. The force, of course, required air, and they believed that all the methods of introducing air somehow changed it so it could not interact with the force.

The proponents of spontaneous generation were finally defeated mainly by the work of the French chemist Louis Pasteur and the English physicist John Tyndall. When the French Academy of Science sponsored a competition in 1859 "to try by well-performed experiment to throw new light on the question of spontaneous generation," Pasteur entered the competition.

During the years Pasteur worked in the wine industry, he had established that alcohol was produced in wine only if yeast was present, and he had learned a lot about the growth of microorganisms. Pasteur's experiment for the competition involved his famous "swan-necked" flasks (Figure 1.7). He boiled *infusions* (broths of food stuffs) in flasks, heated the glass necks, and drew them out into long, curved tubes open at the end. Air could enter the flasks without being subjected to any of the treatments that critics had claimed destroyed its effectiveness. Airborne microorganisms

FIGURE 1.7 The "swan-necked" flasks that Pasteur used in refuting the theory of spontaneous generation. Although air could enter the flasks, microbes became trapped in the curved necks and never reached the contents. The contents, therefore, remained sterile despite their exposure to the air.

could also enter the necks of the flasks, but they became trapped in the curves of the neck and never reached the infusion. The infusions from Pasteur's experiments remained sterile unless the flasks were tipped so that the infusion flowed into the neck and back into the flask. This manipulation allowed microorganisms trapped in the neck to wash into the infusion, where they could grow and cause the infusion to become cloudy. In another experiment Pasteur filtered air through three cotton plugs. He then immersed the plugs in sterile infusion, demonstrating that growth occurred in the infusions from organisms trapped in the plugs.

Tyndall delivered another blow to the idea of spontaneous generation when he arranged sealed flasks of boiled infusion in an airtight box. After allowing time for all dust particles to settle to the bottom of the box, he carefully removed the covers from the flasks. These flasks, too, remained sterile. Tyndall had shown that air could be sterilized by settling, without any treatment that would prevent the "vital force" from acting.

Both Pasteur and Tyndall were fortunate that the organisms present in their infusions at the time they were boiled were destroyed by heat. Others tried the same experiments and observed that the infusions became cloudy from growth of microorganisms. We now know that the infusions in which growth occurred contained heat-resistant or spore-forming organisms, but at the time the growth of such organisms appeared to provide evidence for spontaneous generation. Nevertheless, the works of Pasteur and Tyndall did succeed in disproving spontaneous generation to most scientists of the time. Recognition that microbes must be introduced into a medium before their growth can be observed paved the way for further development of microbiology, and especially for the development of the germ theory of disease.

FIGURE 1.8 Louis Pasteur in his laboratory. The first rabies vaccine, developed by Pasteur, was made from the dried spinal cords of infected rabbits.

Pasteur's Further Contributions

Louis Pasteur (Figure 1.8) was such a giant among scientists working in microbiology in the nineteenth century that we must consider some of the many contributions he made. Born in 1822, the son of a sergeant in Napoleon's army, Pasteur worked as a portrait painter and a teacher before he began to study chemistry in his spare time. These studies led to posts in several French universities as professor of chemistry and to significant contributions to the wine and silkworm industries. He discovered that carefully selected yeasts made good wine, but that mixtures of other microorganisms competed with the yeast for sugar and made wine taste oily or sour. To combat this problem Pasteur developed the technique of pasteurization (heating wine to 56°C in the absence of oxygen for 30 minutes) to kill unwanted organisms. While studying silkworms, he identified three different microorganisms, each of which caused a different disease. His association of specific organisms with particular diseases, even though in silkworms rather than in humans, was an important first step in proving the germ theory of disease.

In spite of personal tragedy—the deaths of three daughters and a cerebral hemorrhage that left him with permanent paralysis—Pasteur went on to contribute to the development of vaccines. The best known of Pasteur's vaccines is the rabies vaccine, made of dried spinal cord from rabbits infected with rabies, which was tested in animals. When a 9-year-old boy who had been severely bitten by a rabid dog was brought to him, he administered the vaccine. The boy, who had been doomed to die, survived and became the first person to be immunized against rabies.

In 1894 Pasteur became director of the Pasteur Institute, which was built for him in Paris. He spent the remainder of his life until his death in 1895 guiding the training and work of other scientists at the Institute. Today the Pasteur Institute is a thriving research center—an appropriate memorial to its founder.

Koch's Contributions

Robert Koch (Figure 1.9), a contemporary of Pasteur, finished his medical training in 1872 and worked as a physician in Germany throughout most of his career. After he bought a microscope and photographic equipment, he spent most of his time studying bacteria, especially those that cause disease. Koch identified the bacterium that causes anthrax, a highly contagious and lethal disease in cattle and sometimes in humans. He recognized both actively dividing cells and dormant cells (spores) and developed techniques for studying them *in vitro* (outside a living organism).

Koch also found a way to grow bacteria in *pure cultures*—cultures that contained only one kind of or-

FIGURE 1.9 Berlin, 1891: Robert Koch's first postgraduate course in bacteriology. Koch is the bearded man in the center of the front row.

ganism. He tried streaking bacterial suspensions on potato slices and then on solidified gelatin. Finally, the wife of a colleague suggested that he add agar (a thickener used in cooking) to his bacteriological media. This created a firm surface over which microorganisms could be spread very thinly—so thinly that some individual organisms were separated from all others. Each individual organism then multiplied to make a colony of thousands of descendants. Koch's technique of preparing pure cultures is still used today.

Koch's outstanding achievement was the formulation of four postulates to associate a particular organism with a specific disease. **Koch's postulates,** which provided scientists with a method of establishing the germ theory of disease, are as follows:

1. The specific causative agent must be found in every case of the disease.
2. The disease organism must be isolated in pure culture.
3. Inoculation of a sample of the culture into a healthy, susceptible animal must produce the same disease.
4. The disease organism must be recovered from the body of the inoculated animal.

Implied in Koch's postulates is his one organism–one disease concept. The postulates assume that an infectious disease is caused by a single organism, and they are directed toward establishing that fact. This concept also was an important advance in the development of the germ theory of disease.

After obtaining a laboratory post at Bonn University in 1880, Koch was able to devote his full time to studying microorganisms. He identified the bacterium that causes tuberculosis and developed a complex method of staining this organism. He also guided the research that led to the isolation of *Vibrio cholerae,* the bacterium that causes cholera.

In a few years Koch became professor of hygiene at the University of Berlin, where he taught a microbiology course believed to be the first ever offered.

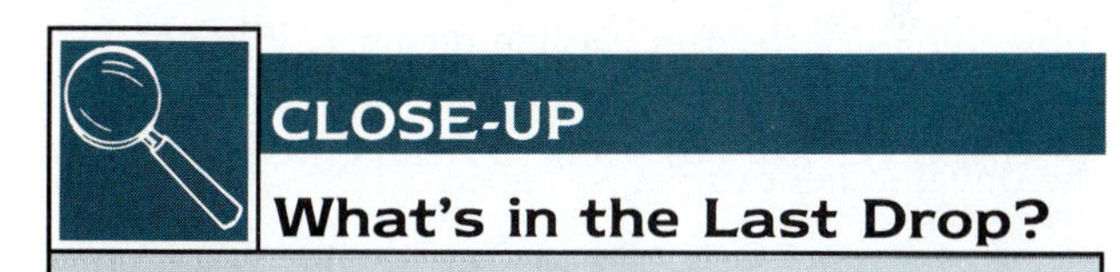

CLOSE-UP

What's in the Last Drop?

During the nineteenth century French and German scientists were fiercely competitive. One area of competition was the preparation of pure cultures. Koch's reliable method of preparing pure cultures from colonies on solid media allowed German microbiologists to forge ahead. The French microbiologists' method of broth dilution, though now often used to count organisms (Chapter 6), hampered their progress. They added a few drops of a culture to fresh broth, mixed it, and added a few drops of the mixture to more fresh broth. After several successive dilutions they assumed that the last broth that showed growth of microbes had contained a single organism. Unfortunately, the final dilution often contained more than one organism, and sometimes the organisms were of different kinds. This faulty technique led to various fiascos, such as inoculating animals with deadly organisms instead of vaccinating them.

(a)

(b)

FIGURE 1.10 Two nineteenth-century pioneers in the control of infections: (a) Ignaz Philipp Semmelweis, who died in an asylum before his innovations were widely accepted, depicted on a 1965 Austrian postage stamp; (b) Joseph Lister, who successfully carried on Semmelweis's work.

He also developed *tuberculin,* a vaccine against tuberculosis. Because he underestimated the difficulty of killing the tubercle organism, use of the vaccine resulted in several deaths from tuberculosis. Although tuberculin was unacceptable as a vaccine, its development laid the groundwork for a skin test to diagnose tuberculosis. After the vaccine disaster, it is not surprising that Koch left Germany. He made several visits to Africa, at least two visits to Asia, and one visit to the United States.

In the remaining 15 years of his life, his accomplishments were many and varied. He conducted research on malaria, typhoid fever, sleeping sickness, and several other diseases. His studies of tuberculosis won him the Nobel Prize in 1905, and his work in Africa and Asia won him great respect on those continents.

Work toward Controlling Infections

Like Koch and Pasteur, two nineteenth-century physicians, Ignaz Philipp Semmelweis of Austria and Joseph Lister of England, were convinced that microorganisms caused infections (Figure 1.10). Semmelweis recognized a connection between autopsies and puerperal (childbirth) fever. Many physicians went directly from performing autopsies to examining women in labor without so much as washing their hands. When Semmelweis attempted to encourage more sanitary practices, he was ridiculed and harassed until he had a nervous breakdown and was sent to an asylum. Ultimately, he suffered the curious irony of succumbing to an infection caused by the same organism that produces puerperal fever. In 1865, Lister, who had read of Pasteur's work on pasteurization and Semmelweis's work on improving sanitation, in 1865 initiated the use of dilute carbolic acid on bandages and instruments to reduce infection. Lister, too, was ridiculed, but with his imperturbable temperament, resolute will, and tolerance of hostile criticism, he was able to continue his work. At age 75, some 37 years after he introduced the use of carbolic acid, Lister was awarded the Order of Merit for his work in preventing the spread of infection.

EMERGENCE OF SPECIAL FIELDS OF MICROBIOLOGY

Pasteur, Koch, and most other microbiologists considered to this point were generalists interested in a wide variety of problems. Certain other contributors to microbiology had more specialized interests, but their achievements were no less valuable. In fact, those achievements helped to establish the special fields of immunology, virology, chemotherapy, and microbial genetics—fields that are today prolific research areas. Selected fields of microbiology are defined in Table 1.2.

Immunology

Disease depends not only on microorganisms invading a host but also on the host's response to that invasion. Today, we know that the host's response is in part a response of the immune system.

The ancient Chinese knew that a person scarred by smallpox would not again get the disease. They took dried scabs from lesions of people who were recovering from the disease and ground them into a powder that they sniffed. As a result of inhaling weakened organisms, they acquired a mild case of smallpox but were protected against subsequent infection.

Smallpox was unknown in Europe until the Crusaders carried it back from the Near East in the twelfth century. By the seventeenth century it was widespread, and in 1717 Lady Montagu, wife of the British ambassador to Turkey, introduced a kind of immunization to England. A thread was soaked in fluid from a smallpox vesicle (blister) and drawn through a small incision in the arm. This technique, called *variolation,* was used at first by only a few prominent people, but eventually it became widespread.

In the late eighteenth century Edward Jenner (Figure 1.11) realized that milkmaids who got cowpox did not get smallpox, and he inoculated his own son with fluid from a cowpox blister. He later similarly inoculated an 8-year-old and subsequently courageously inoculated the same child with smallpox. The child remained healthy. The word *vaccinia* (*vacca,* the Latin name for cow) gave rise both to the name of the virus that causes cowpox and the word *vaccine.* In the early 1800s Jenner received grants amounting to a total of 30,000 British pounds to extend his work on vaccination. Today, those grants would be worth more than $1 million. They may have been the first grants ever made for medical research.

FIGURE 1.11 Edward Jenner vaccinating a child against smallpox.

FIGURE 1.12 Elie Metchnikoff, one of the first scientists to study the body's defenses against invading microorganisms.

Pasteur contributed significantly to the emergence of immunology with his work on vaccines for rabies and cholera. In 1879, when Pasteur was studying chicken cholera, his assistant accidentally used an old chicken cholera culture to inoculate some chickens. The chickens did not develop disease symptoms. When he later inoculated the same chickens with a fresh chicken cholera culture, they remained healthy. Though he hadn't planned to use the old culture first, he did realize that the chickens had been immunized against chicken cholera. Pasteur reasoned that the organisms must have lost their ability to produce disease but retained their ability to produce immunity. This finding led him to look for techniques that would have the same effect on other organisms. His development of the rabies vaccine was a successful attempt.

Along with Jenner and Pasteur, the nineteenth-century Russian zoologist Elie Metchnikoff was a pioneer in immunology (Figure 1.12). In the 1880s many scientists believed immunity was due to noncellular substances in the blood. Metchnikoff discovered that certain cells in the body could ingest microbes, and he named these cells *phagocytes,* which literally means "cell-eating." The identification of phagocytes as cells that defend the body against invading microorganisms was a first step in understanding immunity. Metchnikoff also developed several vaccines. Some were successful, but unfortunately some infected the recipients with the disease against which they were supposedly being immunized. A few of his subjects acquired gonorrhea and syphilis from his vaccines.

CLOSE-UP

A "Thorny" Problem

Metchnikoff's personal life played a role in his discovery of phagocytes. Rather late in life he married and as part of the marriage agreement took in his young wife's brothers and sisters. On one occasion Metchnikoff left a starfish he was studying under his microscope when he went to lunch with his wife. While the starfish was unattended one of the mischievous children poked thorns into it. Metchnikoff was enraged to discover his mutilated starfish, but he took the time to look through the microscope before he discarded his ruined specimen. He was amazed to discover cells of the starfish gathered around the thorns. On further study he identified those cells as *leukocytes* (white blood cells) and found that they could devour foreign particles. Working from what started as a childish prank, Metchnikoff was able to formulate his theory of phagocytosis.

Virology

The science of virology emerged after that of bacteriology because viruses could not be recognized until certain techniques for studying and isolating larger particles such as bacteria had been developed. When Pasteur's collaborator Charles Chamberland developed a porcelain filter to remove bacteria from water in 1884, he had no idea that any kind of infectious agent could pass through the filter. But researchers soon realized that some filtrates (materials that passed through the filters) remained infectious even after the bacteria were filtered out. The Dutch microbiologist Martinus Beijerinck determined why such filtrates were infectious and was thus the first to characterize viruses. The term *virus* had been used earlier to refer to poisons and to infectious agents in general. Beijerinck used the term to refer to specific pathogenic molecules incorporated into host cells. He also believed these molecules could borrow existing metabolic and replicative mechanisms of the host cells for their own use.

Further progress in virology required development of techniques for isolating, propagating, and analyzing viruses. The U.S. scientist Wendell Stanley crystallized tobacco mosaic virus in 1935, showing that an agent with properties of a living organism also behaved as a chemical substance (Figure 1.13). The crystals consisted of protein and ribonucleic acid (RNA). The nucleic acid was soon shown to be important in the infectivity of viruses. Viruses were first observed with an electron microscope in 1939. From that time both chemical and microscopic studies were used to investigate viruses.

By 1952 Alfred Hershey and Martha Chase had demonstrated that the genetic material of some vi-

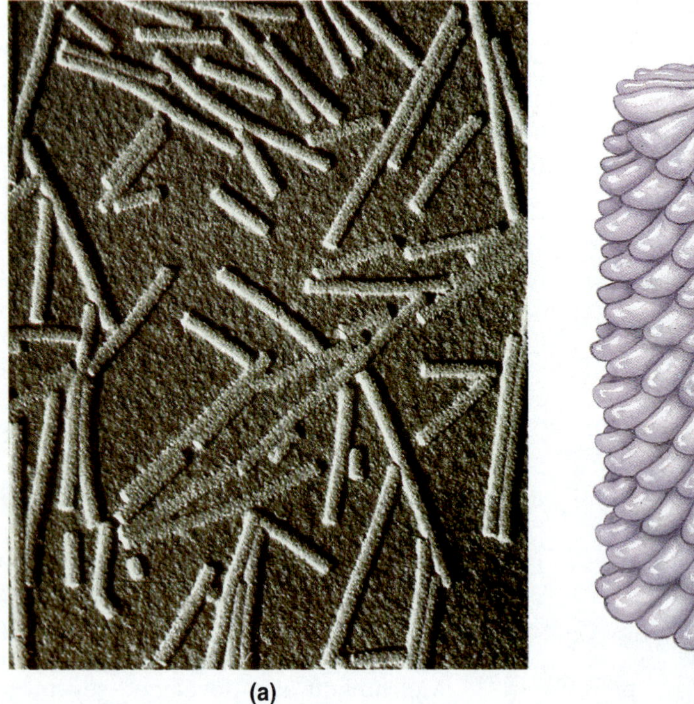

(a)

(b)

FIGURE 1.13 (a) Electron micrograph of tobacco mosaic virus (magnification approx. 500,000×). (b) The structure of the tobacco mosaic virus. A helical core of RNA is surrounded by a coat that consists of repeating protein units. The structure of the particles is so regular that the viruses can be crystallized.

ruses is another nucleic acid, deoxyribonucleic acid (DNA). In 1953 the American postdoctoral student James Watson and the British researcher Francis Crick determined the structure of DNA. The stage was set for rapid advances in understanding how DNA functions as genetic material both in viruses and in cellular organisms. Since the 1950s literally hundreds of viruses have been isolated and characterized. Although much remains to be learned about viruses, tremendous progress has been made in understanding their structure and how they function.

FIGURE 1.14 Paul Ehrlich, pioneer in the development of chemotherapy for infectious disease.

Chemotherapy

The Greek physician Dioscorides compiled *Materia Medica* in the first century A.D. This five-volume work listed a number of substances derived from medicinal plants still in use today—digitalis, curare, ephedrine, and morphine—along with a number of herbal medications. Credit for bringing herbal medicine to the United States is given to many groups of settlers, but Native Americans used many medicinal plants before the arrival of white people in the Americas. Many so-called primitive peoples still use herbs extensively, and some pharmaceutical companies finance expeditions into the Amazon Basin and other remote areas to investigate the uses the natives make of the plants around them.

During the Middle Ages virtually no advances were made in the use of chemical substances to treat diseases. Early in the sixteenth century the Swiss physician Aureolus Paracelsus used metallic chemical elements to treat diseases—antimony for general infections and mercury for syphilis. In the mid-seventeenth century Thomas Sydenham, an English physician, introduced cinchona tree bark to treat malaria. This bark, which we now know contains quinine, had been used to treat fevers in Spain and South America. In the nineteenth century morphine was extracted from the opium poppy and used medicinally to alleviate pain.

Paul Ehrlich, the first serious researcher in the field of chemotherapy (Figure 1.14), received his doc-

PUBLIC HEALTH

Swamp Air or Mosquitoes?

During the American effort to dig the Panama Canal in 1905, yellow fever struck the men as they struggled in the swamps. Yellow fever was a terrible and fatal disease. As Paul de Kruif put it in *Microbe Hunters,* "when folks of a town began to turn yellow and hiccup and vomit black, by scores, by hundreds, every day—the only thing to do was to get up and get out of that town." The entire canal project was in jeopardy because of the disease, and the physician Walter Reed was assigned the task of controlling the disease. Reed listened to the advice of Dr. Carlos Finlay y Barres of Havana, Cuba, who for years had claimed that yellow fever was carried by mosquitoes. Reed ignored those who called Dr. Finlay a theorizing old fool and insisted yellow fever was due to swamp air. Several people, including James Carroll, Reed's longtime associate, volunteered to be bitten by mosquitoes known to have bitten yellow-fever patients. Although Carroll survived after his heart had nearly stopped, most of the other volunteers died. Jesse Lazear, a physician working with Reed, was accidentally bitten while working with patients. He began to show symptoms in 5 days and was dead in 12 days. Thus, it became clear that mosquitoes carried the yellow fever agent. Similar experiments in which volunteers slept on sheets filthy with vomitus of yellow-fever patients demonstrated that bad air, contaminated water, sheets, and dishes were not involved. Later Carroll passed blood from yellow-fever victims through a porcelain filter and used the filtrate to inoculate three people who had not had yellow fever. How he got their cooperation is not known, but it is known that two of them died of yellow fever. The agent that passed through the porcelain filter was eventually identified as a virus.

toral degree in chemistry from the University of Leipzig, Germany, in 1878. His discovery that certain dyes stained microorganisms but not animal cells suggested that the dyes or other chemicals might selectively kill microbial cells. This led him to search for the "magic bullet," a chemical that would destroy specific bacteria without damaging surrounding tissues. Ehrlich coined the term *chemotherapy* and headed the world's first institute concerned with the development of drugs to treat disease.

In the early twentieth century the search for the magic bullet continued, especially among scientists at Ehrlich's institute. After testing hundreds of compounds (and numbering each compound), Ehrlich found compound 418 (arsenophenylglycine) to be effective against sleeping sickness and compound 606 (Salvarsan) to be effective against syphilis. For 40 years Salvarsan remained the best available treatment for this disease. In 1922 Alexander Fleming, a Scottish physician, discovered that lysozyme, an enzyme found in tears, saliva, and sweat, could kill bacteria. Lysozyme was the first body secretion shown to have chemotherapeutic properties.

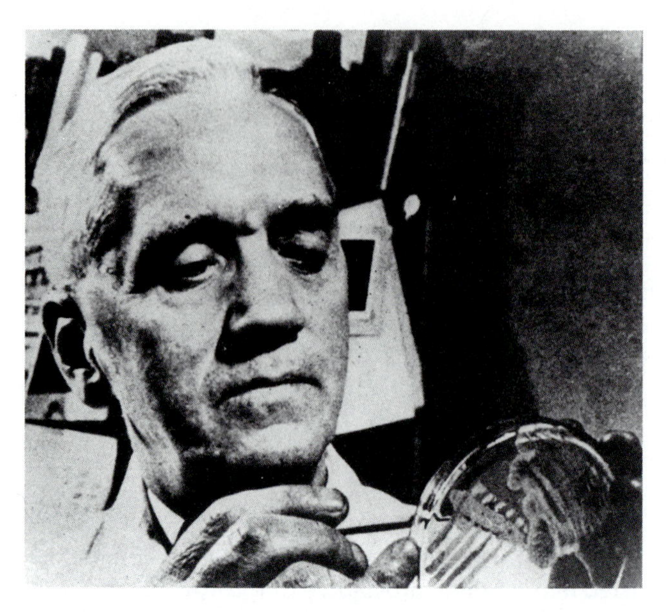

FIGURE 1.15 Alexander Fleming, who discovered the antibacterial properties of penicillin.

The development of antibiotics began in 1917 with the observation that certain bacteria (actinomycetes) stopped the growth of other bacteria. Lieske made a study of antibacterial substances and published a monograph on them in 1921. In 1928 Fleming (Figure 1.15) observed that a colony of *Penicillium* mold contaminating a culture of *Staphylococcus* bacteria had prevented growth of bacteria adjacent to itself. Though not the first to observe this phenomenon, Fleming did recognize its potential for countering infections. However, purification of the substance he called *penicillin* proved to be very difficult. The great need for such a drug during World War II, money from the Rockefeller Institute, and the hard work of biochemist Ernest Chain, pathologist Howard Florey, and researchers at Oxford University accomplished the task. Penicillin became available as a safe and versatile chemotherapeutic agent for use in humans.

While this work was going on, sulfa drugs also were being developed. In 1935, prontosil rubrum, a reddish dye containing a sulfonamide chemical group, was used in treating streptococcal infections. Further study showed that sulfonamides were converted in the body to sulfanilamides; much subsequent work was devoted to developing drugs containing sulfanilamide. The German chemist Gerhard Domagk played an important role in this work, and one of the drugs, prontosil, saved the life of his daughter. In 1939 he was awarded a Nobel Prize for his work, but Hitler refused to allow him to make the trip to receive it. Extensions of Domagk's work led to the development of isoniazid, an effective agent against tuberculosis. Both sulfa drugs and isoniazid are still used today.

The development of antibiotics resumed with the work of Selman Waksman, who was born in the Russian Ukraine and moved to the United States in 1910. Inspired by the French microbiologist Rene Dubos's 1939 discovery of tyrothricin, an antibiotic produced by soil bacteria, Waksman examined soil samples from all over the world for growth-inhibiting microorganisms or their products. He coined the term *antibiotic* in 1941 to describe actinomycin and other products he isolated. Both tyrothricin and actinomycin proved to be too toxic for general use as antibiotics. After repeated efforts, Waksman isolated the less toxic drug streptomycin in 1943. Streptomycin constituted a major breakthrough in the treatment of tuberculosis. In the same decade Waksman and others isolated neomycin, chloramphenicol, and chlortetracycline.

Examining soil samples proved to be a good way to find antibiotics, and explorers and scientists still collect soil samples for analysis. The more common antibiotic-producing organisms are rediscovered repeatedly, but the possibility of finding a new one always remains. Even the sea has yielded antibiotics, especially from the fungus *Cephalosporium acremonium*. The Italian microbiologist Giuseppe Brotzu noted the absence of disease organisms in sea water where sewage entered, and he determined that an antibiotic must be present. Cephalosporin was subsequently purified, and a variety of cephalosporin derivatives are now available for treating human diseases.

The fact that many antibiotics have been discovered does not stop the search for more. As long as there are untreatable infectious diseases, the search will continue. Even when effective treatment becomes available, it is always possible that a better, less toxic, or cheaper treatment can be found. Of the many chemotherapeutic agents currently available, none can cure viral infections. Consequently, much of today's drug research is focused on developing effective antiviral agents.

Genetics and Molecular Biology

Modern genetics began with the rediscovery in 1900 of Gregor Mendel's principles of genetics. Even after this significant event, for nearly three decades little progress was made in understanding how microbial characteristics are inherited. For this reason, microbial genetics is the youngest branch of microbiology. In 1928 British scientist Frederick Griffith discovered that previously harmless bacteria could change their nature so as to become capable of causing disease. The remarkable thing about this discovery was that live bacteria were shown to acquire heritable traits from dead ones. During the early 1940s, Oswald Avery, Maclyn McCarty, and Colin MacLeod of the Rockefeller Institute in New York demonstrated that the change was produced by DNA, thus paving the way for the epoch-making discovery of the structure of DNA by James Watson and Francis Crick. This breakthrough ushered in the modern era of molecular genetics.

About the same time, Edward Tatum and George Beadle used genetic variations in the mold *Neurospora* to demonstrate how genetic information controls metabolism. In the early 1950s Barbara McClintock discovered that some genes (units of inherited information) can move from one location to another on a chromosome. Before McClintock's work, genes were thought to remain stationary; her revolutionary discovery has forced geneticists to revise their thinking about genes.

More recently, scientists have discovered the genetic basis that underlies our ability to make an enormous diversity of antibodies: molecules that the immune system produces to combat invading microbes and their toxic products. Within cells of the immune system, genes are shuffled about and spliced together in various combinations, allowing us to make literally millions of different antibodies, including some that can protect us from threats that the body has never previously encountered.

TOMORROW'S HISTORY

Today's discovery is tomorrow's history. In an active research field such as microbiology, it is impossible to present a complete history. Some of the microbiologists omitted from this discussion are listed in Table 1.3. This period was known as the "Golden Age of Microbiology." Though many terms used to describe their accomplishments will be unfamiliar, you will become familiar with them as you pursue the study of microbiology. Beginning in 1900, Nobel Prizes have been awarded annually to outstanding scientists, many of whom were in the fields of physiology or medicine (Table 1.4). In some years the prize is shared by several scientists, although the scientists may have made independent contributions. Look to see who is listed in Tables 1.3 and 1.4 as you begin to study each new area of microbiology.

FIGURE 1.16 Plant scientists are using knowledge gained from research with microorganisms to produce superior agricultural crops, such as strains that resist insect pests or that have greater productivity due to soil microorganisms that help the plants grow better. The poinsettia on the right has been inoculated with a combination of fungi and bacteria that help the plant grow better.

You can see from Table 1.4 that microbiology has been in the forefront of research in medicine and biology for several decades, and probably never more so than today. One reason is the renewed focus on infectious disease brought about by the advent of AIDS. Another is the dramatic progress in genetic engineering that has been made in the past decade. Microorganisms have been and continue to be an essential part of the genetic engineering revolution. Most of the key discoveries that led to our present understanding of genetics emerged from research with microbes. Today scientists are attempting to redesign microorganisms for a variety of purposes (see Chapter 8 and the related Microbiologist's Notebook feature). Bacteria have been converted into factories that produce drugs, hormones, vaccines, and a variety of biologically important compounds. And microbes, viruses in particular, are often the vehicle by which scientists insert new genes into other organisms. Such techniques may soon also enable us to produce improved varieties of plants and animals such as pest-resistant crops (Figure 1.16) and may even enable us to correct genetic defects in human beings.

TABLE 1.3 The golden age of microbiology: early microbiologists and their achievements

Year	Investigator	Achievement
1874	Billroth	Discovery of round bacteria in chains
1876	Koch	Identification of *Bacillus anthracis* as causative agent of anthrax
1878	Koch	Differentiation of staphylococci
1879	Hansen	Discovery of *Mycobacterium leprae* as causative agent of leprosy
1880	Neisser	Discovery of *Neisseria gonorrhoeae* as causative agent of gonorrhea
1880	Laveran and Ross	Identification of life cycle of malarial parasites in red blood cells of infected humans
1880	Eberth	Discovery of *Salmonella typhi* as causative agent of typhoid fever
1880	Pasteur and Sternberg	Isolation and culturing of pneumonia cocci from saliva
1881	Koch	Animal immunization with attenuated anthrax bacilli
1882	Leistikow and Loeffler	Cultivation of *Neisseria gonorrhoeae*
1882	Koch	Discovery of *Mycobacterium tuberculosis* as causative agent of tuberculosis
1882	Loeffler and Schutz	Identification of actinobacillus that causes the animal disease glanders
1883	Koch	Identification of *Vibrio cholerae* as causative agent of cholera
1883	Klebs	Identification of *Corynebacterium diphtheriae* and toxin as causative agent of diphtheria
1884	Loeffler	Culturing of *Corynebacterium diphtheriae*
1884	Rosenbach	Pure culturing of streptococci and staphylococci
1885	Escherich	Identification of *Escherichia coli* as a natural inhabitant of the human gut
1885	Bumm	Pure culturing of *Neisseria gonorrhoeae*
1886	Flugge	Staining to differentiate bacteria
1886	Fraenckel	*Streptococcus pneumoniae* related to pneumonia
1887	Weichselbaum	*Neisseria meningitidis* related to meningitis
1887	Bruce	Identification of *Brucella melitensis* as causative agent of brucellosis in cattle
1888	Roux and Yersin	Discovery of action of diphtheria toxin
1889	Charrin and Roger	Discovery of agglutination of bacteria in immune serum
1889	Kitasato	*Clostridium tetani* related to tetanus toxin
1890	Pfeiffer	Identification of Pfeiffer bacillus, *Hemophilus influenzae*
1890	von Behring and Kitasato	Immunization of animals with diphtheria toxin
1892	Ivanovski	Discovery of filterability of tobacco mosaic virus
1894	Roux and Kitasato	Identification of *Yersinia pestis* as causative agent of bubonic plague
1894	Pfeiffer	Discovery of bacteriolysis in immune serum
1895	Bordet	Discovery of alexin (complement) and hemolysis
1896	Widal and Grunbaum	Development of diagnostic test based on agglutination of typhoid bacilli by immune serum
1897	van Ermengem	Discovery of *Clostridium botulinum* as causative agent of botulism
1897	Kraus	Discovery of preciptins
1897	Ehrlich	Formulation of sidechain theory of antibody formation
1898	Shiga	Discovery of *Shigella dysenteriae* as causative agent of dysentery
1898	Loeffler and Frosch	Discovery of filterability of virus that causes foot-and-mouth disease
1899	Beijerinck	Discovery of intracellular reproduction of tobacco mosaic virus
1901	Bordet and Gengou	Identification of *Bordetella pertussis* as causative agent of whooping cough; development of complement fixation test
1901	Reed and colleagues	Identification of virus that causes yellow fever
1902	Portier and Richet	Work on anaphylaxis
1903	Remlinger and Riffat-Bey	Identification of virus that causes rabies
1905	Schaudinn and Hoffmann	Identification of *Treponema pallidum* as causative agent of syphilis
1906	Wasserman, Neisser, and Bruck	Development of Wasserman reaction for syphilis antibodies
1907	Asburn and Craig	Identification of virus that causes dengue fever
1909	Flexner and Lewis	Identification of virus that causes poliomyelitis
1915	Twort	Discovery of viruses that infect bacteria
1917	d'Herelle	Independent rediscovery of viruses that infect bacteria (bacteriophages)

In September 1990 a 4-year-old girl became the first gene-therapy patient. She had inherited a defective gene that crippled her immune system. Doctors at the National Institutes of Health (NIH) inserted a normal copy of the gene into some of her white blood cells in the laboratory and then injected these gene-treated cells back into her body, where, it is hoped, they will restore her immune system. Critics are worried that a new gene randomly inserted into her white blood cells could damage other genes and cause cancer. The experiment is underway, and we hope such damage will not occur.

TABLE 1.4 Nobel Prize awards for research involving microbiology

Year of Prize	Prize Winner	Topic Studied
1901	von Behring	Serum therapy against diphtheria
1902	Ross	Malaria
1905	Koch	Tuberculosis
1907	Laveran	Protozoa and the generation of disease
1908	Ehrlich and Metchnikoff	Immunity
1913	Richet	Anaphylaxis
1919	Bordet	Immunity
1928	Nicolle	Typhus exanthematicus
1939	Domagk	Antibacterial effect of prontosil
1945	Fleming, Chain, and Florey	Penicillin
1951	Theiler	Vaccine for yellow fever
1952	Waksman	Streptomycin
1954	Enders, Weller, and Robbins	Cultivation of polio virus
1958	Lederberg	Genetic mechanisms
	Beadle and Tatum	Transmission of hereditary characteristics
1959	Ochoa and Kornberg	Chemical substances in chromosomes that play a role in heredity
1960	Burnet and Medawar	Acquired immunological tolerance
1962	Watson and Crick	Structure of deoxyribonucleic acid
1965	Jacob, Lwoff, and Monod	Regulatory mechanisms in microbial genes
1966	Rous	Viruses and cancer
1968	Holley, Khorana, and Nirenberg	Genetic code
1969	Delbruck, Hershey, and Luria	Mechanism of virus infection in living cells
1972	Edelman and Porter	Structure and chemical nature of antibodies
1975	Baltimore, Temin, and Dulbecco	Interactions between tumor viruses and genetic material of the cell
1976	Blumberg and Gajdusek	New mechanisms for the origin and dissemination of infectious diseases
1978	Smith, Nathans, and Arber	Restriction enzymes for cutting DNA
1980	Benacerraf, Snell, and Dausset	Immunological factors in organ transplants
1984	Milstein, Kohler, and Jerne	Immunology
1987	Tonegawa	Genetics of antibody diversity
1988	Black, Elion, and Hitchings	Principles of drug therapy
1989	Bishop and Varmus	Genetic basis of cancer
1990	Murray, Thomas, and Corey	Transplant techniques and drugs

New information is constantly being discovered and sometimes supersedes earlier findings. Occasionally, new discoveries lead almost immediately to the development of medical applications, as occurred with penicillin and as will most certainly occur when a cure or vaccine for AIDS is discovered. However, old ideas such as spontaneous generation and old practices such as unsanitary measures in medicine can take years to replace. Many new bioethics problems will require considerable thought. Decisions regarding AIDS testing and reporting, transplants, environmental cleanup, and related issues will not come easily or quickly. Because of the wealth of prior knowledge, it is likely that you will learn more about microbiology in a single course than many pioneers learned in a lifetime. Yet, those pioneers still deserve great credit because they worked with the unknown and had few people to teach them.

Human Genome Project

Microbial genetic techniques have made possible the undertaking of a colossal and controversial scientific plan, the Human Genome Project. At a cost of approximately $3 billion over a period of about 15 years, this project will identify the entire location and chemical sequence of the human genome, that is, all the genetic material in the human species. The project is expected to be completed by the year 2005. When finished, it will be like having the "owner's manual" for humankind. This will make possible an incredible array of manipulations of human genetics and functions. Researchers are developing methods of locating and sequencing genes using simple microbes at first and then shifting to human genes as techniques become more efficient. Let us hope we will use the information gained in a wise fashion.

ESSAY

How Microbiologists Investigate Problems

Like other scientists, microbiologists investigate problems by designing and carrying out experiments. Such experiments have provided the information health scientists apply to solve medical problems. Much of this text is devoted to presenting information obtained from experiments and to showing how that information is used in understanding infectious diseases. However, we believe health scientists will be interested in knowing how scientific problems are investigated.

First, a scientific problem must concern some aspect of the natural world because scientific methods can deal only with natural conditions and events. Microbiological problems deal with natural conditions and events involving microbes. Second, scientific problems must be clearly defined and sufficiently limited in scope so that a hypothesis and a prediction can be formulated. A **hypothesis** is a tentative explanation to account for an observed condition or event. The hypothesis in a particular experiment (1) must be an explanation for the defined problem and (2) must be testable. A testable hypothesis is one for which evidence can be collected to support or refute the hypothesis. A **prediction** is an outcome or consequence that will result if the hypothesis is true. Before beginning a scientific experiment one must define the problem and make a hypothesis and a prediction.

A good hypothesis is one that offers the most reasonable explanation and the simplest solution to a problem. The purpose of scientific experiments is to test hypotheses by determining the correctness of predictions derived from the hypotheses. Scientific progress is made by making and testing hypotheses.

For example, suppose a microbiologist has isolated an organism in pure culture and wants to know the effects of temperature on its growth. He or she might (1) hypothesize that the organism's growth rate increases with temperature and (2) predict that the rate of increase in the number of organisms in a culture is proportional to the increase in temperature. After making the hypothesis and a prediction, the investigator designs an experiment to test the hypothesis. The experiment must be designed specifically to test the hypothesis and to collect evidence to determine whether the prediction is true.

To design a good experiment an investigator must consider all variables that might affect the outcome. A **variable** is anything that can change for the purposes of an experiment. An experiment should have only one **experimental variable,** the factor that is purposely changed for the experiment. For example, in the study of the effects of temperature on the growth of an organism, temperature is the experimental variable. The hypothesis and prediction are related to the experimental variable. All other variables are **control variables,** factors that can change but that are prevented from changing for the duration of the experiment. In our example, the control variables include the number and characteristics of the organism, the quantity and properties of the medium, and all environmental factors except temperature.

When all variables have been identified, the investigator establishes the procedures for carrying out the experiment. Once the experiment has been designed it must be carried out exactly as planned, and all observations must be made and recorded accurately and precisely. If problems or unusual situations are encountered, they must be noted carefully. For example, should an incubator fail to maintain certain cultures at the proper temperature for the appropriate length of time, this failure should be noted and taken into consideration in interpreting the experiment. When the experiment is completed, the researcher analyzes and interprets the results in light of the hypothesis and prediction. The analysis of the results of an experiment often involves preparation of tables and graphs and usually compares results obtained under experimental and control conditions. The goal of an experiment is to draw conclusions as to whether the prediction is true. If the experimental results, when analyzed, do not support the hypothesis, they may nevertheless suggest a better alternative hypothesis. The experimenter might then wish to design further experiments to test this new hypothesis. Often, it is the most unexpected experimental results that lead to the most interesting discoveries.

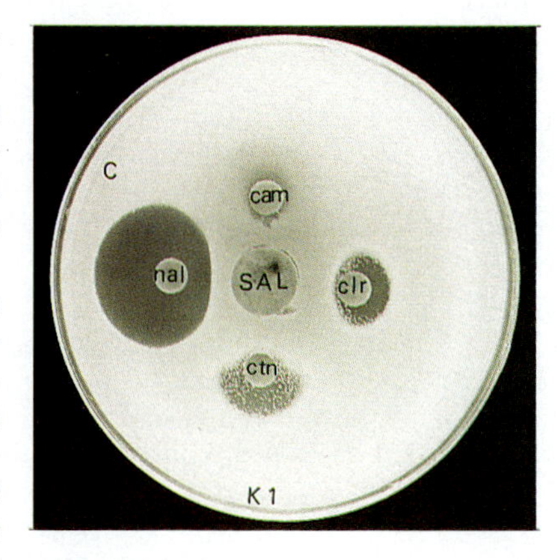

FIGURE 1.17 Petri dish showing clear zones of inhibition of bacterial growth around disks soaked in antibiotics. Areas of growth are tan. Note that zones are not circular. Inhibition is greater on the sides away from the aspirin-soaked disk.

Let us take as an example some scientific research that is currently under way at the laboratories of the National Institutes of Health (NIH). A scientist there, Dr. John Foulds, is studying the effects of aspirin and acetaminophen (Tylenol) on the amounts of antibiotics needed to inhibit the growth of several common disease-causing microorganisms. See if you can plan a set of experiments that will help answer some interesting questions raised by the results of his work.

Bacteria of several species grown in separate Petri dishes with aspirin added to their nutrient medium become resistant to at least four common antibiotics. The amount of antibiotic must then be increased between twofold and eightfold to inhibit bacterial growth (Figure 1.17). Similar results occur when Tylenol is used instead of aspirin. This is medically important because many patients fighting infection with prescription antibiotics also take aspirin or Tylenol for relief from fever or pain. Some arthritis patients take aspirin daily. When these people have a bacterial infection, do they need to take higher doses of antibiotics or, perhaps, different antibiotics? If so, must they do so for all infections or just some infections?

The investigator does not currently have facilities for care of mice or other laboratory animals. When he does get them, how should he plan his experiments? What will they prove or not be able to prove? In the meantime he has repeated his experiments with human blood serum used as the nutrient medium, observing the same results. Does he need to repeat these same experiments with human subjects? Why, or why not?

For those of you wondering what is happening in his original experiments, he has found that the bacteria make fewer pores when grown with aspirin or Tylenol. Some antibiotics enter bacteria through their pores. In a way, it is as if the bacteria respond to a hostile environment by "rolling up the windows as they drive through."

Perhaps you can discuss your experimental designs during lecture or in lab. What other questions occur to you? How would you go about forming and testing hypotheses for them? In addition to performing experiments, scientists also should report their results so that other scientists can verify and use the information. Scientific knowledge increases by the sharing of information. This allows other scientists to repeat experiments and determine whether the results are reproducible. It also allows them to develop new experiments that build on existing information.

CHAPTER SUMMARY

WHY STUDY MICROBIOLOGY?

- Microorganisms are part of the human environment and are therefore important to human health.
- The study of microorganisms provides insight into life processes in all forms of life.

SCOPE OF MICROBIOLOGY

The Microbes

- **Microbiology** is the study of all **microorganisms** (**microbes**) in the microscopic range. These include **bacteria, algae, fungi, viruses,** and **protozoa.**

The Microbiologists

- Immunology, virology, chemotherapy, and genetics are especially active research fields of microbiology.
- Microbiologists work as researchers or teachers in university, clinical, and industrial settings. They do basic research in the biological sciences; help to perform or devise diagnostic tests; develop and test antibiotics and vaccines; work to control infection, protect public health, and safeguard the environment; and play important roles in the food and beverage industries.

HISTORICAL ROOTS

- The Greeks, Romans, and Jews all contributed to early understandings of the spread of disease.
- Diseases such as bubonic plague and syphilis caused millions of deaths because of the lack of understanding of how to control or treat the infections.
- The development of high-quality lenses by Leeuwenhoek made it possible to observe microorganisms and later to formulate the **cell theory.**

THE GERM THEORY OF DISEASE

- The **germ theory of disease** states that microorganisms (germs) can invade other organisms and cause disease.

Early Studies

- Progress in microbiology and acceptance of the germ theory of disease required that the idea of **spontaneous generation** be refuted. Redi and Spallanzani demonstrated that organisms did not arise from nonliving material. Pasteur, with his swan-necked flasks, and Tyndall, with his dust-free air, finally dispelled the idea of spontaneous generation.

Pasteur's Further Contributions

- Pasteur also studied wine making and disease in silkworms and developed the first rabies vaccine. His association of particular microbes with certain diseases furthered the establishment of the germ theory.

Koch's Contributions

- Koch developed four postulates that aided in the definitive establishment of the germ theory of disease. **Koch's postulates** are as follows:

1. The specific causative agent must be found in every case of the disease.
2. The disease organism must be isolated in pure culture.
3. Inoculation with the culture must produce the same disease in a healthy susceptible experimental animal.
4. The disease organism must be recovered from the experimental animal.

Koch also developed techniques for isolating organisms, identified the bacillus that causes tuberculosis, developed tuberculin, and studied various diseases in Africa and Asia.

Work toward Controlling Infections

- Lister and Semmelweis contributed to improved sanitation in medicine by applying the germ theory and using aseptic technique.

EMERGENCE OF SPECIAL FIELDS OF MICROBIOLOGY

Immunology

- Immunization was first used against smallpox; Jenner used fluid from cowpox blisters to immunize against it.
- Pasteur developed techniques to weaken organisms so they would produce immunity without producing disease.

Virology

- Beijerinck characterized viruses as pathogenic molecules that could take over a host cell's mechanisms for their own use.
- Reed demonstrated that mosquitoes carry the yellow fever agent, and several other investigators identified viruses in the early twentieth century. The structure of DNA, the genetic material in many viruses and in all cellular organisms, was discovered by Watson and Crick.
- New techniques for isolating, propagating, and analyzing viruses allowed them to be observed and crystallized and allowed their nucleic acids to be studied.

Chemotherapy

- Substances derived from medicinal plants provided nearly all of the chemotherapeutic agents until Ehrlich began a systematic search for chemically defined substances that would kill bacteria.
- Fleming and his colleagues developed penicillin, and Domagk and others developed sulfa drugs.
- Waksman and others developed streptomycin and other antibiotics derived from soil organisms.

Genetics and Molecular Biology

- Griffith discovered transformation of colony characteristics in pneumococci. This genetic change was shown by Avery, McCarty, and MacLeod to be due to DNA. Tatum and Beadle studied biochemical mutants of *Neurospora* to show how genetic information controls metabolism.

RELATED KEY TERMS

control variable
experimental variable
hypothesis
microbiology prediction
variable

QUESTIONS FOR REVIEW

A.

1. What are the two main reasons for studying microbiology, and why are they important?

B.

2. What kinds of organisms are included among microbes?
3. What kinds of work do microbiologists do?

C.

4. What were the major contributions to microbiology of Greeks, Romans, and Jews?
5. What was Leeuwenhoek's contribution?
6. Which two diseases caused many deaths in the early days of microbiology?

D.

7. What is the germ theory of disease?
8. What is spontaneous generation, and why did it have to be refuted before the germ theory could be accepted?
9. What were the contributions of Redi, Spallanzani, Pasteur, and Tyndall toward dispelling belief in spontaneous generation?

10. What were Pasteur's and Koch's contributions to establishing the germ theory?
11. List Koch's postulates, and explain their importance.
12. What did Semmelweis and Lister contribute to microbiology?

E.

13. How did the use of porcelain filters affect the emergence of virology?
14. What was Beijerinck's contribution to virology?
15. What factors prevented progress in virology before the 1930s?
16. What major events took place in virology between 1930 and 1953?
17. What did chemotherapy consist of before the work of Ehrlich?
18. What were the contributions of Ehrlich, Fleming, Domagk, and Waksman to chemotherapy?
19. How did Griffith, Tatum, and Beadle contribute to microbial genetics?

PROBLEMS FOR INVESTIGATION

1. Suppose Jenner were alive today and had a vaccine he thought would immunize against a dreaded disease. Would he be allowed to give the vaccine, and if so, under what circumstances? Would he be allowed to inoculate a child he had vaccinated with the disease-causing organism? Should he be allowed to do these things? Why or why not?
2. It is well established that sometimes a disease is not produced following the inoculation of a sample of a culture of microorganisms into a healthy, susceptible animal. Does that fact invalidate Koch's postulates? Defend your answer.
3. Select 10 events in the history of microbiology. Explain how each demonstrates the validity of the two reasons for studying microbiology discussed in this chapter.
4. For any one branch of microbiology prepare a short report on accomplishments not mentioned in your text. (If possible, present an oral report to your class.)
5. Cite at least three examples of delays in the development of microbiology because of the lack of proper tools or procedures.

SOME INTERESTING READING

General References

American Society for Microbiology. Professional journals and other publications. Washington, D.C.: American Society for Microbiology. Most professional microbiologists belong to this society and read its journals to keep up to date in their fields.

Baron, S., et al. 1991. *Medical microbiology.* New York: Churchill Livingstone Inc. An excellent medical-school microbiology textbook.

Brooks, G. F., J. S. Butel, and L. N. Ornston. 1991. *Jawetz, Melnick and Adelberg's Medical Microbiology.* East Norwalk, CT: Appleton & Lange. A regularly updated review of microbiology of interest to physicians and health scientists.

Davis, B. D., et al. 1990. *Microbiology.* New York: Harper & Row. A good general reference on most classical topics in microbiology.

Hoeprich, P. D., and M. C. Jordan, eds., 1989. *Infectious diseases.* Philadelphia: Lippincott. An exhaustive survey of clinical microbiology.

Joklik, W. K., et al., eds. 1988. *Zinsser microbiology.* East Norwalk, CT: Appleton & Lange. A classic textbook of medical microbiology, taxonomically arranged.

U.S. Centers for Disease Control. Morbidity and Mortality Weekly Reports. Boston: Massachusetts Medical Society. In addition to providing current statistics on death and disease in the United States, these reports include news items on infectious and some other diseases.

Chapter References

Asimov, I. 1982. *Asimov's biographical encyclopedia of science and technology.* New York: Doubleday.

Baxby, D. 1981. *Jenner's smallpox vaccine: The riddle of vaccinia and its origins.* London: William Heinemann.

Brock, T. D., ed. and trans. 1975. *Milestones in microbiology.* Washington, D.C.: American Society for Microbiology.

Brock, T. D. 1988. *Robert Koch: A life in medicine and bacteriology.* Madison, WI: Science Tech Publishers.

Brown, W. E., and R. P. Williams. 1990. Ignaz Semmelweis and the importance of washing your hands. *The American Biology Teacher,* 52 (May): 291–4.

Cannon, R. E. 1990. *Experiments with writing to teach microbiology. The American Biology Teacher.* 52 (March): 156–58.

Defoe, D. 1722. *A journal of the plague year.* Republished 1966. London: J. M. Dent.

De Kruif, P. 1966. *Microbe hunters.* New York: Harcourt Brace Jovanovich.

Dubos, R. 1988. *Pasteur and modern science.* Madison, WI: Science Tech Publishers.

Gest, H. 1987. *The world of microbes.* Madison, WI: Science Tech Publishers.

Harre, R. 1981. *Great scientific experiments.* England: Oxford University Press.

Karlen, A. 1984. *Napoleon's glands and other ventures in biohistory.* New York: Warner Books.

Morshead, O. F., ed. 1926. *The diary of Samuel Pepys.* New York: Harcourt Brace Jovanovich.

Rosebury, T. 1969. *Life on man.* New York: The Viking Press.

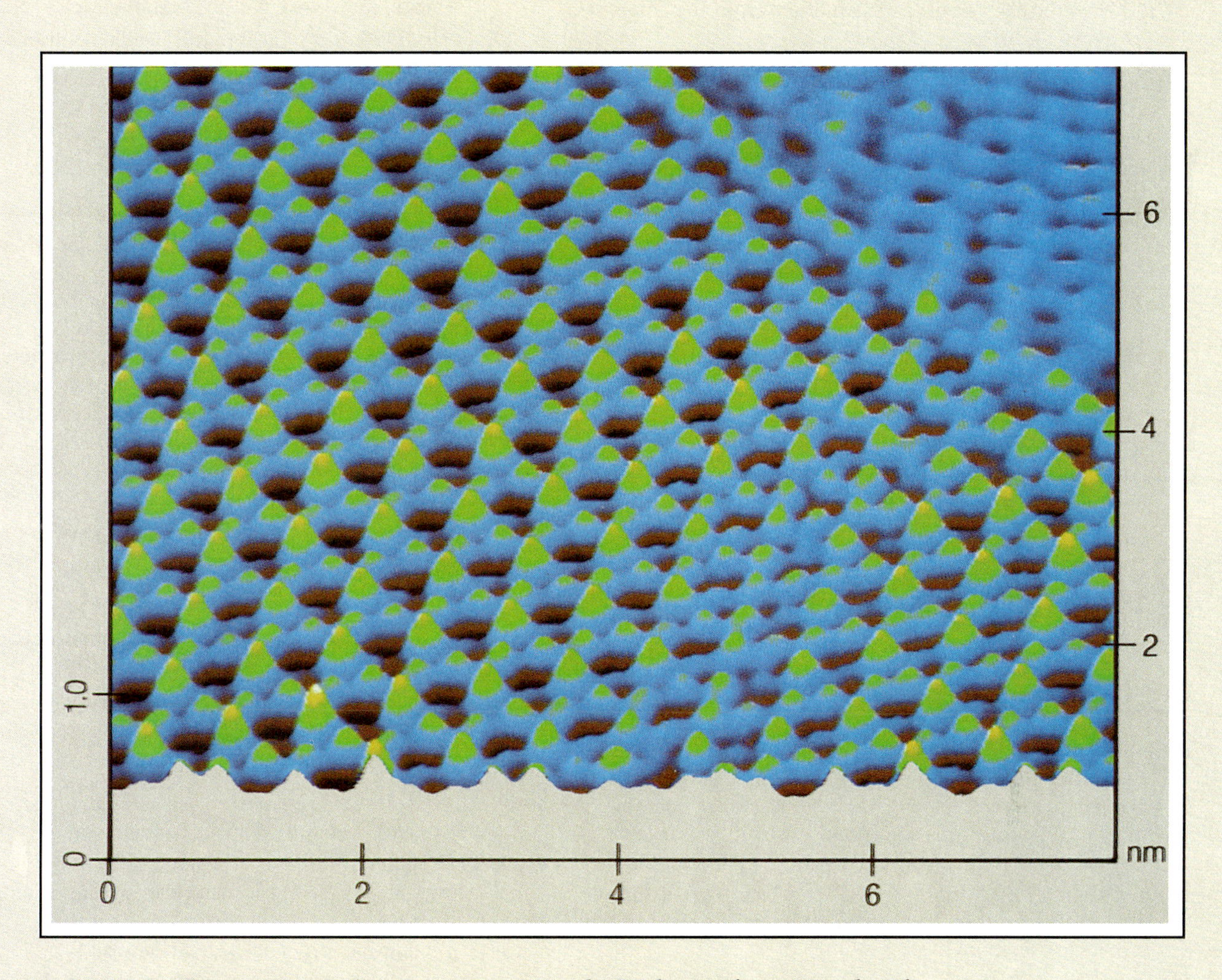

Individual iodine atoms can be seen as green peaks in this surface viewed with a scanning tunneling microscope. It is now possible to monitor some chemical processes in real time, in three dimensions, and with visualization of individual atoms.

2 Fundamentals of Chemistry

This chapter focuses on the following questions:

A. Why is knowledge of basic chemistry necessary to understanding microbiology?

B. What terms describe the organization of matter, and what elements are found in living organisms?

C. What are the properties of chemical bonds and chemical reactions?

D. What properties of water, solutions, colloidal dispersions, acids, and bases make them important in living things?

E. What is organic chemistry, and what are the major functional groups of organic molecules?

F. How do the structures and properties of carbohydrates contribute to their roles in living things?

G. How do the structures and properties of proteins, including enzymes, contribute to their roles in living things?

H. How do the structures and properties of simple lipids, compound lipids, and steroids contribute to their role in living things?

I. How do the structures and properties of nucleotides contribute to their role in living things?

WHY STUDY CHEMISTRY?

All living and nonliving things, including microbes, are composed of matter, so it is not surprising that all properties of microorganisms are determined by the properties of matter. Because chemistry deals with the basic properties of matter, we need to know some chemistry to begin to understand microorganisms. Chemical substances undergo change and interact with one another in chemical reactions. Metabolism, the use of nutrients for energy or for making the substance of cells, consists of many different chemical reactions. This is true regardless of whether the organism is a human or a microorganism. Thus, understanding the basic principles of chemistry is essential to understanding metabolic processes in living things. A microbiologist uses chemistry to understand the structure and function of microorganisms themselves and to understand how they affect humans in disease processes.

CHEMICAL BUILDING BLOCKS AND CHEMICAL BONDS

Chemical Building Blocks

Matter is composed of very small particles that form the basic chemical building blocks. Over the years, chemists have observed matter and deduced the characteristics of these particles. Just as the alphabet can be used to make thousands of words, the chemical building blocks can be used to make thousands of different substances. The complexity of chemical substances greatly exceeds the complexity of words. Words rarely contain more than 20 letters; complex chemical substances can contain as many as 20,000 building blocks!

The smallest chemical unit of matter is the **atom,** and many different kinds of atoms exist. Matter composed of one kind of atom is called an **element.** Each element has specific properties. For example, carbon is an element; a pure sample of carbon consists of a vast number of carbon atoms. Oxygen and nitrogen also are elements; they are found as gases in the earth's atmosphere. Chemists use one- or two-letter symbols to designate elements—C for carbon, O for oxygen, N for nitrogen, Na for sodium (from its Latin name, *natrium*).

Atoms combine chemically in various ways. Sometimes atoms of a single element combine with each other. For example, carbon forms long chains that are important in the structure of living things. Oxygen and nitrogen form paired atoms, O_2 and N_2. More often, atoms of one element combine with atoms of other elements. Carbon dioxide (CO_2) contains one atom of carbon and two atoms of oxygen; water (H_2O) contains two atoms of hydrogen and one atom of oxygen. (The subscripts in these formulas indicate how many atoms of each element are present.)

When two or more atoms combine chemically, they form a **molecule.** Molecules can consist of atoms of the same element, such as N_2, or atoms of different elements, such as CO_2. Molecules made up of atoms of two or more elements are called **compounds.** Thus, CO_2 is a compound, but N_2 is not. Compounds have properties different from those of their component elements. For example, in their elemental state both hydrogen and oxygen are gases at ordinary temperature. They can combine to form water, however, which is a liquid at ordinary temperature.

Living things consist of atoms of relatively few elements, principally carbon, hydrogen, oxygen, and nitrogen, but these are combined into highly complex compounds. A simple sugar molecule, $C_6H_{12}O_6$, contains 24 atoms. Many molecules found in living organisms contain thousands of atoms.

Structure of Atoms

Although the atom is the smallest unit of any element that retains the properties of that element, atoms do contain smaller particles that together account for those properties. Physicists study many such subatomic particles, but we discuss only **protons, neutrons,** and **electrons.** Three important properties of these particles are mass, electrical charge, and location in the atom (Table 2.1). The mass of a proton or a neutron is almost exactly equal to one atomic mass unit (AMU); electrons have a much smaller mass. With respect to electrical charge, electrons are negatively charged, protons are positively charged, and neutrons are neutral, with no charge. Atoms normally have an equal number of protons and electrons and so are electrically neutral. The heavy protons and neutrons are densely packed into the tiny, central nucleus of the atom, whereas the lighter electrons move in orbits around the nucleus.

The atoms of a particular element always have the same number of protons; that number of protons is the **atomic number** of the element. Atomic numbers

TABLE 2.1 Properties of atomic particles

Particle	Relative Mass	Charge	Location
Proton	1	+	Nucleus
Neutron	1	None	Nucleus
Electron	1/1836	−	Orbiting the nucleus

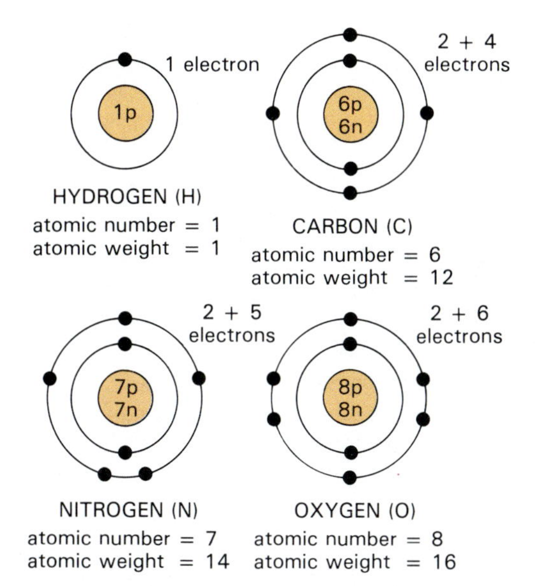

FIGURE 2.1 The structure of four biologically important atoms. Hydrogen, the simplest element, has a nucleus consisting of a single proton and a single electron in the first shell. In carbon, nitrogen, and oxygen the first shell is filled with two electrons and the second shell is partly filled. Carbon, with six protons in its nucleus, has six electrons, four of them in the second shell. Nitrogen has five electrons and oxygen six electrons in the second shell. It is the electrons in the outermost shell that take part in chemical bonding.

range from 1 to over 100. Though the numbers of neutrons and electrons in the atoms of many elements can change, the number of protons—and therefore the atomic number—remains the same.

Protons and electrons are oppositely charged, and consequently attract each other. This attraction keeps the electrons in orbits around the nucleus of an atom. The electrons are in constant, rapid motion, forming an electron cloud around the nucleus. Because some electrons have more energy than others, chemists use a model with concentric circles, or *electron shells,* to suggest different energy levels. Electrons with the least energy are located nearest the nucleus, and those with more energy are farther from the nucleus. Each energy level corresponds to an electron shell (Figure 2.1).

An atom of hydrogen has only one electron, which is located in the innermost shell. An atom of helium has two electrons in this shell, the maximum number of electrons that can be found in the innermost shell. Atoms with more than two electrons always have two electrons in the inner shell and up to eight additional electrons in the second shell. The inner shell is filled before electrons are found in the second shell, the second shell is filled before electrons are found in the third shell, and so on. Very large atoms have several more electron shells of larger capacity, but in elements found in living things, the outer shell is chemically stable if it contains eight electrons. This principle, known as the **rule of octets,** is important for understanding chemical bonding, which we will discuss shortly.

Atoms whose outer electron shells are nearly full (containing six or seven electrons) or nearly empty (containing one or two electrons) tend to form ions. An **ion** is a charged atom produced when an atom gains or loses one or more electrons (Figure 2.2a). When an atom of sodium loses the one electron in its outer shell without losing a proton, it becomes a positively charged ion called a **cation** (kat'i-on). When an atom of chlorine gains an electron to fill its outer shell, it becomes a negatively charged ion called an **anion** (an'i-on). In the ionized state, chlorine is referred to

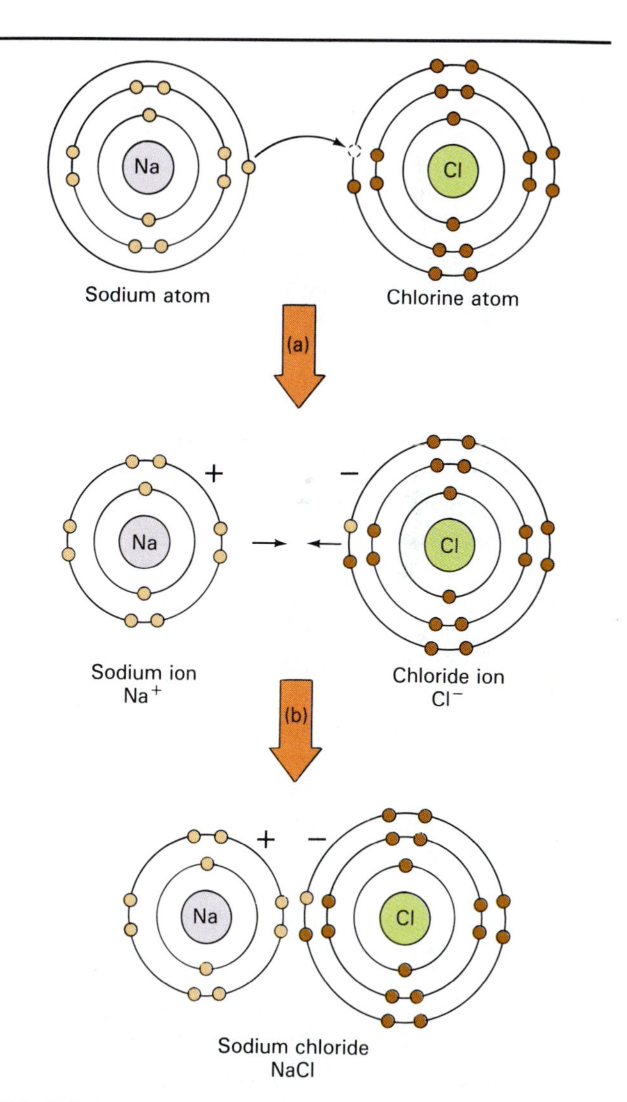

FIGURE 2.2 (a) the formation of ions, or electrically charged atoms. When a neutral sodium atom loses the single electron in its outermost shell, the result is a sodium ion, Na^+. When a neutral chlorine atom gains an extra electron in its outer shell, the result is a chlorine ion, Cl^-.
(b) Oppositely charged ions attract one another. Such attraction creates an ionic bond and results in the formation of an ionic compound, in this case sodium chloride (NaCl).

TABLE 2.2 Some common ions

Ion	Name	Brief Description
Na^+	Sodium	Contributes to salinity of natural bodies of water and body fluids of multicellular organisms.
K^+	Potassium	Important positive ion inside most cells.
H^+	Hydrogen	Responsible for the acidity of solutions.
Ca^{2+}	Calcium	Often acts as a chemical messenger.
Mg^{2+}	Magnesium	Sometimes required for chemical reactions to occur.
Fe^{2+}	Ferrous iron	Carries electrons to oxygen during some chemical reactions that capture energy. Can prevent growth of some microbes that cause human disease.
NH_4^+	Ammonium	Found in animal wastes and degraded by some bacteria.
Cl^-	Chloride	Often found with a positively charged ion, where it usually neutralizes charge.
OH^-	Hydroxyl	Usually present in excess in basic solutions where H^+ is depleted.
HCO_3^-	Bicarbonate	Often neutralizes acidity of bodies of water and body fluids.
NO_3^-	Nitrate	A product of the action of certain bacteria that convert atmospheric nitrogen into a form plants can use.
SO_4^{2-}	Sulfate	Component of sulfuric acid in atmospheric pollutants and acid rain.
PO_4^{3-}	Phosphate	Can be combined with certain other molecules to form high-energy bonds where energy is stored in a form living things can use.

as chloride. Ions of elements such as sodium or chlorine are chemically more stable than atoms of these elements because their outer electron shells are full. Many elements are found in microorganisms or their environments as ions (Table 2.2). Those with one or two electrons in their outer shell tend to lose electrons and form ions with +1 or +2 charges, respectively; those with seven electrons in their outer shell tend to gain an electron and form ions with a charge of −1. Some ions, such as the hydroxyl (hi-drox'l) ion (OH^-), contain more than one element.

Though all atoms of the same element have the same atomic number, they may not have the same atomic weight. **Atomic weight** is the sum of the number of protons and neutrons in an atom. Many elements consist of atoms with differing atomic weights. For example, carbon usually has six protons and six neutrons, giving it an atomic weight of 12. But some naturally occurring carbon atoms have one or two extra neutrons, giving these atoms an atomic weight of 13 or 14. In addition, laboratory techniques are available to create atoms with different numbers of neutrons. Atoms of a particular element that contain different numbers of neutrons are called **isotopes.** The superscript to the left of the symbol for the element indicates the atomic weight of the particular isotope. For example, carbon with an atomic weight of 14, which is often used to date fossils, is written ^{14}C. For an element that has naturally occurring isotopes the atomic weight is the average atomic weight of the natural mixture of isotopes. Thus, atomic weights are not always whole numbers, even though any particular atom contains a specific number of whole neutrons and protons.

A **gram molecular weight,** or **mole,** is the weight of a substance in grams equal to the sum of the atomic weights of the atoms in a molecule of the substance. For example, a mole of glucose, $C_6H_{12}O_6$, weighs 180 grams: 6 carbon atoms × 12 (atomic weight) + 12 hydrogen atoms × 1 (atomic weight) + 6 oxygen atoms × 16 (atomic weight) = 180 grams. Because of the way the mole is defined, a mole of any substance always contains 6.023×10^{23} particles. Properties of elements found in living things are summarized in Table 2.3.

Some isotopes are stable, and others are not. The nuclei of unstable isotopes tend to emit particles and radiation. Such isotopes are said to be *radioactive* and are called **radioisotopes.** Emissions from radioactive nuclei can be detected by radiation counters. Such emissions can be useful in studying chemical processes, but they also can harm living things.

Chemical Bonds

Chemical bonds form between atoms through interactions of electrons in their outer shells. Energy associated with these electrons holds the atoms together. Three kinds of chemical bonds commonly found in living organisms are ionic, covalent, and hydrogen bonds.

Ionic bonds result from the attraction between ions having opposite charges. For example, sodium ions with a positive charge (Na^+) combine with chloride ions with a negative charge (Cl^-) (Figure 2.2b).

Many compounds, especially those that contain carbon, are held together by **covalent bonds.** Instead of gaining or losing electrons, carbon and some other atoms share pairs of electrons (Figure 2.3). One carbon atom, which has four electrons in its outer shell, can

TABLE 2.3 Some properties of elements found in living organisms

Element	Symbol	Atomic Number	Atomic Weight	Electrons in Outer Orbit	Biological Occurrence
Oxygen	O	8	16.0	6	Component of biological molecules; required for aerobic metabolism
Carbon	C	6	12.0	4	Essential atom of all organic compounds
Hydrogen	H	1	1.0	1	Component of biological molecules; H^+ released by acids
Nitrogen	N	7	14.0	5	Component of proteins and nucleic acids
Calcium	Ca	20	40.1	2	Found in bones and teeth; regulator of many cellular processes
Phosphorus	P	15	31.0	5	Found in nucleic acids, ATP, and some lipids
Sulfur	S	16	32.0	6	Found in proteins; metabolized by some bacteria
Iron	Fe	26	55.8	2	Carries oxygen; metabolized by some bacteria
Potassium	K	19	39.1	1	Important intracellular ion
Sodium	Na	11	23.0	1	Important extracellular ion
Chlorine	Cl	17	35.4	7	Important extracellular ion
Magnesium	Mg	12	24.3	2	Needed by some enzymes
Copper	Cu	29	63.6	1	Needed by some enzymes; inhibits growth of some microorganisms
Iodine	I	53	125.9	7	Component of thyroid hormones
Fluorine	F	9	19.0	7	Inhibits microbial growth
Manganese	Mn	25	54.9	2	Needed by some enzymes
Zinc	Zn	30	65.4	2	Needed by some enzymes; inhibits microbial growth

share an electron with each of four hydrogen atoms. At the same time, each of the four hydrogen atoms shares an electron with the carbon atom. Four pairs of electrons are shared, each pair consisting of one electron from carbon and one electron from hydrogen. Such mutual sharing makes a carbon atom stable with eight electrons in its outer shell and a hydrogen atom stable with two electrons in its outer shell. Sometimes a carbon atom and an atom, such as an oxygen atom, share two pairs of electrons to form a double bond, but the octet rule still applies, and each atom has eight electrons in its outer shell and is therefore stable. In structural formulas chemists use a single line to represent a single pair of shared electrons and a double line to represent two pairs of shared electrons (Figure 2.3).

Atoms of four elements, carbon, hydrogen, oxygen, and nitrogen, commonly form covalent bonds that fill their outer electron shells. Carbon shares four electrons, hydrogen one electron, oxygen two electrons, and nitrogen three electrons. Unlike many ionic bonds, covalent bonds are stable in solutions. Because of this stability, covalent bonds are important in molecules that form biological structures.

Hydrogen bonds, though weaker than ionic and covalent bonds, are important in biological structures. Oxygen and nitrogen nuclei attract electrons very strongly, so when hydrogen is covalently bonded to

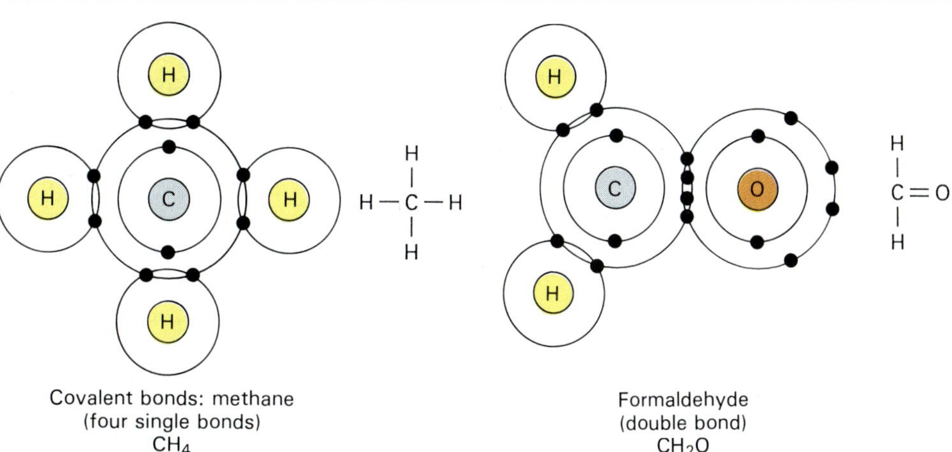

FIGURE 2.3 Covalent bonds are formed by sharing electrons. In methane, a carbon atom, with four electrons in its outermost shell, shares pairs of electrons with four hydrogen atoms. In this way all five atoms acquire stable, filled outer shells. Each shared electron pair constitutes a single covalent bond. In formaldehyde, a carbon atom shares pairs of electrons with two hydrogen atoms and also shares two pairs of electrons with an oxygen atom, forming a double covalent bond.

oxygen or nitrogen, the electrons of the covalent bond are shared unevenly—they are held closer to the oxygen or nitrogen than to the hydrogen. The hydrogen atom then has a partial positive charge, the other atom has a partial negative charge, and the molecule is called a **polar compound** because of its oppositely charged regions. The weak attraction between such partial charges is called a hydrogen bond.

Polar compounds such as water often contain hydrogen bonds. In a water molecule electrons from hydrogen atoms stay closer to the oxygen atom, and the hydrogen atoms lie to one side of the oxygen atom (Figure 2.4). Thus, water molecules are polar molecules having a positive hydrogen region and a negative oxygen region. Covalent bonds between the hydrogen and oxygen atoms hold the atoms together. Hydrogen bonds between the hydrogen and oxygen regions of different water molecules hold the molecules in clusters.

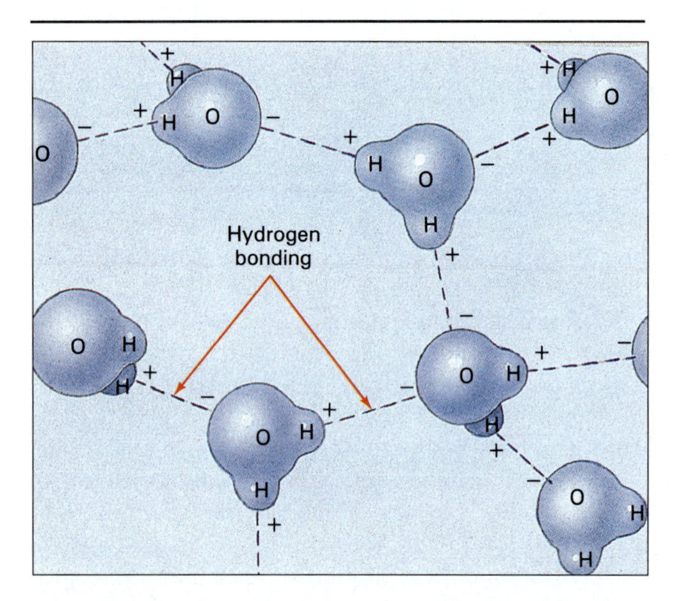

FIGURE 2.4 Water molecules are polar—they have a region with a partial positive charge (the hydrogen atoms) and a region with a partial negative charge (the oxygen atom). Hydrogen bonds, created by the attraction between oppositely charged regions of different molecules, hold the water molecules together in clusters.

Hydrogen bonds also contribute to the structure of large molecules such as proteins and nucleic acids, which contain long chains of atoms. The chains are coiled or folded into a three-dimensional configuration, and the configuration is held together in part by hydrogen bonds.

Chemical Reactions

Chemical reactions in living organisms typically involve using energy to form chemical bonds, and releasing energy as chemical bonds are broken. For example, the food we eat consists of molecules that have much energy stored in their chemical bonds. During **catabolism** (kat-ab'ol-izm), the breakdown of substances, food is degraded and some of that energy is released. Microorganisms use nutrients in the same general way. A catabolic reaction can be symbolized by

$$X{-}Y \longrightarrow X + Y + \text{energy}$$

where X—Y represents a nutrient molecule and where energy was originally stored in the bond between X and Y. Burning also is a catabolic reaction that releases energy in the form of heat. Catabolic reactions release energy and are thus **exergonic.** Conversely, energy is used to form chemical bonds in the synthesis of new compounds. In **anabolism** (an-ab'ol-izm), the buildup, or *synthesis*, of substances, energy is used to create bonds. An anabolic reaction can be symbolized by $X + Y + \text{energy} \rightarrow X{-}Y$, where energy is stored in the new substance X—Y. Anabolic reactions occur in living cells when small molecules are used to synthesize large molecules. Cells can store small amounts of energy for later use or use energy to make new molecules. Anabolic reactions require energy and are thus **endergonic.**

WATER AND SOLUTIONS

Water, one of the simplest of chemical compounds, is also one of the most important to living things. It takes part directly in many chemical reactions. Numerous substances dissolve in water or form colloidal dispersions. Acids and bases exist and function principally in water solutions.

Water

Water is so essential to life that humans can live only a few days without it. Many microorganisms die almost immediately if removed from their normal aqueous environments, such as lakes, ponds, oceans, and moist soil. Yet, some survive for several hours without water, and spores formed by a few microorganisms survive for many years away from water. Several bacteria find the moist, nutrient-rich secretions of human skin glands to be an ideal environment.

Water has several properties that make it important to living things. Because water is a polar compound and forms hydrogen bonds, it can act as a solvent and form thin layers on surfaces.

Water is a good dissolving medium, or **solvent,** for ions because the polar water molecules surround the ions. The positive region of water molecules is attracted to negative ions, and the negative region of

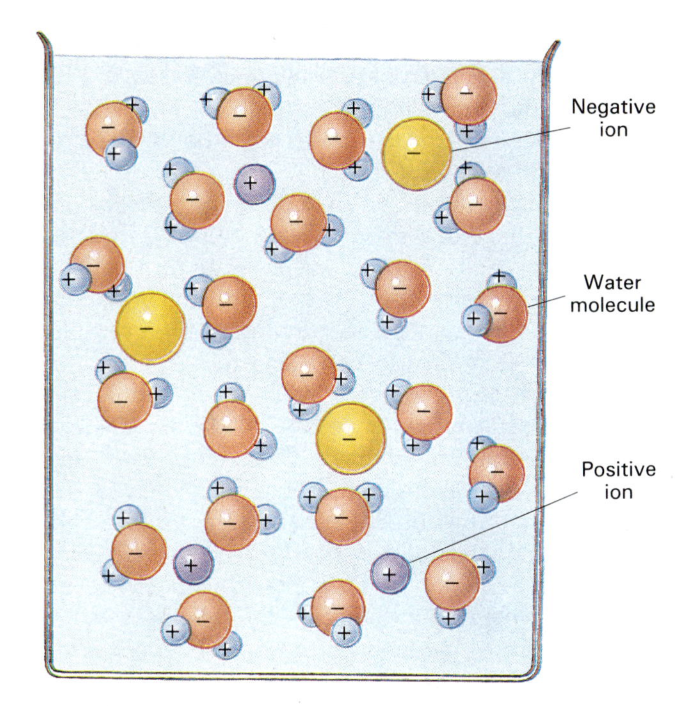

FIGURE 2.5 The polarity of water molecules enables water to dissolve many ionic compounds. The positive regions of the water molecules surround negative ions, and the negative regions of the water molecules surround positive ions, holding the ions in solution.

water molecules is attracted to positive ions. Many different kinds of ions can therefore be distributed evenly through a water medium, forming a **solution** (Figure 2.5).

Water forms thin layers because it has a high surface tension. **Surface tension** (Figure 2.6) is a phenomenon in which the surface of water acts as a thin, invisible, elastic membrane. The polarity of water molecules gives them a strong attraction for one another but no attraction for gas molecules in air at the water's surface. Therefore, surface water molecules cling together, forming hydrogen bonds with other molecules below the surface. In living cells this feature of surface tension allows a thin film of water to cover membranes and keep them moist.

Water has a high *specific heat*, that is, it can absorb or release large quantities of heat energy with little temperature change. This property of water helps to stabilize the temperature of living organisms, which are composed mostly of water, as well as bodies of water where many microorganisms live.

Finally, water provides the medium for most chemical reactions in cells, and it participates in many of these reactions. Suppose, for example, that substance X can gain or lose H^+ and substance Y can gain or lose OH^-. These substances that enter a reaction are called **reactants.** In an anabolic reaction, the components of water (H^+ and OH^-) are removed from the reactants to form a larger product molecule:

$$X{-}H + HO{-}Y \longrightarrow X{-}Y + H_2O$$

This kind of reaction, called **dehydration synthesis,** is involved in the synthesis of complex carbohydrates, some lipids (fats), and proteins. Conversely, in many catabolic reactions, water is added to a reactant to form simpler products:

$$X{-}Y + H_2O \longrightarrow X{-}H + HO{-}Y$$

This kind of reaction, called **hydrolysis,** occurs in the breakdown of large nutrient molecules to release simple sugars, fatty acids, and amino acids.

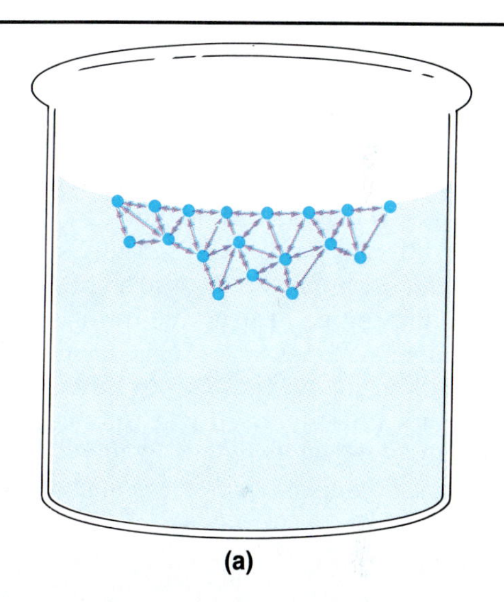

(a)

(b)

FIGURE 2.6 (a) Hydrogen bonding between water molecules creates surface tension, which causes the surface of water to behave like an elastic membrane. (b) Surface tension is strong enough to support the weight of the insects known as water striders.

Solutions and Colloids

Solutions and colloidal dispersions are examples of *mixtures*. Unlike a chemical compound, which consists of molecules whose atoms are present in specific proportions, a **mixture** consists of two or more substances combined in any proportion and not chemically bound. Each of the substances in a mixture contributes its properties to the mixture. For example, a mixture of sugar and salt could be made using any proportions of the two ingredients. The degree of sweetness or saltiness of the mixture would depend on the relative amounts of each substance present, but both sweetness and saltiness would be detectable.

A **solution** is a mixture of two or more substances in which the molecules of the substances are evenly distributed and ordinarily will not separate out upon standing. In a solution the medium in which substances are dissolved is the **solvent.** The substance dissolved in the solvent is the **solute.** Solutes can consist of atoms, ions, or molecules. In cells and in the medium in which cells live, water is the solvent in nearly all solutions. Typical solutes include the sugar glucose, the gases carbon dioxide and oxygen, and many different kinds of ions. Many smaller proteins also can act as solutes in true solutions.

Few living things can survive in highly concentrated solutions. We make use of this fact in preserving several kinds of foods. Can you think of foods that are often kept unrefrigerated and unsealed for long periods of time? Jellies, jams, and candies do not spoil because microorganisms cannot tolerate the high concentration of sugar. Salt-cured meats are too salty to allow growth of microorganisms, and pickles are too acidic for most microbes.

Particles too large to form true solutions can sometimes form *colloidal dispersions,* or **colloids.** Gelatin dessert is an example of a colloid in which the protein gelatin is dispersed in a water medium. Similarly, colloidal dispersions in cells usually are formed from large protein molecules dispersed in water. The fluid or semifluid substance inside living cells is a complex colloidal system. Large particles are suspended by opposing electrical charges, layers of water molecules around them, and other forces. Media for growing microorganisms sometimes are solidified with agar; these media are colloidal dispersions. Some colloidal systems have the ability to change from a semisolid state, such as gelatin that has "set," to a more fluid state, such as gelatin that has melted. Amoebae seem to move, in part, by the ability of the colloidal material within them to change back and forth between semisolid and fluid states.

Acids, Bases, and pH

Chemically speaking, most living things exist in relatively neutral environments, but some microorganisms live in environments that are acidic or basic (alkaline). Understanding acids and bases is important in studying microorganisms themselves and in studying their effects on human cells. An **acid** is a hydrogen ion donor, or proton donor. (A hydrogen ion is a proton.) An acid donates H^+ to a solution. The

CLOSE-UP

Where Have All the Zeros Gone?

The term *pH* is defined as the negative logarithm of the hydrogen ion concentration (in moles per liter). To understand pH we need to understand a little about exponents, scientific notation, and logarithms. The exponent of a number, a superscript to the right of the number, designates the number of times it is to be multiplied by itself. For example, the symbol 2^3 indicates that 2 is to be multiplied by itself 3 times, that is, $2^3 = 2 \times 2 \times 2 = 8$. Similarly, 10^5 indicates that 10 is to be multiplied by itself 5 times. Try it: $10^5 = 100{,}000$.

When scientists work with extremely large or extremely small numbers, they sometimes use *scientific notation.* Instead of writing all the zeros needed to express one billion (1,000,000,000), they observe that a billion is equal to 10 multiplied by itself 9 times, and simply write 1.0×10^9. Likewise, instead of writing one-millionth as 1/1,000,000, they note that the denominator of the fraction is equal to 1.0×10^6, and they write 1.0×10^{-6}. A negative exponent means that the number is less than 1.

When expressing quantities in scientific notation, the exponent is conventionally chosen so that a single digit appears to the left of the decimal value before the exponential term. For example, the expression 22.6×10^6 is not in scientific notation; it should be expressed 2.26×10^7.

The *logarithms* we will use are based on powers of 10. In this system, the logarithm of a number is the power to which 10 must be raised to get the number. For example, the logarithm of 100 is 2, which means that 10 must be multiplied by itself two times to get 100. The logarithm of 1/100 is -2.

Now we can apply the preceding concepts to the definition of pH. The concentration of H^+ in a neutral solution is 1/10,000,000. This can be expressed in exponential notation as 10^{-7}. The logarithm of this number is -7. To determine its negative logarithm, we multiply -7 times -1; the answer is 7. Therefore, the pH of a neutral solution is 7.

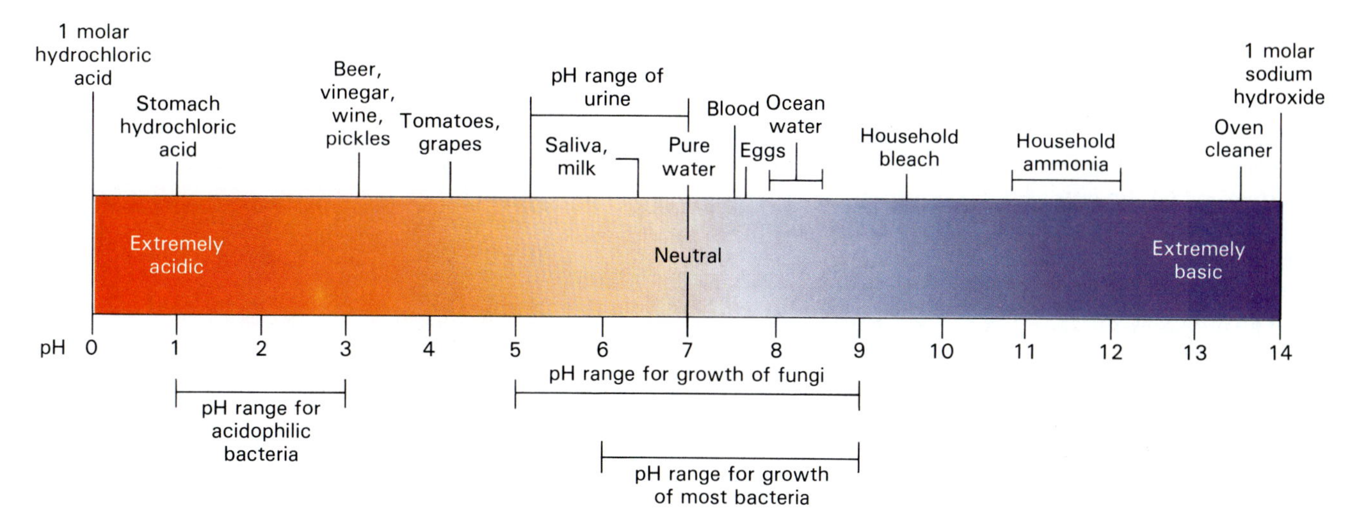

FIGURE 2.7 The pH values of some common substances. Each unit of the pH scale represents a tenfold increase or decrease in the concentration of hydrogen ions. Thus, vinegar, for example, is 10,000 times more acidic than pure water.

acids found in living organisms usually are weak acids such as acetic acid (vinegar). They release H^+ when carboxyl groups (—COOH) ionize to COO^- and H^+. A **base** is a proton acceptor, or a hydroxyl ion donor. It accepts H^+ from the solution or donates OH^- (hydroxyl ion) to it. The bases found in living organisms usually are weak bases such as the amino (NH_2) group, which accepts H^+ to form $NH_3{}^+$.

Chemists have devised the concept of **pH** to specify the acidity or alkalinity of a solution. The pH scale (Figure 2.7), which relates proton concentrations to pH, is a logarithmic scale (see box). This means that the concentration of hydrogen ions (protons) changes by a factor of 10 for each unit of the scale. The practical range of the pH scale is from 0 to 14, and a pH of 7 is neutral—neither **acidic** nor **alkaline** (basic). Pure water has a pH of 7 because the concentrations of H^+ and OH^- in it are equal. Figure 2.7 also shows the pH of some body fluids, selected foods, and other substances.

COMPLEX MOLECULES

The basic principles of general chemistry also apply to **organic chemistry,** the study of compounds that contain carbon. The first organic compound to be synthesized in the laboratory was urea, which is a small molecule excreted as a waste material by many animals. It was made by the German scientist Friedrich Wöhler in 1828. Since that time thousands of organic compounds—plastics, fertilizers, and medicines—have been made in the laboratory. Many organic compounds occur naturally in living things and in the products or remains of living things. They include carbohydrates, lipids, proteins, and nucleic acids. The ability of carbon atoms to form covalent bonds and link up in long chains makes possible the formation of an almost infinite number of organic compounds.

The simplest carbon compounds are the hydrocarbons, chains of carbon atoms with their associated hydrogen atoms. Carbon chains can have not only hydrogen but other atoms such as oxygen and nitrogen bound to them. Some of these atoms form functional groups. A **functional group** is a part of a molecule that generally participates in chemical reactions as a unit and that gives the molecule some of its chemical properties.

Four significant groups of compounds—alcohols, aldehydes, ketones, and organic acids—have functional groups that contain oxygen (Figure 2.8). An

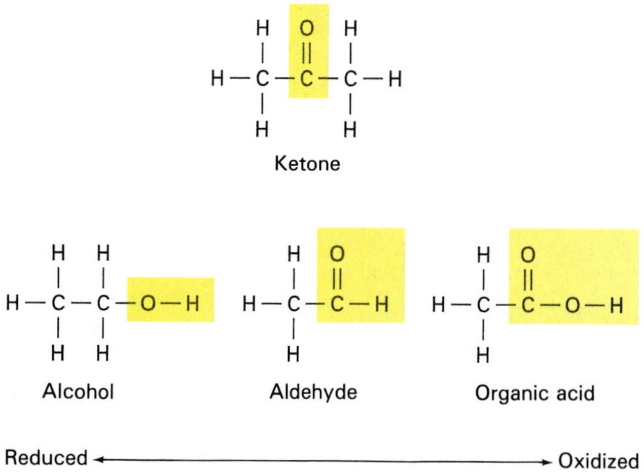

Ketone

Alcohol

Aldehyde

Organic acid

Reduced ⟵ ⟶ Oxidized

FIGURE 2.8 Four classes of organic compounds that incorporate oxygen. Alcohols contain one or more hydroxyl groups (—OH), aldehydes and ketones contain carbonyl groups (=O), and organic acids contain carboxyl groups (—COOH).

alcohol has one or more hydroxyl groups, an aldehyde has a carbonyl group at the end of the carbon chain, a ketone has a carbonyl group within the chain, and an organic acid has one or more carboxyl groups. A functional group that does not contain oxygen is the amino (—NH_2) group. Amino groups are found mainly in amino acids and account for the nitrogen in proteins.

The relative amount of oxygen in different functional groups is significant. Groups with little oxygen, such as alcohol groups, are said to be *reduced;* groups with relatively more oxygen, such as carboxyl groups, are said to be *oxidized.* As we shall see in Chapter 5, *oxidation* is the addition of oxygen or the removal of hydrogen or electrons from a substance. Burning is an example of oxidation. *Reduction* is the removal of oxygen or the addition of hydrogen or electrons to a substance. In general, the more reduced a molecule, the more energy it contains. Hydrocarbons, such as gasoline, have no oxygen and thus represent the extreme in reduced molecules. They also make good fuels because they contain so much energy. Conversely, the more oxidized a molecule, the less energy it contains. Carbon dioxide (CO_2) represents the extreme in an oxidized molecule because two oxygen atoms are the maximum that can bond to a single carbon atom. As we shall see, oxidation releases energy from molecules.

Let us now consider the major classes of large, complex molecules of which all living things, including microbes, are composed.

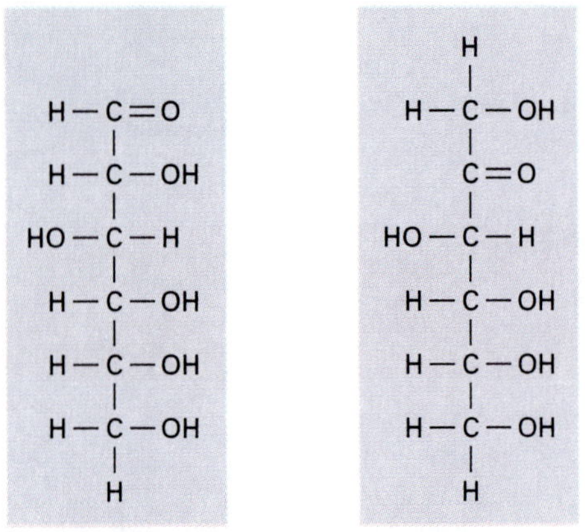

Glucose ($C_6H_{12}O_6$)

Fructose ($C_6H_{12}O_6$)

FIGURE 2.9 Glucose and fructose are isomers: They contain the same atoms, but they differ in structure.

Carbohydrates

Carbohydrates serve as the main source of energy for most living things. Plants make carbohydrates, including structural carbohydrates such as cellulose and energy-storage carbohydrates such as starch. Animals use carbohydrates as food, and many, including humans, store energy in a carbohydrate called glycogen. Many microorganisms use carbohydrates from their environment for energy and also make a variety of carbohydrates. Carbohydrates in the membranes of cells can act as markers that make a cell chemically recognizable. Chemical recognition is important in immunological reactions and other processes in living things.

All carbohydrates contain the elements carbon, hydrogen, and oxygen, generally in the proportion of two hydrogen atoms for each carbon and oxygen atom. The three groups of carbohydrates are monosaccharides, disaccharides, and polysaccharides. **Monosaccharides** consist of a carbon chain or ring with several alcohol groups and one other functional group, either an aldehyde group or a ketone group. Several monosaccharides, such as glucose and fructose, are **isomers**—they have the same molecular formula, $C_6H_{12}O_6$, but different structures and different properties (Figure 2.9). Thus, even at the chemical level we can see that structure and function are related.

Glucose, the most abundant monosaccharide, is represented schematically in Figure 2.10 as a straight chain and a ring and more realistically as a three-

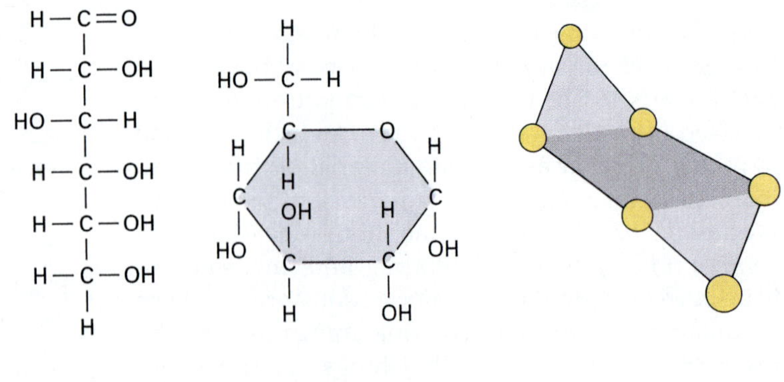

(a) (b) (c)

FIGURE 2.10 Three ways of representing the glucose molecule. In solution, the straight-chain form (a) is rarely found. Instead, the molecule bonds to itself, forming a six-membered ring (b). The ring is conventionally depicted as a flat hexagon, but the actual three-dimensional structure (c) is more complex. The spheres in this depiction represent carbon atoms.

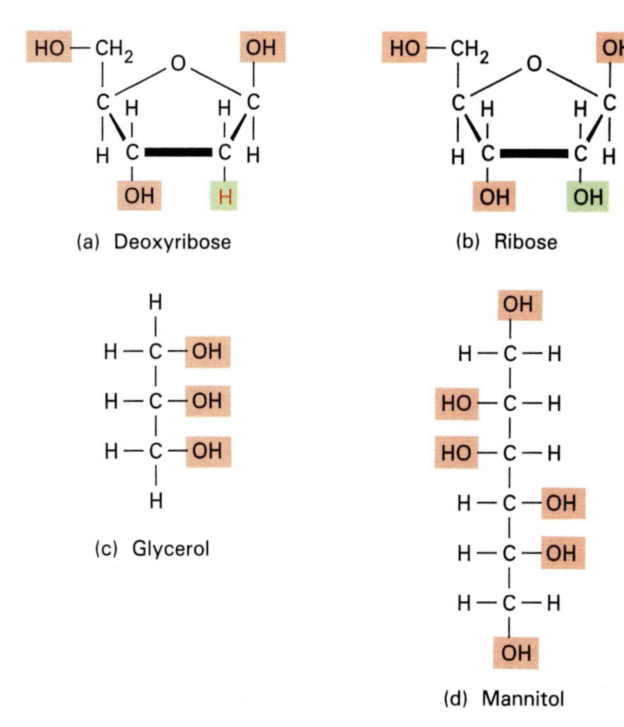

FIGURE 2.11 (a) The deoxy sugar deoxyribose is so called because it lacks a hydroxyl group on one of its carbons, which ribose (b) has. Deoxy refers to having one less oxygen atom. (c) Glycerol, a three-carbon sugar alcohol that is a component of fats (see Figure 2.13). (d) Mannitol, another sugar alcohol used in diagnostic tests for certain microbes.

dimensional structure. The chain structure clearly shows a carbonyl group at carbon 1 (the first carbon in the chain) and alcohol groups on all the other carbons. The second structure shows how a glucose molecule in solution rearranges and bonds to itself to form a closed ring. The three-dimensional projection more nearly shows the actual shape of the molecule. In studying structural formulas it is important to imagine each molecule as a three-dimensional object.

Monosaccharides can be reduced to form deoxy sugars and sugar alcohols (Figure 2.11). The deoxy sugar deoxyribose, which has a hydrogen atom instead of an —OH group on one of its carbons, is a component of DNA. Certain sugar alcohols, which have an additional alcohol group instead of an aldehyde or ketone group, can be metabolized by particular microorganisms. Mannitol and other sugar alcohols are used to identify some organisms in diagnostic tests.

Disaccharides are formed when two monosaccharides are connected by the removal of water and the formation of a glycosidic bond. **Polysaccharides** are formed when many monosaccharides are linked together by **glycosidic bonds** (Figure 2.12). Sucrose, common table sugar, is a disaccharide made of glucose and fructose. Polysaccharides such as starch, glycogen, and cellulose are **polymers** (long chains of repeating units) of glucose. However, the glycosidic bonds in each polymer are arranged differently. Plants and most algae make starch and cellulose. Starch serves as a way to store energy, and cellulose is a structural component of cell walls. Animals make and

APPLICATIONS

Can a Cow Actually Explode?

Cows, as everyone knows, can derive a good deal of nourishment from grass, hay, and other fibrous vegetable matter that humans consider inedible. The fact that you don't find grass in your salad isn't just a matter of taste; grass, hay, and the like are inedible, in that we can't digest the cellulose that is their chief ingredient. If you had to live on hay you would probably starve to death. How, then, do cows and other hooved animals manage on such a diet?

Oddly enough, cows can't digest cellulose either. But they don't need to—it's done for them. Cows and their relatives harbor large populations of microorganisms in their stomachs, and the microorganisms do the actual work of breaking down cellulose into sugars that the animal can use. The same is true of termites: if it weren't for microbes in their guts that help them to digest cellulose, they couldn't dine on the wooden beams in your basement.

Cellulose is very similar to starch—both consist of long chains of glucose molecules. The bonds between these molecules, however, are slightly different in their geometry in the two substances. As a result, the enzymes that animals use to break down a starch molecule into its component glucose units have no effect at all on cellulose. In fact, very few organisms produce enzymes that can attack cellulose. Even the protists (unicellular plants and animals) that live in the stomachs of cows and termites cannot always do it by themselves. Just as cows and termites depend on the protists in their stomachs, the protists frequently depend on certain bacteria that reside permanently within them. It is these bacteria that actually make the essential digestive enzymes.

The activities of the intestinal microorganisms that perform these digestive services are a mixed blessing, both to the cows and to the humans who keep them. The bacteria also produce methane gas, CH_4—as much as 50 to 100 gallons per day from a single cow. (Methane production can be so rapid that a cow's stomach may rupture if the cow can't burp, and some ingenious inventers have actually patented cow safety valves to release the gas buildup directly through the animal's side.) When this gas eventually makes its way out of the cow by one route or another, it rises to the upper atmosphere. There it is suspected of contributing to the "greenhouse effect," trapping solar heat and causing an overall warming of the earth's climate (Chapter 26). Scientists have estimated that the world's cows release 50 million metric tons of methane annually, and that's not counting the sheep, goats, antelope, water buffalo, and other grass eaters.

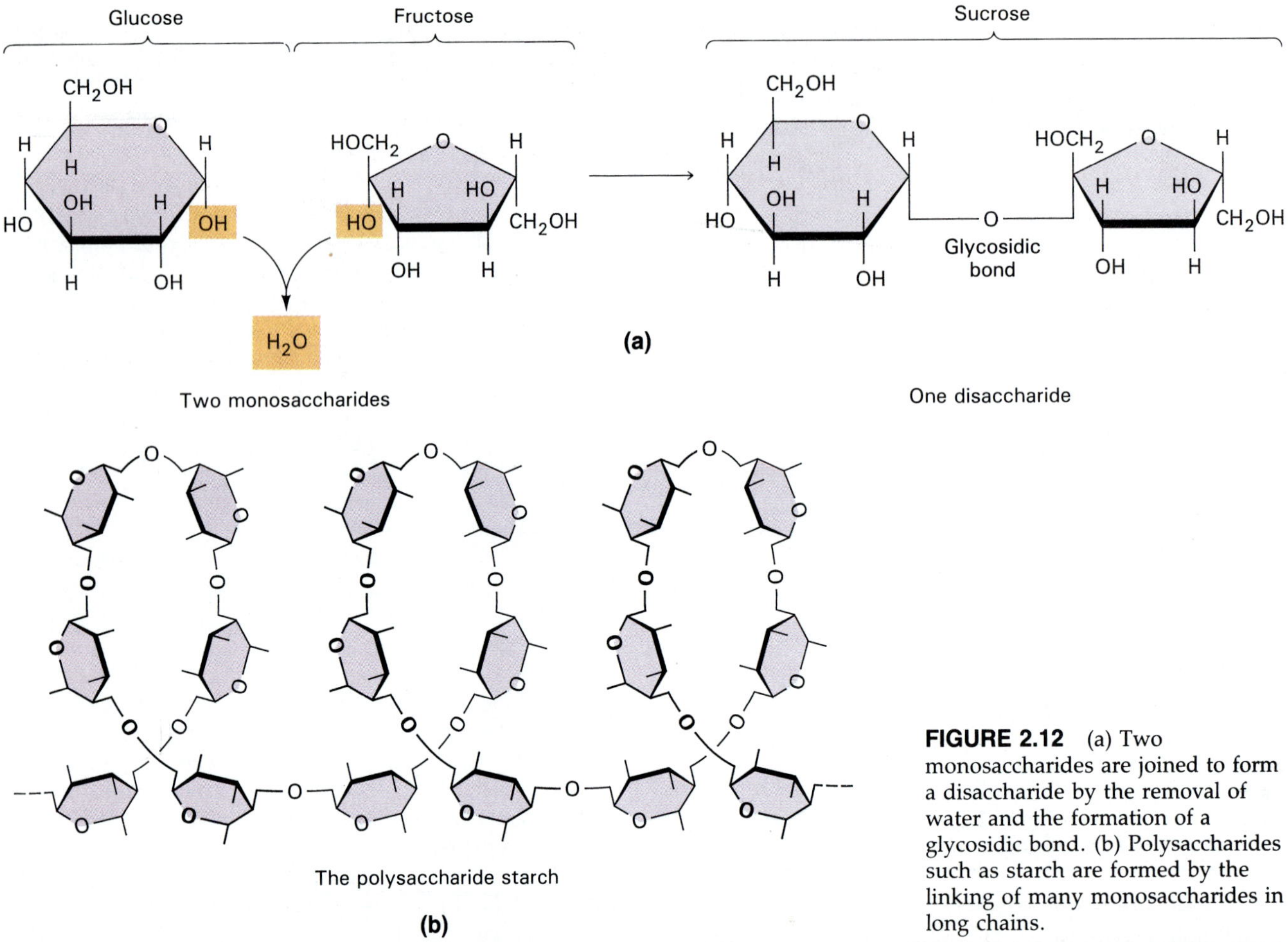

FIGURE 2.12 (a) Two monosaccharides are joined to form a disaccharide by the removal of water and the formation of a glycosidic bond. (b) Polysaccharides such as starch are formed by the linking of many monosaccharides in long chains.

store glycogen, which they can break down to glucose as it is needed for energy. Microorganisms contain several other important polysaccharides, as we shall see in later chapters.

The properties of carbohydrates are summarized in Table 2.4.

Lipids

Lipids constitute a chemically diverse group of substances, which includes fats, phospholipids, and steroids. They are relatively insoluble in water but are soluble in nonpolar solvents such as ether and benzene. Lipids form part of the structure of cells, especially cell membranes, and many can be used for energy. Generally, lipids contain relatively more hydrogen and less oxygen than carbohydrates and therefore contain more energy than carbohydrates. **Fats** contain the three-carbon alcohol glycerol and one or more fatty acids (Figure 2.13). A **fatty acid** consists of a long chain of carbon atoms with associated hydrogen atoms and a carboxyl group at one end of the chain. The synthesis of a fat from glycerol and fatty acids involves removing water and forming an **ester bond** between the carboxyl group of the fatty acid and an

TABLE 2.4 Types of carbohydrates

Class of Carbohydrates	Examples	Description and Occurrence
Monosaccharides	Glucose	Sugar found in most organisms
	Fructose	Sugar found in fruit
	Galactose	Sugar found in milk
	Ribose	Sugar found in RNA
	Deoxyribose	Sugar found in DNA
Disaccharides	Sucrose	Glucose and fructose; table sugar
	Lactose	Glucose and galactose; milk sugar
	Maltose	Two glucose units; product of starch digestion
Polysaccharides	Starch	Polymer of glucose found in plants, digestible by humans
	Glycogen	Polymer of glucose stored in animal liver and skeletal muscles
	Cellulose	Polymer of glucose found in plants, not digestible by humans; digested by some microbes

alcohol group of glycerol. A **triacylglycerol,** formerly called a *triglyceride,* is formed when three fatty acids are connected to glycerol. *Monoacylglycerols* (monoglycerides) and *diacylglycerols* (diglycerides) contain one and two fatty acids, respectively, and usually are formed from the digestion of triacylglycerols.

Fatty acids can be saturated or unsaturated. A **saturated fatty acid** contains all the hydrogen it can have; that is, it is saturated with hydrogen. An **unsaturated fatty acid** has lost at least two hydrogen atoms and contains a double bond between any two carbons that have lost hydrogen atoms. Unsaturated thus means not completely saturated with hydrogen. Oleic acid (Figure 2.13) is an unsaturated fatty acid. Polyunsaturated fats, many of which are vegetable oils that remain liquid at room temperature, contain many unsaturated fatty acid molecules.

Certain important lipids contain one or more other

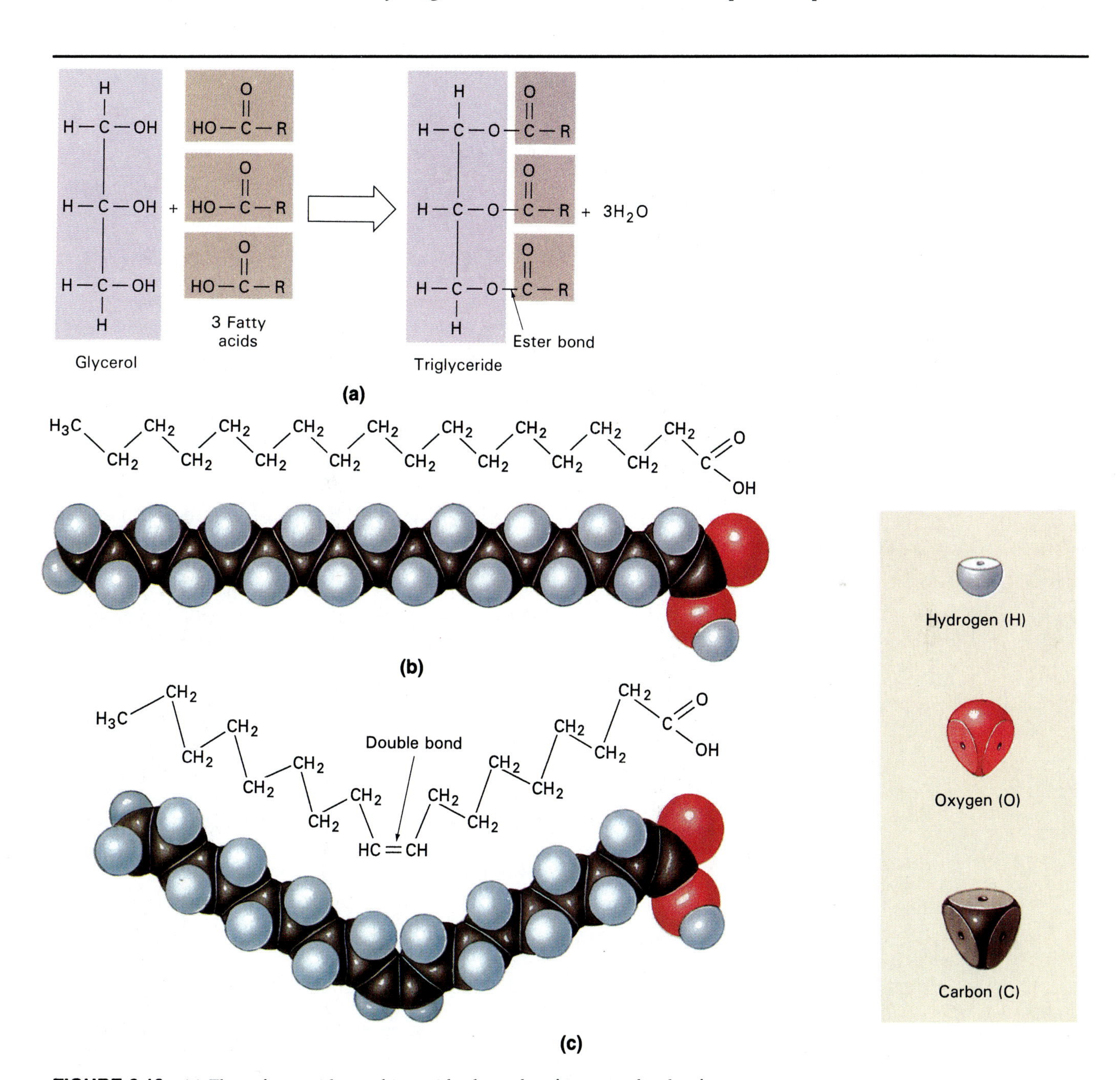

FIGURE 2.13 (a) Three fatty acids combine with glycerol to form a molecule of triacylglycerol, or fat. The group designated R is a long chain and varies in different fatty acids. It may be saturated or unsaturated. (b) Saturated fatty acids have only single bonds between carbon atoms in their carbon chains and can therefore accommodate the maximum possible number of hydrogens. (c) Unsaturated fatty acids such as oleic acid have one or more double bonds between carbons and thus contain fewer hydrogens. The double bond causes a bend in the carbon chain.

$CH_3-CH_2-CH_2-CH_2-CH_2-CH_2-CH_2-CH_2-CH_2-CH_2-CH_2-CH_2-CH_2-C(=O)-O-CH$

$CH_3-CH_2-CH_2-CH_2-CH_2-CH_2-CH_2-CH_2-CH=CH-CH_2-CH_2-CH_2-CH_2-CH_2-C(=O)-O-CH_2$

$H_2C-O-P(=O)(OH)-O^-$

Charged phosphate group

Uncharged fatty acid chains

(a)

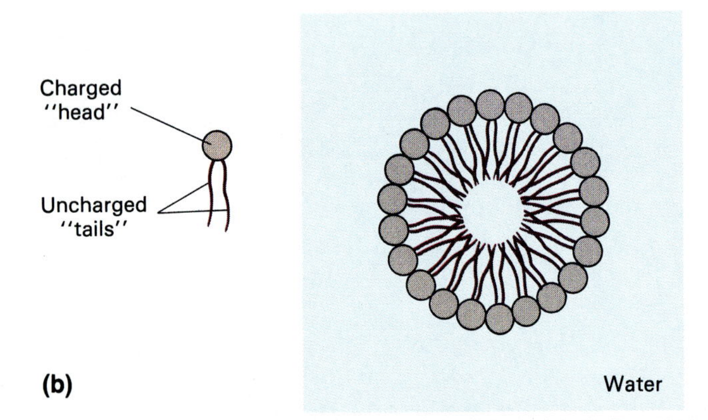

FIGURE 2.14 In phospholipids, (a) one of the fatty acid chains of a fat molecule is replaced by a phosphate group. The charged phosphate group can interact with water molecules, which are polar, but the two long, uncharged fatty acid tails cannot. As a result, phospholipid molecules in water tend to form globular structures with the phosphate groups facing outward and the fatty acids in the interior (b).

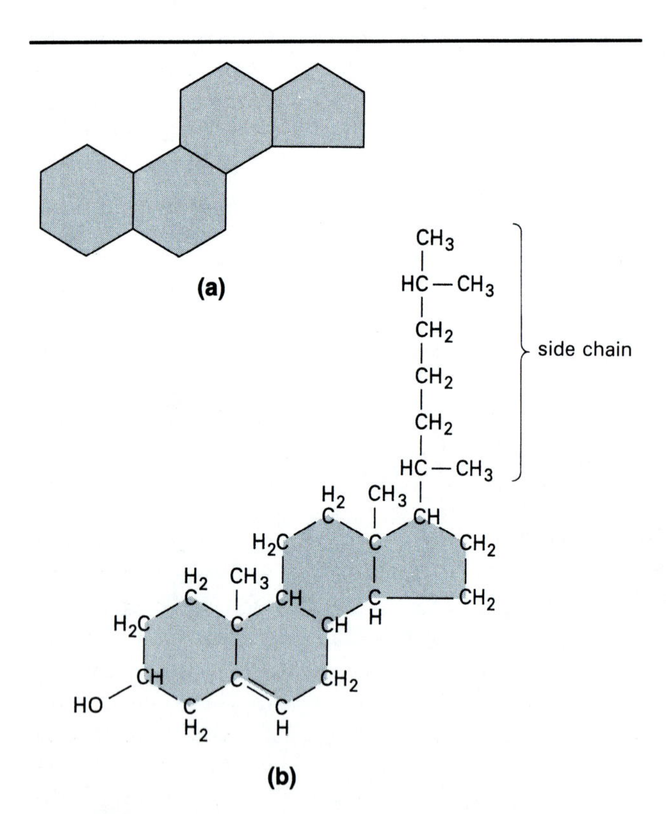

FIGURE 2.15 Steroids are lipids having a characteristic four-ring structure (a). The specific chemical groups attached to the rings determine the properties of different steroids. One of the most biologically important steroids is cholesterol (b), a component of the cell membranes of animal cells and one group of bacteria.

molecules in addition to fatty acids and glycerol. For example, **phospholipids,** which are found in all cell membranes, differ from fats by the substitution of a phosphate group (—HPO_4^-) for one of the fatty acids (Figure 2.14). The charged phosphate end of the molecule can mix with water, but the fatty acid end cannot. This and other properties of the phospholipids are important in determining the characteristics of cell membranes, as we shall see in Chapter 4.

Steroids (Figure 2.15) have a four-ring structure and are quite different from other lipids. They include cholesterol, steroid hormones, and vitamin D. Cholesterol is insoluble in water and is found in the cell membranes of animal cells and a group of bacteria called mycoplasmas. Steroid hormones and vitamin D are important in many animals.

Proteins

Properties of Proteins and Amino Acids

Among the molecules found in living things, proteins have the greatest diversity of structure and function. **Proteins** are composed of building blocks called **amino acids,** which have at least one amino (—NH_2) group and one acidic carboxyl (—COOH) group. The general structure of an amino acid and some of the 20 amino acids found in the proteins are shown in Figure 2.16. Note that each has a different chemical group, called

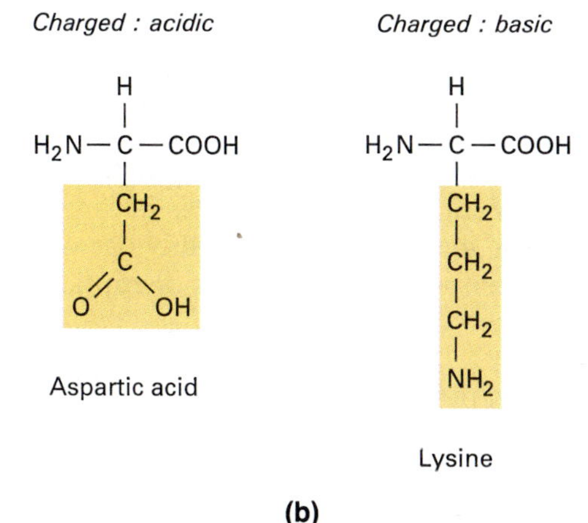

FIGURE 2.16 The general structure of an amino acid, and six representative examples. All amino acids have four groups attached to the central carbon atom: an amino ($—NH_2$) group, a carboxyl (—COOH) group, a hydrogen atom, and a group designated R that is different in each amino acid. The R group determines many of the chemical properties of the molecule; for example, whether it is nonpolar, polar, acidic, or basic.

APPLICATIONS

How To Get Oil and Water To Mix

You probably have washed enough greasy dishes to know that soaps and detergents help to get oily substances off the dishes and into the water. Soap and detergent molecules have both a charged and an uncharged region, like a fatty acid with its ionized carboxyl group at *one end* and its nonionized, hydrogen-saturated carbons at the *other end.* In fact, soap can be made by boiling fat with sodium hydroxide (lye) until the fatty acids break away from the glycerol molecules. The charged ends of soap molecules are attracted to the polar water molecules, and their uncharged ends interact with nonpolar fats and oils, thereby allowing water and oil to mix.

an **R group,** attached to the central carbon atom. Because all amino acids contain carbon, hydrogen, oxygen, and nitrogen and some contain sulfur, proteins also contain these elements. A protein is a polymer of amino acids joined by **peptide bonds** between an amino group of one amino acid and a carboxyl group of another amino acid (Figure 2.17). Two amino acids linked together make a *dipeptide,* three make a *tripeptide,* and many make a **polypeptide.** In addition to the amino and carboxyl groups, some amino acids have a functional group called a sulfhydryl (—SH) group. Sulfhydryl groups in adjacent chains of amino acids can lose hydrogen and form disulfide linkages (—S—S—) from one chain to the other.

Structure of Proteins

Proteins have several levels of structure (Figure 2.18). The **primary structure** of a protein consists of the specific sequence of amino acids in a polypeptide chain. The **secondary structure** of a protein consists of the folding or coiling of amino acid chains into a particular pattern, such as a helix or pleated sheet. Further bending and folding of the protein molecule into globular (irregular spherical) shapes or fibrous threadlike strands produces the **tertiary structure.** Some large proteins such as hemoglobin have **quaternary structure,** formed by the association of several tertiary-shaped polypeptide chains (Figure 2.19). Secondary and higher structures are maintained by disulfide linkages, hydrogen bonds, and other forces between amino acids. The three-dimensional shapes of protein molecules and the nature of sites at which other molecules can bind to them are extremely important in determining how proteins function in living organisms.

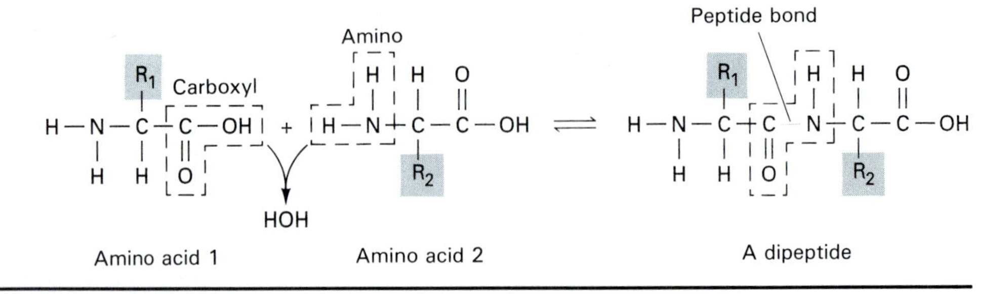

FIGURE 2.17 Two amino acids are joined by the removal of a water molecule and the formation of a peptide bond between the —COOH group of one and the $—NH_2$ group of the other.

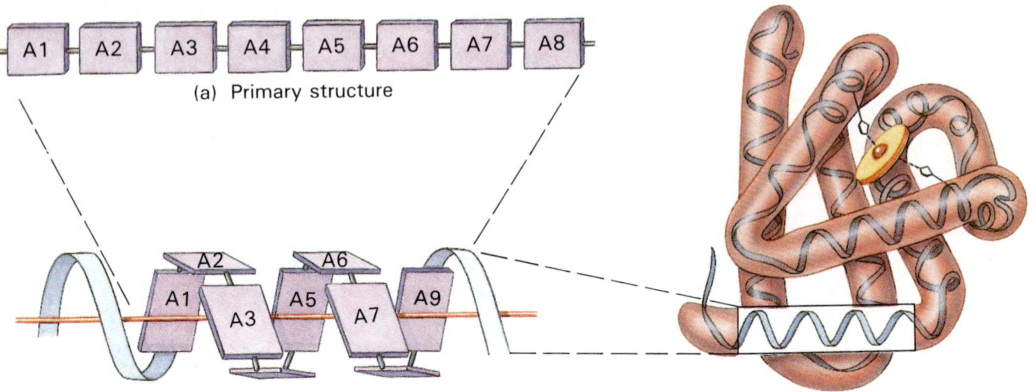

FIGURE 2.18 Three levels of protein structure. (a) Primary structure is simply the sequence of amino acids in a polypeptide chain. Imagine it as a straight telephone cord. (b) Polypeptide chains, especially those of structural proteins, tend to coil or fold into a few simple, regular, three-dimensional patterns called secondary structure. Now imagine the telephone cord as a coiled cord. (c) Polypeptide chains of enzymes and other soluble proteins may also exhibit secondary structure. In addition, the chains tend to fold up into complex, globular shapes that constitute the protein's tertiary structure. Finally, imagine the knot formed when a coiled telephone cord tangles.

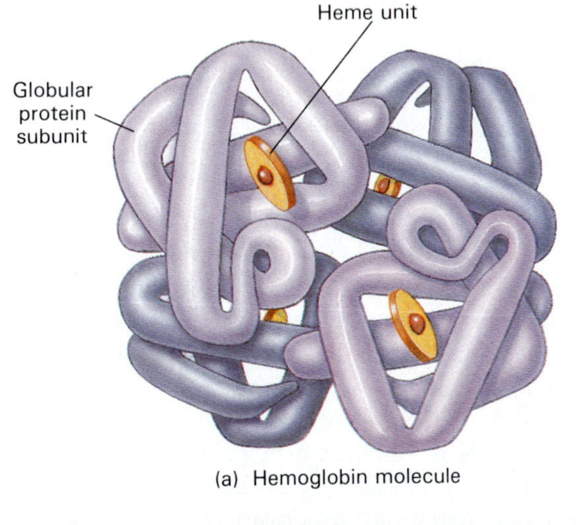

FIGURE 2.19 (a) Many large proteins such as hemoglobin, which carries oxygen in human red blood cells, are made up of several polypeptide chains. The arrangement of these chains makes up the protein's quaternary structure. (b) Some structural proteins such as keratin, a component of human skin and hair, also consist of several polypeptide chains and so have quaternary structure.

Several conditions can disrupt hydrogen bonds and other weak forces that maintain protein structure. They include highly acidic or basic conditions and temperatures above 50°C. Such disruption is called **denaturation.** Sterilization and disinfection procedures often make use of heat or chemicals that kill microorganisms by denaturing their proteins. Also, the cooking of meat tenderizes it by denaturing proteins. Therefore, microbes and cells of larger organisms must be maintained within fairly narrow ranges of pH and temperature to prevent disruption of protein structure.

Classification of Proteins

Most proteins can be classified by their major functions as structural proteins or enzymes. **Structural proteins,** as the name implies, contribute to the three-dimensional structure of cells, cell parts, and membranes. Certain proteins, called *motile proteins,* contribute both to structure and to movement. They account for contraction of animal muscle cells and for some kinds of movement in microbes. **Enzymes** are protein catalysts that control the rate of chemical reactions in cells. A few proteins are neither structural proteins nor enzymes. They include proteins that form receptors for certain substances on cell membranes and antibodies that participate in the body's immune responses (Chapter 18).

Enzymes

Enzymes increase the rate at which chemical reactions take place within living organisms in the temperature range compatible with life. Though enzymes are discussed in more detail in Chapter 5, we summarize their properties here. In general, enzymes speed up reactions by decreasing the energy required to start reactions. They also hold reactant molecules close together in the proper orientation for reactions to occur. Each enzyme has an *active site,* which is the site at which it combines with its *substrate,* the substance on which an enzyme acts. Enzymes have specificity, that is, each acts on a particular substrate or on a certain kind of chemical bond.

Like catalysts in inorganic chemical reactions, enzymes are not permanently affected or used up in the reactions they initiate. Although enzyme molecules will eventually "wear out," probably by losing the shape needed for their catalytic properties, they can be used over and over again to catalyze a reaction. Because enzymes are proteins, they are denatured by extremes of temperature and pH.

Nucleotides and Nucleic Acids

Nucleotides have chemical properties that allow them to perform several essential functions, such as storing energy in high-energy bonds. Nucleotides joined to form nucleic acids are, perhaps, the most remarkable of all biochemical substances. They store information that directs protein synthesis and that can be transferred from parent to progeny.

A **nucleotide** consists of three parts: (1) a nitrogenous base, so named because it contains nitrogen and has alkaline properties, (2) a five-carbon sugar, and (3) one or more phosphate groups (Figure 2.20). The nucleotide adenosine triphosphate (ATP) is the

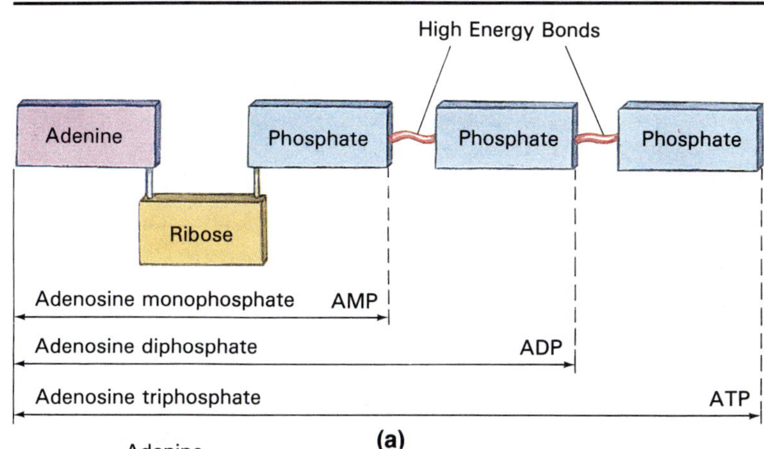

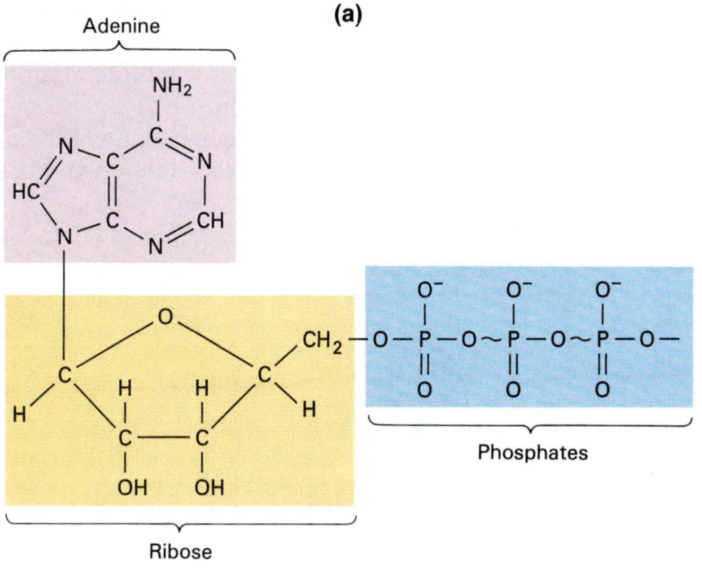

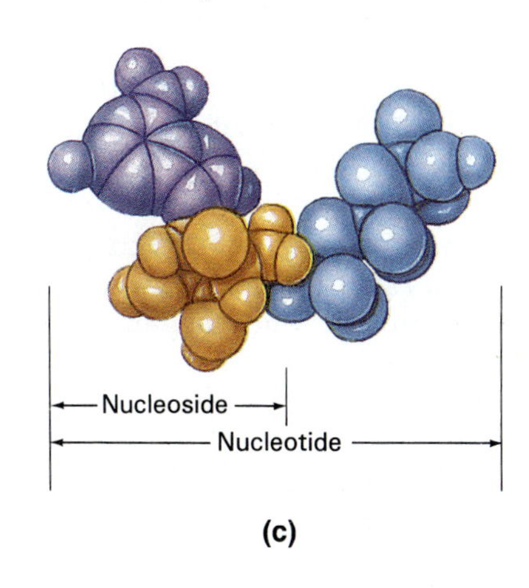

FIGURE 2.20 The nucleotide adenosine triphosphate (ATP), the immediate source of energy for most activities of living cells. A nucleotide consists of a nitrogenous base, a five-carbon sugar, and one or more phosphate groups. (The sugar and base without the phosphates are called a nucleoside.) In ATP the base is adenine, and the sugar is ribose. Adding another phosphate group to adenosine diphosphate greatly increases the energy of the molecule; removing the third phosphate group releases energy that can be used by the cell.

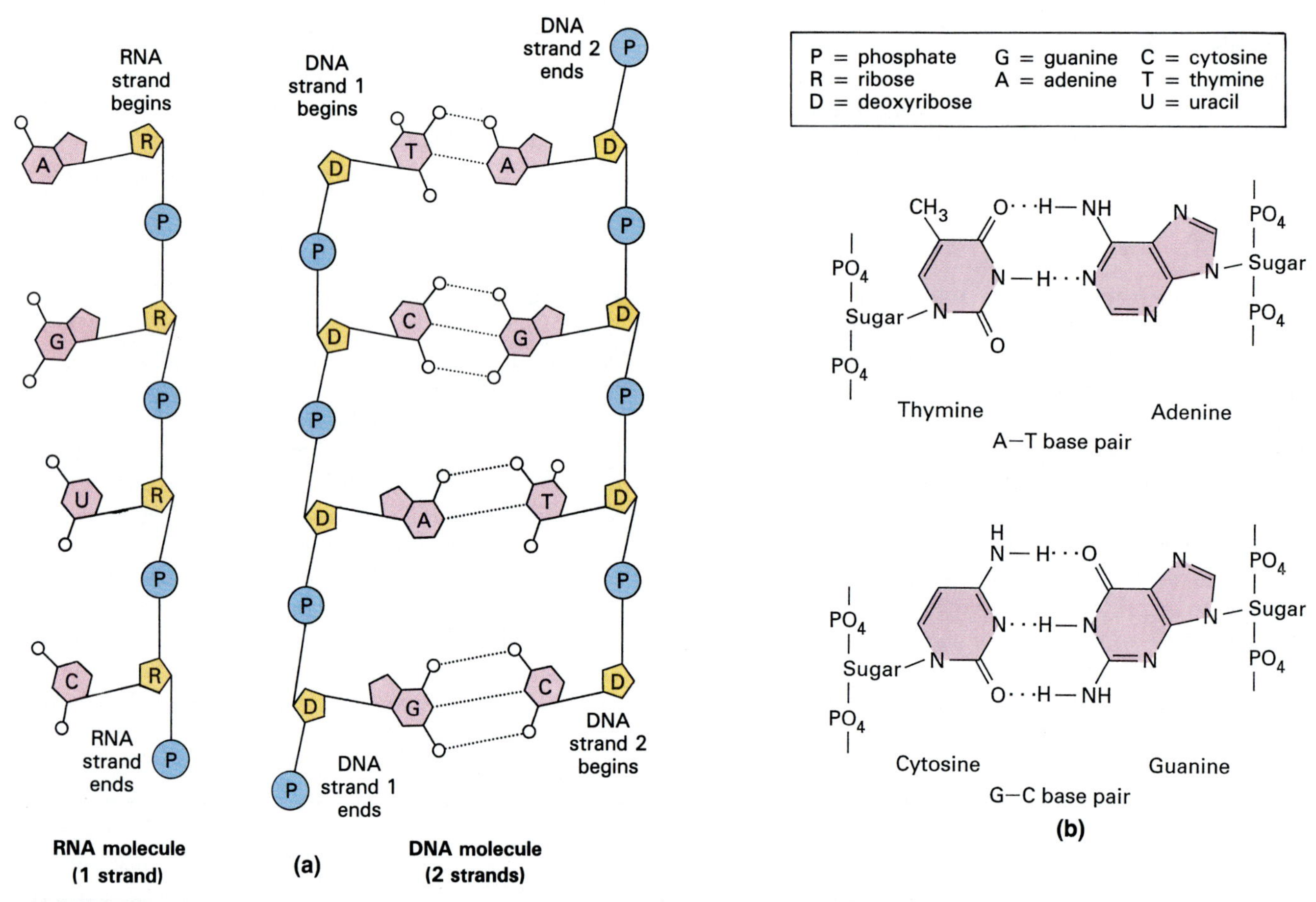

FIGURE 2.21 Nucleic acids consist of a backbone of alternating sugar and phosphate groups to which nitrogenous bases are attached. (a) RNA is usually single-stranded. DNA molecules typically consist of two chains held together by hydrogen bonds between bases. (b) The complementary base pairs in DNA, showing how hydrogen bonds are formed.

main source of energy in cells because it stores chemical energy in a form cells can use. The bonds between phosphates in ATP that are **high-energy bonds** are designated by wavy lines. They contain more energy than most covalent bonds, in that more energy is released when they are broken. Enzymes control the forming and breaking of high-energy bonds so that energy is released as needed within cells. The capture, storage, and use of energy is an important component of cellular metabolism (Chapter 5).

Nucleic acids consist of long polymers of nucleotides. They contain genetic information that determines all the heritable characteristics of a living organism, be it a microbe or a human. Such information is passed from generation to generation and acts in each organism to direct protein synthesis. By directing protein synthesis, nucleic acids determine what structural proteins and enzymes an organism will have. The enzymes determine which other substances the organism can make and which other reactions it can carry out.

The two nucleic acids found in living organisms are **ribonucleic acid (RNA)** and **deoxyribonucleic acid (DNA).** Except in a few viruses, RNA is a single chain of nucleotides, and DNA is a double chain of nucleotides arranged as a double helix. In both nucleic acids the phosphate and sugar molecules form a sturdy but inert chain from which nitrogenous bases protrude. In DNA each chain is connected by hydrogen bonds between the bases, so the whole molecule resembles a ladder with many rungs (Figure 2.21).

DNA and RNA contain somewhat different building blocks (Table 2.5). RNA contains the sugar ribose,

TABLE 2.5 Components of DNA and RNA

Component		DNA	RNA
	Phosphoric acid	X	X
Sugars	Ribose		X
	Deoxyribose	X	X
Bases	Adenine	X	X
	Guanine	X	X
	Cytosine	X	X
	Thymine	X	
	Uracil		X

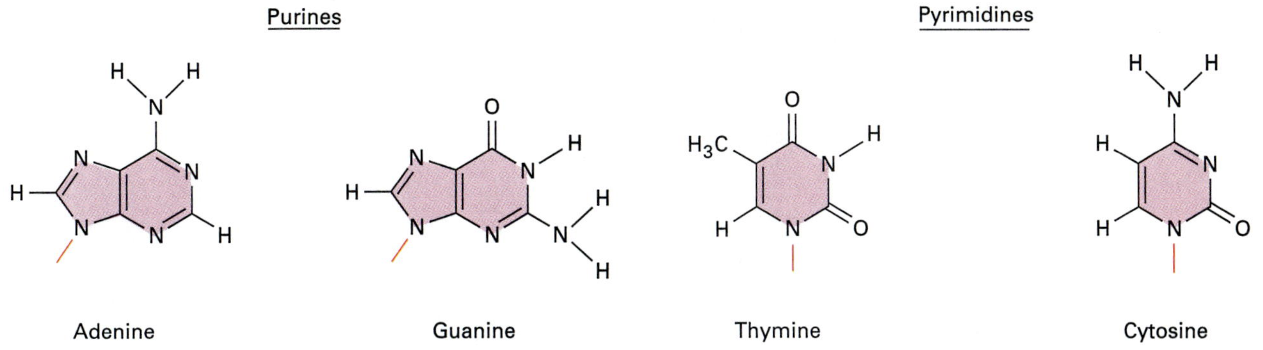

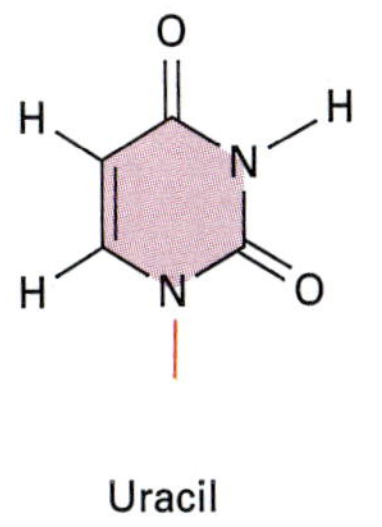

Purines

Adenine

Guanine

Pyrimidines

Thymine

Cytosine

Uracil

FIGURE 2.22 The five bases found in nucleic acids. DNA contains the purines adenine and guanine and the pyrimidines cytosine and thymine. In RNA, thymine is replaced by uracil.

and DNA contains deoxyribose, which has one less oxygen atom than ribose. Three nitrogenous bases, adenine, cytosine, and guanine, are found in both DNA and RNA. In addition, DNA contains the base thymine, and RNA contains the base uracil. Of these bases, adenine and guanine are *purines*, molecules that contain two-ring structures, and thymine, cytosine, and uracil are *pyrimidines*, molecules that contain a single-ring structure (Figure 2.22). All cellular organisms have both DNA and RNA. Viruses have either DNA or RNA but not both.

The two nucleotide chains of DNA are held together by hydrogen bonds between the bases and by other forces. The hydrogen bonds always connect adenine to thymine and cytosine to guanine, as shown in Figure 2.21. This linking of bases is called **complementary base pairing** and is determined by the sizes and shapes of the bases. The same kind of complementary base pairing also occurs when information is transmitted from DNA to RNA at the beginning of protein synthesis (Chapter 7).

DNA and RNA chains contain hundreds or thousands of nucleotides with bases arranged in a particular sequence. This sequence of nucleotides, like the sequence of letters in words and sentences, contains information that determines what proteins an organism will have. As noted earlier, an organism's structural proteins and enzymes, in turn, determine what the organism is and what it can do. Like changing a letter in a word, changing a nucleotide in a sequence can change the information it carries. The number of different possible sequences of bases is almost infinite, so DNA and RNA can contain a great many different pieces of information.

The functions of DNA and RNA are related to their ability to convey information. DNA is transmitted from one generation to the next. It determines the heritable characteristics of the new individual by supplying the information for the proteins its cells will contain. In contrast, RNA carries information from the DNA to the sites where proteins are manufactured in cells. There it directs and participates in the actual assembly of proteins, as is explained in Chapters 7 and 8. The functions of these nucleic acids are discussed in more detail in Chapters 7 and 8.

ESSAY

Bacterial Acids Are Eating the *Last Supper*

Past civilizations built their temples, tombs, and monuments out of stone—so that they would last forever. Indeed, some of these structures have lasted for thousands of years. In the last several decades, however, this situation has been changing rapidly. Automobiles and industrial chimneys have increasingly spewed gaseous pollutants into the atmosphere. People, such as museum directors who help maintain ancient architecture, began to notice a correlation between atmospheric pollutants and damage to stonework. At first people believed this damage to be primarily due to the direct action of the chemical pollutants on the stone. Now, however, a whole new world, a microecosystem of microbes, has been revealed to be living in the

ESSAY Bacterial Acids Are Eating the *Last Supper*

stone itself, supported by the pollutant gases.

The prime villain devouring marble is *Thiobacillus thioparis*, a bacterium that takes pollutant sulfur dioxide (SO_2) gas from the air and converts it into sulfuric acid (H_2SO_4). The sulfuric acid acts upon the calcium carbonate ($CaCO_3$), of which marble is made, to release carbon dioxide (CO_2) and the salt calcium sulfate ($CaSO_4$). *Thiobacilli* utilize the CO_2 as their source of carbon, which is necessary for all the organic molecules they produce. They are literally "eating" the marble. Calcium sulfate is a form of plaster. On many buildings, such as the Parthenon in Greece, this bacterial "epidemic" has already turned a 2-inch-thick layer of the marble surface into plaster. More marble has been destroyed in the last 30 years than in the previous 300 years. The plaster produced by this chemical process is soft and gets washed away by rain, or just crumbles and falls off. Statues lose facial features, and buildings lose their decorative carvings (Figure 2.23). What previous generations have marveled at is being lost now to all future generations.

FIGURE 2.24 A close-up of Leonardo da Vinci's *Last Supper* shows deterioration due to microbial action.

FIGURE 2.23 Greek authorities are trying to prevent further deterioration of caryatid statues on the Acropolis. Note the loss of facial features and other details due to microbial action and air pollutants.

Not all of the damage, however, is due to chemical processes. Some of the damage is due to the physical activities of fungi pushing their thread-like growth (hyphae) into the rock, splitting and reducing it to powder. Other bacteria and fungi produce nitric and nitrous acids, plus organic acids such as acetic acid (vinegar). Dozens of kinds of bacteria, yeast, filamentous fungi, and even algae attack the stone chemically and physically. But, is it just the stone that is being destroyed? No, frescoes painted on walls, such as Leonardo da Vinci's *Last Supper*, are also being consumed: pieces flaking off, brilliant colors being dulled (Figure 2.24). Is there any cure for these microbial "infestations"?

Italian microbiologist Sergio Curri and his colleagues have devised a treatment plan that the Museum of Athens has requested that they undertake on several Greek national treasures. Curri first identifies which microorganisms are attacking the stone, and he then finds which antibiotics will kill them most effectively. However, administering antibiotics to a building is a tricky business. A spraygun must be used rather than a hypodermic needle or pills. Some antibiotics cannot be used in this way, so it is difficult to find a suitable one. Sometimes disinfectants such as isothiazolinone chloride are used.

Unfortunately, this treatment does not reverse the damage already done. Statues cannot grow a new skin, as you might after recovering from an infection. Therefore, researchers at the Museum of Athens are working on a process to harden chemically the plaster layer formed by microbes. To do this, they must carefully bake the affected pieces at high temperatures—hardly feasible for an entire statue or cathedral. Even where this can be done, is it safe to return these statues into the atmosphere as long as the gaseous pollutants remain? Stopgap measures have included taking some statues inside and building protective domes over others. Of course, the real answer lies in cleaning up our environment. Little did we imagine that changes in the chemistry of our air would have such dramatic effects on microbes and that the microbes in turn would cause such damage to the products of Western civilization.

TRY IT

Does Salt Affect Soil Microbes?

A similar situation regarding careless or excessive release of chemicals occurs when winter salt runoff from icy roads affects microbial populations in soil. You might want to investigate this phenomenon in the laboratory using local soils and various anti-icing products. What effects occur? What concentrations of these products are necessary? Do different products have different effects? Are all soil types equally affected? Are all organisms affected in the same way? How far from the edge of the road, median strip, or sidewalk are the effects found? Do melt waters in ditches carry these chemicals to distant areas where their effects are also felt? How long do any effects last? Interestingly, when soil is too alkaline, powdered elemental sulfur is sometimes sprinkled into the soil, where *Thiobacilli* will then colonize the sulfur granules and release sulfuric acid, which adjusts the soil pH back toward normal.

CHAPTER SUMMARY

WHY STUDY CHEMISTRY?

- A knowledge of basic chemistry is needed to understand how microorganisms function and how they affect humans.

RELATED KEY TERMS

metabolism

CHEMICAL BUILDING BLOCKS AND CHEMICAL BONDS

Chemical Building Blocks

- An **element** is a fundamental kind of matter, and the smallest unit of an element is an **atom;** a **molecule** consists of two or more atoms chemically combined, and a **compound** consists of two or more different kinds of atoms chemically combined.
- The most common elements in living things are carbon (C), hydrogen (H), oxygen (O), and nitrogen (N).

Structure of Atoms

- Atoms consist of positively charged **protons** and neutral **neutrons** in the nucleus and very small, negatively charged **electrons** orbiting the nucleus.
- The number of protons in an atom determines the **atomic number.** The total number of protons and neutrons determines the **atomic weight.**
- **Ions** are atoms that have gained or lost one or more electrons.
- **Isotopes** are atoms of the same element that contain different numbers of neutrons; some may be **radioisotopes.**

rule of octets **cation** **anion** **mole** **gram molecular weight**

Chemical Bonds

- Atoms of molecules are held together by chemical bonds.
- **Ionic bonds** involve attraction of oppositely charged ions, **covalent bonds** involve atoms sharing pairs of electrons, and **hydrogen bonds** involve weak attractions between polar regions of hydrogen atoms and oxygen or nitrogen atoms.

polar compound

Chemical Reactions

- Chemical reactions involve breaking or forming chemical bonds and associated energy changes.
- Catabolic reactions break down molecules and release energy. Anabolic reactions require energy to synthesize larger molecules.

anabolism **exergonic** **catabolism** **endergonic**

WATER AND SOLUTIONS

RELATED KEY TERMS

Water

- Water is a **polar compound,** acts as a solvent, and forms thin layers.
- Water also has high specific heat, and it serves as a medium for and participates in many chemical reactions.

surface tension **reactant**
dehydration synthesis
hydrolysis

Solutions and Colloids

- **Solutions** consist of **mixtures** with one or more **solutes** evenly distributed throughout a **solvent.**
- **Colloids** contain particles too large to form true solutions.

Acids, Bases, and pH

- In most solutions containing acids or bases, **acids** release H^+ ions, and **bases** accept H^+ ions (or release OH^- ions).
- The **pH** of a solution is a measure of its acidity or alkalinity. A pH of 7 is neutral, below 7 is **acidic,** and above 7 is **basic,** or **alkaline.**

acidic **alkaline**

COMPLEX MOLECULES

- **Organic chemistry** is the study of carbon-containing compounds.
- Organic compounds such as alcohols, aldehydes, ketones, organic acids, and amino acids can be identified by their **functional groups.**

Carbohydrates

- **Carbohydrates** consist of carbon chains in which most of the carbon atoms have an associated alcohol group and one carbon has either an aldehyde or a ketone group.
- The simplest carbohydrates are **monosaccharides,** which can combine to form **disaccharides** and **polysaccharides.** Long chains of repeating units are called **polymers.**
- The body uses carbohydrates primarily for energy.

glycosidic bond **isomer**

Lipids

- All **lipids** are soluble in nonpolar solvents.
- **Fats** consist of glycerol and **fatty acids.**
- Phospholipids contain phosphoric acid instead of a fatty acid.
- **Steroids** have a complex four-ring structure.

ester bond **triacylglycerol**
saturated fatty acid
unsaturated fatty acid

Proteins

- **Proteins** consist of chains of **amino acids** linked by **peptide bonds.**
- Proteins form part of the structure of cells, act as enzymes, and contribute to other functions such as motility, transport, and regulation.
- **Enzymes** are biological catalysts of great **specificity** that increase the rate of chemical reactions in living organisms. Each enzyme has an **active site** to which its **substrate** binds.

R group **polypeptide**
primary structure
secondary structure
tertiary structure
quaternary structure
denaturation
structural protein

Nucleotides and Nucleic Acids

- A **nucleotide** consists of a niteogenous base, a sugar, and one or more phosphates.
- Some nucleotides contain **high-energy bonds.**
- **Nucleic acids** are important information molecules that consist of chains of nucleotides.

ribonucleic acid (RNA)
deoxyribonucleic acid (DNA)
complementary base pairing

QUESTIONS FOR REVIEW

A.

1. Why study chemistry?

B.

2. Define atom, element, molecule, and compound.
3. What are the most common elements in living things and what are their symbols?
4. Show how protons, electrons, and neutrons are arranged in atoms.
5. How do ions differ from atoms?
6. How do isotopes of an element differ from each other?

C.

7. Use diagrams to illustrate the differences among ionic, covalent, and hydrogen bonds.
8. Distinguish among metabolism, catabolism, and anabolism and between exergonic and endergonic.

D.

9. In what ways is water important to living things?

10. What are the characteristics of a solution?
11. Distinguish between an acid and a base.
12. How is the pH scale used to measure acidity?

E.

13. What is organic chemistry?
14. List five kinds of organic compounds and their associated functional groups.

F.

15. What is the basic structure of a monosaccharide?
16. How are disaccharides and polysaccharides different from monosaccharides?
17. In what ways are carbohydrates used in living organisms?

G.

18. Describe the structure and uses in the body of simple lipids, phospholipids, and steroids.
19. Distinguish between a saturated and an unsaturated fatty acid.

H.

20. Define amino acid, peptide bond, and denaturation.
21. What are the four levels of protein structure and how is each maintained?
22. In what ways do structural proteins and enzymes differ?
23. What are the main properties of enzymes?

I.

24. Briefly describe the structure and function of ATP, DNA, and RNA.

PROBLEMS FOR INVESTIGATION

1. Virtually all living organisms use the nucleotide ATP as a vehicle for chemical energy. Why? What properties of the molecule enable it to perform this function? What do you think this universality tells us about the evolutionary history of life?
2. If a protein were able to control its own replication without using DNA or RNA, should we then call it a "genetic material"? Look ahead to the discussion of prions in Chapter 11 to consider how these proteinaceous infectious particles may replicate.
3. Lipids play an important part in the cell wall structure of some bacteria such as *Mycobacterium tuberculosis* and *Mycobacterium leprae*, the causes of tuberculosis and leprosy. Read about the waxy materials tuberculin and lepromin. What clinical use is made of them?
4. How can weak interactions such as hydrogen bonds hold together large molecules such as proteins and DNA? Find out more about the various forces involved in creating and maintaining the structure of proteins and nucleic acids.

SOME INTERESTING READING

Brown, T. L., H. E. LeMay, and B. E. Bursten. 1991. *Chemistry: the central science.* Englewood Cliffs, NJ: Prentice Hall.

Gibbons, A., and M. Hoffman. 1991. New 3-D protein structures revealed. *Science* 253, no. 5018 (July 26):382–83. (The shape of cholera toxin and the first protein kinase structure.)

Inoue, M., and M. Koyano. 1991. Fungal contamination of oil paintings in Japan. *Int. Biodeterior.* 28(1–4):23–35.

McMurry, J., and M. E. Castellion. 1992. *Fundamentals of general, organic, and biological chemistry.* Englewood Cliffs, NJ: Prentice Hall.

"The molecules of life." 1985. *Scientific American* 253, no. 4 (October). (Special issue, various authors.)

Sharon, N. 1980. "Carbohydrates." *Scientific American* 243, no. 5 (November):90.

Stryer, L. 1988. *Biochemistry.* San Francisco: W. H. Freeman and Co.

Watson, J. D. 1980. *The double helix.* ed. Gunther S. Stent. New York: W. W. Norton.

Wiggins, P. M. 1990. Role of water in some biological processes. *Microbiological Reviews* 54(4):432–49.

A magnificent French gilt-bronze microscope made around 1745 by Juste-Aurèle Meissonnier (1695–1750), official designer to the king. This microscope, which combined scientific and Rococco artistic elements, was no doubt made for someone of eminence, perhaps even a member of the royal family. The original is now included in the collection of the Cleveland Museum of Art.

3 Microscopy and Staining

This chapter focuses on the following questions:

A. How is the evolution of microscopy instruments related to progress in microbiology?

B. Which metric units are most useful for the measurement of microbes?

C. What are the relationships among wavelength, resolution, numerical aperture, and total magnification?

D. How are the following properties of light related to microbiology: transmission, absorption, fluorescence, luminescence, phosphorescence, reflection, refraction, and diffraction?

E. What is the function of each part of a compound microscope?

F. What are the special uses and adaptations of bright-field, dark-field, phase-contrast, differential interference contrast, and fluorescence (UV) microscopes?

G. What are the principles of transmission and scanning electron microscopy? How do the advantages and limitations of electron microscopy compare with those of light microscopy?

H. Which techniques are used to prepare and heighten contrast in specimens to be viewed with a light microscope?

I. What are the uses of the common types of microbial stains?

J. What are the functions and results of each of the steps in the Gram staining procedure?

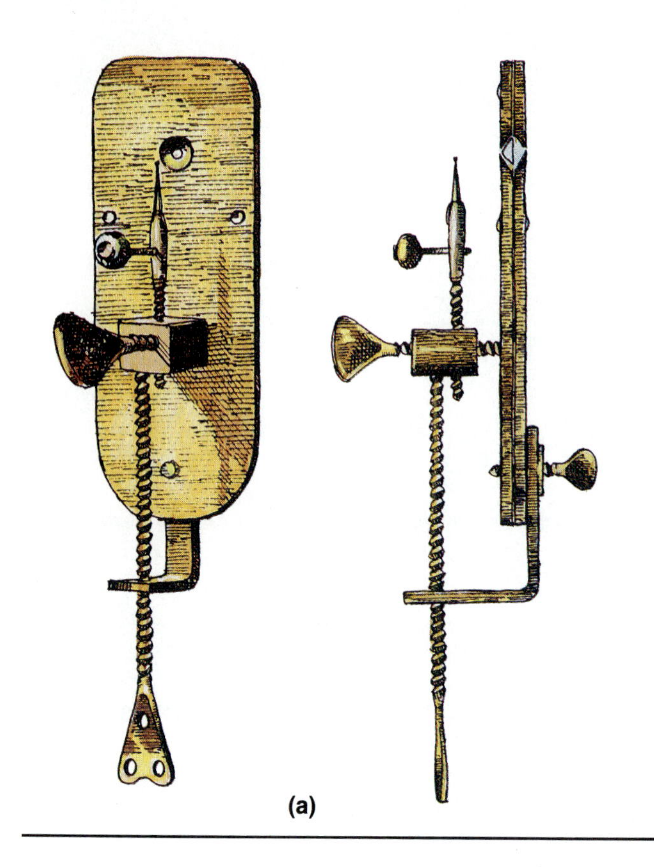

(a)

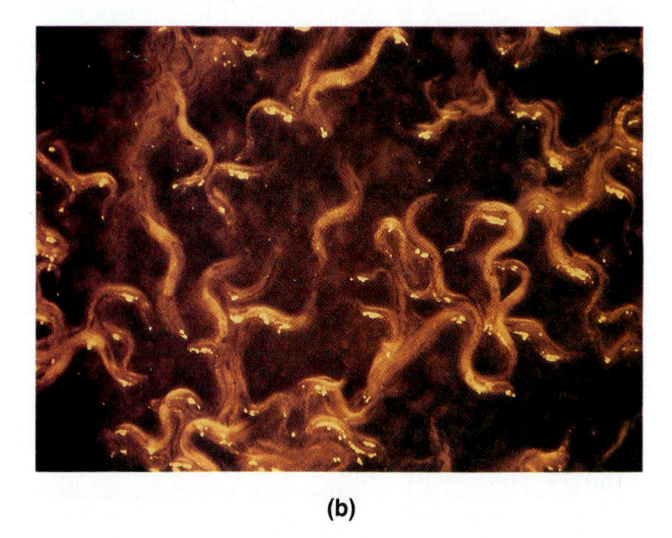

(b)

FIGURE 3.1 (a) Drawing of one of Leeuwenhoek's microscopes. This simple microscope—really a very powerful magnifying glass—made use of a single, tiny, almost spherical lens set into the metal plate. The specimen was mounted on the needlelike end of the vertical shaft and examined through the lens from the opposite side. The various screws were used to position the specimen and bring it into focus—a very difficult process. (b) The vinegar eels (nematodes) that so upset Leeuwenhoek's friends.

DEVELOPMENT OF MICROSCOPY

For thousands of years people were aware of the activities of microorganisms—the rotting of food, fermentations by yeasts, infectious diseases, and the like—without being aware of the existence of the microbes themselves. Until microscopes were invented, no one could attribute these effects to microorganisms.

Anton van Leeuwenhoek (1632–1723), living in Delft, Holland, was almost certainly the first person to see microorganisms. He constructed simple microscopes capable of magnifying objects 100 to 300 times. These instruments were unlike what we commonly think of as microscopes today. Consisting of a single tiny lens, painstakingly ground, they were actually very powerful magnifying glasses (Figure 3.1). It was so difficult to focus one of Leeuwenhoek's microscopes that instead of changing specimens he just built a new microscope for each specimen, leaving the previous specimen and microscope together. When foreign investigators came to Leeuwenhoek's laboratory to look through his microscopes, he made them keep their hands behind their backs, to prevent them from touching the focusing apparatus!

Leeuwenhoek kept his techniques secret. Even today we are not sure of his methods of illumination, although it is likely that he used indirect lighting, with light bouncing off the sides of specimens rather than passing through them. He was also unwilling to part with any of the 419 microscopes he made. It was only near the time of his death that his daughter, at his direction, sent 100 of them to the Royal Society of London.

Following his death, no one came forward to continue the work of perfecting the design and construction of microscopes, and the progress of microbiology slowed. Still, Leeuwenhoek had taken the first steps. In 1676 he reported to the Royal Society the first observations of bacteria and protozoans in water, and in 1683 he described bacteria taken from his own mouth. Through his letters to the Royal Society, the existence of microbes was revealed to the scientific community. However, Leeuwenhoek could see very

CLOSE-UP

The Secret Kingdom Uncovered

"I have had several gentlewomen in my house, who were keen on seeing the little eels in vinegar; but some of 'em were so disgusted at the spectacle, that they vowed they'd never use vinegar again. But what if one should tell such people in future, that there are more animals living in the scum on the teeth in a man's mouth, than there are men in a whole kingdom?"

—Anton van Leeuwenhoek, 1683

TABLE 3.1 Some commonly used units of length

Unit (abbreviation)	Metric Equivalent	English Equivalent
meter (m)		3.28 feet
centimeter (cm)	0.01 m = 10^{-2} m	0.39 inches
millimeter (mm)	0.001 m = 10^{-3} m	0.039 inches
micrometer (μm)	0.000001 m = 10^{-6} m	0.000039 inches
nanometer (nm)	0.000000001 m = 10^{-9} m	0.000000039 inches
Ångström (Å)	0.0000000001 m = 10^{-10} m	0.0000000039 inches

little detail of their structure. Further study required the development of more complex microscopes, as we shall see shortly.

PRINCIPLES OF MICROSCOPY

Metric Units

Microscopy is the technology of making very small things visible to the unaided human eye. Because microorganisms are so small, the units used to measure microorganisms are likely to be unfamiliar to beginning students used to dealing with a macroscopic world. Unfortunately, to complicate matters, there has also been a change in the accepted names of the three units most often used to describe microbes (Table 3.1).

The **micrometer** (μm), formerly called a micron (μ), is equal to 0.000001 m. A micrometer also can be expressed as 10^{-6} m. The second unit, the **nanometer** (nm), formerly called a millimicron (mμ), is equal to 0.000000001 m. It also is expressed as 10^{-9} m. A third unit, the **Ångström** (Å), is found in much of the current and older literature but no longer has any official recognition. It is equivalent to 0.0000000001 m, 0.1 nm, or 10^{-10} m. Figure 3.2 is a scale summarizing the

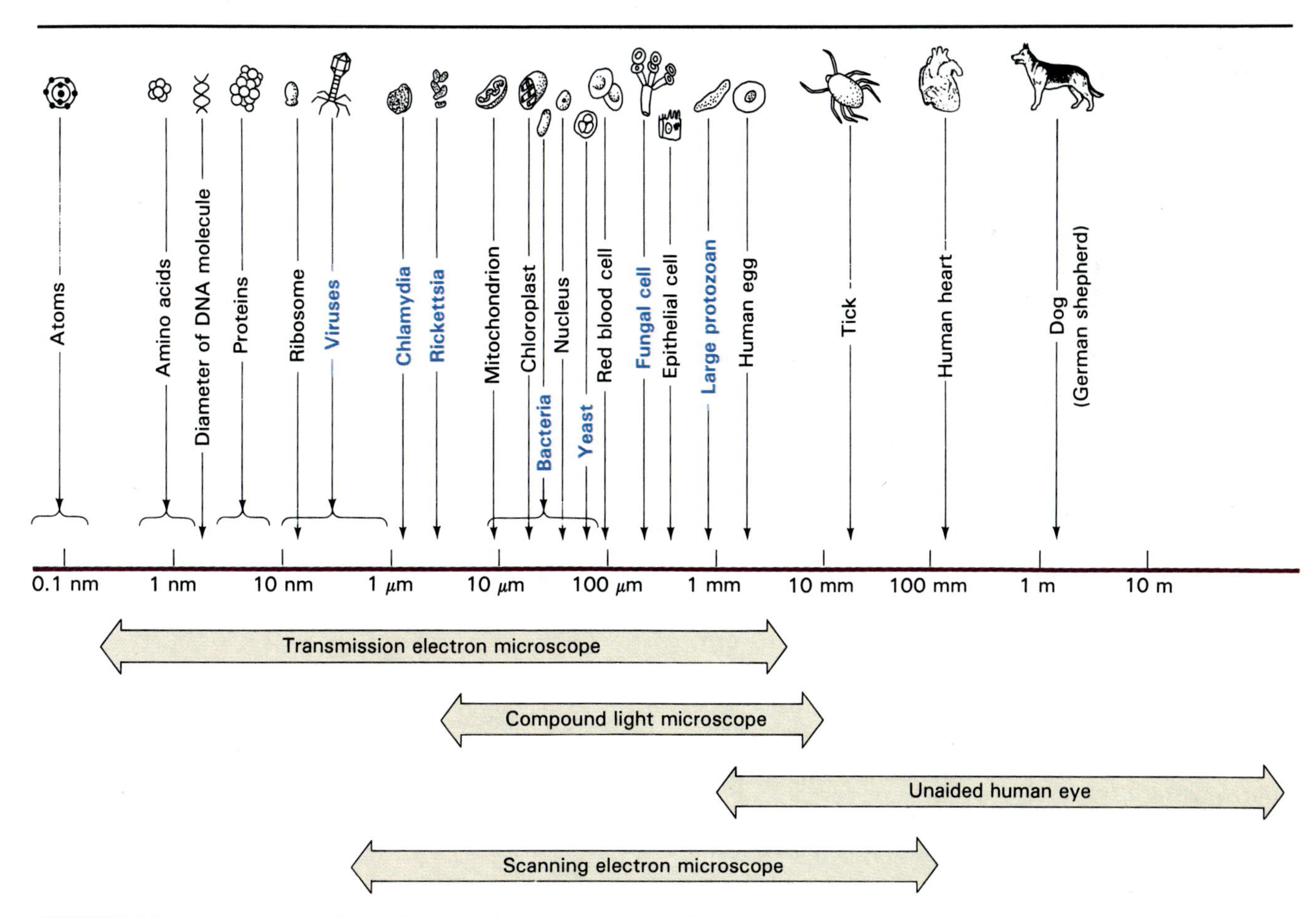

FIGURE 3.2 Relative sizes of microbes. Various organisms and their relation to metric units of measurement are shown; the range of effective use for various instruments is also depicted.

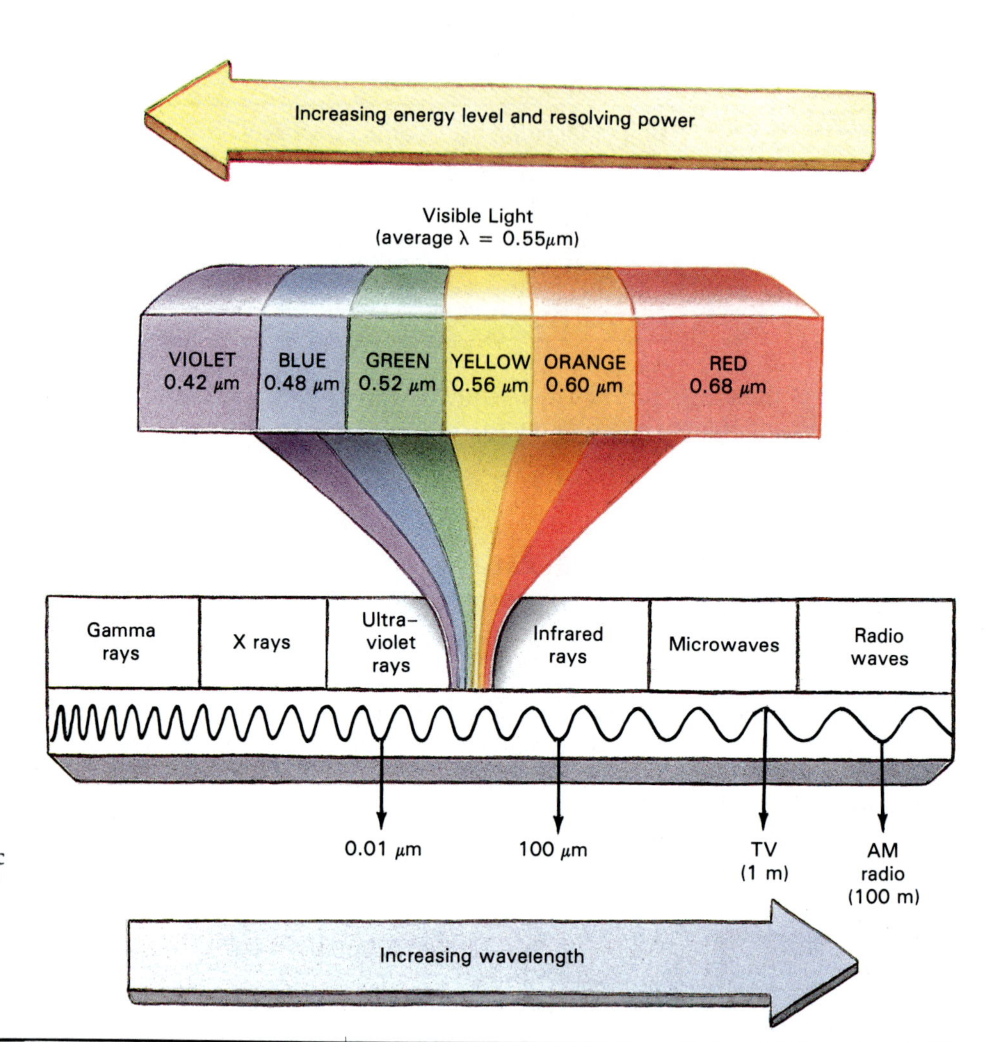

FIGURE 3.3 The electromagnetic spectrum. Only a narrow range of wavelengths—those of visible and ultraviolet light—are used in light microscopy. The shorter the wavelength used, the greater the resolution that can be attained.

metric system unit equivalents, the ranges of sizes that can be detected by the unaided human eye and by various types of microscopy, and examples of where various organisms fall on this scale.

Properties of Light: Wavelength and Resolution

Light has a number of properties that affect our ability to visualize objects, both with the unaided eye and (more crucially) with the microscope. Understanding these properties will allow you to take them into consideration and improve your practice of microscopy.

One of the most important properties of light is its wavelength. The sun produces a continuous spectrum of electromagnetic radiation with waves of various lengths (Figure 3.3). Visible light rays as well as ultraviolet and infrared rays constitute particular parts of this spectrum. The **wavelength,** or the length of a light ray (Figure 3.4), used for observation is crucially related to the resolution that can be obtained. **Resolution** refers to the ability to see two items as separate and discrete units, rather than as a fuzzily overlapped single image (Figure 3.5). One can magnify objects, but if the objects cannot be resolved the magnification is useless.

Light must pass between two objects for them to be seen as separate things. If the wavelength of the light by which we see the objects is too long to pass between them, they will appear as one. So the key to resolution is to get light of a small-enough wavelength to fit between the objects you are trying to see separately. To visualize this (Figure 3.6), imagine throwing ink-covered objects with diameters corresponding to various wavelengths at a target with a foot-high letter *E* on a white background. If the object you throw

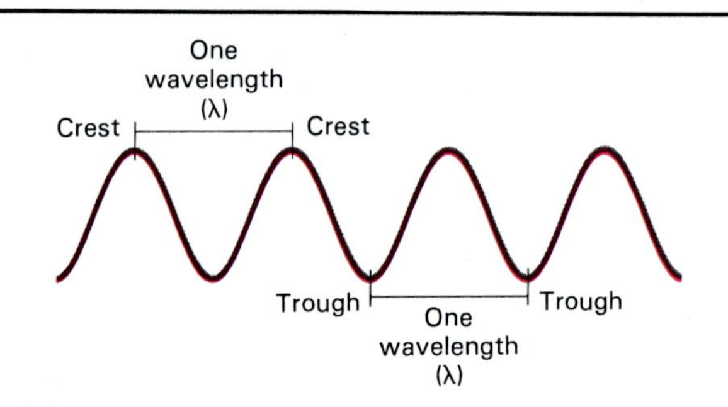

FIGURE 3.4 The distance between two adjacent crests or two adjacent troughs of any wave is defined as one wavelength, designated by the Greek letter lambda (λ).

FIGURE 3.5 Resolution. (a) The two dots are resolved; that is, they can clearly be seen as separate structures. (b) These two dots are not resolved—they appear fused.

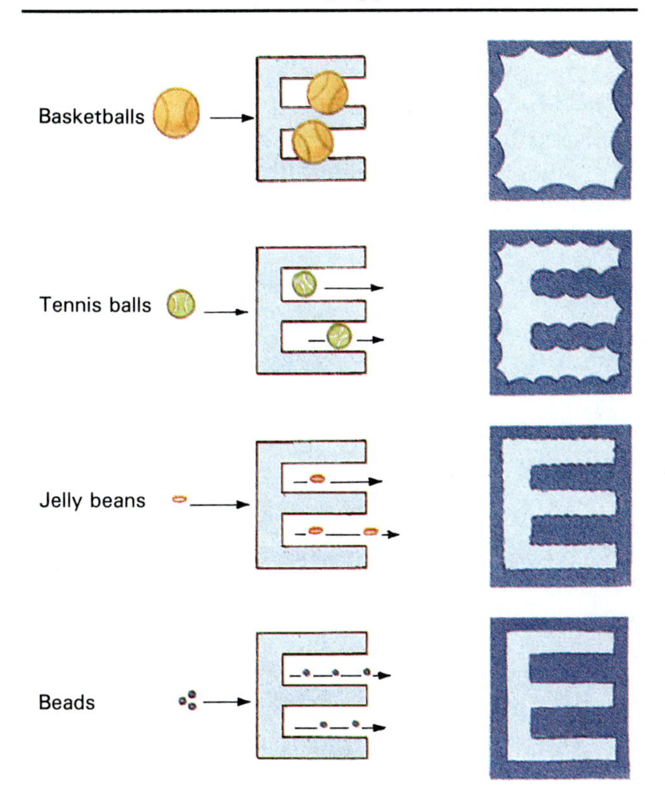

FIGURE 3.6 An analogy for the effect of wavelength on resolution. Smaller objects (corresponding to shorter wavelengths) can pass more easily between the arms of the letter E, defining it more clearly and producing a sharper image.

has a diameter smaller than the distance between the "arms" of the letter *E*, the object will pass between the arms, and they will be distinguishable as separate structures.

First, imagine tossing basketballs. Since they cannot fit between the arms, light rays of that size would give poor resolution. Next toss tennis balls at the target. The resolution will improve. Then try jelly beans and, finally, tiny beads. With each decrease in the diameter of the object thrown, the number of such objects that can pass between the arms of the *E* increases. Resolution improves, and the shape of the letter is revealed with greater and greater precision.

Microscopists use shorter and shorter wavelengths of electromagnetic radiation to improve resolution. Visible light, which has an average wavelength (represented by the Greek letter lambda, λ) of 550 nm, cannot resolve distances less than 220 nm. Ultraviolet light, which has a wavelength of 100 to 400 nm, can resolve distances as small as 110 nm. Thus, microscopes that used ultraviolet light instead of visible light allowed researchers to find out more about the details of cellular structures. But the invention of the electron microscope was the major step in increasing resolving power. Electrons behave both as particles and waves. Their wavelength is about 0.005 nm, which allows resolution of distances as small as 0.2 nm.

The **resolving power** (RP) of a lens refers to the ability to distinguish clearly among objects located close to one another. The smaller the distance between objects that can be distinguished, the greater the resolving power of the lens. We can calculate the RP of a lens if we know its **numerical aperture** (NA), the widest cone of light that can enter a lens. The formula for calculating resolving power is $RP = \lambda/2NA$.

As this formula indicates, the smaller the value of λ and the larger the value of NA, the greater is the resolving power of the lens.

The NA values of lenses differ in accordance with the power of magnification and other properties. The NA is engraved on the side of each objective lens (the lens nearest the stage) of a microscope. Look at those on the one you use the next time you are in the laboratory. Typical values for the objective lenses commonly found on modern microscopes are low power, 0.25; high power, 0.65; and oil immersion, 1.25.

Properties of Light: Light and Objects

Various things can happen to light as it travels through a medium such as air or water and strikes an object (Figure 3.7). We look at some of those things

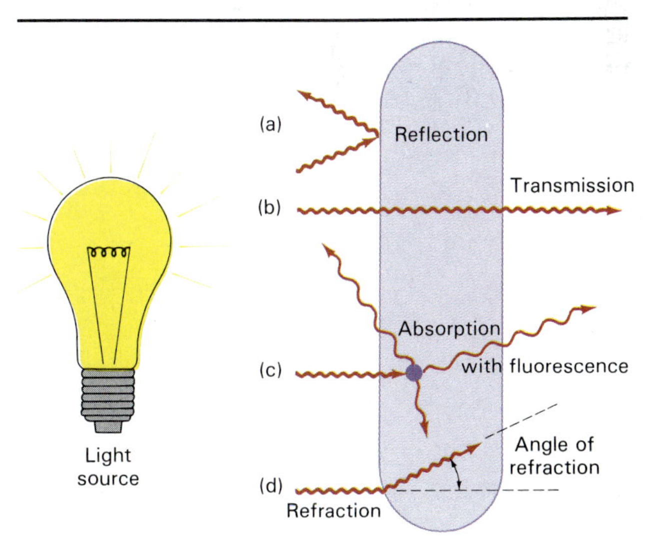

FIGURE 3.7 Various interactions of light with an object that it strikes. (a) Light may be reflected back from the object. The particular wavelengths reflected back to the eye determine the perceived color of the object. (b) Light may be transmitted directly through the object. (c) Light may be taken up, or absorbed, by the object. In some cases, the absorbed light rays are reemitted as longer wavelengths, a phenomenon known as fluorescence. (d) Light passing through the object may be bent, or refracted by it.

now and consider how they can affect your ability to see through a microscope.

Reflection

If the light strikes an object and bounces back (giving the object color), we say that **reflection** has occurred. For example, light rays in the green range of the spectrum are reflected off the surfaces of the leaves of plants, and these reflected rays are responsible for our seeing the leaves as green.

Transmission

Transmission refers to the passage of light through an object. You cannot see through a rock because light cannot pass through it. In order for you to see objects through a microscope, light must either be reflected from the objects or transmitted through them. Because only relatively low powers of magnification are possible with reflected light, most of your observations of microorganisms will make use of transmitted light.

Absorption

If light rays neither pass through nor bounce off an object but are taken up by the object, **absorption** has occurred. Energy in absorbed light rays can be used in various ways. For example, all wavelengths of the sun's light rays except those in the green range are absorbed by a leaf. Some of the energy in these other light rays is captured in photosynthesis and used by the plant to make food. Energy from absorbed light can also raise the temperature of an object. A black object, which reflects no light, will gain heat much faster than a white object, which reflects all light rays. In some cases, absorbed light rays, especially ultraviolet light rays, are changed into longer wavelengths and reemitted. This phenomenon is known as **luminescence.** If luminescence occurs only during irradiation (when light rays are striking an object), the object is said to **fluoresce.** Many fluorescent dyes are important in microbiology, especially in the field of immunology, because they assist in visualizing immune reactions and internal processes in microorganisms. If an object continues to emit light when light rays no longer strike it, the object is said to be **phosphorescent.** A number of bacteria that live deep in the ocean are phosphorescent.

Diffraction

As light passes through a small opening, such as a hole, slit, or space between two adjacent cellular structures, the light waves are broken up into bands of different wavelengths. This phenomenon is called **diffraction.** Diffraction patterns formed when light passes through a small aperture or around the edge of an object are shown in Figure 3.8. You may have seen similar diffraction patterns when water passes through an opening in a breakwater or around the back side of the breakwater.

Diffraction is a problem for microscopists because the lens itself acts as a small aperture through which the light must pass. This then results in diffraction and a blurry image. The higher the magnifying power of a lens, the smaller it must be, and therefore the greater the diffraction and blurring it causes. The oil immersion (100X) lens, with its total magnification capacity of about 1500X, represents the limit of useful magnification with the light microscope, as the small size of higher-power lenses causes such severe diffraction that resolution is impossible.

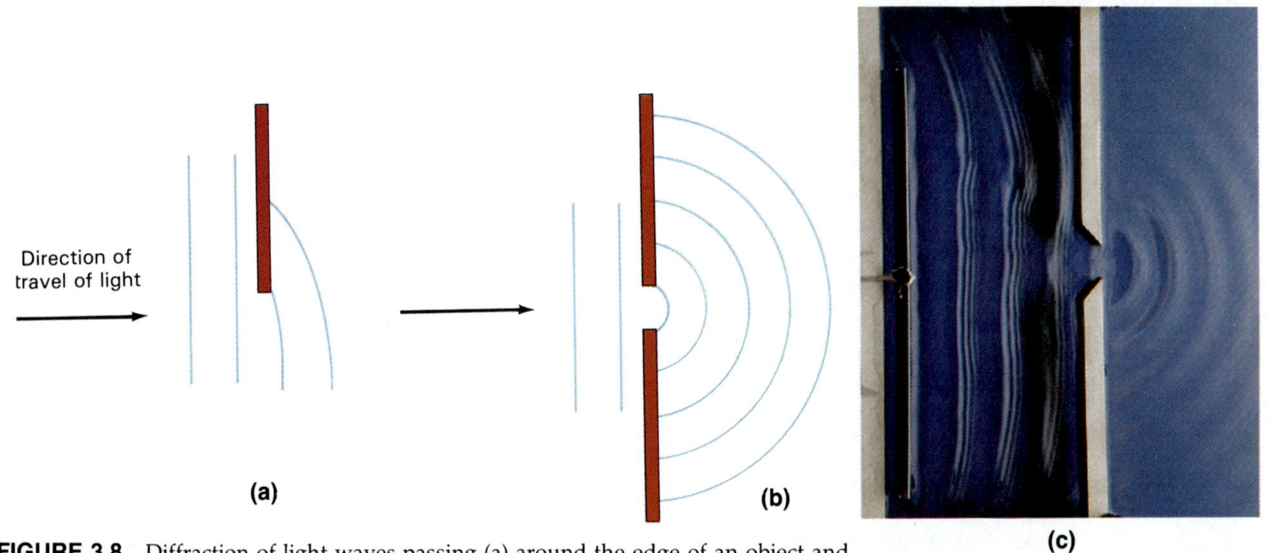

FIGURE 3.8 Diffraction of light waves passing (a) around the edge of an object and (b) through a small aperture. (c) Water waves being diffracted as they pass through an opening in a breakwater.

FIGURE 3.9 Refraction of light rays passing from water into air causes the spoon to appear bent.

Refraction

Refraction is the bending of light as it passes from one medium to another of different density. The bending of the light rays gives rise to an *angle of refraction* (Figure 3.9). You have probably seen how the underwater portion of a pole sticking out of water or a drinking straw in a glass of water seems to bend. When you remove these objects from the water, you can see that they are actually straight. They look bent because light rays deviate, or bend, when they pass from the water into the air. The degree of deviation is called the **index of refraction.**

Light passing through a glass microscope slide, air, and on through a glass lens is refracted each time it goes from one medium to another. To avoid this difficulty, microscopists use **immersion oil,** which has the same index of refraction as glass, to replace the air. The slide and the lens are joined by a layer of oil; there is no refraction to cause blurring of an image (Figure 3.10). If you forget to use oil with the oil immersion lens of a microscope, it will be impossible to focus clearly on a specimen.

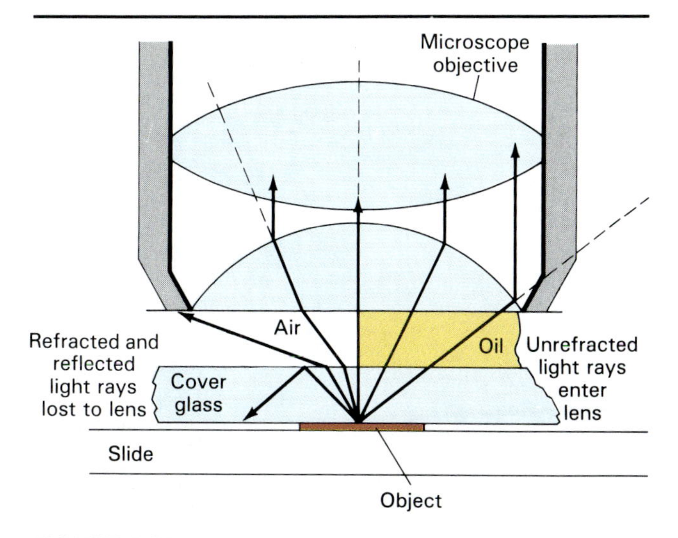

FIGURE 3.10 Use of immersion oil to prevent loss of light due to refraction. The focusing of more light adds to the clarity of the image.

TRY IT

Now You See It, Now You Don't

If you wanted to take some diamonds through customs without declaring them and thus avoid paying duty on them, here's a way you could do it. Obtain an oil with the same refractive index as the diamonds, pour it into a clear glass bottle labeled "Baby Oil," and drop the diamonds in. If you make certain that the surfaces of the diamonds are clean, they will be invisible. This trick works because when light passes from one medium to another with the same index of refraction, it is not bent. Thus, the boundary between the diamonds and the oil will not be apparent. (However, if the customs agent shakes the bottle, you're out of luck, for he or she will hear them rattle.)

If larceny is not in your heart or the price of diamonds not in your pocket, you could try this entertaining, but legal, activity. Clean a glass rod and dip it in and out of a bottle of immersion oil. Watch it disappear and reappear. Try this little experiment to help you understand what is happening when you use the oil immersion lens.

What has happened to the bottom of the glass rod? The rod disappears in immersion oil because glass and immersion oil have the same index of refraction. Therefore, light passing between them is not bent, so we cannot observe the surface of the rod.

LIGHT MICROSCOPY

Light microscopy refers to the use of any kind of microscope that uses light to make specimens observable. The modern microscope is a descendant not of Leeuwenhoek's single lenses but of Hooke's compound microscope (Chapter 1)—a microscope with more than one lens. Single lenses produce two problems: They cannot bring the entire field into focus simultaneously, and there are colored rings around objects in the field. Both of these problems are solved today by use of multiple correcting lenses placed next to the primary magnifying lens (Figure 3.11). Used in modern compound microscopes, these give us nearly distortion-free images.

Over the years several kinds of light microscopes have been developed, each adapted for making certain

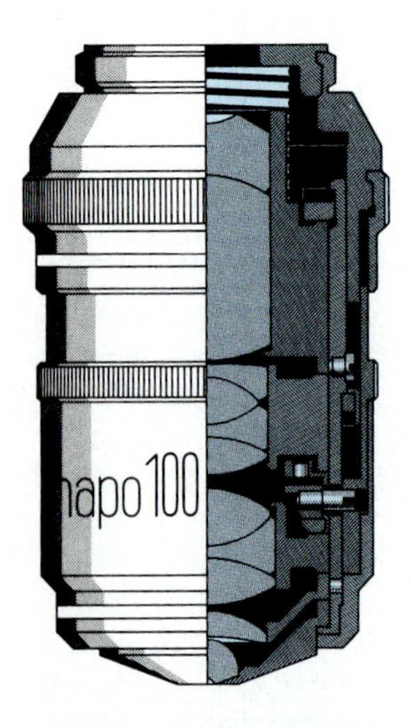

FIGURE 3.11 Cutaway view of a modern microscope objective. Note that what we refer to as a single objective lens is really a series of several lenses, which are necessary to correct aberrations of color and focus. The best objectives may have as many as a dozen or more elements.

kinds of observations. We look first at the standard light microscope and then at some special kinds of microscopes.

The Compound Light Microscope

The **optical microscope,** or *light microscope,* has undergone various improvements since Leeuwenhoek's time and essentially reached its current form shortly before the turn of the twentieth century. This microscope is a **compound light microscope**—that is, it has more than one lens. The parts of a modern compound microscope and the path light takes through it are shown in Figure 3.12. A compound microscope with a single eyepiece is said to be **monocular;** one with two eyepieces is said to be **binocular.**

Light enters the microscope from a source in the **base** and often passes through a blue filter, which filters out the long wavelengths of light, leaving the shorter wavelengths and improving resolution. It then goes through a **condenser,** which converges the light beams so they will pass through the specimen. The **iris diaphragm** controls the amount of light passing through the specimen and into the objective lens. The higher the magnification, the greater the amount of light needed to see the specimen clearly. The **objective lens** magnifies the image before it passes through the **body tube** to the ocular lens in the eyepiece. The **ocular lens** further magnifies the image. The total magnification is the product of the magnifications of the objective and ocular lenses. A **mechanical stage** allows precise control of moving the slide, which is especially useful in the study of microbes.

The focusing mechanism consists of a **coarse adjustment,** which changes the distance between the objective lens and the specimen fairly rapidly, and a **fine adjustment,** which changes the distance very slowly. The coarse adjustment is used to locate the

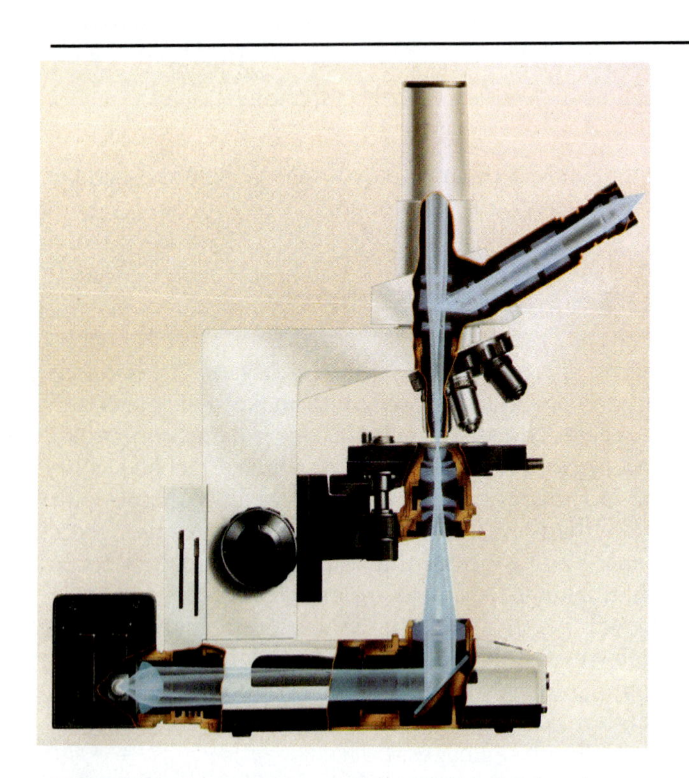

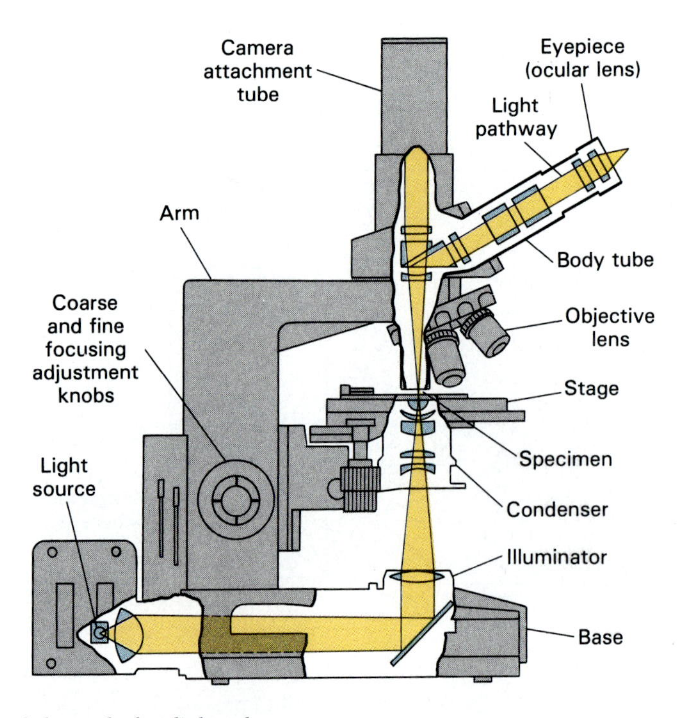

FIGURE 3.12 Parts of a modern compound light microscope and the path that light takes through it.

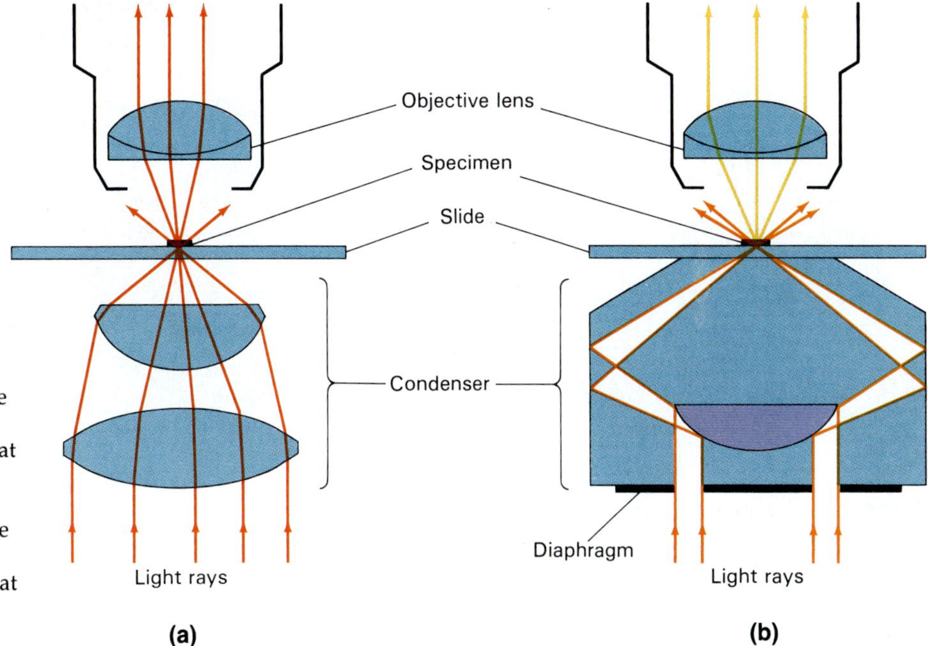

FIGURE 3.13 Comparison of the illumination in (a) bright-field and (b) dark-field microscopy. Note that the condenser of the bright-field microscope concentrates and transmits light directly through the specimen, whereas the dark-field condenser deflects light rays so that they reflect off the specimen at an angle.

specimen; the fine adjustment is used to bring it into sharp focus.

Compound microscopes have three, and sometimes four, interchangeable objective lenses that have different powers of magnification.

The **total magnification** of a light microscope is calculated by multiplying the magnifying power of the objective lens (the lens nearest your specimen) by the magnifying power of the ocular lens (the lens nearest your eye). Typical values using a 10X ocular are:

- scanning (3X) × (10X) = 30X magnification
- low power (10X) × (10X) = 100X magnification
- high "dry" (40X) × (10X) = 400X magnification
- oil immersion (100X) × (10X) = 1000X magnification

Most microscopes are designed so that when the microscopist increases or decreases the magnification by changing from one objective lens to another, the specimen will remain very nearly in focus. Such microscopes are said to be **parfocal.** The development of parfocal microscopes greatly improved the efficiency of microscopes, and reduced the amount of damage to slides and objective lenses. Most student-grade microscopes are parfocal today, but a few older nonparfocal models may still be in use, so be careful the first time you switch objectives on a microscope. Some microscopes are equipped with an **ocular micrometer** for measuring objects viewed. This is a glass disk with a scale marked on it, which is placed inside the eyepiece between its lenses. This scale must first be calibrated with a stage micrometer, which has metric units engraved on it. When these units are viewed through the microscope at various magnifications, the microscopist can determine the corresponding metric values of the divisions on the ocular micrometer for each objective lens. Thereafter, he or she needs only to count the number of divisions covered by an object being observed, and multiply by the calibration factor for that lens in order to determine the actual size of the object.

Dark-Field Microscopy

The condenser used in an ordinary light microscope causes light to be concentrated and transmitted directly through the specimen, as shown in Figure 3.13a. This gives **bright-field illumination.** However, sometimes it is more useful, especially with light-sensitive organisms, to examine specimens that would lack contrast with their background in a bright field under other illumination. Live spirochetes (spi'ro-kets), spiral-shaped bacteria that cause syphilis and other diseases, are just such organisms. In this situation **dark-field illumination** is used. A microscope adapted for dark-field illumination has a condenser that prevents light from being transmitted through the specimen but instead causes the light to reflect off the specimen at an angle (Figure 3.13b). One sees a light object on a dark background (Figure 3.14b).

Phase-Contrast Microscopy

Most living microorganisms are difficult to examine because they cannot be stained by coloring them with

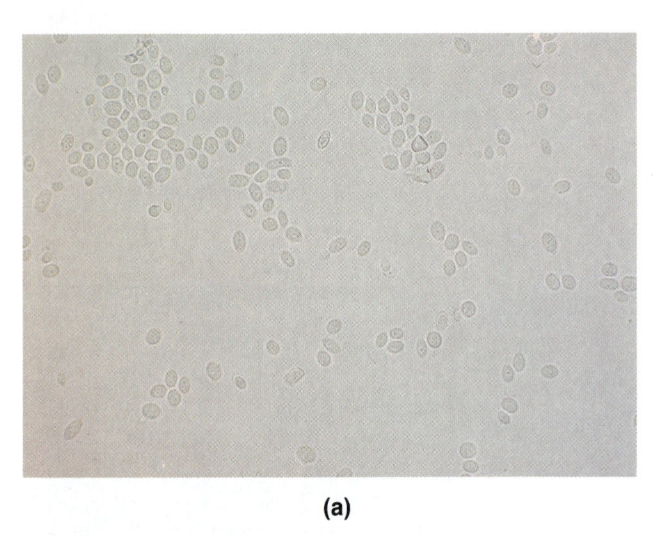

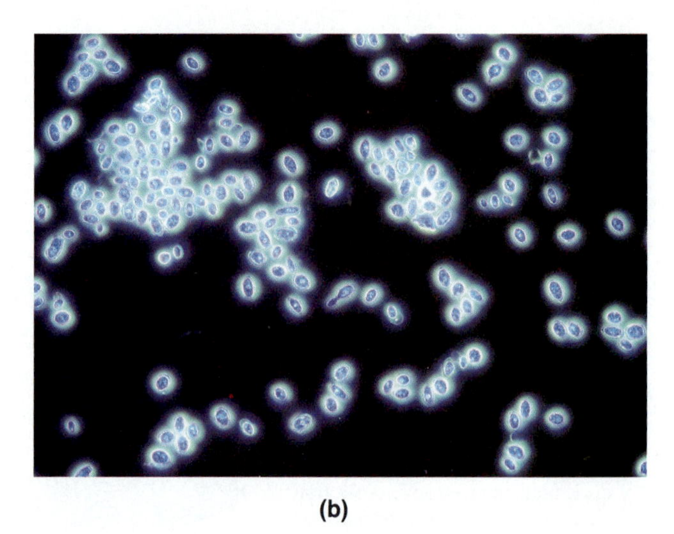

(a) (b)

FIGURE 3.14 (a) Bright-field and (b) dark-field microscope views of *Saccharomyces cerevisiae* (brewer's yeast). Note the enormous increase in contrast provided by the dark-field illumination.

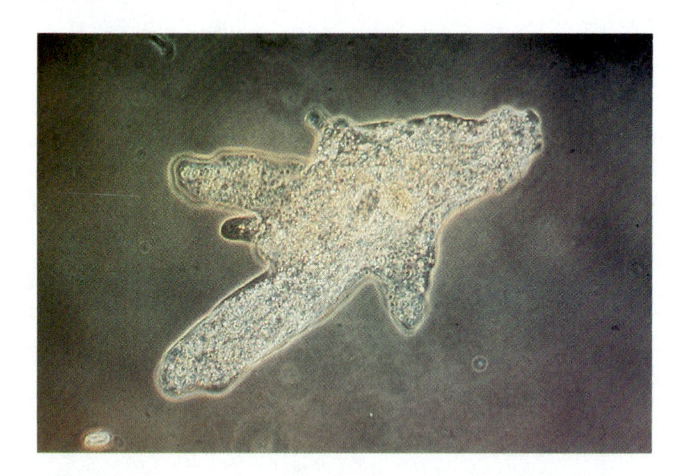

FIGURE 3.15 Phase contrast view of *Amoeba*, a protozoan.

dyes, since stains usually kill the organisms. To observe them alive and unstained requires the use of **phase-contrast microscopy.** A phase-contrast microscope has a special condenser that accentuates small differences in the refractive index of various structures within the organism. Light passing through objects with different refractive indices is slowed down and diffracted. The changes in the speed of light are seen through the microscope as different degrees of brightness (Figure 3.15).

Fluorescence Microscopy

In **fluorescence microscopy** ultraviolet light is used to excite molecules so that they release light of a different wavelength than that originally striking them (see Figure 3.7). The different wavelengths produced are often seen as brilliant shades of orange, yellow, or yellow-green. Some organisms, such as *Pseudomonas*, fluoresce naturally when irradiated with ultraviolet light. Other organisms, such as *Mycobacterium tuberculosis* and *Treponema pallidum* (the spirochete of syphilis), must be treated with a fluorescent dye called a *fluorochrome*, after which they stand out sharply against a dark background (Figure 3.16).

Fluorescent antibody staining is now widely used in diagnostic procedures to determine whether an antigen (a foreign substance such as a microbe) is present. Antibodies—molecules produced by the body as an immune response to an invading antigen—are found in many clinical specimens such as blood and serum. If a patient's specimen contains a particular antigen, that antigen and the antibodies specifically made against it will clump together. However, this ordinarily is not a visible reaction. Therefore, fluo-

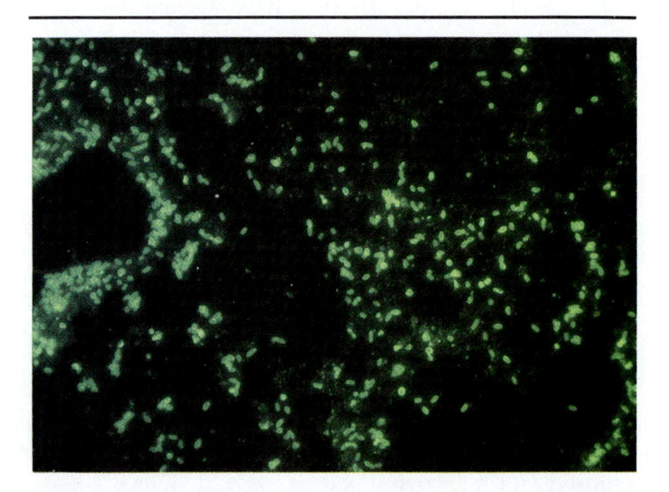

FIGURE 3.16 Fluorescent antibody staining of *Legionella pneumophila*, the bacteria that cause Legionnaire's disease (legionellosis).

FIGURE 3.17 Nomarski interference microscope image of the protozoan *Paracineta*, attached by a long stalk to the green alga *Spongomorpha*.

rescent dye molecules are attached to the antibody molecules. If the dye molecules are retained by the specimen, the antigen is presumed to be present, and a positive diagnosis can be made. Thus, if fluorescent dye–tagged antibodies against syphilis organisms are added to a slide of spirochetes and are seen to bind to the organisms, the organisms can be identified as being syphilis organisms. This technique is especially important in immunology, in which the reactions of antigens and antibodies are studied in great detail (see Chapters 18 and 19, especially Figure 19.28 on technique of fluorescent antibody staining). Often diagnoses can be made in minutes rather than the hours or days it would take to isolate, culture, and identify organisms.

Differential Interference Contrast (Nomarski) Microscopy

Nomarski microscopy, like phase-contrast microscopy, makes use of differences in refractive index to visualize structures. However, the microscope used produces much higher resolution than the standard phase-contrast microscope. It has a very short *depth of field* (the thickness of specimen that is in focus at any one time) and therefore can produce a nearly three-dimensional image (Figure 3.17).

Figure 3.18 contrasts the images produced by four different microscopic techniques.

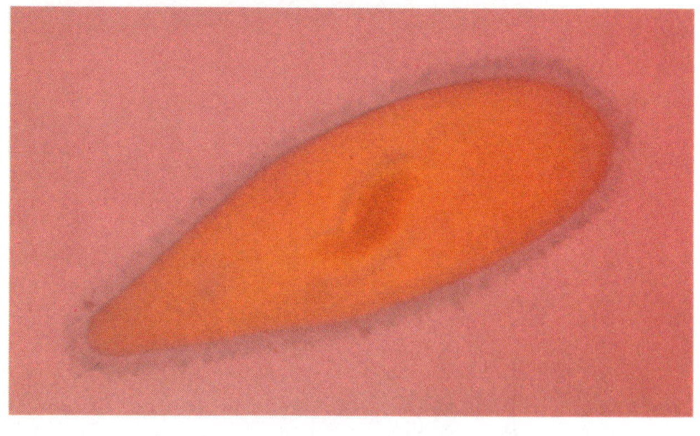

(a)

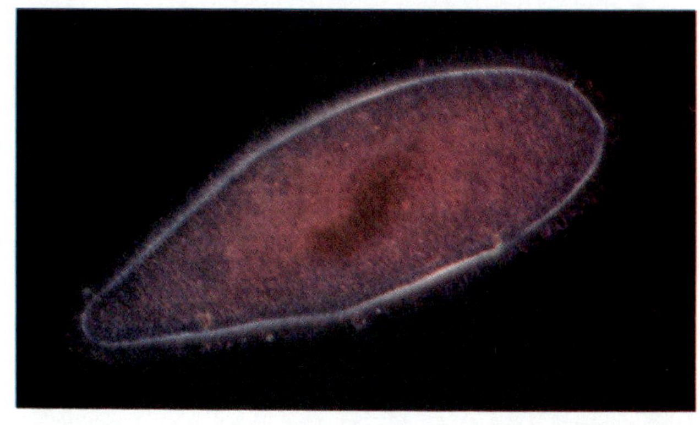

(b)

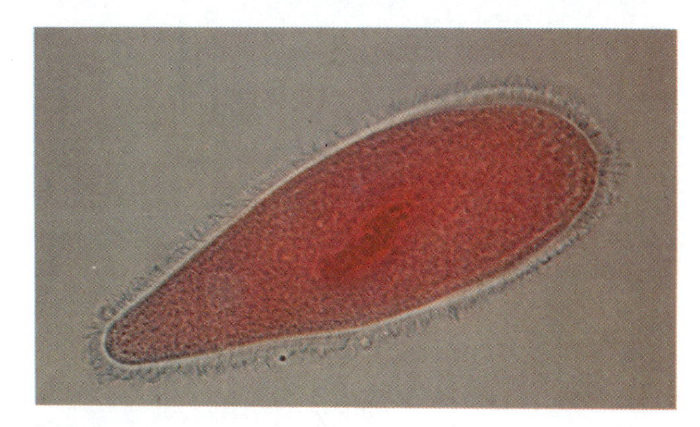

(c)

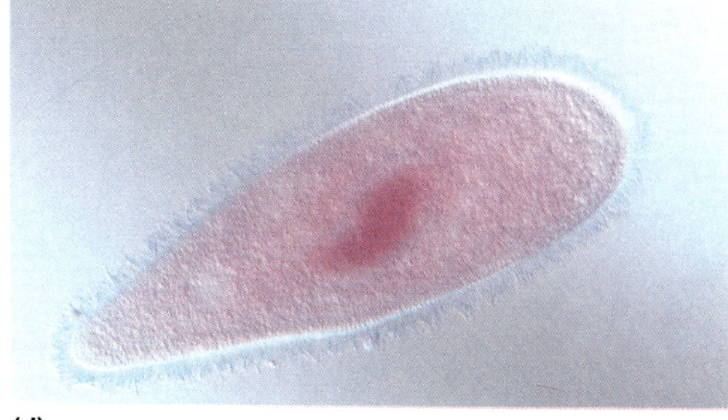

(d)

FIGURE 3.18 Images of the same organism (a paramecium) produced by means of four different techniques: (a) bright-field microscopy, (b) dark-field microscopy, (c) phase-contrast microscopy, (d) Nomarski microscopy.

ELECTRON MICROSCOPY

The light microscope opened doors on the world of microbes. However, the view was limited to observations at the level of whole cells and their arrangements. Few subcellular structures could be seen; nei-

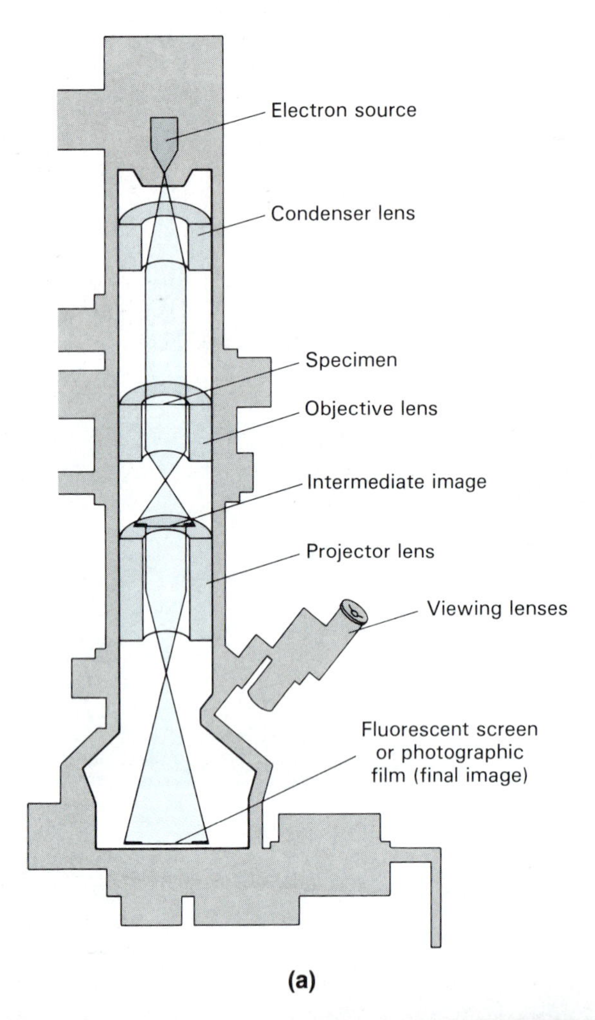

(a)

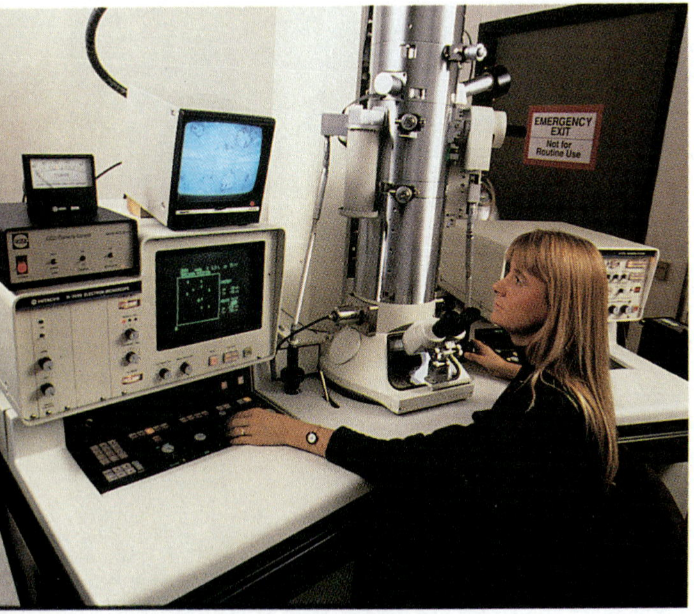

(b)

FIGURE 3.19 (a) Cross-sectional diagram of an electron microscope, showing the pathways of the electron beam as it is focused by electromagnetic lenses. (b) Using a modern electron microscope.

ther could viruses. The advent of the electron microscope allowed these small structures to be visualized and studied. The **electron microscope** (EM) was developed in the 1930s and was in use in many laboratories by the 1940s.

The EM (Figure 3.19) uses a beam of electrons instead of a beam of light, and electromagnets are used to focus the beam. The electrons must travel through a vacuum because collisions with air molecules would distort the image. Electron microscopes are much more expensive than light microscopes; they also take up much more space and require additional rooms for preparation of specimens for viewing and for processing of the photographs, called micrographs. It takes many months to become a proficient electron microscopist, but the time and effort are worthwhile. Nothing else can show us the great detail of minute biological structures that EMs can (Figure 3.20).

The two most common types of electron microscopes are the transmission electron microscope and the scanning electron microscope. Both are used to study microbes.

Transmission Electron Microscopy

The **transmission electron microscope** (TEM) gives better knowledge of the internal structure of microbes than do other types of microscopes. It can resolve objects as close as 1 nm and magnify microbes (and other objects) up to 200,000X. In transmission electron microscopy very thin slices of a specimen are cut, using a glass or diamond knife. These sections are placed on wire grids for viewing so that a beam of electrons will pass directly through the section. The section must be exceedingly thin (0.07 mm) because electrons cannot penetrate very far into materials. Sometimes a heavy metal such as gold or palladium is sprayed at an angle onto the specimen, a technique known as **shadow casting.** Areas behind the specimen that did not receive a coating of metal appear as "shadows," which can give a three-dimensional effect to the image (Figure 3.21). Electron beams are deflected by the densely coated parts of the specimen but, for the most part, pass through the shadows. The specimens can also be "stained" with acids that sit on the surface, scattering electrons and forming an image.

The image formed by the electron beam is made visible as a light image on a televisionlike screen. (The actual image made by the electron beam is not visible and would burn your eyes if you tried to view it directly.) The electrons are used to excite the phosphors (light-generating compounds) coating the screen. However, the electron beam will eventually burn through the specimen; therefore, photographs called **electron micrographs** are made, either by photographing the image on the video screen or by re-

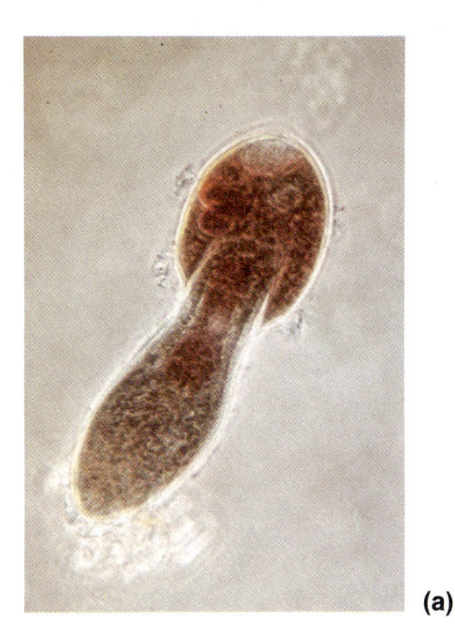

(a)

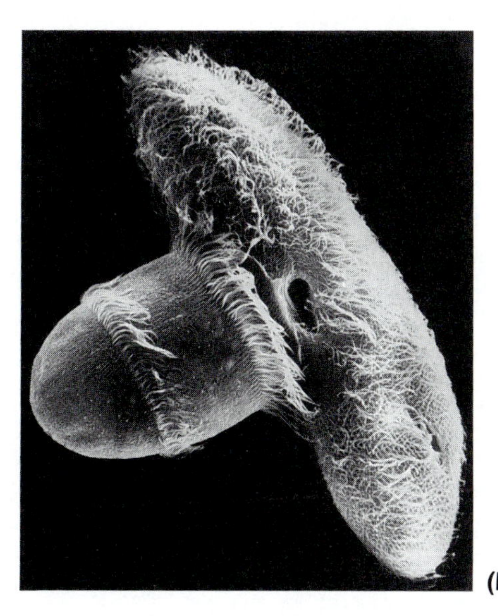

(b)

FIGURE 3.20 Comparison of (a) light and (b) electron microscope images of a *Didinium* eating a *Paramecium*. Notice how much more detail is revealed by the scanning electron micrograph.

placing the screen itself with a photographic plate. These can be enlarged photographically to obtain an image magnified 20 million times! The micrographs are permanent records of structures observed and can be studied at leisure. The study of such micrographs has provided most of our knowledge of the internal structure of microbes.

Scanning Electron Microscopy

The **scanning electron microscope** (SEM) is a more recent invention than the transmission electron microscope and is used to create images of the surfaces of specimens. This microscope can resolve objects as close as 20 nm, giving magnifications up to approximately 10,000X. The SEM gives us wonderful three-dimensional views of the exterior of cells (Figure 3.22).

Preparing a specimen for the SEM involves freeze-drying it and then coating it with a thin layer of a heavy metal, such as gold or palladium. It is also

FIGURE 3.21 Spraying a heavy metal (such as gold or platinum) at an angle over a specimen leaves a "shadow," or darkened area, where metal is not deposited. This technique, known as shadow casting, produces images with a three-dimensional appearance, as in this photograph of polio virus (magnification approximately 300,000X). It is also possible to calculate the height of the organisms from the length of their shadows if you know the angle of the metal spray.

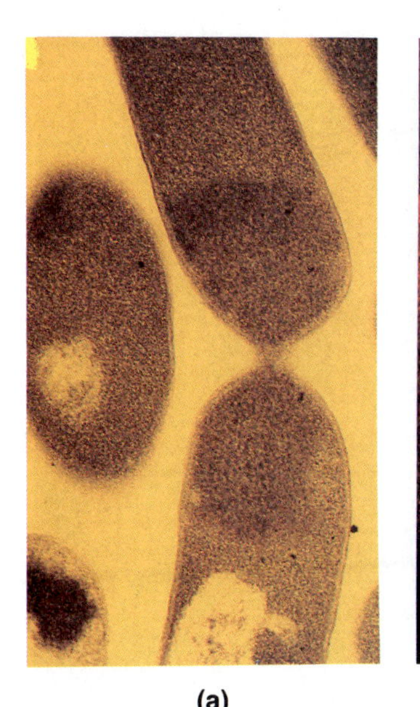

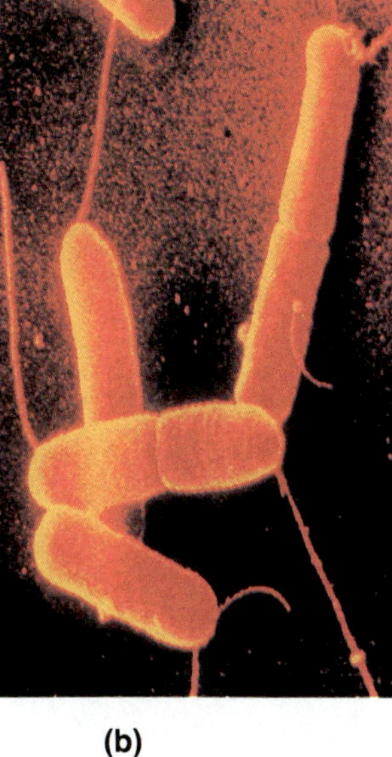

(a) (b)

FIGURE 3.22 Comparison of micrographs of *Escherichia coli* produced by (a) transmission electron microscopy (TEM) and (b) scanning electron microscopy (SEM).

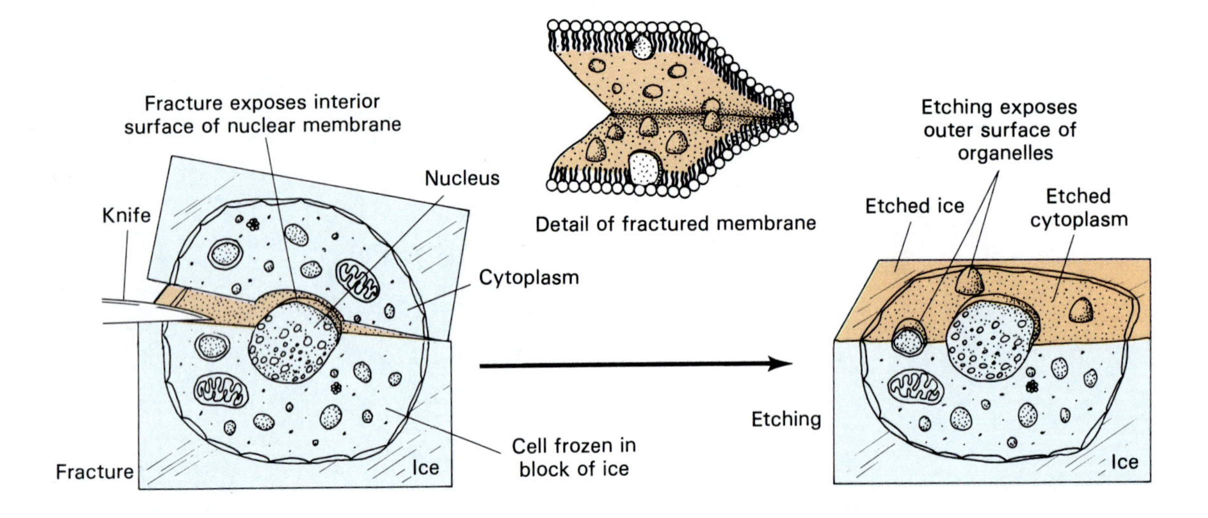

FIGURE 3.23 In freeze-fracture preparation, a specimen is frozen in a block of ice and broken apart with a very sharp knife. The fracture reveals the interiors of cellular structures and often passes through the center of membrane bilayers, exposing their inner faces. In freeze-etching, water is evaporated directly from the ice and frozen cytoplasm of the specimen in vacuum, uncovering additional surfaces for observation.

possible to view the interior of a cell with SEM or TEM using a technique called **freeze-fracturing,** in which the cell is first frozen and then broken with a knife. The fracture reveals the surfaces of structures inside the cell (Figure 3.23). **Freeze-etching,** which involves the evaporation of water from the frozen and fractured specimen, can then be used to expose additional surfaces for examination (Figure 3.24). These surfaces must also be coated with a heavy metal layer before being viewed. Inside the SEM a thin primary electron beam, which moves back and forth across the specimen, displaces secondary electrons as it hits the metal coating of the specimen. As the secondary electrons are emitted, they are collected and form a pattern on an electron detector. This pattern of electrons is then converted to a display on a video monitor. Photographs are made of the screen and enlarged for further study. Views of the three-dimensional world of microbes, as shown in Figure 3.25, are breathtakingly beautiful. The various types of microscopy and their uses are summarized in Table 3.2.

FIGURE 3.24 Photo of the toxic cyanobacterium *Microcystis aeruginosa* prepared by the freeze-etch method, showing details of large spherical gas vacuoles.

TECHNIQUES OF LIGHT MICROSCOPY

Microscopes of any kind are of little use unless one knows how to prepare specimens for viewing. In this section we explain some important techniques used in light microscopy.

Although resolution and magnification are important in microscopy, the degree of contrast between structures to be observed and their backgrounds is equally important. Nothing can be seen without contrast, so special techniques have been developed to enhance contrast.

Preparation of Specimens for the Light Microscope

Wet Mounts

Wet mounts, in which a drop of medium containing the organisms is placed on a microscope slide, can be

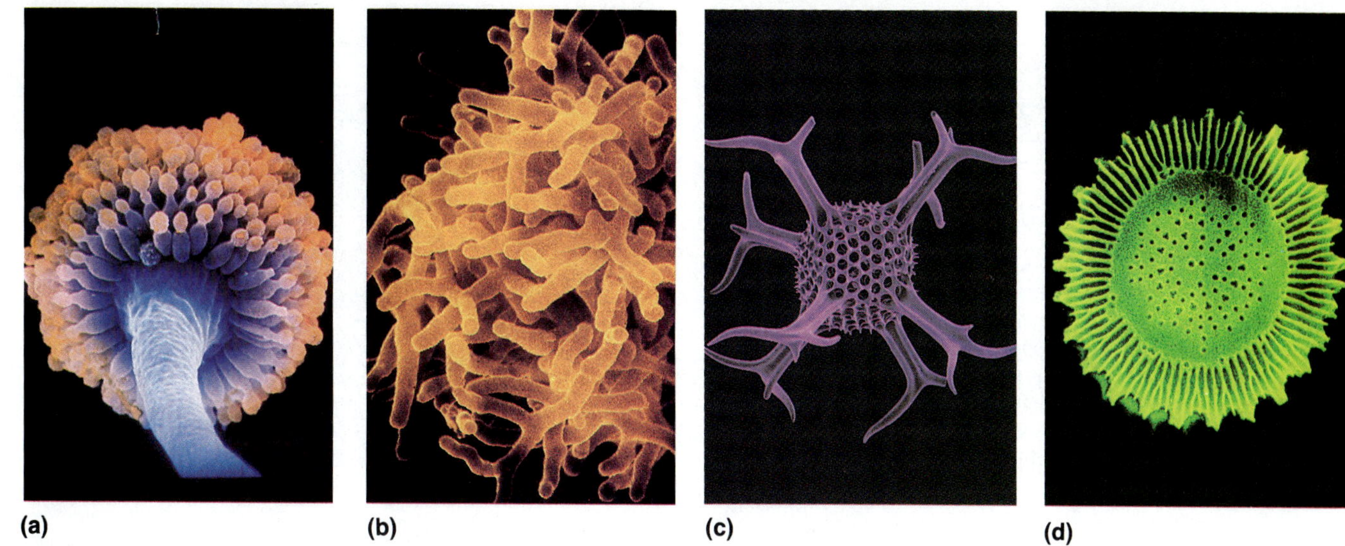

FIGURE 3.25 Color-enhanced scanning electron micrographs of representative microbes: (a) the fungus *Aspergillus*, a cause of human respiratory disease; (b) *Actinomyces*, long though to be a fungus because of its unusual filamentous form, but now known to be a bacterium; (c) a radiolarian from the Indian Ocean; (d) a diatom, one of many that carry on photosynthesis and form the base of many aquatic food chains.

TABLE 3.2 Comparison of types of microscopy

Type	Special Features	Appearance	Uses
Bright-Field	Uses visible light. Simplest to use, least expensive	Colored or clear specimen on light background	Observation of dead stained organisms, or live ones with sufficient natural color contrast
Dark-Field	Uses visible light with a special condenser that causes light rays to reflect off specimen at an angle	Bright specimen on dark background	Observation of unstained living or difficult-to-stain organisms. Allows one to see motion
Phase-Contrast	Uses visible light plus phase-shifting plate in objective with a special condenser that causes some light rays to strike specimen out of phase with each other	Specimen has different degrees of brightness and darkness	Detailed observation of internal structure of living unstained organisms
Nomarski	Uses visible light out of phase, has higher resolution than standard phase-contrast microscope	Produces a nearly three-dimensional image	Observation of finer details of internal structure of living unstained organisms
Fluorescent	Uses ultraviolent light to excite molecules to emit light of different wavelengths, often brilliant colors. UV can burn eyes; therefore, special lens materials are used	Bright, fluorescent, colored specimen on dark background	Diagnostic tool for detection of organisms or antibodies in clinical specimens or for immunologic studies
Transmission Electron	Uses electron beam instead of light rays and electromagnetic lenses instead of glass lenses; image is projected on a videoscreen. Very expensive, techniques require considerable study	Highly magnified, detailed image; not three-dimensional except with shadow casting	Examination of thin sections of cells for details of internal structure, exterior of cells and viruses
Scanning Electron	Uses electron beam and electromagnetic lenses. Expensive, and techniques require considerable study	Three-dimensional view of surfaces	Observation of exterior surfaces of cells, or of internal surfaces when freeze-fracturing is used

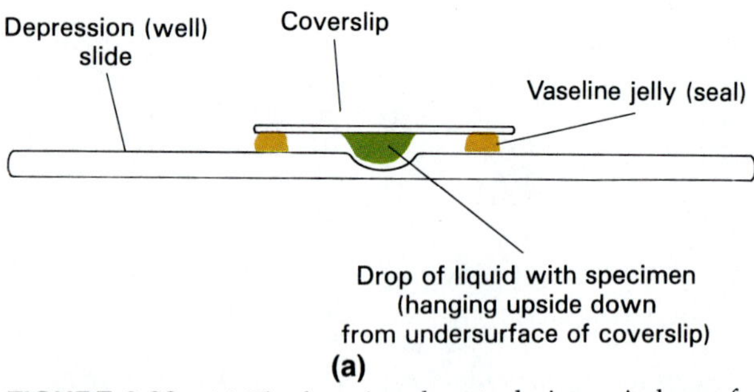

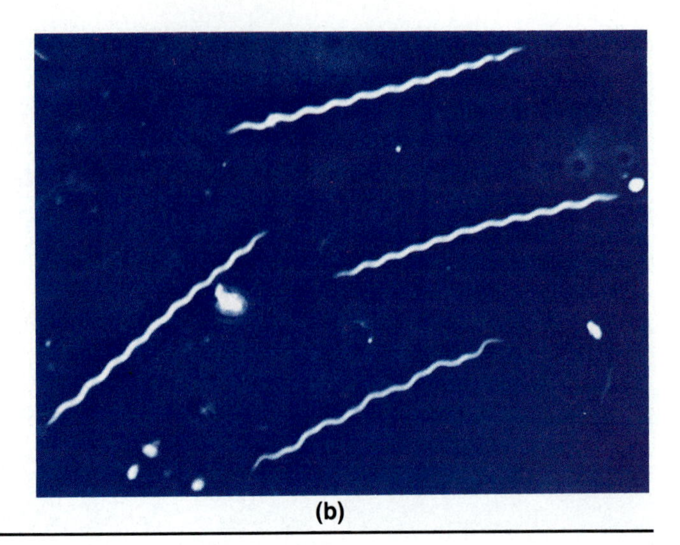

FIGURE 3.26 (a) The hanging-drop technique. A drop of culture is placed on a coverslip ringed with petroleum jelly that is then inverted and placed over the well in a depression slide. The petroleum jelly forms a seal to prevent evaporation. (b) Photograph of a dark-field hanging-drop preparation (1000X) showing the spiral bacterium *Treponema pallidum,* the cause of syphilis.

used to view living microorganisms. Adding a 2 percent solution of carboxymethyl cellulose, a thick syrupy solution, helps to slow fast-moving organisms so they can be studied. A special version of the wet mount, called a **hanging drop,** often is used with dark-field illumination (Figure 3.26). A drop of culture is placed on a coverslip that has a thin ring of petroleum jelly near its edges. This is then inverted over the well of a depression slide, the drop hanging from the coverslip and the petroleum jelly forming a seal. This preparation gives good views of microbial motility.

Smears

Smears, in which microorganisms from a drop of medium are spread onto the surface of a glass slide, can be used to view killed organisms. Though they are living when placed on the slide, the organisms are killed by the techniques used to fix them to the slide. Smear preparation often is difficult for beginners. If you make smears too thick, you cannot see through them; if you make them too thin, you cannot find any organisms. If you stir the drop of medium too much as you spread it on the slide, you will disrupt cell arrangements. You may see organisms that normally appear in tetrads (groups of four) as single or double organisms. You may see organisms that normally appear in simple chains as broken or tangled clusters of chains wrapped around themselves. These variations lead some beginners to imagine that they see more than one kind of organism when, in fact, the organisms are all of the same species.

After a smear is made it is allowed to air-dry completely. Then it is quickly passed two or three times through an open flame. This process is called **heat fixation.** If the slide is not completely dry when passed through the flame, the organisms will be boiled and destroyed and will wash off the slide in subsequent steps. Heat fixation accomplishes three things: (1) It kills the organisms; (2) it causes the organisms to adhere to the slide; and (3) it alters the organisms so that they more readily accept stains (dyes). If you heat-fix too little, the organisms may not adhere to the slide, some may remain alive, and they will stain poorly. When you look at the slide you will find only a few pale organisms. If you heat-fix too much, the organisms may be incinerated, and you will find nothing. The slide should feel warm, but not too hot to touch. Certain structures, such as the capsules found on some microbes, are destroyed by heat-fixing, so heat-fixing is omitted and these microbes are affixed to the slide just by air-drying.

Principles of Staining

A **stain,** or dye, is a molecule that can bind to a cellular structure and give it color. Staining techniques are used to make the microorganisms stand out against their backgrounds. They are also used to help investigators group major categories of microorganisms, examine the structural and chemical differences in cell walls, and look at the parts of the cell.

In microbiology the most commonly used dyes are **cationic** (positively charged), or **basic dyes,** such as methylene blue, crystal violet, safranin, and malachite green. Such basic dyes are best for staining nuclear materials such as chromosomes. Also, the cell membranes of most bacteria have negatively charged surfaces, to which positively charged basic dyes are attracted. Other stains, such as eosin and picric acid, are **anionic** (negatively charged) or **acidic dyes.** They are best for staining cytoplasmic material, rather than nuclear materials.

Two main types of stains, simple stains and differential stains, are used in microbiology. They are compared in Table 3.3. A **simple stain** makes use of a single dye and reveals basic cell shapes and cell arrangements. Methylene blue, safranin, carbolfuch-

TABLE 3.3 Comparison of staining techniques

Type	Examples	Result	Uses
Simple stains			
Use a single dye; do not distinguish organisms or structures by different staining reactions	Methylene blue Safranin Crystal violet	Uniform blue stain Uniform red stain Uniform purple stain	Shows sizes, shapes, and arrangements of cells
Differential stains			
Utilize two or more dyes that react differently with various kinds or parts of bacteria, allowing them to be distinguished	Gram stain	Gram +: purple with crystal violet Gram −: red with safranin counterstain Gram-variable: intermediate or mixed colors (some stain + and some − on same slide) Gram-nonreactive: stain poorly or not at all	Distinguish gram +, gram −, gram-variable, and gram-nonreactive organisms
	Ziehl-Neelsen acid-fast stain	Acid-fast bacteria retain carbolfuchsin and appear red. Non-acid-fast bacteria accept the methylene blue counterstain and appear blue	Distinguishes members of the genera *Mycobacterium* and *Nocardia* from other bacteria
Special stains			
Identify various specialized structures	Flagellar stain	Flagella appear as dark lines with silver, or red with carbolfuchsin	Indicates presence of flagellae by building up layers of stain on their surface
	Schaeffer-Fulton spore stain	Endospores retain malachite green stain. Vegetative cells accept safranin counterstain and appear red	Allows visualization of hard-to-stain bacterial endospores, e.g., in members of genera *Clostridium* and *Bacillus*
	Negative staining	Capsules appear clear against a dark background	Allows visualization of organisms having structures that will not accept most stains, e.g., capsules

sin, and gentian violet are commonly used simple stains. A **differential stain** makes use of two or more dyes and distinguishes between two kinds of organisms or between two different parts of an organism. The Gram stain (see below) is probably the most frequently used differential stain. However, the Schaeffer-Fulton spore stain (also described below) and the **Ziehl-Neelsen acid-fast stain** (Figure 3.27) also are frequently used. The latter is a modification of a staining method developed by Paul Ehrlich in 1882. It can be used to detect tuberculosis- and leprosy-causing organisms of the genus *Mycobacterium*. Slides of organisms are covered with carbolfuchsin and heated, then decolorized with 3 percent hydrochloric acid (HCl) in 95 percent ethanol, and then stained with Loeffler's methylene blue. Most genera of bacteria will lose the red carbolfuchsin stain when decolorized. However, those that are "acid-fast" retain the bright red color. The lipid components of their walls responsible for this characteristic are discussed in Chapter 4. Bacteria that are not acid-fast lose the red color and can therefore be stained blue with the Loeffler's methylene blue counterstain.

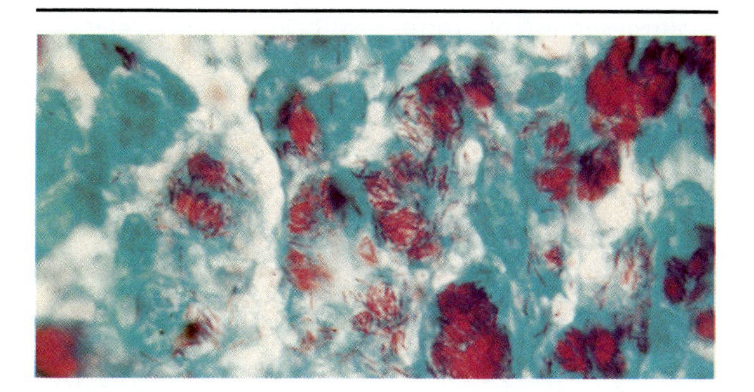

FIGURE 3.27 The Ziehl-Neelsen method of staining produces vivid red color in acid-fast organisms such as *Mycobacterium leprae*, the cause of leprosy, shown here.

The Gram Stain

The **Gram stain** was devised by a Danish physician, Hans Christian Gram, in 1884. Gram was testing new methods of staining biopsy and autopsy materials, and he noticed that with certain methods some bacteria were stained differently than the surrounding tissues. As a result of his experiments with stains, the highly useful Gram stain was developed. The Gram stain is a differential stain, in which certain structures, mainly the cell wall, take up crystal violet. The iodine acts as a **mordant,** a chemical that helps the stain to adhere to the cell. Those structures that cannot retain crystal violet are decolorized and subsequently

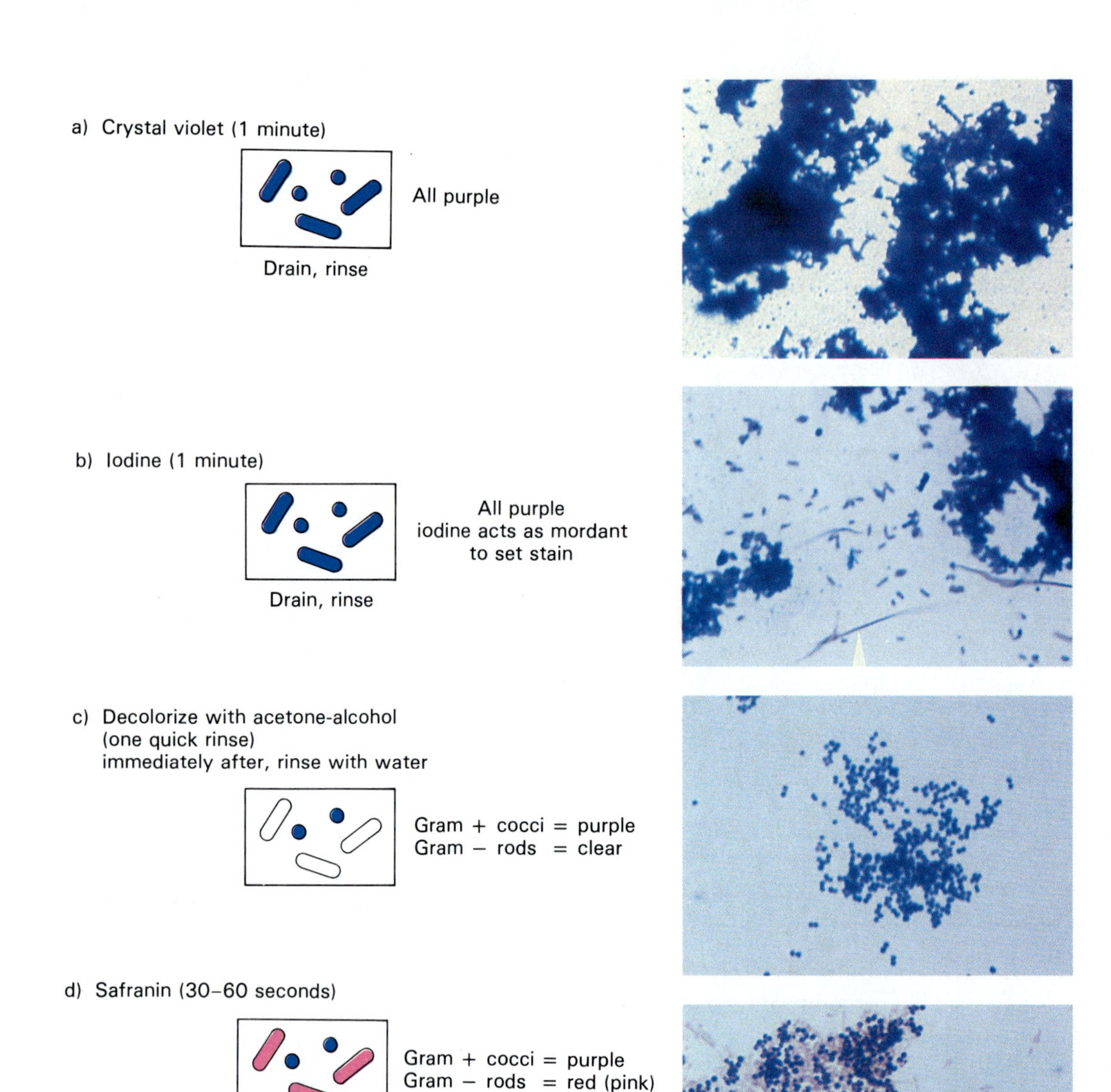

FIGURE 3.28 Steps in Gram staining. Gram-positive cells retain the purple color of crystal violet, whereas gram-negative cells are decolorized with acetone-alcohol and subsequently pick up the red color of the safranin counterstain.

stained (counterstained) with safranin. The steps in the Gram staining procedure are shown in Figure 3.28.

Four groups of organisms can be differentiated with the Gram stain: (1) gram-positive organisms whose cell walls retain crystal violet stain, (2) gram-negative organisms whose cell walls do not retain crystal violet stain, (3) gram-nonreactive organisms that lack cell walls, and (4) gram-variable organisms that stain unevenly. The differentiation between gram-positive and gram-negative organisms reveals a fundamental difference in the nature of the cell walls of bacteria, as is explained in Chapter 4. Furthermore, the reactions of bacteria to the Gram stain have helped in distinguishing gram-positive, gram-negative, and gram-nonreactive groups that belong to radically different taxonomic groups (Chapter 10).

Gram-variable organisms have somehow lost their ability to react distinctively to the Gram stain. Organisms from cultures over 48 hours old (and sometimes only 24 hours old) are often gram-variable, probably because of changes in the cell wall with aging. Therefore, to determine the reaction of an organism to the Gram stain you should use organisms from cultures less than 24 hours old.

Special Stains

Negative Staining **Negative staining** is used when

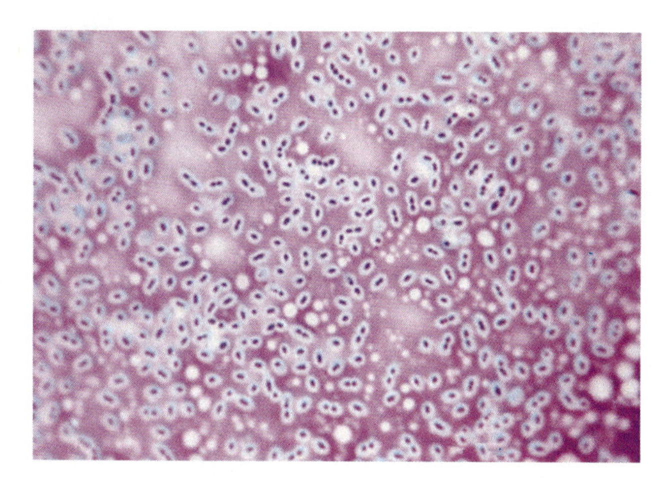

FIGURE 3.29 Negative staining for capsules reveals a clear area (the capsule, which does not accept stain) in a dark background of India ink. The cells themselves are stained deep purple with a counterstain. The bacteria are *Diplococcus pneumoniae* (1000X), which are arranged in pairs.

FIGURE 3.30 Schaeffer-Fulton stain of a species of *Clostridium*, a spore-forming anaerobic bacterium, in which the endospores are visible as green oval structures. The red, rod-shaped cells are vegetative cells, which represent a nonsporulating stage.

a specimen—or a part of it, such as the capsule—resists taking up a stain. The background around the organisms is filled with a stain such as India ink or nigrosin, leaving the organisms themselves as clear unstained objects that stand out against the dark background. A second simple or differential stain can be used to demonstrate the presence of the cell inside the capsule. Thus a typical slide would show a dark background of nigrosin, clear unstained areas of capsular material, and inside these, blue cells stained with methylene blue (Figure 3.29).

Flagellar Staining Flagella are too thin to be seen easily with the light microscope. When it is necessary to determine their presence or arrangement, special painstaking **flagellar staining** procedures are used to coat the surfaces of the flagella with dye or metals such as silver. These techniques are very difficult and time-consuming, and so are usually omitted from the beginning course in microbiology. Look ahead to Figure 4.11 to see some examples of stained flagella.

Endospore Staining A few types of bacteria produce resistant cells called endospores. Endospore walls are very resistant to penetration of ordinary stains. When a simple stain is used, the spores will be seen as clear, glassy, easily recognizable areas within the bacterial cell. Thus, strictly speaking, it is not absolutely necessary to perform an endospore stain to see the spores. However, the **Schaeffer-Fulton spore stain** is a differential stain that makes spores easier to visualize (Figure 3.30). Smears are covered with malachite green and then gently heated until they steam. Approximately 5 minutes of such steaming causes the endospore walls to become more permeable to the dye. The slide is then washed with water for 30 seconds to remove the green dye from all parts of the cell except for the endospores, which retain it. Then a counterstain of safranin is placed on the slide, which stains the nonspore, or vegetative, areas of the cells. Cultures without endospores will appear red; those with endospores will have green spores and red vegetative cells.

ESSAY

Images of Atoms

Being able to look inside of cells, to see the individual atoms that compose them; what a mind-boggling experience it would have been for Leeuwenhoek! In 1980, Gerd Binnig and Heinrich Rohrer invented the first of a series of rapidly improving scanning-probe (tunneling) microscopes. Five years later they received the Nobel Prize for their discovery. Instead of using light, Binnig and Rohrer used a thin wire probe made of platinum and iridium to trace over the surface of a substance, much as you might use your finger to feel the ups and downs in reading Braille. Electron clouds from the surfaces of the probe and the specimen overlap, producing a kind of pathway through which electrons can "tunnel" into each other's clouds. This tunneling sets up an observable electric current. The stronger the current, the closer the top of the atom must be to the

ESSAY Images of Atoms

probe. Running the probe across in a straight line reveals the highs and lows of individual atoms in a surface (Figure 3.31). Even movies can be made this way. The first one ever produced showed individual fibrin molecules coming together to form a clot.

Soon biologists will be able to see actual molecular events that they previously could only imagine. Paul Hansma, a physicist at the University of California at Santa Barbara, has hopes that this technique can speed up the Human Genome Project. Current machines are able to identify and sequence 7000 bases in a DNA molecule each day. Scanning probe microscopes, programmed to recognize labels attached to the four different DNA bases, could sequence 10 bases per second, or nearly a million bases per day. At this rate, the Human Genome Project could be finished years ahead of schedule, for millions instead of billions of dollars, and by a machine that would probably cost around $200,000.

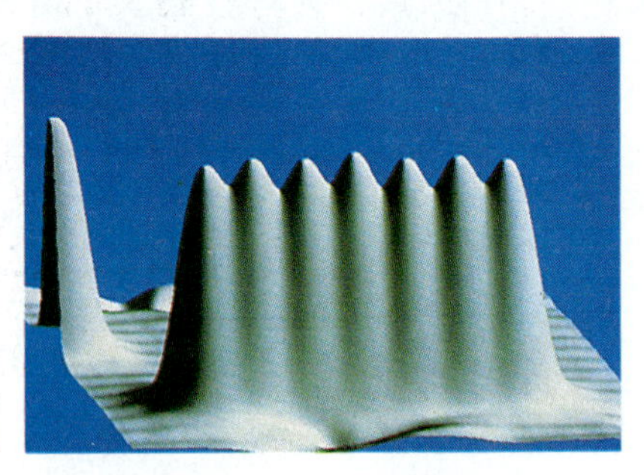

FIGURE 3.31 Individual atoms of the element xenon can be clearly distinguished using a scanning tunneling microscope (STM). This chain of seven xenon atoms was built by IBM scientists moving each atom into position, one at a time. The atoms are 20 billionths of an inch, or 0.5 nm (5 Å) apart. The atoms are bonded together; moving an end atom will relocate up to three of them at a time.

And now, IBM researchers have found a way to use the scanning-tunneling probe not to just look at, but to touch and move atoms around. Their first feat was to spell out "IBM" in letters just five atoms tall! The possibilities of literally constructing or engineering molecules has just taken on new meaning.

CHAPTER SUMMARY

RELATED KEY TERMS

DEVELOPMENT OF MICROSCOPY

- The existence of microorganisms was unknown until the invention of the microscope. Leeuwenhoek (1632–1723) was probably the first to see microorganisms.
- Leeuwenhoek's simple microscopes could reveal little detail of specimens. Today, multiple-lens, compound microscopes give us nearly distortion-free images, enabling us to delve further into the study of microbes.

PRINCIPLES OF MICROSCOPY

microscopy

Metric Units

- The three units most used to describe microbes are the **micrometer** (μm), formerly called a micron, which is equal to 0.000001 m, also written as 10^{-6} m; the **nanometer** (nm), formerly called a millimicron (mm), which is equal to 0.000000001 m, also written as 10^{-9} m; and the **Ångström** (Å) no longer officially recognized, which is equal to 0.0000000001 m, 0.1 nm, or 10^{-10} m.

Properties of Light: Wavelength and Resolution

- The **wavelength,** or the length of light rays, is the limiting factor in resolution.
- **Resolution** is the ability to see two objects as separate and discrete entities.
- Light wavelengths must be small enough to fit between two objects in order for them to be resolved.
- **Numerical aperture** is the widest cone of light that can enter a particular lens. Its value is engraved on the side of each objective lens.
- **Resolving power** can be defined as $RP = \lambda/2NA$, where λ = wavelength of light, and NA = numerical aperture. The smaller the value of λ and the larger the value of NA, the greater is the resolving power of the lens.
- The total magnification of a light microscope is calculated by multiplying the magnifying power of the objective lens by the magnifying power of the ocular lens. Increased magnification is of no value unless good resolution can also be maintained.

Properties of Light: Light and Objects

- If light bounces back, **reflection** (which gives an object its color) occurs.

- **Transmission** is the passage of light through an object. Light must either be reflected from or transmitted through an object in order for it to be seen with a light microscope.
- **Absorption** of light rays occurs when they neither bounce off nor pass through an object but are taken up by that object. Absorbed light energy is used to perform photosynthesis or to raise the temperature of the irradiated body.
- Reemission of absorbed light as light of longer wavelengths is known as **luminescence.** If this occurs only during irradiation, the object is said to **fluoresce.** If reemission continues after irradiation ceases, the object is said to be **phosphorescent.**
- **Diffraction** occurs when light waves are bent as they pass through a small opening, such as a hole, a slit, a space between two adjacent cellular structures, or a small, high-powered, magnifying lens in a microscope. The bent light rays distort the image obtained and limit the usefulness of the light microscope.
- **Refraction** is the bending of light as it passes from one medium to another of different density. **Immersion oil,** which has the same index of refraction as glass, is used to replace air and prevent refraction at a glass-air interface.

RELATED KEY TERMS

LIGHT MICROSCOPY

The Compound Light Microscope

- The major parts of a **compound light microscope** and their functions are as follows:

Base	Supports microscope and usually contains light source.
Condenser	Converges light beams to pass through specimen.
Iris diaphragm	Controls amount of light passing through specimen.
Objective lens	Magnifies image.
Body tube	Conveys light to the ocular lens.
Ocular lens	Set in eyepiece, it further magnifies the image.
Mechanical stage	Allows precise control in moving slide.
Coarse adjustment	Used to locate specimen.
Fine adjustment	Used to bring specimen into sharp focus.

light microscopy **binocular**
total magnification **parfocal**
optical microscope
index of refraction **monocular**
ocular micrometer

Dark-Field Microscopy

- **Bright-field illumination** is used in the ordinary light microscope, with light passing directly through the specimen.
- **Dark-field illumination** utilizes a special condenser that causes light to reflect off the specimen at an angle rather than pass directly through it.

Phase-Contrast Microscopy

- **Phase-contrast microscopy** involves microscopes with special condensers that accentuate small differences in refractive index of structures within the cell, allowing live unstained organisms to be examined.

Fluorescence Microscopy

- **Fluorescence** (UV) **microscopy** uses ultraviolet light instead of white light to excite molecules within the specimen or dye molecules attached to the specimen. These emit different wavelengths, often of brilliant colors.

fluorescent antibody staining

Differential Interference Contrast (Nomarski) Microscopy

- Differential interference contrast, or **Nomarski microscopy,** uses microscopes that operate essentially like phase-contrast microscopes but with a much greater resolution and a very short depth of field. They produce a nearly three-dimensional image.

ELECTRON MICROSCOPY

- The **electron microscope** (EM) uses a beam of electrons instead of a beam of light and electromagnets instead of glass lenses for focusing. They are much more expensive and difficult to use but give magnifications of up to 200,000X and a resolving power of less than 1 nm. Viruses can be seen only using EMs.

Transmission Electron Microscopy

- For the **transmission electron microscope** (TEM), very thin slices of a specimen are cut, revealing the internal structure of microbial cells.

shadow casting
electron micrographs

Scanning Electron Microscopy

- For the **scanning electron microscope** (SEM), a specimen is coated with a metal. The electron beam is scattered by this coating to form a three-dimensional image.

freeze-fracturing
freeze-etching

TECHNIQUES OF LIGHT MICROSCOPY

Preparation of Specimens for the Light Microscope

- **Wet mounts** are used to view living organisms. The **hanging-drop** technique is a special type of wet mount, often used to determine whether organisms are motile.
- **Smears** of appropriate thickness are allowed to air-dry completely and are then passed through an open flame. This process, called **heat fixation,** kills the organisms, causing them to adhere to the slide and more readily accept stains.

Principles of Staining

- A **stain,** or dye, is a molecule that can bind to a structure and give it color.
- Most microbial stains are **cationic** (positively charged), or **basic dyes,** such as methylene blue. Because most bacterial surfaces are negatively charged, these dyes are attracted to them.
- **Simple stains** use one dye and reveal basic cell shapes and arrangements. **Differential stains** use two or more dyes and distinguish various properties of organisms. The **Gram stain,** the **Schaeffer-Fulton spore stain,** and the **Ziehl-Neelsen acid-fast stain** are examples.
- **Negative staining** colors the background around an object that resists taking up stain.
- **Flagellar staining** adds layers of dye or metal to the surfaces of flagella but is too difficult and time-consuming for beginning students.
- In the **Schaeffer-Fulton spore stain,** endospores are colored green due to the uptake of malachite green, whereas vegetative cells are colored red by safranin.

anionic (acidic) dyes

mordant

QUESTIONS FOR REVIEW

A.

1. Leeuwenhoek built simple microscopes. How did these differ from today's modern light microscopes?

B.

2. If a bacterium is 0.5 μm wide and 15 μm long, how many Ångströms wide is it, and how many nanometers long is it?

C.

3. What is resolution, and why is it important in microscopy?
4. As wavelength decreases, does resolving power increase or decrease? Why?

D.

5. Define and contrast: absorption, reflection, and transmission.
6. Define and contrast: fluorescence and phosphorescence.
7. Define and contrast: diffraction and refraction.
8. Explain how and why immersion oil is sometimes used in light microscopy.

E.

9. Where are the condenser, objective lens, and ocular lens located in a modern light microscope? What is the function of each?
10. What is the total magnification of a microscope having:
    a. 20X ocular and 40X objective lenses?
    b. 10X ocular and 99X objective lenses?
11. In the days before microscopes were parfocal, what do you think was one of the greatest problems students faced in using the microscope?
12. What is an ocular micrometer used for?

F.

13. Explain the differences between bright-field and dark-field illumination, and the advantages of the latter.
14. When is phase-contrast microscopy used? What special advantages does it have?
15. What is fluorescence microscopy? What special advantages does it have?

G.

16. What does the electron microscope use instead of light beams and glass lenses?
17. If you were to see an electron micrograph in this book, how would you know whether it was a TEM or an SEM picture?

H.

18. What three things does heat fixation accomplish?

19. Explain the hanging-drop technique and why it is useful.

I.

20. What is the difference between a simple and a differential stain?
21. When would you use the Ziehl-Neelsen acid-fast stain?
22. List some examples of differential stains and the features they can distinguish between.
23. Name three special staining techniques, and explain briefly the uses of each.

J.

24. What is a mordant? Which reagent is a mordant in the Gram-staining procedure?
25. What is the difference between gram-nonreactive organisms and gram-variable organisms?

PROBLEMS FOR INVESTIGATION

1. A student failed to complete the Gram stain on her "unknown" culture in her drawer and plans to do a Gram stain on that same slant next week. What has she overlooked that may prevent her from obtaining proper results?
2. What would happen if a student forgot to do the iodine step in the Gram staining procedure of a mixed culture of gram-positive and gram-negative organisms? What would his results look like? Why?
3. Find out something about the comparative costs of the various types of microscopes and any additional facilities needed for their use, e.g., darkrooms, preparation rooms. Ask your instructor how much the microscope you use in lab would cost if you had to buy it today. Keep this figure in mind as you use it during the semester, treating it kindly and gently.
4. Why do you suppose a depression slide is used for a hanging-drop preparation? Why would it be difficult to use a regular flat slide?
5. What are the advantages and disadvantages of observing living-specimen slides and fixed-specimen slides of microorganisms?

SOME INTERESTING READING

Balows, A., et al., eds. 1991. *Manual of clinical microbiology.* Washington, D.C.: American Society for Microbiology.

Bennig, G. and H. Rohrer. 1985. The scanning tunneling microscope. *Scientific American* 253(2):50.

Dobell, C. 1932. *Anthony van Leeuwenhoek and his "Little Animals."* London: Constable. (Reprinted in paperback 1960 by Dover, New York.)

England, B. M. 1991. The state of the science: scanning electron microscope. *Mineralogical Record* 22, no. 2 (March–April):123–33.

Holt, S. C. and T. J. Beveridge. 1982. "Electron microscopy: Its development and application to microbiology." *Canadian Journal of Microbiology* 28:1.

Lennette, E. H., et al., eds. 1985. *Manual of clinical microbiology,* 4th ed. Washington, D.C.: American Society for Microbiology.

Lillie, R. D. 1977. *H. J. Conn's biological stains: A handbook on the nature and uses of the dyes employed in the biological laboratory,* 9th ed. Baltimore: Williams and Wilkins.

Molina, T. C., H. D. Brown, and R. M. Irbe. 1990. Gram staining apparatus for space station applications. *Applied and Environmental Microbiology* 56(March):601–6.

Powell, C. S. 1990. Science writ small: a microscope builds an atomic-scale billboard. *Scientific American* 262(5):26.

Sieburth, J. M. 1975. *Microbial seascapes.* Baltimore: University Park Press.

Trifiro, S., A.-M. Bourgault, F. Lebel, and P. Rene. 1990. Ghost mycobacteria on gram stain. *Journal of Clinical Microbiology* 28(1):146–47.

Trux, J. 1991. Through the looking glass (objects viewed through a scanning electron microscope). *World Magazine* 58, no. 8 (March):58–64.

Woeste, S. and P. Demchick. 1991. New version of the negative stain. *Applied and Environmental Microbiology* 57(6a):1858–59.

False-color transmission electron micrograph (TEM) of *Streptococcus faecalis* cell dividing by binary fission. This is a gram-positive coccus that forms short chains and lives in the human digestive tract as part of the normal microbial population. Under certain conditions, however, it may cause infection of the heart or urogenital systems. Magnification is 76,000X.

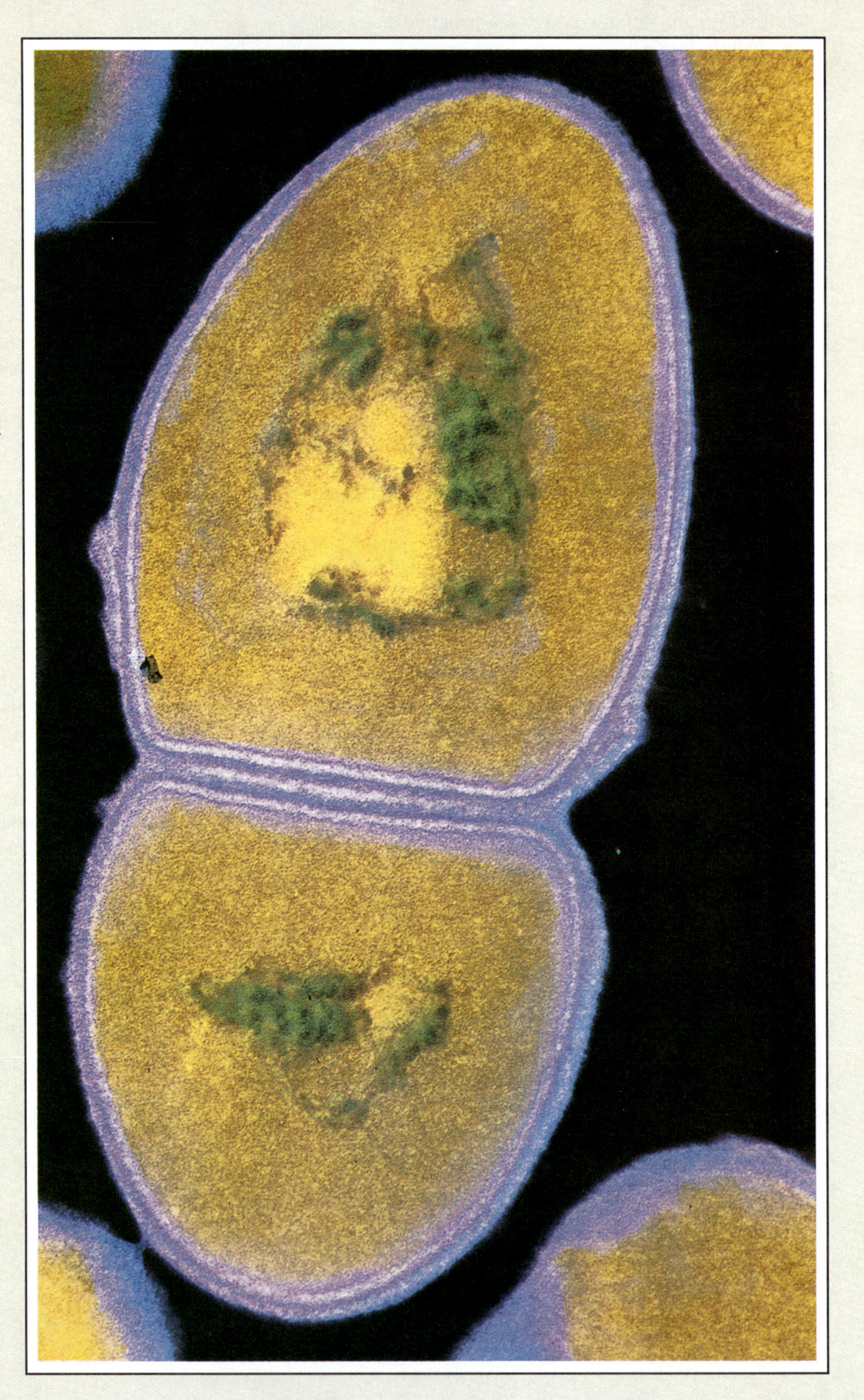

4 Characteristics of Prokaryotic and Eukaryotic Cells

This chapter focuses on the following questions:

A. What are the characteristics of eukaryotic and prokaryotic cells?

B. How do prokaryotic cells differ in size, shape, and arrangement?

C. How are structure and function related in bacterial cell walls and cell membranes?

D. How are structure and function related in other bacterial components?

E. How are structure and function related in eukaryotic cell membranes?

F. How are structure and function related in other eukaryotic components?

G. How do passive transport processes function, and why are they important? How does active transport function, and why is it important?

H. How do exocytosis and endocytosis occur, and why are they important?

BASIC CELL TYPES

Having considered the chemical principles that apply to cells and how to use microscopes and stains to observe cells, we can now look at the structure and function of the cells themselves. All living cells can be classified as prokaryotic or eukaryotic. **Prokaryotic** (pro-kar"e-ot'ik) **cells** lack a nucleus and other membrane-bound structures, and **eukaryotic** (u-kar" e-ot'ik) **cells** have such structures.

Prokaryotic and eukaryotic cells are similar in several ways. Both are surrounded by a cell membrane, or plasma membrane. Although some cells have structures that extend beyond this membrane or surround it, the cell membrane defines the boundaries of the living cell. Both prokaryotic and eukaryotic cells also encode genetic information in DNA molecules.

These two types of cells are different in other important ways. In eukaryotic cells DNA is in a nucleus surrounded by a nuclear membrane, but in prokaryotic cells it is in a nuclear region not surrounded by a membrane. Eukaryotic cells also have a variety of internal structures called **organelles** (or-gan-elz'), or "little organs," also surrounded by a membrane. Prokaryotic cells generally lack such organelles. In this chapter we examine the similarities and differences of prokaryotic cells and eukaryotic cells, as summarized in Table 4.1.

PROKARYOTIC CELLS

All prokaryotic cells are, in fact, whole unicellular organisms. Prokaryotic organisms include two small groups, the primitive archaebacteria and the photosynthetic cyanobacteria, and the large group of true bacteria.

Size, Shape, and Arrangement

Size

Prokaryotes are among the smallest of all organisms. Most prokaryotes range from 0.5 to 2.0 μm in diameter. For comparison, a human red blood cell is about 7.5 μm in diameter. Keep in mind, however, that although we often use diameter to specify cell size, many cells are not spherical in shape. Some spiral

TABLE 4.1 Similarities and differences between prokaryotic and eukaryotic cells

Characteristic	Prokaryotic Cells	Eukaryotic Cells
Genetic information	Found in single chromosome	Found in paired chromosomes
Location of genetic information	Nuclear area (nucleoid)	Membrane-bound nucleus
Nucleolus	Absent	Present
Histones	Absent	Present
Extrachromosomal DNA	In plasmids	In organelles, such as mitochondria and chloroplasts
Mitotic spindle	Absent	Present during cell division
Plasma membrane	Fluid-mosaic structure lacking sterols	Fluid-mosaic structure containing sterols
Internal membranes	Only in photosynthetic organisms	Numerous membrane-bound organelles
Endoplasmic reticulum	Absent	Present
Respiratory enzymes	Cell membrane	Mitochondria
Chromatophores	Present in photosynthetic bacteria	Absent
Chloroplasts	Absent	Present in some
Golgi apparatus	Absent	Present
Lysosomes	Absent	Present
Peroxisomes	Absent	Present
Ribosomes	70S	80S in cytoplasm and on endoplasmic reticulum, 70S in organelles
Cytoskeleton	Absent	Present
Cell wall	Usually peptidoglycan found on most cells	Cellulose, chitin, or both found on plant and fungal cells
External layer	Capsule or slime layer	Pellicle, test, or shell in certain protists
Flagella	When present, consist of fibrils of flagellin	When present, consist of complex membrane-bound structure with "9 + 2" microtubule arrangement
Cilia	Absent	Present as structures shorter than, but similar to, flagella in some eukaryotic cells
Pili	Present as attachment or conjugation pili in some prokaryotic cells	Absent

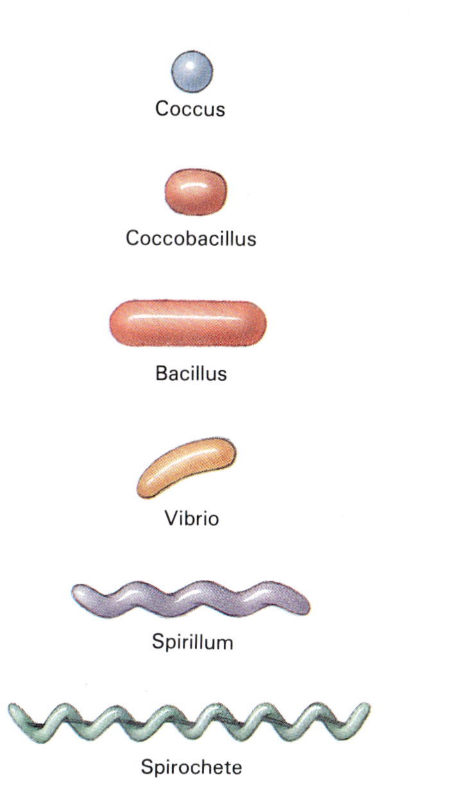

FIGURE 4.1 The most common bacterial shapes.

bacteria have a much larger diameter, and some cyanobacteria are 60 μm long. Because of their small size, bacteria have a large surface-to-volume ratio. For example, bacteria with a diameter of 2 μm have a surface area of about 12 μm^2 and a volume of about 4 μm^3. Their surface-to-volume ratio is 12:4, or 3:1. In contrast, eukaryotic cells with a diameter of 20 μm have a surface area of about 1200 μm^2 and a volume of about 4000 μm^3. Their surface-to-volume ratio is 1200:4000, or 0.3:1—only one-tenth as great. The large surface-to-volume ratio of bacteria means that no internal part of the cell is very far from the surface, and nutrients can easily reach all parts of the cell.

Shapes

Typically, bacteria display three basic shapes: spherical, rodlike, and spiral (Figure 4.1). A spherical bacterium is called a **coccus** (kok'us), and a rodlike bacterium is called a **bacillus** (bas-il'us). The plurals of these terms are *cocci* (kok'se) and *bacilli* (bas-il'e), respectively. Some bacteria, called coccobacilli, are short rods intermediate in shape between cocci and bacilli. Spiral bacteria have a variety of curved shapes. A comma-shaped bacterium is called a **vibrio** (vib're-o); a corkscrew-shaped one, a **spirillum** (spi-ril'um; plural: *spirilli*); and a flexible, wavy-shaped one, a **spirochete** (spi'ro-ket). Some bacteria do not fit in any of the preceding categories but rather have spindle shapes or irregular, lobed shapes. Square bacteria were discovered on the shores of the Red Sea in 1981. They are 2 to 4 μm on a side and sometimes aggregate in wafflelike sheets.

Even bacteria of the same kind sometimes vary in size and shape. When nutrients are abundant in the environment and cell division is rapid, rods are often twice as large as those in an environment with only a moderate supply of nutrients. Although variations in shape within a single species of bacteria are generally small, there are exceptions to this rule. Some bacteria vary widely in form even within a single culture, a phenomenon known as **pleomorphism.** Moreover, in aging cultures where organisms have used up most of the nutrients and have deposited wastes, cells not only are generally smaller, but they often display a great diversity of unusual shapes.

Arrangements

In addition to characteristic shapes, many bacteria also are found in distinctive arrangements of groups of cells (Figure 4.2). Such groups occur when cells divide without separating. Cocci can divide in one or two

FIGURE 4.2 Arrangements of bacteria. (a) Cocci arranged in chains (*Streptococccus*). (b) Cocci arranged in a cluster (*Staphylococcus*). (c) Cyanobacteria in chains. (The rod-shaped cells are *Anabaena*, the hoselike filaments, *Microcoleus*.) (d) *Merisopedia* bacteria form clusters of four cells.

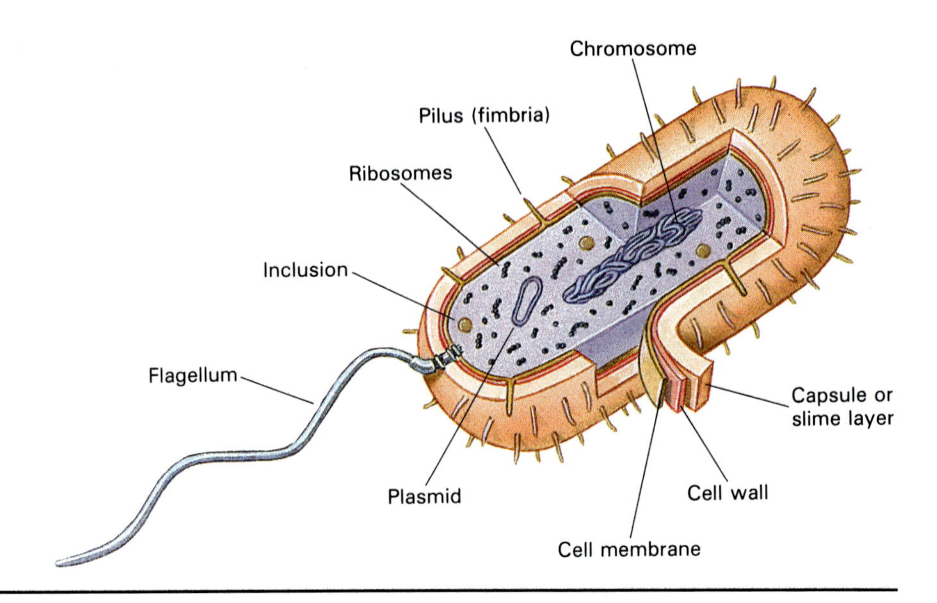

FIGURE 4.3 Diagram of a typical prokaryotic cell. The cell depicted is a bacillus with polar flagellum.

planes. Division in one plane produces cells in pairs (indicated by the prefix *diplo-*) or chains (*strepto-*). Division in two planes produces cells in tetrads (four cells arranged in a cube); random division planes produce grapelike clusters (*staphylo-*). Bacilli divide in only one plane, but they can produce cells connected end-to-end (like train cars) or side by side. Spiral bacteria are not generally found in multicellular arrangements.

Overview of Structure

Structurally, bacterial cells (Figure 4.3) consist of the following:

1. A cell membrane, usually surrounded by a cell wall and sometimes by an outer membrane.
2. An internal cytoplasm with ribosomes, a nuclear region, granules, and vesicles.
3. A variety of external structures, such as capsules, flagella, and pili.

Let us look at each of these kinds of structures in some detail.

Cell Wall

The semirigid **cell wall** lies outside the cell membrane. It is present in nearly all bacteria and performs two important functions. First, it maintains the characteristic shape of the cell. If the cell wall is digested away by enzymes, the cell takes on a spherical shape. Second, it prevents the cell from bursting when fluids flow into the cell by osmosis (described later in this chapter). Though the cell wall surrounds the cell membrane, it is extremely porous and does not play a major role in regulating the entry of materials into the cell.

Components of Cell Walls

Peptidoglycan **Peptidoglycan,** the single most important component of the bacterial cell wall, is a polymer so large it can be thought of as one immense, covalently linked molecule. It forms a supporting net around a bacterium that resembles multiple layers of chain link fence (Figure 4.4). In the peptidoglycan polymer, molecules of *N*-acetylglucosamine (gluNAc) alternate with molecules of *N*-acetylmuramic acid (murNAc). These molecules are cross-linked by tetrapeptides, chains of four amino acids. Amino acids, like many other organic compounds, have stereoisomers—structures that are mirror images of each other, just as a left hand is a mirror image of a right hand. Some of the amino acids in the tetrapeptide chains are mirror images of those most commonly found in living things. These chains are not readily broken down because most organisms lack enzymes that can digest such mirror-image forms. Peptidoglycan usually has teichoic acids attached to it. **Teichoic acids,** which consist of glycerol, phosphates, and the sugar alcohol ribitol, occur in polymers up to 30 units long. These polymers extend beyond the cell wall and even beyond the capsule in encapsulated bacteria. Teichoic acid furnishes attachment sites for bacteriophages, which are viruses that infect bacteria.

Outer Membrane The **outer membrane,** found primarily in gram-negative bacteria, is a typical bilayer membrane (discussed in the next section). It surrounds the cell wall and is attached to the peptidoglycan by an almost continuous layer of small lipoprotein molecules. The lipoproteins are embedded in the outer membrane and covalently bonded to the peptidoglycan. The outer membrane acts as a coarse sieve and exerts little control over the movement of substances into and out of the cell. However, it does

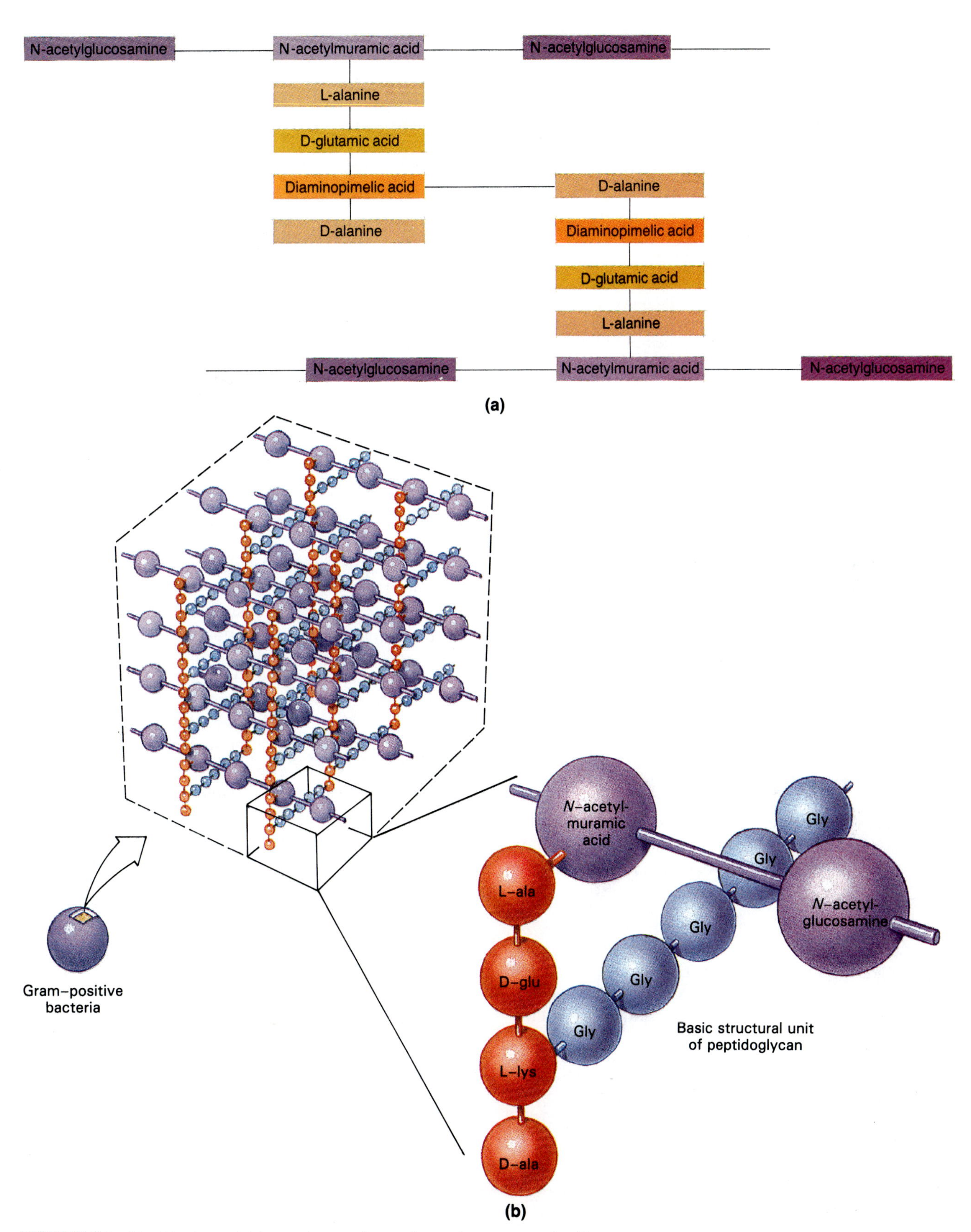

FIGURE 4.4 Peptidoglycan (a) is a polymer of two alternating sugar units, *N*-acetylglucosamine and *N*-acetylmuramic acid, both of which are derivatives of glucose. The sugars are cross-linked by short peptide chains consisting of four amino acids. The resulting structure (b) somewhat resembles a chain link fence surrounding the bacterial cell.

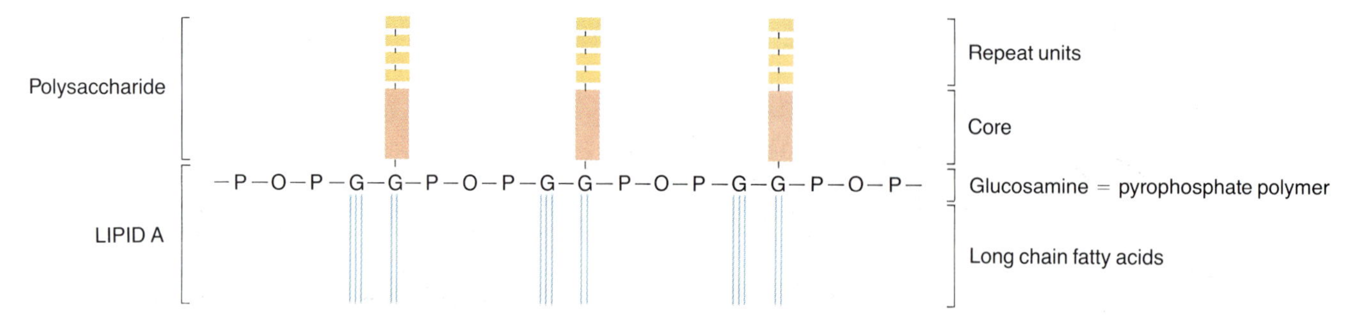

FIGURE 4.5 Lipopolysaccharide, also called endotoxin, an important component of the outer membrane in gram-negative cell walls. The lipid A portion of the molecule consists of a backbone of alternating pyrophosphate units (linked phosphate groups) and glucosamine (another glucose derivative), to which long fatty acid side chains are attached. Lipid A is a toxic substance that contributes to the danger of infection by gram-negative bacteria. Polysaccharide side chains extending outward from the glucosamine units make up the remainder of the molecule.

control the transport of certain proteins from the environment. The outer surface of the outer membrane has surface antigens and receptors. Some receptors bind viruses and thereby help them to infect the bacterium.

Lipopolysaccharide, an important part of the outer membrane, can be used to identify gram-negative bacteria. It is an integral part of the cell wall and is not released until the cell walls of dead bacteria are broken down. Lipopolysaccharide, also known as **endotoxin,** consists of polysaccharides and **lipid A** (Figure 4.5). The polysaccharides are found in repeating side chains that extend outward from the organism. The lipid A portion is responsible for the toxic properties that make any gram-negative infection a potentially serious medical problem. It causes fever and dilates blood vessels so that the blood pressure drops precipitously. Because bacteria release endotoxin mainly when they are dying, killing them may increase the concentration of this very toxic substance.

Periplasmic Space Another distinguishing characteristic of gram-negative bacteria is the **periplasmic space,** the space between the cell membrane and the outer membrane. This space is separated into two parts—the wider part between the peptidoglycan and the cell membrane and the narrower part between the peptidoglycan and the outer membrane. The periplasmic space contains digestive enzymes that help to destroy substances that might harm the bacteria.

Distinguishing Bacteria by Cell Walls

Certain properties of cell walls result in different staining reactions. Gram-positive, gram-negative, and acid-fast bacteria can be distinguished on the basis of these reactions (Table 4.2 and Figure 4.6).

Gram-Positive Bacteria The cell wall in gram-positive bacteria has a relatively dense layer of peptidoglycan 20 to 80 nm thick. The peptidoglycan layer is closely attached to the outer surface of the cell membrane. Chemical analysis shows that 60 to 90 percent of the cell wall of a gram-positive bacterium is peptidoglycan. Except for streptococci, most gram-positive cell walls contain very little protein. If peptidoglycan is digested from their cell walls, gram-positive bacteria become **protoplasts,** or cells with a cell membrane but no cell wall. These are very osmotically fragile and must be kept in isotonic solutions.

The thick cell walls of gram-positive bacteria retain such stains as the crystal violet-iodine dye, but fungal cells, which have thick cellulose walls, also retain them. Thus retention of Gram stain seems to be di-

TABLE 4.2 Characteristics of the cell walls of gram-positive, gram-negative, and acid-fast bacteria

Characteristic	Gram-Positive Bacteria	Gram-Negative Bacteria	Acid-Fast Bacteria
Peptidoglycan	Thick layer	Thin layer	Relatively small amount
Lipids	Very little present	Lipopolysaccharide	Mycolic acid and other waxes and glycolipids
Outer membrane	Absent	Present	Absent
Periplasmic space	Absent	Present	Absent
Cell shape	Always rigid	Rigid or flexible	Rigid or flexible
Effects of enzyme digestion	Protoplast	Spheroplast	Difficult to digest
Sensitivity to dyes and antibiotics	Most sensitive	Moderately sensitive	Least sensitive

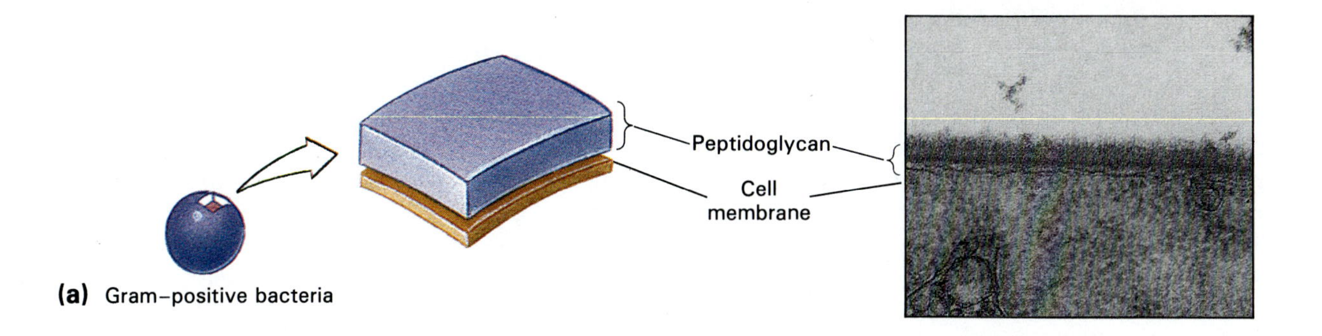

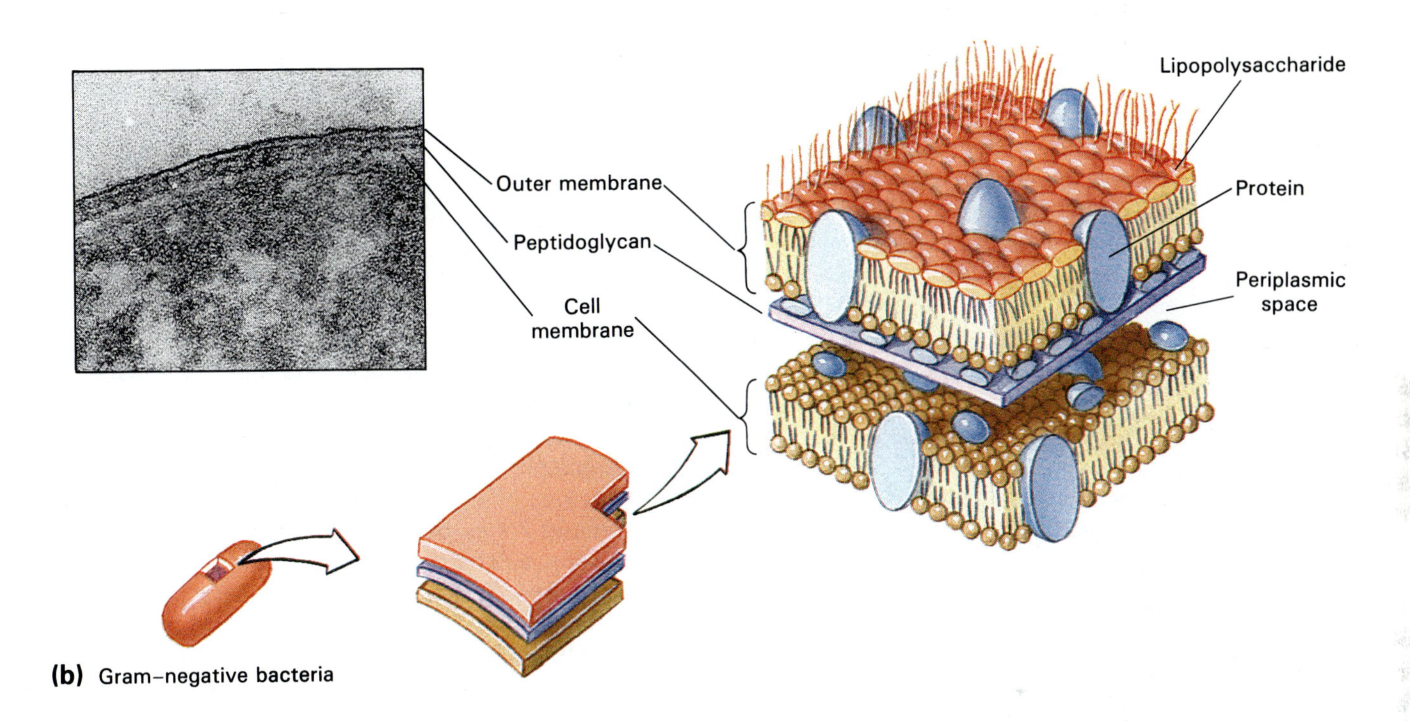

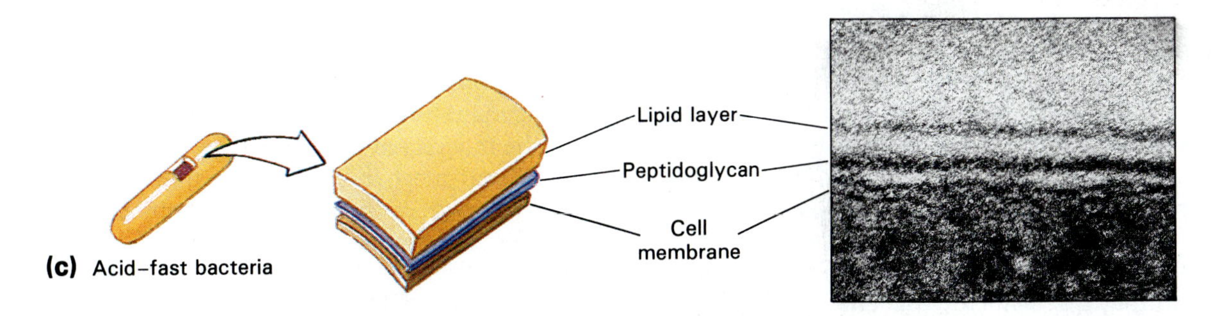

FIGURE 4.6 Schematic representations of bacterial cell walls, paired with TEM photographs of representative bacteria. (a) Gram-positive (*Bacillus fastidosus*). (b) Gram-negative (*Azomonas insignis*). (c) Acid-fast (*Mycobacterium phlei*).

rectly related to wall thickness and not to peptidoglycan. Physiological damage or aging can make a gram-positive cell wall leaky, so that the dye complex escapes. Such organisms can become gram-variable or even gram-negative as they age. Therefore, Gram staining must be performed on cultures less than 24 hours old.

Gram-positive bacteria lack both an outer membrane and a periplasmic space. Therefore, digestive enzymes are released into the environment, where they sometimes become so diluted that the organisms derive no benefit from them.

Gram-Negative Bacteria The cell wall of a gram-negative bacterium is thinner but more complex than that of a gram-positive bacterium. Only 10 to 20 percent of the cell wall is peptidoglycan; the remainder consists of various polysaccharides, proteins, and lipids. The cell wall is surrounded by an outer membrane, which adheres to the outer surface of the wall, leaving only a very narrow periplasmic space. The inner surface of the wall is separated from the cell membrane by a wider periplasmic space. Toxins and enzymes remain in the periplasmic space in sufficient concentrations to help destroy substances that might harm the bacterium, but they do not harm the organism that produced them. If the cell wall is digested away, gram-negative bacteria become **spheroplasts,** which have both a cell membrane and most of the outer membrane. Gram-negative bacteria fail to retain Gram stain during the decolorizing procedure partly because of their thin cell walls and partly because of the relatively large quantities of lipoproteins and lipopolysaccharides in the walls.

Acid-Fast Bacteria The cell wall of acid-fast bacteria, the mycobacteria, is thick, like that of gram-positive bacteria. It is approximately 60 percent lipid and contains much less peptidoglycan. In the acid-fast staining process, carbolfuchsin binds to the lipids and resists removal by an acid-alcohol mixture. (Chapter 3, p. 67) The lipids make acid-fast organisms impermeable to most other stains and protect them from acids and alkalis. The organisms grow slowly because the lipids impede entry of nutrients into cells, and the cells must expend large quantities of energy to synthesize lipids.

Control of Bacteria by Damage to Cell Walls Some methods of controlling bacteria are based on properties of the cell wall. For example, the antibiotic penicillin interferes with cell wall synthesis. It prevents peptide cross linkages in peptidoglycan by inhibiting the formation of particular peptide bonds found only in peptidoglycan. If penicillin is present when bacterial cells are dividing, the cells cannot form complete walls, and they die. Similarly, the enzyme lysozyme, found in tears and other human body secretions, digests peptidoglycan. This enzyme helps to prevent bacteria from entering the body and is the body's main defense against eye infections.

Cell Membrane

The **cell membrane,** or **plasma membrane,** is a living membrane that forms the boundary between a cell and its environment. This dynamic, constantly changing membrane is not to be confused with the cell wall, which is a static structure external to the membrane.

Bacterial cell membranes have the same general structure as the membranes of all other cells. Such membranes, sometimes called unit membranes, consist mainly of phospholipids and proteins. The **fluid-mosaic model** (Figure 4.7) represents the current understanding of the structure of such a membrane. The model's name is derived from the fact that phospholipids in the membrane are in a fluid state and that proteins are dispersed among the lipid molecules in the membrane, forming a mosaic pattern.

Membrane phospholipids form a bilayer, or two adjacent layers. In each layer the phosphate ends of the lipid molecules extend toward the membrane surface, and the fatty acid ends extend inward. The charged phosphate ends of the molecules are **hydrophilic** (water-loving) and thus can interact with the watery environment. The fatty acid ends, consisting largely of nonpolar hydrocarbon chains, are **hydrophobic** (water-fearing) and thus are suitable for forming a barrier between the cell and its environment. Some membranes also contain other lipids. The membranes of mycoplasmas, bacteria that lack a cell wall, include lipids called sterols that make them more rigid.

Interspersed among the lipid molecules are protein molecules. Some extend through the entire membrane and act as carriers or form pores or channels through which materials enter and leave the cell. Others are embedded in, or loosely attached to, the inner or outer surface of the membrane. Proteins on the inner surface are usually enzymes; those on the outer surface include those that make the cell identifiable as a particular organism.

Membranes are dynamic, constantly changing entities. Materials constantly move through pores and through the lipids themselves. Also, both the lipids and the proteins in membranes are continuously changing positions.

The main function of the cell membrane is to regulate the movement of materials into and out of a cell by transport mechanisms, which are discussed later in this chapter. In bacteria the membrane also performs certain functions carried out by other structures in eukaryotic cells. It synthesizes cell wall components, assists with DNA replication, secretes proteins, carries on respiration, and captures energy in ATP. It also contains bases of flagella, the actions of which cause the flagella to move. Finally, the cell membrane sometimes responds to chemical substances in the environment.

Internal Structure

Bacterial cells typically contain ribosomes, a nucleoid region, and a variety of vacuoles within their cyto-

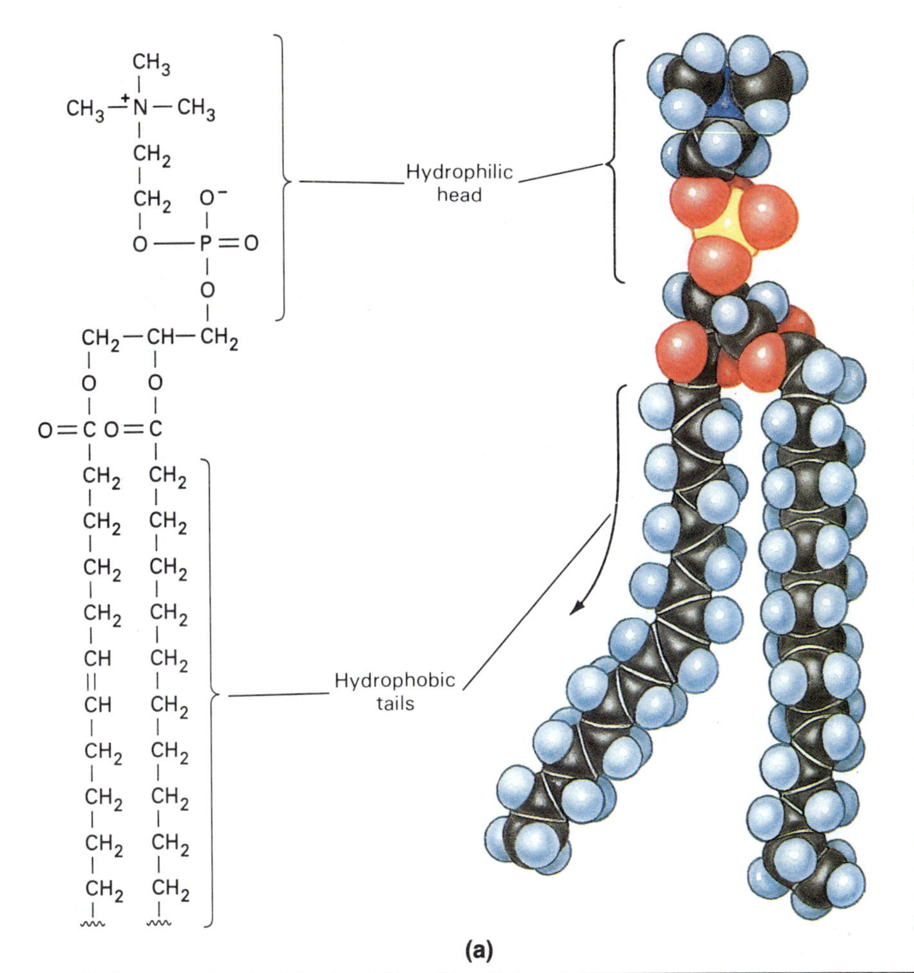

FIGURE 4.7 The fluid mosaic model of the cytoplasmic membrane. (a) The basic structural component of the membrane is the phospholipid molecule. A phospholipid has two long fatty acid "tails" of hydrocarbon. This portion of the molecule is very hydrophobic—that is, it does not interact with water and forms an oily barrier to most water-soluble substances. The "head" of the molecule, however, consists of a charged phosphate group, usually joined to another charged, nitrogen-containing group. This part of the molecule is very hydrophilic—it interacts easily with water. (b) The phospholipids form a bilayer, or double layer, in which the hydrophobic tails form the central core and the hydrophilic heads form the surfaces that face both the interior of the cell and the outside environment. In this fluid bilayer, proteins with a variety of functions float like icebergs in a sea. Some extend through the bilayer; others are anchored to the inner or outer surface. Proteins and membrane lipids that have carbohydrate chains attached to them are called glycoproteins and glycolipids, respectively.

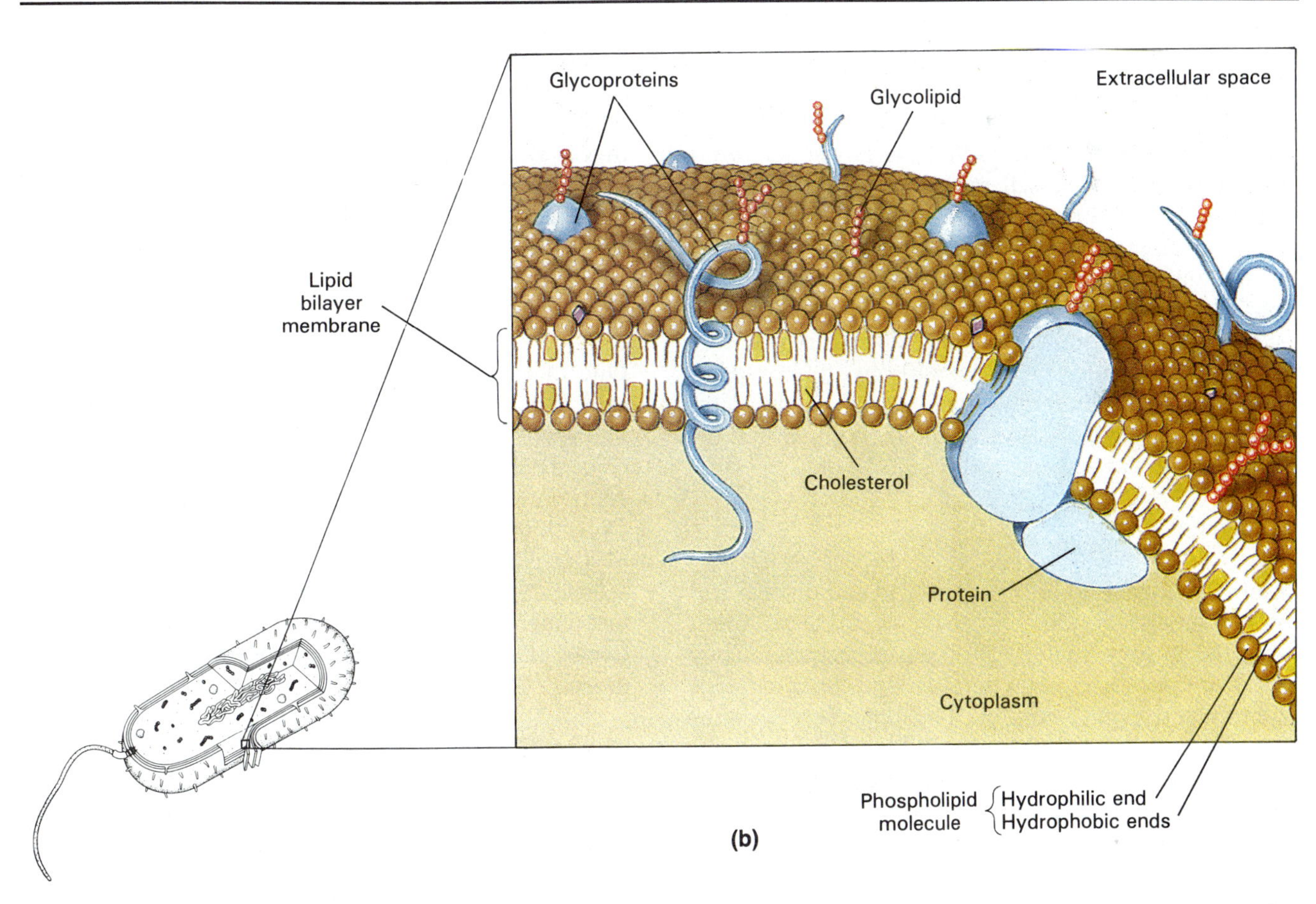

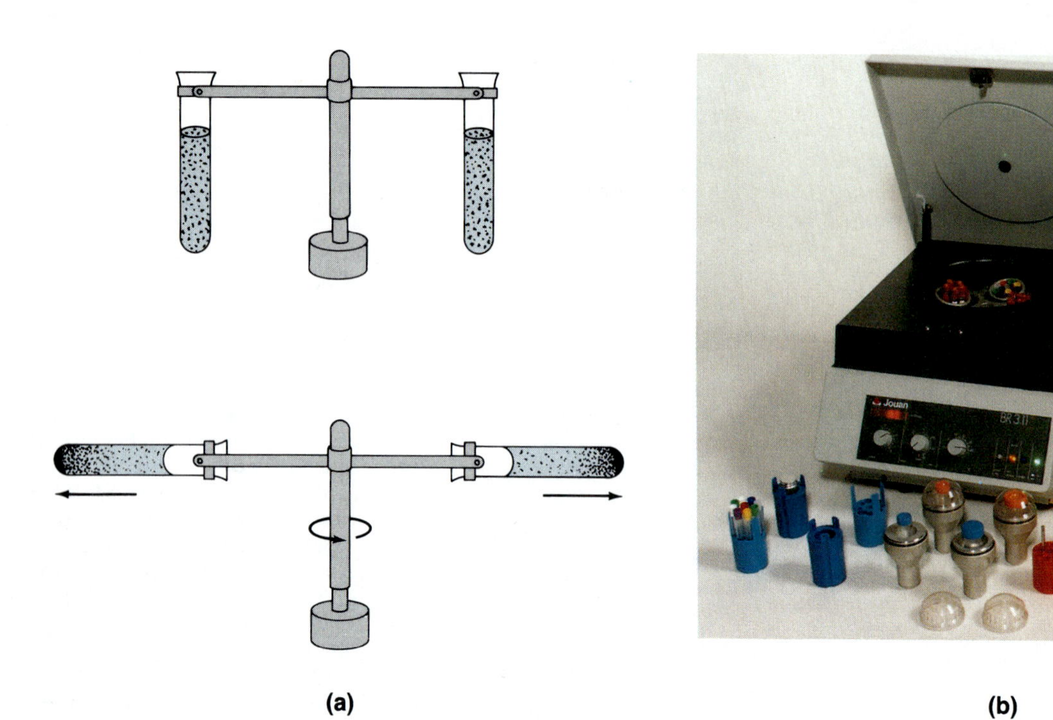

FIGURE 4.8 (a) In a centrifuge, suspended particles in tubes of liquid are whirled around at high speeds, causing them to settle to the bottom of the tubes or to form bands at different levels. The rate of settling or the locations of the bands can be used to determine the size, weight, and shape of the particles. (b) A table-top centrifuge with various containers for holding samples.

plasm. Their locations in a generalized prokaryotic cell are shown in Figure 4.3. Certain bacteria sometimes contain endospores.

Cytoplasm

The **cytoplasm** of prokaryotic cells is the semifluid substance inside the cell membrane. Because these cells typically have only a few clearly defined structures, such as a chromosome and some ribosomes, they consist mainly of cytoplasm. Cytoplasm is about four-fifths water and one-fifth substances dissolved or suspended in the water. These substances include enzymes and other proteins, carbohydrates, lipids, and a variety of inorganic ions. Many chemical reactions, both anabolic and catabolic, occur in the cytoplasm.

Ribosomes

Ribosomes consist of ribonucleic acid and protein and are abundant in the cytoplasm of bacteria, often grouped in long chains called **polyribosomes.** They are nearly spherical, stain densely, and contain a large subunit and a small subunit. Ribosomes serve as sites for protein synthesis (Chapter 7).

The relative size of ribosomes and their subunits can be determined by measuring their sedimentation rates—the rates at which they move toward the bottom of a tube when the tube is rapidly spun in an instrument called a centrifuge (Figure 4.8). Sedimentation rates are expressed in Svedberg (S) units. Whole bacterial ribosomes, which are smaller than eukaryotic ribosomes, have a rate of 70S; their subunits have rates of 30S and 50S. Certain antibiotics, such as streptomycin and erythromycin, bind specifically to 70S ribosomes and disrupt protein synthesis. Because they do not affect the 80S ribosomes found in eukaryotic cells, they kill bacteria without harming host cells.

Nuclear Region

One of the key features differentiating prokaryotic cells from eukaryotic cells is the absence of a nucleus bounded by a nuclear membrane. Instead of a nucleus, bacteria have a **nuclear region,** or **nucleoid.** The centrally located nuclear region consists mainly of DNA but has some RNA and protein associated with it. The DNA is arranged in one large, circular chromosome. Some bacteria also contain small circular molecules of DNA called *plasmids.* Genetic information in plasmids supplements information in the chromosome (Chapter 8).

Internal Membrane Systems

Photosynthetic bacteria contain internal membrane systems, sometimes known as **chromatophores** (Fig-

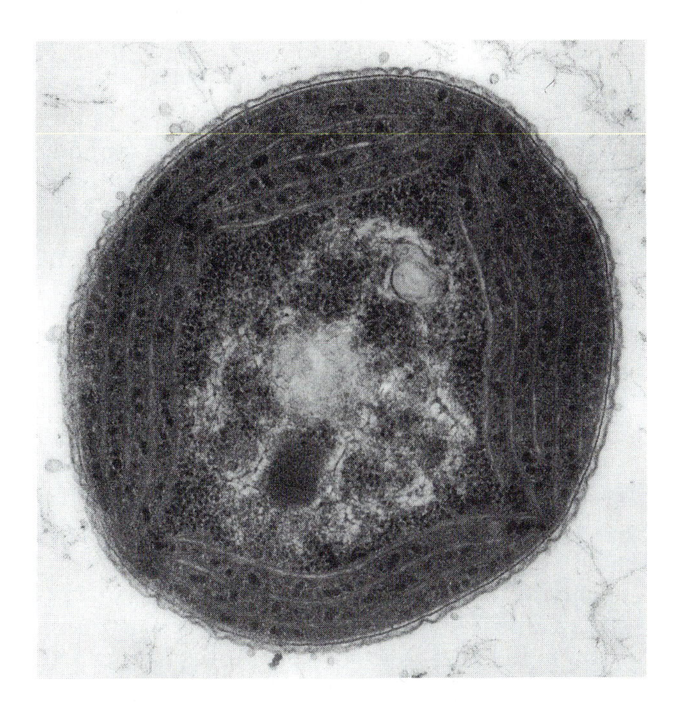

FIGURE 4.9 TEM of the cyanobacterium *Coccochloris elabens*. The outer regions of the cell are filled with photosynthetic membranes. The dark spots between the membranes are granules where carbohydrates produced by photosynthesis are stored.

ure 4.9). The membranes of the chromatophores, derived from the plasma membrane, contain the pigments used to capture light energy for the synthesis of sugars. Nitrifying bacteria, soil organisms that convert nitrogen compounds into forms usable by green plants, also have internal membranes. They house the enzymes used in deriving energy from the oxidation of nitrogen compounds (Chapter 5).

Electron micrographs of bacterial cells often show large infoldings of the plasma membrane, called *mesosomes*. Although these were originally thought to be structures present in living cells, they are now generally believed to be artifacts: that is, they are created by the processes used to prepare specimens for electron microscopy.

Inclusions

Bacteria can have within their cytoplasm a variety of small bodies collectively referred to as **inclusions.** Some are called granules; others are called vesicles.

Granules, though not bounded by membrane, contain substances so densely compacted that they do not easily dissolve in cytoplasm. Each granule contains a specific substance, such as glycogen or polyphosphate. Glycogen, a glucose polymer, is used for energy. Polyphosphate, a phosphate polymer, supplies phosphate for a variety of metabolic processes. Polyphosphate granules are called **volutin** (vo-lu'tin), or **metachromatic granules,** because they display

TRY IT

Living Magnets

Magnetotactic bacteria synthesize magnetite (Fe_3O_4), or lodestone, and store it in membranous vesicles called magnetosomes. (Lodestone was the first substance with magnetic properties to be discovered.) The presence of these magnetic inclusions enables these bacteria to respond to magnetic fields. In the Northern Hemisphere they swim toward the North Pole, in the Southern Hemisphere they swim toward the South Pole, and near the equator some swim north and some swim south. However, they also swim downward in water, because the magnetic force from the earth's poles is deflected through the earth and not over its horizon. This phenomenon is called *magnetotaxis*, and it apparently helps these anaerobic bacteria to move down toward sediments where their food (iron oxide) is abundant and oxygen, which they cannot tolerate, is deficient. Magnetotactic bacteria live in mud and brackish waters, and more than a dozen species have been identified. Most have a single flagellum, but *Aquaspirillum magnetotacticum* has two, one at each end, so it can swim forward or backward. When those that have one flagellum are placed in the field of an electromagnet, they make U-turns as the poles of the magnet are reversed.

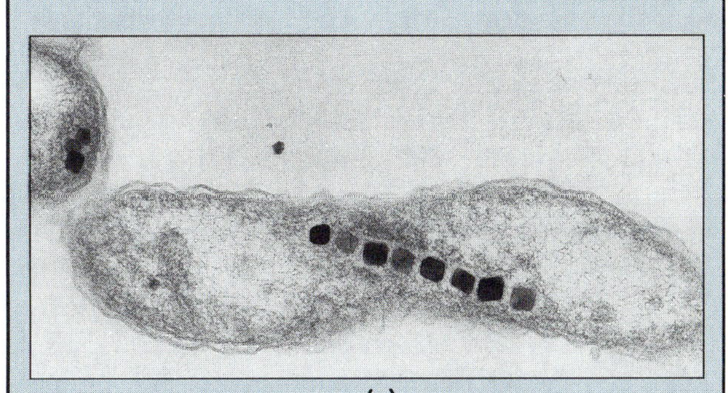

(a)

(b)

(a) Electron micrograph of the magnetotactic bacterium *Aquaspirillum magnetotacticum*. The numerous dark, square inclusions, called magnetosomes, are composed of iron oxide, (Fe_3O_4). (b) The magnetosomes enable these organisms to orient themselves in a magnetic field.

Magnetosomes are nearly constant in size and are oriented in parallel chains like a string of tiny magnets. Experiments are currently under way to use these bacteria in the manufacture of magnets and audio- and videotapes. Because it is difficult to manufacture magnetic particles of the very small, uniform size needed for high-quality tapes, these lowly bacteria may make it possible for all of us to enjoy crisper sound from our magnetic tapes.

TRY IT: You can easily find your own magnetotactic bacteria. These are very common organisms. Bring back a bucket with a few inches of mud from a pond and enough pondwater to fill the rest of the bucket. Try mud from different locations: fresh, salt, and brackish waters. Store the covered bucket in the dark (you don't want to grow algae) for about 1 month. Remove a large beaker of the water, and place a magnet against the outside of the glass. Allow this to stand for a day or 2, until you see a whitish spot in the water near the magnet's end. Which attracts more organisms, the north or south end of the magnet? Pipette off the cloudy liquid and examine it under the microscope. Laying a magnet on the stage will cause the magnetotactic bacteria, if they are present, to orient themselves to the field. They can be purified by streaking on agar, a process described in Chapter 6. Microbiology doesn't exist only in books or labs. It's everywhere around you in the real world—just look for it!

metachromasia. That is, although most substances stained with a simple stain such as methylene blue take on a uniform solid color, metachromatic granules exhibit different intensities of color. Although quite numerous in some bacteria, these granules become depleted during starvation.

Certain bacteria have specialized **vesicles,** or membrane-bound structures. Some aquatic photosynthetic bacteria and cyanobacteria, for example, have rigid gas-filled vesicles (Figure 3.24). These organisms regulate the amount of gas in vesicles and, therefore, the depth at which they float to obtain optimum light for photosynthesis. The membranes that surround vesicles differ from most cellular membranes in being single-layered.

Endospores

The properties of bacterial cells just described pertain to **vegetative cells,** or cells that are metabolizing nutrients. However, vegetative cells of some bacteria, such as *Bacillus* and *Clostridium,* produce **endospores.** Though bacterial endospores are sometimes referred to simply as spores, they should not be confused with fungal spores. A bacterium produces a single endospore, which merely helps that organism to survive and is not a means of reproduction. A fungus produces numerous spores, which both help the organism to survive and provide a means of reproduction.

Endospores, which are formed within cells, contain very little water and are highly resistant to heat, drying, acids, bases, certain disinfectants, and even radiation. The presence of such adverse conditions will usually induce a large number of cells to produce spores. However, many investigators believe that they are part of the normal life cycle and that a few are formed even when nutrients are adequate and environmental conditions are favorable. Thus, sporulation seems to be a means by which some bacteria prepare for the *possibility* of adverse conditions.

Structurally, an endospore (Figure 4.10) consists of a *core,* surrounded by a *cortex, spore coat,* and *exosporium.* The core has an outer core wall, derived from the wall of the vegetative cell, a cell membrane, nuclear region, and other cell components. Unlike vegetative cells, spores contain *dipicolinic acid* and a large quantity of calcium ions (Ca^{2+}). These materials, which are probably stored in the core, appear to contribute to the heat resistance of spores, as does their very low water content.

Endospores are capable of surviving adverse environmental conditions for long periods of time. Some withstand hours of boiling. When conditions become more favorable, endospores germinate and develop into functional vegetative cells. (The process of spore formation and germination is discussed in Chapter 6.) Because of their resistance, special care must be used to kill endospores during sterilization procedures. Otherwise, they germinate and grow in media thought to be sterile. Methods to assure that endospores are killed when bacterial media or foods are sterilized are described in Chapter 13.

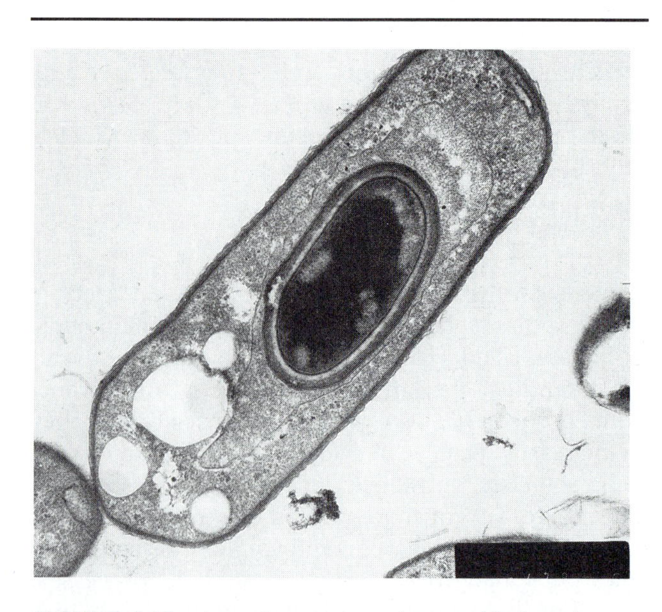

FIGURE 4.10 A nearly mature endospore (dark oval structure) forming within a *Bacillus megaterium* cell.

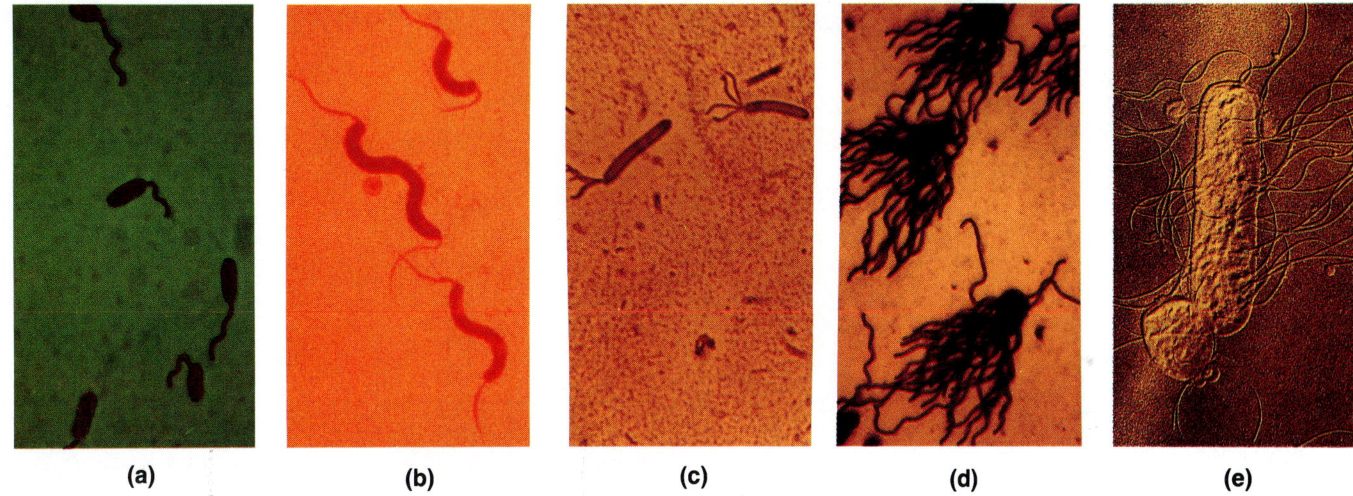

FIGURE 4.11 Arrangements of bacterial flagella. (a) Polar, monotrichous (single flagellum at one end). (b) Polar, amphitrichous (single flagellum at each end). (c) Lophotrichous (tuft of flagella at one or both ends). (d) Peritrichous (flagella distributed all over). (e) SEM of *Legionella* bacterium showing peritrichous flagella.

External Structure

In addition to cell walls, many bacteria have structures that extend beyond or surround the cell wall. Flagella and pili extend from the cell membrane through the cell wall and beyond it. Capsules and slime layers surround the cell wall. These structures have a variety of properties and functions.

Flagella

About half of all known bacteria are *motile*, or capable of movement. They often move with speed and apparent purpose, and they usually move by means of **flagella** (singular: **flagellum**). A bacterium can have one, two, or many flagella (Figure 4.11). The flagella can be **polar,** single and located on one end (**monotrichous**) or at both ends (**amphitrichous**) of the bacterium; **lophotrichous** (lo-fo-trik'us), with two or more at one or both ends; or **peritrichous** (per-e-trik'us), distributed all over the surface of the bacterium.

A flagellum is a long, helical structure with a diameter about one-tenth that of a eukaryotic flagellum. It is made of protein subunits called **flagellin.** Each flagellum is attached to the cell membrane by a basal region consisting of a different protein (Figure 4.12). The basal region has a hooklike structure and a complex basal body. The basal body consists of a central rod and a set of rings surrounding the rod. Gram-positive bacteria have one ring embedded in the cell membrane and another in the cell wall. Gram-negative bacteria have a pair of rings embedded in the cell membrane and another pair of rings associated with the peptidoglycan and lipopolysaccharide layers of the cell wall.

Most flagella rotate like twirling L-shaped hooks, such as a dough hook on a kitchen mixer or the rotating string on a hand-carried grass cutter. It is thought that motion occurs as energy from ATP is used to make one of the rings in the cell membrane rotate with respect to the other. When flagella rotate counterclockwise, the bacteria run, or move in a straight line. When the flagella rotate clockwise, the bacteria twiddle, or tumble randomly. Both runs and twiddles are generally random movements, that is, no one direction of movement is more likely than any other direction.

Chemotaxis Sometimes bacteria move toward or away from substances in their environment by a nonrandom process called **chemotaxis** (Figure 4.13). The concentrations of most substances in the environment vary along a gradient, that is, from high to low concentration. When a bacterium is moving in the direction of increasing concentration of an attractant (such as a nutrient), it tends to lengthen its runs and reduce the frequency of its twiddles. When it is moving away from the attractant, it shortens its runs and increases the frequency of its twiddles. Even though the direction of the individual runs is still random, the net result is movement toward the attractant, or **positive chemotaxis.** Movement away from a repellent, or **negative chemotaxis,** results from the opposite responses: long runs and few twiddles when moving in the direction of lower concentration of the harmful substance, short runs and many twiddles when moving in the direction of higher concentration. The exact mechanism that produces these behaviors is not fully understood, but certain structures on their cell surfaces can evidently detect changes in concentration over time.

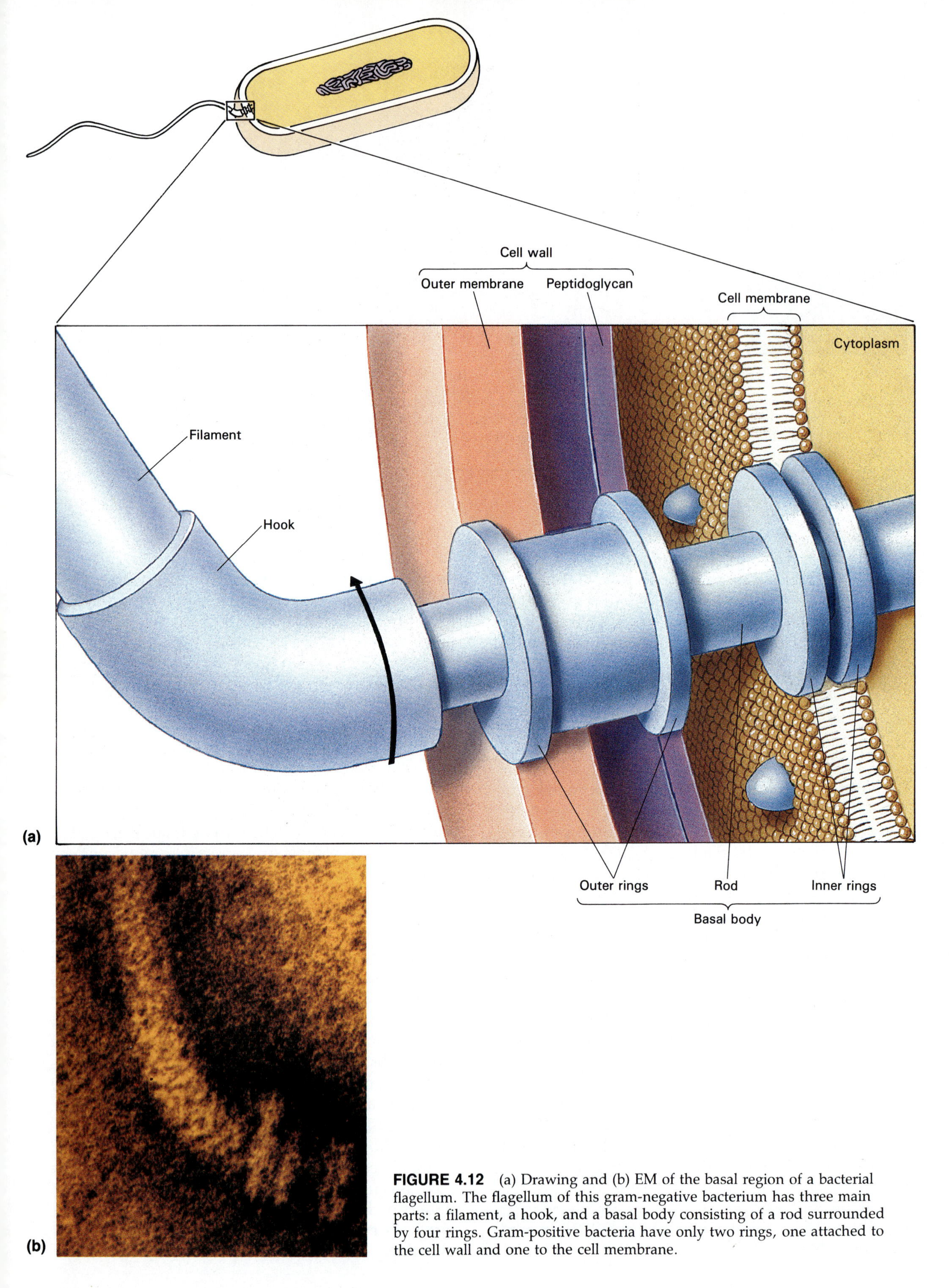

FIGURE 4.12 (a) Drawing and (b) EM of the basal region of a bacterial flagellum. The flagellum of this gram-negative bacterium has three main parts: a filament, a hook, and a basal body consisting of a rod surrounded by four rings. Gram-positive bacteria have only two rings, one attached to the cell wall and one to the cell membrane.

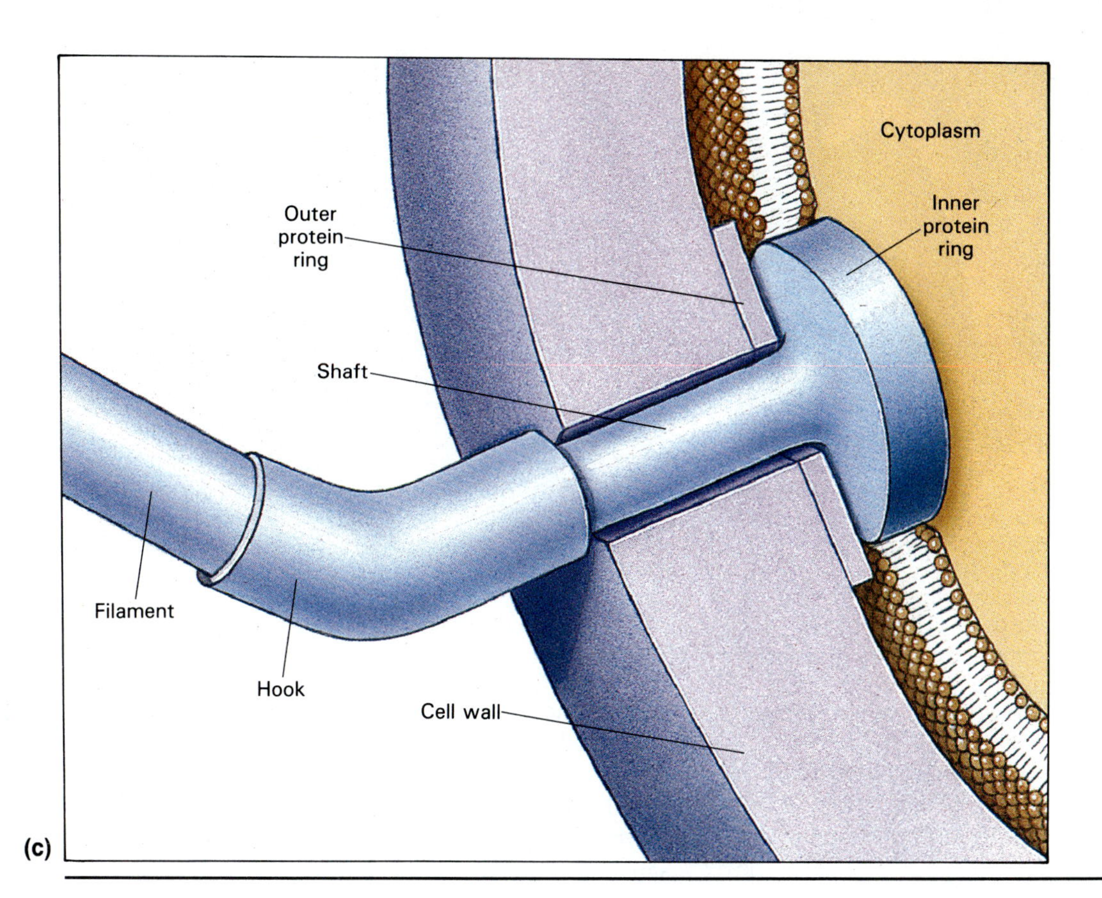

Axial Filaments

Instead of flagella that extend beyond the cell wall, spirochetes have **axial filaments,** or **endoflagella** (Figure 4.14). These filaments are attached near the ends of the cytoplasmic cylinder that forms the body of the spirochete. They cause the rigid spirochete body to rotate like a corkscrew when they twist inside the outer sheath.

Pili

Pili (singular: *pilus*) are tiny hollow projections, somewhat like flagella. They are used to attach to surfaces and are not concerned with movement. A pilus (Figure 4.15) is composed of subunits of a protein called **pilin.** Bacteria can have two kinds of pili: (1) long, conjugation, or F pili and (2) short, attachment pili, or fimbriae.

Conjugation Pili **Conjugation pili,** found only in certain groups of bacteria, attach two cells and furnish a pathway for genetic material (DNA) to be transferred from one to the other. This transfer process is called *conjugation* (Chapter 8). Transfer of DNA furnishes genetic variety for bacteria as sexual reproduction does for many other life forms. Such transfers among bacteria cause problems for humans because antibiotic resistance can be passed on with the DNA transfer. Consequently, more and more bacteria acquire resis-

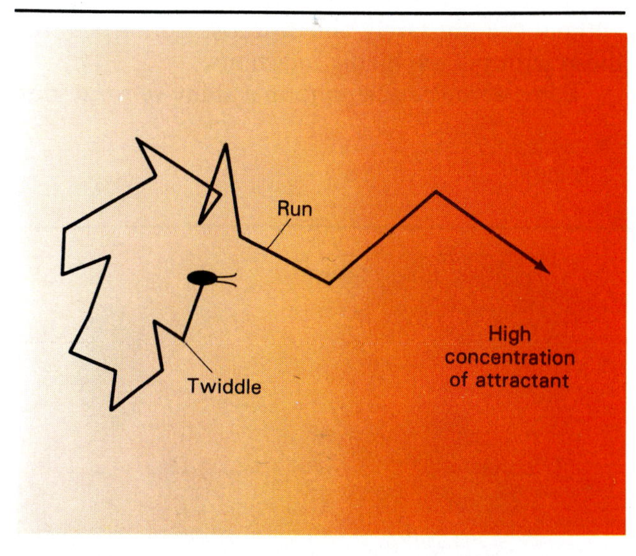

FIGURE 4.13 Movements of a typical bacterium exhibiting chemotaxis. An attractive chemical, such as a nutrient, is concentrated toward the right. When the bacterium finds itself moving away from the attractant, it decreases the length of its runs and increases the frequency of its twiddles. When it eventually finds itself moving toward the attractant, it increases the length of its runs and decreases the number of twiddles.

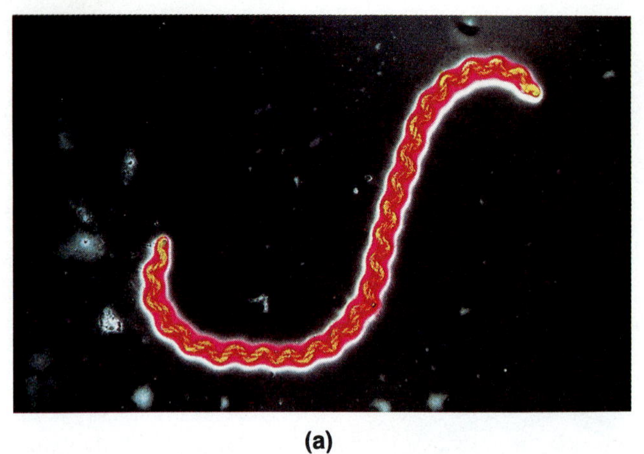

(a)

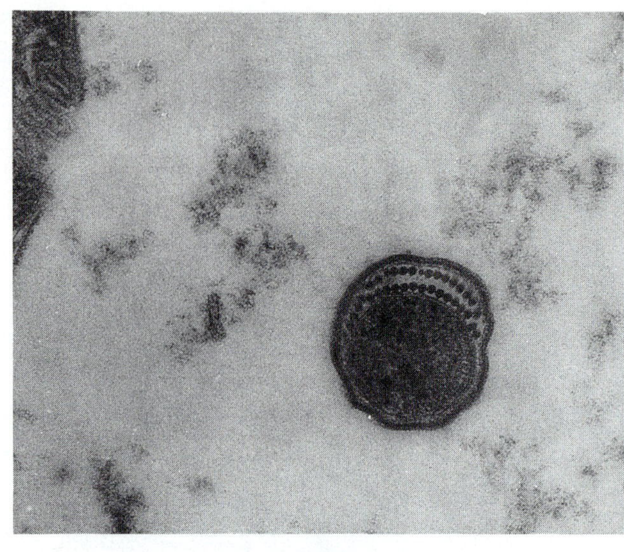

(b)

FIGURE 4.14 (a) Axial filaments made visible by false coloring are clearly seen running along the body of a spirochete. (b) Cross section of a spirochete, showing numerous axial filaments (dark circles). Note that they lie between the outer sheath and the cell wall.

tance, and humans must look for new ways to control these bacteria.

Attachment Pili **Attachment pili,** or **fimbriae** (fim-bre'e), help bacteria adhere to surfaces, such as cell surfaces and the surface at the interface of water and air. They contribute to the pathogenicity of bacteria by enhancing colonization (the development of colonies) on the surfaces of the cells of other organisms. For example, some bacteria adhere to red blood cells by attachment pili and cause the blood cells to clump, a process called *hemagglutination.* In certain species of bacteria, some individuals have attachment pili and others lack them. In *Neisseria gonorrhoeae,* organisms without pili are rarely able to cause gonorrhea, but those with pili are highly infectious.

Some aerobic bacteria form a shiny or fuzzy thin layer at the air-water interface of a broth culture. This layer, called a **pellicle,** consists of many bacteria that adhere to the surface by their attachment pili. Thus the attachment pili allow the organisms to remain in the broth from which they take nutrients and at the same time congregate near air where the oxygen concentration is greatest.

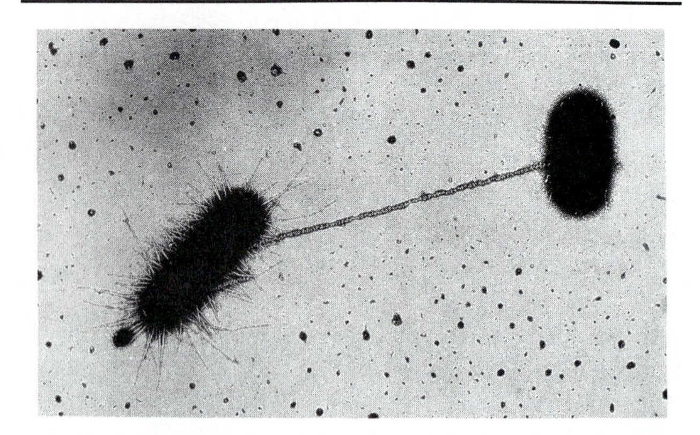

FIGURE 4.15 Electron micrograph of an *Escherichia coli* cell showing two kinds of pili. The shorter ones are fimbriae, used for attachment to surfaces. The long tube reaching to another cell is a conjugation pilus, through which DNA is being transferred.

Glycocalyx

Glycocalyx is the currently accepted term used to refer to all substances containing polysaccharides found external to the cell wall, from the thickest capsules to the thinnest slime layers. All bacteria have at least a thin slime layer.

Capsule A **capsule** is a protective structure outside the cell wall of the organism that secretes it. Only certain bacteria are capable of forming capsules, and not all members of a species have capsules. For example, the bacterium that causes anthrax, a disease found mainly in cattle, does not produce a capsule when it grows outside an organism but does when it infects an animal. Capsules typically consist of complex polysaccharide molecules arranged in a loose gel. However, the chemical composition of each capsule is unique to the strain of bacteria that secreted it. When encapsulated bacteria invade a host, the capsule prevents host defense mechanisms from destroying the organism. If bacteria lose their capsules, they become less infectious and more vulnerable to destruction.

Slime Layer A **slime layer** is less tightly bound to the cell wall and is usually thinner than a capsule. When present, it protects the cell against drying, helps to trap nutrients near the cell, and sometimes binds cells together. Slime layers allow bacteria to adhere to objects in their environments, such as rock surfaces or the root hairs of plants, so that they can remain near sources of nutrients or oxygen (Figure 4.16). Some oral bacteria, for example, adhere by their slime layers and form dental plaque. The slime layer keeps the bacteria in close proximity to tooth surfaces, where they can cause dental caries. Plaque is extremely tightly bound to tooth surfaces. If not removed within 48 hours by brushing, it can be removed only by a dental procedure called scaling.

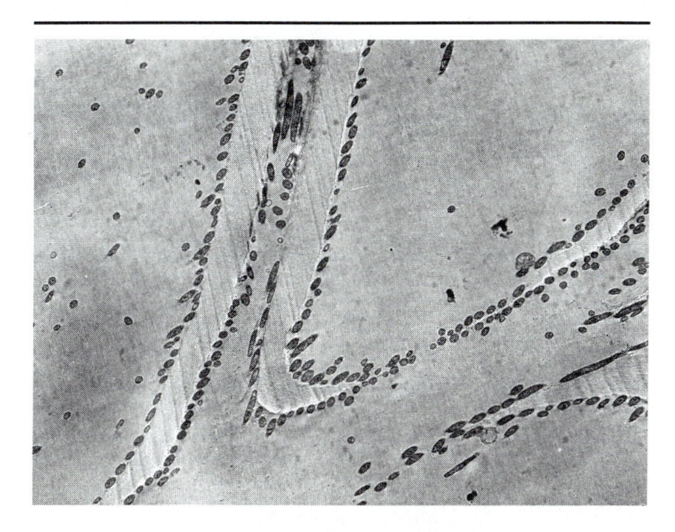

FIGURE 4.16 *Sporocytophaga* bacteria growing on a cellulose fiber, to which they adhere by means of their slime layer.

EUKARYOTIC CELLS

Overview of Structure

Eukaryotic cells are larger and more complex than prokaryotic cells. Most have a diameter of more than 10 μm, and many are much larger. They also contain a variety of highly differentiated structures. These cells are the basic structural unit of all organisms in the kingdoms Protista, Plantae, Fungi, and Animalia (see Chapter 9). Many of these eukaryotic organisms are microscopic protozoa, algae, and fungi and are thus appropriately considered in microbiology. The general structure of the eukaryotic cell is shown diagramatically in Figure 4.17.

Cell Membrane

The cell membrane, or plasma membrane, of a eukaryotic cell has the same fluid-mosaic structure as that described for a prokaryotic cell. In addition to the plasma membrane, eukaryotes also contain several organelles bounded by membranes having the same unit-membrane structure.

Eukaryotic membranes do differ from prokaryotic membranes in some respects, especially in the greater variety of lipids they contain. Eukaryotic membranes contain sterols, found among prokaryotes only in the mycoplasmas. Sterols add rigidity to a membrane, and this may be important in keeping membranes intact in eukaryotic cells. Because of their larger size, eukaryotic cells have a much lower surface-to-volume ratio than prokaryotic cells. As the volume of cytoplasm enclosed by a membrane increases, the membrane is placed under greater stress. The sterols in the membrane may help it to withstand the stress.

Functionally, eukaryotic cell membranes are less versatile than prokaryotic ones. They do not have respiratory enzymes that capture energy in ATP; in the course of evolution that function has been taken over by mitochondria. Also they generally secrete fewer enzymes.

Internal Structure

The internal structure of eukaryotic cells is exceedingly more complex than that of prokaryotic cells. It is also much more highly organized and contains numerous organelles.

Cytoplasm

The cytoplasm makes up a relatively smaller portion of eukaryotic cells than of prokaryotic cells because it contains a nucleus and many organelles. Like the cytoplasm of prokaryotic cells, the cytoplasm of eukaryotic cells is a semifluid substance consisting mainly of water with the same substances dissolved in it. In addition this cytoplasm contains elements of a cytoskeleton that give these larger cells shape and support.

Nucleus

The most obvious difference between eukaryotic and prokaryotic cells is the presence of a nucleus in the eukaryotic cells. The **nucleus** (Figure 4.18) is a distinct organelle with a nuclear envelope, nucleoli, and paired chromosomes. The **nuclear envelope** consists of a double membrane, each layer of which is structurally like the cell membrane. **Nuclear pores** in the envelope allow RNA molecules to leave the nucleus and participate in protein synthesis. Each nucleus has one or more **nucleoli** (singular: *nucleolus*), which contain a significant amount of RNA and serve as sites for the assembly of ribosomes.

Also present in the nucleus of most eukaryotic organisms are paired **chromosomes,** each of which contains DNA and proteins called histones. **Histones** contribute directly to the structure of chromosomes, and other proteins associated with chromosomes

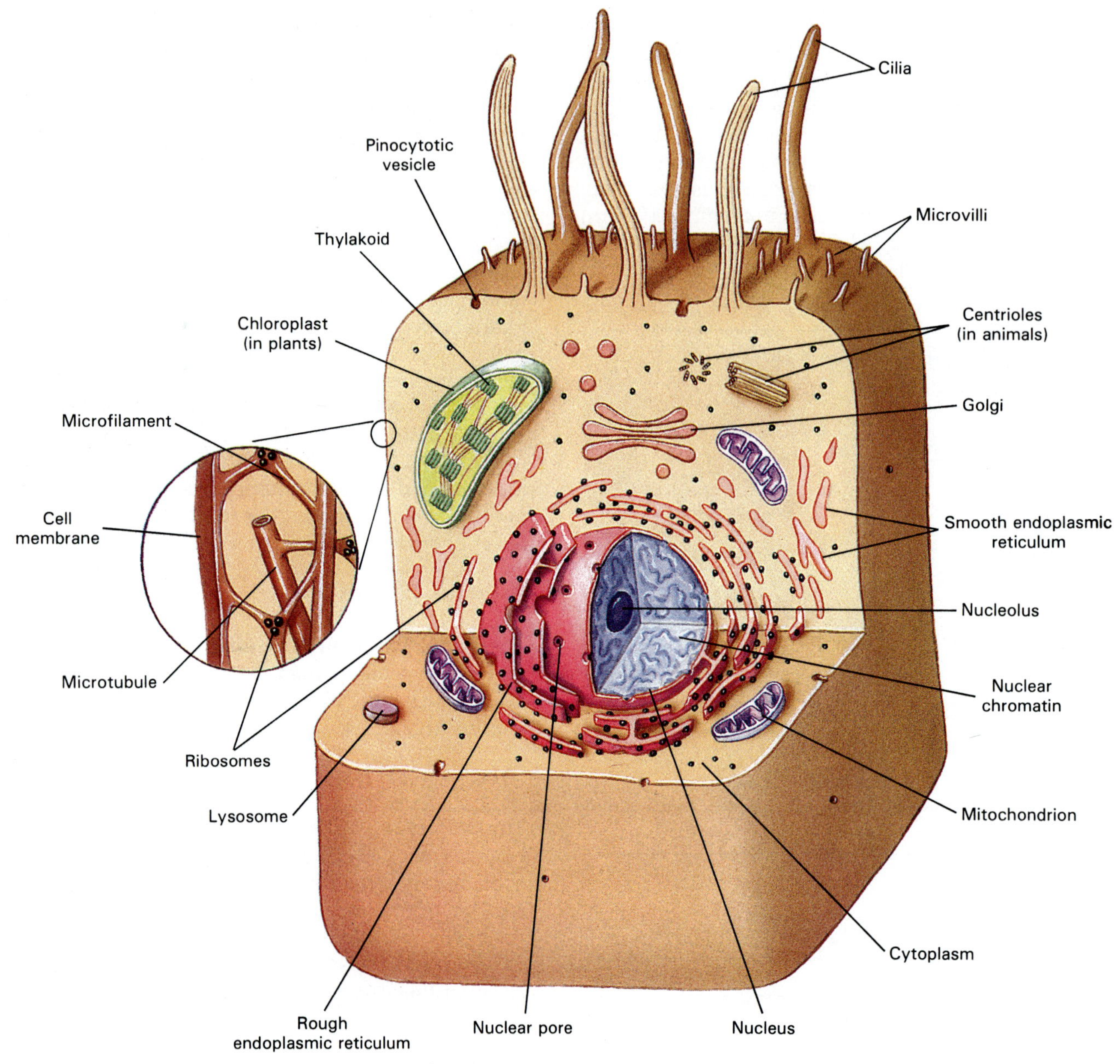

FIGURE 4.17 Diagram of a generalized eukaryotic cell. Most of the features shown are present in nearly all eukaryotic cells, but some (the centrioles) occur only in animal cells and others (the chloroplast) are found only in cells capable of carrying out photosynthesis.

probably regulate their function. During cell division the chromosomes are extensively coiled and folded into compact structures. Between divisions, however, the chromosomes are visible only as a tangle of fine threads called **chromatin** that give the nucleus a granular appearance.

The nuclei of eukaryotic cells divide by a process called **mitosis** (Figure 4.19a). Prior to the actual division of the nucleus the chromosomes replicate but remain attached, forming **dyads.** During mitosis the nuclear envelope dissolves, and a system of tiny fibers called the **spindle apparatus** guides the movement of chromosomes. Dyads aggregate in the center of the spindle and separate into single chromosomes as they move along fibers to the poles of the spindle. Each new cell receives one copy of each chromosome that was present in the parent cell. Because the parent cell contained paired chromosomes, the progeny likewise contain paired chromosomes. Cells with paired chromosomes are said to be *diploid* (2N) cells.

During sexual reproduction the nuclei of sex cells divide by a process called **meiosis** (Figure 4.19b). In this process the chromosomes replicate, forming dyads. Pairs of dyads come together, and in the course of two cell divisions are distributed to four new cells, so that each cell receives only one chromosome from

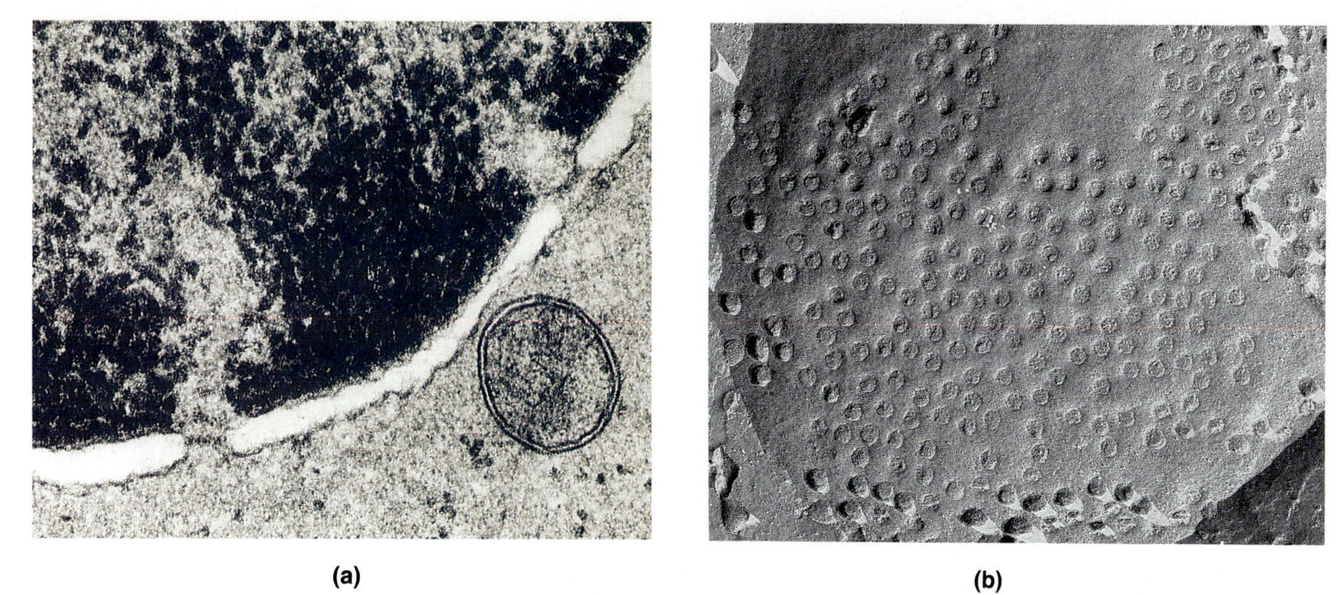

FIGURE 4.18 (a) TEM of a cell nucleus. The dark, granular material is chromatin. The pores in the nuclear membrane are clearly visible. These allow for entry and exit of materials. (b) Freeze-fracture SEM of a nucleus (compare Figure 3.24). The many circular structures are nuclear pores.

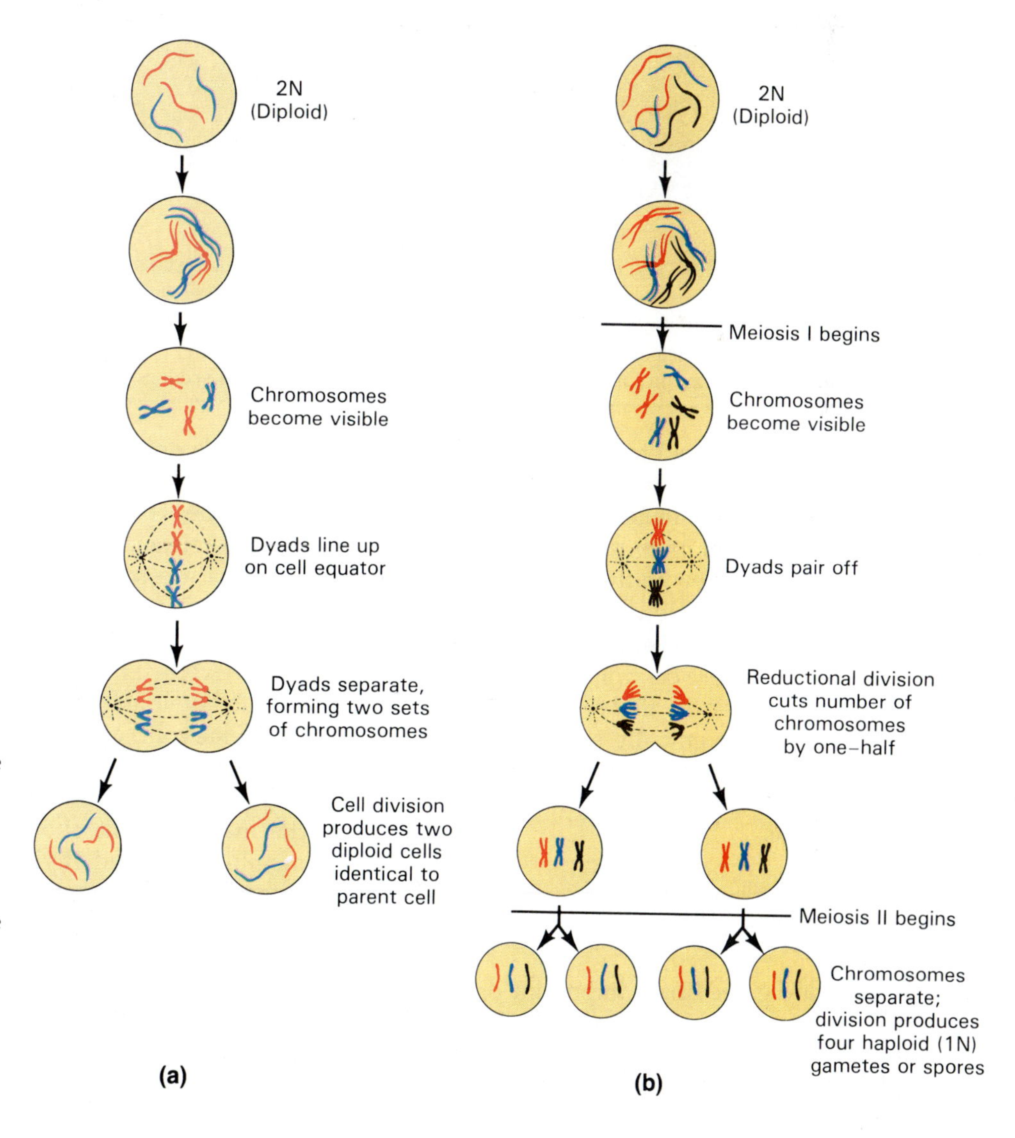

FIGURE 4.19 Comparison of (a) mitosis and (b) meiosis. Both processes begin with duplication of DNA before the chromosomes become visible. In mitosis, two identical daughter cells having the same number and kinds of chromosomes are formed. In meiosis, two divisons give rise to four cells, each with half the number of chromosomes as the original parent cell. For this reason, meiosis is sometimes called "reduction division."

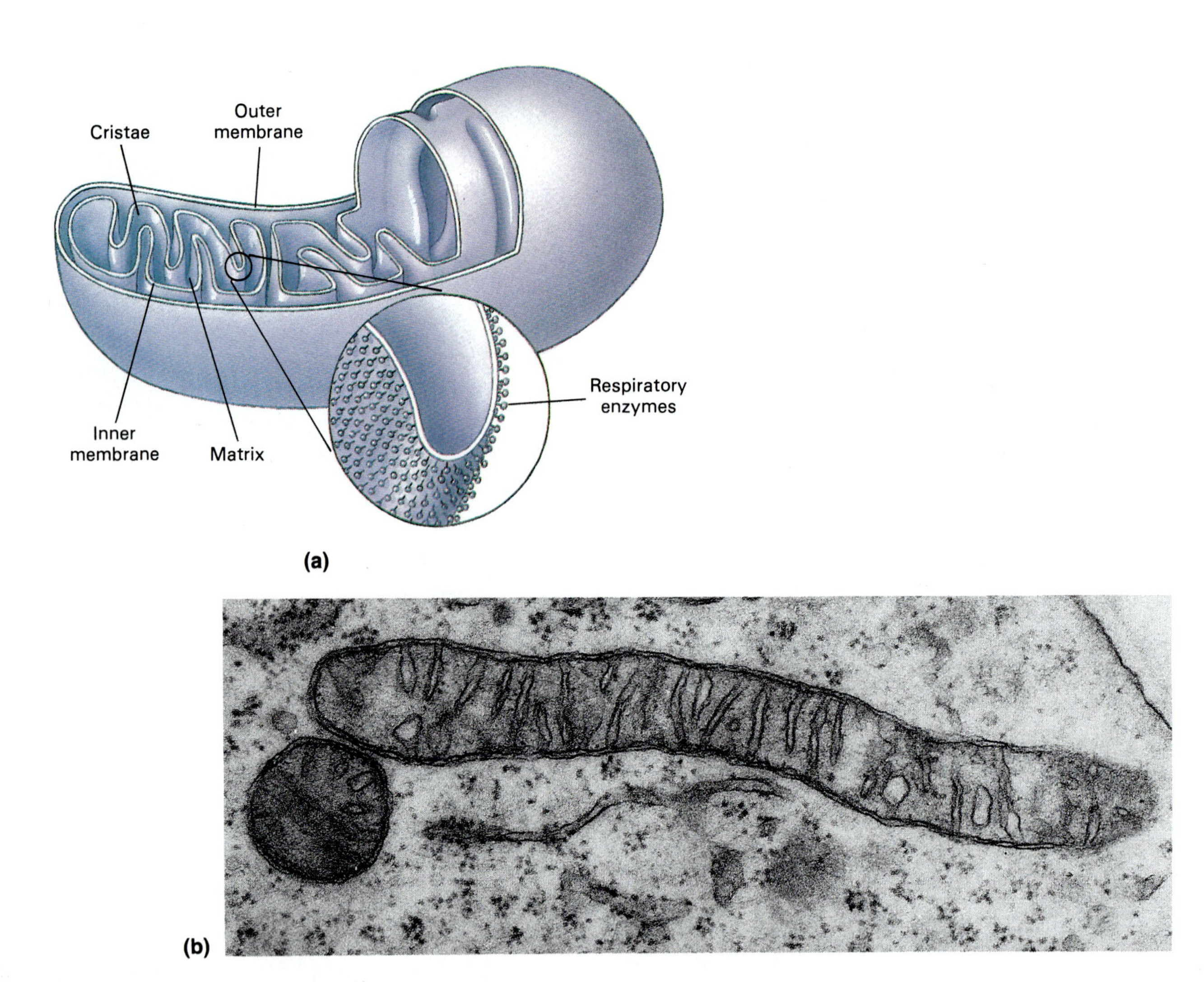

FIGURE 4.20 (a) Diagram of a mitochondrion. Respiratory enzymes that make ATP are located on the surface of the cristae, which are infoldings of the inner membrane. (b) TEM showing two mitochondria, one in longitudinal section and the other in cross section.

each pair. Such cells are said to be *haploid* (1N) cells. Haploid cells can become gametes or spores. **Gametes** participate in sexual reproduction; gametes from each of two parent organisms unite to form a diploid **zygote,** the first cell of a new individual. Some **spores** become dormant, whereas others reproduce by mitosis as haploid vegetative cells. Dormant spores allow for survival during adverse environmental conditions. When conditions improve, these spores germinate and begin to divide. Eventually some of these cells produce gametes, which can unite to form zygotes. Thus, the organism alternates between haploid and diploid generations.

Mitochondria and Chloroplasts

Mitochondria are exceedingly important organelles. They are quite numerous in some cells and can account for up to 20 percent of the cell volume. Mitochondria (Figure 4.20) are complex structures about 1 μm in diameter with an outer membrane, an inner membrane, and a fluid-filled **matrix** inside the inner membrane. The inner membrane is extensively folded to form **cristae,** which extend into the matrix. Mitochondria are sometimes called the powerhouses of eukaryotic cells because they carry out the oxidative reactions that capture energy in adenosine triphosphate (ATP). Energy in ATP is in a form usable by cells for their activities.

Eukaryotic cells capable of carrying out photosynthesis contain **chloroplasts** (Figure 4.21). The inner **stroma** of chloroplasts corresponds structurally with the matrix of mitochondria. The folds of the inner membrane, called **thylakoids,** contain the pigment chlorophyll that captures energy from light during photosynthesis. Both mitochondria and chloroplasts contain DNA and can replicate independently of the cell in which they function. This and other evidence has led many biologists to speculate that these organelles may have originated as free-living organisms. (See the essay at the end of this chapter.)

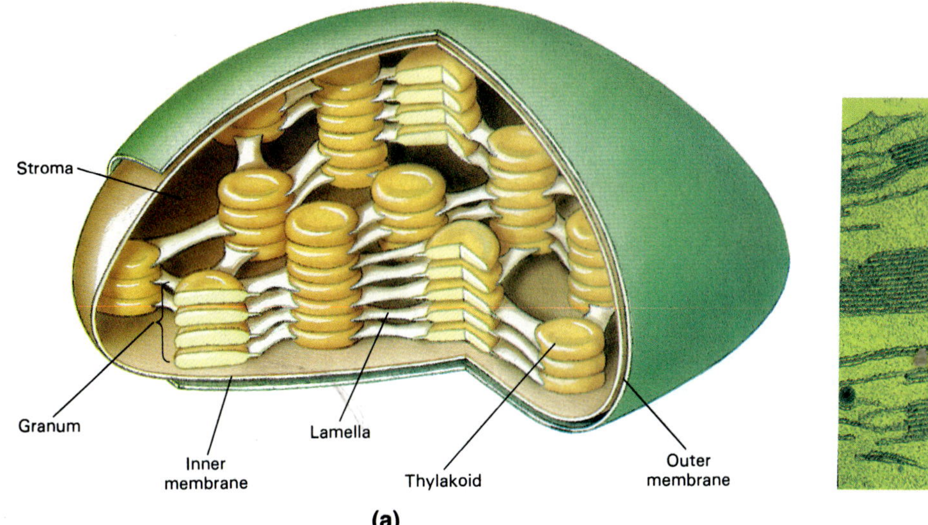

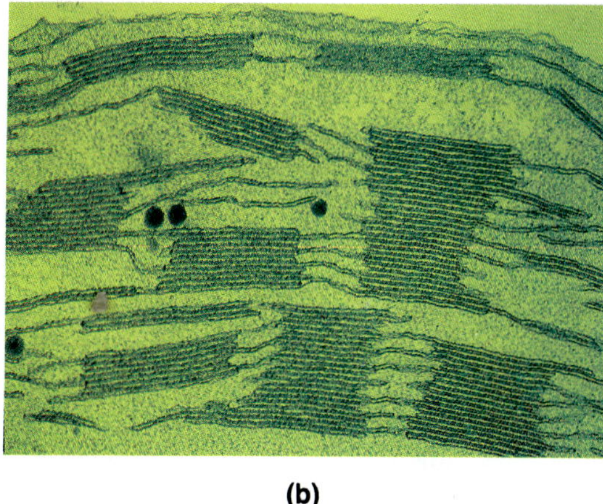

FIGURE 4.21 (a) Diagram of a eukaryotic chloroplast. The thylakoid membranes contain chlorophyll and other pigments and enzymes needed for photosynthesis. (b) TEM of a chloroplast from corn.

Ribosomes

Ribosomes of eukaryotic cells, which are larger than those in prokaryotic cells, are about 60 percent RNA and 40 percent protein. They have a sedimentation rate of 80S, and their subunits have sedimentation rates of 60S and 40S. All ribosomes provide sites for protein synthesis, and some are arranged in chains called polyribosomes. Those attached to an organelle called the endoplasmic reticulum usually make proteins for secretion from the cell; those free in the cytoplasm usually make proteins for use in the cell.

Endoplasmic Reticulum

The **endoplasmic reticulum** (Figure 4.17) is an extensive system of membranes that forms numerous tubes and vesicles in the cytoplasm. The vesicles store substances synthesized on the membrane. Endoplasmic reticulum can be smooth or rough. Smooth endoplasmic reticulum contains enzymes that synthesize lipids, especially those to be used in making membranes. Rough endoplasmic reticulum has ribosomes bound to its surface, which give it a rough texture.

Golgi Apparatus

The **Golgi** (gol'je) **apparatus** (Figure 4.17) consists of interconnected, stacked, flattened vesicles made of unit membrane. The Golgi apparatus receives substances synthesized on the endoplasmic reticulum, stores them, and sometimes alters their chemical structure. It packages these substances in small segments of membrane called **secretory vesicles.** The secretory vesicles fuse with the plasma membrane and release secretions to the exterior of the cell. The Golgi apparatus also helps to form plasma membranes and membranes of other organelles.

Lysosomes

Lysosomes (Figure 4.17) are extremely small membrane-covered organelles probably made by the Golgi apparatus. They contain digestive enzymes that could destroy a cell if freed into the cytoplasm. Lysosomes fuse with vacuoles formed as a cell ingests substances and release enzymes that digest the substances in the vacuoles.

Peroxisomes

The **peroxisome** is a fairly recently discovered, small, membrane-bound organelle filled with enzymes. Peroxisomes are found in both plant and animal cells but appear to have different functions in the two kinds of cells. In animal cells their enzymes oxidize amino acids, whereas in plant cells they typically oxidize fats. However, they are named peroxisomes because their enzymes convert hydrogen peroxide to water in both plant and animal cells. If hydrogen peroxide were to accumulate in cells, it would kill them just as it kills bacteria when humans use it as an antiseptic.

Vacuoles

In eukaryotic cells **vacuoles** are membrane-bound structures that store materials such as starch, glycogen, or fat to be used for energy. Some vacuoles are formed when cells engulf food particles. Such vacuoles are eventually digested by lysosomal enzymes as already noted.

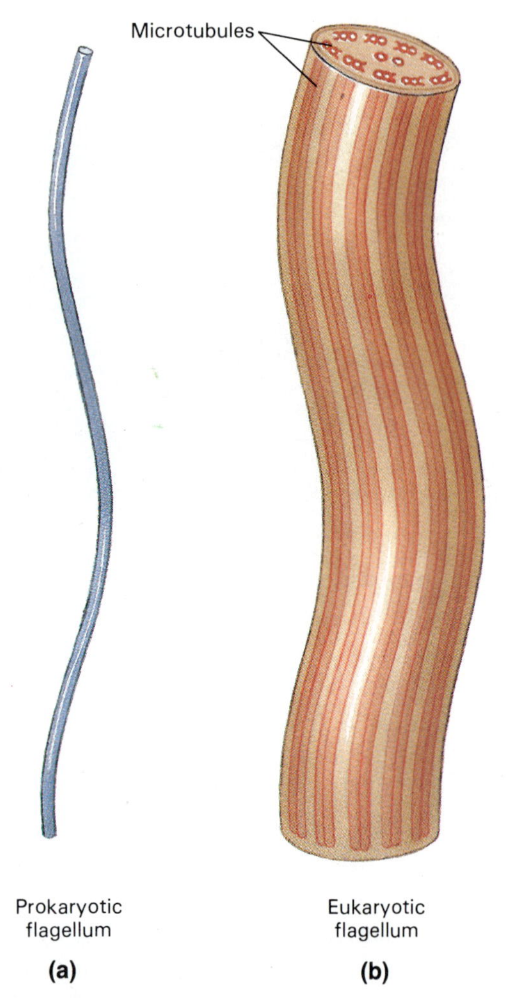

FIGURE 4.22 Comparison of (a) prokaryotic and (b) eukaryotic flagella. Note the substantial difference in the diameter of these two structures.

Flagella

Flagella in eukaryotes (Figure 4.22), which are larger and more complex than in prokaryotes, consist of two central fibers and nine pairs of peripheral fibers surrounded by a membrane. Each fiber is a microtubule made of the protein tubulin. One of these fibers is about the same size as an entire prokaryotic flagellum. Associated with each pair of fibers are small molecules of the protein dynein, which plays a role in converting chemical energy in ATP to mechanical energy that makes the flagellum move. Eukaryotic flagella move like a whip and not like prokaryotic flagella, which move like a rotating hook. The mechanism of eukaryotic flagellar movement seems to be analogous to the sliding of protein filaments in muscles. Fibers of the flagellum are thought to slide toward or away from

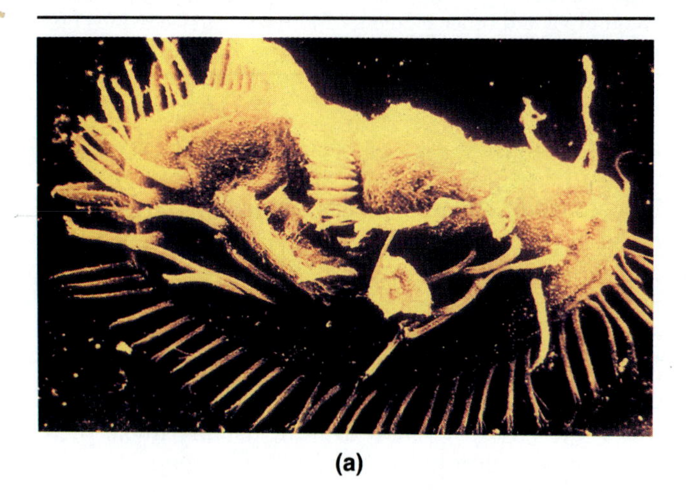

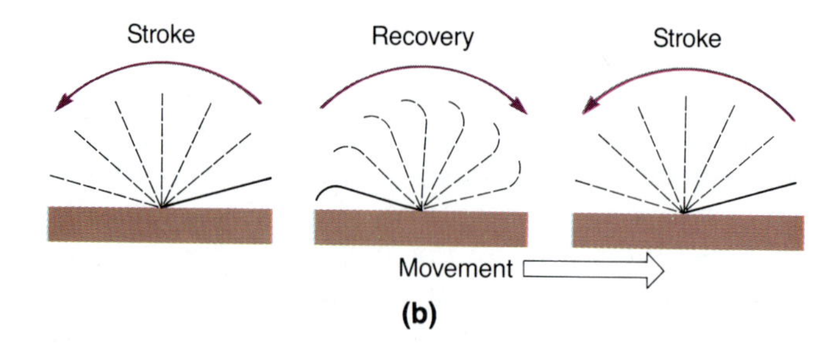

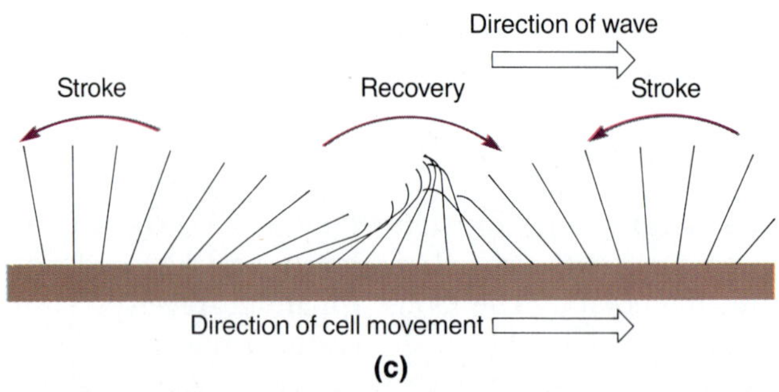

FIGURE 4.23 (a) SEM of the ciliated protozoan *Oxytrichia*. (b) The forward and recovery strokes of a cilium. (c) Cilia on an organism move in a synchronized fashion, creating a wave that propels the organism forward.

Cytoskeleton

The **cytoskeleton** is a network of protein fibers called **microtubules** and **microfilaments** that support and give rigidity and shape to a cell. They also are involved in cell movements, such as those that occur when cells engulf substances or when they make amoeboid movements.

External Structure

As with prokaryotic cells, external structures of eukaryotic cells either assist with movement or provide a protective covering for the cell membrane. These structures include flagella, cilia, and cell walls and other coverings. Though pseudopodia are not, strictly speaking, external structures, they do achieve movement and so are discussed here. Cells of algae and macroscopic green plants have cell walls and some protozoa have special body coverings.

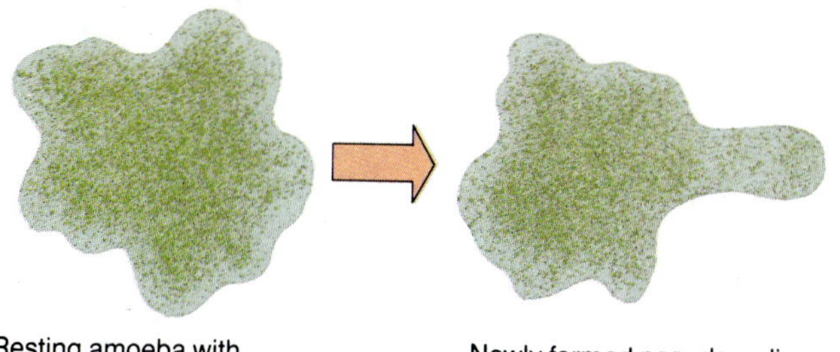

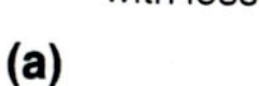

FIGURE 4.24 (a) Formation of a pseudopodium, a cytoplasmic extension that allows organisms such as amoebae to move and capture food. (b) Micrographs of an amoeba engulfing food.

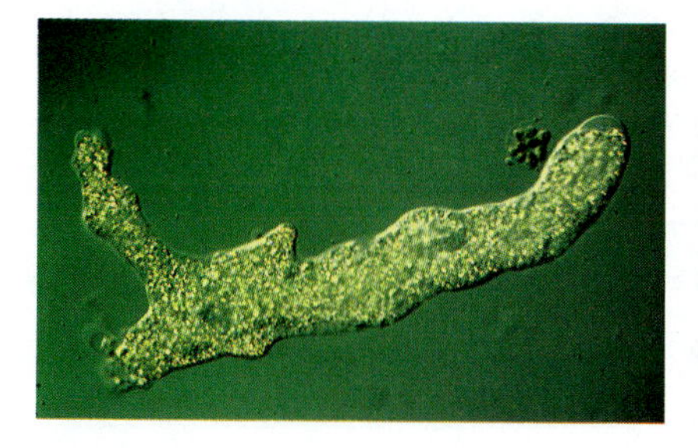

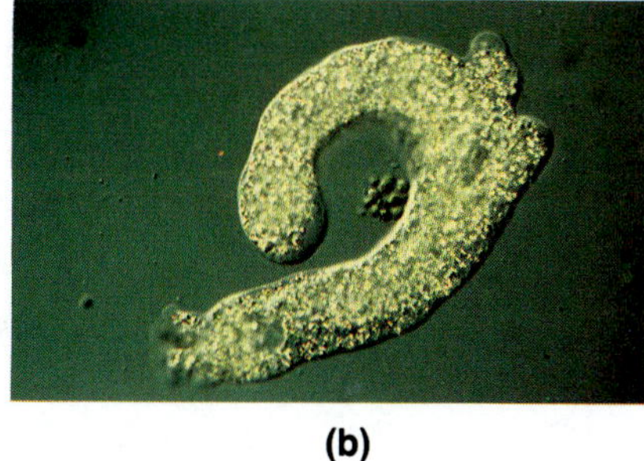

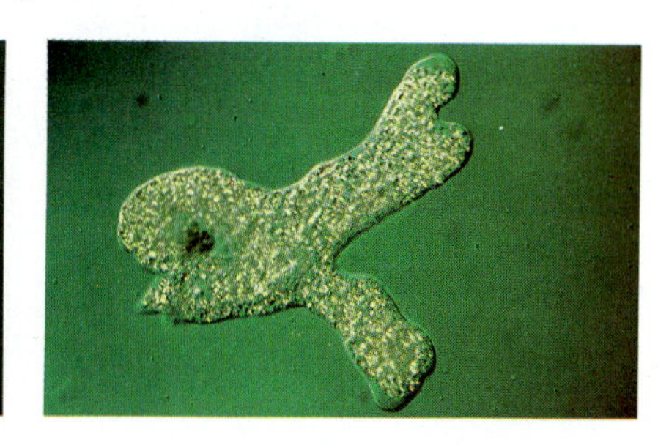

(b)

the base of the cell and thereby cause the whole flagellum to move.

Flagella are most common among protozoa but are found among algae as well. Most flagellated organisms have one flagellum, but some have two or more.

Cilia

Cilia are shorter and more numerous than flagella, but they have the same basic pattern of fibers. They are found mainly among ciliated protozoa, which have 10,000 or more cilia distributed over their bodies (Figure 4.23). Each cilium passes through a stroke and recovery cycle as it beats. Together the cilia of an organism beat in a coordinated pattern, which creates a wave that passes from one end of the organism to the other. The large number of cilia and their coordinated beating allow ciliated organisms, such as paramecia, to move much more rapidly than those with flagella.

Pseudopodia

Pseudopodia ("false feet") are temporary projections of cytoplasm associated with **amoeboid movement.** This kind of movement occurs only in cells without walls, such as amoebae and slime molds, and only when the organism is resting on a solid surface. For example, when an amoeba first extends a portion of its body to form a pseudopodium, the cytoplasm in the pseudopodium is much less dense than in other areas of the cell. As a result cytoplasm from elsewhere in the organism flows into the pseudopod by **cytoplasmic streaming.** Thus, amoeboid movement is a slow, inching-along process (Figure 4.24).

Cell Walls

A number of unicellular eukaryotic organisms have cell walls, but none contains the peptidoglycan that is characteristic of prokaryotes. Algal cell walls consist mainly of cellulose, but some contain other polysaccharides. Cell walls of fungi consist of cellulose or chitin, or both. Chitin is a structural polysaccharide common in the exoskeletons of arthropods such as insects and crustacea. Regardless of composition, cell walls give cells rigidity and protect them from bursting when water moves into them from the environment.

MOVEMENT OF SUBSTANCES ACROSS MEMBRANES

A living cell, either prokaryotic or eukaryotic, is a dynamic entity. It is separated from its environment by the cell membrane, across which substances constantly move in a carefully controlled manner. Understanding how these movements occur is essential to understanding how a cell functions. Very small polar substances, such as water, small ions, and small water-soluble molecules, probably pass through pores in the membrane. Nonpolar substances, such as lipids and other uncharged particles, dissolve in and pass through the membrane lipids. Still other substances are moved through the membrane by carrier mole-

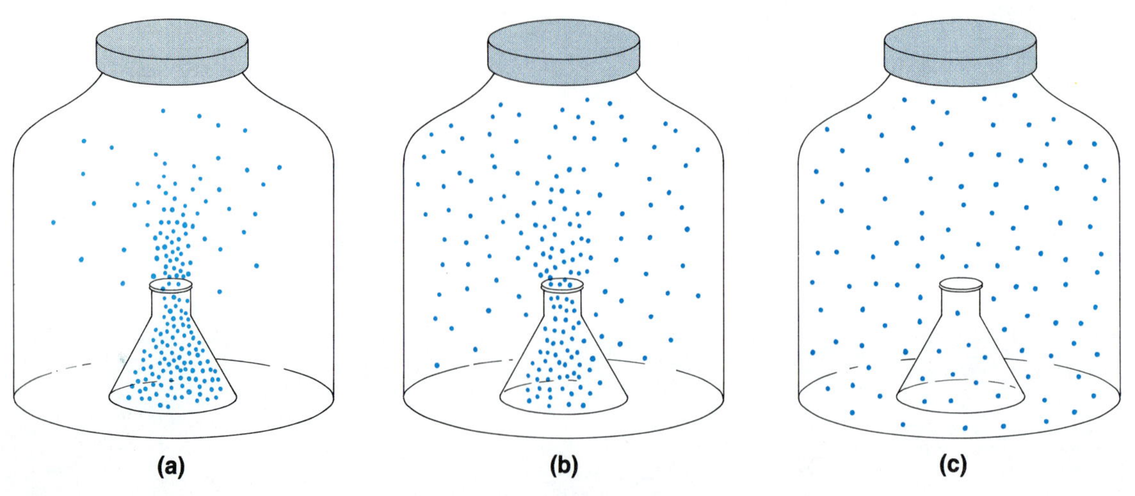

FIGURE 4.25 Simple diffusion. The random movements of molecules causes them to spread out (diffuse) from an area of high concentration to areas of lower concentration until eventually they are equally distributed throughout the available space.

cules. Most large molecules are unable to enter cells without the aid of specific carriers.

The mechanisms by which substances move across membranes can be passive or active. In passive transport the cell expends no energy to move substances down a gradient, that is, from higher to lower concentration. Passive processes include simple diffusion, facilitated diffusion, and osmosis. In active processes the cell expends energy from ATP, enabling it to transport substances against a gradient. These processes include active transport, endocytosis, and exocytosis. Of these, endocytosis and exocytosis occur only in eukaryotic cells.

Simple Diffusion

All molecules have kinetic energy; that is, they are constantly in motion and are continuously redistributed. **Simple diffusion** (Figure 4.25) is the net movement of particles from a region of higher to lower concentration. Suppose, for example, you drop a lump of sugar into a cup of coffee. At first a concentration gradient exists, with the sugar concentration greatest at the lump and least at the rim of the cup. Eventually, though, sugar molecules will become evenly distributed throughout the coffee even without stirring.

Diffusion occurs because of random movement of particles. Although particles move at high velocity, they do not travel far in a straight line before they collide with other randomly moving particles. Even so, some particles from a region of high concentration eventually move toward a region of lower concentration. Fewer particles move in the opposite direction for two reasons: (1) There are fewer of them in regions of low concentration to begin with, and (2) they are likely to be repelled by collision with particles from a region of high concentration.

The length of time required for particles to diffuse across a cell increases with cell diameter. Materials can diffuse throughout small prokaryotic cells very quickly and throughout larger eukaryotic cells fast enough to supply nutrients and remove wastes fairly efficiently. If cells were much larger, diffusion throughout the cell would be too slow to sustain life, so diffusion rates may be responsible in part for limiting the size of cells.

Any membrane severely limits diffusion, but many substances diffuse through the lipids of cell membranes. Diffusion through the lipid bilayer is affected by several factors: (1) the solubility of the diffusing substance in lipid, (2) the temperature, and (3) the difference between the highest and lowest concentration of the diffusing substance. Nonpolar substances such as steroids and gases cross the membrane rapidly by dissolving in the nonpolar fatty acid tails of the membrane phospholipids.

A few substances also diffuse through pores. ∞ (Chapter 1, p. 23) Such diffusion is affected by the size and charge of the diffusing particles and the charges on the pore surface. Pores probably have a diameter less than 0.8 nm, so only water, small water-soluble molecules, and ions such as H^+, K^+, Na^+, and Cl^- pass through them. This is one reason the plasma membrane is said to be **selectively permeable** (*semipermeable*). Some membranes are thought to contain special pores for specific ions. These pores have an arrangement of charges that allows rapid passage of a particular ion.

Facilitated Diffusion

Facilitated diffusion is diffusion down a concentration gradient and across a membrane with the assistance of a carrier molecule. The carrier molecule is a protein embedded in the membrane that binds to one or a few specific molecules and assists in their movement. A possible mechanism for facilitated diffusion is that the carrier acts like a revolving door or shuttle and provides a convenient one-way channel for the movement of substances across a membrane (Figure 4.26). Carrier molecules can become saturated, and similar molecules sometimes compete for the same carrier. Saturation occurs when all the carrier molecules are moving the diffusing substance as fast as they can. Under these conditions the rate of diffusion reaches a maximum and cannot increase further. When a carrier molecule can transport more than one substance, the substances compete for the carrier in proportion to their concentrations. For example, if there is twice as much of substance A as substance B, substance A will move across the membrane twice as fast as substance B.

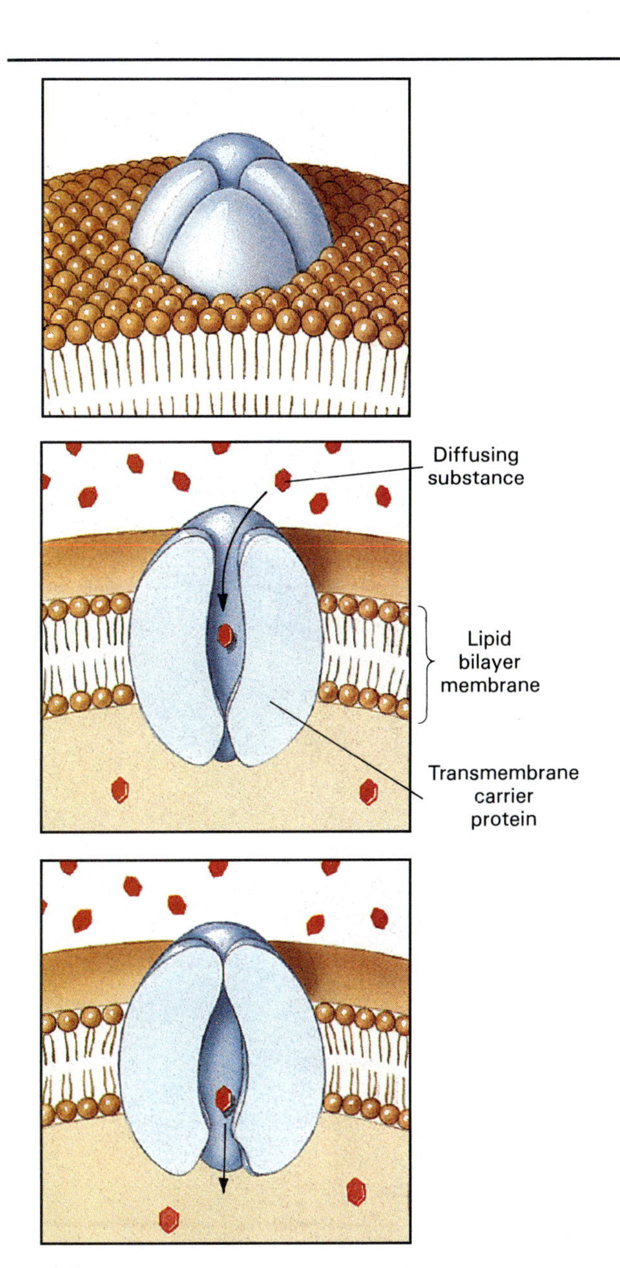

FIGURE 4.26 Facilitated diffusion. Carrier protein molecules aid in the movement of substances through the cell membrane, but only down their concentration gradient (from a region where their concentration is high to one where their concentration is low). This process does not require the expenditure of any energy (ATP) by the cell.

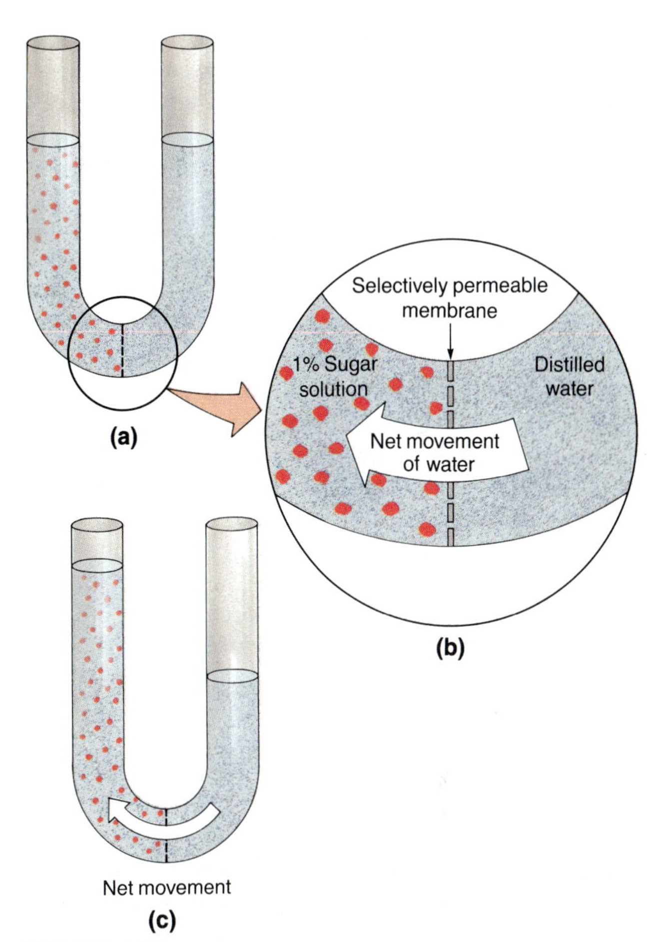

FIGURE 4.27 Osmosis: the diffusion of water from an area of higher water concentration to an area of lower water concentration through a semipermeable membrane. Here the net movement of water is into the sugar solution because the concentration of water there is slightly lower than on the other side of the membrane.

Osmosis

Osmosis is a special case of diffusion—one in which water diffuses across a selectively permeable membrane. To demonstrate osmosis, we start with two compartments separated by a membrane permeable only to water. One compartment contains pure water, and the other compartment contains some large, nondiffusible molecules, such as proteins (Figure 4.27).

Water moves in both directions, but its net movement is from pure water (concentration 100 percent) toward the water that contains other molecules (concentration therefore less than 100 percent). Thus, osmosis is the net flow of water from a region of its higher concentration to a region of its lower concentration.

Osmotic pressure is defined as the pressure required to *prevent* the net flow of water by osmosis. The least amount of hydrostatic pressure required to prevent the movement of water from a given solution into pure water is the osmotic pressure of the solution. The osmotic pressure of a solution is proportional to the number of particles dissolved in a given volume of solution. Thus, NaCl and other salts that form two ions per molecule exert twice as much osmotic pressure as glucose and other substances that do not ionize, provided each compound is present at the same concentration.

For a microbiologist the important thing to know about osmosis and osmotic pressure is how particles dissolved in fluid environments affect microorganisms in those environments. For this purpose tonicity is a useful concept. **Tonicity** is determined by observing the behavior of cells in a fluid environment (Figure 4.28). It is important to note that the cells are the reference point and the fluid environments are compared to the cells. The fluid surrounding cells is **isotonic** to the cells when no change in cell volume occurs. The fluid is **hypertonic** to the cells if the cells shrivel or shrink as water moves out of them into the fluid environment; it is **hypotonic** to the cells if the cells swell or burst as water moves from the environment into the cells. Though bacteria become dehydrated and their cytoplasm shrinks away from the cell wall in a hypertonic environment, their cell walls usually prevent them from swelling or bursting in a hypotonic environment.

Active Transport

In contrast to passive processes, **active transport** moves molecules and ions against concentration gradients from regions of lower concentration to those of higher concentration (Figure 4.29). This process is analogous to rolling something uphill, and it does require the cell to expend energy from ATP. Active transport is important in microorganisms for moving nutrients present in low concentration in the environment into the cells. It requires membrane proteins that act both as carriers and enzymes. These proteins display specificity in that each carrier transports a single substance or a few closely related substances. The results of active transport are to concentrate a substance on one side of a membrane and to maintain that concentration against a gradient. As with facili-

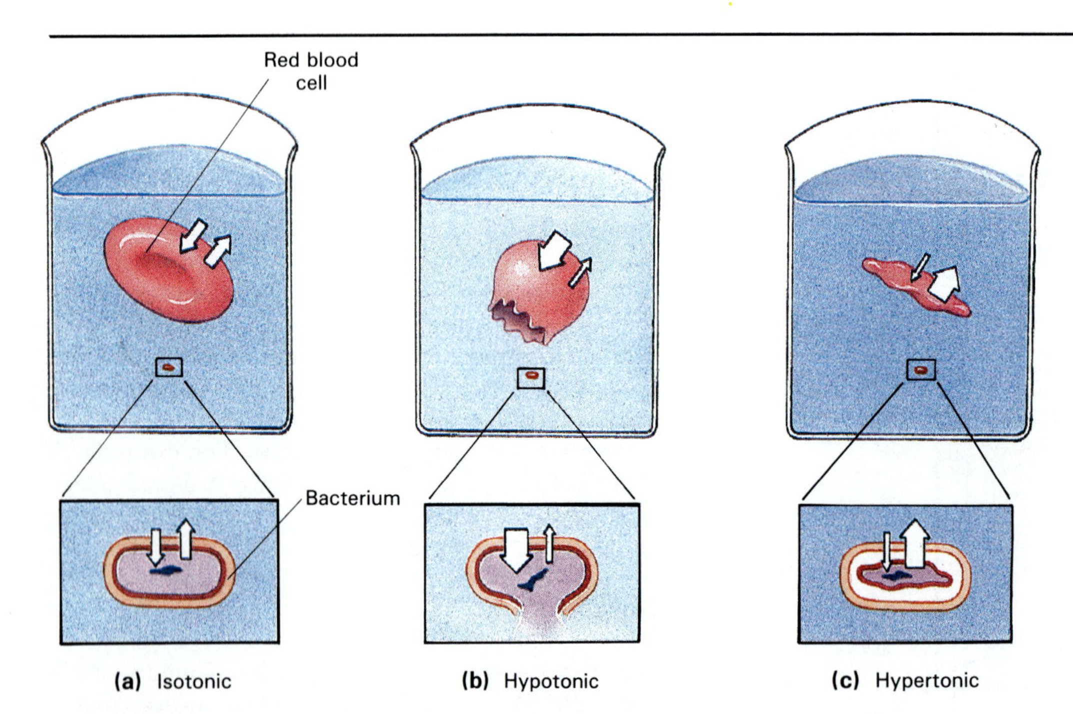

FIGURE 4.28 (a) A cell in an isotonic environment—one that has the same concentration of dissolved material as the interior of the cell—will experience no net gain or loss of water and will retain its original shape. (b) A cell in a hypotonic environment—one with a lower concentration of dissolved material than the interior of the cell—will gain water, swell, and perhaps even burst. (c) A cell in a hypertonic environment—one with a higher concentration of dissolved material than the interior of the cell—will lose water and shrink.

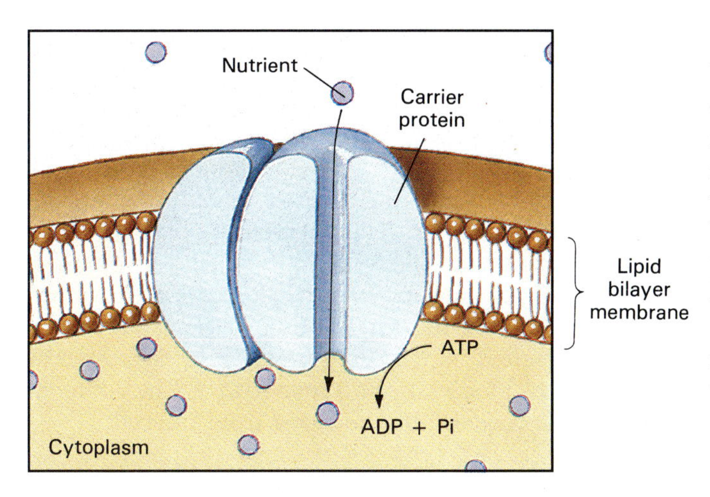

FIGURE 4.29 Active transport. Carrier protein molecules aid in movement of molecules through the cell membrane. This process can take place against a concentration gradient and so requires the use of energy (in the form of ATP) by the cell. The protein to the left of the carrier is an accessory protein that participates in its function.

tated diffusion, active transport carriers also are subject to saturation and competition for binding sites by similar molecules.

A process limited to procaryotes, called **group translocation,** moves a substance from the outside of a cell to the inside, while simultaneously chemically modifying it so that it cannot get back out. This allows molecules such as glucose to be accumulated against a concentration gradient. Because the modified molecule inside the cell is different than those outside, no actual concentration exists. Energy for this process is supplied by phosphoenolpyruvate (PEP), a high-energy phosphate compound.

Endocytosis and Exocytosis

In addition to the processes that move substances directly across membranes, eukaryotic cells move substances by forming membrane-bound vesicles. Such vesicles are made from portions of the cell membrane. If they form by invagination (poking in) and surround substances outside of the cell, the process is called **endocytosis.** These vesicles pinch off from the plasma membrane and enter the cell. If vesicles inside the cell fuse with the plasma membrane and extrude their contents from the cell, the process is called **exocytosis.** Both endocytosis and exocytosis require energy, probably to allow contractile proteins of the cell's cytoskeleton to move vesicles (Figure 4.30).

Endocytosis

Several types of endocytosis occur. In one type, known as *pinocytosis*, or cell drinking (Figure 4.17), a substance outside the cell binds to the plasma membrane to invaginate and surround the substance. The exact mechanisms that trigger binding and invagination are not yet clearly understood, but specific receptor sites on the cell membrane seem to be involved. Once the substance is completely surrounded by plasma membrane to form a vesicle, the vesicle pinches off from the plasma membrane and becomes a vacuole.

Of all the types of endocytosis, only phagocytosis is of special interest to microbiologists. In **phagocytosis** large vacuoles form around microorganisms and debris from tissue injury. These vacuoles enter the cell, taking with them large amounts of the plasma

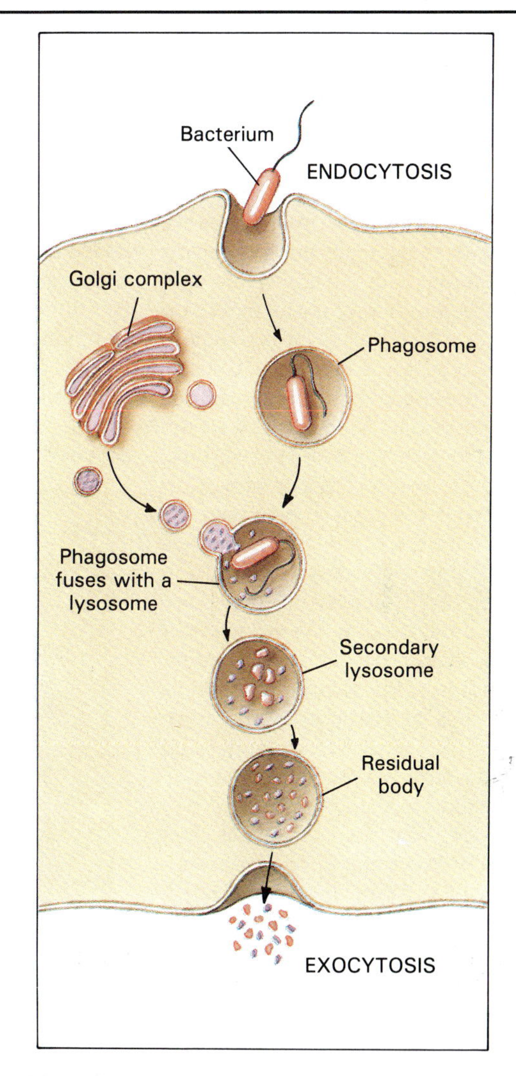

FIGURE 4.30 Endocytosis is the process of taking materials into the cell; exocytosis is the process of releasing materials from the cell. Material taken in by the form of endocytosis called phagocytosis is enclosed in phagosomes. Lysosomes fuse with the phagosomes and release powerful enzymes that degrade their contents. Reusable components are absorbed into the cell, and debris is released by exocytosis.

membrane. Lysosomes fuse with the vacuole membrane and release their enzymes into the vacuole. The enzymes digest the contents of the vacuole and release small molecules into the cytoplasm. Often, undigested particles remain in the vacuole. Much of the membrane that surrounded the original vacuole is returned to and fuses with the plasma membrane when the particles are released from the cell by exocytosis. Certain white blood cells are especially adept at phagocytosis, and they play an important role in defending the body against infection by microorganisms.

Exocytosis

Exocytosis, the mechanism by which cells release secretions, can be thought of as the opposite of endocytosis. Most secretory products are synthesized on ribosomes or smooth endoplasmic reticulum. They are transported through the membrane of the endoplasmic reticulum, packaged in vesicles, and moved to the Golgi apparatus, where their contents are sometimes processed to form the final secretory product. Once secretory vesicles are formed they move toward the plasma membrane and fuse with it. The contents of the vesicles are then released from the cell.

ESSAY

Evolution by Endosymbiosis

Biologists believe that life arose on earth about 4 billion years ago in the form of simple organisms much like the prokaryotic organisms of today. However, fossil evidence suggests that eukaryotic organisms arose only about 1 billion years ago. Why it took so long for living organisms to develop from prokaryotic to eukaryotic form is unknown. Considering the many kinds of eukaryotic organisms that have evolved in the last billion years, we can only surmise that the transition from prokaryote to eukaryote must have been exceedingly complex, involving numerous small steps. How development from prokaryote to eukaryote took place also is unknown, but the *endosymbiont theory* offers a plausible explanation. According to this theory, the various organelles of eukaryotic cells arose from prokaryotic cells. It is suggested that the first eukaryotic cell was an amoebalike cell that somehow had developed a nucleus. Knowing what we now know about the ease with which bits of unit membrane pinch off to form vesicles, it is fairly easy to imagine that a primitive chromosome might have become surrounded by membrane.

The primitive eukaryotic cell probably was a phagocytic cell that obtained nutrients by engulfing particles from its environment. Some of the particles it ingested were prokaryotic cells. Although many prokaryotic cells were probably digested and used to nourish the phagocyte, some apparently survived. These prokaryotes formed a symbiotic relationship with the host eukaryote. **Symbiosis** is a relationship between two organisms living in contact. An *endosymbiont* is an organism that lives within another organism. In the context of the endosymbiont theory, the symbiotic relationship was one in which both organisms benefited. The engulfed prokaryotes were protected by the eukaryote, and the eukaryote acquired some new capabilities through the presence of its symbionts. If the symbiont could capture energy from light and synthesize food, that food became available to both organisms. If the symbiont could use oxygen to obtain more than the usual amount of energy from nutrients, the energy became available to both organisms. And, if the symbiont could move, it helped the eukaryote to move, too.

Evidence for the endosymbiont theory comes from comparing the

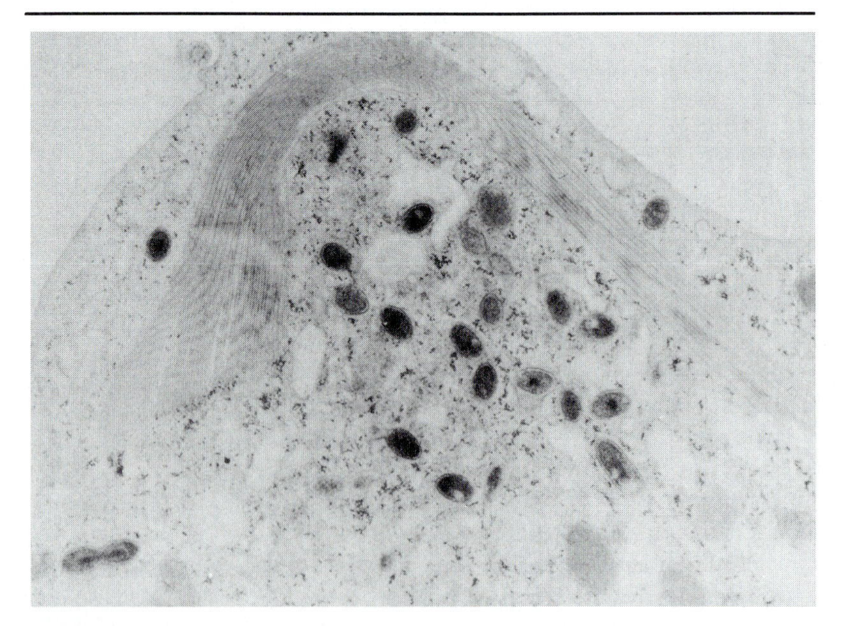

FIGURE 4.31 Inside the cytoplasm of *Pyrsonympha,* a protist that lives symbiotically inside the hindgut of termites, bacteria (dark ovals) that act as mitochondria for the protist. At lower left, one of the bacteria can be seen dividing.

characteristics of eukaryotic organelles with those of prokaryotic organisms. First, mitochondria and chloroplasts are approximately the same size as prokaryotic cells. It seems probable that they once might have been free-living prokaryotes, the mitochondria having been organisms like aerobic bacteria and the chloroplasts having been photosynthetic organisms like cyanobacteria. Furthermore, both the mitochondria and chloroplasts of modern eukaryotic cells contain their own DNA and ribosomes. And, even more important in support of the endosymbiont theory, the DNA resembles that of modern bacteria, and the ribosomes are small like bacterial ribosomes. Finally, organelle DNA and ribosomes carry out protein synthesis as it occurs in bacteria instead of as it occurs when directed by nuclear DNA of modern eukaryotes. Certain eukaryotes living in low-oxygen environments lack mitochondria, yet get along quite well thanks to bacteria that live inside of them and serve as "surrogate mitochondria." Protists living symbiotically in the hindgut of termites are, in turn, colonized by symbiotic bacteria similar in size and distribution to mitochondria (Figure 4.31). The bacteria function better in these low-oxygen conditions than mitochondria would. They oxidize food and provide energy in the form of ATP for their protist partner.

Living in the mud at the bottom of ponds is a giant amoeba, *Pelomyxa palustris* (Figure 4.32). It also lacks mitochondria and has at least two kinds of endosymbiotic bacteria. Killing just the bacteria with antibiotics allows lactic acid to accumulate. This suggests that the bacteria oxidize the end products of glucose fermentation, a function that mitochondria ordinarily perform. What else do mitochondria do? They must perform some function necessary for the formation or functioning of the Golgi apparatus, because all cells that lack mitochondria or bacterial "surrogate mitochondria" do not form Golgi apparati. This group includes all prokaryotes. Perhaps integration of bacterial endosybionts into a cell led to the development of mitochondria and Golgi apparati.

Dr. Lynn Margulis has proposed that eukaryotic flagellae and cilia (she calls them "undulipodia") originated from symbiotic associations of motile bacteria, called spirochetes, with heterotrophic protists. Such associations of present-day species are well known. *Mixotricha paradoxa*, a protist endosymbiont found in the hindgut of the Australian termite, *Mastotermes darwiniensis*, uses the four flagellae at its front end to steer, but depends on the half-million spirochetes covering its surface for driving power. These spirochetes have a natural tendency to coat living or dead surfaces. Once attached, they coordinate their undulations and beat in unison, propelling their host particle along. Margulis has hypothesized that some ancient spirochetes integrated into their host cells to become cilia and flagella. She has further suggested that other spirochetes were drawn down inside the cell (a process that can be observed with modern-day species) and eventually transformed into microtubules.

Although currently available evidence is less convincing, endosymbionts may have given rise to other organelles as well. For example, spirochetes might have attached to the surface of primitive eukaryotes. They would have obtained nutrients that leaked from the eukaryote, while giving the eukaryote motility.

Today, mitochondria and chloroplasts have not been demonstrated to survive outside cells. Nearly all photosynthetic cells have chloroplasts, and eukaryotic cells usually cannot survive without mitochondria. (Human red blood cells lack mitochondria, but they live only a few months.) Otherwise, these organelles show a surprising independence, and they divide independently of their host cells. Demonstrating survival of mitochondria and chloroplasts outside eukaryotic cells would provide strong evidence for the endosymbiont theory. However, the absence of such evidence does not refute the theory because the organelles have almost certainly undergone great evolutionary changes after a billion years as symbionts.

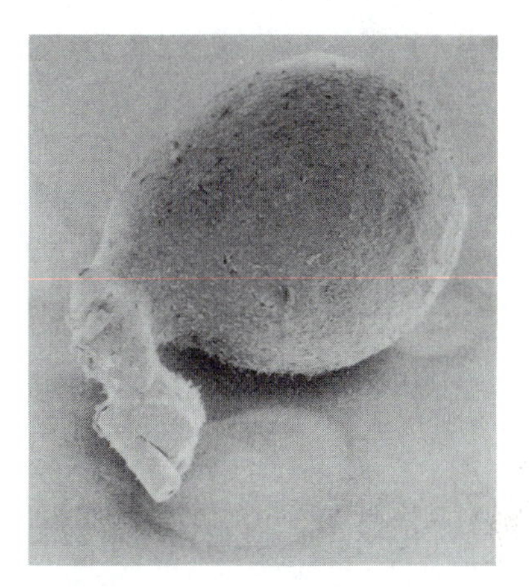

FIGURE 4.32 The free-living, herbivorous giant amoeba *Pelomyxa palustris* feeds on nonmotile algae such as *Spirogyra* and diatoms. Bacteria located close to the nuclear membrane serve the same function as mitochondria.

CHAPTER SUMMARY

BASIC CELL TYPES

RELATED KEY TERMS

- Both **prokaryotic** and **eukaryotic cells** have membranes that define the bounds of the living cell, and both contain genetic information stored in DNA.
- Prokaryotic cells differ from eukaryotic cells in that they lack a defined nucleus and membrane-bound **organelles** (except for a few simple membrane-covered bodies).

PROKARYOTIC CELLS

RELATED KEY TERMS

Size, Shape, and Arrangement

- Prokaryotic cells are the smallest of living things.
- Bacteria are named by shape as **cocci** (spherical), **bacilli** (cylindrical), **spirilli** (spiral), **vibrios** (comma-shaped), and **spirochetes** (flexible, wavy).
- Arrangements include groupings such as pairs, tetrads, grapelike clusters, and long chains.

pleomorphism

Overview of Structure

Cell Wall

- The rigid **cell wall** outside the cell membrane is composed mainly of **peptidoglycan.**
- Cell walls differ in composition and structure. In gram-positive bacteria the wall consists of a thick, dense layer of peptidoglycan, usually with **teichoic acid** attached to it. In gram-negative bacteria the wall has a thin layer of peptidoglycan, separated from the cytoplasmic membrane by the **periplasmic space** and enclosed by an **outer membrane** made of **lipopolysaccharide,** or **endotoxin.** In acid-fast bacteria the wall consists mainly of lipids, some of which are true waxes, and some of which are glycolipids.
- Some bacterial cell walls are damaged by penicillin and lysozyme.

lipid A
protoplasts **spheroplasts**
hydrophilic **hydrophobic**

Cell Membrane

- The **cell membrane** or **plasma membrane** has a **fluid-mosaic** structure with phospholipids forming a bilayer and proteins interspersed in a mosaic pattern.
- The main function of the cell membrane is to regulate the movement of materials into and out of cells.
- Bacterial cell membranes also perform functions usually carried out by organelles of eukaryotic cells.

Internal Structure

- **Ribosomes,** which consist of RNA and protein, serve as sites for protein synthesis.
- The **nuclear region** contains a single, large, circular chromosome, which contains the cell's DNA, and some RNA and protein.
- Bacteria contain a variety of **inclusions,** including **granules** that store glycogen or other substances and **vesicles** filled with gas.
- Some bacteria form resistant **endospores.** The *core* of an endospore contains living material and is surrounded by a *cortex, spore coat,* and *exosporium.*

cytoplasm
polyribosomes
nucleoid
chromatophore
volutin
metachromatic granules
metachromasia
vegetative cells

External Structure

- Motile bacteria have one, two, or many **flagella,** which propel the cell by the action of rings in their basal body.
- Much bacterial movement is random, but some bacteria exhibit **chemotaxis** (movement toward attractants and away from repellents).
- Some bacteria have **pili: Conjugation pili** allow exchange of DNA, whereas **attachment pili** help bacteria to adhere to surfaces.
- The **glycocalyx** includes all polysaccharides external to a bacterial cell. **Capsules** prevent host cells from destroying a bacterium; capsules of any species of bacteria have a specific chemical composition. **Slime layers** protect cells from drying, trap nutrients, and sometimes bind cells together, as in dental plaque.

flagellin **polar**
lophotrichous
peritrichous
monotrichous
amphitrichous
positive chemotaxis
negative chemotaxis
axial filaments **endoflagella**
pilin **fimbriae**
pellicle

EUKARYOTIC CELLS

Overview of Structure

- Eukaryotic cells, generally larger and more complex than prokaryotic cells, are the basic structural unit of microscopic and macroscopic organisms of the kingdoms Protista, Plantae, Fungi, and Animalia.

Cell Membrane

- Cell membranes of eukaryotic cells are almost identical to those of prokaryotic cells, except that they contain sterols, but their function is limited primarily to regulating movement of substances into and out of cells.

Internal Structure

- Eukaryotic cells are characterized by the presence of a membrane-bound **nucleus,** with a **nuclear envelope, nucleoli,** and paired **chromosomes** that contain DNA and proteins called **histones.**

nuclear pores
chromatin **dyads**
spindle apparatus **zygote**

- In cell division by **mitosis,** each cell receives one of each chromosome found in parent cells. In cell division by **meiosis,** each cell receives one member of each pair of chromosomes, and the progeny can be **gametes** or **spores.**
- Photosynthetic cells contain **chloroplasts,** which have membranes called **thylakoids** that contain chlorophyll.
- Eukaryotic ribosomes are larger than those of prokaryotes and can be free or attached to endoplasmic reticulum. Free ribosomes make protein to be used in the cell and those attached to endoplasmic reticulum make proteins to be secreted.
- The **endoplasmic reticulum** is an extensive network of membrane, which without ribosomes (smooth ER), synthesizes lipids and, when combined with ribosomes (rough ER), produces proteins.
- The **Golgi apparatus** is a set of interconnected vesicles that receives, sometimes modifies, and secretes proteins.
- **Lysosomes** are organelles that contain digestive enzymes, which destroy dead cells and digest contents of vacuoles.
- **Peroxisomes** are membrane-bound organelles that convert peroxides to water and oxygen and sometimes oxidize amino acids and fats.
- **Vacuoles** contain various stored substances and materials engulfed by phagocytosis.
- The **cytoskeleton** is a network of **microfilaments** and **microtubules** that support and give rigidity to cells.

RELATED KEY TERMS

mitochondria **matrix** **cristae** **stroma**

secretory vesicles

External Structure

- Most external components of eukaryotic cells are concerned with movement. Eukaryotic flagella are composed of microtubules made of tubulin; sliding of proteins at their bases causes them to move.
- **Cilia** are smaller than flagella and beat in coordinated waves.
- **Pseudopodia** are projections into which cytoplasm flows, causing a creeping movement.
- Eukaryotic cells of the plant and fungi kingdoms have cell walls.

amoeboid movement
cytoplasmic streaming

MOVEMENT OF SUBSTANCES ACROSS MEMBRANES

- All passive processes involved in movement across membranes involve net movement of substances from a region of higher concentration to a region of lower concentration. These processes do not require expenditure of energy by the cell.

Simple Diffusion

- **Simple diffusion** results from the molecular kinetic energy and random movement of particles. The role of diffusion in living cells depends on the size of particles, nature of pores in membranes, and distances substances must move inside cells.

selectively permeable

Facilitated Diffusion

- **Facilitated diffusion** is essentially the same as simple diffusion except that it makes use of a carrier molecule in a cell membrane.

Osmosis

- **Osmosis** is the net movement of water molecules through a semipermeable membrane from a region of higher to a region of lower concentration. The **osmotic pressure** of a solution is the pressure required to prevent such a flow.

tonicity **isotonic** **hypertonic** **hypotonic**

Active Transport

- Active processes involved in movement of substances across membranes generally result in movement from regions of lower concentration to those of higher concentration, and require the cell to expend energy.
- **Active transport** requires a carrier molecule in a cell membrane, a source of ATP, and an enzyme that releases energy from ATP.
- Active transport is important in cell functions because it allows cells to take up substances that are in low concentration in the environment and to concentrate those substances within the cell.

Endocytosis and Exocytosis

- **Endocytosis** and **exocytosis,** which occur only in eukaryotic cells, involve formation of vesicles from fragments of cell membrane.
- In endocytosis the vesicle enters the cell, as in **phagocytosis.**

- In exocytosis the vesicle leaves the cell, as in secretion.
- Endocytosis and exocytosis are important because they allow the movement of relatively large quantities of materials across membranes.

RELATED KEY TERMS

QUESTIONS FOR REVIEW

A.

1. How are prokaryotic and eukaryotic cells alike?
2. How are such cells different?

B.

3. What are the three basic shapes of bacteria?
4. How do planes of division affect the arrangement of bacterial cells?

C.

5. Describe the properties of the cytoplasmic membrane and relate them to the fluid-mosaic model.
6. What is the major function of the cytoplasmic membrane, and what additional functions does a prokaryotic membrane perform?
7. What are the distinguishing characteristics of the cell walls of gram-positive, gram-negative, and acid-fast bacteria?
8. Describe the chemical composition of peptidoglycan.
9. Describe the chemical composition of the outer membrane.
10. How do penicillin and lysozyme affect the cell wall?

D.

11. Describe the composition and function of ribosomes.
12. What structures and functions are associated with the nuclear region?
13. What materials are found in granules and vesicles of bacteria?
14. Name the components of an endospore and describe the function of endospores.
15. Under what conditions do endospores form, and under what conditions do they germinate?
16. Describe the structure of a bacterial flagellum.
17. How do bacteria move in chemotaxis?
18. How do attachment and conjugation pili differ in structure and function?
19. What is glycocalyx?
20. What are capsules, and what properties do they impart to bacteria?
21. What are slime layers, and which functions do they perform?

E.

22. How do eukaryotic membranes differ from prokaryotic membranes?
23. What is the function of sterols in eukaryotic cell membranes?

F.

24. Describe the structure and function of the nucleus.
25. Distinguish between mitosis and meiosis and between gametes and spores.
26. Describe the structure of a chloroplast and briefly describe its function in photosynthesis.
27. Describe the structure, function, and location of ribosomes.
28. What are the distinguishing characteristics of endoplasmic reticulum, Golgi apparatus, peroxisomes, lysosomes, vacuoles, and cytoskeleton?
29. How do flagella in eukaryotes differ from those in prokaryotes?
30. Describe the structure and function of cilia and of pseudopodia.
31. What types of organisms have flagella, cilia, and pseudopodia?
32. What different materials are found in cell walls of eukaryotes, and which groups of organisms have cell walls?

G.

33. What are the general characteristics of passive processes by which materials cross cell membranes?
34. What are the similarities and differences between simple diffusion and facilitated diffusion?
35. What is osmosis?

H.

36. How does active transport differ from passive processes with respect to movement across cell membranes?
37. How does active transport occur, and why is it important?

I.

38. Distinguish between endocytosis and exocytosis.
39. How are endocytosis and exocytosis important in cell function?

PROBLEMS FOR INVESTIGATION

1. Use what you know about the organelles of eukaryotic cells and the general structure of both prokaryotic and eukaryotic cells to explain how prokaryotic cells perform functions of the organelles they are lacking.
2. Predict what would happen to eukaryotic cells and to prokaryotic cells placed in distilled water, 2 percent saline, and isotonic saline.
3. What might have happened if the flasks Pasteur used to disprove spontaneous generation had contained bacterial endospores?
4. Design an experiment that would demonstrate that flagella are responsible for the movement of bacteria in solutions.

SOME INTERESTING READING

Adler, J. 1976. The sensing of chemicals by bacteria. *Scientific American* 235, no. 4 (April):40.

Berg, H. C. 1975. How bacteria swim. *Scientific American* 233, no. 2 (February):36.

Blakemore, R. P., and R. B. Frankel. 1981. Magnetic navigation in bacteria. *Scientific American* 245, no. 6 (December):58.

Bretcher, M. S. 1985. The molecules of the cell membrane. *Scientific American* 253, no. 4 (October):100.

Brock, T. D. 1988. The bacterial nucleus: A history. *Microbiological Reviews* 52:397–411.

Costerton, J. W., R. T. Irvin, and K. J. Cheng. 1981. The bacterial glycocalyx in nature and disease. *Annual Review of Microbiology* 35:299.

Doolittle, R. F. 1985. Proteins. *Scientific American* 253, no. 4 (October):88.

Felsenfeld, G. 1985. DNA. *Scientific American* 253, no. 4 (October):58.

Ferris, F. G., and T. J. Beveridge. 1985. Functions of bacterial cell surface structures. *BioScience* 35:172.

Gray, M. W., and W. F. Doolittle. 1982. Has the endosymbiotic hypothesis been proven? *Microbiological Reviews* 46:1.

Hancock, R. E. W. 1991. Bacterial outer membranes: evolving concepts. *ASM News* 57(4):175–82.

Koch, A. L. 1990. Growth and form of the bacteria growth. *American Scientist* 78, no. 4 (July–August):327–42.

Koppel, T. 1991. Learning how bacteria swim could set new gears in motion. *Scientific American* 265(September):168–69.

Moir, A., and D. A. Smith. 1990. The genetics of bacterial spore germination. *Annual Review of Microbiology* 44:531–53.

Nikaido, H., and M. Vaara. 1985. Molecular basis of bacterial outer membrane permeability. *Microbiological Reviews* 49:1.

Shapiro, J. A. 1991. Multicellular behavior of bacteria. *ASM News* 57(5):247–53.

Unwin, N., and R. Henderson. 1984. The structure of proteins in biological membranes. *Scientific American* 250, no. 2 (February):78.

Vidal, G. 1984. The oldest eukaryotic cells. *Scientific American* 250, no. 2 (February):48.

Fermentation reactions carried out by yeasts living on the skins of these grapes transform grape juice into wine.

5 Essential Concepts of Metabolism

This chapter focuses on the following questions:

A. How do the following terms relate to metabolism: autotrophy, heterotrophy, oxidation, reduction, photoautotrophy, photoheterotrophy, chemoautotrophy, chemoheterotrophy, glycolysis, fermentation, aerobic metabolism, and biosynthetic processes?

B. What are the characteristics of enzymes, and how do those characteristics contribute to their function?

C. What are the main steps and significance of glycolysis and fermentation?

D. What are the main steps and significance of the Krebs cycle?

E. What are the roles of electron transport and oxidative phosphorylation in energy capture?

F. How do microorganisms metabolize fats and proteins for energy?

G. What are the main steps and significance of photosynthesis in microbes?

H. How do photoheterotrophy and chemoautotrophy differ?

I. How do bacteria carry out biosynthetic activities?

J. How do bacteria use energy for membrane transport and for movement?

Until the middle of the nineteenth century people didn't know what caused a fruit juice to become wine or milk to sour. Then, in 1857, Louis Pasteur proved that alcoholic fermentation was due to microorganisms. A few years later he identified specific organisms from samples of fermenting juices and souring milk. Pasteur was one of the first to study chemical processes in a living organism. Since his time much has been learned about such processes.

METABOLISM: AN OVERVIEW

Metabolism (Figure 5.1) is the sum of all the chemical processes carried out by living organisms. It includes **anabolism,** reactions that require energy to synthesize complex molecules from simpler ones, and **catabolism,** reactions that release energy by breaking complex molecules into simpler ones, which can then be reused as building blocks. Anabolism is needed for growth, reproduction, and repair of cellular structures. Catabolism provides the organism with energy for its life processes, including movement, transport, and the synthesis of complex molecules—that is, anabolism.

All catabolic reactions involve electron transfer, which allows energy to be captured in high-energy bonds in ATP and similar molecules. Electron transfer is directly related to oxidation and reduction (Table 5.1). **Oxidation** can be defined as the loss of electrons. Though many substances combine with oxygen and transfer electrons to oxygen, oxygen need not be present if another electron acceptor is available. **Reduction** can be defined as the gain of electrons. When a substance loses electrons, or is oxidized, energy is released, but another substance must gain electrons, or be reduced, at the same time. For example, during the oxidation of organic molecules, hydrogen atoms are removed and used to reduce oxygen to form water:

$$\underset{\text{hydrogen}}{2\ H_2} + \underset{\text{oxygen}}{O_2} \longrightarrow \underset{\text{water}}{2\ H_2O}$$

In this reaction hydrogen is an electron donor, or reducing agent, and oxygen is an electron acceptor, or oxidizing agent. Because oxidation and reduction must occur simultaneously, the reactions in which they occur are sometimes called redox reactions.

Among all living things, microorganisms are particularly versatile in the ways in which they obtain energy. The various ways different microorganisms capture energy can be classified as **autotrophy** (aw-to-trof'e), self-feeding, or **heterotrophy** (het"er-o-trof'e), other-feeding (Figure 5.2). **Autotrophs** use carbon dioxide to synthesize organic molecules. They include **photoautotrophs,** which obtain energy from light, and **chemoautotrophs,** which obtain energy from oxidizing simple inorganic substances such as sulfides and nitrites. **Heterotrophs** use ready-made organic molecules obtained from other organisms, living or dead. They include **photoheterotrophs,** which obtain energy from light, and **chemoheterotrophs,** which obtain energy from breaking down ready-made organic molecules.

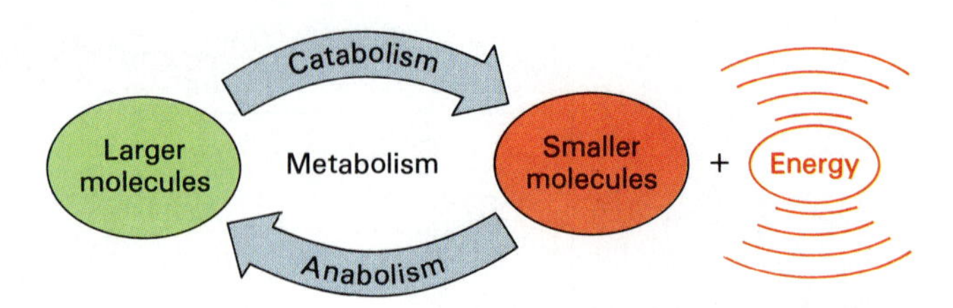

FIGURE 5.1 Large, complex molecules are generally richer in energy than are small, simple ones. Catabolic reactions break down large molecules into smaller ones, releasing energy. Organisms capture some of this energy for their life processes. Anabolic reactions use energy to build larger molecules from smaller components. The molecules synthesized in this way are used for growth, reproduction, and repair.

Autotrophic metabolism, especially photosynthesis, is important as a means of energy capture in many free-living microorganisms. However, such microorganisms do not usually cause disease. We emphasize metabolic processes that occur in chemoheterotrophs because many microorganisms, including nearly all infectious ones, are chemoheterotrophs. These processes include *glycolysis* (oxidation of glucose to pyruvic acid), *fermentation* (conversion of pyruvic acid to ethyl alcohol, lactic acid, or other organic compounds), and *aerobic respiration* (oxidation of pyruvic acid to carbon dioxide and water). Glycolysis and fermentation do not require oxygen and transfer only a small amount of the energy in a glucose molecule to ATP. Aerobic respiration does require oxygen as an electron acceptor and captures a relatively large amount of the energy in a glucose molecule in ATP. Complete oxidation of glucose by glycolysis and aerobic respiration is summarized as follows:

$$\underset{\text{glucose}}{C_6H_{12}O_6} + \underset{\text{oxygen}}{O_2} \longrightarrow \underset{\text{carbon dioxide}}{CO_2} + \underset{\text{water}}{H_2O} + \text{energy}$$

TABLE 5.1 Comparison of oxidation and reduction

Oxidation	Reduction
Loss of electrons	Gain of electrons
Gain of oxygen	Loss of oxygen
Loss of hydrogen	Gain of hydrogen
Loss of energy (liberates energy)	Gain of energy (stores energy in the reduced compound)
Exothermic; exergonic (gives off heat energy)	Endothermic; endergonic (requires energy; e.g., heat)

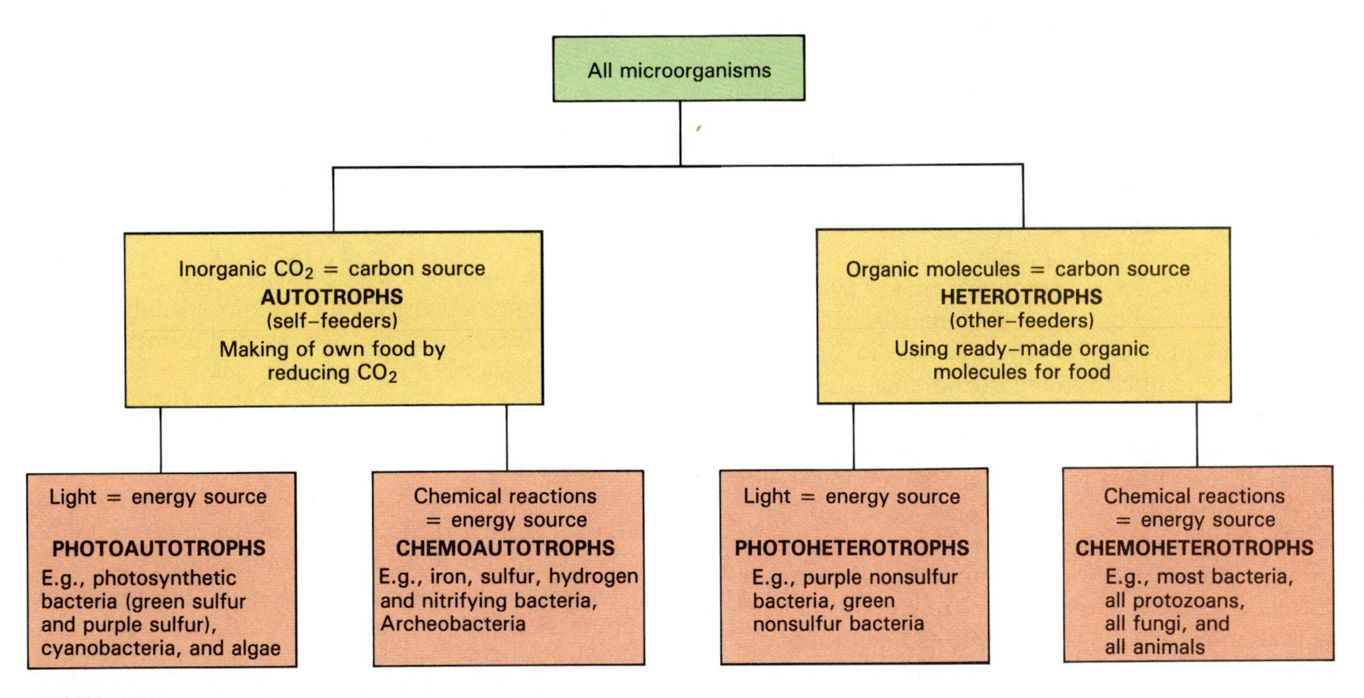

FIGURE 5.2 Kinds of energy-capturing metabolism.

A large number of microorganisms obtain energy by *photosynthesis,* the use of light energy and hydrogen from water or other compounds to reduce carbon dioxide to an organic substance containing more energy. The synthesis of glucose by photosynthesis in cyanobacteria and algae is summarized as follows:

$$\underset{\text{carbon dioxide}}{CO_2} + \underset{\text{water}}{H_2O} \xrightarrow[\text{chlorophyll}]{\text{light energy}} \underset{\text{glucose}}{C_6H_{12}O_6} + \underset{\text{oxygen}}{O_2}$$

(True bacteria, as we shall see later, use a slightly different version of this process.) Photosynthetic organisms then use the glucose or other carbohydrates made in this way for energy. Respiration and photosynthesis are related in Figure 5.3.

Like nearly all chemical processes in living organisms, glycolysis, fermentation, aerobic respiration, and photosynthesis each consist of a series of chemical reactions in which the product of one reaction serves as the substrate for the next: $A \rightarrow B \rightarrow C \rightarrow D \rightarrow E$, and so on. Such a chain of reactions is called a **metabolic pathway.** Each reaction in a pathway is controlled by a particular enzyme.

In addition to **catabolic pathways,** which capture energy in a form cells can use, living things also make

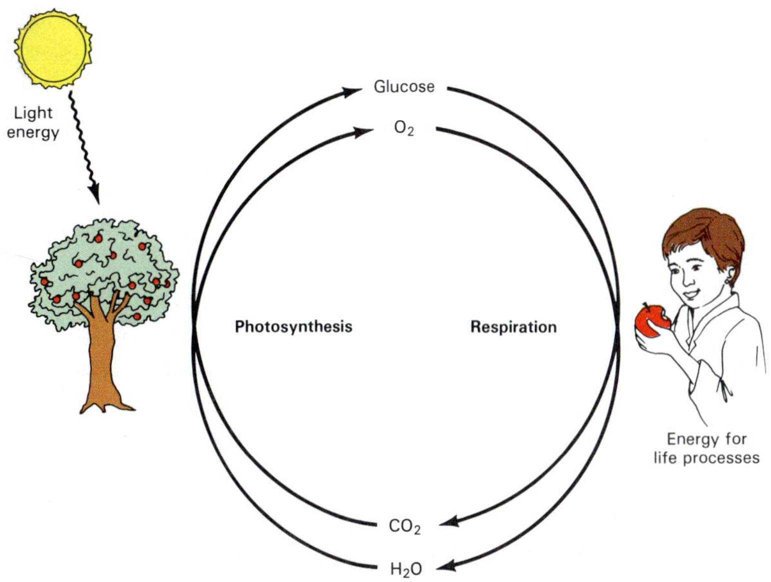

FIGURE 5.3 In photosynthesis, light energy is used to reduce carbon dioxide, forming energy-rich compounds such as glucose and other carbohydrates. In aerobic respiration, energy-rich compounds are oxidized to carbon dioxide and water, and some of the energy released is captured for use in life processes. (The form of photosynthesis depicted here is carried out by cyanobacteria, algae, and higher plants. Green and purple bacteria use compounds other than water as a source of hydrogen atoms to reduce CO_2.)

use of anabolic, or biosynthetic, pathways, which are likewise controlled by enzymes. **Biosynthetic** (*anabolic*) **pathways** make the complex molecules that form the structure of cells, enzymes, and other molecules that control cells. These pathways use building blocks such as sugars, glycerol, fatty acids, amino acids, purines, pyrimidines, and other molecules to make carbohydrates, lipids, proteins, nucleic acids, or combinations such as glycolipids (made from carbohydrates and lipids), glycoproteins (from carbohydrates and proteins), lipoproteins (from lipids and proteins), and nucleoproteins (from nucleic acids and proteins). ATP molecules are the links that couple catabolic and anabolic pathways. Energy released in catabolic reactions is captured and stored in the form of ATP molecules, which are later broken down to provide the energy needed to build up new molecules in biosynthetic pathways. Bacteria transfer approximately 40 percent of the energy in a glucose molecule to ATP molecules during aerobic metabolism and 25 percent during anaerobic fermentation processes.

ENZYMES

Enzymes are a special category of proteins found in all living organisms. In fact, most cells contain hundreds of enzymes, and cells are constantly synthesizing enzymes. Enzymes act as catalysts. Themselves unchanged, they speed up reactions by as much as a million times the uncatalyzed rate, which is ordinarily not sufficient to sustain life. The only other way to speed up the reaction rate would be to increase the temperature: in general, a 10 degree increase in temperature results in a doubling of the reaction rate. However, this would raise the temperature to a level at which a cell would die. Thus, enzymes are necessary for life at temperatures cells can withstand. To explain how enzymes do these things we must consider their properties. (Chapter 2, p. 43)

Properties of Enzymes

In general, reactions that release energy can occur without input of energy from the surroundings. Nevertheless, such reactions often occur at unmeasurably low rates because the molecules lack the energy to start the reaction. For example, though the oxidation of glucose releases energy, it does not occur unless energy to start the reaction is available. The energy required to start such a reaction is called **activation energy** (Figure 5.4). Activation energy can be thought of as a hurdle over which molecules must be raised to get a reaction started. By analogy, a rock resting in a depression at the top of a hill would easily roll down the hill if pushed out of the depression. Activation energy is like the energy required to lift the rock out of the depression.

A common way to activate a reaction is to raise the temperature, thereby increasing molecular movement, as you do when you strike a match. Matches ordinarily do not burst into flame spontaneously. If energy from friction (striking) is added to the reactants on the match head, the temperature increases, and the match bursts into flame. Such a reaction in cells would raise the temperature enough to denature proteins and evaporate liquids. Enzymes lower the activation energy so reactions can occur in living cells.

Enzymes provide a surface on which reactions take place. Each enzyme has a certain area on its surface called the **active site,** a binding site at which it forms a loose association with its **substrate,** the

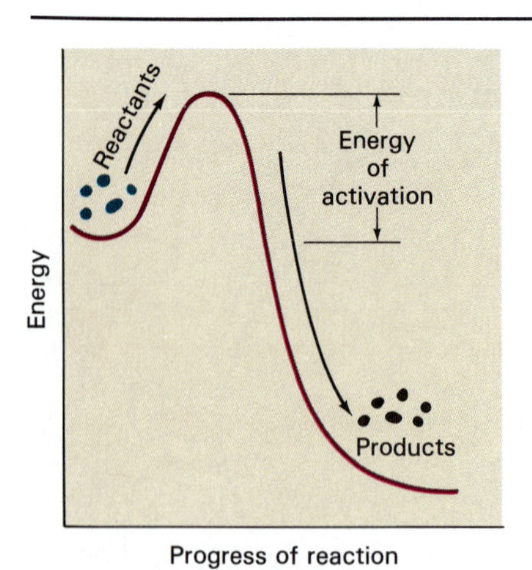

FIGURE 5.4 A chemical reaction cannot take place unless a certain amount of activation energy is available to start it. Enzymes lower the amount of activation energy needed to initiate a reaction. They thus make it possible for biologically important reactions to occur at the relatively low temperatures that living organisms can tolerate.

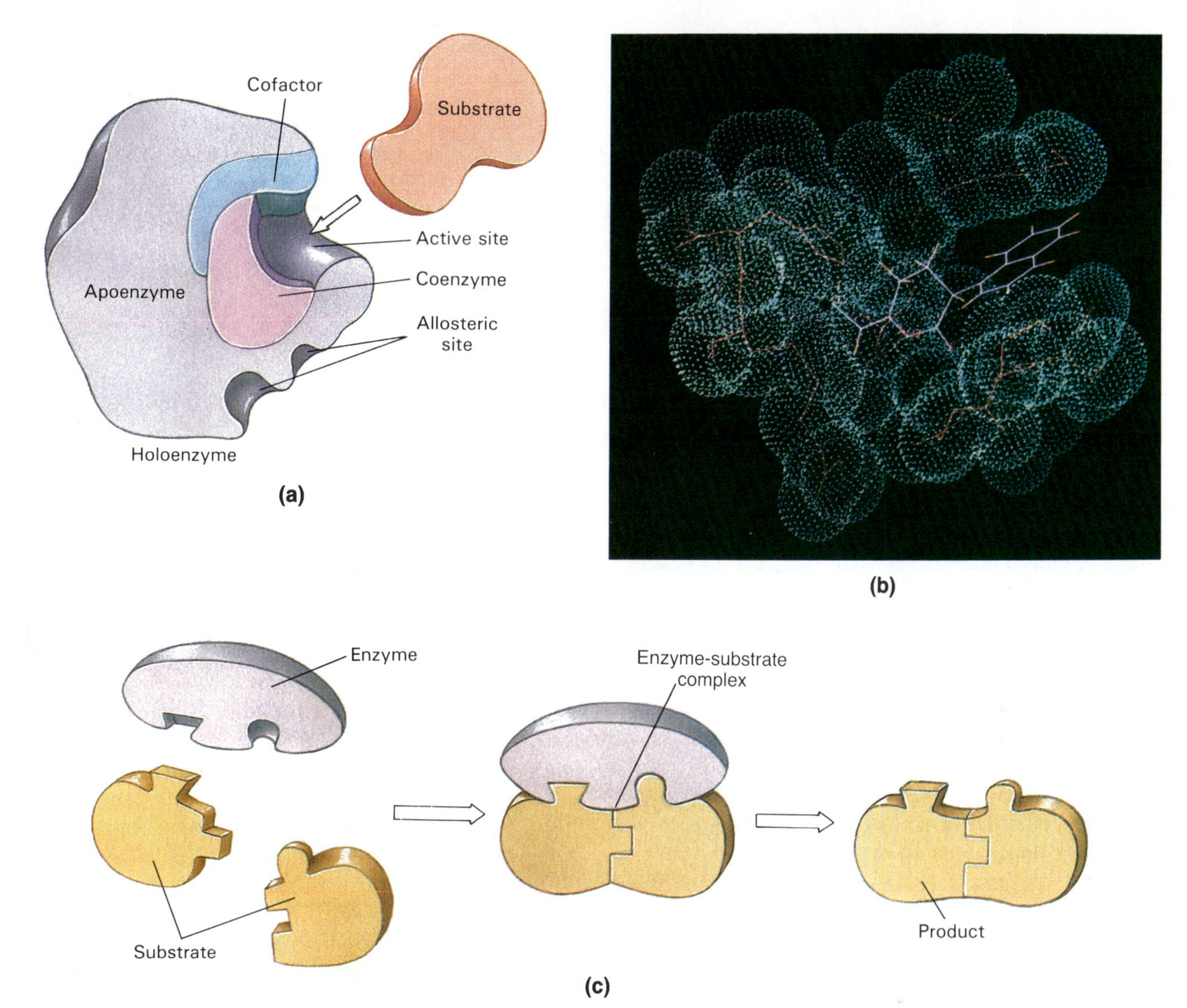

FIGURE 5.5 (a) Many enzymes consist of a protein apoenzyme that must combine with a nonprotein coenzyme (an organic molecule) or cofactor (an inorganic ion) to form the functional holoenzyme. The active site of the enzyme is a cleft or pocket with a shape and chemical composition that enable it to bind the substrate—the molecule on which the enzyme acts. (b) A computer-generated model of an enzyme (blue) with a substrate molecule (purple) bound to the active site. (c) The substrate or substrates bind to the active sites of an enzyme, producing an enzyme-substrate complex. The enzyme helps a chemical reaction to occur, and a product is formed. In this example, the reaction is one that joins two substrate molecules. Other enzyme-catalyzed reactions can involve the splitting of one substrate molecule into two parts or the chemical modification of a substrate.

substance the enzyme acts upon (Figure 5.5a and b). Like all molecules, a substrate molecule has kinetic energy, and it collides with various molecules within a cell. When it collides with the active site of its enzyme, an **enzyme-substrate complex** forms. As a result of binding to the enzyme, some of the chemical bonds in the substrate are weakened. The substrate then undergoes chemical change, and the product or products are formed (Figure 5.5c).

Enzymes generally have a high degree of **specificity;** they catalyze only one type of reaction, and most act only on a particular substrate. An enzyme's structure, especially the shape and electrical charges at its active site, accounts for its specificity. When an enzyme acts on more than one substrate, it usually acts on substrates with the same functional group or the same kind of chemical bond. For example, proteolytic, or protein-splitting, enzymes act on different proteins but always act on peptide bonds.

Enzymes are usually named by adding the suffix *-ase* to the name of the substrate upon which they act. For example, phosphatases act on phosphates, sucrase breaks down the sugar sucrose, lipases break down lipids, and peptidases break peptide bonds. Enzymes are commonly placed in one of six categories according to their functions (Table 5.2). They also can be divided into two categories based on where they act. **Endoenzymes,** or intracellular enzymes, act

TABLE 5.2 Classification of enzymes by their function

Kind of Enzyme	Functions of Enzyme	Examples
Oxidoreductase	Simultaneously oxidizes one substance and reduces another	Lactic dehydrogenase, enzymes that add or remove hydrogen from coenzymes
Transferase	Transfers a functional group from one molecule to another	Phosphatases, transaminases
Hydrolase	Adds water and breaks large molecules into two smaller molecules	Peptidase, lipase, maltase
Lyase	Removes chemical groups from molecules without adding water	Decarboxylase
Isomerase	Rearranges atoms of a molecule	Aconitase
Ligase	Joins two molecules together and usually requires energy from ATP	Acetyl-CoA synthetase

within the cell that produced them. **Exoenzymes,** or extracellular enzymes, are synthesized in a cell but cross the cell membrane to act in the periplasmic space or the cell's immediate environment.

Properties of Coenzymes and Cofactors

Many enzymes can catalyze a reaction only if other substances called *coenzymes,* or *cofactors,* are present. Such enzymes consist of a protein portion called the **apoenzyme** that must combine with a nonprotein coenzyme or cofactor to form an active **holoenzyme** (Figure 5.5a). A **coenzyme** is an organic molecule bound to or loosely associated with an enzyme. Many coenzymes are synthesized from vitamins, which are essential nutrients precisely because they are required to make coenzymes. For example, coenzyme A is made from the vitamin pantothenic acid, and nicotinamide adenine dinucleotide (NAD) is made from the vitamin niacin. A **cofactor** is usually an inorganic ion, such as magnesium, zinc, or manganese. Cofactors often improve the fit of an enzyme with its substrate, and their presence can be essential to allow the reaction to proceed.

Coenzymes carry hydrogen atoms or electrons in many oxidative reactions (Figure 5.6). When a coenzyme receives hydrogen atoms or electrons, it is reduced; when it releases them it is oxidized. The coenzyme FAD (flavin adenine dinucleotide), for example, receives two hydrogen atoms to become $FADH_2$ (reduced FAD). Coenzymes called cytochromes contain an atom of iron (Fe). When a cytochrome molecule gains an electron (is reduced), the iron has a 2^+ charge (Fe^{2+}); when it loses an electron (is oxidized), the iron has a 3^+ charge (Fe^{3+}). The coenzyme NAD (nicotinamide adenine dinucleotide) has a positive charge in its oxidized state (NAD^+). In its reduced state, NADH, it carries a hydrogen atom

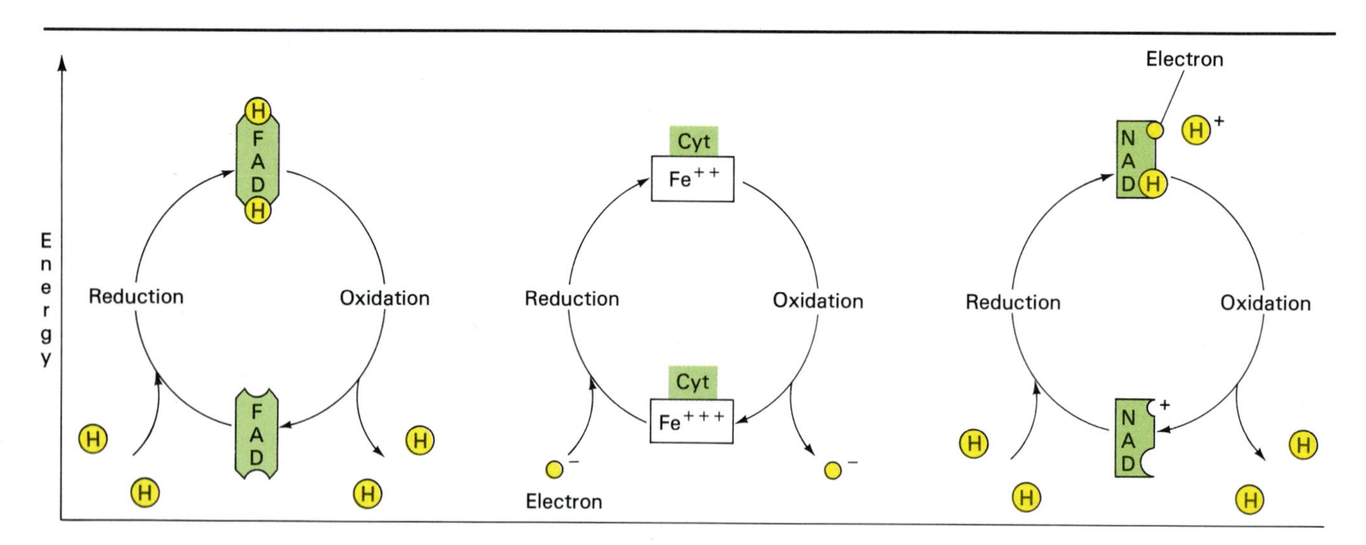

FIGURE 5.6 Coenzymes carry energy in the form of electrons in many biochemical reactions. Some coenzymes, such as cytochromes, carry only electrons. Others, such as FAD, carry whole hydrogen atoms (electrons together with protons). NAD carries one hydrogen atom and one "naked" electron. When coenzymes are reduced (gain electrons), they rise in energy; when they are oxidized (lose electrons), they fall in energy.

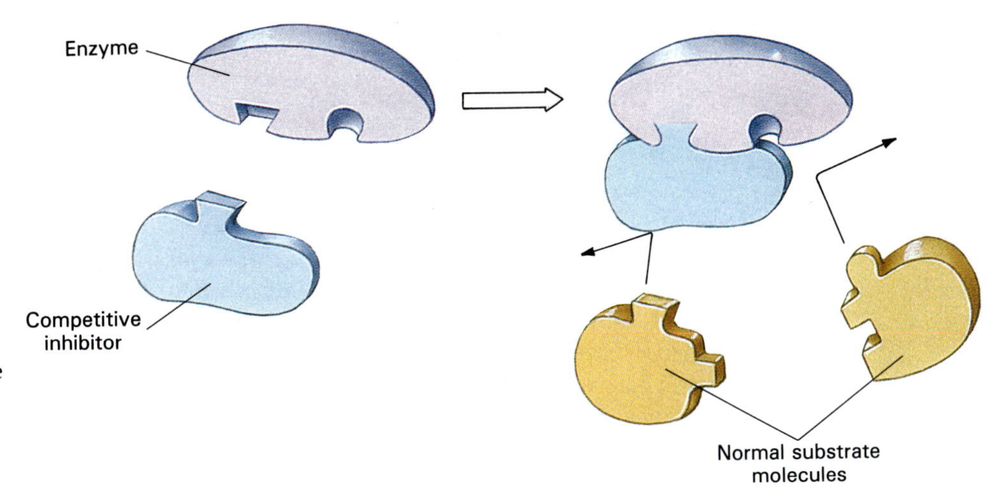

FIGURE 5.7 A competitive inhibitor binds to the active site of an enzyme, preventing the normal substrate from reaching it, but cannot take part in the reaction.

and an electron from another hydrogen atom, the proton of which remains in the cellular fluids. In all such oxidation-reduction reactions, the electron carries the energy that is transferred from one molecule to another. Thus, for simplicity, we will refer to *electron transfer* regardless of whether "naked" electrons or hydrogen atoms (electrons with protons) are transferred.

Enzyme Inhibition

A molecule similar in structure to a substrate can sometimes bind to an enzyme's active site even though it is unable to react. This nonsubstrate molecule is said to act as a **competitive inhibitor** of the reaction because it competes with the substrate for the active site (Figure 5.7). When the inhibitor binds to an active site, it prevents the substrate from binding and thereby inhibits the reaction.

Because the attachment of such an inhibitor is reversible, the degree of inhibition depends on the relative concentrations of substrate and inhibitor. When the concentration of the substrate is high and that of the inhibitor is low, the active sites of only a few enzyme molecules are occupied by the inhibitor, and the rate of the reaction is only slightly reduced. When the concentration of the substrate is low and that of the inhibitor is high, the active sites of many enzyme molecules are occupied by the inhibitor, and the rate of the reaction is greatly reduced.

Enzymes also can be inhibited by substances called **noncompetitive inhibitors** (Figure 5.8). Some

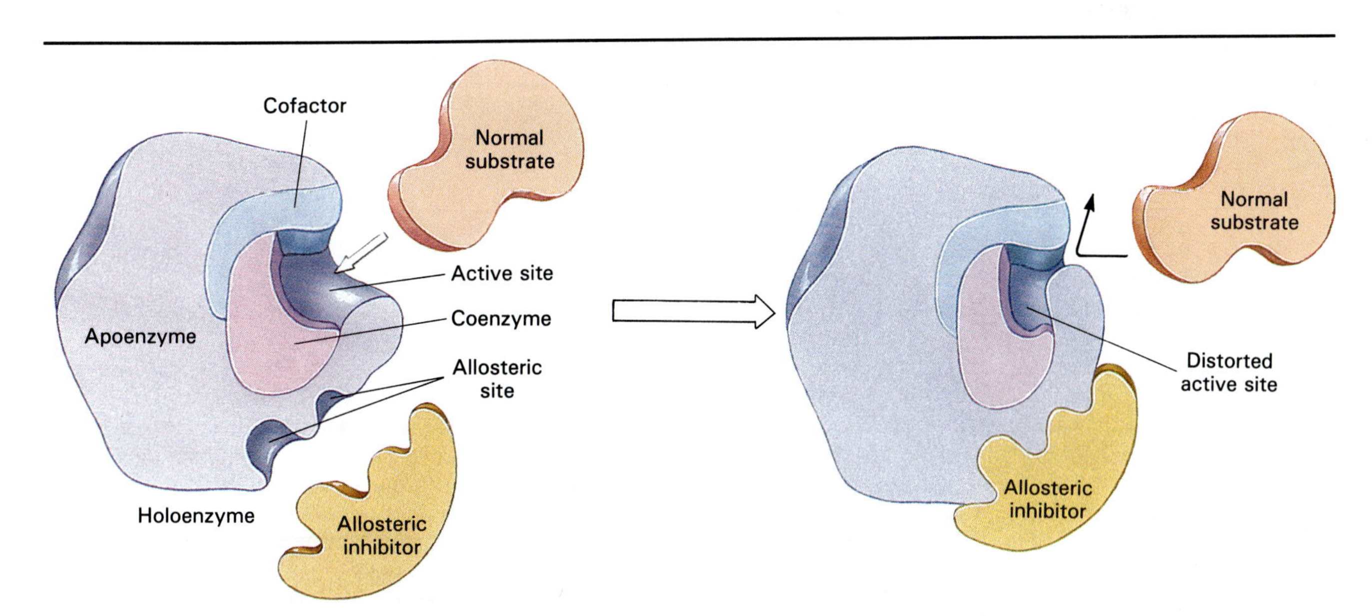

FIGURE 5.8 A noncompetitive (allosteric) inhibitor usually binds at a site other than the active site (an allosteric site). Its presence changes the shape of the enzyme enough to interfere with binding of the normal substrate. Some noncompetitive inhibitors are used in the regulation of metabolic pathways, but others are poisons.

noncompetitive inhibitors attach to the enzyme at an **allosteric site,** or a site other than the active site. They distort the tertiary protein structure and alter the shape of the active site. Any enzyme molecule so affected no longer can bind substrate, so it cannot catalyze a reaction. Although some noncompetitive inhibitors bind reversibly, others bind irreversibly and permanently inactivate enzyme molecules, thereby greatly decreasing the reaction rate. In noncompetitive inhibition, increasing the substrate concentration does not increase the reaction rate as it does in the presence of a competitive inhibitor. Lead and other heavy metals, while not noncompetitive inhibitors, can bind to other sites on the molecule and permanently change its shape, thus inactivating it.

Certain noncompetitive inhibitors temporarily bind to the allosteric site of an enzyme and reduce its activity. **Feedback inhibition,** a kind of reversible noncompetitive inhibition, regulates the rate of many metabolic pathways. For example, when an end product of a pathway accumulates, the product often binds to and inactivates the enzyme that catalyzes the first reaction in the pathway. Feedback inhibition is discussed in more detail in Chapter 7.

Factors That Affect Enzyme Reactions

Factors that affect the rate of enzyme reactions include (1) temperature, (2) pH, and (3) the concentrations of substrate, product, and enzyme. Like other proteins, enzymes are affected by heat and extremes of pH. Even small pH changes can alter the electrical charge on various chemical groups in enzyme molecules, thereby altering the enzyme's ability to bind its substrate and catalyze a reaction.

Most human enzymes have an *optimum temperature,* near normal body temperature, and an *optimum pH,* near neutral, at which they catalyze a reaction most rapidly. Microbial enzymes likewise have optimum temperatures and pHs related to an organism's normal environment. The enzymes of microbes that infect humans have approximately the same optimum temperature and pH as human enzymes.

Changes in enzyme activity, the rate at which an enzyme catalyzes a reaction, are shown in Figure 5.9. Enzyme activity increases with temperature up to the enzyme's optimum temperature. Above 40°C, however, the enzyme is rapidly denatured (Chapter 2), and its activity decreases accordingly. ∞ (p. 43) Activity is maximal at an enzyme's optimum pH and decreases as the pH becomes higher or lower. Like high temperatures, extremely acidic or alkaline conditions also denature enzymes. Such conditions are used to kill or control the growth of microorganisms.

To understand the *effects* of *concentrations* on enzyme-catalyzed reactions, we must first note that all chemical reactions are, in theory, reversible. Enzymes can catalyze a reaction to go in either direction: $AB \rightarrow A + B$ or $A + B \rightarrow AB$. The concentrations of substrates and products determine the direction of a reaction. A high concentration of AB drives the reaction toward formation of A and B. Using A and B in other reactions as fast as they are formed also drives the reaction toward forming more A and B. Conversely, using AB in another reaction so that its concentration remains low drives the reaction toward formation of AB. When neither AB nor A and B are removed from the system, the reaction will ultimately

APPLICATIONS

How To Ruin an Enzyme

If its target enzyme has a vital metabolic function, an enzyme inhibitor acts as a poison. A competitive inhibitor, competing with the normal substrate for the active site, temporarily poisons enzyme molecules and slows the reaction. If it binds to the active site of all molecules of an enzyme at one time, it can stop the reaction. The enzyme itself is unharmed, however, and resumes function if the poison is removed. If the poison forms a covalent bond to the enzyme, distorting the active site so that it can no longer bind its substrate, it is a permanent poison; the enzyme's ability to function is irreversibly destroyed.

Enzyme inhibition has important medical applications. Antibiotics, including penicillin, kill microorganisms by competitively inhibiting one or more enzymes. The drug sulfanilamide is an inhibitor of the reaction in some bacteria in which para-aminobenzoic acid (PABA) is used to make folic acid. Fluoride, which prevents tooth decay, hardens enamel and poisons enzymes. In low concentrations it kills bacteria in the mouth without damaging human cells, but if the concentration is large enough it can kill human cells, too. Many pesticides and herbicides exert their effects through competitive inhibition. Certain chemotherapeutic agents used to treat cancer inhibit enzymes that are most active in rapidly dividing cells, including malignant cells, and have lesser effects on normal cells. Heavy metals inactivate enzymes noncompetitively and permanently and so function as active ingredients in many disinfectants. Though safe to use on inanimate objects, they cause severe toxic effects if ingested or absorbed through the skin.

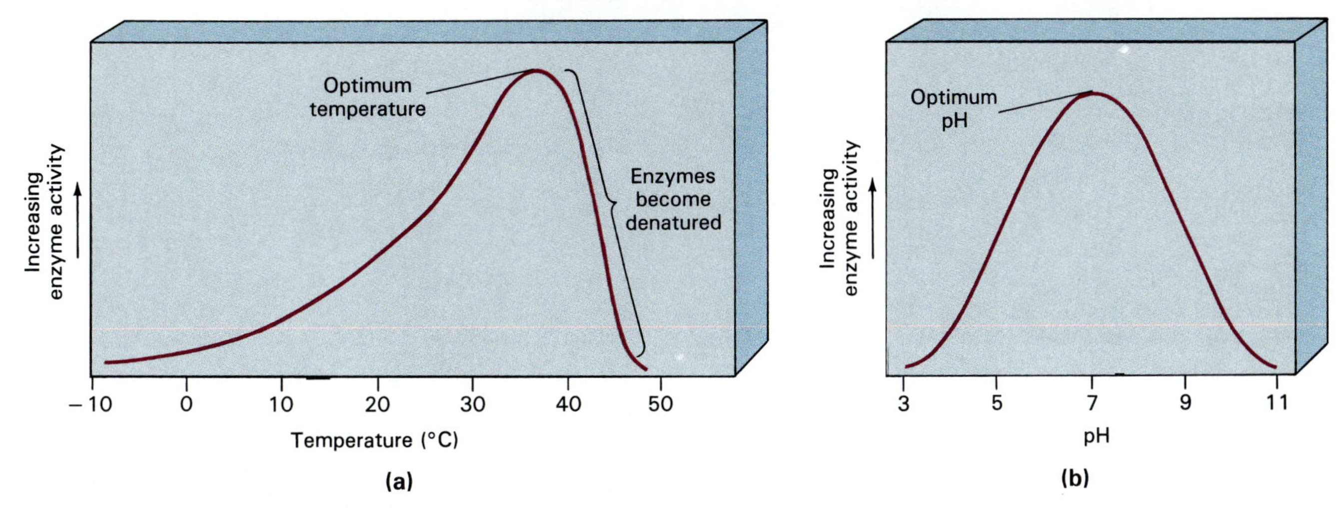

FIGURE 5.9 (a) Enzymes become more active as the temperature rises. Above about 40°C, however, most enzymes become denatured, and their activity falls off sharply. (b) Most enzymes also have an optimal pH at which they function most effectively.

reach a steady state known as **chemical equilibrium.** At equilibrium, no net change in the concentrations of AB, A, or B occurs.

The quantity of enzyme available usually controls the rate of a metabolic reaction. A single enzyme molecule can catalyze only a specific number of reactions per second, that is, can act on only a specific number of substrate molecules. The reaction rate increases with the number of enzyme molecules and reaches a maximum when all available enzyme molecules are working at full capacity. However, if the substrate concentration is too low to keep all enzyme molecules working at capacity, the substrate concentration will determine the rate of the reaction.

With an overview of metabolic processes and an understanding of enzymes and how they work, we are ready to look at metabolic processes in more detail. We begin with glycolysis, fermentation, and aerobic respiration, the processes used by most microorganisms to capture energy.

ANAEROBIC METABOLISM: GLYCOLYSIS AND FERMENTATION

Glycolysis

Glycolysis (gli-kol'is-is) is the metabolic pathway used by most autotrophic and heterotrophic organisms, both aerobes and anaerobes, to begin to break down glucose. It does not require oxygen, but it can occur either in the presence or absence of oxygen. Figure 5.10 shows the ten steps of the glycolysis pathway, within which four important events occur: (1) substrate-level phosphorylation, (2) the breaking of a 6-carbon molecule (glucose) into two 3-carbon molecules, (3) the transfer of two electrons to the coenzyme NAD, and (4) the capture of energy in ATP.

Phosphorylation (fos″for-il-a'shun) is the addition of a phosphate group to a molecule, often from ATP. This addition generally increases the molecule's energy. Thus, phosphate groups commonly serve as energy carriers in biochemical reactions. Early in glycolysis, phosphate groups from two molecules of ATP are added to glucose. This expenditure of two ATPs raises the energy level of glucose, enabling it to participate in subsequent reactions and rendering it incapable of leaving the cell.

After phosphorylation, glucose is broken into two 3-carbon molecules, and each molecule is oxidized as two electrons are transferred from it to NAD. The end products are two molecules of pyruvic acid, or pyruvate (the ionized form), and two molecules of reduced NAD.

Energy is captured in ATP at the substrate level in two separate reactions late in glycolysis. With adenosine diphosphate (ADP) and inorganic phosphate (P_i) available in the cytoplasm, the energy released from substrate molecules is used to form high-energy bonds between ADP and P_i:

$$\text{ADP} + \text{P}_i + \text{energy} \longrightarrow \text{ATP}$$

Glycolysis provides cells with a relatively small amount of energy. Energy is captured in two molecules of ATP during the metabolism of each 3-carbon molecule, and a total of four ATPs are formed as one 6-carbon glucose molecule is metabolized by glycolysis to two molecules of pyruvate. Because energy from two ATPs was used in the initial phosphorylations, glycolysis results in a net energy capture of only two ATPs per glucose molecule. When atmospheric

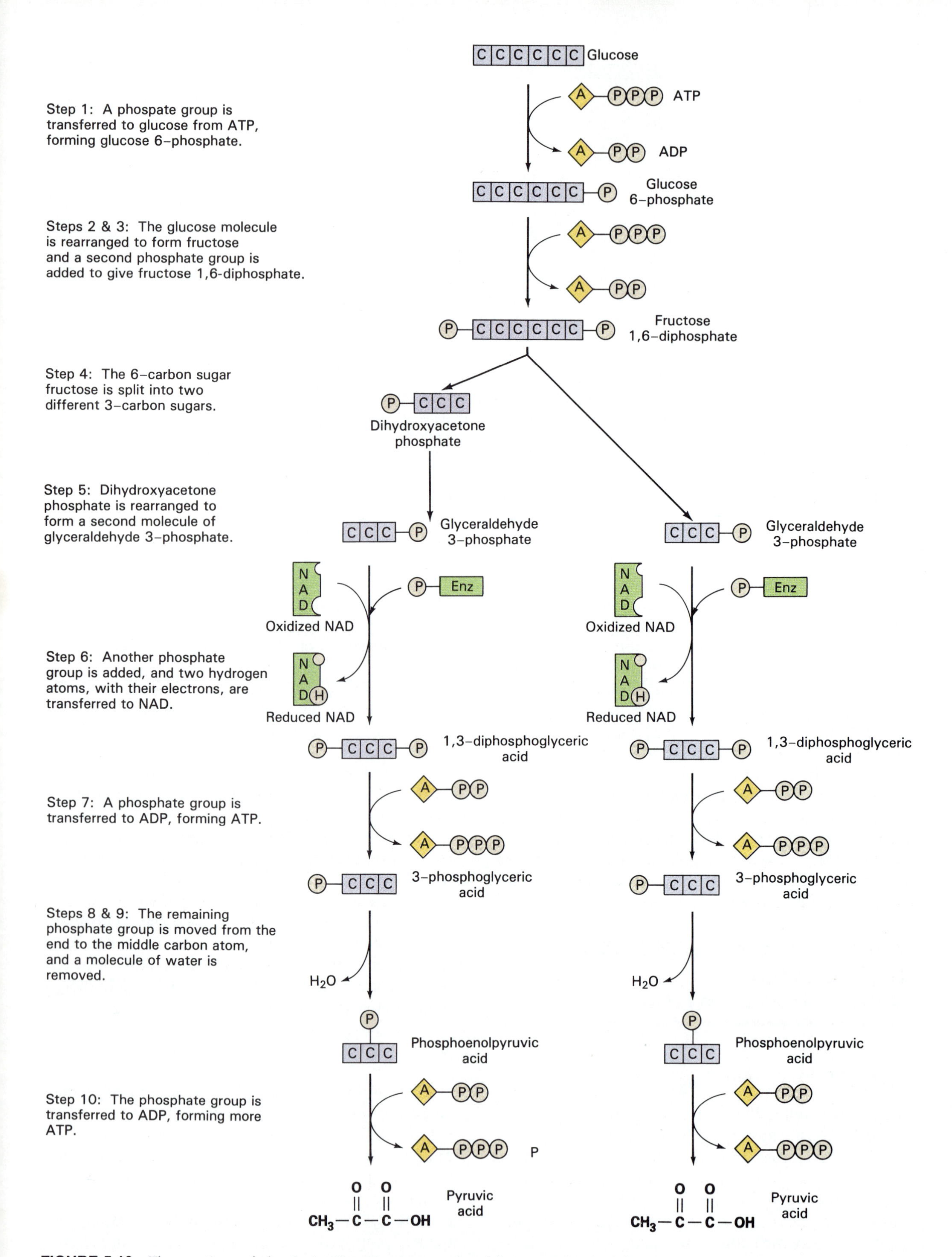

FIGURE 5.10 The reactions of glycolysis. Note that in steps 1 and 3, two molecules of ATP are used. In steps 7 and 10, two molecules of ATP are formed. Because each glucose molecule yields two of the 3-carbon sugars that undergo reactions 7 and 10, four molecules of ATP are actually formed, giving a net yield of two ATP per glucose.

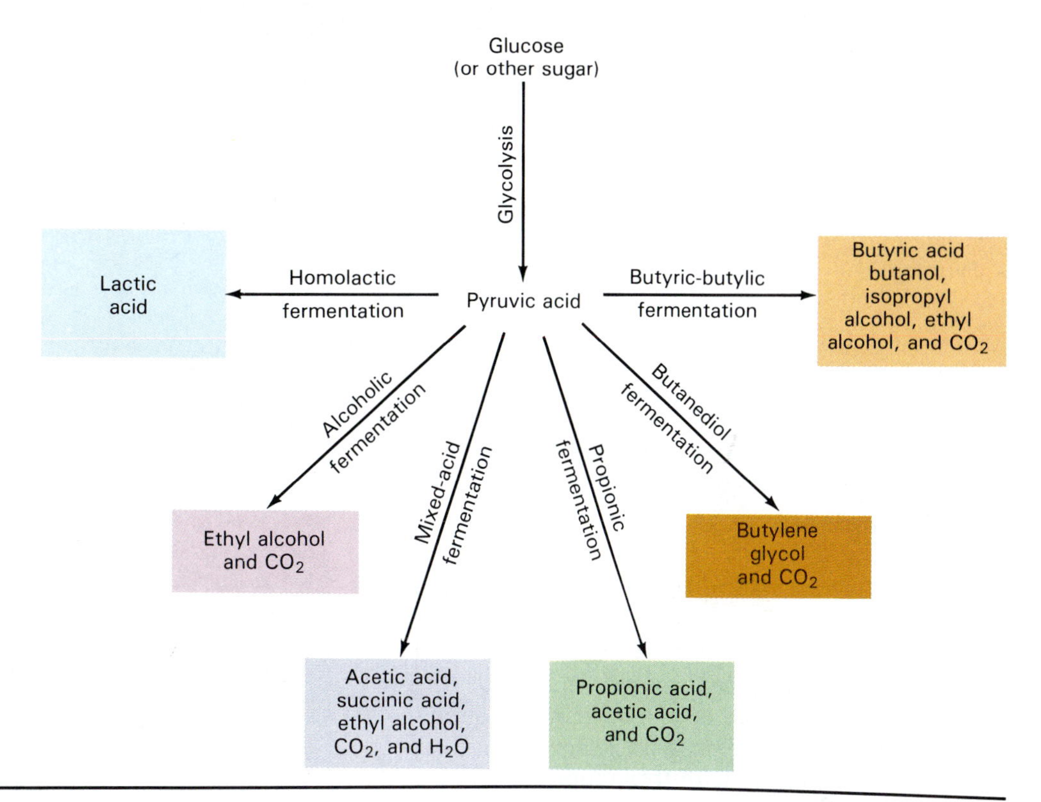

FIGURE 5.11 Some of the many different fermentation pathways found among microorganisms.

oxygen is present and the organism has the enzymes to carry out aerobic metabolism, electrons from reduced NAD are transferred to oxygen during biological oxidation, as is explained later.

Two additional features of glycolysis illustrate principles that apply to metabolic pathways in general:

1. Each reaction is catalyzed by a specific enzyme. Though enzyme names have been omitted in our account, it is important to remember that each reaction in glycolysis and in other pathways we will consider is catalyzed by an enzyme.
2. When electrons are removed from substrates such as glucose, they are transferred to particular coenzymes. In glycolysis, oxidized NAD becomes reduced NAD (NADH). As explained next, electrons are removed from reduced NAD during fermentation, freeing it to remove more electrons from glucose and keep glycolysis operating. Because cells contain limited quantities of both enzymes and coenzymes, the rate at which the reactions of glycolysis and other pathways occur is limited by the availability of these important molecules.

Though glucose is the main nutrient of most microorganisms, some can obtain energy from other sugars. These organisms usually have specific enzymes to convert a sugar to fructose or another intermediate substance in the glycolysis pathway. (All substances in a metabolic pathway between the first and last substance are called intermediate substances, or, simply, intermediates.) Once the sugar has entered glycolysis, it is metabolized to pyruvic acid and then fermented or metabolized aerobically by processes to be described later.

Fermentation

Anaerobic metabolism of glucose or another sugar by glycolysis is a process carried out by nearly all cells. The subsequent metabolism of pyruvate in the absence of oxygen is called fermentation. **Fermentation** is the result of needing to recycle the limited amount of NAD by passing the electrons of reduced NAD off to other molecules, and occurs by many different pathways (Figure 5.11). Two of the most important and commonly occurring pathways are homolactic-acid fermentation and alcoholic fermentation. Neither captures energy in ATP from the metabolism of pyruvate, but both remove electrons from reduced NAD so that it can continue to act as an electron acceptor. Thus, they indirectly foster energy capture by keeping glycolysis going.

Homolactic Acid Fermentation

The simplest pathway for pyruvate metabolism is **homolactic acid fermentation** (Figure 5.12). Pyruvate is converted directly to lactate using electrons from reduced NAD. Unlike other fermentations, this type produces no gas. It occurs in some of the bacteria called lactobacilli, in streptococci, and in mammalian cells. This pathway in lactobacilli is used in making some cheeses.

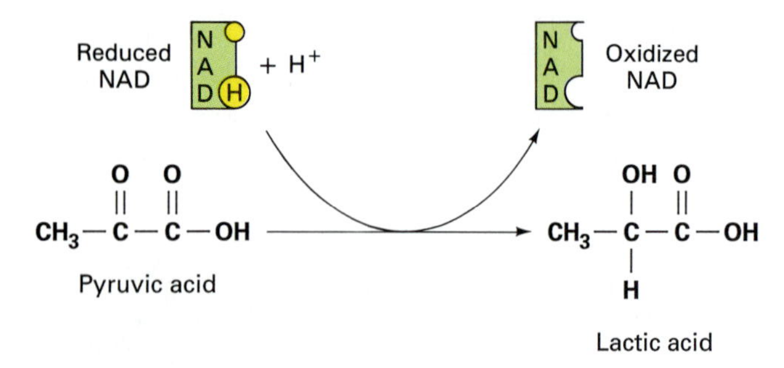

FIGURE 5.12 In homolactic acid fermentation, pyruvic acid is reduced to lactic acid by the NAD from step 6 of glycolysis (Figure 5.10).

Alcoholic Fermentation

In **alcoholic fermentation** (Figure 5.13), carbon dioxide is released from pyruvate to form the intermediate acetaldehyde, which is quickly reduced to ethyl alcohol by electrons from reduced NAD. Alcoholic fermentation, while rare in bacteria, is common in yeasts and is used in making bread and wine. Chapter 27 deals extensively with these topics.

Other Kinds of Fermentation

The other kinds of fermentation summarized in Figure 5.11 are performed by a great variety of microorganisms. For our purposes the most important things about these processes are that they occur in certain infectious organisms and their products are used in diagnosis. For example, the Voges-Proskauer test for acetoin, a product of butanediol fermentation, helps to detect the bacterium *Klebsiella pneumoniae,* which can cause pneumonia. Anaerobic butyric-butylic fermentation occurs in *Clostridium* species that cause tetanus and botulism. This fermentation also produces the unpleasant odors of rancid butter and cheese.

The ability to ferment sugars other than glucose forms the basis of other diagnostic tests. One such test (Figure 5.14) uses the sugar mannitol and the pH indicator phenol red. The pathogenic bacterium *Staphylococcus aureus* ferments mannitol and produces acid,

$$CH_3-\overset{O}{\overset{\|}{C}}-\overset{O}{\overset{\|}{C}}-OH \longrightarrow CO_2 + CH_3-\overset{O}{\overset{\|}{C}}-H$$

Pyruvic acid — Carbon dioxide — Acetaldehyde

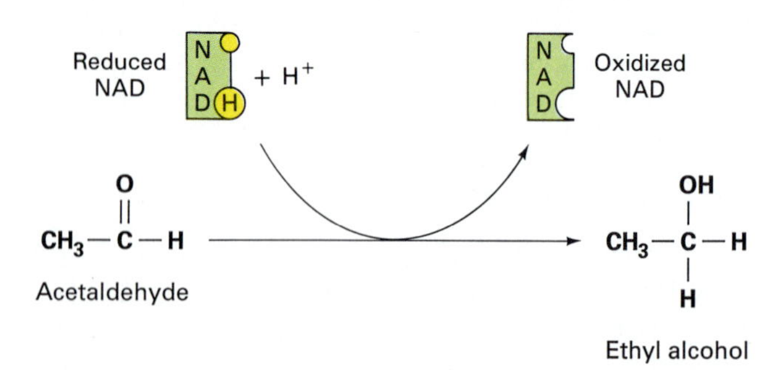

FIGURE 5.13 Alcoholic fermentation is a two-step process. A molecule of carbon dioxide is first removed from pyruvic acid to form acetaldehyde. Acetaldehyde is then reduced to ethyl alcohol by NAD.

APPLICATIONS

Involuntary Drunkenness

A man was arrested in Virginia for drunk driving. He offered a most unusual defense: involuntary drunkenness—due to yeast fermenting in food in his stomach, thereby producing alcohol that was absorbed into his bloodstream. The judge didn't think much of his plea, and appeals are still pending. However, there are documented cases in Japan and the United States of people with stomach infections of peculiar stains of the yeast *Candida albicans,* who were unable to remain sober. *Candida* is commonly found throughout the digestive tract, where it ordinarily causes no problems. However, those odd strains convert any meal or drink containing carbohydrates into alcohol, although usually not enough to raise blood alcohol levels to the legal limits of intoxication (unless a big meal was eaten). Fortunately the infection can be cured, and the victim returned to sobriety. However, until this occurs, it would seem prudent for people with this problem to refrain from driving.

The yeasts that leaven bread also produce alcohol. Why, then, don't you get drunk from eating your dinner rolls? The reason is that alcohol evaporates in the oven during baking.

FIGURE 5.14 The mannitol-fermentation test distinguishes the pathogenic *Staphylococcus aureus* (left) from the nonpathogenic *Staphylococcus epidermidis* (right). *S. aureus* ferments mannitol, producing acid that turns an indicator in the medium a yellowish color.

which causes the medium to turn yellow. The non-pathogenic bacterium *Staphylococcus epidermidis* fails to ferment mannitol and does not change the color of the medium.

AEROBIC METABOLISM: RESPIRATION

As we have noted, most organisms obtain some energy by metabolizing glucose to pyruvate by glycolysis. Among microorganisms, both anaerobes and aerobes carry out these reactions. **Anaerobes** are organisms that do not use oxygen; they include some that are killed by exposure to oxygen. **Aerobes** are organisms that *do* use oxygen; they include some that must have oxygen. In addition, a significant number of species of microorganisms are facultative anaerobes (Chapter 6) that use oxygen if it is available but can function without it. Though aerobic organisms obtain some of their energy from glycolysis, they use glycolysis chiefly as a prelude to a much more productive process, one that allows them to obtain far more of the energy potentially available in glucose. This process is **aerobic respiration** via the Krebs cycle and oxidative phosphorylation.

APPLICATIONS

Why Doesn't All Swiss Cheese Taste Alike?

Fermentation byproducts differ greatly among strains of microorganisms, and such differences account for the distinctive flavors and textures of cheeses, wines, and other fermented foods. For example, some organisms, such as *Propionibacterium* and *Veillonella*, metabolize lactate (lactic acid) instead of pyruvate. Very little energy is available from lactate, so cultures of such organisms grow slowly. Lactate fermentation of milk traps large carbon dioxide bubbles that make the holes in Swiss cheese. It also releases propionic acid, which gives the cheese its characteristic flavor.

A mutation, or change in the DNA, of a strain being used in a commercial process can occur at any time and might alter the byproducts of fermentation for better or worse. To guard against a change for the worse, many commercial users of special strains of organisms preserve dormant cultures in a vault at the American Type Culture Collection (Chapter 9). Then if a mutation occurs, organisms with the original characteristics can be retrieved from the dormant culture.

The Krebs Cycle

The **Krebs cycle,** named for Hans Krebs, who identified its steps in the late 1930s, metabolizes 2-carbon units called acetyl groups to CO_2 and H_2O. It also is called the **tricarboxylic acid (TCA) cycle** because some molecules in the cycle have three carboxyl (COOH) groups, or the **citric acid cycle** because citric acid is an important intermediate.

Before pyruvate, the product of glycolysis, can enter the Krebs cycle, it must first be converted to acetyl-CoA. This complex reaction involves the transfer of electrons to NAD, removal of one molecule of CO_2, and addition of coenzyme A (Figure 5.15).

The Krebs cycle is a sequence of reactions in which acetyl groups are oxidized to carbon dioxide. Hydrogen atoms are also removed, and their electrons are transferred to coenzymes that serve as electron carriers (Figure 5.16). (The hydrogens, as we will see, are eventually combined with oxygen to form water.) Each reaction in the Krebs cycle is controlled by a specific enzyme, and the molecules are passed from one enzyme to the next as they go through the cycle. The reactions constitute a cycle because oxaloacetic acid (oxaloacetate), a first reactant, is regenerated at the end of the cycle. As one acetyl group is metabolized, oxaloacetate combines with another and goes through the cycle again.

Certain events in the Krebs cycle are of special significance: (1) oxidation of carbon, (2) removal of electrons to coenzymes, and (3) substrate-level energy capture. As each acetyl group goes through the cycle, two molecules of carbon dioxide arise from the complete oxidation of its two carbons. Four pairs of electrons are transferred to coenzymes: three pairs to NAD and one pair to FAD. Much energy is derived

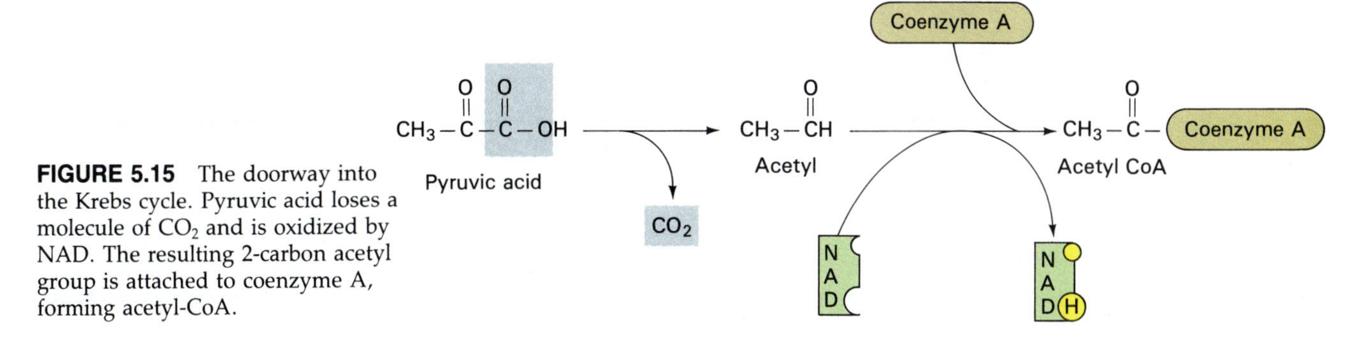

FIGURE 5.15 The doorway into the Krebs cycle. Pyruvic acid loses a molecule of CO_2 and is oxidized by NAD. The resulting 2-carbon acetyl group is attached to coenzyme A, forming acetyl-CoA.

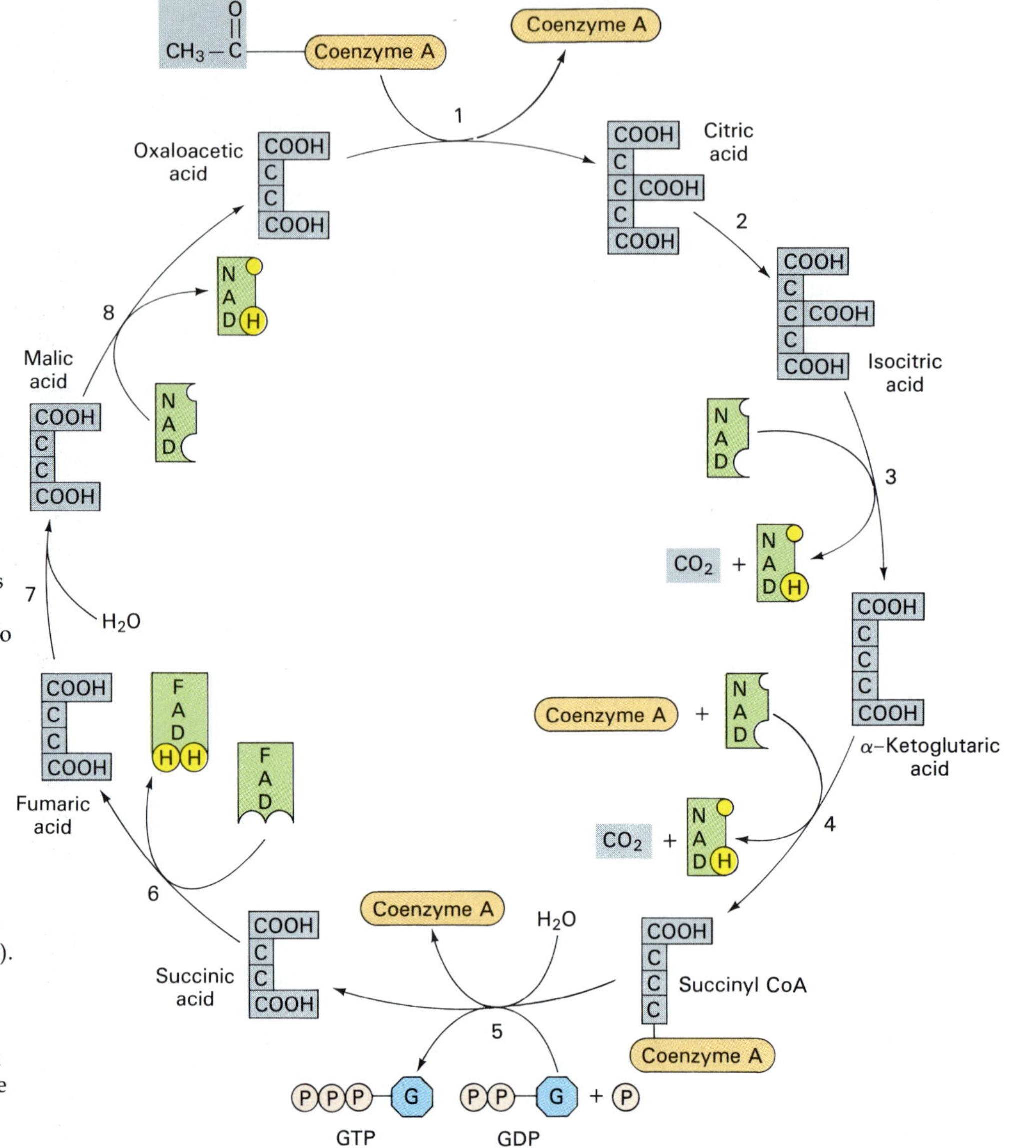

FIGURE 5.16 The reactions of the Krebs cycle. The intermediates are simplified to show only the number of carbon atoms and carboxyl groups for each. A 2-carbon acetyl group enters the cycle as acetyl-CoA in step 1, and two carbon atoms leave the cycle as molecules of CO_2 in steps 3 and 4. Energy is captured in guanosine triphosphate (GTP) in step 5 and eventually transferred to adenosine triphosphate (ATP). In addition, electrons are removed by coenzymes in steps 3, 4, 6, and 8. More energy will be extracted from these electrons when they are subsequently fed into the electron transport chain.

BIOTECHNOLOGY

Putting Microbes To Work

What can you get from *Klebsiella pneumoniae* besides pneumonia? Acrylic plastics, clothing, pharmaceuticals, paints—all products of the parent compound 3-hydroxypropionaldehyde, which *K. pneumoniae* produces by fermentation of glycerol. Glycerol is a common byproduct of processing animal fats and vegetable oils, such as those from soybeans. Thus our nation's surplus farm commodities could eventually help replace costly imported petrochemicals, thanks to microbial fermentations.

Scientists are also exploring the possibility of modifying other microbes such as *Saccharomyces cerevisiae* (baker's yeast) and *S. carlsbergensis* (brewer's yeast) to make them more useful to us. They hope to obtain from yeast a fast-acting enzyme that, when given intravenously, will dissolve blood clots and thus lessen the damage caused by heart attacks and strokes. The biochemical versatility of yeast may also help to clear our air of petrochemical pollutants. The yeast *Pachysolen tannaphilus* converts xylose, a sugar found in woody parts such as corn stalks, directly into ethyl alcohol. The U.S. Department of Agriculture estimates that such yeasts could produce four billion gallons of clean-burning fuel alcohol from agricultural wastes per year.

from these electrons in the next phase of respiration, as we will soon see. Finally, some energy is captured in a high-energy bond in guanosine triphosphate (GTP). This reaction takes place at the substrate level. That is, it occurs directly in the course of a reaction of the Krebs cycle. Energy in GTP is easily transferred to ATP. Note that because each glucose molecule produces two molecules of acetyl-CoA, the quantities of the products just given must be doubled to represent the yield from metabolism of a single glucose molecule.

Electron Transport and Oxidative Phosphorylation

Reduced NAD and reduced FAD from the Krebs cycle and some other metabolic pathways feed their electrons into a series of coenzymes called an **electron transport chain** (Figure 5.17). In **electron transport,** pairs of electrons are relayed from one coenzyme to the next until they reach oxygen, the final electron acceptor. The electrons, reunited with the protons from which they were separated in the course of the Krebs cycle, combine with oxygen to form water. Oxygen must be available to serve as the final electron acceptor for electron transport to occur, and this process must take place in order for the Krebs cycle to continue to operate. Unless electrons are continuously transferred from reduced NAD and FAD to oxygen via the electron transport chain, these enzymes cannot accept more electrons from the Krebs cycle, and the entire process will grind to a halt. From the metabolism of a single glucose molecule 12 pairs of electrons are transported by NAD and FAD to other electron carriers in the biological oxidation pathway. Eight pairs come from the Krebs cycle, 2 pairs from glycolysis, and 2 pairs from the conversion of pyruvate to acetyl-CoA.

At three sites in the electron transport chain, some of the energy of the electrons is captured in high-

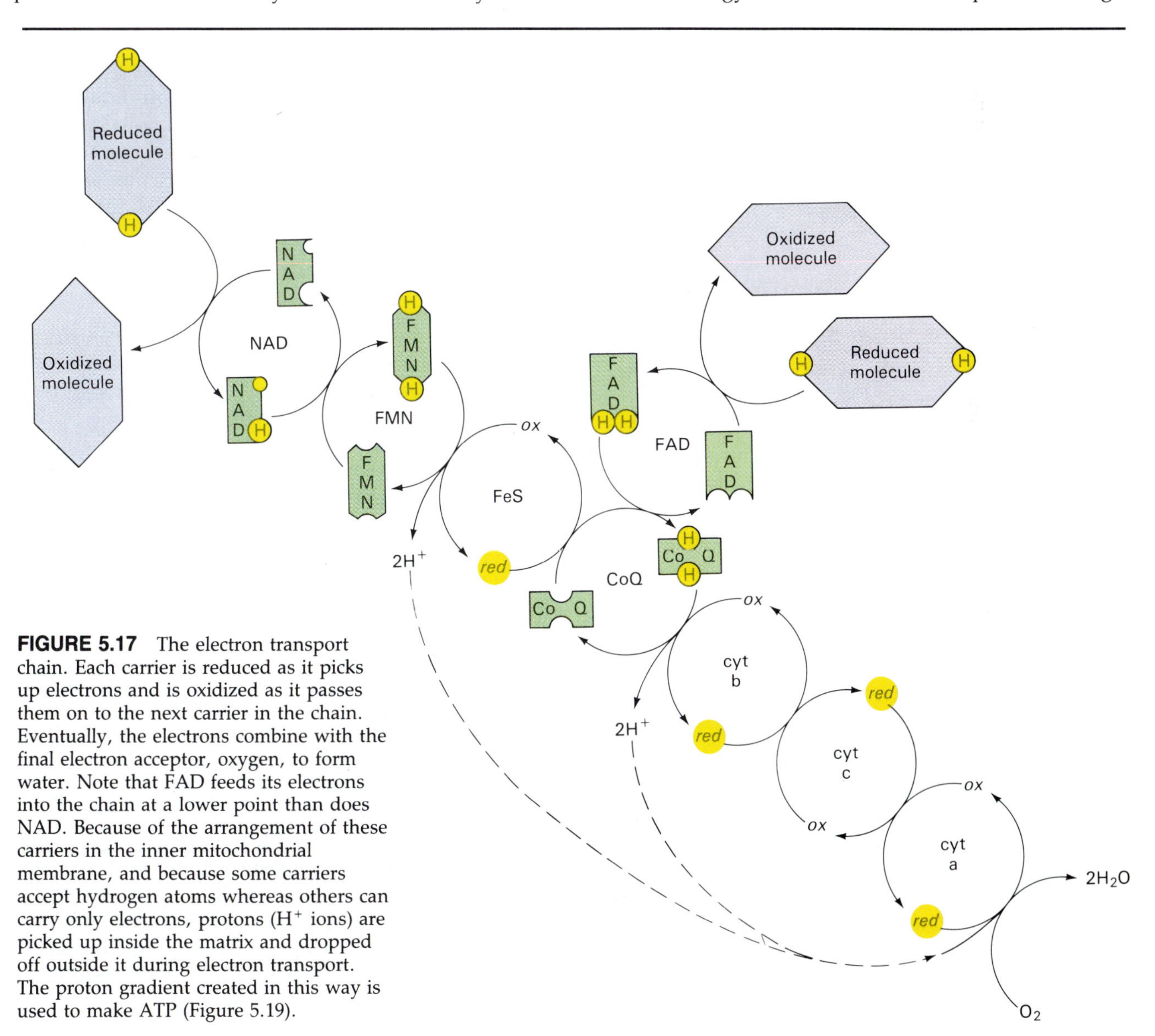

FIGURE 5.17 The electron transport chain. Each carrier is reduced as it picks up electrons and is oxidized as it passes them on to the next carrier in the chain. Eventually, the electrons combine with the final electron acceptor, oxygen, to form water. Note that FAD feeds its electrons into the chain at a lower point than does NAD. Because of the arrangement of these carriers in the inner mitochondrial membrane, and because some carriers accept hydrogen atoms whereas others can carry only electrons, protons (H^+ ions) are picked up inside the matrix and dropped off outside it during electron transport. The proton gradient created in this way is used to make ATP (Figure 5.19).

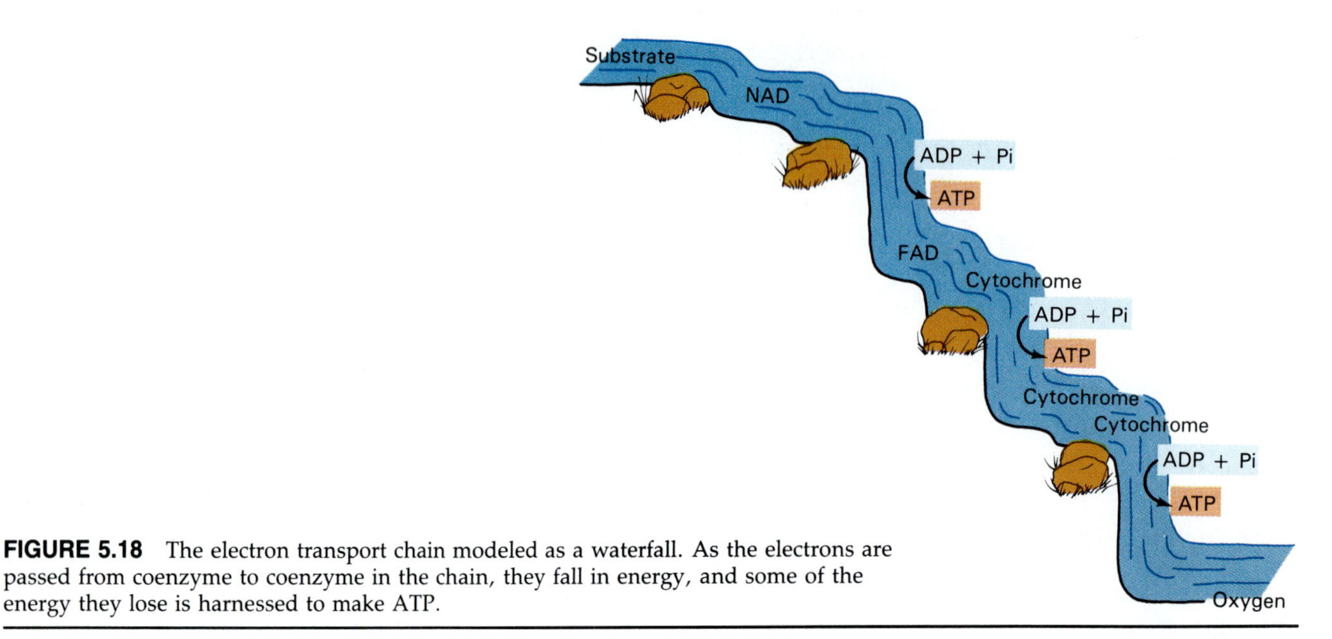

FIGURE 5.18 The electron transport chain modeled as a waterfall. As the electrons are passed from coenzyme to coenzyme in the chain, they fall in energy, and some of the energy they lose is harnessed to make ATP.

energy bonds as P_i combines with ADP to form ATP in the process of **oxidative phosphorylation.** Electron transport and oxidative phosphorylation can be likened to a series of waterfalls where the water makes many small drops and three larger ones (Figure 5.18). In most electron transfers (the small drops) only small amounts of energy are released. At three points (the larger drops) more energy is released, and some is captured in a high-energy bond.

In our waterfall analogy, we can think of water entering the falls at two sites, one higher up the mountain than the other. Water from the higher site falls farther than water entering lower down the mountain. Electrons entering the electron transport chain at NAD start at the top, and their descent releases enough energy to make three ATPs. Electrons entering at FAD start partway down the chain and contribute only enough energy to make two ATPs. Thus, during aerobic metabolism of a glucose molecule, the 10 pairs of electrons from NAD produce 30 ATPs, and 2 pairs from FAD produce 4 ATPs, for a total of 34 ATPs. Four substrate-level ATPs make a total yield of 38 ATPs per glucose molecule.

Chemiosmosis

Energy capture, though not fully understood, probably occurs through a process known as **chemiosmosis** (kem″e-os-mo′sis). Formulation of the chemiosmotic theory was of such significance that it earned British biochemist Peter Mitchell a Nobel Prize in 1978. In chemiosmosis (Figure 5.19) protons (H^+) are pumped across the inner mitochondrial membrane in eukaryotic cells during electron transport. The proton concentration becomes much higher in the space between the inner and outer mitochondrial membranes than in the matrix of the mitochondrion. Protons move from their region of higher concentration through channels in the otherwise impermeable membrane to their region of lower concentration in the matrix. As they cross the membrane, energy is captured in high-energy bonds in ATP, much as the energy of falling water can be captured by a water wheel or turbine. Energy captured in prokaryotic cells, which lack mitochondria, is thought to occur in the cell membrane in a process similar to chemiosmosis in mitochondrial membranes.

Significance of Energy Capture

In glycolysis and fermentation, as we noted earlier, a net of 2 ATPs is usually produced for every glucose molecule metabolized anaerobically. When glycolysis is followed by respiration, in addition each glucose molecule produces 2 ATPs in the Krebs cycle at the substrate level and 34 ATPs by oxidative phosphorylation. Thus, a glucose molecule metabolized aerobically yields 38 ATPs, but when metabolized anaerobically yields only 2 ATPs (Table 5.3). Hence, 19 times as much energy is captured in aerobic metabolism as in anaerobic metabolism! Thus, aerobic microorganisms in environments with ample oxygen generally grow more rapidly than anaerobes, but they will die if oxygen is depleted.

METABOLISM OF FATS AND PROTEINS

For most organisms, including microorganisms, glucose is a major source of energy. However, some microorganism can be found to degrade nearly any organic substance for energy. This attribute of microorganisms accounts for their ability to degrade dead remains and wastes of all organisms.

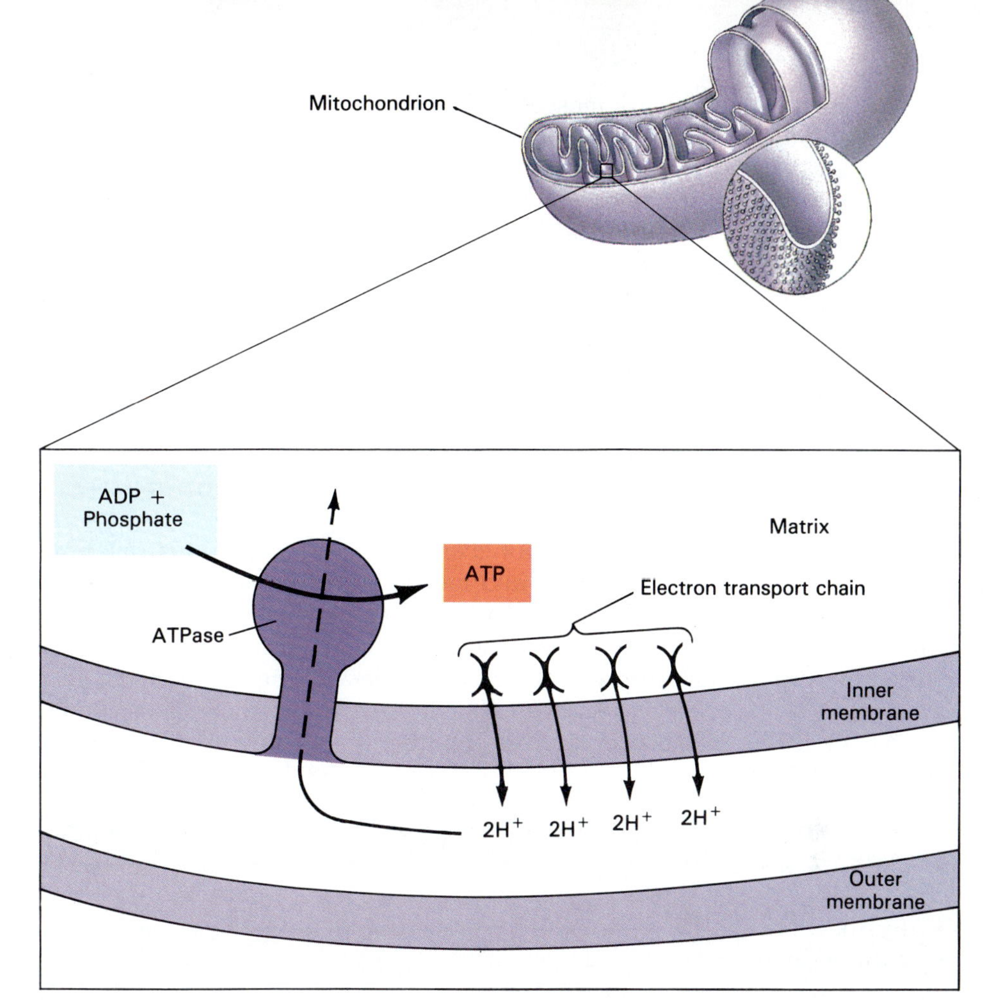

FIGURE 5.19 Energy capture by chemiosmosis. Protons "pumped" out of the matrix during electron transport pass back in through channels in the inner membrane. The channels are associated with enzymes that phosphorylate ADP to make ATP. By some mechanism not yet fully understood, the flow of protons provides the energy to drive this reaction.

TABLE 5.3 Energy captured in ATP molecules from a glucose molecule by anaerobic and aerobic metabolism

	Number of ATP Molecules	
Prokaryotic Metabolic Process	**Anaerobic Conditions**	**Aerobic Conditions**
Glycolysis		
Substrate level	4	4
Hydrogen to NAD	0	6
Pyruvate to acetyl-CoA		
Hydrogen to NAD	0	6
Krebs cycle		
Substrate level	0	2
Hydrogen to NAD	0	18
Hydrogen to FAD	0	4
Less energy for phosphorylation	−2	−2
Total	2	38

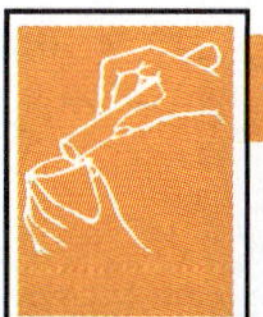

TRY IT

Fueling the Race To Reproduce

Two students, working as a team in the laboratory, observed the growth of yeast in a sugar solution on microscope slides. One student focused her microscope near the edge of the coverslip, where oxygen levels were sufficient for aerobic respiration. The other student focused his microscope on yeast growing under the center of the coverslip. They kept the same fields in focus for the duration of the laboratory, counting the number of yeast cells in the field every 20 to 30 minutes.

Can you predict what results they found? What would explain these results? With your instructor's consent, you could easily try this experiment while completing your assigned laboratory exercises and then share your data with the rest of the class. Do you think this experiment would give the same results with all types of organisms? Why?

Most microorganisms, like most animals, can obtain energy from lipids. The following examples give a general idea of how such processes occur. Fats are hydrolyzed to glycerol and three fatty acids. The glycerol is metabolized by glycolysis. The fatty acids, which usually have an even number of carbons (16, 18, or 20), are broken down into 2-carbon pieces by a metabolic pathway called **beta oxidation.** In this process the beta carbon, the second carbon from the carboxyl group of a fatty acid, is oxidized, and the carbon chain is broken. Coenzyme A is attached to each piece to form acetyl-CoA, which is then metabolized through the Krebs cycle (Figure 5.20) to obtain additional energy.

Proteins also can be metabolized for energy (Figure 5.21). They are first hydrolyzed into individual amino acids by proteolytic (protein-digesting) enzymes. Then the amino acids are **deaminated;** that is, their amino groups are removed. The resulting deaminated molecules enter glycolysis, fermentation, or the Krebs cycle. The metabolism of all major nutrients (fats, carbohydrates, and proteins) for energy is summarized in Figure 5.22.

OTHER METABOLIC PROCESSES

Having considered energy capture in chemoheterotrophs, the most common organisms by nutritional type, we will now briefly consider energy capture in photoautotrophs, photoheterotrophs, and chemoautotrophs.

Photoautotrophy

Organisms called photoautotrophs carry out **photosynthesis,** the capture of energy from light and the use of this energy to manufacture carbohydrates from carbon dioxide. Photosynthesis occurs in green and purple bacteria, in cyanobacteria, in algae, and in higher plants. Photosynthetic bacteria, which probably evolved early in the evolution of living organ-

APPLICATIONS

Microbial Clean-up

A few species of bacteria, such as some members of the genus *Pseudomonas*, can use crude oil for energy. They can grow in seawater with only oil, potassium phosphate, and urea (a nitrogen source) as nutrients. These organisms can clean up oil spills in the ocean acting as "bioremediators." They have also proven useful in degrading oil that remains in the water carried by tankers as ballast after unloading their oil. Then the water pumped from the tankers into the sea in preparation for a new cargo of oil does not pollute. A detergentlike substance has been isolated recently from these organisms. When the detergent is added to a quantity of oil sludge, it converts 90 percent of the sludge into usable petroleum in about 4 days, thereby reducing waste and providing a convenient means of cleaning oil-fouled tanks.

(a)

(b)

(c)

The 1989 Exxon *Valdez* oil spill left great quantities of pooled oil on sites in the Gulf of Alaska such as on Green Island (a). Bioremediation in 1989 (b) by application of nutrients (nitrogen and phosphorus) to the shoreline accelerated the bacterial biodegradation of the oil into carbon dioxide and water. In 1991 the area was surveyed (c) and found to be mostly cleared of oil, with no further treatment recommended.

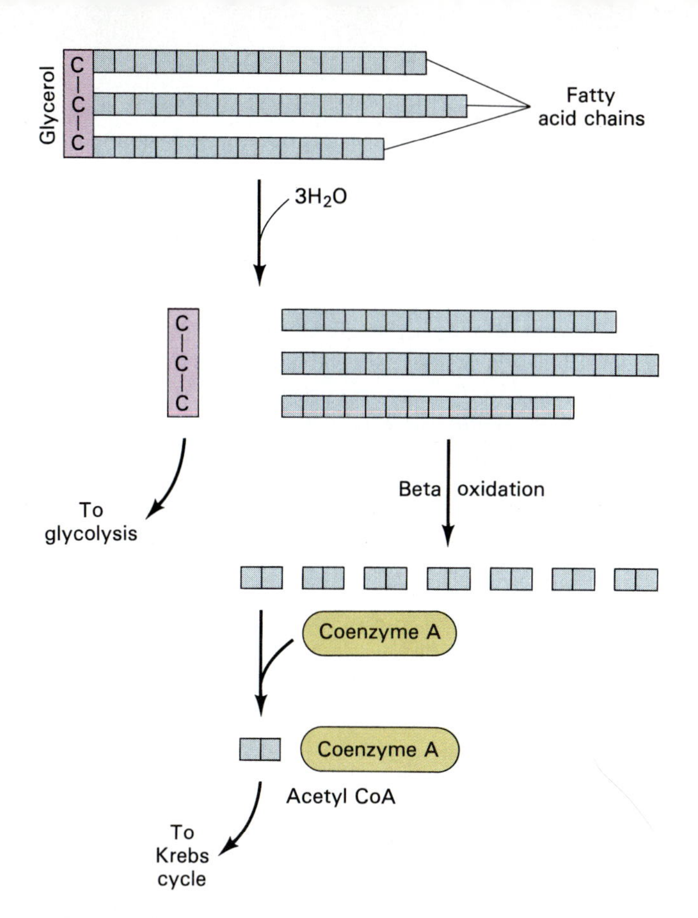

FIGURE 5.20 The catabolism of fats. Triglycerides are hydrolized into glycerol and fatty acids. The glycerol is broken down via glycolysis. The fatty acids are broken down into 2-carbon units and fed into the Krebs cycle, where they are metabolized to produce additional energy.

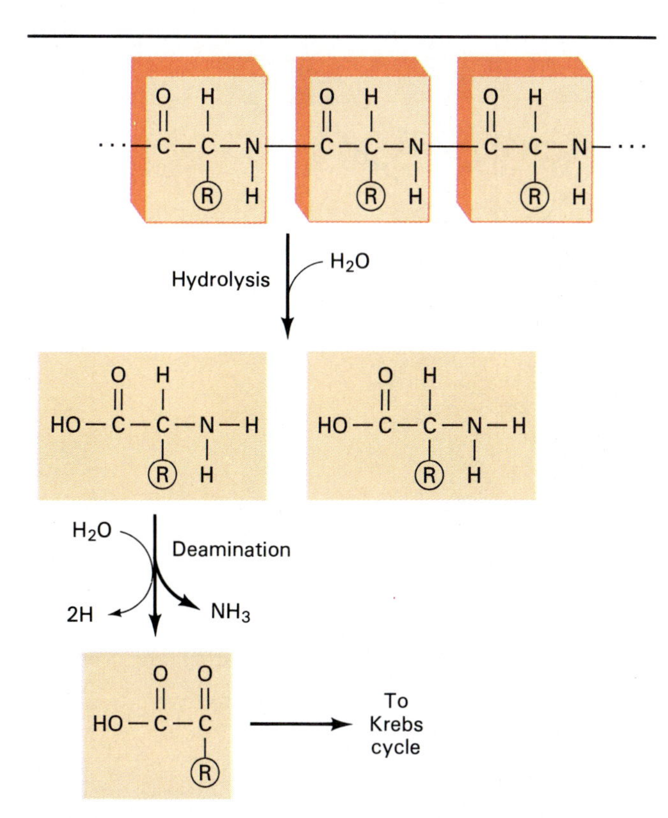

FIGURE 5.21 The catabolism of proteins. Polypeptides are hydrolized to amino acids. The amino acids are deaminated, and the resulting molecules enter pathways leading to the Krebs cycle.

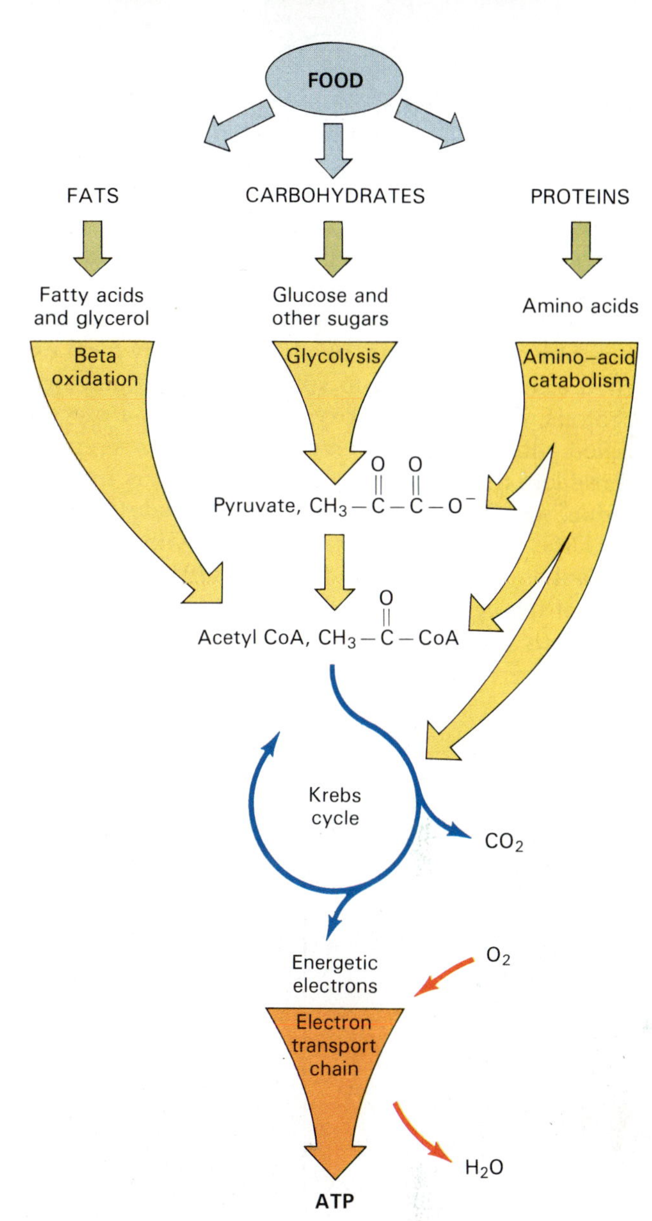

FIGURE 5.22 Metabolism of the major classes of biomolecules: a summary.

isms, perform their own distinctive version of photosynthesis. However, algae and green plants make much more of the world's carbohydrate supply, so we will consider the process in those organisms first and then see how it is modified in bacteria.

In green plants and algae, photosynthesis occurs in two parts—the "photo" part, or the light reactions, in which light energy is converted to chemical energy, and the "synthesis" part, or the dark reactions, in which chemical energy is used to make organic molecules. Each part involves a series of steps.

In the **light reactions,** light strikes the green pigment chlorophyll in thylakoids of chloroplasts (described in Chapter 4). ∞ (p. 94) Electrons in the chlorophyll become excited, that is, raised to a higher energy level. These electrons participate in generating ATP in cyclic photophosphorylation and in noncyclic photoreduction (Figure 5.23). In **cyclic photophos-**

phorylation excited electrons from chlorophyll are passed down an electron transport chain. As they are transferred, energy is captured in ATP by chemiosmosis, as described previously in connection with oxidative phosphorylation. When the electrons return to the chlorophyll, they can be excited over and over again, so the process is said to be *cyclic*.

In noncyclic photoreduction energy also is captured by chemiosmosis. In addition, energy from excited electrons is used to split water molecules into protons, electrons, and oxygen molecules, a process called **photolysis** (fo-tol'eh-sis). The electrons replace those lost from chlorophyll, which are thus freed to reduce nicotinamide dinucleotide phosphate (NADP). NADP is simply the coenzyme NAD with an added phosphate group. ATP and reduced NADP, the products of the light reaction, and atmospheric CO_2 subsequently participate in the dark reaction.

The **dark reactions,** or carbon fixation, occur in the stroma of chloroplasts. Carbon dioxide is *reduced* by electrons from NADP to form various carbohydrate molecules, chiefly glucose. Energy from ATP is required in this synthetic process, but much of the energy is recaptured in the reduced molecules (Figure 5.24).

Photosynthesis in bacteria differs from that in green plants in ways related to the evolution of living organisms. The first photosynthetic organisms probably were purple and green bacteria, which evolved in an atmosphere containing much hydrogen but no oxygen. They differ from green plants in three ways:

1. Their chlorophyll absorbs slightly longer wavelengths of light than does plant chlorophyll.
2. They use hydrogen compounds other than water for reducing carbon dioxide. Electrons from their pigments

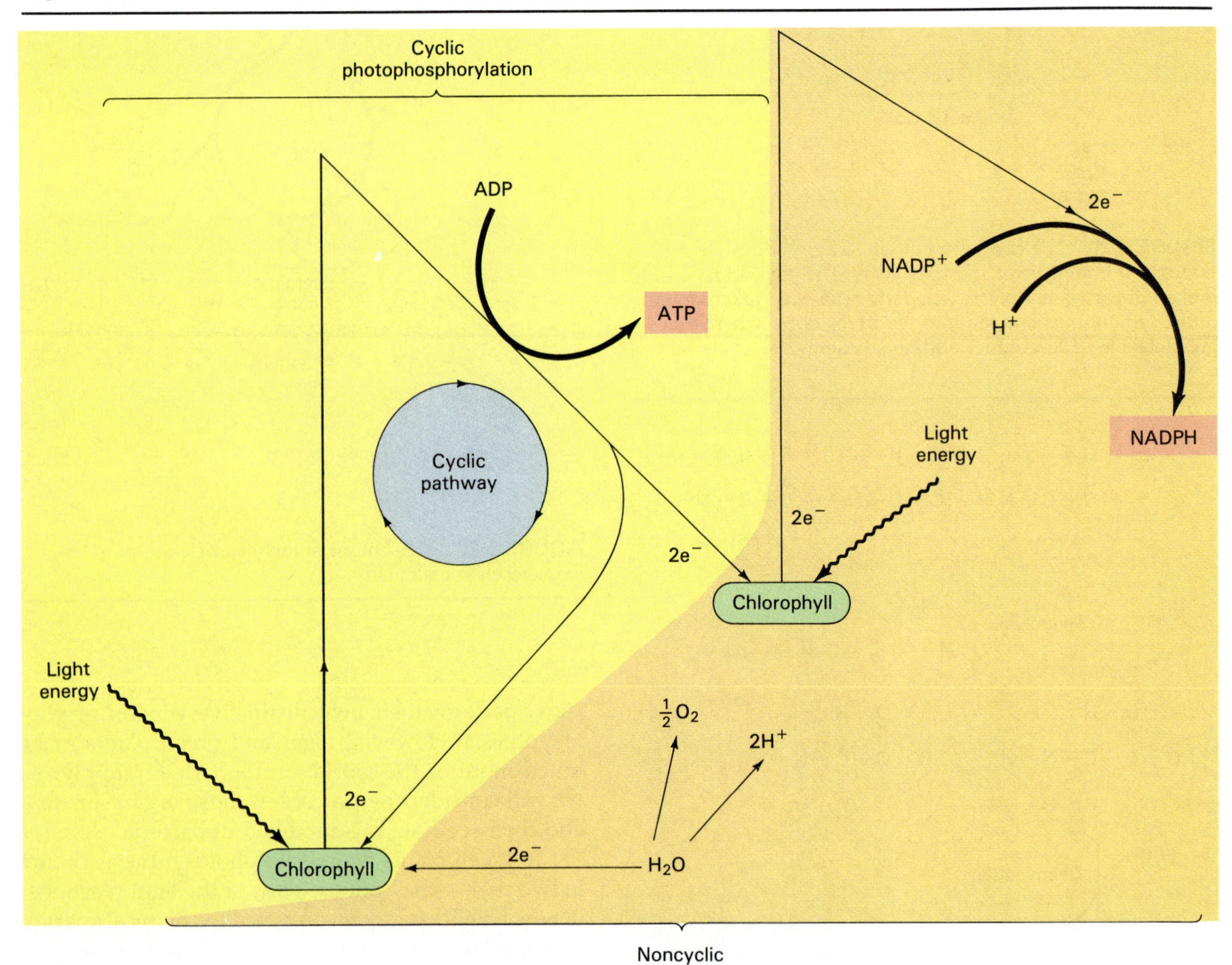

FIGURE 5.23 The light reactions of photosynthesis as performed by cyanobacteria, algae, and higher plants. Electrons in chlorophyll receive a boost in energy from light, and their extra energy is used to make ATP. In the pathway of cyclic photophosphorylation, the electrons return to chlorophyll and thus can be used over and over again. In noncyclic photoreduction, the electrons receive a second boost that gives them enough energy to reduce NADP. The electrons are replaced by the splitting of water.

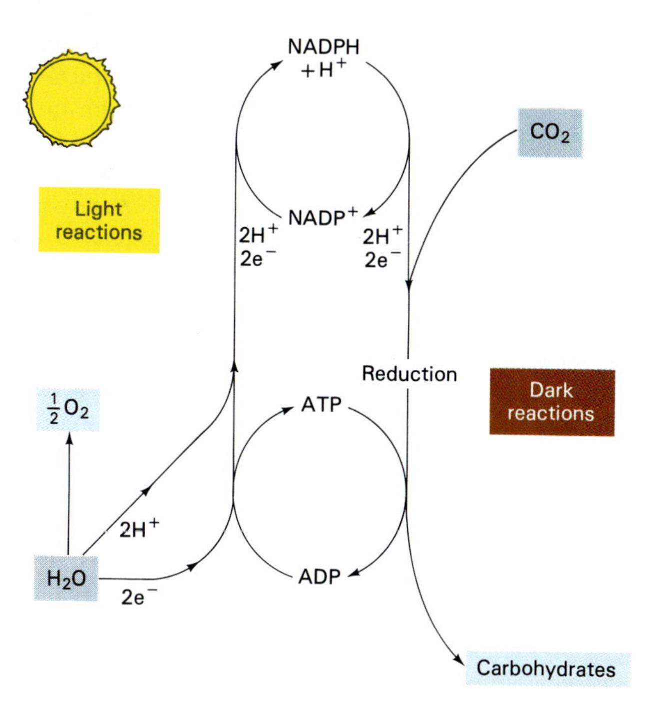

FIGURE 5.24 The relation between the light and dark reactions. In the dark reactions, ATP and NADPH from the light reactions are used to reduce carbon dioxide, forming carbohydrates such as glucose. The dark reactions do not *require* darkness; they are so named because they *can* take place in the dark, so long as the products of the light reactions are available.

reach an energy level high enough to split hydrogen sulfide (H_2S) but not high enough to split water. (Some produce elemental sulfur as a byproduct; a few produce strong sulfuric acid.)

3. They are usually strict anaerobes and can live only in the absence of oxygen. They do not release oxygen as a product of photosynthesis as green plants do.

Characteristics of the groups of bacteria that carry out this primitive form of photosynthesis are summarized in Table 5.4.

The cyanobacteria also are photosynthetic, but they probably evolved after the purple and green bacteria. Though prokaryotic, the cyanobacteria release oxygen during photosynthesis as do green plants. In fact, they are probably responsible for the addition of oxygen to the primitive atmosphere.

TABLE 5.4 Characteristics of photosynthetic bacteria

Group	Family and Representative Genus	Pigments
Green sulfur bacteria	Chlorobiaceae *Chlorobium*	Bacterial chlorophyll
Purple sulfur bacteria	Chromaticeae *Chromatium*	Bacterial chlorophyll and red and purple carotenoid pigments

Photoheterotrophy

A small group of bacteria called the *photoheterotrophs* can capture energy from light but must have an organic substance such as methanol (CH_3OH) in addition to carbon dioxide. These organisms include the nonsulfur, colored (purple or green) bacteria.

Chemoautotrophy

Bacteria called chemoautotrophs, or **chemolithotrophs,** though unable to carry out photosynthesis, can oxidize inorganic substances for energy. With this energy and carbon dioxide as a carbon source, they are able to synthesize a great variety of substances, including carbohydrates, fats, proteins, nucleic acids, and substances that are required as vitamins by many organisms.

The ability to oxidize, and therefore extract energy from, inorganic substances is probably the most outstanding characteristic of these bacteria, but they have some other noteworthy attributes. The nitrifying bacteria are especially important because they increase the quantity of usable nitrogen compounds available to plants and replace nitrogen that plants remove from the soil. *Thiobacillus* and some other sulfur bacteria produce sulfuric acid by oxidizing elemental sulfur or hydrogen sulfide. Acidity greater than pH 1 has been produced by sulfur bacteria. Sulfur is sometimes added to alkaline soil to acidify it, a practice that works because of the numerous thiobacilli present in most soils. Finally, some chemolithotrophic archaebacteria have been found near volcanic vents in the ocean floor, where they grow at extremely hot temperatures and sometimes under very acid conditions. Characteristics of chemolithotrophs are summarized in Table 5.5.

TABLE 5.5 Characteristics of chemolithotrophic bacteria

Group and Representative Genus/Genera	Source of Energy	Products after Oxidizing Reaction
Nitrifying bacteria		
Nitrobacter	HNO_2	HNO_3
Nitrosomonas	NH_3	$HNO_2 + H_2O$
Nonphotosynthetic sulfur bacteria		
Thiothrix	H_2S	$H_2O + 2S$
Thiobacillus	S	H_2SO_4
Iron bacteria		
Siderocapsa	Fe^{2+}	$Fe^{3+} + OH^-$
Hydrogen bacteria		
Hydrogenomonas	H_2	H_2O

Unusual Energy Sources

A small group of bacteria obtain energy from inorganic substances such as hydrogen gas and use inorganic electron acceptors such as nitrate (NO_3^-), sulfate (SO_4^{2-}), or carbon dioxide (CO_2) instead of oxygen. They include denitrifying bacteria, methane bacteria, and members of the genus *Desulfovibrio.* The denitrifying bacteria release nitrogen gas and ammonia. The methane bacteria make methane gas (CH_4) and are found in anaerobic sewage digesters. The *Desulfovibrios* produce hydrogen sulfide (H_2S) and are found in polluted streams, estuaries, and poorly aerated soils.

USES OF ENERGY

Microorganisms use energy for such processes as biosynthesis, membrane transport, movement, and growth. Here we will summarize some biosynthetic activities and some mechanisms for membrane transport and movement. We will consider growth in Chapter 6.

Biosynthetic Activities

Microorganisms share many biochemical characteristics with other organisms. All require the same building blocks to make proteins and nucleic acids. Many of these building blocks (amino acids, purines, pyrimidines, and ribose) can be derived from intermediate products of energy-yielding pathways (Figure 5.25). When the energy-yielding pathways were first discovered, they were thought to be purely catabolic. Now that many of their intermediates are known to be involved in biosynthesis, they are more properly called **amphibolic** (am-fe-bol'ik) **pathways** (*amphi-*, either) because they can yield either energy or building blocks for synthetic reactions.

Some biosynthetic pathways are quite complex. For example, synthesis of amino acids in organisms that can make them often requires many reactions, with an enzyme for each reaction. Tyrosine synthesis requires no less than 10 enzymes, and tryptophan synthesis needs at least 13. The synthetic pathways for making purines and pyrimidines also are complex. The absence of a single enzyme in a synthetic pathway can prevent synthesis of a substance. If the substance is essential to an organism, the inability to synthesize it means the substance must be present in the medium. Missing enzymes thus increase the nutritional needs of organisms.

Microorganisms of many different types also synthesize a variety of carbohydrates and lipids. The rate at which they are synthesized also varies and depends on the availability and activity of enzymes. Some organisms, such as the aerobe *Acetobacter,* synthesize cellulose, which is ordinarily found in plants. As strands of cellulose reach the cell surface, they form

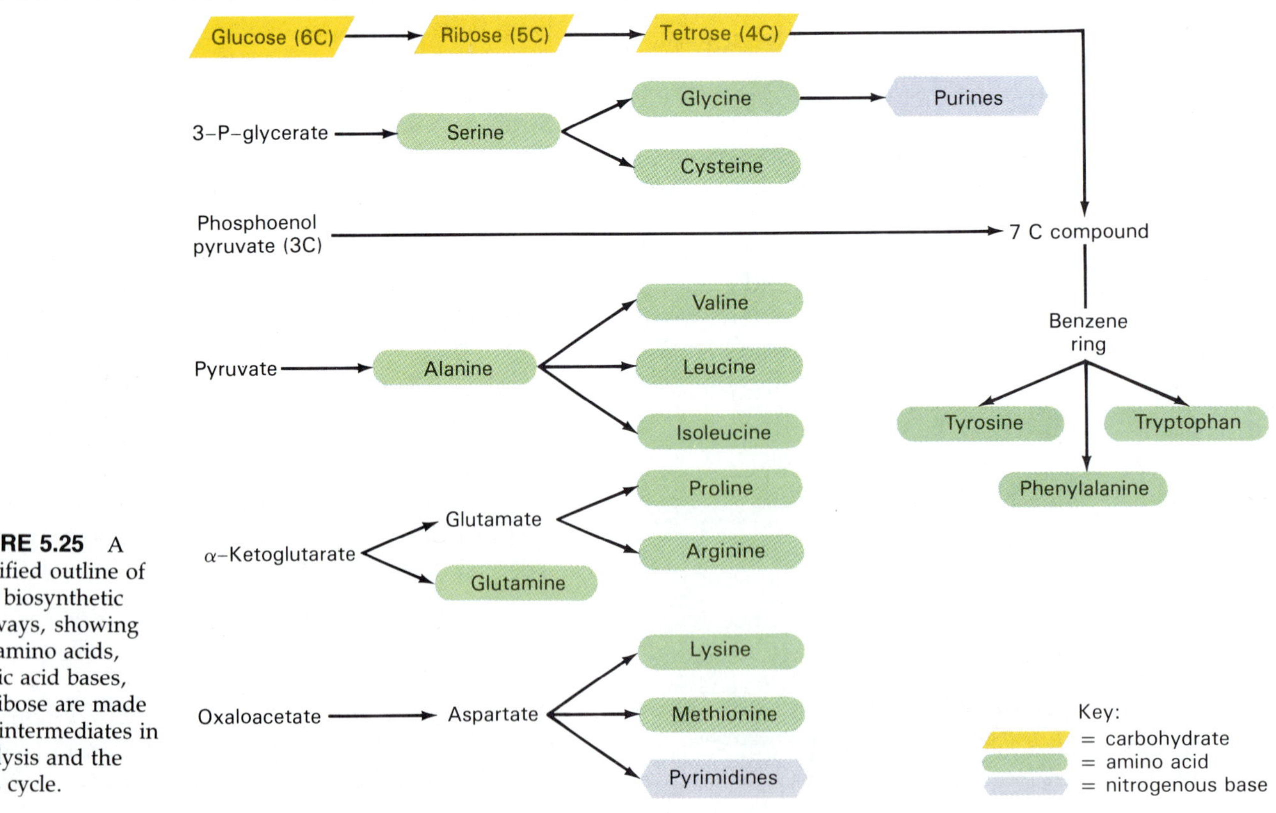

FIGURE 5.25 A simplified outline of some biosynthetic pathways, showing how amino acids, nucleic acid bases, and ribose are made from intermediates in glycolysis and the Krebs cycle.

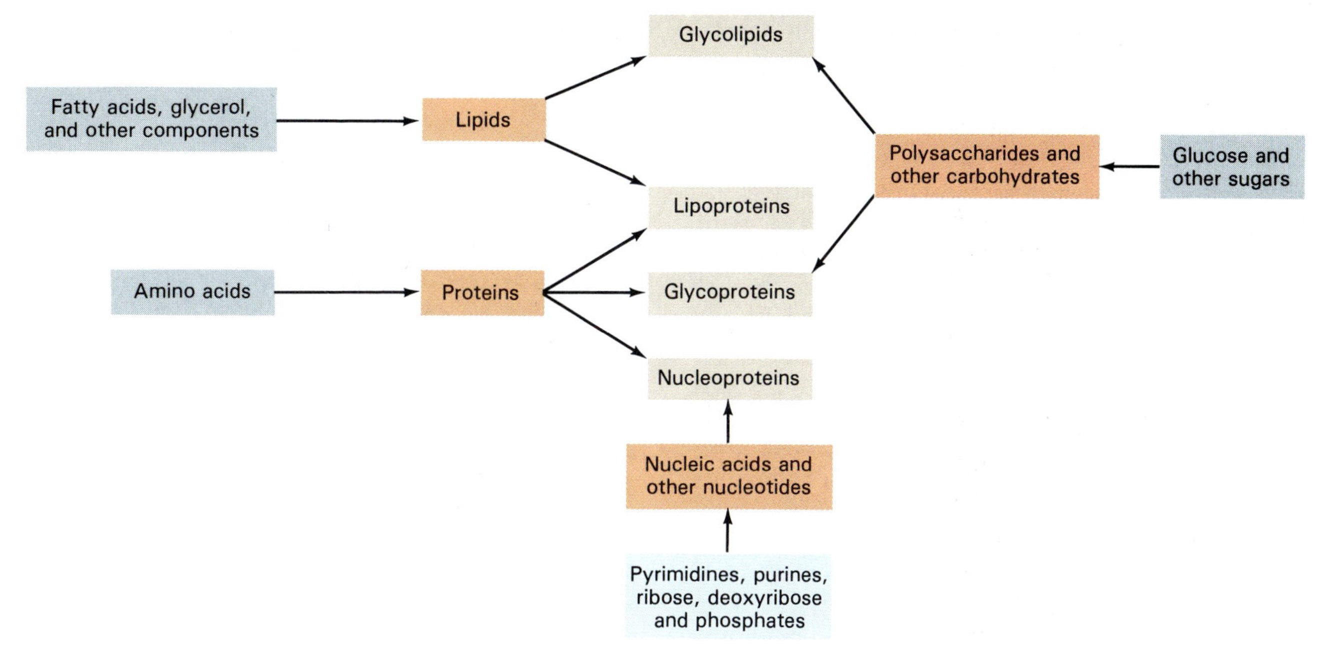

FIGURE 5.26 The formation of complex biomolecules from simpler components.

a mat that traps carbon dioxide bubbles and keeps the cell afloat. Because these organisms must have oxygen, the mat contributes to their survival by keeping them near the surface, where oxygen is plentiful.

Many bacteria synthesize peptidoglycan, lipopolysaccharide, and other polymers associated with cell walls (Chapter 4). (p. 78) Some bacteria form capsules, especially in media that contain serum or large amounts of sugar. Capsules usually consist of polymers of one or more saccharides. However, in *Bacillus anthracis*, the organism that causes anthrax, the capsule is a polypeptide of glutamic acid. Much of what is known about the synthesis of capsule polymers has been learned in investigations of antibiotic actions and immunological phenomena. The biosynthetic processes in microorganisms are summarized in Figure 5.26.

Membrane Transport and Movement

In addition to using energy for biosynthetic processes, microorganisms also use energy for transporting substances across membranes and for their own movement. These energy uses are as important to the survival of the organisms as their biosynthetic activities.

Membrane Transport

Bacteria use energy to move substances across their cell membranes against concentration gradients. For example, they can transport a sugar or an amino acid from a region of low concentration outside the cell to a region of higher concentration inside the cell. This means they accumulate nutrients within cells in concentrations a hundred to a thousand times the concentration outside the cell. They also concentrate certain inorganic ions by the same means.

Two mechanisms exist in bacteria for concentrating substances inside cells, and both require energy. The mechanism of active transport (Chapter 4) involves enzymes called **permeases** (per'me-ās-es) that permeate, or extend through, the cell membrane. (p. 100) When a permease is oriented toward the outside of the membrane, it has a high affinity for the particular molecule or ion it transports. As energy is released from ATP in the cell membrane, the permease changes position and carries its attached molecule to the inside of the cell. As the change in position occurs, the affinity of the permease for the substance it carries is lowered so that the substance is released into the cell.

Another mechanism, called the **phosphotransferase system,** uses energy from phosphoenolpyruvate, a molecule with a high-energy bond. When phosphoenolpyruvate is present in the cytoplasm, it can provide energy and a phosphate group to a carrier protein in the membrane. Then the carrier protein transfers the phosphate to a sugar molecule and at the same time moves the sugar molecule across the membrane. A phosphorylated sugar molecule is thus transported inside the cell and is prepared to undergo metabolism. Both the addition of phosphate and the transfer of the sugar require specific enzymes.

Movement

Most motile bacteria move by means of flagella, but some move by gliding, creeping, or thrashing. Flagellated bacteria move by rotating their flagella (Chapter 4). ∞ (p. 87) The mechanism by which flagella rotate, though not well understood, appears to involve a proton gradient as in chemiosmosis. As the protons move down the gradient, they drive the rotation. Gliding bacteria move only when in contact with a solid surface such as decaying organic matter. Rotation of the cell on its own axis often occurs with gliding. A number of mechanisms have been proposed to explain gliding, but the mechanism that propels the gliding bacterium *Myxococcus* is best understood. This organism uses energy to secrete a substance called a **surfactant** (ser-fak'tant), which lowers surface tension at its posterior end. The difference in surface tension between the organism's anterior and posterior ends (a passive phenomenon) causes it to glide.

Spirochetes expend energy for both creeping and thrashing movements. On a solid surface they creep along in an inchworm fashion by alternately attaching front and rear ends. Suspended in a liquid medium, they thrash (twist and turn) about. Both creeping and thrashing probably occur by waves of contraction within the cell substance that exert force against axial filaments.

CLOSE-UP

Random Motion versus Motility

Bacteria sometimes appear to engage in random, vibrating movements when viewed under the microscope in a hanging-drop preparation. Such nondirectional movements are due to Brownian movement—the bombardment of the bacteria by molecules of water and other substances in the medium. Brownian movement is thus caused by forces external to the bacteria and is not due to energy expenditure by the cells themselves.

ESSAY

Bioluminescence

Bioluminescence, the ability of an organism to emit light, appears to have evolved as a byproduct of aerobic metabolism. Bacteria of the genera *Photobacterium* and *Achromobacter,* fireflies, glowworms, and certain marine organisms living at great depths in the ocean exhibit bioluminescence (Figure 5.27). Many light-emitting organisms have an enzyme *luciferase* (lu-sif'er-as), along with other components of the electron

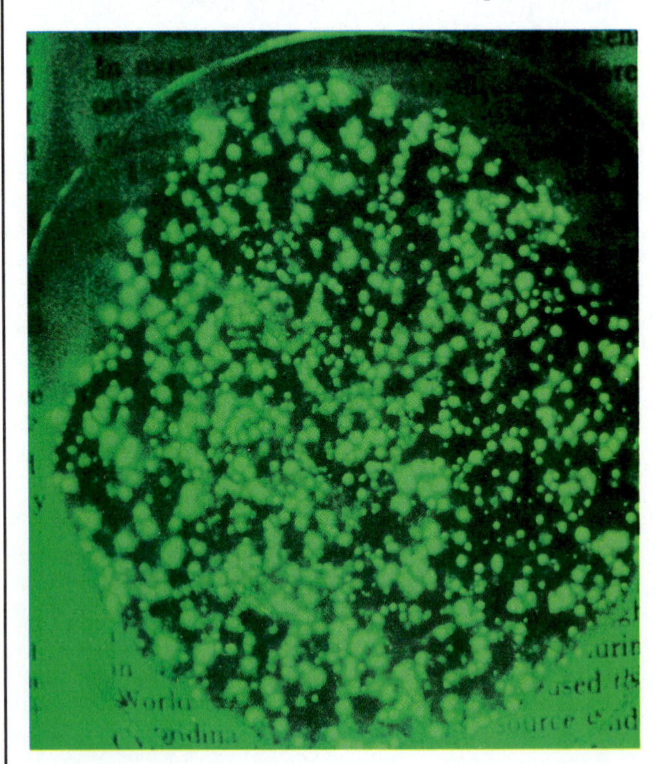

FIGURE 5.27 Bioluminescent bacteria in the Petri dish produce enough light to read by.

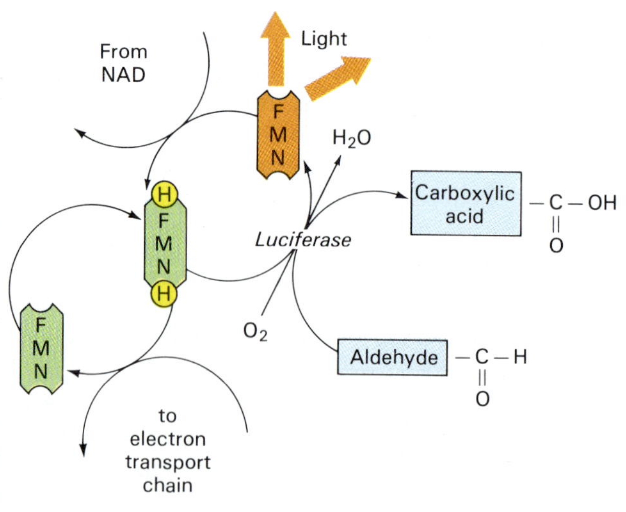

FIGURE 5.28 The reactions that give rise to bioluminescence are catalyzed by the enzyme luciferase. Oxygen is used to oxidize an aldehyde or ketone to a carboxylic acid. In the process, $FMNH_2$ is also oxidized, giving rise to an excited form of FMN that quickly radiates away its extra energy as light. Electrons are thus diverted from the electron transport chain, and light is produced at the expense of generation of ATP.

transport system. (Luciferase derives its name from Lucifer, which means "morning star.") Luciferase catalyzes a complex reaction in which molecular oxygen is used to oxidize a long-chain aldehyde or ketone to a carboxylic acid. At the same time, $FMNH_2$ from the electron transport chain is oxidized to an excited form of FMN, which emits light as it returns to its unexcited state (Figure 5.28). In this process phosphorylation reactions are bypassed, no ATP is generated, and instead energy is released as light.

Luminescent microorganisms often live on the surface of marine organisms such as some squids and fish. More than 300 years ago the Irish chemist Robert Boyle observed that the familiar glow of the skin of dead fish lasted only as long as oxygen was available. At that time the electron transport system and the role of oxygen in it were not understood.

Bioluminescence exhibited by larger organisms has survival value. It is the sole light source for marine creatures that live at great depths, and it helps land organisms such as fireflies to find mates. How bioluminescence arose among microorganisms is less clear. One hypothesis is that early in the evolution of living things, bioluminescence served to remove oxygen from the atmosphere as it was produced by some of the first photosynthetic organisms. Although this is not an advantage to aerobes, it is an advantage to strict anaerobes. Because most of the microorganisms in existence at that time were anaerobes susceptible to the toxic effects of oxygen, bioluminescence would have been beneficial to them. Today, many bioluminescent microbes are beneficiaries of symbiotic relationships with their hosts. They provide light in return for a shelter and nutrients.

CHAPTER SUMMARY

METABOLISM: AN OVERVIEW

- **Metabolism** is the sum of all the chemical processes in a living organism.
- **Autotrophy,** or self-feeding, includes photoautotrophy (*photosynthesis*) and chemoautotrophy.
- **Heterotrophy,** or other-feeding, includes chemoheterotrophy (*glycolysis, fermentation,* and *aerobic respiration*) and photoheterotrophy.
- **Biosynthetic pathways** use energy captured in the **catabolic pathways** for growth, movement, and other activities.

RELATED KEY TERMS

anabolism **catabolism**

oxidation **reduction**
electron acceptor **electron donor**
autotrophs **photoautotrophs**
chemoautotrophs
metabolic pathway
photoheterotrophs
chemoheterotrophs

ENZYMES

Properties of Enzymes

- Enzymes are proteins that catalyze chemical reactions in living organisms by lowering the **activation energy** needed for a reaction to occur.
- Enzymes have an **active site,** where the **substrate** attaches to form an **enzyme-substrate complex.** They typically exhibit a high degree of **specificity** in the reactions they catalyze. Enzymes are named according to their substrates and functions.

endoenzymes **exoenzymes**

Properties of Coenzymes and Cofactors

- Some enzymes require **coenzymes,** which can combine with a protein **apoenzyme** to form a **holoenzyme;** some enzymes also require inorganic ions as **cofactors.**

Enzyme Inhibition

- Enzyme activity can be reduced by **competitive inhibitors,** molecules that compete with the substrate for the active site, or by **noncompetitive inhibitors,** molecules that bind to an **allosteric site.**

feedback inhibition

Factors That Affect Enzyme Reactions

- Factors that affect the rate of enzyme reactions include temperature, pH, and concentrations of substrate, product, and enzyme.

chemical equilibrium

ANAEROBIC METABOLISM: GLYCOLYSIS AND FERMENTATION

Glycolysis

- **Glycolysis** is the metabolism of glucose into pyruvic acid.
- Under anaerobic conditions, glycolysis yields a net of 2 ATPs per molecule of glucose.

phosphorylation

Fermentation

- **Fermentation** includes the anaerobic glycolysis of glucose (or another sugar) to form pyruvate and its subsequent anaerobic metabolism.

- Six pathways of fermentation are summarized in Figure 5.11.
- **Homolactic acid** and **alcoholic fermentations** occur frequently.

RELATED KEY TERMS

AEROBIC METABOLISM: RESPIRATION

The Krebs Cycle

- The **Krebs cycle** involves the metabolism of 2-carbon groups to CO_2 and H_2O, the production of 1 ATP directly from each acetyl group, and the transfer of hydrogen atoms to the electron transport system.
- The significance of the Krebs cycle in energy production is mainly that it processes acetyl-CoA so that hydrogen atoms can be oxidized for energy.

anaerobes **aerobes**
aerobic respiration
tricarboxylic acid cycle
citric acid cycle

Electron Transport and Oxidative Phosphorylation

- **Electron transport** is the transfer of electrons to oxygen (the final electron acceptor).
- **Oxidative phosphorylation,** which occurs at three points in the **electron transport chain,** is an oxidative process that captures energy in ATP.
- The theory of **chemiosmosis** explains how energy is captured in ATP.

electron transport chain

Significance of Energy Capture

- Aerobic (oxidative) metabolism captures 19 times as much energy as anaerobic metabolism.

METABOLISM OF FATS AND PROTEINS

- Though most organisms get energy mainly from glucose, almost any substance can be metabolized by some microorganism.
- Fat metabolism involves glycerol metabolism in glycolysis and fatty acid metabolism by **beta oxidation** to acetyl-CoA, which enters the Krebs cycle.
- The metabolism of proteins involves the breakdown of proteins to amino acids, the deamination of the amino acids, and the subsequent metabolism of them in glycolysis, fermentation, or the Krebs cycle.

deaminated

OTHER METABOLIC PROCESSES

Photoautotrophy

- **Photosynthesis** is the use of light energy to synthesize carbohydrates: (1) **Light reactions** can involve **cyclic photophosphorylation** or **photolysis** accompanied by noncyclic photoreduction of NADP; (2) the **dark reactions** involve the reduction of carbon dioxide to carbohydrate.
- Photosynthesis in microorganisms provides a means of making food as it does in green plants; however, photosynthetic bacteria generally use some substances besides water to reduce carbon dioxide.

chemolithotrophs

Photoheterotrophy

- Photoheterotrophy is the use of light energy to metabolize simple organic molecules.

Chemoautotrophy

- Chemoautotrophs, or **chemolithotrophs,** oxidize inorganic substances to obtain energy. Chemolithotrophs require only carbon dioxide as a carbon source.

USES OF ENERGY

Biosynthetic Activities

- An **amphibolic pathway** is a metabolic pathway that can capture energy or synthesize substances needed by the cell.
- The intermediate products of energy-yielding metabolism and some of the building blocks for synthetic reactions that can be produced from them are summarized in Figure 5.25.
- Bacteria synthesize a variety of cell wall polymers. Units are synthesized in cytoplasm, usually moved across the cell membrane by a membrane transport protein, and incorporated into an existing polymer.

Membrane Transport and Movement

- Membrane transport uses energy derived from the ATP-producing electron transport system in the membrane to concentrate substances against a gradient. It occurs by active transport and the **phosphotransferase system.**
- Movement in bacteria can be by flagella, and by gliding, creeping, and thrashing movements.

permeases **surfactant**

QUESTIONS FOR REVIEW

A.

1. Define metabolism and distinguish between anabolism and catabolism.
2. Explain the differences between autotrophy and heterotrophy, and list the kinds of metabolism that fall into each of those categories.

B.

3. List the main properties of enzymes.
4. How can enzyme-catalyzed reactions be inhibited?
5. Which factors affect the rate of enzyme activity?

C.

6. Summarize the process of glycolysis, and show where energy is produced.
7. Which substances are produced in glycolysis from one molecule of glucose?
8. What is fermentation?
9. What are the products of the two main forms of fermentation, and which organisms carry out each type?
10. How are other fermentations useful in diagnosis?

D.

11. Summarize the main events in the Krebs cycle.
12. What are the end products of the Krebs cycle, including electron carriers?
13. What is the significance of the Krebs cycle in energy capture?

E.

14. What is electron transport, and what are its products?
15. What is oxidative phosphorylation, and what is its significance?
16. What is chemiosmosis?

F.

17. How are fats and proteins used for energy?

G.

18. What is the overall function of photosynthesis?
19. What occurs in the light reaction and in the dark reaction?
20. How does photosynthesis in bacteria differ from photosynthesis in plants?

H.

21. What is chemolithotrophy?
22. What is photoheterotrophy?

I.

23. Define amphibolic, and give an example of an amphibolic process.
24. Which building blocks for biosynthetic activities are derived from intermediates in energy-producing reactions?

J.

25. What is membrane transport?
26. What are two ways bacteria transport substances against concentration gradients?
27. By which mechanisms do bacteria move?

PROBLEMS FOR INVESTIGATION

1. Spontaneous combustion caused by bacteria sometimes sets fire to a barn where damp hay has been stored. How can you explain this?
2. In what sequence do you think the different kinds of metabolism mentioned in this chapter might have evolved? Why didn't just one type of metabolism evolve? Give reasons for your answers.
3. Suppose you had a culture known to contain an *Enterobacter* species and *Escherichia coli*. Devise a way to separate and identify the organisms.
4. Look up the chemical reaction by which certain bacteria change wine into vinegar. Is this an aerobic or anaerobic process?
5. More than 125 human diseases are caused by enzyme deficiencies. Chapters 7 and 8, on bacterial genetics, will explain how bacteria are being used by genetic engineers to remedy these enzyme deficiencies. List some human diseases caused by lack of enzymes, indicating which enzymes are involved.

SOME INTERESTING READING

Bohinski, R. C. 1987. *Modern concepts in biochemistry*. Boston: Allyn and Bacon.

Dickerson, R. E. 1980. Cytochrome c and the evolution of energy metabolism. *Scientific American* 242, no. 3 (March):136.

Gottschalk, G. 1988. *Bacterial metabolism*. New York: Springer-Verlag.

Hinkle, P. C., and R. E. McCarty. 1978. How cells make ATP. *Scientific American* 238, no. 3 (March):104.

Lechtman, M., B. Rookk, and R. Egan. 1979. *The games cells play*. Menlo Park, CA: Benjamin-Cummings.

Mandelstam, J., K. McQuillen, and I. Dawes. 1982. *Biochemistry of bacterial growth*. Oxford: Blackwell Scientific Publications.

Meighen, E. A. 1991. Molecular biology of bacterial bioluminescence. *Microbiological Reviews* 55(1):123–42.

Monastersky, R. 1988. Bacteria alive and thriving at depth. *Science News* 133 (March 5):149.

Pritchard, P. H. 1991. Bioremediation as a technology: experiences with the *Exxon Valdez* oil spill. *J. Hazardous Mater.* 28(1–2):115–30.

Proton pumping. 1985. *BioScience* 35:14.

Trumpower, B. L. 1990. Cytochrome *bc* complexes of microorganisms. *Microbiological Reviews* 54(2):101–29.

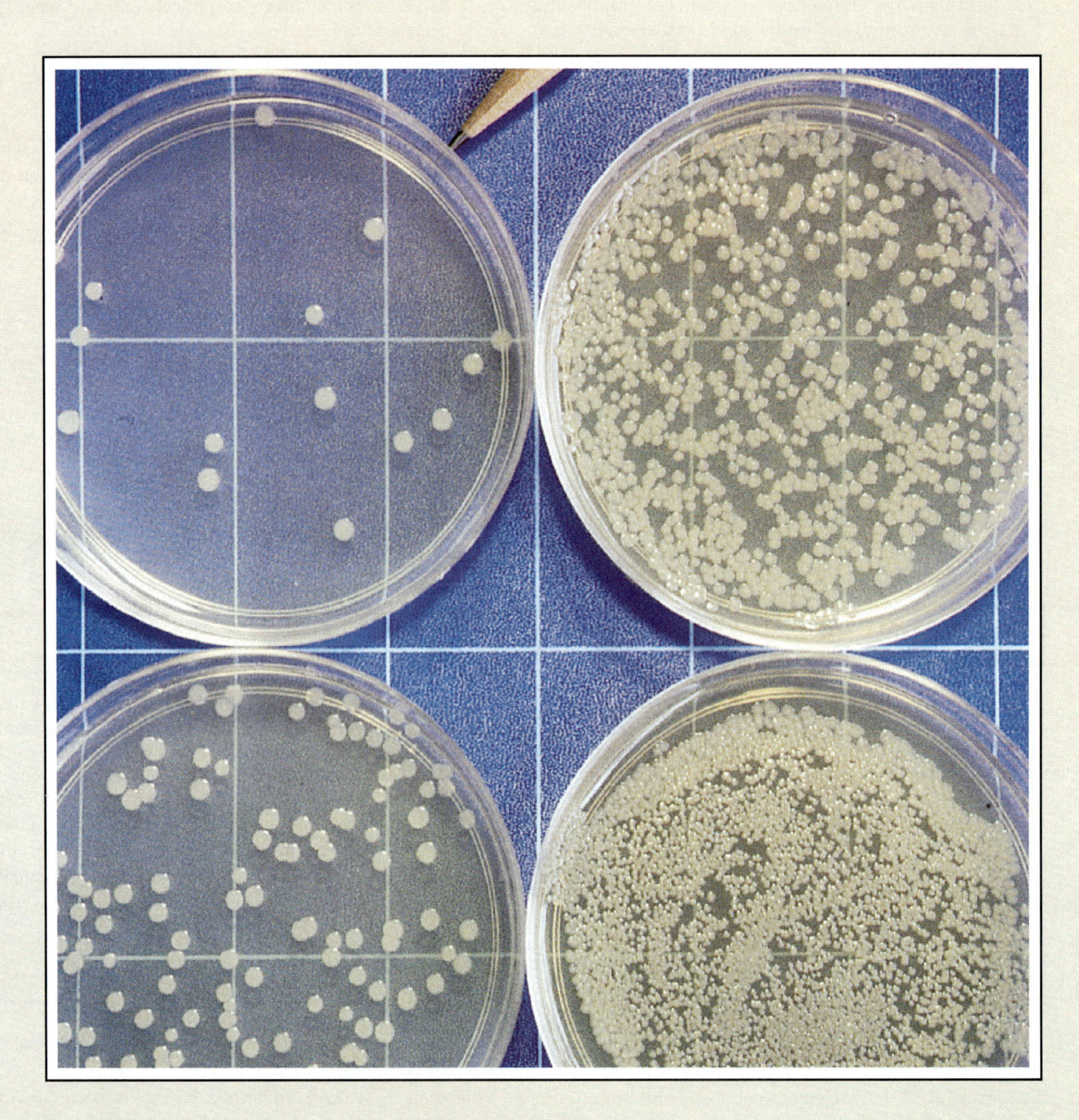

Bacterial colonies growing on culture media in Petri dishes.

6 Growth and Culturing of Bacteria

This chapter focuses on the following questions:

A. How is growth defined in bacteria?

B. How does cell division occur in microorganisms?

C. What are the phases of growth in a bacterial culture?

D. How is bacterial growth measured?

E. How do physical factors affect bacterial growth?

F. How do nutritional factors affect bacterial growth?

G. What occurs in sporulation, and what is its significance?

H. What methods are used to obtain a pure culture of an organism for study in the laboratory?

I. How are different nutritional requirements supplied by various media?

In this chapter we will use what we learned in Chapter 5 about energy in microorganisms to study how to grow them in the laboratory. ∞ (p. 110) Growth in bacteria, which has been more thoroughly studied than growth in other microorganisms, is affected by a variety of physical and nutritional factors. Knowing how these factors influence growth is useful in culturing organisms in the laboratory and in preventing their growth in undesirable places. Furthermore, growing the organisms in pure cultures is essential for making diagnostic tests that are used to identify disease-causing organisms.

GROWTH AND CELL DIVISION

Definition of Growth

In everyday language, growth refers to an increase in size. We are accustomed to seeing children, other animals, and plants grow. Unicellular organisms also grow, but as soon as a cell, called the **mother cell,** has approximately doubled in size, it divides into two **daughter cells.** Then the daughter cells grow, and subsequently they also divide. Because individual cells grow larger only to divide into two new individuals, **microbial growth** is defined not in terms of cell size but as the increase in the number of cells, which occurs by cell division.

Cell Division

Cell division in bacteria, unlike eukaryotes, usually occurs by binary fission or sometimes by budding. In **binary fission,** a cell duplicates its components and divides into two cells (Figure 6.1), which become independent when a septum (partition) grows between them and they break apart. Prokaryotic cells do not have a cell cycle with a specific period of DNA synthesis as eukaryotic cells do. Instead, in continuously dividing cells, DNA synthesis also is continuous and replicates the single bacterial chromosome shortly before the cell divides. The chromosome is attached to the cell membrane, which grows and separates the chromosomes. Replication of the chromosome is completed before cell division, and the cell may temporarily contain two or more nucleoids. In some species incomplete separation of the septa produces linear chains (linked bacilli), **tetrads** (cuboidal groups of four cocci), **sarcinae** (groups of eight cocci in a cube), or grapelike clusters (staphylococci). Some bacilli always form chains or filaments; others form them only under impaired growth conditions. Cell division in yeast and a few bacteria occurs through **budding,** as a small, new cell develops from the surface of an existing cell and subsequently separates from the parent cell.

Phases of Growth

A population of organisms introduced into a fresh nutrient-rich medium displays four major phases of growth: (1) the lag phase, (2) the log (logarithmic) phase, (3) the stationary phase, and (4) the decline, or death, phase. These phases form the standard **bacterial growth curve,** as is illustrated in Figure 6.2.

Lag Phase

In the **lag phase,** the organisms do not increase in number, but they are metabolically active—growing

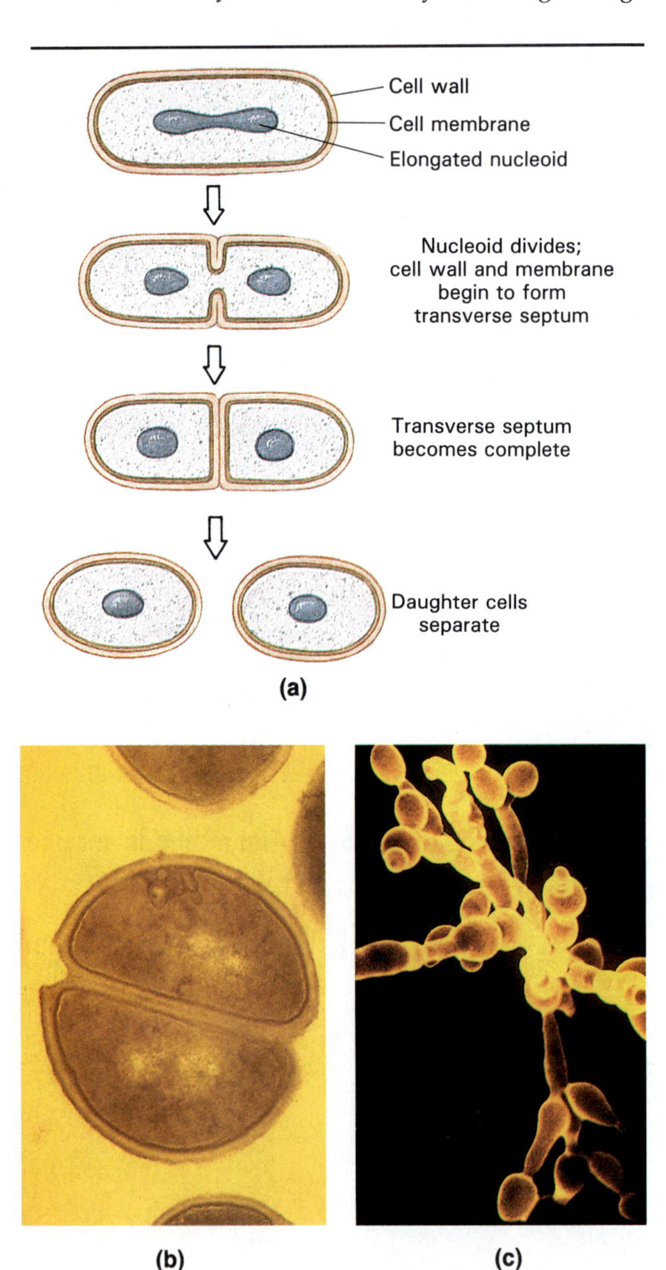

FIGURE 6.1 (a) Binary fission in a bacterium. (b) Electron micrograph of a thin section of the bacterium *Staphylococcus* undergoing binary fission. (c) Budding in the yeast *Candida albicans,* the major cause of vaginal yeast infections.

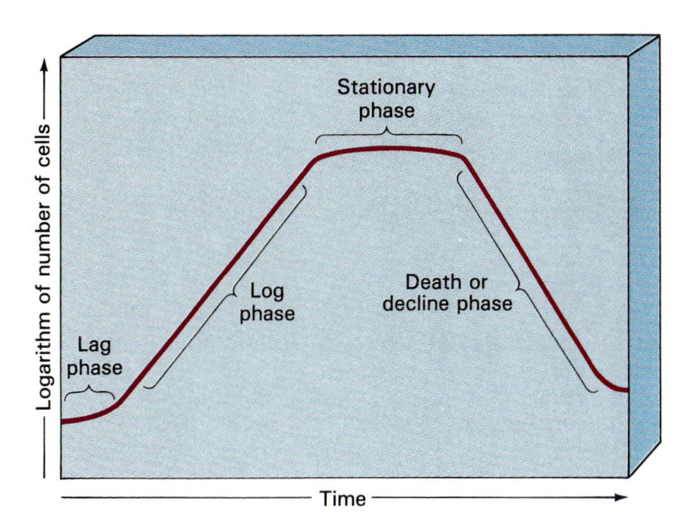

FIGURE 6.2 Phases of growth in a bacterial culture, shown as a standard bacterial growth curve.

in size, synthesizing enzymes, and incorporating various molecules from the medium. During this phase the individual organisms increase in size, and they capture large quantities of energy in ATP.

The length of the lag phase is determined in part by characteristics of the species and in part by conditions in their media—both the medium from which the organisms come and the one to which they are transferred. Some species adapt to the new medium in an hour or two; others take several days. Organisms from old cultures, adapted to limited nutrients and large accumulations of wastes, take longer to adapt to a new medium than those transferred from relatively fresh, nutrient-rich media. Organisms transferred to a minimal nutrient medium take longer to adapt than do those transferred to a rich medium.

Log Phase

Once organisms have adapted to a medium, population growth occurs at an **exponential,** or **logarithmic,** rate. When the scale of the vertical axis is logarithmic, growth in this **log phase** appears on a graph as a straight diagonal line, which represents the size of the bacterial population. (On the base 10 logarithmic scale, each successive unit represents a tenfold increase in the number of organisms.) During the log phase the organisms divide at their most rapid rate, a regular, genetically determined interval called the **generation time.** The generation time for most bacteria is from 20 minutes to 20 hours, and is typically less than 1 hour. Some bacteria, such as those that cause tuberculosis and leprosy, have much longer generation times.

The population of organisms doubles in each generation time. For example, a culture containing 1000 organisms per milliliter with a generation time of 20 minutes would contain 2000 organisms per milliliter after 20 minutes, 4000 organisms after 40 minutes, 8000 after 1 hour, 64,000 after 2 hours, and 512,000 after 3 hours. Such growth is said to be exponential, or logarithmic (Figure 6.2). Some organisms take slightly longer than others to go from the lag phase to the log phase, and they do not all divide precisely together. If they divided together and the generation time were exactly 20 minutes, the number of cells in a culture would increase in a stair-step pattern, exactly doubling every 20 minutes (Figure 6.3)—a hypothetical situation called **synchronous growth.** In an actual culture, each cell divides sometime during the 20-minute generation time, with about 1/20 of them dividing each minute—a natural situation called **nonsynchronous growth.** Nonsynchronous growth appears as a straight line on a logarithmic graph rather than as a series of steps.

Organisms in a tube of culture medium can maintain logarithmic growth for only a limited period of time. As the number of organisms increases, nutrients are used up, metabolic wastes accumulate, living space may become limited, and aerobes suffer from oxygen depletion. Generally, the limiting factor for logarithmic growth seems to be the rate at which energy can be captured in ATP. As the availability of

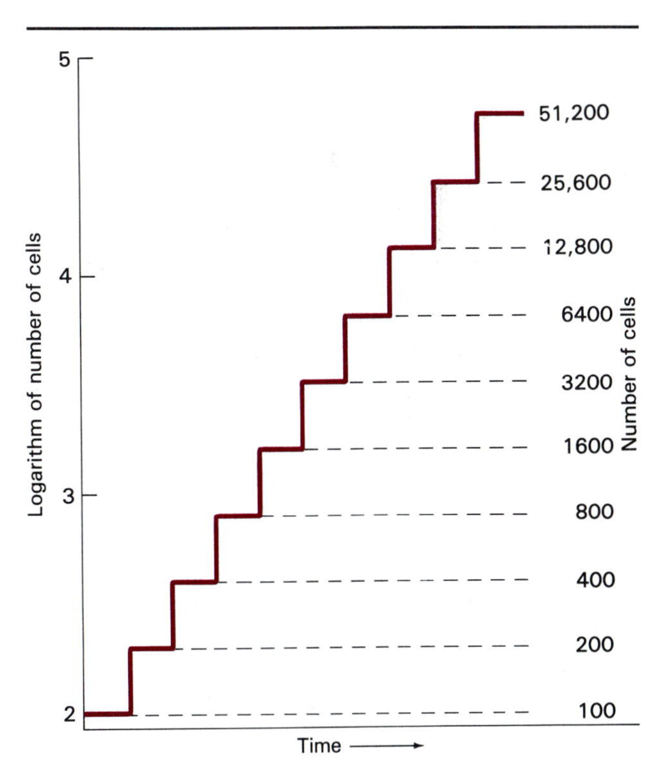

FIGURE 6.3 Bacterial growth curve for the hypothetical situation in which growth is synchronous. The "stair-step" shape of the curve results when all cells in the population divide at the same time, doubling the population with every division.

APPLICATIONS

Determining Generation Time

How do scientists determine generation time? Knowing that the generation time for a particular species of bacteria is constant provides a basis for calculating it. The only information needed is the bacterial population at the beginning and end of a measured time interval during the phase. The calculations are as follows:

Known Information	**Information To Be Obtained**
B_0 = number of bacteria at zero time	n = number of generations
B_t = number of bacteria at the end of a selected period of time, t	G = generation time
t = time period	

Logarithmic growth can be described by the equation $B_t = B_0 \times 2^n$. To solve for n, take the logarithm of both sides of the equation and rearrange terms:

$$\log B_t = \log B_0 \times n \log 2$$

$$n = \frac{\log B_t - \log B_0}{\log 2}$$

Use a table of base 10 logarithms to solve the equation. For example,

$$B_t = 49{,}000{,}000 \qquad \log B_t = 7.690$$

$$B_0 = 12{,}000 \qquad \log B_0 = 4.079$$

$$\log 2 = 0.301$$

Thus, $n = (7.690 - 4.079)/0.301 = 12$ generations. This means that it takes 12 generations—that is, 12 population doublings—for the size of a culture to increase from 12,000 to 49,000,000 cells. Note that the number of generations required for this increase in cells is the same regardless of whether the generation time is long or short.

Next, suppose we observe that for a particular species, the time actually elapsed during this population increase is 4 hr (240 min). We can then calculate the generation time by dividing the total time by the number of generations. In other words, G (generation time) = t (total time) ÷ n (number of generations). Solving this equation, we see that the generation time is 240/12 = 20 min.

nutrients decreases, the cells become less able to generate ATP, and the growth rate decreases. The decrease in growth rate is shown in Figure 6.2 by a gradual leveling off of the growth curve (the curved segment to the right of the logarithmic phase).

Leveling off of growth is followed by the stationary phase unless fresh medium is added or organisms are transferred to fresh medium. Logarithmic growth can be maintained by a device called a **chemostat** (Figure 6.4), which has a growth chamber and a reservoir from which fresh medium is continuously added to the growth chamber as old medium is withdrawn. Alternatively, organisms from a culture in the stationary phase can be transferred to fresh medium. After a brief lag phase, such organisms quickly reenter the log phase of growth.

Stationary Phase

When cell division decreases to the point that new cells are produced at the same rate as old cells die, the number of live cells stays constant. The culture is then in the **stationary phase,** represented by a horizontal straight line in Figure 6.2. The medium contains a limited amount of nutrients and may contain toxic quantities of waste materials. Also, the oxygen supply may become inadequate for aerobic organisms.

Decline Phase

As conditions in the medium become less and less supportive of cell division, many cells lose their ability to divide, and thus the cells die. In this **decline phase,** or **death phase,** the number of live cells decreases at a logarithmic rate, as indicated by the straight, downward sloping diagonal line in Figure 6.2. During the decline phase, many cells undergo **involution,** assuming a variety of unusual shapes, which makes

FIGURE 6.4 A chemostat constantly renews nutrients in a culture, making it possible to grow organisms continuously.

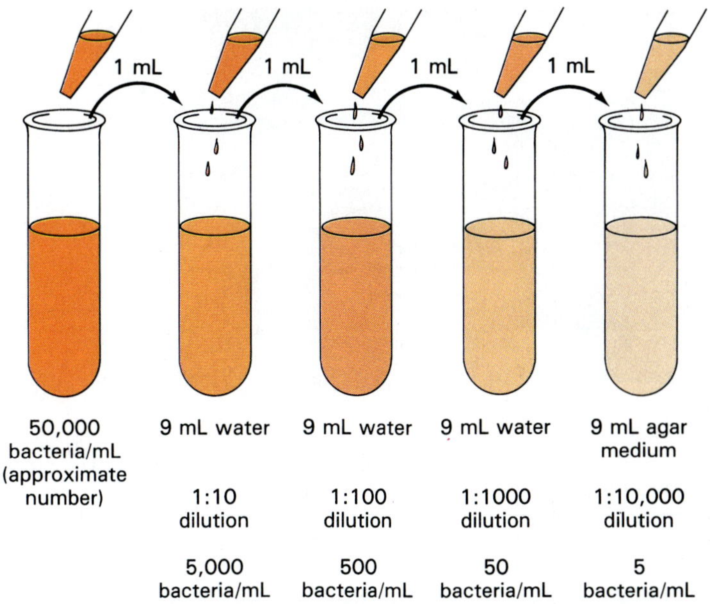

FIGURE 6.5 The technique of serial dilution. One milliliter is taken from a broth culture and added to 9 mL of water, thereby diluting it by a factor of 10. This procedure is repeated until the desired concentration is reached.

them difficult to identify. In cultures of spore-forming organisms, more spores than vegetative (metabolically active) cells survive. The duration of the death phase is as highly variable as the duration of the logarithmic growth phase. Both depend primarily on the genetic characteristics of the organism. Cultures of some bacteria go through all growth phases and die in a few days; others contain a few live organisms after months or even years.

Growth in Colonies

Growth phases are displayed in different ways in colonies growing on solid medium. Typically, a cell divides exponentially, forming a small **colony** of descendants of the original cell. The colony grows rapidly at its edges; cells nearer the center grow more slowly or begin to die because they have smaller quantities of nutrients and are exposed to more waste products. All phases of the growth curve occur simultaneously in a colony.

Measuring Bacterial Growth

Bacterial growth is measured by estimating the number of cells that have arisen by binary fission during a growth phase. This measurement is expressed as the number of viable (living) organisms per milliliter of culture. Several methods of measuring bacterial growth are available.

Serial Dilution and Standard Plate Counts

One method of measuring bacterial growth uses the technique of serial dilution followed by growing colonies of organisms on agar plates. An **agar plate** is a plate of medium solidified with *agar,* a polysaccharide extracted from certain marine algae. To make **serial dilutions** (Figure 6.5), you start with organisms in liquid medium. Adding 1 mL of this medium to 9 mL of sterile water makes a 1:10 dilution; adding 1 mL of the 1:10 dilution to 9 mL of sterile water makes a 1:100 dilution; and so on. The number of bacteria per milliliter of fluid is reduced by 1/10 in each dilution. Subsequent dilutions are made in ratios of 1:1000, 1:10,000, 1:100,000, 1:1,000,000, or even 1:10,000,000 if the original culture contained a very large number of organisms.

From each dilution, usually beginning with the 1:100, 1 mL of the culture is transferred to an agar plate. (A milliliter of the 1:10 dilution usually is not transferred to an agar plate because it contains too many organisms to yield countable colonies.) Wherever a single living bacterium is deposited on an agar plate, it will divide to form a colony. One or more plates should have a small enough number of colonies to distinguish each one clearly and to count them. If dilutions have been made properly, plates with a **countable number** of colonies (30 to 300 per plate) should be available. The plate is placed under the magnifying lens of a colony counter (Figure 6.6), and colonies on the entire plate are counted. The number of colonies on a plate is multiplied by the denominator of the dilution factor, such as 1000 for a plate made from a 1:1000 dilution sample, or 10,000 for a 1:10,000 sample. The result is the number of organisms per milliliter of the original culture (Figure 6.7).

The accuracy of the serial dilution and plate count method depends on homogeneous dispersal of organisms in each dilution. Error can be minimized by shaking each culture before sampling and making sev-

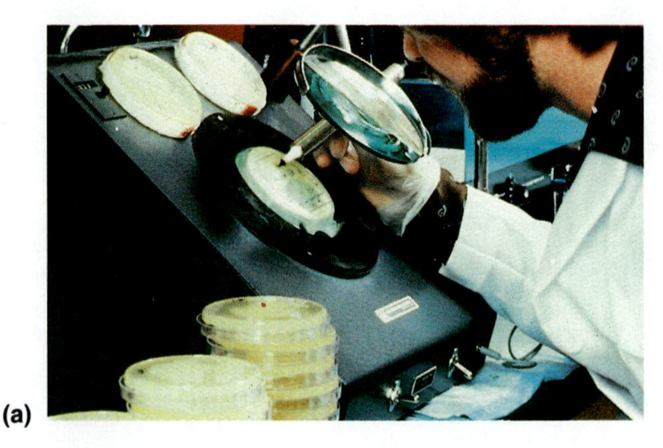

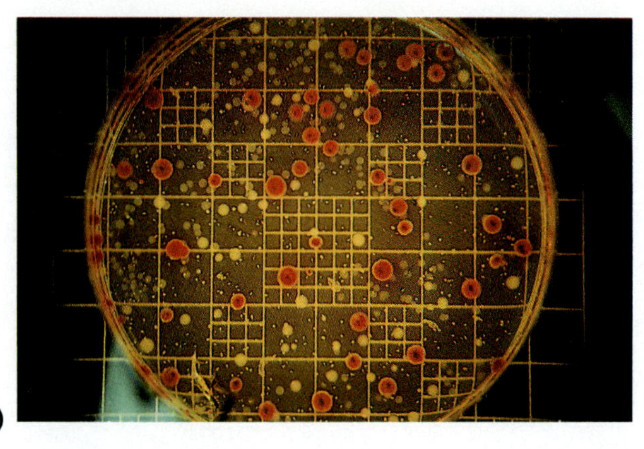

FIGURE 6.6 (a) Technician using a bacterial colony counter. (b) Bacterial colonies viewed through the magnifying glass against a colony-counting grid. The plate shown was produced by the pour plate method described in Figure 6.7.

eral plates from each dilution. Accuracy is also affected by the death of cells. Because the number of colonies counted represents the number of living organisms, it does not include organisms that may have died by the time plating is done, nor does it include organisms that canot grow on the chosen medium. Using young cultures in the log phase of growth minimizes this kind of error.

Direct Microscopic Counts

Bacterial growth can be measured by **direct microscopic counts.** In this method a known volume of medium is introduced into a specially calibrated counting chamber, the Petroff-Hausser counter, on a microscope slide (Figure 6.8). After the organisms that are visible through the microscope in prescribed areas

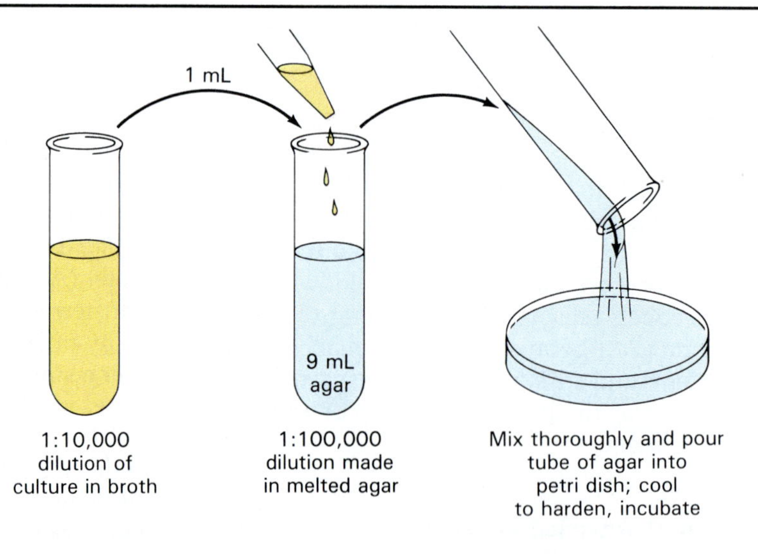

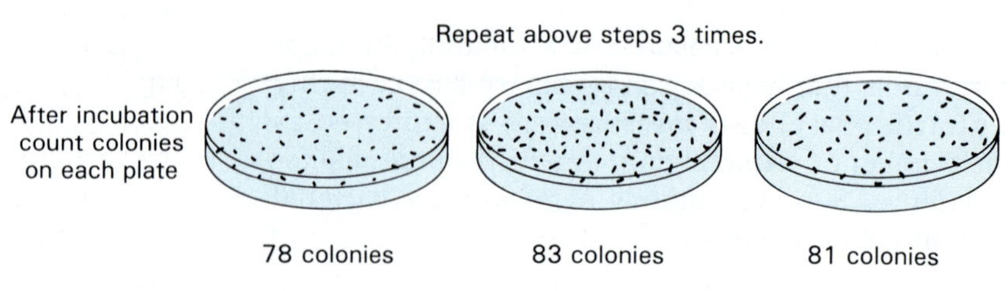

Average number of colonies per plate $= \frac{78 + 83 + 81}{3} = 80.7 \cong 81$

Number of organisms $= 81 \times 100{,}000$ (dilution factor) $= 8.1 \times 10^6$/mL of culture

FIGURE 6.7 Calculation of the number of bacteria per milliliter of culture using serial dilution and pour plate counts. One milliliter of broth is mixed with 9 mL of melted agar, which is warm enough to stay liquid but not hot enough to kill the organisms being mixed into it. After thorough mixing, the warm agar is quickly poured into an empty, sterile Petri dish. When cooled to hardness, it is incubated, and the colonies that develop are counted. A single measurement is not very reliable, so the procedure is repeated at least three times, and the results are averaged. The average number of colonies is multiplied by the dilution factor to ascertain the total number of organisms per milliliter of the original culture.

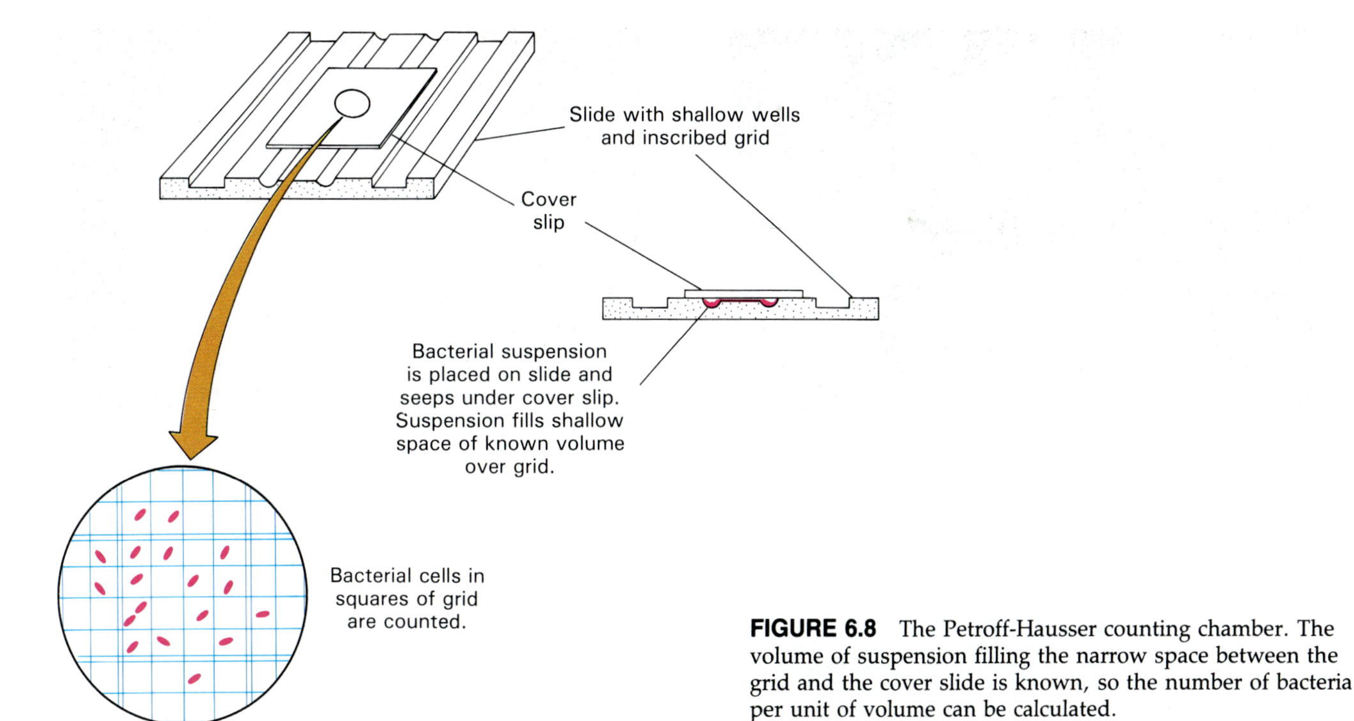

FIGURE 6.8 The Petroff-Hausser counting chamber. The volume of suspension filling the narrow space between the grid and the cover slide is known, so the number of bacteria per unit of volume can be calculated.

of the counting chamber have been counted, the number of bacteria per milliliter of medium can be estimated with a reasonable degree of accuracy. The accuracy of direct microscopic counts depends on the presence of more than 10 million bacteria per milliliter of culture. This is because counting chambers are designed to allow accurate counts only when large numbers of cells are present. An accurate count also depends on the bacteria being homogeneously distributed through the culture. Finally, this technique has the disadvantage of not distinguishing between living and dead cells.

Most Probable Number

When samples contain too few organisms to give reliable measures of population size by the standard plate count method, as in food and water sanitation studies, or when organisms will not grow on agar, the **most probable number** (MPN) method is used. The technician observes the sample, estimates the number of cells in it, and makes a series of progressively greater dilutions. As the dilution factor increases, a point will be reached at which some tubes will contain a single organism and others, none. A typical MPN test consists of five tubes of each of three volumes (10, 5, and 0.1 mL) of a dilution. Those that contain an organism will display growth when incubated. The number of organisms in the original culture is estimated from a most probable number table. The values in the table, which are based on statistical probability, specify that the number of organisms in the original culture has a 95 percent chance of falling within a particular range. A large number of tubes showing growth, especially at higher dilutions, indicates that more organisms were present in the sample than when a smali number of tubes show growth.

Filtration

Another method of estimating the size of small bacterial populations makes use of **filtration.** A known volume of water or air is drawn through a filter with pores too small to allow passage of bacteria. When the filter is placed on a solid medium, each colony that grows represents one organism. Thus, the number of organisms per liter of water or air can be calculated.

Other Methods

Several other methods of monitoring bacterial growth are available. They include simple observation with or without special measurements, detection of gas or acid production, and determination of dry weight of cells.

Turbidity (a cloudy appearance) in a culture tube indicates the presence of organisms (Figure 6.9). Fairly accurate estimates of growth can be obtained by measuring turbidity with photoelectric devices, such as a colorimeter or a spectrophotometer (Figure 6.10). This method is particularly useful in monitoring the rate of growth without disturbing the culture. Measures of bacterial growth based on turbidity are especially subject to error when cultures contain fewer than 1

FIGURE 6.9 Turbidity, or a cloudy appearance, is an indicator of bacterial growth.

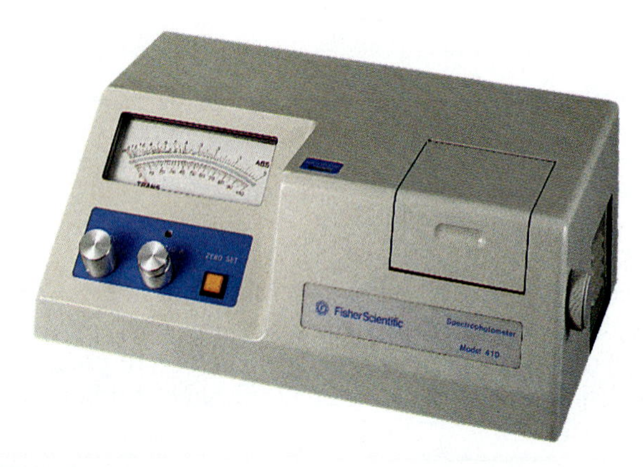

FIGURE 6.10 A spectrophotometer. This instrument can be used to measure bacterial growth by determining the degree of light transmission through the culture. Samples of culture in special optically clear tubes are placed inside the spectrophotometer (inside lid at right of machine) and are measured against standards.

million cells per milliliter. Such cultures can display little or no turbidity even when growth is occurring. Conversely, turbidity can be produced by a high concentration of dead cells in a culture.

Gas production can be detected by observing bubbles in solid media or by capturing gas in small inverted tubes placed inside larger tubes of liquid media. Acid production can be detected by incorporating pH indicators—chemical substances that change color with changes in pH—in media. Finally, cells can be extracted from a medium and dried, and their dry weight can be determined.

FACTORS AFFECTING BACTERIAL GROWTH

Microorganisms are found in nearly every environment on earth, including environments in which no other organisms can survive. They can survive in a great many environments because they are small and easily dispersed, occupy little space, need only small quantities of nutrients, and are remarkably diverse in their nutritional requirements. They also have great capacity for adapting to environmental changes. Some microorganisms can be found to metabolize almost anything as a nutrient or to survive almost any environmental change.

As warm-blooded, air-breathing, land-dwelling mammals, we tend to forget that 72 percent of our planet's surface is water, that 90 percent of that water is salt water, and that environments containing living organisms have an average temperature of about 5°C. Unlike humans, microorganisms live mostly in water, and many are adapted to temperatures above or below those we consider optimum. The organisms of particular interest in the health sciences account for only a fraction of all microorganisms—those that have adapted to conditions found in or on the human body.

Different species of microorganisms can grow in a wide range of environments—from highly acidic to somewhat alkaline conditions, from Antarctic ice to hot springs, in pure spring water or salty marshes, in oceans with or without oxygen, and even under great pressure and in boiling steam vents in the ocean floor. They use a variety of substances to obtain energy, and some require special nutrients.

The kinds of organisms found in an environment and the rates at which they grow can be influenced by a variety of factors, both physical and nutritional. **Physical factors** include pH, temperature, oxygen concentration, moisture, hydrostatic pressure, osmotic pressure, and radiation. **Nutritional factors** include availability of carbon, nitrogen, sulfur, phosphorus, certain other minerals, and, in some cases, vitamins.

Physical Factors

pH

As we saw in Chapter 2, the acidity or alkalinity of a medium is expressed in terms of pH. ∞ (p. 34) Though the pH scale is now widely used in chemistry, it was invented by Danish chemist Søren Sørenson to describe the limits of growth of microorganisms in various media. Microorganisms have an **optimum pH**—the pH at which they grow best. Their optimum pH is usually near neutrality (pH 7), and most do not grow at a pH more than 1 pH unit above or below their optimum pH.

Bacteria are classified as acidophiles, neutrophiles, and alkalinophiles according to the conditions of acidity or alkalinity they can tolerate. However, no single species can tolerate the full pH range of any of these categories, and many tolerate a pH range that overlaps two categories. **Acidophiles** (as-id'o-filz), or acid-loving organisms, exist from pH 0.0 to 5.4. *Lactobacillus*, which produces lactic acid, is an acidophile, but it tolerates only mild acidity. Some bacteria that oxidize sulfur to sulfuric acid, however, can create and tolerate conditions as low as pH 1. **Neutrophiles** (nu'tro-fīlz) exist from pH 5.4 to 8.5. Most of the bacteria that cause disease in humans are neutrophiles. **Alkalinophiles** (al-kah-lin'o-fīlz), alkali- or base-loving organisms, exist from pH 7.0 to 11.5. *Vibrio cholerae*, the causative agent of the disease cholera, grows best at a pH of about 9. *Alcaligenes faecalis*, which sometimes infects humans already weakened by another disease, can create and tolerate alkaline conditions of pH 9 or higher. The soil bacterium *Agrobacterium* grows in alkaline soil of pH 12.

The effects of pH on organisms can be related to the concentration of organic acids in the medium and the protection that bacterial cell walls sometimes provide. *Lactobacillus* and other organisms that produce organic acids during fermentation inhibit their own growth as acids such as lactic acid and pyruvic acid accumulate in the medium. It appears that the acids themselves rather than the hydrogen ions per se inhibit growth. Other organisms have relatively impervious cell walls that prevent the cell membrane from being exposed to an extreme pH in the medium. The organism appears to tolerate environmental acidity or alkalinity because the cell itself is maintained at a nearly neutral pH.

Temperature

Most species of bacteria can grow over a 30°C temperature range, but the minimum and maximum temperatures for different species vary considerably. Sea water remains liquid below 0°C, and organisms living there can tolerate below-freezing temperatures. Bacteria can be classified according to growth temperature ranges as psychrophiles, mesophiles, and thermophiles, although most do not tolerate the whole temperature range of a category and some tolerate a range that overlaps categories. Within these groups bacteria are further classified as obligate or facultative. **Obligate** means that the organism must have the specified environmental condition. **Facultative** means that the organism is able to tolerate the environmental condition, although it can also live in another.

Psychrophiles (si'kro-fīlz), or cold-loving organisms, grow best at temperatures of 15° to 20°C, although some live quite well at 0°C. They can be further divided into **obligate psychrophiles** such as *Bacillus globisporus*, which cannot grow above 20°C, and **facultative psychrophiles** such as *Xanthomonas pharmicola*, which grows best below 20°C but also can grow above 20°C. Psychrophiles live mostly in cold water and soil. None can live in the human body, but they can cause spoilage of refrigerated foods.

Mesophiles (mes'o-fīlz), which include most bacteria, grow best at temperatures somewhere between 25° and 40°C. Human pathogens are included in this category, and most of them grow best near human body temperature (37°C).

Thermophiles (therm'o-fīlz), or heat-loving organisms, grow best at temperatures from 50° to 60°C. Many are found in compost heaps, and a few tolerate temperatures as high as 110°C in boiling hot springs. They can be further divided into **obligate thermophiles**, which can grow only at temperatures above 37°C, and **facultative thermophiles,** which can grow both above and below 37°C. *Bacillus stearothermophilus*, which usually is considered an obligate thermophile, grows at its maximum rate at 65° to 75°C but can display minimal growth and cause food spoilage at temperatures as low as 30°C. Thermophilic sulfur bacteria display zones of optimum growth temperatures in the runoff troughs of geysers (Figure 6.11). Different species collect at various locations along the trough, with the most heat-tolerant near the geyser and those with lesser heat tolerance distributed in regions where the water has cooled to their optimum

FIGURE 6.11 Geyser Hot Springs, Black Rock Desert, Nevada. Bacteria can live and grow in the runoff waters from such geysers despite the near-boiling temperatures.

temperature. In deep channels the most heat-tolerant species are found at the greatest depths and the least heat-tolerant near the surface, where water has cooled. Under laboratory conditions that utilize high pressure to increase water temperature above 100°C, archeobacteria from deep sea vents have grown at 130°C (266°F). See Chapter 10 for more information about these remarkable organisms.

The temperature range over which an organism grows is determined largely by the temperatures at which its enzymes function. Within this temperature range three critical temperatures can be identified:

1. The minimum growth temperature is the lowest temperature at which cells can divide.
2. The maximum growth temperature is the highest temperature at which cells can divide.
3. The optimum growth temperature is the temperature at which cells divide most rapidly, that is, have the shortest generation time.

Regardless of the type of bacteria, growth gradually increases from the minimum to the optimum temperature and decreases very sharply from the optimum to the maximum temperature. Furthermore, the optimum temperature is often very near the maximum temperature (Figure 6.12). These growth properties are due to changes in enzyme activity (Chapter 5). ∞ (p. 116) Enzyme activity generally doubles for every 10°C rise in temperature until the high temperature begins to denature all proteins, including enzymes. The sharp decrease in enzyme activity at a temperature only slightly higher than the optimum temperature occurs as enzyme molecules become so distorted by denaturation that they cannot catalyze reactions.

Temperature is important not only in providing conditions for microbial growth but also in preventing such growth. The refrigeration of food, usually at 4°C, reduces the growth of psychrophiles and prevents the growth of most other bacteria. However, food and other materials such as blood can support growth of some bacteria even when refrigerated. For this reason, materials that can withstand freezing are stored at temperatures of −30°C if they are to be kept for long periods of time. High temperatures also can be used to prevent bacterial growth (Chapter 13). Laboratory equipment and media are often sterilized with heat, and food is frequently preserved by heating and storing in closed containers. Bacteria are more apt to survive extremes of cold than extremes of heat; enzymes are not denatured by chilling but can be permanently denatured by heat.

Oxygen

Bacteria, especially heterotrophs, can be divided into aerobes, which require oxygen to grow, and anaerobes, which do not require it. Among the aerobes, cultures of rapidly dividing cells require more oxygen than do cultures of slowly dividing cells. **Obligate aerobes** such as *Pseudomonas*, which causes many hospital-acquired infections, must have free oxygen, whereas **obligate anaerobes** such as *Bacteroides* are killed by free oxygen. In a culture tube containing nutrient broth, obligate aerobes grow near the surface, where atmospheric oxygen diffuses into the medium; obligate anaerobes grow near the bottom of the tube, where little or no free oxygen reaches them (Figure 6.13).

In the case of aerobic organisms, oxygen is often the environmental factor that limits growth rate. Oxygen is poorly soluble in water, and a variety of methods are sometimes employed to maintain a high O_2 concentration in cultures, including vigorous mixing or forced aeration by bubbling air through a culture, as is done in a fish tank. This is especially important in such commercial processes as the production of antibiotics and in sewage treatment.

Between the preceding extremes are the microaerophiles, the facultative anaerobes, and the aero-

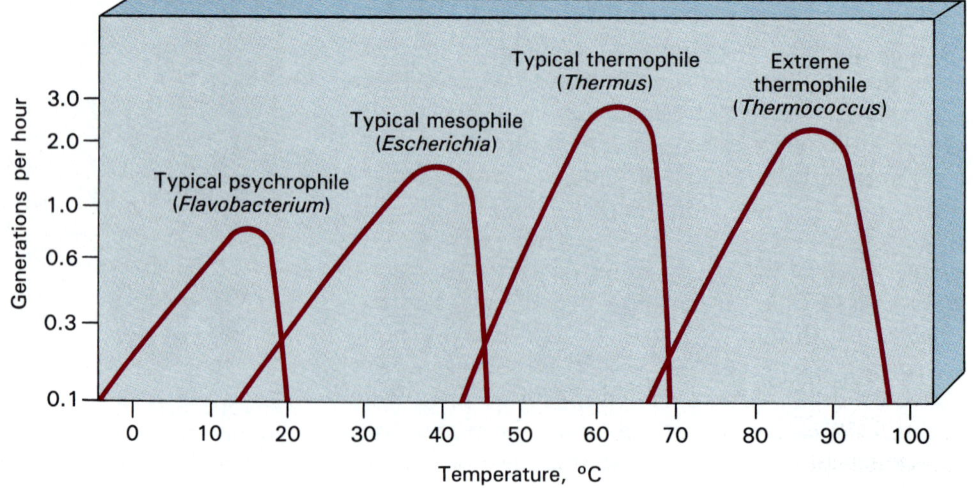

FIGURE 6.12 Comparison of the growth rates of typical psychrophilic, mesophilic, and thermophilic organisms. Note that there is some overlap of ranges at which these organisms can survive, but that rates of growth are much lower at the extreme ends of the ranges.

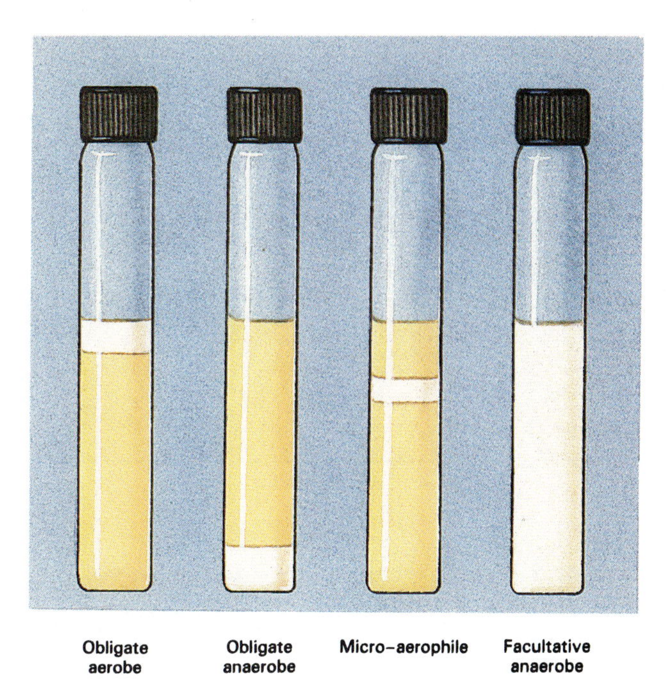

FIGURE 6.13 Different organisms incubated for 24 hours in nutrient broth tubes accumulate in different regions depending on their need for, or sensitivity to, oxygen.

CLOSE-UP

Even in the Coldest, Driest Part of the Planet

A few bacteria live in very cold, dry valleys of Antarctica where very few other organisms can survive. The relative humidity is so low that water passes directly from the frozen to the vapor state, and is rarely found as a liquid. Organisms that live there nevertheless manage to carry out their metabolic activities, either by using water vapor or by melting tiny amounts of ice with their metabolic heat.

Wright Dry Valley in Antarctica, one of the coldest places on earth.

tolerant anaerobes. **Microaerophiles** (mi″kro-a′er-o-filz) appear to grow best in the presence of a small amount of free oxygen. They grow below the surface of the medium in a culture tube at the level where oxygen availability matches their needs. Microaerophiles such as *Campylobacter*, which can cause intestinal disorders, also are **capnophiles** (carbon dioxide–loving organisms). They thrive under conditions of low oxygen and high carbon dioxide concentration. **Facultative anaerobes** ordinarily carry on aerobic metabolism when oxygen is present, but they shift to anaerobic metabolism when oxygen is absent. *Bacillus* and *Staphylococcus* are facultative anaerobes; they often are found in the intestinal and urinary tracts, where only a small amount of oxygen is available. The **aerotolerant anaerobes** can survive in the presence of oxygen but do not use it in their metabolism. *Lactobacillus*, for example, always captures energy by fermentation, regardless of whether the environment contains oxygen.

Compared with other groups of organisms defined according to oxygen requirements, facultative anaerobes have the most complex enzyme systems. They have one set of enzymes that enables them to use oxygen as an electron acceptor and another set that enables them to use another electron acceptor when oxygen is not available. In contrast, the enzymes of the other groups defined here are limited to either aerobic or anaerobic respiration.

Obligate anaerobes are killed not by gaseous oxygen, but by a highly reactive form of oxygen called **superoxide** (O_2^-). Superoxide is formed by certain oxidative enzymes and is converted to molecular oxygen (O_2) and hydrogen peroxide (H_2O_2) by an enzyme called **superoxide dismutase.** Hydrogen peroxide is converted to water and molecular oxygen by the enzyme **catalase.** Obligate aerobes and most facultative anaerobes have both enzymes. Some facultative and aerotolerant anaerobes have superoxide dismutase but lack catalase. Most obligate anaerobes lack both enzymes and succumb to the toxic effects of superoxide and hydrogen peroxide.

Moisture

All actively metabolizing cells generally require a water environment. Unlike larger organisms that have protective coverings and internal fluid environments, single-celled organisms are exposed directly to their environment. Most vegetative cells can live only a few hours without moisture; only the spores of spore-forming organisms can exist in a dry environment.

Hydrostatic Pressure

Water in oceans and lakes exerts **hydrostatic pressure,** pressure exerted by standing water, in proportion to its depth. Such pressure doubles with every 10 m increase in depth. For example, in a lake 50 m deep, the pressure is 32 times the atmospheric pressure. Some ocean valleys have depths in excess of 7000 m, and certain bacteria are the only organisms known to survive the extreme pressure at these depths. Bacteria that live at high pressures die if left in the laboratory for only a few hours at standard atmospheric pressure. It appears that their membranes and enzymes do not simply tolerate pressure but require pressure to function properly.

Osmotic Pressure

As we saw in Chapter 4, all cell membranes, including those of microorganisms, are selectively permeable and allow water to move by osmosis between the cell and the environment. (p. 99) **Environments that** contain dissolved substances exert osmotic pressure, and the pressure can exceed that exerted by dissolved substances in cells. Cells in such hyperosmotic environments lose water and undergo **plasmolysis** (plas-mol'e-sis), or shrinking of the cell with separation of the cell membrane from the cell wall. Conversely, cells in distilled water have a higher osmotic pressure than their environment and, therefore, gain water. The rigid cell wall prevents bacterial cells from swelling and bursting, but the cells fill with water and become turgid (distended).

Most bacterial cells can tolerate a fairly wide range of concentrations of dissolved substances. Their cell membranes contain enzyme systems called permeases (Chapter 5) that regulate the movement of dissolved substances across the membrane. (p. 131) **Yet, if** concentrations outside the cells become too high, water loss can inhibit growth or even kill the cells.

The use of salt as a preservative in curing hams and bacon and in making pickles is based on the fact that high concentrations of dissolved substances exert sufficient osmotic pressure to kill or inhibit microorganisms. The use of sugar as a preservative in making jellies and jams is based on the same principle.

Certain bacteria called **halophiles** (hal'o-fīlz), or salt-loving organisms, require moderate to large quantities of salt. Their membrane transport systems pump sodium ions out of the cells and concentrate potassium ions inside them. Two possible explanations for why halophiles require sodium have been proposed. One is that the cells need sodium to maintain a high intracellular potassium concentration so that their enzymes will function. The other is that they need sodium to maintain the integrity of their cell walls.

Halophiles (Figure 6.14) are typically found in the ocean, where the salt concentration (3.5 percent) is optimum for their growth. Extreme halophiles require salt concentrations of 20 to 30 percent. They are found in exceptionally salty bodies of water such as the Dead Sea and sometimes even in brine vats, where they cause spoilage of pickles being made there.

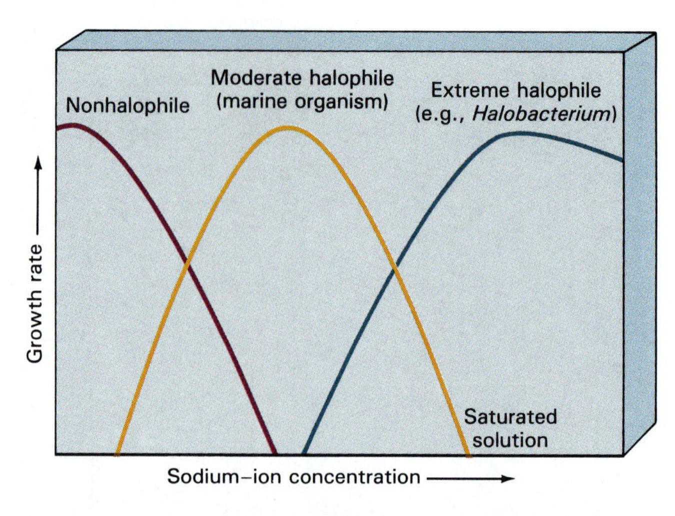

(a)

(b)

FIGURE 6.14 (a) Growth rates of halophilic ("salt-loving") and nonhalophilic organisms are related to sodium ion concentration. (b) The Great Salt Lake in Utah, an example of an environment in which halophilic organisms thrive. Note the white areas of dried salt around the edges of the lake.

Radiation

Radiant energy, particularly ultraviolet light, can cause mutations (changes in DNA) and even kill organisms. However, some microorganisms have pigments that screen radiation and help to prevent DNA damage. Others have enzyme systems that can repair certain kinds of DNA damage.

Nutritional Factors

The growth of microorganisms is affected by nutritional factors as well as physical factors. Nutrients needed by microorganisms include carbon, nitrogen, sulfur, phosphorus, certain trace elements, and vitamins. Although we are concerned with ways microorganisms satisfy their own nutritional needs, we can note that in satisfying such needs they also help to recycle elements in the environment. Activities of microbes in the carbon, nitrogen, sulfur, and phosphorus cycles are described in Chapter 26. A few microbes are **fastidious,** that is, they have many special nutritional needs that can be difficult to meet in the laboratory. Some fastidious organisms, including those that cause gonorrhea, grow quite well in the human body but still cannot be easily grown in the laboratory on an artificial medium.

Carbon Sources

Most bacteria use some carbon-containing compound as an energy source, and many use carbon-containing compounds as building blocks to synthesize cell components. Photosynthetic organisms reduce carbon dioxide to glucose and other organic molecules. Both autotrophic and heterotrophic organisms can obtain energy from glucose by glycolysis, fermentation, and the Krebs cycle. They also synthesize some cell components from intermediates in these pathways.

Nitrogen Sources

All organisms, including microorganisms, need nitrogen to synthesize enzymes and other proteins and nucleic acids. Some microorganisms obtain nitrogen from inorganic sources, and a few even obtain energy by metabolizing inorganic nitrogen-containing substances. Many microorganisms reduce nitrate ions (NO_3^-) to amino groups (NH_2) and use the amino groups to make amino acids. Some can synthesize all 20 amino acids found in proteins, whereas others must have one or a few amino acids provided in their medium. Certain fastidious organisms require all 20 amino acids and some other building blocks in their medium. Many disease-causing organisms obtain amino acids for making proteins and other nitrogenous molecules from the cells of humans and other organisms they invade.

Once amino acids are synthesized or obtained from the medium, they can be used in protein synthesis. Similarly purines and pyrimidines can be used to make DNA and RNA. The processes by which proteins and nucleic acids are synthesized are directly

APPLICATIONS

Picky Eaters

Spiroplasmas, tiny spiral bacteria that lack cell walls, are among the most nutritionally fastidious organisms known. Recently, a U.S. Department of Agriculture scientist devised an exact formula of 80 ingredients, including lipids, carbohydrates, amino acids, salts, vitamins, organic acids, and penicillin (to suppress potential competitors) to meet their needs. In his laboratory, he uses this medium to keep more than 30 species of spiroplasmas alive and well, making it possible for researchers to study them outside the more than 100 insect, tick, and plant species that they normally inhabit. Until now, most have been impossible to keep alive outside their hosts.

The spiroplasmas are responsible for hundreds of crop and animal diseases. Medical researchers are particularly interested in one organism that can experimentally cause tumors in animals. Another species kills honey bees, and a third lives harmlessly in the Colorado potato beetle, an insect that damages potato, eggplant, and tomato plants. Scientists hope to alter this last species genetically so that it will kill its potato beetle host.

USDA scientists are now trying to formulate complex media to grow mycoplasmalike organisms, a related group of bacteria that lack cell walls. These organisms cause hundreds of crop diseases and millions of dollars in economic losses each year. They are spread from plant to plant by infected insects. Another medium being designed would grow the bacterium *Mycoplasma pneumoniae*, which is important as the cause of "walking pneumonia" in humans.

USDA scientist Kevin Hackett working on his microbial "witch's brew"—a mix of some 80 ingredients that will support the growth of nutritionally fastidious spiroplasmas outside their hosts.

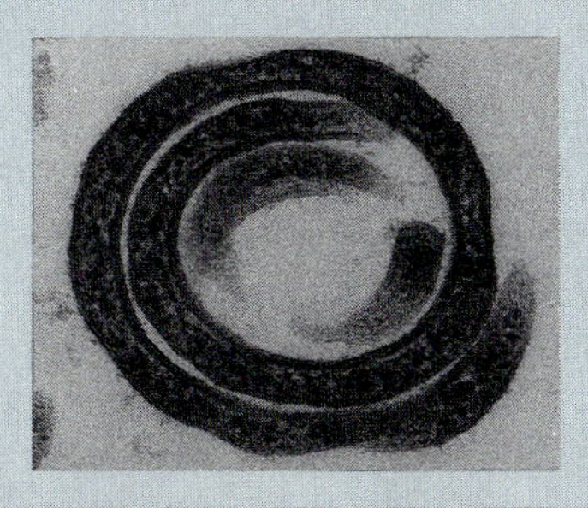

This spiroplasma, which inhabits the Colorado potato beetle, may one day be turned into a weapon against its destructive host.

related to the genetic information contained in a cell. Thus, their synthesis will be discussed in Chapters 7 and 8.

Sulfur and Phosphorus

In addition to carbon and nitrogen, microorganisms need a supply of certain minerals, especially sulfur and phosphorus, which are important cell components. They obtain sulfur from inorganic sulfate salts and from sulfur-containing amino acids. They use sulfur and sulfur amino acids to make proteins, coenzymes, and other cell components. Some organisms can synthesize sulfur amino acids from inorganic sulfur and other amino acids. Microorganisms obtain phosphorus mainly from inorganic phosphate ions (PO_4^{3-}). They use phosphorus (as phosphate) to synthesize ATP, phospholipids, and nucleic acids.

Trace Elements

Many microorganisms require a variety of **trace elements,** tiny amounts of minerals such as copper, iron, zinc, and cobalt, usually in the form of ions. Often they serve as cofactors in enzymatic reactions. All organisms require some sodium and chloride, and halophiles require large amounts of these ions. Potassium, zinc, and manganese are used to activate certain enzymes. Cobalt is required by organisms that can synthesize vitamin B_{12}. Iron is required for the synthesis of heme-containing compounds and for certain enzymes. (The cytochromes of the electron transport system contain heme.) Though little iron is required, a shortage severely retards growth. Calcium is required by gram-positive bacteria for synthesis of cell walls and by spore-forming organisms for synthesis of spore walls.

Vitamins

A substance is a **vitamin** for an organism only if the organism requires the substance and cannot synthesize it. Many microorganisms need no vitamins because they can synthesize any molecules they need from simpler substances. Other microorganisms require several vitamins in their media because they lack the enzymes to synthesize them. Vitamins required by some microorganisms include inositol, choline, folic acid, vitamin B_{12}, and vitamin K. Human pathogens often require a variety of vitamins and thus are able to grow only when they can obtain these substances from the host organism. Growing such organisms in the laboratory requires a complex medium that contains all the nutrients they normally obtain from their hosts. Microbes living in the human intestine manufacture vitamin K, which is necessary for blood clotting, and some of the B vitamins, thus benefiting their host.

APPLICATIONS

Bacteria That Measure

The discovery that certain organisms require vitamins or other special nutrients led to the development of *bioassay techniques*, the use of living organisms to measure the amount of a particular substance in a food or other material. To conduct a bioassay, a medium is prepared that contains all nutrients an organism needs except the one to be assayed. Then a known quantity of the substance to be assayed and the organism that requires it are added to the medium. For example, if you wanted to know how much folic acid a food contained, you could add a known amount of the food to a medium lacking folic acid and inoculate it with an organism with known folic acid requirements. Folic acid would be the limiting factor for growth, and the growth would therefore be proportional to the amount of folic acid present in the medium.

Nutritional Complexity

An organism's **nutritional complexity,** the number of nutrients it must obtain to grow, is determined by the kind and number of its enzymes. The absence of a single enzyme can render an organism incapable of synthesizing a substance. The organism then is dependent on obtaining the substance as a nutrient from its environment. Microorganisms vary in the number of enzymes they possess. Those with many enzymes have simple nutritional needs because they can synthesize nearly all the substances they need. Those with fewer enzymes have complex nutritional requirements because they lack the ability to synthesize many of the substances they need for growth. Thus, *nutritional complexity reflects a deficiency in biosynthetic enzymes.*

Locations of Enzymes

Most microorganisms move a variety of small molecules across their cell membranes and metabolize them. These substances include glucose, amino acids, small peptides, nucleosides, and phosphates, as well as various inorganic ions. Though they cannot move large molecules across membranes, in nature they use large molecules from organisms by digesting them before absorbing them. In addition to the endoenzymes that are produced for use within the cell (Chapter 5), many bacteria (and fungi) produce exoenzymes and release them through the cell membrane. These enzymes include **extracellular enzymes,** usually produced by gram-positive rods, which act in the medium around the organism, and **periplasmic enzymes,** usually produced by gram-negative organisms, which act between the cell wall and the cell membrane. Most

TABLE 6.1 Examples of exoenzymes

Enzymes	Action
Enzymes That Act on Complex Carbohydrates	
Carbohydrases	Break down large carbohydrate molecules into smaller ones
Amylase	Breaks down starch to maltose
Cellulase	Breaks down cellulose to cellobiose
Enzymes That Act on Sugars	
Sucrase	Breaks down sucrose to glucose and fructose
Lactase	Breaks down lactose to glucose and galactose
Maltase	Breaks down maltose to two glucose molecules
Enzymes That Act on Lipids	
Lipases	Break down fats to glycerol and fatty acids
Enzymes That Act on Proteins	
Proteases	Break down proteins to peptides and amino acids
Caseinase	Breaks down milk protein to amino acids and peptides
Gelatinase	Breaks down gelatin to amino acids and peptides

exoenzymes are hydrolases; they add water as they split large molecules of carbohydrate, lipid, or protein into smaller ones that can be absorbed (Table 6.1).

Adaptation to Limited Nutrients

Microorganisms adapt to limited nutrients in several ways:

1. Some synthesize increased amounts of enzymes for uptake and metabolism of limited nutrients. This allows the organisms to obtain and use a larger proportion of the few nutrient molecules that are available.
2. Others have the ability to synthesize enzymes needed to use a different nutrient. For example, if glucose is in short supply, some microorganisms can make enzymes to take up and use a more plentiful nutrient such as lactose.
3. Finally, many organisms adjust the rate at which they metabolize nutrients and the rate at which they synthesize molecules required for growth to fit the availability of the least plentiful nutrient. Metabolism and growth are both slowed, but no energy is wasted on synthesizing products that cannot be used. Growth is as rapid as conditions will allow.

SPORULATION

Sporulation, the formation of endospores, occurs in *Bacillus, Clostridium,* and a few other groups but has been studied most carefully in *B. subtilis* and *B. megaterium.* Bacteria that form endospores do so when nutrients are plentiful and environmental conditions are favorable, so they are prepared at all times for adverse conditions should they arise. In fact, once conditions have seriously deteriorated they are unable to form endospores.

When nutrients such as carbon, nitrogen, or phosphorus become depleted in the environment or when physical factors such as temperature, pressure, or moisture become adverse, highly resistant, dehydrated endospores already exist inside parent cells. Though endospores are not metabolically active, they can survive long periods of drought and are resistant to killing by extreme temperatures, radiation, and some toxic chemicals. Some endospores can withstand about 1 million times more heat than vegetative cells can. The endospore itself cannot divide, and the parent cell can produce only one endospore, so sporulation is a protective mechanism and not a means of reproduction.

As endospore formation begins (Figure 6.15), DNA is replicated and forms a long, compact **axial nucleus.** The two chromosomes formed by replication separate and move to different locations in the cell. In some bacteria the endospore forms at one end of the cell, and in others it forms near the middle (Figure 6.16). The DNA where the endospore will form directs endospore formation. Most of the cell's RNA and some condensed, dehydrated protein molecules gather around the DNA to make the **core,** or living part of the endospore. An **endospore septum,** consisting of a cell membrane but lacking a cell wall, grows around the core, enclosing it in a double thickness of cell membrane. Both layers of this membrane synthesize peptidoglycan and release it into the space between the membranes. Thus, a laminated layer called the **cortex** is formed. The cortex also contains **dipicolinic** (di-pik-o-lin'ik) **acid** and calcium ions, which probably contribute to an endospore's heat resistance. The cortex also protects the core against changes in osmotic pressure such as those that result from drying. A **spore coat** of keratinlike protein, which is impervious to many chemicals, is laid down around the cortex by the mother cell. Finally, in some endospores an **exosporium,** a lipid-protein membrane, also is formed outside the coat by the mother cell. The function of the exosporium is unknown.

Once favorable conditions return, an endospore germinates into a vegetative cell and begins to divide by binary fission in about 90 minutes. **Germination** occurs in three stages. The first stage, *activation,* usually requires some traumatic agent such as low pH or heat, which damages the coat, though some endospores germinate spontaneously in a favorable medium. The second stage, *germination proper,* requires water and a germination agent, such as the amino

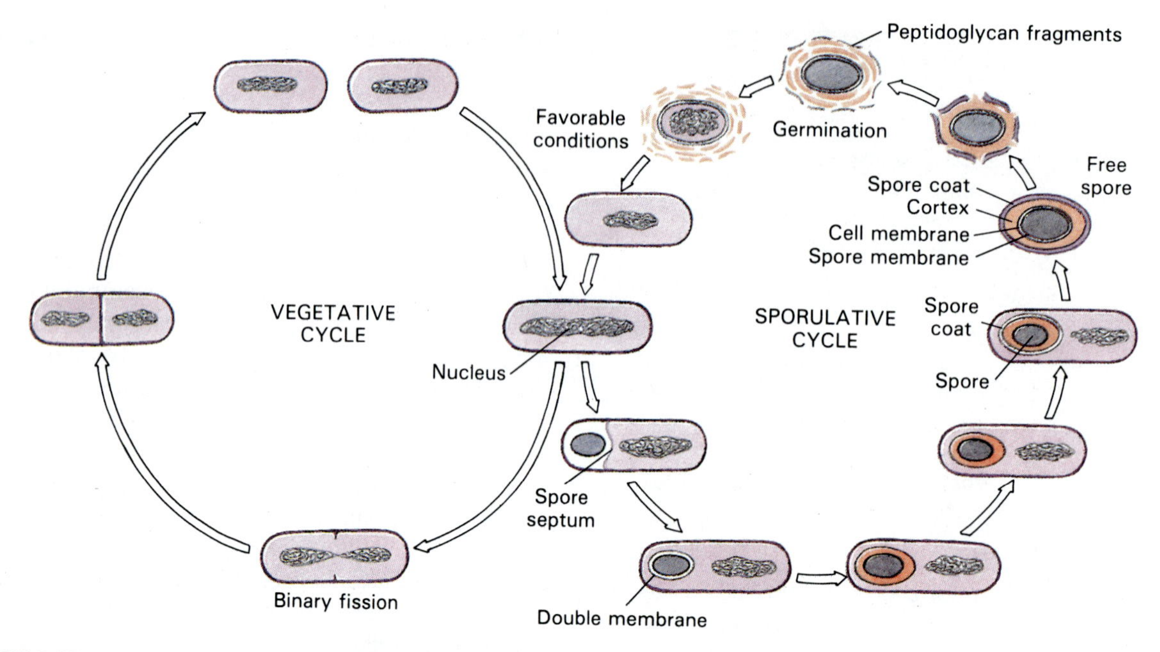

FIGURE 6.15 The vegetative and sporulation cycles in bacteria capable of sporulation.

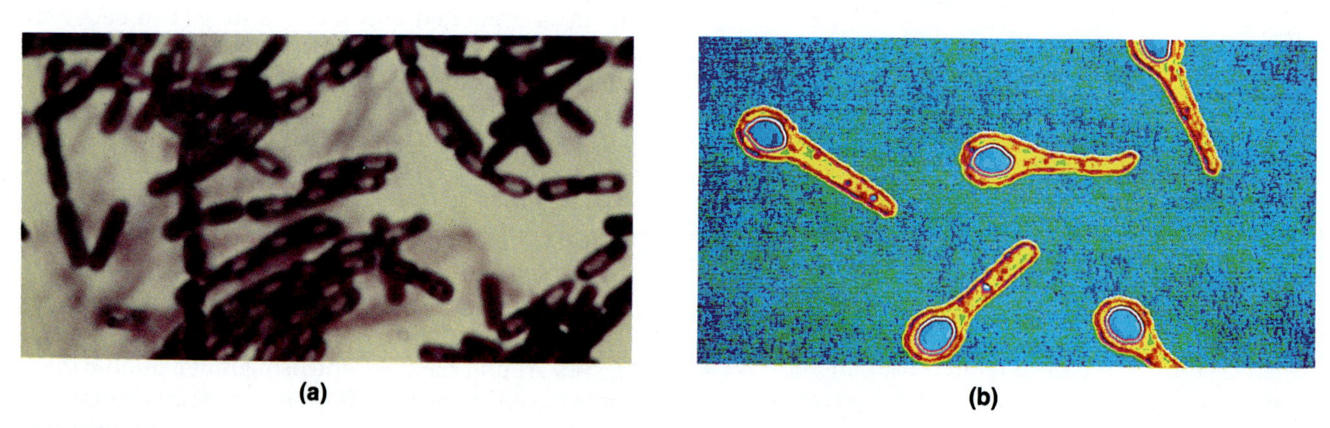

FIGURE 6.16 Bacterial endospores in two *Clostridium* species. (a) Cells with centrally located endospores. (b) False-colored TEM of cells with terminally located endospores, which give the organisms a club-shaped appearance.

acid alanine or certain inorganic ions, which penetrates the damaged coat. During this process, much of the cortical peptidoglycan is broken down, and its fragments are released into the medium. The living cell (which occupied the core) now takes on large quantities of water and loses its resistance to heat and staining, plus, its refractility (ability to bend light rays). Finally, *outgrowth* occurs in a medium with adequate nutrients. Proteins and RNA are synthesized, and in about an hour DNA synthesis begins. The cell is now a vegetative cell and undergoes binary fission.

Thus, bacterial cells capable of sporulation display two cycles—the vegetative cycle and the sporulation cycle (Figure 6.15). The vegetative cycle is repeated at intervals of 20 minutes or more, and the sporulation cycle is initiated periodically. Endospores known to be 300 or more years old have been observed to undergo germination when placed in a favorable medium.

Certain bacteria, such as *Azotobacter*, form resistant **cysts,** or spherical, thick-walled cells, that resemble endospores. Like endospores, cysts are metabolically inactive and resist drying. Unlike endospores they lack dipicolinic acid and have only limited resistance to high temperatures. Cysts germinate into single cells and, therefore, are not a means of reproduction.

Some filamentous bacteria, such as *Micromonospora* and *Streptomyces*, form asexually reproduced **conidia** (ko-nid'e-ah), or chains of aerial spores with thick outer walls. These spores are temporarily dormant but are not especially resistant to heat or drying. When the spores, which are produced in large numbers, are dispersed to a suitable environment, they form new filaments. Unlike endospores, these spores do contribute to reproduction of the species.

CULTURING BACTERIA

Culturing of bacteria in the laboratory presents two problems. First, a pure culture of a single species is needed to study an organism's characteristics. Second, a medium must be found that will support the growth of the desired organism. Let us look at some of the ways these problems are solved.

Methods of Obtaining Pure Cultures

To study bacteria in the laboratory, it is important to obtain a **pure culture,** one that contains only a single species of organism. Today, pure cultures are obtained by isolating the progeny of a single cell. Prior to the development of pure culture techniques, scientists studied mixed cultures, or cultures containing several different kinds of organisms. They could make observations of different shapes and sizes of organisms, but they could find out little about the nutritional needs or growth characteristics of individual species.

Simple as it seems now, the technique of isolating pure cultures was difficult to develop. Attempts to isolate single cells by serial dilution were often unsuccessful because two or more organisms of different species were often present in the highest dilutions. Koch's technique of spreading bacteria thinly over a solid surface was more effective because it deposited a single bacterium at some sites. However, he tried several different solid substances before discovering that agar was an ideal solidifying agent. Only a very few organisms digest it, and in 1.5 percent solution it does not melt below 95°C. Furthermore, after being melted, it remains in the liquid state until it has cooled to about 40°C, a temperature cool enough to allow the addition of nutrients and living organisms that might be destroyed by heat.

Streak Plate Method

Today, the accepted way to prepare pure cultures is the **streak plate method,** which uses agar plates. Bacteria are picked up on a sterile wire loop, and the wire is moved lightly along the agar surface, depositing streaks of bacteria on the surface. The inoculating loop is flamed between streaking different areas (Figure 6.17). Fewer and fewer bacteria are deposited as the streaking continues, and individual organisms are deposited in the region streaked last. After the plate is incubated at a suitable growth temperature for the organism, small colonies derived from a single organism appear. The wire loop is used to pick up a portion of a colony and transfer it to any appropriate sterile medium for further study. If sterile technique is carefully followed, it assures that the new medium will contain organisms of a single species.

Pour Plate Method

Another way to obtain pure cultures, the **pour plate method,** makes use of serial dilutions. A series of dilutions are made such that the final dilution contains about a thousand organisms. Then 1 mL of liquid medium from the final dilution is placed in 9 mL of melted agar medium (45°C), and the medium is quickly poured into a sterile plate. The resulting pour plate will contain a small number of bacteria, some of which will form isolated colonies on the agar. This method allows some organisms to be embedded in the medium. It is particularly useful for growing microaerophiles that cannot tolerate exposure to oxygen in the air at the surface of the medium.

Culture Media

In nature many species of bacteria and other microorganisms are found growing together in oceans, lakes, and soil and on living or dead organic matter. These materials might be thought of as natural **media**

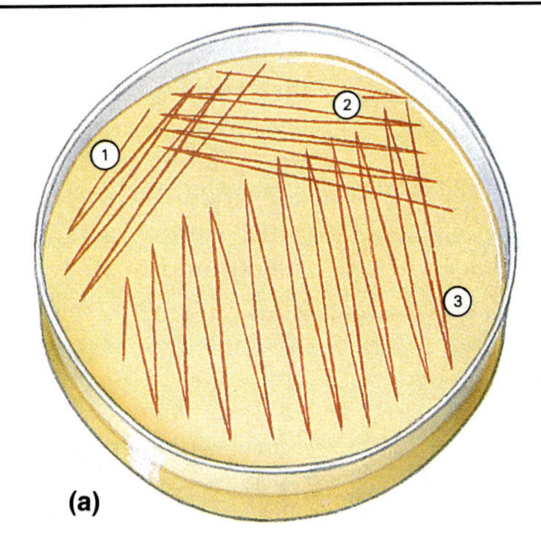

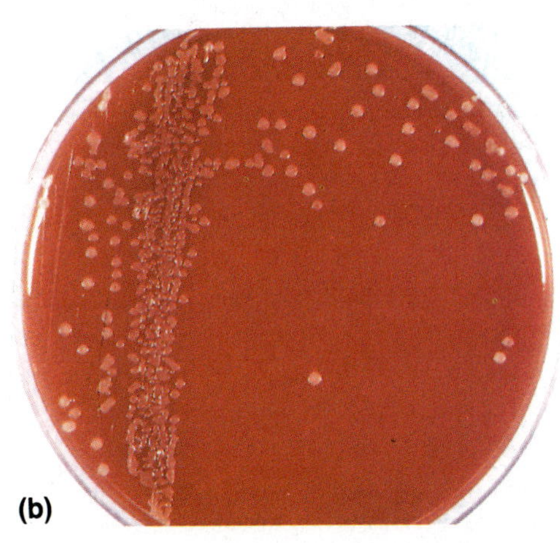

FIGURE 6.17 The streak-plate method of obtaining pure cultures. (a) A drop or bit of culture on a wire inoculating loop is lightly streaked across the top of the agar in region 1. The loop is flamed, the plate is rotated, and a few organisms are picked up from region 1 and streaked out into region 2. The loop is flamed again, and the process is repeated in region 3. The plate is then incubated. (b) A streak plate after incubation. Note the greatly reduced numbers of colonies in each successive region.

(singular: *medium*), that is, substances on or in which microorganisms grow. Though soil and water samples are often brought into the laboratory, organisms from them are typically isolated, and pure cultures are prepared for study.

Growing bacteria in the laboratory requires knowledge of their nutritional needs and the ability to provide the needed substances in a medium. Through years of experience in culturing bacteria in the laboratory, microbiologists have learned what nutrients must be supplied to each of many different organisms. Certain organisms, such as those that cause syphilis and leprosy, still cannot be cultured in laboratory media. They must be grown in cultures that contain living human or other animal cells. Many other organisms whose nutritional needs are reasonably well known can be grown in one or more types of media.

Types of Media

Laboratory media are generally synthetic media, as opposed to the natural media mentioned previously. A **synthetic medium** is a medium prepared in the laboratory from materials of precise or reasonably well-defined composition. A **defined synthetic medium** is one that contains known specific kinds and amounts of chemical substances. Examples of defined synthetic media are given in Tables 6.2 and 6.3. A **complex medium,** or **chemically nondefined medium,** is one that contains certain reasonably familiar materials but that varies slightly in chemical composition from batch to batch. Such media contain blood or extracts from beef, yeasts, soybeans, and other organisms. A common ingredient is **peptone,** a product of enzyme digestion of proteins. It provides small peptides that microorganisms can use. Though the exact concentrations are not known, trace elements and vitamins are present in sufficient quantities in complex media to support the growth of many organisms. Both liquid nutrient broth and solidified agar medium used to culture many organisms are complex media. An example of a complex medium is given in Table 6.4.

TABLE 6.2 A defined synthetic medium for growing *Proteus vulgaris*

Ingredient	Amount	Ingredient	Amount
Water	1 L	K_2HPO_4	1 g
$MgSO_4 \cdot 7H_2O$	200 mg	$FeSO_4 \cdot 7H_2O$	10 mg
$CaCl_2$	10 mg	Glucose	5 g
NH_4Cl	1 g	Nicotinic acid	0.1 mg
Trace elements (Mn, Mo, Cu, Co, Zn as inorganic salts, known quantities of 0.02–0.5 mg each)			

SOURCE: Adapted from R. Y. Stanier, et al. 1986. *The microbial world.* 5th ed. Englewood Cliffs, NJ: Prentice Hall.

TABLE 6.3 A defined synthetic medium for growing a fastidious bacterium, *Leuconostoc mesenteroides*

WATER	1 L		
ENERGY SOURCE			
Glucose	25 g		
NITROGEN SOURCE			
NH_4Cl	3 g		
MINERALS			
KH_2PO_4	600 mg	$FeSO_4 \cdot 7H_2O$	10 mg
K_2HPO_4	600 mg	$MnSO_4 \cdot 4H_2O$	20 mg
$MgSO_4 \cdot 7H_2O$	200 mg	NaCl	10 mg
ORGANIC ACID			
Sodium acetate	20 g		
AMINO ACIDS			
DL-α-Alanine	200 mg	L-Lysine · HCl	250 mg
L-Arginine	242 mg	DL-Methionine	100 mg
L-Asparagine	400 mg	DL-Phenylalanine	100 mg
L-Aspartic acid	100 mg	L-Proline	100 mg
L-Cysteine	50 mg	DL-Serine	50 mg
L-Glutamic acid	300 mg	DL-Threonine	200 mg
Glycine	100 mg	DL-Tryptophan	40 mg
L-Histidine · HCl	62 mg	L-Tyrosine	100 mg
DL-Isoleucine	250 mg	DL-Valine	250 mg
DL-Leucine	250 mg		
PURINES AND PYRIMIDINES			
Adenine sulfate · H_2O	10 mg	Uracil	10 mg
Guanine · HCl · $2H_2O$	10 mg	Xanthine · HCl	10 mg
VITAMINS			
Thiamine · HCl	0.5 mg	Riboflavin	0.5 mg
Pyridoxine · HCl	1.0 mg	Nicotinic acid	1.0 mg
Pyridoxamine · HCl	0.3 mg	p-Aminobenzoic acid	0.1 mg
Pyridoxal · HCl	0.3 mg	Biotin	0.001 mg
Calcium pantothenate	0.5 mg	Folic acid	0.01 mg

SOURCE: H. E. Sauberlich and C. A. Baumann, A factor required for the growth of *Leuconostoc citrovorum*, *J. Biol. Chem.* 176(1948):166.

Commonly Used Media

Most routine laboratory cultures make use of media containing peptone from meat or fish in nutrient broth or solid agar medium. Such media are sometimes enriched with **yeast extract,** which contains a number of vitamins, coenzymes, and nucleosides. **Casein hydrolysate** made from milk protein contains many amino acids and is used to enrich certain media. Because blood contains many nutrients needed by human pathogens, **serum** (the liquid part of the blood after clotting factors have been removed), whole blood, and heated whole blood can be useful in enriching media. **Blood agar** is useful in identifying organisms that can cause hemolysis, or breakdown of red blood cells. Sheep's blood is used because its hemolysis is more clearly defined than when human blood is used in the agar plates. **Chocolate agar,** made with heated blood, is so named because it turns a chocolate brown.

TABLE 6.4 A complex medium suitable for many heterotrophic organisms	
Nutrient Broth Ingredient	**Amount**
Water	1 L
Peptone	5 g
Beef extract	3 g
NaCl	8 g
Solidified Medium	
Agar	15 g
Above ingredients in amounts specified.	

Selective, Differential, and Enrichment Media

To isolate and identify particular microorganisms, especially those from patients with infectious diseases, selective, differential, or enrichment media are often used. Such special media (Table 6.5) are an essential part of modern diagnostic microbiology.

A **selective medium** is one that encourages the growth of some organisms and suppresses growth of others. For example, to identify *Clostridium botulinum* in food samples suspected of being agents of food poisoning, sulfadiazine and polymyxin sulfate (SPS) are added to anaerobic cultures of *Clostridium* species. This culture medium is called SPS agar. It allows growth of *Clostridium botulinum* and inhibits growth of most other *Clostridium* species.

A **differential medium** has a constituent that causes an observable change (a color change or a change in pH) in the medium when a particular chemical reaction occurs. The SPS agar medium also serves as a differential medium. Colonies of *Clostridium botulinum* formed on this medium are black because of hydrogen sulfide made by the organisms from the sulfur-containing additives.

Many media, such as SPS agar, are both selective and differential. MacConkey agar is another selective and differential medium. It contains crystal violet and bile salts, which inhibit growth of gram-positive bacteria and allow growth of gram-negative ones. It also contains the sugar lactose and a pH indicator that makes colonies of lactose fermenters red and leaves colonies of nonfermenters uncolored. Though there are some exceptions, most organisms normally found in the human intestines ferment lactose, whereas most pathogens do not.

An **enrichment medium** contains special nutrients that allow growth of a particular organism, which might not otherwise be present in sufficient numbers to allow it to be isolated and identified. Unlike a selective medium, an enrichment medium does not suppress others. For example, because *Salmonella typhi* organisms may not be sufficiently numerous in a fecal sample to allow positive identification, they are cultured on a medium containing the trace element selenium, which fosters growth of the organism. After incubation in the enrichment medium, the greater numbers of the organisms increase the likelihood of a positive identification.

TABLE 6.5 Selected examples of diagnostic media

Medium	Organism(s) Identified	Selectivity and/or Differentiation Achieved
Brilliant green agar	*Salmonella*	Brilliant green dye inhibits gram-positive bacteria and thus selects gram-negative ones. Differentiates *Shigella* colonies (which do not ferment lactose or sucrose and are red to white) from other organisms that do ferment one of the sugars and are yellow to green.
Desoxycholate agar	Gram-negative enterics	Sodium desoxycholate inhibits gram-positive bacteria. Differentiates organisms that ferment lactose (red colonies) from those that do not (colorless colonies).
Eosin methylene blue agar	Gram-negative enterics	Medium partially inhibits gram-positive bacteria. Eosin and methylene blue differentiate among organisms: *Escherichia* coli colonies have a metallic green sheen, *Enterobacter aerogenes* colonies are pink, indicating that they ferment lactose, and colonies of other organisms are transparent, indicating they do not ferment lactose.
Sodium tetrathionate broth	Enteric pathogens	Sodium tetrathionate inhibits normal inhabitants of the gut and enriches growth of certain pathogens, such as *Salmonella* and *Shigella*.
Triple sugar-iron agar	Gram-negative enterics	Used in agar slants (tubes cooled in slanted position) where differentiation is based on both aerobic surface growth and anaerobic growth in agar in base of tube. Medium contains glucose, sucrose, and lactose and a pH indicator, so relative use of each sugar can be detected.

Controlling Oxygen Content of Media

Obligate aerobes, microaerophiles, and obligate anaerobes require special attention to maintain oxygen concentrations suitable for growth. Most obligate aerobes obtain sufficient oxygen from nutrient broth or solidified agar medium, but some need more. Oxygen gas is bubbled through the medium with filters between the gas source and the medium to prevent contamination of the culture. Microaerophiles can be incubated in nutrient broth tubes or agar plates in a jar in which a candle is lit before the jar is sealed (Figure 6.18). (Scented candles should not be used because oils from them inhibit bacterial growth.) The burning candle uses oxygen from the air in the jar and adds carbon dioxide to it. When the carbon dioxide extinguishes the flame, conditions are optimum for the growth of *Neisseria gonorrhoeae,* which causes gonorrhea.

To culture obligate anaerobes, all molecular oxygen must be removed and kept out of the medium. Adding oxygen-binding agents such as thioglycollate, the amino acid cysteine, or sodium sulfide to the medium prevents oxygen from exerting toxic effects on anaerobes. Such tubes are completely filled to exclude air and are sealed with screw caps. Agar plates are incubated in sealed jars containing chemical substances that remove oxygen from the air and generate carbon dioxide (Figure 6.19). Stab cultures can be made by literally stabbing a straight inoculating wire coated with organisms into a tube of agar-solidified medium. In laboratories where anaerobes are regularly handled, an anaerobic chamber (Figure 6.20) is often used. Equipment and cultures are introduced through an air lock, and the technician uses glove ports to manipulate the cultures.

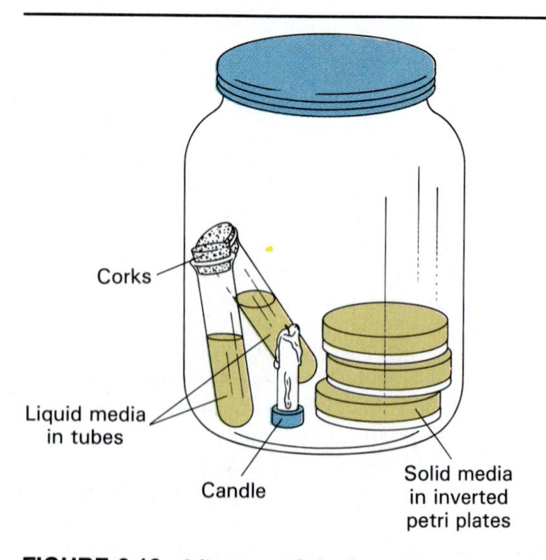

FIGURE 6.18 Microaerophiles are growing in culture tubes and on Petri plates in a sealed jar in which a candle burned until it was extinguished by carbon dioxide accumulation in the atmosphere of the jar. A small amount of oxygen remains.

APPLICATIONS

Special Cave for the Gonococcus

The transport of specimens from patients to the laboratory in a viable condition sometimes presents special problems. The organisms must not be subjected to drying conditions or to too much or too little oxygen. And, it should go without saying, specimen handlers must be protected from infection. Cultures that may contain *Neisseria gonorrhoeae* from patients with gonorrhea pose one such problem—that of providing an atmosphere relatively high in carbon dioxide. Various commercial systems, such as the widely used JEMBEC (John E. Martin Biological Environmental Chamber), are available for this purpose. This system consists of a small plastic plate of selective medium and a tablet of sodium bicarbonate and citric acid. The plate is inoculated, the tablet is placed in the plate, and the plate is placed in a plastic bag and sealed. Moisture from the medium causes the tablet to release carbon dioxide so that an appropriate concentration of 5 to 10 percent is obtained. The culture is incubated for 18 to 24 hours to allow growth to begin before shipment to the laboratory.

The JEMBEC system is used to culture gonorrhea specimens.

Maintaining Cultures

Once an organism has been isolated, it can be maintained indefinitely in a pure culture called a **stock culture.** When needed for study, a sample from a stock culture is inoculated into fresh medium. The stock culture itself is never used for laboratory studies. However, organisms in stock cultures go through growth phases, deplete nutrients, and accumulate wastes just as those in any culture do. As the culture ages, the organisms may acquire odd shapes or other altered characteristics. Stock cultures are maintained by making subcultures in fresh medium at frequent intervals to keep the organisms growing.

The use of careful aseptic techniques is important in all manipulations of cultures. **Aseptic techniques**

FIGURE 6.19 Chemicals which, when activated, remove oxygen can be enclosed with cultures in a sealed jar to create an anaerobic chamber. These are useful for the small laboratory, which has only a few plates needing anaerobic incubation.

minimize the chances that cultures will be contaminated by organisms from the environment or that organisms, especially pathogens, will escape into the environment. Such techniques are especially important in making subcultures from stock cultures. Otherwise an undesirable organism might be introduced, and the stock organism would have to be reisolated. Even with regular transfers of organisms from stock cultures to fresh medium, the organisms can undergo mutations (changes in DNA) and develop altered characteristics.

Preserved Cultures

To avoid the risk of contamination or alteration, stock culture organisms also should be kept in a **preserved culture,** a culture in which organisms are maintained in a dormant state. The most commonly used technique for preserving cultures is lyophilization (freeze-drying), in which cells are quickly frozen, dehydrated while frozen, and sealed in vials under vacuum (Chapter 13). Such cultures can be kept indefinitely at room temperature.

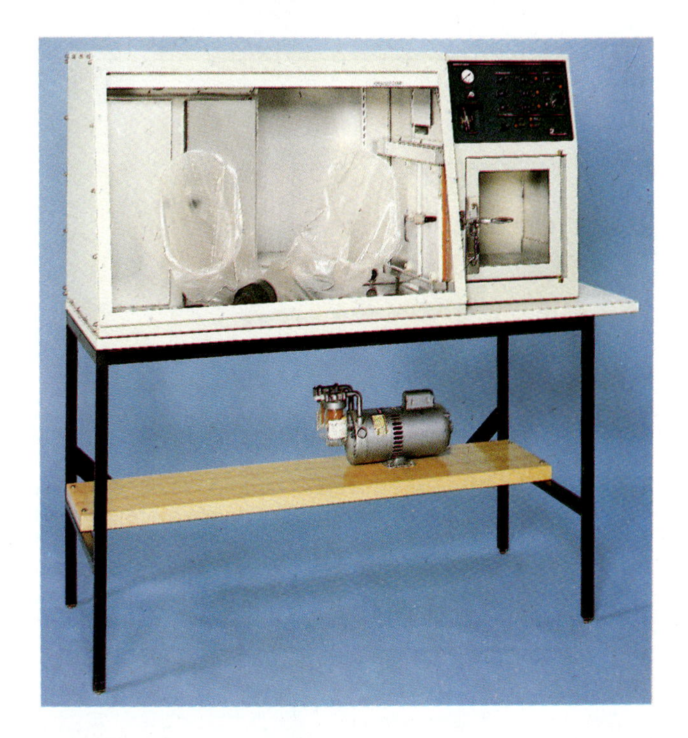

FIGURE 6.20 A large anaerobic transfer chamber with an air lock for introducing equipment and cultures, and glove ports to allow manipulation of the cultures.

Because microorganisms frequently undergo genetic changes, reference cultures are maintained. A **reference culture** is a preserved culture maintained to preserve the organisms with its characteristics as originally defined. Reference cultures of all known species and strains of bacteria and many other microorganisms are maintained in the American Type Culture Collection, and many also are maintained in universities and research centers. Then if stock cultures in a particular laboratory undergo change or if other laboratories wish to obtain certain organisms for study, they are always available.

ESSAY

Methods of Performing Multiple Diagnostic Tests

Many diagnostic laboratories use culture systems that contain a large number of differential and selective media, such as the Enterotube Multitest System (Figure 6.21) or the Analytical Profile Index (API). These systems allow simultaneous determination of an organism's reaction to a variety of carefully chosen diagnostic media from a single inoculation. The advantages of these systems are that they use small quantities of media, occupy little space in an incubator, and provide an efficient and reliable means of making positive identification of infectious organisms.

The Enterotube System® is used to identify enteric pathogens, or organisms that cause intestinal diseases such as typhoid, paratyphoid, shigellosis, gastroenteritis, and some kinds of food poisoning. The causative organisms are all gram-negative rods indistinguishable from one another without biochemical tests. The Enterotube System consists of a tube with compartments, each of which contains a different medium, and a sterile inoculating needle. By touching the tip of the needle to a colony and drawing the needle through the tube, each compartment is inoculated. After the tube has been incubated for 24 hours at 37°C, the results of 15 biochemical tests can be ob-

ESSAY Methods of Performing Multiple Diagnostic Tests

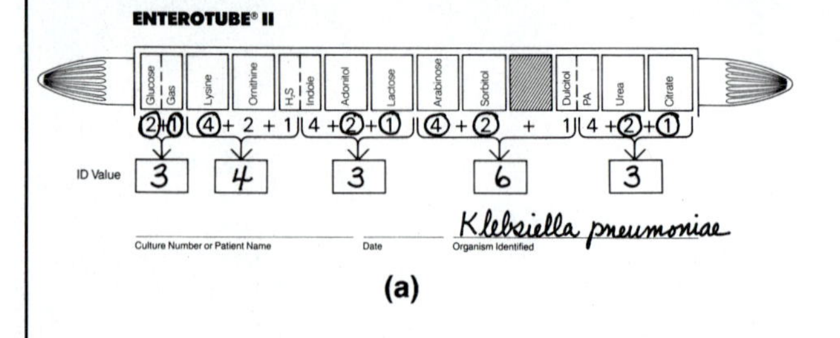

(a)

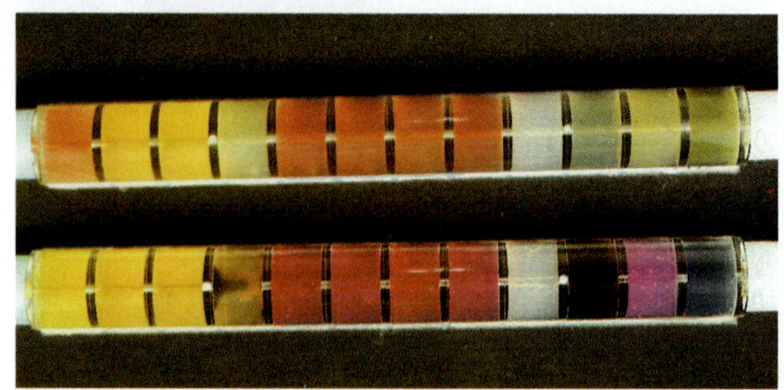

(b)

FIGURE 6.21 The Enterotube Multitest System. After inoculation and incubation, compartments with positive test results are assigned a number. The numbers are summed within zones to get a definitive index number, which identifies an organism on the list in the coding manual. Any necessary confirmatory tests are also noted there. By numbering each test in a zone with a digit equal to a power of 2 (1, 2, 4, 8, and so on), the sum of any set of positive reactions results in a unique number. A given species may, however, be coded for many different numbers, as individual strains of that species will vary somewhat in their characteristics.

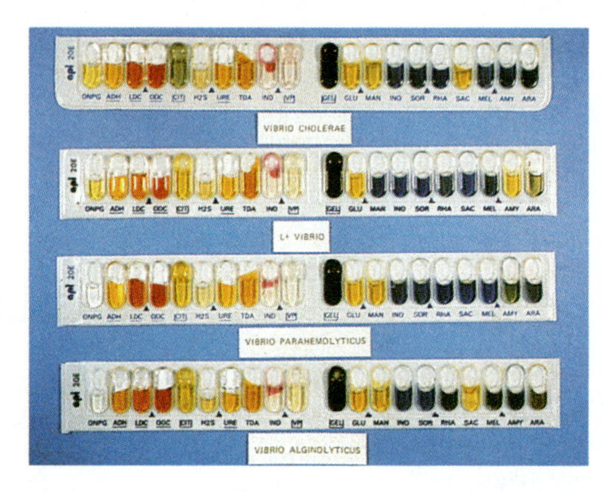

FIGURE 6.22 The Analytical Profile Index (API) 20E System. Various species of the genus Vibrio are shown here, with the differences in reactions that enable them to be identified. This system allows identification to species level of 125 gram-negative intestinal bacilli.

tained by observing (1) whether gas was produced and (2) the color of the medium in each compartment. Tests are grouped in sets of three; within each group tests are assigned a number 1, 2, or 4. The sum of the numbers of positive tests in each group indicates which tests are positive. The sum 3 shows tests 1 and 2 were positive, the sum 5 shows tests 1 and 4 were positive, the sum 6 shows tests 2 and 4 were positive, and the sum 7 shows all tests were positive. The single-digit sums for each of the five sets of tests are combined to form a five-digit identification number for a particular organism. For example, 36601 is *Escherichia coli*, and 70763 is *Klebsiella pneumoniae*. A list of identification numbers and the corresponding organisms is provided with the system.

API consists of a plastic tray with 20 microtubes, each containing a different kind of dehydrated medium. After the medium is rehydrated, bacteria from a colony are mixed with sterile saline solution, and the microtubes are inoculated from the solution. As with Enterotubes, the tray is incubated, test results are determined, and the values 1, 2, and 4 are summed for sets of three tests. The seven-digit profile number identifies the organism (Figure 6.22).

In this brief discussion of diagnostic systems, we have considered only the tip of the iceberg. Of the many other available tests, a large number are based on immunological properties of organisms. Some of these will be considered with immunology or with particular infectious agents. Also, much is known about which organisms are likely to infect certain human organs and tissues, and many diagnostic tests are designed to distinguish among organisms found in respiratory secretions, fecal samples, blood, other tissues, and body fluids.

CHAPTER SUMMARY

GROWTH AND CELL DIVISION

Definition of Growth

- Growth can be defined as the orderly increase in quantity of all components of an organism.
- Because of limited increase in cell size and the frequency of cell division, growth in microorganisms is measured by increase in cell number.

Cell Division

- Most cell division occurs by **binary fission,** in which the nuclear body divides

RELATED KEY TERMS

mother cell **daughter cells**
microbial growth

and the cell envelope forms a transverse septum that separates the original cell into two cells.

- Yeast cells and some bacteria divide by **budding,** in which a small new cell develops from the surface of an existing cell.

Phases of Growth

- In a nutrient-rich medium bacteria divide rapidly; the length of time required for one division is called the **generation time.** Such growth is said to be **exponential,** or **logarithmic.**
- Bacteria introduced into fresh, nutrient-rich medium display four major phases of growth: (1) In the **lag phase,** the organisms are metabolically active—growing and synthesizing various substances, but not increasing in number. (2) In the **log phase,** organisms divide at an exponential, or logarithmic, rate and with a constant generation time. These properties of growth in the log phase can be used to calculate both the number of generations and the generation time. Cultures can be maintained by the use of a **chemostat,** which allows continuous addition of fresh medium. (3) In the **stationary phase,** the number of new cells produced equals the number of cells dying. The medium contains limited amounts of nutrients and may contain toxic quantities of waste materials. (4) In the **death phase,** many cells lose their ability to divide and die. A logarithmic decrease in the number of cells results.
- Growth in colonies parallels that in liquid medium, except that most growth occurs at the edge of the **colony** and all phases of growth occur simultaneously somewhere in the colony.

Measuring Bacterial Growth

- Growth can be measured by transferring a small amount of a known dilution of bacteria onto an **agar plate** and counting the colonies that arise. Each colony represents one live cell from the original sample.
- Growth also can be measured by **direct microscopic counts, most probable number** technique, **filtration,** observation or measurement of **turbidity,** dry weight of cells, and the measurement of products of metabolism.

FACTORS AFFECTING BACTERIAL GROWTH

Physical Factors

- Acidity and alkalinity of the medium affect growth, and most organisms have an **optimum pH** range of no more than one pH unit.
- Particular species may be able to tolerate a pH as low as 0 or as high as 12.
- Temperature affects bacterial growth. (1) Most bacteria can grow over a 30°C temperature range. (2) Bacteria can be classified according to growth temperature into three categories: **psychrophiles** that grow at low temperatures (below 25°C), **mesophiles** that grow best at temperatures between 25° and 40°C, and **thermophiles** that grow at high temperatures (above 40°C). (3) The temperature range of an organism is closely related to the temperature at which its enzymes function best.
- The quantity of oxygen in the environment affects the growth of bacteria. (1) **Obligate aerobes** require relatively large amounts of free molecular oxygen to grow. (2) **Obligate anaerobes** are killed by free oxygen and must be grown in the absence of free oxygen. (3) **Facultative anaerobes** can metabolize substances aerobically if oxygen is available or anaerobically if it is absent. (4) **Aerotolerant anaerobes** metabolize substances anaerobically but are not harmed by free oxygen. (5) **Microaerophiles** must have only limited oxygen to grow.
- Actively metabolizing bacteria require a water environment, except for a few that live in Antarctic dry valleys.
- Some bacteria, but no other living things, can withstand extreme **hydrostatic pressures** in deep valleys in the ocean.
- Osmotic pressure affects bacterial growth, and water can be drawn into or out of cells according to the relative osmotic pressure created by dissolved substances in the cell and the environment. (1) Permeases minimize the effects of high osmotic pressure in the environment. (2) Bacteria called **halophiles** require moderate to large amounts of salt and are found in the ocean or in exceptionally salty bodies of water.

Nutritional Factors

- All organisms require a carbon source: (1) Autotrophs use CO_2 as their carbon source and synthesize other substances they need. (2) Heterotrophs require

RELATED KEY TERMS

tetrad sarcinae
bacterial growth curve

synchronous growth
nonsynchronous growth

decline phase **involution**
serial dilution
countable number

physical factors
nutritional factors
acidophiles
neutrophiles
alkalinophiles

obligate **facultative**
obligate psychrophiles
facultative psychrophiles
obligate thermophiles
facultative thermophiles

superoxide
superoxide dismutase
catalase
capnophiles

plasmolysis

fastidious

glucose or another carbon source from which they obtain energy and intermediates for synthetic processes.

- Microorganisms require an organic or inorganic nitrogen source from which to synthesize proteins and nucleic acids. They also require a source of other elements found within them, including sulfur, phosphorus, potassium, iron, and many **trace elements.**
- Some microorganisms require particular **vitamins** because they lack the enzymes to synthesize them.
- The nutritional requirements of an organism are determined by the kind and number of its enzymes. **Nutritional complexity** reflects a deficiency in biosynthetic enzymes.
- Bioassay techniques use metabolic properties of organisms to determine quantities of vitamins and other compounds in foods and other materials.
- Most microorganisms move low-molecular-weight substances across their cell membranes and metabolize them internally. Some bacteria (and fungi) also produce exoenzymes that digest large molecules outside the cell membrane of the organism.
- Microorganisms adjust to limited nutrient supplies by increasing the quantities of enzymes they produce, by making enzymes to metabolize another available nutrient, or by adjusting their metabolic activities to grow at a rate consistent with availability of nutrients.

RELATED KEY TERMS

extracellular enzymes
periplasmic enzymes

SPORULATION

- **Sporulation,** which occurs in *Bacillus, Clostridium,* and a few other groups, involves the steps summarized in Figure 6.15.
- Sporulation occurs under both favorable and unfavorable environmental conditions and lets the organism withstand long periods of dry conditions and extreme temperatures.
- When more favorable conditions are restored, endospores germinate into vegetative cells.

axial nucleus **core**
endospore septum
cortex **dipicolinic acid**
spore coat **exosporium**
germination **cysts**
conidia

CULTURING BACTERIA

Methods of Obtaining Pure Cultures

- The **streak plate method** of obtaining a **pure culture** involves spreading bacteria across a sterile, solid surface such as an agar plate so that the progeny of a single cell can be picked up from the surface and transferred to a sterile medium.
- The **pour plate method** of obtaining a pure culture involves serial dilution, transferring a sample of a few organisms to melted agar and picking up cells from a colony on the agar.

Culture Media

- In nature, microorganisms grow on natural **media,** or the nutrients available in water, soil, and living or dead organic material.
- In the laboratory, microorganisms are grown in synthetic media: (1) **Defined synthetic media** consist of known quantities of specific nutrients. (2) **Complex media** consist of nutrients of reasonably well-known composition that vary in composition from batch to batch.
- Most routine laboratory cultures make use of **peptones,** or digested meat or fish proteins. Other substances such as **yeast extract, casein hydrolysate, serum,** whole blood, or heated whole blood are sometimes added.
- Diagnostic media are (1) **selective media** if they encourage growth of some organisms and inhibit growth of others, (2) **differential media** if they contain reagents that allow a specific chemical reaction only in the presence of particular organisms, or (3) **enrichment media** if they provide a nutrient that fosters growth of a particular organism.
- Cultures are maintained as **stock cultures** for routine work, as **preserved cultures** to prevent risk of contamination or change in characteristics, and as **reference cultures** to preserve specific characteristics of species and strains.

synthetic medium
chemically nondefined medium

blood agar **chocolate agar**

aseptic techniques

QUESTIONS FOR REVIEW

A.

1. What is growth, and how is it defined in microorganisms?

B.

2. How does cell division occur in microorganisms?

C.

3. What events occur in each of the four phases of bacterial growth?
4. How does growth in a colony differ from growth in a liquid medium?

D.

5. Name three ways of measuring bacterial growth.
6. What kinds of errors are likely to occur in measuring bacterial growth?

E.

7. How does each of the following factors affect growth in bacteria: pH, temperature, moisture, hydrostatic pressure, osmotic pressure, and radiation?
8. Why would you expect facultative organisms to have more different enzymes than other groups of organisms?
9. How do microaerophiles differ from aerotolerant anaerobes?
10. What is superoxide, and how is it destroyed?

F.

11. Which uses do bacteria make of carbon sources?
12. Which other nutrients might bacteria require and why?
13. Under which conditions would bacteria require vitamins?
14. How are nutritional requirements related to an organism's enzymes?

G.

15. What are the steps in the production of endospores?
16. Under which conditions are endospores formed?
17. How and under which conditions do endospores germinate?

H.

18. How are pure cultures obtained in the laboratory?

I.

19. How do natural and synthetic media differ?
20. What kinds of media are used for most routine laboratory cultures?
21. Which substances might be added to simple media for laboratory cultures?
22. How can culture media be used for diagnostic purposes?

PROBLEMS FOR INVESTIGATION

1. An attempt to transfer bacteria to new media during the death phase of a culture resulted in actual growth of the organisms. What is the most likely explanation for this phenomenon?
2. Devise an experiment to determine whether an unknown organism is a psychrophile, a mesophile, or a thermophile. If it is psychrophilic or thermophilic, show how to determine whether it is facultative or obligate.
3. If 100 bacteria having a generation time of 30 min are transferred to new media at 10 A.M., how many organisms will be present by 3 P.M.? How many generations will have been produced by 5 P.M.?
4. Tuberculosis organisms (*Mycobacterium tuberculosis*) are notoriously slow growers and can be overlooked if media are discarded prematurely. They also need special nutrients. Read about the special culture and diagnostic methods for these organisms and other members of this genus.
5. Prepare a library research paper on modern diagnostic media.

SOME INTERESTING READING

Anonymous. 1991. Salty life on Mars. *Discover* 12 (June):12.

Benathen, I. A. 1990. Isolation of pure cultures from mixed cultures: a modern approach. *The American Biology Teacher* 52(1):46–47.

Braunstein, H. 1986. Quality control in microbiology: A review and bibliography. *Clinics in Laboratory Medicine* 6(4)(December):649.

Brock, T. D. 1985. Life at high temperatures. *Science* 230:132.

Burchard, R. P. 1981. Gliding motility of prokaryotes: ultrastructure, physiology, and genetics. *Annual Review of Microbiology* 35:497.

Collins, C. H., and P. M. Lyne. 1976. *Microbiological methods.* London: Butterworth.

Difco manual: dehydrated culture media and reagents for microbiology. 1984. Detroit: Difco Laboratories.

Finegold, S. M., and E. J. Baron. 1986. *Bailey and Scott's diagnostic microbiology,* 7th ed. St. Louis: Mosby.

Gerhardt, P. (ed). 1981. *Manual of methods for general bacteriology.* Washington, D.C.: American Society for Microbiology.

Herring, T. S. 1990. Microbiologists gear up to encourage better clinical laboratory standards—new federal rules for proficiency testing, personnel, and waived tests worry scientists in clinical laboratories. *Scientist* 4(10):2.

Le Rudlier, D., A. R. Strom, A. M. Dandekar, L. T. Smith, and R. C. Valentine. 1984. *Molecular biology of osmoregulation. Science* 224:1064.

Meyer, H. P., et al. 1985. Growth control in microbial cultures. *Annual Review of Microbiology* 39:299.

Pledger, R. J., and J. A. Baross. 1991. Preliminary description and nutritional characterization of a chemoorganotrophic archaeobacterium growing at temperatures of up to 110°C isolated from a submarine hydrothermal vent environment. *The Journal of General Microbiology* 137 (January):203–11.

Robert, F. M. 1990. Impact of environmental factors on populations of soil microorganisms. *The American Biology Teacher* 52(9):364–69.

Washington, J. A. (ed). 1981. *Laboratory procedures in clinical microbiology.* New York: Springer-Verlag.

DNA spills from a ruptured *Escherichia coli* bacterium, revealing the great length of the single circular chromosome as it folds back and forth over itself. Magnification is 7857X.

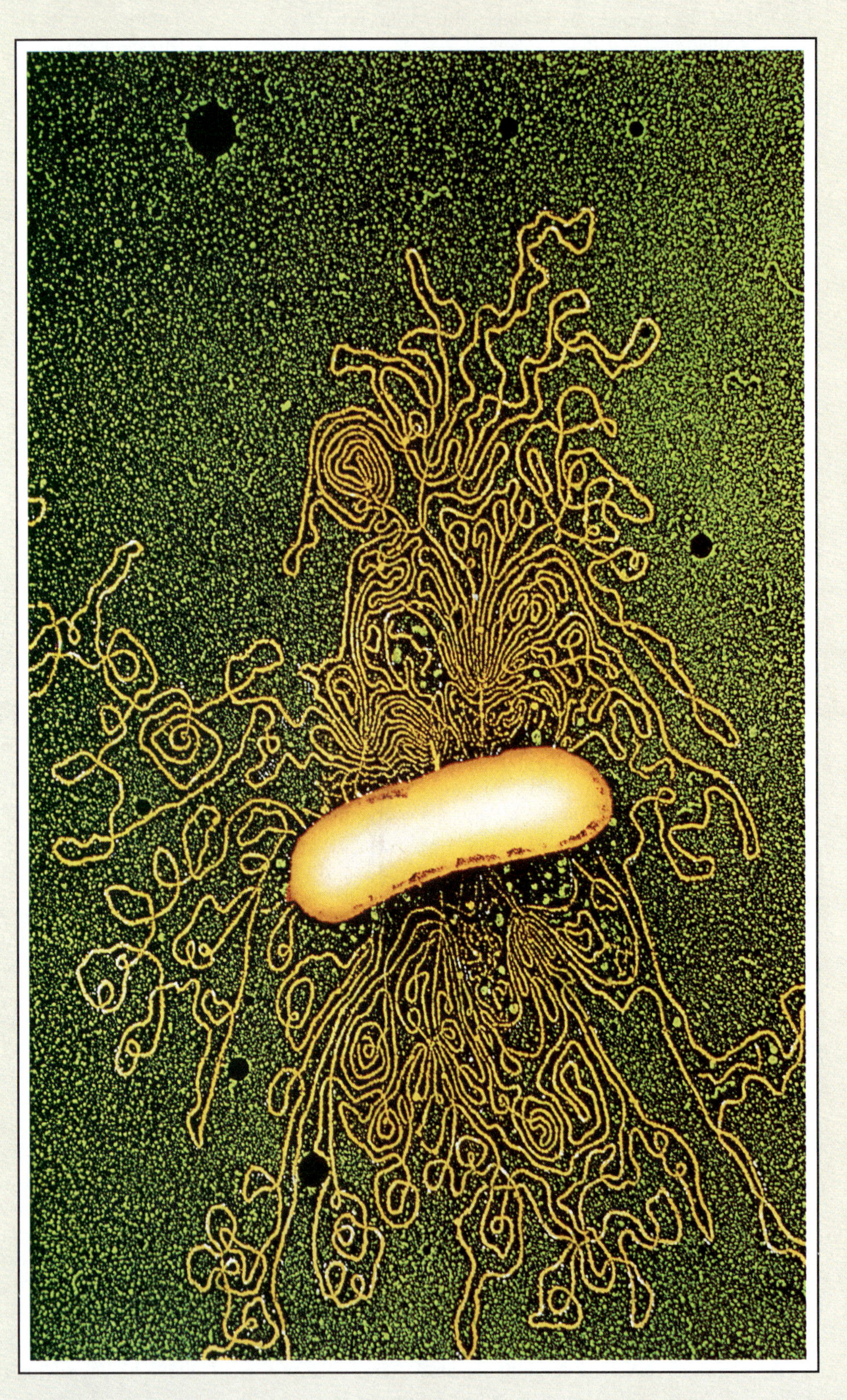

7 Genetics I: Gene Action, Gene Regulation, and Mutation

This chapter focuses on the following questions:

A. How are genes, chromosomes, and mutations involved in heredity in prokaryotic organisms?

B. How do nucleic acids store and transfer information?

C. How is DNA replicated in prokaryotic cells?

D. What are the major steps in protein synthesis?

E. How do mechanisms that regulate enzyme activity differ from those that regulate gene expression?

F. What happens in feedback inhibition, enzyme induction, and enzyme repression?

G. What changes in DNA occur in mutations, and how do they affect organisms?

H. How do spontaneous and induced mutations differ?

I. How do the fluctuation test, replica plating, and the Ames test make use of bacteria in studying mutations?

We have considered many aspects of metabolism and growth, but we have yet to consider the synthesis of nucleic acids and proteins. Synthesis of these complex molecules is the basis of genetics, the topic we will begin to study in this chapter. Genetics of microorganisms is an exciting and active research area; it also has been a rewarding area for microbiologists. Since the inception of the Nobel Prize in physiology or medicine in 1900, more than 25 annual prizes have been awarded to microbiologists. In recent decades, most of those prizes have gone to microbial geneticists. Because of this intensive investigation, much is now known about microbial genetics. We will begin our study of genetics by seeing how bacteria synthesize nucleic acids—DNA and RNA—and how the nucleic acids are involved in the synthesis of proteins. We will also see how genes (specific segments of DNA) act, how they are regulated, and how they are altered in mutations. In the next chapter we will discuss the mechanisms by which genetic information is transferred among microorganisms.

OVERVIEW OF GENETIC PROCESSES

The Basis of Heredity

All information necessary for life is stored in an organism's genetic material, DNA, or, for most viruses, RNA. To explain **heredity,** we must show how this information is transmitted from an organism to its progeny (offspring). To account for such transmission, we must consider the nature of chromosomes and genes.

In prokaryotes a chromosome is a circular threadlike molecule of DNA. As we saw in Chapter 2, DNA consists of a double chain of nucleotides arranged in a helix with the nucleotide base pairs held together by hydrogen bonds (Figure 7.1). ∞ (p. 43) The particular nucleotide sequence in DNA provides information for the synthesis of new DNA and for the synthesis of proteins.

The typical prokaryotic cell contains a single circular chromosome. When a prokaryotic cell reproduces by binary fission, the chromosome reproduces, or *replicates,* itself, and the daughter cells each receive one of the chromosomes. This mechanism provides for the orderly transmission of genetic information from parent to daughter cells.

A **gene** is a linear sequence of nucleotides of DNA that form a functional unit of a chromosome. All information for the structure and function of an organism is coded in the genes. In many cases, a gene determines a single characteristic, but the information in a particular gene, found at a particular **locus,** or location on the chromosome, is not always the same information. Genes with different information at the same locus are **alleles** (al-elz'). For example, in human blood types any one of three genes, A, B, and O, can occupy a certain locus. Allele A causes red blood cells to have a certain glycoprotein, which we will designate as molecule A, on their surfaces. Allele B causes them to have molecule B, and allele O does not cause them to have any molecule on the cell surfaces. People with type AB blood produce both molecules A and B because they have both alleles A and B.

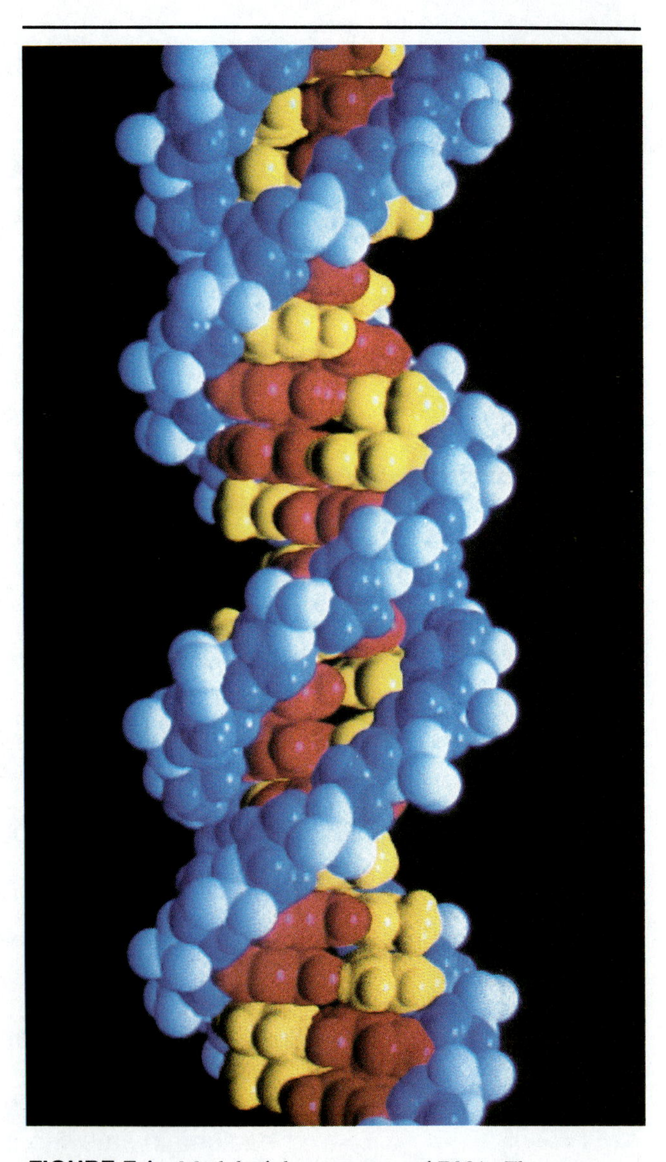

FIGURE 7.1 Model of the structure of DNA. The two strands are held together by hydrogen bonding between complementary bases: Adenine (A) always pairs with thymine (T), and cytosine (C) always pairs with guanine (G). Each strand can thus provide the information needed for the formation of a new DNA molecule.

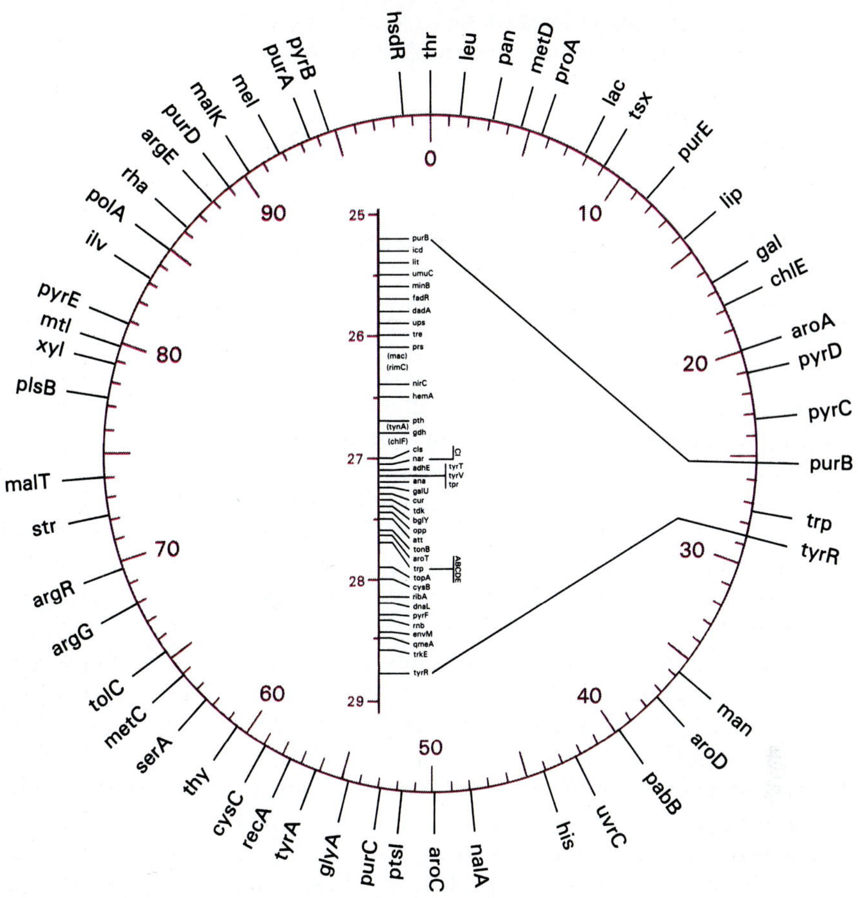

FIGURE 7.2 Part of the chromosome map of *Escherichia coli. E. coli* will be the first organism to have its entire genome (consisting of approximately 3000 genes) mapped. The outer circle is a simplified representation of the chromosome, with a number of the most commonly studied genes marked on it. It takes about 100 minutes to transfer the entire chromosome from a donor to a recipient cell in conjugation (Chapter 8). The numbers marked inside the circle represent the number of minutes of transfer required to reach that point on the chromosome. The insert is a small segment of the *E. coli* map enlarged to show some of the additional genes that have been located within that region (after Bachman).

Heritable variations in the characteristics of progeny can arise from mutations. A **mutation** is a permanent alteration in DNA. Mutations usually change the sequence of nucleotides in DNA and thereby change the information in the DNA. When the mutated DNA is transmitted to a daughter cell, the daughter cell can differ from the parent cell in one or more characteristics. Heritable variations in the characteristics of prokaryotic organisms also can occur by a variety of mechanisms (Chapter 8).

Nucleic Acids in Information Storage and Transfer

Information Storage

All the information for the structure and functioning of a cell is stored in DNA. For example, in the chromosome of the bacterium *Escherichia coli* (*E. coli*), each of the paired strands of DNA contains about 5 million bases arranged in a particular linear sequence. The information in these bases is divided into units of several hundred bases each. Each of these units is a gene. Some of the genes and their locations on the chromosome of *E. coli* are shown in Figure 7.2.

We might think of a gene as a sentence in the language of a cell. Each sentence in this language is constructed from a four-letter alphabet corresponding to the four nitrogenous bases in DNA: adenine (A), thymine (T), cytosine (C), and guanine (G). When these four letters combine to make "sentences" several hundred letters long, the number of possible sentences becomes almost infinite. Likewise, an almost infinite number of possible genes exists. If each gene contained 500 bases, a chromosome containing 5 million bases could contain 10,000 genes. Thus, the information storage capacity of DNA is exceedingly large!

Information Transfer

Information stored in DNA is used both to guide the replication of DNA in preparation for cell division and to direct protein synthesis. In both DNA **replication** and the first step of protein synthesis, the DNA serves as a **template,** or pattern, for the synthesis of a new nucleotide polymer. The sequence of bases in each

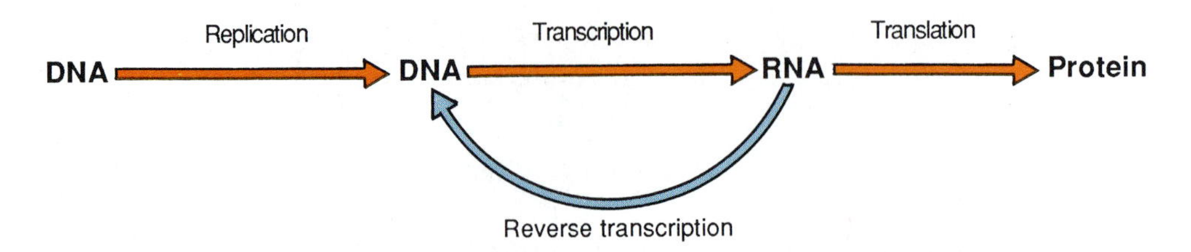

FIGURE 7.3 Information transmission from DNA to protein. As we shall see later, certain viruses, such as the one that causes AIDS, can direct synthesis of DNA from their RNA (reverse transcription).

new polymer is complementary to that in the original DNA. Such an arrangement is accomplished by base pairing. Recall from Chapter 2 that in complementary base pairing in DNA, adenine always pairs with thymine (A-T), and cytosine always pairs with guanine (C-G). Also recall that when DNA serves as a template for synthesis of RNA the base pairing is slightly different because adenine in DNA is paired with uracil (U) in RNA.

In the replication of DNA, the new polymer is also DNA. In protein synthesis the new polymer is a particular RNA called **messenger RNA (mRNA),** which then serves as a second template that dictates the arrangement of amino acids in a protein. Some of these proteins form the structure of a cell and others (enzymes) regulate its metabolism.

In the overall process of protein synthesis, the synthesis of mRNA from a DNA template is called **transcription,** and the synthesis of protein from information in mRNA is called **translation.** By analogy, transcription transfers information from one nucleic acid to another as one might transcribe handwritten sentences to typewritten sentences in the same language. Translation transfers information from the language of nucleic acid to the language of amino acids as one might translate English sentences into another language.

DNA replication, transcription, and translation all transfer information from one molecule to another (Figure 7.3). These processes allow information in DNA to be transferred to each new generation of cells and to be used to control the functioning of cells through protein synthesis.

REPLICATION OF DNA

To understand DNA replication we need to recall from Chapter 2 that pairs of helical DNA strands are held together by base pairing of adenine with thymine and cytosine with guanine. We also need to know that the ends of each strand are different. At one end, called the 3′ (3 prime) end, carbon-3 of deoxyribose is free to bind to other molecules. At the other end, called the 5′ (5 prime) end, carbon-5 of deoxyribose is attached to a phosphate. This structure is somewhat analogous to that of a freight train, with the 3′ end the engine and the 5′ end the caboose. When two strands combine by base pairing, they do so in a head-to-tail, or **antiparallel,** fashion. The arrangement of the strands is somewhat like two trains pointed in opposite directions, and base pairing is like passengers in the two trains shaking hands.

DNA replication begins at a specific location in the circular chromosome of a prokaryotic cell and usually proceeds simultaneously in both directions, creating two moving **replication forks,** the points at which the two strands of DNA separate to allow replication of DNA (Figure 7.4). A variety of enzymes break the hydrogen bonds between the bases in the two DNA strands, unwind the strands from each

CLOSE-UP

If DNA Only Makes Proteins, Who Makes Carbohydrates and Lipids?

If genetic information in DNA is used specifically to determine the structure of proteins, how are the structures of carbohydrates and lipids determined? Stop and think of the kinds of proteins a cell has. Many are enzymes, and, of course, some of those enzymes direct the synthesis of carbohydrates and lipids. The entire cell is controlled by DNA, either directly in DNA replication and synthesis of structural proteins, or indirectly by the synthesis of enzymes that in turn control the synthesis of carbohydrates and lipids.

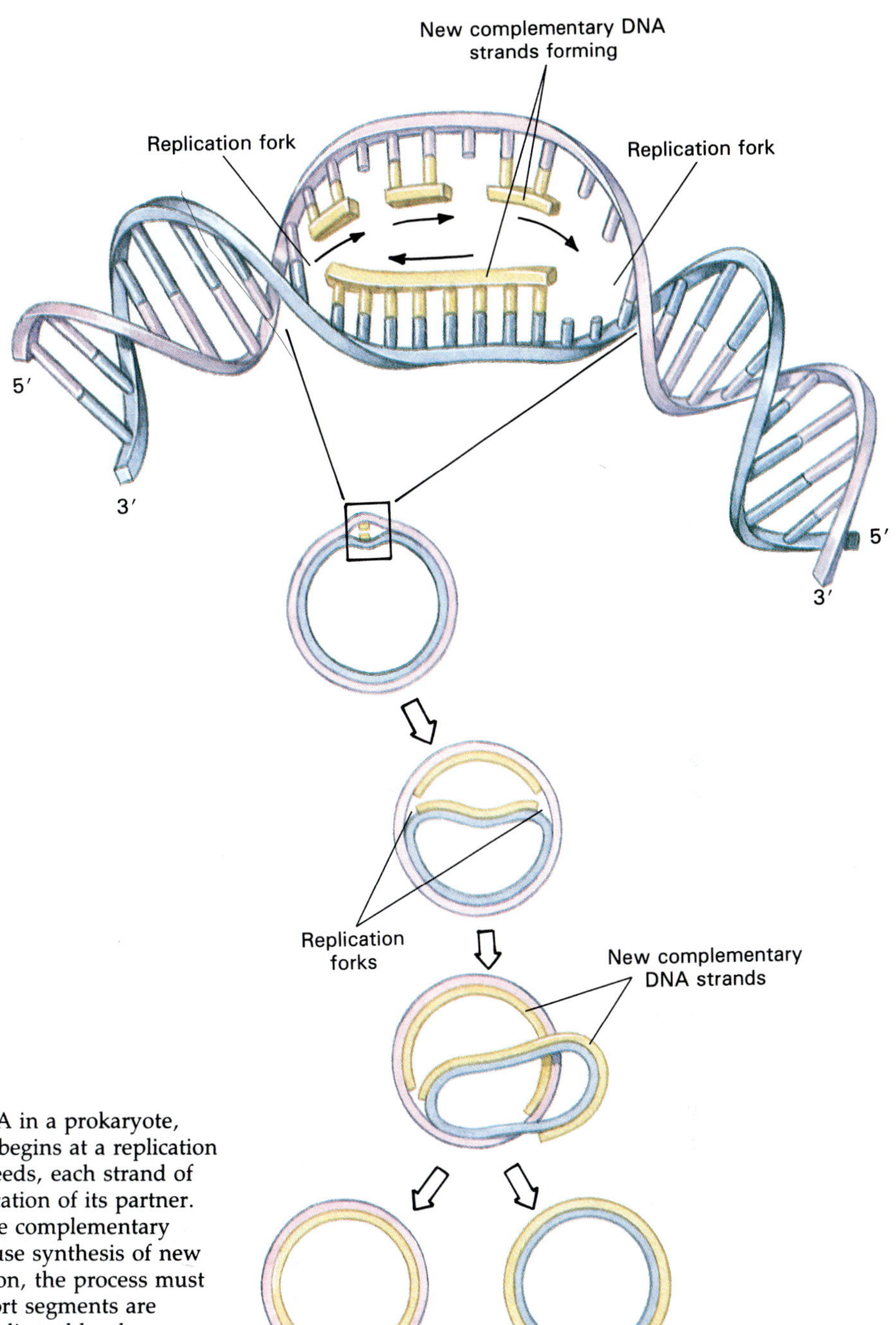

FIGURE 7.4 In the replication of DNA in a prokaryote, DNA strands separate, and replication begins at a replication fork on each strand. As synthesis proceeds, each strand of DNA serves as a template for the replication of its partner. Note the antiparallel arrangement of the complementary strands of the DNA double helix. Because synthesis of new DNA can take place in only one direction, the process must be discontinuous along one strand. Short segments are formed and then spliced together, as indicated by the arrows.

other, and stabilize the exposed single strands. Molecules of the enzyme **DNA polymerase** then move along behind each replication fork, synthesizing new DNA strands complementary to the original ones at a speed of approximately 1000 nucleotides per second.

The polymerase enzyme can add nucleotides only to the 3′ end of a growing DNA strand. Consequently, only one strand, the **leading strand** of original DNA, can serve as template for synthesis of a continuous new strand. Along the other strand, the **lagging strand,** synthesis of new DNA must be **discontinuous;** that is, the polymerase enzyme must continually jump ahead and work backward, making a series of short DNA segments called **Okazaki fragments.** These are then joined together by another enzyme called a **ligase.** Ultimately, two separate chromosomes are formed (Figure 7.4).

In newly synthesized chromosomes, each double helix consists of one strand of old, or parent, DNA and one strand of new DNA. Such replication is called **semiconservative replication** because one strand is always conserved (Figure 7.5). Replication was proven

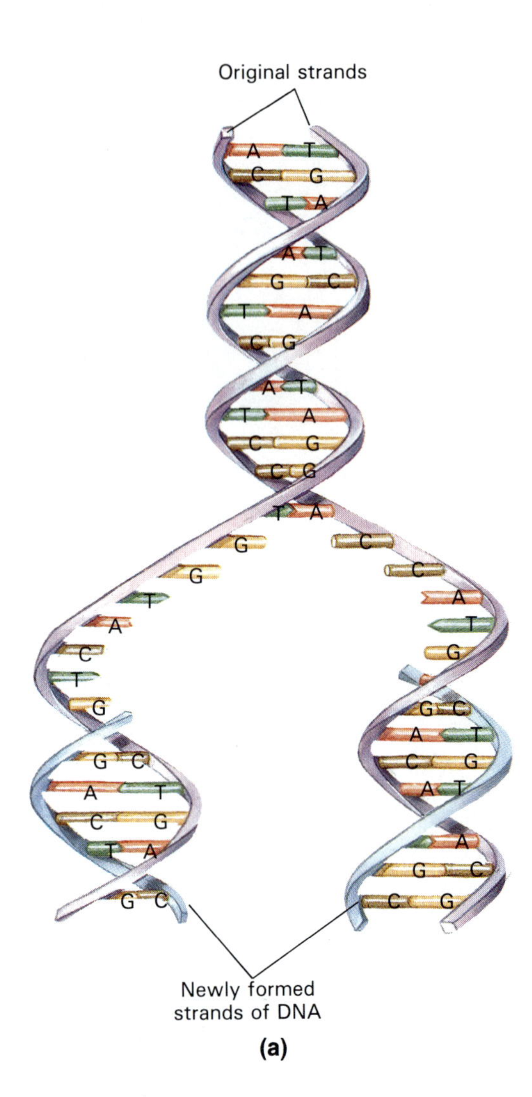

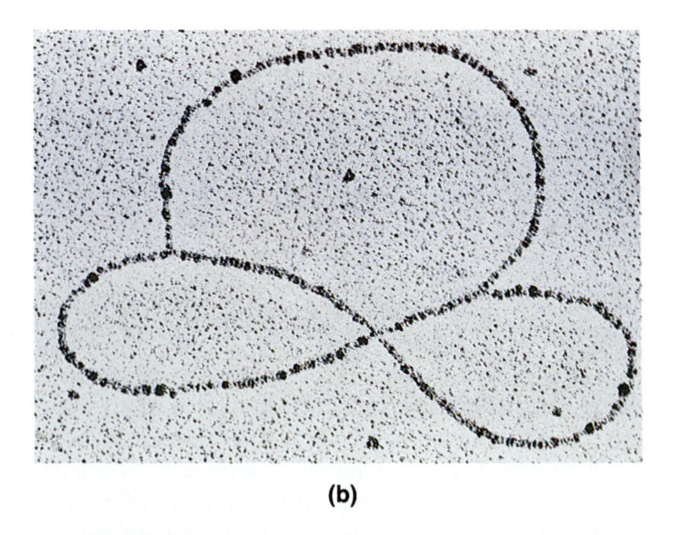

FIGURE 7.5 (a) Semiconservative replication of DNA. Note that the old strands of DNA are not destroyed, but become part of the next generation of DNA molecules. (b) An electron micrograph using the technique of autoradiography (see box) shows that DNA replication is semiconservative. Both chromosomes contain approximately the same amount of radioactive material from which the new DNA was synthesized.

APPLICATIONS

Using Autoradiography

Autoradiography can be used to determine where synthetic reactions take place in cells. For example, when dividing cells are incubated with radioactive thymidine, one of the building blocks of DNA, DNA molecules become radioactive wherever they incorporate the radioactive building block. After incubation some cells are placed on microscope slides, dried, and taken into a darkroom, where they are dipped in melted photographic emulsion like that found on camera film. When the emulsion coating is dry the slides are sealed in a lightproof container, which is kept in refrigerated storage for a period of days or weeks. During this time, radioactive emissions from the DNA strike silver grains nearest them in the emulsion. The exposed grains turn black when the slides are subsequently developed with photographic solutions. Finally the slides are stained and examined under a microscope. Cells are visible under the clear emulsion, with black dots appearing in the emulsion exactly over radioactive cell parts. Thus, it is possible to see just where the radioactive thymidine was incorporated.

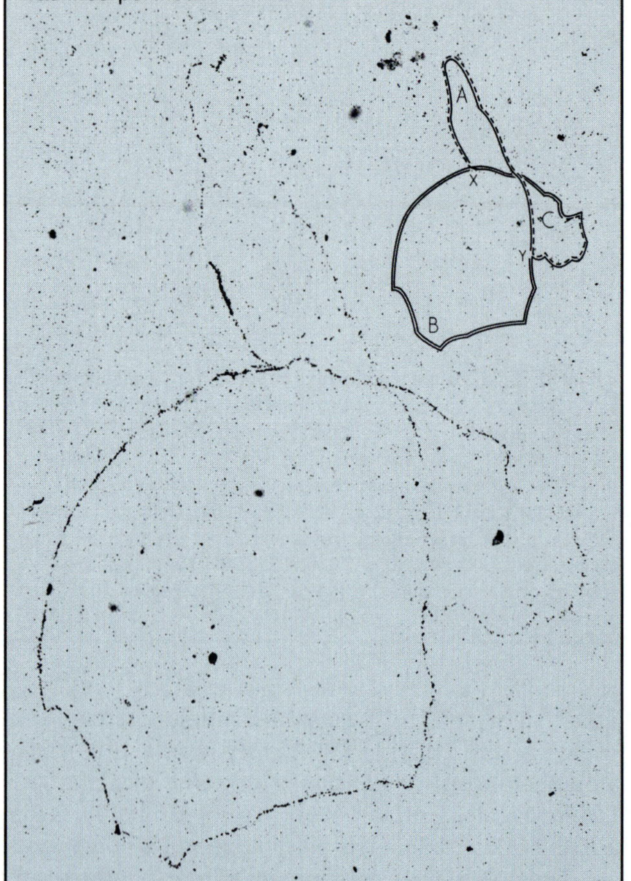

Autoradiograph of the chromosome of *Escherichia coli*, labeled with radioactive thymidine. In the inset (upper right) the same structure is shown diagrammatically and divided into three sections (A, B, and C) that arise at the two forks (X and Y).

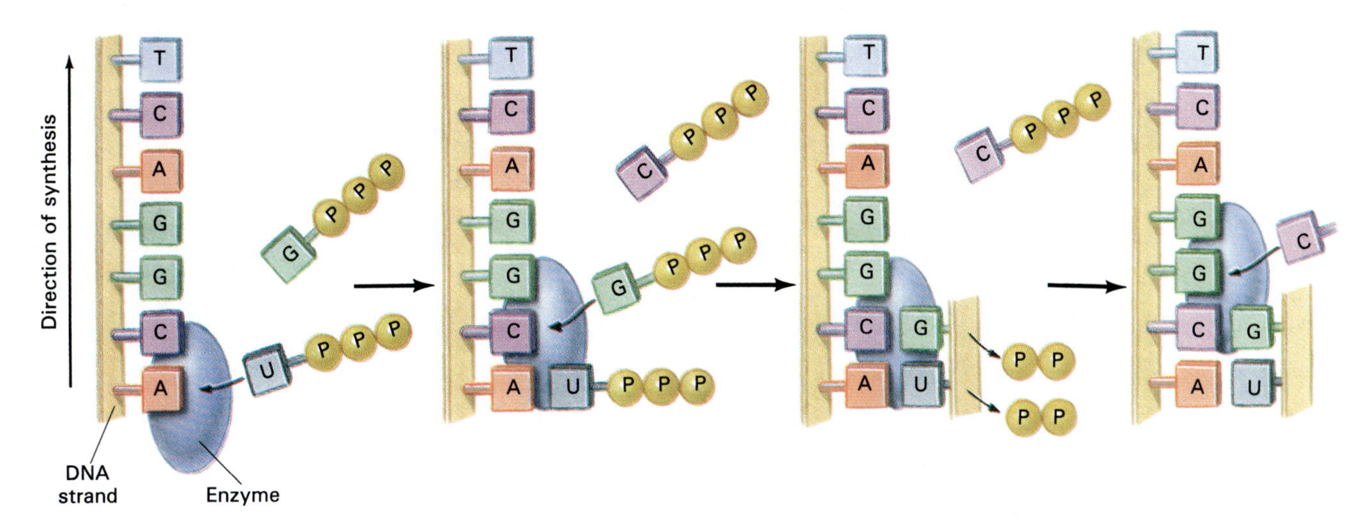

FIGURE 7.6 The transcription of RNA from template DNA. The —PPP represents a triphosphate, and PP represents pyrophosphate. Notice that in RNA, U (rather than T) pairs with A.

to be semiconservative by autoradiography experiments (see the box titled "Using Autoradiography"). When the cells were incubated with a radioactive building block, they incorporated radioactive nucleotides into their DNA. All chromosomes were found to contain approximately the same amount of radioactivity, and none lacked radioactivity. Thus, each chromosome was proved to consist of part old and part new DNA. If half the chromosomes had been made completely of old DNA and half of new DNA, the new ones would have contained more radioactivity, and the old ones would have contained no radioactivity.

PROTEIN SYNTHESIS

Transcription

All cells must constantly synthesize proteins to carry out their life processes: reproduction, growth, repair, and regulation of metabolism. This involves the accurate transfer of linear information of the DNA strands into a linear sequence of amino acids in proteins. To set the stage for protein synthesis, hydrogen bonds between bases in DNA strands are broken enzymatically in certain regions so that the strands separate. Short sequences of unpaired DNA bases are thus exposed to serve as templates in transcription. Only one strand directs the synthesis of mRNA; the complementary strand is used only as a template during DNA replication. Recall from Chapter 2 that RNA contains the base uracil instead of thymine. Thus, when mRNA is transcribed from DNA, uracil pairs with adenine; otherwise, base pairing occurs just as it does in DNA replication.

For transcription to occur, a cell must have sufficient quantities of nucleotides containing high-energy phosphate bonds, which provide energy for the nucleotides to participate in subsequent reactions. The enzyme **RNA polymerase** binds to one strand of exposed DNA. As shown in Figure 7.6, after an enzyme binds to the first base in DNA (adenine, in this case), the appropriate nucleotide joins the DNA base-enzyme complex. The new base then attaches by base pairing to the template base of DNA. The enzyme moves to the next DNA base, and the appropriate phosphorylated nucleotide joins the complex. The phosphate of the second nucleotide is linked to the ribose of the first nucleotide, and pyrophosphate (two attached molecules of phosphate) is released. This forms the first link in a new polymer of RNA. Energy to form this link comes from the hydrolysis of ATP and the release of two more phosphate groups. This process is repeated until the RNA molecule is completed.

Kinds of RNA

At least three kinds of RNA—ribosomal RNA, messenger RNA, and transfer RNA—are involved in protein synthesis. Each RNA consists of a single strand of nucleotides and is synthesized by transcription using DNA as a template. To complete the story of protein synthesis, we will need more information about these RNAs.

Ribosomal RNA (rRNA) binds closely to certain proteins to form two kinds of ribosome subunits. A subunit of each kind combines to form a ribosome. As mentioned in Chapter 4, ribosomes are sites of protein synthesis in a cell. ∞ (p. 84) They serve as binding sites for transfer RNA, and some of their

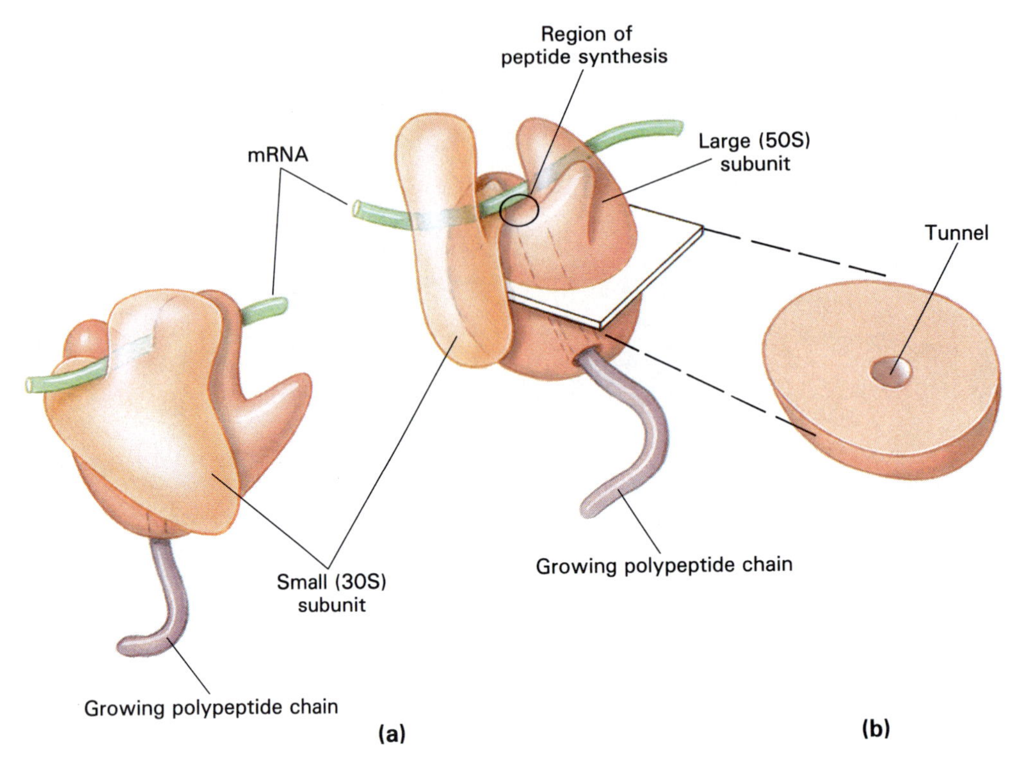

FIGURE 7.7 Ribosomal structure. (a) The small (30S) and large (50S) subunits are shown from two different angles of view. They can be seen to enfold the mRNA strand. The region of peptide synthesis is at the junction of these three components. The growing polypeptide chain passes through a tunnel in the 50S subunit, which can be seen in cross section (b).

proteins act as enzymes that control protein synthesis. Ribosomes are made of a small (30S) and a large (50S) subunit that join together over the strand of mRNA (Figure 7.7). Peptide synthesis occurs at this area of junction, with the newly formed polypeptide chain growing out via a tunnel in the 50S subunit.

Messenger RNA (mRNA) is synthesized in units that contain sufficient information to direct the synthesis of one or more polypeptide chains. It corresponds to a gene, a functional unit of DNA. Each molecule of mRNA becomes associated with one or more ribosomes. On the ribosome, the information coded in mRNA acts in translation to dictate the sequence of amino acids in the protein.

In translation, each triplet (sequence of three bases) in mRNA constitutes a **codon** (ko′don). Codons are the "words" in the language of nucleic acids. Each codon specifies a particular amino acid or acts as a terminator codon. The first codon in a molecule of mRNA acts as a "start" codon. It always codes for the amino acid methionine, even though the methionine may be removed from the protein later. The last codon in a molecule of mRNA is a terminator, or "stop" codon. It acts as a kind of punctuation mark to indicate the end of a protein molecule. Using a sentence as an analogy, the methionine codon is the capital letter at the beginning of the sentence, and the terminator codon is the punctuation mark at the end.

The one-to-one relationship between each codon and a specific amino acid constitutes the **genetic code** (Figure 7.8). At least one codon exists for each of the 20 amino acids found in proteins, and several codons exist for some amino acids. These are called **sense codons.** Early in the study of the code, investigators found some codons that did not code for any amino acid; they therefore called them **nonsense codons.** It was later found that these were terminator codons. Though genetic information is stored in DNA, the genetic code is written in codons of mRNA. Of course, the information in the codons is derived *directly* from DNA by complementary base pairing during transcription.

Comparisons of the codons among different organisms have shown them to be nearly the same in all organisms from bacteria to humans. This universality of the genetic code allows research on other organisms to be applied to understanding information transmission in human cells. Much of what is known about how the genetic code operates has been learned from research on bacteria.

The function of **transfer RNA (tRNA)** is to transfer amino acids from the cytoplasm to the ribosomes for placement in a protein molecule. Many different kinds of tRNAs have been isolated from the cytoplasm of cells. A tRNA molecule consists of 75 to 80 nucleotides folded back on itself to form several loops that are

First position	Second position: U		Second position: C		Second position: A		Second position: G		Third position
U	**UUU**	Phe	**UCU**	Ser	**UAU**	Tyr	**UGU**	Cys	U
U	**UUC**	Phe	**UCC**	Ser	**UAC**	Tyr	**UGC**	Cys	C
U	**UUA**	Leu	**UCA**	Ser	**UAA**	*Stop*	**UGA**	*Stop*	A
U	**UUG**	Leu	**UCG**	Ser	**UAG**	*Stop*	**UGG**	Trp	G
C	**CUU**	Leu	**CCU**	Pro	**CAU**	His	**CGU**	Arg	U
C	**CUC**	Leu	**CCC**	Pro	**CAC**	His	**CGC**	Arg	C
C	**CUA**	Leu	**CCA**	Pro	**CAA**	Gln	**CGA**	Arg	A
C	**CUG**	Leu	**CCG**	Pro	**CAG**	Gln	**CGG**	Arg	G
A	**AUU**	Ile	**ACU**	Thr	**AAU**	Asn	**AGU**	Ser	U
A	**AUC**	Ile	**ACC**	Thr	**AAC**	Asn	**AGC**	Ser	C
A	**AUA**	Ile	**ACA**	Thr	**AAA**	Lys	**AGA**	Arg	A
A	**AUG**	Met	**ACG**	Thr	**AAG**	Lys	**AGG**	Arg	G
G	**GUU**	Val	**GCU**	Ala	**GAU**	Asp	**GGU**	Gly	U
G	**GUC**	Val	**GCC**	Ala	**GAC**	Asp	**GGC**	Gly	C
G	**GUA**	Val	**GCA**	Ala	**GAA**	Glu	**GGA**	Gly	A
G	**GUG**	Val	**GCG**	Ala	**GAG**	Glu	**GGG**	Gly	G

FIGURE 7.8 The genetic code, with standard three-letter abbreviations for amino acids. To find the amino acid for which the mRNA codon AGU codes, go down the left column to the block labeled A, move across to the fourth square labeled G at the top of the figure, and find the first line in the square labeled U on the right side of the figure. There you will find Ser, the abbreviation for serine. *Stop* designates a terminator codon. The *Start* codon is AUG.

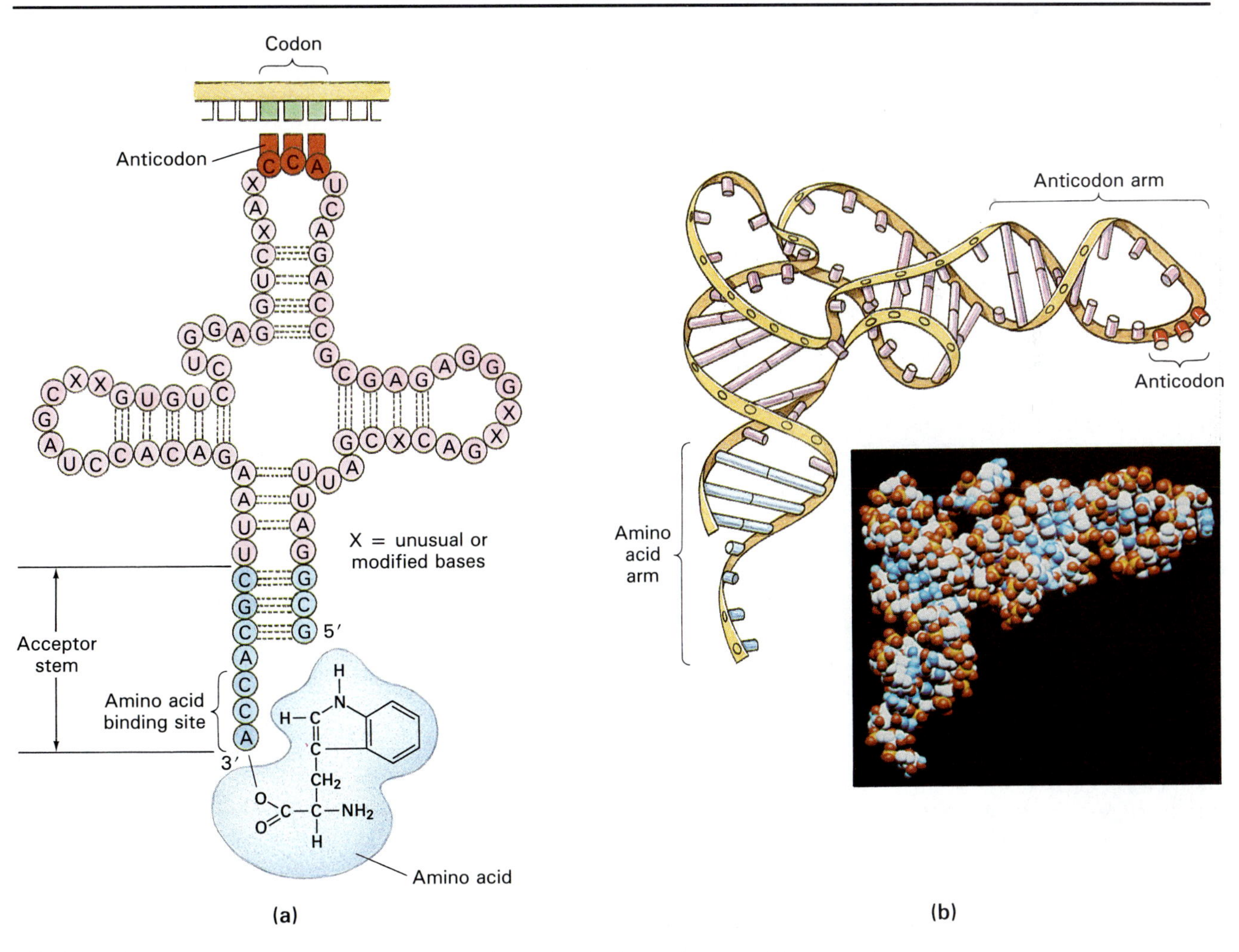

FIGURE 7.9 (a) The two-dimensional structure of a molecule of transfer RNA. The anticodon end will pair up with a codon on a strand of messenger RNA and deliver the desired amino acid, which is bonded to its other end. The molecule is maintained in its cloverleaf pattern by hydrogen bonding between strands that form the arms (dotted lines). (b) Computer-generated model of a tRNA molecule folded into its complex three-dimensional shape.

TABLE 7.1 Properties of kinds of RNA

Kind of RNA	Properties
Ribosomal	Combined with proteins to form ribosomes. Serves as a site for protein synthesis. Associated enzymes function in controlling protein synthesis
Messenger	Molecules usually correspond in length to one gene in DNA. Carries information from DNA for synthesis of a protein. Has base triplets called codons that constitute the genetic code. Attaches to one or more ribosomes.
Transfer	Molecules have a cloverleaf shape with an attachment site for a specific amino acid. Each has a single triplet of bases called an anticodon, which pairs with a corresponding codon in mRNA. Found scattered in cytoplasm, where they pick up amino acids and transfer them to mRNA.

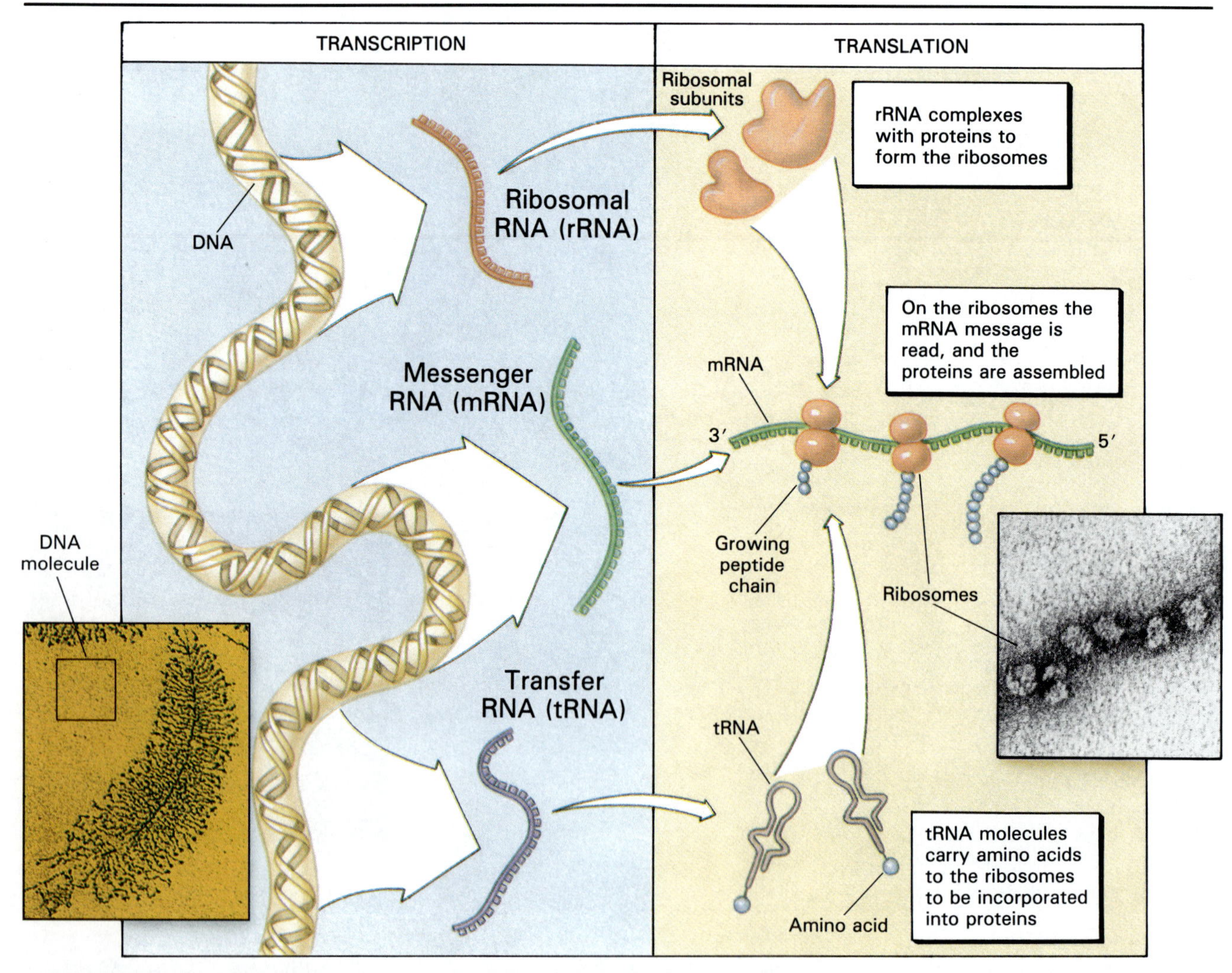

FIGURE 7.10 (left) Transcription from DNA to RNA. (right) Translation from RNA to protein. Many ribosomes riding along, reading the same piece of mRNA, are called a polyribosome.

stabilized by base pairing (Figure 7.9). Each tRNA has a three-base **anticodon** (an"ti-ko'don) that is complementary to a particular mRNA codon. It also has a binding site for an amino acid—the particular amino acid specified by the mRNA codon. (The mRNA codon, of course, got its information directly from DNA.) Thus, the tRNAs are the link between the codons and the corresponding amino acids.

The anticodon attaches by complementary base pairing to the appropriate mRNA codon so that its amino acid is aligned for incorporation into a protein. The accuracy of amino acid placement in protein synthesis depends on this precise pairing of codons and anticodons. The properties of the three types of RNA are summarized in Table 7.1.

Translation

Protein synthesis, an important process in bacterial growth, uses 80 to 90 percent of a bacterial cell's energy. Generally, during protein synthesis (Figure 7.10) the various RNAs and amino acids are available in sufficient quantities. The RNAs can be reused many times before they lose their ability to function. Of the RNAs, mRNA is produced in the most precise quantity in accordance with the cell's need for a particular protein.

Once an mRNA molecule has been transcribed and has combined with a ribosome, the ribosome initiates protein synthesis and provides the site for protein assembly. Several ribosomes can be attached at different points along an mRNA molecule to form a **polyribosome.** Each ribosome first attaches to the end of the mRNA that corresponds to the beginning of a protein. The length of each polypeptide chain extending from a ribosome corresponds to the amount of mRNA it has read.

The main steps in protein synthesis (Figure 7.11) can be summarized as follows: The process begins when a molecule of mRNA becomes properly oriented on a ribosome. As each codon of the mRNA is "read," the appropriate tRNA combines with it and thereby delivers a particular amino acid to the protein assembly site. Matching of codon and anticodon by base pairing allows coded information in mRNA to specify the sequence of amino acids in a protein. Any tRNAs with nonmatching anticodons simply do not bind to the ribosome. As amino acids are delivered one after another and peptide bonds form between them, the length of the polypeptide chain increases. This process continues until the ribosome recognizes a terminator codon. Three codons (designated in Figure 7.8 by the word *stop*) are called **terminator codons** because they signal the end of the information for a particular protein. When the ribosome "reads" a terminator codon, it releases the finished protein.

Any mRNA molecule can direct simultaneous synthesis of many identical protein molecules—one for each ribosome passing along it. Ribosomes, mRNAs, and tRNAs are reusable. The tRNAs shuttle back and forth between the cytoplasm, where they pick up amino acids, and the ribosome, where the amino acids are incorporated into protein.

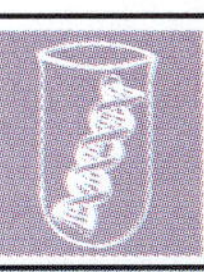

BIOTECHNOLOGY

"Ribozymes"

Dr. Thomas Cech, 1989 co-recipient of the Nobel Prize for Chemistry, has recently patented pieces of RNA that he calls *ribozymes.* These act like enzymes to cut selected pieces out of strands of viral RNA, thereby inactivating the virus. Like enzymes, the ribozymes are not used up in the reaction. Therefore, they can go on to snip up one virus after the next until all the viruses are neutralized. Cech's works has been very successful in laboratory settings. However, the human body is far more complicated than a culture flask of viruses, and it may be some time before ribozymes can be used therapeutically.

REGULATION OF METABOLISM

Significance of Regulatory Mechanisms

Bacteria use most of their energy to synthesize substances needed for growth. These substances include structural proteins that form cell parts and enzymes that control both energy production and synthetic reactions. The survival of bacteria depends on their ability to grow even when conditions are less than ideal—for example, when nutrients are in short supply. In their evolution, cells of bacteria (and all other organisms) have developed mechanisms to turn reactions on and off in accordance with their needs. Energy and materials are too valuable to waste. Also, the cell has a limited amount of space for storing excesses of materials it synthesizes. Thus, cells use energy to synthesize substances in the amounts needed and shut off these processes before wasteful excesses are produced.

All living organisms are presumed to have control mechanisms that regulate their metabolic activities. However, more research on control mechanisms has been done in bacteria than in all other organisms. Bacteria are ideal for such studies for several reasons:

1. They can be grown in large numbers relatively inexpensively under a variety of controlled environmental conditions.
2. They produce many new generations quickly.
3. Because they reproduce so rapidly, a variety of mutations can be observed in a relatively short time.

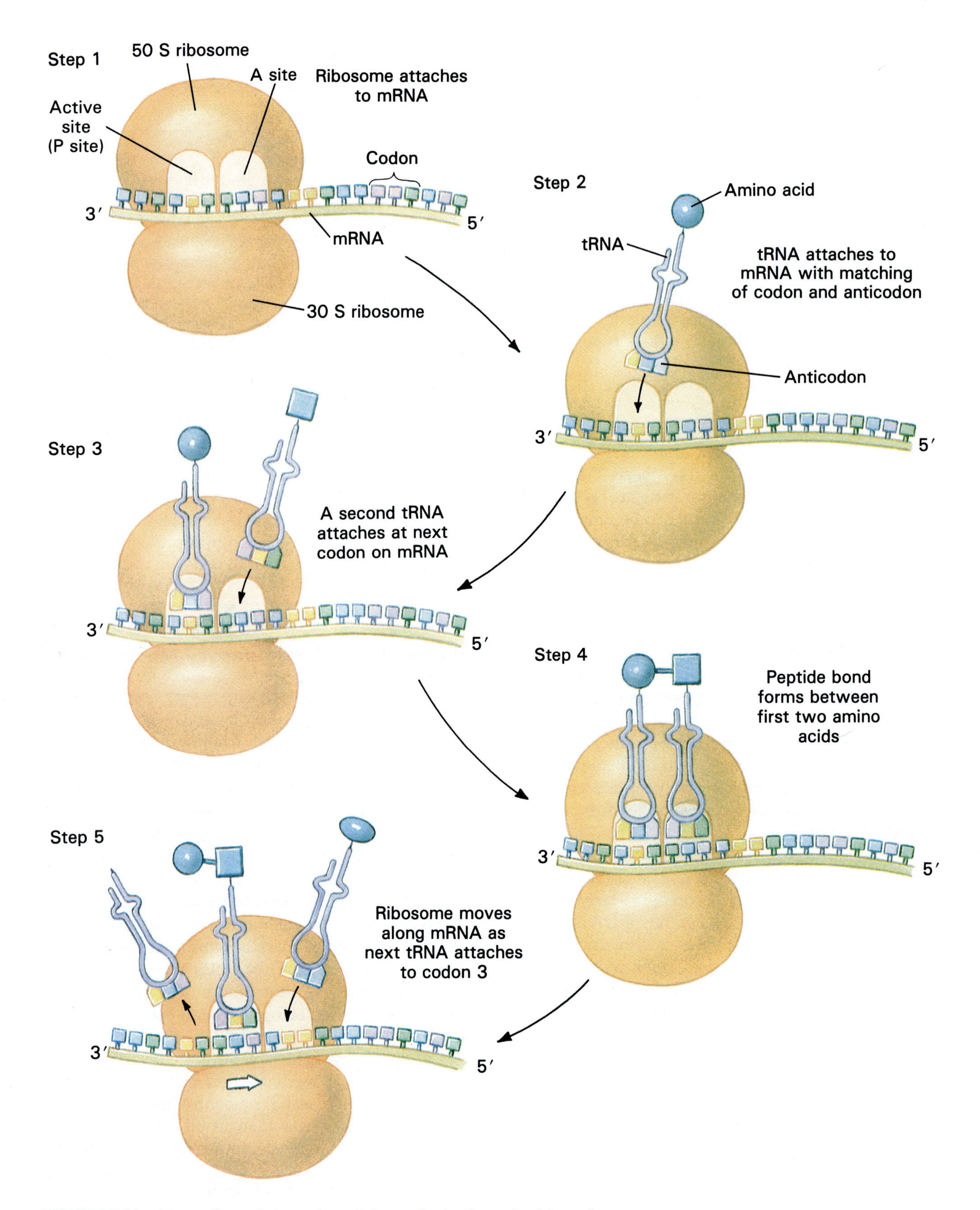

FIGURE 7.11 Above: the main steps in protein synthesis. Opposite: Many ribosomes can "read" the same strand of mRNA simultaneously. The ribosomes are shown moving from left to right.

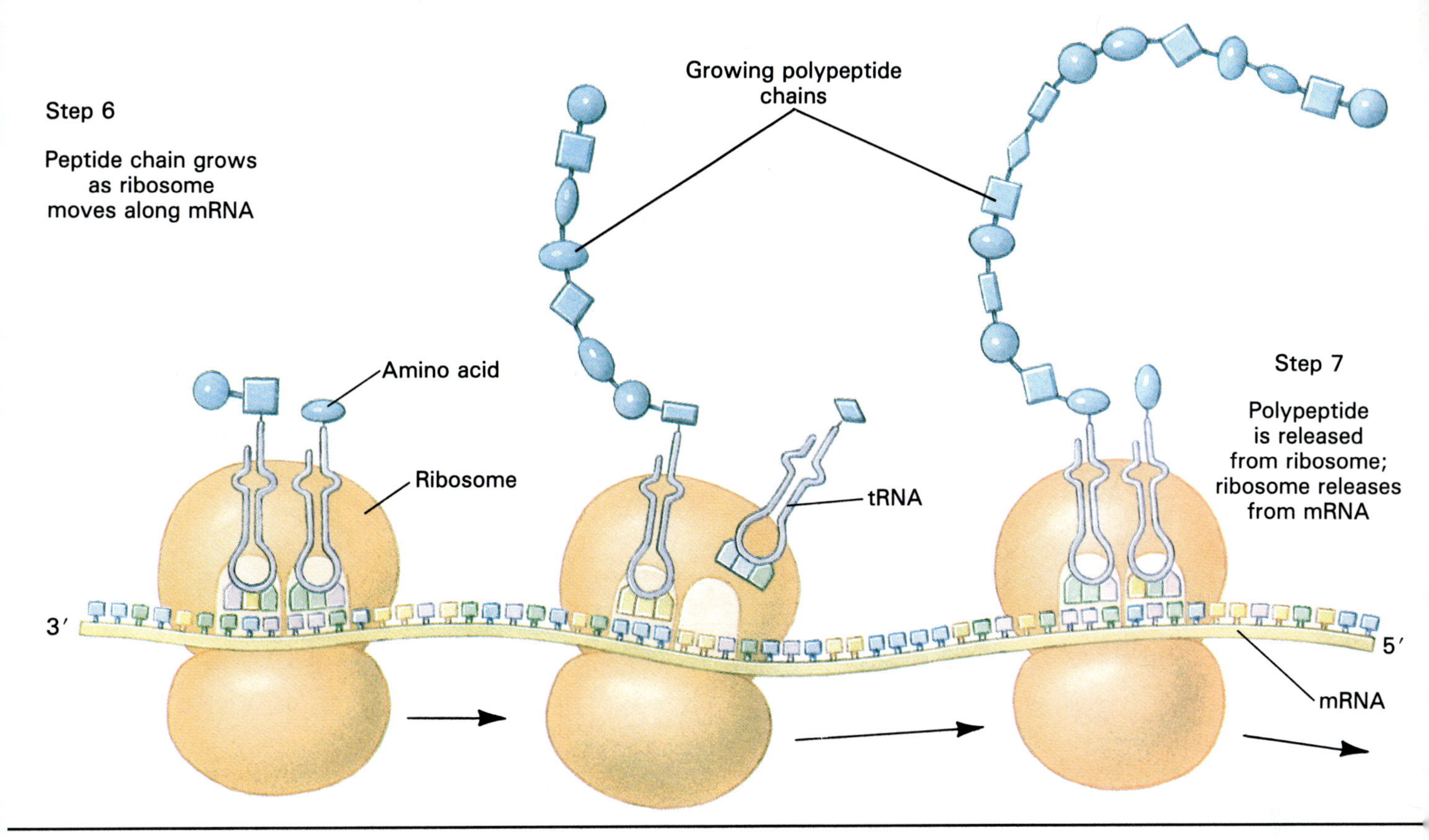

Mutant organisms that have an alteration in their control mechanisms can be isolated and studied along with nonmutated organisms to understand better the operation of control mechanisms.

Categories of Regulatory Mechanisms

The mechanisms that regulate metabolism either regulate enzyme activity directly or regulate enzyme synthesis by regulating gene expression. Where enzyme activity is regulated directly, the control mechanism determines how rapidly enzymes already present will catalyze reactions. Where regulation is by enzyme synthesis, the control mechanism determines which enzymes will be synthesized and in what amounts. Of the various mechanisms that regulate metabolism, three have been extensively investigated in bacteria. Feedback inhibition is an example of regulation of enzyme activity. Enzyme induction and enzyme repression are examples of regulation of gene expression.

Feedback Inhibition

In **feedback inhibition,** also called **end-product inhibition,** the end product of a biosynthetic pathway directly inhibits the first enzyme in the pathway. This mechanism was discovered when it was observed that adding one of several amino acids to a growth medium could cause a bacterium suddenly to stop synthesizing that particular amino acid. Synthesis of the amino acid threonine is regulated by feedback inhibition. Threonine is made from aspartate, and the allosteric enzyme that acts on aspartate is inhibited by threonine (Figure 7.12). (Aspartate is derived from oxaloacetate formed in the Krebs cycle.) When an inhibitor (threonine) attaches to the allosteric site, it alters the enzyme's shape so the substrate (aspartate) cannot attach to the active site (Chapter 5). ∞ (p. 116) Thus, feedback inhibition occurs when the end product of a reaction sequence binds to the allosteric site of the enzyme for the first step in the sequence.

Feedback inhibition regulates the synthesis of various substances besides amino acids (pyrimidines, for example). It also occurs in many organisms other than bacteria. Because feedback inhibition acts quickly and directly on a metabolic process, it allows the cell to conserve energy in two ways:

1. The inhibitor (end product) attaches to the enzyme when it is plentiful and is released from the enzyme when it is in short supply. Thus, the cell expends energy to synthesize the end product only when it is needed.
2. Regulation of enzyme activity requires less energy than the more complex processes involved in the regulation of gene expression.

Enzyme Induction

At one point in the investigation of metabolic regulation, it was discovered that certain organisms always

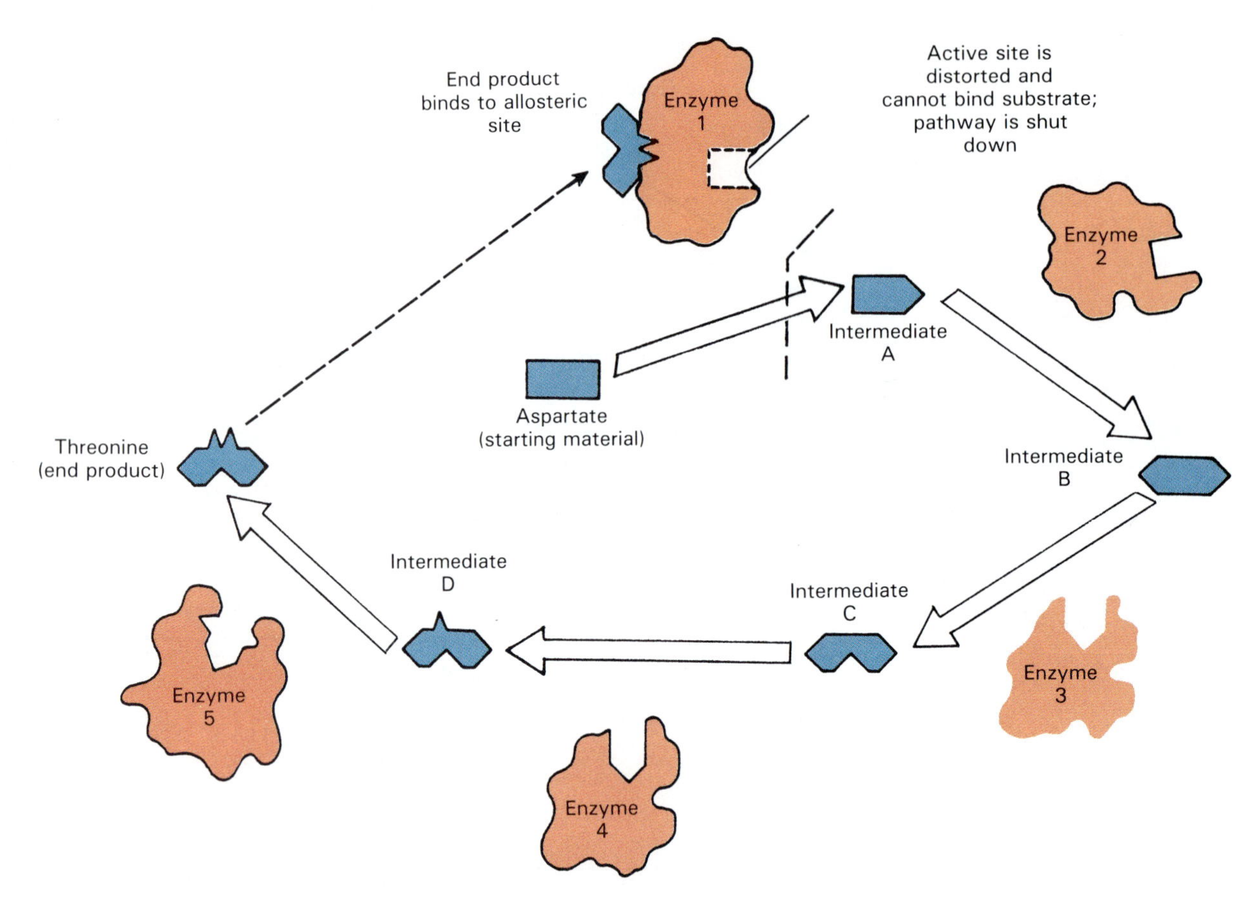

FIGURE 7.12 The synthesis of threonine involves five enzymatically controlled reactions (arrows), and four intermediate products (A, B, C, and D). Threonine (the end product) inhibits an allosteric enzyme (1) that catalyzes the first reaction (1). The allosteric enzyme is functional when its allosteric site is not occupied and is nonfunctional when the end product of a sequence of reactions is bound to that site.

contain active enzymes for glucose metabolism even when glucose is not present in the medium. Such enzymes are called **constitutive enzymes;** they are synthesized continuously regardless of the nutrients available to the organism. The genes that make these enzymes are always active. In contrast, enzymes that are synthesized by genes that are sometimes active and sometimes inactive depending on the presence or absence of substrate are called **inducible enzymes.**

When bacteria such as *E. coli* are grown on a nutrient medium that contains no lactose, the cells do not make any of the enzymes that they would need to utilize lactose as an energy source. When lactose is present, however, the cells are found to synthesize the enzymes needed for its metabolism. This phenomenon is an example of **enzyme induction.** Enzyme induction controls the breakdown of nutrients as they become available in the growth medium. Such a system is turned on when the nutrient is available and turned off when it is depleted.

An explanation of enzyme induction called operon (op'er-on) theory was proposed in 1961 by French scientists Francois Jacob and Jacques Monod, who received a Nobel Prize in 1965 for their work. Though the theory applies to several operons, we will illustrate it with the *lac* operon, which regulates lactose metabolism. An **operon** is a sequence of closely associated genes that include some **structural genes,** which carry information for the synthesis of specific proteins such as enzyme molecules, and **regulator genes,** which control the expression of repressor gene, the genes.

The *lac* operon (Figure 7.13) consists of a repressor gene, the *i* gene, regulator genes called a *promoter* and an *operator,* and structural genes *Z, Y,* and *A,* which direct synthesis of specific enzymes. When lactose is not present in the medium, the *i* gene directs synthesis of a substance called lac repressor. This **repressor** binds to the operator and prevents transcription of the *Z, Y,* and *A* genes. Consequently, the enzymes that metabolize lactose are not synthesized. Note that the *i* gene may be some distance away from the operon and is not under the control of the promoter gene. It is necessary for an RNA polymerase molecule to bind to the promoter before transcription can begin.

When lactose is present in the medium, lactose

acts as the **inducer** by binding to and inactivating the *lac* repressor. The repressor then no longer blocks the operator. The RNAse polymerase then binds to the promoter, causing the operator to initiate transcription of the *Z*, *Y*, and *A* genes in a long strand of mRNA. This mRNA becomes associated with ribosomes and directs synthesis of three enzymes: beta-galactosidase, permease, and transacetylase. Permease transports lactose into cells, and beta-galactosidase breaks down lactose into glucose and galactose. Transacetylase enables the cell to use other molecules similar to galactose. When the available lactose has been broken down, it is no longer available to bind to the repressor. The repressor again binds to the operator, and the operon is turned off.

Enzyme Repression

In contrast to enzyme induction, which typically regulates catabolism, **enzyme repression** typically regulates anabolism. It controls processes in which substances needed for growth are synthesized. Synthesis of the amino acid tryptophan, for example, is regulated by enzyme repression through actions of the *trp* operon. When tryptophan is available, it acts as a repressor by binding to a regulator protein. This complex binds to the promoter and represses synthesis of the enzymes needed to make tryptophan. When tryptophan is not available, the regulator protein remains inactive, and repression does not occur. Structural genes are transcribed, and tryptophan is synthesized.

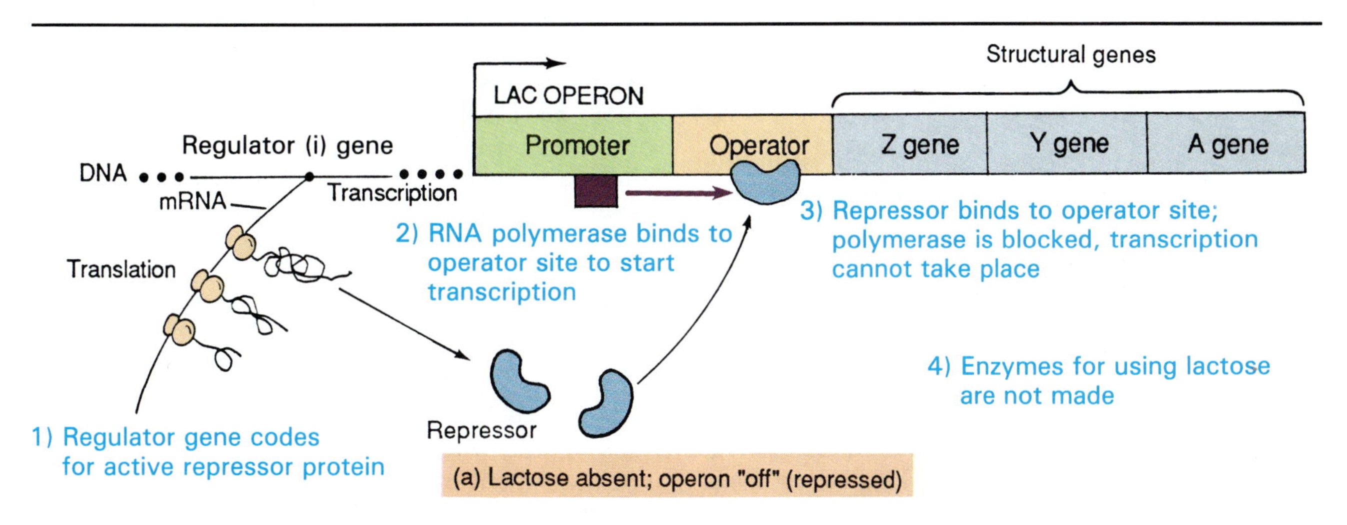

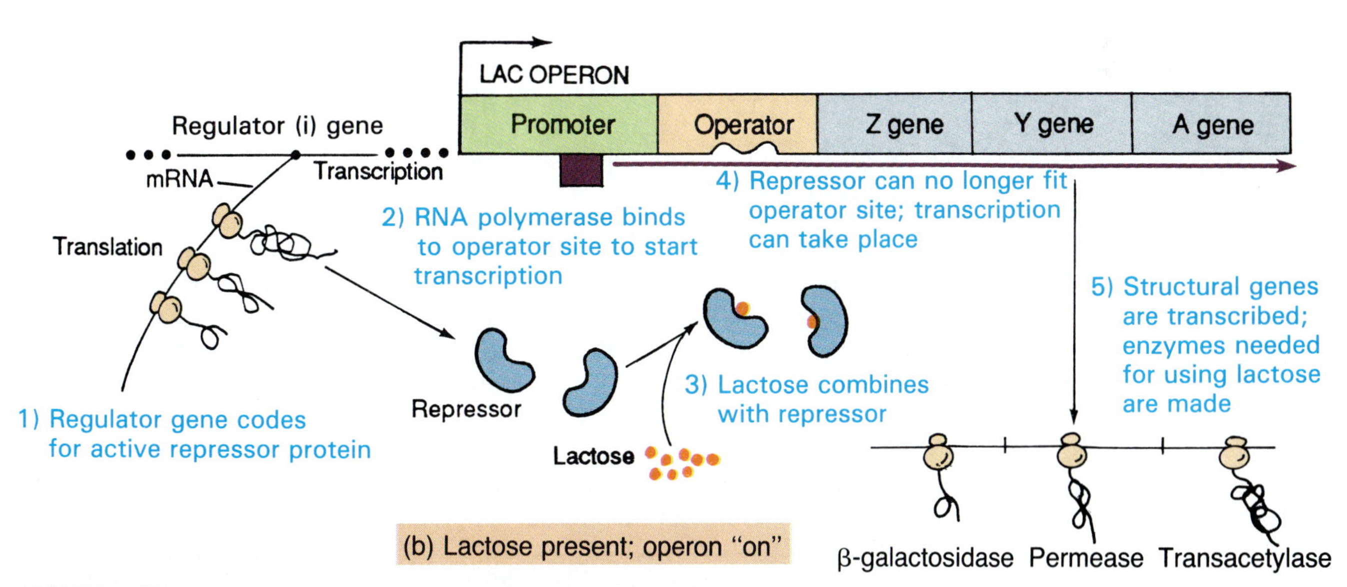

FIGURE 7.13 Enzyme induction: the mechanisms of operation of the *lac* operon. (a) In the absence of lactose, the repressor is synthesized and binds to the operator, preventing transcription of the genes coding for enzymes used to metabolize lactose. (b) When lactose is present, it binds to the repressor and inactivates it. The structural genes of the operon are transcribed, and enzymes for metabolizing lactose are synthesized. Note that the *i* gene may be some distance away from the operon.

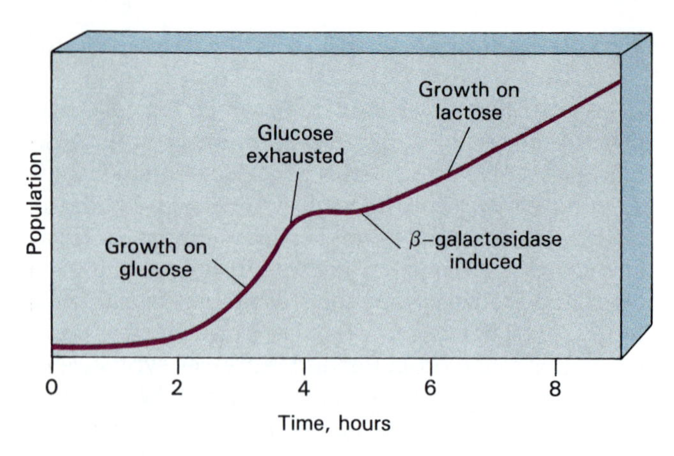

FIGURE 7.14 The growth curve for bacteria in a medium initially containing both glucose and lactose. When glucose is exhausted, growth stops temporarily, but begins again at a slower rate, using lactose as an energy source.

When tryptophan becomes plentiful, it again represses the operon. An even finer control mechanism, called **attenuation,** allows transcription of the *trp* operon to begin but terminates it prematurely by a complex process when sufficient amounts of tryptophan are already present in the cell.

A slightly different kind of repression operates in connection with some catabolic pathways. When certain bacteria (*E. coli,* for example) are grown on a nutrient medium containing both glucose and lactose, they grow at a logarithmic rate as long as glucose is available. When the glucose is depleted, they enter a stationary phase but soon begin to grow again at a logarithmic rate, though not quite as rapidly (Figure 7.14). This time the logarithmic growth rate results from the metabolism of lactose. The stationary phase is the period during which the enzymes needed to utilize lactose are being synthesized.

Why was the synthesis of these enzymes not induced before the glucose was depleted, since lactose was present in the medium from the start? The explanation is that bacteria use glucose as a nutrient with high efficiency. The enzymes for metabolizing glucose, being constitutive, are always present in the cell. Thus when glucose is abundant there is no advantage in making enzymes for metabolizing lactose even if lactose is also available. Consequently, the *lac* operon that we described previously is repressed when glucose is present in adequate quantities, an effect known as **catabolite repression.** In this way the cell saves energy by not making enzymes it doesn't need. When glucose supplies fall, the repression is lifted, the *lac* operon genes are transcribed, and the cell is ready to switch over to using lactose.

Both enzyme induction and enzyme repression are regulatory mechanisms that control enzyme production by altering gene expression. Though these two mechanisms have different effects, they actually represent two examples of the operation of a single mechanism (Table 7.2).

MUTATIONS

Mutations, or changes in DNA, can now be defined more precisely as heritable changes in the sequence of nucleotides in DNA. Mutations account for evolutionary changes in microorganisms (and larger organisms) and for alterations that produce different strains within species of organisms. Here we will consider what kinds of changes occur in DNA during mutations and how these changes affect the organisms.

Types of Mutations and Their Effects

Before we can consider mutations and their effects, we need to distinguish between an organism's genotype and its phenotype. **Genotype** refers to the genetic information contained in the DNA of the organism. **Phenotype** refers to the specific characteristics displayed by the organism. Mutations always change the genotype. This change may or may not be expressed in the phenotype, depending on the nature of the mutation.

Two important kinds of mutations are point mutations, which affect a single base, and frameshift mutations, which affect much larger segments of DNA. Mutations often make an organism unable to synthesize one or more proteins. The absence of a protein often leads to changes in the structure of the organism or in its ability to metabolize a particular substance.

TABLE 7.2 Effects of regulatory systems involving an operon

Regulatory Mechanism	Type of Pathway Regulated	Regulating Substance	Condition That Leads to Gene Expression
Enzyme induction (*lac* operon)	Degradatory and releases energy	Nutrient (lactose)	Presence of nutrient (lactose)
Enzyme repression (*trp* operon)	Biosynthetic and uses energy	End product (tryptophan)	Absence of end product (tryptophan)

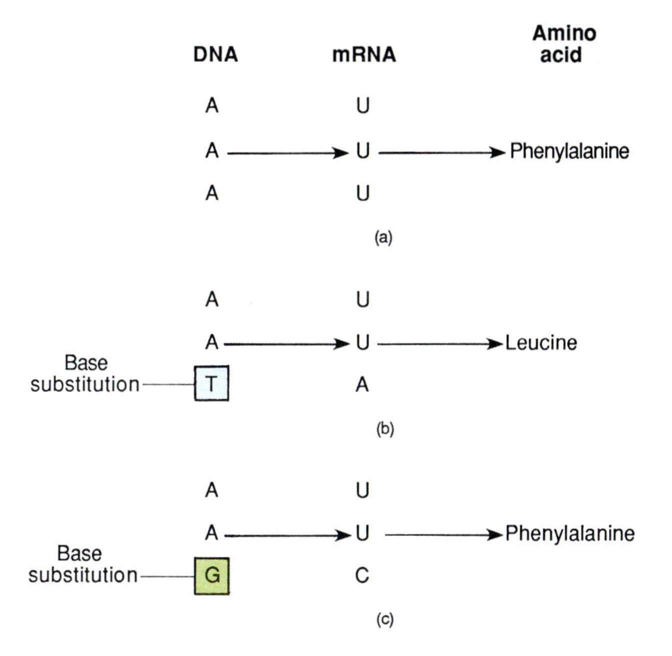

FIGURE 7.15 The effects of nucleotide replacement (a point mutation). The resulting protein may or may not be significantly affected, depending on whether the new codon specifies the same amino acid, one with similar properties, or an entirely different one.

A **point mutation** involves nucleotide replacement, or base substitution, in which one base is substituted for another at a specific location in a gene (Figure 7.15). The mutation changes a single codon in mRNA, and it may or may not change the amino acid sequence in a protein. Let's look at some examples.

Suppose a three-base sequence of DNA is changed from AAA to AAT. During transcription the mRNA codon will change from UUU to UUA. (Recall that uracil in RNA pairs with adenine in DNA.) (Chapter 2, p. 44) When the information in the mRNA is used to synthesize protein, the amino acid leucine will be substituted for phenylalanine in the protein. (To verify this for yourself, refer to the genetic code in Figure 7.8.) Because of the single amino acid substitution, the new protein will be different from the normal protein. The effects on the phenotype of the organism will be negligible if the new protein functions as well as the original one. They will be significant if the new protein functions poorly or not at all. In rare instances the new protein may function better and produce a phenotype that is better adapted to its environment than the original phenotype.

Should the code in DNA be changed from AAA to AAG, the mRNA code becomes UUC instead of UUU. Because the UUC and UUU codons both code for phenylalanine, the mutation has no effect on the protein being synthesized. In this case, though the genotype has changed, the phenotype is unaffected.

Sometimes the substitution of a single base in DNA produces a terminator codon in mRNA. If the terminator codon is introduced in the middle of a molecule of mRNA destined to produce a single protein, synthesis will be terminated part of the way through the molecule. A polypeptide that will most likely be unable to function in the cell will be released, and the appropriate protein will not be synthesized. If the missing protein is essential to cell structure or function, the effect can be lethal.

A **frameshift mutation** is a mutation in which there is a **deletion** or an **insertion** of a single base (Figure 7.16). Such mutations alter all the three-base sequences beyond the deletion or insertion. When mRNA transcribed from such altered DNA is used to synthesize a protein, many amino acids in the sequence are altered. Such mutations also often introduce terminator codons and cause protein synthesis to stop when only a short polypeptide has been made. Frameshift mutations usually prevent synthesis of a particular protein, and they change both the genotype and the phenotype. Their effect on the organism depends on the role of the missing protein in the organism's function. These types of mutations and their effects are summarized in Table 7.3.

Phenotypic Variation

Phenotypic variations frequently seen in mutated bacteria include alterations in colony morphology, colony color, or nutritional requirements. Instead of being a normal smooth, glossy, raised colony, some colonies have a flat, rough appearance. A mutation has impaired synthesis of certain cell surface substances. In organisms that typically form capsules, the mutation can prevent synthesis of capsular polysaccharides. Mutations that alter nutritional requirements generally increase the nutritional needs of an organism, usually by impairing the organism's ability to synthesize one or more enzymes. As a result the organism may require certain amino acids or vitamins in its medium because it can no longer make them itself.

Studies of bacteria that have lost the ability to synthesize a particular enzyme have played an important role in our understanding of metabolic pathways. Such nutritionally deficient organisms are called **auxotrophic** (awks'o-trof"ik) **mutants** (*auxo,* increase, and *trophos,* food) because they require special substances in their medium to maintain growth. In contrast to auxotrophs, normal nonmutant forms are called **prototrophs** (pro'to-trofs), or *wild types.* Comparisons of characteristics of auxotrophs and prototrophs show the effects of a mutation on metabolism. By observing which metabolites accumulate and which nutrients must be added to the medium of

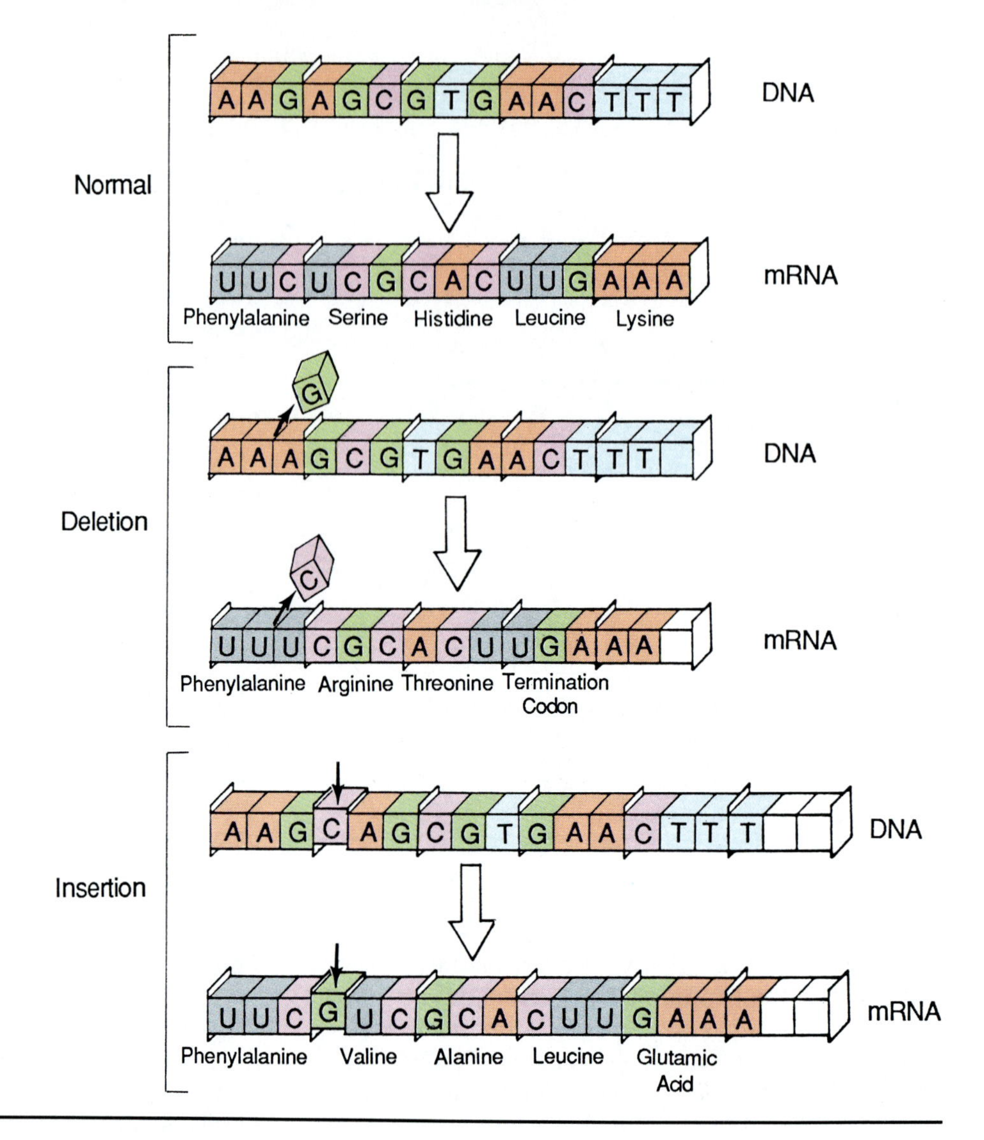

FIGURE 7.16 The effects of frameshift mutations. Adding or deleting a single nucleotide changes the amino acid sequence coded for by the entire gene from that point on. (Note that adding or deleting two nucleotides would cause similar disruption, but adding or deleting three nucleotides might not affect the resulting protein very much. Can you see why?)

TABLE 7.3 Types of mutations and their effects on organisms

Types of Mutations	Effects on Organisms
Point mutation	
Change in DNA with no change in the amino acid specified by the mRNA codon.	No effect on protein; a "silent" mutation.
Change in DNA with change in the amino acid sequence specified by the mRNA codon.	Change in protein by substitution of one amino acid for another; can significantly alter function of protein.
Change in DNA that creates a termination codon in mRNA.	Produces polypeptide of no use to organism and prevents synthesis of normal protein.
Frameshift mutation	
Deletion or insertion of a single base in DNA.	Changes entire sequence of codons and greatly alters amino acid sequence. Can introduce terminator codon and produce useless polypeptides instead of normal proteins.

auxotrophs, the specific steps in the metabolism of certain substances have been determined.

Still another type of phenotypic variation of genetic origin is temperature sensitivity. For example, suppose an organism at one time could grow over a wide range of environmental temperatures. As a result of a mutation, it loses the ability to grow at the higher temperatures in its former range. It can still grow at 25°C, but it can no longer grow at 40°C. This phenomenon may be due to a point mutation that changed a single amino acid in an enzyme. The slightly altered enzyme may function at moderate temperatures but may be easily denatured and inactivated at higher temperatures.

Some phenotypic variations are caused by environmental factors and occur without any change in the genotype (alteration in DNA). For example, large amounts of sugar or irritants in the medium can cause some organisms to form a larger-than-normal capsule. Some organisms, such as the anthrax bacterium, form spores in open air, in spilled blood, or on tissue sur-

CLOSE-UP

Turning Genes On and Off

Variations in organisms not due to changes in DNA are not heritable. For example, humans whose hair loses pigment and turns gray with aging still have the genes for hair color they received from their parents and can pass them on to their offspring. The genes for hair color that are not expressed in pigment formation in the aging adults will be expressed in the children—they will not be born with gray hair. Similarly, genes for making capsules, spores, or pigments in bacteria fail to be expressed under certain conditions. Regulatory mechanisms turn on and off the expression of certain genes during the life of an individual organism, but the genes are transmitted to offspring.

faces but not inside tissues. Variations in environmental temperature can affect pigment synthesis. *Serratia marcescens* usually produces pigment at room temperature but may not do so at higher temperatures. It has the gene for pigment production, but the gene is expressed only at certain temperatures.

Spontaneous versus Induced Mutations

Mutations appear to be random or chance events; it is usually impossible to predict when a mutation will occur or which genes will be altered. Though all mutations result from changes in DNA, they can be spontaneous or induced. **Spontaneous mutations** occur in the absence of any agent known to cause changes in DNA. They arise during the replication of DNA and appear to be due to errors in the base pairing of nucleotides in the old and new strands of DNA. Various genes in the DNA of bacteria have different spontaneous mutation rates ranging from 10^{-3} to 10^{-9} per cell division. In other words, one gene might undergo a mutation once in every thousand ($1/10^3$) cell divisions, whereas another might undergo a mutation only once in every billion ($1/10^9$) cell divisions. **Induced mutations** are produced by agents called **mutagens** that increase the mutation rate. Mutagens include chemical agents and radiation.

Chemical Mutagens

Chemical mutagens act at the molecular level to alter the sequence of bases in DNA. They include base analogs, alkylating agents, deaminating agents, and acridine derivatives.

A **base analog** is a molecule quite similar in structure to one of the nitrogenous bases normally found

Thymine

5-bromouracil

FIGURE 7.17 Similarity of base analog 5-bromouracil structure to structure of normal base thymine allows it to sometimes be taken up in place of thymine. The bromine (Br) group occupies approximately the same-sized area as does the methyl (CH_3) group.

in DNA. Cells incorporate the analog into DNA as it is synthesized. For example, 5-bromouracil can be inserted in DNA instead of thymine. When the DNA containing 5-bromouracil is replicated, the analog can cause an error in base pairing. The 5-bromouracil that replaced thymine may pair with guanine instead of with the adenine that normally pairs with thymine (Figure 7.17). When DNA is replicated in the presence of a significant quantity of 5-bromouracil, the analog can be incorporated at many sites in the DNA molecule. A mutation occurs wherever the analog causes the insertion of guanine instead of adenine in the subsequent replication. Another purine base analog, caffeine, can cause mutations in an unborn child. For this reason pregnant women are advised to avoid or limit their caffeine intake.

Alkylating agents are substances that add alkyl groups (such as a methyl group, CH_3) to other molecules. Adding an alkyl group to a nitrogenous base

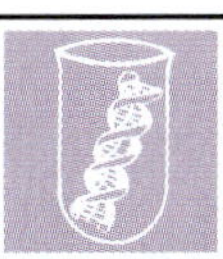

BIOTECHNOLOGY

Anticancer Agents

Certain agents used in cancer therapy create errors in DNA and thereby interfere with the division of malignant cells. The base analogs 6-mercaptopurine and 2-aminopurine act by being erroneously inserted into new DNA. The alkylating agents cyclophosphamide and busulfam cause the addition of a methyl group to the bases of DNA, usually adenine and guanine.

Over several divisions these anticancer agents cause an accumulation of mutations sufficient to kill the cells. Rapidly dividing cells are especially vulnerable to such agents because more divisions mean more DNA synthesis and thus more opportunities for mistakes. Anticancer agents also kill some human cells, especially rapidly dividing cells in bone marrow, hair follicles, and the lining of the digestive tract. Patients receiving such drugs often develop anemia, lose their hair, and suffer from gastrointestinal bleeding.

alters the shape of the base and can cause an error in base pairing. For example, the addition of a methyl group to guanine can cause it to pair with thymine instead of cytosine. Such a change can give rise to a point mutation.

Deaminating agents such as nitrous acid (HNO_2) remove an amino group from a nitrogenous base. Removing an amino group from adenine causes it to resemble guanine, and the deaminated base pairs with cytosine instead of thymine. Nitrates (NO_3^-) and nitrites (NO_2^-) are sometimes added to foods such as hot dogs and cold cuts for coloring, flavoring, or antibacterial action. The hazard of such additives is that in the body they form nitrosamines, deaminating agents known to cause birth defects, cancer, and other mutations in laboratory animals.

In contrast to these alterations, which cause point mutations, **acridine derivatives** cause frameshift mutations. The acridine molecule contains one pyrimidine ring and two benzene rings (Figure 7.18a). This molecule or one of its derivatives can insert in the DNA double helix, replacing both members of a base pair. Such a modification distorts the helix and causes partial unwinding of the DNA strands. The distortion allows one or more bases to be added or deleted, and a frameshift results. The drug quinacrine (Atabrine) is an acridine derivative that was used to treat malaria until other drugs with less unpleasant side effects were developed. It causes mutations in the malarial parasite and possibly in the human host that receives the drug.

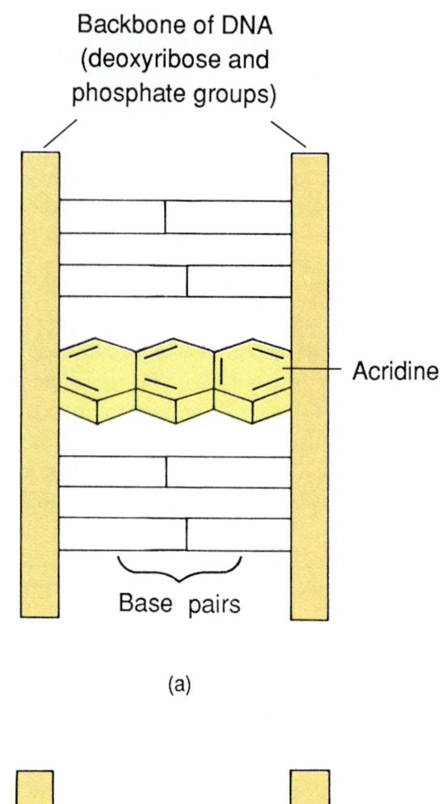

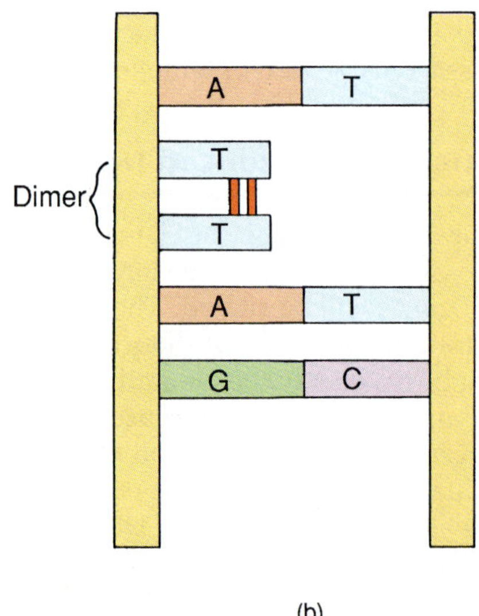

FIGURE 7.18 (a) Insertion of acridine into a DNA helix can produce a frameshift mutation. (b) The formation of a dimer prevents the affected bases from pairing with bases in the complementary chain of DNA, impairing replication and preventing transcription.

Radiation as a Mutagen

Radiation such as X rays and ultraviolet rays can act as mutagens. Ultraviolet rays affect only the skin of humans because the rays lack energy for deeper penetration, but they have significant effects on microorganisms, which they penetrate easily. Ultraviolet lights are sometimes mounted in hospitals and laboratories to kill airborne bacteria. When ultraviolet rays strike DNA, they can cause pyrimidine dimers to form. A **dimer** (di'mer) consists of two adjacent pyrimidines (two thymines, two cytosines, or thymine and cytosine) bonded together in a DNA strand (Figure 7.18b). Binding of pyrimidines to each other prevents base pairing during replication of the adjacent DNA strand, so a gap is produced in the replicated DNA. Transcription of mRNA stops at the gap, and the affected gene fails to transmit information.

X rays and gamma rays, more energetic than ultraviolet (Chapter 3), easily break chemical bonds in molecules. (p. 54) The product is often a free radical, a highly reactive molecular fragment that in turn attacks other cell molecules, including DNA.

Until recently microbiologists had no control over which genes underwent mutation when organisms were treated with mutagens. Now certain enzymes are available that greatly facilitate such studies. **Restriction endonucleases** cut DNA at precise base sequences, and **exonucleases** remove segments of DNA. These enzymes allow individual genes to be isolated and mutated at predetermined sites. The mutated gene can be inserted into a host's chromosome, and the effect of the specific mutation studied.

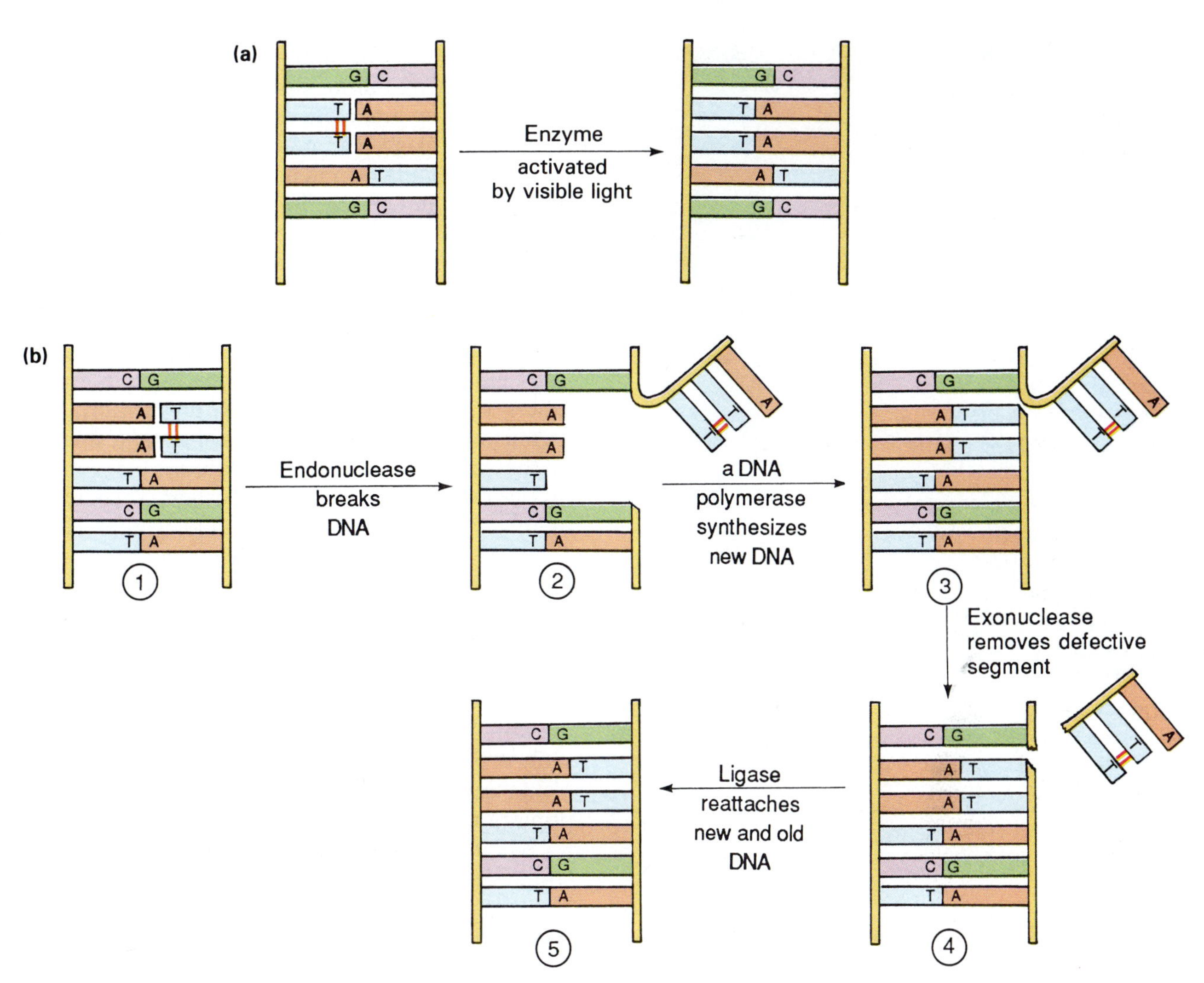

FIGURE 7.19 (a) Light repair of DNA (photoreactivation) removes dimers. (b) In dark repair, a defective segment of DNA is cut out and replaced.

Repair of DNA Damage

Many bacteria, and other organisms as well, have enzymes that can repair certain kinds of damage to DNA. Two mechanisms, light repair and dark repair, are known to repair damage caused by dimers.

Light repair, or **photoreactivation,** occurs in the presence of visible light in organisms previously exposed to ultraviolet light. When organisms containing dimers are kept in visible light, the light activates an enzyme that breaks the bonds between the pyrimidines of a dimer (Figure 7.19a). Thus, mutations that might have been passed along to daughter cells are corrected, and the DNA is returned to its normal state. This mechanism contributes to the survival of the bacteria but creates a problem for microbiologists. Cultures that are irradiated with ultraviolet light to induce mutations must be kept in the dark for the mutations to be retained.

Dark repair, which occurs in some bacteria, requires several enzyme-controlled reactions (Figure 7.19b). First, an endonuclease breaks the defective DNA strand near the dimer. Second, a DNA polymerase synthesizes new DNA to replace the defective segment, using the normal complementary strand as a template. Third, an exonuclease removes the defective DNA segment. Finally, a ligase connects the repaired segment to the remainder of the DNA strand. These reactions were identified in *E. coli* but are now known to occur in many other bacteria. Human cells have similar mechanisms; some human skin cancers such as xeroderma pigmentosum (Figure 7.20) may be due to a defect in the cellular DNA repair mechanism.

The Study of Mutations

Microorganisms are especially useful in studying mutations because of their short generation time and the

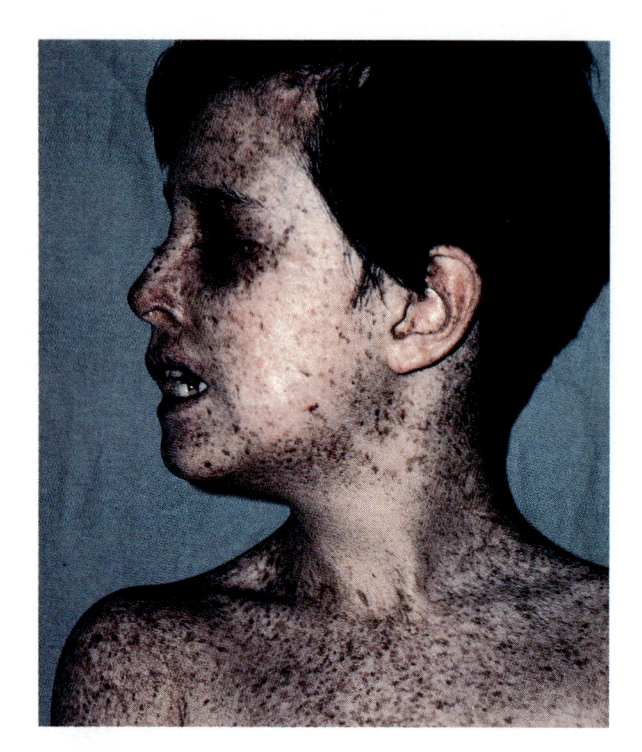

FIGURE 7.20 Xeroderma pigmentosum is a human genetic disease in which the enzymes to repair UV damage to DNA are defective, and exposure to sunlight results in multiple skin cancers.

relatively small expense of maintaining large populations of mutant organisms for study. Comparisons of normal and mutant organisms have led to important advances in the understanding of both genetic mechanisms and metabolic pathways. Microorganisms continue to be important to researchers attempting to further our knowledge of these processes. However, the study of mutations is not without its problems. Two common problems are: (1) distinguishing between spontaneous and induced mutations and (2) isolating particular mutants from a culture containing both mutated and normal organisms. The fluctuation test and the technique of replica plating are used to distinguish between spontaneous and induced mutations; replica plating also is used to isolate mutants.

Why is it important to differentiate between spontaneous and induced mutations? Making this distinction helps us to understand mechanisms in the evolution of microorganisms and presumably other organisms as well. For example, some organisms grow in the presence of penicillin, despite its antibiotic properties. They are penicillin resistant. Theoretically, there are two ways organisms could acquire such resistance: Either the penicillin *induces* a change in the organism that enables it to grow in the presence of penicillin, or a mutation occurs *spontaneously* that will allow the organism to grow if it is later exposed to penicillin. In the latter case, penicillin will kill nonresistant organisms, thereby *selecting* for the resistant mutant. Various experiments, two of which are described below, have shown that the second mechanism, the selection of spontaneous mutants, is the primary means of evolution in microorganisms.

The **fluctuation test,** designed by Salvador Luria and Max Delbruck in 1943, is based on the following hypothesis: If mutations that confer resistance occur spontaneously and at random, we would expect great fluctuation in the number of resistant organisms per culture among a large number of cultures. This would occur regardless of whether the substance to which resistance develops is present. A mutation might occur early in the incubation period, late in that period, or not at all. Cultures with early mutations would

APPLICATIONS

Ozone Biosensors

Because ozone filters out harmful ultraviolet radiation, the discovery of holes in the ozone layer of the earth's atmosphere has raised concern about how much ultraviolet light reaches the earth's surface. Of particular concern are the questions of how deeply into seawater ultraviolet penetrates and how it affects marine organisms, especially plankton (floating microorganisms) and viruses that attack plankton. The photosynthetic plankton utilize carbon dioxide (CO_2) and are believed to have major effects on our planet's temperature and weather.

Now researcher Deneb Karentz of the Laboratory of Radiobiology and Environmental Health at the University of California in San Francisco has devised a simple method for measuring ultraviolet penetration and intensity. Working in the Antarctic Ocean, she submerged to various depths thin plastic bags containing special strains of *E. coli* that are almost totally unable to repair ultraviolet damage to their DNA. Bacterial death rates in these bags were compared with rates in unexposed control bags of the same organism. The bacterial "biosensors" revealed constant significant ultraviolet damage at depths of 10 meters and frequently at 20 and 30 m. Karentz plans additional studies of how ultraviolet may affect seasonal plankton "blooms" in the oceans. These form the base of the marine food chains and may affect weather by their uptake of CO_2 for photosynthesis.

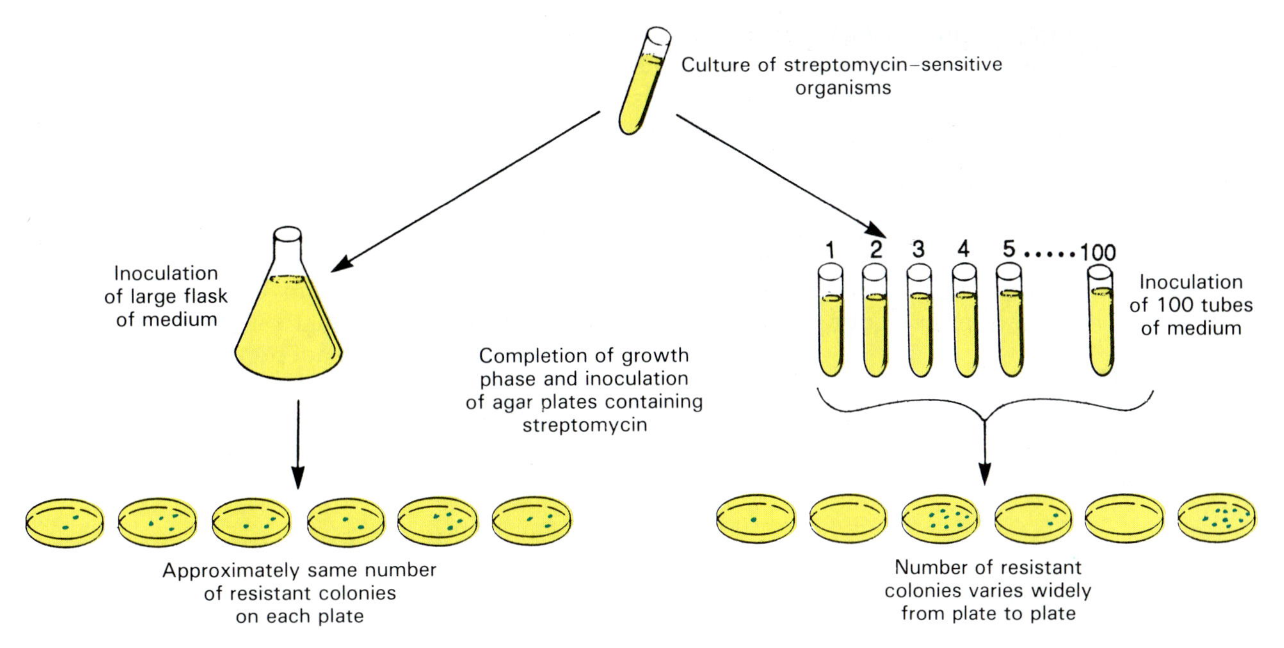

FIGURE 7.21 The fluctuation test of Luria and Delbruck proves that mutations conferring antibiotic resistance are random—they are not induced by exposure to the antibiotic.

contain many mutated progeny, those with late mutations would have few mutated progeny, and those without a mutation would have none. An alternate hypothesis is that mutations conferring resistance to a substance occur only in the presence of the substance. Then cultures containing the substance would be expected to have approximately equal numbers of resistant organisms, whereas cultures lacking the substance would have no resistant organisms.

To test these hypotheses, the investigators inoculated a large flask of liquid medium with an organism that was sensitive to the antibiotic streptomycin. At the same time they inoculated 100 small tubes of liquid medium with the same organism. No streptomycin was present in either the flask or the tubes. Both the flask and the tubes were allowed to reach maximum growth (10^9 organisms per milliliter). One-milliliter samples were then used to inoculate agar plates containing streptomycin; a plate was made from each tube, and many plates were made from the flask. After 24 hours the colonies on each plate were counted. Each colony represented a resistant mutant that could grow in the presence of streptomycin. There was far greater fluctuation in the number of colonies among the plates inoculated from the tubes than among the plates inoculated from the flask (Figure 7.21). Therefore, mutations must have occurred at different times or not at all in the various tubes. Mutations also must have occurred at different times in the flask, but progeny of mutated organisms became distributed through the medium, so that the number of mutants in each sample did not vary greatly. The investigators concluded that resistance was conferred from random mutations occurring at different times among the organisms in the tubes and not from exposure to streptomycin. (Can you predict what results would have been obtained if resistance arose only from exposure to streptomycin?)

The technique of **replica plating,** devised by Joshua and Esther Lederberg in 1952, is also used to study mutations. Based on the same reasoning as the fluctuation test, it hypothesizes that resistance to a substance arises spontaneously and at random without the need for exposure to the substance. In the original replica plating studies (Figure 7.22), organisms from a liquid culture were evenly spread on an agar plate and allowed to grow for 4 to 5 hours. Then a sterile velveteen pad was gently pressed against the surface of the plate to pick up organisms from each colony. The tiny fibers of velveteen acted like hundreds of tiny inoculating needles. The pad was carefully kept in the same orientation and used to inoculate an agar plate containing a substance such as penicillin to which organisms might be resistant. After incubation, the exact positions of corresponding colonies on the two plates were noted. The organisms in colonies found on the penicillin plate had resistance to penicillin without ever having been exposed to it.

Replica plating not only demonstrates spontaneity of mutations that confer resistance, it also provides a means of isolating resistant organisms without exposing them to a substance. By keeping the velveteen pad in perfect alignment during transfer, resistant colonies on the original plate could be identified by their location relative to colonies on the penicillin plate.

Replica plating is now widely used to study changes in the characteristics of many bacteria. The velveteen pads have been replaced by other materials that are easier to sterilize and manipulate. The technique is especially useful to identify mutants whose nutritional needs have changed. Replicas can be transferred to a variety of different media, each deficient in a particular nutrient. Failure of particular colonies to grow on the deficient medium indicates a mutation that caused the organism to require that nutrient.

The Ames Test

Human cancers can be induced by environmental substances that act by altering DNA. Much research effort is now being devoted to determining which substances are **carcinogenic** (cancer-producing). Carcinogenic substances usually also are mutagenic, so determining whether a substance is mutagenic is often a first step in identifying it as a carcinogen. Bacteria, being subject to mutation and being easier and cheaper to study than larger organisms, are ideal organisms to use in screening substances for mutagenic properties. Proving that a substance causes mutations in bacteria does not prove that it does so in human cells. Even proving that a substance causes mutations in human cells does not prove that the mutations will lead to cancer. Additional tests, including tests in animals, are necessary to identify carcinogens, but initial screening using bacteria can eliminate some substances from further study. If a substance induces no

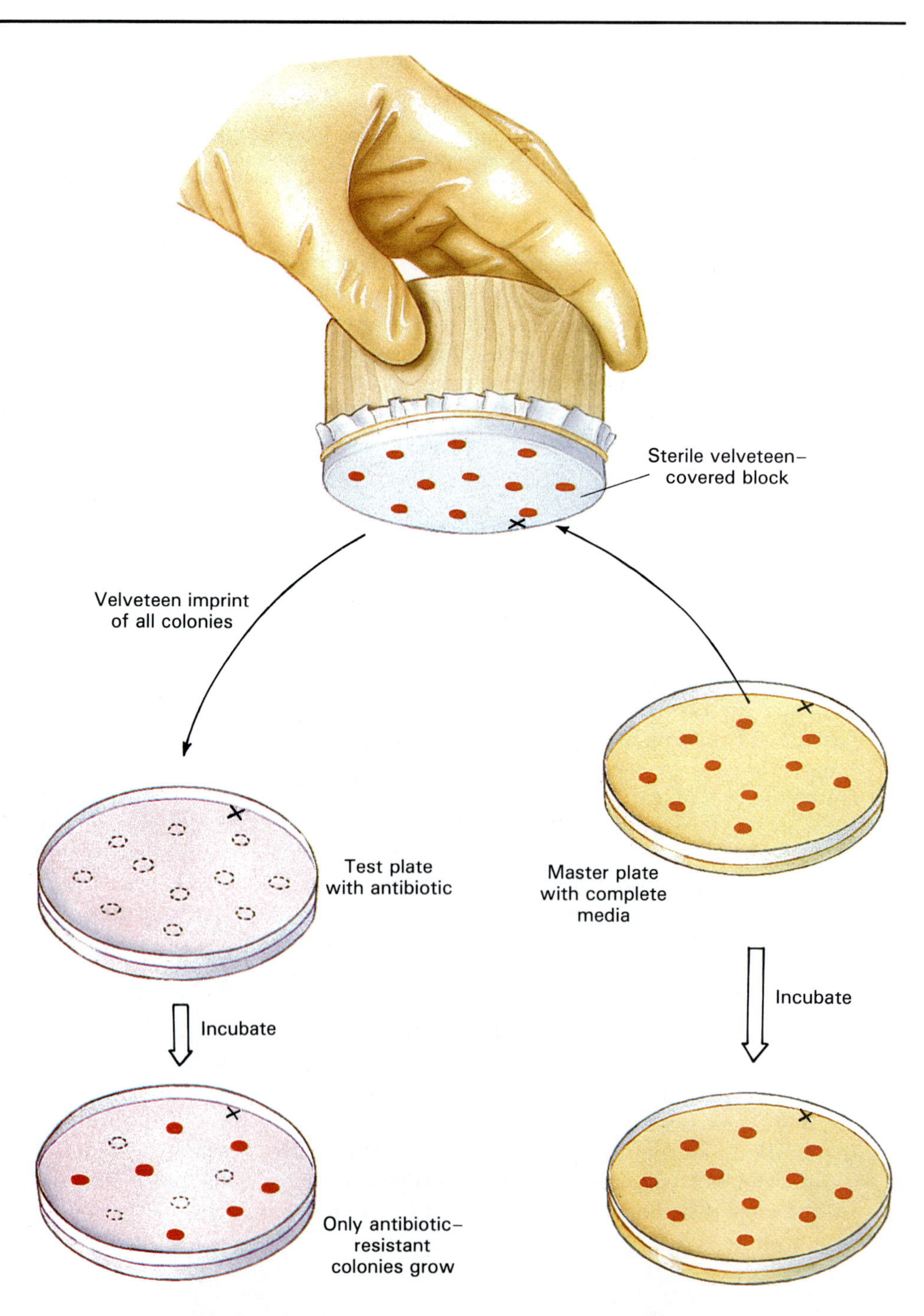

FIGURE 7.22 Replica plating allows detection of antibiotic-resistant organisms. The *x* on the side of the plate provides a reference for identifying colonies from the same organism.

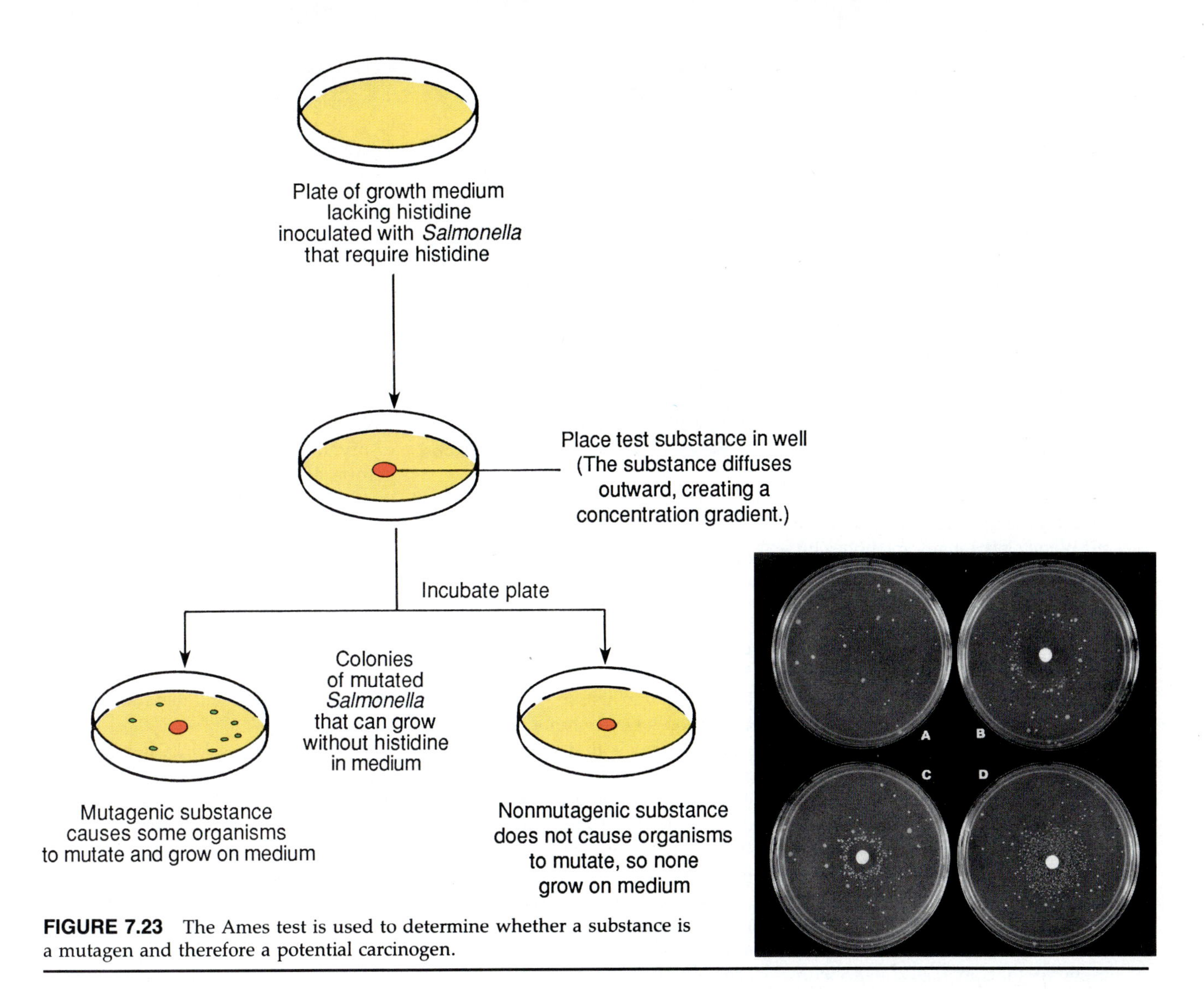

FIGURE 7.23 The Ames test is used to determine whether a substance is a mutagen and therefore a potential carcinogen.

mutations in a large population of bacteria, most researchers believe that it is not likely to be a carcinogen. The **Ames test** (Figure 7.23), devised by Bruce Ames at the University of California at Berkeley, is used to test whether substances induce mutations in certain strains of *Salmonella* that have lost their ability to synthesize the amino acid histidine. These strains easily undergo another mutation that restores their ability to synthesize histidine. The Ames test is based on the hypothesis that if a substance is a mutagen, it will increase the rate at which these organisms revert to histidine synthesizers. Furthermore, the more powerful a substance's mutagenic capacity, the greater the number of reverted organisms it causes to appear. In practice the organism is grown in the presence of a test substance. If any organisms regain the ability to synthesize histidine, the substance is suspected of being a mutagen. The larger the number of organisms regaining the synthetic ability, the stronger the substance's mutagenic capacity is likely to be.

ESSAY

Polymerase Chain Reaction (PCR)— Key to Past and Future Worlds of DNA

Is there any privacy in the grave? Not anymore. Ancient DNA, sometimes as old as 17 million years, is being recovered, and its bases sequenced. Which genes did past organisms have? How many were handed down to us, and how many newer mutant versions do we have? Brains taken from prehistoric Indians, buried and mummified in Florida peat bogs 7500 years ago, have had their DNA extracted. DNA from

ESSAY Polymerase Chain Reaction (PCR)

91 brains is now undergoing analysis thanks to **polymerase chain reaction (PCR),** a technique that first became available in 1985. PCR allows us to produce rapidly (amplify) a billion copies of DNA without needing a living cell. These large quantities are then easily analyzed. Scientists have applied this tool to many questions concerning the past. For example, they have extracted DNA from fossil animals embedded in amber; from fossil leaves embedded in the shale of Idaho 17 million years ago (which turn out to have DNA very similar to that of modern magnolias); and from the blood stains, hair, and bone chips preserved by doctors attending President Lincoln at the time of his assassination. Many people have wondered whether Lincoln had a hereditary disease, Marfan's syndrome, that causes weakened arteries that can rupture and cause death. Most people having Marfan's syndrome would have died before they reached Lincoln's age. Would Lincoln have died soon, even had he not been assassinated? We can now make a library of Lincoln's DNA, and, as the Human Genome Project identifies the sequence of various genes (including those for Marfan's), we can match them to Lincoln's DNA and know with certainty which genes he really had.

A modern forensic problem could soon yield to PCR scrutiny. DNA has been collected from bones buried in Brazil that reputedly belong to the Nazi concentration camp war criminal Dr. Josef Mengele (Figure 7.24). Half of his DNA should have been passed to his son. The other half of his son's DNA should have come from Mengele's first wife. Comparison of the DNA from all three people can tell us unequivocally whether the man in the grave was truly Mengele. Indeed, there will be few secrets left in the grave.

FIGURE 7.24 Is it Josef Mengele? On June 21, 1985, members of the team of forensic specialists at the Coroner's Office in São Paulo, Brazil, exhibit the skull of the man suspected of being the Nazi war criminal Josef Mengele.

Rising from an 11,000-year-old grave in Ohio are cultures of two strains of *Enterobacter cloacae,* recovered alive but in a frozen state of suspended animation from the intestine of a 4-ton mastodon that had been killed by prehistoric hunters, butchered, and then sunk into a peat bog (a primitive form of food preservation) (Figure 7.25). And preserved they are! No mutations have occurred for the past 11,000 years. PCR analysis will reveal how these ancient organisms differ from today's strains of *E. cloacae.* Botanists await the tantalizing analysis of DNA from the mastodon's last meal: pollen grains stuck in his teeth, swamp grass, mosses, leaves, and even a water lily. Chapters of evolutionary history may need to be rewritten.

For the living, PCR-amplified DNA analysis can reveal the presence of organisms that are difficult, dangerous, slow-growing, or require extra skill to culture in standard clinical laboratories. Tuberculosis cultures require 8 weeks to grow; PCR techniques will confirm the presence of the DNA of tuberculosis organisms in just hours. Medical-technology programs will need to train students in this technique of the future, and current personnel will need to be retrained.

What happens in PCR amplification of DNA? A large piece, or mixture of pieces, of DNA is cut up into smaller pieces by restriction endonuclease enzymes. It is necessary to know the base composition of the ends of the exact piece of DNA you wish to replicate. In less than 24 hours automated synthesizing equipment can make oligonucleotide (*oligos,* few) primer molecules that will pair complementarily with the ends of the desired section of DNA. A primer is a molecule that will serve as a starting point for DNA synthesis.

The sequence of events in PCR amplification is shown in Figure 7.26.

(a)

(b)

FIGURE 7.25 *Enterobacter cloacae* (a). These bacteria were isolated and cultured from the remains of an 11,000-year-old American mastodon found in Newark, Ohio, in December 1989 (b). The mastodon's digestive tract could be identified as a darker-colored, discrete cylindrical mass bent into the shape of intestinal loops. No organisms were found in areas sampled adjacent to the intestine.

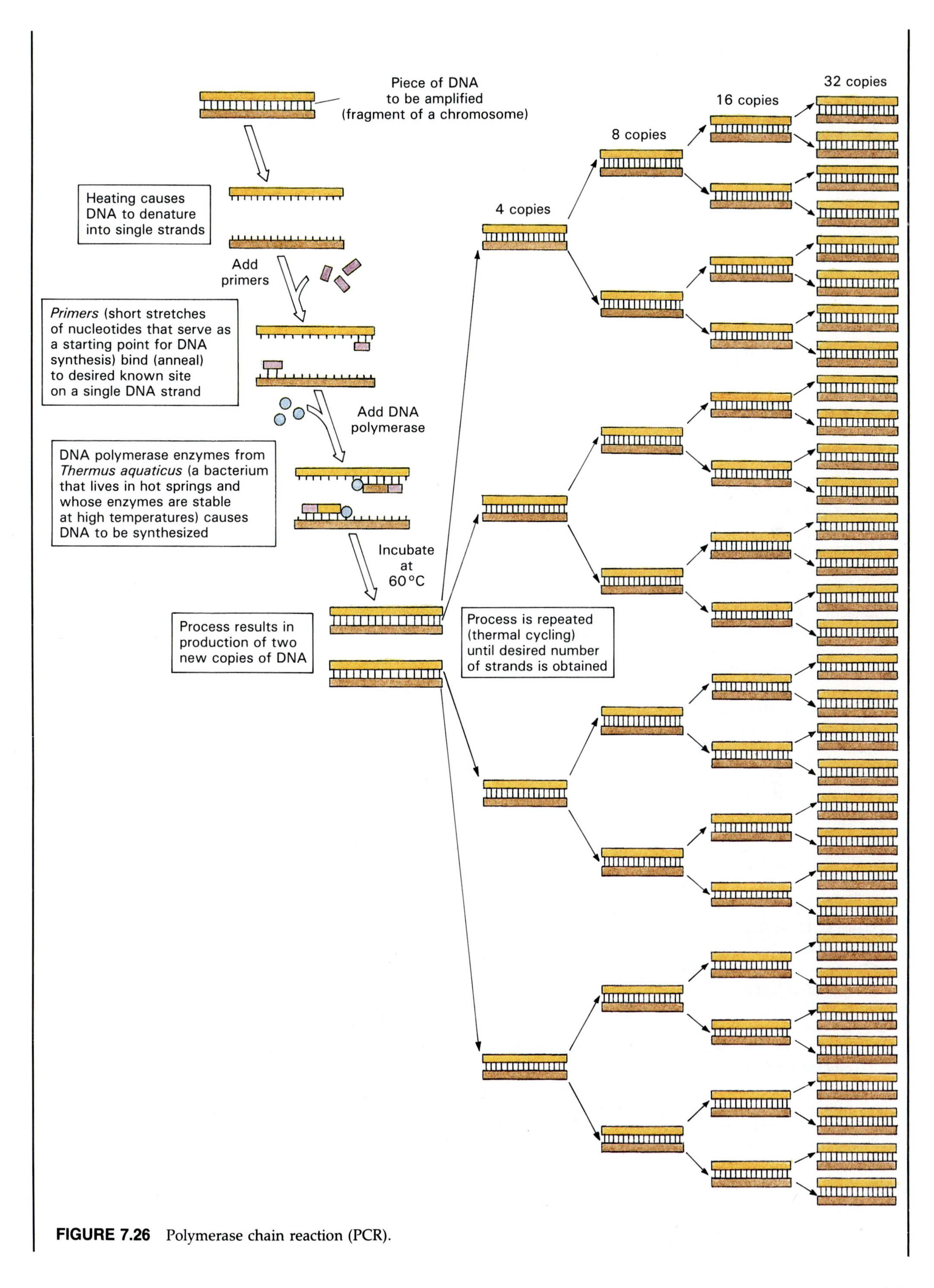

FIGURE 7.26 Polymerase chain reaction (PCR).

ESSAY Polymerase Chain Reaction (PCR)

When the thermal cycling has produced billions of copies of the desired piece of DNA, the DNA is easily detected (as in a clinical diagnostic test) or analyzed for total base sequence. Cutting a large piece of DNA up into smaller pieces, sequencing the PCR-amplified quantities of these, and then looking for overlaps at the ends will allow us finally to determine the sequence of the entire original DNA piece.

CHAPTER SUMMARY

OVERVIEW OF GENETIC PROCESSES

The Basis of Heredity

- **Heredity** involves the transmission of information from an organism to its progeny.
- **Genes** are linear sequences of DNA that carry coded information for the structure and function of an organism.
- Prokaryotic chromosomes are threadlike circular structures made of DNA.
- Transmission of information in prokaryotes typically occurs during asexual reproduction in which the chromosome **replicates,** and each daughter cell receives a chromosome like the one in the parent cell.
- **Mutations** (alterations in DNA) transmitted to progeny account for much of the variation in organisms.

RELATED KEY TERMS: **locus** **alleles**

Nucleic Acids in Information Storage and Transfer

- All information for the functioning of a cell is stored in DNA in a specific sequence of the nitrogenous bases: adenine, thymine, cytosine, and guanine.
- Information stored in DNA is used for two purposes: (1) to replicate DNA in preparation for cell division and (2) to provide information for protein synthesis. In both of these processes information is transferred by base pairing.

REPLICATION OF DNA

- **Replication** of DNA begins at a specific point in the circular chromosome and usually proceeds in both directions simultaneously.
- The main steps in the replication of DNA are summarized in Figure 7.3.
- DNA replication is **semiconservative**—each chromosome consists of one strand of old (parent) DNA and one of newly synthesized DNA.

RELATED KEY TERMS: **antiparallel** **replication forks** **DNA polymerase** **leading strand** **lagging strand** **discontinuous** **Okazaki fragments** **ligase**

PROTEIN SYNTHESIS

Transcription

- In **transcription, messenger RNA (mRNA)** is transcribed from DNA as summarized in Figure 7.6, to serve as a **template** for protein syntheses.

RELATED KEY TERMS: **RNA polymerase**

Kinds of RNA

- Two other kinds of RNA are similarly produced: (1) **Ribosomal RNA (rRNA)** combines with proteins to form ribosomes, the sites for protein assembly. (2) **Transfer RNA (tRNA)** carries amino acids to the assembly site.

Translation

- In the process of **translation,** three-base sequences in mRNA act as **codons** and are matched by base pairing with **anticodons** of tRNA; the mRNA codons constitute the **genetic code**—a code that is essentially the same for all living organisms.
- The process of protein synthesis, once the mRNA and ribosomes are aligned, proceeds as summarized in Figure 7.11.

RELATED KEY TERMS: **sense codon** **nonsense codon** **terminator codon**

REGULATION OF METABOLISM

RELATED KEY TERMS

Significance of Regulatory Mechanisms

- Mechanisms that regulate metabolism turn reactions on and off in accordance with the needs of cells, allowing the cells to use various energy sources and to limit synthesis of substances to the amounts needed.

Categories of Regulatory Mechanisms

- The two basic categories of regulatory mechanisms are: (1) mechanisms that regulate the activity of enzymes already available in the cell and (2) mechanisms that regulate the action of genes, which determine what enzymes will be available.

Feedback Inhibition

- In **feedback inhibition** the end product of a biochemical pathway directly inhibits the first enzyme in the pathway (Figure 7.12).
- Enzymes subject to such regulation are usually allosteric.
- Feedback inhibition regulates the activity of existing enzymes and is a quick-acting control mechanism.

end product inhibition

Enzyme Induction

- In **enzyme induction** (Figure 7.13) the presence of a substrate activates an **operon**—a sequence of closely associated genes: (1) In the absence of lactose, the *i* gene produces a **repressor** that attaches to the operator and prevents transcription of the genes of the *lac* operon. (2) When lactose is present, it inactivates the repressor and allows transcription of the genes of the *lac* operon.

constitutive enzymes
inducible enzymes
structural genes **regulator genes**
inducer

Enzyme Repression

- In **enzyme repression** the presence of a synthetic product inhibits its further synthesis by inactivating an operon: (1) When tryptophan is present, it attaches to the regulator protein and represses genes of the *trp* operon. (2) In the absence of tryptophan, the repressor is not activated, and genes of the *trp* operon are transcribed.
- In **catabolite repression** the presence of a preferred nutrient (often glucose) represses the synthesis of enzymes that would be used to metabolize some alternative substance.
- Both enzyme induction and enzyme repression regulate by altering gene expression. The effect on enzyme synthesis in both cases depends on the presence or absence of the regulator substance—lactose, tryptophan, or glucose in the above examples.

attenuation

MUTATIONS

Types of Mutations and Their Effects

- Mutations cause a change in the **genotype;** the change may or may not be expressed in the **phenotype.**
- Two major classes of mutations are point mutations and frameshift mutations (Table 7.3): (1) **Point mutations** consist of changes in a single nucleotide, and (2) **frameshift mutations** consist of the **insertion** or **deletion** of a single nucleotide.

Phenotypic Variation

- Phenotypic variations produced by mutations can involve alterations in colony morphology, nutritional requirements, and temperature sensitivity.

auxotrophic mutants
prototrophs

Spontaneous versus Induced Mutations

- **Spontaneous mutations** occur in the absence of any known mutagen and appear to be due to errors in base pairing during DNA replication. Various genes have different rates of mutation.
- **Induced mutations** are mutations produced by agents called **mutagens;** mutagens increase the mutation rate.

Chemical Mutagens

- Chemical mutagens include **base analogs, alkylating agents, deaminating agents,** and **acridine derivatives.**

Radiation as a Mutagen

- **Radiation** often causes the formation of **dimers**—adjacent pyrimidine bases bound to each other that interfere with replication.

Repair of DNA Damage

- Many bacteria have enzymes that can repair certain damages to DNA (Figure 7.19). (1) **Light repair** involves an enzyme that is activated by visible light and that breaks bonds between pyrimidines of a dimer. (2) **Dark repair** involves several enzymes that do not require light for activation; they excise defective DNA and replace it with DNA complementary to the normal DNA strand.

The Study of Mutations

- Microorganisms are useful in studying mutations because many generations can be produced quickly and inexpensively.
- The **fluctuation test** (Figure 7.21) demonstrated that resistance to chemical substances occurs spontaneously rather than being induced.
- **Replica plating** (Figure 7.22) likewise demonstrated the spontaneous nature of mutations; it also can be used for isolating mutants without exposing them to the substance to which they are resistant.

The Ames Test

- The **Ames test** (Figure 7.23) is based on the ability of bacteria to mutate by reverting to their original synthetic ability. It is used for screening chemicals for mutagenic properties, which indicate potential carcinogenicity.

RELATED KEY TERMS

restriction endonucleases
exonucleases

photoreactivation

carcinogenic

QUESTIONS FOR REVIEW

A.

1. Define chromosome, gene, allele, mutation, and heredity.

B.

2. How do nucleic acids function in information storage?
3. How do they function in information transfer?
4. Define transcription and translation.

C.

5. What are the steps in the replication of DNA?
6. What are the steps in protein synthesis?
7. What are the three types of RNA, and how do they differ in structure and function?

D.

8. What is the significance of the presence of mechanisms to regulate metabolism?
9. What factors distinguish the two basic regulatory mechanisms?

E.

10. How does feedback inhibition operate, and what does it accomplish?

F.

11. How does enzyme induction operate, and what does it accomplish?
12. How does enzyme repression operate, and what does it accomplish?
13. In what ways are enzyme induction and repression similar, and in what ways are they different?

G.

14. What are the kinds of mutations that can occur?
15. How might each kind of mutation affect an organism?

H.

16. How do spontaneous and induced mutations differ?
17. How do chemical mutagens alter DNA?
18. How might radiation alter DNA?
19. Briefly describe two mechanisms by which bacteria can sometimes repair damage to DNA.

I.

20. Why are bacteria useful in the study of mutations?
21. How is the fluctuation test performed, and what does it show?
22. How is replica plating done, and what can be accomplished by it?
23. What is the Ames test, and what is it used for?

PROBLEMS FOR INVESTIGATION

1. How does the role of the repressor differ in induction and repression?
2. Devise an experiment to produce and isolate a mutant organism. If possible carry out your experiment and perform tests that will confirm your hypothesis that the organism is, in fact, a mutant.
3. During the early stages of development of the earth's atmosphere, the planet was exposed to greater amounts

of ultraviolet radiation than it is today. What do you suppose were the effects of this radiation on the longevity of individual organisms and on the rate of evolution of life forms?

4. Read about and prepare a report on xerodema pigmentosum, the genetic disease in which people lack the ability to repair ultraviolet damage to their DNA.
5. Given the DNA sequence of ATA GCA AAA CCG ATG, what is the amino acid sequence of the polypeptide this would code to produce?

SOME INTERESTING READING

Anonymous. 1991. PCR diagnostics: ownership, technology are changing. *ASM News* 57(10):503–4.

Beardsley, T. 1991. Smart genes. *Scientific American* 265(8):86–95.

Brock, T. D., and M. T. Madigan. 1990. *Biology of microorganisms*. Englewood Cliffs, NJ: Prentice Hall.

Cathcart, R. 1990. Advances in automated DNA sequencing. *Nature* 347(September 20):310.

Cherfas, J. 1991. Ancient DNA: still busy after death. *Science* 253(September 20):1354–56.

Croce, C. M. 1985. Chromosomal translocations, oncogenes, and B-cell tumors. *Hospital Practice* 20:41.

Devoret, R. 1979. Bacterial tests for potential carcinogens. *Scientific American* 241, no. 2 (August):40.

Folger, T. 1992. Oldest living bacteria tell all. *Discover* 13(January):30–32.

Krawiec, S. and M. Riley. 1990. Organization of the bacterial chromosome. *Microbiological Reviews* 54(4):520–39.

Levine, A. J. 1988. Oncogenes of DNA tumor viruses. *Cancer Research* 48 (February 1):493.

Miller, J. A. 1984. Diagnostic DNA. *Science News* 126:104.

Moss, Robert. 1991. Genetic transformation of bacteria. *The American Biology Teacher* 53(March):179–80.

Mullis, K. B. 1990. The unusual origin of the polymerase chain reaction. *Scientific American* 262(4):56–61.

Osawa, S., T. H. Jukes, K. Watanabe, and A. Muto. 1991. Recent evidence for evolution of genetic code. *Microbiological Reviews* 56(1):229–64.

Persing, D. H. 1991. Polymerase chain reaction: trenches to benches. *Journal of Clinical Microbiology* 29(7):1281–85.

Scott, A. 1985. *Pirates of the cell*. Oxford, Eng.: Blackwell.

Sykes, B. 1991. Ancient DNA: the past comes alive. *Nature* 352(August 1):381–82.

Waldrop, M. M. 1989. Did life really start out in an RNA world? *Science* 246:1248.

Weinberg, R. A. 1983. A molecular basis of cancer. *Scientific American* 249, no. 5 (November):126.

Weintraub, H. M. 1990. Antisense RNA and DNA. *Scientific American* 262(1):40–46.

Witkin, E. M. 1976. Ultraviolet mutagenesis and inducible DNA repair in *Escherichia coli*. *Bacteriological Review* 40:869.

Technician working in a DNA research laboratory. Microbiology has provided the techniques for working with DNA, which are being used to understand and solve many human problems.

8

Genetics II: Transfer of Genetic Material and Genetic Engineering

This chapter focuses on the following questions:

A. What are the nature and significance of gene transfer?

B. What are the mechanisms and significance of transformation?

C. What are the mechanisms and significance of transduction?

D. What are the mechanisms and significance of conjugation?

E. What are the characteristics and actions of plasmids?

F. How are the following techniques of genetic engineering used: (a) genetic fusion, (b) protoplast fusion, (c) gene amplification, (d) recombinant DNA, and (e) hybridomas?

G. Why are scientists concerned about uses of recombinant DNA?

Transfer of genetic material from one organism to another can have far-reaching consequences. In microbes it provides ways for viruses to infect bacteria and ways for bacteria to invade humans or to become resistant to certain antibiotics. Information obtained from studying the transfer of genetic material in microorganisms can be applied to agricultural, industrial, and medical problems. In this chapter we will discuss the mechanisms by which genetic transfers occur and the significance of such transfers.

NATURE AND SIGNIFICANCE OF GENE TRANSFER

Gene Transfer

Gene transfer refers to movement of genetic information between organisms. In eukaryotes, it usually occurs by sexual reproduction. Male and female parents produce gametes (reproductive cells) that unite to form the first cell of a new individual. Because each parent produces many genetically different gametes, many different combinations of genetic material can be transferred to offspring. Before the 1920s, bacteria were thought to reproduce only by binary fission and to have no means of genetic transfer comparable to that achieved through sexual reproduction. Since then three mechanisms of gene transfer in bacteria have been discovered, none of which is associated with reproduction. Each mechanism—transformation, transduction, and conjugation—is discussed in this chapter.

Gene transfer is significant because it greatly increases the genetic diversity of organisms. As noted in the last chapter, mutations account for some genetic diversity, but gene transfer between organisms accounts for even more. When organisms are subjected to changing environmental conditions, genetic diversity increases the likelihood that some will adapt to any particular conditions. Such diversity leads to evolutionary changes. Organisms with genes that allow them to adapt to an environment survive and reproduce, while organisms lacking those genes perish. If all organisms were genetically identical, all would survive and reproduce, or all would die.

TRANSFORMATION

Discovery of Transformation

Bacterial **transformation,** a change in an organism's characteristics because of the transfer of genetic information, was first discovered in 1928 by Frederick Griffith, an English physician, while he was studying pneumococcal infections in mice. Pneumococci with capsules produce smooth, glistening colonies. Those lacking capsules produce rough colonies with a coarse, nonglistening appearance. Many types of smooth pneumococci can now be distinguished by the chemical nature of their capsules. Encapsulated pneumococci that infect mice are extremely virulent, that is, they have a strong power to cause disease. A single organism injected into a mouse can multiply rapidly and kill the mouse!

Griffith injected some mice with live smooth pneumococci, some with heat-killed smooth pneumococci, some with live rough pneumococci, and still others with a mixture of live rough and heat-killed smooth pneumococci (Figure 8.1). As expected, mice that received live smooth pneumococci died, whereas those that received either heat-killed smooth pneumococci or live rough pneumococci survived. Surprisingly, those that received the mixture also died, and Griffith isolated live smooth organisms from them. The presence of heat-killed, encapsulated organisms apparently allowed the live unencapsulated ones to develop capsules and become virulent. However, neither Griffith nor his colleagues understood at that time how this transformation occurred.

In subsequent studies of transformation, Oswald Avery discovered that a capsular polysaccharide was responsible for the virulence of pneumococci. In 1944, Avery, Colin MacLeod, and Maclyn McCarty isolated the substance responsible for the transformation of pneumococci and determined that it was DNA. In retrospect, this discovery marked the "birth of molecular genetics," but at the time DNA was not known to carry genetic information. Researchers working with plant and animal chromosomes had isolated both DNA and protein from them, but they thought the genetic information was in the protein. Only when James Watson and Francis Crick determined the structure of DNA did it become clear that DNA encodes genetic information. After this original work with pneumococci (now called *Streptococcus pneumoniae*), transformation was observed in organisms from a wide variety of genera, including *Acinetobacter, Bacillus, Haemophilus, Neisseria,* and *Staphylococcus.*

Mechanism of Transformation

To study the mechanism of transformation, scientists extract DNA from donor organisms by a fairly complex biochemical process that yields hundreds of naked DNA fragments from each bacterial chromosome. (Naked DNA is DNA that is not incorporated into chromosomes or other structures.) When DNA extract is placed in a medium with organisms capable of incorporating it, most organisms can take up a maxi-

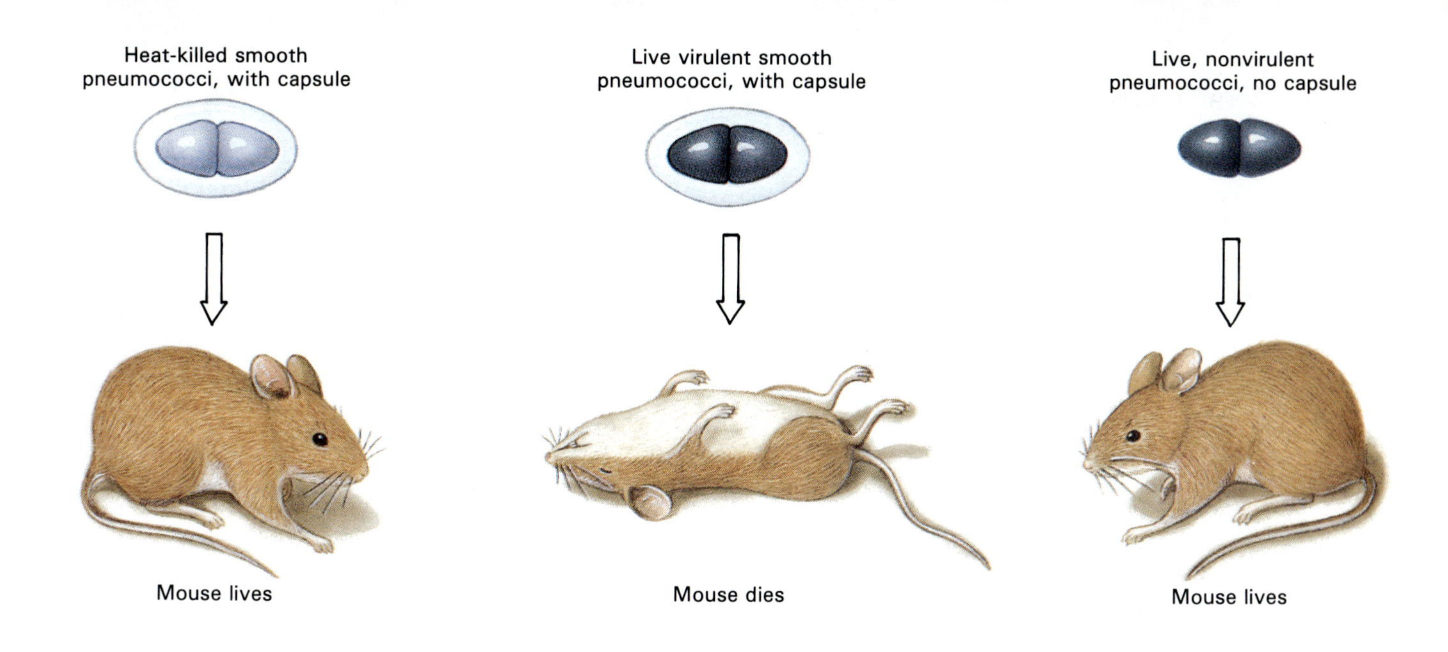

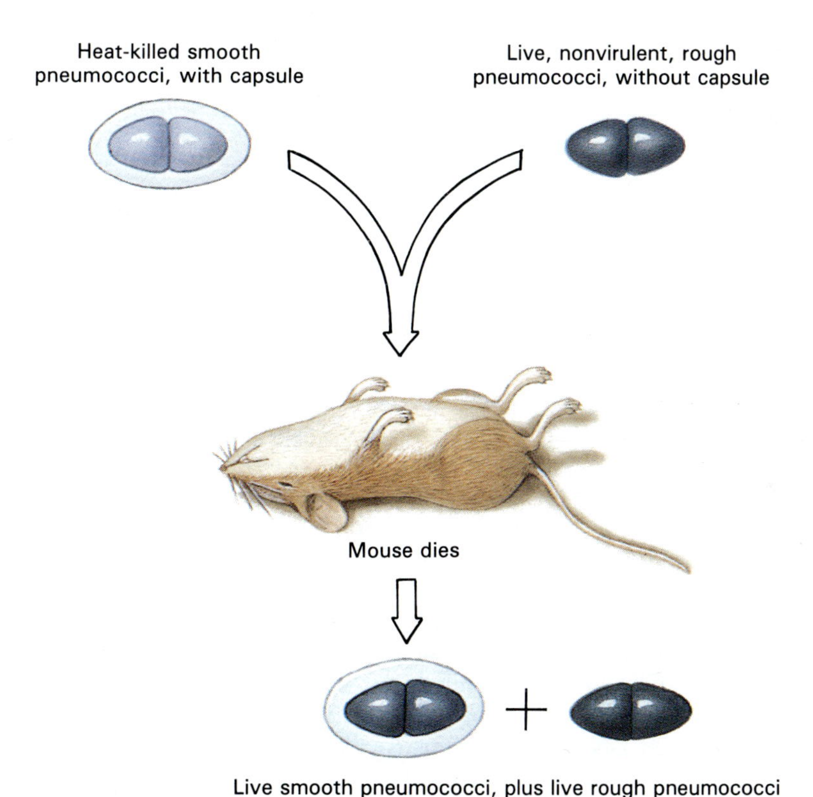

FIGURE 8.1 Griffith's experiment with pneumococcal infections in mice, which led to the discovery of transformation. When S-type pneumococci (which produce smooth-appearing colonies, due to presence of capsules) are injected into mice, the mice die of pneumonia. When R-type pneumococci (which produce rough-appearing colonies, due to lack of capsules) are injected into mice, the mice survive. When heat-killed S-type pneumococci are injected into mice, they also survive. But when a mixture of live R-type and heat-killed S-type pneumococci—neither of which is lethal by itself—are injected into a mouse, it dies, and live S-type organisms as well as R-types are recovered from the dead animal. Griffith had no way of knowing exactly what had happened, but he realized that some cells had clearly been "transformed" from type R to type S. Moreover, the change was heritable. Today we know that R-type bacteria pick up naked DNA, liberated from disintegrated dead S-type bacteria, and incorporate it into their own genome. If they pick up a piece of DNA that has the genes for capsule production, they will be genetically transformed into S-type organisms.

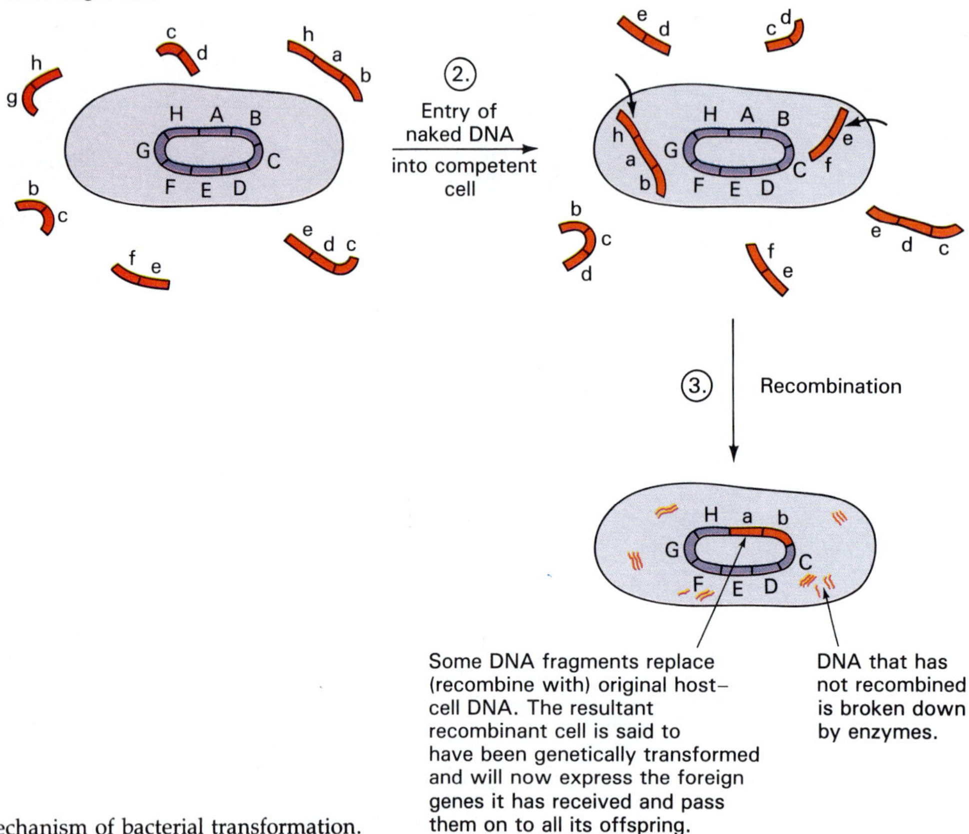

FIGURE 8.2 Mechanism of bacterial transformation.

mum of about 10 fragments, which is less than 5 percent of the amount of DNA normally present in the organism.

Uptake of DNA occurs only at a certain stage in a cell's growth cycle, probably prior to the completion of cell wall synthesis. In this stage, a protein called **competence factor** is released into the medium and apparently facilitates the entry of DNA. When competence factor from one culture is used to treat a culture lacking it, cells in the treated culture become competent to receive DNA. Entry sites on recipient cells appear to recognize different types of DNA. They admit DNA from the same or a closely related species and reject DNA from distantly related species (Figure 8.2).

Once DNA reaches the entry sites, endonucleases cut double-stranded DNA into units of 7000 to 10,000 nucleotides. The strands separate, and only one strand enters the cell. Single-stranded DNA is vulnerable to attack by various nucleases and can enter a cell only if the nucleases on the cell surface somehow have been inactivated. Inside the cell, the donor single-stranded DNA must combine by base pairing with a portion of the recipient chromosome immediately or else be destroyed. In transformation, as well as in other mechanisms of gene transfer, the donor single-stranded DNA is spliced into the recipient DNA. Splicing of a DNA strand involves breaking the strand, removing a segment, inserting a new segment, and attaching the ends. Enzymes in the recipient cell excise (cut out) a portion of the recipient's DNA and replace it with the donor DNA. The leftover recipient DNA is subsequently broken down so that the number of nucleotides in the cell's DNA remains constant.

Significance of Transformation

Though transformation has been observed mainly in the laboratory, it occurs in nature. It probably follows the breakdown of dead organisms in an environment where live ones of the same or a closely related species are present. However, the degree to which transformation contributes to the genetic diversity of organisms in nature is not known. In the laboratory, inducing transformation allows researchers to study the effects of DNA that differs from the DNA that the organism already has. Transformation also can be used to study the locations of genes on a chromosome and to insert DNA from one species into that of another species, thereby producing recombinant DNA.

TRANSDUCTION

Discovery of Transduction

Transduction, like transformation, is a method of transferring genetic material from one bacterium to another. Unlike transformation, in which naked DNA is transferred, in transduction DNA is carried by a **bacteriophage** (bak-te′re-o-faj)—a virus that infects bacteria. The phenomenon of transduction was originally discovered in *Salmonella* in 1952 by Joshua Lederberg and Norton Zinder and has now been observed in many different genera of bacteria.

Mechanisms of Transduction

To understand the mechanisms of transduction, we need to describe briefly the properties of bacteriophages, also called **phages** (faj′ez). Phages, which are described in more detail in Chapter 11, are composed of a core of nucleic acid covered by a protein coat. They infect bacterial cells and reproduce within them, as shown in Figure 8.3. A phage capable of infecting a bacterium attaches to a receptor site on the cell surface. The phage nucleic acid then enters the bacterial cell, leaving its protein coat outside. Once the nucleic acid is in the cell, further events follow one of two

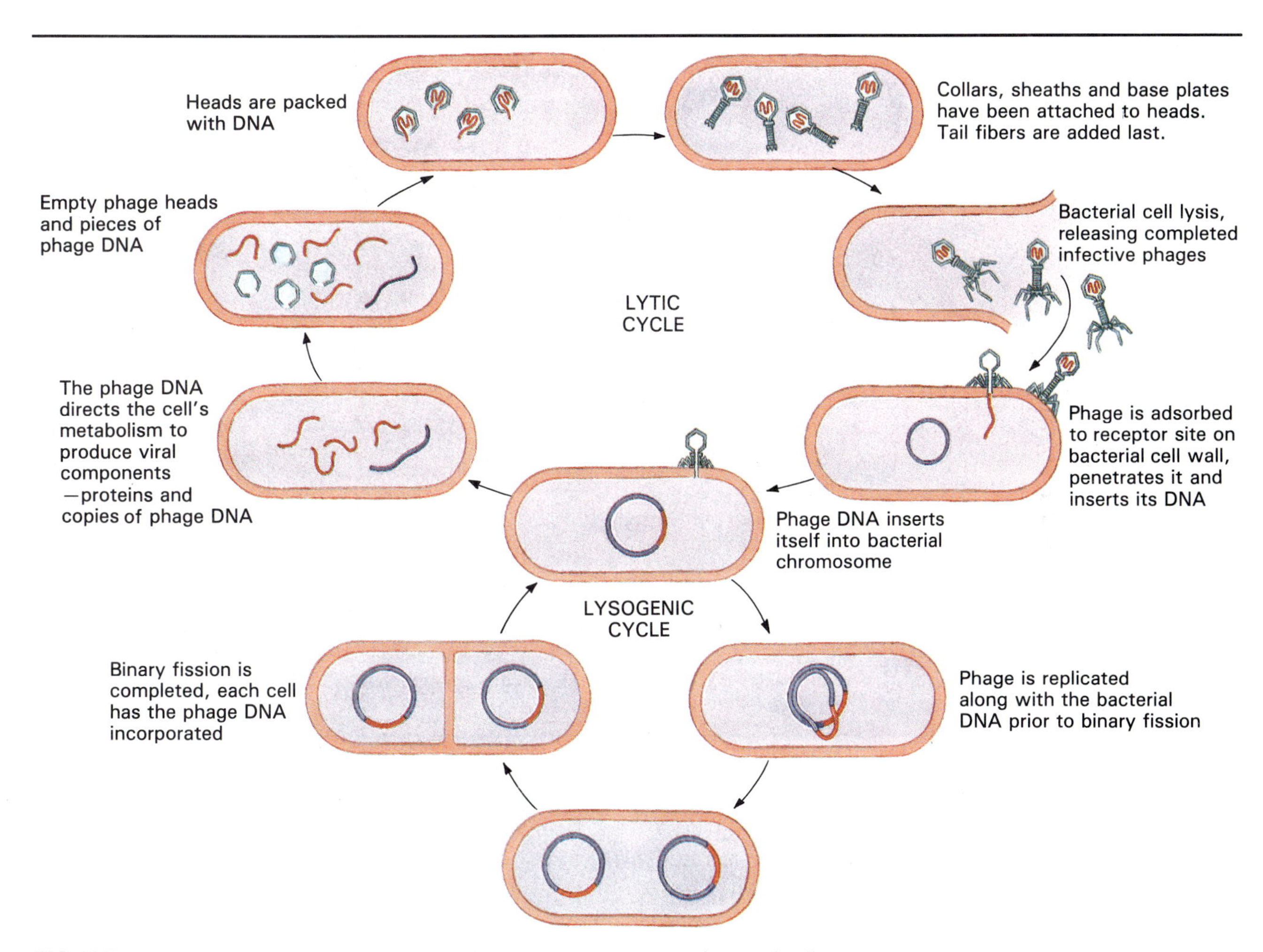

FIGURE 8.3 When a bacteriophage virus injects its DNA into a host bacterial cell, two different outcomes are possible. In the lytic cycle, characteristic of virulent phages, the phage DNA takes control of the cell and causes it to synthesize new viral components, which are assembled into whole viral particles. The cell is lysed to release the infective viruses, which can then enter new host cells. In the lysogenic cycle, the DNA of a temperate phage enters the host cell, becomes incorporated into the bacterial chromosome as a prophage, and replicates along with the chromosome through many cell divisions. However, a lysogenic phage can suddenly revert to the lytic life cycle. A prophage is thus a sort of "time bomb" sitting inside the infected cell.

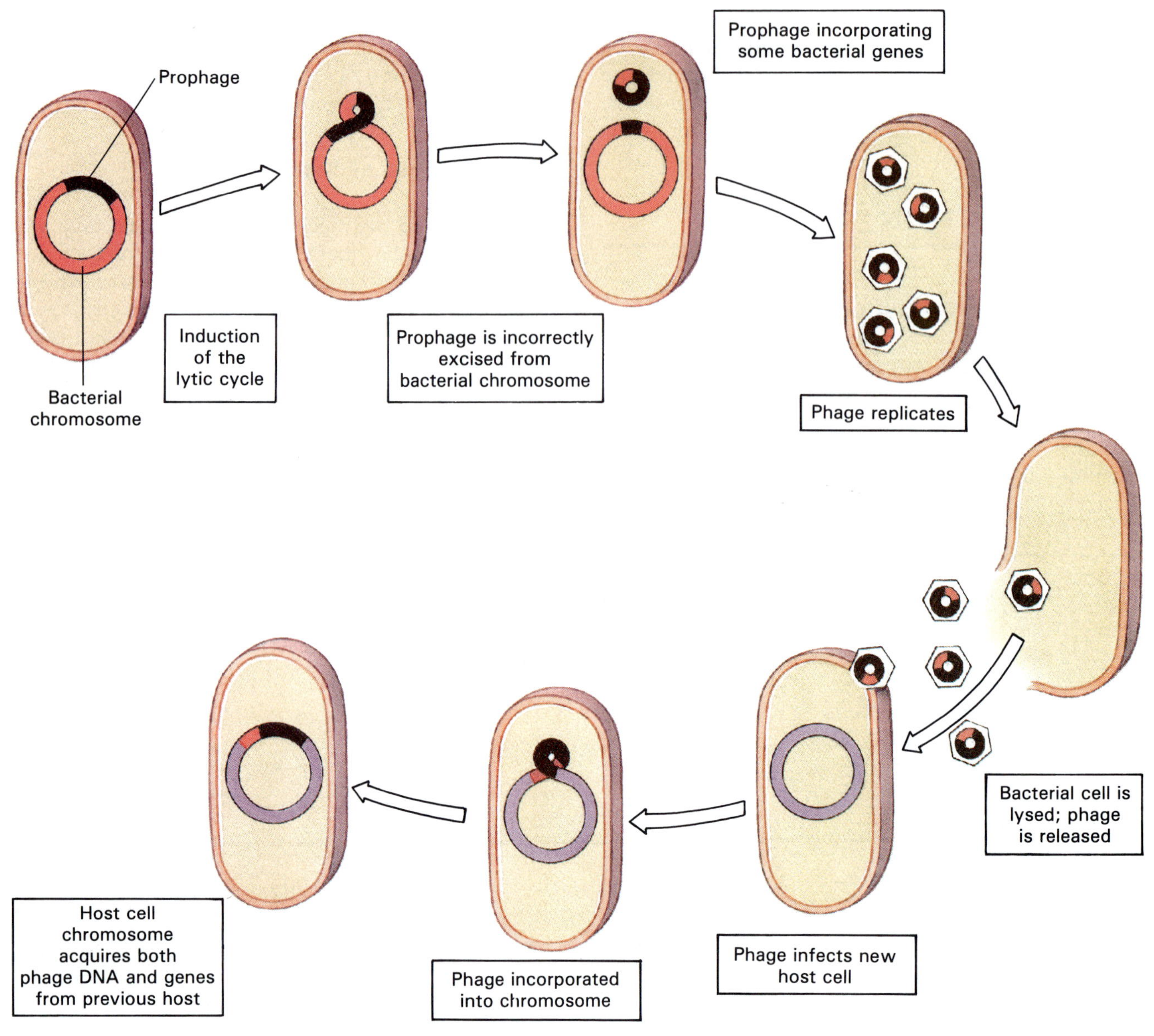

FIGURE 8.4 Specialized transduction by lambda phage in *E. coli.* In this process, phage DNA always inserts itself into the bacterial host chromosome at a particular site. When the phage replicates, it takes bacterial genes from either side of the site and packages them along with its own DNA into new phages. Only genes adjacent to the insertion site are transduced, rather than genes from other parts of the bacterial chromosome. These genes may then be introduced into the phage's next host cell, where they will confer new genetic traits.

pathways depending on whether the phage is virulent or temperate.

A **virulent phage** is one capable of causing severe infection and eventual death of a cell. Once the phage nucleic acid enters the cell, phage genes direct the cell to synthesize phage nucleic acids and protein coats and assemble them into complete phages. When the cell becomes filled with a hundred or more phages, it ruptures and releases the phages, which can then infect other cells. Because this cycle results in **lysis** (li'sis), or destruction, of the infected cell, it is called a **lytic** (lit'ik) **cycle.**

A **temperate phage** is one that ordinarily does not cause a virulent infection. Instead it is incorporated into a bacterium and replicated with it. This phage also produces a repressor substance that prevents the destruction of bacterial DNA, and its DNA does not direct the synthesis of phage particles. Phage DNA that has entered a bacterium is called **prophage** (pro'faj). Persistence of prophage without replication and destruction of the bacterial cell is called **lysogeny** (li-soj'en-e), and cells containing prophages are said to be **lysogenic** (li-so-gen'ik). Several ways to induce such cells to enter the lytic cycle are known, and most involve inactivation of a repressor substance.

Prophage can be incorporated into the bacterial chromosome, or it can exist as a plasmid. A **plasmid** (plas'mid) is a small, circular piece of DNA in a cell that is not part of its chromosome; it is also called *extrachromosomal DNA*. Prophage incorporated into a chromosome participates in specialized transduction, whereas prophage in a plasmid participates in generalized transduction.

Specialized Transduction

Several phages are known to carry out specialized transduction, but lambda phage (λ) in *E. coli* (Figure 8.4) has been extensively studied. Phages usually insert at a specific location when they integrate with a chromosome. Lambda phage inserts into the *E. coli* chromosome between *gal* genes that control galactose use and *bio* genes that control biotin synthesis. The *gal* genes and *bio* genes are operons (Chapter 7). (p. 177) When cells containing lambda phage are induced to enter the lytic cycle, genes of the phage form a loop and are excised from the bacterial chromosome. Lambda phage then directs the synthesis and assembly of new phage particles, and the cell lyses, or ruptures.

In most cases, the new phage particles released contain only phage genes. Occasionally (about one excision in a million) the phage contains one or more bacterial genes, which were adjacent to the phage when it was part of the bacterial chromosome. For example, the *gal* genes might be incorporated into the phage particles. When these particles infect another bacterial cell, they transfer not only the phage genes but also the *gal* genes. Transduction of the *gal* genes from one bacterial cell to another is thus accomplished by a phage particle. In such **specialized transduction** the bacterial DNA transduced is limited to one or a few genes lying adjacent to the prophage.

Generalized Transduction

Generalized transduction occurs when the phage exists as a plasmid (Figure 8.5). When cells containing such a phage enter the lytic cycle, phage enzymes break host cell DNA into many small segments. As the phage directs synthesis and assembly of new phage particles, it packages DNA by the "headful" (enough DNA to fill the head of a virus). This allows a bacterial DNA fragment occasionally to be incorporated into a phage particle. When this phage subsequently infects a bacterial cell, it carries with it a piece of chromosomal DNA, a transfer known as **generalized transduction.** Each bacterial DNA fragment has an equal chance of being incorporated into a phage and transferred (transduced) to another bacterial cell.

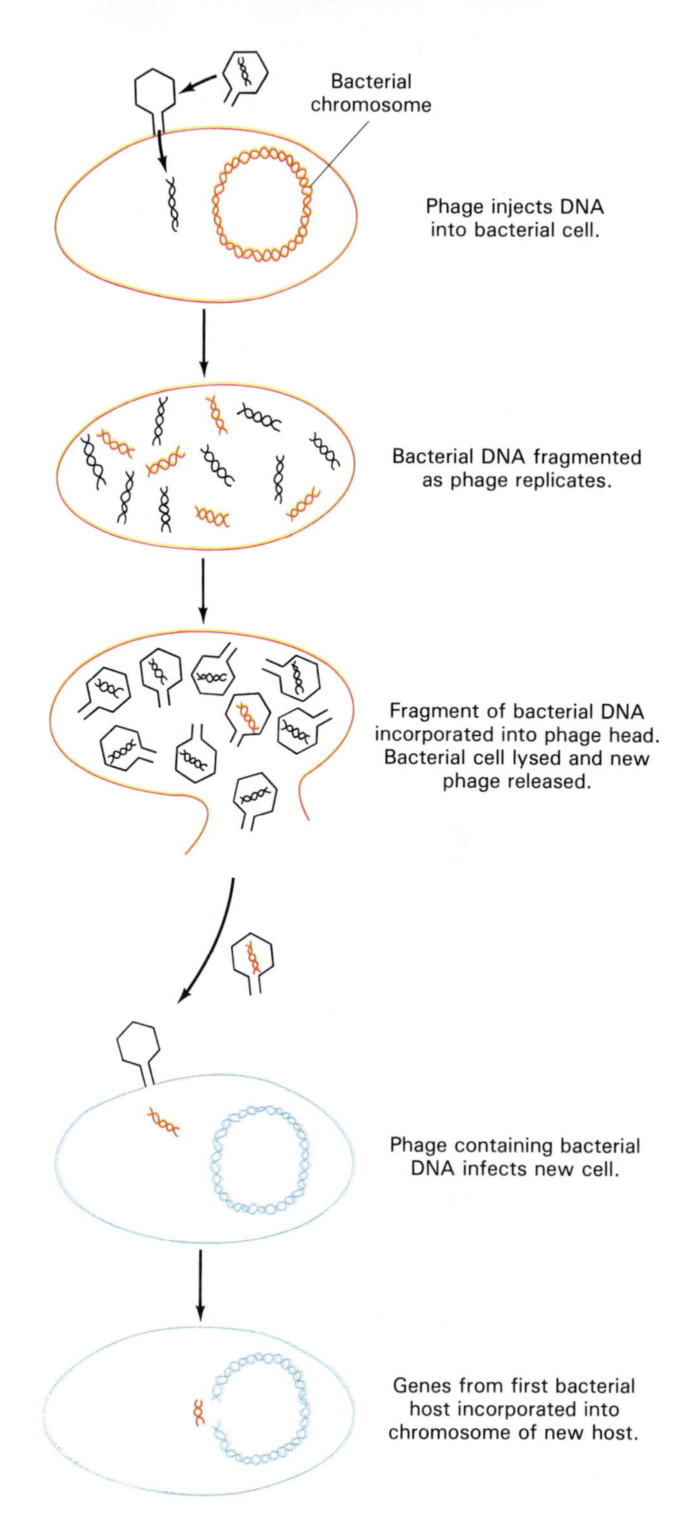

FIGURE 8.5 Generalized transduction. Bacteriophage infection of a host bacterium initiates the lytic cycle. The bacterial chromosome is broken into many fragments, any of which can be picked up and packaged along with phage DNA into new phage particles. When these are released and infect another bacterial cell, the new host acquires the genes that were brought along (transduced) from the previous bacterial host cell.

Significance of Transduction

Transduction is significant for several reasons. First, it transfers genetic material from one bacterial cell to another and alters the genetic characteristics of the recipient cell. As demonstrated by the specialized transduction of the *gal* genes, a cell lacking the ability to metabolize galactose could acquire that ability. Other characteristics also can be transferred either by specialized or generalized transduction.

Second, the incorporation of prophage into a bacterial chromosome demonstrates a close evolutionary relationship between the prophage and the host bacterial cell. The DNA of the prophage and that of the host chromosome must have regions of quite similar base sequences. Otherwise, the prophage would not bind to the bacterial chromosome.

Third, the discovery that prophage can exist in a cell for a long period of time suggests a mechanism for the viral origin of cancer. If a prophage can exist in a cell and at some point alter the expression of the cell's DNA, this could explain how viruses cause malignant changes.

Finally, and of most importance to molecular geneticists, specialized transduction provides a way to study gene linkage. Genes are said to be linked when they are so close together on a DNA segment that they are likely to be transferred together. Different phages can be incorporated into a bacterial chromosome, each kind usually entering at a specific site. By studying many different phage transductions, scientists can determine where they were inserted on the chromosome and which adjacent genes they are capable of transferring. The combined findings of many such studies eventually allow identification of the sequence of genes in a chromosome. This is called **chromosome mapping.**

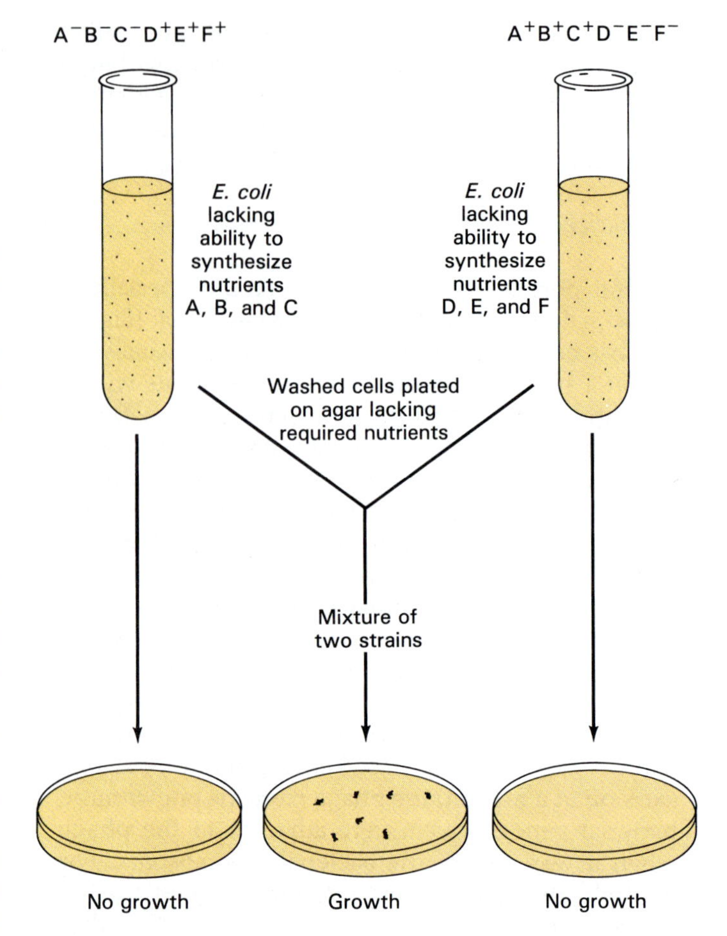

FIGURE 8.6 A schematic diagram of Lederberg's initial experiment that led to the discovery of conjugation.

CONJUGATION

Discovery of Conjugation

Conjugation, like transformation and transduction, transfers genetic information from one bacterial cell to another. It differs from the other mechanisms in two ways: (1) Contact between donor and recipient cells is required, and (2) much larger quantities of DNA (occasionally whole chromosomes) are transferred.

Conjugation was first discovered in 1946 by Joshua Lederberg, who was at that time still a medical student. In his experiments, Lederberg used mutated strains of *E. coli* that were unable to synthesize certain substances. He selected two strains, each defective in a different synthetic pathway, and grew them in a nutrient-rich medium (Figure 8.6). He removed cells from each culture and washed them to remove the residue of the nutrient medium. He then attempted to culture cells of each strain on agar plates that lacked the special nutrients needed by the strain. He also mixed cells from the two strains and plated them on the same medium. Whereas cells from the original cultures failed to grow, some from the mixed cultures did grow. The latter must have acquired the ability to synthesize all the substances they needed. Lederberg and others continued to study this phenomenon and eventually discovered many of the details of the mechanism of conjugation.

Lederberg was indeed fortunate in his choice of organisms, because similar studies of other strains of *E. coli* failed to demonstrate conjugation. In addition to the mutations that led to synthetic deficiencies in Lederberg's organisms, other changes had modified their cell surfaces so that conjugation could occur.

Mechanisms of Conjugation

The mechanisms involved in conjugation were clarified through several important experiments, each of which built on the findings of the preceding one. Of those experiments, we will consider three: transfer of F plasmids, high-frequency recombinations, and transfer of F′ plasmids.

Transfer of F Plasmids

After Lederberg's initial experiment, an important discovery about the mechanism of conjugation was made. Two types of cells, called F^+ and F^-, were found to exist in any population of *E. coli* capable of conjugating. F^+ cells contain extrachromosomal DNA called **F plasmids;** F^- cells lack F plasmids. (Lederberg coined the term *plasmid* in the 1950s to describe these fragments of DNA.) Among the genetic information carried on the F plasmid is information for the synthesis of proteins that make up F pili, sometimes called sex pili (Figure 8.7).

When F^+ and F^- cells conjugate, the F^+ cell makes an **F pilus,** a bridge by which it attaches to the F^- cell, and a copy of the F plasmid is transferred from the F^+ cell to the F^- cell (Figure 8.8). F^+ cells also are called *donor cells,* or male cells, and F^- cells also are called *recipient cells,* or female cells. The DNA is transferred as a single strand through the conjugation bridge, although the exact transfer process remains unknown. Because the sex pilus contains a hole that would permit the passage of single-stranded DNA, it is possible that DNA enters the recipient through this channel. However, there is also evidence to suggest that the mating cells temporarily fuse, during which time the DNA is transferred. Each cell then synthesizes the complementary strand of DNA, so that both have a complete F plasmid. In a culture of both F^+ and F^- cells, all the cells rapidly become F^+ cells, but in a culture of only F^- cells, no transfer occurs, and cells remain F^- cells.

High-Frequency Recombinations

The mechanisms of conjugation were further clarified when the Italian scientist L. L. Cavalli-Sforza isolated

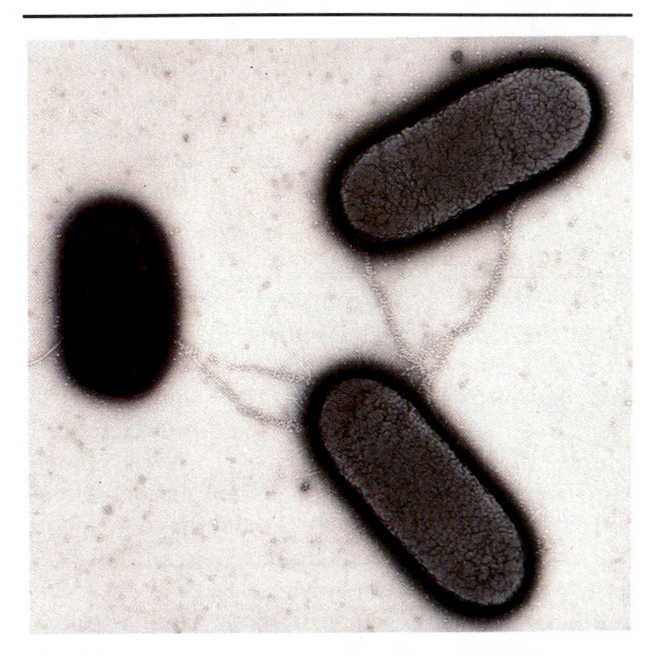

FIGURE 8.7 An F pilus of *E. coli.* Phages along the pilus make it visible. Unlike the shorter common pili (fimbriae), this long type of pilus is used for transfer of genes in conjugation and is often called a sex pilus.

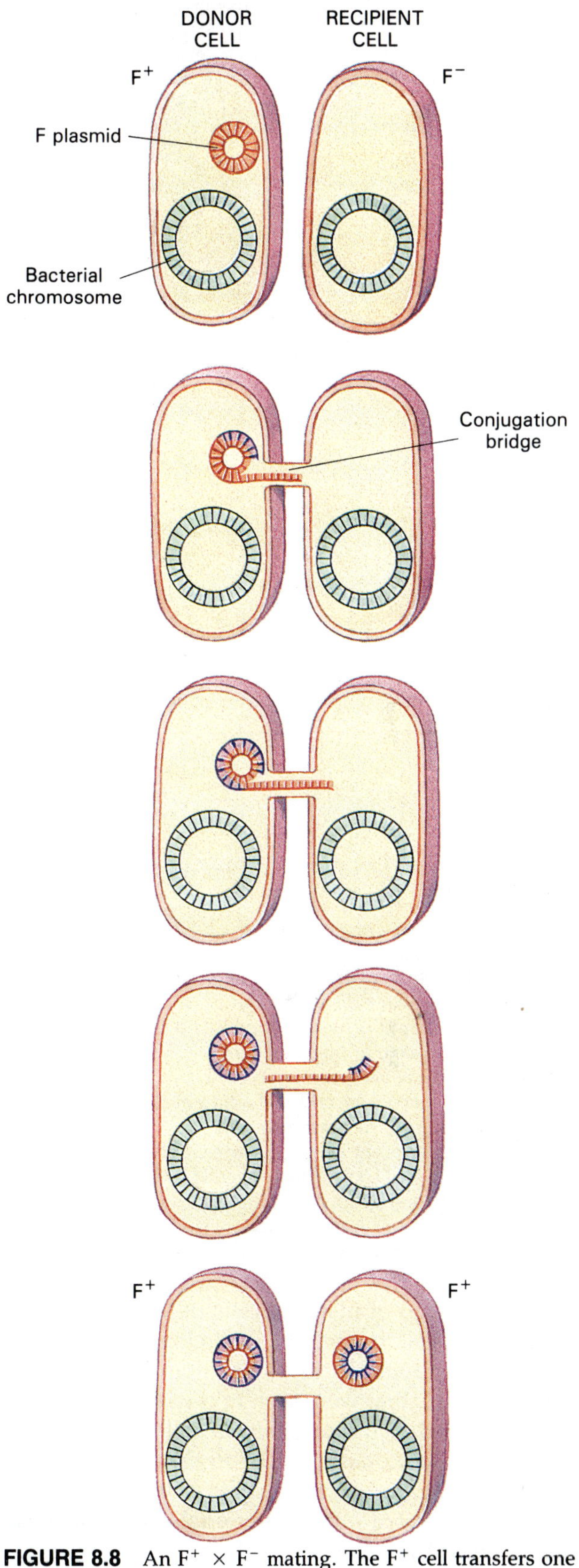

FIGURE 8.8 An $F^+ \times F^-$ mating. The F^+ cell transfers one strand of DNA from its F plasmid to the F^- cell via the conjugation (sex) pilus. As this occurs, the complementary strands of F plasmid DNA are synthesized. Thus, the recipient cell gets a complete copy of the F plasmid, and the donor cell retains a complete copy.

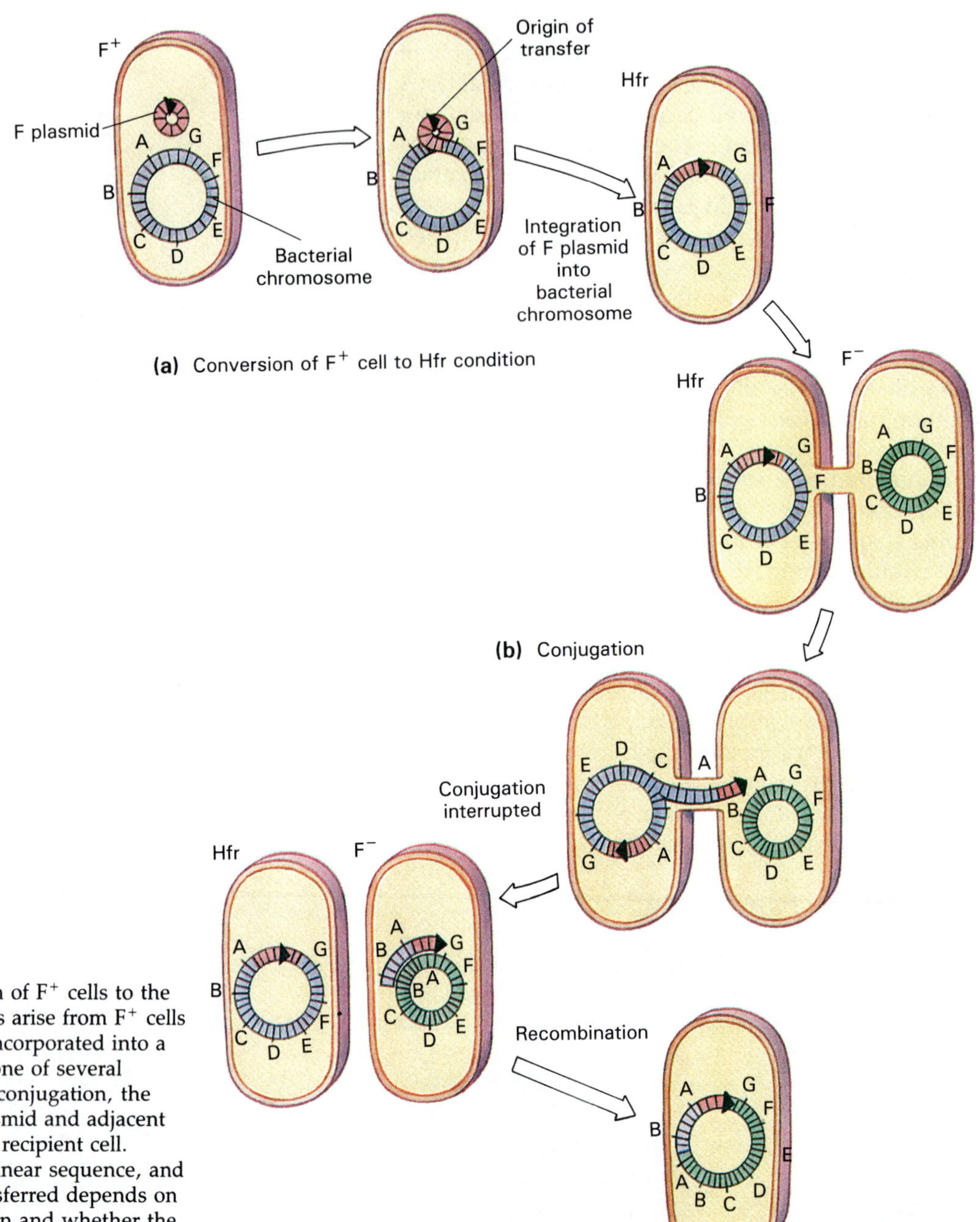

FIGURE 8.9 Conversion of F^+ cells to the Hfr condition. (a) Hfr cells arise from F^+ cells when their F plasmid is incorporated into a bacterial chromosome at one of several possible sites. (b) During conjugation, the initiating site of the F plasmid and adjacent genes are transferred to a recipient cell. Genes are transferred in linear sequence, and the number of genes transferred depends on the duration of conjugation and whether the DNA strand breaks or remains intact.

a **clone,** a group of identical cells descended from a single parent cell, from an F^+ strain that could induce more than a thousand times the number of genetic recombinations that were seen in the F^+ and F^- conjugations. Such a donor strain is called a **high frequency of recombination (Hfr) strain.**

Hfr strains arise from F^+ strains when the F plasmid is incorporated into the bacterial chromosome at one of several possible sites (Figure 8.9a). When an Hfr cell serves as a donor in conjugation, the F plasmid initiates transfer of chromosomal DNA. Usually, only part of the F plasmid, called the **initiating segment,** is transferred, but some adjacent chromosomal genes also are transferred (Figure 8.9b).

In the 1950s French scientists Elie Wollman and Francois Jacob studied this process in a series of interrupted mating experiments. They combined cells of an Hfr strain with cells of an F^- strain and removed samples of cells at short intervals. Each cell sample was subjected to mechanical agitation through vibration or whirling in a blender to disrupt the conjugation process. Cells from each sample were plated on a variety of media, each of which lacked a particular nutrient, to determine their nutrient requirements.

By careful observation of the genetic characteristics of cells from many experiments, the investigators determined that transfer of DNA in conjugation occurred in a linear fashion and according to a precise time schedule. When conjugation was disrupted after 8 minutes, most recipient cells had received one gene. When disrupted after 120 minutes, recipient cells had received a much greater quantity of DNA, sometimes an entire chromosome. At intermediate intervals, the number of donor genes transferred was proportional to the length of time conjugation was allowed to proceed. However, because of a tendency of chromosomes to break during transfer, some cells received fewer genes than would have been predicted by the time allowed. Whatever the number of genes transferred, they were always transferred in linear sequence from the initiation site created by the incorporation of the F plasmid.

Transfer of F′ Plasmids

The process of incorporating an F plasmid into a bacterial chromosome is reversible. In other words, DNA incorporated into a chromosome can separate from it and again become an F plasmid. In some cases this separation occurs imprecisely, and a fragment of the chromosome is carried with the F plasmid, creating what is called an **F′** (F prime) **plasmid** (Figure 8.10). Cells containing such plasmids are called *F′ strains.* When F^+ cells conjugate with F^- cells, the whole plasmid (including the genes from the chromosome) is transferred. Hence, recipient cells have two of some chromosomal genes—one on the chromosome and one associated with the plasmid.

In the transfer of F^+ plasmids, as in all other transfers during conjugation, the donor cell retains all the genes it had prior to the transfer including copies of the F plasmid. Single-stranded DNA is transferred, and both donor and recipient cells synthesize a complementary strand for any single-stranded DNA they contain.

The results of conjugation with respect to F^+, Hfr, and F′ transfers are summarized in Table 8.1.

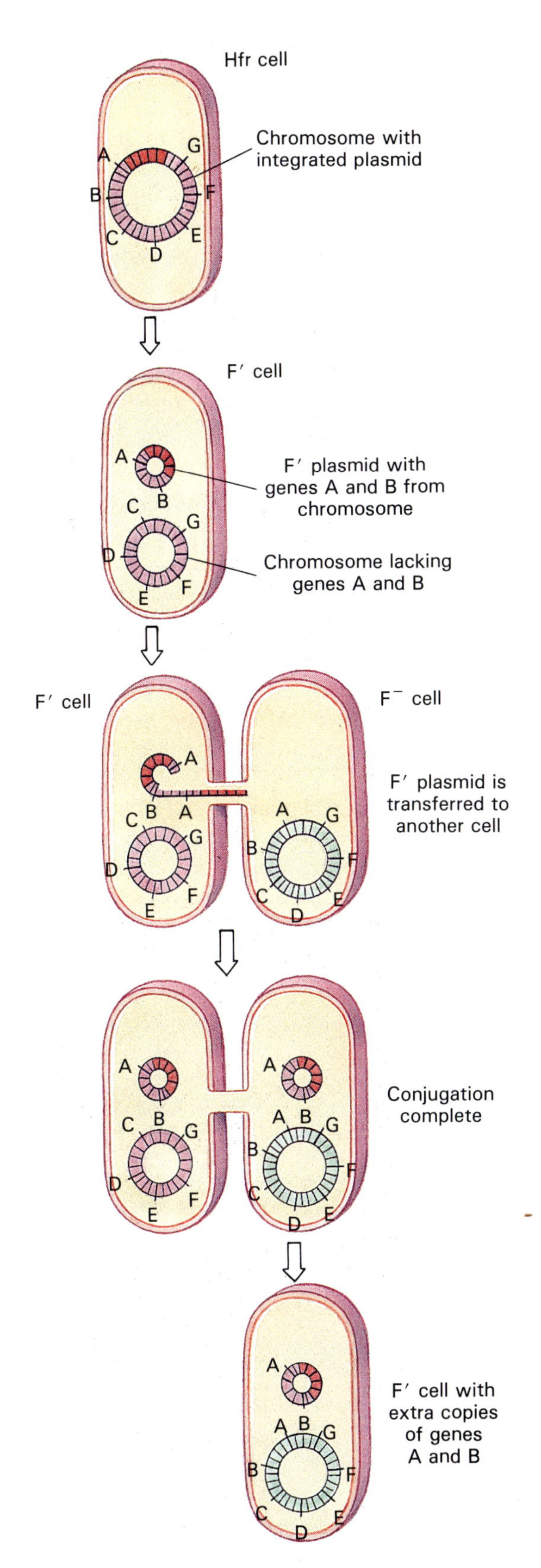

FIGURE 8.10 When the F plasmid in an Hfr cell separates from the bacterial chromosome, it may carry some chromosomal DNA with it. Such an F′ plasmid may then be transferred by conjugation to an F^- cell. The recipient cell will then have two copies of some genes—one on its chromosome and one on the plasmid.

TABLE 8.1 Results of selected conjugations

Donor	Recipient	Molecule(s) Transferred	Product
F^+	F^-	F plasmid	F^+ cells
Hfr	F^-	Initiating segment of F plasmid and variable quantity of chromosomal DNA	F^- with variable quantity of chromosomal DNA
F′	F^-	F plasmid and some chromosomal genes it carries with it	F plasmid and some duplicated chromosomal genes

Significance of Conjugation

Like other mechanisms for gene transfer, conjugation is significant because it contributes to genetic variation. Larger amounts of DNA are transferred in conjugation than in other transfers, so conjugation is especially important in increasing genetic diversity. In fact, conjugation may represent an evolutionary stage between the asexual processes of transduction and transformation and the actual fusion of whole cells (the gametes) that occurs in sexual reproduction. For the microbial geneticist, conjugation is of special significance because precise linear transfer of genes is useful in gene mapping.

COMPARISON OF TRANSFERS OF GENETIC INFORMATION

The most fundamental differences among the major types of transfers of genetic information concern the quantity of DNA transferred and the mechanism by which the transfer takes place. In transformation less than 1 percent of the DNA in one bacterial cell is transferred to another, and the transfer involves only chromosomal DNA.

In transduction, the quantity of DNA transferred varies from a few genes to large fragments of the chromosome, and a bacteriophage is always involved in the transfer. In specialized transduction the phage inserts into a bacterial chromosome and carries a few genes with it when it separates. In generalized transduction the phage causes fragmentation of the bacterial chromosome, and some of those fragments are packed into viruses as they are assembled.

In conjugation, the quantity of DNA transferred is highly variable, depending on the mechanism. A plasmid, or piece of extracellular DNA, is always involved in the transfer. An F plasmid itself can be transferred, as occurs in F^+ to F^- conjugation. An initiating segment of a plasmid and any quantity of chromosomal DNA from a few genes to the whole chromosome is transferred in Hfr conjugation. A plasmid and whatever chromosomal genes it has carried with it from the chromosome are transferred in F′ conjugation. These characteristics are summarized in Table 8.2.

TABLE 8.2 Summary of the effects of various transfers of genetic information

Kind of Transfer	Effects
Transformation	Transfers less than 1% of cell's DNA. Requires competence factor. Changes certain characteristics of an organism depending on which genes are transferred.
Transduction	Transfer is effected by a bacteriophage.
Specialized	A few genes carried by a phage when it leaves one bacterial chromosome and transfers to another.
Generalized	Fragments of DNA of variable length and number packed into the head of a virus.
Conjugation	Transfer is effected by a plasmid.
F^+	A single plasmid is transferred.
Hfr	An initiating segment of a plasmid and a linear sequence of bacterial DNA that follows the initiating segment are transferred.
F′	A plasmid and whatever bacterial genes adhere to it when it leaves a bacterial chromosome.

PLASMIDS

Characteristics of Plasmids

The F plasmid just described was the first plasmid to be discovered. Since its discovery, many other plasmids have been identified. All are ring-shaped, double-stranded extrachromosomal DNA. Being made of DNA, plasmids are self-replicating by the same mechanism that any other DNA uses to replicate itself. Most plasmids have been identified by virtue of some recognizable function that they serve in a bacterium. These functions include the following:

1. F plasmids (fertility factors) direct the synthesis of proteins that self-assemble into sex pili.
2. R plasmids (resistance factors) carry genes that provide resistance to various antibiotics.
3. Other plasmids direct the synthesis of bacteriocidal (bacteria-killing) proteins called bacteriocins.
4. Virulence plasmids, such as those in *Salmonella,* cause disease symptoms.
5. Some plasmids can cause tumors in plants.

Generally, plasmids carry genes that code for functions not essential for cell growth; the chromosome carries the genes that code for essential functions.

Resistance Plasmids

Resistance plasmids, also known as *R plasmids* or *R factors,* were discovered when it was noted that some enteric bacteria, bacteria found in the digestive tract, had acquired resistance to several commonly used antibiotics. We don't know how resistance plasmids arise, but we know that they are not induced by antibiotics. This has been demonstrated by the observation that cultures kept in storage from a time prior to the use of antibiotics exhibited antibiotic resistance on first exposure to the drugs. However, antibiotics contribute to the survival of strains that contain resistance plasmids. That is, when a population of organisms containing both resistant and nonresistant organisms is exposed to an antibiotic, the resistant organisms will survive and multiply, whereas the nonresistant ones will be killed. The resistant organisms are thus said to be *selected* to survive. Such selection is a major force in evolutionary change, as Charles Darwin realized.

According to Darwin, all living organisms are subject to natural selection, the survival of organisms on the basis of their ability to adapt to their environment. After studying many different kinds of plants and animals, Darwin drew two important conclusions. First, living organisms have certain heritable—that is, genetic—characteristics that help them to adapt to their environment. Second, when environmental conditions change, those organisms with characteristics that allow them to adapt to the new environment will survive and reproduce. The organisms lacking such characteristics will perish and leave no offspring. A change in environmental conditions does not directly cause organisms to change. It merely provides a test of their ability to adapt.

Let's look at some examples. Depending on the shapes of their beaks, some birds eat hard seeds, others suck plant juices, and a few capture and eat fish. Should the environment change so that all fish were to die, only the fish-eating birds that can eat seeds, plant juices, or some other available food would survive. Similarly, different kinds of bacteria are adapted to fresh water, salt water, cold lakes, hot geysers, or human bodies containing antibiotics. If the salt content, temperature, or antibiotic content of these various environments changes, only the bacteria that can carry out their life processes under the new conditions will survive.

Resistance plasmids (Figure 8.11) contain two components: a **resistance transfer factor (RTF)** and one or more **resistance genes (R genes).** The DNA in an RTF is similar to that in F plasmids. The resistance transfer factor implements transfer by conjugation of the whole resistance plasmid; it is essential for the transfer of resistance from one organism to another. Each R gene carries information that confers resistance to a specific antibiotic. Such genes usually direct synthesis of an enzyme that inactivates the antibiotic. Some resistance plasmids carry R genes for resistance to four widely used antibiotics: sulfanilamide, chloramphenicol, tetracycline, and streptomycin. Transfer of such a plasmid confers resistance to all four antibiotics to any recipient organism. Other resistance plasmids carry genes for resistance to one or more of

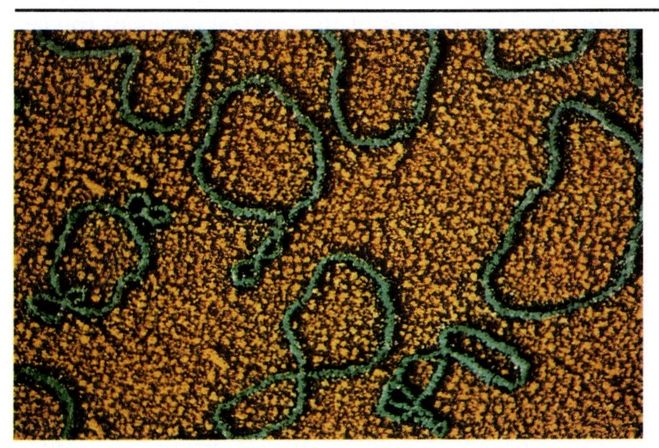

(a)

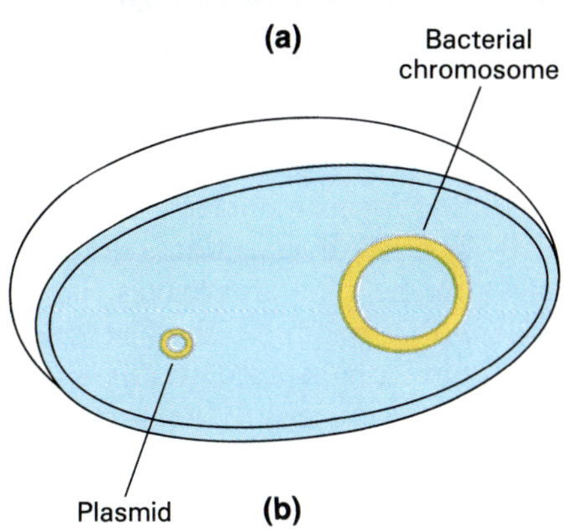

(b)

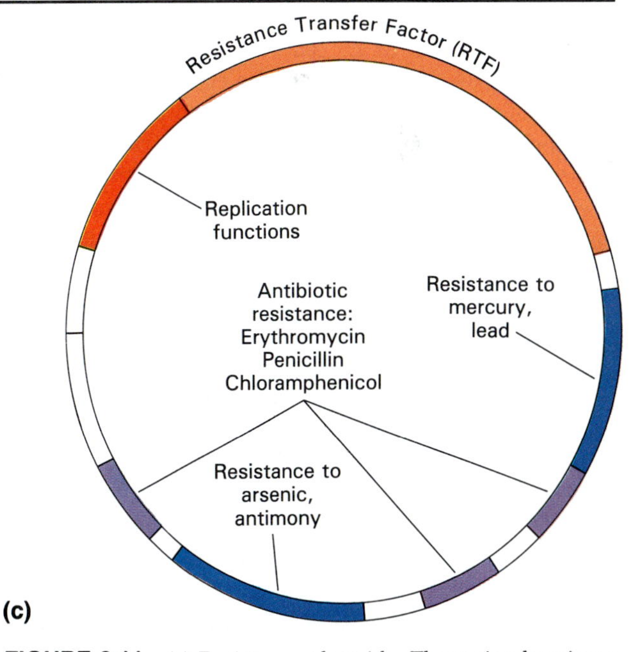

(c)

FIGURE 8.11 (a) Resistance plasmids. These circular pieces of DNA are much smaller than a bacterial chromosome (b). A typical resistance plasmid (c) can carry genes for resistance to various antibiotics and to inorganic toxic substances, sometimes used in disinfectants. The resistance transfer factor includes genes needed for the plasmid to replicate itself.

these antibiotics. A few plasmids carry genes for resistance to even more than four antibiotics.

The transfer of resistance plasmids from resistant to nonresistant organisms is rapid, so that large numbers of previously nonresistant organisms can acquire resistance quickly. Furthermore, transfer of resistance plasmids occurs not only within a species but also between closely related genera such as *Escherichia, Klebsiella, Salmonella, Serratia, Shigella,* and *Yersinia.* Transfer has even been observed between less closely related genera. Transfer of resistance plasmids is of great medical significance because it accounts for increasingly large populations of resistant organisms and reduces the effective use of antibiotics.

As health scientists accumulate information on plasmids and how they confer antibiotic resistance, they become more concerned about the development of resistant strains. As we shall see in Chapter 14, penicillin-resistant strains of *Neisseria gonorrhoeae, Haemophilus influenzae,* and some species of *Staphylococcus* already exist. Other antibiotics must now be used to treat the diseases caused by these strains, and the day may come when no antibiotic will effectively treat them. The more frequently antibiotics are used, the greater the selection is for resistant strains. Therefore, it is extremely important to identify the antibiotic to which an organism is most sensitive before using any antibiotic to treat a disease.

Bacteriocinogens

In 1925 the Belgian scientist André Gratia observed that some strains of *E. coli* released a protein that inhibited growth of other strains of the same organism. This allows them to compete more successfully for food and space against these other strains. About 20 such proteins, called **colicins** (ko'leh-sinz), have been identified in *E. coli,* and similar proteins have been identified in many other bacteria. All these growth-inhibiting proteins are now called **bacteriocins** (bak-te"-re-o'sinz). Typically, bacteriocins inhibit growth only in other strains of the same species or in closely related species.

Bacteriocin production is directed by a plasmid called a **bacteriocinogen** (bak-te"re-o-sin'o-jen). Though in most situations bacteriocinogens are repressed, sometimes the plasmid escapes repression and causes synthesis of its bacteriocin. Ultraviolet radiation can induce formation and release of bacteriocin. When a bacteriocin is released it can have a very potent effect on susceptible cells; one molecule of bacteriocin can kill a bacterium.

The mechanisms of action of bacteriocins are quite variable. Some enter a bacterial cell and destroy DNA. Others arrest protein synthesis by disrupting the molecular structure of enzymes and other molecules required for protein synthesis. Still others act on cell membranes by inhibiting active transport or by increasing membrane permeability to ions.

GENETIC ENGINEERING

Genetic engineering refers to the purposeful manipulation of genetic material to alter the characteristics of an organism in a desired way. Various methods of genetic manipulation allow microbial geneticists to create new combinations of genetic material in microbes. Transfer of genes between different members of the same species occurs in nature and has been done in the laboratory for several decades. Lederberg's experiment (Figure 8.6) provides one example of such a technique. Transfer of genes between different species also is now possible. We have selected five techniques of genetic engineering for discussion here. They are genetic fusion, protoplast fusion, gene amplification, recombinant DNA, and creation of hybridomas.

Genetic Fusion

Genetic fusion allows transposition of genes from one location on a chromosome to another. It can also involve deletion of a DNA segment, resulting in the coupling of portions of two operons. For example, suppose the *gal* operon, which regulates galactose use, and the *bio* operon, which regulates biotin synthesis, lie adjacent to each other on a chromosome (Figure 8.12). Deletion of the control genes of the *bio* operon and subsequent coupling of the operons would constitute genetic fusion. Such fusion would allow the genes that control the use of galactose to control the entire operon, including the making of the enzymes involved in biotin synthesis.

The major applications of genetic fusion within a species, as just described, are in research studies on the properties of microbes. However, the techniques developed for genetic fusion experiments have been extended and modified in the development of other kinds of genetic engineering.

One application of genetic fusion involves *Pseudomonas syringae,* a bacterium that grows on plants. Genetically altered strains have been developed that increase the resistance of plants, such as potatoes and strawberries, to frost damage. Strains of this bacterium naturally occurring on the leaves of plants produce a protein that forms a nucleus for the formation of ice crystals. The ice crystals damage the plants by causing cracks in the cells and leaves. By removing part of the gene that produces the "ice crystal" protein, scientists have engineered strains of *P. syringae* that cannot make the protein. When organisms of this strain are sprayed on the leaves of plants, they crowd out the naturally occurring strain. The treated plants

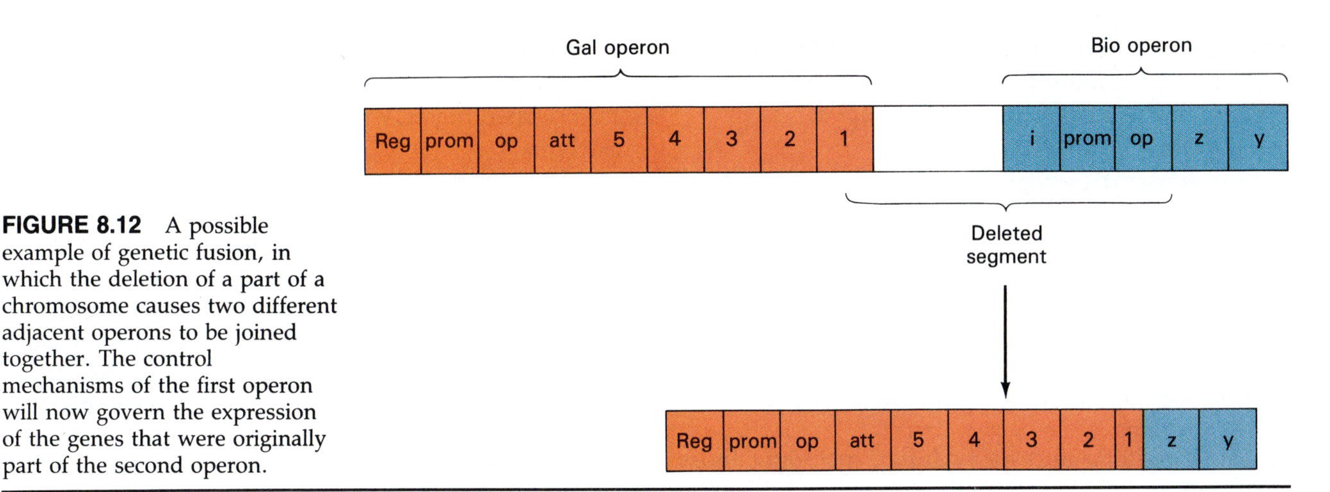

FIGURE 8.12 A possible example of genetic fusion, in which the deletion of a part of a chromosome causes two different adjacent operons to be joined together. The control mechanisms of the first operon will now govern the expression of the genes that were originally part of the second operon.

then become resistant to frost damage at temperatures as low as −5°C.

Protoplast Fusion

A **protoplast** is an organism with its cell wall removed. **Protoplast fusion** (Figure 8.13) is accomplished by removing the cell walls of organisms of two strains and mixing the resulting protoplasts. This allows fusion of genetic material; that is, material from one strain recombines with that from the other strain before new cell walls are produced. Though genetic recombination occurs in nature in about one in a million cells, it occurs in protoplast fusion in as many as one in five cells. Thus, protoplast fusion simply speeds up a process that occurs in a very limited way in nature.

By mixing two strains, each of which has a desirable characteristic, new strains that have both characteristics can be produced. For example, a slow-growing strain that produces large quantities of a desired substance can be mixed with a fast-growing, poor producer. After protoplast fusion, some organisms will probably be fast-growing, good producers of the substance. Other organisms that turn out to be slow-growing, poor producers are discarded. Alternatively, two good producers can be mixed to obtain a super producer. This has been done with two strains of *Nocardia lactamdurans*, which produce the antibiotic cephalomycin. The new strains produced 10 to 15 percent more antibiotic than the best of the parent strains.

Though protoplast fusion works best between strains of the same species, it has been accomplished in molds between two species of the same genus (*Aspergillus nidulans* and *A. rugulosus*) and even between two genera of yeasts (*Candida* and *Endomycopsis*).

Microbiologists have only recently begun to ex-

PUBLIC HEALTH

Genetically Engineered Microbes

The idea of releasing genetically engineered microorganisms, such as the Frostban strains of *Pseudomonas syringae*, has met with great resistance. Farmers worry that the organisms might spread to other crops that benefit from frost. (Frost causes chemical changes in apples, for example, that make them more flavorful.) Other people are fearful that altered genes could be transferred between species of bacteria, producing organisms with new and unpredictable traits. Some citizens have become so frightened and angry about the prospect of releasing engineered organisms that they have sabotaged test fields. Scientists have countered objections by demonstrating that during a growing season the organisms move only a few feet from the fields where they are applied. However, nobody can predict the effects of the engineered organisms over many years or determine whether genetic changes occurred in the crop plants.

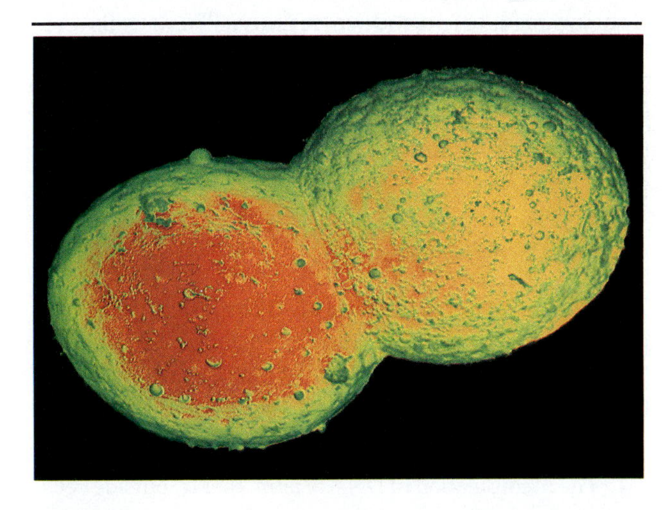

FIGURE 8.13 Protoplast fusion involves the use of enzymes to digest away the cell walls of cells from two different organisms. The cells are then placed together, fuse, and develop a new cell wall around the hybrid cell containing the genes of both organisms. In the photo, two tobacco plant cells are fusing.

plore possible applications of protoplast fusion. It offers great promise for the future as procedures are refined and useful strains are developed.

Gene Amplification

Gene amplification is a process by which plasmids, or sometimes bacteriophages, are induced to reproduce within cells at a rapid rate. If the genes required for the production of a substance are in the plasmids or can be moved to them, increasing the number of plasmids will increase production of the substance by the host cells.

Most bacteria and many fungi, including those that produce antibiotics, contain plasmids. Such plasmids, which often carry genes for antibiotic synthesis, provide many opportunities to use gene amplification to increase antibiotic yields. Even when genes concerned with antibiotic production are in the chromosome, it may be possible to transfer them to plasmids. Increased reproduction of plasmids would then greatly increase the number of copies of genes that act in antibiotic synthesis. This, in turn, would significantly increase the amount of antibiotic such cells could produce.

The possible applications of gene amplification are not limited to increasing antibiotic production. In fact, gene amplification may turn out to be even more effective in increasing production of substances that are synthesized by somewhat simpler pathways. These substances include enzymes and other products such as amino acids, vitamins, and nucleotides.

Rapid reproduction of bacteriophages already can be used to make the amino acid tryptophan. Bacteriophages carrying the *trp* operon (genes that control synthesis of enzymes to make tryptophan) of *E. coli* are induced to reproduce rapidly. Thus, cells containing large numbers of copies of the *trp* operon synthesize large quantities of the enzymes. Subsequent analysis of such cells has shown that half the intracellular proteins are enzymes for tryptophan synthesis.

Recombinant DNA

One of the most useful of all techniques of genetic engineering is the production of **recombinant DNA**—DNA that contains information from two different species of organisms. Making recombinant DNA involves three processes:

1. The manipulation of DNA *in vitro*, that is, outside cells.
2. The recombination of DNA from another organism with bacterial DNA in a phage or a plasmid.
3. The cloning, or production of many genetically identical progeny, of phages or plasmids.

These processes were first carried out in 1972 by Paul Berg and A. D. Kaiser, who inserted other prokaryotic DNA into bacteria, and then by S. N. Cohen and Herbert Boyer, who inserted eukaryotic DNA into bacteria.

DNA from either prokaryotic or eukaryotic cells is removed from the cells and cut into small segments. The donor DNA segments are then incorporated into a **replicon** (rep'leh-kon), a self-replicating carrier such as a phage or a plasmid (Figure 8.14). First, restriction endonucleases make cuts in double-stranded DNA that leave overlapping ends. A particular restriction endonuclease always produces the same complementary ends. Then, donor DNA is incorporated into the phage or plasmid by an enzyme called a ligase that reunites the ends of nucleotide chains. Thus, a replicon containing a new segment of DNA is created.

Once this new segment of DNA is inserted into the replicon, it can be introduced into cells such as *E. coli* that have been rendered competent by heating in a solution of calcium chloride. As the *E. coli* cells divide, the replicons in them also are reproduced by cloning. Such cells can be lysed, and the cloned replicons containing a specific segment of DNA can be retrieved.

Medical Applications of Recombinant DNA

One of the most medically significant applications of recombinant DNA techniques is the creation of bacterial cells that make substances useful to humans. To make bacterial cells produce human proteins, a segment of human DNA with the information for synthesizing the protein is inserted into the replicon. Interferon, a substance that helps cells to resist viral infection (Chapter 17), and the hormone insulin were among the first products made with recombinant DNA. Human growth hormone now can be made that way, and new products—vaccines, blood coagulation proteins for people with hemophilia, and enzymes such as cholesterol oxidase to diagnose disorders in cholesterol metabolism—are being developed.

The use of recombinant DNA technology to make substances useful to humans makes certain treatments safer, cheaper, and available to more patients. Prior to the manufacture of human insulin using recombinant DNA, the insulin for diabetic patients came from slaughtered cattle and pigs. (About half of the insulin in use at this writing still does.) Some patients develop allergies to such insulin, and the number of patients requiring insulin is increasing. Making nonallergenic human insulin and increasing the insulin supply are two important benefits of making insulin by recombinant DNA technology.

Likewise, prior to the manufacture of human growth hormone by recombinant DNA technology, the hormone was obtained from the pituitary glands of cadavers, and several cadavers were needed to obtain a single dose, which cost more than $1000 per

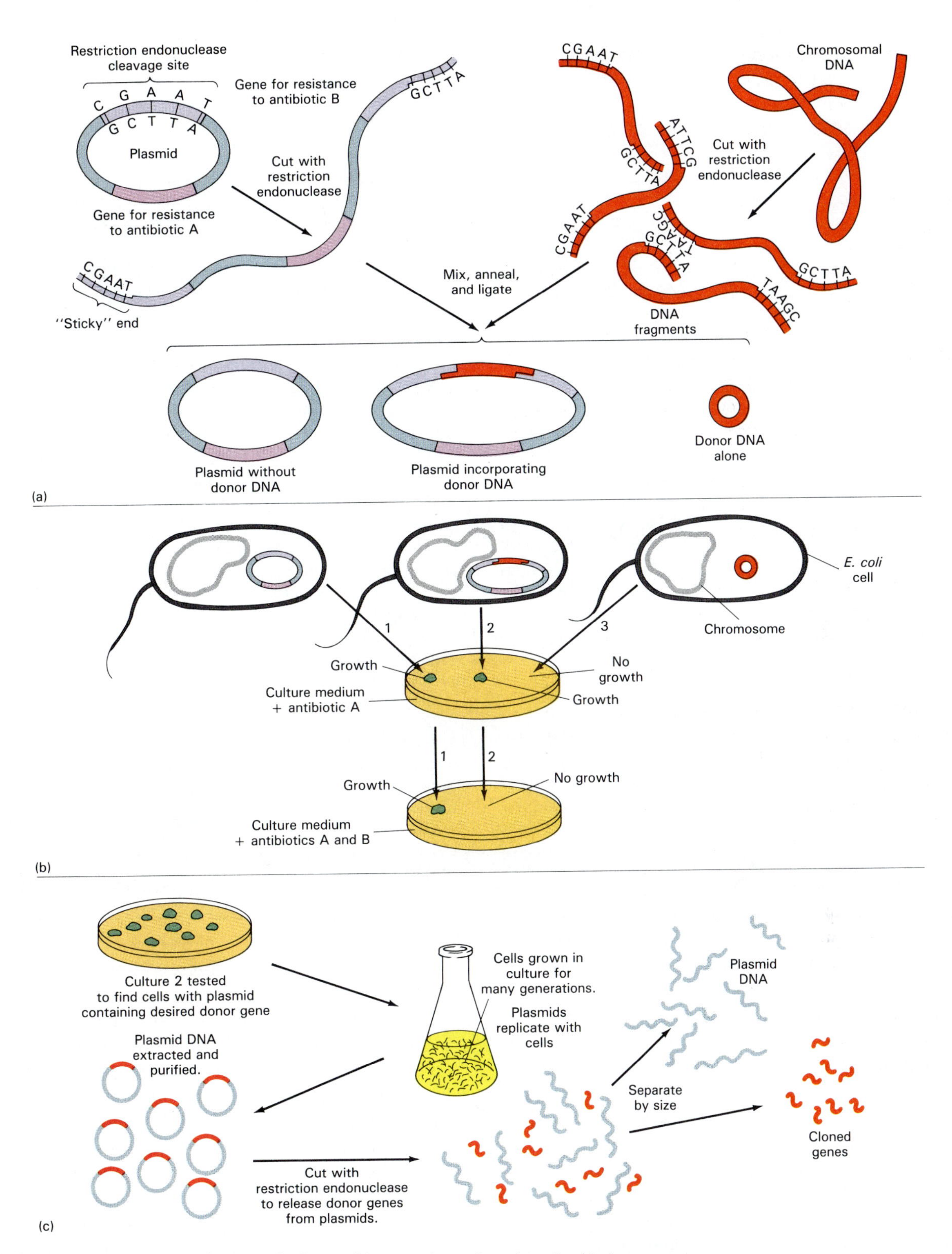

FIGURE 8.14 An example of the methods used in genetic engineering. In this instance, one or more genes from another organism are cloned in bacterial plasmids. The key tools in such work are restriction endonucleases—enzymes produced by bacteria that cut DNA only at specific base sequences. The cuts are staggered—that is, they are made at different spots, a few bases apart, on the two strands of the DNA molecule. This leaves "sticky ends" that can join by hydrogen bonding with similar ends on any segment of DNA cut with the same restriction enzyme. Restriction enzymes are used to cut the donor DNA into short segments (a), some of which become incorporated into plasmids that have been cut with the same enzyme. The plasmids carry two genes for antibiotic resistance, with the cut site in the middle of one of them. If donor DNA is inserted there, that gene will be altered and will no longer confer antibiotic resistance. This provides a means of determining which plasmids have actually incorporated donor DNA (b). Once cells with plasmids containing the desired donor gene have been isolated and cultured, it is easy to obtain many copies of the plasmid, from which the donor DNA can then be extracted using the same restriction enzyme (c).

BIOTECHNOLOGY

Need a Blood Transfusion? Call on a Genetically Altered Pig

By the mid-1990s researchers expect government approval of a blood-substitute product composed primarily of human hemoglobin that is produced by *transgenic* (genetically altered) pigs.

The biotechnology firm DNX, located in Princeton, New Jersey, has succeeded in producing three pigs that have human hemoglobin in about 15 percent of their red blood cells. Eventually DNX hopes to increase that figure to 50 percent. In a kind of "molecular farming," DNX scientists injected thousands of copies of the two human hemoglobin genes into 1-day-old pig embryos that had been removed from their mothers' uteri. The embryos were then implanted into a second pig's uterus to grow until delivery. Only about 0.5 percent of such transfers have succeeded. Once such pigs are obtained and mated to one another, however, all offspring will have the desired gene. To acquire the blood substitute, the pigs are bled, the red blood cells are ruptured, and human hemoglobin is separated from pig hemoglobin on the basis of differing electrical charges.

This substitute product has several advantages over actual human blood:

1. It has a storage life of months instead of weeks.
2. Because naked hemoglobin does not stimulate the immune system to act against it, as do intact red blood cells containing hemoglobin, it can be transfused into anyone without the need for blood typing and matching.
3. It can ensure safety from human pathogens (including AIDS) that might now contaminate human blood.
4. It can serve as an immediate source of oxygen for victims on battlefields and in accidents, which might enable them to survive the trip to the hospital.

The substitute would cost about $50 to $70 more per unit than the $175 to $200 now charged for a unit of whole blood. However, storage would be cheaper, and blood-typing costs would be eliminated.

One drawback to this procedure is that, once transfused, naked hemoglobin lasts only hours or days instead of 6 months, but this might be long enough to treat emergency cases. Another problem with this product is possible contamination with pig molecules or pig pathogens, if purification processes fail. One issue that does *not* appear to be a problem is the use of this product by Jewish people who do not eat pork. The director of the Rabbinical Council of America has said that there would probably be no religious objection because, within Judaism, pigs may be used for purposes other than eating, and kosher rules are suspended in cases of life or death.

This pig has genes for production of human hemoglobin. Such pigs will be bred and bled to collect human hemoglobin to save people's lives by transfusion. This is an example of biotechnologic "pharming" of molecules.

dose. Patients with a deficiency of this hormone require several doses of the hormone per year. Without treatment, these individuals are destined to be pituitary dwarfs because their own pituitary glands do not produce enough growth hormone. Thus, making the hormone by recombinant DNA technology has made the treatment less expensive and available to more patients.

The manufacture of certain blood coagulation proteins by recombinant DNA technology makes these substances more readily available to individuals with hemophilia or other blood disorders. It also assures that the recipient will not be subjected to the risk of acquiring AIDS from a contaminated blood product.

Recombinant DNA is being used to make vaccines more economically and in larger quantities. In this application some microorganisms are used to combat the disease-causing capacity of others. DNA that directs the synthesis of specific substances, called *antigens,* from a disease-causing bacterium, virus, or parasite is inserted in another organism. The organism then makes a pure antigen. When the antigen is introduced into a human, the human immune system makes another specific substance, called an *antibody,* which takes part in the body's defense against the disease-causing organism (Chapter 18).

Procedures for making vaccines for hepatitis B and influenza using recombinant DNA are already available. The vaccines are not only cheaper than conventional ones, they are also purer and more specific, and they cause fewer undesirable side effects. Vaccine recipients receive only the specific antigen they need to make them immune. They do not receive other antigens that might cause allergic reactions, and they do not risk acquiring other diseases from blood products such as gamma globulin. Now that procedures have been developed, it should be reasonably easy to adapt them to making many other vaccines.

Many other applications of recombinant DNA techniques are being developed. An especially important one is the diagnosis of genetic defects in a fetus, which can be done by studying enzymes in fetal cells from amniotic fluid. Such defects are detected by using recombinant DNA with a known nucleotide sequence to find errors in the nucleotide sequence in fetal DNA segments. Such errors in fetal DNA denote genetic defects that can be responsible for absent or defective enzymes. Application of these techniques could greatly improve prenatal diagnosis of many genetic defects. Ultimately, as techniques for preparing recombinant DNA in animal cells improve, it may become possible to insert a missing gene or replace a defective one in human cells (gene therapy). Inserting a functional gene in appropriate cells might cure a genetic disease. Inserting such a gene in a defective gamete (egg or sperm) might prevent offspring from inheriting a genetic disease.

Forensic applications of DNA technology are rapidly coming into use in the courtroom. In paternity cases, for example, experts can now determine with about 99 percent certainty that a given man is the father of a particular child, based on comparison of DNA (refer to the Polymerase Chain Reaction Essay in Chapter 7). (p. 187) Likewise, rapists and murderers can be identified by the "DNA fingerprints" they leave behind at the scene of the crime in the forms of semen, blood, hair, or tissue under their victim's fingernails.

BIOTECHNOLOGY

Virus with a Scorpion's Sting

Caterpillars that devour valuable crop plants are sometimes infected by a naturally occurring virus that is harmless to all other organisms. The caterpillars develop a long illness that culminates in their climbing to the top of a plant and bursting to release a shower of infectious viral particles. But during their long illness they still eat quite a bit of the crops. Now genetic engineers have inserted the gene for scorpion venom into the virus. Caterpillars infected with the recombinant virus manufacture scorpion venom and quickly develop convulsions, paralysis, and death—all without stopping for a bite to eat. Any altered virus that remains uneaten is destroyed in sunlight, and no chemical pesticide residues enter the food chain.

Industrial Applications of Recombinant DNA

Fermentation processes used in making wine, antibiotics, and other substances might be greatly improved by the use of recombinant DNA. For example, addition of genes for the synthesis of amylase to the yeast *Saccharomyces* could allow these organisms to produce alcohol from starch. Malting of grain to make beer would be unnecessary, and wines could be made from juices containing starches instead of sugars. Still other applications might include degradation of cellulose and lignin (plant materials often wasted), manufacture of fuels, cleaning up environmental pollutants, and leaching of metals from low-grade ores. Strains of *Pseudomonas putida*, already known to degrade different components of oil, might be engineered so that one strain degrades all components. Industrial leaching of metals from copper and uranium ores is already carried out by certain bacteria of the genus *Thiobacillus*. If these organisms could be made more resistant to heat and to the toxicity of the metals that they leach, the leaching process could be greatly speeded up.

Agricultural Applications of Recombinant DNA

Certain bacteria are being engineered to control insects that destroy crops. The Monsanto Company has recently modified the genetic makeup of certain bacteria that colonize the roots of corn. These bacteria have been induced to carry genetic information, allowing them to synthesize a protein that kills insects. The modified bacteria are a strain of *Pseudomonas fluorescens*, and the genetic information inserted into them came from *Bacillus thuringiensis* (Figure 8.15). Toxin made by *B. thuringiensis* has been extracted and used for many years as an insecticide. Now the pseu-

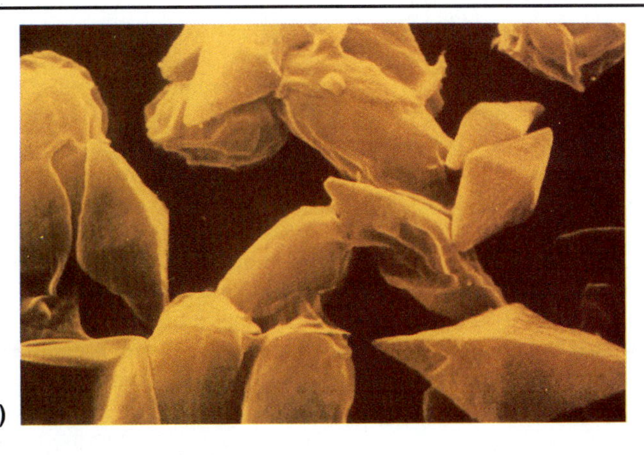

(a)

(b)

FIGURE 8.15 (a) Crystals of a substance toxic to many insects. The genes for production of this toxin are being taken from *Baccillus thuringiensis*, which produces it naturally, and are being incorporated into other organisms through genetic engineering techniques. Imagine the benefits of crop plants that have their own built-in pesticide: no expense to farmers for purchase of chemical pesticides, no danger during application, no buildup in soil or water, and no entry or magnification of pesticides in the food chain. (b) A male gypsy moth, an imported pest that has devastated U.S. forests especially along the East coast and has been very difficult to control.

MICROBIOLOGIST'S NOTEBOOK

Redesigning Bacteria

Little did I imagine, when I first dabbled in horticulture as a kid, that one of my projects would be controversial enough to make the front page of the *New York Times!* But then, in those days, the words "genetic engineering" didn't mean anything, except in a science fiction novel.

My name is Trevor Suslow. I've always liked science, particularly the natural sciences, and ever since middle school I've had a special interest in agriculture. Now it's my vocation—I'm director of microbial pesticides at the DNA Plant Technology Corporation in Oakland, California. My concerns are with food production—specifically, trying to improve the production of food for Third World countries. Alternatives to chemical pesticides are a natural extension of that interest. My research focuses on learning how to use the biological controls that already exist in nature, and on evolving new ones.

Anyone who's ever tried to grow tomatoes in the backyard knows how damaging that first frost can be to the fruits. Beautiful red tomatoes turn into blackened, spongy masses overnight. If that were your main source of food, you would be in real trouble. So the concept of making plants—especially food plants—frost-resistant becomes very attractive.

How can we do that? Well, many people are surprised to learn that frost is partly a microbiological phenomenon—certain bacteria are largely responsible for the formation of ice crystals. These *ice-plus* bacteria produce a protein that has the same structure as that of water molecules in an ice crystal, and so water molecules tend to align themselves on it. As the ice crystals grow, they enter the spaces inside the leaf or flower, causing frost injury. This damage makes the plant susceptible to disease caused by *Pseudomonas syringae,* the predominant frost-causing strain.

Dr. Trevor Suslow in the greenhouse with some of the strawberry plants on which Frostban was tested.

You can use chemicals to control the bacteria that cause frost to form, but this approach has its problems. Antibacterial chemicals are detrimental to the environment, they are expensive, their effectiveness has been shown to be limited, and they may actually lead to other diseases. Because these chemicals kill all bacteria—beneficial ones as well as harmful ones—they destroy the delicate balance of the plant's ecology.

So, we microbiologists stepped in with biocontrol. We wanted to leave the beneficial bacteria untouched, and just control the ice-plus strains. Our strategy was to locate or develop *ice-minus* bacteria to compete successfully with the ice-plus strains. Then if the ice-minus bacteria were applied to a frost-susceptible crop, they might protect the plant by keeping the ice-plus bacteria from becoming established—an effect known as competitive exclusion. [See Chapter 15.]

Our work focused on altering the ice-plus bacteria so that they could no longer produce the ice-building protein. Using genetic engineering techniques, we were able to locate the gene that controls the production of this protein and transfer it to *E. coli* cells. (The fact that the gene will function normally in *E. coli* was very convenient for our work.) With the aid of restriction enzymes, we then removed about one-third of the gene. This was enough, not only to inactivate the gene, but also to ensure that no mutation could ever restore it to functional form. Next, we cloned the defective gene, put it on a plasmid, and inserted the plasmid back into *Pseudomonas syringae* bacteria. Now, a plasmid and a chromosome in the same cell will sometimes exchange similar segments of DNA—genes—that are present on both. So we simply looked for bacteria in which this natural process (called homologous recombination) had taken place, leaving the bacterial cells with a nonfunctional copy of the ice-forming gene in its chromosome. These were our ice-minus bacteria.

It's important to understand that deletion mutations of this kind occur constantly in nature. It's just more accurate, predictable, and cost effective to perform the genetic surgery ourselves in the laboratory. The altered, ice-minus strain—whether naturally or genetically engineered—is identical in every other respect to the unaltered, ice-plus variety.

In 1983, we began our research in the laboratory and greenhouse. We conducted several hundred tests of the ice-minus strain, which we called Frostban. Then it was time to go to field trials. We obtained permits and approvals for field testing from dozens of federal, state, and regional regulatory agencies, from the U.S. Environmental Protection Agency down to local air and water pollution control boards. But, although we were prepared in terms of official permission to conduct field trials of Frostban, we were not at all prepared for the commotion such tests would cause—nor did we expect such publicity.

Our field trials quickly became the focal point of a multilevel controversy. Protests came from several quarters. Many organizations opposed the release of the bacteria into the environment. It didn't matter that Frostban was clearly safe. They had all sorts of objections to genetically engineering a living organism—any organism. Frostban became the symbol in their fight against a future full of recombinant DNA.

Environmentalists were concerned that we were destroying the ecological balance of an increasingly fragile Earth. We also had trouble with local farmers with more immediate concerns. Some simply weren't quite sure what these bacteria were and what they might do. Others worried about potential risks to their own crops from a product that offered them no benefit. These California farmers had no major frost problems—in fact, some of them even *liked* frost on their crops! A good frost will sweeten up certain vegetables, such as carrots or Brussels sprouts. So these farmers were concerned that the Frostban bacteria would drift over onto their fields and keep their crops from getting the frost they needed. Research had in fact shown that there was virtually no likelihood of Frostban spreading in this way. But people with such anxieties weren't easily convinced.

A member of the research team, in full protective clothing complete with "Frostbusters" insignia, sprays Frostban bacteria on plants during field trials. In the background, reporters and photographers in shirtsleeves drink their coffee while covering the historic event.

All this attention had its humorous side. Even though Frostban is not a toxic substance, all of us researchers were required to wear "spacesuits," covering us from head to toe, while spraying the substance on the crops in the field trials. So our team was suited up like Ghostbusters, spraying Frostban on strawberry plants, while several yards downwind, the officials, reporters, and photographers covering the event were standing around in shirtsleeves having coffee and doughnuts!

But the opposition we encountered was serious and very determined. We found ourselves plagued by an increasing number of injunctions—lawsuits—even vandalism. The world just didn't seem ready for strawberries served with a large dollop of recombinant DNA technology. We discontinued our field trials and went back to the laboratory. We decided to concentrate instead on isolating natural ice-minus bacteria—those in which the ice-building gene is already missing. This procedure is more expensive and time-consuming, but it can be done. Although such organisms have the same effect as the bioengineered ones, they're more acceptable to many people because they occur naturally.

Why did the field trials of Frostban cause such controversy? Probably because Frostban was a *first*—the first commercially available, genetically engineered microbial agent. As such, it was bound to attract a great deal of publicity. But more fundamentally, there is a relatively high level of scientific ignorance in our country. People are unaware of the roles that microorganisms play in our lives. If they don't feel comfortable with the naturally occurring organisms, how can they feel comfortable about the bioengineered ones?

Still, there is tremendous potential for genetically engineered biological controls. Some type of genetic manipulation has to take place in order to give growers what they need. And this will happen—but probably not until sometime in the next century. It's likely that we'll see genetically engineered plants first. Until then, we'll continue the educational process that began with Frostban. There will be many more conferences and workshops on the testing of microorganisms. The next time, everyone—the scientific community, the regulatory agencies, and the public—will be better informed. And we scientists, together with governmental officials, are developing protocols to ensure that safety will always be the main concern.

In the meantime, I'm looking forward to some commercial use of naturally occurring Frostban within a couple of years. After that, we'll probably tackle plant fungal diseases. If we involve the public in what we're doing early on, they'll be more supportive—and we'll be able to begin controlling some of the major blights that reduce the world's food supply.

BIOTECHNOLOGY

A Blue Rose for Your Valentine?

Australian scientists from the Calgene genetic engineering company have applied for a patent for the gene that produces delphinidum, the pigment that makes delphiniums and petunias blue. However, this is just the beginning of a very complex road to obtaining a blue rose. Pigment color is affected by pH. Cyanidin, the pigment that makes roses red, is red in an acidic pH and blue in an alkaline pH. It is what makes cornflowers blue. Isolation of the gene for a desired characteristic does not guarantee quick success. Organisms are complex systems of gene interactions which must be taken into account.

domonads, applied to the surface of corn seeds, can make the toxin as they grow around the roots of the corn.

If the pseudomonads survive in corn fields as well as they have in greenhouses, they could replace the use of chemical insecticides to control black cutworm and probably other insect larvae that damage crops. With further research, other bacteria normally present on crops might be modified to control additional pests. Some optimistic scientists believe chemical pesticides may be phased out in favor of these safer and cheaper methods of pest control.

Pilot studies are also under way to develop genetically engineered seeds for crop plants that have high yield and other desirable characteristics and that will resist herbicides that kill weeds. If these studies are successful, farmers will be able to buy seeds that would solve many cultivation problems. Attempts have also been made to introduce nitrogen-fixing genes into nonleguminous plants. (See Chapter 26 for a discussion of nitrogen fixation.) This work has been successful in some plants but not yet in any important crop plants. If it can be extended to crop plants, many agricultural crops could be made to satisfy their nitrogen needs and thrive without commercial fertilizers, which are expensive and tend to pollute ground water. This would be especially beneficial in some developing nations, where famine is an ever-present threat and money for expensive fertilizers is not available.

BIOTECHNOLOGY

Easier Harvest of Molecules

A problem with industrial production of proteins by recombinant DNA methods is the isolation of the product after it is manufactured by a microbe. A particular capability of certain yeast cells seems to offer a solution to the problem. Two kinds of haploid cells of *Saccharomyces cerevisiae* are known to exist and to mate with each other. One produces a *pheromone*, a molecule that emits a chemical signal, which initiates mating. Cells making the pheromone first produce a large protein molecule that is cleaved into four active pheromone, or alpha factor, molecules that are released into the environment. Ways to attach genetic information for other proteins to the pheromone gene are being developed. Such techniques would take advantage of the yeast cell's ability to synthesize a substance and secrete it into the medium, where it is easy to harvest.

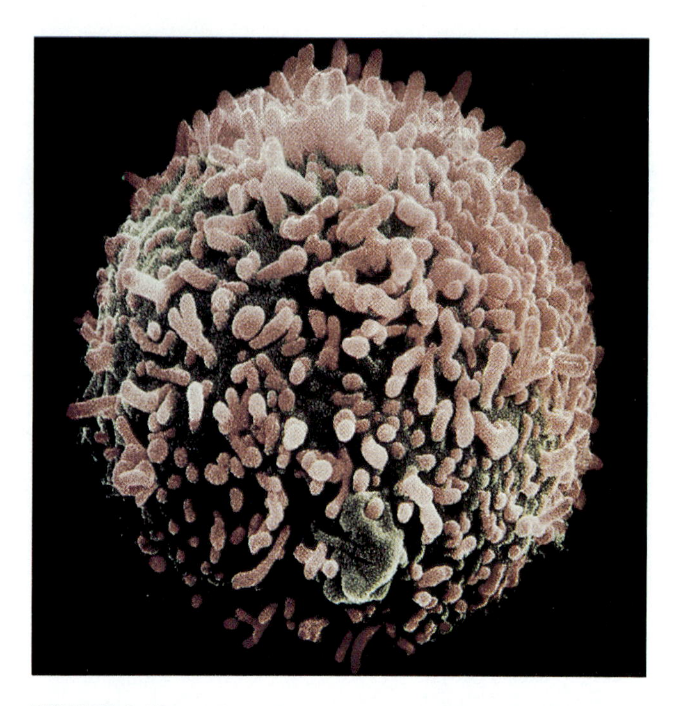

FIGURE 8.16 Hybridomas are often made by fusing an antibody-producing plasma cell and a cancer cell. The latter keeps the culture dividing and growing indefinitely, while the former causes the hybridoma to produce pure antibody against whatever antigen it was sensitized to.

Hybridomas

Along with the study of genetic recombinations in microorganisms came studies of such combinations in higher organisms. The first combination useful in industrial microbiology was the fusion of a myeloma (bone marrow cancer) cell with an antibody-producing white blood cell. Such a fusion of two cells is called a **hybridoma** (hi-brid-o'mah), or hybrid cell (Figure 8.16). This particular hybridoma can be grown in the laboratory, and it produces pure specific antibodies, called **monoclonal** (mon-o-klon'al) **antibodies,** against any antigen to which the white blood cell was previously sensitized. Prior to the production of hybridomas and their monoclonal antibodies, no source of pure antibodies existed. Now many different kinds of monoclonal antibodies are produced commercially, and they represent a major advance in immunology. The production and uses of monoclonal antibodies are discussed in more detail in Chapter 18.

The ability to produce hybridomas may in the future lead to other important advances. Recently, agricultural scientists used this technique to fuse cells from a commercial potato plant with cells from an extremely rare wild strain. The wild strain was selected because it contains the gene for production of a natural insect repellent. The researchers were able to grow the fused cells into a plant that has the necessary commercial properties but also synthesizes the insect repellent in its leaves.

BIOTECHNOLOGY

Viral Insecticides

The baculoviruses, a group of not very well known insect viruses, have been used as natural insecticides for two decades. Recently they have become powerful tools in genetic engineering. One of these viruses was used to make a protein from the AIDS virus, which in turn was used to make the first AIDS vaccine to be approved for human trials in the United States. Baculoviruses also are used to make an insect neurotoxin, which can kill crop pests more rapidly than simple virus infections. Such a toxin may soon become the best available weapon against the gypsy moth. Another application of genetic engineering in baculoviruses is rapid protein synthesis using regulatory signals from the virus and cellular enzymes. The proteins produced can then be used in diagnosis and therapy.

Weighing the Risks and Benefits of Recombinant DNA

In spite of the many potential benefits of recombinant DNA research, at first some scientists working with it were concerned about its hazards. They feared that some recombinants might prove to be new and especially virulent pathogens for which humans would have no natural defenses and no effective treatments. In 1974 they called for a moratorium on certain experiments until the hazards could be assessed. From this assessment emerged the idea of biological containment—the practice of making recombinant DNA only in organisms with mutations that prevent them from surviving outside the laboratory.

In 1981 the constraints on recombinant DNA research were relaxed because of the following observations:

1. No illnesses in laboratory workers could be traced to recombinants.
2. The strain of *E. coli* used in the experiments failed to infect humans who voluntarily received large doses.
3. Incorporation of mammalian genes into *E. coli* was observed in nature, and these genes invariably impaired the organism's ability to adapt to the environment. This suggested that if laboratory organisms did escape, they probably would not survive in the natural environment.
4. Mutants of *E. coli* containing recombinant DNA were subject to control by accepted sanitary practices.

Most scientists now agree that recombinant DNA techniques as currently practiced offer significant benefits and exceedingly small risks to humans.

ESSAY

More about Plasmids

Plasmids come in a wide variety of sizes, and their means of transfer is related to their size. Large plasmids have 60,000 to 120,000 DNA bases, whereas small plasmids have only 1500 to 15,000 DNA bases. Large plasmids usually are transferred by conjugation. DNA from plasmids can be transferred by conjugation or transfection. In *transfection*, DNA released into the medium by lysis of some bacterial cells is taken up by other bacterial cells. This process gets its name from the fact that viruses transferred by this means cause infections in the recipient cells. Plasmids can also be transferred by generalized transduction initiated by a transducing phage.

Another way of classifying plasmids is by the system that regulates their replication. Bacterial cells can contain several different kinds of plasmids. Each kind of plasmid is present in a particular number—1 per cell for large plasmids, and as many as 20 per cell for small plasmids. The number of copies of a plasmid possessed by a cell appears to depend on a replication repressor. For example, when a cell contains a large number of copies of one kind of plasmid, sufficient replication repressor is made to prevent further replication. When that cell divides, the daughter cells receive smaller numbers of the plasmids than the parent cell contained. Daughter cells, in turn, replicate plasmids until the replication repressor stops replication. Though some plasmids have their own specific replication repressor, other plasmids may

ESSAY More about Plasmids

share a common replication repressor (probably because they have extensive common DNA sequences).

Suppose a cell contains two kinds of plasmids, each with its own replication repressor. The numbers of each of these kinds of plasmids are regulated independently. The plasmids are replicated and distributed to daughter cells at each cell division, with approximately the same number of a given kind of plasmid going to each daughter cell. Such plasmids are said to be *compatible*, that is, they can both exist in the same cell.

In contrast, suppose a cell contains two kinds of plasmids that are regulated by the same replication repressor. Such a repressor limits the total number of plasmids but exerts no control over the relative numbers of each kind of plasmid. If the maximum number of plasmids were 20, a cell could contain 10 of each, but it might contain as many as 20 of one kind and none of the other. When such a cell divides, the plasmids are randomly distributed. About half go to each daughter cell, but there is no control over the numbers of each kind being distributed to a daughter cell. Eventually, after many divisions, most cells contain only one kind of plasmid. Such plasmids are said to form an *incompatibility group* because they tend not to be present with other plasmids of the same group for more than a few generations.

In addition to being transferred on resistance plasmids by conjugation, R genes also can move from one plasmid to another in a cell or even become inserted in the chromosome (Figure 8.17). These units of DNA, about the size of a single gene, are called *transposons* (tranz-pos'onz) because they can transpose, or change, their locations. Though transposons replicate only when in a plasmid or a chromosome, their ability to move among plasmids or to chromosomes greatly increases the ways they can affect the genetic makeup of a cell. For this reason transposons play an important role in evolution.

In 1983 Barbara McClintock won the Nobel Prize for her work on transposons, using corn. Transposons were next found in microorganisms and are now considered a universal phenomenon. Transposition is a relatively rare event and is not easily detected in eukaryotes. It is easier to detect in bacteria because we can work with large populations that can be tested more easily for particular characteristics.

Transposons, or "jumping genes" as they are sometimes called, sometimes jump to another location on a piece of DNA and insert themselves somewhere in the middle of a gene. When this occurs, the two ends of the gene lose physical contact with each other, and the functioning of the gene may be disturbed. This constitutes a mutation. Thus, transposons may be useful as a tool to create mutations for study.

Genetic information on plasmids also can cause a bacterium to become capable of producing disease. For example, *Clostridium tetani* produces the neurotoxin that causes tetanus only when it carries a gene for toxin production on a plasmid. Without the plasmid gene the bacterium produces neither toxin nor illness. The same is true for certain strains of *E. coli* that cause diarrhea and for *Staphylococcus aureus* that cause food poisoning. These organisms produce enterotoxin only when particular plasmids are present in the cell.

Even when a toxin gene is present in a bacterium, toxin may not be made. However, iron deficiency facilitates toxin production. Humans suffering from anemia who become infected by a bacterium carrying a gene for toxin are much more likely to suffer from the effects of the toxin.

Much research on plasmids concerns antibiotic resistance, as just described, but other attributes of plasmids are of interest to researchers. One line of research has led to a technique called *plasmid-assisted molecular breeding*. This technique can be used to create strains of bacteria with new genetic characteristics by purposely transferring genes on plasmids. One such strain can use the herbicide Agent Orange as its carbon source. Sunlight initiates breakdown of Agent Orange, and these organisms then completely degrade this toxic substance.

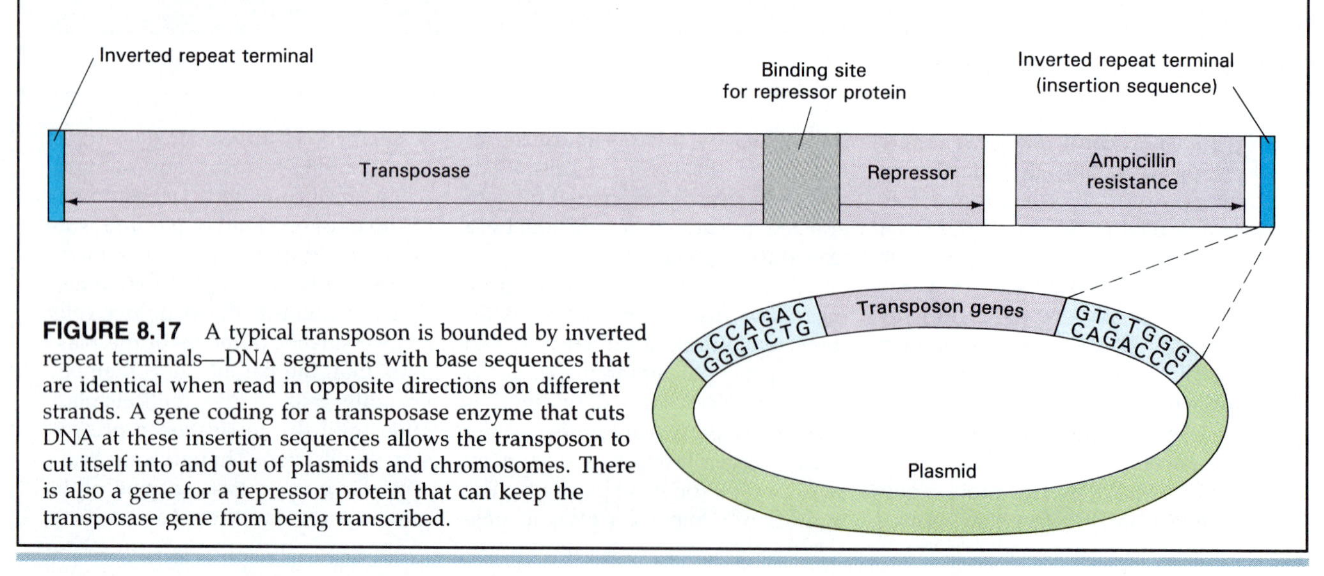

FIGURE 8.17 A typical transposon is bounded by inverted repeat terminals—DNA segments with base sequences that are identical when read in opposite directions on different strands. A gene coding for a transposase enzyme that cuts DNA at these insertion sequences allows the transposon to cut itself into and out of plasmids and chromosomes. There is also a gene for a repressor protein that can keep the transposase gene from being transcribed.

CHAPTER SUMMARY

RELATED KEY TERMS

NATURE AND SIGNIFICANCE OF GENE TRANSFER

- **Gene transfer** refers to movement of genetic information between organisms. It occurs in bacteria by transformation, transduction, and conjugation.
- Gene transfer is significant because it increases genetic diversity within a population, thereby increasing the likelihood that some members of the population will survive environmental changes.

TRANSFORMATION

Discovery of Transformation

- Bacterial **transformation** was discovered in 1928 by Griffith, who showed that live rough and heat-killed smooth pneumococci could produce live smooth pneumococci capable of killing mice.
- Avery later showed that a capsular polysaccharide was responsible for virulence and that DNA was the substance responsible for transformation. Watson and Crick showed that genetic information is coded in DNA.

Mechanism of Transformation

- Transformation involves the release of DNA fragments and their uptake by other cells at a certain stage in their growth cycle: (1) Uptake of DNA requires a protein called **competence factor.** (2) Endonucleases cut double-stranded DNA into units; the strands separate and only one strand is transferred. (3) Ultimately, donor DNA is spliced into recipient DNA. Leftover recipient DNA is broken down so a cell's total DNA remains constant.

Significance of Transformation

- Transformation is significant because (1) it contributes to genetic diversity; (2) it can be used to introduce DNA into an organism, observe its effects, and study gene locations; (3) it can be used to create recombinant DNA.

TRANSDUCTION

Discovery of Transduction

- In **transduction** genetic material is carried by a **bacteriophage** (phage).

Mechanisms of Transduction

- **Phages** can be virulent or temperate. (1) **Virulent phages** destroy a host cell's DNA, direct synthesis of phage particles, and cause lysis of the host cell in the lytic cycle. (2) **Temperate phages (prophages)** produce a repressor substance that prevents destruction of host DNA. They persist in the cell and are replicated with it in **lysogeny.**

  lysis
- Prophage can be incorporated into the bacterial chromosome, or it can exist as a **plasmid,** a piece of extrachromosomal DNA. Cells containing prophage are called **lysogenic** cells because they have the potential to enter the **lytic cycle.**
- Transduction can be specialized or generalized. (1) In **specialized transduction,** the phage is incorporated into the chromosome and can transfer only genes adjacent to the phage. (2) In **generalized transduction,** the phage exists as a plasmid and can transfer any DNA fragment attached to it.

Significance of Transduction

- Transduction is significant because it transfers genetic material and demonstrates a close evolutionary relationship between prophage and host cell DNA. Also, its persistence in a cell suggests a mechanism for the viral origins of cancer, and it provides a mechanism for studying gene linkage.

  chromosome mapping

CONJUGATION

Discovery of Conjugation

- **Conjugation** transfers large quantities of DNA from one organism to another during contact between donor and recipient cells.
- Conjugation was discovered by Lederberg in 1946 when he observed that mixing strains of *E. coli* with different metabolic deficiencies allowed the cells to overcome deficiencies.

Mechanisms of Conjugation

- Three mechanisms of conjugation have been observed: (1) In the transfer of

F plasmids, a piece of extrachromosomal DNA (a plasmid) is transferred, possibly by moving through an **F pilus.** (2) In high-frequency recombinations, parts of F plasmids that have been incorporated into the chromosome are transferred along with the **initiating segment** of the chromosome. (3) An F plasmid incorporated into the chromosome and subsequently separated becomes an **F′ plasmid** and transfers chromosomal genes attached to it.

RELATED KEY TERMS

clone
high frequency of recombination (Hfr) strain

Significance of Conjugation

- The significance of conjugation is that it increases genetic diversity, it may represent an evolutionary stage between asexual and sexual reproduction, and it provides a means of mapping genes in bacterial chromosomes.

COMPARISON OF TRANSFER OF GENETIC INFORMATION PLASMIDS

Characteristics of Plasmids

- Plasmids are ring-shaped, self-replicating, double-stranded extrachromosomal DNA that carry information usually not essential for cell growth.

Resistance Plasmids

- **Resistance plasmids** (R factors) carry genetic information that confers resistance to antibiotics. They consist of a **resistance transfer factor (RTF)** and one or more **resistance genes (R genes).**

Bacteriocinogens

- **Bacteriocinogens** are plasmids that produce **bacteriocins,** which inhibit growth of other strains of the same species or closely related species.

colicins

GENETIC ENGINEERING

- **Genetic engineering** is the manipulation of genetic material to alter the characteristics of an organism.

Genetic Fusion

- **Genetic fusion** involves altering the DNA content of a single species of organism.

Protoplast Fusion

- **Protoplast fusion** combines **protoplasts** (organisms without cell walls) and allows mixing of genetic information.

Gene Amplification

- **Gene amplification** involves addition of plasmids to microorganisms to increase yield of useful substances.

Recombinant DNA

- **Recombinant DNA** consists of addition of DNA from another organism to the normal DNA of a microorganism. It has proven especially useful in medicine, industry, and agriculture.

replicon

Hybridomas

- **Hybridomas** are genetic recombinations involving cells of higher organisms.

monoclonal antibodies

Weighing the Risks and Benefits of Recombinant DNA

- When recombinant DNA techniques were first developed, scientists were concerned that virulent pathogens might be created, and they developed containment procedures. As research proceeded and no illnesses caused by recombinants were observed, most scientists came to believe that the benefits of recombinant DNA techniques outweighed the risks.

QUESTIONS FOR REVIEW

A.

1. What are the main characteristics of gene transfer?
2. What is the significance of gene transfer?

B.

3. How was transformation discovered?
4. What events led to the determination that genetic information was being transferred in transformation?
5. What are the steps in transformation?
6. What is the significance of transformation?

C.

7. What is transduction?
8. What is the role of viruses in transduction?
9. How do generalized and specialized transduction differ?

10. What is the significance of transduction?

D.

11. What is conjugation?
12. Which experiments led to the discovery of conjugation?
13. What are the steps in the process of conjugation?
14. How do the following kinds of conjugation differ: F plasmid, high-frequency recombination, and F′ plasmid?
15. What is the significance of conjugation?

E.

16. What are the characteristics of plasmids?
17. How are plasmids classified?
18. What are resistance plasmids, and what do they do?
19. What are bacteriocins, and what do they do?
20. What are bacteriocinogens, and what do they do?

F.

21. What is genetic engineering?
22. What is genetic fusion, and how has it been used?
23. What is protoplast fusion, and what benefits might be derived from this technique?
24. How is gene amplification accomplished, and what are its applications?
25. How is recombinant DNA produced?
26. What applications have been found for recombinant DNA?
27. What are hybridomas, and what uses have been made of them?

G.

28. Compare the risks and benefits of recombinant DNA techniques.

PROBLEMS FOR INVESTIGATION

1. Hold a class debate concerning the benefits and hazards of gene manipulation. In the case of Frostban, described in the Microbiologist's Notebook, do you feel the opposition's viewpoint and tactics were reasonable? Why or why not?
2. Once the Human Genome Project (described in Chapter 1) has identified and sequenced genes that are important to human health, which types of genetic engineering should be permitted on the human genome? If we allow people to "fix" genes in their own body, should they also be allowed to "fix" their eggs or sperm so as to affect all future generations of their offspring? Should we be allowed to transfer nonhuman genes into humans?
3. You have applied for a job and have had a company-requested physical exam. Should the potential employer have the right to know which genes you have? What constitutes the right to privacy? If the company knows you carry a gene for cancer, will they discriminate against you and give the job to someone without a cancer gene? Have a class discussion of these problems of DNA testing in the workplace.
4. Research how human insulin is now being made in bacteria via genetic engineering techniques. Why is this so important to human diabetics? What kind(s) of insulin had previously been used?
5. Research the connection between plasmids and bacteria that produce toxins harmful to humans.
6. Bring in some news articles to share in class about current developments in genetic engineering.

SOME INTERESTING READING

Anonymous. 1992. Biotechnology federal budget initiative shows growth. *ASM News* 58(5):52–54.

Anonymous. 1991. USDA guidelines drafted, deliberate release tests pending. *ASM News* 57(5):240–41.

Brill, W. J. 1985. Safety concerns and genetic engineering in agriculture. *Science* 227:381.

Cohen, S. N., and J. A. Shapiro. 1980. Transposable genetic elements. *Scientific American* 242, no. 2 (February):40.

Cohrssen, J. J. 1988. United States biotechnology policy. *American Biotechnology Laboratory* 6(1)(January):22.

Crawford, M. 1987. California field test goes forward. *Science* 236 (May 1):511.

Erickson, D. 1991. Gene rush: companies seek profits in the genome project. *Scientific American* 264(January):112–13.

Gilbert, W., and L. Villa-Komaroff. 1980. Useful proteins from recombinant bacteria. *Scientific American* 242, no. 4 (April):74.

Koncz, C., W. H. R. Landridge, O. Olsson, J. Schell, and A. A. Szalay. 1990. Bacterial and firefly luciferase genes in transgenic plants: advantages and disadvantages of a reporter gene. *Developments in Genetics* 11(3):224–32.

Lee, C. J. 1987. Bacterial capsular polysaccharides—biochemistry, immunity, and vaccine. *Molecular Immunology* 24(10) (October):1005.

Macario, E., and A. Macario. 1983. Monoclonal antibodies for bacterial identification and taxonomy. *American Society for Microbiology News* 49:1.

Marx, J. 1987. Assessing the risks of microbial release. *Science* 237 (September 18):1413.

Miller, H. I. 1988. FDA regulation of products of the new biotechnology. *American Biotechnology Laboratory* 6(1) (January):38.

Milstein, C. 1980. Monoclonal antibodies. *Scientific American* 243, no. 4 (October):66.

Stahl, F. W. 1987. Genetic recombination. *Scientific American* 256, no.2 (February):91.

Stewart, G. J., and C. A. Carlson. 1986. The biology of natural transformation. *Annual Review of Microbiology* 40:211.

Watson, J. D., J. Tooze, and D. T. Kurtz. 1983. *Recombinant DNA: a short course.* New York: W. H. Freeman.

Weintraub, H. M. 1990. Antisense RNA and DNA. *Scientific American* 262(January):40–46.

Weiss, R. 1988. Engineered microbes stay close to home. *Science News* 133 (February 20):117.

An assortment of diatoms exemplifies the unity and diversity of organisms.

9 Microbes in the Scheme of Life: An Introduction to Taxonomy

This chapter focuses on the following questions:

A. How are microorganisms named?

B. What did Linnaeus contribute to taxonomy?

C. How is a dichotomous taxonomic key used to identify organisms?

D. What are some problems and developments in taxonomy since Linnaeus?

E. What are the main characteristics of the kingdoms in the five-kingdom system of taxonomy?

F. How are viruses classified?

G. What special methods are needed for determining evolutionary relationships among prokaryotes?

TAXONOMY—THE SCIENCE OF CLASSIFICATION

Humans appear to have an innate need to name things. In many primitive societies, a person who knows the true name of an object or another person is believed to have power over that object or person. Naming helps us to understand our world and to communicate with others about it. In science accurate and standardized names are essential. All chemists must mean the same thing when they talk about an element or a compound, physicists must agree upon terms when they discuss matter or energy, and biologists must agree as to the names of organisms, be they tigers or bacteria. Biologists have created **taxonomy**—the science of classification—to provide an orderly, agreed-upon system of naming organisms. A **taxon** (plural taxa) is a category, hence taxonomy is the science of placing things in their proper categories.

Another important aspect of taxonomy is that it makes use of and makes sense of the fundamental concepts of unity and diversity among living things. Organisms classified in any particular group have certain common characteristics, that is, they have unity with respect to these characteristics. For example, humans walk upright and have a well-developed brain, and *Escherichia coli* have a rod shape and a gram-negative cell wall. The organisms within taxonomic groups exhibit diversity as well. Even members of the same species display variations in size, shape, and other characteristics. Humans vary in height, weight, hair and eye color, and facial features, and certain kinds of bacteria vary somewhat in shape and whether they can form endospores. A basic principle of taxonomy is that members of higher-level groups share fewer characteristics than those in lower-level groups. Humans have backbones like all other vertebrates, but they share fewer characteristics with fish and birds than with other mammals. Likewise, nearly all bacteria have a cell wall, but in some the wall is gram-positive and in others it is gram-negative.

Linnaeus—The Father of Taxonomy

The eighteenth-century Swedish botanist Carolus Linnaeus is credited with founding the science of taxonomy (Figure 9.1). He originated **binomial nomenclature,** the system that is still used today to name all living things. In the binomial, or "two-name," system, the first word designates the **genus** of an organism and is capitalized. The second word is the **specific epithet,** and it is not capitalized even when derived from the name of the person who discovered it. Together the genus and specific epithet identify the **species** to which the organism belongs. Both words are italicized in print, or underlined when handwritten. When there is no danger of confusion, the genus name may be abbreviated to a single letter. Thus *Escherichia coli* is often written *E. coli,* and humans (*Homo sapiens*) may be identified as *H. sapiens.*

The name of an organism often tells something about it, such as its shape, where it is found, what nutrients it uses, who discovered it, or what disease it causes. Some examples of names and their meanings are shown in Table 9.1.

In most cases the members of a species have several common characteristics that distinguish that species from all other species. As a rule, members of the species cannot be divided into significantly different groups on the basis of a particular characteristic, but there are exceptions to this rule. Sometimes members of a species are divided on the basis of a small but permanent difference, such as a need for a particular nutrient, resistance to a certain antibiotic, or the presence of a particular antigen. When organisms in one pure culture of a species differ from the organisms in another pure culture of the species, the organisms in each culture are designated as strains. A **strain** is a subgroup of a species with one or more characteristics that distinguish it from other subgroups of the species. Each strain is identified by a name, number, or letter that follows the specific epithet. For example, *E. coli*

FIGURE 9.1 Carolus Linnaeus (1707–1778), the father of taxonomy.

TABLE 9.1 Meanings of names of some microorganisms

Name of Microorganism	Meaning of Name
Entamoeba histolytica	*Ent* = intestinal, *amoebae* = shape and means of movement, *histo* = tissue, *lytic* = lysing, or digesting, tissue
Escherichia coli	Named after Theodor Escherich in 1888; found in the colon
Haemophilus ducreyi	*Hemo* = blood, *phil* = love; named after Augusto Ducrey in 1889
Neisseria gonorrhoeae	Named after Albert L. Neisser in 1879; causes the disease gonorrhea
Saccharomyces cerevisiae	*Saccharo* = sugar, *myco* = mold, *cerevisia* = beer or ale
Staphylococcus aureus	*Staphylo* = cluster, *kokkus* = berry, *aureus* = golden
Streptococcus lactis	*Strepto* = twisted chain, *kokkus* = berry, *lacto* = milk

strain K12 has been extensively studied because of its plasmids and other genetic characteristics, and *E. coli* strain 0157:H7 is a strain that causes hemorrhagic inflammation of the colon in humans. *Treponema pallidum* Reiter, which may be a strain of the organism that causes syphilis, has been cultured on artificial media, although the disease-causing strain has not. In addition to introducing the binomial nomenclature, Linnaeus also established a taxonomic hierarchy of species, genus, family, order, class, phylum or division, and kingdom. At the highest level Linnaeus divided all living things into two kingdoms—plant and animal. In the taxonomic hierarchy, which is still used today, each organism is assigned a species name, and species of very similar organisms are grouped into a genus. As we proceed up the hierarchy, several similar *genera* (plural of genus) are grouped to form a family, several families to form an order, and so on to the top of the hierarchy. Some hierarchies today have additional levels, such as subphyla. Also, it has become accepted practice to refer to the first categories within the animal kingdom as phyla and those within other kingdoms (we now have five) as divisions. The hierarchy of names designating a human and a bacterium are shown in Table 9.2.

APPLICATIONS

Where Do Stock Cultures Come From?

Particular strains of organisms are often of sufficient value to be preserved because their characteristics are important in research or in industrial applications such as winemaking. Preserving these organisms in a dormant (inactive) state prevents them from undergoing genetic changes that might alter their characteristics. One method of preserving organisms is lyophilization (freeze-drying), as is explained in Chapter 13. Preserved cultures can be deposited in a central type culture collection, in which each culture is directly descended from the original organism to which a particular strain or species designation was first assigned. The American Type Culture Collection (ATCC) in Rockville, Maryland, keeps some dormant cultures in a vault, which also protects them against theft. Such organisms are of value to manufacturers of wine or cheese because they create distinctive flavors or other characteristics of products. If the organisms were lost, the ability to make particular products would also be lost. Stock cultures of many organisms are available to qualified scientists. The existence of such a collection allows different researchers studying a particular strain to be confident that they are really dealing with the same organism and that their results can be meaningfully compared.

The American Type Culture Collection uses lyophilization (freeze-drying) to preserve organisms.

TABLE 9.2 Classification of a human and a bacterium

Taxonomic Category	Human	Bacterium That Causes Syphilis
Kingdom	Animalia	Monera (Prokaryotae)
Division/Phylum	Chordata	Gracilicutes
Subphylum	Vertebrata	
Class	Mammalia	Scotobacteria
Order	Primate	Spirochaetales
Family	Hominidae	Spirochaetaceae
Genus	*Homo*	*Treponema*
Specific epithet	*sapiens*	*pallidum*

Using a Taxonomic Key

Biologists often make use of a taxonomic key to identify organisms according to their characteristics. The most common kind of key is a **dichotomous key,**

1a	Grade point average 3.0 or better	Go to 2
1b	Grade point average less than 3.0	Go to 3
2a	Study at least 20 hours per week	Hardworking good student
2b	Study less than 20 hours per week	Lucky good student
3a	Study at least 25 hours per week	Hardworking not-so-good student
3b	Study less than 25 hours per week	More work might make you a good student!

FIGURE 9.2 A dichotomous key for classifying students.

which has paired statements describing characteristics of organisms. Paired statements present an "either-or" choice, such that only one statement is true. Each statement is followed by directions to go to another pair of statements until finally the name of the organism appears. Figure 9.2 is a dichotomous key that will identify you as a member of a certain group of students. Read statements 1a and 1b and decide which statement applies to you. Look at the number to the right of the statement; that tells you which pair of statements to look at next. Continue in this manner until you reach a group designation. If you have followed the key carefully, that designation will describe you.

1a	Gram-positive	Go to 2
1b	Not gram-positive	Go to 3
2a	Cells spherical in shape	Gram-positive cocci
2b	Cells not spherical in shape	Go to 4
3a	Gram-negative	Go to 5
3b	Not gram-negative (lack cell wall)	Mycoplasma
4a	Cells rod-shaped	Gram-positive bacilli
4b	Cells not rod-shaped	Go to 6
5a	Cells spherical in shape	Gram-negative cocci
5b	Cells not spherical in shape	Go to 7
6a	Cells club-shaped	Corynebacteria
6b	Cells variable in shape	Propionibacteria
7a	Cells rod-shaped	Gram-negative bacilli
7b	Cells not rod-shaped	Go to 8
8a	Cells helical with several turns	Spirochetes
8b	Cells comma-shaped	Vibrioids

FIGURE 9.3 A dichotomous key for classifying major groups of bacteria.

Major groups of bacteria can be identified with the key in Figure 9.3. More detailed keys use staining reactions, metabolic reactions (fermentation of particular sugars or release of different gases), growth at different temperatures, nature of colonies on solid media, and similar characteristics. By proceeding step by step through the key, one should be able to identify an unknown organism, or even a strain, if the key is sufficiently detailed.

Problems in Taxonomy

Among the aims of a taxonomic system are organizing knowledge about living things and establishing standard names for organisms so we can communicate about them. Ideally, we would like to classify organisms according to their evolutionary, or **phylogenetic,** relationships, but this is not always easy. Evolution occurs continuously and at a relatively rapid rate in microorganisms, and our knowledge of the evolutionary history of organisms is incomplete. Taxonomy must change with evolutionary changes and new knowledge. We must remember that it is far more important to have a taxonomic system that reflects our current knowledge than to have a system that never changes.

Creating a taxonomic system that provides an organized overview of all living things and how they are related to each other poses certain problems. Two such problems arise at opposite ends of the taxonomic hierarchy: (1) deciding what constitutes a species and (2) deciding what constitutes a kingdom. In the first, taxonomists try to decide how much diversity can be tolerated within the unity of a species. In the second, taxonomists try to decide how to sort the diverse characteristics of living things into categories that reflect fundamental differences of evolutionary significance. Species that reproduce sexually are distinguished primarily by their reproductive capabilities. A male and a female of the same species can mate successfully and produce fertile offspring, whereas ordinarily members of different species either cannot mate successfully or will have sterile offspring. Morphology (structure characteristics) and geographic distribution also are considered in defining species. Today, biochemical properties, intracellular structures, genetic characteristics, and immunological properties are also used to define species. These properties are especially useful in defining species among organisms, such as bacteria, that reproduce asexually.

Prior to the time that taxonomists turned their attention to microorganisms, the two-kingdom system of plants and animals worked reasonably well. Anyone can tell plants from animals—for example, trees from dogs. Plants make their own food but cannot move, and animals move but cannot make their own food. Simple enough, or is it? In this scheme, how do you classify *Euglena,* a mobile microorganism that makes its own food? How would you classify jellyfishes and sponges that are motile or immotile depending on their stage of life? And how do you classify colorless fungi that neither move nor make their own food? Finally, how do you classify slime molds, organisms that can be unicellular or multicellular and mobile or immobile? Obviously, microorganisms pose a number of problems when one tries to use a two-kingdom system.

Developments Since Linnaeus

The problem of classifying microorganisms was first addressed by the German biologist Ernst H. Haeckel in 1866 when he created a third kingdom, the Protista. He included among the protists all "simple" forms of life such as bacteria, many algae, protozoa, and multicellular fungi and sponges. Haeckel's original term, Protista, is still used in taxonomic schemes today, but it is now limited mainly to unicellular eukaryotic organisms.

Classification of bacteria has posed taxonomic problems over the centuries and still does, as the Essay at the end of this chapter indicates. Until recently many taxonomists regarded bacteria as small plants that lacked chlorophyll. As late as 1957 the seventh edition of *Bergey's Manual of Determinative Bacteriology,* a work devoted to the identification of bacteria, considered bacteria to be unicellular plants. Changes in this viewpoint came as the tools to study bacteria were developed. First, light microscopy and stains were used to describe the basic structure of cells. Second, electron microscopy was used to study the ultrastructure of cells. And, third, biochemical techniques were used to study chemical composition and chemical reactions in cells. From these various studies one of the most important discoveries was that DNA looked and behaved differently during cell division in bacteria than in cells that have their DNA organized into chromosomes within a nucleus.

Studies of the structure and function of cells also led to the recognition of two general patterns of cellular organization, prokaryotic and eukaryotic. Basing taxonomy on these two different patterns of cellular organization was proposed as early as 1937. Various taxonomists such as H. F. Copeland, R. Y. Stanier, C. B. van Niel, and R. H. Whittaker, working in the late 1950s, placed bacteria in a separate kingdom of anucleate (lacking a nucleus) organisms rather than with organisms that have true nuclei. In 1962 Stanier and van Niel stated, "The distinctive property of bacteria is the prokaryotic nature of their cells."

Lynn Margulis and H. F. Copeland proposed a scheme of classifying prokaryotes and eukaryotes by the following four-kingdom system of classification:

1. Monera: all prokaryotes, including true bacteria and blue-green algae
2. Protoctista: all eukaryotic algae, protozoa, and fungi
3. Plantae: all green plants
4. Animalia: all animals derived from a **zygote** (zi'got), a cell formed by the union of an egg and a sperm

These taxonomists also proposed that evolution from prokaryotic to eukaryotic life forms had taken place by endosymbiosis, as described in Chapter 4. ∞ (p. 102)

R. H. Whittaker of Cornell University felt that endosymbiosis could not account for all the differences between prokaryotes and eukaryotes. He also felt that a taxonomic system should give more consideration to the methods organisms used to obtain nourishment. Autotrophic nutrition by photosynthesis and heterotrophic nutrition by ingestion of substances from other organisms had been considered in earlier taxonomies. Absorption as a sole means of acquiring nutrients had been overlooked. To Whittaker, fungi, which acquire nutrients solely by absorption, were sufficiently different from plants to justify placing them in a different kingdom. Also, fungi have certain reproductive processes not shared with any other organisms. Consequently, Whittaker proposed a taxonomic system in 1969 that separated the Protoctista into two kingdoms, Protista (pro-tis'tah) and Fungi, but retained the Monera, Plantae, and Animalia. Finally, through refinements of Whittaker's system by several taxonomists over the past few decades, the five-kingdom system was created. Margulis proposed a very similar five-kingdom system in 1982, but she refers to the kingdom of simple eukaryotes as Protoctista instead of Protista.

THE FIVE-KINGDOM CLASSIFICATION SYSTEM

Before we discuss the five-kingdom classification system and how it applies to microorganisms, we must emphasize that all living organisms, regardless of the kingdom to which they are assigned, display certain characteristics that define the unity of life. All are composed of cells, and all carry out certain functions, such as obtaining nutrients and getting rid of wastes.

The cell is the basic structural and functional unit of all living things. The fact that viruses are not cells is one reason they are not considered to be living organisms. All cells are bounded by a plasma membrane, they carry genetic information in DNA, and they have ribosomes where proteins are made. All cells also contain the same kinds of organic compounds—proteins, lipids, nucleic acids, and carbohydrates. They also selectively transport material between their cytoplasm and their environment. Thus, though organisms may be classified in very diverse taxonomic groups, their cells have many similarities in structure and function.

No single classification system is completely accepted by all biologists. For this text we have elected to use the **five-kingdom system** (Figure 9.4). A major advantage of this system is the clarity with which it deals with microorganisms. It places all **prokaryotes,** microorganisms that lack a nucleus (Chapter 4), in the kingdom Monera. ∞ (p. 76) It places most unicellular **eukaryotes,** organisms whose cells contain a distinct nucleus, in the kingdom Protista. The five-kingdom system also places fungi in the separate kingdom Fungi. Microscopic fungi are of concern to microbiologists. Some taxonomists have recommended creating a sixth kingdom for **archaebacteria** (ar'ke-bak-ter''e-ah). These microbes differ in important ways from the **eubacteria** (u'bak-ter''e-ah), or true bacteria, and may be of very ancient origin. (See the Essay at the end of this chapter.)

The properties and members of each of the five kingdoms are described below and summarized in Table 9.3. A more detailed classification of microorganisms and multicellular parasites is provided in Appendix B.

Kingdom Monera

The kingdom **Monera** (mo-ner'ah) is also called the kingdom Prokaryotae, as suggested by the French marine biologist Edouard Chatton in 1937. As shown in Figure 9.5, it consists of all prokaryotic organisms,

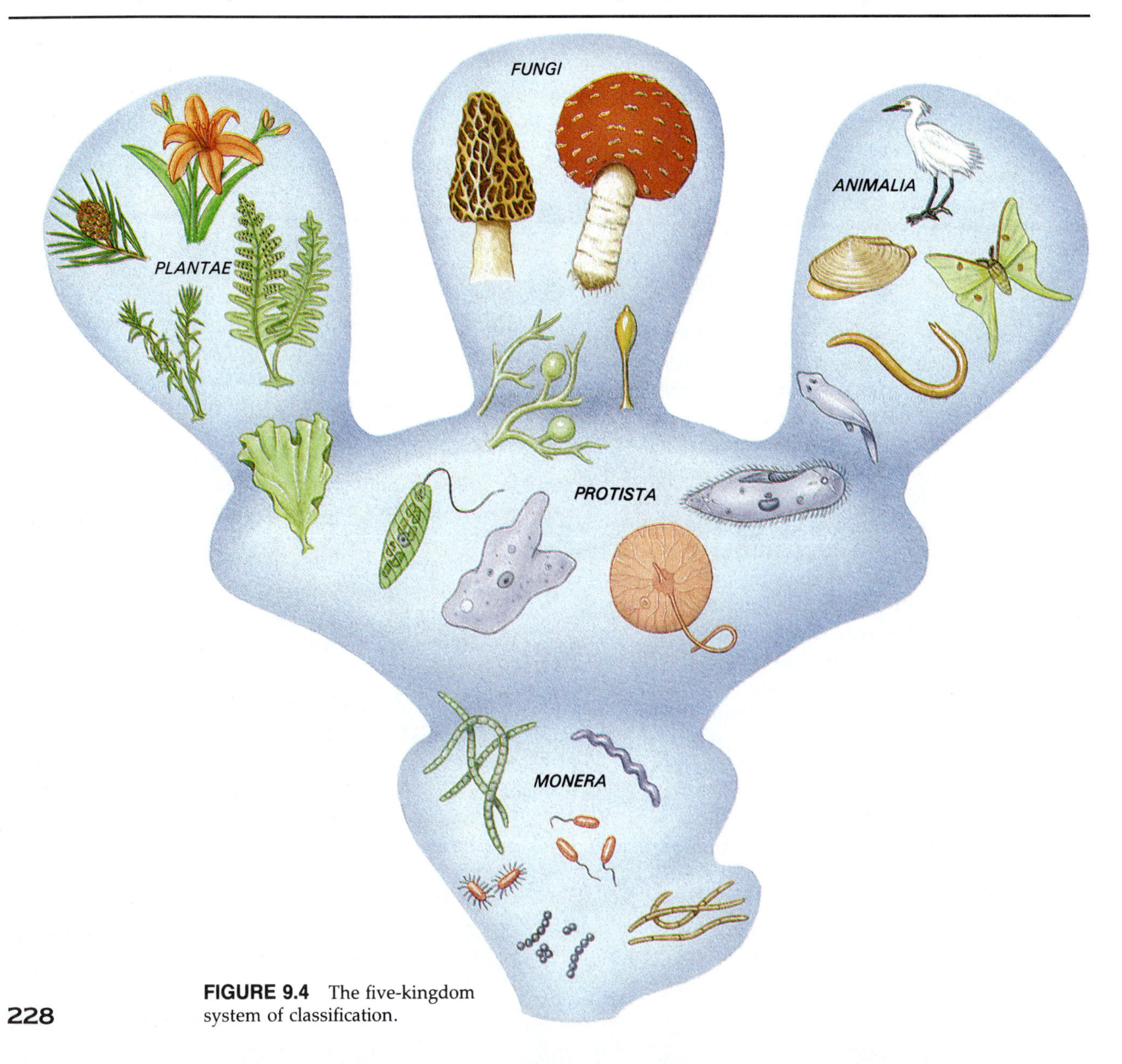

FIGURE 9.4 The five-kingdom system of classification.

TABLE 9.3 The five-kingdom system of classification

Kingdom	Characteristics
Monera	Prokaryotic; unicellular, but sometimes cells are grouped; nutrition by absorption, but in some forms by photosynthesis or chemosynthesis; reproduction asexual, usually by fission.
Protista	Eukaryotic; unicellular but sometimes cells are grouped; nutrition varies among phyla and can be by ingestion, photosynthesis, or absorption; reproduction asexual and in some forms both sexual and asexual.
Fungi	Eukaryotic; unicellular or multicellular, nutrition by absorption; reproduction usually both sexual and asexual and often involves a complex life cycle.
Plantae	Eukaryotic; multicellular; nutrition by photosynthesis.
Animalia	Eukaryotic; multicellular; nutrition by ingestion but in some parasites by absorption; reproduction primarily sexual.

including the true bacteria and the cyanobacteria—and for the present, the archaebacteria, too.

All monerans are unicellular; they lack true nuclei and generally lack organelles. Their DNA has little or no protein associated with it. Reproduction in the kingdom Monera is mainly by binary fission. Of all monerans, the bacteria are of greatest concern in the health sciences and will be considered in detail in several chapters of this book.

The **cyanobacteria** (si''an-o-bak-ter'e-ah), formerly known as blue-green algae, are of special importance in the balance of nature. They are photosynthetic, typically unicellular, organisms, although cells may sometimes be connected together to form threadlike filaments. Being autotrophs, they do not invade other organisms, so they pose no health threat to humans, except for toxins (poisons) some release into water.

Cyanobacteria grow in a great variety of habitats, including anaerobic ones, where they often serve as food sources for more complex heterotrophic organisms. Some "fix" atmospheric nitrogen, converting it to nitrogenous compounds that algae and other organisms can use. Some cyanobacteria also thrive in nutrient-rich water and are responsible for algal blooms—a thick layer of algae on the surface of water that prevents light from penetrating to the water below. Such blooms release toxic substances that can give the water an objectionable odor and even harm fish and livestock that drink the water.

Archaebacteria surviving today are primitive anaerobes adapted to extreme environments. The methanogens reduce carbon to the gas methane. The extreme halophiles live in excessively salty environments, and the thermoacidophiles live in hot acidic environments such as volcanic vents in the ocean floor (Figure 9.6).

Kingdom Protista

While the modern protist group is very diverse, it contains fewer kinds of organisms than when first defined by Haeckel. Organisms now classified in the

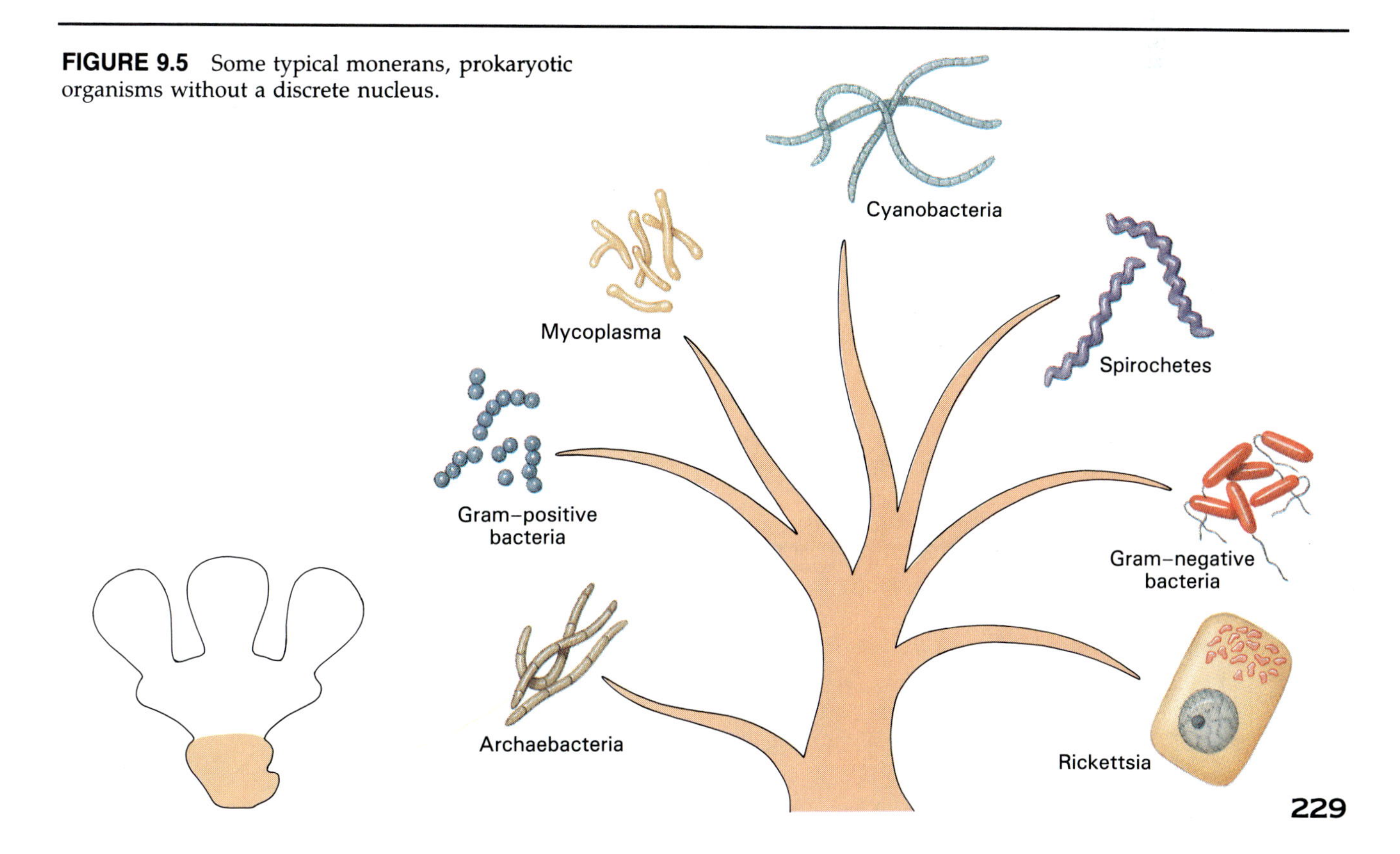

FIGURE 9.5 Some typical monerans, prokaryotic organisms without a discrete nucleus.

FIGURE 9.6 Organisms living at deep ocean vents, where hot volcanic gases are released from the earth's interior, survive in one of the most extreme environments known. Temperature and pressure are extremely high, yet these vents are the most productive ecosystem on our planet.

kingdom **Protista** (Figure 9.7) are all eukaryotic. Most are unicellular, but some are organized into colonies. Protists have a true membrane-bound nucleus and organelles within their cytoplasm, as do other eukaryotes. Many live in fresh water, some live in seawater, and a few live in soil. They are distinguished more by what they don't have than what they have. Protists do not develop from an embryo as plants and animals do, and they do not develop from distinctive spores as the fungi do. Yet, among the protists are the algae that resemble plants, the protozoa that resemble animals, and the euglenoids that have both plant and animal characteristics. The protists of greatest interest to health scientists are the protozoa that can cause human disease.

Kingdom Fungi

The kingdom **Fungi** (Figure 9.8) includes mostly multicellular and some unicellular organisms. Fungi obtain nutrients solely by absorption of organic matter from dead organisms. Even when they invade living tissues, they typically kill cells and then absorb nutrients from them. Though the fungi have some characteristics in common with plants, their filaments are much simpler in organization than true leaves and stems. Fungi form spores but do not form seeds. Many fungi pose no threat to other living things, but some attack plants, animals, and even humans.

Kingdom Plantae

Placing most microscopic eukaryotes with the protists leaves only macroscopic green plants in the kingdom **Plantae.** Most plants live on land and contain chlorophyll in organelles called chloroplasts. Plants are of interest to microbiologists only because some contain medicinal substances such as quinine, which has been used to treat microbial infections.

Kingdom Animalia

The kingdom **Animalia** includes all animals derived from zygotes. Though nearly all members of this king-

FIGURE 9.7 Some typical protists, unicellular eukaryotic organisms.

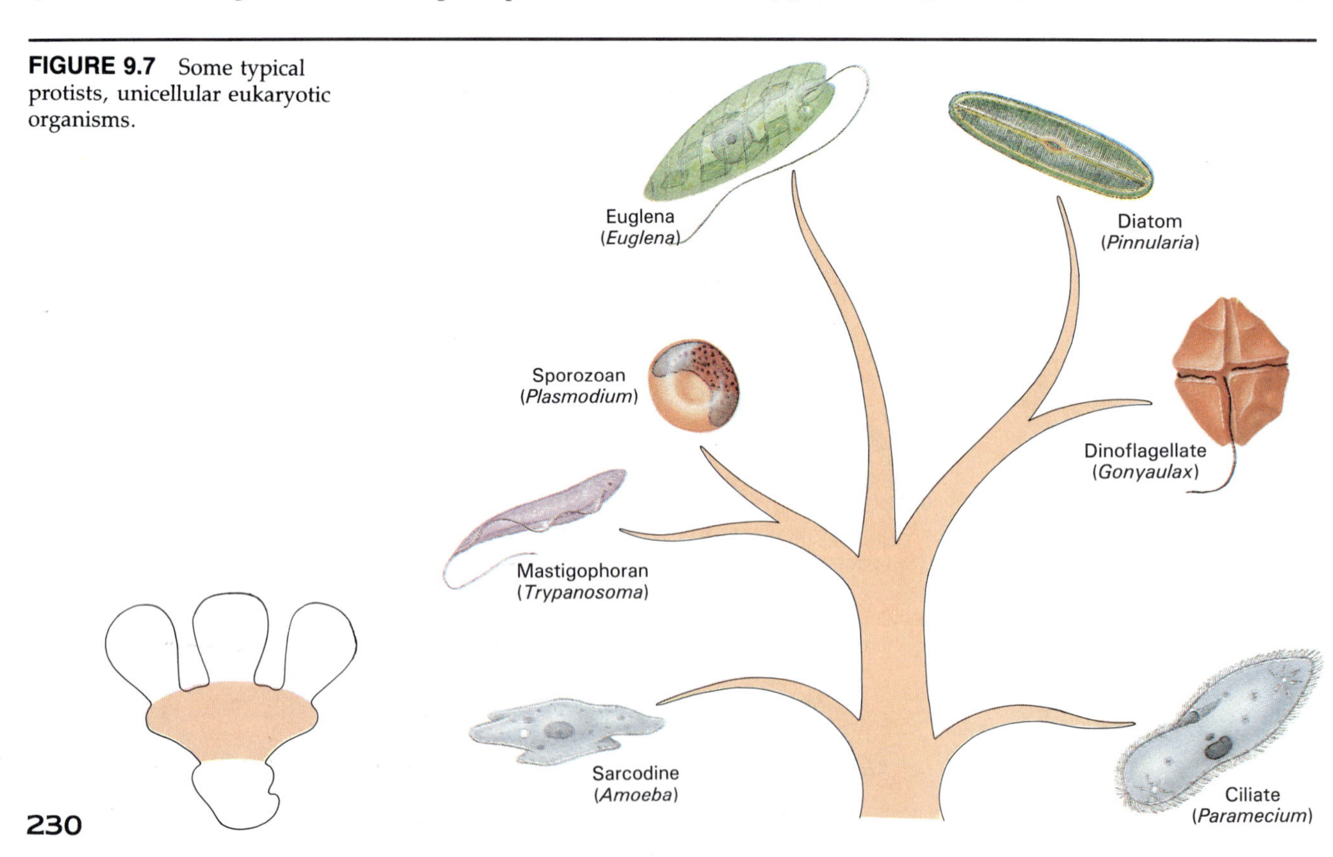

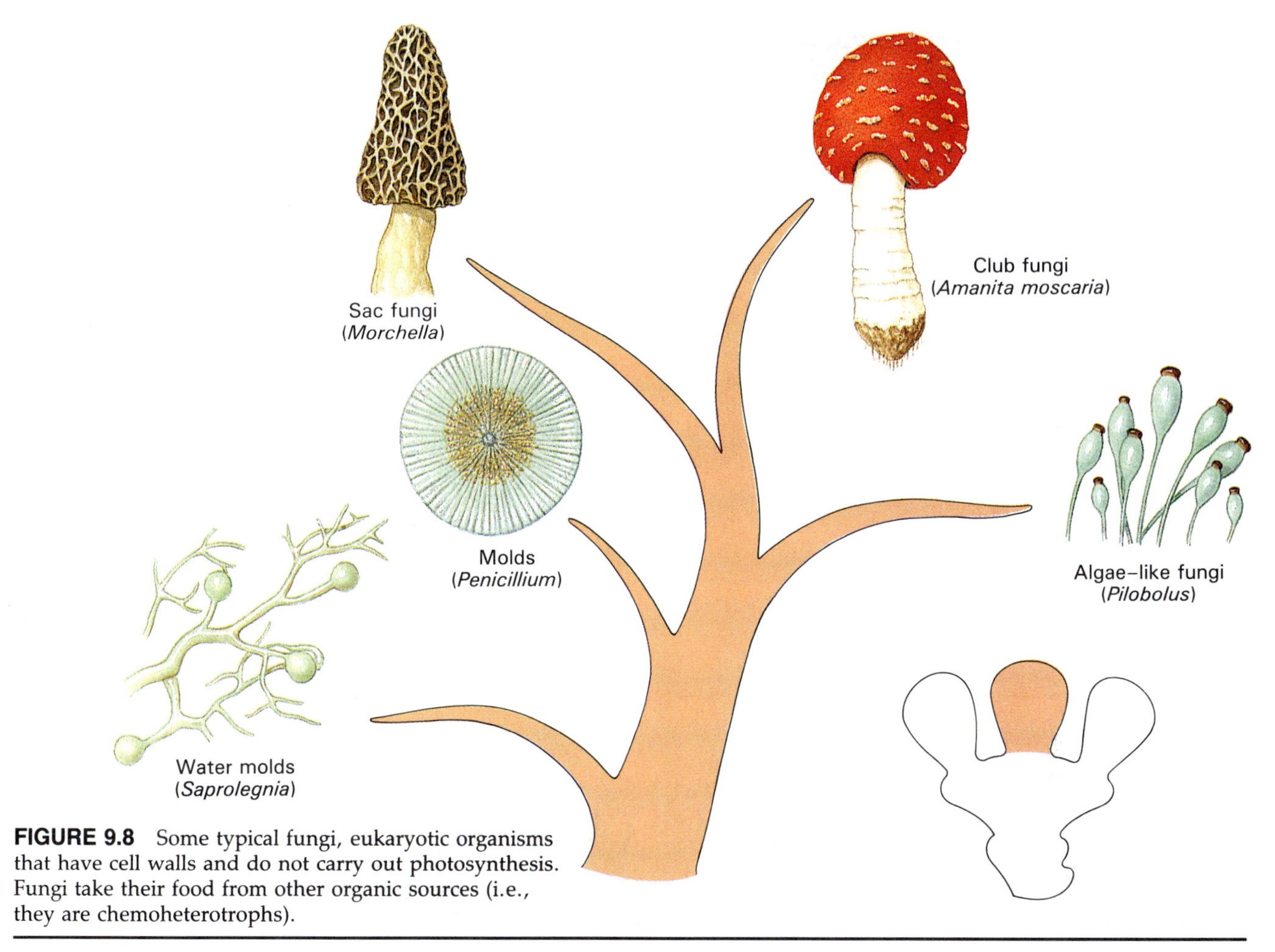

FIGURE 9.8 Some typical fungi, eukaryotic organisms that have cell walls and do not carry out photosynthesis. Fungi take their food from other organic sources (i.e., they are chemoheterotrophs).

dom are macroscopic and, therefore, of no concern to microbiologists, several groups of animals live in or on other organisms, and some serve as carriers of microorganisms (Figure 9.9).

Certain helminths (worms) are parasitic in humans and other animals. They include flukes, tapeworms, and roundworms, which live inside the body of their host. They also include leeches, which live on the surface of their hosts. Microbiologists often are required to identify microscopic forms of helminths.

Certain arthropods live on the surface of their hosts, and some spread disease. Ticks, mites, lice, and fleas live on their hosts for at least part of their lives. Ticks, lice, fleas, and mosquitoes can spread infectious microorganisms from their bodies to those of humans or other animals.

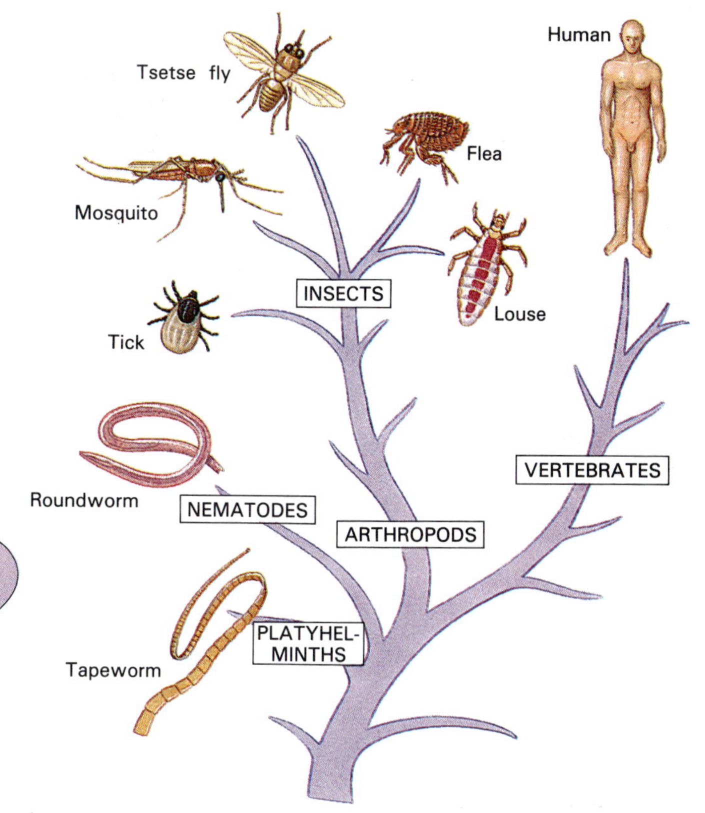

FIGURE 9.9 Groups from the kingdom Animalia that are relevant to microbiology.

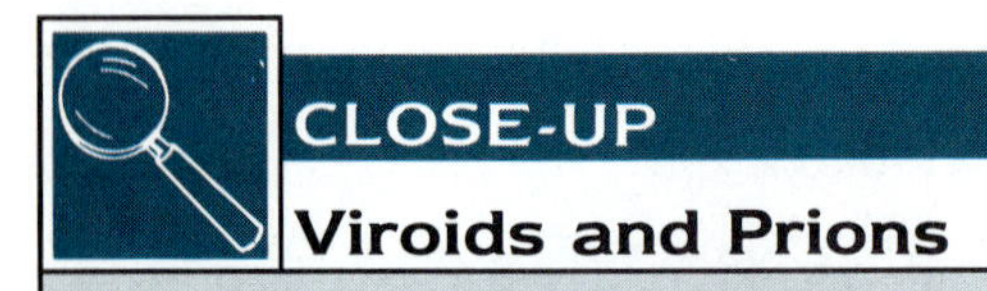

Viroids and Prions

In the past decade particles even smaller than viruses have been discovered, and some appear to serve as infectious agents. They include viroids (vīr'oids) and prions (pre'onz). A viroid consists of a fragment of RNA. One viroid, which causes a disease in potatoes, contains only 359 bases—enough information to specify the location of only 119 amino acids if all the bases function as codons. This amount of nucleic acid is only 1/10 that found in the smallest viruses. Prions, or proteinaceous infectious particles, are only 1/10 the size of a virus and consist of a protein molecule that appears to be self-replicating. (How do you think a protein might replicate itself?) Prions may be responsible for some mysterious human brain infections.

CLASSIFICATION OF VIRUSES

Viruses are acellular particles smaller than cells that contain nucleic acid and are coated with protein. They have not been assigned to a kingdom, and, in fact, they display only a few characteristics associated with living organisms.

Initially viruses were classified according to the hosts they invaded and by the diseases they caused. As more was learned about viruses, the early concept of one virus–one disease used in classification was found to be invalid for many viruses. Today viruses are classified by the type and arrangement of nucleic acids in them, by their cubical or tubular shape, by the symmetry of the protein capsid (covering) surrounding the nucleic acid, and by the presence or absence of such things as a membrane (called an envelope), enzymes, tail structures, or lipids (Figure 9.10). These groupings only reflect common characteristics and are not intended to represent evolutionary relationships.

The study of viruses is extremely important in any microbiology course for two reasons: (1) Virology is a recognized branch of microbiology and techniques to study viruses are derived from microbiological techniques; and (2) viruses are of concern to health scientists because many cause human disease.

THE SEARCH FOR EVOLUTIONARY RELATIONSHIPS

Many biologists are interested in how living things evolved and how they are related to each other. In fact, most people have some curiosity about how life originated and gave rise to the diverse assortment of living things we see today. Though the details of the search for evolutionary relationships are of interest mainly to taxonomists, they are of some significance to health scientists. For example, many of the biochemical properties used to establish evolutionary re-

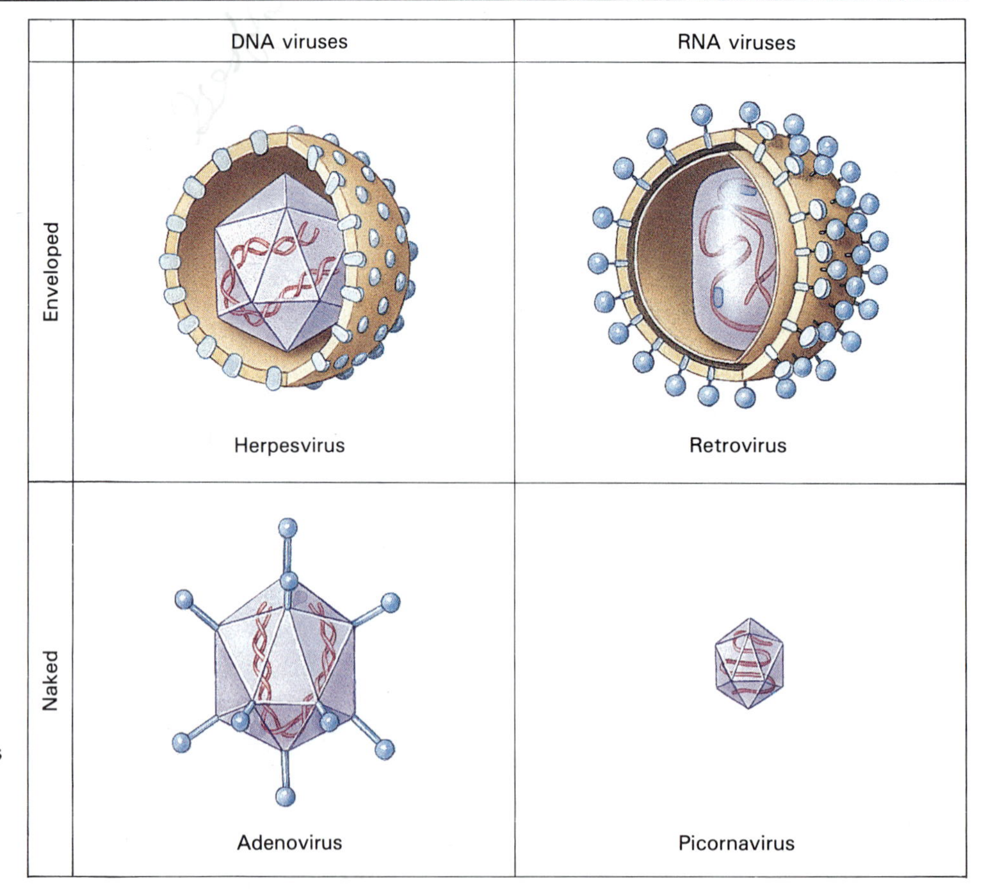

FIGURE 9.10 Some categories of viruses.

lationships also can be used in identifying microorganisms. Also, infectious agents, their hosts, and the relationships between them generally evolve together. Some knowledge of such evolution is useful in understanding the circumstances under which one organism becomes able to infect another and how the disease process occurs.

Special Methods Needed for Prokaryotes

The taxonomy of most eukaryotes is based on morphology (structural characteristics) of living organisms and knowledge of their evolutionary relationships from fossil records. However, morphology and fossil records provide little information about prokaryotes. For one thing, prokaryotes have left few fossil records. Recently some fossilized mats of prokaryotes, or **stromatolites** (stro-mat'o-lītz), have been found mainly at sites where the environment many years ago allowed the deposition of dense layers of algae and bacteria (Figure 9.11a and b). Unfortunately, most bacteria do not form such mats, so most ancestral prokaryotes have disappeared without a trace.

Some rocks containing fossils of microscopic individual cells also have been discovered (Figure 9.11c), but they fail to reveal much information about the organisms. Moreover, prokaryotes have few structural characteristics, and these characteristics are subject to rapid change when the environment changes. Large organisms generally require a fairly long period of time to reproduce, but prokaryotes reproduce rapidly. Rapid reproduction allows for mutations in each generation and much change in a relatively short time.

Because morphology and evolution are of little use in classifying prokaryotes, metabolic reactions and other specialized properties have been used instead. Health scientists use these properties to identify infectious prokaryotes in the laboratory, but such identification does not necessarily reflect evolutionary relationships among the organisms. The methods described next are of use in exploring evolutionary relationships. Although they are particularly appropriate for eukaryotes, they also can be used for prokaryotes.

Numerical Taxonomy

Numerical taxonomy is based on the idea that increasing the number of characteristics of organisms that we observe increases the accuracy with which we can detect similarities among them. Though the idea of numerical taxonomy was developed before computers were available, modern computers allow rapid comparisons among large numbers of organisms according to many different characteristics. In a simple example of numerical taxonomy each characteristic is assigned a value of 1 if present and 0 if not present. Characteristics such as reaction to Gram staining, oxygen requirements, presence or absence of an envelope, properties of nucleic acids and proteins, and the presence or absence of particular enzymes and chemical reactions can be compared. Organisms are then compared, and patterns of similarities and differences are detected (Figure 9.12). If two organisms match on 90 percent or more of the characteristics studied, they are presumed to belong to the same species. Provided the characteristics are genetically determined, organisms sharing a greater number of characteristics have a close evolutionary relationship. Computerized numerical taxonomy offers great promise for improving our understanding of relationships among all organisms.

(a)

(b)

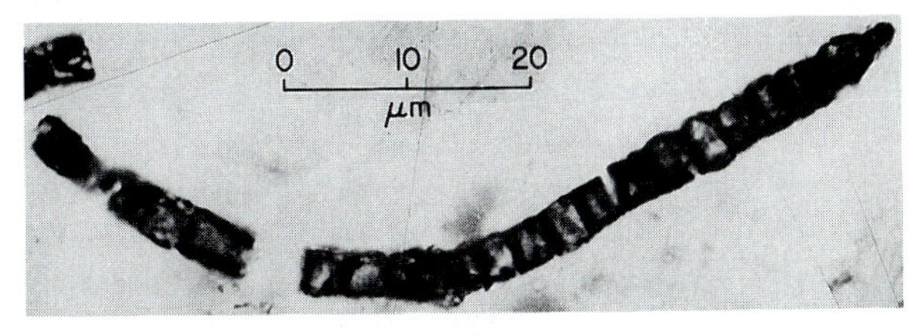

(c)

FIGURE 9.11 (a) Mats of bacteria, growing as stromatolites, are seen in the foreground of a salt lagoon in the Sea of Cortez, Mexico. (b) A cross section through fossil stromatolites from Colorado, showing horizontal layers of bacterial growth. (c) Filamentous cyanobacterial microfossils from the Bitter Springs Formation in Australia. The fossils date from the late Precambrian and are approximately 850 million years old.

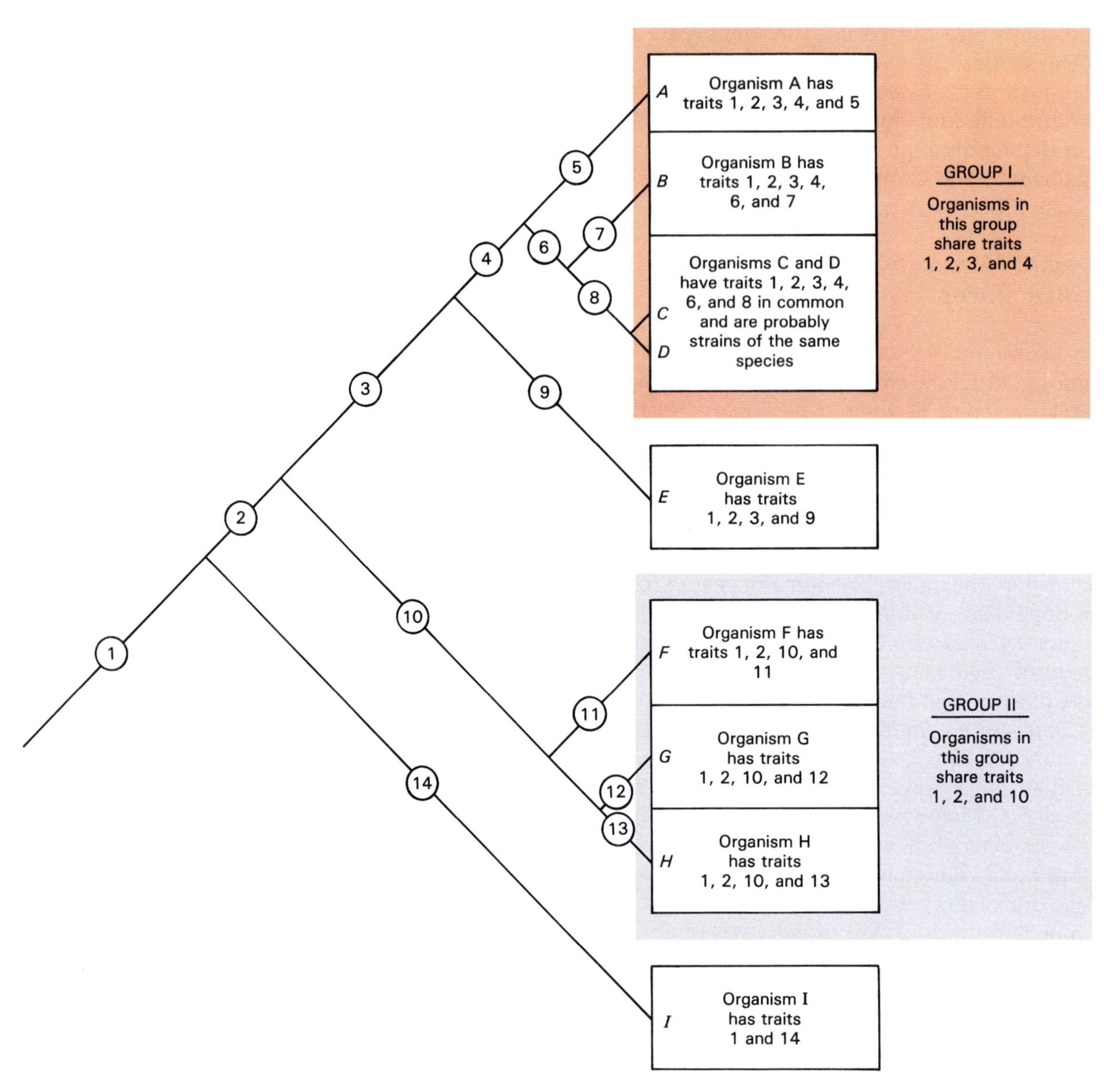

FIGURE 9.12 In numerical taxonomy, organisms are compared on the basis of a large number of characteristics. Those having a high proportion of characteristics in common are presumed to be closely related. Here the numerical taxonomy of selected gram-negative aerobic bacteria, based on 96 characteristics, is displayed in (a) cluster and (b) dendritic (treelike branching) fashion.

Genetic Homology

The discovery of the structure of DNA by James Watson and Francis Crick in 1953 provided new knowledge that was quickly applied by taxonomists, especially those studying taxonomic relationships and evolution of eukaryotes. These scientists began to study the similarity of DNA, or **genetic homology,** among organisms, and several techniques for determining genetic homology are now available. Similarities in DNA can be studied directly by determining the base composition of the DNA, by sequencing the bases in DNA or RNA, and by DNA hybridization. Because the proteins in an organism are determined by its DNA, similarities in DNA can be studied indirectly by preparing protein profiles and analyzing amino acid sequences in proteins.

Base Composition

Organisms can be grouped by comparing the relative percentages of bases present in the DNA of their cells. As explained in Chapter 2, DNA contains four bases, abbreviated by the letters A (adenine), T (thymine), C (cytosine), and G (guanine). Base pairing always occurs between A and T and between C and G. In making base comparisons, the total amount of C and G in a sample of DNA is determined and expressed as a percentage of total DNA. Subtracting this percentage from 100 gives the percent of A and T in the sample. For example, if the DNA is 60 percent C-G, then it is 40 percent A-T.

Various studies of base composition have shown that the C-G percentages vary from 23 to 75 percent in bacteria. These studies also have shown that certain

(a)

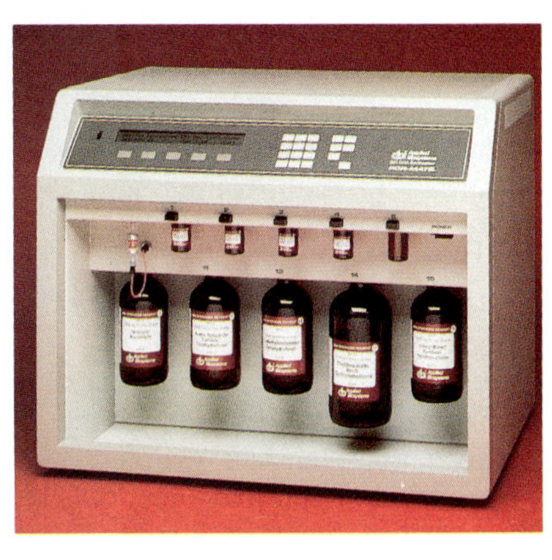

(b)

FIGURE 9.13 (a) DNA sequencer. Automated systems can identify the sequence of nucleotide bases in a piece of DNA. (b) DNA synthesizer. Automated systems can use bottles of reagents supplied to synthesize pieces of DNA with base sequences as desired.

species of bacteria, *Clostridium tetani* and *Staphylococcus aureus*, for example, have very similar DNA composition, but that *Pseudomonas aeruginosa* has a very different DNA composition. Thus, *C. tetani* and *S. aureus* are probably more closely related to each other than either is to *P. aeruginosa*. Similar percentages of bases do not in themselves prove that the organisms are closely related because the sequence of bases may be quite different. (Human beings and *Bacillus subtilis*, for example, have nearly identical C-G percentages.) We can say, however, that if the percentages in two organisms are quite different, they are unlikely to be closely related.

DNA and RNA Sequencing

Automated equipment for identifying the base sequences in DNA or RNA is now available at reasonable cost (Figure 9.13). This makes it possible to go into a culture and look for base sequences known to be unique to certain species. Using PCR techniques (see the Essay in Chapter 7), one can produce a large number of **probes,** DNA fragments that have sequences complementary to those being sought. ∞ (p. 187) A fluorescent dye or a radioactive tag (reporter molecules) can be attached to the probe. When the probe finds its target DNA, it will bind to it, and will not wash off when rinsed. The specimen is then examined for dye or radiation, and the presence or absence of the unique DNA sequence helps in identification of the specimen.

DNA Hybridization

In **DNA hybridization** the double strands of DNA of each of two organisms are split apart, and the split strands from the two organisms are allowed to combine (Figure 9.14). The strands from different organisms will anneal (bond to each other) by base pairing—A with T and C with G. The amount of annealing is directly proportional to the quantity of identical base sequences in the two DNAs. A high degree of homology (similarity) exists when both organisms have long identical sequences of bases. Close DNA homology indicates that the two organisms are closely

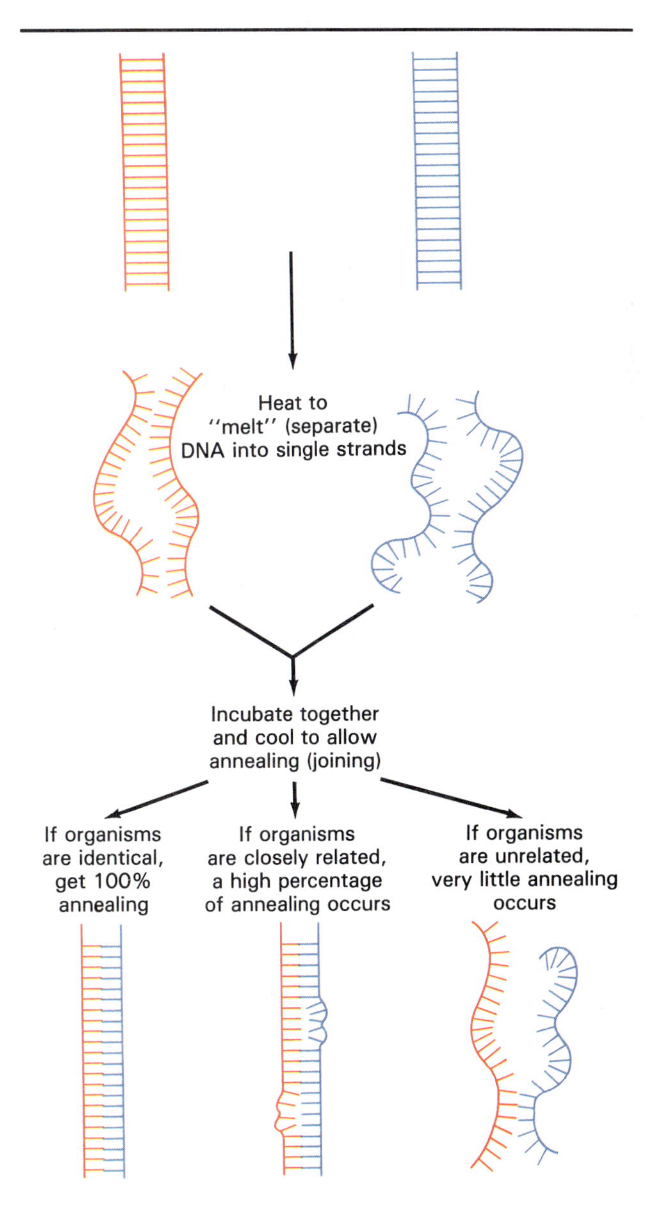

FIGURE 9.14 In DNA hybridization, strands of DNA are separated, and individual strands from two different organisms are allowed to anneal (join by hydrogen bonding at sites where there are many complementary base pairs). The degree of annealing reflects the degree of relatedness between the organisms, based on the assumption that annealing takes place only where genes, or parts of genes, are identical.

related and that they probably evolved from a common ancestor. A small degree of homology indicates the organisms are not very closely related. Ancestors of such organisms probably diverged from each other hundreds of centuries ago and since have evolved along separate lines.

Protein Profiles and Amino Acid Sequences

Every protein molecule consists of a specific sequence of amino acids and has a particular shape with an assortment of surface charges. Modern laboratory methods allow comparison of cells according to these properties of their proteins. Though variations in proteins among cells make these techniques difficult to apply to multicellular organisms, they are quite helpful in studying unicellular organisms.

A **protein profile** is a kind of picture of the proteins a cell contains (Figure 9.15). Protein profiles are obtained by the use of a technique called **polyacrylamide gel electrophoresis (PAGE),** and they are as distinctive for a cell as fingerprints are for humans.

There are four steps in making a protein profile:

1. A thin layer of polyacrylamide gel is prepared in an electrophoresis apparatus. When electric current is applied to the gel, molecules of different sizes and charges migrate at different rates through the gel.
2. Cells are broken apart and dissolved in detergent, and a drop of the resulting solution is placed on the gel.
3. The electric current is turned on until the proteins are separated into bands, as a result of their different migration rates.
4. The gel is stained to make the proteins visible.

The bands in the profile from one kind of cell represent different proteins in that cell. Bands at the same location in profiles from different kinds of cells indicate the same protein is present in the different cells.

Determination of amino acid sequences in proteins—once exceedingly difficult and now done by automated equipment—also identifies similarities and differences among organisms. Certain proteins such as cytochromes, which contribute to oxidative metabolism in many organisms, are commonly used to study amino acid sequences. The sequence of amino acids in the same kind of protein from several organisms is determined. As with DNA hybridization, the extent of matching sequences of amino acids in the proteins indicates the relatedness of the organisms.

The proteins an organism contains are determined directly by the information in that organism's DNA. Thus, both protein profiles and determinations of amino acid sequences are as significant measures of the relatedness of organisms as DNA homologies. All are also related to the evolutionary history of the organisms.

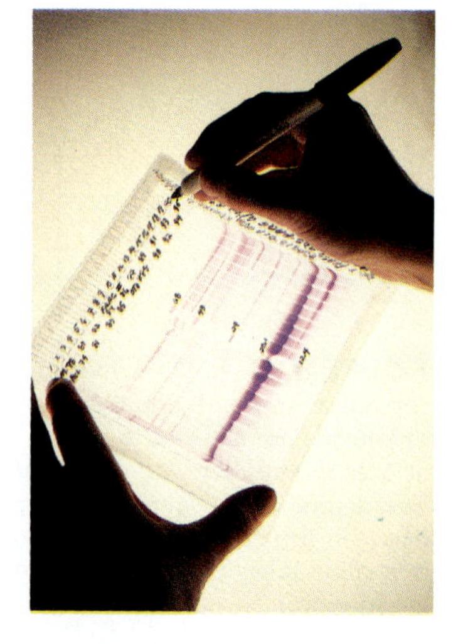

FIGURE 9.15 Protein profiles, which provide a "fingerprint" of the proteins present in particular cells, can be used to compare different organisms to determine their degree of relatedness.

Other Techniques

Other techniques to study evolutionary relatedness include determining properties of ribosomes, immunologic reactions, and phage typing.

Properties of Ribosomes

Ribosomes serve as sites of protein synthesis in both prokaryotic and eukaryotic cells. RNA in the ribosomes can be separated into several types according to the size of the RNA units. A particular RNA unit, the 16S RNA component, has proven especially useful in studying evolutionary relationships for several reasons. It is universally distributed in all bacteria, constant in function, slow to change, and easy to isolate. Moreover, RNA is a simpler molecule than DNA. Studies of the nature of 16S RNA, which constitute an important field of taxonomic research, will likely lead to better understandings of relationships between bacteria.

Immunological Reactions

Immunological reactions also are used to identify and study surface structures and the composition of microorganisms, as explained in Chapter 18. As we shall see, one highly specific and sensitive technique involves proteins called monoclonal antibodies. Monoclonal antibodies can be created so that they will bind to a certain protein, usually a protein found on a cell surface. Binding of the antibodies to the surfaces of more than one kind of organism indicates that the organisms have a protein in common. This technique promises to be particularly useful in identifying specific biochemical properties of microorganisms. Identifying such properties, in turn, will be extremely useful in determining taxonomic relationships.

Phage Typing

Phage typing involves growing an organism, such as a bacterium, known to be attacked by viruses called bacteriophages, or phages. After marking off an agar plate of the bacterial cultures, a drop of a suspension

BIOTECHNOLOGY

Identification by "Breathprints"

Rapid identification of organisms is not always easy. Now a new system is available that provides 96 different kinds of substrates, each in its own well on a plastic plate (see photo). Suspensions of an unknown bacterial culture are added to each well, and the plate is incubated for 4 to 24 hours. In each well there is also an indicator dye, tetrazolium violet. If the organism is able to use the substrate in a given well for its carbon source, respiratory gases released will turn that well violet in color. A quick trip through a plate reader, a computer comparison of the 96 test results with a library of results for known species, and in less than 10 seconds the identity of the organism is known. Think of how long it would take you to run 96 tests on your unknown specimen in the lab—especially if you had to make up all 96 media!

This machine allows rapid diagnosis in the clinical laboratory, but is also very valuable in environmental studies. Many organisms in soil, animals, and other materials are unidentified. A system like this makes it possible to rapidly identify the characteristics of new organisms and to ascertain their possible relationship to known species.

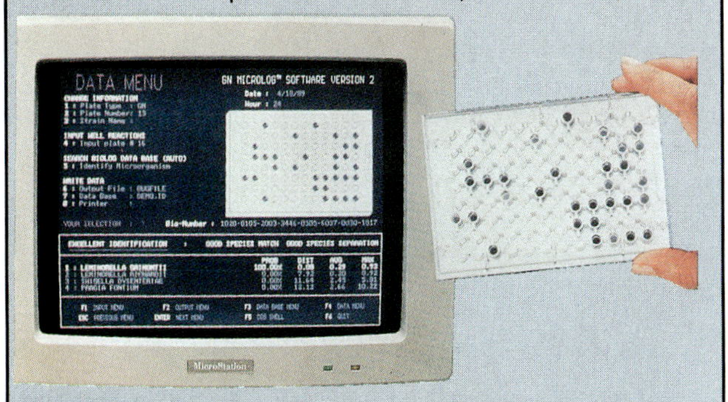

Biolog system of identifying organisms. Organisms grown in different kinds of media, each medium located in a separate well in the large plastic plate at right, are placed in a reading device to assess color changes that result from biochemical reactions. Computerized software analyzes the patterns of these reactions and identifies the organism as one of 300 clinically or environmentally important gram-negative species.

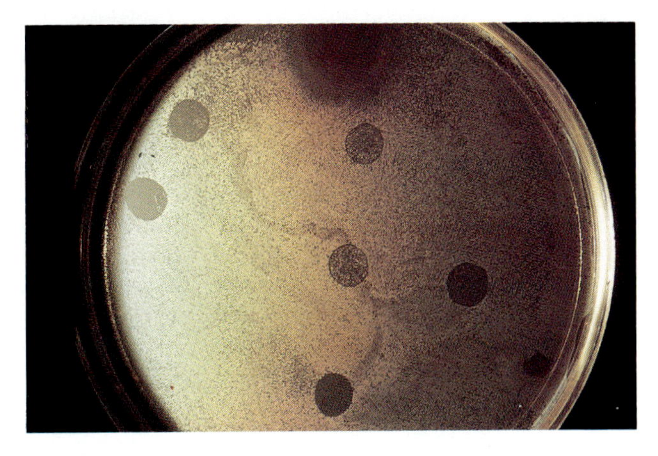

FIGURE 9.16 Phage typing. Receptor sites for bacteriophages are highly specific; certain strains of a species of bacterium are attacked only by particular types of phage. On the basis of which phages have attacked a bacterial culture, leaving clear spots (plaques) where they have killed the cells, one can determine which strain of that bacterial species is present.

of a different phage is placed in each square (Figure 9.16). By observing which phages destroy the bacteria, researchers can identify the strain. Strains lysed by the same phages are presumed to be more closely related than strains that show different patterns of lysis by phages.

Significance of Findings

The main significance of methods of determining evolutionary relationships is that they can be used to group closely related organisms and to separate them from less closely related ones. When groups of closely related organisms are identified, it is presumed that they probably had a common ancestor and that small differences among them have arisen by divergent evolution. **Divergent evolution** occurs as certain members of common-ancestor species undergo sufficient mutation to be identified as separate species.

Within the eubacteria, an early divergence gave rise to two important subgroups, the gram-positive organisms and the gram-negative ones. Subsequent divergence within each group has given rise to many modern species of bacteria. Among the gram-negative bacteria, the purple nonsulfur bacteria gave rise to modern bacteria that inhabit animal digestive tracts. One proposed scheme of divergent evolution among this group is shown in Figure 9.17.

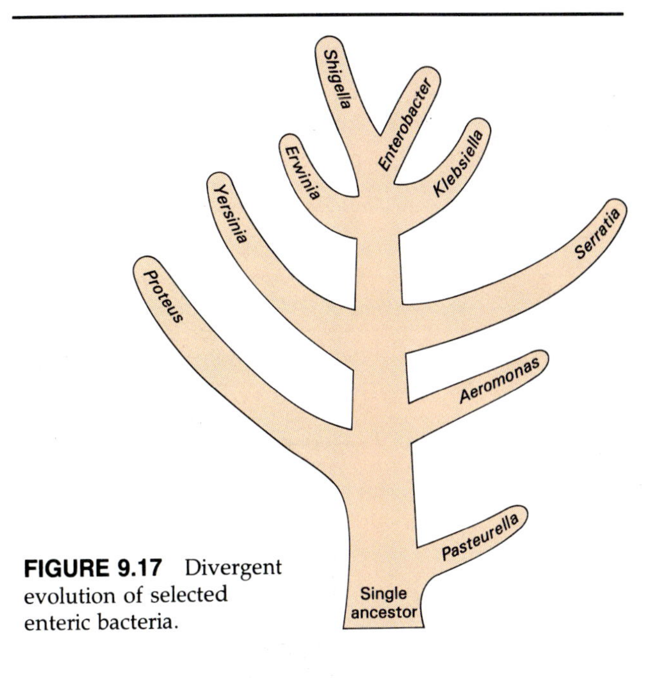

FIGURE 9.17 Divergent evolution of selected enteric bacteria.

ESSAY

Evolution from a Universal Common Ancestor

For centuries humans have pondered the questions of how life arose and how living things are related. Before microorganisms were discovered, living things were divided into plants and animals. When microscopes allowed scientists to observe microorganisms, some scientists tried to classify them as either plant or animal, and others put them in one or more separate categories. As microscopy became more powerful, fundamental differences in cells were recognized, and living organisms were classified as prokaryotes and eukaryotes on the basis of the properties of their cells. The prokaryotes included the then-known bacteria and the cyanobacteria; the eukaryotes included all organisms with nucleated cells. It was presumed that prokaryotes had a universal common ancestor from which all living things evolved and that an ancestral eukaryote arose by endosymbiosis from among various groups of prokaryotes (Figure 9.18).

Then came the discovery of archaebacteria. Studies of these organisms in the late 1970s by C. R. Woese, G. E. Fox, and others suggested that the archaebacteria represented a third cell type (Table 9.4). These investigators proposed another scheme, shown in Figure 9.19, for the evolution of living things from a universal common ancestor. They hypothesized a group of *urkaryotes*, the earliest or original cells, that gave rise to the eukaryotes directly rather than by way of the prokaryotes. They did not, however, dispense with the idea of endosymbiosis. Rather, they proposed that nucleated urkaryotes became true eukaryotes by acquiring organelles by endosymbiosis from certain eubacteria.

At about the same time that

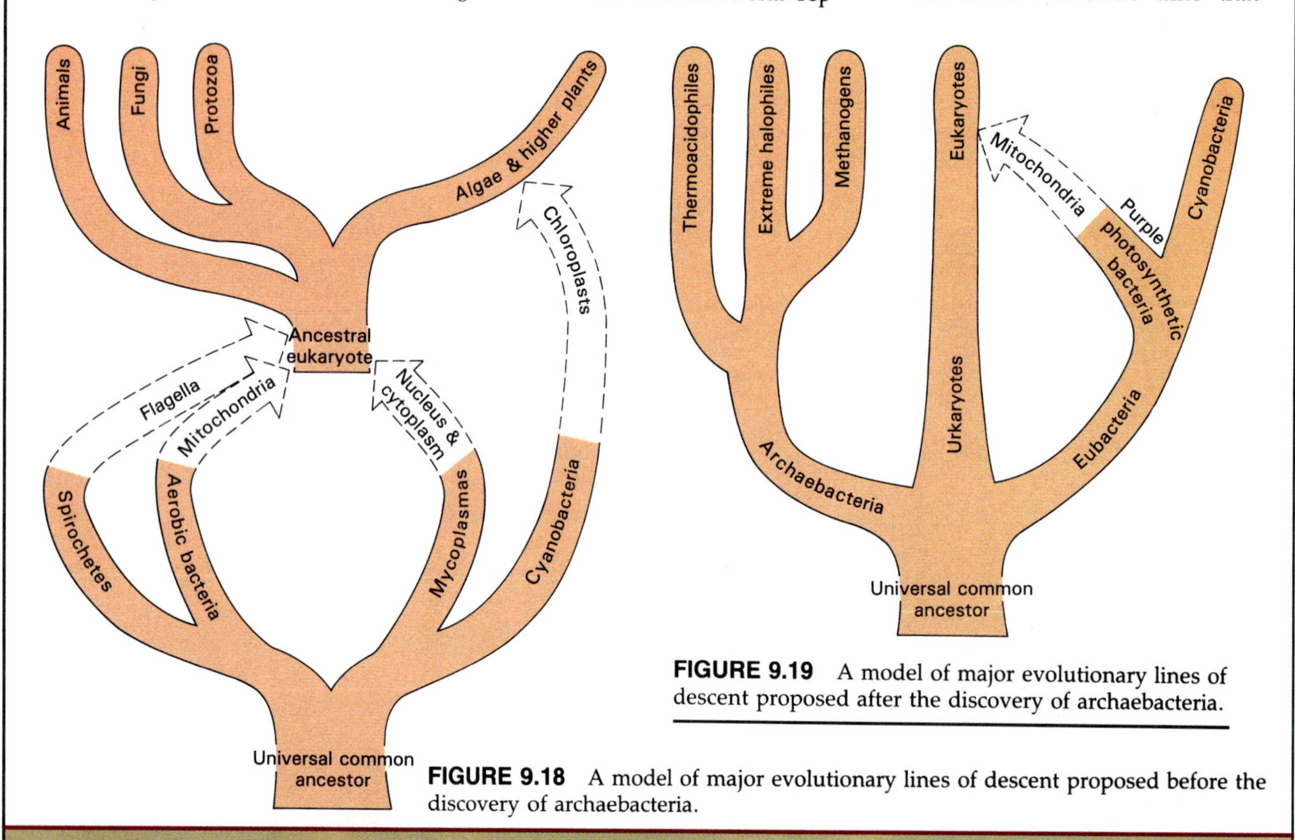

FIGURE 9.19 A model of major evolutionary lines of descent proposed after the discovery of archaebacteria.

FIGURE 9.18 A model of major evolutionary lines of descent proposed before the discovery of archaebacteria.

TABLE 9.4 Comparison of archaebacteria, eubacteria, and eukaryotes

Characteristic	Archaebacteria	Eubacteria	Eukaryotes
Cell wall	Lack peptidoglycan	Contain peptidoglycan	Absent or made of other materials
Lipids of cell membrane	Branched chain fatty acids	Straight chain fatty acids	Straight chain fatty acids and sterols
Protein synthesis	Not impaired by antibiotics such as chloramphenicol	Impaired by antibiotics such as chloramphenicol	Most not impaired by antibiotics
First amino acid in a protein	Methionine	Formylmethionine	Methionine
Habitat	Usually found only in extreme environments	Found in a wide range of environments	Found in a wide range of environments

archaebacteria were first being investigated, studies of stromatolites indicated that life had arisen nearly 4 billion years ago. These studies indicated that an age of microorganisms, in which there were no multicellular living organisms, lasted for about 3 billion years. Combined evidence from studies of archaebacteria and the most ancient stromatolites convinced many scientists that three branches of the tree of life formed during the age of microorganisms and that each branch gave rise to distinctly different groups of organisms.

The three-branch tree was not accepted by all scientists. In 1977 T. Cavalier-Smith of King's College, London, proposed instead that the archaebacteria arose later than the eubacteria by divergent evolution from a group of gram-positive bacteria similar to present-day actinomycetes, which were once thought to be fungi (Figure 9.20a). J. A. Lake of the University of California at Los Angeles proposed in 1988 still another model with two main branches (Figure 9.20b). One branch gave rise to the eubacteria and two groups of archaebacteria, those that live in extremely salty environments and those that release methane. The other branch gave rise to the eukaryotes and a group of archaebacteria he calls the eocytes, which grow in hot, acidic environments.

Which, if any, of these models is correct remains to be seen. Further analysis of nucleotide and amino acid sequences promises to provide much better information about relationships among organisms than has ever before been available. It also may help to clarify the nature and time of origin of the universal common ancestor.

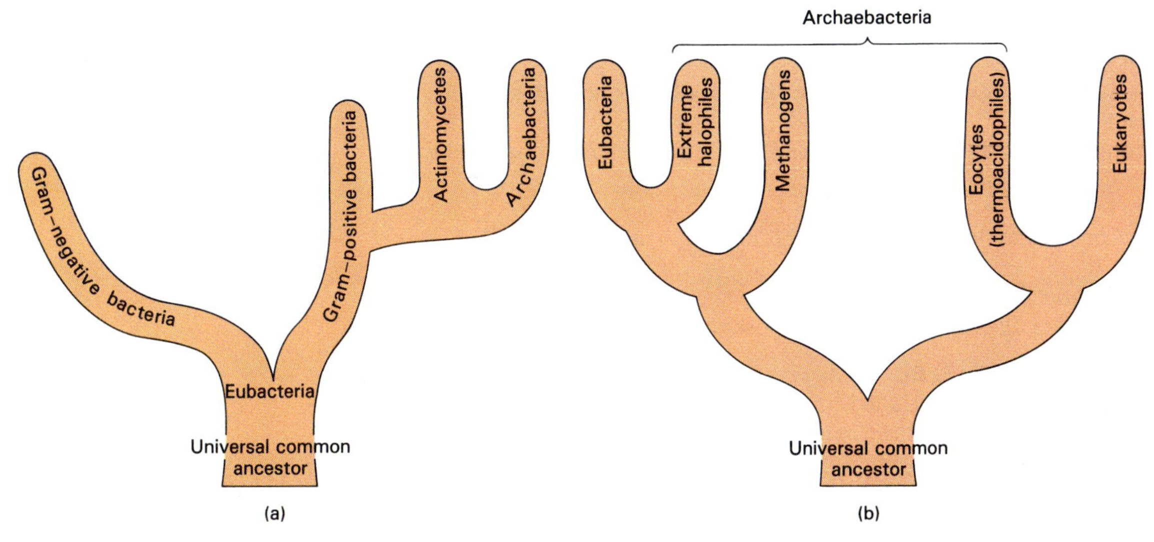

FIGURE 9.20 Other models of evolution, (a) as proposed by Cavalier-Smith, and (b) as proposed by Lake.

CHAPTER SUMMARY

TAXONOMY—THE SCIENCE OF CLASSIFICATION

RELATED KEY TERMS

- Organisms are named according to their characteristics, where they are found, who discovered them, or what disease they cause. — **taxon**

Linnaeus, the Father of Taxonomy

- Linnaeus invented **binomial nomenclature,** which specifically identifies each living organism.
- The **genus** and **specific epithet** of each organism identify the **species** to which it belongs. — **strain**
- Linnaeus also established the hierarchy of **taxonomy** and classified organisms in two kingdoms, Plantae and Animalia.

Using a Taxonomic Key

- A **dichotomous** taxonomic **key** consists of a series of paired statements presented as either-or choices that describe characteristics of organisms. By selecting appropriate statements to progress through the key one can classify organisms and, if the key is sufficiently detailed, identify them by genus and species.

Problems in Taxonomy

- Problems in taxonomy include the rapid pace of evolutionary change in microorganisms and the difficulty in deciding what constitutes a kingdom and what constitutes a species. — **phylogenetic**

Developments since Linnaeus

RELATED KEY TERMS

- Since Linnaeus' time several taxonomists have proposed three- and four-kingdom systems based on various fundamental characteristics of living things. Whittaker proposed a five-kingdom system in 1969.
- Since 1925 *Bergey's Manual of Determinative Biology* has served as an important tool in identifying bacteria.

THE FIVE-KINGDOM CLASSIFICATION SYSTEM

- The kingdoms of the **five-kingdom system** are **Monera** (Prokaryotae), **Protista, Fungi, Plantae,** and **Animalia.**
- The characteristics of members of each kingdom are summarized in Table 9.3.

archaebacteria **eubacteria**

Kingdom Monera

- All monerans are unicellular **prokaryotes;** they generally lack organelles, have no true nuclei, and their DNA has little or no protein associated with it. The **cyanobacteria** are photosynthetic organisms of great ecological importance.

Kingdom Protista

- The protists are a diverse group of mostly unicellular **eukaryotes.**

Protista

Kingdom Fungi

- The fungi include some unicellular and many multicellular organisms that obtain nutrients solely by absorption.

Kingdom Plantae

- Most plants live on land and contain chlorophyll in organelles called chloroplasts.

Kingdom Animalia

- All animals are derived from **zygotes;** most are macroscopic.

CLASSIFICATION OF VIRUSES

- **Viruses,** which are not included in any of the five kingdoms, are classified by their nucleic acids and their morphology.

THE SEARCH FOR EVOLUTIONARY RELATIONSHIPS

Special Methods Needed for Prokaryotes

- Special methods are needed for determining evolutionary relationships among prokaryotes because they have few morphological characteristics and have left only a sparse fossil record.
- Several methods are currently used to determine evolutionary relationships among organisms.

stromatolites

Numerical Taxonomy

- In **numerical taxonomy,** organisms are compared on a large number of characteristics and grouped according to the percent of shared characteristics.

Genetic Homology

- In base comparison, the base composition of DNA is determined, and C-G percentages are compared among organisms.
- In **DNA hybridization,** the degree of matching between strands of DNA is compared among organisms.
- In **protein profiles** and studies of amino acid sequences, similarities among organisms provide a measure of their relatedness.

genetic homology
probe
polyacrylamide gel electrophoresis (PAGE)

Other Techniques

- Other methods make use of properties of ribosomes, immunological reactions, and phage typing.

Significance of Findings

- Evolutionary relationships can be used to group closely related organisms. Small differences among organisms descended from a common ancestor arise by **divergent evolution.** An early divergence gave rise to the two major subgroups of eubacteria, the gram-positive organisms and the gram-negative ones.

QUESTIONS FOR REVIEW

A.

1. How are organisms named?

B.

2. Define taxonomy, and describe Linnaeus' contributions to it.

C.

3. What is a dichotomous key, and how is it used?

D.

4. What are some of the problems associated with developing a good taxonomic system?
5. How have taxonomists since Linnaeus attempted to solve those problems?

E.

6. What are the main characteristics of organisms in each of the kingdoms in the five-kingdom system?

F.

7. How are viruses classified?

G.

8. Why are special methods needed to determine evolutionary relationships among prokaryotes?
9. How do the following methods help to determine evolutionary relationships: numerical taxonomy, genetic homology, protein profiles, and amino acid analysis?

PROBLEMS FOR INVESTIGATION

1. What is a species? How is this defined in organisms that do not reproduce sexually?
2. A series of DNA hybridization experiments were performed in which the DNA of two given organisms were separated into single strands. Then the two organisms' single-stranded DNA was incubated together, and the percentages that hybridized (combined with that of the other species) were determined. From the data given below, which two species are probably most closely related?

Species	Percentage Hybridization
A and B	46
A and C	58
B and C	75

3. When DNA strands are "melted" (separated into single strands by heating), more heat is required to break a guanine-cytosine base pairing that is held together by three hydrogen bonds than is required to break an adenine-thymine base pairing that is held together by only two hydrogen bonds. Four species of organisms' DNA was "melted," and their melting temperatures were recorded.

Species	Melting Temperature (°C)
A	90.2
B	86.3
C	87.1
D	94.7

Which species has the highest G-C content? Which has the lowest? Which two species are the likeliest to be closely related? Why?

4. Cite examples of how the concepts of unity and diversity among living organisms are handled in taxonomic schemes.
5. Select five members of your class to come to the front of the room. Make a dichotomous key to identify them. Which characteristics are good or bad to use in your key? Would height, weight, or hair color be good if you tried to use this key again at your 25th class reunion?

SOME INTERESTING READING

Anonymous. 1991. Germs in space? *Sky and Telescope* 81(April):357.

Bergey's manual of systematic bacteriology. Vol. 1, 1984, R. Krieg and J. G. Holt, eds. Vol. 2, 1986, P. Sneath, ed. Vol. 3, 1989, J. Staley, ed. Vol. 4, 1989, S. Williams, ed. Baltimore: Williams and Wilkins.

Cavalier-Smith, T. 1987. The origin of eukaryotic and archaebacterial cells. *Annals of the New York Academy of Sciences* 503:17.

Doolittle, W. F. 1987. The evolutionary significance of the archaebacteria. Annals of the New York Academy of Sciences 503:72.

Margulis, L. and R. Guerrero. 1991. Kingdoms in turmoil. *New Scientist* 129, no. 1761 (March 23):46–51.

Margulis, L. and K. V. Schwartz. 1988. *Five kingdoms: an illustrated guide to the phyla of life on earth*. New York: Freeman and Co.

Whittaker, R. H. 1969. New concepts of kingdoms of organisms. *Science* 163:150.

Woese, C. R. 1981. Archaebacteria. *Scientific American* 244, no. 6 (June):98.

Zillig, W. 1987. Eukaryotic traits in archaebacteria. *Annals of the New York Academy of Sciences* 503:7.

Scanning electron micrograph (SEM) of a *Methanosarcina mazei* colony.

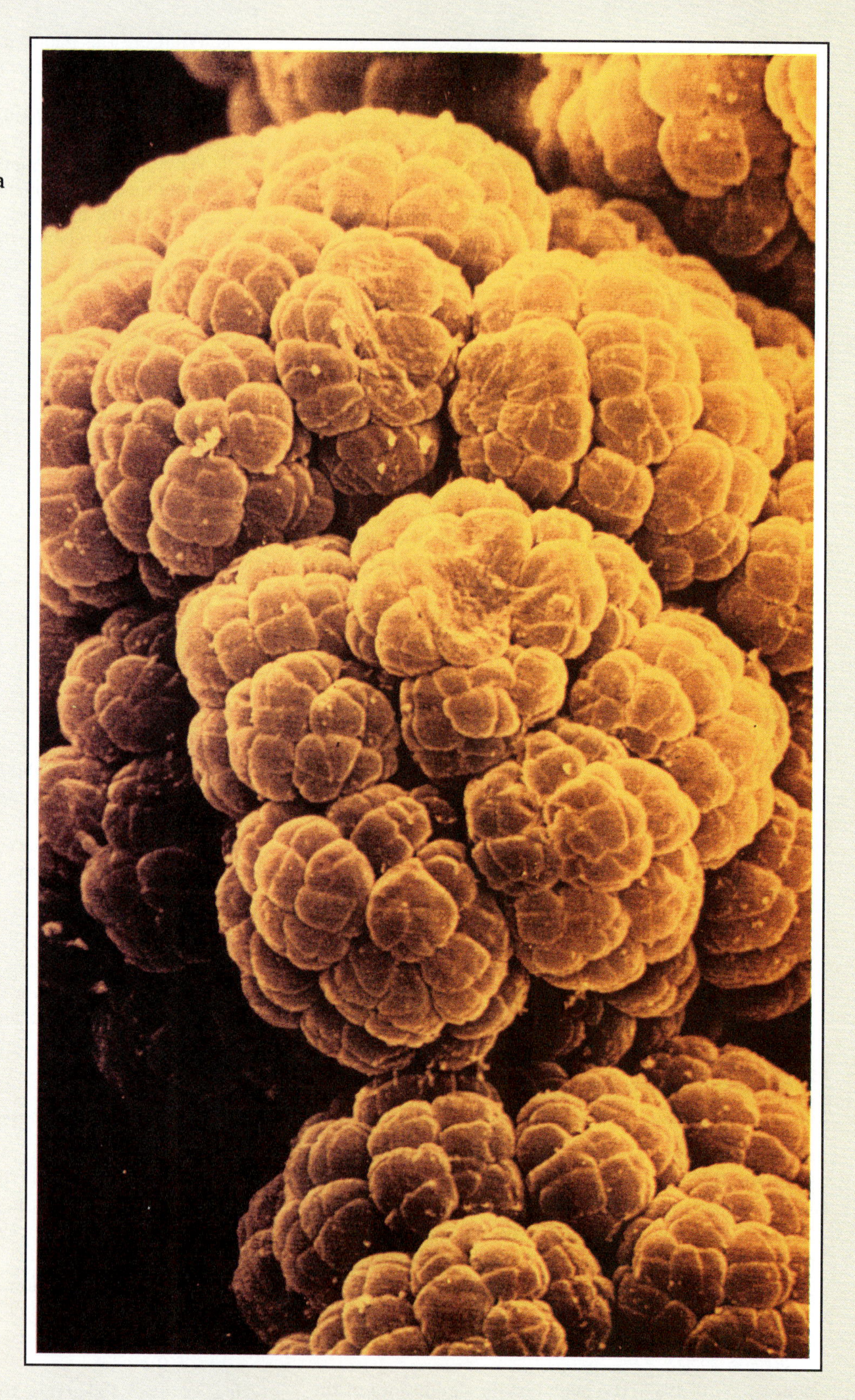

10 The Bacteria

This chapter focuses on the following questions:

A. What criteria are used for classifying bacteria?

B. What problems are associated with bacterial taxonomy?

C. What are the history and significance of *Bergey's Manual*?

D. What are the characteristics of those bacterial genera that have medical significance, and what diseases do they cause?

In Chapter 9 we considered general aspects of classification and the characteristics of members of each kingdom according to the five-kingdom system. Here we will see how bacteria are classified and why there are problems in classifying bacteria.

BACTERIAL TAXONOMY AND NOMENCLATURE

Criteria for Classifying Bacteria

Most macroscopic organisms can be classified according to observable structural characteristics. Classifying microscopic organisms, especially bacteria, is more difficult because many of them have similar structures. Separating them according to cell shape, size, and arrangement does not produce a very useful classification system. Finding flagella, endospores, or capsules still does not allow identification of particular species. Therefore, other criteria must be used. Staining reactions, especially to the Gram stain, were among the first criteria other than morphology to be used in classification. Other criteria now in use include properties of growth, nutrition, physiology, biochemistry, and genetics. These criteria include properties of DNA and proteins, as explained in Chapter 9. Important criteria used in classifying bacteria are summarized in Table 10.1, and biochemical tests used in classifying and identifying them are described in Table 10.2.

By using various classification criteria, an organism can be identified as belonging to a particular genus and species. For bacteria a species is regarded as a collection of strains that share many common features and differ significantly from other strains. A bacterial *strain* consists of descendants of a single isolation in pure culture. Bacteriologists designate one strain of a species as the **type strain.** Usually this is the first one described. It is the name-bearer of the species and is preserved in one or more type culture collections.

TABLE 10.1 Criteria for classifying bacteria

Morphology	Size and shape of cells, arrangement in pairs, clusters or filaments, presence of flagella, pili, endospores, capsules
Staining	Gram-positive, gram-negative, acid-fast
Growth	Characteristics in liquid and solid cultures, colony morphology, development of pigment
Nutrition	Autotrophic, heterotrophic, fermentative with different products, energy sources, carbon sources, nitrogen sources, needs for special nutrients
Physiology	Temperature (optimum and range), pH (optimum and range), oxygen requirements, salt requirements, osmotic tolerance, antibiotic sensitivities and resistances
Biochemistry	Nature of cellular components such as cell wall, RNA molecules, ribosomes, storage inclusions, pigments, antigens
Genetics	Percentages of DNA bases, DNA hybridization

For many strains of bacteria, scientists find it relatively easy to determine that they are members of a particular species. For other strains, however, subjective judgments must be made to decide whether the strain belongs to an existing species or differs sufficiently to be defined as a separate species. In recent years similarities of DNA and proteins among organisms have proven a reliable means of assigning a strain to an existing species or establishing a new species for it.

Curiously, assigning bacterial genera to higher taxonomic levels (families, orders, classes, and divisions) is even more difficult than organizing species and strains within genera. Many macroscopic organisms are classified by establishing their evolutionary relationships to other organisms from fossil records. Efforts are being made to classify bacteria by evolutionary relationships, too, but these efforts are hampered by the absence of a complete fossil record and the limited information gleaned from what fossils there are. Even a complete fossil record would supply only morphological information and would thus be inadequate for determining evolutionary relationships.

Problems of Bacterial Taxonomy

In spite of a tremendous amount of effort spent in classifying bacteria, no complete classification of bacteria from kingdom to species has been established. The plight of bacteriologists looking at the taxonomy of bacteria might be described as follows: Those looking from the top down can propose at least plausible divisions of the kingdom Monera, or kingdom Prokaryotae, as it is commonly designated by bacteriologists. The Prokaryotae include eubacteria, cyanobacteria, and archaebacteria. Those looking from the bottom up can establish strains, species, and genera, and sometimes assign them to higher-order groups. But too little is known about evolutionary relationships to establish clearly defined taxonomic classes and orders.

Taxonomy from the Top Down

Taxonomists generally agree that true bacteria belong in the kingdom Monera, or Prokaryotae. Like other monerans they lack nuclear membranes, and their nucleoproteins lack basic proteins called histones. Basic proteins have a slightly alkaline pH and are associated with DNA in eukaryotic organisms. Taxonomists do not agree, however, on how to separate the true bacteria into divisions or phyla. In 1968 bacteriologist R. G. E. Murray of the University of Western

TABLE 10.2 Specific biochemical tests used in identifying bacteria

Biochemical Test	Nature of Test
Sugar fermentation	Organism is inoculated into a medium containing a specific sugar; growth, end products of fermentation, including gases are noted. Anaerobic fermentations can be detected by inoculating organisms via a "stab" culture into solid medium.
Gelatin liquefaction	Organism is inoculated (stabbed) into a solid medium containing gelatin; liquefaction at room temperature or inability to resolidify at refrigerator temperature indicates the presence of proteolytic enzymes.
Starch hydrolysis	Organism is inoculated onto an agar medium containing starch; clear areas around colonies indicate the presence of starch-digesting enzymes.
Milk digestion	Organism is inoculated into litmus milk medium (10 percent powdered skim milk plus litmus indicator); characteristic changes such as alteration of pH to acid or alkaline, denaturation of the protein casein (curdling), and gas production can be used to help identify specific organisms.
Catalase test	Hydrogen peroxide (H_2O_2) is poured over heavy growth of an organism on an agar slant; release of gas bubbles indicates the presence of catalase, which oxidizes H_2O_2 to H_2O and O_2.
Citrate utilization	Organism is inoculated into citrate agar medium in which citrate is the sole carbon source; an indicator in the medium changes color if citrate is metabolized; use of citrate indicates the presence of the permease that transports citrate into the cell.
Hydrogen sulfide test	Organism is inoculated into peptone iron medium; formation of black sulfides indicates the organism produces hydrogen sulfide (H_2S).
Indole production	Organism is inoculated into a medium containing the amino acid tryptophan; production of indole, a nitrogenous breakdown product of tryptophan, indicates the presence of a set of enzymes that convert tryptophan to indole.
Nitrite test	Organism is inoculated into a medium containing nitrate (NO_3); presence of nitrite (NO_2) indicates the organism has the enzyme nitrase; absence of nitrite indicates either absence of nitrase or presence of nitrite reductase (which reduces nitrite to N_2 or NH_3).
Vogues-Proskauer test	Organism is cultured in MR-VP broth; KOH-creatine and alpha naphthol are added; presence of the enzyme cytochrome oxidase causes color change in an indicator (rose color).
Phenylalanine deaminase test	Organism is inoculated into a medium containing phenylalanine and ferric ions; formation of phenylpyruvate and its reaction with ferric ions produces a color change that demonstrates the presence of the enzyme phenylalanine deaminase.
Urease test	Organism is inoculated into a medium containing urea; production of ammonia, usually detected by an indicator for alkaline pH, indicates the presence of the enzyme urease.
Specific nutrient tests	Organism is inoculated into a medium containing a specific nutrient, such as a particular amino acid (e.g., cysteine) or vitamin (e.g., niacin); growth of an organism that fails to grow in media lacking the specific nutrient can be used to identify some organisms.

Ontario proposed four divisions—gram-negative, gram-positive, gram-variable, and lacking a cell wall. In 1982 Lynn Margulis grouped bacteria as fermenting heterotrophs, respiring heterotrophs, and autotrophs.

Bergey's Manual currently divides the Kingdom Procaryotae into the following four divisions (phyla):

Division I.	Gracilicutes	Prokaryotes with thinner cell walls, implying a gram-negative type of cell wall
Division II.	Firmicutes	Prokaryotes with thick and strong skin, indicative of a gram-positive type of cell wall
Division III.	Tenericutes	Prokaryotes of a pliable, soft nature, indicative of the lack of a rigid cell wall
Division IV.	Mendosicutes	Prokaryotes having faulty cell walls, suggesting the lack of conventional peptidoglycan

Taxonomy from the Bottom Up

Bacteria have been classified into 33 groups, called sections, in *Bergey's Manual,* a widely accepted reference on the identification of bacteria. The sections are based on relatively easily observable characteristics such as shape, staining reactions, presence or absence of a cell wall, motility, budding, and mode of metabolism. Most sections were established many years ago as a practical means of classifying bacteria. The criteria were established before modern techniques for DNA analysis and biochemical studies were available and before any fossil bacteria had been found.

As more and more bacteria are subjected to modern analysis, certain discrepancies in assignments to sections have become apparent. For example, the cell walls of some genera have unusual properties that prevent their being classified as either gram-positive or gram-negative. The organisms assigned to a particular section of *Bergey's Manual* are not necessarily closely related.

CLOSE-UP

Who Cares?

"It also seems worthwhile to emphasize . . . [that] bacterial classifications are devised for microbiologists, not for the entities being classified. Bacteria show little interest in the matter of their classification. For the systematist [taxonomist], this is sometimes a very sobering thought!"

—J. T. Staley and N. R. Krieg (1985)

The Muddle in the Middle

The difficulties of classifying bacteria are greatly magnified as one proceeds up the taxonomic hierarchy. Although families, orders, and classes can be clearly defined for a few bacteria, too little information is available to do this for many. Until discrepancies in section assignments can be resolved and taxonomic levels between genera and divisions can be more precisely determined, it is practical to continue to use section designations. A complete listing of the sections is provided in Appendix B.

Bacterial Nomenclature

In spite of all these taxonomic problems, there is an established nomenclature for bacteria. Bacterial nomenclature refers to the naming of species according to internationally agreed-upon rules. Both taxonomy and nomenclature are subject to change as new information is obtained. Organisms are sometimes moved from one category to another, and their official names sometimes are changed. For example, the bacterium that causes tularemia, a fever acquired by handling infected rabbits, was, for many years, called *Pasteurella tularensis*. Its genus name has been changed to *Francisella*.

CLASSIFICATION USING *BERGEY'S MANUAL*

History and Significance of *Bergey's Manual*

The first edition of *Bergey's Manual of Determinative Bacteriology* was published in 1923 by the American Society for Microbiology with David H. Bergey (Figure 10.1) as chairperson of its editorial board. Over the years eight editions, an abridged version, and several supplements have been published. *Bergey's Manual* has become an internationally recognized reference for bacterial taxonomy. It has also served as a reliable standby for medical workers interested in identifying causative agents of infections.

Bergey's Manual contains names and descriptions of organisms and diagnostic keys and tables for identifying organisms. It is organized into sections, with each section devoted to a group of similar organisms. The term "section" is used rather than established taxonomic terms such as "family" and "order" because evolutionary relationships between sections (and even within sections) are not yet clearly understood.

A new manual, *Bergey's Manual of Systematic Bacteriology,* has a much broader scope than previous manuals. All the kinds of information found in earlier determinative manuals are included, and information on taxonomy, ecology, cultivation, maintenance, and preservation of organisms has been expanded. This manual is a four-volume work. The first volume was published in 1984, the second in 1986, and the final two in 1989. With all four volumes now published, determinative information (information used to identify bacteria) will be collected into a single volume. It will appear as the ninth edition of *Bergey's Manual of Determinative Bacteriology.*

Bergey's Manual is widely accepted as a reference for identifying bacteria. The four-volume work will retain that significance; it will also be of great significance to those interested in bacterial taxonomy. An important motivation for publishing the larger work is to stimulate research to improve the understanding of taxonomic relationships among bacteria.

The beginning student will doubtless find it difficult to remember many characteristics of specific microorganisms that we cover in this course. A four-volume set of *Bergey's Manual* weighs 21 pounds and costs almost $300—not something you could carry to class and back. However, we have used the endpapers, inside the front and back covers of this textbook to help you. If you wish to find out whether a given organism is gram -positive or -negative, its shape, the disease(s) it causes, and so on, look it up by name in the back endpapers. Organisms are grouped as bacteria, viruses, fungi, and parasites. Or, if you are discussing a disease, but can't remember which organism(s) cause it, look in the front endpapers under the name of the disease (again grouped as bacterial, viral, fungal, and parasitic), and you will find the organism's name and some of its characteristics. Page numbers are given to direct you to further information. Thumbing through the book or index can be frustrating when you need some little piece of information and may keep you from learning some facts. But flipping to the cover of your book is easy, and we encourage you to do so often, until the information becomes gradually more and more familiar.

Bergey's Manual of Systematic Bacteriology provides many kinds of information (Figure 10.2), such as de-

FIGURE 10.1 David H. Bergey, originator of the series of *Bergey's Manuals,* the first of which was published in 1923. In 1936 he set up an educational trust to which all rights and royalties from the *Manuals* would be transferred for the purposes of preparing, editing, and publishing future editions, as well as providing funds for research to clarify problems arising in the process. This nonprofit trust ensures that *Bergey's Manual* will be a self-perpetuating publication.

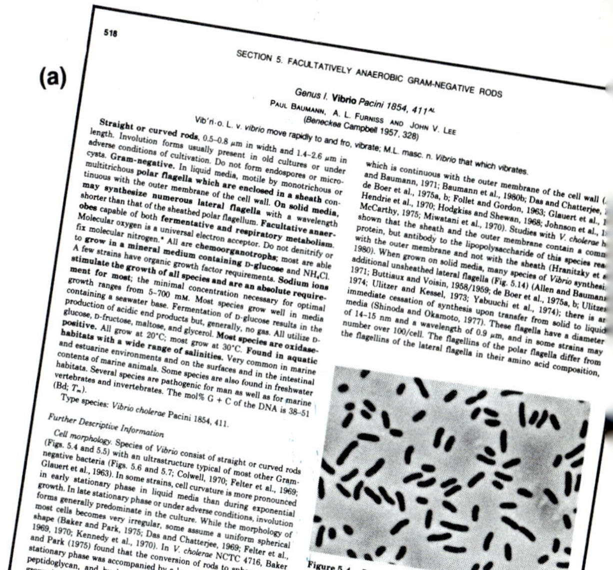

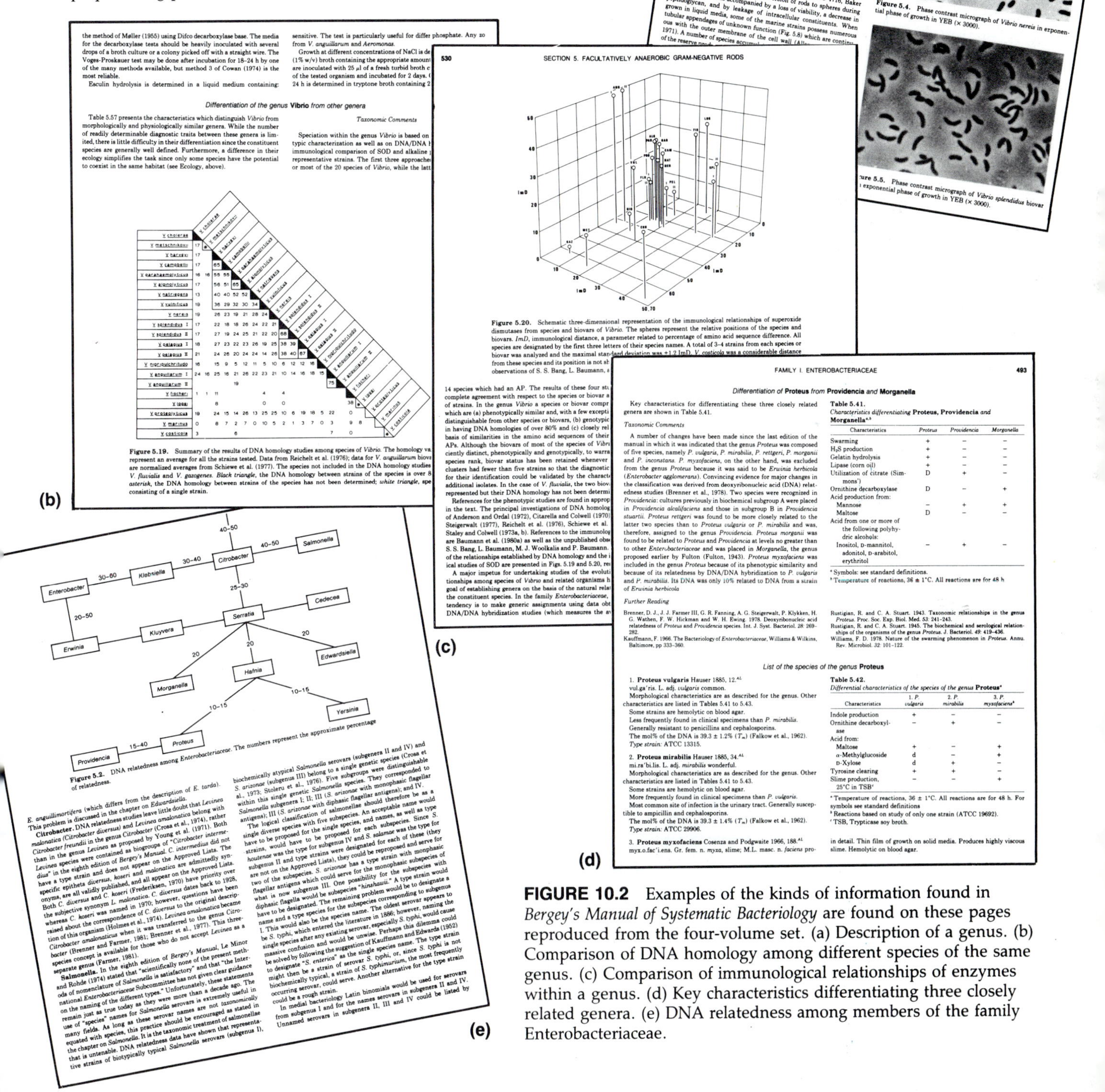

FIGURE 10.2 Examples of the kinds of information found in *Bergey's Manual of Systematic Bacteriology* are found on these pages reproduced from the four-volume set. (a) Description of a genus. (b) Comparison of DNA homology among different species of the same genus. (c) Comparison of immunological relationships of enzymes within a genus. (d) Key characteristics differentiating three closely related genera. (e) DNA relatedness among members of the family Enterobacteriaceae.

scriptions and photos of species, tests to distinguish between genera and species, DNA-relatedness among organisms, and various numerical taxonomy studies. (Chapter 9, p. 233)

Bacteria by Section of *Bergey's Manual*

Of the 33 sections of prokaryotes described in *Bergey's Manual*, some contain only a few organisms, whereas others contain hundreds. Certain sections contain organisms of medical significance, whereas others are of interest mainly to ecologists, taxonomists, or researchers. In the remainder of this chapter we consider characteristics of organisms in the major sections of *Bergey's Manual*, emphasizing those with medical significance. When relevant, order and family names will be noted. Such names have consistent endings: Orders always end in *ales* and families in *aceae*.

Spirochetes (Section 1)

The spirochetes (Figure 10.3) are helically shaped, motile bacteria. They have a multilayered membranous **outer sheath** that surrounds the coiled **protoplasmic cylinder.** Axial filaments (Chapter 4) are structurally similar to other bacterial flagella, but they are enclosed by the outer sheath. Each axial filament is attached at one pole and permanently wound around the cylinder. Filaments vary in number from 2 in some species to as many as 100 in others. In fluid environments spirochetes display three kinds of movement: locomotion, rotation on their longitudinal axis, and flexing. Axial filaments enable spirochetes to move in relatively viscous (syrupy) fluids that prevent movement of bacteria with exposed flagella.

All spirochetes divide by binary fission, but division separates an elongated cell into two shorter ones. None form endospores. Among the spirochetes are aerobic, facultatively anaerobic, and anaerobic organisms. Free-living spirochetes inhabit a variety of aqueous environments, including sewage and muds containing hydrogen sulfide. Pathogenic spirochetes live mainly in body fluids. They belong to the order Spirochaetales.

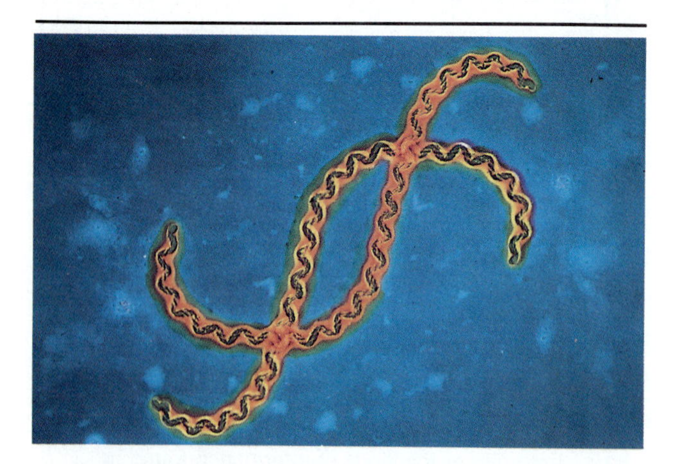

FIGURE 10.3 False-color TEM of *Leptospira interrogans*. Note the tightly coiled shape and axial filaments, both characteristic of spirochetes. This organism causes leptospirosis in humans.

Among the spirochetes are three genera that contain pathogens. The pathogenic **treponemes** have not yet been grown on laboratory media. The type species of the treponemes is *Treponema pallidum*, the causative agent of the venereal disease syphilis. (Other strains of this species cause nonvenereal infections, such as bejel and yaws.) Treponemes such as *T. denticola* and *T. vincentii* are often found in the mouth. Though they usually are not pathogenic, under certain conditions they contribute to periodontal (gum) disease. Several other spirochetes can cause human disease. *Borrelia recurrentis*, which is carried by ticks, causes relapsing fever in which the patient suffers repeated bouts of fever and chills. *B. burgdorferi*, also tick-borne, is responsible for Lyme disease. Several species of *Leptospira* can cause leptospirosis, a disease characterized by fever and liver and kidney damage in humans who ingest contaminated food or water. In laboratory cultures leptospiras oxidize long-chain fatty acids but borrelias do not.

Aerobic/Microaerophilic, Motile, Helical/Vibrioid, Gram-Negative Bacteria (Section 2)

As the title indicates, bacteria of this group are helical, or vibrioid (comma-shaped); as such they are similar to the spirochetes. The helix can have half a turn (comma) or many turns. All these bacteria have flagella and swim in a straight line with a corkscrew motion. Many are small spiral organisms found in soil and in fresh and stagnant water where oxygen to satisfy aerobes or microaerophiles is available; some grow in plant roots. All are gram-negative. Genera representative of this group are *Spirillum*, *Aquaspirillum*, and *Azospirillum*.

Various species of *Campylobacter*, which are slender spiral rods, inhabit the reproductive and intestinal tracts and the mouth of humans. *C. jejuni* (Figure 10.4) causes enteritis (an inflammation of the digestive tract) and bacteremia (bacteria in the blood) in humans.

Gram-Negative Aerobic Rods and Cocci (Section 4)

Among the gram-negative aerobic rods and cocci are a large number of bacteria—eight families and some genera that have not been assigned to families. Many of the genera include significant human pathogens.

The **pseudomonads** (Figure 10.5) are aerobic motile rods with polar flagella. Many species synthesize a soluble yellow-green pigment that fluoresces under ultraviolet light. Most contain an oxidase enzyme and

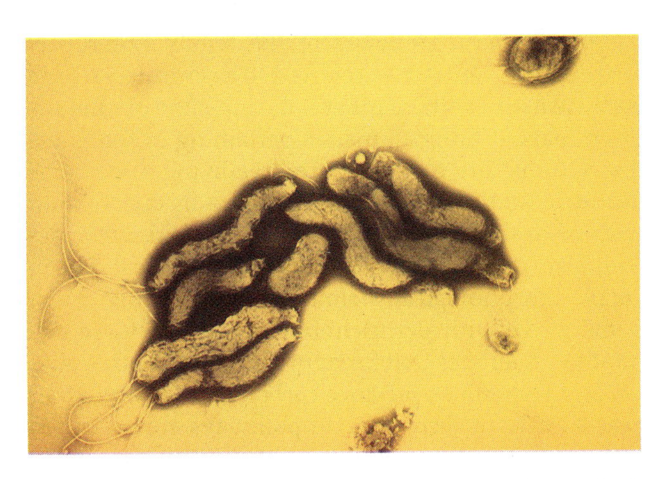

FIGURE 10.4 *Campylobacter jejuni.* False-color TEM (5130X) shows spiral, gram-negative organisms with flagella at one or both ends. *C. jejuni* is part of the normal intestinal flora of birds and other animals. It is a common cause of acute gastroenteritis in humans, causing fever, abdominal pain, and diarrhea.

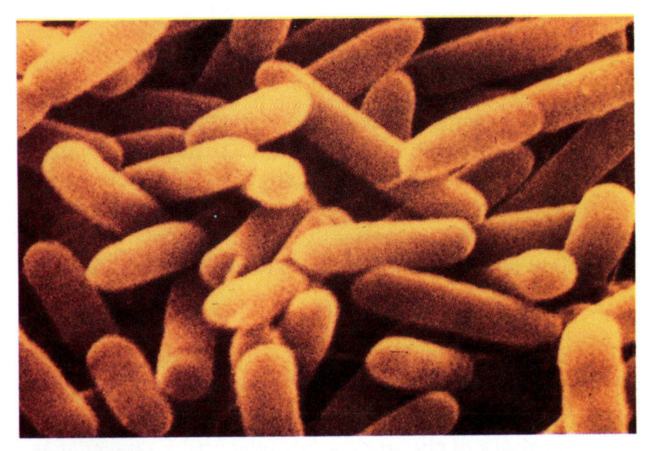

FIGURE 10.5 Pseudomonads, aerobic motile rods, are widely distributed in nature. They are very hardy organisms, and some can even grow in surgical scrub solutions, where they can cause hospital-acquired infections.

give positive oxidase test results (Chapter 6). Free-living forms are found in soils and in fresh water and marine environments, where they are important decomposers of organic material. *Pseudomonas aeruginosa* is the main human pathogen; it is often seen in urinary tract infections and in wounds and burns.

Among the pseudomonads are a variety of animal and plant pathogens. One such animal pathogen causes glanders, a debilitating disease with ulceration of lymph nodes in horses. Another causes melioidosis, a disease similar to pneumonia, in both animals and humans in Southeast Asia. The plant pathogens cause rots, scabs, and wilts on a wide variety of plants, including some common house plants. Pseudomonads that infect plants require quite different growth conditions than those that infect humans, so you needn't worry about such an infection while caring for a sick plant.

Identifying and naming *Legionella* (le″jen-el′la) *pneumophila* as the causative agent of Legionnaire's disease, which affected men staying in one hotel while attending an American Legion convention in Philadelphia in the summer of 1976, required creating a new family. *L. micdadei*, the causative agent of Pittsburgh pneumonia, and several other species have since been added to that family. These organisms had not been previously identified because fastidious nutritional requirements make them difficult to culture and isolate. After **legionellas** were identified in the laboratory, they were found in nature in water and soil, usually associated with other microorganisms such as cyanobacteria, algae, protozoa, and other bacteria.

Members of the Family Neisseriaceae constitute another large group of medically important aerobic gram-negative rods and cocci. Most are found on mucous membranes of humans and animals. *Neisseria gonorrhoeae* thrives in the mucous membranes of the human urogenital tract and causes gonorrhea. *N. meningitidis* first colonizes nasopharyngeal mucous membranes, but it can invade blood and cerebrospinal fluid, where it causes meningitis, an inflammation of the membranes that cover the brain. Several species of the genus *Moraxella* can cause conjunctivitis, an inflammation of membranes of the eyes.

Several genera among the gram-negative, aerobic rods and cocci have been identified but not assigned to a family. Most members of the three genera *Brucella*, *Bordetella*, and *Francisella* are pathogens. All *Brucella* are obligate parasites that usually multiply in phagocytic white blood cells. In humans they cause brucellosis, or undulant fever, in which the patient has daily episodes of fever and chills. They cause similar diseases in domestic animals. *Bordetella* species cause whooping cough. One species, *Francisella tularensis*, causes tularemia in rabbits, some other wild animals, and humans. This pathogen's special need for the amino acid cysteine helps to identify it in laboratory cultures. Because it is highly infectious, specimens suspected of containing *F. tularensis* should be sent to specially equipped laboratories.

Several other gram-negative aerobic organisms are of interest in agriculture and environmental sciences. Some are important in nitrogen metabolism in plants (Figure 10.6); others cause some kinds of tumorlike growths called galls and other plant diseases. A few are specialized to metabolize methane, and many are important decomposers in mineral cycles. Another small group, the halophiles, not only tolerate but require extremely high salt concentrations for growth. Members of the Acetobacteraceae family produce acetic acid (vinegar) and are important in the food industry. They are found on fruits and vegetables and in alcoholic beverages. Some are a nuisance in the brewing industry because they change the alcohol produced by yeasts into acetic acid.

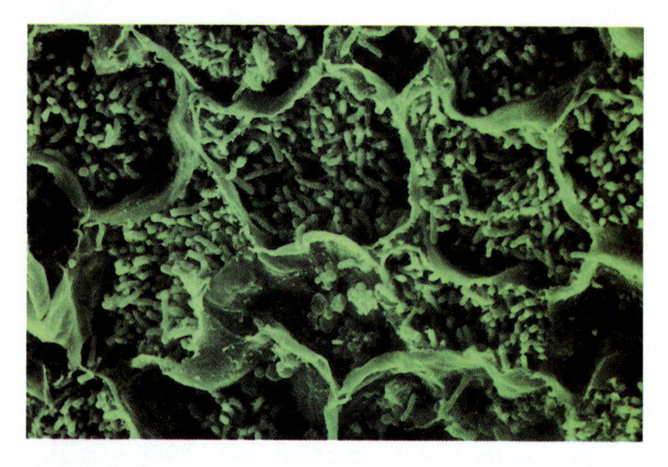

FIGURE 10.6 Nitrogen-fixing bacteria inside nodules on plant roots. These organisms capture N_2 gas from air and make it available to plants in more useful forms.

Facultatively Anaerobic Gram-Negative Rods (Section 5)

The facultatively anaerobic gram-negative rods constitute another large and medically important group of bacteria. The group is divided into three families and includes several genera not assigned to families. Among them are many important human pathogens, including the enterics, the vibrios, and the pasteurella-haemophilus group. The distinguishing characteristics of these organisms, summarized in Figure 10.7, are based on biochemical properties. These properties include kinds of fermentation reactions, percentages of bases in DNA, and presence of certain enzymes (Table 10.2).

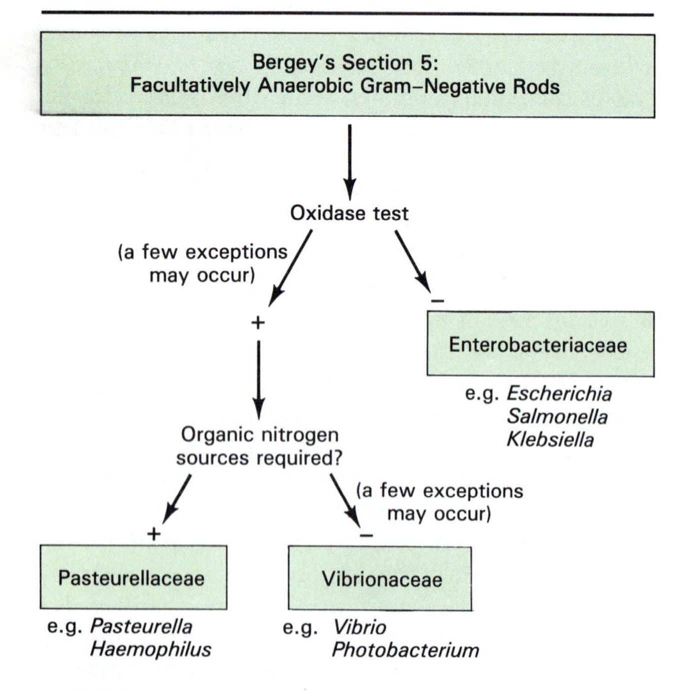

FIGURE 10.7 Distinguishing characteristics of facultatively anaerobic gram-negative rods.

The Enterics Members of the family Enterobacteriaceae are sometimes referred to as **enteric bacteria** because many of them inhabit the intestine of humans or animals. ("Enteric" means pertaining to the intestine.) In addition, some are free-living organisms found in soil and water. Others live in cooperation with or at the expense of their hosts, and a few decompose dead organic matter. Enteric bacteria are small, morphologically similar gram-negative rods. Some move with peritrichous flagella. They are facultative anaerobes that ferment glucose and other sugars and sometimes produce carbon dioxide and other gases. (Various fermentation pathways are described in Chapter 5. (p. 119))

Escherichia coli is an important enteric bacterium for three reasons:

1. *E. coli* is always present among the normal flora of the human intestine. Finding it in water indicates that the water is contaminated with fecal matter and possibly with pathogenic enteric bacteria.
2. *E. coli* can be a pathogen in its own right; it causes opportunistic infections in a variety of sites.
3. *E. coli* is an important organism in genetic engineering. Many important biological products now can be made by recombinant DNA technology, and *E. coli* frequently is the bacterium into which new genetic material is introduced. Consequently, this "lowly" bacterium might be considered the workhorse of today's biotechnology.

Most members of the family Enterobacteriaceae resemble *E. coli* and are sometimes loosely referred to as "coliforms."

Members of the genera *Shigella* and *Salmonella* are closely related to *E. coli* in that all carry out mixed acid fermentation. *Shigella* can be distinguished from other enterics by its inability to produce gas during glucose fermentation. *Salmonella* can be distinguished from *Escherichia* by its inability to produce acid and gas during lactose fermentation. These organisms are responsible for many serious intestinal diseases. *Shigella* species cause a severe diarrhea called bacillary dysentery, and various species of *Salmonella* cause typhoid fever, enteritis, and food poisoning.

Another closely related set of genera are *Klebsiella*, *Enterobacter*, and *Erwinia*, all of which carry out butanediol fermentation. In laboratory cultures, *Klebsiella* can be distinguished from *Enterobacter* by its ability to break down urea. *Klebsiella pneumoniae* causes a bacterial pneumonia, whereas other *Klebsiella* species cause a variety of chronic respiratory diseases. *Enterobacter* species are opportunistic pathogens, that is, they can infect humans with lowered resistance but ordinarily do not cause disease. *Erwinia* species are of no medical significance, but they are plant pathogens and are thus important agriculturally.

Proteus, *Providencia*, and *Yersinia* constitute yet another related set of genera; all are mixed acid fermenters, and they have a much smaller percentage of C and G bases in their DNA than other enterics. They can be distinguished in the laboratory by certain enzyme reactions; *Proteus* has both enzymes, but *Yersinia* lacks phenylalanine deaminase and *Providencia* lacks urease.

Most species of *Proteus* and *Providencia* are associated with urinary tract infections and with burns and wounds. They are especially prominent as causative agents of nosocomial (hospital-acquired) infections. *Proteus mirabilis* is frequently found in urine, blood, and other specimens. Some species of *Proteus* and a few other genera swell and disintegrate into **L forms,** irregularly shaped cells with defective cell walls (Figure 10.8). L forms can arise spontaneously or can be induced by lysozymes or penicillin, which remove the cell wall. They can persist and divide repeatedly or spontaneously revert back to normal-walled cells.

Yersinia pestis has a long history as an agent of human misery. When introduced into the body through the bite of an infected flea, *Y. pestis* causes bubonic plague, which is characterized by buboes, or abscesses of lymph nodes. When it reaches the lungs it causes pneumonic plague, a pneumonialike disease. Other species of *Yersinia* cause mesenteric lymphadenitis, an inflammation of lymph nodes in membranes that support abdominal organs, and a variety of other diseases.

Among enterics, *Serratia marcescens* seems to be in a class by itself. It is very different from common enteric pathogens and is also unusual for its ability to produce pigment. Strains that grow at room temperature are harmless free-living rods that usually produce a red pigment, prodigiosin (Figure 10.9). Strains that grow at human body temperature fail to produce pigment and are less innocuous. In fact, they are avidly opportunistic! They seem to infect almost any system except the digestive tract; they often infect

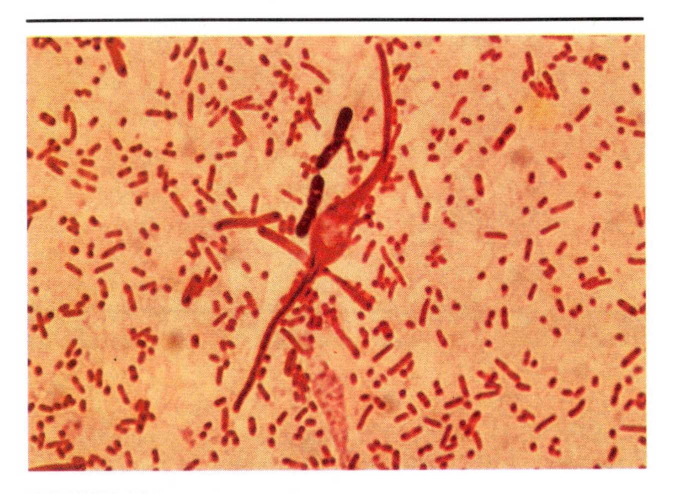

FIGURE 10.8 L forms of *Proteus vulgaris.*

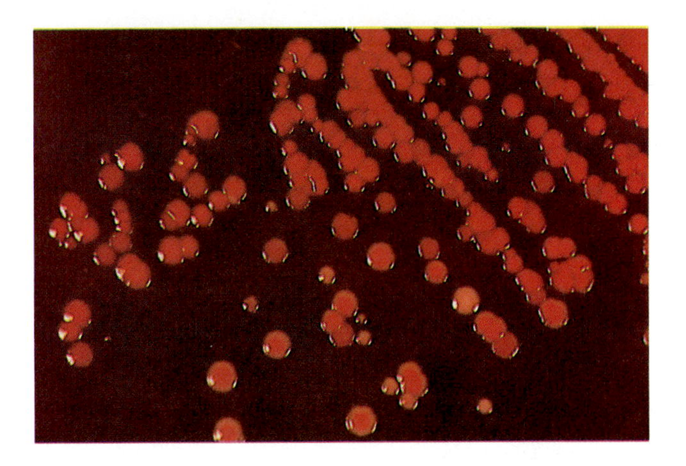

FIGURE 10.9 *Serratia marcescens.* Note the distinctive red pigment produced by this organism.

heart valves and have been known to cause pneumonia, urinary tract infections, meningitis, and wound infections. Many strains are resistant to several antibiotics. Because of their opportunism and antibiotic resistance, they cause formidable problems among surgical patients, especially the debilitated and the elderly.

The Vibrios The vibrios are curved, or comma-shaped, gram-negative facultative anaerobes with polar flagella. Many are found in fresh- or saltwater environments, but some inhabit the intestinal tracts of humans and other animals. Among the vibrios are two important pathogens. *Vibrio cholerae* causes cholera, and *V. parahaemolyticus* causes a foodborne enteritis. Both are waterborne pathogens; the latter is common in coastal marine environments, and humans who eat contaminated shellfish often become infected. Also included with the vibrios are the luminescent marine bacteria of the genus *Photobacterium.*

The Pasteurella-Haemophilus Group Members of the **pasteurella-haemophilus group** are very small gram-negative bacilli and coccobacilli. They lack flagella and are nutritionally fastidious. Most are parasites of animals, and a few are human pathogens, usually attacking mucous membranes of the respiratory tract. The genus *Pasteurella* is named for Louis Pasteur, who identified *P. multocida* as the causative agent of fowl cholera. Most members of the *Pasteurella* genus infect animals. Humans become infected from cat and dog bites and can develop abscesses around the wound and septicemia (blood poisoning). Members of the genus *Haemophilus* ("blood-loving") received their name from the fact that they grow in the laboratory only in media enriched with blood (Figure 10.10). *Haemophilus* species sometimes inhabit the human respiratory membranes as normal flora and can be responsible for opportunistic infections. *H. influen-*

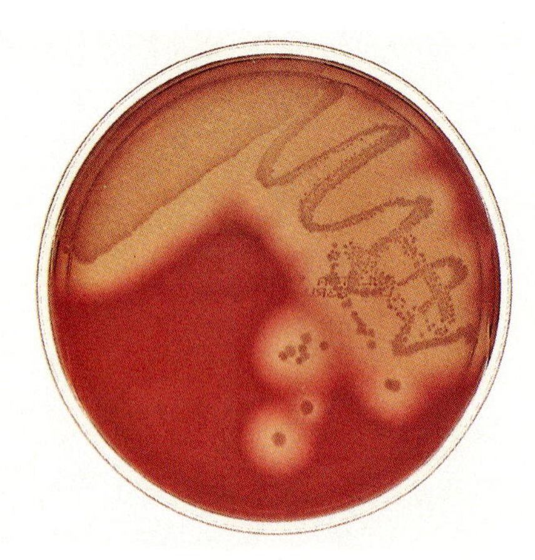

FIGURE 10.10 *Haemophilus* species growing on blood agar. The clear areas surrounding the colonies are due to beta (complete) hemolysis.

zae does not itself cause influenza, but was thought to at the time of its discovery, since viruses could not be seen then. In patients with viral influenza, *H. influenzae* frequently causes a secondary infection (an infection that follows one already established). Some species of *Haemophilus* attack other membranes; *H. ducreyi* causes chancroid, a venereal disease, and *H. aegyptius,* which may be a strain of *H. influenzae* rather than a separate species, causes acute conjunctivitis ("pinkeye").

Anaerobic Gram-Negative Rods (Section 6)

The anaerobic gram-negative rods are members of a single family Bacteroidaceae. They grow only in strictly anaerobic conditions and were cultured in the laboratory only after methods to maintain anaerobic conditions were developed. Although referred to as rods, many of these organisms are pleomorphic; that is, members of the same species have different shapes. Some have peritrichous flagella. Members of this group are found mainly in the intestinal and respiratory tracts of humans and animals. Rod-shaped *Bacteroides* produce propionic acid or succinic acid. Spindle-shaped *Fusobacterium* and comma-shaped *Butyrivibrio* produce butyric acid. *Leptotrichia* produces lactic acid. Anaerobic gram-negative rods cause a variety of opportunistic human infections, not only in the intestine but also in the mouth, upper respiratory tract, and urogenital tract. They are especially likely to be found in deep puncture wounds. *Butyrivibrio* is important in digestion in cows and other animals that chew their cuds. Among members of this section, species of *Bacteroides* are the most frequently encountered in human infections.

Anaerobic Gram-Negative Cocci (Section 8)

Except for shape, the anaerobic gram-negative cocci are quite similar to their rod-shaped cousins described earlier. Although several members of this group are found in animal intestines, only those of the genus *Veillonella* are of medical significance. Certain capnophilic (carbon dioxide-loving) species of *Veillonella* are found in the crevices between teeth and gums, where they often cause tooth abscesses and gum disease. Cocci occur in pairs, chains, or larger masses, and they ferment lactic acid.

Rickettsias and Chlamydias (Section 9)

Rickettsias and **chlamydias** (Figure 10.11) are obligate intracellular parasites. Many are human pathogens. Although once thought to be viruses because they were first found in cells, they are now known to have typical bacterial cell walls and to contain both DNA and RNA. Most lack certain enzymes necessary for life outside host cells. These somewhat simplified organisms probably arose from more typical bacteria by losing certain enzymes and other cell components. As they lost enzymes, they became more dependent on

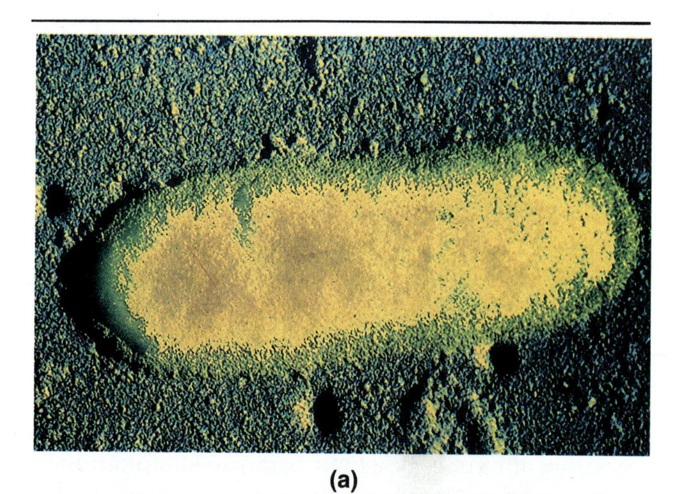

(a)

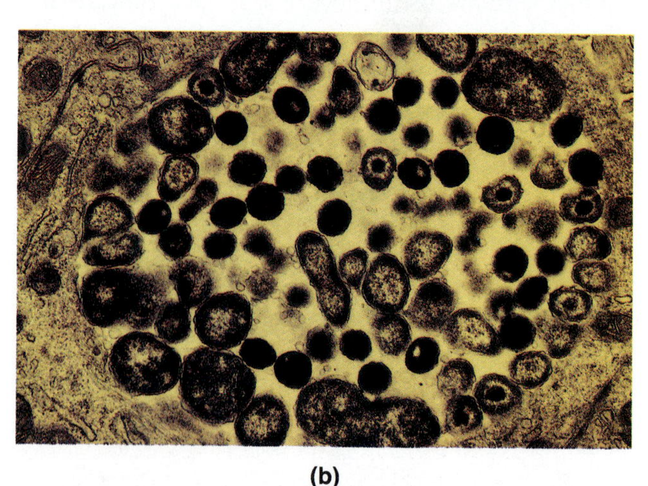

(b)

FIGURE 10.11 (a) TEM of a *Rickettsia* species, an obligate intracellular parasite. (b) TEM of *Chlamydia trachomatis* within an oviduct cell. Damage to the reproductive tract caused by this organism can eventually lead to sterility.

TABLE 10.3 Characteristics of typical bacteria, rickettsias, chlamydias, *Mycoplasma*, *Ureaplasma*, and viruses						
Characteristic	**Typical Bacteria**	**Rickettsias**	**Chlamydias**	***Mycoplasma***	***Ureaplasma***	**Viruses**
Cell wall	Yes	Yes	Yes	No	Sometimes	No
Grow only in cells	No	Yes	Yes	No	No	Yes
Require sterols	No	No	No	Sometimes	Yes	No
Contain DNA and RNA	Yes	Yes	Yes	Yes	Yes	No
Have metabolic systems	Yes	Yes	Yes	Yes	Yes	No

their hosts. These organisms also lack flagella and are nonmotile. Rickettsias and chlamydias each constitute a taxonomic order. Properties of rickettsias, chlamydias, and some other microbes are summarized in Table 10.3.

The rickettsias are small cocci or rods that parasitize cells of mammals and arthropods. Many live in both human cells and cells of ticks, lice, fleas, and mites, which harbor and spread rickettsial diseases. Most human pathogens are members of the genus *Rickettsia*. Variations in their characteristics are related to the diseases they cause. For example, members of the typhus fever group, which cause severe headache, chills, and fever, grow mainly in the nuclei of cells and have an optimum growth temperature of 35°C. In patients they cause hemolysis of red blood cells. Members of the spotted fever group, which cause a more severe but similar disease, grow mainly in cytoplasm and have an optimum growth temperature of 32° to 34°C. They do not cause hemolysis in patients. Other rickettsias cause a variety of fevers. *Rickettsia tsutsugamushi* causes scrub typhus, a fever with rash and lymph node inflammation; it grows mostly in nuclei and has an optimum temperature of 35°C. *Rochalimaea quintana* causes trench fever, which spreads especially among soldiers in trenches. *Rochalimaea* can be grown on artificial media and is an exception to the rule that all rickettsia are obligate intracellular parasites. *Coxiella burnetii* causes Q fever, a pneumonialike disease, and *Bartonella bacilliformis* causes Oroya fever, which is accompanied by a wartlike rash.

The chlamydias are tiny (0.2–1.0 mm) spherical bacteria that invade host cells, where they are protected from immune defense mechanisms. Like rickettsias, chlamydias also were once thought to be viruses because they are so small and because they multiply only in host cells. They have an unusual life cycle (Figure 10.12), in which the organisms appear in two different forms, both of which are nonmotile. When they are multiplying inside the host cell in membrane-bound cytoplasmic vacuoles, chlamydias consist of relatively large, metabolically active **reticulate bodies** with flexible cell walls. Aggregations of reticulate bodies are sometimes referred to as **inclusion bodies.** Reticulate bodies divide many times by binary fission, rupture, and release large numbers of small dense **elementary bodies** with rigid cell walls.

TABLE 10.4 Human disease caused by the trachoma variety of *Chlamydia trachomatis*	
Trachoma	Eyes
Inclusion conjunctivitis	Eyes
Otitis media	Middle ear
Urethral syndrome	Urinary system of females
Nongonococcal urethritis	Urinary system of both males and females
Mucopurulent cervicitis	Reproductive system of females
Salpingitis	Reproductive system of females
Bartholinitis	Reproductive system of females
Epididymitis	Reproductive system of males
Proctitis	Lower intestinal tract
Pneumonia	Respiratory system, especially in the young, the elderly, and the debilitated or immunosuppressed
Peritonitis	Inflammation of the body cavity
Hepatitis	Inflammation of the liver
Endocarditis	Inflammation of the valves of the heart

Elementary bodies are adapted for extracellular existence and are infectious; that is, they can attack other cells of the same host or be transmitted to new hosts.

Two species of chlamydia can cause human infections. *Chlamydia psittaci* causes severe pneumonia in humans and parrot fever in birds. Three varieties of *Chlamydia trachomatis* vary in their pathogenicity. The lymphogranuloma venereum variety infects the genital and anal regions. The mouse variety is not known to cause disease in humans. The trachoma variety can infect many different systems and organs, as summarized in Table 10.4. The ability of the trachoma variety of *C. trachomatis* to attack such a large number of different kinds of cells serves to illustrate an important concept: Although the one organism–one disease concept applies to many pathogens, it has many exceptions, as we shall see in chapters devoted to diseases of particular body systems.

Mycoplasmas (Section 10)

The **mycoplasmas** are one of the few groups of bacteria for which an entire sequence of taxonomic cat-

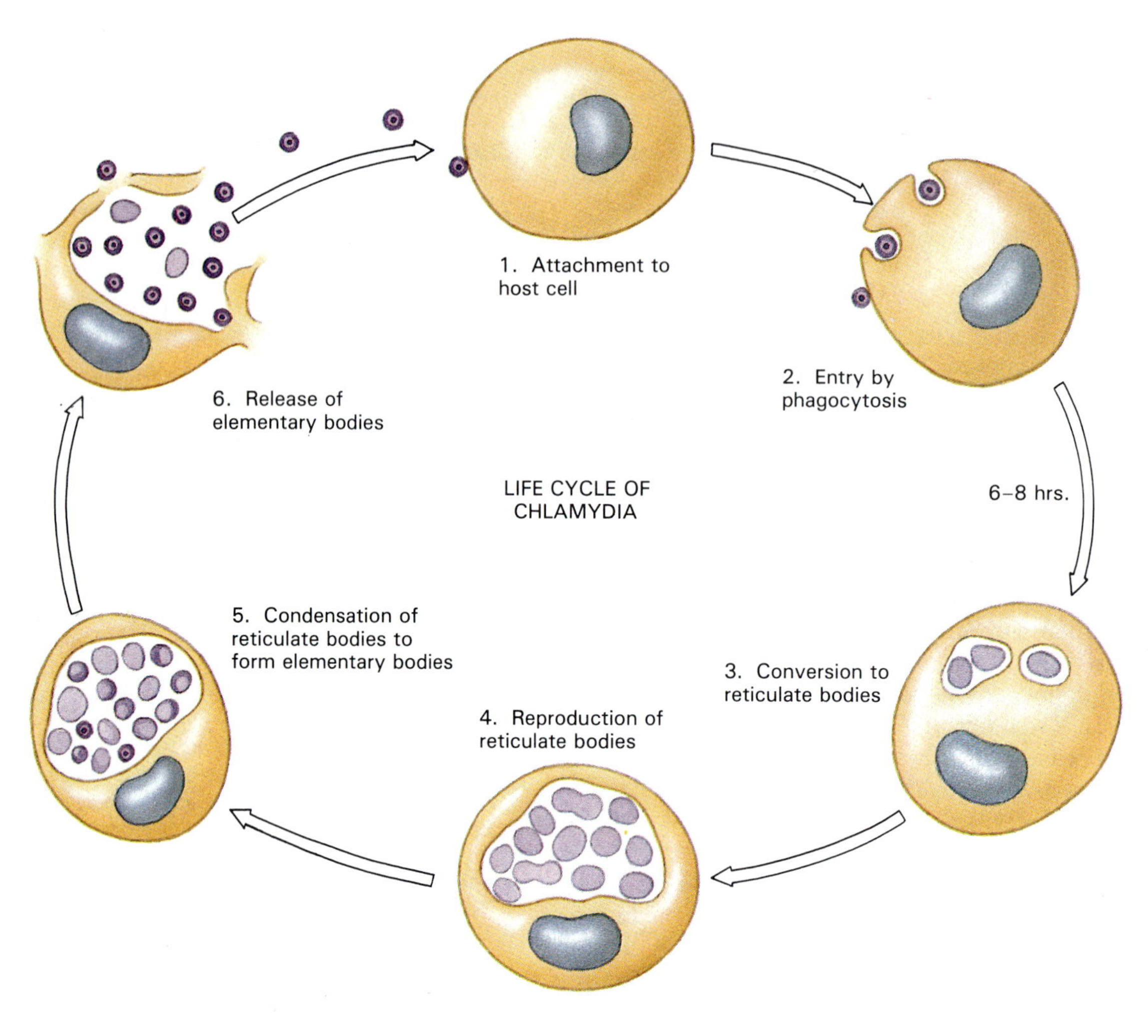

FIGURE 10.12 Life cycle of a *Chlamydia*. Small, dark, elementary bodies (the only infectious stage of the chlamydial life cycle) attach to host cell (1) and enter by phagocytosis (2). The elementary bodies, enclosed within membrane-bound vacuoles, lose their thick walls and enlarge to become reticulate bodies (3). These reproduce by binary fission, rapidly filling the cell (4). The reticulate bodies condense to form infectious elementary bodies (5), which are then released by lysis (6) and are free to attach to a new host cell.

egories from division to species has been defined (Appendix B). These organisms are so very small that they were once classified as viruses, but they are now known to have cell membranes, DNA, and RNA that identify them as bacteria. They are sometimes said to be gram-negative, not because they have the typical gram-negative cell wall layers but because they are entirely lacking a cell wall. Mycoplasmas are pleomorphic because of their flexible cell membrane, and they often form slender branched filaments. Most have sterols in their cell membrane; in this respect they resemble fungi and protists. They lack flagella and are usually nonmotile, but a few display a gliding movement on a wet surface. Most are facultatively anaerobic; a few are obligately anaerobic.

Though mycoplasmas can be grown on agar, where they form colonies with a "fried egg" appearance (Figure 10.13), in nature they have several modes of existence. Some must live at the expense of host organisms, some live on a host without damaging it, and others live on decaying organic matter. Human

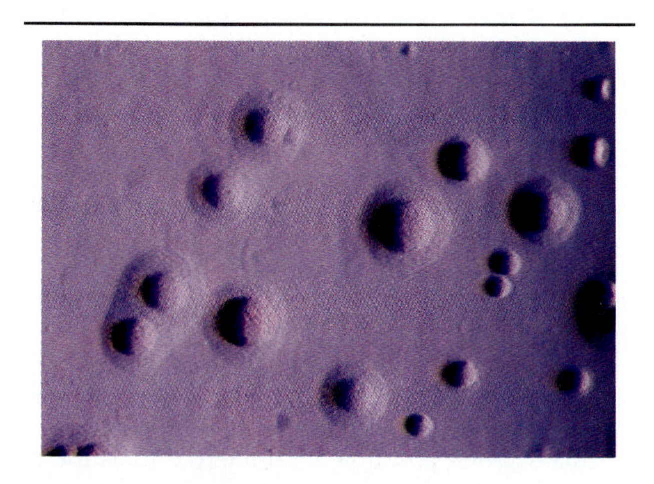

FIGURE 10.13 The "fried egg" appearance of mycoplasma colonies is unique to this group.

pathogens among the mycoplasmas are found in the genera *Mycoplasma* and *Ureaplasma*. *M. pneumoniae* causes primary atypical pneumonia, whereas other species are usually opportunists of the gums, oropharynx, and genitourinary tract. *M. fermentans* may play a role in rheumatoid arthritis. *U. urealyticum* is unusual in that it metabolizes urea; it is found in the urogenital tract, where it sometimes causes opportunistic infections. Other mycoplasmas infect orange trees, which are treated by pumping solutions of the antibiotic tetracycline into the trees.

Gram-Positive Cocci (Section 12)

The gram-positive cocci represent one end of a spectrum of gram-positive organisms distinguishable by shape and whether they form spores (Figure 10.14). As we shall see when we consider the next few sections of bacteria, these organisms have other distinguishing properties.

The gram-positive cocci themselves constitute a relatively large and medically important group of bacteria. These cocci are divided into families on the basis of metabolic properties and cellular arrangements. The **micrococci** are aerobes or facultative anaerobes that form irregular clusters by dividing in two or more planes. The **streptococci** are aerotolerant anaerobes that obtain energy from fermenting sugars to lactic acid. They form pairs, tetrads, or chains by dividing in one or two planes. Most lack the enzyme catalase. The **peptococci** are anaerobes that lack both catalase and the enzymes to ferment lactic acid. They also form pairs, tetrads, or irregular clusters.

Many micrococci are free-living saprophytes found in soil, freshwater, and marine environments. They are easily transmitted on the surfaces of plants and animals to meat, dairy products, and other foods. The micrococci include an aerobic genus, *Micrococcus*, and a facultatively anaerobic genus, *Staphylococcus*. Various species of *Micrococcus* are found on the skin and in the mouth and upper respiratory tract. *Staphylococcus aureus* is a common human pathogen. It is often responsible for skin abscesses and boils. If it invades the blood, it can travel to other tissues and cause pneumonia, meningitis, and osteomyelitis (infection of the marrow cavity of bones). *Staphylococcus epidermidis* normally inhabits skin and mucous membranes; it often causes opportunistic infections.

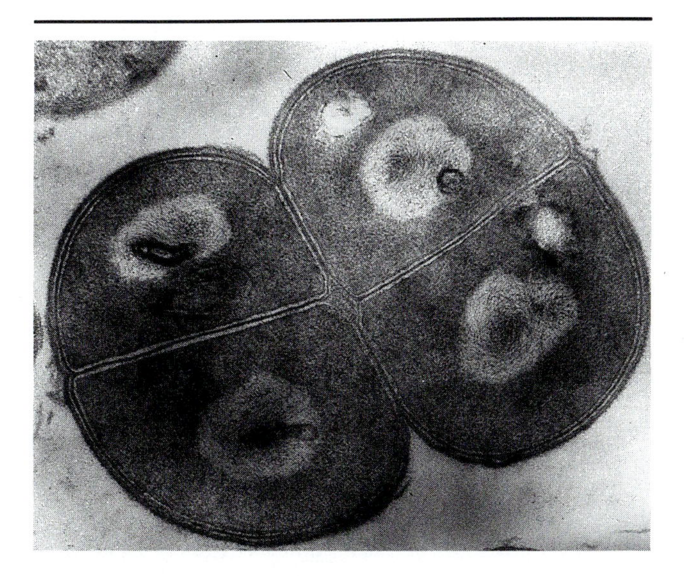

FIGURE 10.14 *Sporosarcina ureae*, gram-positive cocci that produce spores. These cocci are arranged in cubical packets.

The streptococci include a large number of species, which are differentiated by whether they have certain antigens on their cell surfaces and whether they carry out **hemolysis**, the destruction of red blood cells. For example, some strains of *Streptococcus pyogenes* with so-called group-A antigens completely hemolyze red blood cells in laboratory cultures. Such complete hemolysis is called **beta hemolysis.** These strains are the causative agent of strep throat, scarlet fever, rheumatic fever, and a variety of other infections. In contrast, streptococci of the **viridans group,** which often infect the valves and lining of the heart, cause incomplete, or **alpha hemolysis** in laboratory cultures. *S. agalactiae*, long known to cause mastitis (inflammation of the udders) in cattle, is now known to be sexually transmitted and responsible for female urogenital infections in humans. If transmitted to infants during the birth process, it can cause fatal illness.

The peptococci are obligate anaerobes. Though *Sarcina* are saprophytes found in soil and on plants, members of the genera *Peptococcus* and *Peptostreptococcus* are frequently found among the microbes that normally inhabit the digestive, respiratory, and urogenital tracts. These organisms are versatile opportunists; they can cause peritonitis (inflammation of the body cavity and its membranes), postpartum sepsis (blood poisoning after the birth of a child), accumulation of pus in deep wounds, osteomyelitis, vaginitis, and sinus and dental infections.

Endospore-Forming Gram-Positive Rods and Cocci (Section 13)

The endospore-forming gram-positive rods and cocci consist mainly of rod-shaped bacteria of the genera *Bacillus* and *Clostridium*. Cocci of the genus *Sporosarcina* also are included in this group, but they are soil bacteria of no medical significance. The ability to form endospores is the main distinguishing characteristic of these organisms.

All members of the genus *Bacillus* obtain nutrients from dead organic matter; some are obligate aerobes and others are facultative anaerobes. They can be distinguished by the location of spores and by growth temperatures. For example, *B. cereus* and *B. subtilis*

APPLICATIONS

Biological Control of Pests

Bacillus thuringiensis and its toxin are useful as a biological insecticide. Spores of this organism can be added to soil. When the spores germinate, they infect the larval stages of a variety of insects such as moths, butterflies, and mosquitoes. The larvae are thereby killed before they can cause crop damage. It is an important means of controlling the greater wax moth larvae, one of the major predators on honeybee colonies. The spores are sprayed directly onto the hive. Biological insecticides are desirable because they kill undesirable insects without introducing toxic and nonbiodegradable substances into the environment.

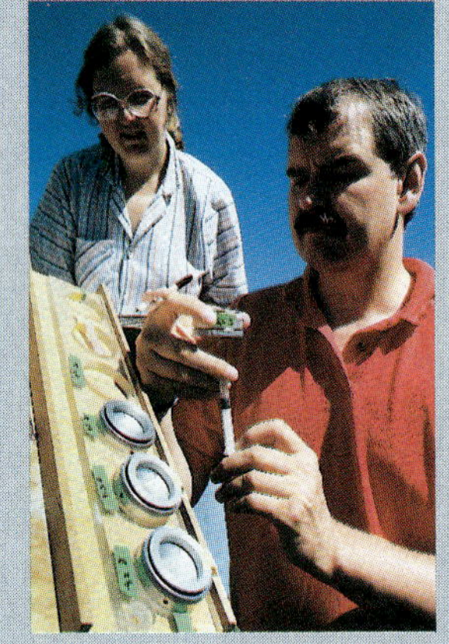

Researchers expose a variety of new strains of *Bacillus thuringiensis* to sunlight to determine how their ability to destroy insect pests may be affected by outdoor conditions.

have centrally located endospores and grow at moderate temperatures. In contrast, *B. stearothermophilus* has terminal spores and grows at temperatures of 65°C or higher.

Several species of *Bacillus*—*B. subtilis, B. licheniformis, B. polymyxa,* and *B. brevis*—produce antibiotics. *B. anthracis* causes anthrax, a severe blood infection of cattle, sheep, and horses that can infect humans. *B. cereus* is sometimes implicated in food poisoning.

Members of the genus *Clostridium* are strictly anaerobic motile rods found in soil, water, and the intestinal tracts of humans and other animals. Several species are important human pathogens that exert their effects mainly through the production of potent toxins. *C. tetani* causes tetanus (lockjaw) and *C. botulinum* causes botulism (a kind of food poisoning). *C. perfringens* also can cause food poisoning. Several species of *Clostridium* produce gas, predominantly hydrogen, under the anaerobic conditions of deep tissue wounds and cause gangrene, or tissue necrosis (death). Gangrene due to an infection is called gas gangrene to distinguish it from dry gangrene, which occurs when a tissue lacks an adequate blood supply. Other clostridia also infect wounds, and some can cause low-grade bacteremias, especially in debilitated patients.

Regular Nonsporing Gram-Positive Rods (Section 14)

The regular nonsporing gram-positive rods are obligate or facultative anaerobes found in fermenting plant and animal products. They include three genera, *Lactobacillus, Listeria,* and *Erysipelothrix.* **Lactobacilli** (Figure 10.15) are found in a wide variety of foods. They are used in the production of cheeses, yogurt, sourdough, and many other fermented foods. *L. acidophilus* and *L. casei* are found among the natural flora of the digestive and urogenital tracts of humans and are generally nonpathogenic. A few lactobacilli have been linked to tooth decay. *Listeria monocytogenes* causes listeriosis, an inflammation of the brain and the membranes that cover it, in humans and animals. *Erysipelothrix rhusiopathiae* causes erysipeloid (red, painful skin lesions) in humans and more severe erysipelas in swine.

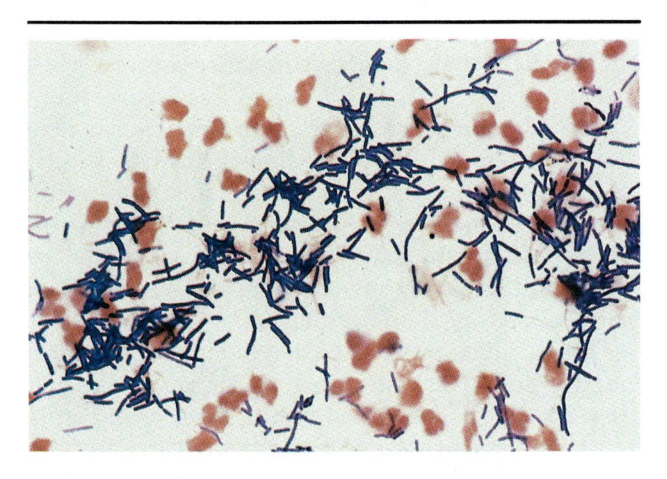

FIGURE 10.15 *Lactobacillus acidophilus* forms the predominant flora in the vagina of sexually mature women. These bacteria produce lactic acid by fermentation of the glycogen in the epithelial cells lining the vagina. The acidic environment that they produce and live in inhibits the growth of many other types of bacteria and thus helps to protect the vagina from infection.

Irregular Nonsporing Gram-Positive Rods (Section 15)

The irregular nonsporing gram-positive rods include club-shaped **corynebacteria,** pleomorphic **propionibacteria,** and filamentous **actinomycetes.** Nearly all are facultative anaerobes, but a few propionibacteria are strict anaerobes.

The genus *Corynebacterium* includes a large and diverse assortment of saprophytes, many of which are found in air, soil, and water or as pathogens in plants and a few animals. *C. xerosis* is a normal inhabitant of human conjunctiva; *C. pseudodiphtheriticum* is a normal inhabitant of the human pharynx. Neither produces damaging toxin. The most medically significant species is *C. diphtheriae.* It produces a potent toxin and causes diphtheria.

The family Propionibacteriaceae includes the genera *Propionibacterium* and *Eubacterium. P. freudenreichii* is found in dairy products. *P. acnes,* which is found in wounds and abscesses, commonly infects but does not cause acne lesions. Most species of *Eubacterium* are obligate anaerobes; they are found in soil and plant products and sometimes infect body cavities of humans and other animals. *E. foedans* is found in dental tartar and in a variety of infections.

Once classified as fungi, the order Actinomycetales, which means ray fungus, includes mainly soil bacteria, which often form branching filaments (Figure 10.16). Some provide nitrogen to plants; a few are normal inhabitants of the human mouth. The most important human pathogens are members of the genus *Actinomyces. A. israelii* causes actinomycosis (jaw abscesses and sometimes lung disease), and several other species are found in dental caries (cavities) and periodontal infections. *A. bovis* causes a disease called lumpy jaw in cattle.

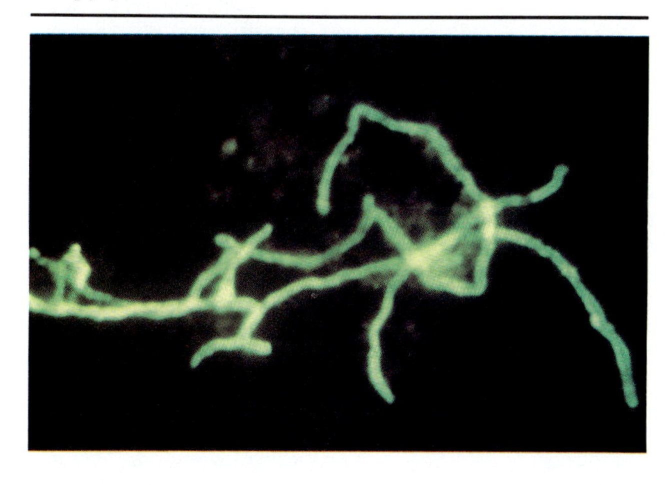

FIGURE 10.16 *Actinomyces israelii* from a brain abscess. Note the branching, filamentous type of growth, which formerly led the actinomycetes to be classified as fungi.

Mycobacteria (Section 16)

The **mycobacteria** are unusual in that their cell envelopes contain large amounts of lipids. The lipids resist basic dyes, and therefore mycobacteria are stained with fuchsin in hot phenol and washed with acid alcohol. The red stain remains; thus, these organisms are said to be acid-fast (refer to Figure 3.27). Mycobacteria are slender rods, usually without clubbed ends, and frequently form filaments. Most are soil saprophytes, but some are human pathogens. *Mycobacterium tuberculosis* causes tuberculosis, and *M. leprae* causes Hansen's disease, formerly known as leprosy. Another organism, *M. bovis,* causes tuberculosis in cattle and can be transmitted to humans.

Nocardioforms (Section 17)

The **nocardioforms** are closely related to corynebacteria, actinomyces, and mycobacteria. They are gram-positive, nonmotile, pleomorphic, aerobic, and usually acid-fast and filamentous. Though many *Nocardia* are found in the soil, *N. asteroides* causes skin abscesses and lung infections. *N. brasiliensis* also causes skin abscesses.

Actinomycetes That Divide in More Than One Plane (Section 27)

The actinomycetes that divide in more than one plane are similar to but distinct from the actinomycetes described earlier. These bacteria nearly always form masses of filaments. Some are free-living in soil; others live in plant root nodules, where they oxidize molecular nitrogen. The medically important species belong to the genus *Dermatophilus,* which causes skin lesions.

Streptomycetes and Their Allies (Section 29)

Like the actinomycetes, the **streptomycetes** and their allies are soil-dwelling organisms that resemble fungi. They develop extensive branching filaments with spores of many different shapes. The characteristics of spores are used to separate the streptomycetes into genera. None are pathogenic, but many are medically significant because of the antibiotics they produce. More than 500 different antibiotic substances have been isolated from various species of the genus *Streptomyces.* Among these antibiotics are substances effective against bacteria, viruses, protozoa, and fungi.

Characteristics of bacteria in 16 medically important sections of *Bergey's Manual* are summarized in Table 10.5.

Bacteria in Ecology and Evolution: The "Nonmedical" Groups

Seventeen of the 33 sections of bacteria defined in *Bergey's Manual* are not known to have medical significance either as pathogens or antibiotic producers. Yet these organisms are of great importance in ecosystems, and they suggest some evolutionary trends. Let us look at some of the unusual and important characteristics of these groups.

The nonmotile, gram-negative curved bacteria (Section 3) are aerobic, free-living organisms found in soil, fresh water, and ocean water. The genus *Spirosoma* is an example. They are a source of food for larger soil organisms.

TABLE 10.5 Characteristics and medically important members of selected sections of bacteria defined in *Bergey's Manual*

Section (number)	Medically Important Members	Diseases
Spirochetes (1): gram-negative, helical, move by axial filaments	*Treponema*	Syphilis
	Borrelia	Relapsing fever, Lyme disease
	Leptospira	Leptospirosis
Aerobic, Motile, Helical Gram-Negative Bacteria (2): move by flagella, helical or comma-shaped	*Campylobacter*	Urogenital and digestive tract infections
Gram-Negative Aerobic Rods and Cocci (4): Some contain pigments or oxidase, some have fastidious nutritional requirements, some obligate parasites.	*Pseudomonas*	Urinary tract infections, burns, and wounds
	Legionella	Pneumonia and other respiratory infections
	Neisseria	Gonorrhea, meningitis, and nasopharyngeal infections
	Moraxella	Conjunctivitis
	Brucella	Brucellosis
	Bordetella	Whooping cough
	Francisella	Tularemia
Facultatively Anaerobic Gram-Negative Rods (5): Some have peritrichous flagella, many can be distinguished by their characteristic fermentation reactions.	*Escherichia*	Opportunistic infections of colon and other sites
	Shigella	Bacillary dysentery
	Salmonella	Typhoid fever, enteritis, and food poisoning
	Klebsiella	Respiratory and urinary tract infections
	Enterobacter	Opportunistic infections
	Serratia	Opportunistic infections
	Proteus	Urinary tract infections (especially nosocomial)
	Providencia	Wound and burn infections, urinary tract infections
	Morganella	Summer diarrhea, opportunistic infections
	Yersinia	Plague, mesenteric lymphadenitis, septicemia
	Vibrio	Cholera, acute gastroenteritis
	Pasteurella	Cat and dog bite wounds
	Haemophilus	Respiratory infections, meningitis, conjunctivitis, chancroid
	Calymmatobacterium	Granuloma inguinale
	Gardnerella	Vaginitis
	Eikenella	Wound infections
	Streptobacillus	Rat-bite fever
Anaerobic Gram-Negative Rods (6): straight, curved, or helical, motile	*Bacteroides* and *Fusobacterium*	Oral, digestive, respiratory, urogenital infections, wounds, and abscesses
Anaerobic Gram-Negative Cocci (8): nonmotile	*Veillonella*	Oral microflora and abscesses

The dissimilatory sulfate- or sulfur-reducing bacteria (Section 7) give off sulfur-containing compounds as wastes. (Assimilation refers to taking in; dissimilation refers to giving off.) These bacteria (Figure 10.17) are anaerobes found mainly in water sediments, sewage, and polluted water. They are very important in the environmental sulfur cycle, and some members of the group can survive at extremes of temperature and high salinity. Although they can be found in the human intestine, they are not known to cause disease. The hydrogen sulfide they produce is toxic and corrosive and smells like rotten eggs. In sufficiently high concentration, it kills fish and plants and corrodes pipes.

The endosymbionts (Section 11) are a diverse as-

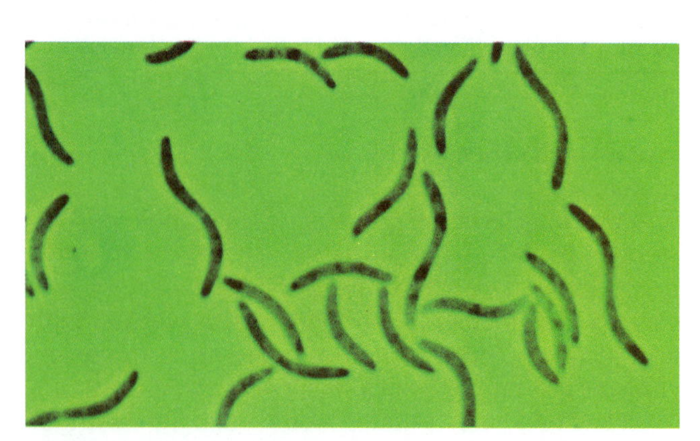

FIGURE 10.17 *Desulfovibrio gigas*, an anaerobic bacterium that reduces sulfate to hydrogen sulfide.

Section (number)	Medically Important Members	Diseases
Rickettsias and Chlamydias (9): intracellular parasites; chlamydias form reticulate and elementary bodies.	*Rickettsia*	Typhus, Rocky Mountain spotted fever, rickettsialpox
	Rochalimaea	Trench fever
	Coxiella	Q fever
	Bartonella	Oroya fever
	Chlamydia	Trachoma, inclusion conjunctivitis, nongonococcal urethritis, lymphogranuloma venereum, parrot fever
Mycoplasmas (10): lack cell walls, extremely small	*Mycoplasma*	Atypical pneumonia, urogenital infections
	Ureaplasma	Opportunistic urogenital infections
Gram-Positive Cocci (12)	*Staphylococcus*	Skin abscesses, opportunistic infections
	Streptococcus	Strep throat and other respiratory infections, skin and other abscesses, puerperal fever, opportunistic infections
	Peptococcus	Postpartum septicemia, visceral lesions
	Peptostreptococcus	Puerperal fever and various pyogenic (pus-forming) infections
Endospore-Forming Gram-Positive Rods and Cocci (13): aerobic to strictly anaerobic; some motile and some nonmotile	*Bacillus*	Anthrax, source of the antibiotic bacitracin
	Clostridium	Tetanus, botulism, gas gangrene, bacteremia
Regular Nonsporing Gram-Positive Rods (14): facultatively or strictly anaerobic; nonmotile	*Lactobacillus*	Microflora of the digestive tract and vagina
	Listeria	Listeriosis
	Erysipelothrix	Erysipeloid
Irregular Nonsporing Gram-Positive Rods (15): club-shaped, pleomorphic, filamentous; aerobic to anaerobic	*Corynebacterium*	Diphtheria and skin opportunists
	Propionibacterium	Wound infections and abscesses
	Eubacterium	Oral and other infections
	Actinomyces	Actinomycoses
Mycobacteria (16): gram-positive, acid-fast	*Mycobacterium*	Tuberculosis, leprosy, and chronic infections
Nocardioforms (17): gram-positive, filamentous, some acid-fast	*Nocardia*	Nocardiosis, mycetoma, abscesses
Actinomycetes That Divide in More Than One Plane (27): gram-positive, filamentous	*Dermatophilus*	Skin lesions
Streptomycetes and Their Allies (29): gram-positive, filamentous	*Streptomyces*	Produce over 500 different antibiotics

sortment of gram-negative bacteria that live in the bodies of protozoa, fungi, algae, other bacteria, and insects. Little is known of how these organisms relate to their hosts. Even less is known about their possible roles in the evolution of microorganisms. Certain endosymbionts of protozoa produce toxins. Some toxins pass from one protozoan to another during conjugation, killing the recipient. Other endosymbionts give toxin resistance to their hosts.

The gliding nonfruiting bacteria (Section 23) glide, or slither, along a solid surface, leaving a slime trail behind them. They are said to be nonfruiting because they do not form cell aggregates called fruiting bodies, as do another group called the gliding fruiting bacteria. Organisms in this group are important in ecological cycles. Members of the genus *Cytophaga* digest cellulose, chitin, agar, and other complex organic material. *Beggiatoa* species oxidize hydrogen sulfide to sulfur, thereby contributing to the sulfur cycle.

As the name suggests, anoxygenic photosynthetic bacteria (Section 18) capture energy from light but do not release oxygen as photosynthetic plants do. These bacteria contain a special bacterial chlorophyll and a variety of other pigments—purple, red, orange, and brown (Figure 10.18). They contribute to ecosystems by providing nutrients for other organisms.

The budding and/or appendaged bacteria (Section 21) reproduce by budding, or dividing unequally. Small buds that contain a nucleus and a small amount of cytoplasm pinch off from the parent cell and grow

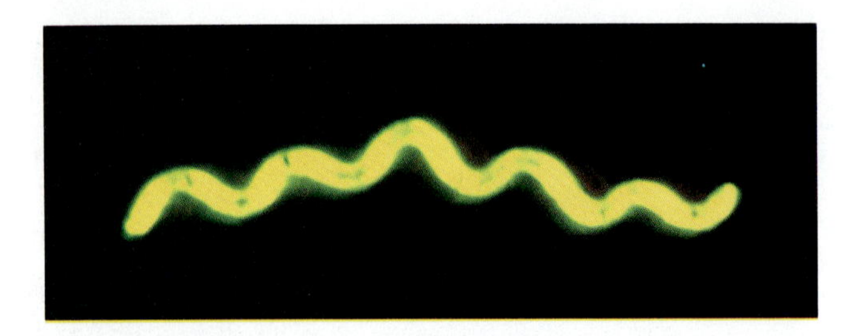

FIGURE 10.18 *Rhodospirillum rubrum*, an anoxygenic photosynthetic bacterium.

to become parent cells themselves. Some of these bacteria have an appendage, or stalk. The stalk usually has a sticky holdfast at its tip, which allows the bacterium to anchor itself on a solid surface (Figure 10.19). These bacteria are found in soil and water. Both budding reproduction and development of stalks are examples of interesting evolutionary developments in this group.

The archaebacteria (Section 25) are a diverse group of anaerobes. They can be gram-positive, gram-negative, motile, nonmotile, rods, or cocci. Some produce methane and are found in soil, water, sewage, and even in the digestive tract of humans and other animals. The curious methanogenic archaebacteria are, in fact, the sole natural source of natural methane, or marsh gas. The extremely halophilic archaebacteria can live in solutions saturated in salt; in some places they can be seen from an airplane as pink patches on salt flats.

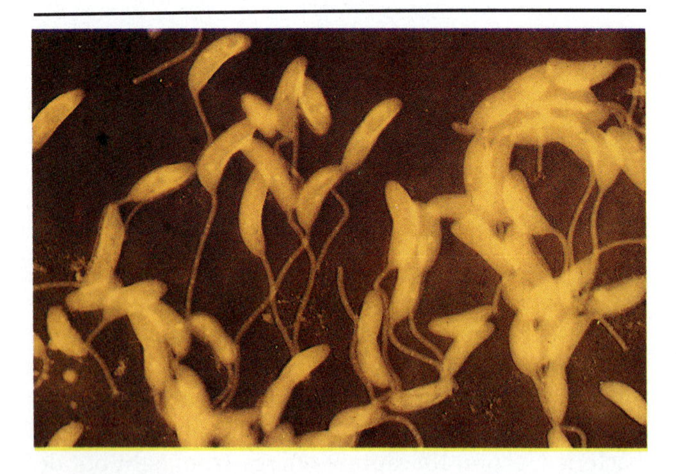

FIGURE 10.19 *Caulobacter* is a nonmotile stalked bacterium anchored to surfaces by a thin filament during part of its life cycle. However, it begins life as a free-swimming "swarmer" cell. Eventually the swarmer cell loses its flagellum and grows an attachment stalk at the site of its former flagellum. After settling down, it grows, elongates, and forms a new flagellum at the end opposite the attachment stalk. It then divides unequally by binary fission to produce a new swarmer cell plus the old attached cell.

Other archaebacteria called the thermoacidophiles metabolize sulfur and live in extremely hot acidic environments. These bacteria and other organisms found near deep sea vents have been studied from the submersible vehicle *Alvin* (Figure 10.20). The organisms are found in surprisingly dense populations. The bacteria provide food for many organisms including mussels, huge clams, and worms 3 m long and as thick as a human arm. Some of the bacteria are free-living and form white, fluffy, tennis ball–sized masses in a bacterial mat. Others live symbiotically in worm tissues. These bacteria make steam vent environments the most productive in terms of energy capture of all known environments.

The sheathed bacteria (Section 22) form filaments and surround themselves with a secreted sheath made of lipoproteins and polysaccharides. If these gram-negative organisms are grown in the presence of min-

FIGURE 10.20 Thermoacidophilic archaebacteria thrive in hot, acidic environments, including regions of the deep ocean floor near volcanic vents. There they serve as the base of food chains, living symbiotically inside the tissues of animals such as these giant tube worms. The worms lack mouths and anuses and thus rely on the bacterial capture of chemical energy to feed their tissues.

erals such as iron or manganese, they incorporate the metals into their sheaths. Sheathed bacteria reproduce by releasing flagellated swarmer cells, which leave the filament and start new colonies. These water dwellers are responsible for the slimy coating on rocks in and along streams and for blooms (dense floating masses of organisms) when the nutrient supply suddenly increases. They provide further evidence of the evolutionary trend toward multicellularity.

The gliding fruiting bacteria (Section 24), also referred to as the slime bacteria, are gram-negative aerobes. Found in well-oxygenated soil, these organisms are important decomposers. Their fruiting bodies consist of large, often highly colored, cell aggregates. Cells released from the fruiting bodies are released and migrate to form new colonies—more evidence for the evolutionary trend toward multicellular structures.

The chemolithotrophs (Section 20) are another diverse group of bacteria that obtain energy from inorganic substances. They include nitrogen bacteria found in soil, such as *Nitrosomonas,* which oxidizes ammonia to nitrite, and *Nitrobacter,* which oxidizes nitrite to nitrate. Also in this group are *Thiobacillus,* which oxidizes sulfur, and certain other bacteria that reduce sulfur. Sulfur bacteria are found in soil, water, sewage, sulfur springs, and acid mine wastes. Still other chemolithotrophs, which oxidize iron and/or manganese, are responsible for large deposits of minerals on the ocean floor. A few of these organisms absorb iron into intracellular crystals, and when placed in a magnetic field, they orient in a north-south direction. Exactly where the chemolithotrophs fit into the evolution of life is not yet clear; they present some interesting taxonomic problems.

The cyanobacteria (Section 19), formerly called blue-green algae and classified with algae, are now known to be prokaryotic. Yet they are remarkably similar to plants: Their chlorophyll is much like plant chlorophyll, and they release oxygen during photosynthesis, as do plants. They may represent a step in the evolution of photosynthetic mechanisms.

The prochlorophytes too (Section 19) are similar to the cyanobacteria, but are placed in a different group because they have two distinctly plantlike chlorophylls found in structures much like chloroplasts. The only two currently known genera of this group are *Prochloron* (Figure 10.21), which lives with marine invertebrates called ascidians, and *Prochlorothrix.* Some scientists believe *Prochloron* represents a "missing link" between photosynthetic prokaryotes and eukaryotic plants.

Some sporangiate actinomycetes (Sections 28 and 30) are soil bacteria that form relatively extensive filaments, complete with sporangia (spore sacs) filled with spores. Though these organisms do not infect

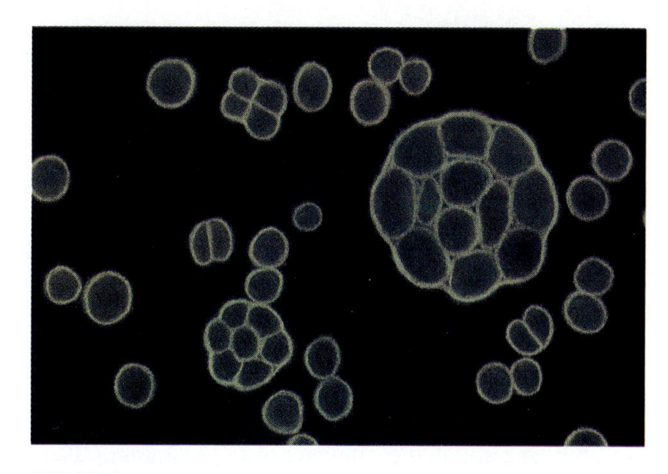

FIGURE 10.21 *Prochloron,* a photosynthetic prokaryote with some similarities to plants. Each cell in a cluster contains chlorophyll arranged on membranous layers.

humans, they are similar to the genera *Actinomyces* and *Dermatophilus,* which do infect humans.

Other conidiate genera (Section 28) have even more complex mycelial structures and more closely resemble fungi than other funguslike bacteria. The species *Micromonospora purpurea* produces the antibiotic gentamicin. If you have found the funguslike bacteria somewhat confusing, you are not alone. For centuries, they confused taxonomists, who placed them with the fungi until they were found to be prokaryotic. Indeed, the editors of *Bergey's Manual* changed their minds between Volumes 2 and 4 as to how to classify some funguslike bacteria, notably the Actinomycetes and Nocardioforms (Sections 17 and 26).

Some bacteria not yet classified have been isolated recently from deep aquifers—porous rock layers that hold underground water. These bacteria produce carbon dioxide that makes some mineral waters effervescent. The carbon dioxide creates carbonic acid that dissolves limestone and increases the size of the aquifers and thereby the quantity of water they can hold. Some such bacteria appear to feed selectively on molecules containing carbon 13, a heavy isotope of carbon, if it is available. Why they should do this is unknown.

We can characterize the "nonmedical" bacteria by saying that they are important in ecosystems and present challenges to biologists interested in evolutionary relationships. In ecosystems they capture energy by photosynthesis and chemolithotrophy; they serve as food for larger organisms, and they decompose the dead bodies and wastes of many organisms. They also provide important links in recycling of nitrogen, sulfur, and other minerals. With respect to evolution they display trends toward multicellularity and a diverse assortment of metabolic processes that may represent some of nature's experiments with the origin of life itself.

ESSAY

Are We Still Discovering New Organisms?

Are there any new worlds to discover, or new creatures in them? Yes! In recent years scientists have discovered living organisms in such diverse environments as submarine hot vents and inside volcanoes. In 1990, a joint U.S. and Soviet team discovered hot vents for the first time ever in fresh water, complete with an associated community of archaebacteria, worms, sponges, and other organisms. The vents lie more than 400 m deep in a most unusual Russian lake, Lake Baikal, which is the deepest lake in the world and holds the greatest quantity of fresh water in the world. Located in central Asia, in Siberia, it lies in a pocket between two continental plates. Asia was formed as a solid mass when several plates collided one after another and remained together. The area of Lake Baikal is now trying to pull apart, forming a rift valley and eventually a new ocean. This is comparable to the sea-floor spreading centers (ridges) of the Pacific, where other vent communities are found. In both locations, hot materials from deep in the earth are emerging. Lake Baikal is a unique treasure for studying evolution of life and microbial forms. Most lakes are only thousands of years old, but Lake Baikal may be 25 million years old. Microbes similar to the early evolving stages of life may still exist in its depths.

Who lives inside a volcano? Studies following the 1980 volcanic eruptions of Mount St. Helens have raised some interesting questions. Archaeobacteria, which had previously been known from the "black smoking" vents located 2200 m below the sea, have been found living on and in Mount St. Helens at temperatures of 100°C. Where did they come from? Some scientists think they may have been present deep down inside volcanoes, just waiting to grow. For that matter, where do the archaeobacteria in the submarine vents come from? Do they point out a linkage between terrestrial and submarine vulcanism? We tend to think of life as being present on the surface of the earth, but perhaps there is a whole different range of life deep *inside* the crust of the earth, which we have not suspected. Daily more evidence accumulates in favor of the idea of "continuous crustal culture."

Various ecologic problems have sent scientists from universities, government, and industry out hunting for new microbes with properties that make them useful in cleaning up the environment. Scientists from the Woods Hole Oceanographic Institution, of Massachusetts, took their search to a depth of more than 1800 m in the Gulf of California, where they have discovered anaerobic bacteria that can degrade naphthalene and possibly other hydrocarbons that might be found in oil spills. Sites needing bioremediation often lack oxygen, making it impossible to utilize aerobic organisms for cleanup; hence the hunt in deep anaerobic environments. General Electric has also found an anaerobic bacterium that it plans to use to destroy polychlorinated biphenyls (PCBs), industrial byproduct chemicals that accumulate in animal tissues and cause damage including cancer and birth defects.

A new bacterium, so far referred to as GS-15, discovered in the Potomac River by U.S. Geological Survey (USGS) scientists, ordinarily changes iron from one form to another. It seems, however, that these bacteria can just as easily feed on uranium, getting twice as much energy in the process and transforming the uranium into an insoluble precipitate. The USGS team plans to use them to remove uranium from contaminated well and irrigation water found in much of the U.S. West and at uranium mining, processing, and nuclear waste sites.

Yes there are many new microbes waiting to be discovered. In addition to the naturally occurring species, scientists will use genetic engineering techniques to design new ones. All these species will need to be classified and named. Clearly, *Bergey's Manual* will never be "finished."

CHAPTER SUMMARY

BACTERIAL TAXONOMY AND NOMENCLATURE

Criteria for Classifying Bacteria

- The criteria used for classifying bacteria are summarized in Table 10.1. These criteria can be used to classify bacteria into species and even into strains within species.
- For many species a particular strain is designated as the **type strain** and is preserved in a type culture collection.

Problems of Bacterial Taxonomy

- Taxonomists do not agree on how members of the kingdom Prokaryotae (Monera) should be divided. Many species of bacteria have been grouped into genera and some, into families. Four *divisions* (phyla) have been established. Much information is needed to determine evolutionary relationships and establish classes and orders.

RELATED KEY TERMS

outer sheath
protoplasmic cylinder
treponemes **pseudomonads**
legionellas **L forms**
enteric bacteria
pasteurella-haemophilus group
rickettsias **chlamydias**
reticulate bodies
inclusion bodies
elementary bodies **mycoplasmas**

CLASSIFICATION USING *BERGEY'S MANUAL*

History and Significance of *Bergey's Manual*

- *Bergey's Manual* was first published in 1923 and has been revised several times; a ninth edition is in preparation.
- It provides definitive information on the identification and classification of bacteria.

Bacteria by Section of *Bergey's Manual*

- The genera of medical significance and their characteristics are summarized in Table 10.5.
- Pathogens and the diseases they cause are summarized in Table 10.5. "Nonmedical" bacteria are important in ecosystems and may represent some of nature's experiments with the origin of life itself.

RELATED KEY TERMS

micrococci **streptococci**
peptococci **hemolysis**
beta hemolysis **alpha hemolysis**
viridans group
lactobacilli **corynebacteria**
propionibacteria **actinomyces**
mycobacteria **nocardioforms**
actinomycetes **streptomycetes**

QUESTIONS FOR REVIEW

A.

1. Describe at least three different kinds of bacteria using the characteristics in Table 10.1.
2. How are the biochemical tests in Table 10.2 used?

B.

3. Why is it difficult to classify bacteria by their evolutionary relationships?

C.

4. When was *Bergey's Manual* first published, and which is the current edition?
5. Of what use is *Bergey's Manual* to medical personnel?

D.

6. For any six diseases, identify the organism and the section of *Bergey's Manual* in which each belongs.
7. Name some organisms that cause disease in more than one body system.
8. Name some sections of *Bergey's Manual* that include organisms responsible for opportunistic infections.
9. Name some sections of *Bergey's Manual* that include organisms normally found on or in the human body.

PROBLEMS FOR INVESTIGATION

1. Identify and describe at least three genera of bacteria in which some species infect humans and others infect animals. Suggest how this situation might have arisen. Assess the likelihood that the animal diseases might someday appear in humans.
2. Use information about bacteria to present an argument to support the thesis that (a) autotrophs evolved before heterotrophs, (b) heterotrophs evolved before autotrophs, (c) bacteria and fungi have common ancestors, (d) bacteria and protists have common ancestors, or (e) bacteria and algae have common ancestors.
3. What benefits might bacteria derive from the antibiotics they produce?
4. Use *Bergey's Manual of Systematic Bacteriology* to look up the following information:
   a) Are members of the genus *Serratia* motile?
   b) What lab test could you use to distinguish *Proteus vulgaris* from *P. mirabilis?*
   c) What percent of DNA to DNA hybridization (homology) exists between *Treponema pallidum* and *T. pertenue?* What conclusion is drawn on the basis of this?
   d) What shape(s) are members of the genus *Mycoplasma?*
   e) Is *Staphylococcus aureus* able to ferment mannitol with production of acid?
5. Explain some of the ways bacteria contribute to recycling in ecosystems.
6. Explain how some particular physical or biochemical characteristic of a given bacterium has helped it adapt to a given specialized niche in the environment.

SOME INTERESTING READING

Buchanan, R. E., and N. E. Gibbons, eds. 1974. *Bergey's manual of determinative bacteriology*. 8th ed. Baltimore: Williams and Wilkins.

Edmond, J. M., and K. Von Damm. 1983. Hot springs on the ocean floor. *Scientific American* 248, no. 4 (April):78.

Holt, J. G., and N. R. Krieg, eds. 1984. *Bergey's manual of systematic bacteriology*. Vol. 1. Baltimore: Williams and Wilkins. (Contains sections 1–11, along with a useful essay on classification of bacteria by J. T. Staley and N. R. Krieg.)

Holt, J. G., P. H. A. Sneath, N. S. Mair, and M. E. Sharpe, eds. 1986. *Bergey's manual of systematic bacteriology*. Vol. 2. Baltimore: Williams and Wilkins. (Contains sections 12–17; volume 3 will contain sections 18–23, and volume 4 will contain sections 27–30.)

Lederberg, J. 1992. Bacterial variation since Pasteur. *ASM News* 58(5):261–65.

Monastersky, R. 1988. Bacteria alive and thriving at depth. *Science News* 133 (March 5):149.

Smith, S. 1990. Afterlife of a whale (whale carcasses may be stepping stones for hydrothermal vent-dwelling creatures). *Discover* 11 (February):46–50.

Stanier, R. Y., et al. 1986. *The microbial world*. Englewood Cliffs, NJ: Prentice Hall.

Weisburd, S. 1986. First fossils of slime bacteria studied. *Science News* 130 (November 29):347.

Viruses bud off into the bloodstream.

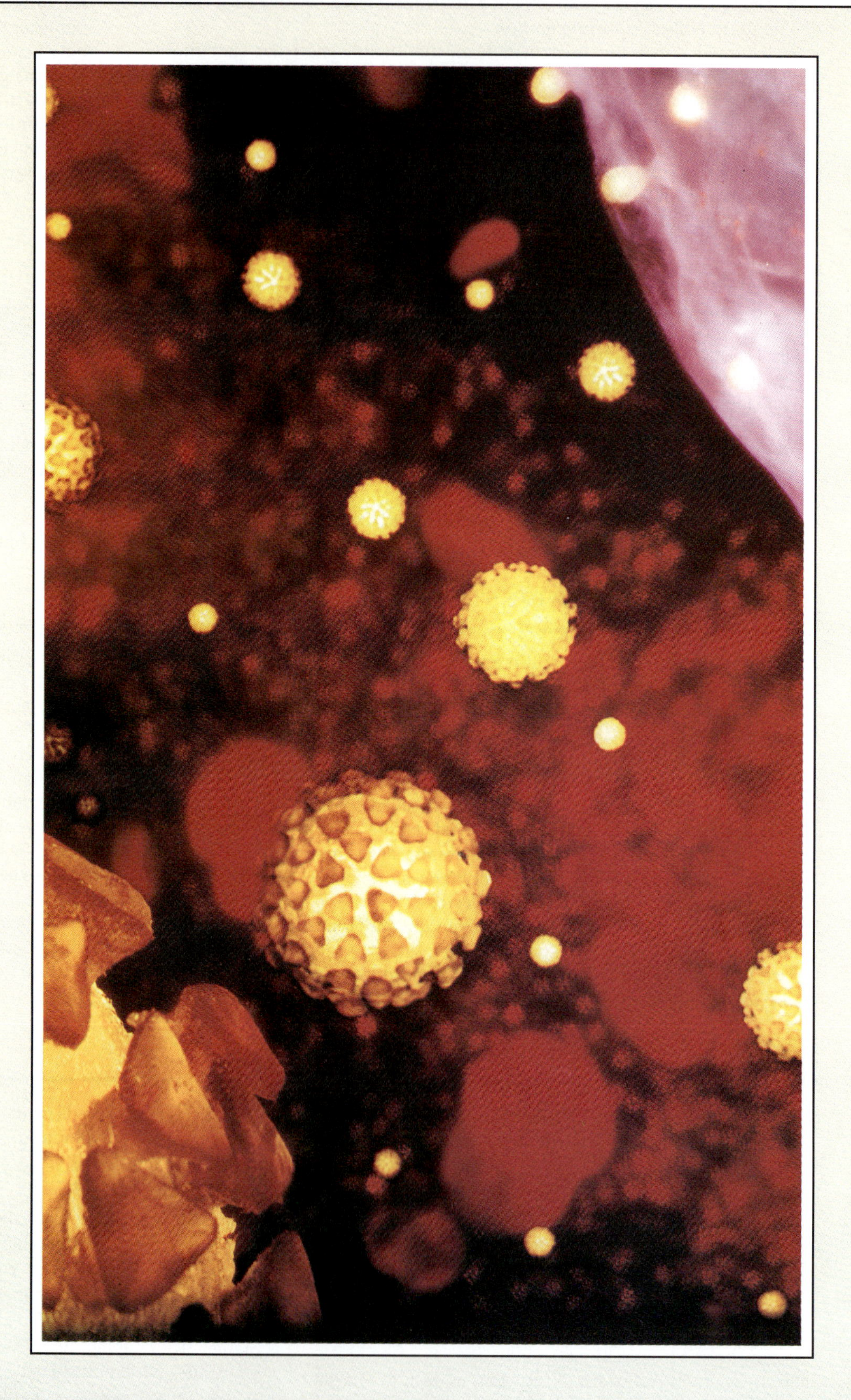

11 Viruses

This chapter focuses on the following questions:

A. What are the general properties of viruses?

B. How are viruses classified?

C. What are the properties of viroids and prions?

D. How do viruses replicate in general?

E. How do lytic and temperate bacteriophages replicate?

F. How do animal viruses replicate?

G. How were methods developed to culture animal viruses?

H. What types of viral cultures are currently in use?

I. What is a teratogen, and how do viruses act as teratogens?

As antibiotics have reduced the incidence of bacterial infections, human viral infections, which usually do not respond to antibiotics, have become increasingly apparent and important to health scientists. Much remains to be learned about controlling viral infections in humans, but research on molecular biology of viruses has provided insight into viral diseases—an important point for health scientists. It has also increased our understanding of the fundamental nature of life—an important point for all biologists.

GENERAL CHARACTERISTICS OF VIRUSES

What Are Viruses?

Viruses are infectious agents too small to be seen with a light microscope. They are not cells. When viruses invade cells, they display some properties of living organisms and so are on the borderline between living and nonliving. Viruses are **obligate intracellular parasites;** they **replicate,** or multiply, only inside a living host cell. They share that distinction with chlamydias and rickettsias. The traditional definition of viruses might have to be reconsidered following the announcement in December 1991 of success in growing virus particles in "cell sap" from lysed cells. No intact living cells were present.

Viruses differ from cells in several important ways. Whereas cells contain both RNA and DNA, grow, and divide, viruses contain only one kind of nucleic acid, either RNA or DNA, and do not grow, and never divide. Viruses multiply by directing synthesis and assembly of viral components inside cells to form new viruses. When not in a host cell, a virus is an inert unit of macromolecules—nucleic acid and proteins. Indeed, they have aptly been described as "a piece of bad news wrapped up in protein." The name *virus* itself comes from the Latin word meaning "poison."

Components of Viruses

The major components of viruses are a central nucleic acid and a protein coat called a **capsid.** Certain viruses also contain enzymes, and some have a bilayer membrane called an **envelope.** A complete virus particle, including its envelope if it has one, is called a **virion.** The components of a typical virus are shown in Figure 11.1.

Nucleic Acids

All viruses contain a nucleic acid, either DNA or RNA, which constitutes its **genome,** or genetic information. The nucleic acid can be single-stranded or double-stranded, and linear, circular, or segmented (existing as several fragments). All genetic information in RNA viruses is carried by RNA. RNA genomes occur only in viruses.

Capsids

Each viral nucleic acid is enclosed within a capsid. This capsid consists of one or more proteins specific to the virus. It protects the nucleic acid from the environment and determines the shape of the virus. Capsids also play a key role in the attachment of some viruses to host cells. Capsids can be helical, icosahedral, or complex in shape. A helical capsid consists of a ribbonlike protein wound to form a cylinder around the nucleic acid. An icosahedral capsid is a regular geometric structure with 20 triangular faces that forms a shell around the nucleic acid. Such a capsid consists of multiple protein subunits called **capsomeres.**

Envelopes

Enveloped viruses have a typical bilayer membrane outside their capsids. Such viruses acquire their envelopes after they are assembled in a host cell as they bud through a membrane. Enveloped viruses are damaged by conditions that damage cell membranes—increased temperature, freezing and thawing, pH below 6 or above 8, lipid solvents, and some chemical disinfectants such as chlorine, hydrogen peroxide, and phenol.

A virus that lacks an envelope is called a **naked virus.** Lacking the delicate membranous envelope, naked viruses are generally more resistant to changes

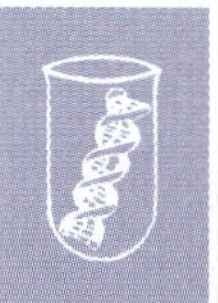

BIOTECHNOLOGY

First Replication of a Virus Outside Living Cells

In December 1991, Dr. E. Wimmer, Dr. A. Molla, and Dr. A. Paul of the State University of New York at Stony Brook reported in the journal *Science* that they had synthesized entire polioviruses in test tubes containing ground-up human cells, but no live cells. RNA from polioviruses was added to the cell-free extract, and about 5 hours later complete new virus particles began to appear. New RNA genomes had been copied, and proteins for viral coatings (capsids) were manufactured and then enclosed the RNA cores. This surprising development in virology will make it easier to study viral replication and perhaps to control it.

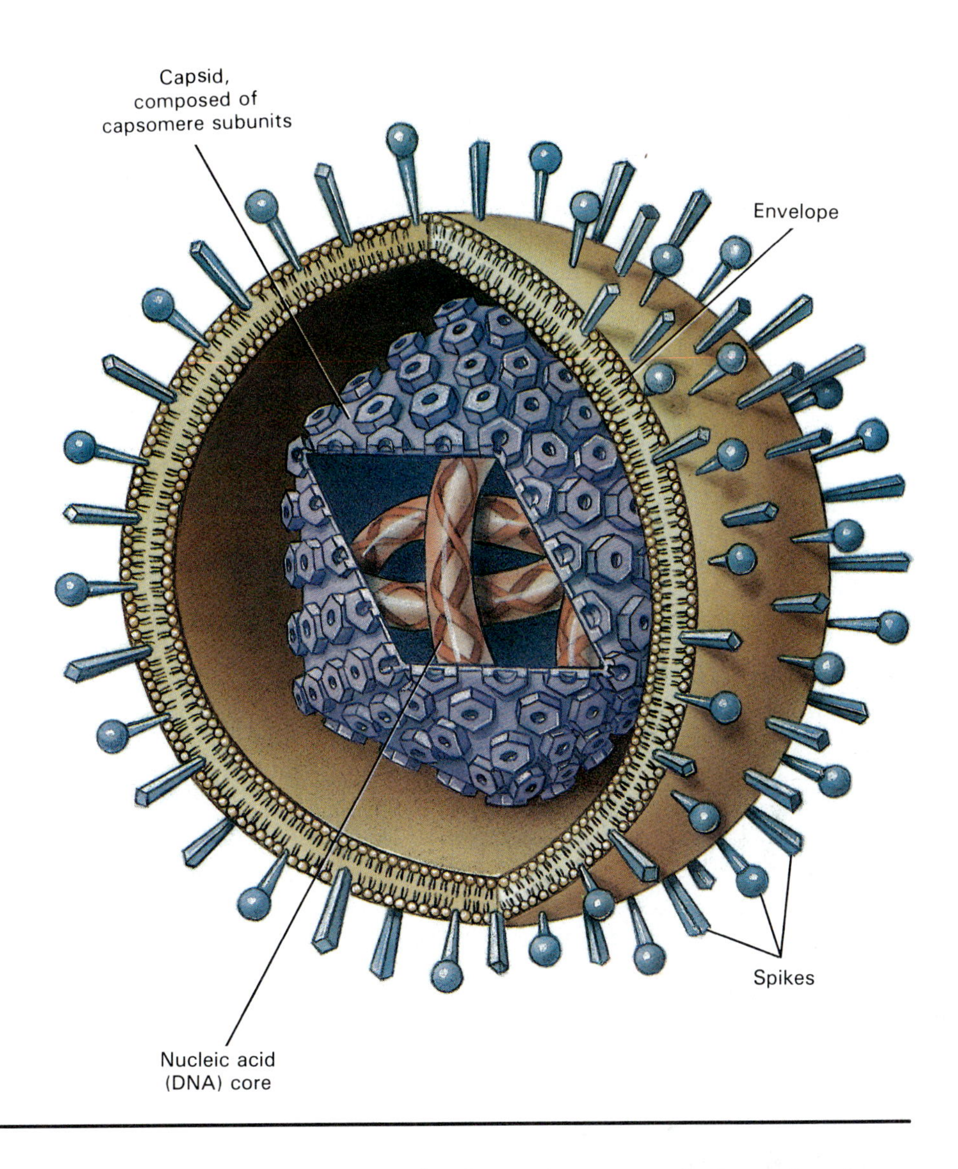

FIGURE 11.1 The components of a typical virus (a herpesvirus).

in temperature or pH and to disinfectants and host defenses than enveloped viruses.

Some proteins in viral envelopes have specialized functions. **Glycoproteins** are long, spikelike molecules, also known as **spikes,** that project beyond the surface of an envelope. Some glycoproteins serve to attach the virus to specific receptor sites on the surface of host cells. Others play a role in the fusion of viral and cellular membranes. Certain proteins, called **matrix proteins,** or M proteins, are found inside the envelope. They contribute to the structure of the envelope and sometimes assist in the assembly of components into new viruses.

Shapes and Sizes

Viruses are helical or icosahedral, according to the shape of their capsids. The virus that causes rabies and the tobacco mosaic virus, which infects tobacco plants, are helical viruses. Polioviruses and herpesviruses are icosahedral viruses. Icosahedral viruses are sometimes seen as crystalline arrays in electron micrographs. Some viruses are called **complex viruses** because they have envelopes or specialized structures such as heads and tails. Most enveloped viruses are spherical, but some are pleomorphic, that is, variable in shape. Bacteriophages are complex viruses with heads, tails, and structures for attaching to hosts and injecting their nucleic acid into them.

Viruses range in size from large brick-shaped poxviruses 250 nm by 350 nm (about 1/10 the size of a red blood cell) to the smallest bacteriophage, only 5 nm in diameter. Viruses that infect humans range from the large poxvirus to the small poliovirus, which is only about 25 nm in diameter. The diversity in sizes and shapes of viruses is illustrated in Figure 11.2.

Host Range and Specificity of Viruses

Although viruses are small and relatively simple, one or more of them are capable of infecting every living organism. The **host range** of a virus refers to the different kinds of organisms it can infect. Though polioviruses can be grown in the laboratory in monkey

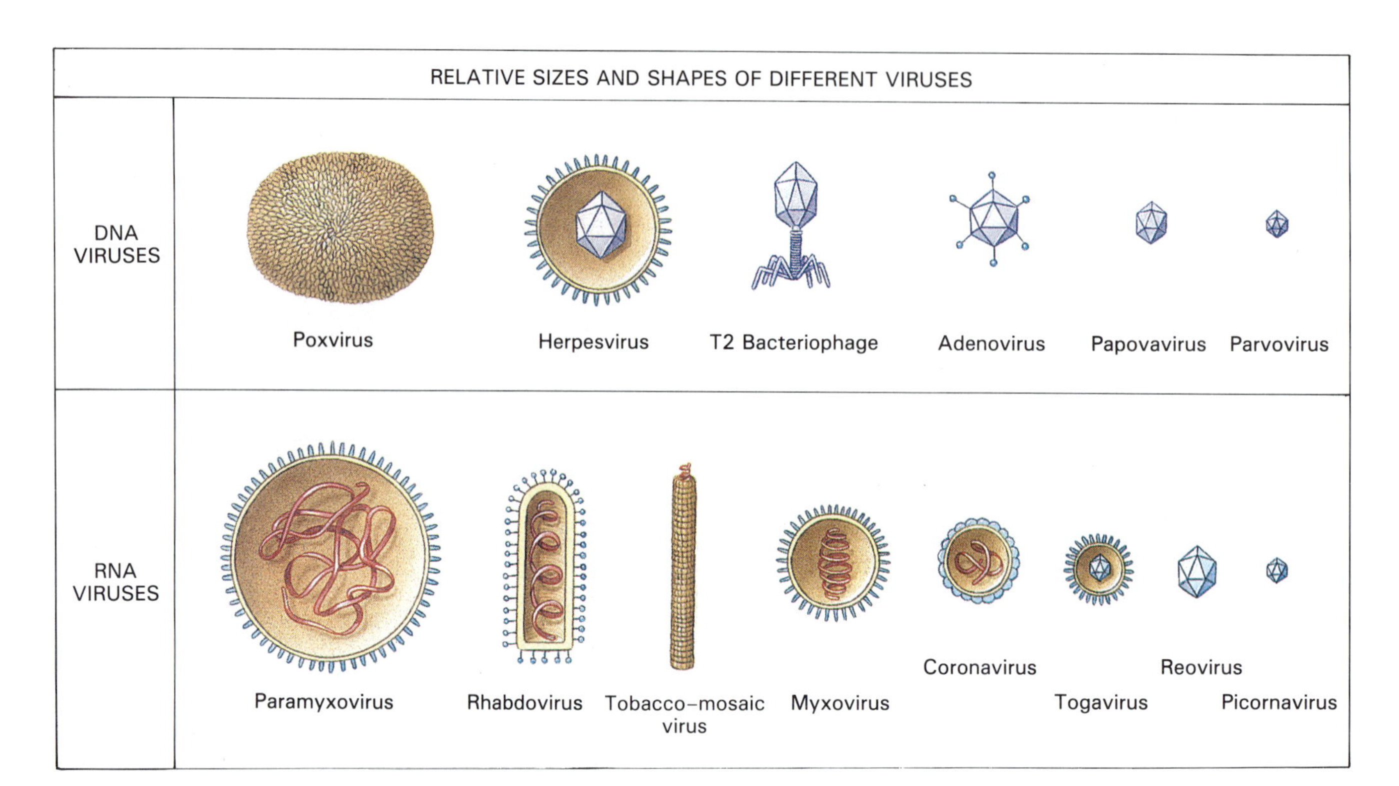

FIGURE 11.2 Variations in shapes and sizes of viruses.

APPLICATIONS

Plant Viruses

Virus specificity extends to viruses that infect plants and insects. Because plant viruses cause serious crop losses, much research has been done on them. As a result of this research tobacco mosaic virus was the first virus to be purified. Other viruses are now known to infect cowpeas and carnations and food crops such as potatoes, beets, cucumbers, tomatoes, lettuce, beans, corn, cauliflower, and turnips. Insects also cause serious crop losses, and researchers now hope to control some crop-destroying insects by infecting them with viruses.

The infectious yellows virus, carried by whiteflies, causes discoloration and stunted growth in lettuce.

The beautiful streaks in these tulips are caused by a viral infection. Unfortunately, the infection (which can spread from plant to plant) also weakens the tulips somewhat.

kidney cells, they have never been observed to cause infection in any species but humans. In contrast, the rabies virus attacks cells of the central nervous system in humans and a wide variety of warm-blooded animals.

The **specificity** of a virus refers to the specific kinds of cells the virus can infect. Wart viruses, for example, are so specific that they infect only skin cells. By contrast, certain viruses called cytomegaloviruses attack cells of salivary glands, the gastrointestinal tract, liver, lungs, and other organs. They can also cross the placenta and attack fetal tissues, especially the central nervous system. The discovery that viruses can cause symptoms in several different body systems made the one virus–one disease concept untenable.

Viral specificity is determined mainly by whether a virus can attach to a cell. Attachment depends on the presence of specific receptor sites on the host cell surface and on specific structures on the capsid or envelope of the virus. Specificity is also affected by whether the appropriate enzymes and other proteins the virus needs to replicate are available inside the cell. Finally, specificity is affected by whether replicated viruses can be released from the cell to spread the infection to other cells.

CLASSIFICATION OF VIRUSES

Before much was known about the structure or chemical properties of viruses, they were classified by where they were found or what organs they infected. Thus, they have been classified as bacterial viruses, plant viruses, or animal viruses. They also have been classified as dermotropic if they infect the skin, neurotropic if they infect nerve tissue, viscerotropic if they infect organs of the digestive tract, or pneumotropic if they infect the respiratory system.

As more has been learned about the structure of viruses at the chemical level, a system reflecting that knowledge has been developed. Today viruses are classified by the type and structure of their nucleic acids, other chemical and physical characteristics, method of replication, and host range. As new viruses were discovered, many competing and conflicting classification systems grew up, resulting in much confusion and some bad feelings. The need for a single, universal taxonomic scheme for viruses led to the establishment in 1966 of a body now called the International Committee on Taxonomy of Viruses (ICTV). This committee established the system that has been built on ever since; it is summarized in Appendix C. Because viruses are so different from other organisms, it is difficult to classify them according to typical taxonomic categories—kingdom, phylum, and the like. Until now, family has been the highest category used by the ICTV, but as we learn more about the evolutionary relationships of viruses, higher taxa will eventually be established.

Currently the ICTV has assigned over 1400 viruses to 61 approved families, with an additional 500 or so viruses placed as "probable" or "possible" members of these groups. Of the 61 families, 21 contain viruses that cause important infections of humans and some animals. Others contain viruses that infect only animals, plants, fungi, algae, or bacteria.

Viral genera have been established but are often new and rarely used by medical virologists. In addition, the problems of defining and naming viral species have not yet been resolved. Controversy cen-

CLOSE-UP

Naming Viruses

Virologists have been creative in naming their charges. The name of the picornaviruses is a combination of *pico*, which means small, and *rna* for the ribonucleic acid. Someone must have thought the envelope of a togavirus resembled a toga, or cloak, and named it accordingly. The coronaviruses too seem to have been named for their appearance in electron micrographs (*corona*, Latin for *crown*). When it became possible to culture viruses, some were isolated that could not be linked to a known disease and that did not cause disease in laboratory animals. These viruses without a disease were dubbed "orphans." Thus, the echoviruses are enteric cytopathic human orphan viruses and the reoviruses are respiratory enteric orphan viruses. Retroviruses (*retro*, Latin for *backward*) use their own RNA to direct synthesis of DNA, reversing the usual process of transcription. Parvoviruses, as their name suggests, are diminutive even for viruses (*parvus*, Latin for *small*). And dependoviruses, not surprisingly, depend on the simultaneous presence of other viruses to replicate.

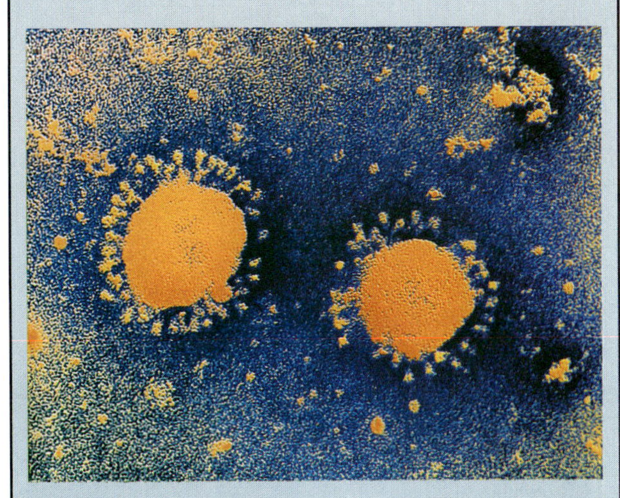

Can you see how the coronavirus acquired its name?

ters on how to distinguish a species from a strain. Meanwhile, the ICTV directs that the English common name be used to designate a species, rather than latinized binomial terms such as *Herpesvirus varicellae*. Thus, the full formal taxonomic designation for a virus would be, for example: family Rhabdoviridae, genus *Lyssavirus*, rabiesvirus.

The major groups of viruses are defined by their nucleic acid as the RNA viruses and the DNA viruses. Further subdivisions are based mainly on other characteristics of nucleic acids, as discussed next. The names of specific viruses often consist of a group name and a number, such as herpesvirus type 1 or type 2. Sometimes viruses are named for the disease they cause, such as poliovirus or measles virus. Classification of viruses is summarized in Table 11.1.

To understand how viruses are classified according to their nucleic acids, we must first distinguish between positive sense and negative sense nucleic acids. A nucleic acid that encodes the information for making proteins needed by a virus is called a **positive sense nucleic acid.** A nucleic acid made up of bases complementary to those of a positive sense nucleic acid is called a **negative sense nucleic acid.** (Recall the discussion of complementary base pairing in Chapter 2.) ∞ (p. 45) Host cell ribosomes reading the base sequence of a negative sense nucleic acid would make the wrong proteins. Viruses with negative sense nucleic acids also contain an enzyme called a **transcriptase.** Inside the host cell this enzyme uses the negative sense nucleic acid as a template (Chapter 7) and makes a complementary positive sense nucleic acid.

RNA Viruses

General Properties of RNA Viruses

Five classes of RNA viruses have been defined according to the properties of their genome (Figure 11.3 and Table 11.1). Class I RNA viruses contain a single piece of positive sense, single-stranded RNA and lack enzymes. Class II RNA viruses contain a single piece of negative sense, single-stranded RNA and a transcriptase enzyme. Class III RNA viruses contain several small segments of negative sense, single-stranded RNA. Class IV RNA viruses have segmented double-stranded RNA (one positive sense and one negative sense). Class V RNA viruses contain two identical positive sense strands of RNA and a special enzyme called a **reverse transcriptase,** which copies RNA into DNA.

Important Groups of RNA Viruses

Picornaviruses **Picornaviruses** are small, spherical (27–30 nm in diameter), naked Class I RNA viruses. They enter cells by a type of phagocytosis and quickly interrupt all functions of DNA and RNA in the host cell. Picornaviruses are divided into two major groups, enteroviruses and rhinoviruses.

TABLE 11.1 Classification of viruses

Class	Nature of Nucleic Acid	Envelope and Shape	Typical Size (nm)	Example	Diseases
RNA Viruses					
Class Ia	Positive, single-stranded RNA	Naked, polyhedral	30	Picornaviruses	Poliomyelitis, common cold
Class Ib	Positive, single-stranded RNA	Enveloped, polyhedral	40–70	Togaviruses	Encephalitis, yellow fever
Class II	Negative, single-stranded RNA	Enveloped, helical	12–15 × 150–300	Paramyxoviruses	Measles, mumps
			70–85 × 130–380	Rhabdoviruses	Rabies
Class III	Negative, segmented, single-stranded RNA	Enveloped, helical	90–120	Orthomyxoviruses	Influenza
			50–300	Arenaviruses	Hemorrhagic fevers
Class IV	Segmented, double-stranded RNA	Naked, polyhedral	60–80	Reoviruses	Respiratory and gastrointestinal infections
Class V	Positive, single-stranded RNA, two strands	Enveloped, helical	80–130	Retroviruses	Leukemia, tumors, and AIDS
DNA Viruses					
Class Ia	Double-stranded, linear DNA	Naked, polyhedral	70–90	Adenoviruses	Respiratory infections
Class Ib	Double-stranded, linear DNA	Enveloped, polyhedral	150–200	Herpesviruses	Oral and genital herpes, chickenpox, shingles, mononucleosis
Class Ic	Double-stranded linear DNA	Enveloped, complex shape	160–260 × 250–450	Poxviruses	Smallpox, cowpox
Class II	Double-stranded circular DNA	Naked, polyhedral	45–55	Papovaviruses	Warts
Class III	Single-stranded linear DNA	Naked, polyhedral	18–26	Parvoviruses	Roseola in children, aggravates sickle cell anemia

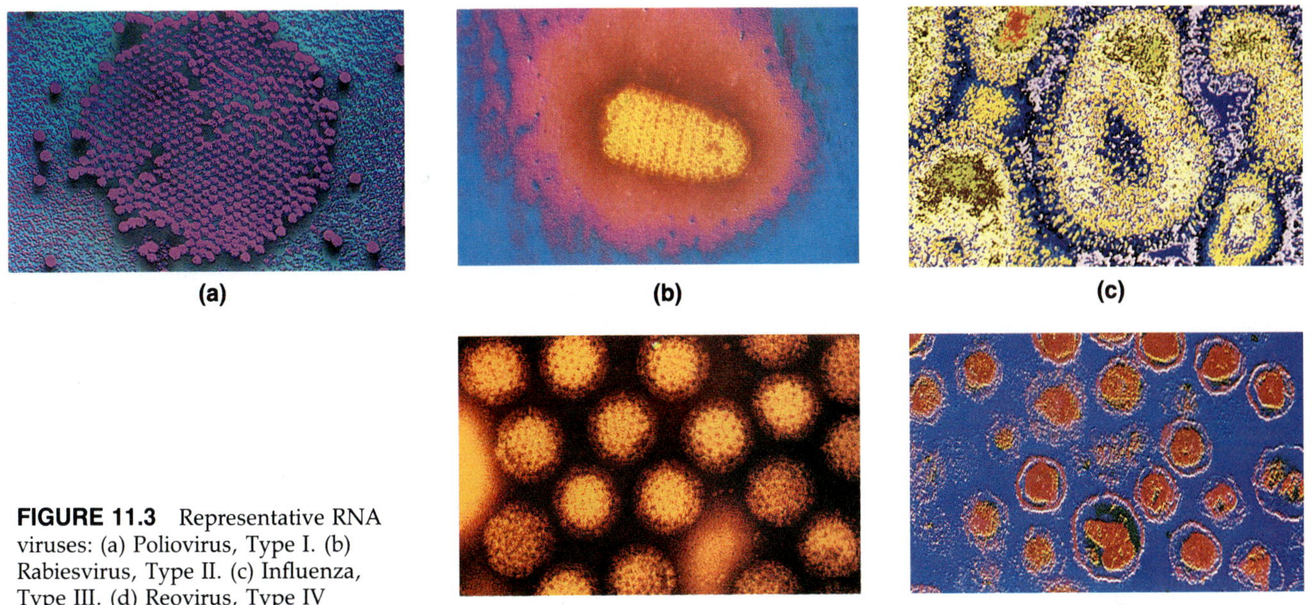

FIGURE 11.3 Representative RNA viruses: (a) Poliovirus, Type I. (b) Rabiesvirus, Type II. (c) Influenza, Type III. (d) Reovirus, Type IV (250,000X). (e) Retrovirus, Type V.

The **enteroviruses** include polioviruses, coxsackieviruses, and echoviruses. Like other naked viruses, enteroviruses are resistant to many chemical substances and can pass through the digestive tract unharmed. They can be inactivated by hydrochloric acid, chlorine and other halogens, drying and heat, light, and certain dyes. These viruses are distributed worldwide and can be isolated year round in the tropics and subtropics, but are most common from midsummer to early fall in temperate climates. Poor sanitation increases the numbers of these viruses, and crowding facilitates their spread. As a result of early and frequent exposure, children living in such conditions develop immunity to them. Enteroviruses replicate first in mucous membranes of the throat and spread to the intestinal tract. Unless inactivated by defense mechanisms, they invade the blood and lymph and spread throughout the body. Enteroviruses can infect nerve and muscle cells, the respiratory tract lining, and skin. They grow readily in cell cultures.

The **rhinoviruses** (*rhin,* Latin for *nose*), of which there are more than 100 types, cause the common cold (Figure 11.4). They replicate in epithelial cells of the upper respiratory tract. They do not cause digestive tract diseases because they cannot survive acid conditions in the stomach. Recently much has been learned about the nucleotide sequences of the genomes of several types of rhinoviruses and their outer structure, making rhinovirus one of the best understood *mammalian viruses.*

Togaviruses **Togaviruses** are small, enveloped Class I RNA viruses that multiply in the cytoplasm of many mammalian and arthropod host cells. Many can be grown in laboratory cultures, including mosquito cells. Like other enveloped viruses, they are inactivated by agents that disrupt membranes. Togaviruses transmitted by arthropods cause several kinds of encephalitis and tropical fevers such as dengue fever and yellow fever. The **rubellavirus,** which causes rubella or German measles, also is a togavirus, but it is not known to be transmitted by arthropods.

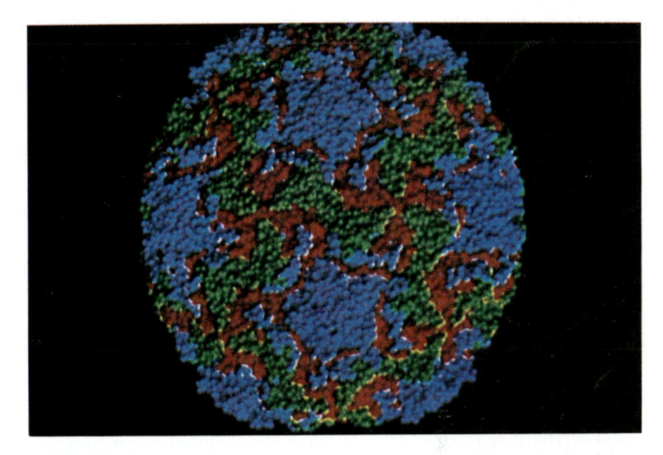

FIGURE 11.4 Computer-generated model of a human rhinovirus, cause of the common cold. The colors represent different protein units of the capsid.

Paramyxoviruses **Paramyxoviruses** are medium-sized, enveloped Class II RNA viruses that have an affinity for mucus (*myxo,* Latin for *mucus*). They cause mumps, measles, croup, viral pneumonia, and bronchitis in children and mild upper respiratory infections in young adults.

Rhabdoviruses Also Class II RNA viruses, rhabdoviruses are medium-sized, enveloped viruses. Although their capsids are helical, with their envelopes they are nearly rod-shaped (*rhabdo*, Latin for *rod*). Rhabdoviruses infect insects, fish, various other animals, and some plants. Many are transmitted by insects. Among the rhabdoviruses, only the rabiesvirus ordinarily infects humans. The Lago virus, which infects bats, and the Mokolo virus, which infects shrews in Africa, are closely related to rabiesviruses and have been isolated from humans on rare occasions.

Orthomyxoviruses **Orthomyxoviruses** are medium-sized, enveloped Class III RNA viruses that vary in shape from spherical to filamentous. They, too, have an affinity for mucus. These viruses cause influenza.

Bunyaviruses and arenaviruses The **bunyaviruses** and **arenaviruses** also are enveloped Class III RNA viruses that display a wide range of sizes. Bunyaviruses have three segments of RNA; arenaviruses have two. Both can be transmitted by arthropods, often have rodents as a principal host, and replicate in the cytoplasm of host cells. Bunyaviruses acquire their envelopes from host cell endoplasmic reticulum, whereas arenaviruses acquire theirs from the cell membrane itself. A bunyavirus causes California encephalitis, an inflammation of the brain. Arenaviruses cause Lassa fever and a variety of other hemorrhagic fevers, in which blood vessels in the skin, mucous membranes, and internal organs can rupture.

Reoviruses **Reoviruses** have a double capsid with no envelope. They are medium-sized Class IV RNA viruses. They replicate in the cytoplasm and form distinctive inclusions that stain with eosin. During replication the virus is never completely uncoated, and replication of the RNA is conservative: one strand of RNA is transcribed and then serves as a template for replication of a new partner strand. Reoviruses include orthoreoviruses (true reoviruses), orbiviruses, and rotaviruses. Some of these viruses are probably responsible for minor upper respiratory and gastrointestinal infections in humans. Some rotaviruses cause severe gastroenteritis in infants and young children.

Retroviruses **Retroviruses** are enveloped Class V RNA viruses with a diameter of 80 to 130 nm. They use their own reverse transcriptase to transcribe their RNA into DNA in the cytoplasm of the host cell. The DNA migrates to the nucleus and becomes circular (like a bacterial chromosome, but smaller). Several molecules of viral DNA are incorporated into host chromosomes; such DNA is referred to as a **provirus.** Proviruses are replicated with the host DNA whenever infected host cells prepare to divide.

As proviruses, retroviruses can remain inactive

PUBLIC HEALTH

AIDS-like Retrovirus Found in Cats

In 1987, Dr. Niels Pedersen, working at the University of California, Davis, isolated a retrovirus from cats. Called *f*eline *i*mmunodeficiency *v*irus (FIV), it is remarkably similar to the human immunodeficiency virus (HIV), including a similar latency period of 3 to 6 years. Unlike HIV, FIV does not seem to be sexually transmitted and does not seem to be transmissible to humans or any animals other than cats. Transmission is mostly through biting, which may account for the fact that the disease is three times more frequent among male cats than females. Approximately 1 to 3 percent of randomly tested U.S. cats are infected, with the disease being found uniformly across the country. Cats that are chronically ill are five times more likely to be carrying the virus than healthy cats. Veterinarians feel that the virus has been around a long time, possibly for decades.

After infection, the virus localizes in the cat's lymph nodes but causes no noticeable symptoms. Gradually the immune system is attacked, and between 3 and 6 years later the animal begins to suffer from frequent infections of the mouth, diarrhea, weight loss, skin and respiratory infections, pneumonia, fever, and neurologic disease. Eventually it either dies of these infections, or the owner has it put to sleep. So far there is no cure for FIV.

A highly reliable test is available for about $40 and should be used on outdoor cats with symptoms and on strays that people plan to adopt, particularly if there are other cats in the house. A cat found to be positive for FIV may have many good years left. However, it should be kept indoors, isolated from other cats.

A virus closely related to FIV has also been found in zoo lions and tigers and in wild panthers. However, the infection in wild felids may be largely asymptomatic. In addition, immunodeficiency viruses are found in cattle, horses, monkeys, goats, and sheep.

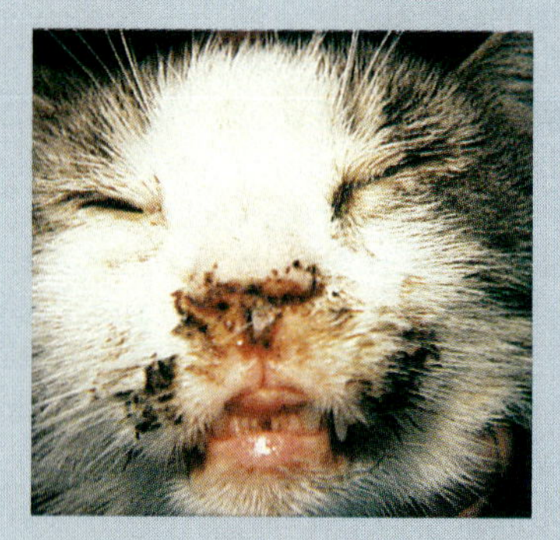

This cat, infected with FIV, has very little immune function left and suffers from herpetic lesions of the nose and oral cavity, as well as feline leukemia. Opportunistic infections are the hallmark of cats with FIV as well as of human patients with AIDS.

and unrecognized in host DNA. If some event occurs that derepresses the provirus, its genes are expressed. The genes are used to make viral genome RNA and viral mRNA, which directs synthesis of viral proteins. Retroviruses that infect humans invade T lymphocytes and are called *h*uman *T* *l*ymphocyte *v*iruses (HTLV). HTLV-I and HTLV-II are associated with malignancies, and HTLV-III, now called the *h*uman *i*mmunodeficiency *v*irus (HIV-1 and HIV-2 strains), causes *a*cquired *i*mmune *d*eficiency *s*yndrome (AIDS). AIDS is discussed in Chapter 19. Other HIV viruses are currently being characterized.

DNA Viruses

General Properties of DNA Viruses

DNA viruses are divided into three classes according to the properties of their DNA (Figure 11.5 and Table 11.1). Class I DNA viruses have double-stranded linear DNA. Among them are naked adenoviruses and enveloped herpesviruses and poxviruses. The poxviruses have more complex shapes than the herpesviruses. Class II DNA viruses have double-stranded circular DNA; papovaviruses are representative of this class. Class III DNA viruses have single-stranded linear DNA; parvoviruses fall in this class.

Important Groups of DNA Viruses

Adenoviruses **Adenoviruses,** discovered in 1953, are medium-sized, naked Class I DNA viruses that are highly resistant to chemical agents and stable from pH 5 to 9 and from 47° to 36°C. Freezing causes little loss of infectivity. Adenoviruses now can be cultured in a variety of human cells. They are icosahedral in shape and have **pentons** (units of the capsomere) with a penton fiber at each "corner." Penton fibers can cause red blood cells to clump, a process called hemagglutination. Different types of adenoviruses can be identified by whether they agglutinate red blood cells from Rhesus monkeys or rats, by the percentage of G + C in their DNA, and by their oncogenic (malignancy-inducing) potential in laboratory animals (Chapter 7).

More than 40 different types of adenoviruses, referred to as **serotypes,** have been identified, and many have been related to human respiratory disease. Two recently identified fastidious adenoviruses (types 40 and 41) cause 10 to 30 percent of all cases of severe diarrhea in babies and young children. Many cases have been traced to day care centers and other institutions. Only half the children carrying the virus in their throats actually become ill.

Adenoviruses can be isolated from half of surgically removed tonsils and adenoids and occasionally from kidneys. They can remain latent in tissues and be shed intermittently, especially from the gut, for years, and maybe for life. Diseases caused by adenoviruses are generally acute (with sudden onset and short duration). Soon after entering the body, the virus appears in the blood and a measleslike rash may develop. Lung infections and meningoencephalitis (inflammation of the brain and membranes around it) can be severe. In rare fatal infections, viruses can be recovered from most of the organs.

Herpesviruses **Herpesviruses** (HV) are relatively large, enveloped Class I DNA viruses. With their envelopes, herpesviruses are 150 to 200 nm in diameter. The capsid itself is about 100 nm in diameter. The core of the virion contains proteins around which the DNA is coiled. Information in the DNA codes for 49 proteins. Of these proteins, 33 form part of the virion; the others probably are enzymes that are released by the virus and act on the host cell.

A universal property of herpesviruses is **latency,** the ability to remain in host cells, usually in neurons, for long periods of time and to retain the ability to replicate. During a period of latency no viruses are found in the blood and no symptoms of infection are seen. Koch's postulates would be difficult or impossible to fulfill while the virus remains sequestered within the cells. However, the latent virus may be reactivated at any time.

Herpesviruses are widely distributed in nature, and most animals are infected with one or more of the 80 that have been characterized. Although many grow slowly and do not kill their hosts, Lucke's virus,

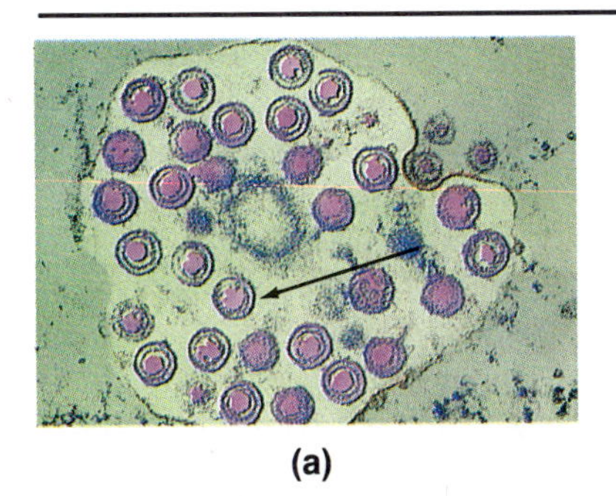

(a)

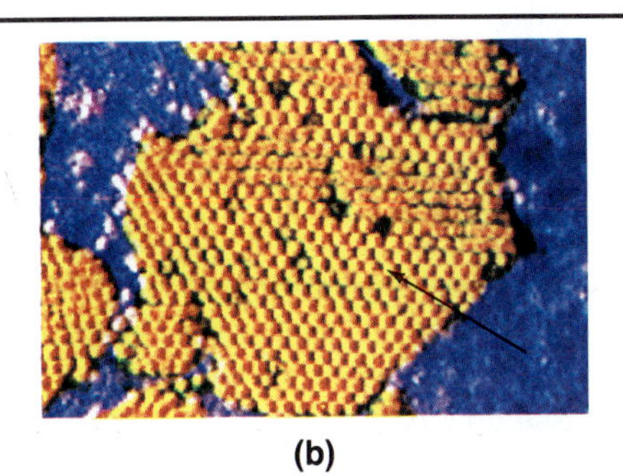

(b)

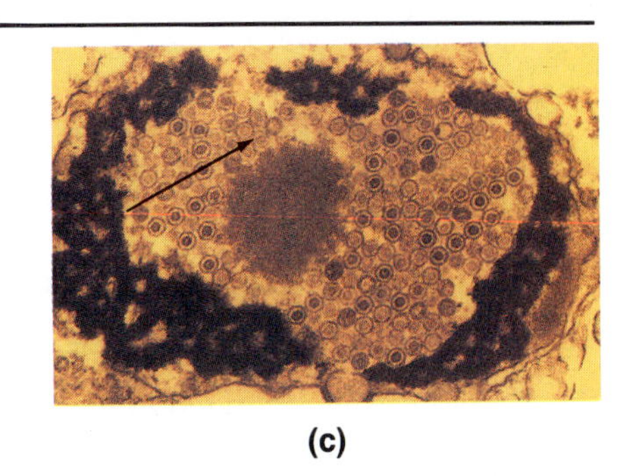

(c)

FIGURE 11.5 Representative DNA viruses: (a) Herpesvirus (pink spheres within the cell), Type I. (b) Papillomavirus, Type II. (c) Parvovirus, Type III.

TABLE 11.2 Herpesviruses that cause human disease

Group	Virus	Disease
Alpha	Herpes simplex type 1	Oral herpes (sometimes genital and neonatal herpes), encephalitis
	Herpes simplex type 2	Genital and neonatal herpes (sometimes oral herpes), meningoencephalitis
	Varicella-zoster	Chickenpox and shingles
Beta	Cytomegaloviruses (salivary gland virus)	Acute febrile illness; infections in AIDS patients, transplant recipients, and others with reduced immune system function; a leading cause of birth defects
Gamma	Epstein-Barr virus	Infectious mononucleosis and Burkitt's lymphoma (a malignancy of the jaw seen mainly in African children); also linked to Hodgkin's disease (a malignancy of lymphocytes) and B cell lymphomas, and to nasopharyngeal cancer in oriental people
Unclassified	Human herpesvirus 6	Exanthema subitum (roseola infantum), a common disease of infancy featuring rash and fever

which causes frog kidney carcinoma, grows very rapidly. Herpesviruses cause a broad spectrum of diseases, possibly because their DNA can assume different arrangements, with each arrangement causing a different disease. Herpesviruses and associated human diseases are summarized in Table 11.2.

Poxviruses **Poxviruses,** also Class I DNA viruses, are the largest and most complex of all viruses. They are widely distributed in nature; nearly every animal can be infected by some poxvirus. They include the orthopoxviruses and the parapoxviruses. **Orthopoxviruses** are large, enveloped, brick-shaped nearly identical viruses 250 to 450 nm long and 160 to 260 nm wide. They replicate in the cytoplasm where they appear as **inclusion bodies** (Figure 11.6). They can cause smallpox and cowpox in humans. **Parapoxviruses** are ovoid structures, approximately 250 nm by 160 nm. They can infect the skin of humans having close contact with infected animals.

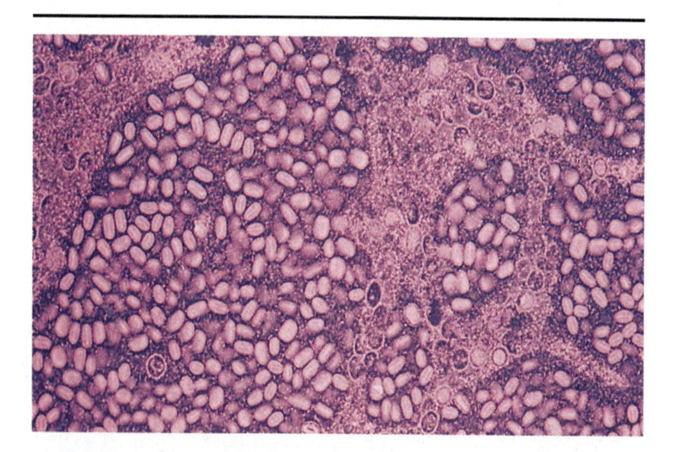

FIGURE 11.6 Oval inclusion bodies, composed of aggregates of poxviruses, inside the cytoplasm of skin cells.

Papovaviruses **Papovaviruses** are named for three related varieties of viruses, papilloma, polyoma, and vacuolating viruses. Papovaviruses are small, naked, Class II DNA viruses that replicate in the nuclei of their host cells. They are widely distributed in nature; more than 25 papillomaviruses and two polyomaviruses have been found in humans. **Papillomaviruses** (55 nm in diameter) are larger than **polyomaviruses** (45 nm in diameter). Papillomaviruses are frequently found in host cell nuclei without being integrated into host DNA; polyomaviruses are nearly always integrated. They cause both benign and malignant warts in humans. The most thoroughly studied **vacuolating virus** is a simian virus, SV-40, which has been used to explore the mechanisms of viral replication, integration, and oncogenesis. Recently it has been proposed that the papovaviruses be divided up into two separate groups, one for the papilloma viruses and another including the polyoma and vacuolating viruses.

Parvoviruses **Parvoviruses** are small, naked Class III DNA viruses 18 to 26 nm in diameter. Their DNA codes for structural proteins but not for any enzymes. To replicate, a parvovirus enlists the aid of an unrelated helper virus or a dividing host cell. Two groups of parvoviruses have been identified in vertebrates, the **parvoviruses proper** and the **dependoviruses.** Parvoviruses proper contain either positive sense DNA or negative sense DNA, but not both. Dependoviruses contain equal amounts of positive sense and negative sense DNA. Dependoviruses require co infection with adenoviruses and also are called adeno-associated viruses. B19, a parvovirus discovered in 1974, is responsible for erythema infectiosum ("fifth disease"), a rash illness of children but a cause of both rash and arthritis in adults. B19 can cross the placenta and damage blood-forming cells in the fetus, leading to anemia, heart failure, and even fetal death. In patients having sickle cell anemia, B19 causes 90 percent of all aplastic crises, events in which the body ceases to produce red blood cells. Canine parvovirus, which first appeared only about 20 years ago, is responsible for severe and sometimes fatal gastroenteritis in dogs and puppies.

VIROIDS AND PRIONS

Viroids

In 1971 plant pathologist O. T. Diener proposed the concept of a **viroid** (Figure 11.7a) to describe the properties of an infectious particle smaller than a virus. Since then viroids have been found to differ from viruses in four ways:

1. Each viroid consists of a single specific RNA.
2. Viroids exist inside cells as particles of RNA without capsids.
3. Viroid particles are not apparent in infected tissues without using special techniques to identify nucleotide sequences in the RNA.
4. Compared with viral nucleic acids, the RNA of viroids is a low-molecular-weight material.

At least three theories have been proposed to account for the origin of viroids. One is that they represent a segment deleted from a retroviral provirus. A second is that they may have originated early in precellular evolution when the primary genetic material probably consisted of RNA. A third is that they may be introns sliced out of mRNA. No one knows whether any of these theories is correct.

Viroids are known to cause several plant diseases, such as potato spindle tuber disease, chrysanthemum stunt, and cucumber pale fruit disease (Figure 11.7b). None of these diseases was recognized before 1922, and several have been identified recently. Some scientists believe that while isolated plants may have contained viroids for an unknown number of years, modern agricultural methods, such as growing large numbers of the same plant in close association and the use of machinery for harvesting, may have allowed viroid diseases to spread and facilitated their recognition. No viroid is presently known to infect animals, but there is no reason to suppose they cannot do so.

Prions

In the 1920s several cases of a progressive dementing illness were observed independently by Hans Gerhard Creutzfeldt and Alfons Maria Jakob. The disease was named Creutzfeldt-Jakob disease. Since that time other neurologic degenerative diseases, such as kuru in humans and scrapie in sheep, have been observed. Some researchers think these diseases, which are discussed in the Essay at the end of Chapter 24, may be caused by an exceedingly small infectious particle called a **prion**, or *pro*teinaceous *in*fectious particle. The agent may be a protein because it is too small to be a nucleic acid and is not destroyed by agents that digest nucleic acids. Other researchers question the existence

FIGURE 11.7 (a) T.O. Diener, plant physiologist who discovered viroids. (b) Viroid particles that cause potato spindle tuber disease (shown as yellow-green rods in this artist's rendition of an electron microscope photograph) are very short pieces of RNA containing a mere 300 to 400 nucleotides. The much larger strand in the micrograph is DNA from a T_7 bacteriophage. Such a comparison makes it easy to see how viroids were overlooked for many years. (c) Most viroids cause plant disease. The tomato on the left is normal, whereas the one on the right is infected with a viroid.

(a)

(b)

(c)

PUBLIC HEALTH

Delta Hepatitis Agent

Hepatitis delta virus (HDV), discovered in the mid-1970s, has not yet been classified. However, sequencing of its genome has shown it to be similar to the viroid and to virusoid RNAs that infect plants. HDV initially was thought to be part of the hepatitis B virus (HBV) because it was never found without the presence of hepatitis B infection. However, it was not present in all cases of hepatitis B, only in especially severe cases. By 1980 it was found to be a separate, defective pathogen that required coinfection with the hepatitis B virus in order to replicate. It can be prevented by vaccinating against HBV, because it cannot infect without its helper virus. HDV has the smallest genome of any known animal virus, with a length of a mere 1679 to 1683 nucleotides. In contrast, HBV is 3000 to 3300, and hepatitis A is about 7500 nucleotide units long. It lacks a well-defined capsid of its own and is surrounded by the portion of the HBV that codes for the surface antigen (formerly called the Australian antigen) of HBV. Further understanding of the HDV is limited because no one has been able to propagate it in cell culture. It is primarily transmitted by blood and blood products. Worldwide, 300 million people who are infected by HBV are at risk for a superinfection with HDV.

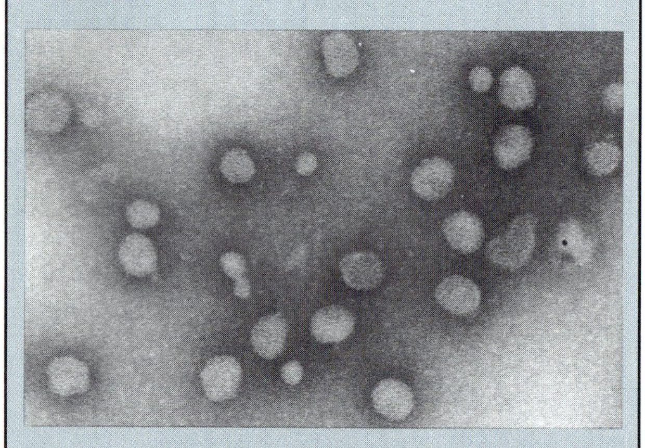

Electron micrograph of the hepatitis delta virus.

of prions and look for other causes of the diseases.

If prions exist, they may be proteins that have escaped from the cells that made them. They may have arisen as endosymbionts, later becoming greatly simplified. Or they may have arisen early in the evolution of life, even before there were nucleic acids.

A major problem with prions is to show how a protein can be self-replicating. Various investigators have proposed that prions might replicate by one of the following mechanisms:

1. Direct pairing of amino acids
2. Reverse translation to RNA
3. Activation of a cellular gene that directs their synthesis
4. Synthesis of a protein directed by a small nucleic acid protected by the protein or activation (again by a nucleic acid) of a cellular gene that accomplishes this

Too little information is presently available to support or refute any of the theories about prions.

VIRAL REPLICATION

General Characteristics of Replication

In general, viruses go through five steps in the replication process: (1) adsorption, (2) penetration, (3) synthesis of viral components, (4) maturation, and (5) release. **Adsorption** refers to the attachment of the virus to the host cell. **Penetration** refers to the entry of the virus (or its nucleic acid) into the host cell. **Synthesis** refers to the making of nucleic acid, coat proteins, and other viral components within the host cell using the cell's synthetic machinery. **Maturation** is the process by which complete virions are assembled from the newly synthesized components. **Release** is the departure from the host cell of new virions; it usually, but not always, kills the host cell.

Replication of Bacteriophages

Bacteriophages, or simply phages, were first observed in 1915 by Frederic Twort in England and in 1917 by Felix d'Herelle in France. D'Herelle named them bacteriophages, or "eaters of bacteria." Sporadic efforts have been made over the years to use phages to fight bacterial infections. Though these efforts were not successful until recently (see the box titled "Viruses instead of Drugs"), they did provide much useful information about bacteriophages. Even today, it is easier to study bacteriophages than other viral infections because they and their bacterial hosts are easier to manipulate in the laboratory than viruses that have multicellular hosts.

Properties of Bacteriophages

Like other viruses, bacteriophages can have either double-stranded or single-stranded RNA or DNA as their genomes. They can be relatively simple, or, as in the case of the T-even phages, complex in structure. T-even phages (Figure 11.8) are relatively well-studied phages designated T2, T4, and T6 (*T* stands for "type"). They have genomes of double-stranded DNA. The most widely photographed phage is the T4 phage, which has a distinct head, neck, and tail.

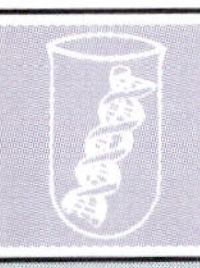

BIOTECHNOLOGY

Viruses Instead of Drugs

Recently, the Polish microbiologist Stefan Slopek and his colleagues successfully used bacteriophages to treat 138 patients with long-term antibiotic-resistant bacterial infections. Each patient received a bacteriophage that specifically attacks the bacterium causing the infection. All patients benefited from the treatment, and 88 percent were completely cured.

Using bacteriophages instead of antibiotics to treat infection has several advantages. The bacteriophages are highly specific. They attack the infectious agent without harming potentially beneficial bacteria that normally inhabit the human digestive tract. They also are cheap and easy to produce, are effective in small doses, and rarely cause side effects. Furthermore, the bacteriophages are effective against bacteria that have developed antibiotic resistance. One potential disadvantage is that bacteria might also develop resistance to bacteriophages.

It is an obligate parasite of the common enteric bacterium *Escherichia coli.*

Replication of T-even Phages

Replication of phages occurs in a series of steps. The steps are illustrated for a T-even phage in Figure 11.9.

Adsorption A T-even phage attaches, or adsorbs, on the host cell surface by its tail fibers. To accomplish this, **recognition factors** in tail fibers bind to specific **receptor sites** on cells. Tail fibers then contract, bringing the tail core into contact with the cell surface.

Penetration An enzyme called **lysozyme** in the phage tail weakens the bacterial cell wall. The tail sheath contracts and injects the viral DNA through the weakened wall into the bacterial cell. The rest of the phage remains outside the bacterium.

Synthesis Once the phage DNA enters the host cell, its genes take control of the cell's synthetic machinery. Phage DNA is transcribed to mRNA, which directs the synthesis of capsid proteins and viral enzymes. Some of these enzymes catalyze reactions that replicate phage DNA. Cellular proteins and nucleic acids are broken down, and their components are used as building blocks in these processes. The host cell's enzymes continue to provide energy for the synthesis of phage components.

Maturation The head of a T-even phage is assembled from capsid proteins. Then viral DNA is packed into the head in an orderly fashion. When the head is properly packed, a specific endonuclease enzyme cuts off excess DNA. Next other components, such as the base plate, sheath, and collar, are assembled and attached to the head. The tail fibers are added last.

Release Enzymes produced by the phages alter host cell membranes so that phages can pass through them. The phage lysozyme lyses the cell wall and progeny viruses escape. Phages that carry out these processes are called *lytic* phages (Chapter 8). (p. 200)

The time from adsorption to release is called the **burst time.** In phages this time varies from 20 to 40 minutes. The number of new virions released is called the **burst size.** In phages it usually varies from 50 to 200, but it can be as many as tens of thousands.

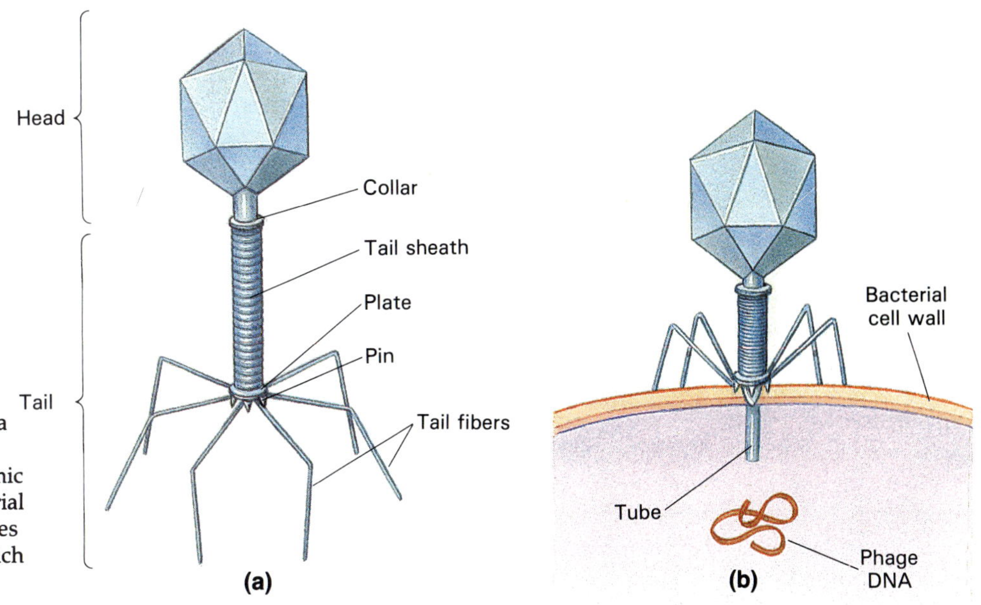

FIGURE 11.8 (a) Structure of a T-even bacteriophage. (b) The phage functions like a hypodermic needle to inject its genetic material into a host bacterial cell. It pierces the cell with a tube through which its nucleic acids will enter.

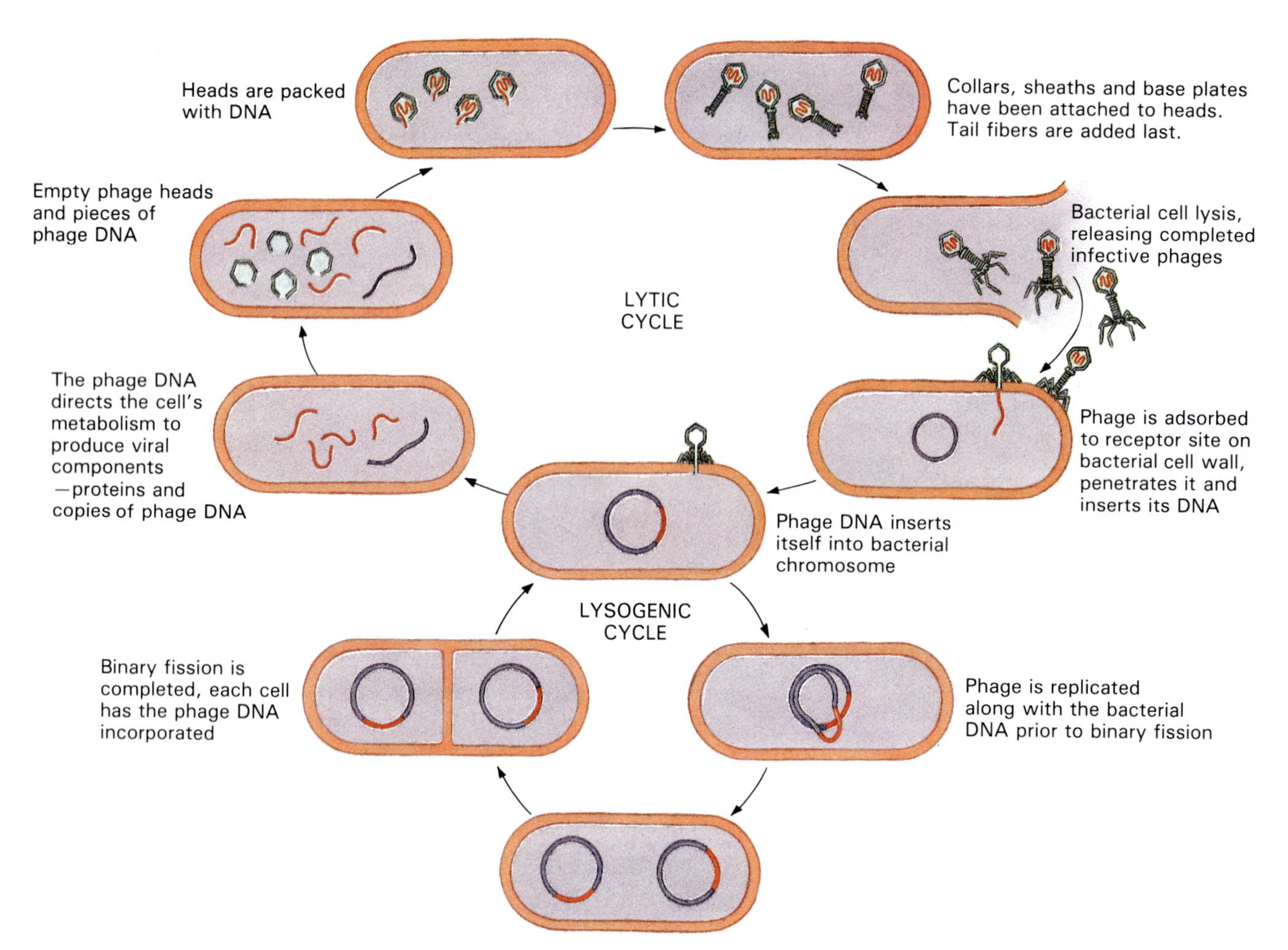

FIGURE 11.9 Replication of a bacteriophage. In the lytic cycle, a virulent phage reproduces within a bacterial cell. The cell is then lysed, releasing new phage particles that can infect other bacteria. Some phages, called temperate phages, exhibit lysogeny—they can exist harmlessly within the host cell as a prophage for long periods of time. Various factors can induce prophages to enter the lytic cycle, and a certain number will do so spontaneously.

Phage Growth

As with bacterial growth, viral growth can be described by a growth curve, which usually is based on observations of laboratory cultures of phage-infected bacteria. The **growth curve** of a phage (Figure 11.10) includes a latent period, when viruses do not increase in number, and a rise period, when they do increase in number. The **latent period** begins when the phages are introduced into a culture and includes an eclipse period and an accumulation period. During the **eclipse period** viruses disappear from the culture because they have adsorbed to and penetrated host cells but cannot yet be detected in cells. During the **accumulation period** the components of viruses are synthesized and assembled within host cells. In the **rise period** the number of phages being released per cell increases to a constant number. This constant number is the **viral yield,** or the average number of phage particles released per infected cell.

The viral yield can be determined by a viral assay, such as a **plaque assay** (Figure 11.11). To perform a plaque assay, serial dilutions of a viral suspension are prepared, such as those described for bacteria in Chapter 6. A sample of each dilution is inoculated onto a plate containing a layer of susceptible bacteria. Such a layer is called a **bacterial lawn.** After a suitable growth period, the cultures are examined for **plaques,** clear areas where viruses have lysed cells. In other parts of the bacterial lawn, bacteria will multiply rapidly and produce a thick, turbid layer.

Ordinarily each plaque represents the progeny of one infectious virus. Sometimes two virions are deposited so close together that they produce a single plaque. Occasionally an infectious virion and a noninfectious one will occupy the same plaque. Thus, counts of the number of plaques on a plate approximate, but do not exactly equal, the number of infectious viruses introduced into the culture. Such counts are reported as **plaque-forming units.**

Similar techniques can be used to study viruses that infect humans. Cultures of human cells are grown in **cell monolayers,** or single layers of cells spread across a medium-covered surface in a culture flask.

FIGURE 11.10 Growth curve for a bacteriophage. During the eclipse period, viruses have penetrated host cells and therefore cannot be detected. During the accumulation period, viral components are being synthesized and assembled into new viruses within cells. During the rise period, extracellular viruses appear once more as the new viral particles start to be released. The number of viruses per infected cell rises and eventually levels off at a value known as the viral yield.

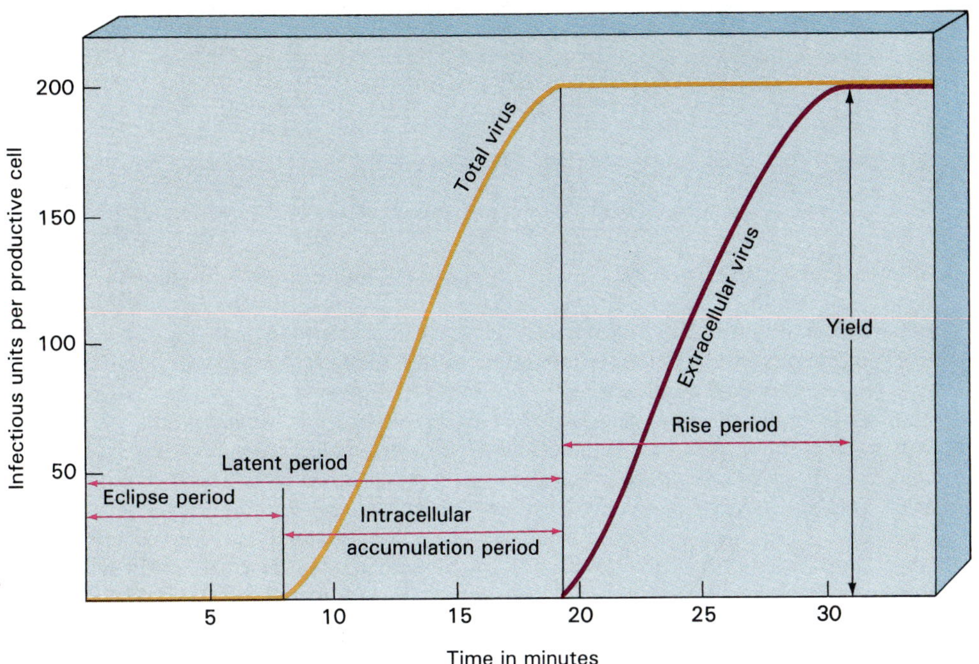

Cultures of susceptible cells are inoculated with viruses and allowed to grow, and the plaques are counted.

Lysogeny

General Properties of Lysogeny

The phages described so far are lytic phages; they are virulent phages that lyse and destroy the bacteria they infect. In contrast, temperate phages, instead of lysing host cells, exhibit lysogeny, a stable long-term relationship between the phage and its host. Lysogeny, which is described in Chapter 8, is of medical significance because the effects of some infectious agents are mediated by temperate phages. (p. 200) Both *Corynebacterium diphtheriae* and *Clostridium botulinum* contain temperate phages with genetic information that codes for the production of a toxin. These toxins are largely responsible for tissue damage that occurs in diphtheria and botulism, respectively. Without the phages the organisms do not cause diseases.

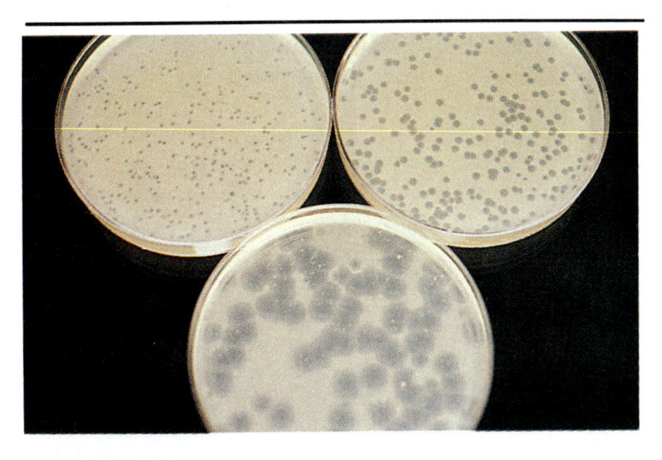

FIGURE 11.11 Plaque assay of the number of bacteriophage particles in a sample is done by spreading the sample out over a "lawn" of solid bacterial growth. When the phages reproduce and kill the bacteria, they leave a hole or clear spot, called a plaque, in the lawn. The number of plaques corresponds roughly to the number of phage particles that were initially present. Different kinds of bacteriophages produce plaques of different size or shape when grown on the same bacterium, in this case *E. coli*. The upper left plate was inoculated by T2 phage, the upper right by T4r phage, and the lower plate by lambda phage.

When bacterial cells containing temperate phages divide, the phages are distributed so that each new cell carries a phage. The combination of a bacterium and a temperate phage is called a **lysogen.** The inactive viral DNA within the lysogen is called a *prophage*. The host cells produce a **protein repressor,** which keeps the virus in an inactive state and also prevents simultaneous infection of the cell by another phage of the same type. It does not protect the bacterium against attack by a different type of phage.

Prophages can revert to the lytic, or vegetative, cycle spontaneously or in response to some outside stimulation. In other words, a temperate phage can become a virulent phage by reversion. The stimulation of a temperate phage to become virulent is called **induction.** In a colony of lysogens about one in a million spontaneously reverts and lyses the host cell.

Occurrence of Lysogeny

Lysogeny is widespread in wild strains of bacteria but occurs only sporadically in strains cultivated in the laboratory. Lysogeny can be recognized by first

MICROBIOLOGIST'S NOTEBOOK

Testing Antiviral Drugs

At times I felt more like a freight manager than a scientist—making arrangements to crate and ship tons of supplies and equipment to the People's Republic of China. It took 40 hours just to fly from my laboratory in Maryland, outside Washington, D.C., to Wuhan, where our hospital laboratory was located. To reach the hospitals where our studies were conducted required another 2 to 12 hours overland. Hauling and transferring all that material, however, proved critical to accomplishing the study.

As the drug developer in a team of virologists, physicians, and technicians, I made several trips to China. We spent 3 to 5 weeks at a time, working 16 to 18 hours per day, 7 days a week each trip. In all, I probably spent more than six months there during a two-year span. I had a sort of home away from home at Hubei Medical University—I'd even leave my quilted, padded jackets hanging in my room, ready to slip into the next time I returned from Maryland.

I'm Dr. John W. Huggins of the United States Army Medical Research Institute of Infectious Diseases (USAMRIID), and you may be wondering why I was keeping up this exhausting commute. It was to study a disease that until recently has received relatively little attention from researchers in the West. It is a disease long recognized in Asia and known by over 80 different names worldwide. Still, it was only during the Korean War, when more than 3000 United Nations troops were affected, that it came to the attention of Western medicine. Intensive clinical and epidemiological studies were conducted during the conflict to define the illness. The disease is now generally known as hemorrhagic fever with renal syndrome (HFRS), but it was formerly called Korean hemorrhagic fever (KHF) or epidemic hemorrhagic fever—the name still used in China. The disease is widespread, especially in Asia. In China alone, there are an estimated 250,000 cases per year. It has also been reported from countries as widely scattered as the former Soviet Union, Japan, Finland, France, Yugoslavia, Greece, and Belgium.

Dr. John Huggins in China.

The virus was first isolated in 1978 and later named Hantaan after the Hantaan River in Korea. Hantaan is the prototype for the Hantaviruses, a subgroup of the Bunyaviridae family. Various strains differ in their virulence and cause infections with mortality rates of 3 to 15 percent. Existing treatment consists of supportive care, because no specific vaccines or therapies are available. Death results primarly from kidney failure or hemorrhage. A few years ago 14 American Marines participating in a joint U.S.-Korean military training exercise in Korea contracted HFRS. Ten required hospitalization, and two died.

The mission of USAMRIID is the development of medical defenses against agents of biological origin, and also against naturally occurring infectious diseases that could impair the deployment of armed forces into an endemic area. We try to develop drugs, vaccines, diagnostic capabilities, and medical management procedures to minimize the effects of any biological attack on U.S. forces. However, in geographical or "tropical" medicine, where research is always underfunded, research that helps to protect our troops also helps protect the civilian community, where these diseases are a daily health problem. The needs of military preparedness, together with the millions of cases among the civilian populations of China and other countries, led us to undertake a joint project with the Chinese government to study HFRS and explore ways to combat it.

I did the preclinical drug development work on HFRS and the early planning for our trial of an experimental antiviral drug, ribavirin. As a member of a multidisciplinary team from our Institute, I was also the study director for the field trials. Chinese physicians did the actual clinical work with 250 patients, mainly from rural areas. Double-blind, placebo-controlled drug trials were not common in China, and the methodology was unfamiliar to the Chinese medical teams. My chief job was to explain the details of the procedures that needed to be followed to assure that the study was conducted as designed and with the required precision. This meant days filled with tending to hundreds of details about the collection and storage of specimens, the conduct of laboratory tests, and the recording of the vast quantities of data for later analysis.

Some of the problems we had in the field stemmed from cultural differences. In China, there is a folk belief that blood lost is not replaced by the body, so some of our experimental subjects were naturally reluctant to give us any of their blood. Of course, we needed to take blood samples for the study. We had to compromise by learning to make do with unusually tiny specimens.

While the most effective means of combatting an infectious threat would be with a preventive vaccine, it is not always possible to produce such a vaccine. In the case of HFRS, we know that the virus does elicit the production of IgM antibodies in humans. [See

Medical personnel who participated in the hemorrhagic fever study at Wuchang County Hospital.

Chapter 18.] Despite the antibodies, though, the virus persists in the blood—we don't know why. Vaccine development for HFRS has been handicapped by the lack of relevant animal models to test experimental vaccines. That is, animals can carry and transmit the virus, but we don't know of any animal species that actually develops a disease like HFRS.

Because of these obstacles, we have been concentrating our efforts with HFRS on finding an effective antiviral drug. Antiviral chemotherapy is still in its infancy—it's at about the same stage as the treatment of bacterial diseases before the development of penicillin. [See Chapter 14.] Developing a new drug typically requires 15 years and may cost 30 to 50 million dollars, and we have yet to find the first really good antiviral drug. Ribavirin—the drug we have been testing in China—may be it! The results against HFRS have been so good that the Chinese have approved the drug for use in their country and are busy manufacturing it right now. Although their supplies are still limited, they are using it today to treat patients. Here in the U.S., it will take 3½ to 5 years of additional testing before the Food and Drug Administration (FDA) reaches a final decision about whether it should be approved for use in this country.

We are not sure of the mechanism by which ribavirin prevents viral replication (nor are we for any other antiviral compounds). So, we are naturally very eager to see if ribavirin is active against other viral diseases. Animal studies show it to be effective against other bunyaviruses, such as the ones that cause Rift Valley fever, sandfly fever, and Crimean Congo hemorrhagic fever. Moreover, it also appears to be effective against the arenavirus that causes Lassa fever, which in the days before AIDS was responsible for 40 percent of medical admissions in parts of Africa. Dr. Joseph McCormick of CDC cooperated in the field trials of ribavirin in Sierra Leone with remarkable results. Unfortunately, there are also other viruses that, based on animal studies, are not affected by ribavirin, including those that cause yellow fever and dengue. We certainly intend to continue our studies with ribavirin, but we hope to develop other new antiviral drugs as well.

Recently, virologists reported finding antibodies to a Hantavirus related to Hantaan in black, inner-city residents of Baltimore. A surprisingly high percentage tested positive for the virus—evidently its presence is a chronic condition. Yet, symptoms typical of HFRS were not observed in these people. What *was* associated with testing positive for antibody to this virus was a higher rate of hypertension—a condition long observed to be particularly prevalent among blacks in this country and which carries an increased risk of heart attack and stroke. Could it be that the Hantavirus is responsible for or contributes to hypertension? Should those persons infected be given ribavirin? More work will have to be done before we know the answers to these questions.

In epidemiological studies it has been noted that the severity of disease caused by the Hantaviruses is correlated with the rodent that transmits it to humans. In Asia, the main vector is a striped field mouse, *Apodemus agrarius,* an animal not found in the U.S. Here, the main source of infection appears to be the rat, *Rattus rattus,* which is unfortunately abundant in inner city areas. Humans can acquire the infection through minor cuts or abrasions that become contaminated with rodent urine or feces, and possibly by exposure to aerosols where virus contamination is heavy. Rats shed virus heavily throughout their lives. Some human cases of HFRS have been due to handling of infected wild or laboratory rodents. Could it be that the strains of Hantavirus carried by American rats produces milder, largely asymptomatic infections?

Clearly there are more questions about HFRS than answers at this point. But that's normal in scientific work. In the meantime, I've recently returned from a trip to Yugoslavia. There have been three especially severe outbreaks of hemorrhagic fever in that country in recent years, and we'd like to find out why. A study is under way.

Thus, a project to protect American troops may have a great impact on civilian populations in many parts of the world.

streaking a phage-sensitive strain of bacteria onto an agar plate and then streaking a strain that might be lysogenized at a right angle to the sensitive strain (Figure 11.12). Narrow clear bands indicating lysis will appear around lysogens that have reverted to the lytic phase.

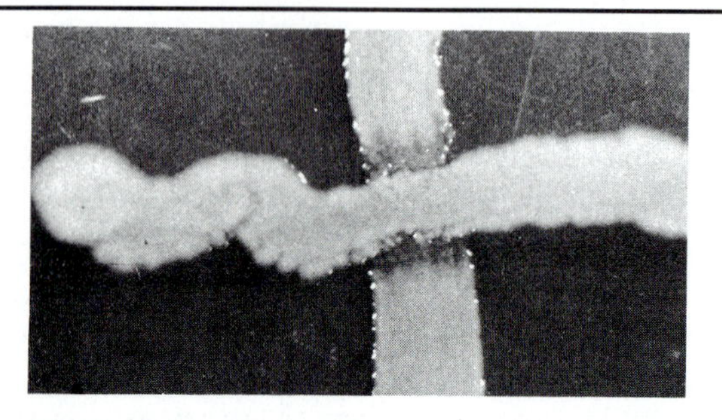

FIGURE 11.12 Demonstration of lysogeny using right-angle (cross) streaking with two strains of *E. coli*. Narrow bands around the lysogenic strain (horizontal streak) mark regions where lysogens have reverted to the lytic phase, liberating phages that have lysed cells of the phage-sensitive strain (vertical streak).

Lysogeny was recognized early in the 1920s but not well understood. In 1950 André Lwoff demonstrated that only a small proportion of lysogens produce phages at any one time. Those that do are lysed. The remaining lysogens do not yield phages and remain immune to attack by phages of the same type that each harbors. Such a phage can adsorb to some immune bacterium and inject its nucleic acid, but the nucleic acid cannot replicate or cause lysis of the bacterium. Some bacteria are so resistant to lysogenization that the phage cannot adsorb to their surfaces.

Lambda Phage

One of the most widely studied prophages is the lambda phage of *Escherichia coli*. Prophages such as lambda insert into the bacterial chromosome at a specific location in a single process called a **recombination event** (Figure 11.13). The lambda phage DNA has cohesive ends that adhere to specific break points in the circular *E. coli* genome. Insertion of a lambda phage into a bacterium alters the genetic characteristics of the bacterium. Such phages are useful in transferring specific genetic information in genetic engineering.

Replication of Animal Viruses

Animal viruses invade and replicate in animal cells as bacteriophages do in bacteria, but they use a greater variety of mechanisms. They go through the processes of adsorption, penetration, synthesis, maturation, and release, but they perform the processes in ways that differ from those employed by bacteriophages and that differ among the animal viruses themselves.

Adsorption

As we have seen, bacteriophages have specialized structures for attaching to bacteria, which have cell walls. Although animal cells lack cell walls, animal viruses have other ways of attaching to host cells. Naked viruses have attachment sites on the surfaces of their capsids that bind to corresponding sites on appropriate host cells. Enveloped viruses such as the orthomyxoviruses and the paramyxoviruses attach to host cells by means of glycoprotein spikes that project from their envelopes. Adenoviruses attach to host cells by penton fibers that project from the corners of their icosahedral structure. In many cases more viruses attach to a cell than can penetrate it, or the attachment may be too brief to allow penetration.

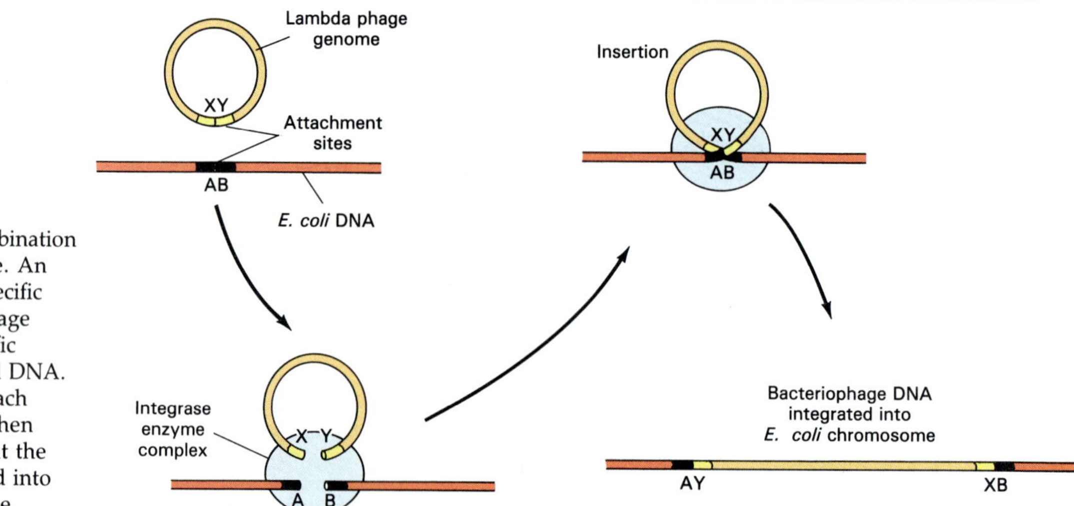

FIGURE 11.13 Recombination event in a lambda phage. An enzyme recognizes a specific base sequence in the phage DNA and another specific sequence in the bacterial DNA. It cuts both strands of each DNA double helix and then joins the cut ends so that the phage DNA is integrated into the bacterial chromosome.

Penetration

Animal viruses, which cannot inject nucleic acid into the host cell, use other methods, which usually allow both the nucleic acid and capsid to penetrate the host cell. Certain naked animal viruses, such as the enteroviruses, are taken into the cell by **viropexis,** a process similar to phagocytosis. Other enveloped viruses, such as influenza and parainfluenza viruses, move to special pitlike regions on the surface of the cell and are incorporated into infoldings of host cell membranes. Viral and cell membranes fuse, and the virus enters the cytoplasm.

The protein coats of animal viruses enter cells, and such viruses undergo a special process of **uncoating.** Uncoating is done by proteolytic enzymes from host cells or from the viruses themselves. For some viruses, such as polioviruses, uncoating begins even before penetration is complete.

Synthesis

Synthesis in DNA Animal Viruses In DNA animal viruses, the DNA is transcribed into mRNA, which directs protein synthesis, and new viral DNA, which is incorporated into new viruses. Except for poxviruses, this takes place in the host cell nucleus. The mRNA travels to the cytoplasm, where it is read by cellular ribosomes, and viral enzymes and other proteins are made. In DNA viruses, synthesis proceeds in a complex series of steps designated as early and late transcription and translation. The early events take place before the synthesis of viral DNA and produce the enzymes and other proteins necessary for viral DNA replication. The late events occur after the synthesis of viral DNA and produce enzymes and structural proteins that will be incorporated into the progeny viruses. Because herpesviruses contain a large number of proteins, their replication requires 8 to 16 hours, a long time when compared with some bacteriophages that replicate in less than 1 hour. The synthesis of a DNA animal virus is summarized in Figure 11.14.

Synthesis in RNA Animal Viruses Synthesis among RNA animal viruses takes place in a greater variety of ways than among DNA animal viruses. In Class I RNA viruses, positive sense, single-stranded RNA

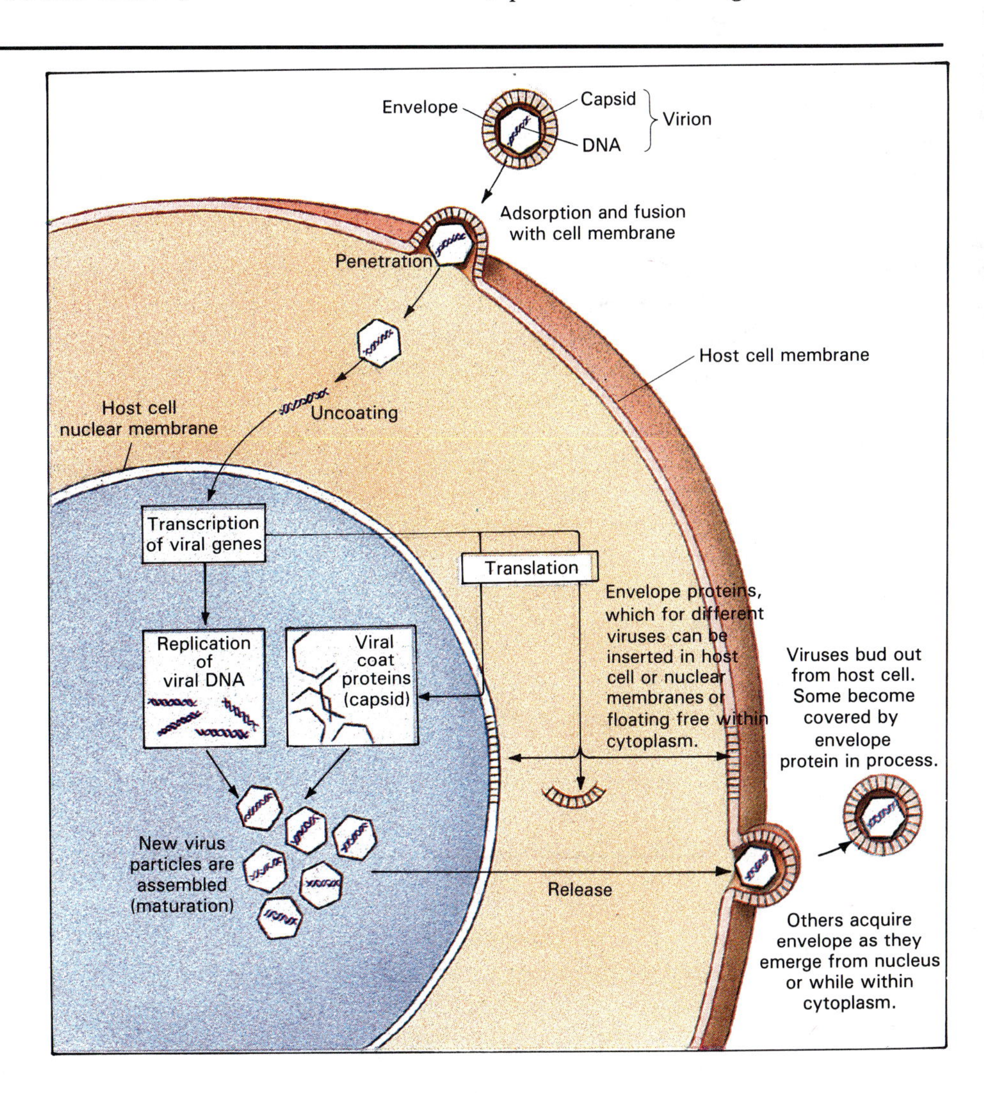

FIGURE 11.14 Replication of a DNA animal virus.

acts as mRNA, and viral proteins are made immediately. The nucleus of the host cell is not involved. Viral proteins also play roles in the synthesis of viruses. One protein inhibits synthetic activities of the host cell. Another, the enzyme transcriptase, uses positive sense RNA as a template to make negative sense RNA. The negative sense RNA in turn acts as a template to produce positive sense mRNA and more viral RNA.

In Class II and Class III RNA animal viruses, an RNA polymerase enzyme uses the negative sense viral RNA to make mRNA and new viral RNA. The process is essentially the same regardless of whether the viral RNA is in a single piece or in segments.

In Class IV RNA animal viruses, each strand of the double-stranded RNA acts as a template for its partner. Replication is semiconservative, so the molecules produced have one strand of old RNA and one strand of new RNA. These viruses have a double-walled capsid that is never completely removed, and replication takes place within the capsid.

In Class V RNA animal viruses the viral RNA can act as mRNA directly or be copied into single-stranded DNA with the help of reverse transcriptase. Single-stranded DNA is replicated to make double-stranded DNA, which forms a circular particle and is inserted into a host cell chromosome. This insertion occurs in much the same way that temperate phages enter bacterial chromosomes. Once viral DNA enters a host cell chromosome it can remain there for an indefinite period of time or it can leave the chromosome and start replication. When the virus replicates, the DNA from the chromosome is transcribed into positive sense RNA for use as mRNA or viral RNA. If the viral DNA remains in the host chromosome while the host cell undergoes division, the viral DNA is copied with the rest of the chromosome. The inactive viral genetic information is thereby passed to progeny cells, where it has the potential of leaving the chromosome and actively directing synthesis of viral particles.

The several mechanisms of replication of RNA animal viruses are summarized in Figure 11.15.

Maturation and Release

Once an abundance of viral genomes, enzymes, and other proteins have been synthesized, assembly of components into complete virions starts. This constitutes maturation of progeny viruses. For naked viruses such as polioviruses, assembly takes place in the cytoplasm. When the cell is filled with progeny virions, the cell membrane ruptures and the progeny are released. Such cell lysis always kills the host cell.

Maturation, or assembly, of enveloped viruses is a longer and more complex process. As we have seen, both the infecting virus and nucleic acids and enzymes made in the host cell participate in synthesizing components. Among the components destined for the progeny viruses, proteins are coded by the virus, lipids are derived from host cells, and glycoproteins get their protein from the virus and their carbohydrate side chains from the cell. If the virus is to have an envelope, the virion is not complete until it buds through a host membrane (Figure 11.16). The budding of new virions through a membrane may or may not kill the host cell.

CULTURING OF ANIMAL VIRUSES

Development of Culturing Methods

Initially, viruses could be grown only in animals. This made it difficult to observe specific effects of the viruses at the cellular level. In the 1930s it was discovered that embryonated (intact, fertilized) chicken eggs could be used to grow herpesviruses, poxviruses, and influenza viruses. Although the chick embryo is a simpler structure than a whole mouse or rabbit, it is still a complex organism. Using it did not completely solve the problem of studying cellular changes caused by viruses. Another problem was that bacteria also grow well in embryos, and the effects of viruses could not be determined in bacterially contaminated embryos. Virology progressed slowly during these years until techniques of cell culture improved.

Two discoveries greatly enhanced the usefulness of cell cultures for virologists and other scientists. Antibiotics made it possible to prevent bacterial contamination. Next, it was found that proteolytic enzymes, particularly trypsin, could be used to free cells from the surrounding tissues without hurting the cells. The cells could be washed, evenly dispersed, counted, and dispensed into flasks, tubes, or Petri dishes. Cells in such suspensions would attach to a solid surface, multiply, and spread to form sheets one cell thick, called monolayers. These monolayers could be treated with trypsin and subcultured. **Subculturing** is the process by which cells from an existing culture are transferred to fresh medium in new containers. A large number of cultures could be made from a single tissue, thereby assuring a reasonably homogeneous set of cultures in which to test the effects of a virus.

The term **tissue culture** remains in widespread use to describe the preceding technique, although the term *cell culture* would be more accurate. Today the majority of cultured cells are in the form of monolayers grown from enzymatically dispersed cells and continuous cultures of cell suspensions. With a wide variety of cell cultures available and with antibiotics to control contamination, virology progressed almost explosively. Within 10 years, more than 400 viruses had

FIGURE 11.15 Two of the mechanisms used for the replication of RNA animal viruses. The one shown at the left is found in Class I RNA viruses, which have positive sense, single-stranded RNA genomes. The viral RNA serves as mRNA—it is translated immediately to produce proteins needed for reproduction of the virus. A negative sense copy is then made, which serves as a template for production of more positive sense viral RNA. The one at the right is found in Class V RNA viruses. The positive sense RNA is copied with the help of reverse transcriptase to make a negative sense strand of DNA, which serves as template for the synthesis of a complementary positive sense strand. The double-stranded DNA is then inserted into the host chromosome, where it can remain for some time. When replication occurs, one strand of the DNA becomes the template for the synthesis of viral RNA.

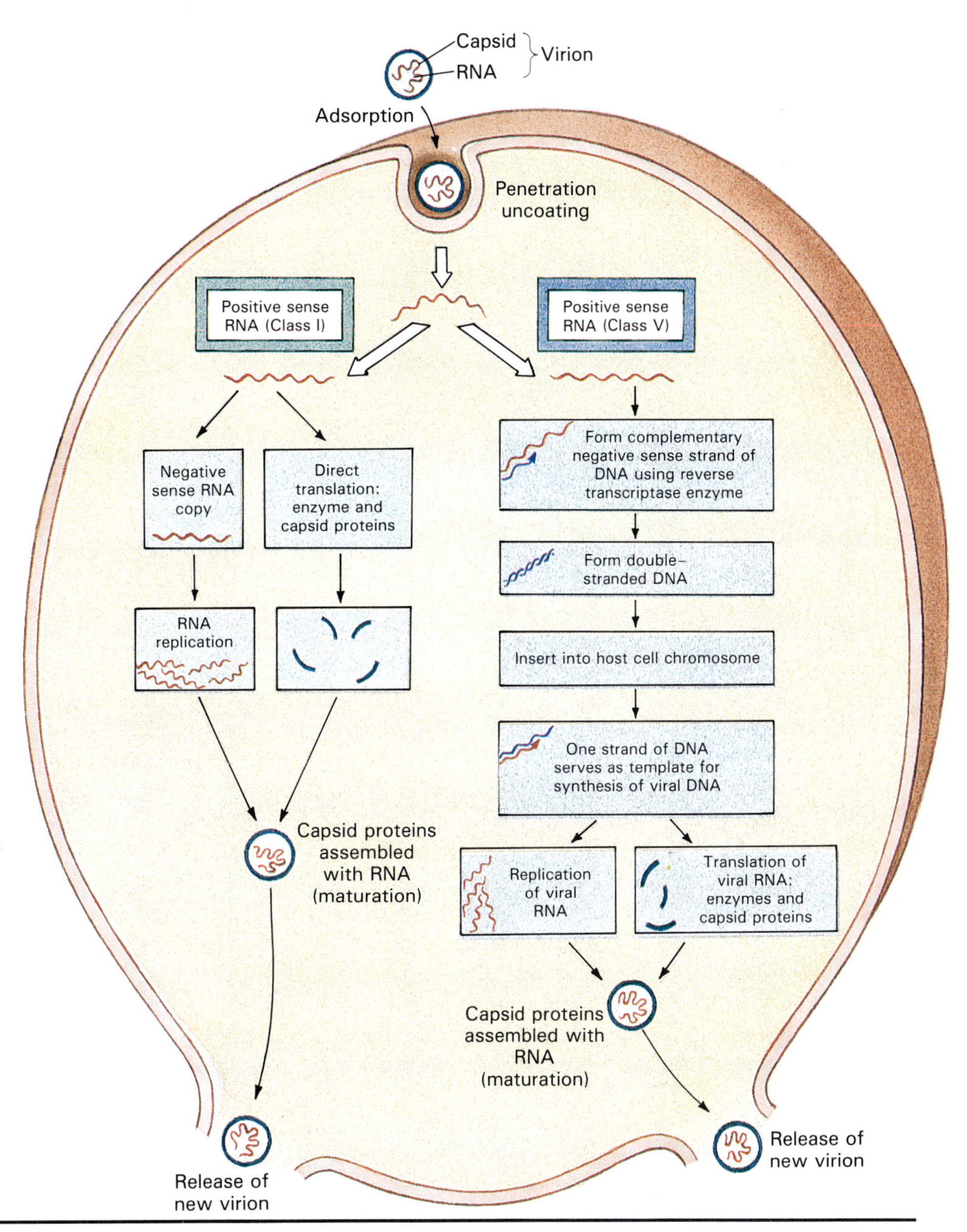

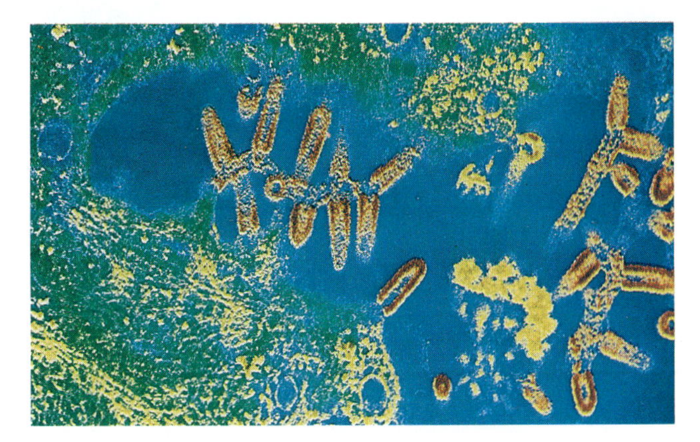

FIGURE 11.16 Colorized SEM of rabiesvirus particles being released from the cytoplasm of an infected cell (green). This Class II RNA virus acquires its envelope when it buds out through the host cell membrane. Note the bullet shape of the rabiesvirus particles.

been isolated and characterized. Although additional new viruses are still being discovered today, emphasis is now on characterizing the viruses in more detail and on determining the precise steps in the processes of viral infection and viral replication.

Types of Cell Cultures

Three basic types of cell cultures are widely used in clinical and research virology: (1) primary cell cultures, (2) diploid fibroblast strains, and (3) continuous cell lines. **Primary cell cultures** (Figure 11.17) come directly from the animal and are not subcultured. The younger the source animal, the longer the cells will survive in culture, but such cells usually do not divide more than a few times. They typically consist of a mixture of cell types, such as muscle and epithelial

FIGURE 11.17 Cultures of viruses in tissue culture flasks.

cells, all of which are diploid. This means they have paired chromosomes, like all animal cells except reproductive cells. Many such cultures support growth of a wide variety of viruses.

If primary cultures are repeatedly subcultured, one cell type will become dominant, and the culture is called a **cell strain.** In cell strains all the cells are very similar to one another. Such strains can be subcultured for several generations with only a very small likelihood that changes in the cells themselves will interfere with determining viral effects.

The visible effect viruses have on cells is called the **cytopathic effect** (CPE). CPE can be so distinctive that an experienced clinical virologist can often use it to identify an infecting virus. Adenoviruses and herpesviruses cause cells to "balloon" because of fluid accumulation, whereas picornaviruses arrest cell functions when they enter and lyse cells when they leave. Paramyxoviruses cause fusion of cell membranes and the aggregation of 4 to 100 nuclei in newly formed giant cells called **syncytia.** Another type of CPE produced by some viruses is transformation: the conversion of normal cells into malignant ones (Figure 11.18).

Among the most widely used cell strains are **diploid fibroblast strains** (Figure 11.19). Fibroblasts are immature cells that produce collagen and other fibers as well as the substance of connective tissues, such as the dermis of the skin. Derived from fetal tissues, these strains retain the fetal capacity for rapid, repeated cell division. Such strains support growth of a wide range of viruses and are usually free of contaminating viruses often found in cell strains from mature animals. For this reason, they are used in making viral vaccines.

The third type of cell culture in extensive use is the **continuous cell line.** A continuous cell line consists of cells that can be propagated over many generations. The most famous continuous cell line is the HeLa line, which has been maintained since 1951. The original cells of the HeLa line came from a woman with cervical cancer; now there are HeLa cells all over the world. Many of the early continuous cell lines used malignant cells because of their rapid growth. Continuous cell lines grow in the laboratory without aging, divide rapidly and repeatedly, and have simpler nutritional needs than normal cells. They are het-

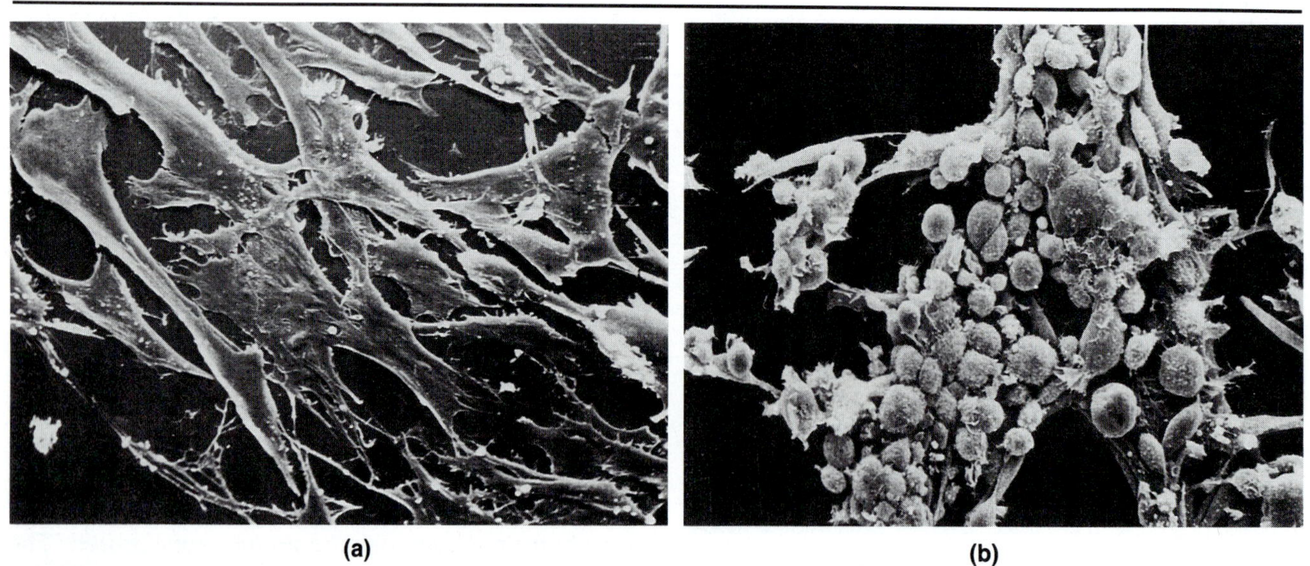

FIGURE 11.18 (a) Normal and (b) transformed (malignant) cells. Such transformation is a cytopathic effect (CPE) caused by infection with certain viruses.

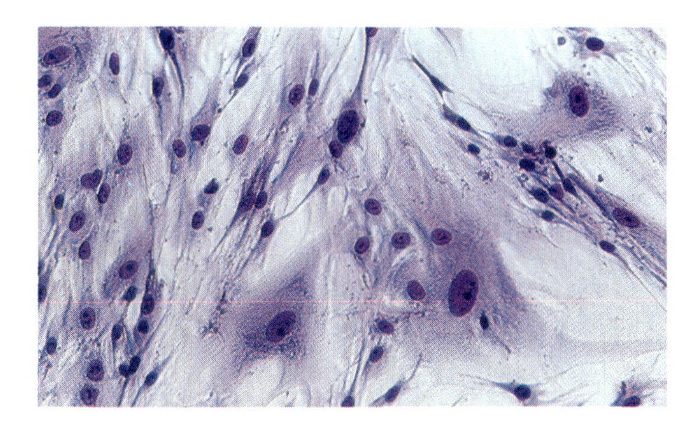

FIGURE 11.19 A culture of diploid fibroblasts. These immature cells, derived from fetal tissue, multiply rapidly and provide an ideal cell culture for many viruses.

eroploid (have different numbers of chromosomes) and are therefore genetically diverse. Unfortunately, continuous cell lines can invade and be invaded by other cell lines. Some research studies have had to be discarded because the cultures had been contaminated with undetected HeLa cells. The results thought to pertain to normal cells were found to be due to malignant ones.

Cell cultures have largely replaced animals and embryonated eggs for studies in animal virology. However, the embryonated chicken egg remains one of the best host systems for influenza A viruses (Figure 11.20). The young albino Swiss mouse is still used to culture arboviruses (*ar*thropod *bo*rne viruses), but mosquito cell lines are coming into use for this purpose.

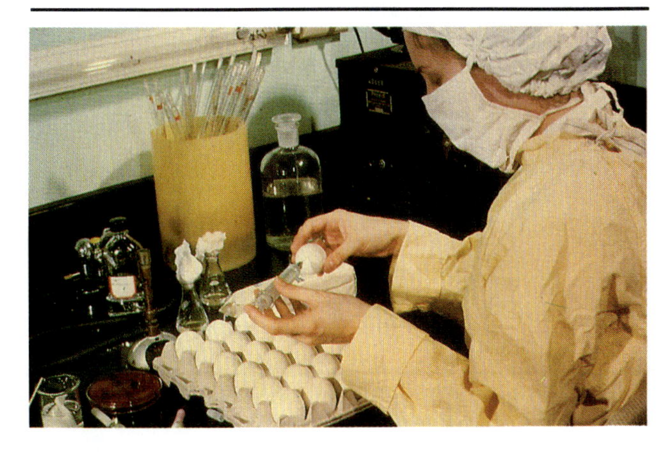

FIGURE 11.20 Some viruses, such as strains of the influenza virus, are grown in embryonated chicken eggs.

VIRUSES AND TERATOGENESIS

Teratogenesis is the induction of defects during embryonic development. A **teratogen** is an agent that induces such defects. Certain viruses are known to be transmitted across the placenta and infect fetuses, and several are known to act as teratogens. The earlier in the gestation period the embryo is infected, the more extensive the damage is likely to be. During the early stages of embryological development, when organs or systems may be represented by only a few cells, damage to those cells can cause damage to an entire organ or system. Later in development, damage to a few cells has a proportionately smaller effect because the total cell population in the fetus has greatly increased, and each organ or system consists of many cells.

Three kinds of viruses—cytomegalovirus, herpes simplex types 1 and 2, and rubella—account for a large number of teratogenic effects. Cytomegalovirus (CMV) infections are found in about 1 percent of live births; of these about 1 in 10 will eventually die of CMV infection. Most of the defects are neurological, and the children have varying degrees of mental retardation. Some also have enlarged spleens, liver damage, and jaundice. Herpes infections usually are acquired at or shortly after birth. Infections acquired before birth are rare. In cases of disseminated infections (those that spread through the body) some infants die, and survivors have permanent damage to the eyes and central nervous system.

Rubella infections in the mother during the first 4 months of pregnancy are most likely to result in fetal defects such as deafness, damage to other sense organs, heart and other circulatory defects, and mental retardation. The degree of impairment is highly variable; some children adapt to their handicaps and live productive lives, but others are severely impaired, and many die as fetuses and are aborted. Congenital rubella is discussed in Chapter 20.

PUBLIC HEALTH

Torch Tests in Pregnancy

A series of serological tests often referred to as the TORCH series is sometimes used to detect diseases in pregnant women and newborn infants. These tests detect antibodies against toxoplasma, rubella, cytomegalovirus, and herpes. *Toxoplasma gondii* (Chapter 23) is a protozoan that can cause blindness and brain damage in fetuses.

ESSAY

Discovering Viruses

Viruses have plagued humans for thousands of years; Egyptian mummies bear the marks of smallpox. But it was not until the last century that we became aware of the true nature of viruses, how they work, and how they can be manipulated. In 1796, Edward Jenner vaccinated his first patient against smallpox, using material from a cowpox lesion. Then in 1885, Louis Pasteur used his experimental rabies vaccine to save the life of a 9-year-old boy who had been bitten by a rabid dog. During the next 15 months, another 2490 people received Pasteur's rabies vaccine. Although Jenner and Pasteur utilized viruses, they could only guess as to their nature.

In 1887, John Buist, a Scottish bacteriologist and surgeon working at the University of Edinburgh, actually saw a virus. He spread fluid from a smallpox vaccination lesion onto a slide, dried, stained and examined it under the microscope. He saw small round dots (Figure 11.21). Cowpox viruses are among the largest of viruses and are just visible under the light microscope. Buist thought that these must be spores of some microbe. He was unable to grow them in his laboratory and eventually gave up the study. He had a virus in his hands and let it slip through his fingers.

Then in 1892, Dimitri Ivanovski, a young Russian botanist, almost closed his hand on the virus. But he, too, failed to realize what he had. It was not until Martinus Beijerinck, a Dutch microbiologist, published similar work in 1897 that Ivanovski awoke to the true significance of his own discoveries. Ivanovski worked with tobacco mosaic disease that causes a mottled (mosaic) pattern of dark and light green areas to develop on leaves of infected plants. He ground up infected leaves, pressed out the sap through a cloth, and then ran the sap through a Chamberlain porcelain filter. This filter has many tiny pores that trap and hold particles, allowing only liquid to emerge at the bottom. Similar, even finer, porcelain filters are still manufactured today for use in purifying drinking water in some Third-World households, and by wilderness campers.

Ivanovski found that the filtered fluid could still cause disease in tobacco plants. He thought that perhaps the filter was defective, but upon examination he found no bacteria in the filtrate, nor could he recover any from his filter. Toxins from bacteria were known to pass through filters and then be able to cause disease symptoms in animals. However, the disease could not be then transmitted to successive animals from the first animal. Always before it had been possible to find the infective bacteria in the filter. This time, however, Ivanovski could not find any, nor could he culture the infectious agent in his test tubes. Ivanovski published his results in a little-read Russian language journal, noting that he would like to be able to conduct more experiments on this puzzling matter. But he abandoned the project for the next 6 years—until he read of Beijerinck's work. Then he wanted credit for the discovery of a virus, overlooking that fact that at the time, he had thought he was dealing with a bacterial toxin. Beijerinck had done much the same work with tobacco mosaic disease and gotten the same results, but unlike Ivanovski, he was able to think of something beyond bacteria. He had not read of Ivanovski's work, but gracefully acknowledged Ivanovski's prior claim and gave up his own work with tobacco mosaic disease. Beijerinck had differed from Ivanovski in his view of his results. He flatly stated that no bacteria were involved in this system. He showed that unlike toxins, this filtrate caused disease that he could transmit serially from plant to plant. This proved to him that the causative agent was reproducing and multiplying, a characteristic he felt indicated life. He treated his filtrate with alcohol and formalin, known to kill bacteria, but without affecting the infectivity of his filtrate. There was still the possibility that resistant spores were present and causing the infection. So, he heated his filtrate to 90°C, just under the 100°C temperature, which kills spores. This time the filtrate could no longer infect, meaning that spores were not the cause. Everything pointed to a nonliving chemical, but one that could reproduce.

Finally, Beijerinck inoculated the top of an agar block with his filtrate. At that time, it was believed that only liquids and soluble compounds—but never cells—were able to move through agar. After 10 days he sampled agar from the lower part of the block and found it to be infectious. He concluded that the nature of this new type of infectious agent was a noncellular, soluble, living germ. Today we call this agent a virus.

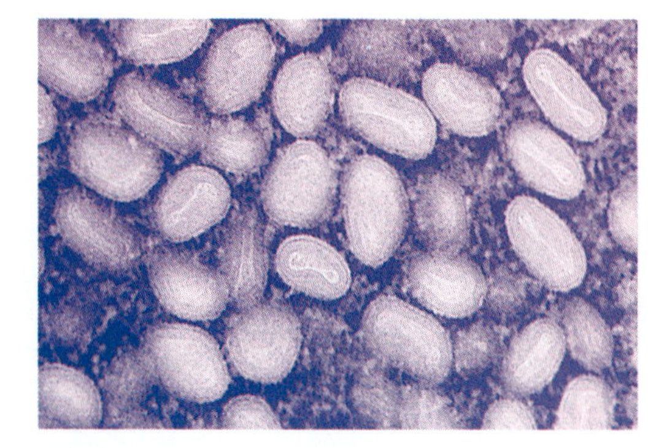

FIGURE 11.21 Pox virus.

CHAPTER SUMMARY

GENERAL CHARACTERISTICS OF VIRUSES

What Are Viruses?

Components of Viruses

- **Viruses** are **obligate intracellular parasites** that consist of a nucleic acid and a protein coat. Some viruses also have a membranous **envelope.**

Shapes and Sizes

- Viruses vary in size from 5 to 300 nm in diameter and have icosahedral, helical, or complex shapes.

Host Range and Specificity of Viruses

- Viruses vary in **host range** and **specificity.** Many viruses infect a single kind of cell in a single host species; others infect several kinds of cells, several hosts, or both.

RELATED KEY TERMS

replicate capsid
virion genome
enveloped viruses capsomeres
naked virus glycoproteins
matrix proteins spikes
complex viruses

CLASSIFICATION OF VIRUSES

- Viruses are classified by structure and type of nucleic acid they contain, other chemical and physical properties, their mode of replication, shape, and host range. Some of these characteristics are summarized in Table 11.1.

positive sense nucleic acid
negative sense nucleic acid
transcriptase

RNA Viruses

- RNA viruses are divided into five classes. Class I is made up of the **picornaviruses** and the **togaviruses;** the picornaviruses include the **enteroviruses,** which cause polio and some intestinal disorders, and the **rhinoviruses,** which cause colds. Class II includes the **paramyxoviruses,** which cause measles, mumps, and several respiratory disorders, and the **rhabdoviruses,** one of which causes rabies. Class III includes the **orthomyxoviruses,** which cause influenza, the **bunyaviruses,** one of which causes encephalitis, and the **arenaviruses,** which cause a variety of hemorrhagic fevers. Class IV includes the **reoviruses,** which cause a variety of upper respiratory and gastrointestinal infections. Class V includes the **retroviruses,** which cause some malignancies and AIDS.

reverse transcriptase
rubellavirus
provirus

DNA Viruses

- DNA viruses are divided into three classes. Class I includes the **adenoviruses,** some of which cause respiratory infections; **herpesviruses,** which cause oral and genital herpes, chickenpox, shingles, and mononucleosis; and **poxviruses,** which cause smallpox and some similar milder infections. Class II includes **papovaviruses,** some of which cause warts. Class III includes **parvoviruses.**

pentons serotypes latency
orthopoxviruses
inclusion bodies
parapoxviruses
papillomaviruses parvovirus
polyomaviruses proper
vacuolating virus dependoviruses

VIROIDS AND PRIONS

Viroids

- **Viroids** are infectious particles smaller than viruses that consist of RNA and are self-replicating. They can infect some plants but how they arose and how they cause disease are not understood.

Prions

- **Prions** are proteinaceous infectious particles believed to cause Creutzfeldt-Jakob disease and kuru in humans. How they originated, how they replicate, and how they cause disease are unknown.

VIRAL REPLICATION

General Characteristics of Replication

- Viruses go through five steps in the replication process: **adsorption, penetration, synthesis, maturation,** and **release.** The nature of these steps is somewhat different in bacteriophages and animal viruses.

Replication of Bacteriophages

- Bacteriophage replication has been thoroughly studied in T-even phages, which are lytic phages (Chapter 8).
- T-phages have **recognition factors** that attach to specific receptors on the surface of bacteria during adsorption. Enzymes weaken the bacterial wall so viral nucleic acid can penetrate it.

receptor sites lysozyme

- During synthesis bacterial DNA directs the making of viral components.
- In maturation the components are assembled: capsid proteins form a head, which is packed with nucleic acid, and other parts are assembled later.
- Release is facilitated by enzymes.
- The **growth curve** of a phage includes an **eclipse period** when viruses first enter cells, an intracellular **accumulation period** when viruses are synthesized and assembled, and a **rise period** when phages are released.

RELATED KEY TERMS: **burst size**, **burst time**, **latent period**, **viral yield**, **plaque assay**, **bacterial yield**, **plaques**, **plaque-forming units**, **cell monolayers**

Lysogeny

- Lysogeny, a stable long-term relationship between certain phages and host bacteria, occurs in temperate phages (Chapter 8). Temperate phage DNA can exist as a prophage or revert to a lytic phage.
- Prophages such as the lambda phage insert in a bacterial chromosome at a specific location.

RELATED KEY TERMS: **lysogen**, **protein repressor**, **induction**, **recombination event**

Replication of Animal Viruses

- Adsorption involves receptors on the surface of viruses instead of in tail fibers; the nucleic acid and coat of animal viruses penetrate the cell.
- Synthesis and maturation differ in DNA and RNA viruses. In DNA viruses, DNA is synthesized in an orderly sequence in the nucleus, and proteins are synthesized in the cytoplasm of the host cell. In RNA viruses, RNA can act as a template for protein synthesis, for making mRNA, or for making DNA by reverse transcription. Viruses are assembled in the cell; sometimes viral DNA is incorporated into cellular DNA.
- Release can be direct lysis of the host cell or can involve budding through the host membrane.

RELATED KEY TERMS: **viropexis**, **uncoating**

CULTURING OF ANIMAL VIRUSES

Development of Culturing Methods

- The development of culturing methods proceeded slowly.
- The discovery of antibiotics to prevent bacterial contamination of cultures and the use of trypsin to separate cells in culture provided great impetus to the study of virology.

RELATED KEY TERMS: **subculturing**, **tissue culture**

Types of Cell Cultures

- **Primary cell cultures** come directly from animals and are not subcultured.
- **Diploid fibroblast strains** from primary cultures of fetal tissues produce stable cultures that can be maintained for years; they are used to produce vaccines, isolate infectious agents, and to study aging.
- **Continuous cell lines,** usually derived from cancer cells, grow in the laboratory without aging, can divide repeatedly, have greatly reduced nutritional needs, and display heteroploidy.

RELATED KEY TERMS: **cell strain**, **syncytia**

VIRUSES AND TERATOGENESIS

- A **teratogen** is an agent that induces defects during embryonic development.
- Viruses act as teratogens by crossing the placenta and infecting embryonic cells. The earlier in development an infection occurs, the more extensive damage is likely to be.
- Rubellaviruses are responsible for the death of many fetuses and severe birth defects in others; cytomegaloviruses and occasionally herpesviruses also act as teratogens.

RELATED KEY TERMS: **teratogenesis**

QUESTIONS FOR REVIEW

A.

1. Describe the shapes of viruses.
2. What is a virion?
3. What are the main characteristics of viruses?

B.

4. How are viruses classified?
5. Summarize the main differences between classes of viruses.

C.

6. How do viroids differ from viruses?
7. What are prions, and what diseases do they cause?

D.

8. What are the general steps in viral replication?

E.

9. How do T-phages attach to and enter cells?
10. Summarize the steps in bacteriophage replication.
11. What happens in each phase of a phage growth curve?
12. What is lysogeny?
13. What is a prophage, and how does it become a lytic phage?

F.

14. In what ways does virus replication in animal cells differ from that in bacteriophages?
15. In what ways does replication of RNA viruses differ from replication of DNA viruses?
16. What are some differences in the replication process among RNA viruses?

G.

17. What problems had to be solved before cell culturing methods could contribute significantly to virology?

H.

18. What is a primary cell culture?
19. What characteristics of diploid fibroblast strains make them useful?
20. What are the characteristics of continuous cell lines?

I.

21. What is a teratogen?
22. What kinds of damage do viruses cause in embryos?

PROBLEMS FOR INVESTIGATION

1. Prepare for, and if possible participate in, a debate on the topic of whether viruses are living or nonliving.
2. Identify places in the DNA and RNA viral replication cycles that might serve as targets for antiviral drugs. Remember, for example, that AZT inhibits reverse transcriptase in the AIDS virus.
3. Which do you think evolved first, viruses or cells? There is a lively controversy among scientists on this point. Try to read something about it.
4. Read about the work of Iwanowski and Beijerinck with tobacco mosaic virus. Prepare a short report.
5. Research the current scientific literature on viroids or prions, and report on your findings to the class.
6. Read about the work of Twort and d'Herelle and the discovery of bacteriophages. Prepare a short report.
7. Research the morphology of the AIDS (HIV) virus and the functions of its various components.
8. Find out how plant viruses are cultured.

SOME INTERESTING READING

Beardsley, T. 1990. Oravske kuru: a human dementia raises the stakes in mad cow disease. *Scientific American* 263 (August):24.

Diener, T.O. 1981. Viroids. *Scientific American* 244 (January):66–73.

Fields, B.N. 1991. *Fundamental virology*. New York: Raven.

Fraenkel-Conrat, H., P. Kimball, and J. Levy. 1988. *Virology*. Englewood Cliffs, NJ: Prentice Hall.

Gallo, R. C. 1987. The AIDS virus. *Scientific American* 256, no. 1 (January):47.

Hogle, J. M., M. Chow, and J. D. Filman. 1987. The structure of poliovirus. *Scientific American* 256, no. 3 (March):42.

Rotbart, H.A. 1991. Nucleic acid detection systems for enteroviruses. *Clinical Microbiology Reviews* 4 (2):156–68.

Simons, K., H. Garoff, and A. Helenuis. 1982. How an animal virus gets into and out of its host cell. *Scientific American* 246, no. 2 (February):58.

Varmus, H. 1987. Reverse transcription. *Scientific American* 257, no. 3 (September):56.

False-color scanning electron micrograph (SEM) of the head (scolex) of the small tapeworm *Acanthrocirrus retrirostris*. The tapeworm spends its early life stages inside the body of a barnacle (*Balanus balanoides*) but reaches maturity inside the intestines of wading birds that eat barnacles. At the top of the head is an extraordinary pistonlike apparatus called the rostellum. It is surmounted by hooks and can be withdrawn into the head or thrust out and buried in the host's intestinal tissue. Beneath this can be seen two of the four suckers that encircle the head and aid in affixing the tapeworm to its host. Magnification is 80X.

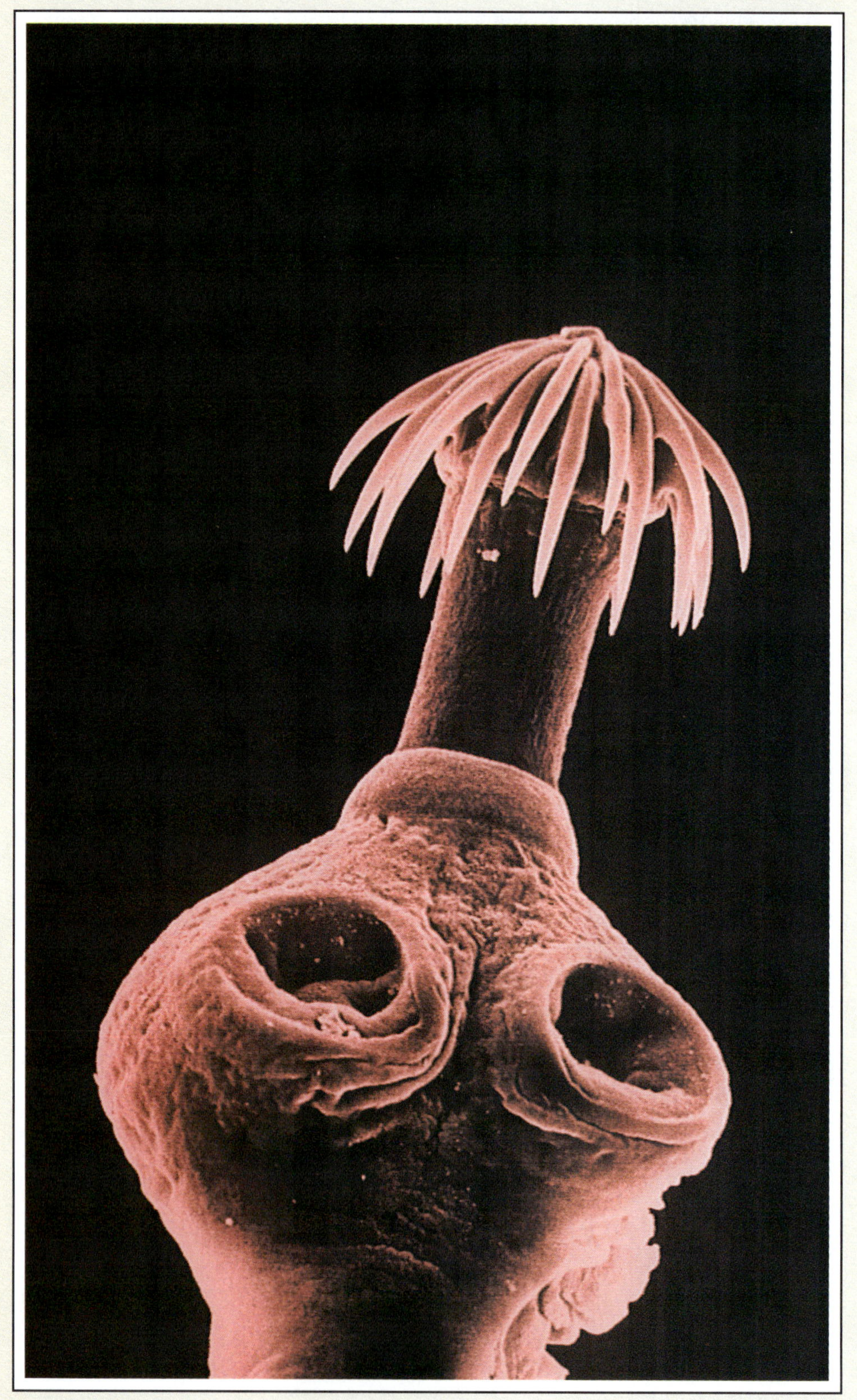

12 Eukaryotic Microorganisms and Parasites

This chapter focuses on the following questions:

A. What is a parasite, and what are the principles of parasitology?

B. What are protists, and why are they important?

C. How do groups of protists differ?

D. What are fungi, and why are they important?

E. How do groups of fungi differ?

F. What are parasitic helminths, and why are they important?

G. How do groups of parasitic helminths differ?

H. What are the characteristics of parasitic and vector arthropods?

I. How do groups of parasitic and vector arthropods differ?

In our survey of microbes we have devoted significant attention to bacteria of the kingdom Monera and to viruses. However, some members of certain eukaryotic kingdoms are also of interest to microbiologists and health scientists. The kingdoms Protista and Fungi contain large numbers of microscopic species, some of which supply food and antibiotics, and some of which cause disease. The kingdom Animalia contains helminths that cause disease and arthropods that cause or transmit diseases. The microscopic eukaryotes, as well as the helminths and arthropods, constitute a significant part of a health scientist's training. Unless health scientists take a course in parasitology, their only opportunity to learn about helminths and arthropods is in conjunction with the study of microscopic infectious agents.

PRINCIPLES OF PARASITOLOGY

A **parasite** is an organism that lives at the expense of another organism, called the **host.** Parasites vary in the degree of damage they inflict on their hosts. Though some cause little harm, others cause moderate to severe diseases. Parasites that cause disease are called **pathogens. Parasitology** is the study of parasites.

Although few people realize it, among all living forms there are probably more parasitic than nonparasitic organisms. Many of these parasites are microscopic throughout their life cycle or at some stage of it. Historically, in the development of the science of biology, parasitology came to refer to the study of protozoa, helminths, and arthropods that live at the expense of other organisms. We will use the term parasite to refer to these organisms. Bacteria and viruses that live at the expense of their hosts also are parasites, but they have been studied by microbiologists.

The manner in which parasites affect their hosts differs in some respects from that described in earlier chapters for bacteria and viruses. Also, some special terms are used to describe parasites and their effects. This introduction to parasitology will make discussions of parasites here and in later chapters more meaningful.

Significance of Parasitism

Parasites have been a scourge throughout human history. In fact, even with modern technology to treat and control parasitic diseases, there are more parasitic infections than there are living humans. This means, of course, that whereas some humans have no significant parasitic infections, many have multiple, chronic, and debilitating infections. Furthermore, it has been estimated that among the 60 million people dying each year, fully one-fourth die of parasitic infections or their complications.

Parasites play an important, though negative, role in the worldwide human economy. For example, less than half the world's cultivatable land is under cultivation, primarily because parasites endemic to (always present in) those lands prevent humans and domesticated animals from inhabiting some of them. As the world population increases, and the need for food with it, cultivation of such lands will become more important. In some inhabited regions many people are near starvation and severely debilitated by parasites (Figure 12.1); much of what they eat goes to nourish their parasites. Furthermore, parasitic infections in wild and domestic animals provide sources of human infection. They also cause debilitation and death among the animals, thus preventing the raising of cattle and other animals for food. Given the many human problems created by parasites, all citizens—and especially health scientists—need to understand the problems associated with controlling and treating parasitic diseases.

FIGURE 12.1 A debilitated, malnourished victim of multiple parasitic infections.

Parasites in Relation to Their Hosts

Parasites can be divided into **ectoparasites,** such as ticks and lice, which live on the surface of other organisms, and **endoparasites,** such as some protozoa and worms, which live within the bodies of other organisms. Most parasites are **obligate parasites:** They must spend at least some of their life cycle in or on a host. For example, the parasite that causes malaria invades red blood cells, and many worms attach to the intestinal lining and feed on tissue fluids. A few parasites are **facultative parasites:** They normally are free-living, such as some soil fungi, but they can obtain nutrients from a host as many fungi do when they cause skin infections. Hosts that parasites invade usually lack effective defenses against them, so such diseases can be serious and sometimes fatal.

Parasites also are classified by the duration of their association with their hosts. **Permanent parasites** such as tapeworms remain in or on a host once they have invaded it. **Temporary parasites** such as many biting insects feed on and then leave their hosts. **Accidental parasites** invade an organism other than their normal host. The ticks that ordinarily attach to dogs or wild animals but sometimes attach to humans are accidental parasites. **Hyperparasitism** refers to a parasite itself having parasites. Some mosquitoes, which are themselves temporary parasites, also harbor malaria or other parasites. Such insects serve as **vectors,** or agents of transmission, of many human parasitic diseases.

If an organism transfers a parasite to a new host it is a vector. A **biological vector** is one in which the parasite goes through part of its life cycle in the vector during transit. The malaria mosquito is both a host and a biological vector. A **mechanical vector** is one in which the parasite does not go through any part of its life cycle in the vector during transit. Flies that carry parasite eggs, bacteria, or viruses on their feet from feces to human food are mechanical vectors.

Hosts can be classified as **definitive hosts** if they harbor a parasite while it reproduces sexually or as **intermediate hosts** if they harbor the parasite during some other developmental stages. The mosquito is the definitive host for the malaria parasite because it reproduces sexually in the mosquito; the human is an intermediate host, even though humans suffer greater damage from the parasite. **Reservoir hosts** are infected organisms that make parasites available for transmission to other hosts. Reservoir hosts for human parasitic diseases usually are wild or domestic animals. **Host specificity** refers to the range of different hosts in which a parasite can mature. Some parasites are quite host specific; they mature in only one host. The malaria parasite matures primarily in *Anopheles* mosquitoes. Others can mature in many different hosts. The worm that causes trichinosis can mature in almost any warm-blooded animal, but the parasite is most often acquired by humans from pigs through the consumption of inadequately cooked pork.

Over thousands of years of evolution, parasites tend to become less injurious to their hosts. Such an arrangement preserves the host so that the parasites are guaranteed a continuous supply of nutrients. A parasite that destroys its host also destroys its own means of support. The adjustment of parasites and hosts to each other is closely related to the host's defense mechanisms. Many parasites have one or more mechanisms for evading host defense mechanisms:

1. Encystment, the formation of an outer covering that protects against unfavorable environmental conditions. These resistant cyst stages also sometimes provide a site for internal reorganization of the organism and cell division, help to attach a parasite to a host, or serve to transmit a parasite from one host to another.
2. Changing the parasite's surface antigens (molecules that elicit immunity) faster than the host can make new antibodies (molecules that recognize and attack antigens).
3. Causing the host's immune system to make antibodies that cannot react with the parasite's antigens.
4. Invading host cells, where the parasites are out of reach of host defense mechanisms.

When parasites successfully evade host defenses, they can cause several kinds of damage. All parasites rob their hosts of nutrients. Some take such a large share of nutrients or damage so much surface area of the host's intestines that the host receives too little nourishment. Many parasites cause significant trauma to host tissues. They cause open sores on the skin, destroy cells in tissues and organs, clog and damage blood vessels, and sometimes even cause internal hemorrhages. Parasites that do not evade defense mechanisms sometimes trigger severe inflammatory and immunological reactions. For example, treatment to rid human hosts of some worm infections effectively kills the worms, but toxins from the dead worms cause more tissue damage than the living parasites.

A hallmark of many parasites is their reproductive capability. Parasitism, although an easy life once the parasite is established, is a hazardous existence during transfers from one host to another. For example, many parasites that leave the human body through feces die from desiccation (drying out) before they reach another host. If several hosts are required to complete the life cycle, the hazards are greatly multiplied. Consequently, many parasites have exceptional reproductive capacities. Some parasites, especially protozoa, undergo **schizogony** (skiz-og'o-ne), or multiple fission, in which one cell gives rise to many cells, all of which are infective. Others, such as various worms, produce large numbers of eggs. Some worms are **hermaphroditic,** that is, one organism has both male and female reproductive systems. In fact, certain worms, such as tapeworms and flukes, lack a digestive tract and consist almost exclusively of reproductive systems.

PROTISTS

Characteristics of Protists

The **protists,** members of the kingdom Protista, are a diverse assortment of organisms that share certain common characteristics. Protists are unicellular (though sometimes colonial), eukaryotic organisms with cells that have true nuclei and membrane-bound organelles. Though most protists are microscopic, they vary in diameter from 5 μm to 5 mm.

Importance of Protists

Protists have captured the fancy of biologists since Leeuwenhoek made his first microscopes. In fact, most of the "animalcules" he observed were protists. Like Leeuwenhoek, many people find protists inherently interesting, and biologists have learned much about life processes from them. Protists also are important to humans for other reasons.

Protists are important in food chains. Autotrophic protists capture energy from sunlight. Some heterotrophic protists ingest autotrophs and other heterotrophs. Others decompose, or digest, dead organic matter, which then can be recycled to living organisms. Protists also serve as food for higher-level consumers. Ultimately some energy originally captured by protists reaches humans. For example, energy from the sun is transferred to protists, protists are eaten by oysters, and the oysters are eaten by humans.

Protists can be economically beneficial or detrimental. Certain protists have shells, or **tests,** of calcium carbonate. Shells deposited in great numbers by organisms that lived in ancient oceans formed the white cliffs of Dover and provided limestone used in building the pyramids of Egypt. Because different test-forming protists gained prominence during different geological eras, identifying the protists in rock layers helps to determine the age of rocks. Certain test-forming protists sometimes appear in rock layers near petroleum deposits, so geologists looking for oil are pleased to find them. Some autotrophic protists produce toxins, which do not harm oysters that eat the protists, but they can cause disease or even death in people who eat the oysters. Oyster beds infected with such protists can cause great economic losses to oyster harvesters. Other autotrophic protists multiply very rapidly in abundant inorganic nutrients and form a "bloom," a thick layer of organisms over a body of water. This process, called **eutrophication** (u"tro-fi-ka'shun), blocks sunlight, killing plants beneath the bloom and causing fish to starve. Microbes that decompose dead plants and animals use large quantities of oxygen, and the lack of oxygen leads to more deaths. Together these events result in great economic losses in the fishing industry.

Finally, some protists are parasitic. They cause debilitation in large numbers of people and sometimes death, especially in poor countries that lack resources to eradicate those protists. Parasitic diseases caused by a group of protists called the protozoa include amoebic dysentery, malaria, sleeping sickness, leishmaniasis, and toxoplasmosis. Together, these diseases account for severe losses in human productivity, incalculable human misery, and many deaths.

Classification of Protists

Like all groups of living things, the protists display great variation, which provides a basis for dividing the kingdom into sections and phyla (Appendix B). However, taxonomists do not agree about how these divisions should be made. We can accomplish our main purpose of illustrating diversity and avoid taxonomic problems by grouping protists according to the kingdom of macroscopic organisms they most resemble (Table 12.1). Thus, we have the protists that resemble plants (Figure 12.2), the protists that resemble fungi (Figure 12.3), and the protists that resemble animals (Figure 12.4).

TABLE 12.1 Properties of protists

Group	Characteristics	Examples
Plantlike protists	Have chloroplasts, live in moist, sunny environments	Euglenoids, diatoms, dinoflagellates
Funguslike protists	Most are saprophytes, may be unicellular or multicellular	True slime molds, cellular slime molds
Animallike protists	Heterotrophs, most are unicellular, most free-living, but some are commensals or parasites	Mastigophorans, sarcodinas, sporozoans, and ciliates

The Plantlike Protists

The plantlike protists have chloroplasts and carry on photosynthesis. They are found in moist, sunny environments. Most have cell walls and one or two flagella, which allow them to move. The **euglenoids** (u-gle'noidz) usually have a single flagellum and a pigmented eyespot called a **stigma.** The stigma may orient flagellar movement so the organism moves toward light. A typical euglenoid, *Euglena gracilis* (Figure

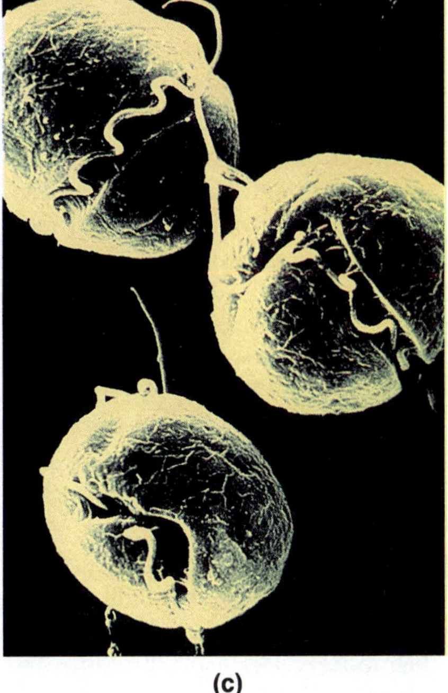

FIGURE 12.2 SEMs of representative examples of plantlike protists. (a) *Euglena*, a euglenoid. (b)4600X magnification of a diatom. (c) *Gonyaulax*, a dinoflagellate that causes red tides.

12.2a), has an elongate, cigar-shaped, flexible body. Instead of a cell wall it has a **pellicle,** or outer membranous cover. Euglenoids usually reproduce by binary fission. Most live in fresh water, but a few are found in the soil.

Another group of plantlike protists have other pigments in addition to chlorophyll. They usually have cell walls surrounded by a loosely attached secreted test, which contains silicon or calcium carbonate. Most reproduce by binary fission. They include the **diatoms** (di'ah-tomz), which lack flagella (Figure 12.2b), and several other groups, which have flagella and are distinguished by their yellow and brown pigments. Diatoms are an especially numerous group and are important as producers in both freshwater and marine environments. Fossil deposits of diatoms, known as diatomaceous earth, are used as filtering agents and abrasives in various industries.

The **dinoflagellates** (di"no-flaj'el-ātz) usually have two flagella, one extending behind the organism like a tail, and the other lying in a transverse groove (Figure 12.2c). They are small organisms that may or may not have a cell wall. Some have a **theca,** a tightly affixed secreted layer that often contains cellulose. Cellulose is an uncommon substance in protists, although it is abundant in plants. Although most dinoflagellates have chlorophyll and are capable of photosynthesis, others are colorless and feed on organic matter. Several dinoflagellates exhibit bioluminescence. The photosynthetic dinoflagellates are second only to the diatoms as producers in marine environments.

PUBLIC HEALTH

Red Tides

Some species of *Gonyaulax* and some other dinoflagellates produce toxins. When these marine organisms appear seasonally in large numbers, they cause a bloom known as a *red tide.* The toxins accumulate in the bodies of shellfish such as oysters and clams that feed on the protists. Though the toxin does not harm the shellfish, it causes paralytic shellfish poisoning in some fish and in humans who eat the infected shellfish. Even animals as large as dolphins have been killed in large numbers by this toxin. Inhaling air that contains small quantities of the toxin can irritate respiratory membranes, so sensitive individuals should avoid the sea and its products during red tides.

A red tide, caused by proliferation of toxin-producing dinoflagellates.

The Funguslike Protists

The funguslike protists, or **slime molds,** have some plant and some animal characteristics. These organisms are commonly found as glistening, viscous masses of slime on rotting logs and other decaying organic matter. Most slime molds are **saprophytes** (sap'ro-fitz), or organisms that feed on dead matter. A few are parasites of algae, fungi, or flowering plants, but not humans. Slime molds occur as true slime molds and cellular slime molds.

True Slime Molds **True slime molds** (Figure 12.3a) form a multinucleate amoeboid mass called a **plasmodium,** which moves about slowly and phagocytizes dead matter. Sometimes a plasmodium stops moving and forms fruiting bodies. Each fruiting body develops sporangia that produce spores. When spores are released, they germinate into flagellated gametes. Two gametes fuse, lose their flagella, and form a new plasmodium. As a plasmodium feeds and grows, it also can divide and produce new plasmodia directly.

Cellular Slime Molds The **cellular slime molds** (Figure 12.3b) produce pseudoplasmodia, fruiting bodies, and spores with quite different characteristics from those of true slime molds. The **pseudoplasmodium** is a slightly motile aggregation of cells with clearly defined membranes. It produces fruiting bodies, which in turn produce spores. Spores germinate into amoeboid phagocytic cells that divide repeatedly, producing more independent amoeboid cells. Depletion of the food supply causes cells to aggregate into loosely organized new pseudoplasmodia.

(a)

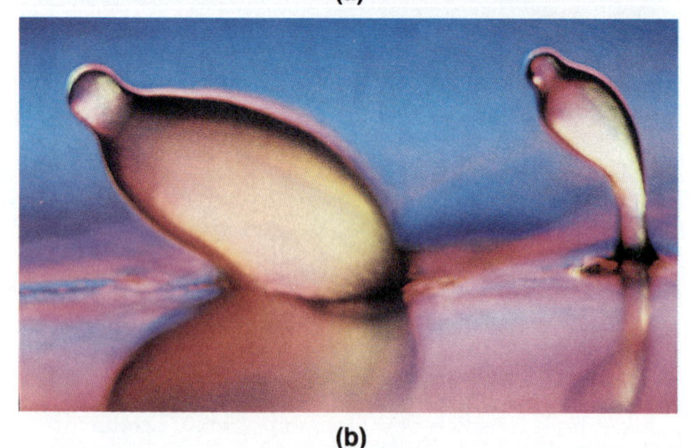

(b)

FIGURE 12.3 (a) A true slime mold, a member of the *Myxomycetes*. (b) A cellular slime mold, *Dictyostelium discoideum*.

The Animallike Protists

The animallike protists, or **protozoa,** are heterotrophic, mostly unicellular organisms, but a few form colonies. Most are free-living. Some are **commensals,** which live in or on other organisms without harming them, and a few are parasites. The parasitic protozoa are of particular interest in the health sciences. Most protozoa are motile and are further classified on the basis of their means of locomotion (Figure 12.4).

Mastigophorans The **mastigophorans** (mast-eh-gof'or-anz) have flagella. A few species are free-living in either fresh water or salt water, but most live in symbiotic relationships with plants or animals. The symbiont *Trichonympha* (Figure 12.4a) lives in the termite gut and contributes enzymes that digest cellulose. Mastigophorans that parasitize humans include members of the genera *Trypanosoma*, *Leishmania*, *Giardia*, and *Trichomonas*. Trypanosomes cause African sleeping sickness, leishmanias cause skin lesions or systemic disease with fever, giardias cause diarrhea, and trichomonads cause vaginal inflammation.

Sarcodinas The **sarcodinas** are usually amoeboid, that is, they move by pseudopodia (Figure 12.4b). A few have flagella at some stage in their life cycles. They feed mainly on other microorganisms, including other protozoa and small algae. Foraminiferans and radiolarians have shells and are found mainly in marine environments.

A total of 45 species of amoebae are capable of inhabiting the human intestinal tract. Most form cysts that help them to withstand adverse conditions. The more commonly observed genera—*Entamoeba*, *Dientamoeba*, *Endolimax*, and *Iodamoeba*—cause amoebic dysenteries of varying degrees of severity. *Entamoeba gingivalis* is found in the mouth. *D. fragilis*, which is unusual in having two nuclei and not forming cysts, is found in the large intestine of about 4 percent of the human population. Its means of transmission is unknown, but it has been found inside the eggs of pinworms. Although usually considered a commensal, it can cause chronic, mild diarrhea.

Sporozoans The **sporozoans** are parasitic and immobile (Figure 12.4c) and usually have complex life cycles. An important example is the life cycle of the malaria parasite, which requires both a human and a mosquito host (Figure 12.5). The parasites, which are

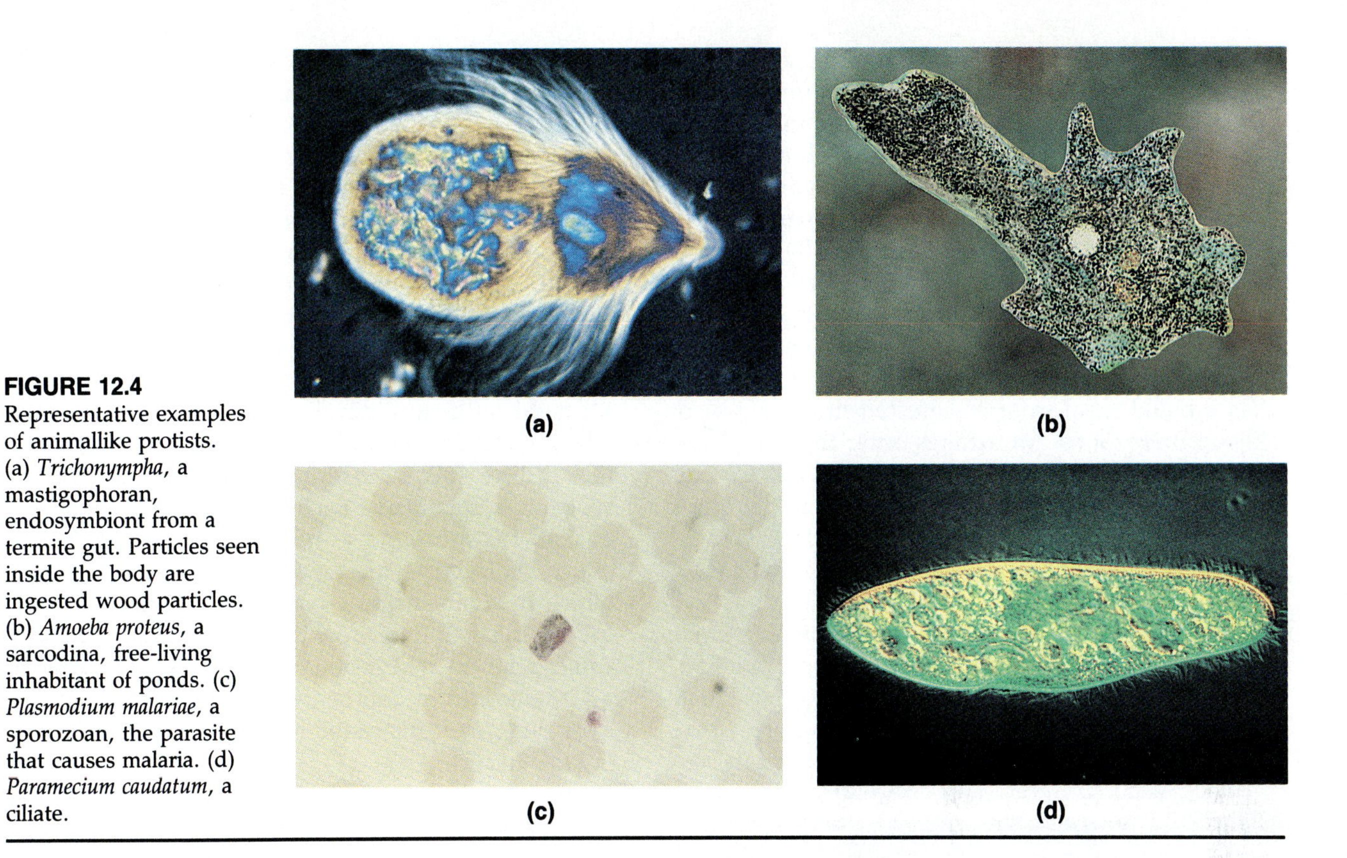

FIGURE 12.4 Representative examples of animallike protists. (a) *Trichonympha*, a mastigophoran, endosymbiont from a termite gut. Particles seen inside the body are ingested wood particles. (b) *Amoeba proteus*, a sarcodina, free-living inhabitant of ponds. (c) *Plasmodium malariae*, a sporozoan, the parasite that causes malaria. (d) *Paramecium caudatum*, a ciliate.

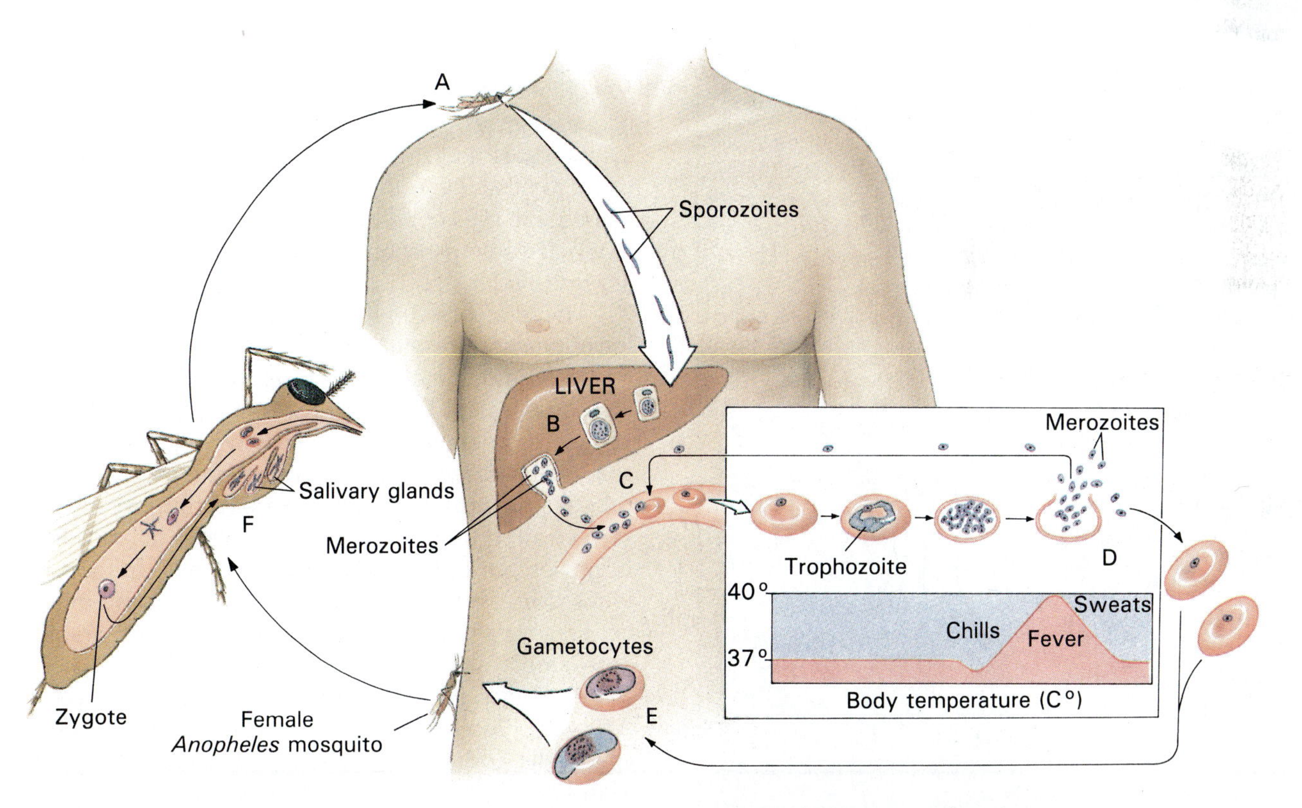

FIGURE 12.5 Life cycle of the malaria parasite, *Plasmodium*. (a) Female *Anopheles* mosquito bites human, transmitting sporozoites from its salivary glands. (b) In the liver the sporozoites multiply and become merozoites, which are shed into the bloodstream when liver cells rupture. (c) The merozoites enter red blood cells and become trophozoites, which feed and eventually form many more merozoites. (d) Merozoites are released by rupture of the red blood cells, accompanied by chills, high fever (40°C), and sweating. They can then infect other red blood cells. (e) After several such asexual cycles, gametocytes (sexual stages) are produced. (f) Upon ingestion by a mosquito, they form a zygote, which gives rise to more infective sporozoites in the salivary glands. These can then infect other people.

present as **sporozoites** (spo-ro-zo'ītz) in the salivary glands of an infected mosquito, enter human blood through the mosquito's bite. The sporozoites migrate to the liver and become **merozoites** (meh-ro-zo'ītz). After about 10 days, they emerge into the blood and invade red blood cells, becoming **trophozoites** (tro-fo-zo'ītz). Trophozoites reproduce asexually, producing many more merozoites, which are released into the blood by the rupture of red blood cells. Multiplication and release of merozoites is repeated several times during a bout with malaria. When a mosquito takes a blood meal from an infected human, it also takes in merozoites. Merozoites enter the sexual reproductive phase and become **gametocytes.** Gametocytes mature and unite to form zygotes in the mosquito's stomach. Zygotes pass through the stomach wall and produce sporozoites as they make their way to the salivary glands.

Several species of *Plasmodium* cause malaria, and each displays slight variations on the life cycle just described or in the particular species of mosquito that serves as a suitable host. Another sporozoan, *Toxoplasma gondii,* causes mild lymphatic infections in adults and severe neurological damage to the fetuses of infected pregnant women.

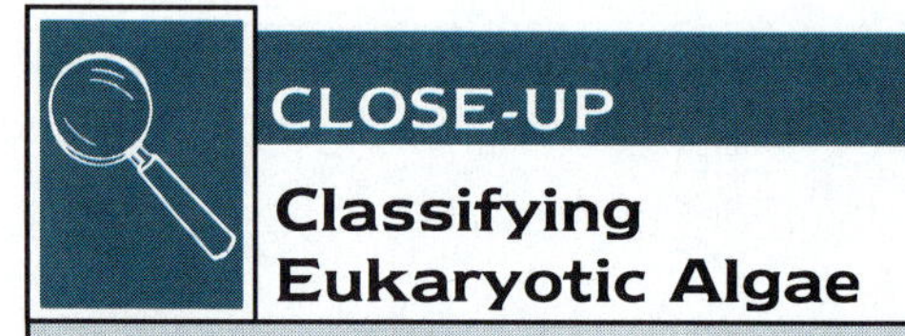

Classifying Eukaryotic Algae

Taxonomists differ in how to classify eukaryotic algae. They are sometimes classified as protists, sometimes classified as plants, and sometimes divided between these two kingdoms. Eukaryotic algae should not be confused with blue-green algae (cyanobacteria), which are prokaryotes. Although both eukaryotic and prokaryotic algae are important as producers in many environments, they are generally not of medical significance. Cyanobacteria of the genus *Prototheca* have lost their chlorophyll and have been reported to cause skin lesions.

The eukaryotic alga *Macrocystis*. This giant kelp can grow several centimeters per day. Air bladders along the stalk keep the photosynthetic leaflike blades afloat, where they can obtain sunlight. This organism is a source of algin, which is used in many foods and other products.

Ciliates The largest group of protozoans, the **ciliates,** have cilia over most of their surfaces. Cilia have a basal body near their origin and extend from the surface of the cell. They allow the organisms to move and in some genera, such as *Paramecium* (Figure 12.4d), assist in food gathering. *Balantidium coli,* the only ciliate that parasitizes humans, causes dysentery.

Ciliates have several highly specialized structures. Most have a well-developed contractile vacuole, which regulates cell fluids. Some have stiff plates in the pellicle (outer covering) that form a kind of skeleton. Others have **trichocysts,** tentacles that can be used to capture prey, or long stalks by which they attach themselves to the substratum, a surface on which an organism grows or to which it attaches itself. Ciliates also undergo **conjugation.** Unlike bacterial conjugation in which one organism receives genetic information from another, conjugation in ciliates allows exchange of genetic information between two organisms.

FUNGI

Characteristics of Fungi

Fungi, studied in the specialty called **mycology,** are a diverse group of heterotrophic organisms. Many are saprophytes that digest dead organic matter and organic wastes. Some are parasites that obtain nutrients from the tissues of other organisms. Most fungi, such as molds and mushrooms, are multicellular, but yeasts are unicellular.

The body of a fungus is called a **thallus.** The thallus of most multicellular fungi consists of a **mycelium** (mi-se'le-um), a loosely organized mass of threadlike structures called **hyphae** (hi'fe; singular: *hypha*). The mycelium is embedded in decaying organic matter, soil, or the tissues of a living organism. Mycelial cells release enzymes that digest the substratum and absorb small nutrient molecules. The cell walls of a few fungi contain cellulose, but those of most fungi contain **chitin** (ki'tin), a polysaccharide also found in the exoskeletons (outer coverings) of insects. Yeasts are an example of a eukaryote that has plasmids. These can be used to clone foreign genes into the yeast cells. All fungi have lysosomal enzymes that digest damaged cells and help parasitic fungi to invade hosts. Many fungi synthesize and store granules of the nutrient polysaccharide glycogen.

Hyphal cells have one or two nuclei and most are separated by cross walls called **septa** (singular: *septum*). Pores in septa allow both cytoplasm and nuclei to pass between cells. Some fungi have septa with so many pores that they are sievelike, and a few lack septa entirely. Certain fungi with a single septal pore have an organelle called a **Woronin** (weh-ro'nin) **body.** When a cell ages or is damaged, the Woronin body moves to and blocks the pore so that materials from a damaged cell cannot enter a healthy cell.

Many fungi reproduce both sexually and asexually, but in a few only asexual reproduction has been observed. Asexual reproduction always involves mitotic cell division and occurs by budding in yeast (Figure 12.6). Sexual reproduction occurs in several ways. In one way, haploid gametes unite, and their cytoplasm mingles in a process called **plasmogamy** (plaz-mog'am-e). However, the nuclei may fail to unite and instead produce a **dikaryotic** ('two-nucleus") **cell**, which can persist for several cell divisions. Eventually, the nuclei fuse in a process called **karyogamy** (kar-e-og'am-e) to produce a diploid cell. Such cells or their progeny later produce new haploid cells. Some fungi also can reproduce sexually during dikaryotic (diploid) phases of their life cycles. Fungi usually go through haploid, dikaryotic, and diploid phases in their life cycles (Figure 12.7).

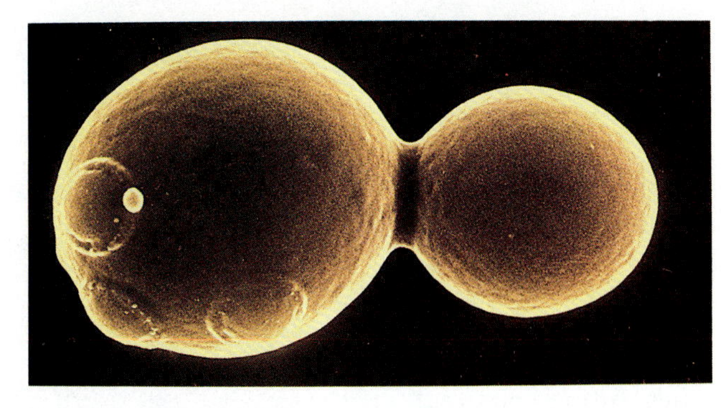

FIGURE 12.6 Budding yeast. Circular scars can be seen on the surface of the cell on the left, representing sites of previous budding.

Fungi can produce spores both sexually and asexually, and spores can have one or several nuclei. Typically, aquatic fungi produce motile spores, and terrestrial fungi produce spores with thick protective walls (Figure 12.8). Germinating spores produce either single cells or germ tubes. Germ tubes are filamentous structures that break through weakened spore walls and develop into hyphae.

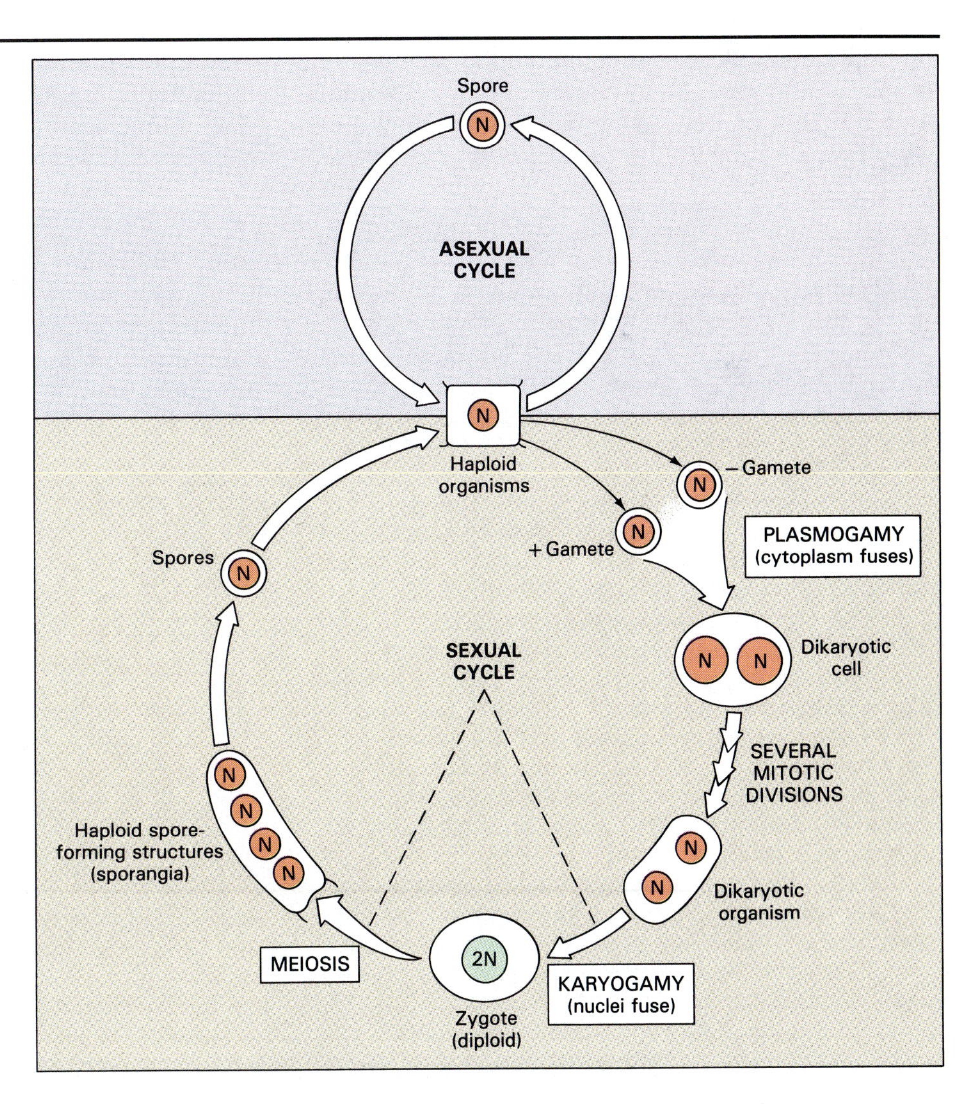

FIGURE 12.7 Sexual reproduction in fungi. Haploid organisms may maintain themselves by asexual budding or spore formation [lavender background]. Alternatively [tan background], they may produce gametes, which initially fuse their cytoplasmic portions. After several mitotic divisions of the still separate nuclei, the two nuclei fuse to form a diploid zygote. The zygote then undergoes meiosis to return to the haploid state and produces reproductive spores.

(a)

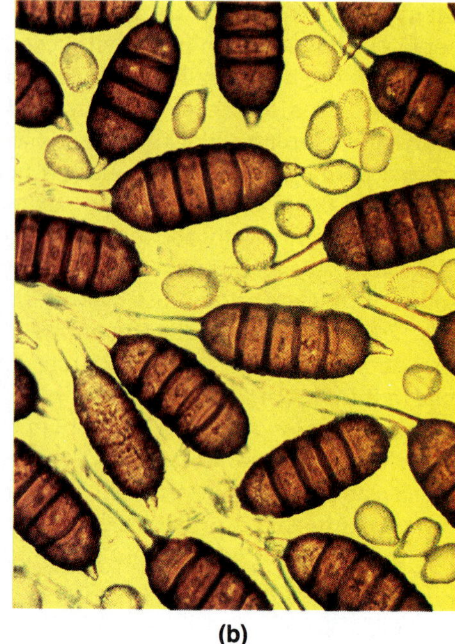

(b)

FIGURE 12.8 (a) Formation of asexual spores (conidiospores) in the fungus *Verticllium*. (b) Spores of the rose rust fungus *Phragmidium*.

Importance of Fungi

In ecosystems, fungi are important decomposers. In the health sciences they are important as facultative parasites—they can obtain nutrients from nonliving organic matter or from living organisms. Fungi are never obligate parasites because all fungi can obtain nutrients from dead organisms. Even when fungi parasitize living organisms, they kill cells and obtain nutrients as saprophytes. Some fungi parasitize nearly every form of life, and some produce antibiotics that lyse bacteria. Parasitic fungi vary in the damage they inflict. Fungi such as those that cause athlete's foot are nearly always present on the skin and rarely cause

CLOSE-UP

Lichens: Dual Plants

A *lichen* consists of a fungus living in symbiosis with either a cyanobacterium or a green alga. The members of the pair can be separated, and each will live a perfectly normal life by itself. Presumably, though, there are advantages to living together. Although there is some debate on the point, most scientists feel that lichens represent a mutualistic relationship, in which each member of the partnership benefits. The fungus obtains food from the photosynthetic organism while providing it with structure and protection from the elements (especially from dehydration).

Lichen organisms take on different and very specific shapes when they grow in association. Crustose lichens grow on surfaces and resemble a crust. Foliose lichens are leaflike, growing in crinkled layers that project up from a substratum, usually a rock or tree trunk. Fruticose lichens are the tallest, sometimes looking like tiny miniature forests. Reindeer moss is not a true moss but rather an example of a fruticose lichen.

Lichens are pioneer organisms, being among the first to colonize bare rock. Gradually they erode the rock with acids they produce. They also accumulate tiny bits of dust and humus around their bases and begin the process of soil formation, making it possible for seeds or spores of other plants to get a start. Lichens are highly sensitive to air pollution, and you may have to leave the city to find them.

Several species of lichens growing on a rock along the shore of Acadia National Park in Maine. At the upper left corner, a gray foliose type can be seen. The pink, purple, and green species are crustose types.

APPLICATIONS

Fungi and Orchids

When explorers first brought orchids from South America back to England, the English were delighted to have such handsome specimens in their conservatories. However, they suffered great disappointment when the plants failed to thrive, no matter how carefully they were potted in fresh soil and new pots. It took several years of experimentation, and perhaps the fortuitous importing of a few orchids in their native medium, to learn to grow orchids out of their natural environment. Eventually, it was discovered that orchids require certain fungi to thrive. These fungi form symbiotic associations with the orchid roots; such associations are called *mycorrhizae* (mi-ko-ri'ze). When medium in which orchids had grown was used to pot fresh specimens, the orchids and the fungi formed mycorrhizae and both thrived.

damage. The fungus that causes histoplasmosis can spread through the lymphatic system to cause fever, anemia, and death.

Saprophytic fungi are beneficial as decomposers and as producers of antibiotics. Their digestive activities provide nutrients not only for the fungi but for other organisms, too. The carbon and nitrogen compounds they release from dead organisms contribute significantly to the recycling of substances in ecosystems. Fungi are essential for decomposing lignins and other woody substances. Some fungi excrete metabolic wastes that are toxic to other organisms, especially soil microorganisms. Production of such toxins is called **antibiosis,** and the toxins are called antibiotics (Figure 12.9). In the soil these toxins presumably help the species that produce them survive. The antibiotics, when extracted and purified, are used to treat human infections (Chapter 14).

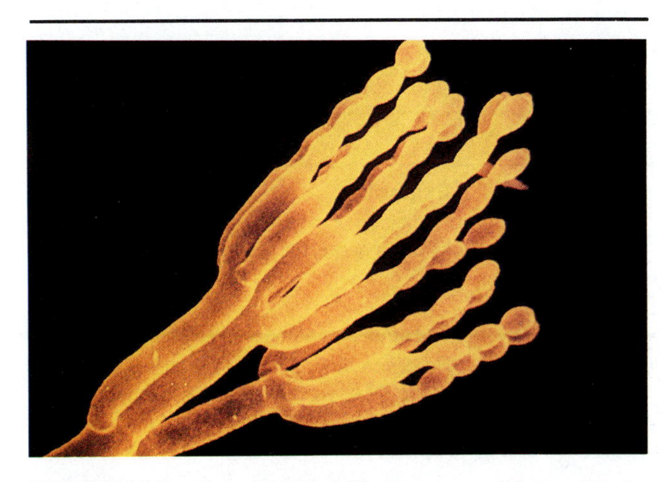

FIGURE 12.9 The mold *Penicillium*, source of the antibiotic penicillin.

Parasitic fungi can be detrimental when they invade other organisms. These fungi have three requirements for invasion: (1) proximity to the host, (2) ability to penetrate the host, and (3) ability to digest and absorb nutrients from host cells. Many fungi reach their hosts by producing spores that are carried by wind or water. Others arrive on the bodies of insects or other animals. For example, wood-boring insects spread spores of the fungal Dutch elm disease (Figure 12.10) throughout the North American continent in the decades following World War I, killing almost all elm trees in some parts of our country. Fungi penetrate plant cells by forming hyphal pegs that press on and push through cell walls. How fungi penetrate animal cells, which lack cell walls, is not fully understood, but lysosomes apparently play an important role. Once fungi have entered cells, they digest cell components and absorb nutrients. As cells die the fungus invades adjacent cells and continues to digest and absorb nutrients.

Fungal parasites in plants cause diseases such as wilts, mildews, blights, rusts, and smuts and thereby produce extensive crop damage and economic losses. Fungal parasites of domestic birds and mammals are also responsible for extensive economic losses. Those that invade humans, either directly or by way of birds and mammals, cause human suffering, decreased productivity, and sometimes long-term medical expenses. Human fungal diseases, or **mycoses,** often are caused by more than one organism. Mycoses can

FIGURE 12.10 Dead American elm (*Ulmus americana*), killed by Dutch elm disease.

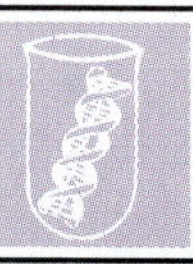

BIOTECHNOLOGY

Roach Trap Uses Fungus

A company specializing in pest control using naturally occurring microbes has recently patented a baited roach trap lined with the soil fungus *Metarhizium anisopliae*. As roaches crawl through the trap the fungus sticks to their back and within 12 hours has penetrated the body wall, grown throughout the interior, and is digesting the tissue. A single roach can spread the fungus throughout a colony, as roaches groom each other by licking. The fungus attacks only roaches and is safer than chemical pesticides.

be classified as superficial, subcutaneous, or systemic. *Superficial* diseases affect only keratinized tissue in the skin, hair, and nails. *Subcutaneous* diseases affect skin layers beneath keratinized tissue and can spread to lymph vessels. *Systemic* diseases invade internal organs. Some fungi are opportunistic; they do not ordinarily cause disease but can do so in individuals whose defenses are impaired.

Culturing and identifying the causative agents of mycoses require special laboratory techniques. Acidic, high-sugar media with antibiotics added help to prevent bacterial growth and allow fungal growth. The medium Sabouraud agar, which was developed nearly a century ago by a French mycologist, is still used in many laboratories. Under the best of conditions, most fungi in cultures grow slowly, and for some it may take 2 to 4 weeks to obtain growth comparable to that obtained with bacteria in 24 hours.

Classification of Fungi

Fungi are classified according to the nature of the sexual stage in their life cycles. Such classification is complicated by two problems: (1) No sexual cycle has been observed for some fungi, and (2) it is often difficult to match the sexual and asexual stages of some fungi. Sometimes one researcher works out an asexual phase and gives the fungus a name; another researcher works out the sexual phase of a fungus and gives it a different name. Because the relationship between the sexual and asexual phases is not always apparent, a particular species of fungi may have two names until someone discovers that the two phases occur in the same organism. Another problem is that many fungi look quite different when growing in tissues (yeastlike) and when growing in their natural habitats (filamentous). The ability of an organism to alter its structure when it changes habitats is called **dimorphism** (di-mor'fizm) (Figure 12.11). Dimorphism in fungi has complicated the problem of identifying causative agents in fungal diseases. We will consider water molds, bread molds, sac fungi, club fungi, and the so-called Fungi Imperfecti, which are believed to have lost their sexual cycle (Table 12.2).

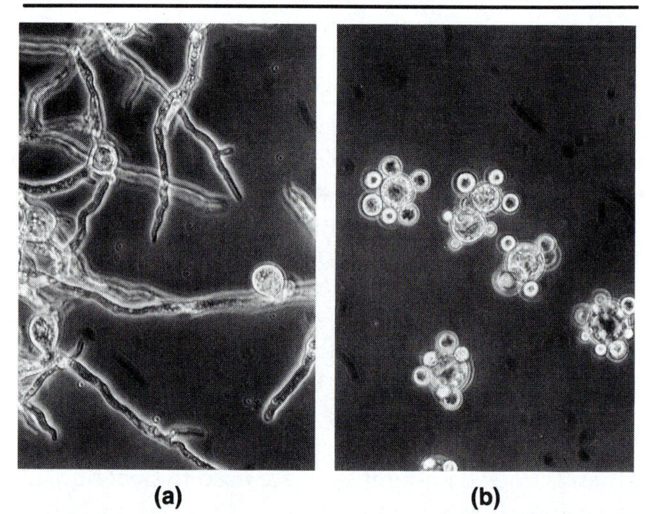

FIGURE 12.11 Dimorphism in fungi, as seen in *Mucor*. (a) Hyphae. (b) Yeast form.

TABLE 12.2 Properties of fungi

Phylum	Common Name	Characteristics	Examples
Oomycota	Water molds	Produce motile spores both sexually and asexually, live in fresh water and as plant parasites	Mildews and plant blights
Zygomycota	Bread molds	Display conjugation	*Rhizopus* and other bread molds
Ascomycota	Sac fungi	Produce asci and ascospores during sexual reproduction	*Neurospora, Penicillin, Saccharomyces,* and other yeasts; *Candida, Trichophyton,* and several other human pathogens
Basidiomycota	Club fungi	Produce basidia and basidiospores	*Amanita* and other mushrooms; *Claviceps* (which produces ergot); *Cryptococcus*
Deuteromycota	Fungi Imperfecti	Sexual stage nonexistent or unknown	Soil organisms, various human pathogens

Water Molds

The **water molds** and related fungi that cause mildew—the **Oomycota**—are so strikingly different from other groups of fungi that some taxonomists classify them as protists. They produce flagellated spores during asexual reproduction and large motile gametes during sexual reproduction. The most prominent phase of their life cycle consists of diploid cells from the union of gametes. These fungi live freely in fresh water and as plant parasites, where they cause such diseases as downy mildew on grapes and sugar beets and late blight in potatoes. A fungus of this group was responsible for the Irish potato famine in the 1840s. With a few exceptions, water molds are not medically significant.

Bread Molds

The **bread molds, Zygomycota** or conjugation fungi, have complex mycelia composed of hyphae with chitinous walls. The black bread mold, *Rhizopus* (Figure 12.12), has hyphae that grow rapidly along a surface and into the substratum. Some hyphae produce spores that are easily carried by air currents. When spores reach an appropriate substratum, they germinate to produce new hyphae. Sometimes short branches of the hyphae of two different strains, called plus and minus strains, grow together. This joining of hyphae gave rise to the name conjugation fungi. How hyphae are attracted to each other is not known, but chemical attractants may be involved. Multinucleate cells form where the hyphae join, and many pairs of plus and minus nuclei fuse to form zygotes. Each zygote is enclosed in a **zygospore,** a thick-walled, resistant structure, which also produces spores. Genetic information in zygospores comes from two strains, that in hyphal spores comes from a single strain. Although bread molds interest mycologists (scientists who study fungi) and frustrate bacteriologists whose cultures they contaminate, they usually do not cause human disease. *Rhizopus*, however, is an opportunistic human pathogen.

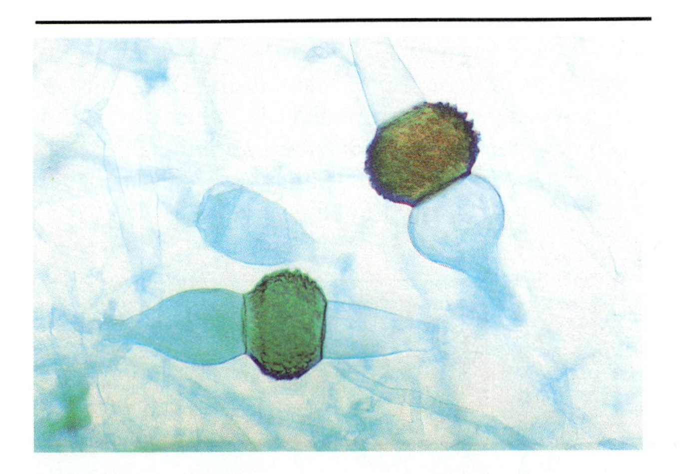

FIGURE 12.12 The black bread mold, *Rhizopus nigricans*. Sexual zygospores (black, spiny structures) are the result of joining and fusion of genetic materials at the tips of special hyphal side branches. The zygospores germinate to produce a sporangium that, in turn, produces many asexual spores. (100X)

Sac Fungi

The **sac fungi** are a diverse group containing over 30,000 species. These fungi have chitin in their cell walls and produce no flagellated cells. With the exception of yeasts, which do not form hyphae, the hyphae have cross walls with a central pore. They are distinguished from other fungi by saclike asci (singular: **ascus**) produced during sexual reproduction and are properly called **Ascomycota** (as-ko-mi'ko-ta). Yeasts are included among the ascomycotes, even though most have no known sexual stage. In species that reproduce both sexually and asexually (Figure 12.13), the asexual phase has **conidia** at the ends of hyphae. In the sexual phase one strain has a large **ascogonium,** and an adjacent strain has a smaller **antheridium.** These structures fuse, their nuclei mingle, and hyphal cells with dikaryotic nuclei grow from the fused mass. Eventually, dikaryotic nuclei fuse to form a zygote, and the zygote nucleus divides to form eight nuclei in each ascus. Each ascus forms eight **ascospores,** sometimes releasing them forcefully.

Several sac fungi are of interest. *Neurospora* is significant because studies of its ascospores have provided important genetic information. *Penicillium notatum* produces the antibiotic penicillin. *P. roquefortii* and *P. camemberti* are responsible for the color, texture, and flavor of Roquefort and Camembert cheeses. Yeasts, especially those of the genus *Saccharomyces*, release carbon dioxide and alcohol as metabolic products and are used to leaven bread and to make alcohol in beer and wine (Chapter 27). A number of sac fungi are human pathogens. *Candida albicans* causes vaginal yeast infections. *Trichophyton* is associated with athlete's foot and *Aspergillus* with opportunistic respiratory infections. Species of *Blastomyces* and *Histoplasma* cause respiratory infections and can spread throughout the body.

Club Fungi

The **club fungi** include mushrooms, toadstools, rusts, and smuts. The rusts and smuts parasitize plants and cause significant crop damage. In addition to having hyphae aggregated to form mycelia, the club fungi have clublike structures called **basidia,** resulting in the name club fungi, or **Basidiomycota** (Figure 12.14). In a typical basidiomycete life cycle, basidiospores germinate to form mycelia, and cells of mycelia unite

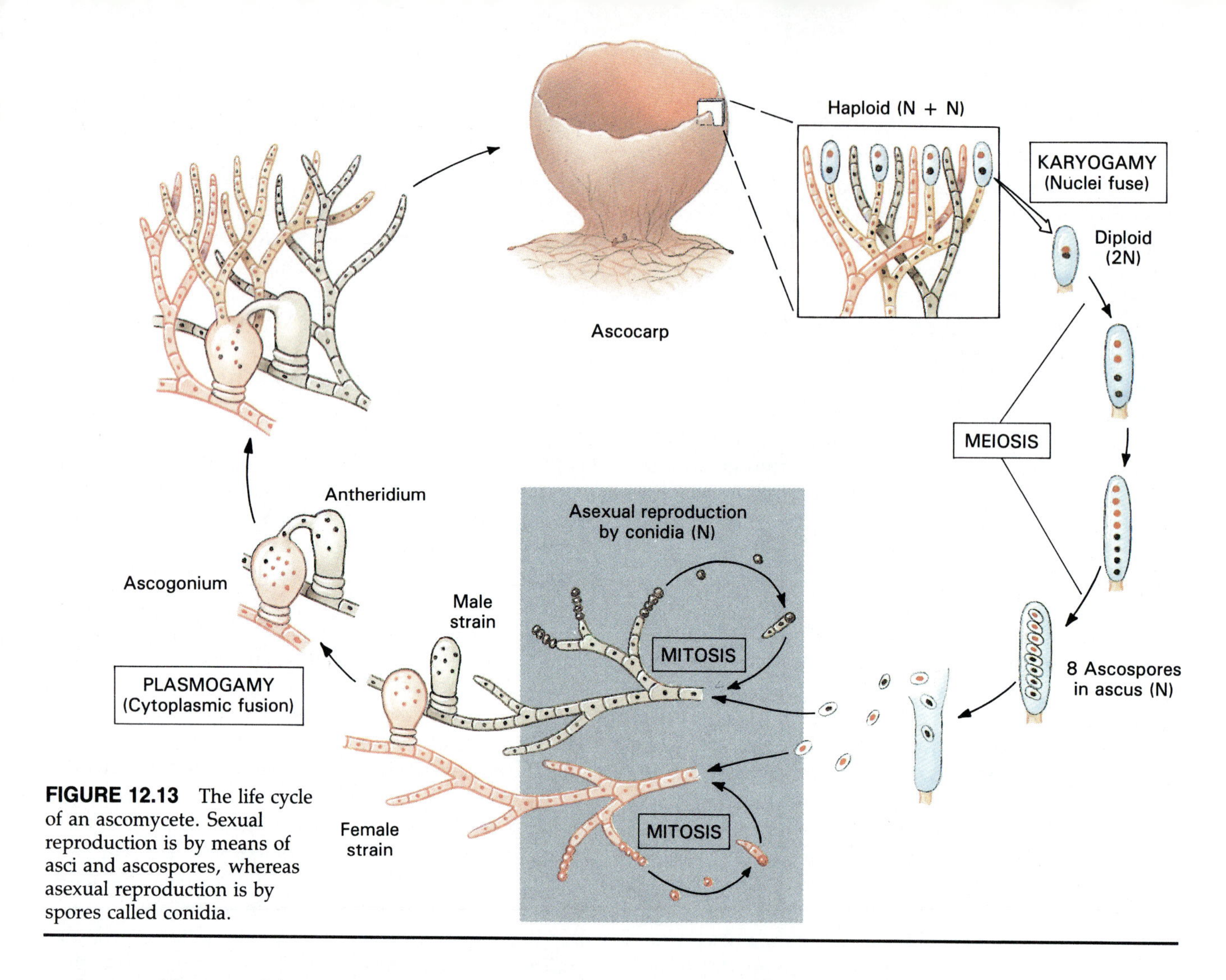

FIGURE 12.13 The life cycle of an ascomycete. Sexual reproduction is by means of asci and ascospores, whereas asexual reproduction is by spores called conidia.

to form a dikaryon. The dikaryon mycelium grows and produces basidia, which produce basidiospores. Some mushrooms, such as *Amanita*, produce toxins that can be lethal to humans. Also, *Claviceps purpura*, a parasite of rye, produces a toxic substance, ergot. This substance can be used in small quantities to treat migraine headaches and induce uterine contractions, but in larger quantities it can kill (Chapter 22). The yeast *Cryptococcus* causes opportunistic respiratory infections, which can be fatal if they spread to the central nervous system causing meningitis and brain infection. It is increasingly being seen in AIDS patients.

Fungi Imperfecti

The **Fungi Imperfecti,** or **Deuteromycota,** are "imperfect" because no sexual stage has been observed in their life cycles. Without information on the sexual cycle they cannot be assigned to a taxonomic group, but by their vegetative characteristics most seem to belong with the sac fungi. Many of the Fungi Imperfecti have recently been placed in other phyla and given new genus names. We have kept the older designations, however, because the new ones are not yet familiar or widely used in clinical work.

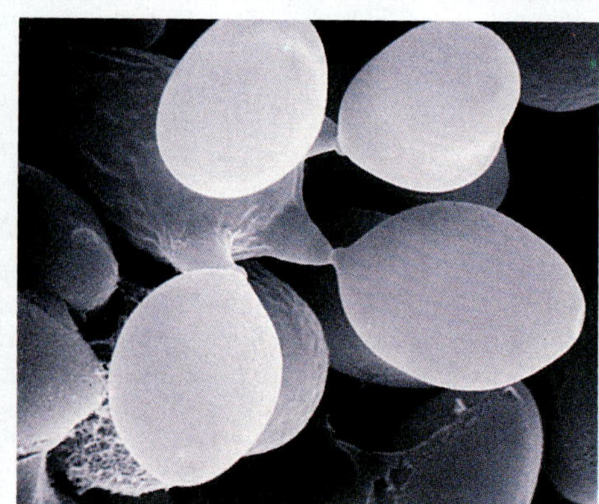

(a) (b)

FIGURE 12.14 The gills on the bottom of a mushroom (a) have microscopic, club-shaped structures called basidia. Each basidium produces four basidiospores, the balloonlike structures in (b).

CLOSE-UP

Are the Biggest and Oldest Organisms on Earth Fungi?

Weighing about 100 tons (more than a blue whale) and extending through nearly 40 acres of soil in a forest near Crystal Falls in upper Michigan is a gigantic individual of *Armillaria bulbosa*. Between 1500 and 10,000 years ago, most likely at the end of the last Ice Age, a single pair of compatible spores germinated and mated, having blown in from parent mushrooms, and began growth that continues today. The fungus grows primarily under the soil, so it is usually not visible to the casual observer. The hyphae of its mycelium probe through the soil, seeking woody debris to decompose and recycle. Experimental measurements of its growth rate through soil enabled scientists to estimate the time required to reach its present size.

DNA analysis of 12 genes from its fruiting structures—commonly called button or or honey mushrooms—and its stringlike underground colonizing structures—called rhizomorphs—reveals the huge fungus to be a giant clone with all parts having identical genetic composition. Although there are minor breaks in its continuity, it is still regarded as a single individual.

Despite the massive size of this fungus, its discoverers, Dr. Myron Smith and Dr. James Anderson of the University of Toronto and Dr. Johann Bruhn of Michigan Technological University, predicted that it might not be the largest organism of its kind. Writing in the journal *Nature* in April 1992, they explained that they found the fungus in a mixed forest containing many kinds of trees. In a single-type forest such as a large stand of birch or aspen, a fungus with a preference for that type of tree could reach even greater size. This one, however, has probably reached its maximum size, as it collides with competing fungi along its borders.

The scientists' prediction quickly proved to be prophetic. About a month after the *Nature* article was published, two forest pathologists—Ken Russell of the State Department of Natural Resources and Terry Shaw of the U.S. Forest Service—announced that they had been studying an even larger fungus near Mount Adams in southwestern Washington. This organism, an individual of *Armillaria ostoyae,* covers 1500 acres (about 2.5 square miles), making it almost 40 times as large as the Michgan fungus. The Washington fungus grows in a region populated largely by a single type of tree—in this case, pine—and therefore enjoys a vast source of nourishment. Although the Washington fungus dwarfs its Michigan counterpart in size, it is actually younger, having an estimated age of 400 to 1000 years. Thus, the Michigan fungus retains the title of "oldest" (at least for now) but not "largest."

These mushrooms are only a small part of the immense 100-ton fungal organism *Armillaria bulbosa* extending through nearly 40 acres of a Michigan forest.

Will scientists eventually discover fungi that are even bigger than the Washington fungus? The answer most likely is yes. In fact, in an interview, Shaw referred to an *Armillaria ostoyae* in Oregon that might be larger than the one he discovered. And still bigger ones may remain to be found. The search for the "biggest and oldest" promises to be an exciting episode in the field of microbiology.

TRY IT

Spore Prints

Mushroom identification often requires knowing something about the spores of your unknown specimen. Some keys to the mushrooms are arranged according to color of spores. How do you obtain such information? Try this simple method of obtaining spores for study. Generations of children, as well as professional scientists, have enjoyed making spore prints.

First, collect fresh mushrooms whose caps are just opening, that is, whose undersides are pulling away from the stem, or that are fully open. Cut off the stem flush with the bottom of the cap. Place the cap, gill-side down, on a piece of paper in a place where it can remain undisturbed overnight or for a few days. When you return, gently lift the cap and see the sunburstlike pattern of spores that have been shed from the surfaces of the gills. If you have two mushrooms of the same sort, place a dark piece of paper under one and a white piece under the other before the drying process, because you won't know which color spores to expect. Spores may range in colors from black to white, tan, or even pink. The spore prints may be made permanent by gently spraying with clear lacquer. Before you spray the spores, you might want to scrape some off onto a wet mount to examine under a microscope.

A spore print made from the mushroom *Psathyrella foenisecii*.

HELMINTHS

Characteristics of Helminths

Helminths, or worms, are bilaterally symmetrical, that is, they have left and right halves that are mirror images. A helminth also has a head and tail end, and its tissues are differentiated into three distinct tissue layers: ectoderm, mesoderm, and endoderm. Helminths that parasitize humans include flatworms and nematodes, or roundworms (Table 12.3).

TABLE 12.3 Properties of helminths

Flatworms	Parasites live in or on hosts	*Taenia* and other tapeworms are internal parasites; flukes can be internal or external parasites.
Nematodes	Most parasites live in the intestine or blood of hosts	Hookworms, pinworms, and several other nematodes live in intestines or lymph.

Flatworms

Flatworms (Platyhelminthes) are primitive worms usually no more than 1 mm thick, but some, such as large tapeworms, can be as long as 10 m. Flatworms lack a **coelom** (se'lom), a cavity between the digestive tract and the body wall, found in higher animals. Most have a simple digestive tract with a single opening, but some parasitic ones, the tapeworms, have lost their digestive tracts. Most flatworms are hermaphroditic, each organism having both male and female reproductive systems. They have an aggregation of neurons in the head end. They lack circulatory systems, and most absorb nutrients and oxygen through their body walls.

More than 15,000 species of flatworms have been identified. They include free-living, mostly aquatic organisms such as planarians and two groups of parasitic organisms, the **flukes** and the **tapeworms.** Both parasitic groups have highly specialized reproductive systems and suckers or hooks by which they attach to their host. The flukes can be internal or external parasites. *Fasciola hepatica* and several other flukes parasitize humans. Tapeworms parasitize the small intestine of animals almost exclusively. *Taenia saginata* and several other tapeworms parasitize humans.

Nematodes

Nematodes, or roundworms, share many characteristics with the flatworms, but they have a **pseudocoelom,** a primitive body cavity that lacks the complete lining found in higher animals. Their cylindrical bodies with tapered ends are covered with a thick protective cuticle. They vary in length from less than 1 mm to more than 1 m. Contractions of strong muscles in the body wall exert pressure on the fluid in the pseudocoelom and stiffen the body. Pointed ends and stiff bodies allow nematodes to move through soil and tissues easily. Nematode females are larger than males. Breeding is enhanced by chemical attractants released by females that attract males. Females can lay as many as 200,000 eggs per day. The large number of eggs, well protected by hard shells, assures that some will survive and reproduce.

Over 80,000 species of nematodes have been described, and as many as a million species may exist. They occur free-living in soil, fresh water, and salt water and as parasites in every plant and animal species ever studied. A single acre of soil can contain billions of nematodes. Many parasitize insects and plants; only a relatively small number of species infect humans, but they cause significant debilitation, suffering, and death. Most nematodes that parasitize humans, such as hookworms and pinworms, live mainly in the intestinal tract, but a few, such as *Wuchereria,* have larval forms that live in blood or lymph. Effects of roundworms on humans were first recorded in ancient Chinese writings and have been recorded by nearly every civilization since then. For an account of modern American experiences with sushi and other forms of raw fish, which may harbor roundworms, see the box in Chapter 22 titled "Sushi."

Parasitic Helminths

We will concern ourselves only with parasitic helminths and consider four groups: flukes, tapeworms, roundworms of the intestine, and roundworm larvae (Figure 12.15). Because helminths have complex life cycles related to their ability to cause diseases, we consider a typical life cycle for each group.

Flukes

Two types of fluke infections occur in humans. One involves tissue flukes, which attach to the bile ducts, lungs, or other tissues; the other involves blood flukes, which are found in blood in some stages of their life cycle. Tissue flukes that parasitize humans include the lung fluke, *Paragonimus westermani,* and the liver flukes, *Clonorchis sinensis* and *Fasciola hepatica.* Blood flukes include various species of the genus *Schistosoma.*

Parasitic flukes have a complex life cycle (Figure 12.16), often involving several hosts. Zygotes, which result from the fusion of male and female gametes, form eggs, and the eggs are encased in tough shells during their passage through the uterus of the female. The eggs pass from the host with the feces, and when they reach water, hatch into free-swimming **miracidia**

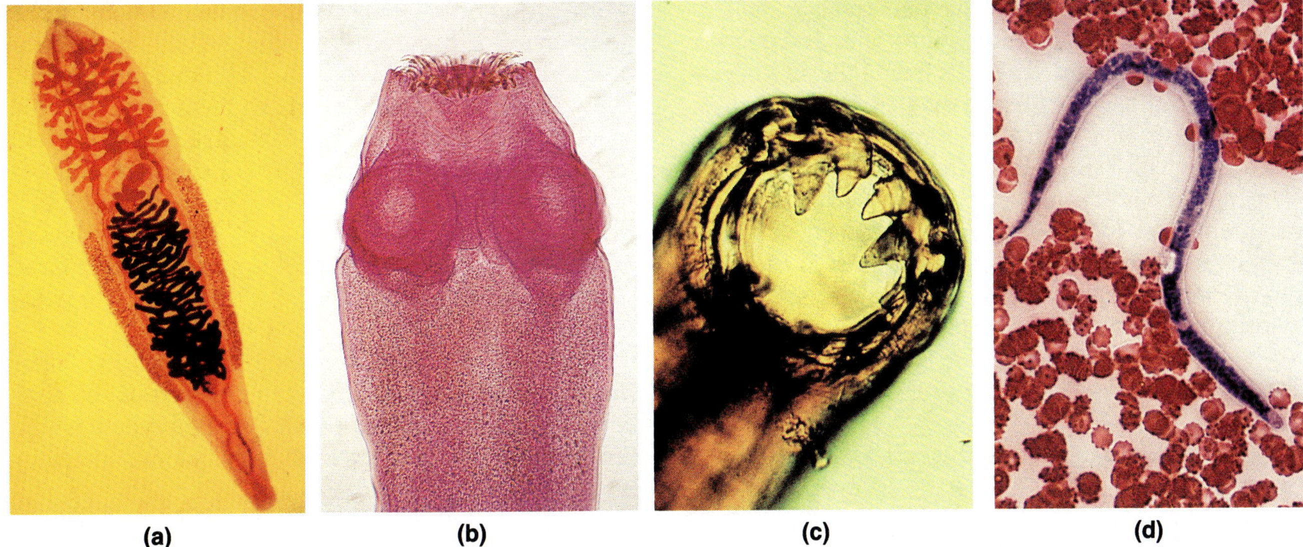

(a) (b) (c) (d)

FIGURE 12.15 Representative examples of helminths. (a) *Clonorchis sinensis*, the Chinese liver fluke, stained to show internal organs. It infests the gallbladder, bile and pancreatic ducts where it causes biliary cirrhosis and jaundice. (b) Head (scolex) of the dog tapeworm *Taenia pisiformis*. Note the hooked spines and suckers used for attachment to intestinal surfaces. (c) Mouth of the Old World hookworm *Ancylostoma duodenale*. The muscular pharynx of this roundworm pumps blood from the intestinal lining of its host. (d) The microfilarial stage of the heartworm *Dirofilaria immitis*, seen in a sample of dog blood, is transmitted by mosquito bites. The larger stages live inside the heart and perforate through its walls.

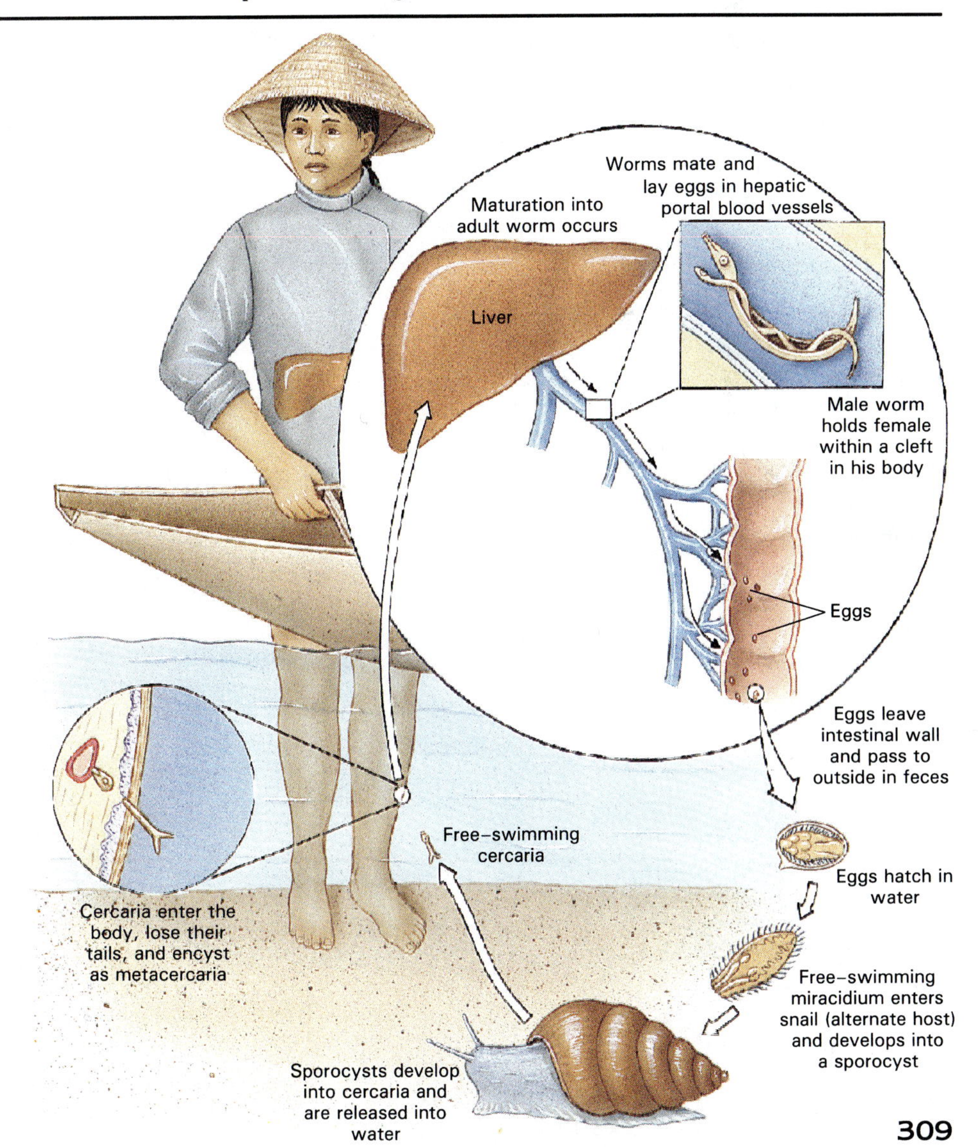

FIGURE 12.16 The life cycle of a fluke, *Schistosoma mansoni*, the cause of schistosomiasis.

(mi-ras-id'e-ah). The miricidia penetrate a snail or other molluscan host, become **sporocysts,** and migrate to the digestive gland. The cells inside the sporocysts divide by mitosis to form **rediae** (re'de-e). Rediae, in turn, give rise to free-swimming **cercariae** (ser-kar'e-e), which escape from the mollusk into water. Cercariae penetrate another host (often an arthropod), using enzymes to burrow through exposed skin, and encyst as **metacercariae.** When this host is eaten by the definitive host, the metacercariae excyst and develop into mature flukes in the host's intestine.

Tapeworms

Tapeworms consist of a **scolex** (sko'lex), or head end, with suckers that attach to the intestinal wall, and a long chain of **proglottids** (pro-glot'tidz), body components that contain mainly reproductive organs. New proglottids develop behind the scolex, and old ones mature, disintegrate, and release eggs at the rear end. Among the tapeworms that can infect humans are beef and pork tapeworms that are species of *Taenia,* dwarf and rat tapeworms that are species of *Hymenolepis,* the hydatid worm *Echinococcus,* the dog tapeworm *Dipylidium,* and the broad fish tapeworm *Diphyllobothrium.*

Though different species display minor variations, the life cycle of tapeworms (Figure 12.17) usually includes the following stages: Embryos develop inside eggs and are released from proglottids; they

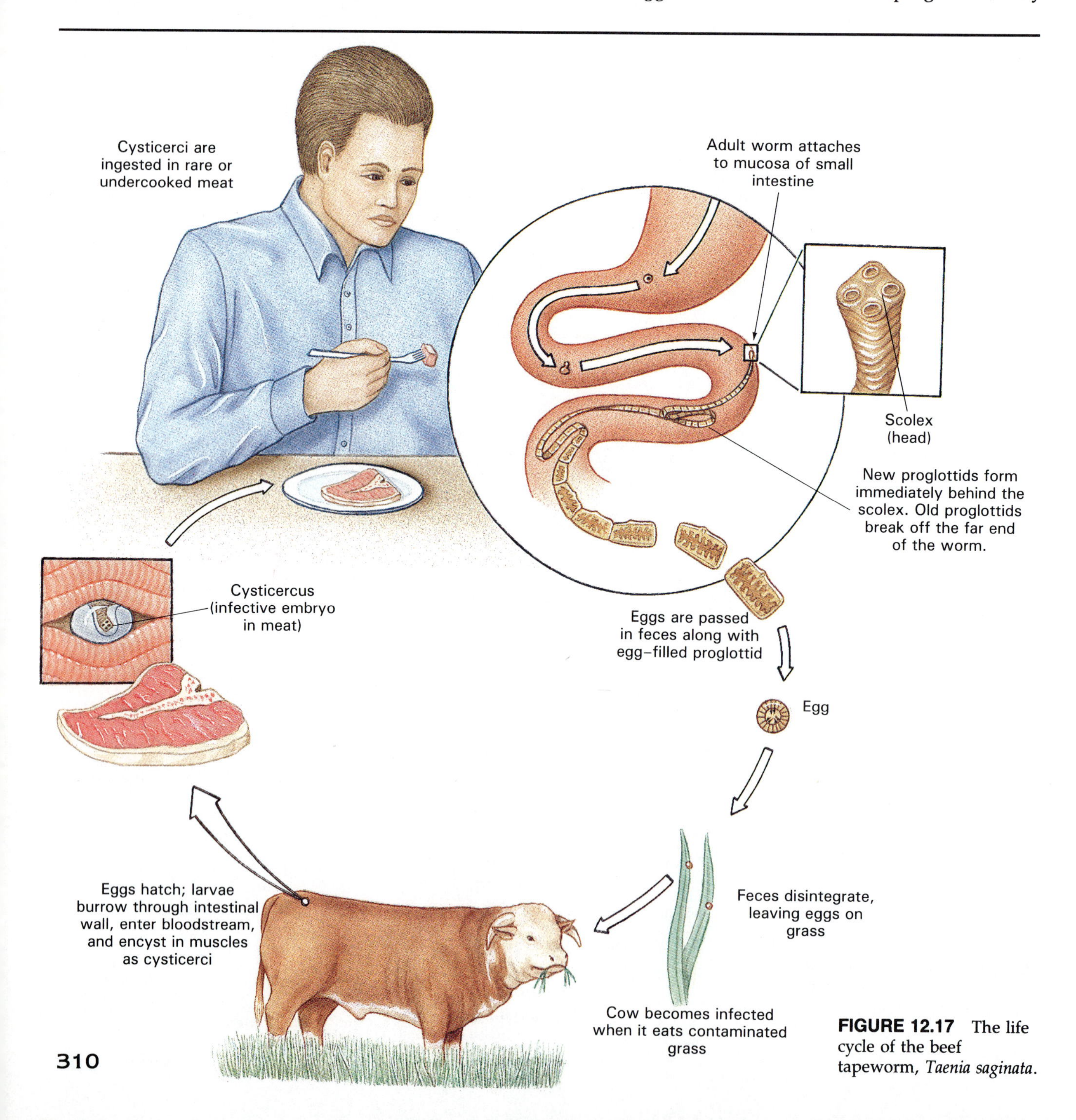

FIGURE 12.17 The life cycle of the beef tapeworm, *Taenia saginata.*

leave the host's body with the feces. When another animal ingests vegetation or water contaminated with eggs, the eggs hatch into larvae, which invade the intestinal wall and can migrate to other tissues. A larva can develop into a **cysticercus** (sis-teh-ser'kus), or bladder worm, or it can form a cyst. A cysticercus can remain in the intestinal wall or migrate through blood vessels to other organs. A cyst can enlarge and develop many tapeworm heads within it, becoming a **hydatid** (hi-da'tid) **cyst** (Chapter 22). If an animal such as a dog eats meat containing such a cyst, each scolex can develop into a new tapeworm.

Adult Roundworms

Most roundworms that parasitize humans live much of their life cycle in the digestive tract. They usually enter the body by ingestion with food or water, but some, such as the hookworm, penetrate the skin. These helminths include the pork roundworm *Trichinella spiralis*, the common roundworm *Ascaris lumbricoides*, the guinea worm *Dracunculus medinensis*, the pinworm *Enterobius vermicularis*, and the hookworms *Ancylostoma duodenale* and *Necator americanus*.

The life cycles of intestinal roundworms show considerable variation. We use the life cycle of *Trichinella spiralis* as an example (Figure 12.18). These worms enter humans as encysted larvae in the muscle of infected pigs when poorly cooked pork is eaten. The cyst walls are digested with the meat, and the larvae are released into the intestine. They mature sexually in about 2 days and then mate. Females burrow into the intestinal wall and produce eggs that hatch inside the adult and emerge as larvae. The larvae migrate to lymph vessels and are carried to the blood. From the blood the larvae burrow into muscles and encyst. These cysts can remain in muscles for years. The same processes occur in the pigs themselves, so cysts are present in their meat.

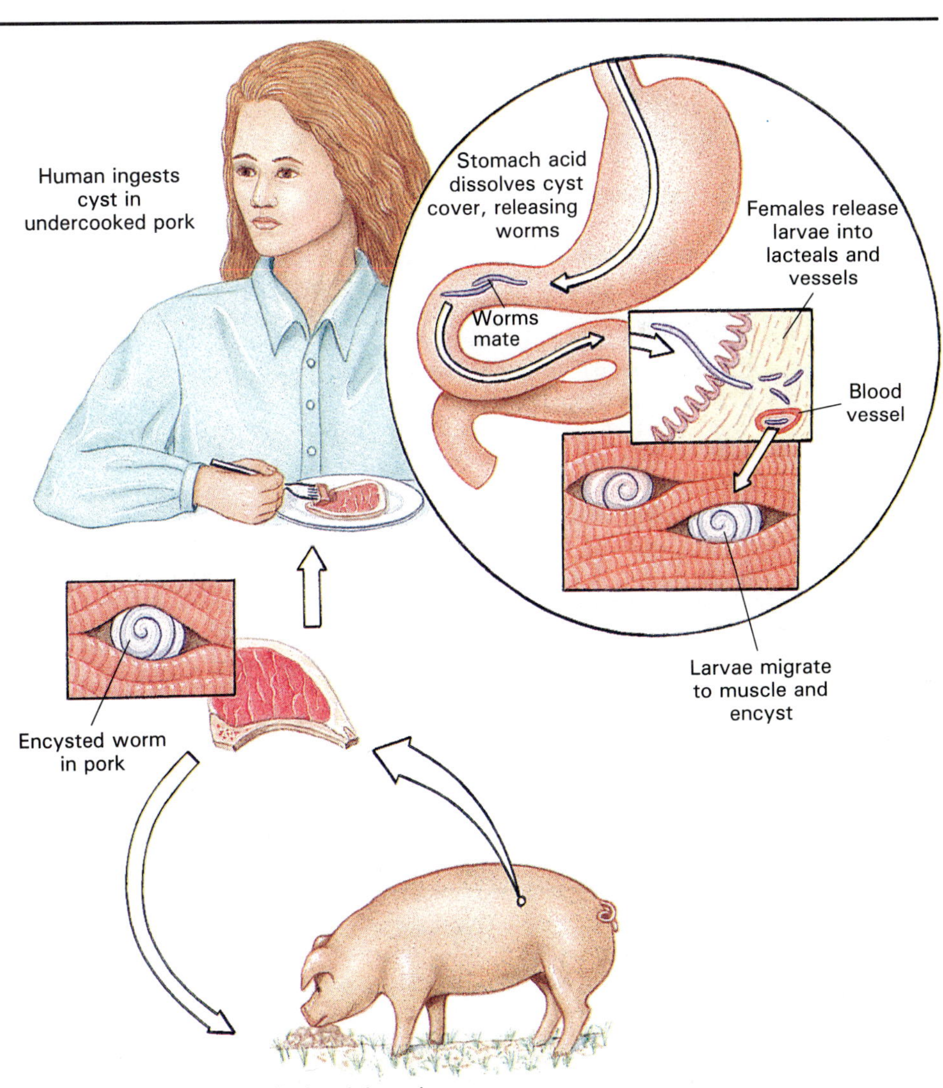

FIGURE 12.18 The life cycle of *Trichinella spiralis*, a roundworm that causes trichinosis.

MICROBIOLOGIST'S NOTEBOOK

Carter Foundation Expects Eradication of Guinea Worm from World by 1996

Former President Jimmy Carter examining a water hole contaminated with guinea worms while on a trip to Africa.

Former President Jimmy Carter: Once you've seen a small child with a 2- or 3-foot-long live Guinea worm protruding from her body, right through her skin, you never forget it. I first saw the devastating effects of Guinea worm disease in two villages near Accra, Ghana, in March 1988. In just a few minutes, Rosalynn and I saw nearly 200 victims, including people with worms coming out of their ankles, knees, groins, legs, arms, and other parts of their bodies. One woman, in great agony, was cradling her breast as if it were an infant. On it was an abscess the size of a fist where a Guinea worm was about to emerge. I have seen people with as many as a dozen or more of these worms emerging at the same time. I was shocked to find that this debilitating disease, which strikes nearly 10 million people each year, could be easily prevented. The disease is caused by contaminated drinking water; prevention is simply a matter of showing people how to make their water supply safe.

We are a compassionate nation, and that is one reason why those people still suffering needlessly from this disease should matter to us. The world is full of difficult problems that we cannot yet solve. This is one we can solve, and we can do it quickly if we put our minds to it. Here in Atlanta, The Carter Center's Global 2000 program is targeting December 31, 1995, as the date for global eradication of Guinea worm. Dr. Donald Hopkins, Senior Consultant for Global 2000's health programs, directs the Center's Guinea worm eradication efforts. He has helped enlist the support of governments and nonprofit organizations in this fight, and he has helped implement eradication programs in several African countries.

Dr. Donald Hopkins: The international community didn't realize until about 1980 that it was possible to eradicate Guinea worm completely. In fact, they didn't even realize the extent of the problem in their own countries, especially in rural areas. Before 1988, Ghana reported an average of 4500 cases of Guinea worm to the World Health Organization each year. Then in 1989, Ghana conducted its first nationwide search for cases and as a result reported more than 170,000 cases. A UNICEF study in southeastern Nigeria estimated that the rice farmers in that area of 1.6 million people are losing $20 million each year in potential profits because so many of them are crippled by Guinea worm at harvest time. In one heavily affected district of Ghana's northern region, the production of yams reportedly increased by 33 percent during the first 9 months of 1991 because so many farmers were restored to full productivity by the sharp reduction of Guinea worm.

Mr. Carter: The value is not just in terms of money. Healthy mothers can take better care of their children. Children can get a better education. Guinea worm disease causes children to miss an average of 12 weeks of school in a year when they are infected, compared to a total loss of 1 to 2 weeks lost due to all other causes combined. Often children are so badly behind when they return to school, especially if this happens several years in a row, that they become permanent drop-outs. Additional income means that people can improve their homes and do all kinds of things they couldn't do before. They become more self-reliant. Tackling the next problem becomes easier when they have pride in the success of this effort.

Guinea worm disease is contracted by drinking from ponds, step wells, cisterns and other sources of stagnant water that have been contaminated by the worm larvae. Guinea worm, *Dracunculis medinensis,* affects only humans, and it actually uses its human host to further its life cycle. Contaminated water contains water fleas that have eaten immature Guinea worm larvae. The larvae escape when the digestive juices in the person's stomach kill the flea. The larvae penetrate the stomach wall, wander around the abdomen, mature in a few months, and mate, after which the male worms die. It is only the female worm that grows to 2 or 3 feet in length and, about a year later, secretes a toxin that causes a blister on the skin. When the blister ruptures, usually when the infected part of the body is immersed in cool water, the worm starts to emerge. This process can take 30 to 100 days before the worm finally finishes making its way out of the body. When an infected person enters the village pond or watering hole, the worm discharges hundreds of thousands of tiny larvae into the water, beginning the cycle again.

Dr. Hopkins: A small incision made before the emergence blister is raised allows the worm to be wound out gradually, wrapped around a stick. This may have been the origin of the symbol of the medical profession, the caduceus, a serpent coiled around a stick. Many scholars think the Guinea worm is the "fiery serpent" of the Bible. It takes several weeks of daily gentle winding to complete the removal of a worm. If the worm breaks and dies, it will decompose inside the host, causing festering and infection. If a portion of it retracts into the tissues, it can carry

Dr. Donald Hopkins examining an extracted guinea worm in an African village.

tetanus spores back with it, leading to fatal tetanus disease. The local practice in some countries of putting cow dung on the wound makes tetanus especially common. In Upper Volta and Nigeria, Guinea worm is the third leading cause of acquiring tetanus. Other types of microorganisms can also enter the wound, and secondary infections are frequent, even if tetanus is avoided. If the worm emerges near a major joint, permanent scarring leads to stiff and crippled joints. One man died of starvation when a worm came out under his tongue and he couldn't eat. Although most worms emerge from the lower limbs, they can be found anywhere: scrotum, scalp, chest, face.

Imagine the suspense of living in an area where Guinea worms abound. Will you have worms again this year? How many will you have? Where will they emerge? Will you become crippled, infected, or die of tetanus? Will you be able to go to work or school? Transmission season is only during wet periods. In sub-Saharan Africa these are the months of June, July, and August. The worms emerge about 12 months later during the next wet season. Years when there are droughts help reduce the number of infections.

Mr. Carter: There is no cure, only prevention. No matter how many times you have been infected with Guinea worms, no immunity to them develops. Prevention involves providing safe sources of drinking water such as borehole wells, which people with emerging worms cannot enter to contaminate. It can also be prevented by teaching villages to boil their water (although many cannot afford enough fuel to do so) or to filter their drinking water through a clean, finely woven cloth. The E. I. DuPont Company, in association with Precision Fabrics Group, donated 1.4 million monofilament nylon cloth filters to The Carter Center for use in the eradication campaign. The American Cyanamid Company has donated over $2 million of the larvicide Abate, which is used to treat water safely, killing the water fleas and Guinea worms but harming nothing else. The company has pledged to continue donating Abate until the goal of eradication is reached. Using the nylon filters and Abate, the two villages I visited in Ghana in March 1988 reduced the number of cases of Guinea worm there by more than 90 percent in one year.

Dr. Hopkins: Money is the limiting factor in our fight against the Guinea worm. We have the training and the manpower. Mr. and Mrs. Carter have been very active, making several trips to Africa to talk with leaders and to go into the villages and explain how the worm is transmitted and to teach people how to break that chain of transmission. This is a problem that affects 19 nations and that will probably require $65 million to eradicate. The countries themselves cannot afford the entire cost themselves, so help has to come from wealthier nations. The Japanese have put in 191 wells. The Swedes, Dutch, and Danes are the only European countries helping.

I was part of the international team that eradicated smallpox. It is exciting to be working on another eradication project that is so close to success. We do expect to make our target date of the end of 1995. We believe we succeeded in totally eliminating Guinea worm from Pakistan by the end of 1991. So far we haven't heard of any new cases there.

Mr. Carter: A while back, I was asked what I felt my top presidential achievements were. Of course there were things such as the Panama Canal treaties, the Camp David accord between Israel and Egypt, and normalization of relations with China. But, if in 1996 we find that where there were 10 million cases of Guinea worm there are none, I would say that eradication of the Guinea worm is more important.

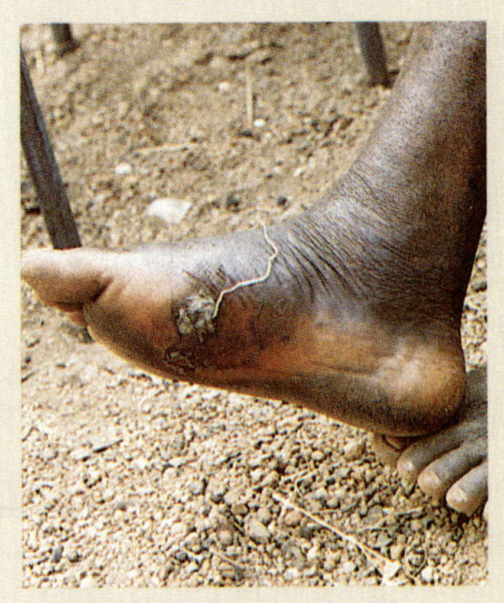

Female guinea worm emerging from a blister on the foot of a victim.

Roundworm Larvae

Whereas most roundworms cause much of their tissue damage as adults in the intestine, some cause their damage mainly as larvae in other tissues. These include *Wuchereria bancrofti*, which lives in lymphatic tissue and causes elephantiasis, *Loa loa*, which infects the eyes and eye membranes, and *Onchocerca volvularis*, which infects both the skin and eyes.

The life cycles of roundworms that parasitize humans as larvae also require a mosquito host (Figure 12.19). These worms enter the human body as immature larvae called **microfilariae** (mi″kro-fil-a′re-i) with the bite of an infected mosquito. The microfilariae migrate through the tissues to lymph glands and ducts and mature and mate as they migrate. Females produce large numbers of new microfilariae, and these enter the blood, usually at night. The microfilariae are

APPLICATIONS

Parasites on Parasites

Parasites can themselves have parasites. A large pinworm, *Heterakis gallinarum*, infects chickens, turkeys, and other birds. Earthworms eat the parasites from bird feces. The parasites cause no disease in the earthworm, but the parasite eggs become infected with a flagellated protozoan, *Histomonas meleagridis*. When a bird eats the earthworm, it becomes infected with both pinworms and the protozoan. The protozoan causes the serious disease histomoniasis in turkeys. Phenothiazine usually is added to chicken and turkey feed to prevent these and other infections. How much of the drug goes into eggs and meat eaten by humans or what its effects might be are unknown.

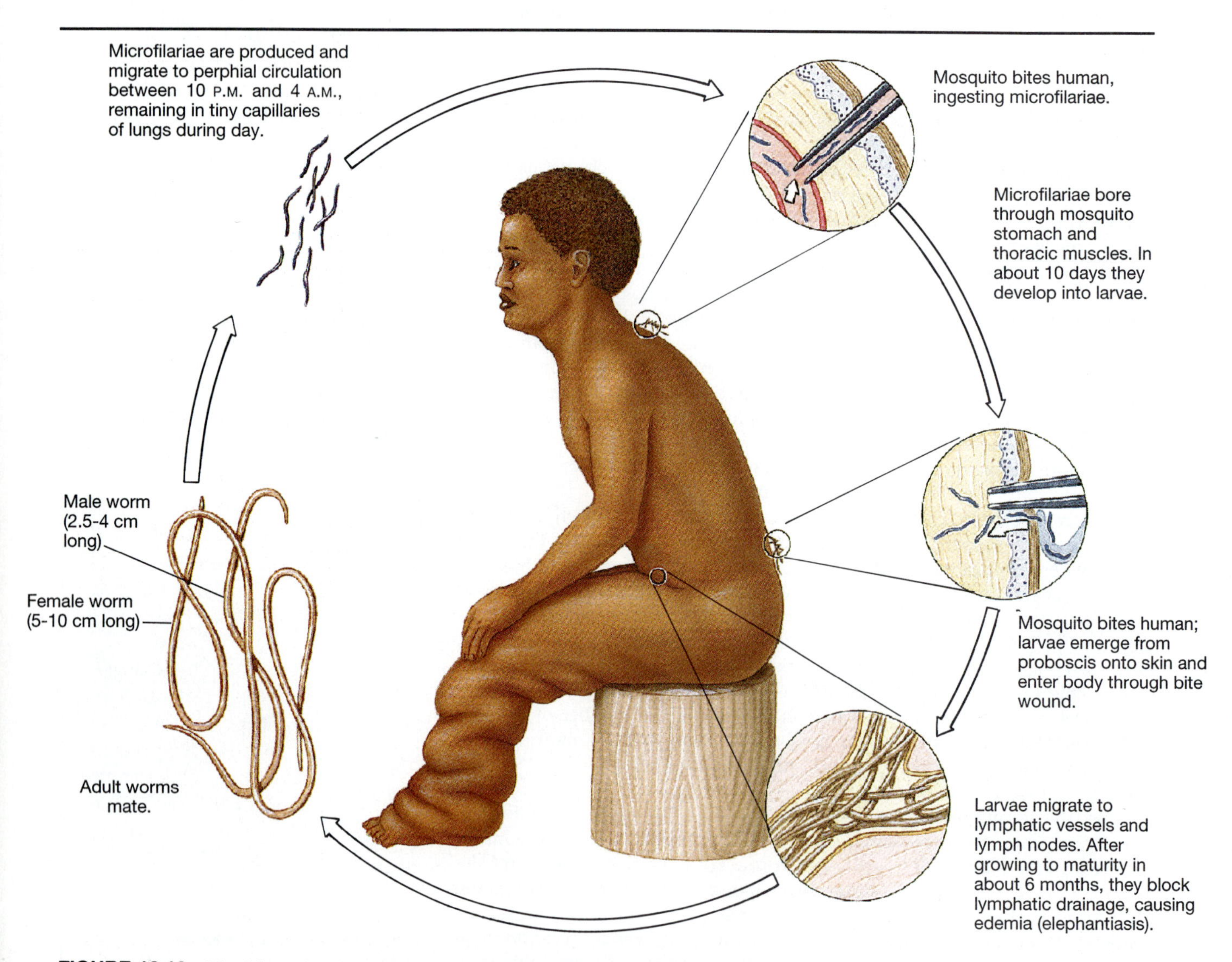

FIGURE 12.19 The life cycle of *Wuchereria bancrofti*, a roundworm that produces microfilariae and causes elephantiasis, especially of the legs and scrotum.

TRY IT

Parasites in Your Yard?

Does your family pet have parasites? Are the neighbors' dogs and cats leaving parasite-laden feces on your lawn? Parasites are not just figures in a textbook. They may be a very real (but not obvious) part of your daily surroundings. Try the following procedure to discover who may be living in your neighborhood.

Using aseptic technique (so that you won't become an alternate host!) and disposable glassware, collect a *fresh* fecal sample from an outdoor pet such as a dog, cat, or horse. Carefully mix about the amount that would fit on a thumbnail into 15 mL of a saturated solution of sodium chloride. Strain through several layers of folded cheesecloth, or gauze, to remove solids. Carefully pour or transfer by bulb pipette some of the strained fluid to a very small test tube, filling it to the top without overflowing. Now place a clean coverslip on the very top of the test tube so that it is in contact with the surface of the liquid. Allow it to sit undisturbed for 15 minutes. During this time, parasite eggs, if present, will float to the top of the tube and adhere to the underside of the coverslip. Using forceps, transfer the coverslip, wet side down, to a slide. Examine this wet mount under low-power (10X) and high-power (40X) objective lenses. How many kinds of helminth eggs can you find? Try to identify these with the help of diagrams in veterinary medicine texts. Carefully dispose of all contaminated materials.

ingested by mosquitoes as they bite infected humans. Any one of several species of mosquitoes can serve as host. When the microfilariae reach the midgut of the mosquito, they penetrate its wall and migrate first to the thoracic muscles and then to the mosquito's mouthparts. Here they can be transferred to a new human host, where the cycle is repeated.

ARTHROPODS

Characteristics of Arthropods

Arthropods constitute the largest group of living organisms; as many as 80 percent of all animal species belong to this phylum. They are characterized by jointed chitinous exoskeletons, segmented bodies, and jointed appendages associated with some or all of the segments. The name arthropod is derived from *arthros*, joint, and *podos*, foot. The exoskeleton both protects the organism and provides sites for the attachment of muscles. These organisms have a true coelom, which is filled with fluid that supplies nutrients as blood does in higher organisms. Arthropods have a small brain and an extensive network of nerves. Various groups have different structures that extract oxygen from air or water environments. The sexes are distinct in arthropods, and females lay many eggs. Arthropods are found in nearly all environments—free-living in soil, on vegetation, in fresh and salt water, and as parasites on many plants and animals.

Classification of Arthropods

Certain members of three subgroups (Classes) of arthropods, the arachnids, insects, and crustaceans (Table 12.4), are important either as parasites or disease vectors (Figure 12.20). The diseases transmitted by arthropods are summarized in Table 12.5.

TABLE 12.4 Properties of three classes of arthropods

Arachnids	Have eight legs	Spiders, scorpions, ticks, mites
Insects	Have six legs	Lice, fleas, flies, mosquitoes, some bugs
Crustacea	A pair of appendages on each body segment	Crabs, crayfish, copepods

Arachnids

Arachnids have two body regions—a cephalothorax and an abdomen; four pairs of legs; and mouthparts that are used in capturing and tearing apart prey. They include spiders, scorpions, ticks, and mites. Spider bites and scorpion stings can produce localized inflammation and tissue death, and their toxins can produce severe systemic effects. Ticks and mites are external parasites on many animals; some also serve as vectors of infectious agents.

Infected ticks transmit several human diseases. Certain species of *Ixodes* carry viruses that cause encephalitis and the spirochete *Borrelia burgdorferi* that causes Lyme disease. The common tick *Dermacentor andersoni*, which also causes tick paralysis, can carry viruses that cause encephalitis and Colorado tick fever, the rickettsias that cause Rocky Mountain spotted fever and typhus, and the bacterium that causes tularemia. Several species of *Amblyoma* ticks also carry the Rocky Mountain spotted fever rickettsia, and *Ornithodorus* ticks transmit the spirochete responsible for relapsing fever. Mites serve as vectors for the rickettsial diseases scrub typhus and Q fever.

TABLE 12.5 Diseases transmitted by arthropods

Disease	Causative Agents	Principal Vectors	Endemic Areas
Plague	*Yersinia pestis*	Fleas	Only sporadic in modern times; reservoir of infection maintained in rodents
Tularemia	*Francisella tularensis*	Fleas and ticks	Western United States
Salmonellosis	*Salmonella* species	Flies	Worldwide
Lyme disease	*Borrelia burgdorferi*	Ticks	Parts of United States, Australia, and Europe
Relapsing fever	*Borrelia* species	Ticks and lice	Rocky Mountains and Pacific Coast of United States and many tropical and subtropical regions
Typhus	*Rickettsia prowazekii*	Lice	Asia, North Africa, and Central and South America
Tick-borne typhus	*Rickettsia conorii*	Ticks	Mediterranean area and parts of Africa, Asia, and Australia
Scrub typhus	*Rickettsia tsutsugamushi*	Mites	Asia and Australia
Murine typhus	*Rickettsia typhi*	Fleas	Tropical and subtropical regions

FIGURE 12.20 Representative examples of arthropods that are parasitic or can serve as disease vectors. (a) False-color SEM of the pubic louse, *Phthirus pubis,* also known as a crab louse, clinging to a human pubic hair. The lice suck blood, feeding about 5 times a day. (b) The housefly, *Musca domestica,* can carry microbes on its feet. (c) The *Aedes* mosquito. (d) A flea, *Ctentocephalides*. (e) A tick, *Dermacentor*.

(a)

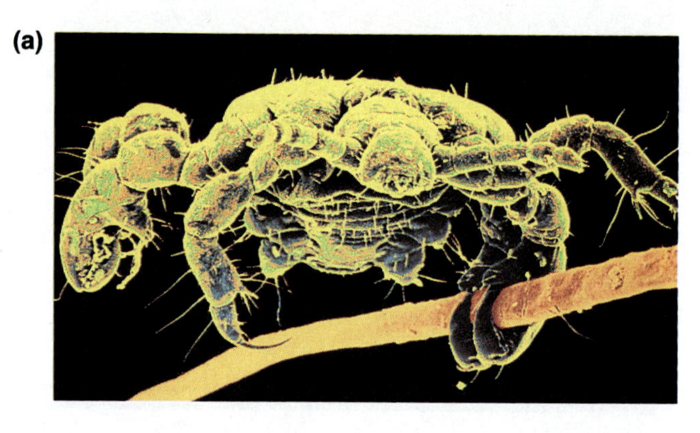

(b)

(c)

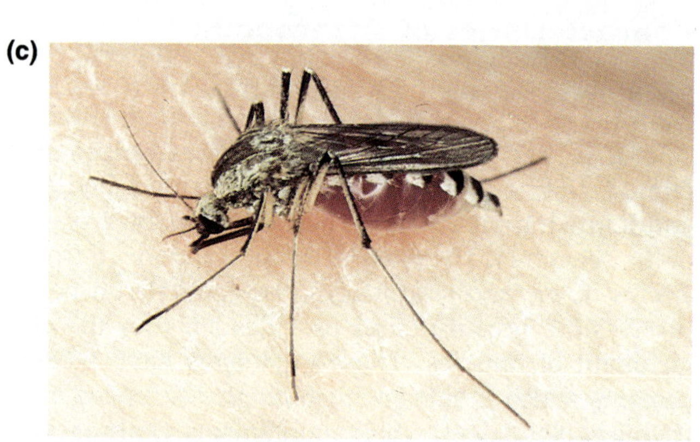

(d)

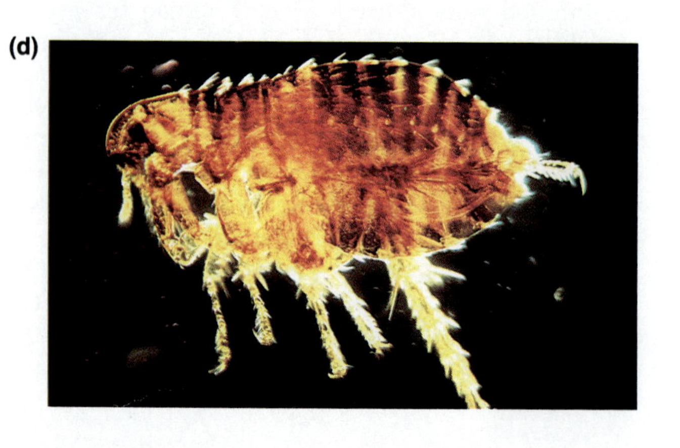

(e)

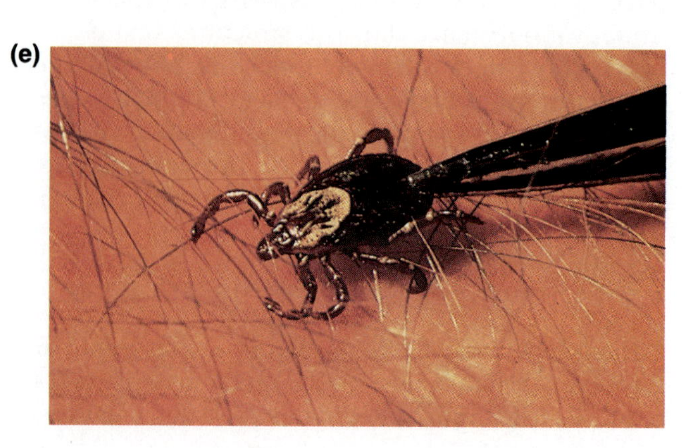

Disease	Causative Agents	Principal Vectors	Endemic Areas
Rocky Mountain spotted fever	*Rickettsia rickettsi*	Ticks	United States, Canada, Mexico, and parts of South America
Q fever	*Coxiella burnetii*	Ticks and mites	Worldwide
Trench fever	*Rochalimaea quintana*	Lice	Known only in fighting armies
Viral encephalitis	Togaviruses	Mosquitoes	Worldwide but varies by virus and vector
Yellow fever	Togavirus	Mosquitoes	Tropics and subtropics
Dengue	Togavirus	Mosquitoes	India, Far East, Hawaii, Caribbean Islands, Africa
Sandfly fever	A virus, probably of bunyavirus family	Female sand fly	Mediterranean region, India, and parts of South America
Colorado tick fever	An orbivirus	Ticks	Western United States
Tick-borne encephalitis	Viruses	Ticks	Europe and Asia
African sleeping sickness	Trypanosomes	Tsetse fly	Africa
Chagas' disease	*Trypanosoma cruzi*	True bug	South America
Kala azar and other leishmaniases	*Leishmania* species	Sand fly	Tropical and subtropical regions

Insects

Insects have three body regions—head, thorax, and abdomen; three pairs of legs; and highly specialized mouthparts. Some insects have specialized mouthparts for piercing skin and sucking blood and can inflict painful bites. Insects that can serve as vectors of disease include all lice and fleas and certain flies, mosquitoes, and true bugs, such as bedbugs and reduviid bugs. Although we often refer to all insects as "bugs," entomologists, scientists who study insects, use the term *true bug* to refer to certain insects that typically have thick, waxy wings.

The body louse is the main vector for the rickettsiae that cause typhus and trench fever and a spirochete that causes relapsing fever. (This spirochete is a different species of *Borrelia* than the one carried by ticks.) Epidemics of all louse-borne diseases usually occur under crowded, unsanitary conditions. All louse-borne disease agents enter the body when louse feces are scratched into bite wounds.

The human flea, *Pulex irritans*, lives on many other hosts and can transmit plague. However, fleas that normally parasitize rats and other rodents are more likely to transmit plague to humans. This bacterial disease still occurs in the United States in individuals who have had contact with wild rodents.

Several kinds of flies feed on humans and serve as vectors for various diseases. The common housefly, *Musca domestica*, is not part of the life cycle of any pathogens, yet it is an important carrier of any pathogens found in feces. This fly is attracted to both human food and human excreta, and it leaves a trail of bacteria, vomit, and feces wherever it goes. Blackflies serve as vectors for *Onchocerca volvulus*. Sand flies serve as vectors for leishmanias, bacteria that cause bartonellosis, and viruses that cause sand fly fever and several other diseases. Tsetse flies are vectors for trypanosomes that cause African sleeping sickness, and deer flies are vectors for the worm that causes loaiasis. Eye gnats look like tiny houseflies. They may be responsible for transmission of bacterial conjunctivitis and the spirochete that causes yaws.

Many species of mosquitoes serve as vectors for diseases. *Culex pipiens*, the common house mosquito, breeds in any water and feeds at night. It is a vector for *Wuchereria*. Another mosquito, *C. tarsalis*, breeds in water in sunny locations and also feeds at night. It is a vector for viruses that cause western equine encephalitis (WEE) and St. Louis encephalitis. Although WEE most often causes severe illness in horses, it also can cause severe encephalitis in children and a milder disease with fever and drowsiness in adults. The latter is sometimes called sleeping sickness. Many species of *Aedes* play a role in human discomfort and disease. *Aedes aegypti* is a vector of a variety of viral diseases, including dengue (breakbone fever), yellow fever, and epidemic hemorrhagic fever. Several species of *Anopheles* serve as vectors for malaria. They have a variety of breeding habits, and thus controlling them requires the application of several different eradication methods.

Several species of reduviid bugs transmit the parasite that causes Chagas' disease, which is a leading cause of cardiovascular disorders in Central and South America. Bedbugs cause dermatitis and may be responsible for spreading one kind of hepatitis, a liver infection.

Crustacea

Crustacea are generally aquatic arthropods that usually have a pair of appendages associated with each segment. Appendages include mouthparts, claws, walking legs, and appendages that aid in swimming and copulation. Those that serve as hosts for diseases that infect humans include some crayfish, crabs, and smaller crustacea called copepods.

ESSAY

Milestones in the Battle Against Malaria

Malaria, caused by protozoa of the genus *Plasmodium,* is one of the most serious parasitic infections. In spite of massive efforts to control the spread of this disease, in any given year it afflicts at least 300 million people and claims the lives of between 2 and 5 million children under the age of 5. Hundreds of U.S. tourists return home with the disease. Written records as early as Egyptian papyrus from 1550 B.C. provide descriptions of a disease with high intermittent fever that must have been malaria. They also refer to the use of tree oils as mosquito repellents, though it was over 30 centuries later that the mosquito was proven to be a vector of the disease. The Greek physician Hippocrates described the fevers and attributed them to bile. Nearly the entire population of certain Greek cities in low areas succumbed to the disease. Only the wealthy could afford to "head for the hills" in late summer to escape the heat, fevers, and mosquitoes. Many of the Crusaders of medieval times also succumbed to malaria, and slave trading in later years contributed significantly to the spread of the disease. The one bright light on this dismal scene was the discovery in the sixteenth century that malaria could be treated with quinine, a drug from the *Cinchona* tree.

The connection between swamps and fevers was recognized early, and the disease was named malaria (*mal,* bad) for the "bad air" thought to cause it. Most of the early investigations into the cause of malaria focused on air and water, and the association between the disease and mosquitoes suggested in Egyptian records was ignored. In 1847 Johann F. Meckel found pigment granules in the blood and spleen during the autopsy of a victim of malaria. By the 1870s some scientists believed they had found the organism that causes malaria and named it *Bacillus malariae.*

Alphonse Laveran, a French army physician in North Africa, was not convinced that the proper organism had been found. Ill-equipped, but determined, he continued the search for the microorganism using unstained preparations and a poor-quality, low-power microscope. He eventually found what are now known to be male sex cells of the malarial parasite in human blood. Most scientists rejected Laveran's findings in favor of the *Bacillus* theory, until he demonstrated the male sex cells to Pasteur in 1884. Only then were protozoa of the genus *Plasmodium* accepted as the cause of malaria.

The Italian physiologist Camillo Golgi added to Laveran's work in 1885 by demonstrating several species of *Plasmodium* and associating them with characteristic periods of fever. The Romanovsky staining procedure (methylene blue and eosin) for malarial parasites in blood smears, developed in Russia in 1891, is still used today with slight modifications.

Though the causative agent of malaria had been found, its transmission was not yet understood. Ronald Ross, a medical officer in India, worked for years in his spare time searching for proof that the malaria parasites were carried by mosquitoes. Eventually, he found these parasites in *Anopheles* mosquitoes, but he never found time to explain fully the mechanism of transmission. His research efforts were neither appreciated nor supported by his superiors, but he was eventually awarded the Nobel Prize in physiology or medicine in 1902 for his work with mosquitoes.

Knowing that mosquitoes transmit malaria was a far cry from being able to prevent them from doing so. Credit for developing the first effective mosquito control measures is attributed to the American physician William Crawford Gorgas, chief medical officer for sanitation during the building of the Panama Canal early in the twentieth century. By draining wet areas and instituting the use of mosquito netting, Gorgas was able to reduce significantly the incidence of both malaria and yellow fever among workers on the canal project.

Knowing the cause of malaria, having a means of controlling mosquitoes, and having the drug quinine to treat it led many people to believe that the disease was no longer a threat. However, some soldiers in World War I thought to have been cured with quinine treatment suffered relapses after they had left malaria-infested areas. Thus, a search began for the sites in which the parasites had been sequestered in human tissues. In 1938 S. P. James and P. Tate discovered the parasite outside red blood cells in certain birds, and in 1948 H. C. Shortt and P. C. C. Garnham made a similar discovery of *Plasmodium vivax* in humans. The parasite is now known to disappear from circulation, out of the reach of drugs, by invading cells of the liver and other tissues. During World War II, several new drugs were developed to treat malaria during initial infections and relapses.

The war against malaria continues. Massive insecticide (DDT) spraying in the 1960s appeared to have eradicated the disease from many regions of the world. Unfortunately, mosquitoes resistant to DDT soon emerged, and the incidence of the disease increased, sometimes to epidemic proportions. Some strains of the parasite also became resistant to chloroquine, one of the best drugs for treating malaria. The parasite even has thwarted efforts to produce a vaccine. Each of its life stages produces different antigens, or molecules that trigger the development of immunity.

To be successful, a malarial vaccine would have to use one of three approaches, which would take effect at different stages in the malarial life cycle (illustrated in Figure 12.5). The first approach is an antisporozoite vaccine that would prevent you from becoming infected. The second approach is a vaccine against stages already in your bloodstream. Such a vaccine would not prevent infection, but it would prevent suffering and damage to you by these stages. The third approach is a vaccine against the stages living in the mosquito. This type of vaccine is called a "love your neighbor" vaccine because it wouldn't protect you against the disease organisms that enter you when you are bitten, but the antibodies the mosquito imbibes (drinks) along with your blood would kill the stages in that mosquito, making her harmless when she subsequently bites your neighbor. Probably the most successful vaccine will use all three approaches simultaneously. If red blood cells are used to grow parasites for vaccine production, public response would probably be poor, as people would fear AIDS contamination in the blood used. Therefore the vaccine will probably be genetically engineered from isolated portions of the malarial parasite's genome. Thus far, the best such vaccine has killed only about 60 percent of the parasites. This is not acceptable, as it takes only one sporozoite to escape and initiate full-blown infection.

Another problem is mosquito migration. A man living near a European airport developed malaria without ever leaving home. The best guess is that an infected mosquito came in with some cargo.

Strangely, a new use has been tried for malaria. Two women infected with Lyme disease, who had failed to get relief with use of antibiotics, were recently purposely infected with malaria. The high fevers of malaria seem to have cured their Lyme disease. Fortunately they did not die from the malaria. This technique of inducing high fever with malaria is an old method for attempting to cure syphilis. Its success rate is not 100 percent, and today's antibiotics have made it obsolete.

The human struggle against malaria is typical of the difficulties encountered in dealing with many parasitic diseases. Much work remains to be done to eradicate this debilitating disease. The efforts of many parasitologists, immunologists, and health professionals will be needed to solve the challenging problems presented by the lowly malarial parasite.

CHAPTER SUMMARY

PRINCIPLES OF PARASITOLOGY

- A **parasite** is an organism that lives at the expense of another organism called its **host**.
- **Parasitology** is the study of parasites, which typically includes protozoa, helminths, and arthropods.

RELATED KEY TERMS: **pathogens**

Significance of Parasitism

- Parasites are responsible for much human disease and death and for extensive economic losses.

Parasites in Relation to Their Hosts

- Parasites can live on or in hosts. They can be **obligate** or **facultative parasites** and **permanent, temporary,** or **accidental parasites.**
- Parasites reproduce sexually in **definitive hosts** and spend other life stages in **intermediate hosts.** They can be transmitted to humans from **reservoir hosts.**

RELATED KEY TERMS: **ectoparasites** **endoparasites** **hyperparasitism** **vectors**

- **Host specificity** refers to the number of different hosts in which a parasite can mature.

RELATED KEY TERMS: **biological vector** **mechanical vector**

- Over time parasites become more adapted to and less destructive of their hosts. Most have mechanisms to evade host defenses and exceptionally adept reproductive capacities.

RELATED KEY TERMS: **schizogony**

PROTISTS

Characteristics of Protists

- **Protists** are eukaryotic, and most are unicellular. They can be autotrophic or heterotrophic, and some are parasitic.

Importance of Protists

- Protists are important in food chains as producers and decomposers; they can be economically beneficial or detrimental.

RELATED KEY TERMS: **tests** **eutrophication**

Classification of Protists

- Protists include plantlike, funguslike, and animallike organisms. The groups of protists are summarized in Table 12.1.

euglenoids **stigma** **slime molds** **true slime molds** **commensals** **mastigophorans** **merozoites**
pellicle **diatoms** **cellular slime molds** **plasmodium** **sarcodinas** **sporozoans** **gametocytes** **ciliates**
dinoflagellates **theca** **pseudoplasmodium** **protozoa** **sporozoites** **trophozoites** **trichocysts** **conjugation**

FUNGI

Characteristics of Fungi

- Fungi are **saprophytes** or parasites that usually have a **mycelium** consisting of **hyphae.** Most reproduce both sexually and asexually, and their sexual stages are used to classify them.

Importance of Fungi

- Fungi are important as decomposers in ecosystems and as parasites in the health sciences.

Classification of Fungi

- Fungi include **water molds, bread molds, sac fungi, club fungi,** and the **Fungi Imperfecti,** which cannot be classified in another group because they lack a sexual stage or it has not been identified. The groups of fungi are summarized in Table 12.2.

RELATED KEY TERMS

thallus **chitin** **septa**
Woronin body **plasmogamy**
dikaryotic cell **karyogamy**
antibiosis

mycoses **dimorphism**
Oomycota **Zygomycota**
zygospore **ascus**
Ascomycota **conidia**
ascogonium **antheridium**
ascospores **basidia**
Basidiomycota **Deuteromycota**

HELMINTHS

Characteristics of Helminths

Parasitic Helminths

- **Helminths** are bilaterally symmetrical and have head and tail ends and differentiated tissue layers.
- Only two groups of helminths, the flatworms and the nematodes (roundworms), contain parasitic species.
- **Flatworms** lack a **coelom,** have a simple digestive tract with one opening, and are **hermaphroditic.** They include **tapeworms** and **flukes.**
- **Nematodes** have a **pseudocoelom,** separate sexes, and a cylindrical body. They include hookworms, pinworms, and other parasites of the intestinal tract and lymphatics.

miracidia **sporocysts**
rediae **cercariae**
metacercariae **scolex**
proglottids **cysticercus**
hydatid cyst **microfilariae**

ARTHROPODS

Characteristics of Arthropods

- **Arthropods** have jointed, chitinous exoskeletons, segmented bodies, and jointed appendages.

Classification of Arthropods

- Parasitic and vector arthropods include some arachnids and insects; a few crustacea also serve as intermediate hosts for human parasites.
- **Arachnids** have eight legs; they include scorpions, spiders, ticks, and mites.
- **Insects** have six legs; they include lice, fleas, flies, mosquitoes, and bugs.
- **Crustacea** are usually aquatic arthropods, usually with a pair of appendages on each segment; they include crayfish, crabs, and copepods.
- Arthropod vectors of disease are summarized in Table 12.5.

QUESTIONS FOR REVIEW

A.

1. Define parasite and parasitology.
2. In what ways can parasites be categorized according to their relationships with their hosts?
3. How do definitive, intermediate, and reservoir hosts differ?
4. What is host specificity?
5. How do host-parasite relationships change over time?
6. How do parasites evade host defenses, and what kinds of damage do they cause?
7. How do parasites assure their own survival?

B.

8. What characteristics distinguish protists from other organisms?
9. In what ways are protists beneficial, and in what ways are they harmful?

C.

10. What are the main characteristics of plantlike protists?
11. What are the two types of slime molds and the characteristics of each?
12. What are the main characteristics of animallike protists?
13. What protists parasitize humans?

D.

14. What characteristics distinguish fungi from other organisms?

15. In what ways are fungi beneficial, and in what ways are they harmful?

E.

16. How do water molds differ from bread molds?
17. How do sac fungi differ from club fungi?
18. Which characteristic distinguishes the Fungi Imperfecti?
19. Which fungi parasitize humans?

F.

20. Which characteristics distinguish helminths from other organisms?
21. How do the two groups of helminths that parasitize humans differ?

G.

22. How do tapeworms and flukes differ?
23. How do adult nematodes differ from larval nematodes in their behavior as parasites?
24. Give some examples of helminths that parasitize humans.

H.

25. Which characteristics distinguish arthropods from other organisms?
26. How would you expect the behavior of a parasitic arthropod to differ from that of a vector arthropod?
27. Do you think an arthropod can be both a parasite and a vector? If so, how?

I.

28. Which groups of arthropods are human parasites?
29. Which groups of arthropods are vectors of disease?

PROBLEMS FOR INVESTIGATION

1. Locate information on the incidence of parasitic diseases in an underdeveloped country and in the United States. Explain which factors contributed to the differences between the two countries, and suggest ways the incidence of parasitic diseases could be decreased in both places.
2. Obtain specimens or photographs of 10 protists, and classify them as plantlike, funguslike, or animallike.
3. Obtain slides or photographs of the eggs of 10 helminths, and devise a classification scheme based on characteristics of the eggs.
4. List several examples of economically important fungi.
5. If you were making champagne, could you use the same strain of yeast you would use to make a Chianti? Why or why not?
6. Find out the proper method to remove a tick that is biting you. Which precautions should be observed? Why?
7. A previously healthy girl, who had never traveled outside South Carolina, developed generalized seizures. A CAT scan revealed a single lesion in the brain. Surgery showed it to be a cysticerus (larval stage) of the pork tapeworm *Taenium solium.* The patient remains asymptomatic on anticonvulsant medication. All members and contacts of the patient were negative when tested for *T. solium.* However, one neighbor who had immigrated from Mexico had other Mexican immigrant friends who often visited him when the girl was also at his house. Three of his friends tested positive for *T. solium* cysticerosis and had proglottids in their stools. Given that *T. solium* cysticerosis is virtually unknown in swine in the United States, transmission through the pig-human cycle was unlikely. In addition the girl denied ever having eaten raw or undercooked pork. How then did she acquire the infection?
(The answer to this question appears in Appendix E.)

SOME INTERESTING READING

Anonymous. 1990. Fungus routs gypsy moth outbreak. *Science News* 138, no. 5 (August 4):77.

Anonymous. 1991. Viruses combat red tide. *USA Today Magazine* 119, no. 2553 (June):15. (Marine viruses may help control phytoplankton.)

Beaver, P. C., and R. J. Jung. 1985. *Animal agents and vectors of human disease.* Philadelphia: Lea and Febiger.

Bold, H. C., and M. J. Wynne. 1985. *Introduction to the algae: Structure and reproduction.* Englewood Cliffs, NJ: Prentice Hall.

Donelson, J. E., and M. J. Turner. 1985. How the trypanosome changes its coat. *Scientific American* 252, no. 2 (February):44.

Farmer, J. N. 1980. *The protozoa: Introduction to protozoology.* St. Louis: C. V. Mosby.

Janerette, C.A. 1991. An introduction to mycorrhizae. *The American Biology Teacher* 53 (January):13–19.

Moore, J. 1984. Parasites that change the behavior of their host. *Scientific American* 250, no. 5 (May):108.

Newhouse, J.R. 1990. Chestnut blight. *Scientific American* 263 (July): 106–11.

Rennie, J. 1991. Proteins 2, malaria 0: malaria-free mice offer clues for developing a human vaccine. *Scientific American* 265 (July):24–25.

Schmidt, G. D., and L. S. Roberts. 1985. *Foundations of parasitology.* St. Louis: Times Mirror/Mosby.

Sieburth, J. M. 1985. *Microbial seascapes.* Baltimore: University Park Press.

Sze, P. 1986. *A biology of the algae.* Dubuque, Iowa: Wm. C. Brown.

Watts, S.J., W.R. Brieger, and M. Yacoob. 1989. Guinea worm: an in-depth study of what happens to mothers, families and communities. *Social Science & Medicine* 29, no. 9 (November 1):1043–50.

Painting by seventeenth-century Flemish artist David Teniers showing a peasant attempting to disinfect a wound on his hand. In the days before good antiseptics and antibiotics, even a small wound such as this could be a death sentence. The fellow has good cause to look worried; he could be dead of "blood poisoning" in a week or two.

13 Sterilization and Disinfection

This chapter focuses on the following questions:

A. How do sterilization and disinfection differ, and which terms are used to describe these processes?

B. What important principles apply to the processes of sterilization and disinfection?

C. What factors affect the potency of antimicrobial chemical agents?

D. How is the effectiveness of an antimicrobial chemical agent assessed?

E. By which mechanisms do antimicrobial chemical agents act?

F. What are the properties of commonly used antimicrobial chemical agents?

G. How are dry heat, moist heat, and pasteurization used to control microorganisms?

H. How are refrigeration, freezing, drying, and freeze-drying used to control and to preserve microorganisms?

I. How are radiation, sonic and ultrasonic waves, filtration, and osmotic pressure used to control microorganisms?

Do you like spicy foods? Perhaps you won't like the original reasons for their popularity. Before modern methods of food preservation such as canning and refrigeration were available, control of microbial growth in foods was a difficult problem. Inevitably after a short while, food began to take on the "off" flavors of spoilage. Spices were used to mask these unpleasant tastes. Some spices were also effective as preservatives. The antimicrobial effects of garlic have long been known. Fortunately we need not eat spoiled food today, and we can use spices solely to enhance our enjoyment of safely preserved foods.

Medical care, especially in the operating room is safer, too, today. As we have seen from the work of Ignaz Semmelweis and Joseph Lister (Chapter 1), careful washing and the use of chemical agents are effective in controlling many infectious microorganisms. (p. 14) In this chapter we will consider properties of various chemical and physical agents used to control microorganisms in laboratories, medical facilities, and homes. Agents called **disinfectants** are typically applied to inanimate objects, and agents called **antiseptics** are applied to living tissue. A few agents are suitable as both disinfectants and antiseptics, although most disinfectants are too harsh for use on delicate skin tissue. *Antibiotics*, though often applied to skin, are considered separately in Chapter 14.

TABLE 13.1 Terms related to sterilization and disinfection

Term	Definition
Sterilization	Killing or removing all microorganisms in a material or on an object.
Disinfection	Reducing the number of pathogenic microorganisms to the point where they pose no danger of disease.
Antiseptic	A chemical agent that can be safely used externally to destroy microorganisms or to inhibit their growth.
Disinfectant	A chemical agent used on inanimate objects to destroy microorganisms. Most disinfectants do not kill spores.
Sanitizer	A chemical agent typically used on food-handling equipment and eating utensils to reduce bacterial numbers so as to meet public health standards. Sanitization may simply refer to thorough washing with only soap or detergent.
Bacteriostatic agent	An agent that inhibits the growth of bacteria.
Germicide	An agent capable of killing microbes rapidly; some such agents are germicidal for certain microorganisms but only inhibitors for others.
Bactericide	An agent that kills bacteria. Most such agents do not kill spores.
Viricide	An agent that inactivates viruses.
Fungicide	An agent that kills fungi.
Sporocide	An agent that kills bacterial endospores or fungal spores.

DEFINITIONS

Sterilization is the killing or removal of all microorganisms in a material or on an object. There are no degrees of sterility—**sterility** means that there are *no* living organisms in or on a material. When properly carried out, sterilization procedures ensure that even highly resistant bacterial endospores and fungal spores are killed. Much of the controversy regarding spontaneous generation in the nineteenth century resulted from the failure to kill resistant cells in materials that were thought to be sterile. In contrast to sterilization, **disinfection** means reducing the number of pathogenic organisms on objects or in materials so that they pose no threat of disease. Terms related to sterilization and disinfection are defined in Table 13.1.

PRINCIPLES OF STERILIZATION AND DISINFECTION

As explained in the discussion of the growth curves in Chapter 6, both the growth and death of microorganisms occur at logarithmic rates. (p. 139) Here we are concerned with the death rate and the effects of antimicrobial agents on it.

Organisms treated with antimicrobial agents obey the same laws regarding death rates as those declining in numbers from natural causes. We will illustrate this principle with heat as the agent because its effects have been most thoroughly studied. When heat is applied to a material, the death rate of the organisms in or on it remains logarithmic, but it is greatly accelerated. Heat acts as an antimicrobial agent. If 20 percent of the organisms die in the first minute, 20 percent of those remaining alive will die in the second minute, and so on. If, at a different temperature, 30 percent die in the first minute, 30 percent of the remaining ones die in the second minute, and so on. From these observations, we can derive the principle that *a definite proportion of the organisms die in a given time interval.*

Consider now what happens when the number of live organisms remaining becomes small, 100 for example. At a death rate of 30 percent per minute, 70 would remain after 1 minute, 49 after 2 minutes, 34 after 3 minutes, and only 1 after 12 minutes. Soon the probability of finding even a single live organism becomes very small. Most laboratories say a sample is sterile if the probability is no greater than one chance in a million of finding a live organism.

The total number of organisms present when disinfection is begun affects the length of time required to eliminate them. We can state a second principle: *The smaller the number of organisms present, the shorter the time needed to achieve sterility.* Thoroughly cleaning objects before attempting to sterilize them is a practical application of this principle. Clearing objects of tissue debris and blood is also important because such organic matter impairs the effectiveness of many chemical agents.

Particular antimicrobial agents have different effects on various species of bacteria and their endospores. Furthermore, any given species may be more susceptible to an antimicrobial agent at one phase of growth than at another. The most susceptible phase for most organisms is the logarithmic growth phase because during that phase many enzymes are actively carrying out synthetic reactions, and interfering with even a single enzyme might kill the organism. From these observations, we can state a third principle: *Microorganisms differ in their susceptibility to antimicrobial agents.*

Application of Principles to Heat Killing

Several measurements have been defined to quantify the killing power of heat. The **thermal death point** is the temperature that kills all the bacteria in a 24-hour-old broth culture at neutral pH in 10 minutes. The **thermal death time** is the time required to kill all the bacteria in a particular culture at a specified temperature. The **decimal reduction time**, also known as the **DRT** or **D value**, is the length of time needed to kill 90 percent of the organisms in a given population at a specified temperature. (The temperature is indicated by a subscript: $D_{80°C}$, for example.)

These measurements have practical significance in industry as well in the laboratory. For example, a food-processing technician wanting to sterilize a food as quickly as possible would determine the thermal death point of the most resistant organism that might be present in the food and employ that temperature. In another situation it might be preferable to make the food safe for human consumption by processing it at the lowest possible temperature. This could be important in processing foods containing proteins that would be denatured (that is, their structure would be altered), thereby altering their flavor or consistency. The processor would then need to know the thermal death time at the desired temperature for the most resistant organism likely to be in the food.

ANTIMICROBIAL CHEMICAL AGENTS

Potency of Chemical Agents

The potency, or power, of an antimicrobial chemical agent is affected by time, temperature, pH, and concentration. The death rate of organisms is affected by the length of time the organisms are exposed to the antimicrobial agent, as was explained earlier for heat. Thus, adequate time should always be allowed for an agent to kill the maximum number of organisms. The death rate of organisms subjected to a chemical agent is accelerated by increasing temperature. Increasing temperature by 10°C roughly doubles the rate of chemical reactions and thereby increases the potency of the chemical agent. Acidic or alkaline pH can increase or decrease potency. A pH that increases the degree of ionization of a chemical agent often increases its ability to penetrate a cell. Such a pH also can alter the contents of the cell itself. Finally, increasing concentration increases the potency of most antimicrobial chemical agents. High concentrations may be **bactericidal** (killing), whereas lower concentrations may be **bacteriostatic** (growth-preventing).

Both ethyl and propyl alcohol are exceptions to the rule regarding increasing concentrations. They have long been believed to be more potent at 70 percent than at higher concentrations, although they are also effective up to 99 percent concentration. Some water must be present for alcohols to disinfect because they act by coagulating (permanently denaturing) proteins, and water is needed for the coagulation reactions. Also, a 70 percent alcohol-water mixture penetrates more deeply than pure alcohol into most materials to be disinfected.

Evaluation of Effectiveness of Chemical Agents

Many factors affect the potency of antimicrobial chemical agents, so evaluation of effectiveness is difficult. No entirely satisfactory method is available. However, we need some way to compare the effectiveness of disinfecting agents, especially as new ones come on the market.

Phenol Coefficient

Since Lister instituted phenol (carbolic acid) as a disinfectant, it has been the standard disinfectant to which other disinfectants are compared under the same conditions. The result of this comparison is called the **phenol coefficient**. Two organisms, *Salmonella typhi*, a pathogen of the digestive system, and *Staphylococcus aureus*, a common wound pathogen, are typically used to determine phenol coefficients. A phenol coefficient of 1.0 means that the disinfectant being compared has the same effectiveness as phenol. A coefficient less than 1.0 means it is less effective; a coefficient greater than 1.0 means it is more effective. Phenol coefficients are reported separately for the different test organisms. Lysol has a coefficient of 5.0 against *Staphylococcus aureus* but only 3.2 when used on *Salmonella typhi*, whereas ethyl alcohol has a value of 6.3 against both.

The phenol coefficient can be determined by the following steps. Prepare several dilutions of a chemical agent, and place the same volume of each in different test tubes. Put the tubes in a 20°C water bath for at least 5 minutes to assure that the contents of all tubes have the same temperature. Transfer 0.5 mL of a culture of a standard test organism to each tube. After 5, 10, and 15 minutes, use a sterile loop to transfer a specific volume of liquid from each tube into a separate tube of nutrient broth, and incubate the tubes. After 48 hours check cultures for cloudiness and find the smallest concentration (highest dilution) of the agent that killed all organisms in 10 minutes but not in 5 minutes. Find the ratio of this dilution to the dilution of phenol that has the same effect. For example, if a 1:1000 dilution of a chemical agent has the same effect as a 1:100 dilution of phenol, the phenol coefficient of that agent is 10 (1000/100). If you performed this test on a new disinfectant and obtained these results, you would have found a very good disinfectant! The phenol coefficient provides an acceptable means of evaluating the effectiveness of chemical agents derived from phenol, but it is less acceptable for other agents. Another problem is that the materials on or in which organisms are found may affect the usefulness of a chemical by complexing with it or inactivating it. These effects are not reflected in the phenol coefficient number.

Filter Paper Method

The **filter paper method** of evaluating a chemical agent uses small filter paper disks, each soaked with a different chemical agent, and is simpler than the determining of a phenol coefficient. The disks are placed on the surface of an agar plate that has been inoculated with a test organism and then incubated. A different plate is used for each test organism. An agent that inhibits growth of a test organism is identified by a clear area around the disk where the bacteria have been killed (Figure 13.1). Note that what is effective against one organism may have little or no effect on the others.

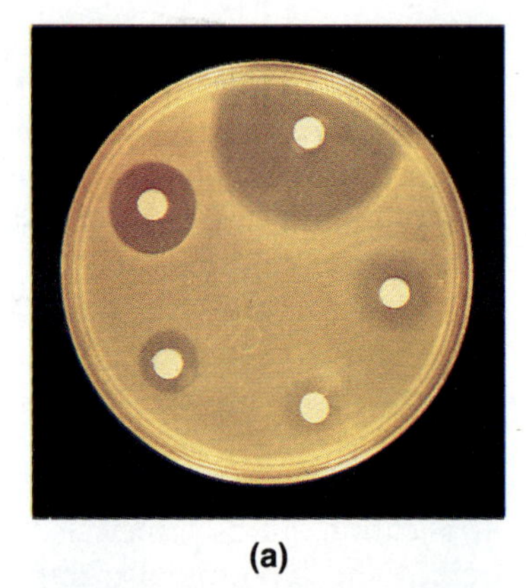

(a)

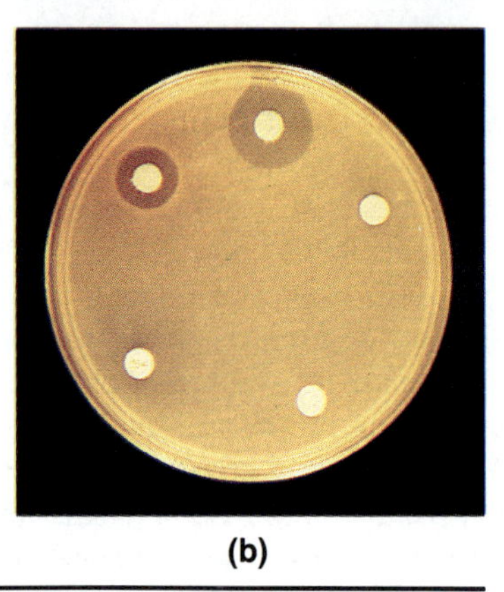

(b)

FIGURE 13.1 Filter paper method of evaluating disinfectants and antiseptics shows the difference in response of (a) *Staphylococcus aureus* (gram positive) and (b) *Escherichia coli* (gram negative) to common products. The greatest inhibition of growth in both cases can be seen near the top of the Petri dish surrounding the paper disk soaked in merthiolate. Moving counterclockwise, the various disks have been soaked in mercurochrome, iodine, Signal® mouthwash, and Listerine® mouthwash respectively before being placed on the surface of the nutrient medium that had first been confluently streaked with one of the test organisms.

Use-Dilution Test

A third way of evaluating chemical agents, the **use-dilution test,** uses standard preparations of certain test bacteria. These bacteria are added to tubes containing different dilutions of a chemical agent, after which the tubes are incubated and then observed for the presence or absence of growth. Agents that prevent growth at the greatest dilutions are considered the most effective. Many people feel that this measurement is more meaningful than is the phenol coefficient.

Disinfectant Selection

Several qualities of disinfectants should be considered in deciding which agent to use. An ideal disinfectant should

1. Be fast acting even in the presence of organic substances
2. Be effective against all types of infectious agents without destroying tissues or acting as a poison if ingested
3. Easily penetrate material to be disinfected without damaging or discoloring the material
4. Be easy to prepare and stable even when exposed to light, heat, or other environmental factors
5. Be inexpensive and easy to obtain and use
6. Not have an unpleasant odor

No disinfectant is likely to satisfy all the criteria, so the agent that meets the greatest number of criteria for the task at hand is chosen.

In actual practice, many agents are tested in a wide

range of situations and are recommended for use where they are most effective. Thus, some agents are selected for sanitizing kitchen equipment and eating utensils and others for rendering pathogenic cultures harmless. Furthermore, certain agents can be used in dilute concentration on the skin and in stronger concentration on inanimate objects.

Mechanisms of Action of Chemical Agents

Antimicrobial chemical agents kill microorganisms by participating in one or more chemical reactions that damage cell components. Although the kinds of reactions are almost as numerous as the agents, they can be grouped by whether they affect proteins, cell membranes, or other cell components.

Reactions That Affect Proteins

Much of a cell is protein, and all its enzymes are proteins. Alteration of protein structure is called *denaturation*, as explained in Chapter 2. (p. 42) In denaturation hydrogen and disulfide bonds are disrupted, and the functional shape of the protein molecule is destroyed. Any agent that denatures proteins prevents them from carrying out their normal functions. When proteins are treated with mild heat or with some dilute acids, alkalis, and other agents for a short time, they are temporarily denatured. After the agent is removed, the proteins can regain their normal structure. However, most antimicrobial agents are used in a strong-enough concentration over a sufficient length of time to permanently denature proteins. Permanent denaturation of the proteins of a microorganism kills the organism. Denaturation is bactericidal if it permanently alters the protein so its normal state cannot be recovered. It is bacteriostatic if it temporarily alters the protein, and its normal structure can be recovered (Figure 13.2).

Reactions that denature proteins include hydrolysis, oxidation, and attachment of atoms or chemical groups. (Recall from Chapter 2 that hydrolysis is the breaking down of a molecule by the addition of water, and oxidation is the addition of oxygen to or the removal of hydrogen from a molecule.) (p. 33) Acids, such as boric acid, and strong alkalis destroy protein by hydrolyzing it. Oxidizing agents (electron acceptors), such as hydrogen peroxide and potassium permanganate, oxidize disulfide linkages (—S—S—) or sulfhydryl groups (—SH). Agents that contain halogens—the elements chlorine, fluorine, bromine, and iodine—also sometimes act as oxidizing agents. Heavy metals, such as mercury and silver, attach to sulfhydryl groups. Alkylating agents, which contain methyl (CH_3—) or similar groups, donate these groups to proteins. Formaldehyde and some dyes are alkylating agents. Halogens can be substituted for hydrogen in carboxyl (—COOH), sulfhydryl, amino (—NH_2), and alcohol (—OH) groups. All these reactions can kill microorganisms.

Reactions That Affect Cell Membranes

Cell membranes contain proteins and so can be altered by all the preceding reactions. They also contain lipids

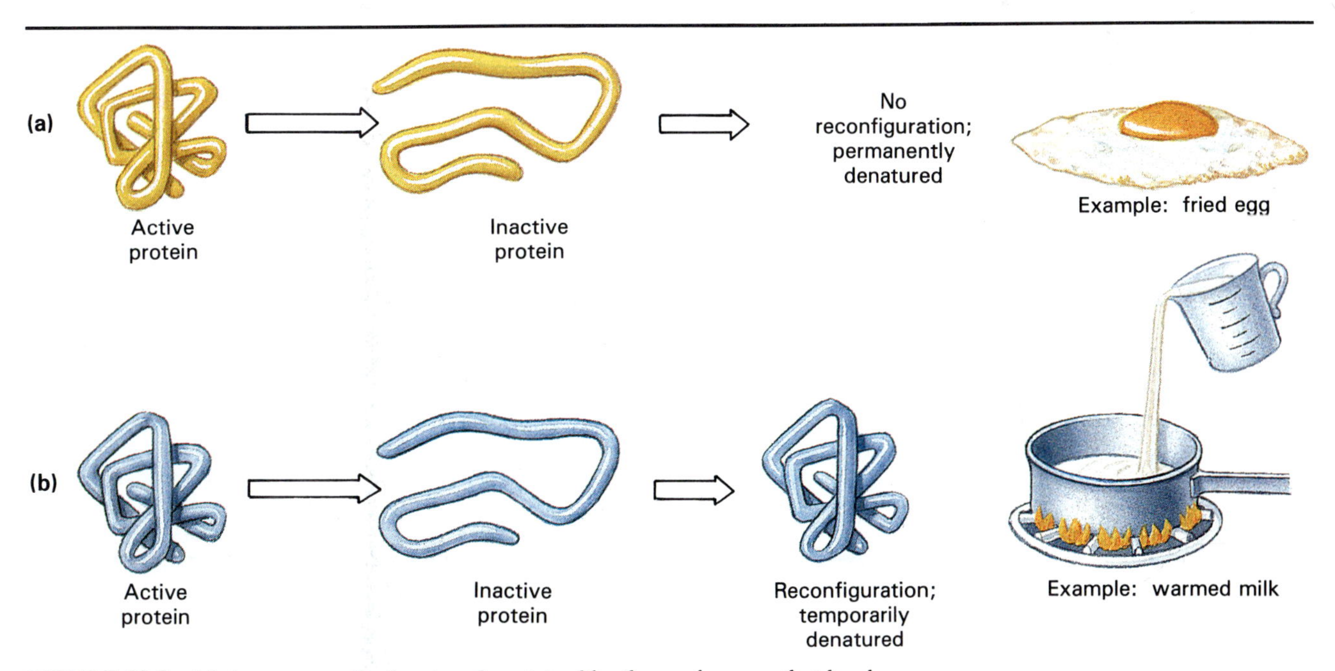

FIGURE 13.2 (a) A permanently denatured protein, like those of an egg that has been fried, cannot return to its original configuration. (b) A temporarily denatured protein, like those in milk that has been warmed, can refold itself into its original configuration.

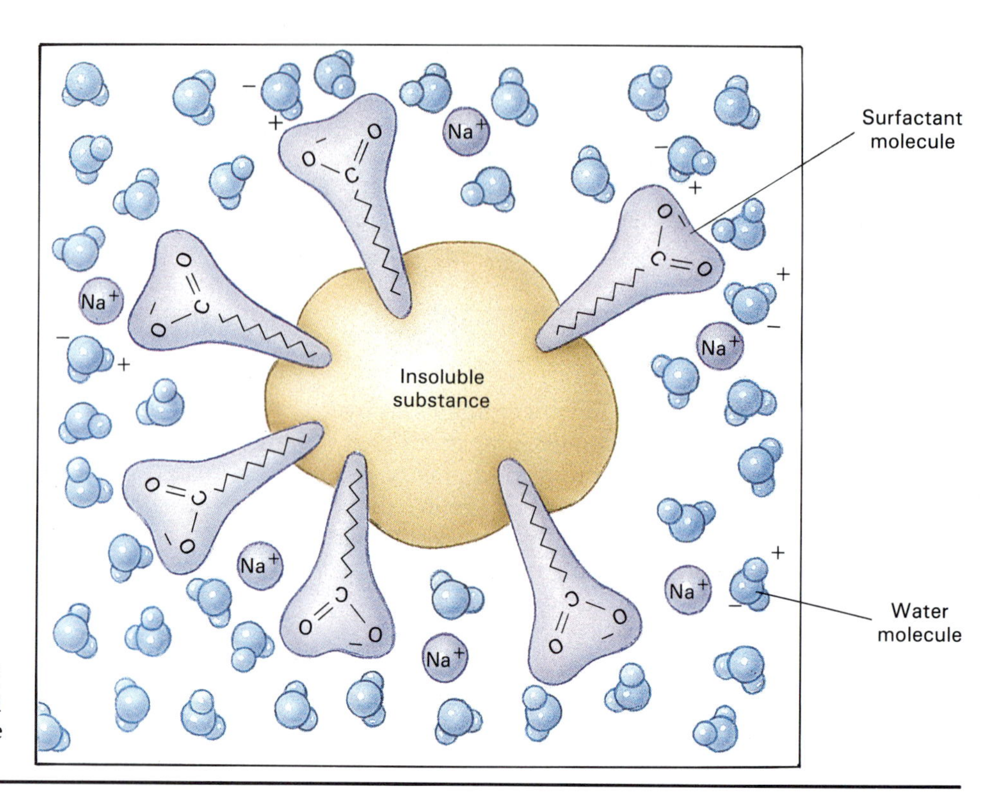

FIGURE 13.3 The action of a surfactant is due to the ability of surfactant molecules to surround an insoluble substance and make it miscible (mixable) with water.

and can be disrupted by substances that dissolve lipids. **Surfactants** (surf-akt'antz) reduce surface tension just as soaps and detergents break up grease particles in dishwater (Figure 13.3), and most also dissolve lipids. Surfactants include alcohols, detergents, and quaternary ammonium compounds, such as benzalkonium chloride, which dissolve lipids. Phenols, which are alcohols, dissolve lipids and also denature proteins. Detergent solutions, also called **wetting agents**, are often used with other chemical agents to help the agent penetrate fatty substances. Though detergent solutions themselves usually do not kill microorganisms, they do help to get rid of lipids and other organic materials so that antimicrobial agents can reach the organisms.

Reactions That Affect Other Cell Components

Other cell components affected by chemical agents include nucleic acids and energy-capturing systems. Alkylating agents can replace hydrogen on amino or alcohol groups in nucleic acids. Certain dyes such as crystal violet probably interfere with cell division, but the mechanism by which they act is not clearly understood. Some substances, such as lactic acid and propionic acid (end products of fermentation), inhibit fermentation and thus prevent energy capture in certain bacteria and molds and some other organisms.

Reactions That Affect Viruses

Like many cellular microorganisms, viruses can cause infections and must be controlled. Control of viruses requires that they be inactivated, that is, rendered permanently incapable of infecting. Inactivation can be effected by destroying either their nucleic acid or their proteins.

Alkylating agents such as ethylene oxide, nitrous acid, and hydroxylamine act as chemical mutagens. They alter DNA or RNA. If the alteration prevents the DNA or RNA from directing synthesis of new viral particles, the alkylating agents are effective inactivators. Detergents, alcohols, and other agents that denature proteins act on bacteria and viruses in the same way. Certain dyes, such as acridine orange and methylene blue, render viruses susceptible to inactivation when exposed to visible light. This process disrupts the structure of the viral nucleic acid.

Viruses sometimes remain infective even after

APPLICATIONS

No More Frogs in Formalin

Formalin, a 37 percent aqueous solution of formaldehyde, was for many years the standard material used to preserve laboratory specimens for dissection. Because formaldehyde is toxic to tissues and may cause cancer, it is rarely used today as a preservative. A variety of other preservatives are now used. Though less toxic to students performing dissections, they also are less effective. Molds growing on the surface of specimens are now a common problem.

their proteins are denatured, so methods used to rid materials of bacteria may not be successful with infectious viruses. Furthermore, using an agent that does not inactivate viruses can lead to laboratory-acquired infections.

Specific Antimicrobial Chemical Agents

Now that we have considered general principles of sterilization and disinfection and the kinds of reactions caused by such agents, we can look at some specific agents and their applications. (The structural formulas of some of the most important compounds discussed are shown in Figure 13.4.)

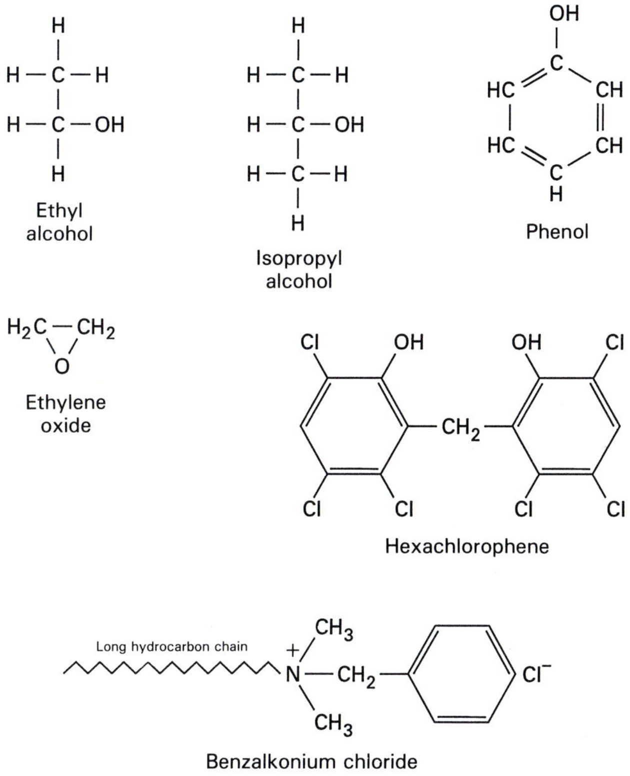

Ethyl alcohol

Isopropyl alcohol

Phenol

Ethylene oxide

Hexachlorophene

Long hydrocarbon chain

Benzalkonium chloride

FIGURE 13.4 Structural formulas of some important disinfectants.

Soaps and Detergents

Soaps and detergents remove microbes, oily substances, and dirt. Mechanical scrubbing greatly enhances their action. In fact, vigorous hand washing is one of the easiest and cheapest means of preventing the spread of disease between patients in hospitals, in medical and dental offices, among employees and patrons in food establishments, and among family members. Germicidal soaps usually are not significantly better disinfectants than ordinary ones.

Soaps contain alkali and sodium and will kill many species of *Streptococcus*, *Micrococcus*, and *Neisseria* and destroy influenza viruses. Many pathogens that survive washing with soap can be killed by a disinfectant applied after washing. A common practice after washing and rinsing hands and inanimate objects is to apply a 70 percent alcohol solution. Even these measures do not necessarily rid hands of all pathogens. Consequently, disposable gloves are used where there is a risk of health care workers becoming infected or transmitting pathogens to other patients.

Detergents, when used in weak concentrations in wash water, allow the water to penetrate into all crevices and cause dirt and microorganisms to be lifted out and washed away. Detergents are said to be *cationic* if they are positively charged and *anionic* if they are negatively charged. Cationic detergents are used to sanitize food utensils. Although not effective in killing endospores, they do inactivate some viruses. Anionic detergents are used for laundering clothes and as household cleaning agents. They are less effective sanitizing agents than cationic detergents, probably because the negative charges on bacterial cell walls repel them.

Many cationic detergents are **quaternary ammonium compounds**, or **quats**, which have four organic groups attached to a nitrogen atom. A variety of quats are available as disinfecting agents; their chemical structures vary according to their organic groups. One problem with quats is that their effectiveness is decreased in the presence of soap, calcium or magnesium ions, or porous substances such as gauze. An even more serious problem with these agents is that they support the growth of some bacteria of the genus *Pseudomonas* rather than killing them. Zephiran (benzalkonium chloride) was once widely used as a skin antiseptic. It is no longer recommended because it is less effective than originally thought and is subject to the same problems as other quats. Quats are now often mixed with another agent to overcome some of these problems and to increase their effectiveness.

Acids and Alkalis

Soap, as just indicated, is a mild alkali, and its alkaline properties help to destroy microbes. A number of organic acids lower the pH of materials sufficiently to inhibit fermentation. Several are used as food preservatives. Lactic and propionic acids retard mold growth in breads and other products. Benzoic acid and several of its derivatives are used to prevent fungal growth in soft drinks, catsup, and margarine. Sorbic acid and sorbates are used to prevent fungal growth in cheeses and a variety of other foods. Boric acid, formerly used as an eyewash, is no longer recommended because of its toxicity.

PUBLIC HEALTH

Soap and Sanitation

Washing and drying clothing in modern public laundry facilities is generally a safe practice because the clothing is almost disinfected if the water temperature is high enough. Soaps, detergents, and bleaches kill many bacteria and inactivate many viruses. Agitation of the clothes in the washer provides good mechanical scrubbing. Many microbes that survive this action are killed by heat in the dryer.

Using bar soap in a public washroom is not such a safe practice, as the soap may be a source of infectious agents. In a study of 84 samples of bar soap taken from public washrooms, every sample contained microorganisms. More than 100 strains of bacteria and fungi were isolated from the soap samples, and some of the organisms were potential pathogens. Most better restaurants and other establishments have installed soap dispensers for this very reason. Many jurisdictions have in fact made the use of bar soap in such facilities illegal.

Heavy Metals

Heavy metals used in chemical agents include selenium, mercury, copper, and silver. Even tiny quantities of such metals can be very effective in inhibiting bacterial growth (Figure 13.5). Silver nitrate was once widely used to prevent gonococcal infection in newborn infants. A few drops of silver nitrate solution were placed in the baby's eyes at the time of delivery to protect against infection by gonococci entering the eyes during passage through the birth canal. Silver nitrate was for a time replaced by antibiotics such as erythromycin in many hospitals. However, the development of antibiotic-resistant strains of gonococci has led some localities to require the use of silver nitrate, to which gonococci do not develop resistance.

APPLICATIONS

For a Clear Aquarium

If you have an aquarium you probably have contended with water that looks like pea soup because of the large numbers of algae growing in it. This problem can be corrected by placing a few pennies in the tank. A quantity of copper sufficient to inhibit algal growth dissolves from the pennies into the water. For this small investment you can greatly increase visibility and enjoyment of your fish.

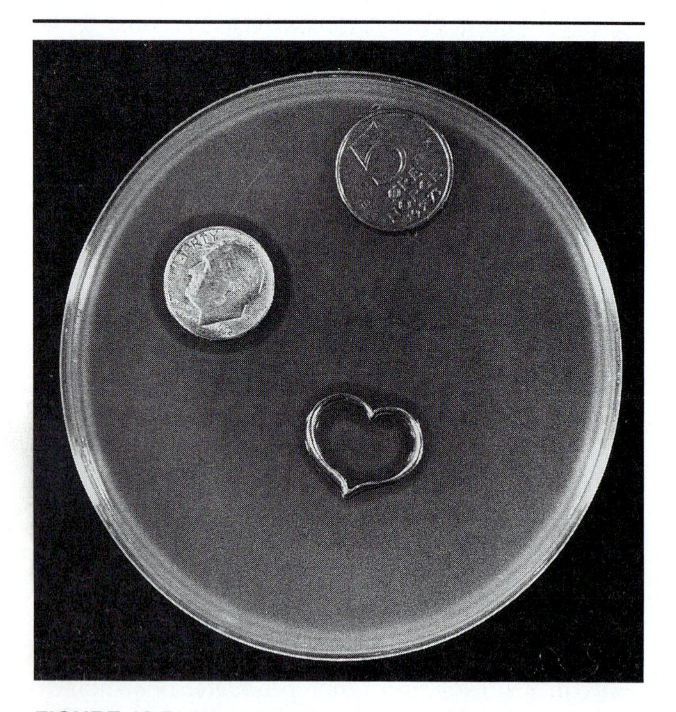

FIGURE 13.5 The inhibitory effects of silver ions (a heavy metal) can be seen as clear zones in which no growth has occurred around the silver heart and silver dime. The nonsilver coin has not inhibited growth of the organisms. A faint circle can be seen where it rested (it has now been pushed aside), and you can see that organisms grew quite well, even beneath it.

Organic mercury compounds, such as merthiolate and mercurochrome, are used to disinfect surface skin wounds. Such agents kill most bacteria in the vegetative state but do not kill spores. They are not effective against *Mycobacterium*. Merthiolate usually is prepared as a **tincture** (tink'tur), that is, dissolved in alcohol. In a tincture the alcohol may have a greater germicidal action than the heavy metal compound. Thimerosal, another organic mercury compound, can be used to disinfect skin and instruments and as a preservative for vaccines. Phenylmercuric nitrate and mercuric naphthenate inhibit both bacteria and fungi and are used as laboratory disinfectants.

Selenium sulfide kills fungi, including spores. Preparations containing selenium are commonly used to treat fungal skin infections. Shampoos that contain selenium are effective in controlling dandruff. Dandruff, a crusting and flaking of the scalp, is often, though not always, caused by fungi.

Copper sulfate is used to control algal growth. Although algal growth usually is not a direct medical problem, it is a problem in maintaining water quality in heating and air-conditioning systems and outdoor swimming pools.

Halogens

Hypochlorous acid, formed by adding chlorine to water (it is also the active ingredient in ordinary household bleach), effectively controls microorganisms in drinking water and swimming pools. It also disinfects food utensils and dairy equipment. It is effective in killing bacteria and inactivating many viruses. However, chlorine itself is easily inactivated by the presence of organic materials. This is why a substance such as copper sulfate is used to control algal growth in water to be purified with chlorine.

Iodine also is an effective antimicrobial agent. Tincture of iodine was one of the first skin antiseptics to come into use. Now iodophors, slow-release compounds in which the iodine is combined with organic molecules, are more commonly used. In such preparations, the organic molecules act as surfactants. Betadine and Isodine are used for surgical scrubs and on skin where an incision will be made. These compounds take several minutes to act and do not sterilize the skin. Betadine in concentrations of 3 to 5 percent destroys fungi, amoebae, and viruses as well as most bacteria, but it does not destroy bacterial endospores. Contamination of Betadine with *Pseudomonas cepacia* has been reported.

Bromine is sometimes used in the form of gaseous methyl bromide to fumigate soil that will be used in propagation of bedding plants.

Alcohols

Alcohols denature protein when mixed with water. They are also lipid solvents and dissolve cell membranes. Ethyl and isopropyl alcohols can be used as skin antiseptics. Isopropyl is more often used because of legal regulation of ethyl alcohol. It is appropriate to disinfect skin where injections will be made or blood drawn. Alcohol disinfects but does not sterilize skin because it evaporates quickly and stays in contact with microorganisms only for a few seconds. It kills vegetative microorganisms on the skin surface but does not kill endospores, resistant cells, or cells deep in skin pores.

Phenols

Phenol and derivatives of it called *phenolics* disrupt cell membranes, denature proteins, and inactivate enzymes. They are used to disinfect surfaces and to destroy discarded cultures because their action is not impaired by organic materials. Amphyl, which contains amylphenol, destroys vegetative forms of bacteria and fungi and inactivates viruses. It can be used on skin, instruments, dishes, and furniture. When used on surfaces it retains its antimicrobial action for several days. The orthophenylphenol in Lysol gives it similar properties. A mixture of phenol derivatives called *cresols* is found in creosote, a substance used to prevent rotting of wooden posts, fences, railroad ties, and such. It is, however, irritating to skin and a carcinogen, and its use is therefore limited. The addition of halogens to phenolic molecules usually increases their effectiveness. The chlorinated phenol derivative chlorhexidine gluconate (Hibiclens) is effective against a wide variety of microbes even in the presence of organic material. It is a good agent for surgical scrubs. Hexachlorophene and dichlorophene, which are also halogenated phenols, inhibit staphylococci and fungi, respectively, on the skin and elsewhere.

PUBLIC HEALTH

Hexachlorophene

Hexachlorophene is an excellent skin disinfectant. In a 3-percent solution, it kills staphylococci and most other gram-positive organisms, and its residue on skin is strongly bacteriostatic. Because staphylococcal skin infections can spread easily among newborn babies in hospitals, this antiseptic was used extensively in the 1960s for daily bathing of infants. The unforeseen price paid for controlling infections was permanent brain damage in infants bathed in it over a period of time. Hexachlorophene is absorbed through the skin and travels in the blood to the brain. Baby powder containing hexachlorophene killed 40 babies in France in 1972. Available only by prescription in the United States today, it is used routinely, though very cautiously, in hospital neonatal units because it is still the most effective agent for preventing the spread of staphylococcal infections.

Oxidizing Agents

Oxidizing agents disrupt disulfide bonds in proteins and thus disrupt the structure of cell membranes and proteins. Hydrogen peroxide, which forms highly reactive superoxide (O_2^-), is used to clean puncture wounds. When it breaks down into oxygen and water, the oxygen kills obligate anaerobes present in the wounds. It is quickly inactivated by enzymes from injured tissues. A new method of sterilization using vaporized hydrogen peroxide (Figure 13.6) can now be used for small rooms or areas such as glove boxes and transfer hoods. Another oxidizing agent, potassium permanganate, is used to disinfect instruments and in low concentrations to clean skin.

Alkylating Agents

Alkylating agents disrupt the structure of both proteins and nucleic acids. Because they can disrupt nu-

FIGURE 13.6 Newly developed biodecontamination equipment uses vaporized hydrogen peroxide to sterilize small sealable enclosures such as glove boxes, hoods, or transfer rooms. It is not sufficient, however, to sterilize a larger space such as an operating room.

FIGURE 13.7 Equipment used for ethylene oxide sterilization. This equipment must be used very carefully, as ethylene oxide is both explosive and carcinogenic. The aerator (at rear) is used to remove all traces of the gas from sterilized material.

cleic acids, these agents may cause cancer and should not be used in situations where they might affect human cells. Formaldehyde, glutaraldehyde, and beta-propiolactone are used in aqueous solutions. Ethylene oxide is used in gaseous form.

Formaldehyde inactivates viruses and toxins without destroying their antigenic properties. Glutaraldehyde kills all kinds of microorganisms including spores and sterilizes equipment exposed to it for 10 hours. Beta-propiolactone destroys hepatitis viruses.

Gaseous ethylene oxide has extraordinary penetrating power. Used at a concentration of 500 mg/L at 50°C for 4 hours, it sterilizes rubber goods, mattresses, plastics, and other materials destroyed by higher temperatures. Special equipment used during ethylene oxide sterilization is shown in Figure 13.7. As will be explained in the discussion of autoclaving, an ampule of endospores should be processed with ethylene oxide sterilization to check the effectiveness of sterilization.

All articles sterilized with ethylene oxide must be well ventilated with sterile air to remove all traces of this toxic gas, which can cause burns if it reaches living tissues and is also highly explosive. Articles such as catheters, intravenous lines, in-line valves, and rubber tubing must be thoroughly flushed with sterile air. Both the toxicity and flammability of ethylene oxide can be reduced by using it in gas containing 90 percent carbon dioxide. *It is exceedingly important that workers be protected from ethylene oxide vapors because they are toxic to skin, eyes, and mucous membranes and may also cause cancer.*

Dyes

Acridine, which interferes with cell replication (Chapter 8), can be used to clean wounds. Methylene blue inhibits growth of some bacteria in cultures. Crystal violet (gentian violet) blocks cell wall synthesis, possibly by the same reaction that causes it to bind to cell wall material in the Gram stain. It effectively inhibits growth of gram-positive bacteria in cultures and in skin infections. It can be used to treat protozoan (*Trichomonas*) and yeast (*Candida albicans*) infections.

Other Agents

Certain plant oils have special antimicrobial uses. Thymol from the herb thyme is used as a preservative,

TABLE 13.2 Properties of antimicrobial chemical agents

Agent	Actions	Uses
Soaps and detergents	Lower surface tension, make microbes accessible to other agents	Hand washing, laundering, sanitizing kitchen and dairy equipment.
Surfactants	Dissolve lipids, disrupt cell membranes, denature proteins, and inactivate enzymes in high concentrations; act as wetting agents in low concentrations	Cationic detergents are used to sanitize utensils; anionic detergents to launder clothes and clean household objects; quaternary ammonium compounds are sometimes used as antiseptics on skin.
Acids	Lower pH and denature proteins	Food preservation.
Alkalies	Raise pH and denature proteins	Found in soaps.
Heavy metals	Denature proteins	Silver nitrate is used to prevent gonococcal infections, mercury compounds to disinfect skin and inanimate objects, copper to inhibit algal growth, and selenium to inhibit fungal growth.
Halogens	Oxidize cell components in absence of organic matter	Chlorine is used to kill pathogens in water and to disinfect utensils; iodine compounds are used as skin antiseptics.
Alcohols	Denature proteins when mixed with water	Isopropyl alcohol is used to disinfect skin; ethylene glycol and propylene glycol can be used in aerosols.
Phenols	Disrupt cell membranes, denature proteins, and inactivate enzymes; not impaired by organic matter	Phenol is used to disinfect surfaces and destroy discarded cultures; amylphenol destroys vegetative organisms and inactivates viruses on skin and inanimate objects; chlorhexidine gluconate is especially effective as a surgical scrub.
Oxidizing agents	Disrupt disulfide bonds	Hydrogen peroxide is used to clean puncture wounds, potassium permanganate to disinfect instruments.
Alkylating agents	Disrupt structure of proteins and nucleic acids	Formaldehyde is used to inactivate viruses without destroying antigenic properties, glutaraldehyde to sterilize equipment, beta-propiolactone to destroy hepatitis viruses, ethylene oxide to sterilize inanimate objects that would be harmed by high temperatures.
Dyes	May interfere with replication or block cell wall synthesis	Acridine is used to clean wounds, crystal violet to treat some protozoan and fungal infections.

and eugenol from oil of cloves is used in dentistry to disinfect cavities. A variety of other agents are used primarily as food preservatives. They include sulfites and sulfur dioxide, used to preserve dried fruits and molasses, sodium diacetate, used to retard mold in bread, and sodium nitrite, used to preserve cured meats and some cold cuts. Foods with nitrites should be eaten in moderation because the nitrites are converted during digestion to substances that may cause cancer.

The properties of antimicrobial chemical agents are summarized in Table 13.2.

ANTIMICROBIAL PHYSICAL AGENTS

Physical agents have been used to preserve food for centuries. Ancient Egyptians dried perishable foods to preserve them. Scandinavians made a supply of dry, crisp bread to hang in their homes during the winter and likewise kept seed grains in a dry place. Otherwise, both flour and grains would have molded during the long, moist winters. Europeans used heat in the food-canning process 50 years before Pasteur's work explained why heating prevented the food from spoiling. Today, physical agents that destroy microorganisms are still used in food preservation and preparation and remain a crucial weapon in the prevention of infectious disease.

Dry Heat, Moist Heat, and Pasteurization

Heat is a preferred agent of sterilization for all materials not damaged by it. It rapidly penetrates thick materials not easily penetrated by chemical agents. Dry heat probably does most of its damage by oxidizing molecules. Moist heat destroys microorganisms mainly by denaturing proteins—the presence of water molecules helps to disrupt the hydrogen bonds and other weak interactions that hold proteins in their three-dimensional shapes (Chapter 4). Moist heat may melt cell membrane lipids as well. Heat also inactivates many viruses, but those that can infect even after their protein coats are denatured require extreme heat treatment, such as steam under pressure, that will disrupt nucleic acids.

FIGURE 13.8 Hot-air oven used for sterilizing metal and glass items.

Dry Heat

Dry (oven) heat penetrates substances more slowly than moist (steam) heat. It is usually used to sterilize metal objects and glassware (Figure 13.8) and is the only suitable means of sterilizing oils and powders. Objects are sterilized by dry heat when subjected to 171°C for 1 hour, 160°C for 2 hours or longer, or 121°C for 16 hours or longer, depending on the volume.

An open flame is a form of dry heat used to sterilize inoculating loops and the mouths of culture tubes by incineration and to dry the inside of pipettes. When flaming objects in the laboratory, it is important to avoid formation of floating ashes and **aerosols** (droplets released into the air). These substances can be a means of spreading infectious agents if the organisms in them are not killed by incineration as intended. For this reason, specially designed loop incinerators with deep throats are often used for sterilizing inoculating loops.

Moist Heat

Moist heat, because of its penetrating properties, is a widely used physical agent. Boiling water destroys vegetative cells of most bacteria and fungi and inactivates some viruses, but it is not effective in killing all kinds of spores. The effectiveness of boiling can be increased by adding 2 percent sodium bicarbonate to the water. However, if water is heated under pressure, its boiling point is elevated so that temperatures above 100°C can be reached. This is normally accomplished by using an **autoclave** (aw'to-klav), as shown in Figure 13.9, and maintaining a pressure of 15 lb/in^2 above atmospheric pressure for 15 to 20 minutes, depending on the volume of the load. At this pressure of 30 lb/in^2 (15 from atmospheric pressure + 15 above), the temperature reaches 121°C and is high enough to kill spores as well as vegetative organisms and to disrupt the structure of nucleic acids in viruses. In this procedure it is the increased temperature, and not the pressure, that kills microorganisms.

Sterilization by autoclaving is invariably success-

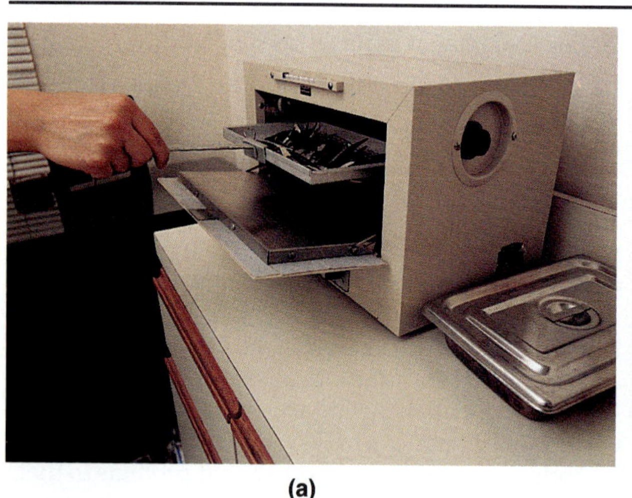

(a)

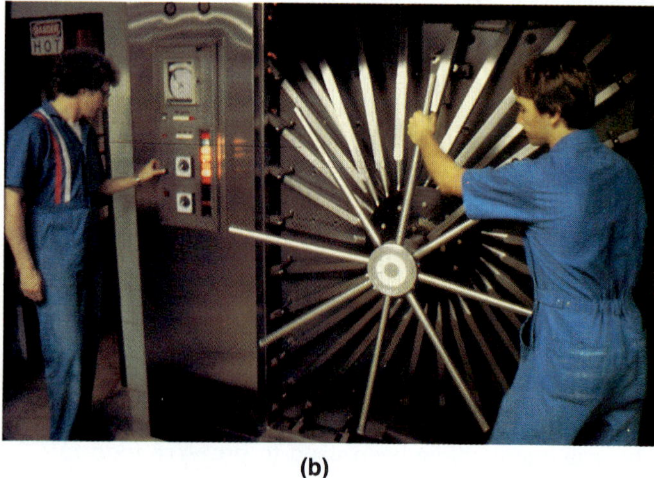

(b)

FIGURE 13.9 (a) A small counter-top autoclave and (b) a large hospital autoclave. Note the recording charts to the left of the door, which keep records of the actual temperatures and pressures reached during the time of operation. These would alert the operator to some malfunctions.

ful if properly done and two commonsense rules are followed: First, articles should be placed in the autoclave so that steam can easily penetrate them, and second, air should be evacuated so that the chamber fills with steam. Wrapping objects in aluminum foil is not recommended because it may interfere with steam penetration. Steam circulates through an autoclave from a steam outlet to an air evacuation port (Figure 13.10). In preparing items for autoclaving, containers should be unsealed and articles should be wrapped in materials that allow steam penetration. Large packages of dressings and large flasks of media require extra time for heat to penetrate them. Likewise, packing many articles close together in an autoclave lengthens the processing time to as much as 60 minutes to assure sterility. It is more efficient and safer to run two separate uncrowded loads than one crowded one.

Several methods are available to assure that autoclaving achieves sterility. Modern autoclaves have devices to maintain proper pressure and record internal temperature during operation. Regardless of the presence of such a device, the operator should check pressure periodically and maintain the appropriate pressure. Tapes impregnated with a substance that causes the word "sterile" to appear when they have been exposed to an effective sterilization temperature can be placed on packages. These tapes are not fully reliable because they do not indicate how long appropriate conditions were maintained. Tapes or other sterilization indicators should be placed inside of and near the center of large packages to determine whether heat penetrated them. This precaution is necessary because when an object is exposed to heat its surface becomes hot much more quickly than its center. (When a large piece of meat is roasted, for example, the surface can be well done while the center remains rare.)

The Centers for Disease Control recommends weekly autoclaving of a culture containing heat-resistant endospores, such as those of *Bacillus stearothermophilus*, to check autoclave performance. Endospore strips (Figure 13.11) are commercially available to make this task easy. The spore strip and an ampule (sealed glass container) of medium are enclosed in a soft plastic vial. The vial is placed in the center of a loaded autoclave and processed; then the inner ampule is broken, releasing the medium, and the whole container is incubated. If no growth appears in the autoclaved culture, sterilization is deemed effective.

In large laboratories and hospitals, where great quantities of materials must be sterilized, special autoclaves, called prevacuum autoclaves, often are used. Instead of simply being emptied of air, a partial vacuum is created in the chamber. The steam enters and heats the chamber much more rapidly than without the vacuum, so the proper temperature is reached quickly. The total sterilization time is cut in half, and the costs of sterilization are greatly decreased.

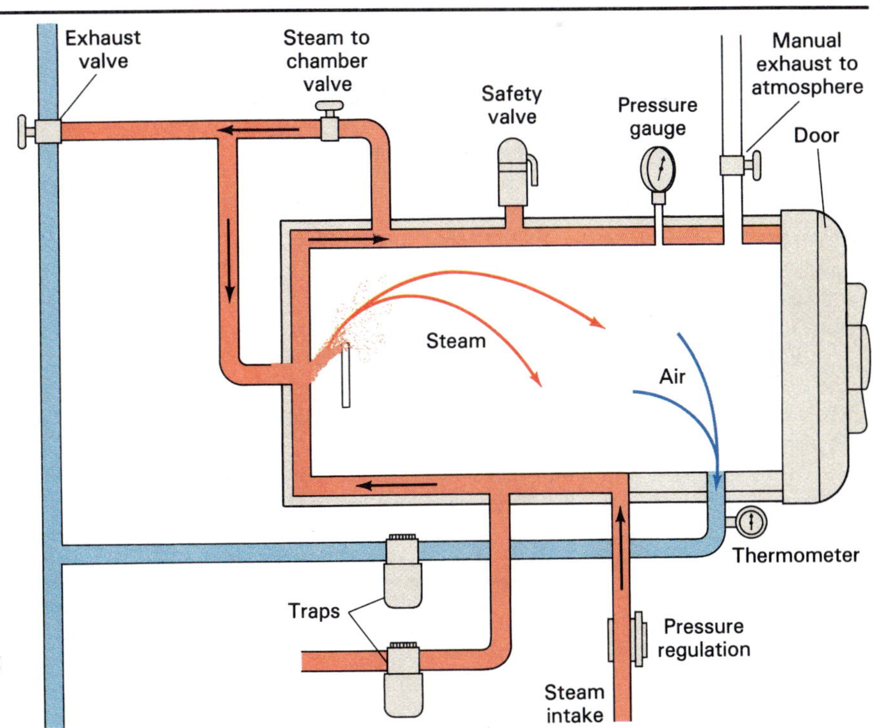

FIGURE 13.10 Steam is heated in the jacket of an autoclave, enters the sterilization chamber through an opening at the upper rear, and is exhausted through an opening at the front bottom.

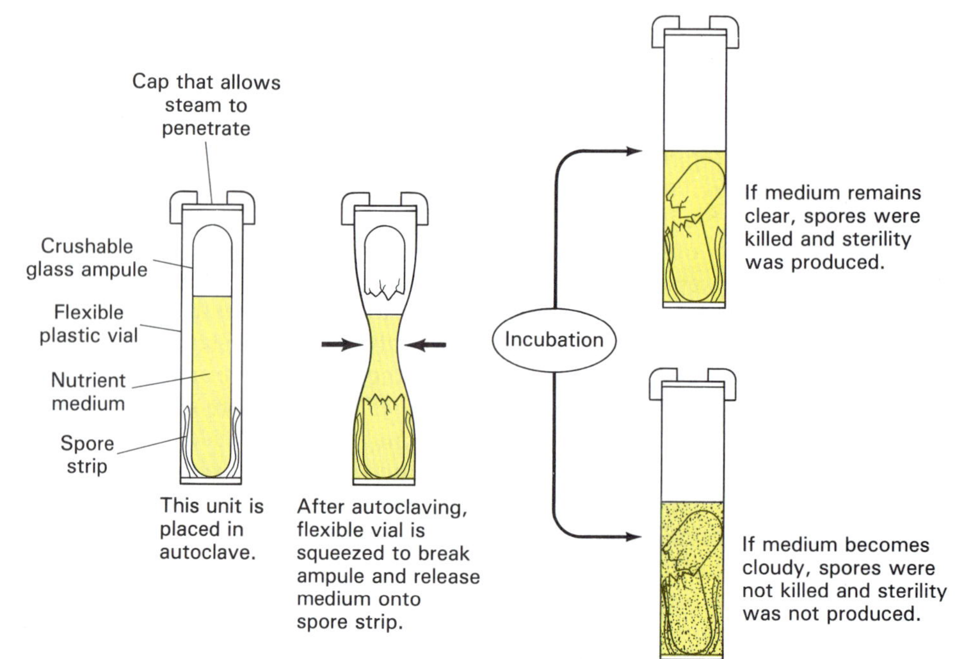

FIGURE 13.11 To be certain that an autoclave is operating properly, a commercially prepared spore test is placed in the autoclave and run through with the rest of the load. Afterwards, the vial is crushed to release medium onto a strip containing spores. If the load was truly sterilized, the spores will have been killed, and growth will not occur in the medium. Sometimes an indicator dye is added to the medium, which will turn color if microbial growth occurs, due to the accumulation of acid byproducts. This is faster than waiting for sufficient growth to turn the medium cloudy.

APPLICATIONS

Home Canning

Home canning is done in an open water bath or in a pressure cooker. Food must be packed loosely in jars with ample fluid to carry heat to the center of the jar. Space also must be left between the jars. Once adequately processed, canned foods will keep indefinitely. Jars of relish in the wreckage of an ironclad Civil War ship, the *Monitor*, were sterile after more than 100 years at the bottom of the Atlantic Ocean. The contents would have been edible if poisonous quantities of lead from the lids had not dissolved in the relish.

The water bath reaches a temperature of 100°C and is adequate for preventing spoilage of acidic foods such as fruits and tomatoes. Acid in these foods inhibits the germination of most spores, should some survive the boiling water treatment. However, meats and alkaline vegetables such as corn and beans must be processed in a pressure cooker. Adding onions or green peppers to a jar of tomatoes increases the pH, so such mixtures also must be cooked under pressure. Because acid-tolerant spores do exist, home canning would be safer if it were all done by pressure cooking.

All commercially canned foods are processed in pressurized equipment. A pressure cooker functions like an autoclave, and foods in jars are processed at least 15 minutes at 15 pounds pressure per square inch. Any kinds of spores that might be present in these foods are killed, so the food is sterile.

Failure to process alkaline foods at the high temperature reached with a pressure cooker can lead to the accumulation of toxin of the bacterium *Clostridium botulinum* while the food is stored. Even a tiny amount of this toxin can be lethal. Any home-canned or commercially canned foods should be discarded if they have an unpleasant odor or if the lids of the containers bulge, because this indicates that gas is being produced by living organisms inside the container. Unfortunately, toxins can be present without lids bulging and without causing any noticeable odor, so great care must be used in home canning to be sure adequate pressure is maintained for a long-enough period of time. As a safety precaution, after a jar of home-canned food is opened, 15 to 20 minutes of vigorous boiling should destroy any botulism toxin. However, the best rule to follow is: When in doubt, throw it out.

Canning peaches at home by the hot water bath method.

Pasteurization

Pasteurization, invented by Pasteur to destroy organisms that caused wine to sour, does not achieve sterility. It does kill pathogens that might be present in milk, other dairy products, and beer, especially *Salmonella* and *Mycobacterium*, which used to cause many cases of tuberculosis among children who drank raw milk. Pasteurization is done by heating the milk to 71.6°C for at least 15 seconds in the *flash method* or by heating the milk to 62.9°C for 30 minutes in the *holding method*. Some years ago certain strains of bacteria of the genus *Listeria* were found in pasteurized milk and cheeses. This pathogen causes diarrhea and encephalitis and can lead to death in pregnant women. Because of a few such infections, revision of standard procedures for pasteurization was considered. However, finding pathogens in pasteurized milk has not become a persistent problem, and no action has been taken.

Though much milk for sale in the United States is pasteurized fresh milk, sterile milk also is available. All evaporated or condensed canned milk is sterile, and some milk packaged in cardboard containers also is sterile. Sterilized milk in cardboard containers is widely available in Europe and can be found in some stores in the United States. The canned milk is subjected to steam under pressure and has a "cooked" flavor. The sterilized milk in cardboard containers is subjected to a process similar to pasteurization using higher temperatures. It too has a "cooked" flavor but can be kept unrefrigerated as long as the container remains sealed. Such milk often is flavored with vanilla, strawberry, or chocolate. **Ultra high temperature** (UHT) raises the temperature from 74°C to 140°C and then falls back to 74°C in less than 5 seconds. A complex cooling process that prevents the milk from ever touching a surface hotter than itself, prevents development of a "cooked" flavor. Some, but not all, small containers of coffee creamer are treated by this method.

APPLICATIONS

Yogurt

Certain foods such as yogurt are made by introducing organisms such as lactobacilli that ferment the milk. The fermented foods often are heat-treated after initial pasteurization to kill the fermenting organisms and increase the shelf life of the products. Information on the labels of such products should indicate whether they contain live fermenting organisms. If you make yogurt at home, purchase a live-culture brand of yogurt to use as your starter.

Refrigeration, Freezing, Drying, and Freeze-drying

Cold temperature retards the growth of microorganisms by slowing the rate of enzyme-controlled reactions but does not kill many of them. Heat is much more effective than cold at killing microorganisms. Refrigeration is used to prevent food spoilage. Freezing, drying, and freeze-drying are used to preserve both foods and microorganisms, but they do not achieve sterilization.

Refrigeration

Many fresh foods eaten by humans can be prevented from spoiling by keeping them at 5°C (ordinary refrigerator temperature). However, storage should be limited to a few days because some bacteria and molds continue to grow at this temperature. To be convinced of this you merely need to recall some of the strange things you have found growing on leftovers in the back of your refrigerator. In rare instances strains of *Clostridium botulinum* have been found growing and producing lethal toxins in a refrigerator if the organisms are deep in a container of food where anaerobic conditions exist.

Freezing

Freezing at −10°C is used to preserve foods in homes and in the food industry. Although freezing does not sterilize foods, it does significantly slow the rate of chemical reactions so that microorganisms do not cause food to spoil. Frozen foods should not be thawed and refrozen. Repeated freezing and thawing of foods causes large ice crystals to form in the foods during slow freezing. Cell membranes in the foods are ruptured, and nutrients leak out. This alters the texture of foods and makes them less palatable. It also

APPLICATIONS

Home Freezing

Home freezing of foods is probably a more common practice today than home canning. Before freezing, fresh fruits and vegetables should be blanched, or immersed for a minute or so in boiling water. This helps to kill microorganisms on the foods, but its main purpose is to denature enzymes in the foods that can cause discoloration or changes in texture even at freezer temperatures. The foods should then be cooled quickly in cold water and placed in clean containers. Finally, they should be placed in the coldest part of the home freezer with space around the containers so that they freeze as quickly as possible.

allows bacteria to multiply while food is thawed and makes the food more susceptible to bacterial degradation.

Freezing can be used to preserve microorganisms, but this requires a much lower temperature than that used for food preservation. Microorganisms usually are suspended in glycerol or protein to prevent the formation of large ice crystals (which could puncture cells), cooled with solid carbon dioxide (dry ice) to a temperature of −78°C, and then held there. Alternatively, they can be placed in liquid nitrogen and cooled to a temperature of −180°C.

Drying

Drying can be used to preserve foods because the absence of water inhibits the action of enzymes. Many foods, including peas, beans, raisins, and other fruits, are often preserved by drying (Figure 13.12). Yeast used in baking also can be preserved by drying. Endospores present on such foods can survive drying, but they do not produce toxins. Dried pepperoni sausage and smoked fish retain enough moisture for microorganisms to grow. Because smoked fish is not cooked, eating it poses a risk of infection. Sealing such fish in plastic bags creates anaerobic conditions that allow botulism toxins to form.

Drying also naturally minimizes the spread of infectious agents. Some bacteria, such as *Treponema pallidum*, which causes syphilis, are extremely sensitive to drying and die almost immediately on a dry surface; thus they can be prevented from spreading by keeping toilet seats and other bathroom fixtures dry. Drying of laundry in dryers or in the sunshine also destroys pathogens.

FIGURE 13.12 Sun drying is an ancient means of preventing growth of microorganisms. These grapes will remain edible as raisins because microbes need more water than remains inside the dried fruit.

Freeze-drying

Freeze-drying, or **lyophilization** (li-of"il-i-za'shun), as shown in Figure 13.13, is the drying of a material from

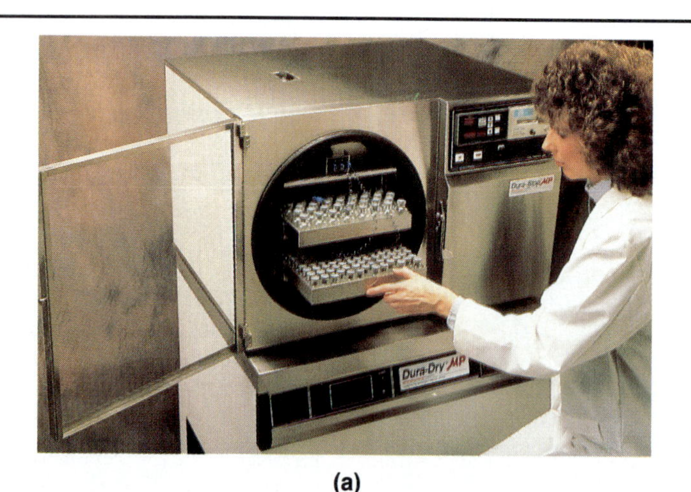

(a)

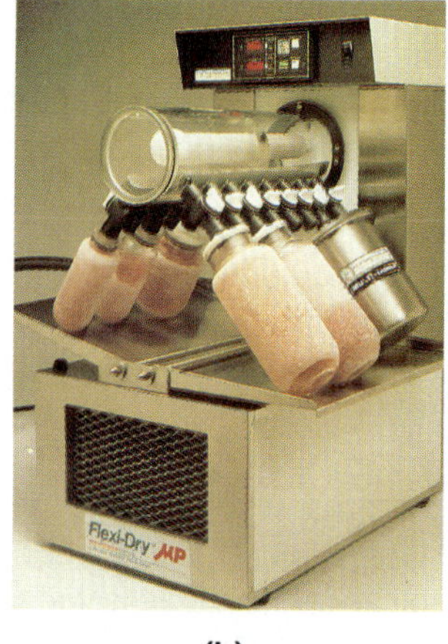

(b)

FIGURE 13.13 Freeze drying (lyophilization) equipment. (a) A stoppering tray dryer, in which the trays supply the heat needed to remove moisture. After completion of the process of lyophilization, which takes about 24 hours, the device automatically stoppers all the vials. (b) A manifold dryer, in which prefrozen samples in vials of different sizes can be attached via ports to the dryer that then supplies the needed heat. The sample will dry in 4 to 20 hours, depending on its initial thickness. Unlike the tray dryer, this device allows samples to be added and removed whenever desired.

the frozen state. This process is used in the manufacture of some brands of instant coffee; freeze-dried instant coffee has a more natural flavor than other kinds. Microbiologists use lyophilization for long-term preservation rather than destruction of cultures of microorganisms. Organisms are rapidly frozen in liquid carbon dioxide or liquid nitrogen and are then subjected to a high vacuum to remove moisture while in the frozen state. Rapid freezing allows only very tiny ice crystals to form in cells, so the organisms survive this process. Organisms so treated can be kept alive for years, stored under vacuum in the freeze-dried state.

Radiation

Four general types of radiation, ultraviolet light, ionizing radiations, plus microwave radiation and strong visible light (under certain circumstances), can be used to control microorganisms and to preserve foods. Refer back to Figure 3.3 of the electromagnetic spectrum to review their wavelengths and positions relative to one another along the spectrum.

Ultraviolet Light

Ultraviolet light consists of light of wavelengths between 40 and 390 nm, but wavelengths in the 200-nm range are most effective in killing microorganisms. Ultraviolet light kills because it is absorbed by proteins and by the purine and pyrimidine bases of nucleic acids. Such absorption can permanently denature these important molecules. It is especially effective in inactivating viruses. However, it kills far fewer organisms than one might expect because of DNA repair mechanisms. Once DNA is repaired, new molecules of RNA and protein can be synthesized to replace the damaged molecules, as described in Chapters 7 and 8.

Ultraviolet light is of limited use because it does not penetrate glass, cloth, paper, or most other materials, and it does not go around corners or under lab benches. It does penetrate air, effectively reducing the number of airborne microorganisms and sterilizing surfaces in operating rooms and rooms that will contain caged animals (Figure 13.14). Ultraviolet lights lose potency and should be monitored frequently. The lights can be turned on when humans are not using the rooms to help sanitize the air without irradiating the humans. Exposure to ultraviolet light can cause burns, as anyone who has had a sunburn knows, and can also damage the eyes. Hanging laundry outdoors in bright sunlight also takes advantage of ultraviolet light, which is present in sunlight. Though the quantity of ultraviolet rays in sunlight is small, these rays may help to kill bacteria on clothing, especially diapers.

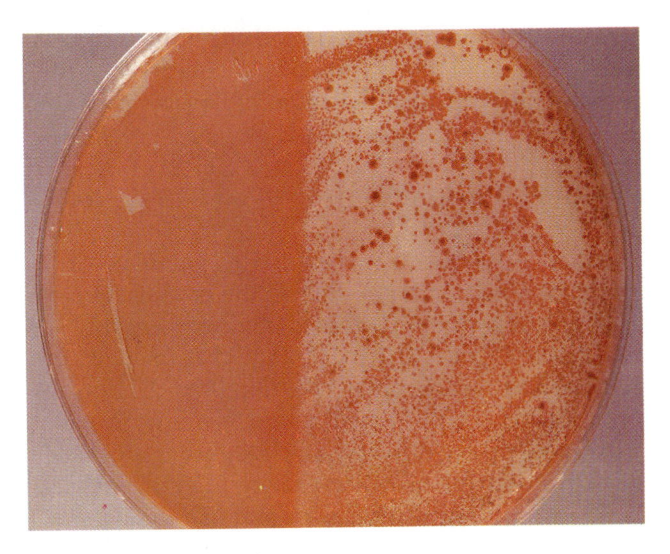

FIGURE 13.14 Effects of ultraviolet radiation can be seen in this Petri plate of *Serratia marcescens;* the right side has been exposed to ultraviolet rays, while the left side was shielded. Most of the organisms on the right side have been killed.

In some communities, ultraviolet light is replacing chlorine in sewage treatment. When chlorine-treated sewage effluent is discharged into streams or other bodies of water, carcinogenic (cancer-causing) compounds form and may enter the food chain. The cost of removing chlorine before discharging treated effluent could add as much as $100 per year to the sewage bills of the average American family, and very few sewage plants do this. Running the sewage effluent under ultraviolet light before discharging it can destroy microorganisms without altering the odor, pH, or chemical composition of the water and without forming carcinogenic compounds.

Ionizing Radiation

X-rays, which have wavelengths of 0.1 to 40 nm, and gamma rays, which have even shorter wavelengths, also kill microorganisms and viruses. Many bacteria are killed by absorbing 0.3 to 0.4 millirads of radiation; polio viruses are inactivated by absorbing 3.8 millirads. (A **rad** is a unit of radiation energy absorbed per gram of tissue, and a millirad is one-thousandth of a rad. Humans usually do not become ill from radiation unless they are subjected to doses greater than 50 rad.) *Ionizing radiation*—so-called because it can dislodge electrons from atoms, creating ions—damages DNA and produces peroxides that act as powerful oxidizing agents in cells. These radiations can also kill or cause mutations in human cells if they reach them. Ionizing radiation is used to sterilize plastic laboratory and medical equipment and pharmaceutical products. It can be used to prevent spoilage in seafoods by doses of 100 to 250 kilorads, in meats and poultry by doses of 50 to 100 kilorads, and in fruits by doses of 200 to 300 kilorads. (A kilorad equals 1000 rads.) Many U.S.

consumers reject irradiated foods fearing they will receive radiation, but such foods are quite safe—free of both pathogens and radiation. In Europe, milk and other foods are often irradiated to achieve sterility.

Microwave Radiation

Microwave radiation, in contrast to gamma, X-ray, and ultraviolet radiation, falls at the end of the electromagnetic spectrum. It has longer wavelengths, approximately 1 mm to 1 m, a range that includes television and police radar wavelengths. Microwave oven frequencies are tuned to match energy levels in water molecules. In the liquid state, water molecules quickly absorb the microwave energy and then release it to surrounding materials as heat. Thus, materials that do not contain water, such as paper, china and plastic plates, remain cool while the moist food on them becomes heated. For this reason the home microwave cannot be used to sterilize items such as bandages and glassware. Conduction of energy in metals leads to problems, such as sparking, which makes most metallic items also unsuitable for microwave sterilization. Moreover, bacterial endospores, which contain almost no water, are not destroyed by microwaves. However, a new style of specialized microwave oven has recently come to the market (Figure 13.15), which can be used to sterilize media in just 10 minutes. It has 12 pressure vessels, each of which holds 100 mL of medium. Microwave energy increases the pressure of the medium inside the vessels until sterilizing temperatures are reached.

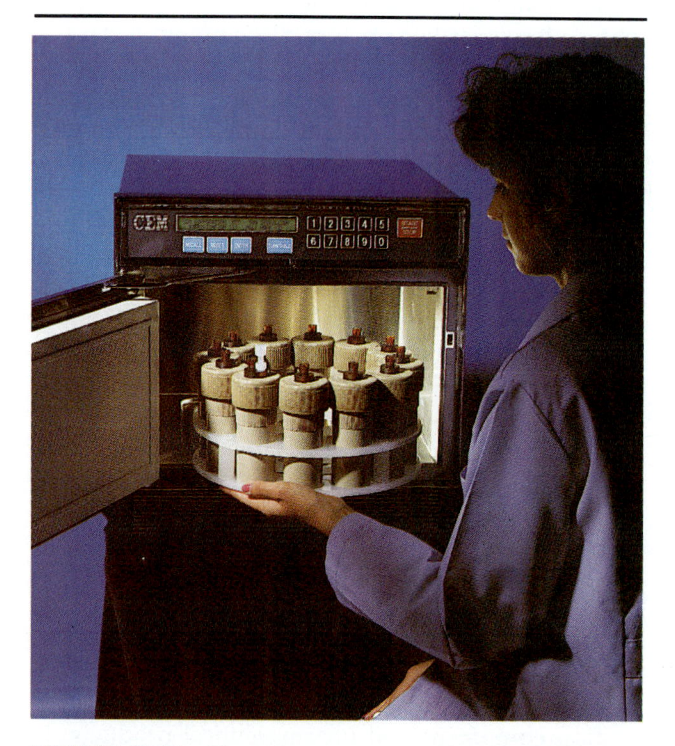

FIGURE 13.15 The MikroClave™ system is specifically designed for rapid sterilization of microbiological media and solutions. Using microwave energy, it can sterilize 1.2 liters of media in 6.5 minutes, or 100 mL in 45 seconds. Agar need not be boiled prior to sterilization.

Caution should be observed in cooking foods in the home microwave oven. Geometry and differences in density of the food being cooked can cause certain regions to become hotter than others, sometimes leaving very cold spots. Consequently, to cook foods thoroughly in a microwave oven it is necessary to rotate the items either mechanically or by hand. For example, pork roasts must be turned frequently and thoroughly cooked to kill cysts of the trichina parasite that might be present (Chapter 22). Failure to kill such cysts could lead to the disease trichinosis, in which cysts of the worm become embedded in human muscles and other tissues. All experimentally infected pork roasts, when microwaved without rotation, showed live worms remaining in some portion at the end of standard cooking time.

Strong Visible Light

Sunlight has been known for years to have a bactericidal effect, but the effect is due primarily to ultraviolet rays in the sunlight. Strong visible light, which contains light of wavelengths from 400 to 700 nm, can have direct bactericidal effects by oxidizing light-sensitive molecules such as riboflavin and porphyrins (components of oxidative enzymes) in bacteria. For that reason, bacterial cultures should not be exposed to strong light during laboratory manipulations. The fluorescent dyes eosin and methylene blue can denature proteins in the presence of strong light because they absorb energy and cause oxidation of proteins and nucleic acids. The combination of a dye and strong light can be used to rid materials of both bacteria and viruses.

Sonic and Ultrasonic Waves

Sonic, or sound, waves in the audible range can destroy bacteria if they are of sufficient intensity. Also, ultrasonic waves, or waves with frequencies above 15,000 cycles per second, can cause bacteria to cavitate. **Cavitation** (kav-eh-ta'shun) is the formation of a partial vacuum in a liquid, in this case, the fluid cytoplasm in the bacterial cell. Bacteria so treated disintegrate, and their proteins are denatured. Enzymes used in detergents are obtained by cavitating the bacterium *Bacillus subtilis*. Disruption of cells by sound waves is called **sonication** (son"eh-ka'shun). Neither sonic nor ultrasonic waves are a practical means of sterilization. They are mentioned here because they are useful in fragmenting cells to study membranes, ribosomes, enzymes, and other components.

Filtration

Filtration is the passage of a material through a filter, or straining device. Sterilization by filtration requires filters with exceedingly small pores. Filtration has been used since Pasteur's time to separate bacteria from media and to sterilize materials that would be destroyed by heat. Over the years, filters have been made of porcelain, asbestos, diatomaceous earth, and sintered glass (glass that has been heated without melting). Membrane filters (Figure 13.16) are widely used today. They usually are made of nitrocellulose and have the great advantage that they can be manufactured with specific pore sizes from 25 nm to less than 0.025 μm. Particles filtered by various pore sizes are summarized in Table 13.3.

Membrane filters have certain advantages and disadvantages. Except for those with the smallest pore sizes, membrane filters are relatively inexpensive, do not clog easily, and can filter large volumes of fluid reasonably rapidly. They can be autoclaved or purchased already sterilized. A disadvantage of membrane filters is that many of them allow viruses and some mycoplasmas to pass through. Other disadvantages are that they may absorb relatively large amounts of the filtrate and may introduce metallic ions into the filtrate.

Membrane filters are used to sterilize materials likely to be damaged by heat sterilization. These materials include media, special nutrients that might be added to media, pharmaceutical products such as drugs, sera, and vitamins. Some filters can be attached to syringes so that materials can be forced through them relatively quickly. Filtration also can be used instead of pasteurization in the manufacture of beer. When using filters to sterilize materials, it is important to select a filter pore size that will prevent any infectious agent from passing into the product.

TABLE 13.3 Pore sizes of membrane filters and particles that pass through them

Pore Size (in μm)	Particles That Pass through Them
10	Erythrocytes, yeast cells, bacteria, viruses, molecules
5	Yeast cells, bacteria, viruses, molecules
3	Some yeast cells, bacteria, viruses, molecules
1.2	Most bacteria, viruses, molecules
0.45	A few bacteria, viruses, molecules
0.22	Viruses, molecules
0.10	Medium-sized to small viruses, molecules
0.05	Small viruses, molecules
0.025	Only the very smallest viruses, molecules
Ultrafilter	Small molecules

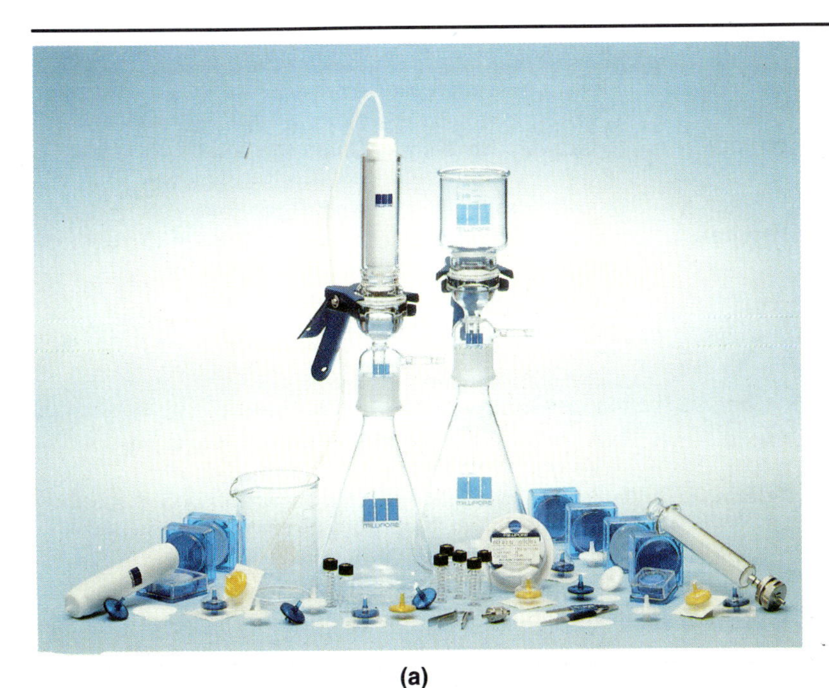

(a)

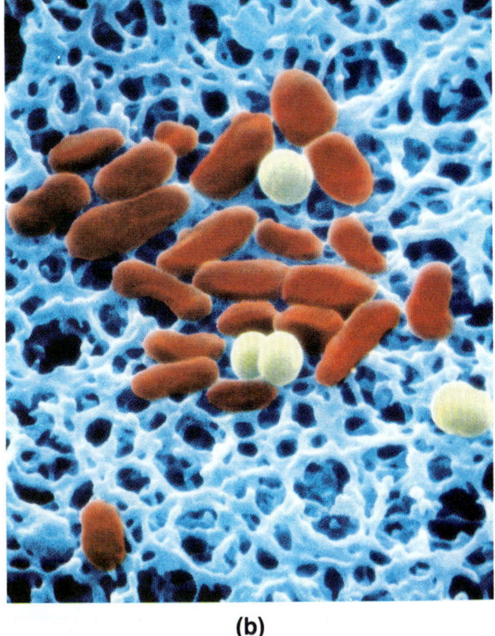

(b)

FIGURE 13.16 (a) Various types of membrane filters are available to sterilize large or small quantities of liquids. Some can be attached to the barrel of a syringe, ensuring that what is forced through the needle will be sterile. (b) Scanning electron micrograph showing red *Serratia marcescens* and yellow *Staphylococcus epidermidis* cells trapped on the surface of a 0.22-μm Millipore membrane filter. Membrane pore size may be selected so as to allow viruses, but not bacteria, to pass through, or to prevent both from passing.

In the manufacture of vaccines that require the presence of live viruses, it is important to select a filter pore size that will allow viruses to pass and prevent bacteria from doing so. By selecting a filter with a proper pore size, scientists can separate polioviruses from the fluid and debris in tissue cultures where they were grown. This procedure simplifies the manufacture of polio vaccine. Cellulose acetate filters with extremely tiny pores are now available and are capable of removing many viruses from liquids (although not the very smallest). However, they are expensive and clog easily.

Membrane filters used to trap bacteria from air and water samples can be transferred directly to agar plates and the quantity of bacteria in the sample determined. Alternately, the filters can be transferred from one medium to another, so organisms with different nutrient requirements can be detected. Filtration also is used to remove microorganisms and other small particles from public water supplies and in sewage treatment facilities. This technique, however, cannot produce sterility but merely reduces contamination.

High-efficiency particulate air (HEPA) filters are used in the ventilation systems of areas where microbial control is especially important, such as in operating rooms, burn units, and laminar flow transfer hoods. They also capture organisms released in rooms occupied by patients with tuberculosis or in laboratories where especially dangerous microbes are studied, such as the maximum containment units shown in Chapter 16. These filters remove almost all organisms larger than 0.3 μm in diameter. Used filters are soaked in formalin before being disposed of.

Osmotic Pressure

High concentrations of salt, sugar, or other substances create a hyperosmotic medium, which draws water from microorganisms by osmosis, as explained in Chapter 4. ∞ (p. 99) The cells **plasmolyze** (plas'mo-līz), or lose water, and collapse. Plasmolysis severely interferes with cell function and eventually leads to cell death. The use of sugar in jellies, jams, and syrups or salt solutions in curing meat and making pickles plasmolyzes most organisms present and prevents growth of new organisms. A few halophilic organisms, however, thrive in these conditions and cause spoilage, especially of pickles, and some fungi can live on the surface of jams.

Properties of antimicrobial physical agents are summarized in Table 13.4.

TABLE 13.4 Properties of antimicrobial physical agents

Agent	Action	Use
Dry heat	Denatures proteins	Oven heat used to sterilize glassware and metal objects; open flame used to incinerate microorganisms.
Moist heat	Denatures proteins	Autoclaving sterilizes media, bandages, and many kinds of hospital and laboratory equipment not harmed by heat and moisture; pressure cooking sterilizes canned foods.
Pasteurization	Denatures proteins	Kills pathogens in milk, dairy products, and beer.
Refrigeration	Slows the rate of enzyme-controlled reactions	Used to keep fresh foods for a few days; does not kill most microorganisms.
Freezing	Greatly slows the rate of enzyme-controlled reactions	Used to keep fresh foods for several months; does not kill most microorganisms; used with glycerol to preserve microorganisms.
Drying	Inhibits enzymes	Used to preserve some fruits and vegetables; sometimes used with smoke to preserve sausages and fish.
Freeze-drying	Dehydration inhibits enzymes	Used to manufacture some instant coffees; used to preserve microorganisms for years.
Ultraviolet light	Denatures proteins and nucleic acids	Used to reduce the number of microorganisms in air in operating rooms, animal rooms, and where cultures are transferred.
Ionizing radiation	Denatures proteins and nucleic acids	Used to sterilize plastics and pharmaceutical products and to preserve foods.
Visible light	Oxidation of light-sensitive materials	Can be used with dyes to destroy bacteria and viruses, may help to sanitize clothing.
Sonic and ultrasonic waves	Cause cavitation	Not a practical means of killing microorganisms but useful in fractionating and studying cell components.
Filtration	Mechanically removes microbes	Used to sterilize media, pharmaceutical products, and vitamins, in manufacturing vaccines, and in sampling microbes in air and water.

ESSAY

Hospital Sanitation and the Infection Control Practitioner

Maintaining sanitation in a modern hospital is becoming an increasingly complex task for several reasons. First, far more patients with infectious diseases are treated in hospitals today than a decade ago. Second, more techniques are available to assist with maintaining sanitation—disposable equipment and supplies, more complex isolation procedures, and advances in sanitizing agents and equipment. Using these techniques improves sanitation, but it also makes more work for the hospital staff. Third, and most important, increasing numbers of pathogens are becoming resistant to antibiotics, and these organisms are especially common in hospitals where antibiotics are in continuous use. Hospital-acquired infections, or *nosocomial infections*, are therefore a constant threat to the lives of already seriously ill patients. The nature of nosocomial infections is discussed in Chapter 16. Here we focus on sterilization and disinfection procedures used to limit the spread of nosocomial and other infections in hospitals.

The most likely mode of transmission of infections in a hospital is by direct contact, as when a health-care worker touches an infected patient and fails to wash his or her hands before touching another patient. Indirect contact, as with contaminated equipment, and airborne transmission of pathogens are less frequent modes of transmission.

Certain characteristics of the hospital environment make it a particularly likely place to acquire an infection. Because people with infections come to hospitals for treatment, the density of pathogens is greater in hospitals than in most other environments. The movements of hospital personnel and visitors and even air currents from elevator shafts tend to spread microbes throughout the hospital.

Several procedures are used to minimize the spread of infection in a hospital. Each room is disinfected after a patient is discharged and before another patient occupies it. Floors are regularly mopped with disinfectant solutions, and carpets are kept dry and vacuumed often. Linens, especially those from patients with infections, are placed in plastic bags before being dropped into a laundry chute. Strong detergents and very hot water (71°C for 25 minutes) are used for laundering hospital linens. Food is heated to an internal temperature of 74°C and kept covered and above 60°C until it is served. Dishes are washed at 60°C for 20 seconds and rinsed at 82°C for 10 seconds. Electronic air filters that ionize airborne microbes are installed in ventilating systems, especially those that serve critical care units, burn units, and nurseries.

Hospital personnel can minimize the risk of infecting themselves and others in several ways. The single most important way is by thoroughly washing their hands between patient contacts. Another way is by receiving appropriate immunizations. Personnel should be immunized against diphtheria and tetanus and against hepatitis B if they will have contact with blood or other potentially infectious fluids. They also should be immunized against measles and mumps if they have not had those diseases and against influenza if they are susceptible to frequent pulmonary infections. Personnel also can learn and practice good aseptic techniques and carry out recommendations of the hospital's employee health programs.

Hospitals are ethically and legally responsible for patients acquiring nosocomial infections. In fact, to maintain accreditation by the American Hospital Association, hospitals must have a program that includes surveillance of nosocomial infections in both patients and staff, a microbiology laboratory, isolation procedures, accepted procedures for the use of catheters and other instruments, general sanitation procedures, and a nosocomial disease education program for staff members. Most hospitals have an infection-control practitioner (ICP) to manage such a program. The goal of the program is to engage all hospital personnel in active measures to prevent infections.

If you are considering a career in health care and you would like to help protect people from the spread of infections in hospitals, you may want to consider a career as an ICP. You will need to qualify for this specialty through a registry examination offered by the Association for Practitioners in Infection Control (APIC). It requires knowledge of both microbiology and patient-care techniques. Among currently registered specialists, some were first trained as microbiologists and many as nurses.

The duties of an ICP include surveillance and identification of infections, supervision of, or collaboration with, the hospital's employee health program, keeping up-to-date on newly available immunizations to determine which ones hospital personnel should receive, assisting with studies of antibiotic use in infection control and detection of resistant organisms, and providing instruction to new staff on aseptic techniques and the hospital's infection-control program, including isolation procedures. Isolation procedures are discussed in more detail in Chapter 16.

Some specific and effective means of infection control are as follows: When physicians, nurses, and other staff members wash their hands thoroughly with soap and water between patients, they can greatly reduce the risk of spreading diseases among patients. Scrupulous care in obtaining sterile equipment and maintaining its sterility while inserting and using catheters and other invasive instruments also are important factors. The use of gloves when drawing blood or handling infectious materials, such as dressings and bedpans, is a third way to prevent infections.

ESSAY Hospital Sanitation

Other techniques are needed to minimize the development of antibiotic-resistant pathogens. Routine use of antimicrobial agents to prevent infections has turned out to be a misguided effort because it contributes to the development of resistant organisms, as is explained in the next chapter. Therefore, some hospitals maintain surveillance of antibiotic use. Antibiotics are given for known infections but are given prophylactically (as a preventive measure) only in special situations. Prophylactic antibiotics are justified in surgical procedures, such as hysterectomies, colorectal surgeries, and repair of traumatic injuries, where the surgical field is invariably contaminated with potential pathogens. They also are justified in immunosuppressed patients and excessively debilitated patients, where natural defense mechanisms may fail.

If all the known techniques for preventing nosocomial infections were rigorously practiced, the incidence of such infections probably could be reduced to half the present level.

CHAPTER SUMMARY

DEFINITIONS

- **Sterilization** refers to the killing or removal of all organisms in any material or on any object.
- **Disinfection** refers to the reduction in numbers of pathogenic organisms on objects or in materials so that the organisms no longer pose a disease threat.
- Important terms related to these processes are defined in Table 13.1.

RELATED KEY TERMS

antiseptics

sterility

PRINCIPLES OF STERILIZATION AND DISINFECTION

- Because of the logarithmic death rate of microorganisms, a definite proportion of organisms die in a given time interval.
- The smaller the number of organisms present, the less time will be needed to achieve sterility.
- Microorganisms differ in their susceptibility to antimicrobial agents.

thermal death point

thermal death time

decimal reduction time (DRT) or D value

ANTIMICROBIAL CHEMICAL AGENTS

Potency of Chemical Agents

- The potency, or power, of an antimicrobial chemical agent is affected by time, temperature, pH, and concentration of the agent.
- Potency increases with length of time organisms are exposed to the agent, increased temperature, acidic or alkaline pH, and usually increased concentration of the agent.

Evaluation of Effectiveness of Chemical Agents

- Evaluation of effectiveness is difficult, and no entirely satisfactory method is available.
- Determining the **phenol coefficient** is used for agents similar to phenol. The phenol coefficient is the ratio of the dilution of the agent to the dilution of phenol that will kill all organisms in 10 minutes but not in 5 minutes.

filter paper method

use-dilution test

Disinfectant Selection

- A variety of criteria are considered in selecting a **disinfectant**. In actual practice, most chemical agents are tested in various situations and used in situations where they produce satisfactory results.

Mechanisms of Action of Chemical Agents

- Actions of antimicrobial chemical agents can be grouped according to their effects on proteins, cell membranes, and other cell components.
- Reactions that alter proteins include hydrolysis, oxidation, and attachment of atoms or chemical groups to protein molecules. Such reactions denature proteins, rendering them nonfunctional.
- Cell membranes can be disrupted by agents that denature proteins and by **surfactants**, which reduce surface tension and dissolve lipids.
- Reactions of other chemical agents damage nucleic acids and energy-capturing systems. Damage to nucleic acids is an important means of inactivating viruses.

Specific Antimicrobial Chemical Agents

- Soaps and detergents aid in the removal of microbes, oils, and dirt but do not sterilize.
- Acids are commonly used as food preservatives; alkali in soap helps to destroy microorganisms.
- Among the agents containing heavy metals, silver nitrate is used to kill gonococci, and mercury-containing compounds are used to disinfect instruments and skin.
- Among the agents containing halogens, chlorine is used to kill pathogens in water, and iodine is a major ingredient in several skin disinfectants.
- Alcohols are used to disinfect skin.
- Phenol derivatives can be used on skin, instruments, dishes, and furniture, and to destroy discarded cultures; they work well in the presence of organic materials.
- Oxidizing agents are particularly useful in disinfecting puncture wounds.
- Alkylating agents can be used to disinfect or to sterilize a variety of materials, but all are carcinogens.
- Some dyes, plant oils, sulfur-containing substances, and nitrates can be used as disinfectants or food preservatives.

RELATED KEY TERMS

wetting agents

quaternary ammonium compounds (quats)

tincture

ANTIMICROBIAL PHYSICAL AGENTS

Dry Heat, Moist Heat, and Pasteurization

- Heat destroys microorganisms by denaturing protein, melting lipids, and when open flame is used, by incineration.
- Dry heat is used to sterilize metal objects and glassware.
- Flame is used to sterilize inoculating loops and the mouths of culture tubes.
- The **autoclave**, which uses moist heat under pressure, is a common instrument for sterilization and is very effective when proper procedures are followed.

Refrigeration, Freezing, Drying, and Freeze-drying

- Refrigeration, freezing, drying, and freeze-drying can be used to retard the growth of microorganisms.
- **Lyophilization**, drying in the frozen state, can be used for long-term preservation of live microorganisms.

Radiation

- Radiation used to control microorganisms includes ultraviolet light, ionizing radiation, and sometimes microwaves and strong sunlight.

Sonic and Ultrasonic Waves

- Sonic and ultrasonic waves can kill microorganisms, but they are used mostly for **sonication**.

Filtration

- **Filtration** can be used to sterilize heat-labile substances, separate viruses, and collect microorganisms from air and water samples.

Osmotic Pressure

- High concentrations of sugar or salt create osmotic pressure that **plasmolyzes** cells and prevents growth of microorganisms in highly sweetened or salted foods.

aerosols

pasteurization

rad

cavitation

QUESTIONS FOR REVIEW

A.

1. How do sterilization and disinfection differ?
2. Define the terms listed in Table 13.1.

B.

3. What are the three principles that apply to the processes of sterilization and disinfection?

C.

4. How do each of the following affect the potency of antimicrobial chemical agents: time, temperature, pH, and concentration of the chemical agent?

D.

5. What is the phenol coefficient, and how is it determined?
6. How is the effectiveness of chemical agents evaluated in actual practice?

E.

7. How do chemical agents affect proteins?
8. How do chemical agents affect cell membranes?
9. What other effects do chemical agents have on cells?

F.

10. How do each of the following agents affect cells: acids and alkalis, heavy metals, halogens, alcohols, phenols, oxidizing agents, alkylating agents, surfactants, and dyes?

G.

11. Compare and contrast the effects of different ways to control microorganisms with heat.

H.

12. What benefits and hazards are associated with refrigerating foods?
13. Explain how freezing, drying, or freeze-drying can be used to control or to preserve organisms.

I.

14. How are cells affected by radiations?
15. What is sonication, and how is it used in microbiology?
16. What are some advantages and disadvantages of membrane filters?
17. What are some applications of membrane filters?
18. What is plasmolysis, and how can it be used to prevent growth of microorganisms?

PROBLEMS FOR INVESTIGATION

1. Suppose you were a member of a committee to maintain the best possible control of microorganisms in a hospital, a laboratory, or a food-handling industry. Show what kinds of problems of controlling microorganisms might exist and how you would effectively solve these problems.
2. Devise a method of evaluating the effectiveness of one of the following groups of chemical agents: skin antiseptics, food-utensil disinfectants, or spore-killing agents.
3. Evaluate the autoclaving procedures used in your laboratory.
4. A technician was testing a new disinfectant to determine its phenol coefficient. The following results were found. What is the coefficient? Is it likely to be a good disinfectant?

EXPOSURE TIME	PHENOL DILUTION				NEW DISINFECTANT DILUTION			
	1:100	1:110	1:120	1:130	1:50	1:60	1:70	1:80
5 min.	+	+	+	–	+	+	–	–
10 min.	+	+	–	–	+	–	–	–

5. Look up the recommended procedures for a surgical scrub. Prepare a report and demonstration for your class.
6. Find out which procedures are used on the area of a patient's skin prior to making a surgical incision.
7. Find out about the industrial use of lasers for sterilization.
8. The central sterilization facilities in a large hospital were located in a basement where air exchange was not very good. One day all employees working in this unit began to complain of headache, burning eyes, nausea, and dizziness. Then two of them fainted. Everyone quickly evacuated the area.
   a) What could have caused this problem?
   b) What further damage may employees have suffered?
   (The answer to this question appears in Appendix E.)

SOME INTERESTING READING

Axnick, K. J., and M. Yarbrough. 1984. *Infection control: An integrated approach.* St. Louis: C. V. Mosby.

Block, S., ed. 1991. *Disinfection, sterilization, and preservation.* 4th ed. Philadelphia: Lea and Febiger.

Brewere, J. H., ed. 1973. *Lectures on sterilization.* Durham, NC: Duke University Press.

Craig, C. P. 1983. *Fundamentals of infection control.* Oradell, NJ: Medical Economics Books.

Jensen, M. M., and D. N. Wright. 1993. *Introduction to medical microbiology.* Englewood Cliffs, NJ: Prentice Hall.

Litsky, W. 1990. Wanted: plastics with antimicrobial properties. *American Journal of Public Health* 80, no. 1 (January):13–16.

McLaughlin, A. J. 1983. *Manual of infection control in respiratory care.* Boston: Little, Brown.

Phillips, G. B., and W. S. Miller, eds. 1973. *Industrial sterilization.* Durham, NC: Duke University Press.

Russell, A. D. 1982. The destruction of bacterial spores. New York: Academic Press.

Russell, A. D., et al., eds., 1982. *Principles and practices of disinfection, preservation, and sterilization.* London: Blackwell.

Spaulding, E. H., and D. H. M. Groschel. 1985. Hospital disinfectants and antiseptics. *Manual of clinical microbiology.* 4th ed. Lennette et al., eds. Washington, D.C.: American Society for Microbiology.

The search for new antibiotics goes on continuously. Perhaps one of these bacterial cultures will yield an antibiotic that you may use in the future.

14 Antimicrobial Therapy

This chapter focuses on the following questions:

A. Which terms are used to discuss chemotherapy and antibiotics, and what do they mean?

B. How have chemotherapeutic agents been developed?

C. How do the terms selective toxicity, spectrum of activity, and modes of action apply to antimicrobial agents?

D. What kinds of side effects are associated with antimicrobial agents?

E. What is resistance to antibiotics, and how do microorganisms acquire it?

F. How are sensitivities of microbes to chemotherapeutic agents determined?

G. What are the attributes of an ideal antimicrobial agent?

H. What are the properties, uses, and side effects of antibacterial agents?

I. What are the properties, uses, and side effects of antifungal agents, antiviral agents, antiprotozoan agents, and antihelminthic agents?

J. How do resistant hospital infections arise, and how can they be treated and prevented?

As you lie in your sickbed, suffering from some infectious disease, how reassuring it is to be able to reach over, swallow some capsules, and look forward to getting well soon. But people have not always been able to do this. Over the ages anxious parents have watched their children die of fevers, diarrhea, infected wounds, and other maladies. Whole villages have been wiped out by plagues. During the middle of the fourteenth century, more than a quarter of the population of Europe died of the Black Death (bubonic plague) in just a few years. More people have died of infection in wartime than from swords or bullets. Until relatively recently, the only defenses against infectious diseases were herbal teas, poultices (soaks), and the like. Modern and effective weapons against microbes—antibiotics and sulfa drugs—were not available until the twentieth century.

Think back through your life. Have there been times when you might have died had it not been for antimicrobial drugs? At what age would you have died? Which of your family members might not be alive now? Happily, we live in a better time with respect to illnesses and deaths caused by infectious organisms. In the United States the life expectancy of a baby born in 1850 was less than 40 years; in 1900, about 50 years; and in 1980 over 70 years. Infectious diseases claimed the lives of about 1 of every 100 U.S. residents per year as late as 1900 but only about 1 out of each 300 in 1980. Antimicrobial agents still don't save all patients, but they have drastically lowered the death rate from infectious disease.

ANTIMICROBIAL CHEMOTHERAPY

The term **chemotherapy** was coined originally by the German chemist Paul Ehrlich to describe the use of chemical substances to kill pathogenic organisms without injuring the host. Today, chemotherapy refers to the use of chemical substances to treat various aspects of disease—aspirin for headache and inflammation, drugs to regulate heart function, and agents to rid the body of malignant cells. Using this modern broad definition of chemotherapy, a **chemotherapeutic agent** is any chemical substance used in medical practice. These agents also are referred to as **drugs.**

In microbiology we are concerned with **antimicrobial agents,** a special group of chemotherapeutic agents used to treat diseases caused by microbes. Thus, in modern terms, an antimicrobial agent is synonymous with a chemotherapeutic agent as Ehrlich defined it. In this chapter we consider a variety of antimicrobial agents and a few agents used to treat helminth (worm) infections.

Antibiosis literally means "against life." In the 1940s Selman Waksman, the discoverer of streptomycin, defined an **antibiotic** as "a chemical substance produced by microorganisms which has the capacity to inhibit the growth of bacteria and even destroy bacteria and other microorganisms in dilute solution." In contrast, agents synthesized in the laboratory are called **synthetic drugs.** Some antimicrobial agents are synthesized by chemically modifying a substance from a microorganism. More often a synthetic precursor different from the natural one is supplied to a microorganism, and the organism completes synthesis of the antibiotic. Antimicrobial agents made partly by laboratory synthesis and partly by microorganisms are called **semisynthetic drugs.**

HISTORY OF CHEMOTHERAPY

Throughout history humans have attempted to alleviate suffering through treating disease—often by taking concoctions of plant substances. Although ancient Egyptians used moldy bread to treat disease, they had no knowledge of the antibiotics it contained. Extracts of willow bark, now known to contain aspirin, were used to alleviate pain. Parts of the foxglove plant were used to treat heart disease in the sixteenth century, although the active ingredient digitalis had not been identified. Likewise, extracts from the cinchona tree, which contain quinine, were used to treat malaria.

In spite of their reputation for using rituals irrelevant to the cure of disease, traditional healers of primitive societies, especially in the tropics, are quite knowledgeable about medicinal properties of plants. What they know has been passed down from generation to generation. Because these healers are disappearing, pharmaceutical companies are attempting to learn from them and make written records of their knowledge, as well as test the plants they use.

In Western civilization, the first systematic attempt to find specific chemical substances to treat infectious disease was made by Paul Ehrlich, as noted in Chapter 1. Although his discovery of Salvarsan to treat syphilis was of great therapeutic benefit, even more important were the concepts he developed in the new science of chemotherapy. He was interested in the mechanisms by which chemical substances bind to microorganisms and to animal tissues. His studies of chemicals that bind to tissues led to histological (tissue) stains that are still used today.

The next advances in chemotherapy were the nearly concurrent development of sulfa drugs and antibiotics. Domagk's discovery of prontosil, which is converted to sulfanilamide in the body, stimulated the development of a group of substances called sulfonamides, or sulfa drugs. As the number of sulfa drugs grew, it was possible to use them to attack directly a variety of pathogens. However, the usefulness of sulfa drugs was limited. They did not attack all pathogens, and they sometimes caused kidney damage and allergies. But they have saved many lives and continue to do so today.

Alexander Fleming's idea that the ability of the mold *Penicillium* to inhibit growth of microorganisms might be exploited led him to identify the inhibitory agent and name it **penicillin.** Fleming had observed the contamination of his bacterial cultures with this fungus many times, as had many other microbiologists. However, instead of grumbling about another contaminated culture and tossing it out, Fleming saw the tremendous potential in this accidental finding. If only the substance (penicillin) could be extracted and collected in large quantities, it could be used to combat infection.

Fleming's idea did not come to fruition until the 1940s, when Ernst Chain and Howard Florey at Oxford University finally isolated penicillin and worked with other researchers to develop methods of mass production. Such mass production occurred during World War II and saved the lives of many wounded people. Supplies of the drug were limited, and it was not readily available to civilians until after the war.

CLOSE-UP

When Doctors Learned To Cure

"We were provided with a thin, pocketsize book called *Useful Drugs*, one hundred pages or so, and we carried this around in our white coats when we entered the teaching wards and clinics . . . , but I cannot recall any of our instructors ever referring to this volume. Nor do I remember much talk about treating disease at any time in the four years of medical school except by the surgeons. . . . Our task for the future was to be diagnosis and explanation. Explanation was the real business of medicine. What the ill patient and his family wanted most was to know the name of the illness, and then, if possible, what had caused it, and finally, most important of all, how it was likely to turn out. . . . It gradually dawned on us that we didn't know much that was really useful, that we could do nothing to change the course of the great majority of the diseases we were so busy analyzing. . . .

Then came the explosive news of sulfanilamide, and the start of the real revolution in medicine. I remember with astonishment when the first cases of pneumococcal and streptococcal septicemia were treated in Boston in 1937. The phenomenon was almost beyond belief. Here were moribund patients, who would surely have died without treatment, improving in their appearance within a matter of hours of being given the medicine and feeling entirely well within the next day or so."

—Lewis Thomas, 1983

Following the war research proceeded rapidly, and new antibiotics were discovered one after another.

The introduction of penicillin and sulfonamides in the 1930s can be said to mark the beginning of modern medicine. As the medical writer Lewis Thomas says, "Doctors could now *cure* disease, and this was astonishing, most of all to the doctors themselves."

GENERAL PROPERTIES OF ANTIMICROBIAL AGENTS

Antimicrobial agents share certain common properties. We can learn much about how these agents work and why they sometimes do not work by considering such properties as selective toxicity, spectrum of activity, mode of action, side effects, and resistance of microorganisms to them.

Selective Toxicity

Some chemical substances have antimicrobial properties but are too toxic to be taken internally and are used only for topical application—application to the skin's surface. For internal use, an antimicrobial drug must have **selective toxicity,** that is, it must harm the microbes without causing significant damage to the host. Some drugs such as penicillin have a wide range between the toxic dosage level, which causes host damage, and the **therapeutic dosage level,** which successfully eliminates the pathogenic organism if the level is maintained over a period of time. The relationship between the toxicity of an agent for the body and its toxicity for an infectious agent is expressed in terms of its **chemotherapeutic index.** For any particular agent the chemotherapeutic index is defined as the maximum tolerable dose per kilogram body weight divided by the minimum dose per kilogram body weight that will cure the disease.

For other drugs, such as those containing arsenic, mercury, and antimony, the dosage must be calculated very precisely because these substances are highly toxic to human and animal hosts as well as to pathogens. Treatment of worm infections is especially difficult because what damages the parasite will also damage the host. In contrast, bacterial pathogens often can be attacked by interfering with metabolic pathways not shared by the host. For example, penicillin interferes with cell wall synthesis; it is not toxic to human cells, which lack walls, though some patients are allergic to it.

Spectrum of Activity

Antimicrobial agents that can attack a great number of microorganisms from a wide range of taxonomic

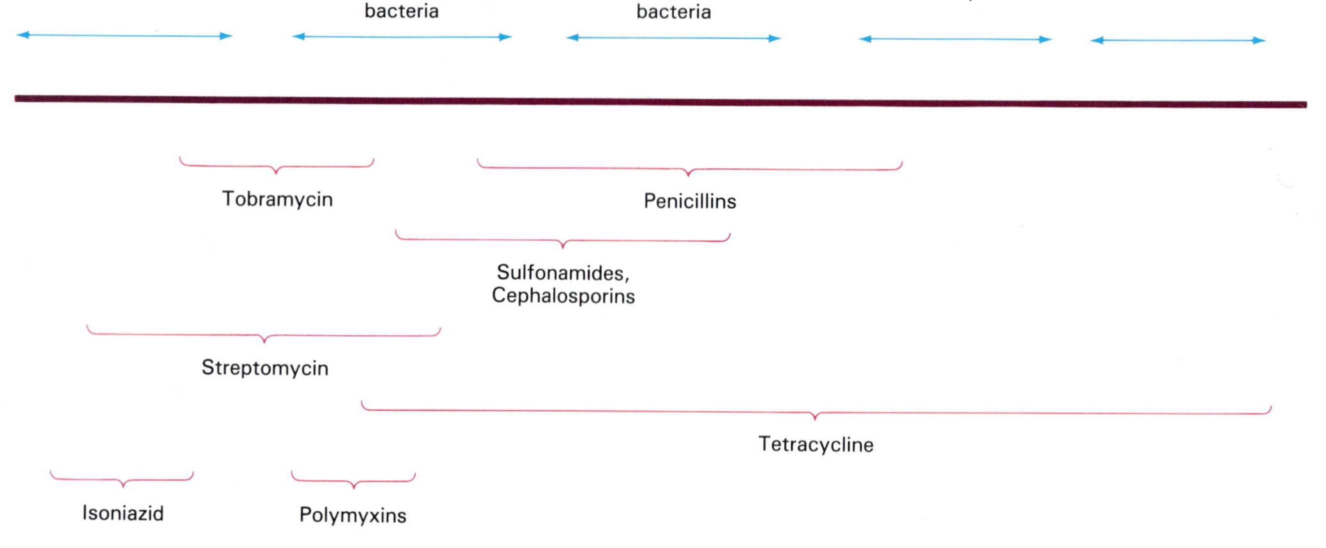

FIGURE 14.1 Broad-spectrum drugs, such as tetracycline, affect a variety of different organisms. Narrow-spectrum drugs, such as isoniazid, affect only a few specific types of organisms.

groups, including both gram-positive and gram-negative bacteria, have a **broad spectrum** of activity. Those that attack only a small number of microorganisms or a single taxonomic group have a **narrow spectrum** of activity (Figure 14.1). Some common antibiotics are classified according to their spectrum of activity in Table 14.1. A broad-spectrum drug is especially useful when a patient is seriously ill with an infection caused by an unidentified organism. Using such a drug increases the chance that the organism will be susceptible to it. However, if the identity of the organism is known, a narrow-spectrum drug should be used. Using such a drug minimizes the destruction of microorganisms normally present in the host that sometimes compete with and help to destroy infectious organisms. Using narrow-spectrum drugs also decreases the likelihood of organisms developing drug resistance.

TABLE 14.1 The spectrum of activity of selected antimicrobial agents

Organisms Affected	Broad-Spectrum Agents*	Narrow-Spectrum Agents
Gram-positive bacteria	Ampicillin	Erythromycin
Bacteroides and other anaerobes	Cephalosporins	Lincomycin
Yeasts	Chloramphenicol	Nystatin
Gram-positive bacteria	Gentamicin	Penicillin
Gram-negative bacteria	Kanamycin	Polymyxins
Streptococci and some gram-negative bacteria	Tetracyclines	Streptomycin
Staphylococci, enterococci, and some clostridia	Tobramycin	Vancomycin

* Broad-spectrum agents affect most bacteria.

Modes of Action

Like other medicines, antimicrobial agents sometimes are used simply because they work, without our always knowing how they work. Many people's lives have been saved by medicines whose actions at the cellular level have never been understood. However, it is always desirable to know the mode of action of an agent. With that knowledge, effects of actions on patients can be better monitored and controlled, and ways of improving them may be found.

Antimicrobial drugs generally act on an important structure or function in microorganisms, which usually differs from its counterpart in animals. This difference is exploited to kill microbes while causing minimal effects on host cells. Five different modes of action of antimicrobials (Figure 14.2) are discussed here: (1) inhibition of cell wall synthesis, (2) disruption of cell membrane function, (3) inhibition of protein synthesis, (4) inhibition of nucleic acid synthesis, and (5) action as an antimetabolite.

Inhibition of Cell Wall Synthesis

Many bacterial and fungal cells have rigid external cell walls, whereas animal cells lack cell walls. Consequently, inhibiting cell wall synthesis selectively

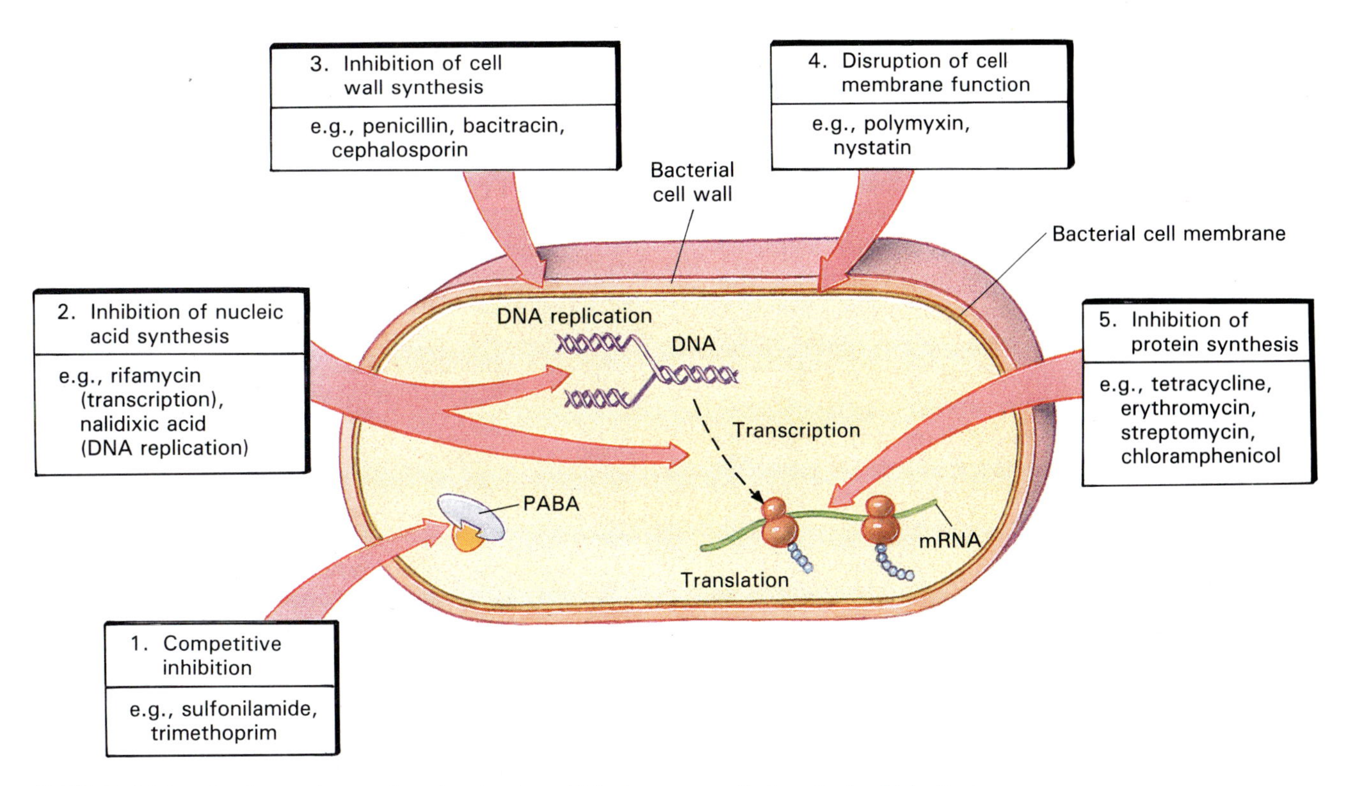

FIGURE 14.2 The five major modes of action by which drugs exert their antimicrobial effects.

damages bacterial and fungal cells. Bacterial cells, especially gram-positive ones, have a high internal osmotic pressure. Without a normal, sturdy cell wall, these cells burst when subjected to the low osmotic pressure of body fluids. (Chapter 4, p. 100) Antibiotics such as penicillin contain a chemical structure called a beta-lactam ring by which they usually attach to components of peptidoglycan. In this way they interfere with synthesis of the cell wall in bacteria (though not in fungi, whose cell walls lack peptidoglycan). (Chapter 4, p. 76)

Disruption of Cell Membrane Function

All cells are bounded by a cell membrane. Although the membranes of all cells are quite similar, those of bacteria and fungi differ sufficiently from those of animal cells to allow selective action of antimicrobial agents. Certain polypeptide antibiotics such as polymyxins act as detergents and dissolve bacterial cell membranes, probably by binding to phospholipids in the membrane. They are especially effective against gram-negative bacteria, which have cell membranes and envelopes rich in phospholipids. (Chapter 4, p. 82) Polyene antibiotics such as amphotericin B bind to particular sterols present in the membranes of fungal cells. Polymyxins do not act on fungi, and polyenes do not act on bacteria.

Inhibition of Protein Synthesis

In all cells protein synthesis requires not only information in DNA and several kinds of RNA, but also ribosomes. Differences between bacterial and animal ribosomes allow antimicrobial agents to attack bacterial ribosomes without significantly damaging animal ribosomes. Aminoglycoside antibiotics, such as streptomycin, derive their name from the amino acids and glycosidic bonds they contain. They act on bacterial ribosomes by interfering with the accurate reading (translation) of the mRNA message—that is, the incorporation of the correct amino acids. (Chapter 7, p. 170) The nonfunctional and incomplete proteins thus formed fail to carry out important cell functions. Aminoglycosides also break up chains of ribosomes and reduce efficiency of protein synthesis. The end result of these various effects is death of the bacterial cell.

Inhibition of Nucleic Acid Synthesis

Differences between the enzymes used to synthesize nucleic acids that exist between bacterial and animal cells provide a means for selective action of antimicrobial agents. Antibiotics of the rifamycin family bind to a bacterial RNA polymerase and inhibit RNA synthesis. (Chapter 7, p. 169)

Action As an Antimetabolite

An **antimetabolite** is a substance that prevents a cell from carrying out a metabolic reaction. Antimetabolites function in two ways: (1) by competitive inhibition of enzymes and (2) by being erroneously incorporated into important molecules such as nucleic acids. Antimetabolites are structurally similar to normally used molecules. Their actions are sometimes called **molecular mimicry** because they mimic, or imitate, the normal molecule, preventing a reaction from occurring or causing it to go awry.

In competitive inhibition (Chapter 5), an enzyme is inhibited by a substrate that binds to its active site but cannot react. (p. 115) The activity of the enzyme is slowed or stopped while the competing substrate occupies its active site. Sulfanilamide and para-aminosalicylic acid (PAS) are chemically very similar to para-aminobenzoic acid (PABA) (Figure 14.3). They competitively inhibit an enzyme that acts on PABA. Many bacteria require PABA as a metabolite, or essential nutrient, to make folic acid, which they use in synthesizing nucleic acids and other metabolic products. When sulfanilamide or PAS is bound to the enzyme instead of PABA, the bacterium cannot make folic acid. Animal cells lack the enzymes to make folic acid and obtain it from their diets, so their metabolism is not disturbed by these competitive inhibitors.

Antimetabolites such as the purine analogue vidarabine and the pyrimidine analogue idoxuridine (Figure 14.4) are erroneously incorporated into nucleic acids. These molecules are very similar to the normal purines and pyrimidines of nucleic acids. When incorporated into a nucleic acid, they garble the information that it encodes because they cannot form the correct base pairs during replication and transcription. Purine and pyrimidine analogues are generally as toxic to animal cells as to microbes because all cells use the same purines and pyrimidines to make nucleotides. These agents are most useful in treating viral

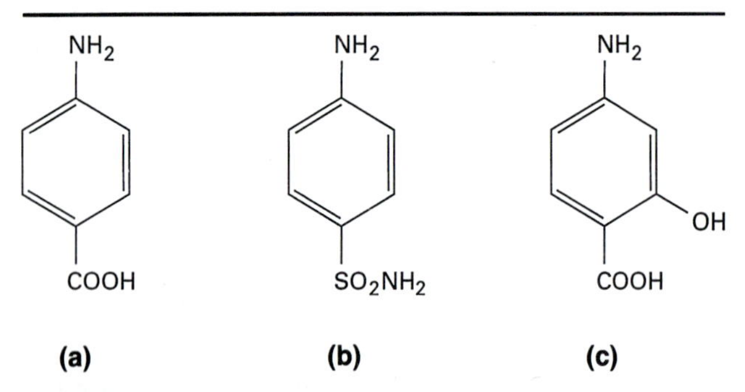

FIGURE 14.3 (a) Para-aminobenzoic acid (PABA), a metabolite required by many bacteria. (b) Sulfanilamide. (c) Para-aminosalicylic acid (PAS). The latter two act as competitive inhibitors to PABA. Note the similarity in the structures of the three compounds.

BIOTECHNOLOGY

Pharmacy of the Future: Triple-stranded DNA

Why struggle to destroy or counteract a harmful microbial product, such as a toxin or a pyrogen, when you could shut it off at its source? Soon pharmacists will dispense preparations containing a third strand of nucleotides. When taken, this strand will nestle between the pair of DNA strands at a specific location on a chromosome. Bonding to both strands, it will prevent them from separating to manufacture messenger RNA, in essence "turning off" the gene. If vital genes are inactivated, organisms will die. This process should work against viruses as well. It would also be possible to turn off defective human genes. Then, with genetic engineering techniques, the patient could be given good genes to replace them. Researchers are currently working with triple-stranded DNA, but the first such products for human use will probably not appear until after the turn of the century.

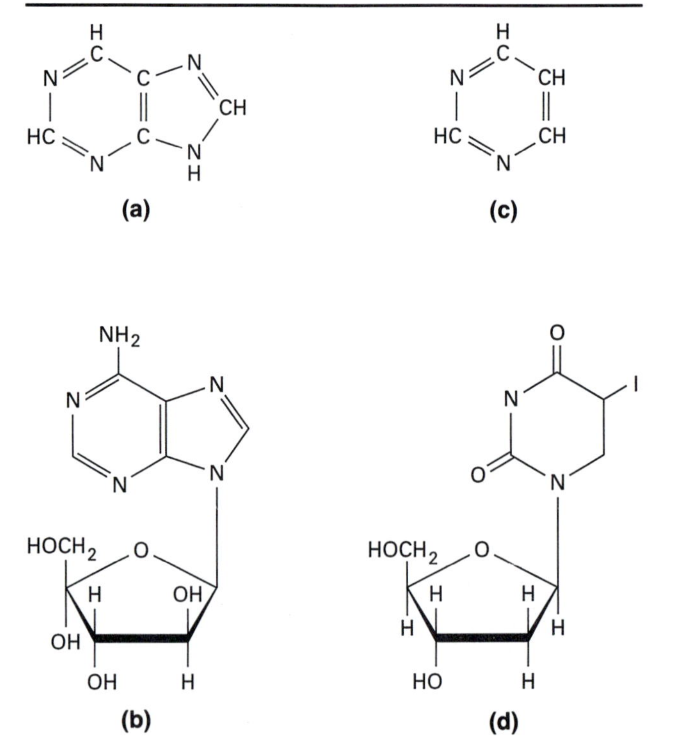

FIGURE 14.4 Nucleic acid bases and their analogues: molecules so similar in structure that they can be incorporated in place of the correct molecule, thus acting as antimetabolites. (a) Basic structure of a purine. (b) The purine analogue vidarabine. (c) Basic structure of a pyrimidine. (d) The pyrimidine analogue idoxuridine.

infections because viruses incorporate analogues more rapidly than cells and are more severely damaged.

Kinds of Side Effects

The side effects of antimicrobial agents on infected persons (hosts) fall in three general categories: (1) toxicity, (2) allergy, and (3) disruption of natural flora. The development of resistance to antibiotics can be thought of as a side effect on the microorganisms. As is explained later, resistance produces infections that can be difficult to treat.

Toxicity

By their selective toxicity and modes of action, antimicrobial agents kill microbes without seriously harming host cells. However, some antimicrobials do exert toxic effects on the patients receiving them. The toxic effects of antimicrobial agents are discussed below in connection with specific agents.

Allergy

An allergy is a condition in which the body's immune system responds to a foreign substance, usually a protein. For example, breakdown products of penicillins combine with proteins in body fluids and form a molecule the body treats as a foreign substance. Allergic reactions can be limited to mild skin rashes and itching, or they can be life-threatening. One kind of life-threatening allergic reaction, called anaphylactic shock (Chapter 19), occurs when a patient is subjected to a foreign substance to which his or her body has already become sensitized.

Disruption of Natural Flora

Antimicrobial agents, especially broad-spectrum antibiotics, exert their adverse effects not only on pathogens but also on indigenous flora, the microorganisms that normally inhabit the skin and the digestive, respiratory, and urinary tracts. When these flora are disturbed, other organisms not susceptible to the antimicrobial agent invade the unoccupied areas and multiply rapidly. Invasion by replacement flora is called **superinfection.** Superinfections are difficult to treat because they are susceptible to few antibiotics.

Although short-term use of penicillins generally does not severely disrupt normal flora, oral ampicillin sometimes allows overgrowth of toxin-producing clostridia. Long-term use of penicillin or aminoglycosides can abolish normal flora and allow colonization of the gut with resistant gram-negative bacteria and fungi such as *Candida*. (A preparation called Lactinex, which contains normal flora, can be given to counteract the effects of antibiotics.) Oral and vaginal superinfections with species of *Candida* yeasts are common after prolonged use of antimicrobial agents such as cephalosporins, tetracyclines, and chloramphenicol. The risk of serious superinfections is greatest in hospitalized patients receiving broad-spectrum antibiotics, for two reasons. First, patients often are debilitated and less able to resist infection. Second, they are in an environment in which drug-resistant pathogens are prevalent.

Resistance of Microorganisms

Resistance of a microorganism to an antibiotic means that a microorganism formerly susceptible to the action of an antibiotic is no longer affected by it. An important factor in the development of drug-resistant strains of microorganisms is that many antibiotics are bacteriostatic, or growth-inhibiting, and not bacteriocidal. Unfortunately, the most resilient microbes evade defenses (Chapter 15) and are likely to develop resistance to the antibiotic.

How Resistance Is Acquired

Microorganisms usually acquire antibiotic resistance by genetic changes, but sometimes resistance is acquired by nongenetic mechanisms. Nongenetic resistance occurs when microorganisms such as those that cause tuberculosis persist in the tissues out of reach of antimicrobial agents. If the sequestered microorganisms start to multiply and release their progeny, the progeny are still susceptible to the antibiotic. This type of resistance might more properly be called evasion. Another type of nongenetic resistance occurs when certain strains of bacteria temporarily change to L-forms that lack most of their cell walls. For several generations, while the cell wall is lacking, these organisms are resistant to antibiotics that act on cell walls. However, when they revert to producing cell walls, they are again susceptible to the antibiotics.

Genetic resistance to antimicrobial agents develops from genetic changes followed by natural selection (Figure 14.5 and Chapter 8). ∞ (p. 207) For example, in most bacterial populations, mutations occur spontaneously at a rate of about 1 per 10 million to 10 billion organisms. Bacteria reproduce so rapidly that billions of organisms, and a few mutants with them, can be produced in a short period of time. If a mutant happens to be resistant to an antimicrobial agent in the environment, that mutant and its progeny will be most likely to survive, whereas the nonresistant organisms die. After a few generations most survivors will be resistant to the antimicrobial agent. Antibiotics do *not* induce mutations, but they can create

FIGURE 14.5 Method for detecting genetic resistance. (a) A mixed population of bacteria of varying resistance to a new antibiotic is present. (b) Antibiotic is added to the Petri dish. Only those organisms with sufficient resistance will survive. Introduction of the antibiotic represents a change in the environment, but it does *not* create the resistant organisms—they were there to begin with.

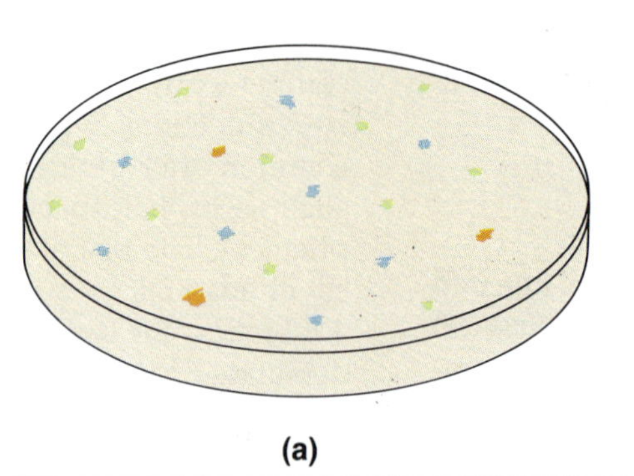

(a)

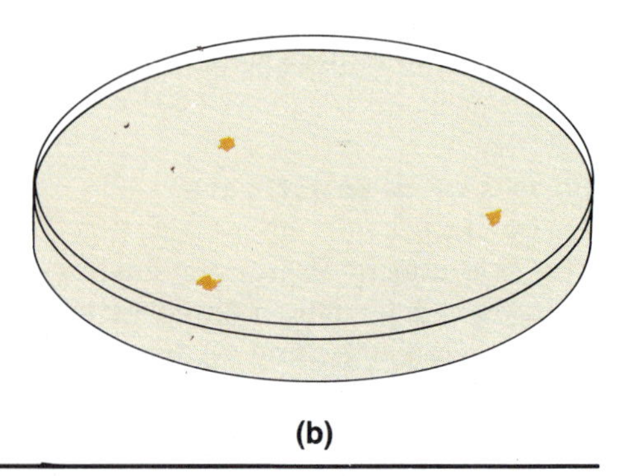

(b)

environments that favor the survival of mutant resistant organisms.

Genetic resistance in bacteria, where it is best understood, can be due to changes in the bacterial chromosome or to the acquisition of extrachromosomal DNA, usually in plasmids. (The mechanisms by which genetic changes occur were described in Chapters 7 and 8.) (p. 196) **Chromosomal resistance** is due to a mutation in chromosomal DNA and will usually be effective only against a single type of antibiotic. Such mutations often alter the DNA that directs the synthesis of ribosomal proteins. **Extrachromosomal resistance** is usually due to the presence of particular kinds of resistance plasmids, called **R factors.** How R factors originated is unknown, but they were first discovered in *Shigella* in Japan in 1959. Since that time many different R factors have been identified; some organisms carry as many as six or seven genes, each of which confers resistance to a different antibiotic. R factors can be transferred from one strain or species of bacteria to another. Most transfers occur by transduction (the transfer of plasmid DNA in a bacteriophage, discussed in Chapter 8), and some occur by conjugation. (pp. 199 and 202)

PUBLIC HEALTH

Antibiotic Resistance I: Sludge

In recent years, some cities have taken to dumping sludge from waste treatment plants on farmland as fertilizer. Although the practice is an economically attractive method of sludge disposal and also helps agricultural productivity, it has certain risks. The EPA requires food plants whose edible parts contact the ground not be grown in sludge-treated soil. Antibiotic-resistant bacteria in the sludge, such as *E. coli*, are released into the environment in large numbers and may later be found on vegetables harvested from the land. Although these organisms are not themselves pathogenic, they are likely to become incorporated into the natural microbial populations of the human gut. There they can serve as a reservoir of antibiotic resistance genes, which could be transferred to pathogens. Other risks include accumulation of heavy metals and toxic compounds in the soil.

Dumping sludge from waste treatment on farm fields in Switzerland.

Mechanisms of Resistance

Five mechanisms of resistance have been identified, each of which involves the alteration of a different microbial structure. One involves the alteration of receptors to which antimicrobial agents bind, a process that usually is caused by a mutation in the bacterial chromosome. The other mechanisms involve alterations in membrane permeability, enzymes, or metabolic pathways, which usually are caused by the acquisition of R factors. These mechanisms are explained below.

(1) Alteration of receptors usually affects bacterial ribosomes. The mutation alters the DNA that produces a ribosomal protein receptor so the antimicrobial agent cannot bind to it. Resistance to streptomycin, several aminoglycosides, and erythromycin has developed by this mechanism.

(2) Alteration of cell membrane permeability occurs when new genetic information changes the nature of proteins in the cell membrane. Such alterations change a membrane transport system or pores in the membrane so an antimicrobial agent can no longer

cross the membrane. Resistance to tetracyclines, polymyxins, and some aminoglycosides has occurred by this mechanism. The presence of a penicillin or cephalosporin can partially overcome resistance because it interferes with cell wall synthesis.

(3) Development of enzymes that can destroy or inactivate antimicrobial agents is a common kind of resistance. One enzyme of this type is beta-lactamase. Several beta-lactamases exist in various bacteria; they are capable of breaking the beta-lactam ring in penicillins and cephalosporins. Other similar enzymes that can destroy various aminoglycosides and chloramphenicol have been found in certain gram-negative bacteria.

(4) Alteration of an enzyme so that a formerly inhibited reaction can occur is exemplified by certain other sulfonamide-resistant bacteria. These organisms have developed an enzyme that has a very high affinity for PABA and a very low affinity for sulfonamide. Consequently, even in the presence of sulfonamide, the enzyme works well enough to allow the bacterium to function.

(5) Alteration of a metabolic pathway to bypass a reaction inhibited by an antimicrobial agent is exemplified by certain sulfonamide-resistant bacteria. These organisms have acquired the ability to use ready-made folic acid from their environment and no longer need to make it from PABA.

First-line, Second-line, and Third-line Drugs

As a strain of microorganism acquires resistance to a drug, another drug must be found to treat resistant infections effectively. If resistance to a second drug develops, a third drug is needed, and so on. Drugs used to treat gonorrhea illustrate this point. Before the 1930s no effective treatment was available for gonorrhea. Then sulfonamides were found to cure the disease. After a few years sulfonamide-resistant strains developed, but penicillin was soon available as a "second-line" drug. Over several decades penicillin-resistant strains developed but were combated with very large doses of penicillin. By the 1970s some strains of gonococci developed the ability to produce a beta-lactamase enzyme, which completely counteracted the effects of penicillin. "Third-line" spectinomycin is now being used. As spectinomycin-resistant strains start to appear, forcing physicians to resort to "fourth-line" drugs, we have to wonder whether the development of new drugs can go on indefinitely.

Drug-resistant organisms have been most frequently encountered within hospitals where seriously ill patients with lowered resistance to infections serve as convenient hosts. However, more and more resistant organisms are being isolated from infections among the general population, and the risk of acquiring a drug-resistant infection is increasing for everyone. Moreover, many organisms are resistant to multiple antibiotics. These are particularly difficult to treat. A new use for genetic probes will be to look for resistance genes in organisms, so that delays in effective treatment are avoided.

Cross-resistance

Cross-resistance is resistance against two or more similar antimicrobial agents because the same mechanisms of resistance are effective against both. The action of beta-lactamases provides a good example of cross-resistance. In many instances an enzyme that will break down one beta-lactam antibiotic also will break down several other beta-lactam antibiotics. The presence of such an enzyme would give a microorganism resistance to all the antibiotics it can break down.

Limiting Drug Resistance

Although, as we have seen, drug resistance is not induced by antibiotics, it is fostered by environments containing antibiotics. The progress of microbes in acquiring resistance can be thwarted in three ways. First, high levels of an antibiotic can be maintained in the bodies of patients long enough to kill all pathogens, including resistant mutants, or to inhibit them so that body defenses can kill them. This is the reason your doctor admonishes you to be sure to take all of an antibiotic prescription and not to stop taking it once you begin to feel better. The development of resistance when medication is discontinued before all pathogens are killed is illustrated in Figure 14.6.

Second, two antibiotics that do not have cross-resistance can be administered simultaneously so that each antibiotic can control the growth of mutants resistant to the other antibiotic. When rifampin and isoniazid are given together in the treatment of tuberculosis, each drug inhibits the growth of microorganisms resistant to the other drug. A variation on this principle is the use of one agent to destroy the resistance of microbes to another agent. When clavulanic acid and a penicillin called amoxicillin are given together, clavulanic acid binds tightly to beta-lactamases and prevents them from inactivating the amoxicillin.

Third, antibiotics can be restricted to essential uses only. Most physicians do not prescribe antibiotics for colds and other viral diseases that fail to respond to antibiotics, except for patients at high risk of secondary bacterial infections. Restrictions could be extended to hospitals, where microbes "just waiting to acquire resistance" lurk in antibiotic-filled environments. In addition, the use of antibiotics in animal feeds could be banned, as is discussed in the box titled "Drugs in Animal Feeds."

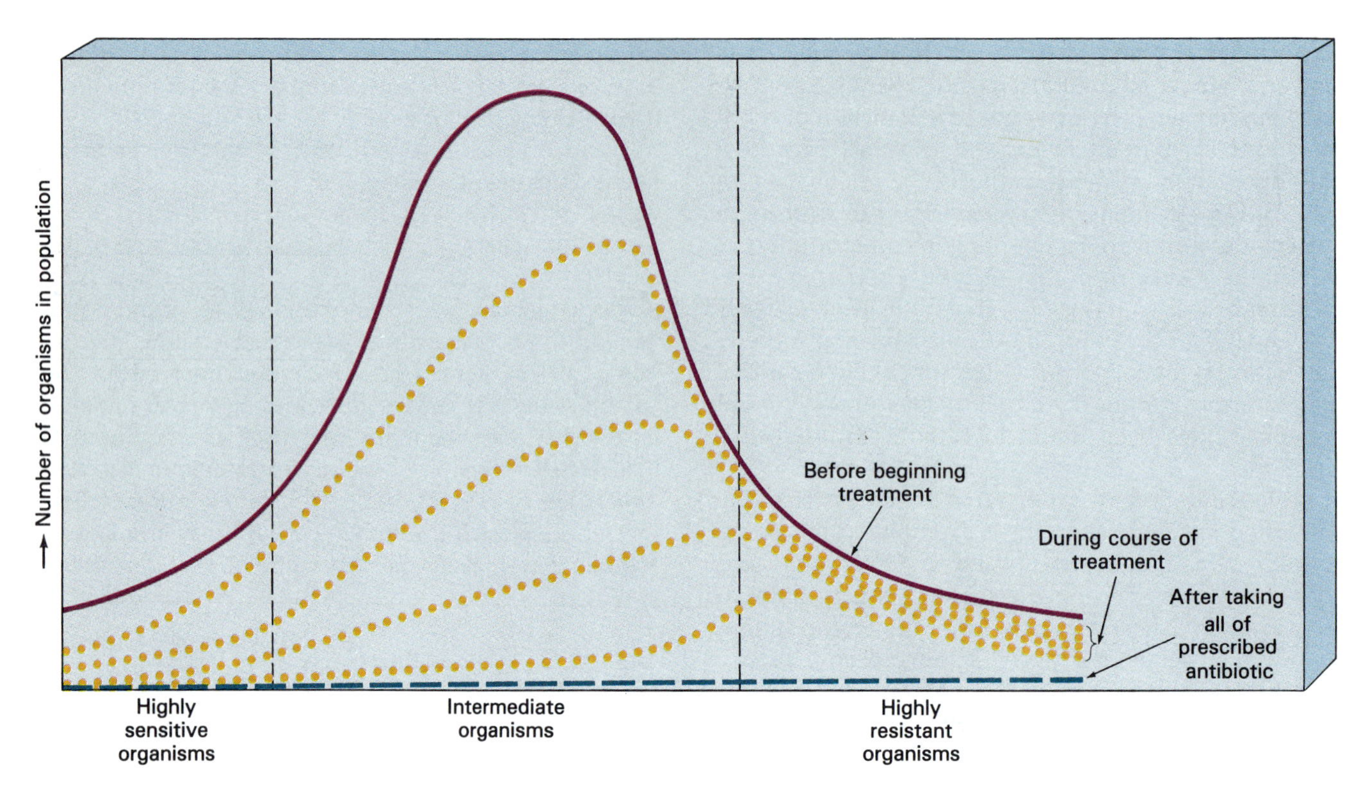

FIGURE 14.6 Why you should finish all the pills in the prescription bottle: This graph illustrates the proportions of a bacterial population showing resistance before, during, and after treatment with an antibiotic. Early in treatment all highly and moderately sensitive organisms are killed off. By the midpoint, a tremendous number of organisms are gone, and the patient feels much better—so much so that he may decide he doesn't need to take the rest of his pills. If he stops taking the antibiotic at this point, there will still be a smaller population, but it now will be made up largely of highly resistant organisms. These organisms will eventually multiply to form a large population again, and the patient will have a relapse. This time, however, many of the organisms will be resistant to the original antibiotic, and it will be useless to treat him or any other people to whom he transmits his infection.

PUBLIC HEALTH

Antibiotic Resistance II: Drugs in Animal Feeds

Soon after the discovery of chlortetracycline in 1948, scientists also discovered that wastes from its manufacture were useful as animal food supplements. Such wastes contained small quantities of the antibiotic. The antibiotic was found to serve as a powerful growth stimulant and therefore was purposely added to animal feeds. By the early 1950s antibiotics were being added to feeds all over the world. Today, penicillin, tetracycline, and many other antibiotics are added to animal feeds. Nearly half the antibiotics manufactured in the United States are used for other purposes than human medicines, and most go into animal feeds.

Unfortunately, the widespread use of antibiotics in animal feeds encouraged the growth of resistant organisms. In response, some European governments have taken steps to limit antibiotic use. For example, a rise in *Salmonella* infections in Britain led to the banning of antibiotics used to treat humans from being used in animal feeds. The proportion of *E. coli* strains resistant to antibiotics subsequently dropped from 31 percent to 18 percent. When the use of tetracycline in animal feeds was banned in the Netherlands, the incidence of tetracycline-resistant *Salmonella* strains dropped from 35 percent to 8 percent.

The danger of widespread use of small doses of an antibiotic is that it provides an ideal environment for the natural selection of mutant organisms resistant to the antibiotic. Microbiologists believe that such organisms are a threat to the health of both humans and farm animals, and the Food and Drug Administration agrees. So far, however, no action to control nonhuman consumption of antibiotics has been taken.

DETERMINATION OF MICROBIAL SENSITIVITIES TO ANTIMICROBIAL AGENTS

Microorganisms vary in their susceptibility to different chemotherapeutic agents, and their susceptibilities can change over time. Ideally, one should determine the appropriate antibiotic to treat any particular infection before any antibiotics are given. Sometimes an appropriate agent can be prescribed as soon as the causative organism is identified from a laboratory culture. Often tests are needed to show which antibiotic kills the organism. Several methods—disk diffusion, dilution, and automated methods—are available to do this.

Disk Diffusion Methods

In the **disk diffusion method,** or **Kirby-Bauer method,** a standard quantity of the causative organism is uniformly spread over an agar plate. Then several filter paper disks impregnated with specific concentrations of selected chemotherapeutic agents are placed on the agar surface (Figure 14.7). Usually only one member of a family of antibiotics—one kind of penicillin or one aminoglycoside—is used in each plate. Finally, the culture with the antibiotic disks is incubated.

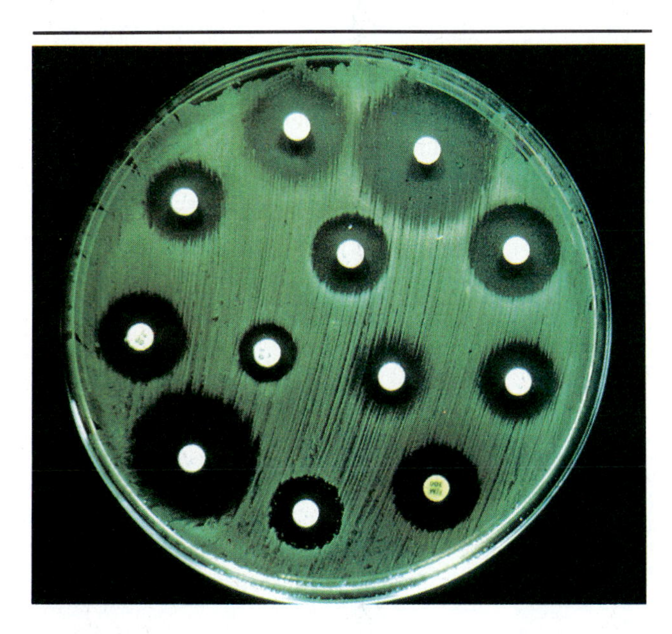

FIGURE 14.7 Disk diffusion (Kirby-Bauer) method of determining sensitivities of an organism to various antibiotics. Each paper disk contains a measured amount of a particular antibiotic that when placed firmly in contact with the agar medium will diffuse outward. Inhibition of the organism that has been prestreaked onto the plate produces clear areas of no growth. Failure to inhibit growth indicates the organism is resistant to that particular antibiotic. The largest zone of inhibition does not always indicate the most effective antibiotic to use, as some molecules diffuse more quickly out into the medium. Also, some drugs do not behave the same way in living organisms as they do on agar. However, this means of testing gives the physician an idea of which drugs are likely to be useful and which to avoid.

During incubation each chemotherapeutic agent diffuses out in all directions from the disk. Agents with lower molecular weights diffuse faster than those with higher molecular weights. Clear areas called **zones of inhibition** appear on the agar around disks where the agents inhibit the organism. The size of a zone of inhibition is not necessarily a measure of the degree of inhibition because of differences in the diffusion rates of chemotherapeutic agents. An agent of large molecular size might be a powerful inhibitor even though it might diffuse only a small distance and produce a small zone of inhibition. Standards for particular media and quantities of organisms have been established and must be used to interpret the inhibitory effects of various agents in a disk diffusion test.

Even when inhibition has been properly interpreted in a disk diffusion test, the most inhibitory chemotherapeutic agent may not cure an infection. The agent probably will inhibit the causative organism, but it may not kill sufficient numbers of the organism to control the infection. A bactericidal agent is often needed to eliminate an infectious organism, and the disk diffusion method does not assure that a bactericidal agent will be identified.

When the disk diffusion method is used to test antibiotic sensitivities, sometimes a clear area is observed between two antibiotic disks (Figure 14.8). In this area the concentrations of each antibiotic are lower (farther from the disk) than their inhibitory concentrations. Such an area indicates that the two antibiotics exert a **synergistic effect,** that is, together they exert an inhibitory effect that neither can achieve alone at low concentration. Moreover, results obtained *in vivo* (in a living organism) often differ from those obtained

FIGURE 14.8 Synergistic action of antibiotics as demonstrated in a disk diffusion test.

in vitro (in a laboratory vessel). Metabolic processes in the body of a living organism may inactivate or inhibit an antimicrobial compound.

Dilution Methods

The **dilution method** of testing antibiotic sensitivity was first performed in tubes of culture broth. In this method a constant quantity of inoculum is introduced into a series of broth cultures containing different known quantities of a chemotherapeutic agent. After incubation (for 16 to 20 hours) the tubes are examined, and the lowest concentration of the agent that prevents growth (turbidity) is noted. This concentration is the **minimum inhibitory concentration (MIC)** for a particular agent acting on a specific microorganism. This test can be done for several agents simultaneously by using several sets of tubes, but it is time-consuming and therefore expensive.

Finding an inhibitory agent by the dilution method does no more to prove that it will kill the infectious organism in the patient than finding one by the disk diffusion method. However, the dilution method allows a second test to distinguish between agents that kill microorganisms and those that merely inhibit their growth. Samples from tubes that show no growth but that might contain inhibited organisms can be used to inoculate broth that contains no chemotherapeutic agent. In this test the lowest concentration of the chemotherapeutic agent that yields no growth following this second inoculation, or subculturing, is the **minimum bactericidal concentration (MBC).** Thus a chemotherapeutic agent and a proper concentration to control an infection can be determined. That concentration should be maintained at the sites of infection because it is the minimum concentration that will cure the disease.

Serum Killing Power

Still another method of determining the effectiveness of a chemotherapeutic agent is to measure **serum killing power.** This test is performed by obtaining a sample of a patient's blood while the patient is receiving an antibiotic. A bacterial suspension is added to a known quantity of the patient's serum. **Serum** is the fluid portion of blood after cells and clotting factors have been removed. If growth (turbidity) occurs after incubation, the patient's serum lacks the proper antibiotic or a sufficient concentration of antibiotic to kill the bacteria. If all organisms are killed by antibiotic in the serum, a more quantitative determination can be made. Various dilutions of the patient's serum are inoculated, and the lowest concentration that prevents growth of the bacteria provides a measure of serum killing power. A lesser concentration will not eradicate the infectious organisms.

Automated Methods

Automated methods are available to identify pathogenic organisms and determine which antimicrobial agents will effectively combat them. One such method makes use of prepared trays with small wells into which a dispenser automatically puts a measured quantity of inoculum. As many as 36 organisms from different patients can be inoculated onto the same tray. Trays containing several kinds of media suitable for identifying members of different groups of organisms such as gram-positive bacteria, gram-negative bacteria, anaerobic bacteria, or yeasts are available. Trays also are available to determine sensitivity of organisms to a variety of antimicrobial agents.

The trays are inserted into a machine that measures microbial growth (Figure 14.9). Some machines

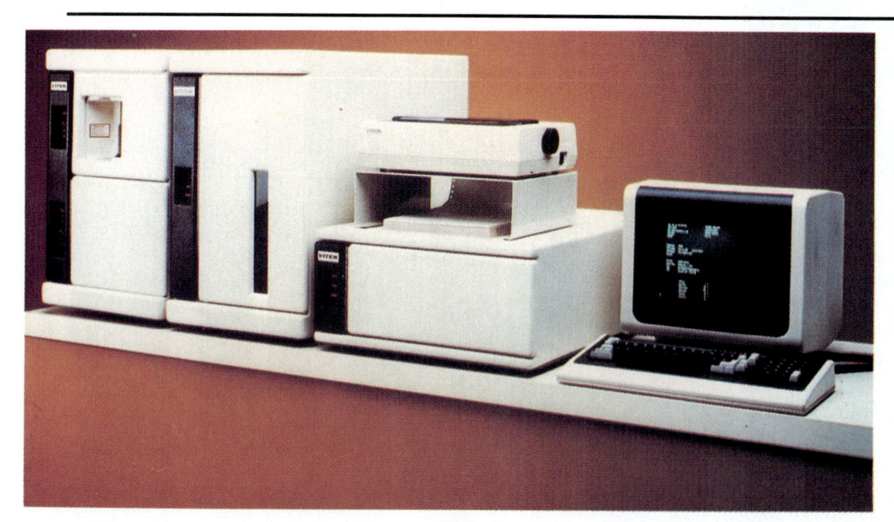

(a)

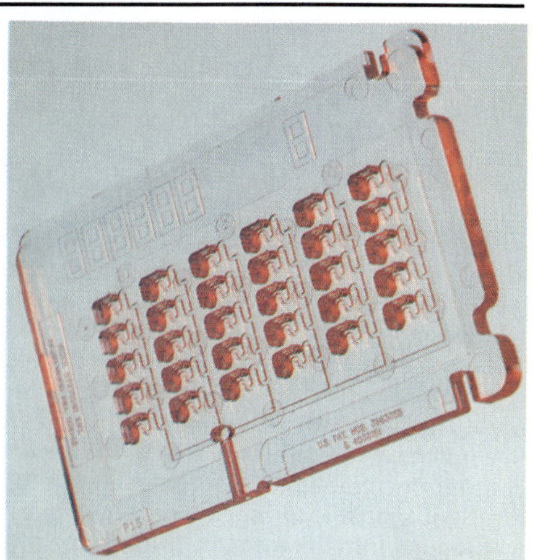

(b)

FIGURE 14.9 (a) An automated system for identifying microorganisms and determining their sensitivity to various antimicrobial agents. A sample containing the organism(s) is automatically inoculated into wells in a card (b), each containing a specific chemical reagent. Tests are carried out in an incubation chamber, and the results are read and recorded by computer. Cards for a wide variety of different tests are available.

measure growth with light and detect turbidity. Others use media containing radioactive carbon. Organisms growing on such media release radioactive carbon dioxide into the air, and a sampling device automatically detects it. Machines vary in their degree of automation and the speed with which results become available. Some require technicians to perform some steps; others provide a computerized printout of results that is relayed to the patient's chart. Some machines provide results in 3 to 6 hours, and most provide them overnight, except for slow-growing organisms, which require 48 hours.

Automated methods make laboratory identification of organisms and their sensitivities to antimicrobials more efficient and less expensive. Once the results of laboratory tests are available, the physician must then choose an appropriate drug based on the nature of the pathogen, the location of the infection, and other factors such as the patient's allergies. Automated methods allow physicians to prescribe an appropriate antibiotic early in an infection rather than prescribing a broad-spectrum antibiotic while waiting laboratory results.

ATTRIBUTES OF AN IDEAL ANTIMICROBIAL AGENT

Having considered various characteristics of antimicrobial agents and methods of determining microbial sensitivities to them, we can now summarize the characteristics of an ideal antimicrobial agent:

1. *Solubility in body fluids.* Agents must dissolve in body fluids to be transported in the body and reach the infectious organisms. Even agents used topically must dissolve in the fluids of injured tissue to be effective.
2. *Selective toxicity.* Agents must be more toxic to microorganisms than to host cells. Ideally, a great difference should exist between the low concentration toxic to microorganisms and the concentration that damages host cells.
3. *Toxicity not easily altered.* The agent should maintain a standard toxicity and not be made more or less toxic by interactions with foods, other drugs, or abnormal conditions such as diabetes and kidney disease in the host.
4. *Nonallergenic.* The agent should not elicit a host allergic reaction.
5. *Maintenance of constant, therapeutic concentration in blood and tissue fluids.* The agent should be sufficiently stable in body fluids to have therapeutic activity over many hours; it should be degraded and excreted slowly.
6. *Resistance by microorganisms not easily acquired.* There should be few, if any, microorganisms with resistance to the agent.
7. *Long shelf life.* The agent should retain its therapeutic properties over a long period of time with a minimum of special procedures such as refrigeration or shielding from light.
8. *Reasonable cost.* The agent should be affordable by patients who need it.

Many antimicrobial agents meet these criteria reasonably well. But few, if any, meet all the criteria for the ideal antimicrobial agents. As long as more ideal drugs might be found, the search for them will continue.

ANTIBACTERIAL AGENTS

Most antimicrobial agents are antibacterial agents, so we will start our "catalog" of antimicrobial agents with them, keeping in mind that some are effective against other microbes. Antibacterial agents can be categorized in several ways; we have chosen their modes of action. Another way of grouping antibiotics is by the microorganism that produces them (Table 14.2).

TABLE 14.2 Selected microbes that serve as sources of anitbiotics

Microbe	Antibiotics
Fungi	
Aspergillus fumigatus	Fumagillin
Cephalosporium species	Cephalosporins
Penicillium griseofulvum	Griseofulvin
Penicillium notatum and *P. chrysogenum*	Penicillin
Streptomycetes	
Streptomyces nodosus	Amphotericin B
Streptomyces venezuelae	Chloramphenicol
Streptomyces erythraeus	Erythromycin
Streptomyces griseus	Steptomycin
Streptomyces kanamyceticus	Kanamycin
Streptomyces fradiae	Neomycin
Streptomyces noursei	Nystatin
Streptomyces antibioticus	Vidarabine
Actinomycetes	
Micronomonospora	Gentamicin
Bacteria	
Bacillus subtilis	Bacitracin
Bacillus polymyxa	Polymyxins
Bacillus brevis	Tyrothricin

Inhibitors of Cell Wall Synthesis

Penicillins

Natural penicillins, such as *penicillin G* and *penicillin V*, are extracted from cultures of the mold *Penicillium notatum*. The discovery in the 1950s that certain strains of *Staphylococcus aureus* were resistant to penicillin provided the impetus to develop semisynthetic penicil-

TRY IT

Intermicrobial Warfare

You, too, can discover an antibiotic. Chances are that it will be missing several of the attributes of an ideal antimicrobial agent and thus would not be of commercial value. Still, it's fun to actually find something yourself instead of just reading about it in books.

Soil microbes are good sources of antibiotics. Pharmaceutical companies pay people to bring back soil samples from all over the world in hopes of finding new and useful antibiotic-producing microbes. Old cemeteries, wildlife refuges, swamps, and airfield landing strips have all been sampled. Try collecting soil samples from some places that interest you. Using aseptic technique, swab the surface of a Petri dish of nutrient agar with a broth culture of an organism such as *E. coli.* Make confluent strokes so that the entire surface is inoculated evenly, like a lawn. You could try different organisms on different plates. Then sprinkle tiny bits of your soil sample over the inoculated plate. Incubate or leave at room temperature, and examine daily. Are there clear areas around some of the colonies that have grown from soil particles? If so, these represent areas where natural antibiotics have diffused out into the agar and prevented the background "lawn" bacteria from growing. Congratulations, you've found some antibiotics!

Next, make stained slides of some of the antibiotic-producing colonies. Are they bacteria or fungi? Fungi may be especially abundant in acid soils. If circumstances permit, subculture one or more of your antibiotic-producing colonies and see if it inhibits all types of microbes equally. Does the antibiotic have a broad spectrum or a narrow spectrum? If you gave your culture to a pharmaceutical company, which further tests would they need to perform to determine whether it has potential commercial value?

lins. The first of these was *methicillin,* which is effective against penicillin-resistant organisms because it is not broken down by beta-lactamase enzymes. Other semisynthetic penicillins emerged in rapid succession. They include *nafcillin, oxacillin, ampicillin, amoxicillin, carbenicillin,* and *ticarcillin.* Each is synthesized by adding a particular side chain to a penicillin nucleus (Figure 14.10).

Penicillin G, the most frequently used natural penicillin, can be administered orally, into muscles, or into veins. When administered orally, some of it is broken down by stomach acids. Penicillin is rapidly absorbed into the blood, reaches its maximum concentration, and is excreted unless it is combined with an agent such as procaine, which slows excretion and prolongs activity.

Penicillin G is the drug of choice in treating infections caused by streptococci, meningococci, pneumococci, spirochetes, clostridia, and aerobic gram-positive rods. It is also suitable for treating infections caused by a few strains of staphylococci and gonococci that are not resistant to it. Because it retains activity in urine, it is suitable for treating some urinary tract infections. Infections caused by organisms resistant to penicillin G can be treated with semisynthetics such as nafcillin, oxacillin, ampicillin, or amoxicillin. Carbenicillin and ticarcillin are especially useful in treating *Pseudomonas* infections. Allergy to penicillin is rare among children but occurs in 1 to 5 percent of adults. Penicillins are generally nontoxic, but large doses can have toxic effects on the kidneys, liver, and central nervous system.

In addition to their use as treatment for infections, penicillins also are used prophylactically, that is, to prevent infection. For example, patients with heart defects or heart disease are especially susceptible to endocarditis, an inflammation of the lining of the heart, caused by a bacterial infection. They often receive penicillin before surgery or dental procedures.

Cephalosporins

Natural cephalosporins, derived from several species of the fungus *Cephalosporium,* have limited antimicrobial action. Their discovery led to the development of a large number of semisynthetic derivatives of natural cephalosporin C. The nucleus of a cephalosporin is quite similar to that of penicillin; both contain beta-lactam rings (Figure 14.10). Semisynthetic cephalosporins, like semisynthetic penicillins, differ in the nature of their side chains. Frequently used cephalosporins include *cephalexin* (Keflex), *cephradine,* and *cefadroxil,* all of which can be administered orally as they are fairly well absorbed from the gut. Other cephalosporins, such as *cephalothin* (Keflin), *cephapirin,* and *cefazolin* must be administered parenterally—by some means other than via the gut, such as intramuscularly or intravenously.

Though cephalosporins usually are not the first drug considered in the treatment of an infection, they are frequently used when allergy or toxicity prevents the use of other drugs. They account for one-fourth to one-third of pharmacy expenditures in American hospitals, mainly because they have a fairly wide spectrum of activity, rarely cause serious side effects, and can be used prophylactically in surgical patients. Unfortunately, they are often used when a less expensive and narrower spectrum agent would be just as effective.

The development of new varieties of cephalosporins seems to be a race against the ability of bacteria to acquire resistance to older varieties. When organisms became resistant to early "first-generation" cephalosporins, new "second-generation" cephalosporins were produced. Now "third-generation" cephalospo-

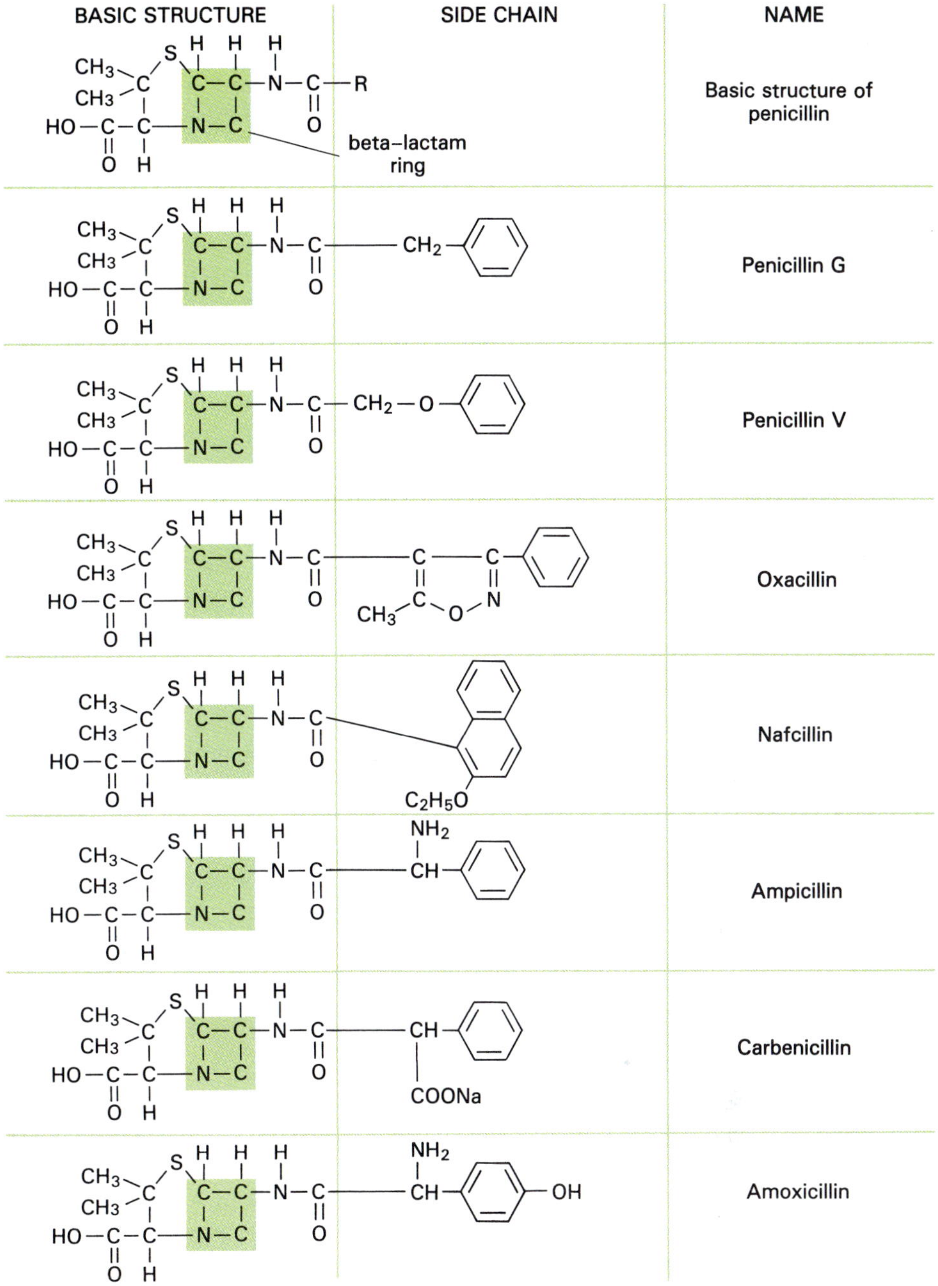

BASIC STRUCTURE	SIDE CHAIN	NAME
R	beta-lactam ring	Basic structure of penicillin
	CH_2	Penicillin G
	CH_2-O	Penicillin V
	C C, CH_3 C O N	Oxacillin
	C_2H_5O	Nafcillin
	NH_2, CH	Ampicillin
	CH, COONa	Carbenicillin
	NH_2, CH, OH	Amoxicillin

FIGURE 14.10 Examples of beta-lactam antibiotics showing the various side chains of penicillin derivatives.

rins, such as *cefotaxime* and *moxalactam,* are used against organisms resistant to older drugs. These drugs are especially effective (for now) in dealing with hospital-acquired infections resistant to many antibiotics. They also are being tried in patients with AIDS and other immunodeficiencies. Do not confuse these with second- and third-line drugs, which were described earlier and are not derivatives of one another.

Adverse effects from cephalosporins are usually local reactions—irritation at the injection site, inflammation of a vein when the drug is administered intravenously, and nausea, vomiting, and diarrhea when the drug is administered orally. Also, 4 to 15 percent of patients allergic to penicillin also are allergic to cephalosporins. Finally, newer cephalosporins have little effect on gram-positive organisms such as staphylococci and enterococci, which can cause superinfections during the treatment of gram-negative infections.

Carbapenems

The carbapenems are a new group of antibiotics with two-part structures. The beta-lactam half of their structure interferes with cell wall synthesis. In Primaxin, a typical carbapenem, the other half of the structure is cilastatin sodium, a compound that prevents degradation of the drug in the kidneys. As a group, the carbapenems have an extremely broad spectrum of activity. Primaxin, for example, is effective against 98 percent of all organisms isolated from patients in hospitals.

Other Antibacterial Agents That Act on Cell Walls

Bacitracin is a small polypeptide derived from the bacterium *Bacillus subtilis.* Because it is poorly absorbed and toxic to the kidneys, it is used only on skin and mucous membrane lesions and wounds. *Vancomycin* is a large, complex molecule produced by the soil actinomycete *Streptomyces orientalis.* It can be used to treat infections caused by staphylococci and enterococci, but given orally, it is effective only against gastrointestinal infections because it is poorly absorbed. Given intravenously it is effective in treating systemic infections, especially those resistant to penicillins and cephalosporins. Vancomycin can cause hearing loss, especially in older patients, and allows superinfections with gram-negative bacteria and fungi if use is prolonged.

Disrupters of Cell Membranes

Polymyxins

Five polymyxins, designated A, B, C, D, and E, have been obtained from the soil bacterium *Bacillus polymyxa.* They are usually applied topically, often with bacitracin, to treat skin infections caused by gram-negative bacteria such as *Pseudomonas.* Used internally, polymyxins can cause numbness in the extremities, serious kidney damage, and respiratory arrest. They are administered by injection when the patient is hospitalized and kidney function can be monitored.

Tyrocidins

Tyrocidins are cyclic polypeptides obtained from *Bacillus brevis.* The first to be discovered was *tyrothricin,* and *tyrocidine* and *gramicidin* (named for Hans Gram, the originator of the Gram stain) have been derived from tyrothricin. All are highly toxic, but they can be used topically to prevent bacterial growth, especially that of gram-positive cocci.

Inhibitors of Protein Synthesis

Aminoglycosides

Aminoglycosides are obtained from various species of the genus *Streptomyces.* The first, *streptomycin,* was discovered in the 1940s and was effective against a variety of bacteria. Since then many bacteria have become resistant to it. Also, it can damage kidneys and the inner ear, sometimes causing permanent ringing in the ears and dizziness. Consequently, streptomycin is now used only in special situations and generally in combination with other drugs. For example, it can be used with tetracyclines to treat plague and tularemia and with isoniazid and rifampin to treat tuberculosis. Other aminoglycosides such as *neomycin, kanamycin, amikacin, gentamicin, tobramycin,* and *netilmicin* also have special uses, and they display varying degrees of toxicity to the kidneys and inner ear. Aminoglycosides usually are administered intramuscularly or intravenously because they are poorly absorbed when given orally.

An important property of aminoglycosides is their ability to act synergistically with other drugs—an aminoglycoside and another drug together often control an infection better than either could do alone. For example, gentamicin and penicillin or ampicillin are effective against penicillin-resistant streptococci. In other synergistic actions, gentamicin or tobramycin work with carbenicillin or ticarcillin to control *Pseudomonas* infections, especially in burn patients, and aminoglycosides work with cephalosporins to control *Klebsiella* infections.

Other applications of aminoglycosides include the treatment of bone and joint infections, peritonitis (inflammation of the lining of the abdominal cavity), pelvic abscesses, and many hospital-acquired infections. In bone and joint infections gentamicin and tobramycin are especially useful because they can penetrate joint cavities. Because peritonitis and pelvic abscesses are severe and are often caused by a mixture of enterococci and anaerobic bacteria, aminoglycoside treatment usually is started before the organisms are identified. Amikacin is especially effective in treating hospital-acquired infections resistant to other drugs. It should not be used in less demanding situations lest organisms also become resistant to it.

Aminoglycosides often damage kidney cells, causing protein to be excreted in the urine, and prolonged use can kill kidney cells. These effects are most pronounced in older patients and those with preexisting kidney disease. Some aminoglycosides damage the eighth cranial nerve: Streptomycin causes dizziness and disturbances in balance, and neomycin causes hearing loss.

Tetracyclines

Several tetracyclines are obtained from species of *Streptomyces,* where they were originally discovered. Commonly used tetracyclines include *tetracycline* itself, *chlortetracycline* (Aureomycin), and *oxytetracycline* (Terramycin). Newer semisynthetic tetracyclines include *minocycline* (Minocin) and *doxycycline* (Vibramycin). All are readily absorbed from the digestive tract and become widely distributed in tissues and body fluids except cerebrospinal fluid.

The fact that tetracyclines have the widest spectrum of activity of any antibiotics is a two-edged sword. They are effective against many gram-positive and gram-negative bacterial infections and are suitable for treating rickettsial, chlamydial, mycoplasma, and some fungal infections. But because they have such a wide spectrum of activity, they destroy the normal intestinal flora and often produce severe gastrointestinal disorders. Recalcitrant superinfections of tetracycline-resistant *Proteus*, *Pseudomonas*, and *Staphylococcus*, as well as yeast infections, also can result.

Tetracyclines can cause a variety of mild to severe toxic effects. Nausea and diarrhea are common, and extreme sensitivity to light is sometimes seen. Effects on the liver and kidneys are more serious. Liver damage can be fatal, especially in patients with severe infections or during pregnancy. Kidney damage can lead to excretion of protein and glucose and acidosis (low blood pH). Anemia can occur, but it is rare.

Staining of the teeth (Figure 14.11) occurs when children under 5 years of age receive tetracycline or when their mothers received it during the last half of their pregnancy. Both deciduous teeth (baby teeth) and permanent teeth will be mottled because the buds of both permanent and deciduous teeth form before birth. Tetracycline taken during pregnancy also can lead to abnormal bone formation in the fetal skull and a permanent abnormal skull shape. The ability of calcium ions to form a complex with tetracycline is responsible for its effects on bones and teeth. Because this reaction destroys the antibiotic effect of the drug, patients should not consume milk or other dairy products with the drug or for a few hours after taking it.

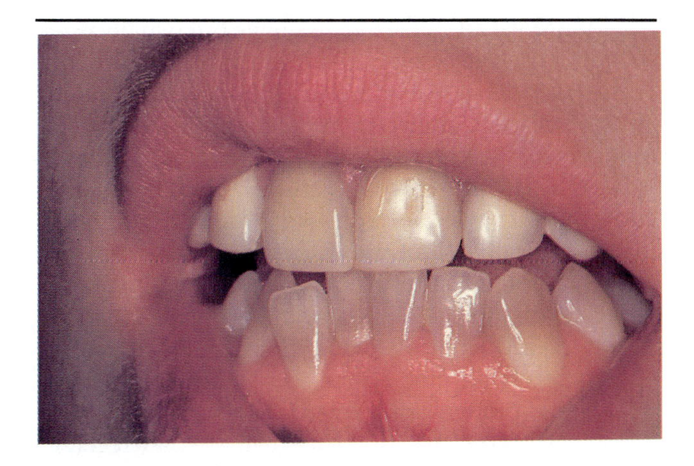

FIGURE 14.11 Staining of teeth caused by tetracycline. If the condition results from ingestion of the antibiotic during pregnancy, both the deciduous (baby) and permanent teeth will be affected, as both sets of tooth buds are forming in the fetus at that time.

Chloramphenicol

Chloramphenicol, originally obtained from cultures of *Streptomyces venezuelae,* is now fully synthesized in the laboratory. Like tetracyclines it is rapidly absorbed from the digestive tract, is widely distributed in tissues, and has a broad spectrum of activity. It is used to treat typhoid fever, infections due to penicillin-resistant strains of meningococci and *Haemophilus influenzae,* brain abscesses, and severe rickettsial infections.

Chloramphenicol damages bone marrow in two ways. It causes a dose-related, reversible anemia in which bone marrow cells produce too few erythrocytes and sometimes too few leukocytes and platelets as well. Terminating use of the drug usually allows the bone marrow to recover normal function. It also causes a non-dose-related, permanent *aplastic anemia* due to destruction of bone marrow. Aplastic anemia appears days to months after treatment is discontinued and is most common in newborn infants. Unless a successful bone marrow transplant can be performed, aplastic anemia usually is fatal. It is seen in only one in 25,000 to 40,000 patients treated with chloramphenicol. Long-term use of chloramphenicol can cause inflammation of the optic and other nerves, mental confusion, delirium, and mild to severe gastrointestinal symptoms.

Macrolides

Erythromycin, the most commonly used macrolide (large-ring) antibiotic, is produced by several strains of *Streptomyces erythreus.* Erythromycin is readily absorbed and reaches most tissues and body fluids except cerebrospinal fluid. It is recommended for infections caused by streptococci, pneumococci, and corynebacteria but also is effective against *Mycoplasma* and some *Chlamydia* and *Campylobacter* infections. Erythromycin is most valuable in treating infections caused by penicillin-resistant organisms or in patients allergic to penicillin. Unfortunately, resistance to erythromycin often emerges during treatment. Dual antibiotic treatment—erythromycin and some other drug—is often used on patients with a pneumonialike disease that might be Legionnaire's disease. Several antibiotics combat other pneumonias, but erythromycin is the only common antibiotic that will combat Legionnaire's disease. Erythromycin is one of the least toxic of commonly used antibiotics. Mild gastrointestinal disturbances are seen in 2 to 3 percent of patients receiving it.

Lincomycin and Clindamycin

Lincomycin is produced by *Streptomyces lincolnensis,* and *clindamycin* is a semisynthetic derivative that is more completely absorbed and less toxic than lincomycin. Lincomycin can be used to treat a variety of infections but is not significantly better than other widely used antibiotics, and organisms become resistant to it quickly. Clindamycin is effective against

PUBLIC HEALTH

Antibiotic Resistance III: Antibiotics and Acne

Low doses of tetracyclines and erythromycin suppress skin bacteria, mostly *Propionibacterium acnes*, and reduce the release of microbial lipases, which contribute to skin inflammation. This therapy is used to treat acne, but its effectiveness has not been proven. Studies intended to assess antibiotic effectiveness in acne therapy have been useless because they lack suitable controls, fail to characterize adequately the type and severity of cases, and make use of other concurrent therapies.

Some studies have shown that low doses of many antibiotics can lead to the appearance of antibiotic-resistant strains. One might ask if the benefits of antibiotics to acne patients are worth the risk of promoting development of resistant organisms.

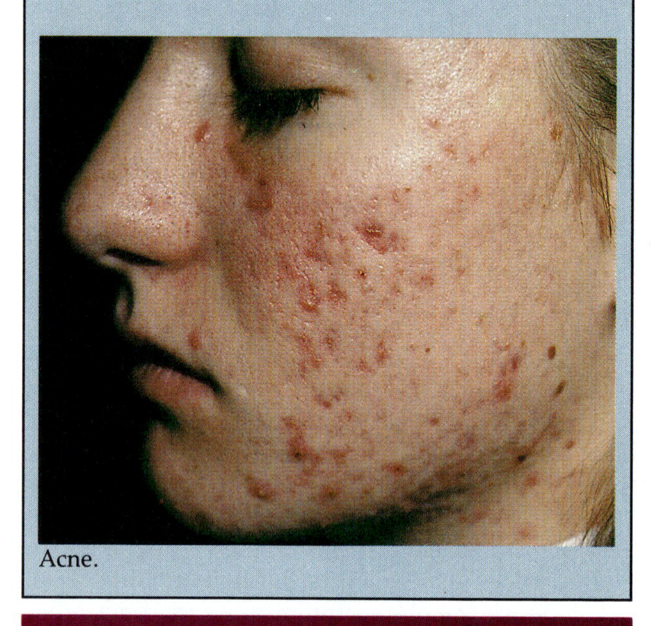

Acne.

Bacteroides and other anaerobes, except *Clostridium difficile,* which often becomes established as a superinfection during clindamycin therapy. Toxins from *C. difficile* can cause a severe, and sometimes fatal, colitis (inflammation of the large intestine) unless diagnosed early and treated with oral vancomycin.

Inhibitors of Nucleic Acid Synthesis

Sulfonamides

The sulfonamides are a large group of entirely synthetic antibacterial agents. Many are derived from *sulfanilamide,* one of the first sulfonamides, or *sulfa drugs.* In general, orally administered sulfonamides are readily absorbed and become widely distributed in tissues and body fluids. They act by blocking the synthesis of folic acid, which is needed to make the nitrogenous bases of DNA. Sulfonamides have now been largely replaced by antibiotics because antibiotics are more specific in their actions and less toxic than sulfonamides. When sulfonamides first came into use in the 1930s, they frequently led to kidney damage. Newer forms of these drugs usually do not damage kidneys, but they do occasionally produce nausea and skin rashes. Certain sulfonamides are still used to suppress intestinal flora prior to colon surgery. They also are used to treat some kinds of meningitis because they enter cerebrospinal fluid more easily than do antibiotics. *Cotrimoxazole* (Septra), a combination of *sulfamethoxazole* and *trimethoprim,* is used to treat urinary tract infections and a few other infections. Unfortunately, both drugs are toxic to bone marrow. Trimethoprine also occasionally causes mild nausea and skin rashes.

Rifampin

From among the *rifamycins* produced by *Streptomyces mediterranei,* only the semisynthetic *rifampin* is currently used. Easily absorbed from the digestive tract except when taken directly after a meal, it reaches all tissues and body fluids, including cerebrospinal fluid. Although it has a wide spectrum of activity, it is approved in the United States only for treating tuberculosis and eliminating meningococci from the nasopharynx of carriers.

Rifampin can cause liver damage but usually does so only when excessive doses are given to patients with preexisting liver disease. It is unusual among antibiotics in its ability to interact with other drugs, and possibilities of such interactions should be considered before the drug is given. Taking rifampin concurrently with oral contraceptives has been implicated in an increased risk of pregnancy and menstrual disorders. Dosages of anticoagulants must be increased while a patient is taking rifampin to achieve the same degree of reduction in blood clotting. Finally, drug addicts receiving methadone sometimes suffer withdrawal symptoms if they are given rifampin without an increase in the quantity of methadone they receive. One explanation for these diverse effects is that rifampin stimulates the liver to produce greater quantities of enzymes that are involved in the metabolism of a variety of drugs.

Quinolones

Quinolones, a new group of synthetic analogues of nalidixic acid, are effective against many gram-positive and gram-negative bacteria. Their mode of action is to inhibit bacterial DNA synthesis by blocking DNA

APPLICATIONS

Red Man Syndrome

Rifampin has been shown to cause the so-called red man syndrome. In this disorder, which occurs with high doses of the antibiotic, colored metabolic products of the drug accumulate in the body and are eliminated through sweat glands. It is characterized by bright orange or red urine, saliva, and tears and skin that looks like a boiled lobster. The red skin secretions can be washed away, but liver damage caused by the drug is only slowly repaired.

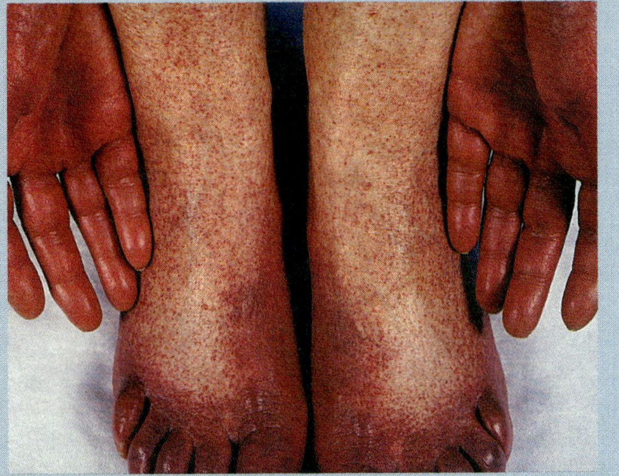

Red man syndrome due to rifampin. The red secretions can be washed away, but they would certainly frighten an unwarned patient.

gyrase, the enzyme that unwinds the DNA double helix preparatory to its replication. *Norfloxacin, ciprofloxacin,* and *enoxacin* are examples of this group of antibiotics. They are especially effective in the treatment of traveler's diarrhea and in urinary tract infections caused by multiply-resistant organisms.

A recent advance has produced a hybrid class of antibiotics. One of these, a quinolone-cephalosporin combination, is currently being tested. When the beta-lactamase enzymes act on the cephalosporin component, the quinolone is released from the hybrid molecule and is available to kill the cephalosporin-resistant organisms. The use of such a dual antibiotic may also prevent or delay development of antibiotic resistance in organisms.

Other Antibacterial Agents

Isoniazid

Isoniazid is an antimetabolite for two vitamins, nicotinamide and pyridoxal (Figure 14.12). It binds to and inactivates the enzyme that converts the vitamins to useful molecules. This synthetic agent, which has little effect on most bacteria, is effective against the mycobacterium that causes tuberculosis. It is completely absorbed from the digestive tract and reaches all tissues and body fluids. Because the mycobacteria present in any such infection usually include some isoniazid-resistant organisms, isoniazid usually is given with another agent such as rifampin or ethambutol (see the following). Dietary supplements of nicotinamide and pyridoxal also should be given with isoniazid.

(a) (b) (c)

FIGURE 14.12 (a) Isoniazid, an antimetabolite for the two vitamins (b) nicotinamide and (c) pyridoxal.

Ethambutol

The synthetic agent *ethambutol* is effective against certain strains of mycobacteria that do not respond to isoniazid. It is well absorbed and reaches all tissues and body fluids. However, mycobacteria acquire resistance to ethambutol fairly rapidly, so it is used with other drugs such as isoniazid and rifampin.

Nitrofurans

Nitrofurans enter susceptible cells and cause the production of highly reactive free radicals, which damage DNA and sensitive enzymes. Several hundred nitrofurans have been synthesized since the first one was made in 1930. Only a few of these are currently used. Oral doses of *nitrofurantoin* (Furdantin) are easily absorbed and quickly metabolized. This drug is especially useful in treating acute and chronic urinary infections. The low incidence of resistance to it makes it an ideal prophylactic agent to prevent recurrences. Unfortunately, 10 percent of patients experience nausea and vomiting as a side effect and must then be treated with an antibiotic instead. Another nitrofuran, *nifuratel,* has been shown to have the same clinical advantages as nitrofurantoin with a much lower incidence of gastrointestinal side effects. Unfortunately, this cheap, effective drug is not currently available.

The chemical structures, uses, and side effects of antibacterial agents are summarized in Figure 14.13.

Agent	Used to Treat	Common Method of Administration*	Side Effects
Agents that inhibit cell wall synthesis			
Penicillin (natural)	Wide variety of infections mostly from gram-positive bacteria	IM,O	Relatively few side effects, but allergies do occur
Penicillin (semisynthetic)	Infections resistant to natural penicillin	O,IV	Same as natural penicillin
Cephalosporins	Wide variety of infections when allergy or toxicity make other agents unsuitable	IV,IM,O	Relatively nontoxic, but can lead to superinfections
Carbapenems	Mixed infections, nosocomial infections, infections of unknown etiology.	IV	Allergic reactions, superinfections, seizures, gastrointestinal disturbances
Vancomycin	Staphylococcal and enterococcal infections resistant to penicillin and cephalosporins	IV	Can cause hearing loss and lead to superinfections
Bacitracin	Skin infections by topical application	T	Internal use toxic to kidneys

Imipenem
(a carbapenem)

Bacitracin

Agent	Used to Treat	Common Method of Administration*	Side Effects
Agents that interfere with cell membrane function			
Polymyxins	Skin infections by topical application, with bacitracin	T,IV	Internal use highly toxic
Tyrocidins	Topically for skin infections caused by gram-positive cocci	T,IV	Internal use highly toxic

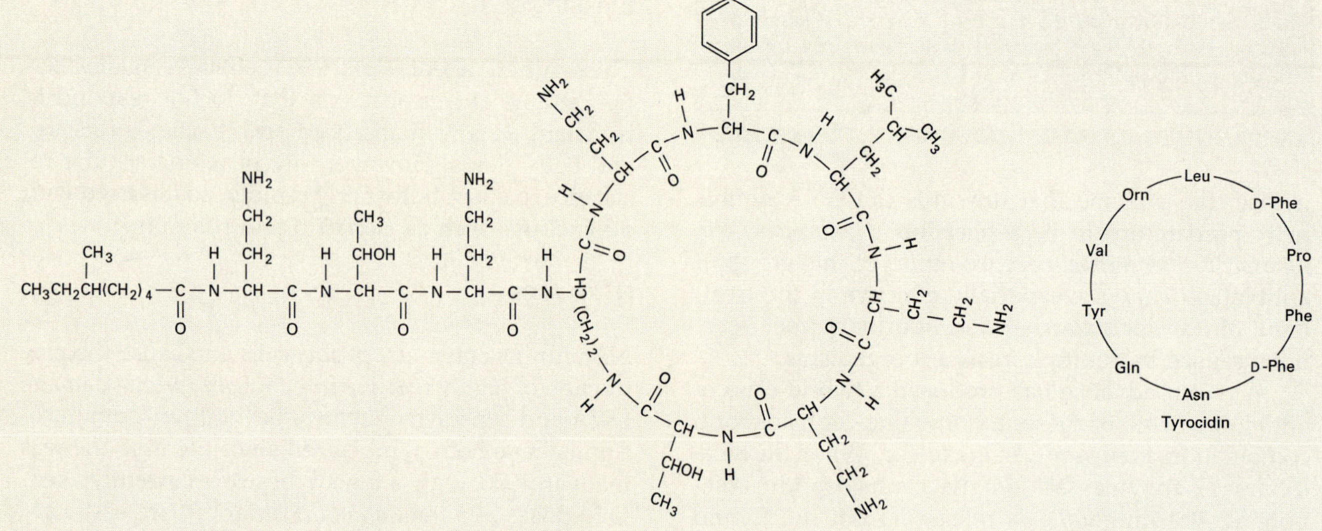

Polymixin B

Tyrocidin

Agent	Used to Treat	Common Method of Administration*	Side Effects
Other Antibacterial Agents			
Isoniazid	Tuberculosis, with ethambutol	O	May cause pyridoxine deficiency
Ethambutol	Tuberculosis, with isoniazid	O	
Nitrofurantoin	Urinary tract infections	O	Nausea and vomiting

Ethambutol

Nitrofurantoin

FIGURE 14.13 Antibacterial drugs.

* IM = intramuscular O = oral
IV = intravenous T = topical

Agent	Used to Treat	Common Method of Administration*	Side Effects
Agents that inhibit protein synthesis			
Streptomycin	Tuberculosis, used with isoniazid and rifampin	IM,O	Damages kidneys and inner ear
Gentamicin and other aminoglycosides	Antibiotic-resistant and hospital-acquired infections, used synergistically with other drugs	IM, T (burns)	Show varying degrees of kidney and inner ear damage
Tetracyclines	A broad spectrum of bacterial infections and some fungal infections	O	Stain teeth; cause gastrointestinal symptoms; can lead to superinfections
Chloramphenicol	A broad spectrum of bacterial infections, brain abscesses and penicillin-resistant infections	O	Can damage bone marrow and cause aplastic anemia
Erythromycin	Gram-positive bacterial infections, some penicillin-resistant infections and Legionnaire's disease	O	One of the least toxic of commonly used antibiotics

Erythromycin

Gentamicin

Tetracycline

Chloramphenicol

Streptomycin

Agent	Used to Treat	Common Method of Administration*	Side Effects
Agents that inhibit nucleic acid synthesis			
Sulfonamides	Some kinds of meningitis and to suppress intestinal flora before colon surgery	O,IV	Early forms caused kidney damage, but ones in use do not
Rifampin	Tuberculosis and to eliminate meningococci from the nasopharynx	O	Bright orange or red urine, saliva, tears, and skin; liver damage; many disorders when used with other agents
Quinolones	Urinary tract infections, traveller's diarrhea; effective against many resistant organisms	O	Nausea; headaches and other nervous system disturbances

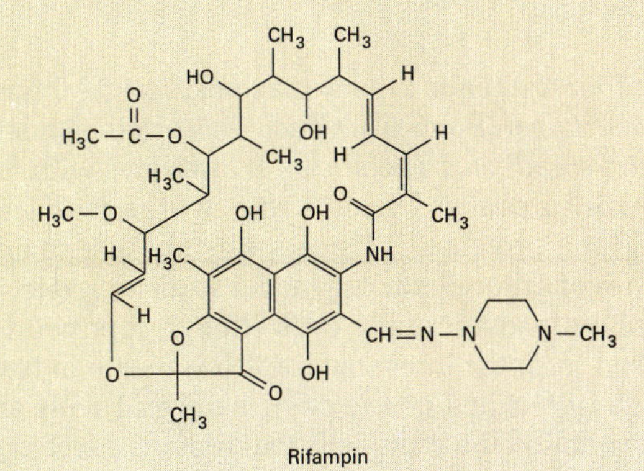

Rifampin

Fleroxacin (a quinolone)

ANTIFUNGAL AGENTS

Antifungal agents are antimicrobial agents used to treat infections caused by fungi. In addition to the particular agents discussed here, the sulfonamides also are effective in treating some fungal infections, as is the simple inorganic salt potassium iodide (KI). Some very promising antifungal drugs, such as hamycin, are under investigation and are not yet commercially available.

Imidazoles

The *imidazoles* are a group of synthetic chemicals developed in the past two decades. Two of them, *clotrimazole* and *miconazole*, are presently available. These agents cause a thickening of the walls of fungal cells and interfere with the uptake of purines and other essential nutrients. The fungi are unable to synthesize adequate quantities of nucleic acids and therefore fail to grow. Imidazoles are most effective when topically applied to treat fungal skin infections and *Candida* yeast infections of the skin, nails, mouth, and vagina. Miconazole has been given intravenously to treat systemic fungal infections, but it is recommended only in instances where other antifungal agents have not been effective.

PUBLIC HEALTH

Nonprescription Vaginal Yeast Infection Remedy

Three out of four women suffer at least one vaginal yeast (*Candida albicans*) infection during their lifetime. Some have several recurrences per year. The overgrowth of this naturally occurring fungus occurs when the natural balance with bacteria is upset due to taking antibiotics or birth control pills, having diabetes, or a number of other factors. Symptoms include a white to yellow vaginal discharge plus vaginal itch and discomfort. Fortunately, women will now be able to get faster, cheaper treatment for this very common condition. Clotrimazole is now available over the counter, without prescription, under the trade name Gyne-Lotrimin. A similar drug, miconazole, available as Monistat, also is effective.

However, some people worry about making antibiotics available on a nonprescription basis. One reason is that symptoms of some sexually transmitted diseases resemble those of a vaginal yeast infection. Women unfamiliar with yeast infection symptoms may mistakenly try to treat other diseases and thus delay proper treatment. Others may overtreat themselves when it is not necessary. But for the majority of women, being able to skip the wait for and cost of an office visit will be most welcome.

Clotrimazole and miconazole often cause local skin irritation when applied topically. When miconazole is given intravenously, one-fifth of patients suffer from itching so severe that the drug must be discontinued. Some patients have nausea, fever, and chills, and a few suffer from thrombophlebitis—inflammation of the veins accompanied by the formation of blood clots.

Amphotericin B

The antibiotic *amphotericin B* (Fungizone), derived from *Streptomyces nodosus*, binds to cell membrane ergosterol in some fungi, algae, and protozoa. It increases membrane permeability so that glucose, potassium, and other essential substances leak from the cell. The drug is poorly absorbed from the digestive tract and so is given intravenously. Even then, only 10 percent of the dose given is found in the blood. Excretion persists for up to 3 weeks after treatment is discontinued, but it is not known where the drug is sequestered in the meantime. Amphotericin B is the drug of choice in treating most systemic fungal infections. Although fungi are not known to develop resistance to this agent, side effects are numerous and sometimes severe. They include abnormal skin sensations, fever and chills, nausea and vomiting, headache, depression, kidney damage, anemia, abnormal heart rhythms, and even blindness. Because some of the fungal infections are fatal without treatment, patients have little choice but to risk these unfortunate side effects.

Nystatin

An antibiotic closely related to amphotericin B, *nystatin* (Mycostatin) has the same mode of action. It is quite effective in the treatment of *Candida* yeast infections. The drug also can be given orally to treat fungal superinfections in the intestine, which often occur after long-term treatment with antibiotics. Nystatin was named for the New York State Health Department, where it was discovered.

Griseofulvin

Another antibiotic effective against fungal infections is *griseofulvin* (Fulvicin), which was originally derived from *Penicillium griseofulvum*. It interferes with fungal growth, probably by impairing synthesis of nucleic acids and proteins. It also appears to cause the membranes of intracellular organelles to disintegrate. Griseofulvin is given orally even though it is poorly absorbed from the intestinal tract. It is useful in treating fungal infections of the skin, hair, and nails and is incorporated into new cells that replace infected cells. Reactions to griseofulvin are usually limited to mild headaches but can include peripheral neuritis (inflammation of nerves) and gastrointestinal disturbances. Skin infections usually are cured within 4 weeks, but recalcitrant infections associated with fingernails and toenails may persist even after a year of treatment.

Flucytosine

The fluorinated pyrimidine *flucytosine* is transformed in the body to fluorouracil, an analog of uracil, which interferes with nucleic acid synthesis. It is useful in treating infections caused by *Candida* and several other fungi. The drug can be given orally and is easily absorbed, but 90 percent of the amount given is found unchanged in the urine within 24 hours. Flucytosine should be given instead of amphotericin B wherever possible because it is less toxic and causes fewer side effects.

ANTIVIRAL AGENTS

Until recent years no chemotherapeutic agents effective against viruses were available. One reason for the difficulty in finding such agents is that the agent must act on viruses within cells without severely affecting the host cells. Currently available antiviral agents act to inhibit some phase of viral replication, but they do not kill the viruses.

Purine and Pyrimidine Analogues

Several purine and pyrimidine analogues are effective antiviral agents. All cause the virus to incorporate erroneous information (the analogue) into a nucleic acid and thereby interfere with the replication of viruses. (Chapter 7, p. 181) The drugs include *idoxuridine, cytarabine, vidarabine, ribavirin,* and *acyclovir.*

Idoxuridine and trifluridine, both analogues of uridine, are administered in eye drops to treat inflammation of the cornea caused by a herpesvirus. They should not be used internally because they suppress bone marrow.

Cytarabine (ARA-C), an analogue of cytosine, is used to treat some kinds of cancer but is of limited use in treating herpesvirus infections. It is more toxic to bone marrow than idoxuridine.

Vidarabine (ARA-A), an analogue of adenine, has been used effectively to treat viral encephalitis, an inflammation of the brain caused by herpes- and cytomegaloviruses. It is not effective against cytomegalovirus infections acquired before birth. Vidarabine is less toxic than either idoxuridine or cytarabine, but it sometimes causes gastrointestinal disturbances.

Ribavirin (Virazole), a synthetic nucleotide analogue of guanine, blocks replication of certain viruses. In an aerosol spray it can combat influenza viruses; in an ointment it can help to heal herpes lesions. Although it has low toxicity, it can induce birth defects and should not be given to pregnant women.

Acyclovir (acycloguanosine), an analogue of guanine, is much more rapidly incorporated into virus-infected cells than into normal cells. Thus, it is less toxic than other analogues. It can be applied topically or given orally or intravenously. It is especially effective in reducing pain and promoting healing of primary lesions in a new case of genital herpes. It is given prophylactically to reduce the frequency and severity of recurrent lesions, which appear periodically after a first attack. It does not, however, prevent the establishment of latent viruses in nerve cells. Acyclovir is more effective than vidarabine against herpes encephalitis and neonatal herpes, an infection acquired at birth, but is not effective against other herpesviruses.

Amantadine

The tricyclic amine *amantadine* prevents influenza A viruses from penetrating cells. Given orally, it is readily absorbed and can be used from a few days before to a week after exposure to influenza A viruses to reduce the incidence and severity of symptoms. Unfortunately, it causes insomnia and ataxia (inability to coordinate voluntary movements), especially in elderly patients, who also are often severely affected by influenza infections. *Rimantadine,* a drug similar to amantadine, may be effective against a wider variety of viruses and be less toxic as well.

Methisazone

The synthetic agent *methisazone* probably acts by inhibiting viral protein synthesis. It is active against viruses in smallpox vaccine and is useful in treating complications of smallpox vaccination.

APPLICATIONS

Drug-resistant Viruses

Evidence is accumulating that viruses, like bacteria, can develop resistance to chemotherapeutic agents. Herpesviruses and cytomegaloviruses with resistance to acyclovir have been observed in AIDS patients. Some laboratory strains of the AIDS virus itself have become resistant to azidothymidine (AZT), the most effective drug currently available to treat the disease. Resistance to chemotherapeutic agents is a greater problem in viruses than in bacteria because there are so very few antiviral agents available. When a bacterium becomes resistant to one antibiotic, another usually can be found to which the bacterium is susceptible. Unfortunately, this is not the case with viruses.

Treatment of AIDS

Several agents are being tested for the treatment of AIDS. New information about AIDS, its complications, and its treatment is becoming available with great rapidity. We consider AIDS, agents used to treat it, and ramifications for health workers in Chapter 19.

Interferon and Immunoenhancers

Cells infected with viruses produce one or more proteins collectively referred to as *interferons* (Chapter 17). When these proteins are released, they induce neighboring cells to produce antiviral proteins, which prevents these cells from becoming infected. Thus, interferons represent a natural defense against viral infection. Some interferons are currently being produced by genetic engineering and tested as antiviral agents. Some positive results have been obtained in controlling viral infections and arresting cancers, especially those associated with viruses.

Because cells produce interferons naturally, a possible way to combat viruses is to induce cells to produce interferons. Synthetic double-stranded RNA has been shown to increase the quantity of interferon in the blood. Experiments with one such substance in virus-infected monkeys have shown sufficient increase in interferon to prevent viral replication.

Two other agents, *levamisole* and *inosiplex*, appear to act by stimulating the immune system to resist viral and other infections. Both appear to stimulate activity of leukocytes called T lymphocytes rather than stimulating interferon release. Levamisole appears to be effective prophylactically in reducing the incidence and severity of chronic upper respiratory infections, which are probably viral in nature. It also reduces symptoms of autoimmune disorders such as rheumatoid arthritis, in which the body reacts against its own tissues. Inosiplex has a more specific action: It stimulates the immune system to resist infection with certain viruses that cause colds and influenza.

Although efforts to improve antiviral therapies by enhancing natural defenses have been somewhat successful, none is yet in widespread use. More research is needed to identify or synthesize effective agents, to determine how they act, and to discover how they can be most effectively used.

ANTIPROTOZOAN AGENTS

Although many protozoa are free-living organisms, a few are parasitic in humans. The parasite that causes malaria invades red blood cells and causes the patient to suffer alternating fever and chills. Other protozoan parasites cause intestinal or urinary tract infections. Chemotherapeutic agents are used to cure or control protozoan infections, just as they are used for other infections.

Quinine

Quinine from the bark of the chinchona tree was used for many years to treat malaria. One of the first chemotherapeutic agents to come into widespread use, it is now used only to treat malaria caused by strains of the parasite resistant to other drugs.

Chloroquine and Primaquine

Currently the most widely used antimalarial agents are the synthetic agents *chloroquine* (Aralen) and *primaquine*. Chloroquine appears to interfere with protein synthesis, especially in red blood cells, which it enters more readily than it does other cells. It may concentrate in vacuoles within the parasite and prevent it from metabolizing hemoglobin. Chloroquine is used to combat active infections. The malarial parasite persists in red blood cells and can cause relapses when it multiplies and is released into the blood plasma. A combination of chloroquine and primaquine can be used prophylactically to protect people who are at risk of becoming infected when visiting or working in regions of the world where malaria occurs. The drugs, however, must be taken both before and after entering a malarial zone. A new prophylactic agent, *mefloquine* (Lariam), has proven effective against resistant strains.

Pyrimethamine

Pyrimethamine (Daraprim) interferes with the synthesis of folic acid, which protozoa need in greater quantities than host cells. It is used with sulfanilamide to treat some protozoan infections and can be used prophylactically to prevent malaria.

Metronidazole

The synthetic imidazole *metronidazole* (Flagyl) is effective in treating trichomonas infections, which typically cause a vaginal discharge and itching. It also is effective against intestinal infections caused by parasitic amoebae and *Giardia*. Though metronidazole controls these infections, it does not prevent overgrowth of *Candida* yeast infections. It also can cause birth defects and cancer, and it can be passed to infants in breast milk. Metronidazole sometimes causes an unusual side effect called "black hairy tongue" or "brown furry tongue" because it breaks down hemoglobin and leaves deposits in papillae (small projections) on the surface of the tongue (Figure 14.14).

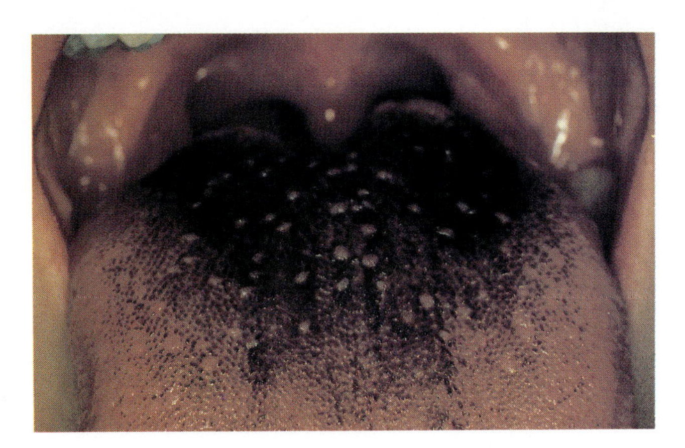

FIGURE 14.14 Black hairy tongue is a reaction to the drug Flagyl. The papilli on the surface of the tongue become elongated and filled with breakdown products of hemoglobin, which give it the dark appearance.

Other Antiprotozoan Agents

A variety of other organic compounds have been found effective in treating certain infections caused by protozoa. *Suramin sodium,* a sulfur-containing compound, can be given intravenously to treat sleeping sickness and other trypanosome infections. A nitrofuran, *nifurtimox,* is used against the trypanosomes that cause Chagas' disease. Pentavalent antimony compounds are effective against a group of protozoans called leishmanias, and pentavalent arsenic compounds can be used to treat intestinal amoeba infections. Both antimony and arsenic compounds are highly toxic.

ANTIHELMINTHIC AGENTS

Various helminths (worms) can infect humans, and a variety of chemotherapeutic agents are available to help rid the body of these unwelcome parasites.

Niclosamide

Niclosamide interferes with carbohydrate metabolism and thereby causes a parasite to release large quantities of lactic acid. This drug also may inactivate products made by the worm to resist digestion by host proteolytic enzymes. It is effective mainly in the treatment of tapeworm infections.

Piperazine

The simple organic compound *piperazine* (Antepar) is a powerful neurotoxin. It paralyzes body wall muscles of roundworms and is useful in treating *Ascaris* and pinworm infections. Although piperazine exerts its effect on worms in the intestine, if absorbed it can reach the human nervous system and cause convulsions, especially in children.

Mebendazole

The imidazole *mebendazole* (Vermox) blocks the uptake of glucose by parasitic worms. It is useful in treating whipworm, pinworm, roundworm, and hookworm infections. However, it can damage a fetus and thus should not be given to pregnant women.

Ivermectin

The compound *ivermectin,* originally developed for the treatment of parasitic nematodes in horses (and widely used to prevent heartworm infections in dogs), has been found to be extremely effective against *Onchocerca volvulus* in humans. Infection with this roundworm, widespread in many parts of Africa, causes a progressive loss of sight known as river blindness.

The chemical structures, uses, and side effects of antifungal, antiviral, antiprotozoan, and antihelminthic agents are summarized in Figure 14.15.

SPECIAL PROBLEMS WITH RESISTANT HOSPITAL INFECTIONS

As soon as antibacterial agents became available, resistant organisms began to appear. One of the first successes in treating bacterial infections was the use of sulfanilamide to treat infections caused by hemolytic streptococci. Then it was discovered that sulfadiazine was useful in preventing recurrent infections in the form of rheumatic fever. Strains of streptococci resistant to sulfonamides soon emerged. Epidemics (mostly in military installations during World War II) caused by resistant strains led to many deaths. These epidemics were brought under control when penicillin became available, but soon penicillin-resistant streptococci were seen.

This chain of events has been repeated again and again. As new antibiotics were developed, strains of streptococci resistant to many of them evolved. Similar events led to the emergence of antibiotic-resistant strains of many other organisms, including staphylococci, gonococci, *Salmonella, Neisseria,* and especially *Pseudomonas. Pseudomonas* infections are now a major problem in hospitals. Many of these organisms are now resistant to several different antibiotics, and new resistant strains are constantly being encountered.

Why are resistant organisms found more often in hospitalized patients than among outpatients? This question can be answered by looking at the hospital environment and the patients likely to be hospitalized. First, in spite of efforts to maintain sanitary condi-

Agent	Used to Treat	Common Method of Administration*	Side Effects
Antifungal Agents			
Clotrimazole	Skin and nail infections	O	Skin irritation
Miconazole	Skin infections and systemic infections resistant to other agents	T,IV	Severe itching, nausea, fever, thrombophlebitis
Amphotericin B	Systemic infections	IV	Fever, chills, nausea, vomiting, anemia, kidney damage, blindness
Nystatin	*Candida* yeast infections, intestinal superinfections	T	
Griseofulvin	Infections of skin, hair, and nails	T,O	Mild headaches, nerve inflammation, gastrointestinal disturbances
Flucytosine	*Candida* and some systemic infections	O	Less toxic than many fungal agents

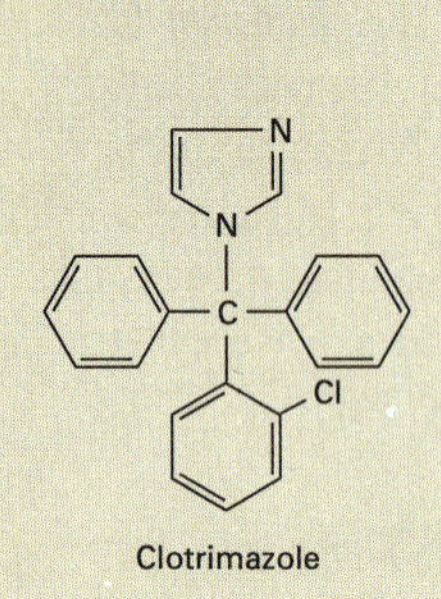

Clotrimazole

Miconazole

Amphotericin B

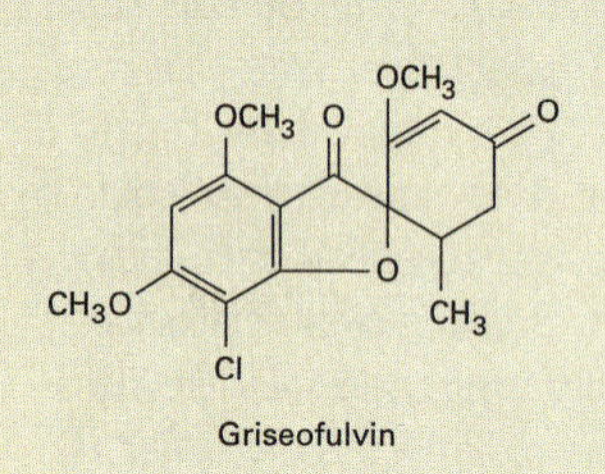

Griseofulvin

Flucytosine

Nystatin

Agent	Used to Treat	Common Method of Administration*	Side Effects
Antihelminthic agents			
Niclosamide	Tapeworm infections	O	Irritation of gut
Piperazine	Pinworm and *Ascaris* infections	O	Can cause convulsions in children
Mebendazole	Whipworm, pinworm, roundworm, and hookworm infections	O	Can damage fetus if given to pregnant woman
Ivermectin	*Onchocerca volvulus* infections (cause of river blindness), heartworm infections in animals	O	Minimal

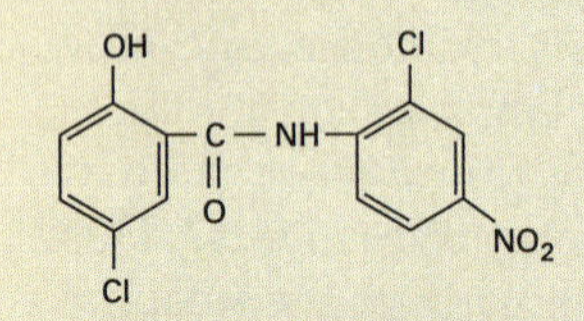

Niclosamide

Piperazine

Mebendazole

FIGURE 14.15 Antifungal, antiviral, antiprotozoan, and antihelminthic drugs.

* IM = intramuscular; IV = intravenous; O = oral; T = topical

Agent	Used to Treat	Common Method of Administration*	Side Effects
Antiviral Agents			
Idoxuridine	Corneal infections	T	Suppresses bone marrow
Cytarabine	More useful in cancer therapy than as an antiviral agent	T	Suppresses bone marrow
Vidarabine	Viral encephalitis	T,IV	Less toxic than other antiviral agents
Ribavirin	Topically on herpes lesions, in aerosol for influenza	T	Can cause birth defects if given to pregnant women
Acyclovir	Herpesvirus infections; lessens severity of symptoms	IV,O,T	Less toxic than other analogs
Amantidine	Prevents influenza A viruses from entering cells	O	Insomnia and ataxia

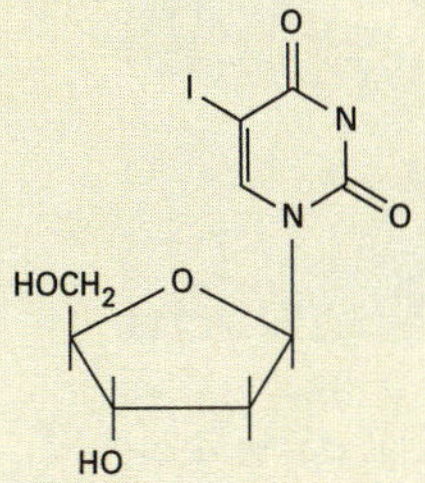

Idoxuridine

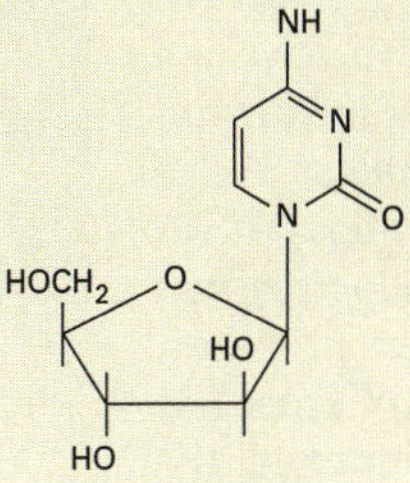

Cytarabine

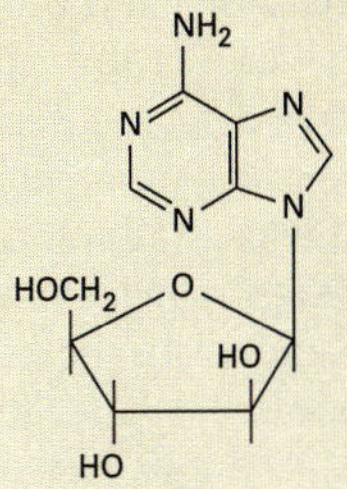

Vidarabine

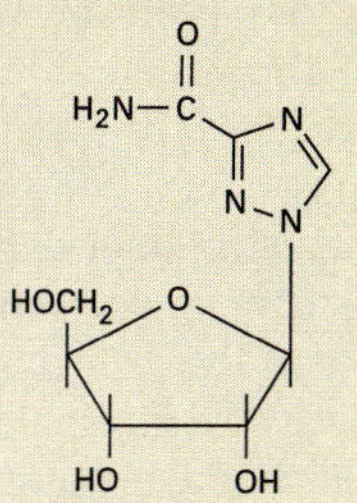

Ribavirin

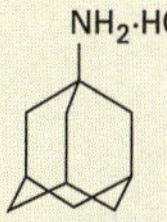

Amantidine

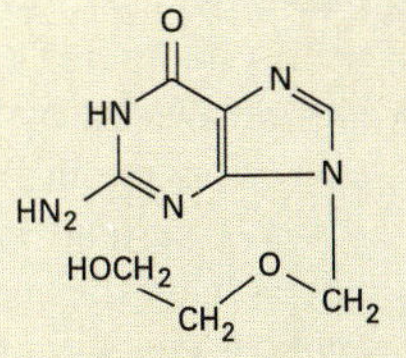

Acyclovir

Agent	Used to Treat	Common Method of Administration*	Side Effects
Antiprotozoan agents			
Quinine	Malaria infections resistant to other agents	O	
Chloroquine	Malaria infections	O	Headache, itching
Primaquine	With chloroquine to prevent relapse of malaria	O	Slight nausea and abdominal pain
Pyrimethamine	Various protozoan infections	O	Large doses damage bone marrow
Metronidazole	*Trichomonas*, *Giardia*, and amoebic infections	O,IV,T	Black hairy tongue

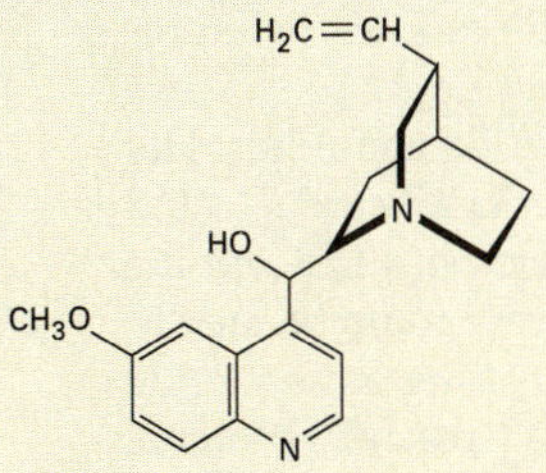

Quinine

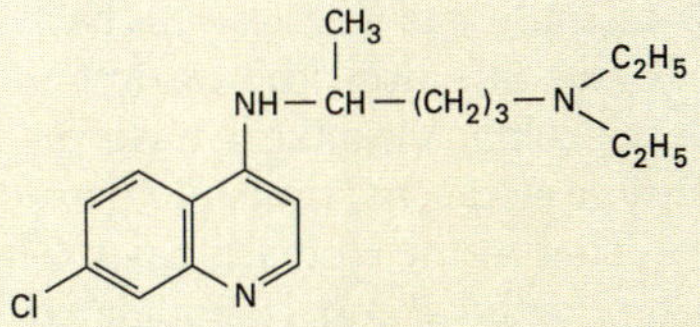

Chloroquine

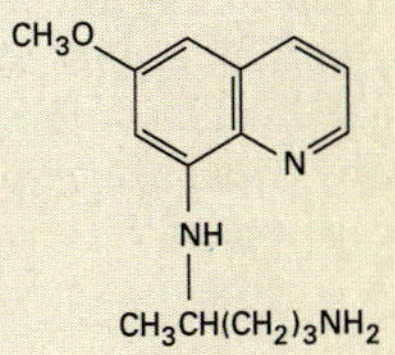

Primaquine

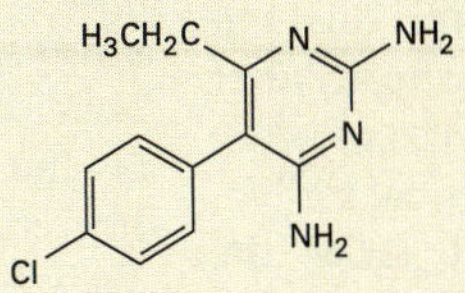

Pyrimethamine

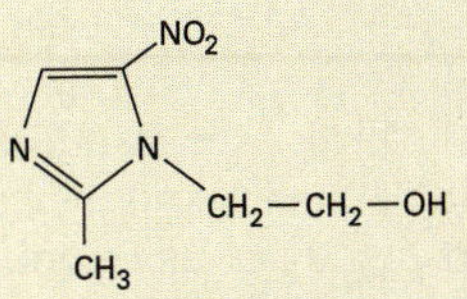

Metronidazole

MICROBIOLOGIST'S NOTEBOOK

Fighting Disease at the Zoo

Rob: As soon as I walked into the monkey house that morning I knew that one of them was really sick. The first thing I noticed was a large amount of liquid stool in the cage. Colobus monkeys have stools like little pellets, so one of them was definitely in trouble.

I looked at the animals to see who it was. We keepers look at each one every day, you know, and we really get to know them. I realized that it was Emily. Her face was very swollen—so much so that her eyes were nearly swollen shut. Her tail was filthy. I was really concerned. Emily is the oldest female in the group, which makes her the core of the females. Also, we suspected that she was pregnant.

I'm Rob Schumaker. I work as a keeper of monkeys and great apes at the Smithsonian's National Zoo in Washington, D.C. I'm studying for a graduate degree in zoology, and I have a particular interest in microbiology. The monkeys I'm talking about are the black and white colobus: *Colobus abyssinicus.* Their range is Africa, from Ethiopia to Tanzania and Zaire. Their natural diet consists of leaves and fruits. Colobus monkeys live in groups of two to 15 or more. The males are somewhat solitary, but the females have a highly structured society with many rules.

I watched Emily some more. The veterinarians here go on rounds every day, but I was concerned about Emily and didn't want to wait for them to come to us. Even though it could be very stressful for her, we decided to put her in her transport cage and take her to the hospital. We had to find out what this illness was. Our groups had always been okay; up until then we had escaped the difficulties that other monkey colonies have experienced.

Dr. Montali: Those other difficulties were outbreaks of dysentery caused by *Shigella* bacteria. [See Chapter 22.] It's a problem in many primate colonies, and even in good zoos, monkeys and apes are often carriers. I'm Dr. Richard Montali, the veterinary pathologist at the National Zoo. Our zoo never had a *Shigella* problem until 1984. Now Dr. Mitchell Bush, the zoo's clinical veterinarian, and I are doing an intensive study of *shigella* infection. Twice a year, we screen cultures of each animal's feces on three consecutive days. We've found *S. flexneri* and *S. sonnei.* We've found carrier animals. So far none of the keepers has caught it—although one of the veterinarians did.

The *Shigella* could have been introduced by humans, or it could have been brought in by a new animal. We really don't know. However, we are considering the possibility of using human vaccine for the animals. We're talking to Johns Hopkins about a *Shigella* vaccine they're developing that might be applicable to our primates. Walter Reed Army Medical Center has also been helpful. We identified the dysentery bacterium in our own labs, but now they do special cultures for us, to characterize organisms by species and even by strain. It's fortunate that these medical centers are close by.

We've also identified other pathogens from the zoo animals in our labs: *Salmonella, Campylobacter,* parasites. Timing is very important. For example, to isolate *Yersinia pseudotuberculosis* and *Y. enterocolitica,* carried by rats, we have to take cultures 3 to 5 days in a row. Even then, we may find the pathogen on only one of those days.

It's terrible when a zoo population falls victim to disease like this. We lost two gibbons to *Shigella.* One was old, the other had other underlying problems. But even if the mortality is low, it's upsetting to the staff *and* to the other animals—their whole social structure can be thrown off.

Once Emily was diagnosed as having a *Shigella* infection, the treatment was clear. Colobus monkeys have ruminant-type stomachs, similar to those of cows. That is, the stomach has several compartments and is populated with symbiotic bacteria that help break down the cellulose in the plant material the monkeys eat. [See Chapter 2.] These very special gastrointestinal microorganisms are acquired shortly after birth from handling the feces of other animals, and they are essential for life. A colobus monkey that lacks the proper microbes cannot get enough nutrients from its food to survive.

When these animals are given antibiotics, the bacterial populations that help them digest their food are destroyed. Saving them from possible death by infection involves exposing them to possible illness from malnutrition. So after the *Shigella* infection is cured, we try to repopulate the gut with stomach juices aspirated from other, healthy animals. Unfortunately, this technique doesn't always work. In Emily's case, it proved effective. But Emily was a very weak animal for a long time afterward. She lost the baby she was carrying, and also miscarried in her next two pregnancies. This year, though, she has a healthy baby.

One of zoo's black and white colobus monkeys—ruminants that depend on gut microbes to help them digest their food.

Rob: Even though we knew what organism was making Emily so sick, we were still very worried about her. Her separation from her family group could be very stressful for both her and the group, and that stress could be the difference in her recovering or not. Being in a hospital is also stressful for animal patients—they don't understand where they are or why. The hospital keepers are wonderful, but the animals don't know them. And they may not always know how to talk to individual animals or know their specific needs. So the primate keepers go up to the hospital and visit their sick charges. They handfeed them, if necessary.

Dr. Montali: The keepers in the zoo and the veterinary staff work closely together. It's important that the staff be educated, and aware of potential problems. For example, we have learned to avoid mixed exhibits of different species of animals, because some may be endangered by the food of others. We saw some losses in the small mammal house because of such a situation. Some tamarins died from *Streptococcus zooepidemicus*, Group C. This is an opportunistic pathogen in humans, and is carried by horses. It is usually present in the respiratory and genital tracts of the horses, and can sometimes cause serious infection even in them. In a small mammal, such as a tamarin, infection begins in the lymph nodes and then develops into septicemia.

How did the tamarins get an equine infection? Probably from the food that was given to the armadillos they lived with. We fed the armadillos a commercially sold raw horsemeat product that we received frozen. In a mixed exhibit such as the one we're talking about, the tamarin may have gone over to the armadillo's food dish and tasted the food. Or it may just have handled the meat and then put its fingers in its mouth. Although the pathogen didn't affect the armadillos, it killed several of the tamarins.

It's easy to understand how the horsemeat could be contaminated in this way. Do many *healthy* horses go to slaughter? Probably not. And the quality controls on food intended for animal consumption are much less stringent than those on food for human consumption. Our solution was to switch to beef, and to reorganize the mixed exhibits.

A pair of squirrel-sized Golden Lion tamarin monkeys becoming acclimated to uncaged life at the National Zoo in Washington, D.C. before being released in Brazil.

Rob Schumaker at the Washington zoo.

Rob: The Zoo has developed a cooperative program to reintroduce golden lion tamarins to their native Brazil. They are an endangered species in the wild but breed well in zoos where a surplus number of animals have accumulated. We run a staging program here in which we keep the tamarins in an outdoor, uncaged setting. This helps them adjust to the conditions they will find in the wild. We want them to survive when they get there.

Dr. Montali: And we want to be sure that they don't bring any diseases into the wild population that might kill what remains of the endangered populaton. In the zoo, animals can pick up parasites and infections not found in their native countries. A cockroach might scurry from the cage of an Asian animal into an adjacent cage where an animal from South America will eat it, thus acquiring a parasite from a different continent. The German cockroach has been found to carry an intestinal nematode parasite from its indigenous host, the slow lorus, to tamarins. Thus, we examine these tamarins *very* carefully before we send them to Brazil. We also eliminate animals with genetic defects from the program.

Rob: We keepers have to be extraordinarily careful to avoid contamination of our animals and of ourselves. Whenever there is illness, we take extra precautions. For example, we always wear waterproof boots around the enclosures—for washing them down and so forth. When Emily got sick, we set up a footbath, containing a diluted bleach solution, at the door to her cage. The keepers stepped in it going into and coming out of the cage. (We didn't want to contaminate anything outside that enclosure.) We kept special tools only for that particular cage. We tended to that cage last and washed all tools in a bleach solution. We took extra care in scrubbing the cage, using a special antiseptic solution.

Those were the precautions we took over and above our regular procedures. We ordinarily clean the inside of the cages daily. They're scrubbed, hosed, and disinfected. We rotate among three products: an antiseptic detergent; bleach, which is a good stain remover; and a deodorizing agent.

As for the keepers, we always wear filter masks so that we don't inhale any aerosol material. We wear gloves, boots, and coveralls on the job, and we don't wear any work clothes out of the building—there are a washer and dryer in the basement of the building that we use. We wear our street clothes in and out.

Just as we check the monkeys to make sure they're healthy on a routine basis, we check ourselves too. All keepers' stools are cultured for three consecutive days twice a year. There are a number of pathogens that can be transmitted back and forth between humans and animals. Early in this century, for example, tuberculosis was a problem among both animals and staff. The keepers gave it to the animals; the animals gave it to the keepers. Even now, mammalian tuberculosis can be a danger to keepers, so we undergo regular TB testing. We have all come to the realization that disease transmission in a zoo can be a two-way problem. And since the animals are totally dependent on us, it is our problem to solve.

tions, a hospital provides an environment where sick people live in close proximity and where many different kinds of infectious agents are constantly present. Second, the patients are usually more severely ill than outpatients; many have lowered resistance to infection because of their illnesses or because they have received immunosuppressant drugs. Finally, and most importantly, hospitals typically make intensive use of a variety of antibiotics. Because many infections are being treated and different antibiotics are used, organisms resistant to one or more of the antibiotics are likely to emerge. The resistant strains can readily spread among patients.

Treatment of resistant infections creates a vicious cycle. If an antibiotic can be found to which an organism is susceptible, it can be used to treat the infection. However, some strains of the organism that are resistant to the new antibiotic may then proliferate. A recurrent cycle of using new antibiotics and the organisms developing resistance to them is established. Preventing infections caused by antibiotic-resistant strains of microorganisms is a difficult task, but several things can be done to reduce their incidence. First, the use of antibiotics should be limited to situations in which the patient is unlikely to recover without antibiotic treatment. Second, sensitivity tests should be done, and the patients should receive only an antibiotic to which the organism is known to be sensitive. Third, when antibiotics are used they should be continued until the organism is completely eradicated from the patient's body. Double antibiotic use, as described earlier under quinolones, is especially useful (Figure 14.16). Finally, any patient with an infectious disease should be completely isolated from other patients.

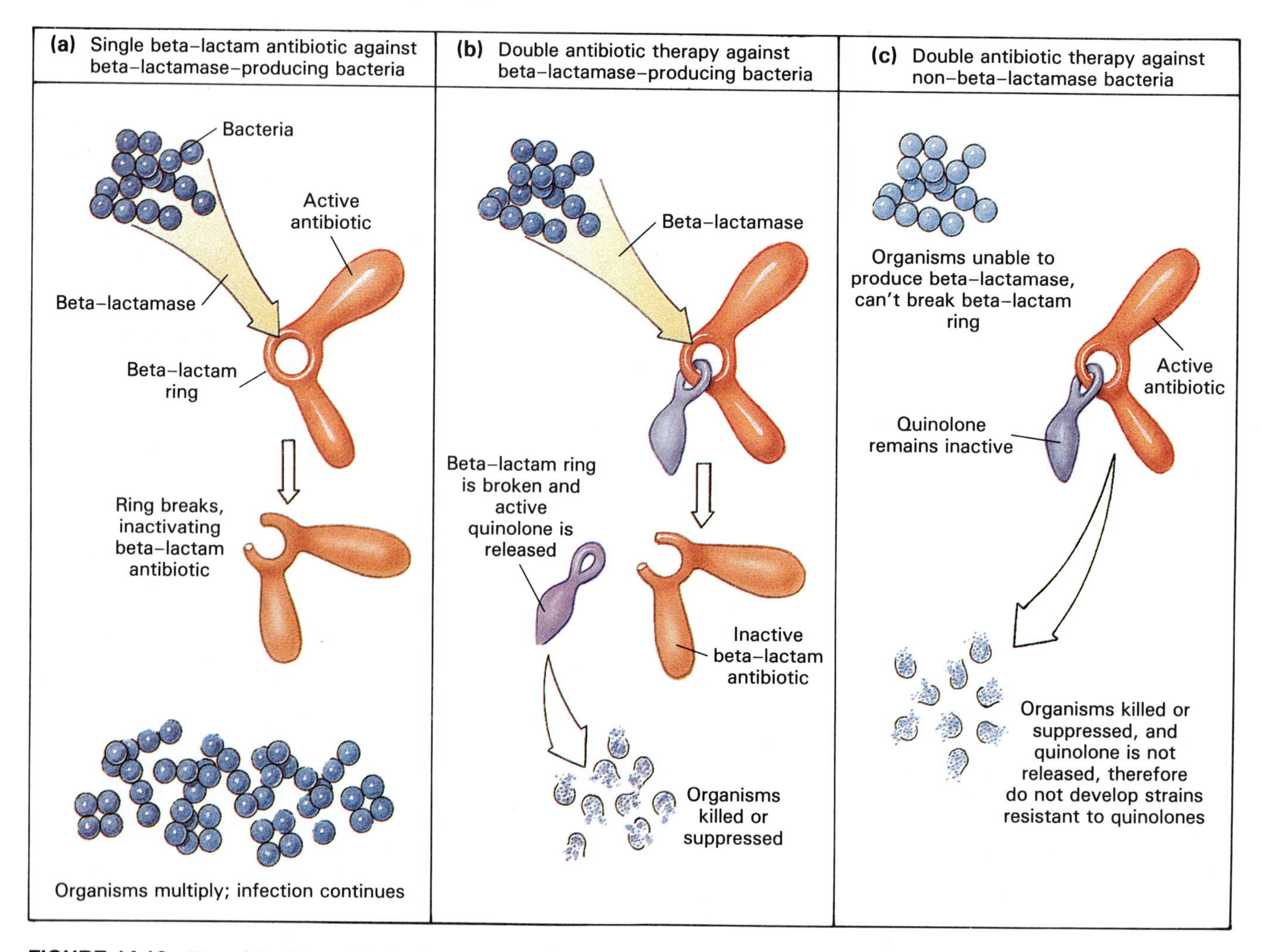

FIGURE 14.16 Use of double antibiotic therapy to eradicate resistant-strain infections.

ESSAY

The Role of *Clostridium difficile* in Antibiotic-associated Intestinal Disease

Antibiotic therapy sometimes has its dark side. *Clostridium difficile* does not cause intestinal disease unless antibiotics have been administered. A few exceptions to this rule existed in the preantibiotic era, and no completely convincing explanation is available to explain these cases. Today, *C. difficile* is the most important intestinal bacterial pathogen in the developed world, in terms of prevalence and severity of disease.

C. difficile was first described in 1935, but it was not definitely associated with disease until 1978. During the 1950s, especially, *Staphylococcus aureus* was blamed for what are now known to have been *C. difficile* infections. *C. difficile* is currently the only organism recognized as a common cause of antibiotic-associated colitis. *C. difficile* is found in 15 to 25 percent of patients having antibiotic-associated diarrhea; 50 to 75 percent of patients with colitis; and over 90 percent of patients with pseudomembranous colitis (PMC) (Figure 14.17). It is rarely found in healthy people except for newborns, among whom it occurs frequently but without harm. It disappears as babies approach 6 to 12 months of age. Older children rarely develop *C. difficile* problems, despite frequent use of antibiotics. Older adults are most likely to develop disease, but this age factor has not yet been explained.

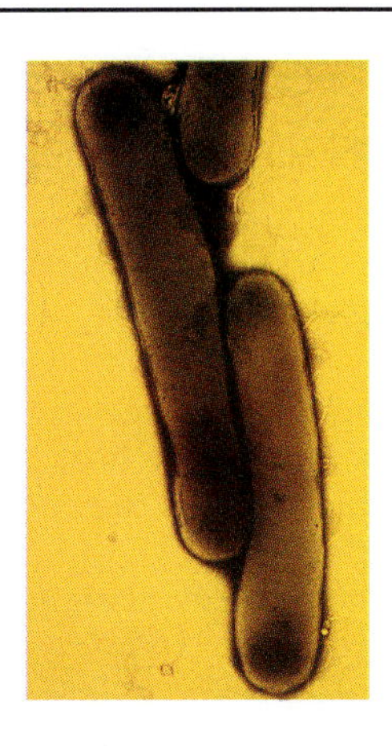

FIGURE 14.17 Scanning electron micrograph of the intestinal lining of a person with pseudomembraneous colitis.

Clindamycin is the antibiotic most frequently associated with *C. difficile* infections, followed by ampicillin and cephalosporins. These drugs are often administered to patients who will undergo abdominal surgery, in order to decrease normal flora. The *C. difficile* organisms are usually very susceptible to these antibiotics and survive by sporulation. After the drug is discontinued, they regenerate and can overgrow the intestinal tract, but they do not invade its tissues. *C. difficile* produces two toxins: toxin A, which has a cytoxic (cell-killing) effect and is responsible for symptoms observed, and toxin B, which does not seem to have any disease-producing effects.

C. difficile is widely distributed in nature, being found in soil and in feces of animals such as cows and horses. It seems to establish itself in an environment and may be difficult to remove. Once established in a hospital ward or nursing home, it may persist for months or years despite vigorous eradication efforts.

Symptoms of disease include abdominal cramps, diarrhea, fever (up to 106°), electrolyte imbalance, toxic megacolon, and even perforation of the colon. The antibiotic vancomycin is the single most effective treatment and is almost 100 percent effective when perforation has not occurred. Fever drops within 24 to 48 hours, and normal bowel action returns in 5 to 7 days. However, relapse is frequent due to the sporulating forms. Vancomycin costs four times as much as an equivalent weight of gold. Cost of the treatment ranges from $200 to $600.

Diagnosis is generally by cell culture assay. Toxin A is so potent that one molecule is enough to cause the changes seen in cells cultured.

CHAPTER SUMMARY

ANTIMICROBIAL CHEMOTHERAPY

- **Chemotherapy** refers to the use of any chemical agent in the practice of medicine.
- A **chemotherapeutic agent,** or **drug,** is any chemical agent used in medical practice.
- An **antimicrobial agent** is a chemical agent used to treat a disease caused by a microbe.
- **Antibiosis** means "against life."
- An **antibiotic** is a chemical substance produced by microorganisms that inhibits the growth of or destroys other microorganisms.
- A **synthetic drug** is one made in the laboratory.

- A **semisynthetic drug** is one made partly by microorganisms and partly by laboratory synthesis.

RELATED KEY TERMS

HISTORY OF CHEMOTHERAPY

- The first chemotherapeutic agents were concoctions from plant materials used by primitive societies.
- Paul Ehrlich's search for the 'magic bullet' was the first systematic attempt to find chemotherapeutic agents.
- Subsequent events included the development of sulfa drugs, **penicillin,** and many other antibiotics.

GENERAL PROPERTIES OF ANTIMICROBIAL AGENTS

Selective Toxicity

- **Selective toxicity** refers to the property of antimicrobial agents that allows them to exert greater toxic effects on microbes than on the host.
- The **therapeutic dosage level** of an antimicrobial agent is the concentration over a period of time required to eliminate a pathogen.
- The **chemotherapeutic index** is a measure of the toxicity of an agent to the body relative to its toxicity for an infectious organism.

Spectrum of Activity

- The spectrum of activity of an antimicrobial agent refers to the variety of microorganisms sensitive to the agent. A **broad-spectrum** agent attacks many different organisms. A **narrow-spectrum** agent attacks only a few different organisms.

Modes of Action

- Agents that inhibit cell wall synthesis allow the cell membrane of the affected microbe to rupture and release the cell contents.
- Agents that disrupt cell membrane function dissolve the membrane or interfere with the movement of substances into or out of cells.
- Agents that inhibit protein synthesis prevent the growth of microbes by disrupting ribosomes or otherwise interfering with the process of translation.
- Agents that inhibit nucleic acid synthesis interfere with synthesis of RNA (transcription) or DNA (replication) or disrupt the information these molecules contain.
- Agents that act as **antimetabolites** compete with a microbial enzyme or disrupt information in a nucleic acid.

molecular mimicry

Kinds of Side Effects

- Side effects of antimicrobial agents on the host include toxicity, allergy, and disruption of natural flora.
- Allergic reactions to antimicrobial agents occur when the body reacts to the agent as a foreign substance.
- Many antimicrobial agents attack not only the infectious organism but also natural flora; **superinfections** with new pathogens can occur when the defensive capacity of natural flora is destroyed.

Resistance of Microorganisms

- **Resistance** to an antibiotic means that a microorganism formerly susceptible to the action of an antibiotic is no longer affected by it.
- Nongenetic resistance occurs when microorganisms are sequestered from antibiotics or when they undergo a temporary change, such as the loss of their cell walls, that renders them nonsusceptible to the action of an antibiotic.
- Genetic resistance occurs when organisms survive exposure to an antibiotic because of their genetic capacity to avoid damage by the antibiotic. As susceptible organisms die, the resistant survivors multiply unchecked and increase in numbers.
- **Chromosomal** genetic **resistance** is due to a mutation in microbial DNA; **extrachromosomal** genetic **resistance** is due to plasmids (**R factors**).
- Mechanisms of resistance include alterations of receptors, cell membranes, enzymes, or metabolic pathways.
- **Cross-resistance** is resistance against two or more similar antimicrobial agents.

Drug resistance can be minimized by: (1) continuing treatment with an appropriate antibiotic until all the organisms causing a disease are removed from the body, (2) using two antibiotics that do not have cross-resistance to treat an infection, and (3) using antibiotics only when they are absolutely necessary.

RELATED KEY TERMS

DETERMINATION OF MICROBIAL SENSITIVITIES TO ANTIMICROBIAL AGENTS

- Sensitivity of microbes to chemotherapeutic agents is determined by exposing them to the agents in laboratory cultures.

Disk Diffusion Methods

- In the **disk diffusion (Kirby-Bauer) method,** organisms causing an infection are cultured on agar plates and subjected to various antibiotics impregnated in filter paper disks, and **zones of inhibition** around disks indicate to which antibiotics the organism is sensitive.

synergistic effect

Dilution Methods

- In the **dilution method,** a constant inoculum is introduced into broth cultures with differing known quantities of chemotherapeutic agents. The **minimum inhibitory concentration** (MIC) of the agent is the lowest concentration in which no growth of the organism is observed. The **minimum bactericidal concentration** (MBC) of the agent is the lowest concentration in which subculturing of broth yields no growth.

Serum Killing Power

- In the **serum killing power** method a bacterial suspension is added to a patient's **serum** drawn while the patient is receiving an antibiotic, and it is noted whether the organisms are killed.

Automated Methods

- Automated methods allow rapid identification of microorganisms and determination of their sensitivities to antimicrobial agents.

ATTRIBUTES OF AN IDEAL ANTIMICROBIAL AGENT

- An ideal antimicrobial agent is soluble in body fluids, selectively toxic, and nonallergenic; can be maintained at a constant therapeutic concentration in blood and body fluids; is unlikely to elicit resistance; has a long shelf life; and is reasonable in cost.
- Most drugs fail to meet all these criteria, so the search for ideal drugs continues.

ANTIBACTERIAL AGENTS

- **Inhibitors of Cell Wall Synthesis**
- **Disrupters of Cell Membranes**
- **Inhibitors of Protein Synthesis**
- **Inhibitors of Nucleic Acid Synthesis**
- **Other Antibacterial Agents**
- Antibacterial agents inhibit cell wall synthesis, disrupt cell membrane functions, inhibit protein synthesis, inhibit nucleic acid synthesis, or act by some other means to kill bacteria.
- The properties of antibacterial agents are summarized in Figure 14.13.

ANTIFUNGAL AGENTS

- **Antifungal agents** thicken walls of fungal cells, increase cell membrane permeability, interfere with nucleic acid synthesis, or otherwise impair cell functions.
- The properties of antifungal agents are summarized in Figure 14.15.

ANTIVIRAL AGENTS

- Antiviral agents have been difficult to find because they must damage intracellular viruses without severely damaging host cells.
- Most antiviral agents are analogues of purine or pyrimidine.
- The properties of antiviral agents are summarized in Figure 14.15.
- Interferon is released by virus-infected cells and stimulates neighboring cells to produce antiviral proteins. It is being made by genetic engineering and tried in treatment of viral infections and cancer.
- Agents to stimulate interferon release and agents to stimulate the immune system are being developed.

ANTIPROTOZOAN AGENTS

- Some antiprotozoan agents interfere with protein synthesis or folic acid synthesis. For others, the mechanism of action is presently not well understood.
- The properties of antiprotozoan agents are summarized in Figure 14.15.

ANTIHELMINTHIC AGENTS

- Antihelminthic agents interfere with carbohydrate metabolism or act as nerve toxins.
- The properties of antihelminthic agents are summarized in Figure 14.15.

SPECIAL PROBLEMS WITH RESISTANT HOSPITAL INFECTIONS

- Resistant hospital infections are due largely to intensive use of a variety of antibiotics, which fosters the growth of resistant strains.
- Treatment and prevention of resistant hospital infections are extremely difficult.

QUESTIONS FOR REVIEW

A.

1. Define and give an example of each of the following: antibiotic, synthetic drug, semisynthetic drug.

B.

2. How were the first chemotherapeutic agents developed?
3. How has chemotherapy developed since Ehrlich's time?

C.

4. Define selective toxicity, and relate it to antimicrobial agents.
5. What is the significance of the terms therapeutic dosage level and chemotherapeutic index?
6. What is meant by the spectrum of activity of an antimicrobial agent?
7. When is the use of a broad-spectrum agent indicated? What are the advantages of a narrow-spectrum agent?
8. Name five modes of action of antimicrobial drugs.
9. Briefly explain how each mode of action affects microbes.

D.

10. Describe at least three common side effects of antimicrobial drugs.

E.

11. What is drug resistance?
12. How do nongenetic and genetic drug resistance differ?
13. What is cross-resistance?
14. How can drug resistance be minimized?

F.

15. How does the disk diffusion method determine sensitivity of an organism to antimicrobial agents?
16. What are the advantages and disadvantages of the tube dilution method of determining sensitivities over the disk diffusion method?
17. How can sensitivity be determined by automated methods?

G.

18. List the attributes of an ideal drug.
19. Evaluate any three drugs according to the criteria for an ideal drug.

H.

20. Summarize the properties, uses, and side effects of antibacterial agents that inhibit cell wall synthesis.
21. Why are cephalosporins usually not the first antibiotic tried in treating an infection?
22. Summarize the properties, uses, and side effects of antibacterial agents that interfere with protein synthesis.
23. Give examples of synergistic drug action.
24. What are some precautions in the use of streptomycin, tetracyclines, chloramphenicol, and sulfonamides?
25. Given that rifampin has a wide spectrum of activity, why is it used in only a few specific situations?

I.

26. Summarize the properties, uses, and side effects of antifungal agents.
27. Summarize the properties, uses, and side effects of antiviral agents.
28. Why has it been more difficult to develop antiviral agents than to develop antibacterial agents?
29. Summarize the properties, uses, and side effects of antiprotozoan agents.
30. Summarize the properties, uses, and side effects of antihelminthic agents.

J.

31. Why do antibiotic-resistant infections so often occur in hospitals?
32. Why are treatment and prevention of such infections so difficult?

PROBLEMS FOR INVESTIGATION

1. Explain how you would design a program to help prevent the emergence of resistance to antimicrobial agents, especially multiply resistant *Staphylococcus aureus*.
2. Design a procedure for the isolation of a microorganism that will produce a new antibiotic useful to humans.
3. Discuss the advantages and disadvantages of using more than one drug simultaneously in treatment.
4. Study the structural formulas of antimicrobial agents given in this chapter, and cite at least three examples of similar structures that have similar modes of action. Also city an example of agents with similar structures that have different actions.
5. Read about the combined efforts of the United States and Britain to develop penicillin production during World War II. What were some of the major problems? Why was the research begun in Britain but later moved to the United States?
6. An outbreak of Legionnaire's disease kills six members of a small rural community. Proper diagnosis is made only at autopsy. What could have prevented this tragedy?
(The answer to this question appears in Appendix E.)

SOME INTERESTING READING

Abraham, E. P. 1981. The beta-lactam antibiotics. *Scientific American* 244, no. 6 (June):76.

Amantea, M. A., D. J. Drutz, and J. R. Rosenthal. 1990. Antifungals: a primary care primer. *Patient Care* 24, no. 18 (November 15):58–74.

Anonymous. 1990. Attacking HIV with antisense and catalytic RNA. *ASM News* 56(2):73–74.

Bean, B. 1992. Antiviral therapy: current concepts and practices. *Clinical Microbiology Reviews* 5(2):146–82.

Davey, P. G. 1990. New antiviral and antifungal drugs. *British Medical Journal* 300, no. 6727 (March 24):793–99.

Hirsch, M. S., and J. C. Kaplan. 1987. Antiviral therapy. *Scientific American* 256, no. 4 (April):76.

Katzung, B. G., ed. 1984. Basic and clinical pharmacology. 2d ed. Palo Alto, Calif.: Lange Medical Publishers.

McNabb, P. C. 1988. Antifungal agents: Which and where to use. *Postgraduate Medicine* 83, no. 1 (January):101.

Neu, H. C. 1984. Changing mechanisms of bacterial resistance. *American Journal of Medicine* Section 77(1B):11.

Newton, R. W., and A. R. W. Forrest. 1975. Rifampicin overdosage—the red man syndrome. *Scottish Medical Journal* 20:55.

Parry, M. F. 1987. The penicillins. *Medical Clinics of North America* 71, no. 6 (November):1093.

Peterson, P. K., and J. Verhoef. 1986. *The antimicrobial agents annual I.* New York: Elsevier.

Physician's Desk Reference. (Annual.) Oradell, NJ: Medical Economics Co.

Pratt, W. B., and R. Fekety. 1986. *The antimicrobial drugs.* New York: Oxford University Press.

Russell, A. D., et al. 1986. Bacterial resistance to antiseptics and disinfectants. *Journal of Hospital Infections* 7, no. 3 (May):213.

Sanders, C. C., and W. E. Sanders. 1985. Microbial resistance to newer generation beta-lactam antibiotics. *Journal of Infectious Diseases* 151:399.

Silberer, J. 1987. Drug resistance: Malaria-cancer similarity? *Science News* 131 (March 7):148.

Skolnick, A. 1991. New insights into how bacteria develop antibiotic resistance. *The Journal of the American Medical Association* 265, no. 1 (January):14–17.

Thomas, L. 1983. *The youngest science.* New York: The Viking Press.

Weiss, R. 1988. Delivering the goods. *Science News* 133 (June 4):360. (Problems with getting drugs to disease-causing agents.)

Wexler, H. M. 1991. Susceptibility testing of anaerobic bacteria: myth, magic, or method? *Clinical Microbiology Reviews* 4(4):470–84.

Young, L. S. 1985. Treatment of infections due to gram-negative bacilli: a perspective of past, present, and future. *Review of Infectious Diseases* 7, Suppl. 4 (November-December):S572.

A manifestation of the disease process: lesions on the arm caused by *Nocardia madurae,* a bacterium belonging to the group of Actinomycetes. These lesions are sometimes called mycetomas.

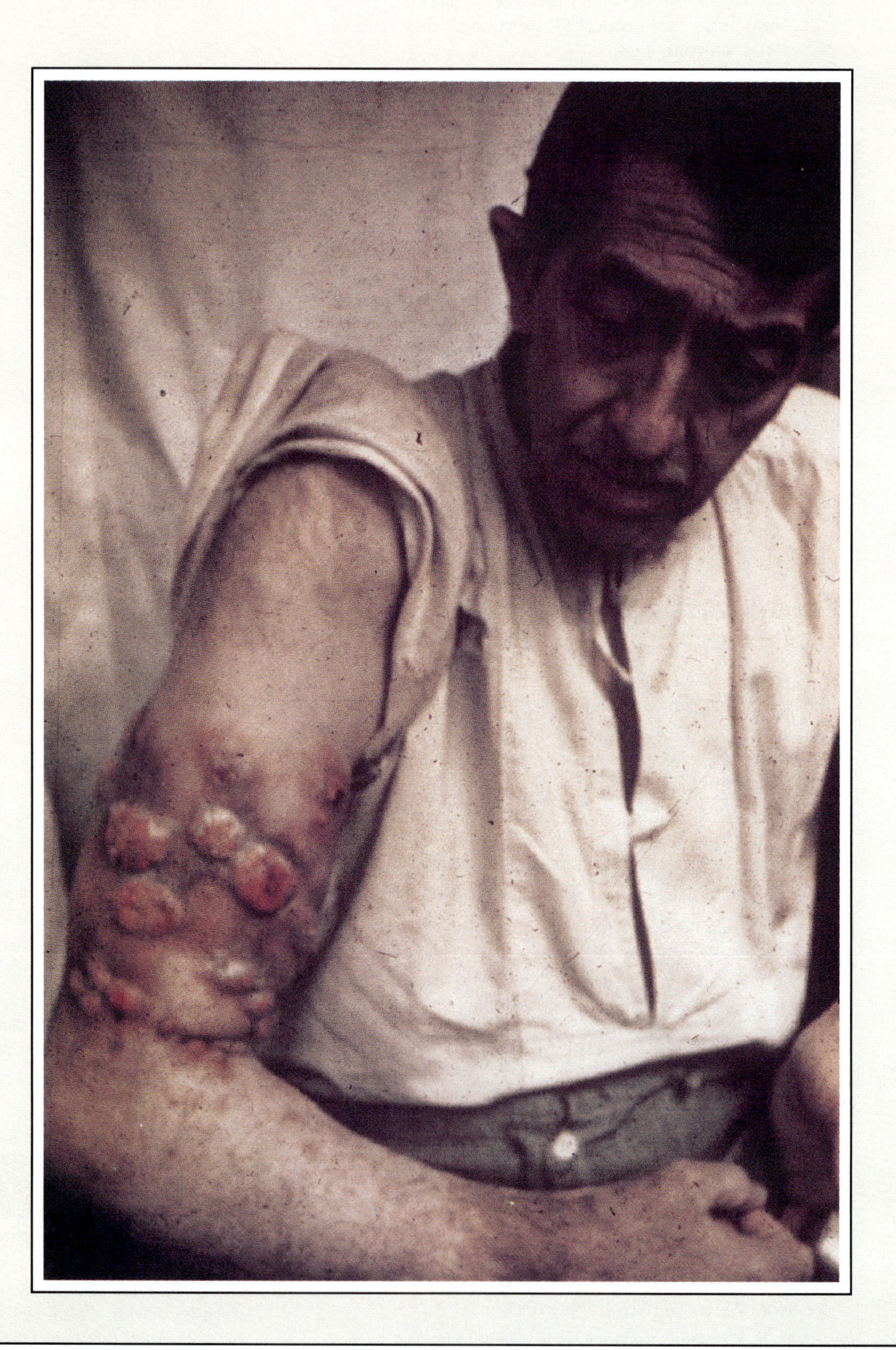

15 Host-Microbe Relationships and Disease Processes

This chapter focuses on the following questions:

A. What terms are used to define host-microbe relationships, and what do they mean?

B. How do Koch's postulates relate to infectious disease?

C. What are the major differences between infectious and noninfectious diseases, between communicable and noncommunicable infectious diseases, and between exogenous and endogenous diseases?

D. How do microbes cause disease?

E. What are the meanings of terms used to describe diseases?

F. What steps occur in the course of an infectious disease?

How is it that every now and then, no matter how careful you are, you "catch" an infectious disease? You become ill. With or without antimicrobial agents, you usually eventually recover from the disease. And in many instances you have developed immunity so that you will not have the disease again.

Whether a pathogen causes disease in a host is a matter of whether the pathogen or the host wins the battle they wage against each other. The pathogen has certain invasive capabilities, and the host has a variety of defenses. For example, in many countries the measles virus is present in some people at all times. Those that are infected release the virus, and it makes its way into the tissues of susceptible individuals. There, it can overcome defenses, invade tissues, and cause disease. However, some potential hosts do not become infected. They may not come in contact with an infected individual.

If the virus does make its way into tissues, the host's defenses may destroy it before it can cause disease. Even when the host's first defenses fail and the disease occurs, the host develops immunity and will not be susceptible to the disease on subsequent exposures.

To begin the study of host-microbe interactions, we will look at a variety of relationships between species and at how some of these relationships result in disease. We will then characterize diseases and look at the disease process brought on by infectious agents.

HOST-MICROBE RELATIONSHIPS

Microorganisms display a variety of complex relationships to other microorganisms and to the larger organisms that serve as hosts for them. A **host** is any organism that harbors another organism.

Symbiosis

All interactions between organisms are examples of **symbiosis,** if we use the literal definition of symbiosis as "living together." Using this definition, symbiosis includes mutualism, commensalism, and parasitism.

In **mutualism** the two species living together both benefit from the relationship. The relationship between certain deep-ocean fish and luminescent bacteria that live on their skin is an example of mutualism. The bacteria acquire nutrients, and the fish use the luminescence to lure prey in their otherwise dark environment. Certain bacteria that live in the human large intestine release metabolic products that we require as vitamins, especially vitamin K, which our bodies need to make certain blood-clotting factors. Although the quantities of these vitamins absorbed by humans are small, they do make a modest contribution to satisfying our vitamin needs. The bacteria, in turn, get free room and board.

In **commensalism** two species of organisms live together in a relationship in which one benefits and the other one neither benefits nor is harmed by the relationship. For example, many microorganisms live on our skin surfaces and make use of metabolic products secreted from pores in the skin (Figure 15.1). Because these products are released regardless of whether they are used by microorganisms, the microorganisms benefit, and ordinarily we are neither benefited nor harmed. By taking up space and utilizing nutrients, however, these organisms may prevent colonization of the skin by other, potentially harmful, microbes—a phenomenon known as **competitive exclusion.** They therefore confer an indirect benefit on the host. This example indicates that the line between commensalism and mutualism is not always a clear one.

In **parasitism** one organism, the parasite, benefits from the relationship, whereas the other organism, the host, is harmed by it. Using this broad definition of the term parasite, bacteria and viruses as well as protozoa, fungi, and helminths are parasites. Animal biologists use the term parasite to refer only to protozoa, fungi, and helminths that live off other organisms. Parasitism encompasses a wide range of relationships, from those in which the host sustains only slight harm to those in which the host is killed. As we saw in Chapter 12, some parasites obtain an easy living by causing only modest harm to their host, and others kill their hosts, thereby rendering themselves homeless. (p. 294) The most successful parasites are those that maintain their own life processes without severely damaging their hosts.

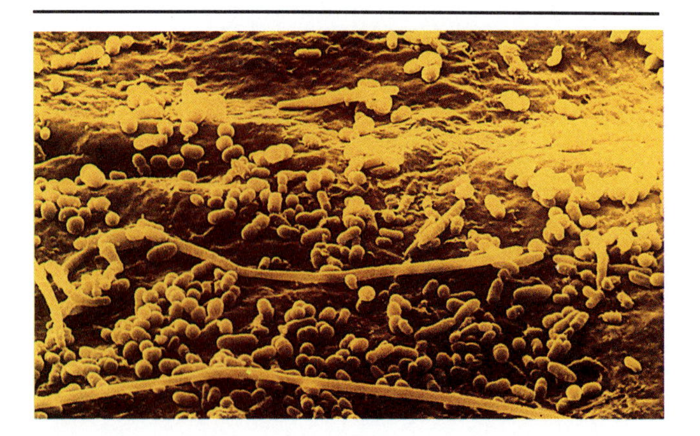

FIGURE 15.1 Bacteria on human skin. Most of these organisms are commensals, which indirectly benefit us by competing with harmful organisms for nutrients and preventing them from finding a site to attach and invade tissues.

Contamination, Infection, and Disease

Contamination, infection, and disease can be viewed as a sequence of conditions in which microorganisms have increasingly significant effects on their hosts. **Contamination** means merely that the microorganisms are present. We have seen that inanimate objects and the surfaces of skin and mucous membranes can be contaminated with a wide variety of microorganisms. Commensals do no harm, but parasites have the capacity to invade tissues. **Infection** refers to the multiplication within or upon the body by any parasitic organism. (Sometimes the term **infestation** is used to refer to the presence of larger parasites, such as worms.) If an infection disturbs the health of the host, disease occurs. **Disease** is a disturbance in the state of health wherein the body cannot carry out all its normal functions.

Infection and disease result from interactions between parasites and their hosts. Sometimes an infection produces no observable effect on the host even though the organisms have invaded tissues. More often an infection produces observable disturbances in health; that is, disease occurs. Even when an infection causes disease, the effects of the disease can range from mild to severe.

Let us look at some examples to distinguish among contamination, infection, and disease. A health care worker who fails to follow aseptic procedures while dressing a skin wound contaminates her hands with staphylococci. However, after she finishes her task she washes her hands and suffers no ill effects. Although her hands were contaminated, she did not get an infection. Another worker performing the same task on another patient fails to wash his hands thoroughly after treating the patient, and the organisms infect a small cut. Soon the skin around the cut becomes reddened for a day or so. This worker was contaminated and infected. In a similar situation, a third worker develops a reddened area on her skin; she ignores this, and in a few days she has a large boil. This worker has experienced contamination, infection, and disease.

Disease, or illness, is characterized by changes in the host that interfere with normal function. These changes can be mild or severe and reversible or irreversible. For example, if you become infected with one of the viruses that cause the common cold, you may have a runny nose for a few days or you may have a severe cold with a sore throat, cough, and fever. However, the disease will run its course in a week or so without any permanent effects. The changes in your state of health are reversible, and you suffer no permanent ill effects. On the other hand, if you develop trachoma, an infection of the eye, scarring of the cornea can lead to permanent vision impairment and sometimes to blindness. Likewise, if you fail to get proper treatment for a streptococcal infection, you might suffer irreversible damage to your heart or kidneys.

Pathogens, Pathogenicity, and Virulence

As we saw in Chapter 12, any parasite capable of causing disease in its host is called a *pathogen*. ∞ (p. 294) Pathogens vary in their abilities to disturb health; that is, they display different degrees of pathogenicity. **Pathogenicity** is the capacity to produce disease. The pathogenicity of an organism depends on its ability to invade a host, multiply in the host, and avoid being damaged by the host's defenses. Some agents, such as *Mycobacterium tuberculosis*, cause disease nearly every time they enter a susceptible host. Other agents, such as *Staphylococcus epidermidis*, cause disease only in rare instances and usually only in hosts with poor defenses. Most infectious agents have a degree of pathogenicity between these extremes. An important factor in pathogenicity is the number of infectious organisms that enter the body. If only a small number enter, the host's defenses may be able to eliminate the organisms before they can cause disease. If a large number enter, they may overwhelm the host's defenses and cause disease.

Virulence refers to the degree of intensity of the disease produced. Virulence varies among different species of pathogen. For example, *Bacillus cereus* causes mild gastroenteritis, whereas the rabiesvirus causes neurological damage that is nearly always fatal. Virulence also varies among members of the same species of pathogen. For example, organisms freshly discharged from a patient usually are more virulent than those from a carrier, who characteristically shows no symptoms.

The virulence of a pathogen can be increased by

CLOSE-UP

Pathogens: Unsuccessful Attempts at Symbiosis

"In real life, however, even in our worst circumstances we have always been a relatively minor interest of the vast microbial world. Pathogenicity is not the rule. Indeed, it occurs so infrequently and involves such a relatively small number of species, considering the huge population of bacteria on earth, that it has a freakish aspect. Disease usually results from inconclusive negotiations for symbiosis, an overstepping of the line by one side or the other, a biological misinterpretation of borders."

—Lewis Thomas, 1974

animal passage, the rapid transfer of the pathogen through animals of a species susceptible to infection by the pathogen. As one animal becomes diseased, organisms from that animal are passed to a healthy animal, and this sequence is repeated several times. Each newly infected animal suffers a more serious case of the disease than the one before it as the microbe becomes better able to invade and damage the host. Sometimes diseases spread through human populations in this fashion, and an epidemic of a severe form of a disease results. Influenza epidemics often proceed in this fashion; the first people to become infected have a mild illness, but those infected later have a much more severe disease.

The virulence of a pathogen can be decreased by **attenuation,** or weakening of the disease-producing ability of the organism. Attenuation can be achieved by repeated subculturing on laboratory media, by treating with chemical substances such as formaldehyde, or by transposal of virulence. **Transposal of virulence** is a laboratory technique in which a pathogen is passed from its normal host to a new host species and then sequentially through many individual members of the new host species. Eventually, the pathogen adapts so completely to the new host that it is no longer virulent for the original host. In other words, virulence has been transposed to another organism. Pasteur made use of transposal of virulence in preparing rabies vaccine. By repeated passage through rabbits, the virus eventually became harmless to humans and was safe to use in a human vaccine. As Chapter 18 shows, attenuation is an important step in the production of some vaccines in use today.

In the past, pathogenicity was viewed largely in terms of the causative agent. Today, pathogenicity is still regarded as a property of a particular infectious agent, but the response of the host is given greater consideration. Chemotherapy, immunization with vaccines, and improved sanitation practices now assist hosts in resisting the invasion of a pathogen.

Normal (Indigenous) Flora

As we have seen, microorganisms found in the several kinds of symbiotic associations with humans do not necessarily cause disease. An adult human body consists of approximately 10^{13} (10 million million) eukaryotic cells. It harbors an additional 10^{14} prokaryotic and eukaryotic microorganisms on the skin surface and on mucous membranes, or inside passageways of the digestive, respiratory, and reproductive systems. Thus, for every human cell there are 10 microbial cells associated with the body!

Before birth a fetus exists in a sterile environment. At birth certain microorganisms become permanently or temporarily associated with it. These organisms are referred to collectively as **normal flora** (Table 15.1).

TABLE 15.1 Major normal flora of the human body (unless otherwise noted, the organisms listed are bacteria.)

Skin
Staphylococcus epidermidis
Staphylococcus aureus
Lactobacillus species
Propionibacterium acnes
Pityrosporon ovale (fungus)
Upper Respiratory Tract
Staphylococcus epidermidis
Staphylococcus aureus
Streptococcus mitis
Streptococcus pneumoniae
Branhamella catarrhalis
Lactobacillus species
Haemophilus influenzae
Mouth
Streptococcus hominis (salivarius)
Streptococcus pneumoniae
Streptococcus mitis
Staphylococcus epidermidis
Staphylococcus aureus
Branhamella catarrhalis
Veillonella alcalescens
Lactobacillus species
Klebsiella species
Haemophilus influenzae
Fusobacterium nucleatum
Treponema denticola
Candida albicans (fungus)
Entamoeba gingivalis (protozoan)
Trichomonas tenax (protozoan)
Intestine
Staphylococcus epidermidis
Staphylococcus aureus
Streptococcus mitis
Enterococcus species
Lactobacillus species
Clostridium perfringens
Clostridium tetani
Eubacterium limosum
Bifidobacterium bifidum
Actinomyces bifidus
Escherichia coli
Enterobacter species
Klebsiella species
Proteus species
Pseudomonas aeruginosa
Bacteroides fragilis
Bacteroides melaninogenicus
Bacteroides oralis
Fusobacterium nucleatum
Fusobacterium necrophorum
Treponema denticola
Endolimax nana (protozoan)
Giardia intestinalis (protozoan)
Urogenital tract
Streptococcus mitis
Staphylococcus epidermidis
Streptococcus species
Lactobacillus species
Clostridium species
Actinomyces bifidus
Candida albicans (fungus)
Trichomonas vaginalis (protozoan)

APPLICATIONS

Can a Bloodhound Find the Correct Identical Twin?

As we walk around, we shed a "dandruff cloud" of skin flakes, which falls to the ground or lands on nearby objects. A bloodhound sniffs these flakes to follow a trail. Normal flora metabolize oils and other secretions into byproducts with particular odors. Identical twins are not colonized by identical normal flora, and so their skin flakes develop slightly different odors which a bloodhound can distinguish, leading him unerringly to the right twin.

Following the unique mixture of odors produced by normal body flora present on skin flakes, which have been shed, a bloodhound trails his quarry.

Most organisms among the normal flora are commensals—they derive their living from waste substances on the surfaces of skin and mucous membranes. Among the normal flora, three categories of organisms can be distinguished: resident flora, transient flora, and opportunists.

Species that are always present on or in the human body comprise the **resident flora.** They are found on the skin and conjunctiva, in the mouth, nose, and throat, in the large intestine, and in passageways of the urinary and reproductive systems, especially near their openings (Figure 15.2). In each of these body regions resident flora are adapted to prevailing conditions. The mouth and the lower part of the large intestine provide warm, moist conditions and ample nutrients. Mucous membranes of the nose, throat, urethra, and vagina also provide warm, moist conditions, although nutrients are in shorter supply. The skin provides ample nutrients but is cooler and less moist.

Other regions of the body lack resident flora either because they provide conditions unsuitable for microorganisms, are protected by host defenses, or are inaccessible to microorganisms. For example, conditions in the stomach are too acidic to permit survival of flora. Blood has no resident flora because it is relatively inaccessible, and host defense mechanisms destroy microorganisms before they become established.

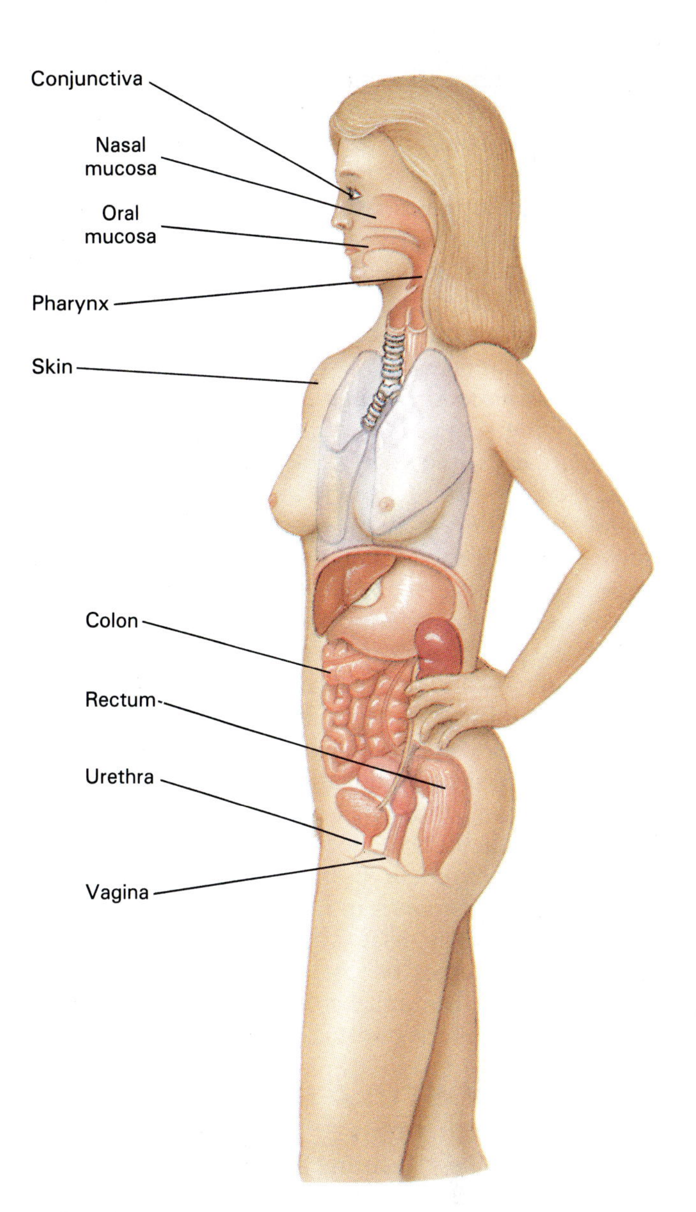

FIGURE 15.2 Locations of resident flora of the human body.

Transient flora can be present under certain conditions in any of the locations where resident flora are found. They persist for hours to months, but only as long as the necessary conditions are met. Transient flora appear on mucous membranes when greater than normal quantities of nutrients are available or on the skin when it is warmer and more moist than usual. Even pathogens can be transient flora. For example, suppose that you come in contact with a child infected with measles, and some of the viruses enter your nose and throat. You had measles years ago and are immune to the disease, so your body's defenses prevent the viruses from invading cells. But the viruses are present as transients for a short period of time.

Finally, among the resident and transient flora are some species of organisms that do not usually cause disease but can do so under certain conditions. These organisms are called **opportunists** because they take advantage of particular opportunities to cause disease. Conditions that create opportunities for these organisms include failure of the host's normal defenses, introduction of the organisms into unusual sites, and disturbances in the normal flora. Failure of host defenses leads to opportunistic infections in patients with AIDS and other immune disorders. The bacterium *Escherichia coli* is a normal resident of the human large intestine, but it can cause disease if it gains entrance to unusual sites such as the urinary tract, surgical wounds, or burns.

Thriving populations of normal flora provide an important benefit to humans by competing with pathogenic organisms and even in some instances actively combating their growth, an effect known as **microbial antagonism.** The normal flora interfere with the growth of pathogens by depleting nutrients the pathogens need or by creating an acidic environment in which the pathogens cannot grow. As we have seen in Chapter 14, antibiotics sometimes destroy normal flora as they bring a pathogen under control. ∞ (p. 355) This allows other potential pathogens, such as yeasts not harmed by the antibiotic, to thrive in the absence of their antagonists, the normal flora.

In later chapters we focus on a few microorganisms that cause human disease. But we must not lose sight of the importance of the many nonpathogenic microorganisms associated with the human body. And we must remember that disease can result from disturbances in the normal ecological balance between resident populations and the host.

KOCH'S POSTULATES

The work of Robert Koch and the role of his postulates in relating causative agents to specific diseases was described briefly in Chapter 1. ∞ (p. 13) Now we can use our understanding of infection and disease to look at those postulates more carefully. For example, we now know that infection with an organism does not necessarily indicate that disease is present. Because of that knowledge we can better appreciate the need for all four of **Koch's postulates** to be met to prove that an organism is the causative agent of a particular disease:

1. A specific causative agent must be observed in every case of a disease.
2. The agent must be isolated from a host displaying the disease and grown in pure culture.
3. When the agent from the pure culture is inoculated into susceptible hosts, it must cause the disease.
4. The agent must be reisolated from the diseased host and identified as the original specific causative agent.

It is relatively easy today to demonstrate that each postulate is met for a variety of diseases caused by bacteria (Figure 15.3). However, intracellular parasites such as viruses and rickettsias cannot be grown in artificial media and must instead be grown in cells. Because cells sometimes harbor latent viruses, it can be difficult to obtain pure cultures of intracellular

CLOSE-UP

Armadillo: Culture Vessel for Leprosy

The organism that causes Hansen's disease (leprosy) is very difficult to culture. Many different methods had been tried and found unsatisfactory until someone tried inoculating the organism into the footpads of the nine-banded armadillo. Here it grows very well; in fact, the organism multiplies faster here than in human tissues. When the organism does infect humans it can have an incubation period of up to 30 years before disease symptoms appear. Before using the armadillo to culture the organism, Koch's third postulate could not be fulfilled. No one wanted to hold out an arm and say, "Here, try to give me leprosy." Also, 30 years was a long time to wait to determine the results of such an experiment. Using the armadillo, it has been possible to confirm Koch's postulates for *Mycobacterium leprae* as the causative agent of leprosy. This organism, seen by Armauer Hansen in 1878, was one of the first infectious agents to be identified and associated with a disease but one of the last to satisfy Koch's postulates. The armadillo was chosen after naturally occurring leprosy infections were found in the armadillo populations of Texas and Louisiana. Happily no cases of human infection arising from contact with armadillos have ever been found.

Armadillos are used to culture the organism that causes Hansen's disease in humans.

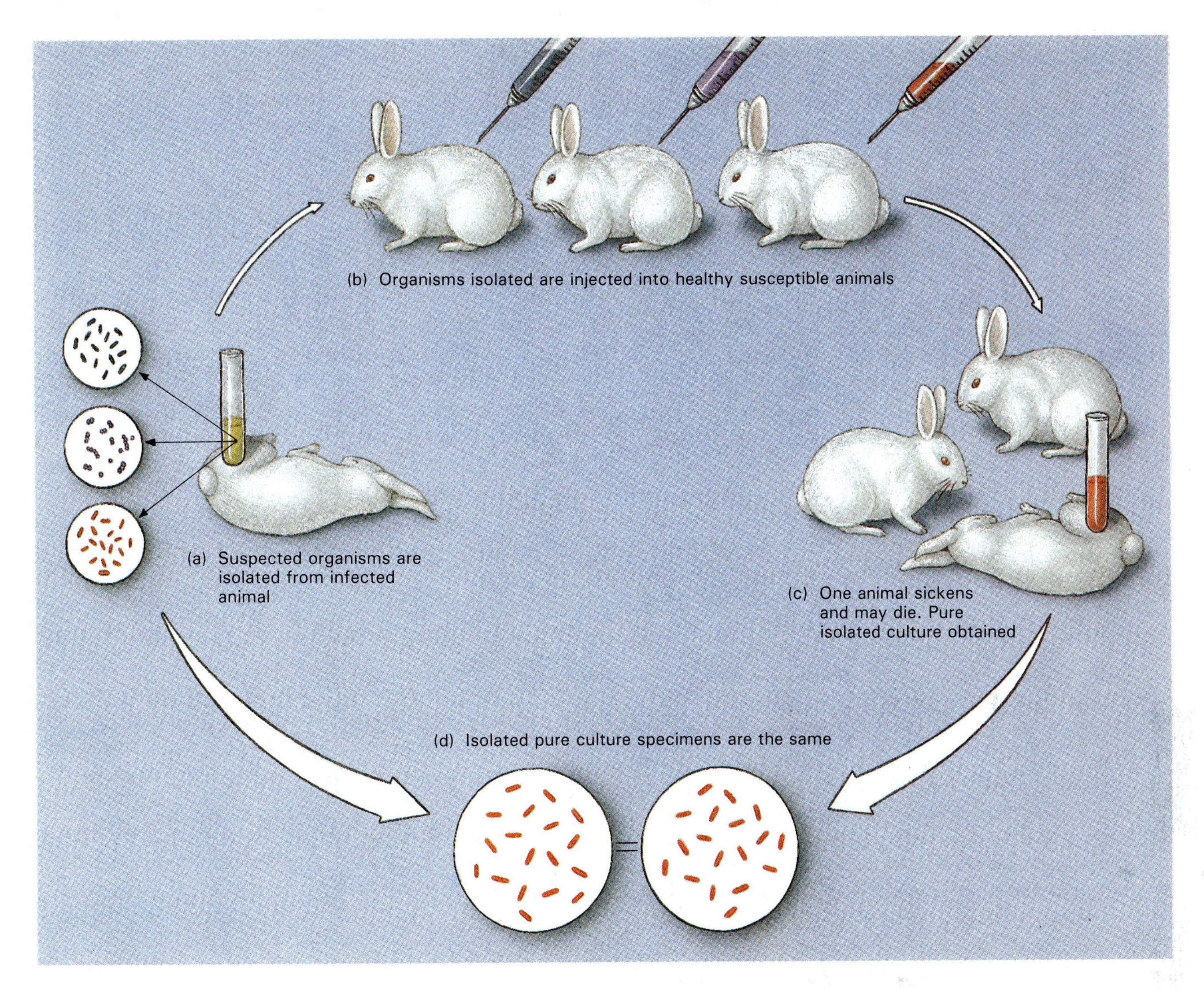

FIGURE 15.3 Demonstration that a bacterial disease satisfies Koch's postulates.

pathogens. Other parasites are difficult to culture because they have fastidious nutritional requirements or other special needs. Although the causative agent of syphilis, *Treponema pallidum*, has been known for many years, it has not been successfully grown on artificial media. For some agents that cause disease in humans, no other host has been found. Consequently, inoculation into a susceptible host is impossible unless human volunteers can be found. And there are ethical problems associated with inoculating humans with infectious agents, even if volunteers might be available.

KINDS OF DISEASES

Human disease can be caused by infectious agents, structural or functional defects, environmental factors, or any combination of these agents. Although in microbiology we are concerned mainly with infectious diseases, it is important to recognize other kinds of diseases and how microbes interact with other factors to cause disease. Among infectious diseases, it is important to understand why some are more communicable, or easily spread, among human populations than others.

Infectious and Noninfectious Diseases

Infectious diseases are diseases caused by infectious agents such as bacteria, viruses, fungi, and parasitic protozoa and helminths. Chapters 20 through 25 of this text are devoted to discussions of infectious agents and the diseases they cause. **Noninfectious diseases** are caused by any factor other than infectious organisms.

Classification of Diseases

Classifying diseases as infectious or noninfectious gives a very limited view of human disease. The following scheme for classifying diseases provides a more comprehensive view of disease and shows how infectious agents interact with other factors to cause disease:

1. **Inherited diseases** are due to errors in genetic information present in the fertilized egg. Some inherited diseases, such as sickle-cell anemia, weaken patients and make them more susceptible to infectious diseases.
2. **Congenital defects** are structural and functional defects present at birth. When a mother has a rubella (German measles) or a syphilis infection, the infectious agent crosses the placenta and often causes congenital defects in the infant. Some medicines, such as the antiwrinkle drug retinoid-A, also cause congenital defects when taken by pregnant women.
3. **Degenerative diseases** are disorders that develop in one or more body systems as aging occurs. Patients with degenerative diseases such as emphysema or impaired kidney function are susceptible to lung and kidney infections, respectively. Conversely, infectious agents can cause tissue damage that leads to degenerative disease, as occurs in bacterial endocarditis, rheumatic heart disease, and some kidney diseases.
4. **Nutritional deficiency diseases** lower resistance to infectious diseases and contribute to the severity of infections. For example, the bacterium that causes diphtheria produces more toxin in people with iron deficiencies than in those with normal amounts of iron. Poor nutrition also increases the severity of measles and contributes to deaths from the disease. Conversely, several helminths severely damage the intestinal lining, causing patients to develop nutritional deficiencies.
5. **Endocrine diseases** are due to excesses or deficiencies of hormones. Viral infection has been linked to pancreatic damage that leads to insulin-dependent diabetes.
6. **Mental disease** can result from brain infections such as congenital syphilis and a mysterious disease called Creutzfeld-Jakob disease. Conversely, emotional stress predisposes toward diseases such as asthma and dermatitis and infections that sometimes occur with these diseases.
7. **Immunological diseases** such as allergies, autoimmune diseases, and immunodeficiencies are caused by malfunction of the immune system. AIDS is a consequence of a viral infection of certain cells of the immune system called T lymphocytes.
8. **Neoplastic diseases** can be benign tumors or malignant, invasive cancers. As we saw in Chapter 11, viruses contribute to the development of some kinds of cancer. ∞ (p. 274)
9. **Iatrogenic** (i-at″ro-gen′ik) **diseases** (*iatros*, Latin for physician) are caused by medical treatment. They include malignancies triggered by medicines or radiation, surgical errors, drug reactions, and infections acquired during hospital treatment. The latter are called nosocomial infections and are discussed in Chapter 16.
10. **Idiopathic** (id-e-o-path′ik) **diseases** are diseases whose cause is unknown. An example is Alzheimer's disease, which causes mental deterioration. Some researchers think an infectious agent may play a role in the development of this disease.

Communicable and Noncommunicable Diseases

Some infectious diseases can be spread from one host to another and are said to be communicable. Some **communicable diseases** are more easily spread than others. Rubeola (red measles) and rubella are highly communicable, or **contagious diseases,** especially among young children. Vaccines protect children in developed countries, but nearly all children in underdeveloped countries still get these diseases. Influenza is highly communicable among adults, especially the elderly. Gonorrhea and genital herpes infections are easily spread among sexual partners. Although also communicable, certain other diseases such as Hansen's disease and *Klebsiella* pneumonia are less contagious. Some diseases that normally affect animals are transmissible to humans (Chapter 16).

Noncommunicable infectious diseases are acquired from the environment but are not spread from one host to another, even though they are caused by infectious agents. An example is tetanus, which is a bacterial infection of a wound acquired from spores in soil. Other noncommunicable infectious diseases, such as a kind of pneumonia called legionellosis, are spread when waterborne organisms are incorporated into aerosols.

Some diseases caused by microbes are not due to infection and invasion of tissues by the organisms. Instead, they are due to ingestion of preformed toxins made by the organisms. For example, botulism food poisoning strikes within hours of ingesting food that contains a toxin produced by *Clostridium botulinum*—too short a time for the microbe to invade tissues and cause disease. The toxins accumulate during the storage of an improperly sterilized jar or can of food and have an immediate and often ultimately lethal effect on the consumer. Such a disease is therefore properly termed an **intoxication** rather than an infection.

Exogenous and Endogenous Diseases

Exogenous diseases are caused by microorganisms from outside the body. They include all diseases transmitted from human and nonhuman hosts and from the environment—for instance from food or soil. In contrast, **endogenous diseases** are caused by microorganisms already present in or on the body. They are caused by opportunistic flora that invade tissues in hosts with lowered resistance.

THE DISEASE PROCESS

How Microbes Cause Disease

Microorganisms act in certain ways that allow them to cause disease. These actions include gaining access to the host, adhering to and colonizing cell surfaces, invading tissues, and producing toxins and other harmful metabolic products. However, host defense mechanisms tend to thwart the actions of microorganisms. Whether disease occurs is determined by whether the pathogen or the host wins this battle.

Most of the parasites to be considered in this text are prokaryotic microorganisms and viruses. However, eukaryotes such as fungi, parasitic protozoa, and multicellular parasites (mostly worms) display pathogenicity similar to that produced by prokaryotic microorganisms. (Chapter 12, p. 308) They can be present in a host without causing disease symptoms, or they can cause severe disease. The extent of damage caused by these parasites, like that caused by prokaryotic infectious agents, is determined by the properties of the parasites and the host's response to them. Thus, it appears that parasitologist Theobald Smith was correct when in 1934 he described parasitism as a universal phenomenon representing normal interactions and interdependency among all living things.

How Bacteria Cause Disease

Direct Actions of Bacteria Bacteria enter the body by penetrating the skin, usually at wound sites, by being ingested with food, by being inhaled in aerosols, or by transmission on a *fomite*, which is any inanimate object contaminated with an infectious agent. If the bacteria immediately are swept out of the body in urine or feces or by coughing or sneezing, they cannot initiate an infection.

A critical point in the production of bacterial disease is the organism's **adherence** to a host cell's surface. Whether certain infections occur depends in part on the interaction between host cell membranes and bacterial adhesins. **Adhesins** are substances that help pathogens attach to the host cell. Although adhesins are only now being identified, *E. coli*, some strains of streptococci, and influenza viruses are known to have specific adhesins.

Each particular adhesin apparently permits the pathogen to adhere only to certain tissues. For example, an adhesin on surface fimbriae of certain strains of *E. coli* attaches to receptors on certain host epithelial cells. (Host leukocytes also have receptors for this adhesin, so the same adhesin that helps the bacterium to attach may also help the host to destroy it.) Adhesins of some streptococci consist of glycoproteins that allow them to adhere to teeth. Adhesins of other streptococci allow them to adhere to respiratory epithelium.

Attachment to a host surface alone is not enough to cause an infection. The microbes also must be able to colonize the surface, penetrate the surface, or produce toxins that can penetrate it.

Colonization refers to the growth of microorganisms on epithelial surfaces, such as skin or mucous membranes. For colonization to occur after adherence, the pathogens must be able to survive and reproduce in spite of host defense mechanisms. Those on the skin's surface must withstand environmental conditions and bacteriostatic skin secretions. Those on respiratory membranes must escape the action of mucus and cilia. Those on membranes in the digestive tract must withstand peristaltic movements, mucus, digestive enzymes, and acid.

Only a few pathogens cause disease by colonizing surfaces; most invade tissues. The degree of **invasiveness** of a pathogen sometimes is related to the severity of disease it produces. Some bacteria, such as pneumococci and streptococci, release digestive enzymes that allow them to invade tissues rapidly and cause severe illnesses. At the other extreme, some viruses can invade tissues without causing any apparent illness. Even the same pathogen can display varying degrees of invasiveness and pathogenicity in different tissues. Both bubonic plague and pneumonic plague are caused by the bacterium *Yersinia pestis*. In bubonic plague the organisms enter the body with a flea bite, migrate through the blood, and infect many organs and tissues. This disease has a mortality rate of about 55 percent. In pneumonic plague, as a plague victim coughs or sneezes, the bacteria are spread by aerosol to the next hapless victim as that victim inhales. They cause a severe infection of the lungs with a mortality rate as high as 98 percent.

Most bacteria that invade tissues damage cells and are found around cells, although some actually enter cells (Figure 15.4). Viruses, rickettsias, and a few other pathogens must invade cells to be able to grow and produce disease. Pathogens that establish themselves within cells are out of reach of most host defense mechanisms. Those that can survive within host phagocytic cells not only escape destruction by the phagocytes, they also obtain free transportation to deeper body tissues.

Actions of Toxins Although the growth of pathogens within tissues causes tissue damage and disease, production of toxins also plays an important role in disease. A **toxin** is a poisonous substance. *Shigella* species and many other bacteria produce toxins, as do some protozoa, fungi, and helminths. Viruses do not produce toxins.

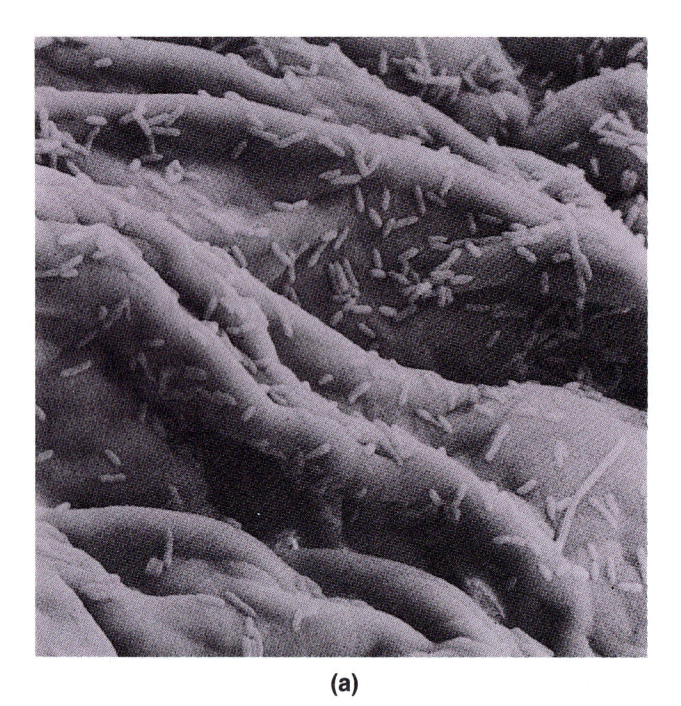

(a)

(b)

FIGURE 15.4 Bacteria can live (a) outside cells, like these bacilli growing on the surface of salamander skin (SEM, 2170X), or (b) inside cells, like these *Mycobacterium avium* intracellular bacilli (stained fuschia by the acid-fast method) seen growing inside the pale blue cytoplasm of skin cells.

Bacterial toxins, which are synthesized inside bacterial cells, are classified according to how they are released. **Exotoxins** are soluble substances secreted into host tissues. **Endotoxins** are incorporated into the cell wall and released into host tissues—sometimes in large quantities—when gram-negative bacteria die. (Chapter 4, p. 80) Giving antibiotics that kill such bacteria can release sufficient toxin to cause the patient to die of severely reduced blood pressure (endotoxic shock). Properties of exotoxins and endotoxins are summarized in Table 15.2.

Exotoxins are powerful toxins produced by several gram-positive and a few gram-negative bacteria. Most are polypeptides, which are relatively easily denatured by heat, ultraviolet light, and chemicals such as formaldehyde. Species of *Clostridium, Bacillus, Staphylococcus, Streptococcus,* and certain other bacteria produce exotoxins and cause intoxications rather than attacking tissues directly. Exotoxins have special affinity for particular tissues. **Neurotoxins** act on tissues of the nervous system. **Enterotoxins** act on tissues of the gut.

TABLE 15.2 Properties of toxins

Property	Exotoxins	Endotoxins
Organisms producing	Usually gram-positive, some gram-negative	Almost all gram-negative
Location in cell	Extracellular, excreted into medium	Bound within bacterial cell wall; released upon death of bacterium
Chemical nature	Mostly polypeptides	Lipopolysaccharide complex
Stability	Unstable; denatured above 60°C and by ultraviolet light	Relatively stable; can withstand several hours above 60°C
Toxicity	Among the most powerful toxins known (some are 100 to 1,000,000 times the strength of strychnine)	Weak, but can be fatal in relatively large doses
Effect on tissues	Highly specific; some act as neurotoxins and cardiac muscle toxins	Nonspecific; ache-all-over systemic effects or local site reactions
Fever production	Little or no fever	Rapid rise in temperature to high fever
Antigenicity	Strong; stimulates antibody production and immunity	Weak; recovery from disease often does not produce immunity
Toxoid conversion and use	By treatment with heat or chemicals; toxoid used to immunize against toxin	Cannot be converted to toxoid; cannot be used to immunize
Examples	Botulism, gas gangrene, tetanus, diphtheria, staphylococcal food poisoning, cholera enterotoxins, plague	Salmonellosis, tularemia, endotoxic shock

Many exotoxins can act as *antigens*, foreign substances against which the immune system reacts. Antigenic exotoxins inactivated by treatment with chemical substances such as formaldehyde are called **toxoids.** A toxoid (*-oid*, Latin for like) is an altered toxin that has lost its ability to cause harm but that retains antigenicity. Toxoids can be used to stimulate development of immunity without causing disease. The roles of bacterial exotoxins in human disease are summarized in Table 15.3. The specific effects of such diseases are discussed in later chapters.

Endotoxins are relatively weak toxins produced by certain gram-negative bacteria. All consist of lipopolysaccharide complexes, the components of which vary among genera. They are relatively stable molecules that do not display affinities for particular tissues. Bacterial endotoxins have nonspecific effects such as fever and a sudden drop in blood pressure. Endotoxins cause tissue damage in human diseases such as typhoid fever and epidemic meningitis, an inflammation of membranes that cover the brain. In addition to the exotoxins and endotoxins just described, some pathogens release proteins such as hemolysins, leukocidins, and certain enzymes that act like toxins.

Hemolysins were first discovered in cultures of bacteria grown on blood agar. Their action is to lyse (rupture) red blood cells, and two kinds were identified from their effects on blood in cultures. **Alpha hemolysins** partially hemolyze blood cells and leave a greenish ring around colonies; **beta hemolysins** completely hemolyze blood cells and leave a clear ring around colonies (Figure 15.5). Streptococci and staphylococci produce hemolysins that are helpful in identifying them in laboratory cultures. Iron is a critical element for growth of all cells, both host and microbe. There is very little free iron within the human body. Most of it is bound in a form such as hemoglobin, and the microbe must release it. Without access to the iron freed by hemolysis, microbes can still cause disease, but not as easily.

Leukocidins, also produced by streptococci and staphylococci, are similar to hemolysins, and some may be identical. Leukocidins damage or destroy cer-

TABLE 15.3 Effects of exotoxins

Bacterium	Name of Toxin or Disease	Action of Toxin	Host Symptoms
Bacillus anthracis	Anthrax	Increases vascular permeability	Hemorrhage and pulmonary edema
Bacillus cereus	Enterotoxin	Causes excessive loss of water and electrolytes	Diarrhea
Clostridium botulinum	Botulism (six serological types)	Blocks release of acetylcholine at nerve endings	Respiratory paralysis, double vision
Clostridium perfringens	Gas gangrene (alpha toxin, a hemolysin)	Breaks down lecithin in cell membranes	Tissue destruction
	Food poisoning (enterotoxin)	Causes excessive loss of water and electrolytes	Diarrhea
Clostridium tetani	Tetanus (lockjaw)	Inhibits antagonists of motor neurons of brain; 1 nanogram can kill 2 tons of cells	Violent skeletal muscle spasms, respiratory failure
Corynebacterium diphtheriae	Diphtheria; produced by virus-infected bacteria	Inhibits protein synthesis	Heart damage can cause death weeks after apparent recovery
Escherichia coli	Traveler's diarrhea (enterotoxin)	Causes excessive loss of water and electrolytes	Diarrhea
Pseudomonas aeruginosa	Exotoxin A	Inhibits protein synthesis	Lethal, necrotizing lesions
Shigella dysenteriae	Bacillary dysentery (enterotoxin)	Cytotoxic effects; as potent as botulinium toxin	Diarrhea, causes paralysis in rabbits from spinal cord hemorrhage and edema
Staphylococcus aureus	Food poisoning	Stimulates brain center that causes vomiting	Vomiting
	Scalded skin syndrome	Causes intradermal separation of cells	Redness and sloughing of skin
Streptococcus pyogenes	Scarlet fever (erythrogenic, or red-producing toxin)	Causes vasodilation	Maculopapular (slightly raised, discolored) lesions
Vibrio cholerae	Cholera (enterotoxin)	Causes excessive loss of water (up to 30 L/day) and electrolytes	Diarrhea; can kill within hours

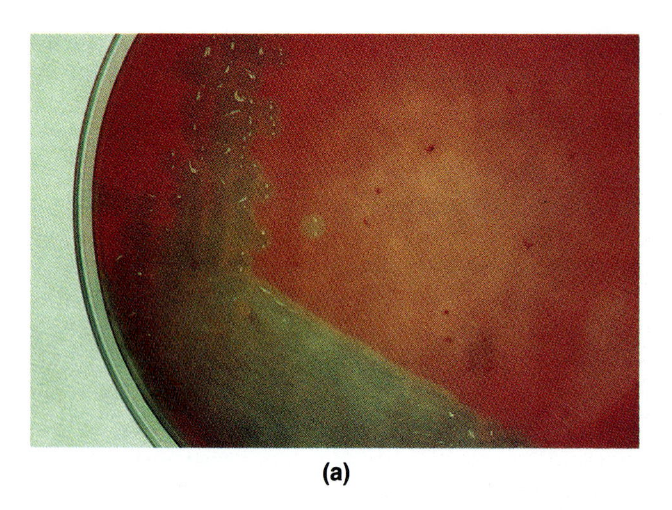

(a)

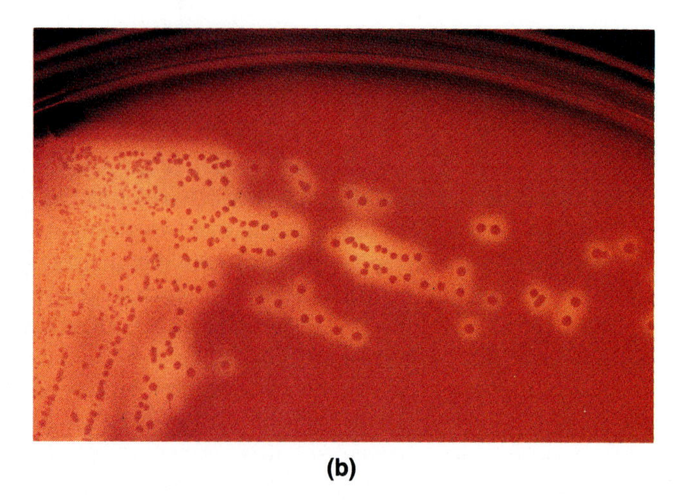

(b)

FIGURE 15.5 (a) Alpha, or partial, hemolysis of red blood cells results in a greenish zone around colonies of *Streptococcus pneumoniae* grown on blood agar. (b) *Streptococcus pyogenes* colonies release beta hemolysins, which produce complete hemolysis, causing clear zones to form around colonies grown on blood agar.

tain kinds of white blood cells called neutrophils. They are most effective when released by microbes that have been engulfed by a neutrophil. Because of their action the number of white blood cells declines in certain diseases, although most infections are characterized by an elevated white cell count. A similar substance, called **leukostatin,** interferes with the ability of leukocytes to engulf microorganisms that release it.

Bacteria produce a variety of enzymes that contribute to tissue damage. They include hyaluronidase and fibrinolysin produced by streptococci, and co-

APPLICATIONS

Clinical Use of Botulinum Toxin

The powerful effects of the exotoxin produced by *Clostriduim botulinum*, best known as a cause of lethal food poisoning, have been harnessed to help victims of dystonia. *Dystonia* refers to a group of neurologic disorders characterized by abnormal, sustained involuntary movements, often twisting, which are of unknown origin. In one form, blepharospasm, the patient's eyes remain tightly closed at all times. Doctors now inject small quantities of botulinum toxin (trade name, Oculinum A) at several sites around each eye. The toxin blocks nerve impulses to muscles, thereby relieving the spasms of the eyelids. Injections are needed every 2 to 3 months. Some people have received such treatment for 4 to 5 years without problems. Other people develop antibodies against the toxin after many injections. Because there are seven different forms of the toxin, it is hoped that patients can switch to a different form once they form antibodies against one version. A round of injections can cost from $400 to $1800.

Other dystonias being helped by botulinum toxin include oromandibular, in which the patient clenches his or her jaws together so tightly that the bones may actually break, causing midfacial collapse. Eating and speaking are difficult, and patients starved to death. Vocal cord spasms, causing a cracked tremulous voice, and "stenographers cramp," causing the middle finger to extend rigidly, are also being treated experimentally with the toxin.

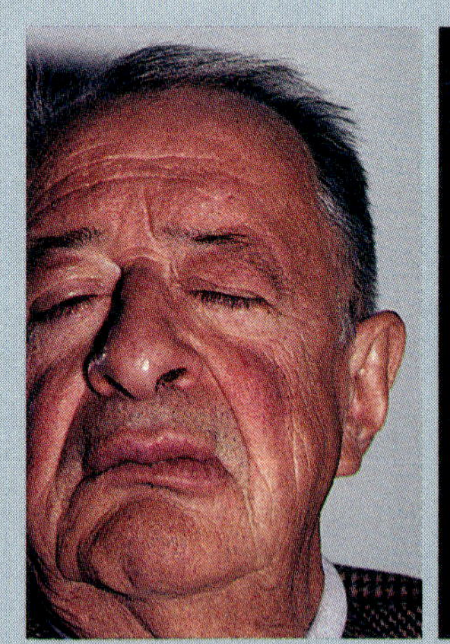

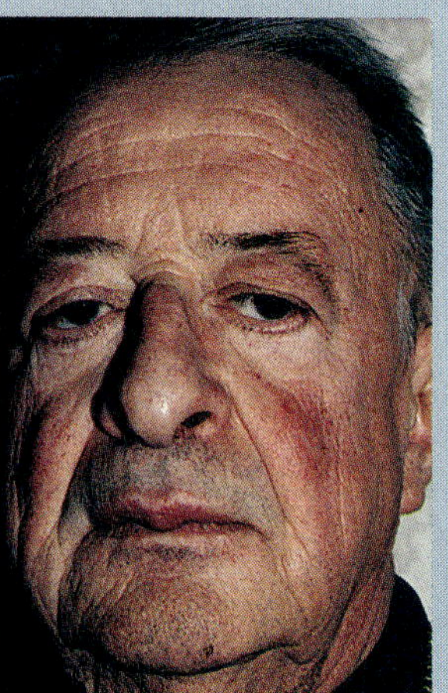

A patient with hemifacial spasm before (left) and two weeks after (right) treatment with botulinum toxin.

Oculinum A is licensed for use in treating adults with strabismus (cross-eye, lazy-eye). Small amounts are injected into the overcontracted eye muscle, which then relaxes and lengthens. The antagonistic muscles on the other side of the eye will contract to take up the slack, and the eye is able to look straight ahead.

PUBLIC HEALTH

Clotbusters

Each year, approximately 1.5 million Americans have heart attacks, with 350,000 dying before they reach the hospital. Another 400,000 don't seek medical aid. Of the 750,000 arriving alive at the hospital, 10 percent subsequently die. However, new "clotbuster" drugs can cut this death rate in half! The streptokinase enzyme produced by streptococci dissolves blood clots in coronary arteries, thereby limiting damage to the heart. When administered within 1 hour after onset of symptoms, the death rate drops 47 percent. Administering the drug in the first 24 hours reduces the death rate by 42 percent. Not all patients are considered candidates for clotbusters. People of advanced age (over 75) or with medical factors such as very high blood pressure or history of strokes are not given these drugs because these "Clotbusters" can cause internal bleeding, leading to brain hemorrhages or strokes. Still, an estimated 400,000 to 600,000 U.S. patients could be given such drugs, saving between 20,000 and 30,000 lives each year. In Europe, 90 percent of heart attack patients routinely receive streptokinase. By contrast, in the United States only 20 percent receive clot-dissolving treatment. Moreover, most U.S. patients receive genetically engineered drugs that cost between $1700 and $2000 per dose, rather than streptokinase at $150 to $200 per dose. Yet streptokinase is just as effective and has a slightly lower probability of causing strokes.

agulase produce by staphylococci. (All of these substances also are produced by other bacteria.) **Hyaluronidase**, or spreading factor, digests hyaluronic acid, a substance that helps to hold the cells of certain tissues together. Digestion of this gluelike substance allows the organism to pass between skin cells and invade deeper tissues. **Coagulase** accelerates the coagulation (clotting) of blood. When blood plasma, the fluid portion of blood, leaks out of vessels into tissues, coagulase causes the plasma to clot. With respect to infections, coagulase is a two-edged sword: It keeps organisms from spreading, but also helps to wall them off from white blood cells that might otherwise destroy them. **Kinases,** the most familiar of which is **streptokinase,** or **fibrinolysin,** digest fibrin and thereby dissolve blood clots. Pathogens trapped in blood clots free themselves to spread to other tissues by releasing fibrinolysin.

How Viruses Cause Disease

Viruses reproduce only when they have invaded host cells. Once inside a cell, viruses cause observable changes collectively called the **cytopathic effect** (CPE). CPE is **cytocidal** when the viruses kill the cell and **noncytocidal** when they do not. Cytocidal viruses can kill cells by causing enzymes from cellular lysosomes to be released or by diverting the host cell's synthetic processes, thereby stopping the synthesis of host proteins and other macromolecules. CPE can be observed in laboratory cultures with a compound microscope (Figure 15.6) and sometimes with the unaided eye (Figure 15.7). Adenoviruses and herpesviruses cause cells to "balloon," and picornaviruses cause cells to swell and lyse. Paramyxoviruses can cause fusion of cell membranes and the aggregation of up to 100 nuclei in a newly formed giant cell. CPE can be so distinctive that an experienced clinical virologist can make a tentative diagnosis by looking at infected cells through the microscope, even though further tests are needed to confirm it.

Many viruses produce CPE in host cells in the form of *inclusion bodies*, consisting of nucleic acids and proteins not yet assembled into viruses, masses of viruses, or remnants of viruses. Rabiesviruses make inclusion bodies called Negri bodies, which are so distinctive they can be used to diagnose rabies. Ret-

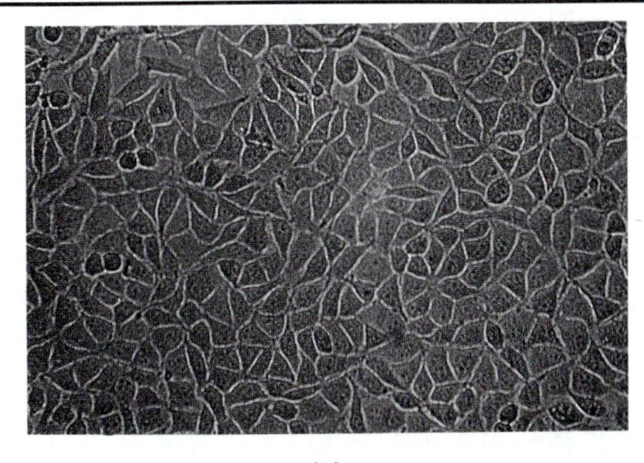

(a)

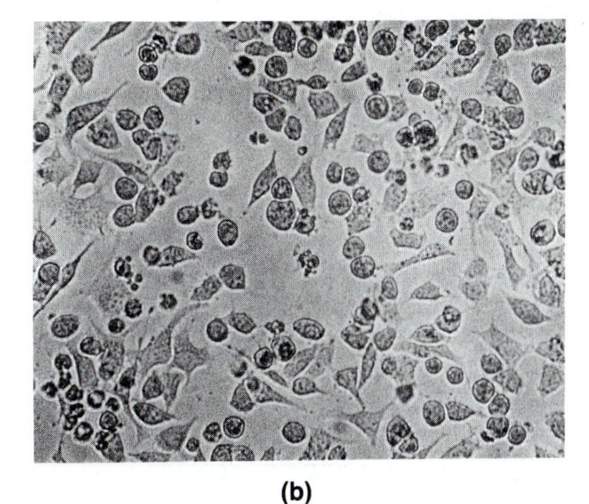

(b)

FIGURE 15.6 An example of cytopathic effect (CPE). (a) Uninfected mouse cells and (b) the same cells 24 hours after infection with vesicular stomatitis virus. Note the large number that have died and others that have rounded up into abnormal shapes.

FIGURE 15.7 Smallpox viruses (pale pink areas) growing on the membranes of a developing chicken egg.

roviruses and oncoviruses integrate into host chromosomes and remain in the cell indefinitely, and they sometimes insert their antigens on host cell surfaces. They are transmitted to daughter cells whenever cell division occurs. Influenza and parainfluenza viruses produce hemagglutinins that cause agglutination, or clumping together, of erythrocytes.

Viral infections can be productive or abortive. A **productive infection** occurs when viruses enter a cell and produce infectious progeny. An **abortive infection** occurs when viruses enter a cell but are unable to express all their genes to make infectious progeny. Productive infections vary in the degree of damage they cause, depending on the kind and number of cells the virus invades. If an enterovirus infects the gut, it can destroy millions of intestinal epithelial cells. Because these cells are rapidly replaced, the infection causes temporary, though sometimes severe, symptoms such as diarrhea but no permanent damage. If a poliovirus infects motor neurons of the central nervous system it can destroy these cells. Because neurons cannot be replaced, permanent paralysis results. Wart viruses infect small numbers of cells in localized areas. In contrast, smallpox and measles viruses replicate and spread throughout the body and damage many tissues.

How Other Pathogens Cause Disease

In addition to bacteria and viruses, infectious diseases also can be caused by fungi, protozoa, and helminths. Even a few algae produce neurotoxins, and one (*Prototheca*) directly invades skin cells.

Fungi damage host tissues by releasing enzymes that attack cells. As the first cells are killed, the fungi progressively digest adjacent cells. Some fungi also release toxins or cause allergic reactions in the host. Certain fungi that parasitize plants produce toxins that cause disease if ingested by humans. Ergot, from a fungus that grows on rye, and aflatoxin, a highly carcinogenic compound that can be found in peanut butter made from moldy peanuts, are fungal toxins (see Chapter 22).

Pathogenic protozoans and helminths cause human disease in several ways. Some pathogens, such as the protozoan *Giardia intestinalis*, attach to tissues and ingest cells and tissue fluids of the host. *Giardia* has an **adhesive disc** by which it attaches to cells that line the small intestine (Figure 15.8). As the parasite burrows into the tissue it uses its flagella to expel tissue fluids. This process creates so strong a suction that the parasite is not disturbed by peristaltic contractions. Many helminths release toxic waste products that often cause allergic reactions in the host. Such infections are difficult to treat because killing the parasites may lead to an allergic reaction much more serious than the infection itself. Some protozoans,

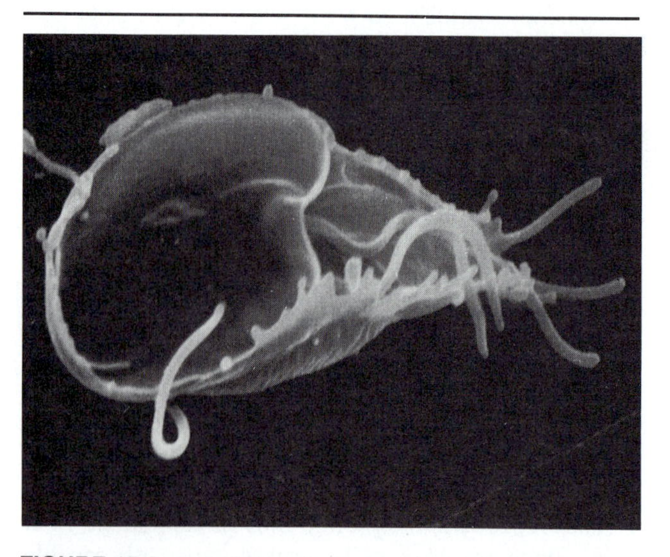

FIGURE 15.8 *Giardia intestinalis* parasite in the intestinal tract. These organisms attach to cells of the intestinal lining by means of an adhesive disk.

APPLICATIONS

Counting Viruses

Several laboratory techniques are available to measure the quantity of viruses present in a specimen. Some techniques make use of cytopathic effects, hemagglutination, and other effects viruses have on cells. Quantitative assays such as plaque assays determine the number of active viruses present in a specimen. Viruses are inoculated onto monolayer cell cultures (cultures that contain a layer of cells one cell deep). As cells are destroyed by lytic viruses such as polioviruses, clear areas appear in the monolayer and can be counted. Viruses that infect but do not lyse cells produce cloudy areas in the culture.

including those that cause malaria, actually invade and reproduce in erythrocytes. (Chapter 12, p. 300)

Signs, Symptoms, and Syndromes

Most diseases are recognized by their signs and symptoms. A **sign** is a characteristic of a disease that can be observed by examining the patient. A **symptom** is a characteristic of a disease that can be observed or felt only by the patient. Signs of disease include such things as swelling, redness, coughing, runny nose, fever, vomiting, and diarrhea. Symptoms include such things as pain, shortness of breath, nausea, and malaise (feeling bad).

A **syndrome** is a combination of signs and symptoms that occur together. For example, most infectious diseases cause the body to mount an acute inflammatory response. This response, which is discussed in Chapter 17, is characterized by a syndrome of fever, malaise, swollen lymph nodes, and leukocytosis (an increase in the numbers of white blood cells circulating in the blood).

In addition to the inflammatory response, many infectious diseases cause other signs and symptoms. Infections of the gut called *enteric infections* often cause nausea, vomiting, and diarrhea. Upper respiratory infections usually cause coughing, sneezing, sore throat, and runny nose. The signs and symptoms usually are related to the tissue damage being done by the pathogen (Table 15.4). Unfortunately, the signs and symptoms of diseases caused by different pathogens may be too similar to allow a specific diagnosis to be made. Thus, laboratory tests to identify infectious agents are an important component of modern medicine.

Even after recovery some diseases leave after-

TABLE 15.4 Correlation of signs and symptoms with tissue damage

Signs and Symptoms	Probable Nature of Tissue Damage
Incubation Period	None
Prodromal Phase	
Local redness and swelling	Pathogen has damaged tissue at site of invasion and caused release of chemical substances that dilate blood vessels (redness) and allow fluid from blood to enter tissues (swelling).
Headache	Chemical substances from tissue injury dilate blood vessels in the brain.
General aches and pains	Chemical substances from tissue injury stimulate pain receptors in joints and muscles.
Invasive Phase	
Fever	Leukocytes release pyrogens that reset the body's thermostat and cause temperature to rise.
Swollen lymph nodes	Leukocytes release other substances that stimulate cell division and fluid accumulation in lymph nodes; lymph nodes themselves release substances that flow to and affect other lymph nodes; some pathogens multiply in lymph nodes.
Skin rashes	Leukocytes release substances that damage capillaries and allow small hemorrhages; some pathogens invade skin cells and cause pox, vesicles, macules, papules, and other skin lesions.
Nasal congestion	Nasal mucosal cells have been damaged by pathogens (usually viruses), which release fluids and increase mucous secretions.
Cough	Mucosal cells of respiratory tract have been damaged by pathogens; excess mucus is released, and neural centers in the brain elicit coughing to remove mucus.
Sore throat	Lymphatic tissue of the pharynx is swollen and inflamed by substances released by pathogens and leukocytes.
Pain at specific sites (earache, local pain at a wound site)	Substances from pathogens or leukocytes have stimulated pain receptors; messages are relayed to the brain, where they are interpreted as pain.
Nausea	Toxins from pathogens have stimulated neural centers; you interpret the stimuli as nausea.
Vomiting	Toxins in food have stimulated the brain's vomiting center; vomiting helps to rid the body of toxins.
Diarrhea	Toxins in food cause fluids to enter the digestive tract; some pathogens directly injure the intestinal epithelium; both toxins and pathogens stimulate peristalsis; frequent watery stools result.
Acme	Full development of all signs and symptoms.
Decline Phase	
Signs and symptoms subside	Host defense mechanisms have overcome the pathogen.
Convalescence	
Patient regains strength	Tissue repair occurs; substances that caused signs and symptoms no longer released.

effects, called **sequelae** (se-kwe'le). Bacterial infections of heart valves often cause permanent valve damage, and poliovirus infections leave permanent paralysis.

Types of Infectious Disease

Infectious diseases vary in duration, location in the body, and other attributes. Several important terms, summarized in Table 15.5, are used to describe these attributes.

An **acute disease** develops rapidly and runs its course quickly. Measles and colds are examples of acute diseases. A **chronic disease** develops more slowly, is usually less severe than an acute disease, and persists for a long, indeterminate period. Tuberculosis and Hansen's disease (leprosy) are examples of chronic diseases. A **subacute disease** is intermediate between an acute and a chronic disease. Gingivitis, or gum disease, can exist as a subacute disease. A **latent disease** is characterized by periods of inactivity either before symptoms appear or between attacks. Herpes and several other viral infections produce latent disease.

TABLE 15.5 Terms used to describe infections

Term	Characteristic of Infection
Acute disease	Disease in which symptoms develop rapidly and disease runs its course quickly
Chronic disease	Disease in which symptoms develop slowly and disease is slow to disappear
Subacute disease	Disease with symptoms intermediate between acute and chronic
Latent disease	Disease in which symptoms appear and/or reappear long after infection
Local infection	An infection confined to a small region of the body, such as a boil or bladder infection
Focal infection	An infection in a confined region from which pathogens travel to other regions of the body, such as an abscessed tooth or infected sinuses
Systemic infection	An infection in which the pathogen is spread throughout the body, often by traveling through blood or lymph
Septicemia	Presence and multiplication of pathogens in blood
Bacteremia	Presence but not multiplication of bacteria in blood
Viremia	Presence but not multiplication of viruses in blood
Toxemia	Toxins in blood
Sapremia	Metabolic products of saprophytes in blood
Primary infection	Infection in a previously healthy person
Secondary infection	Infection that immediately follows a primary infection
Superinfection	Infection that results from destruction of normal flora
Mixed infection	Infection caused by two or more pathogens
Inapparent infection	Infection that fails to produce symptoms

A **local infection** is confined to a specific area of the body. Boils and bladder infections are local infections. A **focal infection** is confined to a specific area, but pathogens from it, or their toxins, can spread to other areas. Abscessed teeth and sinus infections are focal infections. A **systemic infection,** or **generalized infection,** affects the entire body, and the pathogens are widely distributed in many tissues. Typhoid fever is a systemic infection. When focal infections spread, they become systemic infections. For example, organisms from an abscessed tooth can enter the bloodstream and be carried to other tissues, including the kidneys. The organisms can then infect the kidneys and other parts of the urinary tract.

Pathogens can be present in the blood with or without multiplying there. In **septicemia,** once known as blood poisoning, pathogens are present and multiplying in the blood. In **bacteremia** and **viremia,** bacteria and viruses, respectively, are transported in the blood but do not multiply in transit. This often occurs after a cut, abrasion, or teeth cleaning. Pathogens also release toxins into the blood, and their presence in the blood is called **toxemia.** Saprophytes feed on dead tissues. Fungi behave as parasites when they kill cells and as saprophytes when they feed on them. They release metabolic products into the blood, thereby causing a condition called sapremia.

A **primary infection** is an initial infection in a previously healthy person. Most primary infections are acute infections. A **secondary infection** follows a primary infection, especially in patients weakened by the primary infection. A **superinfection** is a secondary

APPLICATIONS

To Squeeze or Not To Squeeze

There is good cause for the common warning not to squeeze pimples and boils. Left alone, the body's defense mechanisms ordinarily confine these lesions to the skin. However, squeezing them can disperse microorganisms into the blood and cause septicemia—a far worse condition. In septicemia the organisms are spread throughout the body and can cause severe infections. If they infect the brain, they can cause permanent brain damage.

infection that results from destruction of normal flora and often follows the use of broad-spectrum antibiotics. Although many infections are caused by a single pathogen, **mixed infections** are caused by several species of organisms present at the same time. Dental caries are due to mixed bacterial and protozoan infections; athlete's foot and certain other skin lesions are due to mixed fungal infections. An **inapparent,** or **subclinical, infection** is one that fails to produce symptoms either because too few organisms are present or because host defenses effectively combat the pathogens. People with inapparent infections, such as carriers of the infectious hepatitis virus, can spread the disease to others.

Steps in the Course of an Infectious Disease

At one time or another all of us have suffered from infectious diseases like the common cold, for which there is no cure. We simply have to let the disease "run its course." Most diseases caused by infectious agents have a fairly standard course, or series of stages. These stages include incubation, the prodromal phase, the invasive phase, the acme, the decline phase, and the convalescence phase. Even when treatment is available to eliminate the pathogen, the disease still passes through all its stages. Treatment lessens the severity of symptoms because pathogens can no longer multiply. It shortens the duration of the disease and the time required for recovery.

Incubation Period

The **incubation period** for an infectious disease is the time between infection and the appearance of signs and symptoms. Although the infected person is not aware of the presence of an infectious agent, he or she can spread the disease to others. Each infectious disease has a typical incubation period. The incubation periods of selected infectious diseases are shown in Figure 15.9. The length of the incubation period is

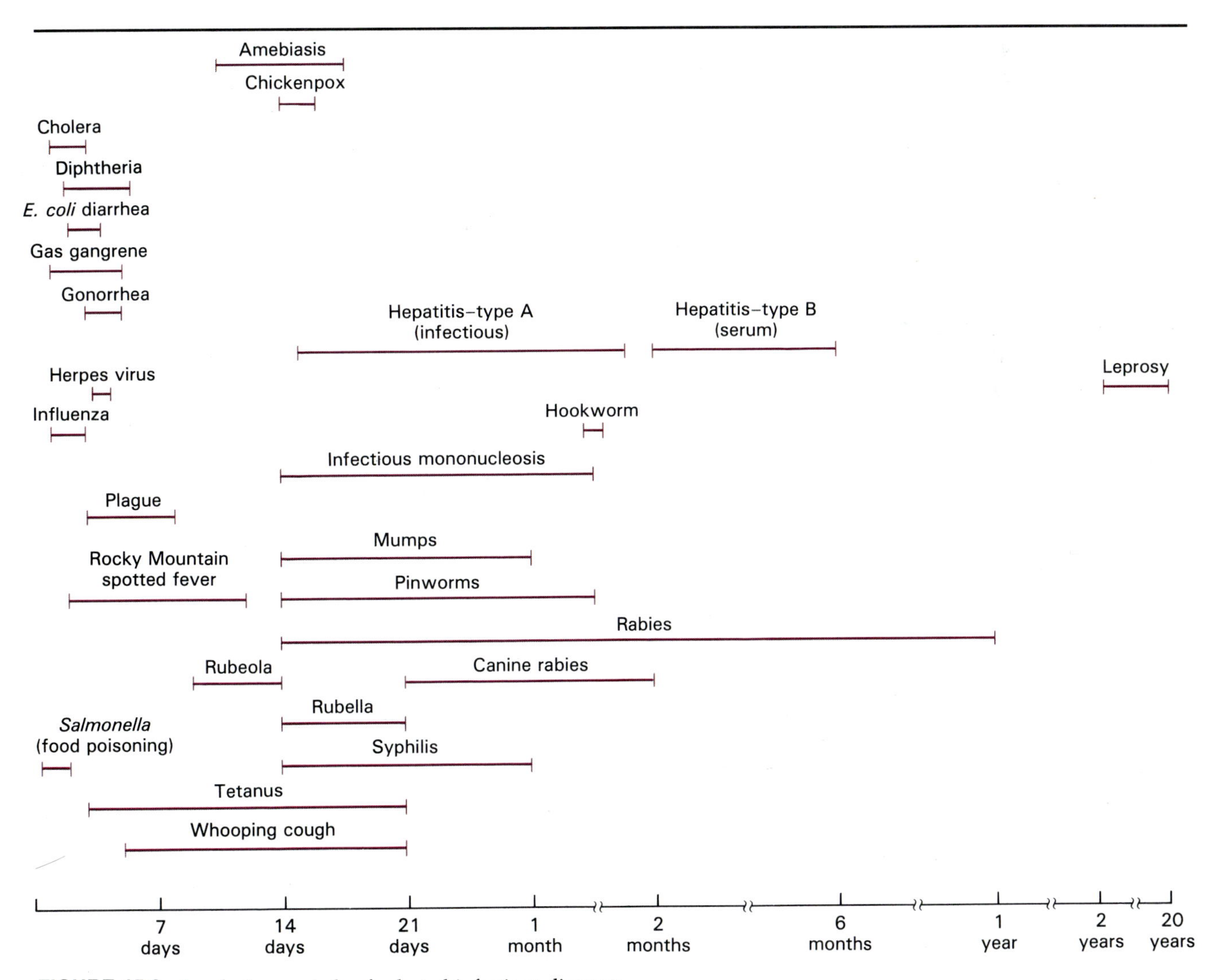

FIGURE 15.9 Incubation periods of selected infectious diseases.

determined by the properties of the infectious organism and the response of the host to the organism.

Properties of organisms that affect the incubation period include the nature of the organism, its virulence, how many organisms enter the body, and where they enter in relation to the tissues they affect. For example, if large numbers of an extremely virulent strain of *Shigella* quickly reach the intestine, profuse diarrhea can appear in a day. In contrast, if only small numbers of a less virulent strain enter the digestive tract with a large quantity of food, the disease will develop more slowly. Host defenses might be able to destroy the small number of organisms so that the disease will not occur at all. In the course of a lifetime, we undoubtedly have many more infections than we have overt diseases. As Chapter 17 shows, host defenses frequently attack pathogens as they start to invade tissues, thus averting potential diseases.

Prodromal Phase

The **prodromal phase** of disease is a short period during which nonspecific symptoms such as malaise and headache sometimes appear. A **prodrome** (*prodromos*, Latin for forerunner) is a symptom indicating the onset of a disease. You wake up one morning feeling bad, and you know you're coming down with something, but you don't know yet whether you will break out in spots, start to cough, develop a sore throat, or experience some other symptoms. Many diseases lack a prodromal phase and begin with sudden, acute symptoms such as fever and chills.

Invasive Phase

The **invasive phase** is the time during which the disease develops its most severe signs and symptoms. This phase corresponds with the period during which pathogens invade and damage tissues. In some diseases, such as some kinds of meningitis, this phase is **fulminating** (*fulmen*, Latin for lightning), or sudden and severe. In other diseases, such as hepatitis, it can be insidious, or appear over several days. A period of chills followed by fever marks the invasive phase of many diseases. Muscle pain, sore throat, vomiting, diarrhea, and swelling of lymph nodes also are common first symptoms of infectious diseases. As your signs and symptoms appear, you discover which form your infection will take.

Fever is an important component of the invasive phase of many diseases. Certain pathogens produce substances called **pyrogens** that act on a center in the hypothalamus sometimes referred to as the body's "thermostat." They set the thermostat at a higher-than-normal temperature, and the body responds with involuntary muscle contraction that generates heat and constriction (narrowing) of blood vessels in the skin that prevents heat loss. Because our bodies are colder than the new set temperature, we feel cold and have chills. We shiver and get "goose flesh" as muscles contract involuntarily, and we crawl under heavy blankets.

As the effects of the pyrogens diminish, the thermostat is reset to a lower temperature, and the body responds to reach that temperature. This response includes sweating and dilation (widening) of blood vessels in the skin to increase heat loss. Because our bodies are warmer than the new set temperature setting, we feel hot and say we have fever. Our skin gets moist from sweat and appears red as more blood circulates near the skin surface. In many infectious diseases repeated episodes of pyrogen release occur, thereby accounting for bouts of fever and chills.

Acme

The **acme,** or **critical stage,** of a disease is the period of most intense symptoms. During this time the battle between the pathogens and the host defenses is at its height.

Decline Phase

As symptoms begin to subside, the disease enters the **decline phase.** The decline phase corresponds to the period during which host defenses finally overcome the pathogen. The body's thermostat gradually returns to and stays at its normal setting. Although the terms *crisis* and *lysis* are not frequently used today, crisis describes symptoms that subside in a day or less, and lysis describes symptoms that subside over several days. High fevers that come on suddenly are likely to disappear quickly by crisis, whereas slowly mounting low-grade fevers usually disappear by lysis.

Convalescence

During **convalescence,** or recovery, tissues are repaired, healing takes place, and the body regains strength. Patients no longer have disease symptoms, but in some diseases, especially those in which scabs form over lesions, patients still can spread the disease to others.

ESSAY

The War on Infectious Diseases—Past, Present, and Future

In all the centuries of human history up to the twentieth century, recovery or death from infectious diseases was determined largely by whether the human host or the pathogen won the war they waged against each other. Assorted potions and palliative (pain-reducing) treatments were available, but none could cure infectious diseases. Sometimes treatment was based on the notion that imbalances in body fluids such as blood and bile caused disease. Depending on which fluid was judged to be in excess, efforts were made to remove some of it. Blood was removed by opening a vein or by applying blood-sucking leeches to the skin of the patient. In eighteenth-century Europe, patients were bled until they lost consciousness. In 1774, when King Louis XV of France came down with smallpox, his desperate physicians bled him for 3 days in a row, each time removing "four large basin-fuls of blood." In other cases, harsh laxatives were given to rid the body of excess bile. In most instances these treatments failed to rid the patient of infectious agents, which at the time were unknown. The best that can be said for such treatments is that they probably reduced suffering merely by hastening death.

Even after microorganisms were generally recognized as agents of disease, many years of painstaking research were required to relate specific diseases with the agents that caused them. More tedious research was needed to find antimicrobial agents that could cure diseases and to develop vaccines that could prevent them. The effects of these medical advances are clearly reflected in changes in death rates in the United States.

The death rate in the United States has dropped from 1600 per 100,000 people in 1900 to 860 per 100,000 people in the 1980s (Figure 15.10). The greatest single factor in decreasing the death rate was the control of infectious diseases by better treatment or immunization. Better sanitation has also helped. In 1900 the proportion of the population dying from infectious diseases was 20 times the current rate. Deaths from typhoid fever, syphilis, and childhood diseases (measles, whooping cough, diphtheria) have been nearly eliminated, and deaths from pneumonia, influenza, and tuberculosis have been greatly reduced. The death rates for selected infectious diseases are shown in Figure 15.11.

Yet we fail to do everything we can to eliminate diseases using technology that is already available. Measles killed nearly 1.5 million children in 1990, although single doses of vaccine that cost less than 12 cents would have saved most of these lives. In fact, each year 14 million children under the age of 5 die from infectious diseases such as measles, whooping cough, tetanus, diarrhea, and pneu-

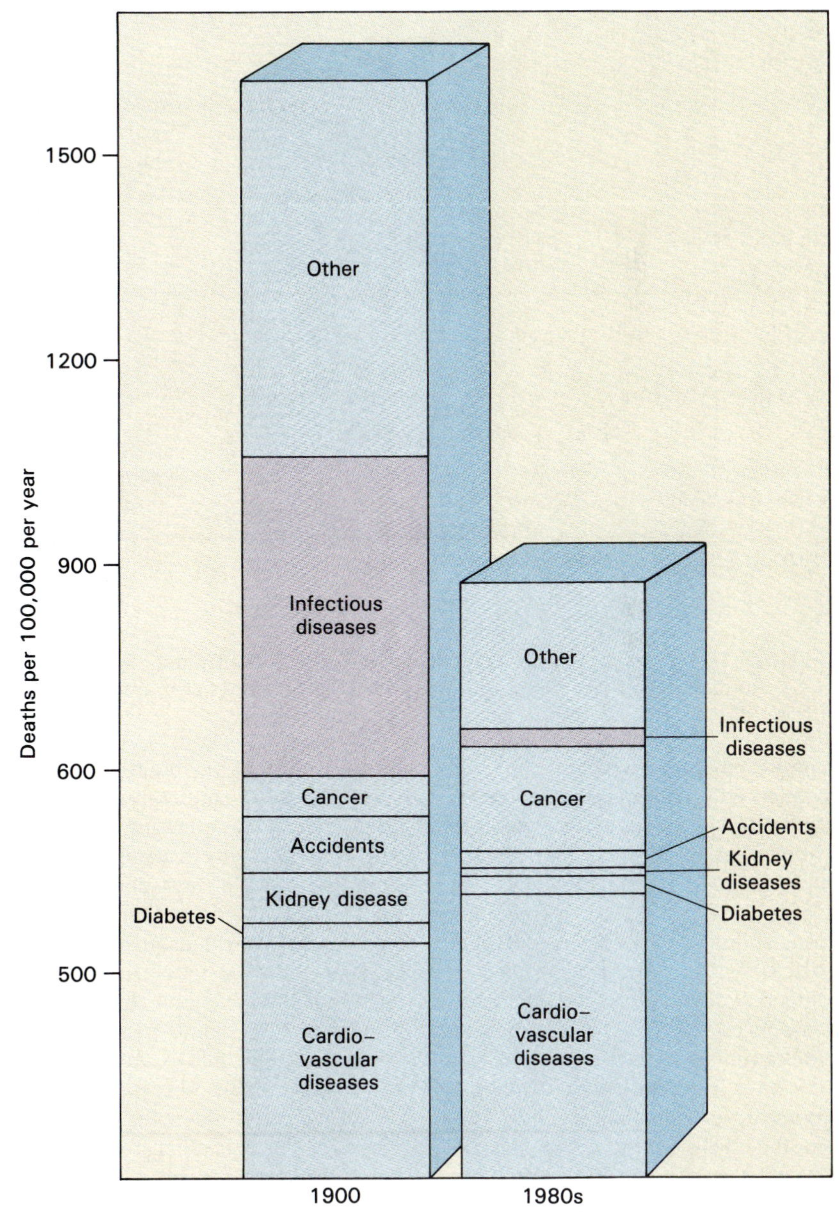

FIGURE 15.10 Changes in the causes of death in the United States from 1900 to the 1980s.

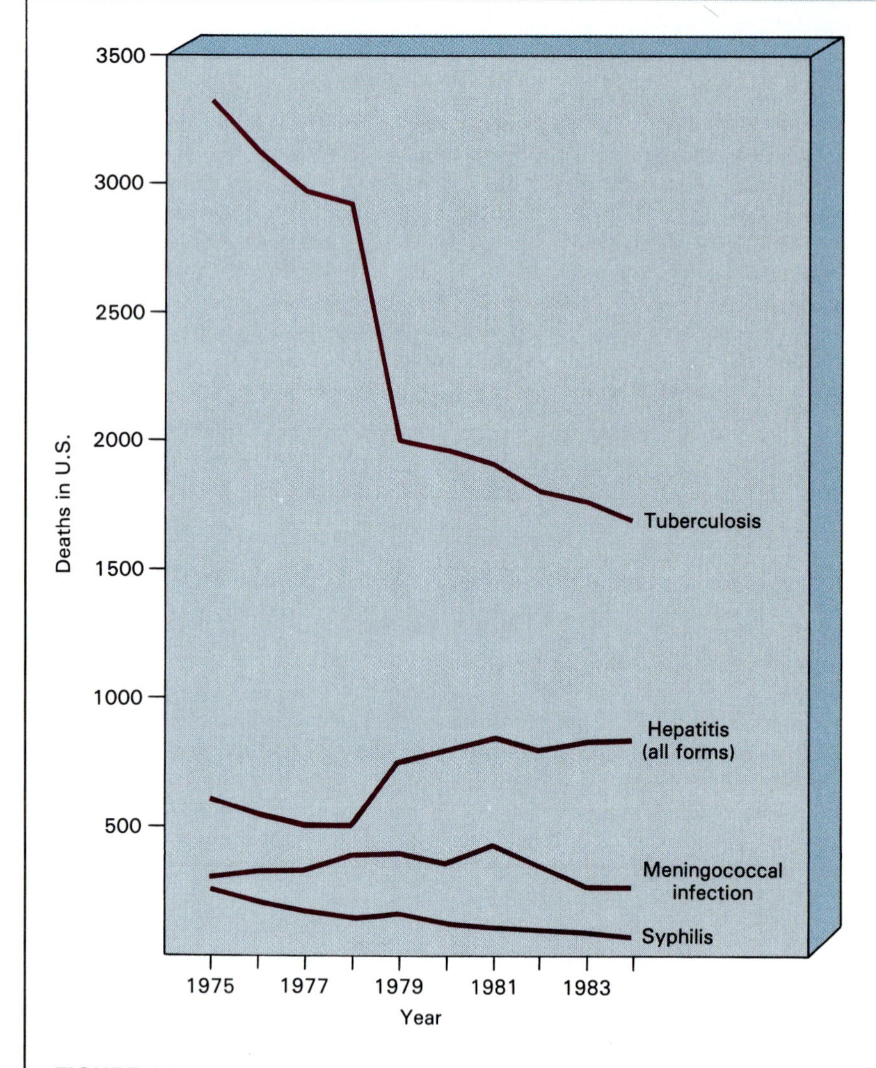

FIGURE 15.11 Death rates from infectious diseases in the United States, 1975–1984. Notifiable diseases with death rates greater than 100 per year are shown. (From *MMWR*, Dec. 26, 1987.)

monia, all of which could be prevented with minimal cost and effort. Still one child dies every 2 seconds from these.

As deaths from infectious diseases have decreased, people live long enough to develop degenerative diseases, and as they age their likelihood of developing a malignant disease increases. Thus, the death rates for cardiovascular diseases and cancer have increased. The death rate from cancer nearly tripled from 1900 to the 1980s (from 68 to 188 per 100,000), and the death rate from cardiovascular diseases increased by about 16 percent (from 360 to 418 per 100,000).

Recent successes in treating infectious diseases suggest that in the future we might be able to eradicate such diseases completely. At least three factors make eradication difficult. First, applying available medical expertise is not always possible. As we just discussed, certain preventable diseases such as measles and mumps continue to occur because parents fail to have their children immunized. Also, some people, both young and old, fail to obtain treatment for curable diseases. These problems could be solved by improved access to health care. Our failure to resolve this problem is "penny-wise but pound-foolish" because the costs of treating advanced disease and the economic losses from untimely deaths are far greater than the costs of providing vaccines and early treatments.

Second, infectious agents are often highly adaptable. Many strains of microorganisms are developing resistance to some of the available antibiotics. The widespread use of antibiotics, which has prevented so many deaths, now appears to foster the development of resistant strains. Treating diseases caused by antibiotic-resistant microorganisms is fast becoming a significant challenge in medical practice. Developing new antimicrobial agents is a corresponding research challenge.

Third, previously unknown or rare diseases can become significant as a result of changes in human activities or social conditions. The epidemic of legionellosis that marred the festivities of a Bicentennial celebration in 1976 was eventually found to be caused by a microorganism previously undiscovered but which had existed and occasionally caused disease in the past. However, this time it was spread through a hotel air-conditioning system—something that could not have happened before air conditioning was invented. In the early 1980s many cases of toxic shock syndrome suddenly appeared. This disease was shown to be caused by staphylococcal toxins that usually reached the blood from organisms growing in certain rough-material, high-absorbency tampons used by women during menstruation. Toxic shock syndrome was very rare before such tampons were invented, and since they have been taken off the market, it has again become rare.

Sometimes, too, a new disease caused by a new agent appears. First recognized in 1981, AIDS is thought to have occurred as early as 1959. The virus responsible for AIDS may have arisen earlier as a mutant strain in remote villages of Africa, but with limited mobility of population would have died out. This time, however, it succeeded in escaping out into the rest of the world. Now thousands of new cases are seen each year, and neither a cure nor a vaccine is yet available. Conquering AIDS is perhaps the greatest medical challenge currently facing the scientific community.

CHAPTER SUMMARY

HOST-MICROBE RELATIONSHIPS

RELATED KEY TERMS

- A **host** is an organism that harbors another organism.

Symbiosis

- **Symbiosis** means "living together" and includes **commensalism,** in which one organism benefits and the other neither benefits nor is harmed, **mutualism,** in which both organisms benefit, and **parasitism,** in which one organism (the parasite) benefits and the other is harmed.

competitive exclusion

Contamination, Infection, and Disease

- **Contamination** refers to the presence of microorganisms; in **infection,** pathogens invade the body; in **disease,** pathogens or other factors disturb health.

infestation

Pathogens, Pathogenicity, and Virulence

- A **pathogen** is a parasite capable of causing disease. **Pathogenicity** is the capacity to produce disease. **Virulence** is the degree of intensity of a disease. **Attenuation** is weakening of a pathogen's disease-producing capacity.

animal passage
transposal of virulence

Normal (Indigenous) Flora

- **Normal flora** are organisms found in or on another organism.
- **Resident flora** are always present. **Transient flora** are present under certain conditions. **Opportunists** are resident or transient flora that can cause disease under certain conditions or in certain locations in the body.

microbial antagonism

KOCH'S POSTULATES

- **Koch's postulates** provide a way to link an organism with a disease.

1. A specific causative agent must be observed in every case of a disease.
2. The agent must be isolated from a host displaying the disease and grown in pure culture.
3. When the agent from the pure culture is inoculated into susceptible hosts, it must cause the disease.
4. The agent must be reisolated from the diseased host and identified as the original specific causative agent.

- When these postulates are met, an organism has been proven to be the causative agent of an infectious disease.

KINDS OF DISEASES

Infectious and Noninfectious Diseases

- **Infectious diseases** are caused by infectious agents; **noninfectious diseases** are caused by other factors. Some diseases involve both infectious agents and other factors.

Classification of Diseases

inherited disease
neoplastic diseases
congenital defects
degenerative diseases
nutritional deficiency diseases
mental disease **endocrine diseases**
immunological diseases
iatrogenic diseases
idiopathic diseases

Communicable and Noncommunicable Diseases

- A **communicable,** or **contagious** infectious **disease** can be spread from one host to another. A **noncommunicable infectious disease** cannot be spread from host to host and usually is acquired from soil or water.

intoxication

Exogenous and Endogenous Diseases

- **Exogenous diseases** are caused by pathogens or other factors from outside the body. **Endogenous diseases** are caused by pathogens (usually opportunists) from within the body.

THE DISEASE PROCESS

How Microbes Cause Disease

- Bacteria cause disease by **adherence** to a host, by **colonization** and/or invasion of host tissues, and sometimes by invasion of cells.

adhesins
invasiveness

- Many bacteria also produce **toxins.** **Exotoxins** are released from bacteria; they are called **neurotoxins** if they affect the nervous system and **enterotoxins** if they affect the digestive system. **Endotoxins** are part of the cell wall of gram-negative bacteria and are released when cells divide or are killed. **Toxoids** are inactivated exotoxins that retain antigenic properties.
- Bacteria release other substances, most of which damage host tissues. **Hemolysins** lyse red blood cells in cultures and may or may not directly cause tissue damage in the host. **Leukocidins** destroy neutrophils. **Hyaluronidase** helps bacteria to make their way through tissues. **Coagulase** accelerates blood clotting. **Fibrinolysin** digests fibrin in blood clots.
- Viruses damage cells and produce a **cytopathic effect (CPE);** they can release digestive enzymes and alter host DNA.
- Fungi progressively digest cells, and some produce toxins.
- Protozoa and helminths damage tissues by ingesting cells and tissue fluids, releasing toxic wastes, and causing allergic reactions.

RELATED KEY TERMS

alpha hemolysins
beta hemolysins
leukostatin
kinase **streptokinase**
cytocidal **noncytocidal**
productive infection
abortive infection
adhesive disc

Signs, Symptoms, and Syndromes

- A **sign** is an observable effect of a disease. A **symptom** is an effect of a disease reported by the patient. A **syndrome** is a group of signs and symptoms that occur together.

Types of Infectious Disease

- Terms used to describe types of diseases are defined in Table 15.5.

Steps in the Course of an Infectious Disease

- The **incubation period** is the time between infection and the appearance of signs and symptoms of a disease.
- The **prodromal phase** is the period during which pathogens begin to invade tissues; it is marked by early nonspecific symptoms.
- The **invasive phase** is the period during which pathogens invade and damage tissues; it is marked by the most severe signs and symptoms.
- The **acme,** or **critical, stage,** is the period of most intense symptoms.
- The **decline phase** is the period during which host defenses overcome pathogens; signs and symptoms subside during this phase.
- **Convalescence** is the period when tissue damage is repaired and the patient regains strength.

acute disease
chronic disease
subacute disease
latent disease **local infection**
focal infection
systemic infection
generalized infection
septicemia
bacteremia **viremia**
toxemia **sapremia**
primary infection
secondary infection
superinfection
mixed infections
inapparent infection
subclinical infection
sequelae **prodrome**
fulminating **fever**
pyrogens

QUESTIONS FOR REVIEW

A.

1. Which characteristics distinguish the various kinds of symbiosis?
2. How does infection differ from contamination?
3. Relate the terms pathogen and disease.
4. How do pathogenicity and virulence differ, and how does attenuation affect virulence?
5. Describe the three classes of normal flora.

B.

6. What are Koch's postulates?
7. In what ways are Koch's postulates sometimes difficult to demonstrate?

C.

8. How are infectious and noninfectious diseases distinguished?
9. What kinds of diseases have both infectious and noninfectious components?
10. How are communicable and noncommunicable diseases distinguished?
11. How are exogenous and endogenous diseases distinguished?

D.

12. Describe the mechanisms used by bacteria to cause disease.
13. How do toxins cause disease?
14. What is a toxoid?
15. What substances other than toxins are released by bacteria?
16. How do other pathogens cause disease?

E.

17. How do signs, symptoms, and syndromes differ?
18. Write a paragraph, correctly using all the terms in Table 15.5.

F.

19. Trace the course of a disease, and relate signs and symptoms to activities of a pathogen.

PROBLEMS FOR INVESTIGATION

1. Which factors lead to the wide variation in the incubation period for rabies (14 days to 1 year)? Would it be better to be bitten on the foot or the shoulder?
2. Identify the type of symbiotic relationship represented in each of the following cases:
   a) bacteria living in the rumen of the cow's digestive system break down the cellulose in grass and hay
   b) tapeworms living in the human intestine
   c) AIDS virus living in a human
3. A microbiology student, clad in shorts, overturns his motorcycle on a patch of gravel. He sustains considerable abrasions and deep lacerations of the right leg. Emergency-room personnel have great difficulty trying to remove embedded gravel and dirt. Many areas of infection develop, and dead tissue can be seen at the sites of abrasion. Are his infections most likely due to pathogens or opportunists? Why is it difficult to determine?
4. Identify which are characteristics of: (EX) exotoxins only, (EN) endotoxins only, or (B) both:
   a) usually found in gram-positive organisms
   b) cause high fever
   c) easily inactivated by heat
   d) lipopolysaccharide complex
   e) enterotoxins
   f) Salmonellosis
   g) toxoid can be used to immunize against
5. Recall the last time you had an infectious disease. Relate the signs and symptoms you experienced to the steps in the course of an infection and the ways in which the pathogen was damaging tissues.
6. Research the topic of opportunistic infections to find out more about what organisms are opportunists, under what conditions they cause infections, and why such infections are difficult to treat.
7. A young mother tried to save money by buying cracked eggs at a local farm. The eggs were labeled "not for human consumption—for pets only." She cooked some of the eggs and fed them to her toddler son, who became intensely ill within hours. He was hospitalized with a diagnosis of salmonellosis (food poisoning). He nearly died and is now permanently retarded.
   a) Did the boy suffer from an infection or from an intoxication? What piece of clinical data supports your answer?
   b) Why didn't cooking the eggs protect him?
   c) Can he be immunized against further occurrences?
   d) Where did the *Salmonella* organisms come from?
   (The answer to this question appears in Appendix E.)

SOME INTERESTING READING

Brubaker, R. R. 1985. Mechanisms of bacterial virulence. *Annual Review of Microbiology* 39:21.

Centers for Disease Control. *Morbidity and mortality reports.* Atlanta, Ga. (A weekly report)

Eisenstein, B. I. 1990. New opportunistic infections—more opportunities. *The New England Journal of Medicine* 323 (December 6):1625–27.

Miller, V. L. 1992. *Yersinia* invasion genes and their products. *ASM News* 58(1):26–33.

Schantz, E. J. and E. A. Johnson. 1992. Properties and use of botulinum toxin and other microbial neurotoxins in medicine. *Microbiological Reviews* 56(1):80–99.

Stephen, J., and R. A. Petrowski. 1986. *Bacterial toxins. Aspects of Microbiology.* Washington, D.C.: American Society for Microbiology.

Many preindustrial peoples attributed diseases to "evil spirits" or "bad air." Industrial societies generally have a very different approach to epidemiology. These modern-day Lassa spiritual healers from Sierra Leone, however, have knowledge that we lack. Western medical scientists, especially those in the pharmaceutical industry, are anxious to learn of herbs and other materials and methods used in disease treatment, before this knowledge is lost.

16 Epidemiology and Nosocomial Infections

This chapter will focus on the following questions:

A. What is epidemiology, and what special terms are used by epidemiologists?

B. How are diseases categorized according to their spread in populations?

C. How do various kinds of reservoirs of infection contribute to human disease?

D. What are the roles of portals of entry and exit and modes of transmission in the spread of human disease?

E. What is a disease cycle, and how is group immunity related to disease cycles?

F. How do the functions of organizations and the reporting of diseases contribute to public health?

G. What are the purposes and methods of epidemiological studies?

H. What methods are used to control communicable diseases?

I. What are nosocomial infections and how are they studied epidemiologically?

J. How can nosocomial infections be prevented and controlled?

EPIDEMIOLOGY

Up to this point we have considered the characteristics of pathogens that lead to infectious diseases and how the disease process occurs in individuals. But individuals with infectious diseases are members of a population—they acquire diseases and transmit them within a population. Therefore, to further our understanding of infectious diseases we must consider their effects on populations.

What Is Epidemiology?

Epidemiology (from Greek: *epi,* among; *demos,* the people) is the study of factors and mechanisms involved in the spread of disease within a population. Thus, **epidemiologists,** scientists who study epidemiology, are concerned with the **etiology** (cause) and transmission of a disease in a population. They use this information to design ways to control and prevent outbreaks of disease.

Epidemiology can be considered a branch of microbiology because many of the diseases of interest to epidemiologists are caused by microorganisms. Some are caused by other infectious agents; a few are caused by other known or unknown factors. Epidemiology also can be considered a branch of ecology because it concerns relationships among pathogens, their hosts, and the environment. Finally, epidemiology is related to public health because it provides information and methods used to understand and control outbreaks of disease in the human population. Agricultural and environmental scientists are often concerned with the epidemiology of animal and plant diseases. When such diseases can be transmitted to humans, they, too, become a public health problem.

The term epidemiology is derived from *epidemios,* a Greek word meaning frequent or prevalent. Epidemiologists are especially concerned with the frequency of diseases in populations and use specific terms to express frequencies. The **incidence** of a disease is the number of new cases seen in a specific period of time. The **prevalence** of a disease is the number of people infected at any one time. It includes old and new cases and is influenced by the duration of the disease. If weekly surveys are conducted in regard to a disease that lasts a month, infected individuals could be counted as many as four times in a prevalence study but only once in an incidence survey. Incidence data are a good measure of the progress of an epidemic—a drop in the incidence indicates the epidemic is coming to an end. On the other hand, prevalence data are a good measure of how seriously and how long the disease is affecting the population.

Frequencies also are expressed as proportions of the total population. The **morbidity rate** is the number of cases in relation to the total number of people in the population. It usually is expressed as the number of cases per 100,000 people per year. The **mortality rate** is the number of deaths in relation to the total population, usually expressed as deaths per 100,000 people per year. In the United States the **Centers for Disease Control (CDC)** carries out epidemiologic studies. At its headquarters in Atlanta, Georgia, CDC publishes the *Morbidity and Mortality Weekly Report* (MMWR), which provides statistics for specific diseases in various parts of the country (Figure 16.1).

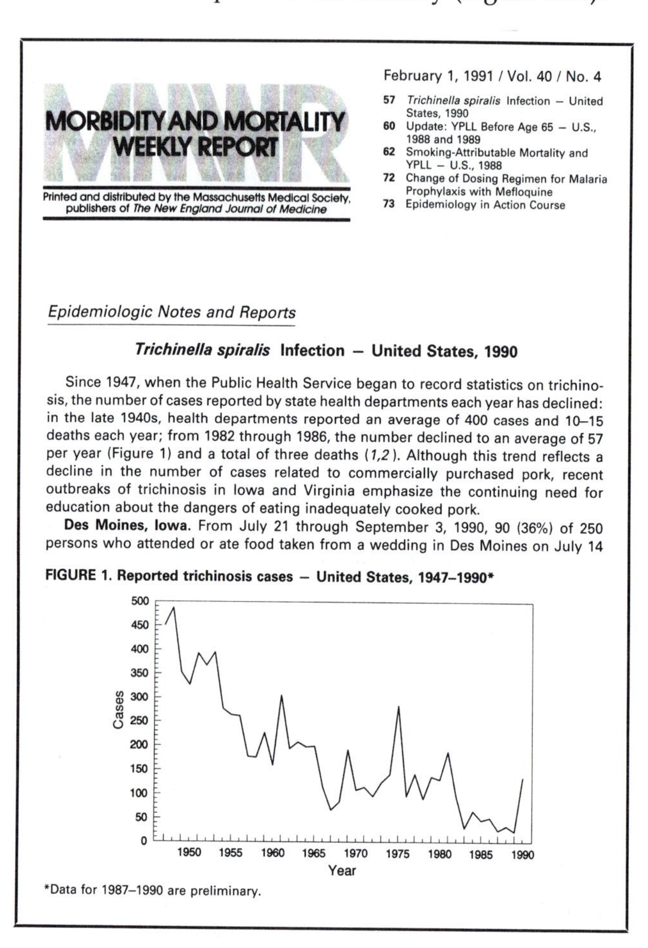
MORBIDITY AND MORTALITY WEEKLY REPORT

Printed and distributed by the Massachusetts Medical Society, publishers of *The New England Journal of Medicine*

February 1, 1991 / Vol. 40 / No. 4

57 *Trichinella spiralis* Infection – United States, 1990
60 Update: YPLL Before Age 65 – U.S., 1988 and 1989
62 Smoking-Attributable Mortality and YPLL – U.S., 1988
72 Change of Dosing Regimen for Malaria Prophylaxis with Mefloquine
73 Epidemiology in Action Course

Epidemiologic Notes and Reports

***Trichinella spiralis* Infection – United States, 1990**

Since 1947, when the Public Health Service began to record statistics on trichinosis, the number of cases reported by state health departments each year has declined: in the late 1940s, health departments reported an average of 400 cases and 10–15 deaths each year; from 1982 through 1986, the number declined to an average of 57 per year (Figure 1) and a total of three deaths (*1,2*). Although this trend reflects a decline in the number of cases related to commercially purchased pork, recent outbreaks of trichinosis in Iowa and Virginia emphasize the continuing need for education about the dangers of eating inadequately cooked pork.

Des Moines, Iowa. From July 21 through September 3, 1990, 90 (36%) of 250 persons who attended or ate food taken from a wedding in Des Moines on July 14

FIGURE 1. Reported trichinosis cases – United States, 1947–1990*

*Data for 1987–1990 are preliminary.

FIGURE 16.1 The *Morbidity and Mortality Weekly Report* is issued weekly by CDC. It reports new and unusual cases and trends. It also lists, by state, the number of cases of notifiable diseases recorded that week and the number recorded for the same week in the previous year. It also compares cumulative totals. It is available from the Massachusetts Medical Society; C.S.P.O. Box 9120; Waltham, MA 02254-9120, at a cost of about $50 per year for 52 issues.

Diseases in Populations

Epidemiologists also consider the frequency of diseases in populations in terms of the size of the geographic areas affected and the degree of harm the diseases cause in the population. On the basis of their findings, they classify diseases as endemic, epidemic, pandemic, or sporadic.

An **endemic disease** is constantly present in the

population of a particular geographic area, but the number of cases and the severity of the disease both remain low enough not to constitute a public health problem. Mumps is endemic to the entire United States, and valley fever is endemic to the southwestern United States. Chickenpox is an endemic disease with seasonal variation; that is, many more cases are seen from late winter through spring than at other times (Figure 16.2). Endemic diseases also vary in incidence in different parts of the endemic region.

An **epidemic** arises when a disease has a sudden very high incidence in a population and when the mortality rate, the degree of harm or morbidity, or both are sufficiently high to pose a public health problem. Endemic diseases can give rise to epidemics, especially when a particularly virulent strain of a pathogen appears or when a large proportion of a population lacks immunity. This phenomenon is illustrated by the incidence of St. Louis encephalitis, an inflammation of the brain, in the United States (Figure 16.3). Poliomyelitis also was said to cause epidemics before immunization became available, not so much because of the large number of cases but because of the severe consequences of the disease. Influenza and childhood diseases such as measles, mumps, and chickenpox also often produce epidemics.

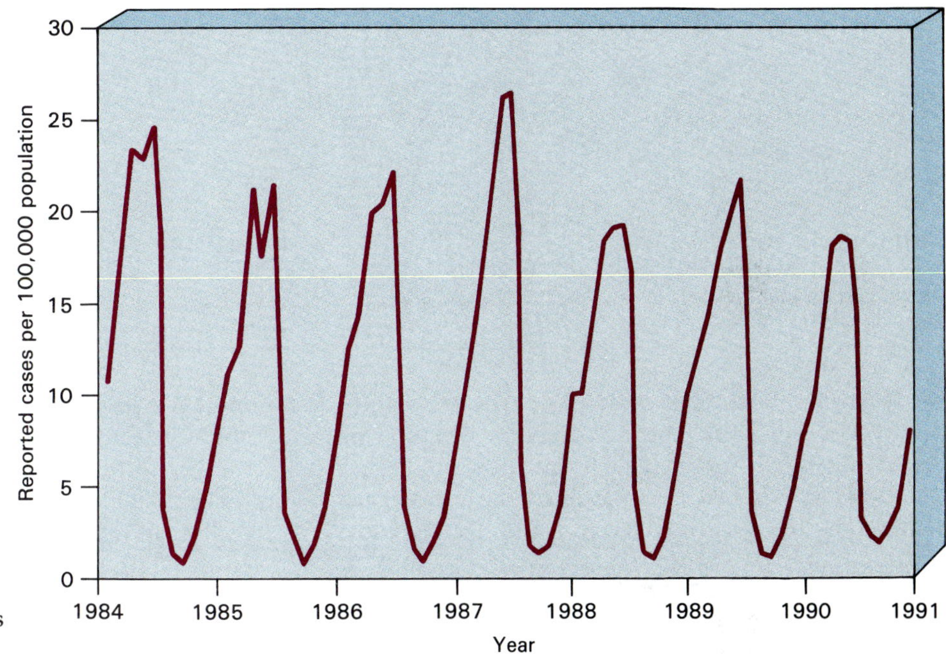

FIGURE 16.2 Incidence of chickenpox, by month, in the United States. Chickenpox, an endemic disease, shows marked seasonal variation, with most cases occurring during the spring.

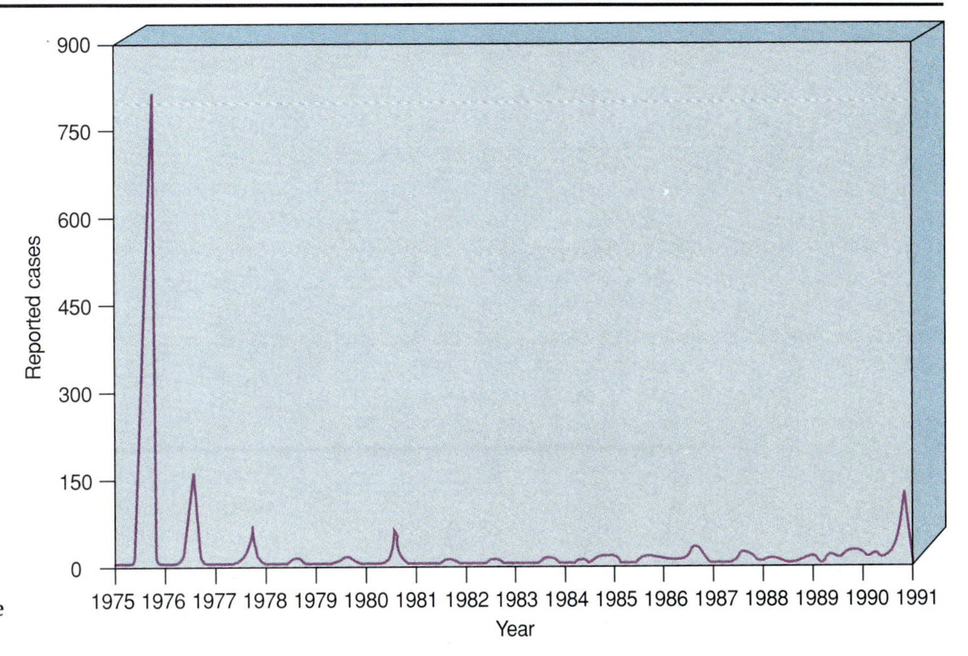

FIGURE 16.3 Incidence of St. Louis encephalitis cases in the United States, showing epidemic outbreaks during late 1975 and late 1990.

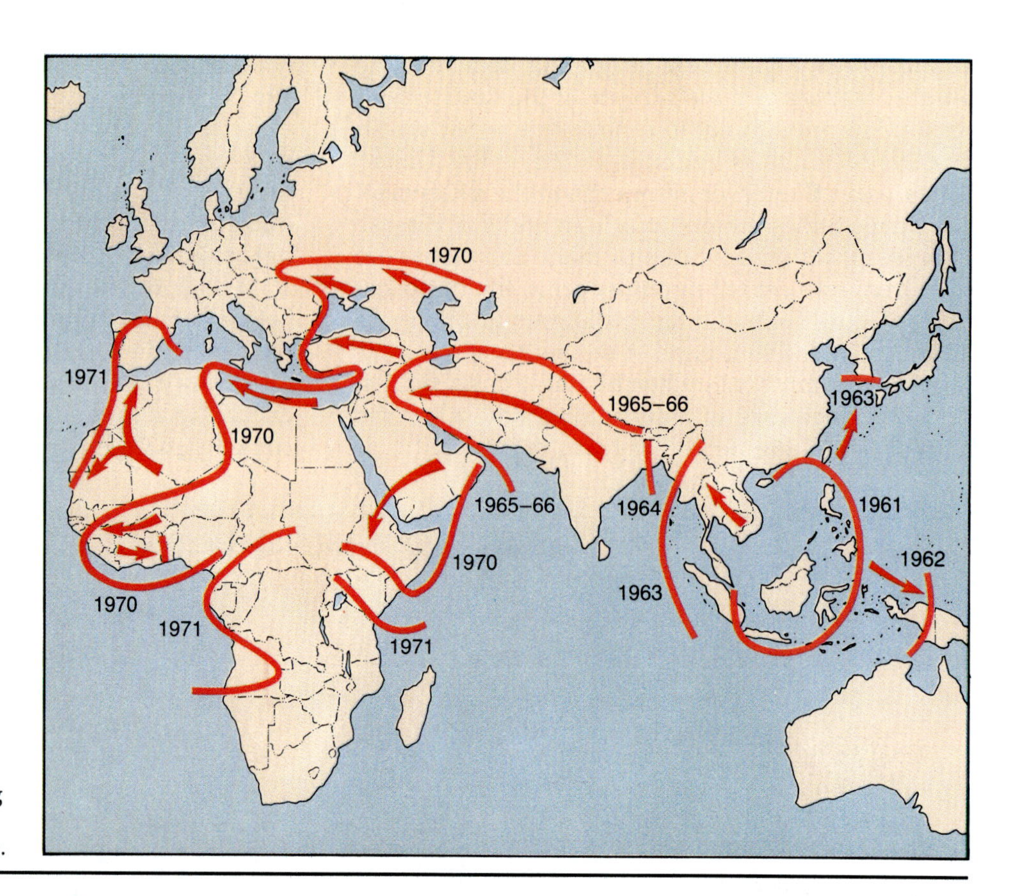

FIGURE 16.4 The cholera pandemic of 1961–1971, showing the geographical spread of the outbreak over the course of time.

A **pandemic** occurs when an epidemic becomes worldwide. Influenza reached pandemic proportions in 1918 during the great swine flu pandemic (Chapter 21). Cholera has been responsible for several pandemics over the centuries. Its spread during the most recent pandemic (1961–1971) is shown in Figure 16.4.

A **sporadic disease** occurs as a small number of isolated cases that do not pose any great threat to the population as a whole. Eastern equine encephalitis is a sporadic disease in the United States. Its sporadic nature is contrasted with endemic and epidemic types of encephalitis in Figure 16.5.

The nature and spread of epidemics can vary according to the source of the pathogen and how it reaches susceptible hosts. A **common-source epidemic** arises from contact with contaminated substances. Such epidemics can often be traced to a water supply contaminated with fecal material or to improperly handled food. Large numbers of people become ill quite suddenly—for example, all the people who ate the potato salad at the school cafeteria on a particular day. The epidemic subsides equally quickly when the source of infection is eradicated. A **propagated epidemic** arises from person-to-person contacts. The pathogen moves from infected people to uninfected, susceptible people. In a propagated epidemic the number of cases rises and falls more slowly, and the pathogen is more difficult to eliminate than in a common-source epidemic. Differences between common-source and propagated epidemics are illustrated in Figure 16.6.

Infectious diseases, either endemic or epidemic, threaten human populations only when they can be spread in some way. Factors in the spread of infection include reservoirs of infection, portals by which organisms enter and leave the body, and mechanisms of transmission. Let's look at each of these factors in more detail.

CLOSE-UP

Cities: Cesspools of Civilization

"Cities are essential to civilization, and until well into the nineteenth century, all cities were the spawning grounds of infectious disease. How could it be otherwise? For centuries all the precautions we now know to be necessary to prevent the spread of bacteria and animal parasites were unthought of. City streets were littered with human and animal filth, water came from contaminated wells, rats, fleas and lice were universal. Crowded together in such filthy environments, every city dweller was inevitably exposed to infection every day of his life. It is no wonder that the population of cities through all history has had to be recruited periodically from the country."

—Macfarlane Burnet and David White, 1972

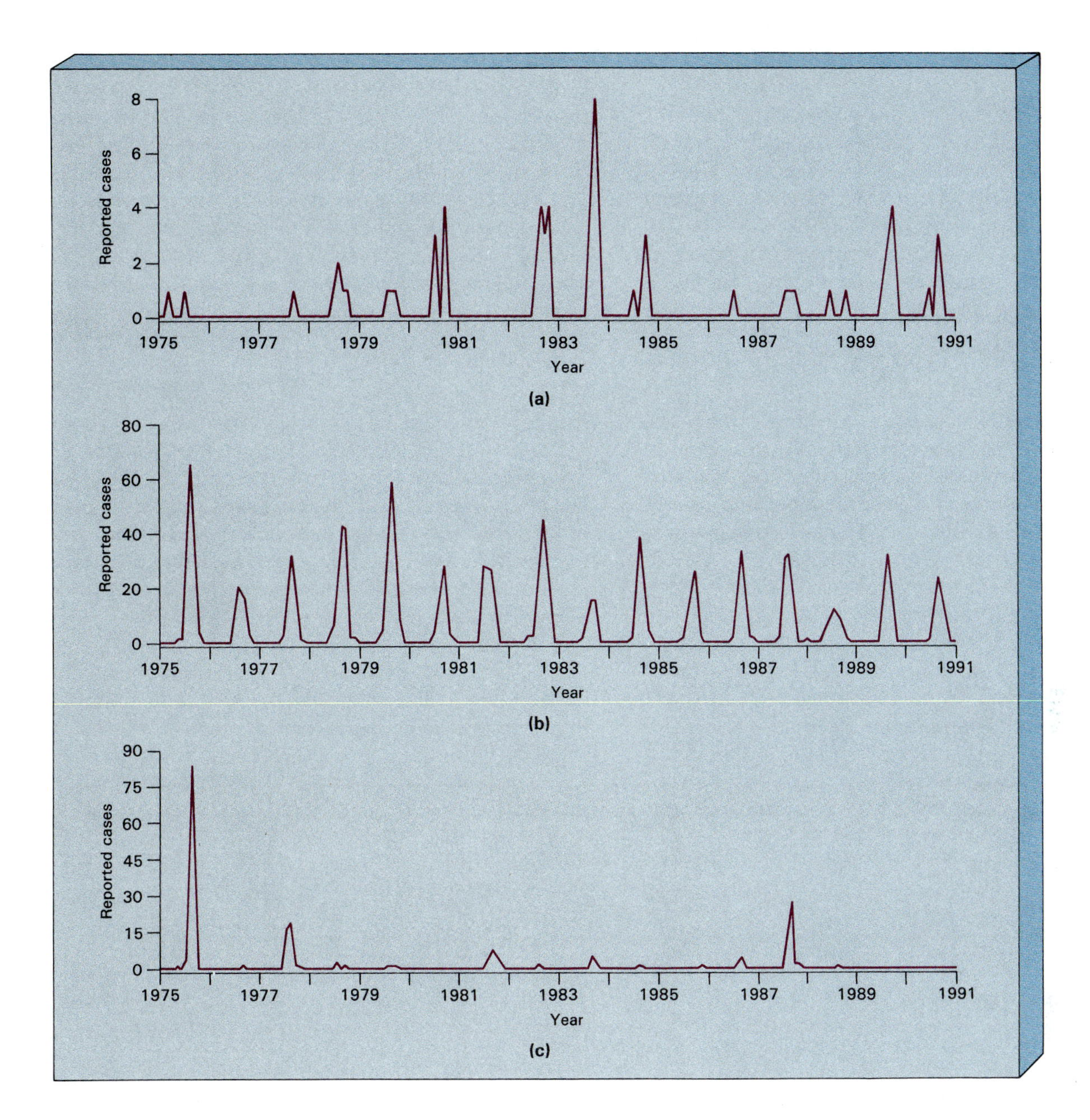

FIGURE 16.5 Incidence of three different types of encephalitis. (a) The sporadic pattern of Eastern equine encephalitis. (b) The endemic pattern of cases due to California serogroup viruses. (c) Epidemic outbreaks of Western equine encephalitis.

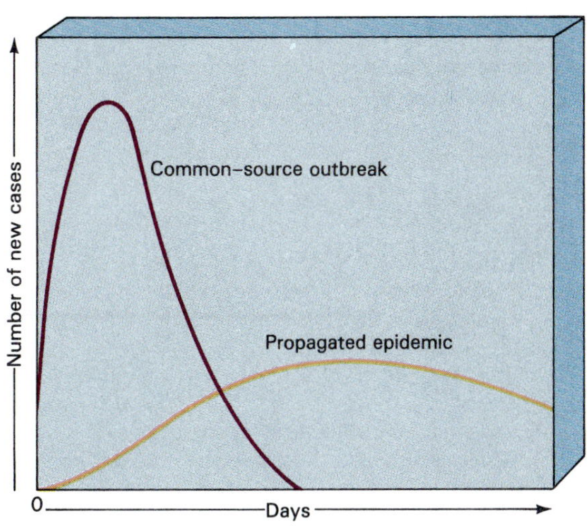

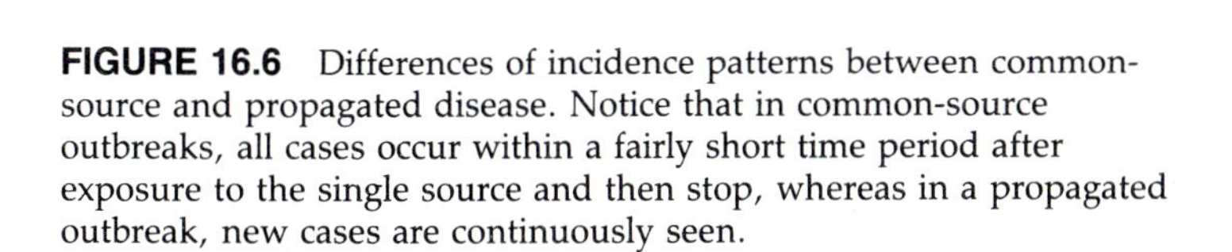

FIGURE 16.6 Differences of incidence patterns between common-source and propagated disease. Notice that in common-source outbreaks, all cases occur within a fairly short time period after exposure to the single source and then stop, whereas in a propagated outbreak, new cases are continuously seen.

Reservoirs of Infection

Most microorganisms capable of infecting humans cannot survive outside the body of a host long enough to serve as a source of infection. Therefore, sites in which organisms can persist and maintain their ability to infect are essential for new human infections to occur. Such sites are called **reservoirs of infection** and include humans, some animals, and certain nonliving media.

Human Reservoirs

Humans with active infections are important reservoirs because they can easily transmit organisms to other humans. Some diseases are communicable during the incubation period (before symptoms are apparent) and during recovery. Often diseases are communicated by people with subclinical infections—those with symptoms too mild to be recognized. Many cases of whooping cough in adults, for example, are never diagnosed. In addition to transmitting a disease during an illness, some humans are healthy *carriers* of disease. **Passive carriers** release organisms without ever having shown signs and symptoms of a disease. **Active carriers** release organisms for a long period of time after recovering from a disease. **Intermittent carriers** periodically release organisms. Depending on the disease, carriers can discharge organisms from the mouth or in urine or feces. Diseases commonly spread by carriers include diphtheria, typhoid fever, amoebic and bacillary dysenteries, hepatitis, streptococcal infections, and pneumonia. International jet travel has greatly increased the risk of introducing infectious agents from reservoirs in one region to populations in another region that have no immunity to them.

Animal Reservoirs

About 150 pathogenic microorganisms can infect both humans and some other animals. In such instances the animals can serve as reservoirs of infection for humans. Animals that are physiologically similar to humans are most likely to serve as reservoirs for human infections. Therefore, monkeys are important reservoirs for malaria, yellow fever, and many other human infections. Once humans are infected, they, too, can serve as reservoirs for the infections.

Diseases that can be transmitted from animals to humans are called **zoonoses** (zo-o-no'-ses; singular: *zoonosis*). Selected zoonoses are summarized in Table 16.1. Of these diseases, rabies is perhaps the greatest

TABLE 16.1 Selected zoonoses (with emphasis on those that occur in pets)

Disease	Animals Infected	Modes of Transmission
Bacterial diseases		
Avian tubercolusis	Birds	Respiratory aerosols
Anthrax	Dogs, cats, and domestic animals	Direct contact with animals, contaminated soil, and hides; ingestion of contaminated milk or meat; inhalation of spores
Brucellosis (undulant fever)	Domestic animals	Direct contact with infected tissues, ingestion of milk from infected animals
Bubonic plague	Rodents	Fleas
Leptospirosis	Primarily dogs, also rodents, and wild animals	Direct contact with urine, infected tissues, and contaminated water
Psittacosis	Parrots, parakeets, and other birds	Respiratory aerosols
Relapsing fever	Rodents	Ticks and lice
Rocky Mountain spotted fever	Dogs, rodents, and wild animals	Ticks
Salmonellosis	Dogs, cats, poultry, and rats	Ingestion of infected tissues and contaminated water
Viral diseases		
Equine encephalitis (several varieties)	Horses, birds, and other domestic animals	Mosquitoes
Rabies	Dogs, cats, bats, skunks, and wolves	Bites, infectious saliva in wounds
Lassa fever, hemorrhagic fevers	Rodents	Urine
Fungal diseases		
Histoplasmosis	Birds	Aerosols of dried infected feces
Ringworm (several varieties)	Cats, dogs, and other domestic animals	Direct contact
Parasitic diseases		
African sleeping sickness	Wild game animals	Tsetse flies
Tapeworms	Cattle, swine, rodents	Ingestion of cysts in meat or via proglottids in feces
Toxoplasmosis	Cats, birds, rodents, and domestic animals	Aerosols, contaminated food and water, and placental transfer

threat in the United States because of the severity of the disease and because both domestic pets and wild animals can serve as reservoirs for it. In the United States, where vaccination of dogs and cats is widespread, humans are more likely to acquire rabies from wild animals such as skunks, raccoons, bats, and foxes. Another zoonosis, malaria, has an exceedingly high incidence in some parts of the world. Controlling malaria is especially difficult because it has been impossible to eradicate it among domestic and wild animals. The larger the animal reservoir, both in number of species and in the total number of susceptible animals, the more unlikely it is that a disease will be eradicated. This is especially true if the reservoir animals are wild; it is impossible to find all of them and to control a disease among them. Even today *Yersinia pestis*, the organism responsible for plague, persists among gophers, ground squirrels, and other wild rodents in the American West and causes occasional human cases. The eradication of smallpox was a difficult task, and this disease was limited to human reservoirs.

Humans, their pets, and domestic animals likewise serve as reservoirs of infection for wild animals. Distemper, an infectious viral disease in dogs, has spread to and killed many black-footed ferrets. This animal, already an endangered species, is now in even greater jeopardy. Thus, pets and domestic animals should not be allowed in wildlife refuges.

Nonliving Reservoirs

Soil and water can serve as reservoirs for pathogens. Soil is the natural environment of most species of *Clostridium* and many fungi. Certain species of *Clostridium* cause tetanus and botulism. Some of the fungi can invade human tissues and cause ringworm, other skin diseases, or systemic infections. Water contaminated by human or animal feces can contain a variety of pathogens. Most of these cause gastrointestinal diseases. Improperly prepared or stored food also can serve as a temporary nonliving reservoir of disease. Poorly cooked meats can be a source of infection with *Salmonella* species and a variety of helminths. Failure to refrigerate foods can lead to growth of microorganisms and production of toxins that cause food poisoning. Even with proper refrigeration, helminth eggs remain infectious unless the foods that contain them are thoroughly cooked.

Portals of Entry

To cause an infection a microorganism must gain access to body tissues. The sites at which microorganisms can enter the body are called **portals of entry.** Common portals of entry include the skin and the mucous membranes of the digestive, respiratory, and genitourinary systems. Microorganisms also can be introduced directly into tissues, and some can cross the placenta to reach a fetus (Figure 16.7). Intact skin usually prevents the entry of microorganisms, but some gain access through the ducts of sweat glands or through hair follicles. Some fungi invade cells on the skin's surface, and a few can pass on to other tissues. Larvae of some parasitic worms such as the hookworm can bore through the skin to enter other tissues.

Body systems that have mucous membrane linings also have openings that communicate with the environment. Openings such as the nose, mouth, eyes, anus, urethra, and vagina allow microbes to enter the body. Organisms that infect the respiratory system typically enter in inhaled air, on dust particles, or in liquid aerosol droplets. Those that infect the

CLOSE-UP

What's in Dust?

Household dust typically contains an amazing assortment of microbes, spores, a few larger organisms, dandruff and other debris from the human body, and other nonliving materials. Spores of *Clostridium perfringens*, which causes gas gangrene in deep wounds, have been found in air conditioner filters. Many different genera of fungi, including *Penicillium*, *Rhizopus* (bread mold), and *Aspergillus*, which can cause swimmer's ear, have been found in dust. Bacterial endospores, fungal spores, plant pollens, insect parts, and hundreds of mites (small organisms related to spiders) also abound in dust. Mites feed on skin particles sloughed from our body surfaces but fortunately do not feed on living skin. Dust also contains large quantities of human and animal hair and sometimes nail clippings as well, all of which have microorganisms on their surfaces. Finally, dust contains nonliving particles from sources as diverse as flaking paint and meteorites.

Household dust provides a home for dust mites.

FIGURE 16.7 Portals of entry for disease-causing organisms.

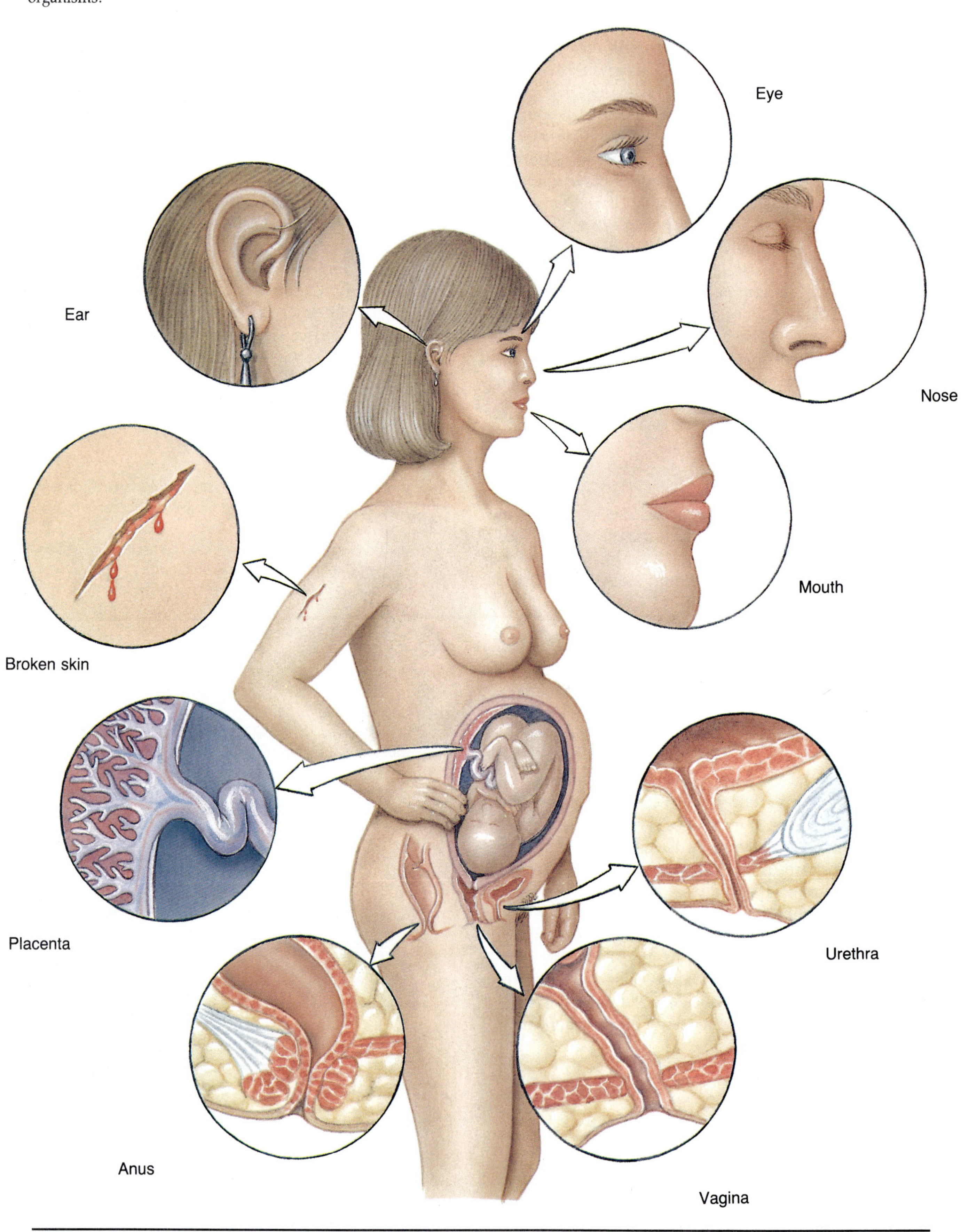

digestive system typically enter in food and water but also can enter from contaminated fingers. Many organisms that infect the genitourinary system enter through sexual contact, but some travel through the urethra and vagina from skin surfaces. Most organisms that enter the body through mucous membranes have the capacity to overcome the effects of mucus and other defenses mounted by these membranes.

Sometimes microorganisms are introduced directly into the damaged tissues created by bites, burns, injections, and accidental or surgical wounds. Insect bites are a common portal of entry of a variety of parasitic diseases carried by insects. *Pseudomonas aeruginosa* infections are especially common in hospitals among burn and surgical patients.

Finally, a few infectious organisms, mostly viruses, can cross the placenta from an infected mother and cause an infection in the fetus. Congenital infectious diseases such as syphilis, AIDS, and rubella (German measles) occur in this way.

Effects of Portal of Entry on Disease

Some organisms can enter the body through only a single portal, but others can enter through several portals. Furthermore, the same organism can have greatly different effects, depending on its portal of entry. For example, many pathogens that cause diseases of the digestive tract cause no disease at all if they happen to enter the respiratory tract. Similarly, most pathogens that cause respiratory diseases do not infect the skin or the tissues of the digestive tract. Yet a few organisms cause quite different illnesses depending on where they enter the body. For example, the plague bacterium causes different symptoms when acquired through a bite than when acquired through inhalation into the lungs. In the former case the infection, called bubonic plague, has a mortality of about 60 percent if untreated. In the latter case, known as pneumonic plague, mortality for untreated patients is very close to 100 percent.

Even when a pathogen enters the body, it may not reach an appropriate site to cause an infection. Almost everyone carries *Klebsiella pneumoniae* in the pharynx at some time during a winter. Why then do we not all come down with pneumonia? The reason is that the organism must enter the lungs before it can cause disease.

Portals of Exit

How infectious agents get out of their hosts is nearly as important to their ability to cause disease as how they get in. The sites at which organisms leave the body are called **portals of exit** (Figure 16.8).

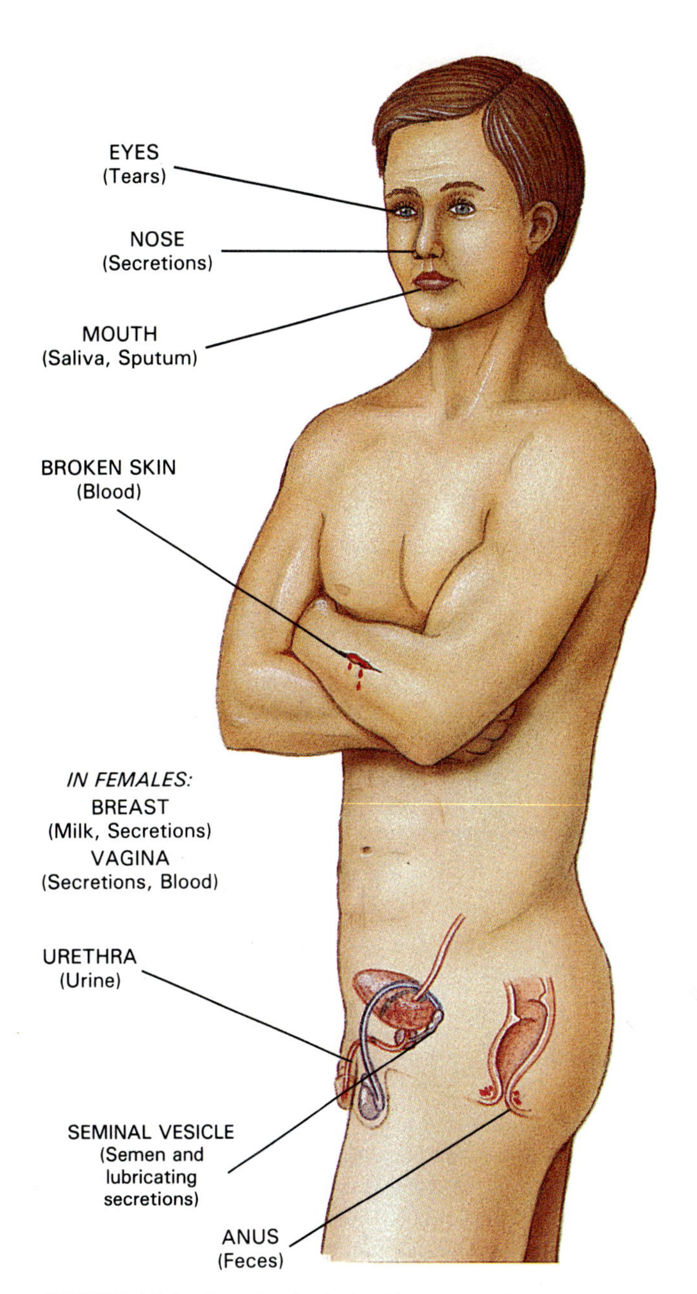

FIGURE 16.8 Portals of exit for disease-causing organisms.

In most cases the pathogens exit with body fluids or feces. Respiratory pathogens exit through the nose or mouth in fluids expelled during coughing, sneezing, or speaking. Pathogens of the gastrointestinal tract exit with fecal material. Some of these pathogens are eggs of helminths, which are exceedingly resistant to drying and other environmental conditions. Urine and, in males, semen from the urethra carry genitourinary pathogens. Semen, which is sometimes ignored as a means by which pathogens exit the body, is an important medium for the viruses that cause AIDS and hepatitis. Blood sometimes contains infectious organisms and can be a source of infection for health workers or others rendering aid to an injured person. Another person's blood should always be considered potentially infectious. Finally, saliva from dogs, cats, insects, and other animals can transmit infectious organisms.

TABLE 16.2 Modes of transmission of selected diseases	
Modes of Transmission	**Examples of Diseases Transmitted**
Contact transmission	
Direct contact	Rat-bite fever, rabies, syphilis, gonorrhea, herpes, staphylococcal infections, cutaneous anthrax, genital warts
Indirect contact by fomites	Hepatitis B, tetanus, rhinovirus and enterovirus infections
Droplets	Common cold, influenza, measles, Q fever, pneumonia, whooping cough
Vehicle transmission	
Foodborne	Intoxication with aflatoxins and botulinum toxin, paralytic shellfish poisoning, staphylococcal food poisoning, thyphoid fever, salmonellosis, listeriosis, toxoplasmosis, tapeworms
Waterborne	Cholera, shigellosis, leptospirosis, *Campylobacter* infections
Airborne, including dust particles	Chickenpox, tuberculosis, coccidioidomycosis, histoplasmosis, influenza, measles
Vector transmission	
Biological	Plague, malaria, yellow fever, typhus, Rocky Mountain spotted fever, Chagas' disease
Mechanical (on insect bodies)	*E. coli* diarrhea, salmonellosis, conjunctivitis

Modes of Transmission of Diseases

For new cases of infectious diseases to occur, pathogens must be transmitted from portals of exit to portals of entry. Transmission can occur by several modes, which we have grouped into three categories: (1) *contact transmission,* (2) *transmission by vehicles,* and (3) *transmission by vectors.* Modes of transmission and diseases transmitted by each mode are shown in Table 16.2.

Contact Transmission

Contact transmission can be direct, indirect, or by droplets. **Direct contact transmission** requires person-to-person body contact. Such transmission can be horizontal or vertical. In **horizontal transmission** pathogens usually are passed by handshaking, kissing, contact with sores, or sexual contact. They also can be spread from one part of the body to another through unhygienic practices. Touching genital herpes lesions and then touching other parts of the body, such as the eyes, can spread the lesions. Pathogens from fecal matter also can be spread by unwashed hands to the mouth; this is **direct fecal-oral transmission.** In **vertical transmission** pathogens are passed from parent to offspring in an egg or sperm, across the placenta, or while traversing the birth canal (as can happen with syphilis and gonorrhea).

Indirect contact transmission occurs through **fomites** (fo'mit-ez), contaminated nonliving objects such as clothing, dishes, eating utensils, bedding, toys, bar soap, and money. (U.S. paper currency is treated with an antimicrobial agent that reduces transmission of microorganisms.)

Droplet transmission, another kind of contact transmission, occurs when a person in close proximity to others coughs, sneezes, or speaks (Figure 16.9). **Droplet nuclei** consist of dried mucus, which protects microorganisms embedded in it. These particles can be inhaled directly, or they can collect on the floor with dust particles and later become part of aerosols (described later). Droplet transmission occurs over a distance of less than 1 m, and such droplets are not considered to be airborne.

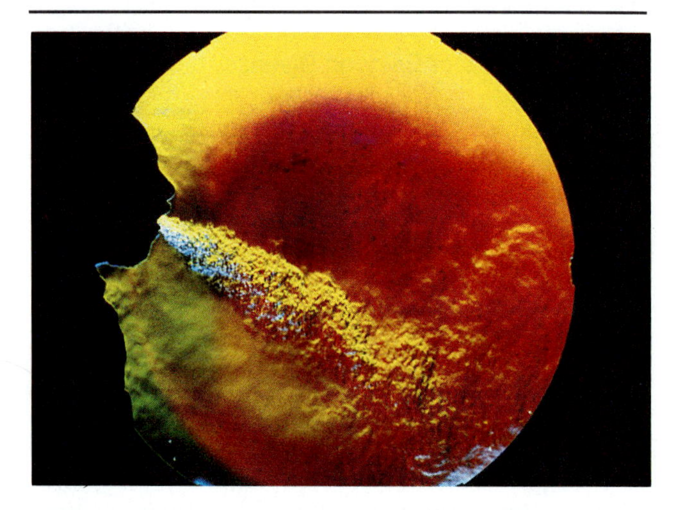

FIGURE 16.9 A cough is photographed with the Schlieren technique that detects air-speed differences as different colored areas. Such dispersal is most important within a radius of about 3 feet. However, the smallest particles can be dispersed for a much greater distance and kept aloft by air currents. Even a surgical mask will not prevent spread of all droplets.

Transmission by Vehicles

Common **vehicles** for transmission of pathogens include water, air, and food. Blood, other body fluids, intravenous fluids, and drugs also can serve as vehicles of disease transmission.

Waterborne Transmission Although waterborne pathogens do not grow in pure water, some survive transit in water with small quantities of nutrients, and many thrive in nutrient-rich water such as sewage and water polluted with fertilizer. Waterborne pathogens usually are transmitted in water contaminated with untreated or inadequately treated sewage. Such **indirect fecal-oral transmission** occurs when pathogens from feces of one organism infect another organism. Pathogens have been isolated from public water supplies, semiprivate water supplies (camps, parks, and hotels that have their own water systems), and private water supplies (springs and wells). Polioviruses and enteroviruses as well as a number of bacteria infect the digestive system and cause gastrointestinal symptoms. Waterborne infections can be prevented by proper treatment of water and sewage, but enteroviruses are especially difficult to eradicate from water.

Airborne Transmission Airborne microorganisms are mainly transients from soil, water, plants, or animals. They do not grow in air, but some reach new hosts through air in spite of dryness, temperature extremes, and ultraviolet radiation. Dry air actually enhances transmission of many viruses. Pathogens are said to be airborne if they travel 1 m or more through an air medium. Both airborne pathogens and those suspended in droplets have the best chance of reaching new hosts when people are crowded together indoors. Increased incidence of airborne infections is associated with nearly sealed modern buildings where temperatures are controlled with heating and air conditioning systems, and little fresh air enters.

Airborne pathogens become suspended in aerosols or fall to the floor and combine with dust particles. An **aerosol** is a cloud of tiny water droplets or fine solid particles suspended in air. Microorganisms in aerosols need not come directly from humans; they also can come from dust particles stirred by dry mopping, changing bedding, or even changing clothing. Flaming a full inoculating loop in the laboratory also can disperse microorganisms into aerosols. Dust particles can harbor many pathogens. Bacteria with sturdy cell walls, such as staphylococci and streptococci, can survive for several months in dust particles. Naked viruses and bacterial and fungal spores can survive for even longer periods. Air currents disturb dust and easily suspend contaminated particles in the air.

Hospitalized patients are at greatest risk of acquiring airborne diseases because they often have lowered resistance and because former patients may have left pathogens deposited in dust particles. Cleaning floors with a wet mop, wiping surfaces with a damp cloth, and carefully unfolding bed linens and towels help to reduce aerosols. Masks and special clothing are used in operating rooms, burn wards, and other areas where patients are at greatest risk of infection. Some hospitals also use ultraviolet lights and special air flow devices to prevent exposure of patients to airborne pathogens.

Foodborne Transmission Pathogens are most likely to be transmitted in foods that have been processed in unsanitary conditions, incompletely cooked, or poorly refrigerated. As with waterborne pathogens, foodborne pathogens are most likely to produce gastrointestinal symptoms.

Transmission by Vectors

As we saw in Chapter 12, *vectors* are living things that transmit disease to humans. (p. 295) Most vectors are arthropods such as ticks, flies, and mosquitoes. *Mechanical vectors* passively transmit pathogens on their feet and bodies; in this mode they are similar to fomites. Houseflies and other insects, for example, frequently feed on animal and, if available, human fecal matter. They then move on to feed on human fare, depositing pathogens in the process. Disease transmission by mechanical vectors can be prevented simply by keeping these vectors out of areas where food is prepared and eaten. The fly that walked across dog feces in the park should not be allowed to walk across your picnic potato salad. Using screened areas and keeping insects out of them also minimize disease transmission by mechanical vectors. Unfortunately, in some poverty-ridden areas of the world, screens are lacking on windows—even those that open into hospital operating rooms!

Biological vectors actively transmit pathogens that complete part of their life cycle in the vector. Compared with direct transmission through bites, the transmission of zoonoses through vectors is much more common. In most vector-transmitted diseases a biological vector is the host for some phase of the life cycle of the pathogen. Control of zoonoses transmitted by biological vectors can often be achieved by controlling or eradicating the vectors. Spraying standing water with oil kills many larval insects. Spraying breeding grounds with pesticides also can be an effective control, at least until the vectors become resistant to the pesticides.

Special Problems in Disease Transmission

Transmission of disease by carriers poses special epidemiologic problems because carriers are often difficult to identify. The carriers themselves usually do not know they are carriers and sometimes cause sudden outbreaks of disease. Depending on the disease they carry, they can transmit it by direct or indirect contact or through vehicles such as water, air, or food; they can even provide a source of pathogens for vectors.

Another special transmission problem arises with people who have so-called sexually transmitted diseases (STDs). Such diseases are most often transmitted by direct sexual contact, including kissing, but some can be transmitted by anal contact or by fomites. STDs present epidemiologic problems because in-

PUBLIC HEALTH

"Dog Germs!"

When Snoopy kisses Lucy, she always screams, "Dog germs!" However, Snoopy is the one who should be worried. Some infections can be transmitted in either direction between children and pets, but dogs are in greater danger of becoming infected from children than the other way around. A human mouth is more likely to contain pathogens than a dog's mouth because dog saliva is much more acidic and is thus a less hospitable environment for microorganisms.

Though kissing endangers the dog more than the human, swimming with man's best friend may not be a good idea. *Leptospira* spirochetes ordinarily infect animal kidneys and are passed out in their urine. Water contaminated with such urine can lead to human leptospirosis, a disease characterized by fever, headache, and kidney damage, as one man discovered when he went swimming in a river with his dog. An infected dog would be even more hazardous in a crowded swimming pool. Though vigilant chlorination of pools and vaccination of pets against leptospirosis might prevent them from transmitting the disease, it is best not to allow pets in pools.

A variation on this chain of transmission occurred in a recent year when some teenagers drove their swamp buggy through water contaminated with urine from leptospirosis-infected deer. Droplets splashed up by the tires apparently infected the passengers.

Is it the child or the dog who should be less affectionate?

fected individuals sometimes have contact with multiple sexual partners and because the incidences of AIDS, genital herpes, syphilis, and *Chlamydia* infections are rapidly increasing.

Zoonoses present still another epidemiologic problem. They can be transmitted by direct contact, as when humans get rabies from the bite of an infected domestic or wild animal. Research is in progress to develop an oral vaccine for rabies to be used in wildlife. Eating poorly cooked meat can lead to the worm infection called trichinosis. Drinking unpasteurized milk from infected animals can lead to tuberculosis and other diseases. Biological vectors of zoonoses, such as mosquitoes, ticks, and fleas, provide still another way to transmit diseases. Epidemiologists find real challenges in identifying the sources of such infections and teaching people how to avoid them.

Disease Cycles

Many diseases have been observed to occur in cycles. Only a few cases are seen over a period of several years, and then many cases suddenly appear. Bubonic plague—or Black Death, as it was called—has occurred in pandemic outbreaks followed by recurrent cycles for centuries. Between A.D. 543 and 548 it ravaged the eastern empire of the Roman Emperor Justinian. In the capital city of Constantinople (now Istanbul, Turkey), plague killed 200,000 people in only 4 months. It came from India or Africa to Egypt and from there via fleas on ship rats to Europe, where it spread from Constantinople throughout the Mediterranean basin. About a half century later it appeared in China, with equally devastating results. After the initial outbreaks it occurred in cycles of 10 to 24 years for the next two centuries.

When no further cases of plague were seen, Europe breathed a sigh of relief and relegated plague to ancient history. But in 1346, the Black Death struck again. This second pandemic was worse than the first, affecting North Africa, the Middle East, and most of Europe. Nearly one-third of the population of Europe died, and three-fourths of the population of many cities lost their lives to the dreaded disease. The disease spread northward from the Mediterranean lands, taking only a few years to reach the British Isles, Scandinavia, Poland, and Russia. Then cyclic recurrences claimed more lives in epidemics such as those of seventeenth-century England and eighteenth-century France. Near the beginning of the twentieth century, a pandemic killed more than a million people in India and spread to many parts of the world, including San Francisco.

Is this sleeping giant of a disease gone forever? Will it return in a few centuries or a few years? If it does strike again, it will become a monumental problem in the rat-infested cities of the world, including those in the United States. The available vaccine is not completely effective, and many rats have become resistant to the best available rat poisons. Furthermore, plague has become endemic to the southwestern United States, where *Yersinia pestis* survives in animal reservoirs such as wild desert rodents.

Cyclic diseases pose special epidemiologic problems. Epidemiologists still cannot predict when one will break out and reach epidemic proportions. It is difficult to be prepared to treat sudden, large increases in the incidence of a disease and nearly impossible to persuade people to be immunized against a disease they have never seen.

Group Immunity

An important factor in cyclic disease is **group immunity,** also called **herd immunity,** the proportion of individuals in a population who are immune to a particular disease. If group immunity is high, that is, if most of the individuals in a population are immune to a disease, then the disease can spread only among the small number of susceptible individuals in the population (Figure 16.10). Even when a member of the population becomes infected, the likelihood of that person transmitting the disease to others is small. Thus, a sufficiently high group immunity protects the entire population, including its susceptible members.

We can illustrate the principle of group immunity by considering the spread of measles (rubeola) in two different hypothetical populations of young people in which the disease commonly occurs. In a city in the United States with 10,000 people under 20, the city health department has mounted a successful campaign to get children immunized, and only 100 have not received the vaccine. Thus, only 100 of 10,000, or 1 percent, are susceptible to measles. If a child with measles moves into the community, that child will encounter a susceptible child in only 1 out of 100 contacts.

Now let us look at a city of the same size in a country where measles vaccine has not been available. Of the 10,000 people under 20, 5000 had measles when the disease last struck the population, which means the other 5000 are susceptible. Should a child with measles move to this city, that child will encounter a susceptible child in 1 out of 2 contacts. He or she is 50 times as likely to transmit the disease as an infected child in the U.S. city.

Another way to look at group immunity is to consider the threshold proportion of susceptible individuals, that is, the proportion of the population that

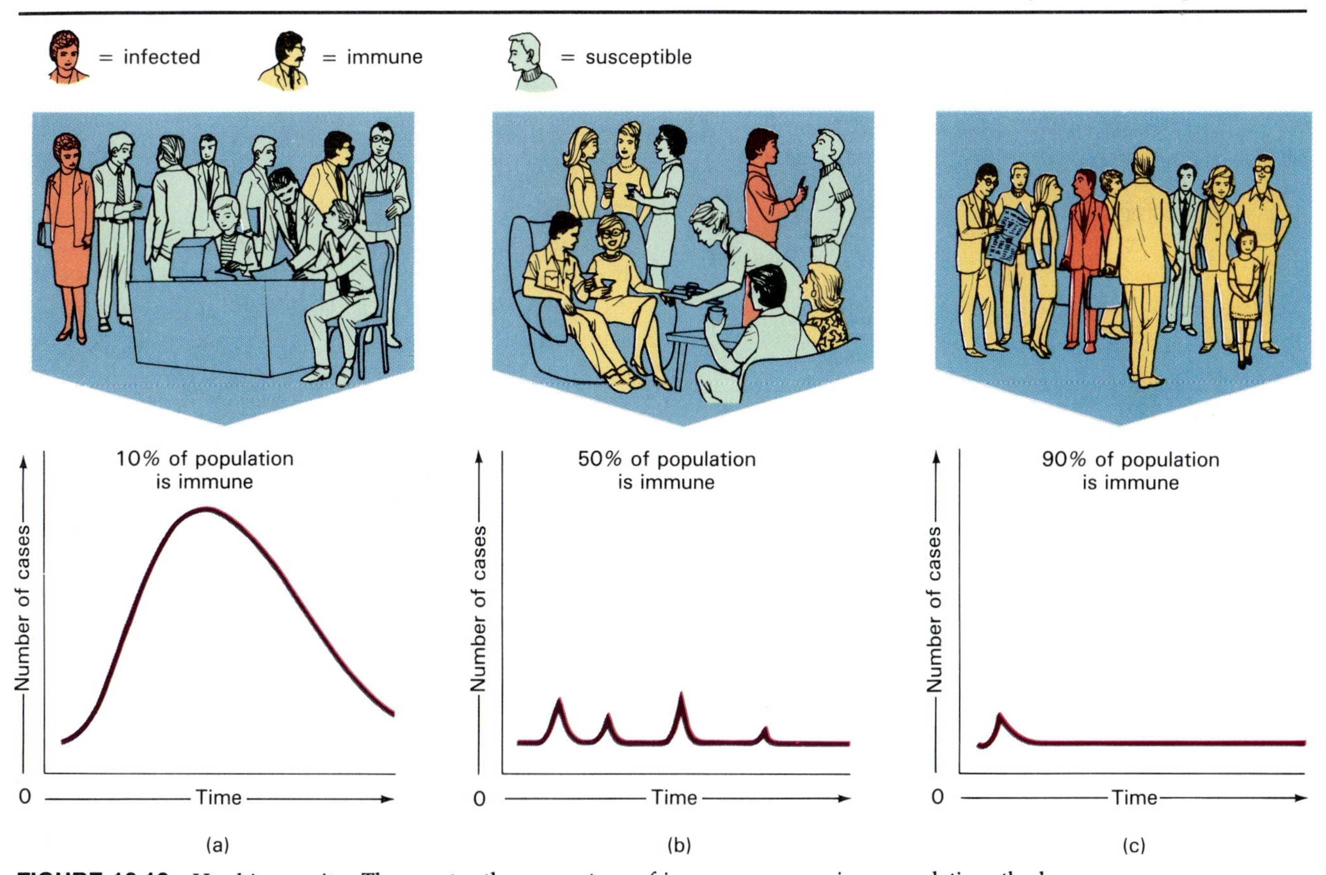

FIGURE 16.10 Herd immunity. The greater the percentage of immune persons in a population, the less likely a susceptible individual is to be exposed to the disease.

must be susceptible for the disease to spread (Figure 16.11). An epidemic can occur when the number of susceptible individuals exceeds the threshold proportion. During an epidemic, the number of susceptible individuals decreases as more and more of them get the disease, develop immunity, and become nonsusceptible. The epidemic subsides when the proportion of susceptible individuals drops below the threshold.

These examples illustrate the advantage of group immunity. A high level of group immunity protects a population from a disease, whereas a low or even moderate level of group immunity leaves susceptible individuals at great risk of becoming infected. It is easy to see, then, why public health officials desire to maintain the highest possible group immunity, especially against common cyclic diseases. They encourage parents to have children immunized against measles and other communicable diseases. Sometimes immunizations are required before children can attend school.

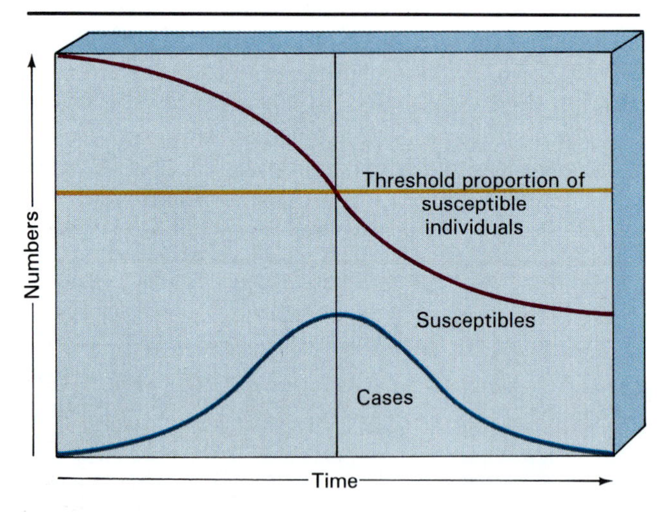

FIGURE 16.11 When the number of susceptible individuals in a population drops, the number of cases of disease also drops.

For a number of years in the United States, many jurisdictions have required children entering school to be immunized against measles. As a result, about 95 percent of elementary-age children are immune to measles; this group immunity has significantly decreased the incidence of measles in the United States. High school and college students who have neither had measles nor received vaccine are susceptible but usually are protected by group immunity. Infected exchange students from countries where measles is common have been responsible for serious outbreaks on college campuses. Now many school systems and some colleges and universities are requiring immunization against measles for all students regardless of age (Figure 16.12).

Immunizing school-age children is essential in order to provide group immunity for the protection of infants under 15 months, in whom the disease is most serious and the vaccine is not effective. Many infants suffer severe complications or die from measles.

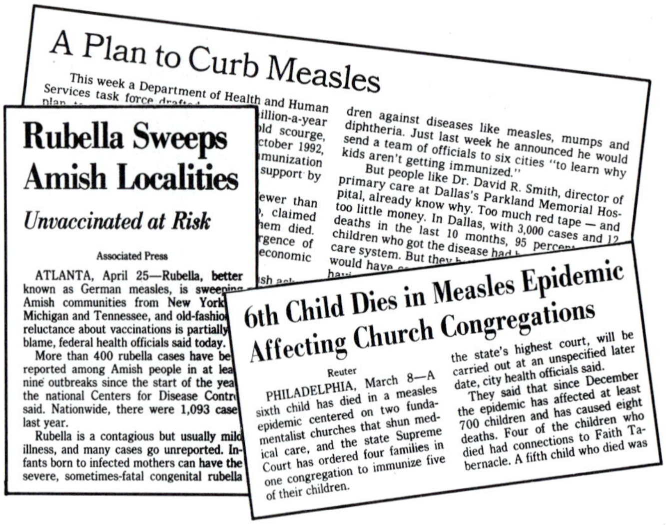

A Plan to Curb Measles

Rubella Sweeps Amish Localities

Unvaccinated at Risk

Associated Press

ATLANTA, April 25—Rubella, better known as German measles, is sweeping Amish communities from New York Michigan and Tennessee, and old-fashio reluctance about vaccinations is partially blame, federal health officials said today. More than 400 rubella cases have be reported among Amish people in at lea nine outbreaks since the start of the yea the national Centers for Disease Contr said. Nationwide, there were 1,093 case last year.

Rubella is a contagious but usually mild illness, and many cases go unreported. Infants born to infected mothers can have the severe, sometimes-fatal congenital rubella

6th Child Dies in Measles Epidemic Affecting Church Congregations

Reuter

PHILADELPHIA, March 8—A sixth child has died in a measles epidemic centered on two fundamentalist churches that shun medical care, and the state Supreme Court has ordered four families in one congregation to immunize five of their children.

the state's highest court, will be carried out at an unspecified later date, city health officials said. They said that since December the epidemic has affected at least 700 children and has caused eight deaths. Four of the children who died had connections to Faith Tabernacle. A fifth child who died was

FIGURE 16.12 Measles has been a recent problem, despite required immunization of children entering school. Older populations of students (high school or college age) have not all been immunized, and many remain susceptible. While the incidence of measles cases in the United States is very low, exchange students or travelers from countries where measles immunization is at a low level can bring the disease and set off an outbreak among these older susceptible populations. Older adults are likely to be immune as a result of having had the disease in childhood.

In contrast to measles control by vaccination, malaria control in the United States was accomplished mainly by controlling mosquito vectors. Consequently, Americans have little group immunity to malaria because most have never had the disease. Since the 1940s new cases of malaria have occurred mainly among people infected in other countries (Figure 16.13). As long as infected individuals do not reintroduce the parasite into the mosquito population, a malaria epidemic in the United States is most unlikely in spite of the low group immunity. In recent years, however, laborers from areas where malaria is still common have brought the parasite back into this country, and the disease has become endemic in some parts of California.

Public Health Organizations

In the United States and many other countries, the importance of controlling infectious diseases and minimizing other health hazards has been recognized since the eighteenth century. Public health agencies have been created at every governmental level—in cities, counties, and states, at the federal level, and even at the world level. City and county health departments provide immunizations and inspect restaurants and food stores. They work with other local agencies to assure that water and sewage are properly

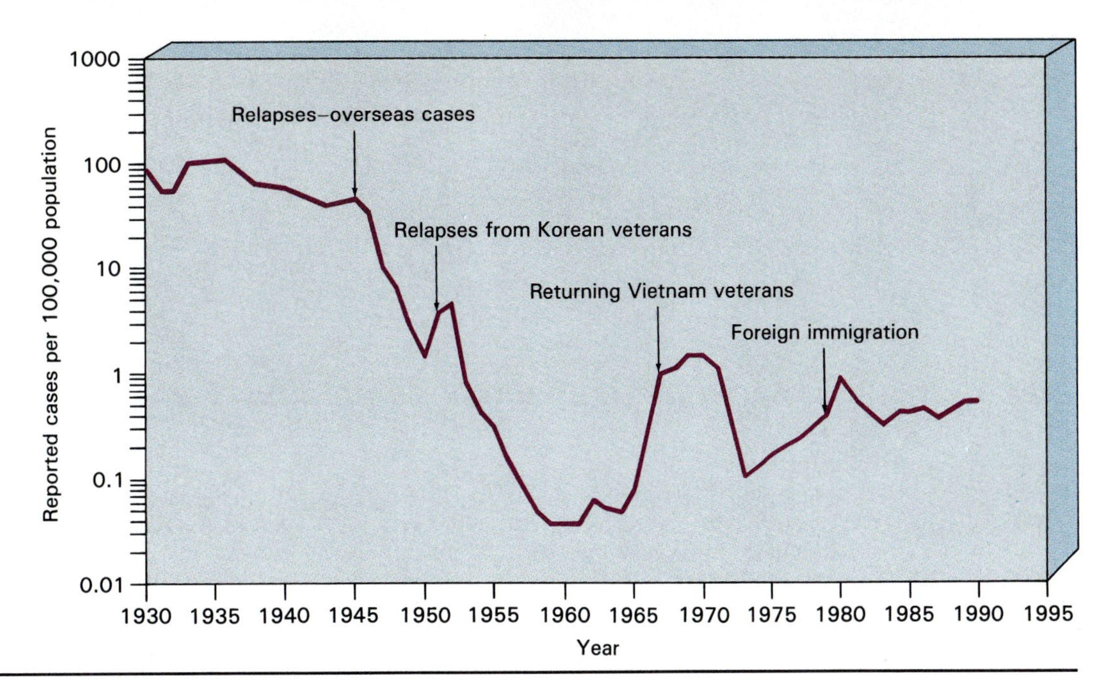

FIGURE 16.13 Incidence of malaria cases, as related to sources of influx into this country.

treated. State health departments deal with problems that extend beyond cities and counties. They often perform laboratory tests, such as the identification of rabies in animals and hepatitis and toxins in water.

The federal government operates the United States Public Health Service (USPHS), which has several branches. Of these branches, the CDC (Figure 16.14) has major responsibilities for the control and prevention of infectious diseases and other preventable conditions. Currently, the study of the epidemiology of AIDS is an important project. CDC also is concerned with occupational health and safety, quarantines, tropical medicine, cooperative activities with national agencies in other countries and with international agencies, and public health education (Figure 16.15). Although licensing and upgrading of clinical laboratories was at one time a responsibility of CDC, the College of American Pathologists now has that responsibility.

Some CDC activities are of special interest to microbiologists. The CDC makes recommendations to the medical community regarding the use of antibiotics, especially for the treatment of diseases caused by antibiotic-resistant organisms such as penicillin-resistant *Neisseria gonorrhoeae*. It also maintains supplies of infrequently used drugs and provides them to physicians who encounter patients with tropical parasitic diseases and other diseases rarely seen in the United States. The CDC also makes recommendations regarding the administration of vaccines—which should be used, who should receive them, and at what ages.

The World Health Organization (WHO) coordinates and implements programs to improve health in more than 100 member countries. Its basic objective is the attainment by all peoples of the highest possible level of health. This aim is embodied in the current theme: "Health for All by the Year 2000." Specific activities are carried out by six regional organizations in Africa, the Eastern Mediterranean, Europe, Southeast Asia, the Western Pacific, and the Americas. WHO works closely with the United Nations on pop-

FIGURE 16.14 CDC headquarters in Atlanta, Georgia.

PUBLIC HEALTH

Monitoring Labs

Many state health departments send numbered cultures to hospital laboratories for identification. If technicians cannot properly identify the organisms in the cultures, the laboratory may lose its certification. Care and skill are critical not only for accurate diagnostic work but also to maintain a safe laboratory environment. Infections with *Shigella dysenteriae* are commonly acquired by laboratory workers using poor aseptic technique. Inept technicians endanger themselves as well as the public, and the state acts as an independent agency to encourage their exclusion from the health care system.

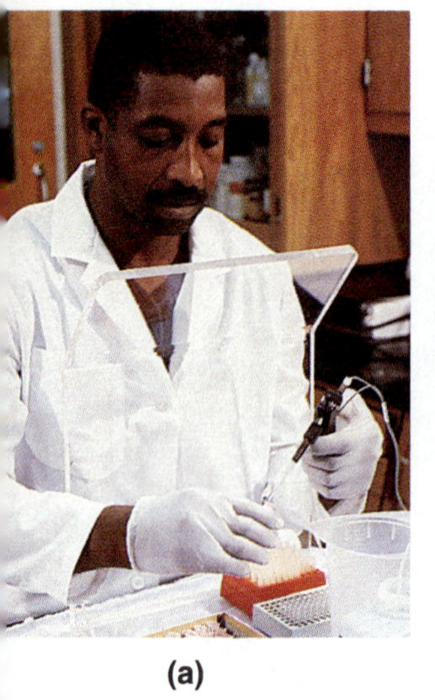

(a)

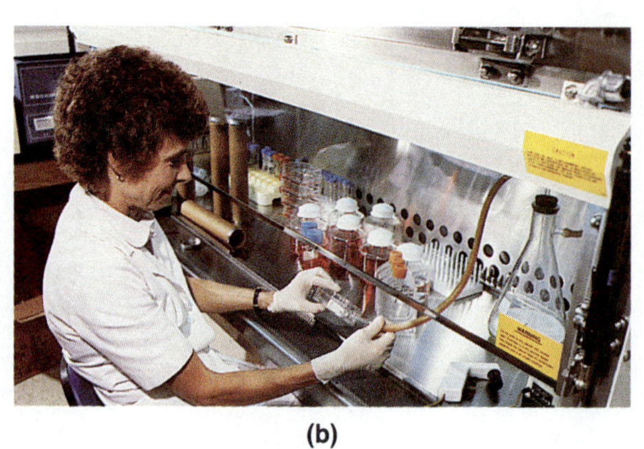

(b)

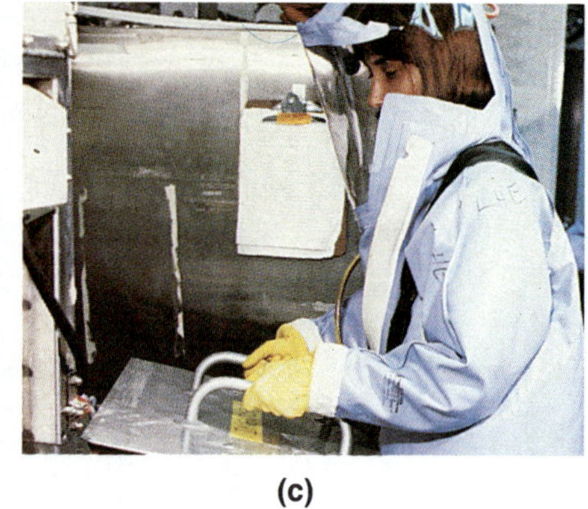

(c)

FIGURE 16.15 Activities of the Centers for Disease Control (CDC) include research in many areas of pathology and epidemiology. (a) Performing enzyme-linked immunosorbent assay (ELISA) on blood samples, to diagnose disease. (b) Processing cell cultures in a biosafety cabinet. (c) Retrieving specimens from a tank of liquid nitrogen where viruses are stored at −83°C.

ulation control, management of food supplies, and various other scientific and educational activities.

WHO aids in international disease control by setting health standards, helping developing countries to mount effective control and immunization programs, and maintaining surveillance of potential epidemics. It collects, analyzes, and distributes health data and provides training and research programs for health personnel and information for individuals (Figure 16.16). WHO is currently working to provide sanitary drinking water and to control diarrhea worldwide. It has assisted more than 100 countries in immunizing against diphtheria, measles, whooping cough, poliomyelitis, tetanus, and tuberculosis and hopes eventually to eradicate measles worldwide. It conducts research and training to combat widespread tropical diseases such as leprosy, malaria, and several diseases caused by worms. WHO also collects data on AIDS worldwide, but efforts to obtain comprehensive information have been hampered by several African countries that refuse to report significant data.

Extensive cooperation between WHO, the USPHS

(a)

(b)

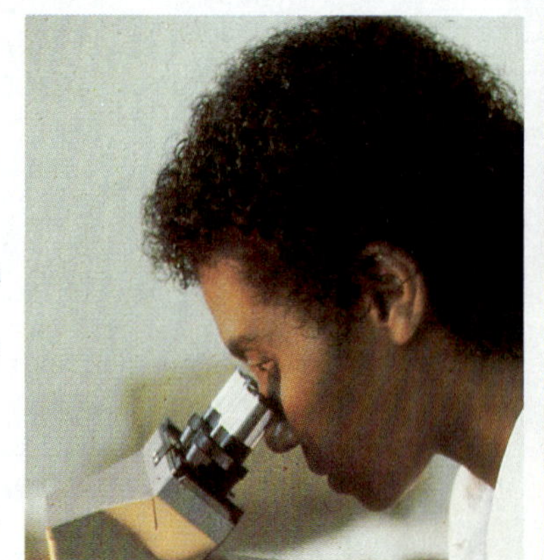

(c)

(d)

FIGURE 16.16 Some typical activities of the World Health Organization. (a) Ambulance boat transporting a patient to a hospital, Burma. (b) Eye tests for the prevention of blindness are given high in the Andes Mountains of Peru. (c) Performing clinical laboratory tests for pathogens, Guyana. (d) Medical assistant travels by horseback to treat patients in remote villages of China.

and its affiliated CDC, and other organizations such as the American Public Health Association has led to major advances in health standards and living conditions all over the world. These efforts have recently led to the eradication of smallpox worldwide.

Notifiable Diseases

Cooperation among health organizations has led to the establishment of a list of **notifiable diseases,** which physicians must report to public health officials. Requiring such reporting is intended to accomplish two things: (1) ensure that public health officials learn of diseases that jeopardize the health of populations and (2) provide consistency and uniformity in the reporting of those diseases. Various kinds of information about notifiable diseases in the United States are available from CDC; a sample of this information is provided in Table 16.3.

Epidemiological Studies

In 1855, some 30 years before Robert Koch isolated the causative agent of Asiatic cholera, the London physician John Snow made what might be the first epidemiological study. He was concerned with the

TABLE 16.3(a) Summary of cases of notifiable diseases, United States, weeks ending December 28, 1991, and December 29, 1990

AIDS	Aseptic Meningitis	Encephalitis		Gonorrhea		Hepatitis (Viral), by type				Legionellosis	Lyme Disease
		Primary	Post-infectious			A	B	NA, NB	Unspecified		
Cum. 1991	Cum. 1991	Cum. 1991	Cum. 1991	Cum. 1991	Cum. 1991	Cum. 1991	Cum. 1991	Cum. 1991	Cum. 1991	Cum. 1991	Cum. 1991
43,389	14,102	923	76	602,577	678,811	22,953	16,790	3,113	1,230	1,222	8,884

Malaria	Measles (Rubeola)					Meningococcal Infections	Mumps		Pertussis			Rubella		
	Indigenous		Imported		Total									
Cum. 1991	1991	Cum. 1991	1991	Cum. 1991	Cum. 1990	Cum. 1991	1991	Cum. 1991	1991	Cum. 1991	Cum. 1990	1991	Cum. 1991	Cum. 1990
1,173	27	9,276	—	212	26,951	1,998	44	4,031	53	2,575	4,450	11	1,372	1,093

Syphilis (Primary & Secondary)		Toxic-shock Syndrome	Tuberculosis		Tularemia	Typhoid Fever	Typhus Fever (Tick-borne) (RMSF)	Rabies, Animal
Cum. 1991	Cum. 1990	Cum. 1991	Cum. 1991	Cum. 1990	Cum. 1991	Cum. 1991	Cum. 1991	Cum. 1991
41,006	48,867	274	23,543	23,973	188	456	635	6,486

TABLE 16.3(b) Notifiable diseases of low frequency, United States

	Cumulative 1991		Cumulative 1991
Anthrax	—	Leptospirosis	59
Botulism: Foodborne	22	Plague	10
Infant	70	Poliomyelitis, paralytic	—
Other	6	Psittacosis	87
Brucellosis	89	Rabies, human	3
Cholera	24	Syphilis, congenital, age < 1 year	1703
Congenital rubella syndrome	36	Tetanus	49
Diphtheria	2	Trichinosis	61
Hansen's disease	140		

PUBLIC HEALTH

AIDS Testing: Mandatory for Health-Care Workers?

If your doctor, dentist, or nurse had AIDS, would you still want to be treated by him or her? If you test positive for the AIDS virus, should you be forced to reveal this? Does it matter whether you are the provider or the recipient of health care?

CDC estimates that between 13 and 128 patients may have been infected with AIDS virus from health-care workers during invasive surgical procedures during the decade ending in 1990. During this same time, 40 health-care workers are known to have acquired the infection on the job, with a few hundred other undetected cases thought to exist. One study estimates that 2000 or more infected surgeons and dental workers are currently practicing in the United States.

In addition, with over 1,000,000 persons in the United States thought to be infected with the AIDS virus, people are increasingly uneasy that the dental drill in their mouth, or some other invasive instrument being used on them, has previously been used on an AIDS-positive individual. Who should be tested, when, how often, and with what consequences?

In July 1991, CDC issued new guidelines recommending that health-care workers performing certain invasive surgical procedures should undergo voluntary (not mandatory) tests for the viruses that cause AIDS and hepatitis B. Those found to be infected should stop performing such procedures until they obtain permission from a panel of experts and inform their patients. The procedures include tooth extractions and other dental procedures, vaginal and cesarean deliveries, hysterectomies, operations on bones, joints, colon, and rectum, plus X-rays and tests of the heart and blood vessels known as cardiac catheterization and angiography.

Reactions to these recommendations have been mixed. Thus far professional societies have refused to cooperate with CDC in agreeing to enforce the new regulations. Some people fear that a strict interpretation would prevent an infected dentist from ever practicing again. Others fear that a hospital's worries of liability would cause individual hospitals to require testing. Insurers might require testing to obtain malpractice insurance. Critics point out that 4 days after CDC made their recommendation, a New York State physician disclosed his positive status for the AIDS virus, and his hospital demanded his resignation which he then gave. In the same week, the U.S. Senate passed a bill sponsored by Senator Jesse Helms of North Carolina that would impose criminal penalties of up to $10,000 and 10 or more years in jail on infected health-care workers who knowingly fail to tell patients before performing invasive procedures. This bill failed to make it through the House.

Although most public health agencies and health professionals societies, such as the American Medical Association, oppose mandatory testing, the majority of health-care workers do not. A July 1991 survey found that 57 percent of the doctors and 63 percent of the nurses questioned favored mandatory testing of health-care workers. Three-quarters of those surveyed also favored mandatory AIDS testing of surgical patients and pregnant women, as well as mandatory reporting of the names of patients who test positive to local health departments.

As you can see, deciding who should be tested is a complex matter. How do you feel about this issue? What does a poll of your classmates reveal?

cause of a cholera epidemic sweeping through the city at that time. Snow eventually traced the source of the epidemic to the Broad Street pump in Golden Square (Figure 16.17). He showed that people became infected by drinking water contaminated with human feces.

Since this landmark study, many other investigators have attempted to conduct **epidemiological studies** to learn more about the spread of disease in populations. Such studies can be descriptive, analytical, or experimental.

Descriptive epidemiological studies note the number of cases of a disease, which segments of the population were affected, where the cases occurred and over what time period. The age, sex, race, marital status, socioeconomic status, and occupation of each patient are determined. From careful analysis of cumulative data from several studies, epidemiologists can determine whether people of a certain age group, males or females, or members of a certain race are particularly susceptible to the disease. Data on marital status or sexual behavior patterns might show whether the disease is sexually transmitted. Data on socioeconomic status might show that a disease is most easily transmitted in undernourished people or those living in substandard conditions. Noting the occupation of infected individuals can help investigators trace diseases to certain factories, slaughterhouses, or hide-processing plants. If most cases are among veterinarians, the disease is probably transmitted by the animals they handle.

The geographic distribution of cases is also important, as Snow's study showed. Modern counterparts of Snow's work sometimes trace an outbreak of a disease to a contaminated water supply, a restaurant where a hepatitis carrier works, or an area where a particular vector is present.

Some investigators think an infectious agent may contribute to the development of the disease multiple sclerosis because most cases are found in clusters, in

FIGURE 16.17 In 1854, Dr. John Snow, a London physician, recorded the locations of cholera cases in the city. He found that they were clustered around the Broad Street pump, which supplied water to the nearby area. He tracked the epidemic's source to contamination of the pump by sewage. When the handle of the pump was removed, the outbreak subsided. Snow's study was the first modern, systematic, scientific epidemiological study.

temperate climates, and among Caucasians. In the United States its incidence increases significantly from Mississippi to Minnesota. However, an important determining factor seems to be where a person spent the first 14 years of life. For example, if you grew up in a northern region with a high incidence of multiple sclerosis, moving south later will not reduce the likelihood of your acquiring the disease. Although genetic and environmental factors also may be involved, the pattern of incidence suggests that an infectious agent may play some role in causing multiple sclerosis.

CLOSE-UP

A Tale of Two Cities

"Hamburg persisted in postponing costly improvements to its water supply. . . . [It] drew its water from the Elbe without special treatment. Adjacent lay the town of Altona . . . where a solicitous government installed a water filtration plant. In 1892, when cholera broke out in Hamburg, it ran down one side of the street dividing the two cities and spared the other completely. . . . A more clearcut demonstration of the importance of the water supply in defining where the disease struck could not have been devised. Doubters were silenced; and cholera has, in fact, never returned to European cities, thanks to systematic purification of urban water supplies from bacteriological contamination."

—William H. McNeill, 1976

Finally, the period of time over which cases appear and the season of the year are important considerations. To study the role of time in an epidemic, epidemiologists define an **index case** as the first case of the disease to be identified. As we have seen, common-source epidemics can be distinguished from propagated epidemics by the rate at which the number of cases increases and time required for the epidemic to subside. The season of the year in which epidemics occur may help to identify the causative agent. Arthropod-borne infections usually occur in relatively warm weather, and certain respiratory infections usually occur in cold weather, when people tend to be crowded together indoors. The seasonal nature of encephalitis, with most cases occurring in fall, was shown in Figure 16.5.

Analytical epidemiological studies focus on establishing cause-and-effect relationships in the occurrence of diseases in populations. Such studies are *retrospective* when factors that preceded an epidemic are considered or *prospective* when factors that occur as an epidemic proceeds are considered. In a retrospective study the investigator might ask patients where they had been and what they had done in the month or so prior to their illness. If most had hiked in a certain wooded area, had contact with horses, or shared some other common activity, that activity might provide a clue to the source of the infection. Several investigations of this sort are described in Berton Roueche's fascinating book *The Medical Detectives* (see Some Interesting Reading at the end of this chapter). In a prospective study, groups in a population—which children get chickenpox, at what age, and under what kinds of living conditions—are used to determine which factors relate to susceptibility and resistance to infection.

Experimental epidemiological studies are designed to test a hypothesis, often about the value of a particular treatment. Such studies are limited to animals or to humans in which participants are not subjected to harm. For example, an investigator might test the hypothesis that a particular treatment is effective in controlling a disease for which no accepted cure is available. One group from a population would receive the treatment, and another group would receive a placebo. A **placebo** (plas-e'bo) is a nonmedicated substance that has no effect on the recipient but that the recipient believes is actually a treatment. From the results of the study, the investigator could determine whether the new treatment was effective.

Control of Communicable Diseases

Several methods are currently available for full or partial control of communicable diseases. They include isolation, quarantine, immunization, and vector control.

In **isolation** a patient with a communicable disease is prevented from contact with the general population. Isolation is generally accomplished in a hospital where appropriate procedures can be carried out. CDC has designated five categories of isolation: strict, protective, respiratory, enteric, and wound and skin. Strict isolation makes use of all available procedures to prevent transmission of organisms to medical personnel and visitors (Figure 16.18). In other kinds of isolation, somewhat less stringent measures are used, as shown in Table 16.4. Isolation minimizes the spread of disease among susceptible individuals. It prevents spread of the disease in the general population because the patient is immune to the disease before returning to the population. It minimizes the chance that a disease will spread to other patients with lowered resistance, while also protecting patients from exposure to new diseases.

Quarantine is the separation of humans or animals from the general population when they have a communicable disease or have been exposed to one (Figure 16.19). Although it is one of the oldest methods of controlling communicable diseases, it is now used mainly for serious diseases such as cholera and yellow fever. It differs from isolation in two ways: (1) It can be applied to healthy people exposed to a disease during the incubation period, and (2) it pertains to limiting the movements of people and not necessarily to precautions during treatment. Quarantine is rarely used today because it is very difficult to implement. To ensure that no infected persons spread a disease, all who had been exposed to it would have to be quarantined for the disease's incubation period. This would mean, for example, that all travelers returning to the United States from a region of the world where cholera is endemic would have to be quarantined for 3 days in accommodations provided at airports and seaports—not likely to be a popular idea.

Large-scale *immunization* programs are an extremely effective means of controlling communicable diseases for which safe vaccines are available. They greatly increase group immunity and thus greatly decrease human suffering and deaths from infectious diseases. In the United States polio, measles, mumps, diphtheria, and whooping cough have been nearly eradicated because of immunizations. Unfortunately, as the incidence of these diseases becomes very small, people become complacent about getting immunizations. Such complacency can lead to a sufficient decrease in group immunity to allow outbreaks of preventable diseases.

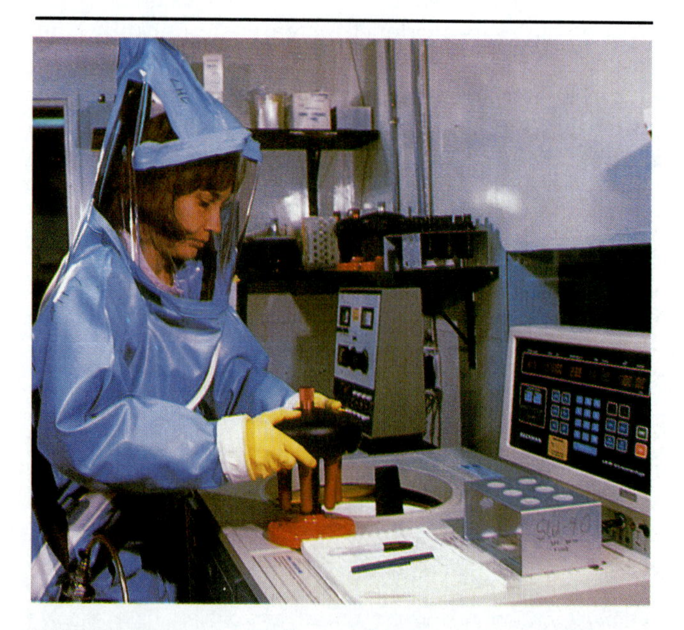

FIGURE 16.18 Scientists working with very dangerous and often easily disseminated organisms are protected by the P5 containment laboratory isolation equipment in this CDC laboratory. This lab represents the highest level of protection, both for the laboratory workers inside the lab and for the population outside. Extreme precautions (such as the use of special ventilation systems) are taken to avoid contact of personnel with microorganisms and to prevent the escape of microorganisms from the facility.

FIGURE 16.19 Quarantine of animals is much easier than attempting to quarantine people. The U.S. Department of Agriculture has long exercised this authority and continues to do so today.

Vector control is an effective means of controlling infectious diseases when a vector, such as an insect or a rodent, can be identified and its habitat, breeding habits, and feeding behavior determined. Places where a vector lives and breeds can be treated with insecticides or rodenticides. Window screens, mosquito netting, insect repellents, and other barriers can be used to protect humans from becoming victims of the bites of feeding vectors. Unfortunately, vectors have their own defenses. Some escape or become resistant to pesticides or make their way through barriers.

In spite of the fact that communicable diseases are theoretically preventable, some communicable diseases still have a high incidence in every human population. In countries with relatively high living standards, the common cold and numerous sexually transmitted diseases occur with great frequency. Countries with lower living standards, especially those in the tropics, are plagued with malaria and a variety of other diseases, including some that have been nearly eradicated in other countries. AIDS has become a worldwide threat that now extends beyond the special high-risk groups with which it was once associated.

NOSOCOMIAL INFECTIONS

A **nosocomial infection** is an infection acquired in a hospital or other medical facility. The term nosocomial

TABLE 16.4 A summary of important isolation procedures

	Kinds of Isolation				
	Strict	**Protective**	**Respiratory**	**Enteric Precautions**	**Wound and Skin Precautions**
Visitors must check in at nursing station before entering patient's room	Yes	Yes	Yes	Yes	Yes
Hands must be washed on entering and on leaving patient's room	Yes	Yes	Yes	Yes	Yes
Gowns must be worn by personnel and visitors	Yes	Yes	No	Only for direct patient contact	Only for direct patient contact
Masks must be worn by personnel and visitors	Yes	Yes	Unless not susceptible to disease	No	While changing dressings
Gloves must be worn by personnel and visitors	Yes	Only for direct patient contact	No	Only for direct contact with patient or feces	Only for direct contact with lesion site
Private room required with door closed	Yes	Yes	Yes	Only for children	Desirable but not required
Examples of diseases	Pneumonic plague, rabies, diphtheria, disseminated herpes zoster, Lassa fever, chickenpox, draining *Staphylococcus aureus* wound, AIDS	Severe noninfected dermatitis, noninfected burns	Measles, mumps, rubella, pertussis, pulmonary tuberculosis	Typhoid fever, cholera, salmonellosis, shigellosis, hepatitis	Bubonic plague, gas gangrene, localized herpes, puerperal sepsis

is derived from two Greek words: *nosos*, disease, and *komeion*, to take care of. Thus, they are literally diseases acquired while the patient is being cared for. Although many such infections occur in patients, infections acquired at work by staff members also are considered nosocomial infections.

Among patients admitted to a hospital, 5 to 10 percent will acquire an infection that increases the risk of death, the duration of the hospital stay, and the cost of treatment. Of these, over 20,000 people per year die of their nosocomial infections. Curiously, nosocomial infections are largely a product of advances in medical treatment. Intravenous, urinary, and other catheters, invasive diagnostic tests, and complex surgical procedures increase the likelihood that pathogens will enter the body. Intensive use of antibiotics contributes to the development of resistant strains of pathogens. And, therapies to minimize rejection of transplanted organs impair the immune response to pathogens. In spite of the risk of nosocomial infections, the medical treatments now available save far more patients than are lost to such infections.

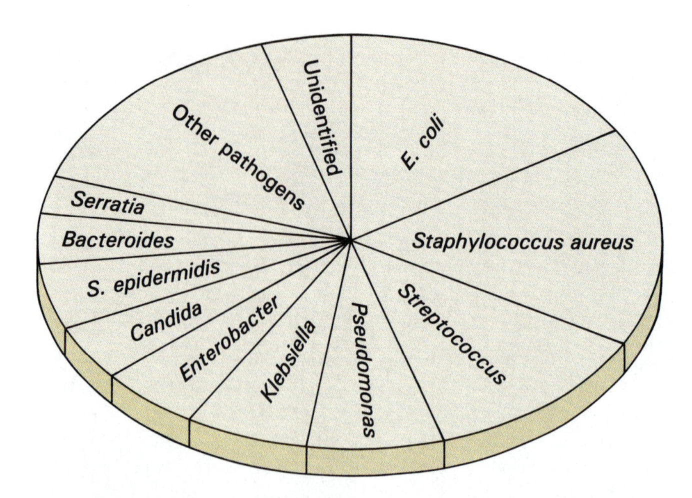

FIGURE 16.20 Common causative agents of nosocomial infections.

Epidemiology of Nosocomial Infections

The epidemiology of nosocomial infections, like the epidemiology of diseases acquired in the community, considers sources of infection, modes of transmission, susceptibility to infection, and prevention and control. In addition, it focuses on medical procedures that increase the risk of infection, the sites at which infections often occur, and the correlation between procedures and sites of infection.

Sources of Infection

Nosocomial infections can be exogenous or endogenous. **Exogenous infections** are caused by organisms that enter the patient from the environment. The organisms can come from other patients, staff members, and visitors. They also can come from inanimate objects, such as equipment used in respiratory or intravenous therapy, bathroom fixtures and soap, and water systems. Some nosocomial infections have even been traced to disinfectants such as quaternary ammonium compounds, to which certain organisms are resistant. Because hospitals and other medical facilities treat patients with infections, the numbers and variety of pathogens are higher than in the general community. **Endogenous infections** are caused by opportunists among the patient's own flora. Opportunists are most likely to cause infection if the patient has lowered resistance or if normal flora that compete with pathogens have been eliminated by antibiotics.

A small group of organisms, *Escherichia coli*, *Streptococcus* group D, *Staphylococcus aureus*, and *Pseudomonas* species are responsible for about half of all nosocomial infections (Figure 16.20). These organisms are particularly likely to cause such infections because they are ubiquitous (present everywhere), have given rise to numerous antibiotic-resistant strains, and can survive outside the body for long periods.

Susceptibility of the Host

Compared with the general population, patients in hospitals are much more susceptible to infection; that is, they are *compromised hosts*. Many patients have breaks in the skin from lesions, wounds (surgical and accidental), or bed sores. Some also have breaks in mucous membranes that line the digestive, respiratory, urinary, and reproductive systems. The lack of intact skin and mucous membranes provides easy access for infectious organisms. Also, most patients are debilitated to some degree, so their resistance to infectious organisms is lower than normal. Patients undergoing organ transplants receive immunosuppressant drugs, and patients with AIDS and other disorders of the immune system also have reduced resistance. Factors that contribute to host resistance are discussed in Chapters 17 and 18.

Modes of Transmission

Theoretically, nosocomial infections can be transmitted by all modes of transmission that occur in the community. However, direct person-to-person transmission, indirect transmission through equipment and supplies, and transmission through air are most common in hospitals (Figure 16.21). Some organisms can be transmitted by more than one route.

As noted earlier, direct person-to-person transmission can occur when an infected patient, staff member, or visitor has direct contact with noninfected patients. The use of isolation procedures described earlier can minimize this kind of transmission. Indirect transmission through equipment and hospital procedures is a common mode of transmission of nosocomial infections.

PUBLIC HEALTH

Infection Control in Dentistry

Infection control in the dental suite is designed to protect both patients and health professionals. Safe practice requires using latex gloves, eye protection, a mask to prevent inhalation of aerosols (which are used extensively with modern drills), and the use of sterile equipment. Between patients, all items that could be contaminated with oral secretions or blood must be disinfected. For example, the gloved hand of the dentist leaves the patient's mouth, moves up to the light to adjust it, pushes away the rim of the instrument tray, and then reaches for the drill handle, leaving organisms behind on all these sites. It is not enough for the dentist merely to change gloves between patients—the clean glove might pick up organisms left on the equipment and transport them into the mouth of the next patient.

In September 1990, Florida dentist Dr. David Acer died of AIDS. Five of his patients have since been found to have the exact same strain of AIDS virus as Dr. Acer and are presumed to have contracted it from him in the dental office. None of the five had any other risk factors. Officials at CDC are at a loss to explain the method of possible transmission, but current suspicion falls on the drill handle. A University of Georgia study has shown that internal parts of the hollow handle, which holds drill bits, polishing brushes and other tools, can become coated with blood, saliva, and tooth fragments. The American Dental Association recommends heat sterilization of drill handles between patients, but this is inconvenient, and many dentists do not do so. Although the drill handles may be chemically disinfected, this is not sufficient to sterilize against the AIDS virus.

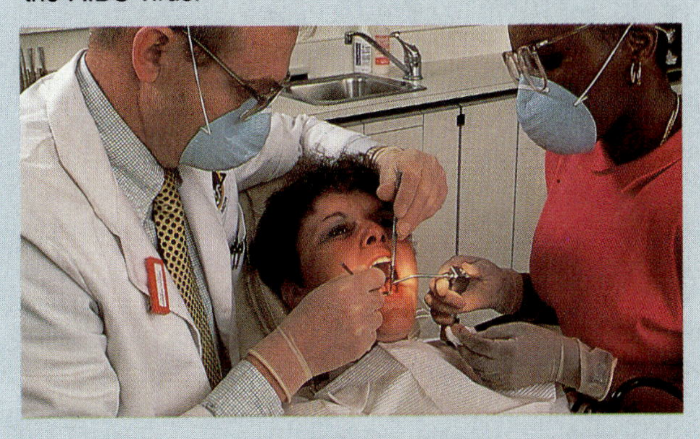

Use of gloves and masks by all dental personnel is essential if the spread of infection is to be avoided, but other measures are also necessary.

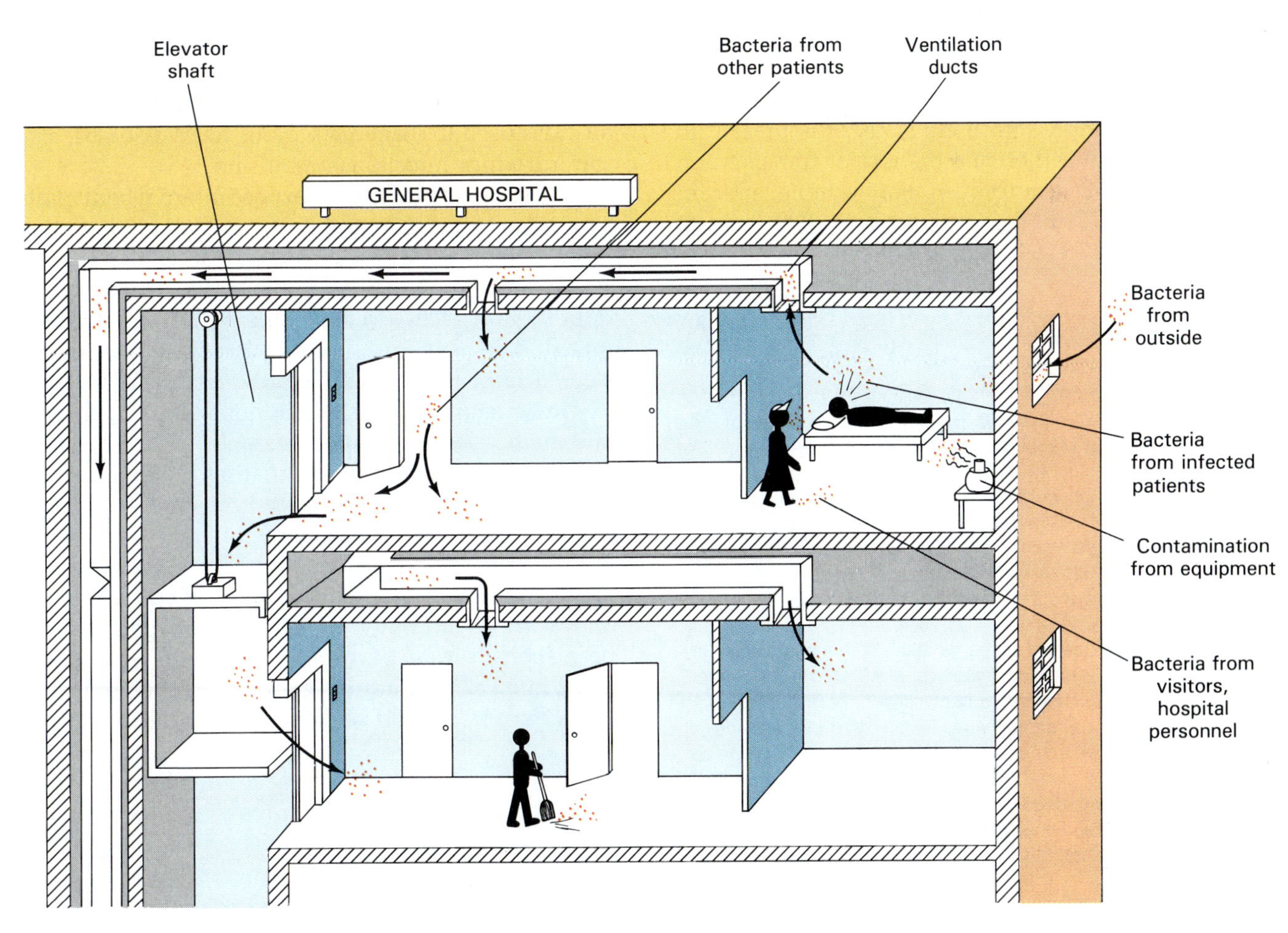

FIGURE 16.21 Some common modes of transmission of nosocomial infections in a hospital setting.

Universal Precautions

In 1988, concerned about the possibilities that the AIDS virus would be transmitted in the health-care setting, CDC issued guidelines to minimize the risks. These guidelines are now called **Universal Precautions.** Later that year, CDC added clarifications of some of these guidelines. Both documents are reproduced in part in Appendix D at the back of this book. Some hospitals and other medical facilities choose to exercise even more caution than CDC recommends. A shortened list of these precautions is given in Table 16.5. Universal Precautions apply to *all* patients, not just those known to be infected with the AIDS virus; hence the term *universal*. These precautions are intended to limit contact with blood or other body fluids that may transmit the AIDS virus. It is important to remember that other pathogens, such as the hepatitis B virus, are also transmitted in these fluids. Thus, even when patients test negative for the AIDS virus, their body fluids may still be dangerous. In fact, each year in the United States, over 12,000 health-care workers acquire hepatitis B infection in the course of their duties. Nearly 200 of these workers die every year.

Universal Precautions apply to the following body fluids: blood, semen, vaginal fluids, tissue, cerebrospinal fluid, synovial fluid, pleural fluid, peritoneal fluid, pericardial fluid, and amniotic fluid. CDC has stated that Universal Precautions do *not* apply to feces, nasal secretions, sputum, sweat, tears, urine, and vomitus, as long as these do not contain visible blood. This is not to imply that there is *no* virus present in these fluids, but rather that the risk of transmission is either very low or unproven. For example, a recent study using PCR methods found the AIDS virus to be present in *all* tested samples of saliva from AIDS patients. However, the levels were so low that it took the very sensitive PCR methods to detect them. AIDS patients are often infected with other disease organisms that may be present in these fluids, such as tuberculosis bacilli in sputum, bacteria such as *Salmonella, Shigella,* and *Cryptosporidium* in feces, and herpesvirus in oral secretions. Therefore, some health-care facilities require their employees to use universal precautions with *all* body fluids.

TABLE 16.5 Some Important Universal Precautions and Recommendations from CDC

1. Gloves and gowns should be worn if soiling of hands, exposed skin, or clothing with blood or body fluids is *likely.*
2. Masks *and* protective eyewear or chin-length plastic face-shields should be worn whenever splashing or splattering of blood or body fluids is *likely.* A mask alone is not sufficient.
3. Wash hands before and after patient contact, and after removal of gloves. Change gloves between *each* patient.
4. Use disposable mouthpiece/airway for cardiopulmonary resuscitation.
5. Contaminated needles and other sharp items should be discarded *immediately* into a *nearby,* special puncture-proof container. Needles must *not* be bent, clipped, or recapped.
6. Spills of blood or contaminated fluids should be cleaned by: (1) put on gloves and any other barriers needed, (2) wipe up with disposable towels, (3) wash with soap and water, (4) disinfect with a 1:10 solution of household bleach and water, allowing it to stand on surface for at least 10 minutes. Bleach solution should not have been prepared more than 24 hours beforehand.

Equipment and Procedures That Contribute to Infection

The use of equipment such as catheters and respiratory devices and surgical procedures are major contributors to nosocomial infections. The smallest abrasion can provide a site of entry for infectious agents. Urinary catheters, which are sometimes left in place for weeks, are used to drain urine from the bladder. Infections can arise from three sources: (1) a contaminated catheter, (2) inadequate cleaning of the site of catheter insertion so that the patient's flora invade the bladder, and (3) movement of organisms from leaky connections or the collection bag up the catheter into the bladder.

Catheters inserted into veins to administer nutrients, fluids, and medicines provide several means by which infections can arise. Needles and skin around the needle insertion site can be contaminated. Tubing, joints, containers of fluids, and the fluids themselves also can be contaminated. Diagnostic catheters, such as those used to diagnose various heart diseases, also can introduce infectious organisms.

All surgical procedures expose internal body parts to air, instruments, surgeons, and other operating room personnel, all of which can be contaminated. These procedures also can allow the patient's own flora to gain access to sites where they can produce infection. For example, bacteria that cause pneumonia can reach the lungs from the pharynx during surgery.

Respiratory devices that administer oxygen or air and medications to expand passageways in the lungs provide a means for disseminating microorganisms deep into the lungs. The nebulizer jet that delivers oxygen and medications can be contaminated, as can the containers and medications themselves. Organisms can grow in the reservoir pans of both cold mist and warm steam humidifiers, and the organisms can be dispersed in an aerosol as the machines operate. Therefore, all respiratory equipment, including humidifier pans, should be disinfected or sterilized daily and, if not disposable, should be disinfected before being moved from one patient to the next.

Other devices and procedures account for smaller, but significant, numbers of nosocomial infections. The embedding in tissues of prostheses, such as heart valves or joint replacements, allows introduction of

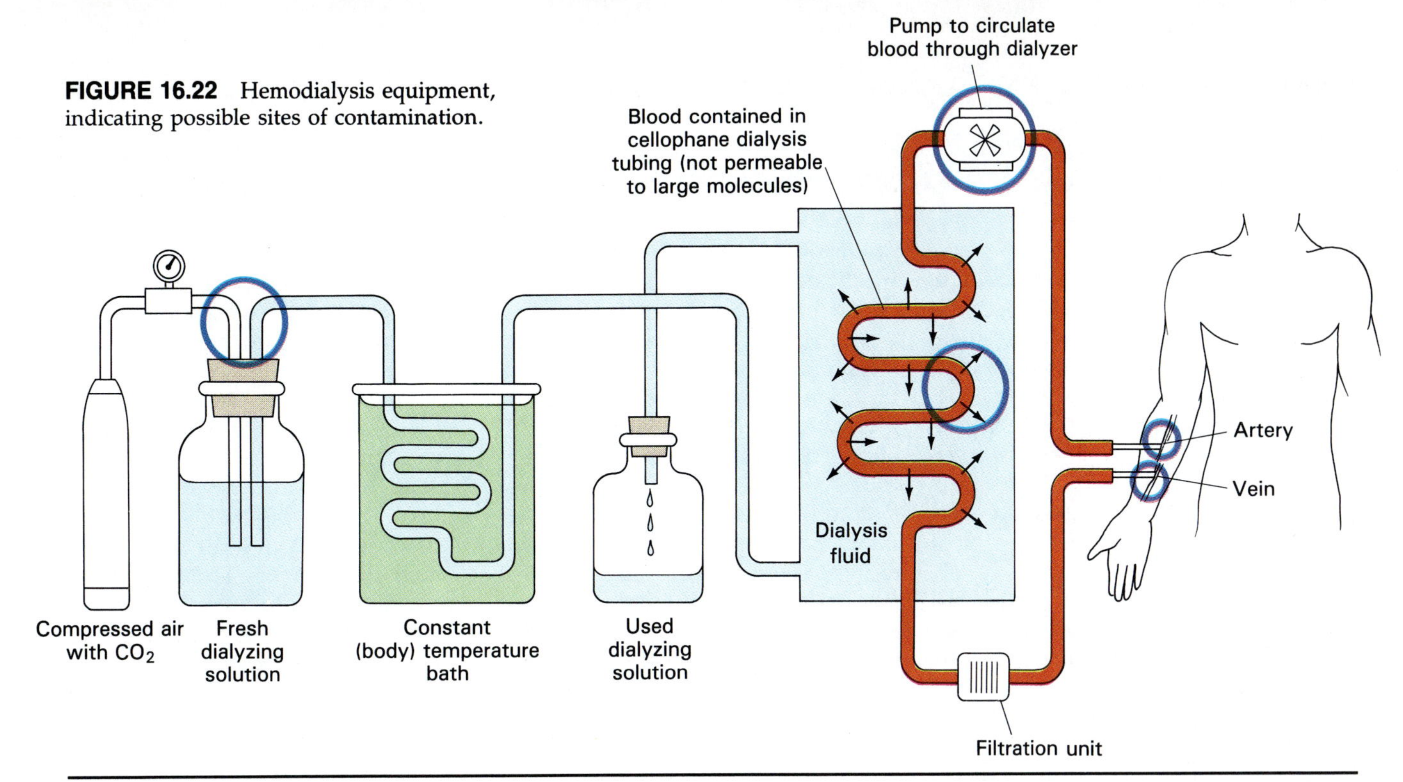

FIGURE 16.22 Hemodialysis equipment, indicating possible sites of contamination.

microorganisms. They can come from contaminated prostheses, surgical equipment, or directly from the surgeon. Hemodialysis, the removal of wastes from blood, provides a variety of means of introducing microorganisms into the body (Figure 16.22). Devices used to monitor blood pressure in the heart or major vessels or cerebrospinal fluid pressure have tubing extending outside the body; they can be contaminated or allow introduction of organisms from the patient or the environment. Endoscopes are introduced through body openings and are used to examine the linings of organs such as the bladder, large intestine, stomach, and respiratory passageways. Being difficult to sterilize, they can transfer microorganisms from one patient to another.

Another important factor that contributes to nosocomial infections is the intensive use of antibiotics, especially in hospital environments. How antibiotics contribute to the development of antibiotic-resistant pathogens and how these pathogens contribute to nosocomial infections were discussed in Chapter 14.

Sites of Infection

The sites of nosocomial infections, in order from most to least common, are as follows: urinary tract, surgical wounds, respiratory tract, skin (especially burns), blood (bacteremia), gastrointestinal tract, and central nervous system (Figure 16.23).

Prevention and Control of Nosocomial Infections

The problem of nonsocomial infections is widely recognized, and nearly all hospitals now have infection-control programs. In fact, to maintain accreditation by the American Hospital Association, hospitals must have programs that include surveillance of nosocomial infections in both patients and staff, a microbiology laboratory, isolation procedures, accepted procedures for the use of catheters and other instruments, general sanitation procedures, and a nosocomial disease education program for staff members. Most hospitals have an infection-control specialist to manage such a program.

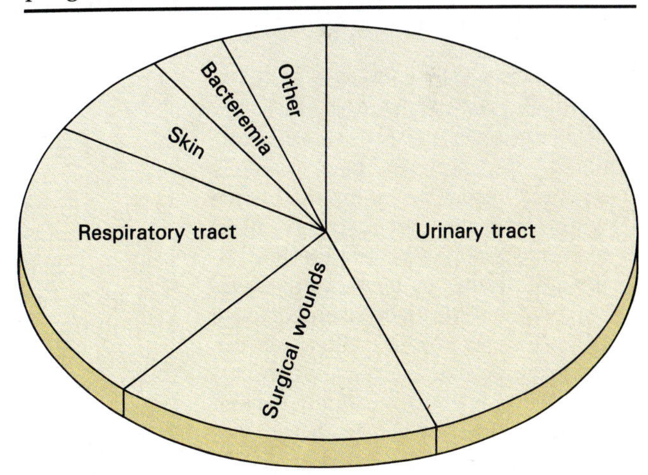

FIGURE 16.23 Relative frequency of sites of nosocomial infections.

Several techniques are available to prevent the introduction and spread of nosocomial infections. Hand washing is the single most important technique. When physicians, nurses, and other staff members wash hands thoroughly with soap and water between patient contacts, they can greatly reduce the risk of spreading diseases among patients. Scrupulous care in obtaining sterile equipment and maintaining its

MICROBIOLOGIST'S NOTEBOOK

Controlling Infection in a Burn Unit

Mrs. Edith Hollan: I remember one time we found a resistant strain of *Staphylococcus* on the Cabbage Patch doll of a little girl who was a patient here. Of course, we didn't want to take the doll away. Instead, the occupational therapist told the girl that the doll too had a burn, and devised a cellophane bandage for it. As play therapy, it probably helped the child emotionally—and it kept her safe from infection.

That kind of vigilance is essential in infection control. You have to use your imagination as well as your experience to anticipate where microorganisms might be found. And today, controlling infection is the most important factor in saving our patients' lives. Dr. Mason can explain why.

Dr. Arthur Mason: The treatment of burn victims has changed a great deal in the past half century. When I first entered the field in 1954, two of the major advances in burn care had recently been made. We had learned that skin grafting could be used to replace the skin destroyed by the burn, and that shock could be averted by the administration of fluids to replace those lost to the burn.

The advent of effective fluid replacement sharply decreased early loss of life due to burn shock. Thus a whole new population of burn patients emerged—severely injured people who would not previously have lived for more than a day or two. But, although these patients now lived longer, many still died. In most of these cases, burn wound infection and bacteremia preceded death. Nevertheless, bacteremia was not generally considered to be the *cause* of death.

Dr. Albert McManus: It's hard to understand from today's perspective, but at the time, bacterial infection of burns was thought to be significant only in so far as it interfered with healing of the burn area itself. People weren't aware of the systemic consequences of these infections. Organisms such as *Pseudomonas aeruginosa* and other gram-negative bacteria were often found in the blood and wounds of burn patients, but because these organisms were not recognized as causes of infection in other surgical situations, they were considered nonpathogens. Moreover, the first-generation antibiotics available at that time, such as penicillin, were effective mostly against streptococci and staphylococci—the traditional agents causing wound infections. They were of little use against *Pseudomonas*, so treating badly burned patients with these drugs usually made no difference in mortality. This fact masked the true role of infection as an often fatal complication in burn cases.

Dr. Mason: During the 1950s, we at the Institute set out on a systematic examination of the process of infection in burns. By the end of the decade, we had developed an animal model of *Pseudomonas* sepsis in which contamination of nonlethal burn wounds with *P. aeruginosa* consistently caused death. In this way we were able to convince the medical community that often it wasn't the burn itself that proved lethal, but rather the generalized bacteremia that developed—especially with gram-negative organisms.

The next step was naturally to explore possible antimicrobial therapies. We soon found that antibiotics in the bloodstream failed in human infections because they did not penetrate the dead tissue of burns, where there was no longer any blood supply. Even newer antibiotics, such as Polymyxin B, that had high activity in the laboratory against strains taken from burned patients, did not reach burn wounds in high enough concentrations to be effective. Dead burned tissue was the site of contamination and growth of organisms; by the time the organisms had penetrated to the circulatory system, bacterial numbers were just too large for effective antibiotic action.

The logical alternative approach was to try and deliver a drug directly to the surface of the burn site. After much experimentation, we eventually found that Sulfamylon, an antibacterial agent used by the German army for wounds during World War II, was highly effective in controlling bacterial growth in burn wounds. The results were remarkable: bacteremia was much less frequent, and mortality decreased sharply.

But this Utopian result—a single, simple, effective therapy—didn't last;

The Institute of Surgical Research, at Brooke Army Medical Center in Fort Sam Houston, Texas, is known throughout the world for its pioneering work in the treatment of burns. An important factor in its success has been its microbiology laboratory, where studies of surgical wound infection have gone on since the Institute was founded.

Dr. Arthur D. Mason is a surgeon who came to the Institute some 35 years ago. As chief of the Laboratory Division, he has been, in large part, responsible for the program of research on burn infections and ways to combat them.

Dr. Albert McManus is chief of the Microbiology Branch and holds a Ph.D. degree in microbiology. He came to the Institute 14 years ago. Working with Dr. Mason and Mrs. Edith Hollan, he has developed a microbiology laboratory that is considered a leader both in research and in the application of its results to the treatment of burn victims.

Mrs. Edith Hollan received her B.S. degree in chemistry, then trained as a nurse at Yale University during World War II. Upon graduation, she was assigned to a Navy burn unit. She plans to continue her work with burn patients and conduct scientific studies of her own. She hopes eventually to write about her experiences—and perhaps try her hand at some detective stories.

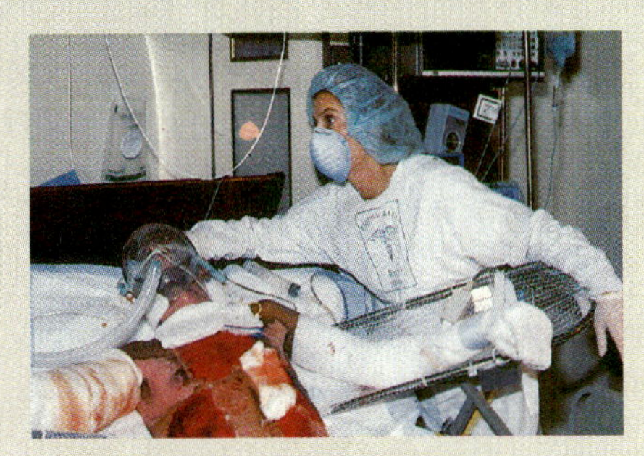

A burn ward nurse adjusting equipment used to support a seriously burned patient. Care of burn victims requires much contact between patients and medical personnel, with consequent danger of cross-contamination of patients.

bacteria are much too versatile for that. In 1969, a strain of *Providencia stuartii* that was resistant to Sulfamylon appeared in our patients. Within a year or two we again had a serious infection problem. This challenge was ultimately met with another topical agent, but it was clear that the topical approach alone was not a complete answer. It fell to Dr. McManus, assisted by Mrs. Hollan, to develop a more complete method for infection control in these patients.

Dr. McManus: During the first few years after the introduction of Sulfamylon, many patients with previously fatal burn injuries survived. This effect, however, slowly eroded, as if this therapeutic net was developing holes; the protective effect of Sulfamylon was waning. Sulfamylon's laboratory activity against *P. aeruginosa* had not changed; rather, new, resistant types of organisms were causing infections.

Clearly, microbial evolution in response to Sulfamylon's use had occurred. There was also the possibility that some of the new resistant organisms had been introduced from outside the burn ward. Since the Institute is a referral center, many of our patients are transferred from other hospitals. With each transfer, there was a chance that the new patient had already been contaminated with resistant nosocomial pathogens. And these strains were probably being maintained somewhere in the ward environment and passed to new patients after admission.

When I was assigned to the Microbiology Branch of the Institute, I found that most of the microbiology cultures were taken to identify an organism that had produced clinical symptoms. This type of information was very useful for describing what infections were occurring, but it helped very little in trying to find the sources and mechanisms of cross-contamination. How many patients carried the same strains but were not cultured because they were not sick? I therefore proposed a culture system that would provide a wider sample of data, reflecting all the patients throughout their hospitalization.

This system came to be known as the Microbial Surveillance System, and it does just what its name suggests. It's based on the premise that a burn wound presents the ideal environment for the colonization of microorganisms. Thus, if the patient survives the initial catastrophe of shock, fluid loss, and the burn itself, infection *will* occur—if not from the environment, then from the patient's own normal flora. To counteract this, we must get as much information as possible before infection manifests itself clinically. So on admission to the Burn Unit, the infection-control nurse enters a wide range of clinical data into our computer: results from cultures of the patient's blood, sputum, stool, and urine. If an organism is present, we group or subtype it according to its antibiotic susceptibility or resistance. If a patient must be transported—for example, a soldier injured overseas—cultures are taken during the flight. In this way, we can know in advance what's coming in—a peculiar strain of a bug, what its sensitivity is—and we are prepared for it.

As long as each patient remains in intensive care, we take cultures on a regular schedule from common sites of infection. We also take cultures from selected sites and equipment in the unit.

Dr. Arthur D. Mason.

Culture results, updated on a daily basis, are made available on computer terminals in the patient care areas and are set up so that they can be easily used even by the "computer-shy." Microbial colonization can thus be followed in each patient, and when an infection is clinically suspected, the surgeon can select antibiotics based on these prior cultures. This, in effect, puts the doctor several days ahead of the old system, in which cultures were taken only after clinical symptoms of infection had developed. Stored data from the Microbial Surveillance System is also used for epidemiological studies.

Within a year or so after starting the system Mrs. Hollan joined our Institute. She had an extensive background and, as she told me soon after we met, was involved in burn care before I was born. (I was 35 at the time!) She immediately saw the utility of the system in her infection-control activities and quickly became an indispensable member of the team. Mrs. Hollan provides most of the data for the Microbial Surveillance System. She and I work together as a team; we count on each other for up-to-the-minute information about each patient.

Mrs. Hollan: Dr. McManus made a huge understatement when he said we work together as a team! I think of my work here at the Burn Unit as a direct

continued

Controlling Infection in a Burn Unit *Continued*

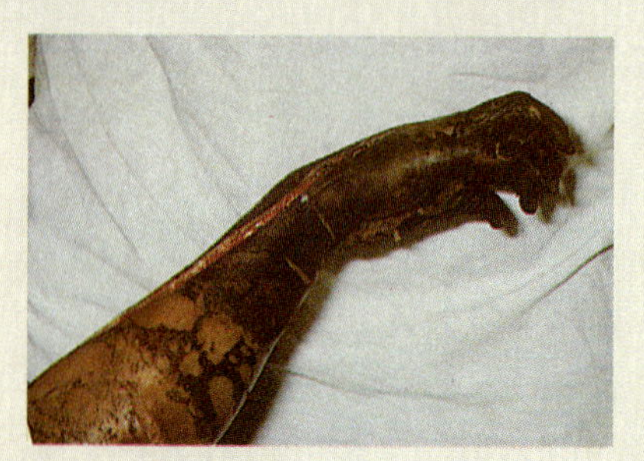

Third-degree burn of hand and forearm shows charred tissue (eschar) in which bacteria can grow. This layer must be removed to prevent infection.

extension of the work done in Dr. McManus's microbiology laboratory. In other words, I apply his research. Conversely, he relies on my input of clinical data.

The Microbial Surveillance System is the main channel for this exchange of information between us. As the infection-control nurse, one of my responsibilities is to augment and update the computer data—hourly if necessary. Our policy generally is to culture specimens obtained from patients at least three times a week. In addition, I monitor the course of therapy, and keep accurate records of each patient's vital signs. The patient's body temperature is often very significant—a decline may be indicative of septicemia.

Our system allows us to study the natural history of microbial infection in each of our cases. Each morning, we print out hard copy from the computer of the data relating to every patient in the ward so that the clinicians can follow the dynamics of microbial colonization over the course of time: shifts in microbial populations, the appearance of new organisms, the emergence of antibiotic resistance.

The availability of such information can greatly reduce the likelihood of a wrong or ineffective drug being given. For instance, if we know that a potential pathogen is present and the patient develops signs of infection, we can assume that the infection is likely to be due to that particular organism. When the computer data has warned us of the presence of a virulent organism and the patient takes a turn for the worse, we are ready with the best available drug.

The key factor is that each patient's microbiological status is monitored constantly, and the information we obtain is available and ready to be used *before* it manifests itself in the form of clinical symptoms. Without our surveillance system, the clinician would not only have far less information, but also far less time in which to make decisions—decisions on which the patient's survival may depend. With patients so seriously ill, there just isn't enough time for trial and error.

Dr. McManus: Of course, knowing what organisms are present or are coming doesn't always guarantee that we have the weapons to deal with them successfully. Usually we do—but there are exceptions. Once, we had a "visitor" so tough that we never really licked it; at best, you could say we fought it to a draw. The episode began when the institute was called on to evacuate three patients from a hospital in South America—American merchant seamen who had been burned in a fire in their ship's engine room. They proved to be infected with a resistant strain of *Providencia stuartii*.

Dr. Albert McManus at work in the microbiology laboratory.

We had done bacterial cultures on the airplane, so we knew the patients had it—but in this case forewarned definitely wasn't forearmed. Within a few days, all three of the seamen developed serious infections with this organism, and despite the use of infection-control measures, the bacteria quickly spread to other patients.

In addition to being highly contagious, the organism proved to be resistant to everything we could throw at it—not only antibiotics, but even many of the disinfectants we use. It harbored a plasmid that was just loaded with resistance genes. It even had several different genes for resistance to some agents. To this day, we haven't fully analyzed everything that was on that plasmid. Worse, within a few weeks, other types of organisms with very similar antibiotic resistance patterns began to appear. It soon became obvious that not only were we having an outbreak of *Providencia* infections, but also the spread of a plasmid capable of transferring multiple antibiotic resistance to other organisms. [See Chapter 8. (pp. 207)]

Mrs. Hollan: An organism like that is the nightmare of every infection-control officer. It spread through the unit very quickly. We soon found that we had a mini-epidemic on our hands. Eventually, we resorted to a kind of quarantine. We simply isolated all newly arriving patients from the patients who had already acquired the organism. Separate teams were set up to work with the new and old patients. The surveillance system continued to find the resistant *Providencia* and its plasmid only in the older group.

We can't say that we ever "cured" the patients infected with the original resistant organism. But *Providencia* is an opportunistic pathogen—it may be present, but if you're reasonably healthy, your immune system will keep it from causing any damage. Only debilitated people, like our burn patients, are seriously affected. So, as our patients improved, they fought off the infection. And as they left the unit one by one, the pathogen left with them. It was a bit like letting a forest fire burn itself out.

Nevertheless, we had to be very

careful. The recovered patients still harbored the bacteria, so even after they were well themselves, we had to keep them away from the other patients. This is a common problem in hospitals. When people recover from an illness, often one of their first impulses is to go back to the ward they've left and visit friends they may have made there, other patients. But if they do, they often bring organisms that are no longer dangerous to them but can be devastating to someone who's still sick. Recovered patients are potential reservoirs of infection. That's one reason we have to be so fussy about visiting.

Maintaining the entire ward as infection-free as possible is a major part of my job. Since our patients are severely immunocompromised by their injuries, they cannot effectively fight invasive organisms—even opportunistic ones that would not harm a healthy person. Unfortunately, though, we must keep the environment of the ward warm and humid. The physical environment is critical because our patients have lost the external layer of protection, and, at all costs, we must prevent shock and any more fluid loss. But these conditions also make the ward an ideal breeding ground for microbes. So, it is vitally important to guard against the introduction of new organisms. That's why we're very cautious about staff members and family members as well as patients. We check the immune status of all staff members at the time they're hired. If they're susceptible to some common ailment such as measles or rubella, we make sure to vaccinate them. Just recently we've started to immunize staff members against hepatitis B as well. Chickenpox remains a bit of a problem, though—there's no vaccine yet.

I constantly think of myself as a detective; I must be alert at all times for any unusual occurrence or strange incident. For example, I once noticed that every patient who had stayed in one particular cubicle seemed to come down with a *Candida* infection. One had a pneumonia, another developed bacteremia. But when we did cultures from objects in the room we could never find any pathogens. Finally, I performed a detailed inspection of the room and noticed a small pull chain to an outside vent. The vent was required by the fire code. Now, there are a lot of pigeons around the hospital—and pigeons are notorious carriers of *Candida*! I got permission to caulk up the vent, and we've had no problems with that room since then. I still have no proof that the vent or the pigeons were the source of the problem. It's always hard to rule out other explanations. The cases we saw might just have been coincidental. But I prefer to err on the side of caution—especially where fungal contamination is concerned. We don't encounter fungal infections very often, but when they occur they're usually the most virulent, the hardest to treat, and the most likely to be fatal.

I devised a lot of the procedures that we use to control infection, and I train all new personnel in our special techniques—for example, in the use of isolation garments. (Our standards here are *very* strict.) One of the biggest problems we have on the unit is getting the staff to adhere to our methods. Handwashing, for example, is probably the single most important defense against infection and cross-contamination. Often, I watch a few cubicles for half an hour and make a list of how often hands should be—or should have been—washed. People forget, even though they know it is important. We have inspections during which I take cultures of fingernails and fingertips of the staff. I also check for abrasions of the skin, which are often produced by some of the harsh chemicals used as disinfectants. These can all be breeding areas for *Staphylococcus* and *Streptococcus*, as well as an ideal means of transport of infectious bacteria from patient to patient.

Another of my responsibilities is approval of new equipment. A piece of apparatus may be quite appropriate for use on the unit in most respects, yet still can present a myriad of problems regarding infection control. For example, we often have to administer heparin, an anti-clotting agent, intravenously. I would prefer to use a heparin bag that does not require changing for 72 hours rather than the 24-hour bag that's now standard. Changing the bag requires a break in the sterile line from bag to bloodstream, and the more often this is done, the more opportunity there is for infection.

Mrs. Edith Hollan examining a culture from the burn ward.

Another example: some years ago we had bedpan "sterilizers" that used a burst of steam to kill microorganisms. These pans have a little bit of water in the bottom, like a toilet. We cultured the water—and found that it was full of bacteria! So now we use disposable bedpans much of the time. A lot of the battle in infection control is simply taking nothing for granted.

Dr. Mason: It's gratifying to us that many hospitals and clinicians from all over the world make use of our research. Indeed, many institutions use the Institute as a model. But the most important thing is that at the start of my career, most patients who sustained serious burns over 50 percent of their body died. Today, for the first time ever, young adults with burns over 80 percent of their body have an even chance of recovery. And our work has contributed to that change.

sterility while in use are also important. In addition, using gloves when handling infectious materials such as dressings and bedpans and when drawing blood prevents the spread of infections.

Opening laundry bags to dump used linen into hospital washing machines can represent a health hazard for laundry workers. Recently, laundry bags made of biodegradable plastic containing about 40 percent starch have been introduced. Bags of hospital laundry are thrown into the washing machine unopened. During the washing cycle they disintegrate as the starch dissolves in the water. Hospital workers need never touch the laundry, and the risk of infection is minimized.

Other techniques are needed to minimize the development of antibiotic-resistant pathogens. Routine use of antimicrobial agents to prevent infections has turned out to be a misguided effort because it contributes to the development of resistant organisms. Therefore, some hospitals maintain surveillance of antibiotic use. Antibiotics are given for known infections but are given prophylactically (as preventive measures) only in special situations. Prophylactic antibiotics are justified in surgical procedures such as those involving the intestinal tract and repair of traumatic injuries, where the surgical field is invariably contaminated with potential pathogens. They also are justified in immunosuppressed and excessively debilitated patients where natural defense mechanisms may fail.

If all the known techniques for preventing nosocomial infections were rigorously practiced, the incidence of such infections probably could be reduced to half the present level.

ESSAY

Microbes and War—Epidemiology in Action

Biological warfare may be defined as the deliberate use of living organisms or their poisonous products to produce death, injury, or disease in man, domestic animals, or plants, or destruction of materials. Although ancient armies knew nothing of microbes, they did know that dumping dead animals or human corpses into wells would poison their enemies' drinking water supplies. Generals ordered bodies of victims of plague and other contagious diseases to be thrown over the walls into besieged cities. Roman gladiators, before entering the arena, first ran their swords or tridents into manure or through decomposing animal bodies. Thus, if the cut itself didn't sever any vital structure, their opponents would still die of infection, often tetanus. Indeed, infections have killed more soldiers throughout the centuries than have guns and swords. For this reason, the Allied discovery and use of antibiotics was very important during World War II.

Changes of monarchs, boundaries, religion, and culture have often been decided by disease rather than by generals. The establishment of the Haitian Republic, for example, was greatly aided by yellow fever, which killed 22,000 of 25,000 French soldiers sent by Napoleon in 1801 to put down the revolt there. However, these disasters were not deliberately created. Although they were biological phenomena that affected the outcomes of wars, they were not biological warfare. White Americans and Europeans who gave or traded blankets containing scabs from smallpox lesions to Native-American tribes *were* indulging in deliberate biological warfare. Prisoners in German concentration camps who threw lice, picked from the bodies of fellow prisoners dying of typhus, onto guards as they walked past were also waging biological warfare in a small way.

Recent events in the Persian Gulf, during Operation Desert Storm, had the world worrying about biological warfare. Would our troops face anthrax spores or bacterial toxins? How surprised some of the most outraged might have been, however, to discover that in 1943 the United States, with British assistance, began work on anthrax bombs. Use of anthrax bombs would have been a violation of the Geneva Convention, which the United States and Great Britain had both signed in 1925.

Fortunately, biological warfare never was used during World War II. But in 1941, fearing that Hitler would use anthrax bombs against them, the British exploded a bomb containing anthrax spores over Gruinard Island (Figure 16.24). This was a small, remote Scottish island, occupied by 60 sheep, all of which promptly died. Inhalation of anthrax spores leads to pulmonary anthrax, with lesions in the lungs, which is usually fatal to both humans and animals. This makes anthrax a favorite choice of those involved in biological weaponry. In addition, the spores of anthrax bacilli remain viable and can cause disease for decades after they are released, a difficult situation for a country to recover from. Gruinard Island was tested 30 years later, in 1971,

FIGURE 16.24 Gruinard Island, Scotland, where anthrax bombs were tested during 13 field trials during 1942–1943. The island remains unsafe to this day, despite efforts of the Ministry of Defence to decontaminate it.

and was still found to have viable anthrax spores on and in the soil. In 1987, sheep were reintroduced on the island to see whether the spores were finally gone. But the island is still not safe, and remains uninhabited today, a warning of what germ warfare would bring.

Feelings about biological warfare research run high, both pro and con. Proponents argue that in order to defend ourselves against attack by biological warfare, we must study it to know how best to protect ourselves. Critics contend that once we have such knowledge, it will be difficult to restrain politicians from using it for attack rather than defense.

In 1979 an outbreak of anthrax in the (then) Soviet Union, near Sverdlovsk, raised deep suspicion that this outbreak was due to an escape of organisms from a biological warfare weapons laboratory located there. This would have been a violation of a 1972 agreement that bans development, production, and possession of biological weapons. All such stockpiles were to have been destroyed at that time. The United States accused, and the Soviets denied; full disclosure was never reached, and the question remained unresolved for many years. Then in June of 1992, Russian president Boris Yeltsin announced that the 1979 outbreak *was* due to escape of anthrax organisms from a biological warfare weapons laboratory. Let us hope that the world never experiences the full horror of biological warfare.

CHAPTER SUMMARY

EPIDEMIOLOGY

RELATED KEY TERMS

What Is Epidemiology?

- **Epidemiology** is the study of factors and mechanisms in the spread of infectious diseases.
- In describing infectious diseases, **epidemiologists** use **incidence** to refer to the number of new cases in a specific time period, **prevalence** to refer to the number of people infected at any one time, **morbidity rate** to indicate the number of cases as a proportion of the population, and **mortality rate** to indicate the number of deaths as a proportion of the population.

etiology

Centers for Disease Control (CDC)

Diseases in Populations

- In a **sporadic disease,** a small number of isolated cases appear in a population.
- In an **endemic disease,** a large number of cases appear in a population but the harm to patients is not sufficient to create a public health problem.
- In an **epidemic** disease, a large number of cases appear in a population, and patients are sufficiently harmed to create a public health problem. **Common-source epidemics** spread from a single contaminated substance, such as a water supply. **Propagated epidemics** spread by person-to-person contact.
- A **pandemic** disease is an epidemic disease that has spread over an exceptionally wide geographic area or several geographic areas.

Reservoirs of Infection

- **Reservoirs of infection** include humans, animals, and nonliving sources from which infectious diseases can be transmitted.
- In human reservoirs carriers often transmit diseases. They are **passive carriers** if they are not known to be infected, **active carriers** if recovering from an infection, and **intermittent carriers** if pathogens are released periodically.
- In animal reservoirs diseases can be transmitted by direct contact with animals or by vectors. Diseases that can be transmitted from animals to humans are called **zoonoses.**
- In nonliving reservoirs diseases are transmitted by water, soil, or wastes.

Portals of Entry

- **Portals of entry** include skin, mucous membranes that line various body systems, direct introduction into tissues, and through the placenta.

Portals of Exit

- **Portals of exit** include the nose, mouth, and openings from which products of the digestive, urinary, and reproductive systems are released. Organisms usually are in body fluids or feces.

Modes of Transmission of Diseases

- Transmission can be by contact, vehicle, or vector. **Direct contact transmission** includes person-to-person **horizontal transmission** and **vertical transmission** from parent to offspring. **Indirect contact transmission** occurs through **fomites** (inanimate objects) and by **droplets. Vehicles** of transmission include water, air, and food. **Vectors** of transmission are usually arthropods, which can transmit disease mechanically or biologically.
- Transmission by carriers, sexual practices, and transmission of zoonoses pose special epidemiologic problems.

contact transmission
direct fecal-oral transmission
droplet nuclei
indirect fecal-oral transmission
aerosol

RELATED KEY TERMS

Disease Cycles

- Some diseases occur in cycles—a few cases for several years and then many cases suddenly appear.
- **Group immunity** (also called **herd immunity**) refers to immunity enjoyed by a large proportion of a population that reduces disease transmission among nonimmune individuals.
- A drop in group immunity can lead to the sudden appearance of cases of a cyclic disease.

Public Health Organizations

- Public health organizations exist at city, county, state, federal, and world levels. They help to establish and maintain health standards, cooperate in the control of infectious diseases, collect and disseminate information, and assist with professional and public education.

Notifiable Diseases

- **Notifiable diseases** are listed in Table 16.3.

Epidemiological Studies

- The purpose of **epidemiological studies** is to learn more about the spread of diseases in populations and how to control them.
- The methods of epidemiological studies are **descriptive, analytical** (retrospective and prospective), and **experimental.**

index case **placebo**

Control of Communicable Diseases

- Methods used to control communicable diseases include isolation, quarantine, active immunization, and vector control.
- **Isolation** procedures are summarized in Table 16.4. **Quarantine** is rarely used but can prevent exposed individuals from infecting others; active immunization prevents many infections. Vector control is effective where vectors can be identified and eradicated.

NOSOCOMIAL INFECTIONS

- A **nosocomial infection** is an infection acquired in a hospital or other medical facility.

Epidemiology of Nosocomial Infections

- Nosocomial infections can be **exogenous** or **endogenous infections;** about half are caused by only four pathogens, of which many strains are antibiotic resistant.
- Host susceptibility is an important factor in the development of such infections.

compromised host

- Medical equipment and procedures, including surgery, are often responsible for infections.

Universal Precautions

- Modes of nosocomial infections are illustrated in Figure 16.21.

Prevention and Control of Nosocomial Infections

- Most hospitals have an extensive infection control program; hand washing, use of gloves, scrupulous attention to maintaining sanitary conditions and sterility where possible, and surveillance of antibiotic use and other hospital procedures help to minimize infections.
- Nosocomial infections could be reduced by half if all known procedures were carefully followed in all medical facilities.

QUESTIONS FOR REVIEW

A.

1. What is epidemiology?
2. Define the terms used by epidemiologists to describe frequencies of diseases.
3. What kind of data could be used to determine when an epidemic is beginning to draw to a close?

B.

4. Distinguish between sporadic, endemic, epidemic, and pandemic diseases.
5. How does a common-source epidemic differ from a propagated epidemic?

C.

6. Describe the properties of the various kinds of reservoirs of disease.
7. By what mechanisms are diseases transmitted to susceptible humans from reservoirs?

D.

8. Name the portals of entry and the portals of exit for pathogens.
9. Define and give examples of direct and indirect transmission of pathogens.

10. How do water, air, and food transmit disease?
11. How do vectors and fomites differ?

E.

12. How can changes in herd immunity contribute to an outbreak of a cyclic disease?

F.

13. What are the major activities of public health organizations at various governmental levels?
14. Why is reporting of diseases important, and why are there different categories of notifiable diseases?

G.

15. What is the main purpose of epidemiological studies?
16. How do methods of epidemiological studies differ?

H.

17. Evaluate the methods of control of communicable diseases.

I.

18. What is a nosocomial infection?
19. Which organisms most often cause nosocomial infections?
20. How does host susceptibility enter into such infections?
21. Give at least three examples of how medical procedures and sites of nosocomial infections are correlated.

J.

22. What methods are used to control nosocomial infections?

PROBLEMS FOR INVESTIGATION

1. Design a procedure to trace the source of an infectious disease in a large city.
2. Why do medical authorities believe it is impossible to prevent all nosocomial infections?
3. Plastic refreezeable ice cubes, filled with water and produced in Hong Kong, were linked to cases of *Salmonellosis* in this country. What is the most likely explanation? What precautions should be taken to prevent similar cases?
4. What are Universal Precautions? Why are they important?
5. Read about quarantine conditions formerly placed on people by health departments. Do older people in your class or family remember seeing quarantine signs? Which diseases were involved? How long did the quarantine last? What conditions were imposed?
6. A 7-year-old girl developed pertussis (whooping cough), followed by her 3- and 2-year-old siblings. Four other children in her apartment were the next to develop the disease. The local health department attempted to find the index case for this outbreak. Exhaustive study revealed no cases of pertussis among her schoolmates or any other known contacts. Her father had a deep cough and had taken a few days off from work. The mother was healthy, and an older brother had a runny nose.
   a) Who is a likely index case?
   b) Why did the local health department know about these cases?
   (The answer to this question appears in Appendix E.)

SOME INTERESTING READING

Anonymous. 1988. Update: universal precautions for prevention of transmission of human immunodeficiency virus, hepatitis B virus, and other bloodborne pathogens in health-care settings. *The Journal of the American Medical Association* 269 (July 22):464–66.

Ayliffe, G. A., and L. J. Taylor. 1990. *Hospital-acquired infections: Principles and prevention.* Litleton, MA: John Wright-PSG.

Benenson, A. L., ed. 1981. *Control of communicable diseases in man.* Washington, D.C.: American Public Health Association.

Bruce, N. G. 1991. Epidemiology and the new public health: implications for training. *Social Science & Medicine* 32, no. 1 (January 1):103–7.

Cliff, A., and P. Haggett. 1984. Island epidemics. *Scientific American* 250, no. 5 (May):138.

Craig, C. P. 1983. *Fundamentals of infection control.* Oradell, NJ: Medical Economics Books.

Dixon, R. E., ed. 1981. *Nosocomial infections.* New York: Yorke Medical Books.

Eagan, J. 1991. Measles: an infection control nightmare. *RN* 54, no. 6 (June):26–30.

Ellerbrock, T. V., T. J. Bush, M. E. Chamberland, and M. J. Oxtoby. 1991. Epidemiology of women with AIDS in the United States, 1981 through 1990. *The Journal of the American Medical Association* 265, no. 22 (June 12):2971–77.

Fleming, D. 1992. OSHA issues rules for controlling disease exposure in the workplace. *ASM News* 58 (3):127–29.

Gerbert, B., T. Bleeker, C. Miyasaki, and B. T. Maguire. 1991. Possible health care professional-to-patient HIV transmission: dentists' reactions to a Centers for Disease Control report. *The Journal of the American Medical Association* 265, no. 14 (April 10):1845–49.

Mann, J. 1991. How AIDS has changed epidemiology. *New Scientist* 129, no. 1755 (February 9):16.

Milliken, J., et al. 1988. Nosocomial infections in a pediatric intensive care unit. *Critical Care Medicine* 16 (3):233.

Pickering, L. K. and R. R. Reves. 1990. Occupational risks for child-care providers and teachers. *The Journal of the American Medical Association* 263, no. 15 (April 18):2096–98.

Salk, D. 1980. Eradication of poliomyelitis in the United States. *Review of Infectious Diseases* 2 (2):228.

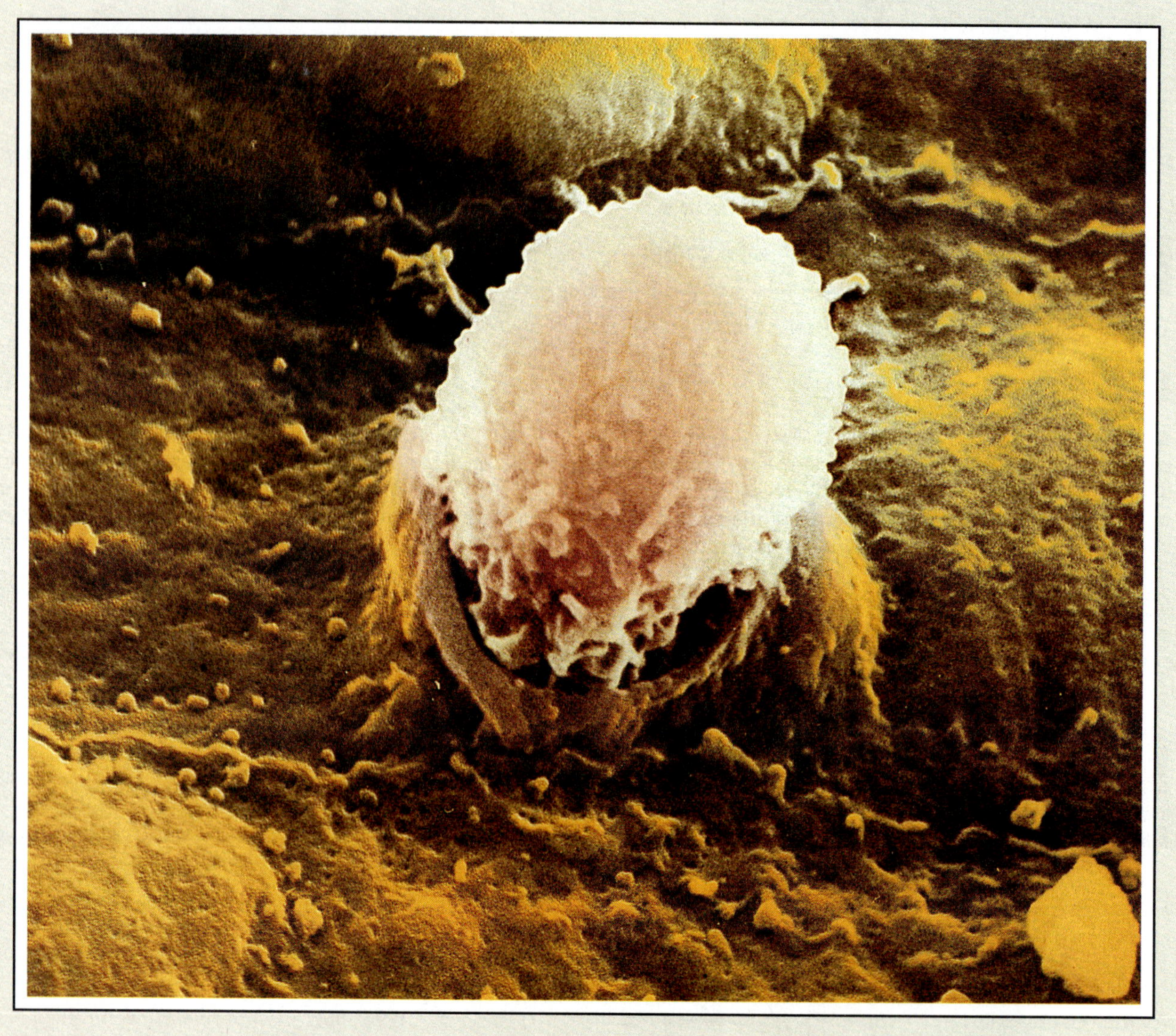

False-color scanning electron micrograph (SEM) of a white blood cell (leukocyte) migrating through the wall of a blood vessel out into tissues, where it will protect the body by performing its immune functions.

17 Host Systems and Nonspecific Host Defenses

This chapter focuses on the following questions:

A. How do nonspecific and specific host defenses differ?

B. What are the important structures, sites of infection, and nonspecific defenses of the skin, eyes and ears, respiratory, digestive, cardiovascular, nervous, urogenital, and lymphatic systems?

C. What are the stages in the process of phagocytosis, and what kinds of cells are involved?

D. What is inflammation?

E. What are the steps in the acute inflammatory process and their functions?

F. How do repair and regeneration occur following acute inflammation?

G. How do the causes and effects of chronic and acute inflammation differ?

H. How does fever function as a nonspecific defense?

I. How do interferon and complement function in nonspecific defenses?

We can look at infectious disease as a battle between the powers of infectious agents to invade and damage the body and the body's powers to resist such invasions. In the last two chapters we considered how infectious agents enter and damage the body and how they leave the body and are spread through populations. In the next several chapters we consider how the body resists invasion by infectious agents. We will begin by distinguishing between nonspecific and specific defenses and follow with a brief review of the structure of body systems. We will look at body systems from the point of view of a microbiologist, emphasizing sites at which infections often occur and the nature of defenses available in each system.

NONSPECIFIC VERSUS SPECIFIC HOST DEFENSES

With potential pathogens ever present, why do we only rarely succumb to them in illness or death? The reason is that our bodies have defenses against the pathogens. In other words, we are **resistant** to many pathogens. If our resistance fails, we are said to be **susceptible** to certain pathogens.

Host defenses that give rise to resistance can be nonspecific or specific. **Nonspecific defenses**—those that operate regardless of the invading agent—constitute the body's first line of defense against pathogens (Figure 17.1). They include:

1. Anatomical barriers such as the skin and mucous membranes.
2. Antimicrobial substances present in body fluids such as saliva and mucus.
3. Phagocytosis, the process by which a cell engulfs an invading microorganism.
4. Inflammation, the reddening and swelling of tissues at sites of infection.

Nonspecific defenses operate to prevent pathogens from entering the body. If pathogens do enter the body, nonspecific defenses act immediately to destroy them before they damage tissues. This chapter focuses on nonspecific defenses.

Specific defenses—those that respond to particular agents—constitute the body's second line of defense against pathogens. They include responses to particular foreign substances, or antigens. These responses are said to be specific because each response is elicited by and tailored to a particular antigen. Specific responses are carried out by the immune system and are more effective against subsequent invasions by the same pathogen than against initial invasions. Chapter 18 focuses on specific defenses of the immune system, and Chapter 19 on immune disorders.

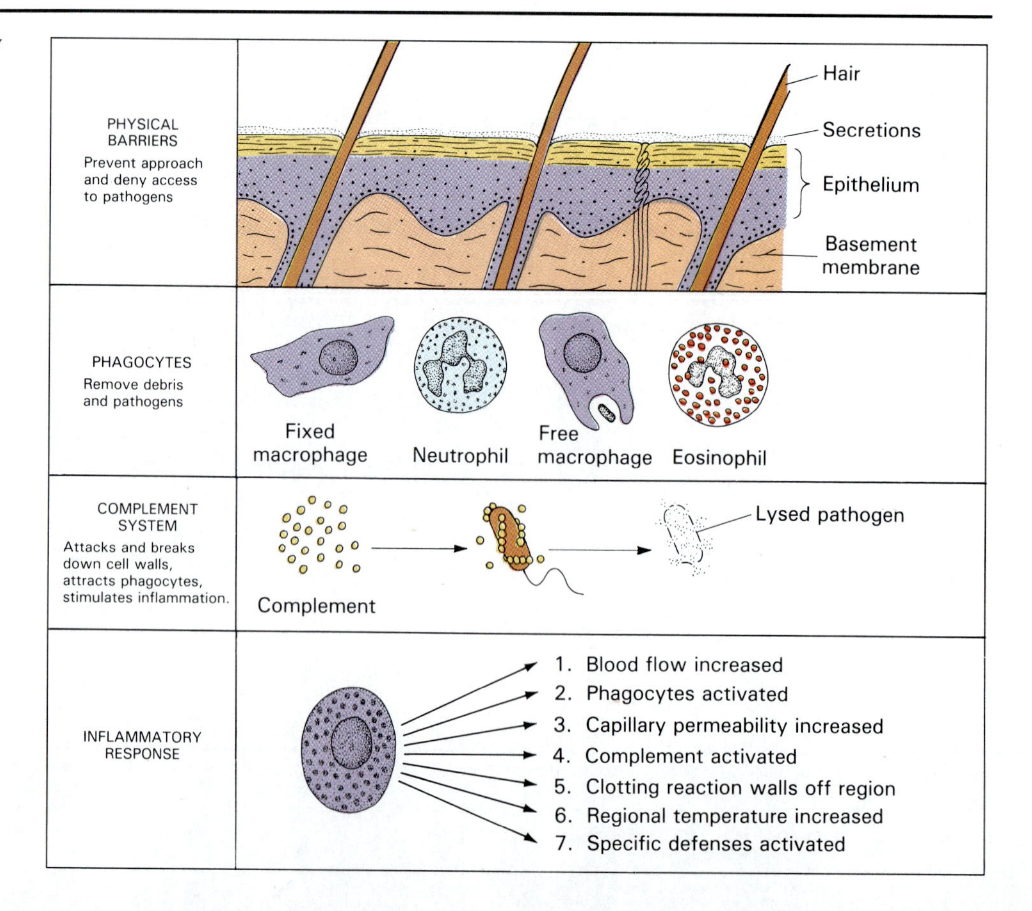

FIGURE 17.1 An overview of the body's nonspecific defenses.

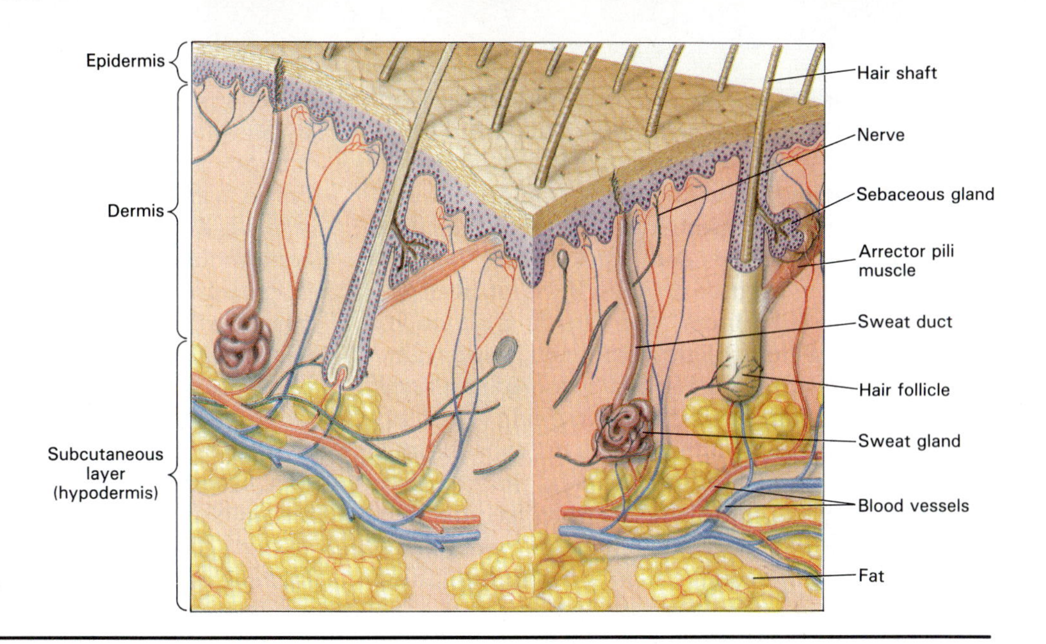

FIGURE 17.2 The structure of the skin.

SYSTEM STRUCTURE, SITES OF INFECTION, AND NONSPECIFIC DEFENSES

Skin

The skin (Figure 17.2) consists of the relatively thin outer **epidermis** and the thicker inner **dermis.** These layers are tightly fastened together, and the dermis is bound to underlying muscles by loose connective tissue. Hair, nails, and glands are called *skin derivatives* because they develop from the epidermis.

The epidermis has several layers of cells. Cells next to the dermis divide throughout life and migrate toward the surface. Outer cells are filled with a waterproofing protein called **keratin.** Old cells are sloughed at the same rate as new cells are produced, so the epidermis is completely renewed every 15 to 30 days. The epidermis lacks blood vessels and is nourished by nutrients that diffuse from blood vessels in the dermis. The nails consist of hardened epidermal cells packed with keratin.

The dermis is the durable part of the skin. In fact, leather is the toughened dermis of an animal skin after the epidermis has been removed. The dermis contains blood vessels and sensory receptors and has epidermal structures such as hair follicles, sebaceous (oil) glands, and sweat glands embedded in it. Most **sebaceous glands** are associated with hair follicles. They produce an oily secretion called **sebum** that consists mainly of organic acids and other lipids. The acids lower the skin pH. Fungi usually prefer acid environments, which explains why most skin infections are fungal in origin. **Sweat glands** are widely distributed over the body and empty a watery secretion through pores in the skin. Sweat glands in the armpits and groin also secrete organic substances that lower the skin pH, provide nutrients for some microorganisms, and inhibit the growth of other microorganisms. Sweat is odorless until bacteria begin to act on it; body odor is due to metabolic products of bacterial action.

The epidermis is inhabited by a variety of normal flora. Although its entire surface is subject to infection, certain sites are more vulnerable than others (Figure 17.3). Cuts, scratches, insect and animal bites, burns,

FIGURE 17.3 Sites of skin infections.

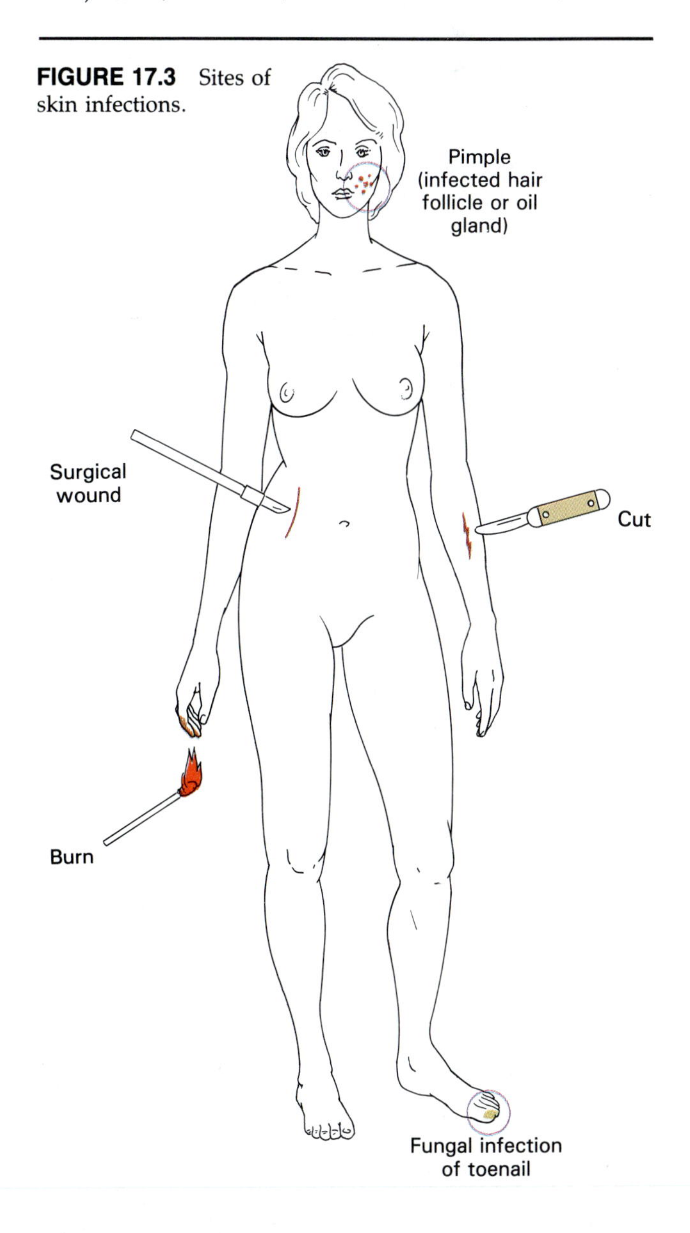

and other wounds disrupt the continuity of the skin and provide sites of infection. Sebaceous glands also can become infected, and pores of glands and hair follicles sometimes provide portals of entry for microbes. Furthermore, infections can spread to the dermis from infected glands and hair follicles.

Your skin is exposed to the effects of microorganisms, toxic substances, and objects that touch, abrade, and tear it: sunlight, heat, cold, wind, rain, even radiation from outer space. It provides several barriers to invasion. The surface of the skin forms the first barrier. Thick skin on the palms of the hands and soles of the feet reduces the likelihood that the skin barrier will be broken. Microorganisms feed on skin secretions, but the intact skin surface prevents these organisms and other foreign substances from entering the body.

Acidic secretions in sebum maintain a pH of 3 to 5 on the skin, which is inhibitory to some microorganisms. The quantities of these secretions increase at puberty and contribute to the development of acne. The high salt concentration in sweat also inhibits many microorganisms. Normal skin flora contribute to skin defenses, as some of their metabolic products inhibit growth of other microbes. Unsaturated fatty acids from *Staphylococcus epidermidis* and *Propionibacterium acnes* are especially toxic to gram-negative organisms. Lactic acid from lactobacilli also contributes to the acid skin pH.

Other skin barriers include waterproof keratin in the outer epidermis and the **basement membrane** between the dermis and the epidermis. Keratin prevents water-soluble substances from entering the body. The basement membrane, a combination of secretions of epithelial and dermal cells, prevents microbes that have evaded other defenses from reaching deeper tissues.

Eyes and Ears

The eyes (Figure 17.4) have several protective external structures. These structures work so well that deep infections are exceedingly rare. Each eye is protected by eyelids, eyelashes, **conjunctiva** (mucous membranes), and the tough **cornea** (the transparent part of the eyeball exposed to the environment). Each eye also has a **lacrimal gland,** which produces tears that continuously flush the cornea and keep it moist. Ducts drain tears from the surface of the eyeball into the nasal cavity. Eye infections usually are limited to the conjunctiva and eyelids, but the cornea too sometimes becomes infected.

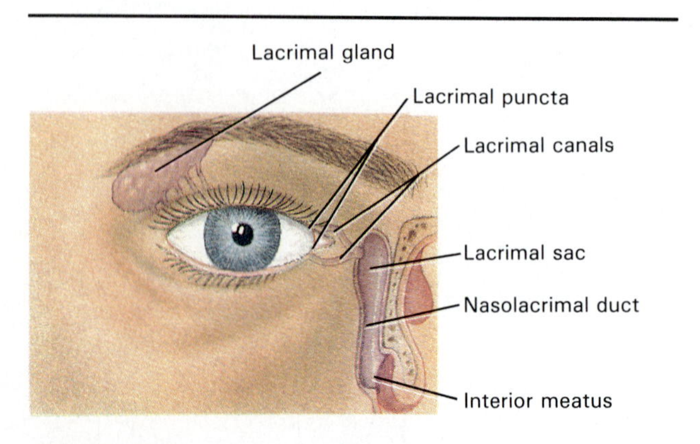

FIGURE 17.4 Structure of the eye.

Eyelashes and eyelids act mechanically to prevent foreign objects from reaching the cornea. Tears contain lysozyme, an enzyme also found in lesser amounts in saliva, nasal secretions, and vaginal secretions. (This enzyme was discovered by Alexander Fleming in 1922, before he discovered penicillin.) Lysozyme helps to destroy bacteria by breaking down bacterial cell walls (Figure 17.5). It is especially effective against gram-positive organisms because the peptidoglycan is nearer the surface. (Not surprisingly, it has no effect on viruses.) Tears, like other body secretions, also contribute to specific defenses—they contain antibodies (Chapter 18) that coat microorganisms and prevent their attachment to tissues. Glandular cells in the conjunctiva add a mucous substance to the tears. This substance may help to trap microorganisms and convey them out of the eye.

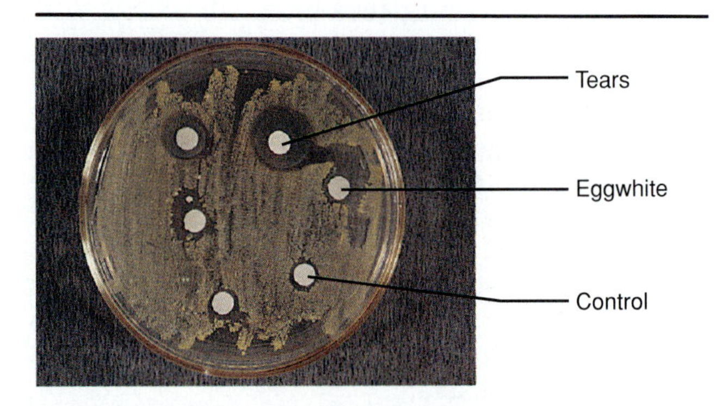

FIGURE 17.5 Lysis of *Micrococcus luteus* bacteria by lysozyme enzymes derived from tears (two upper disks) and from egg white (two disks below uppermost). The two lowermost disks are controls that have been dipped in sterile distilled water.

Like the eyes, the ears have protective structures. Each ear (Figure 17.6) is divided into the outer ear, the middle ear, and the inner ear. Outer and middle ear infections are relatively common, but inner ear infections are rare. The outer ear has a flaplike **pinna** (commonly called the ear) covered with skin and an ear canal lined with skin that has many small hairs and numerous ceruminous glands. The **ceruminous** (se-ru'min-us) **glands** are modified sebaceous glands

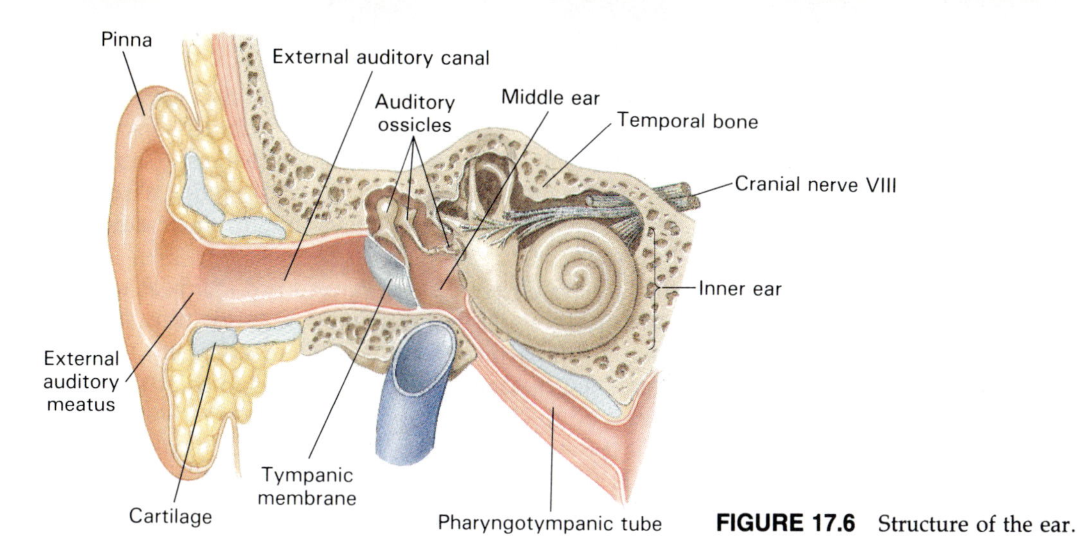

FIGURE 17.6 Structure of the ear.

that secrete **cerumen** (se-ur'men), or earwax. Both hairs and wax help to prevent microorganisms and other foreign objects from entering the ear canal. Ear skin surfaces have most of the same flora as other parts of the skin, and the ear canal often becomes infected with fungi. The **tympanic membrane,** or *eardrum*, separates the outer and middle ear. The **middle ear** is a small cavity in the temporal bone that contains the small bones that transmit sound waves. It has five openings—three that are associated with hearing and two that communicate with the throat and **mastoid area** of the temporal bone. Infectious organisms from the pharynx can infect the middle ear and sometimes also the mastoid area.

Respiratory System

You would live no more than a few minutes without oxygen and only a little longer if your cells had no way to rid themselves of carbon dioxide. Your respiratory system moves oxygen from the atmosphere to the blood and removes carbon dioxide from the blood to the atmosphere. You breathe in microorganisms with every breath you take.

The **respiratory system** (Figure 17.7) consists of the **upper respiratory tract**—the nasal cavity, pharynx, larynx, trachea, bronchi, and larger bronchioles—and the **lower respiratory tract**—thin-walled bronchioles and alveoli where gas exchange occurs. The entire system is lined with moist epithelium. In the upper portion the epithelium contains mucus-secreting cells and is covered with cilia, but in the lower portion these structures are lacking.

Air is warmed and some particles are removed by hairs as the air passes through the **nasal cavity.** Sometimes microorganisms in air enter the sinuses, which are hollow cavities lined with mucous membrane within certain skull bones. Air then passes through the **pharynx** (throat), a common passageway for the respiratory and digestive systems with tubes that con-

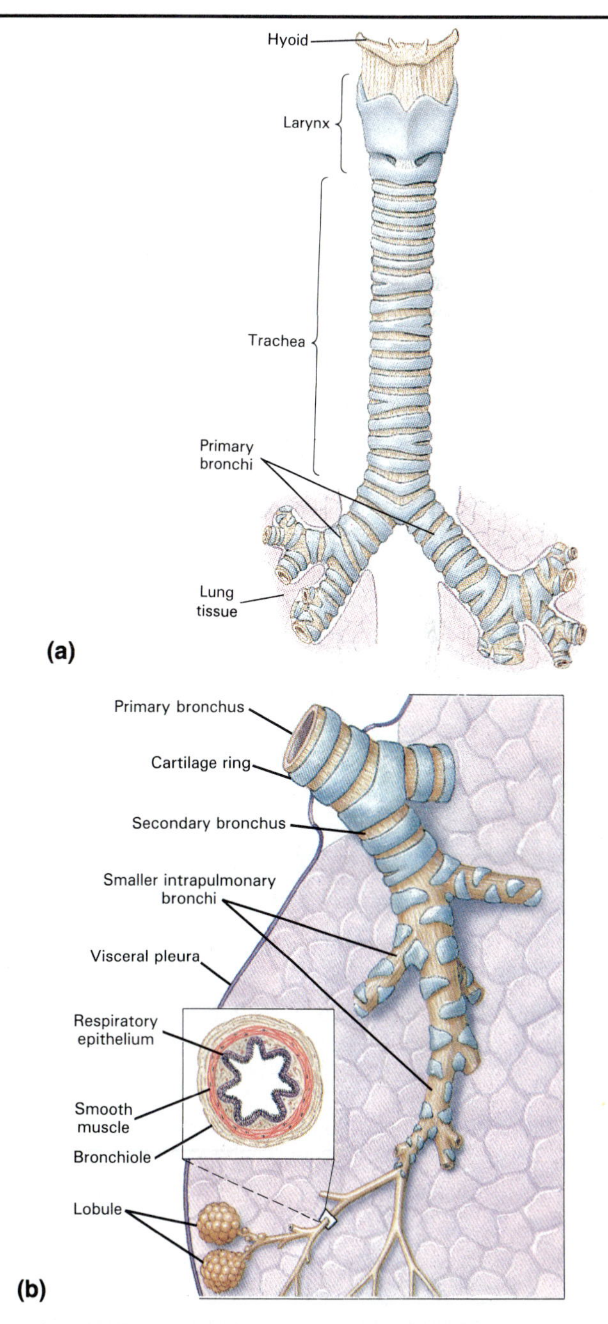

FIGURE 17.7 Structure of the respiratory system.

nect with middle ear chambers. The pharynx is surrounded by a ring of lymphoid tissue (tonsils) that is important in both nonspecific and specific defenses.

From the pharynx, air passes through a series of rigid-walled tubes: the larynx, the trachea, bronchi, and larger bronchioles. The **larynx** (voice box) contains the vocal cords, which produce sound when they vibrate. The **epiglottis** is a flap of tissue that prevents food and fluids from entering the larynx. When the epiglottis fails to operate properly, fluids can be *aspirated*, or sucked, into the lungs. The **trachea** (windpipe), **bronchi,** and **bronchioles** form a branching structure known as the **bronchial tree.** This complex branching arrangement greatly increases the surface area exposed to air flowing into and out of the lungs.

When air reaches the respiratory bronchioles and **alveoli** (saclike structures at the ends of the bronchioles), gas exchange occurs. Oxygen diffuses into the blood, and carbon dioxide from the blood diffuses into the alveoli. Alveoli are arranged in clusters, each of which is surrounded by elastic tissue and well supplied with blood and lymphatic vessels. Arteries carry blood to the capillaries of the alveoli, and veins carry blood from these capillaries back to the heart. Lymphatic vessels drain tissue fluid from around the alveoli.

The bronchial tree, alveoli, and blood and lymph vessels form the bulk of the lungs. The surfaces of the lungs and the cavities they occupy are covered with serous membranes called **pleura. Serous membranes** secrete a watery fluid that lubricates them.

The mucous membranes of the upper respiratory system are common sites of infection, which often spread to the sinuses, the middle ear chambers, and even to the lower respiratory system. Some infections reach the lungs by way of blood and lymph vessels.

Mucus from the membranes that line the nasal cavity and pharynx traps microorganisms and most particles of debris, preventing them from passing beyond the pharynx. Mucus from the lining of the bronchial tree also traps foreign materials that have passed beyond the pharynx. Mucus contains lysozyme, which degrades the cell walls of bacteria. Coughing and sneezing mechanically agitate mucus, increasing exposure of microorganisms to mucus and helping to expel them.

Cilia in the nasal cavity and bronchi beat toward the pharynx, so that mucus with debris trapped in it is moved into the pharynx. This mechanism, the **mucociliary escalator,** allows materials in the bronchi to be lifted to the pharynx and spit out or swallowed.

If these defense mechanisms fail and microorganisms get into bronchi and bronchioles, phagocytes, or cells that engulf microorganisms and other foreign substances, help to remove them. The actions of phagocytes are discussed under phagocytosis. Only when the numbers of organisms exceed the capacity of the phagocytes to destroy them does a lower-respiratory infection occur.

Alcoholics and Pneumonia

Alcoholics are especially likely to get pneumonia. When alcoholics are "sleeping off" a drunken spell, the epiglottis gapes and allows pharyngeal secretions to be aspirated, or drawn into the lungs with breathing. If pneumonia-causing organisms are present in those secretions, they can infect the lungs. Once in the lungs, the microbes are unlikely to be destroyed because alcoholics have "lazy leukocytes." In other words, their white blood cells are less efficient in destroying invading microbes than are those of nonalcoholics.

Digestive System

The **digestive system** (Figure 17.8) consists of an elongated tube, or **tract,** lined with epithelial cells that are constantly sloughed and replaced. Modifications along the tract form various organs—mouth, pharynx, esophagus, stomach, and intestines. Accessory organs such as the salivary glands, liver, and pancreas assist in the digestive process. Together these organs carry out four major functions:

1. Secretion of digestive juices and mucus.
2. Motility, which moves materials through the tract.
3. Digestion of food substances.
4. Absorption of small molecules into blood or lymph.

The mouth, or **oral cavity,** is lined with mucous membrane and contains the tongue, salivary glands, and teeth. The tongue helps to move food and mix it with saliva.

Each tooth (Figure 17.9) has a **crown** covered with **enamel** above the gum and a **root** covered with **cementum** below the gum. Under these coverings is a porous substance called **dentin,** a centrally located **pulp cavity,** and the **root canals,** where blood vessels and nerves are located. Each tooth is held in a tooth socket by fibers running from the cementum to the bone of the socket. Although enamel is the hardest substance in the body, it can be digested by microbial acid and enzymes. Microbes also can infect the gums and form pockets of infection between the gums and the teeth. When the infection spreads to the bone, it causes bone loss and loosening of the teeth.

After food is chewed and mixed with saliva in the mouth, it passes through the pharynx and esophagus to the stomach. Here is mixed with hydrochloric acid and the enzyme pepsin, which together begin the

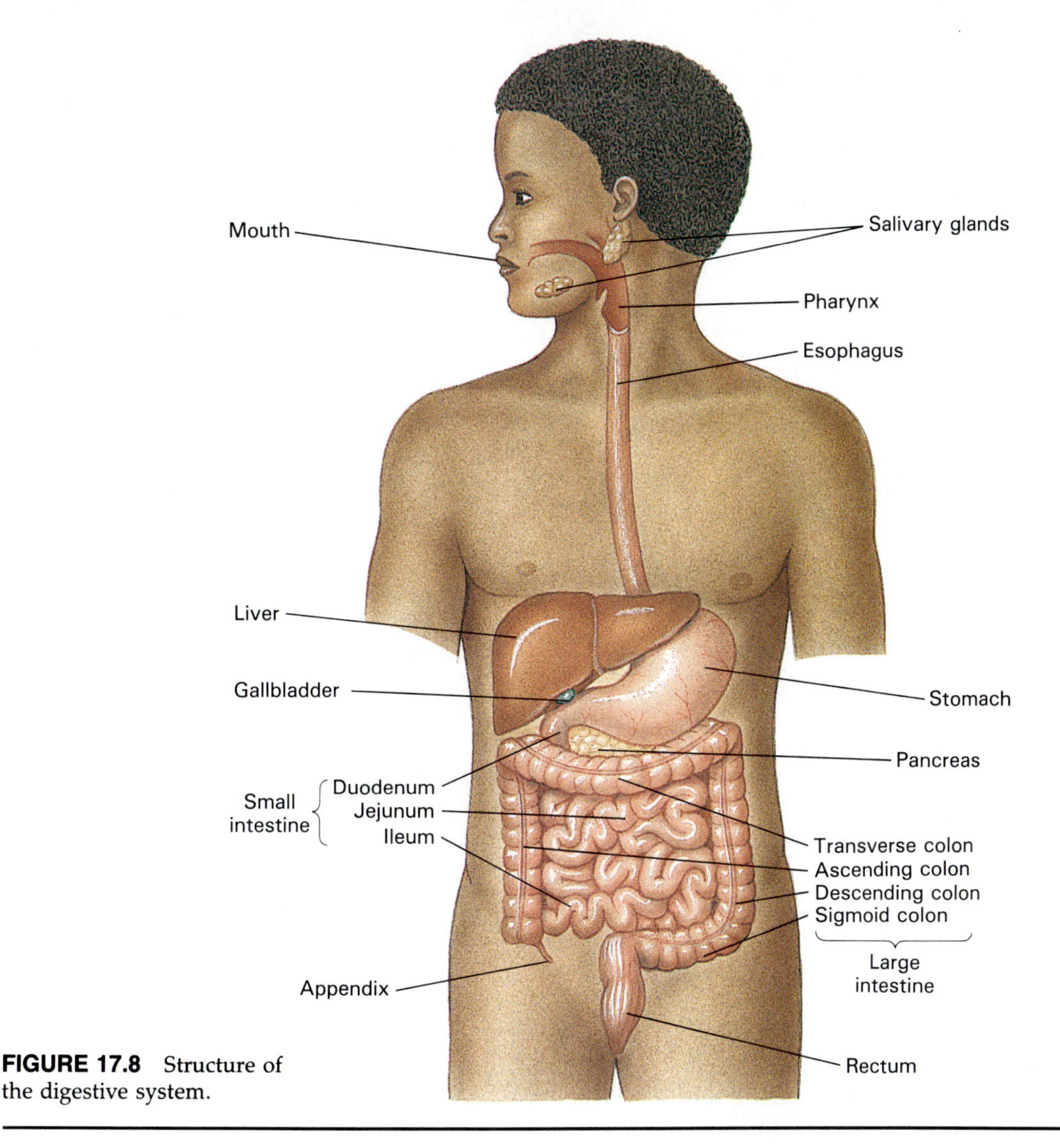

FIGURE 17.8 Structure of the digestive system.

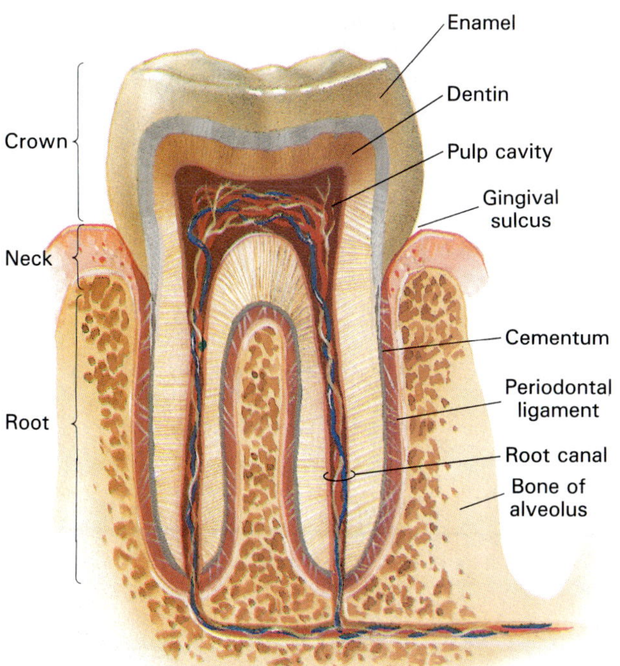

FIGURE 17.9 Tooth and gum structure.

digestion of proteins. The lining of the stomach is protected from the acid by a viscous mucus. Only alcohol, aspirin, and some lipid-soluble drugs that can cross the mucous barrier are absorbed in the stomach.

Food leaving the stomach enters the small intestine and is immediately mixed with secretions from the liver and pancreas. Liver cells secrete bile, a mixture of bile salts, cholesterol, and other lipids. Bile is collected from cells and carried to the small intestine or the gallbladder by various ducts. Bile is released in relatively large quantities after meals and aids in the digestion of fats. In some tissues, especially the liver, capillaries are enlarged to form **sinusoids.** Sinusoids are lined with phagocytic **Kupffer cells,** which remove foreign material from the blood as it passes through the sinusoids. The pancreas releases hormones into the blood and digestive juices into ducts that empty into the small intestine. These digestive juices include enzymes that digest starch, proteins, lipids, and nucleic acids and bicarbonate that neutralizes the acidic materials from the stomach.

The **small intestine** has an extensive internal absorptive area (about $^{1}/_{10}$ the area of a football field) due to folds in the wall itself, fingerlike projections called **villi,** and folds in membranes of mucosal cells called **microvilli.** Villi contain blood and lymph vessels. Digestion is completed in the small intestine; sugars and amino acids are absorbed into blood vessels, and fats into lymph vessels.

The **large intestine,** or **colon,** joins the small intestine near the appendix and merges into the rectum. Whatever digestion occurs in the large intestine is performed by the bacteria from among the normal flora. Byproducts of bacterial metabolism, such as amino acids, thiamine, riboflavin, and vitamins K and B_{12}, are absorbed, but the quantities are too small to satisfy nutritional needs. Much water is absorbed, and undigested food is converted to feces. **Feces,** which consist of about three-fourths water and one-fourth solids, are stored in the rectum until they are eliminated from the body.

Normal flora are found mainly in the mouth and large intestine near portals of entry. Pathogens can be found in all digestive organs except the stomach, where most are killed by acid and pepsin. (Eggs of helminths can pass through the stomach unharmed.) Plaque deposited on teeth by microorganisms contributes to dental caries and gum disease. The mumps virus attacks salivary glands. Enteric pathogens (viruses, bacteria, protozoa, fungi, and helminths) can infect the small and large intestine and sometimes the liver and pancreas. They reach these organs via the mouth, anus, or blood and lymph vessels.

Although food contains many microorganisms, most are killed by various defense mechanisms in the digestive tract. Throughout the digestive tract **mucin** (mu'sin), a glycoprotein in mucus, coats bacteria and prevents their attaching to surfaces. Saliva contains antibodies that likewise coat bacteria and thiocyanate that kills some bacteria. The high acidity of the stomach and bile acids and enzymes in the small intestine kill most microorganisms and inactivate most viruses in food. (People who are deficient in stomach acid frequently get intestinal infections.) A few microorganisms that enter the blood through breaks in the intestinal wall usually are destroyed by Kupffer cells in the liver. Finally, the normal flora of the large intestine compete with pathogens and prevent their growth under normal conditions. Gastrointestinal infections usually occur only when defense mechanisms have been overwhelmed by large numbers of a pathogenic organism. Even when an infection occurs, vomiting and diarrhea help to rid the body of the pathogens and their toxins.

Cardiovascular System

The **cardiovascular system** (Figure 17.10) consists of the heart, blood vessels, and blood. This system sup-

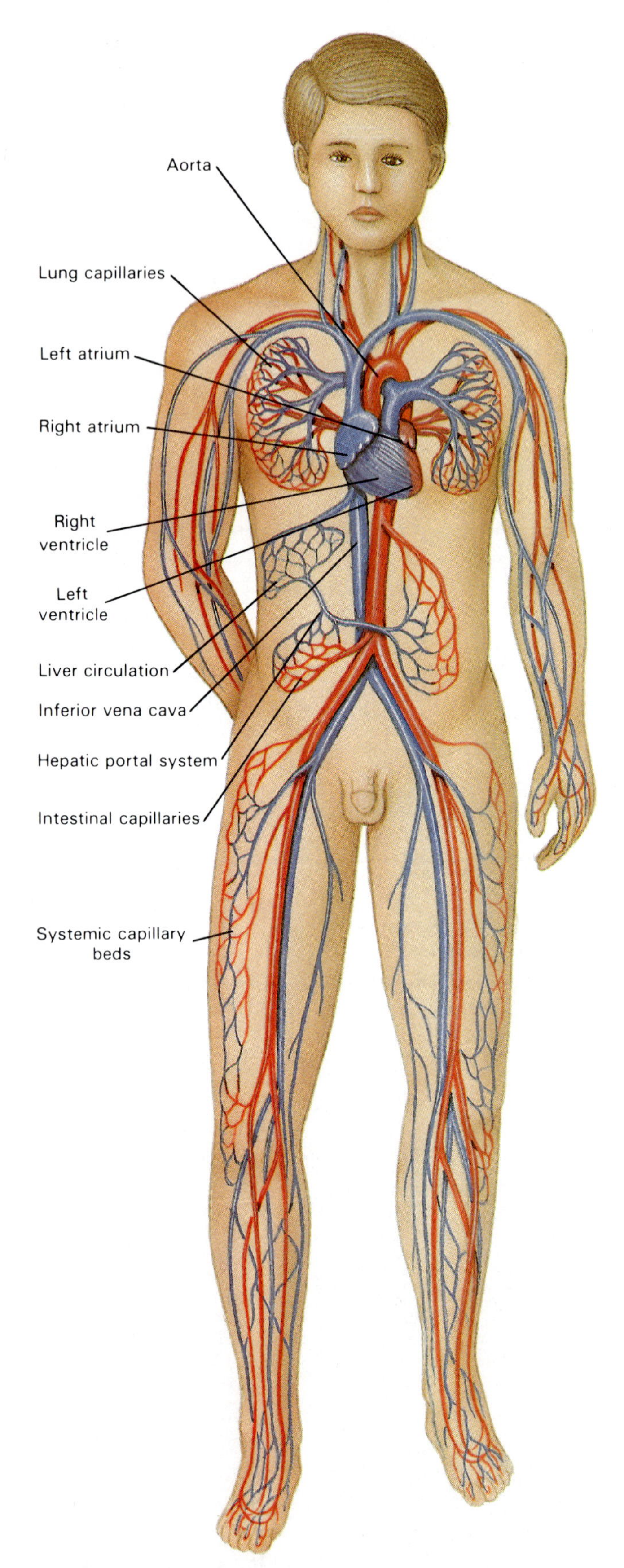

FIGURE 17.10 Generalized structure of the cardiovascular system. Oxygenated blood is shown in red, unoxygenated blood is blue.

plies oxygen and nutrients to all parts of the body and removes carbon dioxide and other wastes from them.

The heart lies between the lungs in a tough, membranous **pericardial sac** lubricated by serous fluid. Its wall consists of a thin internal **endocardium,** a thick muscular **myocardium,** and an outer **epicardium.** The heart has four chambers—two **atria** and two **ventricles** with valves that direct blood flow at the exits of each chamber. The right side of the heart pumps blood to the respiratory portion of the lungs; the left side to all other organs and tissues. The myocardium itself is supplied with blood, not from that flowing through the heart chambers, but from its own vessels, the coronary arteries.

Blood leaving the heart circulates through a closed system of blood vessels and subsequently returns to the heart. Its flow is regulated so that all cells receive nutrients and get rid of wastes according to their needs. The blood vessels include **arteries** that receive blood from the heart, **arterioles** that branch from arteries, **capillaries** that branch from arterioles, **venules** that receive blood from capillaries, and **veins** that receive blood from venules and return it to the heart. Capillary walls are composed of a single layer of cells that allow exchange of materials between the blood and the tissues. White blood cells sometimes force their way between cells of the capillary walls, but red blood cells and large protein molecules normally do not leave the capillaries.

Blood consists of about 60 percent liquid called **plasma** and 40 percent **formed elements** (cells and cell fragments). Plasma is more than 90 percent water and contains proteins such as albumins, globulins, and fibrinogen. Certain globulins are important in defending the body against infection, and fibrinogen is important in blood clotting. Plasma also contains **electrolytes** (ions such as Na^+, K^+, and Cl^-), gases such as oxygen and carbon dioxide, nutrients, and waste products. In contrast to plasma, serum is the fluid that remains after both formed elements and clotting factors have been removed.

Formed elements of the blood include erythrocytes (red blood cells), leukocytes (white blood cells), and platelets (Figure 17.11 and Table 17.1). **Erythrocytes** are the most abundant of the formed elements and account for 40 to 45 percent of the total blood volume. Erythrocyte volume is an important indicator of the oxygen-carrying capacity of the blood because erythrocytes contain the oxygen-binding molecule **hemoglobin. Leukocytes** contribute to various specific and nonspecific defenses. They are divided into two groups according to the cell characteristics: (1) **Granulocytes,** or **polymorphonuclear leukocytes** (PMNLs), have a granular cytoplasm and irregularly shaped, lobed nuclei. (2) **Agranulocytes** lack granules in the cytoplasm and have round nuclei. Granulocytes include neutrophils, basophils, and eosinophils, which are distinguished by their staining reactions in the laboratory. Agranulocytes include monocytes and lymphocytes. **Platelets,** which are short-lived fragments of large cells called **megakaryocytes,** are important components of the blood-clotting mechanism.

TABLE 17.1 Formed elements of the blood

Element	Normal Numbers per Microliter*
Erythrocytes	
Adult male	4.6 to 6.2 million
Adult female	4.2 to 5.4 million
Infant and child	4.5 to 5.0 million
Newborn	5.0 to 5.1 million
Leukocytes	5000 to 9000
Neutrophils	50–70%
Eosinophils	1–4%
Basophils	0.1%
Monocytes	2–8%
Lymphocytes	20–40%
Platelets	250,000 to 300,000

* 1 μL = 1/1,000,000 L.

Under normal conditions the blood, blood vessels, and heart are sterile—they have no normal flora. However, when pathogens evade other defenses they are often transported in the blood, as in bacteremia and viremia. Sometimes they grow in the blood and produce septicemia. Blood cells sometimes harbor pathogens safe from defenses and can be a site for their growth. Certain sites, such as heart valves, are especially susceptible to bacterial infection.

When the skin is broken by any kind of trauma, microorganisms from the environment enter the wound. Blood flowing out of the wound helps to remove the microorganisms. Subsequent constriction of ruptured blood vessels and the clotting of blood seals off the injured area until more permanent repair in the form of a scar can occur.

When microorganisms enter blood through abrasions in internal mucous membranes, such as from dental procedures or breaks in the intestinal lining, several defense mechanisms come into play. **Phagocytes**—cells that engulf foreign organisms—located in liver sinusoids, blood, and tissues engulf and destroy microorganisms before they can invade other tissues. In the case of organisms for which the body has developed immunity, specific antibodies are available to inactivate the organisms before they can cause disease (Chapter 18).

Leukocytes are especially important in body defenses. **Neutrophils,** the most numerous of all leukocytes, are avid phagocytes. During an infection they are released from blood-forming tissues into the blood. Many leave the blood and enter tissues, where they guard skin and mucous membranes against in-

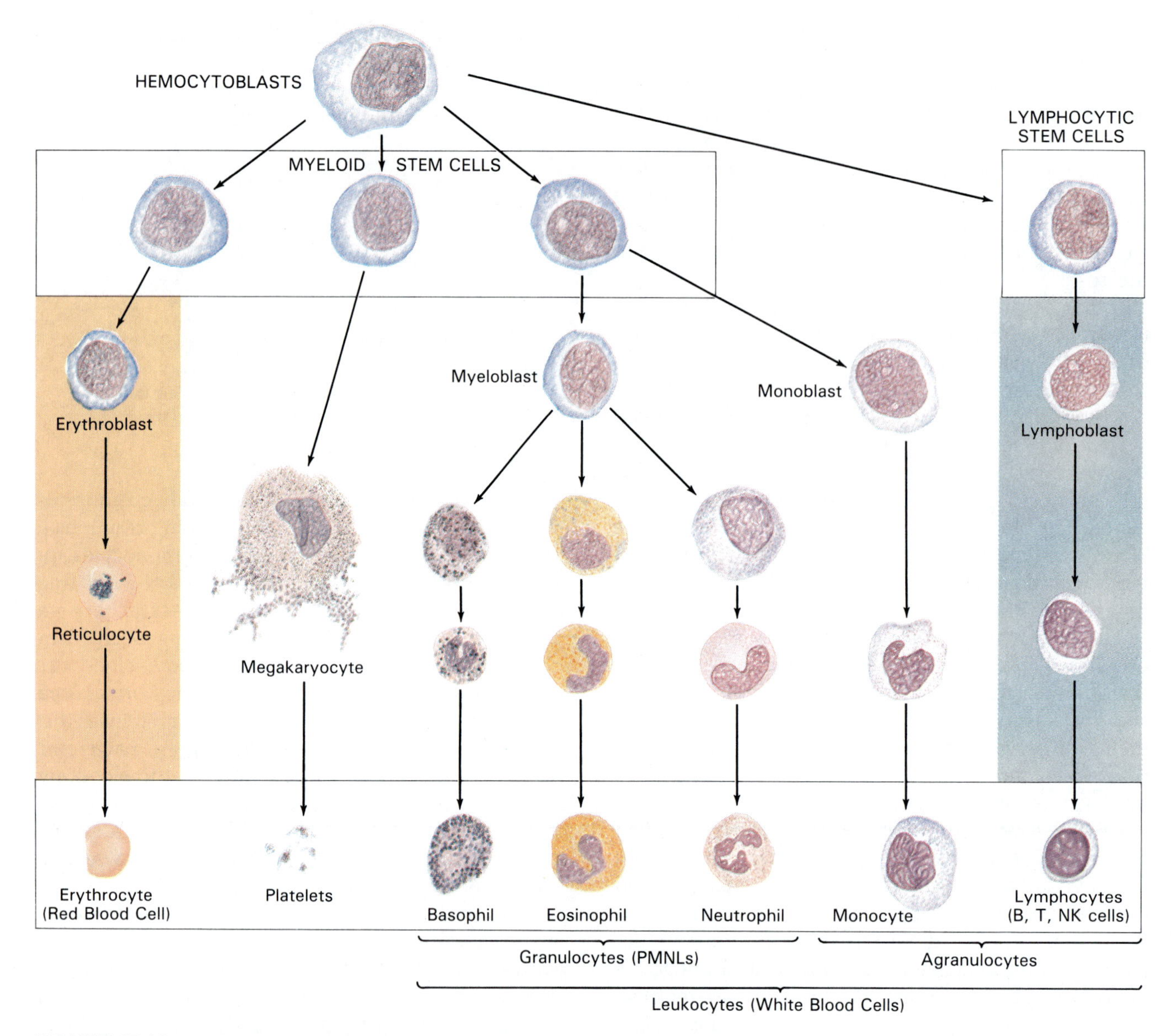

FIGURE 17.11 Formed (cellular) elements of the blood are derived from stem cells of the bone marrow. The myeloblast differentiates to form three kinds of polymorphonuclear leukocytes, neutrophils, basophils, and eosinophils.

vasion by microorganisms and foreign substances. **Eosinophils** are released in large numbers during allergic reactions. They are thought to detoxify foreign substances and turn off inflammatory reactions. They may also act as phagocytes. **Basophils** migrate into tissues, where they are called **mast cells,** and release histamine and heparin. Histamine initiates the inflammatory response, and heparin inhibits blood clotting. **Monocytes** also enter tissues, where they are called **macrophages.** Being ravenous phagocytes, they engulf not only microorganisms but larger particles of debris left from neutrophils that have died after ingesting smaller particles. **Lymphocytes** circulate in the blood and are found in large numbers in lymphoid tissues—lymph nodes, spleen, thymus, and tonsils—where they contribute to specific immunity.

Nervous System

As you read this sentence, millions of neural signals allow you to understand words and at the same time maintain an upright posture and carry on various internal processes such as breathing. The ability of the nervous system to simultaneously control many body functions depends on many **neurons,** or nerve cells, working together. Structurally, the **nervous system** (Figure 17.12) has two components. The **central nervous system,** which consists of the brain and spinal cord, receives and responds to signals from the **peripheral nervous system,** which consists of nerves that supply all parts of the body. The brain and spinal cord are protected by membranes called **meninges.** Hollow chambers in the brain and spinal cord and spaces between meninges are filled with **cerebrospinal fluid.** **Nerves** of the peripheral nervous system consist of fibers that extend from the cell bodies of neurons. Aggregations of these cell bodies are called **ganglia.**

As in the circulatory system, the nervous system is normally sterile and has no normal flora. However, pathogens can enter cerebrospinal fluid from the blood. Infections usually appear first in cerebrospinal fluid and meninges but can spread to nerve cells themselves. Some invade nerve endings in the nasal cavity, the skin, and mucous membranes, and certain viruses invade ganglia. Traumatic injury can give microorganisms access to the nervous system.

The central nervous system is well protected by bone and meninges from invasion by pathogens. Phagocytes of the nervous system can destroy invaders that reach the brain and spinal cord. The brain has special thick-walled capillaries without pores in their walls. These capillaries form the **blood-brain barrier,** which limits entry of substances into brain cells. It protects the cells from microorganisms and toxic substances, but it also prevents them from receiving medications that easily reach other cells.

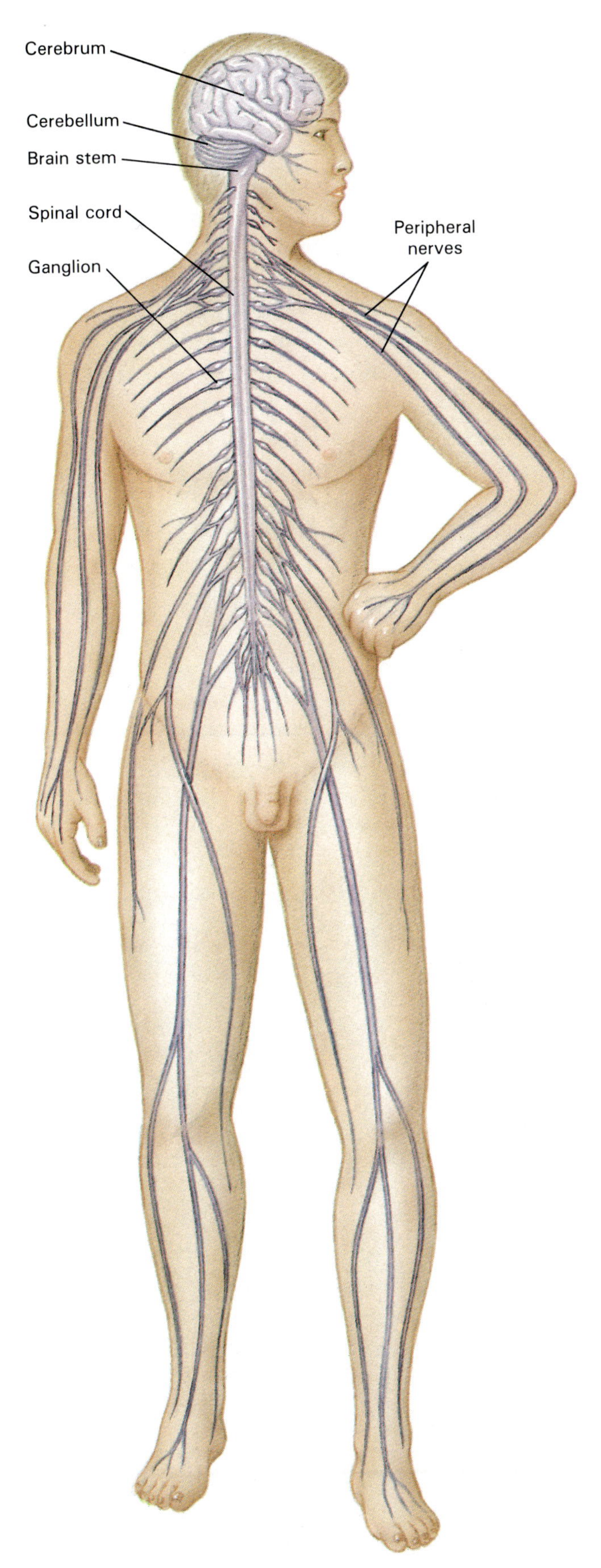

FIGURE 17.12 The structure of the nervous system. Nervous system infections typically affect the meninges (which cover the brain and spinal cord) or the nerve roots.

Urogenital System

The **urogenital system,** or **genitourinary system,** includes the urinary system and the reproductive system. Thus, in considering the urogenital system we will look at the structure and sites of infection in the urinary system, the female reproductive system, and the male reproductive system.

The **urinary system** (Figure 17.13) consists of paired kidneys and ureters, the urinary bladder, and the urethra. It regulates the composition of body fluids and removes nitrogenous and certain other wastes from the body. Each **kidney** contains about 1 million functional units called **nephrons.** In a nephron the fluid part of the blood is filtered from the **glomerulus,** a coiled cluster of capillaries, to the **kidney tubule.** Urine formed in kidney tubules passes through collecting ducts to the **ureter** of each kidney. The ureters carry urine to the **urinary bladder,** where it is stored until released through the **urethra** during micturition (urination). **Urinalysis,** the laboratory analysis of urine specimens, can reveal imbalances in pH or water concentration, the presence of substances such as glucose or proteins, and other conditions associated with infections, metabolic disorders, and other diseases.

The **female reproductive system** (Figure 17.14) consists of the ovaries, uterine tubes, uterus, vagina, and external genitalia. The paired **ovaries** produce **ovarian follicles,** which contain an ovum (egg) and cells that secrete estrogen and progesterone hormones. During a woman's reproductive years an ovum capable of being fertilized is released once each month. The **uterine tubes** receive ova and convey

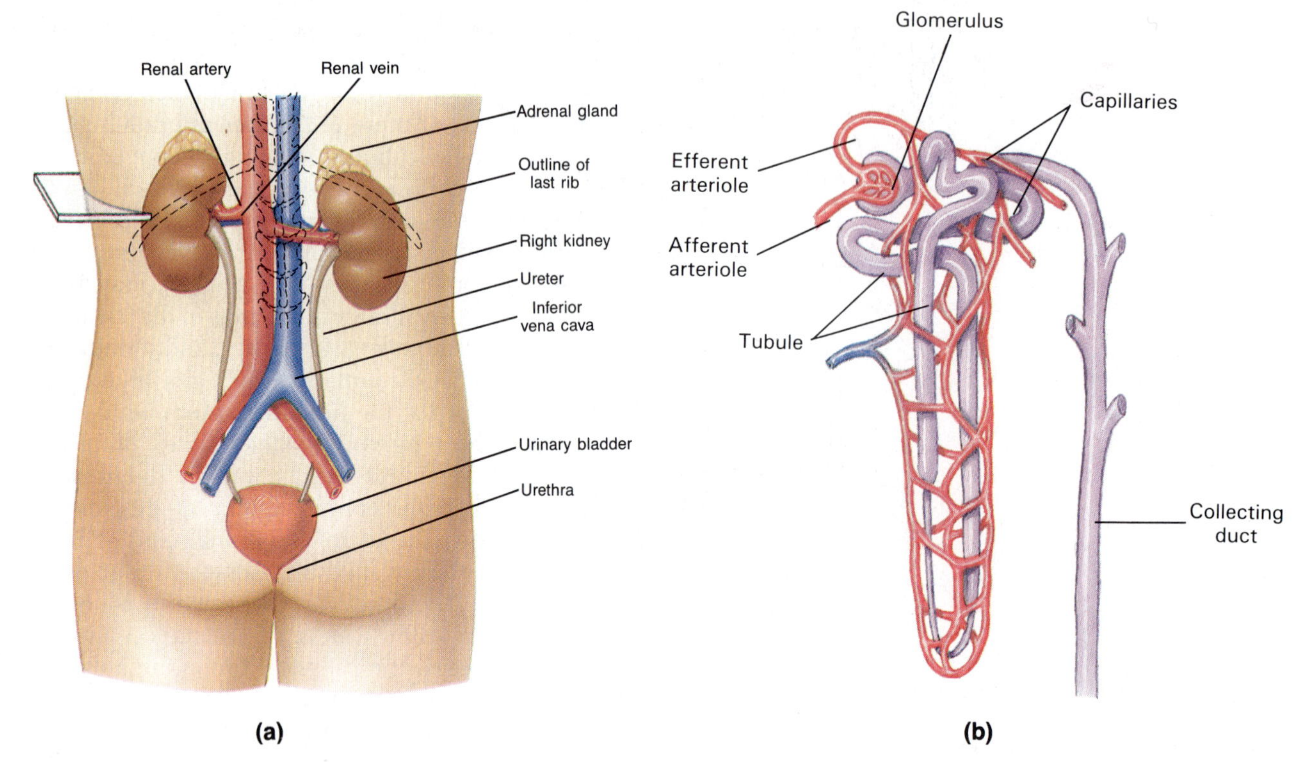

FIGURE 17.13 (a) The structure of the urinary system. (b) Structure of a nephron.

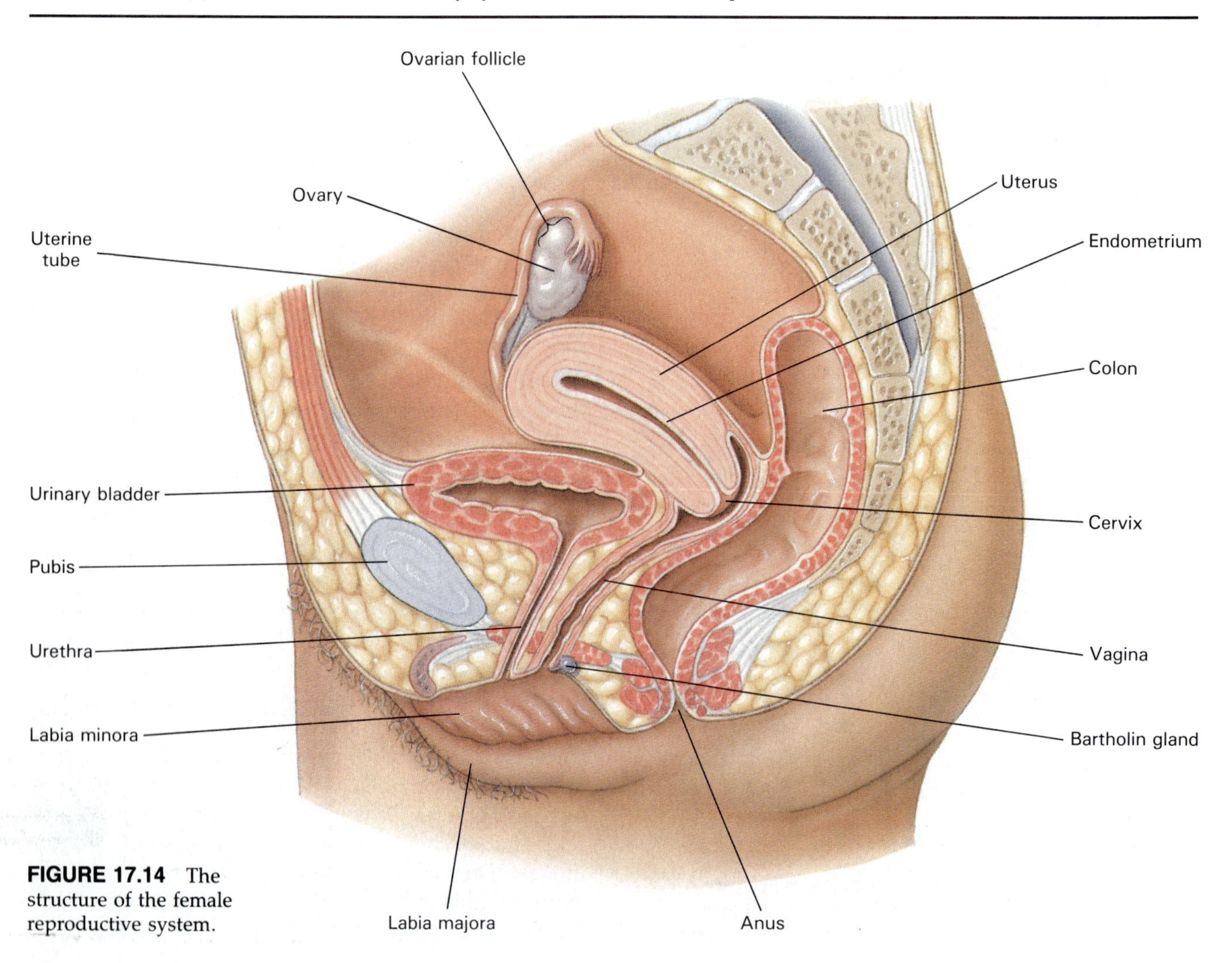

FIGURE 17.14 The structure of the female reproductive system.

them to the uterus. Fertilization usually occurs in the uterine tubes. The **uterus** is a pear-shaped organ in which a fertilized ovum develops. It is lined with a mucous membrane called the **endometrium,** the outer portion of which is sloughed during menstruation. The **vagina,** also lined with mucous membrane, extends from the cervix (an opening at the narrow lower portion of the uterus) to the outside of the body. It allows passage of menstrual flow, receives sperm during intercourse, and forms part of the birth canal. The female **external genitalia** include the sexually sensitive clitoris, two pairs of labia (skin folds), and the mucous-secreting **Bartholin glands.** Because they nourish offspring, **mammary glands** are considered part of the female reproductive system. These modified sweat glands develop at puberty and contain gland cells embedded in fat and ducts that convey milk to the nipple.

The **male reproductive system** (Figure 17.15) consists of testes, ducts, glands, and the penis. The **testes** produce the hormone testosterone and sperm, which are conveyed through a series of ducts to the urethra. Secretions from **seminal vesicles** and the **prostate gland** mix with sperm to form **semen.** Other glands secrete mucus that lubricates the urethra. The **penis** deposits sperm into the female reproductive tract during sexual intercourse.

The normal flora of the urogenital system are found mainly at or near the external opening of the urethra of both sexes and in the vagina of females. Sites of normal flora also are common sites of infections, partly because of their close proximity to the anus and partly because they provide warm, moist conditions suitable for microbial growth. Many of the infectious diseases of the urogenital system are sexually transmitted. Other urogenital organs generally remain sterile, but a breakdown in defense mechanisms can easily lead to infection. A pouch posterior to the uterus often collects pus during pelvic infections.

Defense mechanisms in the genitourinary system are numerous,. Urinary sphincters (muscles that close openings) act as mechanical barriers to microbes and also help to prevent backflow of urine. The flow of urine through the urethra and of mucus through both the urethra and vagina tends to wash away microbes. The low pH within the urethra and, during reproductive years, of the vagina prevents invasion by pathogens. Normal flora compete with pathogens and prevent them from causing disease. In spite of its defenses, the urogenital tract is poorly protected against sexually transmitted diseases. The organisms that cause such diseases tolerate acid conditions and successfully compete with natural flora of the urogenital tract.

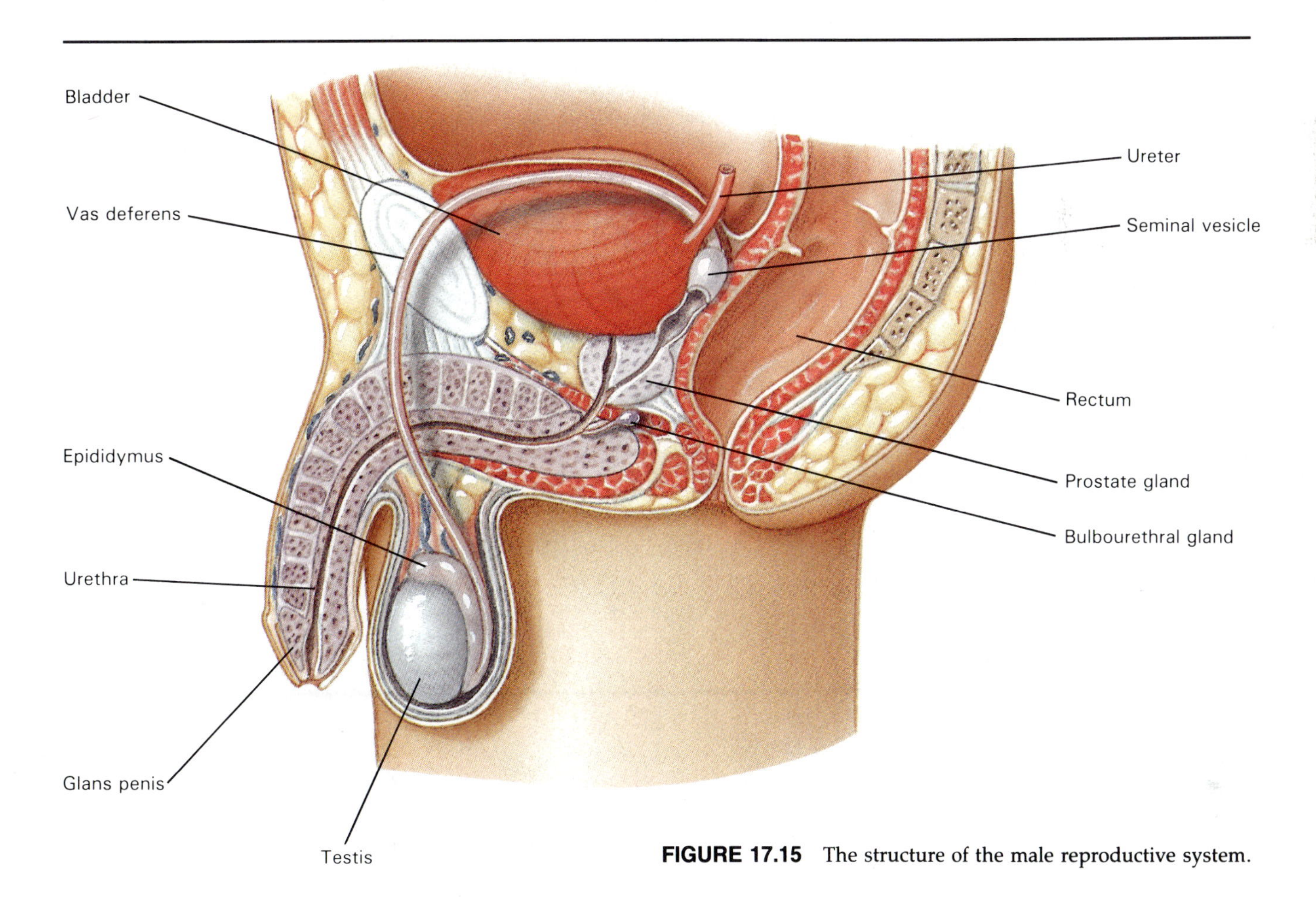

FIGURE 17.15 The structure of the male reproductive system.

Lymphatic System

The **lymphatic system** (Figure 17.16), which is closely associated with the cardiovascular system, consists of a network of vessels, lymph nodes and other lymphatic tissues, and the fluid lymph. Blind-ended **lymph capillaries** accumulate **lymph,** which is composed of fluid and protein molecules draining from the spaces between cells. Lymph, which represents fluid lost through capillary walls, is returned to the blood circulatory system through a system of **lymphatic vessels.** Other lymphatic tissues include the thymus gland, tonsils, spleen, and Peyer's patches of the intestine. These organs perform important functions in both nonspecific host defenses and specific immunity.

Lymphatic tissues contain cells that phagocytize microorganisms. If these cells encounter more pathogens than they can destroy, the lymphatic tissues can become sites of infection. Thus, swollen lymph nodes and tonsillitis are common signs of many infectious diseases. All lymphatic organs contain large numbers of lymphocytes, which originate in bone marrow and are released into blood and lymph. They live from weeks to years, becoming dispersed to various lymphatic organs or remaining in the blood and lymph. Lymphocytes differentiate into functional B lymphocytes (B cells) or T lymphocytes (T cells) and function in specific immunity, as is explained in Chapter 18.

In lymphatic tissues such as **lymph nodes** (Figure 17.17), lymphocytes are tightly packed together in a network of connective tissue fibers. Some lymphocytes are covered with a connective tissue capsule. Lymph vessels carry lymph to and from these tissues. Lymphatic tissues contain wide passageways called **sinuses** that are lined with phagocytic cells, which engulf microorganisms and other foreign material. Lymph nodes are widely distributed throughout the body but are most numerous in the thoracic (chest) region, neck, armpits, and groin. Lymph nodes contain both B cells and T cells.

Small unencapsulated aggregations called **lymph**

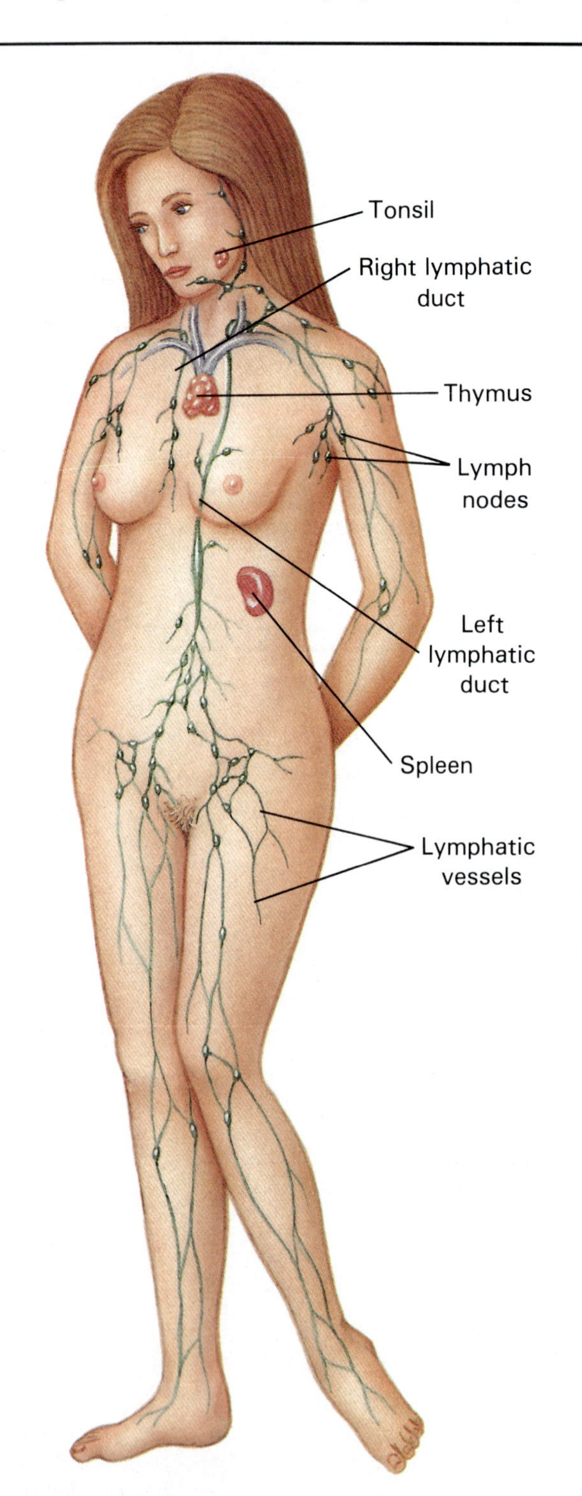

FIGURE 17.16 The structure of the lymphatic system.

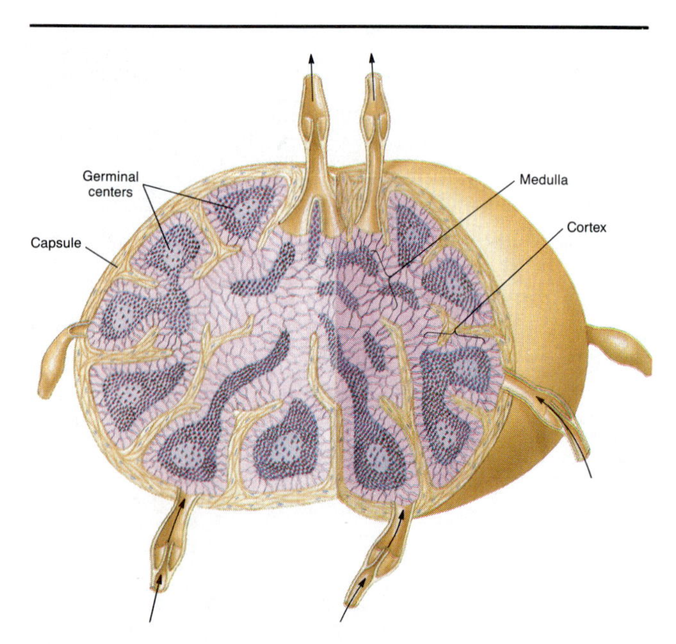

FIGURE 17.17 The structure of a lymph node.

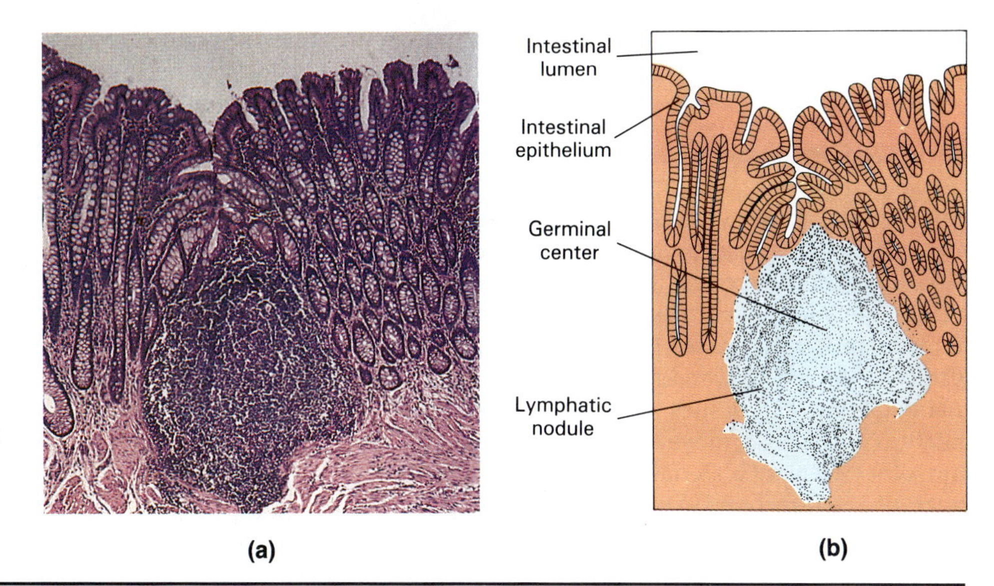

FIGURE 17.18 (a) Micrograph (b) and drawing of a lymph nodule in the gastrointestinal tract.

nodules (Figure 17.18) develop in many tissues in response to microorganisms. (They are not found in fetuses or in animals raised in a germ-free environment.) Lymph nodules are especially numerous in the digestive, respiratory, and urogenital tracts. They include Peyer's patches in the small intestine and the lymphoid tissue ring formed by the tonsils in the pharynx. Collectively, the tissues of lymph nodules are referred to as **gut-associated lymphatic tissue** (GALT); they are the body's main sites of antibody production.

The **thymus gland** (Figure 17.19) is a multilobed lymphatic organ located beneath the sternum (breastbone). It is present at birth, grows until puberty, then atrophies (shrinks) and is replaced by fatty tissue by adulthood. Around the time of birth the thymus begins to process lymphocytes and releases them into the blood as T cells. T cells play several roles in immunity—they regulate the development of B cells into antibody-producing cells, and they kill virus-infected cells directly. The thymus also produces the hormone **thymosin,** which stimulates production of lymphocytes in lymph nodes and other tissues, and probably two other hormones, **thymopoietin I** and **thymopoietin II,** which are believed to stimulate the transformation of lymphocytes into T cells.

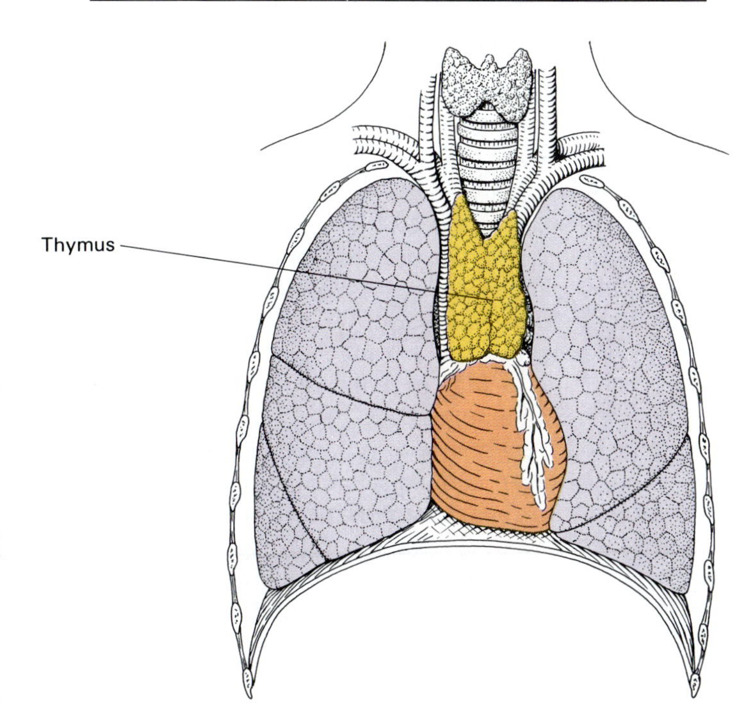

FIGURE 17.19 The location of the thymus gland.

The **spleen,** the largest of the lymphatic organs, is located in the upper left quandrant of the abdominal cavity. It is encapsulated, lobed, and well supplied with blood and lymphatic vessels. Its sinusoids contain many phagocytes, which engulf and digest worn-out erythrocytes and probably microorganisms and other foreign material, too. It also contains B cells and T cells.

In summary, lymphoid tissues contribute to nonspecific defenses by phagocytizing microorganisms and other foreign material. They contribute to specific immunity through the activities of their B and T lymphocytes.

PHAGOCYTOSIS

A phagocyte is a cell that ingests and digests foreign particles. Literally, it is a cell that eats (*phago*, Latin for to eat; *cyte*, Latin for cell). **Phagocytosis** is the process of ingesting and digesting foreign particles. These foreign particles include dead cells and cellular debris that must constantly be removed from the body as cells die and are replaced. But the foreign particles of greatest interest to microbiologists are microorganisms that gain access to host tissues. Phagocytes destroy microorganisms by direct phagocytosis or by a combination of specific immune reactions and phagocytosis.

Kinds of Phagocytic Cells

Neutrophils, monocytes, and macrophages are the major types of phagocytic cells in human tissues. Monocytes that have left the blood and entered the tissues are called macrophages and may have different names in different tissues (Table 17.2). Macrophages also can be fixed or wandering. *Fixed macrophages* remain stationary in tissues; *wandering macrophages* circulate in the blood or move into tissues when microbes and other foreign material are present. Wandering macrophages are found in many tissues (Figure 17.20), but especially in the lung alveoli and in the peritoneum, the membrane that lines the abdominal cavity. Macrophages sometimes are collectively referred to as the macrophage system, or **reticuloendothelial system.**

Neutrophils, also called **microphages,** are best at phagocytizing bacteria and other small particles. Monocytes and macrophages can phagocytize larger particles, such as debris from dead cells and even macrophages that have engulfed bacteria. Eosinophils also are believed to act as phagocytes in some instances. They are particularly important in defending the body against certain parasitic worm infections. In addition to acting as phagocytes, these cells play important roles in specific immunity.

TABLE 17.2 Names of macrophages in various tissues

Name of Macrophage	Tissue
Alveolar macrophage (dust cell)	Lung
Histiocyte	Connective tissue
Kupffer cell	Liver
Microglial cell	Neural tissue
Osteoclast	Bone
Sinusoidal lining cell	Spleen

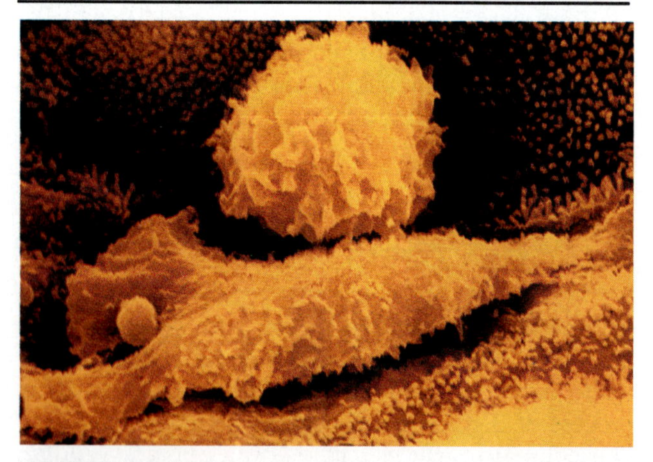

FIGURE 17.20 False-color SEM of two macrophages in a human lung. The macrophage at top is in its normal spherical shape, covered with ruffles. The one beneath it has elongated itself to engulf the small, round particle at left. Macrophages clear the lungs of dust, pollen, bacteria, and some components of tobacco smoke.

The actions of phagocytes constitute an exceedingly important component of nonspecific host defenses because these cells can attack pathogens at many sites in the body. They first attack microbes and other foreign material at portals of entry, such as wounds in skin or mucous membranes. If some microbes escape destruction at the portal of entry and gain access to deeper tissues, blood or lymph phagocytes in these locations mount a second attack on them.

Unfortunately, some pathogens survive phagocytosis and can be spread through the body in the phagocytes that attempted to destroy them. A few pathogens even multiply within phagocytes. Among the phagocytes, macrophages can live for years, whereas neutrophils live for days. Macrophages sometimes provide pathogens with a long-term, stable environment where they can multiply out of the reach of other defense mechanisms. Pathogens known to multiply in macrophages include bacteria that cause Rocky Mountain spotted fever, tuberculosis, and Hansen's disease (leprosy) as well as poxviruses and herpesviruses.

The Process of Phagocytosis

For phagocytes to destroy microorganisms, they must find them, ingest them, and digest them. Let's look at each of these processes in more detail.

Chemotaxis

Phagocytes find microorganisms by chemotaxis, movement toward a higher concentration of an attractant substance. (Chapter 4, p. 87) Both infectious agents and damaged tissues release chemical substances that attract phagocytes. Such chemical attractants are believed to stimulate molecular processes within phagocytes, causing them to move along the gradient of increasing concentration to the source of an attractant. Chemotaxis also can be initiated by immunological reactions in which macrophages "process" the pathogen and cause certain lymphocytes to release lymphokines. **Lymphokines** are chemical substances secreted by T cells when they encounter an antigen. Among the various lymphokines that have been studied, many regulate immunological reactions, but some attract macrophages, neutrophils, and other leukocytes to sites of infection.

Some pathogens can escape phagocytes by interfering with chemotaxis. For example, most strains of the bacterium that causes gonorrhea remain in the urogenital tract, but some strains escape local defenses and enter the blood. It is believed that the invasive strains fail to release the attractants that bring phagocytes to the noninvasive strains. The bacterium that causes tuberculosis is thought to escape phagocytosis by producing a lipid that prevents phagocytes from being attracted to it.

Ingestion

Once a phagocyte has found and attached itself to a microbe, the cell membrane of the phagocyte forms a pouch and eventually completely surrounds the microbe. The cell membrane covering the microbe separates from the rest of the cell membrane and forms a **phagosome,** or vacuole, within the cytoplasm of the phagocyte. Lysosomes, which contain digestive enzymes, fuse with the phagosome membrane, forming a **phagolysosome,** and release their enzymes into the vacuole. This process is called **degranulation** because the lysosome granules disappear. In addition to lysosomal enzymes, phagocytes also release metabolic products that kill ingested microbes. Certain bacteria such as *Mycobacterium* produce a waxy surface material, and others produce protein or polysaccharide surface molecules, such as the capsules of pneumococci, that interfere with ingestion by phagocytes.

Digestion

Lysosomal enzymes released into the vacuole digest the substance of the microbe into small molecules (amino acids, sugars, fatty acids) that the phagocyte can use for energy. This energy compensates for energy used by the phagocyte to reach the microbe and ingest it. Oxygen used by the phagocyte increases significantly during digestion of the contents of a vacuole. Some of the oxygen is used to metabolize the substance of the microbe, but much is used to form hydrogen peroxide and hypochlorite that kill microbes. (Hypochlorite is the ingredient in household bleach that accounts for its antimicrobial action.)

Microbes resist digestion by phagocytes in three ways:

1. Some, such as the bacterium that causes plague, produce capsules. This capsule can be digested by neutrophils, but in macrophages the capsule not only protects the pathogen from digestion but also allows it to multiply.
2. Others, such as the bacteria that cause brucellosis, tuberculosis, and Hansen's disease, are thought to resist digestion by preventing the release of lysosomal enzymes by the phagocyte. These organisms can be spread by neutrophils and can cause chronic infections when they multiply in macrophages.
3. Still other microbes produce toxins that kill phagocytes by releasing the phagocyte's own lysosomal enzymes. Examples of such toxins are **leukocidin,** released by staphylococci, and **streptolysin,** released by streptococci.

INFLAMMATION

Do you remember the last time you cut yourself? If the cut was not too serious, the bleeding soon stopped. You washed the cut and put on a bandage. A few hours later the area around the cut became hot, red, swollen, and painful. It had become inflamed.

Definition and Characteristics of Inflammation

Inflammation is the body's response to tissue damage from injury or infection. It is characterized by (1) an increase in temperature, (2) redness, (3) swelling, and (4) pain. Inflammation is often caused by microbial pathogens, but it also can be caused by mechanical injury (cuts and abrasions), heat and electricity (burns), ultraviolet light (sunburn), chemicals (phenols, acids, and alkalis), and allergies.

Inflammation is the initial step in the repair of injured tissue. Even so, it is often uncomfortable and, in some instances, can itself be harmful. What happens in the inflammatory process, and why? Let's follow the process on the cellular and tissue levels and briefly look at its effects. Figure 17.21 illustrates the steps in the process.

The Acute Inflammatory Process

As with disease, inflammation can be either acute (short-term) or chronic (long-term). In **acute inflammation,** the battle between microbes or other agents of inflammation and host defenses is usually won by the host. Invading microbes are killed, tissue debris is cleared away, and injured tissue is repaired. Let's look at this process more closely.

Inflammation is initiated by cell damage. When cells are damaged, the chemical substance **histamine** is released into tissue fluids, diffuses into nearby capillaries and venules, and causes the walls of these vessels to dilate and to become more permeable. Dilation increases the amount of blood flowing to the area, and around skin wounds it causes the skin to become red and hot to the touch. Fluids leave the blood and accumulate around the injured cells, causing swelling. The blood delivers clotting factors, nutrients, and other substances to the injured area and removes wastes and some of the excess fluids. Although histamine has some unpleasant side effects

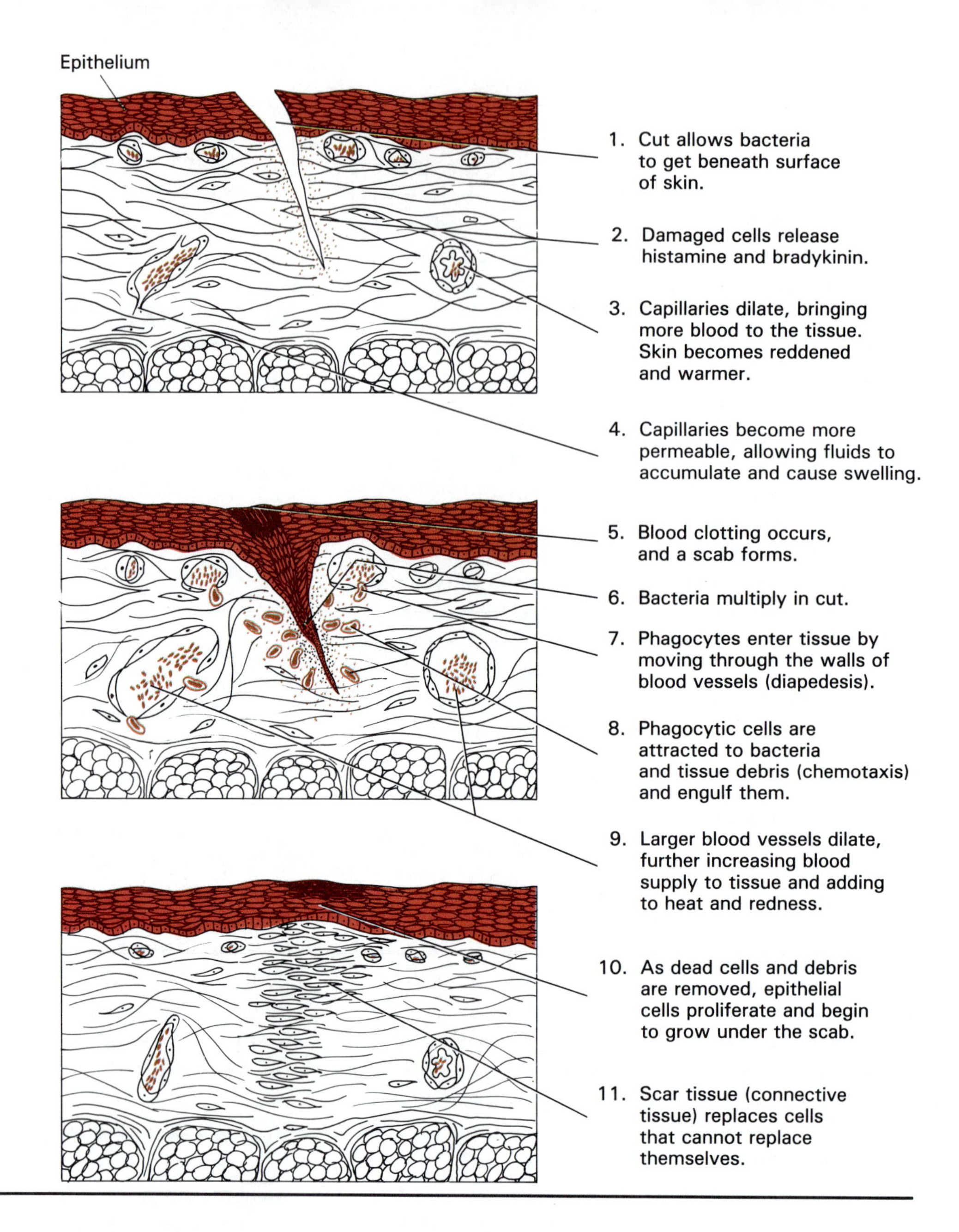

FIGURE 17.21 Steps in the process of inflammation and subsequent healing.

(redness and swelling), it also has beneficial effects that result from increased blood flow and increased permeability of blood vessels. Small peptides called **kinins** are always present in blood in an inactive form. When tissue injury occurs, they are activated and further increase blood flow and vessel permeability. Some also attract neutrophils to the injured tissue.

All kinds of tissue injury—cuts, infections, insect bites, and allergies—cause histamine release. In conjunction with its effects on blood vessels, histamine also causes the red, watery eyes and runny nose of hay fever and the itching of an insect bite. From this information you can probably guess what drugs called **antihistamines** do. They alleviate symptoms caused by histamine by blocking histamine from reaching the receptors. Unfortunately, some antihistamines have unpleasant side effects, such as sleepiness, dizziness, disturbed coordination, digestive disturbances, and thickening of mucous secretions.

The fluid that enters the injured tissue also carries the chemical components of the blood-clotting mechanism. If the injury has caused bleeding, the bleeding is stopped by the formation of a blood clot in the severed blood vessel. The dried clot is called a **scab.** Clotting also takes place in tissue fluids near the injury, where it greatly reduces fluid movement around damaged cells and walls off the injured area from the rest of the body. Pain associated with tissue injury is thought to be due to the release of a particular kinin called **bradykinin.** Injections of bradykinin cause much more pain than similar injections of saline. How bradykinin stimulates pain receptors in the skin is unknown, but cellular regulators called **prostaglandins** seem to intensify its effect. Pain is beneficial because it causes us to protect an injured area and perhaps use it less. This can prevent further injury and allow the tissue to heal more quickly. But pain sometimes lasts longer or is more severe than is needed to warn us. Then drugs such as aspirin can be given to alleviate the pain. Aspirin acts by inhibiting prostaglandins. Certain other *analgesics* (pain relievers) block pain signals before we become conscious of them. No analgesics have been shown to interfere with the release or action of bradykinin.

Inflamed tissues also appear to release another substance called **leukocytosis-promoting factor (LP factor).** Leukocytosis is a condition in which leukocytes increase in numbers in the blood. The LP factor makes more leukocytes available, and many of them pass out of blood between cells of capillary walls in a process called **diapedesis** (di-ah-ped-e'sis). The leukocytes congregate in tissue fluids in the injured region. Phagocytic leukocytes are attracted by chemotaxis to invading microorganisms, foreign substances, and debris that remains from dead cells.

When a phagocyte reaches a microorganism or other foreign substance, it attempts to engulf the particle by phagocytosis. In the process of phagocytizing particles, many of the phagocytes themselves die. Many cells of the tissue damaged by injury or infection also die, leaving an accumulation of tissue debris. The accumulation of dead phagocytes and the materials they have ingested, along with tissue debris, forms the white or yellow fluid called **pus.** Bacteria evoke pus formation, but viruses do not. Pus continues to form until the infection or tissue damage has been brought under control. An accumulation of pus in a cavity hollowed out by tissue damage is called an **abscess.** Pockets of pus that form around the roots of teeth and in boils and pimples are common kinds of abscesses.

Phagocytes usually prevent an infection from spreading or getting worse, but sometimes this natural defense mechanism is overwhelmed. The infectious organisms then invade other parts of the body. Antibiotics are often used to inhibit the growth of microorganisms in injured tissue and to minimize the chance of an infection spreading from the site of an injury. In addition to the actions of phagocytes, lymphocytes carry out specific immune reactions that help to overcome an initial infection and to prevent future infections by the same organism.

On balance the inflammatory process is beneficial, but it can sometimes be harmful, as when it causes brain damage from swelling of meninges. Too much of a good thing like swelling, which delivers phagocytes to injured tissue, also can interfere with breathing if the airway swells shut. Increased blood flow delivers more oxygen and nutrients to injured tissues. Ordinarily this is of greater benefit to host cells than to pathogens, but sometimes it helps the pathogens to thrive as well. Rapid clotting and too-complete walling off of an injured area prevents pathogens from spreading, but it also can prevent natural defenses and antibiotics from reaching the pathogens. Boils must be lanced to deliver therapeutic drugs to them. Tubercles, or pockets that wall off the bacteria that cause tuberculosis, can persist for years. When an anti-inflammatory drug such as cortisone is given, the organisms sequestered in tubercles can be liberated, and symptoms of tuberculosis will reappear. Reactivations of this kind account for most of the newly reported cases of tuberculosis in the United States today.

Attempting to suppress the inflammatory process also can be harmful. It can allow tubercles to form when natural defenses might otherwise have destroyed the bacteria. Giving aspirin to treat symptoms of inflammation is believed by some investigators to contribute to the development of Reye's syndrome, which causes brain damage and degeneration of some internal organs. Reye's syndrome usually appears as a complication after influenza, chickenpox, and some other viral infections, especially in children, and is fatal in about one-fourth of the cases.

Repair and Regeneration

During the entire inflammatory reaction, the healing process is also under way. Once the inflammatory reaction has subsided and most of the debris has been cleared away, healing accelerates. Capillaries grow into the fibrin of a blood clot, and new connective tissue cells called **fibroblasts** replace the fibrin as the clot dissolves. The fragile reddish, grainy tissue made up of capillaries and fibroblasts is called **granulation tissue.** As granulation tissue accumulates fibroblasts and fibers, it replaces nerve and muscle tissues that cannot be regenerated. New epidermis replaces the part that was destroyed. In the digestive tract and other organs that are lined with epithelium, an injured lining can be similarly replaced. Although scar tissue is not able to contract or carry signals, it does provide a strong durable "patch" that allows the remaining normal tissue to function.

Several factors affect the healing process. The tissues of young people heal more rapidly than those of older people. Their cells divide more quickly, their bodies are generally in a better nutritional state, and their blood circulation is more efficient. As you might guess from the many contributions of blood to healing, good circulation is extremely important. Certain vitamins also are important in the healing process. Vitamin A is essential for the division of epithelial cells, and vitamin C is essential for the production of collagen and other components of connective tissue. Vitamin K is required for blood clotting, and vitamin E also may promote healing and reduce the amount of scar tissue formed.

Chronic Inflammation

In contrast to acute inflammation, in chronic inflammation neither the agent of inflammation nor the host is a decisive winner of the battle. What began as an acute inflammation becomes a **chronic inflammation.** The agent causing the inflammation continues to cause tissue damage; the phagocytic cells and other host defenses attempt to destroy or at least confine the region of inflammation. Pus may be formed continuously. Chronic inflammation can persist for years.

Granulomatous inflammation, a special kind of chronic inflammation, is characterized by the presence of monocytes, histiocytes, lymphocytes, and plasma cells. Plasma cells are derived from lymphocytes and are capable of producing antibodies. The monocytes and histiocytes surround a central region of dying tissue. Groups of histiocytes (macrophages of connective tissue) fuse to form giant, multinucleate cells. Lymphocytes and plasma cells carry out specific immune reactions. This collection of necrotic (dead) tissue, phagocytes, and cells of the specific immune system is called a **granuloma** (Figure 17.22). Granulomas may be given special names when seen in syphilis (*gumma*), tuberculosis (*tubercle*), and Hansen's disease (*leproma*). As long as necrotic tissue is present, the inflammatory response will persist. If only a small quantity of necrotic tissue is present, the lesions sometimes become walled off and hardened as calcium is deposited in them. Calcified lesions are common in tuberculosis patients. When larger granulomas rupture they can allow the spread of infection, but they also give body defenses an opportunity to eradicate the infection completely. Ruptured granulomas in the lungs drain into bronchioles or large blood vessels; in peripheral lymph nodes they drain to the skin surface; in postsurgical infections they sometimes drain through a weakened site in the abdominal wall, such as where a drainage tube has been removed.

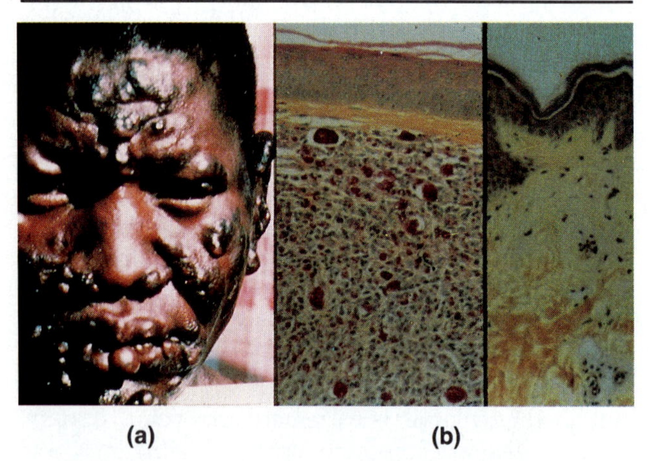

FIGURE 17.22 Granulomatous tissue formed in a Hansen's disease (leprosy) patient. (a) External view of granulomas formed in response to chronic presence of Hansen's disease bacteria. (b) Microscopic views of normal (right) and granulomatous (left) tissue sections. The numerous macrophages that have invaded the granulomatous tissue are filled with bacterial cells (stained red).

FEVER

We have seen that the temperature of infected or injured tissue increases as a part of a local inflammatory reaction. **Fever,** a systemic increase in body temperature, often accompanies inflammation. Although fevers are referred to in the Bible as punishment from God, they were not studied until 1868, when a German physician, Carl Wunderlich, devised a method to measure body temperature. He placed a foot-long thermometer in the armpit of his patients and left it in place for 30 minutes. Using this cumbersome technique he was able to record human body temperatures during febrile (feverish) illnesses.

Normal body temperature is said to be 37°C (98.6°F), although individual variations in normal temperature within the range from 36.1 to 37.5°C are not uncommon. Fever is defined clinically as an oral temperature above 37.8°C (100.5°F) or rectal temperature of 38.4°C (101.5°F). Fever accompanying infectious diseases rarely exceeds 40°C (104.5°F); if it reaches 43°C (109.4°F), death usually ensues.

Body temperature is now known to be maintained within a narrow range by a temperature-regulating center in the hypothalamus, a part of the brain. Fever occurs when the setpoint of this mechanism is raised to a higher temperature. Fever can be caused by many pathogens from among all groups of microbes, by certain specific immunological processes, and by nearly any kind of tissue injury.

In most instances fever is caused by a substance called a **pyrogen. Exogenous pyrogens** include exotoxins and endotoxins from infectious agents, which cause fever by stimulating the release of an endogenous pyrogen. **Endogenous pyrogens** are secreted mainly by monocytes and macrophages. They circulate to the hypothalamus, where they cause the body temperature to begin rising within 20 minutes, as explained in Chapter 15. If you suffer the sensation of chills while having a fever, it means that your body is trying to drive your temperature higher, in response to the pyrogens, by creating heat from the muscle contractions of your shivering.

The two main effects of fever are to increase the level of immune responses and to inhibit growth of microorganisms. Fever increases the immune response by stimulating the activity of T cells. It inhibits microbial growth by decreasing the amount of iron absorbed from the digestive tract and increasing the rate at which it is moved to iron storage deposits, thereby lowering the plasma iron concentration. In

APPLICATIONS

Aspirin

Aspirin acts not only as an antipyretic but also as an anti-inflammatory agent and an anticoagulant in the blood. Therefore, it is often recommended for treating arthritis, where it relieves pain and its anti-inflammatory effect prevents further joint destruction due to chronic inflammation. Nonaspirin pain relievers such as acetaminophen (Tylenol) and ibuprofen (Motrin) reduce arthritis pain but do not halt joint destruction. Aspirin also has recently been recommended in small doses to reduce the likelihood of clots forming within blood vessels and thereby lower the danger of heart attacks and strokes.

addition, fever increases the rate of chemical reactions in the body, raises the body temperature above the optimum temperature for growth of some pathogens, and makes the patient feel ill. Increasing the rate of chemical reactions generally speeds the rate at which defense mechanisms attack pathogens and so can shorten the course of the disease. Raising the temperature above the optimum for a pathogen slows its rate of growth, reducing the number of organisms to be combated. Finally, when fever makes a patient feel ill, the patient is likely to rest. Rest prevents further damage to the body and allows energy to be used to fight the infection.

Knowledge of the benefits of fever has changed the clinical approach to fever. In the past, antipyretics—fever-reducing drugs such as aspirin—were given almost routinely to reduce fever whenever it occurred. Now many physicians recommend allowing fevers to run their course unless they go above 40°C or the patient has a disorder that might be worsened by fever. In those cases, antipyretics are still used. Untreated high fever increases the metabolic rate, makes the heart work harder, increases water loss, alters concentrations of electrolytes, and can cause convulsions, especially in children. Patients with severe heart disease and fluid and electrolyte imbalances and children subject to convulsions usually receive antipyretics. Children should not be given aspirin because it may increase the risk of Reye's syndrome.

MOLECULAR DEFENSES

In addition to phagocytosis, inflammation, and fever, the nonspecific defenses of the human body include some processes that occur at the molecular level. These processes involve the actions of interferon and complement.

Interferon

As early as the 1930s scientists had observed that infection with one virus prevented for a time any infection by another virus. Then, in 1957, a small protein called **interferon** was shown to be released from virus-infected cells. Efforts to purify interferon led to the discovery that many different types of interferon with small differences in molecular structure exist in different species and even in different tissues of the same individual. Three groups of interferons identified in humans are *alpha interferon,* made by leukocytes, *beta interferon,* made by fibroblasts, and *gamma interferon,* made by T cells.

Several studies have focused on determining how interferon acts. It appears that once interferon is released, it binds to surface receptors on adjacent cells. The presence of a single interferon molecule on a receptor can cause a cell to produce many molecules of **antiviral protein.** Although viruses enter cells that are producing antiviral protein, the protein interferes with replication of viruses. In some cases antiviral protein prevents the virus from making nucleic acid, in other cases from making protein. The actions of interferon and antiviral protein are summarized in

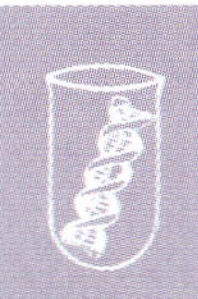

BIOTECHNOLOGY

Alpha Interferon Used in Treatment

Early in 1991, alpha interferon was approved by the U.S. Food and Drug Administration (FDA) for the treatment of hepatitis C (sometimes called hepatitis non-A, non-B). The FDA, starting in 1986, had previously approved alpha interferon for use in treating the AIDS-related cancer Kaposi's sarcoma, genital warts, and hairy cell leukemia, a very rare blood cancer.

Current research has found alpha interferon to be effective in preventing nonsymptomatic patients infected with the AIDS virus from developing a full-blown case of the disease. Interferon is being used together with the drug azidothymidine (AZT). In other research, it has cleared hepatitis B virus from the bloodstream of 10 percent of infected patients and halted viral replication in another 33 percent. This procedure required daily shots of interferon, which caused flulike side effects.

Treatment for hepatitis C requires a dose of 3 million units of interferon to be given three times per week for 6 months, at a total cost of $2400. The drug, however, is a treatment but not a cure for the disease, and needs to be repeated. Sales of alpha interferon in 1990 totaled over $175 million, about $10 million of which was for hepatitis C treatment. The worldwide market for all applications of alpha interferon in 1991 was estimated at $500 million and is expected to reach $1 billion per year by 1994.

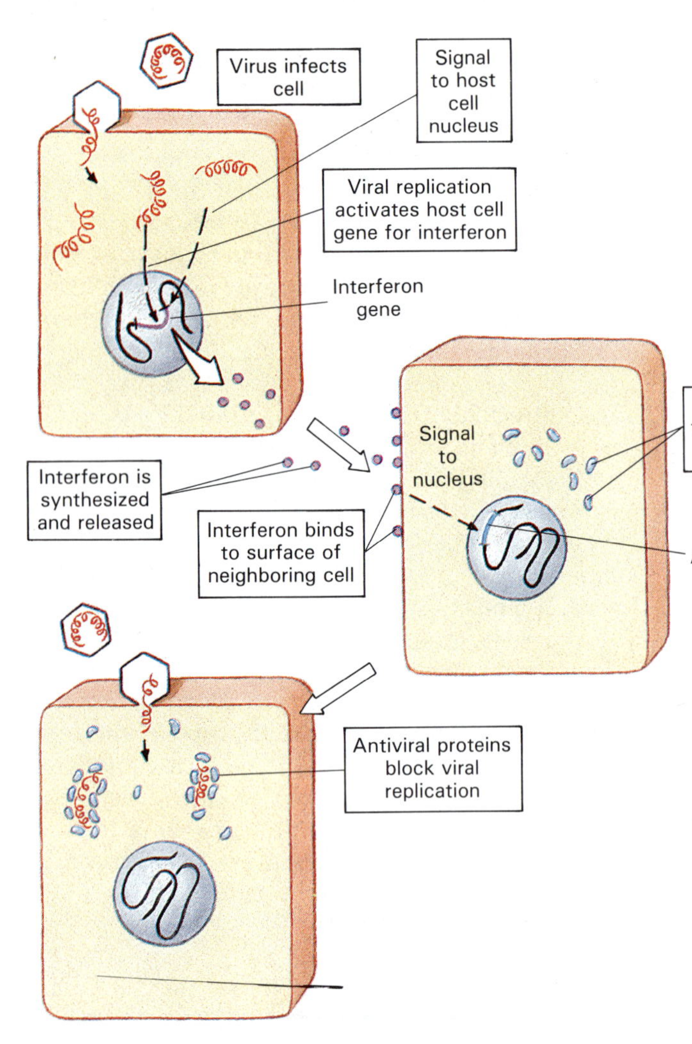

FIGURE 17.23 The mechanism by which interferon acts.

Figure 17.23. Fever increases interferon production and thereby assists the body in fighting viral infections.

Interferon can be produced in laboratory cultures of human fibroblasts and white blood cells and in bacteria by recombinant DNA techniques. Late in 1979 Charles Weissman and his colleagues at the University of Zurich developed such a recombinant DNA technique. They isolated genes from human cells, selecting those most likely to produce interferon, and inserted them into bacterial plasmid DNA. They then allowed the bacterial cells spontaneously to take up this DNA. Some of the cells thereby acquired the human interferon gene and began synthesizing interferon. By a complex sequence of steps these investigators identified and isolated bacterial colonies that could produce interferon. Organisms from these colonies were allowed to multiply so they could produce relatively large quantities of the substance. Interferon produced by this method, called *recombinant interferon*, is relatively cheap and abundant (Figure 17.24).

The ability to produce recombinant interferon spurred research on therapeutic applications for it. Several studies have focused on the use of interferon to treat malignancies. Tests on one form of bone cancer show that after most of the cancerous tissue is removed by surgery or destroyed by radiation, interferon therapy will reduce the incidence of metastasis (spread). How it does this is not yet known. Perhaps it suppresses cancer cells because such cells contain as yet unidentified viruses, or perhaps interferon prevents growth of the cancer cells. Interferon also has been shown to be somewhat effective in treating kidney cell cancer and melanoma (a kind of skin cancer that does not respond to other therapies). It has not been effective against lung, breast, or colon cancer.

Other studies have focused on viral diseases such as influenza, the common cold, hepatitis, AIDS, and genital herpes. Interferon delays the onset of symptoms of influenza but does not effectively treat the disease. It does inhibit rhinoviruses, which cause most colds, but it has unwelcome side effects such as irritation and nasal bleeding. In patients with hepatitis, interferon prevents both relapses and the spread of viruses to others. Finally, interferon sometimes lessens the severity of genital herpes.

The therapeutic usefulness of interferon has not proven to be as great as was hoped when it was first discovered. However, only a few of the many different kinds of interferon have been tested and those on only a small number of diseases. Many possibilities

FIGURE 17.24 Flasks of interferon produced by recombinant DNA techniques.

for therapeutic applications remain, and the search for them continues.

Complement

Complement, or the **complement system,** refers to a set of more than 20 large regulatory proteins produced by the liver that circulate in plasma in an inactive form. They account for about 10 percent (by weight) of all plasma proteins. When complement was named, it was believed to be a single substance that complemented, or completed, certain immunologic reactions. Even though complement can be activated by immune reactions, its effects are nonspecific—it has the same effect on many different microorganisms.

The general functions of the complement system are to enhance phagocytosis, produce inflammation, and directly lyse microorganisms. These functions are nonspecific: They occur regardless of which microorganism has invaded the body. Furthermore, complement goes to work as soon as an invading microbe is detected and constitutes an effective host defense long before the specific immune system is mobilized.

When the complement system is activated, complement proteins participate in a cascade of reactions that trigger an inflammatory response. A **cascade** is a set of reactions in which magnification of effect occurs—that is, a greater quantity of product is formed in the second reaction than in the first, still more in the third, and so on. Of the proteins so far identified in the complement system, 13 participate in the cascade itself, and 7 activate or inhibit reactions in the cascade.

Two pathways have been identified in the sequence of reactions carried out by the complement system. They are referred to as the **classic complement pathway** and the **properdin pathway,** or **alternate pathway.** The steps in the classic pathway and the factors that initiate them are summarized in Figure 17.25. Complement proteins C1 through C9 (C stands

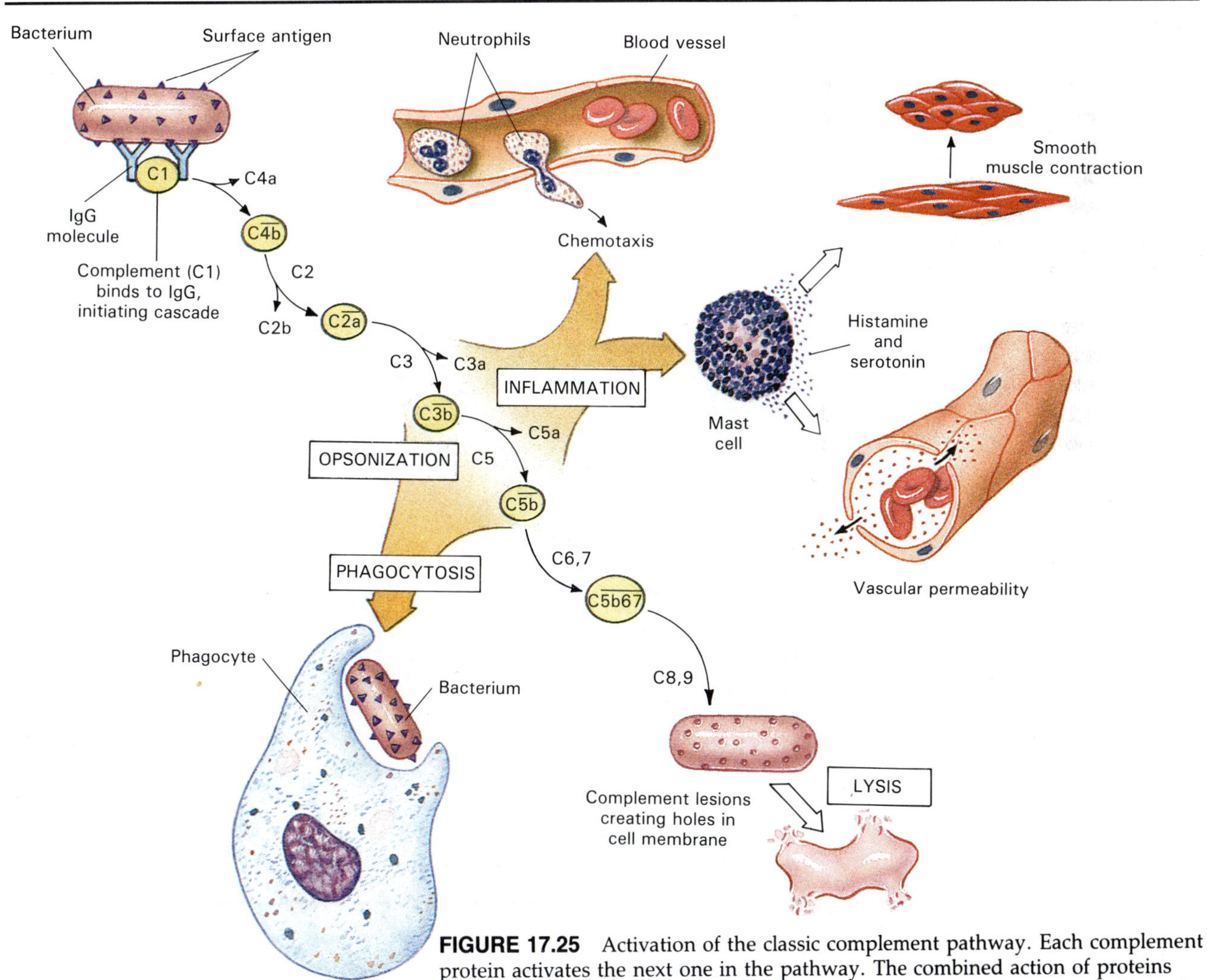

FIGURE 17.25 Activation of the classic complement pathway. Each complement protein activates the next one in the pathway. The combined action of proteins C5–C9 eventually punches a hole in the bacterial cell membrane, causing lysis of the cell. Components of the C3 and C5 proteins have a variety of other defensive effects, including attracting and stimulating phagocytic cells and promoting the inflammatory response.

for complement) participate in the classic pathway. C1 is a large complex molecule made of three smaller proteins, C1q, C1r, and C1s. In the properdin pathway, the proteins **properdin,** factor B, and factor D replace C1, C2, and C4 in the initial steps. The reactions involving C3 through C9, which are common to both pathways, produce the effects of the complement system. Consequently, the effects are the same regardless of the pathway by which C3 is produced.

The contributions of the complement system to nonspecific defenses depend on C3, a key protein in the system. Once C3 is formed, it immediately splits into C3a and C3b. C3b then participates in two kinds of defenses, whereas C3a contributes mainly to the development of inflammation. Let's look at the effects of these molecules in more detail.

Opsonization Some of the C3b binds to the surface of microorganisms that have been coated with antibodies called **opsonins.** C3b so bound reacts with specific C3b receptors on the membranes of phagocytes and promotes phagocytosis. One might say that C3b captures the microorganisms and literally feeds them to the phagocytes. This process is called **opsonization,** or **immune adherence.**

Membrane Attack Some of the C3b initiates the C5 through C9 sequence of reactions, and C5 splits into two components, C5a and C5b, early in the reaction sequence. C5b, C6, C7, C8, and C9, known as the **membrane attack complex,** are responsible for the direct lysis of invading microorganisms. They produce lesions on cell membranes through which the contents of the cells leak out. This process is called **immune cytolysis** (Figure 17.26). The effects of the membrane attack complex form the basis of an important laboratory test, called complement fixation, used to detect antibodies against any one of many microbial pathogens.

Inflammation C3a, assisted by C5a, C5b, C6, and C7, enhances the acute inflammatory reaction by stimulating chemotaxis and thus phagocytosis and by increasing the permeability of blood vessels. C3a and C5a adhere to the membranes of mast cells and platelets and cause them to release histamine and other substances that contribute to the inflammatory response.

One of the great advantages of the complement system as a host defense mechanism is that, once it is activated, the reaction cascade occurs very rapidly. Because it is a cascade, a very small quantity of an activating substance can activate a few molecules of C1 or properdin. They, in turn, activate large quantities of C3; sufficient quantities of C3a and C3b are quickly available to cause opsonization, immune cytolysis, and inflammation.

Unfortunately, complement activity can be impaired by the absence of one or more of its protein components. Impaired complement activity causes

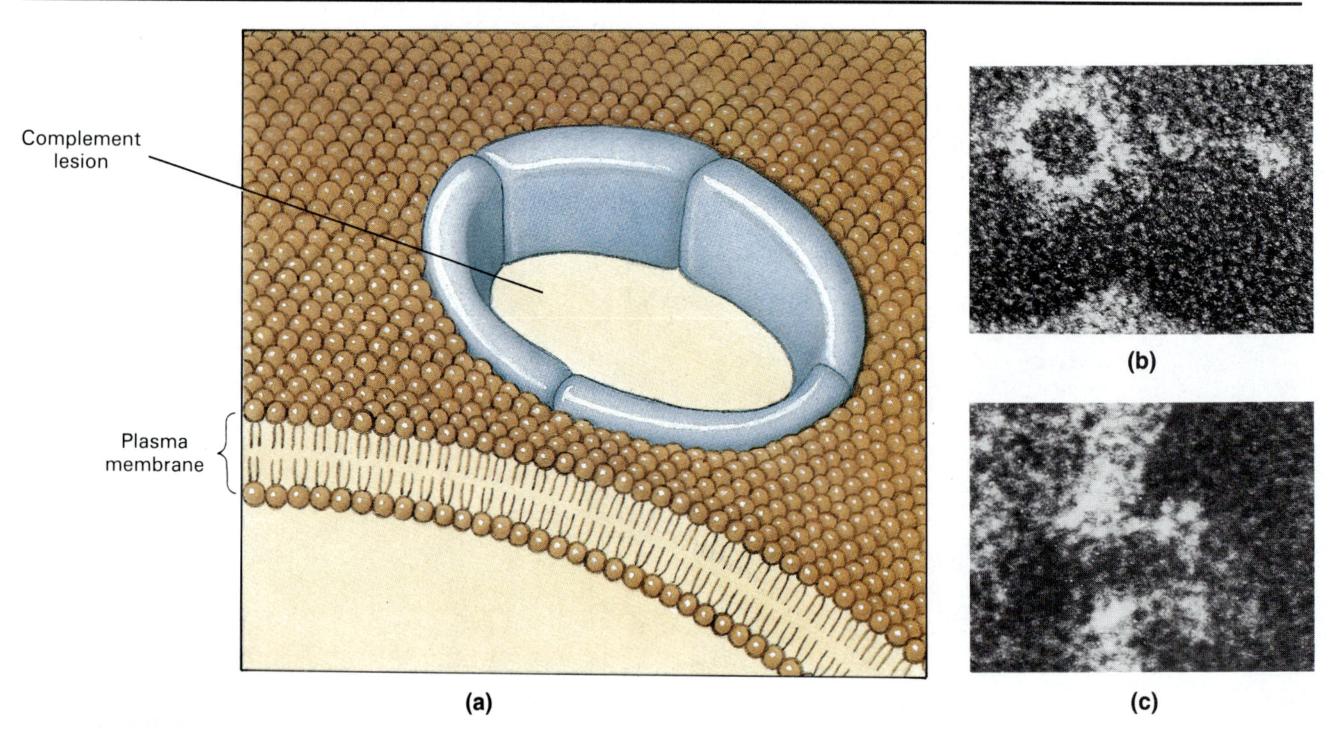

FIGURE 17.26 (a) Complement lyses cells by creating a hole in the cell membrane. (b) Face-on and (c) cross-sectional views of cell membranes showing lesions made by complement.

Disease State	Complement Deficiencies
Severe recurrent infections	C3
Recurrent infections of lesser severity	C1, C2, C5
Systemic lupus erythematosus (a bodywide immunologic disease)	C1, C2, C4, C5, C8
Glomerulonephritis (an immunologic disease of the kidneys)	C1, C8
Gonococcal infections	C6, C8
Meningococcal infections	C6

TABLE 17.3 Disease states related to complement deficiencies

various diseases (Table 17.3), most of which can be either acquired or congenital. Acquired diseases result from temporary depletion of a complement protein; they subside when cells again become able to synthesize the protein. Congenital complement deficiencies are due to permanent genetic defects that prevent synthesis of one or more complement components.

The most significant effect of complement deficiencies is lack of resistance to infection. Although deficiencies in several complement components have been observed, the greatest degree of impaired complement function occurs with a deficiency of C3. This situation is not surprising because, as noted earlier, C3 is the key component in the system. In patients with C3 deficiencies, chemotaxis, opsonization, and cell lysis all are impaired.

Acute Phase Response

Observations of acutely ill patients have led to the characterization of the *acute phase response,* a response to acute illness that involves increased production of certain blood proteins. These proteins include fibrinogen, ceruloplasmin (a metal transport protein), a component of the complement system, certain enzyme inhibitors, and a substance called **C-reactive protein (CRP).** Increased production of CRP and other proteins appears to constitute a body defense mechanism distinct from the inflammatory response and specific immune responses. This mechanism appears to recognize foreign substances before the immune system does and acts early in the inflammatory process before antibodies can be produced. All humans studied thus far have the capacity to produce CRP, whereas some humans lack certain immune responses.

CRP, now known to consist of five identical units composed of 187 amino acids each, is synthesized mainly by liver cells. How its synthesis is accelerated in acute illness is not known, but lymphokines and prostaglandins may be involved. When CRP binds to a cell membrane it can initiate an inflammatory response or accelerate an ongoing one. CRP also can activate the complement system, stimulate the migration of phagocytic cells, especially PMNLs, and initiate platelet aggregation. Finally, CRP sometimes prevents death in certain otherwise fatal bacterial infections, such as those caused by particularly virulent strains of pneumococci. If we knew how to enhance CRP activity, it could be an effective therapy in combating many infections.

ESSAY

Development of the Immune System

Do all organisms have immune systems? In the case of vertebrate animals, the answer is yes. As we shall see in the next chapter, vertebrates have well-developed immune systems. In contrast, most invertebrate animals show no evidence of specific immune response and survive largely because of their enormous reproductive rates. However, they are not totally defenseless in a world full of pathogens and parasites. They have some excellent nonspecific defenses. These mechanisms include actions such as growing tissues to wall off infecting microbes. In fact, in plants one measure of how well a given strain of trees will resist infection after pruning or damage is their ability to grow tissue to wall off the damaged area from the rest of the trunk (see Figure 17.27).

Another nonspecific defense mechanism invertebrates (as well as vertebrates) employ is phagocytosis. How does an invertebrate recognize another cell as foreign (nonself) and something which should be attacked and destroyed? Membrane receptors on invertebrate phagocytes bind either directly to foreign cell surfaces or indirectly to opsonins attached to foreign cell surfaces. Phagocytosis is important to invertebrates in obtaining food, but it is also necessary to prevent sedentary organisms permanently fixed to a surface, and living where space is limited, from being overgrown by neighbors. So, phagocytosis is used to defend one's territory. In lower animals lacking a circulatory system, amoebocytes wander through the body engulfing foreign matter as well as damaged or aged cells of the organism. When your white blood cells phagocytize a bacterium, they are using ancient mechanisms preserved and transformed from lower life forms.

Opsonization is made possible by complement or complementlike

ESSAY Development of the Immune System

FIGURE 17.27 Experimentally damaged areas of tree trunk are walled off in successful trees, thus keeping infection from spreading throughout the entire tree.

components of body fluids, for example, coelomic (body cavity) fluid in sea urchins. These fluids share many characteristics with human complement proteins, which were probably derived from these early versions in invertebrates. Secretion of enzymes having antimicrobial activity is another means of defense present even in simple protozoa.

Almost all invertebrates reject grafts of foreign tissue, as do vertebrates. However, vertebrates will reject them even more vigorously on a second encounter. Invertebrates do not; in fact, the second rejection may be slower than the first. Memory increasing response is lacking. There are many unique and quasi-immunologic responses in invertebrates that vary with particular groups.

The immune system, as we think of it, is found only in vertebrates. Specific antibodies are found in all types of fish. The swiftest and most complex immune responses, however, are found in mammals and birds. Birds have a saclike structure, the Bursa of Fabricius, which is not present in mammals and probably represents a higher state of evolution of the immune system. It produces the B cells, special lymphocytes discussed in Chapter 18. Mammals have tissues that we refer to as "bursal equivalent" tissues, which also produce B cells The height of immune system development culminates in this two-part system of B cells and T cells. B cell immunity evolved later than T-cell immunity. The next chapter will investigate this vertebrate achievement of the specific immune system.

CHAPTER SUMMARY

NONSPECIFIC VERSUS SPECIFIC HOST DEFENSES

- **Nonspecific defenses** operate regardless of the invading agent; they constitute a first line of defense.
- **Specific defenses** respond to particular agents; they constitute a second line of defense.

RELATED KEY TERMS: **resistant** **susceptible**

SYSTEM STRUCTURE, SITES OF INFECTION, AND NONSPECIFIC DEFENSES

Skin

- Skin consists of an outer **epidermis** and an inner **dermis.**
- Skin infections occur anywhere the skin is broken, in ducts of glands, in hair follicles, and sometimes on unbroken skin.
- Nonspecific defenses include the mechanical barrier of intact skin, acidic and salty secretions, waterproofing **keratin,** and **basement membranes.**

RELATED KEY TERMS: **sebaceous glands** **sebum** **sweat glands**

Eyes and Ears

- Protective structures associated with the eyes include the eyelids, eyelashes, **conjunctiva, cornea,** and **lacrimal glands.**
- Protective structures associated with the ears include hairs and **ceruminous glands** in the ear canal.
- Eye infections occur on eyelids, conjunctiva, and cornea; ear infections occur in the ear canal and **middle ear.**
- Eyelashes, eyelids, and hairs and wax in the external ear are mechanical barriers; lysozyme in tears is microbicidal.

RELATED KEY TERMS: **cerumen** **pinna** **tympanic membrane** **mastoid area**

Respiratory System

- The **respiratory system** consists of the **nasal cavity, pharynx, larynx, trachea, bronchi, bronchioles,** and **alveoli.**
- Infections of the mucous membranes of the **upper respiratory tract** are common. These infections sometimes spread to sinuses, the middle ear, and the **lower respiratory tract.**
- Nonspecific defenses include mucus and cilia, including the **mucociliary escalator,** and phagocytic cells.

RELATED KEY TERMS: **epiglottis** **bronchial tree** **pleura** **serous membranes**

Digestive System

- The **digestive system** consists of a tube with modifications to form the mouth, pharynx, esophagus, stomach, and intestines and accessory organs such as the salivary glands, liver, and pancreas. Infections occur in mucous membranes of all organs except the stomach, in accessory glands, and in and around the teeth.
- Nonspecific defenses include mucus, stomach acid, **Kupffer cells** in liver **sinusoids,** patches of submucosal lymphatic tissue, and competition from normal flora.

Cardiovascular System

- The **cardiovascular system** consists of the heart, an extensive system of blood vessels, and blood.
- Though the cardiovascular system is normally sterile, pathogens can be transported in blood, multiply in blood, and infect the heart valves and pericardium.
- Nonspecific defenses include the cleansing effect of blood flow out of wounds, constriction of injured blood vessels, blood clotting, phagocytic actions of certain classes of **leukocytes** (the **neutrophils** and **monocytes,** or **macrophages**) and release of heparin and histamine, which initiates the inflammatory response.

RELATED KEY TERMS

tract **oral cavity** **crown**
root **enamel** **cementum**
dentin **pulp cavity**
root canals **small intestine**
villi **microvilli**
large intestine **colon**
feces **mucin**
pericardial sac **endocardium**
myocardium **epicardium**
atria **ventricles**
arteries **arterioles**
capillaries **venules** **veins**
plasma **formed elements**
electrolytes **erythrocytes**
hemoglobin **granulocytin**
polymorphonuclear leukocytes
agranulocytes **platelets**
megakaryocytes **eosinophils**
basophils **mast cells**
lymphocytes

Nervous System

- The **nervous system** consists of the brain and spinal cord, the peripheral **nerves, meninges,** and **cerebrospinal fluid.**
- The nervous system is normally sterile; when pathogens invade they can attack meninges, invade nerve endings, or even damage brain tissue.
- Nonspecific defenses include macrophages and the **blood-brain barrier** of the brain.

ganglia **neurons**
central nervous system
peripheral nervous system

Urogenital System

- The **urogenital system** includes the urinary system and the reproductive system. The **urinary system** consists of the **kidneys, ureters, urinary bladder,** and **urethra.** The **female reproductive system** consists of the **ovaries, uterine tubes, uterus, vagina, external genitalia,** and **mammary glands.** The **male reproductive system** consists of the **testes,** a system of ducts, glands, and the **penis.**
- Urogenital infections are most common near body openings.
- Nonspecific defenses include urinary sphincters, cleansing action of the outflow of urine, acidity of mucous membranes, and competition from normal flora.

genitourinary system **nephrons**
glomerulus **urinalysis**
ovarian follicles **endometrium**
Bartholin glands
seminal vesicles **prostate gland**
semen

Lymphatic System

- The **lymphatic system** consists of a network of **lymphatic vessels, lymph nodes** and **nodules, thymus gland, spleen,** and **lymph.**
- All lymphatic tissues that filter blood and lymph are susceptible to infection by pathogens they filter when the pathogens overwhelm defenses.
- Nonspecific defenses consist of the actions of phagocytic cells.

lymph capillaries **sinuses**
gut-associated lymphatic tissue
thymosin **thymopoietin I**
thymopoietin II

PHAGOCYTOSIS

- A **phagocyte** is a cell that ingests and digests foreign substances.
- **Phagocytosis** is the process by which ingestion and digestion occurs.

Kinds of Phagocytic Cells

- Phagocytic cells include neutrophils in the blood and in injured tissues, monocytes in the blood, and fixed and wandering macrophages.

reticuloendothelial system
microphages

The Process of Phagocytosis

- The process of phagocytosis occurs as follows: (1) Invading microorganisms are located by chemotaxis, which is aided by the release of **lymphokines** by T lymphocytes. (2) Ingestion occurs as the phagocyte surrounds and incorporates a microbe or other foreign substance into a **phagosome** (vacuole). (3) Digestion occurs as lysosomes surround a vacuole and release their enzymes into it. Enzymes break down the contents of the vacuole and produce substances toxic to microbes.
- Microbes resist phagocytosis by producing capsules, preventing release of lysosomal enzymes, and producing toxins.

phagolysosome **degranulation**
leukocidin **streptolysin**

INFLAMMATION

Definition and Characteristics of Inflammation

- **Inflammation** is the body's response to tissue damage, characterized by increased temperature of the tissue, redness, swelling, and pain.

The Acute Inflammatory Process

- Inflammation is initiated by **histamine** released by damaged tissues, which dilates and increases permeability of blood vessels. Activation of **kinins** also contributes to initiation of inflammation.
- Dilation of blood vessels accounts for redness and increased tissue temperature; increased permeability accounts for swelling.
- Tissue injury also initiates the blood-clotting mechanism.
- **Bradykinin** stimulates pain receptors; **prostaglandins** intensify its effect.
- Inflamed tissues also release **leukocytosis-promoting factor,** which promotes release of leukocytes from blood-forming tissues and their migration to the site of injury.
- Leukocytes and macrophages phagocytize microbes and tissue debris.

Repair and Regeneration

- Repair and regeneration occur as capillaries grow into the site of injury and **fibroblasts** replace the dissolving blood clot.
- The resulting **granulation tissue** is strengthened by connective tissue fibers (from fibroblasts) and the overgrowth of epithelial cells.

Chronic Inflammation

- **Chronic inflammation** is a persistent inflammation in which the inflammatory agent continues to cause tissue injury, and host defenses fail to completely overcome the agent.
- **Granulomatous inflammation** is a chronic inflammation in which monocytes, histiocytes, lymphocytes, and plasma cells surround necrotic tissue to form a **granuloma.**

RELATED KEY TERMS

acute inflammation
antihistamine

scab

diapedesis **pus** **abscess**

FEVER

- **Fever** is an increase in body temperature caused by **pyrogens** increasing the setpoint of the temperature-regulating center in the hypothalamus.
- **Exogenous pyrogens** come from outside the body (usually pathogens), and **endogenous pyrogens** come from inside the body.
- Fever augments the immune response and inhibits the growth of microorganisms by lowering plasma iron. It also increases the rate of chemical reactions, raises the temperature above the optimum growth rate for some pathogens, and makes the patient feel ill (thereby lowering activity).
- Antipyretics are recommended only for high fevers and for patients with disorders that would be exacerbated by fever.

MOLECULAR DEFENSES

Interferon

- **Interferons** are proteins that act nonspecifically to cause cells to produce **antiviral protein.**
- Interferon can be made by recombinant DNA technology and has proven to be therapeutic for certain malignancies; other therapeutic applications are being studied.

Complement

- **Complement** refers to a set of blood proteins that when activated produce a **cascade** reaction. The **complement system** can be activated by the **classic complement pathway** or the **properdin pathway.**
- Action of the complement system is rapid and nonspecific. It promotes **opsonization,** membrane attack and cell lysis, and the inflammatory response.
- Deficiencies in complement reduce resistance to infection.

properdin
alternate pathway **opsonins**
immune adherence
membrane attack complex
immune cytolysis

Acute Phase Response

- Acutely ill patients increase production of certain blood proteins, including **C-reactive protein (CRP),** which is distinct from the inflammatory response and acts early, before antibodies can be made.
- CRP can initiate or accelerate inflammation, activate complement, stimulate migration of phagocytes, and initiate platelet aggregation. No human beings lack CRP.

QUESTIONS FOR REVIEW

A.

1. What are the differences between nonspecific and specific defenses?

B.

2. What are the major components of each of the human body systems?
3. What sites in each body system are most vulnerable to infection?
4. What nonspecific defenses operate in each of the body systems?

C.

5. Define phagocyte and phagocytosis.
6. Which kinds of phagocytes are found in blood and which in tissues?
7. What are the major steps in phagocytosis, and what happens in each step?
8. How do microbes resist phagocytosis?

D.

9. What is inflammation, and what are its characteristics?

E.

10. How is the inflammatory process initiated?
11. What are the effects of dilation of blood vessels? Of increased permeability of blood vessels?
12. How is blood clotting initiated?
13. How is pain related to the inflammatory response?
14. What is the role of leukocytosis-promoting factor in inflammation?
15. How are microbes and tissue debris disposed of during inflammation?

F.

16. What are the first events in repair and regeneration?
17. How is healing completed?

G.

18. What are the distinguishing characteristics of chronic inflammation and granulomatous inflammation?

H.

19. What is fever?
20. What are the roles of exogenous and endogenous pyrogens in fever?
21. What are the benefits of fever?
22. When are antipyretics recommended?

I.

23. What is interferon, and how does it work?
24. What is the therapeutic value of interferon?
25. What is the complement system, and how is it activated?
26. What are the defensive values of complement?
27. What are the effects of complement deficiencies?
28. What are the effects of C-reactive protein?

PROBLEMS FOR INVESTIGATION

1. Recall the last time you suffered from an infectious disease. In which ways do you think your body's nonspecific defenses failed you? Be specific.
2. Patients with cystic fibrosis (a genetic disorder) have very thick mucus and poor mucociliary escalator function. Which types of infections would you expect to find frequently in these patients, and why?
3. Make flow charts to show how nonspecific defenses combat a pathogen that enters the body through (a) a small cut, (b) the upper respiratory tract, (c) the digestive tract, and (d) the urogenital tract.
4. List some ways nonspecific defenses combat a pathogen that has evaded all the defenses considered in Problem 3.
5. Prepare a research paper on the current status of one of the following: (a) therapeutic uses of interferon, (b) effects of complement deficiencies, or (c) the mechanism of action of C-reactive protein.

SOME INTERESTING READING

Anonymous. 1991. Alpha-interferon therapy for chronic hepatitis B. *American Family Physician* 44, no. 3 (September):1027–29.

Atkins, E. 1984. Fever: The old and the new. *Journal of Infectious Disease* 149:339.

Demling, R. H. 1985. Burns. *New England Journal of Medicine* 313 (22):1389.

Edelson, R. L., and J. M. Fink. 1985. The immunologic function of skin. *Scientific American* 252, no. 6 (June):46.

Figueroa, J. E. and P. Densen. 1991. Infectious diseases associated with complement deficiencies. *Clinical Microbiological Reviews* 4 (3):359–95.

Finter, N. B. 1986. The classification and biological functions of interferons. *Journal of Hepatology* 3 (Suppl. 2):S157.

Ho, M. 1987. Interferon for the treatment of infections. *Annual Review of Medicine* 38:51.

Horwitz, M. A. 1982. Phagocytosis of microorganisms. *Review of Infectious Diseases* 4:104.

Joiner, K. A., et al. 1984. Complement and bacteria: Chemistry and biology in host defense. *Annual Review of Immunology* 2:461.

Kotwal, G. J., S. N. Isaacs, R. McKenzie, M. M. Frank, and B. Moss. 1990. Inhibition of the complement cascade by the major secretory protein of vaccinia virus. *Science* 250, no. 4982 (November 9):827–30.

Old, L. J. 1988. Tumor necrosis factor. *Scientific American* 258, no. 5 (May):59.

Reyes-Flores, O. 1986. Granulomas induced by living agents. *International Journal of Dermatology* 25 (3):158.

Youmans, G. P., P. Y. Paterson, and H. M. Sommers. 1985. *The biological and clinical basis of infectious diseases.* Philadelphia: W. B. Saunders.

Scientist working with a flask of mammalian hybridoma cells as part of the process of commercial production of monoclonal antibodies. The initial stage involves the fusion of a normal antibody-producing cell (a lymphocyte) with a rapidly multiplying tumor cell. The resulting hybridoma cells are multiplied in fermenters to produce large numbers of genetically identical copies, each secreting the same antibody produced by the original lymphocyte. In this way, large quantities of specific antibodies can be produced for use in medicine as vaccines, in treatment, and in diagnostic tests.

18 Immunology I: Basic Principles of Specific Immunity and Immunization

This chapter focuses on the following questions:

A. What do immune, immunity, susceptibility, nonspecific immunity, specific immunity, immunology, and immune system mean?

B. How do innate immunity, acquired immunity, and active and passive immunity differ?

C. What are the properties of antigens and antibodies, and how do cells and tissues function in the dual roles of the immune system?

D. How do recognition of self, specificity, heterogeneity, and memory function in the immune system?

E. How do B cells and antibodies function in humoral immunity?

F. What are monoclonal antibodies, and how are they made and used?

G. How does cell-mediated immunity differ from humoral immunity, and how do reactions of cell-mediated immunity occur?

H. What are the special roles of killer cells and activated macrophages?

I. What factors modify immune responses?

J. What are the mechanisms of immunization, recommended immunizations, and the benefits and hazards of immunization?

When you were very young you probably received a variety of immunizations against diphtheria, tetanus, whooping cough, poliomyelitis, and maybe measles, German measles, and mumps as well. Your parents probably became immune to both kinds of measles and to mumps by having the diseases. Either being immunized or having a disease can confer specific immunity to that disease. As we saw in the last chapter, nonspecific host defenses protect against infections in a general way. In this chapter we will see how specific immunity and immunization protect against particular diseases.

IMMUNOLOGY AND IMMUNITY

The word *immune* literally means "free from burden." Used in a general sense, **immunity** refers to the ability of an organism to recognize and defend itself against infectious agents. *Susceptibility,* the opposite of immunity, is the vulnerability of the host to harm by infectious agents.

As we saw in the last chapter, host organisms have many general defenses against invading infectious organisms, regardless of what type of organism invades. (Chapter 17, p. 444) Immunity produced by such defenses is called **nonspecific immunity.** In contrast, **specific immunity** is the ability of a host to mount a defense against particular infectious agents by physiological responses *specific to that infectious agent.*

Immunology is the study of specific immunity and how the immune system responds to specific infectious agents. The **immune system** consists of various cells, especially lymphocytes, and organs such as the thymus gland, that help provide the host with specific immunity to infectious agents.

KINDS OF IMMUNITY

Innate Immunity

Innate immunity, also called **genetic immunity,** exists because of genetically determined characteristics. One kind of innate immunity is **species immunity,** which is common to all members of a species. For example, all humans have immunity to many infectious agents that cause disease in pets and domestic animals, and animals have similar immunity to some human diseases. Humans do not have the appropriate receptor sites and will not therefore become infected with canine distemper no matter how much contact they have with infected puppies. *Mycobacterium avium* causes tuberculosis in birds, but rarely in humans with normal immune systems. (It does often infect AIDS patients.) Some diseases appear only in a few species. Gonococci infect humans and monkeys but usually not other species. *Bacillis anthracis* causes anthrax in all mammals and some birds but not in many other animals.

Acquired Immunity

In contrast to innate immunity, **acquired immunity** is immunity obtained in some manner other than by heredity. It can be naturally acquired or artificially acquired. **Naturally acquired immunity** is most often obtained through having a specific disease. During the course of the disease the immune system responds to molecules called *antigens* on invading infectious agents. It produces molecules called *antibodies* and initiates other specific defenses that protect against future invasions by the same agent. Immunity also can be naturally acquired from antibodies transferred to a fetus across the placenta or to an infant in colostrum and breast milk. **Colostrum** is the first fluid secreted by the mammary glands after childbirth. Although deficient in many nutrients found in milk, colostrum contains large quantities of antibodies that cross the intestinal mucosa and enter the infant's blood.

In contrast, **artificially acquired immunity** is obtained by receiving an antigen by the injection of vaccine or immune serum that produces immunity. Sticking needles full of vaccine or serum into people is not a natural process. Thus, the immunity produced is artificially acquired.

Active and Passive Immunity

Regardless of whether immunity is naturally or artificially acquired, it can be active or passive. **Active immunity** is created when the person's own immune system produces antibodies or other defenses against an infectious agent. It can last a lifetime or for a period of weeks, months, or years, depending on how long the antibodies persist. **Naturally acquired active immunity** is produced when a person is exposed to an infectious agent. **Artificially acquired active immunity** is produced when a person is exposed to a vaccine containing live, weakened, or dead organisms or their toxins. In both types of active immunity the host's own immune system responds specifically to defend the body against an antigen. Furthermore, the immune system generally "remembers" the antigen to which it has responded and will mount another response any time it again encounters the same antigen.

Passive immunity is created when ready-made antibodies are introduced into the body. This immunity is passive because the host's own immune system does not make antibodies. **Naturally acquired passive immunity** is produced when antibodies made by a mother's immune system are transferred to her offspring. New mothers are encouraged to breast-feed for a few days even if they are not planning to continue so that their infants obtain antibodies from colostrum. **Artificially acquired passive immunity** is produced when antibodies made by other hosts are introduced into a new host. In this kind of immunity the host's immune system is not stimulated to respond. Ready-made antibodies and the immunity they confer persist for a few weeks to a few months and are destroyed by the host; the host's immune system cannot make new ones.

Relationships among the various types of immunity are shown in Figure 18.1. The properties of each type of immunity are summarized in Table 18.1.

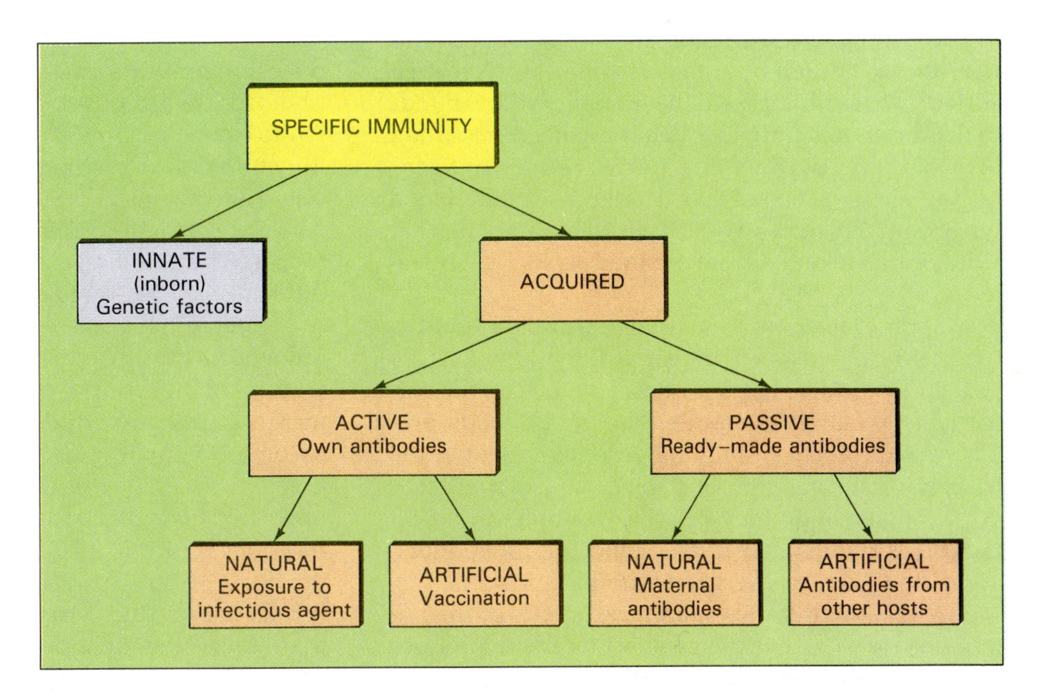

FIGURE 18.1 The various types of immunity.

TABLE 18.1 Characteristics of types of immunity

		Kind of Immunity	
Characteristic	**Innate**	**Actively Acquired**	**Passively Acquired**
Agent	Genetic and physiological factors	Antibodies elicited by antigens	Ready-made antibodies
Source of antibodies	None	Person immunized	Plasma of others, such as mother
How elicited	Genetic expression	**Natural:** by having disease **Artificial:** by receiving vaccine	**Natural:** by receiving antibodies across placenta or in colostrum **Artificial:** by receiving injection of gamma globulin or immune serum
Time to develop immunity	Always present	5 to 14 days after receiving antigen	Immediately after receiving antibodies
Duration of immunity	Lifetime	Months to lifetime	Days to weeks

CHARACTERISTICS OF THE IMMUNE SYSTEM

Antigens and Antibodies

Actions of the immune system are triggered by antigens. An **antigen** is a substance the body identifies as foreign and toward which it mounts an immune response. Most antigens are large protein molecules with complex structures and molecular weights greater than 10,000. Some antigens are polysaccharides, and a few are glycoproteins (carbohydrate and protein) or nucleoproteins (nucleic acid and protein). Proteins usually have greater antigenic strength because they have a more complex structure than polysaccharides. Large, complex proteins can have several **antigenic determinants,** or **epitopes,** areas on the molecule to which antibodies can bind.

Antigens are found on the surfaces of viruses and cells, including bacteria, other microorganisms, and human cells. The exact chemical structure of each of a cell's antigens is determined by genetic information in its DNA. Bacteria can have antigens on capsules, cell walls, and even flagella. Many microorganisms have several different antigens somewhere on their surface. Determining how the human body responds to these different antigenic determinants is important in making effective vaccines. As we shall see, antigens on the surfaces of red blood cells determine blood types, and antigens on other cells determine whether a tissue transplanted from another person will be rejected.

In some instances a small molecule called a **hapten** (hap'ten) can act as an antigen if it binds to a larger protein molecule. Haptens act as epitopes on the surfaces of proteins. Sometimes they bind to body proteins and provoke an immune response. Neither the hapten nor the body protein alone acts as an antigen, but in combination they can. For example, penicillin molecules can act as haptens, bind to protein molecules, and elicit an allergic reaction.

One of the most significant responses of the immune system to any foreign substance is to produce antiantigen proteins, or antibodies. An **antibody** is a protein produced in response to an antigen that is capable of binding specifically to the antigen. Each kind of antibody binds to a specific antigenic determinant. Such binding may or may not contribute to inactivation of the antigen. A typical antigen-antibody reaction is shown diagrammatically in Figure 18.2.

In discussing concentrations of antigens and antibodies, immunologists often refer to titers. A **titer** (ti'ter) is the quantity of a substance needed to produce a given reaction. For example, an antibody titer is the quantity required to bind to and neutralize a particular quantity of an antigen.

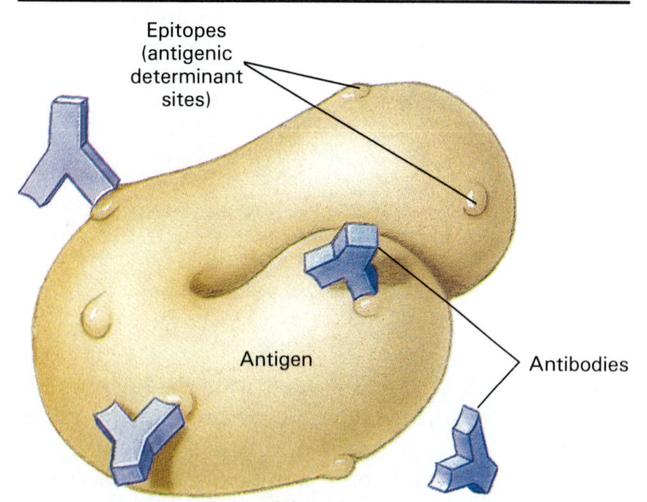

FIGURE 18.2 A typical antigen-antibody reaction. Antibodies bind to specific chemical groups or structures, called epitopes, or antigenic determinants, on the antigen. Large, complex protein molecules may have many epitopes.

Cells and Tissues of the Immune System

Specific immune responses are carried out by lymphocytes, which develop from stem cells as do other white blood cells, red blood cells, and platelets. Early in embryonic development, undifferentiated stem cells from regions in the yolk sac called primitive blood islands proliferate. Later, they migrate into the body through the umbilical cord to various sites where they differentiate into specific types of lymphocytes.

Differentiation of stem cells into lymphocytes is influenced by other tissues of the immune system (Figure 18.3). Lymphocytes that are processed and mature in tissue referred to as bursal-equivalent tissue become **B lymphocytes,** or **B cells.** Differentiation of B cells was first observed in birds, where they are processed in an organ called the *bursa of Fabricius* (Fabris'e-us) (Figure 18.4). Although no site equivalent to the bursa of Fabricius has been identified in humans, B cells are produced in humans. Most investigators think this differentiation takes place in bone marrow or in gut-associated lymphoid tissues (*GALT*)—lymphoid tissues in the digestive tract, including the appendix and Peyer's patches of the small intestine. Regardless of where they differentiate, functional B cells are found in all lymphoid tissues—lymph nodes, spleen, tonsils, adenoids, and gut-associated lymphoid tissues. B cells account for about one-fourth of the lymphocytes circulating in the blood.

Other stem cells migrate to the thymus, where they undergo differentiation into thymus-derived cells called **T lymphocytes,** or **T cells.** In adulthood, when the thymus becomes less active, differentiation of T cells is thought to occur in bone marrow or tissues under the influence of hormones from the thymus. ∞ (Chapter 17, p. 457) T cells are found in all tissues that contain B cells and account for about three-fourths of the lymphocytes circulating in the blood. The distribution of B and T cells in lymphatic tissues is summarized in Table 18.2. Subsequent differentiation of T cells produces four different kinds of cells: (1) cytotoxic (killer) T cells, (2) delayed-hypersensitivity T cells, (3) helper T cells, and (4) suppressor T cells. After differentiation these T cells migrate among lymphatic tissues and the blood.

A few lymphocytes that cannot be identified as either B cells or T cells are found in tissues and circulating in blood. These undifferentiated cells are called **null cells** and include the so-called natural killer cells, discussed later in this chapter.

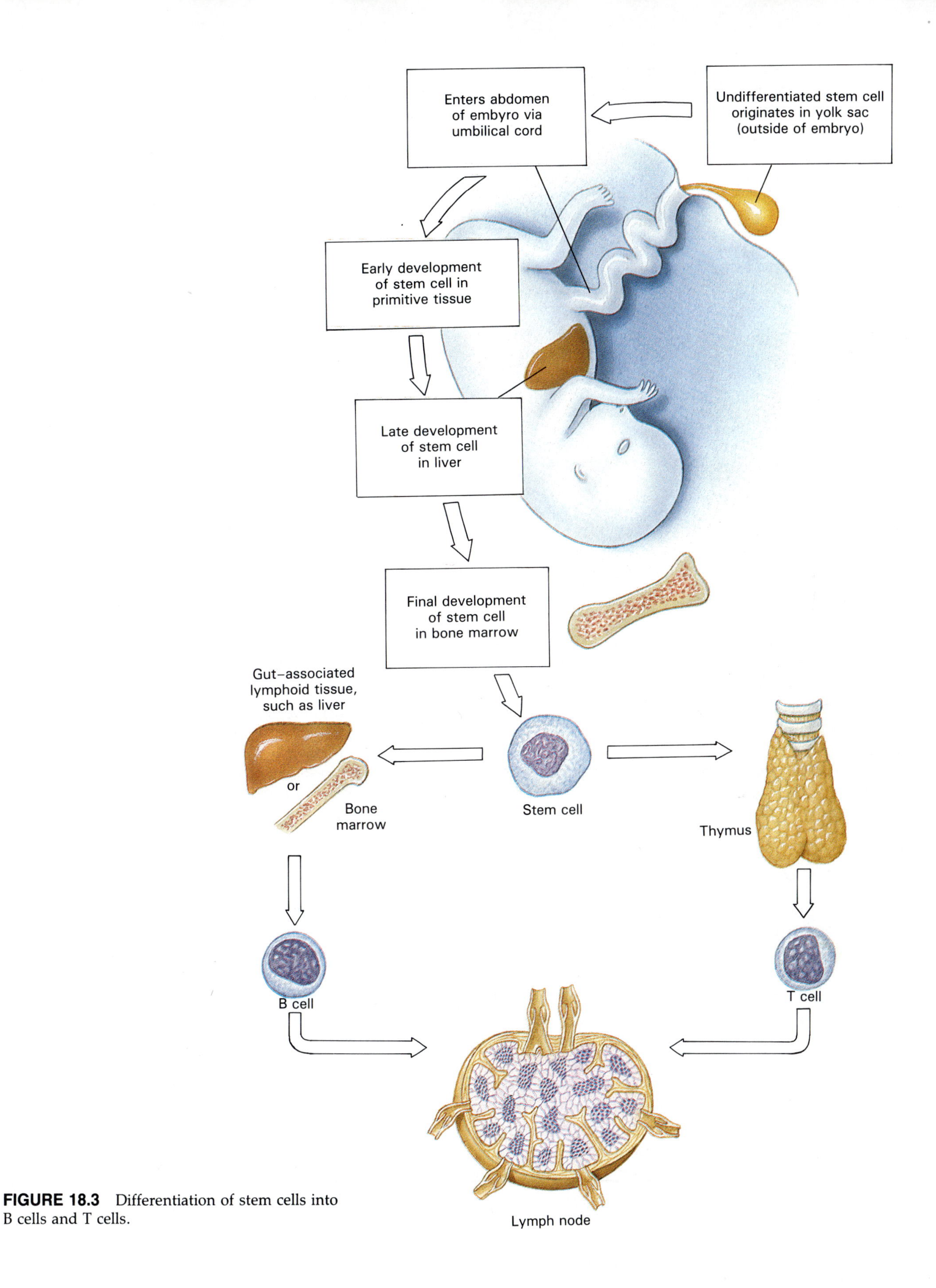

FIGURE 18.3 Differentiation of stem cells into B cells and T cells.

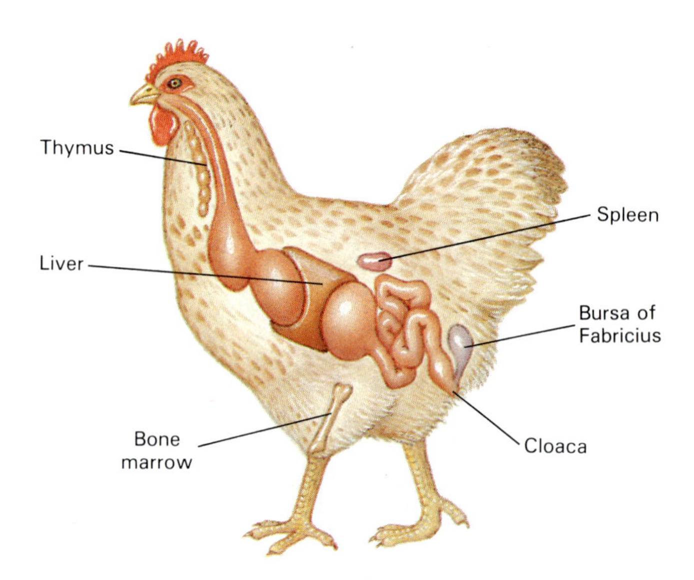

FIGURE 18.4 In chickens, the bursa of Fabricius, where B cells develop, is a pouch located off the cloaca, a chamber into which waste and reproductive materials empty. (Some other organs of importance to the immune system are also shown.)

TABLE 18.2 Proportions of B and T lymphocytes in human lymphoid tissues[a]

Lymphoid Tissue	% B Cells	% T Cells
Peyer's patches and nodules in digestive tract	60	25
Spleen	45	45
Lymph nodes	20	70
Blood	25	75
Thymus	1	99

[a] Where percentages do not add to 100, some lymphocytes are undifferentiated. Based on data from E. J. Moticka. In R. F. Boyd and J. J. Marr, (eds.), *Medical Microbiology*, New York: Little, Brown & Co., 1980.

Dual Nature of the Immune System

Lymphocytes give rise to two major types of immune responses, humoral immunity and cell-mediated immunity. However, the presence of a foreign substance in the body often triggers both kinds of responses.

Humoral (hu'mor-al) **immunity** is carried out by antibodies circulating in the blood. When stimulated by an antigen, B lymphocytes initiate a process that leads to the release of antibodies. Humoral immunity is most effective in defending the body against bacterial toxins, bacteria, and viruses before these agents enter cells.

Cell-mediated immunity is carried out by T cells. It occurs at the cellular level, especially in situations where antigens are embedded in cell membranes or are inside host cells and thus inaccessible to antibodies. It is most effective in clearing the body of virus-infected cells, but it also may participate in defending against fungi and other eukaryotic parasites, cancer, and foreign tissues, such as transplanted organs.

Humoral Immunity: What's In A Name?

Where did the name *humoral immunity* come from? The term obviously is derived from the word *humor,* which originally referred to the four basic body fluids—blood, phlegm, yellow bile, and black bile—that ancient physicians believed must be present in proper proportions for an individual to enjoy good health. If any of these fluids were out of balance, a person was said to be "in bad humor" and likely to be diseased. Because this particular type of immunity involves antibodies that circulate in the blood, the phrase *humoral immunity* seemed logical.

General Properties of Immune Responses

Both humoral and cell-mediated responses have certain common attributes that enable them to confer immunity: (1) recognition of self versus nonself, (2) specificity, (3) heterogeneity, and (4) memory. We will look at each in some detail.

Recognition of Self versus Nonself

For the immune system to respond to foreign substances, it must distinguish between host tissues and substances that are foreign to the host. Immunologists refer to normal host substances as **self** and foreign substances as **nonself.** The **clonal** (klo'nal) **selection hypothesis** (Figure 18.5), first proposed by Frank Macfarlane Burnet in the 1950s, explains one way the immune system might distinguish self from nonself. According to this hypothesis, embryos contain many different lymphocytes, each genetically programmed to recognize a particular antigen and make antibodies to destroy it. If a lymphocyte encounters and recognizes that antigen after development is complete, it divides repeatedly to produce a clone, a group of identical progeny cells that make the same antibody. If, during embryological development, it encounters its programmed antigen as part of a normal host substance (self), the lymphocyte is somehow destroyed or inactivated. This mechanism removes lymphocytes that can destroy host tissues and thereby creates **tolerance** for self. It also selects for survival lymphocytes that will protect the host from foreign antigens.

Tolerance also can be acquired by irradiation dur-

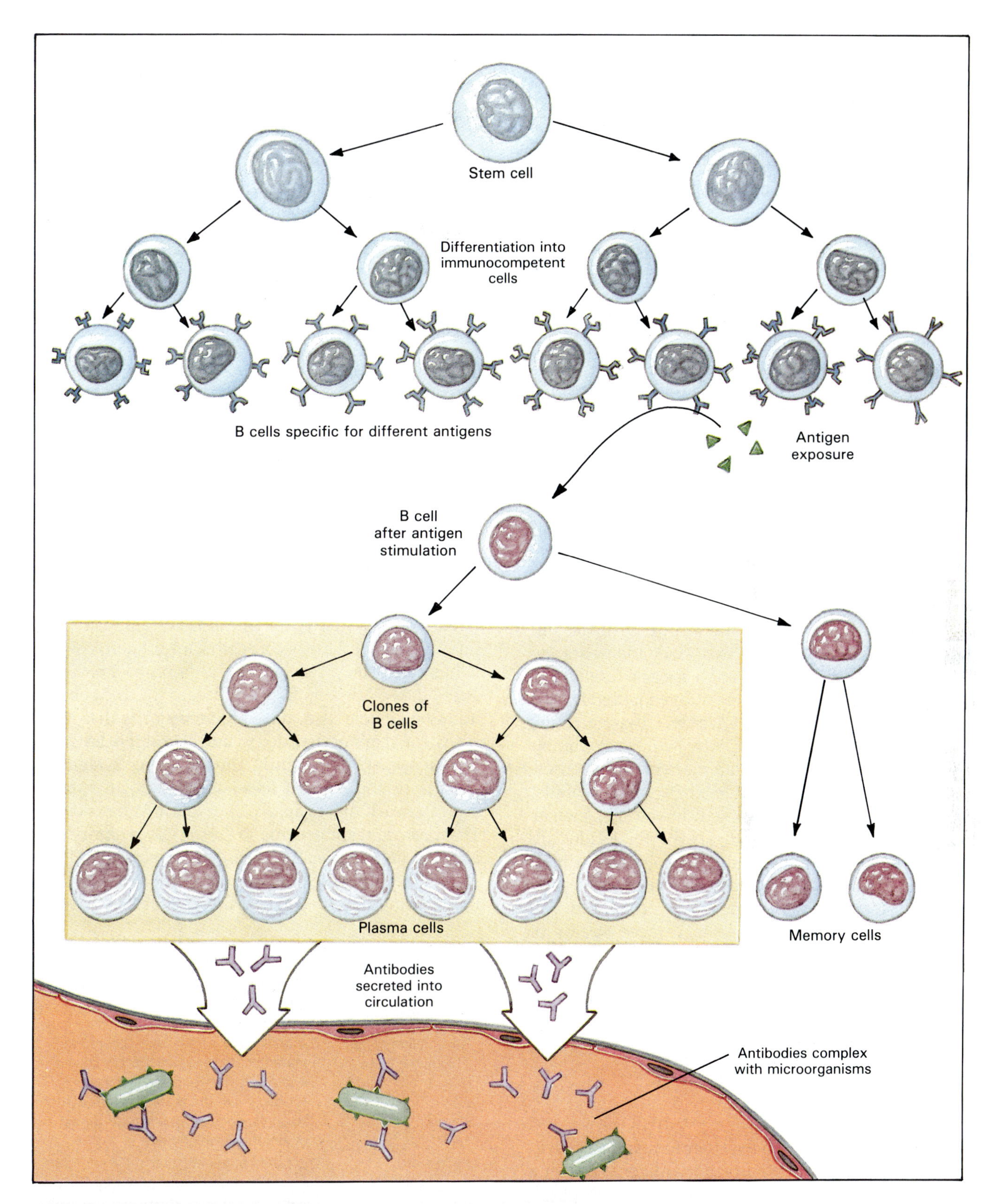

FIGURE 18.5 According to the clonal selection theory, one of many B cells responds to a particular antigen and begins to divide, thereby producing a large population of identical B cells (a clone). All cells of such a clone produce the same antibody against the original antigen. Other B cells that make different antibodies fail to respond.

CLOSE-UP

Immune System: Friend or Foe?

In a lethal viral meningitis in mice, the brain is covered with pus composed entirely of lymphocytes, and brain damage is due to the lymphocytes rather than to the virus. In mice infected with the virus before birth, the maturing immune system learns to recognize the virus as "self" and does not attack it. In the absence of an immune response, the virus invades all tissues but does no harm. However, if the mice subsequently receive transplants of normal lymphoid tissue, which has not acquired such tolerance, the virus elicits an immune response. Lymphocytes from the transplanted tissue then invade and damage the brain. (We will encounter other instances of diseases caused by the body's defense rather than the invading organism in the next chapter.)

ing cancer treatment or the administration of immunosuppressant drugs to prevent rejection of transplanted organs. The host loses the ability to detect and respond to foreign antigens in transplanted organs, but then also fails to respond to infectious organisms.

Specificity

By the time the immune system matures at age 2 to 3, it can recognize a vast number of foreign substances as nonself. Furthermore, it reacts in a different way to each foreign substance. This property of the immune system is called **specificity.** Due to specificity, each reaction is directed toward a specific foreign antigen, and the response to one antigen generally has no effect on other antigens. However, **cross-reactions,** reactions of a particular antibody with very similar antigens, can occur. For example, certain microorganisms, such as the bacterium that causes syphilis, have the same haptens as some human cells, such as heart muscle cells, although the carrier molecules are quite different. This allows antibodies against this particular hapten to react with these otherwise vastly different cells. Cross-reactions also occur between strains of bacteria. For example, if three strains of pneumococci can cause pneumonia, and if each produces a particular antigen, A, B, or C, a person who has recovered from an infection with strain A has anti-A antibodies. The person then may also have some resistance to strains B and C because anti-A antibodies cross (that is, they react with antigens B and C).

Heterogeneity

The ability of the immune system to respond specifically allows it to attack particular antigens. But in a lifetime, the human body encounters hundreds of different foreign antigens. The property of **heterogeneity** (versatility, or flexibility) refers to the ability of the immune system to produce many different kinds of antibodies, each of which reacts with a different epitope (antigenic determinant). When a bacterium or other foreign agent has more than one kind of antigenic determinant, the immune system may make a different antibody against each. And it is capable of producing antibodies even against foreign substances, such as newly synthesized molecules never before encountered by any immune system.

Memory

In addition to its ability to respond specifically to a heterogeneous assortment of antigens, the immune system also has the property of **memory**—that is, it can recognize substances it has previously encountered. Memory allows the immune system to respond rapidly to defend the body against an antigen to which it has previously reacted. In addition to producing antibodies during its first reaction to the antigen, the immune system also makes **memory cells** that stand ready for years or decades to quickly initiate antibody production. Consequently, the immune system responds to second and subsequent exposures to an antigen much more rapidly than to the first exposure. This prompt response due to "recall" by memory cells is called an **anamnestic** (secondary) **response** (Figure 18.6). The attributes of specific immunity are summarized in Table 18.3. With these attributes in mind

TABLE 18.3 Main attributes of specific immunity

Attribute	Description
Recognition of self versus nonself	The ability of the immune system to tolerate host tissues while recognizing and destroying foreign substances. Probably due to the destruction (deletion) of clones of lymphocytes during embryonic development.
Specificity	The ability of the immune system to react in a different and particular way to each foreign substance.
Heterogeneity	The ability of the immune system to respond in a specific way to a great variety of different foreign substances.
Memory	The ability of the immune system to recognize and quickly respond to foreign substances to which it has previously responded.

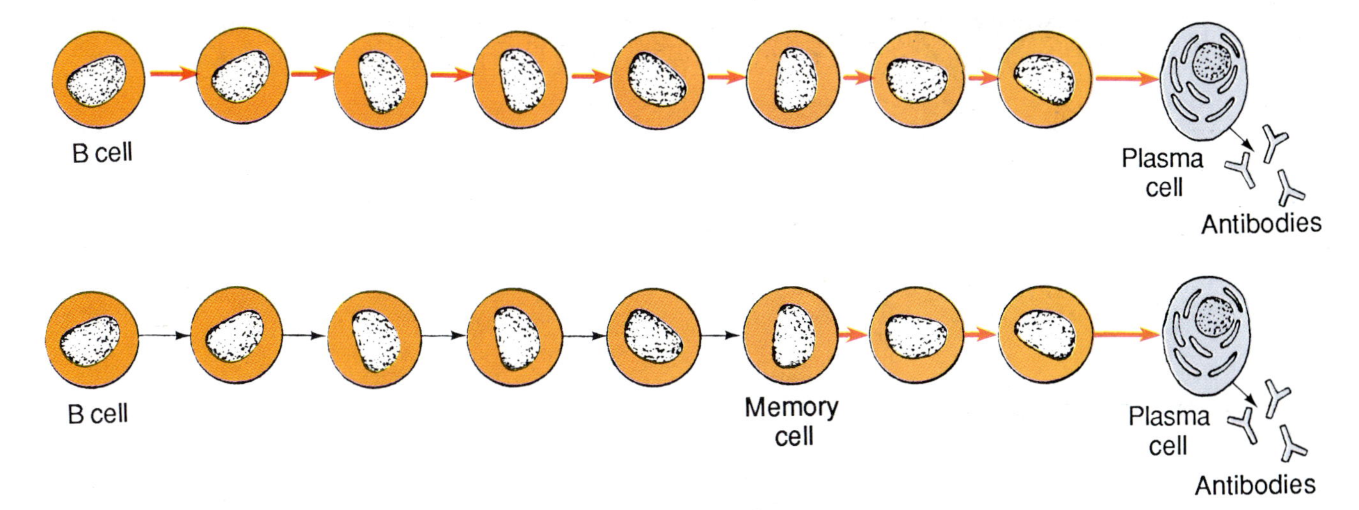

FIGURE 18.6 Through a series of cell divisions, B cells give rise to plasma cells, which produce antibodies. Memory cells need only a few divisions to generate plasma cells. They are thus able to bring about the production of protective levels of antibodies much more quickly than undifferentiated B cells. (Only one progeny cell is shown for each division.)

we will now look in more detail at the two kinds of specific immunity, humoral and cell-mediated.

HUMORAL IMMUNITY

Humoral immunity depends first on the ability of B lymphocytes to recognize specific antigens and second on their ability to initiate responses that protect the body against foreign agents. In most instances the antigens are on the surfaces of infectious organisms. The most common response is the production of antibodies that will inactivate an antigen and lead to destruction of infectious organisms.

Each kind of B cell carries its specific antibody on its membrane and can bind immediately to a specific antigen. The binding of an antigen **sensitizes,** or activates, the B cell and causes it to divide many times. Some of the progeny are memory cells, but most are plasma cells. **Plasma cells** are large lymphocytes that synthesize and release many antibodies like those on their membranes. While it is active a single plasma cell can produce as many as 2000 antibodies per second!

The responses of B cells can be influenced by certain T cells. When B cells encounter very large antigen molecules with many epitopes, they appear to generate plasma cells and antibodies without the aid of T cells. However, most antigens found on infectious agents have relatively few epitopes. In such instances helper T cells assist B cells by "processing" antigens (see the following discussion) so that B cells can respond to them. Macrophages also present antigens to B cells. Suppressor T cells can block activity of helper T cells and so may be important in terminating an antigen-antibody reaction. How T cells carry out these functions will be explained later. Current evidence suggests that helper and suppressor T cells secrete helper and suppressor factors, respectively. Their effect on B cells seems to depend on the relative quantities of the two factors present in the region of an immune reaction. In a typical immune reaction, helper T cells facilitate growth and differentiation of plasma cells. After about a week this reaction reaches a peak and then subsides, largely because suppressor T cells inhibit further antibody production.

Properties of Antibodies (Immunoglobulins)

Antibodies, or **immunoglobulins** (Ig), are Y-shaped protein molecules composed of four polypeptide chains—two identical **light (L) chains** and two identical **heavy (H) chains** (Figure 18.7). The chains, which are held together by disulfide bonds, have constant regions and variable regions. The chemical structure of the *constant regions* determines the particular class that an immunoglobulin belongs to, as described next. The *variable regions* of each chain have a particular shape and charge that enable the molecule to bind a particular antigen. Each of the millions of different immunoglobulins has its own unique pair of identical antigen-binding sites formed from the variable regions at the ends of the L and H chains. These binding sites are identical to the receptors in the membrane of the parent B cell. In fact, the first immunoglobulins made by B cells are inserted into their membranes to form the receptors. When the B cells form plasma cells, they continue to make the same immunoglobulins. When an antibody is cleaved with the enzyme papain at the hinge region, two *Fab* (antibody binding fragment) pieces and one *Fc* (crystallizable fragment) piece result. The Fab fragment binds to the epitope. The Fc region formed by parts of the H chains in the tail of

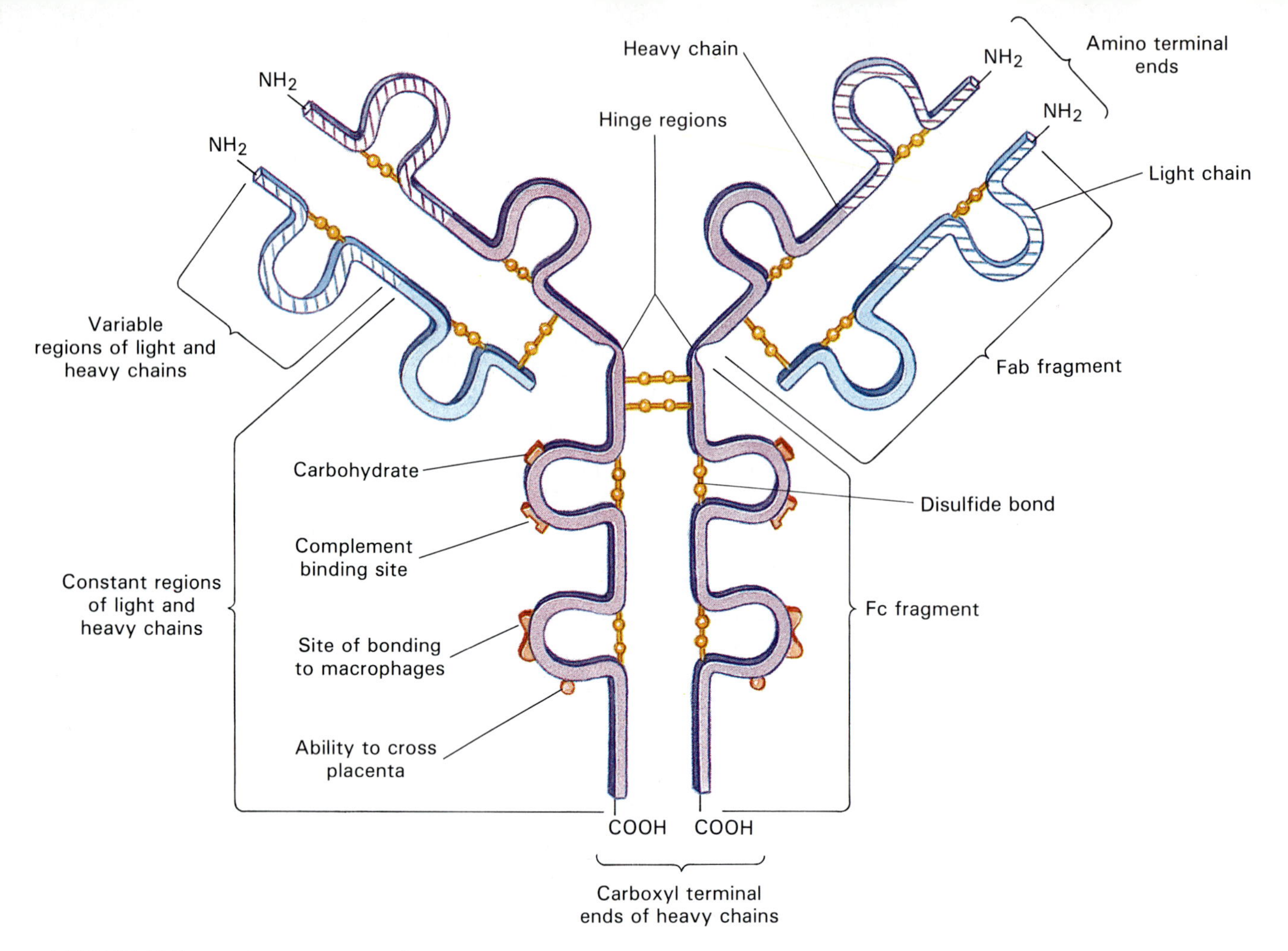

FIGURE 18.7 The basic structure of an antibody molecule comprises two heavy (H) and two light (L) chains, joined by disulfide linkages (-S-S-) to form a "Y" shape. The upper ends of the Y consist of variable regions and so differ from antibody to antibody. These regions, which make up the site that combines with the antigen, are responsible for the specificity of the antibody. The remaining part of the molecule consists of constant regions that are the same in all antibodies of a particular class. These regions determine the role that the antibody plays in the body's immune response. When an antibody is cleaved with the enzyme papain at the hinge region, two Fab (antibody binding fragment) pieces and one Fc (crystallizable fragment) piece result.

the Y has a site that can bind to and activate complement, participate in allergic reactions, and combine with phagocytes in opsonization.

Five classes of immunoglobulins have been identified in humans and other higher vertebrates (Table 18.4). Each class has a particular kind of constant region, which gives that class its distinguishing properties. The five classes are identified as IgG, IgA, IgM, IgE, and IgD (Figure 18.8).

IgG, the main class of antibodies found in the

TABLE 18.4 Properties of immunoglobulins

	Class of Immunoglobulin				
Property	**IgG**	**IgM**	**IgA**	**IgE**	**IgD**
Number of units	1	5	1 or 2	1	1
Activation of complement	Yes	Yes, strongly	Yes, by properdin	No	Yes, by properdin
Crosses placenta	Yes	No	No	No	No
Binds to phagocytes	Yes	No	Yes	No	?
Binds to lymphocytes	Yes	Yes	Yes	Yes	?
Binds to mast cells and basophils	No	No	No	Yes	?
Average survival time (days)	23	5	6	1.5	2.8
Percent of total blood antibodies	75–85	5–10	5–15	0.5	0.2

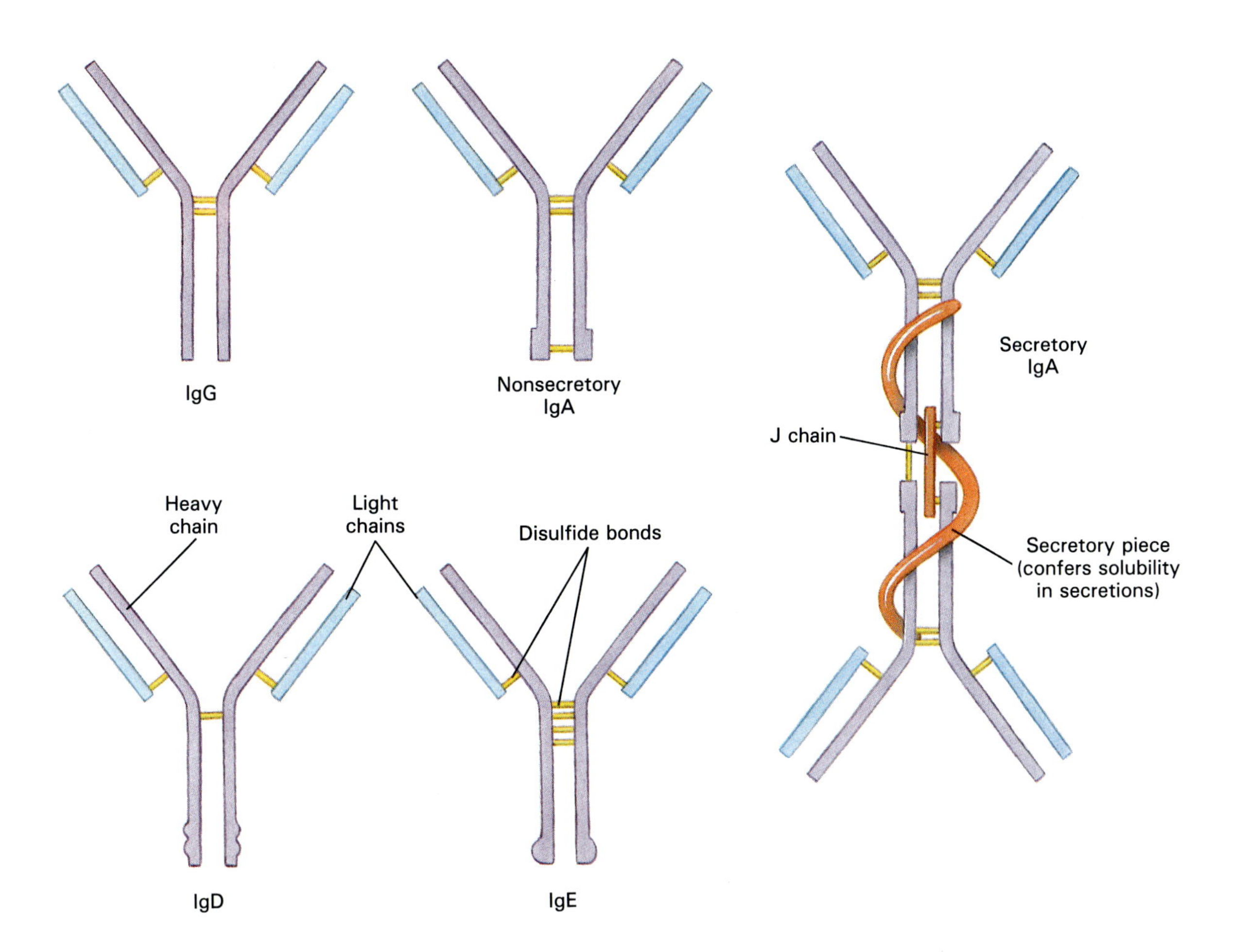

FIGURE 18.8 The structures of the different classes of antibodies.

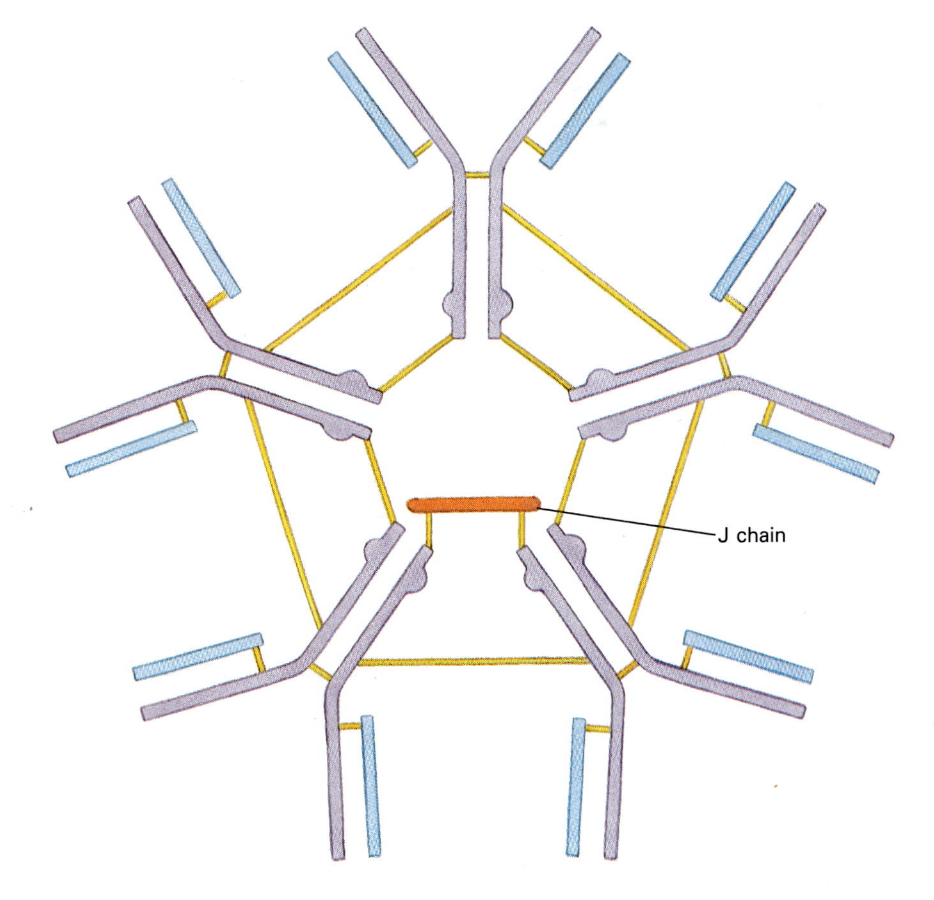

blood, accounts for as much as 20 percent of all plasma proteins. IgG is produced in largest quantities during a secondary response. The antigen-binding sites of IgG attach to antigens on microorganisms, and their tissue-binding sites attach to receptors on phagocytic cells. Thus, as a microorganism is surrounded by IgG, a phagocytic cell is brought into position to engulf the organism. The tail section of the H chains also activates complement. Complement, as explained in Chapter 17, consists of proteins that lyse microorganisms and attract and stimulate phagocytes. ∞ (p. 465) IgG is the only immunoglobulin that can cross the placenta from mother to fetus and provide antibody protection for it.

IgA occurs in small amounts in blood, in larger amounts in body secretions such as tears, milk, saliva, and mucus and attached to the linings of the digestive, respiratory, and genitourinary systems. IgA is secreted into the blood, transported through epithelial cells that line these tracts, and either released in secretions or attached to linings by tissue-binding sites. Blood IgA consists of a single unit of two H and two L chains. Secretory IgA, which consists of two units held together by a J chain (joining chain), has an attached **secretory component,** which protects the IgA from proteolytic (protein-splitting) enzymes and facilitates its transport. The main function of IgA is to bind antigens on microorganisms before they invade tissues. It also activates complement, which helps to kill the microorganisms. IgA does not cross the placenta, but it is abundant in colostrum.

IgM, the first antibody secreted into the blood during the early stages of a primary immune response, can be made by both B cells and plasma cells. IgM consists of five units connected by their tails to a J chain and so has ten peripheral antigen-binding sites. IgM is found mainly on B cell membranes and is rarely secreted. As IgM binds to antigens it also activates complement and causes microorganisms to clump together. These actions probably account for the initial effects the immune system has on infectious agents. It is also the first antibody formed in life, being synthesized by the fetus. In addition it is the antibody of the inherited ABO blood types.

IgE (also called *reagin*) has a special affinity for receptors on the plasma membranes of basophils in the blood or mast cells in the tissues. It binds to these cells by tissue-binding sites, leaving antigen-binding sites free to bind antigens to which humans can develop allergies, such as drugs, pollens, and certain foods. When IgE binds antigens, the associated basophils or mast cells secrete various substances, such as histamine, which produces allergy symptoms. Levels of IgE are elevated in patients with allergies and in those harboring parasites. IgE is found mainly in body fluids and skin and is rare in blood.

Like IgM, **IgD** is found mainly on B-cell membranes and is rarely secreted. Although it can bind to antigens, its function is unknown. It may help initiate immune responses and some allergic reactions.

Primary and Secondary Responses

In humoral immunity the **primary response** to an antigen occurs when the antigen is first recognized by host B cells. After recognizing the antigen, B cells divide to form plasma cells, which begin to synthesize antibodies. In a few days antibodies begin to appear in the blood plasma, and they increase in concentration over a period of 1 to 10 weeks. The first antibodies are IgM, which can attack foreign substances directly. As IgM production wanes, IgG production accelerates. Eventually, it too wanes. The concentrations of both IgM and IgG can become so low as to be undetectable in plasma samples. However, memory cells persist in lymphoid tissues. They do not participate in the initial response, but they retain their ability to recognize a particular antigen. They can survive without dividing for many months to many years.

When an antigen recognized by memory cells enters the blood, a **secondary response** occurs. The presence of memory cells makes the secondary response much faster than the primary response. Some memory cells divide rapidly, producing plasma cells, and others remain as memory cells. Plasma cells quickly synthesize and release large quantities of antibodies. In the secondary response, as in the primary response, IgM is produced before IgG. However, IgM is produced in smaller quantities over a shorter period, and IgG is produced sooner and in much larger quantities than in the primary response. Thus, the secondary response is characterized by a rapid increase in antibodies, most of which are IgG. The primary and secondary responses are compared in Figure 18.9.

Kinds of Antigen-Antibody Reactions

The antigen-antibody reactions of humoral immunity are most useful in defending the body against bacterial infections, but they also neutralize toxins and viruses that have not yet invaded cells. The defensive capability of humoral immunity depends on recognizing antigens associated with pathogens.

For bacteria to colonize surfaces or for viruses to infect cells, these agents first must adhere to surfaces. IgA antibodies in tears, nasal secretions, saliva, and other fluids react with antigens on the microbes. They coat bacteria and viruses and prevent them from adhering to mucosal surfaces.

Microbes that escape IgA invade tissues and encounter IgE in lymph nodes and mucosal tissues. Gut-associated lymphoid tissue releases large quantities of IgE, which bind to mast cells; these cells then release histamine and other substances that initiate and accelerate the inflammatory process. Included in this

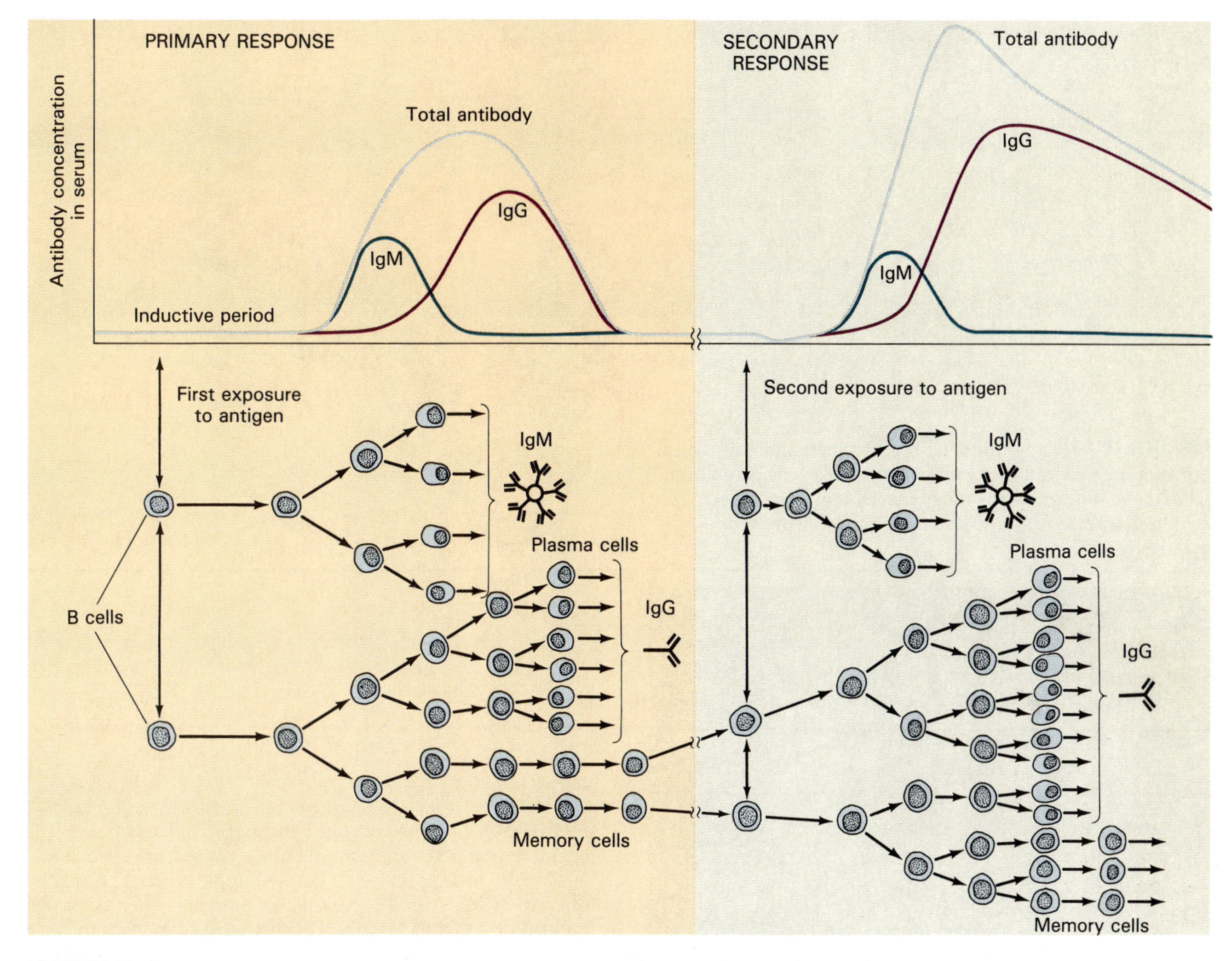

FIGURE 18.9 Primary and secondary responses to an antigen, showing the correlation of antibody concentrations with the activities of B cells.

process is the delivery of IgG and complement to the injured tissue.

Microbes that have reached lymphoid tissue without being recognized by B cells are acted on by macrophages and presented to B cells. B cells then bind the antigens and produce antibodies, usually with the aid of helper T cells. Antibodies binding with antigens on the surfaces of microbes form antigen-antibody complexes.

The formation of antigen-antibody complexes is an important component of the inactivation of infectious agents because it is the first step in removing such agents from the body. However, the means of inactivation varies according to the nature of the antigen and the kind of antibody with which it reacts. Inactivation can be accomplished by such processes as agglutination, opsonization, activation of complement, cell lysis, and neutralization. These reactions occur naturally in the body and can be made to occur in the laboratory. Here we will describe reactions chiefly as they relate to destruction of pathogens. We will discuss their laboratory applications more fully in Chapter 19.

Because bacterial cells are relatively large particles, the particles that result from antigen-antibody reactions also are large. Such reactions result in **agglutination** (ag-lu-tin-a'shun), or sticking together of microbes (Figure 18.10). IgM produces strong, and IgG produces weak, agglutination reactions with certain bacterial cells. Agglutination reactions produce results that are visible to the unaided eye and can be used as the basis of laboratory tests to detect the presence of antibodies or antigens (Figure 18.11). Some antibodies act as opsonins. (Chapter 17, p. 466) That is, they neutralize toxins and coat microbes so that they can be phagocytized, a process called opsonization.

Complement is an important component in inactivating infectious agents, as was discussed in Chapter 17. Both IgG and IgM are powerful activators of the complement system; IgA is less powerful. Sometimes, antibodies, especially IgM, directly lyse cell membranes of infectious agents without the aid of complement.

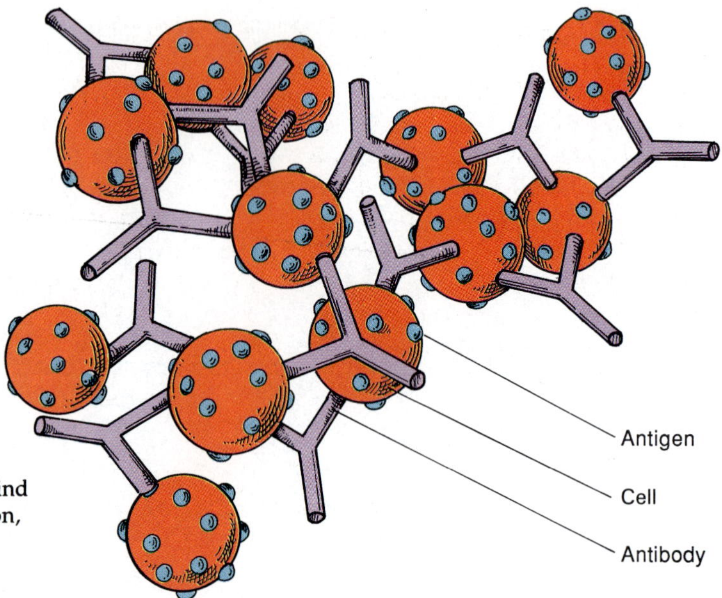

FIGURE 18.10 Agglutination occurs when antibodies bind cells together to form large clumps or lattices. This reaction, which is an important means of inactivating infectious agents, is also useful in the laboratory for a variety of diagnostic tests.

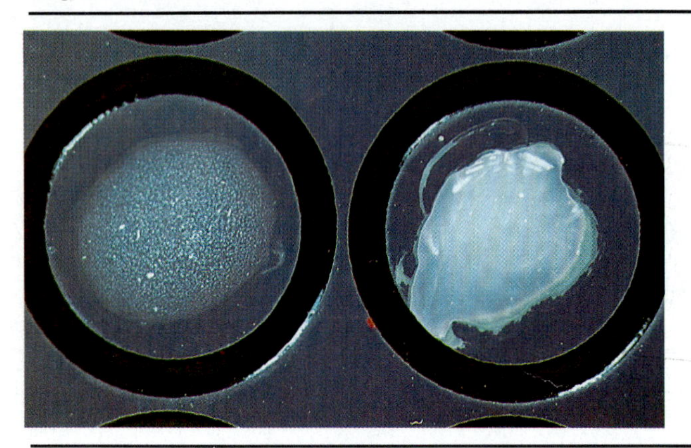

FIGURE 18.11 Positive (left) and negative (right) agglutination tests. Agglutination has caused the sample at left to take on a speckled or curdled appearance. This test is often used to detect the presence of antibodies to a particular antigen or antigens that react with a particular antibody.

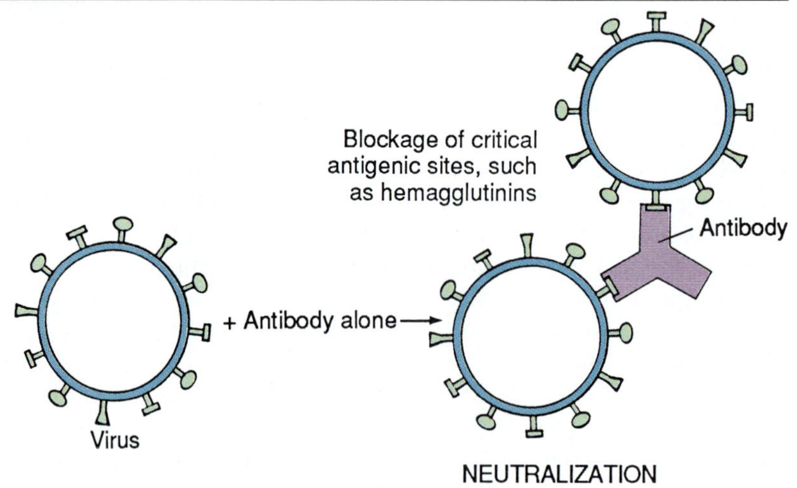

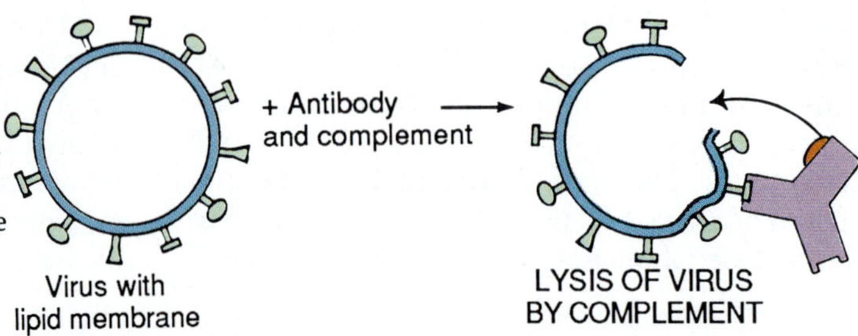

FIGURE 18.12 Neutralization of viruses. (a) Antibodies bind to critical antigenic sites, such as hemagglutination antigens, on the surface of the virus. Blocking these sites prevents the virus from infecting cells. Blockage of less important sites may allow the virus to persist in the body. (b) When a virus has a lipid membrane, antibodies also activate complement, which lyses the virus.

Bacterial toxins, being small molecules secreted from the cell, usually are inactivated simply by the formation of antigen-antibody complexes, or **neutralization.** IgG is the main neutralizer of bacterial toxins. Neutralization effectively stops the toxin from doing further damage to the host. It does not destroy the organisms that produce the toxin—antibiotics are needed to prevent persisting organisms from continuing to produce toxin. Viruses too can be inactivated by neutralization (Figure 18.12a); IgM, IgG, and IgA are all effective neutralizers of viruses. Those viruses that have an envelope may then be lysed by complement (Figure 18.12b).

We have now considered the major characteristics of humoral immunity—how B cells are activated, how antibodies are produced, and how they function. These processes are summarized in Figure 18.13.

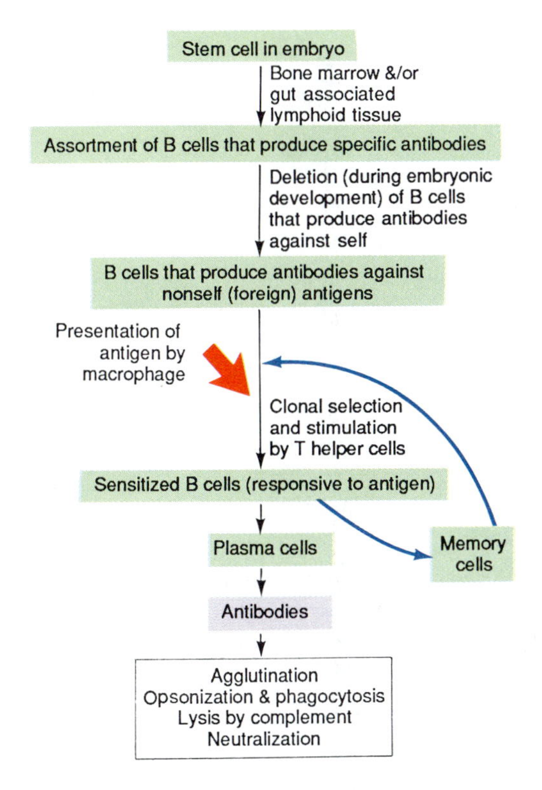

FIGURE 18.13 Summary of humoral immunity.

MONOCLONAL ANTIBODIES

Monoclonal antibodies are antibodies produced in the laboratory by a clone of cultured cells that make one specific antibody. In one method of making monoclonal antibodies, myeloma cells (malignant cells of the immune system) are mixed with sensitized lymphocytes. The malignant cells are used because they will keep dividing indefinitely. The lymphocytes are used because each makes a particular antibody. When the two cell types are mixed in cultures, they can be made to fuse with one another to make a cell called a *hybridoma* (Figure 18.14). (Chapter 8, p. 216) Hybridomas, which contain genetic information from each original cell, divide indefinitely, all the while producing large quantities of antibody. Which antibody a given hybridoma produces is determined by the antigen to which the lymphocytes were sensitized before their progeny were mixed with myeloma cells.

Generally, when a population of lymphocytes is exposed to an antigen, many different clones of B cells will proliferate, each making a different antibody. Many different hybridomas will therefore be produced by this technique. If one specific antibody is wanted, tests must be used to find which hybridomas are synthesizing that antibody, and those cells are then cloned.

Although monoclonal antibodies were first produced in 1975 as research tools, scientists were quick to recognize their practical uses. With experience, techniques for making monoclonal antibodies have improved. Culture media in which hybridomas thrive and produce large quantities of antibodies have been developed, and methods to grow hybridomas in large vat cultures in commercial laboratories are now available.

Theoretically, a monoclonal antibody can be produced for any antigen, provided lymphocytes sensitized to it can be obtained. Large numbers of hybridomas are now available, each producing a specific antibody. In addition to being used in research, many are produced commercially for use in diagnostic tests and in therapy.

Several diagnostic procedures that use monoclonal antibodies are now available. Generally, these procedures are quicker and more accurate than previously used procedures. For example, a monoclonal antibody can be used to detect pregnancy only 10 days after conception. Other monoclonal antibodies allow rapid diagnosis of hepatitis, influenza, and herpes-

BIOTECHNOLOGY

Deodorant Monoclonal Antibodies

The Gillette Company has patented a deodorant that contains monoclonal antibodies that attack *Staphylococcus haemolyticus*, the organism that causes perspiration-associated body odor. These antibodies block a bacterial enzyme necessary for the production of the unpleasant-smelling compounds.

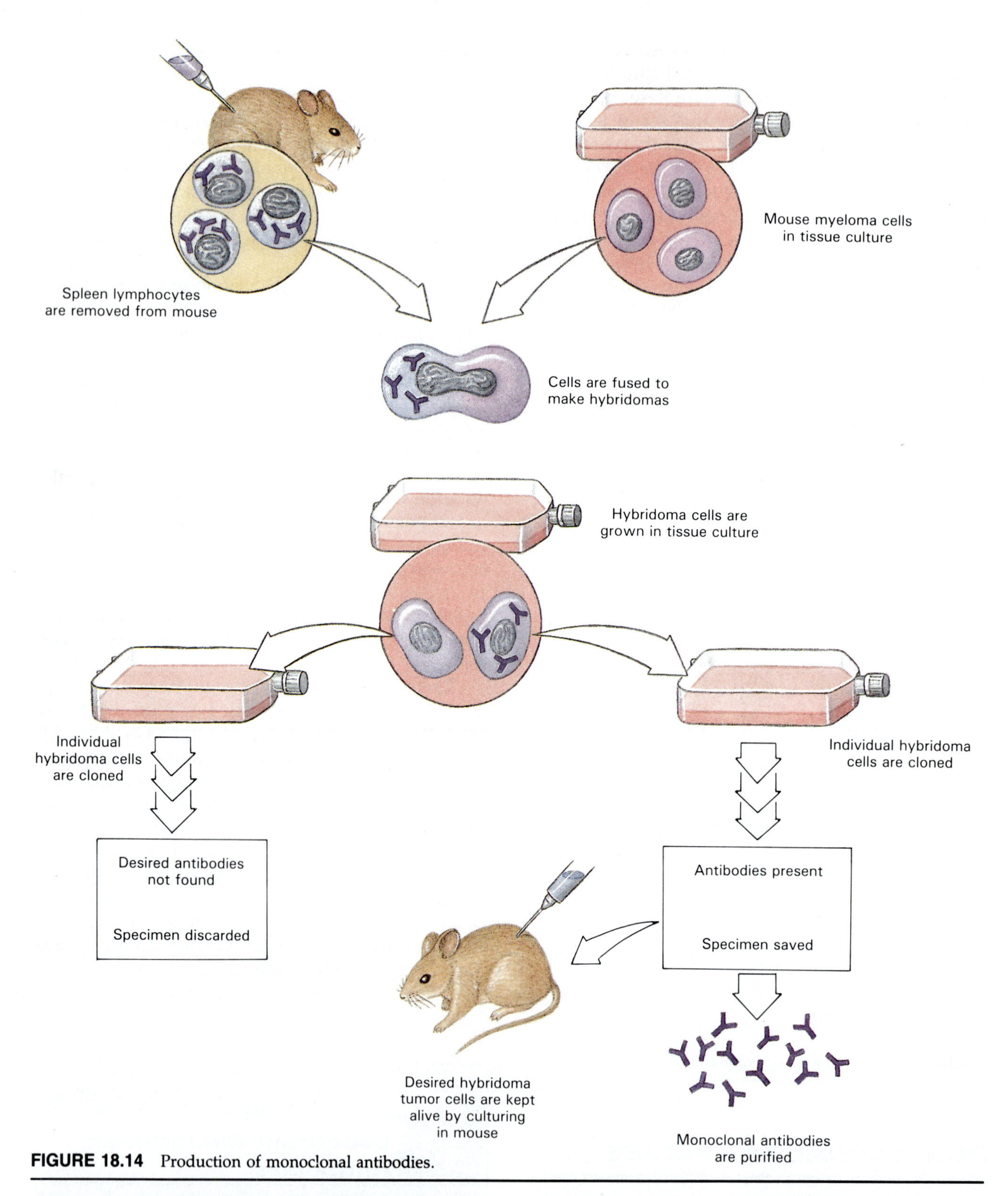

FIGURE 18.14 Production of monoclonal antibodies.

virus and chlamydial infections. Diagnostic tests for other infectious diseases and allergies are being developed at a rapid rate, and progress is being made in using monoclonal antibodies to diagnose various kinds of cancer. Monoclonal antibodies can detect malignant cells before they have multiplied to produce a tumor.

Methods of using monoclonal antibodies in therapy are under development. These methods first require preparing antibodies to infectious agents or malignant cells. Then an appropriate drug or radioactive substance must be attached to the antibodies. If such antibodies are given to a patient, they carry the toxic substance directly to the cells bearing the appropriate antigen. The great advantage of therapeutic monoclonal antibodies is that they selectively damage infected or malignant cells without damaging normal cells.

Monoclonal antibodies against tumor antigens have been tried in a few cancer patients. Unfortunately, the patients often displayed allergic reactions to myeloma proteins that accompany the antibodies. Researchers are now trying to produce monoclonal antibodies that will kill malignant cells without causing allergic reactions in the patients that receive them. Diphtheria exotoxin delivered to cancer cells by monoclonal antibodies is being tried as a therapy for cancer (see the Essay at the end of this chapter).

CELL-MEDIATED IMMUNITY

In contrast to humoral immunity, which involves B cells and immunoglobulins, cell-mediated immunity involves the direct actions of T cells. In cell-mediated immunity, T cells interact directly with other cells that display foreign antigens. These interactions clear the body of viruses and other pathogens that have invaded host cells. They also account for rejection of tumor cells, some allergic reactions, and immunological responses to transplanted tissues.

The cell-mediated immune response involves the differentiation and actions of different types of T cells and the production of chemical mediators called **lymphokines.** Much recent research has been devoted to determining the characteristics, origins, and functions of T cells, including the functions of secreted lymphokines. Much more research is needed to understand fully cell-mediated immunity. What follows is a brief discussion of current knowledge about it.

T cells, as noted earlier, either are processed by the thymus or develop under the influence of thymic hormones. T cells differ from B cells in that they do not make antibodies. However, they do have a particular cell membrane receptor protein that corresponds to the antibodies of B cells and other receptor proteins as well.

The Cell-mediated Immune Reaction

The cell-mediated immune reaction typically begins with the processing of an antigen—usually one associated with a pathogenic organism—by macrophages. When macrophages phagocytize pathogens, they ingest and degrade the pathogen; then they insert some of the pathogen's antigen molecules into their own cell membranes. This constitutes processing the antigen. When a macrophage presents the antigen to T cells having the proper antigen receptor, the antigen and receptor bind. Macrophages also have on their surfaces genetically determined histocompatibility proteins (Chapter 19) that bind to other T cell receptors. These reactions are summarized in Figure 18.15.

Binding with macrophages causes T cells to divide and differentiate into different types of T cells, including memory cells (Figure 18.16). Each cell is sensitized to the antigen that initiated the process, and each type has a different function in cell-mediated immune reactions. Some cells act directly and others release lymphokines, which are chemical substances that trigger certain immunologic reactions. The reactions of cell-mediated immunity are summarized in Figure 18.17. Refer to the figure as you read about the functions of different kinds of T cells.

Macrophages that have processed an antigen secrete the lymphokine interleukin-1 (IL-1), which activates **helper T (T_H) cells.** T_H cells, in turn, secrete lymphokines such as interleukin-2 (IL-2) and gamma interferon. IL-1 from macrophages and IL-2 from T_H cells activate other T cells—**suppressor T (T_S) cells, delayed hypersensitivity T (T_D) cells,** and **cytotoxic (killer) T (T_C) cells.** Also, IL-1, IL-2, and gamma interferon together cause null cells to become **natural killer (NK) cells.**

At the same time that these cells are differentiating, some T memory cells also are formed. As in humoral immunity, the persistence of memory cells in cell-mediated immunity allows the body to recognize antigens to which T cells have previously reacted and to mount more rapid subsequent responses.

As we noted in the discussion of humoral immunity, T_H cells stimulate the growth and differentiation of B cells, and T_S cells suppress T_H cells. Such suppression apparently helps to prevent both humoral and cell-mediated immune processes from getting out of hand.

Activated T_D cells also release various lymphokines. These include:

1. Macrophage chemotactic factor, which helps macrophages to find microbes.
2. Macrophage activating factor, which stimulates phagocytic activity.
3. Migration inhibiting factor, which prevents macrophages from leaving sites of infection.
4. Macrophage aggregation factor, which causes macrophages to congregate at such sites.

T_D cells also participate in delayed hypersensitivity, a kind of allergic reaction explained in Chapter 19.

T_C cells and NK cells kill infected host cells. When pathogens have evaded humoral immunity and established themselves inside cells, they can cause long-term infections unless the infected cells are destroyed by cell-mediated immunity. An agent that infects T cells is especially devastating because it destroys the very cells that might have combated the infection. AIDS is just such a disease. The AIDS virus invades

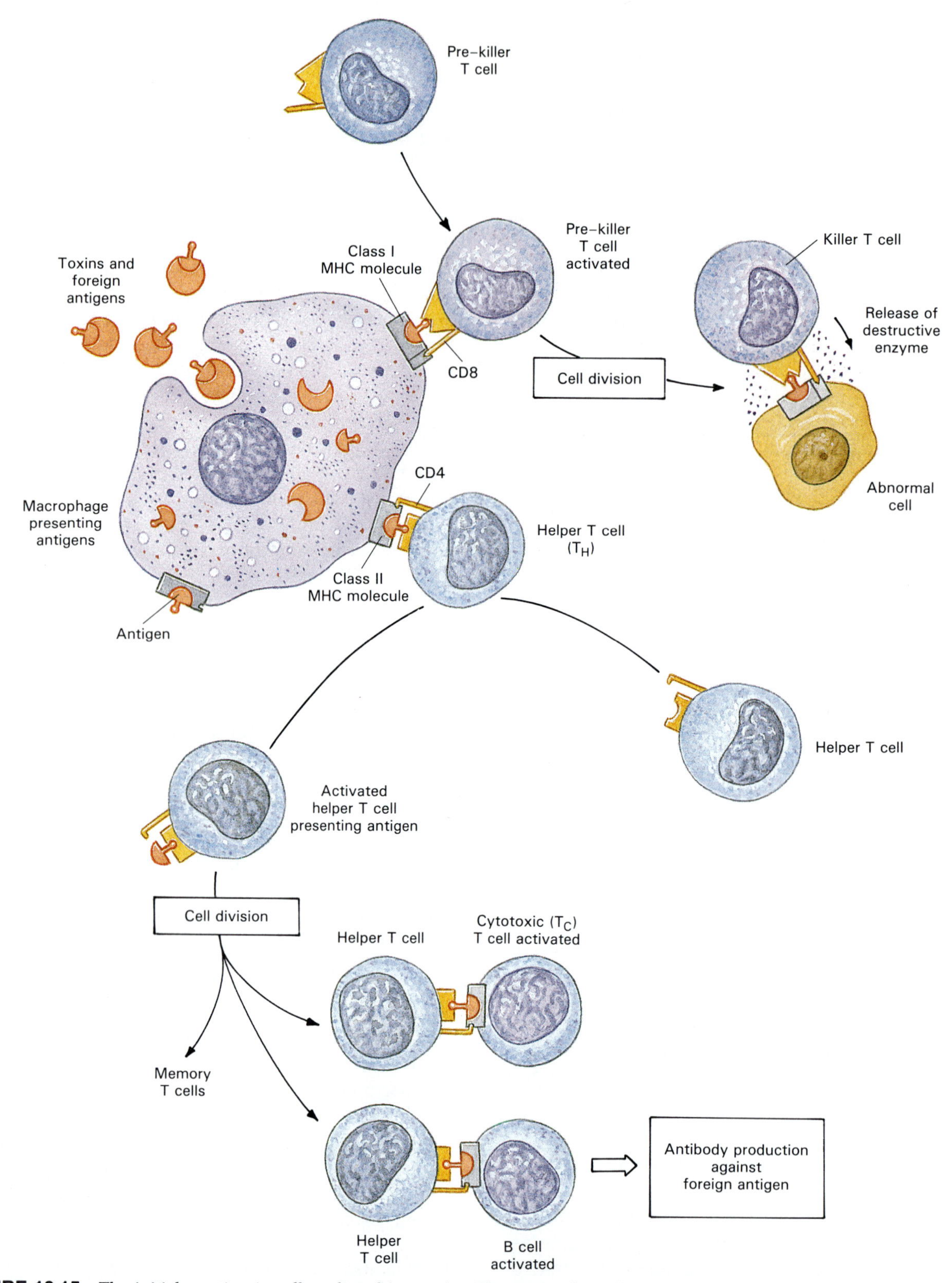

FIGURE 18.15 The initial reaction in cell-mediated immunity. The macrophage has processed an antigen and inserted it into its cell membrane. Presenting the antigen to helper T cells in conjunction with the appropriate histocompatibility antigen causes the T cells to become activated. The activated helper cells will then in turn activate cytotoxic T cells and B cells.

T_H cells, prevents them from carrying out their normal immunological functions, and eventually kills them. The lack of T_H cells impairs both humoral and cell-mediated immune responses, including the destruction of malignant cells. Thus, because of extensive destruction of T_H cells, AIDS patients are susceptible to a host of opportunistic infections and to various malignancies.

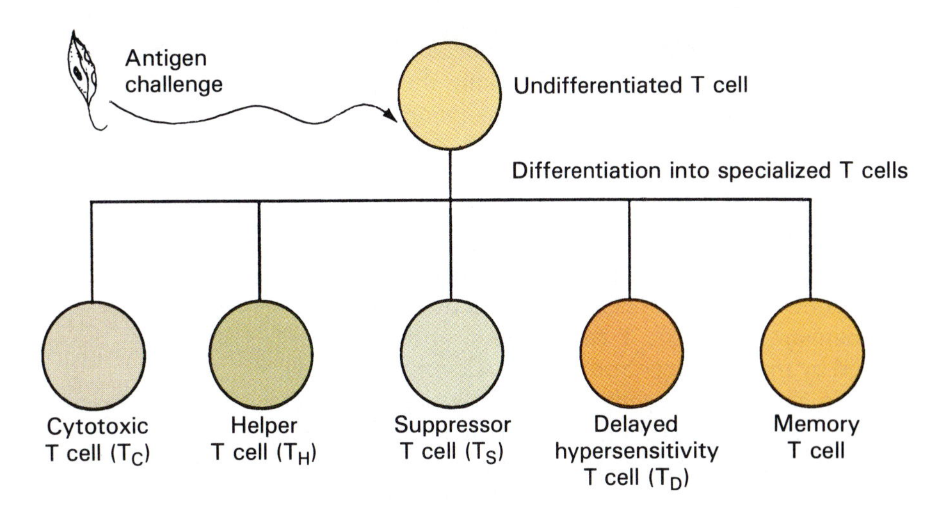

FIGURE 18.16 Differentiation of a T cell, following challenge by an antigen, into one of several types of functioning T cells.

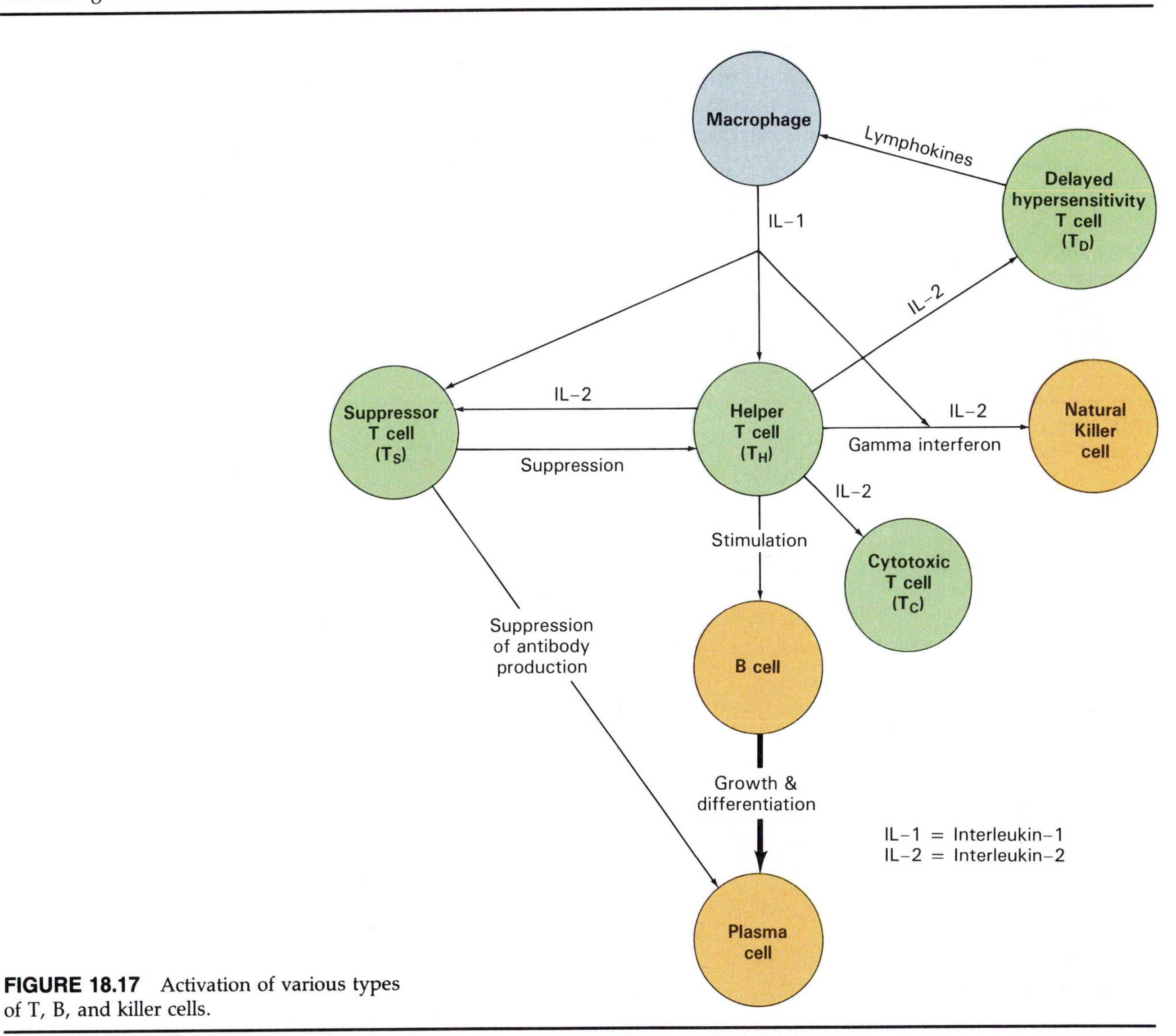

FIGURE 18.17 Activation of various types of T, B, and killer cells.

How Killer Cells Kill

Recent research shows that T_C cells and NK cells kill other cells by making a lethal protein and firing it at target cells. Eosinophils have a similar protein, which they may use to kill certain helminths and other parasites. But lethal proteins are not the sole property of hosts' defensive cells—the amoeba that causes amoebic dysentery and some other parasites and fungi also have them. Learning more about these lethal proteins may one day enable us to treat amoebic dysentery and other parasitic diseases by blocking the action of these proteins or to treat AIDS and malignant diseases by enhancing their actions.

Cytotoxic T cells act mainly on virally infected cells, whereas NK cells act mainly on tumor cells, cells of transplanted tissues, and possibly on cells infected with intracellular agents such as rickettsias and chlamydias. Each kind of killer cell acts by a different mechanism. Cytotoxic T cells bind to antigens presented by macrophages and then attack virus-infected cells. In contrast, NK cells bind directly to malignant or other target cells without the help of macrophages.

Both kinds of killer cells contain granules of a lethal protein, **perforin,** which is released when they bind to a target cell. Perforin bores holes in the target cell membranes so that essential molecules leak out and the cells die. This process is similar to the action of complement. By killing infected cells while they are few in number and before new virus particles are released from them, cytotoxic T cells prevent the spread of infection—but at the expense of destroying host cells. Similarly, NK cells destroy malignant cells before they have a chance to multiply. Both kinds of killer cells can withdraw from cells they have damaged and move on to other target cells.

The discovery of such an efficient mechanism for killing cells raises two important questions: What prevents perforin from killing adjacent uninfected cells, and what prevents it from attacking the membranes of the killer cells themselves? Perforin doesn't kill adjacent cells because it is effective only when secreted at the binding site between the killer and target cells. Why perforin doesn't attack killer-cell membranes is not known, but it has been suggested that killer cells produce a protein called protectin that inactivates perforin.

The Role of Activated Macrophages

Some bacteria, such as those that cause tuberculosis, Hansen's disease (leprosy), listeriosis, and brucellosis, can continue to grow even after they have been engulfed by macrophages. T_D cells combat such infections by releasing the lymphokine macrophage activating factor. This factor causes macrophages to increase production of toxic hydrogen peroxide, along with enzymes that attack the phagocytized organisms and accelerate the inflammatory response. Organisms that survive these defenses are walled off in granulomas. (Chapter 17, p. 462)

We have now completed the discussion of cell-mediated immunity—how it is initiated and how its effects are produced. These processes are summarized in Figure 18.18. The various functions of B and T cells

FIGURE 18.18 Summary of cell-mediated immunity.

Stem cell
Lymphocyte
thymus, or influence of thymic hormones
T cells with receptors for antigens
processing of antigen by macrophage & presentation to T cell; secretion of IL-1
Production of memory cells
Activation of helper T cells
Lymphokine
Activation of Natural Killer cells
Activation of cytotoxic T cells
Activation of helper T cells
Activation of suppressor T cells
Activation of delayed hypersensitivity T cells
Killing of cells
B cell stimulation & suppression humoral immunity

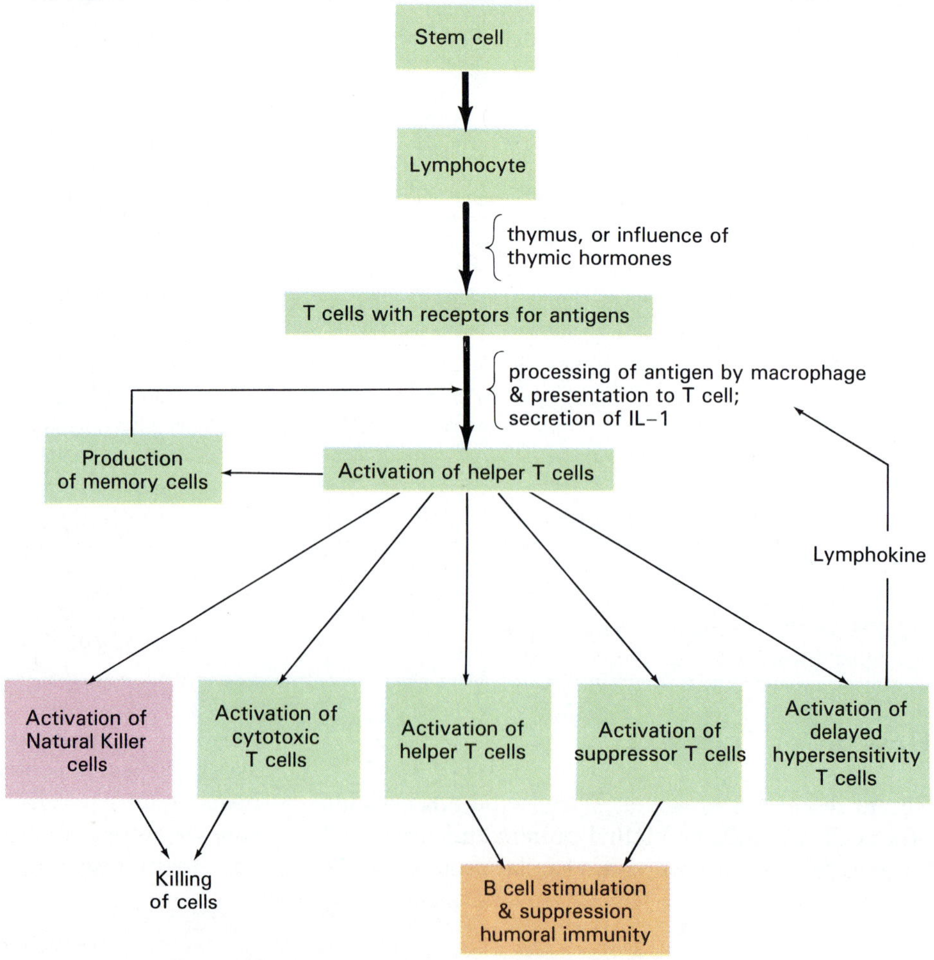

TABLE 18.5 Characteristics of B cells, T cells, and macrophages

Characteristic	B cells	T cells	Macrophages
Site of production	Bursal-equivalent tissues	Thymus or under thymic hormones	
Type of immunity	Humoral	Cell-mediated and assist humoral	Humoral and cell-mediated
Subpopulations	Plasma cells and memory cells	Cytotoxic, helper, suppressor, delayed hypersensitivity, and memory cells	Fixed and wandering
Presence of surface antibodies	Yes	No	No
Presence of foreign surface antigens	No	No	Yes
Presence of receptors for antigens	Yes	Yes	No
Life span	Some long, most short	Long and short	Long
Secretory product	Antibodies	Lymphokines	Interleukin-1
Distribution (% leukocytes)			
Peripheral blood	15–30	55–75	2–12
Lymph nodes	20	75	5
Bone marrow	75	10	10–15
Thymus	10	75	10

are summarized in Table 18.5. Comparing Figures 18.13 and 18.18 and studying Table 18.5 will serve to highlight similarities and differences in humoral and cell-mediated immunity and to provide an overview of specific immunity.

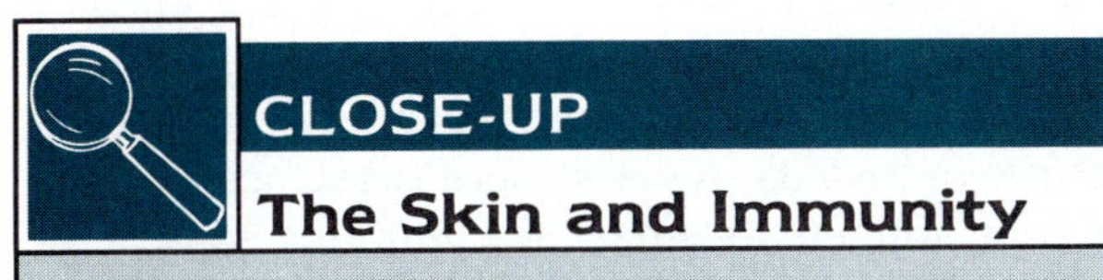

CLOSE-UP

The Skin and Immunity

Although the role of skin in mechanically and chemically resisting microorganisms has been recognized for some time, its role in immunity has been discovered recently. Three kinds of skin cells appear to be involved in immune reactions in the skin: Langerhans cells, Grandstein cells, and keratinocytes. T cells also are present in skin. When an antigenic agent evades other barriers and enters the skin, it can bind to Langerhans cells, Grandstein cells, or both. Langerhans cells present antigens to helper T cells, and Grandstein cells present antigens to suppressor T cells. Langerhans cells, which are much more numerous than Grandstein cells, and helper T cells work together to destroy the invading agent. If Langerhans cells have been damaged or reduced in activity, Grandstein cells and suppressor T cells take over, and the invasion can continue. Keratinocytes, long known to make keratin, are now known to make lymphokines, which stimulate T cells to attack invading agents.

FACTORS THAT MODIFY IMMUNE RESPONSES

The host defenses of young healthy human adults living in an unpolluted environment are capable of preventing nearly all infectious diseases. However, a variety of disorders, injuries, medical treatments, environmental factors, and even age can affect resistance to infectious diseases. An individual with reduced resistance is called a **compromised host.**

In the beginning of this chapter we noted that humans are genetically immune to some diseases. It has also been found that different races have different degrees of resistance and susceptibility to various diseases. When black and white military personnel live under the same conditions (same barracks, food, exercise regime, and such), blacks still develop TB at a higher rate.

Age also affects immune responses. In general, the very young and the elderly are most susceptible to infections, and young adults are least susceptible. The young are susceptible because the immune system is not fully developed until age 2 or 3. Infants can, however, produce some IgM shortly after birth, and they receive maternal IgG passively. The elderly are susceptible to infections and malignancies because the immune system, and especially cell-mediated response, is one of the first to decline in function during the aging process. Thus, it makes sense to take special precautions against unnecessarily exposing infants and elderly people to infectious agents. And it also makes sense to obtain recommended immunizations during infancy and early childhood.

Even seasonal patterns affect the immune system. T cells have a yearly cycle, falling to their lowest level in June. People with Hodgkin's disease are most often diagnosed in spring, leading some researchers to feel there is a link between the two cycles.

Genetic and age factors that modify immunity are beyond our control, but we have some control over diet and environment. Let's see how these factors contribute to resistance—or the lack of it.

An adequate diet, especially adequate protein and vitamin intake, is essential for maintaining healthy intact skin and mucous membranes and phagocytic activity. It is likewise important for lymphocyte production and antibody synthesis. Poor nutrition and poor inflammatory response of alcoholics and drug addicts greatly lower their resistance to infection. In the elderly an inadequate diet can further weaken a declining immunological response.

Regular moderate exercise such as 45 minutes of brisk walking, 5 days per week can produce a 20 percent increase in antibody level, which occurs during the exercise and for about 1 hour afterwards. Natural killer-cell activity is also increased. However, excessive exercise such as running more than 20 miles per week depresses the immune system. Marathoners who ran at their fastest pace for 3 hours experienced a drop in natural killer cell activity of more than 30 percent for about 6 hours. Long-distance runners are more vulnerable to infection, especially of the upper respiratory tract, for about 12 to 24 hours after a race, experiencing six times the rate of illness after a race compared with trained runners who did not race.

Traumatic injuries lower resistance at the same time that they provide easier access to tissues for microbes. Tissue repair competes with immune processes because both require extensive protein synthesis. When normal systems that flush away microbes, such as tears, urinary excretions, and mucous secretions, are impaired by injuries, pathogens have easier access to tissues. Antibiotics destroy commensals that sometimes compete with pathogens. Impaired defenses and use of antibiotics allow opportunistic infections to become established.

Environmental factors such as pollution and exposure to radiation also lower resistance to infection. Air pollutants, including those in tobacco smoke, damage respiratory membranes and reduce their ability to remove foreign substances. They also depress the activities of phagocytes. Excessive exposure to radioactive substances damages cells, including cells of the immune system. These factors can be compounded by induced and inherited immunological disorders. Immunosuppressant drugs used to prevent rejection of transplanted tissue impair the functions of lymphocytes and some phagocytes. Diseases such as AIDS destroy T cells. Finally, genetic defects in the immune system itself can result in the absence of B cells, T cells, or both. How these disorders impair immunity is discussed in Chapter 19.

IMMUNIZATION

Throughout the world each year nearly 3.5 million children, most of them under 5 years of age, die of three infectious diseases for which immunization is available. About 2 million die of measles, 800,000 die of tetanus, and 600,000 die of whooping cough. Another 4 million die of various kinds of diarrhea, against some of which immunization is possible. Most of these deaths occur in underdeveloped countries.

These statistics dramatize three important facts about immunization. First, immunization can prevent significant numbers of deaths. Second, methods of immunization are not yet available for some infectious diseases, such as certain diarrheas. Most organisms that cause diarrhea exert their effects in the digestive tract, where antibodies and other immune defenses cannot reach them. Finally, much greater effort is needed to make immunizations available in underdeveloped countries.

Active Immunization

To develop active immunity, as noted earlier, the immune system must be induced to recognize and destroy infectious agents whenever they are encountered. **Active immunization** is the process of inducing active immunity. It can be conferred by administering vaccines or toxoids. A **vaccine** is a substance that contains an antigen to which the immune system responds. Antigens can be derived from living but attenuated (weakened) organisms, dead organisms, or parts of organisms. A **toxoid** is an inactivated toxin that is no longer harmful, but retains its antigenic properties.

Principles of Active Immunization

Regardless of the nature of the immunizing substance, the mechanism of active immunization is essentially the same. When the vaccine or toxoid is administered, the immune system recognizes it as foreign and produces antibodies, or sometimes cytotoxic T cells, and memory cells. This immune response is the same as the one that occurs during a disease. The disease itself does not occur either because whole organisms are not used or because they have been sufficiently weakened to have lost their virulence. In other words, vaccines retain important antigenic properties but lack the ability to cause disease. In fact, organisms in vaccines sometimes do multiply in the host, but without producing disease symptoms. Similarly, toxoids retain antigenic properties but cannot exert their toxic effects.

An important factor in the longevity of immunity from active immunization is the nature of the immunizing substance. In most instances, vaccines made with live organisms confer longer-lasting immunity than those made with dead organisms, parts of organisms, or toxoids. For example, measles (both rubella and rubeola) vaccines and oral poliomyelitis vaccine, which contain live viruses, usually confer lifelong immunity. Intramuscular polio vaccine, which contains killed viruses, and typhoid fever vaccine, which contains dead bacteria, confer immunity that lasts 3 to 5 years. Tetanus and diphtheria toxoids confer immunity of about 10 years' duration.

Because immunity is not always lifelong, "booster shots" are often needed to maintain immunity. As we have noted, the first dose of a vaccine or toxoid stimulates a primary immune response analogous to that during the course of a disease. Subsequent doses stimulate a secondary immune response analogous to that following exposure to an organism to which immunity has already developed. Thus, booster shots boost immunity by greatly increasing the number of antibodies. This increases the length of time sufficient antibodies are available to prevent disease.

The route of administration of a vaccine can affect the quality of immunity. Compared with injecting vaccines into muscle (Figure 18.19), immunity is more durable when oral vaccines are used against gastrointestinal infections and nasal aerosols are used against respiratory infections.

Vaccines and toxoids, especially those containing live organisms, must be properly stored to retain their effectiveness. Some require refrigeration, and serious failures of measles immunization have resulted from inadequate refrigeration. Others must be used within a certain number of hours or days after a vial of vaccine is opened. Thus, clinics offer some immunizations only on selected days. Immunizations that require expensive vaccines and that are in low demand may be given one day per week or even less frequently.

In general, active immunization cannot be used to prevent a disease after a person has been exposed. This is because the time required for immunity to develop is greater than the incubation period of the disease. Rabies immunization is an exception to this rule. Because rabies typically has a long incubation period, active immunization can be used with some hope that immunity will develop before the rabiesvirus reaches the brain. The farther the virus must travel to reach the brain, the greater the chance of effective immunization. So, a bite on the ankle received while kicking off a rabid animal may be less hazardous than one on the trunk or neck. As we shall see later, passive immunization sometimes is used to prevent or lessen the severity of diseases after exposure to them. Several vaccines and toxoids have been licensed for general use in the United States; their properties are summarized in Table 18.6. Many more are available for people with special needs, such as foreign travel, or for experimental purposes. Properties of some special use vaccines are given in Table 18.7.

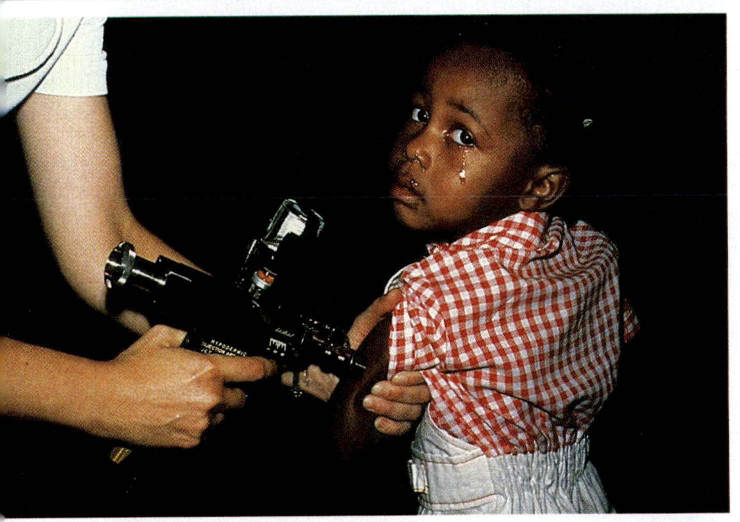

FIGURE 18.19 Mass inoculation using a "gun" rather than individual hypodermic needles and syringes to administer vaccine. Such a method allows large numbers of individuals to be immunized quickly and at low cost and is especially useful in developing countries where sterile materials are in short supply. Some cases of AIDS have already been documented in countries where nonsterile needles are reused in infant-vaccination programs.

Recommended Immunizations

Three vaccines that immunize against seven diseases are currently recommended in the United States for routine immunization of normal infants and children. The **DPT vaccine** contains diphtheria toxoid, killed pertussis (whooping cough) bacteria, and tetanus toxoid. Although most children tolerate DPT vaccine, a few suffer severe complications, as described shortly in connection with hazards of vaccines. The **poliomyelitis vaccine** generally used in this country contains three different types of live polioviruses. Vaccines with similar antigens are available for administration orally or intramuscularly. **MMR vaccine** contains live rubella, rubeola, and mumps viruses. This vaccine can be used to immunize against all three diseases simultaneously, or separate vaccines can be given for each disease.

The recommended age for administering vaccines varies. DPT and polio vaccines can be administered effectively as early as 2 months of age. However, MMR vaccine is not recommended before 15 months of age.

TABLE 18.6 Properties of materials available for active immunization				
Disease	**Nature of Material**	**Route of Administration**[a]	**Use and Comments**	**Duration of Effect**
Cholera	Killed bacteria	SC, IM, ID	2 doses a week or more apart, 50% effective, may be required for travel	6 months
Diphtheria	Toxoid	IM	3 doses 4 weeks apart, and boosters, 90% effective	10 years
Haemophilus infection	Polysaccharide-protein conjugate	IM	2, 4, 6, and 15 months, 75% effective	14–34 years
Hepatitis B	Viral antigen	IM	2 doses 4 weeks apart, booster in 6 months	About 5 years
Influenza (viral)	Inactivated virus	IM	1 or 2 doses, depending on type of virus, recommended for high-risk patients and medical personnel, 75% effective	1–3 years
Measles (rubeola)	Live virus	SC	1 dose at 15 months, revaccination around age 12; 95% effective, may prevent disease given within 48 hours of exposure	Lifelong
Meningococcus	Polysaccharide	SC	1 dose, recommended during epidemics and for high-risk patients	Lifelong if given after age 2
Mumps	Live virus	SC	1 dose given after age 1, 95% effective	Lifelong
Pertussis (whooping cough)	Killed bacteria	IM	Same as diphtheria, not given to children subject to seizures	10 years
Plague	Killed bacteria	IM	3 doses 4 weeks apart, for travel to some parts of the world	6 months
Pneumococcus	Polysaccharide	SC, IM	1 dose before chemotherapy	5 years or more
Poliomyelitis	Live viruses	O	2 doses 6 weeks apart, and booster, 95% effective	Lifelong
Rabies	Killed virus	IM	2 doses 1 week apart with third dose in 2 weeks, 80% effective, used after probable exposure	2 years
Rubella	Live virus	SC	1 dose at 15 months, some recommend second at 12 years	Lifelong
Smallpox	Live vaccinia virus	ID	1 dose, 90% effective, used only by laboratory workers exposed to poxviruses	3 years
Tetanus	Toxoid	IM	3 doses 4 weeks apart and boosters	10 years
Tuberculosis	Attenuated bacteria	ID, SC	1 dose for inadequately treated patients and high-risk groups	Lifelong?
Typhoid	Killed bacteria	SC	2 doses 4 weeks apart, 70% effective, recommended for travel, epidemics, and carriers	3 years
Yellow fever	Live virus	SC	1 dose, recommended for travel to endemic areas	10 years

[a] SC—subcutaneous, IM—intramuscular, ID—intradermal, O—oral.

When it is given to younger infants, the quality of immunity that results usually is not sufficient to protect against infection, probably because of immaturity of the immune system. Vaccines against *Haemophilus influenzae*, type b (Hib), first became available in 1985 but didn't work well in children under 2 years old. The FDA has now approved several new **Hib vaccines** for younger children that could prevent about 10,000 cases per year of meningitis, which kills about 500 children and leaves thousands of survivors mentally retarded, deaf, or otherwise neurologically damaged. Immunizations against Hib are scheduled at 2, 4, 6, and 15 months, or at 2, 4, and 12 months. Children between 15 months and 5 years need only one shot. Immunization is not recommended for those over 5 years old because nearly all children have contracted

PUBLIC HEALTH

BCG Vaccine

Health authorities in the United States have opposed the use of BCG vaccine because they want to use a skin test called the tuberculin test to identify people who have been infected with tuberculosis. BCG vaccine causes its recipients to test positive for tuberculin when, in fact, they are immune to tuberculosis. Thus, if BCG were in use, a positive tuberculin test could mean that the person has active tuberculosis, has had a previous infection, or has received the vaccine.

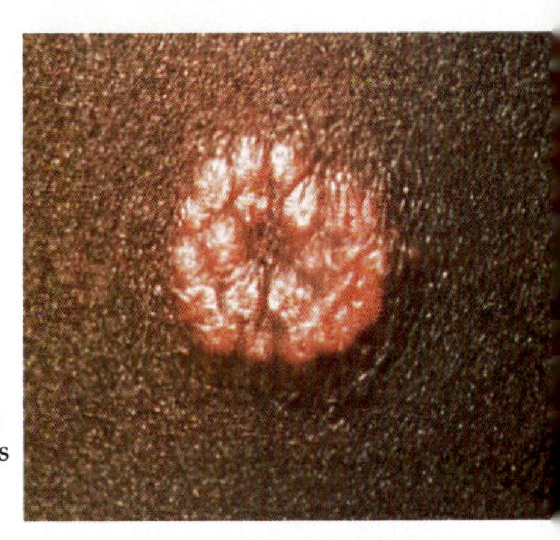

FIGURE 18.20 Vaccination mark from inoculation with BCG vaccine, used in some countries to immunize against tuberculosis. It is introduced under the skin by a cluster of tines that leave a permanent raised mark.

a Hib infection by then and have thus developed natural active immunity. When given early, the Hib vaccine can be combined with DPT. The recommended immunization schedule for normal infants and children in the United States is given in Table 18.8.

Immunization recommendations vary between developed and underdeveloped countries. Most developed countries use approximately the same immunizations as are recommended in the United States. However, several countries use BCG (Bacille Calmette-Guerin) (Figure 18.20) vaccine to protect against tuberculosis. The World Health Organization (WHO) has administered it to over 150,000 people in various countries in the past several decades. It is not used in the United States because serious controversy existed about its safety and efficacy when it was first developed. Now that it has been proven safe and effective, the incidence of tuberculosis is not sufficient to warrant widespread use. WHO recommends immunization at earlier ages in underdeveloped countries to combat serious contagious diseases: BCG and oral polio vaccines at birth, DPT and polio vaccines at 6, 10, and 14 weeks, and measles vaccine at 9 months.

Hazards of Vaccines

There are overwhelming benefits of using vaccines to prevent serious infectious diseases in populations.

TABLE 18.7 Selected examples of materials for special immunization and experimentation

Infectious Agent	Nature of Material	Uses
Adenovirus	Live virus	Military recruits
Anthrax	Antigen extract	Handlers of animals and hides
Campylobacter	Toxoids of *Escherichia coli* and *Vibrio cholera*	Experimentation
Cholera	Attenuated bacteria	Experimental oral administration to obtain more effective immunization
Cytomegalovirus	Live virus	Experimentation, may produce latent infections
Equine encephalitis	Live/inactivated viruses	Laboratory workers and experimentation

TABLE 18.8 Recommended immunizations for normal infants and children in the United States

Disease	Vaccine	Dosage Schedule
Diphtheria	Toxoid	2, 4, and 6 months, 1½ and 4 to 6 years
Tetanus	Toxoid	Same as diphtheria, administered in DPT vaccine
Pertussis	Killed bacteria	Same as diphtheria, administered in DPT vaccine
Poliomyelitis	Live viruses Types I, II, and III	2, 4, and 6 months, 1½ and 4 to 6 years
Haemophilus (Hib) infection	Polysaccharide-protein conjugate	2, 4, 6, and 15 months
Measles	Live virus	15 months, booster at school age, 11–12 years
Mumps	Live virus	15 months, 11–12 years
Rubella	Live virus	15 months, 11–12 years

However, vaccines also pose hazards that must be weighed in deciding whether they should be administered to entire populations, to certain individuals, or not at all. And, of course, the prevalence and severity of diseases also must be weighed in such decisions.

Active immunization often causes fever, malaise, and soreness at the site of injection. Thus, patients already suffering from fever and malaise should not receive immunization because a worsening of their condition might be erroneously attributed to the vaccine. More importantly, the patient's immune system, overburdened by the existing infection, may be unable to mount an adequate response to the antigen in the vaccine. Particular reactions are associated with certain immunizations. For example, joint pain can be caused by rubella vaccine, and convulsions by pertussis vaccine. Allergic reactions sometimes follow the use of influenza and other vaccines that contain egg protein or vaccines that contain antibiotics as preservatives. However, reactions occur in only a small proportion of vaccine recipients and are generally less severe than the disease. An exceedingly small number of vaccine recipients die or suffer permanent damage from vaccines (see the Polio Vaccine Controversy box in Chapter 24).

Live vaccines pose particular hazards to pregnant women, patients with immunological deficiencies, and patients receiving immunosuppressants such as radiation or corticosteroid drugs. In the case of pregnant women, live viruses sometimes cross the placenta and infect the fetus, whose immune system is immature. They also can cause birth defects. In immunodeficient or immunosuppressed patients, the attenuated virus sometimes has sufficient virulence to cause disease. Therefore, patients who test positive for the AIDS virus should not receive live-virus vaccines. This presents a problem for routine immunization of infants, some of whom, unknown to health care personnel, could have become infected with AIDS before birth. It also means that U.S. military and State Department employees and their accompanying families who are being posted abroad must be tested for AIDS to determine whether they can receive the required live-virus vaccines.

Passive Immunization

To develop passive immunity, as noted earlier, ready-made antibodies are introduced into a host. Because antibodies are found in the serum portion of the blood, these products are often called **antisera.** Passive immunity is obtained immediately but lasts only as long as a sufficiently high titer of the antibodies remains. The immune system is not induced to make antibodies or to recognize and destroy infectious agents whenever they are encountered. **Passive immunization** is accomplished by administering a substance such as gamma globulin, hyperimmune serum, or an antitoxin that contains large numbers of ready-made antibodies. However, the specificity and degree of immunization depend on which antibodies and how many of them are administered.

Gamma globulin consists of a pooled sample of the gamma globulin fractions (the portion of serum containing antibodies) from many individuals. If the donors are not specially selected, the gamma globulin will contain assorted antibodies corresponding to the diseases to which the donors are immune. This kind of gamma globulin usually contains sufficient antibodies to provide passive immunity to measles and hepatitis A because most people are immune to these diseases.

If the donors are selected, gamma globulins that have high titers of specific kinds of antibodies can be prepared. These often are referred to as **hyperimmune sera,** or **convalescent sera.** For example, gamma globulin from patients recovering from mumps or from recent recipients of mumps vaccine contains especially large numbers of antimumps antibodies. Similar sera can be collected from donors with high titers of antibodies to other diseases. Hyperimmune sera also can

BIOTECHNOLOGY

Human Antibodies Made by Phage

A variety of antibody genes isolated from human blood samples have been cloned and then inserted into bacteriophages. The phages each produced the specific human antibody coded for by the particular gene they had acquired. The antibodies are displayed on the surface of the phage. Scientists looking for a particular antibody lower an inert material coated with the matching antigen into the culture, where only the desired antibody will complex with it. Antibody-producing phages can be harvested by removing and washing the surface of the material. Multiplication of these phages in bacteria will produce large quantities of the desired antibody. This system can be easily manipulated by genetic engineering.

A unique feature of this system is that no immunization step is needed. Until now, animals such as mice have been inoculated with an antigen and their antibody-producing cells fused with cancer cells. Antibodies of mouse origin are probably not as useful as human antibodies for therapeutic use.

be manufactured by introducing particular antigens into another animal, such as a horse, and subsequently collecting the antibodies from the animal serum. Hyperimmune serum is given immediately after suspected exposure to rabies to inactivate as much of the virus as possible.

Antitoxins are antibodies against specific toxins, such as botulism, diphtheria, and tetanus. Passive immunization against tetanus toxins also can be achieved by the use of tetanus immune globulin, a gamma globulin that contains antibodies against tetanus toxin. The properties of currently available materials used to produce passive immunity are summarized in Table 18.9.

Passive immunization gives a nonimmune person exposed to a disease immediate immunity or lessens the severity of the disease if immunity is not achieved. Usually, a vaccine or toxoid is given at the same time as the passive substance or shortly thereafter to provide active immunity. Prior to the advent of antibiotics, passive immunization was frequently used to prevent or lessen the severity of several kinds of pneumonia and a variety of other diseases. With respect to infectious diseases, the most common current use of passive immunity is to protect people with deep wounds against tetanus toxins. Although the incidence of exposure to rabies, diphtheria, and botulism is lower than that of tetanus, passive immunization also can be used against these diseases.

TABLE 18.9 Properties of materials available for passive immunization

Material	Uses
Human gamma globulin	To prevent recurrent infections in patients with deficiencies in humoral immunity, to prevent or lessen disease symptoms after exposure of nonimmune persons to measles or hepatitis A
Specific gamma globulins	
Varicella-zoster immune globulin	To prevent chickenpox in high-risk children, must be given within 4 days of exposure
Hepatitis-B immune globulin	To prevent hepatitis B after exposure (via blood or needles) and to prevent spread of the disease from mothers to newborn infants
Mumps immune globulin	May prevent orchitis (inflammation of the testes) in adult males exposed to mumps
Pertussis immune globulin	To reduce severity of disease and mortality in children under 3 or in debilitated children
Rabies immune globulin	To prevent rabies after a bite from a possibly rabid animal, applied to the bite if possible and administered intramuscularly
Tetanus immune globulin	To prevent tetanus after injury in nonimmune patients
Vaccinia immune globulin	To halt progress of disease in immunodeficient patients who develop a progressive vaccinia infection after smallpox immunization

Passive immunization also is used to counteract the effects of snake and spider bites and to prevent damage to fetuses from certain immunological reactions. Antivenins, or antibodies to the venom of certain snakes and the black widow spider, are given as emergency treatments. They bind venom molecules that have not already bound to tissues. Thus, the sooner after a bite the antivenin is administered, the more effective it is in counteracting the effects of the venom.

A fairly common fetal immunological reaction occurs when a mother with Rh-negative blood carries her second Rh-positive fetus. As is explained in more detail in Chapter 19, the mother becomes sensitized to the Rh antigen when her first Rh-positive child is born. Her immune system makes anti-Rh antibodies, which will damage the second Rh-positive child. To prevent this serious condition, anti-Rh antibodies are given to the mother when the first child is born. Like the antivenins, the anti-Rh antibodies bind Rh antigens before the mother's immune system can make antibodies to them.

As with active immunization, passive immunization poses some hazards. The most common hazards are allergic reactions. Some antitoxins contain proteins from other animals as a result of their manufacture in eggs or horses. They are particularly likely to cause allergic reactions, especially when the patient receives them for the second time. Thus, vaccines of human origin are safer, at least with respect to the risk of allergic reactions. Allergic reactions to large IgG molecules also can occur if gamma globulins or hyperimmune sera are accidentally given intravenously instead of by their normal intramuscular route. A new immune globulin IV, containing smaller molecules, can be given safely by the intravenous route.

Another hazard, or at least detriment, of passive immunity is that giving ready-made antibodies can interfere with a host's ability to produce its own antibodies. One way this might occur is by ready-made antibodies binding to antigens and preventing them from stimulating the host's immune system. Thus, maternal antibodies, while they protect infants from some infections, may prevent the infant's immune system from making its own antibodies.

Future of Immunization

Whole-cell killed vaccines sometimes produce unwanted side effects due to extraneous cellular materials (leftover bits and pieces). Therefore efforts are being made to identify and obtain cellular subunits that contain only the purified antigenic portion of a microorganism that will produce immunity. *Subunit vaccines* are much safer than *attenuated vaccines,* which use live organisms that are treated so as to eliminate their virulence. There is always the possibility that the organisms may be insufficiently treated or may somehow revert back to virulence. Attenuated vaccines are considered too risky in the hunt for an AIDS vaccine, but they are good for other diseases. In general, live organisms produce higher and longer-lasting immunity than do nonliving organisms. *Recombinant vaccines* are being produced by inserting the genes for pure antigens into the genomes of nonvirulent organisms. The hepatitis B viral antigens have been cloned in yeast cells and form a safe and very effective vaccine. Rabiesvirus antigen has been inserted into harmless vaccinia (cowpox) virus and is being tested for control of rabies in wild animal populations such as raccoons.

IMMUNITY TO VARIOUS KINDS OF PATHOGENS

This chapter has focused on the basic principles of immunity. However, you may find it helpful to note the ways in which the immune responses to various kinds of pathogens differ. We'll begin this summary with bacteria.

Bacteria

As we saw in Chapter 17, nonspecific defenses such as skin, mucus, and stomach acid prevent many bacteria from gaining access to host tissues. ∞ (p. 444) When bacteria do infect a host, most immune responses serve to alter the bacteria so that they can be phagocytized. Once B cells produce antibodies, the antibodies can interfere with any one of several steps in bacterial invasion. They can attack fimbriae and capsules, preventing bacterial attachment to cell surfaces. They can work with complement to opsonize bacteria for later phagocytosis or lysis by other cells of the immune system. Finally, they can neutralize bacterial toxins or inactivate bacterial enzymes.

Viruses

Viruses infect by invading cells. They usually attack cells that line body passages first. They then directly invade target organs such as skin or travel in the blood (viremia) to target organs, such as the liver or nervous system. Polioviruses invade cells that line the digestive tract, but they also can enter nerve endings.

Immune responses can combat viral infections at any of these locations. Interferon, IgA, and some IgG act at the surfaces of lining cells and prevent or minimize entry of viruses. IgG and IgM act in the blood to neutralize viruses directly or to promote their destruction by complement. Finally, cell-mediated immunity is especially important in clearing the body of cells infected with viruses. Although viral antigens are sequestered in infected cells, some viruses, especially enveloped viruses, cause host cells to make antigens and insert the antigens on the cell surfaces. The immune system recognizes these virus-induced host antigens as foreign. Cell-mediated responses to them usually lead to destruction of the host cell. Unfortunately, it does not always cause destruction of the viruses and may, in fact, afford a means for their release. The mechanisms by which the immune system combats virus infections are summarized in Figure 18.21.

Viruses thwart the activities of the immune system by causing a variety of immunological disorders. They can be incorporated into antigen-antibody complexes and deposited in kidneys and other tissues. They also sometimes cause allergic reactions or induce autoimmunity, a lack of tolerance for one's own tissue antigens. Certain viruses destroy cells of the immune system—measles and AIDS viruses destroy T cells, and the Epstein-Barr virus, which causes infectious mononucleosis, destroys B cells.

In addition to specific immune responses to a viral infection, a number of nonspecific responses can limit infection. Fever is an important defense against viruses. Several viruses, such as influenza, parainfluenza, and rhinoviruses, are temperature sensitive. They replicate in the lining cells of the respiratory tract, which normally has a temperature between 33°C and 35°C—lower than the normal body temperature of 37°C because the cells are cooled as atmospheric air moves over their moist surfaces. When a person has a fever of even 1°C to 2°C, the virus's ability to replicate is damaged. Another benefit of fever in resisting viral infection is that temperature increases cause an increase in interferon production.

The fact that some viruses replicate poorly at normal body temperature is used in the manufacture of live viral vaccines. Temperature-sensitive viral mutants can be found for some pathogenic viruses. These viruses elicit antibody production and other immunological phenomena but have limited ability to replicate. The spread of such viruses within the body is limited, but their antigenic properties are intact.

A variety of other factors can contribute to nonspecific immunity or to a lack of it. Phagocytes, particularly macrophages, engulf viruses and have a special appetite for enveloped viruses. Hormones sometimes increase the virulence of a virus. For example, pregnant women are far more apt to develop paralytic poliomyelitis than nonpregnant women; this observation suggests that hormones may be involved.

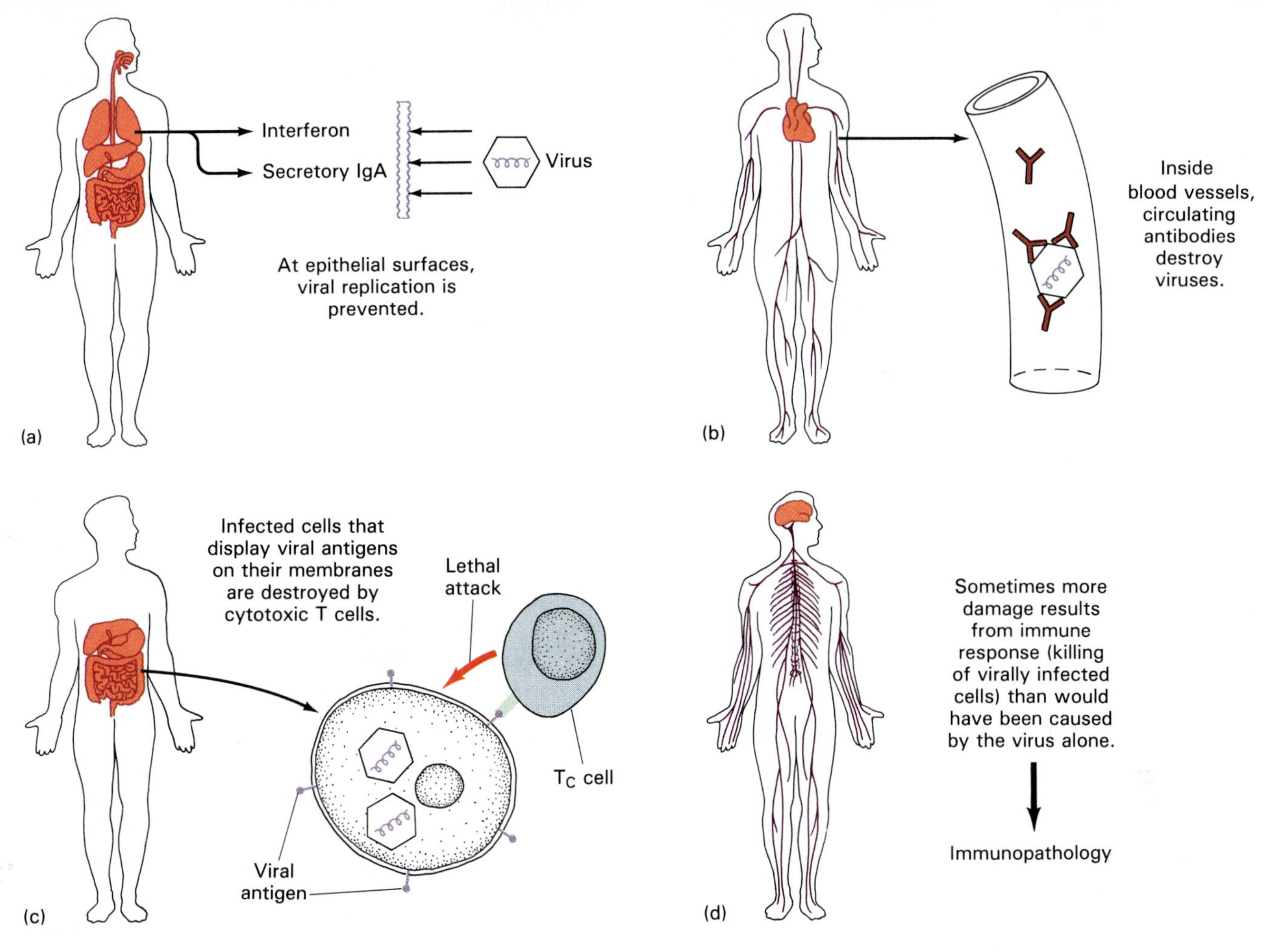

FIGURE 18.21 How the immune system combats viruses.

Diet also can affect viral pathogenesis. Malnutrition, in particular protein deficiency, often increases the severity of viral infections and greatly increases the mortality rate for measles.

Fungi

Fungal infections progress through a tissue as fungal cells invade and destroy one cell after another. Immunity to fungi is poorly understood, but it appears to be mainly cell-mediated. Fungal skin infections probably are combated by T cells releasing lymphokines that activate macrophages. The macrophages, in turn, engulf and digest fungi. Commensal fungi apparently are kept in their place by cell-mediated responses. Evidence for this is derived from studies of hosts with impaired T-cell functions. Such hosts are extremely likely to become infected with opportunistic fungi, such as *Candida albicans*.

Protozoa and Helminths

Protozoa and helminths are quite dissimilar in size and complexity, but they make use of similar methods of invading the body. Host defenses against them also are similar, except that allergic reactions to helminths can be severe enough to cause more damage to host cells than to the parasite. Antigens on the surface of *Ascaris* worms are potent inducers of allergic reactions, and individuals who have developed such allergies can absorb enough antigen through the skin to cause a severe allergic reaction, even just by contacting formalin in which ascaris worms have been stored.

Parasitic protozoa and helminths interact with their hosts in ways that allow host survival and, therefore, parasite survival. They cause chronic, debilitating diseases that usually are not immediately life-threatening. The parasites manage this by a variety of mechanisms that thwart immune responses.

Most parasites are relatively large and difficult to phagocytize. Large parasites, such as heartworms in dogs, can block blood vessels and cause sudden death. When attacked by phagocytes, some parasites release toxins that are more damaging to the host than were the live parasites. Medical intervention also can be hazardous. Giving drugs to kill some parasites causes them to release large quantities of toxic decay products. Thus, once some parasitic infections have been acquired, coexistence with the parasites may be the best course of action.

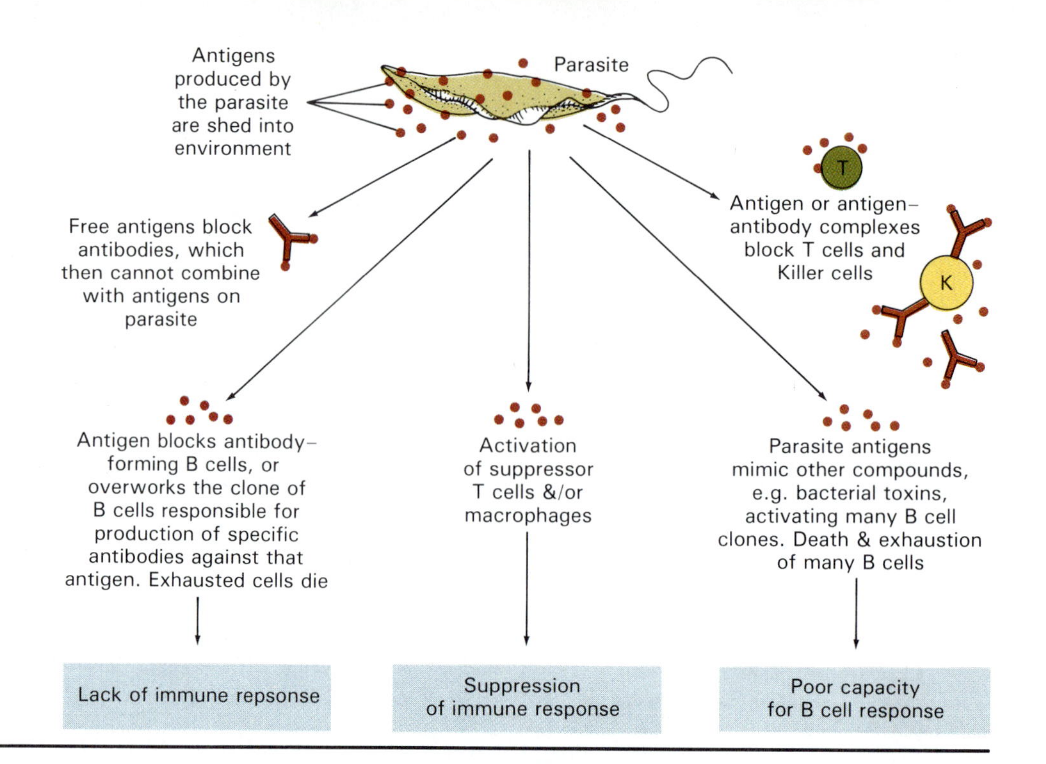

FIGURE 18.22 How parasite antigens thwart the immune system.

Many parasites have complex life cycles with more than one host, and some infect animals that serve as reservoirs for human infection. Thus, they stand ready to take advantage of appropriate conditions in various hosts. Each life stage of a parasite can have several kinds of surface antigens. Although hosts may produce antibodies against such antigens, the parasite's ability to change them provides a way to thwart host defenses. For example, the malaria parasite elicits antibodies before it invades host cells. While it multiplies inside cells it makes different antigens, so the host's original antibodies are not effective against the new parasites when they are released.

The host immune system also combats parasites by cell-mediated processes. Although cytotoxic T cells usually are not effective against parasites, some T cells release lymphokines that activate macrophages. These macrophages can attack malarial parasites, blood flukes, and several kinds of worms. Other lymphokines attract eosinophils to the gut mucosa when parasites are present. Eosinophils apparently help to combat worm infections by amplifying the host's allergic reaction to the worms.

A most ingenious assortment of escape mechanisms have been observed in various protozoa and helminths:

1. Some parasites protect themselves by invading cells, forming protective cases called cysts or otherwise becoming inaccessible to host defenses.
2. Some avoid recognition by changing their surface antigens.
3. Some suppress the host's immune responses by releasing toxins that damage lymphocytes, enzymes that inactivate IgG, and soluble antigens that thwart the immune system in a variety of ways (Figure 18.22).
4. Some suppress the action of phagocytes by inhibiting fusion of lyosomes with vacuoles and resist digestion by lysosomal enzymes.

Given the various methods protozoa and helminths have to evade and thwart immune responses of hosts, it is not surprising that they cause chronic debilitating infections, or that immunization against them is virtually impossible.

ESSAY

Cancer and Immunology

Cancer cells probably arise by mutation of normal cells in response to chemical carcinogens, radiation, or viruses; by expression of previously repressed human oncogenes (tumor-producing genes) within cells; or by a combination of these factors.

Regardless of the means by which they are produced, many cancer cells have certain cell membrane antigens not found in normal cells. These antigens are an ideal target for destruction by the immune system. According to the theory of *immune surveillance*, T cells recognize and destroy these abnormal cells before they develop into cancers (Figure 18.23). If this theory is correct, each of us has within our body many different potentially malignant cells. But we develop cancer only if the T cells fail to identify and destroy the mutated cells.

One way T cells might fail to recognize malignant cells is by the actions of antigens on the malignant cells themselves. The antigens stimulate B cells to make antibodies that bind to the antigens without damaging the malignant cells. Although this binding does not occur before the antigens sensitize T cells, it does block T cells from attacking the malignant cells. Such antibodies are called *enhancement antibodies;* that is, they enhance growth indirectly by protecting malignant cells from sensitized T cells.

The notion that cancer cells can be destroyed by immune reactions has led researchers to work on developing immunotoxins and cancer vaccines. An *immunotoxin* is a monoclonal antibody with an anticancer drug, a microbial toxin such as diphtheria toxin, or a radioactive substance attached to it. The antibody itself is designed to bind to a specific cancer cell antigen; the attached substance is selected for its ability to destroy the cancer cell. Such immunotoxins are expected to seek and destroy specific cells—the cancer cells that have the appropriate antigen. Cancer vaccines containing one or more cancer cell antigens are also being developed. Some of these vaccines elicit immunity to specific cancer cells. Others contain antigens frequently found on certain kinds of cancer cells.

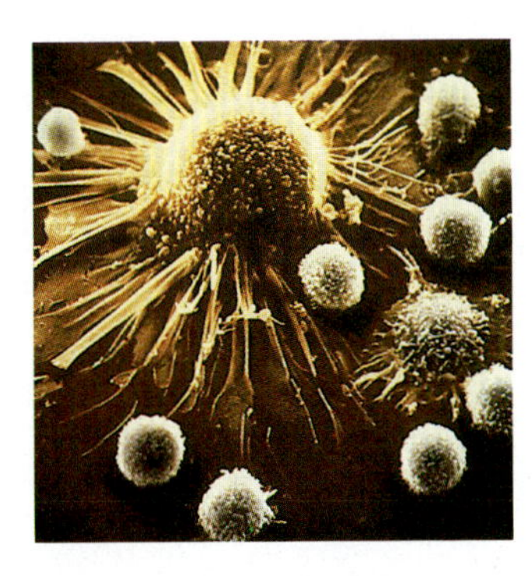

FIGURE 18.23 Cytotoxic T lymphocytes attacking a malignant cell. Artificially colored SEM of a B lymphocyte (red) covered with bacteria (green).

A major difficulty in making immunotoxins and cancer vaccines is that many different kinds of antigens are found on cancer cells. To be effective against an existing cancer, an immunotoxin or vaccine must be specific for the antigens on the cells of that specific cancer. Similarly, to immunize against common forms of cancer, vaccines would need to elicit antibodies or cytotoxic T cells that will destroy cancer cells should they develop. Another difficulty is that some malignant cells produce substances that inhibit T cell production. Immunotoxins and vaccines need to be powerful enough to overcome this inhibition. Finally, some antigens are found on both malignant and normal cells. Great care must be taken to develop immunotoxins and vaccines that will react only with malignant cells.

In October 1991 the first attempt to cure cancer by immunizing patients against their own tumors was made by removing cells from a malignant melanoma tumor, adding genes for tumor necrosis (killing) factor to the cancer cells' genomes, and reinjecting the cells into the patient. There, it is hoped, they will secrete enough of this immune-system hormone to change the regulation of the immune system so as to overcome its tolerance of the tumor and cause it to begin attacking the cancerous cells.

Meanwhile, immunization does offer us today, the ability to prevent about 80 percent of liver cancer cases. Liver cancer is among the most prevalent kinds of cancer. One form, primary hepatocellular carcinoma, accounts for 80 to 90 percent of all cases. It now appears that this form of liver cancer is associated with infection by hepatitis B virus early in life and especially with being a carrier of the virus. Hepatitis B viral DNA has been found in chromosomes of cancerous liver cells, and the intact virus has been isolated from cancerous liver cells. The incidence of liver cancer is highest in regions of Africa and Asia where the incidence of hepatitis B infections also is high.

Vaccination against the hepatitis B virus would protect recipients of the vaccine against infection, prevent them from becoming carriers, and thus protect them against liver cancer. However, the vaccine is still too expensive to make mass immunization feasible.

CHAPTER SUMMARY

IMMUNOLOGY AND IMMUNITY

RELATED KEY TERMS

- **Immune** means free of burden; **immunity** refers to the capacity to recognize and defend against infectious agents and other foreign substances.
- **Susceptibility** is vulnerability to infectious agents. Immunity is **nonspecific** when it acts against any infectious agent and **specific** when it acts against a particular infectious agent. **Immunology** is the study of specific immunity. The **immune system** is the body system that provides the host with specific immunity to particular infectious agents.

KINDS OF IMMUNITY

Innate Immunity

Acquired Immunity

Active and Passive Immunity

- Kinds of immunity are illustrated in Figure 18.1 and their properties are summarized in Table 18.1.

innate immunity
genetic immunity
species immunity
acquired immunity
colostrum
naturally acquired immunity
artificially acquired immunity

active immunity
naturally acquired active immunity
artificially acquired active immunity

passive immunity
naturally acquired passive immunity
artificially acquired passive immunity

CHARACTERISTICS OF THE IMMUNE SYSTEM

RELATED KEY TERMS

Antigens and Antibodies

- An **antigen** is a foreign substance capable of eliciting a specific immune response.
- Most antigens are proteins, but some are polysaccharides, glycoproteins, or nucleoproteins.
- Each antigen has several **antigenic determinant** sites, also called **epitopes**.
- An **antibody** is a protein produced in response to the presence of an antigen. Antibodies are capable of binding specifically to antigenic determinants on the antigen.

hapten **titer**

Cells and Tissues of the Immune System

- Lymphocytes differentiate into **B cells** in bursal-equivalent tissues or into **T cells** in the thymus or under the influence of thymic hormones. **Null cells** remain undifferentiated.

B lymphocytes
T lymphocytes

Dual Nature of the Immune System

- The dual roles of the immune system consist of **humoral immunity**, which is carried out mainly by B cells, and **cell-mediated immunity**, which is carried out mainly by T cells.

General Properties of Immune Responses

- Both humoral and cell-mediated immunity share certain general properties described next.
- The property of recognizing **self** versus **nonself** probably is accomplished by deletion during embryonic development of lymphocytes capable of destroying self.

clonal selection hypothesis
tolerance cross-reactions

- **Specificity** refers to the ability of lymphocytes to respond to each antigen in a different and particular way.
- **Heterogeneity** refers to the ability of lymphocytes to produce many different substances in accordance with the many different antigens they encounter.
- **Memory** refers to the ability of T and B lymphocytes to recognize substances to which the immune system has previously responded.

anamnestic response

HUMORAL IMMUNITY

- B cells are selected to respond to specific antigens in accordance with the particular antibody present on the B cell membrane prior to encountering an antigen.
- When a B cell detects an antigen with which it can react, it binds to the antigen, divides many times to produce a clone of many plasma cells and some memory cells.
- Many B cells require helper T cells to assist them in binding to antigens; suppressor T cells probably limit the duration of antibody production.

sensitizes

- **Plasma cells** synthesize and release large numbers of antibodies.
- **Memory cells** remain in lymphoid tissue ready to respond to subsequent exposure to the same antigen.

Properties of Antibodies (Immunoglobulins)

- Structurally, antibodies consist of two **heavy** and two **light** polypeptide **chains**, each of which has a variable region capable of reacting with a specific antigen.
- Properties of particular kinds of antibodies **(immunoglobulins)** are summarized in Table 18.4.

IgG **IgA** **secretory component**
IgM **IgE** **IgD**

Primary and Secondary Responses

- **Primary** and **secondary responses** to an antigen are summarized in Figure 18.9.

DPT vaccine
MMR vaccine
Hib vaccine

Kinds of Antigen-Antibody Reactions

- Humoral immunity is most effective against bacteria, which are destroyed by **agglutination** (clumping) or lysed by complement after opsonization or directly by IgMs or neutralized. Toxins and some viruses can be inactivated by antibody **neutralization**.

MONOCLONAL ANTIBODIES

- **Monoclonal antibodies** are antibodies produced in the laboratory by hybridomas, cells that contain genetic information from both a myeloma cell and a sensitized lymphocyte.
- Specific monoclonal antibodies can be used in some diagnostic tests, and

methods to use them to treat infectious diseases and cancer are being developed.

CELL-MEDIATED IMMUNITY

- Cell-mediated immunity concerns the direct actions of T cells that defend the body against viral infections, reject tumors and transplanted tissues, and cause delayed hypersensitivity.

The Cell-Mediated Immune Reaction

- Cell-mediated immune responses involve differentiation and activation of several kinds of T cells and the secretion of **lymphokines**.
- T cells have membrane receptors for antigens and for histocompatibility proteins; they do not make antibodies.
- Cell-mediated immune reactions begin with the processing of an antigen by a macrophage. During processing the macrophage inserts antigen molecules into its own cell membrane.
- Processed antigens bind with T cell receptors, and histocompatibility proteins on macrophages bind with other T cell receptors.
- Macrophages then secrete interleukin-1 that activates **helper T(T_H) cells.** IL-1 from macrophages, and IL-2 from helper cells then activate **cytotoxic T(T_C) cells, suppressor T(T_S) cells,** and **delayed hypersensitivity T(T_D) cells.** IL-1, IL-2, and gamma interferon activate **natural killer (NK) cells.**
- T_H and T_S cells regulate B cells; T_S cells also regulate T_H cells.
- T_D cells secrete several lymphokines that stimulate various activities of macrophages.
- T_C cells kill virus-infected cells; NK cells kill tumor cells.
- AIDS destroys T cells, thereby impairing both humoral and cell-mediated immunity.

How Killer Cells Kill

- T_C and NK cells destroy target cells by releasing the lethal protein **perforin**.

The Role of Activated Macrophages

- Certain pathogenic bacteria can grow in macrophages after phagocytosis. The lymphokine macrophage activating factor helps stimulate antimicrobial processes so that macrophages can kill the pathogens.
- When macrophages fail to kill pathogens, the pathogens are walled off in granulomas.

FACTORS THAT MODIFY IMMUNE RESPONSES

- Host defenses in healthy adults in an unpolluted environment prevent nearly all infectious diseases. Individuals with reduced resistance are called **compromised hosts**.
- Factors that reduce host resistance include very young or old age, poor nutrition, traumatic injury, pollution, and radiation. Complement deficiencies, immunosuppressants, infections such as AIDS, and genetic defects impair immune system function.

IMMUNIZATION

Active Immunization

- **Active immunization** occurs by the same mechanism as having a disease; it challenges the immune system to develop specific defenses and memory cells.
- Active immunization is conferred by **vaccines** and **toxoids**: Vaccines can be made from live, attenuated organisms, dead organisms, or parts of organisms. Toxoids are made by inactivating toxins.
- The recommended immunizations for normal infants and children in the United States are summarized in Table 18.8.
- The benefits of active immunization against common severe diseases nearly always outweigh the hazards. Reactions to vaccines can cause death or serious impairments, but their incidence is lower than that for the diseases themselves.

Passive Immunization

- **Passive immunization** occurs by the same mechanism as natural passive transfer of antibodies.

- Passive immunity is conferred by **antisera** such as **gamma globulin**, **hyperimmune**, or **convalescent sera**, and **antitoxins**.
- The benefits of passive immunization are limited to providing immediate but temporary protection; the hazards are mainly allergic reactions.

Future of Immunization

- **Subunit vaccines** produce fewer side effects than **whole cell killed vaccines**, and offer greater safety than do **attenuated vaccines**.
- **Recombinant vaccines** have genes for antigens of pathogens inserted into nonpathogenic organisms' genomes, and are very safe.

IMMUNITY TO VARIOUS KINDS OF PATHOGENS

Bacteria

- Nonspecific defenses are important in preventing bacterial infection. Antibodies produced by B cells are the chief immunologic defense. Most immune responses to bacteria serve to promote phagocytosis of the invading cells.

Viruses

- Viral infection is combated by nonspecific defenses, interferon, and antibodies. In addition, cell-mediated responses are important in destroying virus-infected cells.

Fungi

- Immune responses to fungi are primarily cell-mediated.

Protozoa and Helminths

- Immune responses to parasites are also largely cell-mediated. T cells release lymphokines that activate macrophages and attract other leukocytes. Allergic reactions to helminths can be more damaging to the host than to the parasite.

QUESTIONS FOR REVIEW

A.

1. How do immunity and susceptibility differ?
2. How do nonspecific and specific immunity differ?
3. What is immunology, and how does the immune system contribute to it?

B.

4. How do innate and acquired immunity differ?
5. How do active and passive immunity differ, and how can each be acquired?

C.

6. What are the properties of an antigen?
7. What are the general properties of an antibody?
8. How do lymphocytes become B cells and T cells?
9. What are the dual roles of the immune system?

D.

10. How does the immune system become able to destroy nonself and avoid harming self?
11. What is the difference between specificity and heterogeneity in the immune system?
12. What is the meaning of memory in the immune system?

E.

13. What determines whether a B cell will respond to an antigen?
14. What happens when a B cell does respond to an antigen?
15. How are helper and suppressor T cells involved in humoral immunity?
16. How do the functions of plasma cells and memory cells differ?
17. What are the main structural characteristics of antibodies?
18. What are the distinguishing characteristics of each kind of antibody?
19. How do the primary and secondary responses to the same antigen differ?
20. What is the significance of the above differences in responses?
21. How does humoral immunity defend against bacteria?
22. How does humoral immunity defend against toxins and viruses?

F.

23. What are monoclonal antibodies, and how are they made?
24. What are some current medical applications of monoclonal antibodies?
25. What kinds of medical applications for monoclonal antibodies are being developed?

G.

26. What are the properties of cell-mediated immunity that distinguish it from humoral immunity?
27. What are the two main kinds of products of cell-mediated immune reactions?
28. How do differentiated T cells differ from B cells?
29. What is the initial step in cell-mediated immunity?
30. What role do processed antigens play in cell-mediated immunity?
31. How is each kind of T cell activated?
32. What are the main functions of each kind of T cell?
33. How do NK cells differ from T cells, and how are they activated?
34. Why is AIDS such a devastating disease?

H.

35. How do killer cells attack target cells?
36. What causes these target cells to die?
37. Why do neither adjacent cells nor lymphocytes die when target cells die?
38. How does activation of macrophages contribute to defense against pathogens that have invaded cells?

I.

39. What factors contribute to lowered resistance to infection?
40. What factors contribute to impaired function of the immune system?

J.

41. How is active immunization accomplished?
42. How do vaccines and toxoids differ?
43. What major differences exist among vaccines?
44. What immunizations are currently recommended for normal infants and children in the United States?
45. What are the major benefits and hazards of active immunization?
46. How is passive immunization accomplished?
47. How do gamma globulin, hyperimmune serum, and antitoxin differ?
48. What are the major benefits and hazards of passive immunization?

PROBLEMS FOR INVESTIGATION

1. How would you respond to parents who wish to avoid (a) all vaccines for their infant or (b) pertussis vaccine for their infant?
2. Explain how a person may be a compromised host on one day, week, or month, but not on the next.
3. If you were born without T cells, would you have normal B cell functioning? Why or why not?
4. Identify each of the following types of immunity as (I) innate or acquired, (II) natural or artificial, and (III) active or passive:
   a) People do not get canine distemper
   b) After a case of polio
   c) After polio vaccine
   d) After a shot of gamma globulin
   e) After ingestion of colostrum
5. Select a bacterium, a virus, and a protozoan that infect the human body. Prepare a flowchart to show how the body uses nonspecific defenses and specific immunity to prevent infection. Explain also how immunity develops if an infection occurs.
6. Research current literature on the genetics of antibodies.
7. Research current literature, and describe new medical applications for monoclonal antibodies.
8. Research current literature on the development of vaccines against protozoan and helminth infections.
9. The parents of a 2-month-old infant delighted in taking him with them on their frequent excursions to shopping malls. One of the grandmothers suggested that they leave him at home, as it was flu season and the malls were full of coughing and sneezing people. The mother immediately responded that she was breast-feeding the baby, and with all those antibodies from her, he was surely protected against any diseases in the mall. Later, as she sat at the pediatrician's office holding her sick baby on her lap, she expressed disbelief that he could have caught the flu. If you were the doctor, what fallacies in her beliefs could you have explained to her?
   (The answer to this question appears in Appendix E.)

SOME INTERESTING READING

Ada, G. L., and G. Nossal. 1987. The clonal selection theory. *Scientific American* 257, no. 2 (August):62.

Anonymous. 1990. Changes pending for pertussis, *Haemophilus* vaccines. *ASM News* 56 (12):634–35.

Bellanti, J. A. 1985. Immunology: Basic processes. Philadelphia: W. B. Saunders.

Centers for Disease Control. 1983. General recommendations on immunization; Recommendations of the immunization practices advisory committee. *Annals of Internal Medicine* 98:615.

Hanna, K. E. 1990. Rubella and pertussis vaccines: convincingly safe? *ASM News* 56 (9):470–72.

Kolata, G. 1986. Vaccine compensation bill passed. *Science* 234 (November 7):666.

Lerner, R. A., and A. Tramontano. 1988. Catalytic antibodies. *Scientific American* 258, no. 3 (March):58.

Marrack, P., and J. Kappler. 1986. The T cell and its receptors. *Scientific American* 254, no. 1 (January):36.

Ozanne, G. and M. d'Halewyn. 1992. Secondary immune response in a vaccinated population during a large measles epidemic. *Journal of Clinical Microbiology* 30(7):1778–82.

Robbins, A. 1990. Progress towards vaccines we need and do not have. *Lancet* 335, no. 8703 (June 16):1436–39.

Seifer, H. S. and M. So. 1988. Genetic mechanisms of bacterial antigenic variation. *Microbiological Reviews* 52 (3):327–36.

Smith, K. A. 1990. Interleukin-2. *Scientific American* 262 (March):50–57.

Stites, D. P., J. D. Stobo, and J. V. Wells. 1987. *Basic and clinical immunology*. Los Altos, CA: Appleton and Lange.

Tonegawa, S. 1985. The molecules of the immune system. *Scientific American* 253, no. 4 (October):122.

Young, J. D., and Z. A. Cohn. 1988. How killer cells kill. *Scientific American* 253, no. 4 (October):122.

———. 1988. How killer cells kill. *Scientific American* 258, no. 1 (January):38.

As infection rates soar worldwide, AIDS education becomes more imperative. Sadly, the rates are climbing most rapidly among teenagers, via heterosexual transmission. The 1990s will see a significant increase in symptomatic cases on college campuses. It cannot be avoided, as the students who will exhibit them are already infected.

19 Immunology II: Immunologic Disorders and Tests

This chapter focuses on the following questions:

A. What are the different types of immunologic disorders?

B. What are the causes, mechanism, and effects of immediate hypersensitivity?

C. What are the causes, mechanism, and effects of cytotoxic reactions?

D. What are the causes, mechanism, and effects of immune complex disorders?

E. What are the causes, mechanism, and effects of cell-mediated reactions?

F. How does autoimmunization occur, and how is hypersensitivity involved in it?

G. Why are organ transplants sometimes rejected, and how can rejection be prevented?

H. How is hypersensitivity involved in drug reactions?

I. What are the causes, mechanisms, and effects of immunodeficiency diseases?

J. How can antigens and antibodies be detected and measured?

K. What is AIDS, and why is it causing an epidemic?

L. How can AIDS be diagnosed, treated, and prevented?

In Chapter 18, we emphasized that specific immunity defends the body against harmful substances, but this is not always the case. Sometimes the immune system reacts in ways that are at best unpleasant and at worst life-threatening. Perhaps you or someone you know has a runny nose and watery eyes every time the hay fever season rolls around. Maybe you know of people who have other allergies, who have had transfusion reactions, or who suffer from more severe immunologic disorders—even the dreaded AIDS. In this chapter we will learn more about the immune system by looking at ways it goes awry and reacts inappropriately or inadequately. We will also look at some of the methods used to detect and measure immune reactions.

OVERVIEW OF IMMUNOLOGIC DISORDERS

An **immunologic disorder** is a disorder that results from an inappropriate or inadequate immune response. Most disorders involve either hypersensitivity or immunodeficiency.

In **hypersensitivity,** or **allergy,** the immune system reacts inappropriately, usually by responding to an antigen it normally ignores. Such exaggerated responses can be thought of as "too much of a good thing": The immune system responds by doing harm instead of good. Although allergy is another name for hypersensitivity, many disorders that people call allergies are not due to immunologic reactions. These disorders include toxic responses to drugs, digestive upsets from nonallergic responses to foods, and emotional disturbances.

The four types of hypersensitivity are (1) immediate hypersensitivity (Type I), (2) cytotoxic hypersensitivity (Type II), (3) immune complex hypersensitivity (Type III), and (4) cell-mediated, or delayed, hypersensitivity (Type IV). **Immediate** (Type I) **hypersensitivity,** or *anaphylaxis,* is elicited by a foreign substance called an *allergen,* an antigen that evokes a hypersensitivity response. Allergies to pollen, foods, and insect stings are examples of immediate hypersensitivity. **Cytotoxic** (Type II) **hypersensitivity** is elicited by antigens on cells, especially red blood cells, that the immune system treats as foreign. This occurs in transfusion reactions. **Immune complex** (Type III) **disorders** are elicited by antigens in vaccines, on microorganisms, or on a person's own cells. Large molecules called **immune complexes** (antigen-antibody complexes) form, precipitate on blood vessel walls, and cause tissue injury within hours. **Cell-mediated** (Type IV) **hypersensitivity** is elicited by foreign substances from the environment (such as poison ivy), infectious agents, transplanted tissues, and the body's own malignant cells. T cells react with the foreign substance and cause tissue destruction over a period of several days. Autoimmune diseases, rejection of transplanted organs, and drug sensitivity can involve more than one kind of hypersensitivity.

In **immunodeficiency** the immune system responds inadequately to an antigen, either because of inborn or acquired defects in B cells or T cells. The weak responses in immunodeficiency are "too little of a good thing." They do no direct harm, but they leave the patient susceptible to infections, which can be severe and even life-threatening.

Immunodeficiencies can be primary or secondary. **Primary immunodeficiencies** are genetic or developmental defects in which the patient lacks T cells or B cells or has defective ones. **Secondary immunodeficiencies** result from damage to T cells or B cells after they have developed normally. These disorders can be caused by malignancies, malnutrition, infections such as AIDS, or drugs that suppress the immune system.

APPLICATIONS

Cows and Colic

The mother cow who passes antibodies on to her calf is doing a good thing. But when these same antibodies are passed on to human babies, either in milk-based formula or in the breast milk of mothers who consume dairy products, a bad thing can result. *Colic,* which affects about 20 percent of all babies 1 to 4 months old, causes inconsolable crying, hours on end, despite parents' every effort. Because colic affects breast- and bottle-fed babies, it was originally thought to be related to the parents' behavior. Recent research, however, has found that the average level of cow antibodies in breast milk of mothers with colicky babies was significantly higher than it was in other mothers. Because antibodies are proteins, they should be degraded in the digestive system of the baby. Medical experts suspect, however, that some babies' immature digestive systems are unable to process these antibodies properly. Just how antibodies may cause the symptoms of colic is still unknown.

IMMEDIATE (TYPE I) HYPERSENSITIVITY

Immediate (Type I) hypersensitivity is also called immediate allergy, or **anaphylaxis** (an-a-fi-lak'sis). The term anaphylaxis refers to detrimental effects, as opposed to prophylaxis, which refers to beneficial effects. Early investigators discovered that a substance they called **reagin** (re-a'jin) was responsible for this type of hypersensitivity. We now know that reagin consists of IgE antibodies.

Anaphylaxis is the harmful result of IgE antibodies made in response to allergens. It can be local or systemic. **Localized anaphylaxis** appears as reddening of the skin, watery eyes, hives, asthma, and digestive disturbances. **Generalized anaphylaxis** appears as a systemic life-threatening reaction such as airway constriction or **anaphylactic shock,** a condition resulting from a sudden extreme drop in blood pressure.

Allergens

An **allergen** is an ordinarily innocuous foreign substance that can elicit an adverse immunologic response in a sensitized person. Allergens can enter the body by inhalation, ingestion, or injection. They include airborne substances such as pollen, household dust, molds, and dander—tiny scales from hair, feathers, or skin. The agents in household dust usually are tiny, almost microscopic, mites and their fecal pellets. Other allergens include antibiotics and other drugs, certain foods, and foreign molecules found in vaccines and other diagnostic and therapeutic materials. Some common allergens are listed in Table 19.1.

Mechanism of Immediate Hypersensitivity

Although immediate hypersensitivity is not fully understood, the main characteristics of the reaction are sensitization, production of IgE (antiallergen) antibodies, allergen-antiallergen reactions, and local and systemic effects of those reactions (Figure 19.1). Such reactions occur only in people who have been previously sensitized to an allergen. **Sensitization** is the initial recognition of a foreign substance in which B cells are stimulated to make IgE antibodies against it. What makes some people become sensitized to normally innocuous substances is poorly understood.

TABLE 19.1 Common allergens

Ingested	Inhaled	Injected
Animal proteins, especially from milk and eggs	Cocaine	Antibiotics, especially cephalosporins and penicillins
Aspirin	Danders	Heroin
Fruits	Dust (household and from tree bark)	Hormones (ACTH and animal insulin)
Grains	Face powder	Insect venoms (from bees, hornets, wasps, and yellow jackets)
Hormone preparations	Insecticides	Snake venoms (from vipers and cobras)
Nuts	Mites and their feces	Spider venoms, especially from black widow and brown recluse
Penicillin	Pollen (from grass, trees, and weeds)	
Seafoods	Spores (fungal and bacterial)	

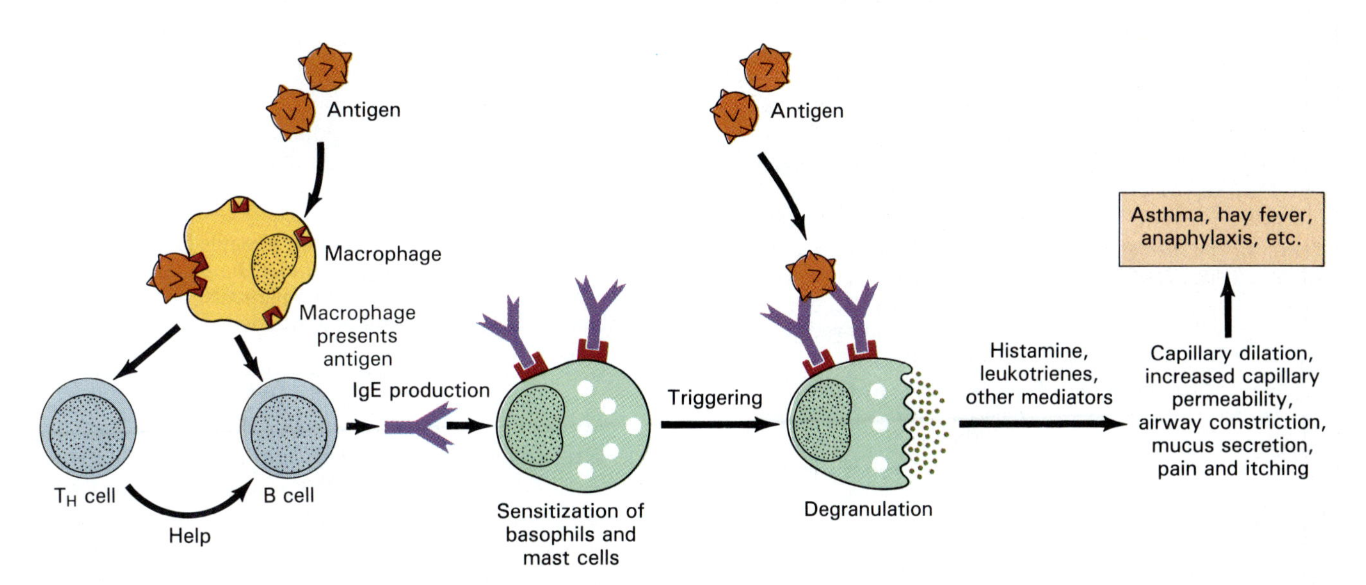

FIGURE 19.1 The mechanism of immediate (Type I) hypersensitivity or anaphylaxis.

In sensitized people, IgE antibodies attach to receptors on mast cells and basophils so that those cells, in turn, become sensitized to the allergen. Thus, the immune system is primed to stage a massive attack on the allergen the next time it is encountered. The sensitizing (first) dose of an allergen must be fairly high, but the eliciting (second) dose can be quite small. When the second encounter occurs, the antigen attaches to sensitized mast cells and basophils. The cells then undergo **degranulation,** releasing histamine and other *mediators* (chemical substances that induce responses) of allergic reactions.

Mediators of allergic reactions have a variety of effects (Table 19.2). *Histamine* dilates capillaries and makes them more permeable. It also contracts bronchial smooth muscle and increases mucus secretion in addition to causing itching. *Prostaglandins,* which are cellular messengers, also contribute to airway constriction. Many years ago, a substance from mast cells was shown to produce slow, long-lasting airway constriction in animals. The substance, called **slow-reacting substance of anaphylaxis** (SRS-A), is now known to consist of three leukotriene mediators. These *leukotrienes* are 100 to 1000 times as potent as histamines and prostaglandins in causing airway constriction and are not inactivated by antihistamines. Both leukotrienes and prostaglandins also dilate and increase the permeability of capillaries, help to form very thick mucus, and stimulate nerve endings that cause pain and itching.

TABLE 19.2 Mediators of immediate hypersensitivity and their effects

Mediator	Effects
Preformed mediators	
Histamine	Vascular dilation and increased capillary permeability, bronchial smooth muscle contraction, edema of mucosal tissues, secretion of mucus, and itching
Neutrophil and eosinophil chemotactic factors	Attraction of neutrophils, eosinophils, and other leukocytes to the site of an allergic reaction
Newly formed mediators	
Leukotrienes of SRS-A	Prolonged bronchial smooth muscle contraction, increased capillary permeability, edema of mucosal tissues, and secretion of mucus
Platelet-activating factor (a prostaglandin)	Formation of minute blood clots, bronchoconstriction, and capillary dilation

Atopy

Atopy (ah'to-pe), which literally means "out of place," refers to localized allergic reactions. These reactions occur first at the site where the allergen enters the body. If the allergen enters the skin, it causes a wheal and flare reaction of redness, swelling, and itching (Figure 19.2). If the allergen is inhaled, mucous membranes or the respiratory tract become inflamed, and the patient has a runny nose and watery eyes. If the allergen is ingested, mucous membranes of the digestive tract become inflamed, and the patient may have abdominal pain and diarrhea. Some ingested allergens, such as foods and drugs, also cause skin rashes.

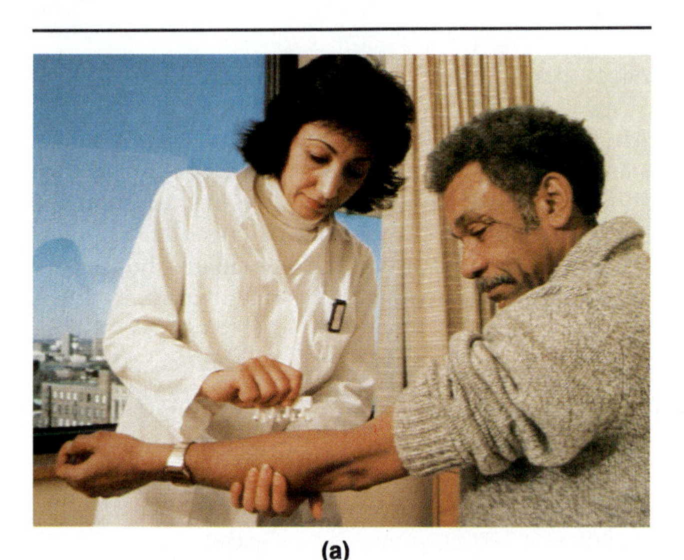

(a)

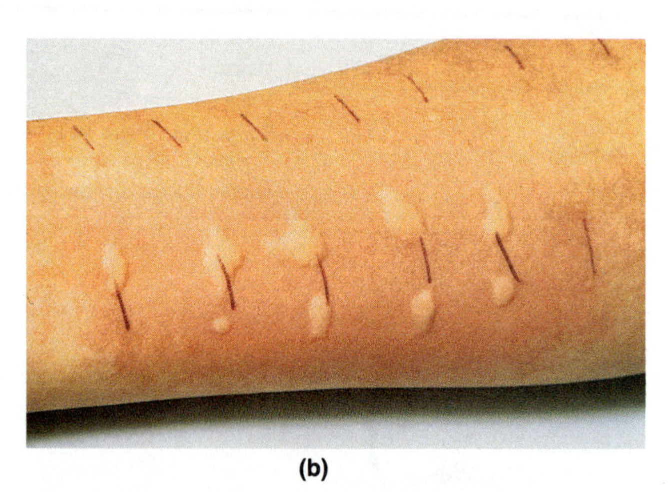

(b)

FIGURE 19.2 (a) In allergy testing, possible allergens are placed on prongs and introduced under the patient's skin. (b) If a patient is hypersensitive to a particular substance, a wheal (white raised area) and flare (reddened area) soon become visible on the skin.

Hay fever, or seasonal allergic rhinitis, is a common kind of atopy. Over 20 million Americans suffer from it. First described in 1819 as resulting from exposure to newly mown hay, it is now known to result from exposure to airborne pollen—tree pollens in the spring, grass pollens in the summer, and ragweed pollen in the fall (Figure 19.3). Some plants, such as

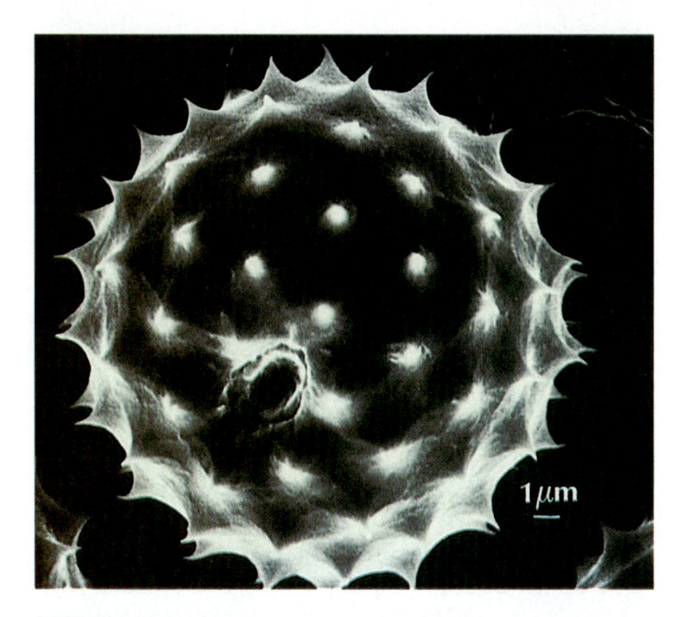

FIGURE 19.3 Ragweed (*Ambrosia*) pollen, a cause of hay fever. Various types of pollen are associated with hay fever. In early spring, the culprits are primarily tree pollens such as oak, elm, birch (especially in Europe), and box elder. In late spring and early summer, grass pollens plus those of some broad-leaved plants are most likely to be involved. In late summer and early autumn, the chief allergens are ragweeds, saltbush, and Russian thistle pollens.

goldenrod and roses, that have been incriminated in hay fever are innocent. Although their flowers are highly visible at hay fever time, their pollens are too heavy to be airborne for any great distance. It is far less conspicuous green ragweed flowers that cause much of the misery for hay fever sufferers. Hay fever can be distinguished from the common cold by the increased numbers of eosinophils in nasal secretions. Finding elevated numbers of eosinophils in blood also suggests allergy or infection with eukaryotic parasites, especially worms.

Generalized Anaphylaxis

Sometimes anaphylactic reactions are generalized, severe, and immediately life-threatening. A generalized reaction begins with sudden reddening of skin, intense itching, and hives, especially over the face, chest, and palms of hands. Then it progresses to respiratory anaphylaxis or anaphylactic shock.

In **respiratory anaphylaxis** the airways become severely constricted and filled with mucous secretions, and the patient may die of suffocation. **Asthma,** which is characterized by these symptoms, is often caused by inhaled or ingested allergens but also can be caused by hypersensitivity to endogenous microorganisms. For example, some patients become sensitized to *Branhamella catarrhalis,* a normal resident of respiratory mucous membranes.

In anaphylactic shock blood vessels suddenly dilate and become more permeable, causing an abrupt and life-threatening drop in blood pressure. Insect bites and stings are a common cause of anaphylactic shock because many people are allergic to insect venoms (Figure 19.4a).

Both airway blockage and shock must be treated immediately. Unless epinephrine (adrenaline) is administered immediately, death can occur. Epinephrine acts by constricting blood vessels and relaxing smooth muscle of respiratory passageways. Patients who have recovered from generalized anaphylactic reactions often carry an emergency anaphylactic kit (Figure 19.4b). The kit contains a tourniquet, benadryl (antihistamine) tablets, and a syringe containing two doses of epinephrine. Having such a kit on hand could easily mean the difference between life and death because of the rapid onset of life-threatening symptoms in patients who already have had anaphylactic reactions.

FIGURE 19.4
(a) Honey bee sting to the face has swollen an eye shut but has not caused the airway to shut, as happens in more severe reactions in which the larynx becomes edematous.
(b) Anaphylactic kit for emergency use, showing syringe loaded with epinephrine. Many individuals with severe insect-sting allergies carry these with them at all times. They are available only by prescription.

(a)

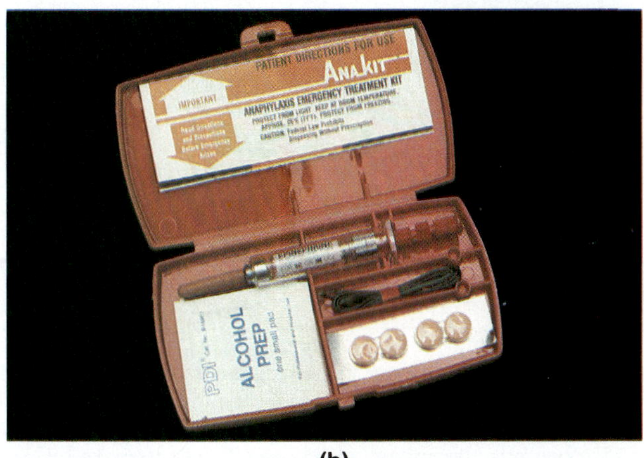

(b)

Genetic Factors in Allergy

Genetic factors are thought to contribute to the development of allergies. Among the population of the United States, 10 to 20 percent have some kind of allergy. In some families the percentage is much higher, although members of such families often have different allergies—asthma in one person, house-dust allergy in another. At least 60 percent of children with atopy have a family history of asthma or hay fever, and half these children later develop other allergies. Thus, allergy probably has a genetic basis, possibly through properties of membranes or performance of phagocytes. Normal membranes screen out all but the tiniest microorganisms and virtually all potential allergens, but membranes of allergic individuals are permeable to larger particles such as pollen grains. Even when allergens pass through membranes, phagocytic cells ordinarily engulf them in normal individuals, but they sometimes fail to do so in allergic individuals.

Treatment of Allergies

Desensitization (Figure 19.5) is the only currently available treatment intended to cure an allergy. The allergic individual receives injections with gradually increasing doses of the allergen, which are presumed to elicit IgG antibodies, called **blocking antibodies,** against the allergen. Blocking antibodies complex with the allergen before it has a chance to react with IgE, so mast cells do not release mediators of allergic symptoms. Suppressor T cells sensitized to the allergen also increase significantly during hyposensitization. Increases in IgG and sensitized suppressor T cells and decreases in IgE work together to make the patient less sensitive to the allergen. Unfortunately, hyposensitization does not alleviate symptoms in many allergies, and the treatment itself can cause anaphylactic shock because the injections contain the very substance to which the patient is allergic. Patients must remain in the physician's office for 20 to 30 min-

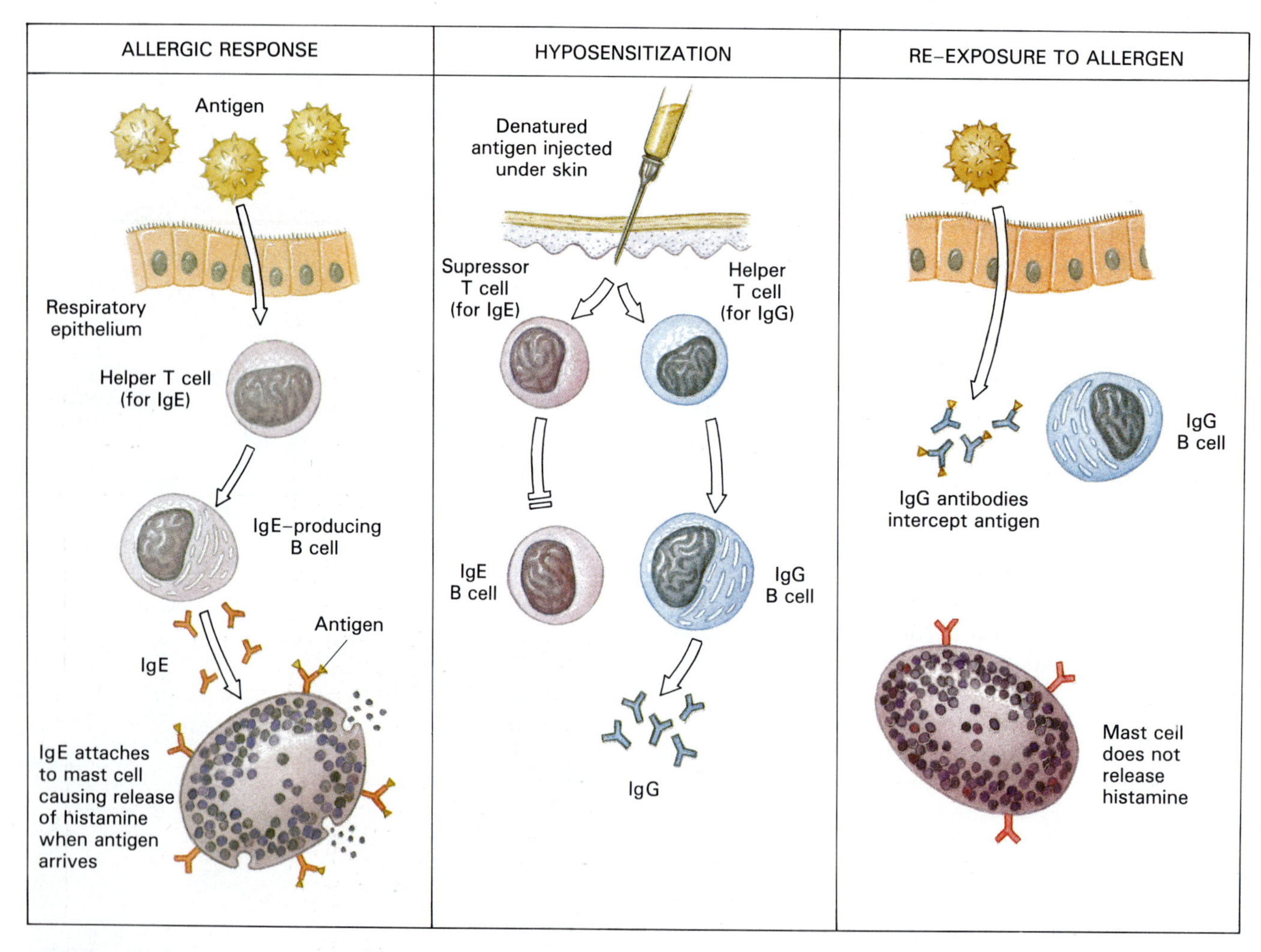

FIGURE 19.5 A proposed mechanism of action for desensitization allergy shots (hyposensitization). (a) Natural exposure to antigen causes helper T cells to stimulate B cells to make IgE antibodies, which attach to mast cells. (b) Denatured antigen is used in treatment, causing suppressor T cells to prevent B cells from making IgE antibodies. (c) Exposure to the antigen also causes B cells to make IgG (blocking antibody), which complexes with the incoming antigen before it reaches the IgE molecules attached to mast cells. Since complexing of antigen with these attached IgE molecules would cause the mast cells to degranulate and release histamine, blocking this step is the key to preventing allergic responses.

utes after the injection so that emergency treatment will be available if a generalized anaphylactic reaction occurs.

Other allergy treatments alleviate symptoms but do not cure the disorder. Antihistamines counteract the swelling and redness due to histamine, and corticosteroid hormones suppress the inflammatory response. However, antihistamines are not effective against SRS-A, which constricts airways. Aspirin taken before ingesting food allergens prevents symptoms such as gastrointestinal upsets and hives. Better methods of treating allergies are greatly needed. As we learn more about the properties of leukotrienes and immunoglobulins (especially IgE and IgD), perhaps this need can be met.

CYTOTOXIC REACTIONS (TYPE II HYPERSENSITIVITY)

In cytotoxic (Type II) hypersensitivity, specific antibodies react with cell-surface antigens interpreted as foreign by the immune system. Antigens that initiate cytotoxic hypersensitivity often enter the body in mismatched blood transfusions or during delivery of an Rh-positive infant to an Rh-negative mother. Antigens on cell surfaces of certain body tissues also can trigger cytotoxic reactions if the immune system erroneously treats them as foreign. In contrast to damage caused by mediators in Type I hypersensitivity, cytotoxic hypersensitivity reactions cause direct cellular damage.

Mechanism of Cytotoxic Reactions

When an antigen on a cell membrane is first recognized as foreign, B cells become sensitized and stand ready to make antibodies when the antigen is again encountered. Such reactions often occur in patients who do not know they have been sensitized to an antigen. At second and subsequent encounters with a cell-surface antigen, antibodies bind to the antigen and activate complement. Then phagocytic cells, such as macrophages and neutrophils, are attracted to the site. The activities of these cells and complement destroy cells, as summarized in Figure 19.6.

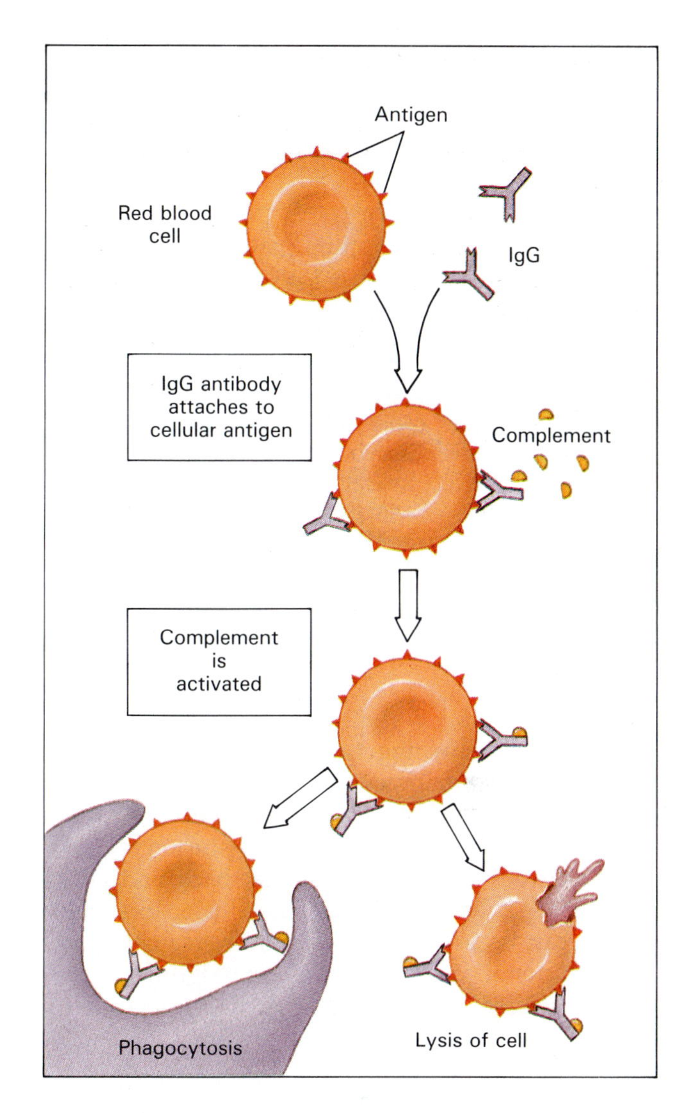

FIGURE 19.6 The mechanisms of cytotoxic (Type II) hypersensitivity. Mismatched red blood cell antigen is bound to IgG or IgM. Complement is activated and results in either opsonization with subsequent phagocytosis, or lysis of the red cell.

Transfusion Reactions

Normal human erythrocytes have genetically determined surface antigens that specify many different blood types. A **transfusion reaction** can occur when matching antigens and antibodies are present in the blood at the same time. Although such reactions can be triggered by any erythrocyte antigens, we will focus on antigens A and B, which determine the ABO blood groups. As shown in Table 19.3, four blood types, A, B, AB, and O, are named according to whether erythrocytes have antigen A, antigen B, both, or neither. Normally, a person's serum has no IgM antibodies against the antigens present on his or her own erythrocytes. If a patient receives erythrocytes with a different antigen during a blood transfusion, IgM antibodies cause a Type II hypersensitivity reaction

TABLE 19.3 Properties of ABO blood types

Blood Type	Antigens on Erythrocytes	Antibodies in Serum
A	A	anti-B
B	B	anti-A
AB	A and B	neither anti-A nor anti-B
O	neither A nor B	anti-A and anti-B

APPLICATIONS

IgA and Transfusion Reactions

Sometimes transfusion reactions can occur even when blood is carefully matched. Some people, for example, lack IgA antibodies of their own and thus do not acquire tolerance to this immunoglobulin. When such individuals receive transfusions they are likely to produce antibodies against IgA in the donated blood, and any subsequent transfusion will then evoke a transfusion reaction. Interestingly, some patients exhibit such reactions on receiving their first transfusion, indicating a previous exposure to IgA. One possible explanation is that they have been sensitized by eating rare beef, the blood of which contains IgA.

against the foreign antigen. The foreign erythrocytes are agglutinated (clumped), complement is activated, and hemolysis (rupture of blood cells) occurs within the blood vessels. Symptoms of a transfusion reaction include fever, low blood pressure, back and chest pain, nausea, and vomiting. Transfusion reactions usually can be prevented by carefully matching antigens of donor and recipient blood. (Figure 19.7).

Transfusion reactions to other erythrocyte antigens, such as Rh (Rhesus), also occur, but they usually are less serious than reactions to foreign A or B antigens because the antigen molecules are less numerous.

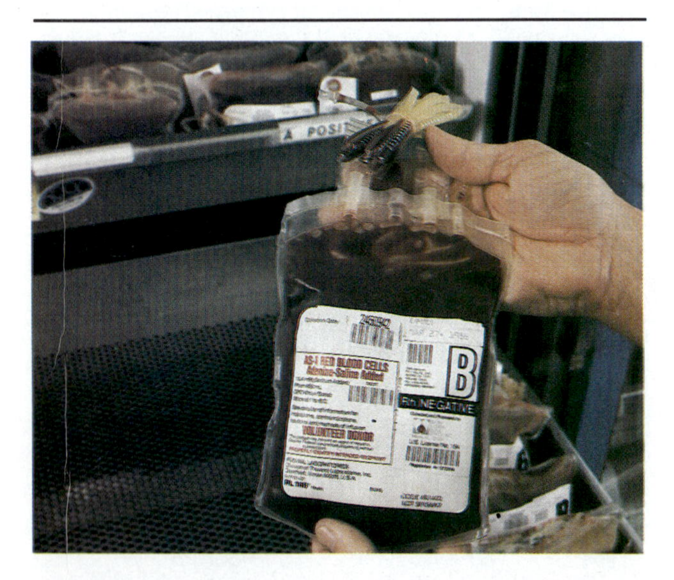

FIGURE 19.7 Careful blood typing and matching of donor and recipient types prevents most transfusion reactions. Persons with type AB blood can safely receive transfusions of any of the four blood types. Persons with type O blood can safely donate blood to recipients regardless of their ABO blood type. (Can you see why? Refer to Table 19.3.)

Hemolytic Disease of the Newborn

In addition to A or B antigens, erythrocytes also can have relatively large numbers of Rh antigens, so named because they were first discovered in Rhesus monkeys. Blood with Rh antigens on erythrocytes is designated Rh-positive; blood lacking Rh antigens on erythrocytes is designated Rh-negative. Anti-Rh antibodies normally are not present in the serum of either Rh-positive or Rh-negative blood. Consequently, sensitization is necessary for an Rh antigen-antibody reaction.

Sensitization typically occurs when an Rh-negative woman carries an Rh-positive fetus, which inherited this blood type from its father. The fetal Rh antigen rarely enters the mother's circulation during pregnancy but can leak across the placenta during delivery, miscarriage, or abortion. The Rh-negative mother's immune system then becomes sensitized to the Rh antigen and can produce anti-Rh antibodies when it again encounters it.

Because sensitization usually occurs at delivery, the first Rh-positive child of an Rh-negative mother rarely suffers from hemolytic disease. But when a sensitized Rh-negative mother carries a second Rh-positive fetus, the mother's anti-Rh antibodies cross the placenta and cause a Type II hypersensitivity reaction in the fetus. Fetal erythrocytes agglutinate, complement is activated, and erythrocytes are destroyed by opsonization and cell membrane attack. The result is **hemolytic disease of the newborn.** The baby is born with an enlarged liver and spleen caused by efforts of these organs to destroy damaged erythrocytes. Such babies exhibit the yellow skin color of **jaundice** due to excessive bilirubin in the blood from the breakdown of erythrocytes. These processes are summarized in Figure 19.8.

Hemolytic disease of the newborn can be prevented by giving Rh-negative mothers intramuscular injections of anti-Rh IgG antibodies (Rhogam) within 72 hours after delivery. The antibodies presumably bind to Rh antigens that have leaked into the mother's blood before they can act to sensitize her immune system. It is essential to treat all Rh-negative women after delivery, miscarriage, or abortion in case the fetus may have been Rh-positive. Nowadays anti-Rh antibodies are often administered to Rh-negative women during pregnancy as well. Such treatment at 3 and 5 months prevents sensitization to the fetus in case fetal antigens leak into the mother's circulation.

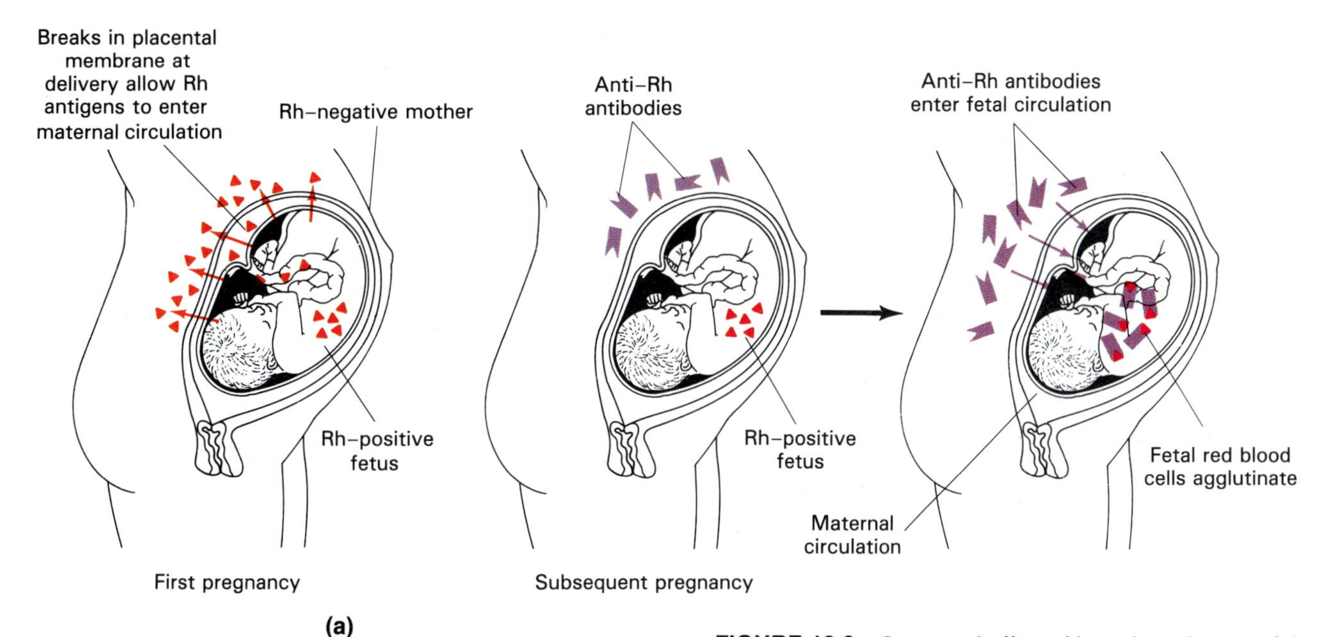

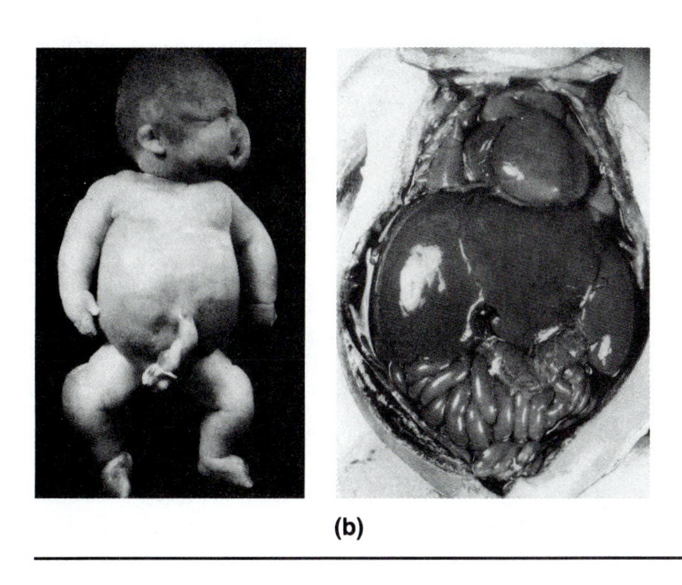

FIGURE 19.8 Cause and effect of hemolytic disease of the newborn. (a) The stage is set for an Rh-incompatibility pregnancy when the mother is Rh^- and the fetus is Rh^+. (This will usually be the case if the father is Rh^+.) Rh^+ antigens may cross the placenta and enter the mother's bloodstream before or at the time of delivery. She responds by making anti-Rh antibodies, which also can cross the placenta. If there has been significant leakage of fetal antigens through the placenta early in the pregnancy (this can result simply from hard coughing and sneezing), the maternal antibodies will be created in time to attack the fetus. Even if antibody production is not stimulated until the time of delivery, the resulting antibodies will persist in the mother's circulation and attack the red blood cells of any subsequent Rh^+ fetus. To prevent this, Rhogam (anti-Rh antibody) is administered to the mother during pregnancy to neutralize any fetal antigens that have entered her bloodstream, and it is given again immediately after delivery, miscarriage, or abortion. This reduces exposure to the antigen, and thus lessens anti-Rh antibody production, making both current and future pregnancies safer. (b) Child affected by hemolytic disease caused by Rh incompatibility. Note the greatly enlarged liver (see text).

Even a violent sneeze can cause leakage of the placenta. Before effective treatment was available, hemolytic disease of the newborn occurred in about 0.5 percent of all pregnancies, and 12 percent of these terminated in stillbirths.

IMMUNE COMPLEX DISORDERS (TYPE III HYPERSENSITIVITY)

Immune complex (Type III) hypersensitivity results from the formation of antigen-antibody complexes. Under normal circumstances these large immune complexes are engulfed and destroyed by phagocytic cells. Hypersensitivity occurs when the complexes persist or are continuously formed.

Mechanism of Immune Complex Disorders

Immune complex disorders are initiated by sensitization. On second exposure to the sensitizing antigen, specific IgG antibodies combine with the antigen in the blood and activate complement. Antibodies bind to live cells or parts of damaged cells in blood vessel walls and other tissues. Such immune complexes are phagocytized by Kupffer cells in the liver only if they are coated with *both* antibodies and complement. Some immune complexes are very small molecules that fail to bind tightly to Kupffer cells and thereby escape destruction. Antigen-antibody complexes and complement, in turn, cause basophils and platelets to release histamine and other mediators of allergic responses, with the effects described earlier. Platelets,

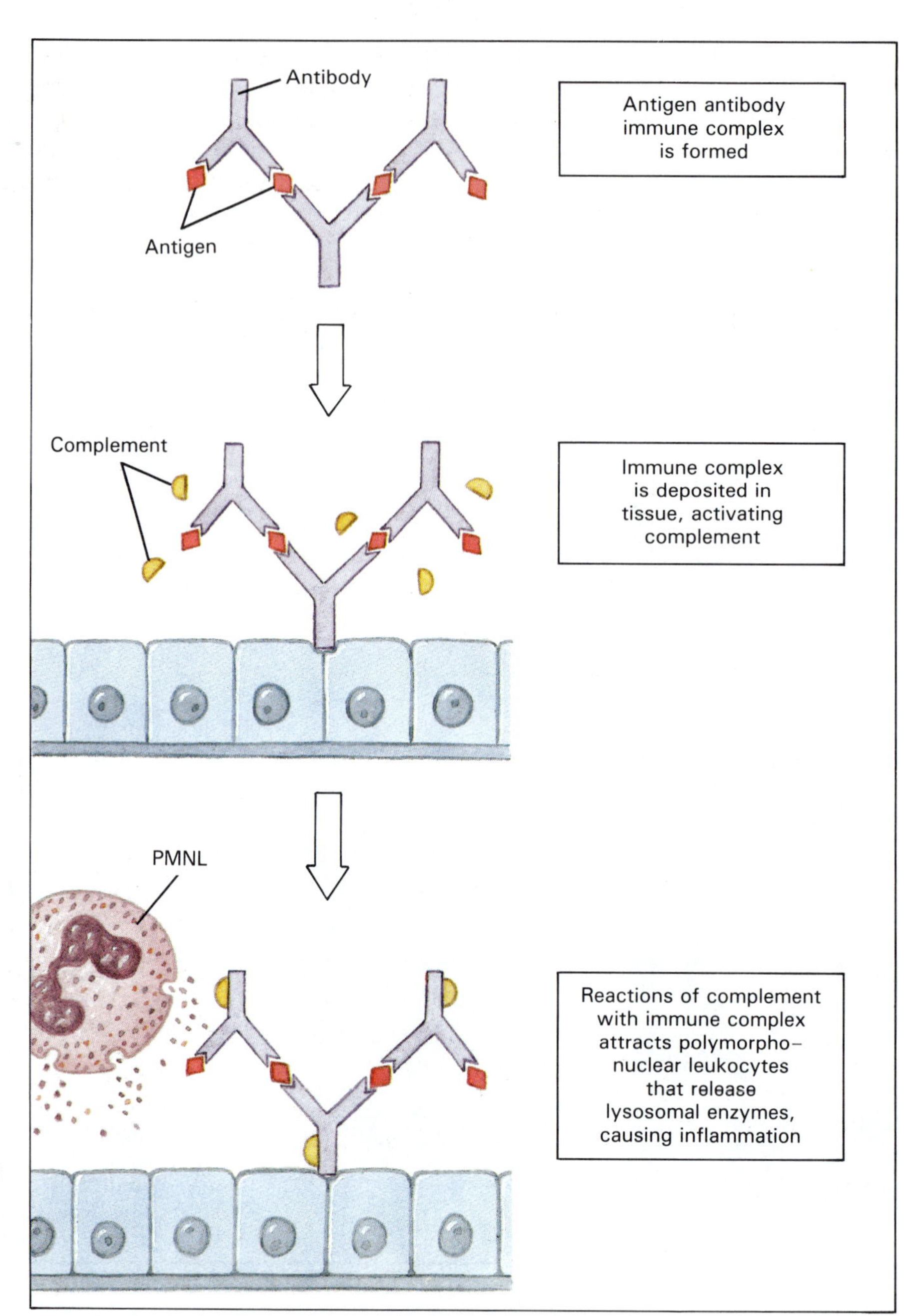

FIGURE 19.9 The mechanism of immune complex (Type III) hypersensitivity. Immune complexes are formed when antigen enters a previously sensitized individual. The resulting IgG or IgM complex is deposited, where it activates complement, producing fever, itching, rash or hemorrhagic areas, joint pain, and acute inflammation. On a systemic basis, this can cause serum sickness, and is observed in acute glomerulonephritis and lupus erythematosus. Localized symptoms are called Arthus reactions.

circulating cell fragments that are involved in blood clotting, form tiny blood clots. As a part of the inflammatory reaction, circulating leukocytes release lysosomal enzymes that damage blood vessel walls. These processes are summarized in Figure 19.9.

Examples of Disorders

We will illustrate immune complex disorders with two phenomena, serum sickness and the Arthus reaction. Other disorders that involve immune complexes, such as glomerulonephritis, rheumatoid arthritis, and systemic lupus erythematosus, are discussed later in connection with autoimmune diseases because the antibodies involved in these disorders react with the patient's own tissues.

Serum sickness was frequently seen in the preantibiotic era when large doses of antisera were used to immunize passively against infectious diseases such as diphtheria. Diphtheria toxin given to horses caused them to make antibodies to it. A patient would receive antibodies against the toxin from horse serum but would also receive foreign horse proteins. On this first exposure to the serum the patient's immune system would make antibodies against horse proteins, resulting in the formation of antigen-antibody complexes upon second exposure. Binding of these complexes to glomeruli of the kidneys impairs the filtration

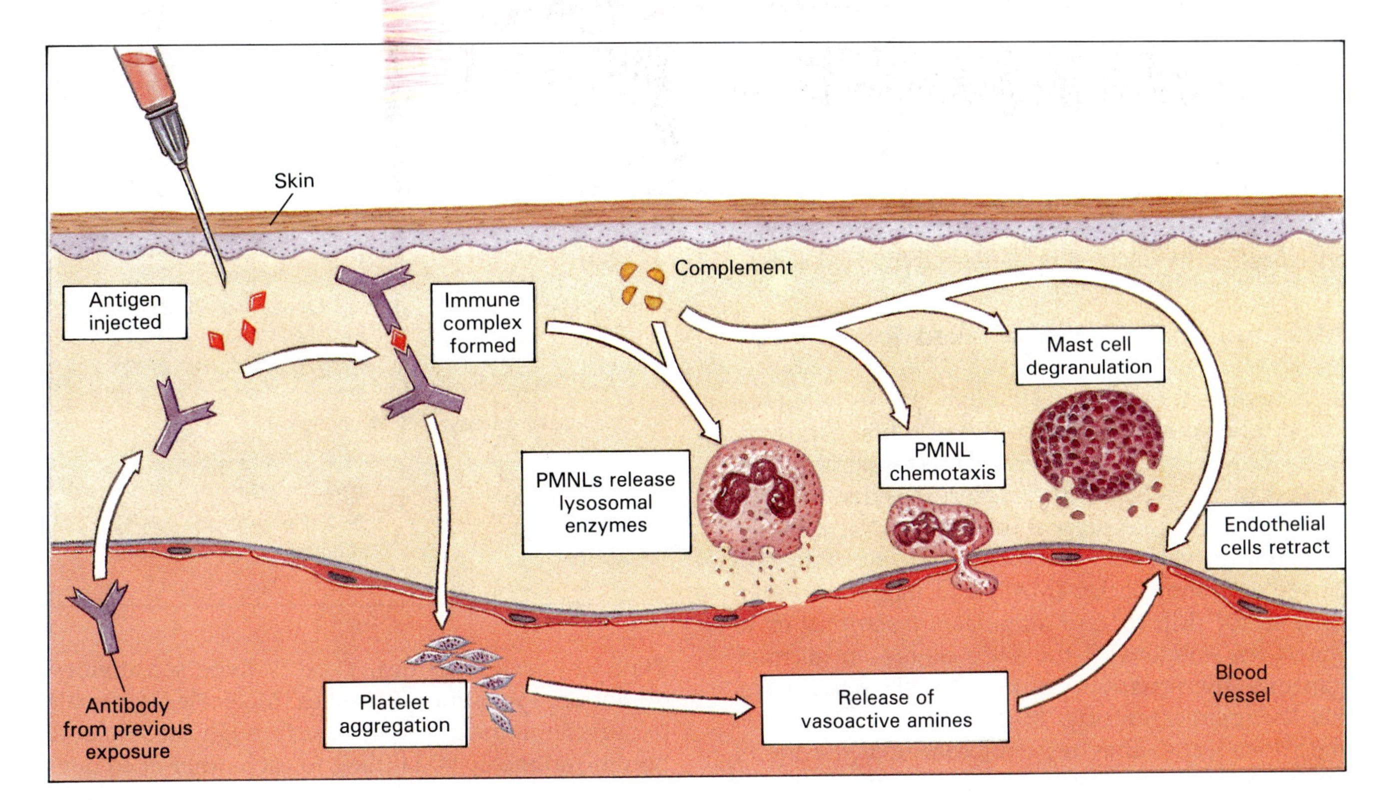

FIGURE 19.10 The mechanism that gives rise to an Arthus reaction and hemorrhagic areas. In severe cases, tissue death and Arthus reactions usually follow readministration of a biological material, such as a vaccine produced in horse serum, to a patient who has previously become sensitized to that material by earlier administration.

capacity of the glomeruli and allows proteins and blood cells to be excreted in the urine.

Patients with serum sickness usually have fever, enlarged lymph nodes, decreased numbers of circulating leukocytes, and swelling at the injection site. Most patients recover from serum sickness as the complexes are cleared from the blood and tissue repair occurs in the glomeruli. However, the disorder became chronic in many diphtheria patients because they received horse serum daily over the course of the disease.

Today, serum sickness is rare and usually is due to second exposure to a foreign substance in a biological product, such as horse serum in a vaccine. An advantage of vaccines made by genetic engineering is that they do not contain such foreign substances. When use of any biological product is contemplated, the patient should be tested for sensitivity first. A small quantity of the product should be given intradermally (within the skin) or intravenously. A wheal and flare reaction or a drop of 20 points or more in blood pressure following intravenous injection indicate hypersensitivity, and the product should not be given.

The **Arthus reaction** is a local reaction seen in the skin after subcutaneous (under the skin) or intradermal injection of an antigenic substance. It occurs in patients who already have large quantities of antibodies (mainly IgG) to the antigen. In 4 to 10 hours edema and hemorrhage develop around the injection site as immune complexes and complement trigger cell damage and platelet aggregation (Figure 19.10). In severe reactions, tiny clots obstruct blood vessels and cells normally nourished by the blocked vessels die (Figure 19.11).

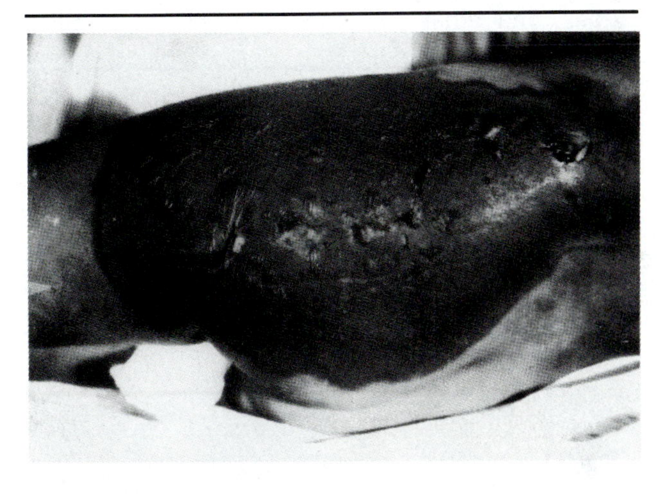

FIGURE 19.11 Arthus reaction in patient showing extensive area of hemorrhagic damage which will result in tissue necrosis.

CELL-MEDIATED REACTIONS (TYPE IV HYPERSENSITIVITY)

Cell-mediated (Type IV) hypersensitivity is also called **delayed hypersensitivity** because reactions take more than 12 hours to develop. It is mediated by T cells—specifically, the delayed hypersensitivity (T_D) cells.

Mechanism of Cell-mediated Reactions

Cell-mediated reactions, although not fully understood, appear to occur as follows. On first exposure, antigen molecules bind to macrophages and are presented to T_D cells, which become sensitized. When macrophages again present the same antigen, the T_D cells release various lymphokines, including macrophage activating factor (MAF) and migration inhibiting factor (MIF). MAF stimulates macrophages to ingest the antigens. If the antigens are on microorganisms, the macrophages usually, but not always, kill the microorganisms. MIF prevents migration of macrophages, so they remain localized at the site of the hypersensitivity reaction. Other lymphokines are presumed to cause the hypersensitivity reaction itself. Such reactions account for patches of raw, reddened skin in eczema, swelling, and granulomatous lesions. These processes are summarized in Figure 19.12.

Examples of Disorders

Three common examples of delayed hypersensitivity—contact dermatitis, tuberculin hypersensitivity, and granulomatous hypersensitivity—illustrate the diversity of cell-mediated reactions.

Contact dermatitis occurs in sensitized individuals on second or subsequent exposure to allergens such as oils from poison ivy, rubber, certain metals, dyes, soaps, cosmetics, some plastics, topical medications, and other substances (Table 19.4). Unlike Type I hypersensitivities, Type IV hypersensitivities do not appear to run in families. Molecules too small to cause immune reactions pass through the skin and become antigenic by binding to each other or to normal proteins on Langerhans cells of the skin. The antigen is then presented to T_D cells and a hypersensitivity reaction occurs, usually as eczema, in about 48 hours.

Urushiol, an oil from the poison ivy plant, is a major cause of contact dermatitis in the United States (Figure 19.13). Most people get poison ivy from direct contact with leaves or other plant parts, but some get it by inhaling smoke from burning brush that contains poison ivy plants. Poison ivy is particularly severe when oil droplets come in contact with respiratory membranes. Sensitivity to poison ivy can develop at any age, even among people who have come in contact with it for years without reacting. One way to minimize a reaction to poison ivy is to wash exposed areas thoroughly with strong soap or detergent within minutes of contact, before much oil has penetrated

TABLE 19.4 Selected contact allergens

Allergen	Common Sources of Contact
Benzocaine	Topical anesthetic
Chromium	Jewelry, watches, chrome-tanned leather, cement
Formaldehyde	Facial tissues, nail hardeners, synthetic fabrics
Nickel	Jewelry, watches, objects made of stainless steel and white gold
Mercaptobenzothiazole	Rubber goods
Methapyrilene	Topical antihistamine
Merthiolate	Topical antiseptic
Neomycin	Topical antibiotic
Oleoresin	Oil from poison ivy and similar plants

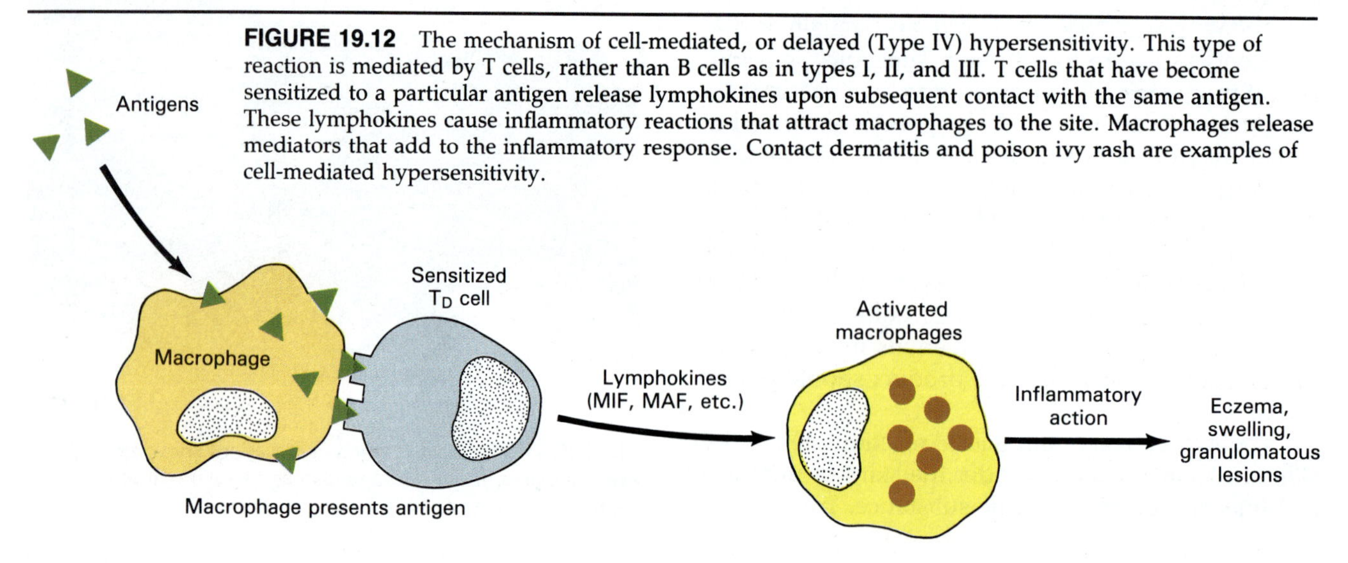

FIGURE 19.12 The mechanism of cell-mediated, or delayed (Type IV) hypersensitivity. This type of reaction is mediated by T cells, rather than B cells as in types I, II, and III. T cells that have become sensitized to a particular antigen release lymphokines upon subsequent contact with the same antigen. These lymphokines cause inflammatory reactions that attract macrophages to the site. Macrophages release mediators that add to the inflammatory response. Contact dermatitis and poison ivy rash are examples of cell-mediated hypersensitivity.

(a)

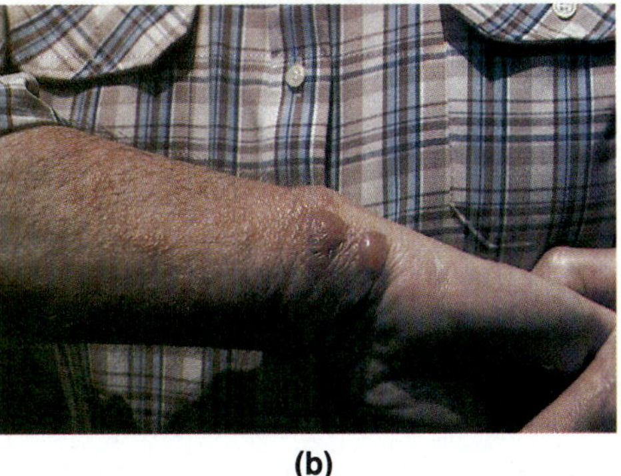

(b)

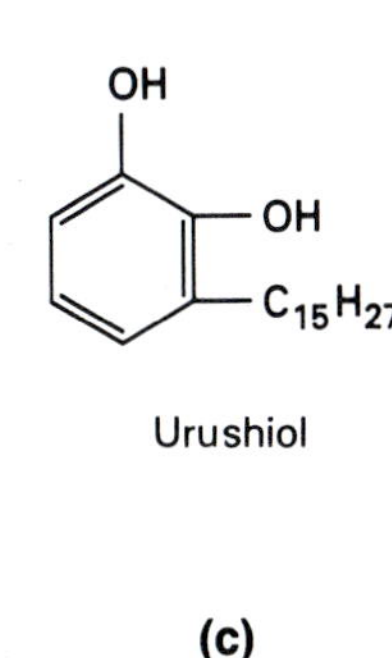

(c)

FIGURE 19.13 (a) Poison ivy (*Toxicodendron radicans*), showing the leaves with their characteristic three leaflets. Poison ivy vines also contain the irritating oil urushiol, so it is important to be able to recognize them in winter. When burning brush or logs, avoid using any with furry-looking vines on them, or you may be hospitalized with poison ivy of the lungs from inhaling the smoke (which contains the volatile oil) or have extensive areas of skin affected by smoke blown over them. (b) Poison ivy dermatitis showing two large fluid-filled vesicles and many smaller ones. (c) The chemical structure of urushiol.

the skin and bound chemically to skin cells. Once a person is sensitized, an exceedingly small quantity of oil will elicit a reaction. Scratching lesions does not spread the oil, but it can lead to infections. Cashews and mangos contain substances chemically similar to urushiol, and some people display delayed hypersensitivity to these substances.

Tuberculin hypersensitivity occurs in sensitized individuals when they are exposed to tuberculin, an antigenic lipoprotein from the tubercle bacillus. Similar antigens from the agents that cause leprosy and leishmaniasis produce similar reactions. The antigen activates T_D cells, which in turn release lymphokines that cause large numbers of lymphocytes and monocytes to infiltrate the dermis. The hypersensitivity reaction usually is limited to the dermis. The normally soft tissues of the skin form a raised, hard, red region called an **induration** (Figure 19.14). In a tuberculin skin test the diameter and height of the induration indicate the degree of hypersensitivity.

Granulomatous hypersensitivity, the most serious of the cell-mediated hypersensitivities, usually occurs when macrophages have engulfed pathogens but have failed to kill them. Inside the macrophages, the protected pathogens survive and sometimes continue to divide. T_D cells sensitized to an antigen of the pathogen elicit the hypersensitivity reaction, and a granuloma develops (see Figure 17.22, leprosy). This kind of hypersensitivity is the most delayed of all, appearing 4 weeks or more after exposure to the antigen.

The characteristics of the types of hypersensitivity are summarized in Table 19.5.

APPLICATIONS

Poison Ivy? But It Doesn't Grow Here!

Contact dermatitis similar to poison ivy was seen in American military personnel in Japan after World War II. Lesions appeared on elbows, forearms, and in a horseshoe shape on the buttocks and thighs. Knowing that poison ivy did not grow in Japan, the medical staff was puzzled. The puzzle was solved when scientists discovered that oils from a Japanese plant containing a small quantity of urushiol were used to manufacture lacquer. Although the quantity of urushiol in lacquer was not sufficient to sensitize Japanese people, it did elicit an allergic response in Americans previously sensitized to poison ivy. Resting the arms on bar tops explained lesions on elbows and forearms; contact with toilet seats explained the horseshoe-shaped lesions.

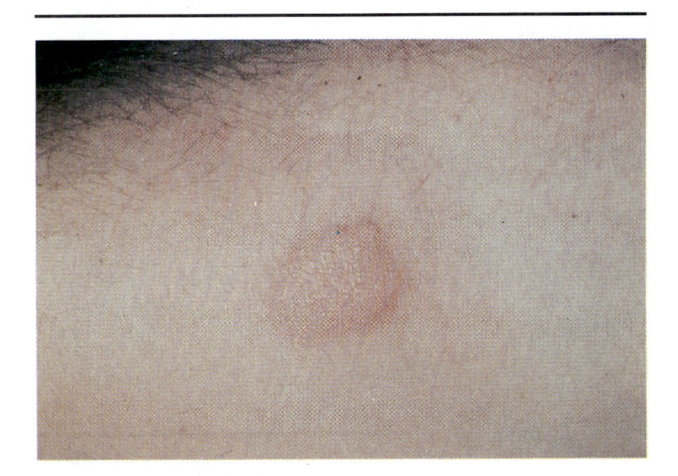

FIGURE 19.14 A positive tuberculin skin test reaction. The raised area of induration should be observed and measured after 48 to 72 hours. A positive reaction will measure 5 mm or more; 2 mm or smaller is negative; 3 and 4 mm are considered doubtful.

TABLE 19.5 Characteristics of the types of hypersensitivity

	Type I	Type II	Type III	Type IV
Characteristic	Immediate	Cytotoxic	Immune Complex	Delayed
Main mediators	IgE	IgG, IgM	IgG, IgM	T cells
Other mediators	Mast cells, anaphylactic factors, eosinophils	Complement	Complement, inflammatory factors, eosinophils, neutrophils	Lymphokines, macrophages
Antigen	Soluble or particulate	On cell surfaces	Soluble or particulate	On cell surfaces
Reaction time	Seconds to 30 minutes	Variable, usually hours	3 to 8 hours	24 to 48 hours
Nature of reaction	Local flare and wheal, airway constriction, anaphylactic shock	Clumping of erythrocytes, cell destruction	Acute inflammation effects	Cell-mediated cell destruction
Therapy	Desensitization, antihistamines, steroids	Steroids	Steroids	Steroids

AUTOIMMUNE DISORDERS

Autoimmune disorders occur when individuals become hypersensitive to antigens on cells of their own bodies, in spite of mechanisms that ordinarily create tolerance to those antigens. The antigens elicit an immune response in which **autoantibodies,** antibodies against one's own tissue, are produced. These disorders are characterized by cell destruction in various types of hypersensitivity reactions. Autoimmune disorders range over a wide spectrum from those that affect a single organ or tissue to those that affect many organs and tissues (Table 19.6).

TABLE 19.6 The spectrum of autoimmune disorders

Disorder	Organ(s) Affected	Nature of Autoantibody
Organ-Specific Disorders		
Addison's disease	Adrenal glands	Antiadrenal gland
Autoimmune hemolytic anemia	Erythrocytes	Antierythrocyte
Glomerulonephritis	Kidneys	Streptococcal cross-reactivity with kidney
Graves' disease	Thyroid gland	Anti-TSH
Hashimoto's thyroiditis	Thyroid gland	Antithyroglobulin
Idiopathic thrombocytopenia purpura	Blood platelets	Antiplatelet
Juvenile diabetes	Pancreas	Antibeta cells and anti-insulin
Myasthenia gravis	Skeletal muscles	Antiacetylcholine
Pernicious anemia	Stomach	Antivitamin B_{12} binding site
Postvaccine/postinfection encephalomyelitis	Myelin	Measles cross-reactivity with myelin
Premature menopause	Ovaries	Anticorpus luteum cells
Rheumatic fever	Heart	Streptococcal cross-reactivity with heart
Spontaneous male infertility	Sperm	Antisperm
Ulcerative colitis	Colon	Anticolon
Systemic (Disseminated) Disorders		
Goodpasture's syndrome	Basement membrane	Antibasement membrane
Mixed connective tissue disease (MCTD)	Connective tissues	Antiribonucleoprotein
Polymyositis/dermatomyositis	Muscles and skin	Antinuclear and other
Rheumatoid arthritis	Basement membranes and other tissues	Antinuclear, antigamma globulins, anti-EBV
Scleroderma	Connective tissues	Antinuclear
Sjögren's syndrome	Lacrimal and salivary glands	Antinuclear
Systemic lupus erythematosus	Many tissues	Antinuclear, antihistone, antilymphocyte, antierythrocyte, antiplatelet, antineuron
Vasculitis	Blood vessels	Circulating immune complexes

Autoimmunization

Autoimmunization is the process by which hypersensitivity to "self" develops. This process is poorly understood and may not be the same for all autoimmune disorders. Autoantibodies produced by B cells or autoreactive T cells have been identified in all kinds of autoimmune disorders, but it is not clear whether autoimmunity causes the disease, the disease causes autoimmunity, or both the disease and the autoimmunity are caused by some other factor.

Genetic factors may predispose toward autoimmune disorders, either directly or by making individuals susceptible to other causative agents. For example, if a parent has autoantibodies to a single organ, the children are likely to develop autoantibodies to the same or a different single organ. As we shall see later, individuals who have genes for certain histocompatibility antigens are at greater risk of developing particular autoimmune disorders.

In addition to predisposing genetic factors, several different mechanisms of autoimmunity probably exist. Mutations might give rise to cells that make autoantibodies. Antigens hidden away in tissues and lacking contact with B or T cells during development could be released and perceived as foreign by the immune system (see the box titled "Sympathetic Blindness"). Diminished suppressor T cell function, which accompanies aging, might allow immune responses to substances normally not antigenic. T cells might attack tissue antigens that are similar to antigens of some pathogens. (Such resemblances are known as antigenic mimicry). Viral components inserted into host cell membranes might act as antigens, or virus-antibody complexes might be deposited in tissues. Defects in stem cells, thymus cells, or macrophages might lead to inappropriate immune responses.

APPLICATIONS
Sympathetic Blindness

Lens proteins are normally sequestered within the lens capsule of the eye. Because they are never exposed to lymphocytes during development, the immune system never acquires tolerance for them. Sometimes an eye injury allows these proteins to leak into the bloodstream, where they elicit an immune response. The antibodies formed in this way then attack proteins in the undamaged eye. The immune response in the healthy eye can actually be more intense than that in the injured eye and sometimes leads to sympathetic blindness, or loss of sight in the uninjured eye.

The sympathetic nervous system, which along with the parasympathetic system controls internal body functions, helps to regulate the immune system. When the sympathetic nervous system is damaged, the number of suppressor T cells decreases. This lack of suppressor T cells might account for the development of autoimmune diseases.

Examples of Disorders

Autoimmune disorders usually are chronic inflammatory disorders with symptoms that can alternately worsen and disappear. They can affect one organ or many. We look at three examples to illustrate the diversity of such disorders.

Myasthenia Gravis

Myasthenia gravis, a rare autoimmune disease (3 per 100,000) specific to skeletal muscle, especially affects muscles of the limbs and those involved in eye movements, speech, and swallowing. It affects women twice as often as men and usually appears in late childhood to middle age. For muscles to contract normally, pores called calcium channels in the membranes of neurons that stimulate the muscles must open. Recent evidence shows that muscle contraction is impaired by autoantibodies blocking the calcium channels. The formation of autoantibodies is still not understood, but their effects are clear. Blocking of the calcium channels prevents neurons from releasing acetylcholine, the substance that initiates contraction in a muscle cell.

Myasthenia gravis is treated with drugs that inhibit cholinesterase, an enzyme that breaks down acetylcholine after it has initiated contraction. By slowing acetylcholine breakdown, the drugs compensate for the small quantity of acetylcholine and allow each molecule to act longer. No means of preventing antibodies from forming or removing them once they have formed is available.

In spite of muscle weakness, many women with myasthenia gravis have children. Their babies have temporary muscular weakness; they are like little rag dolls for the first few weeks of life. Small numbers of autoantibodies from the mother probably cross the placenta and bind to calcium channels in the fetus. Apparently the immune complexes do not permanently damage fetal neurons because the babies soon have normal muscle function.

Rheumatoid Arthritis

In contrast to the single-organ effects of myasthenia gravis, **rheumatoid arthritis** (RA) affects mainly the joints but can extend to other tissues. Of all forms of arthritis, RA is most likely to lead to crippling disabilities and to develop early in life (between the ages of 30 and 40). It does greatest damage to joints of the hands and feet.

Rheumatoid arthritis often starts with a joint inflammation from an infection that causes phagocytic cells to release degradatory enzymes called lysozymes. ∞ (Chapter 17, p. 459) These enzymes attack and alter certain IgG antibodies, causing them to become antigenic. B cells make IgM antibodies to the antigens and cause more joint inflammation. IgM is found in joint fluid and subcutaneous nodules (lumps beneath the skin) that often develop in individuals with RA. IgM also appears in the blood as **rheumatoid factor** in RA patients and their relatives, even when the relatives have no disease symptoms.

The Epstein-Barr virus (EBV), which infects B cells, may cause RA or secondarily infect inflamed joints. Because RA victims have abnormal immune responses, helper T cells fail to prevent spread of the virus, lymphokines are inhibited, and macrophages and cytotoxic T cells mount excessive attacks on tissues, especially cartilage and other joint components (Figure 19.15).

No cure exists for rheumatoid arthritis, but treatment can alleviate symptoms. Hydrocortisone lessens inflammation and reduces joint damage, but long-term use weakens bones and causes other undesirable side effects, including a reduction of normal immune response. Aspirin decreases inflammation and reduces pain with fewer side effects. Physical therapy is used to keep joints movable. In severe cases, surgical replacement of damaged joints can restore movement. Monoclonal antibodies may someday be used to destroy virus–infected cells before they can cause joint destruction.

Systemic Lupus Erythematosus

Systemic lupus erythematosus (SLE) is a widely disseminated, systemic autoimmune disease. The name is derived from the red (erythematose) wolf (lupus) bite, a butterfly-shaped rash over the nose and cheeks

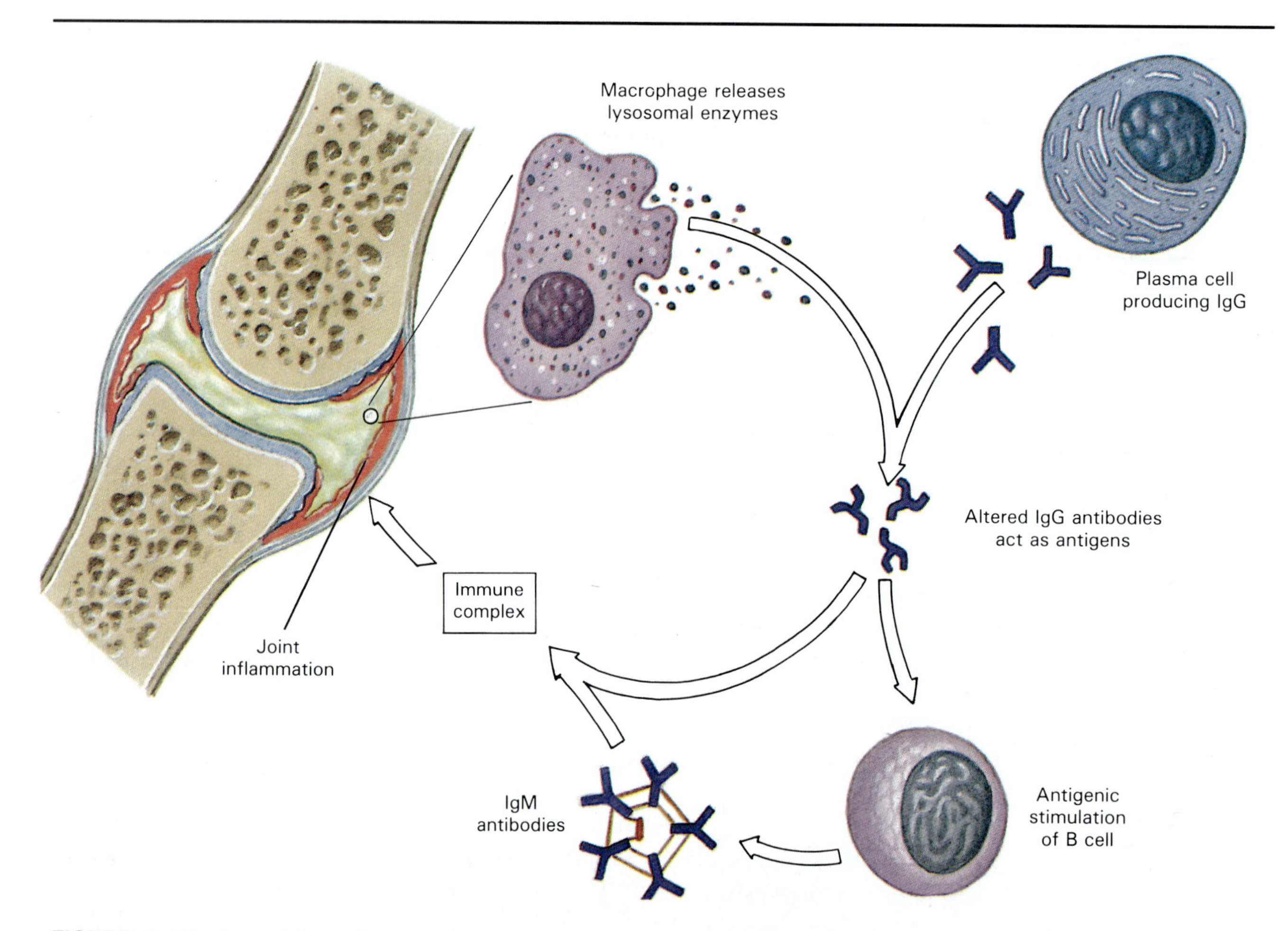

FIGURE 19.15 A possible mechanism of organ damage in rheumatoid arthritis. Joint inflammation triggers release of lysosomal enzymes that damage IgG antibodies. The altered antibodies are perceived as antigens by the body's immune system. B cells respond by making antibodies against the altered IgG, and immune complexes are formed. These in turn cause more joint inflammation. The result is a vicious cycle of continually worsening joint inflammation and destruction.

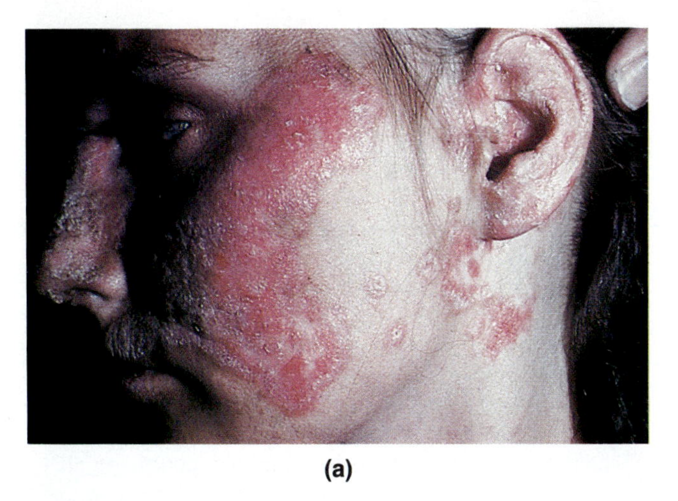

(a)

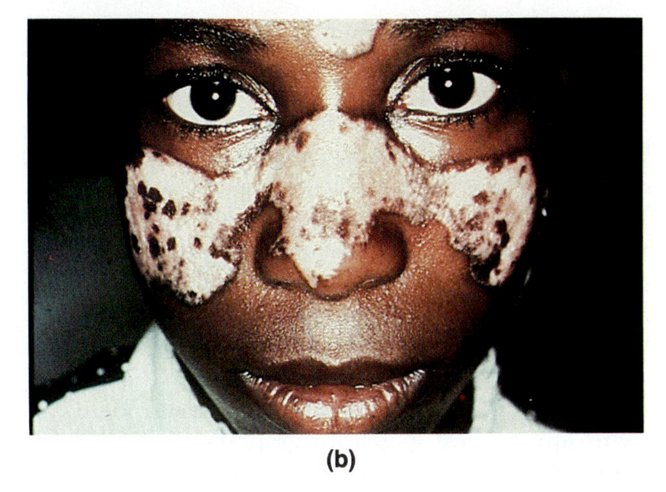

(b)

FIGURE 19.16 The characteristic butterfly-shaped rash of lupus erythematosus has a different appearance in people of different racial origins.

(Figure 19.16). Apparently, someone thought this rash looked like the pattern of colors on a wolf's face. SLE occurs four times as often in women as in men and usually appears during the reproductive years. In SLE autoantibodies are made primarily against components of DNA but can also be made against blood cells, neurons, and other tissues. Immune complexes are deposited between the dermis and epidermis and in blood vessels, joints, glomeruli of the kidneys, and the central nervous system. They cause inflammation and interfere with normal functions wherever they are.

The effects of SLE vary from one patient to another, depending on where the antigen-antibody complexes most interfere with function. Inflammation of blood vessels, heart valves, and joints are common effects. The skin rash appears in 80 percent of patients and is often precipitated by exposure to sunlight. Most SLE patients eventually die from kidney failure as glomeruli fail to remove wastes from the blood. Before fever could be controlled in SLE patients, many died of sudden high fevers early in the disease.

SLE can be diagnosed by finding lupus erythematosus (LE) cells in a patient's tissues. LE cells are neutrophils that have phagocytized breakdown products of cells damaged by the disease. In addition, diagnostic serological tests are available that look for antinuclear antibodies in the patient's serum. SLE cannot be cured but is treated with antipyretics to control fever, corticosteroids to reduce inflammation, and immunosuppressant drugs to prevent or decrease further autoimmune reactions.

TRANSPLANTATION

Transplantation is the moving of tissue, called **graft tissue,** from one site to another. An **autograft** involves grafting tissue from one part of the body to another; for example, the use of skin from a patient's chest to help repair burn damage on a leg. A graft between genetically identical individuals (identical twins in humans, or members of highly inbred animal strains) is called an **isograft.** A graft between two people who are not genetically identical is termed an **allograft.** Most organ transplants fall into this category. Transplants between individuals of different species are known as **xenografts.**

Early transplantation experiments involved grafting skin from one animal to another of the same species. The grafts at first appeared healthy, but in a few days to a few weeks they became inflamed and dropped off the recipient's body. First thought to be due to infection, this reaction, called **transplant rejection,** is now known to be due to the destruction of the grafted tissue by the host immune system. This process accounts for rejection of most organ transplants into humans.

A much less common transplantation effect is **graft-versus-host (GVH) disease,** in which host antigens elicit an immunologic response from graft cells. This response destroys host tissue and is most often seen in immunodeficient patients receiving bone marrow transplants.

Histocompatibility Antigens

All human cells, and those of many other animals, have a set of antigens called **histocompatibility antigens** (*histo,* Latin for tissue). The genes producing these are called the *major histocompatibility complex* (*MHC*). Only identical twins have exactly the same histocompatibility antigens, but family members have some of the same antigens. These antigens are located on the membranes of cells, including the cells of kidneys, hearts, and other organs commonly transplanted. If donor and recipient antigens are different, as they likely would be in randomly chosen donors and recipients, recipient T cells destroy donor tissue.

Like red blood cell antigens, tissue antigens can be identified by laboratory tests so that donor and recipient tissues can be as closely matched as possible. Because lymphocyte antigens are used, histocompatibility antigens are referred to as **human leukocyte antigens (HLAs).** Prospective transplant recipients have their tissues typed. When a donor organ becomes available, it is typed and transplanted into the recipient whose antigens most nearly match. This procedure reduces the risk of tissue rejection. Identical twins are the best match for grafts because all their antigens are the same, but siblings of the same parents can have many matching HLA antigens.

Human HLA antigens are determined by a set of genes designated A, B, C, and D (Figure 19.17). The information in each gene specifies a particular antigen, and some genes are so highly variable that they can specify any one of 49 different antigens. Specific antigens are designated by a number following the letter that designates the gene. The presence of certain HLA antigens is associated with a higher-than-normal risk of developing a particular disease (Figure 19.18), and many of these diseases are autoimmune disorders.

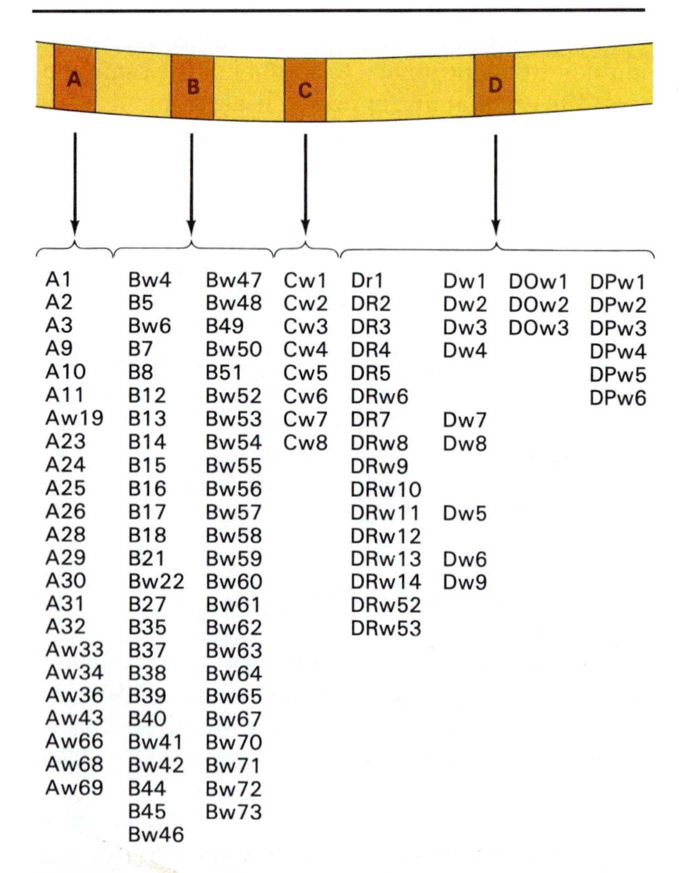

FIGURE 19.17 MHC (HLA) genes and the different antigens that can be produced at each site.

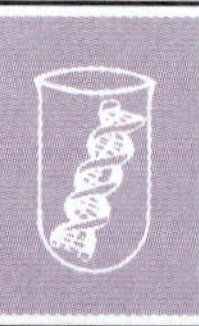

BIOTECHNOLOGY

Babies Conceived To Be Donors

As a teenaged California girl neared death from myelogenous leukemia, her parents, unable to find a compatible donor to replace her cancerous bone marrow, decided to conceive another child. They hoped that it would have the type needed to save the older sister. Their gamble paid off. Before the transplant, 19-year-old Anissa received chemotherapy to destroy her own bone marrow. Healthy marrow taken from the hip of her 13-month-old-sister, Marissa, replaced it. Other families have conceived children to provide donor tissue for a relative; however, this family was the first to state it publicly. Immediate controversy arose as to the morality of such actions. How do you and your classmates feel about this?

Transplant Rejection

Like other immune reactions, transplant rejection displays specificity and memory. Rejection usually is associated with mismatched HLA antigens, and HLA-DR antigens (a subset of D antigens) elicit the strongest rejection reactions. Certain cells that present antigens to phagocytes increase the likelihood of rejection. The fact that DR antigens are found on T cells and macrophages that carry out rejection reactions may explain why DR antigens are so important in graft rejection.

That T cells are responsible for rejection of grafts of solid tissue, such as a kidney, heart, skin, or other organ, has been shown in animal experiments. Grafts are retained by animals lacking T cells and rejected by those lacking B cells. However, how T cells cause rejection is not entirely clear. T_C cells may cause cytotoxic reactions, or lymphokines from T_H cells may stimulate macrophages to cytotoxic action.

The time required for rejection to occur varies from minutes to months. Acute graft rejection, which is caused by a Type II hypersensitivity reaction, occurs when the recipient is already sensitized at the time the graft is done. It causes extensive tissue destruction in minutes but only in grafts, such as kidney transplants where the grafted tissue is immediately supplied with host blood. Corneal transplants are not rejected because the cornea lacks blood vessels, and antibodies cannot reach them. Slower rejection reactions are mainly cell-mediated (Type IV). Rejection occurs within 2 to 5 days of transplantation if the host has been previously sensitized to antigens in the graft, but it takes 7 days to 3 months if sensitization occurs after transplantation. Rejection that begins more than 3 months after transplantation probably involves an

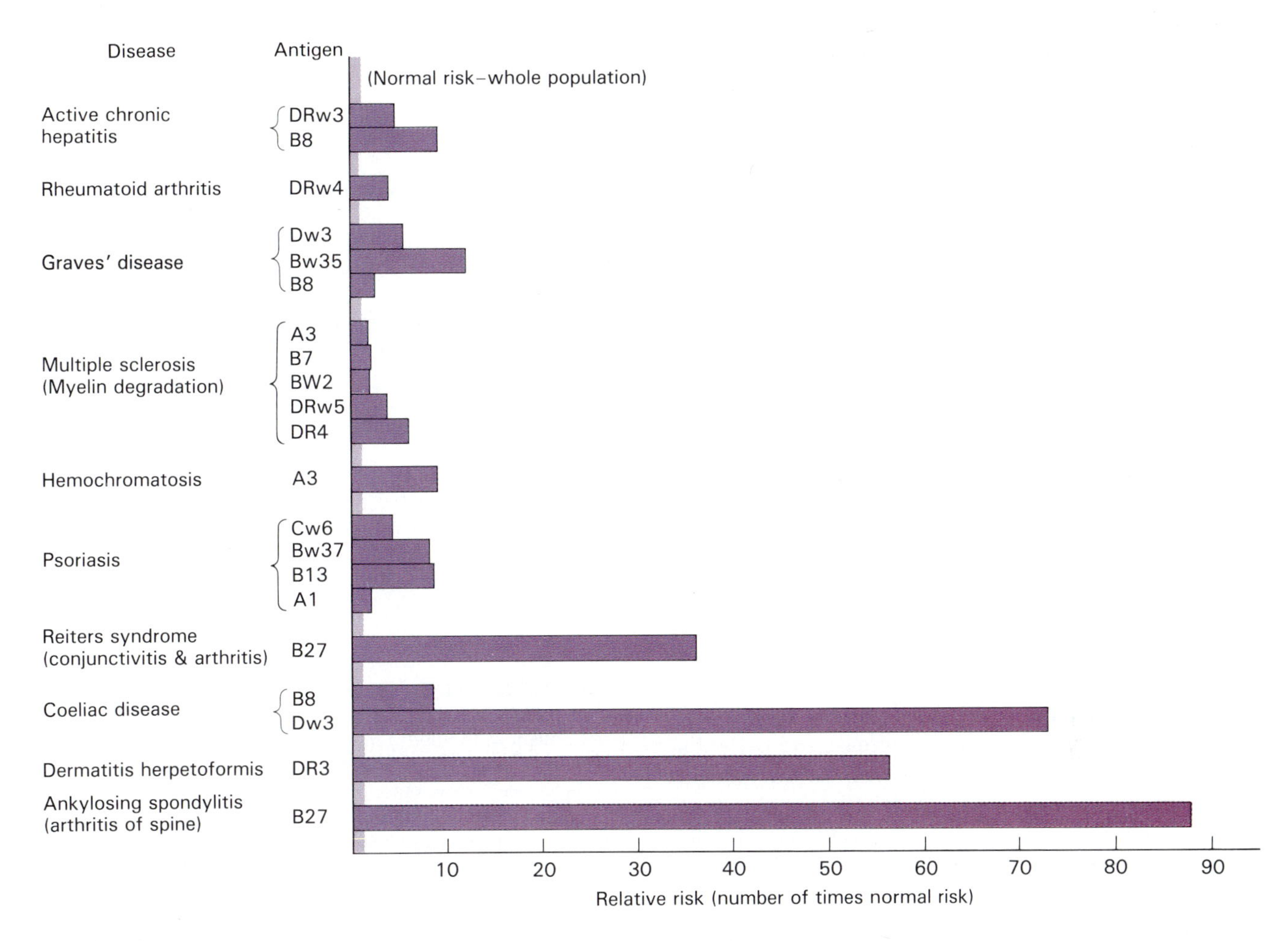

FIGURE 19.18 Correlations between specific HLAs and increased risk of developing certain diseases. Many of these are autoimmune conditions.

interaction between the immune system and the disease process that necessitated the transplant.

Immunosuppression

When a patient is facing an organ transplant, it is important to prevent immune reactions that would destroy the organ. Such minimizing of immune reactions is called **immunosuppression.** Ideally, immunosuppression should be as specific as possible—it should cause the immune system to tolerate only the antigens in transplanted tissue and allow it to continue to respond to infectious agents. This has been accomplished experimentally by introducing an-

BIOTECHNOLOGY

Disguising Tissues Aids in Transplants

Transplanted tissues are rejected because the recipient's immune system recognizes them as foreign. However, scientists at Massachusetts General Hospital have succeeded in disguising foreign cells by covering the HLA class I cell surface proteins that act as antigens and trigger rejection. Ordinarily when antibodies bind to antigens on cells, this starts a process leading to death of the cells. The researchers modified such antibodies so that although they fit tightly to foreign cell surface proteins, they did not destroy the cells. Human pancreatic cells with HLA proteins thus "covered" were transplanted into mice. The mouse immune system ignored the human cells, allowing them to live and produce insulin for more than 6 months. Further research may eventually provide transplants to treat diabetics. An advantage of using "disguised" cells is that immune suppressive drugs now used with transplants will be unnecessary. These drugs, sometimes taken for the rest of the patients' lives, leave patients highly vulnerable to infections and often cause other undesirable side effects.

tibodies to graft antigens into the host, but how the antibodies interfere with the host's immune response is not well understood.

In practice, radiation or cytotoxic drugs, both of which impair the immune responses, are used to minimize rejection reactions. *Radiation* of lymphoid tissues suppresses the immune system and other functions of lymphoid tissues, too. **Cytotoxic drugs,** such as azothioprine and methotrexate, damage many kinds of cells, but they cause most damage to rapidly dividing cells because they interfere with DNA synthesis. As B cells and T cells divide rapidly after sensitization, the drugs exert a somewhat selective effect on the immune system. (Cytotoxic drugs also are used in treating some kinds of cancer because cancer cells too divide with abnormal frequency.)

Radiation and cytotoxic drugs also impair resistance to infections. In contrast, the drug cyclosporine A suppresses, but does not kill, T cells, and it does not affect B cells. It is particularly useful in preventing transplant rejection because it allows T cells to regain function after the drug is stopped and it does not reduce resistance to infections provided by B cells. However, cyclosporine A may increase the patient's risk of developing cancer. Antilymphocytic serum (ALS), made specifically to contain antibodies to a recipient's T cells, also selectively suppresses T cells without suppressing antibody production. The use of immunosuppressive agents, especially cyclosporine A, has greatly increased the success rate of organ transplants.

DRUG REACTIONS

Most drug molecules are too small to act as allergens. If a drug combines with a protein, however, the protein-drug complex sometimes can induce hypersensitivity. All four types of hypersensitivity have been observed in drug reactions.

Type I hypersensitivity can be caused by any drug. Most reactions are atopic, but generalized anaphylactic reactions sometimes occur, especially when drugs are given by injection. Orally administered drugs are less likely to cause hypersensitivity reactions because they are absorbed more slowly. Hypersensitivity reactions require prior sensitization and depend on the production of IgE antibodies. Although penicillin is one of the safest drugs in use, 5 to 10 percent of patients receiving it repeatedly become sensitized. Of the sensitized patients, about 1 percent develop generalized anaphylactic reactions, which account for about 300 deaths per year in the United States.

Type II hypersensitivity (Figure 19.19) can occur when the drug binds to a cell membrane directly, when it binds to a plasma protein and the complex binds to a cell membrane, or when it alters a cell membrane in such a way that cellular antigens elicit autoantibody production. All such reactions involve IgG or IgM and complement, and their targets—erythrocytes, leukocytes, or platelets—are destroyed by complement-dependent cell lysis. Many antibiotics, sulfonamides, quinidine, and methyldopa elicit Type II reactions.

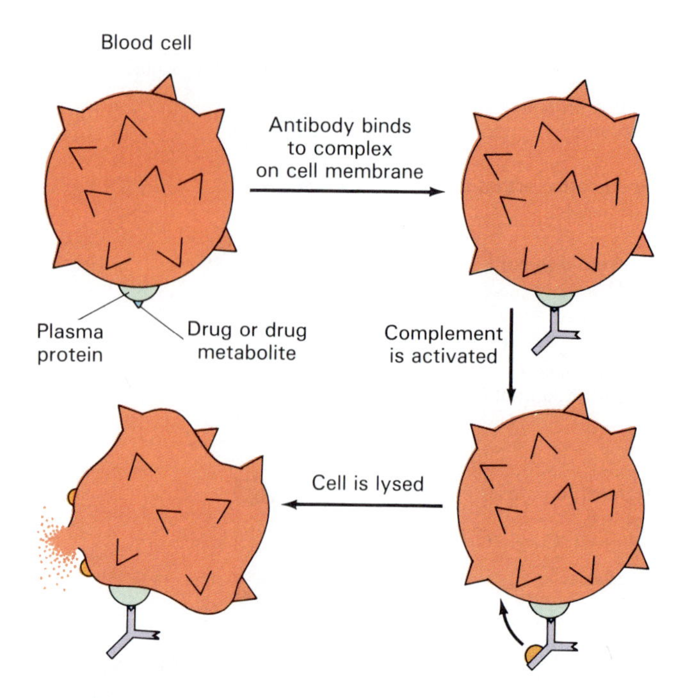

FIGURE 19.19 Drug reactions based on Type II hypersensitivity. A drug (or one of the products of its metabolism in the body) may bind to the membrane of a blood cell, bind to a plasma protein to form a complex that binds to a cell membrane, or alter a cell membrane protein. The result is production of IgG or IgM antibodies, which activate complement to lyse the cell.

Type III hypersensitivity appears as serum sickness and can be caused by any drug that participates in the formation of immune complexes. Symptoms appear several days after administration, when sufficient quantities of immune complexes have accumulated to activate the complement system. A few patients sensitized to penicillin develop serum sickness.

Type IV hypersensitivity usually occurs as contact dermatitis after topical application of drugs. Antibiotics, antihistamines, local anesthetics, and additives such as lanolin are frequent agents of Type IV reactions. Medical personnel who handle drugs sometimes develop Type IV hypersensitivities.

IMMUNODEFICIENCY DISEASES

Immunodeficiency diseases arise from the lack of lymphocytes, defective lymphocytes, or destruction of

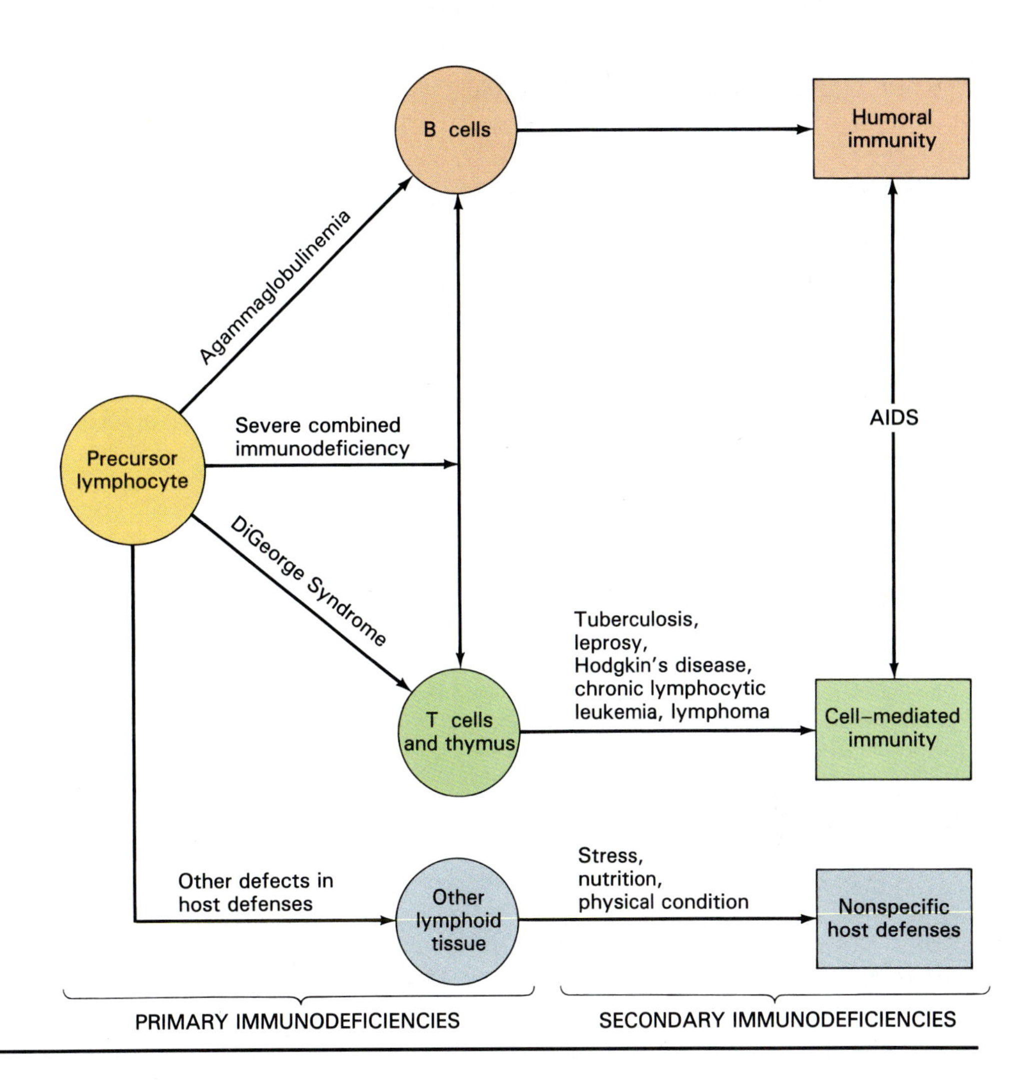

FIGURE 19.20 Kinds of immunodeficiencies.

lymphocytes. Such diseases invariably lead to impaired immunity (Figure 19.20). **Primary immunodeficiency diseases** are caused by abnormalities in embryological development, such as failure of the thymus gland or Peyer's patches to develop normally. **Secondary immunodeficiency diseases** can be caused by infectious agents, such as leprosy, tuberculosis, measles, and AIDS; malignancies, such as Hodgkin's disease or multiple myeloma; or immunosuppressive drugs, some chemotherapeutic drugs, certain antibiotics, and radiation.

Deficiencies of Cells of the Immune System

Agammaglobulinemia, the first immunodeficiency disease to be discovered, is a disorder found primarily in male infants in which B cells, and therefore antibodies, are absent. After maternal antibodies are lost—by about 9 months of age—infants develop severe infections because they cannot produce antibodies. Agammaglobulinemia is treated by giving gamma globulin to replace missing antibodies and antibiotics to prevent infections.

DiGeorge syndrome results from a deficiency of T cells, probably caused by an agent that interferes with embryological development of the thymus gland. Cell-mediated immunity is impaired, so viral diseases pose a greater-than-usual threat. Humoral immunity is not affected, and antibodies can produce immunity to many diseases. Mice lacking a thymus, known as "nude" mice, are bred and raised in germ-free environments for research purposes (Figure 19.21). They are used in the study of DiGeorge syndrome as well as other areas of immunology and genetics.

Severe combined immunodeficiency disease (SCID) is particularly debilitating because both B and T cells are lacking. In SCID, stem cells that give rise to lymphocytes fail to develop properly because of a genetic defect, and both types of lymphocytes are extremely deficient. An infant who inherits this condition is doomed to die unless kept in a germ-free environment until satisfactory treatment can be devised (Figure 19.22).

Several methods for treating SCID are being developed, and early results look good in the first patient who has received a transplant. Bone marrow transplants are sometimes effective if a compatible donor (usually a brother or sister) can be found. If the trans-

FIGURE 19.21 A nude mouse. These animals lack a thymus gland as well as their fur. They are delivered by cesarean section by sterile technique and must be kept in a germ-free environment all their lives, because they completely lack T cells. They are the equivalent of human cases of DiGeorge syndrome. Researchers use them in many types of studies of the immune system.

FIGURE 19.22 David, the boy born without any immune system, shown at age 6 in his self-contained, sterile space suit, a mobile isolation system designed for him by NASA. Other equipment included a pushcart with a battery-powered motor and a seat.

plant is not compatible, it responds immunologically to antigens in host tissue. This response is an example of GVH disease, and it can kill the host. Monoclonal antibodies developed against donor T cells can be used to destroy those cells, minimizing the risk of GVH disease. SCID also can be treated by transplanting immunologically immature fetal liver tissue to the host. This tissue contains stem cells that sometimes persist in the recipient and produce normal T and B cells.

APPLICATIONS

The Boy in the Bubble

David, a child with SCID, lived in a series of specially designed germ-free "bubbles" for most of his life. Lacking both B cells and T cells, he had to be isolated from all sources of infectious agents. At age 12 he received a bone marrow transplant intended to provide him with immune functions. He was watched carefully for signs of graft-versus-host disease after the transplant, and in about 5 months he developed symptoms similar to those of GVH disease and soon died. On autopsy it was learned that he died not of GVH disease, but of a malignancy caused by Epstein-Barr virus, which had contaminated the transplant. Researchers concluded that the lack of immune surveillance of tumor cells (Chapter 18) proved to be the immunodeficiency most responsible for his death.

Acquired Immunodeficiencies

Immunodeficiency diseases are not always inherited; sometimes they are acquired as a result of infections, malignancies, autoimmune diseases, and other conditions. One such disease, acquired immune deficiency syndrome (AIDS), is of such importance that

PUBLIC HEALTH

The Magic Ends

On November 8, 1991, Earvin "Magic" Johnson, star of the Los Angeles Lakers basketball team, announced his retirement from play due to his having contracted the virus that causes AIDS. His career was terminated at age 32, and his life expectancy drastically shortened. He attributed his infection to heterosexual contact—too many women, and failure to practice "safe" sex. Disbelief and shock swept the nation. Up to that time, people hadn't really been worried about getting AIDS from heterosexual activity. Now it had happened to someone they knew and who was a hero to them. In the week that followed, AIDS-testing facilities were swamped with requests for testing. Some health departments used up nearly their entire year's testing budget in just a few weeks.

Since Johnson's announcement, the National Basketball Association (NBA) has taken certain precautions against the possible spread of AIDS. Any player who begins to bleed now must immediately leave the court, and trainers who treat players wear surgical gloves.

Magic Johnson has vowed to use his time and influence now to act as a spokesman for AIDS education and research. Let us hope that people will hear his messages.

it is discussed in a special extended essay at the end of this chapter.

Certain infections sometimes leave the patient with immunodeficiencies. For example, congenital rubella infections can decrease T cell function and antibody production to the extent that infants fail to respond to vaccines. Once patients develop immunodeficiencies, they may suffer from chronic or frequent recurrent infections.

Among malignant diseases that produce immunodeficiencies, those of lymphoid tissues suppress T cell function and those of bone marrow suppress both T cell function and antibody production. Autoimmune diseases (such as lupus erythematosus and rheumatoid arthritis), some kidney disorders, severe burns, malnutrition or starvation, and anesthesia also can cause temporary or permanent immunodeficiencies.

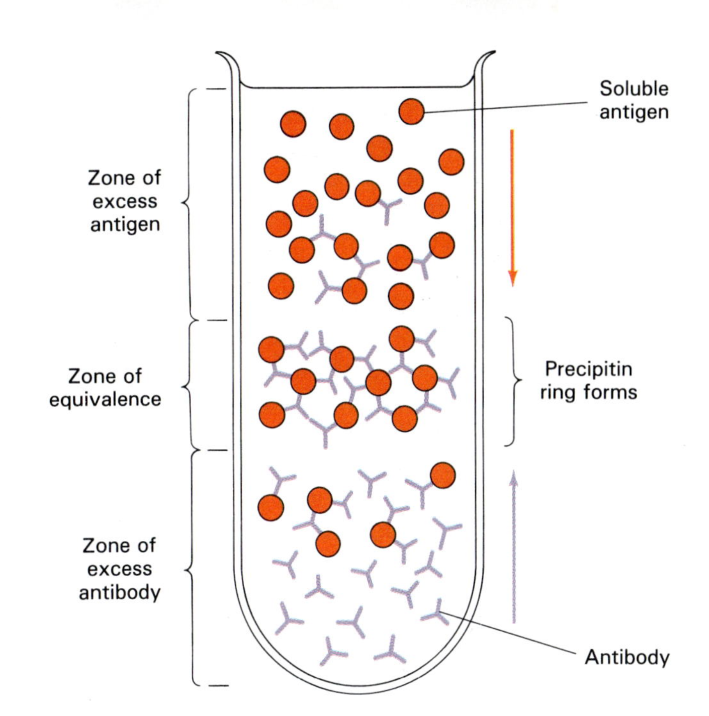

FIGURE 19.23 The precipitin test for antibodies. When IgG or IgM antibodies react with soluble antigens, they quickly form small complexes, which over a longer period of time link up into larger lattices that precipitate from the solution. Such precipitation only occurs, however, when there is an appropriate ratio of antigen to antibody. In the test, antibodies are placed in the bottom of a narrow tube, soluble antigen is added, and the two are allowed to diffuse toward each other. Where the necessary concentration ratio is achieved (the zone of equivalence), precipitation takes place, visible as a hazy ring in the tube.

IMMUNOLOGIC TESTS

In Chapter 18 we considered how certain immunologic reactions—agglutination, cell lysis by complement and by IgM antibodies, and neutralization of viruses and toxins—kill pathogens. Here we will consider how those and other reactions are used as laboratory tests to detect antigens and antibodies. Such laboratory tests make up the branch of immunology called **serology,** so named because many of the tests are performed on serum samples.

Historically, one of the first serological tests to be developed was the **precipitin test** (Figure 19.23), which can be used to detect antibodies. This test is based on **precipitation reactions** in which antibodies called **precipitins** react with antigens and form latticelike networks of molecules that precipitate from solutions. During such reactions antigen-antibody complexes form within seconds, and lattices of these complexes form minutes to hours later.

Immunodiffusion tests are based on the same principle as the precipitin test, but they are carried out in an agar gel medium. They are used to determine whether certain combinations of antigens and antibodies diffusing toward each other in the medium can bind to each other. Small wells are made in the agar, and the antigens and antibodies are placed in separate wells. Antigen-antibody complexes appear as detectable precipitates in the agar. An advantage of immunodiffusion tests is that in a single test medium, several antigens can be reacted with one kind of antibody or several kinds of antibodies with one antigen. After diffusion occurs, one or more bands of precipitation can be detected. The bands can be made more visible by washing the agar surface and applying a stain that colors the antigen-antibody complexes. Figure 19.24 illustrates the results obtained with several combinations of antigens and antibodies.

When samples contain several antigens, **immunoelectrophoresis** can be used to detect antigen-

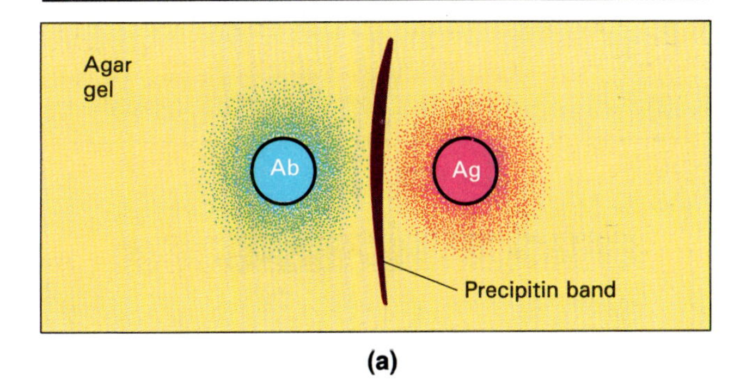

(a)

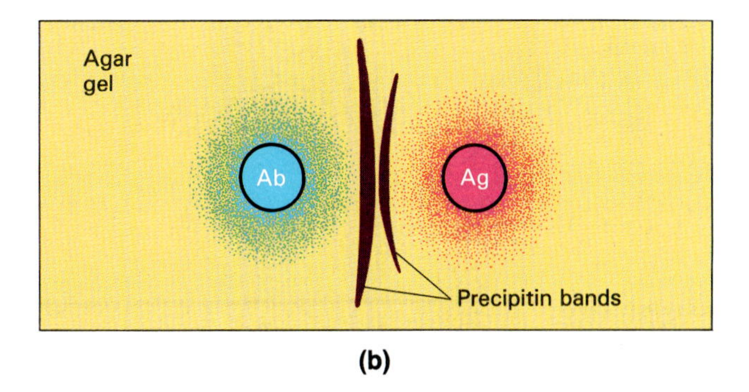

(b)

FIGURE 19.24 In immunodiffusion, wells are punched into sheets of solidified agar gel and filled with test solutions of antigen and antibody. As they diffuse outward from the wells, they meet, react with each other, and precipitate, forming a line called a precipitin band, which is visualized by staining. Two different antigens (or antibodies) will leave two separate bands.

antibody complexes. The antigens are separated by placing the sample on a gel through which an electric current is passed. This process is called **electrophoresis.** The antigen molecules migrate in the electric field at different rates, depending on the size and electrical charges of the molecules. The antibody is then placed in a trough between the antigens, and diffusion is allowed to occur. The results of immunoelectrophoresis are similar to those obtained in other immunodiffusion tests (Figure 19.25).

Radial immunodiffusion provides a quantitative measure of antigen or antibody concentrations. Antigen samples of different concentrations are placed in wells in a gel containing an antibody. After diffusion the antigen concentration is determined by measuring the diameter of the ring of precipitation around the antigen (Figure 19.26). Similarly, antibody concentrations can be determined by placing antibody samples of different concentrations in wells in a gel containing the antigen.

As we saw in Chapter 18, when antibodies react with antigens on cells, they can cause agglutination,

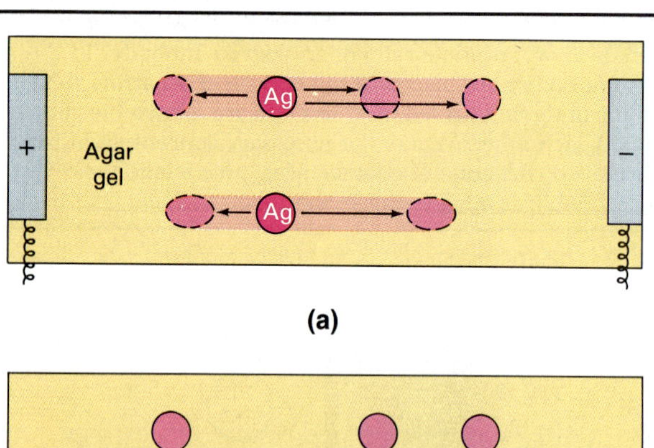

(a)

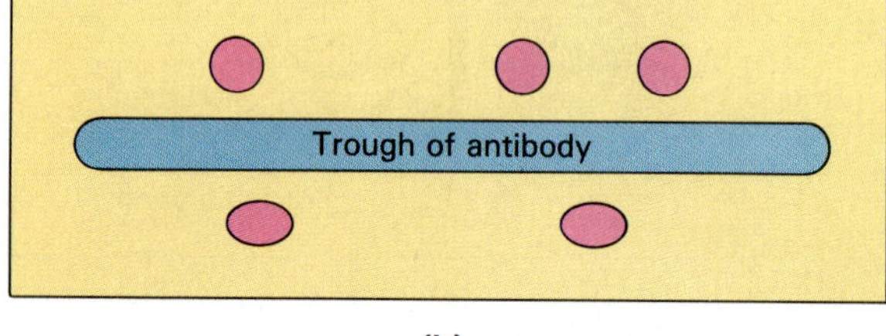

(b)

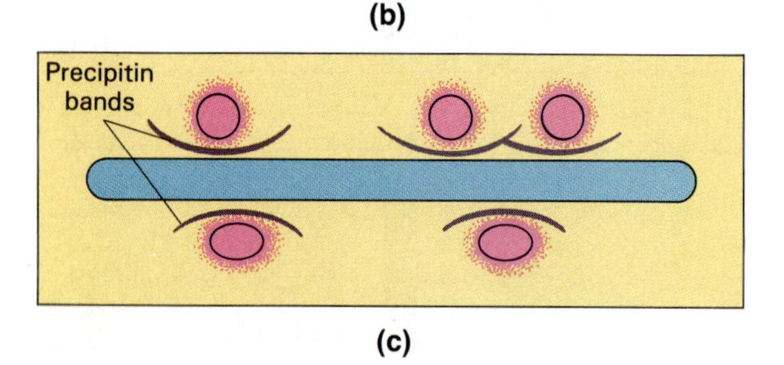

(c)

FIGURE 19.25 Immunoelectrophoresis. (a) Antigens placed in an agar gel are separated by means of an electric current. Positively charged molecules are drawn toward the negative pole, while negatively charged ones move toward the positive pole. (b) A trough is cut into the agar between the wells and filled with antibody. (c) Curved precipitin arcs form where antigens and antibodies diffuse to meet and react.

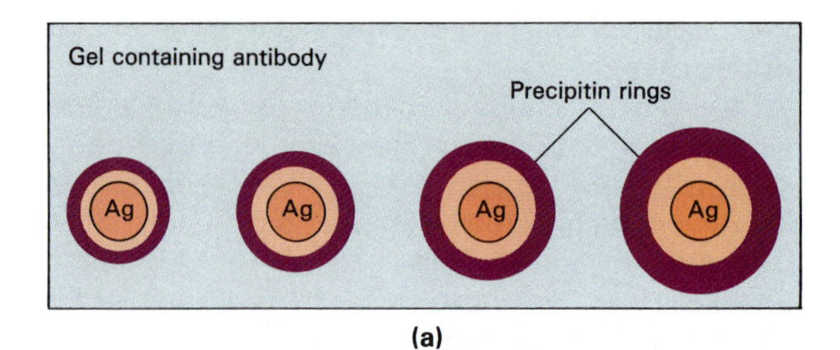

(a)

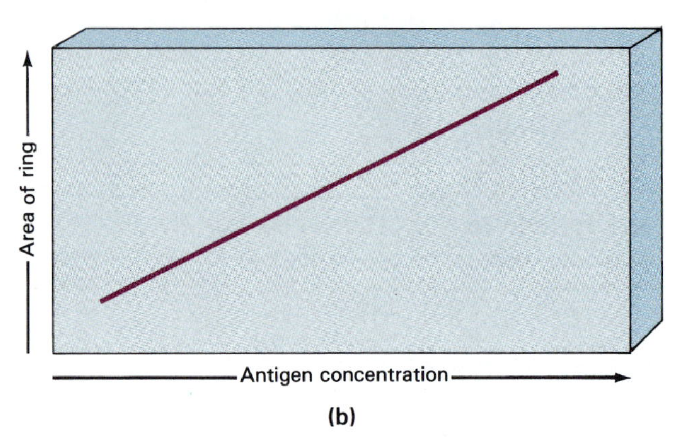

(b)

FIGURE 19.26 In radial immunodiffusion, (a) wells cut into antibody-containing agar sheets are filled with antigen. The antigen diffuses outward, complexing with antibody as it goes. When the amount of antibody is equal to the amount of antigen, the complexes precipitate in a ring. The area within the ring is proportional to the antigen concentration, which can be determined by reference to a standard curve (b). One can also determine the concentration of an antibody by placing it in a well cut into an antigen-containing agar sheet and comparing the size of the resulting ring to a standard curve for that antibody.

or clumping together of large particles. One application of such **agglutination reactions** is to determine whether the quantity of antibodies against a particular infectious agent in a patient's blood is increasing. The quantity of antibodies is called the **antibody titer;** an increase in the antibody titer over time indicates that the patient's immune system is attacking the agent. Diagnosis of the disease agent is possible when it can be shown that the patient's serum had no antibodies against the agent before the onset of disease and that a rise in titer occurred during the course of the disease. This change in titer is called *seroconversion*. The **tube agglutination test** measures antibody titers by comparing various dilutions of the patient's serum against the same known quantity of the antigen. Results are reported as the greatest serum dilution in which agglutination occurs.

Hemagglutination, or agglutination of red blood cells (Figure 19.27), is used in blood typing. However, viruses such as those that cause measles and influenza can bind to red blood cells and cause **viral hemagglutination** without an immunologic reaction. This

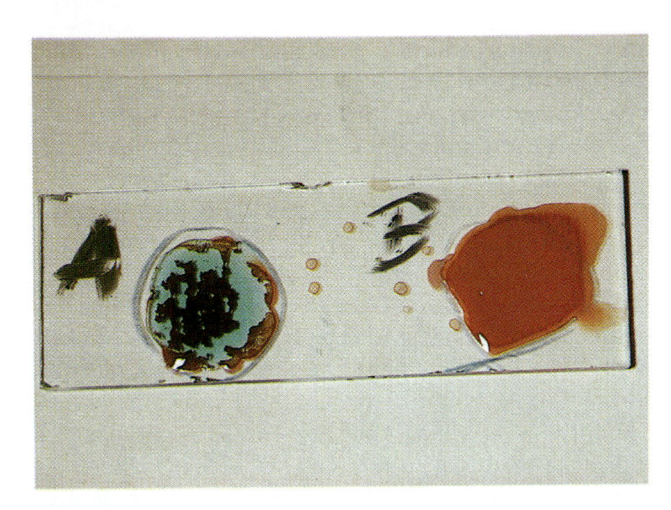

FIGURE 19.27 Hemagglutination. This test, used for matching blood types, is based on the agglutination reaction discussed in Chapter 18 (see Figures 18.10 and 18.11). The presence in the serum of one blood type of antibodies to the red cell antigens of another blood type caused the red cells to clump together.

process is inhibited by antibodies to the viruses because the antibodies bind to the viruses before they can agglutinate red blood cells. Such inhibition is the basis of the **hemagglutination inhibition test,** which can be used to diagnose measles, influenza, and other viral diseases.

The body's natural defenses often make use of the ability of complement to bind to antigen-antibody complexes in the destruction of pathogens. This same ability is used in the laboratory to detect very small quantities of antibodies. The **complement fixation test** is a complex test requiring several steps (Figure 19.28). After complement in a patient's serum has been destroyed by heating, the serum is diluted, and known quantities of complement and an antigen are added. The antigen matches the antibody being sought. This mixture is incubated to allow the antigen to react with any antibody present. Next, cells with antibody attached to them, usually sheep red blood cells, are added to the mixture. These cells can be lysed by complement. If the antibody to the test antigen was present in the serum, the antigen-antibody reaction will have used up the complement, the red blood cells will not be lysed, and the test is positive. If the antibody was not present in the serum, free complement remaining in the mixture will lyse the red blood cells, and the test is negative.

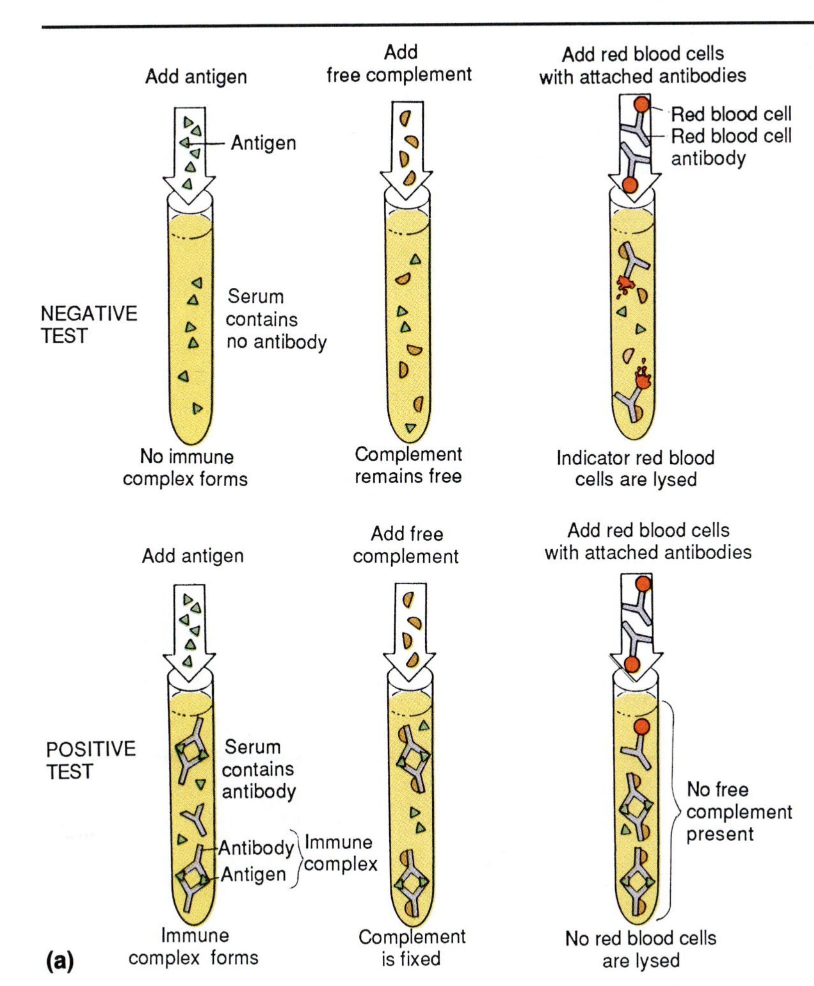

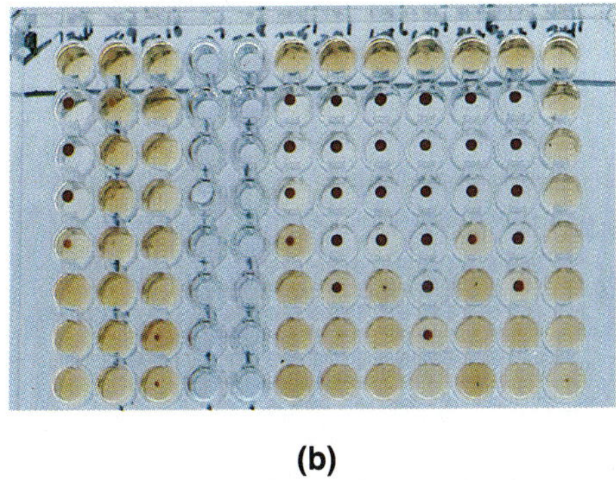

(b)

FIGURE 19.28 (a) The complement fixation test for antibodies. In the first step, the serum to be tested is diluted, and antigen to the antibody being sought is added. If the antibody is present, it reacts with the antigen to form immune complexes. In the second step, free complement is added. If immune complexes have formed, the complement will interact with them and be fixed; if no immune complexes have formed, the complement will remain free. In the third step, sheep red blood cells with bound antibody molecules are added. If free complement is present, it will lyse the red blood cells. This is a negative test result, because it indicates that there was no antibody in the original serum. If all the complement has already been fixed, the red blood cells will not be lysed. This is a positive test result, because it indicates that antibody was present in the original serum. (b) In a positive test, complement has been fixed, and the red blood cells will not be lysed. Instead, they roll down the sides of the well to form a little red blob.

Neutralization reactions can be used to detect antitoxins and antibodies to viruses. Immunity to diphtheria, which depends on the presence of antitoxin, or antibodies to diphtheria toxin, can be detected by the **Schick test.** In this test a person is inoculated with a small quantity of diphtheria toxin. If the person is immune to the disease, antitoxin (antibodies to the toxin) will neutralize the toxin, and no reaction will occur. If the person is not immune and the antitoxin is not present, the toxin will cause tissue damage, detected as a swollen reddened area at the injection site. **Viral neutralization** occurs when antibodies bind to viruses and neutralize them, or prevent them from infecting cells. In the laboratory a patient's serum and a certain virus are added to a cell culture or a developing chick embryo. If the serum contains antibodies to the virus, these antibodies will neutralize the virus and prevent the cells from becoming infected.

Immunofluorescence makes use of antibodies to which a fluorescent substance is bound. For example, the compound fluorescein isothiocyanate fluoresces a bright apple green when exposed to light of a certain wavelength. Fluorescent antibodies can be used to detect antigens, other antibodies, or complement at their locations within tissues (Figure 19.29). A fluorescent antibody that detects another antibody can be thought of as an *anti-antibody;* one that detects complement is an *anti-complement antibody*. Because tissue samples can be examined with a fluorescence microscope, this technique is particularly useful in locating cellular antigens and autoantibodies. It is also a superior diagnostic tool for detecting rabiesviruses.

Radioimmunoassay (RIA) can be used to detect very small quantities (nanograms or billionths of a gram) of antigens and antibodies. Using RIA to measure an antibody (Figure 19.30), a known antigen is placed in a saline (salt) solution and incubated on a plastic surface (tube or plate). Some antigen molecules attach to the plastic, and unattached ones are washed away. The antibody being measured is then allowed to bind with the antigen. Finally, a radioactive antiantibody is applied, and the excess removed. Radioactive material remaining in the specimen is measured with a radiation counter; radiation is proportional to the concentration of the original antibody.

Enzyme-linked immunosorbent assay (ELISA) is a modification of RIA in which the antiantibody, instead of being radioactive, has an enzyme attached to it (Figure 19.31). After the antibody being measured has reacted with the antigen, the antiantibody enzyme is added. Finally, a substrate that the enzyme converts to a colored product is applied. The amount of product is proportional to the concentration of the antibody. An important application of ELISA is the detection of AIDS antibodies. Both RIA and ELISA can be manipulated to detect either antigens or antibodies, depending on which substance is already known.

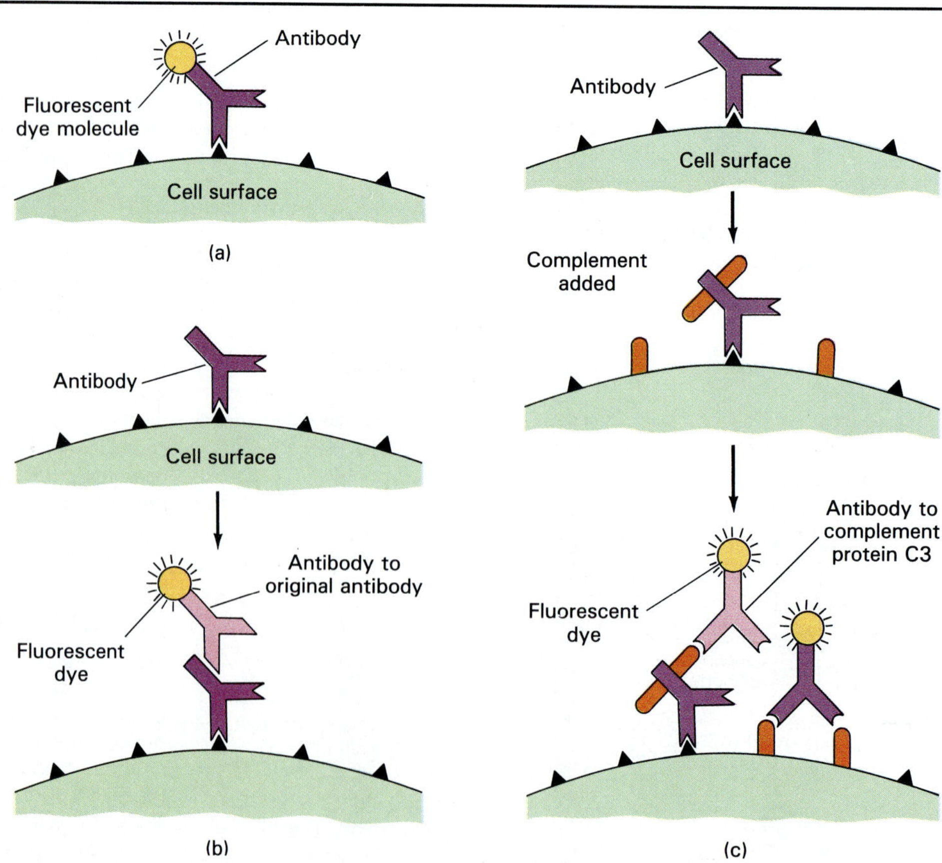

FIGURE 19.29
Immunofluorescence. Flourescein is a fluorescent dye molecule that can be complexed with other molecules. When viewed with ultraviolet light (UV) using a fluorescence microscope it will fluoresce, revealing the presence of the "tagged" molecule. To detect the presence of a specific antigen in a tissue, (a) a solution of fluorescein-tagged antibody to that antigen is prepared, added to a thin section of the tissue, incubated, and then washed. Any tagged antibody that has complexed with antigen in the tissue will fluoresce when viewed under the UV microscope. (b) In indirect testing, the antibody to the antigen being sought is not itself tagged. Instead, its presence is detected by means of a fluorescein-tagged antibody to the original antibody. (c) Complement can be added to the tissue section along with the antibody, and a fluorescein-tagged antibody to one of the complement proteins can then be used to detect the presence of antigen-antibody complexes.

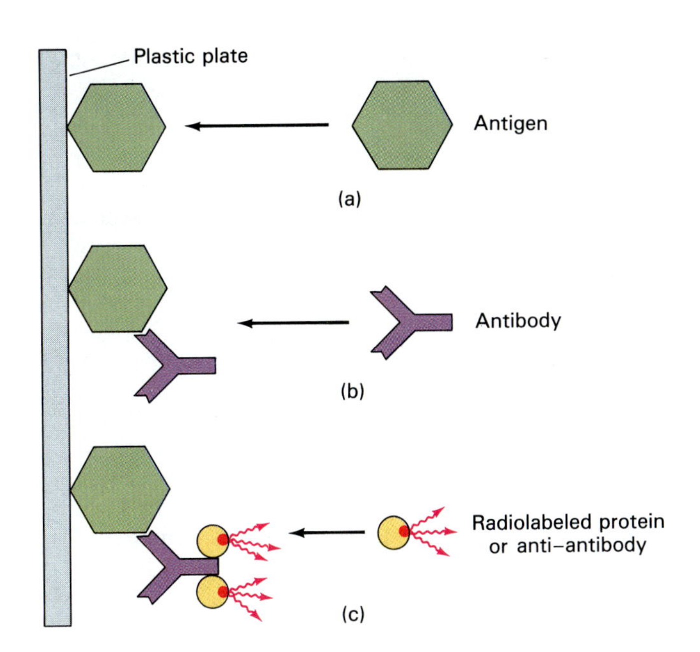

FIGURE 19.30 Radioimmunoassay (RIA) is used to detect very thin quantities of antibody. Antigen to the antibody being sought is first bound to a plastic plate (a) and then allowed to react with the solution being tested. If antibody is present, it reacts with the antigen (b). A radioactively labeled protein or anti-antibody that can bind to the first antibody is then added (c). The amount of bound radioactive label is measured, and is proportional to the concentration of the antibody sought.

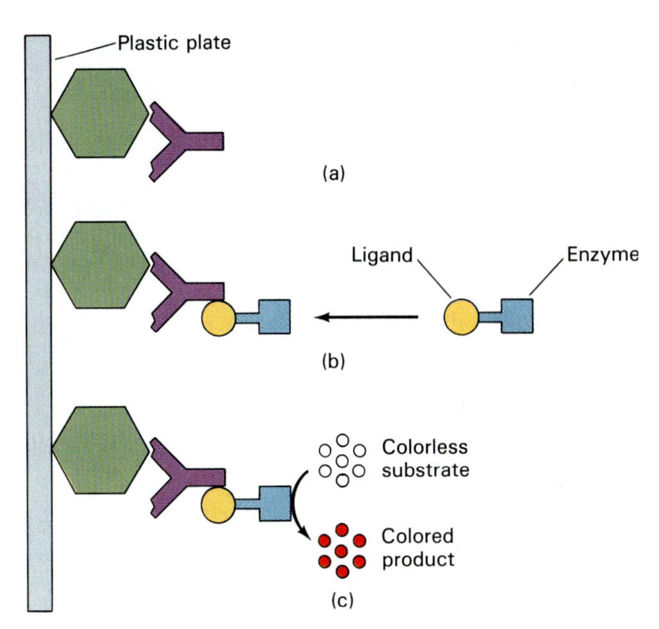

FIGURE 19.31 Enzyme-linked immunosorbent assay (ELISA) is a modification of RIA. As in RIA, antigen bound to a plate reacts with the antibody being sought (a). However, the ligand (the molecule that binds to the antibody) is attached to the enzyme rather than radioactively labeled (b). The enzyme can carry out a reaction that transforms a colorless substrate into a colored product (c). This chemical change, rather than radioactive emission, is used to measure the amount of antibody present.

ESSAY

Acquired Immune Deficiency Syndrome

"Even without mutation, it is always possible that some hitherto obscure parasitic organism may escape its accustomed ecological niche and expose the dense populations that have become so conspicuous a feature of the earth to some fresh and perchance devastating mortality."
—William H. McNeill, 1976

What Is AIDS?

Acquired immune deficiency syndrome (AIDS) is an infectious disease caused by human immunodeficiency virus (HIV). The virus gradually but relentlessly destroys the victim's immune system. The lack of a functional immune system leaves the way open for a variety of malignancies and opportunistic infections, most of which are rarely seen among people not suffering from AIDS. One or a combination of these complications eventually proves fatal.

At least two different types of AIDS virus, designated HIV-1 and HIV-2, are known to exist. Most cases of AIDS in this country are caused by HIV-1. HIV-2, which is most common in certain parts of Africa, has also been shown capable of causing AIDS, though it may be less virulent. Fewer than 100 cases of HIV-2 AIDS have been found in the United States. Only HIV-1 is tested for in screening the nation's blood supplies. Controversy exists now as to whether the additional cost of testing all units of blood for HIV-2 is worthwhile, given its low prevalence. India has recently reported finding that 25 percent of those prostitutes in Bombay infected with AIDS have HIV-2 as the causative virus. Eventually HIV-2 will spread worldwide. The following discussion will focus chiefly on HIV-1 infection.

The Epidemiology of AIDS

AIDS has been called "the epidemic of the century," and certainly few diseases have had such a dramatic impact. By 1991, ending the first decade of reports of the disease, 161,073 cases had been reported in the United States alone, and 100,777 persons had died. One-third of these deaths occurred in 1990. Projections made by the CDC suggest that by the end of 1992 about 263,000 Americans will have died of AIDS.

But even this figure does not give an adequate idea of the magnitude of the public health problem represented by AIDS. Because the latency period for HIV infection is very long, it is hard to estimate how many people have been infected with the virus and what will be their ultimate fate. Publicity about AIDS and education about how to avoid exposure to HIV have probably slowed the spread of the virus. Nevertheless, it is estimated that 1 to 1.5 million people in this country are already infected. It seems likely that without improvements in the means of medical treatment now available, most or all of these people will eventually develop the clinical symptoms we term AIDS, and most or all will die.

ESSAY Acquired Immune Deficiency Syndrome

The official U.S. government definition of AIDS was formulated early in the AIDS epidemic, when most cases were in the male homosexual population. Today the rate of cases is rising most rapidly in the heterosexual population, resulting in many more cases among women. However, these women often cannot meet the criteria for diagnosis as having AIDS because these criteria do not include gynecological conditions such as cervical cancer and pelvic inflammatory disease, commonly seen among women with AIDS. Change in the official definition of AIDS would allow these women (and their dependent children) to qualify for disability and other health and social benefits.

Redefinition would also change statistics being collected. In June 1992, federal health officials announced that they would not meet a July 1, 1992, deadline for changing the definition of AIDS, delaying it indefinitely.

The AIDS pandemic is not confined to the United States. Allowing for the fact that in less developed countries many cases are probably undiagnosed or unreported, World Health Organization officials have recently estimated that the worldwide number of people infected with HIV will exceed 40 million by the year 2000. This estimate is considered by many scientists to be too conservative. The current number of people infected with HIV may be as high as 15 million and is growing rapidly. The regions most seriously affected include Central and East Africa, some countries in South America and the Caribbean, as well as large parts of Europe (Figure 19.32). Parts of Central Africa have infection rates exceeding 15 to 25 percent. In Uganda, 1.5 million people out of a population of 16 million are thought to be infected with HIV. By the year 2002 it is projected that more people will be dying than are being born, plunging Uganda into a negative population growth rate.

To epidemiologists, AIDS offers the rare opportunity to study the spread of a disease from its emergence. Although their findings will be of no consolation to persons with AIDS, what can be learned from such studies may improve our understanding of how diseases spread and how such spread can be prevented.

The Origin of AIDS

The identification of a virus similar to HIV in certain species of African monkeys has led some scientists to suggest that HIV evolved first in nonhuman primates. Although recent studies of the genetic information in these two viruses shows they are only distantly related, there may be other members of this viral family more closely related to HIV. It seems quite possible that a mutation in one such strain could have given rise to the ancestor of HIV.

Regardless of its ultimate origin, most scientists now believe that HIV probably first appeared among humans between 40 and 100 years ago. Early evidence for HIV comes from studies of human blood stored in England and Zaire since 1959, in which HIV-1 antibodies were found. The virus may have existed in relatively isolated regions, perhaps in Central Africa, for decades. Migration of rural people to the rapidly growing cities, where population density was much higher and sexual contacts were more casual and more frequent, could have brought about a great increase in the number of infected individuals. The expansion of international travel in recent years would then have spread the virus to many other parts of the world. The virus probably made multiple entries into this country before becoming established. Evidence indicates that a 16-year-old boy died of AIDS here as early as 1969. The oldest documented case of AIDS has been found in England by polymerase chain reaction (PCR) examination of tissue from a 25-year-old British sailor who died in 1959 of what was then called a "pre-

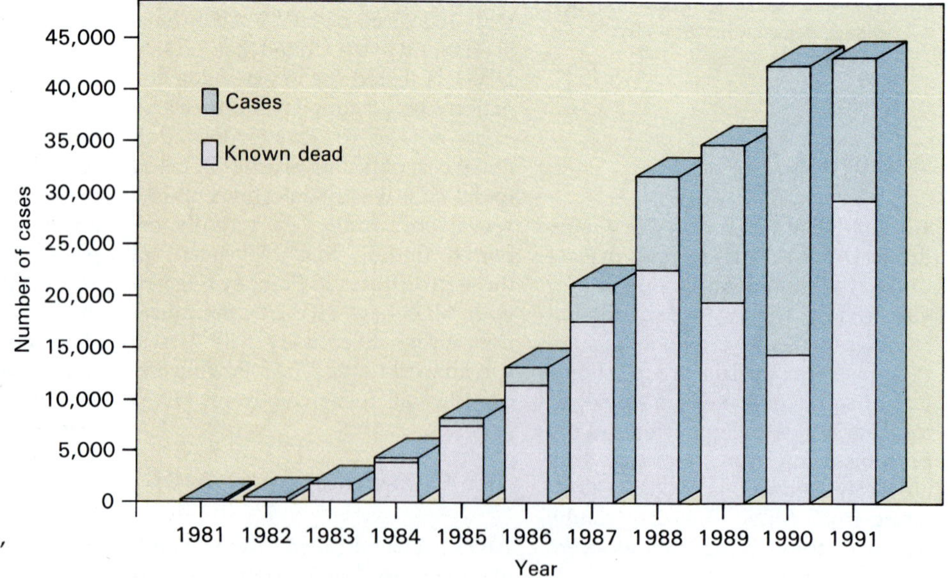

FIGURE 19.32 The total number of AIDS cases and deaths in the United States, by year of report to the CDC, 1981–1991.

viously unknown viral disease." (Chapter 7, p. 187) He had sailed worldwide, and it is not known where he acquired the infection. Researchers are comparing the virus recovered from his tissues with today's virus to see how it may have changed its molecular structure in the last 30 years.

Who Gets AIDS, and How

All available evidence suggests that a person can become infected with the AIDS virus only through intimate contact with the body fluids of an infected individual. It is known that the virus is most commonly transmitted through blood, semen, and vaginal secretions. Transmission by means of saliva and the milk of nursing mothers has also been reported. It seems likely that, however it is transmitted, the virus must make contact with a break or abrasion in the skin or with mucous membranes to cause infection. For this reason, all practices that lead to an exchange of body fluids carry a risk of HIV infection. These include:

1. Sexual contact with an infected individual. In the United States, AIDS has been associated in the popular mind with homosexual practices. It is true that anal intercourse is an important means of transmission because the rectum has less elasticity and is thus more easily torn than the vagina. Small abrasions in the lining of the anus allow the virus to enter the blood, and lining cells also have receptors to which the virus binds. However, it must be emphasized that all forms of sexual intercourse—heterosexual and homosexual, active and passive, vaginal, anal, and oral—carry the risk of HIV infection. (The role of "safe" sexual practices in the avoidance of HIV infection is discussed later in this section.)
2. The sharing of unsterilized needles by users or abusers of intravenous drugs.
3. Receipt of a transfusion of blood or a blood product contaminated with HIV. This means of transmission caused a number of AIDS cases in the early 1980s. Many of the victims were hemophiliacs, who must receive injections of products derived from blood to enable their own blood to clot properly. Transfusions are much less of a threat today, thanks to the testing of donated blood for HIV antibodies. (Antibodies to HIV do not usually appear for some months after infection. Thus, there is still a very slight possibility that blood donated by a recently infected individual could test negative for antibodies while still being infected with the virus.) And, popular fears to the contrary, it is not possible to acquire AIDS by donating blood. New, sterile needles are always used in this procedure.

Finally, it is known that between 30 and 40 percent of infants carried by HIV-positive women can acquire HIV infections from their mothers. Such transmission seems to be possible while the infant is in the uterus, during delivery, and through breast feeding. Which of these routes is the most common, however, has not yet been established. Preliminary studies indicate that more babies are infected during delivery than before birth. First-born of twins are more often infected than the second twin. Perhaps alterations in delivery techniques could prevent infection.

One point that has been clearly established is that it is virtually impossible to become infected with HIV through casual contact—in schools, in offices, or at home. A high rate of AIDS in a Florida town raised the question of whether transmission by mosquitoes had occurred. This was ruled out when it was shown that only people in sexually active age groups were affected. Mosquitoes would have bitten children and very old people, yet there were no cases among these groups.

The mode of transmission of AIDS appears to vary substantially in different parts of the world (Figure 19.33). In Africa it is primarily transmitted by heterosexual contact. By far the highest incidence of AIDS in the United States and other industrialized nations has been among homosexual or bisexual males and intravenous drug users, the majority of whom are males. Thus, the disease is more common among men than among women by a large margin. In much of Africa and the Caribbean, by contrast, the disease is spread largely through heterosexual contact, and so affects roughly equal numbers of men and women. Because the proportion of infected women is higher in these regions, the number of infants at risk is also greater.

Progress of the Disease

In one article, two researchers pointed out that people who ask for a description of the course of AIDS infection "are asking the wrong question. Now that AIDS is known to be caused by a virus . . . the focus should be on the full course of the viral infection, not solely on AIDS. HIV causes a predictable, progressive derangement of immune function, and AIDS is just one, late manifestation of that process."

The sequence of events in HIV infection has now been established in some detail. In some cases infection is followed by a few weeks of fever, fatigue, swollen glands, and perhaps a rash. Most patients, however, experience no symptoms at all at this stage. There follows a long period during which the victim has no symptoms other than chronic lymphadenopathy (swollen lymph nodes). This period varies in length, and precise information is difficult to obtain because most people do not know when they first became infected with HIV. Furthermore, HIV infection is diagnosed by finding antibodies to the virus, and antibodies do not appear until 2 to 36 months after exposure to HIV. Generally, however, this phase lasts from 3 to 5 years.

The progress of the infection evidently depends heavily on how much of the virus a person is exposed to and how often the exposure is repeated (Figure 19.34). (Thus even a person who tests positive for HIV antibodies should avoid behavior that carries the risk of further contact with the virus.) Immune function continues to decline, however, and a variety of symptoms begin to appear. These

ESSAY Acquired Immune Deficiency Syndrome

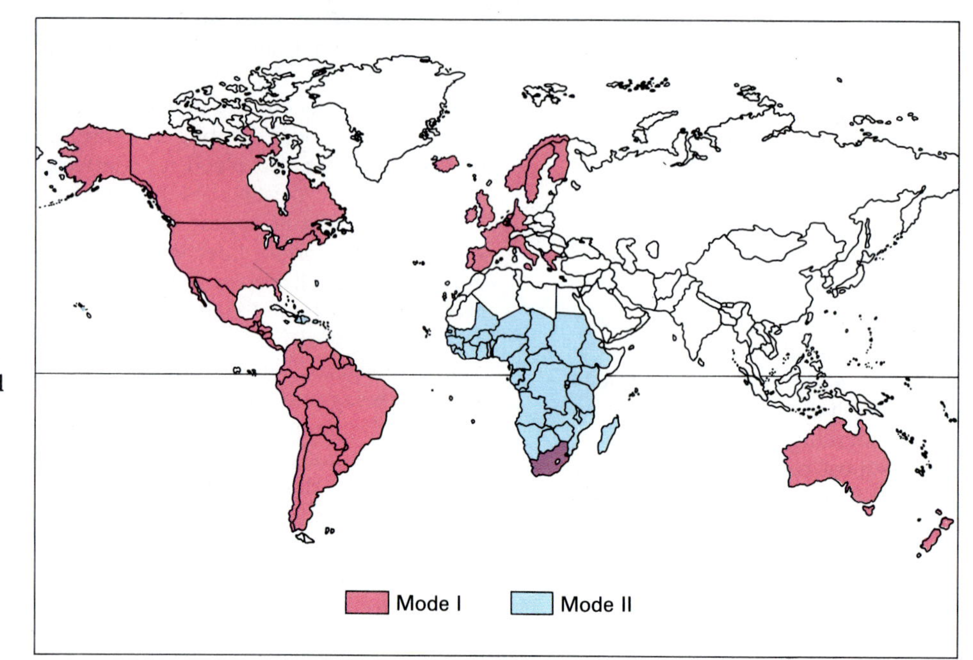

FIGURE 19.33 In Europe, Australia, and the Americas, AIDS has spread most rapidly among homosexual males and intravenous drug users (mode I). In the Caribbean and parts of Africa, however, the infection is spread largely through heterosexual relations and affects men and women about equally (mode II).

symptoms resemble those of many other diseases: fever, night sweats, malaise, nausea, loss of appetite, headache, muscle and joint pain, sore throat, enlarged lymph nodes and spleen, rash, and decreased numbers of lymphocytes and platelets in the blood. Minor secondary infections such as dermatitis, warts, fever blisters, and unexplained diarrhea are sometimes seen. Weight loss with muscle wasting may also occur. These symptoms were often grouped together under the heading of *AIDS-related complex* (*ARC*); however, this term is not much used today. This period, too, is quite variable in length and may last for several years.

Eventually the deterioration of immune function progresses to the point at which severe opportunistic infections, malignancies, and other complications appear (Table 19.7). This is the phase of the infection commonly referred to as AIDS. Latent viral infections such as those caused by herpes simplex and cytomegalovirus, normally kept in check by the immune system, flare up and create a variety of disease symptoms. Severe diarrhea can be caused by opportunistic pathogens such as species of *Cryptosporidium* (Chapter 22), and en-

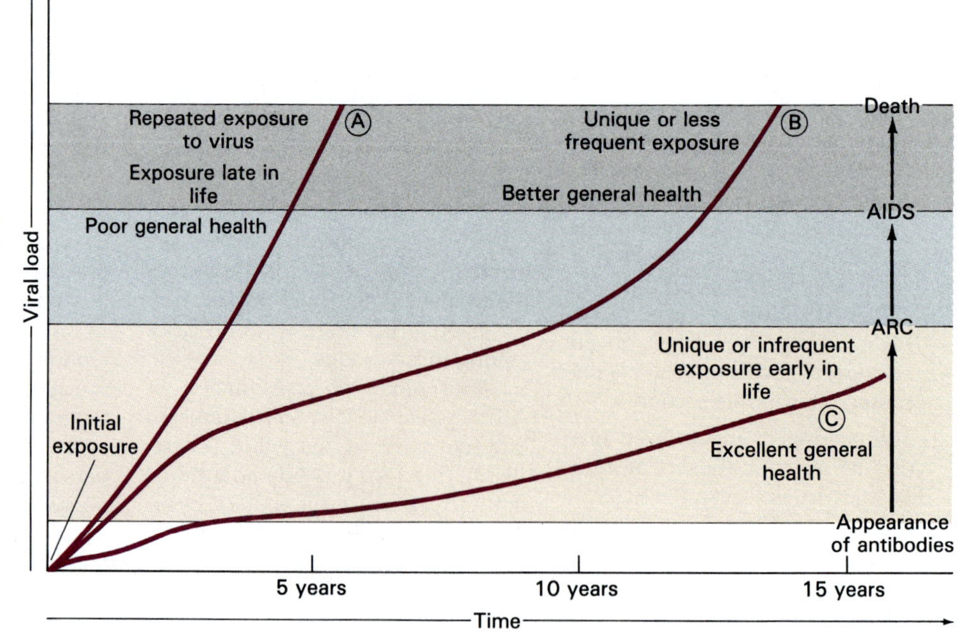

FIGURE 19.34 Hypothetical time scales for the progress of HIV infection. The time for each stage is highly variable, but appears to depend on the frequency of exposure to the virus as well as the general health of the infected person. The greater the frequency of exposure to the virus, the more rapid seems to be the course of the infection. In addition, recent evidence indicates the later in life one is infected, the more rapidly the disease develops.

TABLE 19.7 Infections frequently found in AIDS patients

Pathogen	Disease
Bacteria	
Mycobacterium tuberculosis	Tuberculosis
Mycobacterium avium-intracellulare	Disseminated tuberculosis
Legionella pneumophila	Pneumonia
Salmonella species	Gastrointestinal disease
Viruses	
Herpes simplex	Skin and mucous membrane lesions, pneumonia
Cytomegalovirus	Encephalitis, pneumonia, gastroenteritis, fevers
Epstein-Barr	Oral hairy leukoplakia, possibly lymphoma
Varicella-zoster	Shingles, chickenpox
Fungi	
Pneumocystis carinii	Pneumocystis pneumonia
Candida albicans	Mucous membrane and esophagus infections (thrush)
Cryptococcus neoformans	Meningitis
Histoplasma capsulatum	Pneumonia, disseminated infections, fevers
Protozoans	
Toxoplasma gondii	Encephalitis
Cryptosporidium species	Severe diarrhea

cephalitis can be caused by *Toxoplasma gondii* (Chapter 24). Pneumonia produced by viruses or by *Pneumocystis carinii*, once thought to be a protozoan but now thought to be a fungus, is common in AIDS patients. So is tuberculosis, both the ordinary type caused by *Mycobacterium tuberculosis* and the type caused by *Mycobacterium avium-intracellulare*. The latter, which can affect many organs, is ordinarily so rare that it is now considered to be almost diagnostic for AIDS. New York City health officials reported a 38 percent increase in tuberculosis cases in 1990, attributable to AIDS-related infection. Tests of tuberculosis patients at the city's clinics reveal 35 percent are HIV-infected. Another condition seldom found in the general population but common among sufferers from AIDS is a blood vessel malignancy called Kaposi's sarcoma. This disease causes vessels to grow into tangled masses, filled with blood and easily ruptured, in the skin and viscera (Figure 19.35). Along with *Pneumocystis* pneumonia, it is among the most frequent immediate causes of death.

Many AIDS patients also develop AIDS dementia complex, in which movement, thinking skills, and behavior are altered. Such effects, however, are difficult to separate from effects of other diseases often seen as complications of AIDS.

Most patients who develop full-blown AIDS die within 1 to 3 years. The total course of HIV infection, however, is much longer. A recent study reveals that 8 years after infection, 50 percent of patients will have developed full-blown AIDS. Some of the drugs currently being used or tested against AIDS appear capable of prolonging the life of AIDS victims.

Will everyone with HIV antibodies eventually develop AIDS? Many scientists believe so, but this has not been proven conclusively. One reason for the uncertainty is the existence of a peculiarity sometimes observed in AIDS infections called *seroreversion*. This phenomenon is said to occur when a person who once had HIV antibodies no longer has them and also has no symptoms of the disease. Studies of such individuals have shown that AIDS-virus DNA, which is made by reverse transcription, has been incorporated into the DNA of some body cells. In a few instances even this evidence of the AIDS virus having been present has disappeared. However, this does not preclude the possibility that the virus is hiding in cells that were not tested.

How HIV Attacks the Immune System

AIDS specifically damages a specific group of T lymphocytes, called T4 cells because they bear an antigen called the CD4 marker. T4 cells include helper T cells and certain other T cells that can be recognized by the

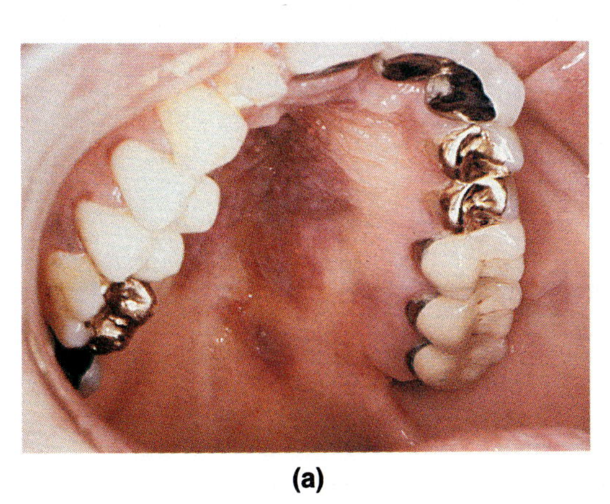

(a)

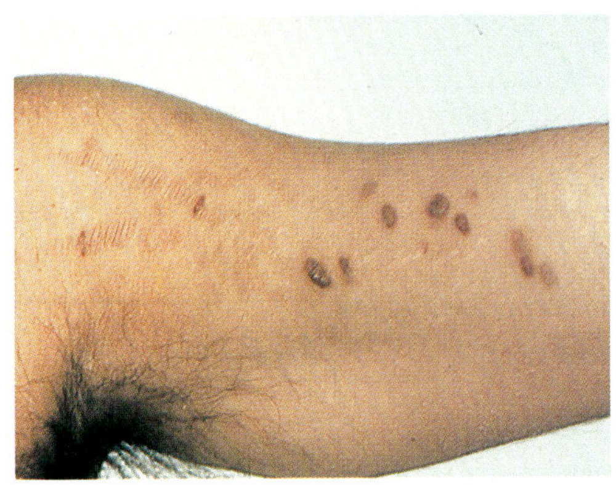

(b)

FIGURE 19.35 Kaposi's sarcoma, a tumor of blood vessels, seen in AIDS patients as (a) three purplish areas on the roof of the mouth, and (b) dark purple areas on the arm.

ESSAY Acquired Immune Deficiency Syndrome

presence in their cell membranes of a particular receptor protein known as CD4. These T4 cells constitute about 70 percent of the total T cell population in a normal person. However, the HIV virus can be harbored in such cells for an indefinite period without replicating or altering cells. Disease symptoms appear only after the virus starts replicating and begins to damage T4 cells. Drop in CD4 cell count can be used to predict the onset of disease symptoms. A normal count is 800 to 1200. If the count remains above 400, only 16 percent of patients will progress to full-blown AIDS within 3 years. With a count of 201 to 400, 46 percent progress, and below 200, 87 percent reach full-blown AIDS within 3 years.

The AIDS virus damages the immune system in the following ways:

1. Being a retrovirus (Chapter 11), HIV synthesizes DNA from RNA when it reproduces itself (Figure 19.36). Some of the DNA thus synthesized is incorporated into the DNA of infected T4 (helper) lymphocytes, causing these cells to make more viruses until they are killed by the infection (Figure 19.37). Nearly all the T4 cells in blood, lymph nodes, and the spleen are destroyed. Death can result from damage to cell membranes as viruses bud out, or from disruption of normal cell metabolism by the large quantities of viral nucleic acid and viral proteins present inside the cell.
2. As T4 cells are destroyed the ratio of T8 (suppressor) cells to T4 cells increases, and the preponderance of suppressor cells depresses other immune functions.
3. Because T4 cells are greatly diminished in number, B cells are not stimulated to produce adequate numbers of antibodies to combat infections. (The anti-HIV antibodies detected early in the course of infection are made before T4-cell populations become too depleted to stimulate B cells.) Similarly, lymphokines are produced in amounts that are insufficient to activate macrophages and cytotoxic T cells.
4. Infected T4 cells also release a substance called *soluble suppressor factor*, which inhibits certain immune responses.
5. Surviving T4 cells lack surface receptors for antigens and are thus incapable of taking the first step in the immune reaction, the recognition of an antigen.

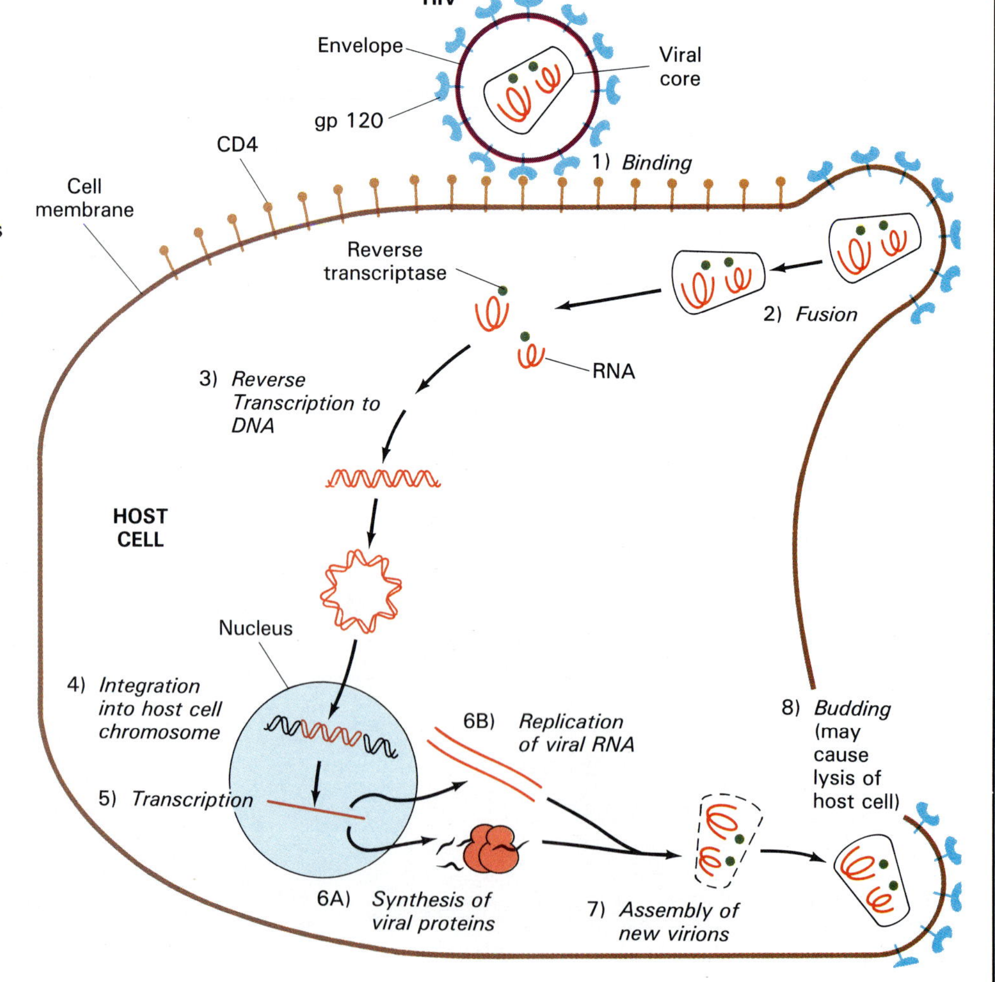

FIGURE 19.36 How the HIV virus enters, replicates in, and destroys lymphocytes. Because HIV is a retrovirus, the process by which it reproduces is complex. Much research has focused on finding ways to interfere with some essential step in the replication of the virus. Other approaches involve attempting to block the binding of the virus's gp 120 protein to the cell's CD4 receptors. The budding of many new virus particles from the infected lymphocyte usually kills the host cell. However, it is suspected that HIV can destroy cells by other means as well—for instance, by causing cells to fuse into syncytia or by provoking an immune response against both infected and uninfected cells.

FIGURE 19.37 False-color TEM of two AIDS viruses (HIV-1) seen as red-coated spheres budding out from the surface of a T4 lymphocyte.

In addition to its assault on the body's immune system, HIV can also damage the nervous system. The virus can infect macrophages as well as T cells. Macrophages can carry HIV particles in cytoplasmic vesicles, rather than at the cell surface, thus allowing the virus to escape detection by the immune system. The virus appears to alter macrophages so as to make them capable of penetrating the blood-brain barrier and entering the brain—something that normal macrophages very rarely do. Within the brain, infected macrophages attack and destroy the myelin coating that insulates many nerve fibers, producing dementia. Recent work shows that loss of mental faculties can occur long before any other symptoms of AIDS appear, and raises concern about employing HIV-positive individuals in jobs that require constant mental alertness and agility.

Diagnosing HIV Infection

The most widely used test for AIDS is the enzyme-linked immunosorbent assay (ELISA). The test, which detects HIV-1 antibodies, was developed to screen the nation's blood supply and protect recipients of blood products from infection. ELISA is a sensitive test, but false positive results do occur at a rate of about 0.4 percent. To confirm the presence of antibodies, another test usually is performed on samples from patients who have a positive ELISA test. Other available tests include the immunofluorescent antibody test and the so-called western blot test. In the latter test, the patient's serum is tested against proteins of HIV-1. The presence of antibodies to viral proteins generally indicates that the patient has been infected with the AIDS virus, but it does not necessarily indicate that the infection is currently active. However, because retroviruses can persist for a lifetime, the presence of any HIV antibodies is considered evidence for an ongoing infection.

The HIV virus can be isolated from most patients with positive immunologic tests by inoculating clinical specimens into human lymphocyte cultures grown under specialized conditions. Although it can take up to 6 weeks for sufficient numbers of viruses to be present in the culture to make a positive identification, the virus can be identified by testing for reverse transcriptase in the cultures, examining cells by electron microscopy, or by immunofluorescent antibody tests.

Many different laboratories are conducting research to develop better AIDS tests. One promising approach is to use genetic engineering techniques to make peptides that mimic key antigens on the AIDS virus with which antibodies will react. Another approach, the PCR test, can detect sequences of DNA in body cells that are derived from reverse transcription of HIV-RNA and subsequently inserted into cellular DNA. This test, already used to study a few cases of seroreversion, requires considerable technical expertise and has not yet become a widely used screening test. It can, however, be done 1 month after a needle stick or sexual contact to avoid waiting 1 year or more to be sure that an ELISA test result will not change. Using ELISA for detection, 70 percent of infected individuals will seroconvert to positive after 3 months, 90 percent after 6 months, and 99 percent after 12 months. PCR testing for AIDS currently costs about $180.

The safety of laboratory workers handling blood samples to be tested for AIDS and assisting with research on the virus has been a concern since the epidemic began. In a study of 265 workers only 1 was found to be HIV positive, possibly through undetected skin contact with fluid from a viral culture. Needle sticks account for 80 percent of all exposures. CDC reports that 3 out of 1000 such sticks transmitted the virus.

Treatment of AIDS

There is currently no cure for AIDS. This is not surprising, for very few antiviral agents of any kind are known, and retroviruses have proven particularly hard to attack. Nevertheless, several drugs are presently being used to prolong the lives of AIDS patients or to alleviate their symptoms, and many others are being investigated.

AZT (3′-azido-2′,3′-dideoxythymidine), the first drug licensed for treatment of AIDS, became available in 1987. This drug, a nucleoside analog (Figure 19.38) which acts by interfering with the synthesis of viral DNA by the enzyme reverse transcriptase, is well tolerated by HIV-infected individuals who have not yet developed symptoms, and is very effective at slowing the progress of the infection and prolonging their lives. Once clinical symptoms of AIDS have developed, however, the drug is less effective, and many patients suffer such severe side effects that they cannot continue treatment. Moreover, it is very expensive: a year's treatment costs about $10,000. Another drug, ddC (2′,3′-dideoxycytidine), is being tested in patients, but it causes side effects ranging from mouth sores and

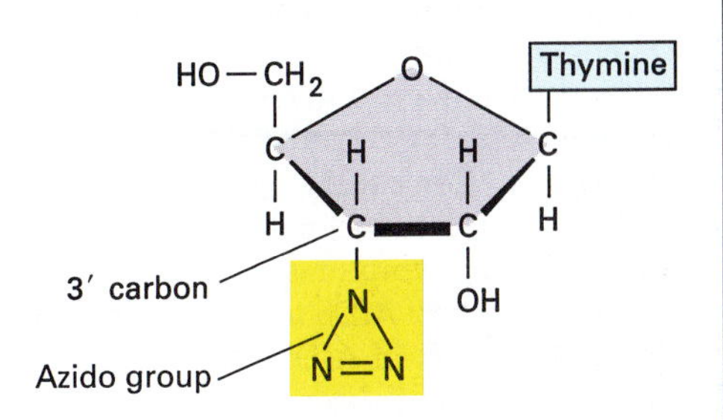

FIGURE 19.38 The structural formula of AZT. This molecule is similar in structure to the DNA nucleotide thymidine, but sufficiently different that its presence halts the process of reverse transcription needed for the virus to reproduce.

dizziness to dementia. Using AZT and ddC alternately appears to alleviate symptoms and prevent neurological disorders with minimal side effects.

The antibiotic fusidic acid, discovered in the 1960s, has been shown to inhibit HIV viruses in cultures and is being tested in clinical trials in Europe. It, too, causes side effects, such as itching and temporary jaundice, and its mechanism of action is unknown. If fusidic acid proves effective, a year's treatment would probably cost only $1000.

Interferon (Chapter 17) has also shown potential, both against HIV infection and against Kaposi's sarcoma. (p. 463) It is being tested, both alone and in combination with other potential anti-HIV agents such as AZT.

Recently, researchers have isolated a protease (protein-cutting) enzyme synthesized by the virus when it reproduces. The function of the protease is to cut another polypeptide made by the virus into two parts, producing two different functional proteins. Because both of these proteins are needed in order for the virus to replicate itself, inactivating the protease should halt viral reproduction. The search is on, therefore, for some compound that might safely inhibit the protease enzyme.

Another promising avenue of research involves attempts to block the binding of HIV to the cells it normally attacks. The first step in binding is an interaction between the CD4 protein on the target cell membrane and a glycoprotein called gp 120 in the outer envelope of the virus. One way to block this interaction might involve the creation of antibodies to gp 120. Another approach would be the use of a soluble form of CD4 to saturate the gp 120 binding sites of the virus, leaving none free to attach to cells. Clinical trials with a genetically engineered soluble CD4 protein are now in progress. Controlling opportunistic infections, treating malignancies, and alleviating symptoms also are important aspects of AIDS treatment. In the short history of AIDS, however, no patient has ever been demonstrably cured of the disease.

Medical personnel treating AIDS patients are at extreme risk of becoming infected. The CDC has recommended the following precautions to minimize that risk (Chapter 16, p. 426):

1. Wear gloves for touching blood, body fluids, mucous membranes, or skin lesions of patients or any objects possibly contaminated by them. Discard gloves, and wash hands immediately and thoroughly after each patient.
2. Wear masks, protective eyewear, and gowns for procedures that might possibly release droplets of body fluids.
3. Scrupulously avoid injury from needles and other sharp objects, and discard them in puncture-proof containers.
4. Use mouthpieces, resuscitation bags, or other ventilator devices for emergency resuscitation.
5. Workers with skin lesions should avoid direct patient care and handling of equipment.
6. Dentists and their technicians should consider blood, saliva, and gingival fluids of all patients potentially infective and use the above procedures to prevent contact with such fluids.

It should be kept in mind that there is danger to medical personnel, not just from the AIDS virus itself, but also from any other infectious disease (such as tuberculosis) that an AIDS patient may have contracted.

What about an AIDS Vaccine?

The outlook for an AIDS vaccine is not promising. If all things went right the first time they were tried, the earliest we could hope for a vaccine would be the mid-1990s. But scientists expect a period of trial and error, lasting beyond the turn of the century, because research with the AIDS virus presents more than ordinary problems. For one thing, HIV has the highest mutation rate of any known virus. The reason seems to be that the reverse transcriptase enzyme is rather imprecise in its operation and often produces variations in the viral DNA during the replication process. Thus, even if we make a successful vaccine against one strain of the virus, another strain would probably develop within a few years, and this new strain might not be affected by the old vaccine. Outbreaks of influenza every few years are caused by similarly high mutation rates. Already several strains of AIDS virus are known. Most cases in the United States are caused by HIV-1, but a few are caused by HIV-2, so already we need a vaccine that immunizes against both viruses as well as the various strains of HIV-1. It is expected, however, that the first vaccines to become available will only protect against one strain of HIV, and with less than 100 percent efficiency.

What can be done to pin down this rapidly changing virus? One hope is to make a vaccine that elicits an immune reaction to structures on the surface of the virus that do not change. Natural viral components, synthetic peptides, and recombinant DNA products are some of the substances now being tried as antigens in AIDS vaccines. Particular attention has been devoted to the viral envelope glycoprotein gp 120. So far, however, the structures tested have not induced a protective immune response.

Manufacturing a vaccine poses more problems. Attenuated viruses cannot be used in a vaccine because they make DNA that can be incorporated into the host's genome, possibly later giving rise to a malignancy. Whole inactivated viruses, which have been used successfully in polio and influenza vaccines, are inappropriate for AIDS vaccine. Such vaccines for AIDS and other viruses of the lentivirus group can predispose the host to severe infections. Recombinant viruses, such as AIDS antigens on a vaccinia virus, are unacceptable because any host might develop disseminated vaccinia and immunocompromised hosts may develop a variety of severe complications. Using adenoviruses instead of vaccinia may avoid these problems. Many problems must be solved before we can have a safe and effective vaccine.

Once we have a safe vaccine that elicits production of antibodies, success is still not guaranteed. People infected with HIV make antibodies early in the course of infection, but because these antibodies are for the most part unable to neutralize the virus (Chapter 18), they do not stop the disease. ∞ (p. 500) Would additional vaccine-induced antibodies do any better? Much may depend upon the answer to this question.

HIV-less AIDS

In July 1992, at an AIDS conference in Amsterdam, reports were made of over 30 cases of AIDS-like symptoms in patients who lacked evidence of infection with HIV. All of these patients had critically low levels of CD4 cells and had opportunistic diseases usually seen in AIDS patients. Some researchers reported other viruses present in a few of these patients. However, we all carry various "silent" viruses, and it will be difficult to prove that the ones reported are actually causative agents. Contamination of cultures is also a common problem. The contaminant might not even be a virus. *Mycoplasma* bacteria can create the impression that a virus is present by altering the results of laboratory tests. Sometimes *Mycoplasma* can even mask the presence of an actual virus, making it appear to be absent.

Another possibility is that there is a third type of HIV strain for which we cannot yet test successfully. Still another explanation may lie in our ignorance of how other chronic illnesses, drugs, heredity, and even environmental factors can affect CD4 cell counts. There may always have been a few people exhibiting these symptoms, but their numbers were too low to be recognized. Now with the recognition of AIDS, we may be identifying these people for the first time.

In August 1992, CDC held a special conference on AIDS-like illnesses. Over 50 cases were reported on. They are distributed evenly across the United States and occur in a somewhat older population than the HIV-infected AIDS cases. No conclusions were reached, however. The studies will continue.

Can AIDS Be Prevented?

In the absence of a vaccine, the main means of preventing AIDS is to educate people about how to avoid becoming exposed to it (Figure 19.39). To this end the U.S. Public Health Service distributed a publication, *Understanding AIDS*, nationwide in the summer of 1988. One of its major recommendations is to avoid "risky behaviors" such as sharing drug needles and syringes and engaging in any sexual activity with persons who are or might be infected with AIDS. (In this connection it is important to keep in mind that all persons who test positive for HIV antibodies have been found to carry the virus, and that HIV infection can be spread by people who do not yet have any clinical symptoms of AIDS or even ARC.) The risks associated with sexual activity can be reduced by the use of condoms, but they cannot be entirely eliminated in this way. Condoms have a significant failure rate (17–50 percent in various recent studies), and not all types of condoms are equally effective in blocking HIV. Natural skin condoms allow passage of the virus; latex is much safer.

The Social Perspective: Economic, Legal, and Ethical Problems

AIDS will have an increasingly significant economic impact in the United States and a catastrophic impact in underdeveloped countries. The lifetime cost of medical care for an AIDS patient in the United States can reach $100,000 or more. Based on the CDC's estimate of cases through 1992, the total cost for all AIDS patients could be many billions. The majority of cases occur in large cities among people lacking health insurance, and public-health resources may not be equal to the demand. If the care of AIDS patients may be a

What Do You Really Know About AIDS?

Are You At Risk?

AIDS And Sex

Why No One Has Gotten AIDS From Mosquitoes

AMERICA RESPONDS TO AIDS

OTIS R. BOWEN, M.D., *Secretary* U.S. Department of Health and Human Services

ROBERT E. WINDOM, M.D., *Assistant Secretary for Health* U.S. Department of Health and Human Services

This brochure has been prepared by the Surgeon General and the Centers for Disease Control, U.S. Public Health Service. The Centers for Disease Control is the government agency responsible for the prevention and control of diseases, including AIDS, in the United States.

☆ U.S. GOVERNMENT PRINTING OFFICE: 1988—533-155

FIGURE 19.39 The Surgeon General's 1988 booklet, sent to all U.S. families.

ESSAY Acquired Immune Deficiency Syndrome

burden in the United States, where income exceeds $12,000 per person per year, imagine what it will be in many underdeveloped countries, where annual income is less than $200 per person.

Laws in the United States protect the confidentiality of medical information, including AIDS test results. They also provide equal opportunity with respect to employment, housing, and education, but many AIDS patients have encountered various kinds of discrimination. Protecting the rights of uninfected citizens and those of health professionals treating AIDS patients also must be considered. Other legal issues concern the responsibility of individuals with AIDS not to transmit the disease and the liabilities of distributors of blood products.

Many ethical issues are related to the legal issues. A major question is how the epidemic can be curtailed without infringing on individual freedoms. Another question weighs the moral obligation of health professionals to care for all patients against the risk of acquiring a fatal disease. Still other questions relate to allocation of scarce medical resources and the rights of terminally ill patients to refuse treatment.

Although many of these legal and ethical questions are raised by any serious disease, they have become more pressing as the AIDS epidemic accelerates. Most have not been answered, and some may be unanswerable.

CHAPTER SUMMARY

OVERVIEW OF IMMUNOLOGIC DISORDERS

- An **immunologic disorder** results from an inappropriate or an inadequate immune response.
- **Hypersensitivity,** or **allergy,** is due to an inappropriate reaction to an antigen.
- **Immunodeficiency** is due to an inadequate immune response.

RELATED KEY TERMS: **primary immunodeficiencies**, **secondary immunodeficiencies**

IMMEDIATE (TYPE I) HYPERSENSITIVITY

Allergens

- An **allergen** is an ordinarily innocuous substance that can elicit a harmful immunologic response in a sensitized person; common allergens are listed in Table 19.1.

Mechanism of Immediate Hypersensitivity

- The mechanism of **immediate hypersensitivity** is summarized in Figure 19.1; mediators of immediate hypersensitivity are summarized in Table 19.2.

RELATED KEY TERMS: **sensitization**, **degranulation**, **slow-reacting substance of anaphylaxis**

Atopy

- **Atopy,** or **localized anaphylaxis,** is a localized reaction to an allergen in which histamine and other mediators elicit flare and wheal reactions and other signs and symptoms of allergy.

Generalized Anaphylaxis

- **Generalized anaphylaxis** is a systemic reaction in which blood pressure is greatly decreased or the airway is occluded.

RELATED KEY TERMS: **reagin**, **anaphylactic shock**, **respiratory anaphylaxis**, **asthma**

Genetic Factors in Allergy

Treatment of Allergies

- Allergy is treated by hyposensitization, as shown in Figure 19.5, and symptoms are alleviated with antihistamines.

RELATED KEY TERMS: **desensitization**, **blocking antibodies**

CYTOTOXIC REACTIONS (TYPE II HYPERSENSITIVITY)

Mechanism of Cytotoxic Reactions

- The mechanism of **cytotoxic hypersensitivity** is summarized in Figure 19.6.

RELATED KEY TERMS: **transfusion reaction**

Transfusion Reactions

- Cytotoxic reactions in transfusions usually are due to the presence of donor erythrocytes with foreign antigens that cause hemolysis.

Hemolytic Disease of the Newborn

- **Hemolytic disease of the newborn** is due to anti-Rh antibodies from a sensitized mother reacting with Rh antigens in an Rh-positive fetus.

RELATED KEY TERMS: **jaundice**

IMMUNE COMPLEX DISORDERS (TYPE III HYPERSENSITIVITY)

Mechanism of Immune Complex Disorders

- The mechanism of **immune complex disorders** is summarized in Figure 19.9.

RELATED KEY TERMS: **immune complex**

Examples of Disorders

- **Serum sickness** occurs when foreign (sometimes animal) antigens in sera cause immune complexes to be deposited in tissues.

- The **Arthus reaction** is a local response to an antigenic substance (usually an injected substance) that causes edema and hemorrhage.

RELATED KEY TERMS

CELL-MEDIATED REACTIONS (TYPE IV HYPERSENSITIVITY)

Mechanism of Cell-mediated Reactions

- The mechanism of **cell-mediated (delayed) hypersensitivity** is summarized in Figure 19.12.

Examples of Disorders

- Contact hypersensitivity **(contact dermatitis)** occurs after second contact with poison ivy, metals, or another substance and usually appears as eczema.
- Other kinds of cell-mediated hypersensitivity include **tuberculin hypersensitivity** and **granulomatous hypersensitivity.**
- Characteristics of the four types of hypersensitivity are summarized in Table 19.5.

induration

AUTOIMMUNE DISORDERS

autoantibodies

Autoimmunization

- **Autoimmunization** is the development of hypersensitivity to self; it occurs when the immune system responds to a body component as if it were foreign.
- Tissue damage from **autoimmune disorders** can be caused by cytotoxic, immune complex, and cell-mediated hypersensitivity reactions.

Examples of Disorders

- Examples of autoimmune diseases include **myasthenia gravis, rheumatoid arthritis,** and **systemic lupus erythematosus.**

rheumatoid factor

TRANSPLANTATION

transplantation
graft tissue **autograft**
isograft **allograft**
xenograft
graft-versus-host disease

Histocompatibility Antigens

- Genetically determined **histocompatibility antigens** are found on the surface membranes of all cells; some are correlated with increased risk of certain diseases.
- **Human leukocyte antigens (HLAs)** in graft tissue are a main cause of **transplant rejection,** but immunocompetent cells in bone marrow grafts sometimes destroy host tissue.

Transplant Rejection

- Acute graft rejection, a Type II hypersensitivity, occurs in an already sensitized recipient. Slower rejection reactions are mainly cell-mediated (Type IV).

Immunosuppression

- **Immunosuppression** is a lowering of the responsiveness of the immune system to materials it recognizes as foreign.
- Immunosuppression is produced by **radiation** and administering **cytotoxic drugs;** it minimizes transplant rejection, but it also can reduce the host's immune response to infectious agents.

DRUG REACTIONS

- All four types of hypersensitivity reactions have been observed in immunologic drug reactions.

IMMUNODEFICIENCY DISEASES

immunodeficiency disease
primary immunodeficiency diseases
secondary immunodeficiency diseases

Deficiencies of Cells of the Immune System

- B cell deficiency, or **agammaglobulinemia,** leads to a lack of humoral immunity.
- T cell deficiency, or **DiGeorge syndrome,** leads to a lack of cell-mediated immunity.
- Deficiencies of both B and T cells, or **severe combined immunodeficiency disease,** leads to a lack of both humoral and cell-mediated immunity.

Acquired Immunodeficiencies

- Acquired immunodeficiencies can be caused by infections, malignancies, and autoimmune disorders.

IMMUNOLOGIC TESTS

precipitin test **precipitins**
immunodiffusion tests
electrophoresis
agglutination reactions
antibody titer
tube agglutination test
hemagglutination

- **Serology** is the use of laboratory tests to detect antigens and antibodies.
- **Precipitation reactions,** including immunodiffusion, **immunoelectrophoresis,** and **radial immunodiffusion,** depend on the formation of antigen-antibody complexes that precipitate from solutions or in agar gels.
- Agglutination tests depend on the formation of large particles that usually contain antigen, antibody, and cells.

- The **complement fixation test** detects antibodies indirectly by determining whether complement added to serum is used up in the reaction.
- **Immunofluorescence** allows detection of products of immune reactions within tissues.
- Assays for antigens and antibodies can be performed with radioactive substances and substances with enzymes coupled to them.

RELATED KEY TERMS

viral hemagglutination
hemagglutination inhibition test
neutralization reactions
Schick test **viral neutralization**
radioimmunoassay
enzyme-linked immunoabsorbent assay

AIDS

- Most AIDS cases are caused by HIV, a virus that probably evolved in the last 40 to 100 years that attacks the immune system. AIDS cases have rapidly increased in number and geographic distribution. The infection is transmitted by contact with infected blood and body fluids.
- The HIV virus destroys T4 cells and eventually impairs nearly all immune functions. Patients die of opportunistic infections and malignancies.
- AIDS infection is detected by immunologic tests for antibodies. The disease is treated with AZT and a few experimental drugs but cannot be cured.
- Efforts to develop a vaccine are under way, but the best means of prevention is to avoid exposure to body fluids of infected persons.
- The AIDS epidemic has raised many legal and ethical questions.

QUESTIONS FOR REVIEW

A.

1. What are the characteristics of an immunologic disorder?
2. What is hypersensitivity?
3. What is immunodeficiency, and how does it differ from hypersensitivity?
4. What are the four types of hypersensitivity?

B.

5. What is an allergen, and what properties make substances likely to be treated as antigens?
6. List examples of common allergens.
7. Explain the mechanism of an immediate hypersensitivity reaction.
8. What substances mediate immediate hypersensitivity reactions, and how do they contribute to the reaction?
9. What events occur in atopy, and what makes them happen?
10. What events occur in generalized anaphylaxis, and what makes them happen?
11. How might allergies be treated or cured?

C.

12. Explain the mechanism of a cytotoxic hypersensitivity reaction.
13. What triggers a transfusion reaction, and what are the effects of such a reaction?
14. What causes hemolytic disease of the newborn?
15. How can cytotoxic hypersensitivity reactions be prevented?

D.

16. What triggers an immune complex hypersensitivity reaction, and what are its effects?
17. What causes serum sickness, and how can it be prevented?
18. What causes an Arthus reaction, and how can it be prevented?

E.

19. Explain the mechanism of cell-mediated hypersensitivity.
20. Why is this reaction called delayed hypersensitivity?
21. What attributes of contact dermatitis identify it as cell-mediated?
22. How does a tuberculinlike reaction differ from granulomatous hypersensitivity?

F.

23. What is autoimmunity?
24. What kind of immunologic defect leads to autoimmunity?
25. For each autoimmune disorder described, how can tissue damage be related to a particular kind of hypersensitivity?

G.

26. What are HLA antigens?
27. How are HLAs related to transplant rejection and disease risk?
28. What is immunosuppression?
29. Under what circumstances would immunosuppressants be used?
30. What are the hazards of immunosuppressants?

H.

31. How are immunologic drug reactions related to hypersensitivity?

I.

32. How are immunodeficiencies related to cells of the immune system?
33. Aside from AIDS infections, how can immunodeficiencies be acquired?

J.

34. Briefly describe the mechanism of each of the immunologic tests discussed in the text.

K.

35. What causes AIDS, and how does the agent affect the body?

36. How is AIDS transmitted, and why is its incidence rising?
37. What are some complications of AIDS, and why do they occur?

L.

38. How is AIDS diagnosed and treated?
39. What problems are associated with making an AIDS vaccine?
40. How can AIDS be prevented?

PROBLEMS FOR INVESTIGATION

1. From members of your family (or another family), obtain as much information as possible about allergies that have been diagnosed over several generations. Use this information to determine whether the allergies are inherited or caused by some other factor.
2. How would you approach the following legal and ethical problems created by AIDS: (a) Should a child with AIDS be allowed to attend school? (b) Should a person with AIDS be allowed to hold a job as long as the illness permits? (c) Should the government be allowed to make testing for AIDS antibodies mandatory? (d) Are there special circumstances in which AIDS testing should be mandatory? (e) Who should have access to the results of a test for AIDS? (f) Should AIDS be a reportable disease? (g) Does a medical worker have the right to refuse to treat an AIDS patient? (h) Should insurance companies be allowed to refuse to provide health or life insurance to an AIDS patient? (i) Must all hospitals accept AIDS patients? (j) Is the use of a large portion of medical resources to care for an AIDS patient justified if it means denying care to indigent mothers and their children? (k) What are the implications to the nation's health-care system of having 90 percent of all hospital beds occupied by AIDS patients? (l) What are the implications to a country such as some in Africa, where cities may have AIDS infection rates of 30 percent?
3. For any immunologic disorder, survey the current literature to determine what has been learned recently about its cause, effects, and treatment.
4. Research and report about immunosuppressive drugs that are currently used in organ transplant cases.
5. Some people suffer hypersensitivity reactions when they wear jewelry. What are they reacting to, and are all such people reacting to the same thing? What type of hypersensitivity is this? What, if anything, can be done?
6. A 53-year-old woman who had smoked since the age of 17 decided to give up smoking. Shortly after stopping she was hospitalized with an autoimmune form of ulcerative colitis. Stress caused her to begin smoking again. Her colitis rapidly disappeared. When she stopped smoking, her colitis returned. Over the years a pattern emerged: no colitis while smoking; return of colitis upon cessation of smoking. What could account for this pattern?
(The answer to this question appears in Appendix E.)

SOME INTERESTING READING

AIDS Special Issue. 1988. *Science* 239 (February 5):573–621.

AIDS Special Issue. 1988. *Scientific American* 259, no. 4 (October).

Barnes, D. M. 1988. Health workers and AIDS: Questions persist. *Science* 241 (July 8):162.

Bartlett, M. S., and J. W. Smith. 1991. *Pneumocystis carinii*, an opportunist in immonocompromised patients. *Clinical Microbiology Reviews* 4(2):137–49.

Brandt, C. D., T. A. Rakusan, A. V. Sison, S. H. Josephs, E. S. Saxena, H. D. Herzog, R. H. Parrott, and J. L. Sever. 1992. Detection of human immunodeficiency virus type 1 infection in young pediatric patients by using polymerase chain reaction and biotinylated probes. *Journal of Clinical Microbiology* 30(1):36–40.

Buisseret, P. D. 1982. Allergy. *Scientific American* 247, no. 2 (August):86.

Cohen, I. R. 1988. The self, the world and autoimmunity. *Scientific American* 258, no. 4 (April):52.

Drew, W. L. 1992. Nonpulmonary manifestations of cytomegalovirus infection in immunocompromised patients. *Clinical Microbiology Reviews* 5(2):204–10.

Findlay, S. 1991. AIDS—the second decade: 500,000 more deaths. *U.S. News & World Report* 110, no. 23 (June 17):20–23.

Gallo, R. C. 1987. The AIDS virus. *Scientific American* 256, no. 1 (January):46.

Herrmann, J. E. 1986. Enzyme-linked immunoassays for the detection of microbial antigens and their antibodies. *Advances in Applied Microbiology* 31:271.

Immunology special issue. 1987. *Journal of the American Medical Association* 258, no. 20 (November 27).

Levine, C., and N. N. Dubler. 1990. Uncertain risks and bitter realities: the reproductive choices of HIV-infected women. *Milbank Quarterly* 68, no. 3 (Fall):321–52.

Loschen, D. J. 1988. Protecting against HIV exposure in family practice. *American Family Physician* 37, no. 1 (January):213.

McFadden, E. R., Jr. 1991. Fatal and near-fatal asthma. *New England Journal of Medicine* 324, no. 6 (February 7):409–11.

Minkoff, H. L., and J. A. DeHovitz. 1991. Care of women infected with the human immunodeficiency virus. *Journal of the American Medical Association* 266, no. 16 (October 23):2253–59.

Nurse's Clinical Library. 1985. *Immune disorders*. Springhouse, PA: Springhouse Corporation.

Rennie, J. 1990. The body against itself. *Scientifc American* 263 (December):106–15.

Rennie, J. 1991. Graft without corruption: antibody treatments could make transplanted organs acceptable. *Scientific American* 265 (September):18.

Rose, N. R. 1981. Autoimmune disease. *Scientific American* 144, no. 2 (February):80.

Rosenberg, S. A. 1990. Adoptive immunotherapy for cancer. *Scientific American* 262 (May):62–69.

Weber, R. W. 1987. Allergens. *Primary Care* 14, no. 3 (September):435.

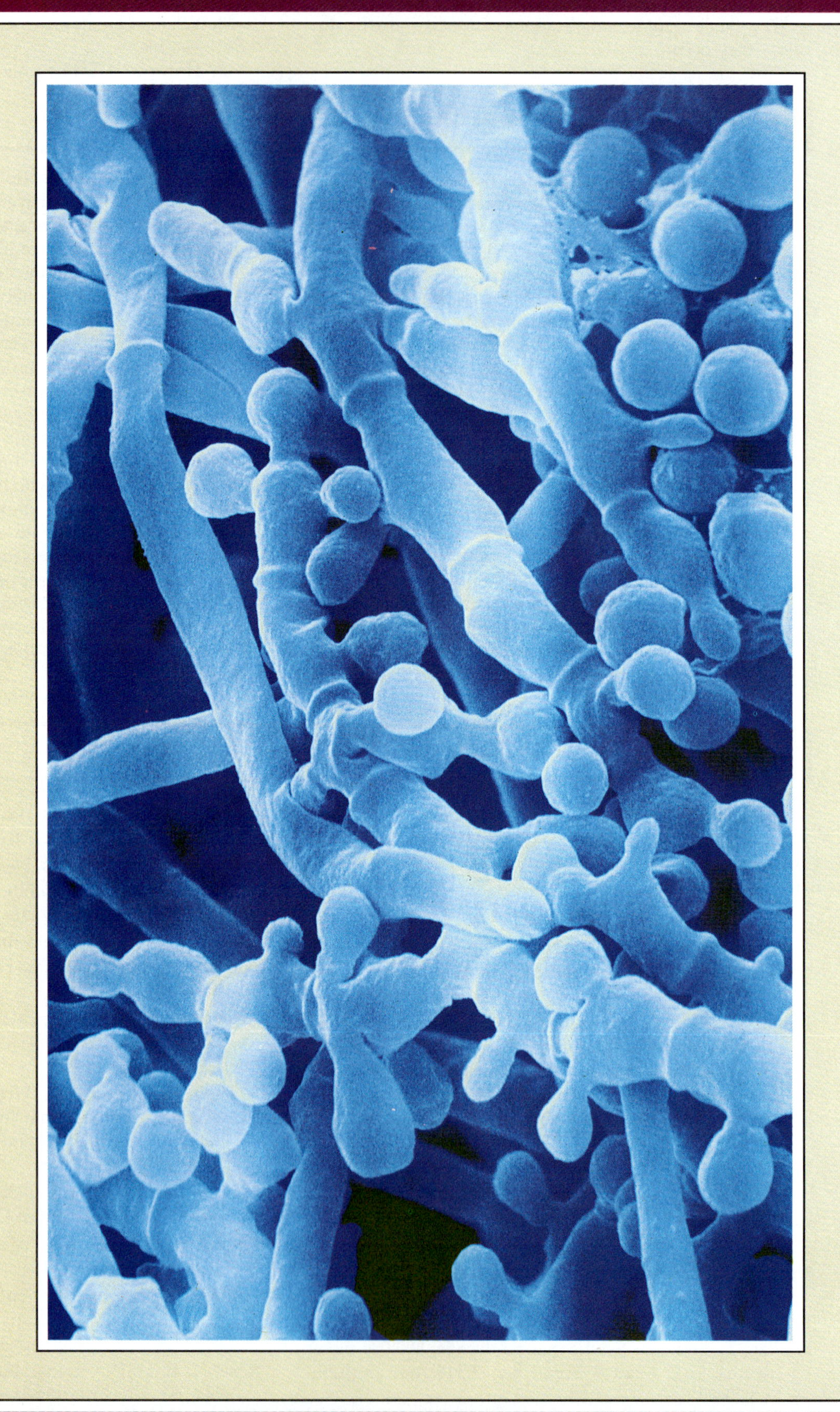

Trichophyton mentagrophytes, a fungal cause of ringworm infections of the skin.

20 Diseases of the Skin and Eyes: Wounds and Bites

This chapter focuses on the following questions:

A. What kinds of pathogens cause skin diseases?

B. What are the important epidemiologic and clinical aspects of skin diseases?

C. What kinds of pathogens cause eye diseases?

D. What are the important epidemiologic and clinical aspects of eye diseases?

E. What kinds of pathogens infect wounds and bites?

F. What are the important epidemiologic and clinical aspects of wound and bite infections?

Having considered the general principles of microbiology in the first five units of this text, we are now ready to apply those principles to understanding infectious diseases in humans. In studying diseases it is important to keep in mind the information on normal flora and disease processes in Chapter 15, epidemiology in Chapter 16, and host systems in Chapter 17. We begin our study of human diseases with diseases of the skin.

DISEASES OF THE SKIN

Your skin covers your body and accounts for 15 percent of your body weight. It provides an effective barrier to invasion by most microbes except when it is damaged. Here we will consider the kinds of skin diseases that occur when the skin surface fails to prevent microbial invasion.

Bacterial Skin Diseases

Many bacteria are found among the normal flora of the skin. Ordinarily they are kept from invading tissues by an intact skin surface and nonspecific defense mechanisms of the skin. (Chapter 17, p. 445) Bacterial and other skin infections usually arise from a failure of these defenses.

Staphylococcal Infections

Folliculitis and Other Skin Lesions Everyone has had a pimple at some time; most likely it was caused by *Staphylococcus aureus*, the most pathogenic of the staphylococci. Staphylococcal skin infections are exceedingly common because the organisms are nearly always present on the skin. Strains of staphylococci colonize the skin and upper respiratory tract of infants within 24 hours of birth. (Half the adult population and virtually all children are nasal carriers of *S. aureus*.) Infection occurs when these organisms invade the skin through a hair follicle, producing **folliculitis** (fol-lik"u-li'tis), also referred to as **pimples** or **pustules.** An infection at the base of an eyelash is called a **sty.** A larger, deeper, pus-filled infection is a **furuncle** (fu'rung-kl), **boil,** or **abscess** (Figure 20.1). An estimated 1.5 million Americans have such infections annually. Further spread of infection, particularly on the neck and upper back, creates a massive lesion called a **carbuncle** (kar'bung-kl). Encapsulation of abscesses prevents them from shedding organisms into the blood, but it also prevents circulating antibiotics from reaching the abscesses in effective quantities. Thus, in addition to antibiotic treatment, it usually is necessary to lance and drain abscesses surgically.

Staphylococcal infections are easily transmitted.

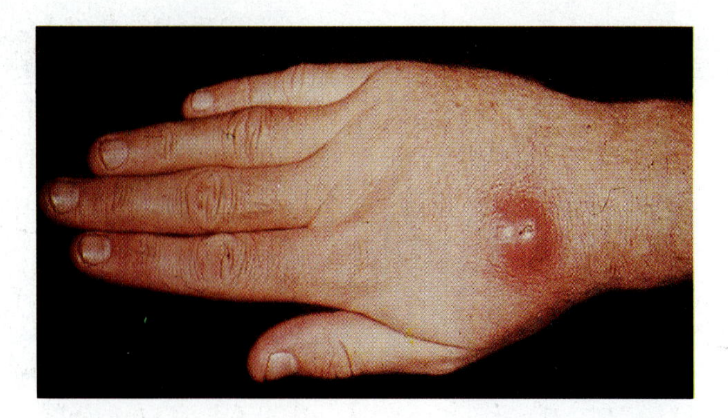

FIGURE 20.1 A furuncle, or deep, pus-filled lesion, caused by *Staphylococcus aureus*.

Asymptomatic carriers, hospital personnel, and hospital visitors often spread staphylococci via the skin as well as by nasal droplets and fomites. Staphylococci usually cause infection in older patients only when foreign bodies such as catheters or splinters are present. Although 5 million organisms must be injected into the skin to cause infection, only 100 are needed if they are soaked into a suture and tied into the skin.

Scalded Skin Syndrome **Scalded skin syndrome,** caused by certain toxin-producing strains of *S. aureus*, begins with a slight reddened area, often around the mouth. Within 24 to 48 hours it spreads to form large, soft, easily ruptured vesicles over the whole body. Skin over vesicles and adjacent reddened areas peels, leaving large, wet scalded-looking areas (Figure 20.2). The lesions dry and scale, and the skin returns to normal in 7 to 10 days. High fever generally is present. Bacteremia is frequent and can lead to septicemia and death within 36 hours.

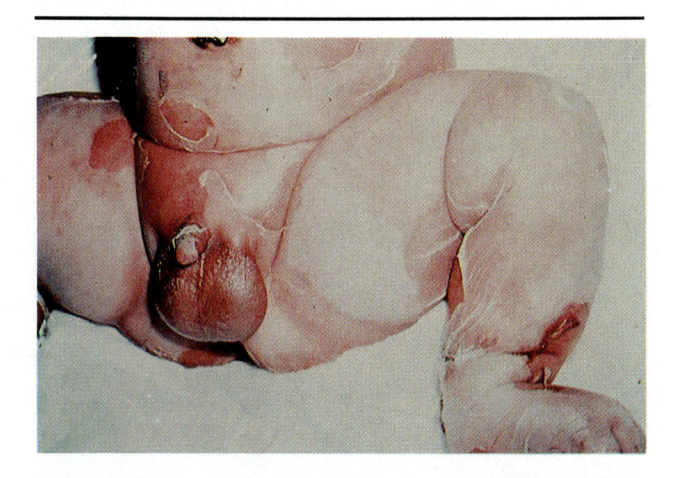

FIGURE 20.2 Scalded skin syndrome in an infant, caused by *Staphylococcus aureus*. Reddened areas of skin peel off, leaving scalded-looking wet areas.

Streptococcal Infections

Scarlet Fever **Scarlet fever,** sometimes called scarlatina, is caused by *Streptococcus pyogenes,* which also causes the familiar strep throat. Strains of the organism that cause scarlet fever have been infected by a temperate phage that enables them to produce an erythrogenic (red-making) toxin that causes the scarlet fever rash. Patients who have previously been exposed to the toxin, and so have antibodies that can neutralize it, can develop strep throat without the scarlet fever rash. However, such people can still spread scarlet fever to others. Three different erythrogenic toxins have been identified, and a person can develop scarlet fever once from each toxin.

In the United States most scarlet fever organisms currently have low virulence. In past decades, when strains were more virulent, it was a much-feared killer. However, even low-virulence strains can cause serious complications, such as glomerulonephritis or rheumatic fever. The use of penicillin has appreciably lowered the mortality rate. Convalescent carriers can shed infective organisms from the nasopharynx for weeks or months after recovery. Fomites also are an important source of streptococcal infections.

Erysipelas **Erysipelas,** also called **St. Anthony's fire,** has been known for over 2000 years and is caused by hemolytic streptococci. Before antibiotics became available, erysipelas often occurred after wounds and surgery and sometimes after very minor abrasions. Mortality was high. Today, it rarely occurs, and mortality is low. The disease begins as a small, bright, raised, rubbery lesion at the site of entry. Lesions spread as streptococci grow at lesion margins and are so sharply defined that they appear to be painted on. The organisms spread through lymphatics and can cause septicemia, abscesses, pneumonia, endocarditis, arthritis, and death if untreated. Curiously, erysipelas tends to recur at old sites. Instead of developing immunity, patients acquire greater susceptibility to future attacks.

Pyoderma and Impetigo

Pyoderma, a pus-producing skin infection, is caused by staphylococci, streptococci, and corynebacteria, singly or in combination. A highly contagious pyoderma, **impetigo,** is caused by staphylococci, streptococci, or both (Figure 20.3). Fluid from early pustules usually contains streptococci, whereas fluid from later lesions contains both. Streptococcal strains that cause skin infections usually differ from those that cause strep throat. Impetigo occurs almost exclusively in children; why adults are not susceptible is unknown. Easily transmitted on hands, toys, and furniture, impetigo can rapidly spread through a day-care center.

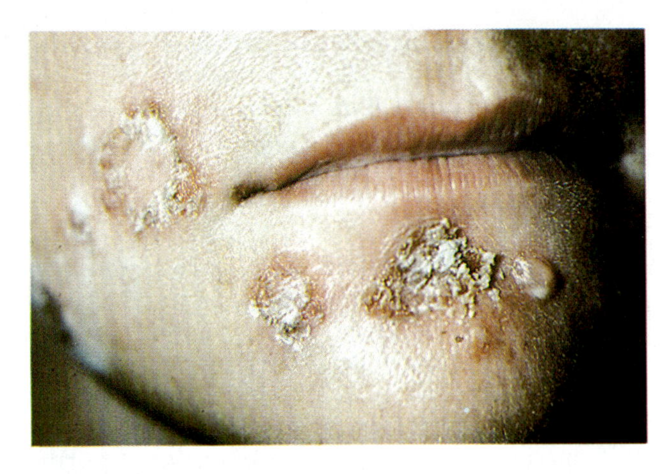

FIGURE 20.3 Impetigo, a highly contagious infection, is caused by staphylococci, streptococci, or both together.

Impetigo rarely produces fever, and it is easily treated with penicillin. Lesions usually heal without scarring, but skin can be discolored for several weeks, and pigment can be permanently lost.

Acne

Acne (acne vulgaris) affects more than 80 percent of teenagers and many adults, too. It is most often caused by male sex hormones that stimulate sebaceous glands to increase in size and secrete more sebum. Acne occurs in both males and females because the hormones are produced by the adrenal glands as well as the testes. Microorganisms feed on sebum, and ducts of the glands and surrounding tissues become inflamed. "Blackheads" are a mild form of acne in which hair follicles and sebaceous glands become plugged with sebum and keratin. In more severe cases (cystic acne) the plugged ducts become inflamed, rupture, and release secretions. Bacteria, especially *Propionibacterium acnes,* infect the area and cause more inflammation, more tissue destruction, and scarring. Such lesions can be widely distributed over the body, and some become encysted in connective tissue.

Acne is treated with frequent cleansing of the skin and topical ointments to reduce the risk of infections. Sometimes acne sufferers are advised to avoid fatty foods, but a connection between diet and acne is not well established. Dermatologists often prescribe oral antibiotics such as tetracycline in low doses to control bacterial infections in the lesions. However, continuous use of antibiotics depletes natural intestinal flora and can contribute to the development of antibiotic-resistant strains of bacteria. ∞ (Chapter 14, p. 355) The drug Accutane, derived from a molecule related to vitamin A, is now used to treat severe and persistent acne. It seems to inhibit sebum production for several months after treatment is stopped, but it can cause serious side effects, such as intestinal bleeding. In most cases acne disappears or decreases in severity as the body adjusts to the hormonal changes of puberty and the functioning of the sebaceous glands stabilizes.

Burn Infections

Severe burns destroy much of the body's protective covering and provide ideal conditions for infection. Burn infections, which are usually nosocomial, account for 80 percent of deaths among burn patients. *Pseudomonas aeruginosa* is the prime cause of life-threatening burn infections, but *Serratia marcescens* and species of *Providencia* also often infect burns. Many strains of these gram-negative bacilli are antibiotic-resistant. See the Microbiologist's Notebook for Chapter 16. ∞ (p. 434)

The thick crust or scab that forms over a severe burn is called **eschar** (es'kar). Bacteria growing in or on eschar pose no great threat, but those growing beneath it cause severe local infections and can move into the blood. Delivering antibiotics to infections under eschar is difficult because the eschar lacks blood vessels. Topical antimicrobials can seep through the eschar, and removal of some of the eschar by a scraping technique called **debridement** (da-bred-maw' or de-brēd'ment) helps them to reach infection sites.

Prevention of burn infections is difficult even when patients are isolated within hospital burn units. Having lost skin, patients have lost the benefit of white blood cells that normally move to sites of infection in the skin. They also have fluid and electrolyte deficiencies because of seepage from burned tissue. Finally, their appetites are depressed at a time when extensive tissue repair increases metabolic needs.

Burn infections are as difficult to diagnose as they are to treat. Early signs of infection can consist of only a mild loss of appetite or increased fatigue. Definitive diagnosis is made by finding more than 10,000 bacteria per gram of eschar. *Pseudomonas aeruginosa* infection should be suspected when a greenish discoloration appears at the burn site and cultures have a grapelike odor. This bacterium produces tissue-killing toxins that erode skin. It is extremely resistant to antimicrobial drugs and has been found growing in surgical scrub solutions.

Viral Skin Diseases

Rubella

Nature of the Disease **Rubella,** or **German measles,** is the mildest of several human viral diseases that cause **exanthema** (ex-an-the'mah), or skin rash. A rash, the main symptom of rubella, appears first on the trunk 16 to 21 days after infection, but the virus spreads in the blood and other tissues before the rash appears. Infected adult women often suffer from temporary arthritis and arthralgia (joint pain) from dissemination of the virus to joint membranes. These complications are seen less frequently in adult men.

Congenital rubella syndrome results from infection of a developing embryo across the placenta. When a woman becomes infected with rubella during the first 8 weeks of pregnancy, severe damage to the embryo is likely because organ systems are developing. After the eighteenth week of pregnancy, damage is rare. The spread of rubella viruses in the infant kills many cells, persistently infects other cells, reduces the rate of cell division, and causes chromosomal abnormalities. Many infants are stillborn, and those that survive may suffer from deafness, heart abnormalities, liver disorders, and low birth weight.

Incidence and Transmission Prior to the development of a vaccine, nearly all humans had rubella at some time, but many cases were not detected. Half the cases in young children and up to 90 percent of those in young adults are not recognized. In the United States the incidence of rubella was nearly 30 cases per 100,000 people in 1969, when a vaccine was licensed. Rubella, including congenital rubella, has now been nearly eliminated (Figure 20.4).

Transmission is mainly by nasal secretions shortly before and for about a week following the appearance of a rash. Many infected individuals do not have a rash and transmit the virus without knowing it. Rubella is highly contagious, especially by direct contact among children aged 5 to 14. Infants infected before birth are rubella carriers; they excrete viruses and expose the hospital staff and visitors, including pregnant women, to the disease.

Diagnosis Rubella can be positively diagnosed by a variety of laboratory tests. Determination of antibody levels is particularly useful in identifying newborn carriers and assessing immunity of pregnant women exposed to rubella.

Immunity and Prevention The currently available rubella vaccine, an attenuated viral vaccine, is the only means of preventing rubella, and many children in the United States have received it. The vaccine produces lower antibody levels than infection, and immunity probably is not as long-lasting as that produced by infection. Also, viruses appear in the nasopharynx a few weeks after immunization, and some individuals experience mild symptoms of the disease. A second immunization is recommended for females before they become sexually active to prevent infection of fetuses. Waiting until a woman is pregnant to immunize may allow viruses from the vaccine to infect the fetus. Similarly, caution must be exercised

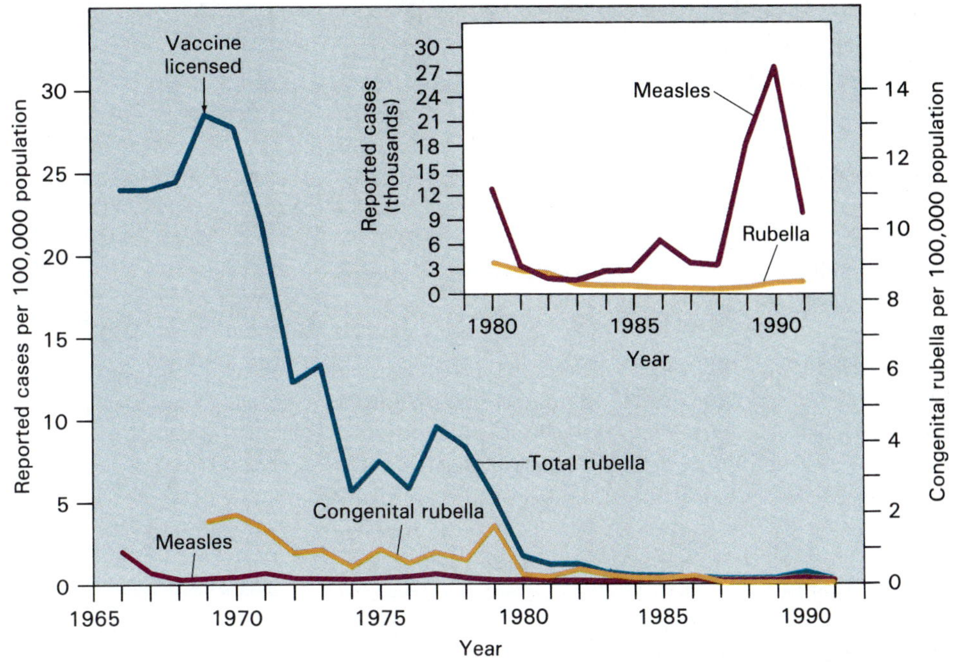

FIGURE 20.4 Incidence rates of rubella (German measles) cases and rubeola (measles) cases dropped rapidly after vaccines were licensed during the 1960s. Failure to vaccinate, however, has led to a recent increase of cases of rubeola.

in immunizing young children whose mothers are pregnant to prevent transmission of the virus from child to mother to fetus.

Measles

Nature of the Disease **Measles,** or **rubeola,** is a febrile (accompanied by fever) disease with a rash caused by the rubeola virus, which invades lymphatic tissue and blood. After the virus enters the body through the nose, mouth, or conjunctiva, symptoms appear in 9 to 11 days in children and in 21 days in adults. **Koplik's spots,** red spots with central bluish specks (Figure 20.5), appear on the upper lip and cheek mucosa 2 or 3 days before other symptoms such as fever, conjunctivitis, and cough. These symptoms persist for 3 or 4 days and sometimes progressively worsen. They are followed by a rash, which spreads during a 3- to 4-day period from the forehead to the upper extremities, trunk, and lower extremities and disappears in the same order several days later. The rash is caused by T cells reacting with virus-infected cells in small blood vessels. Without such reactions, no rash appears, but the virus is free to invade other organs. If the virus invades the lungs, kidneys, or brain, the disease often is fatal.

The most common complications of measles are upper respiratory and middle ear infections. **Measles encephalitis,** a more serious complication, occurs in only 1 or 2 patients per 1000 but has a 30 percent mortality rate and leaves one-third of survivors with permanent brain damage. Another complication, **subacute sclerosing panencephalitis (SSPE),** occurs in only 1 in 200,000 cases but is nearly always fatal. It is

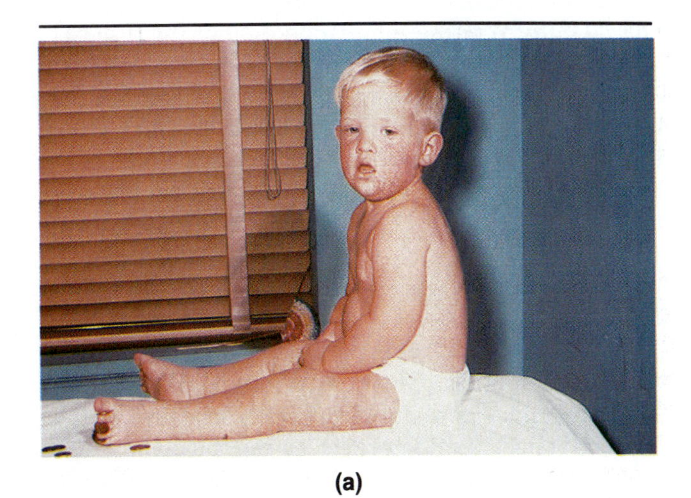

(a)

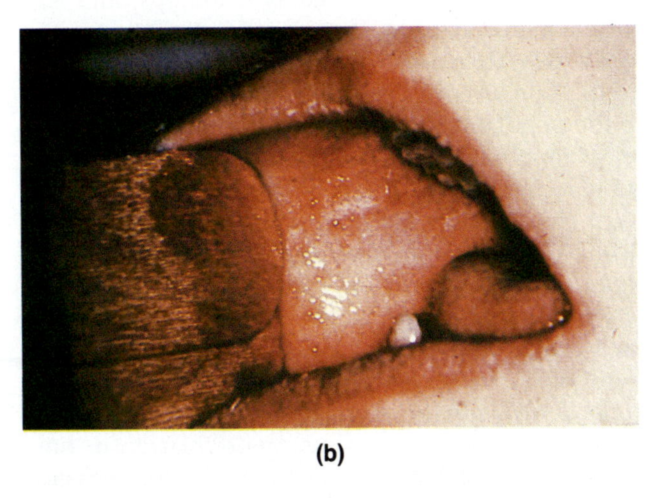

(b)

FIGURE 20.5 Rubeola (measles), showing (a) the typical rash, and (b) Koplik's spots on lip mucosa.

due to the persistence of measles viruses in brain tissue and causes death of nerve cells, with progressive mental deterioration and muscle rigidity. SSPE manifests itself 6 to 8 years after measles, usually in children who had measles before age 3. In poorly nourished children measles causes intestinal inflammation with extensive protein loss and shedding of viruses in stools. More than 15 percent of children infected with measles in developing countries die of the disease or its complications.

Incidence and Transmission The measles virus is highly contagious; a susceptible person has a 99 percent chance of infection if exposed directly to someone who is releasing the virus during coughing and sneezing. Before widespread immunization most children had measles before they were 10 years old. In populations such as that of the United States, where periodic epidemics have occurred and most children are well nourished, the disease is serious but rarely fatal. In populations lacking immunity from periodic epidemics or in malnourished children, measles is a killer. In 1875, when measles was first introduced into Fiji, 30 percent of the population died of the disease.

Diagnosis and Treatment Measles is diagnosed by its symptoms. Treatment is limited to alleviating symptoms and dealing with complications. Secondary bacterial infections can be effectively treated with antibiotics.

Immunity and Prevention In a nonimmunized population, epidemics of large numbers of cases over a 3- or 4-week period generally occur every 2 to 5 years over wide regions. Measles vaccine now prevents such epidemics in many developed countries. In the United States, mandatory measles vaccination has greatly reduced the incidence of the disease. (Figure 20.4) Because the MMR vaccine (Chapter 18) containing attenuated viruses of measles, mumps, and rubella is generally used, the incidence of all three diseases has decreased simultaneously. ∞ (p. 495) Eradication of a disease requires that about 90 percent of the population be immune. Religious groups that reject immunization, immigrants who don't understand its importance, and apathetic parents who don't have their children immunized make eradication difficult. Compounding these problems has been the recent federal government cutbacks in funding used to purchase vaccine for clinics. Parents arrive to have their children immunized and are told that the clinic is out of vaccine. It may take several trips before they are able to obtain the vaccination, and not all parents are diligent about this. Thus, poor children in our country are going without vaccine, and the measles case rate has begun to swing up again.

Immunity acquired from having measles is lifelong. Recent epidemics among college students immunized as children suggests that immunity from the vaccine may not be lifelong or that the children were immunized before 15 months of age. A few cases of SSPE have been attributed to the vaccine, but the incidence of SSPE is much lower than it was prior to the use of the vaccine.

CLOSE-UP

Measles

Humans are the only known hosts of measles, though distemper in dogs is caused by a virus with antigens similar to those of the measles virus. Measles can spread only if a continuous chain exists between infected and susceptible people. This is because the virus doesn't display latency and cannot survive for long outside the body, no other organism serves as a reservoir, and immunity from infection is lifelong. It has been estimated that a world population of at least 300,000 was required to perpetuate the virus, a population size not reached until about 2500 B.C. Measles probably appeared as a human pathogen after that time—possibly as a mutant of a virus like the one that causes distemper.

Chickenpox and Shingles

One Virus—Two Diseases The **varicella-zoster virus** (VZV), a herpesvirus virus, causes both **chickenpox** (varicella) and **shingles** (zoster). Chickenpox is a highly contagious disease that causes skin lesions and usually occurs in children. There are probably over 3 million cases per year in the United States alone. Shingles is a sporadic disease that appears most frequently in older and immunocompromised individuals.

In chickenpox the virus enters the upper respiratory tract and conjunctiva and replicates at the site of entry. New viruses are carried in blood to various tissues, where they replicate several more times. Release of these viruses causes fever and malaise. In 14 to 16 days after exposure, small, irregular, rose-colored skin lesions appear. The fluid in them becomes cloudy, and they dry and crust over in a few days. The lesions appear in cyclic crops over 2 to 4 days as the viruses go through cycles of replication. They start on the scalp and trunk and spread to the face and limbs, and sometimes to the mouth, throat, and vagina, and occasionally to the respiratory and gastrointestinal tracts.

Chickenpox, although sometimes thought of as a mild childhood disease, can be fatal. The viruses invade and damage cells that line small blood vessels and lymphatics. Circulating blood clots and hemorrhages from damaged blood vessels are common. Death from varicella pneumonia is due to extensive blood vessel damage in the lungs and the accumulation of erythrocytes and leukocytes in alveoli. Cells in the liver, spleen, and other organs also die because of damage to blood vessels within them.

In shingles (Figure 20.6) painful lesions like those of chickenpox usually are confined to a single region supplied by a particular nerve. Such eruptions may arise from latent viruses acquired during a prior case of chickenpox. During the latent period these viruses reside in ganglia in the cranium and near the spine. When reactivated, the viruses spread from a ganglion along the pathway of its associated nerve or nerves. Pain and burning and prickling of the skin occur before lesions appear. Symptoms range from mild itching to continuous, severe pain and can include headache, fever, and malaise. Lesions often appear on the trunk in a girdlelike pattern (*zoster*, girdle) but can infect the face and eyes. Shingles is most severe in individuals with malignancies or immune disorders. In these patients lesions may cover wide areas of the skin and sometimes spread to internal organs, where they can be fatal.

The latent virus is activated when cell-mediated immunity drops below a critical minimal level, as can occur in lymphatic cancers, spinal cord trauma, heavy-metal poisoning, or immunosuppression. Release of newly replicated viruses increases antibody production, but the antibodies may fail to stop viral replication. The viruses damage nerve endings, cause intense inflammation, and produce clusters of skin lesions indistinguishable from chickenpox lesions. Recovery from shingles usually is complete, although second and third cases do occur, and depends on development of cell-mediated immunity and local interferon production. Chronic shingles is found in AIDS and other immunocompromised patients. New vesicles constantly erupt, while old ones fail to heal, which can be very debilitating.

Incidence and Transmission Chickenpox is endemic in industrialized societies in the temperate zone, and its incidence is highest in March and April. The primary infection usually occurs between the ages of 5 and 9. In general, chickenpox in adults who have not had it as children is more severe than in children. Shingles also is age-related, with most cases appearing in people over 45 years of age.

Infection can be spread by respiratory secretions and contact with moist lesions but not from crusted lesions. Children experiencing a mild case, with only a few lesions and no other symptoms, often spread the disease. In rare cases, adults with partial immunity can contract shingles from exposure to children with chickenpox, but susceptible children can easily contract chickenpox from exposure to adults with shingles.

Diagnosis and Treatment Chickenpox is diagnosed by a history of exposure and the nature of lesions. Shingles may be impossible to distinguish from other herpes lesions without laboratory tests. Treatment is limited to relieving symptoms, but aspirin should not be given to children because of the risk of Reye's syndrome. Antiviral agents are being tested on infections in immunosuppressed patients and those with disseminated disease.

Immunity and Prevention Having chickenpox as a child confers lifelong immunity in most cases. Recurrences (in the form of shingles) are seen only in individuals with low concentrations of VZV antibodies and waning cell-mediated immunity. Chickenpox is the last of the common communicable childhood diseases for which no vaccine is widely available.

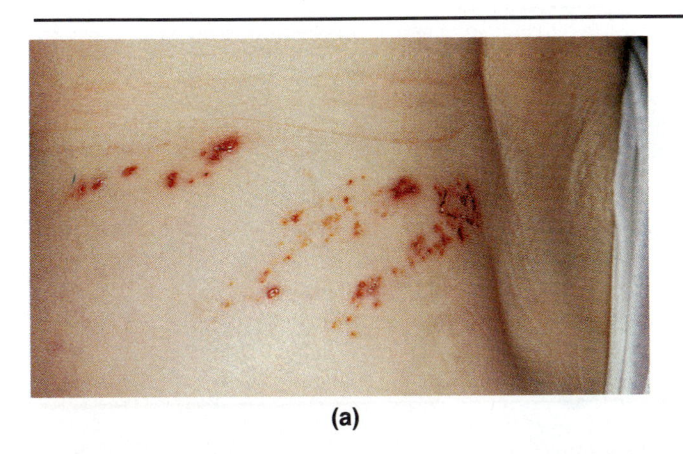

(a)

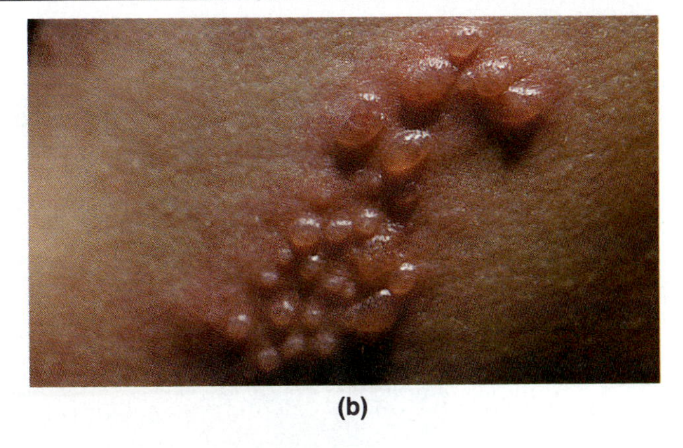

(b)

FIGURE 20.6 Shingles lesions usually result from herpes zoster virus infections acquired during childhood chickenpox. The virus can remain latent within the body for many years before being reactivated in adulthood. (a) Vesicles commonly form a belt around the chest or hips, following the pathway of a nerve. (b) The small yellow vesicles dry up and heal by scabbing, but they can be excruciatingly painful and itchy.

PUBLIC HEALTH

A Vaccine for Varicella?

An attenuated varicella vaccine developed in Japan prevents chickenpox when administered as late as 72 hours after exposure. It is recommended for immunosuppressed children to prevent uncontrollable systemic infection, but not for normal children. If it proves to confer high-quality, lasting immunity it may be used in normal children in the future. However, immunologists have some reservations about administering vaccines containing viruses that can become latent because the mechanism of latency and reactivation is not well understood.

Other Pox Diseases

Smallpox, a formerly worldwide and serious disease, has now been eradicated, as explained in the Essay at the end of this chapter. Other poxviruses cause a variety of diseases.

Cowpox **Cowpox,** caused by the vaccinia virus, causes lesions similar to a smallpox vaccination at abrasion sites, inflammation of lymph nodes, and fever. Vaccinia viruses also can cause a progressive disease, with numerous lesions and symptoms more like those of smallpox, that can be treated with the drug methisazone. Cattle and rodents appear to transmit the disease to humans.

The vaccinia virus was not only used by Jenner to immunize against smallpox, it also was the first animal virus to be obtained in sufficient quantity for chemical and physical analysis. (Chapter 1, p. 15) The modern vaccinia virus may have become attenuated by several centuries of passage on calves' skin, a procedure used in making smallpox vaccine.

Molluscum Contagiosum The **molluscum contagiosum** virus is unusual for several reasons. First, it differs immunologically from both orthopoxviruses and parapoxviruses. Second, it elicits only a slight immune response. Third, although infected cells cease to synthesize DNA, the virus induces neighboring uninfected cells to divide rapidly. Thus, this virus may be intermediate between viruses that cause specific diseases and those that induce tumors. Molluscum contagiosum affects only humans and is distributed worldwide. It causes flesh-colored, painless lesions scattered over the skin, usually in children, where it can persist for years. It is acquired by personal contact or from items such as gym equipment and swimming pools. No treatment is available.

Warts

Human **warts,** or **papillomas,** are caused by **human papillomaviruses** (HPV). HPV specifically attack skin and mucous membranes, and warts grow freely in many sites in the body—the skin, the genital and respiratory tracts, and the oral cavity. Viral infection lasts a lifetime. Even when warts disappear or are removed, the virus remains in surrounding tissue.

Nature of Warts Warts vary in appearance, area of occurrence, and pathogenicity (Figure 20.7). Some are barely visible and self-limiting—that is, they do not grow or spread—and others, such as laryngeal warts, are larger but benign. A few warts are malignant. Genital warts, for example, sometimes become malignant, and some of the viral strains that produce them are strongly associated with cervical cancer.

(a)
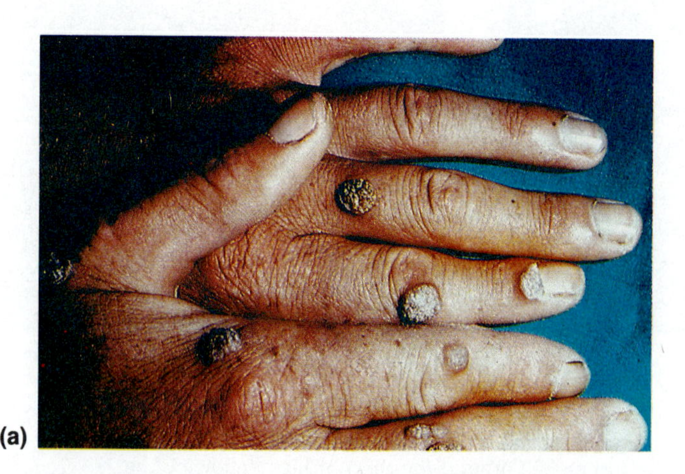

(b)
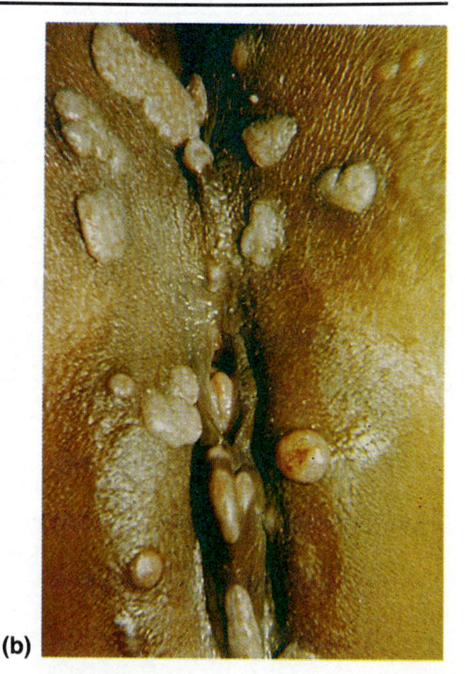

FIGURE 20.7 Warts are caused by human papillomaviruses. (a) Common warts, and (b) genital warts around the vagina, a probable cause of cervical cancer.

APPLICATIONS

Mind Over Body

According to Lewis Thomas, "Warts are wonderful structures. Fully grown, nothing in the body has so much the look of toughness and permanence as a wart, yet, inexplicably and often very abruptly, they come to the end of their lives and vanish without a trace." Thomas goes on to relate how warts "can be ordered off the skin by hypnotic suggestion." In a controlled study, 14 patients with multiple warts were hypnotized and the suggestion made that all warts on one side of the body go away. Within a few weeks, this happened in nine patients, including one who "got mixed up and destroyed the warts on the wrong side."

(Women who have had genital warts removed are advised to have Pap tests twice a year.) Warts of all types grow larger in people with AIDS and other immunodeficiencies.

Transmission of Warts Papillomaviruses are transmitted by direct contact, usually between humans, or by fomites. Dermal warts form when the virus enters the skin or mucous membranes through abrasions. Genital warts are sexually transmitted, and juvenile onset laryngeal warts are acquired during passage through an infected birth canal. The incubation period varies from 1 week to 1 month for dermal warts to 8 months for genital warts. Genital warts and warts acquired at birth will be discussed in Chapter 25.

Dermal Warts Epithelial cells become infected and proliferate to form **dermal warts,** which have distinct boundaries and remain above the basement membrane between the epidermis and the dermis. Children and young adults are more likely than older people to have dermal warts. Only a few warts are present at any one time, and most regress in less than 2 years. Removing one wart often causes regression of all others. In spontaneous regression, all warts usually disappear at the same time. Regression probably is an immunologic phenomenon.

Diagnosis and Treatment Warts can be distinguished by immunological tests and microscopic examination of tissues. Enzyme immunoassay and immunofluorescent antibody tests can detect about three-fourths of the cases in which viruses are found microscopically. These tests sometimes fail because certain papillomas, especially genital and laryngeal papillomas and those progressing toward malignancy, produce only small quantities of antigens.

Available treatments for various kinds of warts are not entirely satisfactory. The most widely used treatment, cryotherapy, involves the freezing of tissue with liquid carbon dioxide and excision of the infected tissue. Caustic agents, such as podophylin, salicylic acid, and glutaraldehyde, surgery, antimetabolites such as 5-fluorouracil, and interferon to block viruses also are used to get rid of warts. Recurrences are still common.

Fungal Skin Diseases

The fungi that invade keratinized tissue are called **dermatophytes** (der-mat'o-fītz), and fungal skin diseases are called **dermatomycoses** (der'mat-o-mi-ko'ses). These diseases can be caused by any one of several organisms, mainly from three genera: *Epidermophyton*, *Microsporum*, and *Trichophyton* (Table 20.1). They cause athlete's foot and ringworm of the skin, nails, scalp, beard, and groin.

Fungi that invade subcutaneous tissues live freely in soil or on decaying vegetation and can be found in bird droppings and as airborne spores. They enter tissue through a wound and sometimes spread to lymph vessels. Subcutaneous fungal infections usually spread slowly and insidiously; response to treatment is likewise slow.

Athlete's Foot In **athlete's foot,** or **tinea pedis,** hyphae invade the skin between the toes and cause dry, scaly lesions. Fluid-filled lesions develop on moist, sweaty feet. Subsequently, the skin cracks and peels and a secondary bacterial infection leads to itchy, soggy, white areas between the toes. Athlete's foot results from an ecological imbalance between normal flora and host defenses. The fungi that cause athlete's

TABLE 20.1 Dermatomycoses and common organisms found in lesions

Dermatomycosis	Epidermophyton floccosum	Microsporum canis	Trichophyton tonsurans	Trichophyton rubrum	Trichophyton mentagrophytes
Body ringworm		X			X
Nail ringworm	X			X	X
Groin ringworm	X			X	X
Scalp ringworm		X	X		
Beard ringworm				X	X
Athlete's foot	X			X	X

foot are always available; they infect tissues when body defenses fail to repel them.

Ringworm **Ringworm** is caused by some of the same organisms as athlete's foot. Being highly contagious, it is easily acquired at hairstyling establishments if strict sanitary practices are not followed. Ringworm infects skin, hair, and nails, and some forms are named for where they are found. **Tinea corporis** (body ringworm) causes ringlike lesions with a central scaly area. (It was the shape of these lesions that originally gave rise to the misleading name "ringworm.") **Tinea cruris** (groin ringworm, or "jock itch") occurs in skin folds in the pubic region (Figure 20.8). **Tinea unguium** (ringworm of the nails) causes hardening and discoloration of fingernails and toenails. In **tinea capitis** (scalp ringworm) hyphae grow down into hair follicles and often leave circular patterns of baldness. **Tinea barbae** (barber's itch) causes similar lesions in the beard.

None of the dermatomycoses cause severe diseases, and they do not usually invade other tissues, but they are unsightly, itchy, and persistent. They grow well at skin temperature, which is slightly below body temperature. Tissue damage caused by dermatophytes can allow secondary bacterial infections.

Diagnosis and Treatment Diagnosis of athlete's foot and ringworm can be made by microscopic examination of scrapings from lesions, but observation of the skin itself is often sufficient. Although fungi in tissues generally do not form spores, those in laboratory cultures often do, and workers must be especially careful not to become infected from escaping spores. Large numbers of people have been infected by the dispersal of fungal spores through a building's ventilation system.

Treatment generally consists of removing all dead epithelial tissues and applying a topical antifungal ointment. If lesions are widespread or difficult to treat

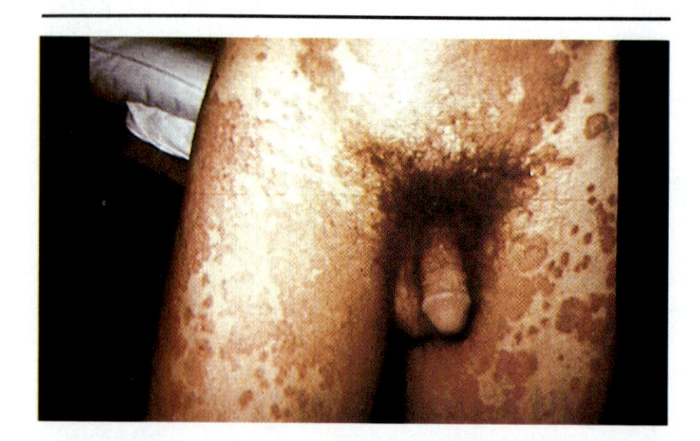

FIGURE 20.8 Tinea cruris, ringworm of the groin ("jock itch"), is caused by the fungus *Trichophyton mentagrophytes.*

APPLICATIONS

Ringworm

Not long after World War II, a specialist in the study and treatment of fungal infections at the start of his career presented himself with the challenge of curing ringworm by inoculating his forearm with appropriate fungi. A lesion soon developed. In spite of his efforts to cure it, he retired from his profession some four decades later with the same ringworm lesion still flourishing. Effective means of treating ringworm are still needed.

topically, as when they infect nailbeds, griseofulvin is administered orally. Preventing athlete's foot depends on maintaining healthy, clean, dry skin that can resist the opportunism of fungi, especially on the feet. Preventing ringworm requires avoiding contaminated objects and spores from laboratory cultures.

Sporotrichosis **Sporotrichosis,** caused by *Sporothrix schenckii,* usually enters the body from plants, especially sphagnum moss and rose and barberry thorns. It also can be acquired from other humans, dogs, cats, horses, and rodents. The disease is most common in the midwestern United States, especially in the Mississippi valley. A lesion appears first at the site of a minor wound as a nodular mass. The mass ulcerates, becomes chronic, granulomatous, and pus-filled, and can spread easily to lymphatic vessels. In rare instances it disseminates to internal organs, especially the lungs. Diagnosis is made by culturing specimens of pus or tissue taken from lesions. The cutaneous and lymphatic forms can be treated with potassium iodide; disseminated infections require amphotericin B. People who work with plants or soil should protect injured skin from materials that might lead to infection.

Blastomycosis North American **blastomycosis** (Figure 20.9), caused by the fungus *Blastomyces dermatitidis,* is most common in the central and southeastern United States. The fungus enters the body through the lungs or wounds, where it causes disfiguring granulomatous, pus-producing lesions, and multiple abscesses in skin and subcutaneous tissue. This condition, which for unknown reasons affects chiefly males in their thirties and forties, is known as **blastomycetic dermatitis.** In some cases, the fungus travels in the blood and invades internal organs, causing **systemic blastomycosis.** The lungs can be infected directly by inhaling spores, and organisms can travel from the lungs to infect other tissues. The disease usually causes relatively mild respiratory symptoms, fever, and general malaise. It can be diagnosed by

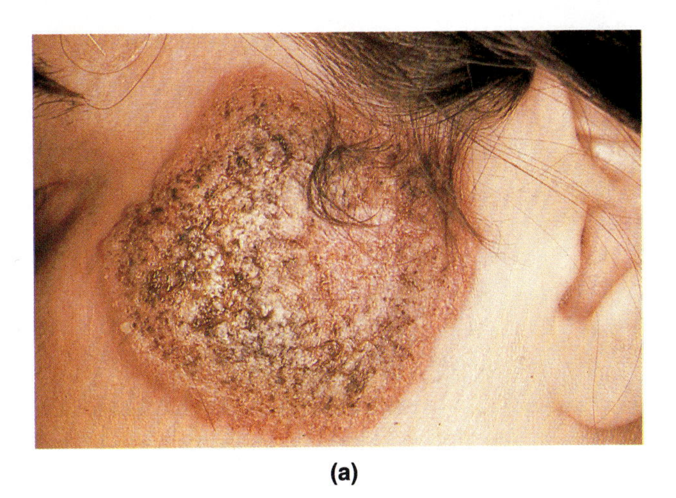

(a)

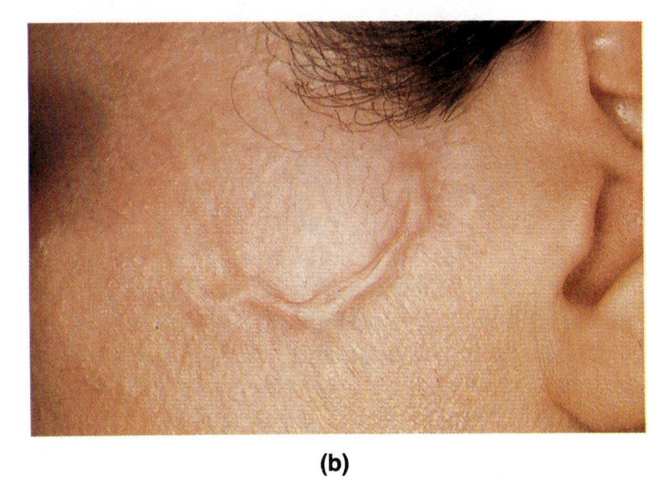

(b)

FIGURE 20.9 Blastomycosis lesion on the face, (a) before, and (b) after treatment with antifungal drugs.

finding budding yeast cells in sputum or pus, and treated with amphotericin B or the less toxic hydroxystilbamidine.

Opportunistic Fungal Infections

Certain fungi, such as some yeasts and black molds, can invade the tissues of humans with impaired resistance. Two yeasts, *Candida albicans* and *Cryptococcus neoformans*, and certain molds, such as *Aspergillus fumigatus*, *A. niger*, and the zygomycetes *Mucor* and *Rhizopus*, account for many opportunistic fungal diseases.

Candidiasis *Candida albicans*, an oval, budding yeast, is present among the normal flora of the digestive, respiratory, and urogenital tracts of humans. In debilitated individuals it can cause **candidiasis,** or **moniliasis,** in one or several tissues. Superficial candidiasis appears as **thrush** (Figure 20.10a), milky patches of inflammation on oral mucous membranes, especially in infants, diabetics, debilitated patients, and those receiving prolonged antibiotic therapy. (Chapter 14, p. 355) It appears as **vaginitis** when vaginal secretions contain large amounts of sugar, as occurs during pregnancy, when oral contraceptives are used, or when diabetes is undiagnosed or poorly controlled. Some pathogenic strains of *Candida* may be sexually transmitted. Cannery workers whose hands are in water for long periods sometimes develop skin and nail lesions (Figure 20.10b). *Candida* can invade the lungs, kidneys, and heart or be carried in the blood, where it causes a severe toxic reaction. Candidiasis, the most common nosocomial fungal infection, is seen in patients with serious diseases such as tuberculosis, leukemia, and AIDS.

Finding budding cells in lesions, sputum, or exudates confirms the diagnosis of candidiasis. Various

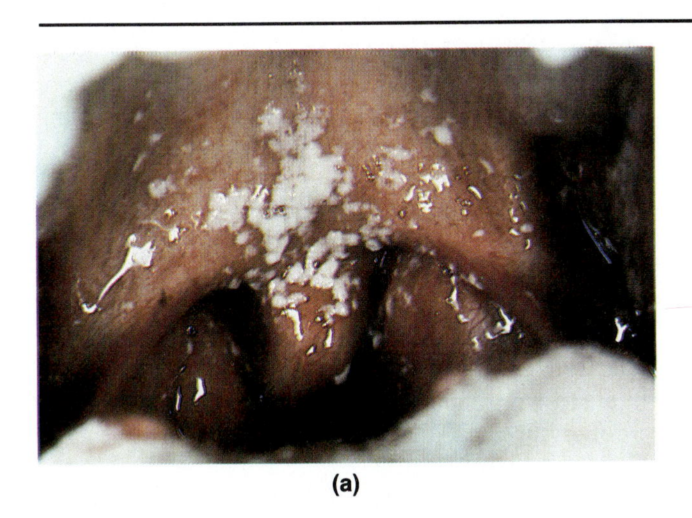

(a)

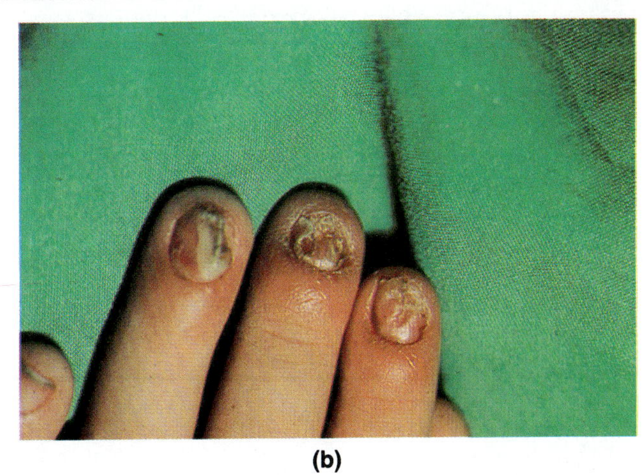

(b)

FIGURE 20.10 *Candida* infections of the oral cavity (thrush), seen as white patches (a), are a common complication of AIDS, diabetes, and prolonged antibiotic therapy. (b) *Candida* infections of the nails are very difficult to eradicate.

antifungal drugs are used to treat it. *Candida* is ubiquitous; infections can be prevented only by preventing debilitating conditions.

Aspergillosis Various species of *Aspergillus*, but especially *A. fumigatus*, which grow on decaying vegetation, can cause **aspergillosis** in humans. This fungus initially invades wounds, burns, the cornea, or the external ear, where it thrives in earwax and can ulcerate the eardrum. In immunosuppressed patients, it can cause severe pneumonia. Diagnosis is made by finding characteristic hyphal fragments in tissue biopsies. Antifungal agents are only modestly successful in treatment. Prevention depends mainly on host defenses because the mold is so ubiquitous that exposure is inevitable.

Zygomycoses Certain zygomycetes of the genera *Mucor* and *Rhizopus* can infect susceptible humans, such as untreated diabetics, and cause **zygomycoses.** Once established, the fungus invades the lungs, central nervous system, and tissues of the eye orbit and can be rapidly fatal, presumably because it evades body defenses. It can be diagnosed by finding broad hyphae in the lumens and walls of blood vessels. Antifungal drugs may or may not be effective in treating zygomycoses.

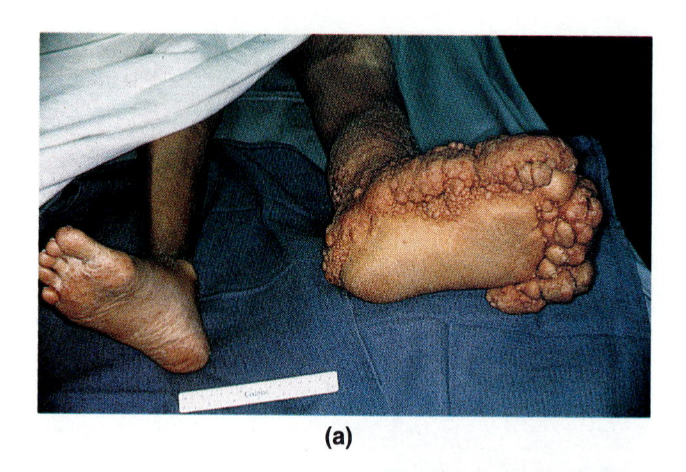

(a)

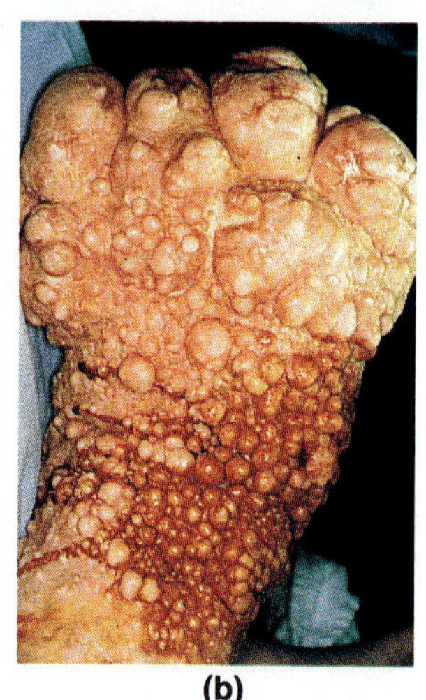

(b)

FIGURE 20.11 Madura foot can be caused by true fungi or by actinomycetes, fungallike bacteria. Lesions filled with pus and pathogens may distort the foot, (a) bottom view, (b) top view, to the point where amputation is necessary.

Other Skin Diseases

Madura foot, or **maduromycosis** (Figure 20.11), occurs mainly in the tropics and is caused by a variety of soil organisms, including fungi of the genus *Madurella* and filamentous actinomycetes, such as *Actinomadura, Nocardia, Streptomyces,* and *Actinomyces*. Organisms enter the body through breaks in the skin, especially in people who do not wear shoes. Initial pus-filled lesions spread and form connected lesions that eventually become chronic and granulomatous. Untreated, the organisms invade muscle and bone, and the foot becomes massively enlarged. Madura foot is diagnosed by finding white, yellow, red, or black granules of intertwined hyphae in pus. So-called sulfur granules in pus are yellow hyphae and not sulfur. Unless antibiotic therapy begins early in the infection and is sufficiently prolonged, amputation may be necessary. Keeping soil particles out of wounds prevents this disease.

Skin reactions to cercariae (larvae) of several genera of schistosomes cause **swimmer's itch.** These cercariae parasitize birds, domestic animals, and primates and can burrow into human skin. Allergic reactions usually prevent them from reaching the blood to cause schistosomiasis. Swimmer's itch occurs throughout the United States but especially in the Great Lakes area.

Dracunculiasis is caused by another parasitic helminth called a guinea worm. This disease is discussed in the Microbiologist's Notebook in Chapter 12.

Skin diseases are summarized in Table 20.2.

APPLICATIONS

Algal Infections

Most algae manufacture their own food and are not parasitic, but some strains of the alga *Prototheca* have lost their chlorophyll and survive by parasitizing other organisms. Found in water and moist soil, they enter the body through skin wounds. By 1987, 45 cases of protothecosis had been reported, 2 from cleaning home aquariums. Protothecosis was first observed on the foot of a rice farmer, and most subsequent cases have occurred on legs or hands. In immunodeficient patients the parasite can invade the digestive tract or peritoneal cavity. A few skin infections have responded to oral potassium iodide or intravenous amphotericin B and tetracycline therapy, but no satisfactory treatment has been found for others.

TABLE 20.2 Summary of skin diseases		
Disease	**Agent**	**Characteristics**
Bacterial Skin Diseases		
Folliculitis	*Staphylococcus aureus*	Skin abscess, encapsulated so not reached by antibiotics
Scalded skin syndrome	*S. aureus*	Vesicular lesions over whole skin and fever; most common in infants
Scarlet fever	*Streptococcus pyogenes*	Sore throat, fever, rash caused by toxin; can lead to rheumatic fever and other complications
Erysipelas	*S. pyogenes*	Skin lesions spread to systemic infection; rare today, but common and fatal before antibiotics
Pyoderma and impetigo	Staphylococci, streptococci	Skin lesions usually in children; easily spread by hands and fomites
Acne	*Propionibacterium acnes*	Skin lesions caused by excess of male sex hormones; infection is secondary. Common in teenagers
Burn infections	*Pseudomonas aeruginosa* and other bacteria	Growth of bacteria under eschar, often a nosocomial infection. Difficult to diagnose and treat, causative agents often antibiotic-resistant
Viral Skin Diseases		
Rubella	Rubella virus	Mild disease with maculopapular exanthema. Infection early in pregnancy can lead to congenital rubella. Vaccine has greatly reduced incidence
Measles	Rubeola virus	Severe disease with fever, conjunctivitis, cough, and rash; encephalitis is a complication. Disease occurs mainly in children; vaccine has greatly reduced incidence
Chickenpox	Varicella-zoster virus	Generalized macular skin lesions; last of childhood communicable diseases for which vaccine is not generally available
Shingles	Varicella-zoster virus	Pain and skin lesions, usually on trunk. Occurs in adults lacking immunity; can be acquired from children with chickenpox
Smallpox	Smallpox virus	Eradicated as a human disease
Other pox diseases	Other poxviruses	Cowpox caused by vaccinia virus from cattle and rodents; all human pox infections are rare
Warts	Human papillomaviruses	Dermal warts are self-limiting; malignant warts occur in immunologic deficiencies
Fungal Skin Diseases		
Dermatomycoses	Dermatophytes	Dry scaly lesions on various parts of the skin; difficult to treat
Sporotrichosis	*Sporothrix schenckii*	Granulomatous, pus-filled lesions; sometimes disseminates to lungs and other organs
Blastomycosis	*Blastomyces dermatitidis*	Granulomatous, pus-filled lesions that develop in lungs and wounds; sometimes disseminates to other organs
Candidiasis	*Candida albicans*	Patchy inflammation of mucous membranes of the mouth (thrush) or vagina (vaginitis). Disseminated nosocomial infections occur in immunodeficient patients
Aspergillosis	*Aspergillus* species	Wound infection in immunodeficient patients; also infects burns, cornea, and external ear
Zygomycosis	*Mucor* and *Rhizopus* species	Occurs mainly in untreated diabetes; begins in blood vessels and can rapidly disseminate
Other Skin Diseases		
Madura foot	Various soil fungi and actinomycetes	Initial lesions spread and become chronic and granulomatous; can require amputation
Swimmer's itch	Cercariae of schistosomes	Itching due to cercariae burrowing into skin; immunological reaction prevents their spread
Dracunculiasis	*Dracunculus medinensis*	Larvae ingested in crustaceans in contaminated water migrate to skin and emerge through lesion; juveniles cause severe allergic reactions

DISEASES OF THE EYES

Like skin diseases, eye diseases frequently result from pathogens that enter from the environment. For that reason we discuss them here.

Bacterial Eye Diseases

Ophthalmia Neonatorum

Ophthalmia neonatorum, or gonococcal conjunctivitis of the newborn, is a pyogenic infection of the eyes caused by *Neisseria gonorrhoeae*. Organisms present in the birth canal enter the eyes as a baby is born. The resulting infection can cause **keratitis,** an inflammation of the cornea, which can progress to perforation and destruction of the cornea and blindness. Early in this century, 20 to 40 percent of children in European institutions for the blind had suffered from this disease. In underdeveloped countries the disease is still prevalent, but the true incidence is unknown because high infant mortality makes it difficult to estimate the number of original infections. Although infections are most common in newborns, adults can transfer organisms by hands or fomites from genitals to the eyes (Figure 20.12).

Penicillin was once the treatment of choice, but resistant strains have developed, and tetracycline is now used. Tetracycline has the advantage of also being effective against chlamydia, with which mothers also may be infected. Preventive measures have almost eradicated the disease in developed countries. A few drops of 1 percent silver nitrate solution in the eyes immediately after birth kills gonococci but can irritate the eyes. Antibiotics such as penicillin, tetracycline, and erythromycin also kill nonresistant gonococci.

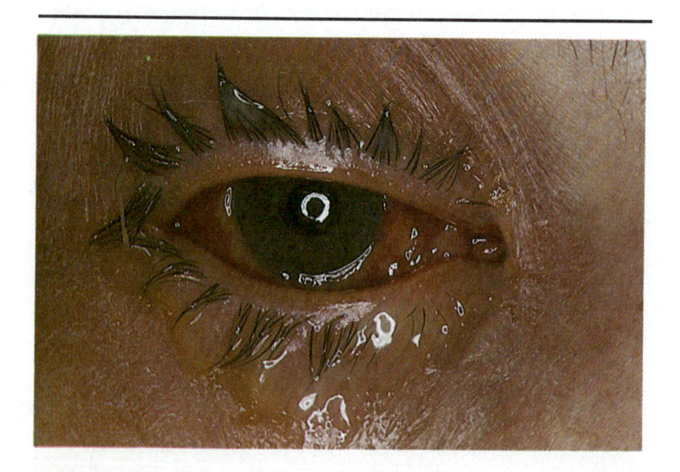

FIGURE 20.12 Gonococcal infection of the eye. Most often this condition is seen in newborns, who are infected as they pass through the birth canal. Adults, however, can also transfer bacteria to the eye from the genitals.

Bacterial Conjunctivitis

Bacterial conjunctivitis, or **pinkeye,** is an inflammation of the conjunctiva caused by organisms such as *Staphylococcus aureus*, *Streptococcus pneumoniae*, *Neisseria gonorrhoeae*, *Pseudomonas* species, and *Haemophilus influenzae* biogroup *aegyptius*. The BPF clone of the latter causes Brazilian pupuric fever, in which young children initially have conjunctivitus and then develop serious, potentially fatal septicemia with extensive hemorrhaging and symptoms of meningitis. Pinkeye is extremely contagious, especially among children, and can spread rapidly through schools and day-care centers. Children rub itchy, running eyes and transfer organisms to their playmates. In warm weather, gnats attracted to the moisture of tears may pick up organisms on their feet and transfer them to other persons' eyes. Topically applied sulfonamide ointment is an effective treatment. Children should not return to school until their infection is completely eliminated.

Trachoma

Trachoma (Figure 20.13), from the Greek word meaning "pebbled," or "rough," is caused by *Chlamydia trachomatis*. It is the leading cause of preventable blindness worldwide; 500 million people have it, and over 20 million are already blind. Although uncommon in the United States, except among American Indians in the Southwest, the disease is widespread in parts of Asia, Africa, and South America, sometimes affecting 90 percent of the population. Flies are important mechanical vectors, and close mother-child contact facilitates human transfer.

FIGURE 20.13 Trachoma, caused by *Chlamydia trachomatis*, is the leading cause of preventable blindness in the world, affecting over 500 million people. Note the pebbled appearance of the vastly swollen conjunctiva.

Viral Eye Diseases

Epidemic Keratoconjunctivitis

Epidemic keratoconjunctivitis (EKC), caused by an adenovirus, is sometimes called shipyard eye because workers are frequently infected from dust particles in the environment. After 8 to 10 days incubation, the conjunctiva become inflamed and eyelid edema, pain, tearing, and photophobia follow. Within 2 days the

TRY IT

What Grows in Your Cosmetics?

Are the shampoos, contact lens fluids, mascara, cold cream, and aerosol spray cans in your house sterile? Or, are you spraying and smearing microorganisms on yourself and into your environment? Try culturing brand new bottles of products as well as ones that have been open for varying times. Sometimes a product will have passed the quality control tests at the factory, but after months on a store shelf the antimicrobial "preservatives" that were originally added will be no longer effective. Single-dose containers such as the little foil packets of shampoo in a hotel room are not required to have the preservatives needed in products that will have repeated reuse with reinoculation of organisms into the product each time. With aerosol cans check both the liquid inside and the nozzle. Sometimes the product is sterile, but a contaminated nozzle can spray organisms into the air. Does the storage case for your contact lenses have a biofilm of organisms on its inner sides? If you make up your own shampoo or contact lens fluids from concentrates, how do they compare as to sterility with store-bought versions?

I am curious to know what you find. Drop me a letter with your findings:

Dr. Jacquelyn Black
Marymount University
2807 N. Glebe Road
Arlington, VA 22207

infection spreads to the corneal epithelium and sometimes to deeper corneal tissue. Clouding of the cornea can last up to 2 years. EKC is often nosocomially acquired in eye clinics and ophthalmologists' offices.

Acute Hemorrhagic Conjunctivitis

Another viral disease, **acute hemorrhagic conjunctivitis** (AHC), caused by an enterovirus, appeared in 1969 in Ghana. Serological studies show it was not prevalent anywhere in the world before then. First seen in the United States in 1981, the disease occurs chiefly in warm humid climates under conditions of crowding and poor hygiene. The disease causes severe eye pain, abnormal sensitivity to light, blurred vision, hemorrhage under conjunctival membranes, and sometimes transient inflammation of the cornea. Onset is sudden, and recovery usually is complete in 10 days. A rare complication is a paralysis resembling poliomyelitis.

Parasitic Eye Diseases

Onchocerciasis—River Blindness

The filarial (threadlike) larvae of the nematode *Onchocerca volvulus* cause **onchocerciasis,** or **river blindness** (Figure 20.14), in many parts of Africa and Central America. Adult worms and microfilariae (very small larvae) accumulate in the skin and are transmitted by blackflies. When the fly bites an infected individual, it ingests microfilariae, which mature in the fly and move to its mouth parts. When the fly bites again, infectious microfilariae enter the skin and invade various tissues, including the eyes. In many small villages where the people depend on water from blackfly–infested rivers, nearly all inhabitants who live past middle age are blind.

Adult worms cause skin nodules that can abscess. Microfilariae cause skin depigmentation and severe dermatitis via immunologic responses to live filariae or toxins from dead ones. The worst tissue damage occurs as the worms invade the cornea and other parts of the eye. Over several years they cause eye blood vessels to become fibrous, and total blindness ensues by about age 40.

The disease can be diagnosed by finding microfilariae in thin skin samples or adult worms visible through the skin. The drug ivermectin kills adult

FIGURE 20.14 Onchocerciasis, or river blindness, is caused by the roundworm *Onchocerca volvulus*, whose microfilarial stages are transmitted by bite of the blackfly. In some African villages, nearly all adults are blind and must be led by children, the only ones in the village who can still see, although they are already infected with the worms that will eventually blind them also.

MICROBIOLOGIST'S NOTEBOOK

Saving Eyes

Probably as many as half of all contact lens wearers store their lenses in contaminated cases. Every time they put their contacts in, they're putting millions of bacteria onto the surface of their eyes—all sorts of gram-positive cocci, and of course *Pseudomonas*, because that's everywhere. Yet not much happens—at most, they may get some irritation. But if the lenses don't fit well, or if the wearer nicks the cornea with a fingernail, it's another story. Once an infection is started, you can lose an eye almost overnight.

Once my colleague, Dr. Louis A. Wilson, and I saw a patient who had gotten a splinter in his eye while taking down a treehouse. A corneal ulcer had developed. At that time, the normal procedure would have been removal of the eye—which would have cost him his job as well, because he was an airline pilot. We tried to find out what was causing the infection, in the hope that some other effective treatment could be devised. It turned out to be *Philaphora verrucosa*, a slow-growing fungus. The slow growth proved to be a key factor—it bought us time. The eye was taped open and antibiotics were administered drop by drop. This didn't work. Then Dr. Wilson stitched conjunctival membranes over the ulcer, forming a sort of natural bandage that brought a blood supply directly to the ulcer. It was no fun for the patient—but he kept his eye, and his job.

Dr. Donald Ahearn and one of his postdoctoral students in the lab, working with the apparatus used for gel electrophoresis.

My name is Don Ahearn, and I'm a research professor of microbiology in the Laboratory for Microbial and Biochemical Sciences at Georgia State University. I've been doing research on microbes in the environment for the past 20 years. One of my special fields is the microbiology of contact lenses, cosmetics, and other products that are applied to the region of the outer eye. My colleagues and I have found out some interesting things, some of which have already influenced the way these products are made and packaged.

The case of the airline pilot prompted us to look more closely at fungi. The first question we were curious about was whether fungi are part of the normal flora of the eye. We found that they aren't. Nevertheless, when soft contact lenses became popular in the 1970s we started to see a number of fungal as well as bacterial infections among their users. The source of the problem was the lens solutions and disinfection systems. Some contact lens wearers were not using the products correctly—but in other cases the products themselves were simply inadequate.

Our research has prompted some changes by manufacturers. For example, saline solutions are preserved and are now usually packaged in smaller bottles—they're used up more quickly and don't sit around long enough to become contaminated.

But contact lens wearers still sometimes have problems. Extended-wear contact lenses cause the most trouble because they're kept on the eye for a longer period of time. They reduce the amount of oxygen that can get to the corneal tissues. If you keep the lenses on the eye for too long a period—say, more than 7 days—then you can cause damage to the eye.

In our research studies, we've come across contact lens users who developed fungal keratitis—a corneal ulcer—because the fungus grew in their hydrogel lens and then attacked the surface of the eye.

Such problems can be caused by members of the genus *Fusarium*, which are omnipresent in the environment—they include the broad group of fungi that can cause infections of tomato plants and flowers and wheat and corn crops. Farmers used to be the most common candidates for *Fusarium* corneal ulcers because they'd get hit in the eye by a stalk of wheat, the cornea would be scratched and the fungus introduced. Nowadays, it's generally contact lens wearers rather than farmers who fall victim to these fungi. I've seen people who complain that they can't get spots off their contacts. They don't realize that those little spots are, in fact, fungal colonies that are growing in their hydrogel lens. Most of the time, this type of fungal contamination is caused by poor hygiene and improper disinfection.

Fungal infections are a problem for two reasons. First, we don't have an extensive arsenal of effective drugs against fungi, as we do against bacteria. Second, because they're slow to develop, they're insidious—people may not pay much attention to them at first. With bacterial infections, we usually have the opposite problem—they can develop so rapidly that a patient can suffer serious eye damage in 24 hours.

Careless use of mascara is one common way to get a corneal ulcer. Women keep and use their cosmetics for too long, and the cosmetics get contaminated. Then, all it takes is for the applicator brush to scratch the surface of the cornea for an infection to start. There's often severe pain and redness

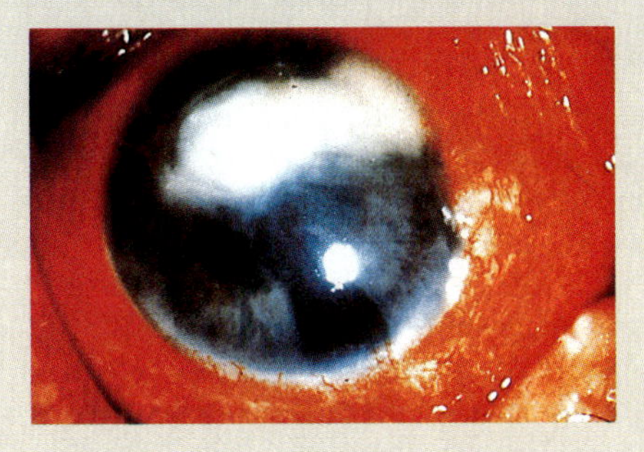

Corneal ulcer caused by the fungus *Fusarium solani*, an organism that can grow in soft contact lenses and attack the eye.

within 3 or 4 hours after the initial trauma, and one can lose part or all of the sight in the infected eye if treatment isn't started promptly. I've seen cases where a mascara-wand injury and subsequent infection have caused severe scarring on the surface of the eye and marked distortion of vision. In severe cases of trauma caused by a mascara wand followed by infection, patients have been hospitalized for several weeks.

Most often the culprit in such cases is *Pseudomonas aeruginosa*. *P. aeruginosa* is a ubiquitious organism that can grow on a great variety of substrates—hydrocarbons, carbohydrates, proteins. You'll find it around the house on many damp areas—around the kitchen sink, for instance, or on the shower curtain. *P. aeruginosa* is an opportunistic pathogen. Ordinarily, you may bathe the eye with high densities of *Pseudomonas* and not cause an infection. But if there's a breach of the body's defense, such as a mascara wand tearing the corneal epithelium and embedding the organism into the eye, infection is likely to occur.

There are other opportunistic bacterial pathogens, such as *Serratia marcescens*, and other ways—sometimes rather surprising ways—of acquiring them from the environment. In one instance, a person who had been in a whirlpool bath subsequently developed a corneal ulcer. The aerosol from a bubbling whirlpool can contain hundreds of thousands or even millions of bacteria per cubic meter of air.

Whirlpools can be quite dangerous if not properly disinfected because the warm water is a perfect breeding ground for organisms. To keep these baths safe, you have to follow a daily regimen of disinfection. You can't just look at the water and say, "It seems okay." We've found that *Pseudomonas aeruginosa* can be back in the water just hours after the disinfection process. Often it's attached to the sides of the whirlpool, or it's in the pipes. The disinfection may kill all the organisms in the water, but once these organisms are attached to a surface and form a biofilm, they're much more resistant to the chemicals used to clean the baths. A person in the bath who vigorously rubbed his eyes could cause a corneal irritation, making him susceptible to bacterial attack. Many people also break out in a body rash in whirlpools—it's a hypersensitivity reaction to being immersed in water filled with *Pseudomonas*.

Really, just about anyplace where you have a large number of people using a product, there's a good chance you'll find bacteria. Testers at cosmetic counters, for instance, can be a breeding ground for potentially pathogenic organisms because so many people use the testers. That's why several states are considering legislation that would outlaw the use of common testers—the cosmetic companies would have to come up with single-use samples.

What should consumers do to guard against these infections? Well, people would be much better off if they used common sense and didn't get careless about the products they use, both inside the home and away from it. Most people know that food sitting out on the counter for a long time is going to be invaded by microorganisms and will spoil. They should realize that the same thing happens to other products like cosmetics. The more times a product is used, the more chance it has of becoming infected by an invading microorganism like *Pseudomonas aeruginosa*. Cosmetics, especially mascara, should be discarded after 3 to 6 months of use. Most products are safe until they're opened because they've been treated with preservatives. Once they're open, however, they're subject to a continual onslaught by exogenous microorganisms, and the preservative may break down.

Many people don't realize that a great number of products in our homes stay fresh and safe because they've been treated with preservatives and microbial inhibitors. Bread would spoil very quickly if it hadn't been treated. Deck furniture and vinyl pillows wouldn't last for very long outside if they hadn't been treated with these substances. The same is true of paints and fabric wall coverings and carpets.

But every story has two sides. The inhibitors do a lot of good, but they can also be harmful to the environment. Some of them just aren't biodegradable. They accumulate in sediment and mud and travel up the food chain. And there is the problem of safety at the manufacturing plants where preservatives are made or incorporated into products. The amount of preservative in your tube of mascara may be so small as to pose no health hazard to you as a user. The same substance, though, might be dangerous to people who work with large quantities of it, day in and day out.

What does all this mean? Well, we're not going to stop using cosmetics or stop putting preservatives into products. What we need to do is to exercise greater care with the substances we use—in industry, at home, and in the environment. And the first step in this process has to be to educate consumers about product safety.

worms quickly and microfilariae over several weeks. Drugs are available to kill microfilariae rapidly, but toxins from large numbers of dying microfilariae can cause anaphylactic shock. Onchocerciasis could be prevented by eliminating blackflies, which congregate along rivers. DDT destroys some flies, but it allows resistant ones to survive, and it accumulates in the environment.

The small size of the pygmies of Uganda is due to onchocerciasis infections. When mothers are infected by *O. volvulus* (and most of them are), the parasite damages the pituitary gland of their fetuses. The result is dwarfism from a deficiency of growth hormone.

Loaiasis

The eye worm *Loa loa*, a filarial worm endemic to African rain forests, is transmitted to humans by deer flies. Adult worms live in subcutaneous tissues and eyes; microfilariae appear in peripheral blood in the daytime and concentrate in the lungs at night. Deer flies feed during the day and acquire microfilariae from infected humans. These parasites develop in the fly, migrate to its mouth parts, and are transferred to another human when the fly feeds again. Microfilariae migrate through subcutaneous tissue, leaving a trail of inflammation, and often settle in the cornea and conjunctiva. Although they usually do not cause blindness, the shock of finding a worm over an inch long in one's eye is undoubtedly a traumatic experience!

PUBLIC HEALTH

Acanthamoeba Keratitis

In the mid-1980s, 24 cases of *Acanthamoeba* keratitis were reported to CDC, 20 of which were in users of contact lenses. Since then, over 100 cases have been reported. The amoeba that causes this condition is common in fresh water and soil and is found in brackish water, seawater, and hot tubs, as well as contact lens cleaning solution. It can also travel via airborne dust. The infection causes severe eye pain and recurrent destruction of corneal epithelium. Some patients have been successfully treated with ketoconazole or miconazole, but others have required corneal transplants. Contamination of water by this amoeba is responsible for the closing of the historic baths in Bath, England, where a girl died from an amoebic brain infection.

Loaiasis is diagnosed by finding microfilariae in blood or actual worms in skin or eyes. It is treated by excising adult worms or by use of suramin or other drugs to eradicate microfilariae. Control could be achieved by eradicating deer flies, but this is an exceedingly difficult task.

Eye diseases are summarized in Table 20.3.

TABLE 20.3 Summary of eye diseases

Disease	Agent	Characteristics
Bacterial Eye Diseases		
Ophthalmia neonatorum	*Neisseria gonorrhoeae*	Infection during birth causes corneal lesions and can lead to blindness. Silver nitrate or antibiotics have nearly eradicated it in developed countries
Bacterial conjunctivitis	*Haemophilus*	Highly contagious inflammation of the conjunctiva in young children
Trachoma	*Chlamydia trachomatis*	Infection and destruction of cornea and conjunctiva; cause of preventable blindness
Keratitis	Bacteria, viruses, and fungi	Ulceration of the cornea; occurs mainly in immunodeficient and debilitated patients
Viral Eye Diseases		
Epidemic keratoconjunctivitis	Adenovirus	Inflammation of the conjunctiva that spreads to the cornea. Called shipyard eye because transmitted in dust particles; also nosocomial
Acute hemorrhagic conjunctivitis	Enterovirus	Severe pain and hemorrhage under conjunctiva; highly contagious under crowded, unsanitary conditions
Parasitic Eye Diseases		
Onchocerciasis	*Onchocerca volvulus*	Microfilariae enter skin through blackfly bite and invade eyes and other tissues; causes dermatitis and blindness. Occurs in tropics
Loaiasis	*Loa loa*	Microfilariae enter skin through deer fly bite; cause inflammation of conjunctiva and cornea

WOUNDS AND BITES

Wound Infections

Gas Gangrene

Associated with deep wounds, **gas gangrene** often is a mixed infection caused by two or more species of *Clostridium*, especially *C. perfringens* (found in 80 to 90 percent of cases), *C. novyi*, and *C. septicum*. Spores of these obligately anaerobic organisms are introduced by injuries or surgery into tissues where circulation is impaired and dead and anaerobic tissue is present. In regions where oxygen concentration is low, spores germinate, multiply, and produce toxins that kill more cells and extend the anaerobic environment.

The onset of gas gangrene occurs suddenly 12 to 48 hours after injury. As the organisms grow they produce gas, mainly hydrogen, and gas bubbles distort and destroy tissue (Figure 20.15). Such tissue is called **crepitant** (rattling) **tissue.** The bubbles audibly "snap, crackle, and pop" when the patient is repositioned. Foul odor is such a prominent feature of gas gangrene that it can be diagnosed without even entering the patient's room. High fever, shock, massive tissue destruction, and blackening of skin accompany this rapidly spreading disease. If left untreated, death occurs swiftly. Diagnosis usually is made on clinical findings, and treatment begun before laboratory test results are available. Dead tissue is removed or limbs are amputated, and penicillin is administered. Gas gangrene often follows illegal abortions performed under unsanitary conditions. It usually necessitates a hysterectomy.

The use of high-pressure oxygen chambers, available in some hospitals, to treat gas gangrene is somewhat controversial. Patients are placed in chambers containing 100 percent oxygen at 3 atmospheres pressure for 90 minutes two or three times a day. The actual mechanism by which high-pressure oxygen aids recovery is not known, but presumably it kills or inhibits obligate anaerobes. Gas gangrene can be prevented by adequate cleansing of wounds, delaying closing of wounds, and providing drainage when they are closed. No vaccine is available.

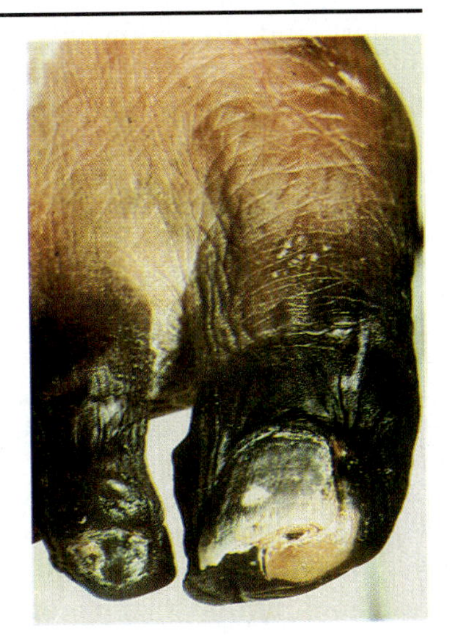

FIGURE 20.15 Gas gangrene has blackened the toes of an infected foot.

Other Anaerobic Infections

In addition to clostridia, certain non-spore-forming anaerobes are associated with some infections. *Bacteroides* and *Fusobacterium* species are normally present in the digestive tract and *Bacteroides* accounts for nearly half of human fecal mass. *Fusobacterium* sometimes causes oral infections. If introduced into the abdominal cavity, genital region, or deep wounds by surgery or human bites, it causes infections there, too. Such infections are difficult to treat because the organisms are resistant to many antibiotics. Abscesses caused by anaerobic organisms must be drained surgically, and appropriate antibiotics used as supportive therapy.

Cat Scratch Fever

Cats are mechanical vectors of **cat scratch fever,** a disease caused by a recently isolated organism that appears to belong in the Chlamydiaceae. The organisms are found mainly in capillary walls or in microabscesses. Presumably cats acquire the organisms from the environment and carry them on claws and in the mouth. They transmit them to humans when they scratch, bite, or lick. After 3 to 10 days a pustule appears at the site of entry, and the patient has a mild fever, headache, sore throat, swollen glands, and conjunctivitis for a few weeks. Diagnosis is based on clinical findings and a history of contact with cats. Treatment with antibiotics has no effect, so only symptomatic relief is available. The only means of prevention is to avoid cats.

Rat Bite Fever

Rat bite fever is caused by *Streptobacillus moniliformis*, which is present in the nose and throat of about half of all wild and laboratory rats. However, only about 10 percent of people bitten by rats develop rat bite fever. Most cases result from bites of wild rats; half are in children under 12 living in overcrowded, unsanitary conditions. It also can result from bites and scratches of mice, squirrels, dogs, and cats.

Rat bite fever begins as a localized inflammation at a bite site that heals promptly. In 1 to 3 days headache begins, and lesions appear elsewhere, especially on palms and soles. Fever is intermittent. By the distribution and appearance of the rash, the disease sometimes is mistaken for Rocky Mountain spotted fever. An arthritis that develops can permanently damage joints.

Another form of rat bite fever, **spirillar fever,** is caused by *Spirillum minor*. First described in Japan as **sodoku** and now known to exist worldwide, this disease is still poorly understood. The initial bite heals easily, but 7 to 21 days later it flares up and occasionally forms an open ulcer. Chills, fever, and inflamed lymph nodes accompany a red or dark purple rash, which spreads out from the wound site. After 3 to 5 days symptoms subside, but they can return after a few days, weeks, months, or even years.

Diagnosis of both rat bite fevers is by dark-field examination of exudates. Treatment is with streptomycin or penicillin; without treatment the mortality rate is about 10 percent. Technicians bitten by rodents should disinfect the bite site, seek medical treatment, and be alert for rat bite fever symptoms.

Arthropod Bites and Infections

A variety of arthropods, including ticks, mites, and insects, cause human disease either directly or as vectors. Many arthropods are ectoparasites that feed on blood. Their bites can be painful, and their toxins can lead to anaphylactic shock. Here we consider the direct effects of injuries caused by arthropods; the diseases they transmit are discussed in other chapters.

Tick Paralysis

As ectoparasites, ticks attach themselves to the skin, where they cause both local and systemic effects. The local effect is a mild inflammation at the bite site. The systemic effects are due to an anticoagulant or toxins secreted into the blood by the tick. The anticoagulant prevents blood clotting while the tick feeds on the host's blood. The toxins can cause **tick paralysis,** especially in children. Although the exact chemical composition of these toxins is not known, the toxins appear to be produced by the ovaries of the ticks. They cause fever and paralysis, which first affects the limbs and eventually affects respiration, speech, and swallowing. Removal of the ticks when symptoms first appear prevents permanent damage. Failure to remove ticks can lead to death by cardiac or respiratory arrest.

Chigger Dermatitis

The term *mites* refers not to a particular species but to an assortment of species. As ectoparasites, adult mites attach to a host long enough to obtain a blood meal and then usually drop off. Chiggers, the larvae of certain species of *Trombicula* mites, burrow into the skin and release proteolytic enzymes that cause host tissue to harden into a tube. The chiggers then insert mouth parts into the tube and feed on blood. They cause itching and inflammation in most people and can cause a violent allergic reaction called **chigger dermatitis** in sensitive individuals. Chiggers are especially prevalent along the southeastern coasts of the United States.

Scabies and House Dust Allergy

Scabies, or **sarcoptic mange,** is caused by the itch mite *Sarcoptes scabiei*. By the time intense itching appears, the lesions usually are quite widespread. Scratching lesions and causing them to bleed provides an opportunity for secondary bacterial infections. Scabies is spread by close human contact. Outbreaks are especially problematic in hospitals and nursing homes. Disinfection of linens and strict isolation are necessary to prevent spread of the infestation. House dust mites are ubiquitous, and we all inhale airborne mites or their excrement. Although this phenomenon may be unappealing, it causes disease only in individuals who have house dust allergy. Two other mites are human commensals. *Demodex folliculorum* lives in hair follicles, and *D. brevis* in sebaceous glands. The incidence of these mites in humans increases with age from 20 percent in young adults to nearly 100 percent in elderly people.

Flea Bites

The sand flea, *Tunga penetrans*, is also sometimes called a chigger because it burrows into the skin and lays its eggs. Sand fleas cause extreme itching, inflammation, and pain and provide sites for secondary infections, including infection with tetanus spores. Surgical removal and sterilization of the wound is used to treat sand flea infections. Fleas in homes and on pets are controlled by insecticides with residual effects for days to weeks after application. Wearing shoes and avoiding sandy beaches helps to protect against sand fleas. Although all fleas encountered by humans are a nuisance, only those that carry infectious agents are a public health hazard.

Pediculosis

Common expressions such as "nit-picking," "going over with a fine-toothed comb," "lousy," and "nitty-gritty" attest to association of lice with humans over the ages. To stay alive, lice must remain on their hosts for all but very short periods of time. They glue their eggs to fibers of clothing and hair. Two varieties of the louse *Pediculus humanus* parasitize humans. One lives mainly on the body and clothing in temperate climates where clothing usually is worn; the other lives on hair in any climate. **Pediculosis,** or lice infestation, results in reddened areas at bites, derma-

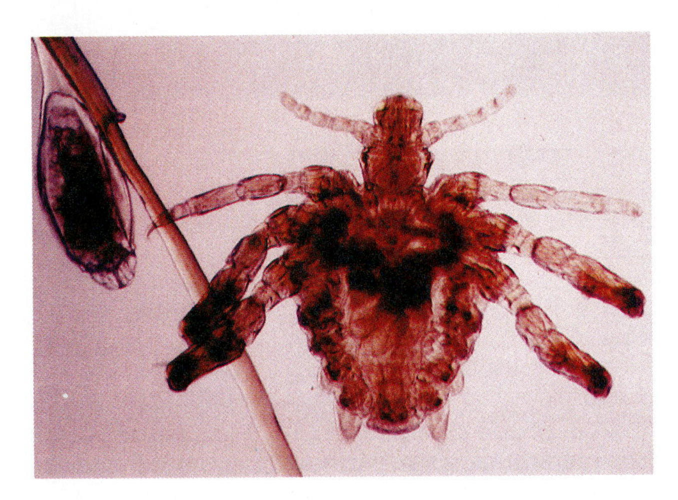

FIGURE 20.16 The human crab louse, *Phthirus pubis,* and its nit (egg), which it has glued to a hair. This parasite typically infests genital hair.

titis, and itching. A lymph exudate from bites provides an ideal medium for secondary fungal infections, especially in the hair. The crab louse, *Phthirus pubis* (Figure 20.16), clings to skin more tightly than the body louse and causes intense itching at bites, especially in the pubic area. It is transmitted between humans by close physical contact, but the louse itself is not known to transmit other diseases. Insecticides can be used to eradicate lice, but sanitary conditions and good personal hygiene must be maintained to prevent their return.

Other Insect Bites

Blackflies inflict vicious wounds, and sensitive individuals who are bitten get **blackfly fever,** characterized by an inflammatory reaction, nausea, and headache. Bloodsucking flies, such as the tsetse fly and the deer fly, are related to horse flies. All inflict painful bites and sometimes cause anemia in domestic animals.

Myiasis is an infection caused by maggots (fly larvae). Wild and domestic animals are susceptible to myiasis in wounds. Human myiasis occurs when larvae of a greenish metallic-looking fly penetrate mucous membranes and small wounds. The congo floor maggot, the only known bloodsucking larva, sucks human blood.

Many mosquitoes, bedbugs, and bloodsucking insects feed on humans and leave painful itchy bites. Bedbugs hide in crevices during the day and come out at night to feed on their sleeping victims. Inflammation at bites results from an allergic reaction to the bug's saliva. Large numbers of bites can lead to anemia, especially in children. Mosquitoes and many other insects can be eradicated by the application of insecticides that have long-term residual effects. They can be kept out of dwellings by the use of tight housing construction and solid roofs instead of thatch and by household cleanliness.

Wound and bite infections are summarized in Table 20.4.

TABLE 20.4 Summary of wound and bite infections

Disease	Agent	Characteristics
Wound Infections		
Gas gangrene	*Clostridium perfringens* and other species	Deep wound infections with gas production in anaerobic tissue. Tissue necrosis and death can result if not treated promptly
Cat scratch fever	Probably a chlamydia	Pustules at scratch site, fever, and conjunctivitis
Rat bite fever	*Streptobacillus moniliformis*	Inflammation at bite site, dissemination of lesions, intermittent fever
Spirillar fever	*Spirillum minor*	Inflammation at rat bite site heals; later reinflammation, fever, and rash
Athropod Bites and Infections		
Tick paralysis	Various ticks	Toxins introduced with tick bite cause fever and ascending motor paralysis
Chigger dermatitis	*Trombicula* mites	Larvae burrow into skin and cause itching and inflammation; can cause violent allergic reactions
Scabies	Itch mites	Widespread lesions with intense itching; similar mites cause house dust allergy if their feces are inhaled by sensitive people
Flea bites	*Tunga penetrans*	Itching and inflammation from adult females in skin
Pediculosis	*Pediculus humanus*	Inflammation at louse bite sites and itching; *Phthirus pubis* found in pubic areas
Other Insect Bites		
Blackfly fever	Blackfly	Bites cause severe inflammatory reaction in sensitive individuals
Myiasis	Fly larvae	Maggots infect wounds in animals; congo floor maggot sucks human blood; screwworms injure cattle
Mosquito and other bites	Mosquitoes, some flies, bedbugs	Painful, itchy bites, several serve as vectors of disease

CLOSE-UP

Screwworm Eradication

An ingenious method of controlling screwworm maggots (larvae), which do great harm to cattle, was developed in the 1930s by Edward Knipling of the U.S. Department of Agriculture. He captured and irradiated male screwworm flies with radioactive cobalt, making them incapable of fertilizing eggs. If females, which mate only once, mate with sterile males, they lay infertile eggs, so far fewer new flies are produced. Irradiated males must be released frequently to maintain a high percentage of sterile males in the population. Another control method uses insects, especially wasp larvae, that are parasitic on other insects. Farmers sometimes purposely release parasitic wasps to infect crop-destroying insects and thereby reduce crop damage. Both the use of sterilized males and parasitic insects make use of natural habits of insects; neither contributes to environmental pollution, as most pesticides do.

ESSAY

Smallpox—The Eradication of a Human Scourge

In 1980 the World Health Organization (WHO) officially proclaimed that smallpox had been eradicated worldwide. This proclamation marked the end of centuries of sickness and death from smallpox. Thus, health professionals are unlikely to encounter patients infected with smallpox. However, the disease and its consequences are of historical interest, and the methods used to eradicate it provide an important lesson in the control of infectious diseases.

History of the Disease Smallpox first appeared sometime after 10,000 B.C. in a small agricultural settlement in Asia or Africa. The mummy of Ramses V, who died in Egypt in 1160 B.C., has smallpox scars on the face, neck, shoulders, and arms. Smallpox ravaged the villages of India and China for centuries, and an epidemic occurred in Syria in A.D. 302. The Persian physician Rhazez clearly described the disease in A.D. 900.

Crusaders brought smallpox to Europe, the Spaniards carried the virus to the West Indies in 1507, and Cortez's army introduced it into Mexico in 1520. In each instance the disease was introduced into a population that lacked immunity, and its spread was rampant. As many as 3.5 million Native Americans may have died from smallpox, and by the eighteenth century more than half the inhabitants of Boston had become infected. Slave traders introduced smallpox into Central Africa in the sixteenth and seventeenth centuries, and immigrants from India brought it to South Africa in 1713. The disease reached Australia in 1789.

History of Immunization The idea of immunization itself originated in Asia with variolation to prevent smallpox. In variolation, threads saturated with fluid from smallpox lesions were introduced into a scratch or dangled in the sleeve of a nonimmune person. This practice immunized some people, but it also started epidemics because live, virulent virus was used. Lady Mary Wortley Montague, wife of the British Ambassador in Turkey, introduced the practice to Britain in 1717. In spite of high morbidity (illness) and even mortality (death), General George Washington ordered all his troops variolated in 1777, and by 1792, 97 percent of the population of Boston had been variolated.

Smallpox immunization was greatly improved by the English physician Edward Jenner, who noticed that milkmaids with cowpox scars never became infected with smallpox. It is now known that immunity to the less severe cowpox confers immunity to smallpox as well. In 1796 Jenner inoculated an 8-year-old boy with cowpox and 6 weeks later inoculated him with smallpox. As Jenner had expected, the boy remained healthy. In the United States in 1799, physician Benjamin Waterhouse introduced the cowpox vaccine by vaccinating his own children and then exposing them to smallpox.

In spite of the knowledge that smallpox could be prevented by vaccination with cowpoxvirus, in 1947 an unimmunized traveler from Mexico infected 12 people in New York City, and 2 of them died. The last eight cases of smallpox in the United States occurred in the Rio Grande Valley in 1949.

Smallpox was still endemic in 33 countries in 1967 when the WHO established its immunization campaign; in 1977 only a single natural case of smallpox occurred worldwide. When smallpox viruses escaped from a laboratory in Birmingham, England, in 1978, an unimmunized medical photographer became infected and died. Her mother suffered a mild case of the disease and recovered. The director of the laboratory from which the virus escaped committed suicide. After this disaster, smallpox stock cultures have been maintained in only two maximum containment laboratories, one of which is the Atlanta division of the Centers for Disease Control (CDC) and the other is in Moscow. However, smallpox surveillance continues and is especially rigorous in West Africa from Zaire to Sierra Leone because of the high incidence of a similar disease, monkeypox. Sufficient vaccine to immunize 300 million people has been stored as an added precaution.

The Disease The smallpox virus enters the throat and respiratory tract and during a 12-day incubation period it infects phagocytic cells and later blood cells. The infection spreads to skin cells, causing pus-filled vesicles. Acute systemic symptoms begin with fever, backache, and headache. In light-skinned individuals, vesicles appear first in the mouth and throat, rapidly spreading to the face, forearms, hands, and, finally, the trunk and legs. Vesicles, which are nearly all the same age because of the speed with which they appear, become opaque and pustular and encrust within about 2 weeks. Lesions are most numerous on the face and generally leave scars (Figure 20.17). Death is most likely to occur 10 to 16 days after onset of the first symptoms.

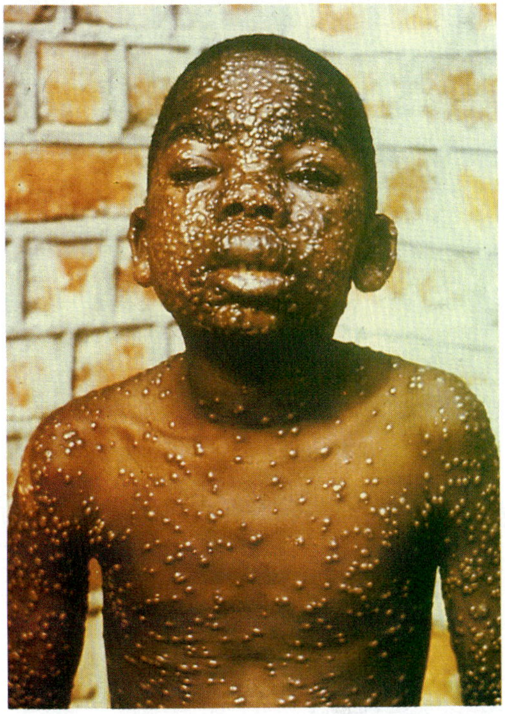

FIGURE 20.17
Smallpox lesions occur in heavier concentration on the face and arms, rather than the body. The reverse is true in chickenpox.

Diagnosis, Immunity, and Control Scrapings from lesions are used to differentiate potential smallpox from leukemia, chickenpox, and syphilis. In a nonimmune population, smallpox is highly contagious; in a largely immunized population it spreads very slowly. Each poxvirus has a hemagglutinin, but there is a large degree of cross-reaction between varieties. Antibodies to one such antigen often protect against more than one pox infection. For example, the vaccinia virus, which causes cowpox, has been used to vaccinate against smallpox. The term vaccination itself was derived from the use of vaccinia virus in early efforts to immunize against smallpox.

In a primary response to vaccination, a vesicle appears at the vaccination site in 4 to 5 days and soon scabs over. The scab drops off in about 3 weeks, leaving a scar. A vesicle must develop at the vaccination site to ensure that viral replication has taken place and the development of immunity initiated.

When working with a specimen from a lesion, a person should first rule out all agents other than the smallpox. If the causative agent cannot be identified, the specimen should be sent to the CDC, which has facilities for working with smallpox virus. Given that smallpox is believed to have been eradicated, just imagine the furor a person might cause by identifying a smallpox virus in a local laboratory!

Why Don't We Vaccinate Anymore? When smallpox infection was an ever-present risk, the risk of complications from vaccination was small compared with the risk of having the disease. Now that smallpox has been eradicated worldwide, complications of vaccination are far more life-threatening than the disease itself. Because of the numerous possible complications—allergic reactions to the vaccine, spreading of lesions over the body or to the fetus of a pregnant women, encephalitis—all but five countries in the world had discontinued vaccination by 1983, although some special groups of U.S. military personnel are still immunized.

The Final Destruction Public health authorities around the world have concluded that the major danger from smallpox now exists in the chance of another accidental escape, such as the one from the laboratory in Birmingham, England. Therefore, in December 1993 (on a date not yet decided upon) officials in both Atlanta and Moscow will place all smallpox stocks into a metal container that will be autoclaved at 130° C for 45 minutes.

Some scientists have expressed concern that valuable knowledge will be lost with the eradication of this very interesting virus. It is one of the largest viruses known and has a very large double-stranded DNA genome that, unlike most viruses, is able to replicate in the cytoplasm rather than in the nucleus of a host cell. To ensure our being able to study this virus in the future, laboratories in the United States and Russia are currently sequencing the bases in the genomes of at least four strains of the virus.

In the unlikely event that smallpox should someday recur, millions of doses of vaccine are stockpiled in readiness. Fortunately, as with rabies, vaccination even after exposure can prevent development of the disease.

CHAPTER SUMMARY

The agents and characteristics of the diseases discussed in this chapter are summarized in Tables 20.2, 20.3, and 20.4. Information in those tables is not repeated in this summary.

DISEASES OF THE SKIN

Bacterial Skin Diseases

- Bacterial skin diseases usually are transmitted by direct contact, droplets, or fomites.
- Most such diseases can be treated with penicillin or other antibiotics; treating **acne** with antibiotics may contribute to the development of antibiotic-resistant organisms; burn infections often are nosocomial and caused by antibiotic-resistant organisms.

RELATED KEY TERMS

folliculitis pimples pustules sty furuncle boil abscess carbuncle scalded skin syndrome scarlet fever erysipelas St. Anthony's fire pyoderma impetigo eschar debridement

Viral Skin Diseases

- **Rubella, measles,** and **chickenpox** viruses usually are transmitted via nasal secretions; pox diseases and **warts** usually are transmitted by direct contact.
- Treatment of viral skin diseases usually is palliative (able to relieve some symptoms but not cure the disease); warts can be excised.

German measles exanthema congenital rubella syndrome rubeola Koplik's spots measles encephalitis subacute sclerosing panencephalitis (SSPE) varicella-zoster virus shingles smallpox cowpox molluscum contagiosum papillomas human papillomaviruses dermal warts

Fungal Skin Diseases

Other Skin Diseases

- Subcutaneous fungal infections often persist in spite of treatment with topical fungicides; systemic infections are even more difficult to eradicate.
- Because of the difficulty in treating fungal skin infections, prevention by maintaining healthy skin and avoiding contamination of wounds is especially important.

dermatophytes dermatomycoses athlete's foot tinea pedis ringworm tinea cruris tinea unguium tinea corporis tinea capitis tinea barbae sporotrichosis blastomycosis blastomycetic dermatitis systemic blastomycosis candidiasis moniliasis thrush vaginitis aspergillosis zygomycosis madura foot maduromycosis swimmer's itch dracunculiasis

DISEASES OF THE EYES

Bacterial Eye Diseases

- Bacterial eye diseases are transmitted in a variety of ways, by direct contact, fomites, insect vectors, and to infants during delivery.
- Most are treated with antibiotics and prevented by good sanitation.

ophthalmia neonatorum keratitis bacterial conjunctivitis pinkeye trachoma

Viral Eye Diseases

- Viral eye diseases are transmitted by dust particles or direct contact; no effective treatments are available, but good sanitation can help to prevent such diseases.

epidemic keratoconjunctivitis acute hemorrhagic conjunctivitis

Parasitic Eye Diseases

- Controlling insect vectors reduces transmission of parasitic eye diseases.

onchocerciasis river blindness loaiasis

WOUNDS AND BITES

Wound Infections

- **Gas gangrene** and other anaerobic infections can be prevented by careful cleaning and draining of deep wounds; they are treated with penicillin, antitoxins, and sometimes with hyperbaric oxygen.

crepitant tissue

Other Anaerobic Infections

- **Cat scratch** and **rat bite fevers** can be prevented by avoiding such injuries; laboratory workers should be aware of dangers of infections from rat bites.

spirillar fever sodoku

Arthropod Bites and Infections

- Arthropod bites and infections can be prevented by good sanitation and hygiene and protecting the skin from bites.

tick paralysis chigger dermatitis scabies sarcoptic mange pediculosis blackfly fever myiasis

QUESTIONS FOR REVIEW

A., B.

1. What causes folliculitis, furuncles, and carbuncles, and how are they treated?
2. How does scarlet fever differ from strep throat, and how can a person have it more than once?
3. What are the causes and effects of erysipelas and scalded skin syndrome?
4. How are pyoderma and impetigo spread?
5. How are microorganisms involved in acne?
6. What organisms are frequently found in burn infections?
7. What are the major problems in diagnosing, treating, and preventing burn infections?
8. In what ways do rubella and rubeola measles differ?
9. What are some possible complications of these diseases?
10. What are the similarities between chickenpox and shingles?
11. What pox diseases are seen in humans?
12. How are warts acquired, and what can be done to get rid of them?
13. What organisms cause dermatomycoses, and how are they treated?
14. How can fungal systemic diseases be treated and prevented?
15. Under what conditions do opportunistic fungal infections occur, and how can they be prevented?
16. What are the causes of madura foot, and how can it be prevented?
17. How are swimmer's itch and dracunculiasis similar, and how are they different?

C., D.

18. How does conjunctivitis differ from keratitis?
19. Which bacteria can infect the eyes, and how do they reach the eyes?
20. What causes ophthalmia neonatorum, and under what circumstances does it occur?
21. How is pinkeye spread, and how is it treated?
22. What causes trachoma, and how is it prevented?
23. How does bacterial conjunctivitis differ from trachoma?
24. What kinds of eye diseases are caused by viruses, and how can they be prevented?
25. How are parasitic eye diseases acquired, and how can they be prevented?

E., F.

26. What causes gas gangrene, and how does the disease progress?
27. How is gas gangrene treated and prevented?
28. What other anaerobic organisms cause wound infections?
29. What is cat scratch fever?
30. What is rat bite fever, and why is it a hazard to laboratory workers?
31. What are the major effects of arthropod bites and infections?
32. What precautions can be taken to prevent arthropod bites?

PROBLEMS FOR INVESTIGATION

1. Study the problem of blindness resulting from infectious diseases, and devise a scheme for reducing its incidence.
2. How can pathogens such as *Acanthamoeba* be detected in the eye-wash stations in your science lab?
3. Why does *Candida albicans* sometimes cause disease when a person takes large amounts of antibiotics?
4. List and describe the skin diseases seen in patients having AIDS.
5. Prepare a report on how the burn unit in your area hospital attempts to prevent burn infection. What is their success rate?
6. A patient, recently arrived from Africa, came to the emergency room late one night complaining of visual problems. Imagine the surprise of the resident on duty when, upon peering through the pupil of the patient's eye, he saw a slender 2.5-cm worm moving vigorously inside the eye. Slides made from peripheral blood showed numerous microfilariae. What was the diagnosis?
(The answer to this question appears in Appendix E.)

SOME INTERESTING READING

Centers for Disease Control. 1986. *Acanthamoeba* keratitis associated with contact lenses—United States. *Morbidity and Mortality Weekly Report* 35, no. 25 (June 27):405.

Edwards, J. E., Jr. 1991. Invasive *Candida* infection: evolution of a fungal pathogen. *New England Journal of Medicine* 324, no. 15 (April 11):1060–63.

Fleiszig, S. M. J., and N. Efron. 1992. Microbial flora in eyes of current and former contact lens wearers. *Journal of Clinical Microbiology* 30(5):1156–61.

Gerber, M. A., et al. 1985. The aetiological agent of cat scratch disease. *Lancet* 1:1236.

Kilvington, S., D. F. P. Larkin, D. G. White, and J. R. Beeching. 1990. Laboratory investigation of *Acanthamoeba* keratitis. *Journal of Clinical Microbiology* 28 (12):2722–25.

Marx, R. 1989. Social factors and trachoma: a review of the literature. *Social Science & Medicine* 29, no. 1 (July 1):23–36.

Radentz, W. H. 1991. Fungal skin infections associated with animal contact. *American Family Physician* 43, no. 4 (April):1253–57.

Schmid, J., F. C. Odds, M. J. Wiselka, K. G. Nicholson, and D. R. Soll. 1992. Genetic similarity and maintenance of *Candida albicans* strains from a group of AIDS patients, demonstrated by DNA fingerprinting. *Journal of Clinical Microbiology* 30(4):935–41.

Tachibana, D. K. 1976. Microbiology of the foot. *Annual Review of Microbiology* 30:351.

Computer graphic representation of the three-dimensional structure of the influenza virus surface glycoprotein hemagglutinin (HA). A portion of the light region at the top of the HA structure binds the virus to sialic acid molecules on the surface of human cells, forming the first step in influenza infection. These binding pockets do not mutate, as do other parts of the HA, giving vaccine makers the hope that they can use molecules shaped similarly to sialic acid in vaccines. These would bind to the HA, blocking attachment of the virus to human cells, thereby preventing influenza infection.

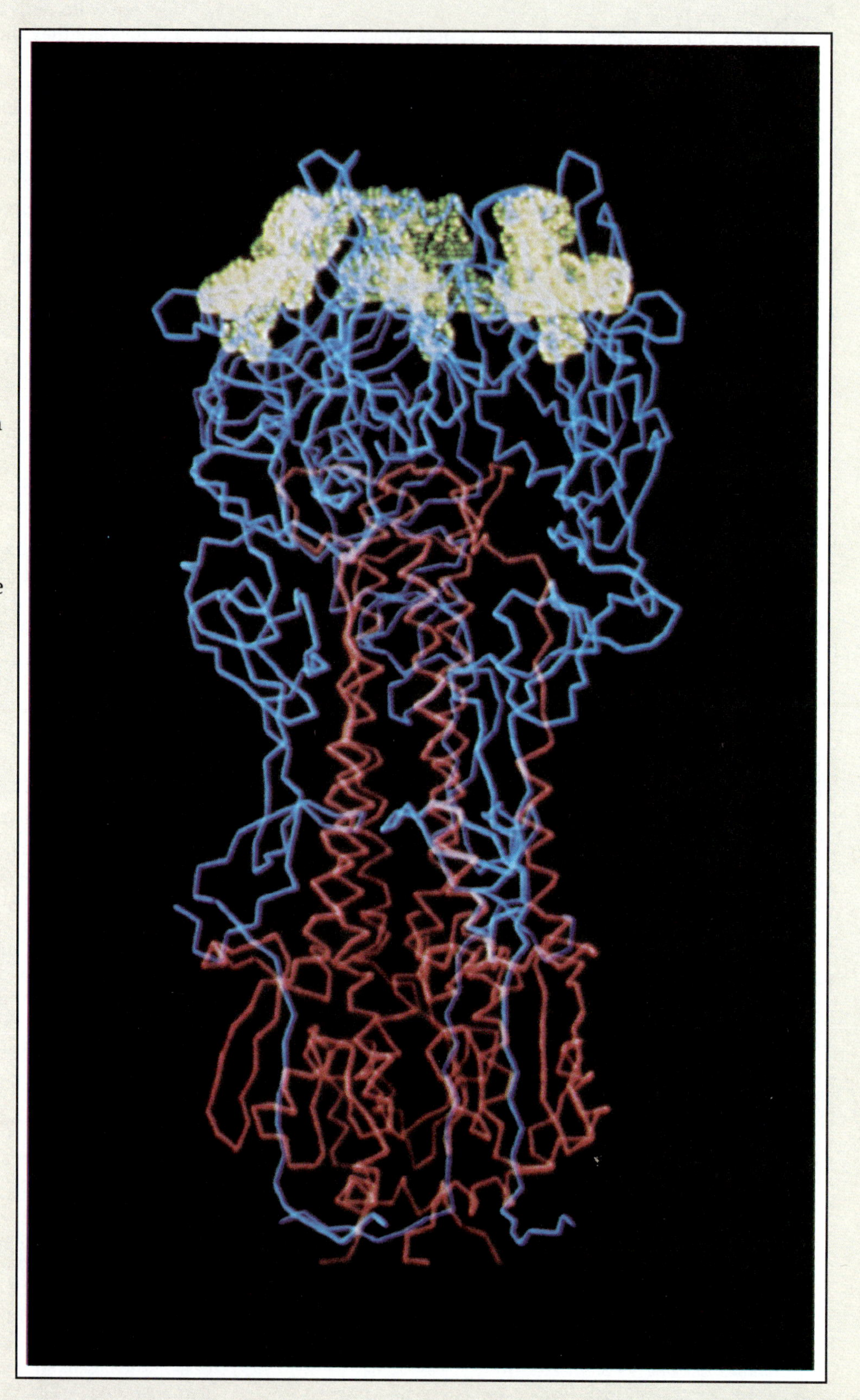

21 Diseases of the Respiratory System

This chapter focuses on the following questions:

A. What bacteria cause upper respiratory infections, and what are the important epidemiologic and clinical aspects of these diseases?

B. What viruses cause upper respiratory infections, and what are the important epidemiologic and clinical aspects of these diseases?

C. What bacteria cause lower respiratory infections, and what are the important epidemiologic and clinical aspects of these diseases?

D. What viruses cause lower respiratory infections, and what are the important epidemiologic and clinical aspects of these diseases?

E. What fungi cause lower respiratory infections, and what are the important epidemiologic and clinical aspects of these diseases?

The human respiratory system can be infected by various bacteria, viruses, fungi, and one helminth. Whether respiratory infections become established depends on host-microbe relationships (Chapter 15) and the condition of the respiratory system and its nonspecific defenses (Chapter 17). Respiratory infections are divided into the less serious upper respiratory infections, including ear infections, and the relatively more serious lower respiratory infections.

DISEASES OF THE UPPER RESPIRATORY SYSTEM

Bacterial Upper Respiratory Diseases

Bacterial infections of the *upper respiratory tract* (you might want to reexamine Figure 17.7) (Chapter 17, p. 447) are exceedingly common and are easily acquired through inhalation of droplets, especially in winter, when people are crowded indoors.

Pharyngitis and Related Infections

Pharyngitis, or sore throat, is an infection of the pharynx, usually caused by a virus but sometimes bacterial in origin. **Laryngitis** is an infection of the larynx, often with loss of voice, and **epiglottitis** is an infection of the epiglottis, which can close the airway and lead to death in just a few hours. Together, infection of the larynx and epiglottis can cause acute obstruction of the larynx, or *croup*. If the infection moves to sinus cavities it is called **sinusitis;** if it moves to the bronchi it is called **bronchitis;** and when it affects the tonsils it is called **tonsillitis.** If the infection spreads to the lungs, it is no longer an upper respiratory disease and is called pneumonia. As will be explained later, diphtheria is both an infection of the pharynx and an intoxication of the whole body.

BIOTECHNOLOGY

Viral Nasal Sprays to Treat Lungs

As currently practiced, gene therapy involves removing blood cells from patients, alterating their genes, and returning the cells to the bloodstream. This systemic form of treatment is not effective against certain organ-specific diseases. For example, it will not work for lung diseases, where the altered gene must act in lung cells. If the lung were stretched out flat, it would occupy an area the size of a tennis court. Reaching all of these cells is a problem—unless you are a cold virus!

Scientists have altered adenoviruses, a common cold virus, to carry genes that counteract cystic fibrosis. These adenoviruses are inserted via a nasal spray into rat lungs, where they preferentially attack infected cells lining the lungs. Researchers believe that the virus entered the cells via surface receptor sites, but they are uncertain whether the new gene was actually inserted into a host chromosome. Nevertheless, this procedure holds hope for eventual cure of diseases such as human cystic fibrosis and hereditary emphysema where organ-specific rather than systemic treatment is needed.

Streptococcal Pharyngitis Less than 10 percent of cases of pharyngitis are caused by beta-hemolytic *Streptococcus pyogenes*. This infection, familiarly known as *strep throat*, is most common in children 5 to 15 years old. It is acquired by inhaling droplets from active cases or healthy carriers. Dogs and other family pets also can be carriers. Detection and elimination of carrier states is often difficult in recurrent or cluster outbreaks. Contaminated food, milk, and water also can spread the disease, so it is important for infected persons and carriers not to handle food. In strep throat, the throat becomes inflamed and tonsils swell and develop white, pus-filled lesions (Figure 21.1).

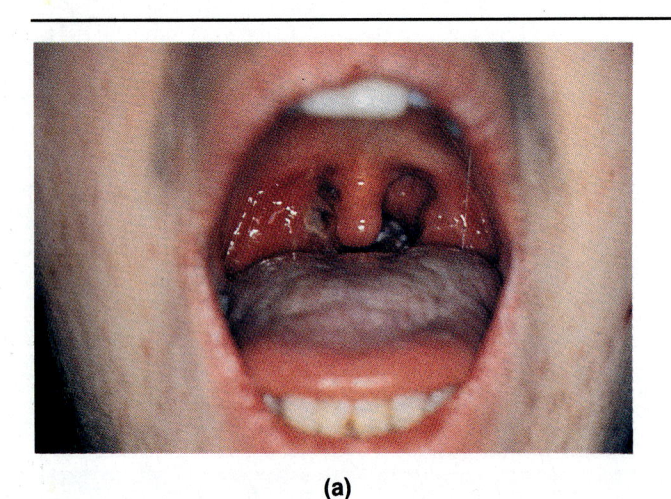

(a)

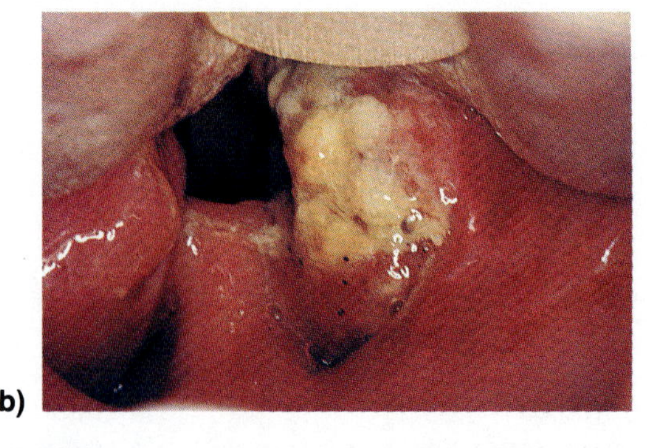

(b)

FIGURE 21.1 "Strep throat," a common form of pharyngitis, showing enlarged and reddened adenoids at the sides of the throat (a), and white, pus-filled lesions on tonsils (b).

Onset is usually quite abrupt, with chills, headache, acute soreness, especially upon swallowing, and often nausea and vomiting. Fever is generally high. The absence of cough and nasal discharge helps to distinguish strep throat from the common cold.

Diagnosis is by positive throat culture. A rapid serological screening test using a throat swab can be completed in the physician's office within minutes. Immediate treatment is important; if treatment is delayed, the organisms sometimes interact with the immune system and give rise to rheumatic fever (Chapter 23), which occurs in 3 percent of untreated cases. For this reason treatment with penicillin or one of its derivatives is often begun even before culture results are available.

Laryngitis and Epiglottitis Laryngitis can be caused by bacteria such as *Haemophilus influenzae* and *Streptococcus pneumoniae*, by viruses alone, or by a combination of bacteria and viruses. Acute epiglottitis is almost invariably caused by *H. influenzae*. Inflammation of the tissues rapidly closes off the airway, causing difficulty in breathing or even death.

Sinusitis More than half the cases of sinusitis are caused by *Streptococcus pneumoniae* and *Haemophilus influenzae*, but some cases are caused by *Staphylococcus aureus* and *Streptococcus pyogenes*. Swelling of sinus cavity linings impedes or prevents drainage and leads to pressure and severe pain. When drainage remains impeded, mucus accumulates and fosters bacterial growth. Secretions consisting of mucus, bacteria, and phagocytic cells then collect in the sinuses. Chronic sinusitis can permanently damage sinus linings and cause smooth pendulous growths called polyps to form. Applying moist heat over the sinuses, instilling drops of vasoconstrictors, such as ephedrine, into the nasal passages, humidifying the air, and holding the head in a position to promote drainage help alleviate symptoms. Treatment with antibiotics such as penicillin or painkillers as strong as codeine or morphine may be necessary. Swimmers and divers often suffer from sinusitis if water is forced into the sinuses, which then become inflamed. Such people can prevent water entering their sinuses by using nose clips or by exhaling as they submerge and continuing to exhale while their heads are under water.

Bronchitis Bronchitis involves the bronchi and bronchioles but does not extend into the alveoli. About 15 percent of the general population has chronic bronchitis. It is most common in older people and is linked to smoking, air pollution, inhalation of coal dust, cotton lint, and other particles, and heredity. Patients cough up sputum containing mucus, organisms, and phagocytic cells. Common causative agents include *Streptococcus pneumoniae*, *Mycoplasma pneumoniae*, and various species of *Haemophilus*, *Streptococcus*, and *Staphylococcus*. Infections can spread to the alveoli of the lungs and cause pneumonia. Diagnosis is from sputum cultures. By the time a patient seeks medical attention, respiratory membranes may have been permanently damaged. Antibiotic treatment can halt further deterioration but cannot reverse damage already done. Eventually severe shortness of breath develops.

Diphtheria

Although fewer than 10 cases per year are now seen in the United States, **diphtheria** was once a feared killer. A century ago 30 to 50 percent of patients died; most deaths occurred by suffocation in children under age 4. *Sequelae*, or symptoms that follow a disease, are common in diphtheria. Myocarditis, an inflammation of the heart muscle, and polyneuritis, an inflammation of several nerves, account for deaths even after apparent recovery. Significant cardiac abnormalities occur in 20 percent of patients, and neurologic problems, including paralysis, can occur after particularly severe cases.

Causative Agent Diphtheria is caused by exotoxin-producing strains of *Corynebacterium diphtheriae*. Club-shaped cells of this gram-positive rod grow side by side in palisades and contain metachromatic granules of phosphates, which turn reddish when stained with methylene blue (Figure 21.2). **Diphtheroids,** which are found in normal throat cultures, fail to produce exotoxin but are indistinguishable by commonly performed laboratory studies from diphtheria-causing organisms. On inoculation into a guinea pig, pathogenic organisms cause disease, but diphtheroids do not. Today, some laboratories also use a gel diffusion test to detect toxin-producing strains.

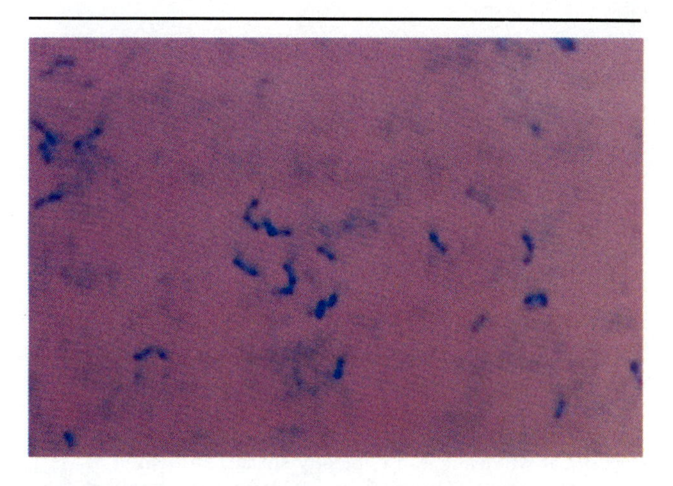

FIGURE 21.2 *Corynebacterium diphtheriae*, stained with methylene blue, showing metachromatic granules of phosphates that have stained a deep reddish blue.

To produce toxin the bacterium must be infected by an appropriate strain of bacteriophage in the lysogenic prophage state. (Chapter 11, p. 279) In other words, the bacteriophage must be integrated into the bacterial chromosome, where its toxin-producing genes are expressed. "Curing" the bacterium of its phage infection removes its ability to produce toxin. The phage-infected cell also fails to produce toxin unless the patient's blood iron concentration drops to a critically low level.

The Disease Humans are the only natural hosts for diphtheria, and the disease usually is spread by droplets of respiratory secretions. Infection usually begins in the pharynx 2 to 4 days after exposure. Bacilli, damaged epithelial cells, fibrin, and blood cells combine to form a **pseudomembrane** (Figure 21.3). Although it is not a true membrane, removing it leaves a raw bleeding surface soon covered by another pseudomembrane. The pseudomembrane can block the airway and cause suffocation. The organisms very rarely invade deeper tissues or spread to other sites, but the extremely potent toxin spreads throughout the body and kills cells by interfering with protein synthesis. The heart, kidneys, and nervous system are most susceptible.

Sometimes diphtheria organisms are found in unusual sites. They can invade the nasal cavities, which have relatively few blood vessels; there they cause a milder disease because less toxin enters the blood. They invade the skin in *cutaneous diphtheria*, but the toxin does not reach the blood. Cutaneous diphtheria is a tropical disease associated with poor hygiene; it is rare in the United States.

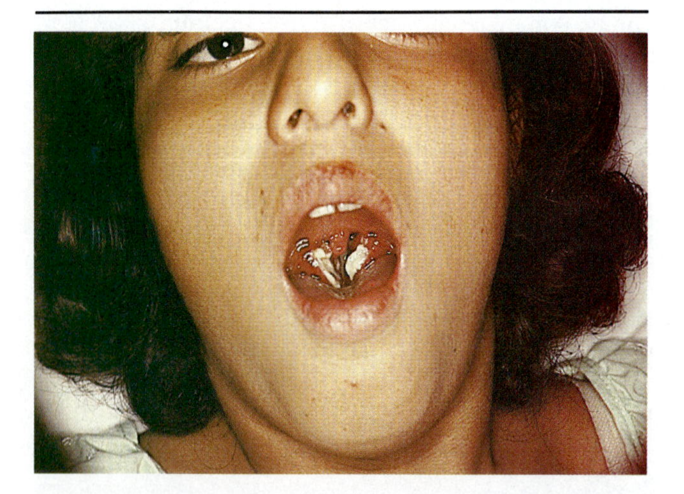

FIGURE 21.3 Pseudomembrane of diphtheria. This is not a true membrane; rather, it adheres to the underlying tissue. If torn off, it will leave a raw, bloody surface and will eventually reform. However, it is necessary to remove it when it blocks the airway.

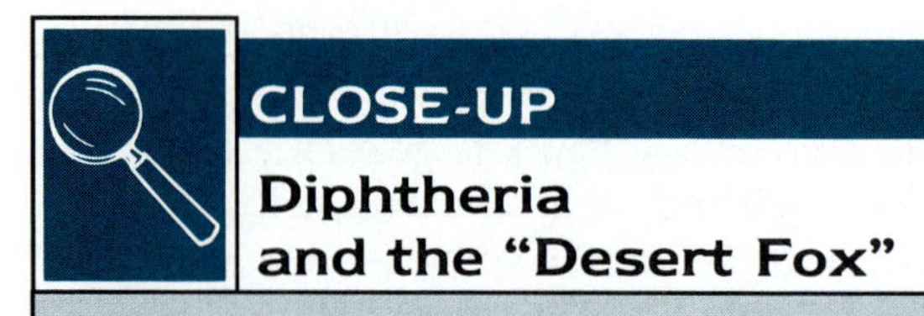

CLOSE-UP

Diphtheria and the "Desert Fox"

Erwin Rommel, the German general known as the "Desert Fox," led the North African tank warfare during World War II. Old newsreels show him with a handkerchief pressed to his nose, not because of desert dust but because of nasal diphtheria. This chronic disease necessitated his being flown back to Berlin periodically for treatment, leaving his troops without his brilliant and charismatic leadership. Battles lost and men disheartened because of his absence might have given the Allies the edge they needed to defeat the Nazis.

Rommel's presence in Berlin led to his joining other generals in a plot to assassinate Hitler. The plot failed, and Rommel's role was discovered. As Gestapo agents were taking him to headquarters, he reportedly died of a heart attack in the car. Whether he was executed, or, knowing the game was up, actually succumbed to a pounding heart on that ride to Berlin is not known. Public trial of a celebrated war hero for treason would not have been politically expedient. On the other hand, his heart, weakened by long exposure to diphtheria toxin, could easily have given out. Had Rommel never contracted diphtheria, who knows how the history of World War II might have been affected?

Treatment and Prevention Diphtheria is treated by administering antitoxin to counteract toxin and antibiotics, such as penicillin, to kill the organisms. It can be prevented with DPT vaccine (diphtheria, pertussis, tetanus), which contains diphtheria toxoid and is administered in a series of shots beginning at age 2 months. Boosters are needed every 10 years. In the past, when diphtheria was endemic in the United States, most of the adult population maintained, or even acquired, immunity by continual exposure to small quantities of the bacterium. Such exposures acted as natural booster shots. Now that diphtheria is no longer endemic, immunity depends on the vaccine. Because diphtheria is not well controlled elsewhere in the world, vaccination must be continued in this country to prevent epidemics should the disease spread from other places.

Ear Infections

Ear infections occur as **otitis media** in the middle ear and as **otitis externa** in the external auditory canal. Because otitis media usually produces a puslike exudate, it often is called otitis media with effusion (OME). *Streptococcus pneumoniae*, *S. pyogenes*, and *Hae-*

mophilus influenzae account for about half of acute cases. Organisms such as species of *Proteus*, *Klebsiella*, and *Pseudomonas* are often responsible for chronic cases. Otitis externa usually is caused by *Staphylococcus aureus* and *Pseudomonas aeruginosa*. Such pseudomonad infections are common in swimmers because the organisms are highly resistant to chlorine.

OME infections arise from pharyngeal organisms passing through the Eustachian tube. Fever and earache, which arise from pus creating pressure in the middle ear, usually are present, but some cases are asymptomatic. The disease is treated with antibiotics, usually penicillin, and it is important to continue treatment until all organisms are eradicated. Even after successful therapy, sterile fluid can remain in the middle ear, impairing vibration of the tiny bones in the middle ear and decreasing sound transmission. Sometimes tubes are inserted to prevent fluid accumulation and repeated middle ear infections (Figure 21.4). If this impairment occurs during speech development, as it often does, speech can be adversely affected. Sometimes children thought to be inattentive or defiant really cannot hear. School can be a nightmare for such children. As children grow older the Eustachian tube changes shape and develops an angle that prevents most organisms from reaching the middle ear—much to the relief of the children and their parents.

Viral Upper Respiratory Diseases

The Common Cold

The common cold (*coryza*) probably causes more misery than any other infectious disease, but it is not life-threatening. Although exact statistics are not available on the number of infections per year, Americans lose more than 200 million days of work and school per year because of colds. Economically, colds are a bane to employers who must deal with lost work time and a boon to manufacturers and retailers of cold remedies. Cold viruses are ubiquitous and present year round, but most infections occur in the early fall or early spring. After an incubation period of 2 to 4 days, symptoms such as sneezing, inflammation of mucous membranes, excessive mucus secretion, and airway obstruction appear. Sore throat, malaise, headache, cough, and occasionally tracheobronchitis also occur. The illness lasts about 1 week, and severity of symptoms is directly correlated with the quantity of viruses released from epithelial cells of the upper respiratory tract.

Causative Agents Different cold viruses predominate at different seasons. In fall and spring, rhinoviruses are in the majority. Parainfluenza virus is present all year, but peaks in late summer. In mid-

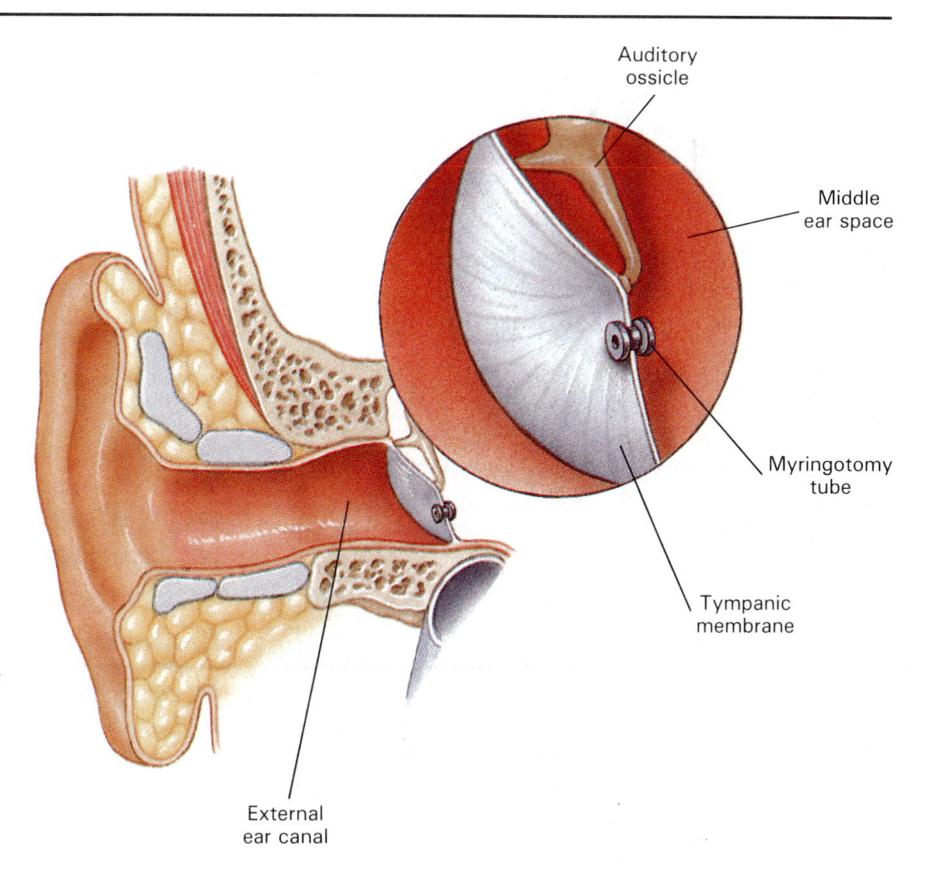

FIGURE 21.4 Tubes to reduce middle ear infections are placed through the tympanic membrane (eardrum) to promote drainage. This can be done in the physician's office. The phlange on the rear end of the tube is designed to hold it in place.

December, the coronaviruses appear. Adenoviruses are present at a lower level all year.

Rhinoviruses are the most common cause of colds. They are resistant to antibiotics, chemotherapeutics, and disinfectants but are quickly inactivated by acid conditions. They grow best at 33 to 34°C and thus replicate in the epithelium of the upper respiratory tract, where air movements lower the tissue temperature. Rhinoviruses can be isolated from nasal secretions and throat washings and identified by their sensitivity to low pH and resistance to ether and chloroform, a unique combination of traits among viruses. At least 113 different rhinoviruses have been identified, all with different antigens. Natural immunity is short-lived, and no effective vaccine has been developed. Even if a person becomes immune to some rhinoviruses, there are always others to cause another cold.

Rhinoviruses account for only about half of all colds; the second most common cause is another group of ubiquitous viruses, the **coronaviruses.** Coronaviruses have clublike projections that give them a halo (*corona*, Latin for halo or crown). In addition to colds, these viruses also cause acute respiratory distress, sometimes mild pneumonia, and acute gastroenteritis. They infect the epithelium of both the respiratory and digestive tracts. In the digestive tract they reduce absorptive capacity and cause diarrhea, dehydration, and electrolyte imbalances.

Transmission Cold viruses are spread by fomites more often than by aerosols. Blowing the nose and handling used tissues contaminates the fingers, so that anything touched becomes contaminated. Tissues impregnated with a disinfectant are effective in reducing viral spread but are harsh-textured and expensive. Conscientiously using tissues once, discarding them immediately, and thoroughly washing hands after each nose wipe can significantly reduce the spread of rhinoviruses.

Diagnosis and Treatment Most people diagnose and treat their own colds with remedies that alleviate some symptoms. Over-the-counter antihistamines are reasonably effective in counteracting inflammatory reactions as the body attempts to defend itself against the viruses. Experiments with certain kinds of human interferon have shown potential in blocking or limiting rhinovirus infections, but they must be given with another agent that helps the interferon to reach epithelial cells beneath a thick blanket of mucus. A quantity of alpha interferon sufficient to block infection causes irritation of membranes and bleeding. Different combinations of interferons and other factors are being explored to control rhinovirus infections.

Parainfluenza

Parainfluenza is characterized by rhinitis (nasal inflammation), pharyngitis, bronchitis, and sometimes pneumonia, mainly in children. **Parainfluenza viruses** initially attack the mucous membranes of the nose and throat. The first symptoms are cough and hoarseness for 2 or 3 days, harsh breathing sounds, and a red throat. Symptoms can progress to a barking cough and a high-pitched, noisy respiration called **stridor** (stri'dor). Recovery is usually quick—within a few days.

Of the four parainfluenza viruses capable of infecting humans, two can cause croup. **Croup** is defined as any acute obstruction of the larynx and can be caused by a variety of infectious agents, including parainfluenza viruses. Both the larynx and the epiglottis become swollen and inflamed; the high-pitched barking cough of croup results from partial closure of these structures. Increased humidity from a cool mist vaporizer or from a hot shower helps to relieve croup symptoms.

By age 10 most children have antibodies to all four parainfluenza viruses whether they have had recognizable illness or not. Thus, the incidence of infection is very high, although the incidence of clinically apparent disease is much lower. Epidemics and smaller outbreaks of parainfluenza infections occur primarily in the fall, but sometimes in early spring after the "flu" season. The viruses are spread by direct contact or by large droplets. They can be inactivated by drying, increased temperature, and most disinfectants, so they do not remain long on surfaces or in the environment. Resistance to infection comes from secretory IgA that defends mucous membranes against infection, and not from bloodborne IgG. Reinfection with parainfluenza viruses is rare, so secretory immunoglobulins must create effective immunity. However, efforts to make a vaccine for parainfluenza viruses have not been successful.

Diseases of the upper respiratory tract are summarized in Table 21.1.

DISEASES OF THE LOWER RESPIRATORY TRACT

Bacterial Lower Respiratory Diseases

Among the bacterial diseases of the lower respiratory tract are two of the great killer infections of history: tuberculosis and pneumonia. Although the advent of antibiotic therapy brought these diseases under control to a considerable extent, both seem to be making comebacks today as a result of the spread of AIDS.

TABLE 21.1 Summary of diseases of the upper respiratory tract

Disease	Agent(s)	Characteristics
Bacterial Upper Respiratory Diseases		
Pharyngitis	*Streptococcus pyogenes*	Inflammation of the throat and fever without cough or nasal discharge
Laryngitis and epiglottitis	*Haemophilus influenzae, Streptococcus pneumoniae*	Inflammation of the larynx and epiglottis, often with loss of voice
Sinusitis	*H. influenzae, S. pneumoniae* and *S. pyogenes, Staphylococcus aureus*	Inflammation of sinus cavities, sometimes with severe pain
Bronchitis	*Streptococcus pneumoniae, Mycoplasma pneumoniae,* and others	Inflammation of bronchi and bronchioles with a mucopurulent cough; shortness of breath in chronic cases
Diphtheria	*Corynebacterium diphtheriae*	Inflammation of the pharynx with pseudomembrane and systemic effects of toxin
Otitis externa	*Staphylococcus aureus, Pseudomonas aeruginosa*	Inflammation of external ear canal; common in swimmers
Otitis media	*Streptococcus pneumoniae, S. pyogenes, Haemophilus influenzae*	Pus-filled infection of the middle ear with pressure and pain
Viral Upper Respiratory Diseases		
Common cold	Rhinoviruses, coronaviruses	Sore throat, malaise, headache, and cough
Parainfluenza	Parainfluenza viruses	Nasal inflammation, pharyngitis, bronchitis, croup, sometimes pneumonia

Whooping Cough

Whooping cough, also called **pertussis,** is a highly contagious disease known only in humans. The word *pertussis* means "violent cough," and the Chinese call it the "cough of 100 days." Although distributed worldwide, strains found in the United States are less virulent than most, but jet travel could bring virulent strains from North Africa or other parts of the world at any time. Whooping cough is a major health problem in developing nations, where lack of vaccination allows 80 percent of those exposed to contract the disease. In the United States, concern and negative publicity concerning vaccine safety discouraged some parents of very young children from having them immunized. As a result, incidence of the disease more than doubled during the 1980s, with almost 4500 cases reported in 1990 (Figure 21.5). The disease tends to occur sporadically, especially in infants or young children, with 50 percent of cases occurring in the first year of life.

Before the development of the vaccine, nearly every child got whooping cough. Adults who contract it today either were not vaccinated or their immunity has declined. Partial immunity lessens the severity of the disease. Many adults may fail to display the characteristic "whooping" sound and so are misdiagnosed. Such people serve as a reservoir to spread the infection.

Causative Agent *Bordetella pertussis*, a small encapsulated gram-negative coccobacillus first isolated in 1906, is the usual causative agent of pertussis. Only about 5 percent of the cases are due to *Bordetella parapertussis* and *B. bronchiseptica*, which usually produce a milder disease. *B. bronchiseptica* is a normal resident of canine respiratory tracts, where it sometimes causes "kennel cough."

Susceptible people become infected by inhaling respiratory droplets, and the organisms colonize cilia lining the respiratory tract. Only active cases of pertussis are known to shed organisms; carriers of the disease are unknown. The pertussis bacillus does not invade tissues or enter the blood, but it does produce several substances that contribute to its virulence. It produces an endotoxin, an exotoxin, and hemagglutinins, which help it adhere to the cilia in the host respiratory tract.

The Disease After an incubation period of 7 to 10 days, the disease progresses through three stages: catarrhal, paroxysmal, and convalescent. The **catarrhal stage** is characterized by fever, sneezing, vomiting, and a mild, dry, persistent cough. A week or two later the **paroxysmal** (par-oks-iz'mal), or intensifying, **stage** begins as mucus and masses of bacteria fill the airway and immobilize the cilia. Strong, sticky, ropelike strings of mucus in the airway elicit violent, paroxysmal coughing. Failure to keep the airway open leads to **cyanosis** (si-an-o'sis), or bluish skin, because too little oxygen gets to the blood. Keeping the airway open is especially difficult in infants, and airway blockage accounts for the high death rate in patients under 1 year of age. Straining to draw in air gives the

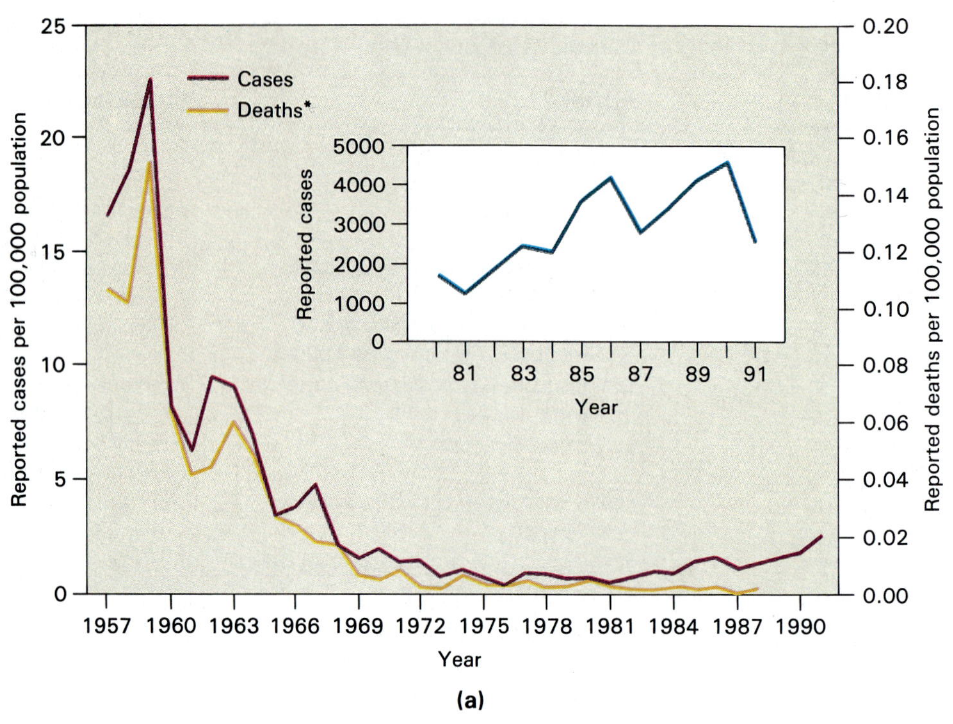

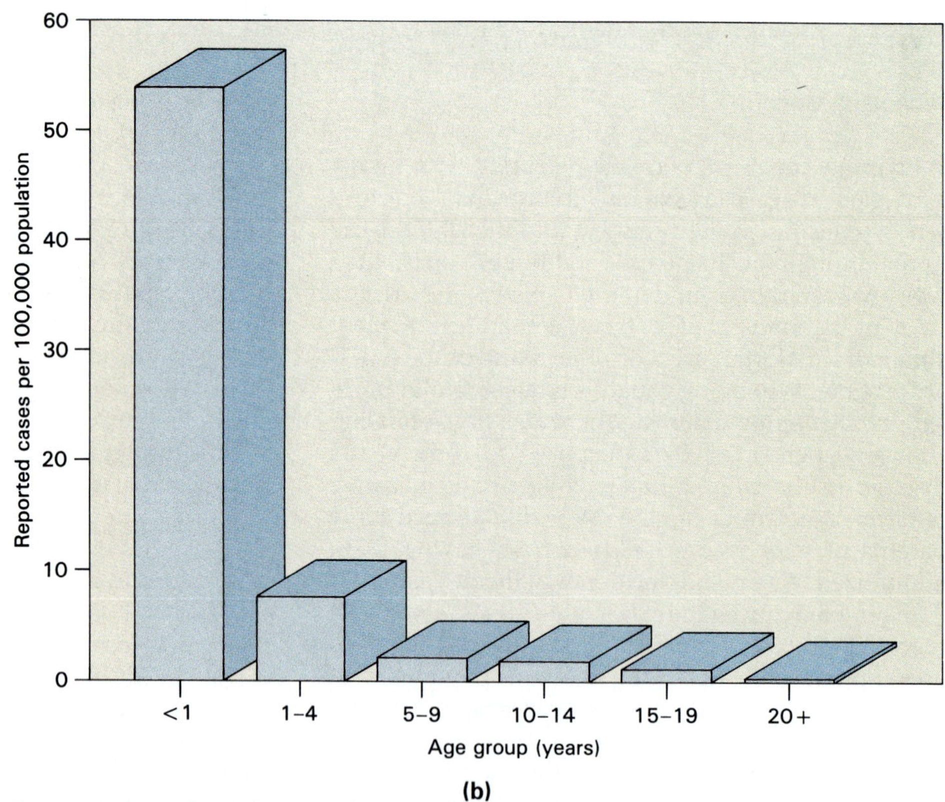

FIGURE 21.5 Incidence of pertussis (whooping cough) in the United States.

characteristic loud "whooping" sound. Coughing sieges occur several times a day and cause exhaustion. Sometimes coughing is so severe that it causes hemorrhage, convulsions, and rib fractures. Vomiting that usually follows a coughing siege leads to dehydration, nutrient deficiency, and electrolyte imbalance—all especially dangerous in infants.

After a paroxysmal stage, usually of 1 to 6 weeks duration but sometimes longer, the patient enters the **convalescent stage**. Milder coughing can continue for several months before it eventually subsides. Secondary infections with other organisms are common at this stage.

Diagnosis and Treatment Whooping cough is diagnosed by obtaining organisms from the posterior

PUBLIC HEALTH

Whooping Cough

In years gone by it was not uncommon to see nannies in Central Park wearing white uniforms covered with blood from holding infants against them during paroxysms of coughing. Of course, the infants also produced plentiful aerosol droplets containing infectious organisms. Very few younger Americans have seen a case of whooping cough. Ask an older family member or friend what they remember about it or how it was when they had it.

nares (Figure 21.6) and culturing them on a potato-blood-glycerol agar called Bordet-Gengou medium. Because penicillin does not kill *B. pertussis*, it is often added to the medium to suppress growth of other organisms. To treat whooping cough, antitoxin is given early in the disease to combat toxin, and erythromycin is given to shorten the second stage. (Chapter 18, p. 499) It cannot entirely eliminate it, but it does reduce the number of viable organisms being shed. Supportive measures such as suctioning, rehydration, oxygen therapy, and attention to nutrition and electrolyte balance are very important, as is appropriate treatment of secondary infections.

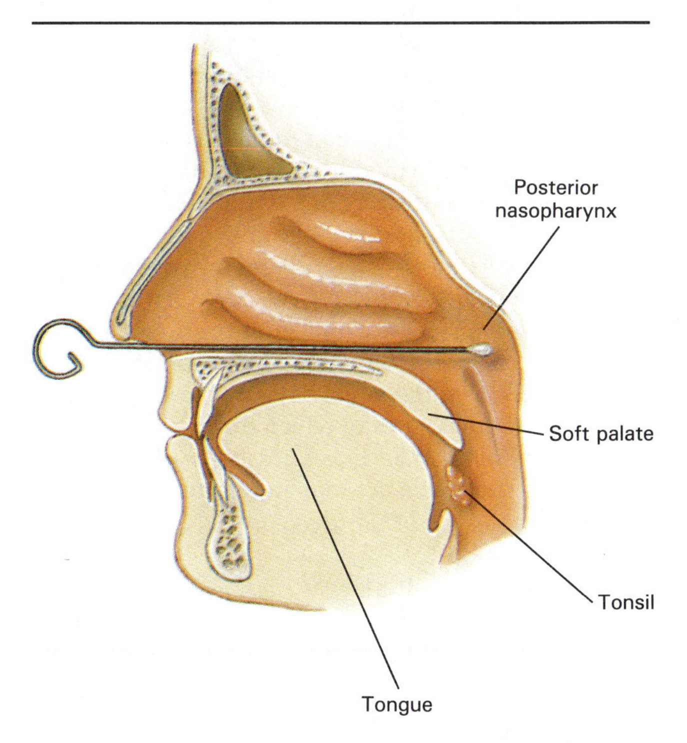

FIGURE 21.6 The technique of culturing for whooping cough. A swab attached to a thin, flexible wire is inserted through the nostrils, and the patient is asked to cough several times.

PUBLIC HEALTH

DPT Vaccine Liability

Deaths and injuries from pertussis and other vaccines have resulted in numerous lawsuits against the manufacturers. After several years of deliberations the National Vaccine Compensation Act was passed by Congress and went into effect January 1, 1988. This new law establishes a trust fund and a minimum compensation of $250,000 per affected child, but it requires parents who accept compensation to sign away their right to pursue further court action. Surcharges of $4.56 on each DPT injection and $4.44 on each MMR (mumps, measles, rubella) injection go into the trust fund. Unfortunate side effects of the surcharges are that some parents in lower middle-income groups cannot afford to have their children immunized, and public health clinics cannot afford to immunize all the children of eligible low-income parents.

Prevention Pertussis vaccine has saved many lives, but it is not entirely safe. In the United States it lowered the number of cases from 227,319 in 1938 to 2575 in 1991, but each year vaccine reactions cause 5 to 20 children to die and another 50 to suffer permanent brain damage. Yet, the number of children hurt by the vaccine is far smaller than the number who would die without it. Vaccination begins with the DPT series at 2 months of age, as soon as the immune system is capable of responding to antigens in the vaccine. Early immunization is important because pertussis antibodies do not cross the placenta, so infants have no passive immunity to whooping cough. Immunity decreases after vaccination, but because of risks with the vaccine, adult immunization is not recommended. Thus, many adults lack protection; 95 percent of those who have not been immunized in the last 12 years would likely contract the disease during an outbreak. Recovery from the disease confers immunity, but not for a lifetime. Second cases are known, especially in adulthood, but they are usually milder.

Classic Pneumonia

In the United States more deaths result from pneumonia than from any other infectious disease. **Pneumonia,** an inflammation of lung tissue, can be caused by bacteria, viruses, or fungi. It sometimes occurs after inhalation of irritating chemicals, such as kerosene or chlorine, or after radiation therapy for lung or breast cancer. Bacterial pneumonias are most often due to *Streptococcus pneumoniae*, also known as pneumococci. *Staphylococcus aureus*, *Klebsiella pneumoniae*, and *My-*

coplasma pneumoniae also account for some cases of pneumonia. *Pseudomonas aeruginosa* also can cause pneumonia, especially in patients with other diseases.

Classification of Pneumonias Pneumonias are classified as lobar or bronchial. **Lobar pneumonia** affects one or more of the five major lobes of the lungs. It is a very serious primary disease, and 95 percent of cases are caused by *Streptococcus pneumoniae*. However, back in 1881 when *Klebsiella pneumoniae* was first discovered, it was thought to be the primary cause. Fibrin deposits are characteristic of lobar pneumonia; when they solidify they cause **consolidation,** or blockage of air spaces. **Pleurisy,** inflammation of the pleural membranes that causes painful breathing, often accompanies lobar pneumonia.

Bronchial pneumonia begins in the bronchi and can spread into the surrounding tissues toward the alveoli in a patchy manner. Although also most likely to be caused by pneumococci, bronchial pneumonia differs from lobar pneumonia in two ways: (1) It often appears as a secondary infection after a primary infection such as a viral influenza, heart disease, or another lung disease, and (2) it lacks the plentiful fibrin deposits seen in lobar pneumonia. Bronchial pneumonia can also follow exposure to chemicals or aspiration of vomitus or other fluids. Sometimes infants aspirate amniotic fluid in the process of being born, especially if delivered by cesarean section. Bronchial pneumonia is common in elderly and debilitated patients. In fact, it is sometimes called the "old man's friend" because it spares an elderly person a possibly longer, more painful death from some other cause.

Transmission Both lobar and bronchial pneumonia are transmitted by respiratory droplets and, in the winter, by carriers—including medical personnel—who have had contact with pneumonia patients. As much as 60 percent of a population can be carriers in closely associated groups such as personnel on military bases or children in preschools.

The Disease After a few days of mild upper respiratory symptoms, the onset of pneumococcal pneumonia is sudden. The patient suffers violent chills and high fever (up to 106°F). Chest pain, cough, and sputum containing blood, mucus, and pus follow. Fever may terminate 5 to 10 days after onset when untreated or within 24 hours after antibiotics are given.

More than a half-million cases of pneumococcal pneumonia occur each year; it is the only infectious disease in the top 10 causes of death in the United States. With immediate and adequate antibiotic therapy, the mortality rate is 5 percent, but without treatment it is 30 percent. This compares favorably with *Klebsiella* pneumonia, which despite best treatment still has a mortality rate of 50 percent. *Klebsiella* pneumonia is an extremely severe pneumonia, which can cause chronic ulcerative lesions in the lungs. Failure to seek medical attention promptly keeps the overall mortality rate for pneumonia in the United States at 25 percent. Conditions that predispose toward pneumonia include old age, chilling, drugs, anesthesia, alcoholism, and a variety of other disease states.

Diagnosis, Treatment, and Prevention Diagnosis of pneumonia is based on clinical observations, X-rays, or sputum culture. Penicillin is the drug of choice for treatment for pneumococcal pneumonia. Following recovery, immunity exists for a few months only against the particular strain that had caused the infection. Thus, a patient can develop case after case of pneumonia from infection with other strains or species. Artificial immunity can be induced with the polyvalent vaccine Pneumovax. This vaccine contains a variety of antigens, so it protects against approximately 80 percent of the strains of *Streptococcus pneumoniae* for a period of from 5 years to life. Vaccination is especially recommended for the elderly and at-risk populations.

Mycoplasma Pneumonia

One of the tiniest bacterial pathogens known, *Mycoplasma pneumoniae,* ordinarily causes mild, and sometimes inapparent, upper respiratory tract infections. In 3 to 10 percent of infections it causes **primary atypical pneumonia,** or mycoplasma pneumonia, usually a mild pneumonia with an insidious onset. The disease is said to be atypical because the symptoms are different from those of classic pneumonia; the causative organisms are typical mycoplasma. Patients often remain ambulatory, so the disease is sometimes called **walking pneumonia.** The mortality rate is less than 0.1 percent. It is unusual in preschool children and is most common among young people 5 to 19 years old.

Transmission is by droplets of respiratory secretions, and onset of symptoms follows an incubation period of 12 to 14 days. Fever lasts 8 to 10 days and gradually declines, during which time the cough and lung symptoms also decline. Unusual features of mycoplasma pneumonia are that alveoli decrease in size by inward swelling of the alveolar walls, and they do not fill with fluid.

Diagnosis by culturing *M. pneumoniae* from sputum takes 2 to 3 weeks due to the slow growth rate of the organisms (Figure 21.7), so treatment usually is based on clinical symptoms. Erythromycin and tetracycline are the drugs of choice; penicillin has no effect. Even untreated cases have a favorable prognosis. No vaccine is currently available, so prevention requires avoiding contact with contaminated secretions.

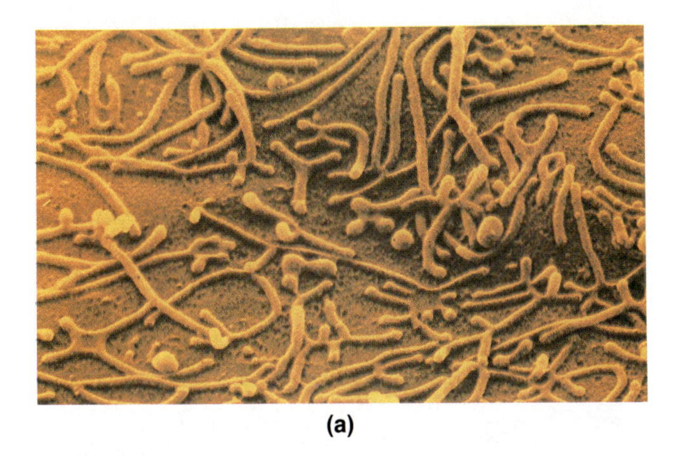

(a)

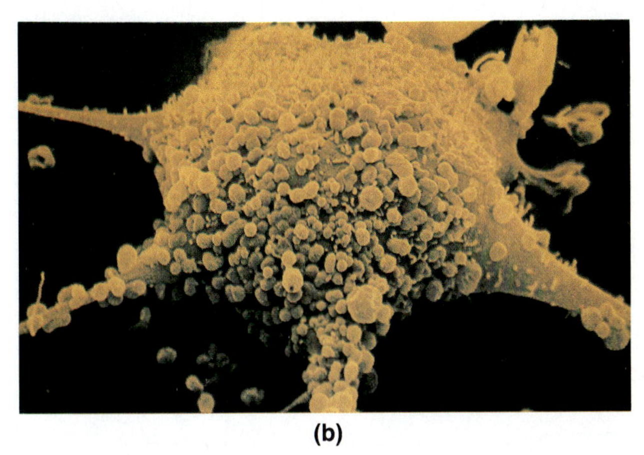

(b)

FIGURE 21.7 (a) *Mycoplasma* cells, lacking a cell wall, assume irregular shapes. (b) A cell infected by numerous *Mycoplasma* cells.

Legionnaire's Disease

In 1976 many war veterans attending a convention in Philadelphia were victims of a mysterious ailment, which became known as **Legionnaire's disease.** After 29 deaths and much frantic investigation, the previously unidentified causative organism, *Legionella pneumophila*, was finally isolated (Figure 21.8). Researchers at CDC later found antibodies to *L. pneumophila* in frozen blood samples from unidentified outbreaks decades ago. One wonders how the organism was overlooked for such a long time. However, it is sufficiently different from previously classified bacteria that a new genus had to be created for it. *Legionella pneumophila* is a weakly gram-negative, strictly aerobic bacillus with fastidious nutritional requirements. It does not ferment sugars and has an obscure life cycle. More than 20 *Legionella* species have been identified; most are free-living in soil or water and do not ordinarily cause disease.

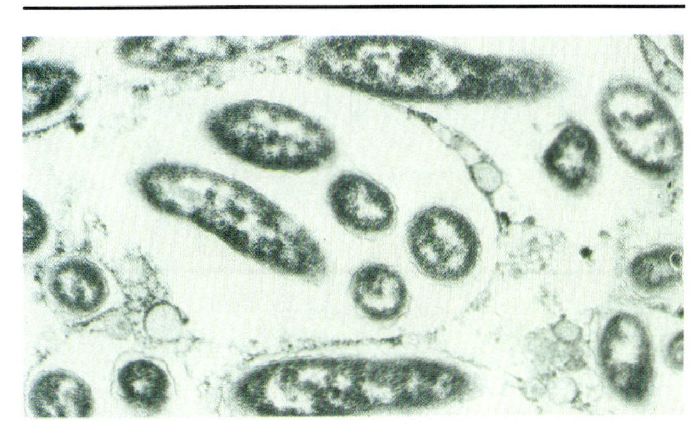

FIGURE 21.8 TEM of *Legionella pneumophila*, the cause of Legionnaire's disease.

Legionellosis is transmitted when organisms growing in soil or water become airborne and enter the patient's lungs as an aerosol. Person-to-person transmission has never been documented. Air conditioners, ornamental fountains, humidifiers, and vaporizers in patient rooms are often implicated in the spread of the disease. Such devices should be regularly disinfected.

After an incubation period of 2 to 10 days, Legionnaire's disease appears with fever, chills, headache, diarrhea, vomiting, fluid in the lungs, pain in the chest and abdomen, and, less frequently, profuse sweating and mental disorders. When death occurs it is usually due to shock and kidney failure. In nonpneumonic legionellosis, after about 48 hours' incubation, the patient suffers 2 to 5 days of flulike symptoms without infiltration of the lungs.

Pontiac fever is a mild legionellosis named for an outbreak in 1968 affecting 144 people—95 percent of the employees of the Pontiac, Michigan, Health Department. None died; all recovered within 3 to 4 days.

Fluorescent microscopy or ELISA tests are used to diagnose *Legionella* infections. Erythromycin is used to treat them; most other antibiotics have no effect. Because it is difficult to distinguish Legionnaire's disease from other pneumonias, erythromycin is frequently given along with a second antibiotic such as penicillin when treating any pneumonialike disease.

Tuberculosis

Tuberculosis (*formerly called consumption*) has plagued humankind since ancient times, as indicated by skeletal damage in Egyptian mummies and earlier human remains. It remains a massive global health problem today. One and one-half billion people (six times the U.S. population) have tuberculosis; each year 600,000 die, and 3 to 5 million new cases arise.

Incidence in the United States In the United States the incidence of tuberculosis in the general population decreased by nearly 6 percent per year from 1953, when uniform national reporting began, to 1986. Temporary increases in the 1980s were caused by Asian and Haitian refugees, many of whom became infected in crowded unsanitary camps and escape boats. Disproportionately large numbers of resident nonwhite groups—blacks, Eskimos, Native Americans, and Hispanics—suffer from tuberculosis, sometimes in epidemic proportions (Figure 21.9). In 1986 the incidence increased by 2.6 percent. Many new cases occurred among AIDS victims; others appear to represent reactivations of old infections triggered by immunodeficiency, crowding, stress, and use of anti-inflammatory drugs.

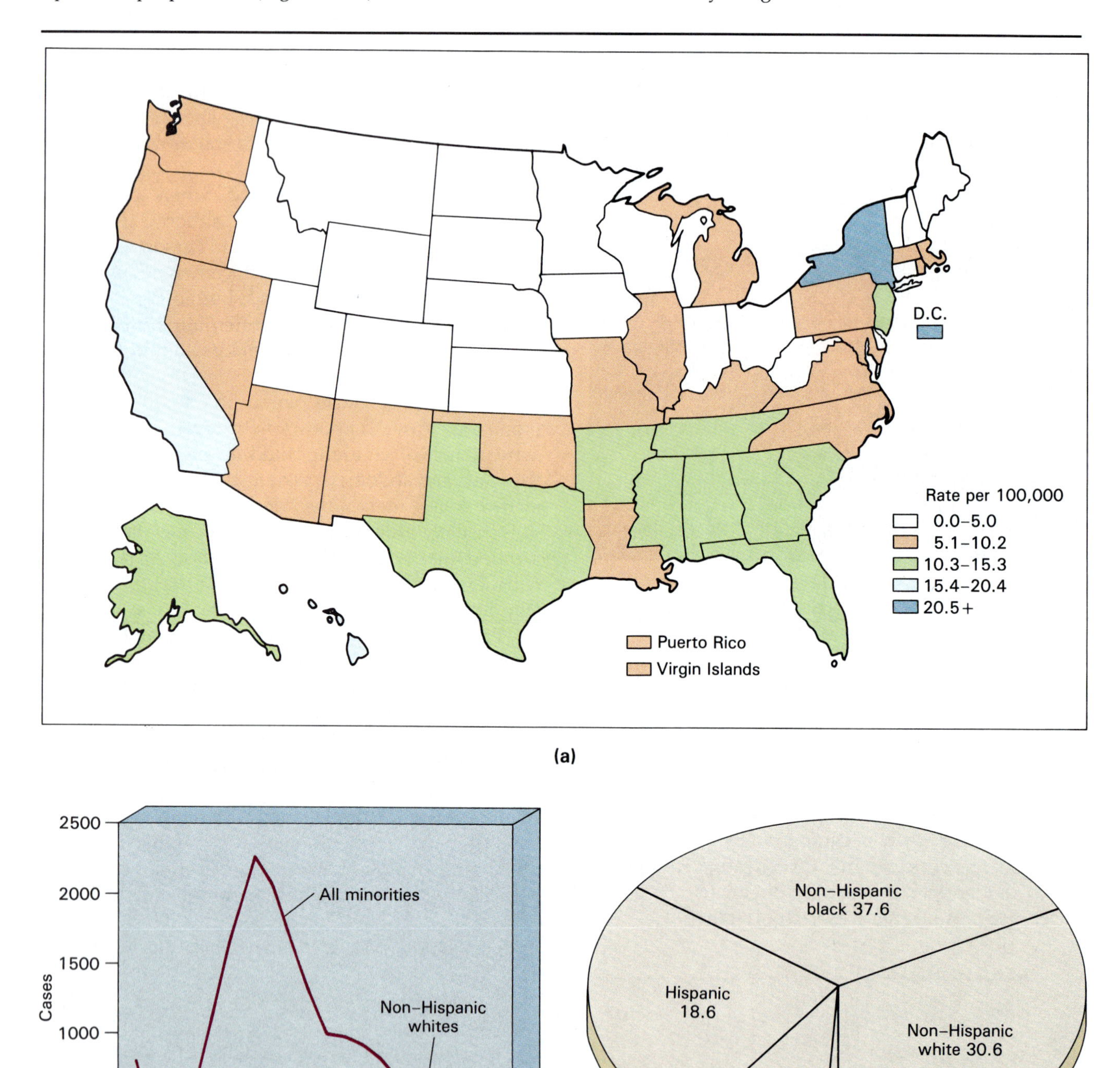

FIGURE 21.9 Incidence of tuberculosis in the United States among white and minority populations, according to age, 1990.

Causative Agents The causative agents of tuberculosis are members of the genus *Mycobacterium*, with *M. tuberculosis* causing the vast majority of cases (Table 21.2). *M. tuberculosis* was discovered by Robert Koch in 1882, when the disease was called the "White Plague" of Europe. Certain other agents, referred to as atypical mycobacteria, also cause tuberculosis, especially *Mycobacterium avium* in AIDS patients. All can be stained with the acid-fast technique, as illustrated earlier in Figure 3.27. (Chapter 3, p. 67)

TABLE 21.2 *Mycobacteria* that cause human disease

Species	Disease
M. tuberculosis	Tuberculosis
M. avium	Tuberculosislike disease in humans, transmitted from birds and swine
M. bovis	Tuberculosis, transmitted from cattle; could be transmitted from subhuman primates
M. fortuitum complex	Wound infections
M. kansasii	Tuberculosislike disease
M. leprae	Hansen's disease (leprosy)
M. marinum	Cutaneous lesions in humans, tuberculosis in fish
M. ulcerans	Ulcerative lesions

Certain properties of mycobacteria are closely associated with their role in tuberculosis. Cell wall lipids that allow mycobacteria to resist Gram staining also protect them from some host defenses. Being obligate aerobes sensitive to slight decreases in oxygen concentration, they grow best in the apical, or upper portions of the lungs, which are the most highly oxygenated. Their exceedingly long generation time (12 to 18 hours, compared with 20 to 30 minutes for most bacteria) contributes to the insidious onset of the disease. They take up to 3 weeks to produce a visible colony on laboratory media and longer to elicit symptoms in patients. Mycobacteria are highly resistant to drying and can remain viable for 6 to 8 months in dried sputum, a property that contributes to public health problems. They are, however, quite sensitive to direct sunlight.

The Disease Tuberculosis is acquired by inhaling droplets of respiratory secretions or particles of dry sputum. Young children and elderly people are particularly at risk, so screening school, day-care, and nursing home workers for tuberculosis is important. After inhalation, organisms multiply very slowly and accumulate within white cells that have phagocytized them. Eventually they kill the phagocyte, and rupture of the dead phagocyte releases infective organisms. No toxins are produced. As additional cells are infected, an acute inflammatory response occurs. A large quantity of fluid is released, especially in lung tissue, where it produces pneumonialike symptoms. Lesions sometimes heal, but more often they produce massive tissue necrosis or solidify to become chronic granulomas, or **tubercles.** Lesions that have a "cheesy" appearance are called **caseous lesions.** Lesions can be walled off from the rest of the lung by encapsulation when the host has sufficient resistance. Lesions near blood vessels can perforate the vessels and cause hemorrhage, which leads to bloody sputum, a major symptom of tuberculosis.

Granulomas, which are produced by inflammatory immune responses to infection, can keep viable organisms walled off for decades. When the immune system becomes impaired by age or other infections, granulomas can open and the disease can be reactivated. People with AIDS often have tuberculosis, either from new infections or reactivation of old ones. Except for initial infections among immigrants, AIDS patients, and residents of certain large cities such as Washington, D.C., most tuberculosis cases in this country are reactivations rather than first-time infections.

Although primary tuberculosis is nowadays most likely to affect the lungs, it can also affect the digestive tract. Before milk pasteurization became widespread, primary digestive tract infections were often seen, especially in children. Most were caused by *Mycobacterium bovis* transmitted in milk. By law, dairy cattle are now regularly screened for tuberculosis, and infected animals are destroyed. However, some cows escape testing, so drinking raw, unpasteurized milk is a health risk. Today tuberculosis of the digestive

When cortisone is used to treat the inflamed arthritic joints of aging patients, the drug also exerts an anti-inflammatory action on granulomatous lesions of tuberculosis. By counteracting factors that keep the lesions walled off, the drug can reactivate the tuberculosis organisms. Most elderly patients today were exposed to tuberculosis in their youth, and many have granulomatous lesions, although their disease was never diagnosed. Bringing relief to their aching joints also can bring active tuberculosis back to their lungs. Those who care for children or spend lots of time with their own grandchildren can unknowingly infect the children. Children, being more susceptible than adults to tuberculosis, soon display signs of primary tuberculosis.

tract usually occurs as a secondary infection site, as when germ-laden sputum coughed up from a primary lung infection is swallowed.

If lesions erode, tuberculosis can spread through blood and lymphatic channels to any part of the body. **Miliary tuberculosis** invades all tissues and forms tiny lesions that resemble millet seeds—hence the name miliary. The prognosis is poor in miliary tuberculosis. Tuberculosis of the bones can cause extensive erosion, especially in the spine (Figure 21.10). The urogenital tract, meninges, lymphatic system, and peritoneum also are prone to develop extrapulmonary tuberculosis. In **disseminated tuberculosis,** now frequently seen in AIDS patients, infected cells become casts: As the tubercle bacillus multiplies in a cell, the cell's organelles and membranes are destroyed, and a clump of microorganisms in the shape, or cast, of the former cell remains. When this process occurs in the intestine, a replica of the intestine consisting of live *Mycobacterium* can be seen.

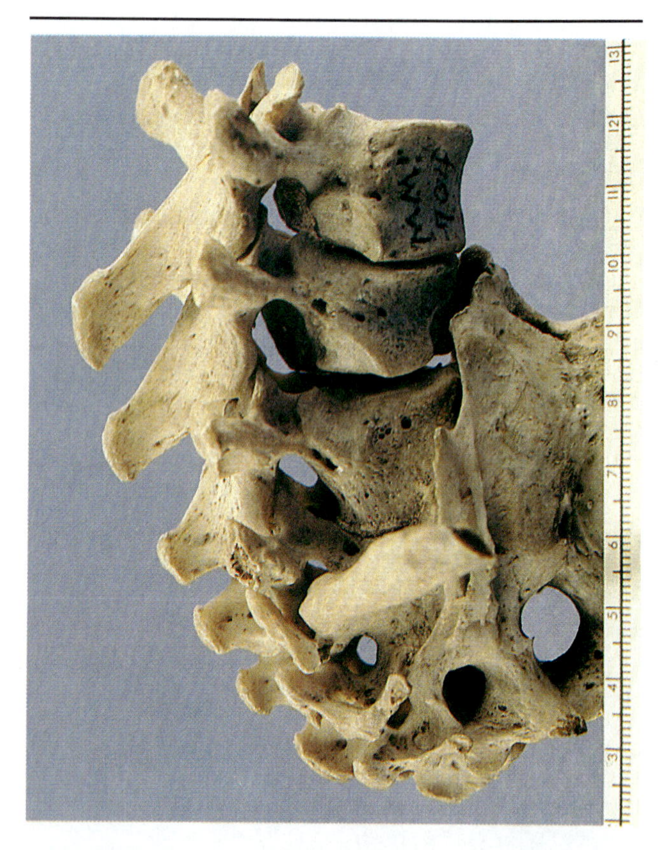

FIGURE 21.10 Tuberculosis of the spinal column. Note fusion of several of the lower vertebrae due to damage by infection.

Diagnosis, Treatment, and Prevention Tuberculosis can be diagnosed by sputum culture, but because the organisms grow very slowly, cultures must be kept for at least 8 weeks before they are declared negative. Chest X-ray examinations fail to reveal lesions outside the lungs and detect only relatively large ones within them. Therefore, screening is now done by skin tests

PUBLIC HEALTH

The New Tuberculosis Epidemic

New York City and the New York State prison system are experiencing severe difficulties with soaring tuberculosis rates in substance abusers and HIV-positive individuals. In November 1991, the front page of the *New York Times* announced, "TB out of control in N.Y. City," and the state prisons began mandatory tuberculosis testing. Thirteen inmates, all HIV-positive, and an HIV-free guard who had been assigned to guard several of them, died of multidrug-resistant tuberculosis (MDR-TB). The guard's immune system was depressed because of cancer. MDR-TB is dangerous even to people with normal immune systems; 40 percent of those given aggressive drug therapy still die. New York City hospitals have recently found that 35 percent of their tuberculosis cases are caused by MDR-TB strains.

New prison inmates show a 25 percent rate of positive tuberculosis skin tests. However, these statistics are not reliable, because the immune systems of people with HIV might be too depressed to respond to the skin test. Mandatory HIV testing is prohibited, so prison officials cannot know whose test results may represent false negatives. Meanwhile, HIV-positive inmates are at high risk of acquiring tuberculosis from such cases. HIV-free prisoners locked in close confinement with people who have tuberculosis also run a significant risk. Guards worry about acquiring tuberculosis and taking it home to their families and other people outside the prisons.

Prison infirmaries are being remodeled to provide quarantine wards. But what about MDR-TB cases among people who are not in prison? Drug-resistant strains require at least 18 months of continuous drug treatment. Patients such as drug abusers, alcoholics, the homeless, and mentally disturbed people cannot be relied on to take their medication as directed. Some New York social workers are now assigned to visit clients at home to supervise their pill swallowing. They may need several pills of as many as four different drugs each day for 2 months, followed by such doses several times a week for at least a year, or more. The cost of supplying such supervision is immense. Worse yet, however, is the cost of letting MDR-TB run rampant. Thus, New York health authorities are discussing reopening tuberculosis sanatoriums, with locked wards for those who cannot be trusted otherwise.

Homeless shelters in some areas have begun refusing HIV-positive clients. They are afraid that many of these clients also are infected with tuberculosis and will spread it to others. They also fear that HIV-positive people who are not infected with tuberculosis will acquire it from undiagnosed cases among other shelter occupants.

CLOSE-UP

Fashionable Tuberculosis

In the late nineteenth century highly intellectual and creative individuals were thought to be especially susceptible to tuberculosis. The merits of poets, artists, and musicians were validated by their deaths in unheated attics. Death from tuberculosis was a prominent theme in opera—the heroines of *La Traviata* and *La Bohème* both die of it. Much art of the period shows pale, lethargic women reclining across sofas, with the bright pink cheeks that marked tuberculosis. Indeed, tuberculosis became a fashionable disease and set the standards for beauty. The buxom, healthy energetic woman was considered less beautiful than the pale, emaciated, languorous woman.

rather than X-ray examinations. In a skin test, a small quantity of **tuberculin,** a waxy substance from the mycobacterium wall, is injected intracutaneously and examined 48 to 72 hours later for an *induration*, or a raised, not necessarily reddened bump. Induration is a delayed hypersensitivity reaction to tuberculin. A positive skin test indicates previous exposure and some degree of immune response. It does not indicate that the person now has or ever has had tuberculosis—the immune response may have prevented infection or eliminated or walled off organisms. In fact, hospital workers, teachers, and other people who have frequent skin tests may eventually develop a positive response due to sensitization to the test material itself.

Treatment is with isoniazid and rifampin for at least 1 year. Many strains of *Mycobacterium*, however, particularly atypical species often found in AIDS patients though rarely in healthy individuals, are now resistant to isoniazid and must be treated with a "second-line" or sometimes even a "third-line" drug.

Tuberculosis can be prevented by vaccination with attenuated organisms in the vaccine **BCG,** or "bacillus of Calmette and Guerin," which these workers developed at the Pasteur Institute in Paris. BCG immunization is widely practiced in other parts of the world but not in the United States, as explained in Chapter 18. (p. 497)

Ornithosis

In the early 1900s, **psittacosis** (sit-ah-ko'sis), or parrot fever, a disease associated with psittacine birds, such as parrots and parakeets, was discovered. Today at least 130 different species of birds, such as ducks, chickens, and turkeys, are found to carry this disease. Because most of these birds are not psittacines, the disease is now called **ornithosis.** Wild and domestic birds are infected, often without showing symptoms. However, stresses such as overcrowding, chilling, or shipping to pet shops can activate the disease. It is caused by *Chlamydia psittaci* and is spread by direct contact, nasal droplets, and feces. Organisms can be found in every organ of an infected bird. The birds have diarrhea and a mucopurulent discharge from nose and mouth.

Humans usually acquire ornithosis from birds, and poultry-plant workers are especially susceptible. Human-to-human transmission also has been documented, and medical personnel caring for patients can contract the disease. Organisms are inhaled and

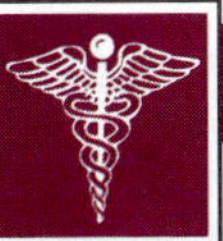

PUBLIC HEALTH

Parrot Smuggling and Ornithosis

Unlike parakeets, parrots do not breed easily in captivity and so are imported from jungles. The cost of capture, shipping to the United States, and holding in quarantine to detect such diseases as ornithosis is high, so a parrot may cost $3000. To increase profits unscrupulous dealers have stolen and smuggled so many parrots that the parrot population has been decimated in many jungle areas. And most birds don't survive the conditions of smuggling—drugged, wrapped so they cannot flap, tied underneath automobile fenders, exposed to the dust of unpaved roads and exhaust fumes. Thousands of parrots die this way each year. A few birds survive and reach pet shops without having been quarantined. If they are infected when they reach the hands of a legitimate pet dealer, they infect the entire stock and all birds must be destroyed.

Bird being held at a U.S. Department of Agriculture quarantine station before being allowed to enter the country. The cage's air supply is being filtered to prevent the spread of any airborne diseases that the animal may have.

spread systemically to the lungs and reticuloendothelial system, especially Kupffer's cells. Most cases of ornithosis are mild and self-limiting, but some patients develop a serious pneumonia. After an incubation time of 1 to 2 weeks, onset of symptoms is sudden, with sore throat, coughing, difficulty in breathing, headache, fever, and chills. This disease is difficult to distinguish from other pneumonias on a clinical basis. Definitive diagnosis is by inoculation into tissue culture. However, the organism is so highly infective that this should be undertaken only by very experienced technicians. With tetracycline treatment the mortality rate of ornithosis pneumonia is about 5 percent; in untreated cases it is around 20 percent. No vaccine is available, so prevention involves strict quarantine regulations for birds entering the United States.

Q Fever

Q fever was first described in Queensland, Australia, but the Q stands not for Queensland but for "query," because determining which organism caused it remained a question for a long time. It is now known to be caused by *Coxiella burnetii*, an organism included among the rickettsias. Unlike other rickettsias, this organism survives long periods outside cells and can be transmitted aerially as well as by ticks. For a long time it was a mystery how Q fever organisms can survive 7 to 9 months in wool, 6 months in dried blood, and over 2 years in water or skim milk. The mystery now appears to have been solved. *C. burnetii* has two forms, called large and small cell variants (Figure 21.11). Electron microscope studies suggest that a type of endospore may be formed at one end of the large cell variant, thereby giving resistance uncharacteristic of other rickettsias.

C. burnetii exist all over the world, especially in cattle- and sheep-raising areas. Wild animals and domestic sheep and cattle are the normal hosts of *C. burnetii*, and it is transmitted among them via tick bites. However, humans become infected by inhaling aerosol droplets from infected domestic animals, which usually do not appear ill. Farmers become infected while attending a cow giving birth or miscarrying if the placenta is laden with organisms. (In Canada recently a dozen cardplayers all contracted Q fever after an infected pet cat had a litter of kittens in the same room.) Slaughterhouse and tannery workers become infected by inhaling dried tick feces from the hides of animals. A third means of transmission to humans is by ingestion of milk from contaminated animals. In areas such as Los Angeles, an estimated 10 percent of the cows shed organisms into their milk. Rates are lower in some other states. Flash pasteurization (Chapter 13) eliminates this hazard. ∞ (p. 337) In fact, flash pasteurization was designed to kill Q fever organisms without giving the milk a "cooked" flavor.

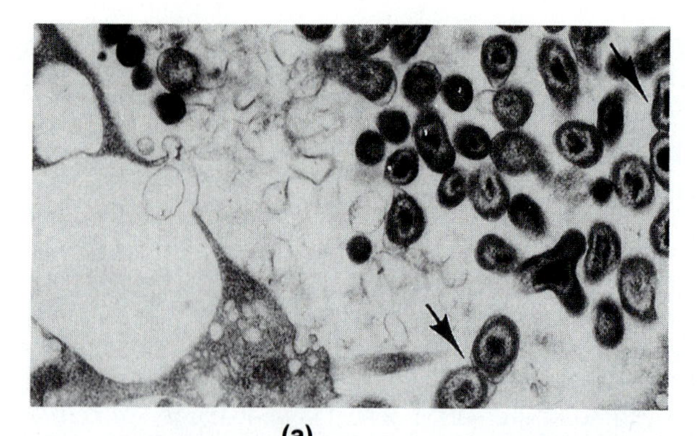

(a)

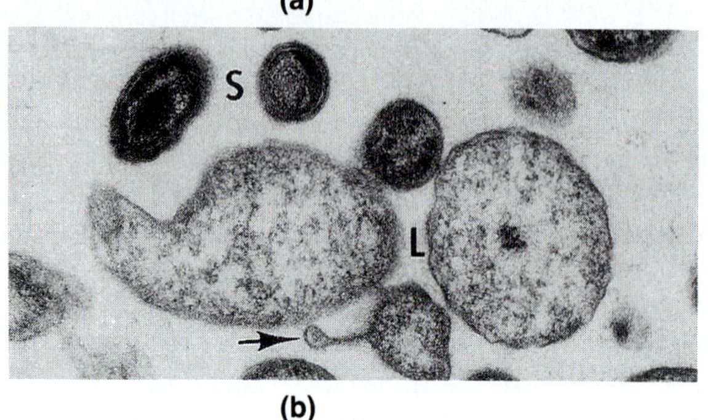

(b)

FIGURE 21.11 (a) *Coxiella burnetii* multiplying inside a phagolysosome (note the dividing cells). (b) Small and large cell forms. The cell walls of the large forms have less peptidoglycan with no cross-links—hence the diversity of cell shapes.

Symptoms of Q fever—chills, fever, headache, malaise, and severe sweats—are very similar to those of primary atypical pneumonia. The incubation period is 18 to 20 days. Diagnosis is by serologic testing or by direct immunofluorescent antibody staining. Treatment is with antibiotics such as tetracycline or fluoroquinolone. Lifelong immunity generally follows an attack of Q fever; a vaccine is available for workers with occupational exposure.

Untreated or inadequately treated Q fever can go into long periods of remission, sometimes lasting years. Cortisone treatment can reactivate it. In chronic cases endocarditis and heart valve infections are sometimes seen. Endocarditis is invariably fatal, as the organisms do not respond to antibiotic therapy.

Nocardiosis

Nocardiosis is characterized by tissue lesions and abscesses. It is caused by an acid-fast staining filamentous bacterium, *Nocardia asteroides* (Figure 21.12), which was first identified in 1888 as the causative agent of a disease in cattle known as farcy. In humans

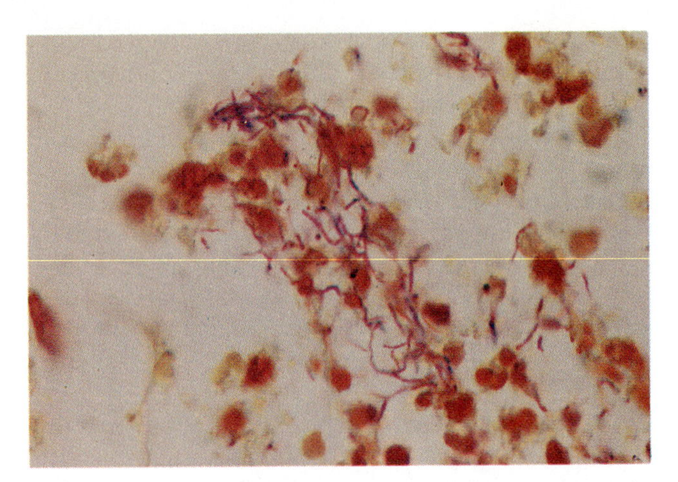

FIGURE 21.12 *Nocardia asteroides* seen as darker, branched purple filaments in this Gram-stained sample from a lung infection.

the primary infection site is the lungs in about three-fourths of all cases, but the disease also originates in the skin and in other organs. The disease usually occurs in immunodeficient patients. Mortality rates as high as 85 percent have been reported, but with early diagnosis and aggressive treatment with sulfonamide in combination with trimethoprim, mortality can be kept below 50 percent.

Viral Lower Respiratory Diseases

Influenza

Influenza is the last great plague from the past. Since Hippocrates described an influenzalike outbreak in 412 B.C., repeated influenza epidemics and pandemics have been recorded, even in recent times (Figure 21.13). One of the greatest killers of all time was the pandemic of swine flu (also known as Spanish flu) of 1918–1919 when 20 to 40 million people died. (See the Essay at the end of this chapter.) The causative virus was unusually virulent, and the crowded, unsanitary conditions created by World War I increased the spread of the virus.

Causative Agents Influenza is caused by **orthomyxoviruses.** These viruses have a surface antigen called hemagglutinin, which is responsible for their infectivity. (Chapter 11, p. 264) It attaches specifically to a receptor on erythrocytes and other host cells. Some influenza viruses also have an enzyme called neuraminidase, which helps the virus to penetrate the mucus layer protecting the respiratory epithelium and also plays a role in the budding of new virus particles from infected cells. These properties

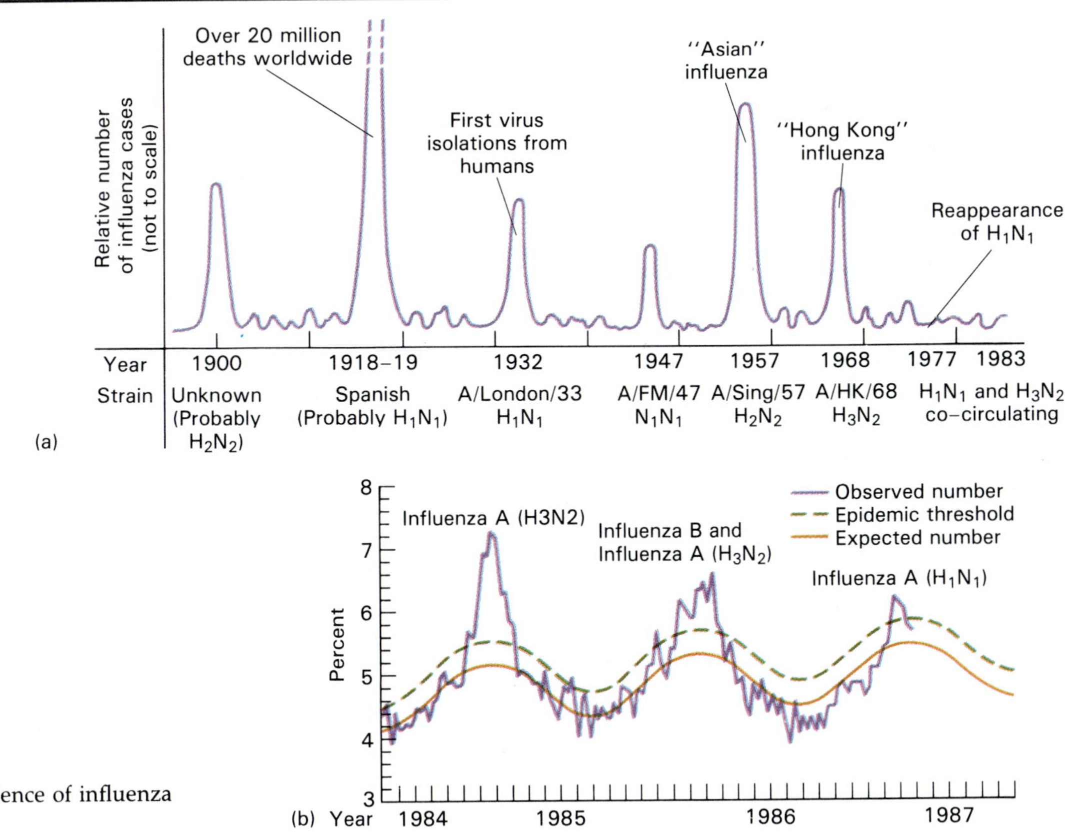

FIGURE 21.13 Incidence of influenza in the United States.

allow identification of influenza viruses in the laboratory (Figure 21.14).

Influenza virus type A, first isolated in 1933, causes epidemics and pandemics at irregular intervals because it periodically undergoes antigenic changes of varying degrees. Various species of birds and mammals are infected with strains of influenza A and may serve as reservoirs of infection.

Influenza B viruses also undergo antigenic changes, but these are less extensive and occur at a slower rate. Epidemics caused by B viruses are limited geographically and tend to center around schools and other institutions. These viruses are found only in humans.

Influenza C viruses are structurally different from the A and B viruses. Infections with C viruses are rarely recognized. When the disease is recognized, it is typically limited to the children in a single family or single classroom. The low infectivity of C viruses, which lack neuraminidase, suggests this enzyme may enhance infectivity in viruses that have it.

Lipid bilayer
from host cell
RNA
Matrix
protein
membrane
from
virus
Hemagglutinin
spike
Neuraminidase
spike

(a)

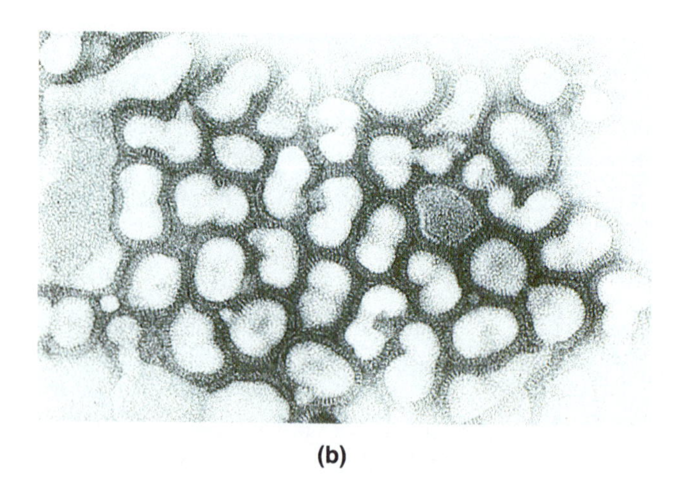

(b)

FIGURE 21.14 (a) A drawing of the influenza virus, showing hemagglutinin and neuraminidase spikes on its outer surface and RNA in the core. (b) EM of influenza virus type A.

Antigenic Variation **Antigenic variation** in influenza viruses occurs by two processes, antigenic drift and antigenic shift. **Antigenic drift** results from mutations in genes that code for hemagglutinin and neuraminidase. Mutations occur continuously, so a diverse assortment of antigens exists. Among recovered patients a similarly diverse assortment of antibodies exists. When the antigens are altered sufficiently, antibodies to their previous forms no longer confer immunity to the virus, and an epidemic can result. **Antigenic shift** probably is caused by reassortment of viral genes, although other explanations have been proposed. However it occurs, it produces viruses so different from their progenitors that antibodies against all previous strains are ineffective. (Chapter 18, p. 500) It is fortunate that antigenic shift is rare because the strains it produces cause severe pandemics. Six have occurred in the last century—in 1890, 1900, 1918, 1957, 1968, and 1977.

Immune responses vary according to antigenic changes. It is as if a contest occurs between the virus's ability to elude the immune system and the immune system's ability to recognize and inactivate the virus. The antigenic change in most viruses is small, and existing antibodies prevent the viruses from causing infections. They cause severe disease in a few people who lack antibodies, mild disease in others who have some effective antibodies, and no disease in others. The diverse assortment of antibodies that hosts acquire from exposure to several different strains tends to protect them from new strains. For a virus to cause an epidemic, it must have undergone sufficient antigenic change to elude most available antibodies.

Influenza viruses replicate in the cytoplasm of nucleated host cells, but not in erythrocytes, which lack a nucleus. The effects of influenza viruses on cells they infect are noteworthy. The viruses acquire their envelopes by budding through host cell membranes and can do so without immediately killing the host cell. (Chapter 11, p. 284) Often a host cell produces thousands of viruses per minute over many hours before the cell's macromolecules are exhausted and it dies.

The Disease Influenza, like many other viral respiratory infections, is a relatively superficial infection. The viruses enter the body through the nose or mouth and immediately invade the oropharyngeal epithelial lining. They destroy cilia of the epithelial cells and then the cells themselves. Although host cells begin regeneration immediately, it takes 10 days or more to restore an intact ciliated epithelium. Loss of the mucociliary blanket, ordinarily a major host defense, allows bacterial invasion and enhanced adherence of bacteria to virus-infected cells. (Chapter 17, p. 448) Impaired phagocytosis and accumulation of fluid in the lungs add to the risk of secondary bacterial infections, especially pneumonia. Death can result from influenza alone, secondary bacterial infection alone, or a combination. Antibiotics reduce the risk of death from bacterial infections, but they have no effect on viral infections. (Chapter 14, p. 371)

In a nonimmune person, disease symptoms begin to appear 36 to 48 hours after infection. Fever, malaise, and muscle soreness are the most common symptoms; cough, nasal discharge, sore throat, and gastroenteritis also occur frequently. The fever usually lasts about 3 days. As systemic symptoms decrease, respiratory symptoms increase. The severity of the disease is directly proportional to the quantity of viruses released from cells. Release of viruses starts as the first malaise appears and peaks about a day before maximum fever and interferon production occur. Viral shedding usually ceases within 8 days of initial exposure. Although the acute phase of the illness is over in about a week, fatigue, cough, and weakness may persist over several weeks.

Incidence and Transmission In northern temperate zones, influenza appears in late November or early December and disappears in April. In most years the greatest number of cases are seen between January and mid-March. In any one area influenza has a 5- to 7-week period of high prevalence. Indoor crowding, poor air circulation, and dry air expedite the spread of the virus. Schools furnish almost ideal conditions for influenza transmission and usually are the focal points of outbreaks.

Diagnosis and Treatment The best specimens for isolation of viruses are throat swabs taken as early in the illness as possible. The viruses can be cultured in embryonated chick eggs and various cell lines and identified by hemagglutination inhibition and immunofluorescent antibody tests. Moderately effective treatment is now becoming available for influenza. The drug amantadine blocks influenza A virus replication, probably by interfering with uncoating. (Chapter 14, p. 371) It is useful for short-term protection of selected individuals but impractical to use for the duration of an influenza epidemic unless patients are especially compromised. Furthermore, it prevents users from developing antibodies to the virus and has some unpleasant side effects. Ribavirin has been demonstrated experimentally to inactivate both A and B viruses.

PUBLIC HEALTH

Flu Vaccines

Each year around June the CDC selects the strains to be used as vaccine for the coming flu season according to what strains are most prevalent and what alterations they have undergone. Strains are named for the geographic location where they were first isolated, a number assigned by the laboratory that isolated them, the year of isolation, and—for type A viruses—the numbers of their hemagglutinin (H) and neuraminidase (N). The 1991 vaccine contained three strains: influenza A/Taiwan/1/86 (H_1N_1), influenza A/Beijing/353/89 (H_3N_2), and influenza B/Panama/45/90. The vaccines are inoculated into ferrets and chickens, and the vaccines and the antibodies they produce are thoroughly tested before they are given to humans.

Immunity and Prevention Because of influenza viruses' mutability, or ability to mutate, annual immunization is recommended. The vaccine provides good protection against influenza, and annual vaccination increases the diversity of the recipient's antibodies. Even less frequent immunization can provide some degree of protection. Although immunized individuals sometimes become infected with flu viruses, they generally have a milder disease of shorter duration than nonimmunized individuals. Furthermore, they shed fewer viruses over a shorter duration of time. Thus, immunization of some people tends to reduce the spread of the disease to other people.

A human gene that confers resistance to influenza has recently been identified at the University of California in Santa Barbara. This gene is turned on by interferon. It produces a specific protein, called the Mx protein, that prevents the virus from making viral RNA and proteins.

Respiratory Syncytial Virus Infection

The **respiratory syncytial virus** (RSV) is the single most important cause of lower respiratory tract infections in children under 1 year and especially in male infants 1 to 6 months old. The virus derives its name from the fact that it causes cells in cultures to lose their cell membranes and independent identity and become multinucleate masses, or **syncytia** (sin-sish'ya). The disease, a kind of **viral pneumonia,** begins with a 3- or 4-day period of fever as the virus infects the respiratory tract; this is followed by hyperventilation, hyperinflation, and infiltration of the lungs with fluid. If an infant with RSV is placed on a rapid pulsing respirator, the infant can shed enough virus to infect an entire intensive care nursery. Major outbreaks of this nature generally result in fatalities. Viruses are released for weeks to months after infection, and reinfection is common. In older children and adults RSV affects mainly the upper respiratory tract, and adults who carry the virus in their noses can be the source of nursery infections. The incidence and transmission of RSV is quite similar to that for parainfluenza viruses, and simultaneous infection by both kinds of viruses is not uncommon. The virus can be identified by enzyme immunoassay performed on nasal secretions or a direct immunofluorescent test on cells from such secretions.

Acute Respiratory Disease

Acute respiratory disease (ARD) ranges from mild to severe. Respiratory symptoms such as sore throat, cough, and other cold symptoms, fever, headache, and malaise are frequently seen. Severe viral pneumonia that lasts about 10 days sometimes occurs in ARD.

Cases of viral ARD have been seen in military training facilities in the United States and in Europe, and epidemics usually start 3 to 6 weeks after the start of training. The epidemic nature of the disease has been attributed to crowding of people from different geographic areas under stressful conditions. However, epidemics have not been observed in colleges and other institutions where people live under similar conditions, so other factors may be involved in such epidemics.

Adenoviruses cause about 5 percent of cases of ARD in children under 5 years old. Usually the symptoms are mild and nonspecific—stuffy nose, cough, and nasal discharge. Sometimes more severe symptoms such as tonsillitis, pharyngitis, bronchitis, bronchiolitis, croup, and often conjunctivitis and abdominal pain also appear. Adenovirus pneumonia accounts for about 10 percent of all childhood pneumonia and is occasionally fatal.

Fungal Respiratory Diseases

Compared with bacteria and viruses, fungi are much less frequent causes of respiratory diseases. Most fungal infections are seen in immunodeficient and debilitated patients. Blastomycosis, usually a skin disease (Chapter 20), can also cause mild respiratory symptoms.

Coccidioidomycosis

The soil fungus *Coccidioides immitis* causes **coccidioidomycosis,** or valley fever (Figure 21.15). This organism is found mainly in warm, arid regions of the southwestern United States and Mexico. Infection usually occurs as a result of inhaling dust particles laden with arthrospores. Dogs, cattle, sheep, and wild rodents deposit spores in feces, and the spores easily become airborne. Susceptible individuals have been known to become infected from merely standing on the platform at a train station for a short period of

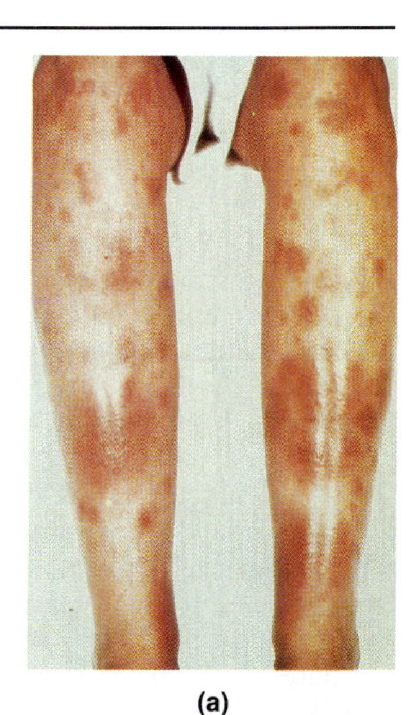

(a)

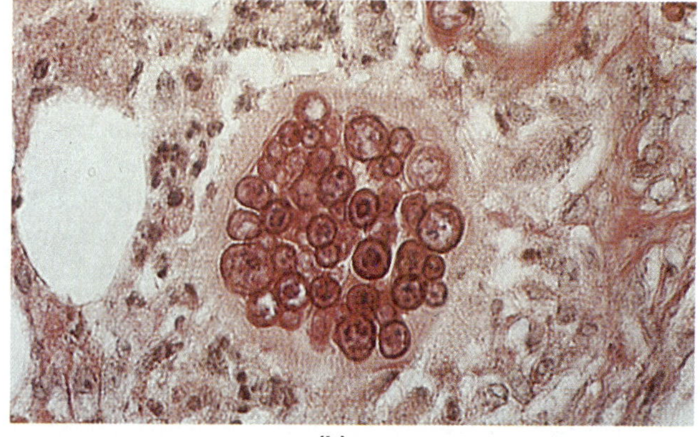

(b)

FIGURE 21.15 (a) The skin rash of coccidioidomycosis on legs. (b) Tissue section showing a spherule containing many endospores.

APPLICATIONS

Student Observers Become Participants

On one occasion a victim of coccidioidomycosis underwent joint surgery to repair damage from the infection. Many members of the medical staff were eager to observe the effect of the surgery, and several students were required to attend. All gathered around the patient while a plaster cast was being removed. Fluid had oozed into the cast, and the fungi in the fluid had reacted to the new environment by forming spores. (These organisms do not form arthrospores in living tissue but only in nonliving materials such as soil or plaster.) As the cast was broken open, airborne spores were released, and most of the spectators were infected with coccidioidomycosis.

time to get a breath of "fresh air." When spores are inhaled by susceptible individuals, an influenzalike illness results. Coccidioidomycosis is always highly infectious, and it can be self-limiting or progressive. In fewer than 1 percent of these individuals, dissemination to the meninges or the bones occurs within 1 year of the initial infection. Dissemination is much more common in blacks than in whites.

Spherules containing yeastlike endospores (reproductive spores that are not ordinarily infectious) can be found in sputum, pus, spinal fluid, or biopsied tissue, and the fungus forms fluffy, white colonies in culture. Various immunological tests are available to aid in diagnosis, but some give false positive results in individuals who have antibodies against other fungal diseases. Intravenous amphotericin B is sometimes successful in treating the disseminated disease; if treatment fails the disease is often fatal. Prevention is difficult, but reducing dust in endemic regions may be helpful. A vaccine is being developed for coccidioidomycosis.

Histoplasmosis

The soil fungus *Histoplasma capsulatum,* which causes **histoplasmosis,** or Darling's disease, is endemic to the central and eastern United States but is found globally in major river valleys. The highest incidence is seen in the Mississippi and Ohio valleys, where 80 percent of the population shows immunological evidence of exposure to the fungus. The organism thrives in soil mixed with feces and especially in chicken houses and in caves containing bat guano (feces). Cave dust is responsible for so many cases that the disease is sometimes referred to as cave sickness.

The fungus enters the body by inhalation of spores. These spores are engulfed, but not killed, by macrophages and thus travel in macrophages throughout the body. Person-to-person spread does not occur. Although most infections do not produce disease symptoms, granulomatous lesions occur in the lungs and spleen of susceptible individuals. Inhaling large numbers of spores can cause pneumonia. In certain individuals, especially the very young, very old, or those receiving immunosuppressants, the organism can invade the spleen, liver, and lymph nodes. Anemia, high fever, and often death result.

Diagnosis is made by microscopic identification of small ovoid cells of the organism inside human cells. When cultured at body temperature, the organisms look like budding yeasts, and at room temperature, they resemble molds. Supportive therapy is used for pulmonary histoplasmosis, and amphotericin B is sometimes effective in treating the disseminated disease. It is very easy for humans to become infected in environments such as chicken houses and caves, where the air is laden with spores. Some employers hire only individuals with positive histoplasmosis skin tests, and therefore immunity to histoplasmosis, to work in high-risk environments. Spraying infected soil and fecal deposits with 3 percent formaldehyde destroys some of the spores.

PUBLIC HEALTH

Histoplasmosis

In the spring of 1984 the Virginia State Health Department sealed off an old Dinwiddie County house after one man died and nine others who had worked on renovating the house were diagnosed as having histoplasmosis. The source of the infection was traced to a pile of old plaster, which had been removed from the house. Similar outbreaks of the disease have been seen among Boy Scouts who had worked industriously to clear their towns of pigeon droppings.

Cryptococcosis

Cryptococcosis is caused by a budding, encapsulated yeast, *Cryptococcus neoformans*. Spores typically enter the body through the skin, nose, or mouth. Birds carry the fungi on their feet and beaks. Although birds do not suffer from cryptococcosis, they do disseminate the spores, and the spores thrive on the nitrogenous waste creatinine, which is present in high concentration in bird feces.

Cryptococcosis is usually characterized by mild symptoms of respiratory infection, but it can become systemic if large quantities of the spores are inhaled by debilitated patients. This fungus often becomes disseminated to meninges, which become thickened and matted, and it can invade brain tissue. As with other fungal opportunists, cryptococcosis is increasing in incidence among AIDS patients.

Observing the organism in body fluids confirms the diagnosis. Flucytosine and amphotericin B can be used in combination to treat systemic disease. Because the organisms thrive in pigeon droppings, some degree of prevention can be achieved by reducing pigeon populations and decontaminating droppings with alkali.

Pneumocystis Pneumonia *Pneumocystis carinii* (Figure 21.16), long thought to be a protozoan of the sporozoan group but now thought to be a fungus, is an opportunistic organism. It invades cells of the lungs and causes alveolar septa to thicken and the epithelium to rupture. Then the parasites and a foamy exudate from cells collect in the alveoli. The disease, called ***Pneumocystis* pneumonia,** occurs in infants, the elderly, and those with compromised immune systems. The marked increase in the incidence of this disease in recent years is due mainly to its ability to infect persons with AIDS.

Diagnosis is made by finding organisms in biopsied lung tissue or bronchial lavage (washings from the bronchial tubes). Because little is known about the life cycle or mode of transmission of this organism, it is impossible to devise effective control measures.

Parasitic Respiratory Disease

Paragonimus westermani (Figure 21.17), a lung fluke, is found in many parts of Asia and the South Pacific. Eggs released in feces into water hatch and parasitize first a snail and then a crab or crayfish, which is eaten by the definitive host. When humans eat these infected animals the metacercariae, or fluke larvae, leave the crab or crayfish as it is digested in the human's small intestine. The larvae then bore through the intestine and embed in the abdominal wall temporarily. Soon they leave it and penetrate the diaphragm and membranes around the lungs to reach the bronchioles. The larvae mature into adults and lay eggs in the bronchioles. When the host coughs the eggs move into the pharynx, are swallowed, and exit the body through the feces. Infected humans have a chronic cough, bloody sputum, and difficulty breathing. Diagnosis can be made by finding eggs in sputum or by any one of several immunologic tests. Several drugs are effective in treating lung fluke infections, but infection can be avoided by cooking crustaceans before eating them.

Diseases of the lower respiratory tract are summarized in Table 21.3.

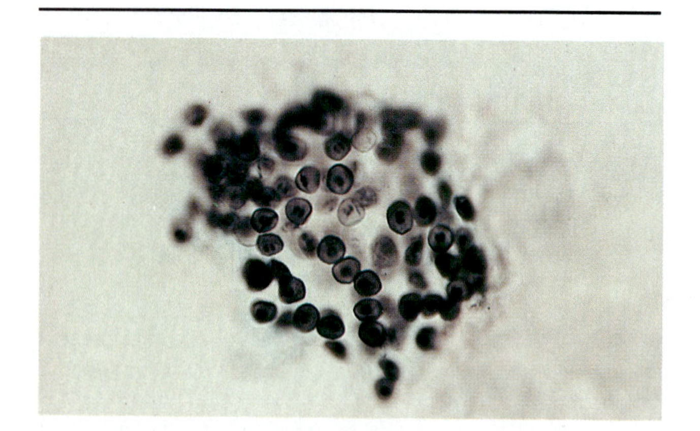

FIGURE 21.16 *Pneumocystis carinii* in sputum. The organism is a frequent cause of pneumonia in AIDS patients.

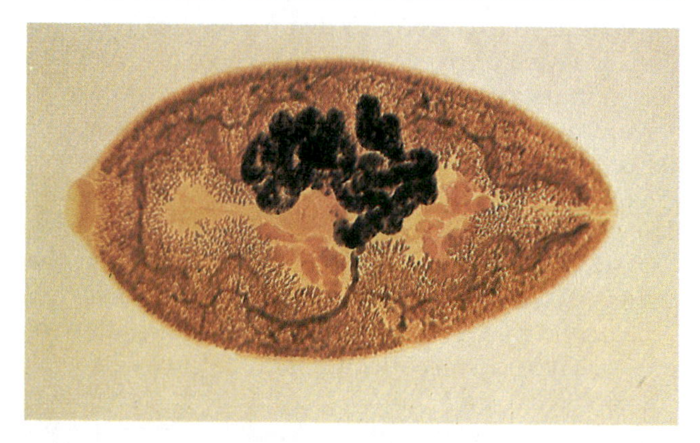

FIGURE 21.17 *Paragonimus westermani*: an adult worm stained to show internal structures.

TABLE 21.3 Summary of diseases of the lower respiratory tract

Disease	Agent(s)	Characteristics
Bacterial lower respiratory diseases		
Whooping cough	*Bordetella pertussis*	Catarrhal stage with fever, sneezing, vomiting, and mild cough. Paroxysmal stage with ropy mucus and violent cough. Convalescent stage with mild cough
Classic pneumonia	*Streptococcus pneumoniae, Staphylococcus aureus, Klebsiella pneumoniae*	Inflammation of bronchi or alveoli of lungs with fluid accumulation and fever
Mycoplasma pneumonia	*Mycoplasma pneumoniae*	Mild inflammation of bronchi or alveoli
Legionnaire's disease	*Legionella pneumophila*	Inflammation of the lungs, fever, chills, headache, diarrhea, vomiting, and fluid in lungs
Tuberculosis	*Mycobacterium tuberculosis*	Tubercles in lungs and sometimes in other tissues; organisms can persist in walled-off lesions and be reactivated
Ornithosis	*Chlamydia psittaci*	Pneumonialike disease transmitted to humans by birds
Q fever	*Coxiella burnetii*	Disease similar to mycoplasma pneumonia transmitted by ticks, aerosols, and fomites
Nocardiosis	*Nocardia asteroides*	Pneumonialike disease seen in immunodeficient patients
Viral lower respiratory diseases		
Influenza	Influenza viruses	Viruses subject to antigenic variation with new strains causing epidemics. Inflammation of oropharyngeal membranes, fever, malaise, muscle pain, cough, nasal discharge, and gastroenteritis
Respiratory syncytial virus infection	Respiratory syncytial virus	Febrile disease of the respiratory tract; can cause viral pneumonia
Acute respiratory disease	Adenoviruses	Mild cough and nasal discharge; can cause viral pneumonia
Fungal respiratory diseases		
Coccidioidomycosis	*Coccidioides immitis*	Influenzalike illness; dissemination to meninges and bones can occur
Histoplasmosis	*Histoplasma capsulatum*	Granulomatous lesions in lungs and spleen in susceptible individuals; can cause pneumonia
Cryptococcosis	*Cryptococcus neoformans*	Usually a mild pulmonary disease; pneumonia and dissemination to meninges can occur
Blastomycosis	*Blastomyces dermatitidis*	Usually a skin disease (see Chapter 20); sometimes disseminated to lungs or acquired directly by inhalation. Mild respiratory symptoms
Pneumocystis pneumonia	*Pneumocystis carinii*	Rupture of alveolar septa, foamy sputum; occurs mainly in immunodeficient patients
Parasitic respiratory diseases		
Lung fluke infection	*Paragonimus westermani*	Larvae mature in bronchioles and cause chronic cough, bloody sputum, and difficulty breathing

ESSAY

The Great Flu Pandemic of 1918

Where could Henry be? He should have been home an hour ago! In normal times, this wouldn't have been terribly alarming—but those weren't normal times. Was his body one of those piled up like stacks of cordwood on some street corner downtown?

Such were the fears that raced through the minds of frantic families during the great Spanish influenza pandemic of 1918. Seemingly healthy people dropped dead in the street, without the slightest warning. Undertakers couldn't begin to answer the demands of a pandemic that killed half a million Americans in only 10 months. (By comparison, the U.S. Civil War claimed about the same number of lives over a period of 4 years.) The city of Philadelphia recorded 528 bodies piled up, awaiting burial, in just one day. Numerous others lay undiscovered in silent houses, apartments, and other buildings. Rescue teams went knocking door to door, seeking victims who were too sick even to call out for aid. Wagons made regular rounds through the streets, stopping at corners to pick up bodies that families, neighbors, storekeepers, or passersby had carried there. No wonder families worried when someone was late getting home. Twenty-five million Americans were infected—who knew which ones would die?

Unlike many diseases, influenza took its heaviest toll among young, healthy individuals. Some died of the high fever; others succumbed to a variety of secondary bacterial infections, chiefly pneumonia. The immediate cause of death for many patients was damage to the lungs. Tremendous pressure built up in the lungs until they degenerated completely (on autopsy they were often found to have the consistency of red pudding). If a patient could relieve this crushing pressure, there was a chance of survival. In one case, a physician was called to the bedside of a teen-aged girl. Little could be done except to comfort the parents as they stood at the foot of her bed. Suddenly a stream of blood shot from her nose, arched across the bed, and soaked the trio. Projectile hemorrhaging had reduced the pressure in her lungs, and she lived. But people who survived cases where the symptoms included high fever often developed Parkinson's disease. Today we think of Parkinson's disease as typical of an older population. But after the 1918 pandemic, it affected large numbers of all ages, including many small children.

What was the cause of this grim experience? It was something too tiny for the researchers of 1918 to see—the swine flu virus. What they did see with their light microscopes were various secondary bacterial invaders. (One of these bacilli was actually named *Haemophilus influenzae* in the mistaken belief that it was the cause of the great influenza pandemic.) Indeed, in one New York City family, all six members were found to be infected with different species of bacteria. No wonder the medical world was confused. Attempts at vaccine production proved futile. Few people even guessed at the existence of the invisible virus.

Where had this virus come from? A new type of influenza had appeared in France during April 1918, as American and European troops fought through the final months of World War I. From France the virus traveled to Spain. That summer it

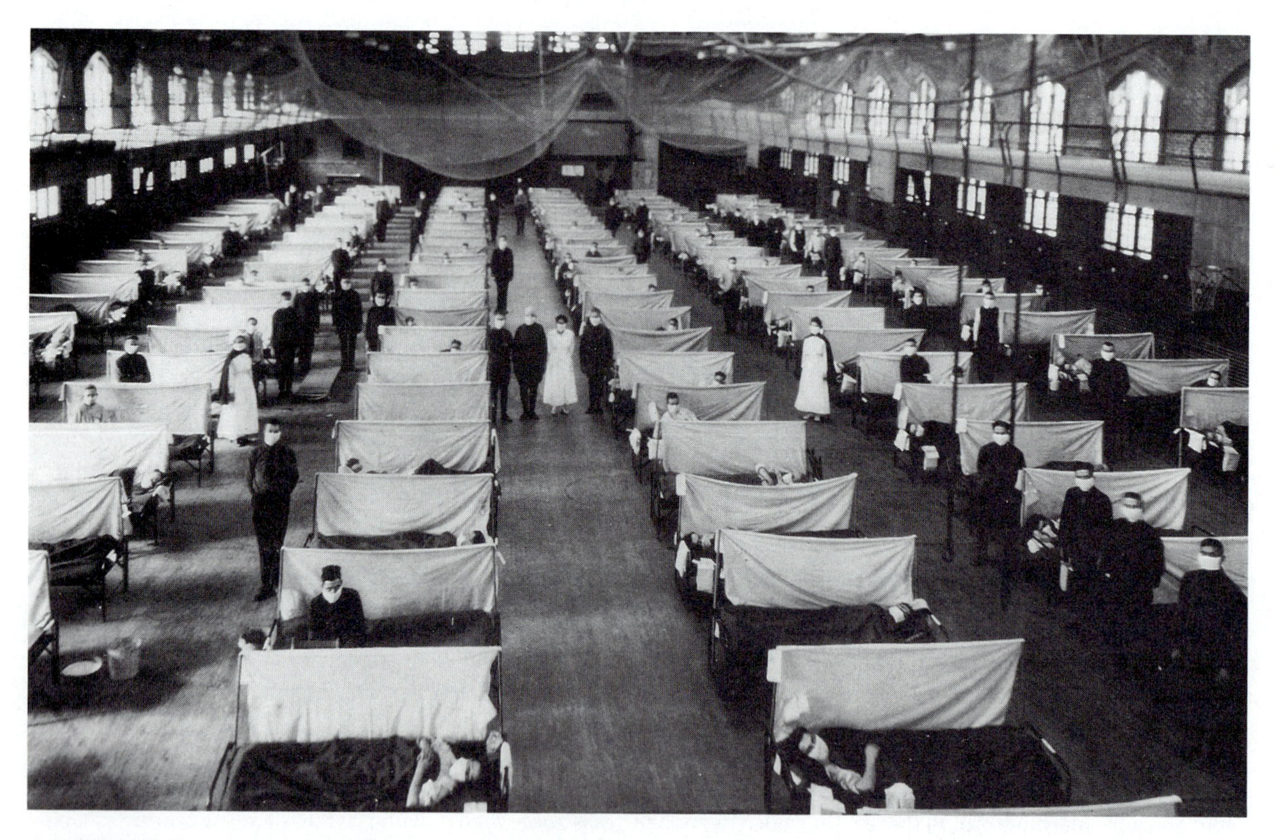

FIGURE 21.18 During the height of the great flu pandemic of 1918, the gymnasium of Iowa State University was temporarily converted into a hospital ward.

spread throughout Europe, and to far-off China and West Africa. By then, the English-speaking countries were calling it the "Spanish Lady." The Lady entered America at Boston in August and had crossed the entire country within a month. Around the world she went, causing a true pandemic. By the time the reverberations ceased in 1920, more than 25 million people had died. The country hardest hit was India, where death wiped out the entire population increase of a decade. In some parts of the world, nearly half the population died. And there is some evidence that the virus took nonhuman victims as well—baboons in South Africa, for example.

It is not certain what brought about this new deadly kind of flu. The influenza virus, which has a long history as a human pathogen, is one of the most rapidly mutating viruses known. But the natural process of mutation may also have had human help. Mustard gas (dichlorodiethyl sulfide), which was widely used in trench warfare in France, is a known mutagen—it is a powerful alkylating agent (see Chapter 7). Could it have interacted with the virus to produce the new, virulent strain? No one can be sure.

The term *swine flu* did not come into use until years after the pandemic was over. Influenza was unknown in pigs before the arrival of the Spanish flu in the United States in 1918. Thereafter, swine were found to suffer from a complex disease, involving both an influenza virus and a bacterium. The virus was believed to be the same one that caused the Spanish flu. Today, the virus that causes influenza in swine is not the same as the one that causes "swine flu" in humans. This fact does not necessarily mean that the name swine flu was a mistake, however, for one would expect the rapidly evolving virus strains to have mutated quite a bit in the years since 1918.

Is the virus that caused the great pandemic still a threat today? In 1976, the U.S. government mounted a massive immunization program against swine flu. Cases had been discovered the previous year in Asia. Flu viruses almost always follow a westward march around the globe, and it is by looking at the viruses that have caused problems to the east of us that our public health officials decide which viral strains to incorporate into next year's vaccine. So, when swine flu was detected, it sent chills up and down medical spines. Not only could there be millions of deaths—there were also the sequelae. How many Parkinson's cases would we have in survivors of a 1976 pandemic?

Nervously the government exhorted all to be immunized. Then came a disturbing development. Some persons who had received swine flu vaccine had subsequently developed Guillain-Barre syndrome resulting in complete paralysis. People pondered which would be worse: swine flu or paralysis? Lines at vaccine clinics shortened. With fewer people immunized, public health officials worried about what would happen when the swine flu arrived. The medical community watched and waited. But swine flu never came.

A reprieve, or just a postponement? No one can say.

CHAPTER SUMMARY

The agents and characteristics of the diseases discussed in this chapter are summarized in Tables 21.1 and 21.3. Information in these tables is not repeated in this summary.

RELATED KEY TERMS

DISEASES OF THE UPPER RESPIRATORY SYSTEM

Bacterial Upper Respiratory Diseases

- Infections related to **pharyngitis** include **laryngitis, epiglottitis, sinusitis,** and **bronchitis.** These diseases usually are transmitted by respiratory droplets. If severe enough they can be treated with penicillin or other antibiotics. — **tonsillitis**
- **Diphtheria,** no longer common in the United States, occurs only in humans. Both the organism and its toxin, produced by a lysogenic prophage, contribute to symptoms, which include formation of a **pseudomembrane** that can block the airway. Diphtheria is spread by respiratory droplets, treated with antitoxin and antibiotics such as penicillin, and prevented by DPT vaccine. — **diphtheroids**
- **Ear infections** occur in the middle and outer ear. Organisms reach the middle ear via the Eustachian tube and usually can be eradicated by penicillin. — **otitis media**, **otitis externa**

Viral Upper Respiratory Diseases

- The common cold is transmitted by fomites and aerosols. Treatment is limited to alleviating symptoms; no vaccine is available. — **coronaviruses**
- **Parainfluenza** infections range from inapparent to severe **croup;** most children have antibodies to these viruses by age 10. — **parainfluenza viruses**, **stridor**

DISEASES OF THE LOWER RESPIRATORY TRACT

Bacterial Lower Respiratory Diseases

- **Whooping cough,** also called **pertussis,** is distributed worldwide, but vaccine has decreased its incidence. It is transmitted by respiratory droplets and treated with antitoxin and erythromycin. Vaccine prevents the disease but also causes some severe complications and a few deaths.
- Classic **pneumonia** can be **lobar** or **bronchial;** it is transmitted by respiratory droplets and carriers. *Klebsiella* pneumonia is more severe than pneumococcal pneumonia. Penicillin is the drug of choice for pneumococcal pneumonia, and vaccine is recommended for high-risk populations. Mycoplasma pneumonia is transmitted by respiratory droplets and is treated with erythromycin or tetracycline.
- **Legionnaire's disease** is transmitted via aerosols from contaminated water and is treated with erythromycin.
- **Tuberculosis** has been a worldwide major health problem for centuries, and its incidence in the United States is increasing. It is transmitted by respiratory droplets; inactive, but viable, organisms can persist for years walled off in **tubercles.** Treatment with isoniazid is effective except for resistant organisms, which must be treated with "second-line" drugs. A vaccine is available, but it is not used in the United States.
- **Ornithosis** is transmitted to humans from infected birds; it is usually mild but can cause a serious pneumonia. The causative organism is dangerous to handle in the laboratory.
- **Q fever** is transmitted by ticks, aerosol droplets, and fomites. It is treated with tetracycline, and a vaccine is available for workers with occupational exposure.

Viral Lower Respiratory Diseases

- **Influenza** viruses display **antigenic variation.** The disease occurs mainly from December through April; it is transmitted in crowded, poorly ventilated conditions and diagnosed by immunologic tests. Vaccine can prevent the disease, but its effectiveness is lessened by antigenic variation.
- **Respiratory syncytial virus** infections and **acute respiratory disease** can lead to pneumonia.

Fungal Respiratory Diseases

- Opportunistic fungal infections occur mainly in immunodeficient and debilitated patients. They usually are transmitted by spores, and some can be treated with amphotericin B.
- In the United States **coccidioidomycosis** occurs in warm, arid regions, **histoplasmosis** is endemic in eastern states, and **cryptococcosis** occurs wherever there are infected birds, especially pigeons.
- ***Pneumocystis* pneumonia,** an opportunistic infection, is a common cause of death among AIDS patients.

Parasitic Respiratory Disease

- Lung fluke infections occur in Asia and the South Pacific where infected shellfish are eaten. The disease can be treated with drugs and prevented by cooking shellfish.

RELATED KEY TERMS

catarrhal stage
paroxysmal stage **cyanosis**
convalescent stage

consolidation **pleurisy**

primary atypical pneumonia
walking pneumonia
Pontiac fever

caseous lesions
miliary tuberculosis
disseminated tuberculosis
tuberculin **BCG**
psittacosis

nocardiosis

antigenic drift **antigenic shift**

syncytia **viral pneumonia**

QUESTIONS FOR REVIEW

A.

1. How is strep throat diagnosed and treated?
2. How do laryngitis, sinusitis, and bronchitis differ from pharyngitis?
3. What agent causes diphtheria, and how does it produce tissue damage?
4. How can diphtheria be diagnosed, treated, and prevented?
5. What are the causes and effects of ear infections?

B.

6. What agents cause colds, and how are they transmitted?
7. Why are colds difficult to treat and to prevent?
8. How are parainfluenza and croup related?

C.

9. What agent causes whooping cough, and why is the disease a world health problem?
10. What are characteristics of each stage of whooping cough, and how does toxin contribute to them?
11. How can whooping cough be transmitted, diagnosed, treated, and prevented?
12. How do lobar and bronchial pneumonia differ?
13. What are the characteristics of pneumococcal pneu-

monia, and how do other typical bacterial pneumonias differ from it?

14. How is classical pneumonia diagnosed, treated, and prevented?
15. What are the special characteristics of primary atypical pneumonia?
16. What are the special characteristics of Legionnaire's disease, and how is it treated?
17. What causes tuberculosis, and why is it a global health problem?
18. How do mycobacteria cause tissue damage, and what tissues can be affected?
19. How is tuberculosis transmitted, and what populations are most susceptible?
20. How is tuberculosis diagnosed, treated, and prevented?
21. What is ornithosis, and how is it diagnosed, treated, and prevented?
22. What is Q fever, and how is it diagnosed, treated, and prevented?
23. What is nocardiosis, and how is it treated?

D.

24. Describe influenza and the viruses that cause it.
25. How can influenza be diagnosed, treated, and prevented?
26. What kinds of viruses can cause viral pneumonia?

E.

27. Under what circumstances do fungal respiratory infections arise?
28. How can coccidioidomycosis, histoplasmosis, and cryptococcosis be distinguished?
29. What is *Pneumocystis* pneumonia, and how is it caused?
30. How do people become infected with lung flukes, and what are their effects?

PROBLEMS FOR INVESTIGATION

1. Devise a plan for reducing the incidence of upper respiratory infections.
2. A patient has *Mycoplasma* pneumonia. Should the physician prescribe penicillin? Why or why not?
3. What do you think the disease consequences would be of having lungs that are constructed in one large unit, rather than in lobes? Why?
4. Prepare a report on current worldwide efforts to control tuberculosis.
5. Look up the CDC statistics for notifiable respiratory diseases in your state. Compare these with other parts of the country. Use recent issues of the *Morbidity and Mortality Weekly Reports* to obtain these.
6. A 34-year-old man was hospitalized with respiratory symptoms of pneumonia. Culture of his sputum revealed infection by *Pneumocystis carinii*. What other condition(s) would you also suspect he has?

(The answer to this question appears in Appendix E.)

SOME INTERESTING READING

Anonymous. 1991. Tuberculosis among HIV-infected persons. *Journal of the American Medical Association* 266, no. 15 (October 16):2058–59.

Bone, R. C. 1991. Chlamydial pneumonia and asthma: a potentially important relationship. *Journal of the American Medical Association* 266, no. 2 (July 10):265.

Cassell, G. H., and B. C. Cole. 1981. Mycoplasmas as agents of human disease. *New England Journal of Medicine* 304:80.

Chazen, G. 1987. Nocardia. *Infection Control* 8(6):260.

Collier, R. 1974. The plague of the Spanish lady. New York: Atheneum.

Cousins, D. V., S. D. Wilton, B. R. Francis, and B. L. Gow. 1992. Use of polymerase chain reaction for rapid diagnosis of tuberculosis. *Journal of Clinical Microbiology* 30(1):255–58.

Edelstein, P. H. 1985. Environmental aspects of *Legionella*. *American Society for Microbiology News* 51:460.

Fields, B. N., ed. 1985. *Virology*. New York: Raven Press.

Fincher, J. 1989. America's deadly rendezvous with the "Spanish lady." *Smithsonian* 20 (January):130.

Fox, J. L. 1992. Coalition reacts to surge of drug-resistant TB. *ASM News* 58(3):135–39.

Fox, J. L. 1990. TB: a grim disease of numbers. *ASM News* 56(7):363–65.

Gilligan, P. H. 1991. Microbiology of airway disease in patients with cystic fibrosis. *Clinical Microbiology Reviews* 4(1):35–51.

Granstrom, M., A. M. Olinder-Nielsen, P. Holmblad, A. Mark, and J. Hanngren. 1991. Specific immunoglobulin for treatment of whooping cough. *Lancet* 338, no. 8777 (November 16):1230–34.

Kaufmann, A. F., J. E. McDade, and C. M. Patton. 1987. Pontiac fever: Isolation of the etiological agent (*Legionella pneumophila*) and demonstration of its mode of transmission. *American Journal of Epidemiology* 114:337.

Middleton, D. B. 1991. An approach to pediatric upper respiratory infections. *American Family Physician* 44, no. 5 (November):33S–40.

Radetsky, P. 1989. Taming the wily rhinovirus; if scientists succeed, the common cold will be a kinder, gentler disease. *Discover* 10, no. 4 (April):38–43.

Rook, G. A. 1987. Progress in the immunology of mycobacteria. *Clinical and Experimental Immunology* 69, no. 1 (July):1.

Thomas, G., and M. Morgan-Witts. 1982. *Anatomy of an epidemic*. New York: Doubleday.

A Peruvian girl paddles her canoe through a canal full of excrement and refuse to reach her home in the village of Belen. The water, used for cooking, bathing, swimming, and fishing, is a perfect breeding ground for cholera, which is now affecting large areas of Peru as well as other South American countries. In the absence of proper water and sewage treatment, it is feared that cholera will become established in South America, to cause deaths for decades to come.

22 Oral and Gastrointestinal Diseases

This chapter focuses on the following questions:

A. What kinds of pathogens cause diseases of the oral cavity, and what are the important epidemiologic and clinical aspects of these diseases?

B. Which bacteria cause gastrointestinal diseases, and what are the important epidemiologic and clinical aspects of these diseases?

C. What other kinds of pathogens cause gastrointestinal diseases, and what are the important epidemiologic and clinical aspects of these diseases?

We all have a wealth of firsthand experience with diseases of the oral cavity and gastrointestinal tract. We've had plaque removed from our teeth and cavities filled. We've had episodes of vomiting and diarrhea—usually of short duration—after eating contaminated food or drinking impure water. Fortunately, for most of us these problems have been minor inconveniences rather than serious illnesses. In this chapter we will consider diseases of the oral cavity and intestinal tract, both minor and serious.

Like the foregoing chapters dealing with diseases of body systems, this chapter presumes a knowledge of disease processes, normal and opportunistic flora, epidemiology, host systems, host defenses, and immunity (Chapters 15 through 19). It also presupposes familiarity with the characteristics of bacteria, viruses, and eukaryotic microorganisms and helminths (Chapters 9 through 12).

DISEASES OF THE ORAL CAVITY

Bacterial Diseases of the Oral Cavity

Plaque

Plaque is a continuously formed coating of microorganisms and organic matter on tooth enamel. Plaque formation, although not a disease itself, is the first step in tooth decay and gum disease. Scrupulous and frequent cleaning of teeth minimizes but does not entirely prevent plaque formation. Unless plaque is removed within 48 hours, it becomes so securely cemented to teeth that it can no longer be removed by home methods. Professional cleaning removes plaque, but it begins to form again before you get home from the dentist's office.

Plaque formation begins as positively charged proteins in saliva adhere to negatively charged enamel surfaces and form a **pellicle** (film) over the tooth surface. Cocci, such as *Streptococcus mutans*, *S. salivarius*, and *S. sanguis*, and some filamentous bacteria among the normal oral flora attach to the pellicle. These organisms hydrolyze sucrose (table sugar) to glucose and fructose. Glucose is used to make the polysaccharide dextran, which forms a polymer over the tooth surface. Plaque consists of microorganisms, dextran, and proteins from saliva (Figure 22.1). If plaque is not removed thoroughly and regularly, streptococci, lactobacilli, and other acid-producing bacteria accumulate within it in layers 300 to 500 cells thick. They metabolize fructose and other sugars that diffuse into the plaque and set the stage for tooth decay. Plaque that accumulates near the gumline offers similar protection to bacteria in the gingival (gum) crevices between the teeth and the gums. These bacteria include species of *Corynebacterium*, *Actinomyces*, and sometimes spirochetes, as well as the streptococci just noted. Some crevices may be anaerobic "pockets." Gumline plaque also mechanically irritates the gums and contributes to inflammation and bleeding.

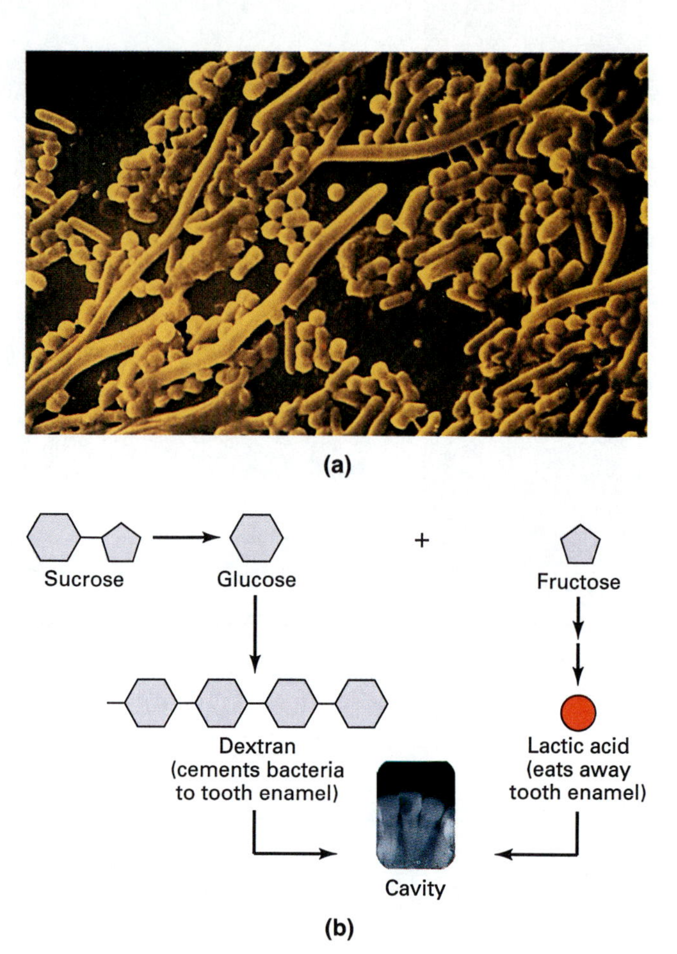

FIGURE 22.1 Dental plaque. (a) A variety of organisms that accumulate in plaque deposits. (b) Production of sticky dextran from sucrose enables bacteria to adhere to tooth surfaces, where the lactic acid produced in this process eats away tooth enamel and forms cavities.

Dental Caries

The Disease Process **Dental caries,** or tooth decay, is the erosion of enamel and deeper parts of teeth. It is the most common infectious disease in developed countries where the diet contains relatively large amounts of refined sugar. Unchecked, it can proceed through enamel, into cementum and dentin, into the pulp cavity, and eventually cause an abscess in the bone that supports the tooth. Sugars easily diffuse through plaque to bacteria embedded in it, but acids produced by bacterial fermentation fail to diffuse out. The acids gradually dissolve enamel and other tooth structures.

The combination of sucrose and the action of *S.*

TRY IT

Testing for Caries Susceptibility

A test for caries susceptibility involves mixing a measured amount of saliva with Snyder test agar, which contains a pH indicator. Bacteria in the saliva produce lactic acid and change the color of the agar from green to yellow. If the color changes after no more than 24 to 48 hours of incubation, the person is said to be susceptible to caries formation.

mutans on it account for much tooth decay. Consequently, the more sucrose one eats and the more frequently one eats it, the greater the risk of dental caries. Saliva helps to rinse sugars from the mouth, but its rinsing efficiency varies with mouth shape, and sugar accumulates in poorly rinsed areas. Starchy foods, which are only partially digested in the mouth, contribute little to tooth decay. Sugar alcohols, such as sorbitol and xylitol used in some chewing gums, do not contribute to tooth decay because the bacteria cannot metabolize them.

Treatment and Prevention Dental caries are treated by removing decay and filling the cavity with amalgam, a mixture of mercury and other metals. Dental caries can be prevented or their incidence greatly reduced by limiting sugar intake, brushing regularly with plaque-removing toothpaste, and flossing between teeth. Vaccines to prevent dental caries are being developed against strains of *S. mutans,* which are most prevalent at decay sites. An injectable vaccine that elicits circulating IgG has been successful in monkeys. An oral vaccine that causes salivary glands to produce secretory IgA has been successful in rats. Neither vaccine has yet been adequately tested in humans. Stress may be a factor in the effectiveness of a vaccine because it appears to depress the immune system. This was noted when dental students were found to secrete less salivary IgA while taking examinations than while on summer vacation. (Chapter 18, p. 484)

The use of **fluoride** has been the most significant factor in reducing tooth decay. It works by poisoning bacterial enzymes and by hardening surface enamel of teeth. Fluoride inhibits certain enzymes that produce phosphates needed by bacteria to capture energy from nutrients. Without the phosphates the bacteria die. This effect of fluoride is easily counteracted. If after brushing your teeth you drink a cola, the phosphates contained in it allow the bacteria to grow as if you had never used fluoride.

To understand how fluoride affects enamel, imagine the tooth surface as consisting of the ends of rods packed together like a fistful of pencils (Figure 22.2). No matter how tightly the rods are packed, channels exist between them. Acids produced by bacteria in plaque seep through channels and dissolve the enamel rods. As channels are enlarged, more acid enters and dissolves more enamel. Eventually, sufficient enamel is eroded to form cavities, or dental caries. Fluoride fills the spaces between rods with a hard mineralized material that strengthens the tooth surface and prevents acid penetration.

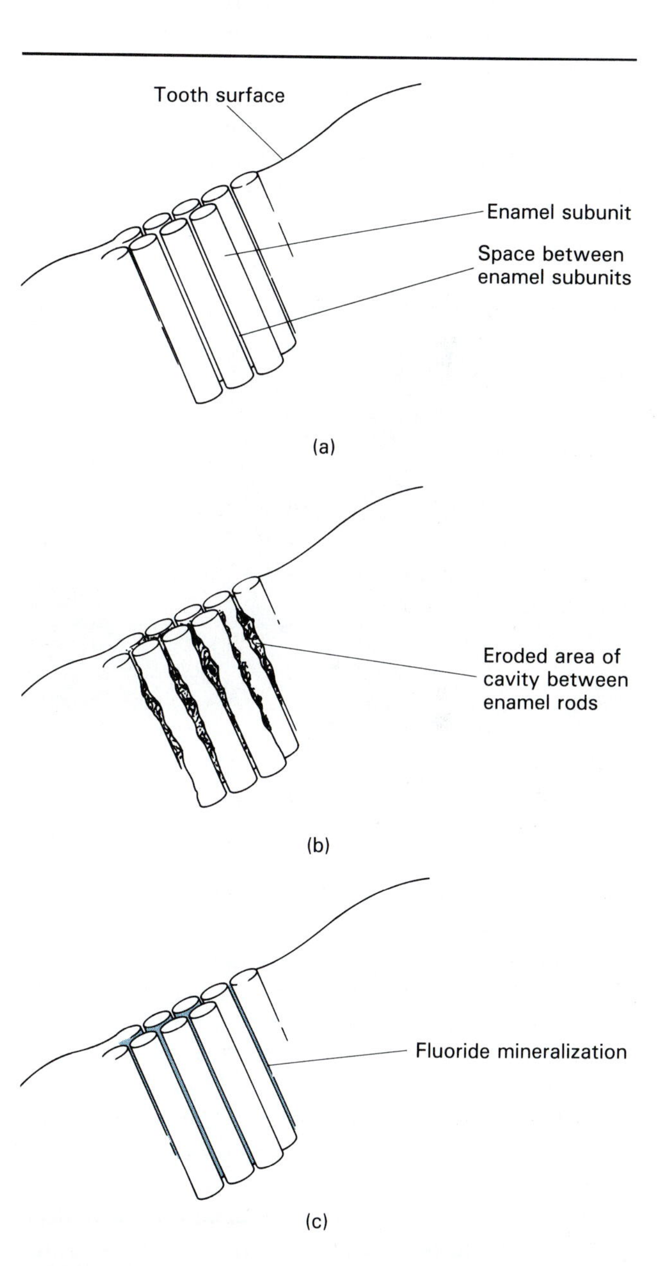

FIGURE 22.2 The effect of fluoride on tooth decay: (a) normal tooth structure; (b) dental caries forming as a result of bacterially produced acid seeping down between rods of enamel; (c) mineralization by application of fluoride, which fills in the spaces between enamel rods, thereby preventing acid from seeping in.

Numerous studies have shown fluoride to be safe and effective in preventing tooth decay. Added to city water supplies in a 1-part-per-million concentration, it reduces tooth decay in children by as much as 60 percent. Because caries occur mainly in childhood and adolescence, a successful program to prevent childhood tooth decay is of major importance. Yet nearly half the U.S. population obtains water from wells or from nonfluoridated community supplies. In some communities that lack fluoridated water, children receive fluoride tablets or gels at school.

Although fluoride is most beneficial before age 20, it provides some benefits at any age even when water is fluoridated and fluoride toothpastes and mouthwashes are used. Fluoride gel treatments should be given every 6 months from about age 4 to adulthood. Each time teeth are cleaned, a little surface is removed, and remineralization with fluoride helps to restore the surface (Figure 22.3).

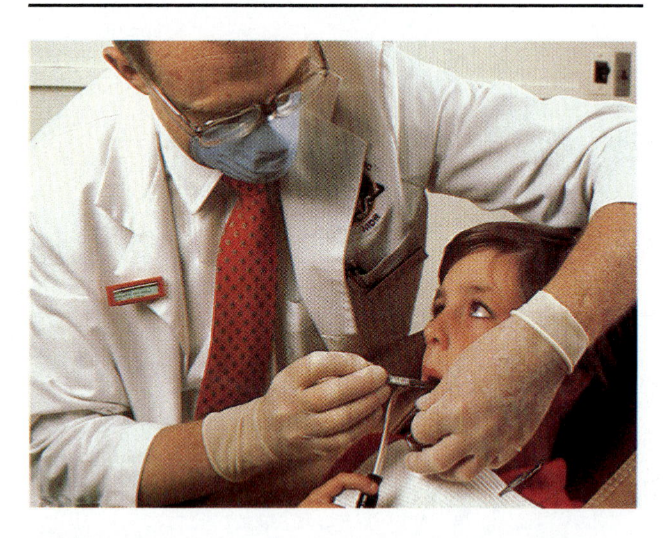

FIGURE 22.3 Visits to the dentist can be minimized by proper oral hygiene and the use of fluoride-containing toothpaste or oral rinse. Fluoride can also be applied in the dentist's office.

Another means of preventing dental caries is sealing of the teeth with **methacrylate** (meth-ak'ril-at). Grinding surfaces of teeth are more susceptible to decay than smooth sides because they have tiny pits and crevices that fill with plaque that cannot be removed with a toothbrush. Sealing teeth with methacrylate provides nearly complete protection against decay as long as the seal remains, which may be 10 years or longer. Applied at about age 6 or 7 after secondary grinding teeth have erupted, seals will last through most of the decay-prone years.

In the sealing process, teeth are thoroughly cleaned and etched with a weak acid to remove the last traces of plaque and then coated with methacrylate. Tiny cavities cease to grow after the sealant is applied and eventually become sterile as the decay organisms die. Sealant can be applied only to teeth without fillings, but it can be used on adults as well as children.

Periodontal Disease

The Disease Process When bacteria become trapped in gingival crevices, they cause both tooth decay and gum inflammation. **Periodontal disease** is a combination of gum inflammation, decay of cementum, and erosion of periodontal ligaments and bone that supports teeth. Exactly how organisms in gingival crevices contribute to periodontal disease is not known, but it appears that new groups of organisms replace previous inhabitants as the disease progresses. If the process is not arrested, the gums recede and can become necrotic, and the teeth loosen as surrounding bone and ligaments are eroded.

Nearly everyone is eventually affected by periodontal disease. In its mildest form, periodontal disease is simply called **gingivitis** (jin-jiv-i'tis); in its most severe form it is called **acute necrotizing ulcerative gingivitis** (ANUG), or trench mouth. The disease got its name because it was common among military men under stress "in the trenches." It is now common in young people, and stress seems to be an important factor in its development. ANUG responds to antibiotics, but more chronic forms of periodontal disease usually do not.

Chronic periodontal disease appears to occur when conditions in the mouth allow overgrowth of potentially virulent bacteria such as species of *Bacteroides, Eikenella, Eubacterium,* and other gram-negative, anaerobic rods. *Fusobacterium nucleatum,* an anaerobic organism normally present in the mouth, is suspected of being a cause of gingivitis. Certain other bacteria such as *Streptococcus mitis, S. sanguis, Veillonella* species, and some *Actinomyces* species may help to control the virulent bacteria.

One of the first thorough studies of organisms associated with periodontal disease, which was done in 1982 by P. H. Keyes, illustrates a method of studying a complex microbial environment. Keyes obtained plaque from individuals with healthy gums, mildly inflamed gums, and severely diseased gums. When he examined them by phase-contrast microscopy, he found distinctly different populations of microorganisms (Figure 22.4). Plaque from healthy gums contains nonmotile filamentous bacteria, a few colonies of cocci, fewer than five leukocytes per microscopic field, and no amoebae or spirochetes. Plaque from mildly inflamed gums shows a greater variety of microbes. Nonmotile organisms form dense masses surrounded by clusters of motile spirochetes and rapidly spinning bacilli too numerous to count. Some spirochetes and leukocytes are present, but amoebae

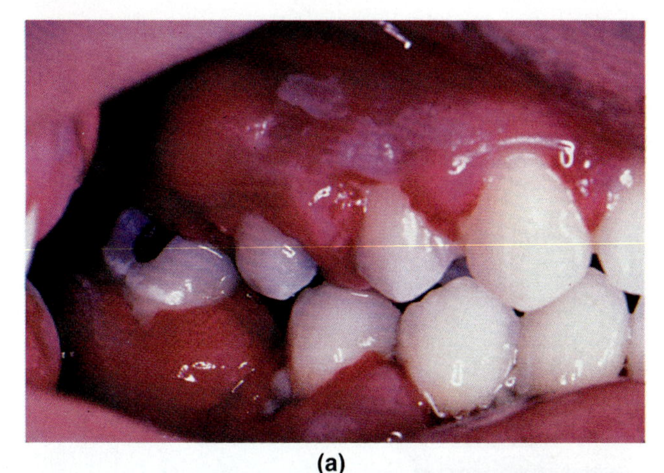

(a)

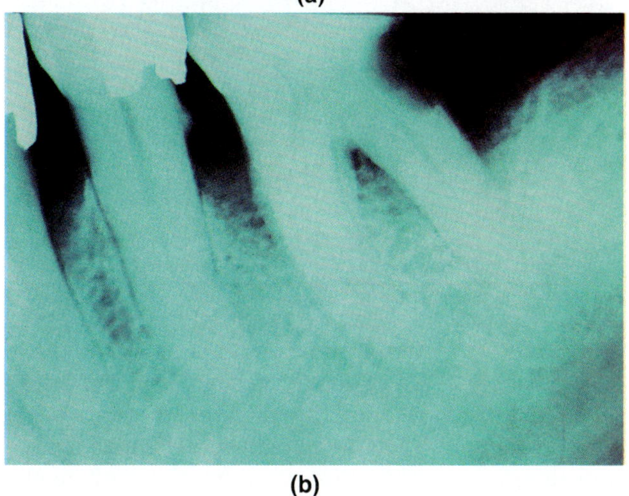

(b)

FIGURE 22.4 Advanced peridontal disease showing (a) severe gum inflammation, which can lead to (b) loss of bone surrounding roots of teeth, causing loosening and eventual loss of teeth.

and trichomonads are absent. Plaque from severely diseased gums contains dense mats of nonmotile filaments, cocci, and numerous motile organisms. Extending from the dense mats are brushlike aggregations of spirochetes and flexible rods that move in rippling waves and migrate from one surface to another. Amoebae always are present, trichomonads sometimes are seen, and leukocytes are numerous.

More recent studies suggest that of 300 different species of bacteria in the mouth *Bacteroides gingivalis* may be a specific cause of some cases of periodontal disease. Researchers at the University of Texas have succeeded in causing a burst of disease in monkeys given doses of the bacterium. They have also had some success treating the monkeys with rifampin.

Prevention and Treatment Chronic periodontal disease can be prevented or its onset delayed by daily thorough cleaning of teeth and frequent professional removal of plaque. Once organisms erode gums and form pockets between the teeth and gums, infections must be kept under control. The treatment of chronic periodontal disease is a somewhat controversial subject in dentistry today, and it appears that patients vary in their responses to different treatments. Keyes has reported success with having patients brush with a mixture of bicarbonate of soda and hydrogen peroxide. Others have tried treating the disease with tetracycline and other antibiotics. Still others recommend antimicrobial mouth rinses.

Viral Diseases of the Oral Cavity

Mumps

Mumps is caused by a paramyxovirus somewhat similar to the measles (rubeola) virus. The virus is transmitted by saliva and invades cells of the oropharynx. After initially replicating in the upper respiratory tract, it travels in the blood to the salivary glands and sometimes to other glands and organs. Swelling of the parotid glands appears 14 to 21 days after initial infection and can persist as long as 7 days. Viruses are

APPLICATIONS

See the Amoebae in Your Mouth

Some dentists now use a television monitor attached to a microscope to show patients their own oral flora, and patients assist the dental hygienist in counting various kinds of microbes. This approach dramatizes the severity of the problem just as stepping on the scales does for overweight people—the patients see why treatment is needed. The effectiveness of the treatment is measured on subsequent visits by tallying changes in flora. These procedures provide high patient motivation and clear evidence of success.

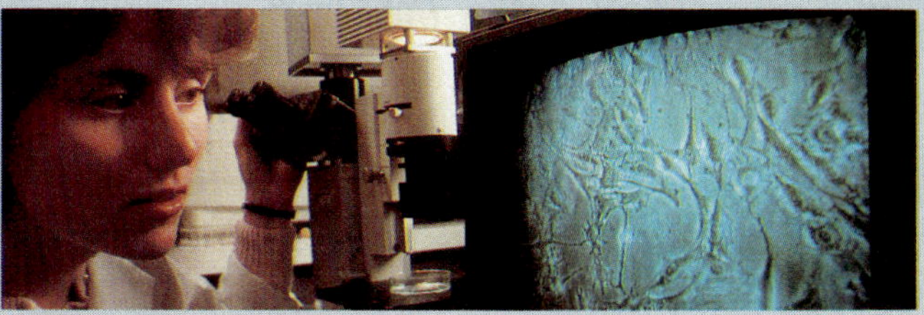

A technician uses a TV monitor to display a patient's oral microorganisms.

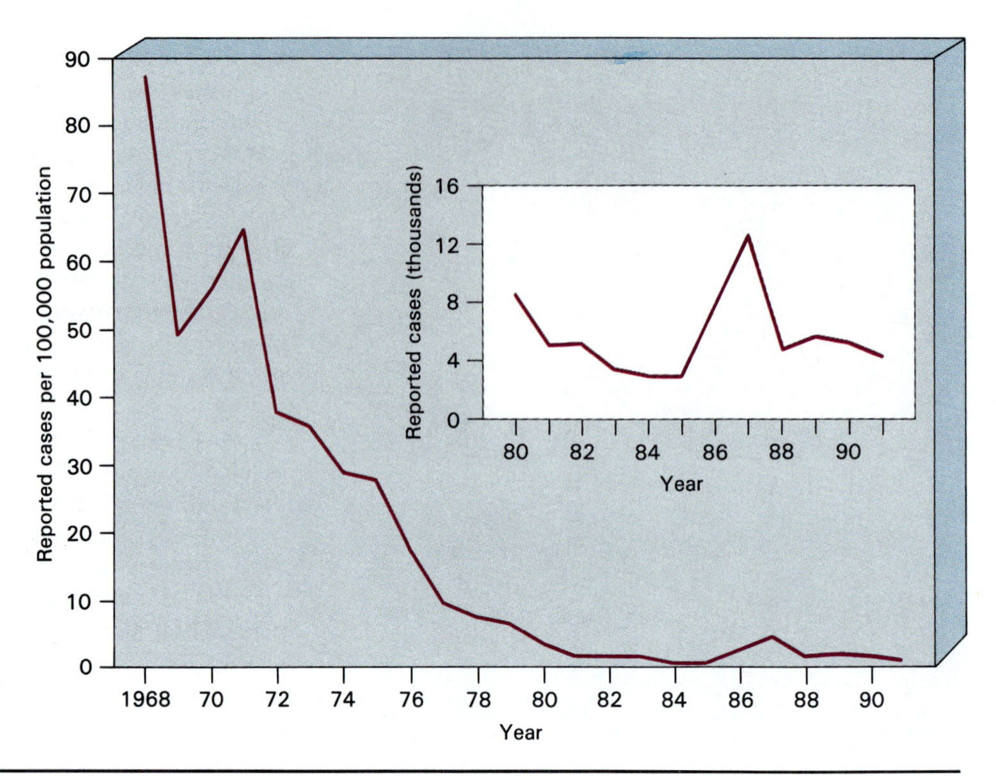

FIGURE 22.5 The incidence of mumps has declined sharply since a vaccine became available.

released, and the patient is contagious 7 days before the glands swell and as long as 9 days after swelling subsides. In contrast to other paramyxoviruses, mumps viruses are excreted in urine as much as 2 weeks after the onset of symptoms.

Humans are the only organisms known to be infected by mumps virus. The virus is found worldwide, especially in the spring, and infections are most frequent in children 6 to 10 years of age. Although as many as 85 percent of a susceptible population will be infected by mumps, 20 to 40 percent will have no symptoms. When the disease appears in postpubertal males, 20 to 30 percent develop **orchitis,** inflammation of the testes. Such infections are capable of causing sterility but rarely do so. Other complications of mumps, regardless of the age or sex of the patient, include meningoencephalitis, eye and ear infections, and inflammation of other glands such as the ovaries and pancreas (where it may be a cause of juvenile onset diabetes). An effective vaccine for mumps contains viruses weakened by passage through a sequence of embryonated egg cultures. It usually is administered in combination with measles and rubella vaccines. The incidence of mumps has decreased dramatically since the vaccine was licensed and put to use in 1967 (Figure 22.5).

Other Diseases

Thrush, an oral infection by the opportunistic fungus *Candida albicans,* was discussed in Chapter 20 and illustrated in Figure 20.10. ∞ (p. 559) Herpes simplex virus infections, a cause of common cold sores in the mouth, are discussed in Chapter 25. Diseases of the oral cavity are summarized in Table 22.1.

TABLE 22.1 Summary of oral diseases

Disease	Agent(s)	Characteristics
Bacterial diseases of the oral cavity		
Dental caries	*Streptococcus mutans* and other species	Erosion of tooth enamel and other structures by acids from microbial metabolism
Periodontal disease	Various bacteria, *Bacteroides gingivalis*	Inflammation and destruction of gums, loosening of teeth, erosion of bone
Viral disease of the oral cavity		
Mumps	Paramyxovirus	Inflammation and swelling of salivary glands and sometimes testes, epididymis, and other tissues

GASTROINTESTINAL DISEASES CAUSED BY BACTERIA

Bacterial Food Poisoning

Food poisoning is caused by the ingestion of food contaminated with preformed toxins. It also can be caused by ingesting foods contaminated with pesticides, heavy metals, or other toxic substances. In food poisoning caused by microbial toxins, organisms that can continue to produce toxin usually are ingested with the toxins. However, tissue damage is due to action of the toxin, so food poisoning is an intoxication rather than an infection. Because the toxin is preformed, the onset of symptoms in intoxication is more rapid than in infection. Food poisoning can be prevented by following proper food-handling procedures, described in Chapter 27. Bacteria that produce toxins responsible for food poisoning include *Staphylococcus aureus, Clostridium perfringens, C. botulinum,* and *Bacillus cereus.* Staphylococci usually enter food by way of infected food handlers. Other organisms that cause food poisoning are ubiquitous soil organisms found in water, feces, sewage, and nearly all foods. We will consider the nature of the intoxication caused by each of these organisms.

Staphylococcal Enterotoxicosis

Certain strains of *Staphylococcus aureus* cause food poisoning, or **enterotoxicosis** (en''ter-o-tox-eh-ko'sis) by releasing an **enterotoxin,** an exotoxin that inflames the intestinal lining. Because the organisms are relatively resistant to heat and drying, foods easily become contaminated with them from food handlers or from the environment. The organisms multiply and release toxin in uncooked or inadequately cooked foods, especially if the foods are unrefrigerated. Nearly any food can be contaminated with *S. aureus,* but cream pies, dairy products, poultry products, and picnic foods such as potato salad are common culprits. Contamination is difficult to detect because it produces no change in the taste or odor of foods. Unlike most exotoxins, the toxin itself is heat-stable and withstands boiling for 30 minutes; cooking foods can kill the organisms but does not destroy the toxin. (Chapter 15, p. 394)

When food contaminated with *S. aureus* and its enterotoxin enters the intestine, the toxin acts directly. The organisms usually continue to produce toxin, although they do not multiply. As toxin comes in contact with the mucosa it is absorbed; it causes tissue damage only after it has entered the blood and has circulated back to the intestine. Symptoms such as abdominal pain, nausea, vomiting, and diarrhea, but usually not fever, appear 1 to 6 hours after ingestion of contaminated food. The time required for symptoms to appear depends on how long it takes for absorption of a sufficient quantity of toxin to produce them. A food loaded with toxin will elicit symptoms quickly; one with only a little toxin will take longer. Once elicited, symptoms usually last about 8 hours. For otherwise healthy adults, no treatment is required because the disease is self-limiting. It can be severe in infants, the elderly, or debilitated patients. Recovery from food poisoning does not confer immunity. The best way of preventing food poisoning by *S. aureus* is to avoid foods that might be contaminated and to use sanitary food-handling procedures.

CLOSE-UP

Food Poisoning

Several strains of *Staphylococcus aureus* cause food poisoning, and the toxins they produce vary in their degree of toxicity. Most strains produce a coagulase enzyme that coagulates blood plasma. Laboratory identification of the enzyme helps to determine whether *S. aureus* was the cause of an outbreak of food poisoning, although the enzyme itself apparently does not contribute to symptoms.

Other Kinds of Food Poisoning

An enterotoxin from *Clostridium perfringens* also causes food poisoning. (Chapter 15, p. 394) The toxin, which is released only during sporulation, is most often produced under anaerobic conditions, as when undercooked meats and gravies are kept warm for a period of time. The main symptom is diarrhea. Compared with *S. aureus* food poisoning, the disease takes longer to appear—10 to 24 hours after ingestion—and lasts longer (about 24 hours). It also is self-limiting and can be prevented by sanitary food handling. Although *C. perfringens* toxin causes food poisoning, the organism itself can infect tissues: It has been incriminated in an intestinal infection, **necrotizing enterocolitis,** and in gas gangrene.

Botulism, which is caused by a neurotoxin of *Clostridium botulinum,* is acquired from eating toxin-contaminated food. Although it is a kind of food poisoning, it has little effect on the digestive system. Its effects on the nervous system are discussed in Chapter 24.

Bacillus cereus secretes two toxins; one causes vomiting, the other causes fluid accumulation in the small intestine and diarrhea. Symptoms occur less than 12 hours after ingestion and last only a short time. Such toxins are often found in dairy products and fried rice.

Food poisoning by *Pseudomonas cocovenenans* occurs in Polynesia and is called **bongkrek disease,** named for its association with a native coconut delicacy called bongkrek. The organism produces a potent, and often fatal, toxin that is frequently found in dishes prepared with coconut.

Bacterial Enteritis and Enteric Fevers

For most of us diarrhea is an unpleasant inconvenience, but it can be deadly. In 1900 in New York City the death rate among infants from diseases grouped together as diarrhea was 5603 per 100,000. Although today the death rate is less than 60 per 100,000, the disease is still life-threatening, and death from diarrhea can occur in hours in infants.

Enteritis is an inflammation of the intestine. **Bacterial enteritis** is an intestinal infection, not an intoxication as in food poisoning; the causative bacteria actually invade and damage the intestinal mucosa or deeper tissues. Enteritis that affects chiefly the small intestine usually causes **diarrhea.** When the large intestine is affected the result is often called **dysentery,** a severe diarrhea that often contains large quantities of mucus and sometimes blood or even pus. Some pathogens spread through the body from the intestinal mucosa and cause systemic infections, such as typhoid fever. Such infections are called **enteric fevers.**

Salmonellosis

Salmonellosis (sal-mo-nel-o'sis) is a common enteritis. The annual reported incidence of salmonellosis (excluding typhoid fever) in the United States has climbed from 20,000 cases in the early 1970s to more than 65,000 cases in 1986. Many more cases go unreported, and the actual annual number of cases has been estimated to exceed 2 million. Salmonellosis can be caused by several species of *Salmonella;* in the United States *S. typhimurium* is the most frequent cause, but *S. enteritidis* is also common. About 1800 strains of salmonellas have been identified by their surface antigens. Identifying strains is sometimes useful in tracing the source of disease outbreaks.

Salmonella species are naturally present on a variety of foods, especially poultry, and they usually are transmitted in contaminated, inadequately cooked food. They can be transmitted in eggs if the hens that laid them were infected. They have also been traced to contaminated water and to food contaminated by carriers. The practice of confining poultry and pigs to small spaces while being reared for market makes feeding and caring for the animals more efficient, but it also facilitates transmission of *Salmonella* and other pathogens among them.

PUBLIC HEALTH

Baby Chicks and Turtles

At one time baby chicks and ducks were frequently purchased to delight small children around Easter. In many localities they are no longer available. Although the little birds often were badly mistreated by children, the ban on their sale probably resulted from local health regulations aimed at preventing salmonellosis. If you or your children do handle chicks, consider all their droppings infectious, and practice careful hand washing.

Similar bans have been passed against the sale of baby turtles, which also carry *Salmonella*. If you handle turtles at all, you should follow the same precautions recommended for handling chicks. Washing the turtle's bowl in the kitchen sink can be hazardous to your health; if you do so, make sure to clean the sink thoroughly. Better yet, whenever possible clean the bowl outdoors.

Symptoms of salmonellosis include abdominal pain, fever, and diarrhea with blood and mucus. They appear 8 to 30 hours after ingestion of organisms and are associated with the organisms invading the mucosa of both the small and large intestines. Fever probably is caused by endotoxins, toxins that are released from a cell only when it is lysed. In otherwise healthy adults salmonellosis lasts 1 to 4 days and is self-limiting. Antibiotics usually are not given because they tend to induce carrier states and to contribute to the development of antibiotic-resistant strains. Infants and elderly or debilitated patients often have more severe and prolonged symptoms, and in such cases antibiotics may be prescribed.

Other species of *Salmonella* also cause disease. Because of their ability to invade intestinal tissue and enter the blood, *S. typhimurium* and *S. paratyphi* cause a somewhat more serious **enterocolitis.** Symptoms and bacteremia appear after an incubation period of 1 to 10 days. Enteric symptoms and fever can last 1 to 3 weeks, and chronic infections of the gallbladder and other tissues are not uncommon. A carrier state becomes established when, after recovery, organisms from chronically infected tissues continue to be excreted with feces. Broad-spectrum antibiotics rid carriers of organisms, but they also activate the disease in some carriers by upsetting the balance of intestinal flora.

Prevention of salmonellosis and enterocolitis depends on maintaining sanitary water and food supplies and eradicating organisms from carriers. The organisms cannot be entirely eradicated because poultry and other animals serve as reservoirs, and no effective vaccine is available.

Typhoid Fever

Typhoid fever, one of the most serious of the epidemic enteric infections, is caused by *Salmonella typhi.* It is rare in places where good sanitation is practiced but is more common where food and water are frequently contaminated. For most of the last 25 years fewer than 500 cases per year have been reported in the United States. The organisms enter the body in food or water and invade the mucosa of the upper small intestine (Figure 22.6). From there they invade lymphoid tissues and are phagocytized and disseminated. The organisms multiply in the phagocytes, emerge, and continue to multiply in the blood.

Bacteremia and septicemia occur at the same time as symptoms appear. During the first week the patient suffers from headache, malaise, and fever, probably due to an endotoxin. During the second week the patient's condition worsens. The organisms invade many tissues, including the intestinal mucosa, and are excreted in the stools. *Salmonella typhi* thrive and multiply in bile; organisms from the gallbladder reinfect the intestinal mucosa and Peyer's patches. Characteristic "rose spots" often appear on the trunk and abdomen for a few days. Abdominal distention and tenderness and enlargement of the spleen are common complaints, but diarrhea usually is absent. Unlike most other infections, where leukocytes increase in number, in typhoid fever they decrease in number. Some patients become delirious and can suffer complications such as internal hemorrhage, perforation of the bowel, and pneumonia. Chloramphenicol is the antibiotic of choice in treating typhoid fever, but some strains of *S. typhi* are resistant to it.

By the fourth week, symptoms subside, convalescence begins, and immunity develops. Cell-mediated immunity provides protection against future infection. Antibodies to surface antigens of the bacterium also are produced, but these are of more use in laboratory diagnosis than in protecting the patient against infection. The Widal test detects antibodies and is used to confirm the diagnosis of typhoid fever. The only vaccine available provides imperfect immunity for a short period of time; it affords some protection to people who must enter areas where typhoid fever is common. The best means of protection against typhoid fever is good sanitation.

PUBLIC HEALTH

Typhoid Mary

"Typhoid Mary," a carrier of typhoid fever named Mary Mallon who lived around the turn of the century, didn't believe in the germ theory of disease, and she refused to stop working as a cook in households. She also refused to have her gallbladder surgically removed, which would have rid her of the typhoid bacilli she carried. After causing many more cases of the disease, a number of them fatal, she was quarantined in a hospital on an island under court order. On arrival at the quarantine facilities, she offered to work in the kitchen—an offer authorities did not accept. Mary remained in quarantine for the rest of her days. Before antibiotics and immunizations were available, quarantine was a common practice. Children with measles and other childhood diseases were quarantined in their homes. With better treatments and more vaccines available, quarantine is rarely used today.

Shigellosis

Shigellosis (shig-el-o'sis), or **bacillary dysentery,** can be caused by several strains, or **serovars,** of *Shigella* (Figure 22.7). They include *S. dysenteriae* (serovar A), *S. flexneri* (serovar B), *S. boydii* (serovar C), and *S. sonnei* (serovar D). Shigellosis was first described in the fourth century B.C. Although less invasive than salmonellosis, it spreads rapidly in overcrowded con-

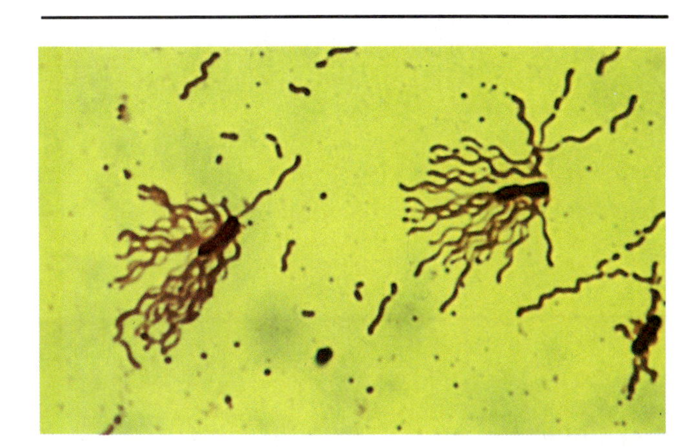

FIGURE 22.6 *Salmonella typhi,* causative agent of typhoid. Note the flagellae.

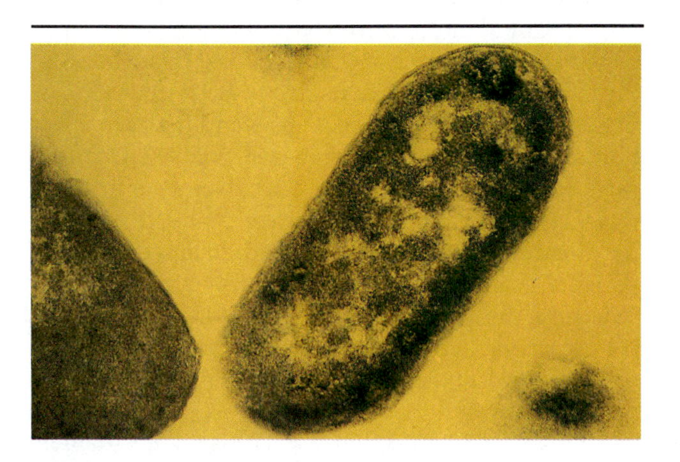

FIGURE 22.7 *Shigella,* causative agent of bacillary dysentery (shigellosis).

ditions with poor sanitation. Humans and higher primates such as chimpanzees and gorillas are the only reservoirs of infection, but the organisms can persist in foods for up to a month. (Chapter 14, p. 376) They also create a hazard for zoo workers.

Children aged 1 to 10 are most susceptible to *Shigella,* which accounts for 15 percent of the infant diarrhea cases in the United States. For the past several decades the total number of annual cases of shigellosis in the United States has ranged between 15,000 and 20,000. In recent years many outbreaks have been recorded in day-care centers. In developing nations it is a major cause of infant mortality, with one-third of all infant deaths due to dehydration from shigellosis and other enteric pathogens. Regardless of the cause, diarrheas are especially life-threatening in infants because their immature kidneys cannot maintain water and electrolyte balance. (Electrolytes are the ions, such as Na^+ and K^+, that are essential for many physiological functions, especially the transmission of nerve impulses.) Essential fluid replacement treatment often is not available in developing countries.

Once ingested, the organisms invade intestinal lining cells. *Shigella* are extremely infective; as few as 100 organisms can cause disease. After an incubation period of 1 to 4 days, abdominal cramps, fever, and profuse diarrhea with blood and mucus suddenly appear. Severity of symptoms ranges from most to least serious in the same order as the serovar designations A to D. In the most serious cases, diarrhea can cause dangerous deficiencies of protein—called *kwashiorkor*—and vitamin B_{12}, which, together with loss of electrolytes, can result in neurological damage. Along with the fever-eliciting endotoxin found in all serovars, *S. dysenteriae* also produces a neurotoxin. What the neurotoxin does in humans is unknown, but it causes paralysis and death in laboratory rabbits and mice. All serovars cause ulceration and bleeding of the intestinal lining and sometimes deeper intestinal layers. Symptoms persist for 2 to 7 days and usually are self-limiting, but they can cause severe dehydration and fluid and electrolyte imbalances.

Specific diagnosis of shigellosis is difficult because the organisms are very sensitive to acids in feces. The organisms die if fecal specimens stand for even a few hours before cultures are prepared. Viable specimens can be obtained by swabbing a bowel lesion during internal examination of the bowel.

Treatment is necessary in children and debilitated patients. A combination of antibiotics—ampicillin, tetracycline, and nalidixic acid—is used. Nalidixic acid is a synthetic agent that inhibits bacterial DNA synthesis, especially in gram-negative intestinal pathogens. Prevention is difficult because many people have inapparent infections. A carrier state, usually lasting less than 1 month, exists and accounts for many new cases by fecal-oral transmission. Any breakdown in sanitation can lead to transmission by way of feces, fingers, and flies. Immunity following recovery from shigellosis is transient. From limited experience in recent years, oral vaccines containing certain strains of *S. flexneri* and *S. sonnei* seem to be safe and effective.

Asiatic Cholera

Asiatic cholera, so named because of its high incidence in Asia, can affect people anywhere sanitation is poor and fecal contamination of water occurs. Worldwide, more than 100,000 cases are reported annually. In the United States fewer than 10 cases are reported per year, and some of those are transmitted to humans from contaminated shellfish in Gulf Coast states. In endemic regions, such as parts of Asia, 5 to 15 percent of patients die; when a seasonal epidemic occurs, as many as 75 percent of patients die.

The causative organism, *Vibrio cholerae* (Figure 22.8), can survive outside the body in cool water. When ingested it invades the intestinal mucosa, multiplies, and releases a potent enterotoxin. The enterotoxin binds to cells of the intestinal mucous membrane and makes the membrane highly permeable to water. Therefore, in addition to sudden severe nausea, vomiting, and abdominal pain, patients have a copious diarrhea that contains large quantities of mucus. The appearance has given it the name "rice water stool." As many as 22 L of fluids and electrolytes can be lost per day, so all patients, regardless of age, are subject to severe dehydration. Special "cholera cots" made of canvas with a hole cut beneath the buttocks have been used. A bucket marked in liters is placed below the hole to measure the fluid lost so that it can be replaced. Most deaths are attributable to shock because of greatly reduced blood volume.

Fluid and electrolyte replacement is the most effective treatment. During the 1971 epidemic in India and Pakistan, medical personnel were able to save

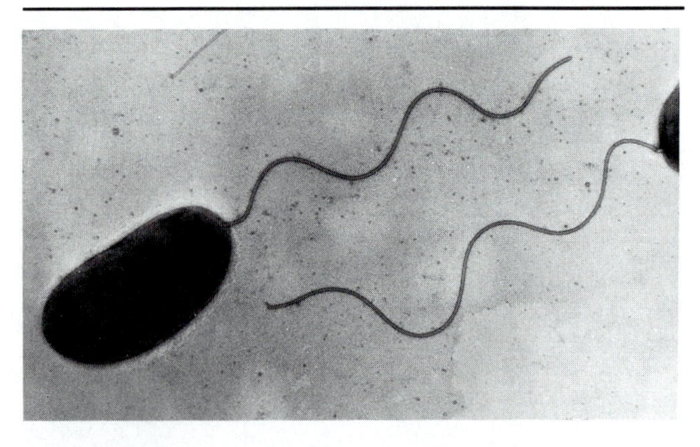

FIGURE 22.8 *Vibrio cholerae,* causative agent of cholera. The cell is slightly curved with a single polar flagellum.

CLOSE-UP
Cholera

"The speed with which cholera killed was profoundly alarming, since perfectly healthy people could never feel safe from sudden death when the infection was anywhere near. In addition, the symptoms were peculiarly horrible: radical dehydration meant that a victim shrank into a wizened caricature of his former self within a few hours, while ruptured capillaries discolored the skin, turning it black and blue. The effect was to make mortality uniquely visible: patterns of bodily decay were exacerbated and accelerated, as in a timelapse motion picture, to remind all who saw it of death's ugly horror and utter inevitability."

—William H. McNeill, 1976

large numbers of victims of the disease largely because of the availability of replacement therapy. Treatment with tetracycline reduces the duration of the symptoms but does not eliminate the organism or the toxin. Recovery from the disease confers only temporary immunity. Many recovered patients remain in a carrier state and can infect others and reinfect themselves. The only available vaccine is not very effective and is not widely used. However, a toxoid-type vaccine is being tested. It also may be possible to develop a vaccine that makes use of secreted IgA against the cholera organism.

A strain of *V. cholerae* known as the El Tor biotype causes a form of cholera that is slower and more insidious in its onset than the classical form of the disease. Cholera outbreaks have often resulted in the imposition of quarantine, halting shipping from port cities. In the past, therefore, countries in South and Central America that were economically dependent on their exports would sometimes report that cholera was not present, only "El Tor." The two diseases, however, are essentially the same. See the Essay at the end of this chapter for a discussion of the recent outbreak of cholera in Peru, which then spread to other Latin American countries.

APPLICATIONS
Oral Rehydration

Oral rehydration therapy, introduced relatively recently, is so simple and effective that one wonders why it was not discovered earlier. The formula for the solution used is:

½ teaspoon salt (NaCl)

¼ teaspoon sodium bicarbonate ($NaHCO_3$)

¼ teaspoon potassium chloride (KCl)

4 tablespoons sugar (sucrose)

Dissolve in 1 L water

The ingredients are extremely inexpensive and easy to obtain, even in undeveloped countries. The solution can be prepared in the field by people with little or no training, and if uncontaminated water is used it is almost certain to be safe. In severe cases of cholera, however, oral rehydration must be used as a supplement to intravenous rehydration, not a substitute.

Vibriosis

An enteritis called **vibriosis** (vib-re-o'sis) is caused by *Vibrio parahaemolyticus.* Although the disease is most common in Japan, where raw fish is considered a delicacy, the organism is widely distributed in marine environments. In the United States infections usually are acquired from contaminated fish and shellfish that have not been thoroughly cooked. Most outbreaks in the United States occur at outdoor festivities where crabs and shrimps are served without thorough cooking and proper refrigeration. Some outbreaks have been traced to the eating of contaminated raw oysters. The organism also can infect skin wounds of people exposed to contaminated water. Once inside the intestine, the organisms colonize the mucosa and release an enterotoxin. Symptoms of nausea, vomiting, diarrhea, and abdominal pain appear about 12 hours after ingesting contaminated food or water and last 2 to 5 days. The disease usually is not treated, and no vaccine is available.

Traveler's Diarrhea

Among the 250 million people who travel between countries each year, it has been estimated that over 100 million suffer from a self-limiting mild to severe diarrhea. Of these travelers, 30 percent are confined to bed, and another 40 percent are forced to curtail their activities. This disorder, officially called **traveler's diarrhea,** also has been called "Delhi belly," "Montezuma's revenge," and some even less attractive names.

The most common causes of traveler's diarrhea are pathogenic strains of *Escherichia coli,* which account for 40 to 70 percent of all cases. Some strains of *E. coli* are normal inhabitants of the human digestive tract, and only certain strains are capable of causing enteritis. **Enteroinvasive strains** have a plasmid with a gene coding for a particular surface antigen called K antigen that enables them to attach to and invade mucosal cells. **Enterotoxigenic strains** contain a plasmid that enables them to make an enterotoxin. They attach to the mucosa by pili or fimbriae. These organisms also cause numerous cases of infant diarrhea. Other causes of traveler's diarrhea include bacteria such as *Shigella*, *Salmonella*, *Campylobacter*, rotaviruses, and protozoa such as *Giardia* and *Entamoeba*. Jet lag and other stresses of travel do not cause diarrhea, but they can lower resistance to infection. Travelers can experience symptoms of diarrhea even when no pathogens are present. Such cases are due to unusual kinds and amounts of dissolved substances in water, which lead to gastrointestinal upsets.

Symptoms of traveler's diarrhea vary from mild to severe and include nausea, vomiting, diarrhea, bloating, malaise, and abdominal pain. A typical case causes 4 to 5 loose stools per day for 3 or 4 days. Fluid loss is greater with invasive strains than with toxigenic strains. The disease is especially hazardous in infants, who are subject to severe dehydration. Bottle-fed infants are much more likely to become infected than breast-fed infants. Newborns are especially at risk of acquiring pathogenic strains from hospital workers before their normal flora are sufficiently established to compete with the pathogens. After infancy, *E. coli* infections usually are acquired during travel in foreign countries, especially those with poor sanitation.

Travelers sometimes medicate themselves with antibiotics before and during a stay in a foreign country. This is not recommended because antibiotics usually are not effective and because such use contributes to the development of antibiotic-resistant strains. A better practice is to keep antidiarrhea medicine available and to use it only after symptoms appear.

Traveler's diarrhea can persist for months or years as a postinfectious irritable bowel syndrome. It also can cause lactose intolerance by damaging intestinal lining cells that normally produce the enzyme lactase, which digests lactose (milk sugar). During bouts of diarrhea, dairy products and other foods containing lactose should be eliminated from the diet. After a few weeks, small quantities of such foods can be reintroduced and gradually increased as long as they are well tolerated. Sometimes normal lactose tolerance is lost forever. *Escherichia coli* and other bacteria, the protozoan *Giardia*, and certain helminths frequently cause lactose intolerance.

Escherichia coli has significance far beyond its ability to cause diarrhea. It is an important indicator organism because it is always present in water contaminated with fecal material. *Escherichia coli* is usually more numerous than other organisms and is easier to isolate. Finding *E. coli* in water means that any pathogens found in feces might also be present.

Escherichia coli also is an extremely versatile opportunistic pathogen—it can infect any part of the body subject to fecal contamination, including the urinary and reproductive tracts and the abdominal cavity after perforation of the bowel. It is present in many bacteremias, causes septicemias, and can infect the gallbladder, meninges, surgical wounds, skin lesions, and lungs, especially in debilitated and immunodeficient patients.

Other Kinds of Bacterial Enteritis

Certain strains of *Campylobacter jejuni* and *C. fetus* can be found in food and water and are becoming increasingly associated with human gastroenteritis, especially in infants and in elderly or debilitated patients. Some strains of *C. fetus* cause infectious abortions in several kinds of domestic animals. Although these organisms apparently do not multiply in foods, they are transmitted passively in undercooked chicken, unpasteurized milk, and poultry held in unchlorinated water during processing. Improper cooking of contaminated poultry can lead to enteritis. In some areas *Campylobacter* surpasses *Salmonella* as the major cause of foodborne enteric disease. Many health departments have begun testing for *Campylobacter* in food handling areas and on equipment.

Campylobacter infections cause copious diarrhea, foul-smelling feces, fever, and abdominal pain. They also cause arthritis in 2 to 10 percent of infected children, but rarely in adults. Because large quantities of fluid can be lost, dehydration and fluid and electrolyte imbalances are common among the populations most affected. The disease is treated with fluid and electrolyte replacement and sometimes with tetracycline and erythromycin.

Yersiniosis (yer-sin-e-o'sis), a severe enteritis, is caused by *Yersinia enterocolitica*. The disease is most common in western Europe, but some cases are seen in the United States. This free-living organism is found mainly in marine environments, but it can survive in many places. Infection can be acquired from water, milk, seafoods, fruits, and vegetables, even when they are refrigerated, because the organism grows more rapidly at refrigerator temperatures than at body temperature. It is most easily identified when cultured at 25°C, a temperature at which the organisms are motile and easily distinguished from other bacteria. Yersiniosis symptoms, which are related to release of an enterotoxin, are similar to other kinds of enteritis, but abdominal pain usually is more severe, and white blood cells increase in number. Yersiniosis sometimes is misdiagnosed as appendicitis because of the similarity of symptoms.

A new source of *Yersinia* infection has caused the CDC to issue a warning that preparers of the Southern delicacy chitterlings (pork intestines, also called chitlins) should avoid touching children or anything used by children until the preparers have very carefully washed their hands. Children should not be allowed to handle raw chitterlings. Fifteen children in Atlanta recently became ill, mainly from contact with preparers, while the adults remained well.

Bacterial Infections of the Stomach, Esophagus, and Duodenum

Peptic Ulcer and Chronic Gastritis

Recent studies have revealed a bacterial cause of *peptic ulcer* and chronic gastritis and a probable cofactor of stomach cancer (Figure 22.9). The organism, *Helicobacter pylori* (formerly called *Campylobacter pylori*), was first cultured in 1982 from gastric biopsy tissues. Peptic ulcers are lesions of the mucous membranes lining the esophagus, stomach, and duodenum caused by sloughing (falling) away of dead inflammatory tissue, exposed to acid, and eventually resulting in an excavation into the surface of the organ. Four million Americans suffer from ulcers each year. Approximately 10 percent of the population will eventually suffer from an ulcer at some point in their lives. Ulcers are responsible for about 46,000 operations and 14,000 deaths per year. *Chronic gastritis* (stomach inflammation) may be so mild as to cause no noticeable symptoms, or it can produce pain and indigestion. It is observed in 70 to 95 percent of people having peptic ulcers. Severe gastritis may lead to ulceration.

Helicobacter pylori is present in 95 percent of patients having duodenal ulcers and in 70 percent of those with gastric ulcers. In developed countries it is uncommon for children to be infected, but the rate of infection increases by about 1 percent per year of age above 20, eventually reaching an average of 20 to 30 percent among U.S. adults. In underdeveloped countries it is commonly found in children and reaches levels of 80 percent of the adult population. In Latin America, stomach cancer rates are the highest in the world, and so is the rate of *H. pylori* infection. Interestingly, U.S. Hispanics also have a 75 percent infection rate. In Japan, rates of stomach cancer and of *H. pylori* infection are dropping together. It may be that just having chronic inflammation of the stomach for a long period of time—which occurs when a person suffers *H. pylori* infection early in life—predisposes toward stomach cancer. Therefore scientists are unwilling to call *H. pylori* a definite cause of stomach cancer but feel it may be only a cofactor because most people infected with *H. pylori* never develop stomach cancer. However, among the various kinds of stomach cancer, the most common type (gastric intestinal) had an 89 percent correlation with *H. pylori* infection.

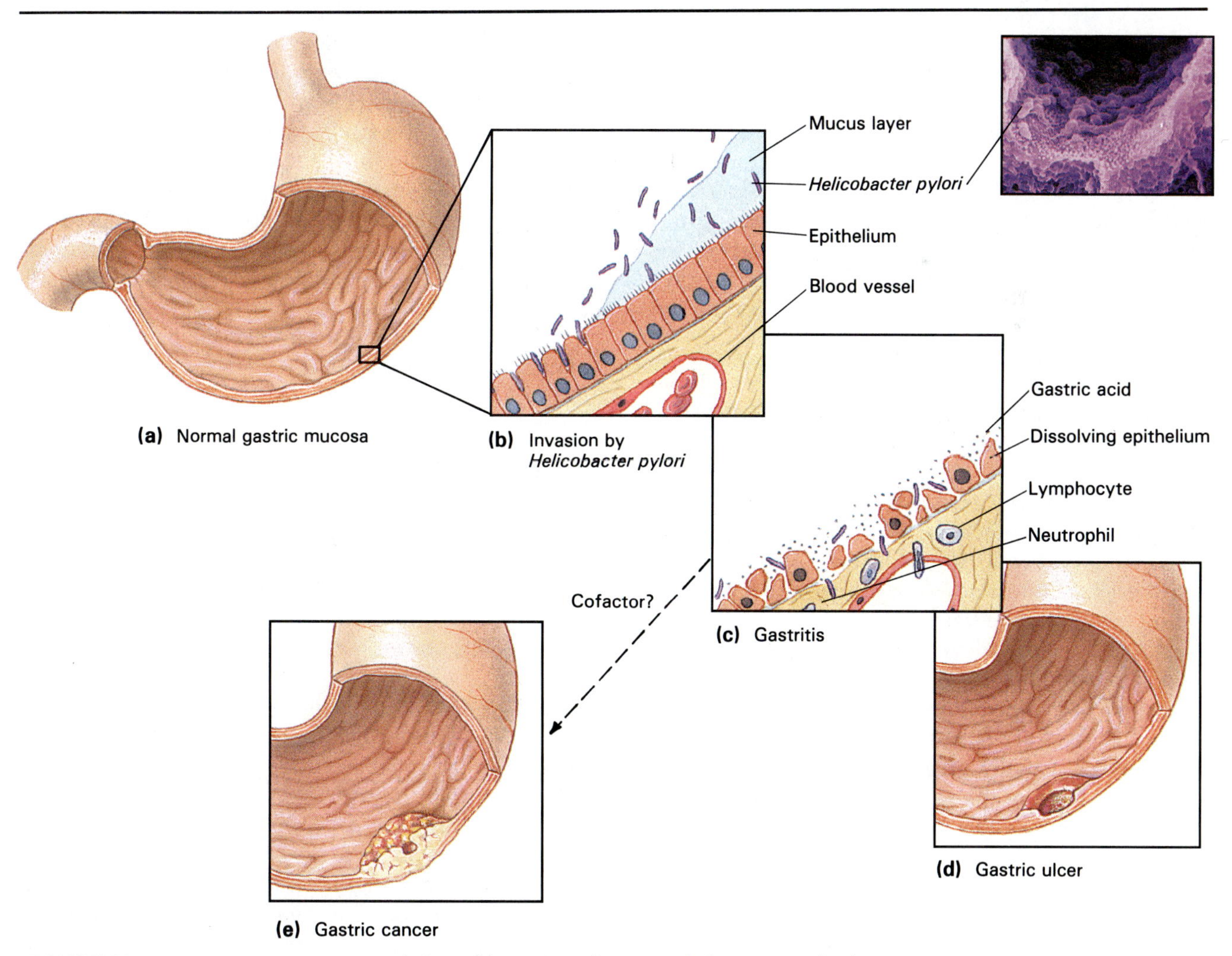

FIGURE 22.9 *Helicobacter pylori*, a spiral-shaped bacterium, has recently been recognized as the cause of peptic ulcers and chronic gastritis, and is involved in stomach cancer. Antibiotic treatment can lead to permanent cure of ulcers, so long as reinfection does not occur.

PUBLIC HEALTH

Fungal Product Alleviates Flatulence

Troubled by intestinal gas? Well, everyone has it. The average person releases about a liter per day. This gas, called *flatus,* is produced by colon bacteria as they metabolize compounds that you have been unable to digest. Complex sugars from foods such as beans, peas, cabbage, brussels sprouts, oat bran, bananas, apples, and high-fiber cereals are broken down by some bacteria, releasing hydrogen gas plus other gases and compounds as byproducts. Other intestinal bacteria consume this hydrogen gas. The balance between these two groups of bacteria partially explains why some people produce more gas than others.

Now, a new commercial product, "Beano," produced by and collected from the fungus *Aspergillus niger,* contains the sugar-digesting enzyme that people lack. The enzyme, an alpha-galactosidase, breaks the alpha linkages in the complex sugars, digesting them before they reach the bacteria in the colon, thus reducing the amount of flatulence.

Clearly more studies need to be done, comparing infected and uninfected populations over long periods of time. These studies also need to be compared with studies of people whose *H. pylori* infections have been eradicated using antibiotics. In underdeveloped countries, such people rapidly become reinfected, but this is not the case in developed countries. No one is certain yet of the route of infection or the portal of exit. It is of great importance, however, if we find that we can prevent or cure ulcers by eliminating *H. pylori* infection. Currently available drugs such as Tagament only control, but cannot cure, ulcers. Experimental treatments with antibiotics have reported cure rates as high as 80 to 90 percent, but these drugs must be used very carefully. Patients treated only with drugs that suppress stomach acid had a relapse rate of 75–95 percent within 2 years. Researchers expect to have better tests, drugs, and treatment plans worked out in a few years.

Bacterial Infections of the Gallbladder and Biliary Tract

The gallbladder and its associated ducts can also be sites of bacterial infection. The liver produces bile, which is stored in the gallbladder and slowly released into the small intestine. Organisms with lipid-containing envelopes are destroyed by the action of bile, which breaks down lipids. Thus, most enteric viruses, poliovirus, and hepatitis A virus (discussed in the next section) lack lipid envelopes. Organisms such as the typhoid bacillus are so resistant to the action of bile that they can actually grow in the gallbladder itself. They are shed from the gallbladder into the intestine and infect feces. People carrying these organisms in the gallbladder are asymptomatic.

Gallstones, formed from crystals of cholesterol and calcium salts, can block the bile ducts, decreasing the flow of bile and predisposing the individual to inflammation of the gallbladder (cholecystitis) or biliary ducts (cholangitis). Distension of the gallbladder due to accumulation of bile fosters the entry of microbes—most commonly *E. coli*—into the bloodstream, probably through minute tears in the gallbladder wall. Infection of the gallbladder can also ascend into the liver.

Diagnosis of gallbladder infections is based on clinical findings of recurring pains (biliary colic), nausea, vomiting, chills, fever, and often jaundice due to absorption of blocked bile into the bloodstream. Blood should be drawn for culture before antibiotic therapy is started, as antibiotics can prevent the growth of the causitive organisms. Despite prompt treatment with antibiotics, gallbladder infections cannot be cured unless the causative obstruction is removed, either by surgery or by the spontaneous passage of the gallstone.

Gastrointestinal diseases caused by bacteria are summarized in Table 22.2.

GASTROINTESTINAL DISEASES CAUSED BY OTHER PATHOGENS

Viral Gastrointestinal Diseases

Viral Enteritis

Most cases of **viral enteritis** are caused by **rotaviruses.** These viruses are transmitted by the fecal-oral route, replicate in the intestine, and cause diarrhea and enteritis within 48 hours. They destroy intestinal epithelial cells, thereby severely impairing absorption. Rotavirus infection is a major cause of infant morbidity and mortality in underdeveloped countries, where 3 to 5 billion cases occur annually, with 5 to 10 million deaths in children under age 5. Rotavirus infections

TABLE 22.2 Summary of gastrointestinal diseases caused by bacteria

Disease	Agent(s)	Characteristics
Bacterial food poisoning		
Staphylococcal enterotoxicosis	*Staphylococcus aureus*	Heat-stable enterotoxin causes tissue damage, abdominal pain, nausea, vomiting
Other kinds of food poisoning	*Clostridium perfringens, C. botulinum, Bacillus cereus*	Diarrhea, and sometimes intestinal infection and gas gangrene
Bacterial enteritis and enteric fevers		
Salmonellosis	*Salmonella typhimurium, S. enteritidis*	Abdominal pain, fever, diarrhea with blood and mucus from toxin; enterocolitis from invasion of organisms; chronic infections and carrier states occur
Typhoid fever	*Salmonella typhi*	Organisms invade mucosa and lymphatics, multiply in phagocytes and other tissues; high fever and "rose spots"; carrier state and life-threatening complications can occur
Shigellosis	*Shigella* strains	Organisms cause intestinal lesions and release toxins; symptoms include cramps, fever, profuse diarrhea with blood and mucus
Asiatic cholera	*Vibrio cholerae*	Organisms invade intestinal lining, release potent toxin that increases lining permeability; symptoms, include nausea, vomiting, copious diarrhea, fluid imbalances
Vibriosis	*Vibrio parahaemolyticus*	Organisms colonize mucosa and release toxin; nausea, vomiting, diarrhea are self-limiting
Traveler's diarrhea	Pathogenic strains of *Escherichia coli*, other bacteria (also viruses and protozoa)	Organisms can invade mucosa and/or produce toxin, cause nausea, vomiting, diarrhea, bloating, malaise, abdominal pain; self-limiting except for postinfection complications; dehydration and death in infants
Other kinds of bacterial enteritis	*Campylobacter jejuni, C. fetus, Yersinia enterocolitica*	*Campylobacter* species cause enteritis in infants and debilitated patients; *Yersinia* releases a toxin that causes enteritis with pain resembling appendicitis
Bacterial infections of the upper gastrointestinal tract		
Cholecystitis, cholangitis	Usually *E. coli*	Blockage of bile ducts by gallstones causes inflammation of gallbladder and bile ducts; accumulation of bile can cause infection to spread to bloodstream or liver
Ulcers, stomach cancer	*Helicobacter pylori*	Definite association with ulcers; probable cofactor in stomach cancer

account for a third of childhood deaths in some countries. Rotavirus infections also occur in infants and young children in developed countries, where they are often nosocomial. Special care should be used with hospitalized children to prevent such infections. The number of cases rises dramatically in the winter in the United States, and this helps to distinguish it from bacterial diarrheas.

Rotaviruses replicate in such great numbers that they can be visualized by electron microscopy in fecal suspensions (Figure 22.10); no effort to concentrate the viruses is required. Immune electron microscopy can be used to identify rotaviruses when antibodies from a patient's serum are reacted with virions in diagnostic specimens. In addition, many hospitals now use ELISA tests that detect rotaviruses in stool specimens. Treatment to restore fluid and electrolyte balance should be prompt.

The recognition of rotavirus infection as a human disease is a relatively recent achievement. Antibodies to the virus have been identified in as many as 90 percent of groups of children tested, even though the

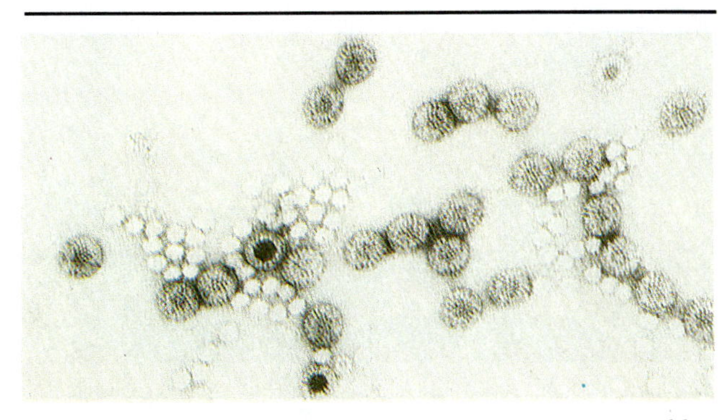

FIGURE 22.10 Rotaviruses, in fecal suspension, resemble little wheels and are the cause of diarrhea.

infection may not have been identified at the time it occurred. Much more research is needed to determine how immunity is produced and how the disease might be controlled.

Enteritis can be caused by other viruses. Echoviruses (*e*nteric *c*ytopathic *h*uman *o*rphan viruses) can cause mild gastrointestinal symptoms and damage intestinal cells, as their name implies. They also sometimes infect other tissues, and with a coxsackie virus can cause meningoencephalitis (inflammation of the brain and meninges).

Patients who have received bone marrow transplants are especially susceptible to infection with rotaviruses, coxsackie viruses, and the bacterium *Clostridium difficile.* As many as 55 percent of such patients succumb to these infections.

The Norwalk virus (named for a 1968 outbreak in Norwalk, Ohio) is responsible for nearly half of all outbreaks in the United States of acute infectious, nonbacterial enteritis. It more often affects older children and adults, rather than preschoolers or infants. Outbreaks occur throughout the year and are common at schools, camps, and nursing homes and on cruise ships. It is the second most common cause of illness (after respiratory disease) among U.S. families and occurs worldwide. It is characterized by 1 to 2 days of diarrhea, vomiting, or both. Immunity does not follow an attack, which makes development of a vaccine unlikely. Careful sanitary practices are the best means of prevention.

Hepatitis

Hepatitis, an inflammation of the liver, usually is caused by viruses (Figure 22.11 and Table 22.3). It also can be caused by an amoeba and various toxic chemicals. The most common viral hepatitis is **hepatitis A,** formerly called **infectious hepatitis;** it is caused by the hepatitis A virus (HAV), a single-stranded RNA virus usually transmitted by the fecal-oral route. **Hepatitis B,** formerly called **serum hepatitis,** is caused by the hepatitis B virus (HBV), a double-stranded DNA virus usually transmitted via blood. A third type of hepatitis transmitted parenterally (by blood) and probably caused by at least two viral agents is diagnosed in the absence of HAV and HBV as **hepatitis C** (HCV), formerly called **non-A, non-B (NANB) hepatitis.** A fourth type of hepatitis, transmitted enterically (fecal-oral route), formerly called **non-A, non-B, non-C hepatitis** has recently been separated out as **hepatitis E** (HEV). An especially severe **hepatitis D,** or **delta hepatitis,** is caused by the presence of both hepatitis D virus (HDV) and HBV. HDV alone does not cause disease. Refer to Chapter 11 for a detailed discussion of delta hepatitis. (p. 276)

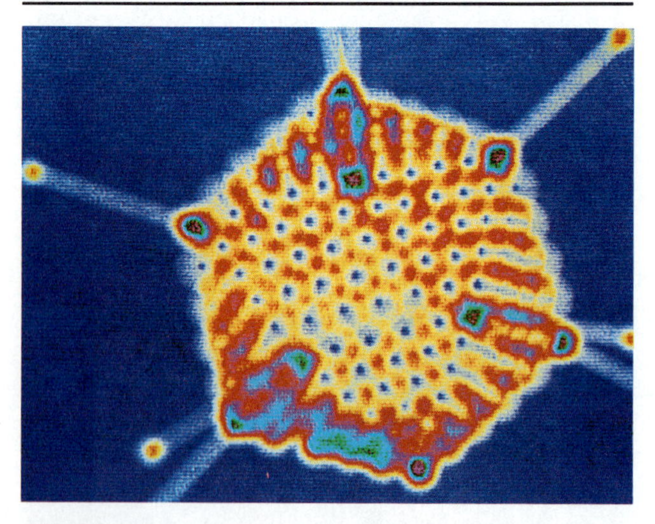

FIGURE 22.11 Hepatitis A virus, computer graphic, 350,000X, showing capsomeres and penton spikes projecting from corners.

Hepatitis A Hepatitis A occurs most often in children and young adults, especially in the autumn and winter. It can occur in epidemics if a population is subjected to water or food, especially shellfish, contaminated with HAV. This disease has an incubation period of 15 to 40 days and begins as an acute febrile illness. After entering the body through the mouth, the viruses replicate in the gastrointestinal tract and spread through the blood to the liver, spleen, and kidneys. Jaundice, a yellowing of the skin common in hepatitis, is caused by impaired liver function: The liver fails to rid the body of a yellow substance called **bilirubin,** which is a product of the breakdown of hemoglobin from red blood cells. Other symptoms of hepatitis are malaise, nausea, diarrhea, abdominal pain, and lack of appetite for a period of 2 days to 3 weeks. Chronic infections are rare, and recovery usually is complete. Immunologic tests are available to detect hepatitis A viruses and host antibodies against them. There is no treatment for hepatitis other than alleviating symptoms. Also, no vaccine for hepatitis A is yet available, but gamma globulin injections provide temporary immunity.

Hepatitis B Hepatitis B occurs in people of all ages with about the same incidence throughout the year. It can be transmitted by intravenous or percutaneous (into the skin) injections, by sexual practices among homosexual males, and by contaminated needles among intravenous drug users. Health-care workers who have routine contact with patients' body fluids (especially blood) have a higher incidence of the disease compared to the general community. Transmission via contaminated semen in artificial insemination has been documented. Hepatitis B has an incubation period of 45 to 180 days, with an average of 90 days. The virus replicates in cells of the liver, lymphoid tissues, and blood-forming tissues, and can persist in

TABLE 22.3 Comparison of types of viral hepatitis

Characteristic	Hepatitis A	Hepatitis B	Hepatitis C	Hepatitis D	Hepatitis E
Alternate names	Infectious hepatitis; epidemic hepatitis; short-term hepatitis	Serum hepatitis	Parenterally transmitted non-A, non-B, posttransfusion hepatitis	Delta hepatitis	Enterically transmitted non-A, non-B, non-C hepatitis
Agent	HAV RNA virus, Picornaviridae	HBV DNA virus, Hepadnaviridae	HCV At least two unclassified RNA viruses; ? Flavivirus, ? Togavirus	HDV Defective RNA virus, resembles plant viroids; has hepatitis B capsid	HEV Unclassified RNA virus; ? Calcivirus
Transmission	Fecal-oral	Blood and other body fluids; crosses placenta with high frequency	Blood and blood products; occasionally crosses placenta	Blood, must coinfect or superinfect with hepatitis B; can cross the placenta	Fecal-oral; more common in adults than in children
Incubation period	15–40 days; average, 28 days	45–180 days; average, 90 days	Short, 2–4 weeks; long, 8–12 weeks	2–12 weeks	2–6 weeks
Severity of disease	Self-limiting; usually mild, rarely severe	Subclinical to severe; most recover completely	Subclinical to severe; most resolve spontaneously	Severe; high mortality rate	Moderate but high mortality in pregnant women
Carrier state	No	Yes, is associated with 80% of liver cancer	Yes, possible association with liver cancer	Yes	No
Chronic liver disease	No	Yes	Yes	Yes	No
Vaccines	None yet; some are in clinical trials	Yes	No	No	No

the blood for years, creating a carrier state. The onset of symptoms is insidious, and fever is uncommon. Otherwise the symptoms are similar to those of hepatitis A, except that chronic active hepatitis B frequently destroys liver cells. Immunologic methods are available to detect the viruses and the host's antibodies. Treatment relieves some symptoms. A vaccine is available, and government regulations require that it be provided by employers for health care workers such as dentists, dental hygienists, dialysis technicians, and emergency room personnel who have contact with blood or body fluids that might contain hepatitis B viruses. The vaccine is safe, having been produced by recombinant genetic means that inserted the appropriate hepatitis genes into a plasmid. The plasmid was then inserted into yeast cells, thereby avoiding all contact with blood or human tissue culture. This yeast-produced vaccine has been given to over 2 million people in the United States and is 95 percent effective. When given to pregnant women who have hepatitis B, it reduces the number of infants who become carriers from 90 percent to 23 percent. This can be reduced to 5 percent when gamma globulin is given along with the vaccine. Given the fact that 40 percent of carriers will die of liver disease and that in some parts of the world nearly 90 percent of mothers are infected, the vaccine clearly could be saving many lives. Currently health officials are urging that all infants be vaccinated at birth. This is expensive and is probably beyond the means of many Third-World countries where it is most important. In the United States it has not been taken seriously, with even many health-care workers remaining unvaccinated.

The hepatitis B virus is unusually stable and resists drying and irradiation. Its double-stranded, circular DNA has a gap in one strand that may help it insert into liver cell DNA. Insertion of viral DNA into liver cell DNA, in turn, may contribute to liver cell carcinoma, a kind of cancer that occurs much more frequently in people who have had hepatitis B than in the general population.

Hepatitis C A virus particle has been recovered from parenterally transmitted non-A, non-B hepatitis and has been only partially characterized, but it is now called hepatitis C. Because this disease has two different incubation periods–2 to 4 weeks and 8 to 12 weeks–some researchers believe there may be two different causative agents. HCV can be distinguished from other kinds of hepatitis by the high blood concentration of a liver enzyme, alanine transferase. Various enzymes are released into the blood from damaged liver cells in all types of hepatitis, but this particular enzyme is unusually elevated in HCV hepatitis. Although the disease usually is mild or even inapparent, the infection becomes chronic in about half those infected. No vaccine is available, and immunity does not follow infection.

Hepatitis E Enterically transmitted through fecally contaminated water supplies, hepatitis E has caused large outbreaks in Asia and Africa. It is more common in adults than in children. Mortality rate is low (1 percent), except in pregnant women, where it is about 20 percent. No vaccine is available, and immunity does not follow infection.

Hepatitis D Hepatitis D has an incubation period of 2 to 12 weeks, but it is shorter when HBV carriers are superinfected with HDV than when individuals are infected with both viruses at the same time. HDV alone fails to cause disease because it requires HBV antigens for replication. HDV and HBV together can result in fatal hepatitis.

Protozoan Gastrointestinal Diseases

Giardiasis

Giardia The flagellated protozoan *Giardia intestinalis* (Figure 22.12) was first observed by Leeuwenhoek in 1681, when he was studying organisms in his own stools. It is, however, a far older organism. Examination of *Giardia*'s DNA has shown it to be the most primitive DNA of any eukaryote, very similar to that of older prokaryotes. On the basis of this, scientists have estimated it to have been around for about 3.5 billion years, long before there were intestines for it to infect.

It infects the small intestine of humans, especially children, and causes a disorder called **giardiasis** (jar-di'ah-sis). Ingestion of cysts from fecal material results in immediate release of motile trophozoites. The parasite has an adhesive disk by which it attaches to the bowel wall. It feeds mainly on mucus, forming cysts that are deposited in the mucus and passed intermittently in mucous stools. Symptoms include inflammation of the bowel, diarrhea, dehydration, and weight loss. Nutritional deficiencies are common in infected children because the parasites can occupy much of the intestinal absorptive area. Fat absorption is greatly reduced, and deficiencies of fat-soluble vitamins are common. Diarrhea is copious and frothy from bacterial action on unabsorbed fats, but it is not bloody because the parasites usually do not invade cells. Some patients experience severe joint inflammation and an itchy rash even before they have diarrhea. This **reactive arthritis of giardia** fails to respond to anti-inflammatory drugs ordinarily used to

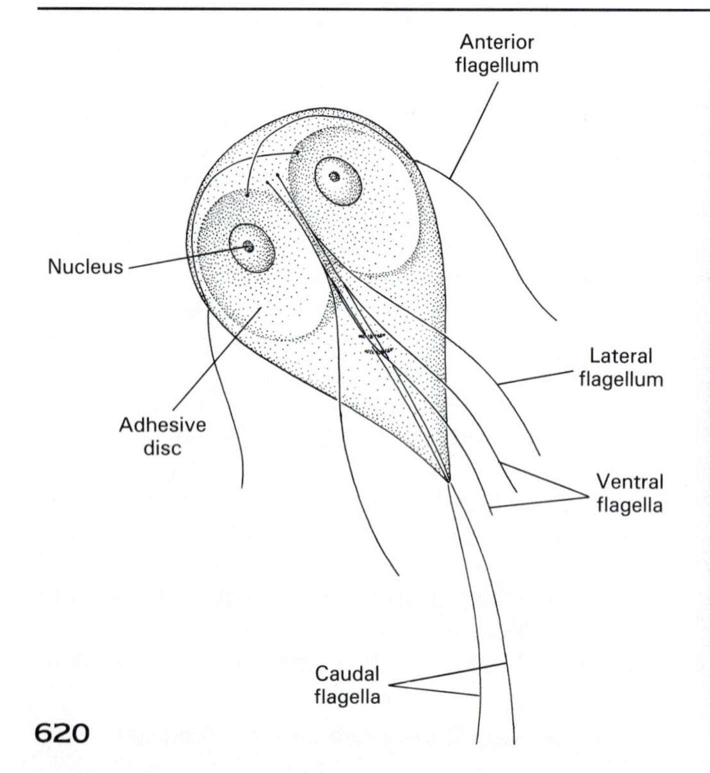

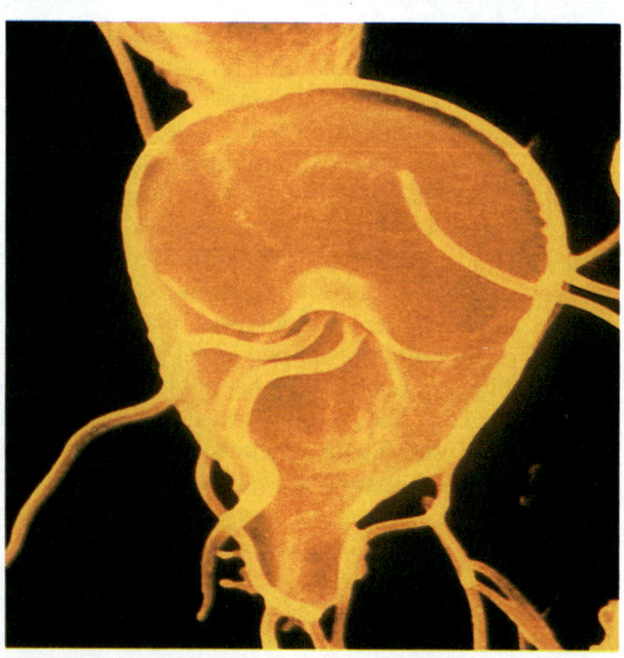

FIGURE 22.12 *Giardia intestinalis*, a protozoan parasite that causes diarrhea.

treat arthritis. The arthritis disappears along with the diarrhea when antiprotozoan drugs are given.

Giardiasis is transmitted through food, water, hands contaminated with fecal matter, and occasionally from wild animals, leading it to be called "beaver fever" by backpackers and hunters in the western United States. Contaminated water supplies in Aspen, Colorado, Leningrad, Russia, and probably many other places have caused large numbers of cases. In some child-care centers up to 70 percent of the children are infected with *Giardia*. *Giardia* cysts are not killed by ordinary sewage treatment chlorination. Some localities in Pennsylvania are being forced to drink bottled water until they can afford to add sand filters to their treatment plants to trap *Giardia* cysts.

Diagnosis is made by finding cysts of the parasite in stools. *Giardia* is found in the mucous sheets that line the intestinal surface. These are released every few days. Samples such as those taken from a bedpan should include some mucus. Because of the intermittent passing of cyst-containing mucus, daily stool samples for several days are needed to increase the likelihood of a positive diagnosis. Trophozoites are sometimes found in watery stools.

Quinacrine (Atabrine) and metronidazole (Flagyl) are used to treat giardiasis. Whenever one member of a family is treated, all should be treated to prevent passage of the infection among them. The disease can be prevented by maintaining pure water supplies uncontaminated by cysts from human or animal wastes.

Amoebic Dysentery and Chronic Amebiasis

Amebiasis is caused by *Entamoeba histolytica* (Figure 22.13), a major pathogenic amoeba. It was isolated from intestinal ulcers of a patient who had succumbed to severe diarrhea, used to infect a dog in which the same disease appeared, and recovered from the dog, thereby demonstrating Koch's postulates. ∞ (Chapter 15, p. 390) Amebiasis can appear as a severe acute disease called **amoebic dysentery** or as **chronic amebiasis** (am-e-bi'as-is), which can suddenly revert to the acute stage. Approximately 400 million people are infected worldwide, most with chronic amebiasis. The proportion of the population varies from 1 percent in Canada to 5 percent in the United States to 40 percent in tropical areas. Humans become infected with the parasite from cysts in food or water contaminated with fecal matter. After ingestion, cysts rupture and release amoeboid trophozoites, which reproduce asexually within the digestive tract. They invade the intestinal mucosa, where they can live indefinitely. As the parasites multiply, they cause more ulceration. Sometimes their proteolytic enzymes digest deep into, or even through, the bowel wall. Thus, the parasites sometimes enter blood vessels and travel to other tissues, or bacteria in fecal material enter the body cavity and cause peritonitis. Patients with amebiasis have abdominal tenderness, 30 or more bowel movements per day, and dehydration from excessive fluid loss. If the parasites invade liver and lung tissue, they can cause abscesses, and bacterial infection of lesions in any tissues can occur.

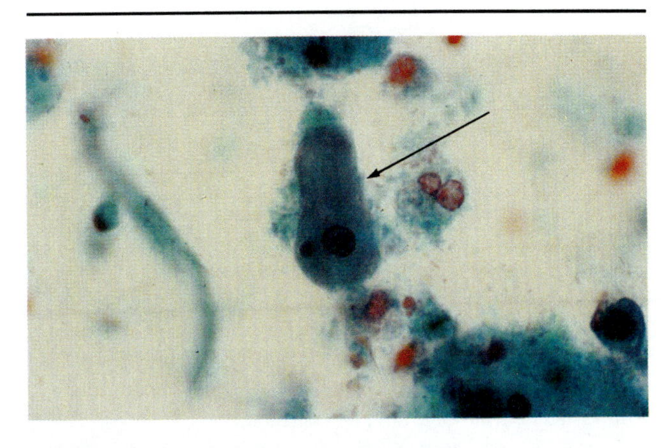

FIGURE 22.13 *Entamoeba histolytica* trophozoite (active amoebic form) and smaller red blood cells.

Because fecal material becomes dehydrated as it passes through the colon, trophozoites tend to encyst, and cysts are passed with the feces. Cysts can survive up to 30 days in a cool, moist environment and are not killed by normal chlorine concentrations in water. When cysts are ingested they pass through the stomach unharmed by acid conditions; the cysts rupture and the parasites multiply in the small intestine. The most common means of transmission is through unsanitary handling of food and water, but cysts can be transmitted through sexual contact among homosexual men.

Amoebic infections can be diagnosed by finding trophozoites or cysts in stools, but several stool samples on consecutive days may be needed to find them. Metronidazole is widely used in treatment, in spite of the fact that it has been found to be mutagenic in bacteria and carcinogenic in rats. Antibiotics also are used to prevent or cure secondary bacterial infections. Such infections can be prevented by sanitary handling of water and food. Although *Entamoeba coli* and *Endolimax nana* are considered commensals, finding them in food or water indicates fecal contamination.

Balantidiasis

Balantidium coli (Figure 22.14) is the only ciliated protozoan that causes human disease. It is distributed worldwide, particularly in the tropics, but human infection is rare except in the Philippines. *Balantidium coli* is transmitted by cysts in fecal matter. After ingestion, cysts rupture and release trophozoites that invade the walls of the small and large intestine, caus-

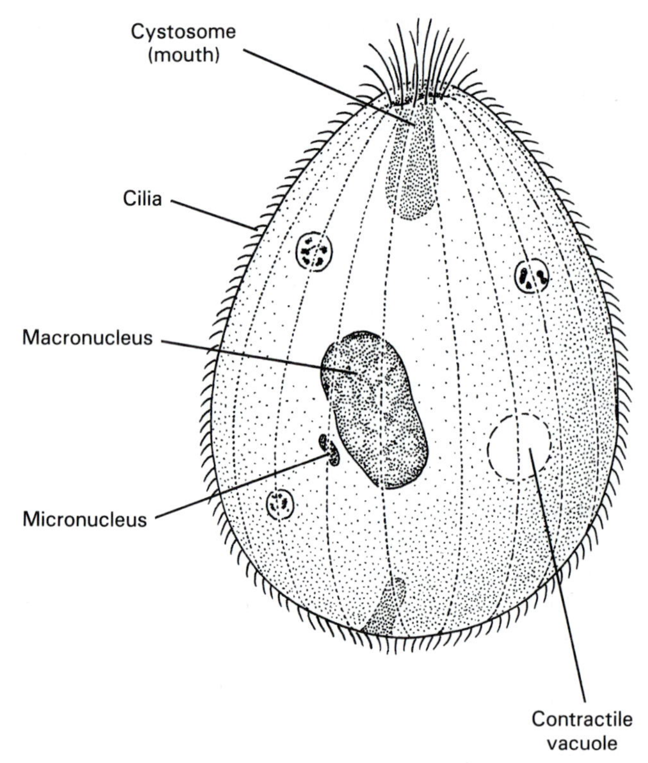

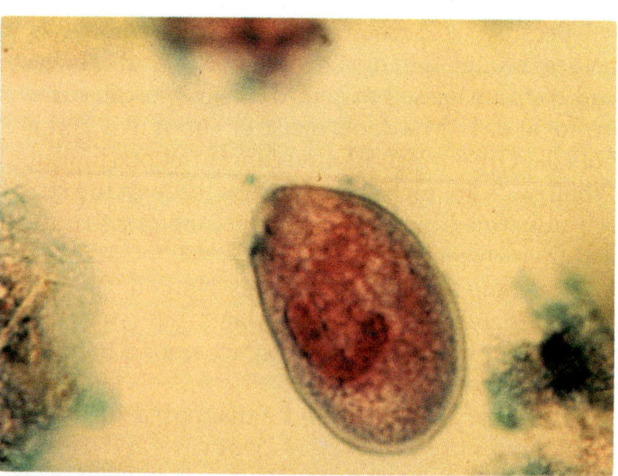

FIGURE 22.14 *Balantidium coli,* seen here in a fecal smear, is a very large ciliated protozoan that causes diarrhea. It is the only ciliate known to infect humans.

ing a dysentery known as **balantidiasis** (bal″an-tid-i′as-is). Symptoms of the disease are similar to those of amoebic dysentery, and, as with amoebic dysentery, perforation of the intestine can lead to fatal peritonitis.

Diagnosis is made by finding trophozoites or cysts in fecal specimens. Tetracycline is used to treat the disease, but some people remain carriers even after treatment. As with other organisms transmitted through fecal matter, infection can be prevented by good sanitation. Pigs serve as a reservoir of infection, so contact with their feces must be avoided.

Cryptosporidiosis

Protozoans of the genus *Cryptosporidium* have recently been observed worldwide to commonly cause opportunistic infections, probably by fecal-oral transmission from kittens and puppies. These organisms live in or under the membrane of cells lining the digestive and respiratory systems. After being swallowed, cysts burst in the intestine, releasing parasites that invade intestinal cells or migrate to other tissues. In immunocompetent individuals the disease is self-limiting, but in immunosuppressed patients it causes severe diarrhea—up to 25 bowel movements per day with a loss of as much as 17 L of fluid. The majority of severe cases of **cryptosporidiosis** (krip″to-spor-id-e-o′sis) have been seen in AIDS patients. No effective treatment has been found.

Effects of Fungal Toxins

Fungi produce a large number of toxins, and most come from members of the genera *Aspergillus* and *Penicillium*. Their various effects on humans include loss of muscle coordination, tremors, and weight loss. Some are carcinogenic.

Aspergillus flavus and other aspergilli produce substances called **aflatoxins** (af-lah-tox′inz). Aflatoxins are the most potent carcinogens yet discovered. Although the effects of the toxins on humans are not fully understood, their presence in foodstuffs may cause cancer of the liver. The toxins reach humans in food made from mold-infested grain and peanuts.

Claviceps purpura, also called **ergot,** is a parasitic fungus of rye and wheat (Figure 22.15). Although most strains of these grains grown in the United States are genetically resistant to the ergot fungus, many strains grown in other parts of the world are not. Ergot causes a variety of effects on humans. It produces lysergic acid, from which the hallucinogen LSD can be made. When the fungus is harvested with rye and incorporated into foodstuffs, it can cause **ergot poi-**

CLOSE-UP

"Hold the Aflatoxin!"

Aflatoxin can be present in peanut butter if any moldy peanuts are used to make it, and 7 percent of peanut butter samples tested in one study contained aflatoxin. Similar toxins can be present in jellies. Even when the moldy top layer of jelly is discarded, some toxins may have diffused down into the jelly. If, in addition, moldy bread is used, you might concoct a potentially highly carcinogenic sandwich from these simple ingredients.

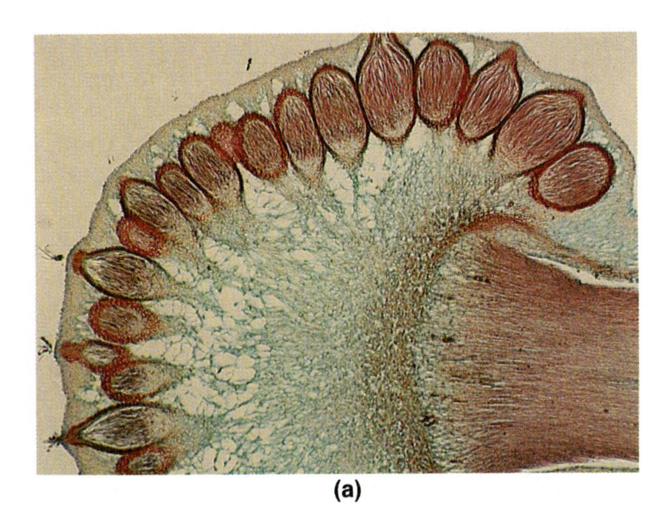

(a)

Ergotamine

(b)

FIGURE 22.15 (a) *Claviceps purpura*. (b) The structure of the toxic ergot alkaloid.

soning—hallucinations, high fever, convulsions, gangrene of the limbs, and ultimately death. The same substance that causes ergot poisoning in large quantities can be used therapeutically in small quantities. Drugs derived from ergot are used in carefully measured doses to control bleeding in childbirth, induce abortions, treat migraine headaches, and lower high blood pressure.

CLOSE-UP
Ergot Poisoning

Ergot poisoning has been implicated in the witchcraft trials of Massachusetts and Connecticut in 1692. The now-known symptoms of ergot poisoning are quite like the visions and fits reported in individuals hanged as witches. Ergot thrives on rye, which was commonly used in bread making in colonial times. Weather records for 1690 to 1692 show that New England summers were especially cool and moist and that winters also were especially cold. Thus, all the conditions necessary for ergot poisoning among the population were present.

Mushroom toxins come mainly from various species of *Amanita*, which are widely distributed around the world (Figure 22.16). The toxins phallotoxin (fal-lo-tox'in) and amatoxin act on liver cells. They cause vomiting, diarrhea, and jaundice; ingestion of a sufficient quantity of the toxin can be lethal or can so damage the liver that an organ transplant is necessary. The toxin muscarin elicits hallucinations.

(a) (b)

FIGURE 22.16 Deadly toxin-producing mushrooms. (a) *Amanita muscaria*. (b) *Amanita virosa*, commonly called the "Destroying Angel." Amanita mushrooms kill by means of a toxin that inhibits RNA polymerase. *Amanita muscaria* was formerly used as an insecticide—it was sprinkled with sugar and set out to attract flies, which died after nibbling on it.

Helminth Gastrointestinal Diseases

A wide variety of helminths can parasitize the human intestinal tract, and some also invade other tissues. Although most are prevalent only in tropical regions, several are endemic to the United States. Health care workers in the United States should be alert to the possibility that patients may have acquired such parasites while traveling in tropical areas.

Liver Fluke Infections

The sheep liver fluke *Fasciola hepatica* (Figure 22.17) is the most thoroughly studied of all flukes. It is found in humans in South America, Cuba, northern Africa, and some parts of Europe. Its intermediate host is a snail; cercaria from the snail mature in water and encyst as metacercaria on water vegetation. When humans eat such vegetation, especially watercress, the metacercaria are released in the intestine, bore through the intestinal wall, and migrate to the liver. There they feed on blood, block bile ducts, and cause

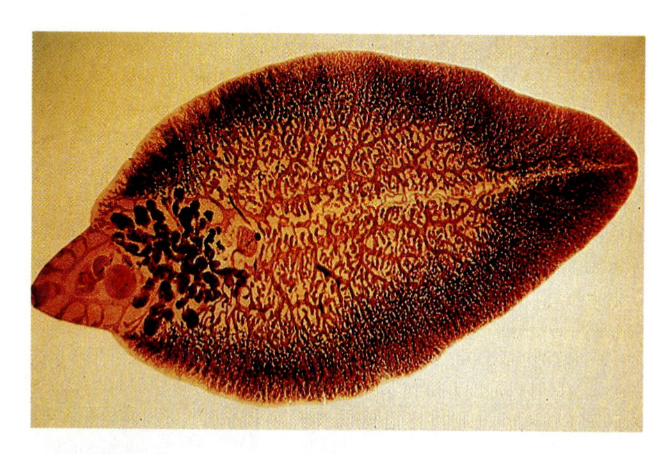

FIGURE 22.17 *Fasciola hepatica*, the liver fluke, stained to reveal inner structures.

inflammation. Sometimes they migrate to the eyes, brain, or lungs. Infections can be diagnosed by finding eggs in stool specimens. Infected humans can be treated with rafoxamide and other antihelminthic agents, but infection can be prevented by not eating water vegetation unless it has been cooked.

The Chinese liver fluke *Clonorchis sinensis* is widely distributed in Asia. As much as 80 percent of the population in rural areas is infected; some travelers and users of raw imported products also become infected. The life cycle is similar to *Fasciola* except that a second intermediate host, usually a fish and sometimes a crustacean, is required. Metacercaria excyst in the duodenum and migrate to the liver, and adult flukes take up residence in bile ducts. They destroy bile duct epithelium, block ducts, and sometimes perforate and damage the liver. The incidence of liver cancer is unusually high in areas where fluke infections are high, but it is unknown whether the fluke is responsible. Finding eggs in feces confirms the diagnosis, but there is no effective treatment. The parasite could be eradicated by cooking fish and crustaceans. However, the cultural habit of eating raw fish and shellfish and the lack of fuel for cooking maintain infections.

Yet another fluke, *Fasciolopsis buski*, is common in pigs and humans in the Orient. It lives in the small intestine and causes chronic diarrhea and inflammation. If several flukes are present, they can cause obstruction, abscesses, and **verminous intoxication,** an allergic reaction to toxins in the flukes' metabolic wastes. Antihelminthic drugs can be used to rid the body of flukes. Human infection can be prevented by controlling snails, avoiding uncooked vegetation, and stopping the use of human excreta for fertilizer.

Tapeworm Infections

Human tapeworm infections can be caused by several species, most of which have worldwide distribution. Illustrations of the tapeworm life cycle were provided in Chapter 12. (p. 310) Humans are most often infected by the eating of uncooked or poorly cooked pork or beef. Infection also can occur through contact with infected dogs or from eating insect-infested grain products or infected raw fish.

The pork tapeworm *Taenia solium* reaches a length of 2 to 7 m, and the beef tapeworm *T. saginata* reaches a length of 5 to 25 m. These worms usually enter the body as cysts in raw or poorly cooked meat, especially pork. Viable cysts can release larvae capable of developing into adult tapeworms. When adult tapeworms develop in the intestine (Figure 22.18a), they absorb large quantities of nutrients and lead to malnutrition even when the person has an adequate diet. Long worms may tangle up into a mass that blocks passage of materials through the intestine.

When humans ingest tapeworm eggs instead of cysts (Figure 22.18b), the egg shells disintegrate in the small intestine, and the embryos penetrate the intestinal wall and enter the blood. In 60 to 70 days the embryo migrates to various tissues and develops into a **cysticercus** (sis-teh-ser'kus), or bladder worm. The bladder worm consists of an oval white sac with the tapeworm head invaginated into it. In humans, cysticerci are frequently found in the brain, where they can reach a diameter of 60 mm. When pigs ingest tapeworm eggs, the cysticerci migrate to muscles and form cysts, which the pigs' defenses cause to be surrounded by calcium deposits. In humans calcification fails to stop the growth of cysticerci, and patients suffer paralysis and convulsions. Death of a cysticercus releases toxins and usually causes a severe, or even fatal, allergic response. Human tapeworms can be prevented by sanitary disposal of human wastes and by thoroughly cooking meats before eating them. Even freezing meats at −5°C for at least a week appears to kill the parasites.

Humans ingest eggs of *Echinococcus granulosus* tapeworms through contact with infected dogs, especially when the dogs lick the faces of small children. Eggs of this tapeworm are especially likely to produce cysts, called **hydatid** (hi-da'tid) **cysts** (Figure 22.18c), in vital tissues such as the liver, lungs, and brain. The cysts, which can contain thousands of tiny immature worm heads and often reach the size of a grapefruit or larger, exert pressure on organs. If a cyst ruptures, as may happen during an attempt at surgical removal, it can release all these infective units. A ruptured cyst can also cause anaphylactic shock and almost certain death.

Humans can ingest *Hymenolepis nana* tapeworm eggs in cereals or other foods that contain parts of infected insects. Such tapeworms in the intestine, especially in children, can cause diarrhea, abdominal pain, and convulsions.

The broad fish tapeworm, *Diphyllobothrium latum*,

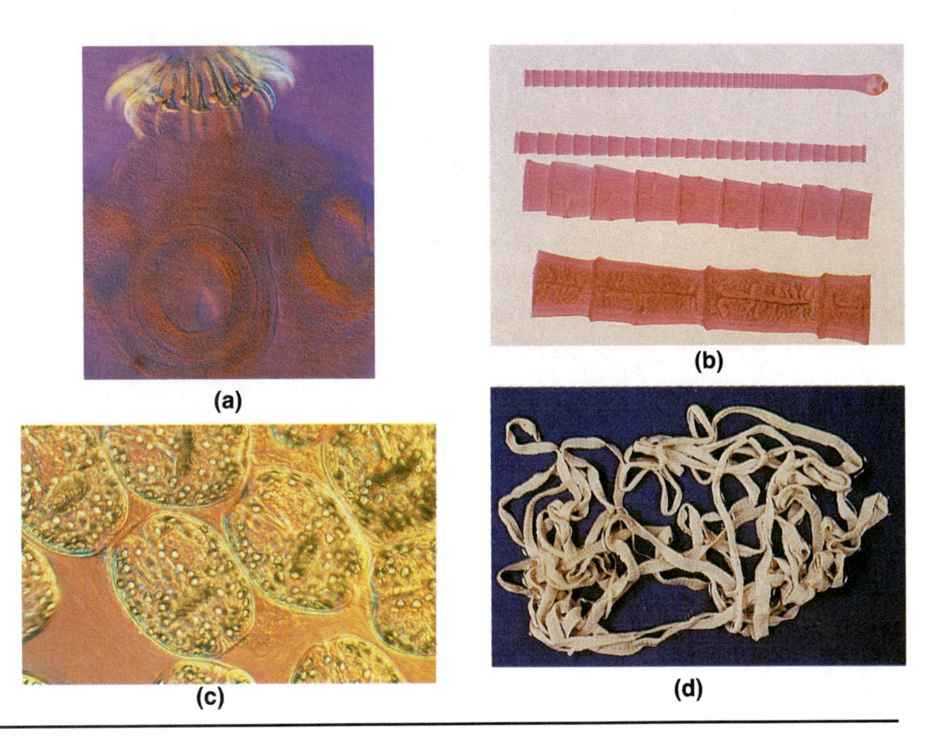

FIGURE 22.18 Tapeworms. (a) The scolex, or head, of *Taenia solium*, showing suction cups and hooks, which aid in attaching to the intestinal lining of its host. (b) Proglottids of *Taenia pisiformis* showing tiny new ones growing behind scolex, which increase in size as they age and move backwards away from the scolex. The last row of proglottids is filled with dark mature eggs. (c) Protoscolices of *Echinococcus granulosus*. Each round structure, the size of a grain of sand, contains the head of a tapeworm and can grow into a complete worm. Many such protoscolices are found in a fluid-filled hydatid cyst, which in extreme cases can contain millions of protoscolices and as much as 15 L of fluid. When even a more moderately sized cyst forms in the brain, it can do extensive damage. (d) Preserved specimen of *Taenia pisiformis* recovered from human intestine.

is common in fish-eating carnivores. It reaches humans through ingestion of raw or poorly cooked fish in sushi and other dishes. Fish tapeworm infections are common in Scandinavia, Russia, and the Baltic countries, approaching 100 percent infestation of the population in some areas and have also occurred in the Great Lakes area of the United States. The worm requires both a small crustacean and a fish to complete its life cycle. When humans ingest an infected fish, the worms coiled up in the fish muscle attach to the intestine and begin producing eggs. The parasites absorb large quantities of vitamin B_{12} and also impair the victim's ability to absorb the vitamin. Vitamin B_{12} deficiency anemia from tapeworm infections is especially high in Finland.

Tapeworm infections are diagnosed by finding eggs or proglottids in feces. Infections can be treated with niclosamide and other antihelminthic agents. Prompt diagnosis and treatment are important to remove worms before they invade tissues beyond the intestine. Human infections could be completely prevented by avoiding raw meats and fish, insect-infested grains, and infected dogs.

Trichinosis

Trichinosis (trik-in-o'sis) is caused by the small nematode *Trichinella spiralis*, also sometimes called *trichina worm*. This parasite, unlike most, is more common in temperate than in tropical climates. (Almost all U.S. adults have antibodies to *Trichinella*, which means that most of us carry around a few worms. But a few will not cause symptoms.) The parasite usually enters the digestive tract as encysted larvae (Figure 22.19) in poorly cooked pork, but infections have been traced to venison and other game animals and to horse meat in France. In the intestine cysts release larvae that develop into adults. The adults mate, the males die, and the females produce living larvae before they too die. The larvae migrate through blood and lymph vessels to the liver, heart, lungs, and other tissues. When they reach skeletal muscles, especially eye, tongue, and chewing muscles, they form cysts.

These parasites cause tissue damage as adults and as migrating and encysted larvae. The adult females penetrate the intestinal mucosa and release toxic wastes that produce symptoms similar to those of food poisoning. Wandering larvae damage blood vessels and any tissues they enter. Death can result from heart failure, kidney failure, respiratory disorders, and reactions to toxins. Encysted larvae cause muscle pain.

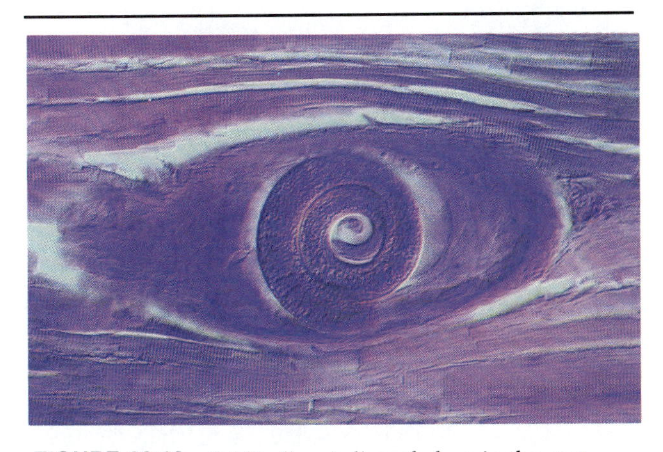

FIGURE 22.19 *Trichinella spiralis* curled up in the cyst stage, embedded in striated muscle fibers.

Trichinosis is difficult to diagnose, but muscle biopsies and immunological tests are sometimes positive. Treatment is directed toward relieving symptoms because the disease cannot be cured. It can be prevented by eating only thoroughly cooked meat. Freezing does not necessarily kill encysted larvae, and microwave cooking is safe only if the internal temperature of the meat reaches 77°C. Microwave cooking depends on geometry. Since pieces of meat are irregularly shaped, they must be rotated during cooking to heat evenly. Studies using pork experimentally infected with *Trichinella* show survival of some worms whenever cooking is done without rotation.

Hookworm Infections

Hookworm can be caused by two species of small roundworms, *Ancylostoma duodenale* and *Necator americanus* (Figure 22.20). Although these parasites have a complex life cycle, that cycle can occur in a single host, and the host is often a human. Eggs in feces quickly hatch in moist soil, releasing free-living larvae that feed on bacteria and organic debris, grow, molt, and become mature parasitic larvae. If these larvae reach the skin, typically of the feet or legs, they burrow through it to reach blood vessels that will carry them to the heart and lungs. The larvae then penetrate lung tissue, and some are coughed up and swallowed. In the intestine the larvae burrow into villi and mature into adult worms. The adult worms mate and start the cycle over again.

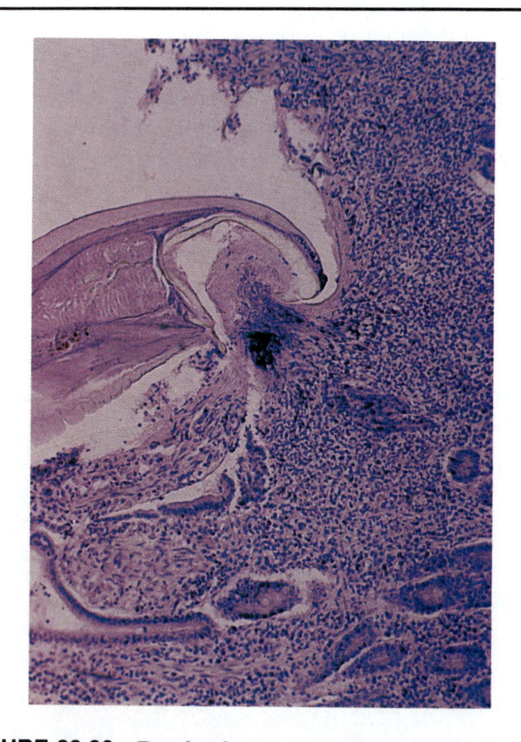

FIGURE 22.20 Dog hookworm, *Ancylostoma canium*, about 1 cm long, attached to intestinal epithelium, where it sucks blood. Note the large piece of intestine taken into its mouth. (Heavy infestations may result in anemia.)

As larvae burrow through the skin, host inflammatory reactions kill many of them, although bacterial infection of penetration sites causes **ground itch.** In the lungs the parasites cause many tiny hemorrhages, but they cause the greatest damage to the lining of the entire small intestine. They feed on blood and cause abdominal pain, loss of appetite, and protein and iron deficiencies. These effects are especially debilitating in people whose diets are barely adequate without the burden of worm infections, so people infected with hookworms often appear to be lazy and shiftless.

Diagnosis is by finding eggs or worms in feces, but samples must be concentrated to find them. Tetrachloroethylene is effective against *Necator* infections and is inexpensive and easy to administer in mass treatment efforts. However, it fails to kill *Ascaris*, a helminth commonly found in patients with hookworm, and can cause *Ascaris* to form clusters that block the intestine. Bephenium hydroxynaphthalate (Alcopar), mebendazole, and several other drugs, although more expensive, kill both species of hookworms and *Ascaris* as well. Dietary supplements should be provided for all hookworm patients. Hookworm is preventable through sanitary disposal of human wastes, but stopping the use of human excreta as fertilizer and getting uneducated people to use latrines can be difficult. Plantation workers often defecate repeatedly in shady areas near fields where they work and in doing so continuously infect themselves and others.

Larvae of species of hookworm for which humans are not the normal host sometimes penetrate the skin and cause **cutaneous larva migrans,** or creeping eruption. Severe skin inflammation results from body defenses that prevent further migration of the parasites. Such infections often are acquired from infected cats and dogs. They can be treated with thiabendazole.

Ascariasis

Ascaris lumbricoides (Figure 22.21) is a large roundworm 25 to 35 cm long that causes **ascariasis** (as-kar-i'a-sis). People become infected by ingesting food or water contaminated with *Ascaris* eggs. Once in the intestine, the eggs hatch, and larvae penetrate the intestinal wall and enter lymph vessels and venules. Although they can invade and cause immunologic reactions in almost any tissue, most move through the respiratory tract to the pharynx and are swallowed. Larvae move to the small intestine, mature, and begin to produce eggs. Eggs are especially resistant to acids and can develop in 2 percent formalin. They also resist drying, and people can be infected by airborne eggs. In some areas of the southern United

States where the soil never freezes, 20 to 60 percent of children are infected.

Ascaris worms cause three kinds of damage:

1. Larvae burrowing through lung tissue cause *Ascaris* pneumonitis, which involves hemorrhage, edema, and blockage of alveoli from worms, dead leukocytes, and tissue debris. Secondary bacterial pneumonia can be fatal.
2. Adult worms cause malnutrition, but because they feed mainly on the contents of the intestine, they do little damage to the mucosa. They also release toxic wastes that elicit allergic reactions. If sufficiently numerous they cause intestinal blockage and sometimes perforation. The peritonitis that follows perforation is nearly always fatal.
3. Wandering worms cause abscesses in the liver and other organs and sometimes traumatize victims by crawling from body openings.

FIGURE 22.21 *Ascaris lumbricoides*, a large roundworm. Females may reach a length of 30 cm and can produce 200,000 eggs per day. The eggs are passed out in feces and can survive in soil for months or even years.

PUBLIC HEALTH

Sushi

In recent years, many Americans have fallen in love with sushi, the Japanese delicacy made of raw fish. If you are one of them, have you ever worried about your chances of ingesting a live worm with that tasty snack—and wondered what the consequences might be if you did?

In Japan, several hundred people each year find out. Typically they consume sushi at dinner and awake in the small hours of the morning with such agonizing pain that they are rushed to the hospital. There, examination of the stomach with a fiber-optic gastroscope reveals a typical bloody ulcer, about 5 cm in diameter, with a 5- to 7.5-cm-long worm in the center, attached by its head. In some cases, the ulcer is so severe that a portion of the stomach wall must be removed. In more fortunate cases, the worm is removed by forceps attached to the gastroscope. The worm is a larval stage of *Anisakis*, a roundworm relative of *Ascaris*. The infestation is called anisakiasis.

The Dutch have had their problems with anisakiasis too, thanks to the relatively recent introduction of a new dish, raw salted herring, which immediately became very popular. The worm is killed by freezing, and Dutch law now requires that all herring served this way must be frozen first.

What about the odds of a sushi dinner giving you anasakiasis in the United States? Only a half dozen or so cases have been reported here, and curiously, ours tend to involve problems somewhat higher up in the digestive tract. A California man had eaten raw white sea bass 10 days before he felt a peculiar sensation at the back of his throat. Coughing, he reached into his mouth and pulled out a lively 7.5-cm-long worm. The other U.S. victims have similarly coughed up and removed their own worms—one as quickly as 4 hours after a meal of cod fillet. Still, considering the number of people in the United States who eat raw or undercooked fish, the total number of diagnosed cases of anisakiasis is so low that there is not much to worry about—so far!

Diagnosis is by finding eggs or worms in feces. The adult worms can be eradicated from the body by piperazine, mebendazole, and several other drugs, but no treatment is available to rid the body of larvae. Infection is fully preventable by good sanitation and personal hygiene.

Toxocara species, relatives of *Ascaris*, ordinarily parasitize cats and dogs. It has been estimated that in the United States 98 percent of puppies, including those from good kennels, are infected. Infection rates may be almost as high in dogs and cats of any age. **Visceral larva migrans** is the migration of larvae of these parasites in human tissues such as liver, lung, and brain, where they cause tissue damage and allergic reactions. The risk of such human infections can be minimized by worming pets periodically, disposing of pet wastes carefully, and keeping children's sandboxes inaccessible to pets.

Trichuriasis

Trichuriasis (trik-u-ri'as-is) is caused by the **whipworm,** *Trichuris trichiura*, which is distributed nearly worldwide. It is estimated that nearly 300 million people are infected, some of them in the southeastern United States. For humans to be infected human feces must be deposited on warm, moist soil in shady areas. Small children, who put dirty hands in their mouths, are particularly susceptible to infection. Eggs deposited with feces contain partially developed embryos. When the eggs are swallowed, they hatch, and juveniles crawl into enzyme-secreting glands of the intestine called crypts of Lieberkühn, where they undergo development. They return to the intestinal lumen, where they reach full maturity within 3 months of initial infection.

Adult worms damage the intestinal mucosa and feed on blood cells. They cause chronic bleeding, anemia, malnutrition, allergic reactions to toxins, and susceptibility to secondary bacterial infection. They can cause rectal bleeding in children. Infection is diagnosed by finding worm eggs in stools. The drug mebendazole is effective in ridding the intestine of parasites, but sanitary disposal of wastes is essential to prevent reinfection.

Strongyloidiasis

Strongyloidiasis (stron"jil-oi-di'as-is) is caused by *Strongyloides stercoralis* (Figure 22.22) and a few closely related species. This parasite is unusual in that females produce eggs by parthenogenesis, that is, without fertilization by a male. Adult females attach to the small intestine, burrow into underlying layers, and release eggs containing partially developed embryos. Many eggs hatch in the intestine and are passed with the feces. Juveniles can become free-living adults or remain as filariae and penetrate the skin of new hosts. Filariae are carried by blood to lungs, where they bore to the trachea, travel to the pharynx, and are swallowed. When they arrive in the small intestine, they develop into adults and restart the life cycle.

The parasites cause itching, swelling, and bleeding at penetration sites, which often become infected with bacteria. Migrating filariae cause immunologic reactions in the host, but the reactions usually do not stop the parasites. Coughing and burning sensations in the chest occur in lung infections, and burning and ulceration occur in intestinal infections. Secondary bacterial infection in any tissue can lead to serious septicemia. The diarrhea and fluid loss associated with intestinal parasites is severe and difficult to control even with electrolyte therapy. Consequently, patients often die of complications such as heart failure or paralysis of respiratory muscles.

Infection in humans usually occurs when filariae are encountered in contaminated soil or water. Diagnosis is difficult because juveniles can be found in fecal smears only in massive infections. Work is in progress to develop a reliable immunological test. The drugs thiabendazole and cambendazole have the greatest effect on the parasites with the fewest undesirable side effects.

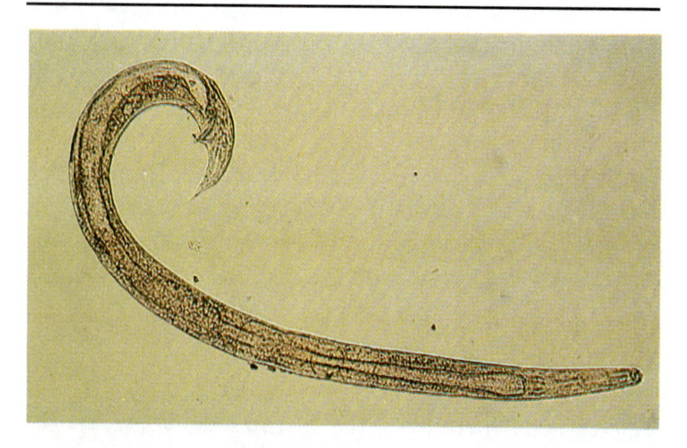

FIGURE 22.22 *Strongyloides stercoralis*, a roundworm parasite of the small intestine.

Pinworm Infections

Pinworm infections are caused by a small roundworm, *Enterobius vermicularis*. Like the hookworm, this parasite can complete its life cycle without an alternate host. Adult worms attach to the epithelium of the large intestine and mate, and the females produce eggs. Egg-laden females migrate toward the anus during the night, release their eggs on the exterior of the anus, and then crawl back in. These eggs are easily transmitted to other people by bedclothes, by debris under fingernails of those who scratch the itchy area around the anus, and even by inhalation of airborne eggs. Ingested eggs hatch in the small intestine and mature and reproduce in the large intestine. The worm population in an infected individual also is increased by **retrofection,** in which eggs hatch in a few hours and larvae reenter the body.

Although pinworm infection usually is not debilitating, it does cause considerable discomfort and can interfere with adequate nutrition, especially in children. Infection with large numbers of the worms can cause the rectum to protrude from the body. Pinworm infections are diagnosed by finding eggs around the anus; at night or right after the host awakens, the pinworms can be picked up with the sticky side of cellophane tape affixed to a wooden tongue depressor. If one member of a family has pinworms, all are presumed infected and are treated with piperazine or another antihelminthic agent. These agents are generally inexpensive and nontoxic. Bed linens, clothing, and towels should be washed and the house thoroughly cleaned at the time of treatment; the treatment and cleaning are repeated in 10 days. Despite these efforts, reinfection in families is very likely.

Gastrointestinal diseases caused by pathogens other than bacteria are summarized in Table 22.4.

TABLE 22.4 Summary of gastrointestinal diseases caused by other pathogens

Disease	Agent(s)	Characteristics
Viral gastrointestinal diseases		
Viral enteritis	Rotaviruses	Viral replication destroys intestinal epithelium; causes diarrhea and dehydration that can be fatal under age 5
Hepatitis A	Hepatitis A virus	Viral replication in intestinal and other cells causes malaise, nausea, diarrhea, abdominal pain, lack of appetite, fever, and jaundice; usually self-limiting
Hepatitis B	Hepatitis B virus	Viral replication and symptoms similar to hepatitis A except onset is insidious and fever usually absent; chronic infections and carrier state occur; transmitted parenterally; vaccine available
Hepatitis C	Unknown, but may involve two agents	Mild symptoms, but disease often chronic
Hepatitis D	Hepatitis D and B viruses	Can result in fatal hepatitis
Hepatitis E	Hepatitis E virus	Moderate disease, but high mortality in pregnant women
Protozoan gastrointestinal diseases		
Giardiasis	*Giardia intestinalis*	Parasite attaches to intestinal wall; feeds on mucus and causes inflammation, diarrhea, dehydration, nutritional deficiencies, and sometimes reactive arthritis
Amoebic dysentery	*Entamoeba histolytica*	Parasites ulcerate mucosa and cause severe acute diarrhea, abdominal tenderness, and dehydration; parasites can live indefinitely in intestine, causing latent amebiasis
Balantidiasis	*Balantidium coli*	Organisms invade walls of intestine; cause dysentery, and sometimes perforation and peritonitis
Cryptosporidiosis	*Cryptosporidium* species	Organisms live in or under mucosal cells and cause severe diarrhea in immunodeficient patients
Helminth gastrointestinal diseases		
Fluke infections	*Fasciola hepatica, Clonorchis sinensis, Fasciolopsis buski*	Organisms excyst in intestine and migrate to liver, block ducts, and damage tissues; *F. buski* causes intestinal obstruction, abscesses, and verminous intoxication
Tapeworm infections	*Taenia solium, T. saginata, Echinococcus granulosis, Hymenolepis nana, Diphyllobothrium latum*	Most infections caused by ingesting encysted larvae, which mature and erode intestinal mucosa; ingestion of eggs allows larval forms to develop in humans as cysticerci or hydatid cysts that can damage the brain and other vital organs
Trichinosis	*Trichinella spiralis*	Larvae excyst in intestine, mature, and produce new larvae, which migrate to various tissues and encyst in muscles where they cause pain; larvae cause tissue damage, and adults release toxins
Hookworm infections	*Ancylostoma duodenale, Necator americanus*	Larvae penetrate skin and migrate via blood vessels to heart and lungs; coughed-up larvae enter digestive tract and burrow into villi; mature worms feed on blood, cause abdominal pain, loss of appetite, and protein and iron deficiencies
Ascariasis	*Ascaris lumbricoides*	Eggs hatch in the intestine, and larvae cause immunologic reactions in many tissues; adults cause malnutrition by feeding on nutrients in gut; wandering worms cause abscesses
Trichuriasis (whipworm)	*Trichuris trichiura*	Eggs hatch in the intestine, and juveniles invade crypts of Lieberkühn and mature; adults damage intestinal mucosa, causing chronic bleeding, anemia, malnutrition, and allergic reactions to toxins
Strongyloidiasis	*Strongyloides stercoralis*	Filariae penetrate the skin, bore to trachea, climb to pharynx and are swallowed; mature worms bore into intestine and release embryos; cause inflammation, bleeding, and immunologic reactions at various sites, severe diarrhea and fluid loss
Pinworm infections	*Enterobius vermicularis*	Ingested eggs mature in intestine and interfere with nutrition, especially in children.

ESSAY

Cholera on Our Doorstep

In 1892, the great German hygienist Dr. Max von Pettenkofer, in the presence of witnesses, raised a broth culture of *Vibrio cholera* to his lips and swallowed approximately 1 billion organisms. His aim was to disprove Robert Koch's studies, done in India, showing that cholera was a contagious disease spread by polluted water. Von Pettenkofer firmly believed that cholera was neither contagious nor related to drinking water, but was due instead to interactions of microbes and soil, with "soil factors" playing the greatest role. Control of disease would rely on removing these "soil factors." Within a few days, von Pettenkofer developed a mild but genuine case of cholera, and *V. cholerae* organisms were recovered from his stool. However, von Pettenkofer insisted that his was *not* a real case of cholera. So, a few days later, his assistant repeated the experiment, fell severely ill with cholera, but recovered. The culture used in these experiments was a weakly virulent strain, supplied by the bacteriologist Georg Gaffky, who had guessed at what von Pettenkofer intended to do and did not want the 74-year-old man to die.

In February 1991 a similar scene was replayed in Peru, with the president of the country, Alberto Fujimori, and his wife being shown repeatedly on television eating raw fish and telling Peruvians that this popular dish was safe to eat. At that time, Peru had over 45,000 reported cholera victims and at least 193 deaths and the Peruvian Health Ministry was insisting that raw fish was likely to be contaminated and therefore unsafe to eat. Doctors all over the world criticized President Fujimori for his foolhardy defense of the Peruvian fisheries industry. Fujimori neglected to make it clear to his viewers that the fish he and his wife ate had been caught far out at sea, away from the sewage-contaminated coastal and river waters where poorer Peruvians caught their fish. The Peruvian minister of fisheries next attempted the same demonstration and promptly came down with cholera.

In April 1991 eight people in New Jersey contracted cholera by eating crab meat illegally imported from Ecuador. And in September of that year, Alabama state health officials closed an oyster reef near Dauphin Island in the Gulf of Mexico, which they said was teeming with the deadly strain of *V. cholerae* that was causing the outbreak in South America. How did this menace arrive at our shores?

Cholera had existed on the Indian subcontinent for centuries before the first Europeans arrived. Portugese explorers described it early in the sixteenth century, but it did not move out to other areas until 1817. We are now experiencing the seventh pandemic to have swept the globe. The United States experienced the second, third, and fourth of these, beginning in 1832, 1849, and 1866 respectively. The fourth pandemic left more than 50,000 Americans dead, with outbreaks continuing until about 1878. In 1887 and 1892 boatloads of infected passengers again arrived in New York as part of the fifth pandemic. However, a drastic shift in knowledge and attitudes in the United States prevented its spread. In the 1832 outbreak, cholera had been viewed as a punishment from God, visited upon the wicked. Indeed, health authorities resisted pleas to clean up streets and water supplies as cholera advanced on their localities, saying that to do so would be to oppose the will of God. And so, herds of pigs wandered the streets of New York City as the sole garbage "collectors," leaving feces and filth in their wake. In contrast, by 1866 most intelligent physicians had come to realize that cholera was a contagious

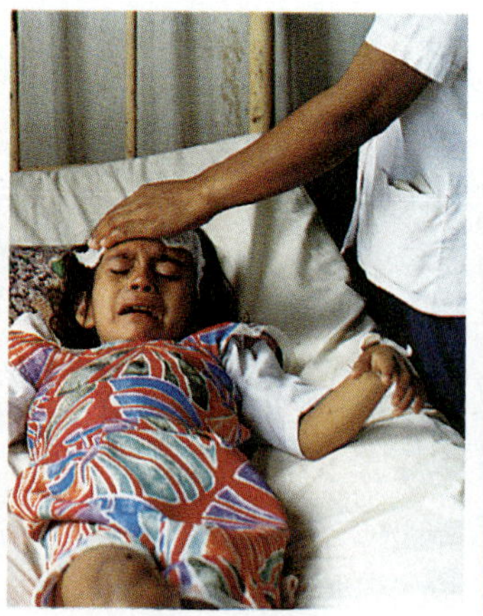

FIGURE 22.23 (a) Contaminated water in Chimbote, Peru, is used for washing, drinking, and bathing. Cholera organisms from excrement of victims also enter this water and spread rapidly through the population. (b) Lima, Peru, April 1991. A young girl cries as she lies in a hospital bed receiving treatment for cholera. Fluid replacement is especially important. (c) Peruvian boys playing in refuse piles hold a puppy while pigs in the background consume edible garbage and add their own urine and feces to the piles—an ideal situation for spread of disease.

disease spread by filth and ignorance. During this pandemic, physicians and city officials learned a great deal about control measures. By implementing these measures, they prevented future pandemics from spreading extensively into the United States.

The seventh pandemic began in Indonesia in 1958 and is of the El Tor variety. It spread to Macao, Hong Kong, and the Philippines, reaching eastern Europe and Africa by 1965. It is still rampant in parts of Africa and Asia. In 1989, sub-Saharan Africa accounted for three-quarters of all cases in the world. Then by ship, cholera reached a port city north of Lima, Peru, in January 1991. It spread through the untreated waters and via infected people and contaminated food to Lima. Within 3 months, some 175,000 people had sickened, and 1258 had died. Because only about one-fourth of infected people show symptoms, a vastly large number of Peruvians must have been infected. Meanwhile, cholera spread into nearby countries: Colombia, Ecuador, Chile, Guatemala, Brazil, and Mexico. Because the current vaccine protects only half of those who receive it, and then only for a 6-month period, mass vaccination campaigns are not being recommended by the World Health Organization. As of August 7, 1991, the CDC reported the following statistics in the Western Hemisphere: 274,768 cases of cholera; 119,644 hospitalized; and 2972 deaths, with 15 epidemic-associated cases in the United States.

With current knowledge of appropriate treatment and hygienic standards, cholera is not expected to pose a major threat to health in the United States. However, there is great fear that cholera will establish itself in South America as an endemic disease, to bring death for decades to poor areas that lack proper sewage- and water-treatment systems. Already the government of Brazil has begun a $1.5 million program of building latrines. Greater efforts than these will be required, however, as is exemplified by an interview with the family of an 18-year-old Mexican girl who had just died of cholera. The family had just finished digging the latrine that the health department had required they build, instead of using the dirt outside their door. When asked if he understood how it was that his sister had gotten cholera and died, her brother nodded yes and explained that it was because of all the dust raised by missiles in Operation Desert Storm the previous summer. The poisonous dust had circled the globe and killed his sister. In Peru, another man refuses to boil his drinking water, saying that boiled water doesn't quench your thirst—it is like drinking nothing. Children continue to play in sewer outfalls. The fight against cholera is also a battle against ignorance.

CHAPTER SUMMARY

The agents and characteristics of the diseases discussed in this chapter are summarized in Tables 22.1, 22.2, and 22.4. Information in those tables is not repeated in this summary.

RELATED KEY TERMS

DISEASES OF THE ORAL CAVITY

Bacterial Diseases of the Oral Cavity

- **Plaque** consists of microorganisms in organic matter on tooth enamel. — **pellicle** **methacrylate**
- Most **dental caries** occur during childhood and early adulthood; they can be prevented by good oral hygiene, **fluoride,** and sealants.
- **Periodontal disease** can be prevented by preventing plaque buildup and treated with peroxide–sodium bicarbonate, mouth rinses, and surgery. — **gingivitis**; **acute necrotizing ulcerative gingivitis**

Viral Diseases of the Oral Cavity

- **Mumps** occurs worldwide mainly in children; it can be prevented by a vaccine. — **orchitis**

GASTROINTESTINAL DISEASES CAUSED BY BACTERIA

Bacterial Food Poisoning

- **Food poisoning** is caused by ingesting food containing preformed toxins; it can be prevented by sanitary handling and proper cooking and refrigeration of foods.
- Staphylococcal **enterotoxicosis** usually occurs from eating poorly refrigerated foods, especially dairy and poultry products. — **enterotoxin**
- Other kinds of food poisoning usually arise from eating undercooked meats and gravies, dairy products, and fried rice. — **necrotizing enterocolitis**; **bongkrek**

Bacterial Enteritis and Enteric Fevers

- **Enteritis** is an inflammation of the intestine; **enteric fever** is a systemic disease caused by pathogens that invade other tissues. All enteritises and enteric fevers are transmitted via the fecal-oral route and can be prevented by good sanitation. — **bacterial enteritis** **dysentery**; **diarrhea**
- **Salmonellosis** is self-limiting and is treated with antibiotics only in high-risk patients. — **enterocolitis**
- **Typhoid fever** is diagnosed by the Widal test, which detects antibodies to

Salmonella typhi, and treated with chloramphenicol. Cell-mediated immunity follows infection, and vaccine elicits temporary partial immunity.

- **Shigellosis,** or **bacillary dysentery,** is treated with antibiotics, and recovery or vaccine elicits transient immunity.
- **Asiatic cholera,** common in Asia and other regions with poor sanitation, is treated with fluid and electrolyte replacement and tetracycline; recovery or vaccine elicits temporary immunity.
- **Vibriosis** is a mild disease common where raw seafood is eaten.
- **Traveler's diarrhea** occurs in more than 1 million travelers each year; it is usually self-limiting but can have complications.
- ***Escherichia coli*** is an indicator of fecal contamination and an opportunistic pathogen.

Bacterial Infections of the Stomach, Esophagus, and Duodenum

- *Helicobacter pylori* is now considered to be the cause of peptic ulcers and chronic gastritis and a probable cofactor of stomach cancer.

Bacterial Infections of the Gallbladder and Biliary Tract

- Bile destroys most organisms that have lipid envelopes. The typhoid bacillus is resistant to bile; it can live in the gallbladder and be shed in feces without causing symptoms. Gallstones blocking bile ducts can cause infections of the gallbladder and ducts, usually by *E. coli*; infection can spread to the bloodstream or ascend to the liver.

RELATED KEY TERMS

serovars

enteroinvasive strains
enterotoxigenic strains
yersiniosis

GASTROINTESTINAL DISEASES CAUSED BY OTHER PATHOGENS

Viral Gastrointestinal Diseases

- Most viral gastrointestinal disorders arise from contaminated water or food and are transmitted via the fecal-oral route, but hepatitis B, C, and D are transmitted parenterally from contaminated blood and other body fluids.
- **Rotavirus** infections kill many children in underdeveloped countries.
- Treatment for **hepatitis** relieves symptoms; gamma globulin gives temporary immunity to **hepatitis A,** and a vaccine is available for **hepatitis B.**

viral enteritis

infectious hepatitis A
serum hepatitis B
non-A, non-B hepatitis **hepatitis C**
hepatitis D **bilirubin** **hepatitis E**

Protozoan Gastrointestinal Diseases

- Protozoan gastrointestinal diseases arise from contaminated food and water via the fecal-oral route and can be prevented by good sanitation.
- Diagnosis usually is by finding cysts in fecal matter; most can be treated with antiprotozoan agents.
- **Giardiasis** is especially common in children; **amoebic dysentery, chronic amebiasis,** and **balantidiasis** occur worldwide but mainly in tropical areas; **cryptosporidiosis** occurs mainly in immunodeficient patients.

reactive arthritis of *Giardia*

cryptosporidium

Effects of Fungal Toxins

- **Aflatoxins** are potent carcinogens produced by fungi of the genus *Aspergillus*; humans ingest them from moldy grain and peanuts.
- **Ergot poisoning** comes from eating grains contaminated with *Claviceps purpura*; small quantities of **ergot** can be used medicinally.
- Mushroom toxins, which come mainly from species of *Amanita*, cause vomiting, diarrhea, jaundice, and hallucinations; in sufficient quantities they are fatal.

verminous intoxication

Helminth Gastrointestinal Diseases

- Helminth gastrointestinal diseases are acquired mainly in tropical regions and include several kinds of fluke, roundworm, and tapeworm infections.
- Helminths that infect humans often have complex life cycles in which meat animals, fish, snails, and crustaceans also may serve as hosts.
- Most such diseases can be diagnosed by finding eggs in fecal specimens. They can be prevented by good sanitation—avoiding contaminated water and soil and thoroughly cooking foods that might be contaminated.

cysticercus **hydatid cysts**
trichinosis **hookworm**
ground itch
cutaneous larva migrans
ascariasis **visceral larva migrans**
trichuriasis **whipworm**
strongyloidiasis **pinworm**
retrofection

QUESTIONS FOR REVIEW

A.

1. What is plaque, and how does it form?
2. What are dental caries, how do they form, and how are they treated?
3. What is periodontal disease, and how is it treated?
4. What causes mumps, and what complications can arise from the disease?
5. Could mumps be eradicated, and if so, how?

B.

6. Why is food poisoning not considered an infection?

7. How is food poisoning acquired, and how can it be prevented?
8. What are the characteristics of staphylococcal enterotoxicosis?
9. How do other kinds of food poisoning differ from that caused by staphylococci?
10. How does enteritis differ from enteric fever?
11. How are enteric infections transmitted, and how can they be prevented?
12. What are the characteristics of salmonellosis, and how does paratyphoid fever differ from it?
13. What causes typhoid fever, and how does the disease progress?
14. How is damage caused by *Salmonella typhi* related to disease symptoms?
15. How is typhoid fever diagnosed, treated, and prevented?
16. How do shigellosis and cholera differ with respect to cause, nature of the illness, and treatment?
17. What causes traveler's diarrhea, and how should travelers prepare to deal with the disease?
18. What other bacteria can cause enteritis?

C.

19. What viruses cause gastroenteritis?
20. How can childhood deaths from viral enteritis be reduced?
21. What are the main differences between the types of hepatitis with respect to cause, nature of the illness, and prevention?
22. What causes giardiasis, and how can it be prevented and treated?
23. How does latent amebiasis differ from amoebic dysentery?
24. What causes balantidiasis, and how can it be prevented and treated?
25. Under what circumstances does cryptosporidiosis occur?
26. Which fungi produce toxins, and how do these toxins affect humans?
27. How can humans avoid fungal toxins?
28. Summarize the variations in the life cycles of flukes, and explain how they affect the ways humans can become infected.
29. How do adult tapeworms and bladder worms differ, and how do they affect humans?
30. What are the causes and effects of trichinosis infections?
31. How can humans avoid helminth infections?
32. Why would improving food and water sanitation not prevent hookworm infections?
33. What are the distinguishing characteristics of ascariasis, trichuriasis, and strongyloidiasis?
34. What are the consequences of pinworm infections, and why are they most likely in children?

PROBLEMS FOR INVESTIGATION

1. Read from other sources about *Clonorchis sinensis*. Although up to 80 percent of rural populations elsewhere are infested with this organism, explain why it is unlikely that Chinese liver fluke would ever become a threat in the United States. Give an example of a parasite that could easily become established in the United States. How does this organism differ from *C. sinensis*?
2. Select an enteric disease that is common in the United States, and devise a public-health program to eradicate it.
3. Select an enteric disease that is common in the tropics, and devise a public-health program to eradicate it.
4. Survey the current literature to determine which vaccines are under development to prevent enteric diseases.
5. Prepare a report on the various kinds of hepatitis vaccines, their methods of manufacture, and the costs and sequence of doses required. How are the latest versions safer than the earlier ones?
6. A veterans hospital in California reported an outbreak of 26 cases of hepatitis B infection in a single ward. All of the affected patients had blood drawn by a hand-held finger-stick instrument. Nurses always changed the lancet, which penetrated the skin, after each use. They sometimes failed to change the disposable prong that is held against the finger to position the device properly. The prong does not penetrate the skin. What is the probable means by which the virus spread from person to person?

(The answer to this question appears in Appendix E.)

SOME INTERESTING READING

Anonymous. 1991. Importation of cholera from Peru. *Journal of the American Medical Association* 265, no. 20 (May 22):2659.

Arends, J., et al. 1986. The nature of early lesions in enamel. *Journal of Dental Research* 65, no. 1 (January):2.

Brown, P. 1991. Cholera under attack from "altered" vaccine. *New Scientist* 131, no. 1783 (August 24):10.

Carpenter, H. A., G. I. Perezperez, and M. J. Blaser. 1991. Gastric adenocarcinoma and *Helicobacter pylori* infection. *Journal of the National Cancer Institute* 83 (23):1734–39.

Diamond, J. 1992. The return of cholera. *Discover* 13(February):60–67.

Luna, A., and A. D. Walling. 1988. Cysticercosis. *American Family Physician* 37, no. 1 (January):105.

Marwich, C. 1990. *Helicobacter*: new name, new hypothesis involving type of gastric cancer. *Journal of the American Medical Association* 264, no. 21 (December 5):2724–26.

Matossian, M. K. 1982. Ergot and the Salem witchcraft affair. *American Scientist* 70 (July–August):355.

Stroh, M. 1992. In the mouths of babes: no more cavities? *Science News* 141, no. 5 (February 1):70.

Togetherness—a male schistosome with a female protruding from the groove in its body.

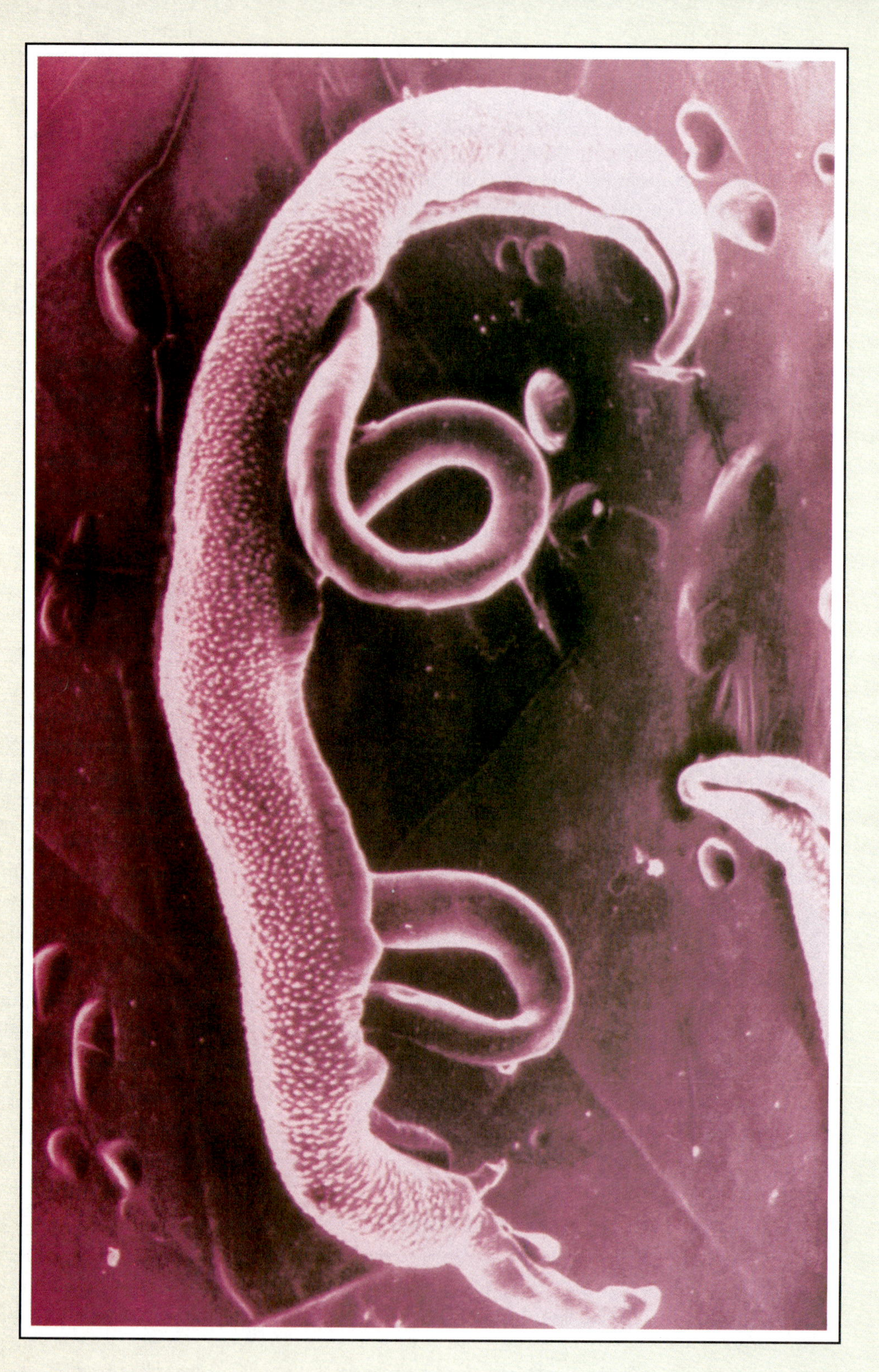

23 Cardiovascular, Lymphatic, and Systemic Diseases

This chapter focuses on the following questions:

A. What pathogens cause bacterial septicemias and related diseases, and what are the important epidemiologic and clinical aspects of these diseases?

B. What pathogens cause parasitic diseases of the blood and lymph, and what are the important epidemiologic and clinical aspects of these diseases?

C. What pathogens cause bacterial systemic diseases, and what are the important epidemiologic and clinical aspects of these diseases?

D. What pathogens cause rickettsial systemic diseases, and what are the important epidemiologic and clinical aspects of these diseases?

E. What pathogens cause viral systemic diseases, and what are the important epidemiologic and clinical aspects of these diseases?

F. What pathogens cause parasitic systemic diseases, and what are the important epidemiologic and clinical aspects of these diseases?

Diseases of the cardiovascular and lymphatic system frequently affect several systems because the infectious agents are easily disseminated through these body fluids. Therefore, diseases that usually affect multiple systems are included in this chapter. As with other chapters on the diseases of organ systems, familiarity with the properties of various groups of organisms and with disease processes, host systems, and host defenses will be helpful in understanding the diseases described here.

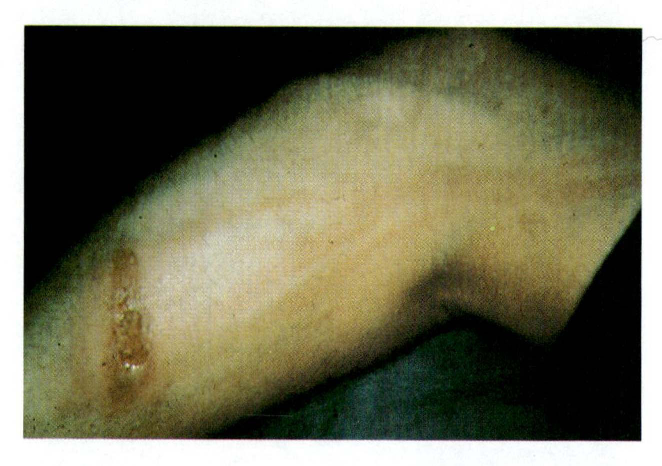

FIGURE 23.1 Lymphangitis from an infected burn. Note the reddened streaks indicating spread of organisms through lymph vessels. The old-fashioned name for this was "blood poisoning."

CARDIOVASCULAR AND LYMPHATIC DISEASES

Bacterial Septicemias and Related Diseases

Ordinarily blood is sterile. When organisms enter the blood from a wound or an infection, bacteremia, in which bacteria circulate without multiplying, can occur. Immune defenses ordinarily eradicate these organisms. If they are not eradicated, bacteremia may progress to septicemia, or *blood poisoning*, as the organisms rapidly multiply.

Septicemias

Before antibiotics, septicemia was often fatal; with antibiotics it is still not easy to treat. Gram-positive organisms such as *Staphylococcus aureus* and *Streptococcus pneumoniae* once caused most septicemias. Today, broad-spectrum antibiotics have made septicemias from these organisms less frequent, but they allow organisms such as *Pseudomonas aeruginosa*, *Bacteroides fragilis*, and species of *Klebsiella*, *Proteus*, *Enterobacter*, and *Serratia* to multiply. These organisms cause **septic shock**, a life-threatening septicemia accompanied by low blood pressure and the collapse of blood vessels. Probably one-third of all septicemias are now gram-negative septic shock, and 10 percent of all septicemias are caused by multiple organisms. Endotoxins produced by these organisms are directly responsible for shock. Antibiotics often worsen the situation; when they kill organisms, the disintegrating organisms release greater quantities of endotoxin, which cause more damage to the host's blood vessels, and blood pressure drops even further. A few cases of septic shock are caused by organisms that produce exotoxins.

Symptoms of septicemia include fever, shock, and **lymphangitis** (lim-fan-ji'tis), or red streaks due to inflamed lymphatics beneath the skin (Figure 23.1). One-third of the cases of septicemia appear within 24 hours after an invasive medical procedure has been performed, and the infection is nosocomial. The transition from bacteremia to septicemia can be sudden or gradual. Therefore, hospital patients who have undergone such procedures should be carefully watched for signs of septicemia. Septicemia has a mortality rate of 50 to 70 percent and accounts for about 100,000 deaths per year in the United States alone.

Diagnosis of septicemia is made by culturing of blood, catheter tips, urine, or other sources of infection. In treating septicemias, blood pressure must be elevated and stabilized; then the infectious organisms must be eliminated by 6 weeks or more of appropriate antibiotic therapy.

Puerperal Fever

Puerperal (pu-er'per-al) **fever,** also called puerperal sepsis or **childbed fever** (Chapter 1), was a common cause of death before antibiotics. It is caused by beta-hemolytic streptococci (*Streptococcus pyogenes*), which are normal vaginal and respiratory flora. They also can be introduced during delivery by medical personnel. Streptococci pass through raw uterine surfaces and invade the blood, giving rise to septicemia. Symptoms of the disease are chills, fever, pelvic distention and tenderness, and a bloody vaginal discharge. Streptococci can be isolated from blood cultures to diagnose puerperal fever. Penicillin is effective except against resistant organisms, and mortality is low with prompt treatment, but recovery usually takes many weeks and relapses are common.

Rheumatic Fever

Rheumatic fever is a multisystem disorder following infection by beta-hemolytic *Streptococcus pyogenes*. That rheumatic fever can follow such infections has been known for decades, but the mechanisms by which it occurs are not yet completely understood. Some form of genetic predisposition is suspected because a certain HLA antigen is present in 75 percent of rheumatic fever patients but in only 12 percent of the general population. Rheumatic fever sometimes is associated with poor hygiene and poor medical care, but it can occur in spite of good hygiene and proper medical attention.

Most rheumatic fever patients are between the ages of 5 and 15. Onset of the disease usually occurs 2 to 3 weeks after a strep throat, but it can occur within 1 week or as late as 5 weeks after the initial infection. Strep throat symptoms have disappeared by the time the rheumatic fever symptoms begin. Classic symptoms include fever, arthritis, and a rash. Evidence of damage to the mitral valve of the heart helps to make a specific diagnosis of rheumatic fever. Weeks or months later subcutaneous nodules appear, especially near the elbows. Approximately 3 percent of untreated strep throat cases progress to rheumatic fever. Culturing of streptococci and antibody tests are helpful in diagnosis, as is a previous history of streptococcal infection.

Heart damage in rheumatic fever results from poorly understood immunological events. Certain streptococcal strains have an antigen that is very similar to heart cell antigens. Antibodies that bind to one antigen will bind to the other; that is, the antibodies are cross-reactive. In the immune reaction, lymphocytes probably become sensitized to the antigen and attack the heart as well as the streptococci. The resulting heart damage can be fatal. Antibiotic therapy will not reverse existing damage, but it can prevent further damage and can be used to prevent recurrences. Once rheumatic fever occurs, mitral valve deformities contribute to eddies in blood flow that predispose toward bacterial colonization of heart valve surfaces. This condition, bacterial endocarditis, is discussed next. Individuals at risk of rheumatic fever should receive a prophylactic antibiotic—usually penicillin—before dental work or other invasive procedures to prevent possible streptococcal infection.

Prompt treatment of beta-hemolytic *S. pyogenes* infections with antibiotics before cross-reactive antibodies can form is the only practical way to prevent rheumatic fever. No effective vaccine exists; vaccines produced so far elicit damaging antibodies, but this problem might some day be solved. Anti-inflammatory drugs such as steroids and aspirin can lessen scarring of heart tissue.

Bacterial Endocarditis

Bacterial endocarditis (en-do-kar-di'tis) is a life-threatening infection and inflammation of the lining and valves of the heart. It can be subacute or acute. Two out of three patients have the subacute type, which manifests itself as fever, malaise, bacteremia, and regurgitating heart murmur sounds usually lasting 2 weeks or more. It occurs primarily in people over age 45 who have a history of valvular disease from rheumatic fever or congenital defects. Many microbes, including fungi, can cause endocarditis, but most cases are due to strains of *Streptococcus* or *Staphylococcus*, many of them normal residents of the mouth or throat. Acute endocarditis is a rapidly progressive disease that destroys heart valves and causes death in a few days.

In bacterial endocarditis, organisms from another site of infection are transported to the heart. A **vegetation** (Figure 23.2) develops, in which exposed collagen fibers on damaged valvular surfaces elicit fibrin deposition. Transient bacteria attach to fibrin and form a bacteria-fibrin mass. Vegetations deform heart valves, decrease their flexibility, and prevent them from closing completely. Blood flows backward from ventricles into atria when the ventricles contract, decreasing the pumping efficiency of the heart. Congestive heart failure, an accumulation of fluids around the heart, is the most common complication and direct cause of death from bacterial endocarditis.

Bacterial endocarditis is diagnosed from blood cultures and is treated with penicillin or other antibiotics, depending on the susceptibilities of causative organisms. Sometimes surgical valve replacement is necessary. Untreated endocarditis results in death. Antibiotics cure about half of all patients, surgery cures another quarter, and one-quarter die. Deaths are most frequent among intravenous drug abusers and others with compromising conditions.

Myocarditis (mi-o-kar-di'tis), an inflammation of the heart muscle (myocardium), and **pericarditis** (per-e-kar-di'tis), an inflammation of the protective membrane around the heart (pericardial sac), also can be caused by microbial infections. Although most such infections are the product of viruses, *Staphylococcus aureus* causes 40 percent of the cases of pericarditis. Untreated cases have a mortality rate of nearly 100 percent, whereas cases receiving modern treatment have a mortality rate between 20 and 40 percent.

Parasitic Diseases of the Blood and Lymph

Schistosomiasis

Three species of blood flukes of the genus *Schistosoma* (Figure 23.3a) cause **schistosomiasis** (skis-to-so-mi'as-is), and each requires a particular snail host to complete its life cycle (Figure 12.15). The parasite was identified by German parasitologist Theodor Bilharz in the 1850s, so the disease also is called **bilharzia** (bil-har'ze-ah). Bilharzia has been known since biblical times; some think Joshua's curse on Jericho was the

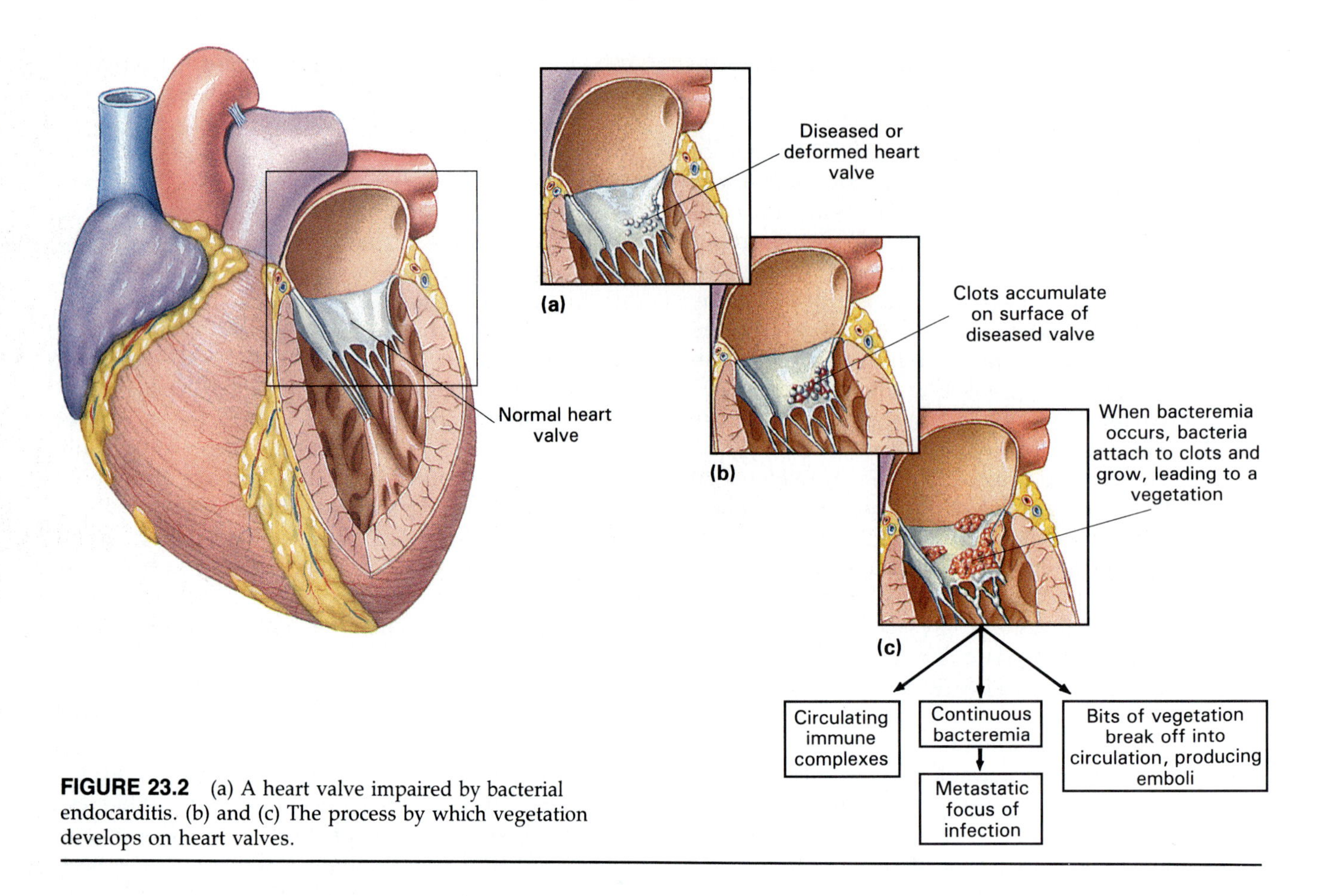

FIGURE 23.2 (a) A heart valve impaired by bacterial endocarditis. (b) and (c) The process by which vegetation develops on heart valves.

placing of blood flukes in communal wells. In fact, the Egypt of the Pharaohs was called "land of the menstruating men" by ancient writers because the prevalence of flukes made bloody urine so common, and we can still find eggs of these parasites in the bladder walls of Egyptian mummies.

Incidence of the disease has increased significantly in Egypt since the building of the Aswan Dam in 1960 because collection of water behind the dam created exceptionally favorable conditions for snail hosts. *Schistosoma japonicum* is found in Asia, *S. haematobium* in Africa, and *S. mansoni* in Africa, South America, and the Caribbean. The last two species probably reached South America during slave trading, but only *S. mansoni* found an appropriate snail host there.

Humans become infected by free-swimming cercaria that have emerged from their snail hosts. Cercaria penetrate the skin when humans wade in snail-infested waters, migrate to blood vessels, and are carried to the lungs and liver. The flukes mature and migrate to veins between the intestine and liver or sometimes the urinary bladder, where they mate and produce eggs. Adults have a special ability to coat themselves with host antigens, thereby evading the immune system. Some eggs become trapped in the tissues and cause inflammation; others penetrate the intestinal wall and are excreted in the feces. Cercariae (Figure 23.3b) cause dermatitis at penetration sites and tissue damage during migration. Metacercariae and adults migrate to and invade the liver (Figure 23.3c), where they cause cirrhosis. These flukes also can invade and damage other organs.

Schistosome eggs are highly antigenic, and allergic reactions to them are responsible for much of the damage caused by blood flukes. If the eggs are released near the spinal cord, the resulting inflammation can cause neurological disorders. The eggs most often damage blood vessels, but which vessels are damaged depends on the species. *Schistosoma japonicum* severely damages blood vessels in the small intestine, *S. mansoni*, those of the large intestine, and *S. haematobium*, those of the urinary bladder. Symptoms of bladder infections include pain on urination, bladder inflammation, and bloody urine.

Diagnosis can be made by finding eggs in feces or urine, but eggs may not be present in chronic cases. Intradermal injection of schistosome antigen and measurement of the area of the wheal or a complement fixation test are good immunologic methods of diagnosis. Until recently toxic antimony compounds were used to treat the disease, but several new drugs, especially praziquantal, seem to be quite effective and less toxic.

Schistosomiasis could be prevented entirely if no human wastes were placed in rivers or if humans

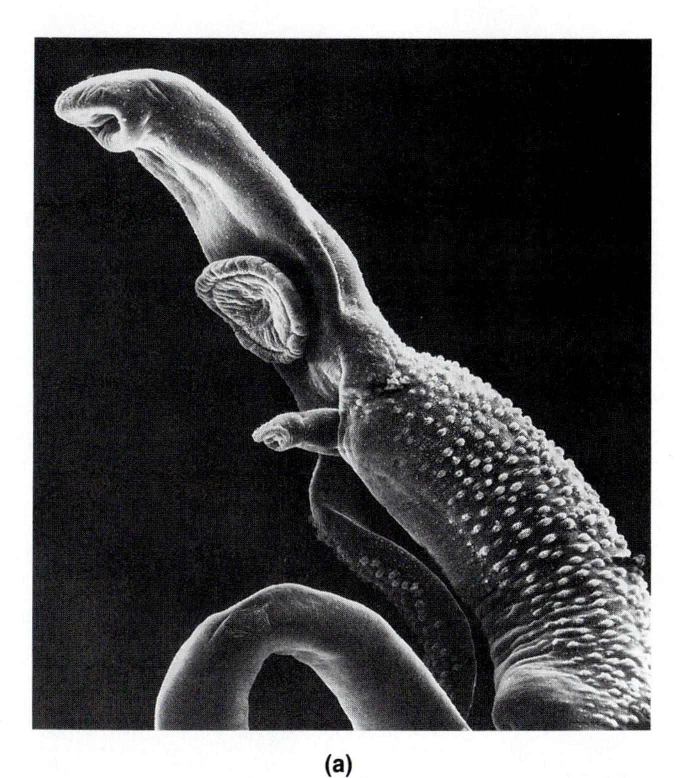

(a)

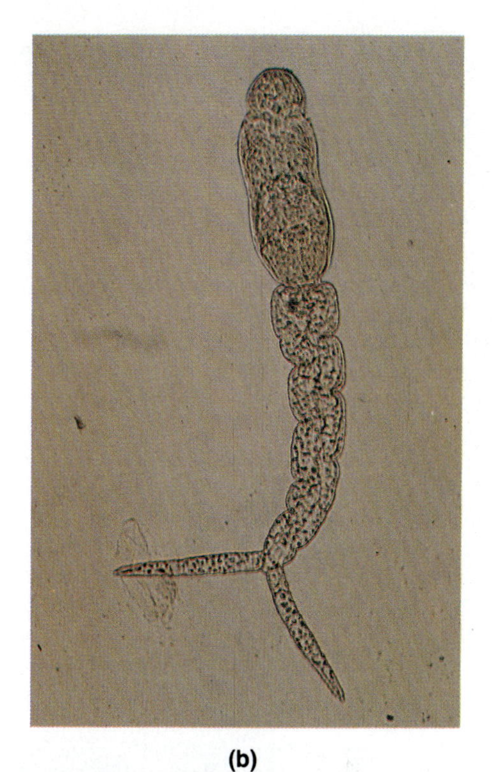

(b)

FIGURE 23.3 *Schistosoma mansoni,* the cause of schistosomiasis. (a) The large worm is the male, with the female's head projecting from the groove in which he holds her. The male attaches itself to the wall of a human blood vessel by means of the round suction cup just below its head. The actual worms, thinner than a cotton fiber, are barely visible to the unaided eye. A mating pair can live in the human body for up to 10 years. (b) The cercarial stage leaves snails and penetrates the skin of humans wading or swimming in water. (c) A child suffering from schistosomiasis. Tracing on her body delineates her enlarged liver, typical of this disease.

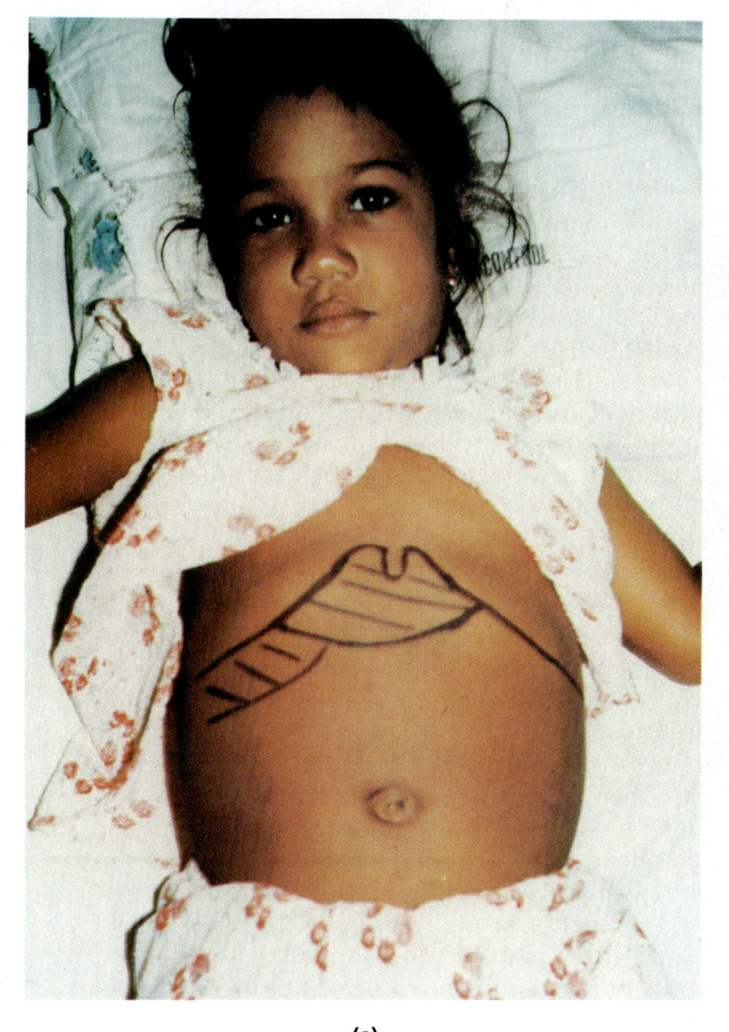

(c)

CLOSE-UP

Heartworm in Dogs

Heartworm infections are common in dogs, and one such infection was recently reported in a Maryland farmer. The heartworm is a nematode, *Dirofilaria immitis*, and is transmitted by mosquitoes. When an infected mosquito bites a dog, it introduces larvae into the blood. The larvae migrate to the skin, mature to about 8 cm in length, and move on to the dog's heart. The worms complete maturation in the heart, mate, and release microfilariae that become larvae only in a mosquito. Adult worms 15 to 30 cm long accumulate in the right side of the heart and also are found in the lungs and liver. The dog becomes weak, tires easily, coughs, and has difficulty breathing. Ultimately it dies of heart damage and circulatory failure. Ivermectin, which has proven so valuable in fighting the nematode that causes river blindness (Chapter 20), was originally developed for veterinary use and is highly effective against heartworm larvae when given in monthly doses. (p. 563) However, it can be administered only to dogs that do not already have adult heartworms. If adult worms are present the drug kills them, but the worms prove more dangerous in death than in life. Disintegration of their corpses releases toxic substances and debris that can clog blood vessels while leaving holes in the wall of the heart where they had resided.

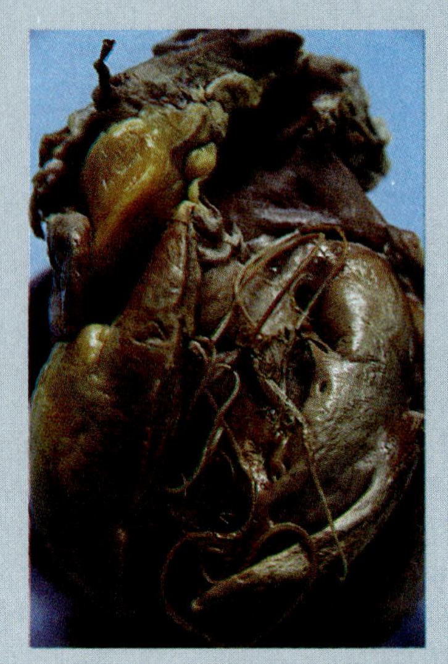

A dog heart that has been opened to show the presence of many adult heartworms, *Dirofilaria immitis*.

never waded in snail-infested water. The practice of wading in the local river to wash orifices after defecation or urination is an important means of transmission wherever the infection occurs. Chemical molluscicides have been used to reduce snail populations, but determining proper concentrations under varying river conditions poses a difficult problem. Recent experiments using predatory snails to destroy schistosome-carrying snails have shown great promise. Finally, work toward developing a vaccine is under way.

Filariasis

Filariasis (fil-ar-i'as-is) can be caused by several different roundworms, but *Wuchereria bancrofti* is a common cause of this tropical disease (Figure 12.18). Adult worms are found in the lymph glands and ducts of humans. Females release embryos called microfilariae. They are present in peripheral blood vessels during the night and retreat to deep vessels, especially those of the lungs, during the day. Mosquitoes also are essential hosts in the life cycle of this parasite, and several species of night feeders among the genera *Culex*, *Aedes*, and *Anopheles* serve as hosts. When a mosquito bites an infected person, it ingests microfilariae that develop into larvae and migrate to the mosquito's mouthparts. When the mosquito bites again, the larvae can infect another person. They enter the blood, develop, and reproduce in the lymph glands and ducts, thereby completing the life cycle. Adults are responsible for inflammation in lymph ducts, fever, and eventual blockage of lymph ducts in filariasis. Repeated infections over a period of years can lead to **elephantiasis** (el″ef-an-ti'as-is), gross enlargement of limbs, scrotum, and sometimes other body parts (Figure 23.4) from accumulation of fluid and connective tissue in blocked lymph ducts.

Filariasis is diagnosed by finding microfilariae in thick blood smears made from blood samples taken at night or by an intradermal test. The drugs diethylcarbamizine (Hetrazan) and metronidazole are effective in treating the disease. Swollen limbs are wrapped in pressure bandages to force lymph from them; if distortion is not too great, nearly normal size can be regained. To control the disease it would be necessary to treat all infected individuals and to eradicate the various species of mosquitoes that carry the parasite. Little progress has been made on this challenging problem.

Cardiovascular and lymphatic diseases are summarized in Table 23.1.

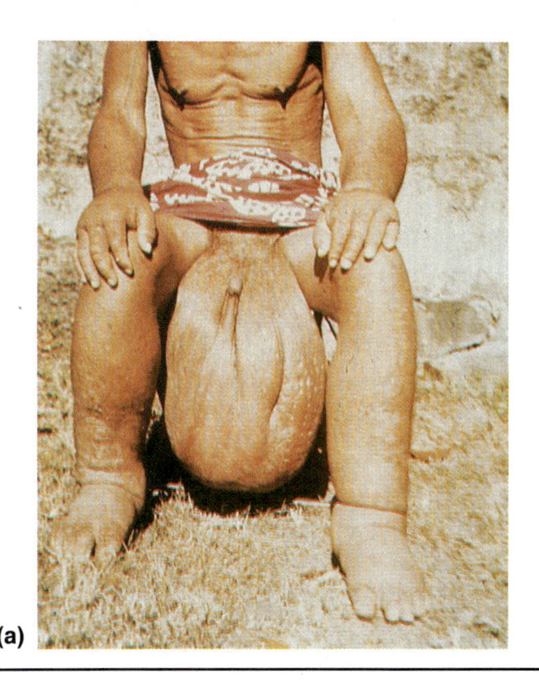

(a)

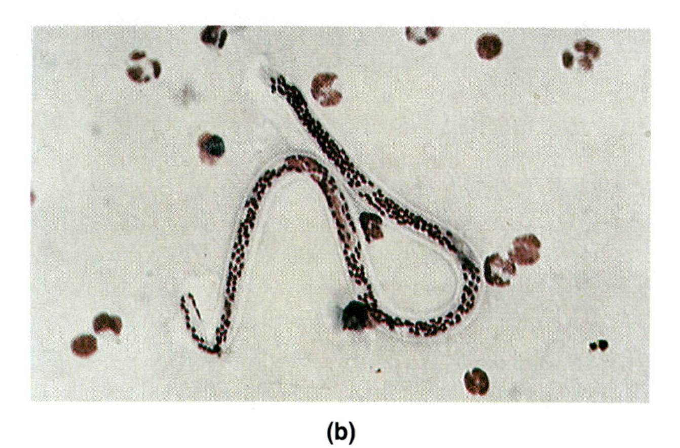

(b)

FIGURE 23.4 (a) Elephantiasis of the scrotum, caused by the roundworm *Wuchereria bancrofti*. Swelling results from blockage of the lymphatic system by adult worms. Another common site of elephantiasis is the leg. (b) The microfilaria stage of the life cycle is transmitted to humans by the bite of mosquitoes.

TABLE 23.1 Summary of cardiovascular and lymphatic diseases

Disease	Agent(s)	Characteristics
Bacterial septicemias and related diseases		
Septicemia	Many bacteria, several per infection	Septic shock due to endotoxins of causative agent(s), fever, lymphangitis
Puerperal fever	*Streptococcus pyogenes*	Organisms from uterus invade blood and cause septicemia, pelvic distention, bloody discharge
Rheumatic fever	*Streptococcus pyogenes*	Fever, arthritis, rash, mitral valve damage due to immunologic reaction
Bacterial endocarditis	*Staphylococcus* or *Streptococcus* strains	Inflammation and vegetation of heart valves and lining, fever, malaise, bacteremia, heart murmur; congestive heart failure can cause death
Parasitic diseases of the blood and lymph		
Schistosomiasis	*Schistosoma haematobium, S. mansoni, S. japonicum*	Dermatitis from cercaria, cirrhosis of liver from eggs, allergic reactions from eggs, tissue damage in intestine and urinary bladder
Filariasis	*Wuchereria bancrofti*	Inflammation and blockage of lymph ducts, leading to elephantiasis, fever

SYSTEMIC DISEASES

Bacterial Systemic Diseases

Anthrax

Anthrax is a zoonosis that affects mostly plant-eating animals, especially sheep, goats, and cattle. Meat-eating animals can acquire the disease by eating infected flesh, but the disease is not spread from animal to animal. Worldwide, on an annual basis, many thousands of animals have anthrax, but only 20,000 to 100,000 humans, mostly in Africa, Asia, and Haiti have it. An anthrax outbreak in Russia in 1979 killed at least 64 people. Health authorities in the United States have made vigorous efforts to eradicate anthrax and to prevent its import from other countries. Only four human cases have occurred in the United States since 1980, and no more than six per year have been reported since 1970.

The Disease The causative agent of anthrax, *Bacillus anthracis*, was discovered in 1877 by Robert Koch. The bacillus is a large, gram-positive, facultatively anaerobic, endospore-forming rod. The endospores form only under aerobic conditions and are not found in tissues or circulating blood. But if an infected animal's blood is spilled during an autopsy, bacilli exposed to air rapidly form endospores. Therefore, veterinarians and farmers should be very careful to avoid contaminating soil or other materials, as endospores can remain viable and ruin a pasture for over 50 years.

Most human anthrax results from contact with endospores during occupational exposure on farms or in industries that handle wool, hides, meat, or bones. Respiratory anthrax, or "woolsorters disease," was such a problem in nineteenth-century England that legislation was enacted to protect textile workers from this occupational disease.

Of human anthrax cases, 90 percent are cutaneous, 5 percent are respiratory, and 5 percent are intestinal. Cutaneous anthrax has a mortality rate of 10 to 20 percent when untreated, but only a 1 percent rate with adequate treatment. Respiratory anthrax is always fatal, regardless of treatment. Clots form inside pulmonary capillaries and lymph nodes, causing swelling, which then obstructs airways. Intestinal anthrax has a fatality rate of 25 to 50 percent. In addition, regardless of the initial site of infection, in about 5 percent of patients septicemia leads to meningitis, which is almost always fatal in 1 to 6 days.

Cutaneous anthrax develops 2 to 5 days after endospores enter epithelial layers of the skin. Lesions 1 to 3 cm in diameter develop at the site of entry (Figure 23.5). Eventually, the center of the lesion becomes black and necrotic; finally, it heals but leaves a scar. Recovery probably confers some—but not total—immunity.

Symptoms of intestinal anthrax closely mimic those of food poisoning. An outbreak in this country resulted from consumption of imported cheese made from unpasteurized goat's milk. The cheese was served at a wine and cheese party to a group of physicians. Some of them traced the source of the infection to a particular cheese factory in France, which as a result was subsequently closed. One wonders, however, what became of the rest of the shipments from that factory, and whether there would ever have been any investigation had those affected not been doctors.

PUBLIC HEALTH

Anthrax Alarm

Skilled detective work unraveled the mystery of a mountain lion's death of anthrax in the Pacific Northwest in the 1970s. One day a loosely cinched pack rubbed a raw spot on the back of a packhorse. The felt pad beneath the pack had been made in North Africa from anthrax-contaminated goat hair. Endospores entered the wound and produced an open lesion. A packhorse that won't pack has a very limited future on a ranch, so it was hustled off to the slaughterhouse. That day, employees of a nearby national forest came to the slaughterhouse to buy carcasses. They took these carcasses, including the packhorse, high into the forested mountains to induce predators to stay there instead of coming into the valley to prey on livestock or pets.

Later, the discovery of a mountain lion found dead of anthrax created a terrible problem in disinfecting the area. Scientists had to determine how the carcass might have been distributed by birds, worms, and other animals and how endospores might have been moved by air and water. Tons of earth were bulldozed, packed up, and disposed of safely; trees were cut down and animals were trapped. Their efforts paid off, and the spread of anthrax was prevented.

Public health agencies maintain constant vigilance against anthrax, recalling mohair yarns from North Africa and banning import from Haiti of voodoo drums made from infected hides. The next time you are tempted to sneak something past customs, consider that you may be the means of importing anthrax or some other serious disease.

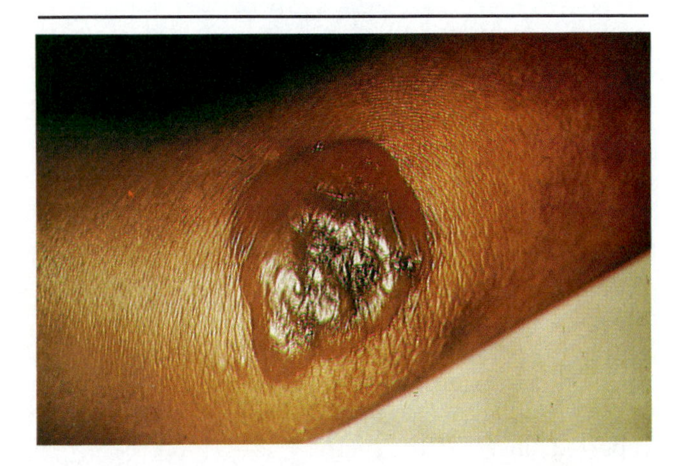

FIGURE 23.5 Cutaneous anthrax lesion can result from contact with infected animals, their hides, or their wool.

Diagnosis, Treatment, and Prevention Anthrax is diagnosed by culturing blood or smears from cutaneous lesions of patients with a history of possible exposure. It is treated with penicillin or tetracycline. Vaccine can be given annually to workers with occupational exposure to anthrax, but industries still should maintain dust-free environments and provide respirators to prevent endospore inhalation. Worker education and on-site employee health services for prompt detection of infections also are important. Unimmunized visitors must be kept away from work areas. Clothing worn by workers should be cleaned at the facility to prevent family members from becoming infected by handling it.

Animal vaccination is an important means of prevention. Farmers also must avoid using bone meal contaminated with anthrax and dispose of infected animals by burying them in deep, lime-lined pits. Lime prevents earthworms from bringing anthrax endospores to the surface, whereas burning can allow bits of contaminated carcass and spores to be spread by the wind. Veterinarians must be especially careful when working with infected animals or giving vaccinations. Accidental vaccination of a human with vaccine intended for animals can cause anthrax!

Plague

From 1937 through 1974 fewer than 10 cases of plague per year were reported in the United States, with no cases reported in some years. Then 20 cases were reported in 1975 and 40 in 1983, mostly in rural settings in the Rocky Mountain states (Figure 23.6). In certain

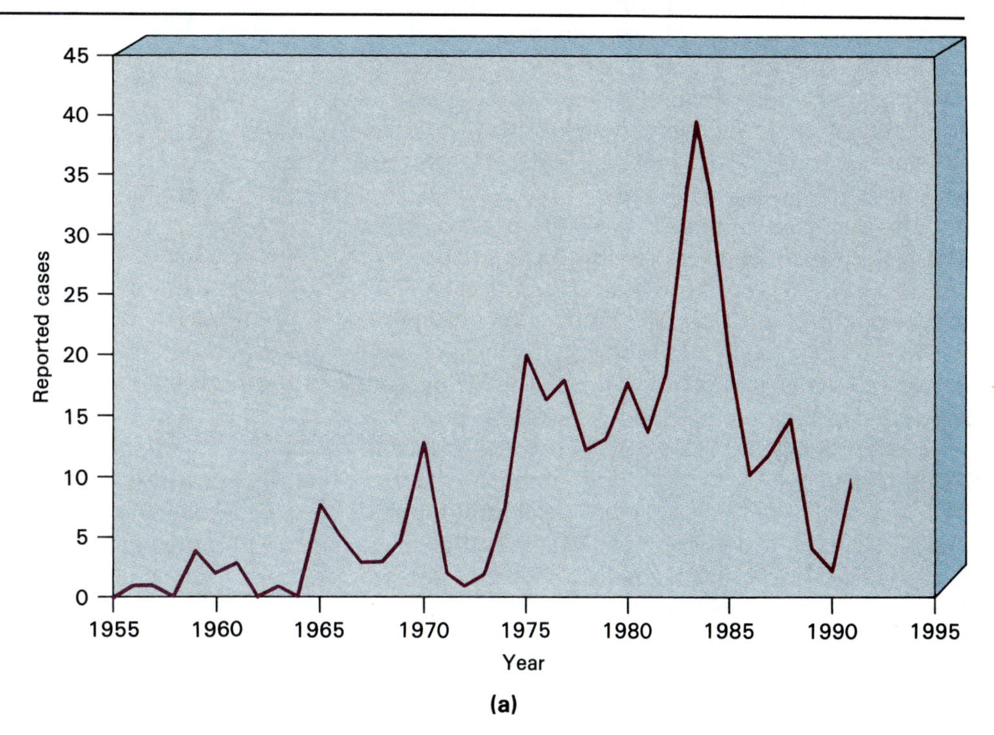

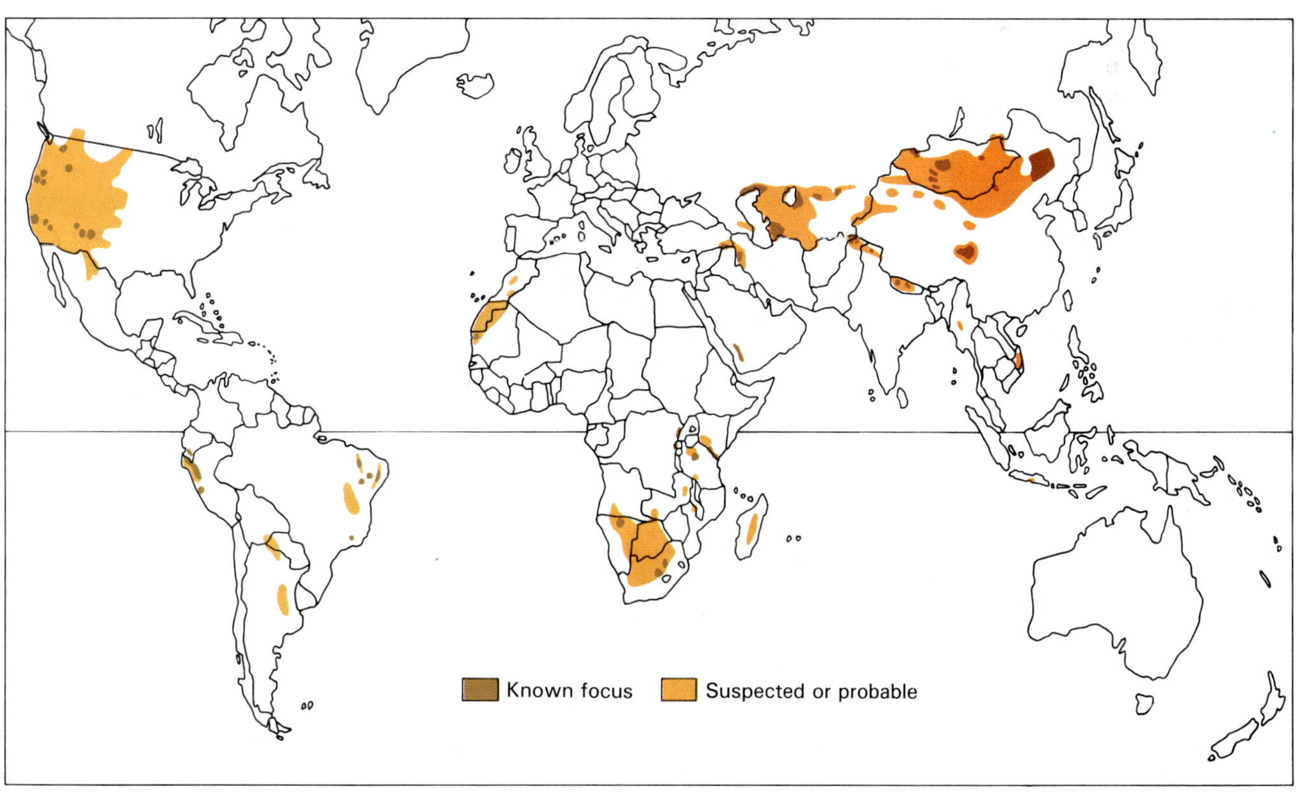

FIGURE 23.6 (a) Incidence of plague in the United States, 1950–1991. (b) Regions of endemic plague.

other areas of the world plague remains an endemic disease, but the number of cases worldwide is much smaller than in the great pandemics mentioned in Chapter 1. ∞ (p. 8)

The Disease The causative agent of plague, *Yersinia pestis*, is a short, fat, gram-negative rod. As a zoonosis, plague infects wild rodents, especially rats, and is transmitted from animal to animal and occasionally to humans by flea bites. As infected rats die, their body temperature drops, and their blood coagulates; hungry fleas jump to nearby sources of warmth and liquid blood. The new host usually is another rat, but in crowded rat-infested living quarters or when curious people poke at a dead carcass, the next host can easily be a human.

The flea itself suffers from plague infection. *Yersinia pestis* organisms ingested from a sick rat multiply and block the flea's digestive tract until food (blood meals) cannot pass through it. The flea gets hungrier, bites more ferociously, and infects new victims as it disgorges plague organisms with each bite. Eventually the flea dies, but this is little consolation to a new human victim, who has a 50 to 60 percent chance of dying if not treated.

Once inside the host, plague bacilli multiply and travel in lymphatics to lymph nodes, where they cause hemorrhages and immense enlargements called **buboes,** especially in the groin and armpit (Figure 23.7). Buboes are characteristic of **bubonic plague** and appear after an incubation time of 2 to 7 days. Hemorrhages turn the skin black, hence the name Black Death. Deaths from bubonic plague can be prevented with adequate timely antibiotic treatment. If organisms move from the lymphatics into the circulatory system, **septicemic plague** develops. It is characterized by hemorrhage and necrosis in all parts of the body, meningitis, and pneumonia. This form of plague is invariably fatal despite the best modern care. **Pneumonic plague** occurs when aerosol droplets from a coughing patient are inhaled. It, too, has a mortality rate approaching 100 percent despite excellent care. Medical personnel working with patients are more likely to acquire pneumonic than bubonic plague.

Diagnosis, Treatment, and Prevention Plague can be diagnosed by identifying organisms from stained smears of sputum or fluid aspirated from lymph nodes, or by fluorescent antibody tests. It is treated with streptomycin, tetracycline, or both. Fortunately no drug-resistant strains have yet appeared.

PUBLIC HEALTH

Plague in the United States

Plague transmitted by rats in cities is sometimes referred to as urban plague, in contrast to sylvan plague carried by wild rodents in rural areas. Sylvan plague is endemic in 15 western states, where it is found in gophers, chipmunks, pack rats, prairie dogs, and ground squirrels. Plague is thought to have persisted for years among similar wild rodent populations of the Central Asian steppe before breaking out to decimate Europe in 1346. So far only sporadic cases have occurred in the American West, often among hunters and Native Americans on reservations. Some years ago a pet-store owner in Los Angeles trapped some desert rodents with the intention of selling them. Unfortunately for him—but perhaps fortunately for the community—he died of plague before he could sell any of the cute pets to unsuspecting customers.

FIGURE 23.7 A Flemish manuscript illumination of the fourteenth century showing a physician lancing a plague-caused bubo. A second patient awaits the lancing of his underarm bubo.

Recovery from a case of plague confers lifetime immunity. Vaccine is available to protect medical personnel and the families of victims. However, during treatment of plague patients in the Vietnamese War, it was discovered that workers, even those protected by vaccine, can become pharyngeal carriers of plague for a short period of time. Plague can be prevented by vaccinating travelers to endemic regions, controlling rat populations, and maintaining surveillance for infections in wild rodent populations. CDC surveys have found plague only among wild rural (usually desert) rodents, which are not likely to come in contact with urban populations. However, plague has moved eastward from the California coast, where it first arrived on a ship from China that docked at San Francisco in 1899. If the disease spreads to city rats, many of which are resistant to rat poisons, the risk to humans will increase.

Tularemia

Tularemia is a zoonosis found in more than 100 mammals and arthropod vectors, especially cottontail rabbits, muskrats, rodents, ticks, and deer flies. It also can be transmitted by body lice and mechanically on claws and teeth of predators that eat infected rodents. It has even been transmitted by the bite of a bullsnake! In ticks, the pathogen is incorporated in eggs as they leave the ovaries in **transovarian transmission** and thereby passes from one generation to the next. Although in about half of human cases the vector is never identified, tularemia is most often associated with cottontail rabbits; the number of cases reported always rises significantly during rabbit-hunting season.

Francisella tularensis, a small gram-negative coccobacillus with worldwide distribution, was first isolated in 1911 from Tulare County, California, from which the species name is derived. The genus is named after Edward Francis, who did much of the early work on this organism and found 20 different routes of infection for it. The annual incidence of tularemia in the United States has dropped from more than 2000 cases in 1939 to fewer than 200 cases in recent years (Figure 23.8). The disease is an occupational hazard for taxidermists.

Tularemia can be acquired in three ways. First, although organisms can probably be transmitted through unbroken skin, they usually enter through minor cuts, abrasions, or bites. Second, organisms can be inhaled, especially from aerosols formed during the skinning of infected animals. Third, organisms can be consumed in contaminated water or meat to cause an intestinal form of the disease. Swimming in a river near a colony of infected river rats and eating undercooked rabbit have been reported as sources of infection. Freezing, even for years, does not destroy the organisms.

Entry through the skin results in the **ulceroglandular** form of disease. After a 48-hour incubation period, symptoms begin with abrupt high fever of 40° to 41°C (104° to 106°F) with chills and shaking. If untreated, fever, severe headache, and buboes, or enlarged regional lymph nodes, can last a month. An ulcer sometimes forms at the site of entry. Handling of animals and skins is most likely to cause ulcers on hands, whereas a bite by an arthropod vector is most likely to cause lymph node lesions in the groin or armpit. The patient is initially disabled for 1 to 2 months and can have frequent relapses. The mortality

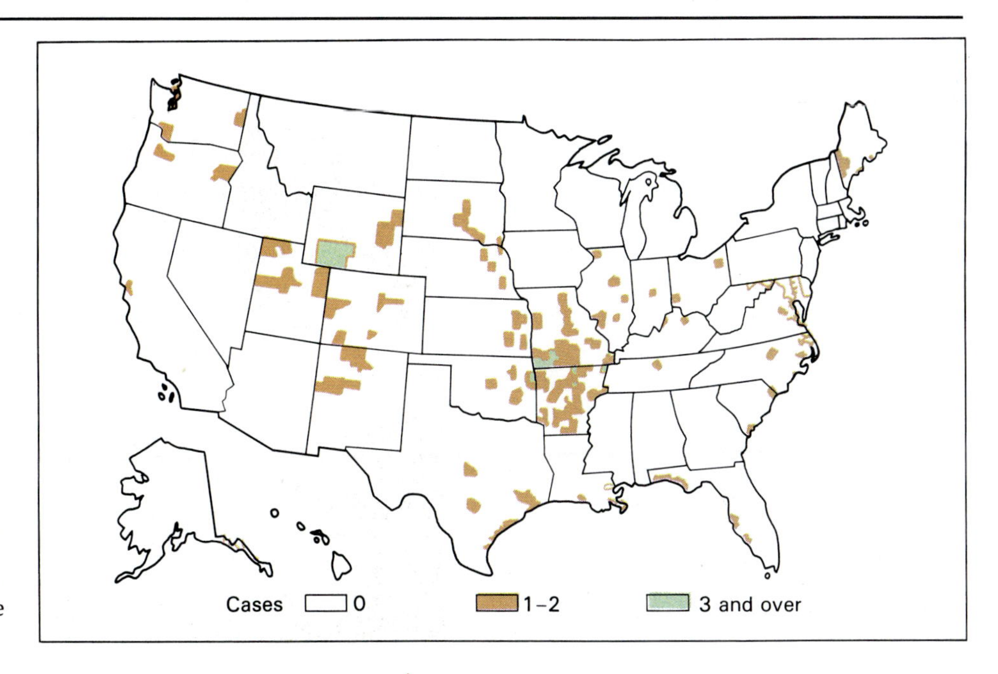

FIGURE 23.8 This map shows the total number of cases of tularemia in the United States in 1990. In addition, two cases were reported in Alaska.

rate is about 5 percent if left untreated. In the days of American pioneers, rabbit was certainly a major part of the diet, and tularemia also must have been a major feature of their lives. Without antibiotic treatment, a rabbit dinner could bring family members long periods of high fever, frequent relapses, and even death. Until the 1960s, when biological warfare was banned, *F. tularensis* was the main organism studied by government scientists for possible use in biological warfare.

Bacteremia from lesions can lead to **typhoidal tularemia,** a septicemia that resembles typhoid fever. Touching the eyes with contaminated hands can lead to conjunctivitis, but this happens in very few cases. Inhalation of organisms or their spread from blood produces a patchy bronchial pneumonia that leads to lung tissue necrosis and 30 percent mortality.

Diagnosis from blood cultures is difficult; the highly infectious organisms are hard to grow on ordinary laboratory media. Only 50 organisms are sufficient to produce a human infection, regardless of the route of administration. Laboratory infections are easily acquired, so culturing of *Francisella tularensis* should be attempted only by very experienced personnel in isolation laboratories equipped with special air-flow hoods and other safety devices (Chapter 6). Animal inoculations should be avoided. Safer and easier hemagglutination tests are the standard method of diagnosis. Streptomycin is an effective treatment, but penicillin is of no value.

Prevention by eliminating the organism from wild reservoir populations is impractical. Therefore, one should avoid handling sickly animals, wear gloves when handling or skinning wild game, and, in tick-infested areas, wear protective clothing and frequently search clothing and skin for ticks. Vaccine is not always protective and must be readministered every 3 to 5 years.

Brucellosis

Brucellosis, also called **undulant fever** or **Malta fever,** is a zoonosis highly infective for humans. It is caused by several species of *Brucella*. *Brucella melitensis,* was isolated in 1887 on the Mediterranean isle of Malta by Sir David Bruce. *Brucella* are small gram-negative bacilli, each of which has a preferred host: *Brucella abortus,* cattle; *B. melitensis,* sheep and goats; *B. suis,* swine; and *B. canis,* dogs. In addition to preferred hosts, each species can infect several other hosts, including humans. The incidence of brucellosis in the United States has dropped sharply from more than 6000 cases per year at the end of World War II to fewer than 200 cases per year since 1978.

Brucella enter hosts through the digestive tract via contaminated dairy products, the respiratory tract via aerosols, or the skin via contact with infected animals on farms or in slaughterhouses. Inside the host, these facultative intracellular parasites multiply and move through the lymph into the blood, where they cause an acute bacteremia within 1 to 6 weeks. Brucellosis has a gradual onset with a daily fever cycle—high in the afternoon and low at night after profuse sweating. The spleen, lymph nodes, and liver can be enlarged, and jaundice can be present, but symptoms may be too mild to diagnose. The initial acute phase lasts from several weeks to 6 months. Recovery usually is spontaneous, but chronic aches and nervousness can develop. If attributed to a psychiatric disturbance, the disease may not be diagnosed or properly treated.

Brucellosis is diagnosed by serological tests and treated with tetracycline and streptomycin. Treatment must be prolonged because the organisms are well sequestered from antibiotics in the blood, but it can prevent fatalities. When death does occur it is usually from endocarditis. The disease can be prevented by pasteurizing dairy products, vaccinating animal herds, and providing education and protective clothing for workers with occupational exposure. No human vaccine is available at present.

Cattle ranchers in the vicinity of Yellowstone, Montana, face a danger from diseased bison who wander outside the park boundaries. Over 50 percent of the Yellowstone herd is infected with brucellosis and can spread it to cattle when they mingle with them, especially during hard winters, when food may be more abundant on ranchers' lands. Hunters have been allowed to kill bison who leave the park. Nearly 600 are shot in some years. Montana has spent millions of dollars in the last decade to eradicate brucellosis from its herds.

Relapsing Fever

Relapsing fever is caused by any of nearly a dozen different species of the genus *Borrelia*. *Borrelia recurrentis,* the most common causative agent, is distinguished from other species by the fact that it alone is transmitted by lice. All other agents are transmitted by ticks and are named according to the type of tick that carries them. Fifteen different species of soft ticks are known to transmit relapsing fever. Louse-borne cases are called **epidemic relapsing fever,** whereas tick-borne cases are called **endemic relapsing fever** (Figure 23.9). Since relapsing fever appeared in ancient Greece, outbreaks have occurred during wars and in unsanitary situations. During epidemics mortality can reach 30 percent of untreated cases.

Borrelia are large spiral organisms distinguished from *Treponema* (syphilis) spirochetes by their coarser, more irregular spirals and their ease of staining. These organisms are so closely related that some taxonomists

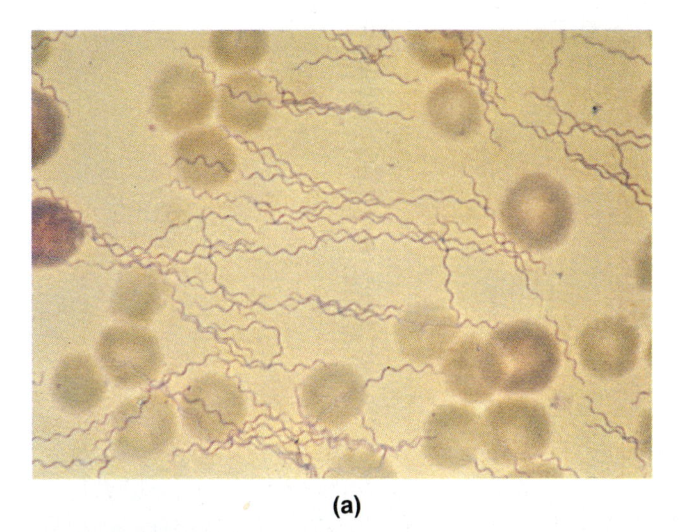

(a)

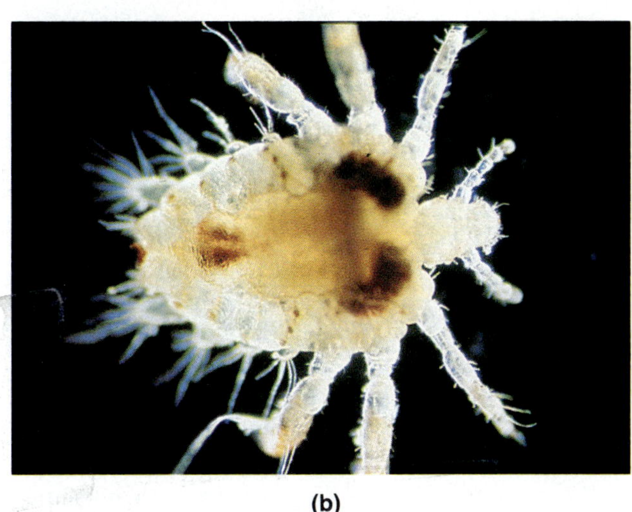

(b)

FIGURE 23.9 Relapsing fever is caused by various *Borrelia* species, such as (a) *Borrelia recurrentis*, 500X, carried by ticks, and (b) lice such as the body louse.

place *Borrelia* with the genus *Treponema* and some put all relapsing fever organisms together in a single species, *Borrelia recurrentis*.

Transmission of relapsing fever varies with the vector involved. Lice must be crushed and their body contents scratched into the skin to transmit the disease, and each louse must acquire the organisms by biting an infected host. Ticks transmit disease organisms in their salivary secretions when biting and by transovarian transmission. They can survive for up to 5 years without a meal and still contain live infectious *Borrelia*. Eradicating these vectors is impossible. The ticks feed mainly at night for only about half an hour; victims such as campers or occupants of houses infested with tick-carrying rodents never realize they have been bitten. Lack of this important piece of clinical history makes diagnosis difficult.

Much of what we know about relapsing fever was learned before the antibiotic era, when syphilis was treated by inducing high fever. Patients were purposely infected with relapsing fever or with malaria; the high fever killed the syphilis organisms and left the patient with a more manageable disease. After an incubation period of 3 to 5 days, relapsing fever begins with a sudden onset of chills and high fever. Fever persists for 3 to 7 days and ends by crisis. About 16 percent of patients never experience a relapse, but most, after a period of 7 to 10 days, have fever for 2 or 3 days. Typically, more respites and more fevers occur, hence the name relapsing fever. The disease is particularly dangerous in pregnant women because the organisms can cross the placenta and infect the fetus.

Relapses are explained by changes in the organism's antigens. During a period of fever, the body's immune response destroys most of the organisms. The few that remain have surface antigens the immune system fails to recognize. These organisms multiply during a period of respite until they are numerous enough to cause a relapse. Each relapse represents a new population of organisms that has evaded the host's defense mechanisms.

Diagnosis is difficult, but the organisms sometimes can be identified in stained blood smears obtained during a rising fever phase. Tetracycline or chloramphenicol are used to treat relapsing fever. Immunity following recovery is usually short-lived. No vaccine is available, so prevention involves tick and louse control, a difficult business at best.

Lyme Disease

Altering ecosystems can give rise to new human diseases or increased incidence and recognition of previously unidentified diseases, as has been demonstrated in the case of **Lyme disease.** More Virginia white-tailed deer, which thrive along borders between forests and clearings, now inhabit the United States than when the Pilgrims landed. Settlers clearing fields created suitable habitats; as hunting of deer for food has declined, their populations have increased to record levels. With this increase has come Lyme disease, first described in 1974 by Allen Steere and his colleagues at Yale University and named after the Connecticut town where the earliest recognized cases occurred. The disease has now been identified on three continents and in more than 46 states in the United States. It is common along with deer from Cape Cod to Virginia, in Wisconsin and Minnesota, and in parts of California, Oregon, Utah, and Nevada (Figure 23.10(a)).

In 1982 Willy Burgdorfer of the National Institutes of Health laboratory in Montana isolated and described the causative organism, *Borrelia burgdorferi*, a

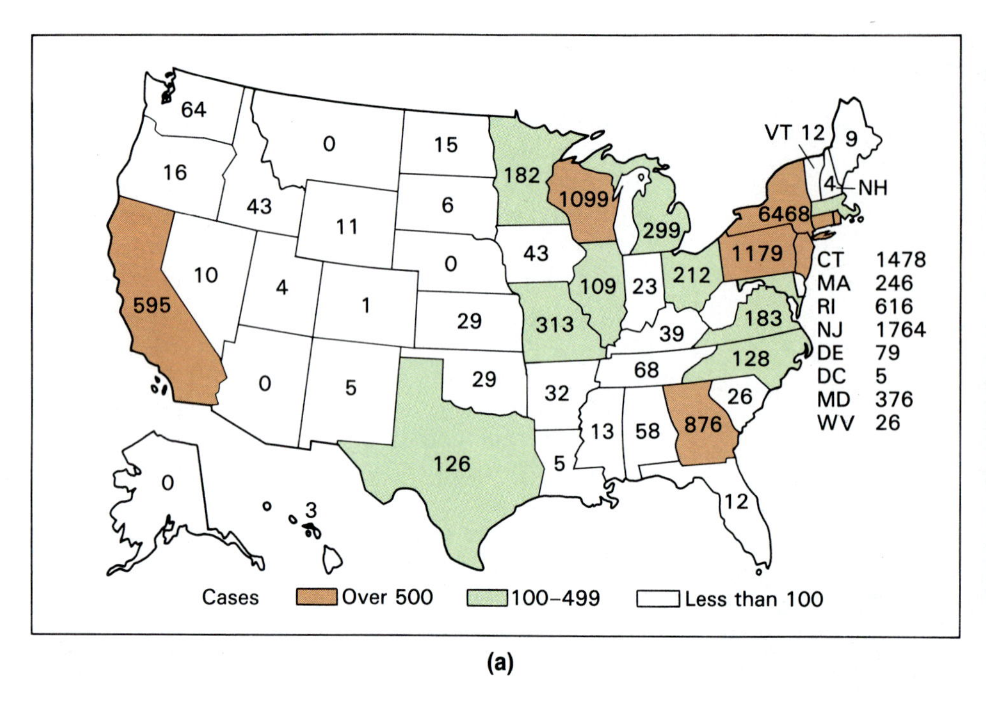

(a)

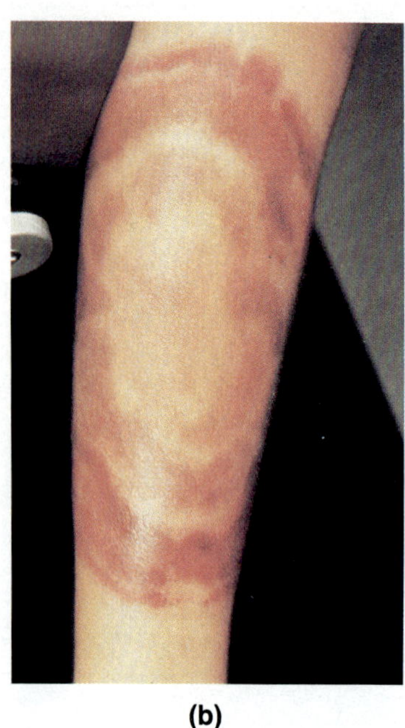

(b)

FIGURE 23.10 (a) Distribution of Lyme disease in the United States. What appears to be spread of the disease may really be better diagnosis due to heightened awareness. However, the disease may also be spreading in some areas. (b) The typical "bull's-eye" rash of Lyme disease, showing concentric rings around the initial site of the tick bite.

previously unknown spirochete (Figure 23.11a). By 1985 Lyme disease was the most commonly reported tick-borne disease in this country. It is carried by *Ixodes dammini*, the deer tick (Figure 23.11b), which feeds on deer and small mammals such as mice. The tick takes three blood meals during its 2-year life cycle. It ingests contaminated blood in one meal and transmits disease during a subsequent one. Dogs, horses, and cows as well as humans can be infected. As the tick spreads to areas of high human density, the risk of tick-borne infections increases.

Symptoms of Lyme disease vary, but most patients develop flulike symptoms shortly after being bitten by an infected tick. A bull's-eye rash at the site

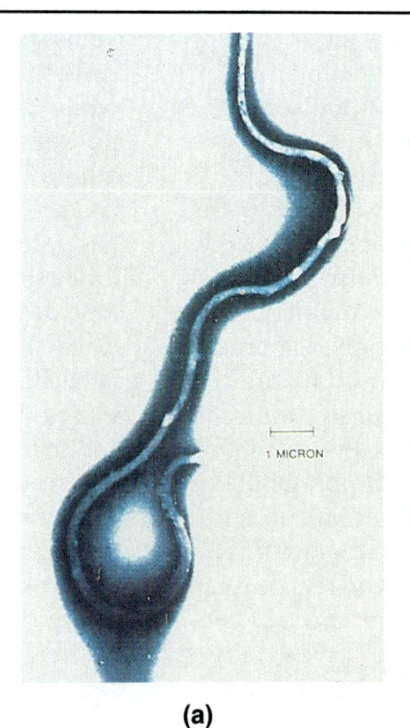

(a)

(b)

FIGURE 23.11 (a) *Borrelia burgdorferi*, the spiral bacterium that causes Lyme disease, and (b) the tick vector *Ixodes dammini*. Three stages of the tick life cycle are shown. The youngest, nymphal stage (top) is about the size of a poppy seed; an unengorged adult (bottom) is about the size of an apple seed; an engorged adult (center) can reach the size of a jelly bean. It is the tiny nymph, however, that is most likely to be carrying disease organisms and transmitting Lyme disease. Very careful inspection is necessary to detect these, which are often found in hairy regions of the body. Cellophane tape wrapped around your hand, sticky side out, can aid in removal when rubbed around ankles and other exposed areas. When walking in tick-infested areas, wear long sleeves and pants with socks pulled up over cuffs. Tick-repellent sprays may also help, but apply them to clothing rather than to skin—they can be toxic.

TABLE 23.2 Summary of bacterial systemic diseases

Disease	Agent(s)	Characteristics
Anthrax	*Bacillus anthracis*	Cutaneous lesions become necrotic, respiratory infections always fatal, intestinal infections similar to food poisoning
Plague	*Yersinia pestis*	Bubonic plague causes buboes in lymph nodes, and hemorrhages turn skin black; septicemic plague occurs when organisms invade blood and cause hemorrhage and necrosis in many tissues; pneumonic plague from inhaling organisms causes pneumonia
Tularemia	*Francisella tularensis*	Ulceroglandular form causes high fever, headache, buboes; bacteremia leads to typhoidal tularemia; inhalation leads to bronchopneumonia; relapses often occur
Brucellosis	*Brucella* species	Gradual onset of symptoms, cyclic fever, enlarged lymph nodes and liver, jaundice
Relapsing fever	*Borrelia recurrentis*	Several days of high fever, respites, and shorter periods of fever due to changes in organisms' antigens; can cross placenta
Lyme disease	*Borrelia burgdorferi*	Rash and flulike symptoms at onset, later arthritis and nerve and heart damage; can cross placenta

of the bite is also seen in about one-half of all cases. Weeks or months later other symptoms occur. Arthritis is the most common symptom, but loss of insulating myelin from nerve cells can cause symptoms resembling Alzheimer's disease and multiple sclerosis. Myocarditis occurs in some patients. Because most patients do not seek medical attention early, Lyme disease usually is diagnosed by clinical symptoms after arthritis appears. An antibody test is available. Lyme disease is treated with antibiotics such as doxycycline and amoxicillin, which are more effective the earlier they are administered in the course of the disease. Unfortunately, the earlier stages are often misdiagnosed. Furthermore, subclinical infections are common in endemic areas.

Although no vaccine for humans currently exists, a vaccine to protect dogs from Lyme disease is now available. Dogs are six times more likely to develop the disease than humans. Cats are less likely to develop Lyme disease than dogs, possibly because of their continual grooming. Other animals such as cattle are also susceptible, as a Wisconsin dairy farmer discovered when three-quarters of his 60 dairy cows went lame from Lyme arthritic enlargements of leg joints, some nearly as large as basketballs. Only about 10 percent of infected people go on to develop crippling Lyme arthritis. Researchers have found a genetic predisposition in those that do, with 89 percent having HLA DR 4 or RLA DR 2 tissue antigens. Antibiotics given in the first 6 weeks after infection generally prevent this chronic arthritis. After a certain point, however, antibiotics are no longer able to prevent the development of arthritis in susceptible individuals.

Because no vaccine for humans is available, control of Lyme disease depends on avoiding tick bites. This is difficult to do, however, because the tick—especially the immature form, which often carries the spirochete—is very tiny. A drastic attempt to control the disease was tried on Great Island, Massachusetts, where an entire 52-member deer herd was killed. Putting antitick eartags on deer and mice, a difficult job, has not been very effective. Investigators also tried saturating cotton balls with insecticide and leaving them around for the mice, who took them back to their nests to sleep on, thereby killing off many of their ticks. Now, a European wasp that parasitizes the ticks is being released in tick-infected areas. This biological control method has given excellent results, almost exterminating the tick in some areas.

Bacterial systemic diseases are summarized in Table 23.2.

Rickettsial Systemic Diseases

Rickettsias are named after Howard T. Ricketts, who identified them as the causative agents of typhus and Rocky Mountain spotted fever. Both he and another investigator, Baron von Prowazek, died of laboratory infections from these highly infectious organisms. Rickettsias are small gram-negative bacteria and are obligate intracellular parasites. Those that cause typhus grow in the cytoplasm; those that cause spotted fever grow in both the nucleus and the cytoplasm. Rickettsias can be cultured only in cells, and embryonated eggs often are used, but they should be handled only by very experienced technicians and only in specially equipped isolation laboratories. Despite improvements in laboratory design and techniques and the availability of vaccine, laboratory infections are still common.

Rickettsial diseases have several common properties. The organisms invade and damage blood vessel linings and cause them to leak. This leakage causes skin lesions and especially **petechiae** (pe-te′ke-e), pinpoint-size hemorrhages most common in skin folds. It also causes necrosis in organs such as the brain and heart. Although each disease produces a particular kind of skin rash, all rickettsial diseases cause fever, headache, extreme weakness, and liver and spleen enlargement. Headaches can be crushing, and the patient may appear confused.

Except for Q fever (Chapter 21), rickettsial diseases are transmitted among vertebrate hosts by an arthropod vector. (p. 590) Humans often are accidental hosts of zoonoses, but they are the sole hosts of epidemic typhus and trench fever. Because of the hazards of culturing organisms, rickettsial diseases usually are diagnosed by clinical findings and serologic tests. The diseases are treated with tetracycline or chloramphenicol; even these antibiotics only inhibit, but do not kill, rickettsias. Therapy must be prolonged until the body's defenses can overcome the infection. Often rickettsias are not totally eliminated; they can remain latent in lymph nodes and have been known to reactivate 20 years after initial infection.

Typhus

Typhus occurs in a variety of forms, including epidemic, endemic (murine), and scrub typhus. Brill-Zinsser disease is a recurrent form of endemic typhus.

Epidemic typhus, also called classic, European, or louse-borne typhus, is caused by *Rickettsia prowazekii* and is most frequently seen during wars and other conditions of overcrowding and poor sanitation. In 1812 typhus helped drive Napoleon from Russia; more recently, during World War I, it infected over 30 million Russians and killed 3 million. The history of warfare is filled with instances of typhus being the "commanding general." These and other examples of the effects of epidemic typhus on human affairs are well documented in Hans Zinsser's *Rats, Lice, and History*. Only after the discovery of the pesticide DDT during World War II were typhus epidemics halted.

Epidemic typhus is transmitted by lice. After a louse feeds on an infected person, rickettsias multiply in its digestive tract and are shed in its feces. When a louse bites it defecates; infected lice deposit organisms next to a bite and themselves die of typhus in a few weeks. As victims scratch bites they inoculate organisms into the wound. Lice become infected by biting an infected human. They abandon dead bodies or people with high fevers, moving to, and infecting new hosts.

After about 12 days' incubation, onset of fever and headache is abrupt and is followed 6 or 7 days later by a rash on the trunk that spreads to the extremities but rarely affects palms or soles. Antibiotic therapy should be started immediately. Without treatment the disease typically lasts up to 3 weeks, and mortality ranges between 3 and 40 percent. The disease can be prevented by eradicating lice with insecticides and by maintaining hygienic living conditions. A vaccine is available. Recovery generally gives lifetime immunity, except when Brill-Zinsser disease occurs.

Brill-Zinsser disease, or **recrudescent typhus,** is a recurrence of a typhus infection; it is named for Nathan Brill and Hans Zinsser, who studied it in the 1930s among New York City's Eastern European immigrants. Compared with first infections, this disease has milder symptoms, is shorter in duration, and often does not cause a skin rash. It is caused by reactivation of latent organisms harbored in lymph nodes, sometimes for years. Lice feeding on patients with Brill-Zinsser disease can transmit the organism and cause initial typhus infections in susceptible individuals. The disease can be prevented by preventing typhus itself and reduced in incidence by adequate antibiotic therapy to those already infected. In spite of rigorous treatment, some victims of typhus still develop it later. Brill-Zinsser disease can be distinguished from epidemic typhus by the type of antibodies formed shortly after onset. Epidemic typhus first elicits IgM and then IgG antibodies, whereas Brill-Zinsser disease, being a second infection, elicits primarily IgG antibodies. (Chapter 18, pp. 482–484).

Endemic or **murine typhus,** named for its association with rats (*murine* refers to rats and mice), is a flea-borne typhus caused by *Rickettsia typhi*. It occurs in isolated pockets around the world, including southeastern and Gulf Coast states, especially Texas. Its incidence in the United States is less than 100 cases

CLOSE-UP
Microbes and War

"Soldiers have rarely won wars. They more often mop up after the barrage of epidemics. And typhus, with its brothers and sisters—plague, cholera, typhoid, dysentery—has decided more campaigns than Caesar, Hannibal, Napoleon, and all the . . . generals of history. The epidemics get the blame for defeat, the generals the credit for victory. It ought to be the other way 'round. . . ."

—Hans Zinsser, 1935

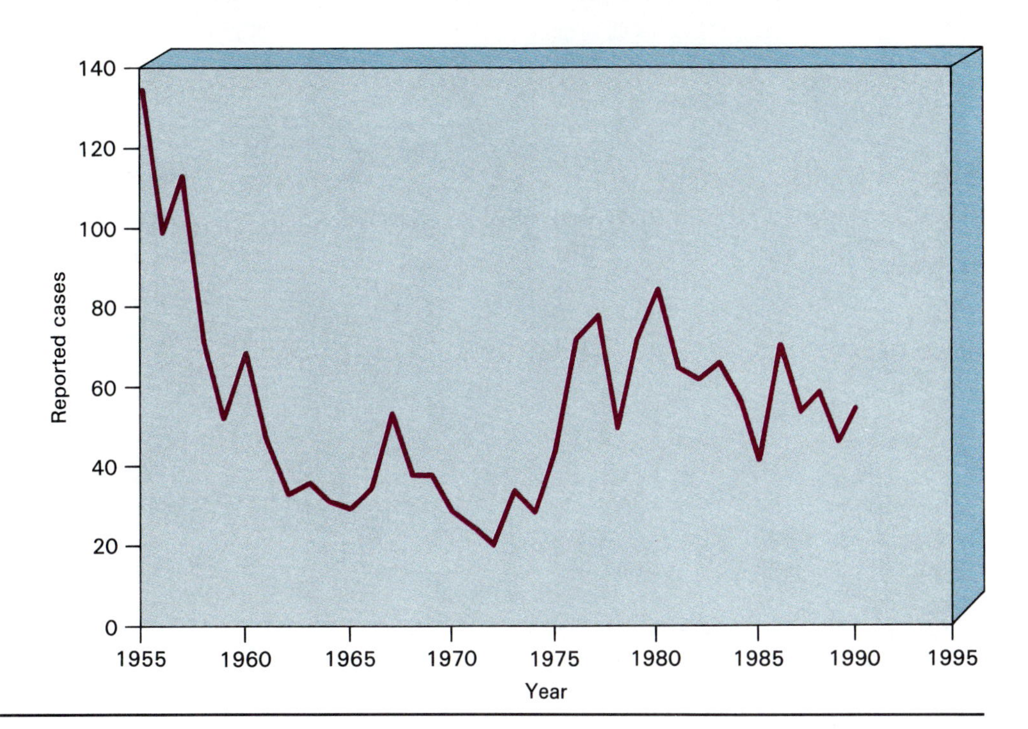

FIGURE 23.12 Incidence of murine typhus in the United States.

per year (Figure 23.12). Fleas from infected rats infect humans when they defecate while biting. The host rubs organisms into the bite wound or transfers them to mucous membranes, another portal of entry. After 10 to 14 days' incubation, onset of fever, chills, and a crushing headache is abrupt, followed by a rash in 3 to 5 days. The disease is self-limiting and lasts about 2 weeks if untreated. Mortality is about 2 percent.

Scrub typhus, or **tsutsugamushi disease,** is caused by *Rickettsia tsutsugamushi.* "Tsutsugamushi" is Japanese for "bad little bug." The "bug" that transmits this disease is a mite, which feeds on rats in Japan, Australia, and parts of Southeast Asia. Mites drop off rodent hosts and infect humans with their bites. Scrub typhus was a problem during World War II and the Vietnamese conflict when soldiers crawled on their bellies through low scrub vegetation to avoid snipers. After 10 to 12 days' incubation, scrub typhus begins abruptly with fever, chills, and headache. Many patients develop sloughing lesions at the bite sites and later a generalized spotty rash. In untreated cases, the fatality rate can reach 50 percent, but with prompt antibiotic treatment, fatalities are rare. Although no vaccine is available, infections can be prevented by controlling mite populations.

Rocky Mountain Spotted Fever

Rocky Mountain spotted fever was first recognized around 1900 in Rocky Mountain states such as Idaho and Montana. However, its geographic distribution (Figure 23.13) suggests that "Appalachian spotted fever" would be a more appropriate name. The incidence has stayed below 1000 cases per year since the 1930s but might increase as more people spend time in tick-infested areas. This disease is caused by *Rickettsia rickettsii* and is transmitted by a tick (Figure 23.14a). After 3 to 4 days' incubation, onset of fever, headache, and weakness is abrupt, followed in 2 to 4 days by a rash (Figure 23.14b). The rash begins on ankles and wrists, is prominent on palms and soles, and progresses toward the trunk—just the reverse of the progression in typhus. Spots are caused by blood leaking out of damaged blood vessels beneath the skin surface; they coalesce as blood leaks from many damage sites. Blood vessels in organs throughout the body are similarly damaged. Strains vary considerably in virulence; likewise, mortality varies from 5 percent to 80 percent, with an average of 20 percent in untreated cases. Prompt antibiotic treatment keeps the mortality rate to between 5 and 10 percent. Rocky Mountain spotted fever can be prevented by wearing protective clothing and by vigilantly inspecting clothing and skin during visits to tick-infested areas. Inspecting children's hair is especially important. The only vaccine is not completely effective.

Worldwide, there are many other types of "spotted fevers," each caused by its own rickettsial species, and often named for its location, such as "Siberian spotted fever."

Rickettsialpox

Rickettsialpox, caused by *Rickettsia akari,* was first discovered in New York City in 1946 and is now known

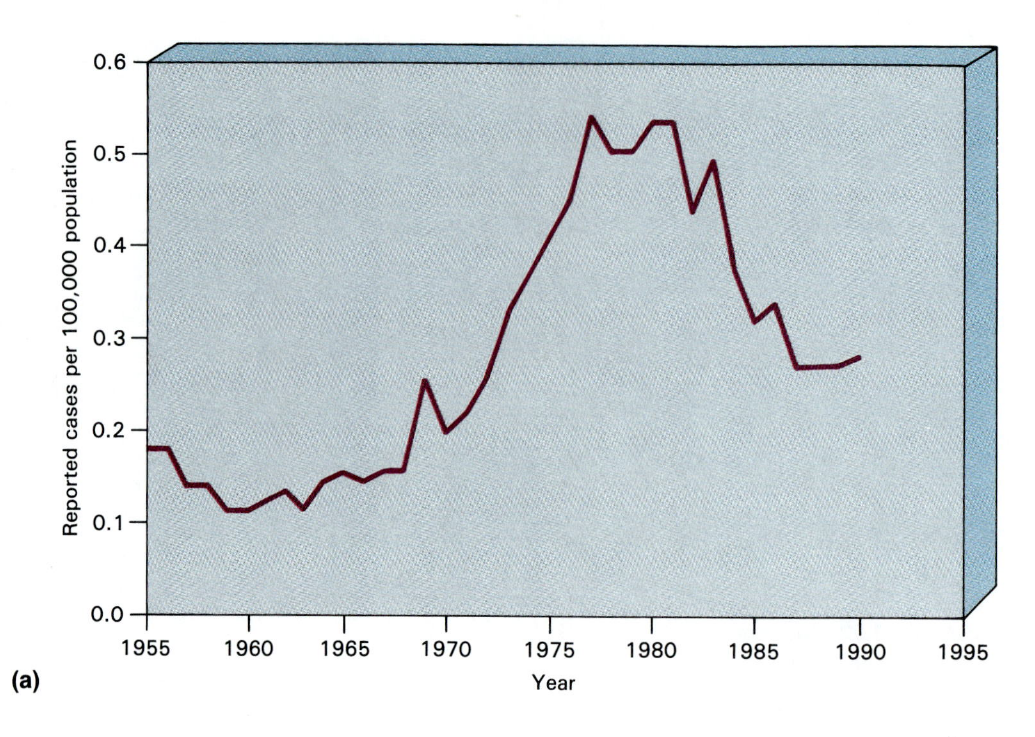

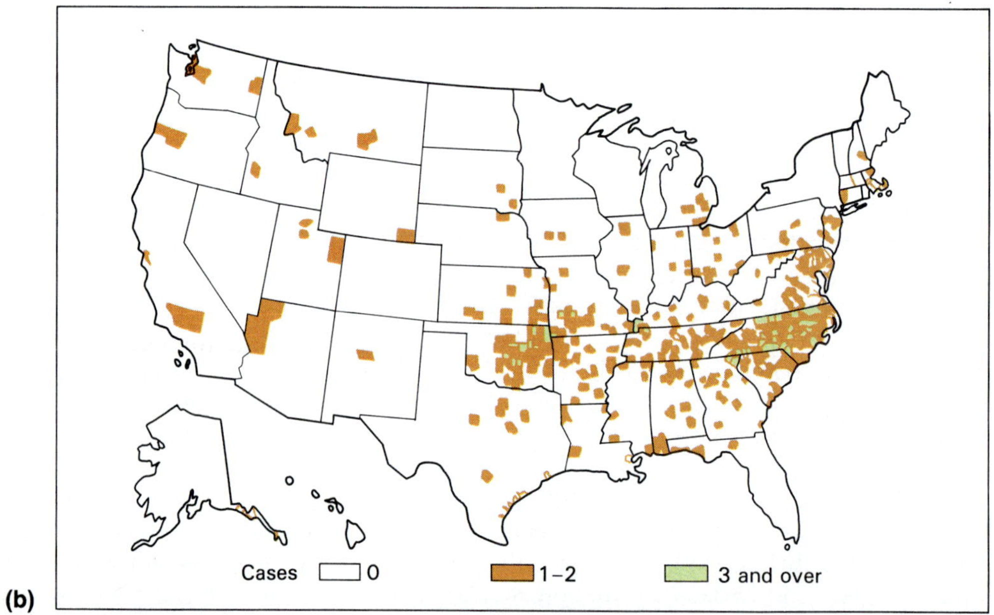

FIGURE 23.13 Distribution of Rocky Mountain spotted fever in the United States.

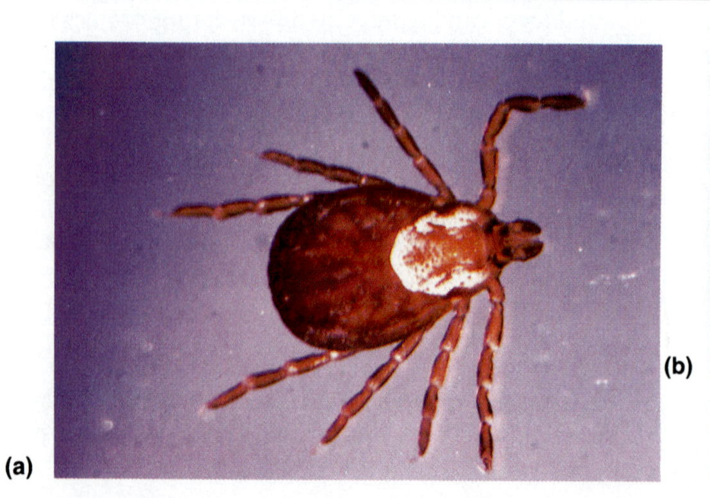

(b)

FIGURE 23.14 (a) Vectors of Rocky Mountain spotted fever include the dog tick *Dermacentor variabilis*. (b) Rash of the disease on a patient's feet.

to occur in Russia and Korea as well. It is carried by mites found on house mice. The disease is relatively mild, and its lesions resemble those of chickenpox. Because of misdiagnosis as chickenpox or other diseases, incidence and mortality data are unreliable, but no fatalities have been reported. It can be prevented by controlling rodents.

Trench Fever

Trench fever, also called **shinbone fever,** resembles epidemic typhus in that it is transmitted among humans by lice and is prevalent during wars and under unsanitary conditions. Stress probably is a predisposing factor. The causative agent is the bacterium *Rochalimaea quintana*, classified with rickettsias even though it is not an obligate intracellular parasite. It can be cultured in artificial media, has worldwide distribution, but only rarely causes disease.

Trench fever was first seen in World War I among soldiers living in trenches and wearing the same clothing day after day. British "trench coats" were developed to protect troops who were without shelter during bad weather, but they were not entirely effective. Soldiers and their clothing, including the trench coats, became infested with body lice; exhausted soldiers crowded together in filth fell victims to the disease. After World War I the disease disappeared and did not reappear until World War II. The symptoms of trench fever are a 5-day fever and severe leg pain, but many soldiers have reported recurrent symptoms, including mental confusion and depression, as long as 19 years after infection. No vaccine is available, and prevention depends on controlling lice.

Bartonellosis

Bartonellosis (bar-to-nel-o'sis) is caused by *Bartonella bacilliformis*, named for the Peruvian physician A. L. Barton, who first described it in 1901. The disease occurs in two forms: **Oroya fever,** or **Carrion's disease,** an acute fatal fever with severe anemia, and **verruga peruana** (ver-oo'gah per-oo-ah'nah), a chronic, nonfatal skin disease (Figure 23.15). Both are found only on the western slopes of the Andes in Peru, Ecuador, and Colombia—the habitat of the sandfly *Phlebotomus*, which transmits the organism. In 1885, Daniel Carrion, a Peruvian medical student, inoculated himself with material from a wartlike verruga peruana lesion to show a connection between it and Oroya fever. His death 39 days later from Oroya fever clearly demonstrated the connection.

After being transmitted to a human host by the bite of an infected sandfly, organisms enter the blood and multiply during an incubation period of a few weeks to 4 months. Little is known of the epidemiology of bartonellosis, but humans appear to be the

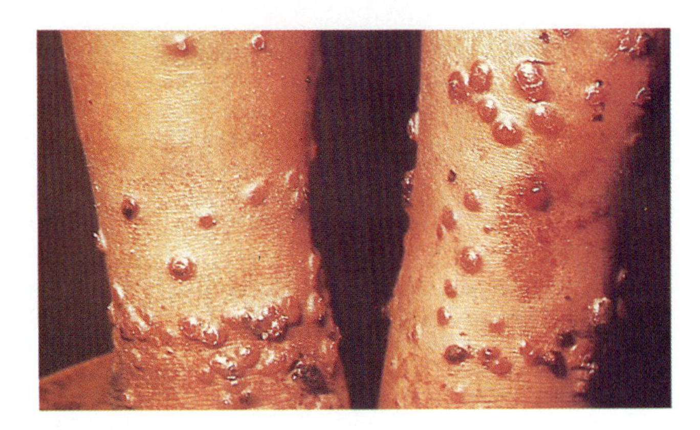

FIGURE 23.15 Lesions of verruga peruana.

sole reservoir. Oroya fever is a severe, febrile, hemolytic anemia. Verruga peruana causes only skin lesions that persist for 1 month to 2 years but usually last for about 6 months. Lesions heal spontaneously but can recur. Oroya fever probably develops in people with no immunity, and verruga peruana occurs in those with partial immunity. Penicillin, tetracyline, or streptomycin can cure Oroya fever but not verruga peruana. No vaccine is available, and prevention depends on controlling sandflies.

The characteristics of rickettsial diseases are summarized in Table 23.3.

Viral Systemic Diseases

Dengue Fever

Dengue (den'ga or den'ge) **fever,** first characterized in 1780 by Benjamin Rush in Philadelphia, has also been called **breakbone fever** because of the severe bone and joint pain it causes. Other symptoms include high fever, headache, loss of appetite, nausea, weakness, and, in some cases, a rash. The disease is self-limiting and runs its course in about 10 days. Four distinct immunologic types of the virus have been identified, and two of them have been correlated with disease symptoms. A first dengue fever infection produces the symptoms just noted. A second dengue fever infection with a different immunologic type of the virus causes a hemorrhagic form of the disease. The hemorrhage occurs while the virus is replicating in circulating lymphocytes. It is due to an immune response made possible by prior infection. Other symptoms are rapid breathing and low blood pressure that may progress to shock. The shock is reversible if treatment is initiated promptly. Serologic tests are available to diagnose dengue fever, and a vaccine against one immunologic type of the virus appears to confer immunity.

Dengue is distributed worldwide in tropical areas, causing over 100 million cases per year, with occasional episodes in the subtropics. Its main vector is

TABLE 23.3 Summary of rickettsial diseases

Disease	Causative Organism	Geographic Area of Prevalance	Arthropod Vector Reservoir	Vertebrate Reservoir
Typhus group				
Epidemic (classic, European) typhus	*R. prowazekii*	Worldwide	Louse	Human
Brill-Zinsser disease (recrudescent typhus)	*R. prowazekii*	Worldwide	(Recurring infection)	Human
Endemic (murine) typhus	*R. typhi*	Worldwide, small scattered foci	Flea	Rodents
Scrub typhus group				
Scrub typhus (tsutsugamushi disease)	*R. tsutsugamushi*	Japan, Southeast Asia	Mite	Rat
Spotted fever group				
Rocky Mountain spotted fever	*R. rickettsii*	Western Hemisphere	Tick	Rodents, dogs
Rickettsialpox	*R. akari*	United States, Korea, Russia	Mite	House mouse
Trench fever	*Rochalimaea quintana*	Worldwide, but disease only during wars	Louse	Human
Bartonellosis	*Bartonella bacilliformis*	Western slopes of the Andes	Sandfly	Humans only known host

Aedes aegypti, although in some areas the Asian tiger mosquito *A. albopictus* can be important. Since 1985, health officials have been concerned about the arrival and spread of *A. albopictus* within the United States. The mosquito apparently came in with used tire casings imported from Asia. In 1980, the first outbreak of dengue fever in the United States in about 40 years occurred. Since then, all four serotypes have arrived in the United States. The rapid spread and aggressive biting habits of *A. albopictus* could bring dengue to areas that previously had been safe. Mosquito control is the primary method of preventing dengue, as vaccines are not available for all viral serotypes.

Yellow Fever

Yellow fever was first studied by Carlos Finley and Walter Reed when infection of workers threatened to disrupt construction of the Panama Canal. Although adequate techniques to identify the causative **flavivirus** were not available, Finley and Reed identified the mosquito vector, *Aedes aegypti*, and instituted control measures that prevented transmission of the disease. The disease is now limited to tropical areas of Central and South America and Africa. Incidence is greatest in remote jungle areas, where monkeys serve as reservoirs of infection and carrier mosquitoes bite both monkeys and people. In recent decades the number of cases reported each year has varied from 12 to 304, but the actual number of cases may be 10 to 20 times as great.

Yellow fever causes fever, nausea, and vomiting, which coincide with viremia. Liver damage from viral replication in liver cells causes the jaundice for which the disease is named. The disease is of short duration: In less than a week the patient has either died or is recovering. In most instances the fatality rate is about 5 percent, but in some epidemics it reaches 30 percent. Two strains of yellow fever viruses are used to produce vaccines. The Dakar strain is scratched into the skin, whereas the 17D strain is administered subcutaneously. Both are effective in establishing immunity.

Infectious Mononucleosis

In 1962 Dennis Burkitt suggested that a virus caused the lymphoid malignancy now called Burkitt's lymphoma, found in children in East Africa. The virus, now called **Epstein-Barr virus** (EBV), is known to cause both Burkitt's lymphoma and **infectious mononucleosis.** EBV infects only human B lymphocytes (Figure 23.16). It replicates like most other herpesviruses and derives its envelope from the inner nuclear membrane of the host cell. The virus has an unusually large number of genes—more than 50 different proteins are produced by complete expression of the DNA in EBV.

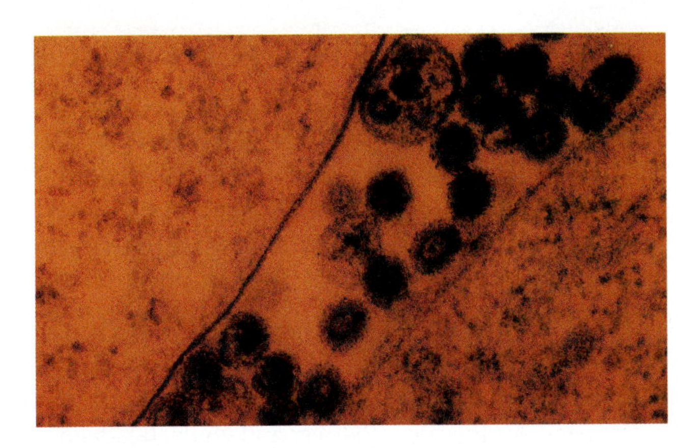

FIGURE 23.16 Epstein-Barr virus, which causes mononucleosis and Burkitt's lymphoma.

EBV enters the body through the oropharynx and establishes a persistent infection in which viruses are shed for months to years. The viruses invade lungs, bone marrow, and lymphoid organs, where they infect certain kinds of mature B lymphocytes. They penetrate B lymphocytes over a 12-hour period, and EBV replication begins within 6 hours after penetration. EBV DNA is replicated much faster than cellular DNA. Viral DNA can exist as circular plasmids or become integrated into the cellular DNA.

EBV exerts three significant effects on lymphocytes:

1. The virus acts on antibody-producing cells and elicits EBV antibodies.
2. Infection and transformation (transfer of viral genes to the cellular DNA) are complex events that occur only in B lymphocytes, which have receptors for EBV. The cells produce a variety of antigens; they also proliferate and account for the excess of lymphocytes seen in infectious mononucleosis.
3. Other antigens are induced on the surface of some infected B cells. They appear to play a role in some B cell and T cell interactions and may account for some symptoms of infectious mononucleosis.

Proliferation of EBV-infected lymphocytes is limited by cytotoxic T cells, T suppressor cells, and cells that make humoral antibodies and complement. (Chapter 18, pp. 489–492) If these defenses fail to limit lymphocyte proliferation, uncontrolled B cell proliferation can lead to B cell cancer or Burkitt's lymphoma.

Infectious mononucleosis is an acute disease that affects many systems. Lymphatic tissues become inflamed, some liver cells become necrotic, and monocytes accumulate in liver sinusoids. In some cases myocarditis and glomerulonephritis are seen. The incubation period for the disease is from 30 to 50 days. Mild symptoms—headache, fatigue, and malaise—occur during the first 3 to 5 days of the disease and worsen as the disease progresses. About 80 percent of patients have a sore throat during the first week. The spleen is enlarged, and cells in lymphoid tissues in the oropharynx multiply. The tonsils are coated with a gray exudate, and the soft palate may be covered with petechiae. Secondary infection with beta-hemolytic streptococci frequently occurs. Although the disease causes great discomfort and requires several weeks of recuperation, fatalities are rare and usually result from underlying immunologic defects.

Diagnosis of EBV infections is complicated by the fact that the disease resembles cytomegalovirus infections, toxoplasmosis, and acute leukemia. The distinguishing symptoms of EBV are the concurrent sore throat, multiplication of lymphocytes, and the presence of antibodies against the antigens on sheep and human erythrocytes. Infectious mononucleosis is treated with bed rest and antibiotics for secondary infections. Ampicillin is not used because it causes a rash in infectious mononucleosis patients. The presence of IgG antibodies to viral capsid protein indicate a past infection, and their numbers furnish an index of immunity. Increasing numbers of IgM antibodies to the protein is evidence of a current infection. No vaccine is available.

In underdeveloped countries the entire population has antibodies to EBV by 1 year of age. Exposure to the virus in infancy produces mild symptoms or no symptoms at all and confers immunity to later infection. Where living standards are higher, a more severe disease is seen later in life. The incidence of infectious mononucleosis in the United States is highest among relatively affluent teenagers and young adults; 10 to 15 percent become infected. The affected age group and the large inoculum required to transmit the disease may be responsible for its being called the "kissing disease."

Chronic Fatigue Syndrome Since 1985 medical researchers have been attempting to determine whether a *chronic EBV syndrome* exists. Patients report fever and persistent fatigue along with a variety of other nonspecific symptoms similar to those of infectious mononucleosis. Some, but not all, have EBV antibodies. Others have had measles or herpesvirus infections. Because a direct relationship between the symptoms and previous EBV infection has not been established, the illness has been renamed **chronic fatigue syndrome.** More study is needed to determine whether the syndrome is associated with one or more previous viral illnesses or whether it might be a psychological disorder. Some recent studies point to possible immune system defects. Herpes human virus number 6 (HHV6), recently discovered, is also a candidate for causative agent of this disorder.

Burkitt's Lymphoma **Burkitt's lymphoma,** a tumor of the jaw, is seen mainly in children (Figure 23.17). It occurs about 6 years after the primary infection with

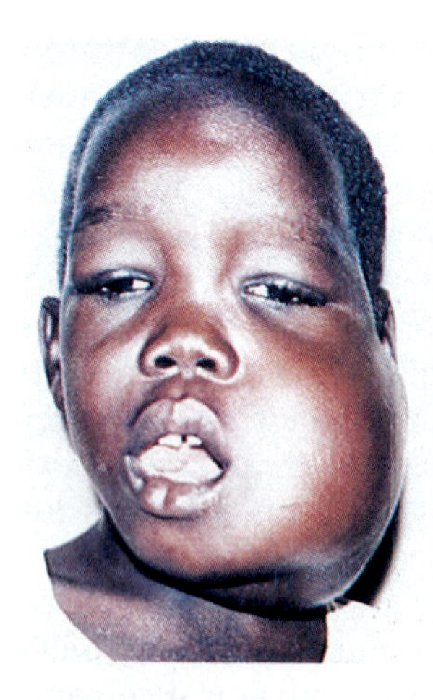
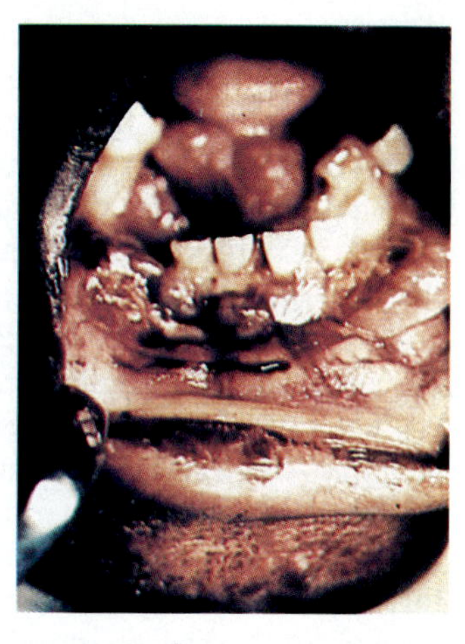

FIGURE 23.17 Burkitt's lymphoma, a form of cancer of the jaw caused by Epstein-Barr virus, is usually seen only in African children who also have malaria.

EBV, and the tumor frequently arises from a single cell. The immune system of affected individuals appears to be normal, but it is incapable of eliminating the tumor cells. Burkitt's lymphoma is found mainly in regions of Africa where malaria is endemic, and infection with malarial parasites may enhance growth of the virus or interfere with the immune response.

Other Effects Another tumor associated with EBV, which also shows a distinct geographic localization, is nasopharyngeal carcinoma, found most often in China and rarely in the Western Hemisphere. Individuals with immunologic defects are especially susceptible to developing lymphomas, presumably because they lack the necessary immune mechanisms to eliminate the malignant cells. Cyclosporine A, used to depress immunity in organ transplant patients, enhances lymphoid cell growth in donor organs. Organs from donors who have ever had EBV infections may contain EBV; immunosuppression from any cause, including AIDS, can release EBV and lead to mononucleosis or malignancy.

Other Viral Infections

Filovirus Fevers **Filoviruses,** or filamentous viruses, display unusual variability in shape. Some are branched, others are fishhook- or U-shaped, and still others are circular. They contain negative-sense RNA in a helical capsid and vary in length from 130 to 4000 nm. Two filoviruses have been associated with human disease. The Ebola virus has caused outbreaks of hemorrhagic fever with a mortality of 88 percent in Zaire and 51 percent in Sudan. Nearly a fifth of the population of rural areas of Central Africa have antibodies to Ebola. Transmission apparently is person to person; neither Ebola viruses nor their antibodies have been found in arthropods, monkeys, rodents, or other animals of the region. The Marburg virus was first recognized in Germany when technicians preparing monkey kidney cell cultures died of a hemorrhagic disease. Nosocomial infections have since been encountered with a mortality rate of about 25 percent. Hemorrhage into skin, mucous membranes, and internal organs, death of cells of the liver, lymph tissue, kidneys, and gonads, and brain edema also have been observed. The virus has been isolated directly from monkeys and from laboratory inoculation of guinea pigs.

Bunyavirus Fevers Infections caused by bunyaviruses begin suddenly with fever and chills, headache, and muscle aches. Although usually not fatal and without permanent effects, the diseases are temporarily incapacitating. When encephalitis occurs, it progresses slowly, either because the viruses replicate slowly in neural tissues or because certain viruses capable of replicating in neural tissue are selected. Rats, bats, and animals with hooves serve as reservoirs of infection. Tropical and temperate forest mosquitoes are vectors.

The LaCrosse bunyavirus has been identified in the northeastern and north-central United States. It causes a mild disease in adults but can cause seizures, convulsions, mental confusion, and paralysis in children. California encephalitis virus, another bunyavirus initially isolated from mosquitoes in the San Joaquin Valley, has similar effects on humans.

Certain bunyaviruses, which are called **phleboviruses** because they are carried by the sandfly *Phlebotomus papatasii*, have been recovered from human infections. The **Rift Valley fever** virus causes epidemics and has unpredictable virulence. It causes sudden vomiting, joint pain, and slowing of the heart rate. One epidemic of Rift Valley fever in 1975 in central Africa infected thousands but left only 4 people dead. Only 2 years later the disease appeared in Egypt, where 200,000 cases and 598 deaths were recorded.

The Hantaan viruses are associated with hemorrhagic fevers, including Korean or epidemic hemorrhagic fever (see Microbiologist's Notebook, Chapter 11), and kidney disease. They are distributed widely over Eurasia and find reservoirs in rodents. They cause capillary leakage, hemorrhage, and cell death in the pituitary gland, heart, and kidneys. Kidney damage can be severe, and low blood pressure can proceed quickly to shock, from which one-third of patients die. Even more die if bleeding occurs in the gastrointestinal tract and the central nervous system or if fluid accumulates in the lungs. The major source of such infections is contact with infected rodents or their excreta. Some rodents shed viruses in feces and saliva for 30 days and into their urine for a year. Humans under 10 or over 60 years are rarely infected, perhaps because they are less likely to come in contact with infected materials.

Arenavirus Fevers Like bunyaviruses, arenaviruses cause hemorrhagic fevers. Of these **Lassa fever** perhaps is the most widely known. It is an African disease that begins with pharyngeal lesions and proceeds to severe liver damage. The prognosis is poor in 20 to 30 percent of cases in which hemorrhage from mucous membranes occurs. Several other arenavirus infections, including **Bolivian hemorrhagic fever,** have been identified in humans, especially in Africa and South America.

Bolivian and other South American hemorrhagic fevers are multisystem diseases with insidious onset and progressive effects. The viruses attack lymphatic tissues and bone marrow and cause vascular damage, bleeding, and shock. However, death, which occurs in about 15 percent of cases, usually results from damage to the central nervous system. How the viruses affect the nervous system is not known.

Colorado Tick Fever **Colorado tick fever** is caused by an **orbivirus,** which is transmitted by dog ticks from reservoir animals such as squirrels and chipmunks to humans. High levels of viremia with infection of immature erythrocytes occur in the disease. The patient suffers from headache, backache, and fever, but recovery usually is complete.

Parvovirus Infections Recently parvovirus B19 has been identified as the probable cause of aplastic crisis in sickle cell anemia. It appears to replicate in rapidly dividing cells of the bone marrow. An **aplastic crisis** is a period during which erythrocyte production ceases. Afflicted children soon are in acute distress. Although normal erythrocytes remain functional in the blood for about 120 days, those of sickle cell anemia patients survive only 10 to 15 days. Under such circumstances severe depletion of functional erythrocytes occurs. Usually a child experiences only one such crisis, possibly because immunity to the virus develops. Another bit of evidence that aplastic crisis is infectious is that, although it occurs only in sickle cell anemia patients, it does occur in 3- to 5-year cycles within particular communities.

CLOSE-UP

Trading One Disease for Another

Bolivian hemorrhagic fever is of interest because of its relationship to efforts to eliminate malaria through improved sanitation. Prior to these efforts, hemorrhagic fever had existed in Bolivia for years without causing a major problem. This changed when a campaign was initiated to control malaria by eradicating mosquitoes. Large quantities of pesticides were used, which killed many village cats. In the absence of the cats, the rodent population multiplied and invaded thatched-roofed human dwellings. The rodents carried arenaviruses, which caused the incidence of hemorrhagic fever to rise to epidemic levels.

Two important parvovirus infections, one in cats and a relatively new one in dogs, are of significance to pet owners. **Feline panleukopenia virus** (FPV) causes severe disease with fever, decreased numbers of white blood cells, and enteritis in cats. The virus replicates in blood-forming and lymphoid tissues and secondarily invades the intestinal mucosa. In 1978 a new virus, the **canine parvovirus,** appeared and infected dogs in widespread geographic areas. It appeared first in North America, Europe, and Australia and rapidly spread worldwide. It causes severe vomiting and diarrhea in dogs of all ages and sudden death with myocarditis in puppies under 3 months old. When the canine parvovirus first appeared in an area, the death rate often exceeded 80 percent. Vaccines are now available for both feline panleukopenia and canine parvovirus.

Fifth Disease Parvovirus B19 destroys the stem cells that give rise to red blood cells. This is not a major problem in healthy adults or children, but it is a serious danger to those who have chronic hemolytic anemias, such as sickle cell anemia, and therefore have difficulty maintaining normal levels of red blood cells. It is also a danger to the fetus if a pregnant woman contracts the virus. The virus can be transmitted across the placenta and cause the fetus to develop fatal anemia. The virus does not, however, cause birth defects. Immunodeficient patients cannot control replication of the virus and may develop chronic anemia.

After parvovirus B19 was recognized as the cause of aplastic crisis in sickle cell anemia, it was also found to be the cause of erythema infectiosum (fifth disease) in normal children. It is common in children aged 5 to 14 and often goes unnoticed. Infected children have a bright red rash on the cheeks, as if someone had slapped them, which may spread to the trunk and extremities. A low-grade fever may accompany this. Often the infection is totally asymptomatic. The virus appears to be spread via the respiratory route. The disease is self-limiting and confers lifetime immunity.

The name "fifth disease" comes from a nineteenth-century list of childhood rash diseases. The first disease on this list is scarlet fever; the second is rubeola; the third, rubella; the fourth, epidemic pseudoscarlatina; and the fifth, erythema infectiosum.

Coxsackie Virus Infections Coxsackie viruses have an affinity for the pericardium and myocardium. They also can cause meningoencephalitis, diarrhea, rashes, pharyngitis, and liver disease. Epidemic muscle pain, diabetes, and inflammation of the pancreas, heart muscle, and the pericardium are associated with coxsackie B viruses. Coxsackie viruses are highly infectious and readily spread among members of a family and in institutions. Most infections probably arise from fecal-oral transmission, but because viruses can be isolated from nasal secretions, infection also may occur by respiratory fomites. Coxsackie virus infections during pregnancy can cause congenital defects, but their incidence is much lower than that for rubellavirus infections, and aborting the fetus is usually not recommended. No effective means are available for treatment, immunization, or prevention.

Parasitic Systemic Diseases

Leishmaniasis

Three species of protozoa of the genus *Leishmania* can cause **leishmaniasis** (lish-man-i'as-is) in humans (Figure 23.18). The protozoa are transmitted by sandflies. When an infected sandfly bites a person, the parasites enter the blood and are phagocytized by macrophages. Inside the macrophages the parasites multiply, and new parasites are released when the macrophage ruptures. Leishmaniasis is endemic to most tropical and subtropical countries, where appropriate species of sandfly vectors are available.

Leishmania donovani causes **kala azar** (Hindi for "black poison"), or visceral leishmaniasis. Symptoms include high irregular fever, progressive weakness, wasting, and protrusion of the abdomen because of extensive liver and spleen enlargement. Extensive damage to the immune system results from parasites destroying large numbers of phagocytic cells. If untreated the disease is usually fatal in 2 to 3 years, and it can be fatal in 6 months in patients with impaired immunity and secondary infections.

Other leishmanias are more localized in their effects and are rarely fatal. *L. tropica* causes a cutaneous lesion, sometimes called an **oriental sore,** at the site of a sandfly bite. *L. braziliensis* causes skin and mucous membrane lesions and sometimes nasal and oral polyps. Parents, observing that people who had oriental sores rarely got kala azar, sometimes purposely have infected children with oriental sores on inconspicuous parts of the body to protect them from the more serious disease.

Diagnosis is made by identifying the parasites in blood smears in kala azar and from scrapings of skin and mucous membrane lesions. Antimony compounds are used to treat both kala azar and skin and mucous membrane lesions. However, such drugs are very toxic. Prevention depends mainly on controlling sandfly breeding and eliminating rodent reservoir infections.

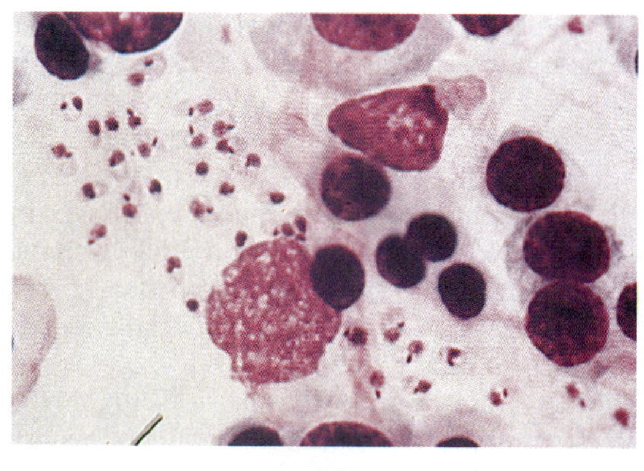

(a)

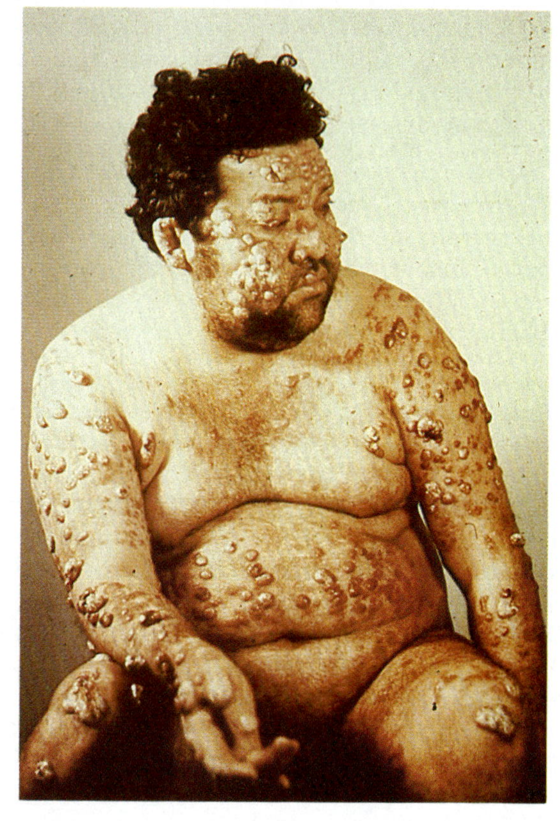

(b)

FIGURE 23.18 (a) *Leishmania* protozoa, seen as small dots between bone marrow cells. (b) A patient suffering from leishmaniasis, probably caused by *L. braziliensis*.

PUBLIC HEALTH

Desert Storm Leishmaniasis

As of June 1992, 37 cases of leishmaniasis have been confirmed in U.S. military personnel who served in the Persian Gulf conflict. Incubation period is 3 years, and the disease is fatal in 90 percent of untreated cases. Currently available treatment lowers the mortality rate to 10 percent. Health officials have banned blood transfusions and organ transplants from Gulf War veterans to prevent the spread of the disease.

Malaria

Several species of the protozoan *Plasmodium* are capable of causing **malaria,** one of the most severe of all parasitic diseases. Malaria is the world's greatest public health problem and is endemic in most tropical areas (Figure 23.19). The annual rate of new infections has been estimated to be between 100 and 300 million cases. Nearly all adults in Africa and India have been infected, and a million people, mostly children, die from the disease each year. At one time malaria was thought to have been eradicated from the United States, but military personnel, travelers, and immigrants have carried the disease to the United States from endemic areas.

Certain individuals, especially blacks, are protected from malaria by carrying the gene for sickle cell anemia. The presence of two such genes causes sickle cell anemia, but the presence of a single gene prevents malarial parasites from growing in erythrocytes. When the parasite enters the cell, the cell sickles (takes on a distorted sickle shape) or dies for lack of nutrients. The spleen removes both sickled and dead cells from the blood.

Members of the genus *Plasmodium* are amoeboid, intracellular parasites that infect erythrocytes and other tissues. They are transmitted to humans through the bite of *Anopheles* mosquitoes. These parasites have a complex life cycle, which was shown in Figure 12.5 in Chapter 12. At least four species, *P. vivax*, *P. malariae*, *P. ovale*, and *P. falciparum*, infect humans. Although they are hard to distinguish morphologically, they can be identified definitively by their effects on red blood cells and the nature of the disease they cause.

The Disease In the pathogenesis of malaria, sporozoites enter the blood from the bite of an infected female mosquito. Male mosquitoes do not suck blood. The parasites disappear from the blood within an hour and invade cells of the liver and other organs. In about a week they begin releasing merozoites that invade and reproduce in red blood cells as trophozoites (Figure 23.20). At intervals of 48 to 72 hours, depending on the infecting species of *Plasmodium*, blood cells rupture and release more merozoites that infect other red blood cells. The release of merozoites soon becomes synchronized and corresponds to intervals of high fever. Among the merozoites are some that have become gametocytes, which can undergo sexual reproduction in mosquitoes should they feed on the patient's blood. Even after the initial disease has subsided, patients are subject to relapses as dormant parasites become activated, emerge from the liver, and initiate a new cycle of disease. Relapses do not occur after infections caused by *P. falciparum* because this species does not remain in the liver.

Of the four species, *P. falciparum* causes the most severe disease because it agglutinates red blood cells and obstructs blood vessels. Such obstruction causes tissue **ischemia,** or reduced blood flow with oxygen and nutrient deficiency and waste accumulation. This species also can cause malignant malaria, an especially virulent, rapidly fatal disease, and a condition called

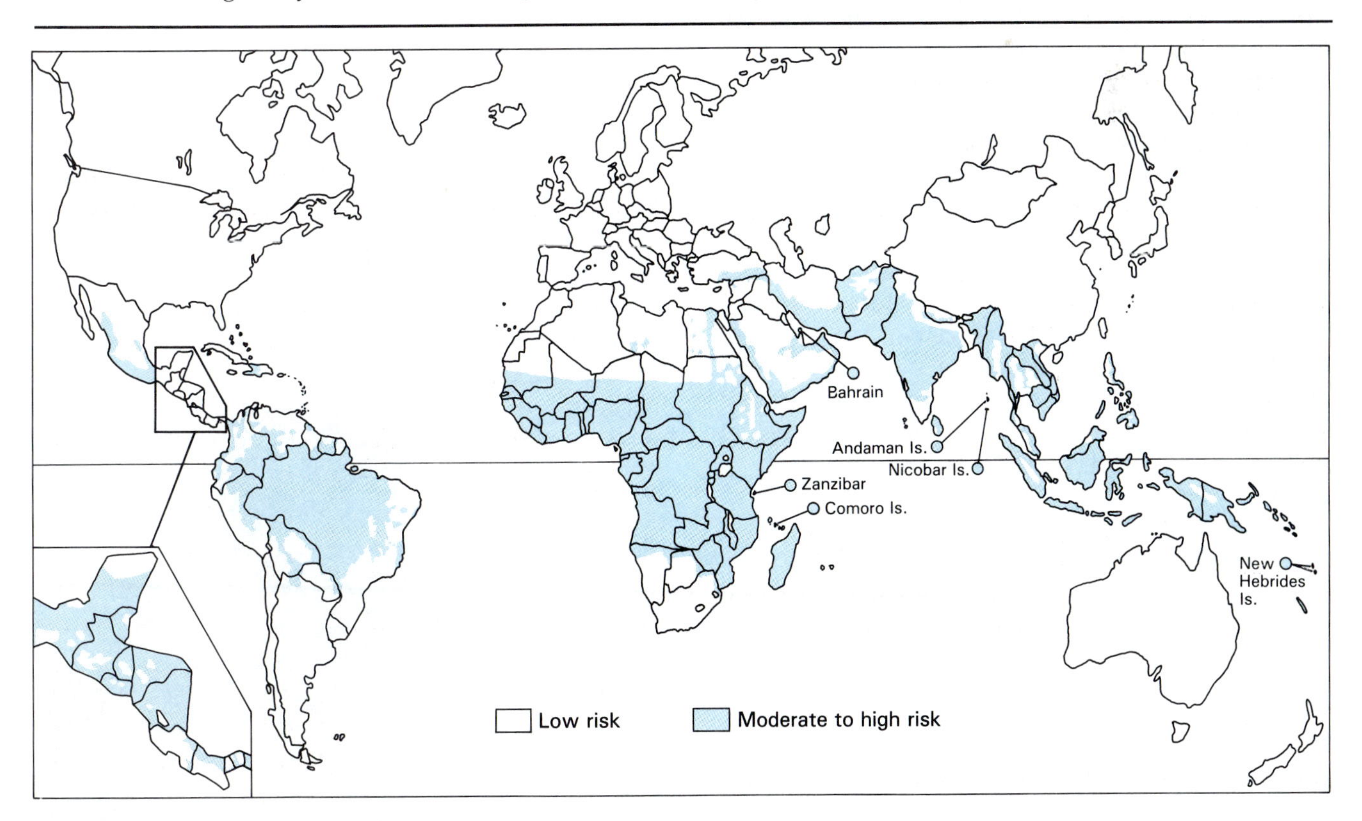

FIGURE 23.19 Regions of endemic malaria.

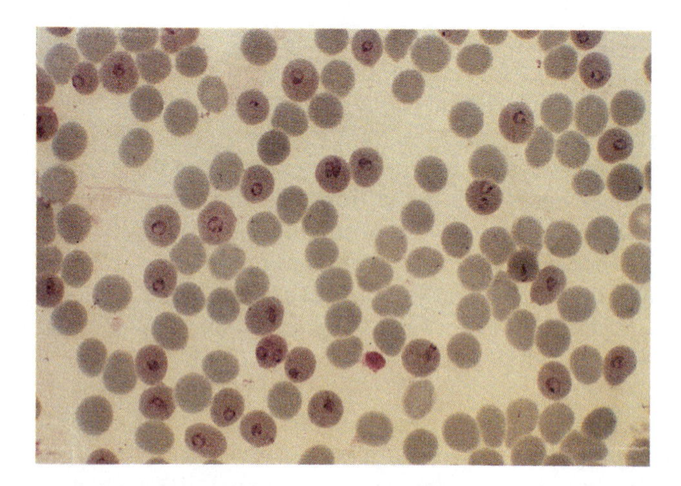

FIGURE 23.20 The "ring" stage of the malarial parasite, *Plasmodium falciparum*, seen as darker circular structures within red blood cells.

blackwater fever. In blackwater fever large numbers of erythrocytes are lysed, probably because of the host's autoimmune reaction to the parasites. Products of hemoglobin breakdown cause jaundice and kidney damage. Pigments from hemoglobin blacken the urine and give the disease its name.

Diagnosis, Treatment, and Prevention The main means of diagnosing malaria is by identifying the parasites in red blood cells. The species of *Plasmodium* responsible for a particular infection can be identified by the distinctive appearance of erythrocytes invaded by the parasite. Chloroquine (Aralen) is the drug of choice for all forms of malaria in the acute stage. A serious problem in the treatment of malaria is that some strains, especially strains of *P. falciparum*, have become resistant to chloroquine. Drugs have recently been found that can be administered with chloroquine to overcome such resistance. This strategy has been tested in monkeys but not yet in humans. Chloroquine can be taken prophylactically for a week before entering a malarial region, during one's stay there, and for 6 weeks after leaving the area. The drug suppresses clinical symptoms of malaria, but it does not necessarily prevent infection. Disease caused by *P. vivax* or *P. ovale* can appear months or years after you leave a malarial area, even when the suppressant drug is taken. Primaquine is the drug of choice for eliminating parasites from the liver and other tissues if they have been infected.

Attempts to destroy mosquitoes that carry malaria have been an important component of malaria control efforts. In the early 1960s the pesticide DDT was used successfully to eradicate malaria-carrying mosquitoes from the United States. It and other insecticides also have been tried in other areas, especially in Africa. Unfortunately, the region in which such mosquitoes thrive in Africa is so vast that no insecticide spraying program has been effective. Some of the mosquitoes, particularly those that carry *P. falciparum* malaria, have now become resistant to DDT and probably to other pesticides. Thus, the use of insecticides has made malaria more deadly by increasing the proportion of mosquitoes that carry the more virulent parasite.

Researchers at the Centers for Disease Control in Atlanta, Georgia, and their colleagues at other institutions have recently developed a strain of *Anopheles gambiae* that is highly resistant to infection with malaria parasites. The researchers believe the resistance is due to a relatively simple genetic change and that it might be possible to induce such resistance in natural vector populations. If large numbers of *A. gambiae*, the most important malaria vector in Africa, could be made resistant, transmission of malaria could be greatly reduced.

Another control effort is directed toward developing a malaria vaccine. One problem has been to identify the stage of the parasite responsible for triggering the immune response in humans. An antigen on the sporozoite has now been identified, and the gene for this antigen can be cloned using recombinant DNA technology. Consequently, the antigen can be manufactured and used to make a vaccine. Thus, great strides have been made toward the development of an effective vaccine, and the vaccine probably will be available in the near future. But even when a vaccine becomes available, administering it to the vast numbers of people living where malaria is endemic will be a monumental task. Mechanisms to distribute the vaccine and ways of gaining cooperation of the population will be needed. The cost of the vaccine in the quantities required will be another major problem.

Toxoplasmosis

Toxoplasma gondii (Figure 23.21) is a widely distributed protozoan that infects many birds and mammals without causing disease. It is an intracellular parasite and can invade many tissues. Humans usually become infected through contact with the feces of domestic cats that forage for natural foods, especially if the cats consume infected rodents (Figure 23.22). Cats kept

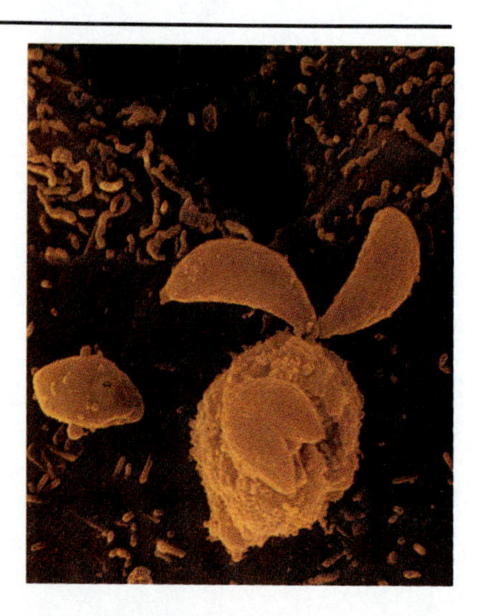

FIGURE 23.21 Crescent-shaped protozoan parasites, *Toxoplasma gondii*, seen leaving an infected cell (beneath them) in which they have multiplied (SEM, 1440X). This organism can be a danger to immunocompromised patients and to pregnant women, causing serious birth defects and miscarriages.

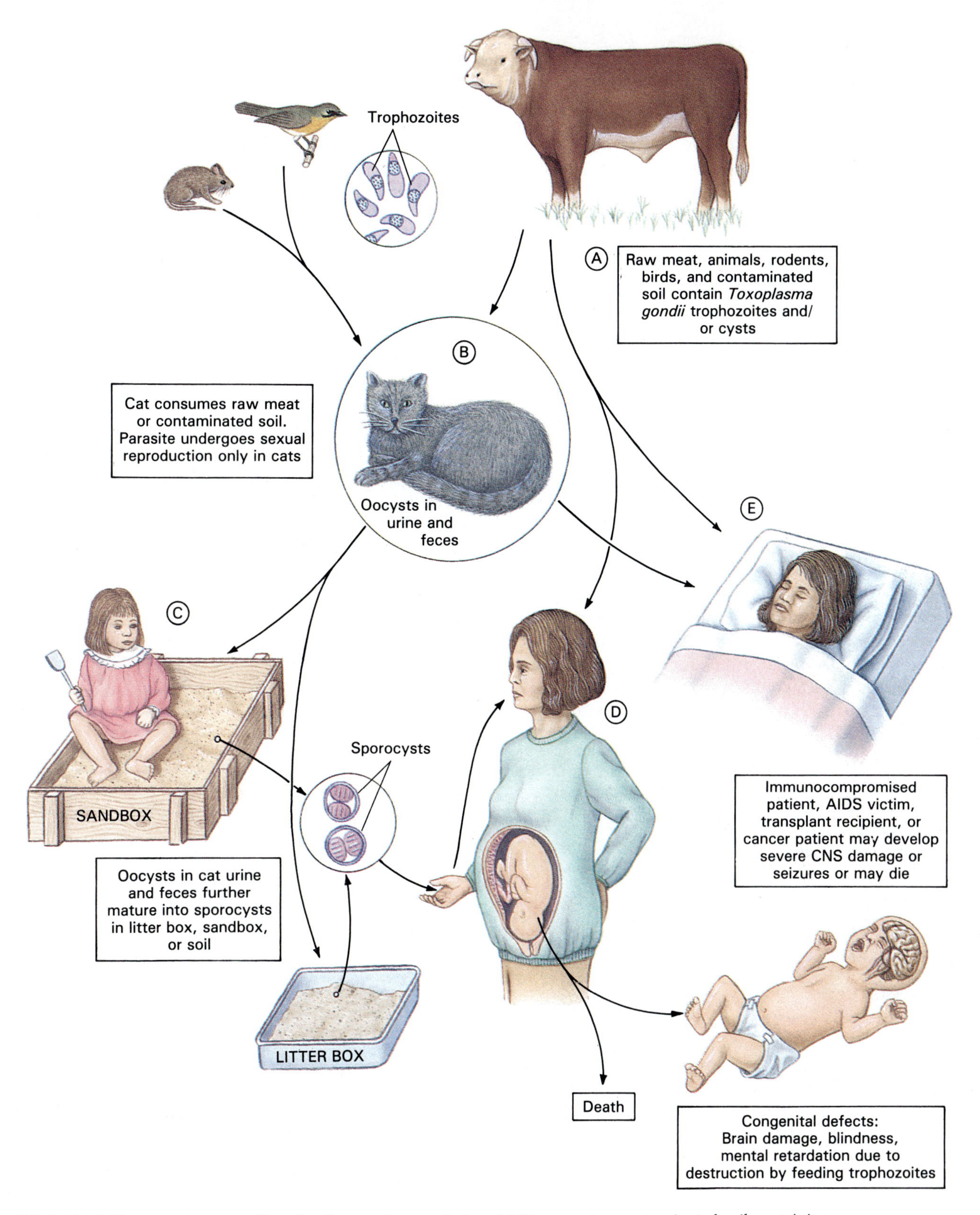

FIGURE 23.22 *Toxoplasma gondii* cycle of natural transmission. (a) Raw meat or contaminated soil containing trophozoites (feeding forms) of *T. gondii* are consumed. (b) In the intestinal tract of cats the trophozoites undergo sexual reproduction, releasing oocysts in urine and feces, which further mature into sporocysts in litter boxes, sandboxes, or soil. (c) If the trophozoites are consumed by older children or adults, a mild flulike syndrome may result. (d) Pregnant women, who themselves suffer only mild or inapparent infection, can transmit this infection to their fetus, which may develop congenital defects or die. (e) Immunocompromised individuals, who may already have the organism in their body, rapidly die of central nervous system destruction.

inside and fed on canned or boiled foods are unlikely to acquire the parasites. Another common means of transmission is by consumption of raw or undercooked meat. The French, who consume large amounts of steak tartare (raw ground beef), have the highest incidence of infection in the world. *T. gondii* causes only mild lymph node inflammation in most humans, but it can cause serious **toxoplasmosis,** especially in developing fetuses, newborn infants, and sometimes in young children. The organism can be transferred across the placenta of an infected mother to the fetus, where it causes serious congenital defects. These defects include accumulation of cerebrospinal fluid, abnormally small head, blindness, mental retardation, and disorders of movement. It also can be responsible for stillbirths and spontaneous abortions. If infection occurs after birth, the symptoms are similar to but less severe than those seen in fetuses. In patients with severe immunosuppression such as AIDS patients, the disease can appear as encephalitis and can also cause dermatologic problems.

Toxoplasmosis can be diagnosed by finding the parasites in the blood, cerebrospinal fluid, or tissues, by animal inoculation with subsequent isolation of the organisms, or by indirect immunofluorescence tests. ∞ (Chapter 19, p. 534). Pyrimethamine and trisulfapyridine are used in combination to treat toxoplasmosis, but no treatment can reverse permanent damage from prenatal infection. To prevent this disease, pregnant women should avoid contact with raw meat and cat feces, and cats should be kept out of sandboxes where children play, especially if a child might carry the organism to a pregnant woman.

Babesiosis

Several species of the sporozoan *Babesia* can cause **babesiosis** (ba-be-se-o'-sis), but *Babesia microti* is most often associated with human infections. The parasites enter the blood via bites of infected ticks and invade and multiply in red blood cells. Although many cases are asymptomatic, when symptoms appear they usually begin with sudden high fever, headache, and muscle pain. Anemia and jaundice can occur as red blood cells are destroyed. The symptoms last for several weeks and are followed by a prolonged carrier state. If babesiosis occurs in a person who has undergone spleen removal, it is usually fatal in 5 to 8 days. This is because the lack of a spleen impairs the body's ability to break down defective red blood cells.

Diagnosis is made from blood smears, but the parasite can be confused with *Plasmodium falciparum*. Chloroquine is the drug of choice for treatment, and avoiding tick bites is the best means of protection.

The properties of nonbacterial systemic diseases are summarized in Table 23.4.

TABLE 23.4 Viral and protozoan systemic diseases

Disease	Agent(s)	Characteristics
Viral systemic diseases		
Dengue fever	Dengue fever virus	Severe bone and joint pain, high fever, headache, loss of appetite, weakness, sometimes rash
Yellow fever	Yellow fever virus	Fever, anorexia, nausea, vomiting, liver damage, jaundice
Infectious mononucleosis	Epstein-Barr virus	Headache, fatigue, malaise, usually sore throat, secondary streptococcal infections common
Other viral fevers	Filoviruses, bunyaviruses, phleboviruses, arenaviruses, orbivirus, and Coxsackie viruses	Some cause hemorrhagic fevers, some cause encephalitis, joint pain, slow heart rate, infection of erythrocytes, diarrhea, rashes, sore throat, liver disease, meningitis, inflammation of the heart and the sac around it
Protozoan systemic diseases		
Kala azar	*Leishmania donovani*	Visceral leishmaniasis with irregular fever, weakness, wasting, enlarged liver and spleen
Localized leishmaniasis	*L. tropica, L. braziliensis*	Oriental sore and skin and mucous membrane lesions
Malaria	*Plasmodium* species	Periods of high fever associated with release of parasites from red blood cells; relapses can occur; one species can cause malignant malaria and blackwater fever
Toxoplasmosis	*Toxoplasma gondii*	Mild lymph node inflammation in adults; can cross placenta and cause serious damage to nervous system of fetus; also causes damage in small children and immunosuppressed patients
Babesiosis	*Babesia microti*	High fever, headache, muscle pain, anemia, and jaundice; fatal in patients whose spleens have been removed

CLOSE-UP

Babesiosis in Cattle

Cattle are affected with babesiosis caused by a tick-borne protozoan, *Babesia bigemina*, so named because two parasites usually are found in each red blood cell. Infected cattle have sudden high fever (106° to 108°F) for a week or more, during which time as many as three-fourths of their erythrocytes may be destroyed. A large quantity of hemoglobin is excreted in the urine, so the disease also is called redwater fever. Among untreated cattle 50 to 90 percent die, but treatment with the same drugs used to treat trypanosome infections reduces the mortality considerably. The disease can be prevented by killing infected ticks and their eggs. The eggs must be killed because they become infected before they are released by the ticks.

ESSAY

Yersinia pestis

Sometimes, details of diseases are so interesting that we forget that the causative microbes themselves can be very interesting. Bubonic plague—*The Black Death*—is certainly a disease unparalled in recorded human history. If there were a "Richter" scale to rate pandemics the way we do earthquakes, the waves of plague that swept around the world would surely rank as the greatest of infectious calamities. The death of perhaps one-third of the world's population during the so-called Black Death of the fourteenth century led to enormous economic, social, religious, and cultural changes. Some experts have estimated that during the sixth century half the population of the Roman Empire died—totaling over 100 million lives snuffed out. Let us take a closer look at the mighty microbe that was able to wreak such havoc (Figure 23.23).

In 1894, the French bacteriologist Alexandre Yersin, working in Hong Kong, isolated the causative organism of plague. For many years it was called *Pasteurella pestis*, the genus named for Pasteur, and the specific epithet from the Latin *pestis* meaning "plague" or "pestilence." In 1971 it was moved to a new genus, *Yersinia* (named for Yersin), in the family Enterobacteriaceae, whose characteristics it more clearly matched (Figure 23.24). Older editions of *Bergey's Manual* divided this family into Tribes, but these arrangements changed greatly as genetic knowledge of the group increased. The current *Manual* no longer uses these designations. DNA studies have shown that *Y. pestis* and *Y. pseudotuberculosis* are 90 percent or more interrelated, and in 1980 scientists proposed that they are really a single species, divided into two subspecies—*Y. pseudotuberculosis* subspecies *pseudotuberculosis* and *Y. pseudotuberculosis* subspecies *pestis*—for taxonomic purposes, but with the old, separate names still used for medical purposes.

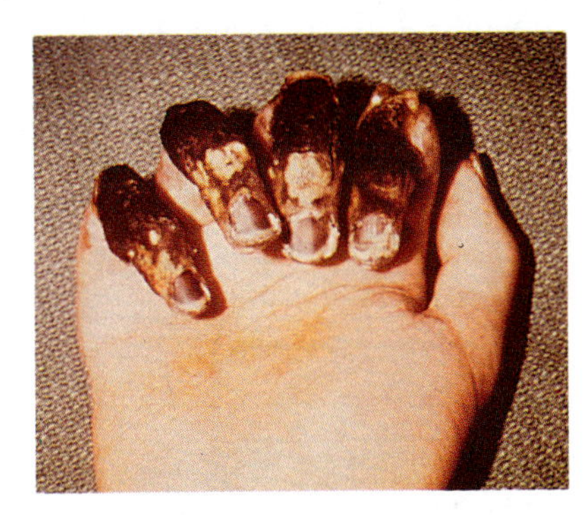

23.23 Discoloration of hemorrhagic fingers of a patient with plague reveals the origin of the name "Black Death." This photo was taken just prior to amputation of the fingers.

Yersinia pestis is a small, gram-negative rod, 0.5 to 0.8 μm in diameter and 1 to 3 μm in length, which often appears as a coccobacillus. The cells exhibit bipolar staining, darker at both ends and lighter in the middle, when Weyson's stain (methylene blue and basic fuchsin in methyl alcohol) is used. No spores are formed. It is nonmotile, a slow grower, and forms tiny colonies on nutrient agar. Fully formed capsules are not present, but when the organism is grown at 37°C or in live cells, an envelope of capsular material forms. Organisms are facultatively anaerobic, having both respiratory and fermentative metabolic pathways. They are oxidase-negative, catalase-positive, and ferment carbohydrates with the formation of acid but no gas. Growth in nutrient broth forms a deposit at the bottom of the tube, leaving a clear supernatant above, with a pellicle at the top. Periodically the pellicle breaks up and falls to the bottom, where it adds to the deposit. Its optimum pH is 7.2 to 7.4, but it can grow in a pH range of 5.0 to 9.6.

Plasmids from *Escherichia coli* have been transferred to *Y. pestis*. The two species have approximately 65 percent of their DNA in common. R factors have been transferred to *Y. pestis* in the laboratory, but they are rare in naturally occurring strains, and few drug-resistant strains have emerged. Conjugation has been observed between *Y. pestis* and *Y. pseudotuberculosis*. *Yersinia pestis* also produces two bacteriocins, Pesticin I and II, which inhibit the growth of *Y. pseudotuberculosis* and *E. coli*.

Several factors contribute to virulence in *Y. pestis*, one of these being Fraction 1 (F1) antigen, which is associated with its envelope or capsular material. When a flea feeds on an infected rat, it ingests *Y. pestis* cells

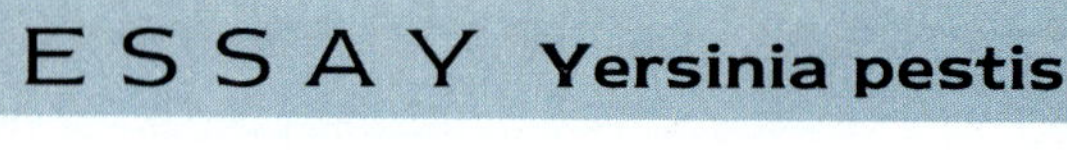

ESSAY Yersinia pestis

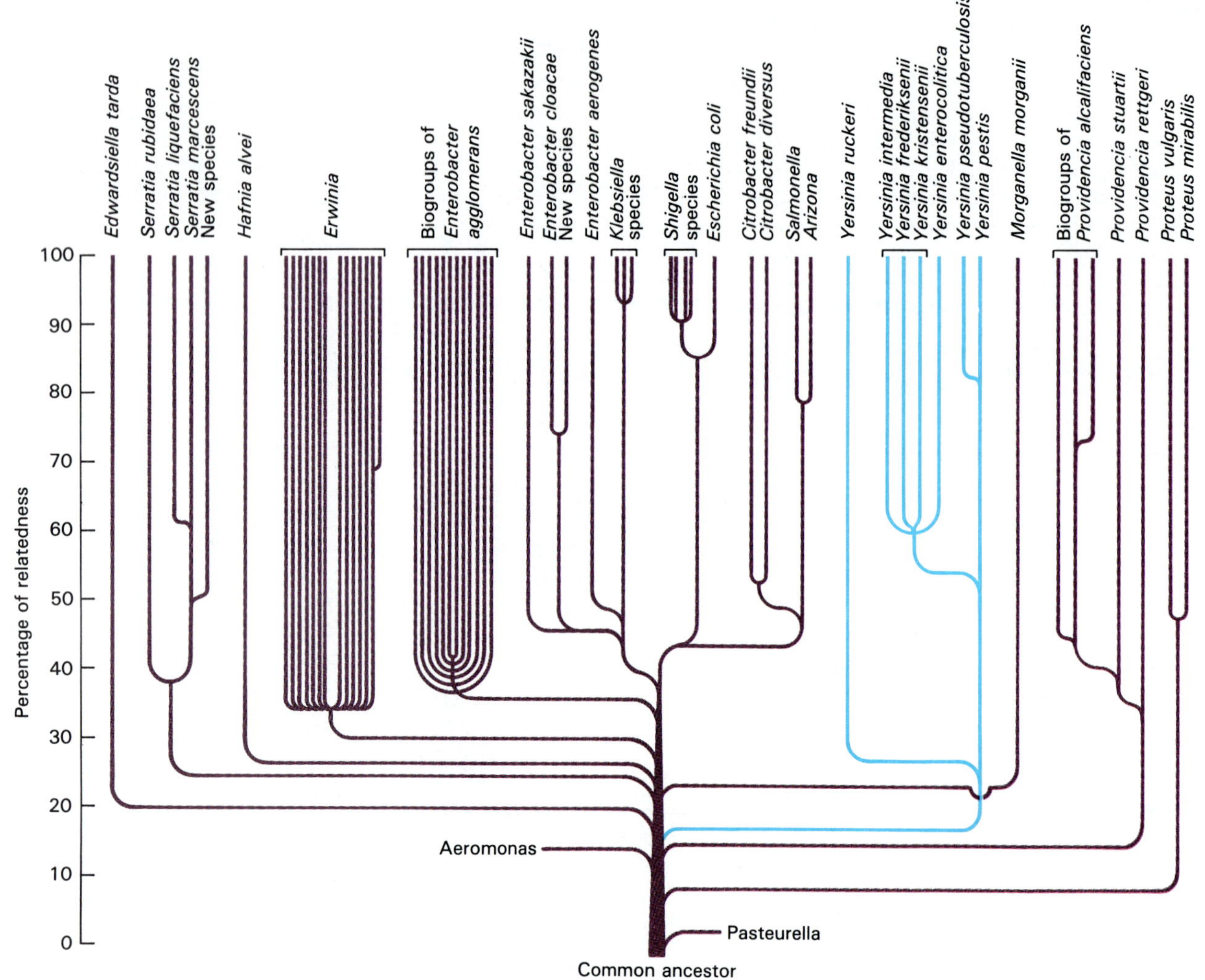

FIGURE 23.24 Position of *Yersinia pestis* within the family Enterobacteriaceae. The percentage number indicates the degree of relatedness of the group to all other groups that have not yet branched off; for example, *Aeromonas* is 15 percent related to all other organisms except *Proteus* and *Pasteurella*. Note that *Yersinia* is most closely related to *Providencia* and *Edwardsiella*. *Yersinia ruckeri*, which causes red mouth disease of trout, is not very closely related to the other *Yersinias*, and may eventually be placed in a new genus of its own.

along with the blood. Inside the flea, the *Y. pestis* loses its envelope layer. Therefore, when the microbes enter the human host during a flea bite, most of the organisms are then easily opsonized and killed by phagocytic polymorphonuclear leukocytes. Some microbes, however, are taken in by tissue macrophages that cannot kill them. Once safely inside these cells, the bacilli regrow their envelope and develop other virulence factors that kill the macrophages. Released from the macrophages, they are now resistant to phagocytosis due to the F1 antigen in their envelopes. Thus, they are able to spread rapidly through the body to lymph nodes, where they produce typical buboes.

Two other antigens, V and W, one a protein and the other a lipoprotein, help the bacilli to survive and multiply inside phagocytes. They are related to the presence of a plasmid that is lost when *Y. pestis* is subcultured at 37°C. To retain virulence, subcultures must be grown at 25° to 28°C. The virulence determinants on the plasmid are controlled by the concentration of calcium ions. Inside the macrophage, the concentration of calcium causes the bacilli to produce outer membrane proteins that protect them from digestion by the macrophage.

Some strains form pigmented colonies due to their ability to utilize iron from hemin. The attenuated live vaccine strain has lost its virulence due to a mutation in its iron metabolism, and it does not produce either V or W antigen despite the presence of the plasmid for them.

Another source of strength for this pathogen is the ability of soil to act as a reservoir for it. Dead infected fleas and rodents contaminate soil deep in burrows. There the organisms can survive for months, awaiting the arrival of noninfected rodents, and the cycle begins again.

CHAPTER SUMMARY

The agents and characteristics of the diseases discussed in this chapter are summarized in Tables 23.1 through 23.4. Information in those tables is not repeated in this summary.

RELATED KEY TERMS

CARDIOVASCULAR AND LYMPHATIC DISEASES

Bacterial Septicemias and Related Diseases

- Septicemia, or **blood poisoning,** involves multiplication of bacteria in the blood.
- Septicemias and related diseases are diagnosed by culturing appropriate samples and are treated with antibiotics.
- **Rheumatic fever** and **bacterial endocarditis** occur most often in patients who have had previous streptococcal infections.

septic shock **lymphangitis**
puerperal fever
childbed fever **vegetation**
myocarditis **pericarditis**

Parasitic Diseases of the Blood and Lymph

- **Schistosomiasis** is acquired by larvae penetrating the skin and is diagnosed by finding eggs in feces. It is treated with praziquantel and could be prevented by eradicating infected snails or avoiding snail-infested water.
- **Filariasis** is transmitted by mosquitoes and is diagnosed by finding microfilariae in blood. It is treated with diethylcarbamizine or metronidazole and could be prevented if infected mosquitoes could be eradicated.

bilharzia
elephantiasis

SYSTEMIC DISEASES

Bacterial Systemic Diseases

- **Anthrax** is acquired through spores from infected domestic animals or their hides. It is diagnosed by culturing blood or from smears from lesions and is treated with penicillin or tetracycline. The disease can be prevented by vaccinating animals and humans exposed in their occupations and carefully burying infected animals.
- **Plague** has occurred in periodic epidemics since the Middle Ages, remains endemic in some regions, and is increasing in incidence in the United States. It is transmitted by fleas from infected rats and is diagnosed by stained smears and antibody tests. The disease is treated with streptomycin or tetracycline and can be prevented by controlling rats and vaccinating people entering endemic areas.
- **Tularemia,** found in more than 100 mammals and arthropod vectors, is transmitted through skin, inhalation, or ingestion. It can be diagnosed by hemagglutination tests and treated with streptomycin. It is prevented by avoiding contact with infected mammals and arthropods; the vaccine is short-lasting and not fully protective.
- **Brucellosis** is transmitted from domestic animals through the skin and by inhalation or ingestion and is diagnosed by serologic tests. It is treated with prolonged antibiotic therapy and can be prevented by avoiding contact with infected animals and contaminated fomites.
- **Relapsing fever** is transmitted by lice and ticks and can sometimes be diagnosed from blood smears. It is treated with tetracycline or chloramphenicol and can be prevented by avoiding or controlling ticks and lice.
- **Lyme disease** is transmitted by ticks from infected deer and other animals and is diagnosed by clinical signs and an antibody test. It is treated with antibiotics and can be prevented by avoiding tick bites.

buboes **bubonic plague**
septicemic plague
pneumonic plague
transovarian transmission
ulceroglandular tularemia
typhoidal tularemia
bronchopneumonia
undulant fever **Malta fever**
epidemic relapsing fever
endemic relapsing fever

Rickettsial Systemic Diseases

- Rickettsias are small gram-negative, highly infectious, intracellular parasites classified among the bacteria. They are widely distributed and transmitted by various arthropods.
- **Typhus** occurs in several forms. **Epidemic typhus,** transmitted by lice, usually occurs in unsanitary, overcrowded conditions. It has a high mortality rate unless treated with antibiotics.
- **Brill-Zinsser disease,** or **recrudescent typhus,** is a recurrence of a latent typhus infection. **Endemic** or **murine typhus** is carried by fleas and **scrub typhus** by mites from infected rats.
- **Rocky Mountain spotted fever,** carried by ticks, damages blood vessels. Strains of the causative organism vary in virulence, and the mortality of the untreated infection can be high.

petechiae
tsutsugamushi disease
shinbone fever

- **Rickettsialpox** is carried by mites that live on house mice. **Trench fever,** transmitted by lice, is prevalent under unsanitary conditions, most often among persons under stress. **Bartonellosis,** transmitted by sandflies, occurs in two forms: **Oroya fever,** an acute fever that causes life-threatening anemia, and **verruga peruana,** a self-limiting skin rash.

RELATED KEY TERMS

Carrion's disease

Viral Systemic Diseases

- **Dengue fever** can be diagnosed by serologic tests; a vaccine is available against one immunologic type of dengue fever virus.
- **Yellow fever** is diagnosed by symptoms and can be prevented with vaccine.
- **Infectious mononucleosis** is diagnosed by symptoms and is treated symptomatically and with antibiotics for secondary infections. In underdeveloped countries babies have mild symptoms and develop antibodies by age 1, but in developed countries patients are teenagers or young adults, who have a much more serious disease.
- **Chronic fatigue syndrome** may be a complication of infection with Epstein-Barr virus.

breakbone fever

flavivirus **Epstein-Barr virus**

Burkitt's lymphoma **filoviruses**

phleboviruses **Rift Valley fever**

Lassa fever **Colorado tick fever** **orbivirus**

aplastic crisis **feline panleukopenia virus** **canine parvovirus**

Parasitic Systemic Diseases

- **Leishmaniasis** occurs in tropical and subtropical countries where appropriate species of sandfly vectors are available. It is diagnosed from blood smears or scrapings from lesions and is treated with antimony. The disease could be prevented by controlling sandfly breeding and eliminating rodent reservoir infections.
- **Malaria** is the world's greatest public health problem; it kills a million people annually, most of them children. Cases in the United States come from people who have been in endemic areas. Malaria is transmitted by female *Anopheles* mosquitoes and is diagnosed by identifying parasites in blood smears. Active disease is treated with chloroquine (except for resistant strains), and latent parasites are eliminated with primaquine. Research is in progress to find a way to control mosquitoes and to develop an effective vaccine.
- **Toxoplasmosis** is usually transmitted from feces of cats that have consumed infected rodents and by human ingestion of raw or undercooked meat. It is diagnosed by finding parasites in body fluids or tissues and is treated with pyrimetamine and trisulfapyridine. The disease can be prevented by avoiding contact with contaminated materials.
- **Babesiosis** is transmitted by ticks and is diagnosed from blood smears. It is treated with chloroquine and can be prevented by avoiding ticks.

kala azar **oriental sore**

ischemia **blackwater fever**

QUESTIONS FOR REVIEW

A.

1. How does septicemia differ from bacteremia?
2. What are the causative agents and symptoms of septicemia?
3. What are the main characteristics of puerperal fever?
4. What are the cause and nature of tissue damage in rheumatic fever?
5. How can rheumatic fever be diagnosed, treated, and prevented?
6. What is bacterial endocarditis, and why is it life-threatening?

B.

7. How is schistosomiasis acquired, and what are its effects?
8. How can schistosomiasis be diagnosed, treated, and prevented?
9. How is filariasis acquired, and what are its effects?
10. How can filariasis be diagnosed, treated, and prevented?

C.

11. What is the cause of anthrax, and how do humans acquire it?
12. What are the forms of anthrax, and how do their effects, treatment, and prevention differ? What are their pathogenic effects?
13. What is the cause of plague, and where is the disease found today?
14. What are the effects of plague, and how is it diagnosed, treated, and prevented?
15. What is the cause of tularemia, and how do humans acquire it?
16. What are the effects of tularemia, and how is it diagnosed, treated, and prevented?
17. What is the cause of brucellosis, and what are its effects?
18. How do humans acquire brucellosis, and how is it diagnosed, treated, and prevented?
19. What are the cause and effects of relapsing fever, and how can it be treated and prevented?
20. What is the cause of Lyme disease, and why is it often misdiagnosed?
21. What are the effects of Lyme disease, and how is it diagnosed, treated, and prevented?

D.

22. What are the general characteristics of pathogenic rickettsias and rickettsial diseases?
23. How are rickettsias transmitted, and how are diseases they cause diagnosed, treated, and prevented?
24. What are the main differences between various kinds of typhus?
25. How does Rocky Mountain spotted fever differ from other rickettsial diseases?
26. What are the main features of rickettsialpox?
27. How does trench fever differ from typhus?
28. In what forms is bartonellosis seen, what causes it, and how can it be treated and prevented?

E.

29. What are the similarities and differences in dengue fever and yellow fever, and how might these diseases be eradicated?
30. What are the special characteristics of the Epstein-Barr virus, what diseases does it cause, and how are they diagnosed and treated?
31. What other viruses cause systemic infections in humans?

F.

32. What are the different kinds of leishmaniases, and how can they be diagnosed, treated, and prevented?
33. Why is malaria the world's greatest public-health problem?
34. How is malaria diagnosed and treated, and how might it be controlled?
35. How do people get toxoplasmosis, and how can it be prevented?
36. How do people get babesiosis, and how can it be prevented?

PROBLEMS FOR INVESTIGATION

1. Using your present knowledge of microbiology and any information you can find in the literature, provide a possible explanation for why Lyme disease and the organism that causes it were not recognized earlier.
2. Discuss the similarities and differences associated with arthropod-borne infections. Are such infections more likely to be systemic than limited to particular tissues? Defend your answer to this question.
3. Why do you suppose there have never been any major outbreaks of bubonic plague in this century? Do you think we could ever have one in the United States? Why or why not?
4. Pretend you are a park employee or a health department official, and devise a way to prevent people from getting arthropod-borne infections. What information would you give to a hiker going into a tick-infested area?
5. What practical advice would you give to a pregnant woman regarding toxoplasmosis?
6. What prophylactic measures should you follow if you needed to travel in a malarial zone? What could you do to lessen the number of mosquito bites you get?
7. A 47-year-old biologist had been studying small wild mammals in a rural area outside of La Paz, Bolivia. She developed chills, sudden high fever, swollen lumps in her right armpit, headache, loss of appetite, and aching back and hip muscles. She immediately returned to Washington, D.C., where her disease was diagnosed, and she was hospitalized in an isolation room. After 6 days of streptomycin therapy she was discharged. Which disease would you diagnose?
(The answer to this question appears in Appendix E.)

SOME INTERESTING READING

Anonymous. 1991. AAP issues policy on treatment of Lyme disease in children. *American Family Physician* 44, no. 1 (July):308–9.

Anonymous. 1991. Malaria in travelers returning to the United States. *American Family Physician* 44, no. 6 (December):2183.

Boyles, S. 1990. Allogenic bone marrow transplantation in adults with Burkitt's lymphoma or acute lymphoblastic leukemia in first complete remission. *NCI Cancer Weekly* (June 4):24–26.

Defoe, D. 1721. (Reprinted 1960). *A journal of the plague year.* New York: Signet Classics.

Grogl, M., M. J. Milhous, W. K. Milhous, E. O. Nuzum, R. K. Martin, J. D. Berman, B. G. Schuster, and C. N. Oster. 1991. Leishmaniasis in Desert Shield–Storm. *Am. J. Trop. Med. Hyg.* 64, no. 3 suppl. (December):179.

Maganrelli, L. A., T. G. Andreadis, K. C. Stafford III, and C. J. Holland. 1991. Rickettsiae and *Borrelia burgdorferi* in ixodid ticks. *Journal of Clinical Microbiology* 29(12):2798–2804.

McEvedy, C. 1988. The bubonic plague. *Scientific American* 258, no. 2 (February):118.

Rybak, L. P. 1990. Deafness associated with Lassa fever. *Journal of the American Medical Association* 264, no. 16 (October):2119.

Suzuki, Y., M. A. Orellana, R. D. Schreiber, and J. S. Remington. 1988. Interferon-gamma: the major mediator of resistance against *Toxoplasma gondii*. *Science* 240, no. 4851 (April 22):516–19.

Taylor, G. C. 1991. Lyme disease: an overview of its public health significance. *Journal of Environmental Health* 54, no. 1 (July–August):24–28.

Tobi, M., and S. E. Strauss. 1988. Chronic mononucleosis—A legitimate diagnosis. *Postgraduate Medicine* 83, no. 1 (January):69.

Vaughan, C. 1988. New name and identity for mysterious Epstein-Barr syndrome. *Science News* 133 (March 12):167.

Young, S. 1991. Dragonflies help to defeat dengue fever. *New Scientist* 130, no. 1766 (April 27):26.

Zinsser, H. 1935. *Rats, lice, and history.* Boston: Little, Brown.

Polio is a preventable disease. We have excellent vaccines against it. Yet each year far more people die of polio than die of AIDS. What are the reasons that vaccine does not reach those who need it? What can be done?

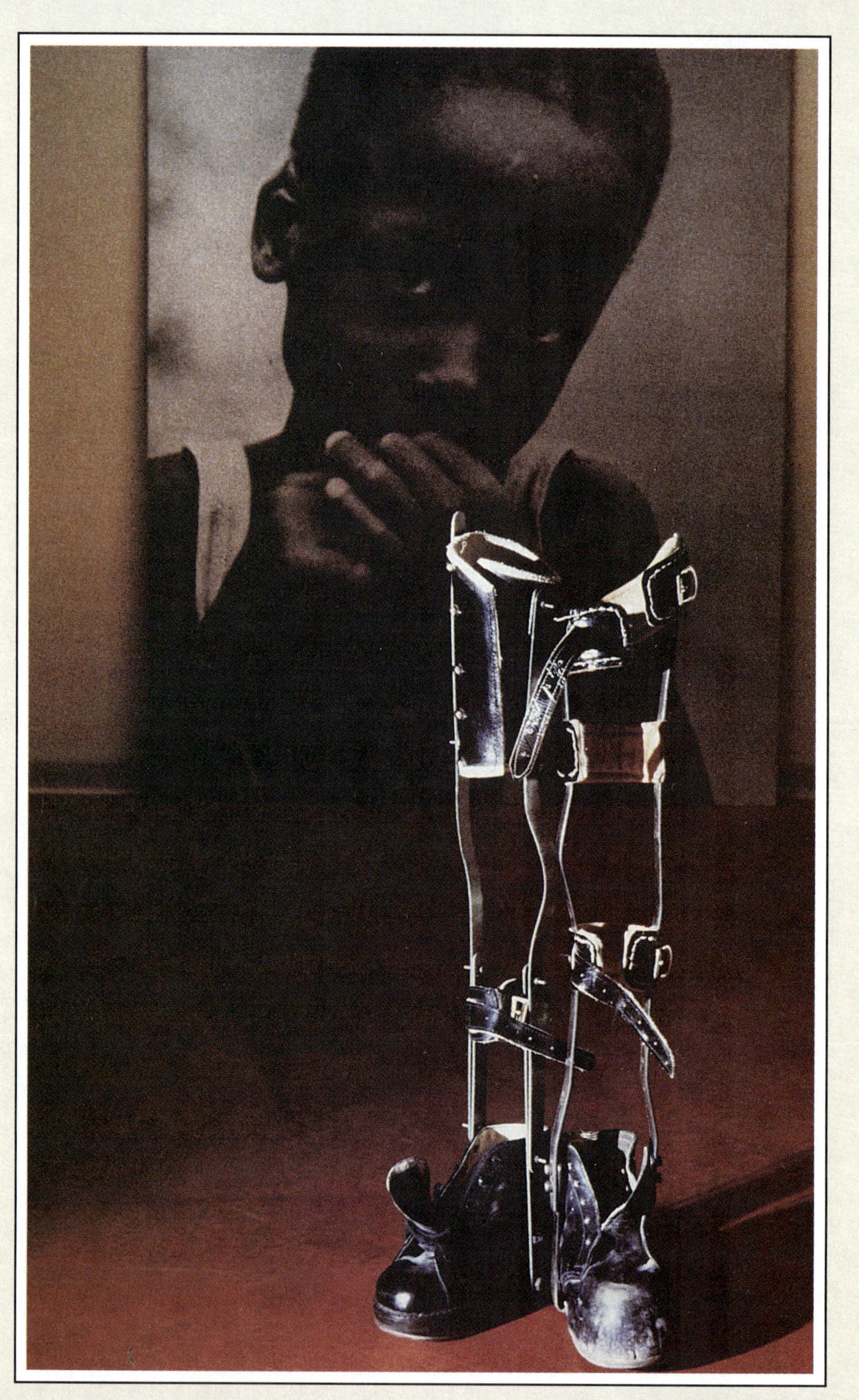

24 Diseases of the Nervous System

This chapter focuses on the following questions:

A. What pathogens cause bacterial diseases of the brain and meninges, and what are the important epidemiologic and clinical aspects of these diseases?

B. What pathogens cause viral diseases of the brain and meninges, and what are the important epidemiologic and clinical aspects of these diseases?

C. What pathogens cause bacterial nerve diseases, and what are the important epidemiologic and clinical aspects of these diseases?

D. What pathogens cause viral nerve diseases, and what are the important epidemiologic and clinical aspects of these diseases?

E. What pathogens cause parasitic diseases of the nervous system, and what are the important epidemiologic and clinical aspects of these diseases?

As with cardiovascular and lymphatic diseases, the diseases of the nervous system also often affect other systems. It is assumed that the reader is familiar with disease processes, host systems, and host defenses.

DISEASES OF THE BRAIN AND MENINGES

Bacterial Diseases of the Brain and Meninges

Bacterial Meningitis

Bacterial meningitis is an inflammation of the *meninges*, the membranes that cover the brain and spinal cord. This life-threatening disease can be caused by several kinds of bacteria, each of which has a prevalence that can be correlated with the age of the host (Table 24.1). Meningitis causes *necrosis* (death of tissues in a given area), clogging of blood vessels, increased pressure within the skull from edema, decreased cerebrospinal fluid flow, and impaired central nervous system function. The early symptoms are headache, fever, and chills. In rare instances seizures develop. Onset can be insidious or fulminating; death can occur from shock and other serious complications within hours of the appearance of symptoms.

Most cases of meningitis are acute, but some are chronic. Acute meningitis is acquired from carriers or endogenous organisms. Organisms gain access to the meninges directly during surgery or trauma, or spread in blood to them from other infections such as pneumonia and otitis media. Host defenses in the arachnoid layer, one of the meninges, ordinarily combat bacteremia, but if organisms overwhelm the defenses, meningitis results. Chronic meningitis occurs with underlying diseases such as syphilis or tuberculosis.

Meningitis is diagnosed by culturing cerebrospinal fluid. The fluid usually is turbid—sometimes so thick with pus that it is difficult to remove with a syringe. Antibiotic treatment varies with the causative organism. If tuberculosis meningitis is suspected, isoniazid therapy is called for immediately.

Meningococcal Meningitis The bacterium *Neisseria meningitidis* has caused about 3000 cases of meningitis per year for the past decade in the United States. Mortality is about 85 percent when untreated but only 1 percent with best treatment. The 15 percent mortality rate in the United States probably reflects delay in seeking treatment. This disease was the leading cause of death from infectious disease among U.S. armed forces during World War II.

In meningococcal meningitis, the organisms colonize the nasopharynx, spread to the blood, and make their way to the meninges, where they grow rapidly (Figure 24.1). In a complication called Waterhouse-Friderichsen syndrome, the meningococci invade all parts of the body, and death occurs within hours from endotoxin shock. The immediate cause of death is usually clotting of blood followed by massive hemorrhage in the adrenal glands (located above the kidneys), leading to fatal deficiency of essential adrenal

TABLE 24.1 Types of bacterial meningitis

Age	Most Frequent Causative Agents	Comments
Newborn (0–2 months)	*Escherichia* coli, other Enterobacteriaceae, *Streptococcus* species	Average mortality about 50%; incidence 40–50/100,000 live births; maternal transmission
Preschool (2 months–5 years)	*Haemophilus influenzae, Neisseria meningitidis*	Maximum incidence 6–8 months; overall incidence 180/100,000 children
Youth and young adult (5–40 years)	*Neisseria meningitidis, Streptococcus pneumoniae*	Sporadic or epidemic
Mature adult (over 40 years)	*Streptococcus pneumoniae, Staphylococcus* species	Sporadic

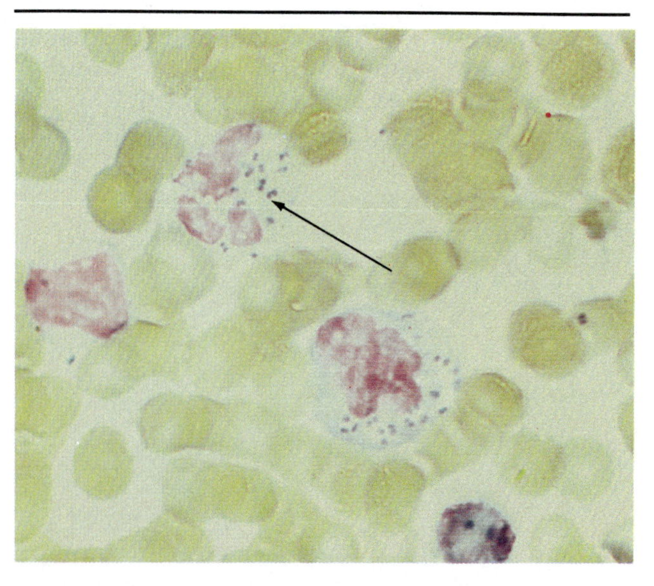

FIGURE 24.1 Meningococci in meninges are a cause of meningitis. These meningococci have been phagocytized by white blood cells in a cerebrospinal fluid specimen.

hormones. A lesser degree of hemorrhaging is sometimes seen in meningitis patients who develop a petechial skin rash.

Penicillin is the drug of choice for treatment—the prevalence of resistant strains makes sulfonamides no longer useful. A vaccine is available, but it is not effective against the most common type B meningococci. The risk of this disease can be decreased by preventing overtiring and overcrowding. The number of inches required between bunks in military barracks is based on experience with meningococcal outbreaks.

Many people become carriers of meningococci without developing the disease. In some closed environments, such as military bases, dormitories, and day-care centers, 90 percent of the population can be carriers yet only one per thousand carriers develops the disease. Among members of patient households, 80 to 90 percent are carriers, compared with only 5 to 30 percent in the general public. Antibiotics can eliminate the carrier state.

APPLICATIONS

Reaching the Brain

The blood-brain barrier (Chapter 17) places some limits on the use of antibiotics to treat diseases of the central nervous system. (p. 453) Beta-lactam antibiotics such as penicillin, for example, do not penetrate the blood-brain barrier easily. When administered orally or intravenously, their average concentration in the cerebrospinal fluid generally reaches only about 15 percent of their concentration in plasma. Some other antibiotics, however, such as chloramphenicol and tetracycline, are lipid-soluble, and so diffuse easily across the blood-brain barrier.

Haemophilus Meningitis During the first year of life roughly two-thirds of bacterial meningitis cases are caused by *Haemophilus influenzae*. Among children, 30 to 50 percent are carriers of this organism; among adults the figure is only 3 percent. Humans are exposed to *H. influenzae* early in life and rapidly acquire immunity, so the disease is rare in adults. Only 10 percent of children between ages 3 and 6 lack antibodies, and those over age 6 all have antibodies. Without treatment this disease is nearly always fatal; even with treatment one-third die. Of those who recover, 30 to 50 percent have serious mental retardation, and 5 percent are permanently institutionalized because of damage to the central nervous system. *Haemophilus* meningitis is the leading cause of mental retardation in the United States and worldwide. Vaccines are available, which all children age 5 and under should receive.

Streptococcus Meningitis Among adults *Streptococcus pneumoniae* is the most common cause of meningitis. Organisms generally spread via the blood from lung, sinus, mastoid, or ear infections. Mortality is 40 percent.

Listeriosis

Another kind of meningitis, **listeriosis,** is caused by *Listeria monocytogenes*, a small gram-positive organism widely distributed in nature. It is sometimes acquired as a zoonosis and is particularly threatening to those with impaired immune systems. Although not an especially significant human disease for several decades, listeriosis is now a leading cause of infection in kidney transplant patients. Transmission by improperly processed milk and cheese has also been documented in recent years. In pregnant women the bacillus can cross the placenta, infect the fetus, and cause abortion, stillbirth, or neonatal death. Listeriosis is responsible for many cases of fetal damage.

Brain Abscesses

Microorganisms that cause **brain abscesses** reach the brain from head wounds or via blood from another site. As would be expected with wounds, multispecies infections are common, and anaerobes are as likely as aerobes to be responsible. Most such abscesses occur in patients under age 40, but two age periods—birth to 20 years and 50 to 70 years—show peak incidences. The infection gradually grows in mass and compresses the brain. The masses can be detected by CAT scans, or X-rays, and causative agents can be identified by serologic tests and by culturing cerebrospinal fluid. In very early stages antibiotic treatment can be sufficient, but later surgical drainage or removal of abscesses is usually necessary. Abscesses in areas of the brain that control the heart or other vital organs cannot be treated surgically. Without treatment half the patients die, but with the current best treatment only 5 to 10 percent die.

Viral Diseases of the Brain and Meninges

Rabies

Rabies was described by Democritus in the fifth century B.C. and by Aristotle in the fourth century B.C. Pasteur made real progress in understanding the disease when he found evidence of the infectious agent in saliva, the central nervous system, and peripheral

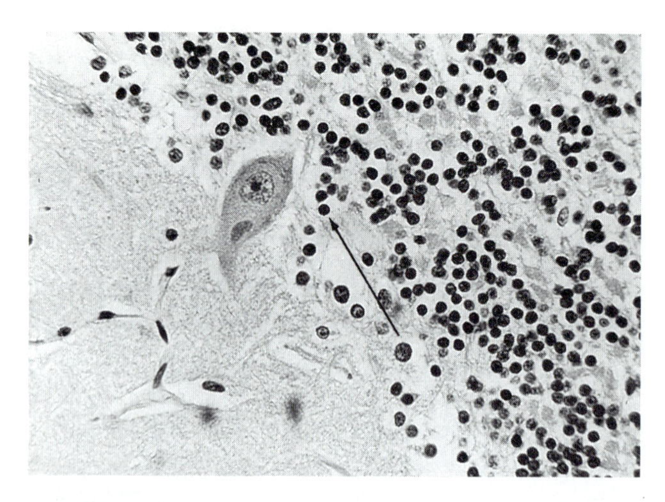

FIGURE 24.2 Negri bodies of rabies in the cerebellum of a human brain.

nerves. He attenuated the agent and proved that a suspension of it could be used to prevent rabies. In 1903 the Italian physician Adelchi Negri found Negri bodies, or clusters of viruses, in neurons (Figure 24.2). Negri bodies were used to diagnose rabies for more than 50 years until an immunofluorescent antibody test (IFAT) was developed in 1958. Still used today, the IFAT is so sensitive that after an animal suspected of having rabies bites a human, it can be killed immediately and its brain examined for rabies antigens. Prior to the availability of IFAT such animals had to be held for 30 days or until symptoms of rabies developed before a search for Negri bodies could be made. Most biting animals are not rabid, so the IFAT has saved many people both the anguish of wondering if they have been exposed to rabies and the discomfort and risk of treatment.

Rabiesvirus has a worldwide distribution. It infects all mammals exposed to it, so the possibilities for reservoir infections are almost limitless. The type of rabies found in different regions of the world varies. In almost all of Asia, Africa, Mexico, and Central and South America, rabies is endemic in dogs. In Canada, the United States, and Western Europe, wildlife rabies predominates, and dog rabies is controlled. The WHO lists 60 countries including England, Australia, Japan, Sweden, and Spain as rabies-free. This success is due to animal vaccination and quarantine programs.

Identifying rabid animals can be a problem. About half of all rabid dogs release viruses into saliva 3 to 6 days before they show symptoms of rabies. In contrast, 90 percent of rabid cats have viruses in their saliva about 1 day before they become symptomatic. Any change in the behavior of an animal can be a warning that it might be rabid. A "friendly" wild animal approaching people or a gentle family animal snapping without provocation sometimes indicates impending rabies symptoms. Small wild animals and some domestic ones, too, will bite if suddenly grabbed and held. Biting is their only defense in such a stressful situation. Much unnecessary anguish could be avoided if small children were taught to appreciate unfamiliar animals without touching them.

PUBLIC HEALTH

Controlling Rabies

Rabies control is fraught with difficulties, as the following incident shows. In 1986 a woman reported to her veterinarian that a fox had had contact with her dog and its four puppies on the front porch of her house. The veterinarian advised destroying or quarantining the dogs but did not report the incident to the health department veterinarian, as is legally required. A few days later the woman delivered the puppies to an animal shelter but neglected to mention the fox contact. The same day two of the puppies were taken to a nursing home and introduced to most of the residents and several visitors.

Within 24 hours the puppies had been adopted by four different families. Within a few days two of the puppies had died of rabies, but not before one of them had ridden in a carriage with two infants and the other had visited the nursing home. A total of 134 people, many of them children, were found to have been exposed. Postexposure rabies treatment consisting of immune globulin and vaccine for these people cost $65,000; the trauma this incident caused is unmeasurable.

Animals vary in their susceptibility to rabies, and susceptibility is directly correlated with the role of the animals in maintaining reservoir infections. Foxes, coyotes, skunks, raccoons, and bats are highly susceptible. No infected fox has ever been known to survive, whereas 10 to 20 percent of dogs and 30 to 40 percent of mongooses do survive. Bats are particularly dangerous because they are asymptomatic and shed viruses into their feces. Two explorers of bat-infested caves in Texas died of rabies. Dogs, cats, cattle, horses, and sheep are less susceptible. Whether a person becomes infected depends mainly on whether an animal is shedding viruses in its saliva at the time it bites the person. Even animals later proven to have rabies may not have been shedding viruses at the time they inflicted bites.

The Disease Rabies is caused by **rabiesvirus,** an RNA-containing rhabdovirus. After entering the body through an animal bite or other break in the skin, the rabiesvirus first replicates in injured tissue for 1 to 4 days and then migrates to nerves, where it replicates slowly until it reaches the spinal cord. It progresses rapidly up the spinal cord to the brain. The time from infection to the appearance of symptoms varies from

13 days to 2 years but is usually between 20 and 60 days. The length of time required for symptoms to appear is proportional to the distance between the wound and the brain and is affected by the accessibility of nerve fibers. Thus a bite on the face, which is well supplied with nerves and close to the brain, produces symptoms much more quickly than one on the leg. The rabiesvirus has a predilection for nervous tissue but also infects salivary glands and the respiratory tract lining.

In humans the first symptoms are headache, fever, nausea, and partial paralysis near the bite site. These symptoms persist for 2 to 10 days and then worsen until the acute neurological phase of the disease ensues. The patient's gait becomes uncoordinated as paralysis becomes more general. Hydrophobia (fear of water) occurs as throat muscles undergo painful spasms, especially during swallowing. Aerophobia (fear of moving air) occurs because the skin is hypersensitive to any sensations. Confusion, hyperactivity, and hallucinations also occur. Within 10 to 14 days of the onset of symptoms, the patient typically goes into a coma and dies. Once symptoms appear, there is no cure for rabies. Of all human patients suffering from clinical rabies, only two have been known to survive and make a complete recovery. Both had some degree of protection from an earlier immunization and it was known that they had been bitten by a rabid animal, so postexposure immunologic treatment was started immediately.

Diagnosis, Treatment, and Prevention A sample from a brain or skin biopsy can be stained by the IFAT to identify rabies antigen before the patient dies. Finding the antigen confirms the diagnosis, but failure to find it does not rule out rabies. Sometimes a diagnosis can be made before death by testing cerebrospinal fluid or serum for neutralizing antibodies, which increase 10 to 12 days after the symptoms appear.

The bite of a rabid animal is treated by first thoroughly cleaning it with soap and flushing it with large amounts of water. Hyperimmune rabies serum is introduced into and around the wound to hopefully neutralize viruses before they reach the nervous system, where they are beyond reach of the antibodies. Interferon also can be applied to the wound. A series of injections of vaccine is given to induce neutralizing antibody formation. The biting animal should be located and confined for examination by IFAT.

The best means of preventing rabies is to immunize pets, and such immunization is required in many countries. Attempts to reduce rabies in raccoons have been made using vaccine in small sponges covered with food bait. When the raccoons eat the bait they also ingest enough vaccine to prevent them from acquiring rabies. Preliminary studies conducted in the United States show promising results. When raccoons from a bait-treated wildlife area were trapped, 15 out of 16 survived a challenge dose of rabiesvirus. All 16 caught in the nonbaited control area died of the same challenge doses. It is not certain whether the animals that died in the experimental group ever ate any bait. It is estimated that a distribution of one bait pellet per acre, at a cost of $1 per pellet, could prove sufficient to reduce rabies dramatically in wild raccoons.

Rabies immunization is recommended for veterinarians and their staffs, hunters who may have contact with wild animals, and technicians who work with the virus. The first vaccine, and the only one for many years, was developed by Pasteur. This vaccine contained viruses modified by 50 transfers, involving drying of infected spinal cords, from one rabbit to another. It was used in all cases of suspected rabies because the disease might develop while waiting for test results on the biting animal. The vaccine was given in 14 daily subcutaneous abdominal injections with unpleasant to serious side effects, including severe abdominal pain, fatigue, fever, and sometimes rabies infection. A relatively new vaccine produced from viruses grown in human diploid fibroblast cultures elicits high levels of neutralizing antibodies with only a few injections and minimal side effects. The old injections were given in the abdomen, where the thick subcutaneous tissue (tissue beneath the skin) slows the rate of absorption. Today's vaccine is given intramuscularly on days 0, 3, 7, 14, and 28. In addition, hyperimmune globulin is placed deep in the wound and infiltrated around the wound.

CLOSE-UP

Rabies in Dogs

Rabies symptoms in a dog begin with the dog acting as if it has a sore throat or has something caught in its throat. As the disease progresses the dog may become groggy or paralyzed (dumb rabies) or agitated and aggressive, biting anything that disturbs it (furious rabies). Throat muscle spasms, difficulty in swallowing, and drooling also are indicative of rabies. Eventually the dog becomes apathetic and stuporous and slides into a final coma.

Encephalitis

Encephalitis is an inflammation of the brain caused by a variety of togaviruses. We will consider the following four diseases, each of which is caused by a different virus: (1) **eastern equine encephalitis** (EEE), seen most often in the eastern United States; (2) **western equine encephalitis** (WEE), seen in the western United States; (3) **Venezuelan equine encephalitis**

(VEE), seen in Florida, Texas, Mexico, and much of South America, and (4) **St. Louis encephalitis** (SLE), seen from east to west in the central United States. The equine varieties are so named because they infect horses more often than humans. The life cycles of these viruses usually involve transmission from a mosquito to a bird, back to a mosquito, and then to a horse, human, or other mammal, and finally back to a mosquito. The St. Louis variety is so named because the first epidemic was identified there in 1933; it appears to be transmitted mostly between English sparrows, mosquitoes, and humans.

The viruses, which are introduced into the body through bites of infected mosquitoes, first multiply in the skin and spread to lymph nodes. Viremia involving especially large numbers of viruses follows, and in a few infections the viruses invade the central nervous system. They cause shrinkage and lysis of neurons. WEE appears every summer, and about a third of the cases occur in children under 1 year of age. Fever and headache are common symptoms, and convulsions sometimes occur. EEE is a much more serious disease; it causes a severe necrotizing infection of the brain. The disease is fatal in 50 to 80 percent of the cases, and survivors often suffer permanent brain damage. Fortunately, because swamp birds are the major reservoir for the virus and swamp mosquitoes the major vectors, very few humans become infected. VEE is mainly a disease of horses; when it occurs in humans, it resembles influenza.

SLE occurs in late summer epidemics about every 10 years and causes the most severe symptoms in elderly patients. The illness starts with malaise, fever, and chills as consequences of viremia. Other common symptoms include anorexia, myalgia (muscle pain), sore throat, and drowsiness. In addition, certain patients have symptoms of a urinary tract infection, neurological disorders, altered states of consciousness, and convulsions. Complications can include secondary bacterial infections, blood clots in the lungs, and gastrointestinal hemorrhage. Most patients escape the complications and recover fully.

Diagnosis, Treatment, and Prevention Encephalitis sometimes can be diagnosed by isolating the causative agent from cell cultures or mice inoculated with blood or spinal fluid. Cultures can be negative when the disease is present because the viremic phase of the disease usually is over before the patient seeks medical attention. Serological methods can be used to identify antibodies at any time during and following the illness. Treatment only alleviates symptoms. Vaccines are available for immunizing horses, but they are rarely used in humans because of the danger of inducing a virulent form of the disease. Prevention by eradicating mosquito vectors is a more appropriate means of decreasing the already low incidence of encephalitis among humans.

Other Viral Diseases

Herpes Meningoencephalitis Herpes simplex virus, which usually is responsible for cold sores, also can cause **herpes meningoencephalitis.** This disease often follows a generalized herpes infection in a newborn infant, child, or adult. The virus reaches the brain by ascending from the trigeminal ganglion. The disease has a rapid onset with fever and chills, headache, convulsions, and altered reflexes. In the middle-aged or elderly, meningoencephalitis causes confusion, loss of speech, hallucinations, and sometimes seizures. Most patients die in 8 to 10 days; survivors usually display neurological damage.

Polyomavirus Infections Polyomaviruses enter the body through the respiratory or gastrointestinal tract. Initial replication takes place in the cells the virus first enters. The viremia that follows allows the viruses to

CLOSE-UP

Foreign Viruses

In Africa, Asia, and South America togaviruses are responsible for various diseases. The chickungunya virus (*chickungunya*, that which bends) causes viremia with high fever, myalgia, and a maculopapular (raised and spotty) rash. Joint pain can persist for months after the infection. The Japanese B encephalitis virus is responsible for a severe, highly fatal form of encephalitis throughout much of Asia. It causes edema and small hemorrhages in blood vessels of the brain. Children suffer abdominal pain, diarrhea, fever, stiff neck, and altered consciousness. Swine are reservoirs of infection, and the Japanese are attempting to reduce incidence of the disease by immunizing domestic pigs. Murray Valley encephalitis, caused by an unusually virulent Australian virus, leads to paralysis and central nervous system damage and usually is fatal. In nonfatal cases motor and intellectual impairment persist. In addition to the preceding mosquito-borne infections, another 14 tick-borne infections cause several varieties of encephalitis and hemorrhagic fevers.

reach target organs, particularly the kidneys, lungs, and brain. Polyomaviruses were first recognized in the 1960s as viral particles in the enlarged nuclei of oligodendrocytes. These are the cells that produce myelin, the lipoprotein that coats nerve fibers in the central nervous system. Infected oligodendrocytes are observed to surround areas that lack myelin in the brains of patients dying from **progressive multifocal leukoencephalopathy.**

One cause of this disease is now known to be the JC virus, named with the initials of a victim from whom it was isolated. Onset is insidious, with vision and speech impairment being the first signs. Typical symptoms of viral infections, such as fever and headache, are absent. Mental deterioration, limb paralysis, and blindness follow. Diagnosis is difficult because cerebrospinal fluid remains normal, and only nonspecific changes are seen in electroencephalograms. The JC virus infects and kills oligodendrocytes but does not affect neurons. An occasional young patient develops this disease as a complication of a primary infection, but most cases result from reactivated latent viruses from childhood infections.

Other polyomaviruses have been isolated from various patients. BK virus was isolated from the urine of a kidney-transplant patient. In one instance a 16-year-old boy with an immunodeficiency had BK viremia and developed kidney inflammation with viruses present in kidney cells. Irreversible kidney failure resulted. The BK virus also has been associated with, but not isolated from, respiratory illness and cystitis, an inflammation of the urinary bladder.

Half the children in the United States have antibodies to JC viruses by age 14; antibodies to BK virus are found by age 4. Although JC and BK viruses apparently persist in most humans for years without causing disease, they sometimes reappear as complications of chronic diseases, immunodeficiencies, and disorders in which lymphocytes proliferate. Pregnancy, diabetes, organ transplantation, antitumor therapy, and immunodeficiency diseases, including AIDS, are among the conditions that can reactivate polyomaviruses. Many kidney-transplant patients, for example, excrete either BK or JC viruses, but the shedding of these viruses rarely has serious consequences. However, unchecked viral multiplication, which is particularly likely with T-cell deficiencies, can sometimes occur and cause clinically apparent disease. Both JC and BK viruses are oncogenic in laboratory animals and possibly also in humans. No diagnostic tests for polyomaviruses are available for routine use, nor is any treatment available for the infections, even if they can be recognized.

Soil amoebae of the genera *Naegleria* and *Acanthamoeba* are opportunistic human pathogens. All are associated with meningoencephalitis. *Naegleria fowleri* usually is seen in swimmers. The amoebae are thought to enter nasal passages and make their way along nerves to the meninges. *Acanthamoeba polyphaga* causes ulceration of eyes or skin. If it invades the central nervous system, death occurs a few weeks after the onset of neurological symptoms. The major source of human infection is contaminated water in hot tubs. While the tubs are covered, the amoebae accumulate on the water surface. Removing the lid disperses cysts among the people as they enter the tub.

OTHER DISEASES OF THE NERVOUS SYSTEM

Bacterial Nerve Diseases

Hansen's Disease

Hansen's disease, the currently preferred name for **leprosy,** has been known since biblical times, when many things, even houses, were said to have "leprosy." Many "leprosy" cases were other skin diseases such as fungal and viral infections, and the houses probably had fungi growing on their walls.

The World Health Organization (WHO) estimates that more than 15 million cases of leprosy exist worldwide today, mainly in Asia, Africa, and South America (Figure 24.3). Hansen's disease also occurs in the United States, where a peak number of 361 cases was reported in 1985, mostly among immigrants from countries where Hansen's disease is endemic. Infected people sometimes show no symptoms when they enter our country. No reliable test is available to disclose all of these subclinical cases, although the **lepromin skin test,** similar to the tuberculin skin test for tuberculosis, detects some of them. Health-care workers should watch for Hansen's disease among immigrants.

The Disease The acid-fast bacillus *Mycobacterium leprae* is found in all cases of Hansen's disease. Although *M. leprae* was the first bacterium to be recognized as a human pathogen, demonstrating that it fulfills Koch's postulates has been slow because the organism is difficult to grow in the laboratory. For one thing, it reproduces very slowly, having a 12-day division cycle. Recently developed methods for growing it in such diverse organisms as nine-banded ar-

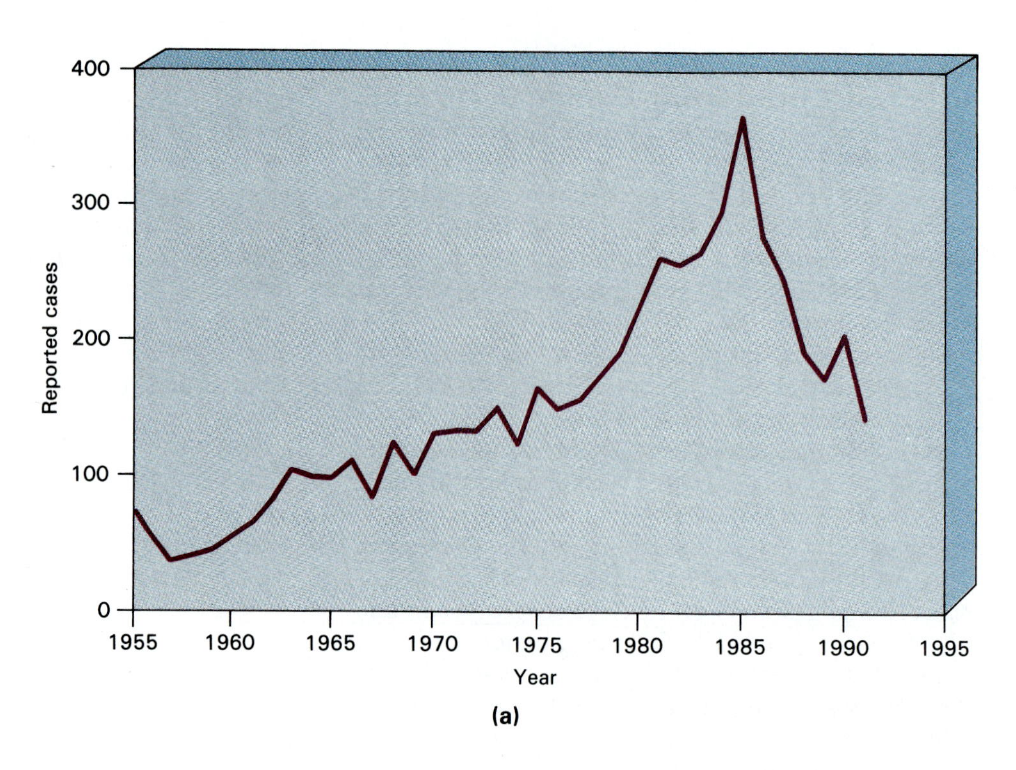

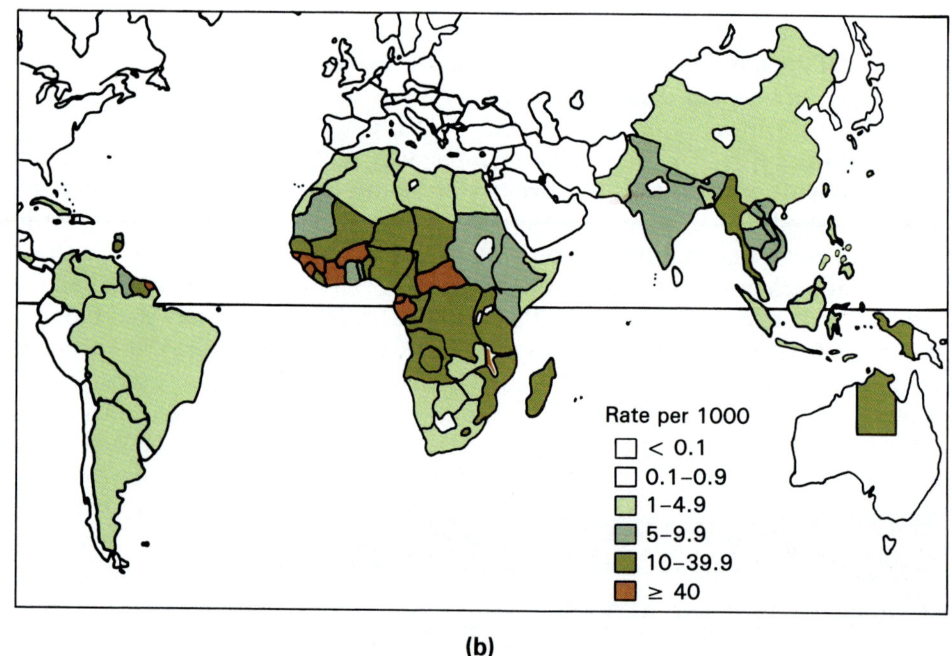

FIGURE 24.3 Incidence of Hansen's disease (leprosy) (a) in the United States, and (b) worldwide.

madillos, chimpanzees, mangabey monkeys, and mice have made the organism available for research. It may now be possible to devise better diagnostic tests and to develop a vaccine.

Clinical forms of Hansen's disease vary along a spectrum from tuberculoid to lepromatous. In the **tuberculoid,** or anesthetic, form (Figure 24.4a), areas of skin lose pigment and sensation; in the **lepromatous** (lep-ro'mat-us), or nodular, form (Figure 24.4b), a granulomatous response causes enlarged, disfiguring skin lesions called **lepromas.** Incubation time averages 2 to 5 years for the tuberculoid form and 9 to 12 years for the lepromatous form.

Mycobacterium leprae is the only bacterium known to destroy peripheral nerve tissue; it also destroys skin and mucous membranes. The organism has a predilection for cooler parts of the human body, such as the nose, ears, and fingers, but large numbers of organisms are seen throughout the body, except for the central nervous system. Continuous bacteremia of 1000 organisms per milliliter of blood has been demonstrated in lepromatous cases. Large numbers of ba-

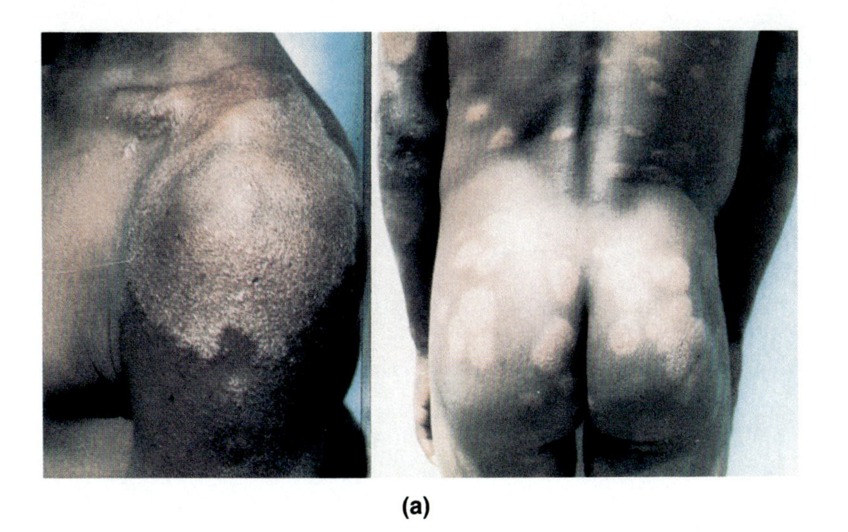

(a)

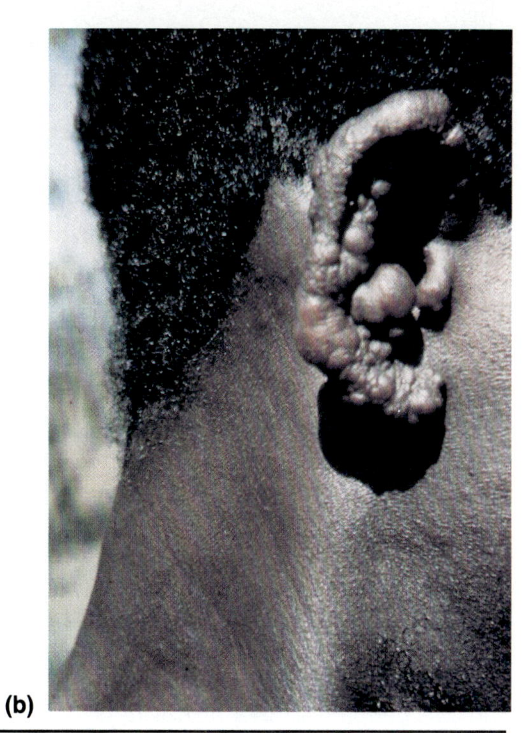

(b)

FIGURE 24.4 Hansen's disease appears in two extremes with gradations and combinations between them. One is the tuberculoid, or anesthetic, form (a), in which areas of skin lose pigment and sensitivity. A pin may be stuck into these "anesthetized" areas and not be felt because of destruction of nerves and nerve endings. The other is the nodular form (b), characterized by disfiguring granulomas called lepromas.

cilli are shed in respiratory secretions and in pus discharged from lesions. Although Hansen's disease is not highly contagious, shedding of organisms probably transmits the disease to those with extensive close contact with patients, such as children of infected parents.

As Hansen's disease progresses it deforms hands and feet (Figure 24.5). Severe lepromatous disease erodes bone: Fingers and toes become needlelike, pits develop in the skull, nasal bones are destroyed, and teeth fall from the jawbone. Surgery sometimes can restore the use of extremely crippled hands and feet. The National Hansen's Disease Center, a U.S. Public Health Service facility in Carville, Louisiana, has pioneered development of these special surgical techniques. In the past decade it was recognized that the changes in the feet of diabetic patients can be helped by these same surgical techniques. Thus, Carville now has an active teaching program for surgeons who will use knowledge gained from one of the most shunned

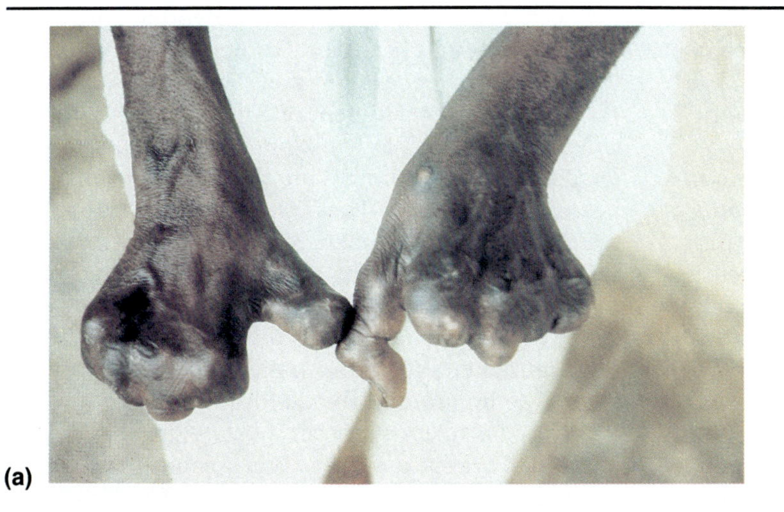

(a)

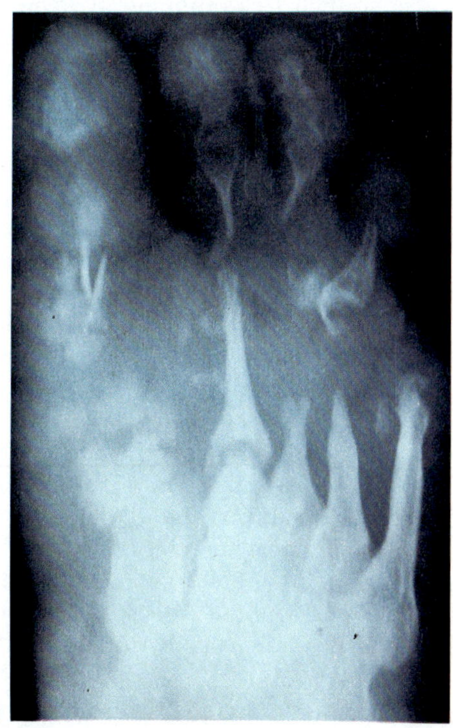

(b)

FIGURE 24.5 (a) Deformed "claw" hand of Hansen's disease. This can be treated surgically in earlier stages, thereby preventing crippling. (b) X-rays of a hand reveal the loss of bone tissue that accompanies Hansen's disease.

diseases to help thousands of victims of diabetes.

Examining ancient skeletons has provided insights into the epidemiology of Hansen's disease in past centuries. It is clear that Hansen's disease traveled from the Old World to the New World and that in the past, even allowing for misdiagnosis, its incidence in Europe was much higher than today. Genetic factors may predispose toward resistance to Hansen's disease; as susceptible individuals have died, resistant ones have made up a greater percentage of the population.

Diagnosis, Treatment, and Prevention Hansen's disease is diagnosed by finding the organism in acid-fast stained smears and scrapings from lesions or biopsies. It is treated with dapsone and rifampin, but dapsone-resistant strains are beginning to appear. Treatment greatly reduces nodules of lepromatous disease, but it cannot restore lost tissue. Until recently victims of Hansen's disease were isolated in special hospitals called leprosariums. Now the disease can be arrested, and the people can live nearly normal lives without infecting others in their community. (They must still sleep in separate bedrooms and use only their own linens and utensils, however, and they cannot live in a household where children are present.)

Immune responses in Hansen's disease are cell-mediated and vary from strong to weak. Strong responses and distinctly positive skin tests are seen in patients with the less serious tuberculoid disease. Weak responses and negative skin tests are seen in patients with rapidly progressing lepromatous disease. However, test results may change over time from positive to negative and vice versa as immune response rises and falls. Lepromatous patients usually have adequate cell-mediated response to other antigens, so their lack of immunity is not due to generalized T-cell absence or dysfunction. The absence of T cells in "nude" mice (Figure 19.20), which lack a thymus, makes them suitable organisms for growing large quantities of organisms.

Vaccine is not available for Hansen's disease. Even if a vaccine became available now, determining its effectiveness would take years because of the disease's long incubation period. Avoiding exposure and receiving prophylactic chemotherapy after exposure are the only means of prevention.

Tetanus

Tetanus is caused by an obligately anaerobic, gram-positive, spore-forming rod, *Clostridium tetani* (Figure 24.6). The organism can be cultured in the laboratory only under strict anaerobic conditions. Spores are exceedingly resistant to drying, disinfectants, and heat. Boiling for 20 minutes does not kill them, and they can survive for years if not exposed to sunlight.

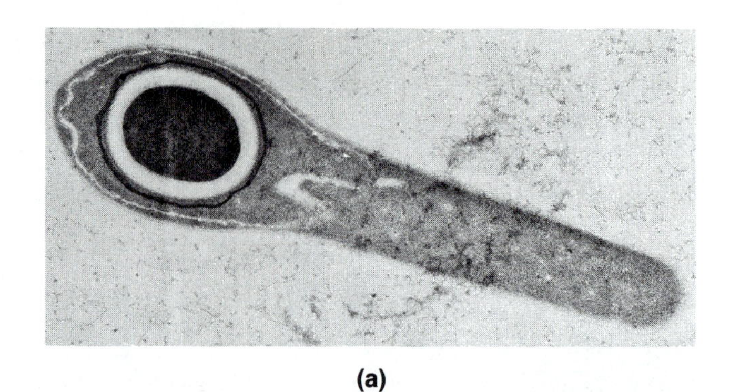

(a)

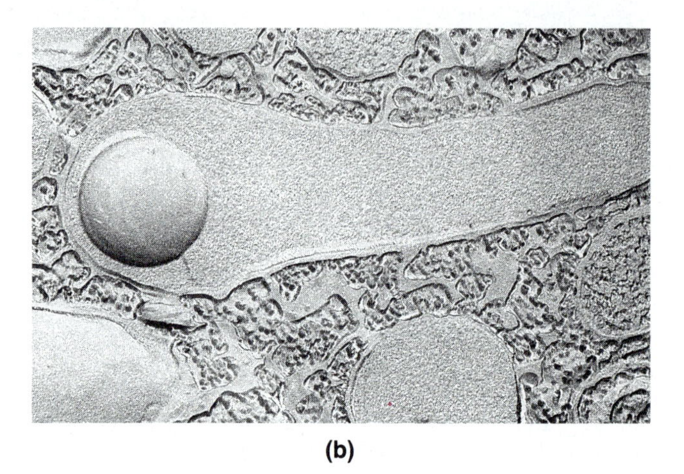

(b)

FIGURE 24.6 (a) *Clostridium tetani* bacillus with large dark terminal endospore. (b) A freeze-etch preparation showing the rounded spore inside a bacillus.

(Chapter 4, p. 86) Spores are found in all soils but especially in those enriched with manure. The organisms are part of the normal bowel flora of horses and cattle and about 25 percent of humans. Therefore, they can be transmitted to persons handling bedpans, dirty diapers, or other objects contaminated with feces if they have any breaks in their skin.

Since the development of tetanus vaccine, the incidence of tetanus in the United States has steadily dropped, with the annual number of cases remaining below 100 since 1975. The highest incidence is in older people, especially women. Vaccine was not available during their childhoods, and they did not receive it during military service as men did. They remain susceptible to tetanus spores as they enjoy gardening in their retirement years. Such elderly family members should be immunized for their protection.

To cause tetanus, spores must be deposited deep in tissues, where oxygen is unavailable. This occurs in deep cuts and puncture wounds. Stepping on a rusty nail has a reputation for leading to tetanus, but it is tetanus endospores, not rust, that cause the disease—a shiny new nail can be just as dangerous if the spores are present. Making puncture wounds bleed helps to flush tetanus spores and other organisms from them. Once inside the host, the noninvasive tetanus organisms stay at the wound site and release a powerful exotoxin, so tetanus is a toxin-mediated disease. After 4 to 10 days' incubation, symptoms begin, with generalized muscle stiffness followed by spasms that affect every muscle. An arched back and clenched fists and jaws (hence the term *lockjaw*), are classic symptoms (Figure 24.7). Spasms can be violent enough to break bones. Eventually, respiratory muscles become paralyzed, heart function is disturbed, and, with rare exceptions, the patient dies. Survivors experience a period of sore muscles but suffer no further sequelae. Before vaccine was available, many soldiers died from tetanus. On battlefields strewn with horses and manure, contamination of wounds with tetanus spores was inevitable. War-related cases were virtually eliminated by vaccinating soldiers; only 12 cases occurred during World War II.

Tetanus toxoid vaccine given prior to injuries protects against the toxin. Antitoxin and antibiotics are given to nonimmunized patients when injuries are treated. Because antitoxin must be administered to inactivate the toxin before the immune system has time to become sensitized to it, infection treated in this way confers no immunity. Patients should receive toxoid immunization after they recover.

Tetanus neonatorum is acquired through the raw stump of the umbilical cord. In some cultures contaminated knives are used to cut umbilical cords after a baby is delivered, and mud is daubed on the cut end. In parts of some developing nations, 10 percent of deaths within a month of birth are due to neonatal tetanus.

Botulism

Botulism derives its name from the Latin word *botulus*, which means sausage. It was coined at a time when the disease was often acquired from eating sausages. Botulism is caused by *Clostridium botulinum*, a spore-forming obligate anaerobe that releases a potent neurotoxin. The disease occurs in three forms: foodborne, infant, and wound. Foodborne botulism accounts for 90 percent of cases and is caused by ingestion of toxin, usually from improperly home-canned nonacid foods, especially green beans and green peppers. Thus, foodborne botulism is an intoxication; the organisms do not infect tissues. Infant botulism and wound botulism involve both infection and intoxication because the organisms grow in tissues and produce toxin.

Endospores of *C. botulinum* are more heat-resistant than those of any other anaerobe; they withstand several hours at 100°C and 10 minutes at 120°C. They also are very resistant to freezing and irradiation. Found in most soils in the Northern Hemisphere, these endospores remain viable for long periods of time and enable the organism to withstand aerobic conditions. Endospores will germinate only in anaerobic conditions.

The ability of *C. botulinum* to form toxin depends on infection with a bacteriophage. If infected with an appropriate bacteriophage, *C. botulinum* produces one of eight different toxins, of which only four cause human disease. The other toxins cause disease in various animals. If a strain of the bacillus is "cured" of its phage infection, it no longer produces toxin. If later

FIGURE 24.7 A soldier dying of tetanus, an all too common cause of death in the days of cavalry troops. Note the extreme contraction of all muscles, from those of the face to those of the toes.

infected with a different phage, it will produce another toxin.

Botulism toxin is the most potent toxin known—even more toxic than *Shigella* and tetanus toxins. As little as 0.000005 μg can kill a mouse! One ounce could kill the entire U.S. population. Originally thought to be an exotoxin, it is now known to be produced inside the cytoplasm and released only upon death and autolysis of the cell. It is activated by proteolytic enzymes, possibly including trypsin in the host's intestine. The toxin is colorless, odorless, and tasteless; people have died from a single taste of affected food. If endospores are not destroyed, they germinate in food during storage under anaerobic conditions and can release large quantities of toxin. Whereas endospores are highly heat-resistant, the toxin can be inactivated by only a few minutes' boiling. Boiling home-canned foods vigorously before serving would eliminate most foodborne botulism.

Botulism is a neuroparalytic disease with sudden onset and rapidly progressing paralysis; it ends in death from respiratory arrest if not treated promptly. The toxin acts at junctions between neurons and muscle cells and prevents the release of acetylcholine, the chemical neurons release to cause muscle cells to contract. The toxin thus paralyzes muscles in a relaxed state, starting with small eye muscles and progressing to the larynx and pharynx and on to the respiratory muscles. This causes double vision, difficulty in speaking and swallowing, and difficulty in breathing. It causes no fever but can cause gastrointestinal disturbances. Although the toxin is an antigen, people who have recovered do not have antibodies, so the amount of antigen required to elicit antibodies must be greater than the lethal dose.

Diagnosis is based on clinical symptoms and history, with confirmation later by demonstrating toxin in serum, feces, or food remains. Although the confirmation test takes 24 to 96 hours, treatment with a polyvalent antitoxin is started immediately. A polyvalent antitoxin is used to assure effectiveness against all toxins that affect humans. Help in maintaining respiration is important and may be continued for up to 2 months. Antibiotics are of no use because foodborne botulism is due to preformed toxin and not to growth of organisms. With proper treatment the mortality rate is less than 10 percent.

Infant botulism was first recognized in 1976, and incidence has ranged between 30 and 100 cases per year, mostly in California. The disease is associated with feeding honey to infants. Studies in California have shown that 10 percent of the jars of honey sold there contain botulism endospores. Endospores germinate and grow in the immature digestive tracts of infants, probably because they lack appropriate competing flora. As toxin is absorbed, the infant becomes lethargic and loses the ability to suck and swallow, so the disease is often called "floppy baby" syndrome. Infant botulism usually occurs in infants under 6 months of age and rarely occurs after 12 months. If parents followed the recommendation that children under 1 year not be given honey, most cases could be prevented. The prognosis is excellent, and death is rare, but the child usually must remain hospitalized for several months.

Wound botulism is the least common form of botulism; fewer than one case per year has been seen in the United States since 1942. It occurs in deep, crushing wounds. Tissue damage impairs circulation and creates anaerobic conditions, so endospores germinate, multiply, and produce toxin. Toxin enters the blood and is distributed throughout the body. It reaches junctions between neurons and muscle cells about a week after injury and causes progressive paralysis. The mortality rate is about 25 percent.

Viral Nerve Disease

Poliomyelitis

Poliomyelitis is a very ancient disease; its effects are clearly depicted in Egyptian wall paintings thousands of years old. As recently as the early 1950s, it was a dreaded disease in the United States, with nearly

CLOSE-UP
Botulism in Waterfowl

Botulism is an important animal disease, especially during ecologic disturbances, and wetland birds are particularly at risk. It is a leading cause of death among ducks in the western United States. After storms uproot marsh vegetation, decomposition uses up available oxygen, so small aquatic invertebrates die. Botulism endospores in their digestive tracts or in mud germinate and produce toxin. Ducks eat the dead invertebrates, die, and fill with toxin. Flies lay eggs on dead ducks; the eggs hatch into toxin-filled maggots. Healthy ducks eat the maggots, die, and provide a site for new fly eggs. The cycle is repeated until thousands of ducks have died.

58,000 cases reported in the peak year of 1952. The coming of summer struck terror in the minds of parents, and the diagnosis of a case of paralytic polio in the community was cause for outright panic. In the United States today, those most likely to become infected are members of religious groups who are opposed to immunization and groups of illegal aliens who are unprotected by a vaccine.

The Disease Poliomyelitis is caused by three strains of polioviruses that have a special affinity for motor neurons of the spinal cord and brain. Although the majority of poliovirus infections are inapparent or mild and nonparalytic, the virus reaches the central nervous system in 1 to 2 percent of cases. High fever, back pain, and muscle spasms result. In less than 1 percent of cases, these symptoms are accompanied by partial or complete paralysis of muscles in a relaxed state. The nature and degree of paralysis depend on which neurons in the spinal cord and brain are infected and how severely they are damaged. Any paralysis remaining after several months is permanent. The very old and the very young are likely to suffer paralysis as a result of poliovirus infection. Malnutrition, physical exhaustion, corticosteroids, radiation, and pregnancy can increase severity of the disease.

Poliovirus infections in small children in impoverished areas may go undetected, whereas teenagers and young adults in affluent areas sometimes acquire severe, paralytic poliovirus infections. Good sanitation in the affluent areas reduces exposure and, therefore, natural immunity to the viruses.

Diagnosis, Treatment, and Prevention Diagnosis of poliomyelitis is by isolating the virus from pharyngeal swabs or feces, culturing it, and noting its cytopathic effects. Methods also are available for identifying antibodies to the virus in serum. Treatment alleviates symptoms, but patients with paralyzed breathing muscles must forever live in an "iron lung" (Figure 24.8).

Before vaccine became available, only nonspecific public health measures were available to prevent the spread of poliomyelitis. Places such as schools and swimming pools where crowds, especially crowds of children, gathered were closed. Large quantities of insecticides were sprayed in the mistaken belief that insect bites somehow played a role in transmission of the disease. Transmission is now known to occur by both the fecal-oral route and from pharyngeal secretions, thus explaining the dangers of fecally contaminated swimming pools in summer. During the first few years that vaccine was available, it could not be made in sufficient quantities to immunize the whole population. Clinics were set up to immunize pregnant women and young children.

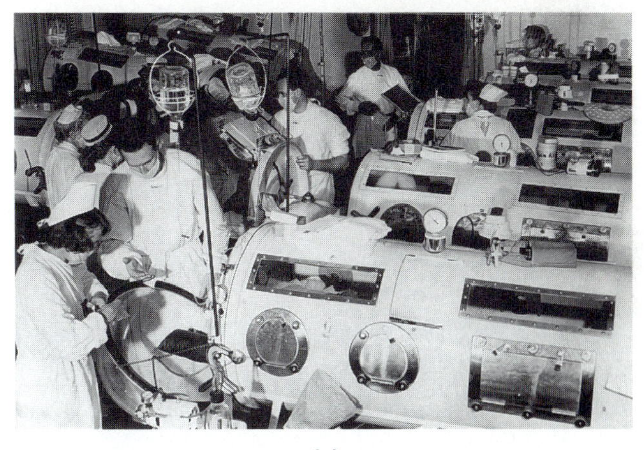

(a)

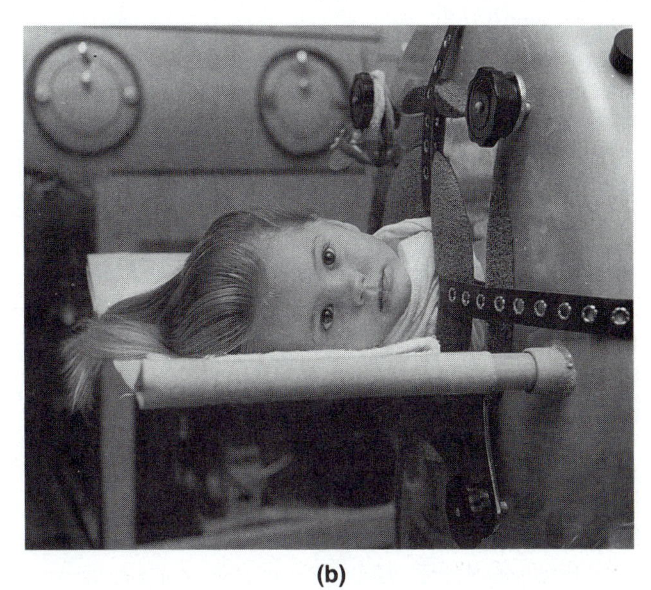

(b)

FIGURE 24.8 During polio epidemics (before 1955) in the United States, row after row of iron lungs (a) were filled with patients like this 2-year-old girl (b). Patients sometimes remained in them for years, or until their death.

Vaccines In 1955 the injectable Salk polio vaccine became available. It contained viruses inactivated by formalin at neutral pH that still retained their antigenic properties. Unfortunately, before the technique was perfected, some batches of vaccine still contained infectious viruses. More than 200 cases of polio and 10 deaths resulted. In 1963 the oral Sabin vaccine, which contained attenuated live viruses, was introduced. In addition to ease of administration, this vaccine has the added advantages of longer-lasting immunity and prevention of fecal-oral transmission by eliminating viruses in the gastrointestinal tract, where they multiply. The use of vaccines has reduced the incidence

Inactivated vaccine

Oral vaccine

Cases per 100,000 population

20.0
10.0
5.0
1.0
0.5
0.1
0.01
0.005

South America

Central America

North America

1955 1958 1961 1964 1967 1970 1973

Year

FIGURE 24.9 Incidence of polio in the Americas. Note that while the incidence has dropped drastically in North America since vaccines became available, the same has not happened in South and Central America. Thus, many people still suffer from what is a preventable disease. It is not generally realized that more people die each year from diseases that could be prevented by vaccines we already have than die from AIDS.

PUBLIC HEALTH

The Polio-vaccine Controversy

In the United States, the attenuated-virus vaccine is almost universally used. Some other countries, however, have returned to the inactivated-virus vaccine. Which vaccine is preferable?

Both vaccines have advantages and disadvantages. The inactivated-virus vaccine can be incorporated with other pediatric vaccines, and at the appropriate dose it can be given to immunodeficient patients. However, it does not produce immunity in all recipients, and recipients may require booster shots. Moreover, virulent viruses are used to make the vaccine; a tragedy could occur if they are not completely inactivated.

The attenuated-virus vaccine induces immunity similar to that produced by a natural infection in that antibodies are produced in the gut (secretory IgA) as well as the bloodstream. This reduces the possibility that people who are themselves immune can serve as reservoirs of infection by carrying the virus in their intestines and passing it on to others. Immunity develops quickly and may be lifelong. Oral administration requires less skill and is more acceptable than injection to many people. Finally, the oral vaccine remains potent without refrigeration for a longer time than the injectable vaccine.

Unfortunately, the live virus used in the oral vaccine does occasionally mutate, and some mutants are virulent. The viruses are shed after immunization and pose a hazard, especially to family members and immunodeficient patients. About 10 cases of polio are caused by the vaccine each year in the United States. Liability costs are now included in the price of the oral vaccine to cover awards that may have to be made to victims of vaccine-induced polio. These costs have made the oral vaccine more expensive than the injected vaccine here, although it is actually cheaper to make and administer. In warm countries with endemic poliomyelitis and other viruses, repeated administration has often failed to induce immunity, probably because immune responses being mounted against other viruses prevent an adequate immune response to the vaccine organisms. Finally, the attenuated-virus vaccine, like all vaccines using live organisms, cannot be administered to immunosuppressed or immunocompromised persons.

Polio has been virtually eliminated in Israel through the use of a combination of inactivated-virus and attenuated-virus vaccines. This procedure shows promise for controlling polio in Third-World countries with warm climates, fecally contaminated water, and endemic polio, but the cost of obtaining and administering both vaccines will delay impiementation.

of polio in the United States from about 29,000 cases in 1955 to 20 cases in 1969 in unimmunized and immunosuppressed individuals (Figure 24.9).

Postpolio syndrome is a condition in which people who survived polio, years before, suffer weakening or paralysis of muscles, which requires them once again to use crutches and braces. It is not infectious, nor is it a recurrence of the disease. It is believed to be due to overuse of compensating muscles that have labored too hard for too many years and now cannot function properly.

Parasitic Diseases of the Nervous System

African Sleeping Sickness

African sleeping sickness, or **trypanosomiasis,** is a disease of equatorial Africa caused by protozoan blood parasites of the genus *Trypanosoma*. Although 100 or more species of this parasite infect various vertebrates and invertebrates, two species, *T. gambiense* and *T. rhodesiense,* cause disease in humans. Typically, trypanosomes have an undulating membrane and a flagellum (Figure 24.10), but in some stages they are shorter and lack a flagellum. For humans to get African sleeping sickness, they must be bitten by an infected tsetse (tset'se) fly. When a tsetse fly bites, it injects infectious trypanosomes, sometimes hundreds in a single bite, into the blood of its victim. The flies serve as vectors and as hosts for part of the life cycle of trypanosomes. Although transmission usually is from one human to another via the fly, game animals serve as natural reservoirs for *T. rhodesiense*.

The Disease African sleeping sickness is a progressive disease characterized according to the tissue in which the parasites congregate during the blood, lymph node, and central nervous system stages. Although the parasites do not actually invade cells, they can damage every tissue and organ in the body.

The bites of tsetse flies cause a local inflammatory reaction. After an incubation period of 2 to 23 days, fever appears initially for about a week while parasites are in the blood and at irregular intervals as the parasites are released from lymph nodes. Patients are able to work through the first and second stages, but they suffer from various symptoms—shortness of breath, cardiac pain, disturbed vision, anemia, and weakness—that become increasingly severe. Invasion of the nervous system by the parasites causes headache, apathy, tremors, and an uncoordinated, shuffling gait. Pain and stiffness in the neck and paralysis occur as the disease progresses. Eventually, the patient cannot be roused to eat, becomes emaciated, has convulsions, sleeps continuously, goes into a profound coma, and dies.

Infection with *T. gambiense* produces a slowly progressive, chronic disease that, if untreated, lasts several years before central nervous system symptoms intensify and death occurs. Infection with *T. rhodesiense* produces a more rapidly progressive disease; it is often fatal within a few months, before central nervous system damage is apparent.

Diagnosis, Treatment, and Prevention Diagnosis of African sleeping sickness depends on finding the parasite in the blood. Until recently, arsenical drugs were used to treat the disease, but they cause eye damage, and the parasites quickly become tolerant to them. Pentamidine, suramin, and melarsoprol are now used, usually in that order. If pentamidine, the least toxic drug, fails to combat the infection, more toxic ones are tried. Results of treatment with any drug are generally more successful if treatment is begun before central nervous system involvement occurs. A com-

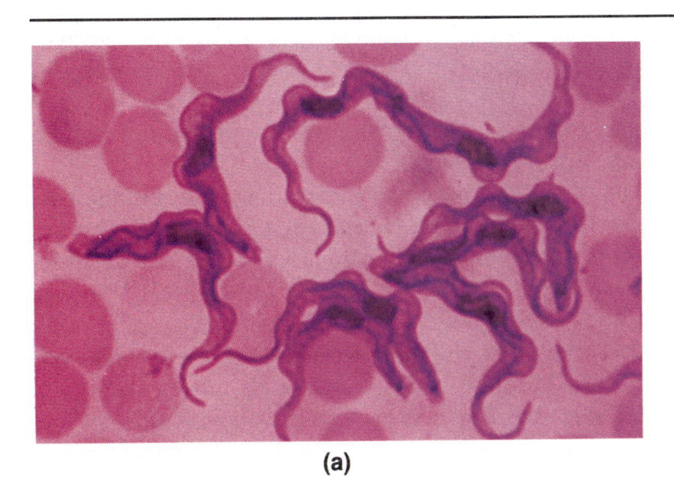

(a)

(b)

FIGURE 24.10 (a) *Trypanosoma gambiense*, 800X, seen in a blood sample, cause African sleeping sickness. (b) Trypanosomes are spread by the bite of the tsetse fly.

bination of Berenil and nitroimidazole is being used to treat the disease after it has progressed to the central nervous system.

Preventing human infection is nearly impossible because of the wide range of tsetse flies (4.5 million square miles) and possibly because of the reservoir of infection in large game animals. Some control has been achieved by clearing brush where flies congregate and by applying aerial pesticides. Another significant means of reducing the tsetse fly population is to release irradiated male flies, which fail to produce viable sperm. The eggs of females that mate with them fail to develop.

Trypanosomes that cause African sleeping sickness have a special means of evading the host's defenses. The intermittent fever of the disease is directly correlated with increasing numbers of parasites in the blood. The unusual thing about these parasites is that each time parasites appear in the blood, they have a glycoprotein coat different from that of previously released parasites. By the time the immune system has developed antibodies to a trypanosome surface antigen, the trypanosome has a different surface antigen. This ability to alter antigens has thwarted efforts to produce a vaccine for African sleeping sickness. When trypanosomes first enter humans, they seem to be able to make only about 15 antigens; later they can make 100 or more. Researchers hope to make a vaccine that will confer immunity against any antigen the parasite might have when it enters the human body. This would allow the body to attack the trypanosomes before an infection becomes established.

Should a vaccine be developed, a means to mount a costly, massive vaccination program would be needed. As available vaccines, such as those for measles and mumps, have not been administered to many of the children in the same region, the prospect for mass immunization is not bright. More than 20,000 people per year are likely to continue to die of African sleeping sickness.

Chagas' Disease

Chagas' disease, caused by *Trypanosoma cruzi*, occurs sporadically in the southern United States and is endemic in Mexico, Central America, and all but the southernmost part of South America. This trypanosome looks like the ones that cause sleeping sickness. It is transmitted by several kinds of reduviid bugs that are hosts for the sexual phase of the trypanosome's life cycle. Each species of bug occupies a particular region, so the ones that transmit Chagas' disease in Mexico are different from those that transmit it in South America. These bugs often bite near the eyes and they defecate as they bite, depositing infectious parasites on the skin. Humans almost automatically rub such a bug bite and they transfer parasites to the eyes or the bite wound.

Chagas' disease begins with subcutaneous inflammation around the bug bite. After 1 to 2 weeks, the parasites have made their way to lymph nodes, where they repeatedly divide and form aggregates called **pseudocysts.** Wherever the pseudocysts rupture, they cause inflammation and tissue necrosis. These parasites enter cells either by invasion or by phagocytosis and can damage lymphatic tissues, all kinds of muscle, and especially supporting tissues around nerve ganglia. Destruction of nerve ganglia in the heart accounts for nearly three-fourths of deaths from heart disease among young adults in endemic areas.

Chagas' disease appears in an acute form and a chronic form. The acute disease, which is most common in children under 2, is characterized by severe anemia, muscle pain, and nervous disorders. In especially virulent acute disease, death can occur in 3 to 4 weeks, but many patients recover after several months of less virulent disease. The chronic disease, which is seen mainly in adults, probably arises from a childhood infection. It is a mild disease and is sometimes asymptomatic but often causes enlargement of various organs. Insidious damage to nerves can cause several severe effects. In the digestive tract it slows or stops muscle contractions, in the heart it can cause irregular heartbeat and accumulation of fluid around the heart, and in the central nervous system it can cause paralysis by destroying motor centers. The parasites also cross the placenta, so chronically infected mothers often give birth to infants with severe acute disease.

Several diagnostic techniques are available for Chagas' disease. The parasites can be found in the blood during fever in acute cases. Small animals, such as guinea pigs and mice, can be inoculated with blood from patients and observed for disease symptoms. This technique is known as **xenodiagnosis.** *Xenos* literally means strange or foreign; in this context it refers to the use of an organism different from a human. Chronic Chagas' disease sometimes can be detected by allowing patients to be bitten by uninfected laboratory-reared reduviid bugs and examining the bugs in 2 to 4 weeks for trypanosomes, which develop in the bug's intestine.

In spite of several ways to diagnose the disease, no effective treatment is available. Drugs used to treat other trypanosome infections are of no use because they fail to reach the parasites inside cells. Work is under way to develop new drugs and a vaccine, but until these are available, control of the reduviid vectors is the only means of reducing misery from this disease. Treating homes with insecticides offers some protection, but the bugs crawl into crevices in walls and thatched roofs and are difficult to eradicate.

The agents and characteristics of the diseases discussed in this chapter are summarized in Table 24.2.

TABLE 24.2 Summary of diseases of the nervous system

Disease	Agent(s)	Characteristics
Bacterial diseases of the brain and meninges		
Bacterial meningitis	See Table 24.1	Tissue necrosis, brain edema, headache, fever, occasionally seizures
Listeriosis	*Listeria monocytogenes*	A kind of meningitis seen in fetuses and immunodeficient patients
Brain abscesses	Various anaerobes	Infection that grows in mass and compresses brain
Viral diseases of the brain and meninges		
Rabies	Rabiesvirus	Invades nerves and brain; headache, fever, nausea, partial paralysis, coma, and death ensue unless patient has immunity
Encephalitis	Several encephalitis viruses	Shrinkage and lysis of neurons of the central nervous system; headache, fever, and sometimes brain necrosis and convulsions
Herpes meningoencephalitis	Herpesvirus	Fever, headache, meningeal irritation, convulsions, altered reflexes
Bacterial nerve diseases		
Hansen's disease	*Mycobacterium leprae*	Range of symptoms from loss of skin pigment and sensation to lepromas and erosion of skin and bone
Tetanus	*Clostridium tetani*	Toxin-mediated disease; muscle stiffness, spasms, paralysis of respiratory muscles, heart damage, and usually death
Botulism	*Clostridium botulinum*	Preformed toxin from food prevents release of acetylcholine; paralysis and death result unless treated promptly; in infants and wounds endospores germinate and produce toxin
Viral nerve diseases		
Poliomyelitis	Several types of polioviruses	Fever, back pain, muscle spasms, partial or complete flaccid paralysis from destruction of motor neurons
Parasitic diseases of the nervous system		
African sleeping sickness	*Trypanosoma gambiense, T. rhodesiense*	Fever, weakness, anemia, tremors, shuffling gait, apathy; as parasites invade nervous system, emaciation, convulsions, and coma ensue
Chagas' disease	*Trypanosoma cruzi*	Subcutaneous inflammation, damage to lymphatic tissues, muscle, and nerve ganglia; muscle pain and paralysis of intestinal, heart, and skeletal muscle

ESSAY

Mysterious Brain Infections

Since the 1920s, when Creutzfeldt-Jakob disease (CJD) was first investigated, several degenerative diseases of the nervous system have been associated with prions. ∞ (Chapter 11, p. 275) These diseases are referred to collectively as transmissible spongiform encephalopathies because they damage neurons so as to give brain tissue a "spongy" appearance (Figure 24.11). They include kuru, Creutzfeldt-Jakob disease, and a special form of CJD called Gerstmann-Strassler disease in humans; scrapie in sheep and goats; transmissible mink encephalopathy; chronic wasting disease of elk and mule deer; and "mad cow" disease affecting English dairy cattle. In addition, in 1991, 29 cases of transmissible spongiform encephalopathies were reported in British cats, and 2 were reported in ostriches in the Berlin Zoo.

A prominent feature of all prion-associated diseases is the lack of any inflammatory response, which is a hallmark of other infectous diseases. There is, however, an increase in the size of astrocytes—the cells that regulate the passage of materials from the blood to neurons—throughout the central nervous system. These cells apparently produce large quantities of a filamentous protein called amyloid, which is characteristic of a

ESSAY Mysterious Brain Infections

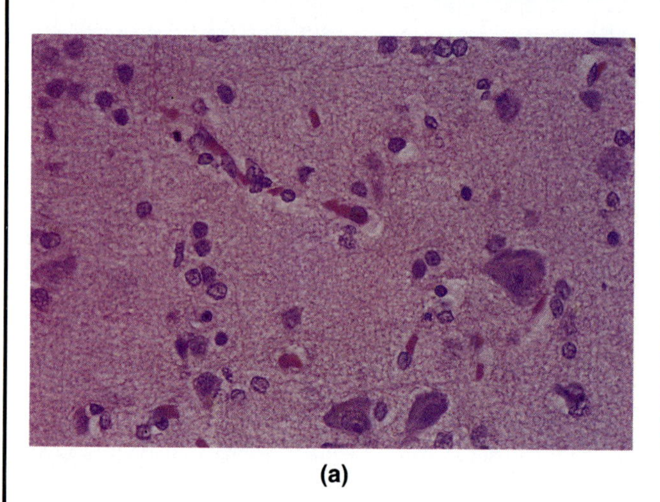

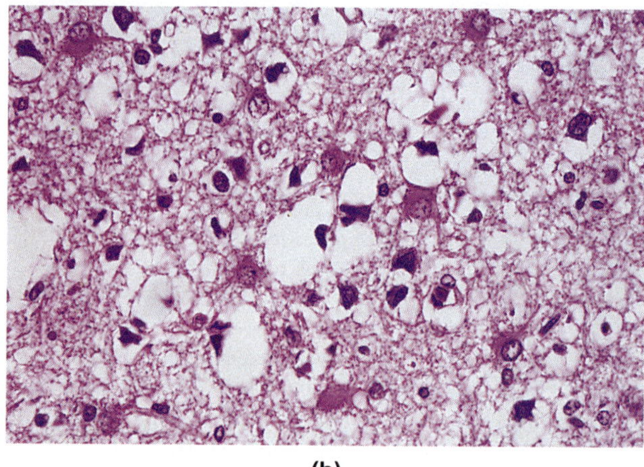

(a) (b)

FIGURE 24.11 (a) A section through the cerebral cortex of a normal human brain (125X) reveals a solid structure, whereas in (b), a brain from a patient with Creutzfeldt-Jakob disease, many holes are evident, making it clear why this is referred to as a subacute *spongiform* encephalopathy.

variety of degenerative diseases of the nervous system.

Like other infectious agents, prions are transmissible, although disease symptoms may not appear until years after initial infection. Kuru, which occurs mainly in New Guinea, is transmitted through small breaks in the skin. Why this disease appeared mainly in women was puzzling until it was discovered that women prepare the bodies of the dead for cannibalistic consumption and smear their bodies with the raw flesh of the corpses. Prions in diseased tissues enter the blood of the women or the children playing at their feet, travel to the brain, and eventually cause kuru. Men and adolescent boys live apart from the women and small children and so are spared infection.

D. Carleton Gajdusek, a U.S. investigator at the National Institutes of Health (NIH), won the Nobel Prize in 1976 for his work on kuru. He found that 1 to 15 years following inoculation with the kuru prion, onset of symptoms began with headaches, minor loss of coordination, and a tendency to giggle at inappropriate times. Three months later, victims needed crutches to walk or stand, and a month after that victims could not move except for spasms. At that point, the swallowing muscles no longer functioned, and malnutrition became a serious problem. Relatives would prechew food and massage it down the esophagus of victims (Figure 24.12), but within 1 year, the patients invariably died.

Some cases of CJD have been traced to inadvertent inoculation with prions present in corneal transplant tissue from patients having undiagnosed Creutzfeldt-Jakob disease, to dwarfed children who received human growth hormone injections made from cadaver pituitaries, and to silver electrodes implanted in the brain during surgical procedures following use in a CJD patient. The exact same electrodes, located 17 months later after numerous supposed sterilizations were implanted into a chimpanzee's brain, where they caused CJD. These findings have caused some newer methods of instrument sterilization to be abandoned. Autoclaving at 121°C at 15 psi of pressure for 60 minutes destroys infectivity, something that years of storage in formaldehyde fails to do.

In one experiment, Gajdusek and Brown (also at NIH) buried scrapie-infected hamster brains in soil and left them undisturbed for 3 years. When they dug up the brain material, it was still infectious. Gajdusek and Brown suspect that infected material may retain its deadliness for more than a decade under such circumstances. Thus, sheep that graze in fields where contaminated carcasses

FIGURE 24.12 Kuru victims have reached the point where they must be fed prechewed food. Since cannibalistic rites have been stopped in New Guinea, the disease has disappeared there.

have been placed have become infected with scrapie (Figure 24.13). Marking burial sites and reexamination of the accepted technique of adding corrosive quicklime to the bodies needs to be done before eradication can be achieved.

In most cases of Creutzfeldt-Jakob disease, no source of prions has been identified. In the most thoroughly studied instance of Gerstmann-Strassler syndrome, CJD developed in every generation of a family line for over 100 years. In this form of CJD a genetic mechanism seems to facilitate prion infection, but how this mechanism works is not clear. Perhaps the presence of a particular gene renders patients susceptible to infection by prions from outside the body, or perhaps the gene activates synthesis of prions within the body.

"Mad cow" disease is currently killing between 400 and 500 British cattle per week (Figure 24.14). It is thought to have existed in Britain since at least the late 1960s, and the number of cases began to mushroom in early 1987. At that time, the method of rendering (boiling down) animal remains for livestock feed was changed so as to omit a solvent extraction step and to increase the number of sheep heads. There is considerable debate as to whether the prions can cross species lines in nature. They definitely can and have been transmitted from one species to another species in laboratory trials. The British government has therefore prohibited addition of beef brains to hamburger, a formerly common practice. They have also changed their butchering methods so that knife or saw blades do not pass through the spinal cord, thus preventing blades from contaminating edible parts. Some countries have banned import of British beef, cattle, and beef products.

The possibility that slow-onset prions play a role in other neurodegenerative diseases such as Alzheimer's and Parkinson's diseases is being investigated. Alzheimer's disease, first described by Alois Alzheimer in 1907, now affects more than 2 million people in the United States—more than 5 percent of the population over age 65. Amyloid proteins arranged in structures called neurofibrillary tangles, or plaques, have been found at autopsy in the brains of patients who suffered from Alzheimer's disease, but the holes found in kuru and other spongiform encephalopathies are missing. Whether prions play a role in the deposition of the proteins is not yet known. So far it has not been possible to transmit these diseases to laboratory animals as has been done with kuru, CJD, and scrapie. Whether this is due to nonsusceptibility of the animals or the absence of an infectious agent also is not yet known. However, recently bits of beta-amyloid protein injected into rat brains have produced a disease similar to Alzheimer's in these rats. Genetic factors may be involved, at least in Alzheimer's disease and CJD, because multiple cases of both diseases have been seen in some families. Much remains to be learned about prions, neurodegenerative diseases, and the relationship between them.

FIGURE 24.13 Sheep infected with "scrapies" will rub or scrape up against fences, poles, or trees, leading to the common name for this disorder. Sometimes they will scrape until they are bloodied. This fatal disease has no cure.

FIGURE 24.14 English cow suffering from "mad cow" disease exhibits weight loss, depressed appearance with lowered head and arched back, and difficulty in walking and turning. At times, such cows will lay their ears back, snort, paw the ground like a bull, and even charge. Eventually, however, their gait becomes very unsteady, and they fall repeatedly.

CHAPTER SUMMARY

The agents and characteristics of the diseases discussed in this chapter are summarized in Table 24.2. Information in that table is not repeated in this summary.

DISEASES OF THE BRAIN AND MENINGES

Bacterial Diseases of the Brain and Meninges

- **Bacterial meningitis,** acquired from carriers or endogenous organisms, can usually be treated with penicillin. A vaccine is available to protect children from *Haemophilus* meningitis.

- **Listeriosis** can be transmitted by improperly processed dairy products and can cross the placenta; it is a great threat to immunodeficient patients.
- **Brain abscesses** arise from wounds or as secondary infections. Antibiotics are effective if given early in an infection, and surgery can be used later unless the abscess is in a vital area of the brain.

Viral Diseases of the Brain and Meninges

- **Rabies** has a worldwide distribution except for a few rabies-free countries. It is difficult to control because of the large number of small mammals that serve as reservoirs.
- Rabies is diagnosed by IFAT and treated by thoroughly cleaning bite wounds and injecting hyperimmune rabies serum and giving vaccine. It can be prevented by immunizing pets and people at risk and by avoidance of contact with wild animals by those who are not immunized.
- **Encephalitis** is transmitted by mosquitoes, often from horses, and can sometimes be diagnosed by culturing blood or spinal fluid in cells or mice; vaccine is available for horses.

Other Viral Diseases

- **Herpes meningoencephalitis** often follows a generalized herpes infection.

RELATED KEY TERMS

rabiesvirus

Eastern equine encephalitis

Western equine encephalitis

Venezuelan equine encephalitis

St. Louis encephalitis

progressive multifocal leukoencephalopathy

OTHER DISEASES OF THE NERVOUS SYSTEM

Bacterial Nerve Diseases

- **Hansen's disease (leprosy)** affects more than 15 million people worldwide. Many patients are asymptomatic for years after infection, and no diagnostic test is available to identify such patients. Dapsone and rifampin are used to treat Hansen's disease, but no vaccine is available.
- **Tetanus** typically follows introduction of spores into deep wounds; it is treated with antitoxin and antibiotics. If all people received the available vaccine, tetanus could be prevented.
- **Botulism** is acquired from ingesting foods containing a potent preformed neurotoxin and is treated with polyvalent antitoxin. In **infant botulism** and **wound botulism,** spores germinate and produce toxin.

lepromin skin test

tuberculoid

lepromatous **leproma**

tetanus neonatorum

Viral Nerve Disease

- Prior to the development of vaccines in the 1950s, **poliomyelitis** was a common and dreaded disease. It can be diagnosed from cultures and by immunologic methods. Treatment is palliative and includes the use of an "iron lung" to maintain breathing when respiratory muscles are paralyzed.
- Both an injectable and an oral vaccine are available; each has advantages and disadvantages. Worldwide use of the vaccine could eradicate poliomyelitis.

Parasitic Diseases of the Nervous System

- **African sleeping sickness** occurs in equatorial Africa and is transmitted by the tsetse fly. It is diagnosed by finding parasites in the blood and can be treated with pentamidine and other drugs.
- **Chagas' disease** occurs from the southern United States to all but the southernmost part of South America and is transmitted by various species of bugs. It is diagnosed by finding parasites in the blood and by **xenodiagnosis;** no effective treatment is available.

trypanosomiasis

pseudocysts

QUESTIONS FOR REVIEW

A.

1. What is meningitis, and why is it life-threatening?
2. What bacteria can cause meningitis, and what is distinctive about the disease caused by each organism?
3. How is meningitis diagnosed, treated, and prevented?
4. What are the special characteristics of listeriosis?
5. Under what circumstances do brain abscesses occur, and how are they treated?

B.

6. What are the characteristics of rabies, and what treatment is available?
7. Why is rabies difficult to control, and what control measures are available?
8. How do humans acquire encephalitis, and in what season of the year is it most common?

9. How can encephalitis be diagnosed, treated, and prevented?
10. Under what conditions does herpes meningoencephalitis occur, and what is its prognosis?

C.

11. What changes have occurred in the incidence of Hansen's disease over the years?
12. How can tuberculoid and lepromatous Hansen's disease be distinguished?
13. How is Hansen's disease diagnosed, treated, and prevented?
14. What is the cause of tetanus, and how do humans acquire it?
15. What are the effects and prognosis for tetanus?
16. How can tetanus be diagnosed, treated, and prevented?
17. What is tetanus neonatorum?
18. How do the forms of botulism differ, and how do humans acquire them?
19. How is botulism diagnosed, treated, and prevented?

D.

20. What are the effects of poliomyelitis?
21. How is poliomyelitis diagnosed and treated?
22. How has vaccine altered the incidence of poliomyelitis?
23. What are the advantages and disadvantages of injectable and oral polio vaccines?

E.

24. Where does African sleeping sickness occur, and how is it transmitted?
25. What agents cause African sleeping sickness, and how does the disease vary according to the causative agent?
26. What efforts have been made to control African sleeping sickness?
27. What causes Chagas' disease, and what are its effects?
28. Where does Chagas' disease occur, and how is it transmitted?

PROBLEMS FOR INVESTIGATION

1. What mechanism would you propose that allows a prion to cause a disease?
2. Survey the problems associated with controlling one of the neurological diseases discussed in this chapter, and propose a better way to control the disease.
3. Read about the U.S. Public Health Service leprosarium at Carville, Louisiana, from its establishment in 1894 to its current research projects.
4. Review the current literature to determine what progress has been made in understanding mysterious brain infections.

SOME INTERESTING READING

Castaño, E. M., and B. Frangione. 1988. Human amyloidosis, Alzheimer's disease and related disorders. *Laboratory Investigation* 58(2):122.

Cohn, J. P. 1989. Leprosy: out of the Dark Ages. *FDA Consumer* 23 (September):24–27.

Fox, J. L. 1990. Rabies vaccine field test undertaken. *ASM News* 56(11):579–83.

Goodfield, J. 1985. *Quest for the killers.* Boston: Birkhauser.

Gray, L. D., and D. P. Fedorko. 1992. Laboratory diagnosis of bacterial meningitis. *Clinical Microbiology Reviews* 5(2):130–45.

Hogle, J. M., M. Chow, and D. J. Filman. 1987. The structure of the poliovirus. *Scientific American* 256, no. 3 (March):42.

Joseph, B. Z., L. J. Yoder, and R. R. Jacobson. 1985. Hansen's disease in native-born citizens of the United States. *Public Health Reports* 100, no. 6 (November–December):666–71.

Marx, J. L. 1987. Leukemia virus linked to nerve disease. *Science* 236 (May 20):1059.

Prusiner, S. B. 1987. Prions and neurodegenerative diseases. *New England Journal of Medicine* 317, no. 25 (December 17):1571.

Whitley, R. J. 1990. Medical progress: viral encephalitis. *New England Journal of Medicine* 323, no. 4 (July 26):242–51.

Wispelwey, B., et al. 1987. Brain abscesses. *Clinical Neuropharmacology* 10(6):483.

False-color TEM of two orange herpes simplex viruses migrating from the nucleus (blue and pink) into the cytoplasm (green) of an infected cell (190,000X). Herpes simplex causes a variety of infections, including whitlows of the finger, cold sores of the mouth, and genital lesions.

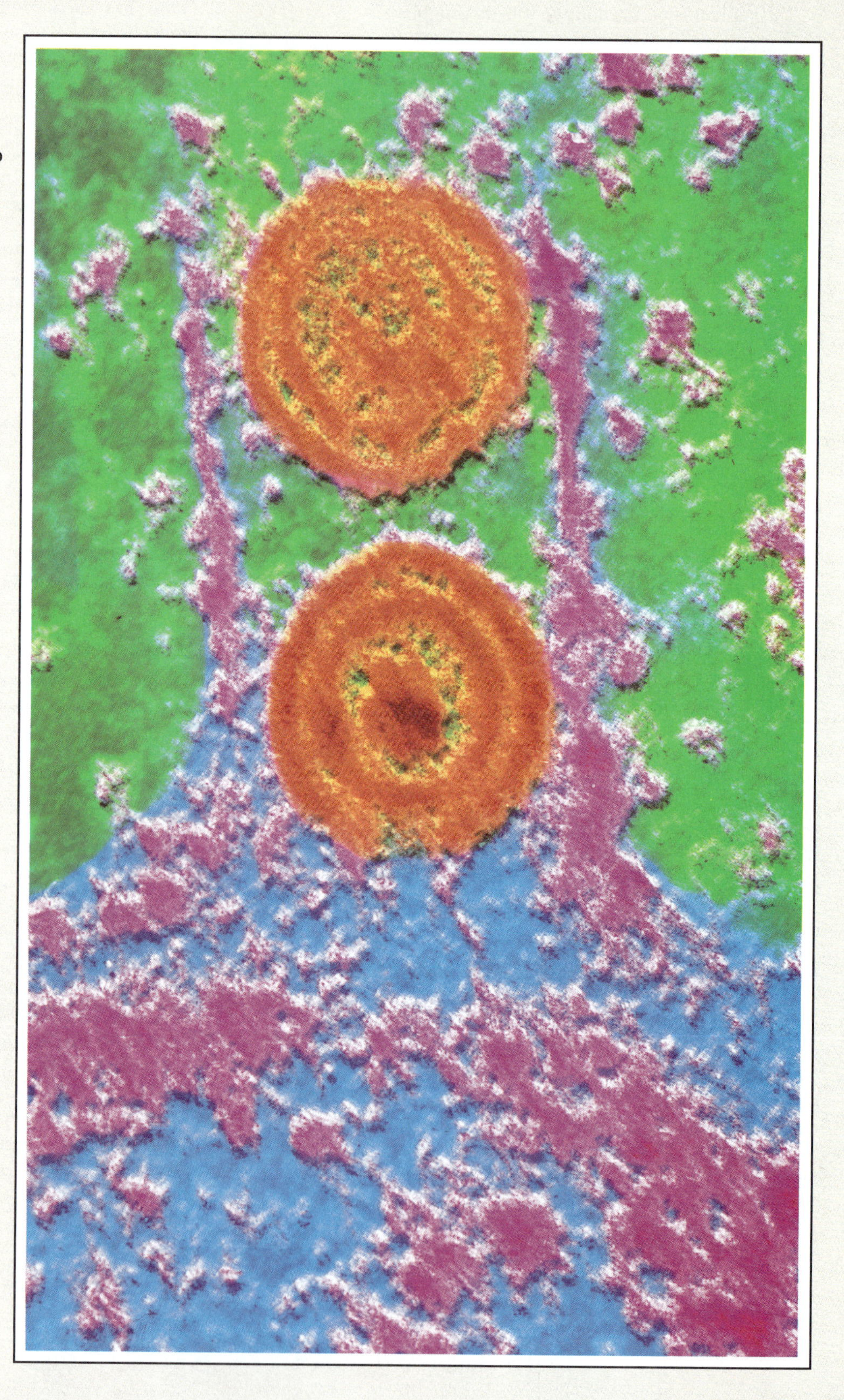

25 Urogenital and Sexually Transmitted Diseases

This chapter focuses on the following questions:

A. What bacteria cause urogenital diseases not usually sexually transmitted, and what are the important epidemiologic and clinical aspects of these diseases?

B. What parasites cause urogenital diseases not usually sexually transmitted, and what are the important epidemiologic and clinical aspects of these diseases?

C. What bacteria cause sexually transmitted urogenital diseases, and what are the important epidemiologic and clinical aspects of these diseases?

D. What viruses cause sexually transmitted urogenital diseases, and what are the important epidemiologic and clinical aspects of these diseases?

In this chapter we conclude our discussion of diseases with diseases of the urogenital system, including sexually transmitted diseases. As with the other chapters on diseases, keep in mind what you have learned about normal flora, disease processes, and body defenses. In studying urogenital diseases it is important to recall that the urinary and reproductive (genital) systems are closely associated. Infections in one system easily spread to the other.

APPLICATIONS

Honeymoon Cystitis

Trauma to the female urethra during sexual intercourse can lead to "honeymoon cystitis," in which damaged tissue easily becomes infected. Urination after intercourse may help to prevent infection. Chronic cystitis, on the other hand, is not particularly associated with sexual intercourse. In many cases no predisposing factor can be found.

UROGENITAL DISEASES USUALLY NOT TRANSMITTED SEXUALLY

Bacterial Urogenital Diseases

Urinary Tract Infections

Urinary tract infections (UTIs) are among the most common of all infections seen in clinical practice. Second only to respiratory infections, they account for 3 million office visits per year in the United States. UTIs cause **urethritis** (u-re-thri'tis), or inflammation of the urethra, and **cystitis** (sis-ti'tis), or inflammation of the bladder. Because infection easily spreads from the urethra to the bladder, most infections are properly named **urethrocystitis** (u-re"thro-sys-ti'tis). Infectious agents reach the bladder more easily through the short (4 cm) female urethra than through the longer (20 cm) male urethra. Thus, women are affected some 40 to 50 times as often as men; by the age of 30, 20 percent of all women have acquired a UTI. In males the prostate gland is closely associated with the urethra and bladder, so **prostatitis** (pros-ta-ti'tis), or inflammation of the prostate gland, often accompanies UTIs.

Each year one out of five women develop **dysuria** (dis-yur'-e-ah), or pain and burning on urination indicative of urethral infection. One-fourth of these infections will develop into chronic cystitis, which will plague the unlucky victims intermittently for years. Elderly women are prone to UTIs, and up to 12 percent of some groups suffer chronically (Table 25.1).

A major cause of UTIs is incomplete emptying of the bladder during urination. Retained urine serves as a reservoir for microbial growth, thereby fostering infection. Any factor that interferes with the flow of urine and the complete emptying of the bladder can therefore predispose an individual to UTIs. The bladder is sometimes compressed by a "sagging" uterus or by the expansion of the uterus in pregnancy. Pregnancy can also result in decreased flow of urine through the ureters. Even the ring of a diaphragm can exert enough pressure on the bladder or ureters to interfere with voiding. In men, the prostate tends to enlarge with age and constrict the urethra. Finally, the problem may be behavorial rather than mechanical: Some people simply do not visit the bathroom often enough. It is important for both females and males to empty the bladder frequently and completely. People with various kinds of paralysis who cannot void completely often have frequent urinary tract infections.

TABLE 25.1 Urinary tract infections by age and sex

Age	Female	Male
First 4 months of life	0.7%	1.3%
Four months to 5 years	4.5%	0.5%
Five years to 60 years	1–4%	<0.1%
Over 60 years	12%	1–4%
Very elderly	up to 30%	up to 30%

UTIs originating in one area often spread throughout the entire urinary tract by "ascending" or "descending." Infections usually begin in the lower urethra and can ascend to cause inflammation of the kidneys, or **pyelonephritis** (pi"e-lo-ne-fri'tis). Less often infections begin in the kidneys and descend to the urethra. Although UTIs are not more frequent during pregnancy, they usually are more serious because they are likely to ascend. Among pregnant women with bacteria present in the urine, 40 percent develop pyelonephritis if the infection is not treated promptly. Descending UTIs originate outside the urinary tract. Organisms that enter the bloodstream from a focal infection, such as an abscessed tooth, can be filtered by the kidneys and cause a single infection or a chronic infection. For this reason it is often suggested that persons with chronic or frequent UTIs visit their dentist to see if some undiagnosed gum infection might be the source.

Escherichia coli is the causative agent in 80 percent of UTIs, but other enteric bacteria from feces can also cause such infections. Poor hygiene, such as wiping from back to front with toilet tissue, especially in females, can introduce fecal organisms into the urethra.

It is important to teach children good toilet habits and the reasons for them. When *Chlamydia* or *Ureaplasma* are responsible for infection, they usually are sexually transmitted and cause nongonococcal urethritis, which is discussed later.

Between 35 and 40 percent of all nosocomial infections are UTIs. Outpatients have a 1 percent chance of developing a UTI following a single catheterization, whereas hospitalized patients have a 10 percent chance. A great many patients with an indwelling catheter develop a UTI in the first week of use as organisms from skin, the lower urethra, or the apparatus colonize the urethra. People with permanent catheters because of paralysis fight a never-ending battle against UTIs. *Staphylococcus epidermidis* often causes infections in patients with indwelling catheters. *Pseudomonas aeruginosa* commonly causes infections that follow the use of instruments to examine the urinary tract. However, *E. coli* causes nearly half of nosocomial UTIs and *Proteus mirabilis* causes about 13 percent of them.

UTIs are diagnosed by identifying organisms in urine cultures (Figure 25.1). Normal urine in the bladder is sterile, but urine is inevitably contaminated by bacteria as it passes through the lower part of the urethra. Even a clean-catch, midstream urine specimen will contain 10,000 to 100,000 organisms per milliliter. Low numbers of organisms do not necessarily rule out infection; in pyelonephritis and acute prostatitis organisms sometimes enter the urine only in small numbers. Generally, however, only a single pathogen will be present at any one time; finding several species in urine almost always means that the specimen was contaminated and the test should be repeated.

UTIs are treated with antibiotics, such as amoxicillin, trimethoprim, and quinolones, or with sulfonamides, according to susceptibilities of their causative agents. Prompt treatment helps prevent the spread of infection. UTIs can be prevented by good personal hygiene and frequent and complete emptying of the bladder.

FIGURE 25.1 Urine samples. (a) Clear specimen, showing no bacterial growth. (b) Cloudy (turbid) specimen, indicating bacterial growth.

CLOSE-UP

Who Is Prone to UTIs?

A simple blood test may soon be used to reveal whether you are an individual prone to UTIs and might benefit from aggressive preventive treatment such as taking antibiotics. Researchers at the Memorial Sloan-Kettering Cancer Center in New York have found that women with one of two specific types of Lewis blood groups have nearly a fourfold higher risk of developing UTIs. The correlation was evident as early as age 2. The Lewis factors in blood are tested for in the same fashion as are A, B, O blood types, but using different antisera. Such testing is available at blood banks across the country. Explanation of the correlation is incomplete but appears to be based on cells lining the urinary tract of people with the two Lewis blood types being more receptive to bacterial attachment, thus leading to infection.

Prostatitis

The symptoms of prostatitis are urgency and frequency of urination, low fever, back pain, and sometimes muscle and joint pain. Most men have had at least one prostate infection by age 40. *Escherichia coli* is the cause of 80 percent of the cases, but it is still uncertain how the bacteria reach the prostate. Four routes of infection are possible: (1) by ascent through the urethra, (2) by backflow of contaminated urine, (3) by fecal organisms from the rectum passing through lymphatics to the prostate, and (4) by descent of bloodborne organisms. Although uncommon, chronic prostatitis is a major cause of persistent UTIs in males and it can cause infertility. Acute prostatitis usually responds well to appropriate antibiotic therapy without leaving sequelae.

Pyelonephritis

Pyelonephritis is an inflammation of the kidney, usually caused by the backup of urine and consequent ascent of microorganisms. Urine backup can be caused by a number of factors, including lower urinary tract blockage or anatomical defects. Young children in particular often have imperfectly formed urinary tract valves that do not prevent urine from backing up. *Escherichia coli* causes 90 percent of outpatient cases and 36 percent of those in hospitalized patients, but yeasts such as *Candida* are occasionally responsible for the infection.

Pyelonephritis and any other urinary tract infections can be asymptomatic. When present, the symptoms are indistinguishable from those of cystitis, except that sometimes chills and fever occur. Dilute urine is another common finding, leading to frequent urination and **nocturia** (nok-tu're-ah), or nighttime urination. Patients must be carefully evaluated to identify underlying, predisposing conditions such as kidney stones or other blockages, which need to be relieved. Pyelonephritis is more difficult to treat than lower UTIs, but nitrofurantoin, sulfonamides, trimethoprim, ampicillin, and quinolones usually are effective. During kidney failure or impaired kidney function, drugs must be used with care to prevent toxic accumulations.

Glomerulonephritis

Glomerulonephritis (glom-er"-u-lo-nef-ri'tis), or Bright's disease, causes inflammation and damage to the glomeruli of the kidneys (Figure 25.2). It is an immune complex disease that sometimes follows a streptococcal or viral infection. Antigen-antibody complexes are filtered out in the kidney, where they cause inflammation of glomerular capillaries. The inflamed vessels leak blood and protein into the urine. Because of the risk of glomerulonephritis, organisms from throat and other infections that might be streptococcal should be cultured and appropriate antibiotics given. (Chapter 21, p. 577) While most peo-

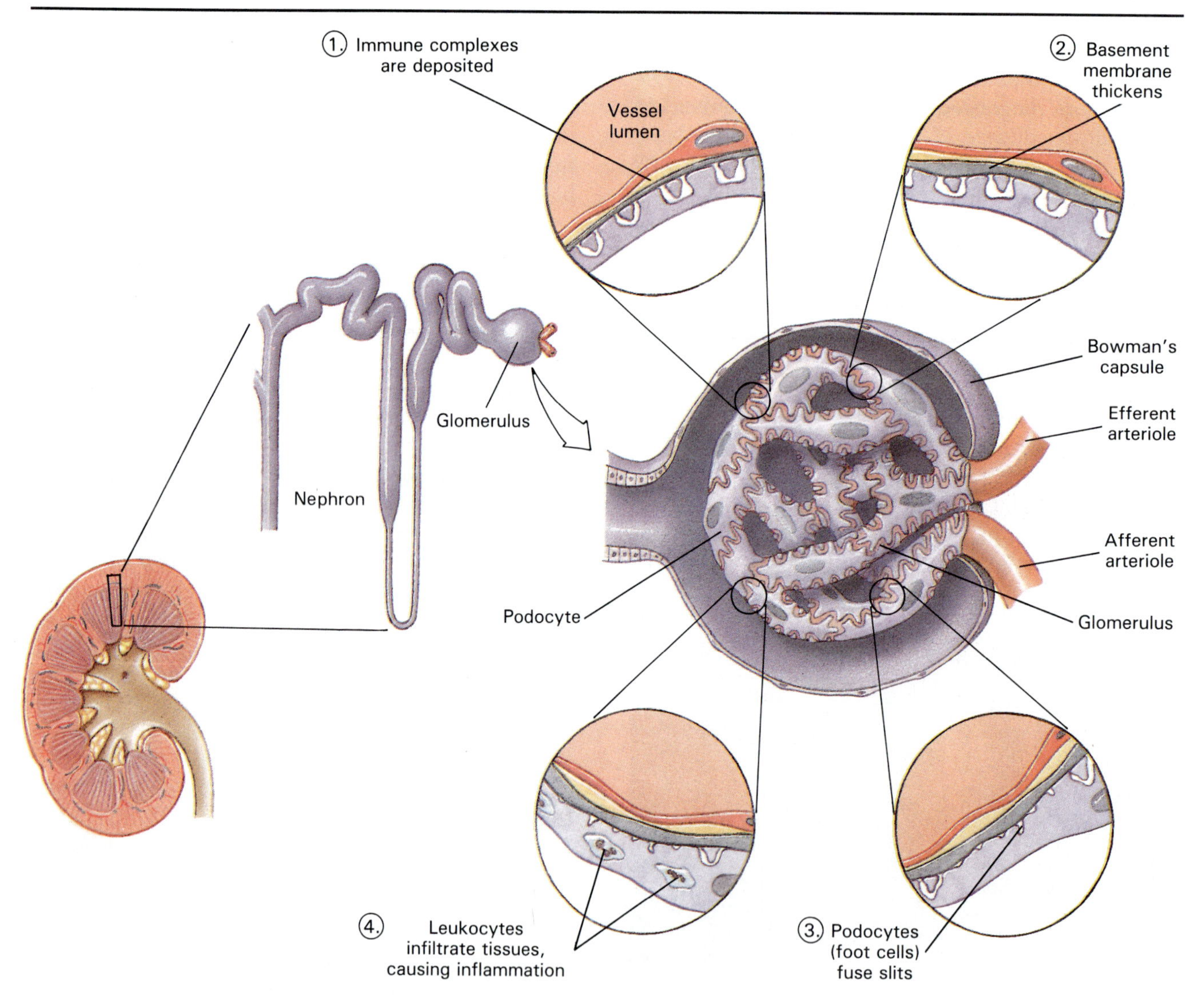

FIGURE 25.2 Glomerulonephritis occurs when immune complexes are deposited on the filtering surfaces of the glomerulus, which is located inside the Bowman's capsule in the cortex of the kidney. Podocytes (foot cells) in the glomerulus, which normally act as filters, fuse their slits; basement membranes of cells thicken, and leukocytes are attracted into the tissues. Kidney function is diminished, sometimes to the point of kidney failure.

ple recover from glomerulonephritis, some have permanent residual kidney damage, and a few die.

Leptospirosis

Leptospirosis is caused by the spirochete *Leptospira interrogans* (Figure 25.3). It is a zoonosis usually acquired by humans through contact with contaminated urine, directly or in water or soil. Dogs, cats, and many wild mammals carry these spirochetes (recall the "Dog Germs" box in Chapter 16). (p. 420) In some parts of the world, more than 50 percent of the rats are carriers of *Leptospira*. The bacteria live within the convoluted tubules of the kidneys and are shed into the urine. Often rain washes them from streets and the soil into natural bodies of water. They die quickly in brackish or acid water but can survive for 3 months or longer in neutral or slightly basic water. Ponds or rivers bordering pastures where animals graze are especially likely to be contaminated.

The organisms enter the body through mucous membranes of eyes, nose, or mouth, or through skin abrasions. Parents of small children often become infected from contact with pets. The children beg for a puppy, promising to take care of it, but the adults frequently clean up the puppy puddles and become infected with *Leptospira*. Dogs, being ignorant of the niceties of toilet training, often account for a cluster infection that includes all who shared a swimming or wading pool with the dogs.

After an incubation period of 10 to 12 days, leptospirosis usually occurs as a febrile and otherwise nonspecific illness. In most cases, an uneventful recovery takes place in 2 to 3 weeks. However, 5 to 30 percent of the untreated cases do result in death. A particularly virulent form of the infection, Weil's disease, is characterized by jaundice and significant liver damage.

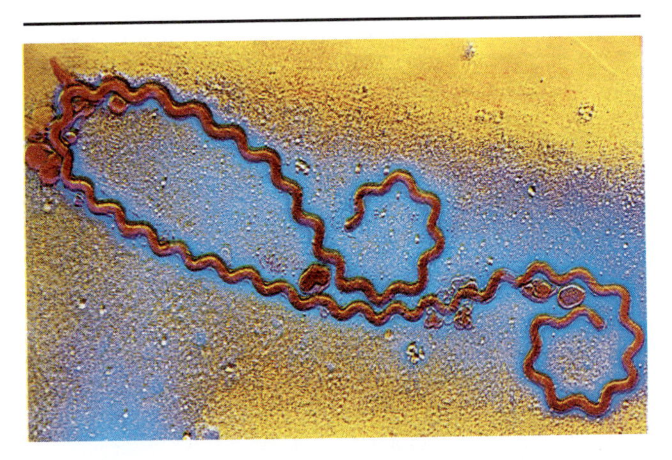

FIGURE 25.3 False color electron micrograph of *Leptospira interrogans*, a spirochete ordinarily found in animals that sometimes causes human urinary tract infections.

Diagnosis can be made by direct microscopic examination of blood, but it is often not even considered because of the nonspecific symptoms and low incidence of leptospirosis. Many cases are asymptomatic, so it is difficult to know the true incidence of the disease. Generally, fewer than 100 cases per year are reported in the United States.

Leptospires are susceptible to almost any antibiotic if it is given in the first 2 or 3 days of the disease, but not after the fourth day. Patients have long-term immunity to the particular strain of *Leptospira* that infected them but not to any other strains. Leptospirosis can be prevented by vaccinating pets and avoiding swimming in or having contact with contaminated water.

Bacterial Vaginitis

Vaginitis, or vaginal infection, usually is caused by opportunistic organisms that multiply when the normal vaginal flora are disturbed by antibiotics or other factors. Predisposing conditions include diabetes mellitus, pregnancy, use of contraceptive pills, menopause, and conditions that result in imbalances of estrogen and progesterone, all of which change the pH and sugar concentration in the vagina. Several organisms each account for a share of vaginitis cases. The bacterium *Gardnerella vaginalis*, in combination with anaerobic bacteria, accounts for about one-third of the cases. *Mobiluncus* is seen in a few infections; it may be a separate organism or a clinical variant of *G. vaginalis*. The protozoan *Trichomonas vaginalis*, which can be opportunistic but usually is transmitted sexually, accounts for about one-fifth of the cases. The fungus *Candida albicans* causes most other cases.

Gardnerella vaginalis This tiny gram-negative rod or coccobacillus is present in the normal urogenital tract in 20 to 40 percent of healthy women. The normal vaginal pH is 3.8 to 4.4 in women of reproductive age and near neutral in young girls and elderly women. When the vaginal pH reaches 5 to 6, *Gardnerella vaginalis* interacts with anaerobic bacteria such as *Bacteroides* and *Peptostreptococcus* to cause vaginitis. None of these organisms alone produces disease. Because different anaerobes interact with *Gardnerella*, this kind of vaginitis sometimes is called "nonspecific vaginitis." *Gardnerella* vaginitis produces a frothy, fishy-smelling vaginal discharge. The discharge, although usually small in volume, contains millions of organisms. Males occasionally get **balantitis** (bal-an-ti'-tis), an infection of the penis that corresponds to female vaginitis. Lesions appear on the penis after sexual contact with a woman who has vaginitis.

Diagnosis can be made when wet mounts of the discharge exhibit "clue cells," vaginal epithelial cells covered with many tiny rods or coccobacilli, as shown

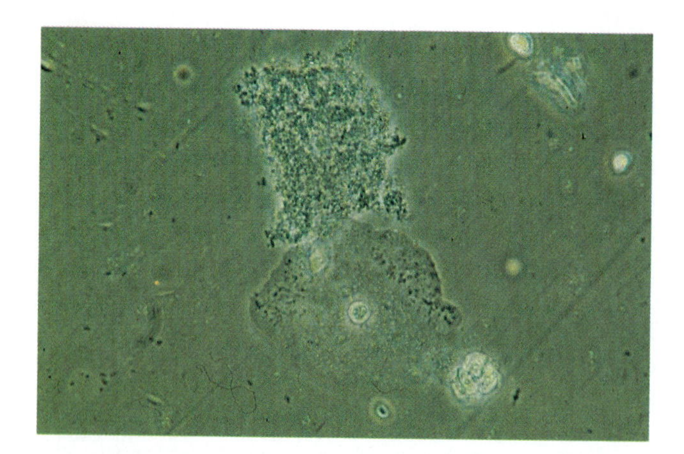

FIGURE 25.4 Clue cell of *Gardnerella* infection (upper), compared with normal epithelial cell (below). Note the many bacilli adhering to the surface of the clue cell.

in Figure 25.4. Metronidazole (Flagyl) suppresses vaginitis by eradicating the anaerobes necessary for continuance of the disease but allows normal lactobacilli to repopulate the vagina. This effect supports the notion that *Gardnerella* vaginitis requires an association with anaerobes. Ampicillin and tetracycline are also sometimes used in treatment. Unflavored (plain) liveculture yogurt used as a douche effectively replaces lactobacilli of normal flora killed by antibiotic treatment.

Toxic Shock Syndrome

Infection with certain toxigenic strains of *Staphylococcus aureus* has produced **toxic shock syndrome** (TSS). Prior to 1977 only 2 to 5 cases were reported each year, but then the incidence rose suddenly, reaching a peak of 955 cases in 1980. Since then the numbers have dropped below 400 per year. The sudden rise was associated with new superabsorbent, but abrasive, tampons, which were left in the vagina for longer than usual periods of time. The tampons caused small tears in the vaginal wall and provided appropriate conditions for bacteria to multiply.

Between 5 and 15 percent of women have *S. aureus* among their vaginal flora, but only a small fraction of these strains cause TSS. Most cases occur in menstruating women, but men, children, and postmenopausal women with focal infections of *S. aureus* occasionally develop TSS. Organisms enter the blood or grow in accumulated menstrual flow in tampons. They produce **exotoxin C,** which enhances the effects of an endotoxin, but how these toxins exert their effects is not clearly understood. Clinical manifestations include fever, low blood pressure (shock), and a red rash, particularly on the trunk, which later peels. Immediate treatment with nafcillin has held the mortality rate to 3 percent. Deaths, when they occur, usually are due to shock. Recurrence is a frequent possibility, especially during subsequent menstrual cycles. Antibiotics can be given prophylactically to prevent recurrences, but women who have had TSS can reduce their risk of recurrence by ceasing to use tampons.

Although infrequent changing of superabsorbent tampons accounts for most cases of TSS, the contraceptive sponge now causes some. Males with boils, furuncles, or other staphylococcal infections also sometimes get TSS.

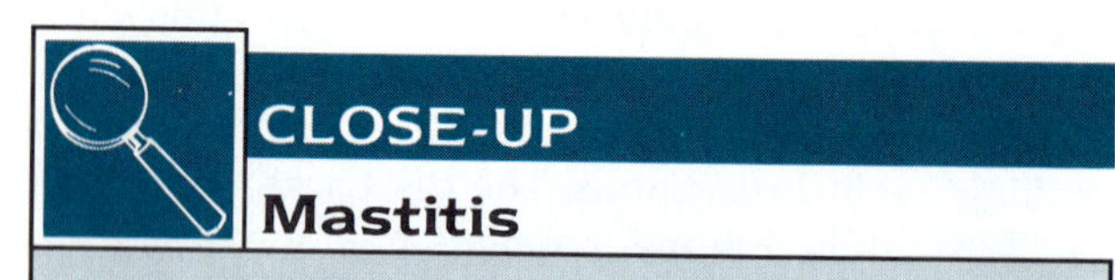

Mastitis

The female breast, another part of the reproductive system, sometimes becomes infected. Mastitis (breast infection), usually caused by *Staphylococcus aureus*, is most likely to occur in nursing mothers. Appropriate antibiotic treatment generally gives prompt results with no sequelae.

Parasitic Urogenital Diseases

Trichomoniasis

Although **trichomoniasis** is transmitted primarily by sexual intercourse, it is discussed here because of its similarity to other kinds of vaginitis and because children can be infected from contaminated linens and toilet seats. At least three species of protozoa of the genus *Trichomonas* can parasitize humans, but only *T. vaginalis* causes trichomoniasis. The others are commensals; *T. hominis* is found in the intestine, and *T. tenax* in the mouth. *Trichomonas vaginalis* is a large flagellate with four anterior flagella and an undulating membrane (Figure 25.5). It infects genitourinary tract

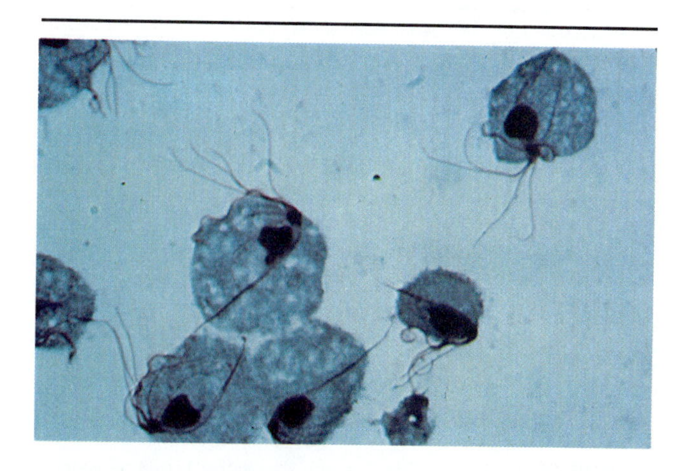

FIGURE 25.5 *Trichomonas vaginalis,* showing characteristic undulating membranes and flagellae.

CLOSE-UP

Vaginal Yeast Infections

In addition to the protozoan *Trichomonas*, fungi and viruses also can cause urogenital infections. In fungal vaginitis, most often caused by *Candida albicans*, lesions are raised gray or white patches surrounded by red areas, and the discharge is scanty, thick, and curdlike. *Candida albicans* is a relatively common organism, being found among the natural flora of 20 percent of nonpregnant women and 30 percent of pregnant ones. *Candida* causes opportunistic infections, especially in patients with AIDS or uncontrolled diabetes, or in those being treated with tetracycline. Diabetes should be regulated and tetracycline therapy stopped so that *Candida* vaginitis can be effectively treated with Nystatin or imidazole. Other fungi that can cause urogenital infections include *Aspergillus*, *Cryptococcus*, and *Histoplasma*. These opportunistic fungi were discussed in Chapter 20. ∞ (p. 559)

surfaces in both males and females and feeds on bacteria and cell secretions. Because the optimum pH for the organism is 5.5 to 6.0, it infects the vagina only when vaginal secretions have an abnormal pH. The symptoms of trichomoniasis are intense itching and a copious white discharge, especially in females.

Trichomoniasis is diagnosed by microscopic examination of smears of vaginal or urethral secretions and treated with metronidazole (Flagyl) and restoration of normal vaginal pH in women. Flagyl cannot be used during pregnancy because it causes abortions, but it is important to get rid of the infection before delivery to prevent infecting the infant. A vinegar douche usually is effective.

The effects of different species of *Trichomonas* illustrate the extreme variation in the degree of damage parasites can do. *Trichomonas hominis* and *T. tenax* are considered commensals. However, *T. foetus* causes serious genital infections in cattle. It is the leading cause of spontaneous abortion in cows and, in fact, is responsible for losses to U.S. cattle breeders of nearly a million dollars per year.

SEXUALLY TRANSMITTED DISEASES

Bacterial Sexually Transmitted Diseases

Sexually transmitted diseases (STDs) have become an increasingly serious public health problem in recent years, in part because of changing sexual behavior. Adding to the problem, some causative agents are becoming resistant to antibiotics, and no vaccines have been developed to control any STDs. Consequently, the only means of preventing these diseases is to avoid exposure to them.

Although AIDS is an STD, we have discussed it in Chapter 19 with disorders of the immune system. ∞ (p. 535)

Gonorrhea

The term **gonorrhea** means "flow of seed" and was coined in A.D. 130 by the Greek physician Galen, who mistook pus for semen. By the thirteenth century the venereal (from Venus, goddess of love in Roman mythology) transfer of this disease was known. But it was not until the mid-nineteenth century that gonorrhea was recognized as a specific disease; until then it was thought to be an early symptom of syphilis. The causative organism, *Neisseria gonorrhoeae*, was first described by Albert Neisser in 1879. It is a gram-negative, spherical or oval diplococcus with flattened adjacent sides and resembles a pair of coffee beans facing each other.

Drying kills the organisms in 1 to 2 hours, but they can survive for several hours on fomites. Cases of *Neisseria* surviving improper laundering in the hospital have been documented, and in dried masses of pus, the bacteria can survive for 6 to 7 weeks! So, while these organisms are ordinarily thought of as being very fragile, some are quite robust.

The infectivity of *Neisseria* is related in several ways to their pili. Sex pili are essential for conjugation, by means of which organisms frequently acquire antibiotic resistance. Moreover, common pili enable gonococci to attach to epithelial cells lining the urinary tract so that they are not swept out with the passage of urine. (In fact, sloughing of epithelial cells is one of the body's defenses against such infections.) These bacteria also use their pili to attach to sperm; conceivably, swimming sperm could carry gonococci into the upper part of the reproductive tract. Strains lacking pili usually are nonvirulent. The search for a vaccine against pili is under way, but success has not yet been achieved.

Gonococci produce an endotoxin that damages the mucosa in fallopian tubes and releases enzymes such as proteases and phospholipases that may be important in pathogenesis. They also produce an extracellular protease that cleaves IgA, the immunoglobulin present in secretions. Gonococci adhere to polymorphonuclear leukocytes (PMNLs) and are phagocytized by them. Phagocytosis kills some of the bacteria, but survivors multiply inside PMNLs. A typical wet mount of a urethral discharge will show PMNLs with diplococci in their cytoplasm (Figure

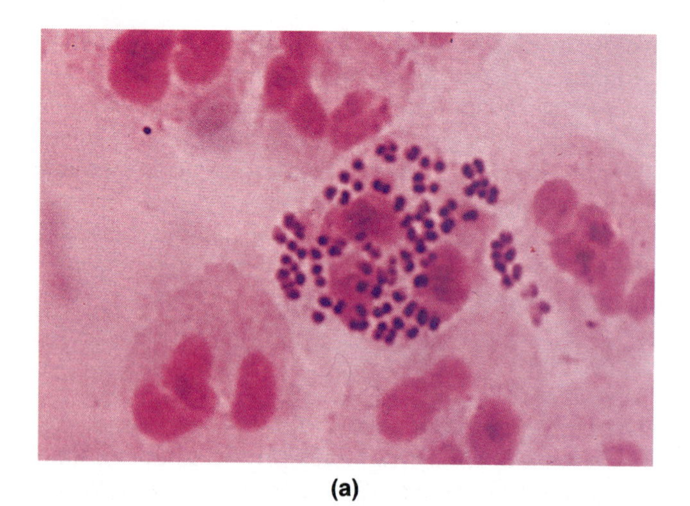

(a)

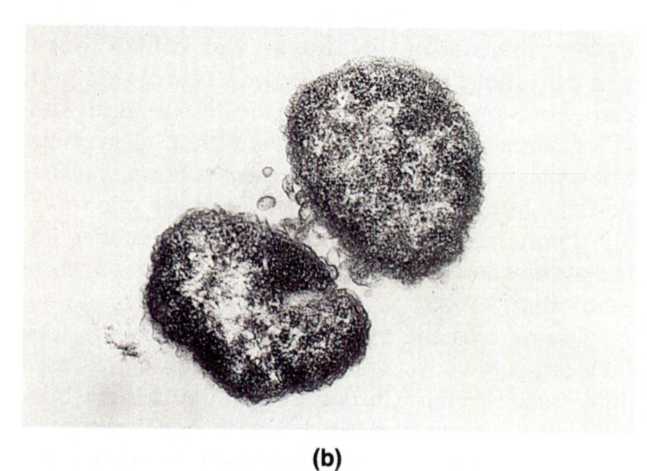

(b)

FIGURE 25.6 Pairs of *Neisseria gonorrhoeae* are seen (a) as small dark purple dots inside the cytoplasm of leukocytes, and (b) magnified 40,000X by TEM, showing the internal structure of the diplococcus pair.

25.6). Gonococci also obtain iron for their own metabolic needs from the iron transport protein transferrin.

The Disease Humans are the only natural hosts for gonococci. Gonorrhea is transmitted by carriers who either have no symptoms or have ignored them. As many as 40 percent of males and 60 to 80 percent of females remain asymptomatic after infection and can act as carriers for 5 to 15 years. Very few organisms are required to establish an infection; half of a group of males given a urethral inoculum of only 1000 organisms developed gonorrhea. Following a single sexual exposure to an infected individual, about one-third of males become infected. After repeated exposures to the same infected person, three-quarters of males eventually become infected. Of males who develop symptoms, 95 percent have pus dripping from the urethra (Figure 25.7) within 14 days, and many develop symptoms sooner. The typical incubation period is 2 to 7 days.

Both the contraceptive pill and IUDs have contributed to the epidemic of gonorrhea now seen in the Western world (Figure 25.8). Their introduction in the 1960s led to increased sexual freedom and decreased use of condoms and spermicides. Women using the contraceptive pill have a 98 percent chance of being infected upon exposure because the pill alters vaginal conditions in favor of gonococcal growth. Although condoms and spermicides are less reliable contraceptives, they do offer some protection from gonorrhea. Gonorrhea spreads into the endometrial cavity and fallopian tubes two to nine times as fast in women who use IUDs as in those who do not use them. The exposure of blood vessels during menstruation allows bacteria to enter the circulation and thus facilitates the development of bacteremia.

Although gonorrhea is thought of as a venereal disease, it also appears in other parts of the body. Pharyngeal infections, which develop in 5 percent of those exposed by oral sex, are most common in women and homosexual men. The patient develops a sore throat, and the infected tissues act as a focal

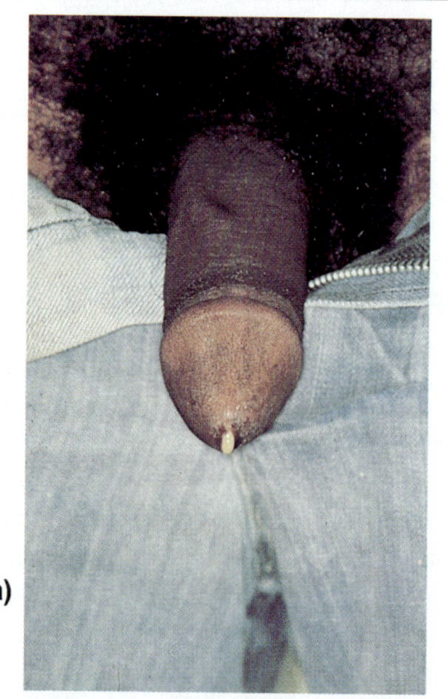

(a)

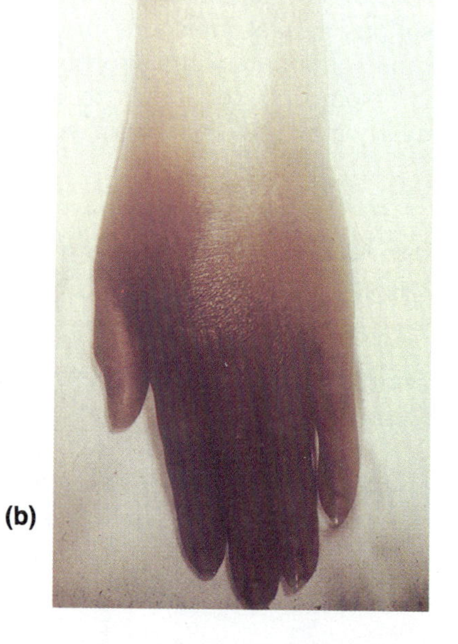

(b)

FIGURE 25.7 Symptoms of gonorrhea sometimes include (a) in males, a urethral "drip" of pus, and (b) an infrequent complication of gonococcally caused arthritis.

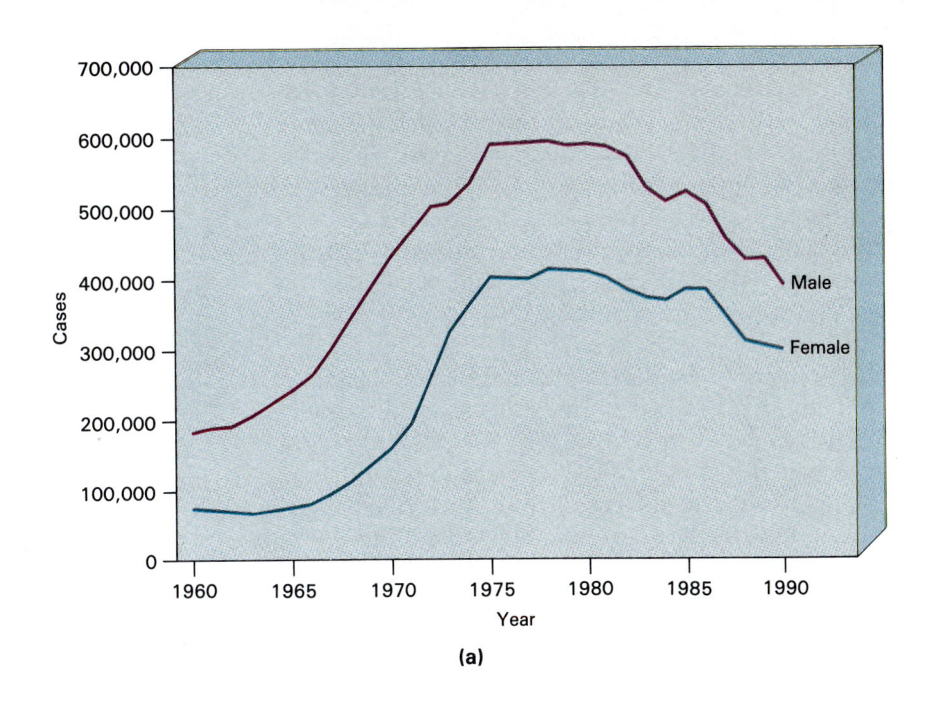

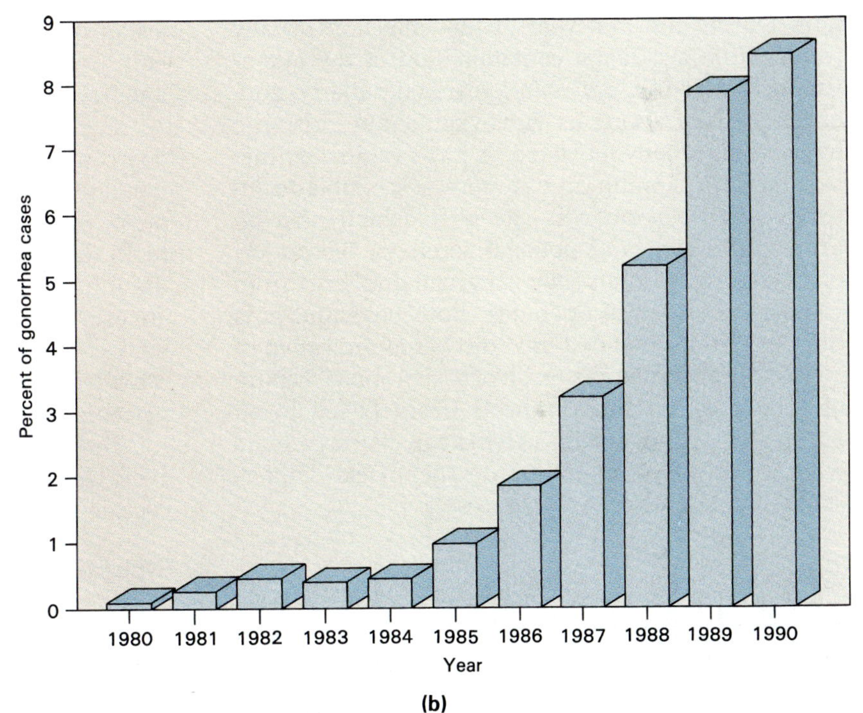

FIGURE 25.8 (a) Incidence of gonorrhea, by sex, in the United States, 1960–1990. Notice the sharp upswing in cases of gonorrhea following the increased use of alternate methods of birth control in place of condoms, beginning in the 1960s. (b) The percentage of cases of gonorrhea caused by penicillin-resistant strains, United States, 1980–1990.

source for bacteremia. Anorectal infection, especially common in homosexual men, occurs in the last 5 to 10 cm of the rectum. It can be either asymptomatic or painful, with constipation, pus, and rectal bleeding. Women with vaginal gonorrhea also often have anorectal infection. Medical personnel doing internal pelvic examinations can avoid spreading the infection by using a different gloved finger for anal examination than was inserted into the vagina.

The urethra is the most common site of gonorrhea infection in males. The most common site of infection in females is the cervix, followed by the urethra, anal canal, and pharynx. Skene's glands in the urethra and Bartholin's glands near the vaginal outlet also can be infected in females. As many as one-half of infected females develop **pelvic inflammatory disease** (PID), in which the infection spreads throughout the pelvic cavity. Studies done in Sweden showed that sterility often follows PID because of tubal occlusion by scarring. The rate of sterility increases with the number of infections—13 percent after one infection, 35 percent after two infections, and 75 percent after three infections.

Disseminated infections, which occur in 1 to 3

percent of cases, produce bacteremia, fever, joint pain, and skin lesions that can be pustular, hemorrhagic, or necrotic. When organisms reach the joints, they can cause arthritis. Gonococcal arthritis is now the most common joint infection in people 16 to 50 years old. Another complication of gonorrhea is infection of the lymphatics that drain the pelvis. Scarring produces tight, inflexible tissue that immobilizes pelvic organs into the condition known as "frozen pelvis."

Transfer of organisms by contaminated hands, or fomites such as towels, can result in eye infections. If untreated, severe scarring of the cornea and blindness can result. Newborns can acquire ophthalmia neonatorum (described in Chapter 20) during passage through the birth canal of an infected mother. ∞ (p. 561) Pus that accumulates behind swollen eyelids can spurt forth with great pressure and infect health-care workers. Such infections have occurred despite immediate treatment.

Age affects the site and type of gonococcal infection. During the first year of life, infection usually results from accidental contamination of the eye or vagina by an adult. Between 1 year and puberty, gonorrhea usually occurs as vulvovaginitis in girls who have been sexually molested. A girl's vaginal epithelium has less keratin and is more susceptible to infection than a woman's. The girls exhibit pain on urination, vulvar and perianal soreness, discomfort on defecation, and a yellowish-green discharge from vaginal and urethral openings. Both boys and girls can develop a pus-filled anal discharge indicative of anorectal gonorrhea as a result of sexual abuse. Failure of a hospital to launder sheets properly led to an outbreak of vulvovaginitis in female pediatric patients who acquired it from sitting on the sheets. Understandably parents were quite upset!

Diagnosis, Treatment, and Prevention Diagnosis is made by identifying *N. gonorrhoeae* in laboratory cultures. Being a rather fastidious organism, it requires high humidity and ambient carbon dioxide for growth. Temperatures of 35° to 37°C and pH of 7.2 to 7.6 are optimal. Many species of *Neisseria* exist. One should not assume that a positive result on a screening test is due to *N. gonorrhoeae;* confirmatory tests should always be done. Samples from patients suspected of having *Neisseria* infections are inoculated from a swab directly into a special medium for transport and incubation in a laboratory (see Chapter 6). ∞ (p. 156)

Sulfonamides were the first agents found to treat gonococcal infections. As some strains became resistant to sulfonamides, penicillin became available. At first, penicillin G was effective in very small amounts but now much larger doses must be used. Some strains of gonococci are entirely resistant to penicillin because they produce the enzyme beta-lactamase, which enables them to break down this antibiotic. Spectinomycin is used for treatment of resistant strains and for patients who are allergic to penicillin. Recently some strains have emerged that are resistant to both penicillin and spectinomycin. Increases in resistance have made combating gonorrhea very difficult. One wonders if it will always be possible to find another antibiotic to combat resistant strains.

PUBLIC HEALTH

Drug-resistant Gonorrhea

The relatively new quinolone antibiotics have not yet been employed to treat gonorrhea in the United States. However, they have been so employed in the Philippines, with the result that quinolone-resistant strains of *Neisseria gonorrhoeae* have already emerged there. If these strains make their way to this country, the quinolones will be obsolete for treating gonorrhea here before they are even used.

Gonorrhea patients often have other STDs. A 7-day course of treatment with tetracycline has the advantage of killing *Chlamydia,* which may be concurrently present. Follow-up cultures should be done 7 to 15 days after completion of treatment to be sure the infection is cured. Additional follow-up cultures are also advised at 6 weeks, as 15 percent of women who are negative at 7 to 14 days are positive again at 6 weeks—possibly because of reinfection. All sexual partners must be treated.

The best means of preventing gonorrhea is to avoid sexual contact with infected individuals. No vaccine is available.

Syphilis

Syphilis is caused by the spirochete *Treponema pallidum,* an active motile organism with fastidious growth requirements. The evolution of this organism has run parallel to human evolution, as discussed in the Essay at the end of this chapter. *Treponema pallidum* eluded discovery until it was finally stained in 1905. Today in the United States syphilis is less common than gonorrhea, but its incidence has recently been increasing (Figure 25.9).

Transmission of syphilis is ordinarily by sexual means, but it can be passed in body fluids such as saliva. It thereby creates a hazard for dentists, dental hygienists, and those who like to kiss. It is not transmitted in food, water, or air or by arthropod vectors. Humans are its only reservoir. Donated blood need not be screened for syphilis because any spirochetes present are killed when the blood is refrigerated.

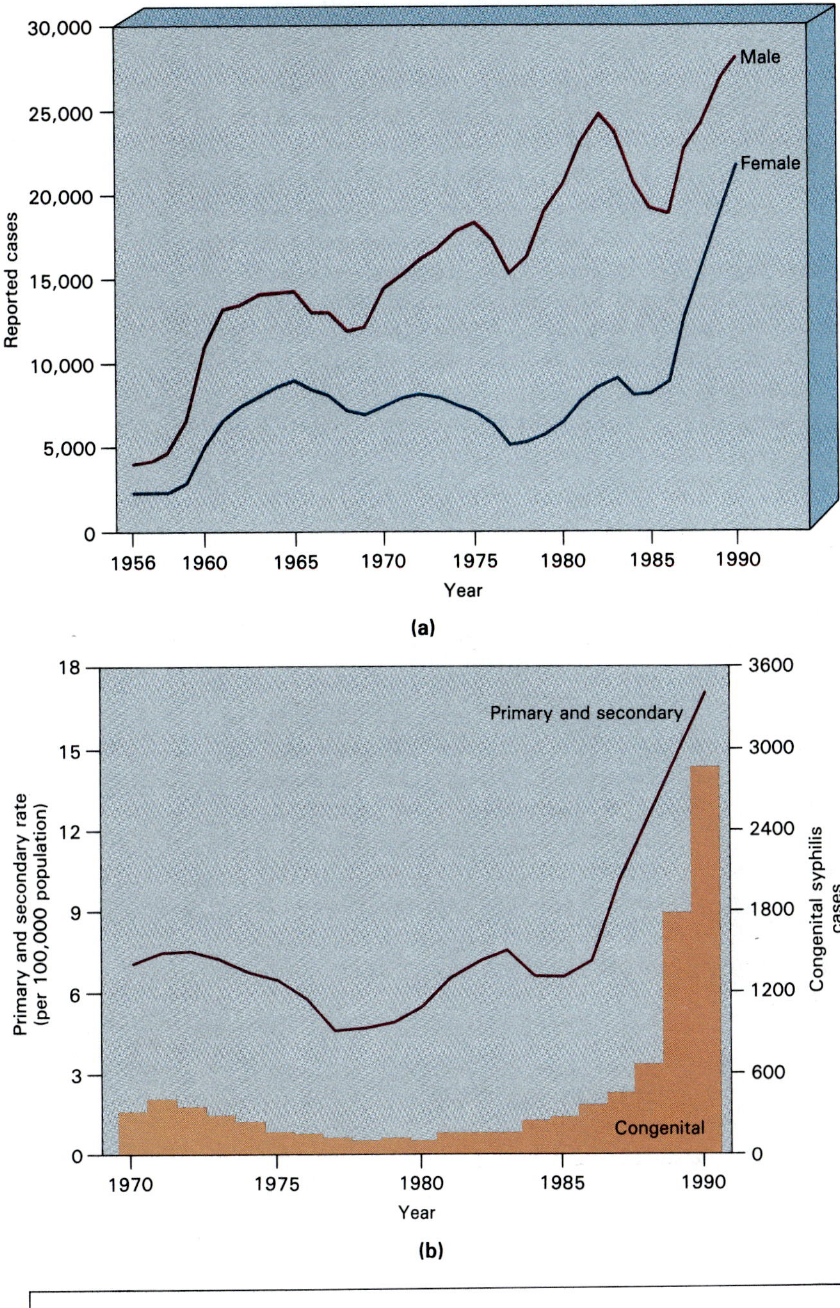

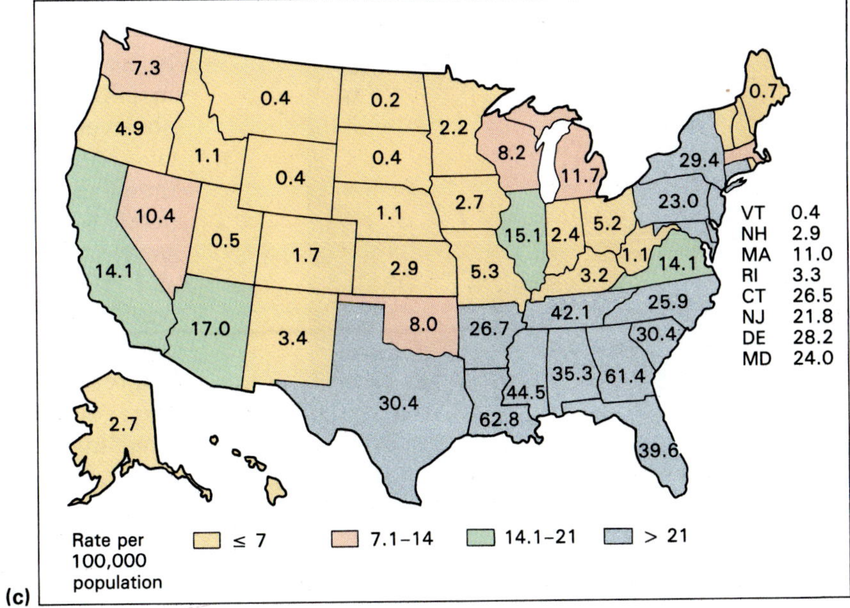

FIGURE 25.9 (a) Incidence of syphilis infections, by sex, United States, 1956–1990. (b) Incidence of syphilis among women and congenital syphilis. As more women contract syphilis, transmission to the fetus becomes more common, leading to a parallel increase in incidence of congenital syphilis. (c) Incidence of syphilis by state.

The Disease Progress of a typical case of syphilis occurs as follows:

1. Incubation stage: Over a period of 2 to 6 weeks after entering the body, the organisms multiply and spread throughout the body.
2. Primary stage: An inflammatory response at the original entry site causes formation of a **chancre** (shang'ker), a hard, painless, nondischarging lesion, about ½ inch in diameter. One or more primary chancres usually develop on the genitals (Figure 25.10) but can develop on lips or hands. In females, chancres on the cervix or another internal location sometimes escape detection. The patient often is embarrassed to seek medical attention for a lesion in a genital location and hopes that it will just "go away." And go away it does after about 4 to 6 weeks, without leaving any scarring. The patient thinks all is well, but the disease has merely entered the next stage.
3. Primary latent period: All external signs of the disease disappear, but blood tests diagnostic for syphilis are positive.
4. Secondary stage: Symptoms appear, disappear, and reappear over a period of about 5 years, during which the patient is contagious. These symptoms include a copper-colored rash, particularly on the palms of the hands and the soles of the feet, and various pustular rashes and skin eruptions. Painful, whitish mucous patches swarming with spirochetes appear on the tongue, cheeks, and gums. Kissing spreads the spirochetes to others. These lesions heal uneventfully, and the patient again thinks all is well. But the disease now has entered the next stage.
5. Secondary latent stage: Again all symptoms disappear, and blood tests can be negative. This stage can persist for life, for a highly variable period, or never occur. Symptoms can reoccur at any time during latency. In some patients syphilis does not progress beyond this stage, but in many patients it progresses to the tertiary stage.
6. Tertiary stage: Permanent damage occurs throughout various systems of the body. A wide assortment of symptoms can appear—syphilis has been called the "great imitator" because its symptoms can mimic those of so many other diseases—but most involve the cardiovascular and nervous systems. Blood vessels and heart valves are damaged. In long-standing cases, calcium deposits in heart valves can be so extensive as to be visible on a chest X-ray. Neurological damage, called **neurosyphilis,** can include thickening of the meninges, ataxia (a-tax'e-ah), or an unsteady gait or inability to walk, and paresis (par'-es-is), or paralysis and insanity. These symptoms often are due to formation of granulomatous inflammations called **gummas** (gum'ahz), as shown in Figure 25.11. Internal gummas typically destroy neural tissue, whereas external gummas destroy skin tissue. Mental illness accompanies neural damage, and in the preantibiotic era as many as half the beds in mental hospitals were occupied by patients with tertiary syphilis.

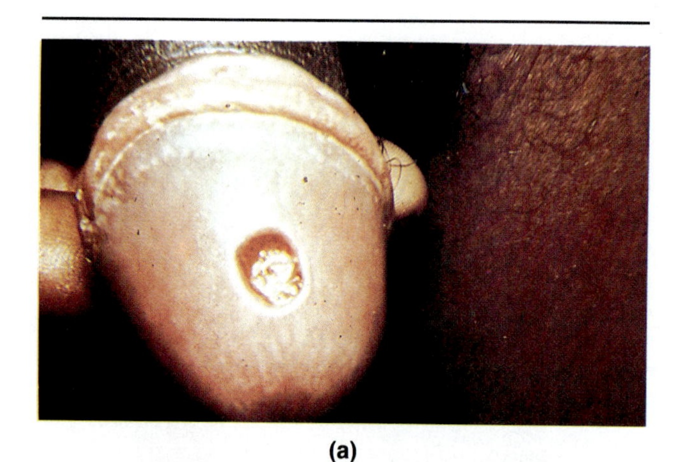

(a)

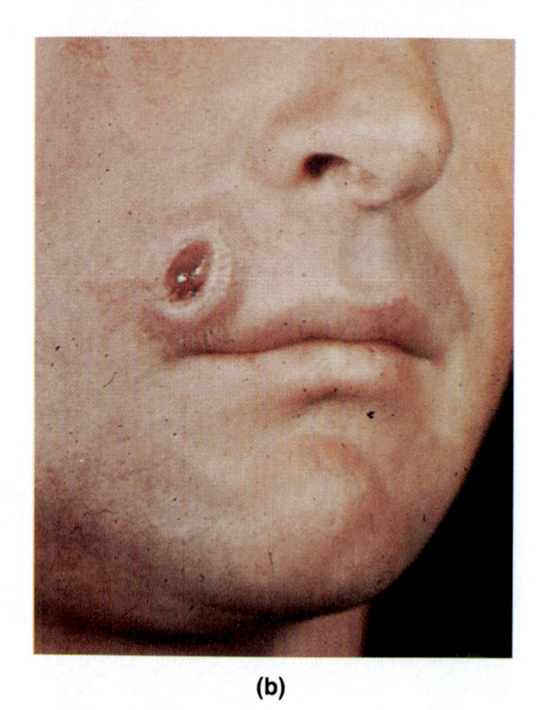

(b)

FIGURE 25.10 Primary chancres of syphilis on (a) genital site (penis) and (b) extragenital site.

Diagnosis, Treatment, and Prevention Diagnostic tests include fluorescent antibody and treponemal immobilization tests. Actively motile organisms can be observed under the dark-field microscope while specific antibody against *Treponema pallidum* is added. Immobilization of organisms by the antibody is 98 percent confirmatory for syphilis. Other blood tests such as VDRL (Venereal Disease Research Laboratory) and Wasserman tests have a high frequency of false positives, as they are based on detecting tissue damage. A severe case of influenza, a myocardial infarction, or an autoimmune disease can cause sufficient damage to elicit a false positive reaction. Therefore, screening tests such as the VDRL must always be followed up by a confirmatory test, such as the fluorescent antibody test.

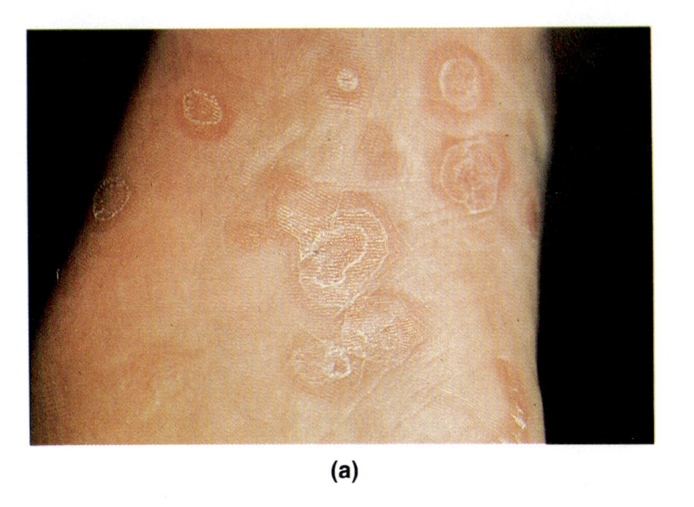

(a)

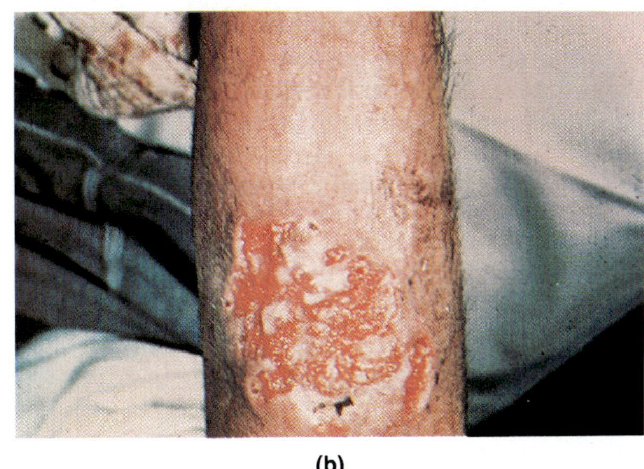

(b)

FIGURE 25.11 Signs of (a) secondary syphilis, a typical papular rash, and (b) of tertiary syphilis, a gumma.

Syphilis usually is treated with penicillin, but tetracycline and erythromycin also are effective. The longer the patient has had syphilis, the more important is continued treatment and testing to assure that organisms have been eradicated. No vaccine is available, and recovery does not confer immunity.

Congenital Syphilis **Congenital syphilis** occurs when treponemes cross the placenta from mother to baby. At birth or shortly thereafter the infant may show such signs as notched incisors, or Hutchinson's teeth (Figure 25.12), a perforated palate, saber shins

FIGURE 25.12 Signs of congenital syphilis. (a) Hutchinson's teeth, showing notched central incisors. (b) Saddle-nose, which causes a snuffling type of breathing. (c) Saber shin, displaying arching of the anterior edge of the tibia. (d) Deformation of hands.

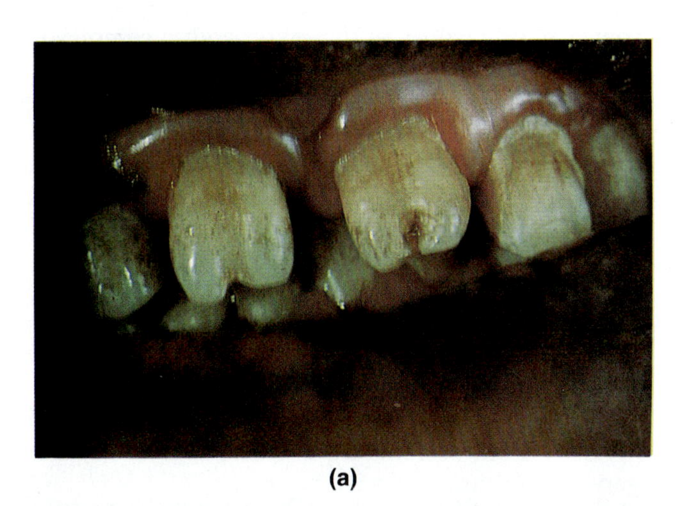

(a)

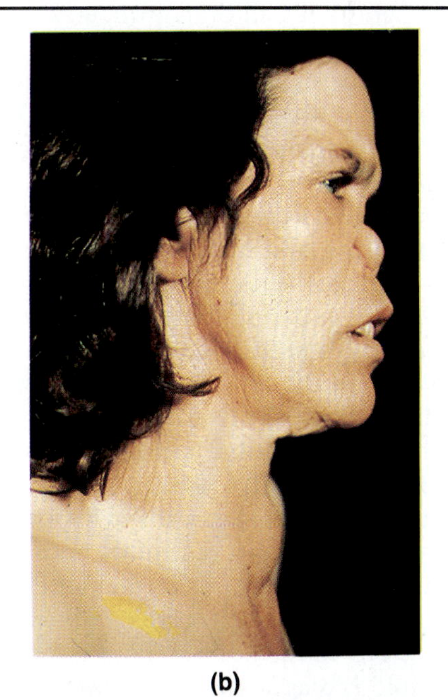

(b)

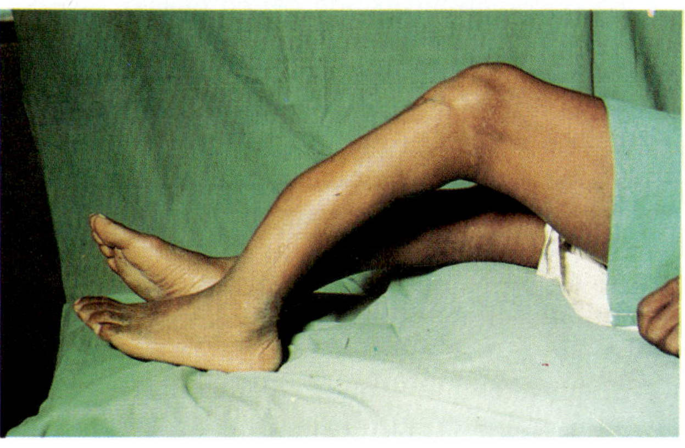

(c)

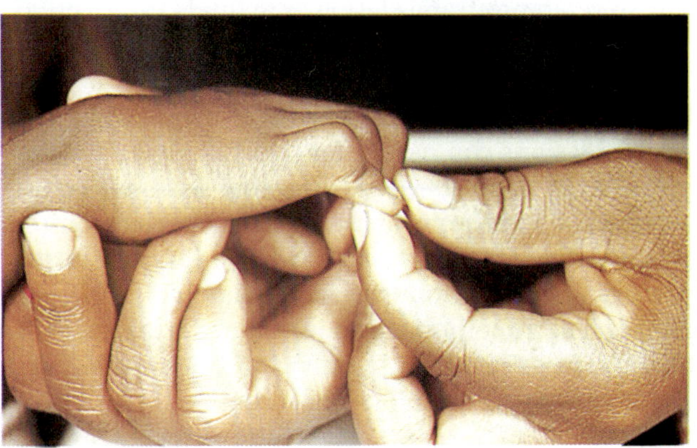

(d)

(where the shin bone projects sharply on the front of the leg), an aged-looking face with a flat, saddle-shaped nose, and a nasal discharge. Women should be tested for syphilis prior to becoming pregnant, and as part of their prenatal testing if they are already pregnant, to prevent such congenital infections.

Chancroid

Chancroid (shang'kroid), called soft chancre to distinguish it from the hard painless chancre of syphilis, is caused by *Haemophilus ducreyi.* Named for Augusto Ducrey, an Italian dermatologist, who first observed it in skin lesions in 1889, the organism is a small gram-negative rod that occurs in strands.

Relatively rare in the United States, with approximately 800 cases reported per year, chancroid is seen most frequently in underdeveloped countries of Africa, the Caribbean, and Southeast Asia. Its worldwide incidence is believed to be greater than that of either gonorrhea or syphilis. Most cases in the United States occur in immigrants and a few in military personnel. In 1982 CDC reported a sudden increase in the number of cases in southern California from an average of 29 per year to over 400 per year. Of these, 90 percent were in Hispanic men, most of whom had recently immigrated from Mexico.

Chancroid begins with the appearance of soft, painful lesions called chancres, which bleed easily, on the genitals 3 to 5 days after sexual exposure. They often occur on the labia and clitoris in females and on the penis in males. However, the infection can be present without apparent lesions, the only symptom being a burning sensation after urination. Chancres also can occur on the tongue and lips. Regardless of their location, chancres are extremely infective. Medical personnel sometimes acquire lesions on the hands merely from contact with chancres. In about one-third of the patients, chancroid spreads to the groin, where it forms enlarged masses of lymphatic tissue called buboes. These appear about 1 week after infection, swell to great size, and can break through the skin, discharging pus to the surface (Figure 25.13).

Chancroid is diagnosed by identifying the organism in scrapings from a lesion or in fluids from a bubo. Patients with chancre also often have syphilis and other STDs. Thus, a patient with a positive diagnosis for one STD should be tested for other STDs. Untreated lesions can persist for months. Often the disease resolves on its own. Infection does not confer permanent immunity, and the disease can be acquired again and again. It is treated with antibiotics such as tetracycline, erythromycin, sulfanilamide, or a combination of trimethoprim and sulfamethoxazole. With treatment lesions heal rapidly, but often leave deep scars with much tissue destruction.

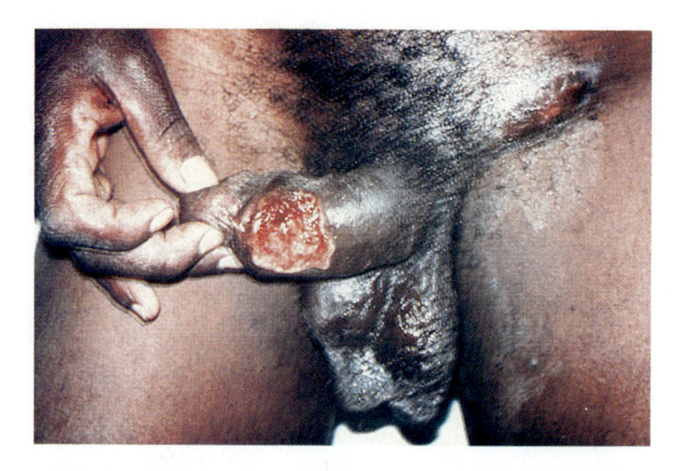

FIGURE 25.13 Lesions of chancroid on the penis, showing draining buboe in the adjacent groin area. Chancroid is caused by *Haemophilus ducreyi.*

Nongonococcal Urethritis

As the name suggests, **nongonococcal urethritis** (NGU) is a gonorrhealike sexually transmitted disease caused by other organisms. Most cases are caused by *Chlamydia trachomatis,* but some are caused by mycoplasmas. The prevalence of chlamydial infections is greater than that of any other sexually transmitted diseases and is increasing dramatically.

Chlamydial Infections *Chlamydia trachomatis* is a tiny spherical bacterium with a complex intracellular life cycle, which was described in Chapter 10. (p. 254) In addition to causing NGU, subspecies strains of *C. trachomatis* cause a wide range of disorders, including conjunctivitis and lymphogranuloma venereum (discussed later). The CDC has estimated that 3 to 5 million Americans have *Chlamydia* NGU each year. The subspecies that causes inclusion conjunctivitis in newborns also causes 30 to 50 percent of NGU cases in males, as well as 30 to 50 percent of vulvovaginitis cases and one-half of cervicitis (inflammation of the cervix) cases in women. The large numbers of humans who are infected and who have organisms in their body secretions makes transmission of chlamydial infections especially easy.

Following an incubation period of 1 to 3 weeks, symptoms of NGU similar to but milder than gonorrhea begin. A scanty watery urethral discharge is observed, especially after passage of first morning urine. Sometimes this is accompanied by tingling sensations in the penis. Many chlamydial NGU infections are asymptomatic, and fortunately most chlamydial STDs produce no significant complications or aftereffects. However, inflammation of the epididymis (the tube through which sperm pass from the testis) can lead to sterility.

Chlamydial infections are difficult to control. In-

PUBLIC HEALTH

Chlamydial Infections

Before 1983 only 200 laboratories in the United States could culture chlamydias. Newly developed diagnostic tests such as enzyme immunoassays and monoclonal antibody tests are revealing a previously unsuspected epidemic of chlamydial infections. Thus, chlamydial infection is now known to be one of the most prevalent diseases in our society. In Britain, where chlamydial infection is a reportable disease, statistics confirm similar high prevalences.

fants can become infected while passing through the birth canal of an infected mother. Silver nitrate used to prevent gonorrheal ophthalmia neonatorum does not protect against *Chlamydia,* but erythromycin protects against both. In venereal disease clinics, penicillin used to treat syphilis and gonorrhea fails to eliminate chlamydial infections. The current increases in chlamydial infections could be stopped with tetracycline and sulfa drugs if all sexual partners were treated.

Adult **inclusion conjunctivitis** can result from self-inoculation with *C. trachomatis* from genitals via fingers or towels, and is especially common in sexually active young adults. It closely resembles trachoma. Prior to the widespread use of chlorine in pools it was called "swimming pool conjunctivitis." Reinfection can occur, but whether by inadequate immune response or by infection with another of eight different infectious strains is unknown.

Each year about 75,000 infants acquire chlamydial **inclusion blennorrhea** (blen-or-e'ah), a name derived from the Greek for mucus flow. This usually benign conjunctivitis begins as a pus-filled mucous discharge 7 to 12 days after delivery and subsides with erythromycin treatment or spontaneously after a few weeks or months. If it persists it becomes indistinguishable from childhood trachoma and can lead to blindness.

Mycoplasmal Infections NGU also can be caused by *Mycoplasma hominis,* which is frequently among the normal urogenital flora, especially in women. Mycoplasmas, also called mollicutes (*mollicute,* Latin for soft-skinned) because they have no cell wall, apparently infect by their cell membranes fusing with the host cell membranes. Such infections are very common; more than half of normal adults have antibodies to *M. hominis.*

Yet another causative agent of NGU is *Ureaplasma urealyticum,* formerly called T-strain (*T* for tiny) mycoplasma. Between 1 and 2.5 million people in the United States are infected. One of the smallest bacteria known to cause human disease, its name is based on the fact that it requires a 10 percent urea medium to grow. Of patients seen in venereal disease clinics, 50 to 80 percent carry *U. urealyticum* in addition to other sexually transmitted pathogens. The organism accounts for more than one-half of all infections that make couples infertile. Low sperm counts and poor sperm mobility have been observed in males; the organisms bind tightly to sperm and can be transmitted by them to sexual partners. They are a major cause of fetal death, recurrent miscarriage, prematurity, and low birth weight—itself a leading cause of neonatal death.

Diagnosis is by culture from urethral and vaginal discharges and from placental surfaces. When both members of an affected couple are treated, pregnancy is achieved in 60 percent of cases, compared with a success rate of only 5 percent in untreated couples infected with this organism. Because of the absence of a cell wall, penicillin is not effective against mycoplasmas. Therefore, people being treated with penicillin for other STDs will not be cured of these infections. Tetracycline is most often used because it also controls *Chlamydia.* The 15 percent of *Mycoplasma* strains resistant to tetracycline can be treated with erythromycin and spectinomycin.

A common complication of NGU and gonorrhea is pelvic inflammatory disease, which can be caused by more than 20 different infectious agents. Chlamydial PID increases the risk of sterility and ectopic pregnancy, a pregnancy in which the embryo begins development outside the uterus (for example, in a fallopian tube or the peritoneal cavity). Surveys of pregnant women reveal that 11 percent have *Chlamydia* in the cervix; postpartum fever is common among infected women. Their infants can develop neonatal chlamydial pneumonia, a rarely fatal disease that accounts for 30 percent of all pneumonias in children under 6 months of age. Because the condition typically does not develop until 2 or 3 months after delivery, its connection with cervical chlamydial infection is often overlooked. Although most often associated with NGU, *M. hominis* sometimes causes PID in women and opportunistic urethritis in men. Mycoplasmas on the cervix during pregnancy probably colonize the placenta and cause spontaneous abortions, premature births, and low birth weight. They also foster ectopic pregnancies.

Lymphogranuloma Venereum

Another sexually transmitted disease, **lymphogranuloma venereum** (LGV), is common in tropical and subtropical regions. Although only about 250 cases per year occur in the United States, mostly in southeastern states, as many as 10,000 cases are treated annually in a single Ethiopian clinic. This nonreportable disease is about 20 times more common in males than in females.

The causative agent of LGV was identified in 1940 as a highly invasive strain of *Chlamydia trachomatis.* Within 7 to 12 days after contact with the organism, lesions appear at the site of infection, usually the genitals but sometimes the oral cavity. In most cases they rupture and heal without scarring. Other symptoms include fever, malaise, headache, nausea and vomiting, and a skin rash. One week to 2 months later, the organisms invade the lymphatic system and cause regional lymph nodes to become enlarged, painful, and pus-filled buboes. Victims have been known to use a razor blade to open the buboes to gain relief (Figure 25.14), but aspiration with a sterile needle is a safer treatment. Lymph node inflammation occasionally obstructs and scars lymph vessels, causing edema of genital skin and elephantiasis (massive enlargement) of the external genitalia in both males and females. Rectal infections often occur in homosexual men. In women lymph from the vagina drains toward the rectum, and lymph nodes in the walls of the rectum become chronically enlarged in 25 percent of the cases. This causes rectal blockage that usually requires surgery. Untreated cases produce a bloody, pus-filled anal discharge and can eventually lead to rectal perforation. Organisms transferred from the hands to the eyes can cause conjunctivitis, and in rare instances the disease progresses to meningitis, arthritis, and pericarditis.

Spontaneous "cures" sometimes represent latent infections. Latency can be prolonged, as is evidenced by males who infect sex partners many years after their own initial infection. The genital tracts and rectums of chronically infected, but at times asymptomatic, persons serve as reservoirs of infection.

LGV is diagnosed by finding chlamydias as inclusions stained with iodine in pus from lymph nodes. Serological tests are available, but they frequently give false positive results. Tetracyline is the drug of choice to treat LGV, but erythromycin and sulfamethoxasole also are effective. Enlarged lymph nodes can take 4 to 6 weeks to subside even after successful antibiotic therapy.

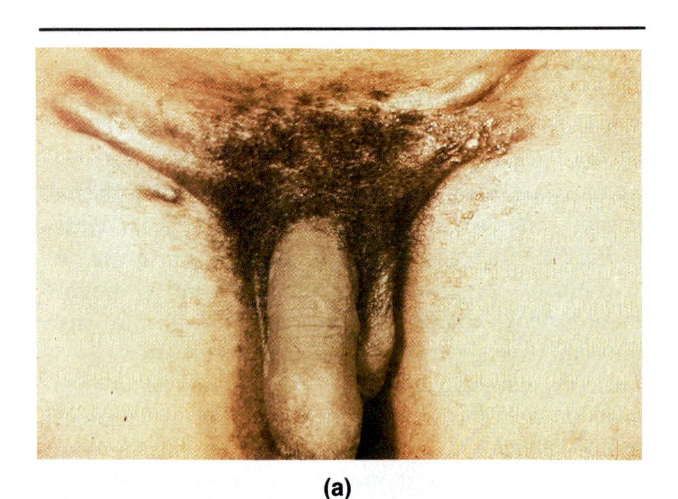

(a)

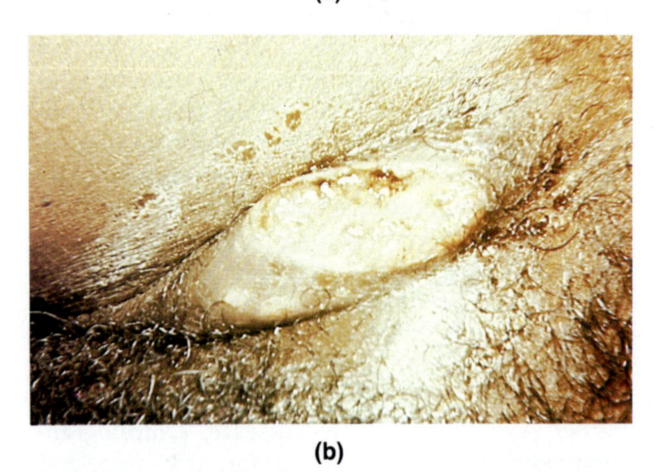

(b)

FIGURE 25.14 Bilateral buboes of lymphogranuloma venereum, caused by *Chlamydia trachomatis.* (a) early in development, (b) after reaching such size that the lesion was opened for drainage.

Granuloma Inguinale

Granuloma inguinale, or donovanosis, is caused by the small gram-negative encapsulated rod, *Calymmatobacterium granulomatis,* which is related to *Klebsiella.* The disease is uncommon in the United States, with about 50 cases reported per year, mostly among male homosexuals. It is common in India, western coastal Africa, South Pacific Islands, and some South American countries, from which it is occasionally brought to this country. The epidemiology of granuloma inguinale is not completely understood. Some cases appear to be sexually transmitted and others not. Even in genital cases, infectivity is low and many partners of infected individuals do not get it.

Granuloma inguinale appears as irregularly shaped, painless ulcers on or around the genitals 9 to 50 days after intercourse. Fever is absent. Ulcers can be spread to other body regions by contaminated fingers. As ulcers heal skin pigmentation is lost, and without treatment, tissue damage can be extensive. Diagnosis from scrapings of lesions is confirmed by finding large mononuclear cells called **Donovan bodies** (Figure 25.15). Antibiotics such as ampicillin, tetracycline, erythromycin, and gentamicin offer effective treatment.

Viral Sexually Transmitted Diseases

Herpesvirus Infections

Two closely related herpesviruses cause disease in humans. **Herpes simplex virus type 1** (HSV-1) typically causes fever blisters (cold sores), and **herpes**

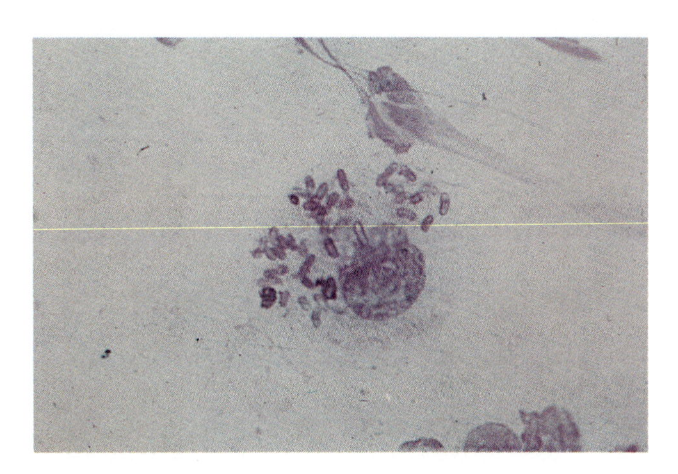

FIGURE 25.15 Scrapings of lesions of granuloma inguinale, caused by *Calymmatobacterium granulomatis*, reveal encapsulated bacterial cells inside the much larger macrophages. These bacterial cells resemble closed safety pins and are called Donovan bodies.

simplex virus type 2 (HSV-2), also sometimes called herpes hominis virus, typically causes **genital herpes.** Both HSV-1 and HSV-2 have an incubation period of 4 to 10 days, cause the same kind of lesions, and have been isolated from skin and mucous membranes of oral and genital lesions. Among oral infections 90 percent are caused by HSV-1 and 10 percent by HSV-2; among genital infections 85 percent are caused by HSV-2 and 15 percent by HSV-1. The presence of HSV-1 in the genital region and HSV-2 in the oral region is most commonly due to the practice of oral sex. Genital herpes is by far the most common and most severe of the herpes simplex viral infections.

Initial HSV-1 or HSV-2 infection can be asymptomatic, especially in children, or cause localized lesions with or without symptoms of acute infection. Most adults have antibodies to herpesviruses, but only 10 to 15 percent have experienced symptoms. In both HSV-1 and HSV-2 infections vesicles form under keratinized cells and fill with fluid from virus-damaged cells, particles of cell debris, and inflammatory cells. Vesicles are painful, but they heal completely in 2 to 3 weeks without scarring unless there is a secondary bacterial infection. Adjacent lymph nodes enlarge and are sometimes tender.

Latency (Chapter 11) is a hallmark of herpes infections. (p. 273) More than 80 percent of the adult population harbors these viruses, but only a small proportion experience recurrent infections. Within 2 weeks of an active infection, the viruses travel via sensory neurons to ganglia (Figure 25.16). Within the ganglia they replicate slowly or not at all, but they can reactivate spontaneously or be activated by fever, ultraviolet radiation, stress, hormone imbalance, menstrual bleeding, a change in the immune system, or trauma. The infection can spread to and kill cells in the adrenal glands, liver, spleen, and lungs. In fatal herpes encephalitis, soft discolored lesions appear in both gray and white matter of the brain.

After reactivation the virus moves along the nerve axon to the epithelial cells where it replicates, causing recurrent lesions. These lesions, which always recur in exactly the same place as the original infection, are smaller, shed fewer viruses, contain more inflammatory cells, and heal more rapidly than primary lesions. Successive recurrences usually become milder until they finally cease. While the virus is in a neuron, neither humoral nor cellular immunity can combat it. Once the virus reaches target epithelial cells and starts to replicate, antibodies can neutralize the viruses and T cells can eliminate virus-infected cells. These immunologic processes that make it difficult to isolate viruses from vesicular fluid in recurrent lesions also reduce their severity and duration. Recurrences can be limited to one or two episodes or can appear periodically for the life of the patient, but typically occur five to seven times. Even if recurrences cease, the viruses remain latent in ganglia, and long-dormant viruses can be reactivated by severe stress, trauma, or impaired immune function (as for example in AIDS patients).

"HSV shedders," people who shed viruses while remaining asymptomatic, pose a significant problem. As many as 1 in 200 women have been shown to shed viruses even when they have no observable lesions and in some cases no knowledge of ever having had a herpes infection. Such women pose a serious threat to any infant they might bear.

Genital Herpes Genital herpes infections usually are acquired after the onset of sexual activity because the virus is transmitted mainly by sexual contact. However, the virus can survive for short periods of time in moist areas such as hot tubs. Changes in sexual practices, especially an increase in oral sex, have increased the incidence of HSV-1 in genital lesions and herpesvirus type 2 in oral lesions. More than 20 million Americans now have genital herpes, and half a million new cases are seen each year.

In females the vesicles appear on the mucous membranes of the labia, vagina, and cervix. Ulcerations sometimes spread over the vulva and can even appear on the thighs. In males tiny vesicles appear on the penis and foreskin and are accompanied by urethritis and a watery discharge. The prostate gland and seminal vesicles also can be affected. In both sexes there is intense pain and itching at the sites of lesions and swelling of lymph nodes in the groin.

A person infected with herpesvirus is contagious any time viruses are being shed. Shedding always occurs when active lesions are present and usually

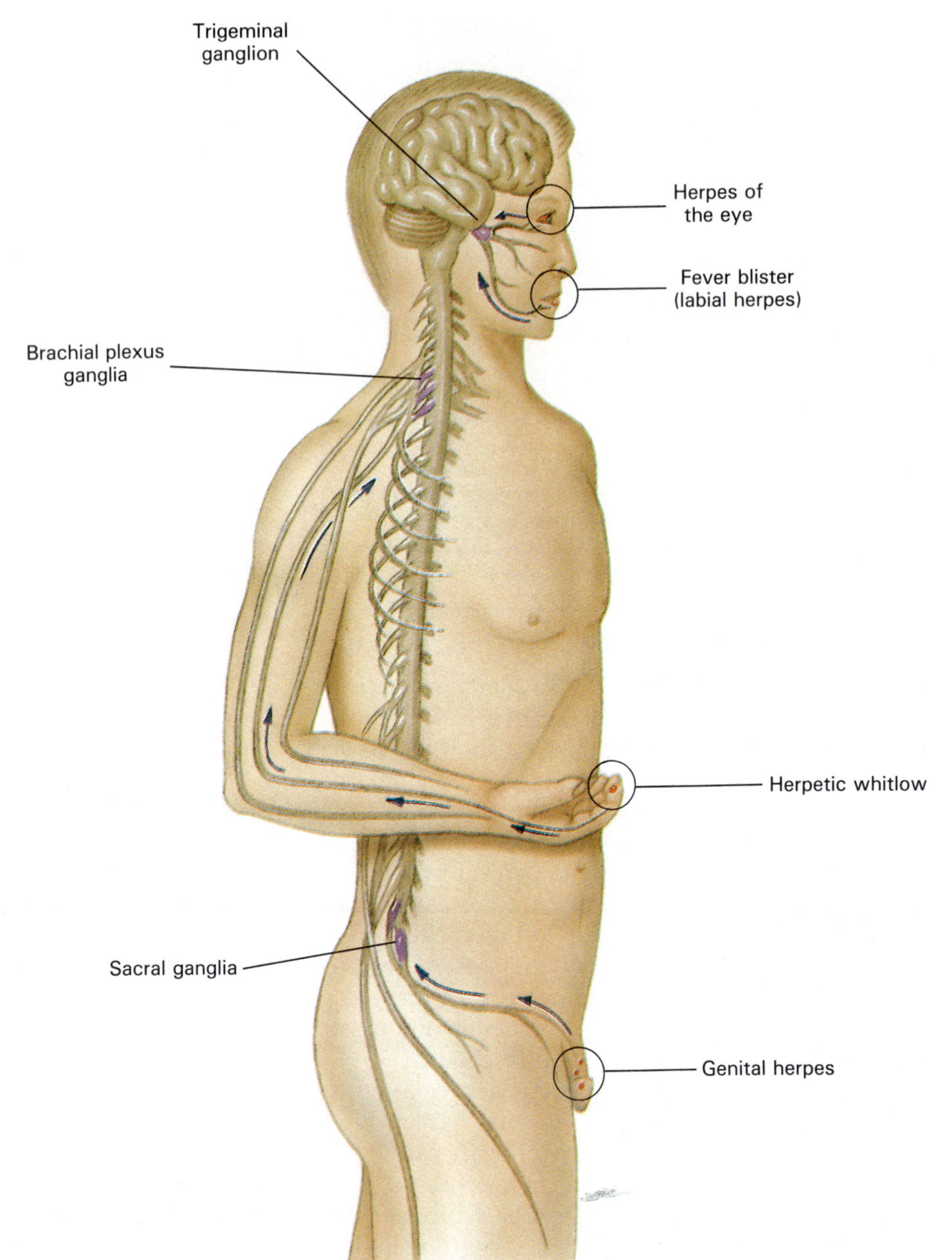

FIGURE 25.16 Herpes simplex viruses are permanent "houseguests" once acquired. After an occurrence of a lesion, they migrate to ganglia, from which they travel back to the site of the original lesion at the time of the next outbreak. An infrequent exception to this pattern sometimes occurs at the trigeminal ganglion, at which three branches of the trigeminal (fifth cranial) nerve join. Virus from a lesion on the lower lip could migrate out along the ophthalmic branch on a subsequent episode, causing the eye to become affected, or even migrate backwards into the brain, causing meningitis and brain damage. It is not clear why this sometimes happens. Fortunately, it is a very rare occurrence.

starts a few days before lesions appear. It can occur continuously even when lesions are not present. Therefore, abstaining from sexual contacts when lesions are present does not always prevent spread of the disease. In recent years promiscuous sexual practices and ignorance of or lack of concern about transmitting the disease have greatly increased the number of cases of genital herpes. This incurable disease has now become one of the most common sexually transmitted diseases.

Women infected with genital herpes may be subject to three other serious problems. First, the incidence of miscarriages among women with genital herpes is higher than that for uninfected women. Second, when infected women become pregnant, the infant must be delivered by cesarean section. Finally, infected women have an increased risk of developing a kind of cancer called cervical carcinoma. Among women with cervical carcinoma, 80 percent have antibodies to HSV-2; they have been infected and probably still harbor the viruses. Most have also had multiple sexual partners. The exact roles of HSV-2, wart viruses (discussed later), or other factors associated with sexual activity in the development of the malignancy are not yet known. Until more information is available, avoiding sexual contact with virus-infected individuals and with multiple partners would seem prudent.

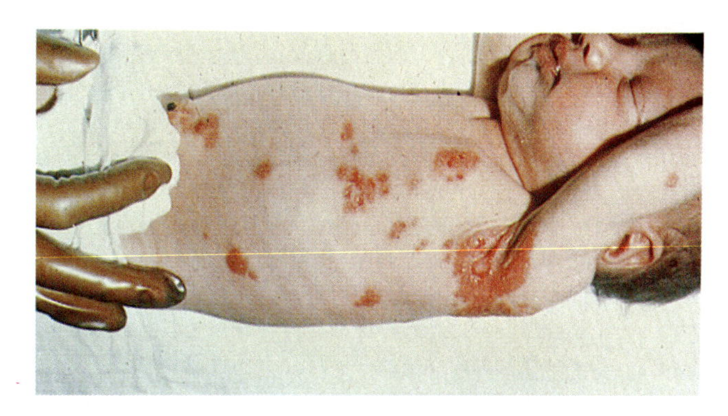

FIGURE 25.17 Neonatal herpes can be acquired when the baby passes through the birth canal of an infected mother. This could be avoided by cesarean section delivery. Sometimes lesions are so extensive that they cover almost the entire skin surface. Such children either suffer profound brain damage or do not survive.

Neonatal Herpes **Neonatal herpes** (Figure 25.17) can appear at birth or up to 3 weeks after birth. Babies most often become infected by delivery through a birth canal contaminated with HSV-2, but they can also become infected through contaminated equipment and hospital procedures. As neonates are highly susceptible to HSV infections, they should not be cared for by individuals with such infections. Infected mothers who must care for their infants must scrupulously follow sanitary procedures.

At diagnosis two-thirds of infected infants have skin vesicles; the others already have disseminated infection with neural or visceral lesions. Skin infections disseminate in 70 percent of infected infants. Infants with disseminated infections display poor appetite, vomiting, diarrhea, respiratory difficulties, and hypoactivity. Some also have neurological symptoms, jaundice, and eye disorders. Neonates with disseminated infections deteriorate rapidly and usually die within 10 days. The few that survive usually have central nervous system and eye damage. Occasionally infants have only a few vesicles, but latent viruses may later cause significant damage. Early diagnosis and treatment of neonatal herpesvirus infections are essential for survival and to reduce the likelihood of neurological damage.

Other Herpes Simplex Infections A variety of manifestations of herpesvirus infections have been seen. Most are thought to be caused by HSV-1, but HSV-2 may be responsible for some. They include gingivostomatitis, herpes labialis, keratoconjunctivitis, herpes meningoencephalitis, herpes pneumonia, eczema herpeticum, traumatic herpes, herpes gladiatorium, and whitlows. HSV-1 is spread by contaminated secretions and through contact with lesions and usually is acquired in childhood from relatives, nurses, or other children. The incidence of the virus is especially high within families, hospitals, and other institutions. Most humans are infected with HSV-1 in the first 18 months of life. Many of these infections are inapparent, and the first apparent lesions result from reactivations.

Gingivostomatitis, lesions of the mucous membranes of the mouth, is most common in children 1 to 3 years old. After an incubation period of 2 to 20 days, small vesicles appear around the mouth over a 7-day period. Each vesicle accumulates fluid, scabs over, and heals in 2 to 3 weeks. Recurrent lesions typically occur in the form of **herpes labialis,** or fever blisters on the lips. If the eyes instead of the mouth are the site of initial infection, vesicles appear on the cornea and eyelids, causing **keratoconjunctivitis.**

The most serious manifestation of HSV infection is herpes meningoencephalitis, which can follow a generalized herpes infection in a newborn infant, child, or adult. How the virus crosses the blood-brain barrier to enter the central nervous system has recently been discovered. The virus remains latent in the ganglion of the trigeminal nerve, which supplies parts of the face and mouth, until some unknown factor reactivates it. Then, instead of moving outward to epithelial sites such as the tongue or lip, it ascends along the nerve into the brain. The disease has a rapid onset, with fever, headache, meningeal irritation, convulsions, and altered reflexes. In the middle-aged and elderly, meningoencephalitis can appear without any preceding symptoms. The patient experiences increasing confusion, hallucinations, and sometimes seizures. Most patients die in 8 to 10 days; survivors usually have permanent neurological damage.

Herpes pneumonia is rare. It usually is seen only in burn patients, alcoholics, and patients with AIDS or other immunodeficiencies. **Eczema herpeticum** is a generalized eruption caused by entry of the virus through the skin. **Traumatic herpes** occurs when the virus enters traumatized skin in the area of a burn or other injury. **Herpes gladiatorium** occurs in skin injuries of wrestlers. A **whitlow** (Figure 25.18) is a herpetic lesion on a finger that can result from exposure to oral, ocular, and probably genital herpes lesions. Dental technicians, nurses, and other medical personnel should therefore use rubber gloves when treating patients with herpes lesions. A person with a whitlow can spread the infection to the mouth, eyes, and genital areas or to other persons.

Diagnosis, Treatment, and Prognosis Herpesviruses are most easily isolated from vesicular fluid and cells from the base of a lesion and can be grown in the laboratory in a variety of cell types. The cytopathic effect of the viruses appears in the culture quickly. The time to obtain a diagnosis has been greatly decreased by rapid immunological tests. Speedy diag-

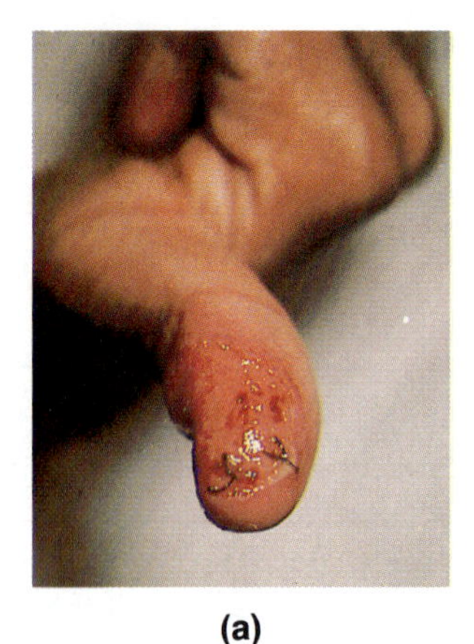

(a)

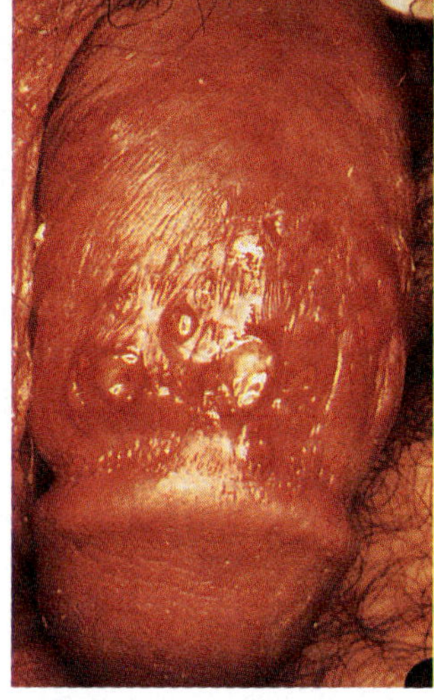

(b)

FIGURE 25.18 Herpes infections include (a) whitlow, a very painful infection of the finger, which can be spread to other areas of the body, for example, by rubbing one's eyes. Health-care workers should be sure to protect themselves by wearing latex gloves. (b) Penile herpes lesions.

nosis is especially important for women in labor or near delivery, as a cesarean delivery can prevent exposure of the infant to the virus.

In recent years several drugs have been used with varying degrees of success in treating HSV-1 and HSV-2 infections. Trifluorothymidine is effective in treating ocular herpes. Although acyclovir fails to prevent recurrences reliably, it does prevent the spread of lesions, decrease viral shedding, and shorten healing time. Best results are obtained with acyclovir when it is used at the beginning of a primary infection. The use of acyclovir, Ara-A, and vidarabine in various combinations early in the infection has increased survival rates in herpes meningoencephalitis and neonatal herpes. The prognosis for alleviating genital herpes symptoms is reasonably good. Most patients have few if any recurrences if they take acyclovir daily, but may have a recurrence if they stop taking it. No treatment can eradicate latent viruses.

Immunity and Prevention Although immunological knowledge of herpesviruses is sufficient to make a vaccine, such a vaccine is not yet available. Even if a vaccine were available and administered to all infants and individuals lacking HSV antibodies, it would take years to eradicate the disease. Large numbers of people already harbor latent viruses, and a vaccine to inactivate the viral genetic information responsible for latency would be needed. A vaccine that elicits antibodies like those from natural infections would probably not be effective; such antibodies apparently have no effect on latent viruses and cannot alone prevent recurrences.

The best available means of preventing HSV infections is to avoid contact with individuals with HSV-1 or HSV-2 lesions. If all infected individuals were to refrain from sexual activity, especially when they have active lesions, some new cases of genital herpes could be prevented. If pregnant women would advise their obstetricians of herpes infections, their infants could be protected from exposure during delivery. Even these precautions will not prevent the spread of genital herpes. Most infected individuals shed viruses a few days before lesions appear, and some shed viruses continuously without having any lesions.

CLOSE-UP

Cortisone Suppression of the Immune System

Cortisone, which is used to treat many inflammatory lesions, should not be used to treat herpes lesions, especially ocular lesions. Because cortisone suppresses the immune system, it allows viral multiplication to increase and cause more extensive cellular injury. The damage can be so severe as to cause perforation of the cornea.

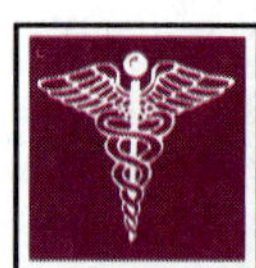

PUBLIC HEALTH

Herpes Singles Clubs

Some herpes victims join "herpes singles clubs" in hope of being able to enjoy sexual activity with other members without inflicting the disease on uninfected persons. Although limiting one's choice of partners in this way may prevent the spread of herpes infections to "outsiders," it can also give herpes sufferers a false sense of security. As several different strains of both HSV-1 and HSV-2 exist, people can infect each other with different strains of HSV. Each new strain can cause a new primary infection and very painful lesions.

Genital Warts

Condylomas (kon-dil-o'mahz), or **genital warts,** caused by the human papillomavirus, most often occur in the sexually promiscuous, young adult population. The incidence of genital warts has increased rapidly in recent years to the extent that such warts are now among the most prevalent sexually transmitted diseases. Two-thirds of the sexual partners of infected individuals also develop warts. The warts can be papillary or flat. In males the warts appear on the penis, anus, and perineum; in females they are found on the vagina, cervix, perineum, and anus.

As with dermal warts (described in Chapter 20), genital warts cause irritation and sometimes intense itching, and they can persist or regress spontaneously. ∞ (p. 556) Genital warts often become infected with bacteria, and those that persist for many years can transform into malignant growths. Some victims suffer psychological damage from the presence of warts. Warts temporarily increase in number and size during pregnancy but decrease after delivery. Infants can be infected during delivery. The number of warts increases in immunosuppressed patients and in those with AIDS and other immunologic deficiencies. (Treatment of warts was discussed in Chapter 20. ∞ (p. 557))

A special hazard of genital warts is their epidemiologic association with cervical carcinoma. Lesions previously diagnosed as cervical dysplasia (dis-pla'se-ah), or abnormal cell growth, are now more often properly recognized as condylomas even though both can be found side by side on the cervix. Distinguishing these conditions from cervical carcinoma is even more difficult. Examining lesions for viral capsid antigens and viral nucleic acids can be helpful, and such examinations suggest that two strains of human papillomaviruses, HPV-16 and HPV-18, are most likely to be found in cervical carcinomas. The extent to which they contribute to such malignancies is not clear, but it has been reported that the virus has been found in 90 percent of patients with cervical carcinoma.

Laryngeal Papillomas **Laryngeal papillomas** (pap-il-lo'maz) are benign growths that can be dangerous if they block the airway. Hoarseness, voice changes, and respiratory distress occur when the airway becomes obstructed. Children are more likely to have laryngeal papillomas than adults. Surgical removal, sometimes every 2 to 4 weeks, is the only treatment for these obstructive growths. Also, there is danger of spreading the virus to the lungs during surgery. Laryngeal papillomas are usually caused by HPV-6 and HPV-11, and they are thought to infect infants during birth to women with active genital warts.

Cytomegalovirus Infections

The **cytomegaloviruses** (CMVs) constitute a widespread and diverse group of herpesviruses. In general, each strain of CMV is capable of infecting a single species. The majority of human CMV infections occur in older children and adults and go unnoticed because they do not produce clinically apparent symptoms. An estimated 80 percent of U.S. adults carry the virus. When there are symptoms, they include malaise, myalgia, protracted fever, abnormal liver function, and lymph node inflammation without swelling. Symptoms are more severe in patients with AIDS and other immunodeficiencies.

Initially the virus can be recovered from the oropharynx; viremia with many virus-infected polymorphonuclear leukocytes can last for months. The viruses replicate and have low pathogenicity but are excreted intermittently over many months, during which time they can infect others. Viruses are shed in all body fluids—saliva, blood, semen, breast milk—but they are found most often and in the largest quantities in urine even a year or more after infection. In a symptomatic primary attack of CMV, cell-mediated immunity is depressed, and the ratio of T helper to T suppressor cells is reversed. However, the immune system returns to normal during convalescence. In subclinical cases cell-mediated immunity remains active. Large quantities of long-lasting antibodies are produced in response to CMV infection, but they do not prevent viral shedding. The virus is usually spread through long, close contact with children who are shedding the virus, but it can be spread by blood transfusions, organ transplants, and sexual intercourse.

Fetal and Infant CMV Infections In fetuses and infants, CMV infections can be life-threatening because the virus disseminates widely to various organs. Fetuses become infected by viruses crossing the placenta from infected mothers. Unlike rubella infection, maternal CMV infections are rarely detected, so the risk to the fetus is unknown. Maternal antibodies also can cross the placenta and inactivate small quantities of virus. Maternal hormones suppress CMV, but their effect diminishes as the pregnancy progresses. Both fetuses and the neonates are dependent on maternal defenses against CMV because their own immune systems are too immature to mount a successful defense. Tests for maternal antibody levels are available. Young children are the most likely source of infection to nonimmune women.

In severe CMV infections where the fetus has been infected with large numbers of viruses (about 4000 cases per year in the United States), intrauterine growth retardation and severe brain damage can occur

FIGURE 25.19 Baby with birth defects due to congenitally acquired cytomegalovirus. The virus is sometimes also called salivary gland virus and can cause grave damage to persons having impaired immune systems.

(Figure 25.19). Many babies have CMV-infected cells in the inner ear and hearing loss due to nerve damage. Some have jaundice with liver damage, and others have impaired vision. Less severe infections (another 4500–6000 per year) cause damage to certain brain areas and mild central nervous system disorders with or without damage to hearing or sight. Mortality can be as high as 30 percent.

CMV infections contracted after birth generally cause fewer permanent defects than those contracted before birth. However, these infections can cause severe illness. Among babies infected with CMV, 5 percent have typical generalized **cytomegalic inclusion disease** (CID), another 5 percent have atypical, less generalized infections, and the rest have subclinical but chronic infections. Many infants with CID infections have significant mental or sensory disorders. These include an abnormally small brain accompanied by intellectual deficits and inflammation of the eyes with impaired vision. In subclinical cases, the prognosis is better, but 10 percent will have deafness or other sensory problems. Some will experience intellectual and behavioral disorders later. Babies who acquire CMV infections from blood transfusions have a gray pallor and symptoms like those of CID. Pneumonia, respiratory deterioration, and death may follow because the infants receiving transfusions usually are premature and debilitated before transfusions are given. Moreover, CMV infections sometimes are not diagnosed in the presence of many other disorders. Finally, CMV sometimes causes pneumonia in infants 1 to 6 months of age in the presence of other infections, such as *Chlamydia trachomatis* and *Ureaplasma urealyticum*.

Disseminated CMV In severe disseminated disease the virus can be present in many organs. When the kidneys become infected immune complexes are deposited in the glomeruli, but renal dysfunction is rare except in renal transplants. Subclinical hepatitis occurs in both children and adults. Lung infections are common in infants and immunosuppressed adults. Brain involvement is rare except in fetuses.

CMV is most virulent when present as a primary infection in immunosuppressed patients. Between 1 and 4 months after an organ transplant the patient develops symptoms like those of infectious mononucleosis, with prolonged fever and enlargement of the liver and spleen. Pneumonia is frequent and severe in bone marrow transplant patients, and mortality can be as high as 40 percent. Hepatitis is usually reversible, but damage to the eyes is likely to be progressive and irreversible.

Diagnosis, Treatment, and Prevention Definitive diagnosis of CMV infections requires identification of the virus from clinical specimens, a procedure that can take up to 6 weeks. New, faster techniques use monoclonal antibodies to detect viral antigens. It is anticipated that nucleic acid probes, which make use of the base sequences unique to CMV nucleic acids, will allow identification of the virus in clinical specimens in less than 24 hours.

No effective treatment for CMV infections in infants is available. The prognosis is poor when infections occur in fetuses because the damage is done before a diagnosis is made in the neonate. Interferon and hyperimmune gamma globulin given before and after organ transplantation reduce the incidence and severity of CMV infections in transplant patients. When blood transfusions are required donor and recipient should be matched for CMV antigens to minimize introduction of CMV to which the recipient is not immune. Also, donors serologically negative for CMV should be used with infants, especially premature infants, to avoid the possibility of transmitting

CLOSE-UP

CMV and Cancer

Because CMV stimulates cellular RNA and DNA synthesis, it also may play a role in the development of cancer. Antigens and other CMV components have been found in human tumors including prostate and cervical cancers and Kaposi's sarcoma. Whether CMV contributes to the cancer or is merely a passenger in the malignant cells is unknown.

CMV through a transfusion. No effective vaccine exists.

The agents and characteristics of the diseases discussed in this chapter are summarized in Table 25.2.

TABLE 25.2 Summary of urogenital and sexually transmitted diseases

Disease	Agent(s)	Characteristics
Bacterial urogenital diseases		
Urinary tract infections	*Escherichia coli, Proteus mirabilis,* and other bacteria	Dysuria, sometimes leads to chronic cystitis, infections often ascend or descend in urinary tract
Prostatitis	*E. coli* and other bacteria	Dysuria, urgency, low fever, back pain, can cause infertility
Pyelonephritis	*E. coli* and other bacteria, sometimes *Candida*	Inflammation of pelvis of kidney, often caused by urinary tract blockage; dysuria, nocturia, sometimes fever
Glomerulonephritis	Streptococcal or viral infections from other sites	Deposition of immune complexes causes inflammation of glomeruli; can cause permanent kidney damage
Leptospirosis	*Leptospira interrogans*	Fever, nonspecific symptoms; can lead to Weil's disease with jaundice and liver damage
Bacterial vaginitis	*Gardnerella vaginalis* with anaerobes	Frothy, fishy-smelling discharge, pain and inflammation
Toxic shock syndrome	*Staphylococcus aureus*	Toxins reach blood and cause fever, rash, and shock that can lead to death
Parasitic urogenital disease		
Trichomoniasis	*Trichomonas vaginalis*	Intense itching, copious white discharge
Bacterial sexually transmitted diseases		
Gonorrhea	*Neisseria gonorrhoeae*	Infectious organisms usually have pili, release endotoxin that damages mucosa. Pus-filled discharge, can cause PID and infect other systems
Syphilis	*Treponema pallidum*	Chancre develops in primary stage, mucous membrane lesions and rash occur in secondary stage, permanent cardiovascular and neurologic damage often occur in tertiary stage
Chancroid	*Haemophilus ducreyi*	Painful, bleeding lesions on genitals, often enlarged lymphatic buboes
Nongonococcal urethritis	*Chlamydia trachomatis* and mycoplasmas	Scanty watery urethral discharge, inflammation, sometimes sterility; can cause fetal death and neonatal infections
Lymphogranuloma venereum	*Chlamydia trachomatis*	Genital lesions, fever, malaise, headache, nausea, vomiting, skin rash; lymph nodes become pus-filled buboes
Granuloma inguinale	*Calymmatobacterium granulomatis*	Painful ulcers on genitals and other sites, loss of skin pigmentation as ulcers heal
Viral sexually transmitted diseases		
Herpes simplex infections	Herpes simplex viruses	Fever blisters usually caused by HSV-1, genital herpes usually caused by HSV-2 (both are latent viruses); recurrent painful vesicular lesions, neonatal herpes and a variety of other manifestations
Genital warts	Human papillomaviruses	Warts on genitals, vagina, and cervix, irritation and sometimes intense itching; may contribute to cervical carcinoma
Cytomegalovirus infections	Cytomegaloviruses	Often asymptomatic but severe in fetuses, neonates, and immunodeficient patients; malaise, myalgia, fever, inflamed lymph nodes, neural damage and death in fetuses and neonates

ESSAY

Homo sapiens and *Treponema pallidum*—An Evolutionary Partnership

"[Infectious disease] is a conflict between man and his parasites which, in a constant environment, would tend to result in a virtual equilibrium, a climax state, in which both species would survive indefinitely. Man, however, lives in an environment constantly being changed by his own activities, and few of his diseases have attained such an equilibrium."
—Macfarlane Burnet and David White, 1972

Historically, four different diseases—syphilis, pinta, yaws, and bejel—have been caused by the same pathogen, *Treponema pallidum*. Studies of the DNA from organisms isolated from all four diseases show the organisms to be extremely similar. Only recently has the causative agent of pinta evolved significantly from *Treponema pallidum;* it has now been given a separate species name, T. *carateum*. Nevertheless the organisms cause different diseases. Evidently they have evolved in response to changing lifestyles of humans. Hackett (1967) suggests the following explanation for that evolution.

The first pathogen infected some wild animal and occasionally infected people as a zoonosis. As such, it caused the set of symptoms known as *pinta*, a mild skin disease that does not reach deeper tissues. Its red, slate blue, and white lesions (Figure 25.20) are disfiguring but not life-threatening. Found in the tropics, it is passed from person to person through direct skin contact. Warm skin, moist with perspiration, provides an ideal environment for the treponeme.

Around 10,000 B.C., probably in Africa, the pathogen mutated and gave rise to *yaws*, a disease that thrived in the somewhat more densely settled human villages. This disease still has its highest incidence in Africa. Deep oozing skin lesions, which can penetrate underlying tissues and erode bone (Figure 25.21) but do not reach the viscera, are typical of yaws. Most victims are children between 4 and 10 years of age.

By 7000 B.C. people carrying the treponeme moved northward. The climate was cooler and drier, required more clothing, and made transmission via skin contact more difficult. The pathogen survived best in warm, moist areas such as the mouth, groin, and armpits. Transfer to a new host was most likely via oral contact such as kissing or by sharing bits of food or dishes. Today the disease is known as nonvenereal syphilis, or *bejel*. The strain responsible for bejel is still more invasive; it causes granulomatous lesions like those of syphilis and attacks tissues of the skeletal, cardiovascular, and nervous systems. It is often transferred from one household to another by young children, because older ones are generally past the infectious stage.

Bejel is almost always acquired prior to puberty and has no relationship whatever to sexual activity. Bejel is never transmitted congenitally, presumably because the mother's infection becomes nontransmissible before she is old enough to become pregnant. Bejel epidemics occurred during the fifteenth century in Africa and in the seventeenth and eighteenth centuries in Scotland and Norway. Today the worldwide incidence of bejel is higher than that of venereal syphilis.

As standards of living improved and people covered more and more of their bodies, the surest means of transmitting the treponeme was via sexual activity and the warm, moist reproductive tracts. The pathogen again evolved in accordance with human lifestyle to cause *venereal syphilis*, in which the primary chancre appears to be analogous to the yaws lesions. The disease is more severe than bejel because it penetrates organs to a greater and more devastating extent.

There is further evidence for the evolution of treponemes with human culture. First, none of the four diseases can be found in the same geographic area as any other. Second, infection with yaws prevents concur-

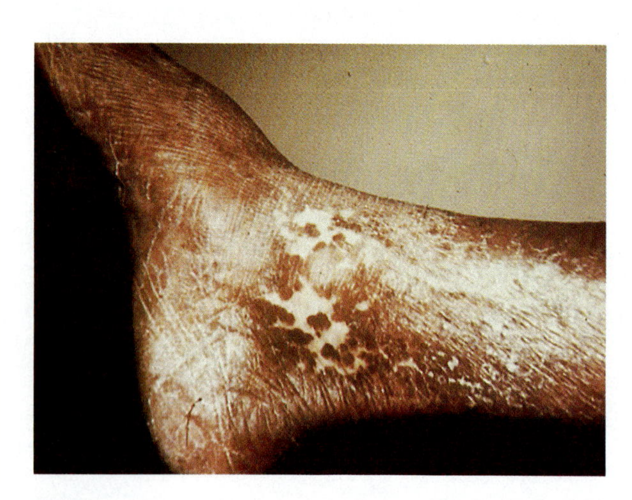

FIGURE 25.20 The rash of pinta.

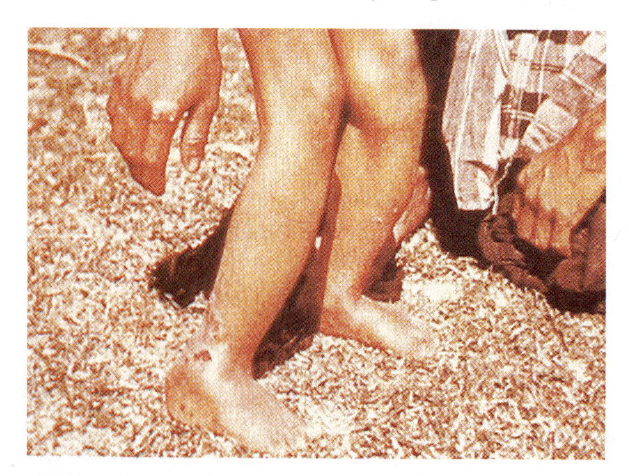

FIGURE 25.21 Bone damage caused by yaws. Compare the very similar saber shin produced by syphilis (Figure 25.12).

rent syphilis. Third, people with yaws who move to cooler climates lose their yaws symptoms and develop those of bejel instead. Evidently, the treponeme is extremely adaptive to its environment. Indeed, syphilis itself seems to have evolved considerably in the past 5 centuries. In the earliest accounts that we have, dating from the 1490s, the disease is described as killing its victims in a matter of weeks. Today it is far less virulent: the incubation period alone is longer than that, and people sometimes survive with the disease for decades.

If the theory of parallel evolution of pathogen and humans is correct, blaming sailors who accompanied Columbus for bringing the organism back from the New World, as historians often did until recently, is a mistake. The fact that the disease appeared in Europe at about the same time as Columbus's voyages is probably just a coincidence. Instead, the treponeme most likely arrived in Europe with slaves from Africa before the discovery of America, eventually evolving to the point where it could cause venereal syphilis. Evidence for this hypothesis has been found in Native-American burial grounds in New England. The remains of individuals who died before the arrival of settlers from the Old World show no trace of syphilis. Signs of syphilis appear first in the skeletons of young native women dating from the period of early European settlement.

CHAPTER SUMMARY

The agents and characteristics of the diseases discussed in this chapter are summarized in Table 25.2. Information in that table is not repeated in this summary.

RELATED KEY TERMS

UROGENITAL DISEASES USUALLY NOT TRANSMITTED SEXUALLY

Bacterial Urogenital Diseases

- **Urinary tract infections,** which are exceedingly common, can ascend or descend to spread through the entire urogenital system. They are diagnosed from urine cultures, treated with various antibiotics, and can be prevented by good personal hygiene and complete emptying of the bladder. (Related key terms: **urethritis**, **cystitis**, **urethrocystitis**)
- **Prostatitis** may result from the spread of a urinary tract infection and usually can be treated with antibiotics. (Related key term: **dysuria**)
- **Pyelonephritis** also may result from the spread of a urinary tract infection; it is somewhat difficult to treat because renal failure allows drugs to accumulate to toxic levels. (Related key term: **nocturia**)
- **Glomerulonephritis** usually results from throat and other infections in which immune complexes deposit in glomeruli. Treating other infections promptly minimizes the risk of glomerulonephritis.
- **Leptospirosis** is acquired through contact with contaminated urine, is diagnosed by microscopic examination of blood. It can be treated with antibiotics and prevented by vaccinating pets, which are usually responsible for transmitting the disease to humans.
- Bacterial and other forms of **vaginitis** occur when normal vaginal flora are disturbed. *Gardnerella* vaginitis is diagnosed by finding "clue cells." It is most often treated with metronidazole to eradicate anaerobes and allow restoration of normal flora. (Related key term: **balantitis**)
- **Toxic shock syndrome** most often arises from the use of high-absorbency tampons. It should be treated promptly with nafcillin and can be prevented by avoiding use of such tampons. (Related key term: **exotoxin C**)

Parasitic Urogenital Diseases

- **Trichomoniasis** usually is transmitted sexually, is diagnosed from smears of secretions, and is treated with metronidazole.

SEXUALLY TRANSMITTED DISEASES

Bacterial Sexually Transmitted Diseases

- **Gonorrhea** is diagnosed by finding organisms in cultures and is treated with penicillin. With penicillin-resistant organisms, another antibiotic is used. (Related key terms: **pelvic inflammatory disease**)
- **Syphilis** is diagnosed by immunological tests and is treated with tetracycline or erythromycin. (Related key terms: **chancre**, **neurosyphilis**, **gumma**, **congenital syphilis**)
- **Chancroid** occurs mainly in underdeveloped countries. It is diagnosed by finding organisms from lesions or buboes and is treated with tetracycline and other antibiotics. (Related key terms: **inclusion conjunctivitis**, **inclusion blennorrhea**)

- **Nongonococcal urethritis** incidence is increasing dramatically. It can be diagnosed by culturing samples from discharges and treated with erythromycin, tetracycline, or other antibiotics, depending on the susceptibility of the causative agent.
- **Lymphogranuloma venereum** occurs mainly in tropical and subtropical regions. It is diagnosed by finding chlamydias as inclusions in pus and is treated with tetracycline or another antibiotic.
- **Granuloma inguinale** is common in India and several other countries; it is diagnosed by finding **Donovan bodies** in scrapings from lesions and can be treated with a variety of antibiotics.
- All the preceding diseases are usually transmitted by sexual contact and can be prevented by avoiding such contacts; no vaccines are available.

Viral Sexually Transmitted Diseases

- **Genital herpes** is the most severe of herpes simplex virus infections. It is diagnosed by finding viruses in vesicular fluids or immunologic tests and can be treated, but not cured, with acyclovir and other antiviral agents.
- **Genital warts** must be distinguished from cervical dysplasia and carcinoma; treatment of warts was discussed in Chapter 20.
- **Cytomegalovirus** infections are of greatest significance in fetuses, newborns, and the immunosuppressed. Rapid diagnosis can be made by monoclonal antibody tests, but no effective treatment is available.

RELATED KEY TERMS

latency **neonatal herpes**
gingivostomatitis
herpes labialis
keratoconjuctivitis
herpes pneumonia
eczema herpeticum
traumatic herpes
herpes gladiatorium **whitlow**
condylomas
laryngeal papillomas
cytomegalic inclusion disease

QUESTIONS FOR REVIEW

A.

1. What are the causative agents of urinary tract infections, and what populations are most susceptible?
2. What is meant by "ascending" and "descending" in urinary tract infections?
3. How are urinary tract infections diagnosed, treated, and prevented?
4. What is prostatitis, and how does it arise?
5. How do glomerulonephritis and pyelonephritis differ in cause, effect, diagnosis, treatment, and prevention?
6. What causes leptospirosis, and how do humans acquire infections?
7. How is leptospirosis diagnosed, treated, and prevented?
8. What kinds of organisms cause vaginitis, and which bacterium is most often responsible?
9. How is vaginitis acquired, diagnosed, and treated?
10. To what was a recent sudden increase in cases of toxic shock syndrome attributed, and how could this have been prevented?

B.

11. What causes trichomoniasis, and how is it acquired?
12. How can trichomoniasis be diagnosed, treated, and prevented?

C.

13. How can sexually transmitted diseases be prevented?
14. What organism causes gonorrhea, and how does it produce its effects?
15. What complications might follow gonorrhea?
16. How can gonorrhea arise in infants and children?
17. How is gonorrhea diagnosed and treated?
18. What organism causes syphilis, and how is it diagnosed and treated?
19. What are the stages of syphilis, and what occurs at each stage?
20. What is chancroid, and what is its cause?
21. How is chancroid diagnosed and treated?
22. What causes nongonococcal urethritis, and what complications arise from it?
23. How is NGU treated, and how can it be prevented?
24. How do lymphogranuloma venereum and granuloma inguinale differ in cause, effect, diagnosis, and treatment?

D.

25. What are the main characteristics of genital herpes, and how do other herpes simplex virus infections (e.g., cold sores) differ from genital herpes?
26. How is genital herpes diagnosed, treated, and prevented?
27. What are the main consequences of genital warts?
28. In what populations are cytomegalovirus infections most likely?
29. Why are fetal and neonatal cytomegalovirus infections of greater concern than those in normal adults?

PROBLEMS FOR INVESTIGATION

1. A college microbiology professor reviewing this chapter remarked that his students were "very sexually active, but very ignorant of sexually transmitted diseases." How true is this statement of your campus? What do you think is the best way to combat ignorance about STDs on campus? How important is the "it'll never happen to me" factor? How can this be combated?
2. Explain how you would educate a patient susceptible to urinary tract infections to minimize the risk of future infections.
3. Select a highly prevalent sexually transmitted disease, and devise a realistic control program for it.
4. Suppose you have volunteered for (or have been assigned) the task of eradicating a sexually transmitted tropical disease. Explain how you would do the job.
5. Is it possible for a person to contract genital herpes from an oral "cold sore"? Explain.
6. Find the package insert from tampons. What warning does it have regarding toxic shock syndrome? Which precautions does it recommend?
7. A man whose first wife had died of cervical cancer remarried. Ten years later his second wife also died of cervical cancer. Is there any possibility that this was not merely a sad coincidence?
   The answer to this question appears in Appendix E.

SOME INTERESTING READING

Anonymous. 1986. Antibiotic-resistant VD on increase. *Science News* 131 (March 28):200.

Aral, S. O., and K. K. Holmes. 1991. Sexually transmitted diseases in the AIDS era. *Scientific American* 264 (February):62–69.

Cates, W., Jr., and A. R. Hinman. 1991. Sexually transmitted diseases in the 1990's. *New England Journal of Medicine* 325, no. 19 (November 7):1368–70.

Collison, M. N-K. 1989. Dramatic increase in genital-warts disease among students worries college health officers. *Chronicle of Higher Education* 35, no. 38 (May 31):A23.

Crum, C. P., S. Barber, and J. K. Roche, 1991. Pathobiology of papillomavirus-related cervical diseases: prospects for immunodiagnosis. *Clinical Microbiology Reviews* 4(3):270–85.

Fletcher, J. L. 1991. Perinatal transmission of human papillovavirus. *American Family Physician* 43, no. 1 (January):143–49.

Frangos, D. N., et al. 1986. Genitourinary fungal infections. *Southern Medical Journal* 79, no. 4 (April):455.

Goldstein, L. C., et al. 1985. Monoclonal antibodies for the diagnosis of sexually transmitted diseases. *Clinical Laboratory Medicine* 5, no. 3 (September):575.

Kirchner, J. T. 1991. Syphilis—an STD on the increase. *American Family Physician* 44, no. 3 (September):843–55.

Lovchik, J. C. 1987. Chlamydial infections. *Maryland Journal of Medicine* 18 (January):54.

Palac, D. M. 1986. Urinary tract infections in women: A physician's perspective. *Laboratory Medicine* 17:25.

Peto, R., and H. zur Hausen. 1986. Viral etiology of cervical cancer. *Banbury Report*, no. 21. Cold Spring Harbor, NY: Cold Spring Harbor Laboratory.

Rice, P. A., and J. Schachter. 1991. Pathogenesis of pelvic inflammatory disease: what are the questions? *Journal of the American Medical Association* 266, no. 18 (November 13):2587–95.

Roizman, B., and A. E. Sears. 1987. An inquiry into the mechanisms of herpes simplex virus latency. *Annual Review of Microbiology* 41:543.

Specter, S., and G. S. Lancz. 1986. *Clinical virology manual.* New York: Elsevier Scientific Publishing Co.

Tolkoff-Rubin, N. E., and R. H. Rubin. 1987. New approaches to the treatment of urinary tract infections. *American Journal of Medicine* 82, Suppl. 4A (April 27):270.

Wicher, K., G. T. Noordhoek, F. Abbruscato, and V. Wicher, 1992. Detection of *Treponema palladium* in early syphilis by DNA amplification. *Journal of Clinical Microbiology* 30(2):497–500.

Antarctic snow colored varying shades of pink and green by bacterial growth. Note penguins for scale.

26 Environmental Microbiology

This chapter focuses on the following questions:

A. What is ecology, and how does energy flow in ecosystems?

B. Why is recycling important, and how are water and carbon atoms recycled?

C. What other biogeochemical cycles exist, and what roles do microorganisms play in them?

D. What kinds of microorganisms are found in air, and how are they detected and controlled?

E. What kinds of microorganisms are found in soil, and what are their roles in biogeochemical cycles and as pathogens?

F. How do freshwater and marine environments, and their microorganisms, differ?

G. How do water pollution and waterborne pathogens affect humans?

H. How is water purified, and how is it tested to determine purity?

I. What is sewage, and what processes are involved in primary, secondary, and tertiary sewage treatment?

Microorganisms are found in every environment—in the air we breathe, the food we eat, the soil where food is grown, and the water we drink. Many benefit humans, and only a few cause human disease. One might consider environmental microbiology of little significance in the health sciences, but that is not the case. Humans, being one of many organisms in an environment, affect and are affected by both living and nonliving components of an environment including its microorganisms. To control disease, health scientists need to know how to control microorganisms in air, food, soil, and water. To do that they need an understanding of their roles in the environment.

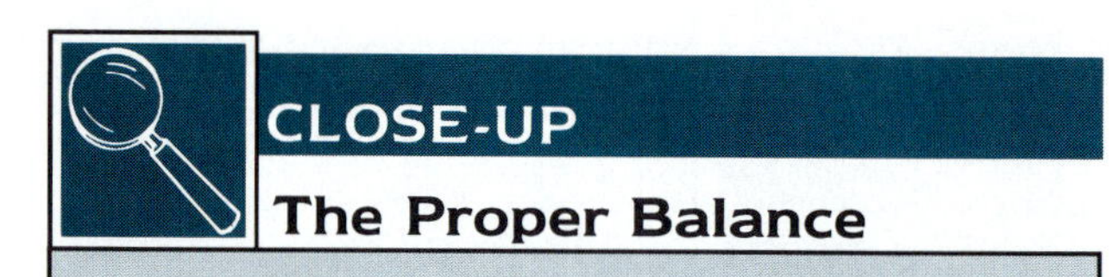

"It begins to appear that man is better off, on balance, for the presence of . . . microbes, just as he is better off for being part of a world of living things larger than microbes."

—Theodor Rosebury, 1969

FUNDAMENTALS OF ECOLOGY

Ecology Defined

Ecology is the study of relationships among organisms and their environment. These relationships include interactions of organisms with physical features—the **abiotic factors**—of the environment, and interactions of organisms with each other—the **biotic factors** of the environment.

The Nature of Ecosystems

An **ecosystem** includes all the biotic (living) components and the abiotic (physical and chemical) components of an environment. Microorganisms can be categorized as indigenous or nonindigenous to an environment. **Indigenous**, or native, **organisms**, are always found in a given environment. They generally are able to adapt to normal seasonal changes or changes in the quantity of available nutrients in the environment. For example, *Spirillum volutans* is indigenous to stagnant water, various species of *Streptomycetes* are indigenous to soil, and *Escherichia coli* is indigenous to the human digestive tract. Regardless of variations in the environment (except for cataclysmic changes), an environment will always support the life of its indigenous organisms. **Nonindigenous organisms** are temporary inhabitants of an environment. They become numerous when growth conditions are favorable for them and disappear when conditions become unfavorable.

An ecological **community** consists of all the kinds of organisms present in a given environment. In general, communities composed of many different species of organisms are more stable than those composed of only a few species. The various species create a kind of system of "checks and balances" so that the numbers of each species remain relatively constant. There are exceptions to this generalization however, especially when humans intervene. A tropical rain forest environment contains the greatest number of species of organisms, yet it is very fragile and can be disturbed quite easily by human attempts to modify it.

Unicellular microorganisms in a community respond to their environment in different ways than do the cells of multicellular organisms. The cells of a large organism exist in a closely regulated internal environment within the organism and display little capacity to adapt to change. However, microorganisms are directly affected by environmental changes. They have great capacity to adapt to such changes by chemotaxis, spore formation, the activity of contractile vacuoles, the production of protective shells, and other mechanisms, as described in earlier chapters.

Energy Flow in Ecosystems

Energy is essential to life, and radiant energy from the sun is the ultimate source of energy for nearly all organisms in any ecosystem. (The chemolithotrophic bacteria that extract energy from inorganic compounds are an exception.) (Chapter 5, p. 129) Organisms called **producers** capture energy from the sun. They use this energy and various nutrients from soil or water to synthesize the substances they need to grow and to support their other activities. Energy stored in the bodies of producers is transferred through an ecosystem when **consumers** (heterotrophs) obtain nutrients by eating the producers. **Decomposers** obtain energy by digesting dead bodies or wastes of producers and consumers. They release substances that producers can use as nutrients. The flow of energy and nutrients in an ecosystem is summarized in Figure 26.1.

Microorganisms can be producers, consumers, or decomposers in ecosystems. Producers include photosynthetic organisms among bacteria, cyanobacteria, protists, and eukaryotic algae. Although on land green plants are the primary producers, in the ocean microorganisms fill this role. Consumers include heterotrophic bacteria, protists, and microscopic fungi.

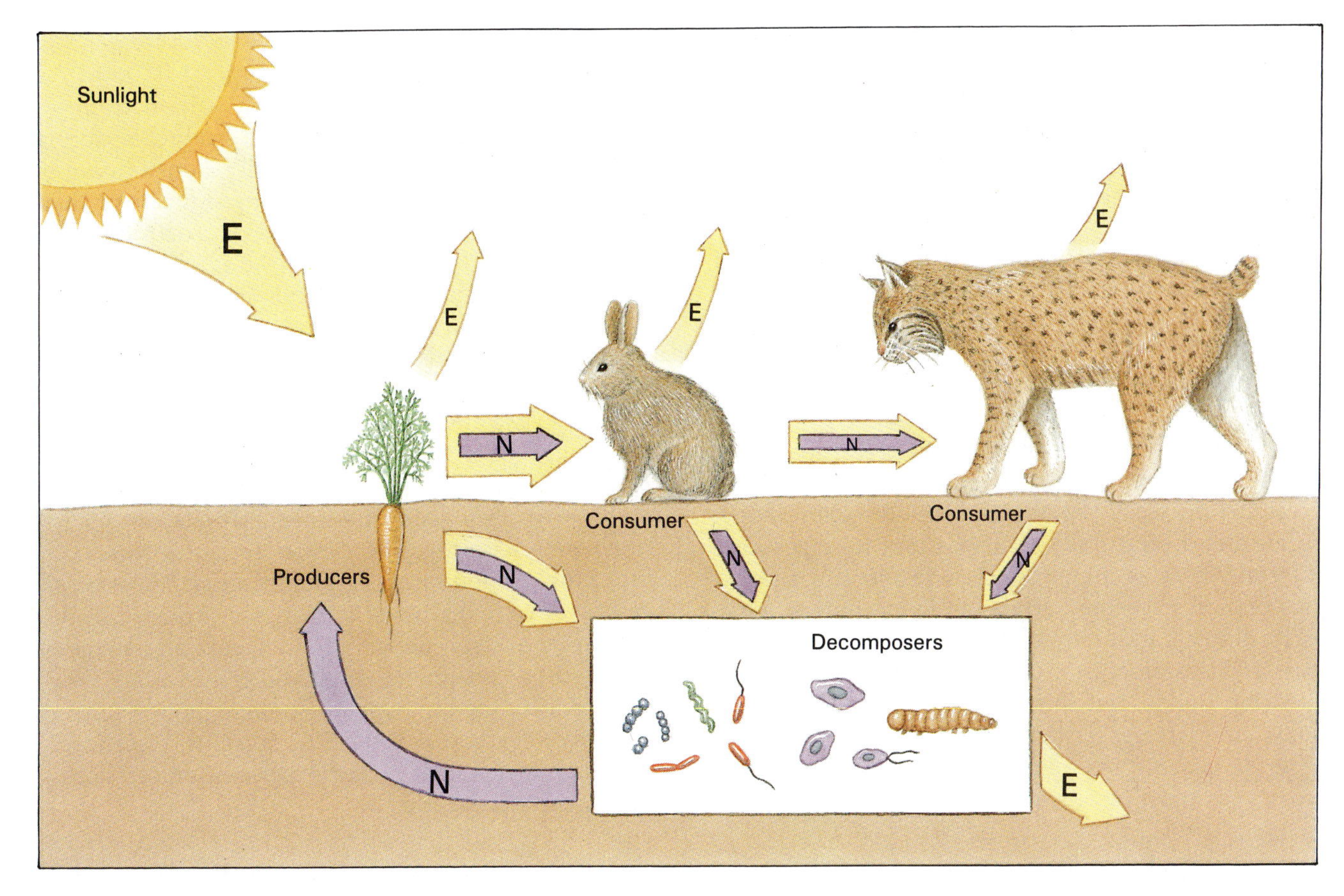

FIGURE 26.1 The flow of energy (E) and nutrients (N) in ecosystems. Note that energy flows through the system (it is obtained continuously from the sun), whereas nutrients must be recycled in order that new life may continue.

(To the extent that viruses divert a cell's energy to the synthesis of new viruses, they too act as consumers.) Many microorganisms act as decomposers; in fact, they play a greater role in decomposition of organic substances than larger organisms.

Heterotrophic microorganisms adapt to limited nutrients in several ways: (1) Some synthesize greater amounts of enzymes and assimilate limited nutrients more efficiently. (2) Others produce inducible enzymes for alternate metabolic pathways to take advantage of more readily available nutrients. (3) Most reduce the rates of all synthetic reactions so that growth occurs at a slower rate, but all substances needed for new organisms are produced.

BIOGEOCHEMICAL CYCLES

Living organisms incorporate water molecules and carbon, nitrogen, and other elements from their environment into their bodies as they carry on essential life processes. Without decomposers to assure the flow of nutrients through ecosystems, much matter would soon be incorporated into bodies and wastes,

CLOSE-UP

Bacteria Make the World Go 'Round

The bacteria particularly . . . are still more important than we. Omnipresent in infinite varieties, . . . they release the carbon and nitrogen held in the dead bodies of plants and animals which would—without bacteria and yeasts—remain locked up forever in useless combinations, removed forever as further sources of energy and synthesis. Incessantly busy in swamp and field, these minute benefactors release the frozen elements and return them to the common stock, so that they may pass through other cycles as parts of other living bodies. . . . Without the bacteria to maintain the continuities of the cycles of carbon and nitrogen between plants and animals, all life would eventually cease. . . . Without them, the physical world would become a storehouse of well-preserved specimens of its past flora and fauna . . . useless for the nourishment of the bodies of posterity. . . .

—Hans Zinsser, 1935

and life would soon become extinct. Although the supply of energy from sunlight is continuously renewed, the supply of water and the chemical elements that serve as nutrients is fixed. These materials must be recycled continuously to make them available to living organisms. The mechanisms by which such recycling occurs are referred to collectively as **biogeochemical cycles,** where *bio* refers to living things and *geo* refers to the earth, the environment of the living things.

The Water Cycle

The **water cycle,** or **hydrologic cycle** (Figure 26.2), recycles water. Water reaches the earth's surface as precipitation from the atmosphere. It enters living organisms during photosynthesis and by ingestion. It leaves them as a byproduct of respiration and by evaporation from the surfaces of living things, such as by transpiration (water loss from pores in plant leaves). Microorganisms use water in metabolism like all living things, but they also live in water or very moist environments. Many form spores or cysts that help them to survive periods of drought, but vegetative cells must have water.

The Carbon Cycle

In the **carbon cycle** (Figure 26.3), carbon from atmospheric carbon dioxide enters producers during photosynthesis or chemosynthesis. Consumers obtain carbon compounds by eating producers or their remains. Carbon dioxide (CO_2) is returned to the atmosphere by respiration and by the actions of decomposers on the dead bodies and wastes of other organisms. Carbon compounds can be deposited in peat, coal, and oil and released from them during burning. A small but significant quantity of carbon dioxide in the atmosphere comes from volcanic activity and from the weathering of rocks, many of which contain the carbonate ion, CO_3^{2-}.

As we have seen in earlier chapters, all microorganisms require some carbon source to maintain life. Most carbon entering living things comes from carbon dioxide dissolved in bodies of water or in the atmosphere. Even the carbon in sugars and starches ingested by consumers is derived from carbon dioxide. Because the atmosphere contains only a limited quantity of carbon dioxide (0.03 percent), recycling is essential to maintain a continuous supply of atmospheric carbon dioxide.

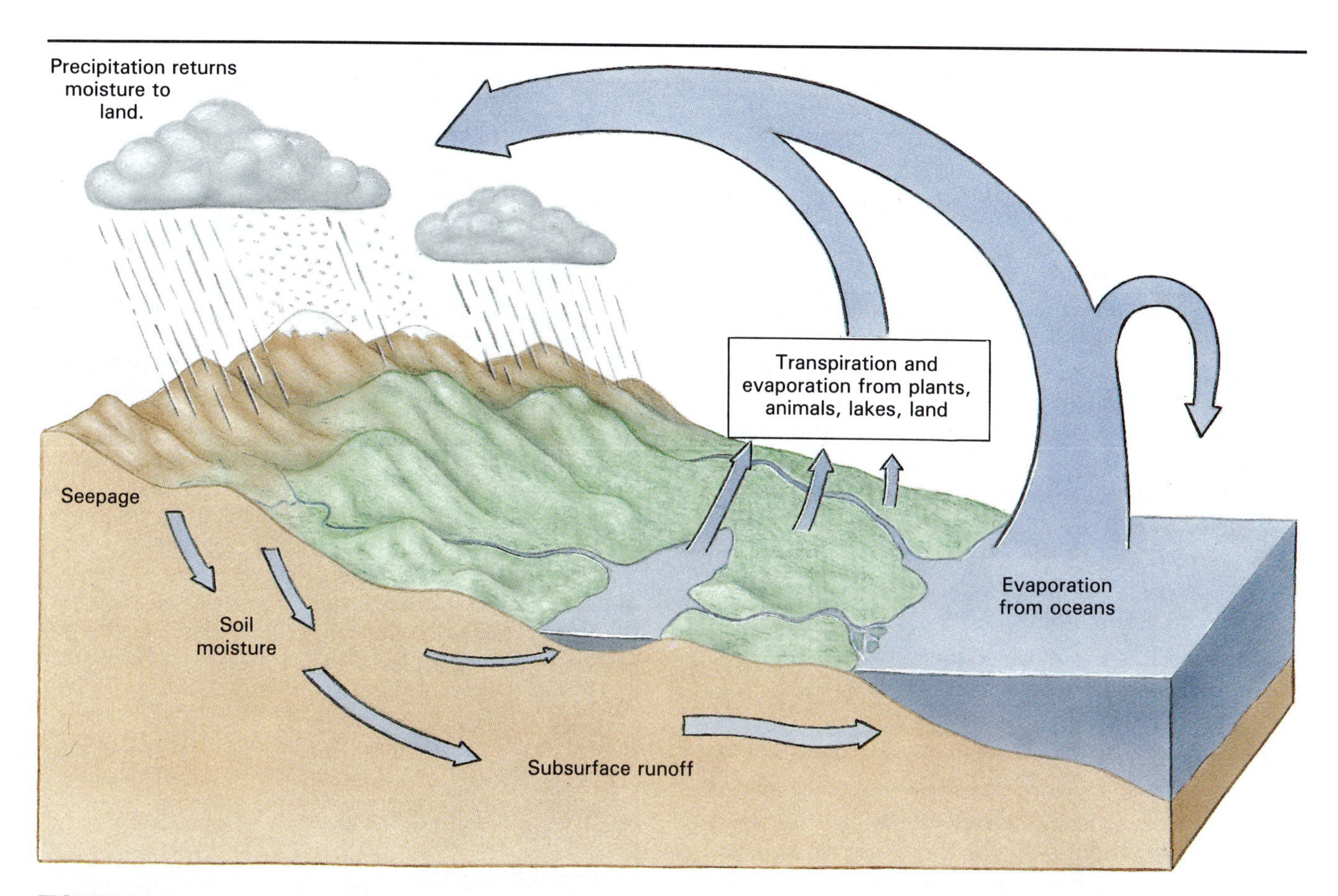

FIGURE 26.2 The water cycle.

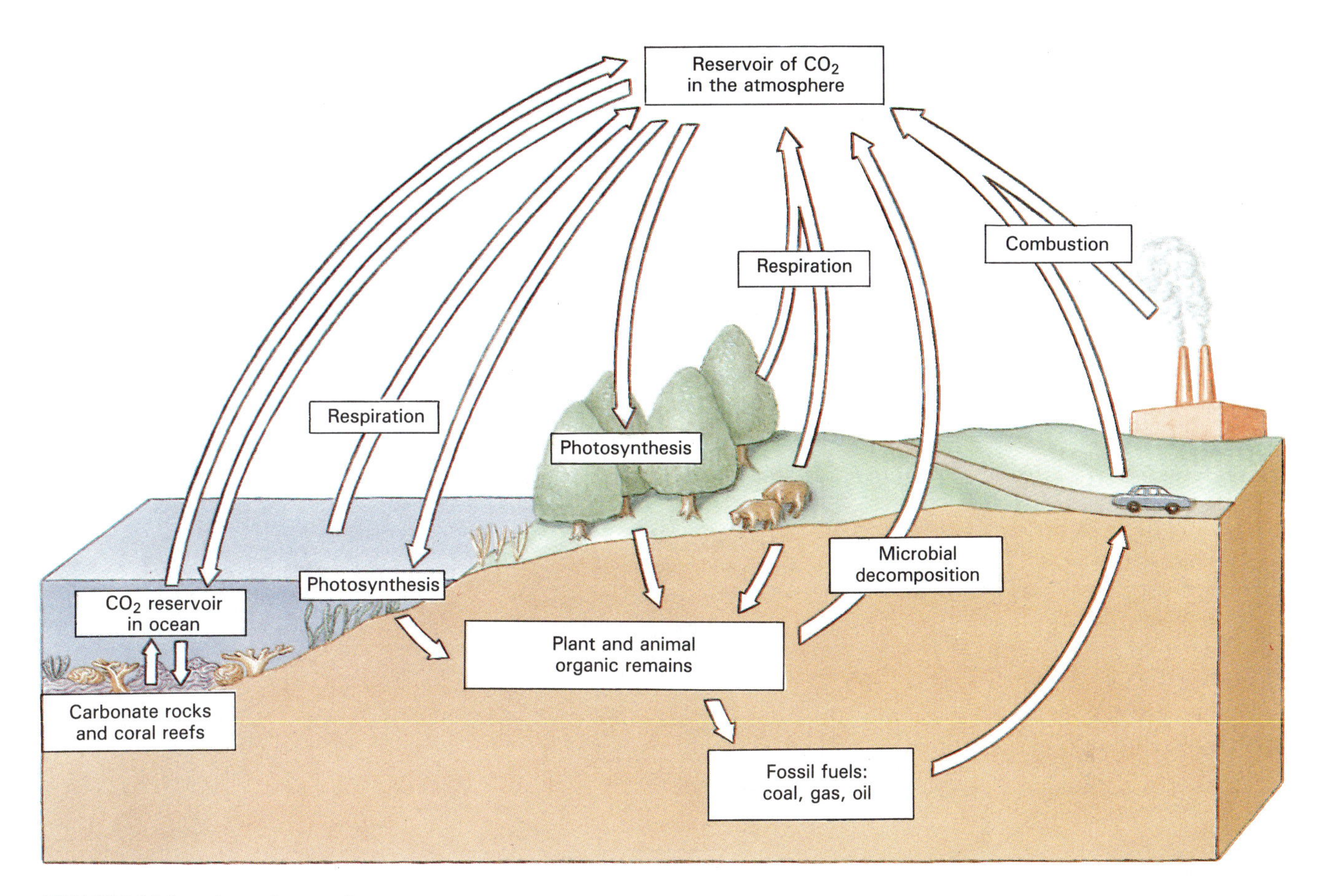

FIGURE 26.3 The carbon cycle.

CLOSE-UP
The Greenhouse Effect

Atmospheric carbon dioxide and water vapor form a blanket over the earth's surface, creating a "greenhouse effect." These gases allow the sun's radiation to penetrate to the earth's surface and warm it. But they trap much of the infrared (heat) radiation produced by the warm surface, reflecting it back to earth. Solar energy is thus captured within the "greenhouse." The overall effect is to reduce temperature variation and raise the temperature near the earth's surface. Over the past century human activities have released relatively large quantities of carbon dioxide from coal and oil combustion. At the same time, the forests that absorb carbon dioxide and replenish the earth's oxygen through photosynthesis have been cut down and burned at an ever-increasing rate. Some scientists think this is causing a general warming trend, increasing average temperatures around the world and possibly changing the balance of organisms in ecosystems. This trend may make now temperate regions too warm to grow wheat and other food crops, create droughts in other areas, and make new deserts. It might even melt polar ice, raising the level of oceans and flooding coastal cities.

The Nitrogen Cycle and Nitrogen Bacteria

In the **nitrogen cycle** (Figure 26.4), nitrogen moves from the atmosphere through various organisms and back into the atmosphere. This cyclic flow depends not only on decomposers but also on various nitrogen bacteria.

Decomposers use several enzymes to break down proteins in dead organisms and their wastes, releasing nitrogen in much the same way they release carbon. Proteinases break large protein molecules into smaller

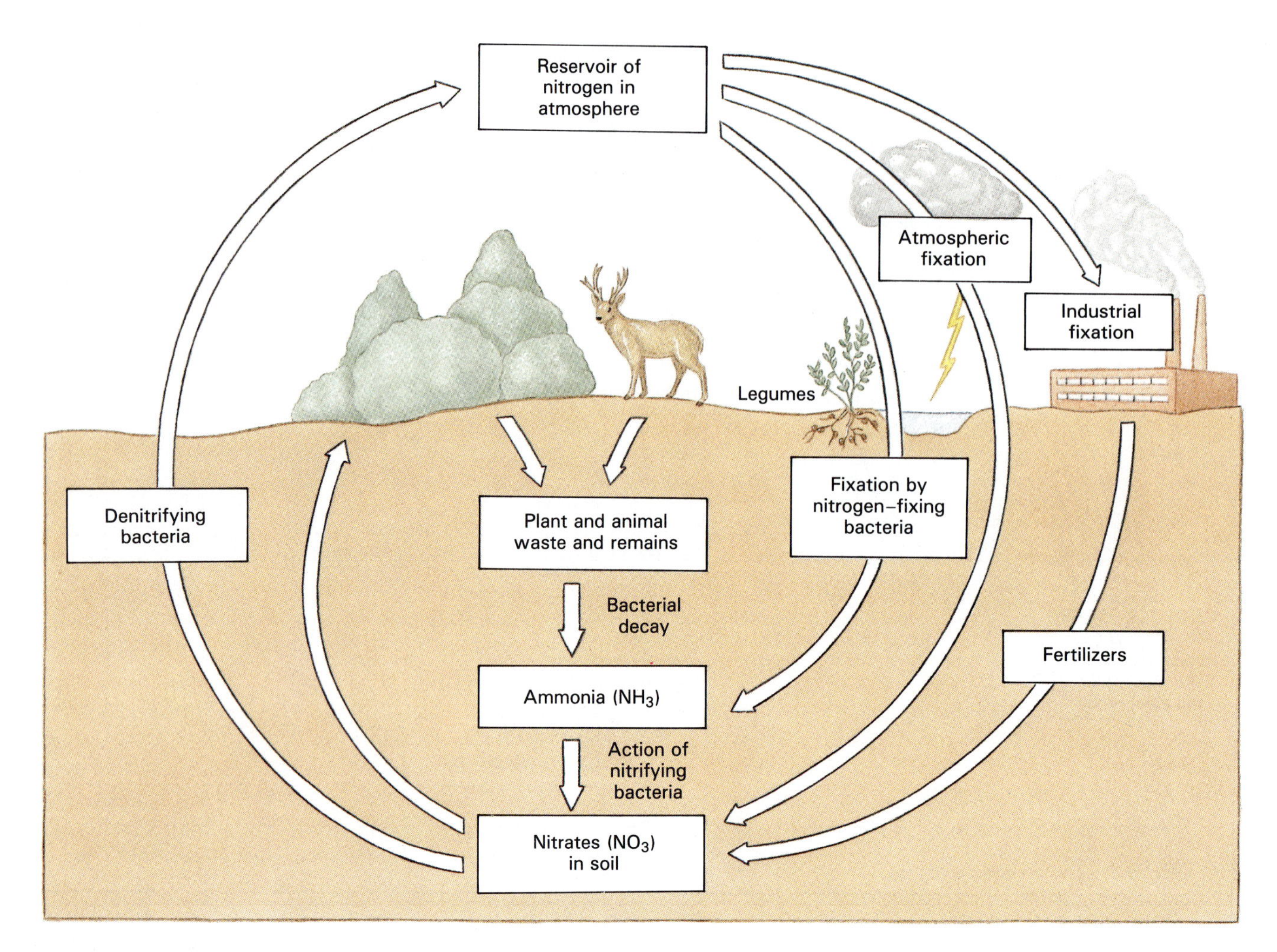

FIGURE 26.4 The nitrogen cycle.

molecules. Peptidases break peptide bonds to release amino acids. Deaminases remove amino groups from amino acids and release ammonia. Eventually free N_2 gas finds its way back into the atmosphere. Many soil microorganisms produce one or more of these enzymes. Clostridia, actinomycetes, and many fungi produce extracellular proteinases that initiate protein decomposition.

Nitrogen bacteria fall into three categories according to the roles they play in the nitrogen cycle:

- nitrogen fixers,
- nitrifying bacteria, and
- denitrifying bacteria.

Nitrogen-fixing Bacteria

Nitrogen fixation is the reduction of atmospheric nitrogen gas (N_2) to ammonia (NH_3). Organisms that can fix nitrogen are essential for maintaining a supply of physiologically usable nitrogen on the planet. About 255 million metric tons of nitrogen are fixed annually—70 percent of it by nitrogen fixers. Bacteria and cyanobacteria (blue-green algae) fix nitrogen in many different environments—from Antarctica to hot springs, acid bogs to salt flats, flooded lands to deserts, in marine and fresh water, and even in the guts of some organisms.

The energy for nitrogen fixation can come from fermentation, aerobic respiration, or photosynthesis. The various organisms that fix nitrogen live independently, in loose associations, or in intimate symbiosis. Regardless of the environment or the associations of the organisms, nitrogen-fixing bacteria must have a functional nitrogen-fixing enzyme called **nitrogenase,** a reducing agent that supplies hydrogen and energy from ATP. In aerobic environments nitrogen fixers must also have a mechanism to protect the oxygen-sensitive nitrogenase from inactivation.

Free-living aerobic nitrogen fixers include several species of the genus *Azotobacter* and some methylotrophic bacteria. Cyanobacteria are also free-living nitrogen fixers. *Azotobacter* is found in soils and is a versatile heterotroph whose growth is limited by the amount of organic carbon available. Methylotrophic

TRY IT

Nitrogen-fixing Bacteria in Root Nodules

It's easy to find and visualize nitrogen-fixing bacteria in root nodules. To do this, first dig up the root system from a leguminous plant such as beans or peas. Carefully wash away the dirt to expose the nodules. Crush a nodule on a clean glass slide. Add a small drop of water, and use the nodule pieces or an inoculating loop to spread the fluid out into a thin film. Remove the nodule pieces. Allow the slide to air-dry, and then quickly pass it through a Bunsen burner flame two or three times to heat-fix the organisms. Flood it with methylene blue stain for 1 minute. Rinse the slide with water, dry it, and examine it under the oil immersion lens. The bacilli you see are likely to belong to the genus *Rhizobium*.

bacteria can fix nitrogen when provided with methane, methanol, or hydrogen from various substrates. Cyanobacteria can fix nitrogen by using hydrogen from hydrogen sulfide, so they increase the availability of nitrogen in sulfurous environments.

Nitrogen-fixing facultative anaerobes include species of *Klebsiella, Enterobacter, Citrobacter,* and *Bacillus.* In addition, a number of obligate anaerobes, including photosynthetic Rhodospirillaceae and bacteria of the genera *Clostridium, Desulfovibrio,* and *Desulfotomaculum,* also fix nitrogen. Several species of *Klebsiella* capable of fixing nitrogen are found in rhizomes of nodulated legumes, such as peas and beans, and in the intestines of humans and other animals. Nitrogen fixation has been observed in about 12 percent of *Klebsiella pneumoniae* oganisms from patients. Various species of *Clostridium* are found in soils and muds. They use a variety of organic substances for energy and withstand unfavorable conditions as spores. They tolerate a range of pH from 4.5 to 8.5 but fix nitrogen best at pH 5.5 to 6.5. *Desulfovibrio* and *Desulfotomaculum,* anaerobic sulfate reducers that live in mud and soil sediments, fix nitrogen at pH 7 to 8.

Some nitrogen fixers are found in symbiotic association with other organisms, which provide them with organic carbon sources. For example, the cyanobacterium *Anabaena* (Figure 26.5a) is found in pores of the leaves of *Azolla* (see Essay), a small water fern found in many parts of the world. The nitrogen-fixing *Anabaena* supplies the necessary nitrogen; the fern supplies substrates for energy capture in ATP and reductants for nitrogen fixation. Together, they fix 100 kg of nitrogen per hectare (the area of about two football fields) per year and are used as "green manure" in rice cultivation in southeast Asia.

Rhizobium is the primary symbiotic nitrogen fixer. It lives in nodules on the roots of certain plants, usually legumes (Figure 26.5b and c). In this symbiotic relationship the plant benefits by receiving nitrogen in a usable form, and the bacteria benefit by receiving nutrients they need for growth. When paired with legumes, *Rhizobium* can fix 150 to 200 kg of nitrogen per hectare of land per year. In the absence of legumes, the bacteria fix only about 3.5 kg of nitrogen per hectare per year—less than 2 percent of that fixed in the symbiotic relationship. Farmers often mix nitrogen-fixing bacteria with seed peas and beans before planting to assure that nitrogen fixation will be adequate for their crops to thrive.

The mechanism by which *Rhizobium* establishes a symbiotic relationship with a legume has been the

(a)

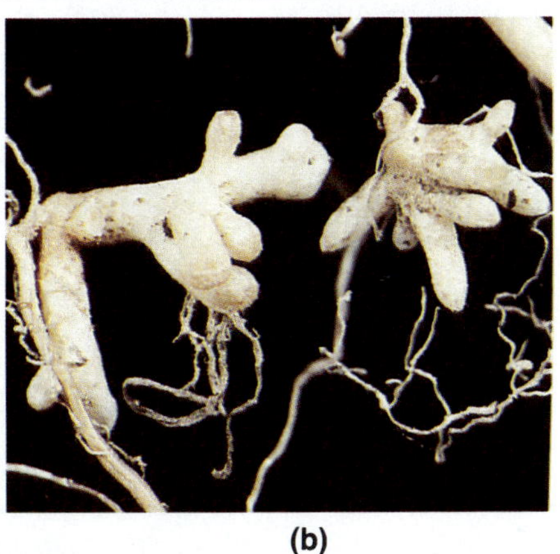

(b)

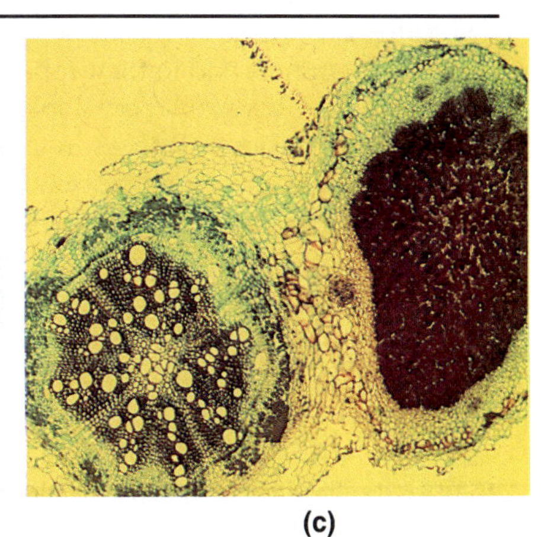

(c)

FIGURE 26.5 (a) SEM of *Anabaena azollae*, a nitrogen-fixing cyanobacterium that lives in a mutualistic relationship with the water fern *Azolla*. (b) Nodules on the roots of a bean plant, resulting from invasion of the plant by nitrogen-fixing bacteria. (c) A cross-section through a root of a leguminous plant shows densely packed *Rhizobium* bacteria inside the nodule (at right).

subject of many research studies. *Rhizobium* multiplies in the vicinity of legume roots, probably under the influence of root secretions. As the rhizobia increase in numbers, they release enzymes that digest cellulose and the substance that cements cellulose fibers together in the root cell walls. The rhizobia then change from their free-living rod shape and become spherical, flagellated cells called **swarmer cells.** These cells are thought to produce indoleacetic acid, a plant growth hormone, that causes curling of root hairs. The swarmer cells then invade the root hairs and form hyphalike networks, killing some root cells and proliferating in others. Swarmer cells become irregularly shaped **bacteroids** tightly packed into root cells, probably under the influence of chemical substances in the plant cells. Accumulations of bacteroids in adjacent root cells form protrusions called nodules on the plant roots.

Bacteroids contain the enzyme nitrogenase, which catalyzes the reaction:

$$\underset{\text{nitrogen gas}}{N_2} + \underset{\text{hydrogen gas}}{3\,H_2} \longrightarrow \underset{\text{ammonia}}{2\,NH_3}$$

However, this enzyme is inactivated by oxygen, so nitrogen fixation can occur only when oxygen is prevented from reaching the enzyme. Nitrogenase is protected from oxygen by a kind of hemoglobin, a red pigment that binds oxygen. This particular hemoglobin is synthesized only in root nodules containing bacteroids because part of the genetic information for its synthesis is in the bacteroids and part in the plant cells. Nitrogenase synthesis is repressed in the presence of excess ammonium (NH_4^+) and derepressed in the presence of free nitrogen. Thus, fixation occurs only when "fixed" nitrogen is in short supply and free nitrogen is available.

Rhizobium species vary in their capacity to invade particular legumes and in their capacity to fix nitrogen once they have invaded. Some species cannot invade any legumes, whereas others invade only certain legumes. This invasive specificity is determined genetically (probably by a single gene or a group of closely related genes) and can be altered by genetic transformation. (Chapter 8, p. 196) Such a transformation might enable a species of *Rhizobium* to invade a group of legumes it previously could not colonize.

Although symbiotic nitrogen fixation occurs mainly in association between rhizobia and legumes, other such associations are known. The alder tree, which grows in soils low in nitrogen, has root nodules similar to those formed by rhizobia. These nodules contain nitrogen-fixing actinomycetes of the genus *Frankia*.

Nitrifying Bacteria

Nitrification is the process by which ammonia or ammonium ions are oxidized to nitrites or nitrates. Nitrification, which is carried out by autotrophic bacteria, is an important part of the nitrogen cycle because it supplies plants with nitrate, the form of nitrogen most usable in plant metabolism. Nitrification occurs in two steps; in each step nitrogen is oxidized and energy is captured by the bacteria that carry out the reaction. Various species of *Nitrosomonas* (Figure 26.6) and related genera, gram-negative rod-shaped bacteria, produce nitrites:

$$\underset{\text{ammonium ion}}{NH_4^+} + \underset{\text{oxygen gas}}{1\tfrac{1}{2}\,O_2} \longrightarrow \underset{\text{nitrite}}{NO_2^-} + \underset{\text{water}}{H_2O} + \underset{\text{hydrogen ions}}{2\,H^+} + \text{energy}$$

Species of *Nitrobacter* and related genera, also gram-negative rods, produce nitrates:

$$\underset{\text{nitrite}}{NO_2^-} + \underset{\text{oxygen}}{\tfrac{1}{2}\,O_2} \longrightarrow \underset{\text{nitrate}}{NO_3^-} + \text{energy}$$

These bacteria use energy derived from the preceding reactions to reduce carbon dioxide in autotrophic metabolism. Because oxygen is required for the nitrification reactions, the reactions occur only in oxygenated water and soils. Furthermore, because nitrite is toxic to plants, it is essential that both of these reactions be carried out in sequence to provide nitrates

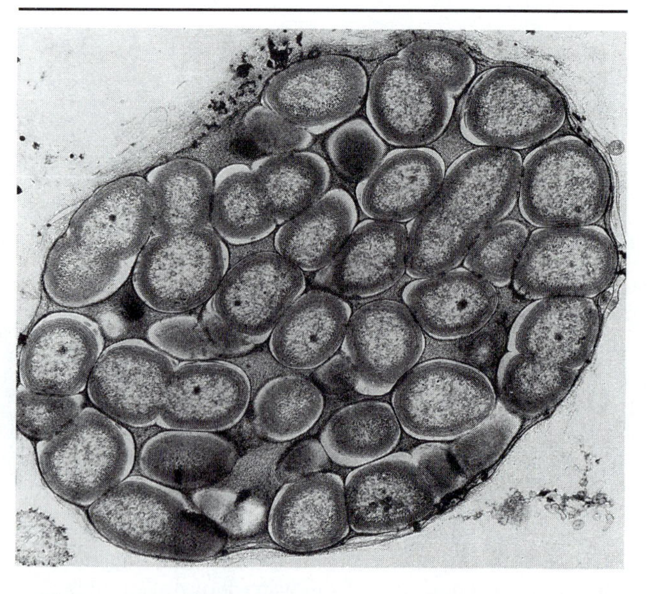

FIGURE 26.6 TEM (28,400X) of a microcolony of the nitrifying bacterium *Nitrosomonas*.

and to prevent the excessive accumulation of nitrites in the soil.

Denitrifying Bacteria

Denitrification is the process by which nitrates are reduced to nitrous oxide or nitrogen gas:

$$\underset{\text{nitrate}}{NO_3^-} \longrightarrow \underset{\text{nitrite}}{NO_2^-} \longrightarrow \underset{\text{nitrous oxide}}{N_2O} \longrightarrow \underset{\text{nitrogen gas}}{N_2}$$

Although this process does not occur to any significant degree in well-oxygenated soils, it does occur in oxygen-depleted, waterlogged soils. Most denitrification is performed by *Pseudomonas* species, but it can also be accomplished by *Thiobacillus denitrificans, Micrococcus denitrificans,* and several species of *Serratia* and *Achromobacter.* These bacteria, although usually aerobic, use nitrate instead of oxygen as a hydrogen acceptor under anaerobic conditions. Another process that reduces soil nitrate is the reduction of nitrate to ammonia. Several anaerobic bacteria carry out this process in a complex reaction that can be summarized as follows:

$$\underset{\text{nitrate}}{HNO_3^-} + \underset{\text{hydrogen gas}}{H_2} \longrightarrow \underset{\text{ammonia}}{NH_3} + \underset{\text{nitrous oxide}}{N_2O}$$

Although this latter process reduces the quantity of available nitrate, it at least retains the nitrogen in the soil in other forms.

Denitrification is a wasteful process because it removes nitrates from the soil and interferes with plant growth. It is responsible for significant losses of nitrogen from fertilizer applied to soils. Another unfortunate effect of denitrification is the production of nitrous oxide (N_2O), which is converted to nitric oxide (NO) in the atmosphere. Nitric oxide, in turn, reacts with ozone (O_3) in the upper atmosphere. Ozone provides a barrier between living things on earth and the sun's ultraviolet radiation. If enough ozone is destroyed, it no longer serves as an effective screen, and living things can be exposed to excessive ultraviolet radiation, which can cause cancer and mutations. (Chapter 7, p. 182)

The Sulfur Cycle and Sulfur Bacteria

The **sulfur cycle** (Figure 26.7), which involves the movement of sulfur through an ecosystem, resembles the nitrogen cycle in several respects. Sulfhydryl (SH) groups in proteins of dead organisms are converted to hydrogen sulfide (H_2S) by a variety of microorganisms. This process is analogous to the release of ammonia from proteins in the nitrogen cycle. Hydrogen sulfide is toxic to living things and thus must be ox-

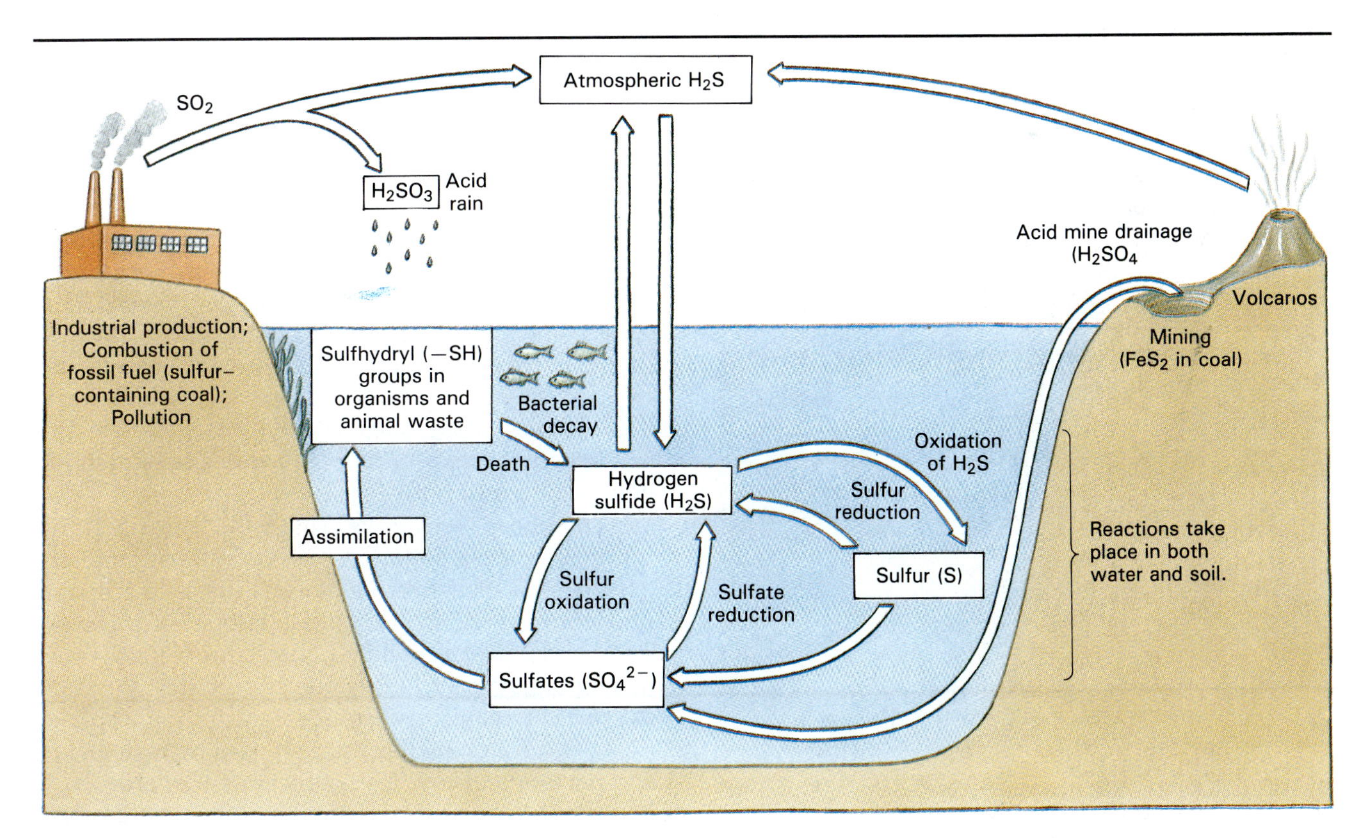

FIGURE 26.7 The sulfur cycle.

idized rapidly. Oxidation to elemental sulfur is followed by oxidation to sulfate (SO_4^{2-}), which is the form of sulfur most usable by both microorganisms and plants. This process is analogous to nitrification. The sulfur cycle is of special importance in aquatic environments, where sulfate is a common ion, especially in ocean water.

The various sulfur bacteria can be categorized according to their roles in the sulfur cycle. These roles include sulfate reduction, sulfur reduction, and sulfur oxidation.

Sulfate-Reducing Bacteria

Sulfate reduction is the reduction of sulfate (SO_4^{2-}) to hydrogen sulfide (H_2S). The sulfate-reducing bacteria are among the oldest life forms, probably more than 3 billion years old. They include the closely related genera of *Desulfovibrio, Desulfomonas,* and *Desulfotomaculum.* In these bacteria sulfate is the final electron acceptor in anaerobic oxidation, as oxygen is the final electron acceptor in aerobic oxidation. By reducing sulfate they produce large amounts of hydrogen sulfide. However, for this process to occur energy from ATP is required to phosphorylate the sulfate and convert it to ADP-SO_4. ADP-SO_4 then can act as an electron acceptor and compete successfully for substrates.

Sulfate-reducing bacteria (Figure 26.8) are strict anaerobes. They are widely distributed and predominate in nearly all anaerobic environments. As discussed in Chapter 6, sulfate-reducing bacteria can be psychrophilic, mesophilic, thermophilic, or halophilic. (p. 145) The variety of organic carbon sources they can metabolize is limited. Most use lactate, pyruvate, fumarate, malate, or ethanol, and some can use glucose and citrate. End products of such metabolism are typically acetate and carbon dioxide. Some sulfate reducers use products derived from anaerobic degradation of plant material by other organisms. *Desulfovibrio, Desulfomonas,* and *Desulfotomaculum* oxidize fatty acids and a variety of other organic acids.

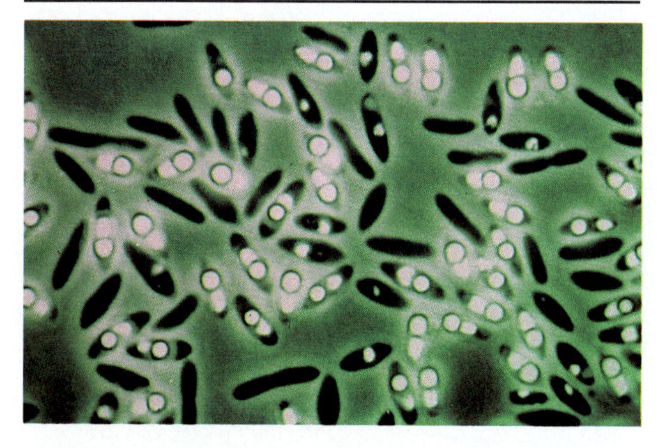

FIGURE 26.8 The sulfate-reducing bacterium *Desulfotomaculum acetoxidans* showing spherical spores and gas vesicles (phase contrast, 2000X).

Sulfur-Reducing Bacteria

Sulfur reduction is the reduction of sulfur to hydrogen sulfide. Like sulfate-reducing bacteria, sulfur-reducing bacteria also are anaerobes. They can use intracellular or extracellular sulfur as an electron acceptor in fermentation. The sulfur can be in elemental form or in disulfide bonds of organic molecules. This process provides energy for the organisms when sunlight is not available for photosynthesis.

Sulfur-Oxidizing Bacteria

Sulfur oxidation is the oxidation of various forms of sulfur to sulfate. *Thiobacillus* and similar bacteria oxidize hydrogen sulfide, ferrous sulfide, or elemental sulfur to sulfuric acid (H_2SO_4). When the acid ionizes it greatly decreases the pH of the environment, sometimes lowering it to a pH of 1 to 2. Sulfur-oxidizing organisms are responsible for oxidizing ferrous sulfide in coal-mining wastes, and the acid they produce is extremely toxic to fish and other organisms in streams fed by such wastes.

Other Biogeochemical Cycles

In addition to the cycling of water, carbon, nitrogen, and sulfur through ecosystems, phosphorus and other elements also move through ecosystems in a cyclic manner. Any element (including trace elements) that appears in the cells of living organisms must be recycled to extract it from dead organisms and make it available to living ones. We will conclude our discussion of biogeochemical cycles with a brief description of the phosphorus cycle.

The **phosphorus cycle** (Figure 26.9) involves the movement of phosphorus between inorganic and organic forms. Soil microorganisms are active in the phosphorus cycle in at least two important ways: (1) They break down organic phosphates from decomposing organisms to inorganic phosphates. (2) They convert inorganic phosphates to orthophosphate (PO_4^{3-}), a water-soluble nutrient used by both plants and microorganisms. These functions are particularly important because phosphorus is often the limiting nutrient in many environments.

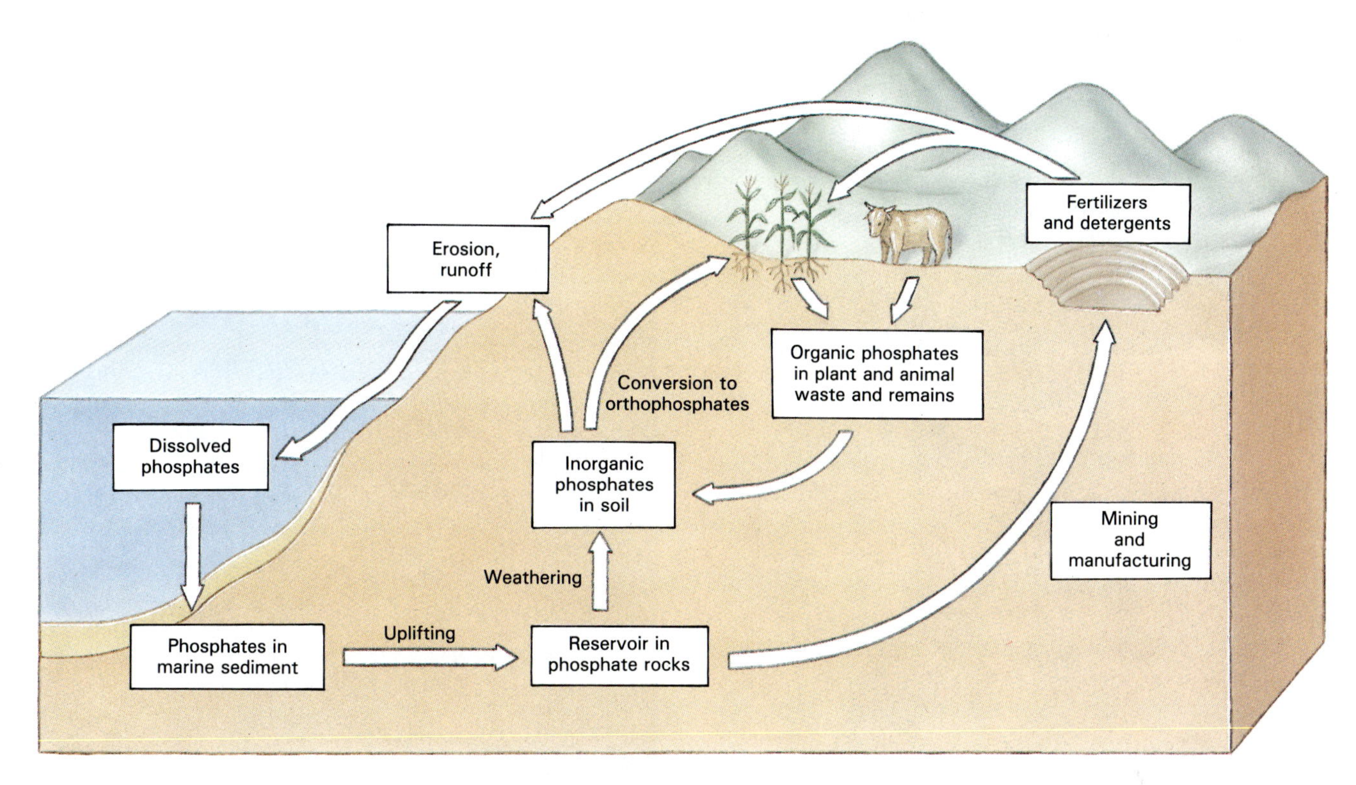

FIGURE 26.9 The phosphorus cycle.

AIR

Having discussed briefly some fundamentals of ecology and of biogeochemical cycles, we now explore different kinds of environments and the microorganisms they contain. We will begin with the air around us and then consider soil and water.

Microorganisms Found in Air

Microorganisms do not grow in air, but spores are carried in air, and vegetative cells can be carried on dust particles and water droplets in air. The kinds and numbers of airborne microorganisms vary tremendously in different environments. Large numbers of many different kinds of microorganisms are present in indoor air where humans are crowded together and ventilation is poor. Small numbers have been detected at altitudes of 3000 m.

Among the organisms found in air, mold spores are by far the most numerous, and the predominant genus is usually *Cladosporium*. Bacteria commonly found in air include both aerobic spore formers such as *Bacillus subtilis* and non-spore formers such as *Micrococcus* and *Sarcina*. Algae, protozoa, yeasts, and viruses also have been isolated from air. Infected humans who cough, sneeze, or even talk can expel pathogenic organisms along with water droplets. Health-care workers should handle wastes from patients carefully to avoid producing aerosols (tiny droplets that remain suspended for some time) of pathogens.

Determining Microbial Content of Air

Airborne microbes can be detected by collecting them on an agar plate or in a liquid medium, but only those that happen to fall into the medium are found. A special centrifugelike instrument provides a better measure of airborne microbes, as shown in Figure 26.10a.

Methods for Controlling Microorganisms in Air

Chemical agents, radiation, filtration, and laminar airflow all can be used to control microorganisms in air. Certain chemical agents, such as triethylene glycol, resorcinol, and lactic acid dispersed as aerosols, kill many if not all microorganisms in room air. These agents are highly bacteriocidal, remain suspended long enough to act at normal room temperature and humidity, are nontoxic to humans, and do not damage or discolor objects in the room.

Ultraviolet radiation has little penetrating power and is bacteriocidal only when rays make direct con-

(a)

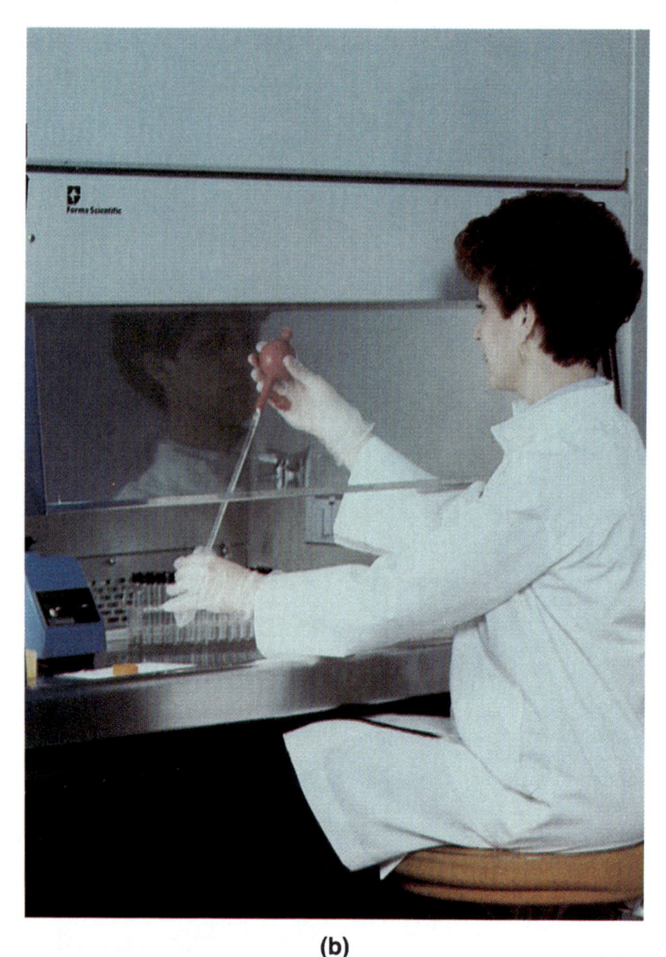

(b)

FIGURE 26.10 (a) Microbial populations in air can be measured by use of equipment such as this instrument, which takes in air and pumps it over culture media. The media are then removed and incubated, and the resulting colonies are counted. (b) Technicians are protected from airborne spread of organisms by use of a laminar flow hood, which suctions air away from the opening and filters it before expelling it.

tact with particles in air. Thus, ultraviolet lamps must be carefully placed to assure treatment of all air in a room. They are most useful in maintaining sterile conditions in rooms only sporadically occupied by humans. They can be turned off while technicians are carrying out sterile techniques and turned on again when the room is not in use. Humans entering a room while ultraviolet lights are on must wear protective clothing and special glasses to protect the eyes from burns. Ultraviolet lamps also can be installed in air ducts to reduce the number of microorganisms entering a room through its ventilation system.

Air filtration involves passing air through fibrous materials, such as cotton or fiberglass. It is useful in industrial processes in which sterile air must be bubbled through large fermentation vats. Cellulose acetate filters can be installed in a laminar airflow system (Figure 26.10b) to remove microorganisms which may have escaped into the air underneath the hood. The air is suctioned away from the opening, filtered, and then returned to the room.

SOIL

We might think of the soil we walk on as an inert substance, but nothing could be further from the truth. Soil, in fact, is teeming with microscopic and small macroscopic organisms, and it receives animal wastes and organic matter from dead organisms. Microorganisms act as decomposers to break down this organic matter into simple nutrients that can be used by plants and by the microbes themselves. (Animals, of course, obtain their nutrients from plants or from other animals.) Soil microorganisms are thus extremely important in recycling substances in ecosystems.

Components of Soil

Soil contains inorganic and organic components. The inorganic components are rocks, minerals, water, and gases. The organic components are **humus** (nonliving organic matter) and living organisms. Soils differ greatly in the relative proportions of these components. Topsoil, the surface layer of soil, contains the greatest number of microorganisms because it is well supplied with oxygen and nutrients. Lower layers of soil and soils depleted of oxygen and nutrients contain fewer organisms.

The most abundant inorganic components of soil are pulverized rocks and minerals. As rocks are exposed to wind, rain, and freezing and thawing of water that seeps into the cracks in the rocks, the rocks break into smaller pieces. Such weathering of rocks releases minerals into the soil. Some microorganisms also act on exposed rock surfaces by secreting acid products that contribute to their disintegration. The most abundant minerals in soil are silicon, aluminum, and iron. Small quantities of other elements such as calcium, potassium, magnesium, sodium, phosphorus, nitrogen, and sulfur also are present in soil.

In addition to rocks and minerals, soil contains water and the gases carbon dioxide, oxygen, and nitrogen. Water molecules adhere to soil particles or are interspersed among them. The amount of water in soil is highly variable and depends on the climate, the quantity of recent rainfall, and the drainage of the soil. Gases are dispersed between soil particles or dissolved in water. The concentration of gases in the soil also varies depending on the metabolic activity of soil organisms. Compared to atmospheric air, soil is lower in oxygen and higher in carbon dioxide.

Humus is constantly changing as organisms die and as decomposers degrade complex molecules to simpler ones. Also soils differ greatly in the amount of humus they contain. Most soils are 2 to 10 percent humus, but a peat bog can be 95 percent humus.

In addition to microorganisms, soil also contains the root systems of many plants, a variety of invertebrate animals (nematodes, earthworms, snails, slugs, insects, centipedes, millipedes, and spiders), and a few burrowing reptiles and mammals. In spite of the great diversity of soil organisms, microorganisms are by far the most numerous both in total numbers and in number of species present. As we have seen, microorganisms convert humus to nutrients usable by other organisms. They also make great demands on soil nutrients. When nutrients are abundant, the number of microorganisms increases rapidly, and as nutrients are depleted, the number of microorganisms decreases accordingly.

Microorganisms in Soil

All the major groups of microorganisms—bacteria, fungi, algae, and protists, as well as viruses—are present in the soil, but bacteria are more numerous than all other kinds of microorganisms (Figure 26.11). Among the bacteria in soil are autotrophs, heterotrophs, aerobes, anaerobes, and, depending on the soil temperature, mesophiles and thermophiles. Soil contains bacteria that digest special substances such as cellulose, protein, pectin, butyric acid, and urea as well as nitrogen-fixing, nitrifying, and denitrifying bacteria.

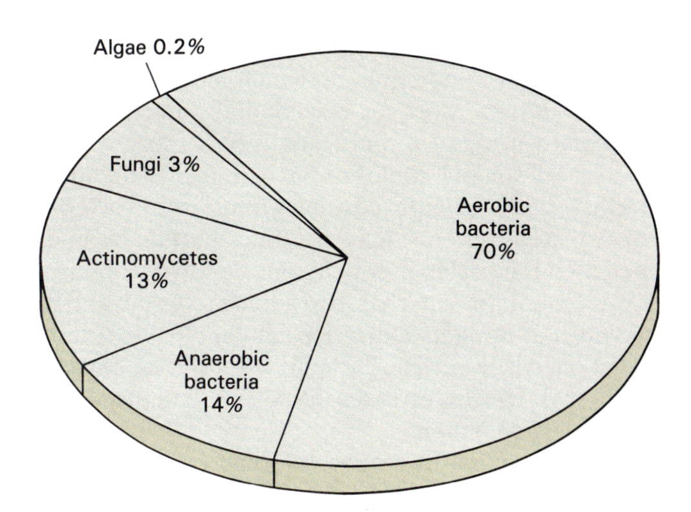

FIGURE 26.11 The relative proportions of various kinds of organisms found in soil.

Soil fungi are mostly molds. Both mycelia and spores are present mainly in the aerobic surface layer of the soil. Fungi serve two functions in soil. They decompose plant tissues such as cellulose and lignin, and their mycelia form networks around soil particles, giving the soil a crumbly texture. In addition to molds, yeasts are abundant in soils where grapes and other fruits are growing.

Small numbers of cyanobacteria, algae, protists, and viruses are found in most soils. Algae are found only on the soil surface, where they can carry on photosynthesis. In desert and other barren soils, algae contribute significantly to the accumulation of organic matter in the soil. Protists, mostly amoebae and flagellated protozoa, also are found in many soils. They feed on bacteria and may help to control bacterial populations. Soil viruses infect mostly bacteria, but a few infect plants or animals.

Factors Affecting Soil Microorganisms

Soil microorganisms, like all organisms, interact with their environment. Their growth is influenced both by physical factors and by other organisms. The microorganisms, in turn, affect the physical characteristics of soil and other organisms in the soil.

Physical factors in soil, as in any environment, include moisture, oxygen concentration, pH, and temperature. The moisture and oxygen content of soil are closely related. Spaces between soil particles ordinarily contain both water and oxygen, and aerobic organisms thrive there. However, in waterlogged soils all the spaces are filled with water, so only anaerobic bacteria grow.

Soil pH, which can vary from 2 to 9, is an important factor in determining which microorganisms will be present. Most soil bacteria have an optimum pH between 6 and 8, but some molds can grow at almost any soil pH. Molds thrive in highly acidic soils partly because of reduced competition from bacteria for available nutrients. Lime neutralizes acidic soil and increases the bacterial population. Fertilizer containing ammonium salts has two effects on soil: (1) It provides a nitrogen source for plants. (2) When it is metabolized by certain bacteria, the bacteria release nitric acid, which decreases the soil pH and increases the mold population.

Soil temperature varies seasonally from below freezing to as high as 60°C at soil surfaces exposed to intense summer sunlight. Mesophilic and thermophilic bacteria are quite numerous in warm to hot soils, whereas cold-tolerant mesophiles (and not true psychrophiles) are present in cold soils. Most soil molds are mesophilic and are found mainly in soils of moderate temperature.

Exceedingly wide variations exist in physical characteristics of soil and in the numbers and kinds of organisms it contains—even in soil samples taken only a few centimeters apart. Such observations led ecologists to develop the concept of **microenvironments**. The interactions among organisms and between the organisms and their environment can be quite different in different microenvironments no matter how close together they are.

Importance of Decomposers in Soil

Soil microorganisms called **decomposers** are important in the carbon cycle because of their ability to decompose organic matter. The decomposition of complex organic substances from dead organisms is a stepwise process that requires the action of several kinds of microorganisms. The organic substances include cellulose, lignins, and pectins in the cell walls of plants, glycogen from animal tissues, and proteins and fats from both plants and animals. Cellulose is degraded by bacteria, especially those of the genus *Cytophaga*, and by various fungi. Lignins and pectins are partially digested by fungi, and the products of fungal action are further digested by bacteria. Protozoa and nematodes can also play a role in the degradation of lignins and pectins. Proteins are degraded to individual amino acids mainly by fungi, actinomycetes, and clostridia.

Under anaerobic conditions in the waterlogged soils of marshes and swamps, methane is the main carbon-containing product. It is produced by three genera of strictly anaerobic bacteria, *Methanococcus*, *Methanobacterium*, and *Methanosarcina*. At the same time they degrade carbon compounds to methane,

BIOTECHNOLOGY

Microbes That Eat Autos

In 1991, manufacture of the East German auto the Trabant ceased. The two-cylinder, two-stroke engine couldn't compete with the Mercedes and BMWs of West Germany. Imagine the pain, after perhaps 10 years of saving to buy one, of finding that what should have been a status symbol was now a costly dinosaur. Customers seeking to trade in their Trabants on a bigger car not only got *no* trade-in allowance but were *charged* $300 more to dispose of their old Trabant. Space in landfills is scarce and costly in Germany.

Biotechnology may soon have a solution to the problem. Trabants are made mostly of plastic. They are so light that two men can lift one into a dumpster. This plastic has a great deal of agricultural waste (cellulose) from the former Soviet Union incorporated into it, which should make it easily biodegradable by microbes. Scientists are searching the soil of landfills for microbes that are degrading the plastic of Trabants already buried there. These organisms could be used to disintegrate the glut of unwanted Trabants more rapidly. One Berlin biotechnology company claims to have developed a bacterium that will eat a Trabant in 20 days, leaving only a small pile of compost.

End of a dream: Dozens of discarded Trabant cars are stacked up on a Berlin junkyard. Scientists are looking for microbes that will eat these cars, which are made mostly of plastic. East Germans once saved for years to afford one of these "dream" cars.

they also get energy from oxidizing hydrogen gas:

$$\underset{\text{hydrogen gas}}{4\,H_2} + \underset{\text{carbon dioxide}}{CO_2} \longrightarrow \underset{\text{methane}}{CH_4} + \underset{\text{water}}{2\,H_2O}$$

In one way or another organic substances are metabolized to carbon dioxide, water, and other small molecules. In fact, all naturally occurring organic compounds can be decomposed by some organisms, so that carbon is continuously recycled. However, certain organic compounds manufactured by humans resist the actions of microorganisms. Accumulations of these synthetic substances create a number of environmental hazards.

Nitrogen enters the soil through the decomposition of proteins from dead organisms and through the actions of nitrogen-fixing organisms. In addition to protein decomposition, which introduces nitrogen into the soil, gaseous nitrogen is fixed by both free-living microorganisms and by symbiotic microorganisms associated with the roots of legumes, as previously described.

Soil Pathogens

Soil pathogens are primarily plant pathogens, many of which have been discussed in earlier chapters. A few soil pathogens can affect humans and other animals. The main human pathogens found in soil belong to the genus *Clostridium*. All are anaerobic spore formers. *Clostridium tetani* causes tetanus and can be introduced easily into a puncture wound. *Clostridium botulinum* causes botulism. Its spores, found on many edible plants, can survive in incompletely processed foods to produce a deadly toxin. *Clostridium perfringens* causes gas gangrene in poorly cleaned wounds. Grazing animals can contract the disease anthrax from spores of *Bacillus anthracis* in the soil. In fact, most soil organisms that infect warm-blooded animals exist as spores because soil temperatures are usually too low to maintain vegetative cells of these pathogens.

WATER

All water—fresh water, ocean water, even rainwater—contains microorganisms as well as inorganic substances. Most of the organisms in aquatic environments have been considered in earlier chapters. Here we will consider properties of the environments, interactions of microorganisms with the environment, transmission of human pathogens in water, and methods for maintaining safe water supplies for humans.

Freshwater Environments

Freshwater environments include surface water, such as lakes, ponds, rivers, and streams, and groundwater that runs through underground strata. Although groundwater contains few microorganisms, surface water contains large numbers of many different kinds of microorganisms. Which organisms are present depends on the temperature and pH of the water, dissolved minerals, and the quantity of nutrients in it. Water temperatures vary from 0°C to nearly 100°C. Most microorganisms thrive in water at moderate temperatures. However, some thermophilic bacteria have been found in geysers having a temperature above 90°C, and psychrophilic fungi and bacteria have been found in water at 0°C. The pH of fresh water varies from 2 to 9. Although most microorganisms grow best in waters having nearly neutral pH, a few have been found in both extremely acidic and extremely alkaline waters. Most natural waters are rich in nutrients, but the quantities of various nutrients in different bodies of water vary considerably. Sometimes nutrients become so abundant that a "bloom," or sudden proliferation of organisms in a body of water, occurs (Figure 26.12).

In water environments oxygen often is the limiting factor in the growth of microorganisms. Because of oxygen's low solubility in water, its concentration never exceeds 0.007 g per 100 g of water. When water contains large quantities of organic matter, decomposers rapidly deplete the oxygen supply as they oxidize the organic matter. Oxygen depletion is much more likely in standing water in lakes and ponds than in running water in rivers and streams because the movement of running water causes it to be continuously oxygenated.

Yet another factor affecting microorganisms in aquatic environments is the depth to which the sunlight penetrates the water. Of minor importance in fresh water except for deep lakes, it is very important

FIGURE 26.12 An overabundance of nutrients has allowed an algal bloom to develop on this pond. Bubbles of oxygen can be seen collecting underneath some of the algae. However, algal blooms often lead to increased biological oxygen demand (BOD) as they die off and decompose, using up oxygen needed by other pond organisms such as fish.

in the ocean. Photosynthetic organisms are limited to locations with adequate sunlight.

Varieties of microorganisms are found in all freshwater environments. Aerobic bacteria are found where oxygen supplies are adequate, whereas anaerobic bacteria are found in oxygen-depleted waters. Eukaryotic algae, cyanobacteria, and sulfur bacteria are limited to water that receives adequate sunlight. *Desulfovibrio* and methane bacteria are found in lake sediments. Protists are found in many different freshwater environments.

Marine Environments

The marine, or ocean, environment covers about 70 percent of the earth's surface and is therefore larger than all other environments combined. Compared to fresh water, the ocean is much less variable in both temperature and pH. Except in the vicinity of volcanic vents in the sea floor, where water reaches 250°C (Chapter 9), ocean water ranges from 30°C to 40°C at the surface near the equator to 0°C in polar regions and in the lowest depths. (p. 229) At any particular location and depth the temperature is nearly constant. The pH of ocean water ranges from nearly neutral to slightly alkaline (pH 6.5 to 8.3). This pH range is suitable for growth of many microorganisms, and sufficient carbon dioxide dissolves in the water to support photosynthetic organisms.

Ocean water, which is about seven times as salty as fresh water, displays a remarkably constant concentration of dissolved salts—3.3 to 3.7 g per 100 g of water. Thus organisms that live in marine environments must be able to tolerate high salinity but need not tolerate variations in salinity.

Marine environments display a far greater range of hydrostatic pressure than do fresh waters. Hydrostatic pressure increases with depth at a rate of approximately 1 atmosphere per 10 m, so the pressure at a depth of 1000 m is 100 times that at the surface. A few microorganisms have been isolated from Pacific trenches at depths greater than 1000 m!

Other factors that vary with the depth of ocean water are penetration of sunlight and oxygen concentration. Sunlight of sufficient intensity to support photosynthesis penetrates ocean water to depths of only 50 to 125 m, depending on the season, latitude, and transparency of the water. Oxygen diffuses into surface water and is released by photosynthetic organisms in sunlit water. However, deep water lacks both sunlight and oxygen.

Nutrient concentrations vary in ocean water, depending on depth and proximity to the shore (Figure 26.13). Nutrients are most plentiful near the mouths of rivers and near the shore, where runoff from land enriches the water. However, ocean water is generally lower in phosphates and nitrates than fresh water.

FIGURE 26.13 Nutrients flow into the ocean with the influx of water from rivers. The lighter-colored area in this photograph is silt-bearing, nutrient-rich river water.

Nutrients in the waters of the open ocean are relatively dilute. Photosynthetic organisms near the surface serve as food for the heterotrophic organisms at the same or deeper levels. Decomposers are usually found in bottom sediments, where they release nutrients from dead organisms.

Large numbers of many different kinds of microorganisms live in the ocean. Yet, much remains to be learned about the exact numbers and the particular species living in various parts of the ocean. The primary producers of the ocean are photosynthetic microorganisms. They are motile and contain oil droplets or other devices to give them buoyancy and allow them to remain in sunlit waters. Producers of the ocean include cyanobacteria, diatoms, dinoflagellates,

CLOSE-UP

Microbial Determinants of Weather

The oceanic phytoplankton (tiny floating plants and cyanobacteria) absorb large amounts of carbon dioxide as they perform photosynthesis. Atmospheric carbon dioxide is one of the *greenhouse gases*, which trap heat and raise global temperature. Seasonal fluctuations of phytoplankton populations affect the earth's temperature more than we had previously realized. Now, further complexity has been revealed by recent research showing that massive populations of marine viruses may be at the base of this. Earlier, the ocean was thought to contain relatively few viruses. Viral infections destroy phytoplankton and thus may be the prime determinants of our earth's weather.

chlamydomonads, and a variety of other protists and eukaryotic algae (Figure 26.14).

Many of the consumers of the ocean are heterotrophic bacteria. Which particular species inhabit a given region is determined by the temperature and pH of the water and the available nutrients. Members of the genera *Pseudomonas*, *Vibrio*, *Achromobacter*, and *Flavobacterium* are common in ocean water. Protozoa, especially radiolarians and foraminiferans, and a variety of fungi also feed on producers in the open ocean. However, most of these organisms inhabit a stratum of water beneath the region of intense sunlight.

Between the stratum of these consumers and the bottom of the ocean is a relatively uninhabited stratum. In the sediments at the bottom of the ocean, the number of microorganisms again increases. The bottom dwellers are generally strict or facultative anaerobic decomposers. Many of them contribute significantly to the maintenance of biogeochemical cycles and produce such substances as ammonium ions, hydrogen sulfate, and nitrogen gas.

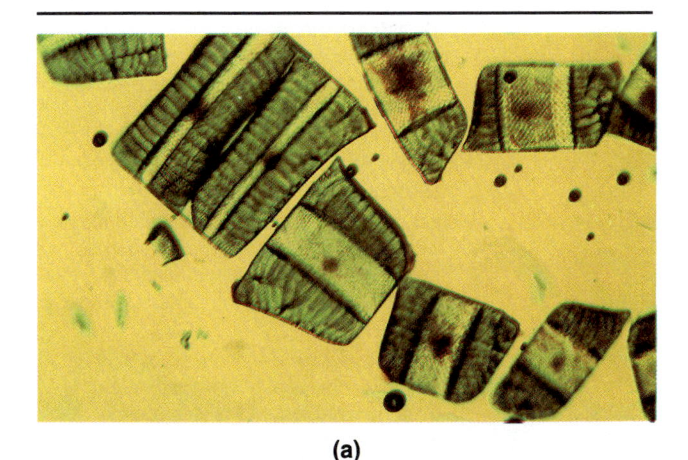

(a)

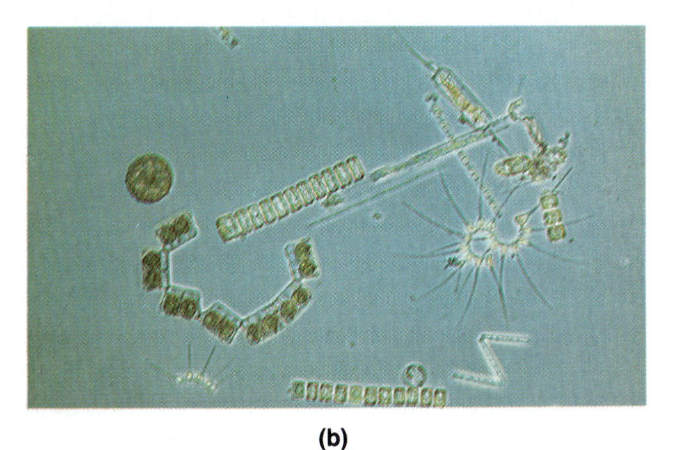

(b)

FIGURE 26.14 Producers of the ocean are called phytoplankton. They include organisms such as (a) the marine diatom *Isthmia* and (b) these diatoms collected in a plankton tow from Rhode Island Sound.

Water Pollution

Water is considered polluted if a substance or condition is present that renders the water useless for a particular purpose. Thus, the concept of water pollution is a relative one, depending on both the nature of the pollutants and the intended uses of the water. For example, human drinking water is considered polluted if it contains pathogens or toxic substances. Water that is too polluted for drinking may be safe for swimming; water that is too polluted for swimming may be acceptable for boating, industrial uses, or generating electrical power. The U.S. Environmental Protection Agency (EPA) has established drinking water standards and methods for water testing. Water that is fit for human consumption is termed **potable water**.

Pollutants

The major types of water pollutants are organic wastes such as sewage and animal manures, industrial wastes, oil, radioactive substances, sediments from soil erosion, and heat. Organic wastes suspended in water are usually decomposed by microorganisms provided the water contains sufficient oxygen for oxidation of the substances. The oxygen required for such degradatory activities is called **biological oxygen demand** (BOD). When BOD is high the water can be depleted of oxygen rapidly. Anaerobes increase in number, while populations of aerobic decomposers dwindle, leaving behind large quantities of organic wastes. Organic wastes also can contain pathogenic organisms from among the bacteria, viruses, and protozoa.

Industrial wastes contain metals, minerals, inorganic and organic compounds, and some synthetic chemicals. The metals, minerals, and other inorganic substances can alter the pH and osmotic pressure of the water, and some are toxic to humans and other organisms. Synthetic chemicals can persist in water because most decomposers lack enzymes to degrade them. Oil is another important water pollutant. Radioactive substances released into water persist as a hazard to living organisms until they have undergone natural radioactive decay.

Soil particles, sand, and minerals from soil erosion enter water from agricultural, mining, and construction activities. Nitrates, phosphates, and other nutrients enter the water from detergents, fertilizers, and animal manures. Such nutrient enrichment of water, called **eutrophication**, leads to excessive growth of algae and other plants. Eventually the plants become so dense that sunlight cannot penetrate the water. Many algae and other plants die, leaving large quantities of dead organic matter in the water. The high BOD of this organic matter leads to oxygen depletion and the persistence of undecomposed matter in the water.

Even heat can act as a water pollutant when large quantities of heated water are released into rivers, lakes, or oceans. Increasing the temperature of water decreases the solubility of oxygen in it. The altered temperature and decreased oxygen supply cause significant changes in the ecological balance of the aquatic environment.

The effects of water pollution are summarized in Table 26.1.

Pathogens in Water

Human pathogens in water supplies usually come from contamination of the water with human feces. When water is contaminated with fecal material, any pathogens that leave the body through the feces—many bacteria and viruses and some protozoa—can be present. The most common pathogens transmitted in water are listed in Table 26.2. Water usually is tested for fecal contamination by isolating *E. coli* from a water sample. *Escherichia coli* is called an **indicator organism**

TABLE 26.1 Effects of water pollution

Pollutant	Effects	Comment
Organic wastes (sewage, decaying plants, animal manures, wastes from food processing plants, oil refineries, and leather, paper, and textile plants)	Increase biological oxygen demand of water.	If adequate oxygen is available, these substances can be degraded by microorganisms usually present in water. If oxygen becomes depleted, decomposition is limited to what can be done by anaerobic decomposers. Water plants may be killed, and animals may be killed or caused to migrate.
Pathogenic organisms	Cause disease in humans who drink the water.	Most bacterial agents are well controlled in public drinking water, but viruses, especially those that cause hepatitis, still cause human disease. More effective means of removing viruses during purification are needed.
Inorganic chemicals and minerals	Increase the salinity and acidity of water and render it toxic.	Such chemicals should be removed during waste treatment. Heavy metals such as mercury, which are toxic to humans, should be prevented from entering water supplies.
Synthetic organic chemicals (herbicides, pesticides, detergents, plastics, wastes from industrial processes)	Can cause birth defects, cancer, neurological damage, and other illness.	Because these substances are not biodegradable, chemical or physical means must be used to remove them during waste treatment. Many such substances become magnified (increased in concentration) as they are passed along food chains.
Plant nutrients	Cause excessive and sometimes uncontrolled growth of aquatic plants (eutrophication). Impart undesirable odors and tastes to drinking water.	Removal of excess phosphates and nitrates from water during waste treatment is costly and difficult.
Sediments from land erosion	Cause silting of waterways and destruction of hydroelectric equipment near dams. Reduce light reaching plants in water and oxygen content of water.	
Radioactive wastes	Can cause cancer, birth defects, radiation sickness in large doses.	Effects can be magnified through food chains. Because such wastes are difficult to remove from water, preventing them from reaching water is exceedingly important.
Heated water	Reduces oxygen solubility in water. Alters habitats and kinds of organisms present. Encourage growth of some aquatic life, but can decrease growth of desired organisms such as fish.	

TABLE 26.2 Human pathogens transmitted in water

Organisms	Diseases Caused
Salmonella typhi	Typhoid fever
Other *Salmonella* species	Salmonellosis (gastroenteritis)
Shigella species	Shigellosis (bacillary dysentery)
Vibrio cholerae	Asiatic cholera
Vibrio parahaemolyticus	Gastroenteritis
Escherichia coli	Gastroenteritis
Yersinia enterocolitica	Gastroenteritis
Campylobacter fetus	Gastroenteritis
Hepatitis A virus	Hepatitis
Poliovirus	Poliomyelitis
Giardia intestinalis	Giardiasis
Balantidium coli	Balantidiasis
Entamoeba histolytica	Amoebic dysentery

because it is a natural inhabitant of the human digestive tract, and its presence in water indicates that the water is contaminated with fecal material.

Water Purification

Purification Procedures

Purification procedures for human drinking water are determined by the degree of purity of the water at its source. Water from deep wells or from reservoirs fed by clean mountain streams requires very little treatment to make it safe to drink. In contrast, water from rivers that contain industrial and animal wastes and even sewage from upstream towns requires extensive treatment before it is safe to drink. Such water is first allowed to stand in a holding reservoir until some of the particulate matter settles out. Then alum (aluminum potassium sulfate) is added to cause **flocculation**, or precipitation of suspended colloids such as clay. Many microorganisms are also removed from the water by flocculation.

Following flocculation treatment, water is filtered through beds of sand. This **filtration** removes nearly all the remaining microorganisms. Charcoal can be used instead of sand for filtration; it has the advantage of removing organic chemicals that are not removed by sand. Finally, water is chlorinated. **Chlorination** readily kills bacteria but is less effective in destroying viruses and cysts of pathogenic protozoa. The amount of chlorine required to destroy microorganisms is increased by the presence of organic matter in the water. Chlorine may combine with some organic molecules to form carcinogenic substances. Although current water chlorination procedures have not been proven to increase the risk of cancer in humans, the long-term effects of the interaction of chlorine and organic compounds are difficult to assess.

Tests for Water Purity

Water purity is usually tested by looking for coliform bacteria. The **coliform bacteria**, which include *E. coli*, are gram-negative, non-spore-forming, aerobic or facultative anaerobic bacteria that ferment lactose and produce acid and gas. Most municipal water supplies are regularly tested for the presence of coliform bacteria. The presence of any significant number of coliforms is evidence that the water may not be safe for drinking. Two methods of testing for coliform bacteria currently in use are the multiple-tube fermentation method and the membrane filter method.

The **multiple-tube fermentation method** (Figure 26.15) involves three stages of testing: the presumptive test, the confirmed test, and the completed test. In the **presumptive test** a water sample is used to inoculate lactose broth tubes. Each tube receives a water volume of 10 mL, 1 mL, or 0.1 mL. The tubes are incubated at 35°C and observed after 24 and 48 hours for evidence of gas production. Gas production provides presumptive evidence that coliform bacteria are present.

Because certain noncoliform bacteria also produce gas, additional testing is necessary to confirm the presence of coliforms. In the **confirmed test**, samples from the highest dilution showing gas production are streaked onto eosin-methylene blue (EMB) agar plates. Coliforms produce acid, and under acid conditions these dyes are absorbed by the organisms of a colony. Thus, after 24-hour incubation, coliform colonies have dark centers and may also have a metallic greenish sheen. Observing such colonies confirms the presence of coliforms. In the **completed test**, organisms from dark colonies are used to inoculate lactose broth and agar slants. The production of acid and gas in the lactose broth and the microscopic identification of gram-negative, non-spore-forming rods from slants constitute a positive completed test.

In the **membrane filter method** (Figure 26.16), a 100-mL water sample is drawn through a membrane filter having pores about 0.45 μm in diameter. This membrane, which traps bacteria on its surface, is then incubated on the surface of an absorbant pad previously saturated with an appropriate growth medium. After incubation, colonies form on the filter where bacteria were trapped during filtration. The presence of more than one colony per 100 mL of water indicates that the water may be unsafe for human consumption. If colonies are observed, additional tests can be performed to identify them specifically.

In addition to contamination with coliforms, drinking water also can contain other organisms sometimes referred to as "nuisance" organisms. Al-

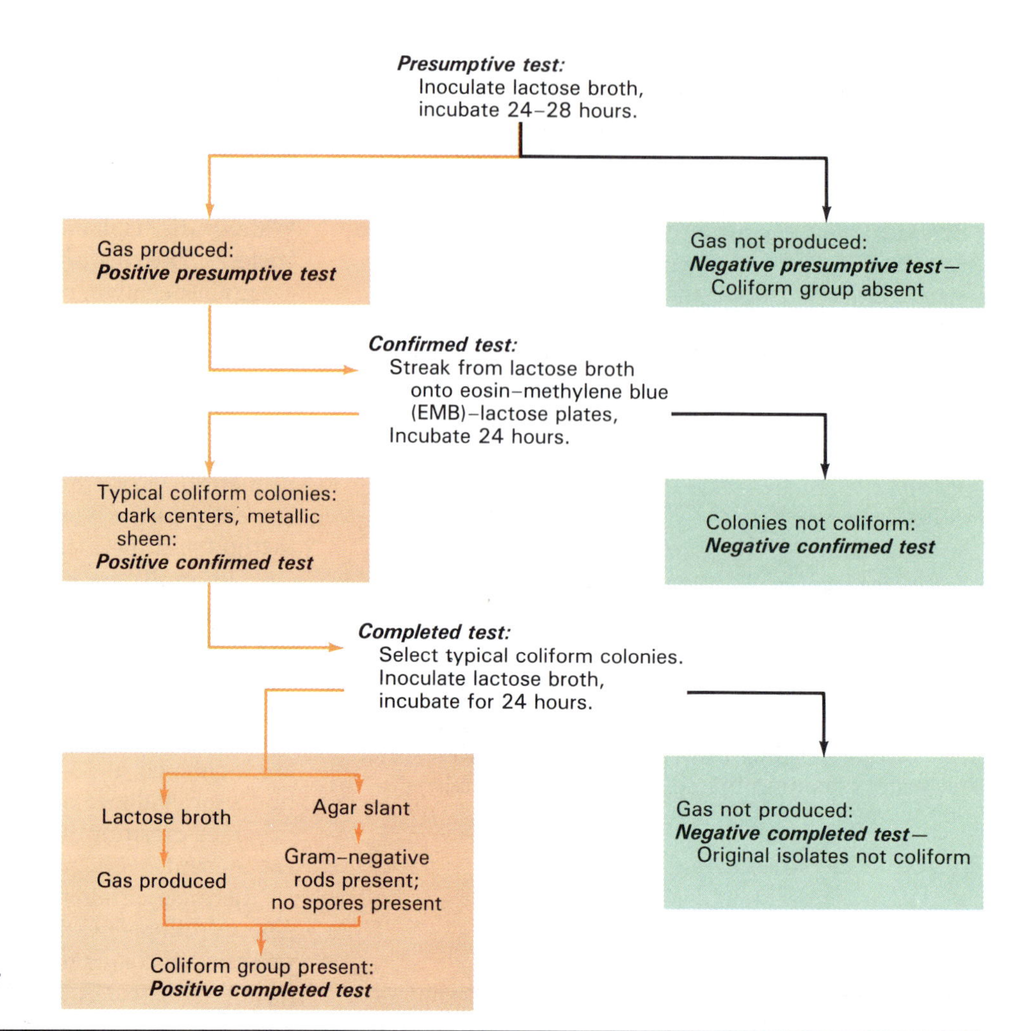

FIGURE 26.15 The multiple-tube fermentation test for water purity.

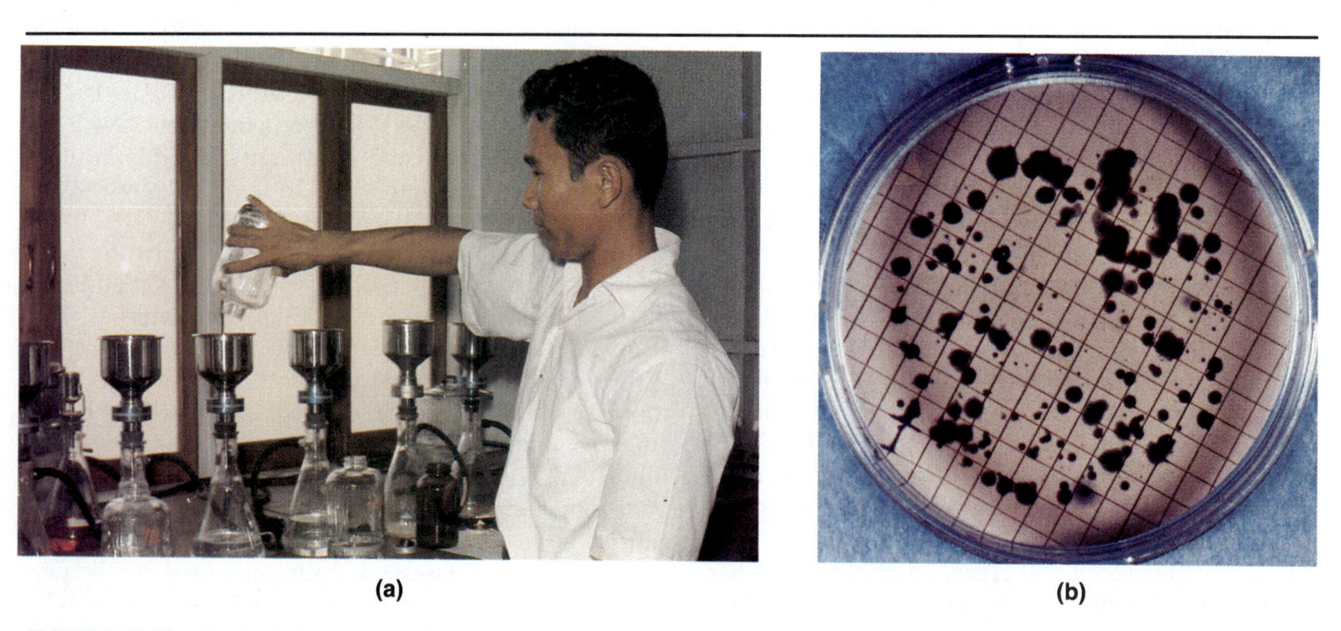

FIGURE 26.16 The membrane filter test for water purity involves (a) filtering water samples and catching organisms on filter pads, which are then incubated. (b) After incubation, the coliform colonies that appear are counted.

though these organisms do not produce human disease, they can affect the taste, color, or odor of the water. Some also can form insoluble precipitates inside water pipes. Nuisance organisms include sulfur bacteria, iron bacteria, slime-forming bacteria, and algae. Among the sulfur bacteria are *Desulfovibrio*, which produces hydrogen sulfide, and *Thiobacillus*, which produces sulfuric acid that can corrode pipes. Iron bacteria deposit insoluble iron compounds that can obstruct water flow in pipes. Eukaryotic algae, diatoms, and cyanobacteria reproduce rapidly in water exposed to sunlight (as in reservoirs); they can become so numerous that they clog filters used in the purification process. Identifying these various nuisance organisms in water is a tedious task because different tests are required for each kind of organism. None of the tests are performed routinely but may be done when citizens complain of objectionable tastes, odors, or colors in water.

Water also can be contaminated with various organic and inorganic substances, especially when water supplies are drawn from rivers that have received upstream effluents of industrial wastes. Although it is at least theoretically possible to detect these substances through chemical analyses, such tests are rarely performed.

SEWAGE TREATMENT

Sewage is used water and the wastes it contains. It is about 99.9 percent water and about 0.1 percent solid or dissolved wastes. These wastes include household wastes (human feces, detergents, grease, and anything else people put down the drain or garbage disposal unit), industrial wastes (acids and other chemical wastes and organic matter from food-processing plants), and wastes carried by rainwater that enters sewers.

Sewage treatment is a relatively modern practice. Until recent years many large U.S. cities dumped untreated sewage into rivers and oceans; many cities along the Mediterranean Sea still do! When small amounts of sewage are dumped into fast-flowing, well-oxygenated rivers, the natural activities of decomposers in the river will purify the water. However, large amounts of sewage overload the purification capacity of the rivers. Cities downstream are then forced to take their water supplies from rivers that contain wastes of the cities upstream. Fortunately, most U.S. cities now have some form of sewage treatment.

Complete sewage treatment consists of three steps: primary, secondary, and tertiary treatment (Figure 26.17). In **primary treatment** physical means are used to remove solid wastes from sewage. In **secondary treatment** biological means (actions of decomposers) are used to remove solid wastes that remain after primary treatment. In **tertiary treatment** chemical and physical means are used to produce an effluent of water pure enough to drink. Let us look at each of these processes in more detail.

Primary Treatment

As raw sewage enters a sewage-treatment plant, several physical processes are used to remove wastes in primary treatment. Screens remove large pieces of floating debris, and skimmers remove oily substances. Water is then directed through a series of sedimentation tanks, where small particles settle out. The solid matter removed by these procedures accounts for about one-half of the total solid matter in sewage. Flocculating substances can be used to increase the amount of solids that settle out and thus the proportion of solids removed by primary treatment. Sludge is removed from the sedimentation tanks intermittently or continuously, depending on the design of the treatment plant.

Secondary Treatment

The effluent from primary treatment flows into secondary treatment systems. These systems are of two types: trickling filter systems and activated sludge systems. Both systems make use of the decomposing activity of aerobic microorganisms. The BOD is high in secondary treatment systems, so the systems provide for continuous oxygenation of the wastewater.

In a **trickling filter system** (Figure 26.18), sewage is spread over a bed of rocks about 2 m deep. The individual rocks are 5 to 10 cm in diameter and are coated with a slimy film of aerobic organisms. Spraying oxygenates the sewage so that the aerobes can decompose organic matter in it. Such a system is less efficient but less subject to operational problems than an activated sludge system. It removes about 80 percent of the organic matter in the water.

In an **activated sludge system**, the effluent from primary treatment is constantly agitated, aerated, and added to solid material remaining from earlier water treatment. This **sludge** contains large numbers of aerobic organisms that digest organic matter in wastewater. However, filamentous bacteria multiply rapidly in such systems and cause some of the sludge to float on the surface of the water instead of settling out. This phenomenon, called **bulking**, allows the floating matter to contaminate the effluent. The sheathed bacterium *Sphaerotilus* (Figure 26.19), which sometimes proliferates rapidly on decaying leaves in small streams and causes a bloom, can interfere with the operation of sewage systems in this way. Its fil-

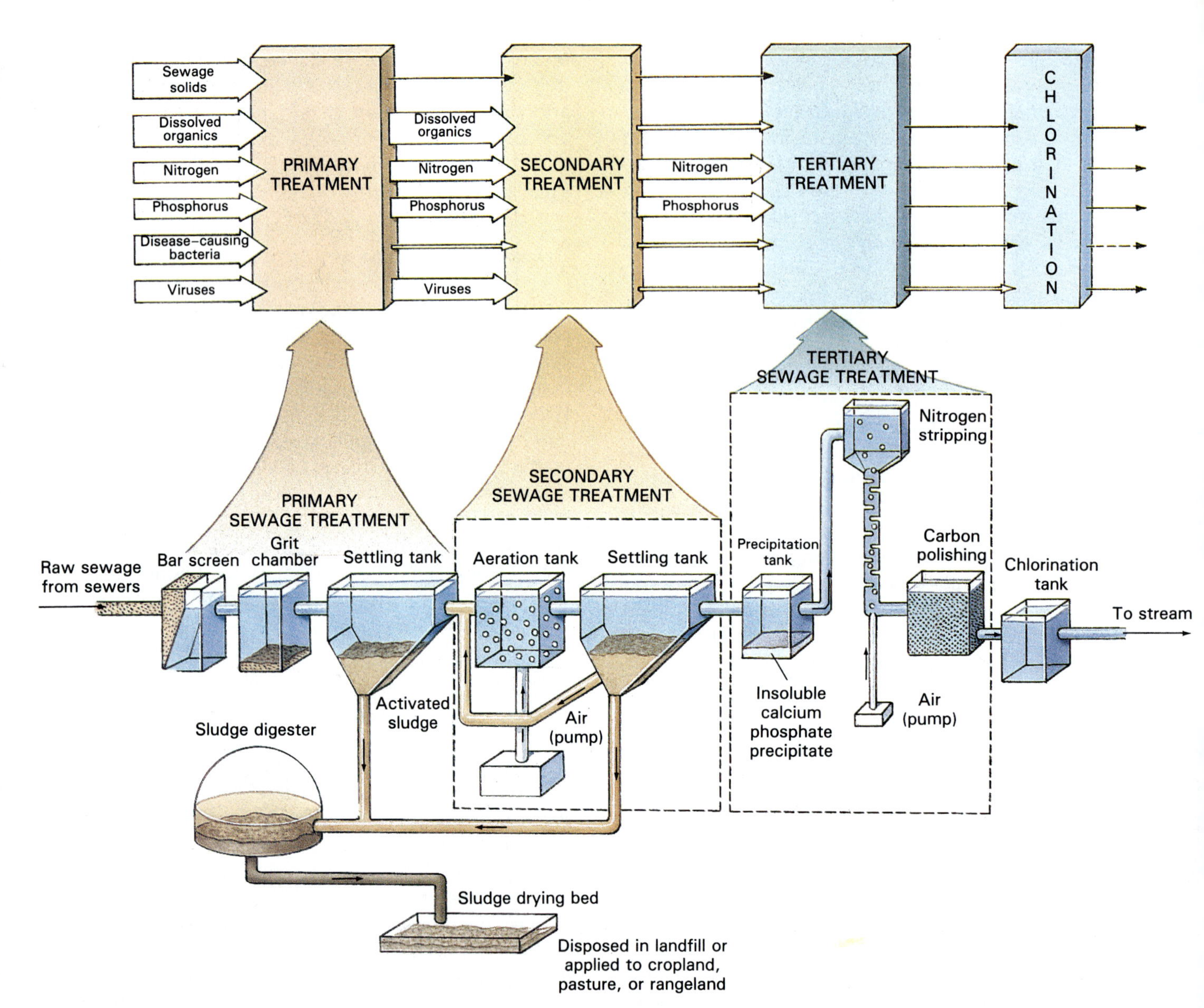

FIGURE 26.17 An overview of a sewage treatment plant showing primary and secondary treatment facilities.

FIGURE 26.18 Trickling filters used in secondary sewage treatment.

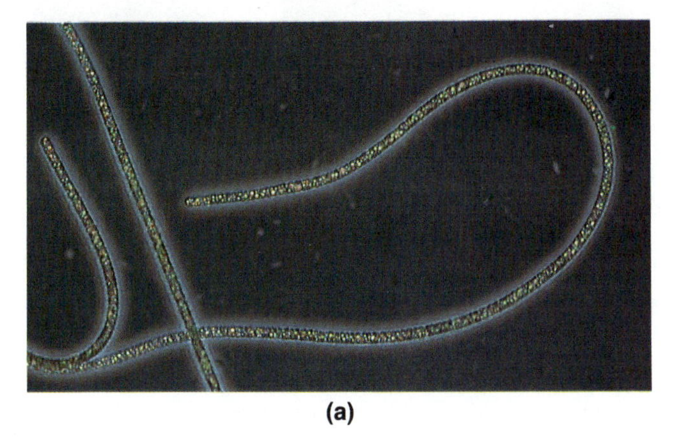

(a)

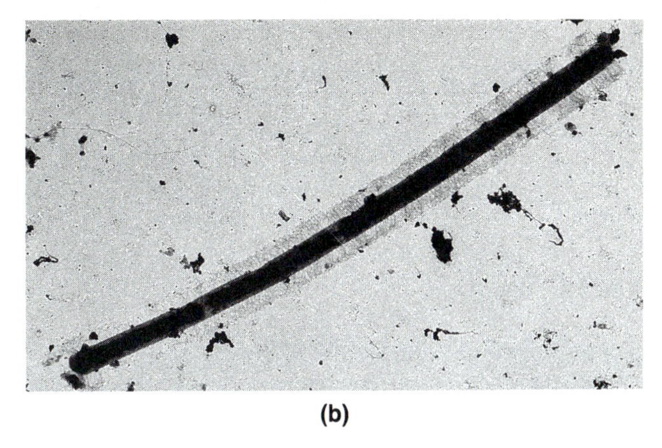

(b)

FIGURE 26.19 Two sheathed bacteria used in trickling filters are (a) *Beggiatoa* and (b) *Sphaerotilus*.

FIGURE 26.20 Injection of municipal sewage sludge into farmland is designed to return nutrients to the soil; but if treatment is not complete, it may also add pathogens to the soil.

aments clog filters and create floating clumps of undigested organic matter.

Sludge from both primary and secondary treatments can be pumped into **sludge digesters**. Here, oxygen is virtually excluded, and anaerobic bacteria partially digest the sludge to simple organic molecules and the gases carbon dioxide and methane. The methane can be used for heating the digester and providing for other power needs of the treatment plant. Undigested matter can be dried and used as a soil conditioner or as landfill (Figure 26.20).

Tertiary Treatment

The effluent from secondary treatment contains only 5 to 20 percent of the original quantity of organic matter and can be discharged into flowing rivers without causing serious problems. However, this effluent can contain large quantities of phosphates and nitrates, which can increase the growth rate of plants in the river. Tertiary treatment is an extremely costly process that involves physical and chemical methods. Fine sand and charcoal are used in filtration. Various flocculating chemicals precipitate phosphates and particulate matter. Denitrifying bacteria convert nitrates to nitrogen gas. Finally, chlorine is used to destroy any remaining organisms. Therefore, water that has received tertiary treatment can be released into any body of water without danger of eutrophication. Such water is pure enough to be recycled into a domestic water supply. However, the chlorine-containing effluent, when released into streams and lakes, can react to produce carcinogenic compounds that may enter the food chain or be ingested directly by humans in their drinking water. It would be safer to remove the chlorine before releasing the effluent, but this is rarely done today, although the cost is not great. As was mentioned in Chapter 13, ultraviolet lights are now replacing chlorination as the final treatment of effluent. (p. 339) They destroy microbes without adding carcinogens to our streams and waters.

Septic Tanks

About 50 million rural families in the United States do not have access to city sewer connections or their treatment facilities. These homes rely on backyard **septic tank** systems (Figure 26.21).

Homeowners must be careful not to flush or put materials such as poisons and grease down the drain, as these might kill the beneficial microbes in the septic tank that decompose sludge solids that accumulate there. This will necessitate immediate pumping of the tank by a vehicle known as the "honey wagon" to prevent sewage from backing up into the house. Even with normal operation, it is occasionally necessary to pump the sludge from the tank and haul it to a sewage treatment plant.

Soluble components of the sewage continue out of the septic tank into the drain field, where they seep through perforated pipe, past a gravel bed, into the soil, which filters out bacteria and some viruses, and binds phosphate. Soil bacteria decompose organic materials. It is, of course, necessary to place drain fields where they will not allow seepage into wells, a difficult problem on hills or in densely populated areas. Drain fields cannot be used where the water table is too high or the soil is insufficiently permeable, such as in rocky areas. After about 10 years, the average drain field clogs up and can no longer be used.

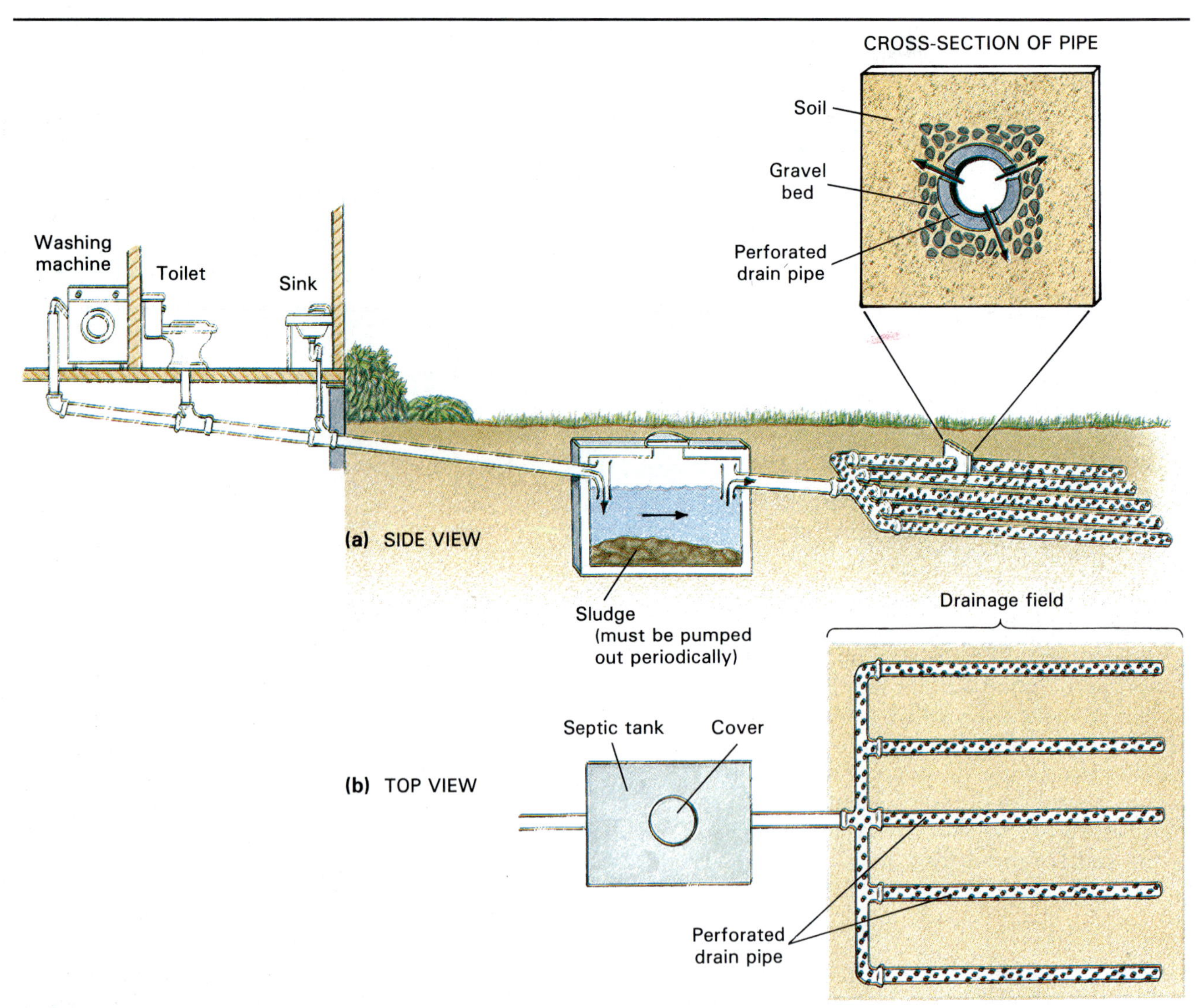

FIGURE 26.21 Sewage disposal by means of a septic tank and drain field system. (a) Solid materials settle out as sludge, which undergoes bacterial decay, while soluble materials continue to the drain field where (b) they seep out into the soil, which filters out bacteria and some viruses and binds phosphate. Soil bacteria decompose organic materials.

ESSAY

The Need for Nitrogen: The *Azolla-Anabaena* Symbiosis

Although 80 percent of our atmosphere consists of nitrogen gas (N_2) and all organisms need nitrogen for proteins, no higher organism—plant, animal, or fungus—can use the N_2 molecule, as is. This is because the triple bond between the two nitrogen atoms is extremely inert. Only some prokaryotes have the nitrogenase enzymes necessary to "fix" atmospheric nitrogen, changing it into ammonia (NH_3), a form that can be used directly or indirectly by all other organisms.

Worldwide, our agricultural soils require about 150 million metric tons of fixed nitrogen per year. As our population grows, these needs will increase. About one-third of the nitrogen used today is supplied by industrially manufactured fertilizers from the Haber chemical process, which produces ammonia directly from atmospheric nitrogen and hydrogen. Another third comes from the symbiotic relationship of *Rhizobium* bacteria in nodules on roots of leguminous plants such as clover and alfalfa. The last third comes from free-living cyanobacteria plus an assortment of other symbiotic associations. A very interesting example of this last category is the symbiosis between the water fern *Azolla* (Figure 26.22a) and the cyanobacterium *Anabaena azollae* (Figure 26.22b). It is the only plant-cyanobacterium symbiosis that has had any practical applications to human affairs.

Exactly when the Chinese began to use *Azolla* as a fertilizer, or "green manure," and as food for poultry and livestock is uncertain. We do know, however, that it is defined in an ancient Chinese dictionary over 2000 years old, and poems from the same period describe gathering of the plants. *Azolla* is also used as a Chinese medicine. The Chinese name for *Azolla*, "Man Chiang Hung," means "whole river is red," referring to the reddish color *Azolla* takes on at extremes of temperature, pH, or nutrient shortages (Figure 26.23).

The genus *Azolla* was established by Lamarck in 1783 after he examined specimens from Chile. Today we recognize six species within this genus: four in the New World, and two Old World species. *Azolla* is the only fern to have a cyanobacterium endosymbiont, and all six species contain the same one species of cyanobacterium, *Anabaena azollae*, a member of the family Nostocaceae. It is interesting that all cyanobacterial symbionts, whether they associate with algae in lichens, or with diatoms, liverworts, ferns, or cycads, are always members of the family Nostocaceae.

(a)

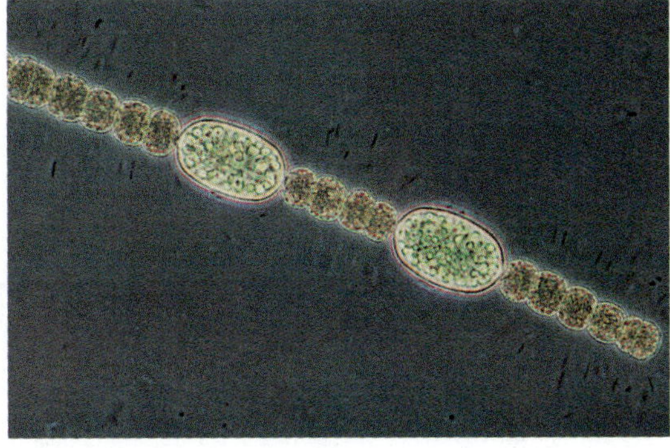

(b)

FIGURE 26.22 (a) Tiny floating fern, *Azolla*, is usually found floating on water with thin roots hanging below the many small leaves. Inside the *Azolla*, in a symbiotic relationship, lives (b) the cyanobacterium *Anabaena*, seen here with two large heterocysts that contain nitrogen-fixing enzymes (phase contrast, 160X).

The *Anabaena azollae* filaments (trichomes) have three types of cells. In symbiotic forms there are approximately 61 percent vegetative cells, 23 percent heterocysts, and 16 percent akinetes. The heterocysts contain the nitrogen-fixing enzymes. Free-living *A. azollae* have only 5 to 8 percent heterocysts. In fact, symbiotic forms are found to have 12 to 20 times greater nitrogen-fixing activity. This gives them a competitive advantage over other types of floating plants.

The *Azolla* fern (*azo*: to dry; *ollyo*: to kill) dies in dry conditions and is usually found floating on water. Its diameter is commonly about 1 cm, although one Old World species, *A. nilotica* can reach over 40 cm. The plants have branched, floating, multileaved stems (rhizomes) with small roots hanging below. Each leaf is deeply bilobed. The lower lobe is thin, colorless, and floats on the water's surface. The upper lobe extends up into the air, is green, and has a hollow chamber within it that contains the symbiotic *Anabaena* plus small populations of the bacteria *Pseudomonas* and *Azotobacter*. At the center of growth (meristem) for the *Azolla* leaves, there is a permanent cluster of undifferentiated, non-nitrogen-fixing *Anabaena* filaments. As each new leaf forms, *Anabaena* filaments grow into the upper lobe cavity and are cut off inside as the leaf separates from the meristem and moves away on the elongating rhizome. As this happens, the *Anabaena* filaments mature, and heterocysts rapidly form and begin fixing nitrogen, with the *Azolla* absorbing most of the ammonia formed. Both the *Azolla* and the *Anabaena* carry on photosynthesis at a very rapid rate—twice that of which *Azolla* is capable alone. Thus, this association is a highly efficient biologic converter of solar energy and fixer of nitrogen.

ESSAY Need for Nitrogen: The *Azolla-Anabaena* Symbiosis

In China, *Azolla* is grown in a separate pond early in the spring. After rice seedlings have been planted, *Azolla* is added to the rice paddies, where it grows on the water surface. Just before the rice enters its elongation stage, the *Azolla* is incorporated into the soil as fertilizer. During the 1970s, Chinese researchers found a hardier strain of *Azolla* that sprouts earlier in the spring. Thus, the Chinese were able to extend *Azolla* cultivation into more northern parts of the country, as well as improve the rate of growth in the southern regions. This process caught the attention of other Far Eastern countries, who thought to extend its use to their rice paddies.

FIGURE 26.23 *Azolla caroliniana* growing on the surface of a pond turns it a deep red color.

Other researchers, however, saw this as a possible basis for a new biotechnology that would use microbial cells as bioreactors for solar energy conversion and nitrogen fixation. They developed plastic foams that were cut into cubes, sterilized, immersed in flasks of nutrient media, and inoculated with *A. azollae*. Evidently the pores inside the foam resemble the leaf cavities of *Azolla*. *Anabaena* filaments grow into them, develop 10 to 18 percent heterocysts, and carry on nitrogen fixation. And, very importantly, these filaments are able to release the ammonia they form from their cells into the surrounding medium—something that free-living *A. azollae* cannot do. This technique of growing cells in synthetic frameworks is called *immobilization*. The *Azolla* probably serves as a natural immobilization system. Research on artificial immobilization of various kinds of cells holds out hope for man-made bioreactors that can supply the needs of this planet, or eventually of humanity in other parts of the universe.

CHAPTER SUMMARY

RELATED KEY TERMS

FUNDAMENTALS OF ECOLOGY

Ecology Defined

- **Ecology** is the study of relationships among organisms and their environment.

The Nature of Ecosystems

- An **ecosystem** includes all the **biotic** and **abiotic factors** of an environment. It will always support life of **indigenous organisms** and sometimes supports **nonindigenous** ones. All the living organisms in an **ecosystem** make up a **community**.

Energy Flow in Ecosystems

- Energy in an ecosystem flows from the sun to **producers** to **consumers. Decomposers** obtain energy from digesting dead bodies and wastes of other organisms. The nutrients they release can be recycled.

BIOGEOCHEMICAL CYCLES

- Although energy is continuously available, nutrients must be recycled from dead organisms and waste to make them available for other organisms. — **biogeochemical cycles**

The Water Cycle

- The **water cycle** is summarized in Figure 26.2. — **hydrologic cycle**

The Carbon Cycle

- The **carbon cycle** is summarized in Figure 26.3.

The Nitrogen Cycle and Nitrogen Bacteria

RELATED KEY TERMS

- The **nitrogen cycle** is summarized in Figure 26.4.
- **Nitrogen fixation** is the reduction of atmospheric nitrogen to ammonia. It is accomplished by certain free-living aerobes and anaerobes but mostly by *Rhizobium*.
- *Rhizobium* cells accumulate around legume roots and change to **swarmer cells**, which invade root cells and become bacteroids. **Nitrogenase** in **bacteroids** catalyzes nitrogen-fixation reactions.
- **Nitrification** is the conversion of ammonia to nitrites and nitrates; nitrites are formed by *Nitrosomonas*, and nitrates by *Nitrobacter*.
- **Denitrification** is the conversion of nitrates to nitrous oxide and nitrogen gas. It is accomplished by a variety of organisms, especially in waterlogged soils.

The Sulfure Cycle and Sulfur Bacteria

- The **sulfur cycle** is summarized in Figure 26.7.
- Various primitive bacteria carry out **sulfate reduction, sulfur reduction,** or **sulfur oxidation**.

Other Biogeochemical Cycles

- The **phosphorus cycle** is summarized in Figure 26.9
- All elements found in living organisms must be recycled.

AIR

Microorganisms Found in Air

- Microorganisms are transmitted by air but do not grow in air. Air is analyzed by exposing agar plates to air and by drawing air over the surface of agar or into a liquid medium and subsequently studying the organisms found.

Methods for Controlling Microorganisms in Air

- Microorganisms in air are controlled by chemical agents, radiation, filtration, and unidirectional airflow.

SOIL

Components of Soil

- Soil contains inorganic rocks, minerals, water, and gases.
- It also contains organic matter (**humus**) and many microorganisms.

Microorganisms in Soil

- Microorganisms of all taxonomic groups are found in soil.
- Physical factors that affect soil microorganisms include moisture, oxygen concentration, pH, and temperature.

microenvironments

- Organisms also alter the characteristics of their environment as they use nutrients and release wastes.
- Soil microorganisms are important as **decomposers** in the carbon cycle and in all phases of the nitrogen cycle.

Soil Pathogens

- Soil pathogens affect mainly plants and insects.
- Various species of *Clostridium* are important human pathogens in soil.

WATER

Freshwater Environments

- Freshwater environments are characterized by low salinity and variability in temperature, pH, and oxygen concentration.
- Microorganisms from all taxonomic groups are found in freshwater environments. Bacteria—aerobes where oxygen is plentiful and anaerobes where oxygen is depleted—are especially abundant.

Marine Environments

- Marine environments are characterized by high salinity, smaller variability in temperature, pH, and oxygen concentration. As depth increases, pressure increases and sunlight penetration decreases. Microorganisms from all taxonomic groups are also found in marine environments. Photosynthetic organisms are found nearest the surface, heterotrophs in surface and lower strata, and decomposers in bottom sediments.

Water Pollution

- Water is polluted if a substance or condition is present that renders the water useless for a particular purpose.

- The effects of water pollution are summarized in Table 26.1.
- Many human pathogens can be transmitted in water.
- Such pathogens are listed in Table 26.2.

Water Purification

- Water purification involves **flocculation** of suspended matter, **filtration**, and **chlorination.**
- Tests for water purity are designed to detect **coliform bacteria**; they include the **multiple-tube fermentation** and the **membrane filter methods**.

RELATED KEY TERMS

potable water
biological oxygen demand (BOD)
eutrophication
indicator organism
presumptive test
confirmed test **completed test**

SEWAGE TREATMENT

- **Sewage** is used water and the wastes it contains.

Primary Treatment

- **Primary treatment** is the removal of solid wastes by physical means.

Secondary Treatment

- **Secondary treatment** is the removal of organic matter by the action of aerobic bacteria.

Tertiary Treatment

- **Tertiary treatment** is the removal of all organic matter, nitrates, phosphates, and any surviving microorganism by physical and chemical means.

trickling filter system
activated sludge system
sludge
bulking
sludge digesters **septic tank**

QUESTIONS FOR REVIEW

A.

1. What is ecology?
2. What are some important properties of an ecosystem?
3. How does energy flow through an ecosystem?

B.

4. Why is recycling of substances essential in an ecosystem?
5. Summarize the water cycle.
6. Summarize the carbon cycle.

C.

7. Summarize the nitrogen cycle.
8. What is nitrogen fixation, and how does *Rhizobium* contribute to it?
9. What is nitrification, and how is it accomplished?
10. What is denitrification, and how is it accomplished?
11. Summarize the sulfur cycle and the roles of microorganisms in it.
12. Briefly summarize the phosphorus cycle.
13. Why are there cycles for all elements found in living things?

D.

14. What kinds of microorganisms are found in air?
15. How is air analyzed for microorganisms?
16. How are microorganisms in air controlled?

E.

17. What are the typical components of soil?
18. What are the properties of soil components?
19. What kinds of microorganisms are found in soil?
20. In what ways do soil microorganisms interact?
21. How do soil microorganisms participate in biogeochemical cycles?
22. What human pathogens are found in soil?

F.

23. What are the main characteristics of freshwater environments?
24. What kinds of microorganisms are found in fresh water?
25. What are the main characteristics of marine environments?
26. What kinds of microorganisms are found in ocean water?

G.

27. What is water pollution?
28. How does water pollution affect microorganisms and humans?
29. What human pathogens are transmissible in water?

H.

30. What processes are used to purify water?
31. What tests are used to determine water purity?

I.

32. What is sewage?
33. What processes are involved in primary sewage treatment?
34. What processes are involved in secondary sewage treatment?
35. What processes are involved in tertiary sewage treatment?

PROBLEMS FOR INVESTIGATION

1. Determine the nature of air or water pollution in your locale. Explain how such pollution affects microorganisms and humans. If possible, suggest ways to reduce its effects.
2. How would you protect workers in sewage disposal plants from being infected with waterborne diseases, and from transmitting such diseases to people outside the treatment plants?
3. Propose a method of controlling photosynthetic bacteria in water supplies.
4. Visit a local water-purification plant or a sewage-treatment plant. Determine which microbiological techniques are used in that plant and the purposes for which they are used.
5. Read about alternative water treatment methods that can be used by campers or in times of emergency.
6. What roles do microbes play in compost piles that recycle nutrients?

SOME INTERESTING READING

American Public Health Association. 1992. *Standard methods for the examination of water and waste water*. Washington, D.C.

Anonymous. 1989. *Valdez* "bugs" chomp away. *Science News* 136, no. 3 (July 15):38.

Atlas, R. M., and R. Bartha. 1987. *Microbial ecology: Fundamentals and Applications*. Menlo Park, CA: Benjamin-Cummings.

Brill, W. J. 1981. Agricultural microbiology. *Scientific American* 245, no. 3 (September):198.

Campbell, R. E. 1984. *Microbial ecology*. Oxford: Blackwell Scientific Publications.

Childress, J. J., H. Felbeck, and G. N. Somero. 1987. Symbiosis in the deep sea. *Scientific American* 256, no. 5 (May):114.

Chollar, S. 1990. The poison eaters: some cultivated bacteria are finally getting down to business, gobbling up hazardous pollutants. *Discover* 11(April):76–79.

Hanselmann, K. W. 1991. Microbial energetics applied to waste repositories. *Experientia* 47(7):645–87.

Margulis, L., D. Chase, and R. Guerero. 1986. Microbial communities. *BioScience* 36:160.

McCabe, A. 1990. The potential significance of microbial activity in radioactive waste disposal. *Experientia* 46(8):779–87.

Nebel, B. J. 1993. *Environmental science: The way the world works*. Englewood Cliffs, NJ: Prentice Hall.

Olson, B. H. 1991. Tracking and using genes in the environment. *Environmental Science & Technology* 25(April):604–11.

Composite SEM (1375X) of a 0.1-mm gold structure panned from Lillian Creek, Alaska, thought to be a low-temperature pseudomorph (chemical replacement or accretion form) of a *Pedomicrobium*-like budding bacterium. Gold has accumulated in or on cell surfaces, revealing the morphology of hyphae and budding cells. The gold most likely originated in nearby rocks that break down, releasing the gold into solution in streams and water-logged soil. There, bacteria may act as scavengers of the soluble gold complexes. Examination of single cells suggests that the bacterioform structures may not be solid gold, but a thin (less than 100 nm) covering built up on the original cells. This is similar to the encrusting with iron and manganese oxides of other hyphal budding bacteria in the genus *Pedomicrobium*. Bacterioform gold has also been found in China and South Africa, indicating that budding bacteria may have played a role in the geochemistry of gold for more than half the Earth's history.

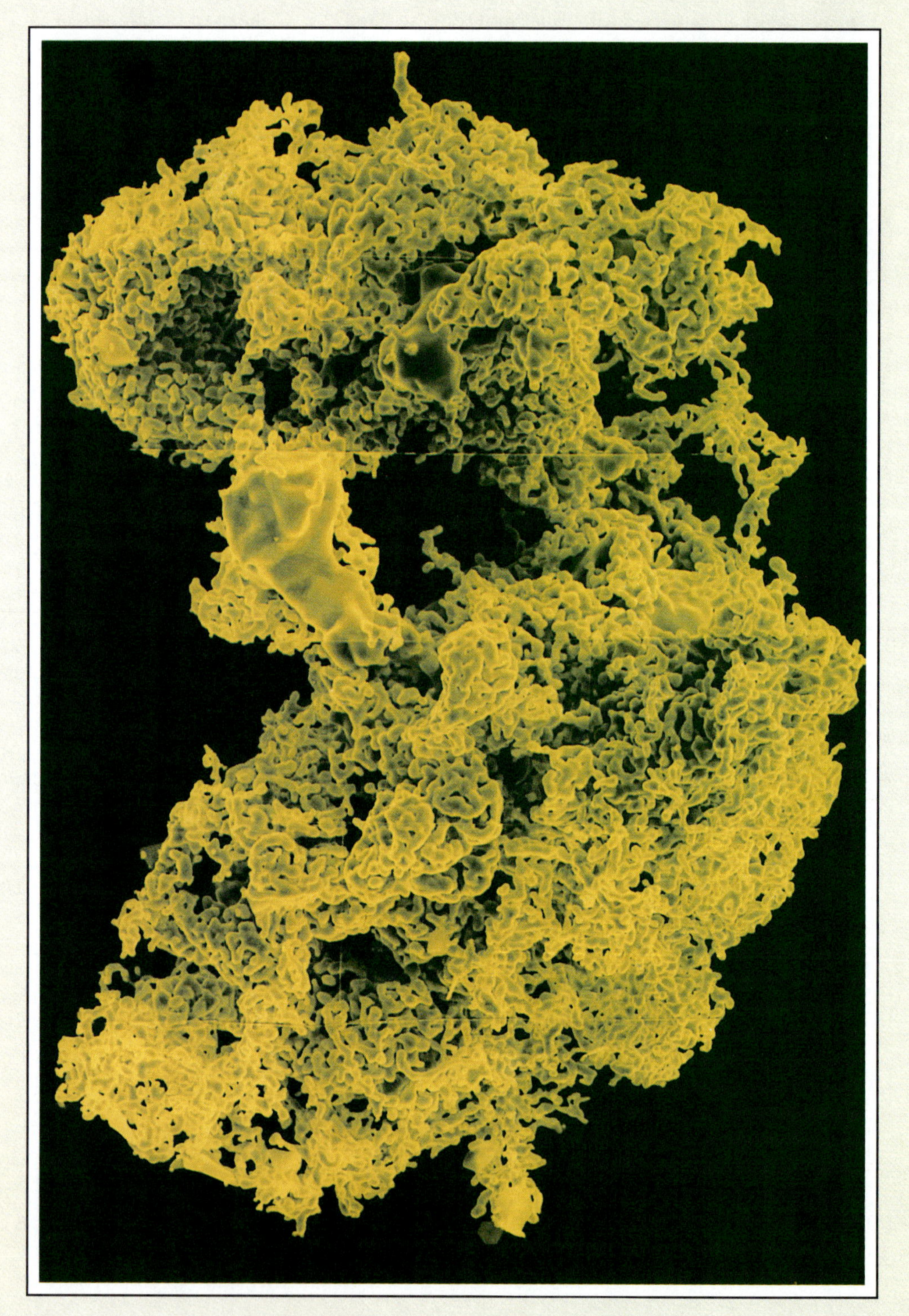

27 Applied Microbiology

This chapter focuses on the following questions:

A. What kinds of microorganisms are found in different categories of food?

B. How are diseases transmitted in food?

C. How can food spoilage and disease transmission be prevented, and what standards relate to these problems?

D. How are microorganisms used as food or in the making of food products?

E. How can microbes be used in industry, and what problems are associated with their use?

F. How are microbes used in the manufacture of beer, wines, and spirits?

G. What is the role of microorganisms in the manufacture of simple organic compounds, antibiotics, enzymes, and other biologically useful substances?

H. How are microorganisms used in mining?

I. How are microorganisms used in waste disposal?

MICROORGANISMS FOUND IN FOOD

Anything that people eat or drink is food for microorganisms, too (Figure 27.1). Most substances consumed by humans are derived from plants, which, of course, grow in soil, or from animals, which live in contact with soil and so have soil organisms on them. Although soil organisms usually are not human pathogens, many can cause food spoilage. The handling of foods from harvest or slaughter to human consumption provides many more opportunities for the foods to become contaminated with microorganisms. Unsanitary practices by food handlers and unsanitary working conditions frequently lead to contamination of foods with pathogens. Improper storage and preparation procedures at home and especially in restaurants can lead to further contamination with pathogens. Improper refrigeration of prepared foods is a major source of food poisoning.

FIGURE 27.1 Microorganisms utilize the same foods that humans do, as can be seen from this moldy peach.

Grains

When properly harvested, the various kinds of edible grains such as rye and wheat are dry. Because of the lack of moisture, few microorganisms thrive on them. However, if stored under moist conditions, they can easily become contaminated with molds and other microorganisms. (Recall the discussion of ergot and aflatoxin in Chapter 22.) ∞ (p. 622) Insects, birds, and rodents also transmit microbial contaminants to grains.

Most grains are used to make breads and cereals. Humans have made breads for thousands of years, and some loaves of bread on display in the British Museum are 4000 years old. Although the first instance of leavening bread with yeast probably occurred by accident, special strains of *Saccharomyces cerevisiae* that produce large amounts of carbon dioxide are now added to bread dough to make it rise. This process is described in more detail later.

PUBLIC HEALTH

Grain Products

Unprocessed grain products, such as granola-type foods, can have rodent feces in them. These rodent feces can contain viable tapeworm eggs. ∞ (Chapter 12, p. 310) As the consumption of such products has increased in the United States in recent years, so has the incidence of mouse tapeworm infections in humans. In spite of federal government standards that limit the amounts of rodent and insect body parts and wastes that can be present in grain products, mouse tapeworm eggs apparently reach consumers. Even the fumigation of grains and grain products in railroad cars fails to kill helminth eggs.

As with raw grain, bread is susceptible to contamination and spoilage by various molds. *Rhizopus nigricans* is the most common bread mold, but several species of *Penicillium*, *Aspergillus*, and *Monilia* also thrive in bread. Contamination of bread with *Monilia sitophila*, a pink bread mold, is especially dreaded by bakers because it is almost impossible to eliminate from a bakery once it has become established. Rye bread is particularly likely to become contaminated with *Bacillus* species, which hydrolyze proteins and starch and give the bread a stringy texture. If not killed by baking, *Bacillus* spores germinate and cause rapid and extensive damage to freshly baked bread.

Fruits and Vegetables

Millions of commensal bacteria, especially *Pseudomonas fluorescens*, are found on the surfaces of fruits and vegetables. These foods also easily become contaminated with organisms from soil, animals, air, irrigation water, and equipment used to pick, transport, store, or process them. Pathogens such as *Salmonella*, *Shigella*, *Entamoeba histolytica*, *Ascaris*, and a variety of viruses can be transmitted on the surfaces of fruits and vegetables. However, the skins of most plant foods contain waxes and release antimicrobial substances, both of which tend to prevent microbial invasion of internal tissues.

Certain vegetables are particularly vulnerable to microbial attack and spoilage. Leafy vegetables and potatoes are susceptible to bacterial soft rot by *Erwinia carotovora*. The fungus *Phytophthora infestans* caused the Irish potato famine of 1846. Tomatoes, cucumbers, and melons can be damaged by the fungus *Fusarium*, which causes soft rot and cracking of tomato skins.

Fruit flies pick up the fungus from infected tomatoes and transit it to healthy ones as they deposit eggs in the cracks. Other insects pierce tomatoes to feed on them and at the same time introduce *Rhizopus*, which breaks down pectin and turns the tomato into a bag of water.

Fruits are likewise susceptible to spoilage by microbial action. Fresh fruit juices, because of their high sugar and acid content, provide an excellent medium for growth of molds, yeasts, and bacteria of the genera *Leuconostoc* and *Lactobacillus*. Grapes and berries are damaged by a wide variety of fungi, and large numbers of stone fruits, such as peaches, can be destroyed overnight by brown rot due to *Monilia fructicola*. *Penicillium expansum*, which grows on apples, produces a toxin, patulin, that can easily contaminate cider. Other *Penicillium* species produce blue and green mold on citrus fruits.

Meats, Poultry, Fish, and Shellfish

Meat animals arrive at slaughterhouses with large numbers of many kinds of microorganisms in the gut and feces, on hides and hoofs, and sometimes in tissues. At least 70 pathogens have been identified among these organisms. Nearly all animal carcasses in slaughterhouses in the United States are inspected by a veterinarian or a trained inspector, and those found to be diseased are condemned and discarded (Figure 27.2). Some of the most common diseases identified in slaughterhouses are abscesses, pneumonia, septicemia, enteritis, toxemia, nephritis, and pericarditis. Lymphadenitis is especially common in sheep and lambs.

Even after animals are slaughtered and the carcasses are hung in refrigerated rooms to age, microorganisms sometimes damage the meat. Several molds grow on refrigerated meats, and *Cladosporium herbarum* can grow on frozen meats. Mycelia of *Rhizopus* and *Mucor* produce a fluffy, white growth referred to as "whiskers" on the surfaces of hanging carcasses. The bacterium *Pseudomonas mephitica* releases hydrogen sulfide and causes green discoloration on refrigerated meat under low oxygen conditions. Several species of *Clostridium* cause putrifaction called bone stink deep in the tissues of large carcasses.

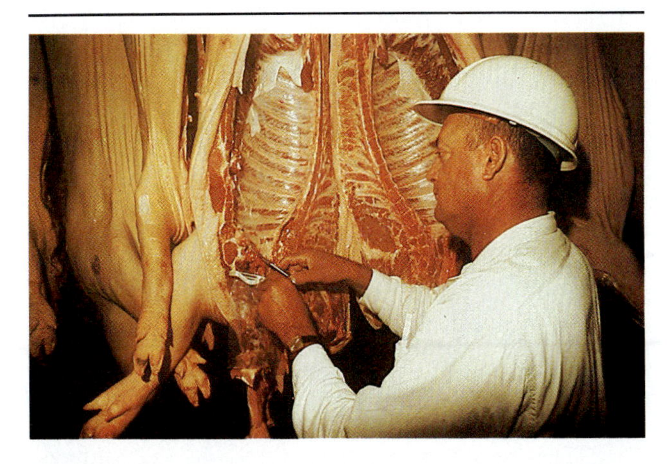

FIGURE 27.2 Meat inspection at the slaughterhouse helps provide, but cannot guarantee, a safe supply of meat for consumption.

PUBLIC HEALTH

Meat Inspection

Although meat inspection is an important means of decreasing the spread of disease from animals to humans, it is not completely effective. It does not guarantee that meat is free of parasites. The inspector examines the carcass surface and the heart (where parasites are often concentrated) but cannot look inside each cut of meat. How could an inspector know what parasites might lurk inside the roast you just bought? Such insight could occur only by microscopic examination of tissues, and that is obviously impractical. The best solution to the problem is for you, the consumer, to cook all meat thoroughly.

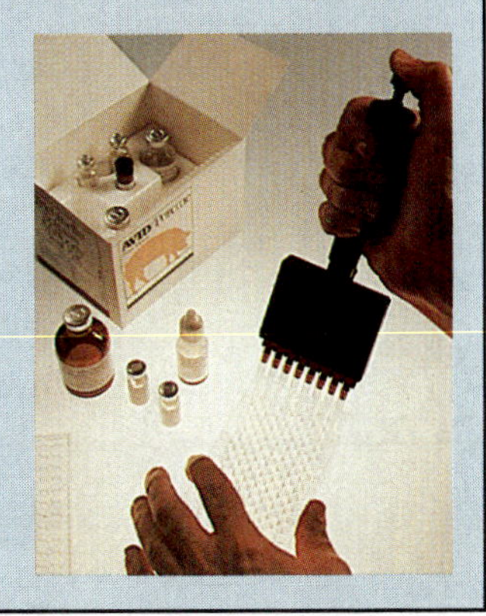

This new blood test kit can detect trichinosis infection in pigs.

Ground meats sometimes contain helminth eggs, and they always contain large numbers of lactobacilli and molds. In most areas butcher shops are required to maintain two separate meat-grinding machines—one for pork and one for other meats. This practice is encouraged because it is difficult to clean a machine thoroughly enough to completely eliminate the possibility of transmitting raw pork bits, which may carry trichinosis, to other meats. (Chapter 12, p. 311) Lactobacilli in ground meats produce acids that retard the growth of enteric pathogens. Ground meats still are subject to spoilage even when refrigerated and should be frozen if not to be used in a day or two.

More than 20 bacterial genera have been found on dressed poultry, and mishandling of poultry in restaurants accounts for many foodborne infections. Nearly half these infections have been traced to *Salmonella* and a fourth each to *Clostridium perfringens* and

CLOSE-UP

Big or Little End Up When Storing Eggs?

Does it make a difference whether you store eggs in your refrigerator with the big end or the little end up? Big end is the recommendation of food scientists. Why? The object is to keep the egg yolk and any embryo residing on it as close to the center of the egg as possible. This maximizes the distance an invading microbe would have to swim from the shell to the yolk. The guardian ocean of egg white is filled with chemical hazards. Lysozymes attack the bacterial cell walls, killing many bacteria as they rupture open. Nutrients, vitamins, and metal ions of iron, copper, and zinc are wrapped tightly by proteins and other substances in the egg white, thus making them unavailable to the bacteria. Deprived of nutrients, chewed up by lysozymes, bacteria do not survive the swim, and the embryo remains safe.

But why big end up? For one thing, the air cell in the big end could deteriorate with the weight of the egg pressing down on it. However the natural tendency of the lipid-rich yolk is to rise, just like oil floating to the top of water. Bird eggs have two ropy cords (look for them next time you crack a raw egg) called chalaza cords, which act sort of like the ropes of a hammock to suspend the yolk from the shell lining. The larger chalaza is at the small end of the egg and can hold the yolk down better, preventing it from rising too close to the edge of the protective egg white.

Staphylococcus aureus. Freezing fails to rid poultry of *Salmonella*. Pseudomonads and several other gram-negative bacteria are common contaminants of poultry, where they cause slime and offensive odors.

You might suppose that eggs, with their hard shells, would be free of microbial contamination. Most are, but pseudomonads and some other bacteria, as well as fungi such as *Penicillium*, *Cladosporum*, and

PUBLIC HEALTH

Poultry and Eggs

Salmonella species are found in many animal reservoirs, especially poultry. Recent investigations of *Salmonella* in the poultry industry were not reassuring, and consumers can assume that poultry products will be heavily contaminated. What steps can you take to maneuver defensively in the sea of *Salmonella?*

First, assume that the shells of all eggs are contaminated, and wash your hands after handling them. (But do not wash the eggs—washing removes a protective surface coating that helps prevent microbes from entering the eggs.) *Salmonella* are part of the intestinal flora of poultry, and eggs surely come in contact with feces, feathers, and contaminated surfaces. Cracked eggs cannot be sold for human consumption, but in some areas they can be sold as pet food. When organisms enter an egg they find a superb, nutrient-rich culture medium in which they can rapidly multiply.

Second, when baking, control your urge to lick raw batter from a bowl. If the batter contains eggs, even powdered eggs in mixes, it might contain *Salmonella.* Pieces of shell dropped into batter, even when promptly retrieved, can inoculate it with organisms. Also, raw eggs can be contaminated internally if laid by an infected hen or if the eggs have undetected cracks or have been immersed in water.

Third, handle raw poultry carefully. Poultry is naturally contaminated with the fecal organisms, and contamination is exacerbated by the industry's practice of holding poultry in water baths during certain phases of the plucking and evisceration processes. The water becomes a *Salmonella* broth, and the surfaces of the birds become uniformly contaminated. Thus, you should use poultry promptly and discard contaminated wrappings and juices carefully. If you have laid poultry on a counter or cutting board, scrub those surfaces thoroughly with hot soapy water before placing other foods on them. Avoid wooden cutting boards, as scratches and seams in them allow *Salmonella* growth unless the board is scrubbed with a stiff brush in hot soapy water. And don't touch other foods until you have washed your hands thoroughly.

Contaminated eggs being destroyed.

Sporotrichum, grow on eggshells and can pass through pores in shells to infect the inside of eggs. *Salmonella* also survives on eggshells and can enter broken eggs or be deposited with bits of shell in foods. Hens infected with *Salmonella pullorum* lay infected eggs. The CDC reports that 1 out of every 10,000 eggs has *Salmonella* inside the shell. Any pathogens on eggshells or in eggs can be transmitted to humans unless eggs and foods that contain them are thoroughly cooked.

Fresh fish abound with microorganisms. Several species of enteric bacteria and clostridia, enteroviruses, and parasitic worms are commonly found on fresh fish. Many of these organisms survive shipment of fish packed in crushed ice, especially if the fish are packed too tightly together or are pressed against the slats of contaminated crates.

Shellfish such as oysters and clams carry many of the same organisms as fish, and raw oysters are particularly likely to carry *Salmonella typhimurium* and sometimes *Vibrio cholerae*. Clams are especially likely sources of human infection because they are filter feeders—that is, they obtain food by filtering water and extracting microbes. If clams are exposed to increased levels of sewage, red tides, and other sources of large numbers of pathogens or toxin producers, clamming may be prohibited until the numbers of organisms decrease. Scallops are less likely to transmit diseases because only the muscular part of the organism, and not its digestive tract, is eaten.

Among crustaceans, shrimp are extremely likely to be contaminated. Some studies have shown that over half the breaded shrimp on the market contain in excess of 1 million bacteria (and more than 5000 coliforms) per gram. Such high bacterial counts probably result from growth of bacteria during processing before the shrimp were frozen. Lobsters and crabs are even more perishable than shrimp and can carry a variety of enteric pathogens. Along the U.S. Gulf Coast improperly cooked crabs have transmitted cholera. Crabs also carry *Clostridium botulinum* and the pathogenic fungi *Cryptococcus* and *Candida*. It should be kept in mind, however, that the mere presence of microorganisms in seafoods—or any other foods—does not necessarily mean that the foods are spoiled or contaminated with pathogens. In fact, *Lactobacillus bulgaricus*, which produces hydrogen peroxide, can be used to inhibit other organisms found on seafoods.

PUBLIC HEALTH

Seafood

Seafood is highly perishable and should always be as fresh as possible when eaten. If you buy a lobster to cook, make sure that it is alive when you bring it home. If it is dead when you are ready to cook it, don't. Similarly, make sure that the shells of clams or oysters are tightly closed, not only when you buy them but also, if you are serving them uncooked, when you are ready to serve them. If you find a shell open when you are about to serve a raw clam or oyster, don't serve it. Better yet, remember that eating raw seafood is a calculated risk.

Milk

Modern mechanized milking and milk handling has greatly reduced the microbial content of raw milk (Figure 27.3). However, breeding dairy cattle for increased milk production has resulted in exceptionally large udders and teats that easily admit bacteria. The first few milliliters of milk drawn from such cows can contain as many as 15,000 bacteria per milliliter, whereas the last milk drawn is free of microorganisms. Most microorganisms in freshly drawn milk are *Staphylococcus epidermidis* and *Micrococcus*, but *Pseudomonas*, *Flavobacterium*, *Erwinia*, and some fungi also can be present.

Microorganisms have many opportunities to enter milk before it is consumed. Hand milking, as opposed to mechanical milking, allows organisms from the body of the cow to enter the milk. These include *Escherichia coli*, which give milk a fecal flavor, and *Acinetobacter johnsoni* (formerly known as *Alcaligenes viscolactis*), which is especially abundant in summer months and causes a viscous slime to form in milk. Storing, transporting, and processing milk allow for contamination with any organisms in containers and growth of those already present. Infectious organisms in milk usually come either from infected cows or from unsanitary practices of milk handlers. Diseased cattle can transmit *Myocobacterium bovis* and *Brucella* species

FIGURE 27.3 Modern mechanized methods of milking and milk handling, typified by this carousel milking parlor, have greatly reduced the microbial content of raw milk.

PUBLIC HEALTH

Dried Milk

Dried milk usually is made from pasteurized liquid milk. The milk is sprayed in a fine mist; water evaporates as the milk drips into a hot pan. Contamination must be avoided as the milk dries because it is not sterilized. Dried milk contains less than 5 percent water and should be kept sealed to keep the powder from taking up atmospheric moisture. When the water content reaches about 10 percent, bacteria and molds will begin to grow in it.

to their milk. Because dairy herds are tested for tuberculosis (and infected animals removed from the herd) and vaccinated against brucellosis, the risk of transmitting these diseases to humans is small. *Staphylococcus aureus*, *Salmonella* species, and other enteric bacteria can enter milk through unsanitary handling.

Some microorganisms, such as certain *Pseudomonas* species and some soil organisms, grow in refrigerated milk. These organisms are psychrophilic; although they normally grow at higher temperatures, they can grow at 5°C (refrigerator temperature). They also survive the concentration of chlorine normally used to purify drinking water.

Organisms that sour milk include *Streptococcus lactis* and species of *Lactobacillus*. When these microbes release enough lactic acid to bring the pH below 4.8, the milk proteins coagulate, and the milk is said to have soured. Souring of milk does not mean that the milk is unsafe for human consumption, but it does, of course, greatly alter the taste and appearance of the milk.

Other Edible Substances

In addition to major nutrients, humans also consume sugar, spices, condiments, tea, coffee, and cocoa, all of which are subject to microbial contamination.

PUBLIC HEALTH

Candy

Candy is generally safe from microbial contamination because its high sugar concentration creates an osmotic pressure too high for most organisms to survive. Occasionally, the cream filling in a batch of chocolates becomes contaminated with *Salmonella* or *Clostridium*. Chocolate lovers have been surprised to discover that the gas produced by clostridia has caused their favorite chocolates to explode.

Fresh, dry, refined sugar is sterile, but raw sugarcane juice supports growth of fungi such as *Aspergillus*, *Saccharomyces*, and *Candida* and several species of *Bacillus* and *Micrococcus*. Most are removed by filtration, and the remainder are killed by heat during evaporation of the juice. Foods to which sugar is added are especially susceptible to spoilage because the sugar is an excellent nutrient for many organisms. *Bacillus stearothermophilus*, a facultative anaerobe that grows best at 55 to 60°C, can multiply rapidly during processing of foods. Another thermophilic anaerobe, *Clostridium thermosaccharolyticum*, is often responsible for the gas production that causes cans to bulge. On the other hand, sugar in high concentration acts as a preservative. The high sugar concentration in such foods as jellies, jams, candies, and candied fruits creates sufficient osmotic pressure to inhibit microbial growth.

Maple trees are tapped and sap is collected in the early spring. The sap becomes increasingly contaminated as the weather becomes warmer. Organisms that thrive in maple sugar sap include species of *Leuconostoc*, *Pseudomonas*, and *Enterobacter*. Although these organisms may consume large quantities of sugar, they are killed as the sap is evaporated to syrup or sugar.

Honey can contain poisonous toxins if it is made of nectars from such plants as *Rhododendron* and *Datura*. It also can contain spores of *Clostridium botulinum* (Chapter 24). (p.679) Although the spores do not germinate in the honey, in infants they can germinate after the honey is ingested, and their toxin can produce floppy baby syndrome.

Spices have been used in food preservation and embalming for centuries and have thereby gained a reputation as antimicrobials. This reputation is undeserved because spices more often mask odors of putrefaction than prevent spoilage. Leeuwenhoek, who first observed bacteria in spices, reported that water containing whole peppercorns teemed with them. Among the many microorganisms found in spices (Table 27.1), most are not pathogens. The small amounts of spices used in cooking probably are not a health hazard.

Condiments such as salad dressings, catsup, pickles, and mustard usually are markedly acidic. Although low pH prevents growth of many microorganisms, some molds are able to grow in these foods if they are not refrigerated.

Americans consume huge quantities of carbonated beverages and coffee and lesser quantities of tea and cocoa. The automated equipment in modern plants can prepare carbonated beverages aseptically, but syrups can become contaminated with molds if there is a mechanical failure. Syrups sold in bulk to restaurants also can become contaminated. Freshly harvested coffee beans are subject to contamination

TABLE 27.1 Numbers of bacteria in spices

Type of Spice	Number of Microorganisms per Gram of Dry Sample				
	Total Aerobes	Coliforms	Yeasts and Molds	Aerobic Spores	Anaerobic Spores
Bay leaves	520,000	0	3,300	9,200	<2
Cloves	3,000	0	5	<2	<2
Curry	>7,500,000	0	70	>240,000	>240,000
Marjoram	370,000	0	18,000	54,000	>24,000
Paprika	>5,500,000	600	2,300	>240,000	620
Pepper	>2,000,000	0	15	>240,000	>24,000
Sage	6,800	0	10	7	1,700
Thyme	1,900,000	0	11,000	160,000	>24,000
Turmeric	1,300,000	50	70	>110,000	430

Adapted from Karlson and Gunderson, 1965, *Food Technology* 1986, with permission from the Institute of Food Technologies.

by various molds and by microorganisms transmitted to them by insects. The coffee rust *Hemileia vastatrix*, a fungus, has devastated coffee plantations in Asia and is now a serious problem in South America. Tea leaves allowed to become moist are susceptible to contamination by *Aspergillus* and *Penicillium* molds, which impart an unpleasant aroma to tea brewed from them.

CLOSE-UP

Tea or Coffee for Britain

When the coffee plantations of Ceylon (now Sri Lanka) were destroyed in the 1860s by rust, the fields were replanted with tea. The British, who had depended on Ceylon for their coffee supplies, were forced to switch to tea. Thus, the lowly fungus played a major role in changing the British into a nation of tea drinkers. The Irish, 2 decades earlier, had not been so lucky. When a fungal disease destroyed the potato crop in the late 1840s, no replacement capable of feeding the population was available. A million people starved to death, and over a million emigrated.

Could such catastrophes occur today? The answer, unfortunately, is probably "Yes." Most of the food crops grown today are potentially vulnerable to diseases and pests because they are monotypes, or pure strains, lacking the genetic diversity that is typical of wild plants. An outbreak of plant disease can therefore spread very rapidly through an entire crop. Agricultural scientists are using genetic engineering techniques to develop strains that are resistant to diseases and insects—rust-resistant strains of coffee, for example (as well as varieties that are naturally low in caffeine). However, there is always the danger that such strains will turn out to be vulnerable to a new species of microorganism—either a mutation or one to which previously cultivated varieties had not been susceptible. This happened to the U.S. corn crop in the 1970s and resulted in large economic losses.

Microorganisms also are useful in preparing coffee and cacao beans for marketing. The bacterium *Erwinia dissolvens* is used to digest pectin in outer coverings of coffee beans. Other bacteria are used to dissolve the covering of cacao beans, the beans from which cocoa and chocolate are made. These beans are subsequently treated with fermenting bacteria before they are roasted.

DISEASES TRANSMITTED IN FOOD

Diseases acquired from food are due mainly to the direct effects of microorganisms or their toxins (Table 27.2), but they can result from microbial action on food substances. Industrialization has increased the spread of foodborne pathogens. Large processing plants provide opportunities for contamination of great quantities of food unless sanitation is strictly practiced. In institutions that feed large numbers of people, contaminated food will cause many cases of disease. Also, the increased popularity of convenience foods, especially warm delicatessen foods, raises the risk of infection.

In addition to the enteric diseases described in Chapter 20, several other diseases can be transmitted in food. *Klebsiella pneumoniae* is commonly found in the human digestive tract. Although it is thought of mainly as a pathogen of the respiratory system, it can cause diarrhea in infants, abscesses, and nosocomial wound and urinary tract infections. Tuberculosis can be transmitted by fomites in food, unpasteurized milk and cheese, and meats from infected animals.

Several diseases can be transmitted to humans through eating infected meat. They include anthrax, brucellosis (undulant fever), Q fever, and listeriosis. Because of meat inspection procedures, animal and meat handlers are much more often exposed to these

TABLE 27.2 Pathogenic organisms transmitted in food and milk

Organism	Disease	Vector
Staphylococcus aureus	Food poisoning	Infected food handlers, unrefrigerated foods, milk from infected cows
Clostridium perfringens	Food poisoning	Unrefrigerated foods
Bacillus cereus	Food poisoning	Unrefrigerated foods
Clostridium botulinum	Botulism	Inadequately processed canned goods
Salmonella species	Salmonellosis	Infected food handlers, poor sanitation, seafood from contaminated water
Shigella species	Shigellosis	Infected food handlers, poor sanitation
Enteropathogenic *Escherichia coli*	"Montezuma's revenge" and other diseases	Infected food handlers (sometimes asymptomatic), poor sanitation
Campylobacter	Gastroenteritis	Undercooked chicken and raw milk
Vibrio cholerae	Cholera	Poor sanitation
Vibrio parahaemolyticus	Asian food poisoning	Undercooked fish and shellfish

diseases than are consumers. Adirondack disease, caused by *Yersinia enterocolitica,* is thought to be transmitted by eating infected meat, but it also may be transmitted through milk and water. The disease has many forms, including mild to severe gastroenteritis, arthritis, glomerulonephritis, a fatal typhoidlike septicemia, and fatal ileitis, an inflammation of the small intestine that can be mistaken for appendicitis. People have also become infected with *Erysipelothrix rhusiopathiae* by eating infected pork. The resulting disease, called erysipelas in animals and erysipeloid in humans, infects swine, sheep, and turkeys, and is most likely to infect farmers and packing plant workers through skin wounds. The organism can infect skin, joints, and the respiratory tract.

Viral infections are frequently transmitted through food. Enteroviruses are often spread during unsanitary handling of food, especially by asymptomatic food handlers. Droplet infection of food can transmit echovirus and coxsackie respiratory infections. Poliomyelitis viruses can be transmitted through milk and other foods. And herpesviruses responsible for hepatitis can be transmitted through shellfish from contaminated waters. The virus that causes lymphocytic choriomeningitis, a flulike illness, can be spread to foods by mice.

Milk is an ideal medium for the growth of many pathogens. In addition to toxin producers and pathogens found in other foods, milk can contain organisms from cows. These organisms include *Mycobacterium bovis, Brucella* species, *Listeria monocytogenes,* and *Coxiella burnetii.* Spore-producing organisms such as *Bacillus anthracis* enter milk from infected cows or from soil.

Milk also contains certain antibacterial substances. They include lysozyme, agglutinins, leukocytes, and lactenin. Lactenin is a combination of thiocyanate, lactoperoxidase, and hydrogen peroxide. It also is present in human milk and other body secretions and may help to prevent enteric infections in newborn infants. Fermented milk contains bacteria such as *Leuconostoc cremoris* that kill pathogens.

PUBLIC HEALTH

Mushrooms and Mercury

Certain other disorders, such as mushroom poisoning and methyl mercury poisoning resemble microbial intoxications. Mushroom poisoning results from the ingesting of neurotoxins from mushrooms of the genus *Amanita.* It causes vomiting, liver damage, and neurological impairment, and is often fatal. Methyl mercury poisoning results from ingestion of fish from waters contaminated with methyl mercury. It causes numbness and blurred vision at low concentrations and coma at high concentrations.

Salts of heavy metals, such as mercuric chloride, are sometimes used to control microorganisms in industrial wastes. If ingested by producers, these toxic substances are passed along food chains and eventually can reach humans. Furthermore, they are concentrated in the tissues of organisms that consume them so that organisms at each successive level of the food chain receive larger doses. Instances have been reported of farmers mistakenly using seed grain treated with mercury salts to feed hogs. The chemicals become concentrated in the hogs' tissues. When these animals were eaten by the farm families, some members of the families died of mercuric poisoning.

PREVENTION OF SPOILAGE AND DISEASE TRANSMISSION

A crucial factor in preventing spoilage and disease transmission in food and milk is cleanliness in handling. Other commonsense practices—prompt use of fresh foods, careful refrigeration, and prompt and ad-

BIOTECHNOLOGY

Antisense RNA Prevents Ripening of Tomatoes

Natural release of ethylene gas by tomatoes causes them to redden and ripen, a problem if they must be shipped long distances or stored without refrigeration, as is the case in many Third-World countries. Now researchers have prevented 99.5 percent of ethylene production in tomato plants by inserting an altered gene. This gene ordinarily makes messenger RNA that produces an enzyme needed to make ethylene. They redesigned the gene so that its genetic code read backwards, and they used a bacterium to inoculate it into the tomato plants. Recipient plants were unable to make ethylene, and tomatoes remained unripened on the vine for 5 months. The added time on the vine allowed sugars and acids important to "vine-ripened" flavor to accumulate. When ripening is desired, ethylene gas is pumped into rooms where the green tomatoes are held, producing red, soft, *and* flavorful tomatoes. These techniques can be used with other perishable types of produce such as bananas and melons. It can also retard wilting in some cut flowers.

equate processing of foods to be stored—also help to control spoilage and disease transmission.

Many procedures used to preserve foods are based on practices begun early in human civilization. The ability to maintain a stable year-round food supply was essential in allowing humans to abandon a nomadic lifestyle, which followed the food supply, for a more settled village lifestyle. These methods were most likely based on simple observations like the following: Grain that was kept dry did not mold. Dried and salted foods remained edible over a long period of time. And milk allowed to sour or made into cheese was usable over a much longer period of time than was fresh milk. Modern methods of food and milk preservation still make use of some of these early methods, but they also use heat and cold and other specialized procedures.

Food Preservation

Many of the methods of food preservation have been described under antimicrobial physical agents in Chapter 10. They include canning using moist heat; refrigeration, freezing, and lyophilization; drying; and the use of ionizing radiation. A number of chemical food additives also are used to retard spoilage.

Canning

The most common method of food preservation is the use of moist heat under pressure. This method, analogous to laboratory autoclaving, is used to preserve fruits, vegetables, and meats in metal cans or glass jars (Figure 27.4). If properly carried out, canning destroys all harmful spoilage microorganisms, including most heat-resistant spores, prevents spoilage, and avoids any hazard of disease transmission. Foods so treated may remain edible for years. However, because some thermophilic anaerobic spores such as those of *Bacillus stearothermophilus* may remain alive, canned food should not be stored in a hot environment such as a car trunk.

At such elevated temperatures, the spores can germinate, grow, and cause spoilage. Usually this produces gas, which causes the ends of cans to bulge so that the ends can be pressed up and down. It also usually produces acid, which gives a sour flavor. Such changes are called *thermophilic anaerobic spoilage*. Sometimes, however, spoilage due to growth of such spores does not cause cans to bulge with gas and is called *flat sour spoilage*. Cans can also bulge due to *mesophilic spoilage*, which occurs when canning procedures were improperly followed or the seal has been broken. This type of spoilage will take place at room temperature, unlike flat sour and thermophilic spoilage, which occur only in properly processed and sealed cans that have been stored at high temperatures. Because of the danger of botulism and other kinds of spoilage in improperly processed canned foods, people who do home canning should carefully follow instructions in a good, up-to-date canning handbook. Any can, whether commercially or home-produced, that has bulging ends should be discarded.

Although many condiments are heat processed, some of their properties help preserve them. Sugar in jellies and jams increases osmotic pressure and retards growth of microorganisms. (Note that sac-

FIGURE 27.4 Commercial canning equipment (of the sort being used here) to can tomatoes destroys all microbes and their spores, thereby preventing food spoilage and disease transmission.

PUBLIC HEALTH

Home Canning

With respect to home canning, what was good enough for your grandparents may not be good enough for you. Better understanding of potential pathogens and toxins, together with changes in the acidity of certain foods by plant breeding activities, has led to changes in recommended home-canning methods. The U.S. Department of Agriculture now recommends that all low-acid foods be processed by pressure cooking. Such foods include low-acid tomatoes, in which spores can persist after waterbath processing. The U.S.D.A. also recommends that jellies and jams, formerly packed hot and sealed with wax, be processed in a boiling water bath and sealed with lids like other canned goods to prevent accumulation of mold toxins.

charine does not have this effect, so artificially sweetened products may require more stringent precautions to prevent spoilage than those with a high sugar content.) Likewise, the high acidity of pickles and other sour foods helps to prevent microbial growth.

Refrigeration, Freezing, and Lyophilization

Refrigeration at temperatures slightly above freezing (about 4°C) is suitable for preserving foods only for a few days. It does not prevent growth of psychrophilic organisms that can cause food poisoning. Freezing, another common method of food preservation, involves storage of foods at temperatures below freezing (about −10°C in most home freezers). All kinds of foods can be preserved by freezing for several months and some for much longer periods of time. An advantage of freezing is that it preserves the natural flavor of foods better than does canning. However, freezing has two disadvantages: (1) It causes some foods, especially fruits and watery vegetables, to become somewhat soft on thawing and detracts from their appearance. (2) Although freezing prevents growth of most microorganisms, it does not destroy them. As soon as food begins to thaw, microorganisms begin to grow. In fact, freezing and thawing of foods actually promotes growth of microorganisms. Ice crystals puncture cell walls and allow nutrients to escape from foods. These nutrients are then readily available to support microbial growth. Consequently, it is important to use foods that have been preserved by freezing quickly and never to thaw and refreeze foods.

Currently, lyophilization (freeze-drying) is employed in the food industry almost exclusively for the preparation of instant coffee and dry yeast for bread making. (The technique is also used to preserve bacterial cultures—see Chapter 13.) (p. 338)

Ionizing Radiation

Although ionizing radiation can be used to sterilize foods as thoroughly as does canning, it does not destroy enzymes. The flavor and appearance of such foods can be altered through autodigestion by their own enzymes unless they are subsequently heat-treated. Radiation has been used successfully to control microbial growth on fresh fish during transport to market. It is also effective in reducing spoilage of fresh fruits and vegetables (Figure 27.5).

Unfortunately, public concern about the hazards of radiation engendered by events such as the Three Mile Island and Chernobyl disasters have caused skepticism about radiation of foods. Thus, it should be emphasized that irradiating foods kills microorganisms bombarded by the radiation, but it does not cause the food itself to become radioactive.

PUBLIC HEALTH

Combating *Salmonella* in Poultry

Salmonella infection in poultry has been a serious source of food spoilage and causes food poisoning in as many as 4 million people in the United States per year. Now microbiologists have found that the sugar D-mannose blocks the ability of *Salmonella* bacteria to adhere to chicken intestinal tissue. Birds fed 2.5 percent solutions of D-mannose the first 10 days of their lives resisted an oral dose of 100 million *Salmonella typhimurium* bacteria on their third day of life. At age 50 days, when broilers are usually sent to market, only one bird had become colonized with *Salmonella*, 99 percent fewer than among those that had received only plain water. Mannose has no side effects, is a natural sugar, and will be metabolized as an energy source by the bird as part of its regular food.

A U.S. Department of Agriculture scientist tests the ability of mannose to prevent *Salmonella* infection in chicks.

FIGURE 27.5 Radiation can be used to sterilize food, as these two boxes of strawberries demonstrate. Irradiating food does not make it radioactive; it is safe to eat.

Chemical Additives

A large number of chemical compounds are added to various foods to kill microorganisms or retard their growth. A few examples and their uses will be described here.

Organic acids, some of which occur naturally in some foods, lower the pH of foods enough to prevent growth of human pathogens and toxin-producing bacteria. Acids such as benzoic, sorbic, and propionic inhibit growth of yeast and other fungi in margarine, fruit juices, breads, and other baked goods.

Alkylating agents such as ethylene oxide and propylene oxide are used only in nuts and spices. Sulfur dioxide, which is most effective at acidic pH, is used in the United States only to bleach dried fruits and to eliminate bacteria and undesired yeasts in wineries. Ozone, a highly reactive form of oxygen, is used to kill coliform bacteria in shellfish and to treat water used in making beverages. It has the advantage of leaving no residue. The disadvantages of ozone are that it tends to give foods a rancid taste by oxidizing fats and it can damage molecules, especially lung lysozyme, if inhaled.

Sodium chloride, perhaps one of the first food additives, increases osmotic pressure in foods so that most microorganisms cannot grow. Salt used to cure meats is especially useful in preventing growth of clostridia deep in tissues, although fungi eventually grow on the surfaces of salted foods. Salt dehydrates bacteria and makes it difficult for them to take in water and nutrients. A recent discovery suggests that salt, in addition to increasing osmotic pressure, also may create electrical charges on the surface of meats that prevent bacteria from adhering to the surfaces.

Still other chemical additives have special uses. Halogen compounds, such as sodium hypochlorite, disinfect water and food surfaces. Gaseous chlorine prevents growth of microorganisms on food-processing equipment. Nitrates and nitrites suppress microbial growth in meats, especially ground meats and cold cuts, but during cooking they can be converted to nitrosamines, which are carcinogenic and toxic to the liver. We continue to use nitrates and nitrites because we have no good alternatives, particularly for sausage. The name botulism comes from the Latin word for sausages, so common was food poisoning from sausage in the days before the use of nitrites. Another reason nitrites are used is to keep colors bright and fresh-looking, especially the red of meat. Carbon dioxide kills microorganisms, except for some fungi, in carbonated beverages. It also retards maturation of fruits and decreases spoilage during shipment. Finally, quaternary ammonium compounds can be used to sanitize many objects—utensils, cows' udders, fresh vegetables, and the surfaces of eggshells, which they do not penetrate.

Antibiotics

Antibiotics are added to foods and milk in some countries. In the United States only the anticlostridial agent nisin, a bacteriocin which is produced naturally during fermentation of milk by *Streptococcus lactis*, can be used. Antibiotic use in foods and milk is prohibited for the following good reasons:

1. The antibiotics might be relied on instead of good sanitation.
2. Pathogenic microorganisms may develop resistance to the antibiotics, so that treatment of the diseases they cause would become difficult or impossible.
3. Humans might be sensitized to the antibiotics and subsequently suffer allergic reactions.
4. The antibiotics might interfere with the activities of microorganisms essential for fermenting milk and making cheese.

Pasteurization of Milk

Prevention of spoilage and disease transmission in milk begins with the maintenance of health in both dairy animals and in milk handlers. In the past, bovine tuberculosis was sometimes transmitted to humans from cows' milk. Many children were infected early in life and usually died by age 15. Establishment of mandatory testing of dairy herds for tuberculosis every 3 years (the time it takes an infection to progress to a transmissible stage) has virtually eradicated the disease in the United States.

Milk is collected under clean, but not sterile, conditions and is usually subjected to pasteurization. Two methods of pasteurization are currently in use: (1) In flash pasteurization milk is heated to 71.6°C for at least 15 seconds. (2) In the low-temperature holding (LTH) method of pasteurization milk is heated to 62.9°C for at least 30 minutes. Both methods destroy all pathogenic organisms likely to be found in milk and decrease the numbers of organisms that can cause souring. Following pasteurization, milk is quickly cooled and refrigerated in sealed containers until it is used.

Milk can be preserved and kept safe to drink by means other than pasteurization. In Europe and some parts of the United States, milk is often sterilized rather than simply pasteurized. Such milk can be kept in sealed paper containers unrefrigerated for long periods of time. Milk can also be preserved as canned condensed milk, which also is sterilized. It is reconstituted by adding an equal volume of water. Although sterilized milk is completely free of microorganisms, the heat treatment necessary to render it sterile alters its flavor.

Various chemical additives are sometimes used in milk. Addition of hydrogen peroxide to milk reduces the temperature required to destroy most pathogens. However, it fails to kill mycobacteria, and its use in milk has been banned in the United States for this reason.

Standards Regarding Food and Milk Production

Because food and milk production are carefully regulated by federal, state, and local laws, consumers are better protected in the United States than in many other countries. In spite of regulations, some hazards remain because of the use of food additives or heat treatment. The Food and Drug Administration regulates inspection of meat and poultry, accurate labeling, and quality standards for products being shipped across state lines. Other similar regulations are imposed by many state and local agencies within their own jurisdictions. Many food producers maintain quality checks on their products. For example, in canning factories counts of microorganisms in samples of foods are made during processing in an effort to minimize the number of organisms in the foods. Milk, because it is an extremely good growth medium for microorganisms, is subjected to several tests (Table 27.3). The use of these tests virtually guarantees high-quality milk to consumers.

APPLICATIONS

No More "Genuine" Camembert Cheese?

The European Community headquarters in Brussels are producing new rules for the borderless Europe that will come into existence in 1993, including standards for what Europeans will eat or drink. Among these standards is a ban on the use of raw milk in cheese making. The French in particular are upset. They use unpasteurized milk to make cheeses with flavors unique to small regions. They argue that the new ban on using raw milk in cheesemaking will lead to a dull sameness of flavor of all cheese because pasteurization will kill good bacteria that give special aroma and flavors, as well as the harmful bacteria. Some producers also insist that heating the milk to pasteurization temperatures will ruin the quality of their cheese by changing its texture as well as flavor. Some see it as an attempt of northern countries "obsessed" with hygiene to impose their standards on the southern parts of Europe. Everyone agrees that it will add to the expense of production. Lobbyists for the French cheese industry are begging Brussels to reconsider.

As this book goes to press, the president of the European Commission announced that this was all a great misunderstanding and that the Commission had always intended to draw up different standards for raw-milk products. The French immediately cried victory!

PUBLIC HEALTH

Raw Milk

The current interest in natural foods has led some consumers to insist on obtaining raw milk. Although federal regulations prevent shipment of such milk across state lines, local processors, especially in California, have acquiesced to consumer demands and sell it within the state. The Centers for Disease Control have objected strenuously to this practice because of its potential for disease transmission. Unfortunately, as this is being written, no way has yet been found to prevent in-state sale of raw milk—and its accompanying hazards.

MICROORGANISMS AS FOOD AND IN FOOD PRODUCTION

Algae and Fungi As Food

The rapid rise in world population is greatly increasing the demand for human food. Given present birthrates, the earth's population is expected to double in about 40 years, from 5 billion to 10 billion. To put these large numbers in perspective, the world population increases by about 156 people per minute, 225,000 per day, or 9 million (the population of New York City) every 40 days. Even now, 25,000 people in developing countries are dying of starvation every day, and many more are suffering from malnutrition. Clearly this situation calls for expanding the human food supply.

Among microorganisms, yeasts show great prom-

TABLE 27.3 Tests used to determine milk quality

Test	Description	Purpose and Significance
Phosphatase test	Detects the presence of phosphatase, an enzyme destroyed during pasteurization.	To determine whether adequate heat was used during pasteurization. If active phosphatase remains, pathogens also might be present.
Reductase test	Indirectly measures the number of bacteria in milk. The rate at which methylene blue is reduced to its colorless form is directly proportional to the number of bacteria in a milk sample.	To estimate the number of bacteria in a milk sample. High-quality milk contains so few bacteria that a standard concentration of methylene blue will not be reduced in 5½ hours. Low-quality milk has so many that methylene blue is reduced in 2 hours or less.
Standard plate count	Directly measures viable bacteria. Diluted milk is mixed with nutrient agar and incubated 48 hours; colonies are counted, and number of bacteria in original sample is calculated.	To determine the number of bacteria in a milk sample. The number per milliliter must not exceed 100,000 in raw milk before pooled with other milk or 20,000 after pasteurization.
Test for coliforms	Same as the test used for water. (See Chapter 26.)	To determine whether coliforms are present. A positive coliform test indicates contamination with fecal material.
Test for pathogens	Detect the presence of pathogens. Methods vary depending on the suspected pathogens.	To identify pathogens. Usually not required but can help to locate the source of infectious agents that may appear in milk.

ise for increasing our food supplies. Yeasts are a good source of protein and vitamins, and they can be grown on a variety of waste materials—grain husks, corncobs, citrus rinds, paper, and sewage. Each kilogram of yeast introduced into one of these media can produce 100 kg of protein—1000 times that obtained from 1 kg of soybeans and 100,000 times that obtained from 1 kg of beef. Yeast food products have been manufactured in Africa, Australia, Puerto Rico, Hawaii, Florida, and Wisconsin. Taiwan makes about 73,000 tons of yeast food per year, most of which is shipped to the United States for adding to processed foods. Growing yeast on wastes could increase the food supply in the United States by 50 to 100 percent, but it has some drawbacks. Expensive equipment is required to begin production, and, more important, a way must be found to make people accept yeast as a desirable food. At present, yeast is used chiefly in animal feeds. Only small amounts of single-cell foodstuffs (yeast or algal) can be handled well by the human digestive tract. Large quantities of nucleic acid derivatives can aggravate gout; therefore yeast must be processed to reduce their levels.

Algal culture is another promising avenue for increasing human food supplies (Figure 27.6). The alga *Spirulina* has been grown for centuries in alkaline lakes in Africa, Mexico, and by the Incas in Peru. The algae are harvested, sun-dried, washed to remove sand, and made into cakes for human consumption. Other algae such as *Scenedesmus* and *Chlorella* also have been cultivated in Asia, Israel, Central America, several European countries, and even the western United States.

The use of algae as human food shortens the food chain. In other words, if humans eat algae directly rather than eating fish that have been nourished on algae, the algae will feed more people than will the fish. Approximately 100,000 kg of algae are required to make 1 kg of fish. Each acre of pond used to cultivate algae can produce 40 tons of dried algae—40 times the protein per acre from soybeans and 160 times that from beef. However, algal culture has so far proven economically feasible only in urban areas where large quantities of sewage on which to grow the algae are available. In addition to the problem of getting people to accept algal products as food, growing algae on sewage creates a potential health hazard because the products may contain viral pathogens.

If technical difficulties could be overcome and the products made acceptable as human food, yeasts and algae could increase the world food supply. But, at

FIGURE 27.6 Algal culture can produce food for human consumption. Especially in the Orient, many red algae are grown in mariculture farms. Nori, the dried sheets of pressed red algae used to wrap some sushi rolls, are produced in this way.

best, the use of microorganisms as food can only buy a little time to allow humans to control their own numbers.

Food Production

The use of microorganisms in making bread, cheese, and wine is as old as civilization itself. Long before any microorganisms were identified, milk was being made into cheese and fermented beverages, and bread was being leavened by microbes. In modern food production, specific organisms are purposely used to make a variety of foods.

Bread

In making bread, yeast is used as a leavening agent—that is, to produce gas that makes the dough rise. A particular strain of *Saccharomyces cerevisiae* is added to a mixture of flour, water, salt, sugar, and shortening. The mixture is allowed to ferment at about 25°C for several hours. During fermentation, yeast cells produce a little alcohol and much carbon dioxide. As the carbon dioxide bubbles become trapped in the dough, they cause the dough to increase in bulk and acquire a lighter, finer texture. When the dough is baked the alcohol and carbon dioxide are driven off. The bread becomes light and porous because of spaces created by carbon dioxide bubbles. Home bakers often use activated dry yeast, a product that is prepared by the lyophilization of yeast cells.

APPLICATIONS

Rising Expectations

A genetically engineered strain of fast-acting yeast cuts the rising time of most breads in half. Some old-time bread bakers complain that the flavor is not quite as "full" because the longer the rising time, the better the flavor. However, bread you didn't bake because of lack of time has *no* flavor. Otherwise, these new yeasts do everything the old ones do—but twice as fast. Using them, however, requires slightly different techniques: Mix the dry yeast directly with the flour rather than dissolving it in water first, and use very hot water (the temperature of which would kill regular yeast strains). Then sit back and wait for the yeast to start "blowing bubbles." Gluten, a protein found in flour, forms elastic sheets in the dough that blow up, much like bubble gum, when yeasts release carbon dioxide gas during their fermentation. The amount of gluten varies with types of flour: Rye has none, and whole wheat has very little. Therefore, rye will not rise unless it is mixed with white, and whole wheat will form a very dense loaf. And so failure of bread to rise well should not always be blamed on the yeast! It may be due to the type of flour. Even the best yeast can't blow bubbles if there's no gluten layer to inflate.

Dairy Products

Microorganisms are used in making a wide variety of dairy products. Cultured buttermilk, popular in the United States, is made by adding *Streptococcus cremoris* to pasteurized skim milk and allowing fermentation to occur until the desired consistency, flavor, and acidity are reached. Other organisms—*Streptococcus lactis, S. diacetylactis* and *Leuconostoc citrovorum, L. cremoris,* or *L. dextranicum*—make buttermilk with different flavors because of variations in fermentation products. Sour cream is made by adding one of these organisms to cream. Yogurt is made by adding *Streptococcus thermophilus* and *Lactobacillus bulgaricus* to milk. These organisms release still other products, so yogurt has a different texture and flavor.

Fermented milk beverages have been made for centuries in various countries, especially Eastern European countries, by adding specific organisms or groups of organisms to milk. The products vary in acidity and alcohol content (Table 27.4). Acidophilus milk is made by adding *Lactobacillus acidophilus* to sterile milk. Sterilization prevents uncontrolled fermentation by organisms that might already be present in nonsterilized milk. Bulgarian milk is made by *L. bulgaricus;* it is similar to buttermilk except that it is more acidic and lacks the flavor imparted by the leuconostocs. In kefir and koumiss, *Streptococcus lactis, Lactobacillus bulgaricus,* and yeasts are responsible for the production of lactic acid, alcohol, and other products. These products are usually made by continuous fermentation—fresh milk is added as fermented product is removed. The Balkan product kefir is made from the milk of cows, goats, or sheep, and the fermentation is usually carried out in goatskin bags. The Russian product koumiss is made from mare's milk.

APPLICATIONS

Combating Lactose Intolerance

Certain strains of *Lactobacillus acidophilus* metabolize cholesterol in milk fat. Acidophilus milk made by appropriate strains might be enjoyed by individuals who must restrict their cholesterol intake. *Lactobacillus acidophilus* also adds the enzyme lactase to milk. Such milk might be tolerated by many people who cannot drink milk because they lack this enzyme, which is needed to digest the lactose present in milk.

TABLE 27.4 Fermented milk beverages

Characteristics	Beverages and Countries Where Made
Less than 1% lactic acid	Sour cream, cultured buttermilk (United States) Filmjolk (Finland)
2–3% lactic acid	Yogurt (United States) Called leben in Egypt, matzoon in Armenia, naja in Bulgaria, and dahi in India Tarho (Hungary) Kos (Albania) Fru-fru (Switzerland) Kaimac (Yugoslavia) Acidophilus milk (United States)
Alcoholic (1–3%)	Koumiss, kefir, and araka (former Soviet Union) Fuli and puma (Finland) Taette (Norway) Lang (Sweden)

Cheeses

The first step in making almost any cheese is to prepare a curd by adding lactic acid bacteria and either rennin (an enzyme from the calf stomach) or bacterial enzymes to milk. The bacteria sour the milk, and the enzymes coagulate the milk protein casein. The products are the solid curd used to make cheese and the liquid whey (Figure 27.7). Whey is a waste product of cheese making, but sometimes lactic acid is extracted from it. In separating the curd and the whey, different amounts of moisture are removed according to the kind of cheese being made. For soft cheeses the whey is simply allowed to drain from the curd; for harder cheeses heat and pressure are used to extract more moisture. Nearly all cheeses are salted. Salting helps to remove water, prevents growth of undesired microorganisms, and contributes to the flavor of the cheese.

A few cheeses, such as cream cheese, cottage cheese, and ricotta, are unripened, but most cheeses

(a)

(b)

FIGURE 27.7 Making gouda cheese. (a) Lactobacilli and the enzyme rennin are added to pasteurized milk. The bacteria sour the milk, and the rennin coagulates the milk protein casein. (b) The milk curdles into the solid curd and the liquid whey. Curds are drained and placed into "hoops," which are then squeezed in a press. (c) The pressed cheeses are removed from their hoops and floated in a tank of brine (salt solution). The high salt concentration extracts even more moisture from the cheese by osmosis (recall Chapter 4) ∞ (p. 99) and thus hardens it. The salt also adds flavor to the cheese and prevents the growth of undesired microorganisms.

(c)

(a)

(b)

FIGURE 27.8 Ripening of gouda cheese involves the action of microorganisms on the curd after it is pressed into its form. (a) A plastic coating is added to the cheese, which will keep it from drying out. (b) The cheese is placed on shelves in the aging room, where the microorganisms that are distributed throughout the curd will ripen it. Because gouda is a hard cheese, it will have to ripen for a relatively long period—3 months to a year—depending on the size of the individual cheese.

are ripened. Ripening involves the action of microorganisms on the curd after it is pressed into a particular form. Soft cheeses are ripened by the action of microorganisms that occur naturally or are inoculated onto the surface of the pressed curd. Because the enzymes that ripen the cheese must diffuse from the surface into the center of the cheese, soft cheeses are relatively small. In contrast, hard cheeses are ripened by the action of microorganisms distributed through the curd. Because microbial action does not depend on diffusion, these cheeses can be quite large (Figure 27.8). Cheese is ripened by microorganisms in a cool, moist environment. Many modern factories have environmentally controlled rooms, but some still use natural caves similar to those where cheeses were first ripened.

Cheeses can be classified by their consistency (soft to hard), by the kind of microorganisms involved in the ripening process, and by the length of time required for ripening (Table 27.5). The ripening period is shorter for soft cheeses (1 to 5 months) than for hard cheeses (2 to 16 months).

Several microbial actions, such as decomposition of the curd and fermentation, occur during the ripening of cheeses. Prior to ripening, the curd consists of protein, lactose, and, if the cheese is made from whole milk, fat. As microorganisms act on the curd, they first break down the lactose to lactic acid and other products such as alcohols and volatile acids. Proteolytic enzymes break down protein, more extensively in soft than in hard cheese. *Brevibacterium linens* and the mold *Penicillium camemberti* are especially adept at releasing proteolytic enzymes. Lipase, especially in *Penicillium roqueforti*, releases relatively short-chain fatty acids such as butyric, caproic, and caprylic acids. These acids and their oxidation products contribute significantly to the flavor of the cheeses. The effects of fermentation in cheese ripening are most easily seen in Swiss cheese. Bacteria of the genus *Propionibacterium* ferment lactic acid and produce propionic acid, acetic acid, and carbon dioxide. The acids flavor the cheese, and the carbon dioxide, which becomes trapped in the curd, produces the characteristic holes in the cheese.

TABLE 27.5 Classification of ripened cheeses

Consistency and Ripening Period	Examples	Organisms Associated with Ripening
Soft (1–2 months)	Limburger	*Streptococcus lactis, S. cremoris, Brevibacterium linens*
Soft (2–5 months)	Brie and Camembert	*S. lactis, S. cremoris, Penicillium camemberti,* and *P. candidum*
Semihard (1–8 months)	Muenster and Brick	*S. lactis, S. cremoris, Brevibacterium linens*
Semihard (2–12 months)	Roquefort and Blue	*S. lactis, S. cremoris,* and *P. roqueforti* or *P. glaucum*
Hard (3–12 months)	Cheddar and Colby	*S. lactis, S. cremoris, S. durans,* and *Lactobacillus casei*
	Edam and Gouda	*S. lactis* and *S. cremoris*
	Gruyere and Swiss	*S. lactis, S. thermophilus, S. helveticus, Propionibacterium shermani,* or *L. bulgaricus* and *P. freudenreichii*
Hard (12–16 months)	Parmesan	*S. lactis, S. cremoris, S. thermophilus,* and *L. bulgaricus*

Other Products

Throughout the world a great variety of fermented foods and food products are made; we consider only a few of them. (The making of beers, wines, and spirits is discussed later.)

Vinegar Vinegar is made from ethyl alcohol by acetic acid bacteria (Figure 27.9), which oxidize the alcohol to acetic acid. Commercially produced vinegar contains about 4 percent acetic acid. Cider vinegar is made from alcohol in fermented apple cider; wine vinegar is made from alcohol in wine.

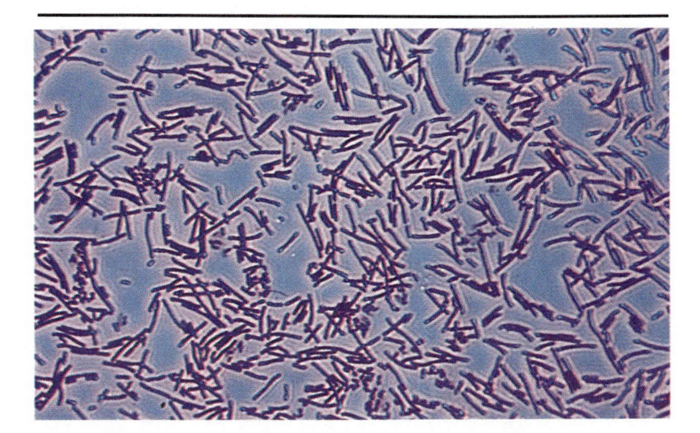

FIGURE 27.9 Manufacture of vinegar makes use of the bacterium *Acetobacter aceti*. This organism oxidizes ethyl alcohol to acetic acid.

Sauerkraut Sauerkraut was first made in Europe in the sixteenth century. Bacteria naturally present on cabbage leaves act on shredded cabbage placed in layers in large crocks. Enough dry salt is placed between the layers to make a 2- to 3-percent salt solution as the salt draws water out of the cabbage. The cabbage is firmly packed and weighted down to create an anaerobic environment. Although many organisms cannot tolerate such an environment, anaerobic, halophilic species of *Lactobacillus* and *Leuconostoc* can carry out fermentation under these conditions. They produce lactic acid, acetic acid, carbon dioxide, alcohol, and small amounts of other substances. After 2 to 4 weeks at room temperature, fermentation changes the cabbage to sauerkraut, which can be refrigerated until eaten or preserved by canning.

Pickles Pickles are made by essentially the same process as that used to make sauerkraut. Fresh cucumbers, either whole, sliced, or ground, are packed in brine and allowed to ferment for a few days to a few weeks. *Leuconostoc mesenteroides* is the major fermenting organism in low-salt brines (less than 5 percent salt), whereas species of *Pediococcus* are more active in high-salt brines (5 percent or higher salt concentration). A problem in pickle making is to prevent a yeast film from forming on the surface of vats. Direct sunlight and ultraviolet light help solve this problem. After fermentation, vinegar and spices are added to sour pickles; sugar also is added to sweet pickles. Most pickles are pasteurized. Some pickles, such as sweet pickles and pickle relish, are made without fermentation. After the pickles have been soaked in brine for a few hours, they are seasoned with vinegar and spices, heat-processed, and sealed. The entire operation can be done in a day in any kitchen.

Olives Olives are treated with lye to hydrolyze the bitter oleuropein they contain. Green olives are fermented in a 5 to 8 percent salt solution by *Leuconostoc mesenteroides* and *Lactobacillus plantarum*. They are packed in water and pasteurized. Ripe olives are picked when reddish in color but not fully ripe. They are oxidized with tannins to blacken them, fermented in dilute (less than 5 percent) salt solution, packed in water in cans, and processed at 116°C for 60 minutes.

Poi Poi, a common food in the South Pacific, is made from ground roots of the taro plant (Figure 27.10). A paste of ground roots is allowed to ferment through the action of a succession of naturally occurring organisms. Pseudomonads and coliforms begin the fermentation process, and lactobacilli later take over. Yeasts add alcohol to the mixture.

FIGURE 27.10 The Hawaiian specialty poi is made by grating the root of the taro plant and then allowing it to ferment for 2 or 3 days. The longer it is fermented, the sharper the flavor. Although many tourists find it reminiscent of wallpaper paste, it is considered to be an acquired taste.

FIGURE 27.11 Fermentation of soy and wheat is carried out in large steel tanks.

Soy Sauce Soy sauce is made in a stepwise process (Figure 27.11). A salted mixture of crushed soybeans and wheat is treated with the mold *Aspergillus oryzae* to break down starch into fermentable glucose. The product, called koji, is mixed with an equal quantity of salt solution to make a mixture called moromi. Moromi is fermented for 8 to 12 months at low temperature and with occasional stirring. The fermenting organisms are mainly the bacterium *Pediococcus soyae* and the yeasts *Saccharomyces rouxii* and *Torulopsis* species. Lactic acid, other acids, and alcohol are produced. Upon completion of fermentation, the liquid and solid parts of the moromi are separated. The liquid part is bottled as soy sauce, and the solid part is sometimes used as animal food.

Soy Products Other soy products include miso, tofu, and sufu. Miso is a fermented paste of soybeans made like soy sauce. Tofu is a soft curd of soybeans. It is made from soybeans ground to make a milk, which is boiled to inactivate enzymes. The curd is precipitated with calcium or magnesium sulfate and pressed into a soft mass similar in consistency to soft cheese. Sufu is made by action of a fungus on soy curd. Cubes of curd are dipped in a mixture of salt and citric acid, inoculated with *Mucor*, and incubated until coated with mycelia. The fungus-coated cubes are aged 6 weeks in a rice-wine brine.

INDUSTRIAL AND PHARMACEUTICAL MICROBIOLOGY

Industrial microbiology deals with the use of microorganisms to assist in the manufacture of useful products or to dispose of waste products. **Pharmaceutical microbiology** is a special branch of industrial microbiology that concerns the manufacture of products used in treating or preventing disease. Today many industrial and pharmaceutical processes make use of genetic engineering, as we saw in Chapter 8. ∞ (pp. 210–213)

Industrial microbiology, albeit in primitive form, had its beginnings more than 8000 years ago when the Babylonians fermented grain to make beer. However, little was understood about fermentation until Pasteur studied the process in the nineteenth century. Over the next several decades, various researchers studied fermentation and its products, but their findings were largely ignored until shortages of materials for making explosives in World War I created a use for them. The Germans developed a process for making glycerol; the British used acetone-butanol fermentation by *Clostridium acetobutylicum* to make acetone. Both glycerol and acetone are used to make explosives and other materials. An important sidelight of the acetone-butanol fermentation was the development of techniques to maintain pure cultures in industrial fermentation vats.

The serendipitous discovery by Fleming in 1928 that *Penicillium notatum* killed *Staphylococcus aureus* was the beginning of the development of the antibiotic industry. Today the manufacture of antibiotics is an immense component of pharmaceutical microbiology. Concurrent with the development of antibiotics was the development and industrial production of a variety of vaccines. The manufacture of antibiotics, vaccines, and many other pharmaceuticals all require pure culture technology.

In recent years genetic engineering has been used to cause cells to synthesize products they otherwise would not make or to increase the yield of products they normally make. Genetic engineering may make it possible to program organisms to carry out specific industrial and pharmaceutical processes. Such processes might well be more efficient and more profitable than any others now available.

Today hundreds of different substances are manufactured with the aid of microorganisms. Various species of yeasts, molds, bacteria, and actinomycetes are used in manufacturing processes. The organisms themselves are sometimes useful, as when they can serve as a source of protein. Animal feed consisting of microorganisms is sometimes called *single-cell protein*. More often the valuable substance is a product of microbial metabolism.

Metabolic Processes Applicable in Industry

Production of complex molecules and metabolic end products in commercially profitable quantities requires the manipulation of microbial processes. In nature microbes have regulatory mechanisms such as induction and repression that cause them to make substances only in the amounts the organisms require. An important task of industrial microbiologists is to manipulate regulatory mechanisms so that large quantities of useful substances are produced. They do this in several ways: (1) by altering nutrients available to the microbes, (2) by altering environmental conditions, (3) by isolating mutant microbes that produce excesses of a substance because of a defective regulatory mechanism, and (4) by using genetic engineering to program organisms to display particular synthetic capabilities.

In some instances the efforts of industrial microbiologists have been remarkably successful. The industrial strain of a mold *Ashbya gossypii* produces 20,000 times as much of the vitamin riboflavin as it uses. Industrial strains of *Propionibacterium shermanii* and *Pseudomonas denitrificans* produce 50,000 times as much cobalamin (vitamin B_{12}) as they use.

Problems of Industrial Microbiology

It is one thing to cause a microbe to carry out a useful process in a test tube. It is quite another to adapt the process so that it will be profitable in a large-scale industrial setting. In the past most industrial processes have been carried out in large fermentation vats. Many new processes simply do not work in large vats, so smaller vats must be used (Figure 27.12). Moreover, many of today's industrial microorganisms have been so extensively modified by selection of mutants or by genetic engineering that their products, useful to humans, may be useless or even toxic to the organisms.

Isolating and purifying the product, with or without killing the organisms, often presents technical difficulties. When the product remains within the cell, the cell membrane must be disrupted to obtain the product. Membranes can be disrupted by spraying a medium containing the organisms through a nozzle under high pressure or by putting the organisms in alcohol or salt solutions that cause the product to form a precipitate. Product molecules from disrupted cells sometimes can be collected on resin beads of appropriate size and electrical charge. Secreted products can be collected relatively easily, sometimes without killing the organisms. This can be done by using a continuous reactor (Figure 27.13), where fresh medium is introduced at one end and medium containing the product is withdrawn at the other end.

(a)

(b)

FIGURE 27.12 Some fermentations do not progress satisfactorily in large vats (a) and therefore must be conducted in small-scale containers like this one (b).

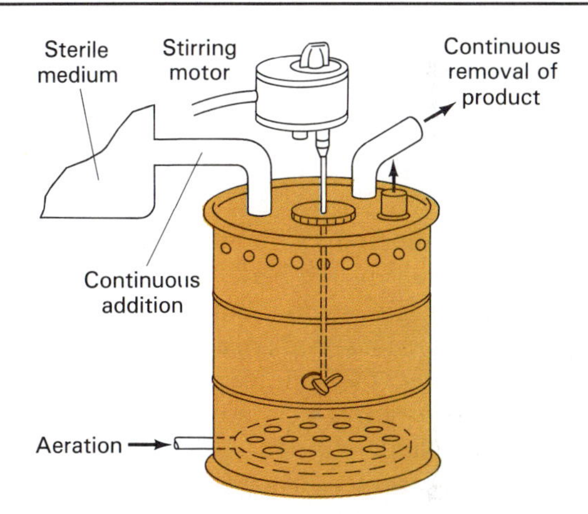

FIGURE 27.13 A continuous reactor, in which fresh medium is introduced at one end and medium containing product is withdrawn at the other end.

BEER, WINE, AND SPIRITS

Beer and wine are made by fermenting sugary juices; spirits, such as whiskey, gin, and rum, are made by fermenting juices and distilling the fermented product. Distillation separates alcohol and other volatile substances from solid and nonvolatile substances. Strains of *Saccharomyces* are the fermenters for all alcoholic beverages. Many different strains have been developed, each having distinctive characteristics, and both the organisms and how they are used are carefully guarded brewers' secrets.

To make beer, cereal grains (usually barley) are malted (partially germinated) to increase the concentration of starch-digesting enzymes that provide the sugar for fermentation. Malted grain is crushed and mixed with hot water (about 65°C), and after a few hours a liquid malt wort is extracted. Hops (flower cones from the hop plant) are added to the wort for flavoring, and the mixture is boiled to stop enzyme action and precipitate proteins. A strain of *Saccharomyces* is added, and fermentation produces ethyl alcohol, carbon dioxide, and other substances. The other substances include amyl and isoamyl alcohols and acetic and butyric acids, which add to the flavor of the beer. After fermentation, the yeast is removed, and the beer is filtered, pasteurized, and bottled.

Most wine is made from juice extracted from grapes (Figure 27.14), although it can be made from any fruit—and even from nuts or dandelion blossoms. Juice is treated with sulfur dioxide to kill any wild yeasts that may already be present. Sugar and a strain of *Saccharomyces* are then added, and fermentation proceeds. Although ethyl alcohol is the main product of fermentation, other products similar to those in beer add to the flavor of the wine. In both beer and wine, the particular characteristics of the juice and the yeast strain determine the flavor of the final product. When fermentation is completed, liquid wine is siphoned to separate it from yeast sediment, and, if necessary, cleared with agents such as charcoal to remove suspended particles. Finally, it is bottled and aged in a cool place.

Spirits are made from the fermentation of a variety of foods, including malted barley (Scotch whiskey), rye (rye whiskey, gin), corn (bourbon), wine or fruit juice (brandy), potatoes (vodka), and molasses (rum). After fermentation, distilling separates alcohol and other volatile substances that impart flavor from the solid and nonvolatile substances. Because of distillation, the alcohol content of spirits ranges from 40 to 50 percent—much higher than the typical 12 percent for wine and 6 percent for beer. (Wines do not contain more alcohol because when the alcohol concentration reaches 12 to 15 percent it poisons the yeasts carrying out the fermentation. To produce fortified wines such as sherry and cognac, extra alcohol is added after fermentation.)

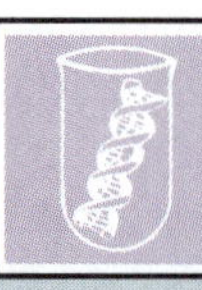

BIOTECHNOLOGY

Vitamins from Wastes

Fermentation wastes from breweries contain significant quantities of vitamins, proteins, and carbohydrates. Releasing these wastes into streams is prohibited in some areas because it enriches the waters, causing overgrowth of algae. Some breweries dispose of these wastes by drying them and marketing them as supplements to animal feed.

Yeasts also can be grown specifically for the vitamins they produce. Certain *Candida* species can make and release into the medium up to 0.1 mg of riboflavin per gram of dry weight of yeast. This process has limited commercial value because the organisms are poisoned by traces of iron in the medium, and iron is nearly always present from equipment. Higher fungi, such as some that parasitize coffee and other plants, are not harmed by traces of iron and can be used commercially to produce riboflavin.

Saccharomyces uvarum produces ergosterol, a sterol that can be converted to vitamin D by ultraviolet radiation. This process is commercially feasible when carbon sources are adequate and vats are aerated.

USEFUL ORGANIC PRODUCTS

Simple Organic Compounds

Simple organic compounds such as solvents and organic acids can be manufactured with the aid of microorganisms. The solvents include ethanol (ethyl alcohol), butanol, acetone, and glycerol. The acids include acetic, lactic, and citric acids. Although microorganisms are not now regularly used to make

(a)

(b)

(c)

FIGURE 27.14 Fermentation of wine begins with the rapid bubbling of the grape juice and yeast mixture (a), which is stored in two-story high stainless steel fermentation vats (b) until the fermentation process is complete. The wine is then transferred to wooden barrels (c), where it is aged, sometimes for many years. During this time, the flavor mellows and develops fully.

these substances, microbial synthesis will become economically feasible as the cost of raw materials from petroleum increases, especially if organisms can be genetically programmed to increase their productivity.

Ethanol, an industrial chemical with annual sales of about $300 million, is used not only as a solvent but also in the manufacture of antifreeze, dyes, detergents, adhesives, pesticides, explosives, cosmetics, and pharmaceuticals. In addition, ethanol is used as a fuel, either alone or mixed with gasoline. Microbes are employed to produce ethanol for industrial purposes in much the same way they are used to make alcoholic beverages, but microbiologists are developing new methods. In one method cellulose is extracted from wood, digested to sugars, and fermented by thermophilic clostridia. Because these organisms act at high temperatures, their metabolic rates are higher than those of other fermenters, and they produce alcohol at a more rapid rate. Also, because the effluent is already heated, less energy is needed to distill and purify the product. In another method the bacterium *Zymomonas mobilis*, which ferments sugar twice as fast as yeast, is used.

The yeast *Pachysolen tannophilus* produces relatively large quantities of alcohol from five-carbon sugars. This is significant because grain products contain both five-carbon and six-carbon sugars; thus, making alcohol with yeasts that use only six-carbon sugars fails to extract a significant amount of energy from the grain. Producing alcohol for fuel is currently almost a break-even activity; the energy obtained from the six-carbon sugars in the alcohol is approximately equal to the energy required to produce it. Extracting energy from both five- and six-carbon sugars would be much more economical.

Clostridium acetobutylicum acting on starch or *C. saccharoacetobutylicum* acting on sugar produces both butanol and acetone. Butanol is used in the manufacture of brake fluid, resins, and gasoline additives. Acetone is used mainly as a solvent.

Glycerol is made by adding sodium sulfite to a yeast fermentation. The sodium sulfite shifts the metabolic pathway so that glycerol, and not alcohol, is the main product. Glycerol is used as a lubricant and as an emollient in a variety of foods, toothpaste, cosmetics, and paper.

Of the organic acids made by microorganisms, acetic acid (vinegar) is used in the greatest volume in the manufacture of rubber, plastics, fabrics, insecticides, photographic materials, dyes, and pharmaceuticals (Figure 27.15). Acetic acid bacteria oxidize ethanol to make vinegar. But thermophilic bacteria can make it from cellulose and *Acetobacterium woodii* and *Clostridium aceticum* can make it from hydrogen and carbon dioxide. Other acids made by microorganisms include lactic and citric acids. *Lactobacillus delbrueckii* metabolizes glucose to lactic acid, which is used to acidify foods, to make synthetic textiles and plastics, and in electroplating. The mold *Aspergillus niger* makes citric acid and does so with great efficiency when molasses is used as the fermentation substrate. Citric acid is widely used to acidify and improve the flavor of foods.

FIGURE 27.15 Equipment for the industrial production of vinegar.

Antibiotics

The antibiotic industry came into being in the 1940s with the manufacture of penicillin (Figure 27.16a), and about 100 antibiotics have been manufactured in quantity. The market value of antibiotics worldwide now exceeds $5 billion annually.

Industrial microbiologists work diligently to find ways to cause organisms that make a particular antibiotic to make it in large quantities. A most effective way is to induce mutations and screen progeny to find strains that produce more antibiotic than the parent strain. Such methods, combined with improved fermentation procedures, have been quite successful. For example, a strain of *Penicillium chrysogenum* that once produced 60 mg of penicillin per liter of culture now produces 20 g/L. Genetic engineering may lead to more effective methods.

The reason only 2 percent of all known antibiotics have been marketed is that many antibiotics are either too toxic to use therapeutically or of no greater benefit than antibiotics already in use. Because pathogens continually develop resistance to antibiotics, indus-

(a)

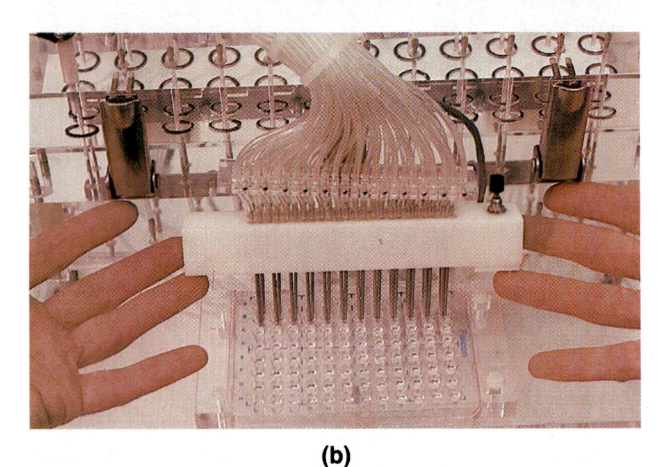

(b)

FIGURE 27.16 (a) The amber-colored droplets seen on the surface of this culture of the mold *Penicillium notatum* are the antibiotic penicillin, which it has produced. (b) Synthetic and semisynthetic antibiotics are now made in the laboratory for use alongside those produced by microorganisms.

trial microbiologists not only seek new antibiotics, they also try to make available antibiotics more effective. They look for ways to increase potency, improve therapeutic properties, and make antibiotics more resistant to inactivation by microorganisms. These efforts have led to the development of semisynthetic antibiotics (Figure 27.16b). Semisynthetic antibiotics, as explained in Chapter 14, are made partly by microorganisms and partly by chemists. (p. 350) This collaboration is illustrated by the way chemists have modified beta-lactam antibiotics (penicillins and cephalosporins), which some microorganisms destroy with an enzyme called beta-lactamase. Binding a molecule such as clavulanic acid to the beta-lactam ring prevents the enzyme from inactivating the antibiotic. The antibiotic augmentin, for example, is a semisynthetic antibiotic consisting of amoxicillin and clavulanic acid.

Enzymes

All enzymes used in industrial processes are synthesized by living organisms. With a few exceptions, such as the extraction of the meat tenderizer papain from papaya fruit, industrial enzymes are made by microorganisms. Enzymes are extracted from microorganisms rather than being synthesized in the laboratory because so far laboratory synthesis is too complicated to be practical. Enzymes, like other proteins, consist of complex chains of amino acids in specific sequences. Organisms use genetic information to synthesize enzymes easily, but chemists in the laboratory find this a tedious and expensive process. Enzymes are especially useful in industrial processes because of their specificity. They act on a certain substrate and yield a certain product, thus minimizing problems of product purification.

The methods industrial microbiologists use to produce enzymes include screening for mutants and gene manipulation. Screening for mutants that produce large quantities of an enzyme has been an effective technique. Gene manipulation has been somewhat less effective. Compared to synthetic pathways for antibiotics, those for enzymes are simpler, and fewer genes are involved. Thus, gene manipulation shows great promise for increasing the yield of some enzymes and making industrial production of others feasible. As only about 200 of some 2000 known enzymes are now produced commercially, there is room for significant progress in this area.

Of the commercially available enzymes, proteases and amylases are produced in greatest quantities. Proteases, which degrade proteins and are added to detergents to increase cleaning power, are made industrially by molds of the genus *Aspergillus* and bacteria of the genus *Bacillus*. Amylases, which degrade starch into sugars, also are made by *Aspergillus* species. Other useful degradative enzymes are lipase from the

BIOTECHNOLOGY

Mowing the Grass Can Be Rewarding

A Danish enzyme researcher returned from his summer vacation to find a compost heap of grass cuttings piled up by the brother-in-law who had thoughtfully mowed his lawn for him. Recognizing this pile as a potential treasure trove, he took samples from its center and isolated fungi that had been growing there. One fungus caused milk to coagulate. Analysis in his lab revealed an enzyme, now trade-named Rennilase, that is used in place of rennet, a mixture containing the enzyme rennin, formerly collected from stomachs of lambs and calves at the slaughterhouse. Rennilase is used to coagulate milk in making cheese. One ounce can curdle 50 gallons of milk into curds and whey.

In the United States over $100 million is spent each year on rennet, which is of uncertain purity and whose cost has risen sharply in recent years. Therefore the dairy industry welcomes industrially produced rennin. The U.S. Food and Drug Administration (FDA) has approved a version of rennin that was bioengineered by inserting the cow gene for rennin into *E. coli*. The altered *E. coli* are grown in large fermentation tanks, and the enzyme is collected.

yeast *Saccharomycopsis* and lactase from the mold *Trichoderma* and the yeast *Kluyveromyces*. Another important industrial enzyme is invertase (glucose isomerase) from *Saccharomyces;* it converts glucose to fructose, which is used as a sweetener in many processed foods.

Pancreatic enzymes were first used more than 70 years ago to remove blood stains from butchers' aprons without weakening the cloth. These enzymes were tried as laundry aids, but it was found that they were inactivated by soap. In the 1970s proteolytic enzymes from bacteria, which retain their activity in the presence of detergents and hot water, were added to detergents. When this was first tried, workers in the detergent factories developed respiratory problems and skin irritations. The enzymes were removed from the detergents when the illnesses were traced to airborne enzyme molecules. Today, enzymes are incorporated in coated granules that dissolve in the wash instead of in the environment of workers.

Proteolytic enzymes also have been added to drain cleaners, where they are especially useful in degrading hair, which often clogs bathroom drains. A drain cleaner with a lipase should do a good job on kitchen drains.

To make high-quality paper much of the lignin, a coarse material in wood, is removed by expensive chemical means to produce a purer cellulose wood pulp. The fungus *Phanerochaete chrysosporium* secretes enzymes that digest both lignin and cellulose. If the enzymes can be separated and purified, one might be used selectively to digest lignin, leaving cellulose unchanged. When perfected, this process would provide a cheap way to prepare wood pulp for high-quality paper. A byproduct of the research may benefit humans. Another lignin-digesting enzyme has been identified in a strain of *Streptomyces* bacteria. This enzyme modifies lignin into a molecule that enhances antibody production in mice. Might it someday do the same for humans?

Sometimes whole organisms are added to fermentation tanks, where their special propensities for producing certain enzymes are used to advantage. The amino acids lysine and glutamic acid are made in this way by *Corynebacterium glutamicum*. Lysine, an essential amino acid in the diet of humans and nonruminant animals, is added to animal feed and is sold in health-food stores for human consumption. Glutamic acid is used to make the flavor enhancer monosodium glutamate (MSG).

Other Biological Products

Vitamins, hormones, and single-cell proteins are the major categories of other biologically useful products of industrial microbiology. The ability of microorganisms to make vitamin B_{12} and riboflavin was cited earlier as an example of highly successful amplification of microbial synthesis. Vaccines, of course, are exceedingly useful products (Figure 27.17). The development of hepatitis vaccine was described in Chapter 8. (p. 212)

FIGURE 27.17 Hepatitis B vaccine is being produced using recombinant DNA. Technicians use a chromatography column to separate key proteins from batches of yeast cells.

Microorganisms also are used in the production of steroid hormones. The process used is called **bioconversion,** a reaction in which one compound is converted to another by enzymes in cells. The first application of bioconversion to hormone synthesis was the use of the mold *Rhizopus nigricans* to hydroxylate progesterone. The use of this microbial step simplified the chemical synthesis of cortisone from bile acids from 37 to 11 steps. It reduced the cost of cortisone from $200 per gram to $6 per gram. (Subsequent improvements in procedures have brought the price under $.70 per gram.) Other hormones that can now be produced industrially include insulin, human growth hormone, and somatostatin. They are made by recombinant DNA technology using modified strains of *Escherichia coli*.

Single-cell proteins, as noted earlier, are whole organisms rich in protein. Currently used in animal feed, they may someday feed humans, too. An important advantage of single-cell protein is that it can be made from substances of less value than the protein produced. Certain *Candida* species make protein from paper pulp wastes, *Saccharomycopsis* makes it from petroleum wastes, and the bacterium *Methylophilus* makes it from methane or methanol.

MICROBIOLOGICAL MINING

As the availability of mineral-rich ores decreases, methods are needed to extract minerals from less concentrated sources. At present copper can be profitably mined by the chemolithotrophic acidophile *Thiobacillus ferrooxidans*. Copper in low-grade ores is often present as copper sulfide. When acidic water is poured on such ore, *T. ferrooxidans* obtains energy as it uses oxygen from the atmosphere to oxidize the sulfur atoms in sulfide ores to sulfate. The bacterium doesn't use the copper, it merely converts it to a water-soluble form that can be retrieved and used by humans (Figure 27.18a).

Other minerals also can be degraded by microbes. *Thiobacillus ferrooxidans* releases iron from iron sulfide by the same process (Figure 27.18b). Combinations of *T. ferrooxidans* and a similar organism, *T. thiooxidans*, degrade some copper and iron ores more rapidly than either one alone. Another combination of organisms, *Leptospirillum ferrooxidans* and *T. organoparus*, degrades pyrite (FeS_2) and chalcopyrite ($CuFeS_2$), although neither organism can degrade the minerals alone. Other bacteria can be used to mine uranium, and bacteria may eventually be used to remove arsenic, lead, zinc, cobalt, and nickel.

(a)

(b)

FIGURE 27.18 Mining is made easier in many places by the activity of microorganisms. (a) Copper sulfide ores are converted to soluble sulfates by the action of acidophilic bacteria. (b) Microorganisms in these ponds reclaim metals from low-grade ores. Ions generated by microbial activity include ferric ions (orange) and ferrous ions (green).

MICROBIOLOGICAL WASTE DISPOSAL

Sewage-treatment plants (Chapter 26) are prime examples of microbiological waste-disposal systems, but sewage disposal is a relatively simple problem compared with the problems associated with disposing of chemical pollutants and toxic wastes. (p. 732) Some wastes persist in the environment and contaminate water supplies for wildlife and humans. Microorganisms have proven beneficial in disposing of some of these wastes, (bioremediations), and further research may find additional ways microorganisms can be used to avert a monumental environmental disaster from accumulation of toxic wastes.

Three strains of microorganisms have been found to deactivate Arochlor 1260, one of the most highly toxic of polychlorinated biphenyl (PCB) compounds. Other organisms have been shown to detoxify chemical substances such as cyanide and dioxin and degrade oil spilled in the ocean. Genetic engineering has been used to develop a bacterium capable of detoxifying the defoliant Agent Orange, and work is under way to modify bacteria to detoxify other toxins.

One of the problems in developing microorganisms that can degrade toxic substances is the limited information available on the genetic characteristics of microorganisms found in wastes. Many researchers in this area focus their efforts on organisms found in wastes because these organisms probably already have degradatory capabilities. The researchers think it should be somewhat easier to modify them to metabolize other wastes than to use better known organisms that have no known capacity to degrade wastes.

ESSAY

Visit to a Mushroom Farm

"Did you know that mushrooms are the number-one cash crop in Pennsylvania?," asked Jim Angelucci, general manager of Phillips Mushroom Farms in Kennett Square, Pennsylvania, as we walked toward the mushroom-growing sheds. I hadn't realized what a big industry it is—and it's one that definitely requires a good knowledge of microbiology. (Chapter 12, p. 305) Over 735 million pounds of mushrooms are grown each year in the United States, 47 percent of these in Pennsylvania. And 44 percent of

Pennsylvania mushrooms are grown right here in Chester County, the largest production area in the United States.

The mushroom industry in Pennsylvania was started in the late 1800s by Quaker farmers. Later, Italian immigrant quarry workers joined together to work cooperatively at different members' farms. Mushroom farming follows a cycle that includes periods of peak labor intensity, interspersed with low-labor periods. Rotating crews made it possible to grow amounts of crops that one or two men would have difficulty trying to manage. Mushroom growing used to be seasonal, from October until April or May, and was used as a "fill-in" crop for farmers. Then, the advent of refrigeration and air conditioning made it possible to farm year-round.

Americans prefer the white button mushroom, *Agaricus bisporus*, even though the brown strain is larger and meatier and has better flavor, longer shelf life, and higher productivity (more crop per square foot). Genetic engineers are trying to combine the admirable qualities of the brown strains with the desired white color. In fact, almost all varieties of button mushrooms grown today are genetically engineered strains and may have an off-white color. Phillips is also experimenting with growing small quantities of specialty mushrooms such as Portobello, Nameko, winecaps, and mitake. These are new to Americans and are appearing in gourmet restaurants and specialty produce stores. Once the cultivation methods are better worked out, production will increase, and more Americans will be able to enjoy these delicacies. Most of the specialty mushrooms must be cooked, although some can be eaten raw in salads.

I wanted to see everything there was about growing mushrooms starting from the very bottom. So we hopped into Jim Angelucci's Trooper and drove over to the co-op that produces compost for 13 growers. *Compost* is the nutrient material on which mushrooms are grown. It is a 50:50 mixture of horse manure and field hay (timothy, alfalfa, clover) plus corn cobs, coco hulls, shredded hardwood bark, and chicken manure, with a little gypsum added to balance the pH. The ingredients are combined on a schedule that varies according to the materials used. The compost mixture is laid out in long *ricks* or rows (Figure 27.19) that measure 10 ft × 8 ft × 400 ft. Bacterial and fungal decomposition of the materials raises the core temperature to 160°F. This temperature is monitored three times per day. The process should be kept aerobic, and so the ricks are picked up and turned by huge specially designed machines at least four times within a 10-day period. Because their core temperature is so high, the ricks emit steam as they are turned, even on hot summer days. After 10–14 days the compost is ready to be delivered to one of the member growers. Another thing I learned about composting is that when you step in the liquid that drains from the ricks, the odor (megaton strength!) stays on your shoes or clothes for weeks. In fact, the waitress at the restaurant where I stopped for dinner on the way home hesitated for a moment and then firmly led me to a table on an empty porch. But I'd had such fun squishing along an area the size of several football fields, filled with mountains of microbes in action!

In some places the mushrooms are grown in caves or old mines. At Phillips, the mushroom houses are set like stairsteps, going down the side of a hill. Before the new compost is delivered, all old material is removed from the growing house, and the cypress boards that form the framework of the beds are dismantled, sprayed with fungicide, and—along with the rooms—are sterilized with a steam boiler. Otherwise all the wood in the house would be colonized and digested by the fungi.

FIGURE 27.19 Ricks of mushroom compost.

Fresh compost moves via conveyor belt down to the lowest room, where men reassemble the wooden beds and shovel compost into them, filling the bottom ones first, then assembling and filling a second and then third bed above these. It takes 240 cubic yd of compost to fill a house. It is very dark in the rooms, with only occasional tubular fluorescent bulbs hanging straight down from the ceiling (Figure 27.20). The walls are thick and well caulked with hardened foam. Temperature and air control are very important here. Four air and eight bed probes in each room constantly monitor temperature and air movement.

A complete cycle of growth from one filling of a house to the next filling takes 14 weeks. Of this cycle, 30 to 35 days are spent on harvesting. Some other growers harvest for 60 to 90 days. These people grow their mushrooms at cooler temperatures that retard the growth of competing organisms but also slow down the rate at which the mushrooms grow. During harvest times, pickers may arrive at work as early as 4 A.M., since it is necessary to pick and package that day's crop in time for morning air freight shipment to markets. This allows delivery of fresh mushrooms to restaurants in time for dinner.

A typical growth cycle begins with delivery of the compost on a Monday, followed by a 10- to 14-day cycle that pasteurizes the compost. By Tuesday the bed temperature should be 135°–140°F, and air temperature 90°–100°F. On Friday the steam boilers raise the beds to a minimum of 140°F and the air to 140°F for a period of 4 hours. The air temperature is then reduced to 100°F. When the air has cooled, the temperature of the compost is reduced by 2°F every 24 hours until conversion is completed. This treatment kills nematodes and mushroom flies. Fly larvae eat the mycelium, which is the underground growth of threadlike hyphae that eventually produces the fruiting structure, or mushroom, on top of the compost (Figure 27.21). One type of mushroom fly brings in fungi and spreads disease. In any

ESSAY Visit to a Mushroom Farm

FIGURE 27.20 White button mushrooms growing in bunk-style beds. Note fluorescent bulbs hanging at intervals.

mushroom house you will always see a few flies attracted to the fluorescent bulbs.

The best bed temperature for conversion of the compost by thermophilic microbes into protein useable to mushrooms is 128°–135°F. This process releases ammonia. When the ammonia is gone (below 0.05 percent), the conversion is over, and the compost has a soft, pliant texture and the proper moisture content (72–74 percent). The second phase of the pasteurization cycle is known as the "cookout" and brings the moisture content down to 64–66 percent. During this phase the bed temperature drops to 75°–80°F, and the air to 70°F. Now the compost is ready to be inoculated with the desired mushroom

FIGURE 27.21 Cross section through a mushroom bed showing the threadlike mycelium growing through the compost, with mushrooms beginning to fruit at the surface.

strain. Later problems with nematodes, or fungi such as *Verticillium* or *Aspergillus*, usually mean that something went wrong in phase one or two of compost treatment. Subsequent growth of inky cap mushrooms indicates pockets of high ammonia, which should be gone by this time. Mushrooms are also very sensitive to viruses, which are present in them at all times and can be triggered to destroy a crop if conditions do not remain properly balanced.

The mushroom culture inoculum, called *spawn*, is purchased by growers from a commercial supplier, such as the Campbell Soup Company, or any number of smaller suppliers. The spawn is produced by sterilizing a grain such as wheat, rye, or millet, inoculating it, and then incubating it at room temperature for a few weeks until the grain is fully colonized. Then it is refrigerated until sold. The grower sows the spawn by hand, along with a growth enhancer such as denatured soybean preparation that is encapsulated in wax or formaldehyde for slow release of nutrients when the mushrooms are ready for them. The spawn is mixed into the compost by machine, and the bed is covered with plastic to keep moisture in and flies out. It takes 14 to 21 days to colonize the compost completely.

After the compost has been colonized, the plastic is removed, and a casing layer of 1.75 in. of Canadian peat is added to the top, where it serves as a moisture reservoir but provides no nutrients. About 10 days later, the mycelium grows up through the casing. Shocking by manipulation of air temperature and CO_2 concentration induces the vegetative mycelium to begin fruiting, that is, producing mushrooms. Room temperature is dropped to 60°F; CO_2 is dropped to 1000 ppm from 5000 to 10,000 ppm, and water is added. This causes tiny "pinhead" mushrooms to form. Fruiting can be stopped by raising the CO_2 concentration again. First picking can begin 25 to 30 days from the time casing was added. The initial 5 to 7 days are the best picking, known as "first flush." The next 5 to 7 days are slower, followed by a second flush for 5 to 7 subsequent days. Picking ends after 30 to 35 days. Mushrooms are picked by hand. Mushrooms can be taken at any size, although growers don't want the caps opening in the beds because the spores they release can spread disease. Closed mushrooms also have a longer shelf life. When a diseased mushroom is found during the daily inspection of beds, it is first coated with Premisan, a phenolic compound, and then cut and removed. This keeps spores from dropping elsewhere on the bed while it is en route to the trash. Bacterial blotch and bacterial wilt invariably claim a few mushrooms from every crop.

Specialty mushrooms require special growth and media conditions. For example, shiitake and other mushrooms ordinarily grow on logs outdoors. Now they are grown indoors on artificial logs composed of sawdust (red oak is best) and other nutrients, which are formulated inside plastic bags. For shiitakes the logs are removed from the bag and inoculated with spawn. The spawn takes 6 weeks to colonize the logs completely. Another 2 to 3 weeks are required until fruiting begins (Figure 27.22). This is followed by about 6 weeks of harvest. The logs gradually shrink due to degradation of the cellulose.

Other types of specialty mush-

FIGURE 27.22 Shiitake mushrooms being grown on an artificial log made of chipped red oak wood plus other nutrients.

rooms, such as oyster mushrooms, are grown on artificial logs that are left in their bags. Light entering slits made in the sides of the bag causes fruiting bodies to emerge through the slits (Figure 27.23).

If you would like to try growing mushrooms yourself, you can find preinoculated kits advertised in many garden catalogs. Buying spawn is not recommended, as amateurs are not likely to have the facilities for proper production of their own compost.

Another interesting use of mushrooms is as decorative accents in the garden. Some mushrooms have bright colors and unusual forms that make them very attractive. Unfortunately some of the prettiest mushrooms are also deadly.

Identification of the wild mushrooms is a tricky business, definitely not for the beginner. There is a saying: "There are old mushroom eaters, and there are bold mushroom eaters, but there are no old, bold mushroom eaters." I'm content to eat those supplied by commercial mushroom growers.

FIGURE 27.23 Yellow oyster mushrooms growing out of slits in the black plastic bag that covers their artificial log. Oyster mushrooms also come in blue- and tan-colored varieties.

CHAPTER SUMMARY

MICROORGANISMS FOUND IN FOOD

- Anything that humans eat is food for microorganisms, too.
- Many organisms in food are commensals, some cause spoilage, and a few cause human disease.

Grains

- Grains stored in moist areas become contaminated with various molds. Flour is purposely inoculated with yeasts to make bread.

Fruits and Vegetables

- Fruits and vegetables are subject to soft rot and mold damage.

Meats, Poultry, Fish, and Shellfish

- Meats, poultry, fish, and seafoods contain many kinds of microorganisms, some of which cause zoonoses. Poultry often is contaminated with *Salmonella*, *Clostridium perfringens*, and *Staphylococcus aureus*; fish and shellfish can be contaminated with several kinds of bacteria and viruses.

Milk

- Milk can contain organisms from cows, milk handlers, and the environment.

Other Edible Substances

- Sugars support growth of various organisms, which are killed during refining.
- Spices contain large numbers of microorganisms; condiments may support mold growth.
- Syrups for carbonated beverages can be contaminated with molds.
- Tea, coffee, and cocoa also are subject to mold growth if not kept dry.

DISEASES TRANSMITTED IN FOOD

- Common diseases transmitted in food and milk are listed in Table 27.2.
- Good sanitation reduces the chance of getting foodborne diseases.

PREVENTION OF SPOILAGE AND DISEASE TRANSMISSION

- The most important factor in preventing spoilage and disease transmission in food and milk is cleanliness in handling.

Food Preservation

- Methods of food preservation include canning, refrigeration, freezing, lyophilization, drying, ionizing radiation, and use of chemical additives.

Pasteurization of Milk

- Methods of preserving milk include pasteurization and sterilization.

Standards Regarding Food and Milk Production

- Certain standards for food and milk production are maintained by federal, state, and local laws.
- Tests for milk quality are summarized in Table 27.3.

MICROORGANISMS AS FOOD AND IN FOOD PRODUCTION

RELATED KEY TERMS

Algae and Fungi As Food

- Rapid human population growth has created a need for new food sources.
- Yeasts can be grown on a variety of wastes and are good sources of cheap protein and vitamins. Equipment to grow them is expensive, and some effort will be needed to persuade humans to consider them acceptable food.
- Algae can be grown in lakes and on sewage and also are good sources of protein. Problems with growing them include the danger of viral contamination and lack of acceptability as food.

Food Production

- Yeast is used to leaven bread.
- The fermentative capabilities of certain bacteria are used to make dairy products such as buttermilk, sour cream, yogurt, a variety of fermented beverages, and cheeses.
- In cheese making the whey of milk is discarded, and microorganisms ferment the curd and impart flavor and texture to a cheese.
- Other foods produced by microbial fermentation include vinegar, sauerkraut, pickles, olives, poi, soy sauce, and other soy products.

INDUSTRIAL AND PHARMACEUTICAL MICROBIOLOGY

- **Industrial microbiology** deals with the use of microorganisms to assist in the manufacture of useful products or to dispose of waste products.
- **Pharmaceutical microbiology** deals with use of microorganisms in the manufacture of medically useful products.

Metabolic Processes Applicable in Industry

- Modifications of microbial processes in industry include altering nutrients available to microbes, altering environmental conditions, isolating mutants that produce excesses of useful products, and modifying the organisms by genetic engineering.

Problems of Industrial Microbiology

- One problem in industrial microbiology is the development of small-scale processes; other problems concern techniques for recovering products.

BEER, WINE, AND SPIRITS

- Beer is made from malted grains; hops are added, and the mixture is fermented.
- Wine is made by fermenting fruit juices.
- Spirits are made by fermenting various substances and distilling the products.

USEFUL ORGANIC PRODUCTS

Simple Organic Compounds

- Microorganisms can make alcohols, acetone, glycerol, and organic acids. Although few microbial processes are currently used to make these products, such processes may become economically feasible in the future.

Antibiotics

- Antibiotics are derived from *Streptomyces*, *Penicillium*, *Cephalosporin*, and *Bacillus* species.
- Many antibiotics have a beta-lactam ring; some semisynthetic antibiotics are made by modifying that ring so microorganisms cannot degrade it.

Enzymes

- Enzymes extracted from microorganisms include proteases, amylases, lactases, lipases, and invertase. They are used in detergents, drain cleaners, enrichment of food, and manufacture of paper.

bioconversion

Other Biological Products

- Vitamins and hormones usually are made by manipulating organisms so that they produce excessive amounts of these useful products.
- Single-cell proteins consist of whole organisms rich in proteins, which are used mainly as animal feeds.

MICROBIOLOGICAL MINING

- Microbes are currently used to extract copper from low-grade ores. Other minerals that can be extracted by microorganisms include iron, uranium, arsenic, lead, zinc, cobalt, and nickel.

MICROBIOLOGICAL WASTE DISPOSAL

- Sewage-treatment plants (Chapter 26) use microorganisms in waste disposal.
- A few organisms have been found that degrade toxic wastes, and research is under way to identify or develop others.

QUESTIONS FOR REVIEW

A.

1. What types of microorganisms are found in food and milk?
2. List organisms that cause food spoilage.
3. List organisms that cause disease.

B.

4. What problems prevent development of food uses of microorganisms?

C.

5. For the major diseases transmitted in food, list the causative organism(s) and explain how they enter foods.
6. How do organisms get into milk, and what diseases do they cause?
7. What is the single most important factor in preventing spoilage and disease transmission in food and milk?
8. List the methods of preserving foods, and describe how they are carried out.
9. What are the differences among the holding method of pasteurization, flash pasteurization, and sterilization of milk?
10. How are standards for food and milk production maintained?
11. What do each of the tests for milk quality accomplish?

D.

12. How are microorganisms used as food?
13. How are microorganisms used in bread making?
14. How are microorganisms used in making dairy products?
15. How are microorganisms used in making other foods?

E.

16. What are some common industrial uses of microorganisms?
17. What kinds of problems often arise in adapting microbiological processes for industrial use?

F.

18. Distinguish among the methods of making beer, wine, and spirits.

G.

19. What simple organic compounds can microorganisms make?
20. How do natural and semisynthetic antibiotics differ?
21. Why does the search for antibiotics and improvements in antibiotics continue?
22. How are enzymes obtained, and how are they used?
23. What other biologically useful products are made in industrial microbiology?

H.

24. What benefits can be derived from mining with microorganisms?

I.

25. What are some advantages of using microbes in waste disposal?

PROBLEMS FOR INVESTIGATION

1. Even though enzymes are used in cheese production, why are microorganisms also important?
2. Which waste materials in your community could be used to grow single-cell protein?
3. Can you think of a way to use yeast as a new type of food acceptable to people in the United States?
4. Devise methods to test samples of spices for bacterial content, and, if laboratory facilities are available, carry out your tests.
5. Read about home canning techniques. What precautions are taken in the processing directions?

SOME INTERESTING READING

Ayres, J. C., J. O. Mundt, and W. E. Sandine. 1980. *Microbiology of foods*. San Francisco: W. H. Freeman.

Biotechnology special issue. 1983. *Science* 219.

Brierley, C. L. 1982. Microbiological mining. *Scientific American* 247, no. 2 (August):44.

Broome, M. C., and M. W. Hickey. 1990. Comparison of fermentation-produced chymosin and calf rennet in cheddar cheese. *Aust. J. Dairy Technol.* 45(2):53–59.

Crueger, W., and A. Crueger. 1984. *Biotechnology: A textbook of industrial microbiology*. Madison, WI: Science Tech.

Erickson, D. 1991. Industrial immunology: antibodies may catalyze commercial chemistry. *Scientific American* 265(September):174–75.

Fox, P. F., and J. Law. 1991. Enzymology of cheese ripening. *Food Biotechnology* 5(3):239–62.

Hesseltine, C. W. 1983. Microbiology of oriental fermented foods. *Annual Review of Microbiology* 37:575.

———. 1985. Fungi, people, and soybeans. *Mycologia* 77:505.

Industrial Microbiology special issue. 1981. *Scientific American* 245, no. 3 (September).

Kosikowski, F. V. 1985. Cheese. *Scientific American* 252, no. 5 (May):88.

Lommi, H., A. Gronqvist, and E. Pajunen. 1990. Immobilized yeast reactor speeds beer production. *Food Technology* 44, no. 5 (May):128–30.

Markle, S. 1990. In the dough: try these simple explorations with yeast and get a rise out of science! *Instructor* 99, no. 5 (January):84.

Veld, J. H. I., and H. Hofstra. 1991. Biotechnology and the quality assurance of foods. *Food Biotechnology* 5(3):313–22.

Webb, A. D. 1984. The science of making wine. *American Scientist* 72:360.

APPENDIX A

Metric System Measurements

METRIC SYSTEM PREFIXES

pico (p) = 10^{-12}
nano (n) = 10^{-9}
micro (μ) = 10^{-6}
milli (m) = 10^{-3}
centi (c) = 10^{-2}
deci (d) = 10^{-1}
kilo (k) = 10^{4}

LENGTH

1 kilometer (km) = 0.62 miles
1 meter = 39.3700 inches = 3.2808 feet
1 meter (m) = 100 centimeters (cm) = 1,000 millimeters (mm)
1 centimeter = 10 millimeters (mm) = 0.394 inch (in)
1 millimeter = 0.0394 in
1 micrometer (μm) = 10^{-6} meter
1 nanometer (nm) = 10^{-9} meter
1 angstrom (Å) = 10^{-10} meter

VOLUME

1 liter = 1.0567 quarts
1 liter = 1,000 milliliters (ml)
1 milliliter = 1 cm^3 = 0.061 cubic in.
1 mm^3 = 10^{-3} cm^3 = 10^{-6} liter

MASS

1 kilogram (kg) = 1,000 grams = 2.205 pounds
1 pound = 453.60 g
1 gram (g) = 1,000 milligrams = 0.0353 ounce
1 ounce = 28.35 g
1 milligram (mg) = 10^{-3} g
1 microgram (μg) = 10^{-6} g

TEMPERATURE

degrees Fahrenheit (°F) = 9/5(°C + 32)
degrees Centigrade (°C) = 5/9(°F − 32)

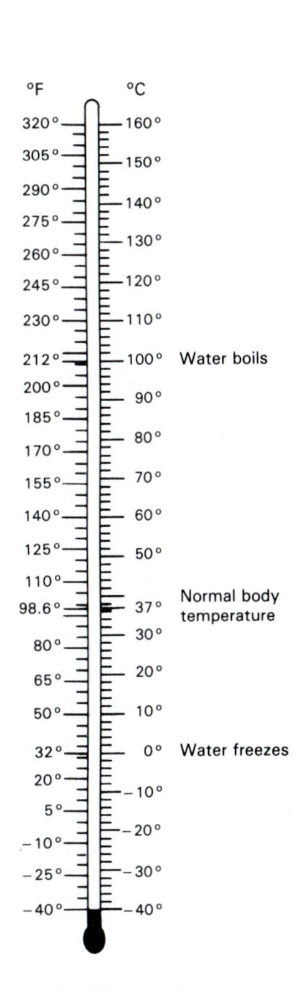

0°C = 32°F (freezing point of water)
100°C = 212°F (boiling point of water)
37°C = 98.6°F (normal body temperature)

EXPONENTIAL NOTATION

Numbers that are either very large or very small are usually represented in *exponential notation* as a number between 1 and 10 multiplied by a power of 10. In this kind of expression, the small raised number to the right of the 10 is the *exponent*.

Number	Exponential Form	Exponent
1,000,000	1×10^6	6
100,000	1×10^5	5
10,000	1×10^4	4
1,000	1×10^3	3
100	1×10^2	2
10	1×10^1	1
1		
0.1	1×10^{-1}	−1
0.01	1×10^{-2}	−2
0.001	1×10^{-3}	−3
0.000 1	1×10^{-4}	−4
0.000 01	1×10^{-5}	−5
0.000 001	1×10^{-6}	−6
0.000 000 1	1×10^{-7}	−7

Numbers greater than 1 have *positive* exponents, which tell how many times a number must be *multiplied* by 10 to obtain the correct value. For example, the expression 5.2×10^3 means that 5.2 must be multiplied by 10 three times:

$$5.2 \times 10^3 = 5.2 \times 10 \times 10 \times 10 = 5.2 \times 1000 = 5200$$

Note that doing this means moving the decimal point three places to the right:

5 2 0 0.

1 2 3

The value of a positive exponent indicates *how many places to the right the decimal point must be moved* to give the correct number in ordinary decimal notation.

Numbers less than 1 have *negative* exponents, which tell how many times a number must be *divided* by 10 (or multiplied by one-tenth) to obtain the correct value. Thus, the expression 3.7×10^{-2} means that 3.7 must be divided by 10 two times:

$$3.7 \times 10^{-2} = \frac{3.7}{10 \times 10} = \frac{3.7}{100} = 0.037$$

Note that doing this means moving the decimal point two places to the left:

0. 0 3 7

2 1

The value of a negative exponent indicates *how many places to the left the decimal point must be moved* to give the correct number in ordinary decimal notation.

CONVERTING DECIMAL NUMBERS TO EXPONENTIAL NOTATION

To convert a number greater than 1 from decimal notation to exponential notation, first move the decimal point to the *left* until there is only a single digit to the left of the decimal point. The *positive* exponent needed for exponential notation is the same as *the number of places the decimal point was moved.*

$$6\,3\,5\,7\,8\,1. = 6.35781 \times 10^5$$

5 4 3 2 1

To convert a number smaller than 1 from decimal notation to exponential notation, first move the decimal point to the *right* until there is *a single nonzero digit* to the left of the decimal point. The *negative* exponent needed for exponential notation is the same as *the number of places the decimal point was moved.*

$$0.\,0\,0\,0\,4\,2\,6 = 4.26 \times 10^{-4}$$

1 2 3 4

MULTIPLYING EXPONENTIAL NUMBERS

To multiply two numbers in exponential form, the exponents are *added*. For example:

$$\begin{aligned}(3.5 \times 10^3) \times (4.2 \times 10^4) &= 3.5 \times 4.2 \times 10^{(3+4)} \\ &= 14.7 \times 10^7 \\ &= 1.47 \times 10^8 = 1.5 \times 10^8 \\ &\quad \text{(rounded off)}\end{aligned}$$

$$\begin{aligned}(5.2 \times 10^4) \times (4.6 \times 10^{-3}) &= 5.2 \times 4.6 \times 10^{[4+(-3)]} \\ &= 23.92 \times 10^1 \\ &= 2.392 \times 10^2 = 2.4 \times 10^2 \\ &\quad \text{(rounded off)}\end{aligned}$$

DIVIDING EXPONENTIAL NUMBERS

To divide two numbers in exponential form, the exponents are *subtracted*. For example:

$$\begin{aligned}\frac{4.1 \times 10^4}{6.2 \times 10^6} &= \frac{4.1}{6.2} \times 10^{(4-6)} = 0.6613 \times 10^{-2} \\ &= 6.6 \times 10^{-3}\end{aligned}$$

$$\begin{aligned}\frac{6.6 \times 10^3}{8.4 \times 10^{-2}} &= \frac{6.6}{8.4} \times 10^{[3-(-2)]} = 0.7857 \times 10^5 \\ &= 7.9 \times 10^4\end{aligned}$$

APPENDIX B

Classification of Bacteria and Viruses

CLASSIFICATION OF BACTERIA ACCORDING TO *BERGEY'S MANUAL OF SYSTEMATIC BACTERIOLOGY*

The classification of bacteria changes as our understanding of them improves. The classification scheme currently accepted by microbiologists is set forth in the four volumes of *Bergey's Manual of Systematic Bacteriology* (Baltimore: Williams and Wilkins): Volume 1, 1984, R. Krieg; Volume 2, 1986, P. Sneath; Volume 3, 1989, J. Staley; and Volume 4, 1989, S. Williams.

This is a working model only, and changes were made even as the volumes were being published. For example, Section 26 in Volume 4 is an updated, slightly enlarged version of Section 17 in Volume 2. Some organisms present special difficulties, and will, therefore, be found listed in two different places until further research can determine their proper position.

KINGDOM PROCARYOTAE

Division I.	**Gracilicutes**	Procaryotes with thinner cell walls, implying a gram-negative type of cell wall
Division II.	**Firmicutes**	Procaryotes with thick and strong skin, indicative of a gram-positive type of cell wall
Division III.	**Tenericutes**	Procaryotes of a pliable, soft nature, indicative of the lack of a rigid cell wall
Division IV.	**Mendosicutes**	Procaryotes having faulty cell walls, suggesting the lack of conventional peptidoglycan

VOLUME 1

SECTION 1: The Spirochetes

Order I	*Spirochaetales*
Family I.	*Spirochaetaceae*
Genus I.	*Spirochaeta*
Genus II.	*Cristispira*
Genus III.	*Treponema*
Genus IV.	*Borrelia*
Family II.	*Leptospiraceae*
Genus I.	*Leptospira*

Other Organisms
Hindgut spirochetes of termites and *Cryptocercus punctulatus* (wood-eating cockroach)

SECTION 2: Aerobic/Microaerophilic, Motile, Helical/Vibrioid Gram-Negative Bacteria

Genus	*Aquaspirillum*
Genus	*Spirillum*
Genus	*Azospirillum*
Genus	*Oceanospirillum*
Genus	*Campylobacter*
Genus	*Bdellovibrio*
Genus	*Vampirovibrio*

SECTION 3: Nonmotile (or Rarely Motile), Gram-Negative Curved Bacteria

Family I.	*Spirosomaceae*
Genus I.	*Spirosoma*
Genus II.	*Runella*
Genus III.	*Flectobacillus*

Other Genera

Genus	*Microcyclus*
Genus	*Meniscus*
Genus	*Brachyarcus*
Genus	*Pelosigma*

SECTION 4: Gram-Negative Aerobic Rods and Cocci

Family I. *Pseudomonadaceae*
Genus I. *Pseudomonas*
Genus II. *Xanthomonas*
Genus III. *Frateuria*
Genus IV. *Zoogloea*
Family II. *Azotobacteraceae*
Genus I. *Azotobacter*
Genus II. *Azomonas*
Family III. *Rhizobiaceae*
Genus I. *Rhizobium*
Genus II. *Bradyrhizobium*
Genus III. *Agrobacterium*
Genus IV. *Phyllobacterium*
Family IV. *Methylococcaceae*
Genus I. *Methylococcus*
Genus II. *Methylomonas*
Family V. *Halobacteriaceae*
Genus I. *Halobacterium*
Genus II. *Halococcus*
Family VI. *Acetobacteraceae*
Genus I. *Acetobacter*
Genus II. *Gluconobacter*
Family VII. *Legionellaceae*
Genus I. *Legionella*
Family VIII. *Neisseriaceae*
Genus I. *Neisseria*
Genus II. *Moraxella*
Genus III. *Acinetobacter*
Genus IV. *Kingella*

Other Genera
Genus *Beijerinckia*
Genus *Derxia*
Genus *Xanthobacter*
Genus *Thermus*
Genus *Thermomicrobium*
Genus *Halomonas*
Genus *Alteromonas*
Genus *Flavobacterium*
Genus *Alcaligenes*
Genus *Serpens*
Genus *Janthinobacterium*
Genus *Brucella*
Genus *Bordetella*
Genus *Francisella*
Genus *Paracoccus*
Genus *Lampropedia*

SECTION 5: Facultatively Anaerobic Gram-Negative Rods

Family I. *Enterobacteriaceae*
Genus I. *Escherichia*
Genus II. *Shigella*
Genus III. *Salmonella*
Genus IV. *Citrobacter*
Genus V. *Klebsiella*
Genus VI. *Enterobacter*
Genus VII. *Erwinia*
Genus VIII. *Serratia*
Genus IX. *Hafnia*
Genus X. *Edwardsiella*
Genus XI. *Proteus*
Genus XII. *Providencia*
Genus XIII. *Morganella*
Genus XIV. *Yersina*

Other Genera of Enterobacteriacae
Genus *Obesumbacterium*
Genus *Xenorhabdus*
Genus *Kluyvera*
Genus *Rahnella*
Genus *Cedecea*
Genus *Tatumella*
Family II. *Vibrionaceae*
Genus I. *Vibrio*
Genus II. *Photobacterium*
Genus III. *Aeromonas*
Genus IV. *Plesiomonas*
Family III. *Pasteurellaceae*
Genus I. *Pasteurella*
Genus II. *Haemophilus*
Genus III. *Actinobacillus*

Other Genera
Genus *Zymomonas*
Genus *Chromobacterium*
Genus *Cardiobacterium*
Genus *Calymmatobacterium*
Genus *Gardnerella*
Genus *Eikenella*
Genus *Streptobacillus*

SECTION 6: Anaerobic Gram-Negative Straight, Curved and Helical Rods

Family I. *Bacteroidaceae*
Genus I. *Bacteroides*
Genus II. *Fusobacterium*
Genus III. *Leptotrichia*
Genus IV. *Butyrivibrio*
Genus V. *Succinimonas*
Genus VI. *Succinivibrio*
Genus VII. *Anaerobiospirillum*
Genus VIII. *Wolinella*
Genus IX. *Selenomonas*
Genus X. *Anaerovibrio*
Genus XI. *Pectinatus*
Genus XII. *Acetivibrio*
Genus XIII. *Lachnospira*

SECTION 7: Dissimilatory Sulfate- or Sulfur-Reducing Bacteria

Genus *Desulfuromonas*
Genus *Desulfovibrio*
Genus *Desulfomonas*
Genus *Desulfococcus*
Genus *Desulfobacter*
Genus *Desulfobulbus*
Genus *Desulfosarcina*

SECTION 8: Anaerobic Gram-Negative Cocci

Family I.	*Veillonellaceae*
Genus I.	*Veillonella*
Genus II.	*Acidaminococcus*
Genus III.	*Megasphaera*

SECTION 9: The Rickettsias and Chlamydias

Order I.	*Rickettsiales*
Family I.	*Rickettsiaceae*
Tribe I.	*Rickettsieae*
Genus I.	*Rickettsia*
Genus II.	*Rochalimaea*
Genus III.	*Coxiella*
Tribe II.	*Ehrlichieae*
Genus IV.	*Ehrlichia*
Genus V.	*Cowdria*
Genus VI.	*Neorickettsia*
Tribe III.	*Wolbachieae*
Genus VII.	*Wolbachia*
Genus VIII.	*Rickettsiella*
Family II.	*Bartonellaceae*
Genus I.	*Bartonella*
Genus II.	*Grahamella*
Family III.	*Anaplasmataceae*
Genus I.	*Anaplasma*
Genus II.	*Aegyptianella*
Genus III.	*Haemobartonella*
Genus IV.	*Eperythrozoon*
Order II.	*Chlamydiales*
Family I.	*Chlamydiaceae*
Genus I.	*Chlamydia*

SECTION 10: The Mycoplasmas

Division *Tenericutes*

Class I.	*Mollicutes*
Order I.	*Mycoplasmatales*
Family I.	*Mycoplasmataceae*
Genus I.	*Mycoplasma*
Genus II.	*Ureaplasma*
Family II.	*Acholeplasmataceae*
Genus I.	*Acholeplasma*
Family III.	*Spiroplasmataceae*
Genus I.	*Spiroplasma*

Other Genera

Genus	*Anaeroplasma*
Genus	*Thermoplasma*

Mycoplasma-like organisms of plants and invertebrates

SECTION 11: Endosymbionts

A. Endosymbionts of Protozoa
Endosymbionts of ciliates
Endosymbionts of flagellates
Endosymbionts of amoebas
Taxa of endosymbionts:

Genus I.	*Holospora*
Genus II.	*Caedibacter*
Genus III.	*Pseudocaedibacter*
Genus IV.	*Lyticum*
Genus V.	*Tectibacter*

B. Endosymbionts of Insects
Blood-sucking insects
Plant sap-sucking insects
Cellulose and stored grain feeders
Insects feeding on complex diets
Taxon of endosymbionts:

Genus	*Blattabacterium*

C. Endosymbionts of Fungi and Invertebrates other than Anthropods
Fungi
Sponges
Coelenterates
Helminthes
Annelids
Marine worms and mollusks

VOLUME 2

SECTION 12: Gram-Positive Cocci

Family I.	*Micrococcaceae*
Genus I.	*Micrococcus*
Genus II.	*Stomatococcus*
Genus III.	*Planococcus*
Genus IV.	*Staphylococcus*
Family II.	*Deinococcaceae*
Genus I.	*Deinococcus*

Other Genera

Genus *Streptococcus*
- Pyogenic Hemolytic Streptococci
- Oral Streptococci
- Lactic Acid Streptococci
- Anaerobic Streptococci
- Other Streptococci

Genus	*Leuconostoc*
Genus	*Pediococcus*
Genus	*Aerococcus*
Genus	*Gemella*
Genus	*Peptococcus*
Genus	*Peptostreptococcus*
Genus	*Ruminococcus*
Genus	*Coprococcus*
Genus	*Sarcina*

SECTION 13: Endospore-forming Gram-Positive Rods and Cocci

Genus	*Bacillus*
Genus	*Sporolactobacillus*
Genus	*Clostridium*
Genus	*Desulfotomaculum*
Genus	*Sporosarcina*
Genus	*Oscillospira*

SECTION 14: Regular, Nonsporing, Gram-Positive Rods

Genus *Lactobacillus*
Genus *Listeria*
Genus *Erysipelothrix*
Genus *Brochothrix*
Genus *Renibacterium*
Genus *Kurthia*
Genus *Caryophanon*

SECTION 15: Irregular, Nonsporing, Gram-Positive Rods

Genus *Corynebacterium*
Plant Pathogenic Species of Corynebacterium
Genus *Gardnerella*
Genus *Arcanobacterium*
Genus *Arthrobacter*
Genus *Brevibacterium*
Genus *Curtobacterium*
Genus *Caseobacter*
Genus *Microbacterium*
Genus *Aureobacterium*
Genus *Cellulomonas*
Genus *Agromyces*
Genus *Arachnia*
Genus *Rothia*
Genus *Propionibacterium*
Genus *Eubacterium*
Genus *Acetobacterium*
Genus *Lachnospira*
Genus *Butyrivibrio*
Genus *Thermoanaerobacter*
Genus *Actinomyces*
Genus *Bifidobacterium*

SECTION 16: The Mycobacteria

Family *Mycobacteriaceae*
Genus *Mycobacterium*

SECTION 17: Nocardioforms

(These organisms are considered again in Section 26.)

Genus *Nocardia*
Genus *Rhodococcus*
Genus *Nocardioides*
Genus *Pseudonocardia*
Genus *Oerskovia*
Genus *Saccharopolyspora*
Genus *Micropolyspora*
Genus *Promicromonospora*
Genus *Intrasporangium*

VOLUME 3

SECTION 18: Anoxygenic Phototrophic Bacteria

I. Purple Bacteria
Family I. *Chromatiaceae*
Genus I. *Chromatium*
Genus II. *Thiocystis*
Genus III. *Thiospirillum*
Genus IV. *Thiocapsa*
Genus V. *Lamprobacter*
Genus VI. *Lamprocystis*
Genus VII. *Thiodictyon*
Genus VIII. *Amoebobacter*
Genus IX. *Thiopedia*
Family II. *Ectothiorhodospiraceae*
Genus *Ectothiorhodospira*
Purple nonsulfur bacteria
Genus *Rhodospirillum*
Genus *Rhodopila*
Genus *Rhodobacter*
Genus *Rhodopseudomonas*
Genus *Rhodomicrobium*
Genus *Rhodocyclus*
II. Green Bacteria
Green sulfur bacteria
Genus *Chlorobium*
Genus *Prosthecochloris*
Genus *Pelodictyon*
Genus *Ancalochloris*
Genus *Chloropherpeton*
Addendum to the green sulfur bacteria
Multicellular filamentous green bacteria
Genus *Chloroflexus*
Genus *Heliothrix*
Genus *"Oscillochloris"*
Genus *Chloronema*
III. Genera Incertae Sedis
Genus *Heliobacterium*
Genus *Erythrobacter*

SECTION 19: Oxygenic Photosynthetic Bacteria

Group I. Cyanobacteria
Preface
Taxa of the Cyanobacteria
Subsection I. Order *Chroococcales*
1. Genus I. *Chamaesiphon*
2. Genus II. *Gloeobacter*
3. *Synechococcus*-group
4. Genus III. *Gloeothece*
5. *Cyanothece*-group
6. *Gloeocapsa*-group
7. *Synechocystis*-group

Subsection II. Order *Pleurocapsales*
1. Genus I. *Dermocarpa*
2. Genus II. *Xenococcus*
3. Genus III. *Dermocarpella*
4. Genus IV. *Myxosarcina*
5. Genus V. *Chroococcidiopsis*
6. *Pleurocapsa*-group

Subsection III. Order *Oscillatoriales*
Genus I. *Spirulina*
Genus II. *Arthrospira*
Genus III. *Oscillatoria*
Genus IV. *Lyngbya*
Genus V. *Pseudanabaena*
Genus VI. *Starria*

Genus VII. *Crinalium*
Genus VIII. *Microcoleus*

Subsection IV. Order *Nostocales*

Family I. *Nostocaceae*
Genus I. *Anabaena*
Genus II. *Aphanizomenon*
Genus III. *Nodularia*
Genus IV. *Cylindrospermum*
Genus V. *Nostoc*
Family II. *Scytonemataceae*
Genus I. *Scytonema*
Family III. *Rivulariaceae*
Genus I. *Calothrix*

Subsection V. Order *Stigonematales*

Genus I. *Chlorogloeopsis*
Genus II. *Fischerella*
Genus III. *Stigonema*
Genus IV. *Geitleria*

Group II. Order *Prochlorales*

Family I. *Prochloraceae*
Genus *Prochloron*
Other taxa
Genus *"Prochlorothrix"*

SECTION 20: Aerobic Chemolithotrophic Bacteria and Associated Organisms

Nitrifying Bacteria
Family *Nitrobacteraceae*
Nitrite-oxidizing bacteria
Genus I. *Nitrobacter*
Genus II. *Nitrospina*
Genus III. *Nitrococcus*
Genus IV. *Nitrospira*
Ammonia-oxidizing bacteria
Genus V. *Nitrosomonas*
Genus VI. *Nitrosococcus*
Genus VII. *Nitrosospira*
Genus VIII. *Nitrosolobus*
Genus IX. *"Nitrosovibrio"*

Colorless Sulfur Bacteria
Genus *Thiobacterium*
Genus *Macromonas*
Genus *Thiospira*
Genus *Thiovulum*
Genus *Thiobacillus*
Genus *Thiomicrospira*
Genus *Thiosphaera*
Genus *Acidiphilium*
Genus *Thermothrix*

Obligately Chemolithotrophic Hydrogen Bacteria
Genus *Hydrogenobacter*

Iron- and Manganese-Oxidizing and/or Depositing Bacteria
Family *"Siderocapsaceae"*
Genus I. *"Siderocapsa"*
Genus II. *"Naumanniella"*
Genus III. *"Siderococcus"*
Genus IV. *"Ochrobium"*

Magnetotactic Bacteria
Genus *Aquaspirillum* (*A. magnetotacticum*)
Genus *"Bilophococcus"*

SECTION 21: Budding and/or Appendaged Bacteria

Prosthecate Bacteria
Budding bacteria
Buds produced at tip of prostheca
Genus *Hyphomicrobium*
Genus *Hyphomonas*
Genus *Pedomicrobium*
Buds produced on cell surface
Genus *Ancalomicrobium*
Genus *Prosthecomicrobium*
Genus *Labrys*
Genus *Stella*
Bacteria that divide by binary transverse fission
Genus *Caulobacter*
Genus *Asticcacaulis*
Genus *Prosthecobacter*

Nonprosthecate Bacteria
Budding bacteria
Lack peptidoglycan
Genus *Planctomyces*
Genus *"Isosphaera"*
Contain peptidoglycan
Genus *Ensifer*
Genus *Blastobacter*
Genus *Angulomicrobium*
Genus *Gemmiger*
Nonbudding, stalked bacteria
Genus *Gallionella*
Genus *Nevskia*
Other bacteria
Nonspinate bacteria
Genus *Seliberia*
Genus *"Metallogenium"*
Genus *"Thiodendron"*
Spinate bacteria

SECTION 22: Sheathed Bacteria

Genus *Sphaerotilus*
Genus *Leptothrix*
Genus *Haliscomenobacter*
Genus *"Lieskeella"*
Genus *"Phragmidiothrix"*
Genus *Crenothrix*
Genus *"Clonothrix"*

SECTION 23: Nonphotosynthetic, Nonfruiting Gliding Bacteria

Order I. *Cytophagales*
Family I. *Cytophagaceae*
Genus I. *Cytophaga*
Genus II. *Capnocytophaga*
Genus III. *Flexithrix*
Genus IV. *Sporocytophaga*

Other Genera

Genus	*Flexibacter*
Genus	*Microscilla*
Genus	*Chitinophaga*
Genus	*Saprospira*
Order II.	*Lysobacterales*
Family I.	*Lysobacteraceae*
Genus I.	*Lysobacter*
Order III.	*Beggiatoales*
Family I.	*Beggiatoaceae*
Genus I.	*Beggiatoa*
Genus II.	*Thiothrix*
Genus III.	*Thioploca*
Genus IV.	*"Thiospirillopsis"*

Other Families and Genera

Family	*Simonsiellaceae*
Genus I.	*Simonsiella*
Genus II.	*Alysiella*
Family	*"Pelonemataceae"*
Genus I.	*"Pelonema"*
Genus II.	*"Achroonema"*
Genus III.	*"Peloploca"*
Genus IV.	*"Desmanthos"*

Other Genera

Genus	*Toxothrix*
Genus	*Leucothrix*
Genus	*Vitreoscilla*
Genus	*Desulfonema*
Genus	*Achromatium*
Genus	*Agitococcus*
Genus	*Herpetosiphon*

SECTION 24: Fruiting Gliding Bacteria: The Myxobacteria

Order	*Myxococcales*
Family I.	*Myxococcaceae*
Genus	*Myxococcus*
Family II.	*Archangiaceae*
Genus	*Archangium*
Family III.	*Cystobacteraceae*
Genus I.	*Cystobacter*
Genus II.	*Melittangium*
Genus III.	*Stigmatella*
Family IV.	*Polyangiaceae*
Genus I.	*Polyangium*
Genus II.	*Nannocystis*
Genus III.	*Chondromyces*

SECTION 25: Archaeobacteria

Group I. Methanogenic Archaeobacteria

Order I.	*Methanobacteriales*
Family I.	*Methanobacteriaceae*
Genus I.	*Methanobacterium*
Genus II.	*Methanobrevibacter*
Family II.	*Methanothermaceae*
Genus	*Methanothermus*
Order II.	*Methanococcales*
Family	*Methanococcaceae*
Genus	*Methanococcus*
Order III.	*Methanomicrobiales*
Family I.	*Methanomicrobiaceae*
Genus I.	*Methanomicrobium*
Genus II.	*Methanospirillum*
Genus III.	*Methanogenium*
Family II.	*Methanosarcinaceae*
Genus I.	*Methanosarcina*
Genus II.	*Methanolobus*
Genus III.	*Methanothrix*
Genus IV.	*Methanococcoides*
Other Taxa	
Family	*Methanoplanaceae*
Genus	*Methanoplanus*
Other Genus	*Methanosphaera*

Group II. Archaeobacterial Sulfate Reducers

Order	*"Archaeoglobales"*
Family	*"Archaeoglobaceae"*
Genus	*Archaeglobus*

Group III. Extremely Halophilic Archaeobacteria

Order	*Halobacteriales*
Family	*Halobacteriaceae*
Genus I.	*Halobacterium*
Genus II.	*Haloarcula*
Genus III.	*Haloferax*
Genus IV.	*Halococcus*
Genus V.	*Natronobacterium*
Genus VI.	*Natronococcus*

Group IV. Cell Wall-less Archaeobacteria

Genus	*Thermoplasma*

Group V. Extremely Thermophilic S^0-Metabolizers

Order I.	*Thermococcales*
Family	*Thermococcaceae*
Genus I.	*Thermococcus*
Genus II.	*Pyrococcus*
Order II.	*Thermoproteales*
Family I.	*Thermoproteaceae*
Genus I.	*Thermoproteus*
Genus II.	*Thermofilum*
Family II.	*Desulfurococcaceae*
Genus	*Desulfurococcus*
Other Bacteria	
Genus	*Staphylothermus*
Genus	*Pyrodictium*
Order III.	*Sulfolobales*
Family	*Sulfolobaceae*
Genus I.	*Sulfolobus*
Genus II.	*Acidianus*

VOLUME 4

SECTION 26: Nocardioform Actinomycetes

(This is an updated, slightly enlarged version of Section 17 in Volume 2.)

Genus	*Nocardia*
Genus	*Rhodococcus*
Genus	*Nocardioides*
Genus	*Pseudonocardia*
Genus	*Oerskovia*

Genus	*Saccharopolyspora*
Genus	*Faenia*
Genus	*Promicromonospora*
Genus	*Intrasporangium*
Genus	*Actinopolyspora*
Genus	*Saccharomonospora*

SECTION 27: Actinomycetes with Multilocular Sporangia

Genus	*Geodermatophilus*
Genus	*Dermatophilus*
Genus	*Frankia*

SECTION 28: Actinoplanetes

Genus	*Actinoplanes*
Genus	*Ampullariella*
Genus	*Pilimelia*
Genus	*Dactylosporangium*
Genus	*Micromonospora*

SECTION 29: Streptomycetes and Related Genera

Genus	*Streptomyces*
Genus	*Streptoverticillium*
Genus	*Kineosporia*
Genus	*Sporichthya*

SECTION 30: Maduromycetes

Genus	*Actinomadura*
Genus	*Microbispora*
Genus	*Microtetraspora*
Genus	*Planobispora*
Genus	*Planomonospora*
Genus	*Spirillospora*
Genus	*Streptosporangium*

SECTION 31: Thermomonospora and Related Genera

Genus	*Thermomonospora*
Genus	*Actinosynnema*
Genus	*Nocardiopsis*
Genus	*Streptoalloteichus*

SECTION 32: Thermoactinomycetes

Genus	*Thermoactinomyces*

SECTION 33: Other Genera

Genus	*Glycomyces*
Genus	*Kibdelosporangium*
Genus	*Kitasatosporia*
Genus	*Saccharothrix*
Genus	*Pasteuria*

ADDENDUM

Some novel Actinomycete Taxa that have been validly published or validated since Volume 4 of *Bergey's Manual of Systematic Bacteriology* went to press

Placement of certain genera has presented difficulties as indicated by the following examples:

(a) The genus *Gardnerella*. The organisms of this genus have had a checkered taxonomic history and it is still not entirely clear whether they should be placed in Volume 1 with Gram-negative bacteria or in Volume 2 with Gram-positive bacteria.

(b) The genus *Butyrivibrio*. Although the cells stain Gram-negative, the ultrastructure of the cell wall is of the Gram-positive type. It is not clear whether the genus should be placed in Volume 1 or Volume 2.

(c) The genus *Xanthobacter*. The cells stain Gram-positive or Gram-variable, yet the cell wall structure and composition, as well as nucleic acid hybridization data, indicate that the organisms are of the Gram-negative type.

(d) The genus *Chromobacterium*. Although 80 percent of the strains attack glucose fermentatively and grow well anaerobically, the remainder attack glucose oxidatively and grow slowly under anaerobic conditions. It is consequently difficult to assign the organisms definitely to either Section 5 (Facultatively Anaerobic Gram-Negative Rods) or Section 4 (Gram-Negative Aerobic Rods and Cocci). Nucleic acid hybridization studies indicate a relationship to certain genera of aerobic rods.

(e) The genus *Zymomonas*. Although the organisms are facultatively anaerobic (a few obligately anaerobic), they are related genetically, phenotypically and ecologically to the acetic acid bacteria, which are aerobic. Moreover, the occurrence of the Entner-Doudoroff pathway is typical of aerobic bacteria.

(f) The genus *Thermoplasma*. The lack of a cell wall makes this genus compatible with Section 10 (The Mycoplasmas); however, studies of the ribosomal RNA, as well as various phenotypic characteristics, indicate that the genus is related to the archaeobacteria, covered in Volume 3 of the *Manual*.

(g) The genera *Halobacterium* and *Halococcus*. Although these extreme halophiles are compatible with Section 4 (Gram-Negative Aerobic Rods and Cocci), nucleic acid studies and certain phenotypic characteristics indicate the genus is related to the archaeobacteria, covered in Volume 3.

As an interim solution to some of these problems, some taxa are described, not only in Volume 1, but in an appropriate subsequent volume as well.

CLASSIFICATION OF VIRUSES

Viral classification is in a greater state of change than is bacterial taxonomy. Many viruses have not even been classified. The concept of viral "species" is being explored, and in the near future we will probably see classifications that include species names for viruses.

The classification presented here is based on those presented in *Fields Virology*, 2nd ed., Volume 1, by Bernard N. Fields and David M. Knipe (1990, Raven Press), and *Introduction to Modern Virology* by N. J. Dimmock and S. B. Primrose, 3rd ed. (1987, Blackwell Scientific Publications, Ltd).

The 21 families of viruses listed here are only a small part of the more than 61 families recognized by the International Committee on Taxonomy of Viruses (ICTV). These families were chosen because they contain viruses that are pathogenic for humans and animals.

1. Family: Picornaviridae

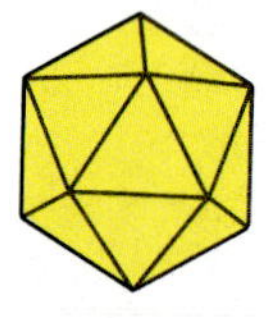

Genera:

Enterovirus (primarily viruses of gastrointestinal tract, polioviruses, coxsackieviruses, echoviruses)
Cardiovirus (EMC virus of mice)
Rhinovirus (mainly viruses of upper respiratory tract)
Aphtovirus (foot-and-mouth disease)
[Unnamed] (proposed, for hepatitis A virus)

Nonenveloped, icosahedral, positive-sense, ssRNA. Replication and assembly take place in cytoplasm, and virus is released via cell destruction.

2. Family: Calciviridae

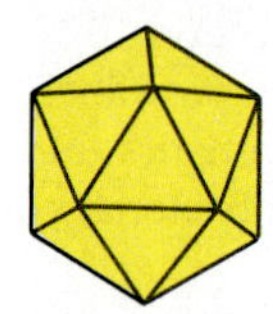

Genera:

Calcivirus (possibly Norwalk virus and similar viruses, which are proven causes of gastroenteritis)

Nonenveloped, icosahedral, positive-sense ssRNA. Replication and assembly take place in cytoplasm, and virus is released via cell destruction.

3. Family: Togaviridae

Genera:

Alphavirus (eastern equine encephalitis virus, western equine encephalitis virus, Venezuelan equine encephalitis virus)
Rubivirus (rubella virus)
Pestivirus (mucosal disease viruses, hog cholera virus)
Artevirus (equine arteritis virus)

Lipid-containing envelope, positive-sense, ssRNA. Replication takes place in cytoplasm, and assembly involves budding through host cell membranes. Many replicate in arthropods as well as in vertebrates.

4. Family: Flaviridae

Genera:

Flavivirus (yellow fever virus, denque viruses, St. Louis encephalitis virus)

Lipid-containing envelope, unknown symmetry, positive-sense, ssRNA. Replication takes place in cytoplasm, and assembly involves budding through host cell membranes, at which time virions become enveloped by an unknown process. Most replicate in arthropods as well.

5. Family: Coronaviridae

Genera:

Coronavirus (common cold, upper respiratory tract infection, probably pneumonia)

Pleomorphic lipid-containing envelope, positive-sense, ssRNA. Replication takes place in cytoplasm and assembly involves budding, usually through intracytoplasmic membranes; release occurs via membrane fusion and exocytosis and by cell destruction.

6. Family: Rhabdoviridae

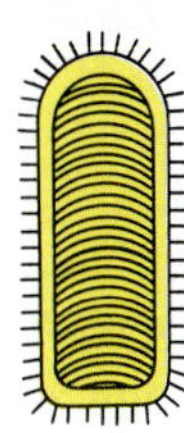

Genera:

Vesiculovirus (vesicular stomatitislike viruses)
Lyssavirus (rabieslike viruses)
[Unnamed] (proposed, for bovine ephemeral feverlike viruses)
[Unnamed] (proposed, for the many ungrouped rhabdoviruses of mammals, birds, fish, arthropods, and plants)

Bullet-shaped (plant rhabdoviruses are often bacilliform), lipid-containing envelope, negative-sense ssRNA. Replication takes place in cytoplasm, and assembly occurs via budding upon plasma (vesicular stomatitis viruses) or intracytoplasmic (rabies viruses) membranes. Many replicate in, and are transmitted by, arthropods.

7. Family: Filoviridae

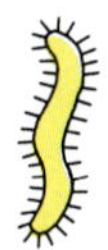

Genera:

Filovirus (Marburg and Ebola viruses)

Long filamentous forms, sometimes with branching, and sometimes as U-shaped, 6-shaped, or circular forms. Lipid-containing envelope, negative-sense ssRNA. Replication takes place in the cytoplasm, and assembly involves envelopment via budding of preformed nucleocapsids. Marburg and Ebola viruses are "Biosafety Level 4" pathogens; they must be handled in the laboratory under maximum containment conditions.

8. Family: Paramyxoviridae

Genera:

Paramyxovirus (paramyxoviruses, mumps virus)
Morbillivirus (measleslike viruses, measles virus, canine distemper virus)
Pneumovirus (respiratory syncytial viruses)

Pleomorphic lipid-containing envelope, negative-sense ssRNA. Replication takes place in cytoplasm, and assembly occurs via budding through the plasma membrane. Morbilliviruses may cause persistent infections.

9. Family: Orthomyxoviridae

Genera:

Influenzavirus (influenza A and B viruses)
[Unnamed] (influenza C virus)

Pleomorphic lipid-containing envelope. Replication takes place in the nucleus and cytoplasm, and assembly occurs via budding upon plasma membrane. The viruses can reassort genes during mixed infections.

10. Family: Bunyaviridae

Genera:

Bunyavirus (Bunyamwera supergroup)
Phlebovirus (sandly fever viruses)
Nairovirus (Nairobi sheep diseaselike viruses)
Unkuvirus (Unkuniemilike viruses)
Hantavirus (hemorrhagic fever, Korean hemorrhagic fever)

Lipid-containing envelope, negative-sense, and in the genus *Phlebovirus* ambisense, ssRNA. Replication takes place in the cytoplasm, and assembly occurs via budding through smooth membranes of the Golgi apparatus. Closely related viruses can reassort genes during mixed infections.

11. Family: Arenaviridae

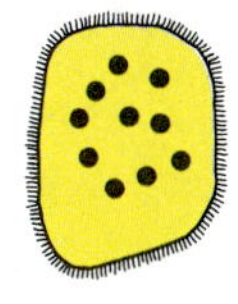

Genera:

Arenavirus (Lassa fever)

Pleomorphic lipid-containing envelope, negative, and ambisense ssRNA. Replication takes place in cytoplasm, and assembly occurs via budding from plasma membrane. The human pathogens Lassa, Machupo, and Junin viruses are "Biosafety Level 4" pathogens and can be worked with in the laboratory only under maximum containment conditions.

12. Family: Reoviridae

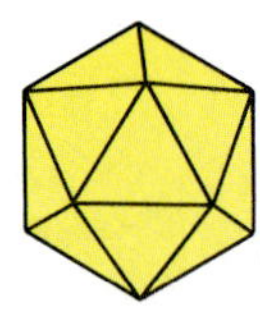

Genera:

Reovirus (reovirus 1, 2, and 3)
Orbivirus (Orungo virus)
Rotavirus (human rotaviruses)
Phytoreovirus (plant reovirus subgroup 1)
Fijivirus (plant reovirus subgroup 2)
Cypovirus (cytoplasmic polyhedrosis viruses)
[Unnamed] (proposed, for Colorado tick fever virus)

Each genus differs in morphologic and physicochemical details. In general, virions are nonenveloped, icosahedral symmetry, linear dsRNA. Replication and assembly take place in cytoplasm, often in association with granular of fibrillar inclusion bodies.

13. Family: Birnaviridae

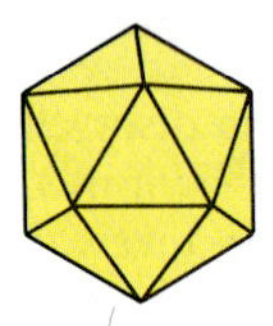

Genera:

Birnavirus

Nonenveloped, icosahedral, two segments of linear dsRNA. Replication and assembly take place in cytoplasm and virus release occurs via cell destruction.
Human pathogens: none known.
Animal pathogens: infectious bursal disease virus of chickens, infectious pancreatic necrosis virus of fish.

14. Family: Retroviridae

Subfamily:

Oncovirinae (the RNA tumor viruses, human T-lymphotropic virus HTLV-I, possibly associated with hairy-cell leukemia, HTLV-II)
Spumavirinae (the foamy viruses)
Letivirinae (the HIV-like viruses; the Maedi/visnalike viruses, human immunodeficiency viruses, HIV-1, HIV-2, HTLV-III, FIV)

Lipid-containing envelope, icosahedral, two identical molecules of positive-sense ssRNA. Virion assembly occurs via budding through plasma membranes. Most of the viruses are oncogenic, causing leukemias, lymphomas, carcinomas, and sarcomas.

15. Family: Hepadnaviridae

Genera:

Hepadnavirus (hepatitis B-like viruses)

Lipid-containing envelope (containing hepatitis B surface antigen [HBsAg]). One molecule of DNA, which is circular, nicked, and mainly double-stranded with a large single-stranded gap. Replication takes place in the nucleus of hepatocytes, HBsAg production occurs in the cytoplasm. Persistence is common and is associated with chronic disease and neoplasia.

16. Family: Parvoviridae

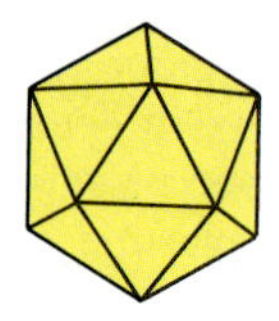

Genera:

Parvovirus (human parvovirus B-19, the cause of aplastic crisis in hemolytic anemias and in sickle-cell disease, erythema infectiosum-fifth disease, and spontaneous abortion, fetal death, and hydrops fetalis, feline panleukopenia virus, canine parvovirus)
Dependovirus (adeno-associated viruses)
Densovirus (insect parvoviruses)

Nonenveloped, icosahedral symmetry, one molecule of ssDNA. Viruses of the genus *Parvovirus* preferentially encapsidate negative-sense DNA, whereas members of the other two genera encapsidate positive- and negative-sense DNA equally. Replication and assembly take place in the nucleus.

17. Family: Papovaviridae

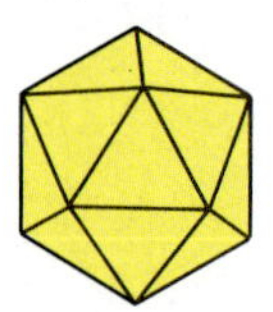

Genera:

Papillomavirus (papillomaviruses, genital condylomas)
Polyomavirus (polyomaviruses)
JC, SV40 (progressive multifocal leukoencephalopathy)

Nonenveloped, icosahedral, one molecule of circular dsDNA. The genome may persist in infected cells in an integrated form (polymaviruses) or in an episomal form (papillomaviruses). Replication and assembly occur in the nucleus and virions are released via cell destruction.

18. Family: Adenoviridae

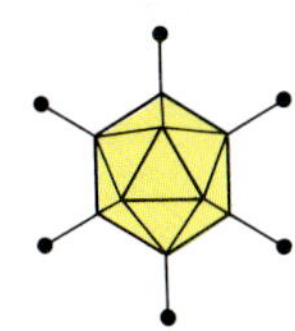

Genera:

Mastadenovirus (infectious canine hepatitis)
Aviadenovirus (Avian adenoviruses)

Nonenveloped, icosahedral. Replication and assembly occur in the nucleus and virions are released via cell destruction.

19. Family: Herpesviridae

Subfamily:
Alphaherpesvirinae

Genera:
Simplexvirus (herpes simplex virus 1 and 2)
Varicellavirus (varicella-zoster virus)

Subfamily
Betaherpesvirinae

Genera:
Cytomegalovirus (human cytomegalovirus)
Muromegalovirus

Subfamily:
Gammaherpesvirinae

Genera:
Lymphocryptovirus (EB-like viruses)
[Unnamed] (Marek's disease-like viruses)
Rhadinovirus (saimiri-ateles-like viruses)
Unclassified: human herpesvirus 6

Envelope, icosahedral, one molecule of dsDNA. Replication takes place in the nucleus, and capsids acquire their envelopes via budding through the inner lamella of the nuclear envelope. Virions are released via transport across the cytoplasm in membranous vesicles which then fuse with plasma membrane.

20. Family: Poxviridae

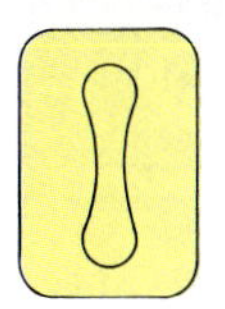

Subfamily:
Chordopoxvirinae (the poxviruses of vertebrates)

Genera:
Orthopoxvirus (vaccina and smallpox)
Parapoxvirus (orf, milker's nodule)
Avipoxvirus (fowlpox)
Capripoxvirus (sheeppox)
Leporipoxvirus (myxoma)
Suipoxvirus (swinepox)
[Yatapoxvirus] (yabapox and tanapox)
[Molliscipoxvirus] (molluscum contagiosum)

Subfamily:
Entomopoxvirinae (the poxviruses of insects)

Large, brick-shaped (or ovoid), external envelope. Replication and assembly occur in cytoplasm in viroplasm ("viral factories"), and virions are released by budding (enveloped virions) or by cell destruction (nonenveloped virions).

21. Family: Iridoviridae

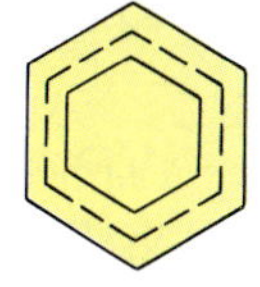

Genera:

Iridovirus (small iridescent insect viruses)
Chloriridovirus (large iridescent insect viruses)
Ranavirus (frog viruses)
Lymphocystivirus (lymphocystis viruses-fish)

Family:
[Unnamed] (for African swine fever virus)

Lipid-containing envelope (missing on some insect viruses), icosahedral. Replication occurs in the cytoplasm (although the nucleus is needed for viral DNA synthesis), and virons are released by budding or cell destruction.
Human pathogens: none known.

APPENDIX C

Word Roots Commonly Encountered in Microbiology

a-, an- not, without, absence   abiotic, not living; anaerobic, in the absence of air

acantho- thorn or spinelike   *Acanthamoeba*, an amoeba with spinelike projections

actino- having rays   *Actinomyces*, a bacterium forming colonies that look like sunbursts

aero- air   aerobic, in the presence of air

agglutino- clumping or sticking together   hemagglutination, clumping of blood cells

albo- white   *Candida albicans*, a white fungus

amphi- around, doubly, both   Amphitrichous describes flagella found at both ends of a bacterial cell.

ant-, anti- against, versus   Antibacterial compounds kill bacteria.

archae- ancient   Archaebacteria are thought to resemble ancient forms of life.

arthro- joint   arthritis, inflammation of joints

asco- sac, bag   Ascospores are held in a saclike container, the ascus.

-ase denotes enzyme   lipase, an enzyme attacking lipids

aureo- golden   *Staphylococcus aureus* has gold-colored colonies.

auto- self   Autotrophs, self-feeding organisms.

bacillo- rod   bacillus, a rod-shaped bacterium

basid- base, foundation   basidium, fungal cell bearing spores at its end

bio- life   biology, the study of living things

blast- bud   blastospore, spore formed by budding

bovi- cow   *Mycobacterium bovis*, bacterium causing tuberculosis in cattle

brevi- short   *Lactobacillus brevis*, a bacterium with short rod-shaped cells

butyr- butter   Butyric acid gives rancid butter its unpleasant odor.

campylo- curved   *Campylobacter*, a curved bacterium

carcino- cancer   A carcinogen causes cancer.

caryo-, karyo- center, kernel   Procaryotic cells lack a true, discrete nucleus.

caseo- cheese   caseous, cheeselike lesions

caul- stalk, stem   *Caulobacter*, a stalked bacterium

ceph-, cephalo- of the head or brain   encephalitis, inflammation of the brain

chlamydo- cloaked, hidden *Chlamydia* are difficult bacteria to detect.

chloro- green chlorophyll, a green pigment

chromo- colored Metachromatic granules stain various colors within a cell.

chryso- golden *Streptomyces chryseus*, a bacterium forming golden colonies

-cide to kill Fungicide kills fungi.

co-, con- with, together Congenital, existing from birth

cocc- berry *Streptococcus*, spherical bacteria in chains

coeno- shared in common coenocytic, many nuclei not separated by septa

col-, colo- colon coliform bacteria, found in the colon (large intestine)

conidio- dust conidia, tiny dustlike spores produced by fungi

coryne- club *Corynebacterium diphtheriae*, a club-shaped bacterium

-cul little, tiny molecule, a tiny mass

cut-, -cut skin cutaneous, of the skin

cyan- blue cyanobacteria, formerly called the blue-green algae

cyst-, -cyst bladder cystitis, inflammation of the urinary bladder

cyt-, -cyte cell leukocyte, white blood cell

de- lack of, removal decolorize, to remove color

dermato- skin dermatitis, inflammation of the skin

di-, diplo- two, double diplococci, pairs of spherical cells

dys- bad, faulty, painful dysentery, a disease of the enteric system

ec-, ecto-, ex- outside, outer ectoparasite, found on the outside of the body

em-, en- in, inside encapsulated, inside a capsule

-emia of the blood pyemia, pus in the blood

endo- inside endospore, spore found inside a cell

entero- intestine enteric, bacteria found in the intestine

epi- atop, over epidemic, a disease spreading over an entire population at one time

erythro- red lupus erythematosus, disease with a red rash

etio- cause etiology, study of the causes of disease

eu- true, good, normal eukaryote, cell with a true nucleus

exo- outside exotoxin, toxin released outside of a cell

extra- outside, beyond extracellular, outside of a cell

fil- thread filament, thin chain of cells

flav- yellow flavivirus, cause of yellow fever

-fy to become, make solidify, to become solid

galacto- milk galactose, monosaccharide from milk sugar

gamet- marriage gamete, a reproductive cell, such as egg or sperm

gastro- stomach gastroenteritis, inflammation of the stomach and intestines

gel- to stiffen, congeal gelatinous, jellylike

gen-, -gen to give rise to pathogen, microbe that causes disease

-genesis origin, development pathogenesis, development of disease

germ, germin- bud germination, process of growing from a spore

-globulin protein immunoglobulins, proteins of the immune system

haem-, hem- blood hemagglutination, clumping of blood cells

halo- salt halophilic, organisms that thrive in salty environments

hepat- liver hepatitis, inflammation of the liver

herpes creeping herpes zoster, or shingles, in which vesicles erupt sequentially along a nerve pathway

hetero- different, other heterotroph, organism deriving nutrition from other sources

histo- tissue histology, the study of tissues

homo- same homologous, having the same structure

hydro- water hydrologic cycle, water cycle

hyper- over, above hyperbaric oxygen, higher than atmospheric pressure oxygen

hypo- under, below hypodermic, going beneath the skin

im-, -in not insoluble, cannot be dissolved

inter- between intercellular, between cells

intra- inside intracellular, inside a cell

io- violet iodine, element purple-colored in gaseous state

iso- same, equal isotonic, having the same osmotic pressure

-itis inflammation of meningitis, inflammation of the meninges

kin- moving kinetic energy, energy of movement

leuko- white leukocyte, white blood cell

lip-, lipo- fat, lipid lipoprotein, molecule having both fatty and proteinaceous parts

-logy, -ology study of microbiology, study of microbes

lopho- tuft lophotrichous, having a tuft or group of flagella

luc-, luci- light luciferase, enzyme that catalyzes a light-producing reaction

luteo- yellow *Micrococcus luteus*, bacterium producing yellow colonies

lys-, -lysis splitting cytolysis, rupture of a cell

macro- large macroconidia, large spores

meningo- membrane meninges, membranes of the brain

meso- middle mesophile, organism growing best at medium temperatures

micro- small, tiny microbiology, study of tiny forms of life

mono- one, single monosaccharide, a single sugar unit

morph- shape, form pleiomorphic, having many different shapes

multi- many multicellular, having many cells

mur- wall muramic acid, a component of cell walls

muri-, mus- mouse murine, in or of mice

mut-, -mute to change mutagen, agent that causes genetic change

myc-, -myces fungus *Actinomyces*, a bacterium that resembles a fungus

myxo- slime, mucus myxomycetes, slime molds

necro- dead, corpse necrotizing toxin, causes death of tissue

nema-, -nema thread *Treponema*, nematode, threadlike organisms

nigr- black *Rhizopus nigricans*, a black mold

oculo- eye binocular, microscope with two eyepieces

-oid like, resembling toxoid, harmless molecule that resembles a toxin

-oma tumor carcinoma, tumor of epthelial cells

onco- mass, tumor oncogenes, genes that cause tumors

-osis condition of brucellosis, condition of being infected with Brucella

pan- all, universal pandemic, a disease affecting a large part of the world

para- beside, near, abnormal parainfluenza, a disease resembling influenza

patho- abnormal pathology, study of abnormal diseased states

peri- around peritrichous, flagella located all around an organism

phago- eating phagocytosis, cell eating by engulfing

philo-, -phil loving, preferring capnophile, organism needing higher than normal levels of carbon dioxide

-phore bearing, carrying electrophoresis, technique in which ions are carried by an electric current

-phyte plant dermatophyte, fungus that attacks skin

pil- hair pilus, hairlike tube on bacterial surface

-plast formed part chloroplast, green-colored body inside plant cell

pod-, -pod foot podocyte, foot cell of kidney

poly- many polyribosomes, many ribosomes on the same piece of messenger RNA

post- afterwards, behind post-streptococcal glomerulonephritis, kidney damage following a streptococcal infection

pre-, pro- before, towards prepubertal, before puberty

pseudo- false pseudopod, projection resembling a foot, false foot

psychro- cold psychrophilic, preferring extreme cold

pyo- pus pyogenic, producing pus

pyro- fire, heat pyrogen, fever-producing compound

rhin- nose rhinitis, inflammation of nasal membranes

rhizo- root mycorrhiza, symbiotic growth of fungi and roots

rhodo- red *Rhodospirillum*, a large red spiral bacterium

-rrhea flow diarrhea, abnormal flow of liquid feces

rubri- red *Rhodospirillum rubrum*, a large, red, spiral bacterium

saccharo- sugar polysaccharide, many sugar units linked together

sapro- rotten, decaying saprophyte, organism living on dead matter

sarco- flesh sarcoma, tumor made up of muscle or connective tissue

schizo- to split schizogony, a type of fission in malarial parasites

-scope, -scopy to see, examine microsocpy, use of the microscope to examine small things

sept-, septo- partition, wall septum, wall between cells

septi- rotting septic, exhibiting decomposition due to bacteria

soma-, -some body chromosome, colored body (when stained)

spiro- coil spirochete, spiral-shaped bacterium

sporo- spore sporocidal, spore killing

staphylo- in bunches, like grapes staphylococci, spherical bacteria growing in clusters

-stasis, stat- stopping, not changing bacteriostatic, able to stop the growth of bacteria

strepto- twisted *Streptobacillus*, twisted chains of bacilli

sub- under, below subclinical, signs and symptoms not clinically apparent

super- above, more than superficial mycosis, fungal infection of the surface tissues

sym-, syn- together symbiosis, living together

tact-, -taxis touch chemotaxis, orientation or movement in response to chemicals

tax-, taxon- arrangement taxonomy, the classification of organisms

thermo- heat thermophile, organism preferring or needing high temperatures

thio- sulfur *Thiobacillus*, organism that oxidizes hydrogen sulfide to sulfates

tox- poison toxin, a harmful compound

trans- through, across transduction, movement of genetic information from one cell to another

trich- hair monotrichous, having a single hairlike flagellum

-troph feeding, nutrition phototroph, organism that makes its own food using energy from light

uni- one, singular unicellular, composed of one cell

undul- waving undulant fever, disease in which fever rises and falls

vac-, vaccin- cow vaccine, disease-preventing product originally produced by inoculating it onto skin of calves

vacu- empty vacuole, empty-appearing structure in cytoplasm

vesic- blister, bladder vesicle, small blisterlike lesions

vitr- glass in vitro, grown in laboratory glassware

xantho- yellow *Xanthomonas oryzae*, bacterium producing yellow colonies

xeno- strange, foreign xenograft, graft from a different species

zoo- animal protozoan, first animal

zygo- yoke, joining zygote, fertilized egg

-zyme ferment enzymes, biological catalysts some of which are involved in fermentation

APPENDIX D

Safety Precautions in the Handling of Clinical Specimens

Concern for maintaining safe conditions in school and hospital laboratories, in other work settings, and especially during patient interactions, has led the federal government to formulate various regulations and recommendations. These are far too extensive to reproduce here in their entirety, inasmuch as several are nearly 200 pages in length. As an introduction to the kinds of safety measures that should be taken, a few of the guidelines set forth in these publications are provided below. Some key references and sources are also listed, with the hope that these will stimulate the interested reader to investigate further.

In 1983, CDC published a document entitled "Guidelines for Isolation Precautions in Hospitals" that contained a section headed "Blood and Body Fluid Precautions." The recommendations in this section specified precautions to be observed regarding contact with the blood or body fluids of any patient known or suspected to be infected with blood-borne pathogens.

In August 1987, CDC published a document entitled "Recommendations for Prevention of HIV Transmission in Health-Care Settings." In contrast to the 1983 document, the 1987 publication recommended that blood and body fluid precautions be consistently used for *all* patients regardless of their blood-borne infection status. These blood and body fluid precautions as they pertain to all patients are referred to as "Universal Blood and Body Fluid Precautions," or more simply as "Universal Precautions."

Following publication of this document there were many requests for clarification—for example, as to exactly which bodily fluids these precautions should apply to. This led to a CDC publication on June 24, 1988 in *MMWR* entitled "Update: Universal Precautions for Prevention of Transmission of Human Immunodeficiency Virus, Hepatitis B Virus, and Other Bloodborne Pathogens in Health-Care Settings."

Copies of these two most recent reports (CDC August, 1987 and June, 1988) are available through the National AIDS Information Clearinghouse, P.O. Box 6003, Rockville, MD 20850.

In the publications cited above, CDC makes the following recommendations regarding sharp instruments:

1. Take care to prevent injuries when using needles, scalpels, and other sharp instruments or devices; when handling sharp instruments after procedures; when cleaning used instruments; and when disposing of used needles. Do not recap used needles by hand; do not remove used needles from disposable syringes by hand; and do not bend, break, or otherwise manipulate used needles by hand. Place used disposable syringes and needles, scalpel blades, and other sharp items in puncture-resistant containers for disposal. Locate the puncture-resistant containers as close to the use area as is practical.
2. Use protective barriers to prevent exposure to blood, body fluids containing visible blood, and other fluids to which universal precautions apply. The type of protective barrier(s) should be appropriate for the procedure being performed and the type of exposure anticipated.
3. Immediately and thoroughly wash hands and other skin surfaces that are contaminated with blood, body fluids containing visible blood, or other body fluids to which universal precautions apply.

Glove use is recommended during phlebotomy (drawing blood samples), but of course cannot protect against penetrating injuries. Some institutions have relaxed recommendations for using gloves for phlebotomy procedures by skilled phlebotomists in settings where the prevalence of blood-borne pathogens is known to be very low (e.g., volunteer blood-donation centers). Such institutions should periodically reevaluate their policy. Gloves should always be available to health-care workers who wish to use them for phlebotomy. In addition, the following general guidelines apply:

1. Use gloves for performing phlebotomy when the health-care worker has cuts, scratches, or other breaks in his/her skin.
2. Use gloves in situations where the health-care worker judges that hand contamination with blood may occur, for example, when performing phlebotomy on an uncooperative patient.
3. Use gloves for performing finger and/or heel sticks on infants and children.
4. Use gloves when persons are receiving training in phlebotomy.

In March, 1989, the Environmental Protection Agency (EPA), published a new set of "Standards for the Tracking and Management of Medical Waste," in part designed to prevent the deplorable pollution of our nation's beaches by medical wastes (*Federal Register*, Mar. 24, 1989, pp. 12325–12395). In May 1989, the Occupational Safety and Health Administration (OSHA) published a new set of rules for "Occupational Exposure to Bloodborne Pathogens" (*Federal Register*, May 30, 1989, pp. 23041–23139). Both of these publications provide very detailed information about procedures that must be followed.

Another useful and quite detailed publication, issued by CDC in February, 1989, and reprinted in *MMWR* for June 23, 1989, is entitled "Guidelines for Prevention of Transmission of Human Immunodeficiency Virus and Hepatitis B Virus to Health-Care and Public-Safety Workers."

For those concerned primarily with the teaching laboratory, we recommend the publication by Gary Ballman entitled "Handling Infectious Materials in the Education Setting" (*American Clinical Laboratory*, July, 1989, pp. 10–11).

APPENDIX E

Answers to Selected Problems for Investigation

Chapter 3

6. Tuberculosis. Very few organisms exhibit a positive acid-fast staining reaction; *Mycobacterium tuberculosis* is one that does. Clinical symptoms are compatible with the diagnosis.

Chapter 12

7. Cysticercosis is acquired by ingesting tapeworm eggs shed in human feces, not by ingesting pork. It is widely endemic in rural areas of Latin America, Asia, and Africa. It is probable that she obtained her infection from one of her neighbor's friends via the fecal-oral route.

Chapter 13

8. **a)** A leak was discovered in the lines supplying ethylene oxide to the ethylene oxide sterilizer.
   b) Ethylene oxide is a carcinogen.

Chapter 14

6. Pneumonialike diseases should be given dual antibiotic therapy, with one being erythromycin, the only commonly used antibiotic that is effective against Legionnaire's disease.

Chapter 15

7. **a)** An intoxication. Such swift onset of symptoms was due to presence of preformed toxin in the eggs. Bacteria could not have multiplied that quickly so as to cause an infection.
   b) Endotoxins can withstand prolonged heating. Perhaps the eggs were soft and runny.
   c) No, endotoxins are weak antigens and cannot be converted to toxoids for immunization.
   d) *Salmonella* are part of the natural flora in the chicken's gut and feces. Contamination of the shell is unavoidable. The bacteria entered through the crack and grew inside.

Chapter 16

6. **a)** The father is. Adults with pertussis lack the characteristic "whooping" cough. It usually presents as a heavy chest cold would. That means that adults are the unidentified cases who spread the disease to children where it *is* recognized.
   b) Pertussis is a notifiable disease.

Chapter 18

9. She can only give her baby antibodies against diseases that she has had. New strains of flu virus arise frequently. She would not have antibodies against these strains. Also, some kinds of antibodies such as pertussis do not cross the placenta at all. She also might not secrete as much antibody in her milk as she imagines. The bulk of antibodies received by a baby are in colostrum, not milk.

Chapter 19

6. Autoimmune diseases improve when the immune system is suppressed. Smoking suppresses the immune system. When the working woman was smoking, her immune system was sufficiently suppressed for her to remain asymptomatic. Rebound of her immune system caused symptoms to return when she stopped smoking.

Chapter 20

7. Loa Loa worm, which is endemic to Africa.

Chapter 21

6. An impaired immune system—probably due to HIV infection.

Chapter 22

6. The prong probably became contaminated with blood, which could then enter the wound made in a subsequent patient's finger when blood was next drawn.

Chapter 23

7. Bubonic plague. Plague is endemic in Bolivia, and the buboes in the armpit are highly diagnostic. CDC confirmed the diagnosis from samples aspirated from one of the buboes.

Chapter 25

7. Yes. The genital wart virus is associated with 90 percent of all cases of cervical cancer. The virus can cause inapparent infections that last for years. It is transmitted sexually, and the husband may have transmitted it to both wives, or he may have acquired it from the first wife and then passed it on to the second wife. (Moral of the story: Ask a widower what his first wife died of.)

Glossary

abiotic factors The physical features of the environment that interact with organisms.

abortive infection Viral infection in which viruses enter a cell but are unable to express all their genes to make infectious progeny.

abscess An accumulation of pus in a cavity hollowed out by tissue damage.

absorption Process in which light rays are neither passed through nor reflected off an object but are retained and transformed to another form of energy.

accidental parasite A parasite that invades an organism other than its normal host.

accumulation period Period during which components of viruses are synthesized and assembled within host cells.

acid A substance that releases hydrogen ions when dissolved in water.

acidic Condition caused by an abundance of hydrogen ions (H^+) resulting in a pH of less than 7.0.

acidophiles Acid-loving organisms that grow best from pH 0.0 to 5.4.

acme In disease, the period of most intense symptoms. (Also called **critical stage.**)

acne Skin condition caused by bacterial infection of hair follicles and the ducts of sebaceous glands.

acquired immunity Immunity obtained in some manner other than by heredity.

acridine derivatives Chemical mutagens that can be inserted between bases of the DNA double helix, causing frameshift mutations.

Actinomyces Filamentous, irregular, nonsporing, gram-positive rods, formerly considered to be fungi.

actinomycetes Gram-positive bacteria that tend to form filaments.

activated sludge system Procedure in which the effluent from primary treatment is agitated, aerated, and added to sludge containing aerobic organisms that digest organic matter.

activation energy The energy required to start a chemical reaction.

activation The first stage in germination, requiring some traumatic agent.

active carrier A human or animal that releases disease microorganisms for a long period of time after recovering from a disease.

active immunity Immunity created when an organism's own immune system produces antibodies or other defenses against an infectious agent.

active immunization Use of vaccines to control disease by increasing group immunity by stimulating immune response.

active site Area on the surface of an enzyme to which its substrate binds.

active transport Movement of molecules or ions across a membrane against a concentration gradient; requires expenditure of energy from ATP.

acute disease A disease that develops rapidly and runs its course quickly.

acute hemorrhagic conjunctivitis Eye disease caused by an enterovirus.

acute necrotizing ulcerative gingivitis (ANUG) A severe form of periodontal disease. (Also called trench mouth.)

acute respiratory disease (ARD) Viral disease that occurs in epidemics with cold symptoms, as well as fever, headache and malaise; sometimes causes viral pneumonia.

adenoviruses Medium-sized, naked Class I DNA viruses that are highly resistant to chemical agents and often cause respiratory infections or diarrhea.

adherence The attachment of a microorganism to a host cell surface.

adhesin A substance that helps a microorganism attach to a host cell.

adhesive disc Structure by which the protozoan parasite *Giardia intestinalis* attaches to cells that line the small intestine.

adsorption The attachment of the virus to the host cell in the replication process.

aerobes Organisms that use oxygen, including some that must have oxygen.

aerobic respiration Process in which aerobic organisms gain energy from the catabolism of organic molecules via the Krebs cycle and oxidative phosphorylation.

aerosol A cloud of tiny liquid droplets suspended in air.

aerotolerant anaerobes Bacteria that can survive in the presence of oxygen but do not use it in their metabolism.

aflatoxins Fungal toxins that are potent carcinogens; found in food made from grain or peanuts infested with *Aspergillus flavus* and other aspergilli.

African sleeping sickness Disease of equatorial Africa caused by protozoan blood parasites of the genus *Trypanosoma*. (Also called **trypanosomiasis.**)

agammaglobulinemia Primary immunodeficiency disease caused by failure of B cells to develop, resulting in lack of antibodies.

agar plate A plate of medium solidified with agar, a polysaccharide extracted from certain marine algae.

agglutination The sticking together of materials such as microbes or blood cells.

agglutination reactions Reactions of antibodies with antigens that result in agglutination, the clumping together of cells or other large particles.

agranulocyte A leukocyte lacking granules in the cytoplasm and having rounded nuclei.

alcoholic fermentation Fermentation in which pyruvate is reduced to ethyl alcohol by electrons from reduced NAD.

algae Photosynthetic, eukaryotic organisms, in kingdoms Protista and Plantae.

alkaline Condition caused by an abundance of hydroxyl ions (OH^-), resulting in a pH of greater than 7.0.

alkalinophiles Base-loving organisms that grow best from pH 7.0 to 11.5.

alkylating agents Chemical mutagens that can add alkyl groups to DNA bases, altering their shapes and causing errors in base pairing.

alleles Genes that occupy the same locus but carry different information.

allergen An ordinarily innocuous foreign substance that can elicit an adverse immunologic response in a sensitized person.

allergy Disorder in which the immune system reacts inappropriately, usually by responding to an antigen it normally ignores. (Also called **hypersensitivity.**)

allograft A graft of tissue between two organisms of the same species that are not genetically identical.

allosteric site The site at which a noncompetitive inhibitor binds.

alpha hemolysin A type of hemolysin that partially lyses red blood cells, leaving a greenish ring in the blood agar medium around the colonies.

alpha hemolysis Incomplete hemolysis of red blood cells by bacterial enzymes.

alternate pathway See **properdin pathway.**

alveoli Sac-like structures arranged in clusters at ends of respiratory bronchioles, having walls 1-cell layer thick, where gas exchange occurs.

Ames test Test used to determine whether a particular substance is mutagenic, based on its ability to induce mutations in bacteria.

amino acid An organic acid containing an amino group and a carboxyl group.

amoebic dysentery Severe, acute form of amebiasis, caused by *Entamoeba histolytica*.

amoeboid movement Movement by means of pseudopodia that occurs in cells without walls, such as amoebas and slime molds.

amphibolic pathways Metabolic pathways that can yield either energy or building blocks for synthetic reactions.

anabolic pathways Chains of chemical reactions in which energy is used to synthesize biologically important molecules.

anabolism Chemical reactions in which energy is used to synthesize large molecules from simpler components.

anaerobes Organisms that do not use oxygen, including some that are killed by exposure to oxygen.

analytical epidemiological study A study that focuses on establishing cause-and-effect relationships in the occurrence of diseases in populations.

anamnestic response Prompt immune response due to "recall" by memory cells. (Also called **secondary response.**)

anaphylactic shock Condition resulting from a sudden extreme drop in blood pressure caused by an allergic reaction.

anaphylaxis Type of allergy caused by a foreign substance known as an allergen. (Also called type I or **immediate hypersensitivity.**)

Ångström (Å) Unit of measurement equal to 0.0000000001 m or 10^{-10} m. No longer officially recognized.

animal passage The rapid transfer of a pathogen through animals of a species susceptible to infection by the pathogen.

anion A negatively charged ion.

anionic dye An ionic compound, used for staining bacteria, in which the negative ion imparts the color.

antheridium The male reproductive structure in a sac fungus.

anthrax A zoonosis caused by *Bacillus anthracis* that exists in cutaneous, respiratory ("woolsorters disease"), or intestinal forms; transmitted by endospores.

antibiosis The natural production of an antimicrobial agent by a bacterium or fungus; literally, "against life."

antibiotic A chemical substance produced by microorganisms that can inhibit the growth of or destroy other microorganisms.

antibody A protein produced in response to an antigen that is capable of binding specifically to that antigen. (Also called **immunoglobulin.**)

antibody titer The quantity of antibodies in a patient's blood, often measured by means of agglutination reactions.

anticodon A three-base sequence in tRNA that is complementary to one of the mRNA codons, forming a link between each codon and the corresponding amino acid.

antigen A substance that the body identifies as foreign and toward which it mounts an immune response.

antigenic determinants Areas on an antigen molecule to which antibodies bind.

antigenic drift Process of antigenic variation that results from mutations in genes coding for hemagglutinin and neuraminidase.

antigenic shift Process of antigenic variation probably caused by a reassortment of viral genes.

antigenic variation Mutations in influenza viruses that occur by antigenic drift and antigenic shift.

antihistamines Drugs that alleviate symptoms caused by histamine.

antimetabolite A substance that prevents a cell from carrying out an important metabolic reaction.

antimicrobial agent A chemotherapeutic agent used to treat diseases caused by microbes.

antiparallel The head-to-tail arrangement of DNA strands in the double helix.

antiseptic A chemical agent that can be safely used externally on tissues to destroy microorganisms or to inhibit their growth.

antitoxin An antibody against a specific toxin.

antiviral protein Protein that interferes with the replication of viruses.

aplastic crisis A period during which erythrocyte production ceases.

apoenzyme The protein portion of an enzyme.

arachnids Arthropods with two body regions, four pairs of legs, and mouth parts that are used in capturing and tearing apart prey.

archaebacteria Prokaryotic organisms lacking peptidoglycan in their cell walls, and differing from eubacteria in many ways.

arenaviruses Enveloped Class III RNA viruses that cause Lassa fever and certain other hemorrhagic fevers.

arteriole A blood vessel that branches from an artery.

artery A blood vessel that receives blood from the heart.

arthropods The largest group of living organisms, characterized by jointed chitinous exoskeletons, segmented bodies, and jointed appendages associated with some or all of the segments.

Arthus reaction A local reaction seen in the skin after subcutaneous or intradermal injection of an antigenic substance, an immune complex (type III) reaction.

artificially acquired active immunity Active immunity acquired when an organism is exposed to a vaccine containing live, weakened, or dead organisms or their toxins.

artificially acquired immunity Immunity obtained by receiving an injection of vaccine or immune serum.

artificially acquired passive immunity Immunity produced when antibodies made by other hosts are introduced into a new host by injection or transfusion.

ascariasis Disease caused by a large roundworm, *Ascaris lumbricoides*, acquired by ingestion of food or water contaminated with eggs.

ascogonium The female reproductive structure in a sac fungus.

Ascomycota See **sac fungi.**

ascospore One of the eight sexual spores produced in each ascus of a sac fungus.

ascus (pl: **asci**) Saclike structure produced by sac fungi during sexual reproduction.

aseptic techniques Techniques used to minimize chances that cultures will be contaminated by organisms from the environment.

Asiatic cholera Severe gastrointestinal disease caused by *Vibrio cholerae*; common in areas of poor sanitation and fecal contamination of water.

aspergillosis Infection caused by various species of *Aspergillus*, a fungus that grows on decaying vegetation.

asthma Respiratory anaphylaxis caused by inhaled or ingested allergens or by hypersensitivity to endogenous microorganisms.

athlete's foot Fungal skin disease in which hyphae invade the skin between the toes, causing dry, scaly lesions. (Also called **tinea pedis.**)

atom The smallest chemical unit of matter.

atomic number The number of protons in an atom of a particular element.

atomic weight The sum of the number of protons and neutrons in an atom.

atopy Localized allergic reactions that occur first at the site where an allergen enters the body.

atrium (pl. **atria**) One of the two upper chambers of the heart.

attachment pili Type of pili that help bacteria adhere to surfaces. (Also called **fimbriae.**)

attenuation (1) The weakening of the disease-producing ability of an organism. (2) A genetic control mechanism that terminates transcription of an operon prematurely when the gene products are not needed.

autoantibodies Antibodies against one's own tissues.

autoclave An instrument for sterilization by means of moist heat under pressure.

autograft A graft of tissue from one part of the body to another.

autoimmune disorders Disorders in which individuals are hypersensitive to antigens on cells of their own bodies.

autoimmunization The process by which hypersensitivity to "self" develops; it occurs when the immune system responds to a body component as if it were foreign.

autotrophy Self-feeding—the use of CO_2 as a source of carbon atoms for the synthesis of biomolecules.

auxotrophic mutants Organisms that have lost the ability to synthesize one or more metabolically important enzymes through mutation.

axial filaments Subsurface filaments attached near the ends of the cytoplasmic cylinder of spirochetes that cause the spirochete body to rotate like a corkscrew. (Also called endoflagella.)

axial nucleus A long, compact nucleus formed during sporulation.

B cell See **B lymphocyte.**

B lymphocyte A lymphocyte that is processed and matures in bursal-equivalent tissue, it gives rise to antibody-producing plasma cells. (Also called **B cell.**)

babesiosis Parasitic protozoan disease caused by the sporozoan *Babesia microti* and other species of *Babesia*.

bacillary dysentery See **shigellosis.**

bacillus (plural: **bacilli**) A rodlike bacterium.

bacteremia An infection in which bacteria are transported in the blood but do not multiply in transit.

bacteria All prokaryotic organisms.

bacterial conjunctivitis A highly contagious inflammation of the conjunctiva caused by various bacterial species. (Also called **pinkeye.**)

bacterial endocarditis A life-threatening infection and inflammation of the lining and valves of the heart.

bacterial enteritis An intestinal infection caused by bacterial invasion of intestinal mucosa or deeper tissues.

bacterial lawn A uniform layer of susceptible bacteria grown on a petri dish, and used to culture viruses.

bacterial meningitis An inflammation of the meninges, the membranes that cover the brain and spinal cord, caused by any one of several bacterial species.

bacteriocinigen A plasmid that directs production of a bacteriocin.

bacteriocins Proteins released by some bacteria that inhibit the growth of other strains of the same or closely related species.

bacteriophage A virus that infects bacteria. (Also called **phage.**)

bacteroids Tightly packed, irregularly shaped cells that develop from *Rhizobium* swarmer cells and form nodules in the roots of leguminous plants.

balantidiasis Type of dysentery caused by a ciliated protozoan, *Balantidium coli*.

balantitis An infection of the penis.

bartonellosis Rickettsial disease, caused by *Bartonella bacilliformis*, that occurs in two forms: Oroya fever, or Carrion's disease, and verruga peruana.

base analog A chemical mutagen similar in molecular structure to one of the nitrogenous bases found in DNA.

base (1) A substance that absorbs hydrogen ions or donates hydroxyl ions. (2) Supporting structure of a microscope.

basement membrane Nonliving skin barrier comprising secretions of epithelial and dermal cells.

Basidiomycota See **club fungi.**

basidium (pl: **basidia**) Clublike structure in club fungi bearing four external spores on short, slender stalks.

basophils Leukocytes that migrate into tissues, becoming mast cells.

BCG Bacillus of Calmette and Guerin, a vaccine for tuberculosis.

beta hemolysin A type of hemolysin that completely hemolyses red blood cells, leaving a clear ring in the blood agar medium around the colonies.

beta hemolysis Complete hemolysis of red blood cells by bacterial enzymes.

beta oxidation A metabolic pathway that breaks down fatty acids into 2-carbon pieces.

bilharzia See **schistosomiasis.**

bilirubin A yellow substance, the product of the breakdown of hemoglobin from red blood cells.

binary fission Process in which a bacterial cell duplicates its components and divides into two cells.

binocular Refers to a light microscope having two eyepieces.

binomial nomenclature The system of taxonomy originated by Linnaeus in which each organism is assigned a genus and species name.

bioconversion A reaction in which one compound is converted to another by enzymes in cells.

biogeochemical cycles Mechanisms by which water and elements that serve as nutrients are recycled.

biological oxygen demand (BOD) The oxygen required to degrade organic wastes suspended in water.

biological vector A vector that actively transmits pathogens that complete part of their life cycle within the vector.

biosynthetic pathways See **anabolic pathways.**

biotic factors The interactions of organisms with one another.

blackfly fever Illness resulting from bites by blackflies, characterized by an inflammatory reaction, nausea, and headache.

blackwater fever Malaria caused by *Plasmodium falciparum* that results in jaundice and kidney damage.

blastomycetic dermatitis Fungal skin disease caused by *Blastomyces dermatitidis* characterized by disfiguring, granulomatous, pus-producing lesions.

blastomycosis Fungal skin disease caused by *Blastomyces dermatitidis* that enters the body through wounds.

blocking antibodies IgG antibodies, elicited in allergy patients by increasing doses of allergen, that complex with allergen before it can react with IgE.

blood agar Type of medium containing sheep blood, used to identify organisms that cause hemolysis, or breakdown of red blood cells.

blood-brain barrier Formation in the brain of special thick-walled capillaries without pores in their walls that limits entry of substances into brain cells.

blood poisoning See **septicemia.**

body tube Microscope part that conveys an image from the objective to the eyepiece.

boil See **furuncle.**

Bolivian hemorrhagic fever A multisystem disease caused by an arenavirus with insidious onset and progressive effects.

bongkrek disease Type of food poisoning caused by *Pseudomonas cocovenenans*, named for a native Polynesian coconut dish.

botulism Disease caused by *Clostridium botulinum*. The most common form, foodborne botulism, results from ingestion of preformed toxin and is, therefore, an intoxication rather than an infection.

bradykinin A kinin thought to cause the pain associated with tissue injury.

bread molds Group of fungi with complex mycelia composed of hyphae with chitinous cross walls. (Also called conjugation fungi or **Zygomycota.**)

breakbone fever See **dengue fever.**

bright-field illumination Illumination produced by the condenser of an ordinary light microscope.

Brill-Zinsser disease A recurrence of an epidemic typhus infection caused by reactivation of latent organisms harbored in the lymph nodes. (Also called **recrudescent typhus.**)

broad spectrum The range of activity of an antimicrobial agent that attacks a wide variety of microorganisms.

bronchial pneumonia Type of pneumonia that begins in the bronchi and can spread through surrounding tissue toward the alveoli.

bronchial tree A branching structure conveying air to and from the lungs, formed by the trachea, bronchi, and bronchioles.

bronchioles Finer subdivisions of the air-conveying bronchi.

bronchitis An infection of the bronchi.

bronchus (pl. **bronchi**) A subdivision of the trachea that conveys air to and from the lungs.

brucellosis A zoonosis highly infective for humans, caused by any of several species of *Brucella*. (Also called **undulant fever** or **Malta fever.**)

buboes Enlargements of infected lymph nodes, especially in the groin and armpit, due to accumulation of pus; characteristic of bubonic plague and other diseases.

bubonic plague A bacterial disease, caused by *Yersinia pestis* and transmitted by flea bites, that spreads in the blood and lymphatic system.

budding Process that occurs in yeast and a few bacteria in which a small new cell develops from the surface of an existing cell.

bulking Phenomenon in which filamentous bacteria multiply, causing sludge to float on the surface of water rather than settling out.

bunyaviruses Enveloped Class III RNA viruses that cause some forms of encephalitis and hemorrhagic fever.

Burkitt's lymphoma A tumor of the jaw, seen mainly in African children; caused by Epstein-Barr virus.

burst size The number of new virions released (in the replication process).

burst time The time from absorption to release in replication of phages.

candidiasis A fungal infection caused by *Candida albicans* that appears as thrush (in mouth) or vaginitis (yeast infection). (Also called **moniliasis.**)

canine parvovirus A parvovirus that causes severe disease in dogs.

capillary A blood vessel that branches from an arteriole.

capsid The protein coating of a virus, which protects the nucleic acid core from the environment and determines the shape of the virus.

capsomeres Multiple protein subunits that make up a viral capsid.

capsule A protective structure outside the cell wall, secreted by the organism.

carbohydrates Compounds composed of carbon, hydrogen, and oxygen that serve as the main source of energy for most living things.

carbon cycle Process by which carbon from atmospheric carbon dioxide enters living and nonliving things and is recycled through them.

carbuncle A massive pus-filled lesion resulting from an infection, particularly of the neck and upper back.

cardiovascular system Body system that supplies oxygen and nutrients to all parts of the body and removes carbon dioxide and other wastes from them.

Carrion's disease One form of bartonellosis; an acute fatal fever with severe anemia. (Also called **Oroya fever.**)

cascade A set of reactions in which magnification of effect occurs, as in the complement system.

casein hydrolysate A substance made from milk protein that contains many amino acids; used to enrich certain media.

caseous lesions Lesions with a "cheesy" appearance that form in lung tissue of patients with tuberculosis.

cat scratch fever A disease probably caused by a chlamydia transmitted in cat scratches and bites.

catabolic pathways Chains of chemical reactions that capture energy by breaking down large molecules into simpler components.

catabolism The chemical breakdown of molecules in which energy is released.

catabolite repression Process by which the presence of a preferred nutrient (often glucose) represses the genes coding for enzymes used to metabolize some alternative nutrient.

catalase An enzyme that converts hydrogen peroxide to water and molecular oxygen.

catarrhal stage Stage of whooping cough characterized by fever, sneezing, vomiting, and a mild, dry, persistent cough.

cation A positively charged ion.

cationic dye An ionic compound, used for staining bacteria, in which the positive ion imparts the color.

cavitation The formation of a cavity inside the cytoplasm of a cell.

cell cultures Cultures in the form of monolayers from dispersed cells and continuous cultures of cell suspensions.

cell-mediated hypersensitivity (type IV hypersensitivity) Type of allergy elicited by foreign substances from the environment, infectious agents, transplanted tissues, and the body's own malignant cells. (Also called **delayed hypersensitivity.**)

cell-mediated immunity Immune response carried out at the cellular level by T cells.

cell membrane A living lipoprotein bilayer that forms the boundary between a cell and its environment. (Also called **plasma membrane.**)

cell monolayers Single layers of cells spread across a medium in a culture flask.

cell strain Dominant cell type resulting from subculturing.

cell theory Theory formulated by Schleiden and Schwann that cells are the fundamental units of all living things.

cell wall Outer layer of most bacterial, algal, fungal, and plant cells that maintains the shape of the cell.

cellular slime mold Funguslike protist consisting of amoeboid, phagocytic cells that aggregate to form a pseudoplasmodium.

cementum The hard, bony covering of the tooth below the gumline.

Centers for Disease Control (CDC) Branch of the United States Public Health Service responsible for the control and prevention of infectious disease.

central nervous system The brain and spinal cord.

cercaria Free-swimming fluke larvae that emerge from the snail or mollusk host.

cerebrospinal fluid Liquid that fills the hollow chambers in the brain and spinal cord as well as the spaces between meninges.

cerumen Earwax.

ceruminous glands Modified sebaceous glands that secrete cerumen.

Chagas' disease Disease caused by *Trypanosoma cruzi* that occurs in the southern United States and is endemic to Mexico; transmitted by several kinds of reduviid bugs.

chancre A hard, painless, nondischarging lesion; a symptom of primary stage syphilis.

chancroid Sexually transmitted disease caused by *Haemophilus ducreyi* that causes soft, painful skin lesions on the genitals, which bleed easily.

chemical equilibrium A steady state in which there is no net change in the concentrations of products or reactants.

chemically nondefined medium See **complex medium.**

chemiosmosis Process of energy capture in which a proton gradient is created by means of electron transport and then used to drive the synthesis of ATP.

chemoautotrophs Autotrophs that obtain energy by oxidizing simple inorganic substances such as sulfides and nitrites.

chemoheterotrophs Heterotrophs that obtain energy from breaking down ready-made organic molecules.

chemolithotrophs See **chemoautotrophs.**

chemostat A device for maintaining logarithmic growth of a culture by the continuous addition of fresh medium.

chemotaxis Process by which bacteria move toward or away from substances in their environment.

chemotherapeutic agent Any chemical substance used to treat disease.

chemotherapeutic index The maximum tolerable dose of a particular drug per kilogram body weight divided by the minimum dose per kilogram body weight that will cure the disease.

chemotherapy The use of chemical substances to treat various aspects of disease.

chickenpox A highly contagious disease, characterized by skin lesions, caused by varicella-zoster herpesvirus; usually occurs in children.

chigger dermatitis A violent allergic reaction caused by chiggers, the larvae of *Trombicula* mites.

childbed fever See **puerperal sepsis.**

chitin A polysaccharide found in the cell walls of most fungi and the exoskeletons of insects.

chlamydias Tiny, nonmotile, spherical bacteria; all are obligate intracellular parasites with a complex life cycle.

chlorination The addition of chlorine to water to kill bacteria.

chloroplasts Chlorophyll containing organelles found in eukaryotic cells that carry out photosynthesis.

chocolate agar Type of medium made with heated blood, so named because it turns a chocolate brown.

chromosomal resistance Drug resistance of a microorganism due to a mutation in chromosomal DNA.

chromosome mapping The identification of the sequence of genes in a chromosome.

chromosomes Structures that contain the DNA of organisms.

chronic amebiasis Chronic infection caused by *Entamoeba histolytica.*

chronic disease A disease that develops more slowly than an acute disease, is usually less severe, and persists for a long, indeterminate period.

chronic EBV syndrome See **chronic fatigue syndrome.**

chronic fatigue syndrome Disease of uncertain origin similar to mononucleosis with symptoms including persistant fatigue and fever. Previously called **chronic EBV syndrome.**

chronic inflammation A condition in which there is a persistent, indecisive standoff between an inflammatory agent and the phagocytic cells and other host defenses attempting to destroy it.

cilia Short cellular projections used for movement that beat in coordinated waves.

ciliates Protozoans that move by means of cilia over most of their surfaces.

citric acid cycle See **Krebs cycle.**

classic complement pathway One of the two sequences of reactions by which proteins of the complement system are activated.

clonal selection hypothesis Hypothesis that explains (1) how exposure to an antigen stimulates a lymphocyte capable of making antibodies against that particular antigen to proliferate, giving rise to a clone of identical antibody-producing cells, and (2) how the immune system acquires tolerance for self antigens by eliminating certain lymphocyte clones during embryological development.

clone A group of identical cells descending from a single parent cell.

club fungi Group of fungi, including mushrooms, toadstools, rusts, and smuts, that produce spores on basidia. (Also called **Basidiomycota.**)

coagulase Bacterially produced enzyme that accelerates the coagulation of blood.

coarse adjustment Focusing mechanism of a microscope that changes the distance between the objective lens and the specimen rapidly.

coccidioidomycosis Fungal respiratory disease caused by the soil fungus *Coccidioides immitis*. (Also called valley fever.)

coccus (plural: **cocci**) A spherical bacterium.

codon A sequence of three bases in mRNA that specifies a particular amino acid in the translation process.

coelom The body cavity between the digestive tract and body wall in higher animals.

coenzyme An organic molecule bound to or loosely associated with an enzyme.

cofactor An inorganic ion necessary for the function of an enzyme.

colicins Proteins released by some strains of *E. coli* that inhibit growth of other strains of the same organism.

coliform bacteria Gram-negative, non-spore-forming, aerobic or facultatively anaerobic bacteria that ferment lactose and produce acid and gas; significant numbers may indicate water pollution.

colloids A mixture formed by particles too large to form a true solution dispersed in a liquid.

colon The large intestine.

colonization Growth of microorganisms on epithelial surfaces such as skin or mucous membranes.

colony A group of descendants of an original cell.

Colorado tick fever Disease caused by an orbivirus carried by dog ticks, characterized by headache, backache, and fever.

commensal An organism that lives in or on other organisms without harming them.

commensalism A symbiosis in which one organism benefits and the other one neither benefits nor is harmed by the relationship.

common-source epidemic An epidemic that arises from contact with contaminated substances.

communicable disease Infectious disease that can be spread from one host to another. (Also called **contagious disease.**)

community All the kinds of organisms present in a given environment.

competence factor A protein released into the medium of a cell that facilitates the uptake of DNA.

competitive exclusion The phenomenon whereby one type of organism is prevented from colonizing a particular environment by the presence of another type of organism that preempts the available space and nutrients.

competitive inhibitor A molecule similar in structure to a substrate that competes with a substrate by binding to the active site.

complement fixation test A complex serological test used to detect small quantities of antibodies.

complement A set of more than 20 large regulatory proteins that circulate in plasma and when activated form a nonspecific defense mechanism against many different microorganisms. (Also called **complement system.**)

complement system See **complement.**

completed test The final test for coliforms in multiple-tube fermentation in which organisms from colonies grown on EMB agar are used to inoculate lactose broth and agar slants.

complex medium A medium that contains certain reasonably well-defined materials but that varies slightly in chemical composition from batch to batch. (Also called **chemically nondefined medium.**)

complex viruses Viruses that have envelopes or specialized structures such as heads and tails.

compound A chemical substance made up of atoms of two or more elements.

compound light microscope A light microscope with more than one lens.

compromised host An individual with reduced resistance.

condenser Device in a microscope that converges light beams so that they will pass through the specimen.

condylomas See **genital warts.**

confirmed test Second stage of testing for coliforms in multiple-tube fermentation in which samples from the highest dilution showing gas production are streaked onto EMB agar.

congenital rubella Complication of German measles causing death or damage to a developing embryo infected by virus crossing the placenta.

congenital syphilis Syphilis passed to a fetus when treponemes cross the placenta from mother to child before birth.

conidia Chains of asexually produced aerial spores with thick outer walls.

conidium (pl: **conidia**) A small asexual spore formed from the tip of a hypha in certain fungi.

conjugation Transfer of genetic information from one bacterial cell to another by means of pili.

conjugation pili Type of pili that attach two bacteria together and furnish a pathway for the exchange of genetic material.

conjunctiva Mucous membranes of the eye.

consolidation Blockage of air spaces as a result of fibrin deposits in lobar pneumonia.

constitutive enzymes Enzymes that are synthesized continuously regardless of the nutrients available to the organism.

consumers Organisms that obtain nutrients by eating producers.

contact dermatitis Cell-mediated hypersensitivity disorder that occurs in sensitized individuals on second exposure of the skin to allergens.

contact transmission A mode of disease transmission effected directly, indirectly, or by droplets.

contagious disease See **communicable disease.**

contamination The presence of microorganisms.

continuous cell line Cell culture consisting of cells that can be propagated over many generations.

convalescence Recovery phase of an illness.

convalescent sera Preparations of gamma globulins having high titers of specific kinds of antibodies. (Also called **hyperimmune sera.**)

convalescent stage Stage of whooping cough in which the disease subsides but secondary infections with other organisms are common.

core The living part of an endospore.

cornea The transparent part of the eyeball exposed to the environment.

coronaviruses Viruses with clublike projections that cause colds and acute upper respiratory distress.

cortex A laminated layer of peptidoglycan between the membranes of the endospore septum.

corynebacteria Club-shaped, irregular, nonsporing, gram-positive rods.

countable number A number of colonies on an agar plate small enough so that one can clearly distinguish and count them (30 to 300 per plate).

covalent bonds Bonds between atoms created by the sharing of pairs of electrons.

cowpox Disease caused by the vaccinia virus characterized by lesions, inflammation of lymph nodes, and fever; virus is used to make vaccine against smallpox.

crepitant tissue Distorted tissue caused by gas bubbles in gas gangrene.

cristae Folds of the inner mitochondrial membrane.

critical stage In disease, the period of most intense symptoms. (Also called **acme.**)

cross-reactions Immune reactions of a single antibody with different antigens that are similar in structure.

cross-resistance Resistance against two or more similar antimicrobial agents through a common mechanism.

croup Acute obstruction of the larynx that produces a characteristic high-pitched, barking cough.

crown The part of the tooth above the gumline, covered with enamel.

crustacea Generally aquatic arthropods that usually have a pair of appendages associated with each body segment.

cryptococcosis Fungal respiratory disease caused by a budding, encapsulated yeast, *Cryptococcus neoformans.*

cryptosporidiosis Disease caused by protozoans of the genus *Cryptosporidium,* common in AIDS patients.

Cryptosporidium Protozoan genus that has been observed to cause opportunistic infections, probably by fecal-oral transmission.

cutaneous larva migrans Severe skin inflammation near the site of penetration of the skin by hookworm larvae. (Also called creeping eruption.)

cyanobacteria Photosynthetic, prokaryotic, typically unicellular organisms, members of kingdom Monera.

cyanosis Bluish skin characteristic of oxygen-poor blood.

cyclic photophosphorylation Pathway in which excited electrons from chlorophyll are used to generate ATP without the splitting of water or fixation of CO_2.

cysticercus An oval white sac with a tapeworm head invaginated into it. (Also called bladder worm.)

cystitis Inflammation of the bladder.

cysts Spherical, thick-walled cells that resemble endospores, formed by certain bacteria.

cytocidal Lethal to its host cell.

cytomegalic inclusion disease A disease common among babies infected with cytomegaloviruses that results in mental and sensory disorders.

cytomegaloviruses A widespread and diverse group of herpesviruses that often produce no symptoms in normal adults but can severely affect AIDS patients and congenitally infected children.

cytopathic effect (CPE) The visible effect viruses have on cells.

cytoplasm The semifluid substance inside the cell membrane, excluding the nucleus if one is present.

cytoplasmic streaming Process by which cytoplasm flows from one part of a eukaryotic cell to another.

cytoskeleton A network of protein fibers that support and give rigidity and shape to a cell.

cytotoxic drugs Drugs that interfere with DNA synthesis, used to suppress the immune system and prevent the rejection of transplants.

cytotoxic (killer) T cells (T_C) Lymphocytes that destroy virally infected cells.

cytotoxic (type II) **hypersensitivity** Type of allergy elicited by antigens on cells, especially red blood cells, that the immune system treats as foreign.

dark-field illumination Illumination that is reflected from an object rather than passing through it, resulting in a light image on a dark background.

dark reactions Part of photosynthesis in which carbon dioxide is reduced by electrons from NADP to form various carbohydrate molecules, chiefly glucose. (Also called carbon fixation.)

dark repair Mechanism for repair of damaged DNA by several enzymes that do not require light for activation; they excise defective nucleotide sequences and replace them with DNA complementary to the unaltered DNA strand.

daughter cells The two identical products of cell division.

deaminating agents Chemical mutagens that can remove an amino acid from a nitrogenous base, causing a point mutation.

deamination The removal of an amino group from a molecule such as an amino acid.

death phase See **decline phase.**

debridement A scraping technique used to remove eschar from a burn and reach infection sites.

decimal reduction time (DRT, D value) The length of time needed to kill 90 percent of the organisms in a given population at a specified temperature.

decline phase (1) In disease, the period during which the host defenses finally overcome the pathogen and symptoms begin to subside. (2) The fourth of four major phases of the bacterial growth curve in which cells lose their ability to divide (due to less supportive conditions in the medium) and thus die. (Also called **death phase.**)

decomposers Organisms that obtain energy by digesting dead bodies or wastes of producers and consumers.

defined synthetic medium A synthetic medium that contains known specific kinds and amounts of chemical substances.

definitive host An organism that harbors the adult, sexually reproducing form of a parasite.

degranulation (1) Release of histamine and other mediators of allergic reactions by sensitized mast cells and basophils after a second encounter with an allergen. (2) The process by which lysosomes fuse with phagosomes, releasing their enzymes and becoming phagolysosomes.

delayed hypersensitivity See **cell-mediated hypersensitivity.**

delayed hypersensitivity T cells (T_D) Lymphocytes that produce lymphokines in delayed (type IV) hypersensitivity reactions.

deletion The removal of one or more bases from DNA, usually producing a frameshift mutation.

delta hepatitis See **hepatitis D.**

denaturation The disruption of hydrogen bonds and other weak forces that maintain the structure of a globular protein, resulting in the loss of its biological activity.

dengue fever Viral systemic disease that causes severe bone and joint pain. (Also called **breakbone fever.**)

denitrification The process by which nitrates are reduced to nitrous oxide or nitrogen gas.

dental caries The erosion of enamel and deeper parts of teeth. (Also called cavities.)

dentin The porous substance forming the bulk of the tooth, covered by enamel or cementum.

deoxyribonucleic acid (DNA) Nucleic acid that carries hereditary information from one generation to the next.

dependoviruses Parvoviruses that contain both positive sense and negative sense DNA and require coinfection with adenoviruses.

dermal warts Warts resulting from the viral infection of epithelial cells.

dermatomycoses Fungal skin diseases.

dermatophytes Fungi that invade keratinized tissue of the skin and nails.

dermis The thick inner layer of the skin.

descriptive epidemiological study A study that notes the number of cases of a disease, which segments of the population are affected, where the cases have occurred, and over what time period.

desensitization Treatment designed to cure allergies by means of injections with gradually increasing doses of allergen.

Deuteromycota See **Fungi Imperfecti.**

diapedesis The process in which leukocytes pass out of blood into inflamed tissues by squeezing between cells of capillary walls.

diatoms Plantlike protists that lack flagella and have a glass-like outer shell.

dichotomous key Taxonomic key used to identify organisms; composed of paired (either-or) statements describing characteristics.

differential medium A medium with a constituent that causes an observable change (in color or pH) in the medium when a particular chemical reaction occurs, making it possible to distinguish organisms.

differential stain Use of two or more dyes to distinguish various structures of an organism, e.g., Gram stain.

diffraction Phenomenon in which light waves, as they pass through a small opening, are broken up into bands of different wavelengths.

DiGeorge syndrome Primary immunodeficiency disease caused by failure of the thymus to develop properly, resulting in deficiency of T cells.

digestive system Body system that converts ingested food into material suitable for the liberation of energy or for assimilation into body tissues.

dikaryotic cell A fungal cell with two nuclei, produced by plasmogamy in which the nuclei fail to unite.

dilution method Method of testing antibiotic sensitivity in which organisms are incubated in a series of tubes containing known quantities of a chemotherapeutic agent.

dimer Two adjacent pyrimidines bonded together in a DNA strand, usually as a result of exposure to ultraviolet rays.

dimorphism The ability of an organism to alter its structure when it changes habitats.

dinoflagellates Plantlike protists, usually with two flagella.

diphtheria A severe upper respiratory disease caused by *Corynebacterium diphtheriae;* can produce subsequent myocarditis and polyneuritis.

diphtheroids Organisms found in normal throat cultures that fail to produce exotoxin but are otherwise indistinguishable from diphtheria-causing organisms.

dipicolinic acid Acid found in the cortex of an endospore that contributes to its heat resistance.

diploid fibroblast strains Cultures derived from fetal tissues that retain fetal capacity for rapid, repeated cell division.

direct contact transmission Mode of disease transmission requiring person-to-person body contact.

direct fecal-oral transmission Direct contact transmission of disease in which pathogens from fecal matter are spread by unwashed hands to the mouth.

direct microscopic counts Method of measuring bacterial growth by counting cells in a known volume of medium that fills a specially calibrated counting chamber on a microscope slide.

disaccharide A carbohydrate formed by the joining of two monosaccharides.

disease A disturbance in the state of health wherein the body cannot carry out all its normal functions.

disinfectant A chemical agent used on inanimate objects to destroy microorganisms.

disinfection Reducing the number of pathogenic organisms on objects or in materials so that they pose no threat of disease.

disk diffusion method (Kirby-Bauer method) Method used to determine microbial sensitivity to antimicrobial agents in which antibiotic disks are placed on an inoculated petri dish, incubated, and observed for inhibition of growth.

disseminated tuberculosis Type of tuberculosis spread throughout body; now seen in AIDS patients, usually caused by *Mycobacterium avium.*

divergent evolution Process in which descendants of a common ancestor species undergo sufficient change to be identified as separate species.

DNA hybridization Process in which the double strands of DNA of each of two organisms are split apart and the split strands from the two organisms are allowed to combine.

DNA polymerase An enzyme that moves along behind each replication fork, synthesizing new DNA strands complementary to the original ones.

Donovan bodies Large mononuclear cells found in scrapings of lesions that confirms the presence of granuloma inguinale.

DPT vaccine A vaccine used to prevent diphtheria, pertussis, and tetanus.

dracunculiasis Skin disease caused by a parasitic helminth, the guinea worm, *Dracunculus medinensis.*

droplet nuclei Particles consisting of dried mucus in which microorganisms are embedded.

droplet transmission Contact transmission of disease through small liquid droplets.

drug See chemotherapeutic agent.

dyads Paired chromosomes that are prepared to divide by mitosis.

dye A molecule that can bind to a cellular structure and give it color. (Also called **stain.**)

dynein Protein associated with each pair of fibers found in eukaryotic flagella that uses ATP to make the flagellum move.

dysentery A severe diarrhea which often contains mucus and sometimes blood or pus.

dysuria Pain and burning on urination.

ear canal Passage leading from the outer ear to the middle ear, lined with skin that has many small hairs and ceruminous glands.

eastern equine encephalitis Type of viral encephalitis seen most often in the eastern United States; infects horses more frequently than humans.

eclipse period Period during which viruses have absorbed to and penetrated host cells but cannot yet be detected in cells.

ecology The study of relationships among organisms and their environment.

ecosystem All the biotic and abiotic components of an environment.

ectoparasite A parasite that lives on the surface of another organism.

eczema herpeticum A generalized eruption caused by entry of the herpesvirus through the skin; often fatal.

electrolyte A substance that is ionizable in solution.

electron acceptor An oxidizing agent in a chemical reaction.

electron donor A reducing agent in a chemical reaction.

electron micrographs Photographs of images from an electron microscope.

electron microscope Microscope that uses a beam of electrons rather than a beam of light and electromagnets instead of glass lenses.

electron transport chain A series of coenzymes that pass electrons to oxygen (the final electron acceptor).

electron transport Process in which pairs of electrons are transferred from coenzyme to coenzyme.

electrons Negatively charged subatomic particles that surround the nucleus of an atom.

electrophoresis Process used to separate large molecules such as antigens by passing an electrical current through a sample on a gel.

element Matter composed of one kind of atom.

elementary bodies An infectious stage in the life cycle of chlamydias.

elephantiasis Gross enlargement of limbs, scrotum, and sometimes other body parts from accumulation of fluid due to blockage of lymph ducts by *Wucheria bancrofti* worms.

enamel The hard substance covering the crown of the tooth.

encephalitis An inflammation of the brain caused by a variety of viruses or bacteria.

end product inhibition See **feedback inhibition.**

endemic disease A disease that is constantly present in a specific population.

endemic relapsing fever Tick-borne cases of relapsing fever caused by *Borrelia* sp.

endemic typhus A flea-borne typhus caused by *Rickettsia typhi.* (Also called **murine typhus.**)

endergonic reaction A chemical reaction that absorbs energy.

endocardium The thin, internal membrane lining the heart.

endocytosis Process in which vesicles form by invagination to move substances into eukaryotic cells.

endoenzyme An enzyme that acts within the cell producing it. (Also called intracellular enzymes.)

endogenous infection An infection caused by opportunistic microorganisms already present in the body.

endogenous pyrogen Pyrogen secreted mainly by monocytes and macrophages that circulates to the hypothalamus and causes an increase in body temperature.

endometrium The mucous membrane lining the uterus.

endoparasite A parasite that lives within the body of another organism.

endoplasmic reticulum An extensive system of membranes that form tubes and vesicles in the cytoplasm of eukaryotic cells; involved in synthesis and storage of protein and lipids.

endospore septum A cell membrane without a cell wall that grows around the core of an endospore.

endospore A resistant, dormant structure, formed inside some bacteria such as *Bacillus* and *Clostridium,* that can survive adverse conditions.

endotoxin A toxin incorporated in gram-negative bacterial cell walls and released when the bacterium dies.

enrichment medium A medium that contains special nutrients that allow growth of a particular organism.

enteric bacteria Members of the family Enterobacteriaceae, many of which are intestinal; small facultatively anaerobic gram-negative rods with peritrichous flagella.

enteric fevers Systemic infections, such as typhoid fever, spread throughout the body from the intestinal mucosa.

enteritis An inflammation of the intestine.

enterocolitis Disorder caused by *Salmonella typhimurium* and *S. paratyphi* that invade intestinal tissue and produce bacteremia.

enteroinvasive strains Strains of *Escherichia coli* with a plasmid-borne gene for a surface antigen (K antigen) that enables them to attach to and invade mucosal cells.

enterotoxicosis See **food poisoning.**

enterotoxigenic strains Strains of *Escherichia coli* carrying a plasmid that enables them to make an enterotoxin.

enterotoxin An exotoxin that acts on tissues of the gut.

enteroviruses One of two major groups of picornaviruses that can infect nerve and muscle cells, the respiratory tract lining, and skin.

envelope A bilayer membrane found outside the capsid of some viruses, acquired as the virus buds through the host cell's membrane.

enveloped viruses Viruses with a bilayer membrane outside their capsids.

enzyme induction A mechanism whereby the genes coding for enzymes needed to metabolize a particular nutrient are activated by the presence of that nutrient.

Enzyme-linked immunoabsorbent assay (ELISA) Modification of RIA in which the antiantibody, instead of being radioactive, is attached to an enzyme that causes a color change in its substrate.

enzyme repression Mechanism by which the presence of a particular metabolite represses the genes coding for enzymes used in its synthesis.

enzyme-substrate complex A loose association of an enzyme with its substrate.

enzymes Protein catalysts that control the rate of chemical reactions in cells.

eosinophils Leukocytes released in large numbers during allergic reactions.

epicardium The outer membrane covering the heart.

epidemic Arises when a disease has a very high incidence in a population over a relatively short period of time.

epidemic keratoconjunctivitis Eye disease caused by an adenovirus.

epidemic relapsing fever Louse-borne cases of relapsing fever caused by *Borrelia* sp.

epidemic typhus Louseborne rickettsial disease caused by *Rickettsia prowazekii,* seen most frequently in conditions of overcrowding and poor sanitation.

epidemiological study A study conducted in order to learn more about the spread of a disease in a population.

epidemiologist A scientist who studies epidemiology.

epidemiology The study of factors and mechanisms involved in the spread of disease within a population.

epidermis The thin outer layer of the skin.

epiglottis Flap of tissue that prevents food and fluids from entering the larynx.

epiglottitis An infection of the epiglottis.

Epstein-Barr virus (EBV) Virus that causes Burkitt's lymphoma and infectious mononucleosis.

ergot A parasitic fungus of rye and wheat, *Claviceps purpura,* that causes ergot poisoning when ingested by humans.

ergot poisoning Disease caused by ingestion of ergot, *Claviceps purpura,* a fungus of rye and wheat.

erysipelas Infection caused by hemolytic streptococci that spreads through lymphatics, resulting in septicemia and other diseases. (Also called **St. Anthony's fire.**)

erythrocyte A red blood cell.

eschar The thick crust or scab that forms over a severe burn.

ester bond Chemical bond between the carboxyl group of an organic acid and the hydroxyl group of an alcohol.

etiology The assignment or study of causes and origins of a disease.

eubacteria True bacteria.

euglenoid A plantlike protist, usually with a single flagellum and a pigmented eyespot (stigma).

eukaryotes Organisms composed of eukaryotic cells (see below).

eukaryotic cells Cells that have a distinct nucleus and other membrane-bound structures.

eutrophication The nutrient enrichment of water from detergents, fertilizers, and animal manures, which causes overgrowth of algae and subsequent depletion of oxygen.

exanthema A skin rash.

exergonic reaction A chemical reaction resulting in the release of energy.

exocytosis Process by which vesicles inside the cell fuse with the plasma membrane and release their contents from the cell.

exoenzyme An enzyme that is synthesized in a cell, but crosses the cell membrane to act in the periplasmic space or the cell's immediate environment. (Also called **extracellular enzyme.**)

exogenous infection An infection caused by microorganisms that enter the body from the environment.

exogenous pyrogen Exotoxins and endotoxins from infectious agents that cause fever by stimulating the release of an endogenous pyrogen.

exonucleases Enzymes that remove segments of DNA.

exosporium A lipid-protein membrane formed outside the coat of some endospores by the mother cell.

exotoxin C A chemical produced by organisms that grow in accumulated menstrual flow in tampons, causing toxic shock syndrome.

exotoxin A soluble toxin secreted by microbes into their surroundings, including host tissues.

experimental epidemiological study Study designed to test a hypothesis about an outbreak of disease, often about the value of a particular treatment.

exponential growth Growth of a bacterial culture characterized by doubling of the population in a fixed interval of time.

external genitalia The female reproductive organs located outside the body, including the clitoris, labia, and Bartholin glands.

extracellular enzymes Exoenzymes produced by bacteria, which act in the medium around the organism.

extrachromosomal resistance Drug resistance of a microorganism due to the presence of resistance plasmids called R factors.

F′ plasmid An F plasmid that has been imprecisely separated from a chromosome so that it carries a fragment of the chromosome.

F pilus A bridge between F^+ and F^- cells during conjugation.

F plasmids Extrachromosomal DNA found in F^+ cells.

facilitated diffusion Diffusion (down a concentration gradient and) across a membrane (from an area of higher concentration to lower concentration) with the assistance of a carrier molecule, but not requiring ATP.

facultative anaerobes Bacteria that carry on aerobic metabolism when oxygen is present but shift to anaerobic metabolism when oxygen is absent.

facultative Able to tolerate the presence or absence of a particular environmental condition.

facultative parasite A parasite that can live either on a host or freely.

facultative psychrophiles Organisms that grow best at temperatures below 20°C but can also grow at temperatures above 20°C.

facultative thermophiles Organisms that can grow both above and below 37°C.

FAD Flavin adenine dinucleotide, a coenzyme.

fastidious Having many special nutritional needs that are difficult to meet in the laboratory.

fat A molecule formed from glycerol and one or more fatty acids.

fatty acid A long chain of carbon atoms and their associated hydrogens with a carboxyl group at one end.

feces Solid waste produced in the large intestine and stored in the rectum until eliminated from the body.

feedback inhibition Regulation of a metabolic pathway by the concentration of one of its intermediates or, typically, its end product, which inhibits an enzyme in the pathway.

feline panleukopenia virus A parvovirus that causes severe disease in cats.

fermentation Anaerobic metabolism of the pyruvate produced in glycolysis.

fever A body temperature that is abnormally high.

fibrinolysin Enzyme produced by streptococci that digests fibrin and thereby dissolves blood clots. (Also called **streptokinase.**)

fibroblasts New connective tissue cells that replace fibrin as a blood clot dissolves, forming granulation tissue.

filariasis Disease of the blood and lymph caused by any of several different roundworms carried by mosquitoes.

filoviruses Filamentous viruses that display unusual variability in shape. Two filoviruses, the Ebola virus and the Marburg virus, have been associated with human disease.

filter paper method Method of evaluating the antimicrobial properties of a chemical agent using filter paper disks placed on an inoculated agar plate.

filtration (1) A method of estimating the size of bacterial populations in which a known volume of air or water is drawn through a filter with pores too small to allow passage of bacteria. (2) The filtering of water through beds of sand to remove most of the remaining microorganisms after flocculation in water treatment plants.

fimbriae Also called **attachment pili.**

fine adjustment Focusing mechanism of a microscope that changes the distance between the objective lens and the specimen very slowly.

five-kingdom system System of classification of organisms into five kingdoms: Monera, Protista, Fungi, Animalia, and Plantae.

fixed macrophages Phagocytic cells that remain stationary in tissues.

flagella Long, thin, helical appendages of certain cells that provide means of locomotion.

flagellar staining Technique of coating surfaces of flagella with a dye or a metal such as silver.

flagellin Protein sub-units that make up a flagellum.

flatworms Primitive, unsegmented, hermaphroditic often parasitic worms. (Also called Platyhelminthes.)

flavivirus Type of virus that causes yellow fever.

flocculation The addition of alum to cause precipitation of suspended colloids, such as clay, in the water purification process.

fluctuation test Test that demonstrates that resistance to chemical substances occurs spontaneously rather than being induced.

fluid-mosaic model Current model of the structure of cellular membranes in which proteins are thought to be dispersed in a phospholipid bilayer.

flukes Flatworms with complex life cycles; all internal or external parasites.

fluoresce To reemit absorbed light of one color when exposed to light of another color.

fluorescence microscopy Use of ultraviolet light in a microscope to excite molecules so that they release light of different colors.

fluorescent antibody staining Procedure in fluorescence microscopy using fluorochrome-stained antibodies to detect the presence of an antigen.

fluoride Chemical that helps in reducing tooth decay by poisoning bacterial enzymes and hardening the surface enamel of teeth.

fluorochrome A fluorescent dye used in fluorescence microscopy, which causes organisms to stand out sharply against a dark background.

focal infection An infection confined to a specific area from which pathogens can spread to other areas.

folliculitis Local infection produced when hair follicles are invaded by pathogenic bacteria. (Also called **pimples** or **pustules.**)

fomites Non-living substances capable of transmitting disease, such as clothing, dishes, paper money, etc.

food poisoning A gastrointestinal disease caused by ingestion of foods contaminated with preformed toxins or other toxic substances.

formed elements Cells and cell fragments comprising about 40 percent of the blood.

frameshift mutation Mutation resulting from the deletion or insertion of one or more bases.

freeze-etching Scanning electron micrography technique in which water is evaporated under vacuum from the freeze-fractured surface of a specimen.

freeze-fracturing Scanning electron microscopy technique in which a cell is first frozen and then broken with a knife so that the fracture reveals structures inside the cell.

fulminating Of a disease, sudden and severe.

functional group A part of a molecule that generally participates in chemical reactions as a unit and gives the molecule some of its chemical properties.

fungi Nonphotosynthetic, eukaryotic organisms that absorb nutrients from their environment.

Fungi Imperfecti Group of fungi termed "imperfect" because no sexual stage has been observed in their life cycles. (Also called **Deuteromycota.**)

furuncle A large, deep, pus-filled infection. (Also called a **boil** or **abscess.**)

gamete A male or female reproductive cell.

gametocyte A male or female protozoan cell.

gamma globulin A pooled sample of antibody-containing fractions of serum from many individuals.

ganglia Aggregations of neuron cell bodies.

gas gangrene A deep wound infection, destructive of tissue, often caused by a combination of two or more species of *Clostridium*.

gene amplification Technique of genetic engineering in which plasmids or bacteriophages are induced to reproduce within cells at a rapid rate.

gene A linear sequence of DNA nucleotides that form a functional unit of a chromosome.

gene transfer Movement of genetic information between organisms by transformation, transduction, or conjugation.

generalized anaphylaxis Type I hypersensitivity that appears as a systemic life-threatening reaction such as airway constriction or anaphylactic shock.

generalized infection An infection that affects the entire body. (Also called **systemic infection.**)

generalized transduction Type of transduction in which a fragment of DNA from the degraded chromosome of an infected bacterial cell is accidentally incorporated into a new phage particle during replication and thereby transferred to another bacterial cell.

generation time Time required for a population of organisms to double in number.

genetic code The one-to-one relationship between each codon and a specific amino acid.

genetic engineering The purposeful manipulation of genetic material to alter the characteristics of an organism in a desired way.

genetic fusion A technique of genetic engineering that allows transposition of genes from one location on a chromosome to another or the coupling of genes from two different operons.

genetic homology Study of similarity of DNA among organisms.

genetic immunity See **innate immunity.**

genital herpes See **herpes, genital.**

genital warts A sexually transmitted viral disease having very high association rate with cervical cancer. (Also called **condylomas.**)

genitourinary system See **urogenital system.**

genome The genetic information in an organism.

genotype The genetic information contained in the DNA of an organism.

genus A taxon consisting of one or more species; the first name of an organism in binomial nomenclature, e.g., *Escherichia* in *Escherichia coli*.

germ theory of disease Theory that microorganisms (germs) can invade other organisms and cause disease.

German measles See **rubella.**

germination The process of a spore or an endospore starting to grow.

germination proper The second stage in germination, requiring water and a germination agent that penetrates the damaged spore coat.

giardiasis A gastrointestinal disorder caused by the flagellated protozoan, *Giardia intestinalis*.

gingivitis The mildest form of periodontal disease, characterized by inflammation of the gums.

gingivostomatitis Lesions of the mucous membranes of the mouth.

glomerulonephritis Inflammation of and damage to the glomeruli of the kidneys. (Also called Bright's disease.)

glomerulus A coiled cluster of capillaries in the nephron.

glycocalyx Term used to refer to all substances containing polysaccharides found external to the cell wall.

glycolysis The metabolic pathway used by most heterotrophic organisms, both aerobes and anaerobes, to break down glucose.

glycoproteins Long spikelike molecules that project beyond the surface of a (viral) envelope, some attach the virus to receptor sites on host cells; others aid fusion of viral and cellular membranes.

Golgi apparatus An organelle that stores substances synthesized on the endoplasmic reticulum.

gonorrhea A sexually transmitted disease caused by *Neisseria gonorrhoeae*.

graft tissue Tissue that is transplanted from one site to another.

graft-versus-host (GVH) disease Disease in which host antigens elicit an immunologic response from graft cells that destroys host tissue.

gram stain A differential stain. Gram-positive bacteria stain dark purple, gram-negative ones stain pink/red.

granulation tissue Fragile, reddish, grainy tissue made up of capillaries and fibroblasts that appears with the healing of an injury.

granules Inclusions that are not bounded by membrane and contain compacted substances that do not dissolve in cytoplasm.

granulocyte (or **polymorphonuclear leukocyte**) A leukocyte with granular cytoplasm and irregularly shaped, lobed nuclei.

granuloma In an inflammation, a collection of dead tissue, phagocytes, and cells of the specific immune system.

granuloma inguinale A sexually transmitted disease caused by *Calymmatobacterium granulomatis*. (Also called donovanosis.)

granulomatous hypersensitivity Cell-mediated hypersensitivity reaction that occurs when macrophages have engulfed pathogens but have failed to kill them.

granulomatous inflammation A special kind of chronic inflammation characterized by the presence of monocytes, histiocytes, lymphocytes, and plasma cells.

ground itch Bacterial infection of sites of penetration by hookworms.

group immunity See **herd immunity.**

growth curve The different growth periods of a bacterial or phage population.

gummas Granulomatous inflammations, symptomatic of syphilis, that destroy tissue.

gut-associated lymphatic tissue (GALT) Collective name for the tissues of lymph nodules; main site of antibody production.

halophiles Salt-loving organisms that require moderate to large concentrations of salt.

hanging drop A special type of wet mount often used with dark-field illumination to study motility of organisms.

Hansen's disease The preferred name for leprosy; caused by *Mycobacterium leprae*, it exhibits various clinical forms ranging from tuberculoid to lepromatous.

hapten A small molecule that can act as an antigenic determinant when combined with a larger molecule.

heat fixation Technique in which air-dried smears are passed through an open flame so that organisms are killed, adhere better to the slide, and take up dye more easily.

heavy (H) chains Larger of the two identical pairs of chains comprising immunoglobulin molecules.

helminths Worms, e.g., roundworms and flatworms.

helper T cells (T_H) Lymphocytes that stimulate the growth, differentiation, and immune response of B cells.

hemagglutination Agglutination (clumping) of red blood cells; used in blood typing.

hemagglutination inhibition test Serological test used to diagnose measles, influenza, and other viral diseases, based on the ability of antibodies to viruses to prevent viral hemagglutionation.

hemoglobin The oxygen-binding compound found in erythrocytes.

hemolysin An enzyme that lyses red blood cells.

hemolytic disease of the newborn Disease in which a baby is born with enlarged liver and spleen caused by efforts of these organs to destroy erythrocytes damaged by maternal antibodies; mother is Rh^- and baby, Rh^+.

hepatitis A Common form of viral hepatitis caused by a single-stranded RNA virus transmitted by the fecal-oral route. Formerly called **infectious hepatitis.**

hepatitis B Type of hepatitis caused by a double-stranded DNA virus usually transmitted in blood or semen. Formerly called **serum hepatitis.**

hepatitis D Severe type of hepatitis caused by presence of both hepatitis D and hepatitis B viruses. (Also called **delta hepatitis.**)

hepatitis An inflammation of the liver, usually caused by viruses but sometimes by an amoeba or various toxic chemicals.

herd immunity (or **group immunity**) The proportion of individuals in a population who are immune to a particular disease.

heredity The transmission of genetic traits from an organism to its progeny.

hermaphroditic Having both male and female reproductive systems in one organism.

herpes, genital Sexually transmitted disease caused primarily by herpes simplex virus type 2, or less often by type 1.

herpes gladiatorium Herpes that occurs in skin injuries of wrestlers; transmitted by contact or on mats.

herpes labialis Fever blisters (cold sores) on lips.

herpes meningoencephalitis A serious disease caused by a herpesvirus that can cause permanent neurological damage or death and that sometimes follows a generalized herpes infection or ascends from the trigeminal ganglion.

herpes pneumonia A rare form of herpes infection seen in burn patients, alcoholics, and AIDS patients.

herpes simplex virus type 1 (HSV-1) A virus that most frequently causes fever blisters (cold sores) and other lesions of the oral cavity, and less often causes genital lesions.

herpes simplex virus type 2 (HSV-2) A virus that typically causes genital herpes, but which can also cause oral lesions.

herpesviruses Relatively large, enveloped Class I DNA viruses that can remain latent in host cells for long periods of time.

heterogeneity The ability of the immune system to produce many different kinds of antibodies, each specific for a different antigenic determinant.

heterotrophy Other-feeding—use of carbon atoms from organic compounds for the synthesis of biomolecules.

high-energy bonds Bonds that release more energy than most covalent bonds when broken.

high frequency of recombination (Hfr) strain A strain of F^+ bacteria in which the F plasmid is incorporated into the bacterial chromosome.

histamine Amine released by tissues in allergic reactions.

histocompatibility antigens Antigens found in the membranes of all human cells that are unique in all individuals except identical twins.

histones Proteins that contribute directly to the structure of eukaryotic chromosomes.

histoplasmosis Fungal respiratory disease endemic to the Central and Eastern United States, caused by the soil fungus *Histoplasma capsulatum*. (Also called Darling's disease.)

holoenzyme An enzyme consisting of an apoenzyme and a coenzyme or cofactor.

homolactic-acid fermentation A pathway in which pyruvate is converted directly to lactate using electrons from reduced NAD.

hookworm A disease caused by two species of small roundworms, *Ancyclostoma duodenale* and *Necator americanus*, whose larvae burrow through skin of feet, enter the blood vessels, and penetrate lung and intestinal tissue.

horizontal transmission Direct contact transmission of disease in which pathogens are usually passed by handshaking, kissing, contact with sores, or sexual contact.

host Any organism that harbors another organism.

host range The different types of organisms that a microbe can infect.

host specificity The range of different hosts in which a parasite can mature.

human leukocyte antigens Lymphocyte antigens used in laboratory tests to determine compatibility of donor and recipient tissues for transplants.

human papillomaviruses Viruses that attack skin and mucous membranes, causing papillomas or warts.

humoral immunity Immune response carried out by antibodies circulating in the blood.

humus The nonliving organic components of soil.

hyaluronidase Bacterially-produced enzyme that digests hyaluronic acid, which holds cells together, thereby making tissues more accessible to microbes. (Also called spreading factor.)

hybridoma A hybrid cell resulting from the fusion of a cancer cell with another cell, usually one that produces antibodies.

hydatid cyst An enlarged cyst containing many tapeworm heads.

hydrogen bonds A relatively weak attraction between a hydrogen atom carrying a partial positive charge and an oxygen or nitrogen atom carrying a partial negative charge.

hydrologic cycle See **water cycle.**

hydrophilic Water-loving.

hydrophobic Water-repelling.

hydrostatic pressure Pressure exerted by standing water.

hyperimmune sera Preparations of gamma globulins having high titers of specific kinds of antibodies. (Also called **convalescent sera.**)

hyperparasitism The phenomenon of a parasite itself having parasites.

hypersensitivity Disorder in which the immune system reacts inappropriately, usually by responding to an antigen it normally ignores. (Also called **allergy.**)

hypertonic solution A solution containing a concentration of dissolved material greater than that within a cell.

hyphae (sing: **hypha**) Long, threadlike structures of cells in fungi or actinomycetes.

hypotonic solution A solution containing a concentration of dissolved material lower than that within a cell.

IgA Class of antibody found in small amounts in the blood and secretions.

IgD Class of antibody found mainly on B cell membranes and rarely secreted; function unknown.

IgE Class of antibody that binds to receptors on basophils in the blood or mast cells in the tissues; responsible for severe allergic reactions.

IgG The main class of antibodies found in the blood; produced in largest quantities during secondary response.

IgM The first class of antibody secreted into the blood during the early stages of a primary immune response; a rosette of five immunoglobulin molecules.

immersion oil Substance used to avoid refraction at a glass-air interface when examining objects through a microscope.

immune adherence See **opsonization.**

immune complex disorders (type III **hypersensitivity**) Disorders caused by antigen-antibody complexes that precipitate in the blood and injure tissues; elicited by antigens in vaccines, on microorganisms, or on a person's own cells.

immune complexes Antigen-antibody complexes that are normally destroyed by phagocytic cells.

immune cytolysis Process in which the membrane attack complex of complement produces lesions on cell membranes through which the contents of the cells leak out.

immune Not susceptible to a disease caused by an infectious agent.

immune system Body system that provides the host organism with specific immunity to infectious agents.

immunity The ability of an organism to recognize and defend itself against infectious agents.

immunodeficiency diseases Diseases that impair immunity, caused by lack of lymphocytes, defective lymphocytes, or destruction of lymphocytes.

immunodeficiency Disorder in which the immune system responds inadequately to an antigen because of inborn or acquired defects in B or T cells.

immunodiffusion tests Serological tests similar to the precipitin test but carried out in an agar gel medium.

immunoelectrophoresis Serological test in which antigens are first separated by gel electrophoresis and then allowed to react with antibody placed in a trough in the gel.

immunofluorescence Serological test that utilizes antibodies to which a fluorescent substance is bound to detect antigens, other antibodies, or complement within tissues.

immunoglobulin (Ig) Y-shaped protein molecule, composed of four polypeptide chains, which can complex with a specific antigen. (Also called **antibody.**)

immunologic disorder Disorder that results from an inappropriate or inadequate immune response.

immunology The study of specific immunity and how the immune system responds to specific infectious agents.

immunosuppression Minimizing of immune reactions using radiation or cytotoxic drugs.

impetigo A highly contagious pyoderma caused by staphylococci, streptococci, or both.

inapparent infection An infection that fails to produce symptoms, either because too few organisms are present or because host defenses effectively combat the pathogens. (Also called **subclinical infection.**)

incidence The number of new cases of a particular disease seen in a specific period of time.

inclusion blennorrhea A mild chlamydial infection of the eyes in infants.

inclusion bodies (1) Aggregations of reticulate bodies within chlamydias. (2) A form of cytopathic effect consisting of viral components, masses of viruses, or remnants of viruses.

inclusion conjunctivitis A chlamydial infection that can result from self-inoculation with *Chlamydia trachomatis*.

inclusions Small nonliving bodies found in the cytoplasm, such as granules and vesicles.

incubation period In a disease, the time between infection and the appearance of signs and symptoms.

index case The first case of a disease to be identified.

index of refraction A measure of the amount that light rays bend when passing from one medium to another.

indicator organism An organism such as *E. coli* whose presence indicates the contamination of water by fecal matter.

indigenous Native to a given environment.

indirect contact transmission Transmission of disease through fomites.

indirect fecal-oral transmission Transmission of disease in which pathogens from feces of one organism infect another organism.

induced mutations Mutations produced by agents called mutagens that increase the mutation rate.

inducer A substance that binds to and inactivates the repressor of an operon.

inducible enzymes Enzymes coded for by genes that are sometimes active and sometimes inactive.

induction The stimulation of a temperate phage to become virulent.

induration A raised, hard, red region on the skin resulting from tuberculin hypersensitivity.

industrial microbiology Branch of microbiology concerned with the use of microorganisms to assist in the manufacture of useful products or disposal of waste products.

infant botulism Form of botulism in infants associated with ingestion of honey. (Also called "floppy baby syndrome.")

infection The multiplication of a parasite, usually microscopic, within the body.

infectious disease Disease caused by infectious microbial or parasitic agents.

infectious hepatitis Former name for **hepatitis A.**

infectious mononucleosis An acute disease that affects many systems, caused by the Epstein-Barr virus.

infestation The presence of protozoan, helminth, or arthropod parasites in or on a living host.

influenza A viral respiratory infection caused by orthomyxoviruses that appears as epidemics.

initiating segment Part of the F plasmid that is transferred in conjugation of an Hfr bacterium.

innate immunity Immunity to infection that exists in an organism because of genetically determined characteristics. (Also called **genetic immunity.**)

insects Arthropods with three body regions, three pairs of legs, and highly specialized mouthparts.

insertion The addition of one or more bases to DNA, usually producing a frameshift mutation.

interferon A small protein released from virus-infected cells which binds to adjacent cells, causing them to produce a protein that interferes with viral replication.

intermediate host An organism that harbors a sexually immature stage of a parasite.

intermediate (Type I) **hypersensitivity** Type of allergy elicited by foreign substances (allergens) such as pollen, foods, and insect venoms. (Also called **anaphylaxis.**)

intermittent carrier A human or animal that releases disease microorganisms periodically.

invasive phase The time during which a disease develops its most severe signs and symptoms.

invasiveness The ability of a microorganism to take up residence in a host.

involution The assumption of unusual shapes by cells in the decline phase of the bacterial growth curve.

ion An electrically charged atom produced when an atom gains or loses one or more electrons.

ionic bonds Bonds between atoms resulting from attraction of ions with opposite charges.

iris diaphragm Adjustable device in a microscope that controls the amount of light passing through the specimen.

ischemia Reduced blood flow to tissues with oxygen and nutrient deficiency and waste accumulation.

isograft A graft of tissue between genetically identical individuals.

isolation Situation in which a patient with a communicable disease is prevented from contact with the general population.

isomers Two molecules with the same molecular formula but different structures.

isotonic Fluid containing same concentration of dissolved materials as is in cell; causes no change in cell volume.

isotopes Atoms of a particular element that contain different numbers of neutrons.

jaundice Yellow skin color due to excessive bilirubin in the blood from the breakdown of erythrocytes; caused by impaired liver function and common in hepatitis.

kala azar Visceral leishmaniasis caused by *Leishmania donovani.*

karyogamy Process by which nuclei fuse to produce a diploid cell.

keratin A water-proofing protein found in epidermal cells.

keratitis An inflammation of the cornea.

keratoconjunctivitis Condition in which vesicles appear on the cornea and eyelids.

kidney tubule Part of the kidney where urine is formed.

kidneys A pair of organs responsible for the formation of urine.

kinases Bacterially produced enzymes that digest fibrin, thereby dissolving blood clots.

kingdom Animalia Kingdom that includes all animals derived from zygotes.

kingdom Fungi Kingdom consisting of mostly multicellular, nonphotosynthetic organisms that obtain nutrients solely by absorption of organic matter from dead organisms.

kingdom Monera Kingdom consisting of all prokaryotic organisms, including the true bacteria, the cyanobacteria, and the archaebacteria.

kingdom Plantae Kingdom consisting of macroscopic green plants.

kingdom Protista Kingdom consisting of eukaryotic, mostly unicellular organisms.

kinins Small peptides always present in the blood in an inactive form that increase blood flow and vessel permeability when activated by injury.

Koplik's spots Red spots with central bluish specks that appear on the upper lip mucosa in early stages of measles.

Koch's postulates Four postulates formulated by Robert Koch, in the 19th century; used to prove that a particular organism causes a particular disease.

Krebs cycle A sequence of enzyme-catalyzed chemical reactions that metabolizes 2-carbon units called acetyl groups to CO_2 and H_2O. Also called the tricarboxylic acid cycle or the **citric acid cycle.**

Kupffer cells Phagocytic cells that remove foreign matter from the blood as it passes through sinusoids.

L forms Irregularly shaped naturally occurring bacteria with defective cell walls.

lacrimal gland Tear-producing gland of the eye.

lactobacilli Type of regular, nonsporing, gram-positive rods found in many foods; used in production of cheeses, yogurt, sourdough, and other fermented foods.

lag phase First of four major phases of the bacterial growth curve, in which organisms grow in size but do not increase in number.

large intestine The lower area of the intestine that absorbs water and converts undigested food into feces.

laryngeal papillomas Benign growths caused by herpesviruses that can be dangerous if they block the airway; infants are often infected during birth by mothers having genital warts.

laryngitis An infection of the larynx, often with loss of voice.

larynx The voicebox.

Lassa fever Hemorrhagic fever caused by arenaviruses that begins with pharyngeal lesions and proceeds to severe liver damage.

latency The ability of a virus to remain in host cells for long periods of time while retaining the ability to replicate.

latent disease A disease characterized by periods of inactivity either before symptoms appear or between attacks.

latent period Period on the growth curve after the bacteria or phages are introduced into a culture, but before their numbers increase.

legionellas The causative bacterial agent in Legionnaire's disease, *Legionella pneumophila.*

Legionnaire's disease Disease caused by *Legionella pneumophila*, transmitted by airborne bacteria.

leishmaniasis A parasitic systemic disease caused by three species of protozoa of the genus *Leishmania* and transmitted by sandflies.

lepromas Enlarged, disfiguring skin lesions that occur in the lepromatous form of Hansen's disease.

lepromatous leprosy The nodular form of leprosy in which a granulomatous response causes enlarged, disfiguring skin lesions called lepromas.

lepromin skin test Test used to detect Hansen's disease; similar to the tuberculin test.

leptospirosis A zoonosis caused by the spirochete *Leptospira interrogans*, which enters the body through mucous membranes or skin abrasions.

leukocidin Toxin produced by staphylococci that kills phagocytes.

leukocydins Substance produced by streptococci and staphylococci that damages or destroy neutrophils.

leukocyte A white blood cell.

leukocytosis-promoting factor (LP factor) A substance released by inflamed tissues that makes more leukocytes available.

leukostatin Substance produced by streptococci and staphylococci that interferes with the ability of leukocytes to engulf microorganisms that release it.

ligase An enzyme that joins together DNA segments.

light (L) chains Smaller of the two identical pairs of chains comprising immunoglobulin molecules.

light microscopy The use of any type of microscope that uses light to make specimens observable.

light reactions The part of photosynthesis in which light energy is used to excite electrons from chlorophyll, which are then used to generate ATP and NADPH.

light repair Repair of DNA dimers by a light-activated enzyme.

lipid A Toxic substance found in the cell wall of gram-negative bacteria. (Also called **endotoxin.**)

lipids A diverse group of water-insoluble compounds.

lipopolysaccharide Part of the outer layer of the cell wall in gram-negative bacteria.

listeriosis A type of meningitis caused by *Listeria monocytogenes* that is threatening to those with impaired immune systems.

loaiasis Tropical eye disease caused by the filarial worm *Loa loa.*

lobar pneumonia Type of pneumonia that affects one or more of the five major lobes of the lungs.

local infection An infection confined to a specific area of the body.

localized anaphylaxis Allergy that appears as reddening of the skin, watery eyes, hives, asthma, and digestive disturbances.

locus The location of a gene on a chromosome.

log phase Second of four major phases of the bacterial growth curve, in which cells divide at an exponential or logarithmic rate.

logarithmic growth See **exponential growth.**

lophotrichous Having two or more flagella at one or both ends of a bacterial cell.

lower respiratory tract Thin-walled bronchioles and alveoli where gas exchange occurs.

luminescence Process in which absorbed light rays are reemitted at longer wavelengths.

Lyme disease Disease caused by *Borrelia burgdorferi*, carried by the deer tick.

lymph capillaries Very small blind-ended vessels that surround tissue cells and accumulate lymph.

lymph nodes Encapsulated globular structures located along the routes of the lymphatic vessels that help clear the lymph of microorganisms.

lymph nodules Small, unencapsulated aggregations of lymphatic tissue that develop in many tissues, especially in the digestive, respiratory and urogenital tracts, collectively called gut-associated lymphatic tissue (or GALT), they are the body's main sites of antibody production.

lymphangitis Symptom of septicemia in which red streaks due to inflamed lymphatics appear beneath the skin.

lymphatic system Body system, closely associated with the cardiovascular system, which transports lymph through body tissues and organs; performs important functions in host defenses and specific immunity.

lymphatic vessels Vessels that form a system for returning lymph to the blood circulatory system.

lymphocytes Leukocytes (white blood cells) found in large numbers in lymphoid tissues that contribute to specific immunity.

lymphogranuloma venereum A sexually transmitted disease, caused by *Chlamydia trachomatis*, that attacks the lymphatic system.

lymphokines Chemical substances secreted by T cells when they encounter an antigen.

lyophilization The drying of a material from the frozen state; freeze-drying.

lysis The destruction of an infected cell by rupture and release of phages.

lysogen The combination of a bacterium and a temperate phage.

lysogenic State of a cell containing prophages.

lysogeny Persistence of prophage in an infected bacterium without replication and destruction of the cell.

lysosomes Extremely small membrane-bound organelles that contain digestive enzymes.

lysozyme An enzyme that acts on polysaccharides to weaken the bacterial cell walls.

lytic cycle The sequence of events in which a bacteriophage infects a bacterial cell, replicates, and eventually causes rupture of the cell.

macrophages Ravenously phagocytic leukocytes found in tissues.

madura foot Tropical disease caused by a variety of soil organisms (fungi and actinomycetes) that often enter the skin through bare feet. (Also called **maduromycosis.**)

maduromycosis See **madura foot.**

malaria A severe parasitic disease caused by several species of the protozoan *Plasmodium* and transmitted by mosquitoes.

Malta fever See **brucellosis.**

mast cells Leukocytes that release histamine and heparin in tissues.

mastigophorans (Flagellates) protozoans with flagella, e.g. *Giardia.*

mastoid area Portion of the temporal bone prominent behind the ear opening.

matrix Fluid-filled inner portion of mitochondria.

matrix proteins M proteins found inside the envelope that contribute to its structure and assist in assembly of components into new viruses.

maturation The process by which complete virions are assembled from newly synthesized components in the replication process.

measles encephalitis A serious complication of measles that leaves many survivors with permanent brain damage.

measles A febrile disease with rash caused by the rubeola virus, which invades lymphatic tissue and blood. (Also called **rubeola.**)

mechanical stage Attachment to a microscope stage that holds the slide and allows precise control in moving it.

mechanical vector A vector in which the parasite does not complete any part of its life cycle during transit.

megakaryocyte Large cell normally present in bone marrow that gives rise to platelets.

meiosis Division process in eukaryotic cells that reduces chromosome number by one-half.

membrane attack complex A set of proteins in the complement system that lyses invading microorganisms by producing lesions in their cell membranes.

membrane filter method Method of testing for coliform bacteria in water in which bacteria are filtered through a membrane and then incubated on the membrane surface in growth medium.

memory cells Long-lived B or T lymphocytes able to make an anamnestic or secondary response.

memory Ability of the immune system to recognize certain substances to which it has previously responded.

meninges Three layers of membrane that protect the brain and spinal cord.

merozoite A malaria trophozoite found in infected red blood or liver cells.

mesophiles Organisms that grow best at temperatures between 25° and 40°C, including most bacteria.

messenger RNA (mRNA) A type of RNA that carries the information from DNA to dictate the arrangement of amino acids in a protein.

metabolic pathway A chain of chemical reactions in which the product of one reaction serves as the substrate for the next.

metabolism The sum of all chemical processes carried out by living organisms.

metacercaria The post-cercarial encysted stage in the development of a fluke, prior to transfer to the final host.

metachromasia Property of exhibiting a variety of colors when stained with a simple stain.

metachromatic granules Polyphosphate granules that exhibit metachromasia. (Also called **volutin.**)

methacrylate Sealant that provides protection against tooth decay.

microaerophiles Bacteria that grow best in the presence of a small amount of free oxygen.

microbial antagonism Production by one species of microbial organism of substances that help it to compete with other species by interfering with their growth.

microbial growth Increase in the number of cells, due to cell division.

microbiology The study of microorganisms.

micrococci Aerobes or facultative anaerobes that form irregular clusters by dividing in two or more planes.

microenvironments Distinct environmental conditions existing in close proximity.

microfilaments Protein fibers that make up part of the cytoskeleton of eukaryotic cells.

microfilaria An immature microscopic roundworm larva.

micrometer (μm) Unit of measure equal to 0.000001 m or 10^{-6} m; formerly called a micron (μ).

microphages Phagocytic cells that digest bacteria and other small particles. (Also called neutrophils.)

microscopy The technology of making very small things visible to the unaided eye.

microtubules Protein tubules that make up cilia, flagella, and part of the cytoskeleton of eukaryotic cells.

microvillus (pl. **microvilli**) A minute projection from the surface of a single cell.

middle ear Small cavity in the temporal bone containing the small bones that transmit sound waves.

miliary tuberculosis Type of tuberculosis that invades all tissues, producing tiny lesions.

minimum bactericidal concentration (MBC) The lowest concentration of an antimicrobial agent that kills microorganisms, as indicated by absence of growth following subculturing in the dilution method.

minimum inhibitory concentration (MIC) The lowest concentration of an antimicrobial agent that prevents growth in the dilution method of determining antibiotic sensitivity.

miracidia Ciliated, free-swimming first-stage fluke larvae that emerge from eggs.

mitochondria Organelles that carry out oxidative reactions that capture energy in ATP.

mitosis Process by which the nucleus of a eukaryotic cell divides to form identical daughter nuclei.

mixed infection An infection caused by several species of organisms present at the same time.

mixture Two or more substances combined in any proportion and not chemically bound.

mole The weight of a substance in grams equal to the sum of the atomic weights of the atoms in a molecule of the substance.

molecular mimicry Imitation of the behavior of a normal molecule by an antimetabolite.

molecule Two or more atoms chemically bonded together.

molluscum contagiosum A viral infection characterized by flesh-colored, painless lesions.

moniliasis See **candidiasis.**

monoclonal antibody A single, pure antibody produced in the laboratory by a clone of cultured cells.

monocular Refers to a light microscope having one eyepiece.

monocytes Ravenously phagocytic leukocytes, called macrophages after they migrate into tissues.

monosaccharides A simple carbohydrate, consisting of a carbon chain or ring with several alcohol groups and aldehyde or ketone group.

morbidity rate The number of persons contracting a specific disease in relation to the total population.

mortality rate The number of deaths from a specific disease in relation to the total population.

most probable number (MPN) A statistical method of measuring bacterial growth, used when samples contain two few organisms to give reliable measures by the plate count method.

mother cell A cell that has approximately doubled in size and is about to divide into two daughter cells.

mucin A glycoprotein in mucus that coats bacteria and prevents their attaching to surfaces.

mucociliary escalator Mechanism involving ciliated cells that allows materials in the bronchi, trapped in mucus, to be lifted to the pharynx and spit out or swallowed.

multiple-tube fermentation method Three-step method of testing for coliform bacteria in drinking water.

mumps Disease caused by a paramyxovirus that is transmitted by saliva and invades cells of the oropharynx.

murine typhus A flea-borne typhus caused by *Rickettsia typhi*. (Also called **endemic typhus.**)

mutagens Agents that increase the rate of mutations.

mutation A permanent alteration in an organism's DNA.

mutualism A symbiosis benefiting both organisms.

myasthenia gravis Autoimmune disease specific to skeletal muscle, especially muscles of the limbs and those involved in eye movements, speech, and swallowing.

mycelium In fungi, a mass of long, threadlike structures (hyphae) that branch and intertwine.

mycobacteria Slender, acid-fast rods, often filamentous; include organisms that cause tuberculosis, leprosy, and chronic infections.

mycoplasmal pneumonia A mild pneumonia with an insidious onset and symptoms different from those of classic pneumonia. (Also known as **primary atypical pneumonia** or **walking pneumonia.**)

mycoplasmas Very small bacteria with cell membranes, RNA and DNA, but no cell walls.

mycoses Human fungal diseases.

myiasis An infestation caused by maggots (fly larvae).

myocarditis An inflammation of the heart muscle.

myocardium The thick, muscular portion of the heart wall between the endocardium and the epicardium.

NAD Nicotinamide adenine dinucleotide, a coenzyme.

naked virus A virus that lacks an envelope.

nanometer (nm) Unit of measure equal to 0.000000001 m or 10^{-9} m; formerly called a millimicron (mμ).

narrow spectrum The range of activity of an antimicrobial agent that attacks only a few kinds of microorganisms.

nasal cavity Part of the upper respiratory tract where air is warmed and particles are removed by hairs as it passes through.

natural killer (NK) cells Lymphocytes that destroy malignant tumor cells and cells of transplanted tissues.

naturally acquired active immunity Active immunity produced when an organism is exposed to an infectious agent.

naturally acquired immunity Immunity conferred either by one's own antibodies, produced as a result of having a disease, or by maternal antibodies, acquired across the placenta or in colostrum.

naturally acquired passive immunity Immunity produced when antibodies made by a mother's immune system are transferred to her offspring across the placenta or in colostrum.

necrotizing enterocolitis An intestinal infection caused by *Clostridium perfringens*.

negative chemotaxis Movement of an organism away from a chemical.

negative sense nucleic acid A nucleic acid made up of bases complementary to those of a positive sense nucleic acid.

negative staining Technique of staining the background of a specimen when a specimen or part of it resists taking up a stain.

nematodes Often parasitic worm of animals and plants with long, cylindrical, unsegmented body and a heavy cuticle. (Also called roundworms.)

neonatal herpes Herpes infection in infants, usually with HSV-2, most often acquired during passage through a birth canal contaminated with the virus.

nephron A functional unit of the kidney.

nerves Bundles of neuron fibers that relay sensory and motor signals throughout the body.

nervous system The body system that coordinates the body's activities in relation to the environment.

neuron A conducting nerve cell.

neurosyphilis Neurological damage, including thickening of the meninges, ataxia, paralysis, and insanity, that results from syphilis.

neurotoxin A toxin that acts on nervous system tissues.

neutralization Inactivation of microbes or their toxins by incorporation into antigen-antibody complexes.

neutrons Uncharged subatomic particles in the nucleus of an atom.

neutrophiles Organisms that grow best from pH 5.4 to 8.5.

neutrophils Numerous, phagocytic leukocytes. (Also called **microphages.**)

nitrification The process by which ammonia or ammonium ions are oxidized to nitrites or nitrates.

nitrogen cycle Process by which nitrogen moves from the atmosphere through various organisms and back into the atmosphere.

nitrogen fixation The reduction of atmospheric nitrogen gas to ammonia.

nitrogenase Enzyme in nitrogen-fixing bacteroids that catalyzes the reaction of nitrogen gas and hydrogen gas to form ammonia.

nocardioforms Gram-positive, nonmotile, pleomorphic, aerobic bacteria, often filamentous and acid-fast; include some skin and respiratory pathogens.

nocardiosis Respiratory disease characterized by tissue lesions and abscesses; caused by the filamentous bacterium *Nocardia asteroides*.

nocturia Nighttime urination, often a result of urinary tract infections.

Nomarski microscopy Differential interference contrast microscopy; utilizes differences in refractive index to visualize structures, producing a nearly three-dimensional image.

non-A, non-B (NANB) hepatitis Type of hepatitis caused by multiple viral agents that is diagnosed in the absence of hepatitis A and B viruses.

noncommunicable infectious disease Disease caused by infectious agents but not spread from one host to another.

noncompetitive inhibitor A molecule that attaches to an enzyme at an allosteric site (a site other than the active site), distorting the shape of the active site so that the enzyme can no longer function.

noncytocidal Nonlethal to its host cell.

nongonococcal urethritis A gonorrhealike sexually transmitted disease most often caused by *Chlamydia trachomatis* and mycoplasmas.

nonindigenous Temporarily found in a given environment.

noninfectious disease Disease caused by any factor other than infectious agents.

nonself Antigens recognized as foreign by an organism.

nonspecific defense Defense against pathogens that operates regardless of the invading agent.

nonspecific immunity Immunity produced by defenses against invading organisms in general.

nonsynchronous growth Natural pattern of growth during the log phase in which every cell in a culture divides at some point during the generation time, but not simultaneously.

normal flora Microorganisms commonly found in or on another organism.

nosocomial infection An infection acquired in a hospital or other medical facility.

notifiable disease A disease that a physician is required to report to public health officials.

nuclear envelope The double membrane surrounding the nucleus in a eukaryotic cell.

nuclear pores Pores in the nuclear envelope that allow RNA molecules to leave the nucleus and participate in protein synthesis.

nuclear region Central location of DNA, RNA, and some proteins in bacteria; not a true nucleus.

nucleic acids Long polymers of nucleotides that encode genetic information.

nucleoid Nuclear region in bacteria.

nucleoli Areas in the nucleus of a eukaryotic organism that contain RNA and serve as sites for the assembly of ribosomes.

nucleotide A compound consisting of a nitrogenous base, a five-carbon sugar, and one or more phosphate groups.

nucleus Organelle that contains the genetic material of a eukaryotic cell.

null cell A lymphocyte that cannot be identified as a B cell or a T cell.

numerical aperture The widest cone of light that can enter a lens.

numerical taxonomy Comparison of organisms based on quantitative assessment of a large number of characteristics.

nutritional complexity The number of nutrients an organism must obtain to grow.

nutritional factors The availability of various nutrients that influence the kind of organisms found in an environment and their growth.

objective lens Lens in a microscope closest to the specimen that creates an enlarged image of the object viewed.

obligate aerobes Bacteria that must have free oxygen to grow.

obligate anaerobes Bacteria that are killed by free oxygen.

obligate Requiring a particular environmental condition.

obligate intracellular parasites Organisms that can live or multiply only inside a living host cell.

obligate parasite A parasite that must spend some or all of its life cycle in or on a host.

obligate psychrophiles Organisms that cannot grow at temperatures above 20°C.

obligate thermophiles Organisms that can grow only at temperatures above 37°C.

ocular lens Lens in microscope that further magnifies the image created by the objective lens.

ocular micrometer A glass disk with an inscribed scale that is placed inside the eyepiece of a microscope; used to measure the actual size of an object being viewed.

onchocerciasis An eye disease caused by the filarial larvae of the nematode *Onchocerca volvulus*, transmitted by blackflies; common in Africa and Central America. (Also known as **river blindness.**)

Oomycota See **water molds.**

operon A sequence of closely associated genes that includes both structural genes and the regulator genes that control their transcription.

opportunists Species of resident or transient flora that do not ordinarily cause disease but can do so under certain conditions.

opsonins Antibodies that promote phagocytosis when bound to the surface of microorganisms.

opsonization The process by which microorganisms are rendered more attractive to phagocytes by being coated with antibodies (opsonins) and C3b complement protein. (Also called **immune adherence.**)

ophthalmia neonatorum Pyogenic infection of the eyes caused by *Neisseria gonorrhoeae*. (Also known as gonococcal conjunctivitis of the newborn.)

optical microscope A light microscope.

optimum pH The pH at which microorganisms grow best.

oral cavity The mouth.

orbivirus Type of virus that causes Colorado tick fever.

orchitis Inflammation of the testes; a symptom of mumps in postpubertal males.

organelles Internal membrane-bound structures of eukaryotic cells.

organic chemistry The study of compounds that contain carbon.

oriental sore A cutaneous lesion at the site of a sandfly bite caused by *Leishmania tropicana*.

ornithosis Disease with pneumonialike symptoms, caused by *Chlamydia psittaci* and acquired from birds. Previously called **psittacosis** and parrot fever.

Oroya fever One form of bartonellosis; an acute fatal fever with severe anemia. (Also called **Carrion's disease.**)

orthomyxoviruses Medium-sized, enveloped Class III RNA viruses that vary in shape from spherical to filamentous and have an affinity for mucus.

orthopoxviruses Large, enveloped, brick-shaped poxviruses that cause smallpox and cowpox.

osmosis A special type of diffusion in which water moves from an area of higher concentration to one of lower concentration across a selectively permeable membrane.

osmotic pressure The pressure required to prevent the net flow of water by osmosis.

otitis externa Infection of the external ear canal.

otitis media Infection of the middle ear.

outer membrane A bilayer membrane, surrounding the cell wall of gram-negative bacteria.

outer sheath A multilayered membranous sheath that surrounds the protoplasmic cylinder in spirochetes.

outgrowth The final stage of germination, in which proteins and RNA are synthesized and DNA synthesis begins.

ovarian follicle An aggregation of cells in the ovary containing an ovum.

ovaries In the female, paired glands that produce ovarian follicles, which contain an ovum and hormone-secreting cells.

oxidation The loss of electrons.

oxidative phosphorylation Process in which energy of electrons is captured in high-energy bonds as phosphate groups combine with ADP to form ATP.

pandemic An epidemic that has become worldwide.

papillomas See **warts.**

papillomaviruses Wart and tumor-causing viruses frequently found in host cell nuclei without being integrated into host DNA.

papovaviruses Small, naked, Class II DNA viruses that cause both benign and malignant warts in humans.

parainfluenza Viral disease characterized by nasal inflammation, pharyngitis, bronchitis, and sometimes pneumonia, mainly in children.

parainfluenza viruses Viruses that initially attack the mucous membranes of the nose and throat.

paramyxoviruses Medium-sized, enveloped Class II RNA viruses having an affinity for mucus.

parapoxviruses Poxviruses that can infect the skin of humans having close contact with infected animals.

parasite An organism that lives at the expense of another organism, the host.

parasitism A symbiosis in which one organism, the parasite, benefits from the relationship, whereas the other organism, the host, is harmed by it.

parasitology The study of parasites.

parfocal Microscope that remains in approximate focus when objective lenses are changed.

paroxysmal stage Stage of whooping cough in which mucus and masses of bacteria fill the airway, causing violent coughing.

parvoviruses Small, naked, Class III DNA viruses.

parvoviruses proper Parvoviruses that contain either positive sense DNA or negative sense DNA, but not both.

passive carrier A human or animal that releases disease microorganisms without ever having shown signs and symptoms of the disease.

passive immunity Immunity created when ready-made antibodies are introduced into, rather than created by, an organism.

passive immunization The process of inducing immunity by introducing ready-made antibodies into a host.

***Pasteurella-Haemophilus* group** Very small gram-negative bacilli and coccobacilli that lack flagella and are nutritionally fastidious.

pasteurization Mild heating to destroy pathogens and other organisms that cause spoilage.

pathogen Any parasite capable of causing disease in its host.

pathogenicity The capacity to produce disease.

pediculosis Lice infestation, resulting in reddened areas at bites, dermatitis, and itching.

pellicle (1) A thin layer of bacteria adhering to the air-water interface of a broth culture by their attachment pili. (2) An outer membranous cover of a protozoan cell. (3) Film over the surface of a tooth at the beginning of plaque formation.

pelvic inflammatory disease An infection of the pelvic cavity in females, caused by any of several organisms including *Neisseria gonorrhoeae* and *Chlamydia*.

penetration The entry of the virus (or its nucleic acid) into the host cell in the replication process.

penicillin An antimicrobial agent made by the mold *Penicillium*.

pentons Units of the capsomere that cause hemagglutination.

peptide bond A bond joining the amino group of one amino acid and the carboxyl group of another amino acid.

peptidoglycan A structural polymer in the bacterial cell wall that forms a supporting net.

peptococci Anaerobes that form pairs, tetrads, or irregular clusters; they lack both catalase and the enzymes to ferment lactic acid.

peptone A product of enzyme digestion of proteins that contains many small peptides; a common ingredient of a complex medium.

perforin A lethal protein produced by cytotoxic killer and natural killer cells that bores holes in target cell membranes.

pericardial sac The tough, membranous bag containing the heart.

pericarditis An inflammation of the protective membrane around the heart.

periodontal disease A combination of gum inflammation, decay of cementum, and erosion of periodontal ligaments and bone that support teeth.

peripheral nervous system All nerves outside the central nervous system.

periplasmic enzymes Exoenzymes produced by gram-negative organisms, which act between the cell wall and the cell membrane.

periplasmic space The space between the cell membrane and the outer membrane; contains toxins and digestive enzymes.

peritrichous Having flagella distributed all over the surface of a bacterial cell.

permanent parasite A parasite that remains in or on a host once it has invaded the host.

permeases Enzymes involved in active transport that extend through the cell membrane.

peroxisome An organelle filled with enzymes that in animal cells oxidize amino acids and in plant cells oxidize fats.

pertussis See **whooping cough.**

petechiae Pinpoint-size hemorrhages, most common in skin folds, that often occur in rickettsial diseases.

pH A means of expressing the hydrogen-ion concentration, and thus the acidity, of a solution.

phage See **bacteriophage.**

phagocyte A cell that ingests and digests foreign particles.

phagocytosis Ingestion of solids into cells by means of the formation of vacuoles.

phagolysosome A structure resulting from the fusion of lysosomes and a phagosome.

phagosome A vacuole that forms around a microbe within the phagocyte that engulfed it.

pharmaceutical microbiology A special branch of industrial microbiology concerned with the manufacture of products used in treating or preventing disease.

pharyngitis An infection of the pharynx, usually caused by a virus but sometimes bacterial in origin; a sore throat.

pharynx The throat, a common passageway for the respiratory and digestive systems with tubes connecting to the middle ear.

phase contrast microscopy Use of a light microscope having a condenser that accentuates small differences in the refractive index of various structures within the cell.

phenol coefficient A numerical expression for the effectiveness of a disinfectant relative to that of phenol.

phenotype The specific observable characteristics displayed by an organism.

phleboviruses Bunyaviruses that are carried by the sandfly *Phlebotomus papatsii*.

phospholipids A lipid composed of glycerol, two fatty acids, and a phosphate group, found in all cell membranes.

phosphorescence Continued emission of light by an object when light rays no longer strike it.

phosphorus cycle The cyclic movement of phosphorus between inorganic and organic forms.

phosphorylation The addition of a phosphate group to a molecule, often from ATP, generally increasing the molecule's energy.

phosphotransferase system A mechanism that uses energy from phosphoenolpyruvate to move sugar molecules into cells.

photoautotrophs Autotrophs that obtain energy from light.

photoheterotrophs Heterotrophs that obtain energy from light.

photolysis Process in which energy from excited electrons is used to split water molecules into protons, electrons, and oxygen molecules.

photosynthesis The capture of energy from light and use of this energy to manufacture carbohydrates from carbon dioxide.

physical factors Factors in the environment, such as temperature, moisture, pressure, and radiation, that influence the kinds of organisms found and their growth.

picornaviruses Small, spherical, naked Class I RNA viruses that enter cells by phagocytosis and interrupt functions of DNA and RNA.

pili Tiny hollow projections used to attach bacteria to surfaces (fimbriae) or for conjugation (sex pili).

pilin A protein that makes up sub-units of pili.

pimples See **folliculitis.**

pinkeye See **bacterial conjunctivitis.**

pinna Flaplike external structure of the ear.

pinworm infections Parasitic disease caused by a small roundworm, *Enterobius vermicularis.*

placebo An unmedicated, usually harmless substance given to a recipient as a substitute for or to test the efficacy of a medication or treatment.

plaque assay A viral assay used to determine viral yield by culturing viruses on a bacterial lawn and counting plaques, clear areas where viruses have lysed cells.

plaque-forming units Plaques counted on a bacterial lawn give only an approximate number of phages present, since a given plaque may have been due to more than one phage.

plaque A continuously formed coating of microorganisms and organic matter on tooth enamel.

plaques Clean areas in bacterial lawn cultures where viruses have lysed cells.

plasma cells Large lymphocytes that synthesize and release antibodies like those on their membranes.

plasma Liquid portion of the blood, excluding the formed elements.

plasma membrane A living, selectively permeable, membrane enclosing a cell. (Also called **cell membrane.**)

plasmid A small circular piece of DNA in a cell that is not part of its chromosome. (Also called extrachromosomal DNA.)

plasmogamy Sexual reproduction in fungi in which haploid gametes unite and their cytoplasm mingles.

plasmolysis Shrinking of a cell, with separation of the cell membrane from the cell wall, resulting from loss of water in a hypertonic solution.

platelet A short-lived fragment of large cells called megakaryocytes, important components of the blood-clotting mechanism.

pleomorphism Phenomenon in which bacteria vary widely in form, even within a single culture under optimal conditions.

pleura Serous membranes covering the surfaces of the lungs and the cavities they occupy.

pleurisy Inflammation of pleural membranes that causes painful breathing; often accompanies lobar pneumonia.

***Pneumocystis* pneumonia** A fungal respiratory disease caused by *Pneumocystis carinii.*

pneumonia An inflammation of lung tissue caused by bacteria, viruses, or fungi.

pneumonic plague Usually fatal form of plague transmitted by aerosol droplets from a coughing patient.

point mutation Mutation in which one base is substituted for another at a specific location in a gene.

polar compound A molecule with an unequal distribution of charges.

polar Having one flagellum located at one or both end of a bacterial cell.

poliomyelitis Disease caused by any of several strains of polioviruses that attack motor neurons of the spinal cord and brain.

polymorphonuclear leukocyte (PMNL or **granulocyte**) A leukocyte with granular cytoplasm and irregularly shaped, lobed nuclei.

polyomaviruses Wart and tumor-causing viruses that are usually integrated into host cell DNA.

polypeptide A chain of many amino acids.

polyribosome A long chain of ribosomes attached at different points along an mRNA molecule.

polysaccharide A carbohydrate formed when many monosaccharides are linked together by glycosidic bonds.

Pontiac fever A mild variety of legionellosis.

portal of entry A site at which microorganisms can gain access to body tissues.

portal of exit A site at which microorganisms can leave the body.

positive chemotaxis Movement of an organism toward a chemical.

positive sense nucleic acid A nucleic acid that encodes information for making proteins needed by a virus.

pour plate method Method used to prepare pure cultures using serial dilutions, each of which is mixed with melted agar and poured into a sterile petri plate.

poxviruses Class I DNA viruses that are the largest and most complex of all viruses.

precipitation reactions Reactions in which antibodies called precipitins react with antigens to form lattice-like networks of molecules that precipitate from solution.

precipitins Antibodies that react with antigens in precipitation reactions.

preserved culture A culture in which organisms are maintained in a dormant state.

presumptive test First stage of testing in multiple-tube fermentation in which gas production in lactose broth provides presumptive evidence that coliform bacteria are present.

prevalence The number of people infected with a particular disease at any one time.

primary atypical pneumonia See **mycoplasmal pneumonia.**

primary cell cultures Cultures that come directly from an animal and are not subcultured.

primary immunodeficiencies Immunodeficiencies resulting from genetic or developmental defects in which T cells or B cells are lacking or defective.

primary infection An initial infection in a previously healthy person.

primary response Humoral immune response that occurs when an antigen is first recognized by host B cells.

primary structure The specific sequence of amino acids in a polypeptide chain.

primary treatment Physical treatment to remove solid wastes from sewage.

prion An exceedingly small infectious particle alleged to consist of protein without any nucleic acid; their existence has been challenged by some investigators.

prodomal phase In a disease, the short period during which nonspecific symptoms such as malaise and headache sometimes appear.

prodome A symptom indicating the onset of a disease.

producers Organisms that capture energy from the sun and synthesize food.

productive infection Viral infection in which viruses enter a cell and produce infectious progeny.

proglottid One of the segments of a tapeworm, containing the reproductive organs.

progressive multifocal leukoencephalopathy Disease caused by the JC polyomavirus with symptoms including mental deterioration, limb paralysis, and blindness.

Prokaryotae Organisms in kingdom Monera, which lack a discrete nucleus; the bacteria and cyanobacteria.

prokaryotes Microorganisms that lack a nucleus; all bacteria are prokaryotes.

prokaryotic cells Cells that lack a nucleus; includes all bacteria.

propagated epidemic An epidemic that arises from person-to-person contacts.

properdin A protein that participates in the reactions carried out by the complement system, initiating the alternate pathway.

properdin pathway One of the two sequences of reactions by which proteins of the complement system are activated. (Also called **alternate pathway.**)

prophage Phage DNA that has entered a bacterium.

propionibacteria Pleomorphic, irregular, nonsporing, gram-positive rods.

prospective analytical epidemiological study Study in which factors that occur, as an outbreak of disease proceeds, are considered.

prostaglandins Substances that act as cellular regulators, often intensifying pain.

prostate gland Gland located at the beginning of the male urethra whose milky fluid discharge forms a component of semen.

prostatitis Inflammation of the prostate gland.

protein A polymer of amino acids joined by peptide bonds.

protein profile A method of visualizing the proteins a cell contains; obtained by the use of polyacrylamide gel electrophoresis.

protein repressor Substance produced by host cells that keeps a virus in an inactive state and prevents the infection of the cell by another phage of the same type.

protists A diverse assortment of unicellular eukaryotic organisms, members of the kingdom Protista.

protons Positively charged subatomic particles in the nucleus of an atom.

protoplasmic cylinder Coiled cylinder surrounded by the outer sheath in spirochetes.

protoplast fusion A technique of genetic engineering in which genetic material is combined by removing the cell walls of two different types of cells and allowing the resulting protoplasts to fuse.

protoplasts A cell from which the cell wall has been removed.

prototrophs Normal, nonmutant organisms. (Also called **wild types.**)

protozoa Single-celled, microscopic, eukaryotic organisms in kingdom Protista.

protozoan An animallike protist.

provirus Viral DNA that is incorporated into a host-cell chromosome.

pseudocoelom A primitive body cavity, typical of nematodes, that lacks the complete lining found in higher animals.

pseudocysts Aggregates of trypanosome parasites that form in lymph nodes in Chagas' disease.

pseudomembrane A combination of bacilli, damaged epithelial cells, fibrin, and blood cells resulting from infection with diphtheria that can block the airway, causing suffocation.

pseudomonads Aerobic motile rods with polar flagella.

pseudoplasmodium A slightly motile aggregation of slime mold cells with clearly defined membranes.

pseudopodia Temporary footlike projections of cytoplasm associated with amoeboid movement.

psittacosis See **ornithosis.**

psychrophiles Cold-loving organisms that grow best at temperatures of 15° to 20°C.

puerperal fever Disease caused by beta-hemolytic streptococci, normal vaginal and respiratory flora that can be introduced during child delivery by medical personnel. Also called **puerperal sepsis** or **childbed fever.**

puerperal sepsis See **puerperal fever.**

pulp cavity Relatively soft area of the tooth beneath the dentin.

pure culture A culture that contains only a single species of organism.

pus Fluid formed by the accumulation of dead phagocytes, the materials they have ingested, and tissue debris.

pustules See **folliculitis.**

pyelonephritis Inflammation of the kidneys.

pyoderma A pus-producing skin infection caused by staphylococci, streptococci, and corynebacteria, singly or in combination.

pyrogen A substance that acts on the hypothalamus to raise the body's "thermostat" to a higher-than-normal temperature.

Q fever Pneumonialike disease caused by *Coxiella burnetii,* a rickettsia that survives long periods outside cells and can be transmitted aerially as well as by ticks.

quarantine The separation of humans or animals from the general population when they have a communicable disease or have been exposed to one.

quaternary ammonium compounds (quats) Cationic detergents that have four organic groups attached to a nitrogen atom.

quaternary structure The three-dimensional structure of a protein molecule formed by the association of several polypeptide chains.

R factors Drug resistance plasmids.

R group A chemical group attached to the central carbon atom in an amino acid.

rabies A viral disease that affects the brain and nervous system with symptoms including hydrophobia and aerophobia; transmitted by animal bites.

rabiesvirus An RNA-containing rhabdovirus that is transmitted through animal bites.

rad A unit of radiation energy absorbed per gram of tissue.

radial immunodiffusion Serological test used to provide a quantitative measure of antigen or antibody concentration by measuring the diameter of the ring of precipitation around an antigen.

radiation Light rays, such as X-rays and ultraviolet rays, that can act as mutagens.

radioactivity The ability of certain isotopes of some elements to emit radiation.

radioimmunoassay (RIA) Technique using a radioactive antiantibody to detect very small quantities of antigens or antibodies.

radioisotopes Isotopes with unstable nuclei that tend to emit particles and radiation.

rat bite fever A disease caused by *Streptobacillus moniliformis* transmitted by bites from wild and laboratory rats.

reactants Substances that enter a reaction.

reactive arthritis of *Giardia* Joint inflammation that does not respond to antiinflammatory drugs ordinarily used to treat arthritis, a symptom of giardiasis in some patients.

reagin IgE antibodies that are responsible for anaphylaxis.

receptor sites Sites in cells that bind to specific molecules or structures.

recognition factors Structures on the tail fibers of bacteriophages that bind to receptor sites on bacterial cells.

recombinant DNA DNA that contains information from two different species.

recombination event Process in which prophages insert into the bacterial chromosome at a specific location.

recrudescent typhus A recurrence of an epidemic typhus infection caused by reactivation of latent organisms harbored in the lymph nodes. (Also called **Brill-Zinsser disease.**)

redia The development stage of the fluke immediately following the sporocyst stage.

reduction The gain of electrons.

reference culture A preserved culture used to maintain an organism with its characteristics as originally defined.

reflection The bouncing of light off an object.

refraction The bending of light as it passes from one medium to another medium of different density.

regulator genes Genes that control the expression of structural genes.

relapsing fever Disease caused by various species of *Borrelia,* most commonly by *Borrelia recurrentis;* transmitted by lice.

release The departure from the host cell of new virions, usually killing the host cell.

reoviruses Medium-sized Class IV RNA viruses that have a double capsid with no envelope; cause upper respiratory and gastrointestinal infections in humans.

replica plating A technique used to transfer colonies from one medium to another.

replication forks The site at which the two strands of the DNA double helix are separated during replication and new complementary DNA strands formed.

replication Process by which an organism or structure (especially a DNA molecule) duplicates itself.

replicon A self-replicating carrier such as a phage or a plasmid.

reservoir host An infected organism that makes parasites available for transmission to other hosts.

reservoirs of infection Sites where microorganisms can persist and maintain their ability to infect.

resident flora Species of microorganisms that are always present on or in an organism.

resistance genes (R genes) Components of resistance plasmids that confer resistance to specific antibiotics.

resistance The ability of a microorganism to remain unharmed by an antimicrobial agent.

resistance plasmids Plasmids that carry genes that provide resistance to various antibiotics. (Also called R plasmids or **R factors.**)

resistance transfer factor (RTF) Component of a resistance plasmid that implements transfer by conjugation of the plasmid.

resistant Able to ward off pathogens.

resolution The ability of an optical device to show two items as separate and discrete entities, rather than as a fuzzily overlapped image.

resolving power A numerical measure of the resolution of an optical instrument.

respiratory anaphylaxis Life-threatening allergy in which airways become constricted and filled with mucous secretions.

respiratory syncytial virus (RSV) Cause of lower respiratory infections affecting children under one year old; causes cells in cultures to lose their cell membranes and become multinucleate masses.

respiratory system Body system that moves oxygen from the atmosphere to the blood and removes carbon dioxide and other wastes from the blood.

restriction endonucleases Enzymes that cut DNA at precise base sequences.

reticulate bodies An intracellular stage in the life cycle of chlamydias.

reticuloendothelial system Collective name for macrophages.

retrofection Reinfection by larvae of pinworms that hatch and reenter the body.

retrospective analytical epidemiological study A study in which factors that preceded an outbreak of disease are considered.

retroviruses Enveloped Class V RNA viruses that use their own reverse transcriptase to transcribe their RNA into DNA in the cytoplasm of the host cell.

reverse transcriptase An enzyme found in Class V RNA viruses that copies RNA into DNA.

rhabdoviruses Rod-shaped, enveloped, Class II RNA viruses that infect insects, fish, various other animals, and some plants.

rheumatic fever A multisystem disorder following infection by beta hemolytic *Streptococcus pyogenes* that can cause heart damage.

rheumatoid arthritis Autoimmune disease that affects mainly the joints but can extend to other tissues.

rheumatoid factor IgM found in the blood of patients with rheumatoid arthritis and their relatives.

rhinoviruses Viruses that replicate in cells of the upper respiratory tract and cause the common cold.

ribonucleic acid (RNA) Nucleic acid that carries information from DNA to sites where proteins are manufactured in cells and that directs and participates in the assembly of proteins.

ribosomal RNA (rRNA) A type of RNA that, together with certain proteins, makes up the ribosomes.

ribosomes Sites for protein synthesis consisting of RNA and protein, located in the cytoplasm.

rickettsialpox Mild rickettsial disease with symptoms resembling those of chickenpox; caused by *Rickettsia akari* and carried by mites found on house mice.

rickettsias Small, nonmotile, gram-negative organisms; obligate intracellular parasites of mammalian and arthropod cells.

Rift Valley fever Disease caused by bunyaviruses that occurs in epidemics.

ringworm A highly contagious fungal skin disease that can cause ringlike lesions.

rise period Period during which the number of phages being released per host cell increases to a constant number.

river blindness See **onchocerciasis.**

RNA polymerase An enzyme that binds to one strand of exposed DNA during transcription and catalyzes the synthesis of mRNA from the DNA template.

Rocky Mountain spotted fever Disease caused by *Rickettsia rickettsii* and transmitted by a tick.

root canal Channel in the tooth where blood vessels and nerves are located.

root The part of the tooth below the gumline, covered with cementum.

rotaviruses Viruses transmitted by the fecal-oral route that replicate in the intestine, causing diarrhea and enteritis.

rubella Viral disease characterized by a skin rash; can cause severe congenital damage. (Also called **German measles.**)

rubellavirus A togavirus that causes rubella (German measles).

rubeola See **measles.**

rule of octets Principle that an element is chemically stable if it contains eight electrons in its outer shell.

sac fungi A diverse group of fungi that produce saclike asci during sexual reproduction. (Also called **Ascomycota.**)

salmonellosis A common enteritis characterized by abdominal pain, fever, and diarrhea with blood and mucus; caused by *Salmonella* species.

sapremia A condition caused by saphrophytes releasing metabolic products into the blood.

saprophytes Organisms that feed on dead organic matter.

sarcodinas Amoeboid protozoans.

sarcoptic mange See **scabies.**

saturated fatty acid A fatty acid containing only carbon-hydrogen single bonds.

scab A dried blood clot.

scabies Highly contagious skin disease caused by the itch mite *Sarcoptes scabiei*. (Also called **sarcoptic mange.**)

scalded skin syndrome Infection caused by staphylococci consisting of large, soft vesicles over the whole body.

scanning electron microscope Type of electron microscope used to study the surfaces of specimens.

scarlet fever Infection caused by *Streptococcus pyogenes* that produce an erythrogenic toxin.

Schick test A serological test based on neutralization reactions which is used to detect the presence of antitoxin or antibodies to diphtheria toxin.

schistosomiasis Disease of the blood and lymph caused by blood flukes of the genus *Schistosoma*. (Also called **bilharzia.**)

schizogony Multiple fission, in which one cell gives rise to many cells.

scolex Head end of a tapeworm, with suckers and sometimes hooks that attach to the intestinal wall.

scrub typhus A typhus caused by *Rickettsia tsutsugamushi*; transmitted by mites that feed on rats. (Also called **tsutsugamushi disease.**)

sebaceous gland Epidermal structure, associated with hair follicles, that secretes an oily substance called sebum.

sebum Oily substance secreted by the sebaceous glands.

secondary immunodeficiencies Immunodeficiencies resulting from damage to T cells or B cells after they have developed normally.

secondary infection Infection that follows a primary infection, especially in patients weakened by the primary infection.

secondary response Humoral immune response that occurs when an antigen is recognized by memory cells; more rapid and stronger than a primary response.

secondary structure The folding or coiling of a polypeptide chain into a particular pattern, such as a helix or pleated sheet.

secondary treatment Treatment of sewage by biological means to remove remaining solid wastes after primary treatment.

secretory vesicles Small membrane-bound structures that store substances in the Golgi apparatus.

selective medium A medium that encourages growth of some organisms and suppresses growth of others.

selective toxicity The ability of an antimicrobial agent to harm microbes without causing significant damage to the host.

selectively permeable Able to prevent the passage of certain specific molecules and ions while allowing others through.

self Molecules that are not recognized as antigenic or foreign by an organism.

semen The male fluid discharge at the time of ejaculation, containing sperm and various glandular and other secretions.

semiconservative replication Replication in which a new DNA double helix is synthesized from one strand of parent DNA and one strand of new DNA.

seminal vesicles Saclike structures whose secretion forms a component of semen.

semisynthetic drug An antimicrobial agent made partly by laboratory synthesis and partly by microorganisms.

sensitization Initial exposure to an antigen, which causes the host to mount an immune response against it.

septic shock A life-threatening septicemia with low blood pressure and blood-vessel collapse, caused by endotoxins.

septicemia An infection caused by rapid multiplication of pathogens in the blood; formerly called **blood poisoning.**

septicemic plague Fatal form of plague that occurs when bubonic plague bacteria move from the lymphatics to the circulatory system.

septum (pl: **septa**) A crosswall separating two fungal cells.

serial dilutions Method of measurement of bacterial growth in which successive 1:10 dilutions are made of the original liquid culture and grown on agar plates.

serology Branch of immunology dealing with laboratory tests to detect antigens and antibodies.

serotypes Strains of an organism.

serous membranes Membranes covering the surfaces of viscera and the cavities that contain them and secreting a watery fluid that lubricates them.

serovars Strains; a subspecies category.

serum The liquid part of blood after cells and clotting factors have been removed.

serum hepatitis Former name for **hepatitis B.**

serum killing power Test used to determine effectiveness of an antimicrobial agent in which a bacterial suspension is added to the serum of a patient who is receiving an antibiotic and incubated.

serum sickness Immune complex disorder that occurs when foreign antigens in sera cause immune complexes to be deposited in tissues.

Severe combined immunodeficiency disease (SCID) Primary immunodeficiency disease caused by failure of stem cells to develop properly, resulting in deficiency of both B and T cells.

sewage Used water and the wastes it contains.

shadow casting The coating of electron microscopy specimens with a heavy metal, such as gold or palladium, to create a three-dimensional effect.

Shaeffer-Fulton spore staining A differential stain used to make spores easier to visualize.

shigellosis Gastrointestinal disease caused by several strains of *Shigella* that invade intestinal lining cells. (Also called **bacillary dysentery.**)

shinbone fever See **trench fever.**

shingles Sporadic disease caused by reactivation of varicella-zoster herpesvirus that appears most frequently in older and immunocompromised individuals.

sign A characteristic of disease that can be observed by examining the patient, such as swelling or redness.

simple diffusion The net movement of particles from a region of higher to one of lower concentration; does not require energy from a cell.

simple stain A single dye used to reveal basic cell shapes and arrangements.

sinuses Large passageways in tissues, lined with phagocytic cells.

sinusitis An infection of the sinus cavities.

sinusoids Enlarged capillaries.

slime layer A thin protective structure loosely bound to the cell wall that protects the cell against drying, helps trap nutrients, and sometimes binds cells together.

slime molds Funguslike protists having both animal and plant characteristics.

slow-reacting substance of anaphylaxis (SRS-A) A mediator of allergic reactions that produces slow, long-lasting airway constriction in animals; consists of three leukotrienes.

sludge digesters Large fermentation tanks in which sludge is digested by anaerobic bacteria into simple organic molecules, carbon dioxide, and methane gas.

sludge Solid matter remaining from water treatment that contains aerobic organisms that digest organic matter.

small intestine The upper area of the intestine where digestion is completed.

smallpox A formerly worldwide and serious disease that has now been eradicated.

smears A thin layer of liquid specimen spread out on a microscope slide.

sodoku See **spirillar fever.**

solute The substance dissolved in a solvent to form a solution.

solution A mixture of two or more substances in which the molecules are evenly distributed and will not separate out upon standing.

solvent The medium in which substances are dissolved to form a solution.

sonication The disruption of cells by sound waves.

specialized transduction Type of transduction in which the bacterial DNA transduced is limited to one or a few genes lying adjacent to a prophage that are accidentally included when the prophage is excised from the bacterial chromosome.

species A group of organisms with many common characteristics; the narrowest taxon.

species immunity Innate immunity that is common to all members of a species.

specific defense Defense that operates in response to a particular invading pathogen.

specific epithet The second name of an organism in binomial nomenclature, following that of the genus, e.g., *coli* in *E. coli*.

specific immunity The ability of an organism to mount a specific defense against a particular infectious agent.

specificity (1) The property of an enzyme that allows it to accept only certain substrates and catalyze only one particular reaction. (2) The property of a virus that restricts it to certain specific types of host cells. (3) The ability of the immune system to mount a unique immune response to each antigen it encounters.

spheroplasts Gram-negative bacteria that have both a cell membrane and most of the outer membrane, but have had the cell wall removed.

spindle apparatus A system of tiny fibers in the cytoplasm that guide the movement of chromosomes during mitosis.

spirillar fever A form of rat bite fever, caused by *Spirillum minor*, first described as **sodoku** in Japan.

spirillum (pl: **spirilli**) A corkscrew-shaped bacterium.

spirochete A flexible, wavy-shaped bacterium.

spleen The largest lymphatic organ; acts as a blood filter.

spontaneous generation Theory that living organisms can arise from nonliving things.

spontanteous mutations Mutations that occur in the absence of any agent known to cause changes in DNA; usually caused by errors during replication.

sporadic disease A disease that is limited to a small number of isolated cases posing no great threat to a large population.

spore coat A keratinlike protein material that is laid down around the cortex of an endospore by the mother cell.

spore A resistant reproductive structure formed by fungi and actinomycetes; different from bacterial endospores.

sporocyst Larval form of fluke that develops in the body of its snail or mollusk host.

sporotrichosis Fungal skin disease caused by *sporothrix shenckii* that often enters the body from plants.

sporozoans Parasitic, immobile protozoans that generally have complex life cycles.

sporozoite A malaria trophozoite present in the salivary glands of infected mosquitoes.

sporulation The formation of endospores.

St. Anthony's fire See **erysipelas.**

St. Louis encephalitis Type of viral encephalitis most often seen in humans in the central United States.

stain A molecule that can bind to a structure and give it color. (Also called **dye.**)

stationary phase The third of four major phases of the bacterial growth curve in which new cells are produced at the same rate that old cells die, leaving the number of live cells constant.

sterility The state in which there are no living organisms in or on a material.

sterilization The killing or removal of all microorganisms in a material or on an object.

steroids Lipids having a four-ring structure, including cholesterol, steroid hormones, and vitamin D.

stigma An eyespot of a euglenoid.

stock culture A reserve culture used to store an isolated organism in pure condition for use in the laboratory.

strain A subgroup of a species with one or more characteristics that distinguish it from other subgroups of that species.

streak plate method Method used to prepare pure cultures in which bacteria are lightly spread over the surface of agar plates, resulting in isolated colonies.

streptococci Aerotolerant anaerobes that form pairs, tetrads, or chains by dividing in one or two planes; most lack the enzyme catalase.

streptokinase Enzyme produced by streptococci that digests fibrin and thereby dissolves blood clots. (Also called **fibrinolysin.**)

streptolysin Toxin produced by streptococci that kills phagocytes.

streptomycetes Gram-positive, filamentous, sporing, soul-dwelling bacteria, producers of many antibiotics.

stridor High-pitched, noisy respiration, a symptom of parainfluenza and similar infections.

stroma Fluid-filled inner portion of the chloroplast.

stromatolites Fossilized mats of prokaryotes.

strongyloidiasis Parasitic disease caused by the roundworm *Strongyloides stercoralis* and a few closely related species.

structural genes Genes that carry information for the synthesis of specific proteins.

structural proteins Proteins which contribute to the structure of cells, cell parts, and membranes.

sty An infection at the base of an eyelash.

subacute disease A disease that is intermediate between an acute and a chronic disease.

subacute sclerosing panencephalitis (SSPE) A complication of measles, nearly always fatal, that is due to the persistance of measles viruses in brain tissue.

subclinical infection An infection that fails to produce symptoms, either because too few organisms are present or because host defenses effectively combat the pathogens. (Also called **inapparent infection.**)

subculturing The process by which cells from an existing culture are transferred to fresh medium in new containers.

substrate The substance upon which an enzyme acts.

sulfate reduction The reduction of sulfate ions to hydrogen sulfide.

sulfur cycle The cyclic movement of sulfur through an ecosystem.

sulfur oxidation The oxidation of various forms of sulfur to sulfate.

sulfur reduction The reduction of elemental sulfur to hydrogen sulfide.

superinfection Invasion of digestive, respiratory, or urinary tracts by resistant replacement flora when normal flora are disturbed.

superoxide dismutase An enzyme that converts superoxide to molecular oxygen and hydrogen peroxide.

superoxide A highly reactive form of oxygen that kills obligate anaerobes.

supressor T cells (T_S) Lymphocytes that inhibit immune response.

surface tension A phenomenon in which the surface of water behaves like a thin, invisible, elastic membrane.

surfactant A substance that reduces surface tension.

susceptibility The vulnerability of an organism to harm by infectious agents.

susceptible Unable to ward off pathogens.

swarmer cells Spherical, flagellated *Rhizobium* cells that invade the root hairs of leguminous plants, eventually to form nodules.

sweat gland Epidermal structure that empties a watery secretion through pores in the skin.

swimmer's itch Skin reaction to cercariae of some species of the helminth *Schistosoma*.

symbiosis Two different kinds of organisms living together.

symptom A characteristic of a disease that can be observed or felt only by the patient, such as pain or nausea.

synchronous growth Hypothetical pattern of growth during the log phase in which all the cells in a culture divide at the same time.

synctia Multinucleate masses in cell cultures, e.g., caused by the respiratory syncytial virus.

syndrome A combination of signs and symptoms that occur together.

synergistic effect An inhibitory effect produced by two antibiotics working together that is greater than either can achieve alone.

synthesis The making of nucleic acid, coat proteins, and other viral components using the host cell's synthetic machinery in the replication process.

synthetic drug An antimicrobial agent synthesized chemically in the laboratory.

synthetic medium A medium prepared in the laboratory from materials of precise or reasonably well-defined composition.

syphilis A sexually transmitted disease, caused by the spirochete *Treponema pallidum*, characterized by a chancre at the site of entry and often eventual neurological damage.

systemic blastomycosis Disease resulting from invasion by *Blastomyces dermatitidis* of internal organs, especially the lungs.

systemic infection An infection that affects the entire body. (Also called **generalized infection.**)

systemic lupus erythematosus A widely disseminated, systemic autoimmune disease resulting from production of antibodies against DNA and other body components.

T cell See **T lymphocyte.**

T lymphocyte Thymus-derived cell of the immune system agent of cellular responses. (Also called **T cell.**)

tapeworms Flatworms that live in the adult stage as parasites in the small intestine of animals.

taxonomy The science of classification.

teichoic acids Polymers attached to peptidoglycan in gram-positive cell walls.

temperate phage A phage that ordinarily does not cause a virulent infection but rather is incorporated into a bacterium and replicated with it.

template DNA used as a pattern for the synthesis of a new nucleotide polymer in replication or transcription.

temporary parasite A parasite that feeds on and then leaves its host (such as a biting insect).

teratogen An agent that induces defects during embryonic development.

teratogenesis The induction of defects during embryonic development.

terminator codon A codon that signals the end of the information for a particular protein.

tertiary structure The folding of a protein molecule into globular shapes or fibrous threadlike strands.

tertiary treatment Chemical and physical treatment of sewage to produce an effluent of water pure enough to drink.

testes (sing: **testis**) Paired male reproductive glands that produce testosterone and sperm.

tetanus Disease caused by *Clostridium tetani* in which muscle stiffness progresses to eventual paralysis and death. (Also called "lockjaw.")

tetanus neonatorum Type of tetanus acquired through the raw stump of the umbilical cord.

thallus The body of a fungus.

theca A tightly affixed, secreted outer layer of dinoflagellates that often contains cellulose.

therapeutic dosage level Level of drug dosage that successfully eliminates a pathogenic organism if maintained over a period of time.

thermal death point The temperature that kills all the bacteria in a 24-hour-old broth culture at neutral pH in 10 minutes.

thermal death time The time required to kill all the bacteria in a particular culture at a specified temperature.

thermophiles Heat-loving organisms that grow best at temperatures from 50° to 60°C.

thrush Milky patches of inflammation on oral mucuous membranes; a symptom of candidiasis, caused by *Candida albicans*.

thylakoids Folds of the inner membrane of chloroplasts that contain chlorophyll.

thymopoietin I and **II** Two hormones produced by the thymus that are believed to stimulate the transformation of lymphocytes into T cells.

thymosin Hormone produced by the thymus gland.

thymus gland Multilobed lymphatic organ located beneath the sternum that processes lymphocytes into T cells.

tincture An alcoholic solution.

tinea barbae Barber's itch; a type of ringworm that causes lesions in the beard.

tinea capitis Scalp ringworm, a form of ringworm in which hyphae grow in hair follicles, often leaving circular patterns of baldness.

tinea corporis Body ringworm, a form of ringworm that causes ringlike lesions with a central scaly area.

tinea cruris Groin ringworm, a form of ringworm that occurs in skin folds in the pubic region. (Also called jock itch.)

tinea pedis See **athlete's foot.**

tinea unguium A form of ringworm that causes hardening and discoloration of fingernails and toenails.

tissue culture Culture made from a single tissue, assuring a reasonably homogenous set of cultures in which to test the effects of a virus, or to culture an organism.

titer The quantity of a substance needed to produce a given reaction.

togaviruses Small, enveloped Class I RNA viruses that multiply in many mammalian and arthropod cells.

tolerance A state in which antigens no longer elicit immune response.

tonicity The osmotic pressure of a solution relative to that of the interior of a cell.

total magnification Obtained by multiplying the magnifying power of the objective lens by the magnifying power of the ocular lens.

toxemia The presence of pathogen-released toxins in the blood.

toxic shock syndrome Condition caused by infection with certain toxigenic strains of *Staphylococcus aureus*; often associated with superabsorbent but abrasive tampons.

toxin A poisonous substance.

toxoid An exotoxin inactivated by chemical treatment which retains its antigenicity and therefore can be used to vaccinate against the toxin.

toxoplasmosis Disease caused by the protozoan *Toxoplasma gondii* that can cause congenital defects in newborns.

trachea The windpipe.

trachoma Eye disease caused by *Chlamydia trachomatis* that can result in blindness.

tract An elongated tube.

transcriptase An enzyme found in viruses with negative sense nucleic acids, which it uses as a template to make a complementary positive sense nucleic acid.

transcription The synthesis of mRNA from a DNA template.

transduction The transfer of genetic material from one bacterium to another by a bacteriophage.

transfer RNA (tRNA) Type of RNA that transfers amino acids from the cytoplasm to the ribosomes for placement in a protein molecule.

transformation A change in an organism's characteristics through the transfer of naked DNA.

transfusion reaction Reaction that occurs when matching antigens and antibodies are present in the blood at the same time.

transient flora Microorganisms that may be present in or on an organism under certain conditions and for certain lengths of time.

translation The synthesis of protein from information in mRNA.

transmission by vehicles Transmission of disease through water, air, food, drugs, blood, or other body fluids.

transmission electron microscope Type of electron microscope used to study internal structures of microbes; very thin slices of specimens are used.

transmission The passage of light through an object.

transovarian transmission Passing of a pathogen from one generation of ticks to the next as eggs leave the ovaries.

transplant rejection Destruction of grafted tissue or a transplanted organ by the host immune system.

transplantation The moving of tissue from one site to another.

transposal of virulence A laboratory technique in which a pathogen is passed from its normal host sequentially through many individual members of a new host species, resulting in a lessening or even total loss of its virulence for the original host.

traumatic herpes Type of herpes infection in which the virus enters traumatized skin in the area of a burn or other injury.

traveler's diarrhea Gastrointestinal disorder generally caused by pathogenic strains of *Escherichia coli*.

trench fever Rickettsial disease, caused by *Rochalimaea quintana*, resembling epidemic typhus in that it is transmitted by lice and is prevalent during wars and under unsanitary conditions. (Also called **shinbone fever.**)

treponemes Spirochetes belonging to the genus *Treponema*.

triacylglycerol A molecule formed from three fatty acids bonded to glycerol.

tricarboxylic acid cycle See **Krebs cycle.**

trichinosis A disease caused by a small nematode, *Trichinella spiralis*, that enters the digestive tract as encysted larvae in poorly cooked meat.

trichocyst Tentacle-like structure on ciliates for catching prey or for attachment.

trichomoniasis A parasitic urogenital disease, transmitted primarily by sexual intercourse, that causes intense itching and a copious white discharge, especially in females.

trichuriasis Parasitic disease caused by the whipworm, *Trichuris trichiura*, that damages intestinal mucosa and causes chronic bleeding.

trickling filter system Procedure in which sewage is spread over a bed of rocks coated with aerobic organisms that decompose the organic matter in it.

trophozoite Vegetative form of a protozoan such as *Plasmodium*.

true slime mold Funguslike protist consisting of a multinucleate amoeboid mass, or plasmodium, that moves about slowly and phagocytizes dead matter.

trypanosomiasis See **African sleeping sickness.**

tsutsugamushi disease A typhus caused by *Rickettsia tsutsugamushi*. (Also called **scrub typhus.**)

tube agglutination test Serological test that measures antibody titers by comparing various dilutions of the patient's serum against known quantities of an antigen.

tubercles Solidified lesions or chronic granulomas that form in the lungs in patients with tuberculosis.

tuberculin A waxy substance from the *Mycobacterium* cell wall that is injected intracutaneously to test for tuberculosis.

tuberculin hypersensitivity Cell-mediated hypersensitivity reaction that occurs in sensitized individuals when they are exposed to tuberculin.

tuberculoid leprosy The anesthetic form of Hansen's disease in which areas of skin lose pigment and sensation.

tuberculosis Disease caused mainly by *Mycobacterium tuberculosis*.

tubulin Protein that makes up fibers found in the flagella of eukaryotes.

tularemia Zoonosis caused by *Francisella tularensis*, most often associated with cottontail rabbits.

turbidity A cloudy appearance in a culture tube indicating the presence of organisms.

tympanic membrane Membrane separating the outer and middle ear. (Also called the eardrum.)

type strain Original reference strain of a bacterial species, descendants of a single isolation in pure culture.

typhoid fever An epidemic enteric infection caused by *Salmonella typhi*; uncommon in areas with good sanitation.

typhoidal tularemia Septicemia that resembles typhoid fever, caused by bacteremia from tularemia lesions.

typhus Rickettsial disease that occurs in a variety of forms including epidemic, endemic (murine), and scrub typhus.

ulceroglandular tularemia Form of tularemia caused by entry of *Francisella tularensis* through the skin; characterized by ulcers on the skin and enlarged regional lymph nodes.

uncoating Process in which protein coats of animal viruses which have entered cells are removed by proteolytic enzymes.

undulant fever See **brucellosis.**

unsaturated fatty acid A fatty acid that contains at least one double bond between adjacent carbon atoms.

upper respiratory tract The nasal cavity, pharynx, larynx, trachea, bronchi, and larger bronchioles.

ureter Tube that carries urine from the kidney to the urinary bladder.

urethra Tube through which urine passes from the bladder to the outside during micturition (urination).

urethritis Inflammation of the urethra.

urethrocystitis Common term used to describe urinary tract infections involving the urethra and the bladder.

urinalysis The laboratory analysis of urine specimens.

urinary bladder Storage area for urine.

urinary system Body system that regulates the composition of body fluids and removes nitrogenous and other wastes from the body.

urinary tract infections (UTIs) Bacterial urogenital infections that cause urethritis and cystitis.

urogenital system (or **genitourinary system**) Body system that (1) regulates the composition of body fluids and removes certain wastes from the body and (2) enables the body to participate in sexual reproduction.

use-dilution test Method of evaluating the antimicrobial properties of a chemical agent using standard preparations of certain test bacteria.

uterine tubes Tubes that convey ova from the ovaries to the uterus. (Also called Fallopian tubes or oviducts.)

uterus The pear-shaped organ in which a fertilized ovum implants and develops.

vaccine A substance that contains an antigen to which the immune system responds.

vacuolating virus A simian virus, SV-40, which has been used to explore the mechanisms of viral replication, integration, and oncogenesis.

vacuoles Membrane-bound structures that store materials such as food or gas in the cytoplasm of eukaryotic cells.

vagina The female genital canal, extending from the cervix to the outside of the body.

vaginitis Vaginal infection, often caused by opportunistic organisms that multiply when the normal vaginal flora are disturbed by antibiotics or other factors.

varicella-zoster virus A herpesvirus that causes both chickenpox and shingles.

vector control A means of controlling disease in which the disease vector is discovered and destroyed or treated.

vector An organism that transmits a disease-causing organism from one host to another.

vegetation A growth that forms on damaged heart valve surfaces in bacterial endocarditis; exposed collagen fibers elicit fibrin deposits, and transient bacteria attach to the fibrin.

vegetative cells Cells that are actively metabolizing nutrients.

vein A blood vessel that receives blood from venules and returns it to the heart.

Venezuelan equine encephalitis Type of viral encephalitis seen in Florida, Texas, Mexico, and South America; infects horses more frequently than humans.

ventricle One of the two large, lower chambers of the heart.

venule A blood vessel that receives blood from a capillary.

verminous intoxication An allergic reaction to toxins in the metabolic wastes of liver flukes.

verruga peruana One form of bartonellosis; a chronic nonfatal skin disease.

vertical transmission Direct contact transmission of disease in which pathogens are passed from parent to offspring in an egg or sperm, across the placenta, or while traversing the birth canal.

vesicles In prokaryotes, inclusions bounded by a single-layered membrane.

vibrio A comma-shaped bacterium.

vibriosis An enteritis caused by *Vibrio parahaemolyticus*, acquired from eating contaminated fish and shellfish that have not been thoroughly cooked.

villus (pl. **villi**) A multicellular projection from the surface of a mucous membrane, functioning for absorption.

viral enteritis Gastrointestinal disease caused by rotaviruses, characterized by diarrhea.

viral hemagglutination Hemagglutination caused by binding of viruses, such as those that cause measles and influenza, to red blood cells.

viral pneumonia Disease caused by viruses such as respiratory syncytial virus.

viral yield The average number of phage particles released per infected cell.

viremia An infection in which viruses are transported in the blood but do not multiply in transit.

viridans group A group of streptococci that often infect the valves and lining of the heart and cause incomplete (alpha) hemolysis of red blood cells in laboratory cultures.

virion A complete virus particle, including its envelope if it has one.

viroids Infectious RNA particles smaller than viruses and lacking capsids that cause various plant diseases.

viropexis Process similar to phagocytosis in which certain naked animal viruses are taken into the cell.

virulence The degree of intensity of the disease produced by a pathogen.

virulent phage A phage that enters the lytic cycle when it infects a cell, causing eventual death of the cell.

viruses Submicroscopic, parasitic, acellular entities composed of a nucleic acid core inside a protein coat.

visceral larva migrans The migration of larvae of *Toxocara* species in human tissues, where they cause damage and allergic reactions.

volutin Polyphosphate granules. (Also called **metachromatic granules.**)

walking pneumonia See **mycoplasmal pneumonia.**

wandering macrophages Phagocytic cells that circulate in the blood or move into tissues when microbes and other foreign material are present.

warts Growths on the skin and mucous membranes caused by infection with papillomavirus. (Also called **papillomas.**)

water cycle Process by which water is recycled through precipitation, ingestion by organisms, respiration, and evaporation. (Also called the **hydrologic cycle.**)

water molds Group of fungi that cause mildew and are strikingly different from other fungi. (Also called **Oomycota.**)

wavelength The distance between successive crests or troughs of a light wave.

western equine encephalitis Type of viral encephalitis seen most often in the western United States; infects horses more frequently than humans.

wet mounts Microscopy technique in which a drop of fluid containing the organisms (often living) is placed on a slide.

wetting agent A detergent solution often used with other chemical agents to penetrate fatty substances.

whipworm *Trichuris trichiura*, a worm that causes trichuriasis infestation of the intestine.

whitlow A herpetic lesion on a finger that can result from exposure to oral, ocular, and probably genital herpes.

whooping cough A highly contagious respiratory disease, caused primarily by *Bordetella pertussis*, that is prevalent among children in developing countries. (Also called **pertussis.**)

wild types See **prototrophs.**

Woronin body Organelle found in certain fungi, which blocks a septal pore, preventing materials from a damaged cell from entering a healthy cell.

wound botulism Rare form of botulism that occurs in deep wounds when tissue damage impairs circulation and creates anaerobic conditions in which endospores can multiply.

xenodiagnosis A technique for diagnosing disease in which small animals are inoculated with blood from patients and then observed for symptoms.

xenograft A graft between individuals of different species.

yeast extract Substance from yeast containing vitamins, coenzymes, and nucleosides; used to enrich media.

yellow fever Viral systemic disease found in tropical areas, carried by the mosquito *Aedes aegypti*.

yersiniosis Severe enteritis caused by *Yersinia enterocolitica*.

Ziehl-Neelsen acid-fast stain A differential stain for organisms that are not decolorized by acid in alcohol, such as the bacteria that cause leprosy and tuberculosis.

zones of inhibition Clear areas that appear on agar in the disk diffusion method, indicating where the agents have inhibited growth of the organism.

zoonoses (sing: **zoonosis**) Diseases that can be transmitted from animals to humans.

zygomycosis Disease in which certain zygomycetes of the genera *Mucor* and *Rhizopus* invade lungs, the central nervous system, and tissues of the eye orbit.

Zygomycota See **bread molds.**

zygospore In bread molds, a thick-walled, resistant, spore-producing structure enclosing a zygote.

zygote A cell formed by the union of gametes (egg and sperm).

Index

Photo Credits

Chapter 1

Chapter Opening 1 National Library of Medicine
1.1 Dr. Jacquelyn G. Black
1.2(a) CNRI/Science Photo Library/Science Source/Photo Researchers
1.2(b) R. B. Taylor/Science Photo Library/Science Source/Photo Researchers
1.2(c) CBC/Phototake NYC
1.2(d) Thomas Broker/CNRI/Phototake
1.2(e) M. Abbey/Visuals Unlimited
1.2(f) Cabisco/Visuals Unlimited
1.3(a) U.S. Department of Agriculture
1.3(b) U.S. Department of Agriculture
1.3(c) U.S. Department of Agriculture
1.3(d) U.S. Department of Agriculture
1.3(e) U.S. Department of Agriculture
1.3(f) U.S. Department of Agriculture
1.4 Scala/Art Resource
Box, page 9 Marcel Miranda
Box, page 9 Courtesy Coulter Corporation
1.5 The Bettmann Archive
1.7 Institut Pasteur, Paris
1.8 The Granger Collection
1.9 National Library of Medicine
1.10(a) The Granger Collection
1.10(b) National Library of Medicine
1.11 The Granger Collection
1.12 New York Public Library Picture Collection
1.13(a) Omikron/Science Source/Photo Researchers
1.14 National Library of Medicine
1.15 National Library of Medicine
1.16 U.S. Department of Agriculture
1.17 Courtesy of John Folds

Chapter 2

Chapter Opening 2 Digital Instruments Inc.
2.6(b) Biophoto Associates/Photo Researchers
2.23 Susan Perry/Woodfin Camp & Associates
2.24 Scala/Art Resource

Chapter 3

Chapter Opening 3 The Cleveland Museum of Art, Purchase from the J. H. Wade Fund, 74.15
3.1(b) John D. Cunningham/Visuals Unlimited
3.8(c) Runk/Schoenberger from Grant Heilman
3.9 E. R. Degginger/Earth Scenes
Box, page 57 Dr. Jacquelyn G. Black
3.12(a) Courtesy Carl Zeiss, Oberkochen, Germany
3.14(a) Jim Solliday/Biological Photo Service
3.14(b) Jim Solliday/Biological Photo Service
3.15 Biophoto Associates/Authenticolor/Photo Researchers
3.16 Centers for Disease Control
3.17 Paul W. Johnson/Biological Photo Service
3.18(a) David M. Phillips/Visuals Unlimited
3.18(b) David M. Phillips/Visuals Unlimited
3.18(c) David M. Phillips/Visuals Unlimited
3.18(d) David M. Phillips/Visuals Unlimited
3.19(b) SIU Biomedical Communications/Science Source/Photo Researchers
3.20(a) Eric V. Grave/Photo Researchers, Inc.
3.20(b) Biophoto Associates/Photo Researchers, Inc.
3.21 J. J. Cardamone, Jr. and B. A. Phillips, University of Pittsburgh/BPS
3.22(a) David M. Phillips/Visuals Unlimited
3.22(b) David M. Phillips/Visuals Unlimited
3.24 G. T. Cole, University of Texas, Austin/BPS
3.25(a) David M. Phillips/Visuals Unlimited
3.25(b) David M. Phillips/Visuals Unlimited
3.25(c) Manfred Kage/Peter Arnold, Inc.
3.25(d) Veronika Burmeister/Visuals Unlimited
3.26(b) A. M. Siegelman/Visuals Unlimited
3.27 John D. Cunningham/Visuals Unlimited
3.28(a) Raymond B. Otero/Visuals Unlimited
3.28(b) Raymond B. Otero/Visuals Unlimited
3.28(c) Raymond B. Otero/Visuals Unlimited
3.28(d) Raymond B. Otero/Visuals Unlimited
3.29 Jack M. Bostrack/Visuals Unlimited
3.30 Raymond B. Otero/Visuals Unlimited
3.31 Courtesy IBM, Thomas J. Watson Research Center

Chapter 4

Chapter Opening 4 CNRI/Science Photo Library/Photo Researchers, Inc.
4.2(a) David M. Phillips/Visuals Unlimited
4.2(b) CNRI/Science Photo Library/Photo Researchers
4.2(c) Dr. Tony Brain/Science Photo Library/Science Source/Photo Researchers
4.2(d) Cabisco/Visuals Unlimited

4.6(a) S. C. Holt, University of Texas Health Science Center, San Antonio/BPS
4.6(b) T. J. Beveridge, University of Guelph/BPS
4.6(c) T. J. Beveridge and T. Paul/Visuals Unlimited
4.8(b) Jouan, Inc.
4.9 Paul W. Johnson/Biological Photo Service
Box, page 85 D. Blackwell, D. Maratea/Scientific American
Box, page 85 R. Blackemore/Scientific American
4.10 H. S. Pankratz, T. C. Beaman, and P. Gerhardt, Michigan State University/BPS
4.11(a) E. C. S. Chan/Visuals Unlimited
4.11(b) Jack M. Bostrack/Visuals Unlimited
4.11(c) E. C. S. Chan/Visuals Unlimited
4.11(d) John D. Cunningham/Visuals Unlimited
4.11(e) Fred Hossler/Visuals Unlimited
4.12(b) J. Adler/Visuals Unlimited
4.14(a) CNRI/Science Photo Library/Photo Researchers, Inc.
4.14(b} Max A. Listgarten/*J. Bacteriol:* Vol. 88
4.15 Charles C. Brinton, Jr. and Judith Cernehan
4.16 B. Berg, B. V. Hofstein, G. Petterson
4.18(a) J. R. Factor/Photo Researchers, Inc.
4.18(b) Don W. Fawcett/Photo Researchers, Inc.
4.20(b) N. G. Flickinger
4.21(b) Dr. Kenneth R. Miller/Science Photo Library/Photo Researchers, Inc.
4.23(a) K. G. Murti/Visuals Unlimited
4.24(b-1) M. Abbey/Visuals Unlimited
4.24(b-2) M. Abbey/Visuals Unlimited
4.24(b-3) M. Abbey/Visuals Unlimited
4.31 Micrograph by David Chase from *Early Life* by Lynn Margulis. Copyright © 1984 by Jones and Bartlett Publishers Inc., Fig. 4–8, pg. 90.
4.32 Micrograph by E. W. Daniels from *Biology of the Amoeba,* edited by Jeon. Copyright 1973 by Academic Press, Fig. 2, pg. 128.

Chapter 5

Chapter Opening 5 U.S. Department of Agriculture
5.5(b) Tripos Associates, St. Louis SYBYL molecular modeling systems
5.14 Dr. Jacquelyn G. Black
Box, page 126 Exxon Company, U.S.A.
Box, page 126 Exxon Company, U.S.A.
Box, page 126 Exxon Company, U.S.A.
5.27 John D. Cunningham/Visuals Unlimited

Chapter 6

Chapter Opening 6 cytoGraphics Inc./Visuals Unlimited
6.1(b) George Musil/Visuals Unlimited
6.1(c) David M. Phillips/Visuals Unlimited
6.4 Courtesy Wheaton Science Products, Inc.
6.6(a) Science VU/SIM/Visuals Unlimited
6.6(b) Barbara J. Miller/Biological Photo Service
6.9 Dr. Jacquelyn G. Black
6.10 Courtesy Fisher Scientific, Educational Materials Division
6.11 Stephen Trimble/DRK Photo
Box, page 147 Kevin Schafer/Tom Stack & Associates
6.14(b) Barry Rokeach/The Image Bank
Box, page 149 U.S. Department of Agriculture
Box, page 149 Jack M. Bostrack/Visuals Unlimited
6.16(a) Soad Tabapchali/Visuals Unlimited
6.16(b) Alfred Pasieka/SPL/Science Source/Photo Researchers, Inc.
6.17(b) Raymond B. Otero/Visuals Unlimited
Box, page 156 Courtesy Becton Dickinson Microbiology Systems
6.19 Jack M. Bostrack/Visuals Unlimited
6.20 Forma Scientific
6.21(a) Roche Diagnostic Systems/Hoffmann–La Roche Inc.
6.21(b) Christine L. Case/Visuals Unlimited
6.22 Courtesy API

Chapter 7

Chapter Opening 7 Science Photo Library/Science Source/Photo Researchers, Inc.
7.1 Richard Feldman/NIH
7.5(b) NIH/Kakefuda/Science Source/Photo Researchers, Inc.
Box, page 168 Micrograph by Dr. John Cairns from Cold Spring Harbor Symposium (Volume XXVIII), 1963, Fig. 2, pg. 44.
7.9(b) Tripos Associates, St. Louis, Missouri
7.10(a) © O. L. Miller, B. R. Beatty, D. W. Fawcett/Visuals Unlimited
7.10(b) Barbara Hamkalo, *Int. Rev. Cytology* 33(1972), 7.
7.20 Ken Greer/Visuals Unlimited
7.23 Dr. Bruce N. Ames, Div. of Biochemistry and Molecular Biology, University of California, Berkeley
7.24 Reuters/Bettmann
7.25(a) Licking County Archeological and Landmarks Society. Courtesy, Dr. Gerald Goldstein
7.25(b) Licking County Archeological and Landmarks Society. Courtesy, Dr. Gerald Goldstein

Chapter 8

Chapter Opening 8 Hank Morgan/Science Source/Photo Researchers, Inc.
8.7 Prof. L. Caro/Photo Researchers, Inc.
8.11(a) K. G. Murti/Visuals Unlimited
8.13 Dr. Jeremy Burgess/Science Photo Library/Science Source/Photo Researchers, Inc.
Box, page 212 © Mark Lyons
8.15(a) J. R. Adams/Science VU/Visuals Unlimited
8.15(b) U.S. Department of Agriculture
Notebook, page 214 Cesare Bonazza; courtesy, Dr. Trevor Suslow, DNAP
Notebook, page 215 Courtesy, Dr. Trevor Suslow, DNAP
8.16 Dr. Jeremy Burgess/Science Photo Library/Photo Researchers, Inc.

Chapter 9

Chapter Opening 9 Robert Brons/Biological Photo Service
9.1 The Bettmann Archive
Box, page 225 ATCC/Complete Photographic Services
9.6 Dudley Foster/Woods Hole Oceanographic Institution
9.11(a) Kevin Schafer/Tom Stack & Associates
9.11(b) Doug Sokell/Visuals Unlimited
9.11(c) S. W. Schopf/UCLA
9.13(a) Applied Biosystems
9.13(b) Applied Biosystems
9.15 Amgen
Box, page 237 Courtesy Biolog, Inc.
9.16 Dr. Edward J. Bottone/Mt. Sinai Hospital

Chapter 10

Chapter Opening 10 Ralph Robinson/Visuals Unlimited
10.1 Photography Collection, ASM Archives, Center for the History of Microbiology, University of Maryland, Baltimore County

10.2 © 1984, the Williams & Wilkins Co., Baltimore
10.3 CNRI/Science Photo Library/Photo Researchers, Inc.
10.4 Heather Davies/Science Photo Library/Science Source/Photo Researchers, Inc.
10.5 David M. Phillips/Visuals Unlimited
10.6 C. P. Vance/Visuals Unlimited
10.8 Dr. Edward J. Bottone/Mt. Sinai Hospital
10.9 Christine Case/Visuals Unlimited
10.10 Dr. Jacquelyn G. Black
10.11(a) H. Pol/CNRI/Science Photo Library/Photo Researchers, Inc.
10.11(b) David M. Phillips/Visuals Unlimited
10.13 Michael Gabridge/Visuals Unlimited
10.14 T. J. Beveridge, University of Guelph/BPS
Box, page 256 Tim McCabe/U.S. Department of Agriculture
10.15 John Durham/Science Photo Library/Photo Researchers, Inc.
10.16 Science VU/Visuals Unlimited
10.17 F. Widdel/Visuals Unlimited
10.18 Science VU-CBS-NIH/Visuals Unlimited
10.19 J. Poindexter/Science VU/Visuals Unlimited
10.20 Frederick Grassle/Woods Hole Oceanographic Institution
10.21 Centers for Disease Control

Chapter 11

Chapter Opening 11 Yoav-Simon/Phototake NYC
Box, page 268 Jack Dykinga/U.S. Department of Agriculture
Box, page 268 Science VU/Wayside/Visuals Unlimited
Box, page 269 Dr. Steve Patterson/Science Photo Library/Science Source/Photo Researchers, Inc.
11.3(a) Omikron/Science Source/Photo Researchers, Inc.
11.3(b) Textoff-RM/CNRI/Science Photo Library/Science Source/Photo Researchers, Inc.
11.3(c) Herbert Wagner/Phototake NYC
11.3(d) K. G. Murti/Visuals Unlimited
11.3(e) CNRI/Science Photo Library/Custom Medical Stock Photo
11.4 Dr. Michael Rossman/Pursue University
Box, page 272 Dr. Niels C. Pedersen, D.V.M., Ph.D., University of California at Davis
11.5(a) CDC/Science Source/Photo Researchers, Inc.
11.5(b) CNRI/Science Photo Library/Science Source/Photo Researchers, Inc.
11.5(c) G. Musil/Visuals Unlimited
11.6 G. Musil/Visuals Unlimited
11.7(a) Agricultural Research Service/U.S. Department of Agriculture
11.7(b) Agricultural Research Service/U.S. Department of Agriculture
11.7(c) U.S. Department of Agriculture
Box, page 276 Courtesy of J. L. Gerin, Georgetown University Medical Center
11.11 Bruce Iverson, BSc.
Notebook, page 280 Courtesy Dr. John Huggins, USAMRIID
Notebook, page 281 Courtesy Dr. John Huggins, USAMRIID
11.12 Renato Dulbecco
11.16 Textoff-RM/CNRI/Science Photo Library/Science Source/Photo Researchers, Inc.
11.17 U.S. Department of Agriculture
11.18(a) G. Steven Martin
11.18(b) G. Steven Martin
11.19 M. G. Gabridge, cytoGraphics Inc./BPS
11.20 Centers for Disease Control
11.21 G. Musil/Visuals Unlimited

Chapter 12

Chapter Opening 12 Cath Ellis, Dept. of Zoology, University of Hull/Science Photo Library/Science Source/Photo Researchers, Inc.
12.1 United Nations
12.2(a) Carolina Biological Supply Co.
12.2(b) Stanley Flegler/Visuals Unlimited
12.2(c) David M. Phillips/Visuals Unlimited
Box, page 297 Sanford Berry/Visuals Unlimited
12.3(a) Carolina Biological Supply Co.
12.3(b) Cabisco/Visuals Unlimited
12.4(a) Eric Grave/Science Source/Photo Researchers, Inc.
12.4(b) Runk/Schoenberger from Grant Heilman
12.4(c) Centers for Disease Control
12.4(d) M. Abbey/Visuals Unlimited
Box, page 300 Daniel Gotshall/Visuals Unlimited
12.6 J. Forsdyke/Science Photo Library/Science Source/Photo Researchers, Inc.
12.8(a) H. C. Huang/Visuals Unlimited
12.8(b) Bruce Iverson/Visuals Unlimited
Box, page 302 Dwight R. Kuhn
12.9 Stanley L. Flegler/Visuals Unlimited
12.10 Richard Thom/Visuals Unlimited
12.11(a) Dr. Michael Orlowski, Dept. of Microbiology, Louisiana State University
12.11(b) Dr. Michael Orlowski, Dept. of Microbiology, Louisiana State University
12.12 Bruce Iverson, BSc.
12.14(a) U.S. Department of Agriculture
12.14(b) Stanley Flegler/Visuals Unlimited
Box, page 307 top Dr. Johann N. Bruhn, School of Forestry and Wood Products, Michigan Technological University
Box, page 307 bottom Dwight R. Kuhn
12.15(a) John D. Cunningham/Visuals Unlimited
12.15(b) Runk/Schoenberger from Grant Heilman
12.15(c) Science VU/Fred Marsik/Visuals Unlimited
12.15(d) George J. Wilder/Visuals Unlimited
Notebook, page 312 The Carter Center
Notebook, page 313 The Carter Center
Notebook, page 313 The Carter Center
12.20(a) Cath Wadforth/Science Photo Library/Science Source/Photo Researchers, Inc.
12.20(b) Runk/Schoenberger from Grant Heilman
12.20(c) Richard Walters/Visuals Unlimited
12.20(d) Arthur M. Siegelman/Visuals Unlimited
12.20(e) Dr. Jacquelyn G. Black

Chapter 13

Chapter Opening 13 National Library of Medicine
13.1(a) Jack M. Bostrack/Visuals Unlimited
13.1(b) Jack M. Bostrack/Visuals Unlimited
13.5 Richard Humbert/Biological Photo Service
13.6 Amsco Scientific
13.7 Raymond B. Otero/Visuals Unlimited
13.8 Ulrich Sapountis/OKAPIA/Photo Researchers, Inc.
13.9(a) Yoav Levy/Phototake NYC
13.9(b) John D. Cunningham/Visuals Unlimited
Box, page 336 Barry L. Runk from Grant Heilman
13.12 U.S. Department of Agriculture
13.13(a) FTS Systems
13.13(b) FTS Systems
13.14 Grant Heilman
13.15 CEM Corporation
13.16(a) Debbie Della Piana/Millipore Corp.
13.16(b) Debbie Della Piana/Millipore Corp.

Chapter 14

Chapter Opening 14 Geoff Tompkinson/Science Photo Library/Science Source/Photo Researchers, Inc.
Box, page 356 Paolo Koch/Photo Researchers, Inc.
14.7 Miles/Science VU/Visuals Unlimited
14.8 Victor Lorian/Bronx Lebanon Hospital
14.9(a) Vitek Systems
14.9(b) Vitek Systems
14.11 Forsyth Dental Center
Box, page 366 Ken Greer/Visuals Unlimited
Box, page 367 New York University, Skin and Cancer Unit
14.14 Forsyth Dental Center
Notebook, page 376 Dr. Jacquelyn G. Black
Notebook, page 377 Jessie Cohen/Office of Graphics and Exhibits/National Zoological Park/Smithsonian Institution
Notebook, page 377 Jessie Cohen/Office of Graphics and Exhibits/National Zoological Park/Smithsonian Institution
14.17 Heather Davies/Science Photo Library/Science Source/Photo Researchers

Chapter 15

Chapter Opening 15 Centers for Disease Control
15.1 David M. Phillips/Visuals Unlimited
Box, page 389 L. Migdale/Stock Boston
Box, page 390 William J. Weber/Visuals Unlimited
15.4(a) S. Murphree, Auburn University/BPS
15.4(b) Frederick C. Skvara M.D.
15.5(a) L. M. Pope and D. R. Grote, University of Texas, Austin/BPS
15.5(b) Michael Abbey/Science Source/Photo Researchers, Inc.
Box, page 396 left Dr. Albert W. Biglan, University of Pittsburgh Medical Center
Box, page 396 right Dr. Albert W. Biglan, University of Pittsburgh Medical Center
15.6(a) Gail W. T. Wertz, University of Alabama/BPS
15.6(b) Gail W. T. Wertz, University of Alabama/BPS
15.7 Copr. James Webb/Phototake NYC
15.8 J. J. Paulin, University of Georgia/BPS

Chapter 16

Chapter Opening 16 Centers for Disease Control
Box, page 415 Slough Laboratory/M.A.F.F.
16.9 Dr. Gary Settles/Science Source/Photo Researchers, Inc.
Box, page 420 Dr. Jacquelyn G. Black
16.14 Centers for Disease Control
16.15(a) Centers for Disease Control
16.15(b) Centers for Disease Control
16.15(c) Centers for Disease Control
16.16(a) Ko San Win, China/World Health Organization
16.16(b) D. Espinoza, Peru/World Health Organization
16.16(c) Carlos Gaggero
16.16(d) Chang Hongen, China/World Health Organization
16.18 SIU/Visuals Unlimited
16.19 U.S. Department of Agriculture
Box, page 431 NIDR
Notebook, page 435 Dr. Albert McManus
Notebook, page 435 Dr. Albert McManus
Notebook, page 436 Mike English, M.D./Custom Medical Stock Photo
Notebook, page 436 Dr. Albert McManus
Notebook, page 437 Dr. Albert McManus
16.24 AP/Wide World Photo

Chapter 17

Chapter Opening 17 NIBSC/Science Photo Library/Science Source/Photo Researchers, Inc.
17.5 Jack M. Bostrack/Visuals Unlimited
17.18(a) Dr. Frederic Martini
17.20 Dr. Arnold Brody/Science Photo Library/Custom Medical Stock Photo
17.22(a) Centers for Disease Control
17.22(b) Centers for Disease Control
17.24 Science VU/Visuals Unlimited
17.26(b) R. Dourmashkin/Middlesex Hospital Medical School, London
17.26(c) R. Dourmashkin/Middlesex Hospital Medical School, London
17.27 Agricultural Research Service/U.S. Department of Agriculture

Chapter 18

Chapter Opening 18 James Holmes/Cell Tech Ltd./SPL/Science Source/Photo Researchers, Inc.
18.11 Leon J. Le Beau/Biological Photo Service
18.19 Centers for Disease Control
18.20 Centers for Disease Control
18.23 Lennart Nilsson/Ville

Chapter 19

Chapter Opening 19 National Library of Medicine
19.2(a) U.S. Department of Agriculture
19.2(b) SIU/Custom Medical Stock Photo
19.3 Richard Kessel, C. Y. Shih/Rex Educational Resources Co.
19.4(a) Scott Camazine/Photo Researchers, Inc.
19.4(b) W. Ober
19.7 Larry Mulvehill/Science Source/Photo Researchers, Inc.
19.8(b) left From *Rh,* by Edith Potter (1947) Chicago: Yearbook Medical Publishers
19.8(b) right From *Rh,* by Edith Potter (1947) Chicago: Yearbook Medical Publishers
19.11 From Top, F. H., Sr.: *Communicable and Infectious Diseases,* ed. 6, St. Louis, 1968, The C. V. Mosby Co.)
19.13(a) Ed Reschke/Peter Arnold, Inc.
19.13(b) Dr. Jacquelyn G. Black
19.14 Ken Greer/Visuals Unlimited
19.16(a) Ken Greer/Visuals Unlimited
19.16(b) Ken Greer/Visuals Unlimited
19.21 Sprague/Darley/Taconic
19.22 NASA
19.27 George Whiteley/Photo Researchers, Inc.
19.28(b) Leon J. Le Beau/Biological Photo Service
19.32 Centers for Disease Control
19.35(a) Centers for Disease Control
19.35(b) Centers for Disease Control
19.37 Institut Pasteur/CNRI/Phototake NYC

Chapter 20

Chapter Opening 20 Manfred Kage/Peter Arnold, Inc.
20.1 University of Virginia (Dermatology Department)
20.2 Ken Greer/Visuals Unlimited
20.3 Ken Greer/Visuals Unlimited
20.5(a) Science VU/Visuals Unlimited
20.5(b) Armed Forces Institute of Pathology
20.6(a) Mike Devlin/SPL/Science Source/Photo Researchers, Inc.

20.6(b) N. Martin Hauprich/Science Source/Photo Researchers, Inc.
20.7(a) Ken Greer/Visuals Unlimited
20.7(b) Centers for Disease Control
20.8 Everett S. Beneke/Visuals Unlimited
20.9(a) NIH
20.9(b) NIH
20.10(a) Ken Greer/Visuals Unlimited
20.10(b) University of Virginia (Dermatology Department)
20.11(a) Ed Rottinger/Custom Medical Stock Photo
20.11(b) Ed Rottinger/Custom Medical Stock Photo
20.12 Leon J. Le Beau/Biological Photo Service
20.13 Science VU/Armed Forces Institute of Pathology/Visuals Unlimited
20.14 World Bank
Notebook, page 564 Courtesy Dr. Arnold Ahearn
Notebook, page 565 Courtesy Dr. Arnold Ahearn
20.15 Science VU/Armed Forces Institute of Pathology/Visuals Unlimited
20.16 R. Calentine/Visuals Unlimited
20.17 Centers for Disease Control

Chapter 21

Chapter Opening 21 Dr. Don C. Wiley, Howard Hughes Medical Institute, Harvard University, Cambridge, MA.
21.1(a) Biophoto Associates/Science Source/Photo Researchers, Inc.
21.1(b) SIU/Peter Arnold, Inc.
21.2 Fred Marsik/Visuals Unlimited
21.3 J. J. Ellen
21.7(a) David M. Phillips/Visuals Unlimited
21.7(b) David M. Phillips/Visuals Unlimited
21.8 Centers for Disease Control
21.10 Armed Forces Institute of Pathology
Box, page 589 U.S. Department of Agriculture
21.11(a) Jim C. Williams
21.11(b) Jim C. Williams
21.12 Frederick C. Skvara, M.D.
21.14(b) Centers for Disease Control
21.15(a) NIAID/NIH
21.15(b) Centers for Disease Control
21.16 G. W. Willis, Ochsner Medical Institution/BPS
21.17 ASCP
21.18 Iowa State University Library, Department of Special Collections

Chapter 22

Chapter Opening 22 Anibal Solimano/Reuters/Bettmann
22.1(a) Fred Hossler/Visuals Unlimited
22.1(b) inset Irving C. Puls, D.D.S.
22.3 National Institute of Dental Research
22.4(a) Dr. R. Gottsegen/Peter Arnold, Inc.
22.4(b) Science VU/Max Listgarten/Visuals Unlimited
Box, page 607 National Institute of Dental Research
22.6 Arthur M. Siegelman/Visuals Unlimited
22.7 London School of Hygiene and Tropical Medicine/SPL/Science Source/Photo Researchers, Inc.
22.8 Leodocia M. Pope/University of Texas, Austin/BPS
22.9 inset Veronika Burmeister/Visuals Unlimited
22.10 Centers for Disease Control
22.11 Alfred Pasieka/Peter Arnold, Inc.
22.12 Jerome Paulin/Visuals Unlimited
22.13 Centers for Disease Control
22.14 Larry Jensen/Visuals Unlimited
22.15(a) John D. Cunningham/Visuals Unlimited
22.16(a) W. Ormerod/Visuals Unlimited
22.16(b) Forest W. Buchanan/Visuals Unlimited
22.17 ASCP
22.18(a) Jim Solliday/Biological Photo Service
22.18(b) Bruce Iverson, BSc.
22.18(c) Larry Jensen/Visuals Unlimited
22.18(d) R. Calentine/Visuals Unlimited
22.19 Jim Solliday/Biological Photo Service
22.20 R. Calentine/Visuals Unlimited
22.21 Dr. Daniel H. Connor, Armed Forces Institute of Pathology
22.22 ASCP
22.23(a) Alejandro Balaguer/AP/Wide World Photos
22.23(b) Anibal Solimano/Reuters/Bettmann
22.23(c) Alejandro Balaguer/Sygma

Chapter 23

Chapter Opening 23 Science VU/Visuals Unlimited
23.1 University of Virginia (Burn Department)
23.3(a) NIAID/NIH
23.3(b) Centers for Disease Control
23.3(c) NIAID/NIH
Box, page 640 Renee Stockdale/Animals Animals
23.4(a) ASCP
23.4(b) ASCP
23.5 University of Virginia (Dermatology Department)
23.7 The Granger Collection
23.9(a) Eric Grave/Phototake NYC
23.9(b) National Medical Slide Bank/Custom Medical Stock Photo
23.10(b) Centers for Disease Control
23.11(a) Centers for Disease Control
23.11(b) Centers for Disease Control
23.14(a) R. Calentine/Visuals Unlimited
23.14(b) Ken Greer/Visuals Unlimited
23.15 Armed Forces Institute of Pathology
23.16 G. Musil/Visuals Unlimited
23.17(a) Science VU/Centers for Disease Control/Visuals Unlimited
23.17(b) Science VU/Centers for Disease Control/Visuals Unlimited
23.18(a) ASCP
23.18(b) Science VU/Armed Forces Institute of Pathology/Visuals Unlimited
23.20 Biophoto Associates/Science Source/Photo Researchers, Inc.
23.21 Moredun Animal Health Ltd/SPL/Custom Medical Stock Photo
23.23 Centers for Disease Control

Chapter 24

Chapter Opening 24 WHO photo by Jean-François Chrétien with the collaboration of Terre des Hommes, Lausanne
24.1 Dr. Edward V. Bottone/Mt. Sinai Hospital
24.2 F. A. Murphy, Centers for Disease Control/Fred English Photographs
24.4(a) left Centers for Disease Control
24.4(a) right Centers for Disease Control
24.4(b) Ken Greer/Visuals Unlimited
24.5(a) University of Virginia (Dermatology Department)
24.5(b) R. Pfaltzgraff/Visuals Unlimited
24.6(a) T. J. Beveridge/University of Guelph/BPS
24.6(b) T. J. Beveridge/University of Guelph/BPS

24.7 By courtesy of The Royal College of Surgeons of Edinburgh
24.8(a) March of Dimes Birth Defects Foundation
24.8(b) March of Dimes Birth Defects Foundation
24.10(a) Arthur M. Siegelman/Visuals Unlimited
24.10(b) R. Ashley/Visuals Unlimited
24.11(a) Frederick C. Skvara, M.D.
24.11(b) Frederick C. Skvara, M.D.
24.12 NIAID/NIH
24.13 U.S. Department of Agriculture
24.14 Dr. Gerald Wells, Neuropathology Unit, Central Veterinary Laboratory, Weybridge, England

Chapter 25

Chapter Opening 25 CNRI/Science Photo Library/Science Source/Photo Researchers, Inc.
25.1 Dr. Jacquelyn G. Black
25.3 CNRI/Science Photo Library/Science Source/Photo Researchers, Inc.
25.4 Mike Rein
25.5 Mike Rein
25.6(a) VU/A. M. Siegelman/Visuals Unlimited
25.6(b) Centers for Disease Control
25.7(a) Mike Rein
25.7(b) Centers for Disease Control
25.10(a) Centers for Disease Control
25.10(b) University of Virginia (Dermatology Department)
25.11(a) University of Virginia (Dermatology Department)
25.11(b) University of Virginia (Dermatology Department)
25.12(a) University of Virginia (Dermatology Department)
25.12(b) University of Virginia (Dermatology Department)
25.12(c) University of Virginia (Dermatology Department)
25.12(d) Centers for Disease Control
25.13 Centers for Disease Control
25.14(a) Centers for Disease Control
25.14(b) Centers for Disease Control
25.15 CDC/Biological Photo Service
25.17 Centers for Disease Control
25.18(a) University of Virginia (Dermatology Department)
25.18(b) Centers for Disease Control
25.19 Centers for Disease Control
25.20 Centers for Disease Control
25.21 Centers for Disease Control

Chapter 26

Chapter Opening 26 Jack Stein Grove/Tom Stack & Associates
26.5(a) Prof. David Hall/Science Photo Library/Photo Researchers, Inc.
26.5(b) John D. Cunningham/Visuals Unlimited
26.5(c) Arthur M. Siegelman/Visuals Unlimited
26.6 Paul W. Johnson/Biological Photo Service
26.8 F. Widdel/Visuals Unlimited
26.10(b) Science VU-Forma/Visuals Unlimited
Box, page 732 AP/Wide World Photos
26.12 Dr. Jacquelyn G. Black
26.13 Bruce F. Molnia/TERRAPHOTOGRAPHICS
26.14(a) Arthur M. Siegelman/Visuals Unlimited
26.14(b) Paul W. Johnson/Biological Photo Service
26.16(a) Leon J. Le Beau/Biological Photo Service
26.16(b) Raymond B. Otero/Visuals Unlimited
26.18 R. F. Ashley/Visuals Unlimited
26.19(a) Paul W. Johnson/Biological Photo Service
26.19(b) T. J. Beveridge, University of Guelph/BPS
26.20 Larry Lefever from Grant Heilman
26.22(a) Gregory G. Dimijian, M.D./Photo Researchers, Inc.
26.22(b) Paul W. Johnson/Biological Photo Service
26.23 William J. Weber/Visuals Unlimited

Chapter 27

Chapter Opening 27 John R. Watterson, U.S. Geological Survey
27.1 A. J. Copley/Visuals Unlimited
27.2 U.S. Department of Agriculture
Box, page 751 Tim McCabe/U.S. Department of Agriculture
Box, page 752 U.S. Department of Agriculture
27.3 Grant Heilman, Grant Heilman Photography
27.4 Larry Lefever from Grant Heilman
Box, page 758 Chris Keith/U.S. Department of Agriculture
27.5 International Atomic Energy Association
27.6 Biophoto Associates/Photo Researchers, Inc.
27.7(a) John Colwell from Grant Heilman
27.7(b) John Colwell from Grant Heilman
27.7(c) John Colwell from Grant Heilman
27.8(a) John Colwell from Grant Heilman
27.8(b) John Colwell from Grant Heilman
27.9 David M. Phillips/Visuals Unlimited
27.10 Camera Hawaii, Inc.
27.11 Bill Frantz
27.12(a) Joseph Nettis/Photo Researchers, Inc.
27.12(b) Science VU-SIM/Visuals Unlimited
27.14(a) Earl Roberge/Photo Researchers, Inc.
27.14(b) Earl Roberge/Photo Researchers, Inc.
27.14(c) Dr. Jacquelyn G. Black
27.15 Heinrich Frings/GmbH & Co. KG
27.16(a) A. McClenaghan/SPL/Science Source/Photo Researchers, Inc.
27.16(b) Dan McCoy from Rainbow
27.17 Hank Morgan/Science Source/Photo Researchers, Inc.
27.18(a) Corale L. Brierlev/Visuals Unlimited
27.18(b) Kennecott
27.19 Dr. Jacquelyn G. Black
27.20 Dr. Jacquelyn G. Black
27.21 Dr. Jacquelyn G. Black
27.22 Dr. Jacquelyn G. Black
27.23 Dr. Jacquelyn G. Black